AUTOMOTIVE ENGINE PERFORMANCE

FIFTH EDITION

James D. Halderman

PEARSON

Boston Columbus Indianapolis New York San Francisco Amsterdam
Cape Town Dubai London Madrid Milan Munich Paris Montreal Toronto Delhi
Mexico City São Paulo Sydney Hong Kong Seoul Singapore Taipei Tokyo

Editor-in-Chief: Andrew Gilfillan
Product Manager: Anthony T. Webster
Program Manager: Holly Shufeldt
Project Manager: Rex Davidson
Editorial Assistant: Nancy Kesterson
Team Lead Project Manager: Bryan Pirrmann
Team Lead Program Manager: Laura Weaver
Director of Marketing: David Gesell
Senior Product Marketing Manager: Darcy Betts
Field Marketing Manager: Thomas Hayward
Procurement Specialist: Deidra M. Skahill
Creative Director: Andrea Nix
Art Director: Diane Y. Ernsberger
Cover Designer: Cenveo
Full-Service Project Management: Abinaya Rajendran, Integra Software Services, Pvt. Ltd.
Composition: Integra Software Services, Pvt. Ltd.
Printer/Binder: R.R. Donnelley & Sons
Cover Printer: Phoenix Color
Text Font: Helvetica Neue

Library of Congress Cataloging-in-Publication Data

Halderman, James D.,
Automotive engine performance/James D. Halderman.—Fifth edition.
pages cm
Includes index.
ISBN 978-0-13-407491-7 (alk. paper)—ISBN 0-13-407491-2 (alk. paper) 1. Automobiles—Motors.
2. Automobiles—Performance. 3. Automotive sensors. I. Title.
TL210.H289 2017
629.25—dc23

2015025866

10 9 8 7 6 5 4 3 2 1

ISBN 10: 0-13-407491-2
ISBN 13: 978-0-13-407491-7

PREFACE

PROFESSIONAL TECHNICIAN SERIES Part of Pearson Automotive's Professional Technician Series, the fifth edition of *Automotive Engine Performance* represents the future of automotive textbooks. The series is a full-color, media-integrated solution for today's students and instructors. The series includes textbooks that cover all eight areas of ASE certification, plus additional titles covering common courses.

The series is also peer reviewed for technical accuracy.

UPDATES TO THE FIFTH EDITION Text was updated with the following features:

- Many new full color line drawings and photos were added to this edition to help bring the subject to life.
- Updated throughout and correlated to the latest ASE/NATEF tasks.
- New Case Studies included in this edition that includes the "three Cs" (Complaint, cause and correction).
- New OSHA hazardous chemical labeling requirements added to Chapter 2.
- Atkinson Cycle engine design added to Chapter 3.
- Scope testing of MAF sensors added to Chapter 23.
- Additional content on gasoline direct injection (GDI) added to Chapter 28.
- Fiat Chrysler Multiair System information added to Chapter 30.
- Tier 3 Emission Standards added to Chapter 31.
- Unlike other textbooks, this book is written so that the theory, construction, diagnosis, and service of a particular component or system are presented in one location. There is no need to search through the entire book for other references to the same topic.

ASE AND NATEF CORRELATED NATEF-certified programs need to demonstrate that they use course material that covers NATEF and ASE tasks. All Professional Technician textbooks have been correlated to the appropriate ASE and NATEF task lists. These correlations can be found in two locations:

- As an appendix to each book.
- At the beginning of each chapter in the Instructor's Manual.

A COMPLETE INSTRUCTOR AND STUDENT SUPPLEMENTS PACKAGE All Professional Technician textbooks are accompanied by a full set of instructor and student supplements. Please see page vi for a detailed list of supplements.

A FOCUS ON DIAGNOSIS AND PROBLEM SOLVING The Professional Technician Series has been developed to satisfy the need for a greater emphasis on problem diagnosis. Automotive instructors and service managers agree that students and beginning technicians need more training in diagnostic procedures and skill development. To meet this need and demonstrate how real-world problems are solved, "Real World Fix" features are included throughout and highlight how real-life problems are diagnosed and repaired.

The following pages highlight the unique core features that set the Professional Technician Series book apart from other automotive textbooks.

IN-TEXT FEATURES

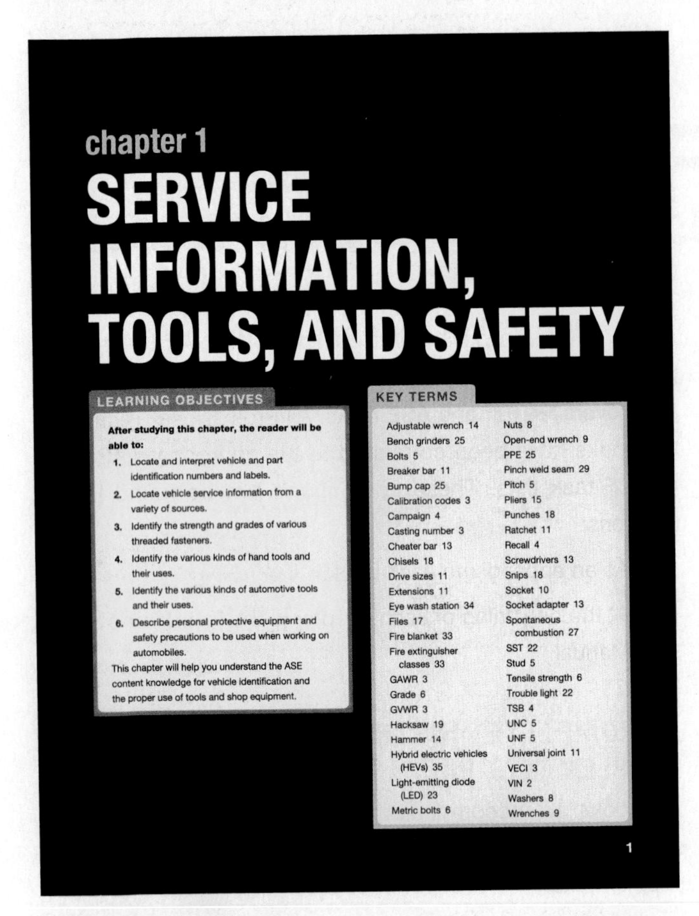
chapter 1

SERVICE INFORMATION, TOOLS, AND SAFETY

LEARNING OBJECTIVES

After studying this chapter, the reader will be able to:

1. Locate and interpret vehicle and part identification numbers and labels.
2. Locate vehicle service information from a variety of sources.
3. Identify the strength and grades of various threaded fasteners.
4. Identify the various kinds of hand tools and their uses.
5. Identify the various kinds of automotive tools and their uses.
6. Describe personal protective equipment and safety precautions to be used when working on automobiles.

This chapter will help you understand the ASE content knowledge for vehicle identification and the proper use of tools and shop equipment.

KEY TERMS

Adjustable wrench 14
Bench grinders 25
Bolts 5
Breaker bar 11
Bump cap 25
Calibration codes 3
Campaign 4
Casting number 3
Cheater bar 13
Chisels 18
Drive sizes 11
Extensions 11
Eye wash station 34
Files 17
Fire blanket 33
Fire extinguisher classes 33
GAWR 3
Grade 6
GVWR 3
Hacksaw 19
Hammer 14
Hybrid electric vehicles (HEVs) 35
Light-emitting diode (LED) 23
Metric bolts 6
Nuts 8
Open-end wrench 9
PPE 25
Pinch weld seam 29
Pitch 5
Pliers 15
Punches 18
Ratchet 11
Recall 4
Screwdrivers 13
Snips 18
Socket 10
Socket adapter 13
Spontaneous combustion 27
SST 22
Stud 5
Tensile strength 6
Trouble light 22
TSB 4
UNC 5
UNF 5
Universal joint 11
VECI 3
VIN 2
Washers 8
Wrenches 9

1

LEARNING OBJECTIVES AND KEY TERMS appear at the beginning of each chapter to help students and instructors focus on the most important material in each chapter. The chapter objectives are based on specific ASE and NATEF tasks.

SAFETY TIP

Shop Cloth Disposal

Always dispose of oily shop cloths in an enclosed container to prevent a fire. ● **SEE FIGURE 1–69.** Whenever oily cloths are thrown together on the floor or workbench, a chemical reaction can occur, which can ignite the cloth even without an open flame. This process of ignition without an open flame is called **spontaneous combustion**.

SAFETY TIPS alert students to possible hazards on the job and how to avoid them.

CASE STUDY

There Is No Substitute for a Thorough Visual Inspection

An intermittent "check engine" light and a random-misfire diagnostic trouble code (DTC) P0300 was being diagnosed. A scan tool did not provide any help because all systems seemed to be functioning normally. Finally, the technician removed the engine cover and discovered a mouse nest. ● **SEE FIGURE 30-24.**

Summary:

- **Complaint—**Customer stated that the "Check Engine" light was on.
- **Cause—**A stored P0300 DTC was stored indicating a random misfire had been detected caused by an animal that had partially eaten some fuel injector wires.
- **Correction—**The mouse nest was removed and the wiring was repaired.

CASE STUDY present students with actual automotive scenarios and show how these common (and sometimes uncommon) problems were diagnosed and repaired.

TECH TIP

It Just Takes a Second

Whenever removing any automotive component, it is wise to screw the bolts back into the holes a couple of threads by hand. This ensures that the right bolt will be used in its original location when the component or part is put back on the vehicle.

TECH TIPS feature real-world advice and "tricks of the trade" from ASE-certified master technicians.

FREQUENTLY ASKED QUESTION

How Many Types of Screw Heads Are Used in Automotive Applications?

There are many, including Torx, hex (also called Allen), plus many others used in custom vans and motor homes. ● **SEE FIGURE 1–9.**

FREQUENTLY ASKED QUESTIONS are based on the author's own experience and provide answers to many of the most common questions asked by students and beginning service technicians.

NOTE: Most of these "locking nuts" are grouped together and are commonly referred to as *prevailing torque nuts*. This means that the nut will hold its tightness or torque and not loosen with movement or vibration.

NOTES provide students with additional technical information to give them a greater understanding of a specific task or procedure.

CAUTION: *Never* use hardware store (nongraded) bolts, studs, or nuts on any vehicle steering, suspension, or brake component. Always use the exact size and grade of hardware that is specified and used by the vehicle manufacturer.

CAUTIONS alert students about potential damage to the vehicle that can occur during a specific task or service procedure.

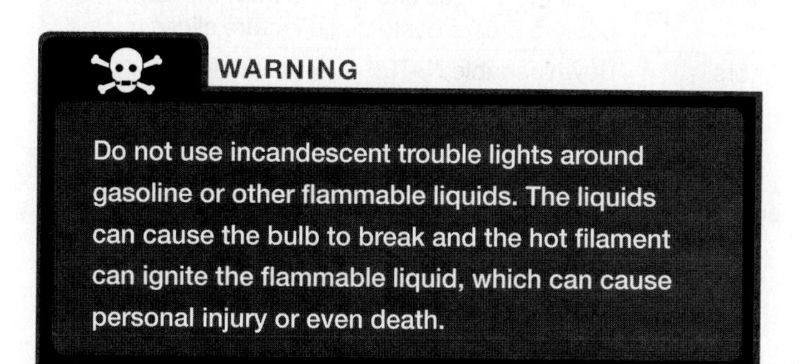

WARNING

Do not use incandescent trouble lights around gasoline or other flammable liquids. The liquids can cause the bulb to break and the hot filament can ignite the flammable liquid, which can cause personal injury or even death.

WARNINGS alert students to potential dangers to themselves during a specific task or service procedure.

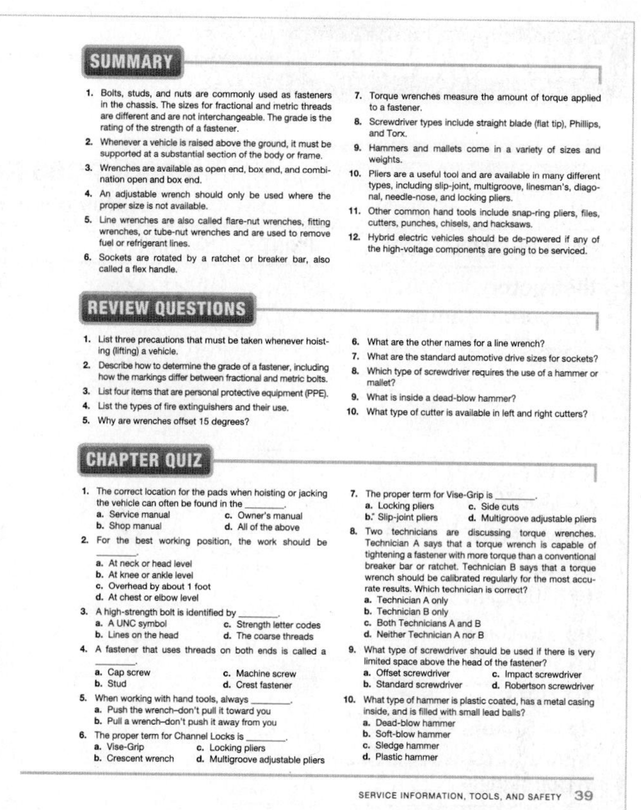

SUMMARY

1. Bolts, studs, and nuts are commonly used as fasteners in the chassis. The sizes for fractional and metric threads are different and are not interchangeable. The grade is the rating of the strength of a fastener.
2. Whenever a vehicle is raised above the ground, it must be supported at a substantial section of the body or frame.
3. Wrenches are available as open end, box end, and combination open and box end.
4. An adjustable wrench should only be used where the proper size is not available.
5. Line wrenches are also called flare-nut wrenches, fitting wrenches, or tube-nut wrenches and are used to remove fuel or refrigerant lines.
6. Sockets are rotated by a ratchet or breaker bar, also called a flex handle.
7. Torque wrenches measure the amount of torque applied to a fastener.
8. Screwdriver types include straight blade (flat tip), Phillips, and Torx.
9. Hammers and mallets come in a variety of sizes and weights.
10. Pliers are a useful tool and are available in many different types, including slip-joint, multigroove, linesman's, diagonal, needle-nose, and locking pliers.
11. Other common hand tools include snap-ring pliers, files, cutters, punches, chisels, and hacksaws.
12. Hybrid electric vehicles should be de-powered if any of the high-voltage components are going to be serviced.

REVIEW QUESTIONS

1. List three precautions that must be taken whenever hoisting (lifting) a vehicle.
2. Describe how to determine the grade of a fastener, including how the markings differ between fractional and metric bolts.
3. List four items that are personal protective equipment (PPE).
4. List the types of fire extinguishers and their use.
5. Why are wrenches offset 15 degrees?
6. What are the other names for a line wrench?
7. What are the standard automotive drive sizes for sockets?
8. Which type of screwdriver requires the use of a hammer or mallet?
9. What is inside a dead-blow hammer?
10. What type of cutter is available in left and right cutters?

CHAPTER QUIZ

1. The correct location for the pads when hoisting or jacking the vehicle can often be found in the ________.
   a. Service manual
   b. Shop manual
   c. Owner's manual
   d. All of the above
2. For the best working position, the work should be ________.
   a. At neck or head level
   b. At knee or ankle level
   c. Overhead by about 1 foot
   d. At chest or elbow level
3. A high-strength bolt is identified by ________.
   a. A UNC symbol
   b. Lines on the head
   c. Strength letter codes
   d. The coarse threads
4. A fastener that uses threads on both ends is called a ________.
   a. Cap screw
   b. Stud
   c. Machine screw
   d. Crest fastener
5. When working with hand tools, always ________.
   a. Push the wrench–don't pull it toward you
   b. Pull a wrench–don't push it away from you
6. The proper term for Channel Locks is ________.
   a. Vise-Grip
   b. Crescent wrench
   c. Locking pliers
   d. Multigroove adjustable pliers
7. The proper term for Vise-Grip is ________.
   a. Locking pliers
   b. Slip-joint pliers
   c. Side cuts
   d. Multigroove adjustable pliers
8. Two technicians are discussing torque wrenches. Technician A says that a torque wrench is capable of tightening a fastener with more torque than a conventional breaker bar or ratchet. Technician B says that a torque wrench should be calibrated regularly for the most accurate results. Which technician is correct?
   a. Technician A only
   b. Technician B only
   c. Both Technicians A and B
   d. Neither Technician A nor B
9. What type of screwdriver should be used if there is very limited space above the head of the fastener?
   a. Offset screwdriver
   b. Standard screwdriver
   c. Impact screwdriver
   d. Robertson screwdriver
10. What type of hammer is plastic coated, has a metal casing inside, and is filled with small lead balls?
   a. Dead-blow hammer
   b. Soft-blow hammer
   c. Sledge hammer
   d. Plastic hammer

SERVICE INFORMATION, TOOLS, AND SAFETY 39

THE SUMMARY, REVIEW QUESTIONS, AND CHAPTER QUIZ at the end of each chapter help students review the material presented in the chapter and test themselves to see how much they've learned.

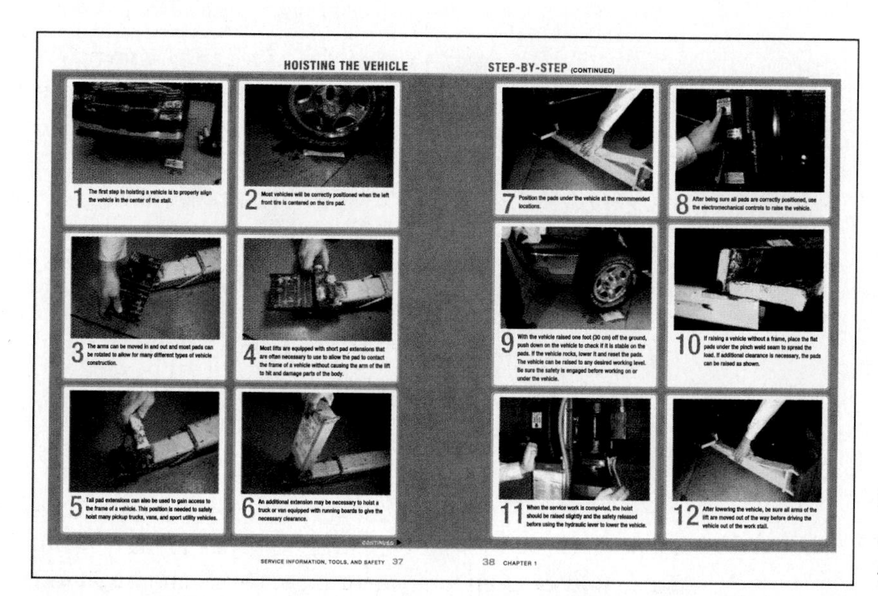

HOISTING THE VEHICLE STEP-BY-STEP (CONTINUED)

1 The first step in hoisting a vehicle is to properly align the vehicle in the center of the stall.

2 Most vehicles will be correctly positioned when the left front tire is centered on the tire pad.

3 The arms can be moved in and out and most pads can be rotated to allow for many different types of vehicle construction.

4 Most lifts are equipped with short pad extensions that are often necessary to use to allow the pad to contact the frame of a vehicle without causing the arm of the lift to hit and damage parts of the body.

5 Tail pad extensions can also be used to gain access to the frame of a vehicle. This position is needed to safely hoist many pickup trucks, vans, and sport utility vehicles.

6 An additional extension may be necessary to hoist a truck or van equipped with running boards to give the necessary clearance.

CONTINUED

7 Position the pads under the vehicle at the recommended locations.

8 After being sure all pads are correctly positioned, use the electromechanical controls to raise the vehicle.

9 With the vehicle raised one foot (30 cm) off the ground, push down on the vehicle to check if it is stable on the pads. If the vehicle rocks, lower it and reset the pads. The vehicle can be raised to any desired working level. Be sure the safety is engaged before working on or under the vehicle.

10 If raising a vehicle without a frame, place the flat pads under the pinch weld seam to spread the load. If additional clearance is necessary, the pads can be raised as shown.

11 When the service work is completed, the hoist should be raised slightly and the safety released before using the hydraulic lever to lower the vehicle.

12 After lowering the vehicle, be sure all arms of the lift are moved out of the way before driving the vehicle out of the work stall.

SERVICE INFORMATION, TOOLS, AND SAFETY 37 38 CHAPTER 1

STEP-BY-STEP photo sequences show in detail the steps involved in performing a specific task or service procedure.

SUPPLEMENTS

RESOURCES IN PRINT AND ONLINE

Automotive Engine Performance

NAME OF SUPPLEMENT	PRINT	ONLINE	AUDIENCE	DESCRIPTION
Instructor Resource Manual 0134067002		✔	Instructors	NEW! The Ultimate teaching aid: Chapter summaries, key terms, chapter learning objectives, lecture resources, discuss/demonstrate classroom activities, and answers to the in text review and quiz questions.
TestGen 0134067053		✔	Instructors	Test generation software and test bank for the text.
PowerPoint Presentation 0134067045		✔	Instructors	Slides include chapter learning objectives, lecture outline of the test, and graphics from the book.
Image Bank 0134067010		✔	Instructors	All of the images and graphs from the text-book to create customized lecture slides.
NATEF Correlated Task Sheets - For Instructors 0134066960		✔	Instructors	Downloadable NATEF task sheets for easy customization and development of unique task sheets.
NATEF Correlated Task Sheets - For Students 0134067061	✔		Students	Study activity manual that correlates NATEF Automobile Standards to chapters and pages numbers in the text. Available to students at a discounted price when packaged with the text.
CourseSmart eText 0134067037		✔	Students	An alternative to purchasing the print text-book, students can subscribe to the same content online and save up to 50% off the suggested list price of the print text. Visit **www.coursesmart.com**

All online resources can be downloaded from the Instructor's Resource Center: **www.pearsonighered.com/irc**

ACKNOWLEDGMENTS

A large number of people and organizations have cooperated in providing the reference material and technical information used in this text. The author wishes to express sincere thanks to the following individuals and organizations for their special contributions:

Carl Borsani, Graphic Home

Tom Birch

Mark Warren

Dr. John Kershaw, Criterion Technical Education Consultants

Tony Martin

Richard Krieger

James (Mike) Watson, Watson Automotive LLC

Bill Fulton, Ohio Automotive Technology

Jim Morton, Automotive Training Center (ATC)

Jim Linder, Linder Technical Services, Inc.

John Thornton, Autotrain

Dave Scaler, Mechanic's Education Association

TECHNICAL AND CONTENT REVIEWERS The following people reviewed the manuscript before production and checked it for technical accuracy and clarity of presentation. Their suggestions and recommendations were included in the final draft of the manuscript. Their input helped make this textbook clear and technically accurate while maintaining the easy-to-read style that has made other books from the same author so popular.

Matt Dixon
Southern Illinois University

Greg Pfahl
Miami-Jacobs Career College

Jeff Rehkopf
Florida State College

Omar Trinidad
Southern Illinois University

Jim Morton
Automotive Training Center (ATC)

Jim Anderson
Greenville High School

Victor Bridges
Umpqua Community College

Dr. Roger Donovan
Illinois Central College

Mike Elder
Pittsburg State University

Dale Eldridge
Iowa Central Community College

Eric Fenske
Gateway Community College, Phoenix

Joseph Gumina
City College of San Francisco & Solano Community College

Trent Lindbloom
Pittsburg State University

A. C. Durdin
Moraine Park Technical College

Herbert Ellinger
Western Michigan University

Al Engledahl
College of Dupage

Larry Hagelberger
Upper Valley Joint Vocational School

Oldrick Hajzler
Red River College

Betsy Hoffman
Vermont Technical College

Steven T. Lee
Lincoln Technical Institute

Carlton H. Mabe, Sr.
Virginia Western Community College

Roy Marks
Owens Community College

Tony Martin
University of Alaska Southeast

Kerry Meier
San Juan College

Fritz Peacock
Indiana Vocational Technical College

Dennis Peter
NAIT (Canada)

Kenneth Redick
Hudson Valley Community College

Mitchell Walker
St. Louis Community College at Forest Park

Jennifer Wise
Sinclair Community College

Special thanks to instructional designer **Alexis I. Skriloff James**.

PHOTO SEQUENCES The author wishes to thank Mike Garblik and Chuck Taylor of Sinclair Community College in Dayton, Ohio, and James (Mike) Watson who helped with many of the photos. Thanks also to Carl Borsani for his help with many of the new figures used in this edition.

Most of all, I wish to thank Michelle Halderman for her assistance in all phases of manuscript preparation.

—James D. Halderman

ABOUT THE AUTHOR

JIM HALDERMAN brings a world of experience, knowledge, and talent to his work. His automotive service experience includes working as a flat-rate technician, a business owner, and a professor of automotive technology at a leading U.S. community college for more than 20 years.

He has a Bachelor of Science degree from Ohio Northern University and a master's degree in education from Miami University in Oxford, Ohio. He also holds a U.S. Patent for an electronic transmission control device. He is an ASE-certified Master Automotive Technician and Advanced Engine Performance (L1) ASE certified.

He is the author of many automotive textbooks, all published by Pearson Education.

He has presented numerous technical seminars to national audiences including the California Automotive Teachers (CAT) and the Illinois College Automotive Instructor Association (ICAIA). He is also a member and presenter at the North American Council of Automotive Teachers (NACAT). He was also named Regional Teacher of the Year by General Motors Corporation and an outstanding alumnus of Ohio Northern University.

He and his wife, Michelle, live in Dayton, Ohio. They have two children. You can reach him at

jim@jameshalderman.com

BRIEF CONTENTS

CONTENTS

chapter 1
SERVICE INFORMATION, TOOLS, AND SAFETY 1

chapter 2
ENVIRONMENTAL AND HAZARDOUS MATERIALS 41

chapter 3
GASOLINE ENGINE OPERATION, PARTS, AND SPECIFICATIONS 54

chapter 4
DIESEL ENGINE OPERATION AND DIAGNOSIS 69

SERVICE INFORMATION, TOOLS, AND SAFETY

LEARNING OBJECTIVES

After studying this chapter, the reader will be able to:

1. Locate and interpret vehicle and part identification numbers and labels.
2. Locate vehicle service information from a variety of sources.
3. Identify the strength and grades of various threaded fasteners.
4. Identify the various kinds of hand tools and their uses.
5. Identify the various kinds of automotive tools and their uses.
6. Describe personal protective equipment and safety precautions to be used when working on automobiles.

This chapter will help you understand the ASE content knowledge for vehicle identification and the proper use of tools and shop equipment.

KEY TERMS

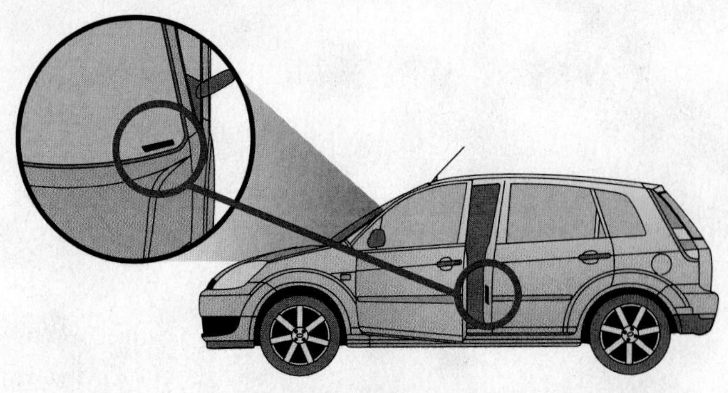

FIGURE 1–1 The vehicle identification number (VIN) is visible through the base of the windshield and on a decal inside the driver's door.

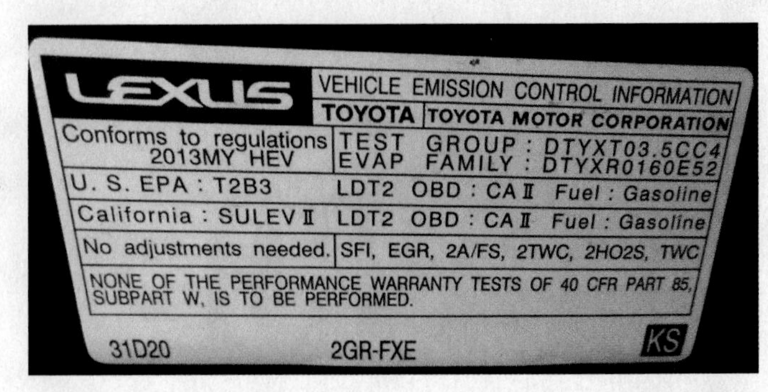

FIGURE 1–2 The vehicle emissions control information (VECI) sticker is placed under the hood.

VEHICLE IDENTIFICATION

MAKE, MODEL, AND YEAR All service work requires that the vehicle and its components be properly identified. The most common identification is the make, model, and year of manufacture of the vehicle.

Make: e.g., Chevrolet

Model: e.g., Impala

Year: e.g., 2008

VEHICLE IDENTIFICATION NUMBER The year of the vehicle is often difficult to determine exactly. A model may be introduced as the next year's model as soon as January of the previous year. Typically, a new model year starts in September or October of the year prior to the actual new year, but not always. This is why the **vehicle identification number,** usually abbreviated **VIN,** is so important. ● **SEE FIGURE 1–1.**

Since 1981, all vehicle manufacturers have used a VIN that is 17 characters long. Although every vehicle manufacturer assigns various letters or numbers within these 17 characters, there are some constants, including:

- The first number or letter designates the country of origin. ● **SEE CHART 1–1.**
- The fourth and fifth characters represent the vehicle line/series.
- The sixth character is the body style.
- The seventh character is the restraint system.
- The eighth character is often the engine code. (Some engines cannot be determined by the VIN.)
- The tenth character represents the year on all vehicles. ● **SEE CHART 1–2.**

1 = United States	J = Japan	T = Czechoslovakia
2 = Canada	K = Korea	U = Romania
3 = Mexico	L = China	V = France
4 = United States	M = India	W = Germany
5 = United States	N = Turkey	X = Russia
6 = Australia	P = Philippines	Y = Sweden
8 = Argentina	R = Taiwan	Z = Italy
9 = Brazil	S = England	

CHART 1–1

The first number or letter in the VIN identifies the country where the vehicle was made.

A = 1980/2010	L = 1990/2020	Y = 2000/2030
B = 1981/2011	M = 1991/2021	1 = 2001/2031
C = 1982/2012	N = 1992/2022	2 = 2002/2032
D = 1983/2013	P = 1993/2023	3 = 2003/2033
E = 1984/2014	R = 1994/2024	4 = 2004/2034
F = 1985/2015	S = 1995/2025	5 = 2005/2035
G = 1986/2016	T = 1996/2026	6 = 2006/2036
H = 1987/2017	V = 1997/2027	7 = 2007/2037
J = 1988/2018	W = 1998/2028	8 = 2008/2038
K = 1989/2019	X = 1999/2029	9 = 2009/2039

CHART 1–2

The pattern repeats every 30 years for the year of manufacture.

FIGURE 1–3 A typical calibration code sticker on the case of a controller. The information on the sticker is often required when ordering parts or a replacement controller.

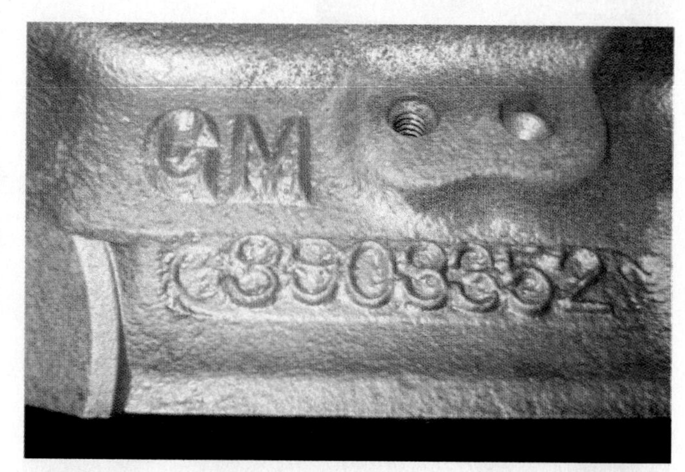

FIGURE 1–4 Casting numbers on major components can be either cast or stamped.

VEHICLE SAFETY CERTIFICATION LABEL A vehicle safety certification label is attached to the left side pillar post on the rearward-facing section of the left front door. This label indicates the month and year of manufacture as well as the **gross vehicle weight rating (GVWR),** the **gross axle weight rating (GAWR),** and the VIN.

VECI LABEL The **vehicle emissions control information (VECI)** label under the hood of the vehicle shows informative settings and emission hose routing information. ● **SEE FIGURE 1–2.**

The VECI label (sticker) can be located on the bottom side of the hood, the radiator fan shroud, the radiator core support, or on the strut towers. The VECI label usually includes the following information.

- Engine identification
- Emissions standard that the vehicle meets
- Vacuum hose routing diagram
- Base ignition timing (if adjustable)
- Spark plug type and gap
- Valve lash
- Emission calibration code

CALIBRATION CODES **Calibration codes** are usually located on powertrain control modules (PCMs) or other controllers. Whenever diagnosing an engine operating fault, it is often necessary to use the calibration code to be sure that the vehicle is the subject of a technical service bulletin or other service procedure. ● **SEE FIGURE 1–3.**

CASTING NUMBERS When an engine part such as a block is cast, a number is put into the mold to identify the casting. ● **SEE FIGURE 1–4.** These **casting numbers** can be used to identify the part and to check dimensions, such as the cubic inch displacement, and other information, such as the year of manufacture. Sometimes changes are made to the mold, yet the casting number is not changed. Most often the casting number is the best piece of identifying information that the service technician can use for identifying an engine.

SERVICE INFORMATION

SERVICE MANUALS Service information is used by the service technician to determine specifications and service procedures, and any needed special tools.

Factory and aftermarket service manuals contain specifications and service procedures. While factory service manuals cover just one year and one or more models of the same vehicle, most aftermarket service manufacturers cover multiple years and/or models in one manual.

FIGURE 1–5 Electronic service information is available from aftermarket sources, such as All-Data and Mitchell-on-Demand, as well as on websites hosted by vehicle manufacturers.

Included in most service manuals are the following:

- Capacities and recommended specifications for all fluids
- Specifications including engine and routine maintenance items
- Testing procedures
- Service procedures including the use of special tools when needed

ELECTRONIC SERVICE INFORMATION Electronic service information is available mostly by subscription and provides access to an Internet site where service manual–type information is available. ● **SEE FIGURE 1–5**. Most vehicle manufacturers also offer electronic service information to their dealers and to most schools and colleges that offer corporate training programs.

TECHNICAL SERVICE BULLETINS **Technical service bulletins,** often abbreviated **TSBs,** sometimes called *technical service information bulletins (TSIBs)*, are issued by the vehicle manufacturer to notify service technicians of a problem and include the necessary corrective action. Technical service bulletins are designed for dealership technicians but are

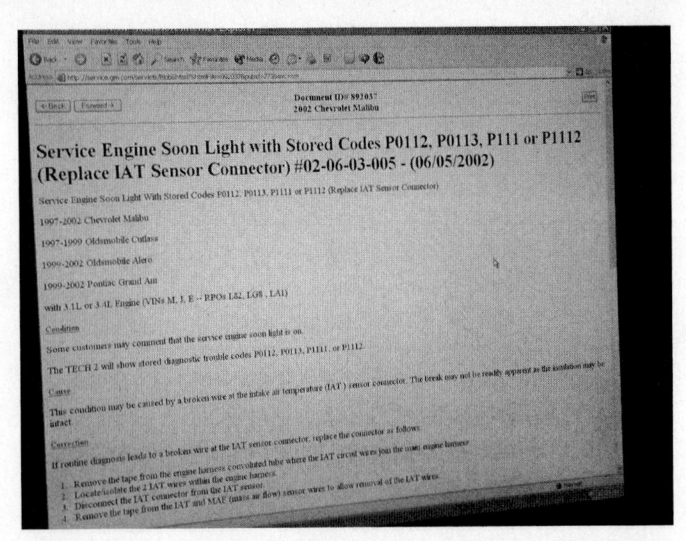

FIGURE 1–6 Technical service bulletins (TSBs) are issued by vehicle manufacturers when a fault occurs that affects many vehicles with the same problem. The TSB then provides the fix for the problem including any parts needed and detailed instructions.

republished by aftermarket companies and made available along with other service information to shops and vehicle repair facilities. ● **SEE FIGURE 1–6.**

INTERNET The Internet has opened the field for information exchange and access to technical advice. One of the most useful websites is the International Automotive Technician's Network at **www.iatn.net**. This is a free site but service technicians must register to join. For a small monthly sponsor fee, the shop or service technician can gain access to the archives, which include thousands of successful repairs in the searchable database.

RECALLS AND CAMPAIGNS A **recall** or **campaign** is issued by a vehicle manufacturer and a notice is sent to all owners in the event of a safety-related fault or concern. Although these faults may be repaired by shops, they are generally handled by a local dealer. Items that have created recalls in the past include potential fuel system leakage problems, exhaust leakage, or electrical malfunctions that could cause a possible fire or the engine to stall. Unlike technical service bulletins whose cost is only covered when the vehicle is within the warranty period, a recall or campaign is always done at no cost to the vehicle owner.

? FREQUENTLY ASKED QUESTION

What Should Be Included on a Work Order?

A work order is a legal document that should include the following information.

1. Customer information
2. Identification of the vehicle including the VIN
3. Related service history information
4. The "three Cs":
   - Customer concern (complaint)
   - Cause of the concern
   - Correction or repairs that the vehicle required to return it to proper operation

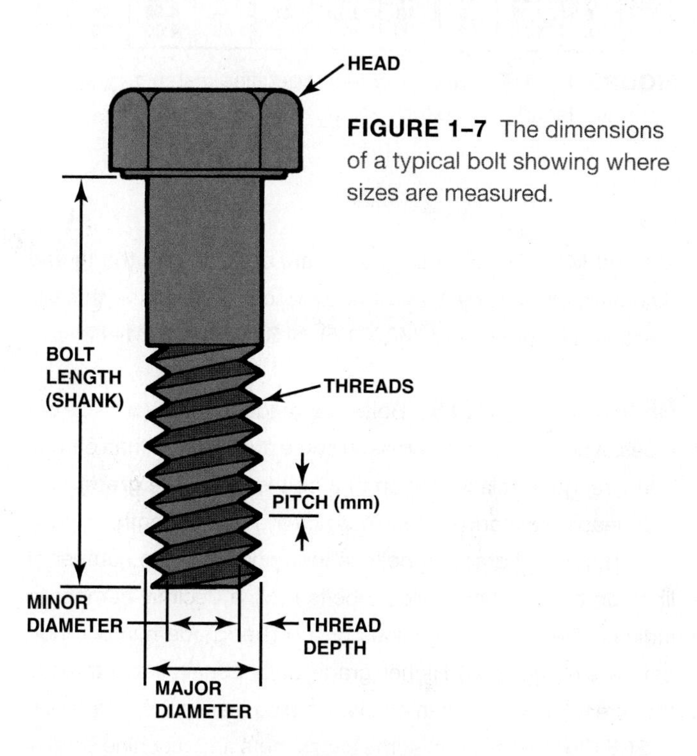

FIGURE 1–7 The dimensions of a typical bolt showing where sizes are measured.

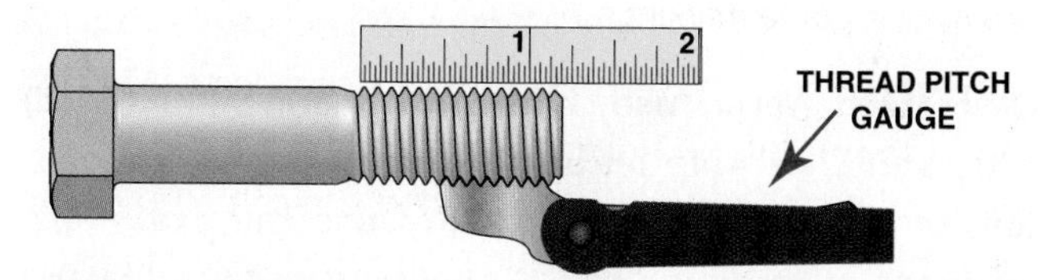

FIGURE 1–8 Thread pitch gauge used to measure the pitch of the thread. This bolt has 13 threads to the inch.

THREADED FASTENERS

BOLTS AND THREADS Most of the threaded fasteners used on vehicles are **bolts.** Bolts are called *cap screws* when they are threaded into a casting. Automotive service technicians usually refer to these fasteners as bolts, regardless of how they are used. In this chapter, they are called bolts. Sometimes, studs are used for threaded fasteners. A **stud** is a short rod with threads on both ends. Often, a stud will have coarse threads on one end and fine threads on the other end. The end of the stud with coarse threads is screwed into the casting. A nut is used on the opposite end to hold the parts together.

The fastener threads *must* match the threads in the casting or nut. The threads may be measured either in fractions of an inch (called fractional) or in metric units. The size is measured across the outside of the threads, called the major diameter or the *crest* of the thread. ● **SEE FIGURE 1–7.**

FRACTIONAL BOLTS Fractional threads are either coarse or fine. The coarse threads are called **Unified National Coarse (UNC),** and the fine threads are called **Unified National Fine (UNF).** Standard combinations of sizes and number of threads per inch (called **pitch**) are used. Pitch can be measured with a thread pitch gauge as shown in ● **FIGURE 1–8.** Bolts are identified by their diameter and length as measured from below the head, not by the size of the head or the size of the wrench used to remove or install the bolt.

Fractional thread sizes are specified by the diameter in fractions of an inch and the number of threads per inch. Typical

FREQUENTLY ASKED QUESTION

How Many Types of Screw Heads Are Used in Automotive Applications?

There are many, including Torx, hex (also called Allen), plus many others used in custom vans and motor homes. ● **SEE FIGURE 1–9.**

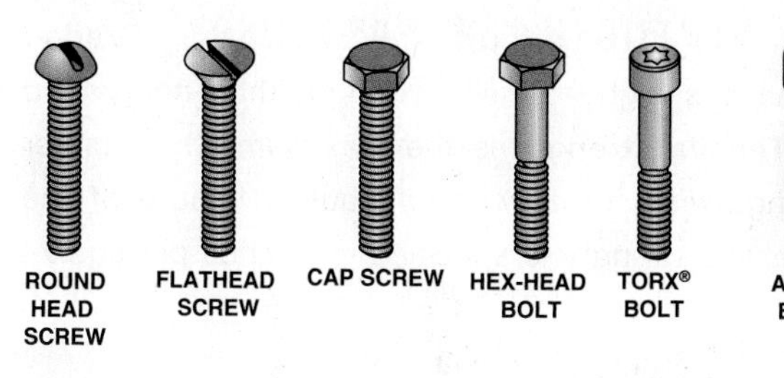
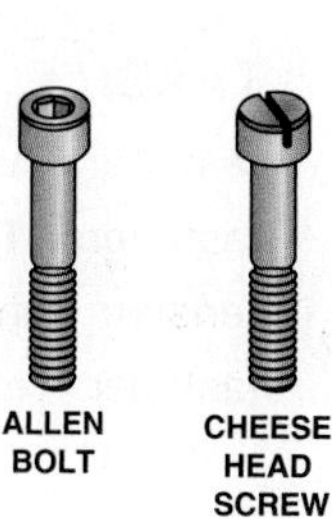

FIGURE 1–9 Bolts and screws have many different heads. The head determines what tool is needed.

SIZE	THREADS PER INCH NC UNC	NF UNF	OUTSIDE DIAMETER INCHES
0	..	80	0.0600
1	64	..	0.0730
1	..	72	0.0730
2	56	..	0.0860
2	..	64	0.0860
3	48	..	0.0990
3	..	56	0.0990
4	40	..	0.1120
4	..	48	0.1120
5	40	..	0.1250
5	..	44	0.1250
6	32	..	0.1380
6	..	40	0.1380
8	32	..	0.1640
8	..	36	0.1640
10	24	..	0.1900
10	..	32	0.1900
12	24	..	0.2160
12	..	28	0.2160
1/4	20	..	0.2500
1/4	..	28	0.2500
5/16	18	..	0.3125
5/16	..	24	0.3125
3/8	16	..	0.3750
3/8	..	24	0.3750
7/16	14	.	0.4375
7/16	..	20	0.4375
1/2	13	..	0.5000
1/2	..	20	0.5000
9/16	12	..	0.5625
9/16	..	18	0.5625
5/8	11	..	0.6250
5/8	..	18	0.6250
3/4	10	..	0.7500
3/4	..	16	0.7500
7/8	9	..	0.8750
7/8	..	14	0.8750

CHART 1–3

American Standard is one method of sizing fasteners.

UNC thread sizes would be 5/16–18 and 1/2–13. Similar UNF thread sizes would be 5/16–24 and 1/2–20. ● **SEE CHART 1–3.**

METRIC BOLTS The size of a **metric bolt** is specified by the letter *M* followed by the diameter in millimeters (mm) across the outside (crest) of the threads. Typical metric sizes would be

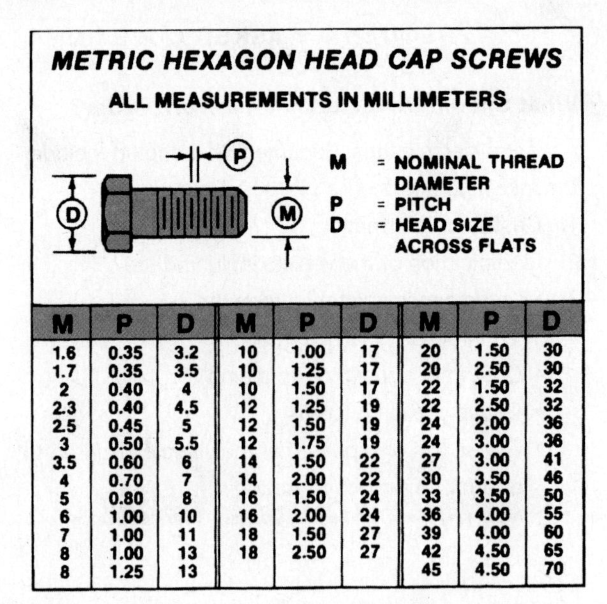

M	P	D	M	P	D	M	P	D
1.6	0.35	3.2	10	1.00	17	20	1.50	30
1.7	0.35	3.5	10	1.25	17	20	2.50	30
2	0.40	4	10	1.50	17	22	1.50	32
2.3	0.40	4.5	12	1.25	19	22	2.50	32
2.5	0.45	5	12	1.50	19	24	2.00	36
3	0.50	5.5	12	1.75	19	24	3.00	36
3.5	0.60	6	14	1.50	22	27	3.00	41
4	0.70	7	14	2.00	22	30	3.50	46
5	0.80	8	16	1.50	24	33	3.50	50
6	1.00	10	16	2.00	24	36	4.00	55
7	1.00	11	18	1.50	27	39	4.00	60
8	1.00	13	18	2.50	27	42	4.50	65
8	1.25	13				45	4.50	70

FIGURE 1–10 The metric system specifies fasteners by diameter, length, and pitch.

M8 and M12. Fine metric threads are specified by the thread diameter followed by X and the distance between the threads measured in millimeters (M8 X 1.5). ● **SEE FIGURE 1–10.**

GRADES OF BOLTS Bolts are made from many different types of steel, and for this reason some are stronger than others. The strength or classification of a bolt is called the **grade.** The bolt heads are marked to indicate their grade strength.

The actual grade of bolts is two more than the number of lines on the bolt head. Metric bolts have a decimal number to indicate the grade. More lines or a higher grade number indicate a stronger bolt. Higher grade bolts usually have threads that are rolled rather than cut, which also makes them stronger. ● **SEE FIGURE 1–11**. In some cases, nuts and machine screws have similar grade markings.

CAUTION: *Never* **use hardware store (nongraded) bolts, studs, or nuts on any vehicle steering, suspension, or brake component. Always use the exact size and grade of hardware that is specified and used by the vehicle manufacturer.**

TENSILE STRENGTH OF FASTENERS Graded fasteners have a higher tensile strength than nongraded fasteners. **Tensile strength** is the maximum stress under tension (lengthwise force) without causing failure of the fastener. Tensile strength is specified in pounds per square inch (psi).

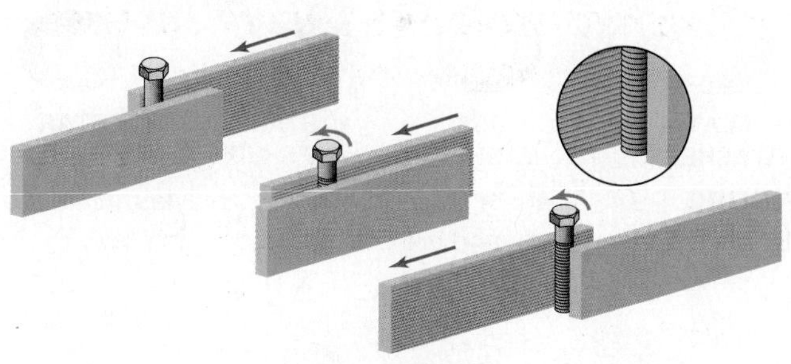

FIGURE 1–11 Stronger threads are created by cold rolling a heat-treated bolt blank instead of cutting the threads, using a die.

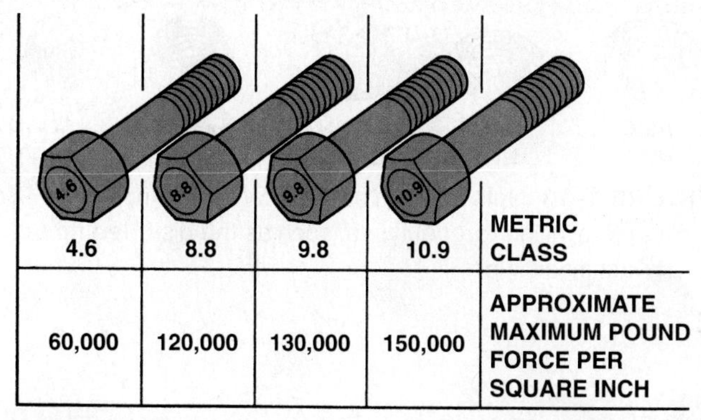

FIGURE 1–12 Metric bolt (cap screw) grade markings and approximate tensile strength.

SAE BOLT DESIGNATIONS

SAE GRADE NO.	SIZE RANGE	TENSILE STRENGTH, PSI	MATERIAL	HEAD MARKING
1	1/4 through 1 1/2	60,000	Low or medium carbon steel	
2	1/4 through 3/4	74,000		
	7/8 through 1 1/2	60,000		
5	1/4 through 1	120,000	Medium carbon steel, quenched and tempered	
	1-1/8 through 1 1/2	105,000		
5.2	1/4 through 1	120,000	Low carbon martensite steel,* quenched and tempered	
7	1/4 through 1 1/2	133,000	Medium carbon alloy steel, quenched and tempered	
8	1/4 through 1 1/2	150,000	Medium carbon alloy steel, quenched and tempered	
8.2	1/4 through 1	150,000	Low carbon martensite steel,* quenched and tempered	

CHART 1–4

The tensile strength rating system as specified by the Society of Automotive Engineers (SAE).

* Martensite steel is steel that has been cooled rapidly, thereby increasing its hardness. It is named after a German metallurgist, Adolf Martens.

The strength and type of steel used in a bolt is supposed to be indicated by a raised mark on the head of the bolt. The type of mark depends on the standard to which the bolt was manufactured. Most often, bolts used in machinery are made to SAE standard J429. ● **CHART 1–4** shows the grade and specified tensile strength.

Metric bolt tensile strength property class is shown on the head of the bolt as a number, such as 4.6, 8.8, 9.8, and 10.9; the higher the number, the stronger the bolt. ● **SEE FIGURE 1–12.**

FIGURE 1–13 Nuts come in a variety of styles, including locking (prevailing torque) types, such as the distorted thread and nylon insert type.

TECH TIP

A 1/2 Inch Wrench Does Not Fit a 1/2 Inch Bolt

A common mistake made by persons new to the automotive field is to think that the size of a bolt or nut is the size of the head. The size of the bolt or nut (outside diameter of the threads) is usually smaller than the size of the wrench or socket that fits the head of the bolt or nut. Examples are given in the following table.

Wrench Size	Thread Size
7/16 inch	1/4 inch
1/2 inch	5/16 inch
9/16 inch	3/8 inch
5/8 inch	7/16 inch
3/4 inch	1/2 inch
10 mm	6 mm
12 or 13 mm*	8 mm
14 or 17 mm*	10 mm

* European (Systeme International d'Unites, or SI) metric

NUTS **Nuts** are the female part of a threaded fastener. Most nuts used on cap screws have the same hex size as the cap screw head. Some inexpensive nuts use a hex size larger than the cap screw head. Metric nuts are often marked with dimples to show their strength. More dimples indicate stronger nuts. Some nuts and cap screws use interference-fit threads to keep them from accidentally loosening. This means that the shape of the nut is slightly distorted or that a section of the thread is deformed. Nuts can also be kept from loosening with a nylon washer fastened in the nut or with a nylon patch or strip on the threads. ● **SEE FIGURE 1–13.**

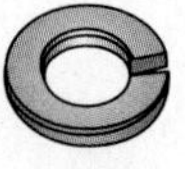

FIGURE 1–14 Washers come in a variety of styles, including flat and star (serrated), and are often used to help prevent a fastener from loosening.

TECH TIP

It Just Takes a Second

Whenever removing any automotive component, it is wise to screw the bolts back into the holes a couple of threads by hand. This ensures that the right bolt will be used in its original location when the component or part is put back on the vehicle. Often, the same diameter of fastener is used on a component, but the length of the bolt may vary. Spending just a couple of seconds to put the bolts and nuts back where they belong when the part is removed can save a lot of time when the part is being reinstalled. Besides making certain that the right fastener is being installed in the right place, this method helps prevent bolts and nuts from getting lost or kicked away. How much time have you wasted looking for that lost bolt or nut?

NOTE: Most of these "locking nuts" are grouped together and are commonly referred to as *prevailing torque nuts*. This means that the nut will hold its tightness or torque and not loosen with movement or vibration. Most prevailing torque nuts should be replaced whenever removed to ensure that the nut will not loosen during service. Always follow the manufacturer's recommendations. Anaerobic sealers, such as Loctite, are used on the threads where the nut or cap screw must be both locked and sealed.

WASHERS **Washers** are often used under cap screw heads and under nuts. ● **SEE FIGURE 1–14.** Plain flat washers are used to provide an even clamping load around the fastener. Lock washers are added to prevent accidental loosening. In some accessories, the washers are locked onto the nut to provide easy assembly.

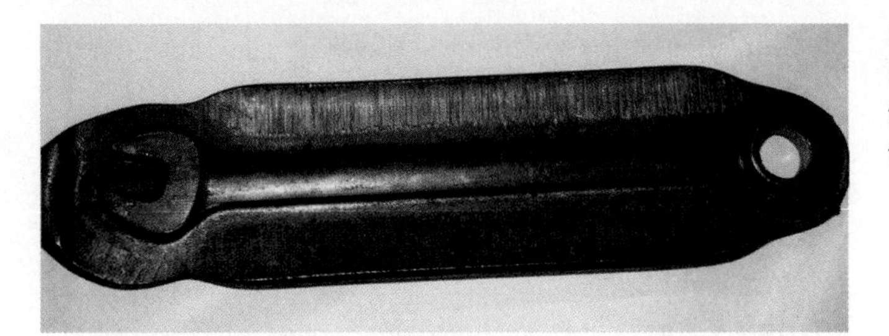

FIGURE 1–15 A wrench after it has been forged but before the flashing (extra material around the wrench) has been removed.

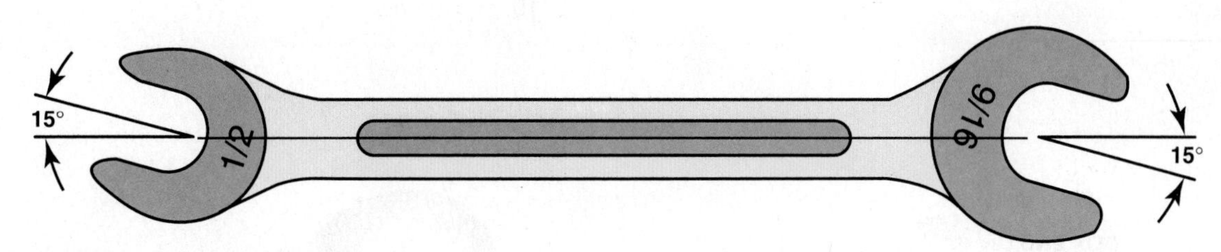

FIGURE 1–16 A typical open-end wrench. Note the size difference on each end and that the head is angled 15 degrees at the end.

HAND TOOLS

WRENCHES Wrenches are the most used hand tool by service technicians. **Wrenches** are used to grasp and rotate threaded fasteners. Most wrenches are constructed of forged alloy steel, usually chrome-vanadium steel. ● **SEE FIGURE 1–15.**

After the wrench is formed, it is hardened, tempered to reduce brittleness, and then chrome-plated. There are several types of wrenches.

- An **open-end wrench** is often used to loosen or tighten bolts or nuts that do not require a lot of torque. Because of the *open* end, this type of wrench can be easily placed on a bolt or nut with an angle of 15 degrees, which allows the wrench to be flipped over and used again to continue to rotate the fastener. The major disadvantage of an open-end wrench is the lack of torque that can be applied due to the fact that the open jaws of the wrench only contact two flat surfaces of the fastener. An open-end wrench has two different sizes, one at each end. ● **SEE FIGURE 1–16.**
- A box-end wrench, also called a *closed-end wrench,* is placed over the top of the fastener and grips the points of the fastener. A box-end wrench is angled 15 degrees to allow it to clear nearby objects.

  Therefore, a box-end wrench should be used to loosen or tighten fasteners because it grasps around the entire head of the fastener. A box-end wrench has two different sizes, one at each end. ● **SEE FIGURE 1–17.**

  Most service technicians purchase *combination wrenches,* which have the open end at one end and the same size box end on the other end. ● **SEE FIGURE 1–18.**

  A combination wrench allows the technician to loosen or tighten a fastener using the box end of the wrench, turn it around, and use the open end to increase the speed of rotating the fastener.
- An **adjustable wrench** is often used where the exact size wrench is not available or when a large nut, such as a wheel spindle nut, needs to be rotated but not tightened. An adjustable wrench should not be used to loosen or tighten fasteners because the torque applied to the wrench can cause the movable jaws to loosen their grip on the fastener, causing it to become rounded. ● **SEE FIGURE 1–19.**
- Line wrenches, also called *flare-nut wrenches, fitting wrenches,* or *tube-nut wrenches,* are designed to grip almost all the way around a nut used to retain a fuel, brake, or refrigerant line, and yet be able to be installed over the line. ● **SEE FIGURE 1–20.**

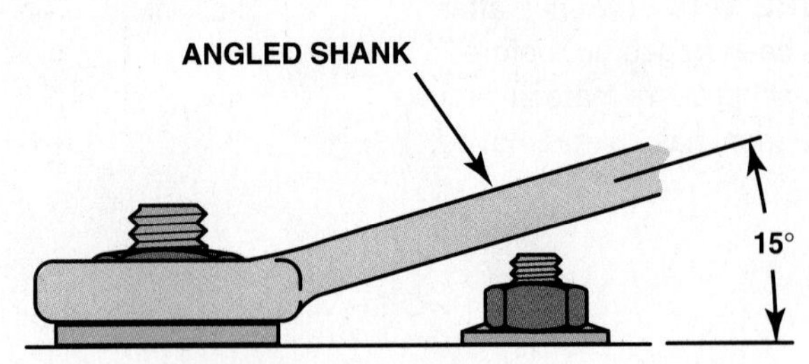

FIGURE 1–17 The end of a box-end wrench is angled 15 degrees to allow clearance for nearby objects or other fasteners.

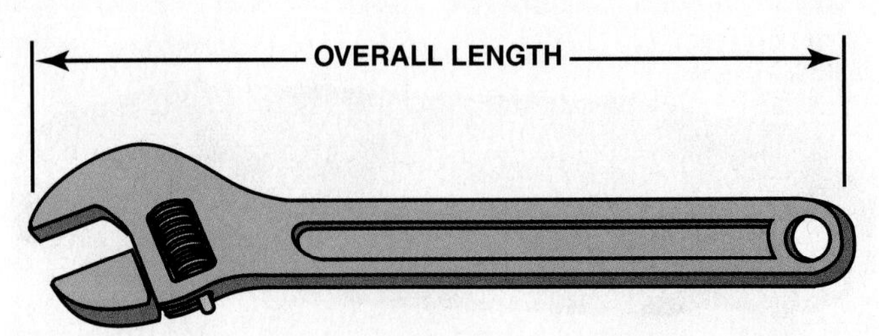

FIGURE 1–19 An adjustable wrench. Adjustable wrenches are sized by the overall length of the wrench, not by how far the jaws open. Common sizes of adjustable wrenches include 8, 10, and 12 inch.

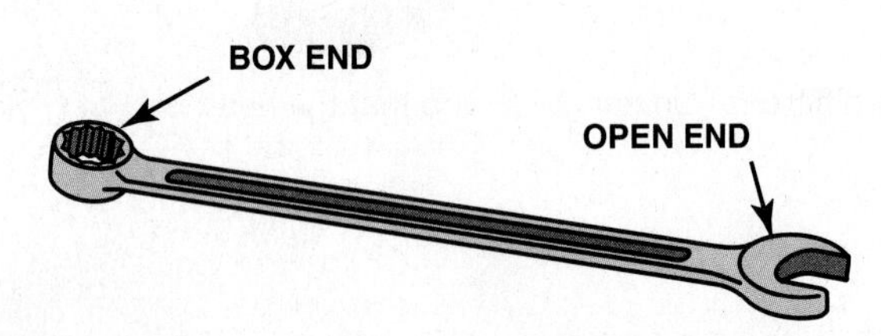

FIGURE 1–18 A combination wrench has an open end at one end and a box end at the other end.

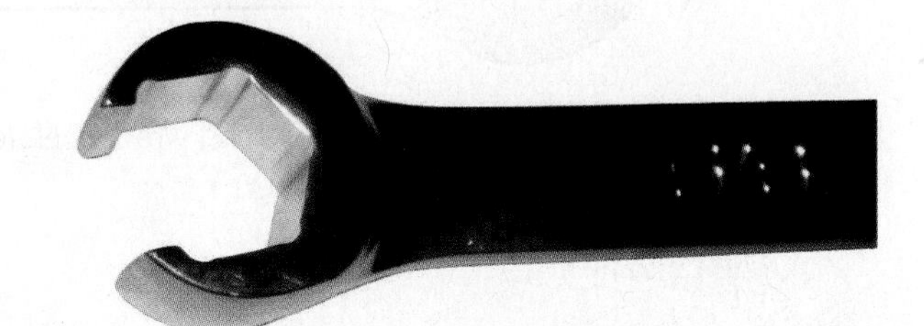

FIGURE 1–20 The end of a typical line wrench, which shows that it is capable of grasping most of the head of the fitting.

TECH TIP

Hide Those from the Boss

An apprentice technician started working for a dealership and put his top tool box on a workbench. Another technician observed that, along with a complete set of good-quality tools, the box contained several adjustable wrenches. The more experienced technician said, "Hide those from the boss." The boss does not want any service technician to use adjustable wrenches. If any adjustable wrench is used on a bolt or nut, the movable jaw often moves or loosens and starts to round the head of the fastener. If the head of the bolt or nut becomes rounded, it then becomes much more difficult to remove.

SAFE USE OF WRENCHES Wrenches should be inspected before use to be sure they are not crack, bent, or damaged. All wrenches should be cleaned after use before being returned to the tool box. Always use the correct size of wrench for the fastener being loosened or tightened to help prevent the rounding of the flats of the fastener. When attempting to loosen a fastener, pull a wrench–do not push it. If you push a wrench, your knuckles may be hurt when forced into another object if the fastener breaks loose or if the wrench slips. Always keep wrenches and all hand tools clean to help prevent rust and to allow for a better, firmer grip. Never expose any tool to excessive heat. High temperatures can reduce the strength ("draw the temper") of metal tools.

Never use a hammer on any wrench unless you are using a special *staking face wrench* designed to be used with a hammer. Replace any tools that are damaged or worn.

RATCHETS, SOCKETS, AND EXTENSIONS A **socket** fits over the fastener and grips the points and/or flats of the bolt or nut. The socket is rotated (driven) using either a long bar called

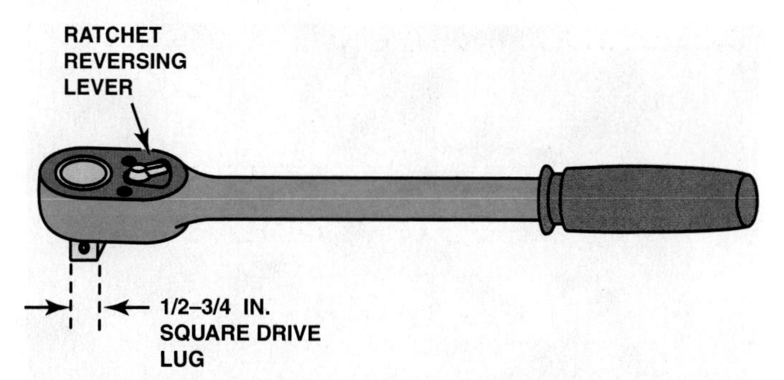

FIGURE 1–21 A typical ratchet used to rotate a socket. A ratchet makes a ratcheting noise when it is being rotated in the opposite direction from loosening or tightening. A knob or lever on the ratchet allows the technician to switch directions.

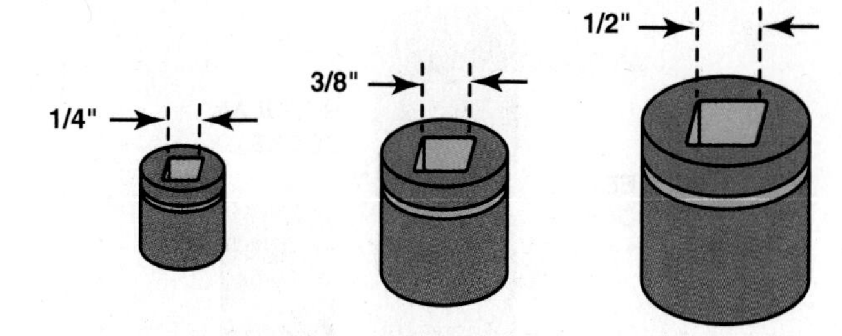

FIGURE 1–23 The most commonly used socket drive sizes include 1/4, 3/8, and 1/2 inch drives.

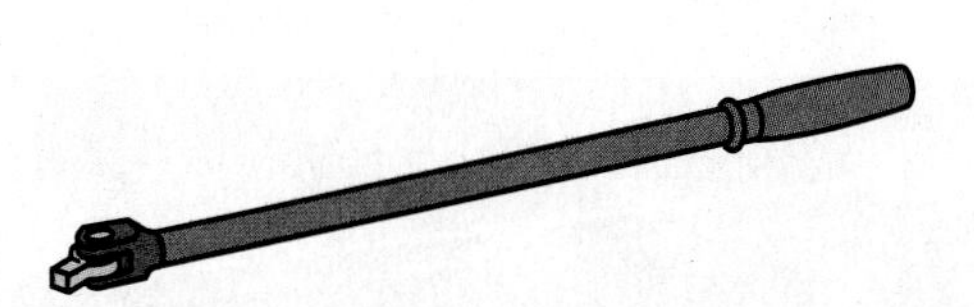

FIGURE 1–22 A typical flex handle used to rotate a socket; also called a breaker bar, because it usually has a longer handle than a ratchet and, therefore, can be used to apply more torque to a fastener than a ratchet.

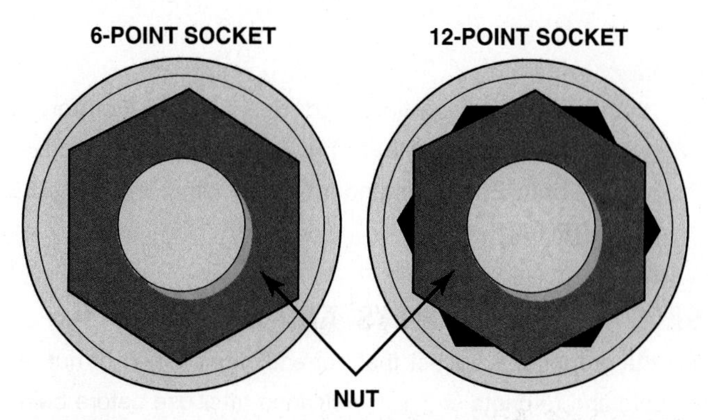

FIGURE 1–24 A 6-point socket fits the head of a bolt or nut on all sides. A 12-point socket can round off the head of a bolt or nut if great force is applied.

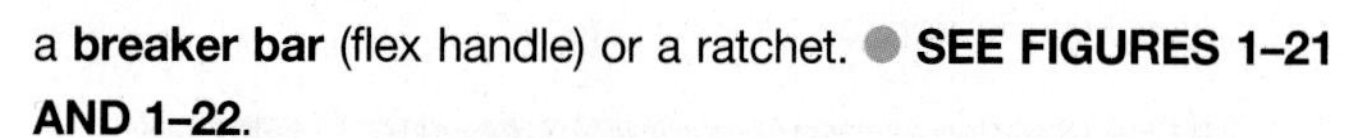

a **breaker bar** (flex handle) or a ratchet. ● **SEE FIGURES 1–21 AND 1–22.**

A **ratchet** is a tool that turns the socket in only one direction and allows the rotating of the ratchet handle back and forth in a narrow space. Socket **extensions** and **universal joints** are also used with sockets to allow access to fasteners in restricted locations.

DRIVE SIZE. Sockets are available in various **drive sizes,** including 1/4, 3/8, and 1/2 inch sizes for most automotive use. ● **SEE FIGURES 1–23 AND 1–24.**

TECH TIP

Right to Tighten

It is sometimes confusing which way to rotate a wrench or screwdriver, especially when the head of the fastener is pointing away from you. To help visualize while looking at the fastener, say "righty tighty, lefty loosey."

Many heavy-duty truck and/or industrial applications use 3/4 and 1 inch sizes. The drive size is the distance of each side of the square drive. Sockets and ratchets of the same size are designed to work together.

Regular and deep well sockets are available in regular lengths for use in most applications or in a deep well design that allows for access to a fastener that uses a long stud or other similar conditions. ● **SEE FIGURE 1–25.**

TORQUE WRENCHES Torque wrenches are socket turning handles designed to apply a known amount of force to the fastener. The two basic types of torque wrenches include:

1. **Clicker type.** This type of torque wrench is first set to the specified torque and it then "clicks" when the set torque value has been reached. When force is removed from the torque wrench handle, another click is heard. The setting on a clicker-type torque wrench should be set back to zero after use and checked for proper calibration regularly. ● **SEE FIGURE 1–26.**
2. **Beam or dial type.** This type of torque wrench is used to measure torque, but instead of presenting the value, the actual torque is displayed on the dial of the wrench as

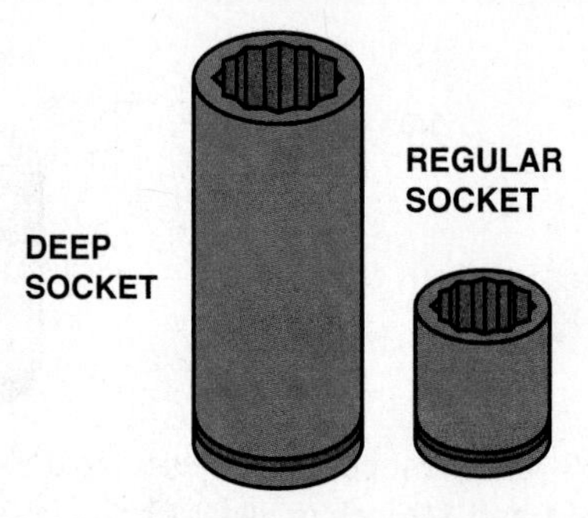

FIGURE 1–25 Allows access to the nut that has a stud plus other locations needing great depth, such as spark plugs.

the fastener is being tightened. Beam or dial-type torque wrenches are available in 1/4, 3/8, and 1/2 inch drives and in both English (standard) and metric units. ● **SEE FIGURE 1–27.**

SAFE USE OF SOCKETS AND RATCHETS Always use the proper size socket that correctly fits the bolt or nut. All sockets and ratchets should be cleaned after use before being placed back into the tool box. Sockets are available in short and deep well designs. Never expose any tool to excessive heat. High temperatures can reduce the strength ("draw the temper") of metal tools.

Never use a hammer on a socket handle unless you are using a special *staking face wrench* designed to be used with a hammer. Replace any tools that are damaged or worn.

Also select the appropriate drive size. For example, for small work, such as on the dash, select a 1/4 inch drive. For most general service work, use a 3/8 inch drive and for suspension and steering and other large fasteners, select a 1/2 inch drive. When loosening a fastener, always pull the ratchet toward you rather than push it outward.

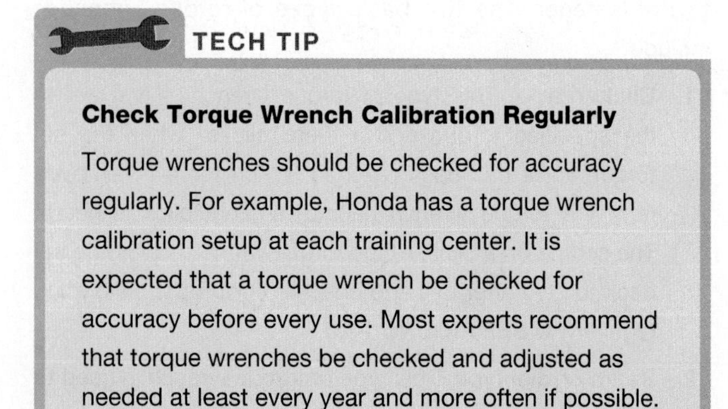

TECH TIP

Check Torque Wrench Calibration Regularly

Torque wrenches should be checked for accuracy regularly. For example, Honda has a torque wrench calibration setup at each training center. It is expected that a torque wrench be checked for accuracy before every use. Most experts recommend that torque wrenches be checked and adjusted as needed at least every year and more often if possible. ● **SEE FIGURE 1–28.**

FIGURE 1–26 Using a torque wrench to tighten connecting rod nuts on an engine.

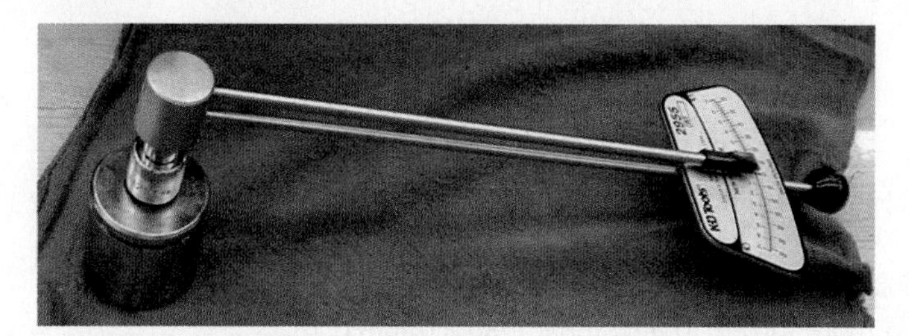

FIGURE 1–27 A beam-type torque wrench that displays the torque reading on the face of the dial. The beam display is read as the beam deflects, which is in proportion to the amount of torque applied to the fastener.

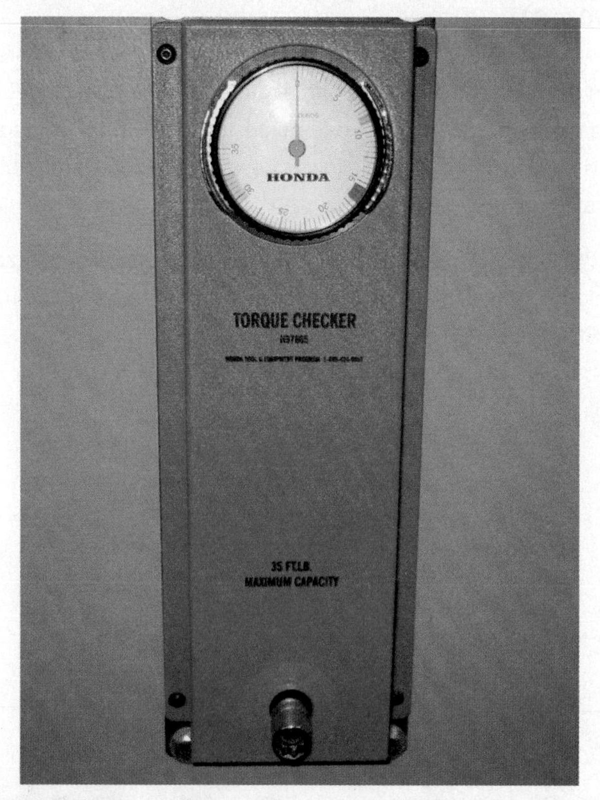

FIGURE 1–28 Torque wrench calibration checker.

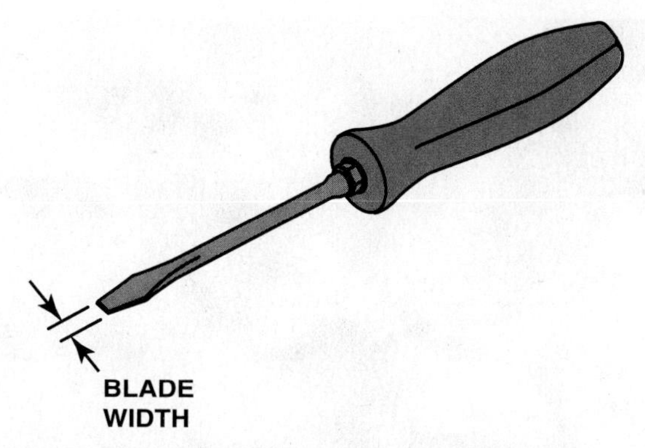

FIGURE 1–29 A flat-tip (straight-blade) screwdriver. The width of the blade should match the width of the slot in the fastener being loosened or tightened.

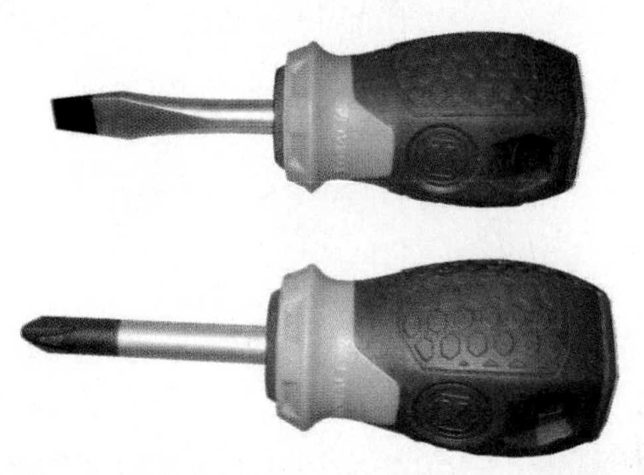

FIGURE 1–30 Two stubby screwdrivers used to access screws that have limited space above. A straight blade is on top and a #2 Phillips screwdriver is on the bottom.

TECH TIP

Use Socket Adapters with Caution

A **socket adapter** allows the use of one size of socket and another drive size ratchet or breaker bar. Socket adapters can be used for different drive size sockets on a ratchet. Combinations include:

1/4 inch drive – 3/8 inch sockets
3/8 inch drive – 1/4 inch sockets
3/8 inch drive – 1/2 inch sockets
1/2 inch drive – 3/8 inch sockets

Using a larger drive ratchet or breaker bar on a smaller size socket can cause the application of too much force to the socket, which could crack or shatter. Using a smaller size drive tool on a larger socket will usually not cause any harm, but would greatly reduce the amount of torque that can be applied to the bolt or nut.

SCREWDRIVERS

- **Straight-blade screwdriver.** Many smaller fasteners are removed and installed using a **screwdriver.** Screwdrivers are available in many sizes and tip shapes. The most commonly used screwdriver is called a *straight blade* or *flat tip.*

  Flat-tip screwdrivers are sized by the width of the blade, and this width should match the width of the slot in the screw. ● **SEE FIGURE 1–29.**

CAUTION: Do not use a screwdriver as a pry tool or chisel. Screwdrivers use hardened steel only at the tip and are not designed to be pounded on or used for prying because they could bend easily. Always use the proper tool for each application.

TECH TIP

Avoid Using "Cheater Bars"

Whenever a fastener is difficult to remove, some technicians will insert the handle of a ratchet or a breaker bar into a length of steel pipe sometimes called a **cheater bar**. The extra length of the pipe allows the technician to exert more torque than can be applied using the drive handle alone. However, the extra torque can easily overload the socket and ratchet, causing them to break or shatter, which could cause personal injury.

- **Phillips screwdriver.** Another type of commonly used screwdriver is the Phillips screwdriver, named for Henry F. Phillips, who invented the crosshead screw in 1934. Due to the shape of the crosshead screw and screwdriver, a Phillips screw can be driven with more torque than can be achieved with a slotted screw.

  A Phillips head screwdriver is specified by the length of the handle and the size of the point at the tip. A #1 tip has a sharp point, a #2 tip is the most commonly used, and a #3 tip is blunt and is only used for larger sizes of Phillips head fasteners. For example, a #2 × 3 inch Phillips screwdriver would typically measure 6 inches from the tip of the blade to the end of the handle (3 inches long handle and 3 inches long blade) with a #2 tip.

  Both straight-blade and Phillips screwdrivers are available with a short blade and handle for access to fasteners with limited room. ● **SEE FIGURE 1–30.**

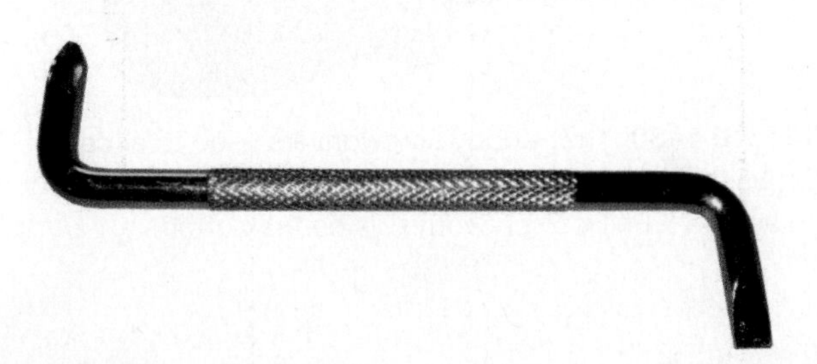

FIGURE 1–31 An offset screwdriver is used to install or remove fasteners that do not have enough space above to use a conventional screwdriver.

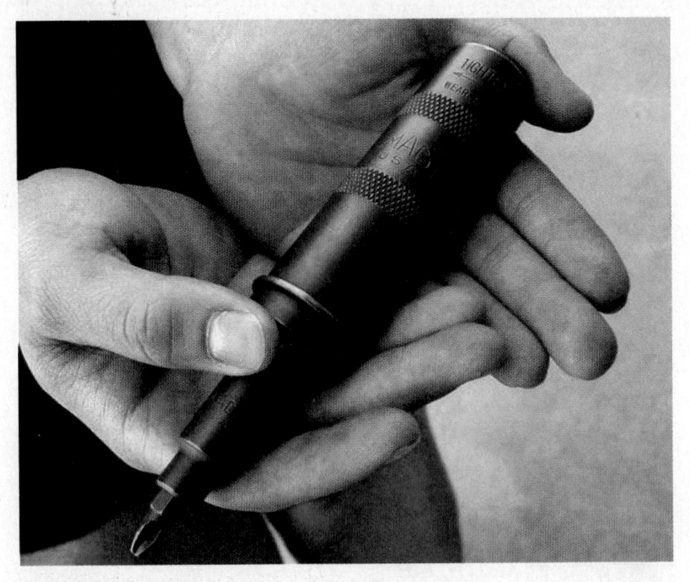

FIGURE 1–32 An impact screwdriver used to remove slotted or Phillips head fasteners that cannot be broken loose using a standard screwdriver.

- **Offset screwdriver.** Offset screwdrivers are used in places where a conventional screwdriver cannot fit. An offset screwdriver is bent at the ends and is used similar to a wrench. Most offset screwdrivers have a straight blade at one end and a Phillips head at the opposite end. ● **SEE FIGURE 1–31**.
- **Impact screwdriver.** An *impact screwdriver* is used to break loose or tighten a screw. A hammer is used to strike the end after the screwdriver holder is placed in the head of the screw and rotated in the desired direction. The force from the hammer blow does two things: It applies a force downward holding the tip of the screwdriver in the slot and then applies a twisting force to loosen (or tighten) the screw. ● **SEE FIGURE 1–32**.

SAFE USE OF SCREWDRIVERS Always use the proper type and size screwdriver that matches the fastener. Try to avoid pressing down on a screwdriver because if it slips, the screwdriver tip could penetrate your hand, causing serious personal injury. All screwdrivers should be cleaned after use. Do not use a screwdriver as a prybar; always use the correct tool for the job.

HAMMERS AND MALLETS **Hammers** and mallets are used to force objects together or apart. The shape of the back of the hammer head (called the *peen*) usually determines the name. For example, a ball-peen hammer has a rounded end like a ball and is used to straighten oil pans and valve covers, using the hammer head, and to shape metal, using the ball peen. ● **SEE FIGURE 1–33**.

FREQUENTLY ASKED QUESTION

What Is a Torx?

A Torx is a six-pointed star-shaped tip that was developed by Camcar (formerly Textron) to offer higher loosening and tightening torque than is possible with a straight (flat tip) or Phillips. Torx is very commonly used in the automotive field for many components. Commonly used Torx sizes from small to large include T15,T20,T25, and T30.

Some Torx fasteners include a round projection in the center requiring that a special version of a Torx bit be used. These are called security Torx bits that have a hole in the center to be used on these fasteners. External Torx fasteners are also used mostly as engine fasteners and are labeled E instead of T, plus the size, such as E45.

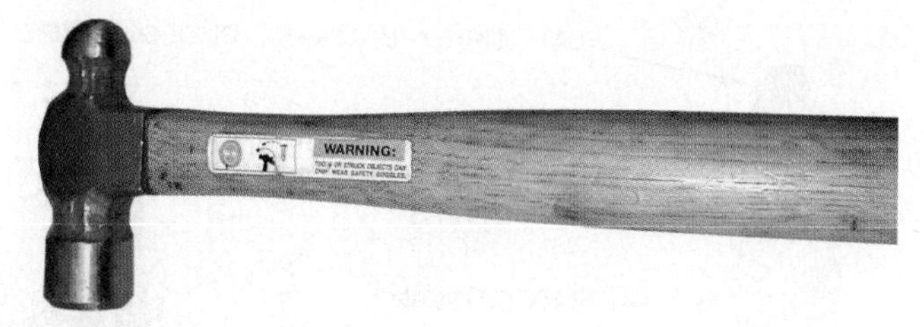

FIGURE 1–33 A typical ball-peen hammer.

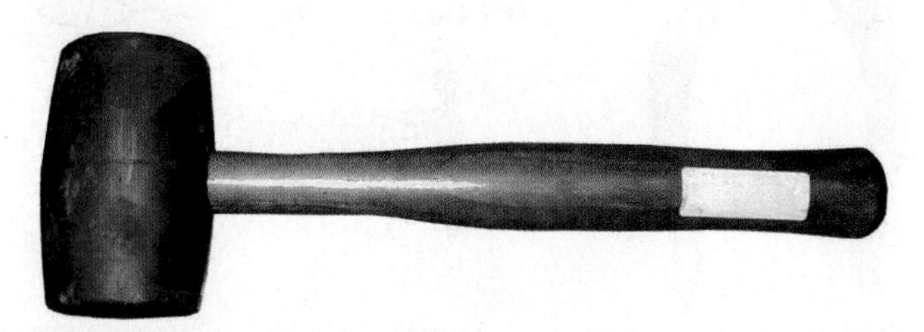

FIGURE 1–34 A rubber mallet used to deliver a force to an object without harming the surface.

FIGURE 1–35 A dead-blow hammer that was left outside in freezing weather. The plastic covering was damaged, which destroyed this hammer. The lead shot is encased in the metal housing and then covered.

NOTE: A claw hammer has a claw used to remove nails; therefore, it is not for automotive service.

A hammer is usually sized by the weight of the hammer's head and the length of the handle. For example, a commonly used ball-peen hammer has an 8 oz head and an 11 inch handle.

- **Mallets.** *Mallets* are a type of hammer with a large striking surface, which allows the technician to exert force over a larger area than a hammer, so as not to harm the part or component being hammered. Mallets are made from a variety of materials including rubber, plastic, or wood. ● **SEE FIGURE 1–34.**
- **Dead-blow hammer.** A shot-filled plastic hammer is called a *dead-blow hammer.* The small lead balls (shot) inside a plastic head prevent the hammer from bouncing off of the object when struck. ● **SEE FIGURE 1–35.**

SAFE USE OF HAMMERS AND MALLETS All mallets and hammers should be cleaned after use and not exposed to extreme temperatures. Never use a hammer or mallet that is damaged in any way and always use caution to avoid doing damage to the components and the surrounding area. Always follow the hammer manufacturer's recommended procedures and practices.

PLIERS

- **Slip-joint pliers. Pliers** are capable of holding, twisting, bending, and cutting objects and form an extremely useful classification of tools. The common household pliers are called the *slip-joint pliers.* There are two different positions where the junction of the handles meets to achieve a wide range of sizes of objects that can be gripped. ● **SEE FIGURE 1–36.**

TECH TIP

Pound with Something Softer

If you must pound on something, be sure to use a tool that is softer than what you are about to pound on to avoid damage. Examples are given in the following table.

The Material Being Pounded	What to Pound with
Steel or cast iron	Brass or aluminum hammer or punch
Aluminum	Plastic or rawhide mallet or plastic-covered dead-blow hammer
Plastic	Rawhide mallet or plastic dead-blow hammer

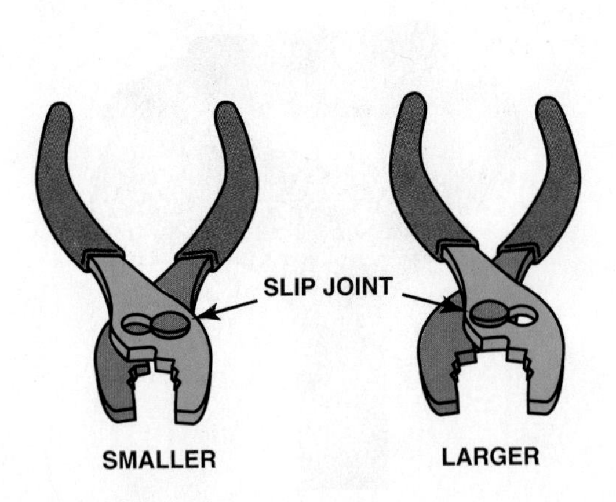

FIGURE 1–36 Typical slip-joint pliers are common household pliers. The slip joint allows the jaws to be opened to two different settings.

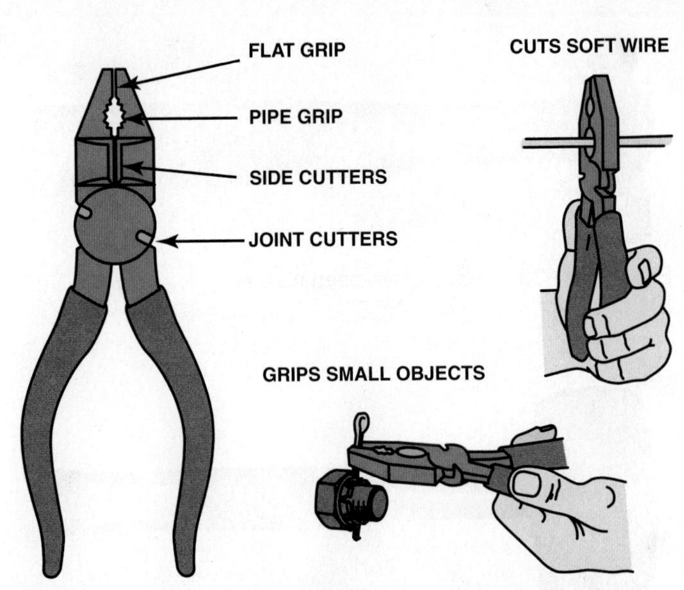

FIGURE 1–38 Linesman's pliers are very useful because they can help perform many automotive service jobs.

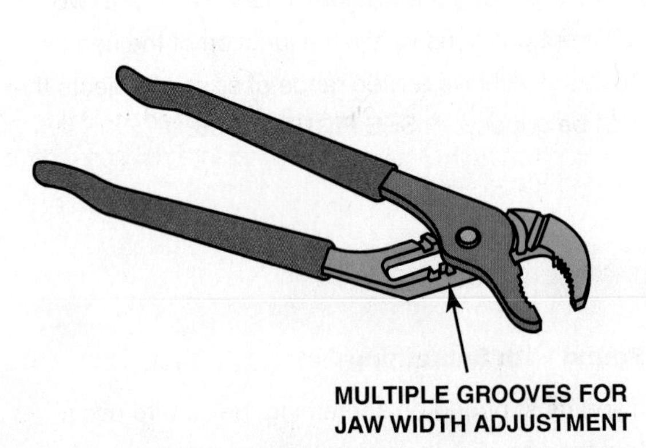

FIGURE 1–37 Multigroove adjustable pliers are known by many names, including the trade name Channel Locks®.

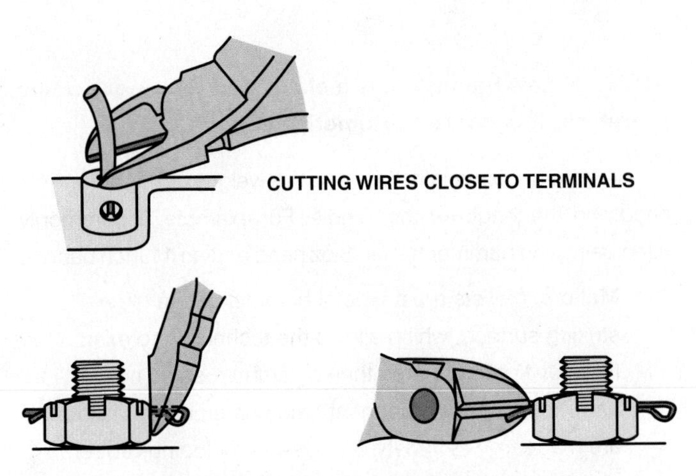

FIGURE 1–39 Diagonal pliers are another common tool that has many names.

- **Multigroove adjustable pliers.** For gripping larger objects, a set of *multigroove adjustable pliers* is a common choice of many service technicians. Originally designed to remove the various size nuts holding rope seals used in water pumps, the name *water pump pliers* is also used. ● **SEE FIGURE 1–37.**
- **Linesman's pliers.** *Linesman's pliers* are specifically designed for cutting, bending, and twisting wire. While commonly used by construction workers and electricians, linesman's pliers are a very useful tool for the service technician who deals with wiring. The center parts of the jaws are designed to grasp round objects, such as pipe or tubing, without slipping. ● **SEE FIGURE 1–38.**
- **Diagonal pliers.** *Diagonal pliers* are designed only for cutting. The cutting jaws are set at an angle to make it easier to cut wires. Diagonal pliers are also called *side cuts* or *dikes.* These pliers are made of hardened steel and they are used mostly for cutting wire. ● **SEE FIGURE 1–39.**
- **Needle-nose pliers.** *Needle-nose pliers* are designed to grip small objects or objects in tight locations. Needle-nose pliers have long, pointed jaws to allow the tips to reach into narrow openings or groups of small objects. ● **SEE FIGURE 1–40.**

Most needle-nose pliers have a wire cutter located at the base of the jaws near the pivot. There are several variations of needle-nose pliers, including right angle jaws or slightly angled jaws, to allow access to certain cramped areas.

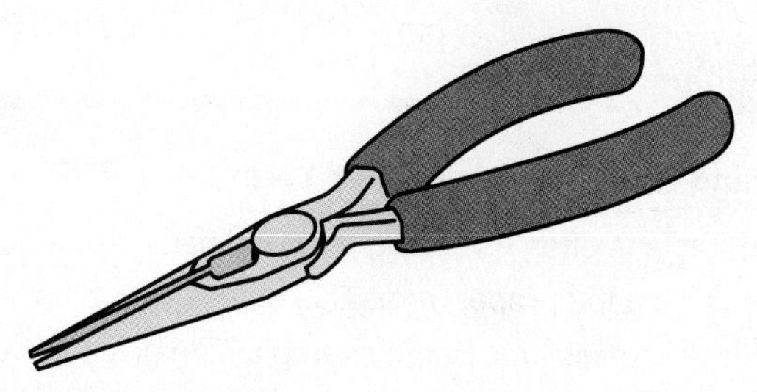

FIGURE 1–40 Needle-nose pliers are used where there is limited access to a wire or pin that needs to be installed or removed.

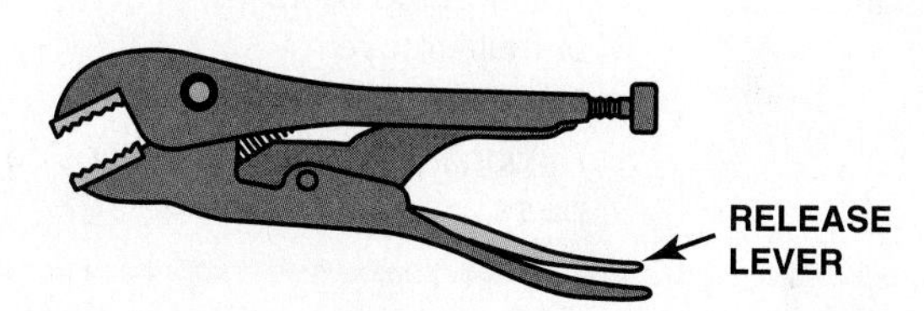

FIGURE 1–41 Locking pliers are best known by the trade name Vise-Grip®.

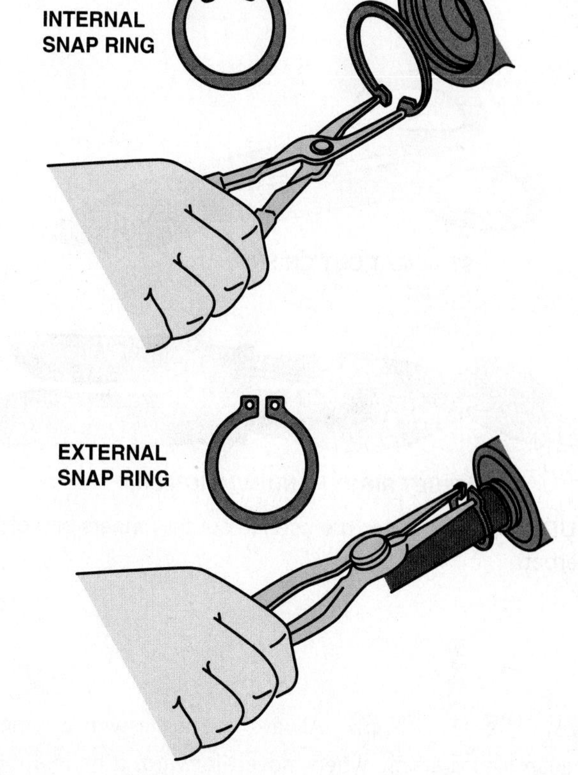

FIGURE 1–42 Snap-ring pliers are also called lock-ring pliers, and they are designed to remove internal and external snap rings (lock rings).

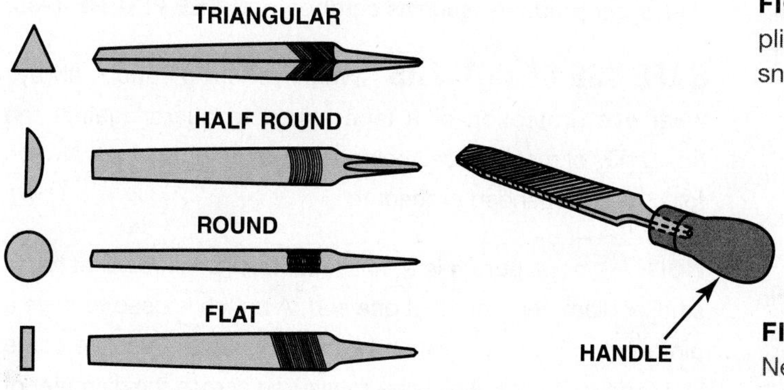

FIGURE 1–43 Files come in many different shapes and sizes. Never use a file without a handle.

- **Locking pliers.** *Locking pliers* are adjustable pliers that can be locked to hold objects from moving. Most locking pliers also have wire cutters built into the jaws near the pivot point. Locking pliers come in a variety of styles and sizes and are commonly referred to by the trade name Vise-Grip®. The size is the length of the pliers, not how far the jaws open. ● **SEE FIGURE 1–41.**
- **Snap-ring pliers.** *Snap-ring pliers* are used to remove and install snap rings. Many snap-ring pliers are designed to be able to remove and install both inward and outward expanding snap rings. Snap-ring pliers can be equipped with serrated-tipped jaws for grasping the opening in the snap ring, while others are equipped with points, which are inserted into the holes in the snap ring. ● **SEE FIGURE 1–42.**

SAFE USE OF PLIERS Pliers should not be used to remove any bolt or other fastener. Pliers should only be used when specified for use by the vehicle manufacturer.

FILES **Files** are used to smooth metal and are constructed of hardened steel with diagonal rows of teeth. Files are available with a single row of teeth called a *single cut file,* as well as two rows of teeth cut at an opposite angle called a *double cut file.* Files are available in a variety of shapes and sizes including small flat files, half-round files, and triangular files. ● **SEE FIGURE 1–43.**

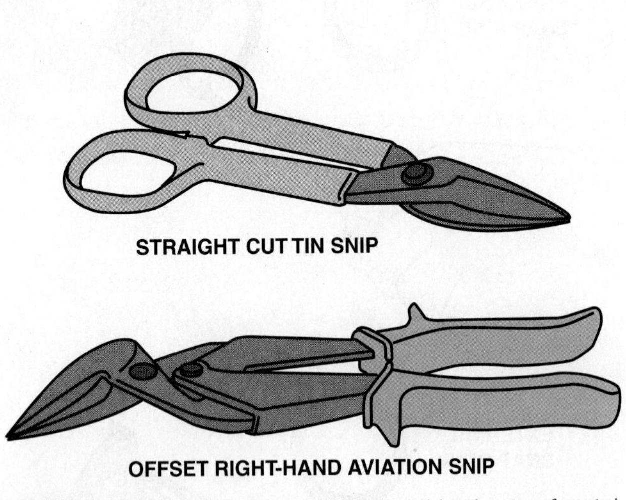

FIGURE 1–44 Tin snips are used to cut thin sheets of metal or carpet.

TECH TIP

Brand Name versus Proper Term

Technicians often use slang or brand names of tools rather than the proper term. This results in confusion for new technicians. Some examples are given in the following table.

Brand Name	Proper Term	Slang Name
Crescent wrench®	Adjustable wrench	Monkey wrench
Vise-Grip®	Locking pliers	Pump pliers
Channel Locks®	Water pump pliers or multigroove adjustable pliers	
	Diagonal cutting pliers	Dikes or side cuts

SAFE USE OF FILES Always use a file with a handle. Because files cut only when moved forward, a handle must be attached to prevent possible personal injury. After making a forward strike, lift the file and return to the starting position; avoid dragging the file backward.

SNIPS Service technicians are often asked to fabricate sheet metal brackets or heat shields and need to use one or more types of cutters available called **snips.** The simplest cutter is called *tin snips,* designed to make straight cuts in a variety of materials such as sheet steel, aluminum, or even fabric. A variation of the tin snips is called the *aviation tin snips.* There are three designs of aviation snips including one designed to cut straight (called a *straight cut aviation snip*), one designed to cut left (called an *offset left aviation snip*), and one designed to cut right (called an *offset right aviation snip*). The handles are color-coded for easy identification. These include yellow for straight, red for left, and green for right. ● **SEE FIGURE 1–44.**

UTILITY KNIFE A *utility knife* uses a replaceable blade and is used cut a variety of materials, such as carpet, plastic, wood, and paper products, such as cardboard. ● **SEE FIGURE 1–45.**

SAFE USE OF CUTTERS Whenever using cutters, always wear eye protection or a face shield to guard against the possibility of metal pieces being ejected during the cut. Always follow recommended procedures.

PUNCHES A **punch** is a small diameter steel rod that has a smaller diameter ground at one end. A punch is used to drive a pin out that is used to retain two components. Punches come in a variety of sizes, which are measured across the diameter of the machined end. Sizes include 1/16, 1/8, 3/16, and 1/4 inch ● **SEE FIGURE 1–46.**

CHISELS A **chisel** has a straight, sharp cutting end that is used for cutting off rivets or to separate two pieces of an assembly. The most common design of chisel used for automotive service work is called a *cold chisel.*

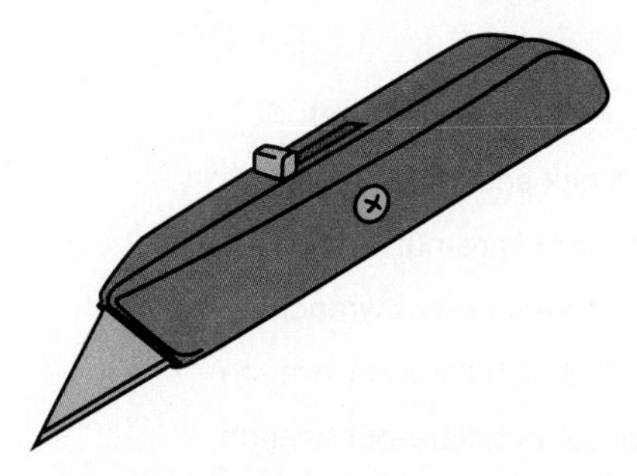

FIGURE 1–45 A utility knife uses replaceable blades and can cut carpet and other materials.

FIGURE 1–47 Warning stamped on the side of a punch that goggles should be worn when using this tool. Always follow safety warnings.

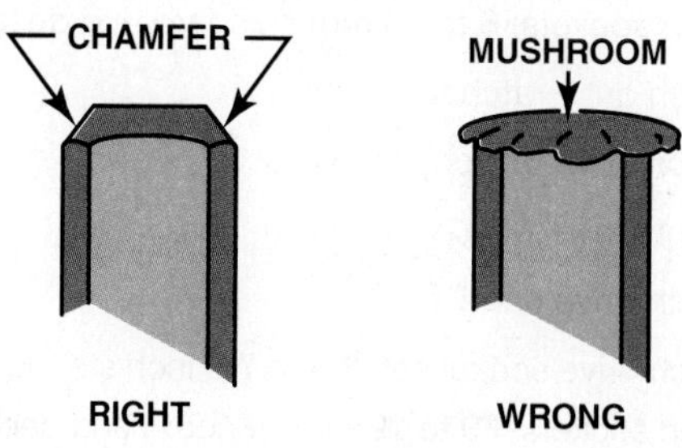

FIGURE 1–48 Use a grinder or a file to remove the mushroom material on the end of a punch or chisel.

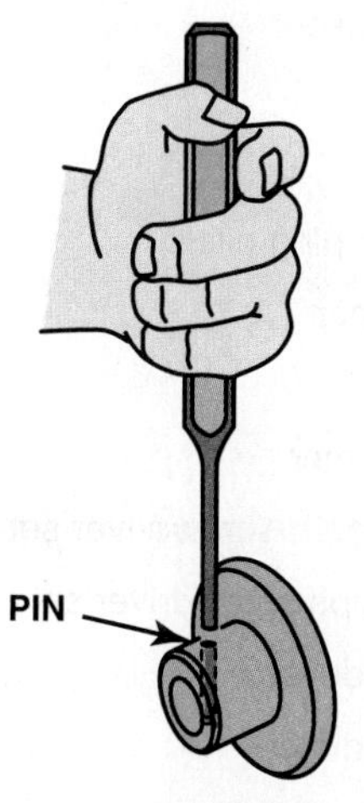

FIGURE 1–46 A punch is used to drive pins from assembled components. This type of punch is also called a pin punch.

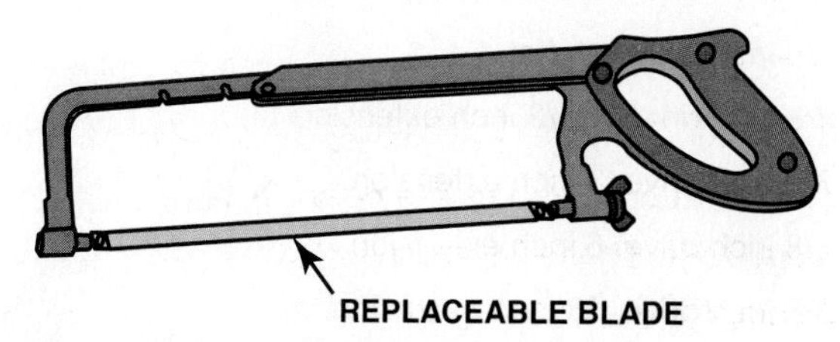

FIGURE 1–49 A typical hacksaw that is used to cut metal. If cutting sheet metal or thin objects, then use a blade with more teeth.

SAFE USE OF PUNCHES AND CHISELS Always wear eye protection when using a punch or a chisel because the hardened steel is brittle and parts of the punch could fly off and cause serious personal injury. See the warning stamped on the side of the automotive punch in ● **FIGURE 1–47**.

The tops of punches and chisels that become rounded off from use are referred to as being "mushroomed." This material must be ground off to help prevent the overhanging material from becoming loosened and airborne during use. ● **SEE FIGURE 1–48**.

HACKSAWS A **hacksaw** is used to cut metals such as steel, aluminum, brass, or copper. The cutting blade of a hacksaw is replaceable and the sharpness and number of teeth can be varied to meet the needs of the job. Use 14 or 18 teeth per inch (TPI) for cutting plaster or soft metals such as aluminum and copper. Use 24 or 32 TPI for steel or pipe. Hacksaw blades should be installed with the teeth pointing away from the handle. This means that a hacksaw only cuts while the blade is pushed in the forward direction. ● **SEE FIGURE 1–49**.

SAFE USE OF HACKSAWS Check that the hacksaw is equipped with the correct blade for the job and that the teeth are pointed away from the handle. When using a hacksaw, move the hacksaw slowly away from you, then lift slightly and return for another cut.

BASIC HAND TOOL LIST

The following is a list of hand tools every automotive technician should possess. Specialty tools are not included.

- Safety glasses
- Tool chest
- 1/4 inch drive socket set (1/4 to 9/16 inch standard and deep sockets; 6 to 15 mm standard and deep sockets)
- 1/4 inch drive ratchet
- 1/4 inch drive, 2 inch extension
- 1/4 inch drive, 6 inch extension
- 1/4 inch drive handle
- 3/8 inch drive socket set (3/8 to 7/8 inch standard and deep sockets; 10 to 19 mm standard and deep sockets)
- 3/8 inch drive Torx set (T40, T45, T50, and T55)
- 3/8 inch drive, 13/16 inch plug socket
- 3/8 inch drive, 5/8 inch plug socket
- 3/8 inch drive ratchet
- 3/8 inch drive, 1 1/2 inch extension
- 3/8 inch drive, 3 inch extension
- 3/8 inch drive, 6 inch extension
- 3/8 inch drive, 18 inch extension
- 3/8 inch drive universal
- 1/2 inch drive socket set (1/2 to 1 inch standard and deep sockets)
- 1/2 inch drive ratchet
- 1/2 inch drive breaker bar
- 1/2 inch drive, 5 inch extension
- 1/2 inch drive, 10 inch extension
- 3/8 to 1/4 inch adapter
- 1/2 to 3/8 inch adapter
- 3/8 to 1/2 inch adapter
- Crowfoot set (fractional inches)
- Crowfoot set (metric)
- 3/8 to 1 inch combination wrench set
- 10 to 19 mm combination wrench set
- 1/16 to 1/4 inch hex wrench set
- 2 to 12 mm hex wrench set
- 3/8 inch hex socket
- 13 to 14 mm flare-nut wrench
- 15 to 17 mm flare-nut wrench
- 5/16 to 3/8 inch flare-nut wrench
- 7/16 to 1/2 inch flare-nut wrench
- 1/2 to 9/16 inch flare-nut wrench
- Diagonal pliers
- Needle-nose pliers
- Adjustable-jaw pliers
- Locking pliers
- Snap-ring pliers
- Stripping or crimping pliers
- Ball-peen hammer
- Rubber hammer
- Dead-blow hammer
- Five-piece standard screwdriver set
- Four-piece Phillips screwdriver set
- #15 Torx screwdriver
- #20 Torx screwdriver
- Awl
- Mill file
- Center punch
- Pin punches (assorted sizes)
- Chisel
- Utility knife
- Valve core tool
- Filter wrench (large filters)
- Filter wrench (smaller filters)
- Test light
- Feeler gauge
- Scraper
- Pinch bar
- Magnet

FIGURE 1–50 A typical beginning technician tool set that includes the basic tools to get started.

FIGURE 1–51 A typical large tool box, showing just one of many drawers.

TOOL SETS AND ACCESSORIES

A beginning service technician may wish to start with a small set of tools before purchasing an expensive tool set. ● SEE FIGURES 1–50 AND 1–51.

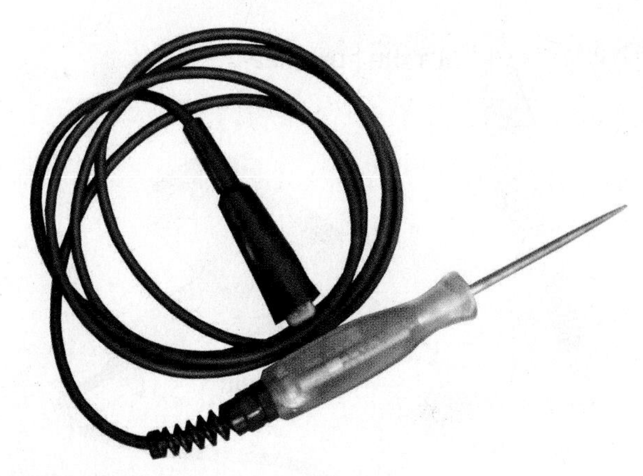

FIGURE 1–52 A typical 12 volt test light.

TECH TIP

Need to Borrow a Tool More Than Twice? Buy It!

Most service technicians agree that it is okay for a beginning technician to borrow a tool occasionally. However, if a tool has to be borrowed more than twice, then be sure to purchase it as soon as possible. Also, whenever you borrow a tool, be sure that you clean the tool and let the technician you borrowed the tool from know that you are returning it. These actions will help in any future dealings with other technicians.

ELECTRICAL WORK HAND TOOLS

TEST LIGHT A test light is used to test for electricity. A typical automotive test light consists of a clear plastic screwdriver-like handle that contains a lightbulb. A wire is attached to one terminal of the bulb, which the technician connects to a clean metal part of the vehicle. The other end of the bulb is attached to a point that can be used to test for electricity at a connector or wire. When there is power at the point and a good connection at the other end, the lightbulb lights. ● SEE FIGURE 1–52.

ELECTRIC SOLDERING GUNS This type of soldering gun is usually powered by 110 volt AC and often has two power settings expressed in watts. A typical electric soldering gun will produce from 85 to 300 watts of heat at the tip, which is more than adequate for soldering.

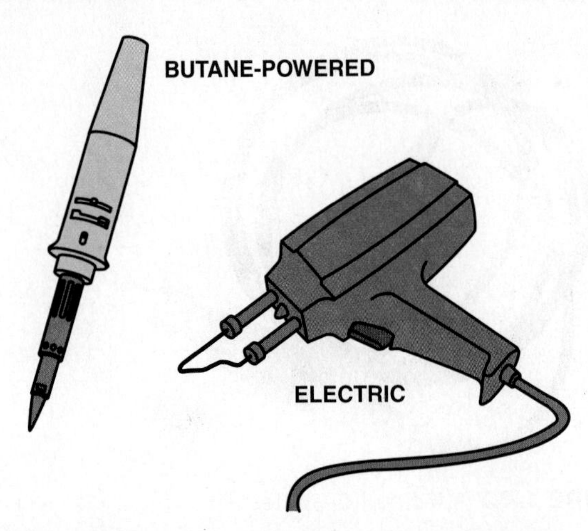

FIGURE 1–53 Electric and butane-powered soldering guns used to make electrical repairs. Soldering guns are sold by the wattage rating: The higher the wattage, the greater the amount of heat created. Most solder guns used for automotive electrical work usually fall within the 60 to 160 watt range.

- **Electric soldering pencil.** This type of soldering iron is less expensive and creates less heat than an electric soldering gun. A typical electric soldering pencil (iron) creates 30 to 60 watts of heat and is suitable for soldering smaller wires and connections.
- **Butane-powered soldering iron.** A butane-powered soldering iron is portable and very useful for automotive service work because an electrical cord is not needed. Most butane-powered soldering irons produce about 60 watts of heat, which is enough for most automotive soldering. ● **SEE FIGURE 1–53.**

ELECTRICAL WORK HAND TOOLS In addition to a soldering iron, most service technicians who do electrical-related work should have the following tools:

- Wire cutters
- Wire strippers
- Wire crimpers
- Heat gun for heat shrink tubing

DIGITAL METER A digital meter is a necessary tool for electrical diagnosis and troubleshooting. A digital multimeter, abbreviated DMM, is usually capable of measuring the following units of electricity.

- DC volts
- AC volts
- Ohms
- Amperes

? FREQUENTLY ASKED QUESTION

What Is an SST?

Vehicle manufacturers often specify a **special service tool (SST)** to properly disassemble and assemble components such as transmissions. These tools are also called special tools and are available from the vehicle manufacturer or its tool supplier, such as Kent-Moore or Miller Tools. Many service technicians do not have access to special service tools so they use generic versions that are available from aftermarket sources.

HAND TOOL MAINTENANCE

Most hand tools are constructed of rust-resistant metals but they can still rust or corrode if not properly maintained. For best results and long tool life, the following steps should be taken.

- Clean each tool before placing it back into the tool box.
- Keep tools separated. Moisture on metal tools will start to rust more readily if the tools are in contact with another metal tool.
- Line the drawers of the tool box with a material that will prevent the tools from moving as the drawers are opened and closed. This helps to quickly locate the proper tool and size.
- Release the tension on all clicker-type torque wrenches.
- Keep the tool box secure.

TROUBLE LIGHTS

INCANDESCENT *Incandescent lights* use a filament that produces light when electric current flows through the bulb. This was the standard **trouble light,** also called a *work light,* for many years until safety issues caused most shops to switch to safer fluorescent or LED lights. If incandescent lightbulbs are used, try to locate bulbs that are rated "rough service," which is designed to withstand shock and vibration more than conventional lightbulbs.

FLUORESCENT A trouble light is an essential piece of shop equipment, and for safety, should be fluorescent rather than incandescent. Incandescent lightbulbs can scatter or break if

FIGURE 1–54 A fluorescent trouble light operates cooler and is safer to use in the shop because it is protected against accidental breakage where gasoline or other flammable liquids would happen to come in contact with the light.

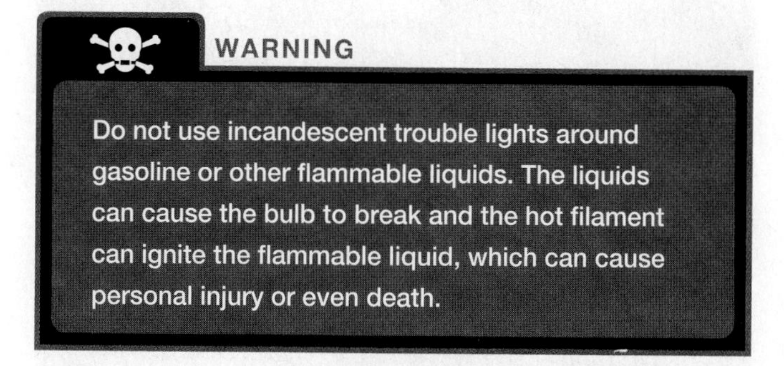

WARNING

Do not use incandescent trouble lights around gasoline or other flammable liquids. The liquids can cause the bulb to break and the hot filament can ignite the flammable liquid, which can cause personal injury or even death.

gasoline were to be splashed onto the bulb, creating a serious fire hazard. Fluorescent light tubes are not as likely to be broken and are usually protected by a clear plastic enclosure. Trouble lights are usually attached to a retractor, which can hold 20 to 50 feet of electrical cord. ● **SEE FIGURE 1–54.**

LED TROUBLE LIGHT **Light-emitting diode (LED)** trouble lights are excellent to use because they are shock resistant, long lasting, and do not represent a fire hazard. Some trouble lights are battery powered and therefore can be used in places where an attached electrical cord could present problems.

FIGURE 1–55 A typical 1/2 inch drive air impact wrench. The direction of rotation can be changed to loosen or tighten a fastener.

FIGURE 1–56 A typical battery-powered 3/8 inch drive impact wrench.

AIR AND ELECTRICALLY OPERATED TOOLS

IMPACT WRENCH An impact wrench, either air or electrically powered, is used to remove and install fasteners. The air-operated 1/2 inch drive impact wrench is the most commonly used unit. ● **SEE FIGURE 1–55.**

Electrically powered impact wrenches commonly include:

- Battery-powered units. ● **SEE FIGURE 1–56.**
- 110 volt AC-powered units. This type of impact is very useful, especially if compressed air is not readily available.

FIGURE 1–57 A black impact socket. Always use an impact-type socket whenever using an impact wrench to avoid the possibility of shattering the socket, which could cause personal injury.

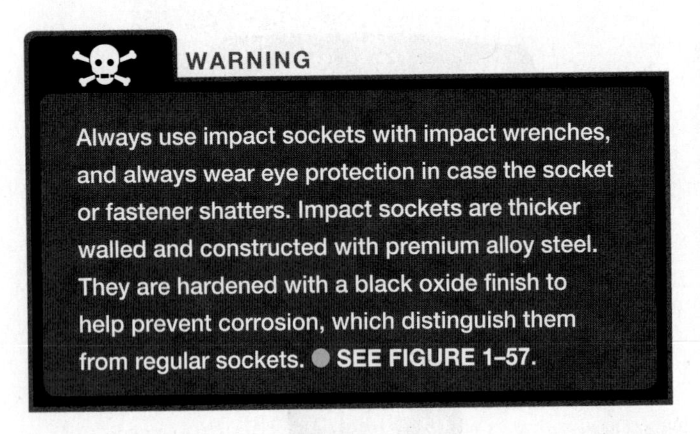

WARNING

Always use impact sockets with impact wrenches, and always wear eye protection in case the socket or fastener shatters. Impact sockets are thicker walled and constructed with premium alloy steel. They are hardened with a black oxide finish to help prevent corrosion, which distinguish them from regular sockets. ● **SEE FIGURE 1–57.**

AIR RATCHET An air ratchet is used to remove and install fasteners that would normally be removed or installed using a ratchet and a socket. ● **SEE FIGURE 1–58.**

DIE GRINDER A die grinder is a commonly used air-powered tool which can also be used to sand or remove gaskets and rust. ● **SEE FIGURE 1–59.**

BENCH-MOUNTED OR PEDESTAL-MOUNTED GRINDER These high-powered grinders are equipped with a wire brush wheel and/or a stone wheel.

- **Wire brush wheel**—This type is used to clean threads of bolts and to remove gaskets from sheet metal engine parts.
- **Stone wheel**—This type is used to grind metal and to remove the mushroom from the top of punches or chisels. ● **SEE FIGURE 1–60.**

FIGURE 1–58 An air ratchet is a very useful tool that allows fast removal and installation of fasteners, especially in areas that are difficult to reach or do not have room enough to move a hand ratchet or wrench.

FIGURE 1–59 This typical die grinder surface preparation kit includes the air-operated die grinder and a variety of sanding disks for smoothing surfaces or removing rust.

FIGURE 1–60 A typical pedestal grinder with a wire wheel on the left side and a stone wheel on the right side. Even though this machine is equipped with guards, safety glasses or a face shield should always be worn whenever using a grinder or wire wheel.

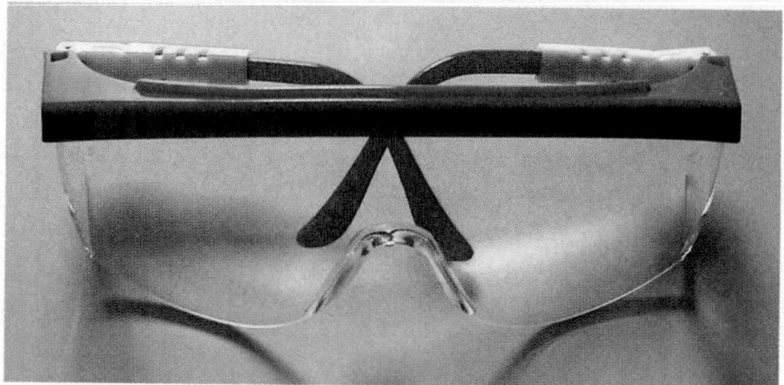

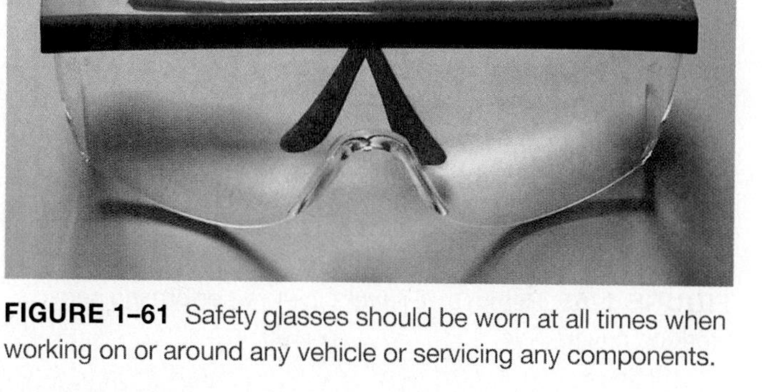

FIGURE 1–61 Safety glasses should be worn at all times when working on or around any vehicle or servicing any components.

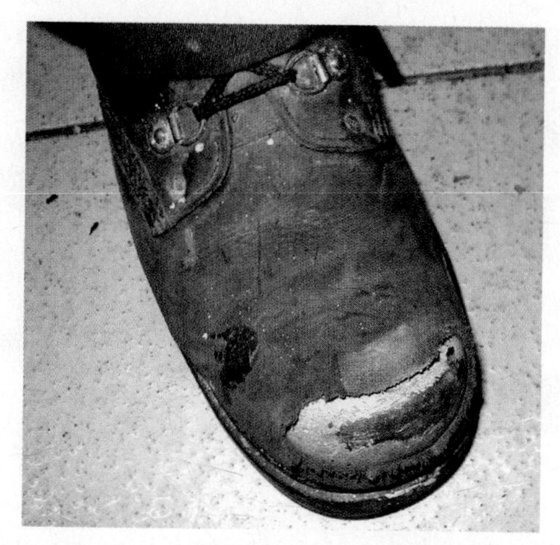

FIGURE 1–62 Steel-toed shoes are a worthwhile investment to help prevent foot injury due to falling objects. Even these well-worn shoes can protect the feet of this service technician.

> **WARNING**
>
> Always wear a face shield when using a wire wheel or a grinder.

Most **bench grinders** are equipped with a grinder wheel (stone) on one end or the other of a wire brush. A bench grinder is a useful piece of shop equipment and the wire wheel end can be used for the following:

- Cleaning threads of bolts
- Cleaning gaskets from sheet metal parts, such as steel valve covers

CAUTION: Use a steel wire brush only on steel or iron components. If a steel wire brush is used on aluminum or copper-based metal parts, it can remove metal from the part.

The grinding stone end of the bench grinder can be used for the following purposes:

- Sharpening blades and drill bits
- Grinding off the heads of rivets or parts
- Sharpening sheet metal parts for custom fitting

PERSONAL PROTECTIVE EQUIPMENT

Service technicians should wear protective devices to prevent personal injury. The **personal protective** devices include the following equipment.

SAFETY GLASSES Be sure that safety glasses meet standard ANSI Z87.1. They should be worn at all times while servicing any vehicle. ● **SEE FIGURE 1–61**.

STEEL-TOED SAFETY SHOES Steel-toed safety shoes help prevent foot injury due to falling objects. ● **SEE FIGURE 1–62**. If safety shoes are not available, then leather-topped shoes offer more protection than canvas or cloth covered shoes.

BUMP CAP Service technicians working under a vehicle should wear a **bump cap** to protect the head against under-vehicle objects and pads of the lift. ● **SEE FIGURE 1–63**.

HEARING PROTECTION Hearing protection should be worn if the sound around you requires that you raise your voice (sound level higher than 90 dB). For example, a typical lawnmower produces noise at a level of about 110 dB. This means that everyone who uses a lawnmower or other lawn or garden equipment should wear ear protection.

GLOVES Many technicians wear gloves not only to help keep their hands clean but also to help protect their skin from the effects of dirty engine oil and other possibly hazardous materials.

FIGURE 1–63 One version of a bump cap is a molded plastic insert worn inside a regular cloth cap.

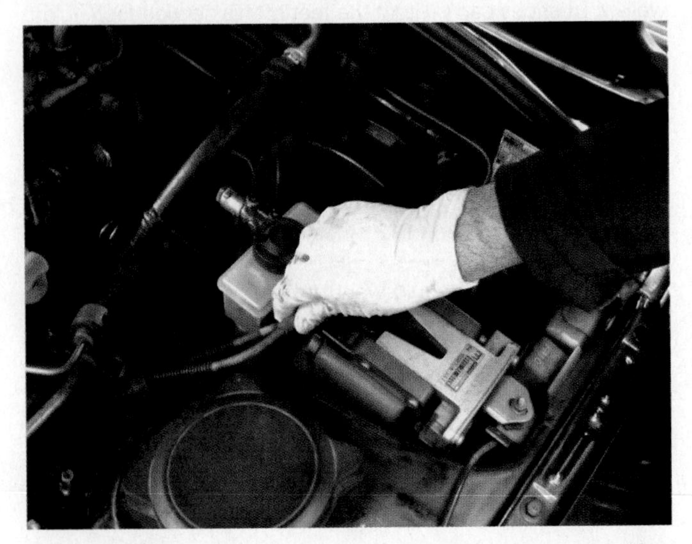

FIGURE 1–64 Protective gloves are available in several sizes and materials.

Several types of gloves and their characteristics are as follows:

- **Latex surgical gloves.** These gloves are relatively inexpensive, but tend to stretch, swell, and weaken when exposed to gas, oil, or solvents.
- **Vinyl gloves.** These gloves are also inexpensive and are not affected by gas, oil, or solvents.
- **Polyurethane gloves.** Polyurethane gloves are more expensive, yet strong. These gloves are not affected by gas, oil, or solvents, but they tend to be slippery.
- **Nitrile gloves.** These gloves are exactly like latex gloves, but are not affected by gas, oil, or solvents, yet they tend to be expensive.
- **Mechanic's gloves.** These gloves are usually made of synthetic leather and spandex and provide thermo protection, as well as protection from dirt and grime.

● **SEE FIGURE 1–64.**

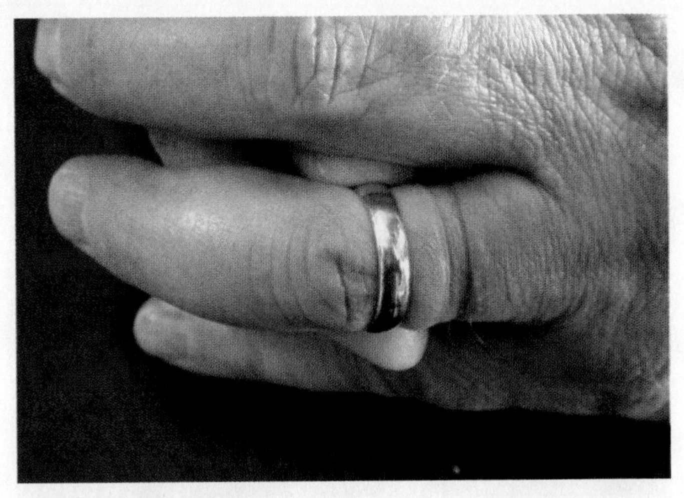

FIGURE 1–65 Remove all jewelry before performing service work on any vehicle.

SAFETY PRECAUTIONS

Besides wearing personal safety equipment, the following actions should be performed to keep safe in the shop.

- Remove jewelry that may get caught on something or act as a conductor to an exposed electrical circuit. ● **SEE FIGURE 1–65.**
- Take care of your hands. Keep your hands clean by washing with soap and hot water that is at least 110°F (43°C).
- Tie back long hair to keep from getting it caught in moving components.
- Avoid loose or dangling clothing.
- When lifting any object, get a secure grip with solid footing. Keep the load close to your body to minimize the strain. Lift with your legs and arms, not your back.
- Do not twist your body when carrying a load. Instead, pivot your feet to help prevent strain on the spine.
- Ask for help when moving or lifting heavy objects.
- Push a heavy object rather than pull it. (This is opposite to the way you should work with tools–never push a wrench! If you do and a bolt or nut loosens, your entire weight is used to propel your hand(s) forward. This usually results in cuts, bruises, or other painful injury.)
- Always connect an exhaust hose to the tailpipe of any running vehicle to help prevent the buildup of carbon monoxide inside a closed garage space. ● **SEE FIGURE 1–66.**
- When standing, keep your objects, parts, and tools between chest height and waist height. If seated, work at tasks that are at elbow height.
- Always be sure the hood is securely held open.

FIGURE 1–66 Always connect an exhaust hose to the tailpipe of a vehicle to be run inside a building.

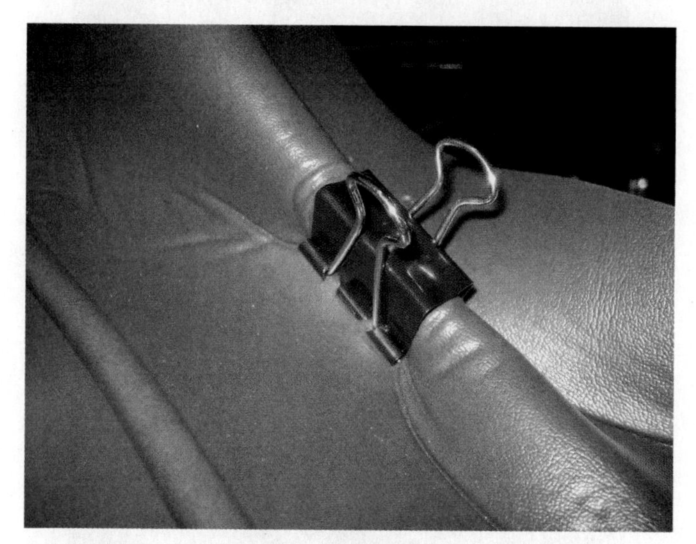

FIGURE 1–67 A binder clip keeps a fender cover from falling off.

FIGURE 1–68 Covering the interior as soon as the vehicle comes in for service helps improve customer satisfaction.

SAFETY TIP

Shop Cloth Disposal

Always dispose of oily shop cloths in an enclosed container to prevent a fire. ● **SEE FIGURE 1–69.** Whenever oily cloths are thrown together on the floor or workbench, a chemical reaction can occur, which can ignite the cloth even without an open flame. This process of ignition without an open flame is called **spontaneous combustion.**

VEHICLE PROTECTION

FENDER COVERS Whenever working under the hood of any vehicle be sure to use fender covers. They not only help protect the vehicle from possible damage but also provide a clean surface to place parts and tools. The major problem with using fender covers is that they tend to move and often fall off the vehicle. To help prevent the fender covers from falling off, secure them to a lip of the fender using a *binder clip* available at most office supply stores. ● **SEE FIGURE 1–67.**

INTERIOR PROTECTION Always protect the interior of the vehicle from accidental damage or dirt and grease by covering the seat, steering wheel, and floor with a protective covering. ● **SEE FIGURE 1–68.**

FIGURE 1–69 All oily shop cloths should be stored in a metal container equipped with a lid to help prevent spontaneous combustion.

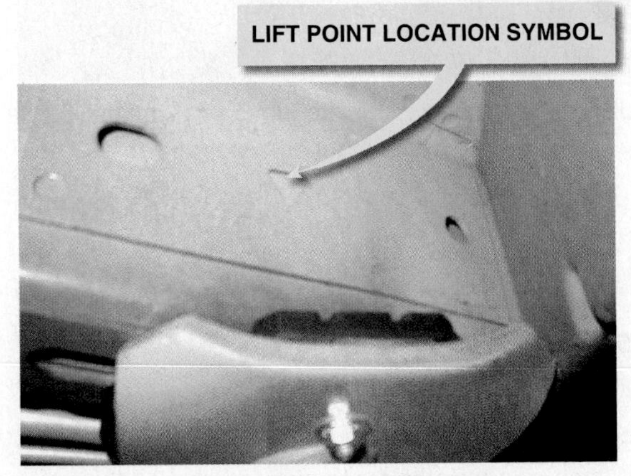

FIGURE 1–70 Most newer vehicles have a triangle symbol indicating the recommended hoisting lift location.

SAFETY IN LIFTING (HOISTING) A VEHICLE

Many chassis and underbody service procedures require that the vehicle be hoisted or lifted off the ground. The simplest methods involve the use of drive-on ramps or a floor jack and safety (jack) stands, whereas in-ground or surface-mounted lifts provide greater access.

Setting the pads is a critical part of this hoisting procedure. Owner's, shop, and service manuals include recommended locations to be used when hoisting (lifting) a vehicle. Newer vehicles have a triangle decal on the driver's door indicating the recommended lift points. The recommended standards for the lift points and lifting procedures are found in SAE standard JRP-2184. ● **SEE FIGURE 1–70**.

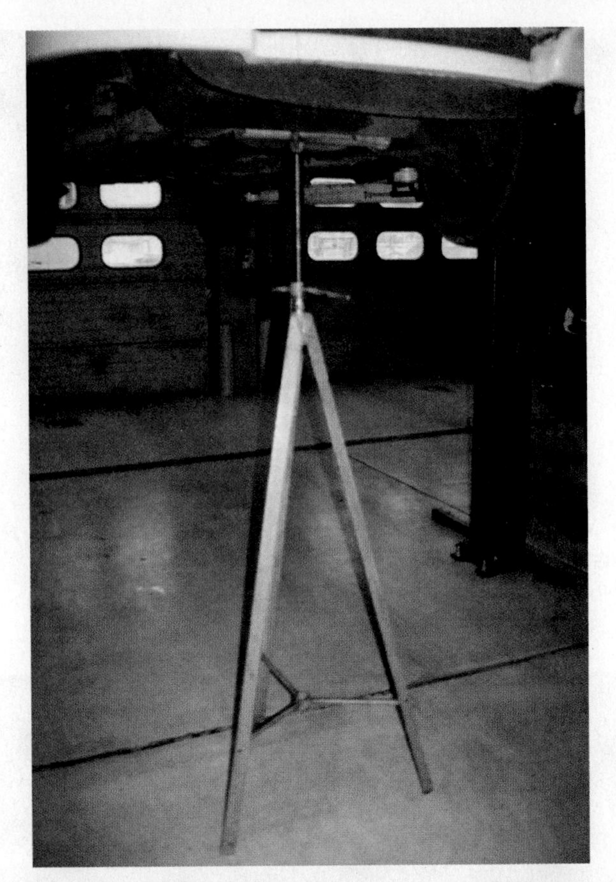

(a)

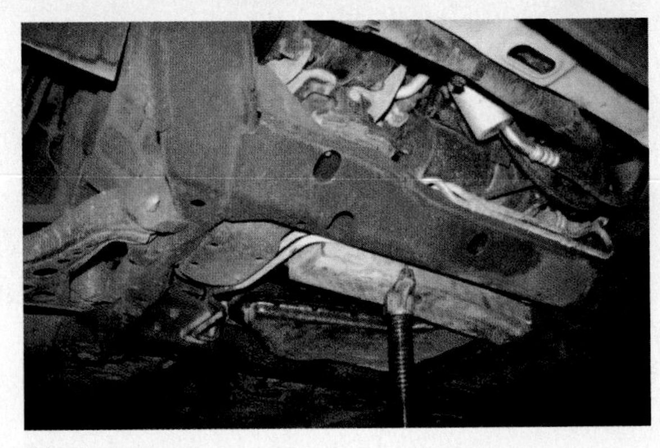

(b)

FIGURE 1–71 (a) Tall safety stands can be used to provide additional support for the vehicle while on the hoist. (b) A block of wood should be used to avoid the possibility of doing damage to components supported by the stand.

These recommendations typically include the following points.

1. The vehicle should be centered on the lift or hoist so as not to overload one side or put too much force either forward or rearward. ● **SEE FIGURE 1–71**.
2. The pads of the lift should be spread as far apart as possible to provide a stable platform.

FIGURE 1–72 This training vehicle fell from the hoist because the pads were not set correctly. No one was hurt, but the vehicle was damaged.

(a)

(b)

FIGURE 1–73 (a) An assortment of hoist pad adaptors that are often needed to safely hoist many pickup trucks, vans, and sport utility vehicles (SUVs). (b) A view from underneath a Chevrolet pickup truck showing how the pad extensions are used to attach the hoist lifting pad to contact the frame.

3. Each pad should be placed under a portion of the vehicle that is strong and capable of supporting the weight of the vehicle.
   a. Pinch welds at the bottom edge of the body are generally considered to be strong.

CAUTION: Even though pinch weld seams are the recommended location for hoisting many vehicles with unitized bodies (unit-body), care should be taken not to place the pad(s) too far forward or rearward. Incorrect placement of the vehicle on the lift could cause the vehicle to be imbalanced, and the vehicle could fall. This is exactly what happened to the vehicle in ● FIGURE 1–72.

   b. Boxed areas of the body are the best places to position the pads on a vehicle without a frame. Be careful to note whether the arms of the lift might come into contact with other parts of the vehicle before the pad touches the intended location. Commonly damaged areas include the following:
      (1) Rocker panel moldings
      (2) Exhaust system (including catalytic converter)
      (3) Tires or body panels (● **SEE FIGURES 1–73 AND 1–74.**)

(a)

(b)

FIGURE 1–74 (a) The pad arm is just contacting the rocker panel of the vehicle. (b) The pad arm has dented the rocker panel on this vehicle because the pad was set too far inward underneath the vehicle.

4. The vehicle should be raised about 1 foot (30 centimeters [cm]) off the floor, then stopped and shaken to check for stability. If the vehicle seems to be stable when checked at a short height from the floor, continue raising it with attention until it has reached the desired height. The hoist should be lowered onto the mechanical locks, and then raised off of the locks before lowering.

CAUTION: Do not look away from the vehicle while it is being raised (or lowered) on a hoist. Often one side or one end of the hoist can stop or fail, resulting in the vehicle being slanted enough to slip or fall, creating physical damage not only to the vehicle and/or hoist but also to the technician or others who may be nearby.

NOTE: Most hoists can be safely placed at any desired height. For ease while working, the area where you are working should be at chest level. When working on brakes or suspension components, it is not necessary to work on them down near the floor or over your head. Raise the hoist so that the components are at chest level.

5. Before lowering the hoist, you must release the safety latch(es) and reverse the direction of the controls. The speed downward is often adjusted to be as slow as possible for additional safety.

FLOOR JACKS

DESCRIPTION Floor jacks are used to lift one side or end of a vehicle. They are portable and relatively inexpensive and must be used with safety (jack) stands.

OPERATING PRINCIPLES A floor jack uses a hydraulic cylinder to raise a vehicle. ● **SEE FIGURE 1–75.**

A jack operates as follows:

- When the jack handle is twisted clockwise, the release valve is closed.
- When the jack handle is moved upward, hydraulic oil is drawn from the reservoir into the pump assembly.
- When the jack handle is moved downward, the oil is forced into the hydraulic cylinder, which forces the ram out and the lifting pad upward.
- When the cylinder ram reaches its maximum height, a bypass valve opens, which directs the oil back into the reservoir.
- When the jack handle is twisted counterclockwise, the release valve opens and allows the oil to flow back into the reservoir.

CAUTION: The valve must be closed to allow the jack handle to remain in the upright position. If the release valve is opened, the jack handle will drop toward the floor.

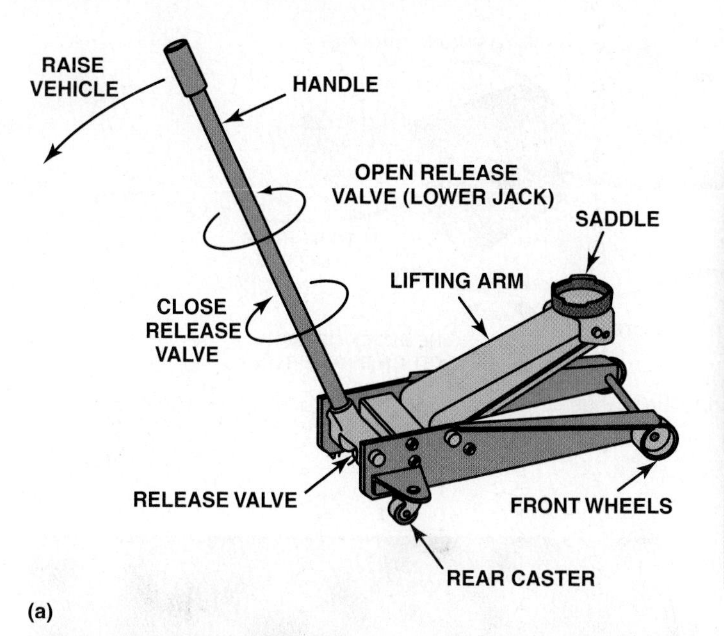

FIGURE 1–75 (a) A typical 3 ton (6,000 lb) capacity hydraulic jack. (b) Whenever a vehicle is raised off the ground, a safety stand should be placed under the frame, axle, or body to support the weight of the vehicle.

SAFE USE OF A FLOOR JACK

To safely use a floor jack, perform the following steps.

STEP 1 Read, understand, and follow all operating and safety items listed in the instructions.

STEP 2 Be sure the vehicle is on a flat, level, and hard surface.

STEP 3 Chock (block) the wheels of the vehicle to prevent it from moving during the lifting operation.

STEP 4 Check vehicle service information to determine the specified lifting point under the vehicle.

STEP 5 Place the lifting pad of the jack under the specified lifting point.

STEP 6 Close the release valve of the jack by rotating the jack handle clockwise. Move the jack handle downward until the lifting pad contacts the vehicle lifting point. Double-check that the jack is located in the specified location.

STEP 7 Continue to move the jack handle downward, and then up and down again until the vehicle has been raised to the desired height.

STEP 8 Place safety (jack) stand(s) under the vehicle.

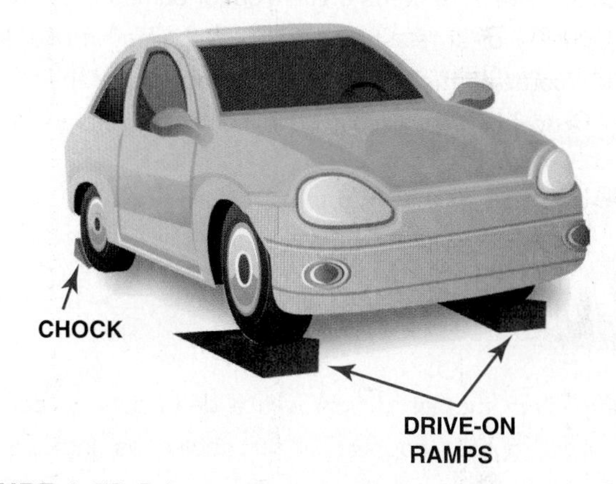

FIGURE 1–76 Drive-on ramps are dangerous to use. The wheels on the ground level must be chocked (blocked) to prevent accidental movement down the ramp.

STEP 9 To lower the vehicle, raise the vehicle just enough to remove the safety stands and then rotate the jack handle *slowly* counterclockwise.

DRIVE-ON RAMPS Ramps are an inexpensive way to raise the front or rear of a vehicle. ● **SEE FIGURE 1–76.** Ramps are easy to store, but may be dangerous because they can "kick out" when driving the vehicle onto the ramps.

CAUTION: Professional repair shops do not use ramps because they are dangerous. Use ramps only with extreme care.

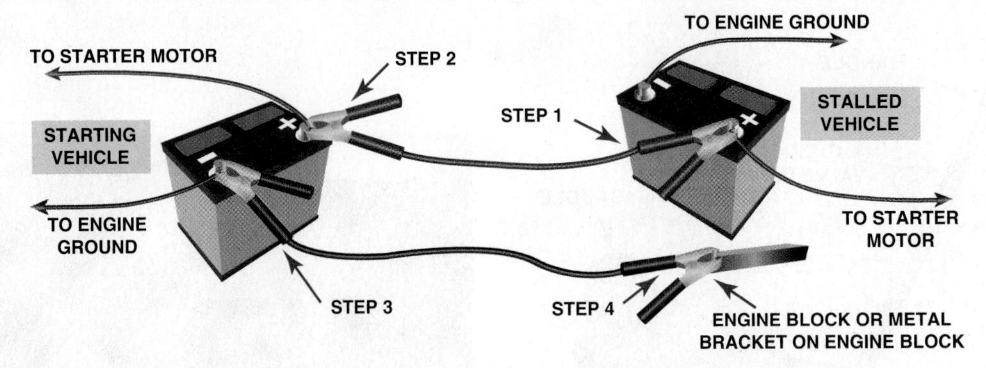

FIGURE 1–77 Jumper cable usage guide. Follow the same connections if using a portable jump box.

ELECTRICAL CORD SAFETY

Use correctly grounded three-prong sockets and extension cords to operate power tools. Some tools use only two-prong plugs. Make sure these are double insulated and repair or replace any electrical cords that are cut or damaged to prevent the possibility of an electrical shock. When not in use, keep electrical cords off the floor to prevent tripping over them. Tape the cords down if they are placed in high foot traffic areas.

JUMP STARTING AND BATTERY SAFETY

To jump start another vehicle with a dead battery, connect either good-quality copper jumper cables as indicated in ● **FIGURE 1–77** or a jump box. The last connection made should always be on the engine block or an engine bracket on the dead vehicle as far from the battery as possible. It is normal for a spark to be created when the jumper cables finally complete the jumper cable connections, and this spark could cause an explosion of the gases around the battery. Many newer vehicles have special ground connections built away from the battery just for the purpose of jump starting. Check the owner's manual or service information for the exact location.

Batteries contain acid and should be handled with care to avoid tipping them greater than a 45-degree angle. Always remove jewelry when working around a battery to avoid the possibility of electrical shock or burns, which can occur when the metal comes in contact with a 12 volt circuit and ground, such as the body of the vehicle.

FIGURE 1–78 The air pressure going to the nozzle should be reduced to 30 psi or less to help prevent personal injury.

SAFETY TIP

Air Hose Safety

Improper use of an air nozzle can cause blindness or deafness. Compressed air must be reduced to less than 30 psi (206 kPa). ● **SEE FIGURE 1–78.** If an air nozzle is used to dry and clean parts, make sure the air stream is directed away from anyone else in the immediate area. Always use an OSHA-approved nozzle with side slits that limit the maximum pressure at the nozzle to 30 psi. Coil and store air hoses when they are not in use.

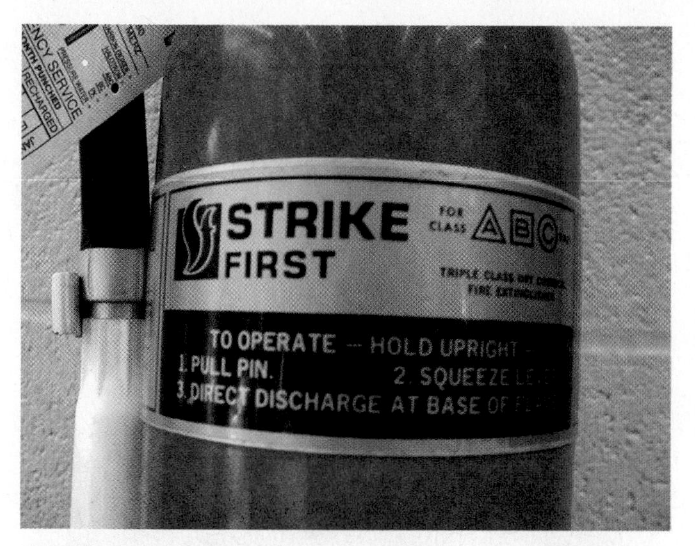

FIGURE 1–79 A typical fire extinguisher designed to be used on class A, B, or C fires.

FIGURE 1–80 A CO_2 fire extinguisher being used on a fire set in an open drum during a demonstration at a fire training center.

FIRE EXTINGUISHERS

There are four **fire extinguisher classes.** Each class should be used on specific fires only.

- **Class A** is designed for use on general combustibles, such as cloth, paper, and wood.
- **Class B** is designed for use on flammable liquids and greases, including gasoline, oil, thinners, and solvents.
- **Class C** is used only on electrical fires.
- **Class D** is effective only on combustible metals such as powdered aluminum, sodium, or magnesium.

The class rating is clearly marked on the side of every fire extinguisher. Many extinguishers are good for multiple types of fires. ● **SEE FIGURE 1–79.**

When using a fire extinguisher, remember the word "PASS."

P = Pull the safety pin.

A = Aim the nozzle of the extinguisher at the base of the fire.

S = Squeeze the lever to actuate the extinguisher.

S = Sweep the nozzle from side to side.

● **SEE FIGURE 1–80.**

TYPES OF FIRE EXTINGUISHERS Types of fire extinguishers include the following:

- **Water.** A water fire extinguisher, usually in a pressurized container, is good to use on Class A fires by reducing the temperature to the point where a fire cannot be sustained.
- **Carbon dioxide (CO_2).** A carbon dioxide fire extinguisher is good for almost any type of fire, especially Class B or Class C materials. A CO_2 fire extinguisher works by removing the oxygen from the fire and the cold CO_2 also helps reduce the temperature of the fire.
- **Dry chemical (yellow).** A dry chemical fire extinguisher is good for Class A, B, or C fires by coating the flammable materials, which eliminates the oxygen from the fire. A dry chemical fire extinguisher tends to be very corrosive and will cause damage to electronic devices.

FIGURE 1–81 A treated wool blanket is kept in an easy-to-open, wall-mounted holder and should be placed in a central location in the shop.

FIRE BLANKETS

Fire blankets are required to be available in the shop areas. If a person is on fire, a fire blanket should be removed from its storage bag and thrown over and around the victim to smother the fire. ● **SEE FIGURE 1–81** showing a typical fire blanket.

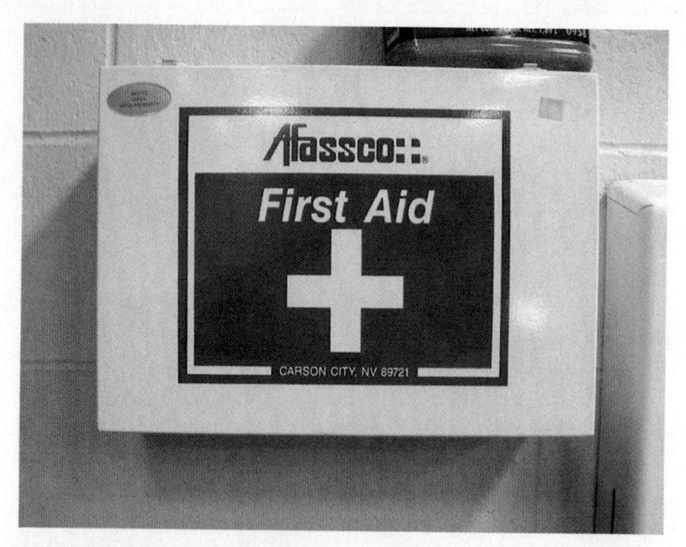

FIGURE 1–82 A first aid box should be centrally located in the shop and kept stocked with the recommended supplies.

FIRST AID AND EYE WASH STATIONS

All shop areas must be equipped with a first aid kit and an eye wash station that are centrally located and kept stocked with emergency supplies. ● **SEE FIGURE 1–82.**

FIRST AID KIT A first aid kit should include:

- Bandages (variety)
- Gauze pads
- Roll gauze
- Iodine swab sticks
- Antibiotic ointment
- Hydrocortisone cream
- Burn gel packets
- Eye wash solution
- Scissors
- Tweezers
- Gloves
- First aid guide

Every shop should have a person trained in first aid. If there is an accident, call for help immediately.

EYE WASH STATION An **eye wash station** should be centrally located and used whenever any liquid or chemical gets into the eyes. If such an emergency does occur, keep eyes in a constant stream of water and call for professional assistance. ● **SEE FIGURE 1–83.**

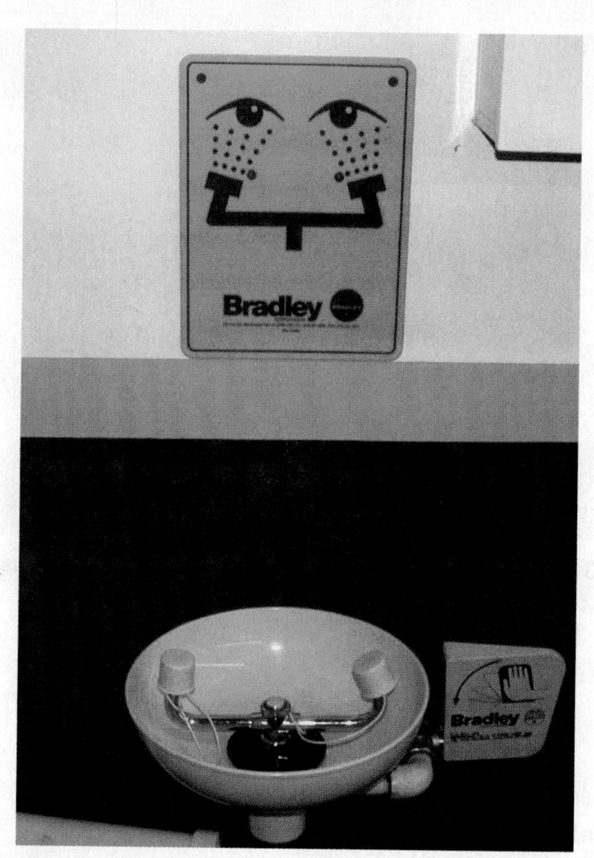

FIGURE 1–83 A typical eye wash station. A thorough flushing of the eyes with water is the first and often the best treatment in the event of eye contamination.

SAFETY TIP

Infection Control Precautions

Working on a vehicle can result in personal injury including the possibility of being cut or hurt enough to cause bleeding. Some infections such as hepatitis B, HIV (which can cause acquired immunodeficiency syndrome, or AIDS), and hepatitis C are transmitted through blood. These infections are commonly called blood-borne pathogens. Report any injury that involves blood to your supervisor and take the necessary precautions to avoid coming in contact with blood from another person.

FIGURE 1–84 A warning label on a Honda hybrid warns that a person can be killed due to the high-voltage circuits under the cover.

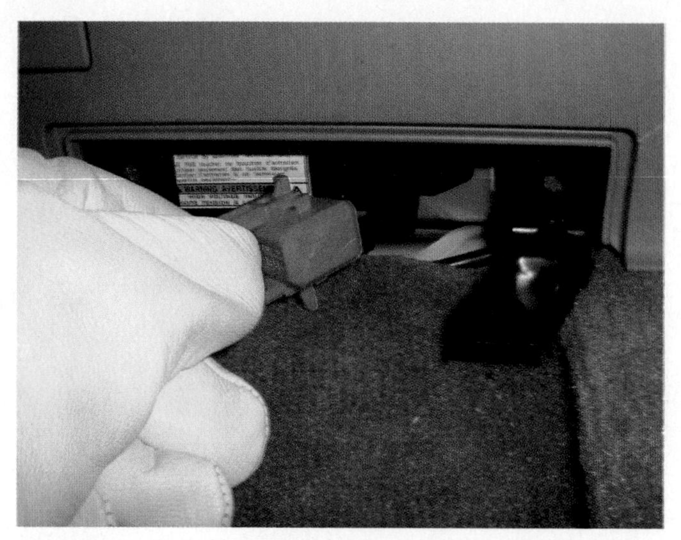

FIGURE 1–85 The high-voltage disconnect switch is in the trunk area on a Toyota Prius. High-voltage lineman's gloves should be worn when removing this plug.

HYBRID ELECTRIC VEHICLE SAFETY ISSUES

Hybrid electric vehicles (HEVs) use a high-voltage (HV) battery pack and electric motor(s) to help propel the vehicle. ● **SEE FIGURE 1–84** for an example of a typical warning label on a hybrid electric vehicle. The gasoline or diesel engine is also equipped with a generator or a combination starter and an integrated starter generator (ISG) or integrated starter alternator (ISA). To safely work around a hybrid electric vehicle, the high-voltage battery and circuits should be shut off following these steps:

STEP 1 Turn off the ignition key (if equipped) and remove the key from the ignition switch. (This will shut off all high-voltage circuits if the relay[s] is[are] working correctly.)

STEP 2 Disconnect the high-voltage circuits.

> **WARNING**
>
> Some vehicle manufacturers specify that rubber-insulated *lineman's gloves* be used whenever working around the high-voltage circuits to prevent the danger of electrical shock.

TOYOTA PRIUS The cutoff switch is located in the trunk on the Toyota Prius. To gain access, remove three clips holding the upper left portion of the trunk side cover. To disconnect the high-voltage system, pull the orange handled plug while wearing insulated rubber lineman's gloves. ● **SEE FIGURE 1–85**.

FORD ESCAPE AND MERCURY MARINER Ford and Mercury specify that the following steps should be included when working with the high-voltage (HV) systems of a hybrid vehicle.

- Four orange cones are to be placed at the four corners of the vehicle to create a buffer zone.
- High-voltage insulated gloves are to be worn with an outer leather glove to protect the inner rubber glove from possible damage.
- The service technician should also wear a face shield, and a fiberglass hook should be in the area and used to move a technician in the event of electrocution.

The high-voltage shut-off switch is located in the rear of the vehicle under the right side carpet. ● **SEE FIGURE 1–86**. Rotate the handle to the "service shipping" position, lift it out

FIGURE 1–86 The high-voltage shut-off switch on a Ford Escape hybrid. The switch is located under the carpet at the rear of the vehicle.

FIGURE 1–87 The shut-off switch on a GM parallel hybrid truck is green because this system uses 42 volts instead of higher, and possible fatal, voltages used in other hybrid vehicles.

to disable the high-voltage circuit, and wait five minutes before removing high-voltage cables.

HONDA CIVIC To totally disable the high-voltage system on a Honda Civic, remove the main fuse (labeled number 1) from the driver's side underhood fuse panel. This should be all that is necessary to shut off the high-voltage circuit. If this is not possible, then remove the rear seat cushion and seat back. Remove the metal switch cover labeled "up" and remove the red locking cover. Move the "battery module switch" down to disable the high-voltage system.

CHEVROLET SILVERADO AND GMC SIERRA PICKUP TRUCK The high-voltage shut-off switch is located under the rear passenger seat on the Chevrolet and GMC vehicles. Remove the cover marked "energy storage box" and turn the green service disconnect switch to the horizontal position to turn off the high-voltage circuits. ● **SEE FIGURE 1–87**.

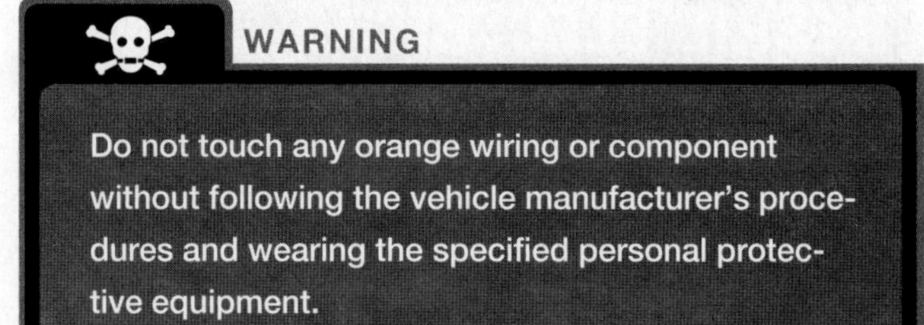

WARNING

Do not touch any orange wiring or component without following the vehicle manufacturer's procedures and wearing the specified personal protective equipment.

1 The first step in hoisting a vehicle is to properly align the vehicle in the center of the stall.

2 Most vehicles will be correctly positioned when the left front tire is centered on the tire pad.

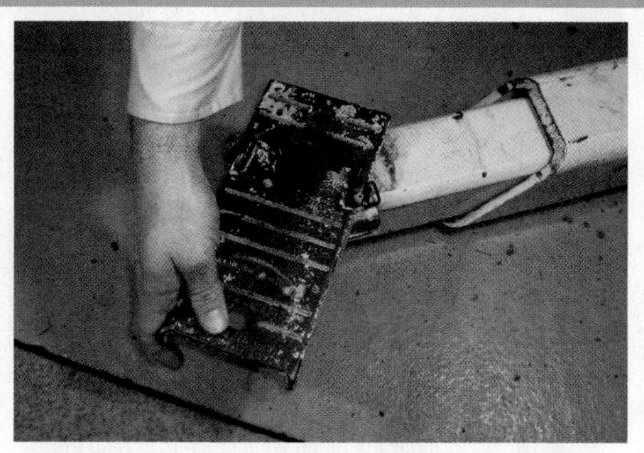

3 The arms can be moved in and out and most pads can be rotated to allow for many different types of vehicle construction.

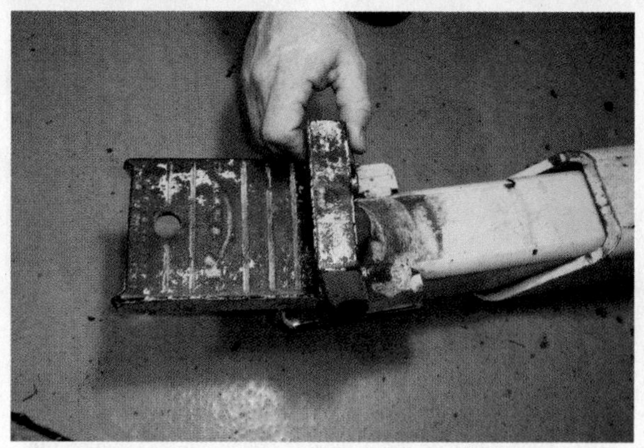

4 Most lifts are equipped with short pad extensions that are often necessary to use to allow the pad to contact the frame of a vehicle without causing the arm of the lift to hit and damage parts of the body.

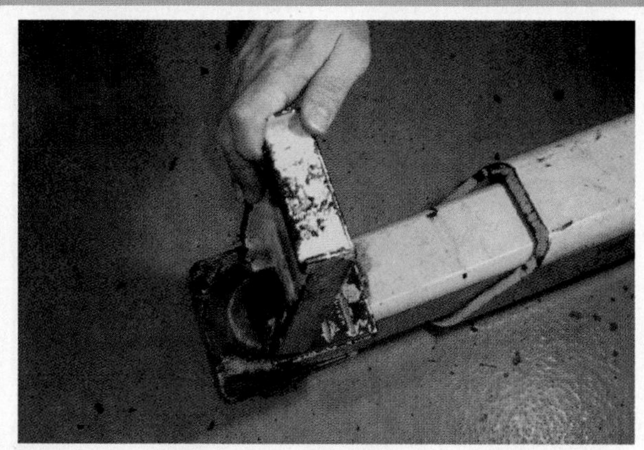

5 Tall pad extensions can also be used to gain access to the frame of a vehicle. This position is needed to safely hoist many pickup trucks, vans, and sport utility vehicles.

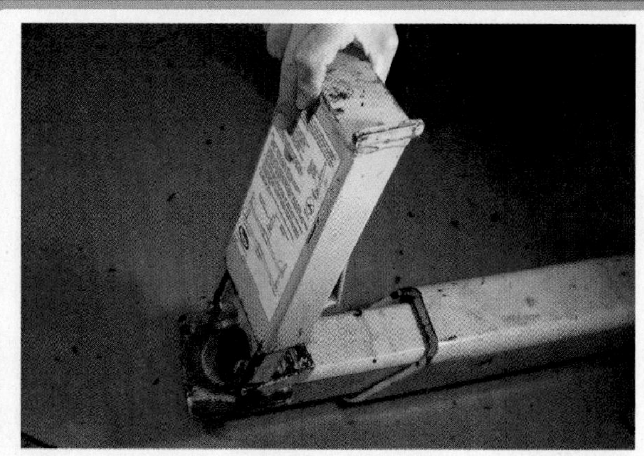

6 An additional extension may be necessary to hoist a truck or van equipped with running boards to give the necessary clearance.

CONTINUED ▶

STEP-BY-STEP (CONTINUED)

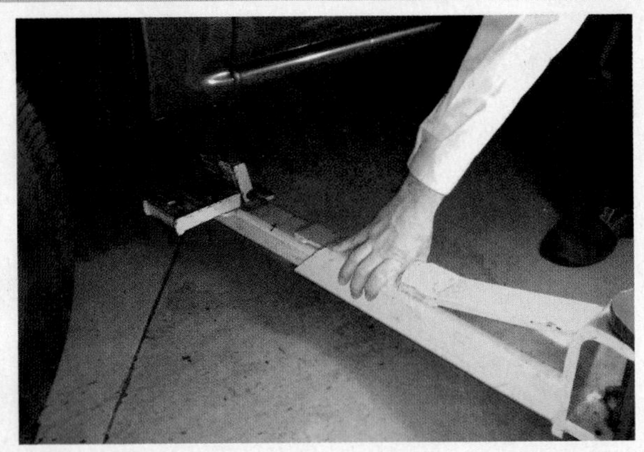

7 Position the pads under the vehicle at the recommended locations.

8 After being sure all pads are correctly positioned, use the electromechanical controls to raise the vehicle.

9 With the vehicle raised one foot (30 cm) off the ground, push down on the vehicle to check if it is stable on the pads. If the vehicle rocks, lower it and reset the pads. The vehicle can be raised to any desired working level. Be sure the safety is engaged before working on or under the vehicle.

10 If raising a vehicle without a frame, place the flat pads under the pinch weld seam to spread the load. If additional clearance is necessary, the pads can be raised as shown.

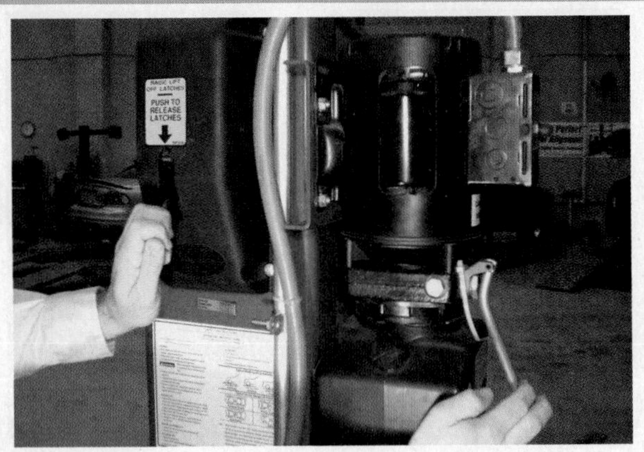

11 When the service work is completed, the hoist should be raised slightly and the safety released before using the hydraulic lever to lower the vehicle.

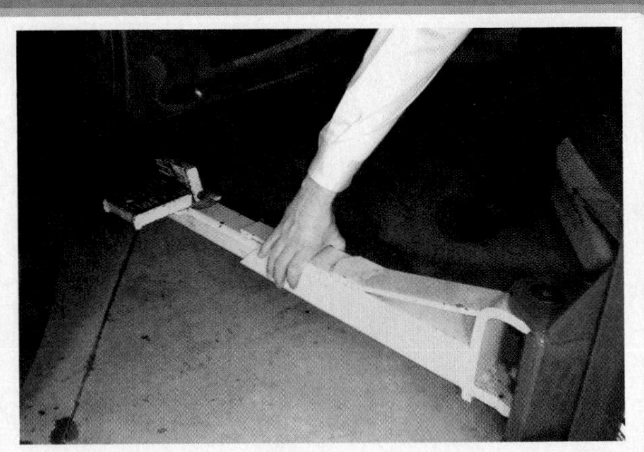

12 After lowering the vehicle, be sure all arms of the lift are moved out of the way before driving the vehicle out of the work stall.

SUMMARY

1. Bolts, studs, and nuts are commonly used as fasteners in the chassis. The sizes for fractional and metric threads are different and are not interchangeable. The grade is the rating of the strength of a fastener.
2. Whenever a vehicle is raised above the ground, it must be supported at a substantial section of the body or frame.
3. Wrenches are available as open end, box end, and combination open and box end.
4. An adjustable wrench should only be used where the proper size is not available.
5. Line wrenches are also called flare-nut wrenches, fitting wrenches, or tube-nut wrenches and are used to remove fuel or refrigerant lines.
6. Sockets are rotated by a ratchet or breaker bar, also called a flex handle.
7. Torque wrenches measure the amount of torque applied to a fastener.
8. Screwdriver types include straight blade (flat tip), Phillips, and Torx.
9. Hammers and mallets come in a variety of sizes and weights.
10. Pliers are a useful tool and are available in many different types, including slip-joint, multigroove, linesman's, diagonal, needle-nose, and locking pliers.
11. Other common hand tools include snap-ring pliers, files, cutters, punches, chisels, and hacksaws.
12. Hybrid electric vehicles should be de-powered if any of the high-voltage components are going to be serviced.

REVIEW QUESTIONS

1. List three precautions that must be taken whenever hoisting (lifting) a vehicle.
2. Describe how to determine the grade of a fastener, including how the markings differ between fractional and metric bolts.
3. List four items that are personal protective equipment (PPE).
4. List the types of fire extinguishers and their use.
5. Why are wrenches offset 15 degrees?
6. What are the other names for a line wrench?
7. What are the standard automotive drive sizes for sockets?
8. Which type of screwdriver requires the use of a hammer or mallet?
9. What is inside a dead-blow hammer?
10. What type of cutter is available in left and right cutters?

CHAPTER QUIZ

1. The correct location for the pads when hoisting or jacking the vehicle can often be found in the ________.
   - a. Service manual
   - b. Shop manual
   - c. Owner's manual
   - d. All of the above
2. For the best working position, the work should be ________.
   - a. At neck or head level
   - b. At knee or ankle level
   - c. Overhead by about 1 foot
   - d. At chest or elbow level
3. A high-strength bolt is identified by ________.
   - a. A UNC symbol
   - b. Lines on the head
   - c. Strength letter codes
   - d. The coarse threads
4. A fastener that uses threads on both ends is called a ________.
   - a. Cap screw
   - b. Stud
   - c. Machine screw
   - d. Crest fastener
5. When working with hand tools, always ________.
   - a. Push the wrench–don't pull it toward you
   - b. Pull a wrench–don't push it away from you
6. The proper term for Channel Locks is ________.
   - a. Vise-Grip
   - b. Crescent wrench
   - c. Locking pliers
   - d. Multigroove adjustable pliers

7. The proper term for Vise-Grip is ________.
   a. Locking pliers
   b. Slip-joint pliers
   c. Side cuts
   d. Multigroove adjustable pliers
8. Two technicians are discussing torque wrenches. Technician A says that a torque wrench is capable of tightening a fastener with more torque than a conventional breaker bar or ratchet. Technician B says that a torque wrench should be calibrated regularly for the most accurate results. Which technician is correct?
   a. Technician A only
   b. Technician B only
   c. Both Technicians A and B
   d. Neither Technician A nor B
9. What type of screwdriver should be used if there is very limited space above the head of the fastener?
   a. Offset screwdriver
   b. Standard screwdriver
   c. Impact screwdriver
   d. Robertson screwdriver
10. What type of hammer is plastic coated, has a metal casing inside, and is filled with small lead balls?
   a. Dead-blow hammer
   b. Soft-blow hammer
   c. Sledge hammer
   d. Plastic hammer

chapter 2

ENVIRONMENTAL AND HAZARDOUS MATERIALS

LEARNING OBJECTIVES

After studying this chapter, the reader will be able to:

1. Identify hazardous waste materials in accordance with state and federal regulations and follow safety precautions while handling and disposing of hazardous waste materials.

This chapter will help you prepare for the ASE assumed knowledge content required by all service technicians to adhere to environmentally appropriate actions and behavior.

KEY TERMS

HAZARDOUS WASTE

When handling hazardous waste material, one must always wear the proper protective clothing and equipment detailed in the right-to-know laws. This includes respirator equipment. All recommended procedures must be followed accurately. Personal injury may result from improper clothing, equipment, and procedures when handling hazardous materials. **Hazardous waste materials** are chemicals, or components, that the shop no longer needs and that pose a danger to the environment and people if disposed of in ordinary garbage cans or sewers. However, no material is considered hazardous waste until the shop has finished using it and is ready to dispose of it.

FEDERAL AND STATE LAWS

OCCUPATIONAL SAFETY AND HEALTH ACT The U.S. Congress passed the **Occupational Safety and Health Act (OSHA)** in 1970. This legislation was designed to assist and encourage the citizens of the United States in their efforts to ensure the following:

- Safe and healthful working conditions by providing research, information, education, and training in the field of occupational safety and health
- Safe and healthful working conditions for working men and women by authorizing enforcement of the standards developed under the act

Because about 25% of workers are exposed to health and safety hazards on the job, the OSHA standards are necessary to monitor, control, and educate workers regarding health and safety in the workplace.

EPA The **Environmental Protection Agency (EPA)** publishes a list of hazardous materials that is included in the **Code of Federal Regulations (CFR)**. The EPA considers waste hazardous if it is included on their list of hazardous materials, or it has one or more of the following characteristics.

- **Reactive—**Any material that reacts violently with water or other chemicals is considered hazardous.
- **Corrosive—**If a material burns the skin or dissolves metals and other materials, a technician should consider it hazardous. A pH scale is used, with number 7 indicating neutral. Pure water has a pH of 7. Lower numbers indicate an acidic solution and higher numbers indicate a caustic solution. If a material releases cyanide gas, hydrogen sulfide gas, or similar gases when exposed to low pH acid solutions, it is considered hazardous.
- **Toxic—**Materials are hazardous if they leak one or more of eight different heavy metals in concentrations greater than 100 times the primary drinking water standard.
- **Ignitable—**A liquid is hazardous if it has a flash point below 140°F (60°C), and a solid is hazardous if it ignites spontaneously.
- **Radioactive—**Any substance that emits measurable levels of radiation is radioactive. When individuals bring containers of a highly radioactive substance into the shop environment, qualified personnel with the appropriate equipment must test them.

WARNING

Hazardous waste disposal laws include serious penalties for anyone responsible for breaking these laws.

RIGHT-TO-KNOW LAWS The **right-to-know laws** state that employees have a right to know when the materials they use at work are hazardous. The right-to-know laws started with the Hazard Communication Standard published by the Occupational Safety and Health Administration in 1983. Under the right-to-know laws, the employer has responsibilities regarding the handling of hazardous materials by their employees. All employees must be trained about the types of hazardous materials they will encounter in the workplace. The employees must be informed about their rights under legislation regarding the handling of hazardous materials.

FIGURE 2–1 Material safety data sheets (MSDS), now called *safety data sheets* (SDS), should be readily available for use by anyone in the area who may come into contact with hazardous materials.

FIGURE 2–2 Tag identify that the power has been removed and service work is being done.

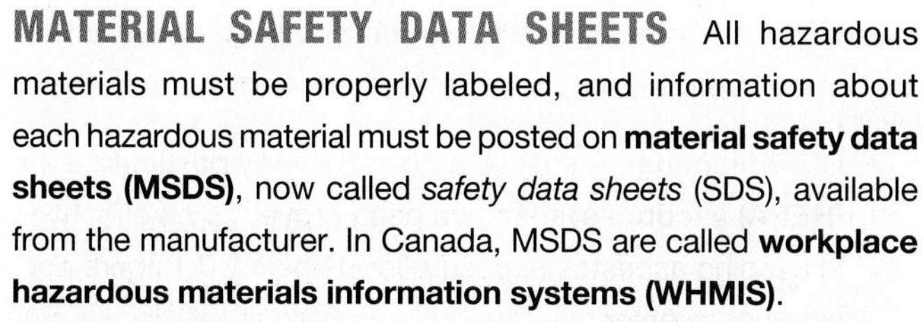

MATERIAL SAFETY DATA SHEETS All hazardous materials must be properly labeled, and information about each hazardous material must be posted on **material safety data sheets (MSDS)**, now called *safety data sheets* (SDS), available from the manufacturer. In Canada, MSDS are called **workplace hazardous materials information systems (WHMIS)**.

The employer has a responsibility to place SDS where they are easily accessible by all employees. These data sheets provide the following information about the hazardous material: chemical name, physical characteristics, protective handling equipment, explosion/fire hazards, incompatible materials, health hazards, medical conditions aggravated by exposure, emergency and first aid procedures, safe handling, and spill/leak procedures.

The employer also has a responsibility to ensure that all hazardous materials are properly labeled. The label information must include health, fire, and reactivity hazards posed by the material, as well as the protective equipment necessary to handle the material. The manufacturer must supply all warning and precautionary information about hazardous materials. This information must be read and understood by the employee before handling the material. ● **SEE FIGURE 2–1**.

RESOURCE CONSERVATION AND RECOVERY ACT Federal and state laws control the disposal of hazardous waste materials and every shop employee must be familiar with these laws. Hazardous waste disposal laws include the **Resource Conservation and Recovery Act (RCRA)**. This law states that hazardous material users are responsible for hazardous materials from the time they become a waste until their proper disposal. Many shops hire an independent hazardous waste hauler to dispose of hazardous waste material. The shop owner, or manager, should have a written contract with the hazardous waste hauler. The RCRA controls the following types of automotive waste.

- Paint and body repair products waste
- Solvents for parts and equipment cleaning
- Batteries and battery acid
- Mild acids used for metal cleaning and preparation
- Waste oil and engine coolants or antifreeze
- Air-conditioning refrigerants and oils
- Engine oil filters

LOCKOUT/TAGOUT According to OSHA Title 29, Code of Federal Regulations, part 1910.147, machinery must be locked out to prevent injury to employees when maintenance or repair work is being performed. Any piece of equipment that should not be used must be tagged and the electrical power disconnected to prevent it from being used. Always read, understand, and follow all safety warning tags. ● **SEE FIGURE 2–2**.

CLEAN AIR ACT Air-conditioning (A/C) systems and refrigerants are regulated by the **Clean Air Act (CAA)**, Title VI, Section 609. Technician certification and service equipment is also regulated. Any technician working on automotive A/C systems must be certified. A/C refrigerants must not be released or vented into the atmosphere, and used refrigerants must be recovered.

ASBESTOS HAZARDS

Friction materials such as brake and clutch linings often contain asbestos. While asbestos has been eliminated from most original equipment friction materials, the automotive service technician cannot know whether the vehicle being serviced is or is not equipped with friction materials containing asbestos. It is important that all friction materials be handled as if they contain asbestos.

Asbestos exposure can cause scar tissue to form in the lungs. This condition is called **asbestosis**. It gradually causes increasing shortness of breath, and the scarring to the lungs is permanent.

Even low exposures to asbestos can cause *mesothelioma*, a type of fatal cancer of the lining of the chest or abdominal cavity. Asbestos exposure can also increase the risk of *lung cancer* as well as cancer of the voice box, stomach, and large intestine. It usually takes 15 to 30 years or more for cancer or asbestos lung scarring to show up after exposure. (Scientists call this the *latency period*.)

Government agencies recommend that asbestos exposure be eliminated or controlled to the lowest level possible. These agencies have developed recommendations and standards that the automotive service technician and equipment manufacturer should follow. These U.S. federal agencies include the National Institute for Occupational Safety and Health (NIOSH), Occupational Safety and Health Administration (OSHA), and Environmental Protection Agency (EPA).

ASBESTOS OSHA STANDARDS The Occupational Safety and Health Administration has established three levels of asbestos exposure. Any vehicle service establishment that does either brake or clutch work must limit employee exposure to asbestos to less than 0.2 fibers per cubic centimeter (cc) as determined by an air sample.

If the level of exposure to employees is greater than specified, corrective measures must be performed and a large fine may be imposed.

NOTE: Research has found that worn asbestos fibers such as those from automotive brakes or clutches may not be as hazardous as first believed. Worn asbestos fibers do not have sharp flared ends that can latch on to tissue, but rather are worn down to a dust form that resembles talc. Grinding or sawing operations on unworn brake shoes or clutch discs *will* contain *harmful* asbestos fibers. To limit health damage, always use proper handling procedures while working around any component that may contain asbestos.

ASBESTOS EPA REGULATIONS The federal Environmental Protection Agency has established procedures for the removal and disposal of asbestos. The EPA procedures require that products containing asbestos be "wetted" to prevent the asbestos fibers from becoming airborne. According to the EPA, asbestos-containing materials can be disposed of as regular waste. Only when asbestos becomes airborne it is considered to be hazardous.

ASBESTOS HANDLING GUIDELINES The air in the shop area can be tested by a testing laboratory, but this can be expensive. Tests have determined that asbestos levels can easily be kept below the recommended levels by using a liquid, like water, or a special vacuum.

NOTE: The service technician cannot tell whether the old brake pads, shoes, or clutch discs contain asbestos. Therefore, to be safe, the technician should assume that all brake pads, shoes, or clutch discs contain asbestos.

- **HEPA vacuum.** A special **high-efficiency particulate air (HEPA) vacuum** system has been proven to be effective in keeping asbestos exposure levels below 0.1 fibers per cubic centimeter.
- **Solvent spray.** Many technicians use an aerosol can of brake cleaning solvent to wet the brake dust and prevent it from becoming airborne. A **solvent** is a liquid used to dissolve dirt, grime, or solid particles. Commercial brake cleaners are available that use a concentrated cleaner mixed with water. ● **SEE FIGURE 2–3.**

The waste liquid is filtered, and when dry, the filter can be disposed of as solid waste.

CAUTION: Never use compressed air to blow brake dust. The fine talclike brake dust can create a health hazard even if asbestos is not present or is present in dust rather than fiber form.

- **Disposal of brake dust and brake shoes.** The hazard of asbestos occurs when asbestos fibers are airborne. Once the asbestos has been wetted down, it is then considered to be solid waste, rather than hazardous waste. Old brake shoes and pads should be enclosed, preferably in a plastic bag, to help prevent any of the brake material from becoming airborne. *Always follow current federal and local laws concerning disposal of all waste.*

FIGURE 2–3 All brakes should be moistened with water or solvent to help prevent brake dust from becoming airborne.

USED BRAKE FLUID

Most brake fluid is made from polyglycol, is water soluble, and can be considered hazardous if it has absorbed metals from the brake system.

STORAGE AND DISPOSAL OF BRAKE FLUID

- Collect brake fluid in containers clearly marked to indicate that it is dedicated for that purpose.
- If your waste brake fluid is hazardous, manage it appropriately and use only an authorized waste receiver for its disposal.
- If your waste brake fluid is nonhazardous (such as old, but unused), determine from your local solid waste collection provider what should be done for its proper disposal.
- Do not mix brake fluid with used engine oil.
- Do not pour brake fluid down drains or onto the ground.
- Recycle brake fluid through a registered recycler.

USED OIL

Used oil is any petroleum-based or synthetic oil that has been used. During normal use, impurities such as dirt, metal scrapings, water, or chemicals can get mixed in with the oil. Eventually, this used oil must be replaced with virgin or re-refined oil. The EPA's used oil management standards include a three-pronged approach to determine if a substance meets the definition of *used oil*. To meet the EPA's definition of used oil, a substance must meet each of the following three criteria.

- **Origin.** The first criterion for identifying used oil is based on the oil's origin. Used oil must have been refined from crude oil or made from synthetic materials. Animal and vegetable oils are excluded from the EPA's definition of used oil.
- **Use.** The second criterion is based on whether and how the oil is used. Oils used as lubricants, hydraulic fluids, heat transfer fluids, and for other similar purposes are considered used oil. Unused oil, such as bottom clean-out waste from virgin fuel oil storage tanks or virgin fuel oil recovered from a spill, does not meet the EPA's definition of used oil because these oils have never been "used." The EPA's definition also excludes products used as cleaning agents, as well as certain petroleum-derived products like antifreeze and kerosene.
- **Contaminants.** The third criterion is based on whether or not the oil is contaminated with either physical or chemical impurities. In other words, to meet the EPA's definition, used oil must become contaminated as a result of being used. This aspect of the EPA's definition includes residues and contaminants generated from handling, storing, and processing used oil.

NOTE: The release of only 1 gallon of used oil (a typical oil change) can make 1 million gallons of freshwater undrinkable.

If used oil is dumped down the drain and enters a sewage treatment plant, concentrations as small as 50 to 100 parts per million (PPM) in the waste water can foul sewage treatment processes. Never mix a listed hazardous waste, gasoline, waste water, halogenated solvent, antifreeze, or an unknown waste material with used oil. Adding any of these substances will cause the used oil to become contaminated, which classifies it as hazardous waste.

STORAGE AND DISPOSAL OF USED OIL Once oil has been used, it can be collected, recycled, and used over and over again. An estimated 380 million gallons of used oil are recycled each year. Recycled used oil can sometimes be used again for the same job or can take on a completely different task. For example, used engine oil can be re-refined and sold at some discount stores as engine oil or processed for furnace fuel oil. After collecting used oil in an appropriate container, such as a 55 gallon steel drum, the material must be disposed of in one of two ways:

1. Shipped offsite for recycling
2. Burned in an onsite or offsite EPA-approved heater for energy recovery

Used oil must be stored in compliance with an existing **underground storage tank (UST)** or an **aboveground storage tank (AGST)** standard, or kept in separate containers. ● **SEE FIGURE 2–4**. Containers are portable receptacles, such as a 55 gallon steel drum.

- **Keep used oil storage drums in good condition.** This means that they should be covered, secured from vandals, properly labeled, and maintained in compliance with local fire codes. Frequent inspections for leaks, corrosion, and spillage are an essential part of container maintenance.
- **Never store used oil in anything other than tanks and storage containers.** Used oil may also be stored in units that are permitted to store regulated hazardous waste.
- **Follow used oil filter disposal regulations.** Used oil filters contain used engine oil that may be hazardous. Before an oil filter is placed into the trash or sent to be recycled, it must be drained using one of the following hot draining methods approved by the EPA.
  - Puncture the filter antidrain back valve or filter dome end and hot drain for at least 12 hours
  - Hot drain and crush
  - Dismantle and hot drain
  - Use another hot draining method to remove all used oil from the filter

After the oil has been drained from the oil filter, the filter housing can be disposed of in any of the following ways.

- Sent for recycling
- Picked up by a service contract company
- Disposed of in regular trash

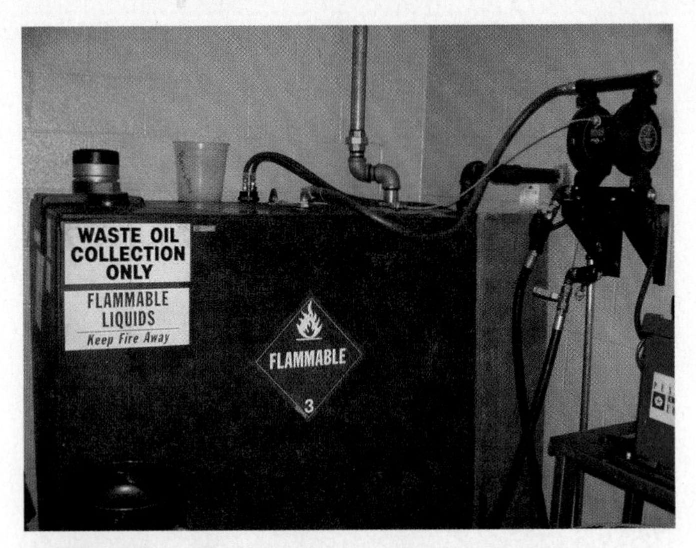

FIGURE 2–4 A typical aboveground oil storage tank.

SOLVENTS

The major sources of chemical danger are liquid and aerosol brake cleaning fluids that contain chlorinated hydrocarbon solvents. Several other chemicals that do not deplete the ozone, such as heptane, hexane, and xylene, are now being used in nonchlorinated brake cleaning solvents. Some manufacturers are also producing solvents they describe as environmentally responsible, which are biodegradable and noncarcinogenic (not cancer causing).

There is no specific standard for physical contact with chlorinated hydrocarbon solvents or the chemicals replacing them. All contact should be avoided whenever possible. The law requires an employer to provide appropriate protective equipment and ensure proper work practices by an employee handling these chemicals.

EFFECTS OF CHEMICAL POISONING The effects of exposure to chlorinated hydrocarbon and other types of solvents can take many forms. Short-term exposure at low levels can cause one or more of the following symptoms.

- Headache
- Nausea
- Drowsiness
- Dizziness
- Lack of coordination
- Unconsciousness

It may also cause irritation of the eyes, nose, and throat, and flushing of the face and neck. Short-term exposure to higher concentrations can cause liver damage with symptoms such as yellow jaundice or dark urine. Liver damage may not become evident until several weeks after the exposure.

FIGURE 2–5 Washing hands and removing jewelry are two important safety habits all service technicians should practice.

FIGURE 2–6 Typical fireproof flammable storage cabinet.

SAFETY TIP

Hand Safety

Service technicians should wash their hands with soap and water after handling engine oil or differential or transmission fluids, or wear protective rubber gloves. Another safety hint is that the service technician should not wear watches, rings, or other jewelry that could come in contact with electrical or moving parts of a vehicle. ● **SEE FIGURE 2–5.**

SOLVENT HAZARDOUS AND REGULATORY STATUS

Most solvents are classified as hazardous wastes. Other characteristics of solvents include the following:

- Solvents with flash points below 140°F (60°C) are considered flammable and, like gasoline, are federally regulated by the Department of Transportation (DOT).
- Solvents and oils with flash points above 140°F (60°C) are considered combustible and, like engine oil, are also regulated by the DOT. All flammable items must be stored in a fireproof container. ● **SEE FIGURE 2–6.**

It is the responsibility of the repair shop to determine if its spent solvent is hazardous waste. Solvents that are considered hazardous waste have a flash point below 140°F (60°C). Hot water or aqueous parts cleaners may be used to avoid disposing of spent solvent as hazardous waste. Solvent-type parts cleaners with filters are available to greatly extend solvent life and reduce spent solvent disposal costs. Solvent reclaimers are available that clean and restore the solvent so it lasts indefinitely.

USED SOLVENTS Used or spent solvents are liquid materials that have been generated as waste and may contain xylene, methanol, ethyl ether, and methyl isobutyl ketone (MIBK). These materials must be stored in OSHA-approved safety containers with the lids or caps closed tightly. These storage receptacles must show no signs of leaks or significant damage due to dents or rust. In addition, the containers must be stored in a protected area equipped with secondary containment or a spill protector, such as a spill pallet. Additional requirements include the following:

- Containers should be clearly labeled "Hazardous Waste" and the date the material was first placed into the storage receptacle should be noted.
- Labeling is not required for solvents being used in a parts washer.
- Used solvents will not be counted toward a facility's monthly output of hazardous waste if the vendor under contract removes the material.

FIGURE 2–7 Using a water-based cleaning system helps reduce the hazards from using strong chemicals.

- Used solvents may be disposed of by recycling with a local vendor, such as SafetyKleen®, to have the used solvent removed according to specific terms in the vendor agreement. ● SEE FIGURE 2–7.
- Use aqueous-based (nonsolvent) cleaning systems to help avoid the problems associated with chemical solvents.

COOLANT DISPOSAL

Coolant is a mixture of antifreeze and water. New antifreeze is not considered to be hazardous even though it can cause death if ingested. Used antifreeze may be hazardous due to dissolved metals from the engine and other components of the cooling system. These metals can include iron, steel, aluminum, copper, brass, and lead (from older radiators and heater cores).

- Coolant should be recycled either onsite or offsite.
- Used coolant should be stored in a sealed and labeled container. ● SEE FIGURE 2–8.
- Used coolant can often be disposed of into municipal sewers with a permit. Check with local authorities and obtain a permit before discharging used coolant into sanitary sewers.

FIGURE 2–8 Used antifreeze coolant should be kept separate and stored in a leakproof container until it can be recycled or disposed of according to federal, state, and local laws. Note that the storage barrel is placed inside another container to catch any coolant that may spill out of the inside barrel.

LEAD–ACID BATTERY WASTE

About 70 million spent lead–acid batteries are generated each year in the United States alone. Lead is classified as a toxic metal and the acid used in lead–acid batteries is highly corrosive. The vast majority (95% to 98%) of these batteries are recycled through lead reclamation operations and secondary lead smelters for use in the manufacture of new batteries.

BATTERY DISPOSAL Used lead–acid batteries must be reclaimed or recycled in order to be exempt from hazardous waste regulations. Leaking batteries must be stored and transported as hazardous waste. Some states have more strict regulations, which require special handling procedures and transportation. According to the **Battery Council International (BCI)**, battery laws usually include the following rules.

1. Lead–acid battery disposal is prohibited in landfills or incinerators. Batteries are required to be delivered to a battery retailer, wholesaler, recycling center, or lead smelter.
2. All retailers of automotive batteries are required to post a sign that displays the universal recycling symbol and indicates the retailer's specific requirements for accepting used batteries.
3. Battery electrolyte contains sulfuric acid, which is a very corrosive substance capable of causing serious personal injury, such as skin burns and eye damage. In addition, the battery plates contain lead, which is highly poisonous. For this reason, disposing of batteries improperly can cause environmental contamination and lead to severe health problems.

BATTERY HANDLING AND STORAGE

Batteries, whether new or used, should be kept indoors if possible. The storage location should be an area specifically designated for battery storage and must be well ventilated (to the outside). If outdoor storage is the only alternative, a sheltered and secured area with acid-resistant secondary containment is strongly recommended. It is advisable that acid-resistant secondary containment be used for indoor storage also. In addition, batteries should be placed on acid-resistant pallets and never stacked!

FUEL SAFETY AND STORAGE

Gasoline is a very explosive liquid. The expanding vapors that come from gasoline are extremely dangerous. These vapors are present even in cold temperatures. Vapors formed in gasoline tanks on many vehicles are controlled, but vapors from gasoline storage may escape from the can, resulting in a hazardous situation. Therefore, place gasoline storage containers in a well-ventilated space. Although diesel fuel is not as volatile as gasoline, the same basic rules apply to diesel fuel and gasoline storage. These rules include the following:

1. Use storage cans that have a flash arresting screen at the outlet. These screens prevent external ignition sources from igniting the gasoline within the can when someone pours the gasoline or diesel fuel.
2. Use only a red approved gasoline container to allow for proper hazardous substance identification. ● **SEE FIGURE 2–9.**
3. Do not fill gasoline containers completely full. Always leave the level of gasoline at least 1 inch from the top of the container. This action allows expansion of the gasoline at higher temperatures. If gasoline containers are completely full, the gasoline will expand when the temperature increases. This expansion forces gasoline from the can and creates a dangerous spill. If gasoline or diesel fuel containers must be stored, place them in a designated storage locker or facility.
4. Never leave gasoline containers open, except while filling or pouring gasoline from the container.
5. Never use gasoline as a cleaning agent.
6. Always connect a ground strap to containers when filling or transferring fuel or other flammable products from one container to another to prevent static electricity that could result in explosion and fire. These ground wires prevent the buildup of a static electric charge, which could result in a spark and disastrous explosion.

FIGURE 2–9 This red gasoline container holds about 30 gallons of gasoline and is used to fill vehicles used for training.

AIRBAG DISPOSAL

Airbag modules are pyrotechnic devices that can be ignited if exposed to an electrical charge or if the body of the vehicle is subjected to a shock. Airbag safety should include the following precautions.

1. Disarm the airbag(s) if you will be working in the area where a discharged bag could make contact with any part of your body. Consult service information for the exact procedure to follow for the vehicle being serviced. The usual procedure is to deploy the airbag using a 12 volt power supply, such as a jump start box, using long wires to connect to the module to ensure a safe deployment.
2. Do not expose an airbag to extreme heat or fire.
3. Always carry an airbag pointing away from your body.
4. Place an airbag module facing upward.
5. Always follow the manufacturer's recommended procedure for airbag disposal or recycling, including the proper packaging to use during shipment.
6. Wear protective gloves if handling a deployed airbag.
7. Always wash your hands or body well if exposed to a deployed airbag. The chemicals involved can cause skin irritation and possible rash development.

USED TIRE DISPOSAL

Used tires are an environmental concern because of several reasons, including the following:

1. In a landfill, they tend to "float" up through the other trash and rise to the surface.
2. The inside area traps and holds rainwater, which is a breeding ground for mosquitoes. Mosquito-borne diseases include encephalitis, malaria, and dengue fever.
3. Used tires present a fire hazard and, when burned, create a large amount of black smoke that contaminates the air.

Used tires should be disposed of in one of the following ways.

1. Used tires can be reused until the end of their useful life.
2. Tires can be retreaded.
3. Tires can be recycled or shredded for use in asphalt.
4. Derimmed tires can be sent to a landfill. (Most landfill operators will shred the tires because it is illegal in many states to landfill whole tires.)
5. Tires can be burned in cement kilns or other power plants where the smoke can be controlled.
6. A registered scrap tire handler should be used to transport tires for disposal or recycling.

FIGURE 2–10 Air-conditioning refrigerant oil must be kept separated from other oils because it contains traces of refrigerant and must be treated as hazardous waste.

AIR-CONDITIONING REFRIGERANT OIL DISPOSAL

Air-conditioning refrigerant oil contains dissolved refrigerant and is therefore considered to be hazardous waste. This oil must be kept separated from other waste oil or the entire amount of oil must be treated as hazardous. Used refrigerant oil must be sent to a licensed hazardous waste disposal company for recycling or disposal. ● **SEE FIGURE 2–10.**

WASTE CHART All automotive service facilities create some waste, and while most of it is handled properly, it is important that all hazardous and nonhazardous waste be accounted for and properly disposed. ● **SEE CHART 2–1** for a list of typical wastes generated at automotive shops, plus a checklist for keeping track of how these wastes are handled.

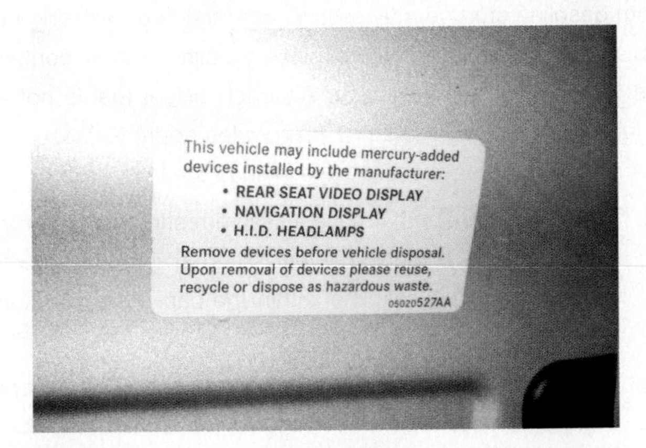

FIGURE 2–11 Placard near driver's door, including what devices in the vehicle contain mercury.

TECH TIP

Remove Components That Contain Mercury

Some vehicles have a placard near the driver's side door that lists the components that contain the heavy metal mercury. **Mercury** can be absorbed through the skin and is a heavy metal that once absorbed by the body does not leave. ● **SEE FIGURE 2–11.**

These components should be removed from the vehicle before the rest of the body is sent to be recycled to help prevent releasing mercury into the environment.

	TYPICAL WASTES		
WASTE STREAM	TYPICAL CATEGORY IF NOT MIXED WITH OTHER HAZARDOUS WASTE	IF DISPOSED IN LANDFILL AND NOT MIXED WITH A HAZARDOUS WASTE	IF RECYCLED
Used oil	Used oil	Hazardous waste	Used oil
Used oil filters	Nonhazardous solid waste, if completely drained	Nonhazardous solid waste, if completely drained	Used oil, if not drained
Used transmission fluid	Used oil	Hazardous waste	Used oil
Used brake fluid	Used oil	Hazardous waste	Used oil
Used antifreeze	Depends on characterization	Depends on characterization	Depends on characterization
Used solvents	Hazardous waste	Hazardous waste	Hazardous waste
Used citric solvents	Nonhazardous solid waste	Nonhazardous solid waste	Hazardous waste
Lead–acid automotive batteries	Not a solid waste if returned to supplier	Hazardous waste	Hazardous waste
Shop rags used for oil	Used oil	Depends on used oil characterization	Used oil
Shop rags used for solvent or gasoline spills	Hazardous waste	Hazardous waste	Hazardous waste
Oil spill absorbent material	Used oil	Depends on used oil characterization	Used oil
Spill material for solvent and gasoline	Hazardous waste	Hazardous waste	Hazardous waste
Catalytic converter	Not a solid waste if returned to supplier	Nonhazardous solid waste	Nonhazardous solid waste
Spilled or unused fuels	Hazardous waste	Hazardous waste	Hazardous waste
Spilled or unusable paints and thinners	Hazardous waste	Hazardous waste	Hazardous waste
Used tires	Nonhazardous solid waste	Nonhazardous solid waste	Nonhazardous solid waste

CHART 2–1

Typical wastes generated at auto repair shops and typical category (hazardous or nonhazardous) by disposal method.

TECH TIP

What Every Technician Should Know

OSHA has adopted new hazardous chemical labeling requirements making it agree with global labeling standards established by the United Nations. As a result, workers will have better information available on the safe handling and use of hazardous chemicals, allowing them to avoid injuries and possible illnesses related to exposures to hazardous chemicals.

● **SEE FIGURE 2–12.**

HEALTH HAZARD

- CARCINOGEN
- MUTAGENICITY
- REPRODUCTIVE TOXICITY
- RESPIRATORY SENSITIZER
- TARGET ORGAN TOXICITY
- ASPIRATION TOXICITY

FLAME

- FLAMMABLES
- PYROPHORICS
- SELF-HEATING
- EMITS FLAMMABLE GAS
- SELF-REACTIVES
- ORGANIC PEROXIDES

EXCLAMATION MARK

- IRRITANT (SKIN AND EYE)
- SKIN SENSITIZER
- ACUTE TOXICITY
- NARCOTIC EFFECTS
- RESPIRATORY TRACT IRRITANT
- HAZARDOUS TO OZONE LAYER (NON-MANDATORY)

GAS CYLINDER

- GASES UNDER PRESSURE

CORROSION

- SKIN CORROSION/BURNS
- EYE DAMAGE
- CORROSIVE TO METALS

EXPLODING BOMB

- EXPLOSIVES
- SELF-REACTIVES
- ORGANIC PEROXIDES

FLAME OVER CIRCLE

- OXIDIZERS

ENVIRONMENT (NON-MANDATORY)

- AQUATIC TOXICITY

SKULL AND CROSSBONES

- ACUTE TOXICITY (FATAL OR TOXIC)

FIGURE 2–12 The OSHA global hazardous materials labels.

SUMMARY

1. Hazardous materials include common automotive chemicals, liquids, and lubricants, especially those whose ingredients contain *chlor* or *fluor* in their name.
2. Right-to-know laws require that all workers have access to safety data sheets (SDS).
3. Asbestos fibers should be avoided and removed according to current laws and regulations.
4. Used engine oil contains metals worn from parts and should be handled and disposed of properly.
5. Solvents represent a serious health risk and should be avoided as much as possible.
6. Coolant should be disposed of properly or recycled.
7. Batteries are considered to be hazardous waste and should be discarded to a recycling facility.

REVIEW QUESTIONS

1. List five common automotive chemicals or products that may be considered hazardous.
2. The Resource Conservation and Recovery Act (RCRA) controls what types of automotive waste?

CHAPTER QUIZ

1. Hazardous materials include all of the following except ________.
   a. Engine oil
   b. Asbestos
   c. Water
   d. Brake cleaner
2. To determine if a product or substance being used is hazardous, consult ________.
   a. A dictionary
   b. An MSDS
   c. SAE standards
   d. EPA guidelines
3. Exposure to asbestos dust can cause what condition?
   a. Asbestosis
   b. Mesothelioma
   c. Lung cancer
   d. All of the above are possible
4. Wetted asbestos dust is considered to be ________.
   a. Solid waste
   b. Hazardous waste
   c. Toxic
   d. Poisonous
5. An oil filter should be hot drained for how long before disposing of the filter?
   a. 30 to 60 minutes
   b. 4 hours
   c. 8 hours
   d. 12 hours
6. Used engine oil should be disposed of by all *except* one of the following methods.
   a. Disposed of in regular trash
   b. Shipped offsite for recycling
   c. Burned onsite in a waste oil-approved heater
   d. Burned offsite in a waste oil-approved heater
7. All of the following are the proper ways to dispose of a drained oil filter *except* ________.
   a. Sent for recycling
   b. Picked up by a service contract company
   c. Disposed of in regular trash
   d. Considered to be hazardous waste and disposed of accordingly
8. Which act or organization regulates air-conditioning refrigerant?
   a. Clean Air Act (CAA)
   b. MSDS
   c. WHMIS
   d. Code of Federal Regulations (CFR)
9. Gasoline should be stored in approved containers that include what color(s)?
   a. Red container with yellow lettering
   b. Red container
   c. Yellow container
   d. Yellow container with red lettering
10. What automotive devices may contain mercury?
    a. Rear seat video displays
    b. Navigation displays
    c. HID headlights
    d. All of the above

chapter 3

GASOLINE ENGINE OPERATION, PARTS, AND SPECIFICATIONS

LEARNING OBJECTIVES

After studying this chapter, the reader will be able to:

1. Describe how an internal combustion engine works.
2. Describe how turbocharging or supercharging increases engine power.
3. Explain how a four-stroke cycle gasoline engine operates.
4. List the various characteristics by which vehicle engines are classified.
5. Discuss how a compression ratio is calculated.

This chapter will help you prepare for Engine Repair (A1) ASE certification test content area "A" (General Engine Diagnosis).

KEY TERMS

FIGURE 3–1 The rotating assembly for a V-8 engine that has eight pistons and connecting rods and one crankshaft.

FIGURE 3–2 A cylinder head with four valves per cylinder, two intake valves (larger) and two exhaust valves (smaller) per cylinder.

ENERGY AND POWER

Energy is used to produce power. The chemical energy in fuel is converted to heat by the burning of the fuel at a controlled rate. This process is called **combustion**. If engine combustion occurs within the power chamber, the engine is called an **internal combustion engine**.

NOTE: An external combustion engine is an engine that burns fuel outside of the engine itself, such as a steam engine.

Engines used in automobiles are internal combustion heat engines. They convert the chemical energy of the gasoline into heat within a power chamber that is called a **combustion chamber**. Heat energy released in the combustion chamber raises the temperature of the combustion gases within the chamber. The increase in gas temperature causes the pressure of the gases to increase. The pressure developed within the combustion chamber is applied to the head of a piston or to a turbine wheel to produce a usable **mechanical force**, which is then converted into useful **mechanical power**.

ENGINE CONSTRUCTION OVERVIEW

BLOCK All automotive and truck engines are constructed using a solid frame, called a **block**. A block is constructed of cast iron or aluminum and provides the foundation for most of the engine components and systems. The block is cast and then machined to very close tolerances to allow other parts to be installed.

ROTATING ASSEMBLY Pistons are installed in the block and move up and down during engine operation. Pistons are connected to *connecting rods*, which connect the pistons to the crankshaft. The crankshaft converts the up-and-down motion of the piston to rotary motion, which is then transmitted to the drive wheels and propels the vehicle. ● **SEE FIGURE 3–1**.

CYLINDER HEADS All engines use a cylinder head to seal the top of the cylinders, which are in the engine block. The cylinder head also contains valves that allow air and fuel into the cylinder, called intake valves and exhaust valves, which open after combustion to allow the hot gases left over to escape from the engine. Cylinder heads are constructed of cast iron or aluminum and are then machined for the valves and other valve-related components. Cooling passages are formed during the casting process and coolant is circulated around the combustion chamber to keep temperatures controlled. ● **SEE FIGURE 3–2**.

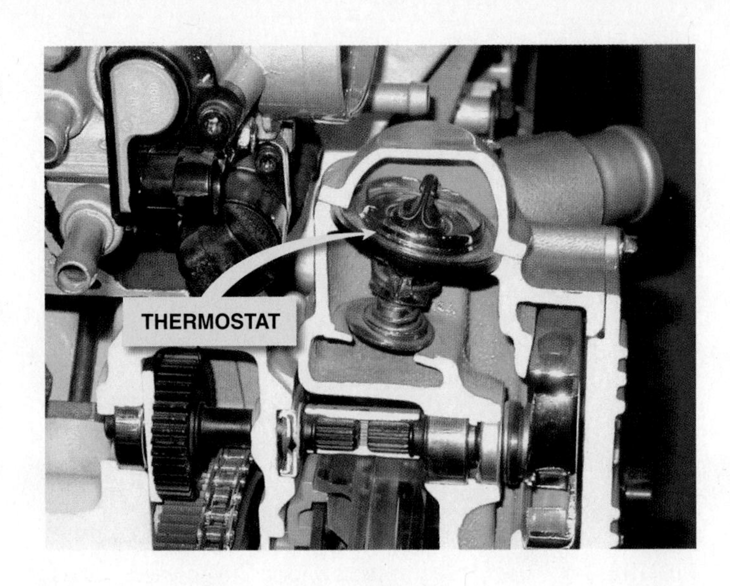

FIGURE 3–3 The coolant temperature is controlled by the thermostat, which opens and allows coolant to flow to the radiator when the temperature reaches the rating temperature of the thermostat.

INTAKE AND EXHAUST MANIFOLDS Air and fuel enter the engine through an intake manifold and exit the engine through the exhaust manifold. Intake manifolds operate cooler than exhaust manifolds and are therefore constructed of nylon reinforced plastic or aluminum. Exhaust manifolds must be able to withstand hot exhaust gases and therefore most are constructed from cast iron.

COOLING SYSTEM All engines must have a cooling system to control engine temperatures. While some older engines were air cooled, all current production passenger vehicle engines are cooled by circulating antifreeze coolant through passages in the block and cylinder head. The coolant picks up the heat from the engine and after the thermostat opens, the water pump circulates the coolant through the radiator where the excess heat is released to the outside air, cooling the coolant. The coolant is continuously circulated through the cooling system and the temperature is controlled by the thermostat. ● **SEE FIGURE 3–3**.

LUBRICATION SYSTEM All engines contain moving and sliding parts that must be kept lubricated to reduce wear and friction. The oil pan, bolted to the bottom of the engine block, holds 4 to 7 quarts (liters) of oil. An oil pump, which is driven by the engine, forces the oil through the oil filter and then into passages in the crankshaft and block. These passages are called **oil galleries**. The oil is also forced up to the valves and then falls down through openings in the cylinder head and block back into the oil pan. ● **SEE FIGURE 3–4**.

FUEL SYSTEM AND IGNITION SYSTEM All engines require a fuel and an ignition system to ignite the fuel–air mixture in the cylinders. The fuel system includes the following components:

- Fuel tank where fuel is stored
- Fuel filter and lines
- Fuel injectors, which spray fuel into the intake manifold or directly into the cylinder, depending on the type of system used

The ignition system is designed to take 12 volts from the battery and convert it to 5,000 to 40,000 volts needed to jump the gap of a spark plug. Spark plugs are threaded into the cylinder head of each cylinder, and when the spark occurs, it ignites the air–fuel mixture in the cylinder creating pressure and forcing the piston down in the cylinder. The components included on the ignition system include:

- Spark plugs
- Ignition coils
- Ignition control module (ICM)
- Associated wiring

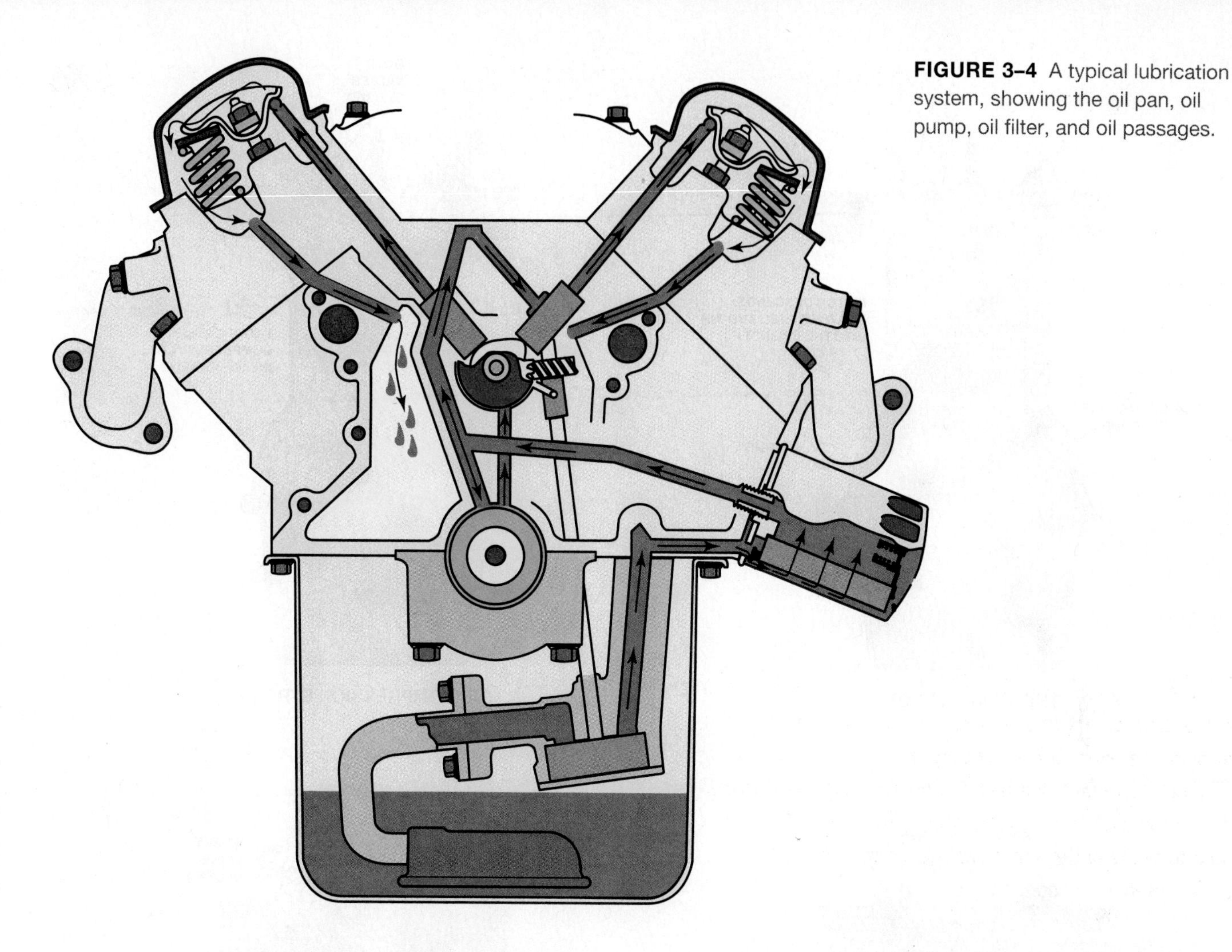

FIGURE 3–4 A typical lubrication system, showing the oil pan, oil pump, oil filter, and oil passages.

FOUR-STROKE CYCLE OPERATION

Most automotive engines use the four-stroke cycle of events, begun by the starter motor, which rotates the engine. The four-stroke cycle is repeated for each cylinder of the engine. ● **SEE FIGURE 3–5**.

- **Intake stroke.** The **intake valve** is open and the piston inside the cylinder travels downward, drawing a mixture of air and fuel into the cylinder.
- **Compression stroke.** As the engine continues to rotate, the intake valve closes and the piston moves upward in the cylinder, compressing the air–fuel mixture.
- **Power stroke.** When the piston gets near the top of the cylinder (called **top dead center [TDC]**), the spark at the spark plug ignites the air–fuel mixture, which forces the piston downward.

  The combustion pressure developed in the combustion chamber at the correct time will push the piston downward to rotate the crankshaft.
- **Exhaust stroke.** The engine continues to rotate, and the piston again moves upward in the cylinder. The exhaust valve opens, and the piston forces the residual burned gases out of the **exhaust valve** and into the exhaust manifold and exhaust system.

This sequence repeats as the engine rotates. To stop the engine, the electricity to the ignition system is shut off by the ignition switch.

A piston that moves up and down, or reciprocates, in a **cylinder** can be seen in this illustration. The piston is attached to a **crankshaft** with a **connecting rod**. This arrangement allows the piston to reciprocate (move up and down) in the cylinder as the crankshaft rotates. ● **SEE FIGURE 3–6**.

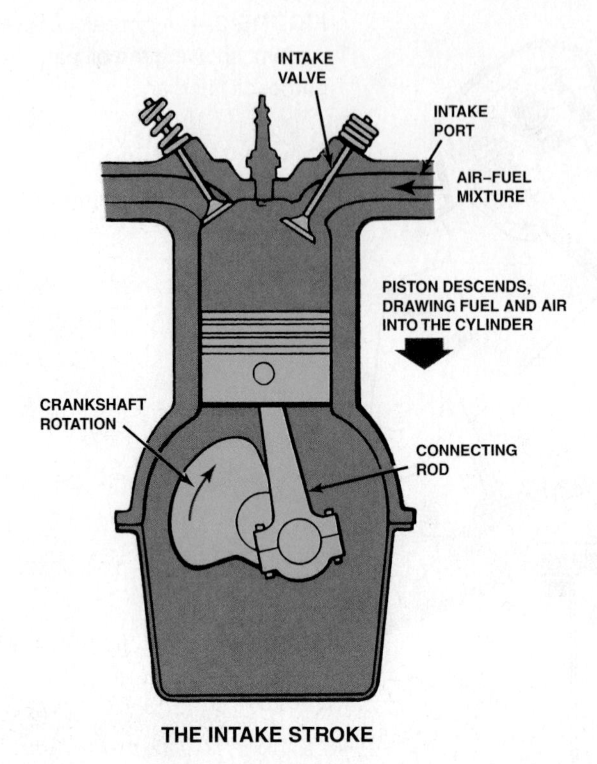

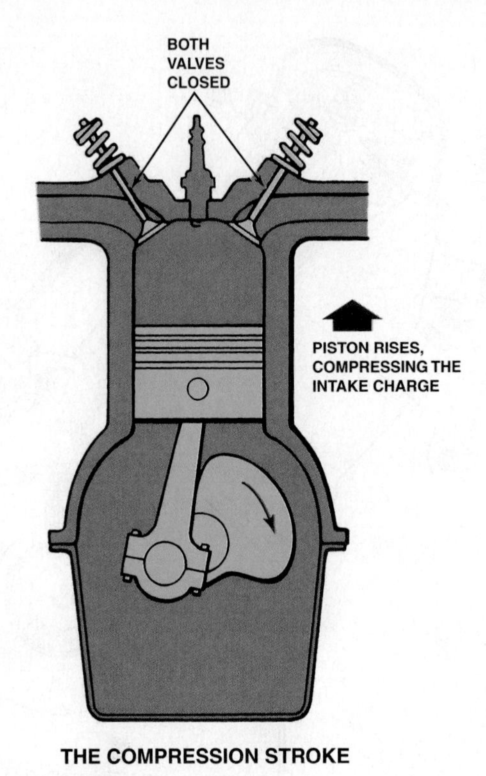

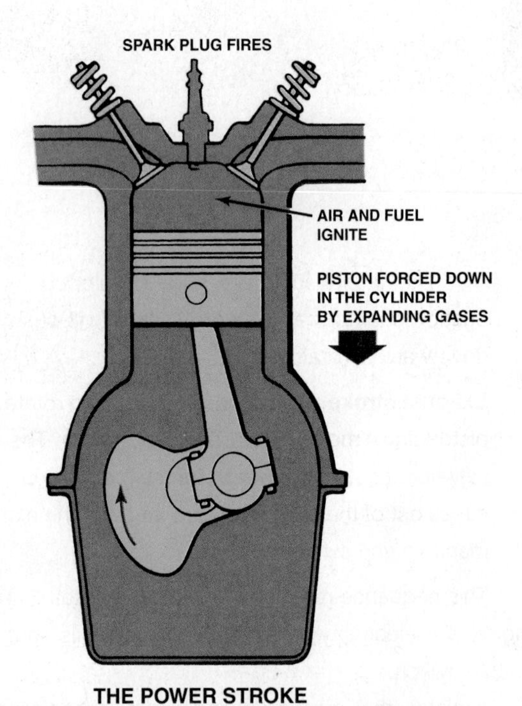

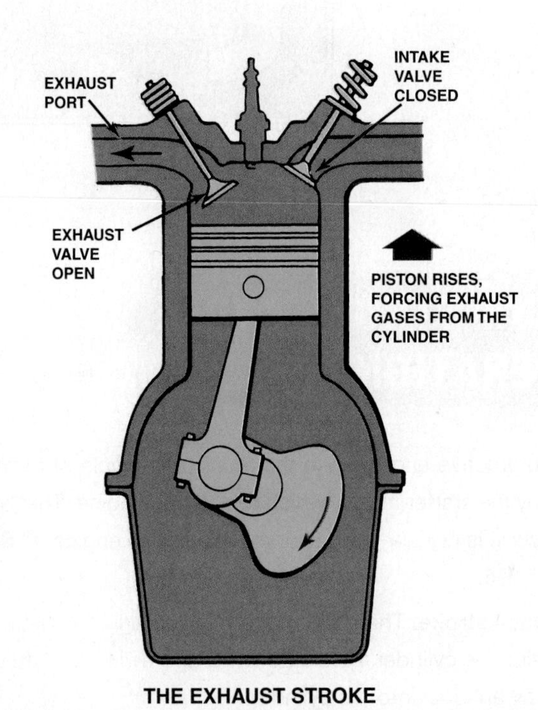

FIGURE 3–5 The downward movement of the piston draws the air–fuel mixture into the cylinder through the intake valve on the intake stroke. On the compression stroke, the mixture is compressed by the upward movement of the piston with both valves closed. Ignition occurs at the beginning of the power stroke, and combustion drives the piston downward to produce power. On the exhaust stroke, the upward-moving piston forces the burned gases out the open exhaust valve.

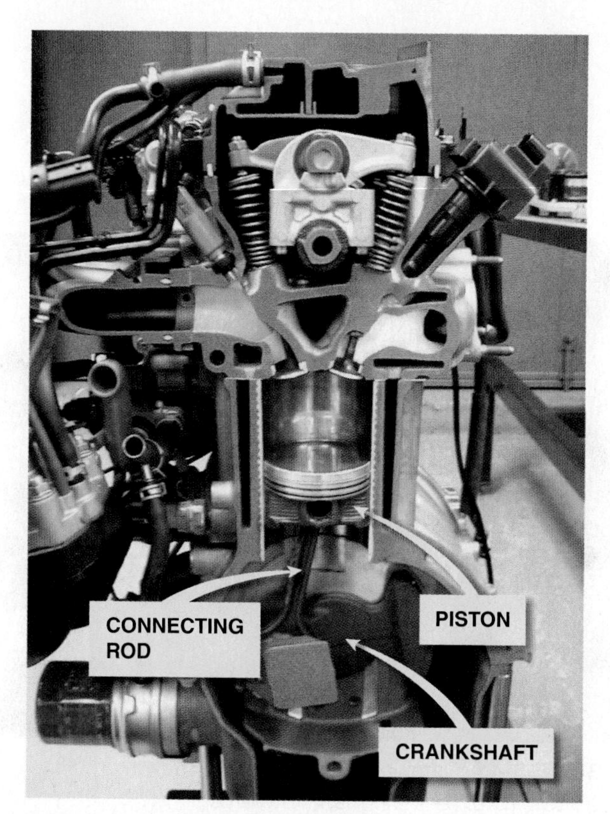

FIGURE 3–6 Cutaway of an engine showing the cylinder, piston, connecting rod, and crankshaft.

THE 720° CYCLE

Each cycle of events requires that the engine crankshaft make two complete revolutions or 720° (360° × 2 = 720°). The greater the number of cylinders, the closer together the power strokes occur. To find the angle between cylinders of an engine, divide the number of cylinders into 720°.

Angle with three cylinders = 720°/3 = 240°

Angle with four cylinders = 720°/4 = 180°

Angle with five cylinders = 720°/5 = 144°

Angle with six cylinders = 720°/6 = 120°

Angle with eight cylinders = 720°/8 = 90°

Angle with 10 cylinders = 720°/10 = 72°

This means that in a four-cylinder engine, a power stroke occurs at every 180° of the crankshaft rotation (every 1/2 rotation). A V-8 is a much smoother operating engine because a power stroke occurs twice as often (every 90° of crankshaft rotation).

Engine cycles are identified by the number of piston strokes required to complete the cycle. A **piston stroke** is a one-way piston movement between the top and bottom of the cylinder. During one stroke, the crankshaft revolves 180° (1/2 revolution). A **cycle** is a complete series of events that continually repeat. Most automobile engines use a **four-stroke cycle**.

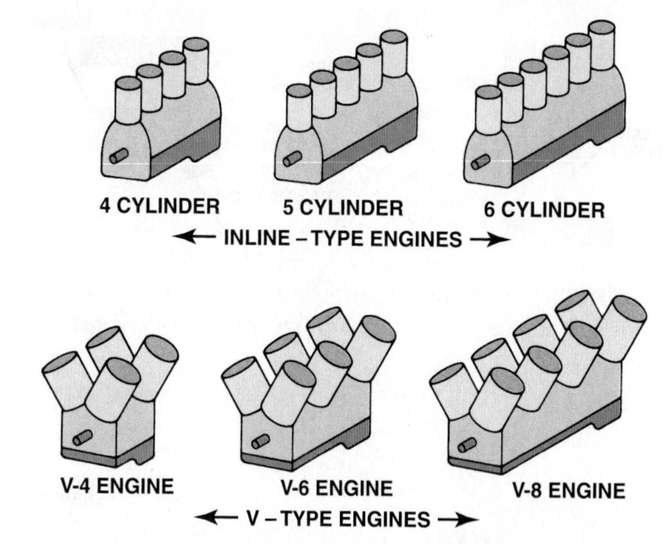

FIGURE 3–7 Automotive engine cylinder arrangements.

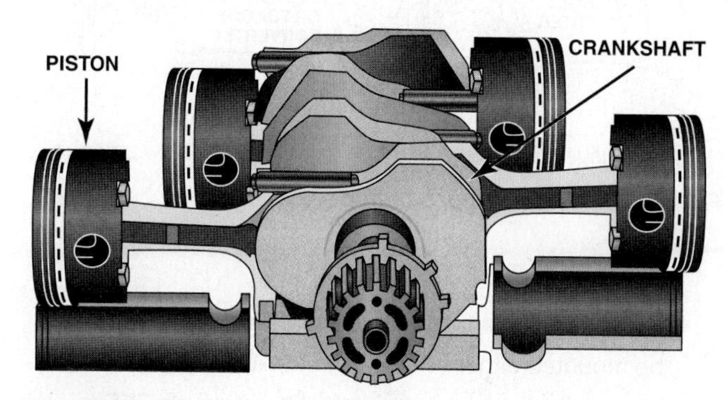

FIGURE 3–8 A horizontally opposed engine design helps to lower the vehicle's center of gravity.

ENGINE CLASSIFICATION AND CONSTRUCTION

Engines are classified by several characteristics including:

- **Number of strokes.** Most automotive engines use the four-stroke cycle.
- **Cylinder arrangement.** An engine with more cylinders is smoother operating because the power pulses produced by the power strokes are more closely spaced. An inline engine places all cylinders in a straight line. Four-, five-, and six-cylinder engines are commonly manufactured inline engines. A V-type engine, such as a V-6 or V-8, has the number of cylinders split and built into a V-shape. ● **SEE FIGURE 3–7.** Horizontally opposed four- and six-cylinder engines have two banks of cylinders that are horizontal, resulting in a low engine. This style of engine is used in Porsche and Subaru engines and is often called the **boxer** or **pancake engine** design. ● **SEE FIGURE 3–8.**

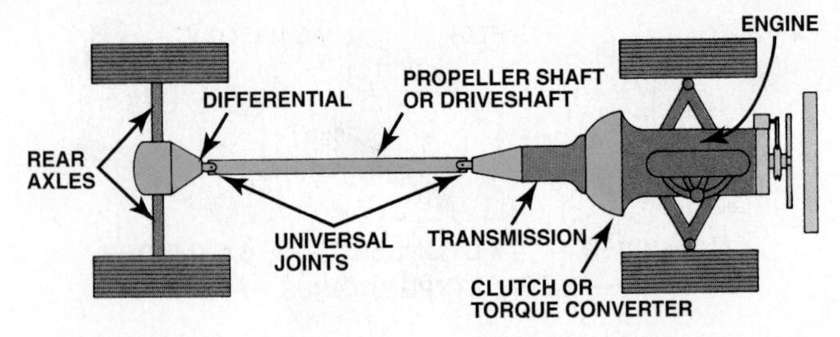

FIGURE 3–9 A longitudinally mounted engine drives the rear wheels through a transmission, driveshaft, and differential assembly.

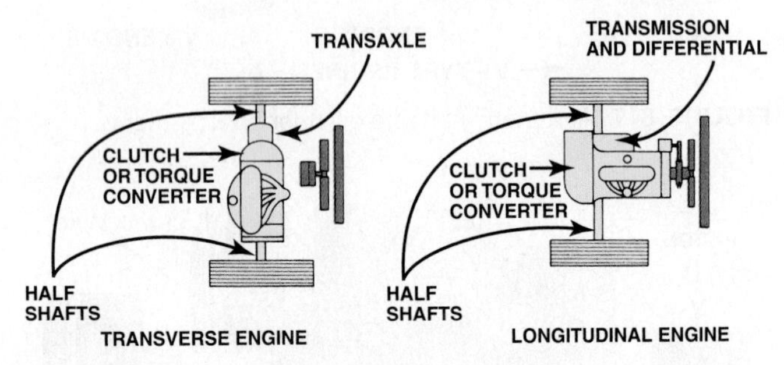

FIGURE 3–10 Two types of front-engine, front-wheel drive.

FIGURE 3–11 Cutaway of a V-8 engine showing the lifters, pushrods, roller rocker arms, and valves.

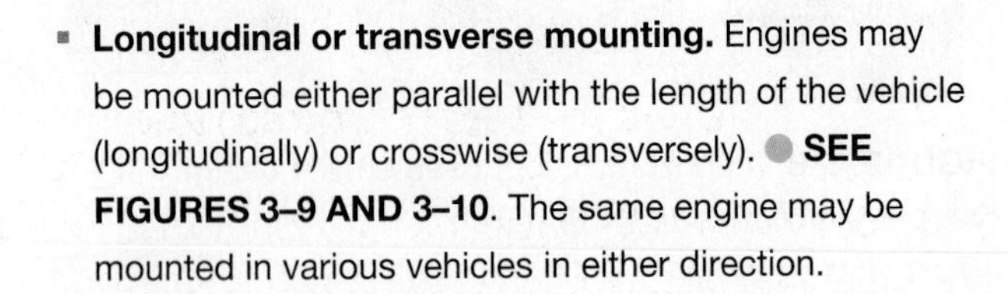

- **Longitudinal or transverse mounting.** Engines may be mounted either parallel with the length of the vehicle (longitudinally) or crosswise (transversely). ● **SEE FIGURES 3–9 AND 3–10.** The same engine may be mounted in various vehicles in either direction.

NOTE: Although it might be possible to mount an engine in different vehicles both longitudinally and transversely, the engine component parts may *not* be interchangeable. Differences can include different engine blocks and crankshafts, as well as different water pumps.

- **Valve and camshaft number and location.** The number of valves and the number and location of camshafts are a major factor in engine operation. A typical older-model engine uses one intake valve and one exhaust valve per cylinder. Many newer engines use two intake and two exhaust valves per cylinder. The valves are opened by a **camshaft**. For high-speed engine operation, the camshaft should be overhead (over the valves). Some engines use one camshaft for the intake valves and a separate camshaft for the exhaust valves. When the camshaft is located in the block, the valves are operated by lifters, pushrods, and rocker arms. ● **SEE FIGURE 3–11.** This type of engine is called a **pushrod engine** or **cam-in-block design**.

  An overhead camshaft engine has the camshaft above the valves in the cylinder head. When one overhead camshaft is used, the design is called a **single overhead camshaft (SOHC)** design. When two overhead camshafts are used, the design is called a **double overhead camshaft (DOHC)** design. See ● **SEE FIGURES 3–12 AND 3-13.**

NOTE: A V-type engine uses two banks or rows of cylinders. An SOHC design therefore uses two camshafts, but only one camshaft per bank (row) of cylinders. A DOHC V-6, therefore, has four camshafts, two for each bank.

- **Type of fuel.** Most engines operate on gasoline, whereas other engines are designed to operate on methanol, natural gas, propane, or diesel fuel.

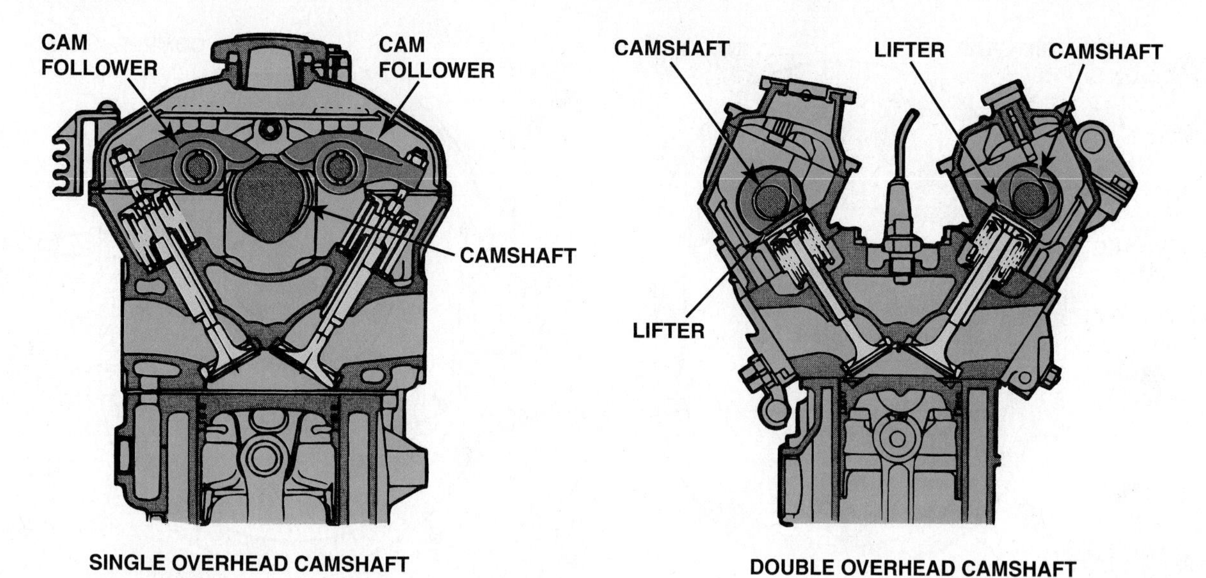

FIGURE 3–12 SOHC engines usually require additional components such as a rocker arm to operate all of the valves. DOHC engines often operate the valves directly.

FIGURE 3–13 A DOHC engine uses a camshaft for the intake valve and a separate camshaft for the exhaust valves in each cylinder head.

- **Cooling method.** Most engines are liquid cooled, but some older models were air cooled.
- **Type of induction pressure.** If atmospheric air pressure is used to force the air–fuel mixture into the cylinders, the engine is called **naturally aspirated**. Some engines use a **turbocharger** or **supercharger** to force the air–fuel mixture into the cylinder for even greater power.

? FREQUENTLY ASKED QUESTION

What Is a Rotary Engine?

A successful alternative engine design is the **rotary engine**, also called the **Wankel engine** after its inventor. The Mazda RX-7 and RX-8 represent the only long-term use of the rotary engine. The rotating combustion chamber engine runs smoothly, and it produces high power for its size and weight.

The basic rotating combustion chamber engine has a triangular-shaped rotor turning in a housing. The housing is in the shape of a geometric figure called a two-lobed epitrochoid. A seal on each corner, or apex, of the rotor is in constant contact with the housing, so the rotor must turn with an eccentric motion. This means that the center of the rotor moves around the center of the engine. The eccentric motion can be seen in ● Figure 3–14.

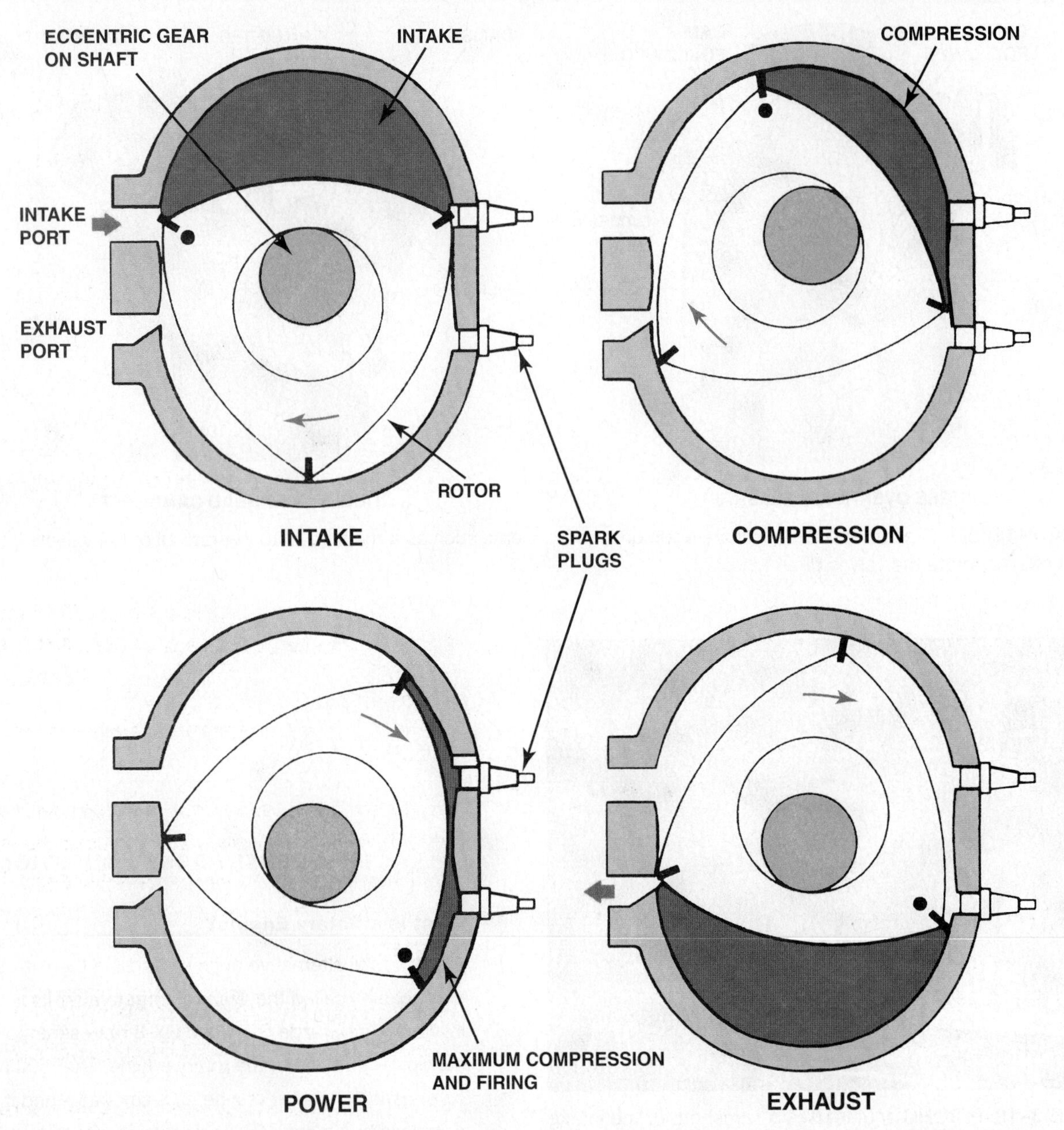

FIGURE 3–14 Rotary engine operates on the four-stroke cycle but uses a rotor instead of a piston and crankshaft to achieve intake, compression, power, and exhaust stroke.

ENGINE ROTATION DIRECTION

The SAE standard for automotive engine rotation is counterclockwise (CCW) as viewed from the flywheel end (clockwise as viewed from the front of the engine). The flywheel end of the engine is the end to which the power is applied to drive the vehicle. This is called the **principal end** of the engine. The **nonprincipal end** of the engine is opposite the principal end and is generally referred to as the ***front*** of the engine, where the accessory belts are used. ● **SEE FIGURE 3–15.**

In most rear-wheel-drive vehicles, therefore, the engine is mounted longitudinally with the principal end at the rear of the engine. Most transversely mounted engines also adhere to the same standard for direction of rotation. Many Honda engines and some marine applications may differ from this standard.

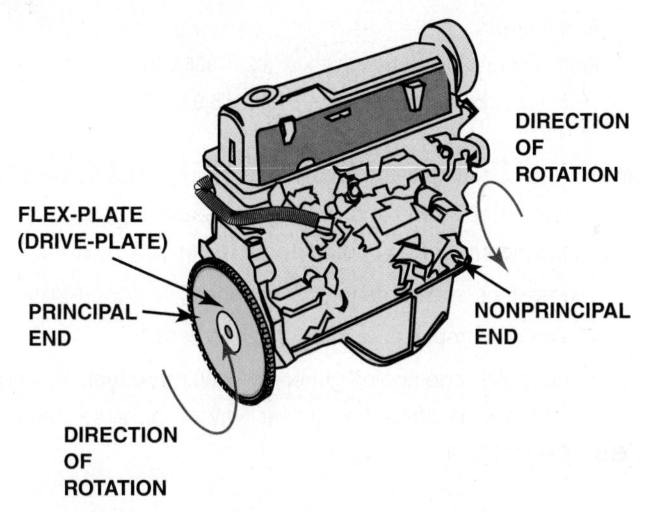

FIGURE 3–15 Inline four-cylinder engine showing principal and nonprincipal ends. Normal direction of rotation is clockwise (CW) as viewed from the front or accessory belt end (nonprincipal end).

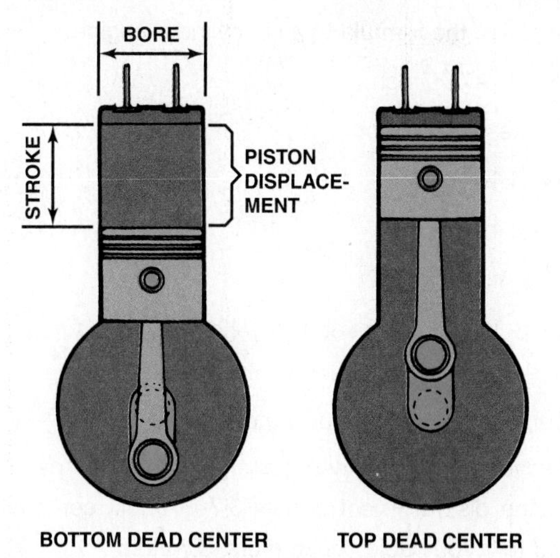

FIGURE 3–16 The bore and stroke of pistons are used to calculate an engine's displacement.

BORE

The diameter of a cylinder is called the **bore**. The larger the bore, the greater the area on which the gases have to work. Pressure is measured in units, such as pounds per square inch (psi). The greater the area (in square inches), the higher the force exerted by the pistons to rotate the crankshaft. ● **SEE FIGURE 3–16.**

FREQUENTLY ASKED QUESTION

What Is the Atkinson Cycle?

In 1882, James Atkinson, a British engineer, invented an engine that achieved a higher efficiency than the Otto cycle but produced lower power at low engine speeds. The Atkinson cycle engine was produced in limited numbers until 1890, when sales dropped, and the company that manufactured the engines finally went out of business in 1893.

However, the one key feature of the Atkinson cycle that remains in use today is that the intake valve is held open longer than normal to allow a reverse flow into the intake manifold. This reduces the effective compression ratio and engine displacement and allows the expansion to exceed the compression ratio while retaining a normal compression pressure. This is desirable for good fuel economy because the compression ratio in a spark ignition engine is limited by the octane rating of the fuel used, while a high expansion delivers a longer power stroke and reduces the heat wasted in the exhaust. This increases the efficiency of the engine because more work is being achieved. The Atkinson cycle engine design is commonly used in hybrid electric vehicles.

STROKE

The distance the piston travels down in the cylinder is called the **stroke**. The longer this distance is, the greater the amount of air–fuel mixture that can be drawn into the cylinder. The more air–fuel mixture inside the cylinder, the more force will result when the mixture is ignited.

ENGINE DISPLACEMENT

Engine size is described as displacement. **Displacement** is the cubic inch (cu. inch) or cubic centimeter (cc) volume displaced or swept by all of the pistons. A liter (L) is equal to 1,000 cubic centimeters; therefore, most engines today are identified by their displacement in liters.

1 L = 1,000 cc

1 L = 61 cu. inch

1 cu. inch = 16.4 cc

The formula to calculate the displacement of an engine is basically the formula for determining the volume of a cylinder multiplied by the number of cylinders.

The formula is:

$$\textbf{Cubic inch displacement} = \pi\textbf{(p)} \times \textbf{R}^2 \times \textbf{Stroke} \times \textbf{Number of cylinders}$$

R = Radius of the cylinder or one-half of the bore.

The πR^2 part is the formula for the area of a circle.

Applying the formula to a six-cylinder engine:

- Bore = 4.000 inch
- Stroke = 3.000 inch
- $\pi = 3.14$
- R = 2 inches
- $R^2 = 4$ (2^2 or 2×2)

Cubic inches = 3.14×4 (R^2) $\times 3$ (stroke) $\times 6$ (number of cylinders).

Cubic inches = 226 cubic inches

Because 1 cubic inch equals 16.4 cubic centimeters, this engine displacement equals 3,706 cubic centimeters or, rounded to 3,700 cubic centimeters, 3.7 liters.

How to convert cubic inches to liters: 61.02 cubic inches = 1 liter

Example:

From liter to cubic inch—5.0 L × 61.02 = 305 CID

From cubic inch to liter—305 ÷ 61.02 = 5.0 L

ENGINE SIZE VERSUS HORSEPOWER The larger the engine, the more power the engine is capable of producing. Several sayings are often quoted about the engine size:

"There is no substitute for cubic inches."

"There is no replacement for displacement."

Although a large engine generally uses more fuel, making an engine larger is often the easiest way to increase power. ● **SEE CHART 3–1**.

TECH TIP

All 3.8-Liter Engines Are Not the Same!

Most engine sizes are currently identified by displacement in liters. However, not all 3.8-liter engines are the same. See, for example, the following table:

Engine	Displacement
Chevrolet-built 3.8-L, V-6	229 cu. inch
Buick-built 3.8-L, V-6 (also called 3,800 cc)	231 cu. inch
Ford-built 3.8-L, V-6	232 cu. inch

The exact conversion from liters (or cubic centimeters) to cubic inches is 231.9 cubic inches. However, due to rounding of exact cubic-inch displacement and rounding of the exact cubic- centimeter volume, several entirely different engines can be marketed with the exact same liter designation. To reduce confusion and reduce the possibility of ordering incorrect parts, the vehicle identification number (VIN) should be noted for the vehicle being serviced. The VIN should be visible through the windshield on all vehicles. Since 1980, the engine identification number or letter is usually the eighth digit or letter from the left.

Smaller, four-cylinder engines can also cause confusion because many vehicle manufacturers use engines from both overseas and domestic manufacturers. Always refer to service manual information to be assured of correct engine identification.

LITERS	CUBIC INCHES	LITERS	CUBIC INCHES
1.0	61	4.3	260/262/265
1.3	79	4.4	267
1.4	85	4.5	273
1.5	91	4.6	280/281
1.6	97/98	4.8	292
1.7	105	4.9	300/301
1.8	107/110/112	5.0	302/304/305/307
1.9	116	5.2	318
2.0	121/122	5.3	327
2.1	128	5.4	330
2.2	132/133/134/135	5.7	350
2.3	138/140	5.8	351
2.4	149	5.9	360
2.5	150/153	6.0	366/368
2.6	156/159	6.1	370
2.8	171/173	6.2	381
2.9	177	6.4	389/390/391
3.0	181/182/183	6.5	396
3.1	191	6.6	400
3.2	196	6.9	420
3.3	200/201	7.0	425/427/428/429
3.4	204	7.2	440
3.5	215	7.3	445
3.7	225	7.4	454
3.8	229/231/232	7.5	460
3.9	239/240	7.8	475/477
4.0	241/244	8.0	488
4.1	250/252	8.4	510
4.2	255/258	8.8	534

CHART 3–1

Engine size conversion chart—liters to cubic inches.

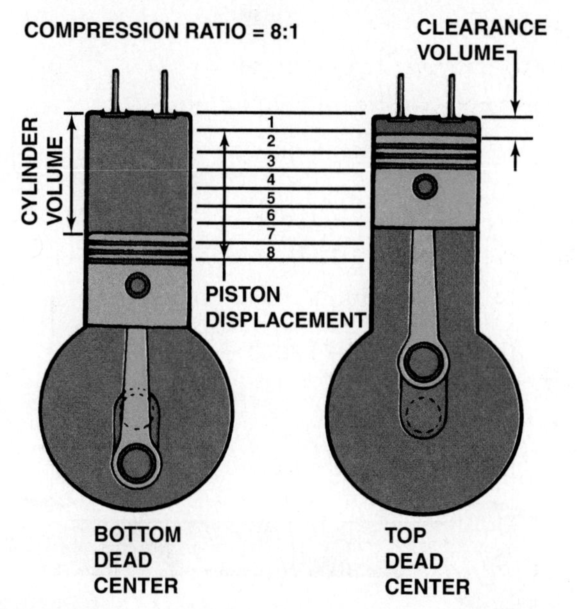

FIGURE 3–17 Compression ratio is the ratio of the total cylinder volume (when the piston is at the bottom of its stroke) to the clearance volume (when the piston is at the top of its stroke).

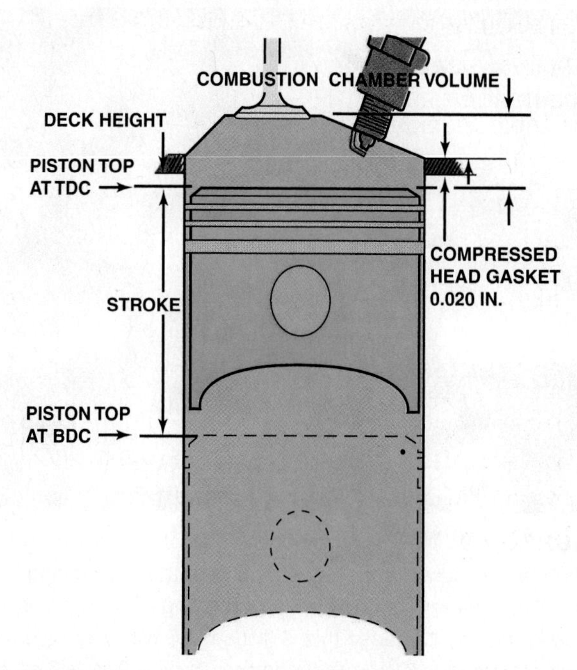

FIGURE 3–18 Combustion chamber volume is the volume above the piston with the piston at top dead center.

IF COMPRESSION IS LOWER	IF COMPRESSION IS HIGHER
Lower power	Higher power possible
Poorer fuel economy	Better fuel economy
Easier engine cranking	Harder to crank engine, especially when hot
More advanced ignition timing possible without spark knock (detonation)	Less ignition timing required to prevent spark knock (detonation)

CHART 3–2

A comparison between compression ratio and how it affects engine operation.

COMPRESSION RATIO

The compression ratio of an engine is an important consideration when rebuilding or repairing an engine. **Compression ratio (CR)** is the ratio of the volume in the cylinder above the piston when the piston is at the bottom of the stroke to the volume in the cylinder above the piston when the piston is at the top of the stroke. ● **SEE FIGURE 3–17 AND CHART 3–2.**

$$\text{CR} = \frac{\textbf{Volume in cylinder with piston at bottom of cylinder}}{\textbf{Volume in cylinder with piston at top center}}$$

● **SEE FIGURE 3–18.**

For example: What is the compression ratio of an engine with 50.3 cu. inch displacement in one cylinder and a combustion chamber volume of 6.7 cu. inch?

$$\text{CR} = \frac{50.3 + 6.7\,\text{cu.inch} = 57.0 = 8.5}{6.7\,\text{cu.inch}}$$

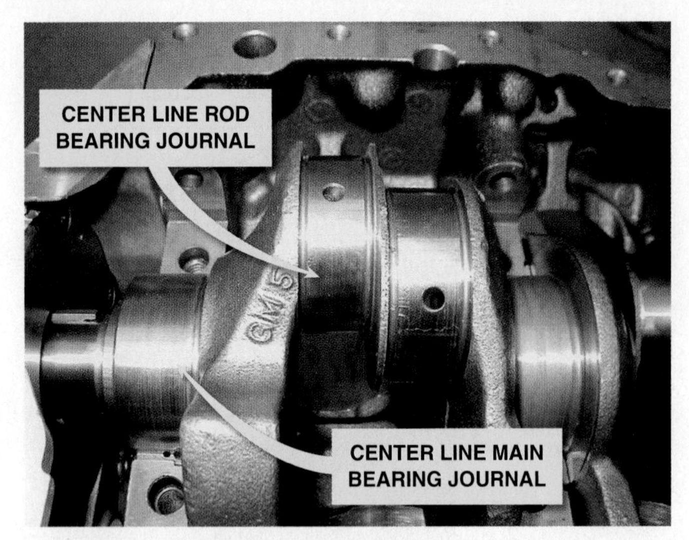

FIGURE 3–19 The distance between the centerline of the main bearing journal and the centerline of the connecting rod journal determines the stroke of the engine. This photo is a little unusual because this is from a V-6 with a splayed crankshaft used to even out the impulses on a 90°, V-6 engine design.

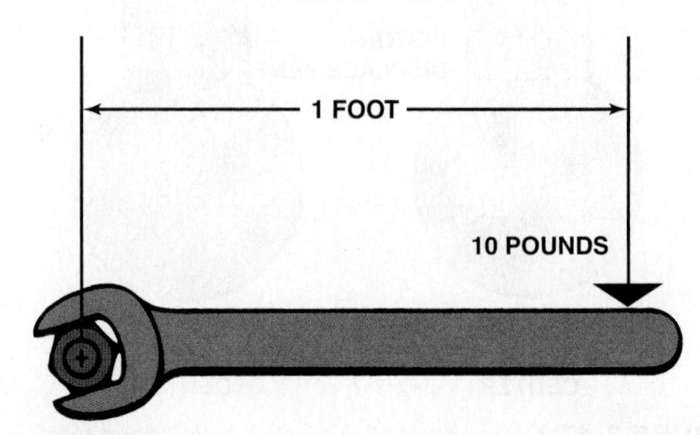

FIGURE 3–20 Torque is a twisting force equal to the distance from the pivot point times the force applied expressed in units called pound-feet (lb-ft) or Newton-meters (N-m).

THE CRANKSHAFT DETERMINES THE STROKE

The stroke of an engine is the distance the piston travels from top dead center (TDC) to bottom dead center (BDC). This distance is determined by the throw of the crankshaft. The throw is the distance from the centerline of the crankshaft to the centerline of the crankshaft rod journal. The throw is one-half of the stroke. ● **SEE FIGURE 3–19** for an example of a crankshaft as installed in a GM V-6 engine.

If the crankshaft is replaced with one with a greater stroke, the pistons will be pushed up over the height of the top of the block (deck). The solution to this problem is to install replacement pistons with the piston pin relocated higher on the piston. Another alternative is to replace the connecting rod with a shorter one to prevent the piston from traveling too far up in the cylinder. Changing the connecting rod length does *not* change the stroke of an engine. Changing the connecting rod only changes the position of the piston in the cylinder.

TORQUE

Torque is the term used to describe a rotating force that may or may not result in motion. Torque is measured as the amount of force multiplied by the length of the lever through which it acts. If a one-foot-long wrench is used to apply 10 pounds of force to the end of the wrench to turn a bolt, then you are exerting 10 pound-feet of torque. ● **SEE FIGURE 3–20**.

The metric unit for torque is Newton-meters because Newton is the metric unit for force and the distance is expressed in meters.

one pound-foot = 1.3558 Newton-meters

one Newton-meter = 0.7376 pound-foot

POWER

The term *power* means the rate of doing work. Power equals work divided by time. Work is achieved when a certain amount of mass (weight) is moved a certain distance by a force. It does not make a difference in the amount of work accomplished if the object is moved in 10 seconds or 10 minutes, but it does affect the amount of power needed. Power is expressed in units of foot-pounds per minute.

HORSEPOWER AND ALTITUDE

Because the density of the air is lower at high altitude, the power that a normal engine can develop is greatly reduced at high altitude. According to SAE conversion factors, a non-supercharged or nonturbocharged engine loses about 3% of its power for every 1,000 feet (300 meters [m]) of altitude.

Therefore, an engine that develops 150 brake horsepower at sea level will only produce about 85 brake horsepower at the top of Pike's Peak in Colorado at 14,110 feet (4,300 meters). Supercharged and turbocharged engines are not as greatly affected by altitude as normally aspirated engines. "Normally aspirated," remember, means engines that breathe air at normal atmospheric pressure.

? FREQUENTLY ASKED QUESTION

What's with These Kilowatts?

A watt is the electrical unit for power, the capacity to do work. It is named after a Scottish inventor, James Watt (1736-1819). The symbol for power is P. Electrical power is calculated as amperes times volts:

$$P \text{ (power)} = I \text{ (amperes)} \cdot E \text{ (volts)}$$

Engine power is commonly rated in watts or kilowatts (1,000 watts equal 1 kilowatt), because 1 horsepower is equal to 746 watts. For example, a 200 horsepower engine can be rated in the metric system, as having the power equal to 149,200 watts or 149.2 kilowatts (kW).

SUMMARY

1. The four strokes of the four-stroke cycle are intake, compression, power, and exhaust.
2. Engines are classified by number and arrangement of cylinders and by number and location of valves and camshafts, as well as by type of mounting, fuel used, cooling method, and induction pressure.
3. Most engines rotate clockwise as viewed from the front (accessory) end of the engine. The SAE standard is counterclockwise as viewed from the principal (flywheel) end of the engine.
4. Engine size is called displacement and represents the volume displaced by all of the pistons.

REVIEW QUESTIONS

1. Name the strokes of a four-stroke cycle.
2. If an engine at sea level produces 100 horsepower, how many horsepower would it develop at 6,000 feet of altitude?

CHAPTER QUIZ

1. All overhead valve engines ________.
   a. Use an overhead camshaft
   b. Have the overhead valves in the head
   c. Operate by the rotary cycle
   d. Use the camshaft to close the valves
2. How many camshafts does a SOHC V-8 engine have?
   a. One
   b. Two
   c. Three
   d. Four
3. The coolant flow through the radiator is controlled by the ________.
   a. Size of the passages in the block
   b. Thermostat
   c. Cooling fan(s)
   d. Water pump
4. Torque is expressed in units of ________.
   a. Pound-feet
   b. Foot-pounds
   c. Foot-pounds per minute
   d. Pound-feet per second
5. Horsepower is expressed in units of ________.
   a. Pound-feet
   b. Foot-pounds
   c. Foot-pounds per minute
   d. Pound-feet per second
6. A normally aspirated automobile engine loses about ________ power per 1,000 feet of altitude.
   a. 1%
   b. 3%
   c. 5%
   d. 6%
7. One cylinder of an automotive four-stroke-cycle engine completes a cycle every ________.
   a. 90°
   b. 180°
   c. 360°
   d. 720°
8. How many rotations of the crankshaft are required to complete each stroke of a four-stroke cycle engine?
   a. One-fourth
   b. One-half
   c. One
   d. Two
9. A rotating force is called ________.
   a. Horsepower
   b. Torque
   c. Combustion pressure
   d. Eccentric movement
10. Technician A says that a crankshaft determines the stroke of an engine. Technician B says that the length of the connecting rod determines the stroke of an engine. Which technician is correct?
    a. Technician A only
    b. Technician B only
    c. Both Technicians A and B
    d. Neither Technician A nor B

chapter 4

DIESEL ENGINE OPERATION AND DIAGNOSIS

LEARNING OBJECTIVES

After studying this chapter, reader will be able to:

1. Explain how a diesel engine works.
2. List the advantages and disadvantages of a diesel engine.
3. Describe the difference between direct injection (DI) and indirect injection (IDI) diesel engines.
4. List the parts of the typical diesel engine fuel system.
5. Explain how glow plugs work.
6. Describe how diesel fuel is rated and tested.

This chapter will help you prepare for Engine Performance (A8) ASE certification test content area "C" (Fuel, Air Induction and Exhaust Systems Diagnosis and Repair)

KEY TERMS

API gravity 79
Cloud point 79
Diesel exhaust fluid (DEF) 84
Diesel oxidation catalyst (DOC) 82
Differential pressure sensor (DPS) 83
Direct injection (DI) 71
Flash point 79
Glow plug 77
Heat of compression 70
High-pressure common rail (HPCR) 74
Hydraulic electronic unit injection (HEUI) 74
Indirect injection (IDI) 70
Injection pump 70
Lift pump 73
Opacity 87
Pop tester 87
Regeneration 82
Selective catalytic reduction (SCR) 84
Urea 84
Water–fuel separator 73

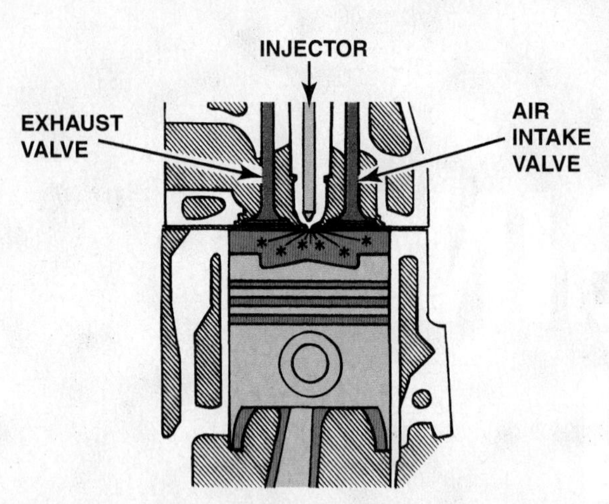

FIGURE 4–1 Diesel combustion occurs when fuel is injected into the hot, highly compressed air in the cylinder.

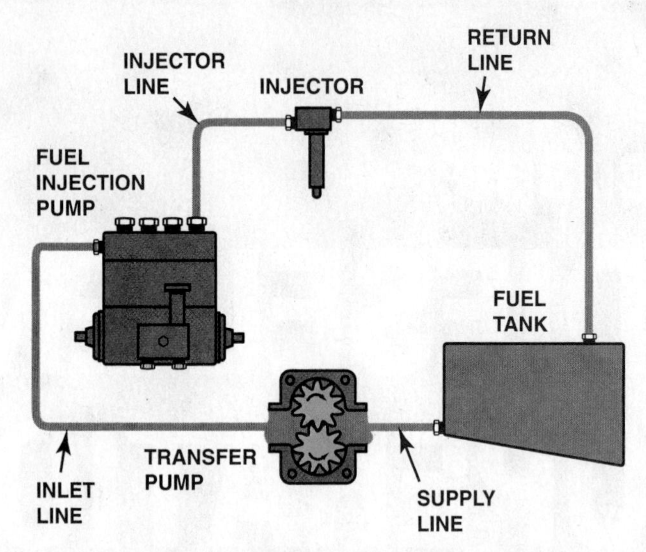

FIGURE 4–2 A typical injector-pump-type automotive diesel fuel injection system.

DIESEL ENGINES

In 1892, a German engineer, Rudolf Diesel, perfected the compression-ignition engine that bears his name. The diesel engine uses heat created by compression to ignite the fuel, so it requires no spark ignition system.

The diesel engine requires compression ratios of 16:1 and higher. Incoming air is compressed until its temperature reaches about 1000°F (540°C). This is called **heat of compression**. As the piston reaches the top of its compression stroke, fuel is injected into the cylinder, where it is ignited by the hot air. ● **SEE FIGURE 4–1**.

As the fuel burns, it expands and produces power. Because of the very high compression and torque output of a diesel engine, it is made heavier and stronger than the same size gasoline-powered engine.

A common diesel engine uses a fuel system precision **injection pump** and individual fuel injectors. The pump delivers fuel to the injectors at a high pressure and at timed intervals. Each injector sprays fuel into the combustion chamber at the precise moment required for efficient combustion. ● **SEE FIGURE 4–2**.

In a diesel engine, air is not controlled by a throttle as in a gasoline engine. Instead, the amount of fuel injected is varied to control power and speed. The air–fuel mixture of a diesel can vary from as lean as 85:1 at idle to as rich as 20:1 at full load. This higher air–fuel ratio and the increased compression pressures make the diesel more fuel-efficient than a gasoline engine in part because diesel engines do not suffer from throttling losses. Throttling losses involve the power needed in a gasoline engine to draw air past a closed or partially closed throttle.

In a gasoline engine, the speed and power are controlled by the throttle valve, which controls the amount of air entering the engine. Adding more fuel to the cylinders of a gasoline engine without adding more air (oxygen) will not increase the speed or power of the engine. In a diesel engine, speed and power are not controlled by the amount of air entering the cylinders because the engine air intake is always wide open. Therefore, the engine always has enough oxygen to burn the fuel in the cylinder and will increase speed (and power) when additional fuel is supplied.

Diesel engines are built in both two-stroke and four-stroke versions. The most common two-stroke diesels were the truck and industrial engines made by Detroit Diesel. In these engines, air intake is through ports in the cylinder wall. Exhaust is through poppet valves in the head. A blower pushes air into the air box surrounding liner ports to supply air for combustion and to blow the exhaust gases out of the exhaust valves.

INDIRECT AND DIRECT INJECTION In an **indirect injection (IDI)** diesel engine, fuel is injected into a small prechamber, which is connected to the cylinder by a narrow opening. The initial combustion takes place in this

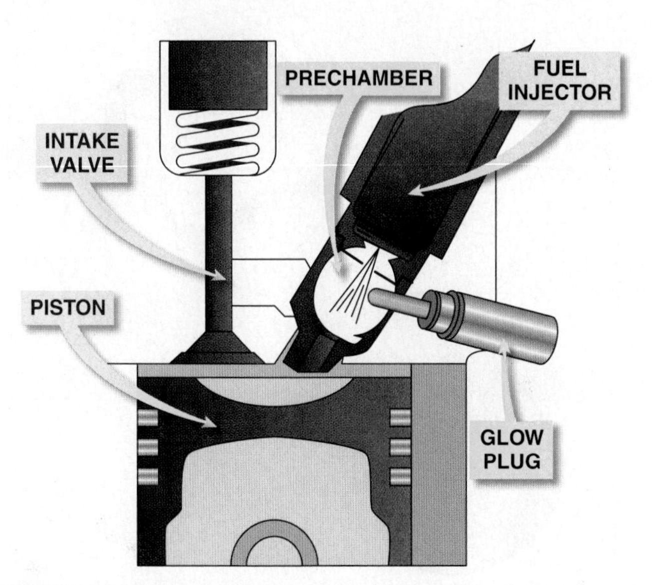

FIGURE 4–3 An indirect injection diesel engine uses a prechamber and a glow plug.

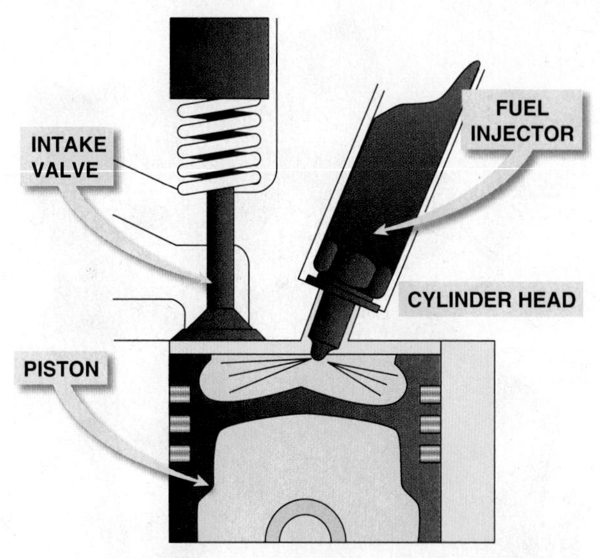

FIGURE 4–4 A direct injection diesel engine injects the fuel directly into the combustion chamber. Many designs do not use a glow plug.

prechamber. This has the effect of slowing the rate of combustion, which tends to reduce noise. ● **SEE FIGURE 4–3.**

All indirect diesel injection engines require the use of a glow plug.

In a **direct injection (DI)** diesel engine, fuel is injected directly into the cylinder. The piston incorporates a depression where initial combustion takes place. Direct injection diesel engines are generally more efficient than indirect injection engines, but have a tendency to produce greater amounts of noise. ● **SEE FIGURE 4–4.**

While some direct injection diesel engines use glow plugs to help cold starting and to reduce emissions, some direct injection diesel engines do not use glow plugs.

DIESEL FUEL IGNITION Ignition occurs in a diesel engine by injecting fuel into the air charge, which has been heated by compression to a temperature greater than the ignition point of the fuel or about 1000°F (538°C). The chemical reaction of burning the fuel liberates heat, which causes the gases to expand, forcing the piston to rotate the crankshaft. A four-stroke diesel engine requires two rotations of the crankshaft to complete one cycle. On the intake stroke, the piston passes top dead center (TDC), the intake valve(s) open, the fresh air is admitted into the cylinder, and the exhaust valve is still open for a few degrees to allow all of the exhaust gases to escape. On the compression stroke, after the piston passes bottom dead center (BDC), the intake valve closes and the piston travels up to TDC (completion of the first crankshaft rotation). On the power stroke, the piston nears TDC on the compression stroke, the diesel fuel is injected by the injectors, and the fuel starts to burn, further heating the gases in the cylinder. During this power stroke, the piston passes TDC and the expanding gases force the piston down, rotating the crankshaft. On the exhaust stroke, as the piston passes BDC, the exhaust valves open and the exhaust gases start to flow out of the cylinder. This continues as the piston travels up to TDC, pumping the spent gases out of the cylinder. At TDC, the second crankshaft rotation is complete.

FIGURE 4–5 The common rail on a Cummins diesel engine. A high-pressure pump (up to 30,000 PSI) is used to supply diesel fuel to this common rail, which has cubes running to each injector. Note the thick cylinder walls and heavy-duty construction.

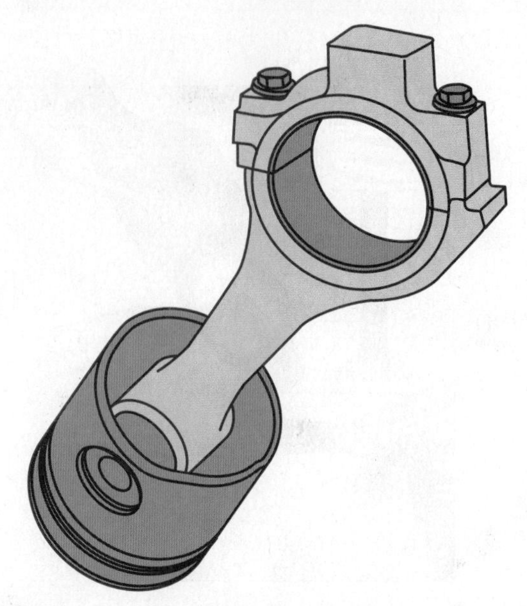

FIGURE 4–6 A rod/piston assembly from a 5.9 liter Cummins diesel engine used in a Dodge pickup truck.

THREE PHASES OF COMBUSTION

There are three distinct phases or parts to the combustion in a diesel engine:

1. **Ignition delay.** Near the end of the compression stroke, fuel injection begins, but ignition does not begin immediately. This period is called delay.
2. **Rapid combustion.** This phase of combustion occurs when the fuel first starts to burn, creating a sudden rise in cylinder pressure. It is this rise in combustion chamber pressure that causes the characteristic diesel engine knock.
3. **Controlled combustion.** After the rapid combustion occurs, the rest of the fuel in the combustion chamber begins to burn and injection continues. This is an area near the injector that contains fuel surrounded by air. This fuel burns as it mixes with the air.

DIESEL ENGINE CONSTRUCTION

Diesel engines must be constructed heavier than gasoline engines because of the tremendous pressures that are created in the cylinders during operation. The torque output of a diesel engine is often double or more than the same size gasoline powered engines. ● **SEE CHART 4–1.**

SYSTEM OR COMPONENT	DIESEL ENGINE	GASOLINE ENGINE
Block	Cast iron and heavy ● SEE FIGURE 4–5.	Cast iron or aluminum and as light as possible
Cylinder head	Cast iron or aluminum	Cast iron or aluminum
Compression ratio	17:1–25:1	8:1–12:1
Peak engine speed	2000–2500 RPM	5000–8000 RPM
Pistons and connecting rods	Aluminum with combustion pockets and heavy-duty rods ● SEE FIGURE 4–6.	Aluminum, usually flat top or with valve relief but no combustion pockets

CHART 4–1

A comparison between the features of a diesel engine to a gasoline engine.

FIGURE 4–7 Using an ice bath to test the fuel temperature sensor.

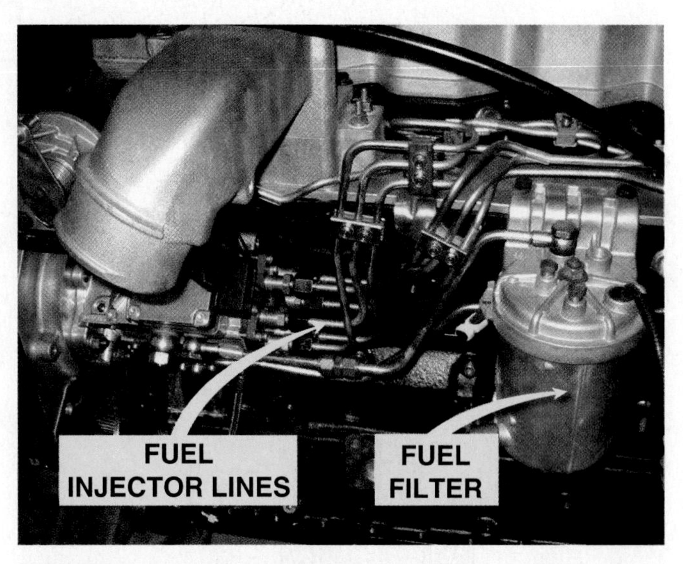

FIGURE 4–8 A typical distributor-type diesel injection pump showing the pump, lines, and fuel filter.

FUEL TANK AND LIFT PUMP

A fuel tank used on a vehicle equipped with a diesel engine differs from the one used with a gasoline engine in several ways, including:

- A larger filler neck for diesel fuel. Gasoline filler necks are smaller for the unleaded gasoline nozzle.
- No evaporative emission control devices or charcoal (carbon) canister. Diesel fuel is not as volatile as gasoline and, therefore, diesel vehicles do not have evaporative emission control devices.

The diesel fuel is drawn from the fuel tank by a **lift pump** and delivers the fuel to the injection pump. Between the fuel tank and the lift pump is a **water–fuel separator**. Water is heavier than diesel fuel and sinks to the bottom of the separator. Part of normal routine maintenance on a vehicle equipped with a diesel engine is to drain the water from the water–fuel separator. A float is usually used inside the separator, which is connected to a warning light on the dash that lights if the water reaches a level where it needs to be drained.

NOTE: Water can cause corrosive damage as well as wear to diesel engine parts because water is not a good lubricant. Water cannot be atomized by a diesel fuel injector nozzle and will often "blow out" the nozzle tip.

Many diesel engines also use a fuel temperature sensor. The computer uses this information to adjust fuel delivery based on the density of the fuel. ● **SEE FIGURE 4–7**.

INJECTION PUMP

A diesel engine injection pump is used to increase the pressure of the diesel fuel from very low values from the lift pump to the extremely high pressures needed for injection.

Injection pumps are usually driven by a gear off the camshaft at the front of the engine. As the injection pump shaft rotates, the diesel fuel is fed from a fill port to a high-pressure chamber. If a distributor-type injection pump is used, the fuel is forced out of the injection port to the correct injector nozzle through the high-pressure line. ● **SEE FIGURE 4–8**.

NOTE: Because of the very tight tolerances in a diesel engine, even the smallest amount of dirt can cause excessive damage to the engine and to the fuel injection system.

DISTRIBUTOR INJECTION PUMP A distributor diesel injection pump is a high-pressure pump assembly with lines leading to each individual injector. The high-pressure lines between the distributor and the injectors must be the exact same length to ensure proper injection timing. The injection pump itself creates the injection advance needed for engine speeds above idle and the fuel is discharged into the lines. The high-pressure fuel causes the injectors to open. Due to the internal friction of the lines, there is a slight delay before fuel pressure opens the injector nozzle. ● **SEE FIGURE 4–9**.

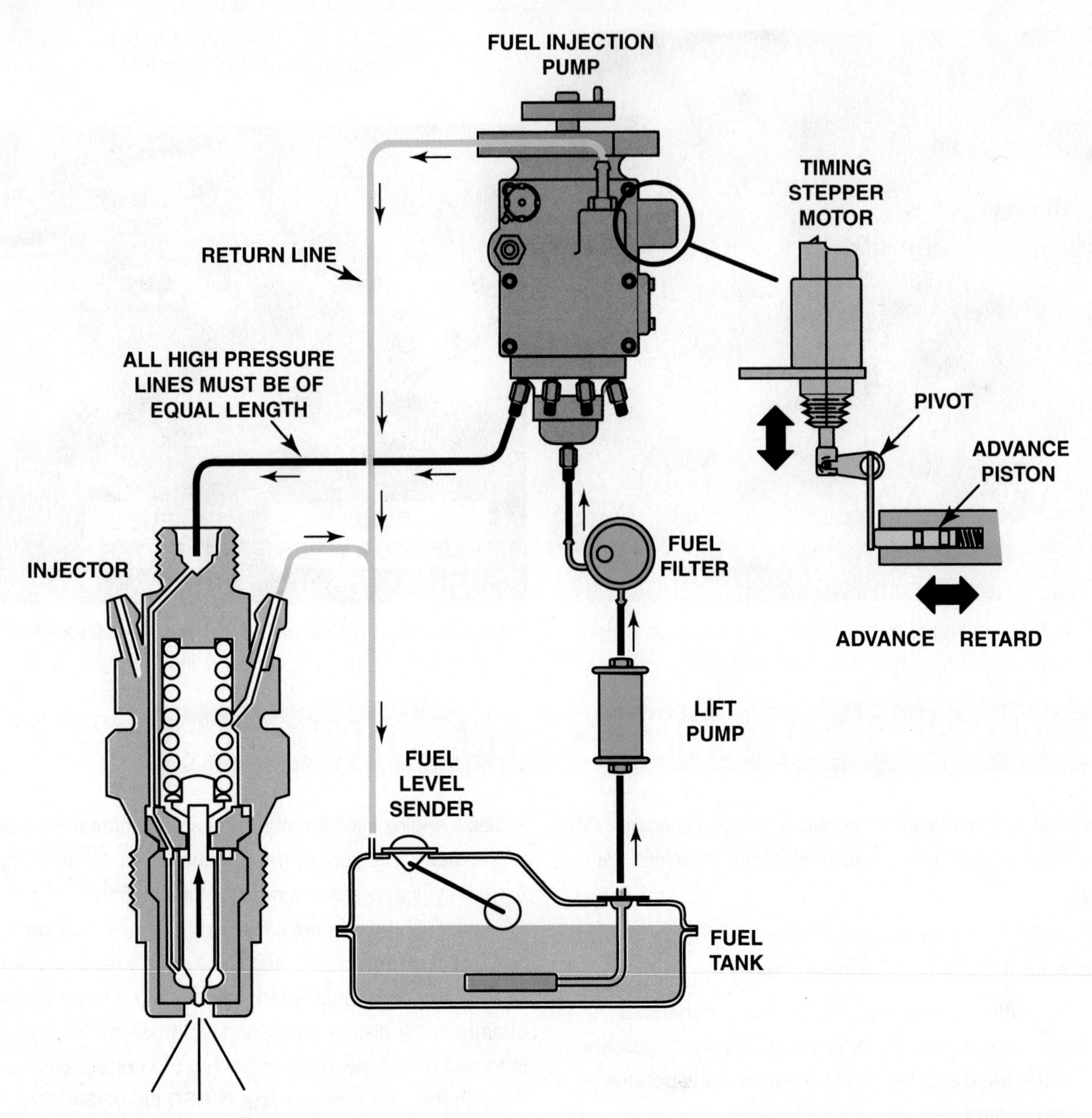

FIGURE 4–9 A schematic of a Stanadyne diesel fuel injection pump assembly showing all of the related components.

NOTE: The lines expand some during an injection event. This is how timing checks are performed. The pulsing of the injector line is picked up by a probe used to detect the injection event similar to a timing light used to detect a spark on a gasoline engine.

HIGH-PRESSURE COMMON RAIL Newer diesel engines use a fuel delivery system referred to as a **high-pressure common rail (HPCR)** design. Diesel fuel under high pressure, over 20,000 PSI (138,000 kPa), is applied to the injectors, which are opened by a solenoid controlled by the computer. Because the injectors are computer controlled, the combustion process can be precisely controlled to provide maximum engine efficiency with the lowest possible noise and exhaust emissions. ● **SEE FIGURE 4–10.**

HEUI SYSTEM

Ford 7.3 and 6.0 liter diesels use a system called a **Hydraulic Electronic Unit Injection (HEUI)** system. The components that replace the traditional mechanical injection pump include a high-pressure oil pump and reservoir, pressure regulator for the oil, and passages in the cylinder head for flow of fuel to the injectors.

Fuel is drawn from the tank by the tandem fuel pump, which circulates fuel at low pressure through the fuel filter/water separator/fuel heater bowl and then fuel is directed back to the fuel pump where fuel is pumped at high pressure into the cylinder head fuel galleries. The injectors, which are hydraulically actuated by the oil pressure from the high-pressure oil pump, are then fired by the Powertrain Control Module (PCM). The control system for

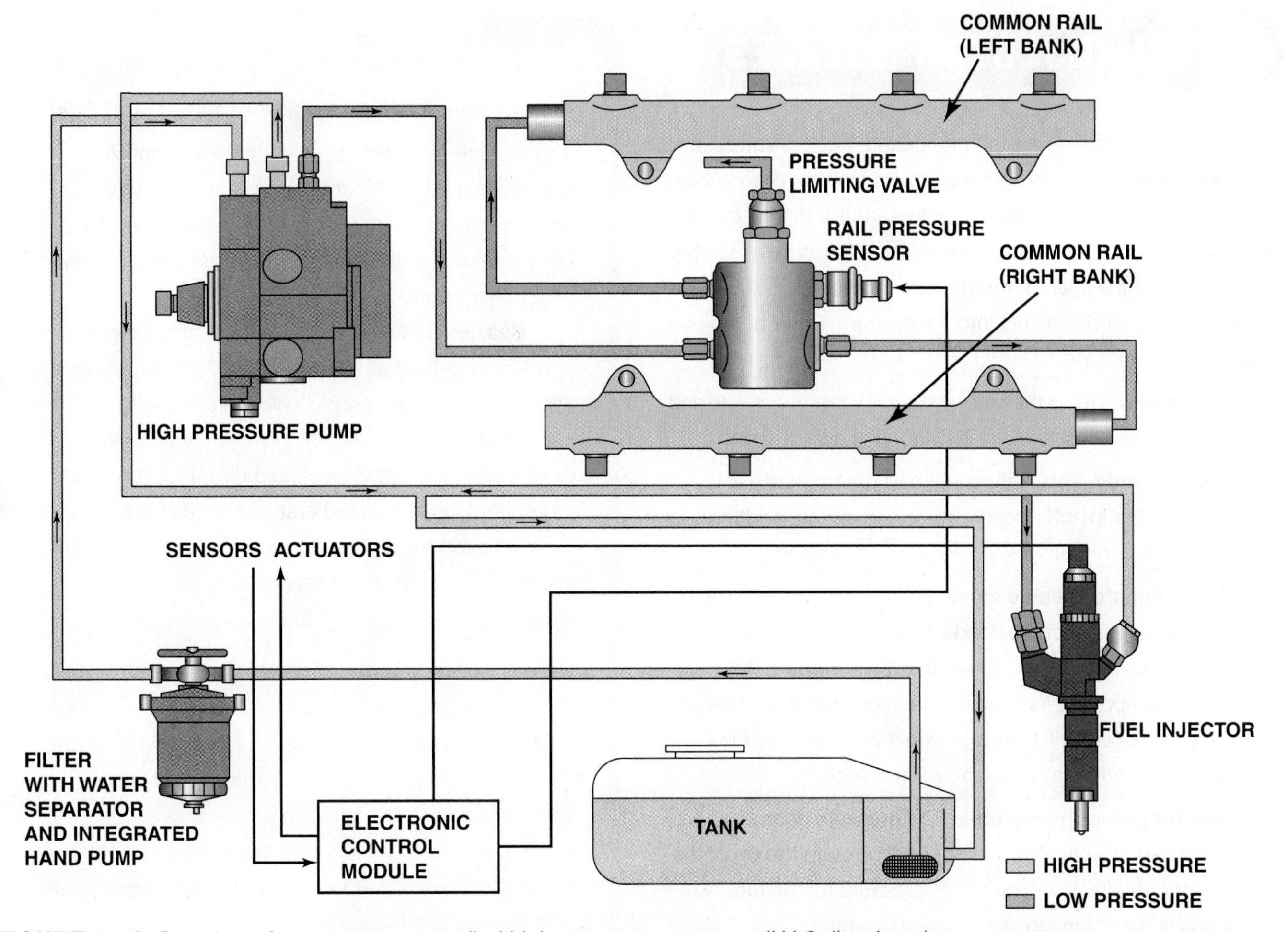

FIGURE 4–10 Overview of a computer-controlled high-pressure common rail V-8 diesel engine.

the fuel injectors is the PCM and the injectors are fired based on various inputs received by the PCM.

HEUI injectors rely on O-rings to keep fuel and oil from mixing or escaping, causing performance problems or engine damage. HEUI injectors use five O-rings. The three external O-rings should be replaced with updated O-rings if they fail. The two internal O-rings are not replaceable and if these fail, the injector or injectors must be replaced. ● **SEE FIGURE 4–11.** The most common symptoms of injector O-ring trouble include:

- Oil getting in the fuel
- The fuel filter element turning black
- Long cranking times before starting
- Sluggish performance
- Reduction in power
- Increased oil consumption often accompanies O-ring problems or any fault that lets fuel in the oil

FIGURE 4–11 A HEUI injector from a Ford PowerStroke diesel engine. The grooves indicate the location of the O-rings.

DIESEL INJECTOR NOZZLES

Diesel injector nozzles are spring-loaded closed valves that spray fuel directly into the combustion chamber or precombustion chamber. Injector nozzles are threaded into the cylinder head, one for each cylinder, and are replaceable as an assembly.

The top of the injector nozzle has many holes to deliver an atomized spray of diesel fuel into the cylinder. Parts of a diesel injector nozzle include:

- **Heat shield.** This is the outer shell of the injector nozzle and has external threads where it seals in the cylinder head.
- **Injector body.** This is the inner part of the nozzle and contains the injector needle valve and spring, and threads into the outer heat shield.
- **Diesel injector needle valve.** This precision machined valve and the tip of the needle seal against the injector body when it is closed. When the valve is open, diesel fuel is sprayed into the combustion chamber. This passage is controlled by a solenoid on diesel engines equipped with computer-controlled injection.
- **Injector pressure chamber.** The pressure chamber is a machined cavity in the injector body around the tip of the injector needle. Injection pump pressure forces fuel into this chamber, forcing the needle valve open.

TECH TIP

Never Allow a Diesel Engine to Run Out of Fuel

If a gasoline-powered vehicle runs out of gasoline, it is an inconvenience and a possible additional expense to get some gasoline. However, if a vehicle equipped with a diesel engine runs out of fuel, it can be a major concern.

Besides adding diesel fuel to the tank, the other problem is getting all of the air out of the pump, lines, and injectors so the engine will operate correctly.

The procedure usually involves cranking the engine long enough to get liquid diesel fuel back into the system while keeping cranking time short enough to avoid overheating the starter. Consult service information for the exact service procedure if the diesel engine is run out of fuel.

NOTE: Some diesel engines such as the first generation General Motors *Duramax* V-8 are equipped with a priming pump located under the hood on top of the fuel filter. Pushing down and releasing the priming pump with a vent valve open will purge any trapped air from the system. Always follow the vehicle manufacturer's instructions.

TECH TIP

Change Oil Regularly in a Ford Diesel Engine

Ford 7.3 and 6.0 liter diesel engines pump unfiltered oil from the sump to the high-pressure oil pump and then to the injectors. This means that not changing oil regularly can contribute to accumulation of dirt in the engine and will subject the fuel injectors to wear and potential damage as particles suspended in the oil get forced into the injectors.

DIESEL INJECTOR NOZZLE OPERATION

The electric solenoid attached to the injector nozzle is computer controlled and opens to allow fuel to flow into the injector pressure chamber. ● **SEE FIGURE 4–12.**

The diesel injector nozzle is mechanically opened by the high-pressure fuel delivered to the nozzle by the injector pump. The fuel flows down through a fuel passage in the injector body and into the pressure chamber. The high fuel pressure in the pressure chamber forces the needle valve upward, compressing the needle valve return spring and forcing the needle valve open. When the needle valve opens, diesel fuel is discharged into the combustion chamber in a hollow cone spray pattern.

Any fuel that leaks past the needle valve returns to the fuel tank through a return passage and line.

FIGURE 4–12 Typical computer-controlled diesel engine fuel injectors.

GLOW PLUGS

Glow plugs are always used in diesel engines equipped with a precombustion chamber and may be used in direct injection diesel engines to aid starting. A **glow plug** is a heating element that uses 12 volts from the battery and aids in the starting of a cold engine. As the temperature of the glow plug increases, the resistance of the heating element inside increases, thereby reducing the current in amperes needed by the glow plugs.

Most glow plugs used in newer vehicles are controlled by the Powertrain Control Module (PCM), which monitors coolant temperature and intake air temperature. The glow plugs are turned on or pulsed on or off depending on the temperature of the engine. The PCM will also keep the glow plug turned on after the engine starts to reduce white exhaust smoke (unburned fuel) and to improve idle quality after starting. ● **SEE FIGURE 4–13**.

The "wait to start" lamp will light when the engine and the outside temperature is low to allow time for the glow plugs to get hot. The "wait to start" lamp will not come on when the glow plugs are operating after the engine starts.

NOTE: The glow plugs are removed to test cylinder compression using a special high-pressure reading gauge.

? FREQUENTLY ASKED QUESTION

What Are Diesel Engine Advantages and Disadvantages?

A diesel engine has several advantages compared to a similar size gasoline-powered engine including:

1. More torque output
2. Greater fuel economy
3. Long service life

A diesel engine has several disadvantages compared to a similar size gasoline-powered engine including:

1. Engine noise, especially when cold and/or at idle speed
2. Exhaust smell
3. Cold weather startability
4. A vacuum pump is needed to supply the vacuum needs of the heat, ventilation, and air-conditioning system
5. Heavier than a gasoline engine. ● **SEE FIGURE 4–14**.
6. Fuel availability

ENGINE-DRIVEN VACUUM PUMP

Because a diesel engine is unthrottled, it creates very little vacuum in the intake manifold. Several engine and vehicle components operate using vacuum, such as the exhaust gas recirculation (EGR) valve and the heating and ventilation blend and air doors. Most diesels used in cars and light trucks are equipped with an engine-driven vacuum pump to supply the vacuum for these components.

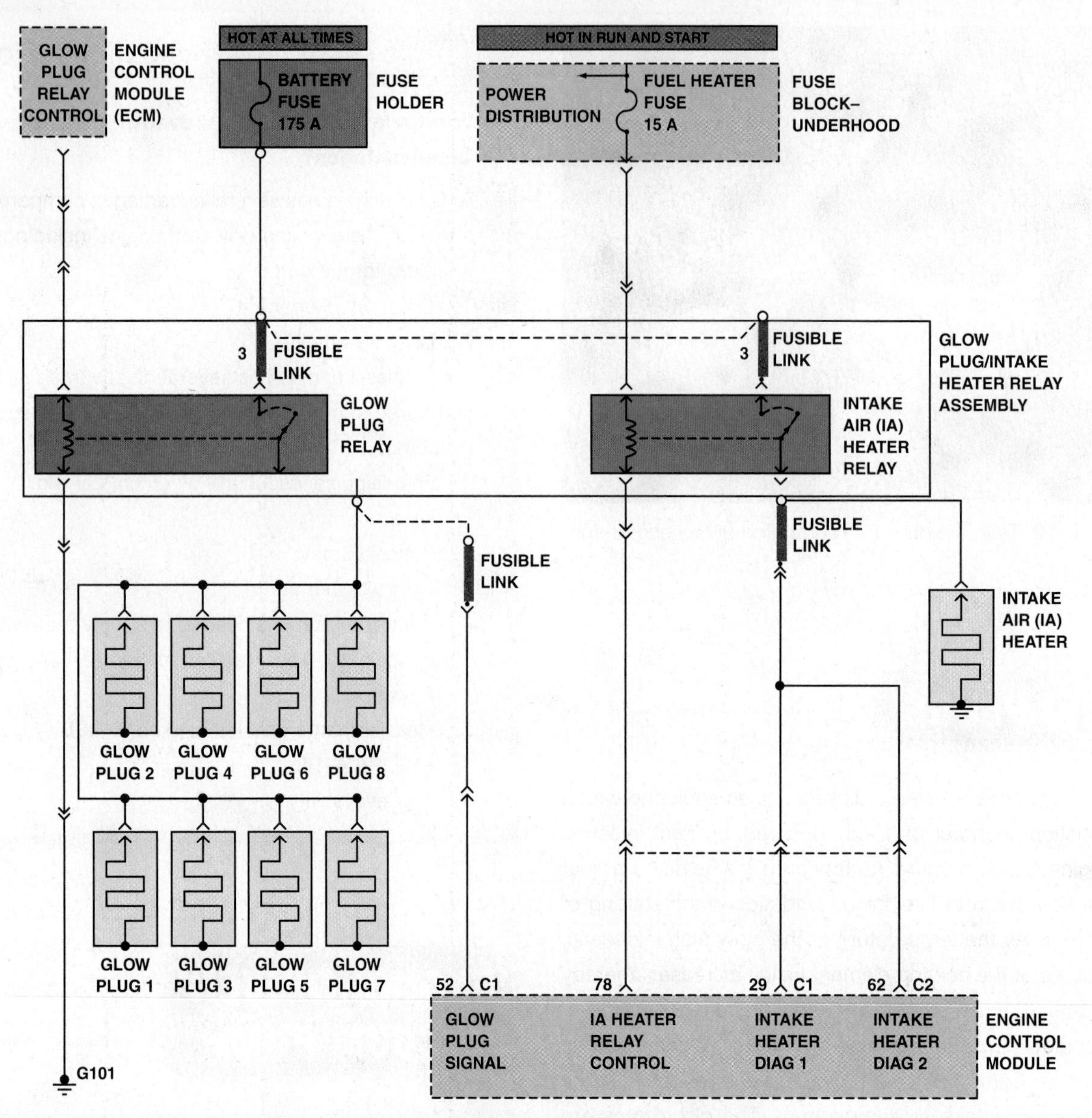

FIGURE 4–13 A schematic of a typical glow plug circuit. Notice that the relay for the glow plug and intake air heater relay are both computer controlled.

FIGURE 4–14 Roller lifter from a GM Duramax 6.6 liter V-8 diesel engine. Notice the size of this lifter compared to a roller lifter used in a gasoline engine.

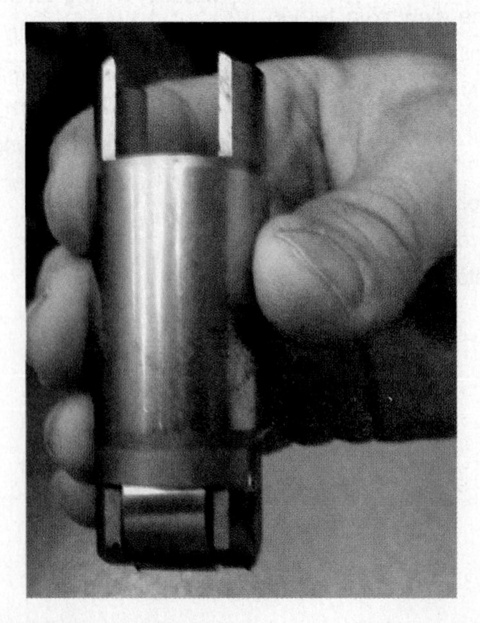

DIESEL FUEL

Diesel fuel must meet an entirely different set of standards than gasoline. The fuel in a diesel engine is not ignited with a spark, but is ignited by the heat generated by high compression. The pressure of compression (400 to 700 PSI or 2,800 to 4,800 kPa) generates temperatures of 1200°F to 1600°F (700°C to 900°C), which speeds the preflame reaction to start the ignition of fuel injected into the cylinder.

All diesel fuel must be clean, be able to flow at low temperatures, and be of the proper cetane rating.

- **Cleanliness.** It is imperative that the fuel used in a diesel engine be clean and free from water. Unlike the case with gasoline engines, the fuel is the lubricant and coolant for the diesel injector pump and injectors. Good-quality diesel fuel contains additives such as oxidation inhibitors, detergents, dispersants, rust preventatives, and metal deactivators.
- **Low-temperature fluidity.** Diesel fuel must be able to flow freely at all expected ambient temperatures. One specification for diesel fuel is its "pour point," which is the temperature below which the fuel would stop flowing. **Cloud point** is another concern with diesel fuel at lower temperatures. Cloud point is the low-temperature point at which the waxes present in most diesel fuel tend to form wax crystals that clog the fuel filter. Most diesel fuel suppliers distribute fuel with the proper pour point and cloud point for the climate conditions of the area.
- **Cetane number.** The cetane number for diesel fuel is the opposite of the octane number for gasoline. The cetane number is a measure of the ease with which the fuel can be ignited. The cetane rating of the fuel determines, to a great extent, its ability to start the engine at low temperatures and to provide smooth warm-up and even combustion. The cetane rating of diesel fuel should be between 45 and 50. The higher the cetane rating, the more easily the fuel is ignited, whereas the higher the octane rating, the more slowly the fuel burns.

Other diesel fuel specifications include its flash point, sulfur content, and classification. The **flash point** is the temperature at which the vapors on the surface of the fuel will ignite if exposed to an open flame. The flash point does ***not*** affect diesel engine operation. However, a lower than normal flash point could indicate contamination of the diesel fuel with gasoline or a similar substance.

The sulfur content of diesel fuel is very important to the life of the engine. Since 2007, all diesel fuel has to have less than 15 parts per million (PPM) of sulfur and is called ultra low sulfur diesel (ULSD). This is way down from the previous limit for low sulfur diesel of 500 PPM. Sulfur in the fuel creates sulfuric acid during the combustion process, which can damage engine components and cause piston ring wear. Federal regulations are getting extremely tight on sulfur content. High-sulfur fuel contributes to acid rain.

ASTM, formerly known as the American Society for Testing and Materials, also classifies diesel fuel by volatility (boiling range) into the following grades:

GRADE #1 This grade of diesel fuel has the lowest boiling point and the lowest cloud and pour points; it also has a lower BTU content—less heat per pound of fuel. As a result, grade #1 is suitable for use during low-temperature (winter) operation. Grade #1 produces less heat per pound of fuel compared to grade #2 and may be specified for use in diesel engines involved in frequent changes in load and speed, such as those found in city buses and delivery trucks.

GRADE #2 This grade has a higher boiling range, cloud point, and pour point as compared with grade #1. It is usually specified where constant speed and high loads are encountered, such as in long-haul trucking and automotive diesel applications.

DIESEL FUEL SPECIFIC GRAVITY TESTING

The density of diesel fuel should be tested whenever there is a driveability concern. The density or specific gravity of diesel fuel is measured in units of **API gravity**. API gravity is an arbitrary scale expressing the gravity or density of liquid petroleum

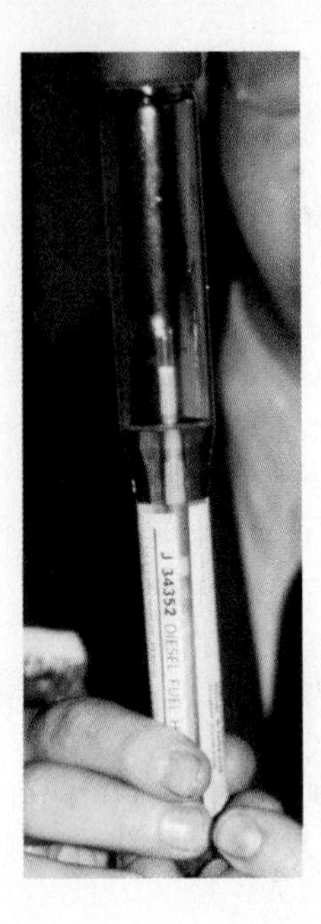

FIGURE 4–15 A hydrometer is used to measure the American Petroleum Institute (API) specific gravity of diesel fuel. The unit of measure is usually the API scale.

products devised jointly by the American Petroleum Institute and the National Bureau of Standards. The measuring scale is calibrated in terms of degrees API. Oil with the least-specific gravity has the highest API gravity. The formula for determining API gravity is as follows:

$$\text{Degrees API gravity} = (141.5/\text{specific gravity at } 60^\circ\text{F}) - 131.5$$

The normal API gravity for #1 diesel fuel is 39 to 44 (typically 40). The normal API gravity for #2 diesel fuel is 30 to 39 (typically 35). A hydrometer calibrated in API gravity units should be used to test diesel fuel. ● **SEE FIGURE 4–15 AND CHART 4–2.**

DIESEL FUEL HEATERS

Diesel fuel heaters help prevent power loss and stalling in cold weather. The heater is placed in the fuel line between the tank and the primary filter. Some coolant heaters are thermostatically controlled, which allows fuel to bypass the heater once it has reached operating temperature.

HEATED INTAKE AIR

Some diesels, such as the General Motors 6.6 liter Duramax V-8, use an electrical heater wire to warm the intake air to help in cold weather starting and running. ● **SEE FIGURE 4–16.**

VALUES FOR API SCALE OIL

API GRAVITY SCALE	SPECIFIC GRAVITY S	WEIGHT DENSITY, IB/FT P	POUNDS PER GALLON
10	1.0000	62.36	8.337
12	0.9861	61.50	8.221
14	0.9725	60.65	8.108
16	0.9593	59.83	7.998
18	0.9465	59.03	7.891
20	0.9340	58.25	7.787
22	0.9218	57.87	7.736
24	0.9100	56.75	7.587
26	0.8984	56.03	7.490
28	0.8871	55.32	7.396
30	0.8762	54.64	7.305
32	0.8654	53.97	7.215
34	0.8550	53.32	7.128
36	0.8448	52.69	7.043
38	0.8348	51.06	6.960
40	0.8251	51.46	6.879
42	0.8155	50.86	6.799
44	0.8030	50.28	6.722
46	0.7972	49.72	6.646
48	0.7883	49.16	6.572
50	0.7796	48.62	6.499
52	0.7711	48.09	6.429
54	0.7628	47.57	6.359
56	0.7547	47.07	6.292
58	0.7467	46.57	6.225
60	0.7389	46.08	6.160
62	0.7313	45.61	6.097
64	0.7238	45.14	6.034
66	0.7165	44.68	5.973
68	0.7093	44.23	5.913
70	0.7022	43.79	5.854
72	0.6953	43.36	5.797
74	0.6886	42.94	5.741
76	0.6819	42.53	5.685
78	0.6754	41.72	5.631
80	0.6690	41.32	5.577
82	0.6628	41.13	5.526
84	0.6566	40.95	5.474
86	0.6506	40.57	5.424
88	0.6446	40.20	5.374
90	0.6388	39.84	5.326
92	0.6331	39.48	5.278
94	0.6275	39.13	5.231
96	0.6220	38.79	5.186
98	0.6116	38.45	5.141
100	0.6112	38.12	5.096

CHART 4–2

Diesel fuel specification comparison chart.

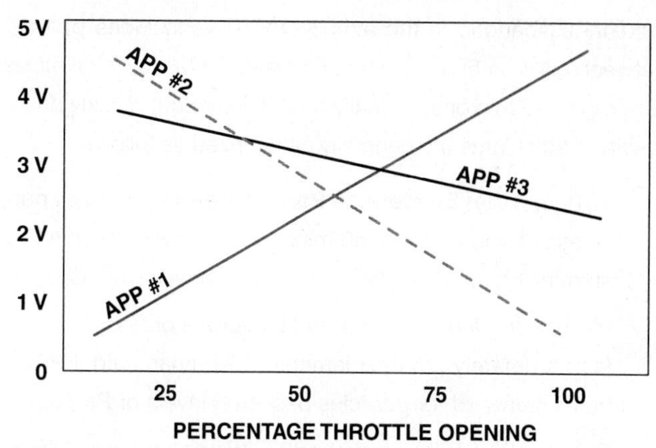

FIGURE 4–17 A typical accelerator pedal position (APP) sensor uses three different sensors in one package with each creating a different voltage as the accelerator is moved.

FIGURE 4–16 A wire wound electrical heater is used to warm the intake air on some diesel engines.

? FREQUENTLY ASKED QUESTION

How Can You Tell If Gasoline Has Been Added to the Diesel Fuel by Mistake?

If gasoline has been accidentally added to diesel fuel and is burned in a diesel engine, the result can be very damaging to the engine. The gasoline can ignite faster than diesel fuel, which would tend to increase the temperature of combustion. This high temperature can harm injectors and glow plugs, as well as pistons, head gaskets, and other major diesel engine components. If contaminated fuel is suspected, first smell the fuel at the filler neck. If the fuel smells like gasoline, then the tank should be drained and refilled with diesel fuel. If the smell test does not indicate a gasoline (or any rancid smell), then test a sample for proper API gravity.

NOTE: Diesel fuel designed for on-road use should be green in color. Red diesel fuel should only be found in off-road or farm equipment.

ACCELERATOR PEDAL POSITION SENSOR

Some light truck diesel engines are equipped with an electronic throttle to control the amount of fuel injected into the engine. Because a diesel engine does not use a throttle in the air intake, the only way to control engine speed is by controlling the amount of fuel being injected into the cylinders. Instead of a mechanical link from the accelerator pedal to the diesel injection pump, a throttle-by-wire system uses an accelerator pedal position sensor. To ensure safety, it consists of three separate sensors that change in voltage as the accelerator pedal is depressed. ● SEE FIGURE 4–17.

The computer checks for errors by comparing the voltage output of each of the three sensors inside the APP and compares them to what they should be if there are no faults. If an error is detected, the engine and vehicle speed are often reduced.

SOOT OR PARTICULATE MATTER

Soot particles may come directly from the exhaust tailpipe or they can also form when emissions of nitrogen oxide and various sulfur oxides chemically react with other pollutants suspended in the atmosphere. Such reactions result in the formation of ground-level ozone, commonly known as smog. Smog is the most visible form of what is generally referred to as particulate matter. Particulate matter refers to tiny particles of solid or semisolid material suspended in the atmosphere. This includes particles between 0.1 and 50 microns in diameter. The heavier particles, larger than 50 microns, typically tend to settle out quickly due to gravity. Particulates are generally categorized as follows:

- **TSP, or Total Suspended Particulate**—Refers to all particles between 0.1 and 50 microns. Up until 1987, the EPA standard for particulates was based on levels of TSP.
- **PM10**—Particulate matter of 10 microns or less (approximately 1/6 the diameter of a human hair). EPA has a standard for particles based on levels of PM10.
- **PM2.5**—Particulate matter of 2.5 microns or less (approximately 1/20 the diameter of a human hair), also called "fine" particles. In July 1997, the EPA approved a standard for PM2.5.

In general, soot particles produced by diesel combustion fall into the categories of fine, that is, less than 2.5 microns, and ultrafine, less than 0.1 micron. Ultrafine particles make up about 80% to 95% of soot.

DIESEL OXIDATION CATALYST (DOC)

Diesel oxidation catalyst (DOC) consists of a flow-through honeycomb-style substrate structure that is washcoated with a layer of catalyst materials, similar to those used in a gasoline engine catalytic converter. These materials include the precious metals platinum and palladium, as well as other base metals catalysts. Catalysts chemically react with exhaust gas to convert harmful nitrogen oxide into nitrogen dioxide, and to oxidize absorbed hydrocarbons. The chemical reaction acts as a combustor for the unburned fuel that is characteristic of diesel compression ignition. The main function of the DOC is to start a regeneration event by converting the fuel-rich exhaust gases to heat.

The DOC also reduces carbon monoxide, hydrocarbons, plus odor-causing compounds such as aldehydes and sulfur, and the soluble organic fraction of particulate matter. During a regeneration event, the Catalyst System Efficiency test will run. The engine control module (ECM) monitors this efficiency of the DOC by determining if the exhaust gas temperature sensor (EGT Sensor 1) reaches a predetermined temperature during a regeneration event.

? FREQUENTLY ASKED QUESTION

What Is the Big Deal for the Need to Control Very Small Soot Particles?

For many years soot or particulate matter (PM) was thought to be less of a health concern than exhaust emissions from gasoline engines. It was felt that the soot could simply fall to the ground without causing any noticeable harm to people or the environment. However, it was discovered that the small soot particulates when breathed in are not expelled from the lungs like larger particles but instead get trapped in the deep areas of the lungs where they accumulate.

DIESEL EXHAUST PARTICULATE FILTER (DPF)

The heated exhaust gas from the DOC flows into the diesel particulate filter (DPF), which captures diesel exhaust gas particulates (soot) to prevent them from being released into the atmosphere. This is done by forcing the exhaust through a porous cell which has a silicon carbide substrate with honeycomb-cell-type channels that trap the soot. The channels are washcoated with catalyst materials similar to those in the DOC filter. The main difference between the DPF and a typical catalyst filter is that the entrance to every other cell channel in the DPF substrate is blocked at one end. So instead of flowing directly through the channels, the exhaust gas is forced through the porous walls of the blocked channels and exits through the adjacent open-ended channels. This type of filter is also referred to as a "wall-flow" filter.

Soot particulates in the gas remain trapped on the DPF channel walls where, over time, the trapped particulate matter will begin to clog the filter. The filter must therefore be purged periodically to remove accumulated soot particles. The process of purging soot from the DPF is described as **regeneration**. ● **SEE FIGURE 4–18.**

EXHAUST GAS TEMPERATURE SENSORS There are two exhaust gas temperature sensors that function in much the same way as engine temperature sensors. EGT Sensor 1 is positioned between the DOC and the DPF where it can measure the temperature of the exhaust gas entering the DPF. EGT

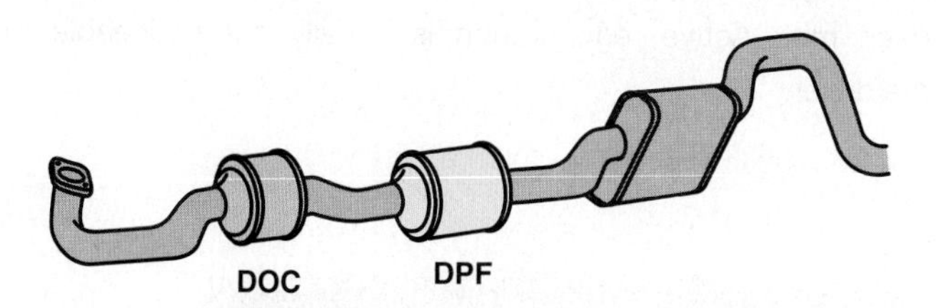

FIGURE 4–18 After treatment of diesel exhaust is handled by the DOC and DPF.

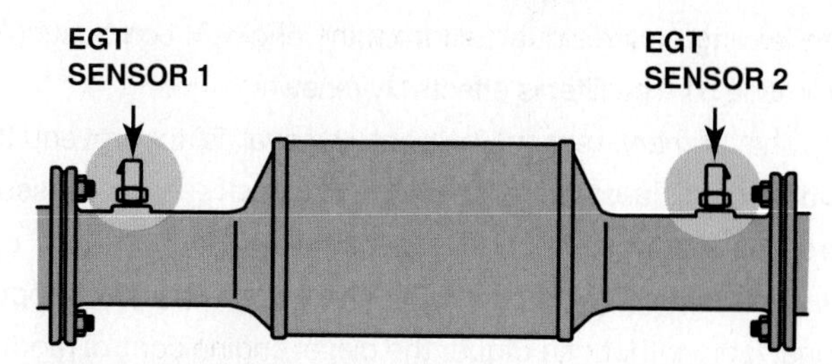

FIGURE 4–19 EGT 1 and EGT 2 are used by the PCM to help control after treatment.

Sensor 2 measures the temperature of the exhaust gas stream immediately after it exits the DPF.

The engine control module (ECM) monitors the signals from the EGT sensors as part of its calibrations to control DPF regeneration. The ECM supplies biased 5 volts to the signal circuit and a ground on the low reference circuit to EGT Sensor 1. When the EGT Sensor 1 is cold, the sensor resistance is high. As the temperature increases, the sensor resistance decreases. With high sensor resistance, the ECM detects a high voltage on the signal circuit. With lower sensor resistance, the ECM detects a lower voltage on the signal circuit. Proper exhaust gas temperatures at the inlet of the DPF are crucial for proper operation and for starting the regeneration process. Too high a temperature at the DPF will cause the DPF substrate to melt or crack. Regeneration will be terminated at temperatures above 1470°F (800°C). With too low a temperature, self-regeneration will not fully complete the soot-burning process. ● **SEE FIGURE 4–19.**

DPF DIFFERENTIAL PRESSURE SENSOR The DPF **differential pressure sensor (DPS)** has two pressure sample lines:

- One line is attached before the DPF, labeled P1
- The other is located after the DPF, labeled P2

The exact location of the DPS varies by vehicle model type (medium duty, pickup, or van). By measuring the exhaust supply pressure (P1) and the post DPF pressure (P2) the ECM can determine differential pressure, also referred to as "delta" pressure, across the DPF. Data from the DPF differential pressure sensor is used by the ECM to calibrate for controlling DPF exhaust system operation.

? FREQUENTLY ASKED QUESTION

What Is an Exhaust Air Cooler?

An exhaust air cooler is simply a length of tubing with a narrower center that acts as a venturi. The cooler is attached to the tailpipe with a bracket that provides a gap between the two. As hot exhaust rushes past the gap, the venturi effect draws surrounding air into the cooler and reduces the exhaust temperature. The cooler significantly lowers exhaust temperature at the tailpipe from a potential 788°F to 806°F (420°C to 430°C) to approximately 520°F (270°C).

DIESEL PARTICULATE FILTER REGENERATION

Soot particulates in the gas remain trapped on the DPF channel walls where, over time, the buildup of trapped particulate matter will begin to clog the filter. The filter must therefore be purged periodically to remove accumulated soot particles. The process of purging soot from the DPF by incineration is described as regeneration. When the temperature of the exhaust gas is increased sufficiently, the heat incinerates the soot particles trapped in the filter, leaving only residual ash from the engine's combustion of lubrication oil. The filter is effectively renewed.

The primary reason for soot removal is to prevent the buildup of exhaust back pressure. Excessive back pressure increases fuel consumption, reduces power output, and can potentially cause engine damage. There are a number of operational factors that can trigger the diesel engine control module to initiate a DPF regeneration sequence. The ECM monitors:

- Distance since last DPF regeneration
- Fuel used since last DPF regeneration
- Engine run time since last DPF regeneration
- Exhaust differential pressure across the DPF

DPF REGENERATION PROCESS A number of engine components are required to function together for the regeneration process to be performed. ECM controls that impact DPF regeneration include late post-injections, engine speed, and adjusting fuel pressure. Adding late post-injection pulses provides the engine with additional fuel to be oxidized in the DOC which increases exhaust temperatures entering the DPF to about 900°F (500°C) and higher. The intake air valve acts as a restrictor that reduces air entry to the engine, which increases engine operating temperature. The intake air heater may also be activated to warm intake air during regeneration.

The variable vane turbocharger also plays a role in achieving regeneration temperatures by reducing or increasing boost depending on engine load.

TYPES OF DPF REGENERATION DPF regeneration can be initiated in a number of ways, depending on the vehicle application and operating circumstances. The two main regeneration types are:

- Passive
- Active

PASSIVE REGENERATION. During normal vehicle operation when driving conditions produce sufficient load and exhaust temperatures, passive DPF regeneration may occur. This passive regeneration occurs without input from the ECM or the driver. A passive regeneration may typically occur while the vehicle is being driven at highway speed or towing a trailer.

ACTIVE REGENERATION. Active regeneration is commanded by the ECM when it determines that the DPF requires it to remove excess soot buildup and conditions for filter regeneration have been met. Active regeneration is usually not noticeable to the driver.

SELECTIVE CATALYTIC REDUCTION

PURPOSE AND FUNCTION **Selective catalytic reduction (SCR)** is a method used to reduce NO_x emissions by injecting urea into the exhaust stream. Instead of using large amounts of exhaust gas recirculation (EGR), the SCR system uses urea.

Urea is used as a nitrogen fertilizer. It is colorless, odorless, and nontoxic. Urea is called **diesel exhaust fluid (DEF)** in North America and AdBlue in Europe. Urea is injected into the catalyst where it sets off a chemical reaction, which converts nitrogen oxides (NO_x) into nitrogen (N_2) and water (H_2O). Vehicle manufacturers size the onboard urea storage tank so that it needs to be refilled at about each scheduled oil change or every 7,500 miles (12,000 km). A warning light alerts the driver when the urea level needs to be refilled. If the warning light is ignored and the diesel exhaust fluid is not refilled, current EPA regulations require that the operation of the engine be restricted and may not start unless the fluid is refilled. This regulation is designed to prevent the engine from being operated without the fluid, which, if not, would greatly increase exhaust emissions. ● **SEE FIGURE 4–20.**

ADVANTAGES OF SCR Using urea injection instead of large amounts of EGR results in the following advantages:

- Potential higher engine power output for the same size engine
- Reduced NO_x emissions up to 90%
- Reduced HC and CO emissions up to 50%
- Reduced particulate matter (PM) by 50%

DISADVANTAGES OF SCR Using urea injection instead of large amounts of EGR results in the following disadvantages:

- Onboard storage tank required for the urea
- Increased costs to the vehicle owner due to having to refill the urea storage tank

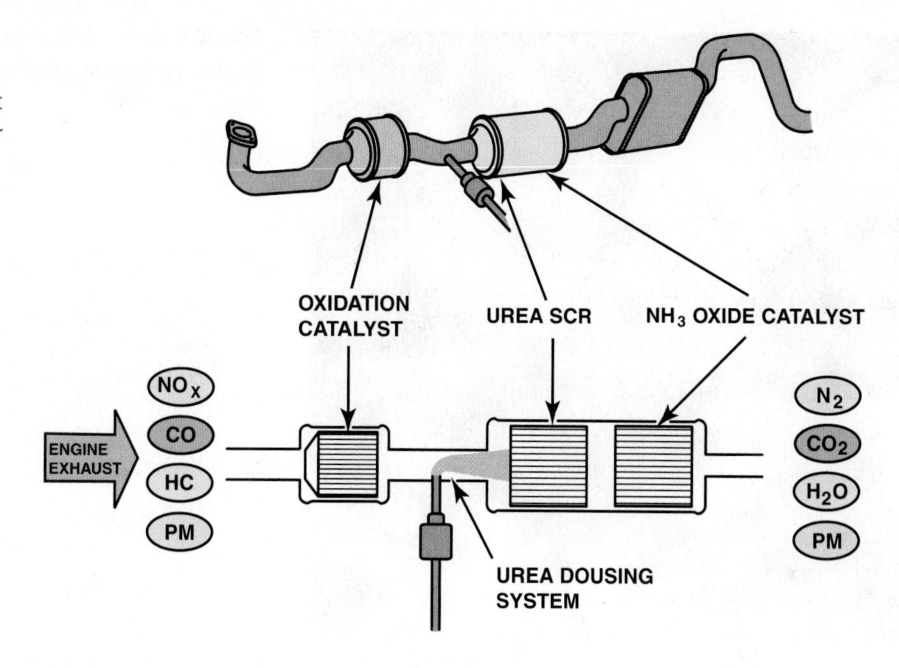

FIGURE 4–20 Urea (diesel exhaust fluid) injection is used to reduce NO_X exhaust emissions. It is injected after the diesel oxidation catalyst (DOC) and before the diesel particulate filter (DPF).

WARNING

Tailpipe outlet exhaust temperature will be greater than 572°F (300°C) during service regeneration. To help prevent personal injury or property damage from fire or burns, keep vehicle exhaust away from any object and people.

ASH LOADING Regeneration will not burn off ash. Only the particulate matter (PM) is burned off during regeneration. Ash is a noncombustible by-product from normal oil consumption. Ash accumulation in the DPF will eventually cause a restriction in the particulate filter. To service an ash loaded DPF, the DPF will need to be removed from the vehicle and cleaned or replaced. Low ash content engine oil (API CJ-4) is required for vehicles with the DPF system. The CJ-4 rated oil is limited to 1% ash content.

DIESEL EXHAUST SMOKE DIAGNOSIS

While some exhaust smoke is considered normal operation for many diesel engines, especially older units, the cause of excessive exhaust smoke should be diagnosed and repaired.

BLACK SMOKE Black exhaust smoke is caused by incomplete combustion because of a lack of air or a fault in the injection system that could cause an excessive amount of fuel in the cylinders. Items that should be checked include the following:

- Check the fuel specific gravity (API gravity).
- Perform an injector balance test to locate faulty injectors using a scan tool.
- Check for proper operation of the engine coolant temperature (ECT) sensor.
- Check for proper operation of the fuel rail pressure (FRP) sensor.
- Check for restrictions in the intake or turbocharger.
- Check to see if the engine is using oil.

WHITE SMOKE White exhaust smoke occurs most often during cold engine starts because the smoke is usually condensed fuel droplets. White exhaust smoke is also an indication of cylinder misfire on a warm engine. The most common causes of white exhaust smoke include:

- Inoperative glow plugs
- Low engine compression
- Incorrect injector spray pattern
- A coolant leak into the combustion chamber

GRAY OR BLUE SMOKE Blue exhaust smoke is usually due to oil consumption caused by worn piston rings, scored cylinder walls, or defective valve stem seals. Gray or blue smoke can also be caused by a defective injector(s).

FIGURE 4–21 A compression gauge designed for the higher compression rate of a diesel engine should be used when checking the compression.

DIESEL PERFORMANCE DIAGNOSIS

Always start the diagnosis of a diesel engine by checking the oil. Higher than normal oil level can indicate that diesel fuel has leaked into the oil. Diesel engines can be diagnosed using a scan tool in most cases, because most of the pressure sensors' values can be displayed. Common faults include:

Hard starting

No start

Extended cranking before starting

Low power

Using a scan tool, check the sensor values to help pin down the source of the problem. Also check the minimum pressures that are required to start the engine if a no-start condition is being diagnosed.

COMPRESSION TESTING

A compression test is fundamental for determining the mechanical condition of a diesel engine. Worn piston rings can cause low power and excessive exhaust smoke. A diesel engine should produce at least 300 PSI (2,068 kPa) of compression pressure and all cylinders should be within 50 PSI (345 kPa) of each other. ● **SEE FIGURE 4–21.**

GLOW PLUG RESISTANCE BALANCE TEST

Glow plugs increase in resistance as their temperature increases. All glow plugs should have about the same resistance when checked with an ohmmeter. A similar test of the resistance of the glow plugs can be used to detect a weak cylinder. This test is particularly helpful on a diesel engine that is not computer controlled. To test for even cylinder balance using glow plug resistance, perform the following on a warm engine:

1. Unplug, measure, and record the resistance of all of the glow plugs.
2. With the wires still removed from the glow plugs, start the engine.
3. Allow the engine to run for several minutes to allow the combustion inside the cylinder to warm the glow plugs.
4. Measure the plugs and record the resistance of all of the glow plugs.
5. The resistance of all of the glow plugs should be higher than at the beginning of the test. A glow plug that is in a cylinder that is not firing correctly will not increase in resistance as much as the others.
6. Another test is to measure exhaust manifold temperature at each exhaust port. Misfiring cylinders will run cold. This can be done with a contact or noncontact thermometer.

FIGURE 4–22 A typical pop tester used to check the spray pattern of a diesel engine injector.

INJECTOR POP TESTING

A **pop tester** is a device used for checking a diesel injector nozzle for proper spray pattern. The handle is depressed and pop off pressure is displayed on the gauge. ● **SEE FIGURE 4–22.**

The spray pattern should be a hollow cone. This will vary depending on design. The nozzle should also be tested for leakage—dripping of the nozzle while under pressure. If the spray pattern is not correct, cleaning, repairing, or replacing of the injector nozzle may be necessary.

TECH TIP

Always Use Cardboard to Check for High-Pressure Leaks

If diesel fuel is found on the engine, a high-pressure leak could be present. When checking for a high-pressure leak, wear protective clothing including safety glasses and face shield plus gloves and long-sleeved shirt. Then use a piece of cardboard to locate the high-pressure leak. When a diesel is running, the pressure in the common rail and injector tubes can reach over 20,000 PSI. At these pressures the diesel fuel is atomized and cannot be seen but can penetrate the skin and cause personal injury. A leak will be shown as a dark area on the cardboard. When a leak is found, shut off the engine and trace the exact location of the leak without the engine running.

CAUTION: Sometimes a leak can actually cut through the cardboard, so use extreme care.

DIESEL EMISSION TESTING

The most commonly used diesel exhaust emission test used in state or local testing programs is called the **opacity** test. Opacity means the percentage of light that is blocked by the exhaust smoke.

- A 0% opacity means that the exhaust has no visible smoke and does not block light from a beam projected through the exhaust smoke.
- A 100% opacity means that the exhaust is so dark that it completely blocks light from a beam projected through the exhaust smoke.
- A 50% opacity means that the exhaust blocks half of the light from a beam projected through the exhaust smoke.

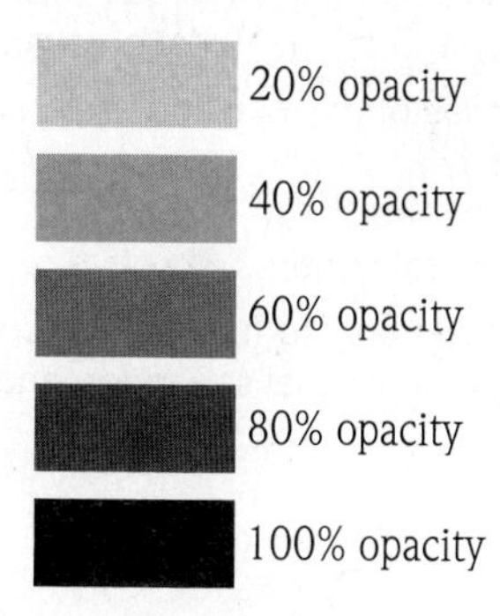

TECH TIP

Do Not Switch Injectors

In the past, it was common practice to switch diesel fuel injectors from one cylinder to another when diagnosing a dead cylinder problem. However, most high-pressure common rail systems used in new diesels use precisely calibrated injectors that should not be mixed up during service. Each injector has its own calibration number. ● **SEE FIGURE 4–23.**

FIGURE 4–23 The letters on the side of this injector on a Cummins 6.7 liter diesel indicate the calibration number for the injector.

SNAP ACCELERATION TEST In a snap acceleration test, the vehicle is held stationary with wheel chocks in place and the brakes released as the engine is rapidly accelerated to high idle with the transmission in neutral while smoke emissions are measured. This test is conducted a minimum of six times and the three most consistent measurements are averaged together for a final score.

ROLLING ACCELERATION TEST Vehicles with a manual transmission are rapidly accelerated in low gear from an idle speed to a maximum governed RPM while the smoke emissions are measured.

STALL ACCELERATION TEST Vehicles with automatic transmissions are held in a stationary position with the parking brake and service brakes applied while the transmission is placed in "drive." The accelerator is depressed and held momentarily while smoke emissions are measured.

The standards for diesels vary according to the type of vehicle and other factors, but usually include a 40% opacity or less.

SUMMARY

1. A diesel engine uses heat of compression to ignite the diesel fuel when it is injected into the compressed air in the combustion chamber.
2. There are two basic designs of combustion chambers used in diesel engines. Indirect injection (IDI) uses a precombustion chamber, whereas a direct injection (DI) occurs directly into the combustion chamber.
3. The three phases of diesel combustion are:
   a. Ignition delay
   b. Rapid combustion
   c. Controlled combustion
4. The typical diesel engine fuel system consists of the fuel tank, lift pump, water–fuel separator, and fuel filter.
5. The engine-driven injection pump supplies high-pressure diesel fuel to the injectors.
6. The two most common types of fuel injection used in automotive diesel engines are:
   a. Distributor-type injection pump
   b. Common rail design where all of the injectors are fed from the same fuel supply from a rail under high pressure
7. Injector nozzles are either opened by the high-pressure pulse from the distributor pump or electrically by the computer on a common rail design.
8. Glow plugs are used to help start a cold diesel engine and help prevent excessive white smoke during warm-up.

9. The higher the cetane rating of diesel fuel, the more easily the fuel is ignited.
10. Most automotive diesel engines are designed to operate on grade #2 diesel fuel in moderate weather conditions.
11. The API specific gravity of diesel fuel should be 30 to 39 with a typical reading of 35 for #2 diesel fuel.
12. Diesel engines can be tested using a scan tool, as well as measuring the glow plug resistance or compression reading to determine a weak or nonfunctioning cylinder.

REVIEW QUESTIONS

1. What is the difference between direct injection and indirect injection?
2. What are the three phases of diesel ignition?
3. What are the most commonly used types of automotive diesel injection systems?
4. Why are glow plugs kept working after the engine starts?
5. What is the advantage of using diesel fuel with a high cetane rating?
6. How is the specific gravity of diesel fuel tested?

CHAPTER QUIZ

1. How is diesel fuel ignited in a warm diesel engine?
   - **a.** Glow plugs
   - **b.** Heat of compression
   - **c.** Spark plugs
   - **d.** Distributorless ignition system
2. Which type of diesel injection produces less noise?
   - **a.** Indirect injection (IDI)
   - **b.** Common rail
   - **c.** Direct injection
   - **d.** Distributor injection
3. Which diesel injection system requires the use of a glow plug?
   - **a.** Indirect injection (IDI)
   - **b.** High-pressure common rail
   - **c.** Direct injection
   - **d.** Distributor injection
4. The three phases of diesel ignition include ________.
   - **a.** Glow plug ignition, fast burn, slow burn
   - **b.** Slow burn, fast burn, slow burn
   - **c.** Ignition delay, rapid combustion, controlled combustion
   - **d.** Glow plug ignition, ignition delay, controlled combustion
5. What fuel system component is used in a vehicle equipped with a diesel engine that is not usually used on the same vehicle when it is equipped with a gasoline engine?
   - **a.** Fuel filter
   - **b.** Fuel supply line
   - **c.** Fuel return line
   - **d.** Water–fuel separator
6. The diesel injection pump is usually driven by a ________.
   - **a.** Gear off the camshaft
   - **b.** Belt off the crankshaft
   - **c.** Shaft drive off of the crankshaft
   - **d.** Chain drive off of the camshaft
7. Which diesel system supplies high-pressure diesel fuel to all of the injectors all of the time?
   - **a.** Distributor
   - **b.** Inline
   - **c.** High-pressure common rail
   - **d.** Rotary
8. Glow plugs should have high resistance when ________ and lower resistance when ________.
   - **a.** Cold/warm
   - **b.** Warm/cold
   - **c.** Wet/dry
   - **d.** Dry/wet
9. Technician A says that glow plugs are used to help start a diesel engine and are shut off as soon as the engine starts. Technician B says that the glow plugs are turned off as soon as a flame is detected in the combustion chamber. Which technician is correct?
   - **a.** Technician A only
   - **b.** Technician B only
   - **c.** Both Technicians A and B
   - **d.** Neither Technician A nor B
10. What part should be removed to test cylinder compression on a diesel engine?
    - **a.** An injector
    - **b.** An intake valve rocker arm and stud
    - **c.** An exhaust valve
    - **d.** A glow plug

chapter 5

GASOLINE

LEARNING OBJECTIVES

After studying this chapter, the reader will be able to:

1. Explain the chemical composition of gasoline and the process of refining gasoline.
2. Discuss how volatility affects driveability.
3. Discuss the benefits of using gasoline additives, reformulating gasoline, and blending gasoline
4. Discuss safety precautions when working with gasoline.

This chapter will help you prepare for Engine Repair (A8) ASE certification test content area "A" (General Engine Diagnosis).

KEY TERMS

Air–fuel ratio 95
Antiknock index (AKI) 96
ASTM 93
British thermal unit (BTU) 95
Catalytic cracking 91
Cracking 91
Detonation 96
Distillation 91
Distillation curve 93
Driveability index (DI) 93
E10 99
Ethanol 99
Fungible 92
Gasoline 91
Hydrocracking 91
Octane rating 96
Oxygenated fuels 98
Petroleum 91
Ping 96
Reformulated gasoline (RFG) 101
RVP 93
Spark knock 96
Stoichiometric ratio 95
Tetraethyl lead (TEL) 96
Vapor lock 93
Volatility 93
WWFC 102

GASOLINE

DEFINITION **Gasoline** is a term used to describe a complex mixture of various hydrocarbons refined from crude petroleum oil for use as a fuel in engines. Gasoline and air burn in the cylinder of the engine and produce heat and pressure, which is transferred to rotary motion inside the engine and eventually powers the drive wheels of a vehicle. When combustion occurs, carbon dioxide and water are produced if the process is perfect and all of the air and all of the fuel are consumed in the process.

CHEMICAL COMPOSITION Gasoline is a combination of hydrocarbon molecules that have between 5 and 12 carbon atoms. The names of these various hydrocarbons are based on the number of carbon atoms and include:

- **Methane**—one carbon atom
- **Ethane**—two carbon atoms
- **Propane**—three carbon atoms
- **Butane**—four carbon atoms
- **Pentane**—five carbon atoms
- **Hexane**—six carbon atoms
- **Heptane**—seven carbon atoms (used to test octane rating—has an octane rating of zero)
- **Octane**—eight carbon atoms (a type of octane is used as a basis for antiknock rating)

REFINING

TYPES OF CRUDE OIL Refining is a complex combination of interdependent processing units that can separate crude oil into useful products such as gasoline and diesel fuel. As it comes out of the ground, **petroleum** (meaning "rock oil") crude can be as thin and light colored as apple cider or as thick and black as melted tar. A barrel of crude oil is 42 gallons, not 55 gallons as commonly used for industrial barrels. Typical terms used to describe the type of crude oil include:

- Thin crude oil has a high American Petroleum Institute (API) gravity, and therefore, is called *high-gravity* crude.
- Thick crude oil is called *low-gravity* crude. High-gravity-type crude contains more natural gasoline and its lower sulfur and nitrogen content makes it easier to refine.
- Low-sulfur crude oil is also known as "sweet" crude.
- High-sulfur crude oil is also known as "sour" crude.

DISTILLATION In the late 1800s, crude was separated into different products by boiling, in a process called **distillation**. Distillation works because crude oil is composed of hydrocarbons with a broad range of boiling points.

In a distillation column, the vapor of the lowest-boiling hydrocarbons, propane and butane, rises to the top. The straight-run gasoline (also called naphtha), kerosene, and diesel fuel cuts are drawn off at successively lower positions in the column.

CRACKING **Cracking** is the process where hydrocarbons with higher boiling points could be broken down (cracked) into lower-boiling hydrocarbons by treating them to very high temperatures. This process, called *thermal cracking*, was used to increase gasoline production starting in 1913.

Instead of high heat, today cracking is performed using a catalyst and is called **catalytic cracking**. A catalyst is a material that speeds up or otherwise facilitates a chemical reaction without undergoing a permanent chemical change itself. Catalytic cracking produces gasoline of higher quality than thermal cracking.

Hydrocracking is similar to catalytic cracking in that it uses a catalyst, but the catalyst is in a hydrogen atmosphere. Hydrocracking can break down hydrocarbons that are resistant to catalytic cracking alone, and it is used to produce diesel fuel rather than gasoline.

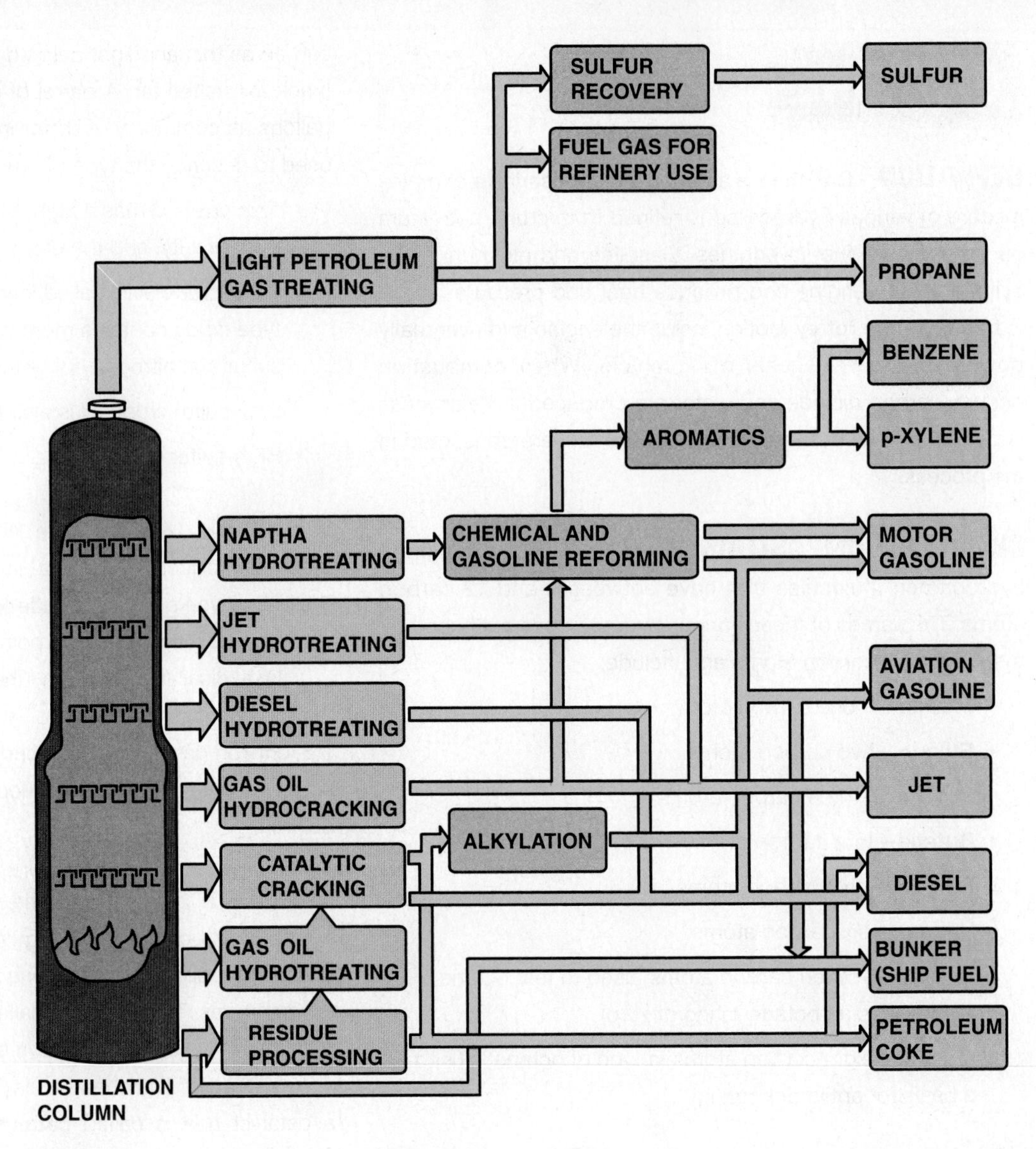

FIGURE 5–1 The crude oil refining showing most of the major steps and processes.

Other types of refining processes include:

- Reforming
- Alkylation
- Isomerization
- Hydrotreating
- Desulfurization

● **SEE FIGURE 5–1.**

SHIPPING The gasoline is transported to regional storage facilities by tank railway car or by pipeline. In the pipeline method, all gasoline from many refiners is often sent through the same pipeline and can become mixed. All gasoline is said to be **fungible**, meaning that it is capable of being interchanged because each grade is created to specification so there is no reason to keep the different gasoline brands separated except for grade. Regular grade, mid-grade, and premium grades are separated in the pipeline and the additives are added at the regional storage facilities and then shipped by truck to individual gas stations.

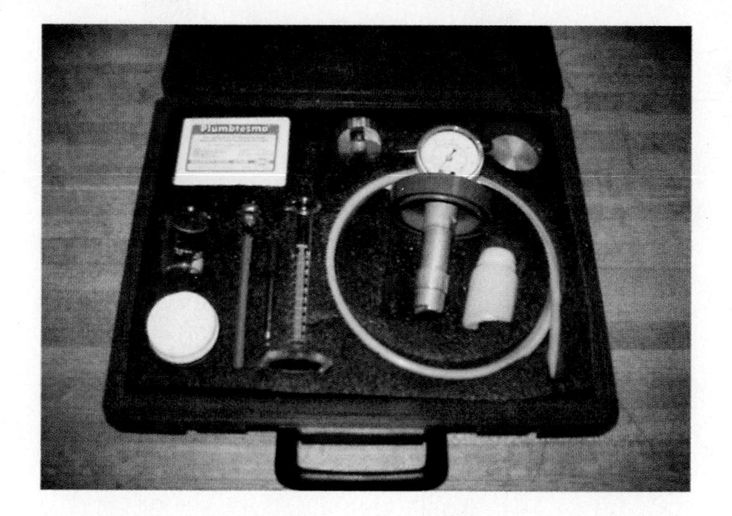

FIGURE 5–2 A gasoline testing kit, including an insulated container where water at 100°F is used to heat a container holding a small sample of gasoline. The reading on the pressure gauge is the Reid vapor pressure (RVP).

VOLATILITY

DEFINITION OF VOLATILITY **Volatility** describes how easily the gasoline evaporates (forms a vapor). The definition of volatility assumes that the vapors will remain in the fuel tank or fuel line and will cause a certain pressure based on the temperature of the fuel.

REID VAPOR PRESSURE (RVP) **Reid vapor pressure (RVP)** is the pressure of the vapor above the fuel when the fuel is at 100°F (38°C). Increased vapor pressure permits the engine to start in cold weather. Gasoline without air will not burn. Gasoline must be vaporized (mixed with air) to burn in an engine. ● **SEE FIGURE 5–2.**

SEASONAL BLENDING Cold temperatures reduce the normal vaporization of gasoline; therefore, winter-blended gasoline is specially formulated to vaporize at lower temperatures for proper starting and driveability at low ambient temperatures. The **American Society for Testing and Materials (ASTM)** standards for winter-blend gasoline allow volatility of up to 15 pounds per square inch (PSI) RVP.

At warm ambient temperatures, gasoline vaporizes easily. However, the fuel system (fuel pump, carburetor, fuel-injector nozzles, etc.) is designed to operate with liquid gasoline. The volatility of summer-grade gasoline should be about 7.0 PSI RVP. According to ASTM standards, the maximum RVP should be 10.5 PSI for summer-blend gasoline.

DISTILLATION CURVE Besides Reid vapor pressure, another method of classifying gasoline volatility is the **distillation curve**. A curve on a graph is created by plotting the temperature at which the various percentage of the fuel evaporates. A typical distillation curve is shown in ● **FIGURE 5–3.**

DRIVEABILITY INDEX A distillation curve shows how much of a gasoline evaporates at what temperature range. To predict cold-weather driveability, an index was created called the **driveability index**, also called the *distillation index (DI).*

The DI was developed using the temperature for the evaporated percentage of 10% (labeled T10), 50% (labeled T50), and 90% (labeled T90). The formula for DI is:

$$\boldsymbol{DI = 1.5 \times T10 + 3 \times T50 + T90}$$

The total DI is a temperature and usually ranges from 1000°F to 1200°F. The lower values of DI generally result in good cold-start and warm-up performance. A high DI number is less volatile than a low DI number.

NOTE: Most premium-grade gasoline has a higher (worse) DI than regular-grade or mid-grade gasoline, which could cause poor cold-weather driveability. Vehicles designed to operate on premium-grade gasoline are programmed to handle the higher DI, but engines designed to operate on regular-grade gasoline may not be able to provide acceptable cold-weather driveability.

VOLATILITY-RELATED PROBLEMS At higher temperatures, liquid gasoline can easily vaporize, which can cause **vapor lock**. Vapor lock is a *lean* condition caused by vaporized

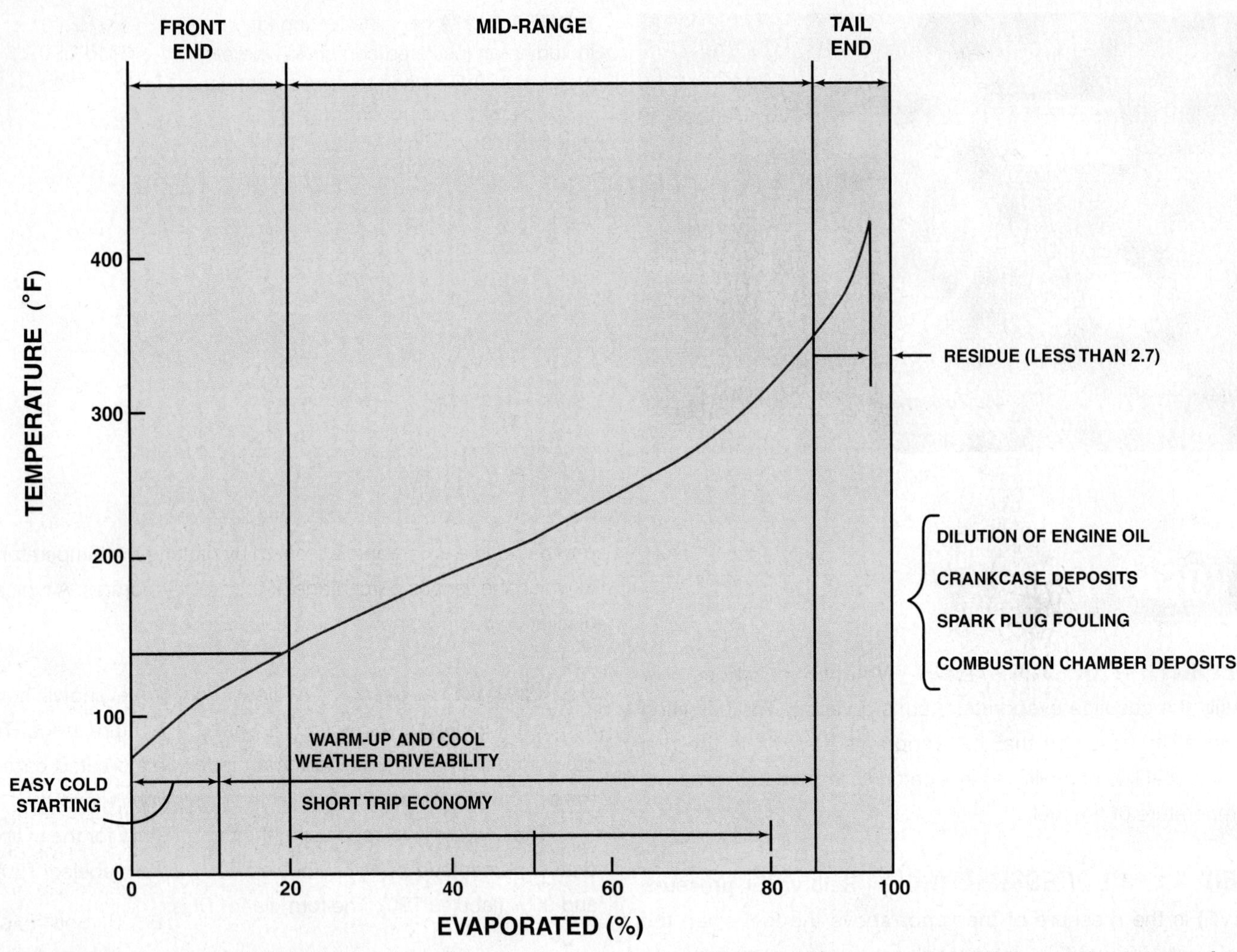

FIGURE 5–3 A typical distillation curve. Heavier molecules evaporate at higher temperatures and contain more heat energy for power, whereas the lighter molecules evaporate easier for starting.

fuel in the fuel system. This vaporized fuel takes up space normally occupied by liquid fuel. Bubbles that form in the fuel cause vapor lock, preventing proper operation of the fuel-injection system.

Heat causes some fuel to evaporate, thereby causing bubbles. Sharp bends cause the fuel to be restricted at the bend. When the fuel flows past the bend, the fuel can expand to fill the space after the bend. This expansion drops the pressure, and bubbles form in the fuel lines. When the fuel is full of bubbles, the engine is not being supplied with enough fuel and the engine runs lean. A lean engine will stumble during acceleration, will run rough, and may stall. Warm weather and alcohol-blended fuels both tend to increase vapor lock and engine performance problems.

If winter-blend gasoline (or high-RVP fuel) is used in an engine during warm weather, the following problems may occur:

1. Rough idle
2. Stalling
3. Hesitation on acceleration
4. Surging

FREQUENTLY ASKED QUESTION

Why Do I Get Lower Gas Mileage in the Winter?

Several factors cause the engine to use more fuel in the winter than in the summer, including:

- Gasoline that is blended for use in cold climates is designed for ease of starting and contains fewer heavy molecules, which contribute to fuel economy. The heat content of winter gasoline is lower than summer-blended gasoline.
- In cold temperatures, all lubricants are stiff, causing more resistance. These lubricants include the engine oil, as well as the transmission and differential gear lubricants.
- Heat from the engine is radiated into the outside air more rapidly when the temperature is cold, resulting in longer run time until the engine has reached normal operating temperature.
- Road conditions, such as ice and snow, can cause tire slippage or additional drag on the vehicle.

GASOLINE COMBUSTION PROCESS

CHEMICAL REACTIONS The combustion process involves the chemical combination of oxygen (O_2) from the air (about 21% of the atmosphere) with the hydrogen and carbon from the fuel. In a gasoline engine, a spark starts the combustion process, which takes about 3 ms (0.003 sec) to be completed inside the cylinder of an engine. The chemical reaction that takes place can be summarized as follows: hydrogen (H) plus carbon (C) plus oxygen (O_2) plus nitrogen (N) plus spark equals heat plus water (H_2O) plus carbon monoxide (CO) (if incomplete combustion) plus carbon dioxide (CO_2) plus hydrocarbons (HC) plus oxides of nitrogen (NO_X) plus many other chemicals. In an equation format it looks like this:

$$\mathbf{H + C + O_2 + N + Spark = Heat + CO_2 + HC + NO_x + H_2O}$$

HEAT ENERGY The heat produced by the combustion process is measured in **British thermal units (BTUs)**. One BTU is the amount of heat required to raise one pound of water one Fahrenheit degree. The metric unit of heat is the *calorie*

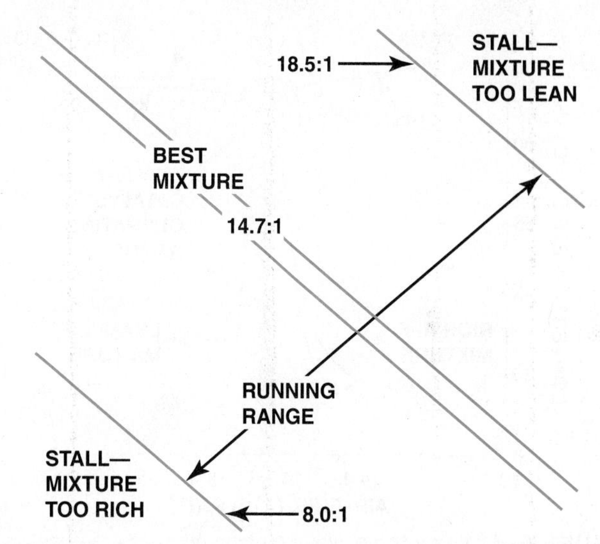

FIGURE 5–4 An engine will not run if the air–fuel mixture is either too rich or too lean.

(cal). One calorie is the amount of heat required to raise the temperature of one gram (g) of water one Celsius degree.

Gasoline—About 114,000 BTUs per gallon

AIR–FUEL RATIOS Fuel burns best when the intake system turns it into a fine spray and mixes it with air before sending it into the cylinders. In fuel-injected engines, the fuel becomes a spray and mixes with the air in the intake manifold. There is a direct relationship between engine airflow and fuel requirements; this is called the **air–fuel ratio.**

The air–fuel ratio is the proportion by weight of air and gasoline that the injection system mixes as needed for engine combustion. The mixtures, with which a gasoline engine can operate without stalling, range from 8 to 1 to 18.5 to 1. ● **SEE FIGURE 5–4.**

These ratios are usually stated by weight, such as:

- 8 parts of air by weight combined with 1 part of gasoline by weight (8:1), which is the richest mixture that an engine can tolerate and still fire reliably.
- 18.5 parts of air mixed with one part of gasoline (18.5:1), which is the leanest practical ratio. Richer or leaner air–fuel ratios cause the engine to misfire badly or not run at all.

STOICHIOMETRIC AIR–FUEL RATIO The ideal mixture or ratio at which all of the fuel combines with all of the oxygen in the air and burns completely is called the **stoichiometric ratio**, a chemically perfect combination. In theory, this ratio for gasoline is an air–fuel mixture of 14.7 to 1. ● **SEE FIGURE 5–5.**

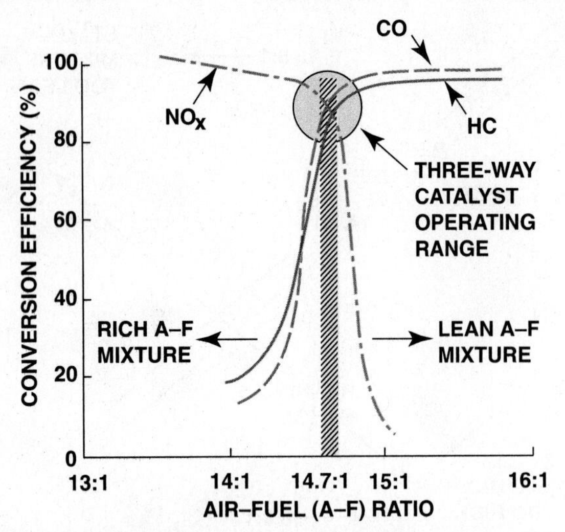

FIGURE 5–5 With a three-way catalytic converter, emission control is most efficient with an air–fuel ratio between 4.65 to 1 and 14.75 to 1.

In reality, the exact ratio at which perfect mixture and combustion occurs depends on the molecular structure of gasoline, which can vary. The stoichiometric ratio is a compromise between maximum power and maximum economy.

NORMAL AND ABNORMAL COMBUSTION

The **octane rating** of gasoline is the measure of its antiknock properties. *Engine knock* (also called **detonation**, **spark knock**, or **ping**) is a metallic noise an engine makes, usually during acceleration, resulting from abnormal or uncontrolled combustion inside the cylinder.

Normal combustion occurs smoothly and progresses across the combustion chamber from the point of ignition. ● **SEE FIGURE 5–6.**

Normal flame-front combustion travels between 45 and 90 mph (72 and 145 km/h). The speed of the flame front depends on air–fuel ratio, combustion chamber design (determining amount of turbulence), and temperature.

During periods of spark knock (detonation), the combustion speed increases by up to 10 times to near the speed of sound. The increased combustion speed also causes increased temperatures and pressures, which can damage pistons, gaskets, and cylinder heads. ● **SEE FIGURE 5–7.**

One of the first additives used in gasoline was **tetraethyl lead (TEL)**. TEL was added to gasoline in the early 1920s to reduce the tendency to knock. It was often called ethyl or high-test gasoline.

OCTANE RATING

The antiknock standard or basis of comparison was the knock-resistant hydrocarbon isooctane, chemically called trimethylpentane (C_8H_{18}), also known as 2-2-4 trimethylpentane. If a gasoline tested had the exact same antiknock characteristics as isooctane, it was rated as 100-octane gasoline. If the gasoline tested had only 85% of the antiknock properties of isooctane, it was rated as 85 octane. Remember, octane rating is only a comparison test.

The two basic methods used to rate gasoline for antiknock properties (octane rating) are the *research method* and the *motor method*. Each uses a model of the special cooperative fuel research (CFR) single-cylinder engine. The research method and the motor method vary as to temperature of air, spark advance, and other parameters. The research method typically results in readings that are 6 to 10 points higher than those of the motor method. For example, a fuel with a research octane number (RON) of 93 might have a motor octane number (MON) of 85.

The octane rating posted on pumps in the United States is the average of the two methods and is referred to as $(R + M) \div 2$, meaning that, for the fuel used in the previous example, the rating posted on the pumps would be

$$\frac{RON + MON}{2} = \frac{93 + 85}{2} = 89$$

The pump octane is called the **antiknock index (AKI)**.

GASOLINE GRADES AND OCTANE NUMBER The posted octane rating on gasoline pumps is the rating achieved by the average of the research and the motor methods. ● **SEE FIGURE 5–8.**

Except in high-altitude areas, the grades and octane ratings are given in ● **CHART 5–1.**

> **? FREQUENTLY ASKED QUESTION**
>
> **What Grade of Gasoline Does the EPA Use When Testing Engines?**
>
> Due to the various grades and additives used in commercial fuel, the government (EPA) uses a liquid called indolene. Indolene has a research octane number of 96.5 and a motor method octane rating of 88, which results in an $(R + M) \div 2$ rating of 92.25.

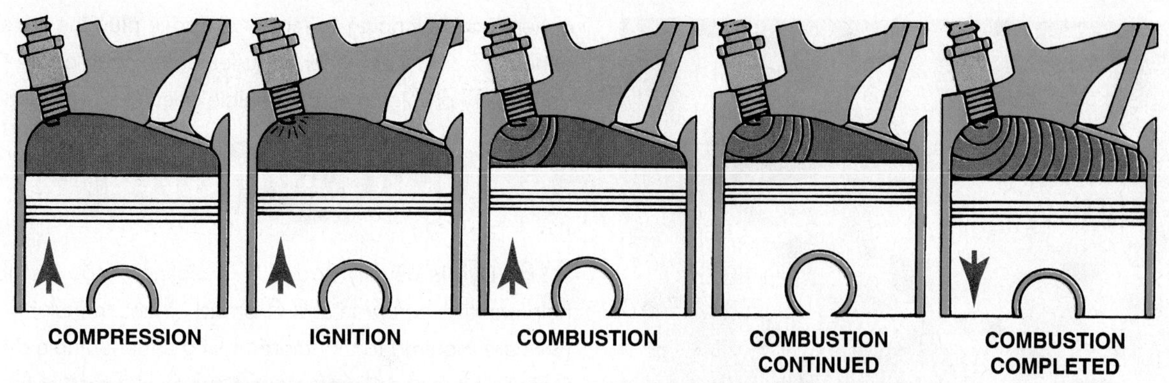

FIGURE 5–6 Normal combustion is a smooth, controlled burning of the air–fuel mixture.

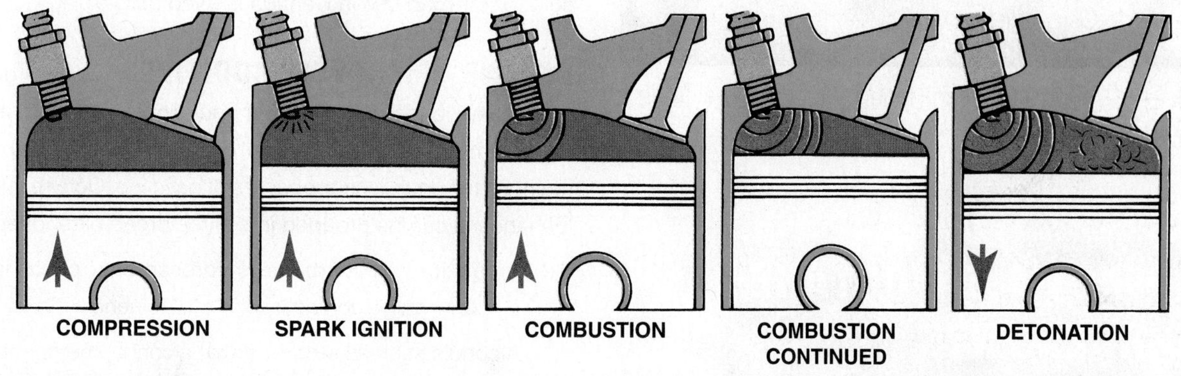

FIGURE 5–7 Detonation is a secondary ignition of the air–fuel mixture. It is also called spark knock or pinging.

Horsepower and Fuel Flow

To produce 1 hp, the engine must be supplied with 0.50 pounds of fuel per hour (lb/hr). Fuel injectors are rated in pounds per hour. For example, a volt-8 engine equipped with 25 lb/hr fuel injectors could produce 50 hp per cylinder (per injector) or 400 hp. Even if the cylinder head or block is modified to produce more horsepower, the limiting factor may be the injector flow rate.

The following are flow rates and resulting horsepower for a V-8 engine:

30 lb/hr: 60 hp per cylinder or 480 hp
35 lb/hr: 70 hp per cylinder or 560 hp
40 lb/hr: 80 hp per cylinder or 640 hp

Of course, injector flow rate is only one of many variables that affect power output. Installing larger injectors without other major engine modifications could decrease engine output and drastically increase exhaust emissions.

FIGURE 5–8 A pump showing regular with a pump octane of 87, plus rated at 89, and premium rated at 93. These ratings can vary with brand as well as in different parts of the country.

GRADES	OCTANE RATING
Regular	87
Mid-grade (also called Plus)	89
Premium	91 or higher

CHART 5–1

Typical octane ratings for gasoline in most parts of the country.

FIGURE 5–9 The posted octane rating in most high-altitude areas shows regular at 85 instead of the usual 87.

HIGH-ALTITUDE OCTANE REQUIREMENTS

As the altitude increases, atmospheric pressure drops. The air is less dense because a pound of air takes more volume. The octane rating of fuel does not need to be as high because the engine cannot take in as much air. This process will reduce the combustion (compression) pressures inside the engine. In mountainous areas, gasoline $(R + M) \div 2$ octane ratings are two or more numbers lower than normal (according to the SAE, about one octane number lower per 1,000 feet or 300 meter in altitude). ● **SEE FIGURE 5–9.**

A secondary reason for the lowered octane requirement of engines running at higher altitudes is the normal enrichment of the air–fuel ratio and lower engine vacuum with the decreased air density. Some problems, therefore, may occur when driving out of high-altitude areas into lower-altitude areas where the octane rating must be higher. Most computerized engine control systems can compensate for changes in altitude and modify air–fuel ratio and ignition timing for best operation.

Because the combustion burn rate slows at high altitude, the ignition (spark) timing can be advanced to improve power. The amount of timing advance can be about 1 degree per 1,000 feet over 5,000 feet. Therefore, if driving at 8,000 feet of altitude, the ignition timing can be advanced 3 degrees.

High altitude also allows fuel to evaporate more easily. The volatility of fuel should be reduced at higher altitudes to prevent vapor from forming in sections of the fuel system, which can cause driveability and stalling problems. The extra heat generated in climbing to higher altitudes plus the lower atmospheric pressure at higher altitudes combine to cause possible driveability problems as the vehicle goes to higher altitudes.

GASOLINE ADDITIVES

DYE Dye is usually added to gasoline at the distributor to help identify the grade and/or brand of fuel. In many countries, fuels are required to be colored using a fuel-soluble dye. In the United States and Canada, diesel fuel used for off-road use and not taxed is required to be dyed red for identification. Gasoline sold for off-road use in Canada is dyed purple.

OCTANE IMPROVER ADDITIVES When gasoline companies, under federal EPA regulations, removed tetraethyl lead from gasoline, other methods were developed to help maintain the antiknock properties of gasoline. Octane improvers (enhancers) can be grouped into three broad categories:

1. Aromatic hydrocarbons (hydrocarbons containing the benzene ring) such as xylene and toluene
2. Alcohols such as ethanol (ethyl alcohol), methanol (methyl alcohol), and tertiary butyl alcohol (TBA)
3. Metallic compounds such as methylcyclopentadienyl manganese tricarbonyl (MMT)

NOTE: MMT has been proven to be harmful to catalytic converters and can cause spark plug fouling. However, MMT is currently one of the active ingredients commonly found in octane improvers available to the public and in some gasoline sold in Canada. If an octane boost additive has been used that contains MMT, the spark plug porcelain will be rust colored around the tip.

Propane and butane, which are volatile by-products of the refinery process, are also often added to gasoline as octane improvers. The increase in volatility caused by the added propane and butane often leads to hot-weather driveability problems.

OXYGENATED FUEL ADDITIVES **Oxygenated fuels** contain oxygen in the molecule of the fuel itself. Examples of oxygenated fuels include methanol, ethanol, methyl tertiary butyl ether (MTBE), tertiary-amyl methyl ether (TAME), and ethyl tertiary butyl ether (ETBE).

Oxygenated fuels are commonly used in high-altitude areas to reduce carbon monoxide (CO) emissions. The extra oxygen in the fuel itself is used to convert harmful CO into carbon dioxide (CO_2). The extra oxygen in the fuel helps ensure that there is enough oxygen to convert all CO into CO_2 during the combustion process in the engine or catalytic converter.

FIGURE 5–10 E10 is 10% ethanol and 90% gasoline.

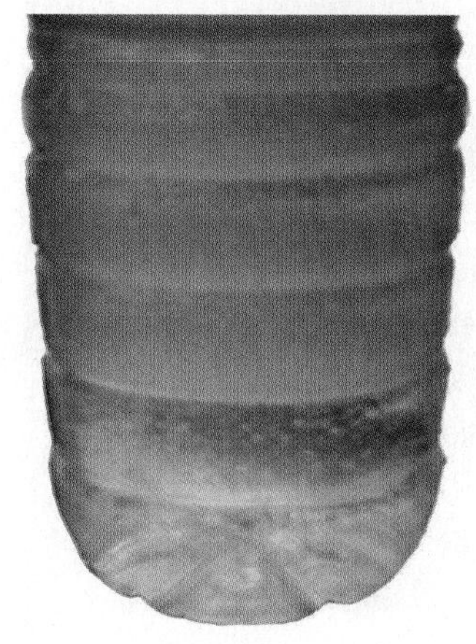

FIGURE 5–11 A container with gasoline containing alcohol. Notice the separation line where the alcohol-water mixture separated from the gasoline and sank to the bottom.

? FREQUENTLY ASKED QUESTION

Can Regular-Grade Gasoline Be Used If Premium Is the Recommended Grade?

Maybe. It is usually possible to use regular-grade or mid-grade (plus) gasoline in most newer vehicles without danger of damage to the engine. Most vehicles built since the 1990s are equipped with at least one knock sensor. If a lower octane gasoline than specified is used, the engine ignition timing setting will usually cause the engine to spark knock, also called detonation or ping. This spark knock is detected by the knock sensor(s), which sends a signal to the computer. The computer then retards the ignition timing until the spark knock stops.

NOTE: Some scan tools will show the "estimated octane rating" of the fuel being used, which is based on knock sensor activity.

As a result of this spark timing retardation, the engine torque is reduced. While this reduction in power is seldom noticed, it will reduce fuel economy, often by four to five miles per gallon. If premium gasoline is then used, the PCM will gradually permit the engine to operate at the more advanced ignition timing setting. Therefore, it may take several tanks of premium gasoline to restore normal fuel economy. For best overall performance, use the grade of gasoline recommended by the vehicle manufacturer.

METHYL TERTIARY BUTYL ETHER (MTBE). MTBE is manufactured by means of the chemical reaction of methanol and isobutylene. Unlike methanol, MTBE does not increase the volatility of the fuel, and is not as sensitive to water as are other alcohols. The maximum allowable volume level, according to the EPA, is 15% but is currently being phased out due to health concerns, as well as MTBE contamination of drinking water if spilled from storage tanks.

TERTIARY-AMYL METHYL ETHER. Tertiary-amyl methyl ether (TAME) contains an oxygen atom bonded to two carbon atoms and is added to gasoline to provide oxygen to the fuel. It is slightly soluble in water, very soluble in ethers and alcohol, and soluble in most organic solvents including hydrocarbons.

ETHYL TERTIARY BUTYL ETHER. ETBE is derived from ethanol. The maximum allowable volume level is 17.2%. The use of ETBE is the cause of much of the odor from the exhaust of vehicles using reformulated gasoline.

ETHANOL. **Ethanol**, also called *ethyl alcohol*, is drinkable alcohol and is usually made from grain. Adding 10% ethanol (ethyl alcohol or grain alcohol) increases the $(R + M) \div 2$ octane rating by three points. The alcohol added to the base gasoline, however, also raises the volatility of the fuel about 0.5 PSI. Most automobile manufacturers permit up to 10% ethanol if driveability problems are not experienced.

The oxygen content of a 10% blend of ethanol in gasoline, called **E10**, is 3.5% oxygen by weight. ● **SEE FIGURE 5–10.**

Keeping the fuel tank full reduces the amount of air and moisture in the tank. ● **SEE FIGURE 5–11.**

FREQUENTLY ASKED QUESTION

What Is Meant by "Phase Separation"?

All alcohols absorb water, and the alcohol-water mixture can separate from the gasoline and sink to the bottom of the fuel tank. This process is called phase separation. To help avoid engine performance problems, try to keep at least a quarter tank of fuel at all times, especially during seasons when there is a wide temperature span between daytime highs and nighttime lows. These conditions can cause moisture to accumulate in the fuel tank as a result of condensation of the moisture in the air.

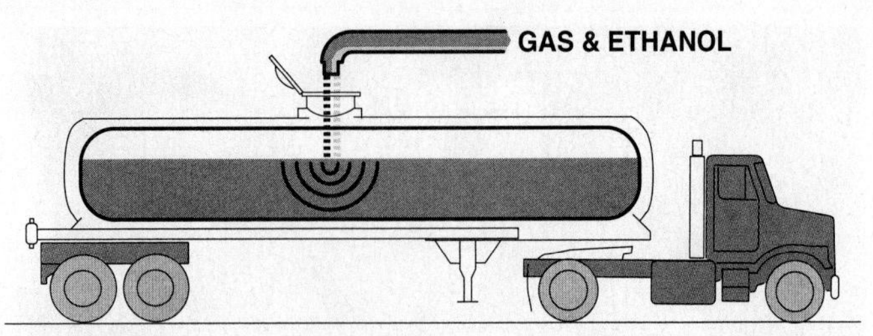

FIGURE 5–12 In-line blending is the most accurate method for blending ethanol with gasoline because computers are used to calculate the correct ratio.

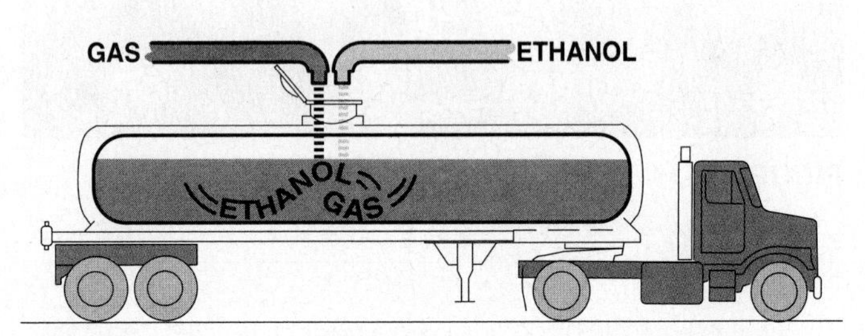

FIGURE 5–13 Sequential blending uses a computer to calculate the correct ratio as well as the prescribed order in which the products are loaded.

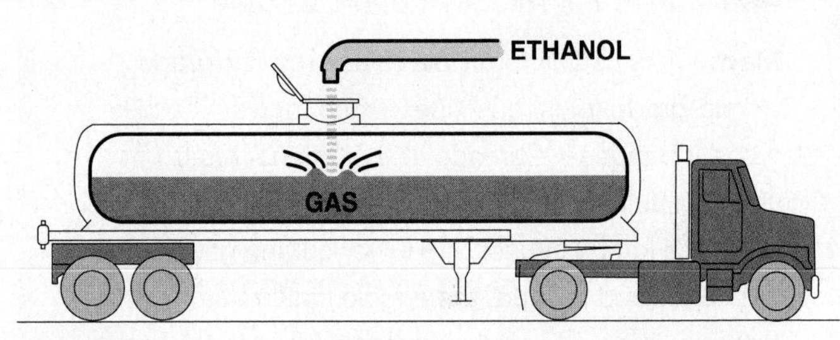

FIGURE 5–14 Splash blending occurs when the ethanol is added to a tanker with gasoline and is mixed as the truck travels to the retail outlet.

GASOLINE BLENDING

Gasoline additives, such as ethanol and dyes, are usually added to the fuel at the distributor. Adding ethanol to gasoline is a way to add oxygen to the fuel itself. Gasoline containing an additive that has oxygen is called *oxygenated fuel*. There are three basic methods used to blend ethanol with gasoline to create E10 (10% ethanol, 90% gasoline).

1. **In-line blending**—Gasoline and ethanol are mixed in a storage tank or in the tank of a transport truck while it is being filled. Because the quantities of each can be accurately measured, this method is most likely to produce a well-mixed blend of ethanol and gasoline. ● **SEE FIGURE 5–12.**
2. **Sequential blending**—This method is usually performed at the wholesale terminal and involves adding a measured amount of ethanol to a tank truck followed by a measured amount of gasoline. ● **SEE FIGURE 5–13.**
3. **Splash blending**—Splash blending can be done at the retail outlet or distributor and involves separate purchases of ethanol and gasoline. In a typical case, a distributor can purchase gasoline, and then drive to another supplier and purchase ethanol. The ethanol is then added (splashed) into the tank of gasoline. This method is the least-accurate method of blending and can result in ethanol concentration for E10 that should be 10%, and ranges from 5% to over 20% in some cases. ● **SEE FIGURE 5–14.**

REFORMULATED GASOLINE

Reformulated gasoline (RFG) is manufactured to help reduce emissions. The gasoline refiners reformulate gasoline by using additives that contain at least 2% oxygen by weight and reducing the additive benzene to a maximum of 1% by volume. Two other major changes done at the refineries are as follows:

1. **Reduce light compounds.** Refineries eliminate butane, pentane, and propane, which have a low boiling point and evaporate easily. These unburned hydrocarbons are released into the atmosphere during refueling and through the fuel tank vent system, contributing to smog formation. Therefore, reducing the light compounds from gasoline helps reduce evaporative emissions.
2. **Reduce heavy compounds.** Refineries eliminate heavy compounds with high boiling points such as aromatics and olefins. The purpose of this reduction is to reduce the amount of unburned hydrocarbons that enter the catalytic converter, which makes the converter more efficient, thereby reducing emissions.

Because many of the heavy compounds are eliminated, a drop in fuel economy of about 1 mpg has been reported in areas where reformulated gasoline is being used. Formaldehyde is formed when RFG is burned, and the vehicle exhaust has a unique smell when reformulated gasoline is used.

FREQUENTLY ASKED QUESTION

Is Water Heavier than Gasoline?

Yes. Water weighs about 8 pounds per gallon, whereas gasoline weighs about 6 pounds per gallon. The density as measured by specific gravity includes:

Water = 1.000 (the baseline for specific gravity)
Gasoline = 0.730 to 0.760

This means that any water that gets into the fuel tank will sink to the bottom.

TESTING GASOLINE FOR ALCOHOL CONTENT

Take the following steps when testing gasoline for alcohol content:

WARNING

Do not smoke or run the test around sources of ignition!

1. Pour suspect gasoline into a graduated cylinder.
2. Carefully fill the graduated cylinder to the 90 mL mark.
3. Add 10 mL of water to the graduated cylinder by counting the number of drops from an eyedropper.
4. Put the stopper in the cylinder and shake vigorously for one minute. Relieve built-up pressure by occasionally removing the stopper. Alcohol dissolves in water and will drop to the bottom of the cylinder.
5. Place the cylinder on a flat surface and let it stand for two minutes.
6. Take a reading near the bottom of the cylinder at the boundary between the two liquids.
7. For percent of alcohol in gasoline, subtract 10 from the reading and multiply by 10.

   For example,

   The reading is 20 mL: 20 − 10 = 10% alcohol

If the increase in volume is 0.2% or less, it may be assumed that the test gasoline contains no alcohol. ● **SEE FIGURE 5–15**. Alcohol content can also be checked using an electronic tester. See the step-by-step sequence at the end of the chapter.

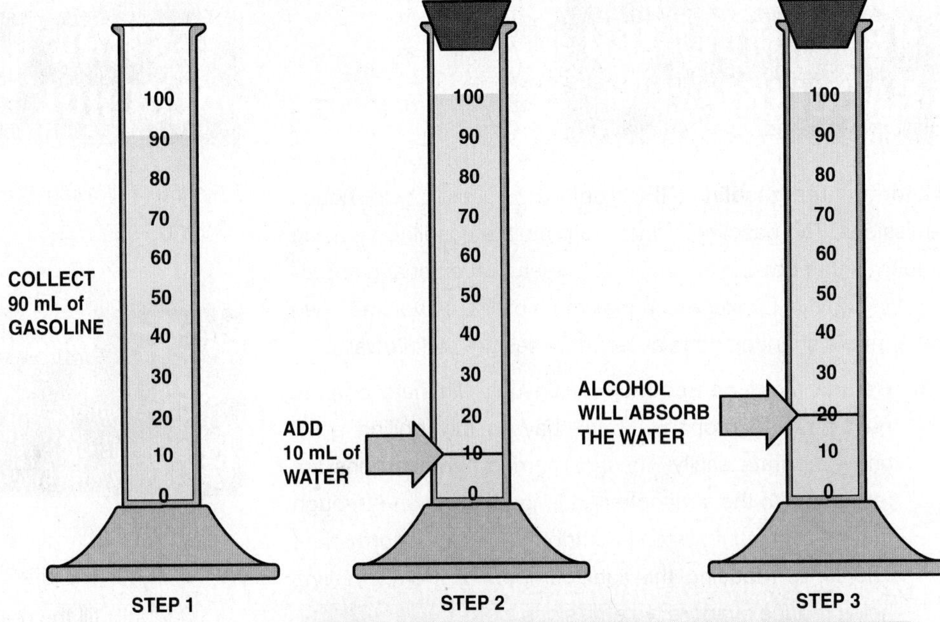

FIGURE 5–15 Checking gasoline for alcohol involves using a graduated cylinder and adding water to check if the alcohol absorbs the water.

FIGURE 5–16 The gas cap on a Ford vehicle notes that BP fuel is recommended.

? FREQUENTLY ASKED QUESTION

How Does Alcohol Content in the Gasoline Affect Engine Operation?

In most cases, the use of gasoline containing 10% or less of ethanol (ethyl alcohol) has little or no effect on engine operation. However, because the addition of 10% ethanol raises the volatility of the fuel slightly, occasional rough idle or stalling may be noticed, especially during warm weather. The rough idle and stalling may also be noticeable after the engine is started, driven, then stopped for a short time. Engine heat can vaporize the alcohol-enhanced fuel causing bubbles to form in the fuel system. These bubbles in the fuel prevent the proper operation of the fuel injection system and result in a hesitation during acceleration, rough idle, or in severe cases repeated stalling until all the bubbles have been forced through the fuel system, replaced by cooler fuel from the fuel tank.

? FREQUENTLY ASKED QUESTION

What Is "Top-Tier" Gasoline?

Top-tier gasoline is gasoline that has specific standards for quality, including enough detergent to keep all intake valves clean. Four automobile manufacturers—BMW, General Motors, Honda, and Toyota—developed the standards. Top-tier gasoline exceeds the quality standards developed by the **World Wide Fuel Charter (WWFC)** that was established in 2002 by vehicle and engine manufacturers. The gasoline companies that agreed to make fuel that matches or exceeds the standards as a top-tier fuel include ChevronTexaco, Shell, and ConocoPhillips. Ford has specified that BP fuel, sold in many parts of the country, is the recommended fuel to use in Ford vehicles. For additional information and a list of all stations that are top tier gas stations, visit www.toptiergas.com.  **SEE FIGURE 5–16.**

GENERAL GASOLINE RECOMMENDATIONS

The fuel used by an engine is a major expense in the operation cost of the vehicle. The proper operation of the engine depends on clean fuel of the proper octane rating and vapor pressure for the atmospheric conditions.

To help ensure proper engine operation and keep fuel costs to a minimum, follow these guidelines:

1. Purchase fuel from a busy station to help ensure that it is fresh and less likely to be contaminated with water or moisture.
2. Keep the fuel tank above one-quarter full, especially during seasons in which the temperature rises and falls by more than 20°F between daytime highs and nighttime lows. This helps to reduce condensed moisture in the fuel tank and could prevent gas line freeze-up in cold weather.

   NOTE: Gas line freeze-up occurs when the water in the gasoline freezes and forms an ice blockage in the fuel line.

3. Do not purchase fuel with a higher octane rating than is necessary. Try using premium high-octane fuel to check for operating differences. Most newer engines are equipped with a detonation (knock) sensor that signals the vehicle computer to retard the ignition timing when spark knock occurs. Therefore, an operating difference may not be noticeable to the driver when using a low-octane fuel, except for a decrease in power and fuel economy. In other words, the engine with a knock sensor will tend to operate knock free on regular fuel, even if premium, higher-octane fuel is specified. Using premium fuel may result in more power and greater fuel economy. The increase in fuel economy, however, would have to be substantial to justify the increased cost of high-octane premium fuel. Some drivers find a good compromise by using mid-grade (plus) fuel to benefit from the engine power and fuel economy gains without the cost of using premium fuel all the time.
4. Avoid using gasoline with alcohol in warm weather, even though many alcohol blends do not affect engine driveability. If warm-engine stumble, stalling, or rough idle occurs, change brands of gasoline.

FIGURE 5–17 Many gasoline service stations have signs posted warning customers to place plastic fuel containers on the ground while filling. If placed in a trunk or pickup truck bed equipped with a plastic liner, static electricity could build up during fueling and discharge from the container to the metal nozzle, creating a spark and possible explosion. Some service stations have warning signs not to use cell phones while fueling to help avoid the possibility of an accidental spark creating a fire hazard.

5. Do not purchase fuel from a retail outlet when a tanker truck is filling the underground tanks. During the refilling procedure, dirt, rust, and water may be stirred up in the underground tanks. This undesirable material may be pumped into your vehicle's fuel tank.
6. Do not overfill the gas tank. After the nozzle clicks off, add just enough fuel to round up to the next dime. Adding additional gasoline will cause the excess to be drawn into the charcoal canister. This can lead to engine flooding and excessive exhaust emissions.
7. Be careful when filling gasoline containers. Always fill a gas can on the ground to help prevent the possibility of static electricity buildup during the refueling process. ● **SEE FIGURE 5–17.**

TECH TIP

The Sniff Test

Problems can occur with stale gasoline from which the lighter parts of the gasoline have evaporated. Stale gasoline usually results in a no-start situation. If stale gasoline is suspected, sniff it. If it smells rancid, replace it with fresh gasoline.

NOTE: If storing a vehicle, boat, or lawnmower over the winter, put some gasoline stabilizer into the gasoline to reduce the evaporation and separation that can occur during storage. Gasoline stabilizer is frequently available at lawnmower repair shops or marinas.

FREQUENTLY ASKED QUESTION

Why Should I Keep the Fuel Gauge Above One-Quarter Tank?

The fuel pickup inside the fuel tank can help keep water from being drawn into the fuel system unless water is all that is left at the bottom of the tank. Over time, moisture in the air inside the fuel tank can condense, causing liquid water to drop to the bottom of the fuel tank (water is heavier than gasoline—about 8 pounds per gallon for water and about 6 lb per gallon for gasoline). If alcohol-blended gasoline is used, the alcohol can absorb the water and the alcohol-water combination can be burned inside the engine. However, when water combines with alcohol, a separation layer occurs between the gasoline at the top of the tank and the alcohol-water combination at the bottom. When the fuel level is low, the fuel pump will draw from this concentrated level of alcohol and water. Because alcohol and water do not burn as well as pure gasoline, severe driveability problems can occur such as stalling, rough idle, hard starting, and missing.

TECH TIP

Do Not Overfill the Fuel Tank

Gasoline fuel tanks have an expansion volume area at the top. The volume of this expansion area is equal to 10% to 15% of the volume of the tank. This area is normally not filled with gasoline, but rather is designed to provide a place for the gasoline to expand into, if the vehicle is parked in the hot sun and the gasoline expands. This prevents raw gasoline from escaping from the fuel system. A small restriction is usually present to control the amount of air and vapors that can escape the tank and flow to the charcoal canister.

This volume area could be filled with gasoline if the fuel is slowly pumped into the tank. Since it can hold an extra 10% (2 gallons in a 20-gallon tank), some people deliberately try to fill the tank completely. When this expansion volume is filled, liquid fuel (rather than vapors) can be drawn into the charcoal canister. When the purge valve opens, liquid fuel can be drawn into the engine, causing an excessively rich air–fuel mixture. Not only can this liquid fuel harm vapor recovery parts, but overfilling the gas tank could also cause the vehicle to fail an exhaust emission test, particularly during an enhanced test when the tank could be purged while on the rollers.

TESTING FOR ALCOHOL CONTENT IN GASOLINE

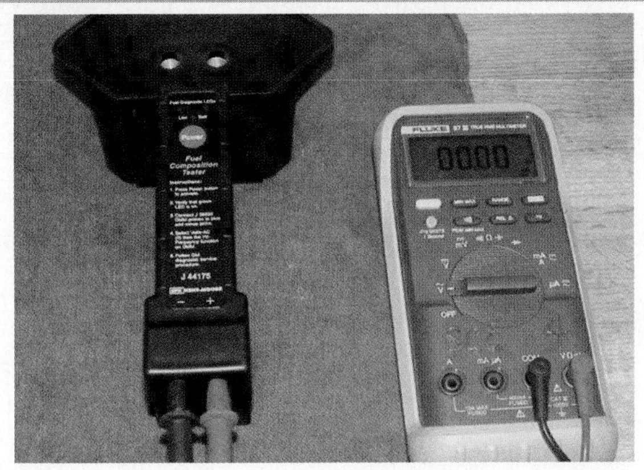

1 A fuel composition tester (SPX Kent-Moore J-44175) is the recommended tool, by General Motors, to use to test the alcohol content of gasoline.

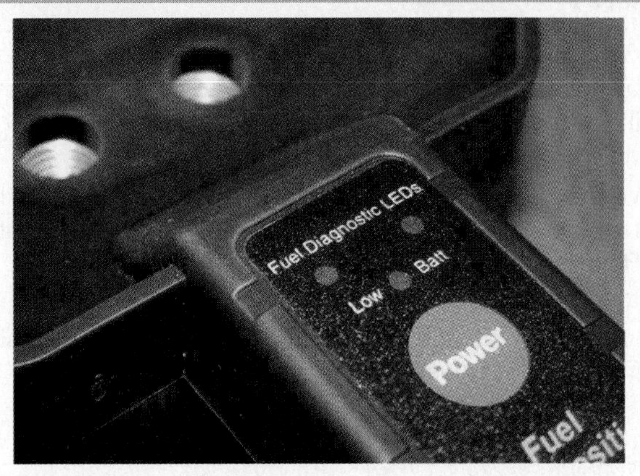

2 This battery-powered tester uses light-emitting diodes (LEDs), meter lead terminals, and two small openings for the fuel sample.

3 The first step is to verify the proper operation of the tester by measuring the air frequency by selecting AC hertz on the meter. The air frequency should be between 35 and 48 Hz.

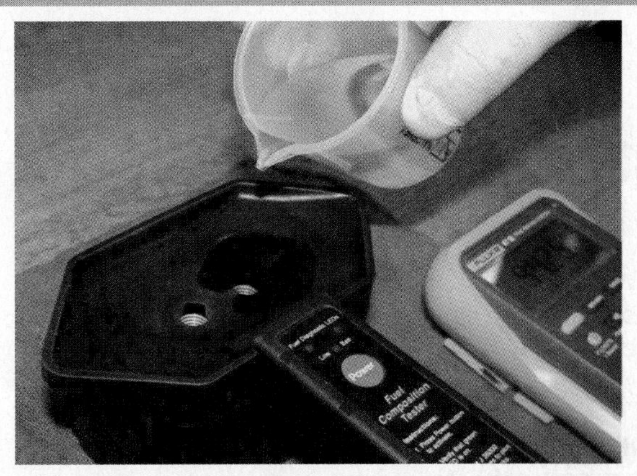

4 After verifying that the tester is capable of correctly reading the air frequency, gasoline is poured into the testing cell of the tool.

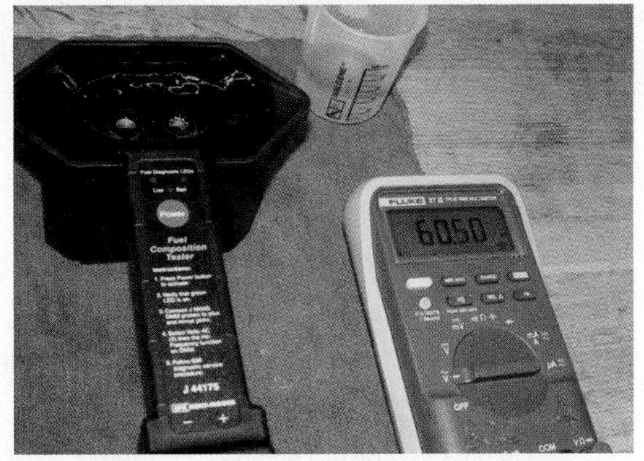

5 Record the AC frequency as shown on the meter and subtract 50 from the reading (e.g., 60.50 − 50.00 = 10.5). This number (10.5) is the percentage of alcohol in the gasoline sample.

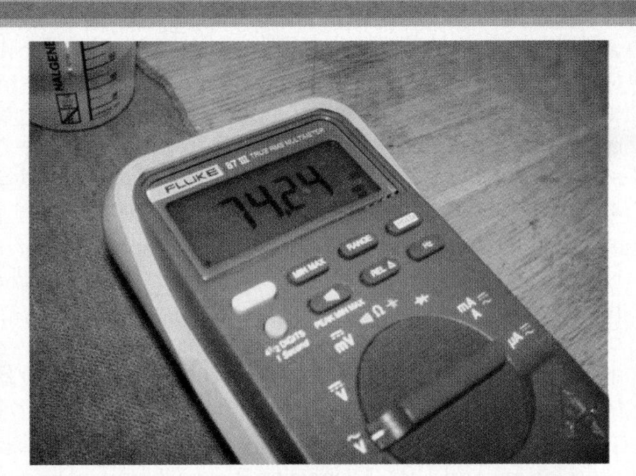

6 Adding additional amounts of ethyl alcohol (ethanol) increases the frequency reading.

SUMMARY

1. Gasoline is a complex blend of hydrocarbons. Gasoline is blended for seasonal usage to achieve the correct volatility for easy starting and maximum fuel economy under all driving conditions.
2. Winter-blend fuel used in a vehicle during warm weather can cause a rough idle and stalling because of its higher Reid vapor pressure (RVP).
3. Abnormal combustion (also called detonation or spark knock) increases both the temperature and the pressure inside the combustion chamber.
4. Most regular-grade gasoline today, using the $(R + M) \div 2$ rating method, is 87 octane; mid-grade (plus) is 89; and premium grade is 91 or higher.
5. Oxygenated fuels contain oxygen to lower CO exhaust emissions.
6. Gasoline should always be purchased from a busy station, and the tank should not be overfilled.

REVIEW QUESTIONS

1. What is the difference between summer-blend and winter-blend gasoline?
2. What is Reid vapor pressure?
3. What is vapor lock?
4. What does the $(R + M) \div 2$ gasoline pump octane rating indicate?
5. What are the octane improvers that may be used during the refining process?
6. What is stoichiometric ratio?

CHAPTER QUIZ

1. Winter-blend gasoline _____.
   a. Vaporizes more easily than summer-blend gasoline
   b. Has a higher RVP
   c. Can cause engine driveability problems if used during warm weather
   d. All of the above
2. Vapor lock can occur _____.
   a. As a result of excessive heat near fuel lines
   b. If a fuel line is restricted
   c. During both a and b
   d. During neither a nor b
3. Technician A says that spark knock, ping, and detonation are different names for abnormal combustion. Technician B says that any abnormal combustion raises the temperature and pressure inside the combustion chamber and can cause severe engine damage. Which technician is correct?
   a. Technician A only
   b. Technician B only
   c. Both Technicians A and B
   d. Neither Technician A nor B
4. Technician A says that the research octane number is higher than the motor octane number. Technician B says that the octane rating posted on fuel pumps is an average of the two ratings. Which technician is correct?
   a. Technician A only
   b. Technician B only
   c. Both Technicians A and B
   d. Neither Technician A nor B
5. Technician A says that in going to high altitudes, engines produce lower power. Technician B says that most engine control systems can compensate the air–fuel mixture for changes in altitude. Which technician is correct?
   a. Technician A only
   b. Technician B only
   c. Both Technicians A and B
   d. Neither Technician A nor B
6. Which method of blending ethanol with gasoline is the most accurate?
   a. In-line
   b. Sequential
   c. Splash
   d. All of the above
7. What can be used to measure the alcohol content in gasoline?
   a. graduated cylinder
   b. electronic tester
   c. scan tool
   d. either a or b
8. To avoid problems with the variation of gasoline, all government testing uses ______ as a fuel during testing procedures.
   a. MTBE (methyl tertiary butyl ether)
   b. Indolene
   c. Xylene
   d. TBA (tertiary butyl alcohol)
9. Avoid topping off the fuel tank because _____.
   a. It can saturate the charcoal canister
   b. The extra fuel simply spills onto the ground
   c. The extra fuel increases vehicle weight and reduces performance
   d. The extra fuel goes into the expansion area of the tank and is not used by the engine
10. Using ethanol-enhanced or reformulated gasoline can result in reduced fuel economy.
    a. True
    b. False

chapter 6

ALTERNATIVE FUELS

LEARNING OBJECTIVES

After studying this chapter, the reader will be able to:

1. List alternatives to gasoline.
2. Discuss how alternative fuels affect driveability.
3. Explain how alternative fuels can reduce CO exhaust emissions.
4. Discuss safety precautions when working with alternative fuels.

This chapter will help you prepare for Engine Repair (A8) ASE certification test content area "A" (General Engine Diagnosis).

KEY TERMS

AFV 110
Anhydrous ethanol 108
Biomass 114
Cellulose ethanol 109
Cellulosic biomass 109
Coal to liquid (CTL) 120
Compressed natural gas (CNG) 116
E85 109
Ethanol 108
Ethyl alcohol 108
FFV 110
Fischer-Tropsch 119
Flex Fuels 110
FTD 119
Fuel compensation sensor 110
Gas to liquid (GTL) 119
Grain alcohol 108
Liquefied petroleum gas (LPG) 115
LP-gas 115
M85 115
Methanol 114
Methanol to gasoline (MTG) 120
NGV 116
Propane 115
Switchgrass 109
Syncrude 120
Syn-gas 114
Synthetic fuel 119
Underground coal gasification (UCG) 120
V-FFV 111
Variable fuel sensor 110

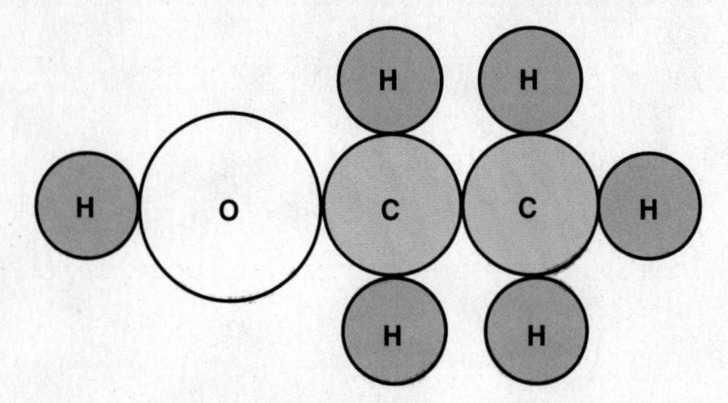

FIGURE 6–1 The ethanol molecule showing two carbon atoms, six hydrogen atoms, and one oxygen atom.

> **? FREQUENTLY ASKED QUESTION**
>
> **Does Ethanol Production Harm the Environment?**
>
> The production of ethanol is considered *carbon neutral* because the amount of CO_2 released during production is equal to the amount of CO_2 that would be released if the corn or other products were left to decay.

ETHANOL

ETHANOL TERMINOLOGY **Ethanol** is also called **ethyl alcohol** or **grain alcohol**, because it is usually made from grain and is the type of alcohol found in alcoholic drinks such as beer, wine, and distilled spirits like whiskey. Ethanol is composed of two carbon atoms and six hydrogen atoms with one added oxygen atom. ● **SEE FIGURE 6–1.**

ETHANOL PRODUCTION Conventional ethanol is derived from grains, such as corn, wheat, or soybeans. Corn, for example, is converted to ethanol in either a dry or wet milling process. In dry milling operations, liquefied cornstarch is produced by heating cornmeal with water and enzymes. A second enzyme converts the liquefied starch to sugars, which are fermented by yeast into ethanol and carbon dioxide. Wet milling operations separate the fiber, germ (oil), and protein from the starch before it is fermented into ethanol.

The majority of the ethanol in the United States is made from:

- Corn
- Grain
- Sorghum
- Wheat
- Barley
- Potatoes

In Brazil, the world's largest ethanol producer, it is made from sugarcane. Ethanol can be made by the dry mill process in which the starch portion of the corn is fermented into sugar and then distilled into alcohol.

The major steps in the dry mill process include:

1. **Milling.** The feedstock passes through a hammer mill that turns it into a fine powder called *meal*.
2. **Liquefaction.** The meal is mixed with water and then passed through cookers where the starch is liquefied. Heat is applied at this stage to enable liquefaction. Cookers use a high-temperature stage of about 250°F to 300°F (120°C to 150°C) to reduce bacteria levels and then a lower temperature of about 200°F (95°C) for a holding period.
3. **Saccharification.** The mash from the cookers is cooled and a secondary enzyme is added to convert the liquefied starch to fermentable sugars (dextrose).
4. **Fermentation.** Yeast is added to the mash to ferment the sugars to ethanol and carbon dioxide.
5. **Distillation.** The fermented mash, now called beer, contains about 10% alcohol plus all the nonfermentable solids from the corn and yeast cells. The mash is pumped to the continuous-flow, distillation system where the alcohol is removed from the solids and the water. The alcohol leaves the top of the final column at about 96% strength, and the residue mash, called *silage*, is transferred from the base of the column to the co-product processing area.
6. **Dehydration.** The alcohol from the top of the column passes through a dehydration system where the remaining water will be removed. The alcohol product at this stage is called **anhydrous ethanol** (pure, no more than 0.5% water).
7. **Denaturing.** Ethanol that will be used for fuel must be denatured, or made unfit for human consumption, with a small amount of gasoline (2% to 5%), methanol, or denatonium benzoate. This is done at the ethanol plant.

CELLULOSE ETHANOL

TERMINOLOGY **Cellulose ethanol** can be produced from a wide variety of cellulose biomass feedstock, including:

- Agricultural plant wastes (corn stalks, cereal straws)
- Plant wastes from industrial processes (sawdust, paper pulp)
- Energy crops grown specifically for fuel production

These nongrain products are often referred to as **cellulosic biomass**. Cellulosic biomass is composed of cellulose and lignin, with smaller amounts of proteins, lipids (fats, waxes, and oils), and ash. About two-thirds of cellulosic materials are present as cellulose, with lignin making up the bulk of the remaining dry mass.

REFINING CELLULOSE BIOMASS As with grains, processing cellulose biomass involves extracting fermentable sugars from the feedstock. But the sugars in cellulose are locked in complex carbohydrates called polysaccharides (long chains of simple sugars). Separating these complex structures into fermentable sugars is needed to achieve the efficient and economic production of cellulose ethanol.

Two processing options are employed to produce fermentable sugars from cellulose biomass:

- Acid hydrolysis is used to break down the complex carbohydrates into simple sugars.
- Enzymes are employed to convert the cellulose biomass to fermentable sugars. The final step involves microbial fermentation, yielding ethanol and carbon dioxide.

NOTE: Cellulose ethanol production substitutes biomass for fossil fuels. The greenhouse gases produced by the combustion of biomass are offset by the CO_2 absorbed by the biomass as it grows in the field.

? FREQUENTLY ASKED QUESTION

What Is Switchgrass?

Switchgrass (*Panicum virgatum*) can be used to make ethanol and is a summer perennial grass that is native to North America. It is a natural component of the tall-grass prairie, which covered most of the Great Plains, but was also found on the prairie soils in the Black Belt of Alabama and Mississippi. Switchgrass is resistant to many pests and plant diseases, and is capable of producing high yields with very low applications of fertilizer. This means that the need for agricultural chemicals to grow switchgrass is relatively low. Switchgrass is also very tolerant of poor soils, flooding, and drought, which are widespread agricultural problems in the southeast.

There are two main types of switchgrass:

- **Upland types**—usually grow five to six feet tall
- **Lowland types**—grow up to 12 feet tall and are typically found on heavy soils in bottomland sites

Better energy efficiency is gained because less energy is used to produce ethanol from switchgrass.

E85

WHAT IS E85 Vehicle manufacturers have available vehicles that are capable of operating on gasoline plus ethanol or a combination of gasoline and ethanol, called **E85**. E85 is composed of 85% ethanol and 15% gasoline.

Pure ethanol has an octane rating of about 113. E85, which contains 35% oxygen by weight, has an octane rating of about 100 to 105. This compares to a regular unleaded gasoline, which has a rating of 87. ● **SEE FIGURE 6–2.**

NOTE: The octane rating of E85 depends on the exact percent of ethanol used, which can vary from 81% to 85%. It also depends on the octane rating of the gasoline, which is used to make E85.

HEAT ENERGY OF E85 E85 has less heat energy than gasoline.

FIGURE 6–2 E85 has 85% ethanol mixed with 15% gasoline.

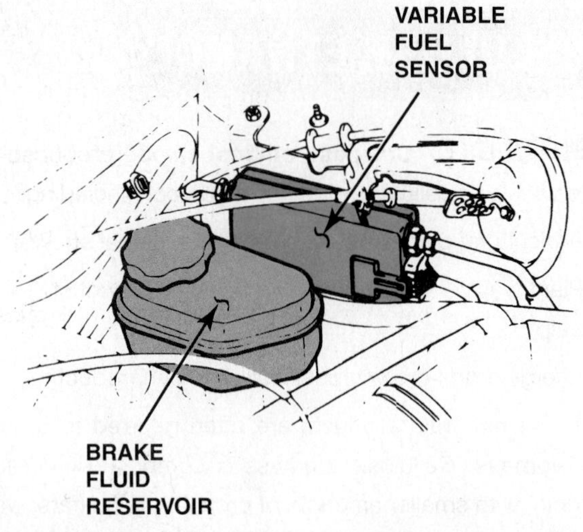

FIGURE 6–3 The location of the variable fuel sensor can vary, depending on the make and model of vehicle, but it is always in the fuel line between the fuel tank and the fuel injectors.

Gasoline = 114,000 BTUs per gallon

E85 = 87,000 BTUs per gallon

This means that the fuel economy is reduced by 20% to 30% if E85 is used instead of gasoline.

Example: A Chevrolet Tahoe 5.3-liter V-8 with an automatic transmission has an EPA rating of 15 mpg in the city and 20 mpg on the highway when using gasoline. If this same vehicle was fueled with E85, the EPA fuel economy rating drops to 11 mpg in the city and 15 mpg on the highway.

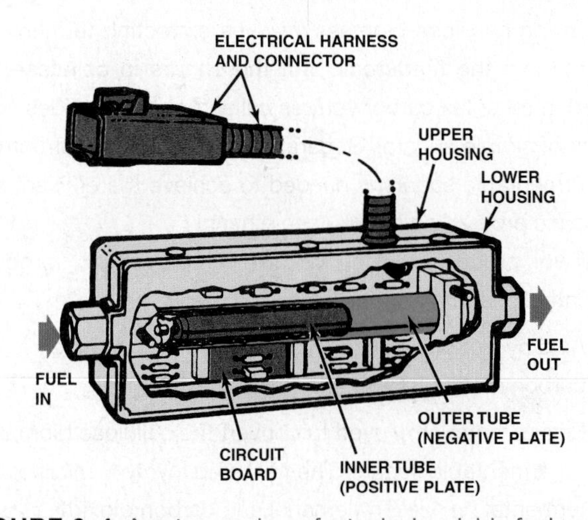

FIGURE 6–4 A cutaway view of a typical variable fuel sensor.

ALTERNATIVE-FUEL VEHICLES

The 15% gasoline in the E85 blend helps the engine start, especially in cold weather. Vehicles equipped with this capability are commonly referred to as **alternative-fuel vehicles (AFVs)**, **Flex Fuels**, and **flexible fuel vehicles (FFVs)**. Using E85 in a flex-fuel vehicle can result in a power increase of about 5%. For example, an engine rated at 200 hp using gasoline or E10 could produce 210 hp if using E85.

NOTE: E85 may test as containing less than 85% ethanol if tested in cold climates because it is often blended according to outside temperature. A lower percentage of ethanol with a slightly higher percentage of gasoline helps engines start in cold climates.

These vehicles are equipped with an electronic sensor in the fuel supply line that detects the presence and percentage of ethanol. The PCM then adjusts the fuel injector on-time and ignition timing to match the needs of the fuel being used.

E85 contains less heat energy, and therefore will use more fuel, but the benefits include a lower cost of the fuel and the environmental benefit associated with using an oxygenated fuel.

General Motors, Ford, Chrysler, Mazda, and Honda are a few of the manufacturers offering E85 compatible vehicles. E85 vehicles use fuel system parts designed to withstand the additional alcohol content, modified driveability programs that adjust fuel delivery and timing to compensate for the various percentages of ethanol fuel, and a **fuel compensation sensor** that measures both the percentage of ethanol blend and the temperature of the fuel. This sensor is also called a **variable fuel sensor**. ● **SEE FIGURES 6–3 AND 6–4.**

TECH TIP

Purchase a Flex-Fuel Vehicle

If purchasing a new or used vehicle, try to find a flex-fuel vehicle. Even though you may not want to use E85, a flex-fuel vehicle has a more robust fuel system than a conventional fuel system designed for gasoline or E10. The enhanced fuel system components and materials usually include:

- Stainless steel fuel rail
- Graphite commutator bars instead of copper in the fuel pump motor (ethanol can oxidize into acetic acid, which can corrode copper)
- Diamond-like carbon (DLC) corrosion-resistant fuel injectors
- Alcohol-resistant O-rings and hoses

The cost of a flex-fuel vehicle compared with the same vehicle designed to operate on gasoline is a no-cost or a low-cost option.

FIGURE 6–5 A warning sticker on an E85 pump warning to only use this fuel in vehicles designated as flexible fuel vehicles (FFV).

FREQUENTLY ASKED QUESTION

How Does a Sensorless Flex-Fuel System Work?

Many General Motors flex-fuel vehicles do not use a fuel compensation sensor and instead use the oxygen sensor to detect the presence of the lean mixture and the extra oxygen in the fuel.

The Powertrain Control Module (PCM) then adjusts the injector pulse-width and the ignition timing to optimize engine operation to the use of E85. This type of vehicle is called a **virtual flexible fuel vehicle (V-FFV)**. The virtual flexible fuel vehicle can operate on pure gasoline or blends up to 85% ethanol.

E85 FUEL SYSTEM REQUIREMENTS Most E85 vehicles are very similar to non-E85 vehicles. Fuel system components may be redesigned to withstand the effects of higher concentrations of ethanol. In addition, since the stoichiometric point for ethanol is 9:1 instead of 14.7:1 as for gasoline, the air-fuel mixture has to be adjusted for the percentage of ethanol present in the fuel tank. In order to determine this percentage of ethanol in the fuel tank, a compensation sensor is used. The fuel compensation sensor is the only additional piece of hardware required on some E85 vehicles. The fuel compensation sensor provides both the ethanol percentage and the fuel temperature to the PCM. The PCM uses this information to adjust both the ignition timing and the quantity of fuel delivered to the engine. The fuel compensation sensor uses a microprocessor to measure both the ethanol percentage and the fuel temperature. This information is sent to the PCM on the signal circuit. The compensation sensor produces a square wave frequency and pulse width signal. The normal frequency range of the fuel compensation sensor is 50 hertz, which represents 0% ethanol, to 150 hertz, which represents 100% ethanol. The pulse width of the signal varies from 1 to 5 milliseconds. One millisecond would represent a fuel temperature of −40°F (−40°C), and 5 milliseconds would represent a fuel temperature of 257°F (125°C). Since the PCM knows both the fuel temperature and the ethanol percentage of the fuel, it can adjust fuel quantity and ignition timing for optimum performance and emissions.

The benefits of E85 vehicles are less pollution, less CO_2 production, and less dependence on oil. ● **SEE FIGURE 6–5.**

Ethanol-fueled vehicles generally produce the same pollutants as gasoline vehicles; however, they produce less CO

FIGURE 6–6 A flex-fuel vehicle often has a yellow gas cap, which is labeled "E85/gasoline".

FIGURE 6–7 A vehicle emission control information (VECI) sticker on a flexible fuel vehicle.

and CO_2 emissions. While CO_2 is not considered a pollutant, it is thought to lead to global warming and is called a greenhouse gas.

FLEX-FUEL VEHICLE IDENTIFICATION Flexible fuel vehicles (FFVs) can be identified by:

- Emblems on the side, front, and/or rear of the vehicle
- Yellow fuel cap showing E85/gasoline (● **SEE FIGURE 6–6**)
- Vehicle emission control information (VECI) label under the hood (● **SEE FIGURE 6–7**)
- Vehicle identification number (VIN)

● **SEE CHART 6–1.**

Chrysler

2004+
- 4.7L Dodge Ram Pickup 1500 Series
- 2.7L Dodge Stratus Sedan
- 2.7L Chrysler Sebring Sedan
- 3.3L Caravan and Grand Caravan SE

2003–2004
- 2.7L Dodge Stratus Sedan
- 2.7L Chrysler Sebring Sedan

2003
- 3.3L Dodge Cargo Minivan

2000–2003
- 3.3L Chrysler Voyager Minivan
- 3.3L Dodge Caravan Minivan 3.3L Chrysler Town and Country Minivan

1998–1999
- 3.3L Dodge Caravan Minivan
- 3.3L Plymouth Voyager Minivan
- 3.3L Chrysler Town & Country Minivan

Ford Motor Company

*Ford offers the flex fuel capability as an option on select vehicles—see the owner's manual.

2004+
- 4.0L Explorer Sport Trac
- 4.0L Explorer (4-door)
- 3.0L Taurus Sedan and Wagon

2002–2004
- 4.0L Explorer (4-door)
- 3.0L Taurus Sedan and Wagon

2002–2003
- 3.0L Supercab Ranger Pickup 2WD

2001
- 3.0L Supercab Ranger Pickup 2WD
- 3.0L Taurus LX, SE, and SES Sedan

1999–2000
- 3.0L Ranger Pickup 4WD and 2WD

General Motors

*Select vehicles only—see your owner's manual.

2005+
- 5.3L Vortec-Engine Avalanche
- 5.3L Vortec-Engine Police Package Tahoe

CHART 6–1

Flexible fuel vehicles by year, make, and model.

2003–2005

- 5.3L V8 Chevy Silverado* and GMC Sierra* Half-Ton Pickups 2WD and 4WD
- 5.3L Vortec-Engine Suburban, Tahoe, Yukon, and Yukon XL

2002

- 5.3L V8 Chevy Silverado* and GMC Sierra* Half-Ton Pickups 2WD and 4WD
- 5.3L Vortec-Engine Suburban, Tahoe, Yukon, and Yukon XL
- 2.2L Chevy S10 Pickup 2WD
- 2.2L Sonoma GMC Pickup 2WD

2000–2001

- 2.2L Chevy S10 Pickup 2WD
- 2.2L GMC Sonoma Pickup 2WD

Isuzu

2000–2001

- 2.2L Hombre Pickup 2WD

Mazda

1999–2003

- 3.0L Selected B3000 Pickups

Mercedes-Benz

2005+

- 2.6L C240 Luxury Sedan and Wagon

2003

- 3.2L C320 Sport Sedan and Wagon

Mercury

2002–2004

- 4.0L Selected Mountaineers

2000–2004

- 3.0L Selected Sables

Nissan

2005+

- 5.6L DOHC V8 Engine

*Select vehicles only—see the owner's manual or VECI sticker under the hood.

CHART 6–1

(Continued)

TECH TIP

Avoid Resetting Fuel Compensation

Starting in 2006, General Motors vehicles designed to operate on E85 do not use a fuel compensation sensor, but instead use the oxygen sensor and refueling information to calculate the percentage of ethanol in the fuel. The PCM uses the fuel level sensor to sense that fuel has been added and starts to determine the resulting ethanol content by using the oxygen sensor. However, if a service technician were to reset fuel compensation by clearing long-term fuel trim, the PCM starts the calculation based on base fuel, which is gasoline with less than or equal to 10% ethanol (E10). If the fuel tank has E85, then the fuel compensation cannot be determined unless the tank is drained and refilled with base fuel. Therefore, avoid resetting the fuel compensation setting unless it is known that the fuel tank contains gasoline or E10 only.

HOW TO READ A VEHICLE IDENTIFICATION NUMBER

The vehicle identification number (VIN) is required by federal regulation to contain specific information about the vehicle. The following chart shows the character in the eighth position of the VIN number from Ford Motor Company, General Motors, and Chrysler that designates their vehicles as flexible fuel vehicles. ● **SEE CHART 6–2.**

NOTE: For additional information on E85 and for the location of E85 stations in your area, go to www.e85fuel.com.

FREQUENTLY ASKED QUESTION

How Long Can Oxygenated Fuel Be Stored Before All of the Oxygen Escapes?

The oxygen in oxygenated fuels, such as E10 and E85, is not in a gaseous state like the CO_2 in soft drinks. The oxygen is part of the molecule of ethanol or other oxygenates and does not bubble out of the fuel. Oxygenated fuels, just like any fuel, have a shelf life of about 90 days.

Ford Motor Company	
VEHICLE	8TH CHARACTER
Ford Crown Victoria	V
Ford F-150	V
Ford Explorer	K
Ford Ranger	V
Ford Taurus	2
Lincoln Town Car	V
Mercury Mountaineer	K
Mercury Sable	2
Mercury Grand Marquis	V
General Motors	
VEHICLE	8TH CHARACTER
Chevrolet Avalanche	Z
Chevrolet Impala	K
Chevrolet Monte Carlo	K
Chevrolet S-10 Pickup	5
Chevrolet Sierra	Z
Chevrolet Suburban	Z
Chevrolet Tahoe	Z
GMC Yukon and Yukon XL	Z
GMC Silverado	Z
GMC Sonoma	5
Chrysler	
VEHICLE	8TH CHARACTER
Chrysler Sebring	T
Chrysler Town & Country	E, G or 3
Dodge Caravan	E, G or 3
Dodge Cargo Minivan	E, G or 3
Dodge Durango	P
Dodge Ram	P
Dodge Stratus	T
Plymouth Voyageur	E, G or 3
Mazda	
VEHICLE	8TH CHARACTER
B3000 Pickup	V
Nissan	
VEHICLE	4TH CHARACTER
Titan	B
Mercedes Benz	
Check owner's manual or the VECI sticker under the hood.	

CHART 6–2

Flexible fuel vehicles by make, model, and VIN.

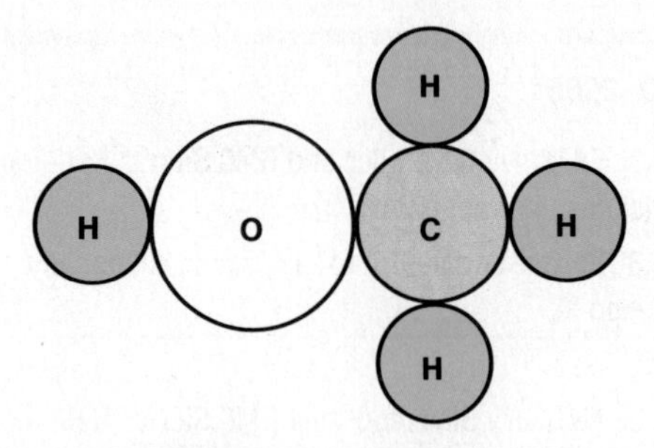

FIGURE 6–8 The molecular structure of methanol showing the one carbon atom, four hydrogen atoms, and one oxygen atom.

METHANOL

METHANOL TERMINOLOGY **Methanol**, also known as *methyl alcohol*, *wood alcohol*, or *methyl hydrate,* is a chemical compound that includes one carbon atom, four hydrogen atoms, and one oxygen atom. ● **SEE FIGURE 6–8.**

Methanol is a light, volatile, colorless, tasteless, flammable, poisonous liquid with a very faint odor. It is used as an antifreeze, a solvent, and a fuel. It is also used to denature ethanol. Methanol burns in air, forming CO_2 (carbon dioxide) and H_2O (water). A methanol flame is almost colorless. Because of its poisonous properties, methanol is also used to denature ethanol. Methanol is often called wood alcohol because it was once produced chiefly as a by-product of the destructive distillation of wood. ● **SEE FIGURE 6–9.**

PRODUCTION OF METHANOL The biggest source of methanol in the United States is coal. Using a simple reaction between coal and steam, a gas mixture called **syn-gas** (*synthesis gas*) is formed. The components of this mixture are carbon monoxide and hydrogen, which, through an additional chemical reaction, are converted to methanol.

Natural gas can also be used to create methanol and is re-formed or converted to synthesis gas, which is later made into methanol.

Biomass can be converted to synthesis gas by a process called partial oxidation, and later converted to methanol. **Biomass** is organic material, such as:

- Urban wood wastes
- Primary mill residues
- Forest residues
- Agricultural residues
- Dedicated energy crops (e.g., sugarcane and sugar beets) that can be made into fuel

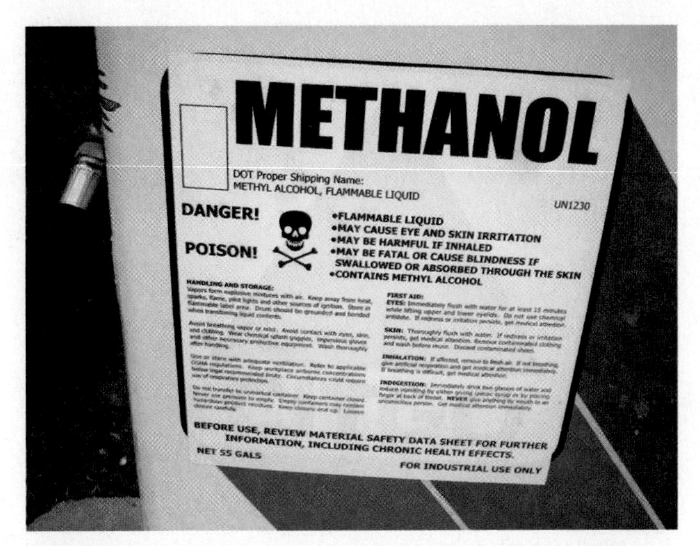

FIGURE 6–9 Sign on methanol pump shows that methyl alcohol is a poison and can cause skin irritation and other personal injury. Methanol is used in industry as well as being a fuel.

FIGURE 6–10 Propane fuel storage tank in the trunk of a Ford taxi.

Electricity can be used to convert water into hydrogen, which is then reacted with carbon dioxide to produce methanol.

Methanol is toxic and can cause blindness and death. It can enter the body by ingestion, inhalation, or absorption through the skin. Dangerous doses will build up if a person is regularly exposed to fumes or handles liquid without skin protection. If methanol has been ingested, a doctor should be contacted immediately. The usual fatal dose is 4 fl oz (100 to 125 mL).

M85 Some flexible fuel vehicles are designed to operate on 85% methanol and 15% gasoline called **M85**. Methanol is very corrosive and requires that the fuel system components be constructed of stainless steel and other alcohol-resistant rubber and plastic components. The heat content of M85 is about 60% of that of gasoline.

PROPANE

Propane is the most widely used of all of the alternative fuels. Propane is normally a gas but is easily compressed into a liquid and stored in inexpensive containers. When sold as a fuel, it is also known as **liquefied petroleum gas (LPG)** or **LP-gas** because the propane is often mixed with about 10% of other gases such as butane, propylene, butylenes, and mercaptan to give the colorless and odorless propane a smell. Propane is nontoxic, but if inhaled can cause asphyxiation through lack of oxygen. Propane is heavier than air and lays near the floor if released into the atmosphere. Propane is commonly used in forklifts and other equipment used inside warehouses and factories because the exhaust from the engine using propane is not harmful. Propane is a by-product of petroleum refining of natural gas. In order to liquefy the fuel, it is stored in strong tanks at about 300 PSI (2,000 kPa). The heating value of propane is less than that of gasoline; therefore, more is required, which reduces the fuel economy. ● **SEE FIGURE 6–10.**

FIGURE 6–11 The blue sticker on the rear of this vehicle indicates that it is designed to use compressed natural gas.

FIGURE 6–12 A CNG storage tank from a Honda Civic GX shown with the fixture used to support it while it is being removed or installed in the vehicle. Honda specifies that three technicians be used to remove or install the tank through the rear door of the vehicle due to the size and weight of the tank.

COMPRESSED NATURAL GAS (CNG)

CNG VEHICLE DESIGN Another alternative fuel that is often used in fleet vehicles is **compressed natural gas (CNG)**, and vehicles using this fuel are often referred to as **natural gas vehicles (NGVs)**. Look for the blue CNG label on vehicles designed to operate on compressed natural gas. ● **SEE FIGURE 6–11.**

Natural gas has to be compressed to about 3,000 PSI (20,000 kPa) or more, so that the weight and the cost of the storage container is a major factor when it comes to preparing a vehicle to run on CNG. The tanks needed for CNG are typically constructed of 0.5 inch-thick (3 mm) aluminum reinforced with fiberglass. ● **SEE FIGURE 6–12**. The octane rating of CNG is about 130 and the cost per gallon is about half of the cost of gasoline. However, the heat value of CNG is also less, and therefore more is required to produce the same power and the miles per gallon is less.

CNG COMPOSITION Compressed natural gas is made up of a blend of:

- Methane
- Propane
- Ethane
- N-butane
- Carbon dioxide
- Nitrogen

Once it is processed, it is at least 93% methane. Natural gas is nontoxic, odorless, and colorless in its natural state. It is odorized during processing, using ethyl mercaptan ("skunk"), to allow for easy leak detection. Natural gas is lighter than air and will rise when released into the air. Since CNG is already a vapor, it does not need heat to vaporize before it will burn, which improves cold start-up and results in lower emissions during cold operation. However, because it is already in a gaseous state, it does displace some of the air charge in the intake manifold. This leads to about a 10% reduction in engine power as compared to an engine operating on gasoline. Natural gas also burns slower than gasoline; therefore, the ignition timing must be advanced more when the vehicle operates on natural gas. The stoichiometric ratio, the point at which all the air and fuel are used or burned, is 16.5:1 compared to 14.7:1 for gasoline. This means that more air is required to burn one pound of natural gas than is required to burn one pound of gasoline. ● **SEE FIGURE 6–13.**

The CNG engine is designed to include:

- Increased compression ratio
- Strong pistons and connecting rods
- Heat-resistant valves
- Fuel injectors designed for gaseous fuel instead of liquid fuel

? FREQUENTLY ASKED QUESTION

What Is the Amount of CNG Equal to in Gasoline?

To achieve the amount of energy of one gallon of gasoline, 122 cubic feet of compressed natural gas (CNG) is needed. While the octane rating of CNG is much higher than gasoline (130 octane), using CNG instead of gasoline in the same engine would result in a reduction 10% to 20% of power due to the lower heat energy that is released when CNG is burned in the engine.

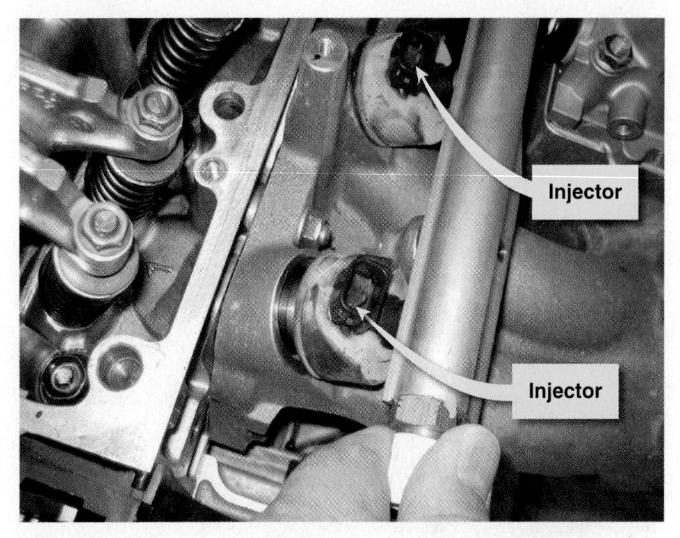

FIGURE 6–13 The fuel injectors used on this Honda Civic GX CNG engine are designed to flow gaseous fuel instead of liquid fuel and cannot be interchanged with any other type of injector.

CNG FUEL SYSTEMS When completely filled, the CNG tank has 3,600 PSI of pressure in the tank. When the ignition is turned on, the alternate fuel electronic control unit activates the high-pressure lock-off, which allows high-pressure gas to pass to the high-pressure regulator. The high-pressure regulator reduces the high-pressure CNG to approximately 170 PSI and sends it to the low-pressure lock-off. The low-pressure lock-off is also controlled by the alternate fuel electronic control unit and is activated at the same time that the high-pressure lock-off is activated. From the low-pressure lock-off, the CNG is directed to the low-pressure regulator. This is a two-stage regulator that first reduces the pressure to approximately 4 to 6 PSI in the first stage and then to 4.5 to 7 inches of water in the second stage. Twenty-eight inches of water is equal to 1 PSI, therefore, the final pressure of the natural gas entering the engine is very low. From here, the low-pressure gas is delivered to the gas mass sensor/mixture control valve. This valve controls the air-fuel mixture. The CNG gas distributor adapter then delivers the gas to the intake stream.

CNG vehicles are designed for fleet use that usually have their own refueling capabilities. One of the drawbacks to using CNG is the time that it takes to refuel a vehicle. The ideal method of refueling is the slow fill method. The slow filling method compresses the natural gas as the tank is being fueled. This method ensures that the tank will receive a full charge of CNG; however, this method can take three to five hours to accomplish. If more than one vehicle needs filling, the facility will need multiple CNG compressors to refuel the vehicles.

FIGURE 6–14 This CNG pump is capable of supplying compressed natural gas at either 3,000 PSI or 3,600 PSI. The price per gallon is higher for the higher pressure.

There are three commonly used CNG refilling station pressures:

P24 —2,400 PSI

P30 —3,000 PSI

P36 —3,600 PSI

Try to find and use a station with the highest refilling pressure. Filling at lower pressures will result in less compressed natural gas being installed in the storage tank, thereby reducing the driving range. ● SEE FIGURE 6–14.

The fast fill method uses CNG that is already compressed. However, as the CNG tank is filled rapidly, the internal temperature of the tank will rise, which causes a rise in tank pressure. Once the temperature drops in the CNG tank, the pressure in the tank also drops, resulting in an incomplete charge in the CNG tank. This refueling method may take only about five minutes; however, it will result in an incomplete charge to the CNG tank, reducing the driving range.

LIQUEFIED NATURAL GAS (LNG)

Natural gas can be turned into a liquid if cooled to below –260°F (–127°C). The natural gas condenses into a liquid at normal atmospheric pressure and the volume is reduced by about 600 times. This means that the natural gas can be more efficiently transported over long distances where no pipelines are present when liquefied.

Because the temperature of liquefied natural gas (LNG) must be kept low, it is only practical for use in short haul trucks where they can be refueled from a central location.

P-SERIES FUELS

P-series alternative fuel is patented by Princeton University and is a non-petroleum- or natural gas-based fuel suitable for use in flexible fuel vehicles or any vehicle designed to operate on E85 (85% ethanol, 15% gasoline). P-series fuel is recognized by the U.S. Department of Energy as being an alternative fuel, but is not yet available to the public. P-series fuels are blends of the following:

- Ethanol (ethyl alcohol)
- Methyltetrahydrofuron (MTHF)
- Natural gas liquids, such as pentanes
- Butane

COMPOSITION OF P-SERIES FUELS (BY VOLUME)

COMPONENT	REGULAR GRADE	PREMIUM GRADE	COLD WEATHER
Pentanes plus	32.5%	27.5%	16.0%
MTHF	32.5%	17.5%	26.0%
Ethanol	35.0%	55.0%	47.0%
Butane	0.0%	0.0%	11.0%

CHART 6–3

P-series fuel varies in composition, depending on the octane rating and temperature.

? **FREQUENTLY ASKED QUESTION**

What Is a Tri-Fuel Vehicle?

In Brazil, most vehicles are designed to operate on ethanol or gasoline or any combination of the two. In this South American country, ethanol is made from sugarcane, is commonly available, and is lower in price than gasoline. Compressed natural gas (CNG) is also being made available so many vehicle manufacturers in Brazil, such as General Motors and Ford, are equipping vehicles to be capable of using gasoline, ethanol, or CNG. These vehicles are called tri-fuel vehicles.

The ethanol and MTHF are produced from renewable feedstocks, such as corn, waste paper, biomass, agricultural waste, and wood waste (scraps and sawdust). The components used in P-type fuel can be varied to produce regular grade, premium grade, or fuel suitable for cold climates. ● **SEE CHART 6–3** for the percentages of the ingredients based on fuel grade.

● **SEE CHART 6–4** for a comparison of the most frequently used alternative fuels.

ALTERNATIVE FUEL COMPARISON CHART					
CHARACTERISTIC	PROPANE	CNG	METHANOL	ETHANOL	REGULAR UNLEADED GAS
Octane	104	130	100	100	87-93
BTU per gallon	91,000	N.A.	70,000	83,000	114,000-125,000
Gallon equivalent	1.15	122 cubic feet—1 gallon of gasoline	1.8	1.5	1
On-board fuel storage	Liquid	Gas	Liquid	Liquid	Liquid
Miles/gallon as compared to gas	85%	N.A.	55%	70%	100%
Relative tank size required to yield driving range equivalent to gas	Tank is 1.25 times larger	Tank is 3.5 times larger	Tank is 1.8 times larger	Tank is 1.5 times larger	
Pressure	200 PSI	3,000-3,600 PSI	N.A.	N.A.	N.A.
Cold weather capability	Good	Good	Poor	Poor	Good
Vehicle power	5-10% power loss	10-20% power loss	4% power increase	5% power increase	Standard
Toxicity	Nontoxic	Nontoxic	Highly toxic	Toxic	Toxic
Corrosiveness	Noncorrosive	Noncorrosive	Corrosive	Corrosive	Minimally corrosive
Source	Natural gas/ petroleum refining	Natural gas/ crude oil	Natural gas/coal	Sugar and starch crops/biomass	Crude oil

CHART 6–4

The characteristics of alternative fuels compared to regular unleaded gasoline shows that all have advantages and disadvantages.

SYNTHETIC FUELS

Synthetic fuels can be made from a variety of products, using several different processes. Synthetic fuel must, however, make these alternatives practical only when conventional petroleum products are either very expensive or not available.

FISCHER-TROPSCH Synthetic fuels were first developed using the **Fischer-Tropsch** method and have been in use since the 1920s to convert coal, natural gas, and other fossil fuel products into a fuel that is high in quality and clean-burning. The process for producing Fischer-Tropsch fuels was patented by two German scientists, Franz Fischer and Hans Tropsch, during World War I. The Fischer-Tropsch method uses carbon monoxide and hydrogen (the same synthesis gas used to produce hydrogen fuel) to convert coal and other hydrocarbons to liquid fuels in a process similar to hydrogenation, another method for hydrocarbon conversion. The process using natural gas, also called **gas-to-liquid (GTL)** technology, uses a catalyst, usually iron or cobalt, and incorporates steam re-forming to give off the by-products of carbon dioxide, hydrogen, and carbon monoxide. ● **SEE FIGURE 6–15.**

Whereas traditional fuels emit environmentally harmful particulates and chemicals, namely sulfur compounds, Fischer-Tropsch fuels combust with no soot or odors and emit only low levels of toxins. Fischer-Tropsch fuels can also be blended with traditional transportation fuels with little equipment modification, as they use the same engine and equipment technology as traditional fuels.

The fuels contain a very low sulfur and aromatic content and they produce virtually no particulate emissions. Researchers also expect reductions in hydrocarbon and carbon monoxide emissions. Fischer-Tropsch fuels do not differ in fuel performance from gasoline and diesel. At present, Fischer-Tropsch fuels are very expensive to produce on a large scale, although research is under way to lower processing costs. Diesel fuel created using the Fischer-Tropsch diesel (**FTD**) process is often called *GTL diesel*. GTL diesel can also be combined with petroleum diesel to produce a GTL blend. This fuel product is currently being sold in Europe and plans are in place to introduce it in North America.

FIGURE 6–15 A Fischer-Tropsch processing plant is able to produce a variety of fuels from coal.

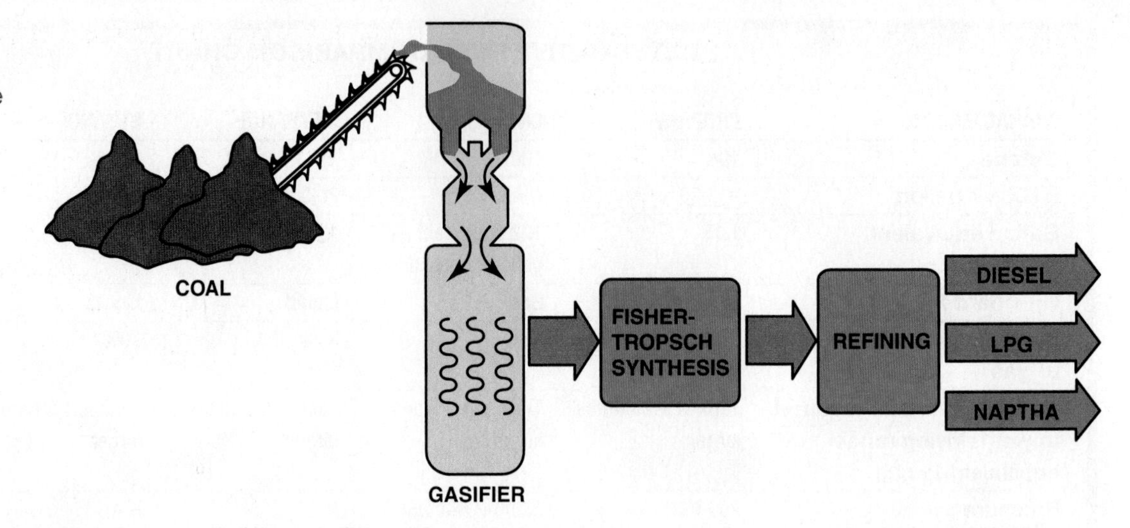

COAL TO LIQUID (CTL) Coal is abundant in the United States and can be converted to a liquid fuel through a process called **coal to liquid (CTL)**. The huge cost is the main obstacle to these plants. The need to invest $1.4 billion per plant before it can make product is the reason no one has built a CTL plant yet in the United States. Investors need to be convinced that the cost of oil is going to remain high in order to get them to commit this kind of money.

A large plant might be able to produce 120,000 barrels of liquid fuel a day and would consume about 50,000 tons of coal per day. However, such a plant would create about 6,000 tons of CO_2 per day. These CO_2 emissions, which could contribute to global warming, and the cost involved make CTL a technology that is not likely to expand.

Two procedures can be used to convert coal-to-liquid fuel:

1. **Direct** —In the direct method, coal is broken down to create liquid products. First, the coal is reacted with hydrogen (H_2) at high temperatures and pressure with a catalyst. This process creates a synthetic crude, called **syncrude**, which is then refined to produce gasoline or diesel fuel.
2. **Indirect** —In the indirect method, coal is first turned into a gas and the molecules are reassembled to create the desired product. This process involves turning coal into a gas called syn-gas. The syngas is then converted into liquid, using the Fischer-Tropsch (FT) process.

Russia has been using CTL by injecting air into the underground coal seams. Ignition is provided and the resulting gases are trapped and converted to liquid gasoline and diesel fuel through the Fischer-Tropsch process. This underground method is called **underground coal gasification (UCG)**.

METHANOL TO GASOLINE Exxon Mobil has developed a process for converting methanol (methyl alcohol) into gasoline in a process called **methanol-to-gasoline (MTG)**. The MTG process was discovered by accident when a gasoline additive made from methanol was being created. The process instead created olefins (alkenes), paraffins (alkenes), and aromatic compounds, which in combination are known as gasoline. The process uses a catalyst and is currently being produced in New Zealand.

FUTURE OF SYNTHETIC FUELS Producing gasoline and diesel fuels by other methods besides refining from crude oil has usually been more expensive. With the increasing cost of crude oil, alternative methods are now becoming economically feasible. Whether or not the diesel fuel or gasoline is created from coal, natural gas, or methanol, or created by refining crude oil, the transportation and service pumps are already in place. Compared to using compressed natural gas or other similar alternative fuels, synthetic fuels represent the lowest cost.

SAFETY PROCEDURES WHEN WORKING WITH ALTERNATIVE FUELS

All fuels are flammable and many are explosive under certain conditions. Whenever working around compressed gases of any kind (CNG, LNG, propane, or LPG), always wear personal protective equipment (PPE), including at least the following items:

1. Safety glasses and/or face shield.
2. Protective gloves.
3. Long-sleeved shirt and pants to help protect bare skin from the freezing effects of gases under pressure in the event that the pressure is lost.
4. If any fuel gets on the skin, the area should be washed immediately.
5. If fuel spills on clothing, change into clean clothing as soon as possible.
6. If fuel spills on a painted surface, flush the surface with water and air dry. If simply wiped off with a dry cloth, the paint surface could be permanently damaged.
7. As with any fuel-burning vehicle, always vent the exhaust to the outside. If methanol fuel is used, the exhaust contains *formaldehyde*, which has a sharp odor and can cause severe burning of the eyes, nose, and throat.

WARNING

Do not smoke or have an open flame in the area when working around or refueling any vehicle.

SUMMARY

1. Flexible fuel vehicles (FFVs) are designed to operate on gasoline or gasoline-ethanol blends up to 85% ethanol (E85).
2. Ethanol can be made from grain, such as corn, or from cellulosic biomass, such as switchgrass.
3. E85 has fewer BTUs of energy per gallon compared with gasoline and will therefore provide lower fuel economy.
4. Older flexible fuel vehicles used a fuel compensation sensor but newer models use the oxygen sensor to calculate the percentage of ethanol in the fuel being burned.
5. Methanol is also called methyl alcohol or wood alcohol and, while it can be made from wood, it is mostly made from natural gas.
6. Propane is the most widely used alternative fuel. Propane is also called liquefied petroleum gas (LPG).
7. Compressed natural gas (CNG) is available for refilling in several pressures, including 2,400 PSI, 3,000 PSI, and 3,600 PSI.
8. P-series fuel is recognized by the U.S. Department of Energy as being an alternative fuel. P-series fuel is a non-petroleum-based fuel suitable for use in a flexible fuel vehicle. However, P-series fuel is not commercially available.
9. Synthetic fuels are usually made using the Fischer-Tropsch method to convert coal or natural gas into gasoline and diesel fuel.
10. Safety procedures when working around alternative fuel include wearing the necessary personal protective equipment (PPE), including safety glasses and protective gloves.

REVIEW QUESTIONS

1. By what other terms ethanol is also known?
2. The majority of ethanol in the United States is made from what farm products?
3. How is a flexible fuel vehicle identified?
4. By what other terms methanol is also known?
5. What other gases are often mixed with propane?
6. Why is it desirable to fill a compressed natural gas (CNG) vehicle with the highest pressure available?
7. P-series fuel is made of what products?
8. The Fischer-Tropsch method can be used to change what into gasoline?

CHAPTER QUIZ

1. Ethanol can be produced from what products?
   a. Switchgrass
   b. Corn
   c. Sugarcane
   d. Any of the above
2. E85 means that the fuel is made from ________.
   a. 85% gasoline, 15% ethanol
   b. 85% ethanol, 15% gasoline
   c. Ethanol that has 15% water
   d. Pure ethyl alcohol
3. A flex-fuel vehicle can be identified by ________.
   a. Emblems on the side, front, and/or rear of the vehicle
   b. VECI
   c. VIN
   d. Any of the above
4. Methanol is also called ________.
   a. Methyl alcohol
   b. Wood alcohol
   c. Methyl hydrate
   d. All of the above
5. Which alcohol is dangerous (toxic)?
   a. Methanol
   b. Ethanol
   c. Both ethanol and methanol
   d. Neither ethanol nor methanol
6. Which is the most widely used alternative fuel?
   a. E85
   b. Propane
   c. CNG
   d. M85
7. Liquefied petroleum gas (LPG) is also called ________.
   a. E85
   b. M85
   c. Propane
   d. P-series fuel
8. How much compressed natural gas (CNG) does it require to achieve the energy of one gallon of gasoline?
   a. 130 cubic feet
   b. 122 cubic feet
   c. 105 cubic feet
   d. 91 cubic feet
9. When refueling a CNG vehicle, why is it recommended that the tank be filled to a high pressure?
   a. The range of the vehicle is increased
   b. The cost of the fuel is lower
   c. Less of the fuel is lost to evaporation
   d. Both a and c
10. Producing liquid fuel from coal or natural gas usually uses which process?
    a. Syncrude
    b. P-series
    c. Fischer-Tropsch
    d. Methanol to gasoline (MTG)

chapter 7

DIESEL AND BIODIESEL FUELS

LEARNING OBJECTIVES

After studying this chapter, the reader will be able to:

1. Explain diesel fuel specifications.
2. List the advantages and disadvantages of biodiesel.
3. Discuss diesel fuel additives.
4. Explain E-diesel specifications.

This chapter will help you prepare for Engine Repair (A8) ASE certification test content area "A" (General Engine Diagnosis).

KEY TERMS

DIESEL FUEL

FEATURES OF DIESEL FUEL Diesel fuel must meet an entirely different set of standards than gasoline. Diesel fuel contains 12% more heat energy than the same amount of gasoline. The fuel in a diesel engine is not ignited with a spark, but is ignited by the heat generated by high compression. The pressure of compression (400 to 700 PSI or 2,800 to 4,800 kilopascals) generates temperatures of 1,200°F to 1,600°F (700°C to 900°C), which speeds the preflame reaction to start the ignition of fuel injected into the cylinder.

DIESEL FUEL REQUIREMENTS All diesel fuel must have the following characteristics:

- **Cleanliness.** It is imperative that the fuel used in a diesel engine be clean and free from water. Unlike the case with gasoline engines, the fuel is the lubricant and coolant for the diesel injector pump and injectors. Good-quality diesel fuel contains additives such as oxidation inhibitors, detergents, dispersants, rust preventatives, and metal deactivators.
- **Low-temperature fluidity.** Diesel fuel must be able to flow freely at all expected ambient temperatures. One specification for diesel fuel is its "pour point," which is the temperature below which the fuel would stop flowing.
- **Cloud point** is another concern with diesel fuel at lower temperatures. Cloud point is the low-temperature point when the waxes present in most diesel fuels tend to form crystals that can clog the fuel filter. Most diesel fuel suppliers distribute fuel with the proper pour point and cloud point for the climate conditions of the area.

(a)

(b)

FIGURE 7–1 **(a)** Regular diesel fuel on the left has a clear or greenish tint, whereas fuel for off-road use is tinted red for identification. **(b)** A fuel pump in a farming area that clearly states the red diesel fuel is for off-road use only (non-taxed).

CETANE NUMBER The cetane number for diesel fuel is the opposite of the octane number for gasoline. The cetane number is a measure of the ease with which the fuel can be ignited. The cetane rating of the fuel determines, to a great extent, its ability to start the engine at low temperatures and to provide smooth warm-up and even combustion. The cetane rating of diesel fuel should be between 45 and 50. The higher the cetane rating, the more easily the fuel is ignited.

SULFUR CONTENT The sulfur content of diesel fuel is very important to the life of the engine. Sulfur in the fuel creates sulfuric acid during the combustion process, which can damage engine components and cause piston ring wear. Federal regulations are getting extremely tight on sulfur content to less than 15 parts per million (PPM). High-sulfur fuel contributes to acid rain.

DIESEL FUEL COLOR Diesel fuel intended for use on the streets and highways is clear or green in color. Diesel fuel to be used on farms and off-road use is dyed red. ● **SEE FIGURE 7–1.**

FIGURE 7–2 Many diesel fuel additives increase the cetane rating which results in improved fuel economy.

GRADES OF DIESEL FUEL **American Society for Testing Materials (ASTM)** also classifies diesel fuel by volatility (boiling range) into the following grades:

GRADE #1 This grade of diesel fuel has the lowest boiling point and the lowest cloud and pour points, as well as a lower BTU content—less heat per pound of fuel. As a result, grade #1 is suitable for use during low-temperature (winter) operation. Grade #1 produces less heat per pound of fuel compared to grade #2 and may be specified for use in diesel engines involved in frequent changes in load and speed, such as those found in city buses and delivery trucks.

GRADE #2 This grade has a higher boiling point, cloud point, and pour point as compared with grade #1. It is usually specified where constant speed and high loads are encountered, such as in long-haul trucking and automotive diesel applications. Most diesel is grade #2.

DIESEL FUEL HEATERS Diesel fuel heaters, either coolant or electric, help prevent power loss and stalling in cold weather. The heater is placed in the fuel line between the tank and the

? FREQUENTLY ASKED QUESTION

What Are Diesel Fuel Additives?

There are several types and many brands of additives that are designed to be added to diesel fuel. These types of additives include:

1. **Winter Conditioners**—Winter conditioners are designed to reduce the Cold Filter Plugging Point (CFPP). CFPP is lowest temperature at that a specified volume of diesel type of fuel can pass through a standardized filtration device in a specified time when cooled under certain conditions.
2. **Multi-functional Conditioners**—Many multifunctional additives increase the cetane rating of the fuel and helps keep injectors clean. By raising the cetane rating of the diesel fuel engine power and fuel economy is improved. This type of additive is designed to be used year-round.
3. **Microbicide**—Microbes can grow in diesel fuel at the junction between the water and the diesel. Water is heavier than the diesel fuel and is near the bottom of the tank. Water in the fuel can be caused by condensation of moist air in the tank and during transport and storage. A microbicide is designed to kill microorganisms including bacteria and fungi.

Always follow the vehicle manufacturer's recommended service procedures and for best results always use additives from a known brand and use according to the instructions on the product label. ● **SEE FIGURE 7–2.**

> ## FREQUENTLY ASKED QUESTION
>
> **How Can You Tell If Gasoline Has Been Added to the Diesel Fuel by Mistake?**
>
> If gasoline has been accidentally added to diesel fuel and is burned in a diesel engine, the result can be very damaging to the engine. The gasoline can ignite faster than diesel fuel, which would tend to increase the temperature of combustion. This high temperature can harm injectors and glow plugs, as well as pistons, head gaskets, and other major diesel engine components. If contaminated fuel is suspected, first smell the fuel at the filler neck. If the fuel smells like gasoline, then the tank should be drained and refilled with diesel fuel. If the smell test does not indicate a gasoline smell (or any rancid smell), then test a sample for proper API gravity.
>
> **NOTE: Diesel fuel designed for on-road use should be green in color. Red diesel fuel (high sulfur) should only be found in off-road or farm equipment.**

primary filter. Some coolant heaters are thermostatically controlled, which allows fuel to bypass the heater once it has reached operating temperature. ● **SEE FIGURE 7–3.**

ULTRA-LOW-SULFUR DIESEL FUEL Diesel fuel is used in diesel engines and is usually readily available throughout the United States, Canada, and Europe, where many more cars are equipped with diesel engines. Diesel engines manufactured to 2007 or newer standards must use ultra-low-sulfur diesel fuel containing less than 15 parts per million (PPM) of sulfur compared to the older, low-sulfur specification of 500 PPM. The purpose of the lower sulfur amount in diesel fuel is to reduce emissions of sulfur oxides (SO_x) and particulate matter (PM) from heavy-duty highway engines and vehicles that use diesel fuel. The emission controls used on 2007 and newer diesel engines require the use of ultra-low-sulfur diesel (ULSD) for reliable operation.

Ultra-low-sulfur diesel (ULSD) will eventually replace the current highway diesel fuel, low-sulfur diesel, which can have as much as 500 PPM of sulfur. ULSD is required for use in all model year 2007 and newer vehicles equipped with advanced emission control systems. ULSD looks lighter in color and has less smell than other diesel fuel.

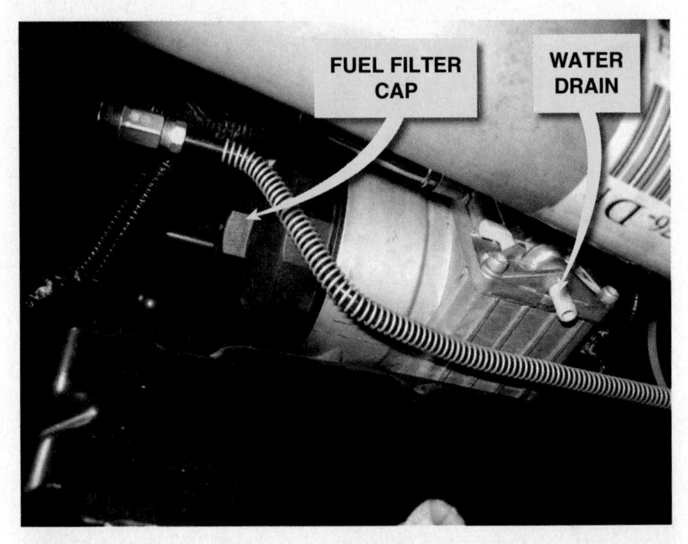

FIGURE 7–3 A fuel heater is part of the fuel filter and water separator located on the frame rail of a Ford pickup truck equipped with a PowerStroke 6.0 liter V-8 diesel engine.

BIODIESEL

DEFINITION OF BIODIESEL **Biodiesel** is a domestically produced, renewable fuel that can be manufactured from vegetable oils, animal fats, or recycled restaurant greases. Biodiesel is safe, biodegradable, and reduces serious air pollutants such as particulate matter (PM), carbon monoxide, and hydrocarbons. Biodiesel is defined as mono-alkyl esters of long-chain fatty acids derived from vegetable oils or animal fats that conform to ASTM D6751 specifications for use in diesel engines. Biodiesel refers to the pure fuel before blending with diesel fuel. ● **SEE FIGURE 7–4.**

BIODIESEL BLENDS Biodiesel blends are denoted as “BXX” with “XX” representing the percentage of biodiesel contained in the blend (e.g., **B20** is 20% biodiesel, 80% petroleum diesel). Blends of 20% biodiesel with 80% petroleum diesel (B20) can generally be used in unmodified diesel engines; however, users should consult their OEM and engine warranty statement. Biodiesel can also be used in its pure form (B100), but it may require certain engine modifications to avoid maintenance and performance problems and may not be suitable for wintertime use. Most diesel engine or vehicle manufacturers of diesel vehicles allow the use of B5 (5% biodiesel). For example, Cummins, used in Dodge trucks, allows the use of B20 only if the optional extra fuel filter has been installed. Users should consult their engine warranty statement for more information on fuel blends of greater than 20% biodiesel.

FIGURE 7–4 A pump decal indicating that the biodiesel fuel is ultra-low-sulfur diesel (ULSD) and must be used in 2007 and newer diesel vehicles.

In general, B20 costs 30 to 40 cents more per gallon than conventional diesel. Although biodiesel costs more than regular diesel fuel, often called **petrodiesel**, fleet managers can make the switch to alternative fuels without purchasing new vehicles, acquiring new spare parts inventories, rebuilding refueling stations, or hiring new service technicians.

FEATURES OF BIODIESEL Biodiesel has the following characteristics:

1. Purchasing biodiesel in bulk quantities decreases the cost of fuel.
2. Biodiesel maintains similar horsepower, torque, and fuel economy.
3. Biodiesel has a higher cetane number than conventional diesel, which increases the engine's performance.
4. It is nontoxic, which makes it safe to handle, transport, and store. Maintenance requirements for B20 vehicles and petrodiesel vehicles are the same.
5. Biodiesel acts as a lubricant and this can add to the life of the fuel system components.

NOTE: For additional information on biodiesel and the locations where it can be purchased, visit www.biodiesel.org.

FREQUENTLY ASKED QUESTION

I Thought Biodiesel Was Vegetable Oil?

Biodiesel is vegetable oil with the glycerin component removed by means of reacting the vegetable oil with a catalyst. The resulting hydrocarbon esters are 16 to 18 carbon atoms in length, almost identical to the petroleum diesel fuel atoms. This allows the use of biodiesel fuel in a diesel engine with no modifications needed. Biodiesel-powered vehicles do not *need* a second fuel tank, whereas vegetable-oil-powered vehicles do. The following are three main types of fuel used in diesel engines:

- Petroleum diesel, a fossil hydrocarbon with a carbon chain length of about 16 carbon atoms.
- Biodiesel, a hydrocarbon with a carbon chain length of 16 to 18 carbon atoms.
- Vegetable oil is a triglyceride with a glycerin component joining three hydrocarbon chains of 16 to 18 carbon atoms each, called straight vegetable oil (**SVO**). Other terms used when describing vegetable oil include:
  - Pure plant oil (**PPO**)—a term most often used in Europe to describe SVO
  - Waste vegetable oil (**WVO**)—this oil could include animal or fish oils from cooking
  - Used cooking oil (**UCO**)—a term used when the oil may or may not be pure vegetable oil

Vegetable oil is not liquid enough at common ambient temperatures for use in a diesel engine fuel delivery system designed for the lower-viscosity petroleum diesel fuel. Vegetable oil needs to be heated to obtain a similar viscosity to biodiesel and petroleum diesel. This means that a heat source needs to be provided before the fuel can be used in a diesel engine. This is achieved by starting on petroleum diesel or biodiesel fuel until the engine heat can be used to sufficiently warm a tank containing the vegetable oil. It also requires purging the fuel system of vegetable oil with petroleum diesel or biodiesel fuel prior to stopping the engine to avoid the vegetable oil thickening and solidifying in the fuel system away from the heated tank. The use of vegetable oil in its natural state does, however, eliminate the need to remove the glycerin component. Many vehicle and diesel engine fuel system suppliers permit the use of biodiesel fuel that is certified as meeting testing standards. None permit the use of vegetable oil in its natural state.

E-DIESEL FUEL

DEFINITION OF E-DIESEL **E-diesel**, also called **diesohol** outside of the United States, is standard No. 2 diesel fuel that contains up to 15% ethanol. While E-diesel can have up to 15% ethanol by volume, typical blend levels are from 8% to 10%.

CETANE RATING OF E-DIESEL The higher the cetane number, the shorter the delay between injection and ignition. Normal diesel fuel has a cetane number of about 50. Adding 15% ethanol lowers the cetane number. To increase the cetane number back to that of conventional diesel fuel, a cetane-enhancing additive is added to E-diesel. The additive used to increase the cetane rating of E-diesel is ethylhexylnitrate or ditertbutyl peroxide.

E-diesel has better cold-flow properties than conventional diesel. The heat content of E-diesel is about 6% less than conventional diesel, but using E-diesel reduces the particulate matter (PM) emissions by as much as 40%, with 20% less carbon monoxide, and a 5% reduction in oxides of nitrogen (NO_X).

Currently, E-diesel is considered to be experimental and can be used legally in off-road applications or in mass-transit buses with EPA approval. For additional information, visit www.e-diesel.org.

SUMMARY

1. Diesel fuel produces 12% more heat energy than the same amount of gasoline.
2. Diesel fuel requirements include cleanliness, low-temperature fluidity, and proper cetane rating.
3. Emission control devices used on 2007 and newer engines require the use of ultra-low-sulfur diesel (ULSD) that has less than 15 parts per million (PPM) of sulfur.
4. The density of diesel fuel is measured in a unit called API gravity.
5. The cetane rating of diesel fuel is a measure of the ease with which the fuel can be ignited.
6. Biodiesel is the blend of vegetable-based liquid with regular diesel fuel. Most diesel engine manufacturers allow the use of a 5% blend, called B20, without any changes to the fuel system or engine.
7. E-diesel is a blend of ethanol with diesel fuel up to 15% ethanol by volume.

REVIEW QUESTIONS

1. What is meant by the cloud point?
2. What is ultra-low-sulfur diesel?
3. Biodiesel blends are identified by what designation?

CHAPTER QUIZ

1. What color is diesel fuel dyed if it is for off-road use only?
   a. Red
   b. Green
   c. Blue
   d. Yellow
2. What clogs fuel filters when the temperature is low on a vehicle that uses diesel fuel?
   a. Alcohol
   b. Sulfur
   c. Wax
   d. Cetane
3. Diesel fuel additives are designed to
   a. Improve cetane rating
   b. Kill microbes
   c. Reduce cold filter plugging point (CFPP)
   d. All of the above
4. What rating of diesel fuel indicates how well a diesel engine will start?
   a. Specific gravity rating
   b. Sulfur content
   c. Cloud point
   d. Cetane rating
5. Ultra-low-sulfur diesel fuel has how much sulfur content?
   a. 15 PPM
   b. 50 PPM
   c. 500 PPM
   d. 1,500 PPM
6. E-diesel is diesel fuel with what additive?
   a. Methanol
   b. Sulfur
   c. Ethanol
   d. Vegetable oil
7. Biodiesel is regular diesel fuel with vegetable oil added.
   a. True
   b. False
8. B20 biodiesel has how much regular diesel fuel?
   a. 20%
   b. 40%
   c. 80%
   d. 100%
9. What grade is most diesel fuel?
   a. Grade #1
   b. Grade #2
   c. Grade #3
   d. Grade #4
10. Most manufacturers of vehicles equipped with diesel engines allow what type of biodiesel?
    a. B100
    b. B80
    c. B20
    d. B5

chapter 8

COOLING SYSTEM OPERATION AND DIAGNOSIS

LEARNING OBJECTIVES

After studying this chapter, the reader will be able to:

1. Diagnose high- and low-temperature engine problems.
2. Describe how coolant flows through an engine.
3. Discuss the operation of the thermostat.
4. Explain the purpose and function of the radiator pressure cap.
5. Describe the operation and service of the water pump.
6. Describe the various types of antifreeze and how to recycle and discard used coolant.
7. Inspect and test cooling system, and perform necessary action.

This chapter will help you prepare for Engine Repair (A1) ASE certification test content area "D" (Lubrication and Cooling Systems Diagnosis and Repair).

KEY TERMS

Bar 140
Bypass 133
Cavitation 140
Centrifugal pump 143
Coolant recovery system 141
Core tubes 137
Ethylene glycol 135
Fins 137
Impeller 143
Reverse cooling 143
Scroll 143
Silicone coupling 145
Surge tank 140
Thermostat 132
Thermostatic spring 145

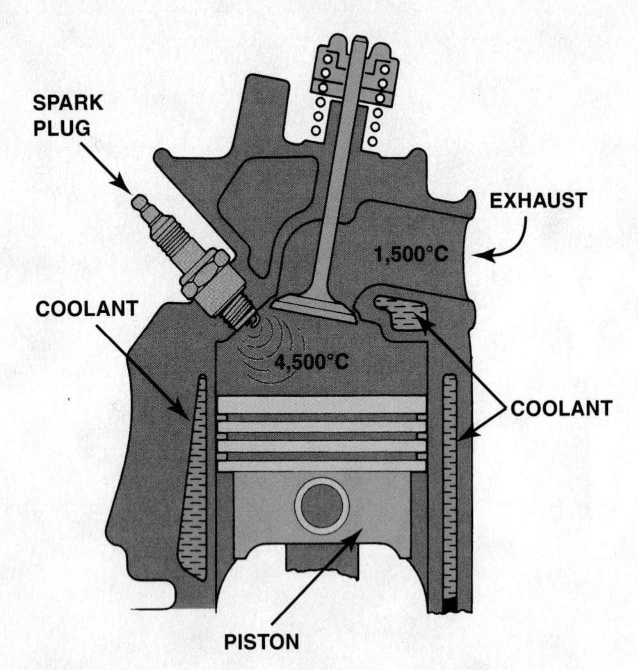

FIGURE 8–1 Typical combustion and exhaust temperatures.

TECH TIP

Overheating Can Be Expensive

A faulty cooling system seems to be a major cause of engine failure. Engine rebuilders often have nightmares about seeing their rebuilt engine placed back in service in a vehicle with a clogged radiator. Most engine technicians routinely replace the water pump and all hoses after an engine overhaul or repair. The radiator should also be checked for leaks and proper flow whenever the engine is repaired or replaced. Overheating is one of the most common causes of engine failure.

COOLING SYSTEM

PURPOSE AND FUNCTION Satisfactory cooling system operation depends on the design and operating conditions of the system. The design is based on heat output of the engine, radiator size, type of coolant, size of water pump (coolant pump), type of fan, thermostat, and system pressure. Unfortunately, the cooling system is usually neglected until there is a problem. Proper routine maintenance can prevent problems.

The cooling system must allow the engine to warm up to the required operating temperature as rapidly as possible and then maintain that temperature. It must be able to do this when the outside air temperature is as low as –30°F (–35°C) and as high as 110°F (45°C).

Peak combustion temperatures in the engine run from 4,000°F to 6,000°F (2,200°C to 3,300°C). The combustion temperatures will *average* between 1,200°F and 1,700°F (650°C and 925°C). Continued temperatures as high as this would weaken engine parts, so heat must be removed from the engine. The cooling system keeps the head and cylinder walls at a temperature that is within the range for maximum efficiency. ● **SEE FIGURE 8–1**.

LOW-TEMPERATURE ENGINE PROBLEMS

Engine operating temperatures must be above a minimum temperature for proper engine operation. Gasoline combustion is a rapid oxidation process that releases heat as the hydrocarbon fuel chemically combines with oxygen from the air. *For each gallon of fuel used, moisture equal to a gallon of water is produced*. It is a part of this moisture that condenses and gets into the oil pan, along with unburned fuel and soot, and causes sludge formation. The condensed moisture combines with unburned hydrocarbons and additives to form carbonic acid, sulfuric acid, nitric acid, hydrobromic acid, and hydrochloric acid. These acids are responsible for engine wear by causing corrosion and rust within the engine. Rust occurs rapidly when the coolant temperature is below 130°F (55°C). High cylinder wall wear rates occur whenever the coolant temperature is below 150°F (65°C).

To reduce cold-engine problems and to help start engines in cold climates, most manufacturers offer block heaters as an option. These block heaters are plugged into household current (110 volts AC) and the heating element warms the coolant.

TECH TIP

Engine Temperature and Exhaust Emissions

Many areas of the United States and Canada have exhaust emission testing. Hydrocarbon (HC) emissions are simply unburned gasoline. To help reduce HC emissions and be able to pass emission tests, be sure that the engine is at normal operating temperature. Vehicle manufacturers' definition of "normal operating temperature" includes the following:

1. Upper radiator hose is hot and pressurized.
2. Electric cooling fan(s) cycles twice.

Be sure that the engine is operating at normal operating temperature before testing for exhaust emissions. For best results, the vehicle should be driven about *20 miles* (32 kilometers) to be certain that the catalytic converter and engine oil, as well as the coolant, are at normal temperature. This is particularly important in cold weather. Most drivers believe that their vehicle will "warm-up" if allowed to idle until heat starts flowing from the heater. The heat from the heater comes from the coolant. Most manufacturers recommend that idling be limited to a maximum of five minutes and that the vehicle should be warmed up by driving slowly after just a minute or two to allow the oil pressure to build.

HIGH-TEMPERATURE ENGINE PROBLEMS

Maximum temperature limits are required to protect the engine. High temperatures will oxidize the engine oil. This breaks the oil down, producing hard carbon and varnish. If high temperatures are allowed to continue, the carbon that is produced will plug piston rings. The varnish will cause the hydraulic valve lifter plungers to stick. High temperatures lower the viscosity of the oil, thus making the oil thin. Metal-to-metal contact within the engine will occur when the oil is too thin. This will cause high friction, loss of power, and rapid wear of the parts. Thinned oil will also get into the combustion chamber by going past the piston rings and through valve guides to cause excessive oil consumption.

High coolant temperatures raise the combustion temperatures to a point that may cause detonation and preignition to occur. These are common forms of abnormal combustion. If they are allowed to continue for any period of time, the engine will get damaged.

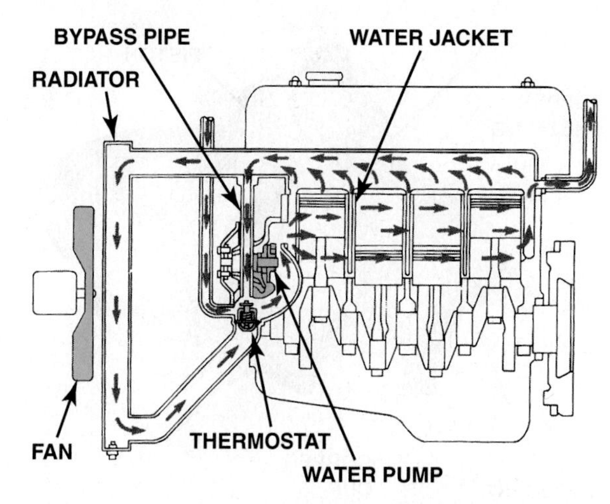

FIGURE 8–2 Coolant flow through a typical engine cooling system.

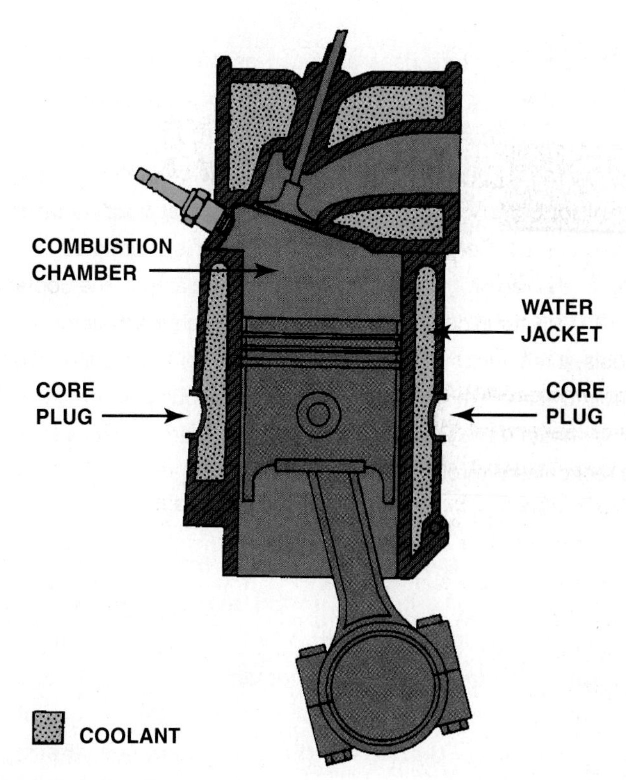

FIGURE 8–3 Coolant circulates through the water jackets in the engine block and cylinder head.

COOLING SYSTEM DESIGN

Coolant flows through the engine, where it picks up heat. It then flows to the radiator, where the heat is given up to the outside air. The coolant continually recirculates through the cooling system, as illustrated in ● **FIGURES 8–2 AND 8–3.**

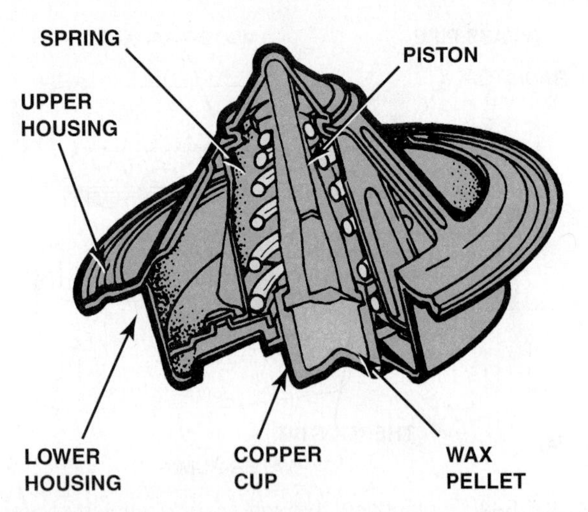

FIGURE 8–4 A cross section of a typical wax-actuated thermostat showing the position of the wax pellet and spring.

Its temperature rises as much as 15°F (8°C) as it goes through the engine; then it recools as it goes through the radiator. *The coolant flow rate may be as high as 1 gallon (4 liters) per minute for each horsepower the engine produces*.

Hot coolant comes out of the thermostat housing on the top of the engine. The engine coolant outlet is connected to the top of the radiator by the upper hose and clamps. The coolant in the radiator is cooled by air flowing through the radiator. As it cools, it moves from the top to the bottom of the radiator. Cool coolant leaves the lower radiator area through an outlet and lower hose, going into the inlet side of the water pump, where it is recirculated through the engine.

NOTE: Some newer engine designs such as Chrysler's 4.7 L, V-8 and General Motors, 4.8, 5.3, 5.7, and 6.0 L V-8s place the thermostat on the inlet side of the water pump. As the cooled coolant hits the thermostat, the thermostat closes until the coolant temperature again causes it to open. Placing the thermostat in the inlet side of the water pump therefore reduces thermal cycling by reducing the rapid temperature changes that could cause stress in the engine, especially if aluminum heads are used with a cast-iron block.

Much of the cooling capacity of the cooling system is based on the functioning of the radiator. Radiators are designed for the maximum rate of heat transfer using minimum space. Cooling airflow through the radiator is aided by a belt- or electric motor-driven cooling fan.

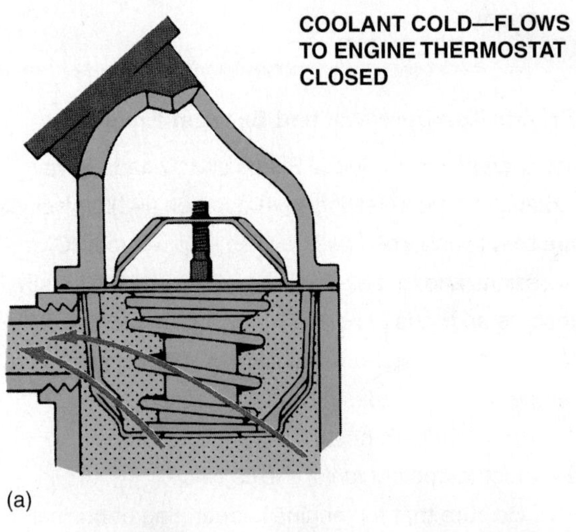

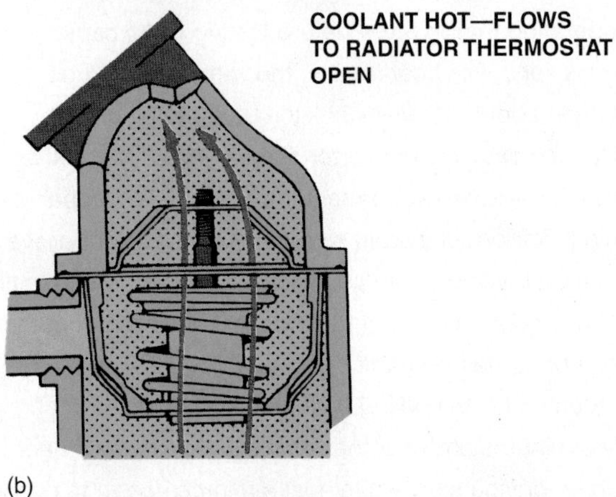

FIGURE 8–5 (a) When the engine is cold, the coolant flows through the bypass. (b) When the thermostat opens, the coolant can flow to the radiator.

THERMOSTAT TEMPERATURE CONTROL

There is a normal operating temperature range between low-temperature and high-temperature extremes. The thermostat controls the minimum normal temperature. The **thermostat** is a temperature-controlled valve placed at the engine coolant outlet. An encapsulated, wax-based, plastic-pellet heat sensor is located on the engine side of the thermostatic valve. As the engine warms, heat swells the heat sensor. ● **SEE FIGURE 8–4.**

A mechanical link, connected to the heat sensor, opens the thermostat valve. As the thermostat begins to open, it allows some coolant to flow to the radiator, where it is cooled. The remaining part of the coolant continues to flow through the bypass, thereby bypassing the thermostat and flowing back through the engine. ● **SEE FIGURE 8–5.**

The rated temperature of the thermostat indicates the temperature at which the thermostat starts to open. The

TECH TIP

Do Not Take Out the Thermostat!

Some vehicle owners and technicians remove the thermostat in the cooling system to "cure" an overheating problem. In some cases, removing the thermostat can *cause* overheating—not stop overheating. This is true for three reasons:

1. Without a thermostat the coolant can flow more quickly through the radiator. The thermostat adds some restriction to the coolant flow, and therefore keeps the coolant in the radiator longer. The presence of the thermostat thus ensures a greater reduction in the coolant temperature before it returns to the engine.
2. Heat transfer is greater with a greater difference between the coolant temperature and air temperature. Therefore, when coolant flow rate is increased (no thermostat), the temperature difference is reduced.
3. Without the restriction of the thermostat, much of the coolant flow often bypasses the radiator entirely and returns directly to the engine.

If overheating is a problem, removing the thermostat will usually not solve the problem. Remember, the thermostat controls the temperature of the engine coolant by opening at a certain temperature and closing when the temperature falls below the minimum rated temperature of the thermostat. If overheating occurs, two basic problems could be the cause:

1. The engine is producing too much heat for the cooling system to handle. For example, if the engine is running too lean or if the ignition timing is either excessively advanced or excessively retarded, overheating of the engine can result.
2. The cooling system has a malfunction or defect that prevents it from getting rid of its heat.

thermostat is fully open at about 20°F higher than its opening temperature. ● **SEE CHART 8–1.**

If the radiator, water pump, and coolant passages are functioning correctly, the engine should always be operating within the opening and fully open temperature range of the thermostat. ● **SEE FIGURE 8–6.**

FIGURE 8–6 A thermostat stuck in the open position caused the engine to operate too cold. The vehicle failed an exhaust emission test because of this defect. If a thermostat is stuck closed, this can cause the engine to overheat.

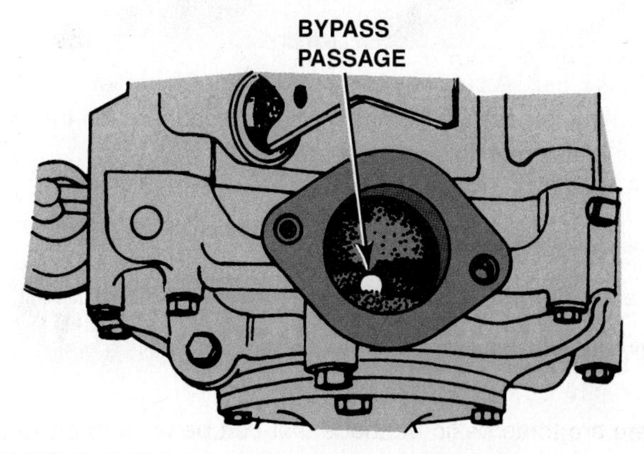

FIGURE 8–7 This internal bypass passage in the thermostat housing directs cold coolant to the water pump.

THERMOSTAT TEMPERATURE RATING	STARTS TO OPEN	FULLY OPEN
180°F	180°F	200°F
195°F	195°F	215°F

CHART 8–1

The temperatures that a thermostat starts to open depends on the rating of the thermostat.

BYPASS A **bypass** around the closed thermostat allows a small part of the coolant to circulate within the engine during warm-up. It is a small passage that leads from the engine side of the thermostat to the inlet side of the water pump. It allows some coolant to bypass the thermostat even when the thermostat is open. The bypass may be cast or drilled into the engine and pump parts. ● **SEE FIGURES 8–7 AND 8–8.**

The bypass aids in uniform engine warm-up. Its operation eliminates hot spots and prevents the buildup of excessive coolant pressure in the engine when the thermostat is closed.

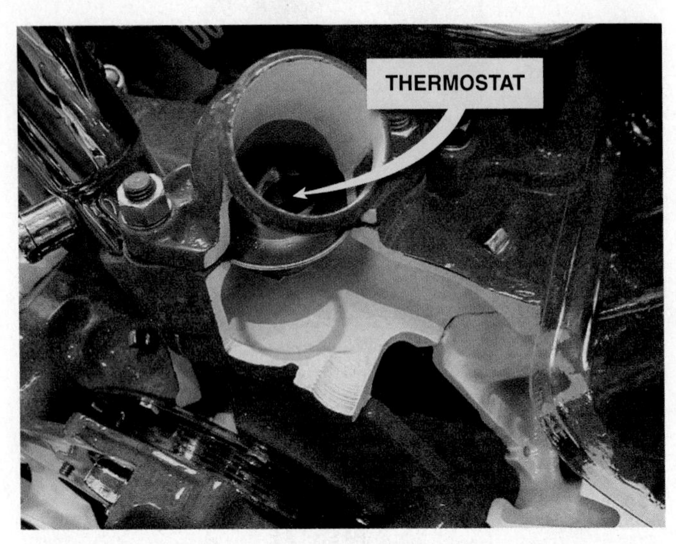

FIGURE 8–8 A cutaway of a small block Chevrolet V-8 showing the passage from the cylinder head through the front of the intake manifold to the thermostat.

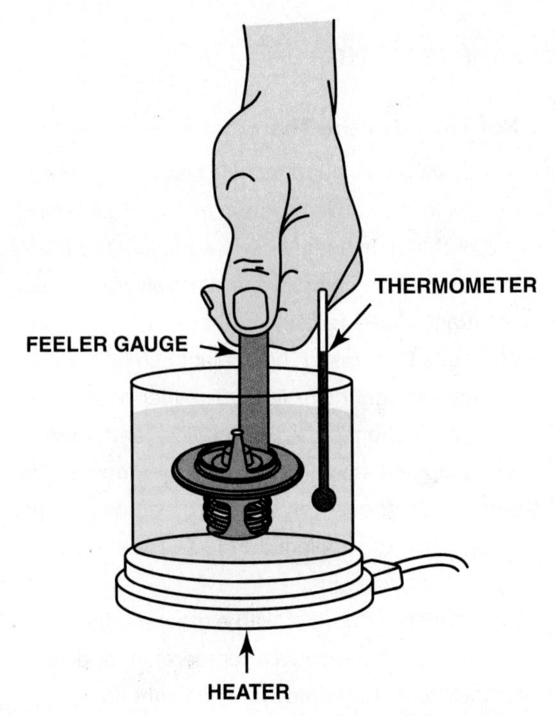

FIGURE 8–9 Setup used to check the opening temperature of a thermostat.

TESTING THE THERMOSTAT

There are three basic methods that can be used to check the operation of the thermostat.

1. **Hot water method.** If the thermostat is removed from the vehicle and is closed, insert a 0.015 inch (0.4 mm) feeler gauge in the opening so that the thermostat will hang on the feeler gauge. The thermostat should then be suspended by the feeler gauge in a bath along with a thermometer. ● **SEE FIGURE 8–9**. The bath should be heated until the thermostat opens enough to release and fall from the feeler gauge. The temperature of the bath when the thermostat falls is the opening temperature of the thermostat. If it is within 5°F (4°C) of the temperature stamped on the thermostat, the thermostat is satisfactory for use. If the temperature difference is greater, the thermostat should be replaced.
2. **Infrared thermometer method.** An infrared thermometer (also called pyrometer) can be used to measure the temperature of the coolant near the thermostat. The area on the engine side of the thermostat should be at the highest temperature that exists in the engine. A properly operating cooling system should cause the pyrometer to read as follows:
   - **a.** As the engine warms, the temperature reaches near thermostat-opening temperature.
   - **b.** As the thermostat opens, the temperature drops just as the thermostat opens, sending coolant to the radiator.
   - **c.** As the thermostat cycles, the temperature should range between the opening temperature of the thermostat and 20°F (11°C) above the opening temperature.

   NOTE: If the temperature rises higher than 20°F (11°C) above the opening temperature of the thermostat, inspect the cooling system for a restriction or low coolant flow. A clogged radiator could also cause the excessive temperature rise.
3. **Scan tool method.** A scan tool can be used on many vehicles to read the actual temperature of the coolant as detected by the engine coolant temperature (ECT) sensor. Although the sensor or the wiring to and from the sensor may be defective, at least the scan tool can indicate what the computer "thinks" the engine coolant temperature is.

FIGURE 8–10 Some thermostats are an integral part of the housing. This thermostat and radiator hose housing is serviced as an assembly. Some thermostats simply snap into the engine radiator fill tube underneath the pressure cap.

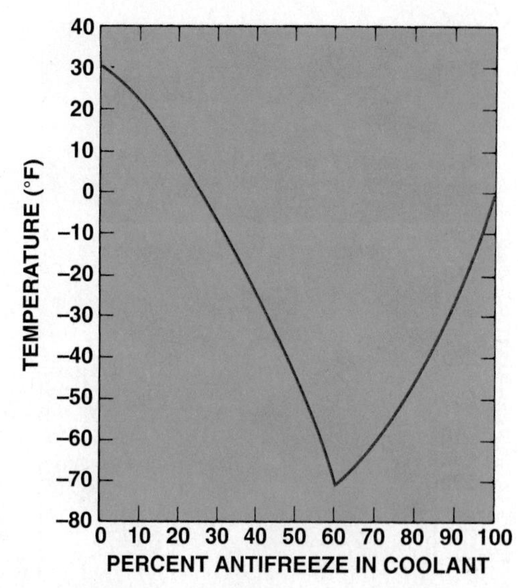

FIGURE 8–11 Graph showing the relationship between the freezing point of the coolant and the percentage of antifreeze used in the coolant.

THERMOSTAT REPLACEMENT

An overheating engine may result from a faulty thermostat. An engine that does not get warm enough always indicates a faulty thermostat.

To replace the thermostat, coolant will have to be drained from the radiator drain petcock to lower the coolant level below the thermostat. It is not necessary to completely drain the system. The upper hose should be removed from the thermostat housing neck; then the housing must be removed to expose the thermostat. ● **SEE FIGURE 8–10.**

The gasket flanges of the engine and thermostat housing should be cleaned, and the gasket surface of the housing must be flat. The thermostat should be placed in the engine with the sensing pellet *toward* the engine. Make sure that the thermostat position is correct, and install the thermostat housing with a new gasket or O-ring.

CAUTION: Failure to set the thermostat into the recessed groove will cause the housing to become tilted when tightened. If this happens and the housing bolts are tightened, the housing will usually crack, creating a leak.

The upper hose should then be installed and the system refilled. Install the proper size of radiator hose clamp.

ANTIFREEZE/COOLANT

Coolant is a mixture of antifreeze and water. Water is able to absorb more heat per gallon than any other liquid coolant. Under standard conditions, water boils at 212°F (100°C) and freezes at 32°F (0°C). *When water freezes, it increases in volume about 9%.* The expansion of the freezing water can easily crack engine blocks, cylinder heads, and radiators. All manufacturers recommend the use of **ethylene glycol**–based antifreeze mixtures for protection against this problem.

A curve depicting freezing point as compared with the percentage of antifreeze mixture is shown in ● **FIGURE 8–11.**

It should be noted that the freezing point increases as the antifreeze concentration is increased above 60%. The normal mixture is 50% antifreeze and 50% water. Ethylene glycol antifreezes contain anticorrosion additives, rust inhibitors, and water pump lubricants.

At the maximum level of protection, an ethylene glycol concentration of 60% will absorb about 85% as much heat as will water. Ethylene glycol–based antifreeze also has a higher boiling point than water. ● **SEE FIGURE 8–12.**

If the coolant boils, it vaporizes and does not act as a cooling agent because it is not in liquid form and in contact with the cooling surfaces.

All coolants have rust and corrosion inhibitors to help protect the metals in the engine and cooling systems. Most conventional green antifreeze contains inorganic salts such as sodium silicate and phosphates. Organic additive technology (OAT) coolant contains inorganic acid salt (carboxylates)

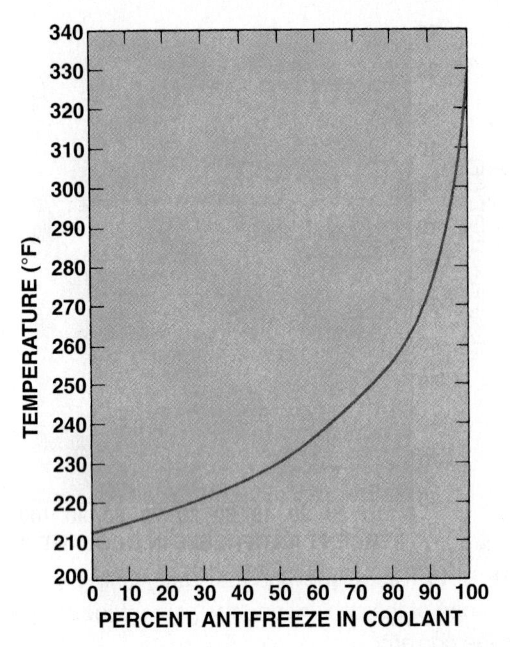

FIGURE 8–12 Graph showing how the boiling point of the coolant increases as the percentage of antifreeze in the coolant increases.

additives that are not abrasive to water pumps. Hybrid organic additive technology (HOAT) coolant contains inorganic acids and some silicate but is phosphate free.

FREQUENTLY ASKED QUESTION

What Is "Pet Friendly" Coolant?

Similar to ethylene glycol, propylene glycol is a type of coolant that is less harmful to pets and animals because it is not sweet tasting, although it is still harmful if swallowed. This type of coolant should not be mixed with ethylene glycol coolant.

NOTE: Some vehicle manufacturers do not recommend the use of propylene glycol coolant. Check the recommendations in the owner's manual or service manual before using it in a vehicle.

	FREEZING POINT
Pure water	32°F (0°C)
Pure antifreeze*	0°F (–18°C)
50/50 mixture	–34°F (–37°C)
70% antifreeze/30% water	–84°F (–64°C)

*Pure antifreeze is usually 95% ethylene glycol, 2% to 3% water, and 2% to 3% additives.

CHART 8–2

The freezing point depends on the concentration of antifreeze in the coolant.

CASE STUDY

If 50% Is Good, 100% Must Be Better

A vehicle owner said that the cooling system of his vehicle would never freeze or rust. He said that he used 100% antifreeze (ethylene glycol) instead of a 50/50 mixture with water.

However, after the temperature dropped to –20°F (–29°C), the radiator froze and cracked. (Pure antifreeze freezes at about 0°F [–18°C].) After thawing, the radiator had to be repaired. The owner was lucky that the engine block also did not crack.

For best freeze protection with good heat transfer, use a 50/50 mixture of antifreeze and water. A 50/50 mixture of antifreeze and water is the best compromise between temperature protection and the heat transfer that is necessary for cooling system operation.

Summary:

- **Complaint**—The engine block cracked in cold weather.
- **Cause**—The coolant used was 100% antifreeze without any water causing the it to freeze when the temperature dropped to −20 degrees F
- **Correction**—The engine was replaced the coolant used was of the proper 50/50 mixture of antifreeze and water.

ANTIFREEZE CAN FREEZE

An antifreeze and water mixture is an example wherein the freezing point differs from the freezing point of either pure antifreeze or pure water. ● SEE CHART 8–2.

Depending on the exact percentage of water used, antifreeze, as sold in containers, freezes between –8°F and +8°F (–13°C and –22°C). Therefore, it is easiest just to remember that most antifreeze freezes at about 0°F (–18°C).

The boiling point of antifreeze and water is also a factor of mixture concentrations. ● SEE CHART 8–3.

	BOILING POINT AT SEA LEVEL	BOILING POINT WITH 15 PSI PRESSURE CAP
Pure water	212°F (100°C)	257°F (125°C)
50/50 mixture	218°F (103°C)	265°F (130°C)
70/30 mixture	225°F (107°C)	276°F (136°C)

CHART 8–3

The boiling point depends on the concentration of antifreeze in the coolant.

FIGURE 8–13 Checking the freezing and boiling protection levels of the coolant using a hydrometer.

TECH TIP

Ignore the Windchill Factor

The windchill factor is a temperature that combines the actual temperature and the wind speed to determine the overall heat loss effect on exposed skin. Because it is the heat loss factor for exposed skin, the windchill temperature is *not* to be considered when determining antifreeze protection levels.

Although moving air does make it feel colder, the actual temperature is not changed by the wind and the engine coolant will not be affected by the windchill. Not convinced? Try this. Place a thermometer in a room and wait until a stable reading is obtained. Now turn on a fan and have the air blow across the thermometer. The temperature will not change.

HYDROMETER TESTING

Coolant can be checked using a coolant hydrometer. The hydrometer measures the density of the coolant. The higher the density, the more concentration of antifreeze in the water. Most coolant hydrometers read the freezing point and boiling point of the coolant. ● **SEE FIGURE 8–13.**

If the engine is overheating and the hydrometer reading is near –50°F (–45.555°C), suspect that pure 100% antifreeze is present. For best results, the coolant should have a freezing point lower than –20°F and a boiling point above 234°F.

RECYCLING COOLANT

Coolant (antifreeze and water) should be recycled. Used coolant may contain heavy metals, such as lead, aluminum, and iron, which are absorbed by the coolant during its use in the engine.

Recycle machines filter out these metals and dirt and reinstall the depleted additives. The recycled coolant, restored to be like new, can be reinstalled into the vehicle.

CAUTION: Most vehicle manufacturers warn that antifreeze coolant should not be reused unless it is recycled and the additives restored.

DISPOSING OF USED COOLANT

Used coolant drained from vehicles can usually be disposed of by combining it with used engine oil. The equipment used for recycling the used engine oil can easily separate the coolant from the waste oil. Check with recycling companies authorized by local or state governments for the exact method recommended for disposal in your area.

RADIATOR DESIGN AND FUNCTION

Two types of radiator cores are in common use in domestic vehicles—the serpentine fin core and the plate fin core. In each of these types the coolant flows through oval-shaped **core tubes**. Heat is transferred through the tube wall and soldered joint to the **fins**. The fins are exposed to airflow, which removes heat

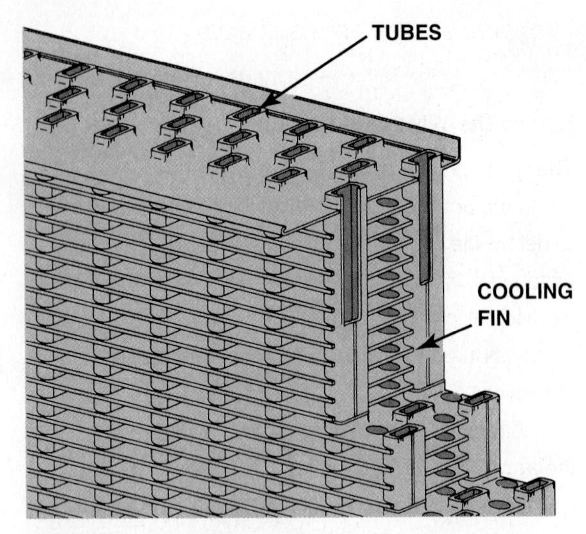

FIGURE 8–14 The tubes and fins of the radiator core.

from the radiator and carries it away. ● **SEE FIGURES 8–14 THROUGH 8–16.**

Older automobile radiators were made from yellow brass. Since the 1980s, most radiators have been made from aluminum. These materials are corrosion resistant, have good heat-transferring ability, and are easily formed.

Core tubes are made from 0.0045 to 0.012 inch (0.1 to 0.3 mm) sheet brass or aluminum, using the thinnest possible materials for each application. The metal is rolled into round tubes and the joints are sealed with a locking seam.

The main limitation of heat transfer in a cooling system is in the transfer from the radiator to the air. Heat transfers from the water to the fins as much as seven times faster than heat transfers from the fins to the air, assuming equal surface exposure. The radiator must be capable of removing an amount of heat energy approximately equal to the heat energy of the power produced by the engine. *Each horsepower is equivalent to 42 BTU (10,800 calories) per minute*. As the engine power is increased, the heat-removing requirement of the cooling system is also increased.

With a given frontal area, radiator capacity may be increased by increasing the core thickness, packing more material into the same volume, or both. The radiator capacity may also be increased by placing a shroud around the fan so that more air will be pulled through the radiator.

NOTE: The lower air dam in the front of the vehicle is used to help direct the air through the radiator. If this air dam is broken or missing, the engine may overheat, especially during highway driving due to the reduced airflow through the radiator.

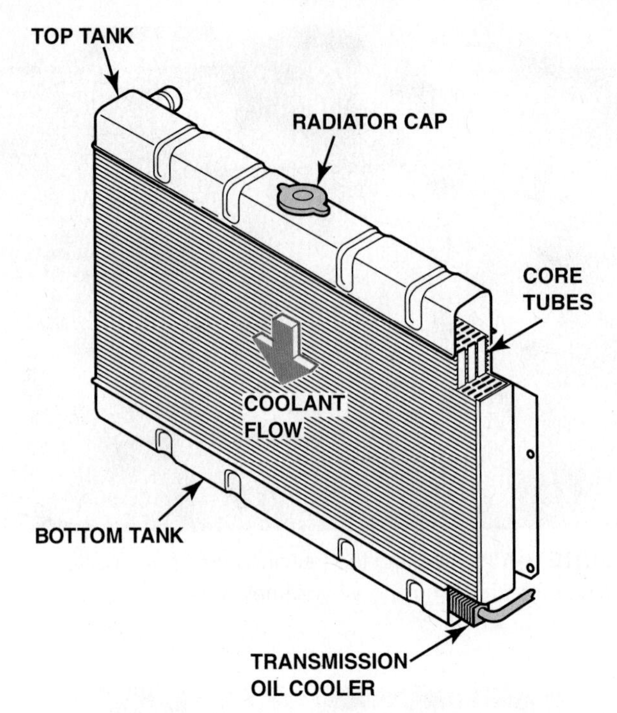

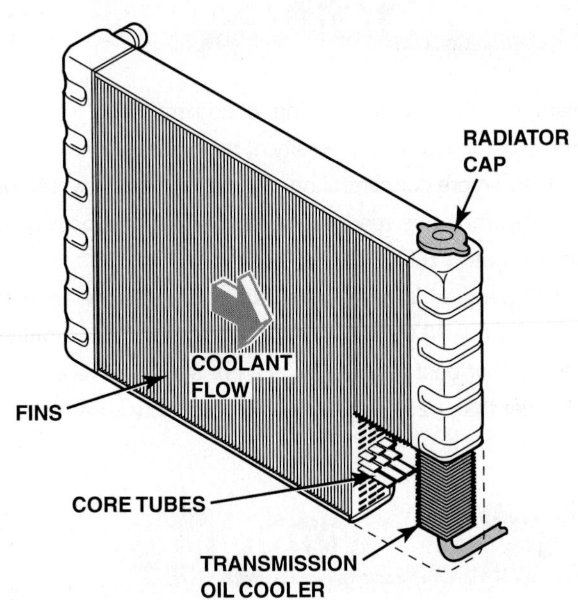

FIGURE 8–15 A radiator may be either a down-flow or a cross-flow type.

Radiator headers and tanks that close off the ends of the core were made of sheet brass 0.020 to 0.050 inch (0.5 to 1.25 mm) thick and now are made of molded plastic. When a transmission oil cooler is used in the radiator, it is placed in the outlet tank, where the coolant has the lowest temperature. ● **SEE FIGURE 8–17.**

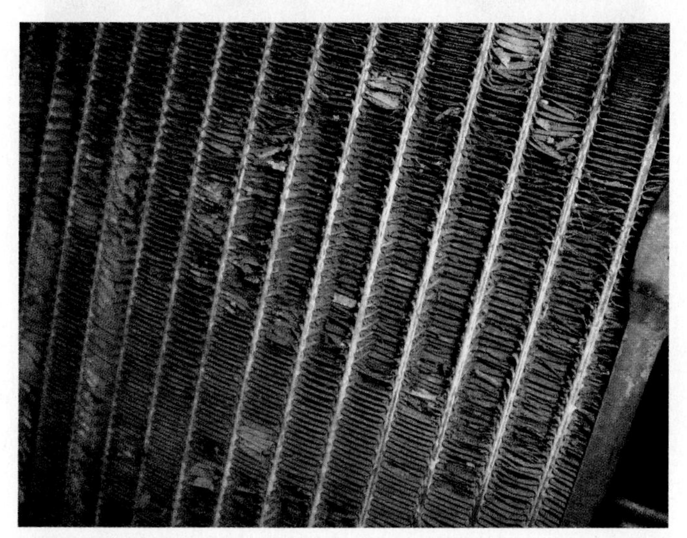

FIGURE 8–16 A heavily corroded radiator from a vehicle that was overheating. A visual inspection discovered that the corrosion had eaten away many of the cooling fins, yet did not leak. This radiator was replaced and it solved the overheating problem.

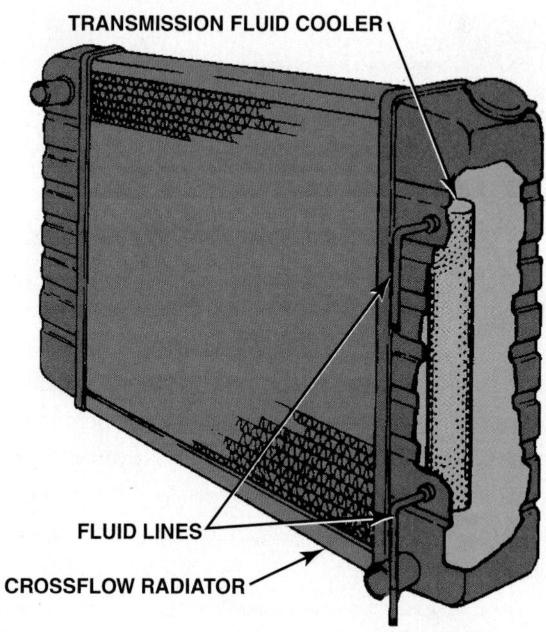

FIGURE 8–17 Many vehicles equipped with an automatic transmission use a transmission fluid cooler installed in one of the radiator tanks.

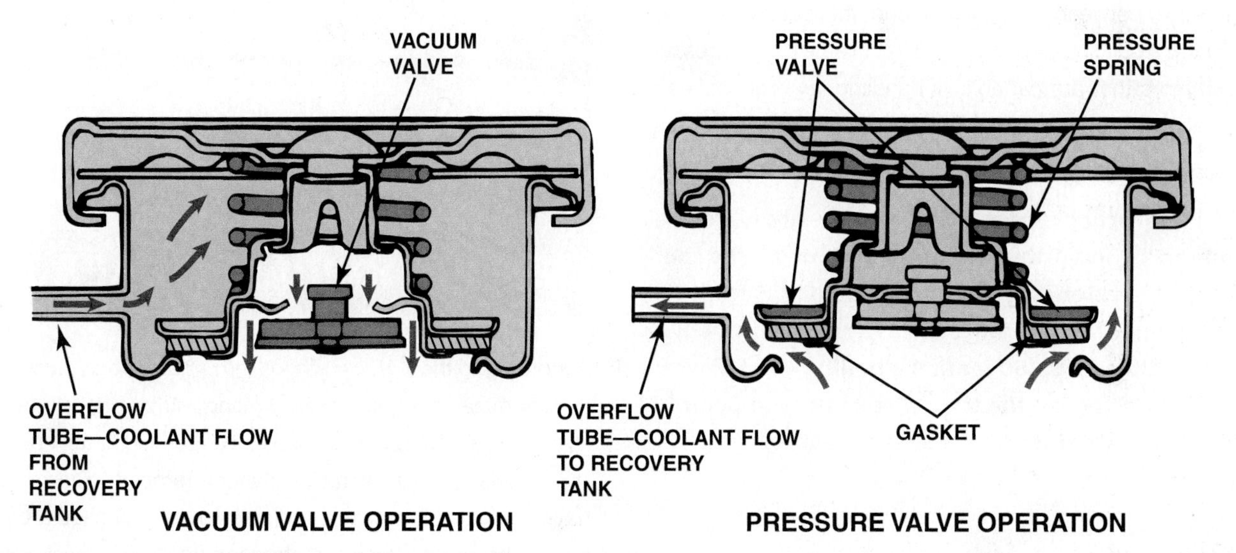

FIGURE 8–18 The pressure valve maintains the system pressure and allows excess pressure to vent. The vacuum valve allows coolant to return to the system from the recovery tank.

PRESSURE CAP

The filler neck is fitted with a pressure cap. The cap has a spring-loaded valve that closes the cooling system vent. This causes cooling pressure to build up to the pressure setting of the cap. At this point, the valve will release the excess pressure to prevent system damage. ● **SEE FIGURE 8–18**.

Engine cooling systems are pressurized to raise the boiling temperature of the coolant. *The boiling temperature will increase by approximately 3°F (1.6°C) for each pound of increase in pressure*. At standard atmospheric pressure, water will boil at 212°F (100°C). With a 15 PSI (100 kPa) pressure cap, water will boil at 257°F (125°C), which is a maximum operating temperature for an engine.

The high coolant system temperature serves two functions:

1. It allows the engine to run at an efficient temperature, close to 200°F (93°C), with no danger of boiling the coolant.

TECH TIP

Working Better Under Pressure

A problem that sometimes occurs with a high-pressure cooling system involves the water pump. For the pump to function, the inlet side of the pump must have a lower pressure than its outlet side. If inlet pressure is lowered too much, the coolant at the pump inlet can boil, producing vapor. The pump will then spin the coolant vapors and not pump coolant. This condition is called pump **cavitation**. Therefore, a radiator cap could be the cause of an overheating problem. A pump will not pump enough coolant if not kept under the proper pressure for preventing vaporization of the coolant.

2. The higher the coolant temperature, the more heat the cooling system can transfer. The heat transferred by the cooling system is proportional to the temperature difference between the coolant and the outside air. This characteristic has led to the design of small, high-pressure radiators that are capable of handling large quantities of heat. For proper cooling, the system must have the right pressure cap correctly installed.

NOTE: The proper operation of the pressure cap is especially important at high altitudes. The boiling point of water is lowered by about 1°F for every 550-foot increase in altitude. Therefore in Denver, Colorado (altitude 5,280 feet), the boiling point of water is about 202°F, and at the top of Pike's Peak in Colorado (14,110 feet) water boils at 186°F.

SURGE TANK

Some vehicles use a **surge tank**, which is located at the highest level of the cooling system and holds about 1 quart (1 liter) of coolant. A hose attaches to the bottom of the surge tank to the inlet side of the water pump. A smaller bleed hose attaches to the side of the surge tank to the highest point of the radiator. The bleed line allows some coolant circulation through the surge tank, and air in the system will rise below the radiator cap and be forced from the system if the pressure in the system exceeds the rating of the radiator cap. ● **SEE FIGURE 8–19.**

FIGURE 8–19 Some vehicles use a surge tank, which is located at the highest level of the cooling system, with a radiator cap.

BAR OR ATMOSPHERES	POUNDS PER SQUARE INCH (PSI)
1.1	16
1.0	15
0.9	13
0.8	12
0.7	10
0.6	9
0.5	7

CHART 8–4

Conversion from BAR to PSI rating of radiator caps.

METRIC RADIATOR CAPS

According to the *SAE Handbook*, all radiator caps must indicate their nominal (normal) pressure rating. Most original equipment radiator caps are rated at about 14 to 16 PSI (97 to 110 kPa).

However, many vehicles manufactured in Japan or Europe have the radiator pressure indicated in a unit called a **bar**. One bar is the pressure of the atmosphere at sea level, or about 14.7 PSI. The following conversion can be used when replacing a radiator cap to make certain it matches the pressure rating of the original. ● **SEE CHART 8–4.**

NOTE: Many radiator repair shops use a 7 PSI (0.5 bar) radiator cap on a repaired radiator. A 7 PSI cap can still provide boil protection of 21°F (3°F × 7 PSI = 21°F) above the boiling point of the coolant. For example, if the boiling point of the antifreeze coolant is 223°F, 21°F is added for the pressure cap, and boilover will not occur until about 244°F (223°F + 21°F = 244°F). Even though this lower-pressure radiator cap does provide some protection and will also help protect the radiator repair, the coolant can still boil *before* the "hot" dash warning light comes on and therefore, should not be used.

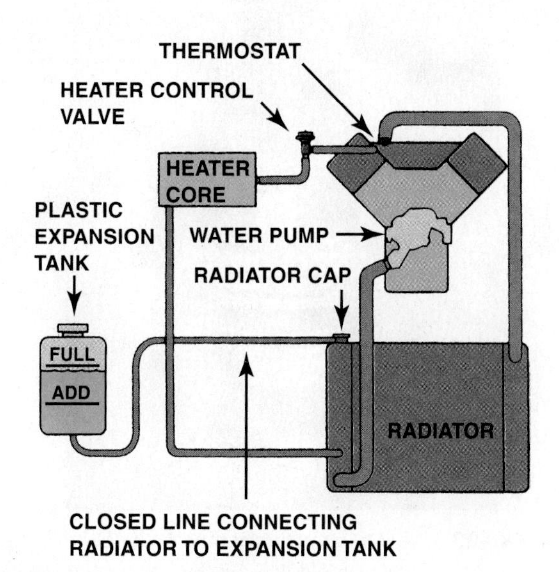

FIGURE 8–20 The level in the coolant recovery system raises and lowers with engine temperature.

FIGURE 8–21 Pressure testing the cooling system. A typical hand-operated pressure tester applies pressure equal to the radiator cap pressure. The pressure should hold; if it drops, this indicates a leak somewhere in the cooling system. An adapter is used to attach the pump to the cap to determine if the radiator can hold pressure, and release it when pressure rises above its maximum rated pressure setting.

COOLANT RECOVERY SYSTEM

Excess pressure usually forces some coolant from the system through an overflow. Most cooling systems connect the overflow to a plastic reservoir to hold excess coolant while the system is hot. ● **SEE FIGURE 8–20**.

When the system cools, the pressure in the cooling system is reduced and a partial vacuum forms. This pulls the coolant from the plastic container back into the cooling system, keeping the system full. Because of this action, this system is called a **coolant recovery system**. The filler cap used on a coolant system without a coolant saver is fitted with a vacuum valve. This valve allows air to reenter the system as the system cools so that the radiator parts will not collapse under the partial vacuum.

TESTING THE COOLING SYSTEM

PRESSURE TESTING Pressure testing using a hand-operated pressure tester is a quick and easy cooling system test. The radiator cap is removed (engine cold!) and the tester is attached in the place of the radiator cap. By operating the plunger on the pump, the entire cooling system is pressurized. ● **SEE FIGURE 8–21**.

CAUTION: Do not pump up the pressure beyond that specified by the vehicle manufacturer. Most systems should not be pressurized beyond 14 PSI (100 kPa). If a greater pressure is used, it may cause the water pump, radiator, heater core, or hoses to fail.

If the cooling system is free from leaks, the pressure should stay and not drop. If the pressure drops, look for evidence of leaks anywhere in the cooling system including:

1. Heater hoses
2. Radiator hoses
3. Radiator
4. Heat core
5. Cylinder head
6. Core plugs in the side of the block or cylinder head

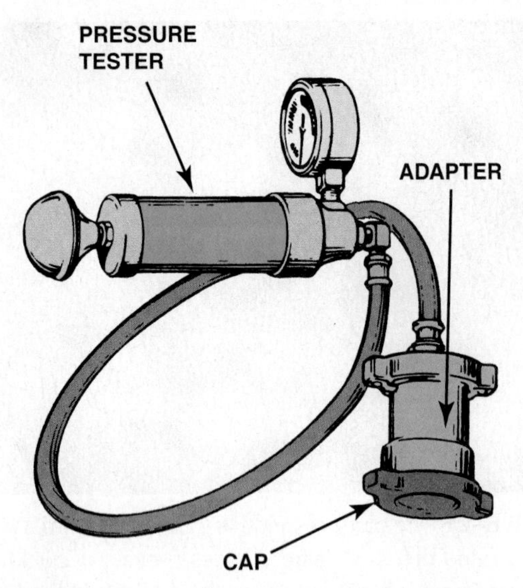

FIGURE 8–22 The pressure cap should be checked for proper operation using a pressure tester as part of the cooling system diagnosis.

FIGURE 8–23 Use dye specifically made for coolant when checking for leaks using a black light.

TECH TIP

Use Distilled Water in the Cooling System

Two technicians are discussing refilling the radiator after changing antifreeze. One technician says that distilled water is best to use because it does not contain minerals that can coat the passages of the cooling system. The other technician says that any water that is suitable to drink can be used in a cooling system. Both technicians are correct. If water contains minerals, however, it can leave deposits in the cooling system that could prevent proper heat transfer. Because the mineral content of most water is unknown, distilled water, which has no minerals, is better to use. Although the cost of distilled water must be considered, the amount of water required (usually about 2 gallons [8 liters] or less of water) makes the expense minor in comparison with the cost of radiator or cooling system failure.

Pressure testing should be performed whenever there is a leak or suspected leak. The pressure tester can also be used to test the radiator cap. An adapter is used to connect the pressure tester to the radiator cap. Replace any cap that will not hold pressure. ● **SEE FIGURE 8–22.**

COOLANT DYE LEAK TESTING One of the best methods to check for a coolant leak is to use a fluorescent dye in the coolant. Use a dye designed for coolant. Operate the vehicle with the dye in the coolant until the engine reaches normal operating temperature. Use a black light to inspect all areas of the cooling system. When there is a leak, it will be easy to spot because the dye in the coolant will be seen as bright green. ● **SEE FIGURE 8–23.**

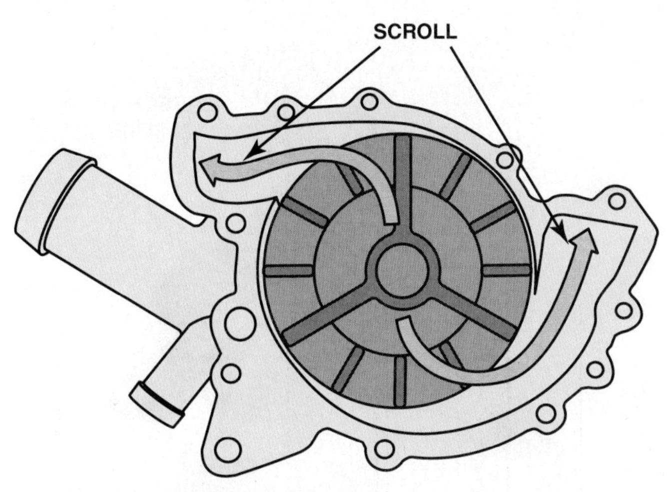

FIGURE 8–24 Coolant flow through the impeller and scroll of a coolant pump for a V-type engine.

FIGURE 8–25 A demonstration engine showing the amount of water that can be circulated through the cooling system.

WATER PUMP

OPERATION The water pump (also called a coolant pump) is driven by a belt from the crankshaft or driven by the camshaft. Coolant recirculates from the radiator to the engine and back to the radiator. Low-temperature coolant leaves the radiator through the bottom outlet. It is pumped into the warm engine block, where it picks up some heat. From the block, the warm coolant flows to the hot cylinder head, where it picks up more heat.

NOTE: Some engines use reverse cooling. This means that the coolant flows from the radiator to the cylinder head(s) before flowing to the engine block.

Water pumps are not positive displacement pumps. The water pump is a **centrifugal pump** that can move a large volume of coolant without increasing the pressure of the coolant. The pump pulls coolant in at the center of the **impeller**. Centrifugal force throws the coolant outward so that it is discharged at the impeller tips. This can be seen in ● **FIGURE 8–24**.

As engine speeds increase, more heat is produced by the engine and more cooling is required. The pump impeller speed increases as the engine speed increases to provide extra coolant flow at the very time it is needed.

Coolant leaving the pump impeller is fed through a **scroll**. The scroll is a smoothly curved passage that changes the fluid flow direction with minimum loss in velocity. The scroll is connected to the front of the engine so as to direct the coolant into the engine block. On V-type engines, two outlets are usually used, one for each cylinder bank. Occasionally, diverters are necessary in the water pump scroll to equalize coolant flow between the cylinder banks of a V-type engine to equalize the cooling.

> **? FREQUENTLY ASKED QUESTION**
>
> **How Much Coolant Can a Water Pump Pump?**
>
> A typical water pump can move a maximum of about 7,500 gallons (28,000 liters) of coolant per hour, or recirculate the coolant in the engine over 20 times per minute. This means that a water pump could be used to empty a typical private swimming pool in an hour! The slower the engine speed, the less power is consumed by the water pump. However, even at 35 miles per hour (56 kilometers per hour), the typical water pump still moves about 2,000 gallons (7,500 liters) per hour or 1/2 gallon (2 liters) per second! ● **SEE FIGURE 8–25.**

SERVICE A worn impeller on a water pump can reduce the amount of coolant flow through the engine. ● **SEE FIGURE 8–26.** If the seal of the water pump fails, coolant will leak out of the hole as seen in ● **FIGURE 8–27**. The hole allows coolant to escape without getting trapped and forced into the water pump bearing assembly.

If the bearing is defective, the pump will usually be noisy and will have to be replaced. Before replacing a water pump

FIGURE 8–26 This severely corroded water pump could not circulate enough coolant to keep the engine cool. As a result, the engine overheated and blew a head gasket.

FIGURE 8–27 The bleed weep hole in the water pump allows coolant to leak out of the pump and not be forced into the bearing. If the bearing failed, more serious damage could result.

that has failed because of a loose or noisy bearing, be sure to do all of the following:

1. Check belt tension
2. Check for bent fan
3. Check fan for balance

If the water pump drive belt is too tight, excessive force may be exerted against the pump bearing. If the cooling fan is bent or out of balance, the resulting vibration can damage the water pump bearing. ● **SEE FIGURE 8–28.**

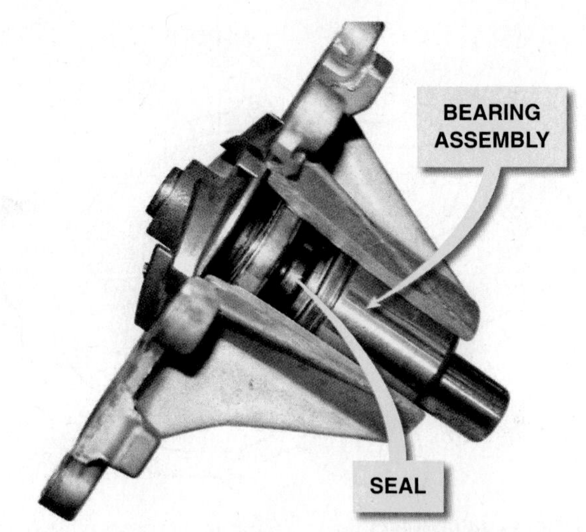

FIGURE 8–28 A cutaway of a typical water pump showing the long bearing assembly and the seal. The weep hole is located between the seal and the bearing. If the seal fails, then coolant flows out of the weep hole to prevent the coolant from damaging the bearing.

COOLING FANS

Air is forced across the radiator core by a cooling fan. On older engines used in rear-wheel-drive vehicles, it is attached to a fan hub that is pressed on the water pump shaft. ● **SEE FIGURE 8–29.**

Many installations with rear-wheel drive and all transverse-engines drive the fan with an electric motor. ● **SEE FIGURE 8–30.**

NOTE: Most electric cooling fans are computer controlled. To save energy, most cooling fans are turned off whenever the vehicle is traveling faster than 35 mph (55 km/h). The ram air from the vehicle's traveling at that speed should be enough to keep the radiator cool. Of course, if the computer senses that the temperature is still too high, the computer will turn on the cooling fan, to "high," if possible, in an attempt to cool the engine to avoid severe engine damage.

The fan is designed to move enough air at the lowest fan speed to cool the engine when it is at its highest coolant temperature. The fan shroud is used to increase the cooling system efficiency. The horsepower required to drive the fan increases at a much faster rate than the increase in fan speed. Higher fan speeds also increase fan noise. Fans with flexible plastic or flexible steel blades have been used. These fans have high blade angles that pull a high volume of air when turning at low speeds. As the fan speed increases, the fan blade angle flattens, reducing the horsepower required to rotate the blade at high speeds. ● **SEE FIGURE 8–31.**

FIGURE 8–29 A typical engine-driven cooling fan.

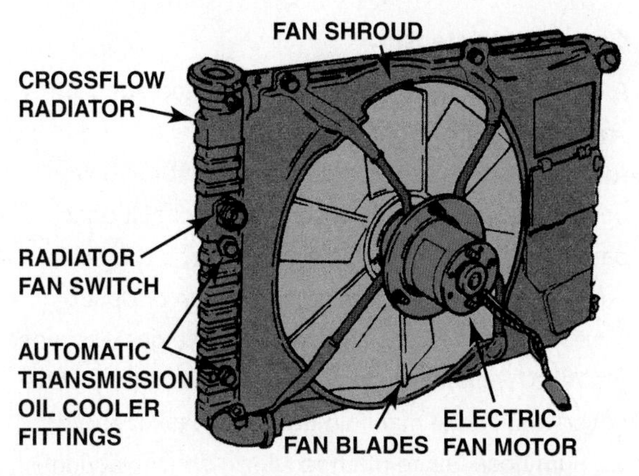

FIGURE 8–30 A typical electric cooling fan assembly showing the radiator and related components.

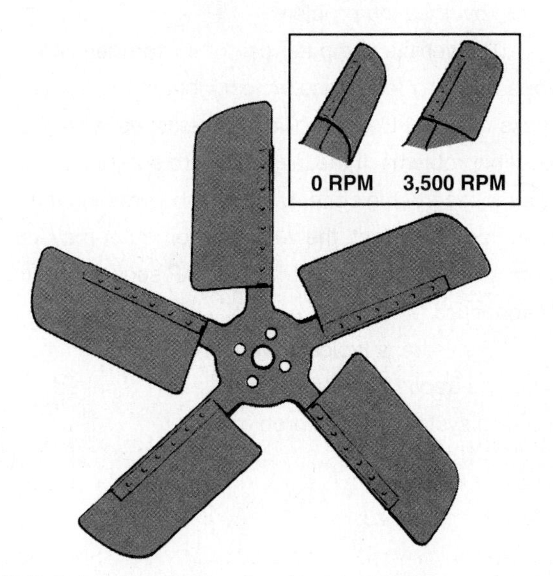

FIGURE 8–31 Flexible cooling fan blades change shape as the engine speed changes.

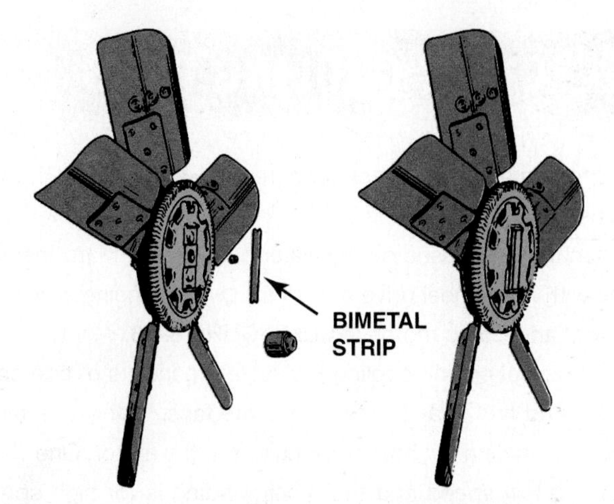

FIGURE 8–32 The bimetallic temperature sensor spring controls the amount of silicone that is allowed into the drive unit, which controls the speed of the fan.

THERMOSTATIC FANS

On some rear-wheel-drive vehicles, a thermostatic cooling fan is driven by a belt from the crankshaft. It turns faster as the engine turns faster. Generally, the engine is required to produce more power at higher speeds. Therefore, the cooling system will also transfer more heat. Increased fan speed aids in the required cooling. Engine heat also becomes critical at low engine speeds in traffic where the vehicle moves slowly.

The thermal fan is designed so that it uses little power at high engine speeds and minimizes noise. The thermal fan has a **silicone coupling** fan drive mounted between the drive pulley and the fan.

NOTE: Whenever diagnosing an overheating problem, look carefully at the cooling fan. If silicone is leaking, then the fan may not be able to function correctly and should be replaced.

A second type of thermal fan has a **thermostatic spring** added to the silicone coupling fan drive. The thermostatic spring operates a valve that allows the fan to freewheel when the radiator is cold. As the radiator warms to about 150°F (65°C), the air hitting the thermostatic spring will cause the spring to change its shape. The new shape of the spring opens a valve that allows the drive to operate like the silicone coupling drive. When the engine is very cold, the fan may operate at high speeds for a short time until the drive fluid warms slightly. The silicone fluid will then flow into a reservoir to let the fan speed drop to idle. ● **SEE FIGURE 8–32.**

ELECTRIC COOLING FANS

Air is forced across the radiator core by a cooling fan. On older engines used in rear-wheel-drive vehicles, it is attached to a fan hub that is pressed on the water pump shaft. Many installations with rear-wheel drive and all transverse engines drive the fan with an electric motor. ● **SEE FIGURE 8–33.**

A typical engine cooling fan system consists of one cooling fan and two relays or two separate fans. If only one fan is used, the cooling fan has two windings in the motor. One winding is for low speed and the other winding is for high speed. When the cooling fan 1 relay is energized, voltage is sent to the cooling fan low-speed winding. The ECM controls the high-speed fan operation by grounding the cool fan 2 relay control circuit. When the cooling fan 2 relay is energized, voltage is sent to the cooling fan high-speed winding. The cooling fan motor has its own ground circuit.

The ECM commands Low Speed Fans ON under the following conditions:

- Engine coolant temperature (ECT) exceeds approximately 223°F (106°C).
- A/C refrigerant pressure exceeds 190 PSI (1,310 kPa).

After the vehicle is shut off, if the ECT at key-off is greater than 284°F (140°C) and system voltage is more than 12 volts, the fans will stay on for approximately 3 minutes.

The ECM commands High Speed Fans ON under the following conditions:

- ECT reaches 230°F (110°C).
- A/C refrigerant pressure exceeds 240 PSI (1,655 kPa).
- When certain diagnostic trouble codes (DTCs) set.

To prevent a fan from cycling ON and OFF excessively at idle, the fan may not turn OFF until the ignition switch is moved to the OFF position or the vehicle speed exceeds approximately 10 mph.

NOTE: To save energy and to improve fuel economy, most cooling fans are turned off whenever the vehicle is traveling faster than 35 mph (55 km/h). The ram air from the vehicle's traveling at that speed should be enough to keep the radiator cool. Of course, if the computer senses that the temperature is still too high, the computer will turn on the cooling fan, to "high," if possible, in an attempt to cool the engine to avoid severe engine damage. Some engines, such as the General Motors NorthStar engine, can disable four of the eight cylinders from firing to air cool the engine in the event of a severe overheating condition.

FIGURE 8–33 A typical electric cooling fan assembly after being removed from the vehicle.

TECH TIP

Cause and Effect

A common cause of overheating is an inoperative cooling fan. Most front-wheel-drive vehicles and many rear-wheel-drive vehicles use electric motor–driven cooling fans. A fault in the cooling fan circuit often causes overheating during slow city-type driving.

Even slight overheating can soften or destroy rubber vacuum hoses and gaskets. The gaskets most prone to overheating damage are rocker cover (valve cover) and intake manifold gaskets. Gasket and/or vacuum hose failure often results in an air (vacuum) leak that leans the air–fuel mixture. The resulting lean mixture burns hotter in the cylinders and contributes to the overheating problem.

The vehicle computer can often compensate for a minor air leak (vacuum leak), but more severe leaks can lead to driveability problems; especially idle quality problems. If the leak is severe enough, a lean diagnostic trouble code (DTC) may be present. If a lean code is not set, the vehicle's computer may indicate a defective or out-of-range MAP sensor code in diagnostics.

Therefore, a typical severe engine problem can often be traced back to a simple, easily repaired, cooling system–related problem.

COOLANT TEMPERATURE WARNING LIGHT

Most vehicles are equipped with a heat sensor for the engine operating temperature. If the "hot" light comes on during driving (or the temperature gauge goes into the red danger zone), then the coolant temperature is about 250°F to 258°F (120°C to 126°C), which is still *below* the boiling point of the coolant (assuming a properly operating pressure cap and system). If this happens, follow these steps:

STEP 1 Shut off the air conditioning and turn on the heater. The heater will help rid the engine of extra heat. Set the blower speed to high.

STEP 2 If possible, shut the engine off and let it cool. (This may take over an hour.)

STEP 3 Never remove the radiator cap when the engine is hot.

STEP 4 Do *not* continue to drive with the hot light on, or serious damage to your engine could result.

STEP 5 If the engine does not feel or smell hot, it is possible that the problem is a faulty hot light sensor or gauge. Continue to drive, but to be safe, stop occasionally and check for any evidence of overheating or coolant loss.

COMMON CAUSES OF OVERHEATING

Overheating can be caused by defects in the cooling system. Some common causes of overheating include:

1. Low coolant level
2. Plugged, dirty, or blocked radiator
3. Defective fan clutch or electric fan
4. Incorrect ignition timing
5. Low engine oil level
6. Broken fan belt
7. Defective radiator cap
8. Dragging brakes
9. Frozen coolant (in freezing weather)
10. Defective thermostat
11. Defective water pump (the impeller slipping on the shaft internally)

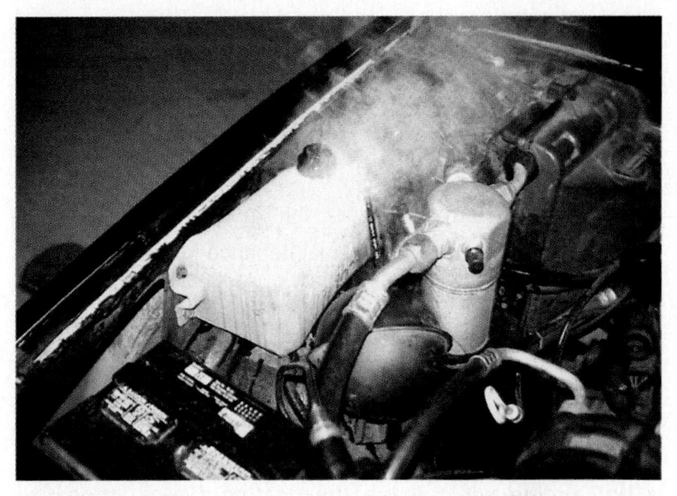

FIGURE 8–34 When an engine overheats, often the coolant overflow container boils.

CASE STUDY

Highway Overheating

A vehicle owner complained of an overheating vehicle, but the problem occurred only while driving at highway speeds. The vehicle, ● **SEE FIGURE 8–34**, would run in a perfectly normal manner in city-driving situations.

The technician flushed the cooling system and replaced the radiator cap and the water pump, thinking that restricted coolant flow was the cause of the problem. Further testing revealed coolant spray out of one cylinder when the engine was turned over by the starter with the spark plugs removed.

A new head gasket solved the problem. Obviously, the head gasket leak was not great enough to cause any problems until the engine speed and load created enough flow and heat to cause the coolant temperature to soar.

The technician also replaced the oxygen sensor (O2S), because some coolants contain phosphates and silicates that often contaminate the sensor. The deteriorated oxygen sensor could have contributed to the problem.

Summary:

- **Complaint**—The engine overheated when driving at highway speed.
- **Cause**—The head gasket was found to defective allowing coolant to enter the combustion chamber.
- **Correction**—The head gasket was replaced and the oxygen sensor as a precaution against possible coolant contamination.

COOLING SYSTEM MAINTENANCE

The cooling system is one of the most maintenance-free systems in the engine. Normal maintenance involves an occasional check on the coolant level. It should also include a visual inspection for signs of coolant system leaks and for the condition of the coolant hoses and fan drive belts.

CAUTION: The coolant level should only be checked when the engine is cool. Removing the pressure cap from a hot engine will release the cooling system pressure while the coolant temperature is above its atmospheric boiling temperature. When the cap is removed, the pressure will instantly drop to atmospheric pressure level, causing the coolant to boil immediately. Vapors from the boiling liquid will blow coolant from the system. Coolant will be lost, and someone may be injured or burned by the high-temperature coolant that is blown out of the filler opening.

The coolant–antifreeze mixture is renewed at periodic intervals. Some vehicle manufacturers recommend that coolant system stop-leak pellets be installed whenever the coolant is changed.

CAUTION: General Motors recommends the use of these stop-leak pellets in only certain engines. Using these pellets in some engines could cause a restriction in the cooling system and an overheating condition.

Drive belt condition and proper installation are important for the proper operation of the cooling system.

FLUSH AND REFILL

Manufacturers recommend that a cooling system be flushed and that the antifreeze be replaced at specified intervals. Draining coolant when the engine is cool eliminates the danger of being injured by hot coolant. The radiator is drained by opening a petcock in the bottom tank, and the coolant in the block is drained into a suitable container by opening plugs located in the lower part of the cooling passage.

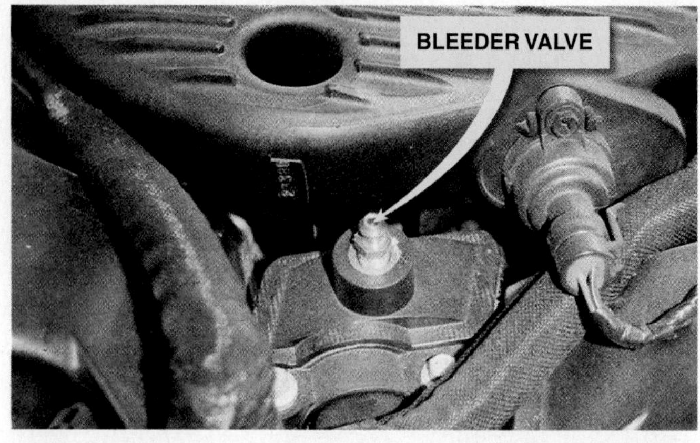

(a)

(b)

FIGURE 8–35 (a) Chrysler recommends that the bleeder valve be opened whenever refilling the cooling system. (b) Chrysler also recommends that a clear plastic hose (1/4" ID) be attached to the bleeder valve and directed into a suitable container to keep from spilling coolant onto the ground and on the engine and to allow the technician to observe the flow of coolant for any remaining oil bubbles.

Water should be run into the filler opening while the drains remain open. Flushing should be continued until only clear water comes from the system.

The volume of the cooling system must be determined. It is specified in the owner's manual and in the engine service manual. The antifreeze quantity needed for the protection desired is shown on a chart that comes with the antifreeze. Open the bleeder valves and add the correct amount of the specified type of antifreeze followed by enough water to completely fill the system. ● **SEE FIGURE 8–35**. The coolant recovery reservoir should be filled to the "level-cold" mark with the correct antifreeze mixture.

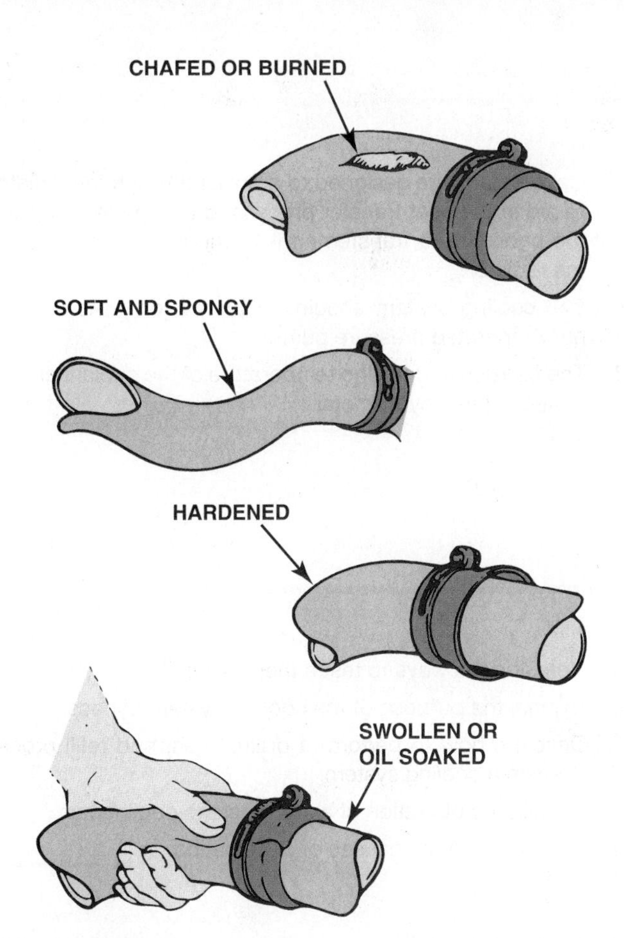

FIGURE 8–36 All cooling system hoses should be checked for wear or damage.

BURPING THE SYSTEM

In most systems, small air pockets can occur. The engine must be thoroughly warmed to open the thermostat. This allows full coolant flow to remove the air pockets. The heater must also be turned to full heat.

NOTE: The cooling system will not function correctly if air is not released (burped) from the system after a refill. An easy method involves replacing the radiator cap after the refill, but only to the first locked position. Drive the vehicle for several minutes and check the radiator level. Without the radiator cap tightly sealed, no pressure will build in the cooling system. Driving the vehicle helps circulate the coolant enough to force all air pockets up and out of the radiator filler. Top off the radiator after burping and replace the radiator cap to the fully locked position. Failure to burp the cooling system to remove all the air will often result in lack of heat from the heater and may result in engine overheating.

HOSES

Coolant system hoses are critical to engine cooling. As the hoses get old, they become either soft or brittle and sometimes swell in diameter. Their condition depends on their material and on the engine service conditions. If a hose breaks while the engine is running, all coolant will be lost. A hose should be replaced anytime it appears to be abnormal. ● **SEE FIGURE 8–36.**

NOTE: To make hose removal easier and to avoid possible damage to the radiator, use a utility knife and slit the hose lengthwise. Then simply peel the hose off.

Care should be taken to avoid bending the soft metal hose neck on the radiator. The hose neck should be cleaned before a new hose is slipped in place. The clamp is placed on the hose; then the hose is pushed fully over the neck. The hose should be cut so that the clamp is close to the bead on the neck. This is especially important on aluminum hose necks to avoid corrosion. When the hoses are in place and the drain petcock is closed, the cooling system can be refilled with the correct coolant mixture.

CLEANING THE RADIATOR EXTERIOR

Overheating can result from exterior radiator plugging as well as internal plugging. External plugging is caused by dirt and insects. This type of plugging can be seen if you look straight through the radiator while a light is held behind it. It is most likely to occur on off-road vehicles. The plugged exterior of the radiator core can usually be cleaned with water pressure from a hose. The water is aimed at the *engine side* of the radiator. The water should flow freely through the core at all locations. If this does not clean the core, the radiator should be removed for cleaning at a radiator shop.

TECH TIP

Quick and Easy Cooling System Problem Diagnosis

If overheating occurs in slow, stop-and-go traffic, the usual cause is low airflow through the radiator. Check for airflow blockages or cooling fan malfunction. If overheating occurs at highway speeds, the cause is usually a radiator or coolant circulation problem. Check for a restricted or clogged radiator.

SUMMARY

1. The purpose and function of the cooling system is to maintain proper engine operating temperature.
2. The thermostat controls engine coolant temperature by opening at its rated opening temperature to allow coolant to flow through the radiator.
3. Most antifreeze coolant is ethylene glycol based.
4. Used coolant should be recycled whenever possible.
5. Coolant fans are designed to draw air through the radiator to aid in the heat transfer process, drawing the heat from the coolant and transferring it to the outside air through the radiator.
6. The cooling system should be tested for leaks using a hand-operated pressure pump.
7. The freezing and boiling temperature of the coolant can be tested using a hydrometer.

REVIEW QUESTIONS

1. Explain why the normal operating coolant temperature is about 200°F to 220°F (93°C to 104°C).
2. Explain why a 50/50 mixture of antifreeze and water is commonly used as a coolant.
3. Explain the flow of coolant through the engine and radiator.
4. Why is a cooling system pressurized?
5. Describe the ways to test a thermostat.
6. Explain the purpose of the coolant system bypass.
7. Describe how to perform a drain, flush, and refill procedure on a cooling system.
8. Explain the operation of a thermostatic cooling fan.
9. List five common causes of overheating.

CHAPTER QUIZ

1. Permanent antifreeze is mostly _________.
   a. Methanol
   b. Glycerin
   c. Kerosene
   d. Ethylene glycol
2. As the percentage of antifreeze in the coolant increases, _________.
   a. The freeze point decreases (up to a point)
   b. The boiling point decreases
   c. The heat transfer increases
   d. All of the above
3. A stuck open thermostat can cause _________.
   a. Lower fuel economy
   b. Increased exhaust emissions
   c. Failure of a state emission test
   d. All of the above
4. A water pump is a positive displacement-type pump.
   a. True
   b. False
5. The weep hole on a water pump is located where?
   a. Between the impeller and the seal.
   b. Between the seal and the bearing.
   c. Between the bearing and the engine.
   d. Any of the above depending on engine.
6. Technician A says that a bleeder valve is located on many engines to allow air to escape when refilling the cooling system. Technician B says that a hose should be attached to the bleeder valve to allow any escaping coolant to be directed to a suitable container. Which technician is correct?
   a. Technician A only
   b. Technician B only
   c. Both Technicians A and B
   d. Neither Technician A nor B
7. Which statement is *true* about thermostats?
   a. The temperature marked on the thermostat is the temperature at which the thermostat should be fully open.
   b. Thermostats often cause overheating.
   c. The temperature marked on the thermostat is the temperature at which the thermostat should start to open.
   d. Both a and b.
8. Used coolant should be _________.
   a. Reused
   b. Recycled
   c. Disposed of properly
   d. Either b or c

9. An engine fails to reach normal operating temperature. Which is the most likely fault?
   a. Defective thermostat
   b. Low coolant level
   c. Wrong antifreeze coolant
   d. Partially clogged radiator

10. The normal operating temperature (coolant temperature) of an engine equipped with a 195°F thermostat is ________.
   a. 175°F to 195°F
   b. 185°F to 205°F
   c. 195°F to 215°F
   d. 175°F to 215°F

chapter 9

LUBRICATION SYSTEM OPERATION AND DIAGNOSIS

LEARNING OBJECTIVES

After studying this chapter, the reader will be able to:

1. Describe basic lubrication principles.
2. Describe how an oil pump and engine lubrication work.
3. Explain how to inspect an oil pump for wear.

This chapter will help you prepare for Engine Repair (A1) ASE certification test content area "D" (Lubrication and Cooling Systems Diagnosis and Repair).

KEY TERMS

Boundary lubrication 153
Hydrodynamic lubrication 153
Longitudinal header 158
Oil gallery 158
Positive displacement pump 154
Pressure regulating valve 155
Viscosity 153
Windage tray 160

Engine oil is the lifeblood of any engine. The purposes of engine oil include the following:

1. *Lubricating* all moving parts to prevent wear
2. Helping to *cool* the engine
3. Helping to *seal* piston rings
4. *Cleaning* and holding dirt in suspension in the oil until it can be drained from the engine
5. *Neutralizing* acids that are formed as the result of the combustion process
6. *Reducing* friction
7. *Preventing* rust and corrosion

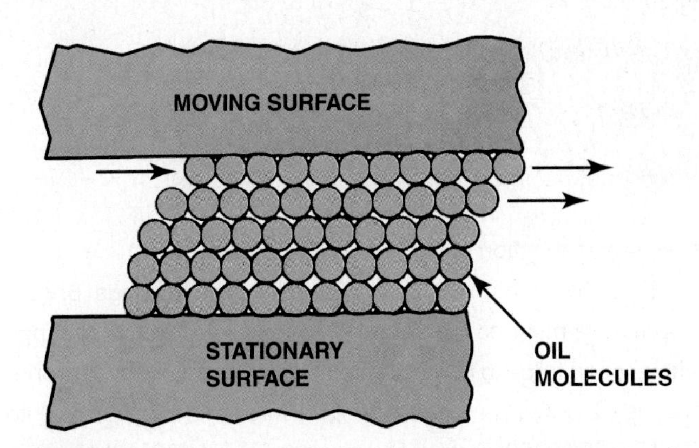

FIGURE 9–1 Oil molecules cling to metal surfaces but easily slide against each other.

LUBRICATION PRINCIPLES

Lubrication between two moving surfaces results from an oil film that separates the surfaces and supports the load. ● **SEE FIGURE 9–1.**

Although oil does not compress, it does leak out around the oil clearance between the shaft and the bearing. In some cases, the oil film is thick enough to keep the surfaces from seizing, but can allow some contact to occur. This condition is called **boundary lubrication.** The specified oil viscosity and oil clearances must be adhered to during service to help prevent boundary lubrication and wear from occurring, which usually happens when the engine is under a heavy load and low speeds. The movement of the shaft helps prevent contact with the bearing. If oil were put on a flat surface and a heavy block were pushed across the surface, the block would slide more easily than if it were pushed across a dry surface. The reason for this is that a wedge-shaped oil film is built up between the moving block and the surface, as illustrated in ● **FIGURE 9–2**. This wedging action is called **hydrodynamic lubrication**. The wedging action depends on the force applied, the speed of difference between objects, and the thickness of the oil. Thickness of oil is called the **viscosity** and defined as the ability of the oil to resist flow. High-viscosity oil is thick and low-viscosity oil is thin. The prefix *hydro* refers to liquids, as in hydraulics, and *dynamic* refers to moving materials. Hydrodynamic lubrication occurs when a wedge-shaped film of lubricating oil develops between two surfaces that have relative motion between them. ● **SEE FIGURE 9–3.**

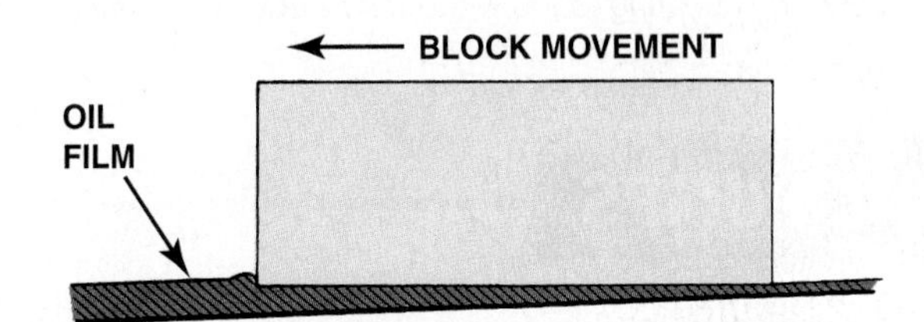

FIGURE 9–2 Wedge-shaped oil film developed below a moving block.

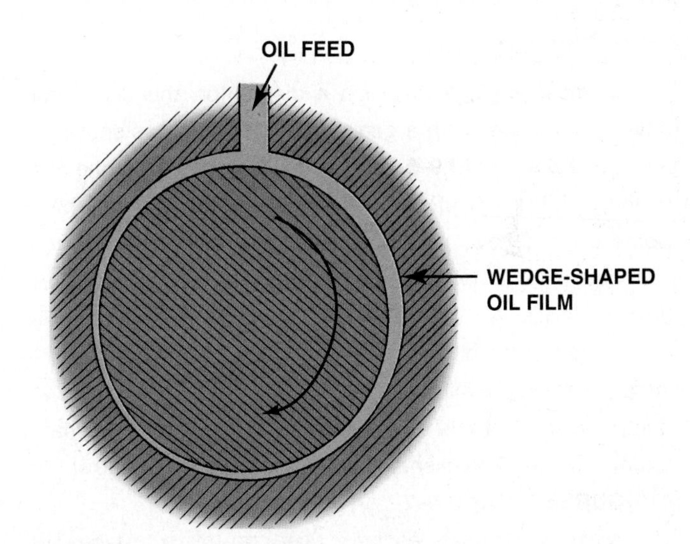

FIGURE 9–3 Wedge-shaped oil film curved around a bearing journal.

The engine oil pressure system feeds a continuous supply of oil into the lightly loaded part of the bearing oil clearance. Hydrodynamic lubrication takes over as the shaft rotates in the bearing to produce a wedge-shaped hydrodynamic oil film that is curved around the bearing. This film supports the bearing and reduces the turning effort to a minimum when oil of the correct viscosity is used.

Most bearing wear occurs during the initial start-up. Wear continues until a hydrodynamic film is established.

ENGINE LUBRICATION SYSTEMS

The primary function of the engine lubrication system is to maintain a positive and continuous oil supply to the bearings. Engine oil pressure must be high enough to get the oil to the bearings with enough force to cause the oil flow that is required for proper cooling. The normal engine oil pressure range is from 10 to 60 PSI (200 to 400 kPa) (10 PSI per 1000 engine RPM). However, hydrodynamic film pressures developed in the high-pressure areas of the engine bearings may be over 1,000 PSI (6,900 kPa). The relatively low engine oil pressures obviously could not support these high bearing loads without hydrodynamic lubrication.

OIL PUMPS

All production automobile engines have a full-pressure oil system. The oil is drawn from the bottom of the oil pan and is forced into the lubrication system under pressure.

NOTE: The oil pump is the only engine component that uses unfiltered oil.

In most engines that use a distributor, the distributor drive gear meshes with a gear on the camshaft, as shown in ● **FIGURES 9-4 AND 9–5**. The oil pump is driven from the end of the distributor shaft, often with a hexagon-shaped shaft. Some engines have a short shaft gear that meshes with the cam gear to drive both the distributor and oil pump. With these drive methods, the pump turns at one-half engine speed. In other engines, the oil pump is driven by the front of the crankshaft, in a setup similar to that of an automatic transmission pump, so that it turns at the same speed as the crankshaft. Examples of a crankshaft-driven oil pump are shown in ● **FIGURES 9-6 AND 9–7**.

Most automotive engines use one of two types of oil pumps: *gear* or *rotor*. All oil pumps are called **positive displacement pumps**, and each rotation of the pump delivers the same volume of oil; thus, everything that enters must exit. The gear-type oil pump consists of two spur gears in a close-fitting housing—one gear is driven while the other idles. As the gear teeth come out of mesh, they tend to leave a space, which is filled by oil drawn through the pump inlet. When the pump is pumping, oil is carried around the *outside* of each gear in the space between the gear teeth and the housing, as shown in ● **FIGURE 9–8**.

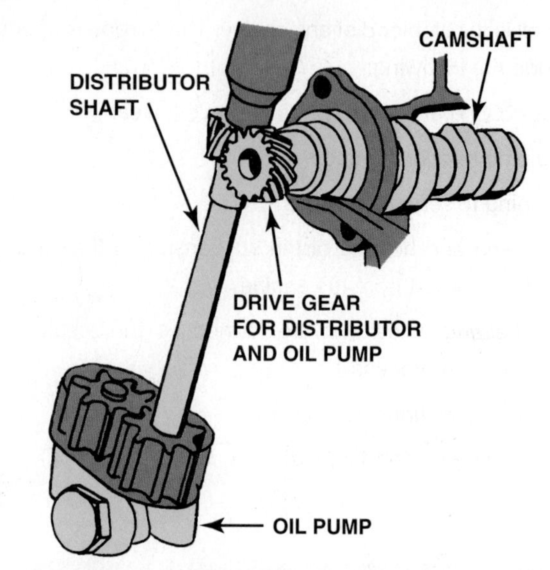

FIGURE 9–4 An oil pump driven by the camshaft.

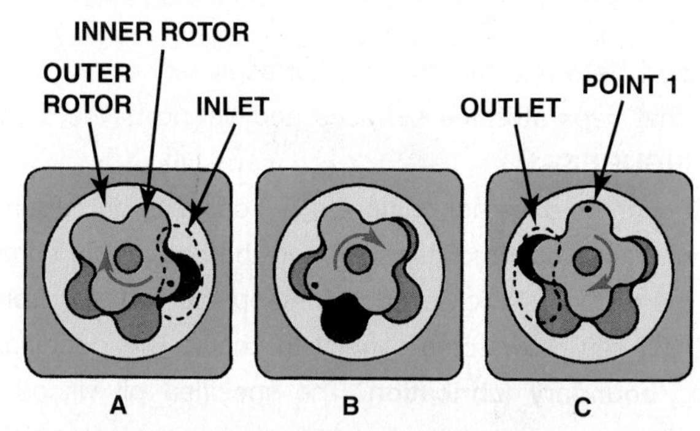

FIGURE 9–5 The operation of a rotor-type oil pump.

FIGURE 9–6 A typical oil pump mounted in the front cover of the engine that is driven by the crankshaft.

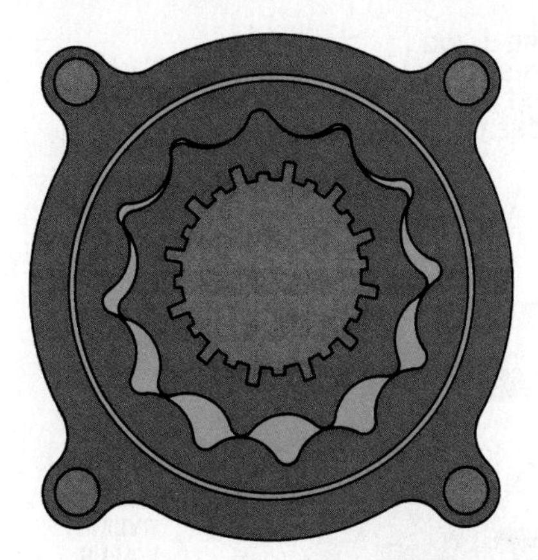

FIGURE 9–7 Gerotor-type oil pump driven by the crankshaft.

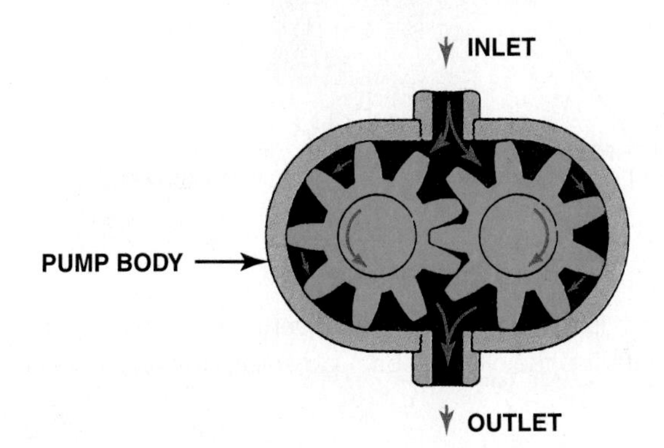

FIGURE 9–8 In a gear-type oil pump, the oil flows through the pump around the outside of each gear. This is an example of a positive displacement pump, where everything entering the pump must leave the pump.

As the teeth mesh in the center, oil is forced from the teeth into an oil passage, thus producing oil pressure. The rotor-type oil pump consists essentially of a special lobe-shaped gear meshing with the inside of a lobed rotor. The center lobed section is driven and the outer section idles. As the lobes separate, oil is drawn in just as it is drawn into gear-type pumps. As the pump rotates, it carries oil around and between the lobes. As the lobes mesh, they force the oil out from between them under pressure in the same manner as the gear-type pump. The pump is sized so that it will maintain a pressure of at least 10 PSI (70 kPa) in the oil gallery when the engine is hot and idling. Pressure will increase by about 10 PSI for each 1000 RPM as the engine speed increases, because the engine-driven pump also rotates faster.

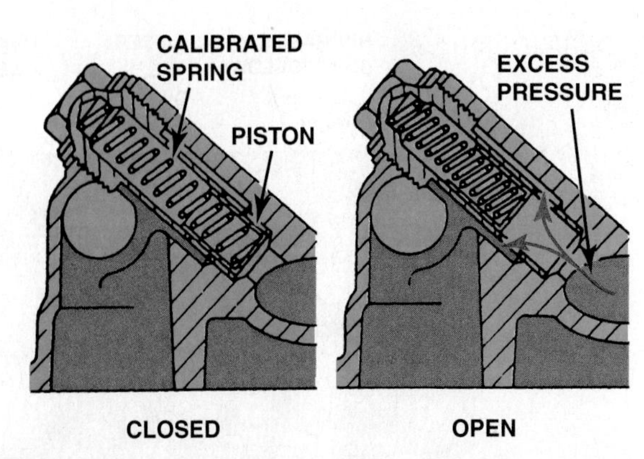

FIGURE 9–9 Oil pressure relief valves are spring loaded. The stronger the spring tension, the higher the oil pressure.

OIL PRESSURE REGULATION

In engines with a full-pressure lubricating system, maximum pressure is limited with a pressure relief valve. The relief valve (sometimes called the **pressure regulating valve**) is located at the outlet of the pump. The relief valve controls maximum pressure by bleeding off oil to the inlet side of the pump. ● **SEE FIGURE 9–9.**

The relief valve spring tension determines the maximum oil pressure. If a pressure relief valve is not used, the engine oil pressure will continue to increase as the engine speed increases. Maximum pressure is usually limited to the lowest pressure that will deliver enough lubricating oil to all engine parts that need to be lubricated. *Three to six gallons per minute are required to lubricate the engine.* The oil pump is made so that it is large enough to provide pressure at low engine speeds and small enough so that cavitation will not occur at high speed. Cavitation occurs when the pump tries to pull oil faster than it can flow from the pan to the pickup. When it cannot get enough oil, it will pull air. This puts air pockets or cavities in the oil stream. A pump is cavitating when it is pulling air or vapors.

NOTE: The reason for sheet-metal covers over the pickup screen is to prevent cavitation. Oil is trapped under the cover, which helps prevent the oil pump from drawing in air, especially during sudden stops or during rapid acceleration.

After the oil leaves the pump, it is delivered to the oil filter and then to the moving parts through drilled oil passages. ● **SEE FIGURE 9–10.** It needs no pressure after it reaches the parts that are to be lubricated. The oil film between the parts is developed and maintained by hydrodynamic lubrication. Excessive oil pressure requires more horsepower and provides no better lubrication than the minimum effective pressure.

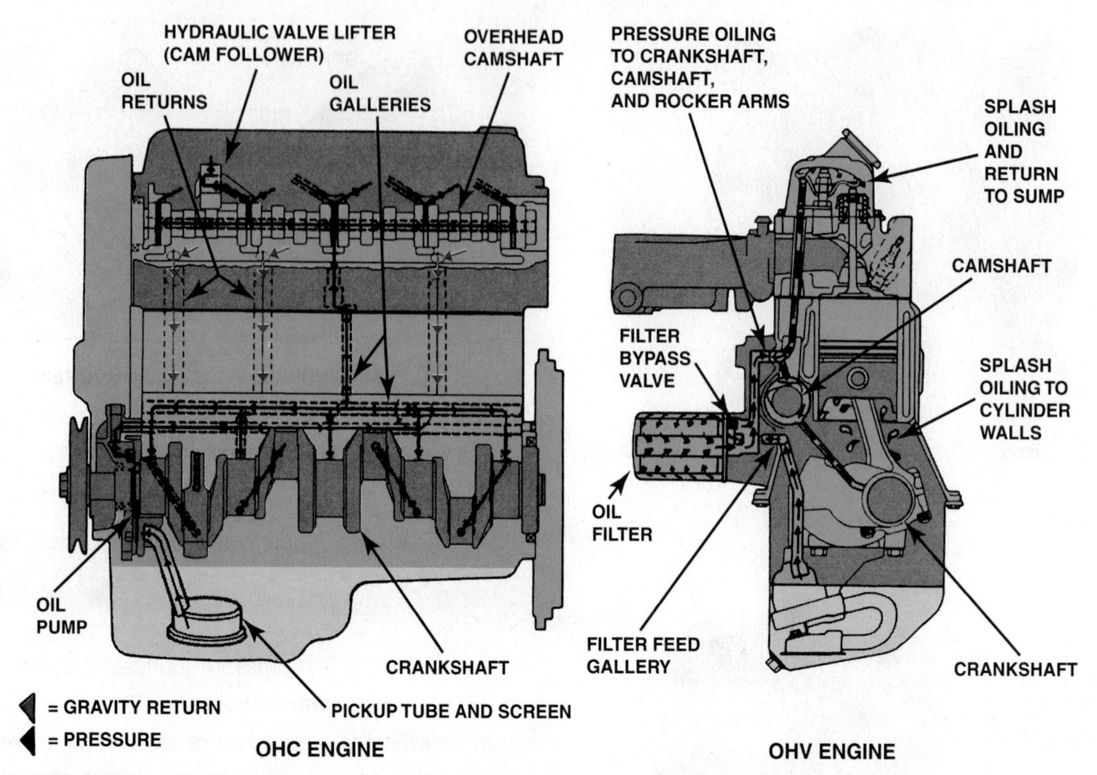

FIGURE 9–10 A typical engine design that uses both pressure and splash lubrication. Oil travels under pressure through the galleries (passages) to reach the top of the engine. Other parts are lubricated as the oil flows back down into the oil pan or is splashed onto parts.

FACTORS AFFECTING OIL PRESSURE

LEAKS Oil pressure can only be produced when the oil pump has a capacity larger than all the "leaks" in the engine. The leaks are the clearances at end points of the lubrication system. The end points are at the edges of bearings, the rocker arms, the connecting rod spit holes, and so on. These clearances are designed into the engine and are necessary for its proper operation.

OIL PUMP CAPACITY As the engine parts wear and clearance becomes greater, more oil will leak out. The oil pump *capacity* must be great enough to supply extra oil for these leaks. The capacity of the oil pump results from its size, rotating speed, and physical condition. If the pump is rotating slowly as the engine is idling, oil pump capacity is low. *If the leaks are greater than the pump capacity, engine oil pressure is low*. As the engine speed increases, the pump capacity increases and the pump tries to force more oil out of the leaks. This causes the pressure to rise until it reaches the regulated maximum pressure.

VISCOSITY OF ENGINE OIL The viscosity of the engine oil affects both the pump capacity and the oil leakage. Thin oil or oil of very low viscosity slips past the edges of the pump and flows freely from the leaks. Hot oil has a low viscosity, and therefore, a hot engine often has low oil pressure. Cold oil is more viscous (thicker) than hot oil. This results in higher pressures, even with the cold engine idling. High oil pressure occurs with a cold engine, because the oil relief valve must open farther to release excess oil than is necessary with a hot engine. This larger opening increases the spring compression force, which in turn increases the oil pressure. Putting higher-viscosity oil in an engine will raise the engine oil pressure to the regulated setting of the relief valve at a lower engine speed.

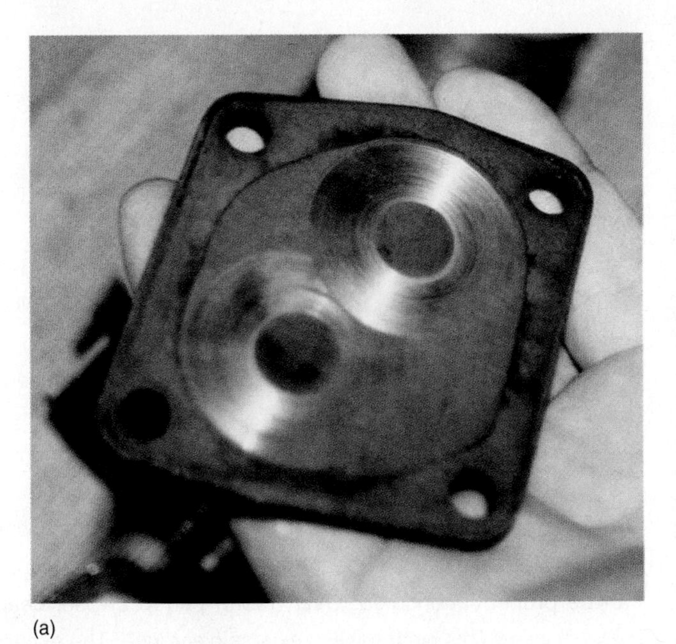

(a)

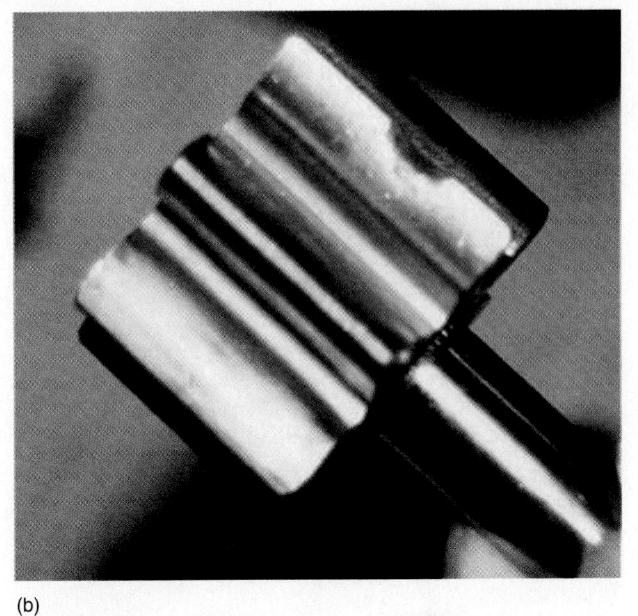

(b)

FIGURE 9–11 (a) A visual inspection indicated that this pump cover was worn. (b) An embedded particle of something was found on one of the gears, making this pump worthless except for scrap metal.

OIL PUMP CHECKS

The cover is removed to check the condition of the oil pump. The gears and housing are examined for scoring. If the gears and housing are heavily scored, the entire pump should be replaced. If they are lightly scored, the clearances in the pump should be measured. These clearances include the space between the gears and housing, the space between the teeth of the two gears, and the space between the side of the gear and the pump cover. A feeler gauge is often used to make these measurements. Gauging plastic can be used to measure the space between the side of the gears and the cover. The oil pump should be replaced when excessive clearance or scoring is found. ● **SEE FIGURE 9–11**.

On most engines, the oil pump should be replaced as part of any engine work, especially if the cause for the repair is lack of lubrication.

NOTE: The oil pump is the "garbage pit" of the entire engine. Any and all debris is often forced through the gears and housing of an oil pump. ● SEE FIGURE 9–12.

Always refer to the manufacturer's specifications when checking the oil pump for wear. Typical oil pump clearances include the following:

1. End plate clearance: 0.0015 inch (0.04 mm)
2. Side (rotor) clearance: 0.012 inch (0.30 mm)
3. Rotor tip clearance: 0.010 inch (0.25 mm)
4. Gear end play clearance: 0.004 inch (0.10 mm)

All parts should also be inspected closely for wear. Check the relief valve for scoring and check the condition of the spring. When installing the oil pump, coat the sealing surfaces with engine assembly lubricant. This lubricant helps draw oil from the oil pan on initial start-up.

(a)

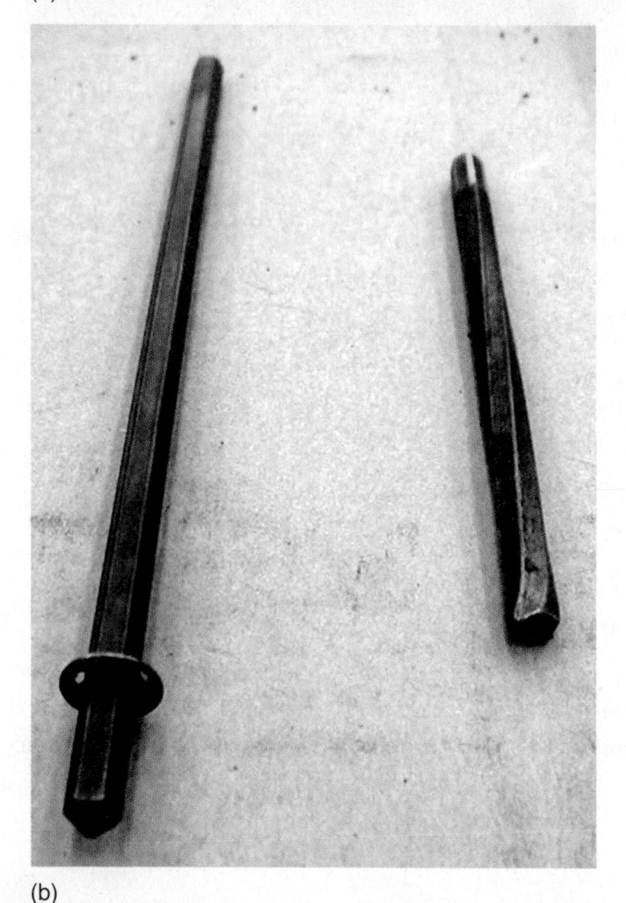

(b)

FIGURE 9–12 (a) The oil pump is the only part in an engine that gets unfiltered engine oil. The oil is drawn up from the bottom of the oil pan and is pressurized before flowing to the oil filter. (b) If debris gets into an oil pump, the drive or distributor shaft can twist and/or break. When this occurs, the engine will lose all oil pressure.

OIL PASSAGES IN THE BLOCK

From the filter, oil goes through a drilled hole that intersects with a drilled main **oil gallery** or **longitudinal header**. This is a long hole drilled from the front of the block to the back. Inline engines use one oil gallery; V-type engines may use two or three galleries. Passages drilled through the block bulkheads allow the oil to go from the main oil gallery to the main and cam bearings. ● **SEE FIGURE 9–13**. In some engines, oil goes to the cam bearings first, and then to the main bearings.

It is important that the oil holes in the bearings match with the drilled passages in the bearing saddles so that the bearing can be properly lubricated. Over a long period of use, bearings will wear. This wear causes excess clearance. The excess clearance will allow too much oil to leak from the side of the bearing. When this happens, there will be little or no oil left for bearings located farther downstream in the lubricating system. This is a major cause of bearing failure. If a new bearing were installed in place of the oil-starved bearing, it, too, would fail unless the bearing having the excess clearance was also replaced.

VALVE TRAIN LUBRICATION

The valve train components are the last parts to get oil from the oil pump. The oil gallery may intersect or have drilled passages to the valve lifter bores to lubricate the lifters. When hydraulic lifters are used, the oil pressure in the gallery keeps refilling them. On some engines, oil from the lifters goes up the center of a hollow pushrod to lubricate the pushrod ends, the rocker arm pivot, and the valve stem tip. ● **SEE FIGURE 9–14**. In other engines, an oil passage is drilled from either the gallery or a cam bearing to the block deck, where it matches with a gasket hole and a hole drilled in the head to carry the oil to a rocker arm shaft. Some engines use an enlarged bolt hole to carry lubrication oil around the rocker shaft cap screw to the rocker arm shaft. Holes in the bottom of the rocker arm shaft allow lubrication of the rocker arm pivot. Mechanical loads on the valve train hold the rocker arm against the passage in the rocker arm shaft. This prevents excessive oil leakage from the rocker arm shaft. Often, holes are drilled in cast rocker arms to carry oil to the pushrod end and to the valve tip. Rocker arm assemblies need only a surface coating of oil, so the oil flow to the rocker assembly is minimized using restrictions or metered openings.

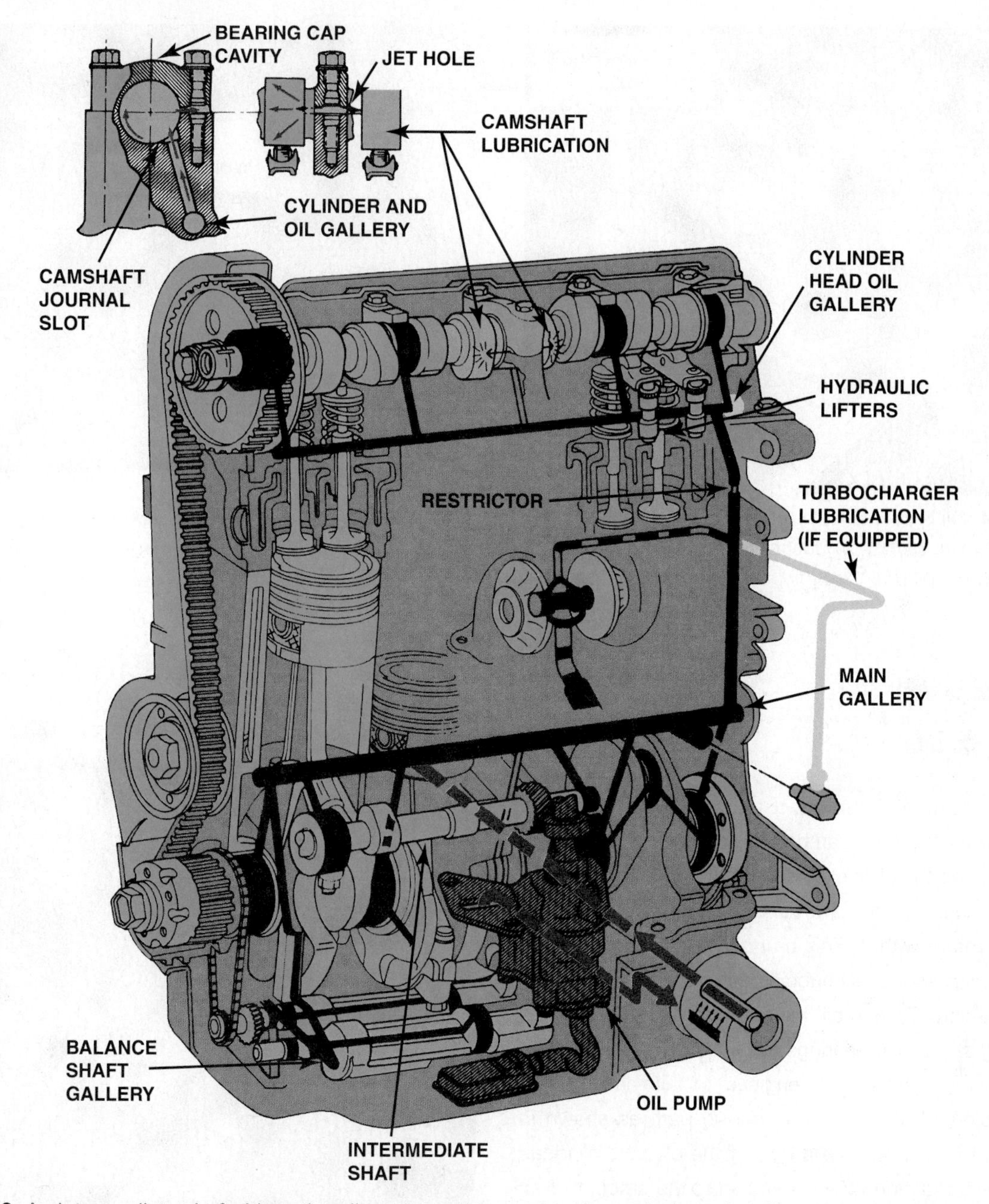

FIGURE 9–13 An intermediate shaft drives the oil pump on this overhead camshaft engine. Note the main gallery and other drilled passages in the block and cylinder head.

The restriction or metering disk is in the lifter when the rocker assembly is lubricated through the pushrod. Cam journal holes that line up with oil passages are often used to meter oil to the rocker shafts.

Oil that seeps from the rocker assemblies is returned to the oil pan through drain holes. These oil drain holes are often so placed that the oil drains on the camshaft or cam drive gears to lubricate them.

Some engines have means of directing a positive oil flow to the cam drive gears or chain. This may be a nozzle or a chamfer on a bearing parting surface that allows oil to spray on the loaded portion of the cam drive mechanism.

FIGURE 9–14 Oil is sent to the rocker arms on this Chevrolet V-8 engine through the hollow pushrods. The oil returns to the oil pan through the oil drainback holes in the cylinder head.

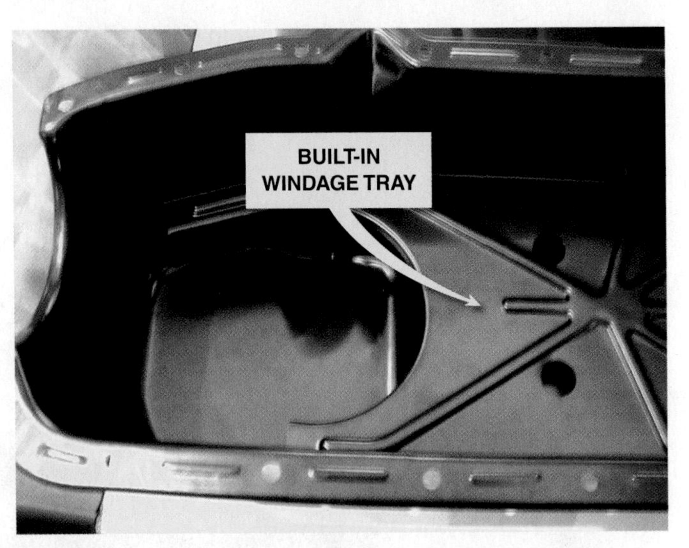

FIGURE 9–15 A typical oil pan with a built-in windage tray used to keep oil from being churned up by the rotating crankshaft.

OIL PANS

As the vehicle accelerates, brakes, or turns rapidly, the oil tends to move around in the pan. Pan baffles and oil pan shapes are often used to keep the oil inlet under the oil at all times. As the crankshaft rotates, it acts like a fan and causes air within the crankcase to rotate with it. This can cause a strong draft on the oil, churning it so that air bubbles enter the oil, which then causes oil foaming. Oil with air will not lubricate like liquid oil, so oil foaming can cause bearings to fail. A baffle or **windage tray** is sometimes installed in engines to eliminate the oil churning problem. This may be an added part, as shown in ● **FIGURE 9–15**, or it may be a part of the oil pan. Windage trays have the good side effect of reducing the amount of air disturbed by the crankshaft, so that less power is drained from the engine at high crankshaft speeds. Oil pans on many engines are a structural part of the engine. ● **SEE FIGURE 9–16.**

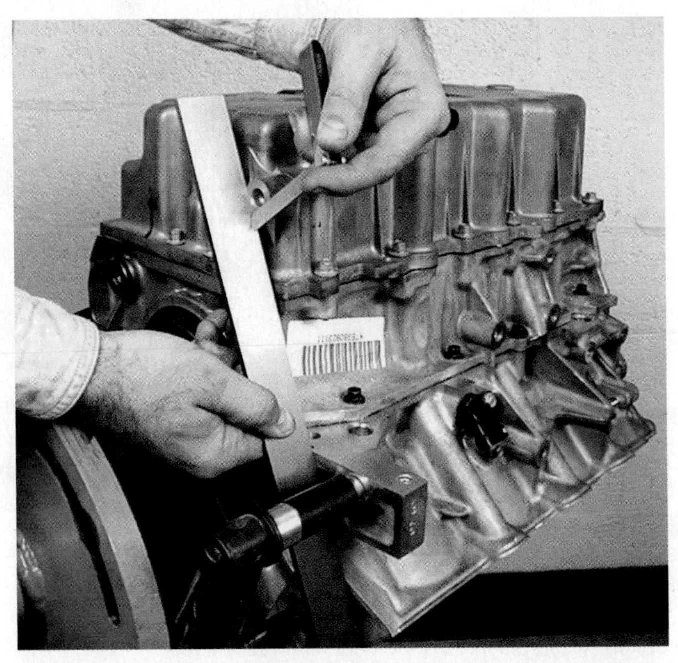

FIGURE 9–16 A straightedge and a feeler gauge are being used to check that the oil pan has been correctly installed on the 5.7 liter Chevrolet V-8 engine. The oil pan is part of the engine itself and must be properly installed to ensure that other parts attached to the engine are not being placed in a bind.

OIL COOLERS

Oil temperature must also be controlled on many high-performance or turbocharged engines. ● **SEE FIGURE 9–17** for an example of an engine oil cooler used on a production high-performance engine. A larger-capacity oil pan also helps to control oil temperature. Coolant flows through the oil cooler to help warm the oil when the engine is cold and cool the oil when the engine is hot. Oil temperature should be above 212°F (100°C) to boil off any accumulated moisture, but it should not exceed about 280° to 300°F (138° to 148°C).

FIGURE 9–17 A typical engine oil cooler. Engine coolant flows through the cooler adjuster that fits between the engine block and the oil filter.

FIGURE 9–19 A typical oil pressure sending unit on a Ford V-8.

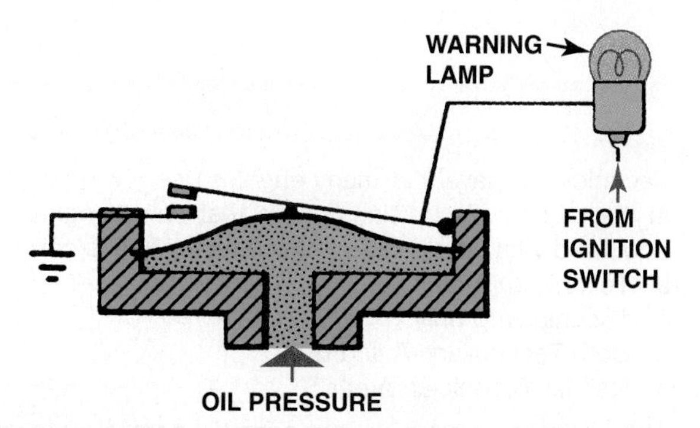

FIGURE 9–18 The oil pressure switch is connected to a warning lamp that alerts the driver of low oil pressure.

OIL PRESSURE WARNING LAMP

All vehicles are equipped with an oil pressure gauge or a warning lamp. The warning lamp comes on whenever the engine oil pressure has dropped to 3 to 7 PSI. Normal oil pressure is considered to be 10 PSI per 1000 RPM. An electrical switch is used to convert the ground circuit of the oil pressure warning lamp if the oil pressure is below the rating of the sending unit. ● **SEE FIGURES 9–18 AND 9–19.**

? FREQUENTLY ASKED QUESTION

What Is Acceptable Oil Consumption?

There are a number of opinions regarding what is acceptable oil consumption. Most vehicle owners do not want their engine to use any oil between oil changes even if they do not change it more often than every 7,500 miles (12,000 km). Engineers have improved machining operations and piston ring designs to help eliminate oil consumption.

Many stationary or industrial engines are not driven on the road; therefore, they do not accumulate miles, yet they still may consume excessive oil.

A general rule for "acceptable" oil consumption is that it should be about 0.002 to 0.004 pounds per horsepower per hour. To figure, use the following:

$$\frac{\textbf{1.82} \times \textbf{quarts used}}{\textbf{Operating hp} \times \textbf{total hours}} = \textbf{lb/hp/hr}$$

Therefore, oil consumption is based on the amount of work an engine performs. Although the formula may not be usable for vehicle engines used for daily transportation, it may be usable by the marine or industrial engine builder. Generally, oil consumption that is greater than 1 quart for every 600 miles (1,000 kilometers per liter) is considered to be excessive with a motor vehicle.

SUMMARY

1. Normal engine oil pump pressure ranges from 10 to 60 PSI (200 to 400 kPa) or 10 PSI for every 1000 engine RPM.
2. Hydrodynamic oil pressure around engine bearings is usually over 1,000 PSI (6,900 kPa).
3. The oil pump is driven directly by the crankshaft or by a gear or shaft from the camshaft.
4. The last components to get oil from the oil pump are the valve train parts.
5. Some engines use an oil cooler.

REVIEW QUESTIONS

1. What causes a wedge-shaped film to form in the oil?
2. What is hydrodynamic lubrication?
3. Explain why internal engine leakage affects oil pressure.
4. Describe how the oil flows from the oil pump, through the filter and main engine bearings, to the valve train.
6. What is the purpose of a windage tray?

CHAPTER QUIZ

1. Normal oil pump pressure in an engine is ________.
   a. 3 to 7 PSI
   b. 10 to 60 PSI
   c. 100 to 150 PSI
   d. 180 to 210 PSI
2. The oil pump pressure relief valve is also called ________.
   a. Oil pump valve
   b. Pressure valve
   c. Pressure regulating valve
   d. Pressure dump valve
3. What type of pump is a typical oil pump?
   a. Positive displacement
   b. Centrifugal
   c. Piston-type
   d. Hydraulically driven
4. Engine oil passages in an engine block are called ________.
   a. Oil passages
   b. Oil galleries
   c. Weep holes
   d. Oil holes
5. Technician A says that the oil pump draws unfiltered oil from the bottom of the oil pan. Technician B says that the oil pump is driven from the front of the crankshaft in some engines. Which technician is correct?
   a. Technician A only
   b. Technician B only
   c. Both Technicians A and B
   d. Neither Technician A nor B
6. Technician A says that oil pressure is affected by the amount of main and rod bearing clearance. Technician B says that the oil pressure is lower when the oil gets hot than when it is cold. Which technician is correct?
   a. Technician A only
   b. Technician B only
   c. Both Technicians A and B
   d. Neither Technician A nor B
7. Technician A says that many engines use a windage tray in the oil pan. Technician B says that some engines are equipped with an oil cooler. Which technician is correct?
   a. Technician A only
   b. Technician B only
   c. Both Technicians A and B
   d. Neither Technician A nor B
8. The oil pressure warning light normally comes on to warn the driver if the oil pressure drops below ________.
   a. 50 PSI
   b. 30 PSI
   c. 10 PSI
   d. 3 to 7 PSI
9. A typical oil pump can pump how many gallons per minute?
   a. 3 to 6 gallons
   b. 6 to 10 gallons
   c. 10 to 60 gallons
   d. 50 to 100 gallons
10. In typical engine lubrication systems, what components are the last to receive oil?
   a. Main bearings
   b. Rod bearings
   c. Valve train components
   d. Oil filters

chapter 10

INTAKE AND EXHAUST SYSTEMS

LEARNING OBJECTIVES

After studying this chapter, the reader will be able to:

1. Describe the air intake filtration system.
2. Describe the purpose and operation of intake air heating systems.
3. Discuss the purpose and function of intake manifolds.
4. Describe the operation of the exhaust gas recirculation system in the intake manifold.
5. List the materials used in exhaust manifolds and exhaust systems.

This chapter will help you prepare for Engine Repair (A8) ASE certification test content area "C" (Fuel, Air Induction and Exhaust Systems Diagnosis and Repair).

KEY TERMS

Annealing 170
Exhaust gas recirculation (EGR) 169
Hangers 173
Helmholtz resonator 166
Micron 164
Plenum 170

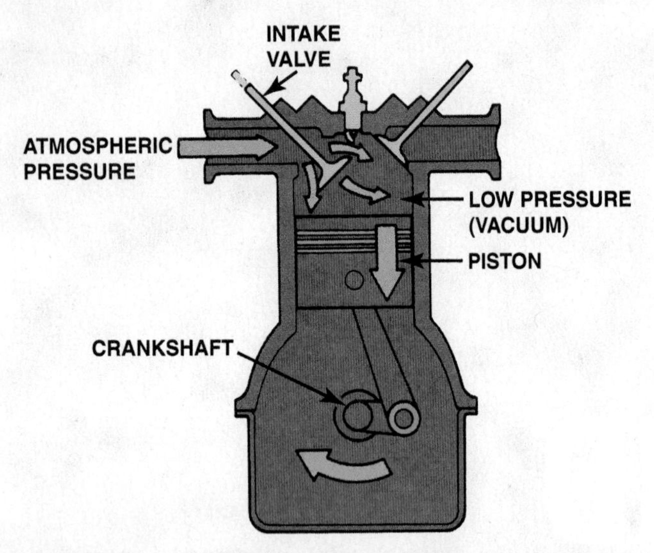

FIGURE 10–1 Downward movement of the piston lowers the air pressure inside the combustion chamber. The pressure differential between the atmosphere and the inside of the engine forces air into the engine.

FIGURE 10–2 Dust and dirt in the air are trapped in the air filter so they do not enter the engine.

AIR INTAKE FILTRATION

PURPOSE AND FUNCTION Gasoline must be mixed with air to form a combustible mixture. Air movement into an engine occurs due to low pressure (vacuum) being created in the engine. ● **SEE FIGURE 10–1.**

Like gasoline, air contains dirt and other materials, which cannot be allowed to reach the engine. Just as fuel filters are used to clean impurities from gasoline, an air cleaner and filter are used to remove contaminants from the air. The three main jobs of the air cleaner and filter are to:

- Clean the air before it is mixed with fuel
- Silence intake noise
- Act as a flame arrester in case of a backfire

The automotive engine uses about 9,000 gallons (34,069 liters) of air for every gallon of gasoline burned at an air–fuel ratio of 14.7 to 1. Without proper filtering of the air before intake, dust and dirt in the air seriously damage engine parts and shorten engine life.

While abrasive particles can cause wear at any place inside the engine where two surfaces move against each other, they first attack piston rings and cylinder walls. Contained in the blowby gases, they pass by the piston rings and into the crankcase. From the crankcase, the particles circulate throughout the engine with the oil. Large amounts of abrasive particles in the oil can damage other moving engine parts.

The filter that cleans the intake air is in a two-piece air cleaner housing made either of stamped steel or composite materials. The air cleaner housing is located on top of the throttle-body injection (TBI) unit or is positioned to one side of the engine. ● **SEE FIGURE 10–2.**

FILTER REPLACEMENT Manufacturers recommend cleaning or replacing the air filter element at periodic intervals, usually listed in terms of distance driven or months of service. The distance and time intervals are based on so-called normal driving. More frequent air filter replacement is necessary when the vehicle is driven under dusty, dirty, or other severe conditions.

It is best to replace a filter element before it becomes too dirty to be effective. A dirty air filter passes contaminants that cause engine wear.

AIR FILTER ELEMENTS The paper air filter element is the most common type of filter. It is made of a chemically treated paper stock that contains tiny passages in the fibers. These passages form an indirect path for the airflow to follow. The airflow passes through several fiber surfaces, each of which traps microscopic particles of dust, dirt, and carbon. Most air filters are capable of trapping dirt and other particles larger than 10 to 25 microns in size. One **micron** is equal to 0.000039 inch.

NOTE: A person can only see objects that are 40 microns or larger in size. A human hair is about 50 microns in diameter.

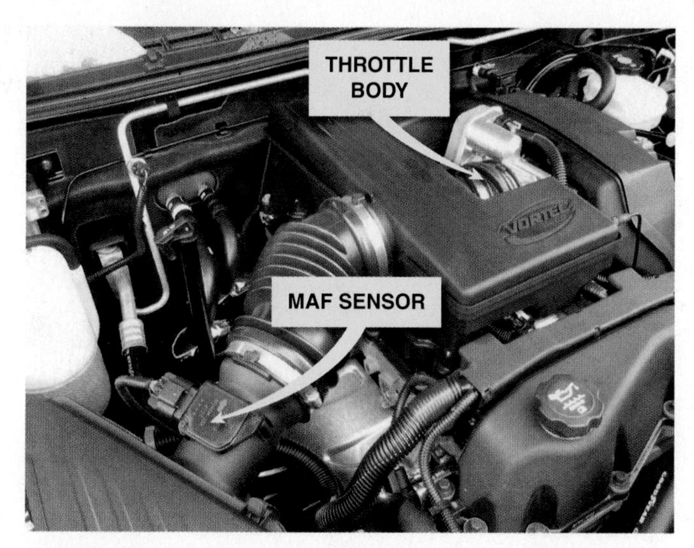

FIGURE 10–3 Most air filter housings are located on the side of the engine compartment and use flexible rubber hose to direct the airflow into the throttle body of the engine.

NOTE: Do not attempt to clean a paper element filter by rapping it on a sharp object to dislodge the dirt, or blowing compressed air through the filter. This tends to clog the paper pores and further reduce the airflow capability of the filter.

REMOTELY MOUNTED AIR FILTERS AND DUCTS Air cleaner and duct design depend on a number of factors such as the size, shape, and location of other engine compartment components, as well as the vehicle body structure.

Port fuel-injection systems generally use a horizontally mounted throttle body. Some systems also have a mass airflow (MAF) sensor between the throttle body and the air cleaner. ● **SEE FIGURE 10–3**. Because placing the air cleaner housing next to the throttle body would cause engine and vehicle design problems, it is more efficient to use this remote air cleaner placement.

Turbocharged engines present a similar problem. The air cleaner connects to the air inlet elbow at the turbocharger. However, the tremendous heat generated by the turbocharger makes it impractical to place the air cleaner housing too close to the turbocharger. For better protection, the MAF sensor is installed between the turbocharger and the air cleaner in some vehicles. Remote air cleaners are connected to the turbocharger air inlet elbow or fuel-injection throttle body by composite ducting, which is usually retained by clamps. The ducting used may be rigid or flexible, but all connections must be airtight.

(a)

(b)

FIGURE 10–4 (a) Note the discovery as the air filter housing was opened during service on a Pontiac Bonneville. The nuts were obviously deposited by squirrels (or some other animal). (b) Not only was the housing filled with nuts, but also this air filter was extremely dirty, indicating that this vehicle had not been serviced for a long time.

TECH TIP

Always Check the Air Filter

Always inspect the air filter and the air intake system carefully during routine service. Debris or objects deposited by animals can cause a restriction to the airflow and can reduce engine performance. ● **SEE FIGURE 10–4.**

FREQUENTLY ASKED QUESTION

What Does This Tube Do?

What is the purpose of the odd-shaped tube attached to the inlet duct between the air filter and the throttle body, as seen in ● **FIGURE 10–5?**

The tube shape is designed to dampen out certain resonant frequencies that can occur at certain engine speeds. The length and shape of this tube are designed to absorb shock waves that are created in the air intake system and to provide a reservoir for the air that will then be released into the airstream during cycles of lower pressure. This resonance tube is often called a **Helmholtz resonator**, named for the discoverer of the relationship between shape and value of frequency Herman L. F. von Helmholtz (1821–1894) of the University of Hönizsberg in East Prussia. The overall effect of these resonance tubes is to reduce the noise of the air entering the engine.

FIGURE 10–5 A resonance tube, called a Helmholtz resonator, is used on the intake duct between the air filter and the throttle body to reduce air intake noise during engine acceleration.

ENGINE AIR TEMPERATURE REQUIREMENTS

Some form of thermostatic control has been used on vehicles equipped with a throttle-body fuel injection to control intake air temperature for improved driveability. In a throttle-body fuel injection system, the fuel and air are combined above the throttle plate and must travel through the intake manifold before reaching the cylinders. Air temperature control is needed under these conditions to help keep the gas and air mixture combined.

Heat radiating from the exhaust manifold is retained by the heat stove and sent to the air cleaner inlet to provide heated air to the throttle body.

An air control valve or damper permits the air intake of:

- Heated air from the heat stove
- Cooler air from the snorkel or cold-air duct
- A combination of both

While the air control valve generally is located in the air cleaner snorkel, it may be in the air intake housing or ducting of remote air cleaners. Most fuel-injection systems do not use temperature control.

THROTTLE-BODY INJECTION INTAKE MANIFOLDS

The *intake manifold* is also called the *inlet manifold*. Smooth operation can only occur when each combustion chamber produces the same pressure as every other chamber in the engine. For this to be achieved, each cylinder must receive a charge exactly like the charge going into the other cylinders in quality and quantity. The charges must have the same physical properties and the same air–fuel mixture.

A throttle-body fuel injector forces finely divided droplets of liquid fuel into the incoming air to form a combustible air–fuel mixture. ● **SEE FIGURE 10–6** for an example of a typical throttle-body injection (TBI) unit. These droplets start to evaporate as soon as they leave the throttle-body injector nozzles. *The droplets stay in the charge as long as the charge flows at high velocities*. At maximum horsepower, these velocities may reach 300 feet per second. Separation of the droplets from the charge as it passes through the manifold occurs when the velocity drops below 50 feet per second. Intake charge velocities at idle speeds are often below this value. When separation occurs—at low engine speeds—extra fuel must be supplied to the charge in order to have a combustible mixture reach the combustion chamber.

FIGURE 10–6 A throttle-body injection (TBI) unit used on a GM V-6 engine.

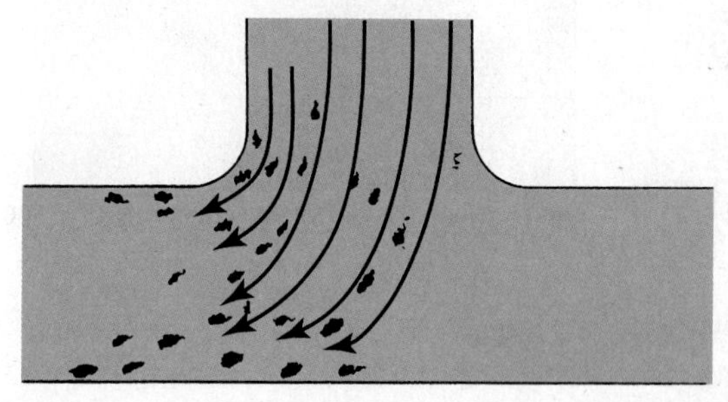
FIGURE 10–7 Heavy fuel droplets separate as they flow around an abrupt bend in an intake manifold.

Manifold sizes represent a compromise. They must have a cross section large enough to allow charge flow for maximum power. The cross section must be small enough that the flow velocities of the charge will be high enough to keep the fuel droplets in suspension. This is required so that equal mixtures reach each cylinder. Manifold cross-sectional size is one reason why engines designed especially for racing will not run at low engine speeds. Racing manifolds must be large enough to reach maximum horsepower. This size, however, allows the charge to move slowly, and the fuel will separate from the charge at low engine speeds. Fuel separation leads to poor accelerator response. ● **SEE FIGURE 10–7**. Standard passenger vehicle engines are primarily designed for economy during light-load, partial-throttle operation. Their manifolds, therefore, have a much smaller cross-sectional area than do those of racing engines. This small size will help keep flow velocities of the charge high throughout the normal operating speed range of the engine.

TECH TIP

Check the Intake If an Exhaust Noise Is Heard

Because many V-type engines equipped with a throttle-body injection and/or EGR valve use a cross-over exhaust passage, a leak around this passage will create an exhaust leak and noise. Always check for evidence of an exhaust leak around the intake manifold whenever diagnosing an exhaust sound.

PORT FUEL-INJECTION INTAKE MANIFOLDS

The size and shape of port fuel-injected engine intake manifolds can be optimized because the only thing in the manifold is air. The fuel injection system is located in the intake manifold about 3 to 4 inches (70 to 100 mm) from the intake valve. Therefore, the runner length and shape are designed for tuning only. There is no need to keep an air–fuel mixture homogenized throughout its trip from the TBI unit to the intake valve. Typically, long runners build low-RPM torque, while shorter runners provide maximum high-RPM power. ● **SEE FIGURES 10–8 AND 10–9**. Some engines with four valve heads utilize a dual or variable intake runner design. At lower engine speeds, long intake runners provide low-speed torque. At higher engine speeds, shorter intake runners are opened by means of a computer-controlled valve to increase high-speed power.

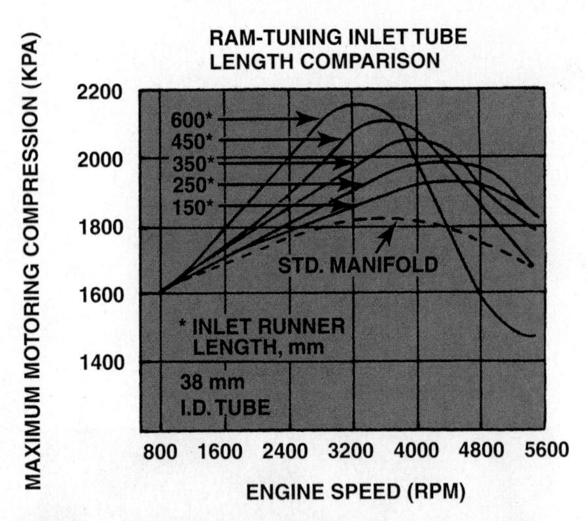

FIGURE 10–8 The graph shows the effect of sonic tuning of the intake manifold runners. The longer runners increase the torque peak and move it to a lower RPM. The 600 mm-long intake runner is about 24 inches long.

FIGURE 10–10 The air flowing into the engine can be directed through long or short runners for best performance and fuel economy.

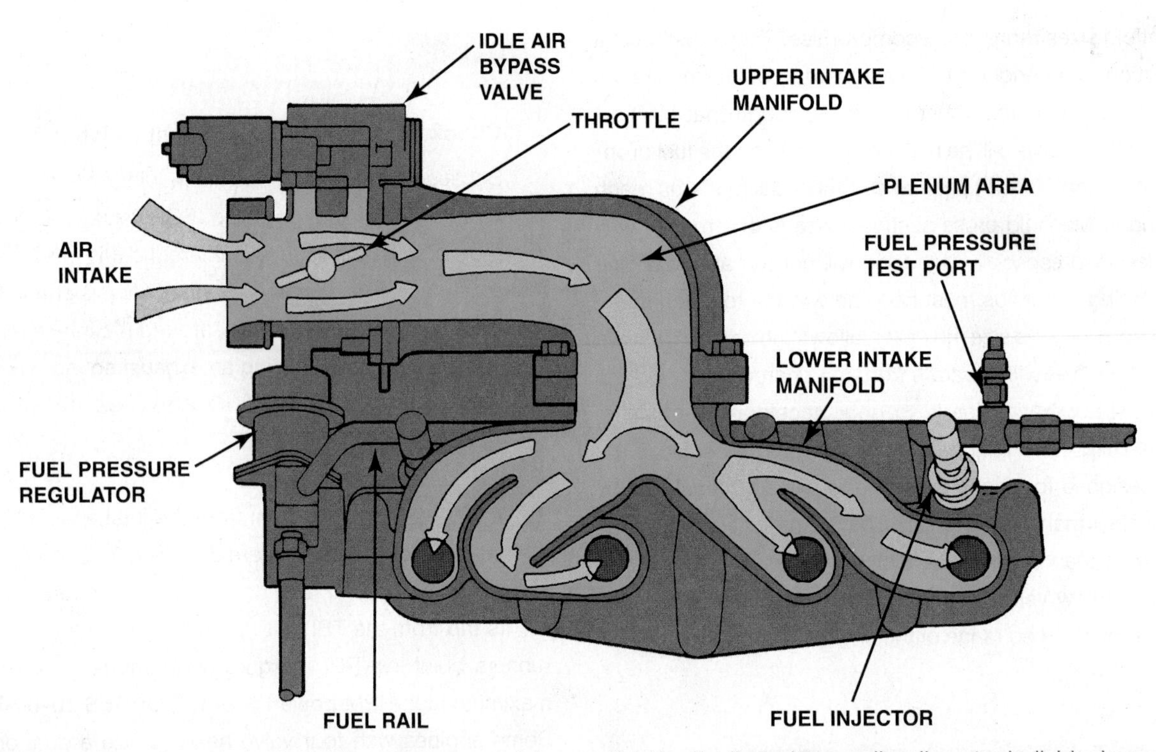

FIGURE 10–9 Airflow through the large diameter upper intake manifold is distributed to smaller diameter individual runners in the lower manifold in this two-piece manifold design.

VARIABLE INTAKES

Many intake manifolds are designed to provide both short runners, best for higher engine speed power, and longer runners, best for lower engine speed torque. The valve(s) that control the flow of air through the passages of the intake manifold are computer controlled. ● **SEE FIGURE 10–10.**

PLASTIC INTAKE MANIFOLDS

Most thermoplastic intake manifolds are molded from fiberglass-reinforced nylon. The plastic manifolds can be cast or injection molded. Some manifolds are molded in two parts and bonded together. Plastic intake manifolds are lighter than aluminum manifolds and can better insulate engine heat from the fuel injectors.

Plastic intake manifolds have smoother interior surfaces than do other types of manifolds, resulting in greater airflow. ● **SEE FIGURE 10–11**.

FIGURE 10–11 Many plastic intake manifolds are constructed using many parts glued together to form complex passages for airflow into the engine.

EXHAUST GAS RECIRCULATION PASSAGES

To reduce the emission of oxides of nitrogen (NO_X), engines have been equipped with **exhaust gas recirculation (EGR)** valves. From 1973 until recently, they were used on almost all vehicles. Because of the efficiency of computer-controlled fuel injection, some newer engines do not require an EGR system to meet emission standards. Some engines use intake and exhaust valve overlap as a means of trapping some exhaust in the cylinder as an alternative to using an EGR valve.

The EGR valve opens at speeds above idle on a warm engine. When open, the valve allows a small portion of the exhaust gas (5% to 10%) to enter the intake manifold. Here, the exhaust gas mixes with and takes the place of some of the intake charge. This leaves less room for the intake charge to enter the combustion chamber. The recirculated exhaust gas is inert and does *not* enter into the combustion process. The result is a lower peak combustion temperature. As the combustion temperature is lowered, the production of oxides of nitrogen is also reduced.

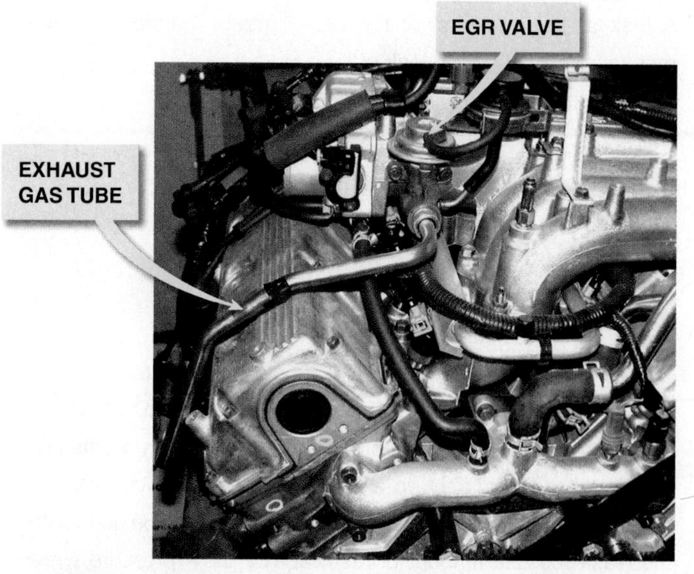

FIGURE 10–12 The exhaust gas recirculation system is more efficient at controlling NO_X emissions if the exhaust gases are cooled. A long metal tube between the exhaust manifold and the intake manifold allows the exhaust gases to cool before entering the engine.

The EGR system has some means of interconnecting the exhaust and intake manifolds. The interconnecting passage is controlled by the EGR valve. On V type engines, the intake manifold crossover is used as a source of exhaust gas for the EGR system. A cast passage connects the exhaust crossover to the EGR valve. On inline type engines, an external tube is generally used to carry exhaust gas to the EGR valve. The exhaust gases are more effective in reducing NO_X emissions if the exhaust is cooled before being drawn into the cylinders. This tube is often designed to be long so that the exhaust gas is cooled before it enters the EGR valve. ● **FIGURE 10–12** shows a typical long EGR tube.

UPPER AND LOWER INTAKE MANIFOLDS

Many intake manifolds are constructed in two parts.

- A lower section attaches to the cylinder heads and includes passages from the intake ports.
- An upper manifold, also called the **plenum**, connects to the lower unit and includes the long passages needed to help provide the ram effect that helps the engine deliver maximum torque at low engine speeds. The throttle body attaches to the upper intake.

The use of a two-part intake manifold allows for easier manufacturing as well as assembly, but can create additional locations for leaks. If the lower intake manifold gasket leaks, not only could a vacuum leak occur affecting the operation of the engine, but a coolant leak or an oil leak can also occur. A leak at the gasket(s) of the upper intake manifold usually results in a vacuum (air) leak only.

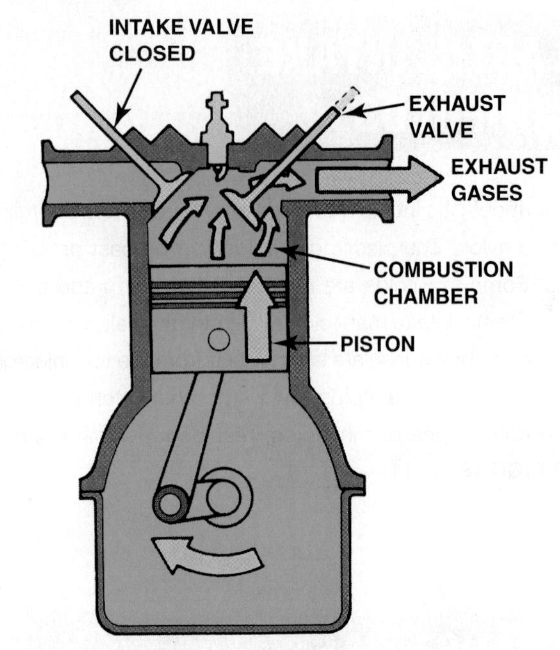

FIGURE 10–13 The exhaust gases are pushed out of the cylinder by the piston on the exhaust stroke.

EXHAUST MANIFOLD DESIGN

The exhaust manifold is designed to collect high-temperature spent gases from the head exhaust ports. ● **SEE FIGURE 10–13.** The hot gases are sent to an exhaust pipe, then to a catalytic converter, to the muffler, to a resonator, and on to the tailpipe, where they are vented to the atmosphere. This must be done with the least-possible amount of restriction or back pressure while keeping the exhaust noise at a minimum.

Exhaust gas temperature will vary according to the power produced by the engine. The manifold must be designed to operate at both engine idle and continuous full power. Under full-power conditions, the exhaust manifold will become red-hot, causing a great deal of expansion.

NOTE: The temperature of an exhaust manifold can exceed 1,500°F (815°C).

At idle, the exhaust manifold is just warm, causing little expansion. After casting, the manifold may be annealed. **Annealing** is a process that reheats the manifold and then it is allowed to cool slowly. This is performed to remove stress in the casting to reduce the chance of cracking from the temperature changes. During vehicle operation, manifold temperatures usually reach the high-temperature extremes. Most exhaust manifolds are made from cast iron to withstand extreme and rapid temperature changes. The manifold is bolted to the head in a way that will allow expansion and contraction. In some cases, hollow-headed bolts are used to maintain a gas-tight seal while still allowing normal expansion and contraction.

The exhaust manifold is designed to allow the free flow of exhaust gas. Some manifolds use internal cast-rib deflectors or dividers to guide the exhaust gases toward the outlet as smoothly as possible.

Some exhaust manifolds are designed to go above the spark plug, whereas others are designed to go below. The spark plug and carefully routed ignition wires are usually shielded from the exhaust heat with sheet-metal deflectors. Many exhaust manifolds have heat shields as seen in ● **FIGURE 10–14.**

Exhaust systems are especially designed for the engine–chassis combination. The exhaust system length, pipe size, and silencer are designed, where possible, to make use of the tuning effect of the gas column resonating within the exhaust system. Tuning occurs when the exhaust pulses from the cylinders are emptied into the manifold between the pulses of other cylinders. ● **SEE FIGURE 10–15.**

FIGURE 10–14 This exhaust manifold has a heat shield to help retain the heat and help reduce exhaust emissions.

FIGURE 10–15 Many exhaust manifolds are constructed of pressed steel and are free flowing to improve engine performance.

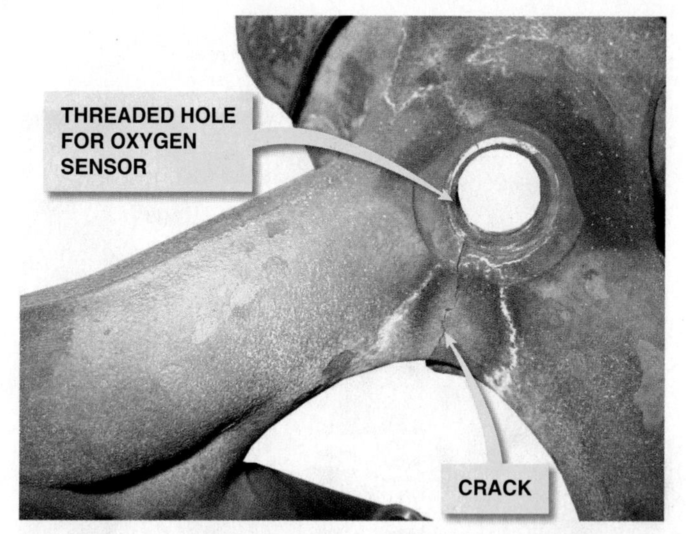

FIGURE 10–16 A crack in an exhaust manifold is often not visible. A crack in the exhaust manifold upstream of the oxygen sensor can fool the sensor and affect engine operation.

? FREQUENTLY ASKED QUESTION

How Can a Cracked Exhaust Manifold Affect Engine Performance?

A crack in an exhaust manifold will not only allow exhaust gases to escape and cause noise but the crack can also allow air to enter the exhaust manifold. ● **SEE FIGURE 10–16.**

Exhaust flows from the cylinders as individual puffs or pressure pulses. Behind each of these pressure pulses, a low pressure (below atmospheric pressure) is created. Outside air at atmospheric pressure is then drawn into the exhaust manifold through the crack. This outside air contains 21% oxygen and is measured by the oxygen sensor (O2S). The air passing the O2S signals the engine computer that the engine is operating too lean (excess oxygen) and the computer, not knowing that the lean indicator is false, adds additional fuel to the engine. The result is that the engine will be operating richer (more fuel than normal) and spark plugs could become fouled causing poor engine operation.

EXHAUST MANIFOLD GASKETS

Exhaust heat will expand the manifold more than it will expand the head. It causes the exhaust manifold to slide on the sealing surface of the head. The heat also causes thermal stress. When the manifold is removed from the engine for service, the stress is relieved and this may cause the manifold to warp slightly. Exhaust manifold gaskets are included in gasket sets to seal slightly warped exhaust manifolds. These gaskets *should* be used, even if the engine did not originally use exhaust manifold gaskets. When a perforated core exhaust manifold gasket has facing on one side only, put the facing side against the head and put the manifold against the perforated metal core. The manifold can slide on the metal of the gasket just as it slid on the sealing surface of the head.

FIGURE 10–17 Typical exhaust manifold gaskets. Note how they are laminated to allow the exhaust manifold to expand and contract due to heating and cooling.

FIGURE 10–18 An exhaust manifold spreader tool is a tool that is absolutely necessary to use when reinstalling exhaust manifolds. When they are removed from the engine, they tend to warp slightly even though the engine is allowed to cool before being removed. The spreader tool allows the technician to line up the bolt holes without doing any harm to the manifold.

Gaskets are used on new engines with tubing- or header-type exhaust manifolds. The gaskets often include heat shields to keep exhaust heat from the spark plugs and spark plug cables. They may have several layers of steel for high-temperature sealing. The layers are spot-welded together. Some are embossed where special sealing is needed. ● **SEE FIGURE 10–17.**

Many new engines do not use gaskets with cast exhaust manifolds. The flat surface of the new cast-iron exhaust manifold fits tightly against the flat surface of the new head.

TECH TIP

The Correct Tools Save Time

When cast-iron exhaust manifolds are removed, the stresses built up in the manifolds often cause the manifolds to twist or bend. This distortion even occurs when the exhaust manifolds have been allowed to cool before removal. Attempting to reinstall distorted exhaust manifolds is often a time-consuming and frustrating exercise.

However, special spreading jacks can be used to force the manifold back into position so that the fasteners can be lined up with the cylinder head. ● **SEE FIGURE 10–18.**

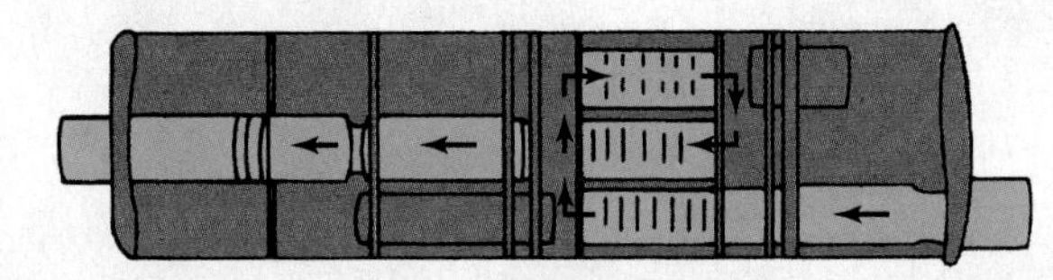

FIGURE 10–19 Exhaust gases expand and cool as they travel through the passages in the muffler.

MUFFLERS

When the exhaust valve opens, it rapidly releases high-pressure gas. This sends a strong air pressure wave through the atmosphere, which produces a sound we call an explosion. It is the same sound produced when the high-pressure gases from burned gunpowder are released from a gun. In an engine, the pulses are released one after another. The explosions come so fast that they blend together in a steady roar.

Sound is air vibration. When the vibrations are large, the sound is loud. The muffler catches the large bursts of high-pressure exhaust gas from the cylinder, smoothing out the pressure pulses and allowing them to be released at an even and constant rate. It does this through the use of perforated tubes within the muffler chamber. The smooth-flowing gases are released to the tailpipe. In this way, the muffler silences engine exhaust noise. Sometimes resonators are used in the exhaust system and the catalytic converter also acts as a muffler. They provide additional expansion space at critical points in the exhaust system to smooth out the exhaust gas flow. ● **SEE FIGURE 10–19.**

FIGURE 10–20 A hole in the muffler allows condensed water to escape.

FREQUENTLY ASKED QUESTION

Why Is There a Hole in My Muffler?

Many mufflers are equipped with a small hole in the lower rear part to drain accumulated water. About 1 gallon of water is produced in the form of steam for each gallon of gasoline burned. The water vapor often condenses on the cooler surfaces of the exhaust system unless the vehicle has been driven long enough to fully warm the muffler above the boiling point of water (212°F [100°C]). ● **SEE FIGURE 10–20.**

Most mufflers have a larger inlet diameter than outlet diameter. As the exhaust enters the muffler, it expands and cools. The cooler exhaust is more dense and occupies less volume. The diameter of the outlet of the muffler and the diameter of the tailpipe can be reduced with no decrease in efficiency.

The tailpipe carries the exhaust gases from the muffler to the air, away from the vehicle. In most cases, the tailpipe exit is at the rear of the vehicle, below the rear bumper. In some cases, the exhaust is released at the side of the vehicle, just ahead of or just behind the rear wheel.

The muffler and tailpipe are supported with brackets called **hangers**. The hangers are made of rubberized fabric with metal ends that hold the muffler and tailpipe in position so that they do not touch any metal part. This helps to isolate the exhaust noise from the rest of the vehicle.

FIGURE 10–21 A high-performance aftermarket air filter often can increase airflow into the engine for more power.

TECH TIP

More Airflow = More Power

One of the most popular high-performance modifications is to replace the factory exhaust system with a low-restriction design and to replace the original air filter and air filter housing with a low-restriction unit as shown in ● **FIGURE 10–21.**

The installation of an aftermarket air filter not only increases power, but also increases air induction noise, which many drivers prefer. The aftermarket filter housing, however, may not be able to effectively prevent water from being drawn into the engine if the vehicle is traveling through deep water.

Almost every modification that increases performance has a negative effect on some other part of the vehicle, or else the manufacturer would include the change at the factory.

SUMMARY

1. All air entering an engine must be filtered.
2. Engines that use throttle-body injection units are equipped with intake manifolds that keep the airflow speed through the manifold at 50 to 300 feet per second.
3. Most intake manifolds have an EGR valve that regulates the amount of recirculated exhaust that enters the engine to reduce NO_x emissions.
4. Exhaust manifolds can be made from cast iron or stainless steel.
5. The exhaust system also contains a catalytic converter, exhaust pipes, and muffler. The entire exhaust system is supported by rubber hangers that isolate the noise and vibration of the exhaust from the rest of the vehicle.

REVIEW QUESTIONS

1. Why is it necessary to have intake charge velocities of about 50 feet per second?
2. Why can port fuel-injected engines use larger (and longer) intake manifolds and still operate at low engine speed?
3. What is a tuned runner in an intake manifold?
4. How does a muffler reduce exhaust noise?

CHAPTER QUIZ

1. Intake charge velocity has to be ________ to prevent fuel droplet separation.
   - a. 25 feet per second
   - b. 50 feet per second
   - c. 100 feet per second
   - d. 300 feet per second
2. The intake manifold of a port fuel-injected engine ________.
   - a. Uses a dual heat riser
   - b. Contains a leaner air–fuel mixture than does the intake manifold of a TBI system
   - c. Contains only fuel (gasoline)
   - d. Contains only air
3. Why are the EGR gases cooled before entering the engine on some engines?
   - a. Cool exhaust gas is more effective at controlling NO_X emissions
   - b. To help prevent the exhaust from slowing down
   - c. To prevent damage to the intake valve
   - d. To prevent heating the air–fuel mixture in the cylinder
4. A heated air intake system is usually necessary for proper cold-engine driveability on ________.
   - a. Port fuel-injection systems
   - b. Throttle-body fuel-injection systems
   - c. Both a port-injected and throttle-body-injected engine
   - d. Any fuel-injected engine
5. Air filters can remove particles and dirt as small as ________.
   - a. 5 to 10 microns
   - b. 10 to 25 microns
   - c. 30 to 40 microns
   - d. 40 to 50 microns
6. Why do many port fuel-injected engines use long intake manifold runners?
   - a. To reduce exhaust emissions
   - b. To heat the incoming air
   - c. To increase high-RPM power
   - d. To increase low-RPM torque
7. Exhaust passages are included in some intake manifolds. Technician A says that the exhaust passages are used for exhaust gas recirculation (EGR) systems. Technician B says that the exhaust heat is used to warm the intake charge on some engines equipped with a throttle-body-type fuel-injection system. Which technician is correct?
   - a. Technician A only
   - b. Technician B only
   - c. Both Technicians A and B
   - d. Neither Technician A nor B
8. The upper portion of a two-part intake manifold is often called the ________.
   - a. Housing
   - b. Lower part
   - c. Plenum
   - d. Vacuum chamber
9. Technician A says that a cracked exhaust manifold can affect engine operation. Technician B says that a leaking lower intake manifold gasket could cause a vacuum leak. Which technician is correct?
   - a. Technician A only
   - b. Technician B only
   - c. Both Technicians A and B
   - d. Neither Technician A nor B
10. Technician A says that some intake manifolds are plastic. Technician B says that some intake manifolds are constructed in two parts or sections: upper and lower. Which technician is correct?
    - a. Technician A only
    - b. Technician B only
    - c. Both Technicians A and B
    - d. Neither Technician A nor B

chapter 11

VARIABLE VALVE TIMING SYSTEMS

LEARNING OBJECTIVES

After studying this chapter, the reader will be able to:

1. List the reasons for variable valve timing.
2. Discuss the various types of variable valve timing.
3. Explain how to diagnose variable valve timing faults.

This chapter will help you prepare for Engine Repair (A8) ASE certification test content area "A" (General Engine Diagnosis).

KEY TERMS

Active fuel management (AFM) 184
Cylinder cutoff system 184
Displacement on demand (DOD) 184
EVCP 178
Ground side switching 183
MDS 184
Oil control valve (OCV) 176
Power side switching 183
PWM 181
Spline phaser 178
Vane phaser 178
Variable displacement system 184
VTEC 182
VVT 177

FIGURE 11–1 Camshaft rotation during advance and retard.

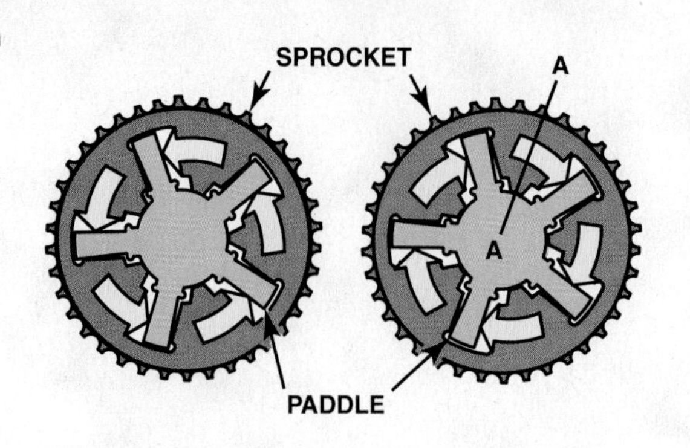

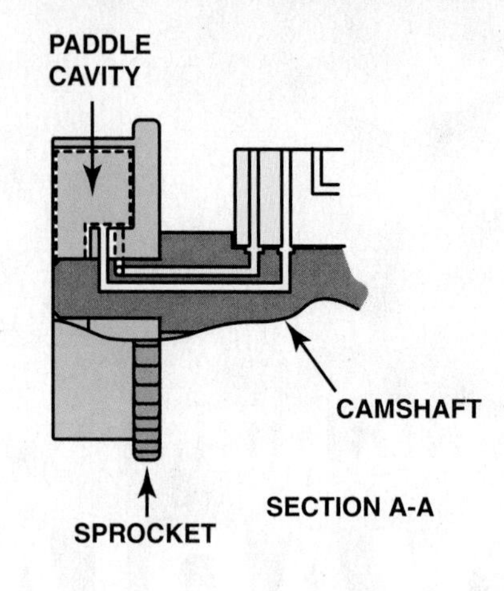

CAMSHAFT POSITION CHART			
DRIVING CONDITION	**CHANGE IN CAMSHAFT POSITION**	**OBJECTIVE**	**RESULT**
Idle	No change	Minimize valve overlap	Stabilize idle speed
Light engine load	Retard valve timing	Decrease valve overlap	Stable engine output
Medium engine load	Advance valve timing	Increase valve overlap	Better fuel economy with lower emissions
Low to medium RPM with heavy load	Advance valve timing	Advance intake valve closing	Improve low to midrange torque
High RPM with heavy load	Retard valve timing	Retard intake valve closing	Improve engine output

CHART 11–1

An overview of how variable valve timing is able to improve engine performance and reduce exhaust emissions.

PRINCIPLES OF VARIABLE VALVE TIMING

PURPOSE OF VARIABLE VALVE TIMING Conventional camshafts are permanently synchronized to the crankshaft so that they operate the valves at a specific point in each combustion cycle. In an engine, the intake valve opens slightly before the piston reaches the top of the cylinder and closes about 60 degrees after the piston reaches the bottom of the stroke on every cycle, regardless of the engine speed or load.

Variable cam timing allows the valves to be operated at different points in the combustion cycle, to improve performance.

There are three basic types of variable valve timing used on vehicles:

1. Exhaust camshaft variable action only on overhead camshaft engines, such as the inline 4.2 liter engine used in Chevrolet Trailblazers.
2. Intake and exhaust camshaft variable action on both camshafts used in many General Motors engines.
3. Overhead valve, cam-in-block engines use variable valve timing by changing the relationship of the camshaft to the crankshaft. ● **SEE CHART 11–1.**

PARTS AND OPERATION The camshaft position actuator **oil control valve (OCV)** directs oil from the oil feed in the head to the appropriate camshaft position actuator oil passages. There is one OCV for each camshaft position actuator. The OCV is sealed and mounted to the front cover. The ported end of the OCV is inserted into the cylinder head with a sliding fit. A filter screen protects each OCV oil port from any contamination in the oil supply.

The camshaft position actuator is mounted to the front end of the camshaft and the timing notch in the nose of the camshaft aligns with the dowel pin in the camshaft position actuator to ensure proper cam timing and camshaft position actuator oil hole alignment. ● **SEE FIGURE 11–1.**

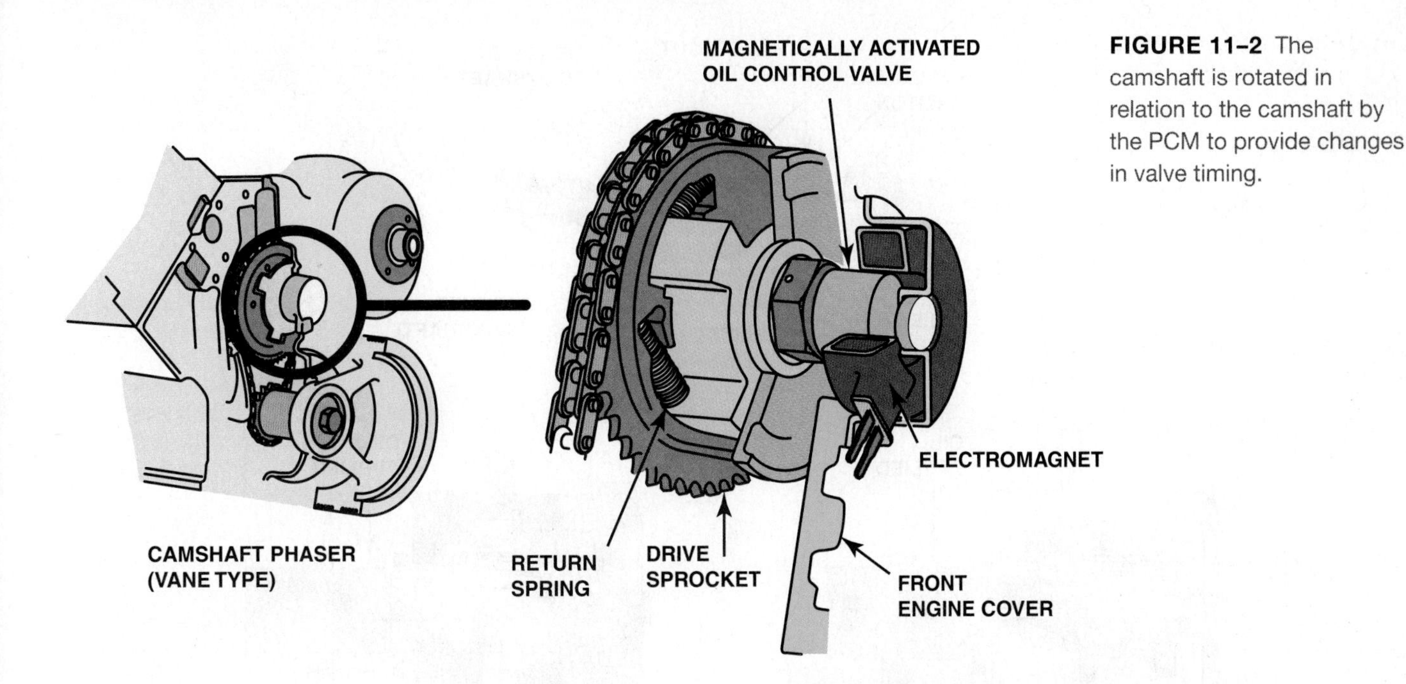

FIGURE 11–2 The camshaft is rotated in relation to the camshaft by the PCM to provide changes in valve timing.

FREQUENTLY ASKED QUESTION

What Are the Various Names Used for Variable Valve Timing Systems?

- **BMW-**VANOS (Variable Nockenwellen Steuerung)
- **Ford-**VVT (Variable Valve Timing)
- **GM-**DCVCP (Double Continuous Variable Cam Phasing), if used for both intake and exhaust camshafts
- **Honda-**VTEC (Variable valve Timing and lift Electronic Control)
- **Hyundai-**MPI CVVT (Multiport Injection Continuously Variable Valve Timing)
- **Mazda-**S-VT (Sequential Valve Timing)
- **Mitsubishi-**MIVECC (Mitsubishi Innovative Valve timing Electronic Control system)
- **Nissan-**N-VCTT (Nissan Variable Control Timing)
- **Nissan-**VVL (Variable Valve Lift)
- **Porsche-**variocam (Variable camshaft timing)
- **Suzuki-**VVT (Variable Valve Timing)
- **Subaru-**AVCS (Active Valve Control System)
- **Toyota-**VVT-i (Variable Valve Timing-intelligent)
- **Toyota-**VVTL-i (Variable Valve Timing and Lift-intelligent)
- **Volkswagen-**VVT (Variable Valve Timing)
- **Volvo-**VVT (Variable Valve Timing)

OHV VARIABLE TIMING

The GM 3900 V-6 engine is an example of an overhead-valve (OHV) cam-in-block engine that uses variable valve timing (VVT) and active fuel management (displacement on demand—DOD). Engine size was increased from 3.5 liters to 3.9 liters because the larger displacement was needed to obtain good performance in the three-cylinder mode.

The variable valve timing system uses electronically controlled, hydraulic gear-driven cam phaser that can alter the relationship of the camshaft from 15 degrees retard to 25 degrees advance (40 degrees overall) relative to the crankshaft. By using **variable valve timing (VVT)**, engineers were able to eliminate the EGR valve and were still able to meet the standards for oxides of nitrogen (NO_X). The VVT also works in conjunction with an active manifold that gives the engine a broader torque curve.

A valve in the intake manifold creates a longer path for intake air at low speeds, improving combustion efficiency and torque output. At higher speed the valve opens, creating a shorter air path for maximum power production.

● **SEE FIGURE 11–2.**

Varying the exhaust and/or the intake camshaft position allows for reduced exhaust emissions and improved performance. ● **SEE CHART 11–2.**

By varying the exhaust cam phasing, vehicle manufacturers are able to meet newer NO_X reduction standards and eliminate the exhaust gas recirculation (EGR) valve. By using exhaust cam phasing, the PCM can close the exhaust valves sooner than usual, thereby trapping some exhaust gases in the combustion chamber. General Motors uses one or two actuators that allow

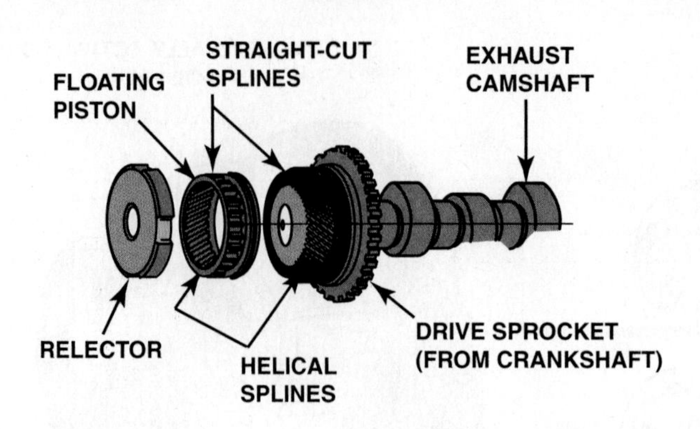

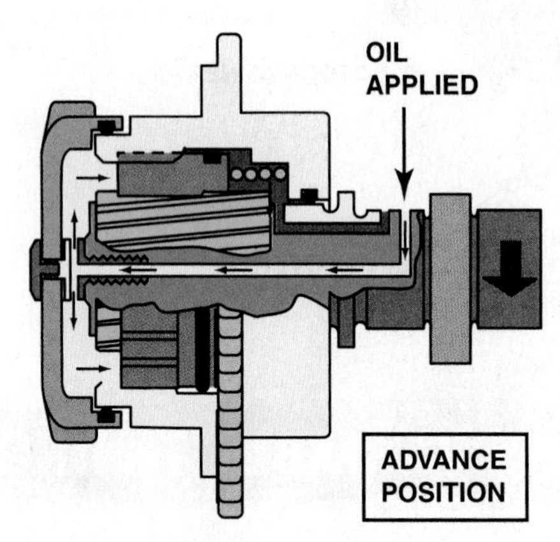

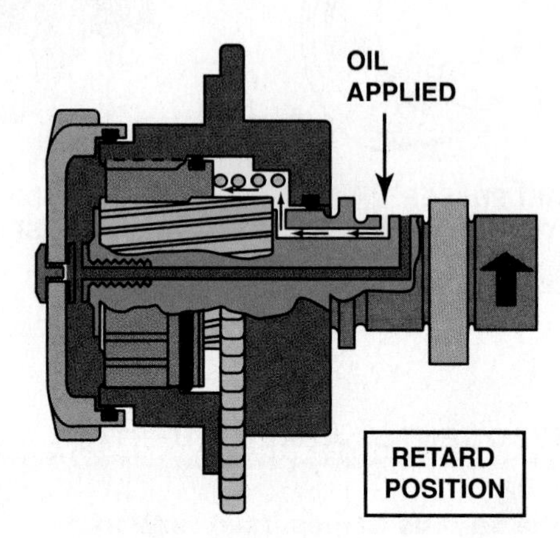

FIGURE 11–3 Spline cam phaser assembly.

INTAKE AND EXHAUST CAMSHAFT PHASING CHART	
CAMSHAFT PHASING CHANGED	**IMPROVES**
Exhaust cam phasing	Reduces NO_X exhaust emissions
Exhaust cam phasing	Increases fuel economy (reduced pumping losses)
Intake cam phasing	Increases low-speed torque
Intake cam phasing	Increases high-speed power

CHART 11–2

By varying the intake camshaft timing, engine performance is improved. By varying the exhaust camshaft timing, the exhaust emissions and fuel consumption are reduced.

the camshaft piston to change by up to 50 degrees in relation to the crankshaft position.

There are two types of cam phasing devices used on General Motors engines:

- **Spline phaser**—used on overhead camshaft (OHC) engines
- **Vane phaser**—used on overhead camshaft (OHC) and overhead valve (OHV) cam-in-block engines

SPLINE PHASER SYSTEM The spline phaser system is also called the **exhaust valve cam phaser (EVCP)** and consists of the following components:

- Engine control module (ECM)
- Four-way pulse-width-modulated (PWM) control valve
- Cam phaser assembly
- Camshaft position (CMP) sensor

● **SEE FIGURE 11–3.**

SPLINE PHASER SYSTEM OPERATION On the 4200 inline six-cylinder engine used in the Chevrolet Trailblazer, the pulse-width-modulated (PWM) control valve is located on the front passenger side of the cylinder head. Oil pressure is regulated by the control valve and then directed to the ports in the cylinder head leading to the camshaft and cam phaser position. The cam phaser is located on the exhaust cams and is part of the exhaust cam sprocket. When the ECM commands an increase in oil pressure, the piston is moved inside the cam phaser and rides along the helical splines, which compresses the coil spring. This movement causes the cam phaser gear and the camshaft to move in an opposite direction, thereby retarding the cam timing. ● **SEE FIGURE 11–4.**

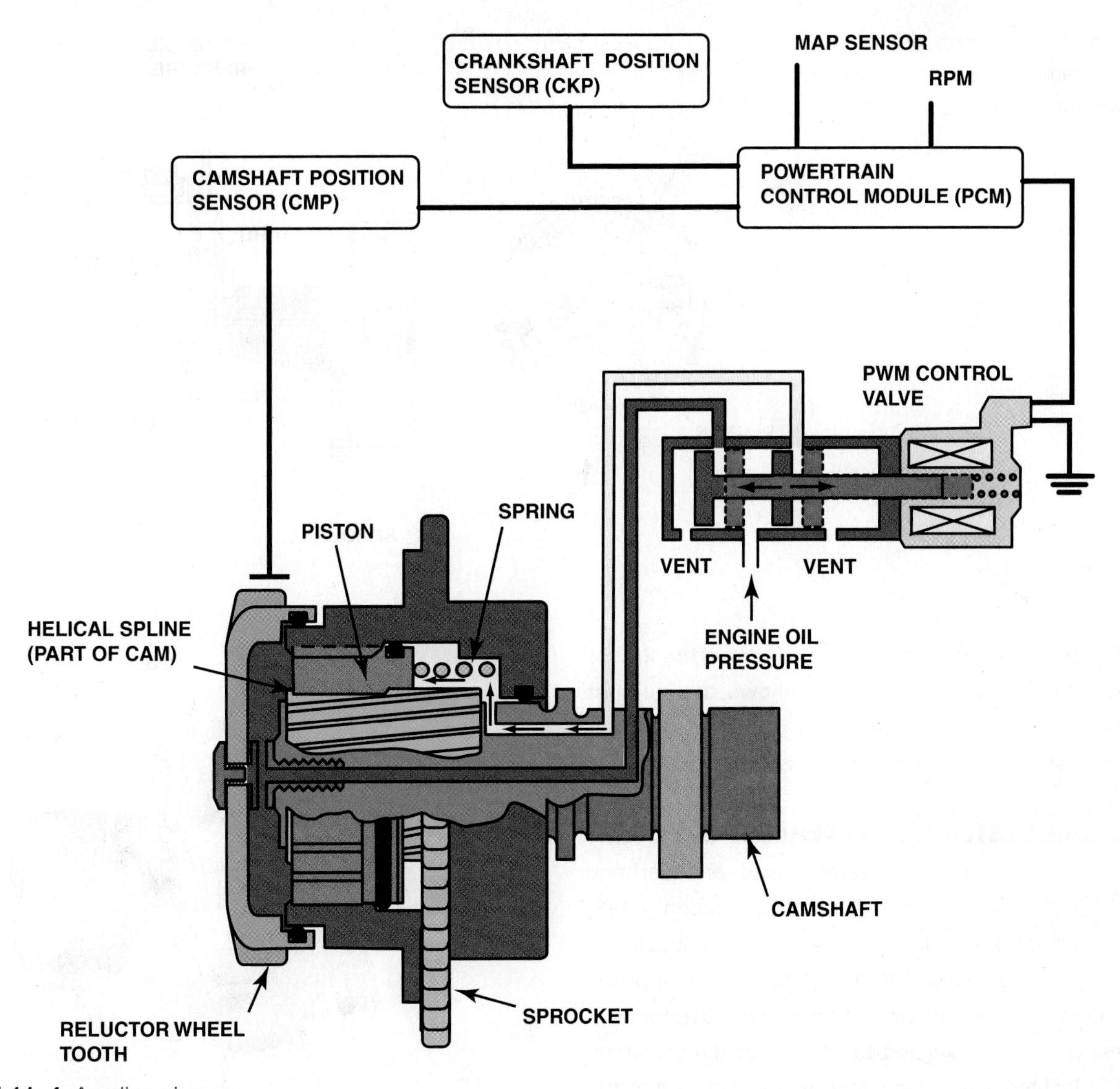

FIGURE 11–4 A spline phaser.

TECH TIP

Check the Screen on the Control Valve if There Are Problems

If a NO_x emission failure at a state inspection occurs or a diagnostic trouble code is set related to the cam timing, remove the control valve and check for a clogged oil screen. A lack of regular oil changes can cause the screen to become clogged, thereby preventing proper operation. A rough idle is a common complaint because the spring may not be able to return the camshaft to the idle position after a long highway trip. ● **SEE FIGURE 11–5.**

FIGURE 11–5 The screen(s) protect the solenoid valve from dirt and debris that can cause the valve to stick. This fault can set a P0017 diagnostic trouble code (crankshaft position–camshaft position correlation error).

FIGURE 11–6 A vane phaser is used to move the camshaft using changes of oil pressure from the oil control valve.

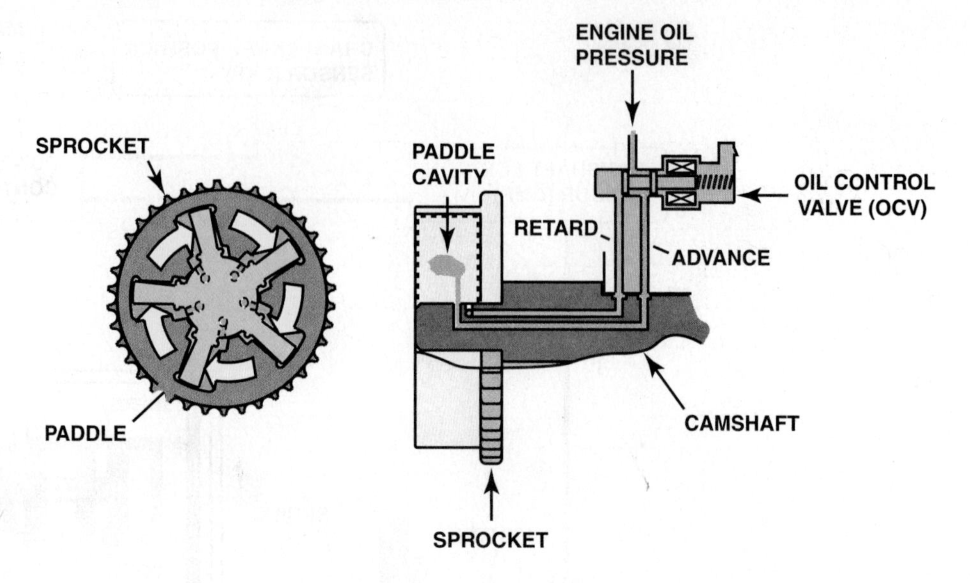

NOTE: A unique cam-within-a-cam is used on the 2008+ Viper V-10 OHV engine. This design allows the exhaust lobes to be moved by up to 36 degrees to improve idle quality and reduction of exhaust emissions.

VANE PHASER SYSTEM ON AN OVERHEAD CAMSHAFT ENGINE The vane phaser system used on overhead camshaft (OHC) engines uses a camshaft piston (CMP) sensor on each camshaft. Each camshaft has its own actuator and its own oil control valve (OCV). Instead of using a piston along a helical spline, the vane phaser uses a rotor with four vanes, which is connected to the end of the camshaft. The rotor is located inside the stator, which is bolted to the cam sprocket. The stator and rotor are not connected. Oil pressure is controlled on both sides of the vanes of the rotor, which creates a hydraulic link between the two parts. The oil control valve varies the balance of pressure on either side of the vanes and thereby controls the position of the camshaft. A return spring is used under the reluctor of the phaser to help return it to the home or zero degrees position. ● **SEE FIGURE 11–6.**

MAGNETICALLY CONTROLLED VANE PHASER A magnetically controlled vane phaser is controlled by the ECM by using a 12 volt pulse-width-modulated (PWM) signal to an electromagnet, which operates the oil control valve (OCV). A magnetically controlled vane phaser is used on many General Motors engines that use overhead camshafts on both the intake and exhaust. The OCV directs pressurized engine oil to either advance or retard chambers of the camshaft actuator to change the camshaft position in relation to the crankshaft position. ● **SEE FIGURE 11–7.**

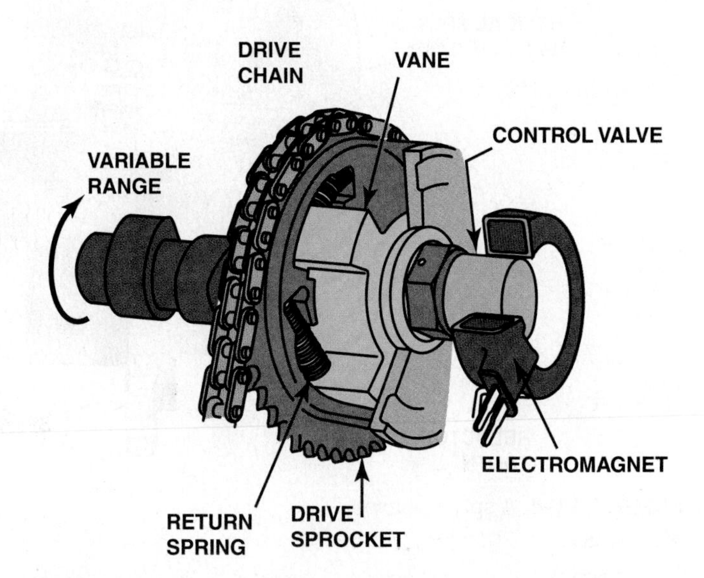

FIGURE 11–7 A magnetically controlled vane phaser.

The following occurs when the pulse width is changed:

- **0% pulse width—**The oil is directed to the advance chamber of the exhaust camshaft actuator and the retard chamber of the intake camshaft activator.
- **50% pulse width—**The PCM is holding the cam in the calculated position based on engine RPM and load. At 50% pulse width, the oil flow through the phaser drops to zero. ● **SEE FIGURE 11–8.**
- **100% pulse width—**The oil is directed to the retard chamber of the exhaust camshaft actuator and the advance chamber of the intake camshaft actuator.

The cam phasing is continuously variable with a range from 40 degrees for the intake camshaft and 50 degrees for

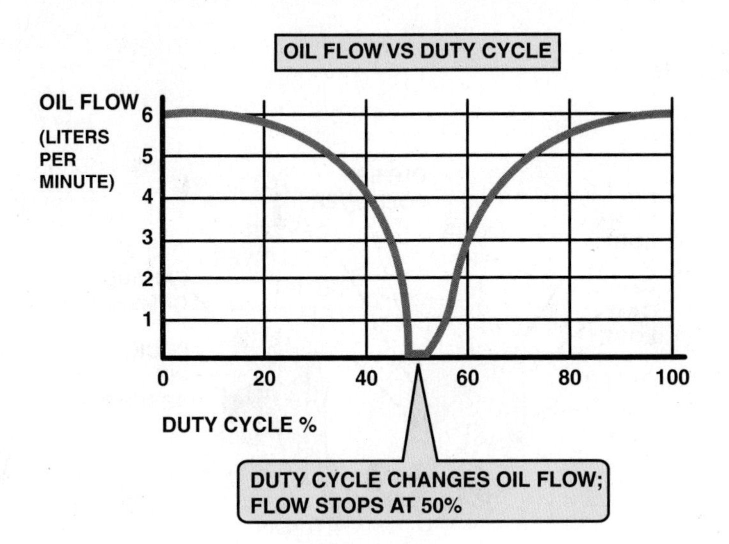

FIGURE 11–8 When the PCM commands 50% duty cycle, the oil flow through the phaser drops to zero.

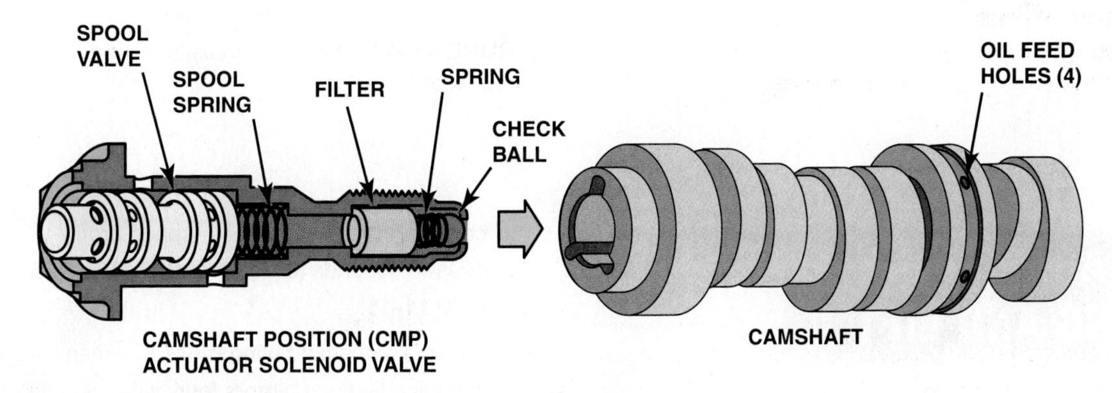

FIGURE 11–9 A camshaft position actuator used in a cam-in-block engine.

the exhaust camshaft. The PCM uses the following sensors to determine the best position of the camshaft for maximum power and lowest possible exhaust emissions:

- Engine speed (RPM)
- MAP sensor
- Crankshaft position (CKP) sensor
- Camshaft position (CMP) sensor
- Barometric pressure (BARO) sensor

CAM-IN-BLOCK ENGINE CAM PHASER Overhead valve engines that use a cam-in-block design use a magnetically controlled cam phaser to vary the camshaft in relation to the crankshaft. This type of phaser is not capable of changing the duration of valve opening or valve lift.

Inside the camshaft actuator is a rotor with vanes that are attached to the camshaft. Oil pressure is supplied to the vanes, which causes the camshaft to rotate in relation to the crankshaft. The camshaft actuator solenoid valve directs the flow of oil to either the advance or retard side vanes of the actuator. ● **SEE FIGURE 11–9.**

The ECM sends a **pulse-width-modulated (PWM)** signal to the camshaft actuator magnet. The movement of the pintle is used to direct oil flow to the actuator. The higher the duty cycle, the greater the movement in the valve position and change in camshaft timing.

? FREQUENTLY ASKED QUESTION

What Happens When the Engine Stops?

When the engine stops, the oil pressure drops to zero and a spring-loaded locking pin is used to keep the camshaft locked to prevent noise at engine start. When the engine starts, oil pressure releases the locking pin.

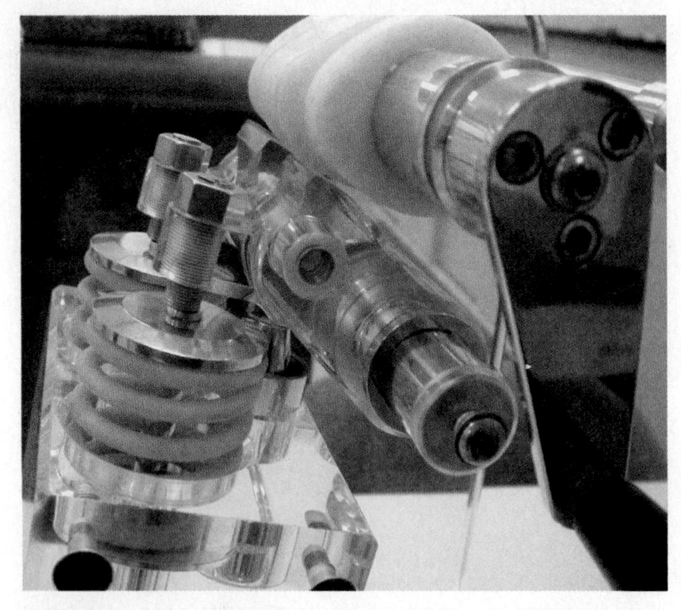

FIGURE 11–10 A plastic mock-up of a Honda VTEC system that uses two different camshaft profiles: one for low-speed engine operation and the other for high speed.

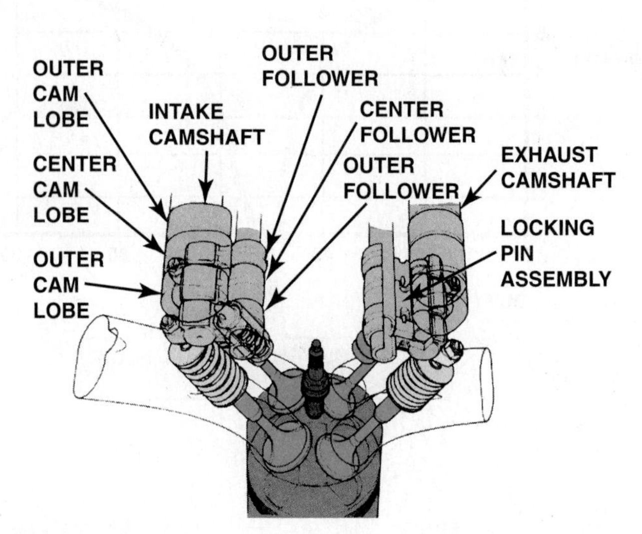

FIGURE 11–11 Engine oil pressure is used to switch cam lobes on a VTEC system.

VARIABLE VALVE TIMING AND LIFT

Many engines use variable valve timing in an effort to improve high-speed performance without the disadvantages of a high-performance camshaft at idle and low speeds. There are two basic systems including:

- Variable camshafts such as the system used by Honda/Acura called **Variable Valve Timing and Lift Electronic Control** or **VTEC**. This system uses two different camshafts for low and high RPM. When the engine is operating at idle and speeds below about 4000 RPM, the valves are opened by camshafts that are optimized by maximum torque and fuel economy. When engine speed reaches a predetermined speed, depending on the exact make and model, the computer turns on a solenoid, which opens a spool valve. When the spool valve opens, engine oil pressure pushes against pins that lock the three intake rocker arms together. With the rocker arms lashed, the valves must follow the profile of the high RPM cam lobe in the center. This process of switching from the low-speed camshaft profile to the high-speed profile takes about 100 milliseconds (0.1 sec). ● **SEE FIGURES 11–10 AND 11–11.**
- Variable camshaft timing is used on many engines including General Motors four-, five-, and six-cylinder engines, as well as engines from BMW, Chrysler, and Nissan. On a system that controls the intake camshaft only, the camshaft timing is advanced at low engine speed, closing the intake valves earlier to improve low RPM torque. At high engine speeds, the camshaft is retarded by using engine oil pressure against a helical gear to rotate the camshaft. When the camshaft is retarded, the intake valve closing is delayed, improving cylinder filling at higher engine speeds. ● **SEE FIGURE 11–12.** Variable cam timing can be used to control exhaust cam timing only. Engines that use this system, such as the 4.2 liter GM inline six-cylinder engines, can eliminate the EGR valve because the computer can close the exhaust valve sooner than normal, trapping some exhaust gases in the combustion chamber and therefore eliminating the need for an EGR valve. Some engines use variable camshaft timing on both intake and exhaust cylinder cams.

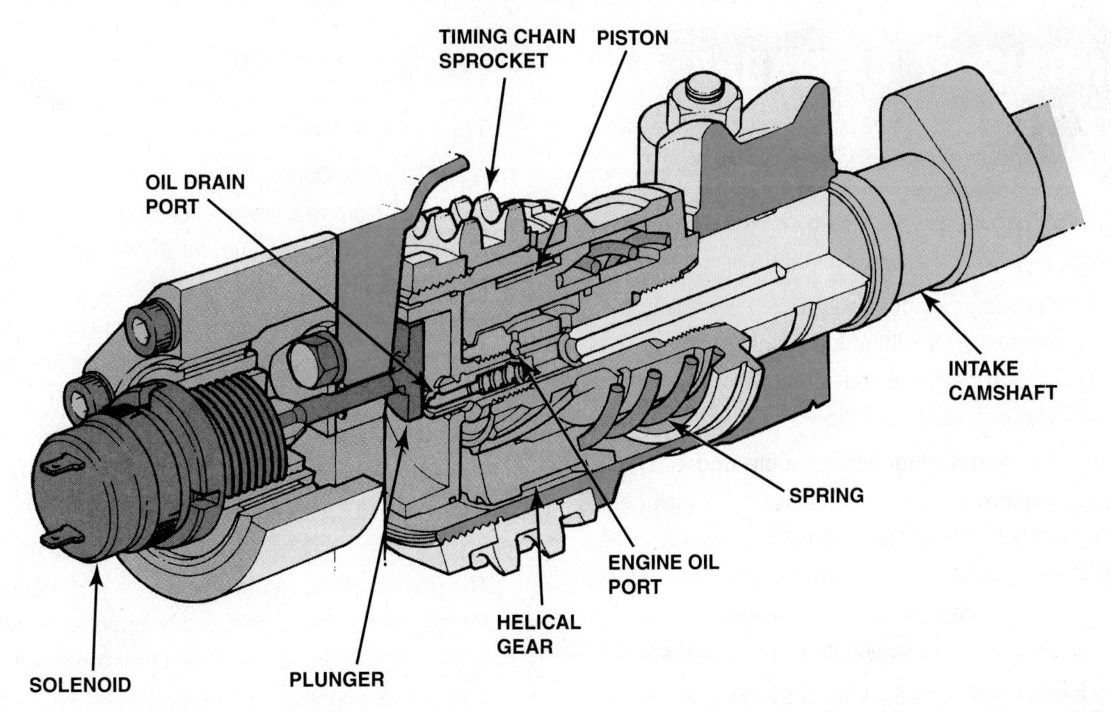

FIGURE 11–12 A typical variable cam timing control valve. The solenoid is controlled by the engine computer and directs engine oil pressure to move a helical gear, which rotates the camshaft relative to the timing chain sprocket.

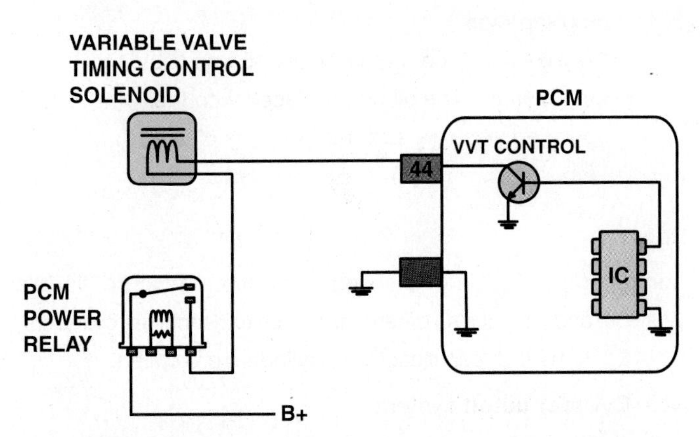

FIGURE 11–13 The schematic of a variable valve timing control circuit, showing that battery power (+) is being applied to the variable valve timing (VVT) solenoid and pulsed to ground by the PCM.

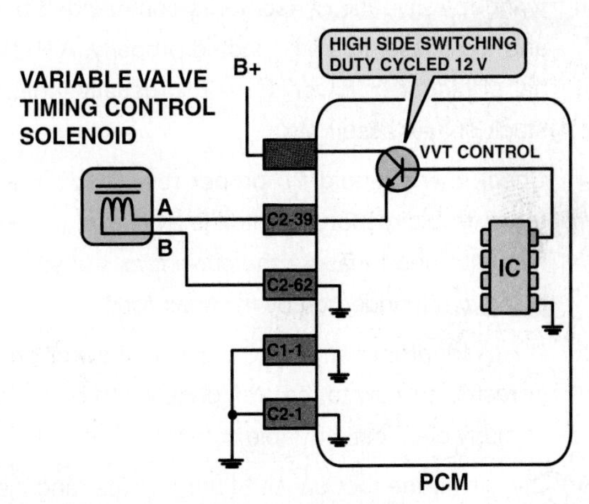

FIGURE 11–14 A variable valve timing solenoid being controlled by applying voltage from the PCM.

COMPUTER CONTROL OF VARIABLE VALVE TIMING

Variable valve timing is controlled by the Powertrain Control Module (PCM) and can be one of two different circuits:

- **Ground side switching** is the most commonly used. ● **SEE FIGURE 11–13**. The variable valve timing (VVT) solenoid usually has 3 to 6 ohms of resistance and therefore requires 2 to 4 amperes of current to operate.
- **Power side switching** is commonly found on General Motors vehicles and the solenoid has 8 to 12 ohms of resistance requiring 1.0 to 1.5 amperes of current to operate. ● **SEE FIGURE 11–14** for an example of a circuit showing high-side switching of the variable valve timing solenoid.

DIAGNOSIS OF VARIABLE VALVE TIMING SYSTEMS

The diagnostic procedure as specified by most vehicle manufacturers usually includes the following steps:

STEP 1 Verify the customer concern. This will usually be a check engine light (malfunction indicator light or MIL), as the engine performance effects would be minor under most operating conditions.

STEP 2 Check for stored diagnostic trouble codes (DTCs). Typical variable valve timing–related DTCs include:
P0011—Intake cam position is over advanced bank 1
P0021—Intake cam position is over advanced bank 2
P0012—Intake cam position is over retarded bank 1
P0022—Intake cam position is over retarded bank 2

STEP 3 Use a scan tool and check for duty cycle on the cam phase solenoid while operating the vehicle at a steady road speed. The commanded pulse width should be 50%. If the pulse width is not 50%, then the PCM is trying to move the phaser to its commanded position and the phaser has not reacted properly. A PWM signal of higher or lower than 50% usually indicates a stuck phaser assembly.

STEP 4 Check the solenoid for proper resistance. If a scan tool with bidirectional control is available, connect an ammeter and measure the current as the solenoid is being commanded on by the scan tool.

STEP 5 Check for proper engine oil pressure. Low oil pressure or restricted flow to the cam phaser can be the cause of many diagnostic trouble codes.

STEP 6 Determine the root cause of the problem and clear all DTCs.

STEP 7 Road-test the vehicle to verify the fault has been corrected.

CASE STUDY

The Case of the Wrong Oil

A 2007 Dodge Durango was in the shop for routine service, including a tire rotation and an oil change. Shortly after, the customer returned and stated that the "check engine" light was on. A scan tool was used to retrieve any diagnostic trouble codes. A P0521, "oil pressure not reaching specified value at 1250 RPM" was set. A check of service information showed that this code could be set if the incorrect viscosity engine oil was used. The shop had used SAE 10W-30 but the 5.7 liter Hemi V-8 with multiple displacement system (MDS) required SAE 5W-20 oil. The correct oil was installed and the DTC cleared. A thorough test-drive confirmed that the fault had been corrected and the shop learned that the proper viscosity oil is important to use in all vehicles.

Summary:

- **Complaint—**The check engine light was on after an oil change.
- **Cause—**The incorrect viscosity oil was used
- **Correction—**The oil was replaced with the specified viscosity and the DTC was cleared

VARIABLE DISPLACEMENT SYSTEMS

PURPOSE AND FUNCTION Some engines are designed to be operated on four of eight or three of six cylinders during low-load conditions to improve fuel economy. The powertrain computer monitors engine speed, coolant temperature, throttle position, and load. It also determines when to deactivate cylinders.

Systems that can deactivate cylinders are called:

- **Cylinder cutoff system**
- **Variable displacement system**
- **Displacement on Demand (DOD)** (now called **Active Fuel Management**) for General Motors
- **Multiple Displacement System (MDS)** for Chrysler

PARTS AND OPERATION The key to this process is the use of two-stage hydraulic valve lifters. In normal operation, the inner and outer lifter sleeves are held together by a pin and operate as an assembly. When the computer determines that the cylinder can be deactivated, oil pressure is delivered to a passage, which depresses the pin and allows the outer portion of the lifter to follow the contour of the cam while the inner portion remains stationary, keeping the valve closed. The electronic operation is achieved through the use of lifter oil manifold containing solenoids to control the oil flow, which is used to activate or deactivate the cylinders. ● **SEE FIGURES 11–15 AND 11–16.**

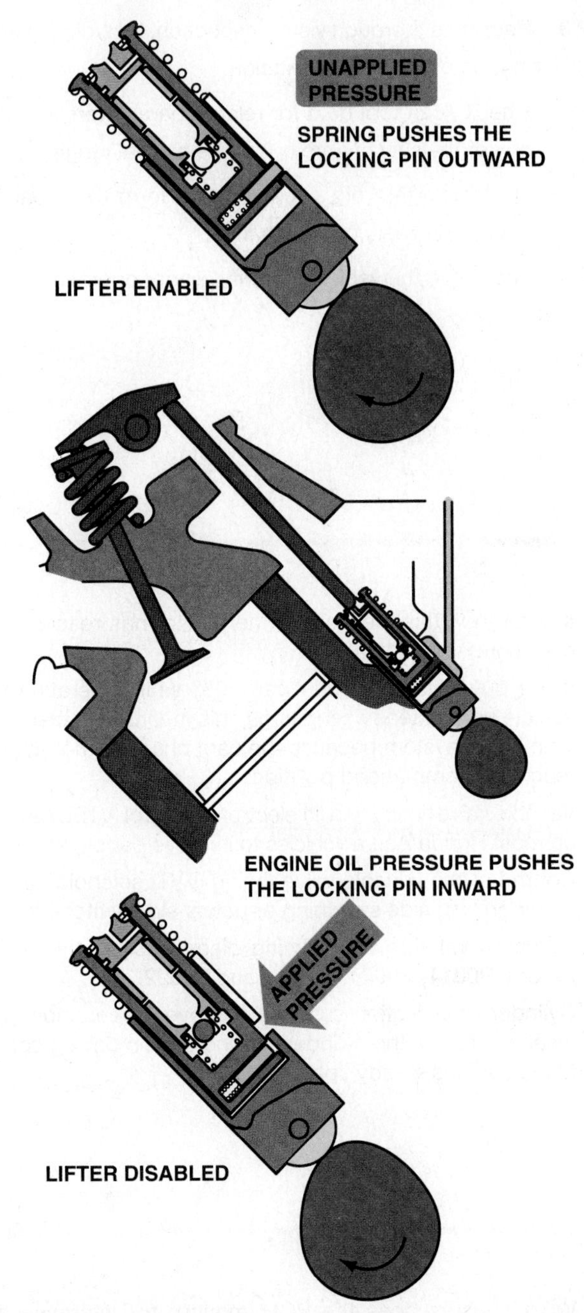

FIGURE 11–15 Oil pressure applied to the locking pin causes the inside of the lifter to freely move inside the outer shell of the lifter, thereby keeping the valve closed.

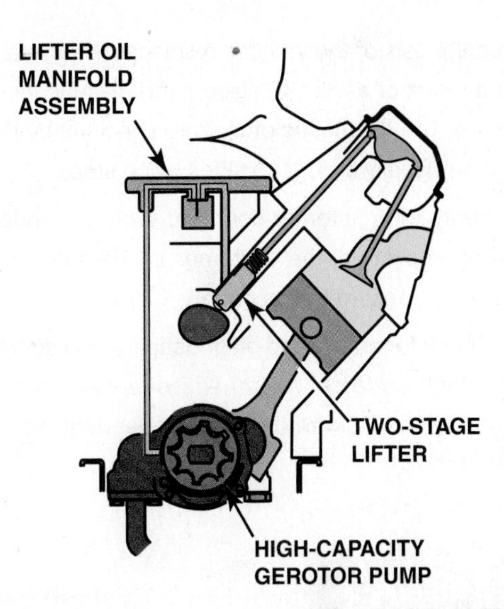

FIGURE 11–16 Active fuel management includes many different components and changes to the oiling system, which makes routine oil changes even more important on engines equipped with this system.

FIGURE 11–17 The driver information display on a Chevrolet Impala with a 5.3 liter V-8 equipped with active fuel management. The transition between four-cylinder mode and eight-cylinder mode is so smooth that most drivers are not aware that the switch is occurring.

CYLINDER DEACTIVATION SYSTEM DIAGNOSIS

A cylinder deactivation system, also called cylinder cutoff system or variable displacement system, often displays when the system is active on the driver information display. ● **SEE FIGURE 11–17.**

The diagnosis of the variable displacement system usually starts as a result of a check engine light (malfunction indicator lamp or MIL). The diagnostic procedure specified by the vehicle manufacturer usually includes the following steps:

STEP 1 Verify the customer concern. With a cylinder deactivation system, the customer concern could be lower than expected fuel economy.

STEP 2 Check for any stored diagnostic trouble codes (DTCs). A fault code set for an emission-related fault could cause the PCM to disable cylinder deactivation.

STEP 3 Perform a thorough visual inspection, including checking the oil level and condition.

STEP 4 Check scan tool data for related parameters to see if any of the sensors are out of the normal range.

STEP 5 Determine the root cause and perform the repair as specified in service information.

STEP 6 Test-drive the vehicle to verify proper operation.

SUMMARY

1. Variable valve timing is used to improve engine performance and reduce exhaust emissions.
2. Intake cam phasing is used to improve low-speed torque and high-speed power.
3. Exhaust cam phasing is used to reduce exhaust emissions and increase fuel economy by reducing pumping losses.
4. Variable valve timing on overhead valve, cam-in-block engines are used to reduce NO_X emissions.
5. Variable valve timing faults are often the result of extended oil change intervals, which can clog the screen on the cam phaser. As a result of the clogged screen, oil cannot flow to and adjust the valve timing, thereby setting valve timing-related diagnostic trouble codes (DTCs).
6. Oil flow to the phasers is controlled by the Powertrain Control Module (PCM). If a 50% duty cycle is shown on a scan tool, this means that the phaser has reached the commanded position.
7. If the duty cycle is other than 50% while operating the vehicle under steady conditions, this means that there is a fault in the system because the cam phaser is not able to reach the commanded position.
8. Variable valve timing and lift electronic control (VTEC) is used on most Honda/Acura vehicles to improve performance.
9. Control of the variable valve timing (VVT) solenoid can be either ground side switching or power side switching.
10. Common variable valve timing diagnostic trouble codes include P0011, P0021, P0012, and P0022.
11. Cylinder deactivation systems improve fuel economy by disabling half of the cylinders during certain driving conditions, such as steady speed cruising.

REVIEW QUESTIONS

1. What is the advantage of varying the intake camshaft timing?
2. What is the advantage of varying the exhaust camshaft timing?
3. Why must the engine oil be changed regularly on an engine equipped with variable valve timing?
4. What sensors does the PCM monitor to determine the best camshaft timing?
5. What diagnostic trouble codes are associated with the variable valve timing (VVT) system?

CHAPTER QUIZ

1. Variable valve timing can be found on which type of engines?
   a. Cam-in-block
   b. SOHC
   c. DOHC
   d. All of the above
2. To reduce oxides of nitrogen (NO_X) exhaust emissions, which camshaft is varied?
   a. Exhaust camshaft only
   b. Intake camshaft only
   c. Both the intake and exhaust camshaft
   d. The exhaust camshaft is advanced and the intake camshaft is advanced.
3. To increase engine performance, which camshaft is varied?
   a. Exhaust camshaft only
   b. Intake camshaft only
   c. Both the intake and exhaust camshaft
   d. The exhaust camshaft is advanced and the intake camshaft is advanced.
4. What is the commanded pulse width of the camshaft phaser that results in the desired position?
   a. 0%
   b. 25%
   c. 50%
   d. 100%
5. What sensors are used by the PCM to determine the best position of the camshafts for maximum power and lowest possible exhaust emissions?
   a. Engine speed (RPM)
   b. Crankshaft position (CKP) sensor
   c. Camshaft position (CMP) sensor
   d. All of the above
6. How is the camshaft actuator controlled?
   a. On only when conditions are right
   b. Pulse-width-modulated (PWM) signal
   c. Spring-loaded to the correct position based on engine speed
   d. Vacuum-controlled valve
7. If the engine oil is not changed regularly, what is the most likely fault that can occur to an engine equipped with variable valve timing (VVT)?
   a. Low oil pressure diagnostic trouble code (DTC)
   b. A no-start condition because the camshaft cannot rotate
   c. The filter screens on the actuator control valve become clogged
   d. Any of the above
8. How quickly can the rocker arms be switched from the low-speed camshaft profile to the high-speed camshaft profile on a Honda equipped with a VTEC system?
   a. 50 milliseconds
   b. 100 milliseconds
   c. 250 milliseconds
   d. 500 milliseconds
9. Which diagnostic trouble code (DTC) may be set if there is a problem with the bank 1 intake cam position?
   a. P0300
   b. P0420
   c. P0011 or P0012
   d. P0021 or P0022
10. If the incorrect grade of engine oil is used in an engine equipped with a variable displacement engine, what diagnostic trouble code (DTC) could be set?
    a. P0521
    b. P0300
    c. P0420
    d. P0011

chapter 12

TURBOCHARGING AND SUPERCHARGING

LEARNING OBJECTIVES

After studying this chapter, the reader will be able to:

1. Describe airflow requirements in automobile engines.
2. Explain the difference between a turbocharger and a supercharger.
3. Describe how the boost levels are controlled.
4. Discuss maintenance procedures for turbochargers and superchargers.

This chapter will help you prepare for Engine Repair (A8) ASE certification test content area "C" (Fuel, Air Induction and Exhaust Systems Diagnosis and Repair).

KEY TERMS

Bar 190
Boost 190
Bypass valve 192
Compressor bypass valve (CBV) 197
Dump valve 197
Intercooler 195
Naturally (normally) aspirated 190
Positive displacement 192
Roots-type Subcharger 192
Supercharging 190
Turbo lag 195
Turbocharger 193
Volumetric efficiency 189
Wastegate 195

FIGURE 12–1 A supercharger on a Ford V-8.

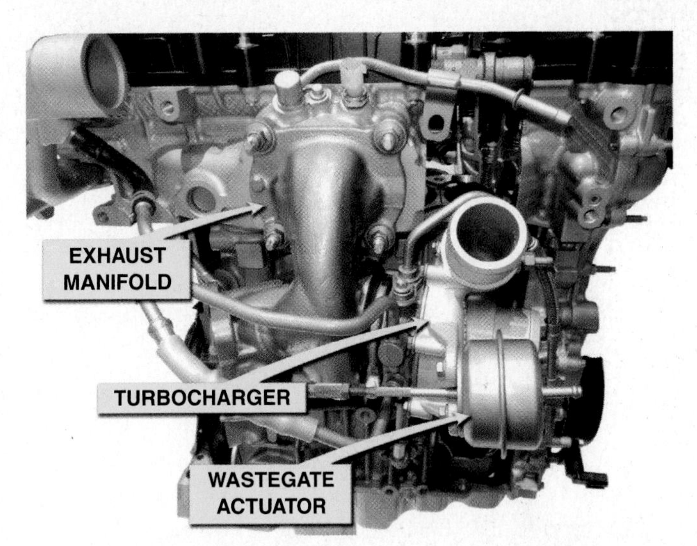

FIGURE 12–2 A turbocharged Ford three-cylinder 1.0 liter Eco Boost engine.

AIRFLOW REQUIREMENTS

Naturally aspirated engines with throttle bodies rely on atmospheric pressure to push an air–fuel mixture into the combustion chamber vacuum created by the downstroke of a piston. The mixture is then compressed before ignition to increase the force of the burning, expanding gases. The greater the mixture compression, the greater the power resulting from combustion.

All gasoline automobile engines share certain air–fuel requirements. For example, a four-stroke engine can take in only a fixed amount of air, and how much fuel it consumes depends on how much air it takes in. Engineers calculate engine airflow requirements using these three factors:

- Engine displacement
- Engine revolutions per minute (RPM)
- Volumetric efficiency

VOLUMETRIC EFFICIENCY **Volumetric efficiency** is a comparison of the actual volume of air–fuel mixture drawn into an engine to the theoretical maximum volume that could be drawn in. Volumetric efficiency is expressed as a percentage, and changes with engine speed. For example, an engine might have 75% volumetric efficiency at 1000 RPM. The same engine might be rated at 85% at 2000 RPM and 60% at 3000 RPM.

If the engine takes in the airflow volume slowly, a cylinder might fill to capacity. It takes a definite amount of time for the airflow to pass through all the curves of the intake manifold and valve port. Therefore, volumetric efficiency decreases as engine speed increases. At high speed, it may drop to as low as 50%.

The average street engine never reaches 100% volumetric efficiency. With a street engine, the volumetric efficiency is about 75% at maximum speed, or 80% at the torque peak. A high-performance street engine is about 85% efficient, or a bit more efficient at peak torque. A race engine usually has 95% or better volumetric efficiency. These figures apply only to naturally aspirated engines, however, and turbocharged and supercharged engines easily achieve more than 100% volumetric efficiency. Many vehicles are equipped with a supercharger or a turbocharger to increase power. ● **SEE FIGURES 12–1 AND 12–2.**

ENGINE COMPRESSION Higher compression increases the thermal efficiency of the engine because it raises compression temperatures, resulting in hotter, more complete combustion. However, a higher compression can cause an increase in NO_X emissions and would require the use of high-octane gasoline with effective antiknock additives.

FIGURE 12–3 The more air and fuel that can be packed in a cylinder, the greater the density of the air–fuel charge.

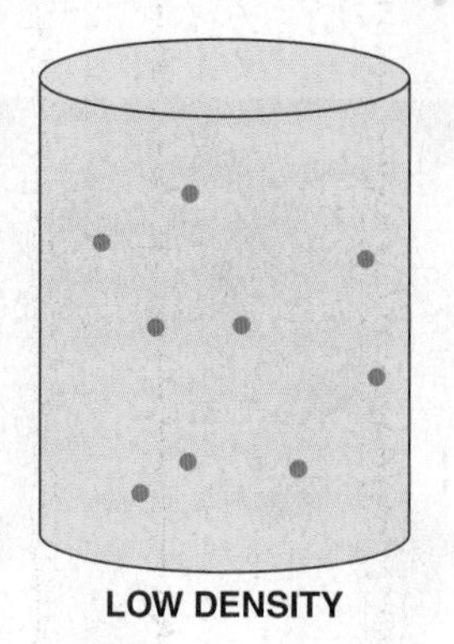

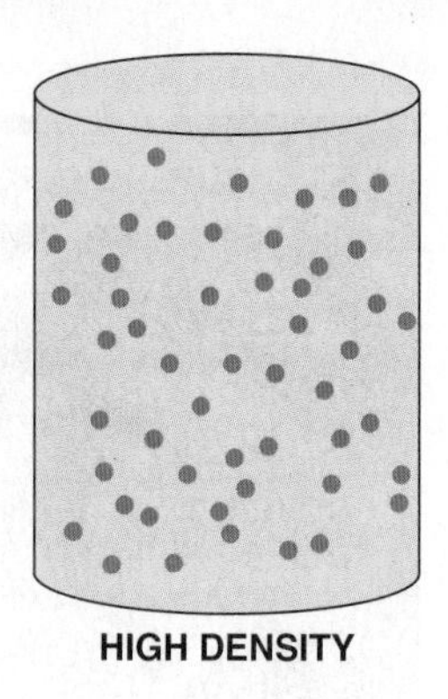

SUPERCHARGING PRINCIPLES

The amount of force an air–fuel charge produces when it is ignited is largely a function of the charge density. Density is the mass of a substance in a given amount of space. ● **SEE FIGURE 12–3.**

The greater the density of an air–fuel charge forced into a cylinder, the greater the force it produces when ignited, and the greater the engine power.

An engine that uses atmospheric pressure for intake is called a **naturally (normally) aspirated** engine. A better way to increase air density in the cylinder is to use a pump.

When air is pumped into the cylinder, the combustion chamber receives an increase of air pressure known as **boost** and is measured in pounds per square inch (PSI), atmospheres (ATM), or **bar.** While boost pressure increases air density, friction heats air in motion and causes an increase in temperature. This increase in temperature works in the opposite direction, decreasing air density. Because of these and other variables, an increase in pressure does not always result in greater air density.

Another way to achieve an increase in mixture compression is called **supercharging.** This method uses a pump to pack a denser air–fuel charge into the cylinders. Since the density of the air–fuel charge is greater, so is its weight—and power—is directly related to the weight of an air–fuel charge consumed within a given time period. The result is similar to that of a high-compression ratio, but the effect can be controlled during idle and deceleration to avoid high emissions.

Air is drawn into a naturally aspirated engine by atmospheric pressure forcing it into the low-pressure area of the intake manifold. The low pressure or vacuum in the manifold results from the reciprocating motion of the pistons. When a piston moves downward during its intake stroke, it creates an empty space, or vacuum, in the cylinder. Although atmospheric pressure pushes air to fill up as much of this empty space as possible, it has a difficult path to travel. The air must pass through the air filter, the throttle body, the manifold, and the intake port before entering the cylinder. Bends and restrictions in this pathway limit the amount of air reaching the cylinder before the intake valve closes; therefore, the volumetric efficiency is less than 100%.

Pumping air into the intake system under pressure forces it through the bends and restrictions at a greater speed than it would travel under normal atmospheric pressure, allowing more air to enter the intake port before it closes. By increasing the airflow into the intake, more fuel can be mixed with the air while still maintaining the same air–fuel ratio. The denser the air–fuel charge entering the engine during its intake stroke, the greater the potential energy released during combustion. In addition to the increased power resulting from combustion, there are several other advantages of supercharging an engine including:

- It increases the air–fuel charge density to provide high-compression pressure when power is required, but allows the engine to run on lower pressures when additional power is not required.
- The pumped air pushes the remaining exhaust from the combustion chamber during intake and exhaust valve overlap.
- The forced airflow and removal of hot exhaust gases lowers the temperature of the cylinder head, pistons, and valves, and helps extend the life of the engine.

A supercharger pressurizes air to greater than atmospheric pressure. The pressurization above atmospheric pressure, or boost, can be measured in the same way as atmospheric pressure. Atmospheric pressure drops as altitude increases, but boost pressure remains the same. If a supercharger develops

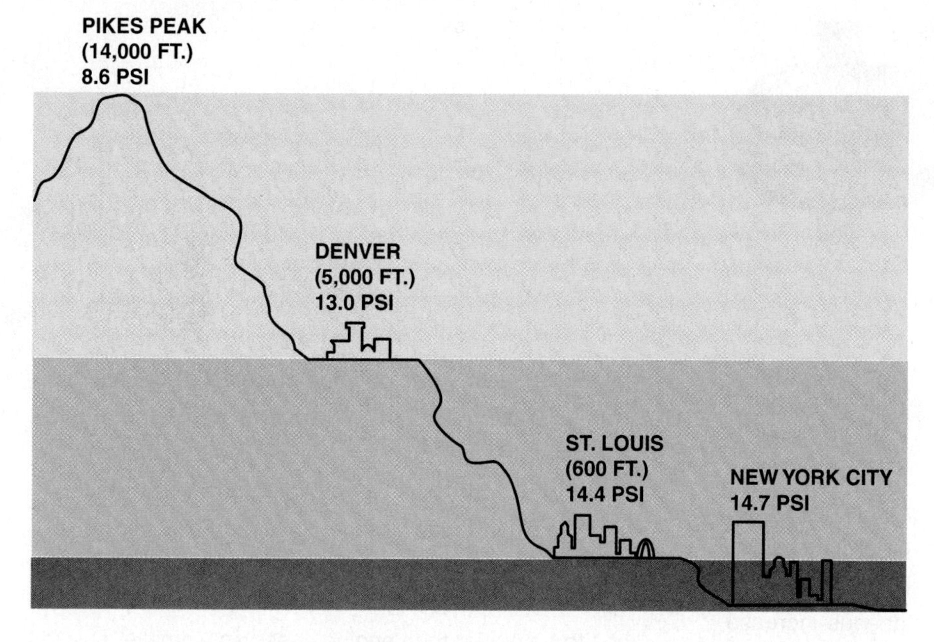

FIGURE 12–4 Atmospheric pressure decreases with increase in altitude.

FINAL COMPRESSION RATIO CHART AT VARIOUS BOOST LEVELS

	BLOWER BOOST (PSI)									
COMP RATIO	2	4	6	8	10	12	14	16	18	20
6.5	7.4	8.3	9.2	10	10.9	11.8	12.7	13.6	14.5	15.3
7	8	8.9	9.9	10.8	11.8	12.7	13.6	14.5	15.3	16.2
7.5	8.5	9.5	10.6	11.6	12.6	13.6	14.6	15.7	16.7	17.8
8	9.1	10.2	11.3	12.4	13.4	14.5	15.6	16.7	17.8	18.9
8.5	9.7	10.8	12	13.1	14.3	15.4	16.6	17.8	18.9	19.8
9	10.2	11.4	12.7	13.9	15.1	16.3	17.6	18.8	20	21.2
9.5	10.8	12.1	13.4	14.7	16	17.3	18.5	19.8	21.1	22.4
10	11.4	12.7	14.1	15.4	16.8	18.2	19.5	20.9	22.2	23.6

CHART 12–1

Equivalent compression ratios for listed boost levels.

12 PSI (83 kPa) boost at sea level, it will develop the same amount at a 5,000-foot altitude because boost pressure is measured inside the intake manifold. ● **SEE FIGURE 12–4.**

BOOST AND COMPRESSION RATIOS Boost increases the amount of air drawn into the cylinder during the intake stroke. This extra air causes the effective compression ratio to be greater than the mechanical compression ratio designed into the engine. The higher the boost pressure, the greater the compression ratio. ● **SEE CHART 12–1** for an example of how much the effective compression ratio is increased compared to the boost pressure.

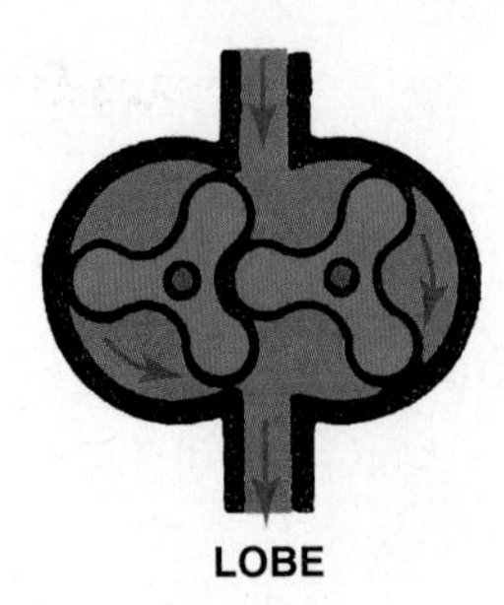

FIGURE 12–5 A roots-type supercharger uses two lobes to force the air around the outside of the housing and forces it into the intake manifold.

SUPERCHARGERS

PURPOSE AND FUNCTION A supercharger is an engine-driven air pump that supplies more than the normal amount of air into the intake manifold and boosts engine torque and power. A supercharger provides an instantaneous increase in power without the delay or lag often associated with turbochargers. However, a supercharger, because it is driven by the engine, does require horsepower to operate and is not as efficient as a turbocharger.

In basic concept, a supercharger is nothing more than an air pump mechanically driven by the engine itself. Gears, shafts, chains, or belts from the crankshaft can be used to turn the pump. This means that the air pump or supercharger pumps air in direct proportion to engine speed.

TYPES OF SUPERCHARGERS There are two general types of superchargers:

- **Roots-type supercharger.** Named for Philander and Francis Roots, two brothers from Connersville, Indiana, who patented the design in 1860 as a type of water pump to be used in mines. Later it was used to move air and is used today on two-stroke cycle Detroit diesel engines and other supercharged engines. The **roots-type supercharger** is called a **positive displacement** design because all of the air that enters is forced through the unit. Examples of a roots-type supercharger include the GMC 6-71 (used originally on GMC diesel engines that had six cylinders each with 71 cu. inch) and Eaton used on supercharged 3800 V-6 General Motors engines. ● **SEE FIGURE 12–5.**
- **Centrifugal supercharger.** A centrifugal supercharger is similar to a turbocharger but is mechanically driven by the engine instead of being powered by the hot exhaust gases. A centrifugal supercharger is not a positive displacement pump and all of the air that enters is not forced through the unit. Air enters a centrifugal supercharger housing in the center and exits at the outer edges of the compressor wheels at a much higher speed due to centrifugal force. The speed of the blades has to be higher than engine speed so a smaller pulley is used on the supercharger and the crankshaft overdrives the impeller through an internal gear box achieving about seven times the speed of the engine. Examples of centrifugal superchargers include Vortech and Paxton.

SUPERCHARGER BOOST CONTROL Many factory-installed superchargers are equipped with a **bypass valve** that allows intake air to flow directly into the intake manifold bypassing the supercharger. The computer controls the bypass valve actuator. ● **SEE FIGURE 12–6.**

The airflow is directed around the supercharger whenever any of the following conditions occurs:

- The boost pressure, as measured by the MAP sensor, indicates that the intake manifold pressure is reaching the predetermined boost level.
- During deceleration.
- Whenever reverse gear is selected.

SUPERCHARGER SERVICE Superchargers are usually lubricated with synthetic engine oil inside the unit. This oil level should be checked and replaced as specified by the vehicle or supercharger manufacturer. The drive belt should also be inspected and replaced as necessary.

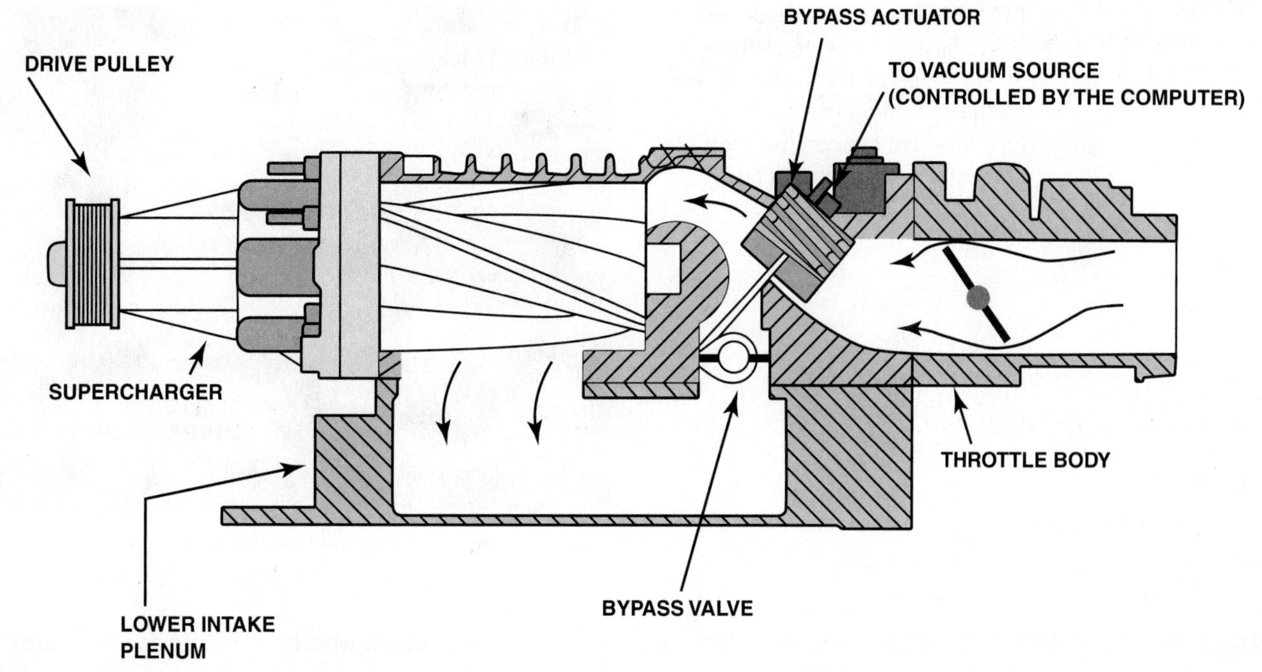

FIGURE 12–6 The bypass actuator opens the bypass valve to control boost pressure.

TURBOCHARGERS

PURPOSE AND FUNCTION The major disadvantage of a supercharger is its reliance on engine power to drive the unit. In some installations, as much as 20% of the engine's power is used by a mechanical supercharger. However, by connecting a centrifugal supercharger to a turbine drive wheel and installing it in the exhaust path, the lost engine horsepower is regained to perform other work and the combustion heat energy lost in the engine exhaust (as much as 40% to 50%) can be harnessed to do useful work. This is the concept of a **turbocharger**.

The turbocharger's main advantage over a mechanically driven supercharger is that the turbocharger does not drain power from the engine. In a naturally aspirated engine, about half of the heat energy contained in the fuel goes out the exhaust system. ● **SEE FIGURE 12–7**. Another 25% is lost through radiator cooling. Only about 25% is actually converted to mechanical power. A mechanically driven pump uses some of this mechanical output, but a turbocharger gets its energy from the exhaust gases, converting more of the fuel's heat energy into mechanical energy.

A turbocharger turbine looks much like a typical centrifugal pump used for supercharging. ● **SEE FIGURE 12–8**. Hot exhaust gases flow from the combustion chamber on to the turbine wheel. The gases heat and expand as they leave the engine. It is not the force of the exhaust gases that makes the turbine wheel to turn, as is commonly thought, but the expansion of hot gases against the turbine wheel's blades.

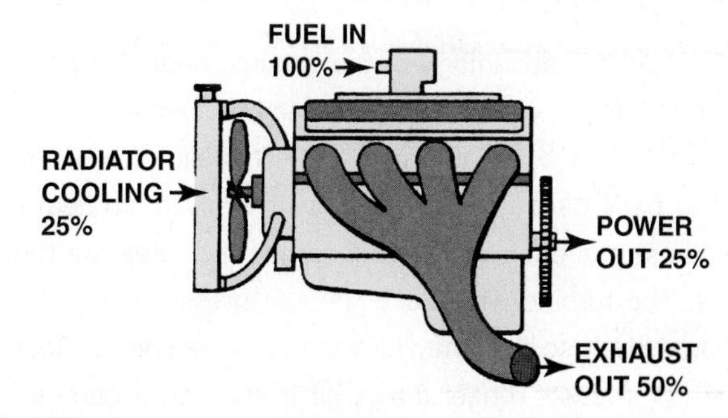

FIGURE 12–7 A turbocharger uses some of the heat energy that would normally be wasted.

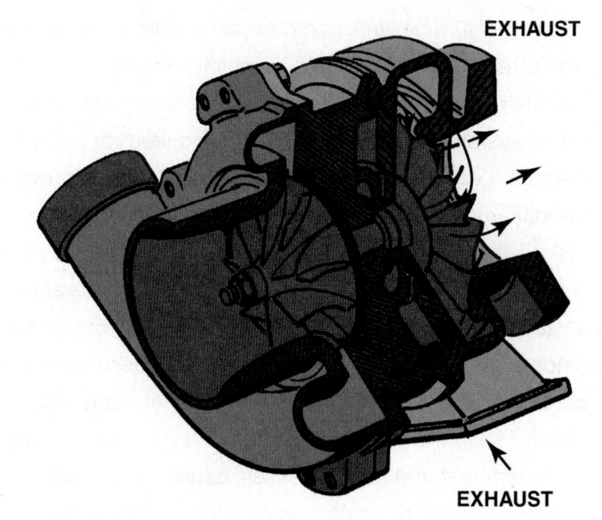

FIGURE 12–8 A turbine wheel is turned by the expanding exhaust gases.

FIGURE 12–9 The exhaust drives the turbine wheel on the left, which is connected to the impeller wheel on the right through a shaft. The bushings that support the shaft are lubricated with engine oil under pressure.

TURBOCHARGER DESIGN AND OPERATION A turbocharger consists of two chambers connected by a center housing. The two chambers contain a turbine wheel and a compressor wheel connected by a shaft that passes through the center housing.

To take full advantage of the exhaust heat that provides the rotating force, a turbocharger must be positioned as close as possible to the exhaust manifold. This allows the hot exhaust to pass directly into the unit with a minimum heat loss. As exhaust gas enters the turbocharger, it rotates the turbine blades. The turbine wheel and the compressor wheel are on the same shaft so that they turn at the same speed. Rotation of the compressor wheel draws air in through a central inlet and centrifugal force pumps it through an outlet at the edge of the housing. A pair of bearings in the center housing support the turbine and compressor wheel shaft, and are lubricated by engine oil. ● **SEE FIGURE 12–9**.

Both the turbine and compressor wheels must operate with extremely close clearances to minimize possible leakage around their blades. Any leakage around the turbine blades causes a dissipation of the heat energy required for compressor rotation. Leakage around the compressor blades prevents the turbocharger from developing its full boost pressure.

When the engine is started and runs at low speed, both exhaust heat and pressure are low and the turbine runs at a low speed (approximately 1000 RPM). Because the compressor does not turn fast enough to develop boost pressure, air simply passes through it and the engine works like any naturally aspirated engine. As the engine runs faster or load increases, both exhaust heat and flow increases, causing the turbine and compressor wheels to rotate faster. Since there is no brake and very little rotating resistance on the turbocharger shaft, the turbine and compressor wheels accelerate as the exhaust heat energy increases. When an engine is running at full power, the typical turbocharger rotates at speeds between 100000 and 150000 RPM.

Engine deceleration from full power to idle requires only a second or two because of its internal friction, pumping resistance, and drivetrain load. The turbocharger, however, has no such load on its shaft, and is already turning many times faster than the engine at top speed. As a result, it can take as much as a minute or more after the engine has returned to idle speed before the turbocharger also has returned to idle. If the engine is decelerated to idle and then shut off immediately, engine lubrication stops flowing to the center housing bearings while the turbocharger is still spinning at thousands of RPM. The oil in the center housing is then subjected to extreme heat and can gradually "coke" or oxidize. The coked oil can clog passages and will reduce the life of the turbocharger.

The high rotating speeds and extremely close clearances of the turbine and compressor wheels in their housings require equally critical bearing clearances. The bearings must keep radial clearances of 0.003 to 0.006 inches (0.08 mm to 0.15 mm). Axial clearance (end play) must be maintained at 0.001 to 0.003 inches (0.025 mm to 0.08 mm). If properly maintained, the turbocharger also is a trouble-free device. However, to prevent problems, the following conditions must be met:

- The turbocharger bearings must be constantly lubricated with clean engine oil—turbocharged engines should have regular oil changes at half the time or mileage intervals specified for nonturbocharged engines.
- Dirt particles and other contamination must be kept out of the intake and exhaust housings.

FIGURE 12–10 The unit on top of this Subaru that looks like a radiator is the intercooler, which cools the air after it has been compressed by the turbocharger.

- Whenever a basic engine bearing (crankshaft or camshaft) has been damaged, the turbocharger must be flushed with clean engine oil after the bearing has been replaced.
- If the turbocharger is damaged, the engine oil must be drained and flushed and the oil filter replaced as part of the repair procedure.

Late-model turbochargers all have liquid-cooled center bearings to prevent heat damage. In a liquid-cooled turbocharger, engine coolant is circulated through passages cast in the center housing to draw off the excess heat. This allows the bearings to run cooler and minimize the probability of oil coking when the engine is shut down.

TURBOCHARGER SIZE AND RESPONSE TIME A time lag occurs between an increase in engine speed and the increase in the speed of the turbocharger. This delay between acceleration and turbo boost is called **turbo lag.** Like any material, moving exhaust gas has inertia. Inertia also is present in the turbine and compressor wheels, as well as the intake airflow. Unlike a supercharger, the turbocharger cannot supply an adequate amount of boost at low speed.

Turbocharger response time is directly related to the size of the turbine and compressor wheels. Small wheels accelerate rapidly; large wheels accelerate slowly. While small wheels would seem to have an advantage over larger ones, they may not have enough airflow capacity for an engine. To minimize turbo lag, the intake and exhaust breathing capacities of an engine must be matched to the exhaust and intake airflow capabilities of the turbocharger.

BOOST CONTROL

Both supercharged and turbocharged systems are designed to provide a pressure greater than atmospheric pressure in the intake manifold. This increased pressure forces additional amounts of air into the combustion chamber over what would normally be forced in by atmospheric pressure. This increased charge increases engine power. The amount of "boost" (or pressure in the intake manifold) is measured in pounds per square inch (PSI), in inches of mercury (in. Hg), in bar, or in atmospheres. The following values will vary due to altitude and weather conditions (barometric pressure).

1 atmosphere = 14.7 PSI

1 atmosphere = 29.50 inch Hg

1 atmosphere = 1.0 bar

1 bar = 14.7 PSI

The higher the level of boost (pressure), the greater the horsepower potential. However, other factors must be considered when increasing boost pressure:

1. As boost pressure increases, the temperature of the air also increases.
2. As the temperature of the air increases, combustion temperatures also increase, which increases the possibility of detonation.
3. Power can be increased by cooling the compressed air after it leaves the turbocharger. *The power can be increased about 1% per 10°F by which the air is cooled.* A typical cooling device is called an **intercooler** and is similar to a radiator, wherein outside air can pass through, cooling the pressurized heated air. ● **SEE FIGURE 12–10.**

   Some intercoolers use engine coolant to cool the hot compressed air that flows from the turbocharger to the intake.
4. As boost pressure increases, combustion temperature and pressures increase, which, if not limited, can do severe engine damage. The maximum exhaust gas temperature must be 1,550°F (840°C). Higher temperatures decrease the durability of the turbocharger *and* the engine.

WASTEGATE A turbocharger uses exhaust gases to increase boost, which causes the engine to make more exhaust gases, which in turn increases the boost from the turbocharger. To prevent overboost and severe engine damage, most turbocharger systems use a wastegate. A **wastegate** is a valve similar to a door that can open and close. The wastegate is a bypass valve at the exhaust inlet to the turbine. It allows all of

FIGURE 12–11 A wastegate is used on the first-generation Duramax diesel to control maximum boost pressure.

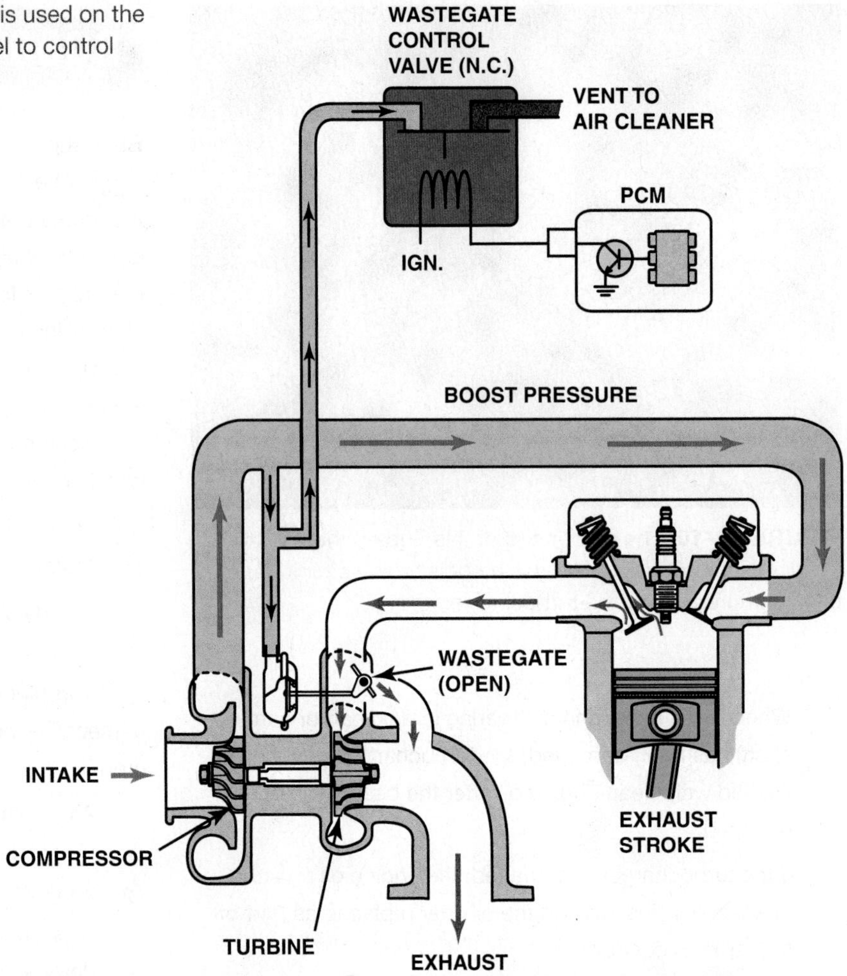

the exhaust into the turbine, or it can route part of the exhaust past the turbine to the exhaust system. If the valve is closed, all of the exhaust travels to the turbocharger. When a predetermined amount of boost pressure develops in the intake manifold, the wastegate valve is opened. As the valve opens, most of the exhaust flows directly out the exhaust system, bypassing the turbocharger. With less exhaust flowing across the vanes of the turbocharger, the turbocharger decreases in speed and boost pressure is reduced. When the boost pressure drops, the wastegate valve closes to direct the exhaust over the turbocharger vanes and again allow the boost pressure to rise. Wastegate operation is a continuous process to control boost pressure.

The wastegate is the pressure control valve of a turbocharger system. The wastegate is usually controlled by the onboard computer through a boost control solenoid. ● **SEE FIGURE 12–11.**

TECH TIP

Boost Is the Result of Restriction

The boost pressure of a turbocharger (or supercharger) is commonly measured in pounds per square inch. If a cylinder head is restricted because of small valves and ports, the turbocharger will quickly provide boost. Boost results when the air being forced into the cylinder heads cannot flow into the cylinders fast enough and "piles up" in the intake manifold, increasing boost pressure. If an engine had large valves and ports, the turbocharger could provide a much greater *amount* of air into the engine at the same boost pressure as an identical engine with smaller valves and ports. Therefore, by increasing the size of the valves, a turbocharged or supercharged engine will be capable of producing much greater power.

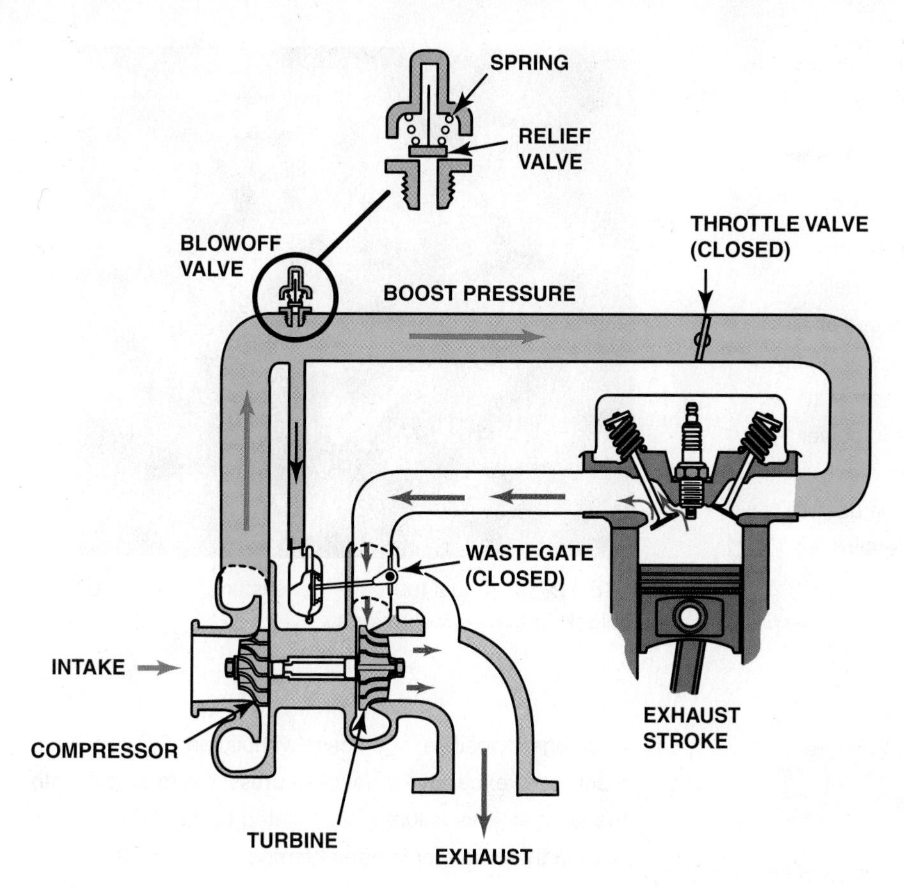

FIGURE 12–12 A blow-off valve is used in some turbocharged systems to relieve boost pressure during deceleration.

RELIEF VALVES A wastegate controls the exhaust side of the turbocharger. A relief valve controls the intake side. A relief valve vents pressurized air from the connecting pipe between the outlet of the turbocharger and the throttle whenever the throttle is closed during boost, such as during shifts. If the pressure is not released, the turbocharger turbine wheel will slow down, creating a lag when the throttle is opened again after a shift has been completed. There are two basic types of relief valves including:

- **Compressor bypass valve (CBV)** This type of relief valve routes the pressurized air to the inlet side of the turbocharger for reuse and is quiet during operation.
- **Blow-off valve (BOV)** This is also called a **dump valve** or vent valve and features an adjustable spring design that keeps the valve closed until a sudden release of the throttle. The resulting pressure increase opens the valve and vents the pressurized air directly into the atmosphere. This type of relief valve is noisy in operation and creates a whooshing sound when the valve opens. ● **SEE FIGURE 12–12**.

TECH TIP

If One Is Good, Two Are Better

A turbocharger uses the exhaust from the engine to spin a turbine, which is connected to an impeller inside a turbocharger. This impeller then forces air into the engine under pressure higher than is normally achieved without a turbocharger. The more air that can be forced into an engine, the greater the power potential. A V-type engine has two exhaust manifolds and so two small turbochargers can be used to help force greater quantities of air into an engine, as shown in **FIGURE 12–13**.

FIGURE 12–13 A dual turbocharger system installed on a small block Chevrolet V-8 engine.

TURBOCHARGER FAILURES

When turbochargers fail to function correctly, a drop in power is noticed. To restore proper operation, the turbocharger must be rebuilt, repaired, or replaced. It is not possible to simply remove the turbocharger, seal any openings, and still maintain decent driveability. Bearing failure is a common cause of turbocharger failure, and replacement bearings are usually only available to rebuilders. Another common turbocharger problem is excessive and continuous oil consumption resulting in blue exhaust smoke. Turbochargers use small rings similar to piston rings on the shaft to prevent exhaust (combustion gases) from entering the central bearing. Because there are no seals to keep oil in, excessive oil consumption is usually caused by:

1. A plugged positive crankcase ventilation (PCV) system resulting in excessive crankcase pressures forcing oil into the air inlet. This failure is not related to the turbocharger, but the turbocharger is often blamed.
2. A clogged air filter, which causes a low-pressure area in the inlet, which can draw oil past the turbo shaft rings and into the intake manifold.
3. A clogged oil return (drain) line from the turbocharger to the oil pan (sump), which can cause the engine oil pressure to force oil past the turbocharger's shaft rings and into the intake *and* exhaust manifolds. Obviously, oil being forced into both the intake and exhaust would create lots of smoke.

SUMMARY

1. Volumetric efficiency is a comparison of the actual volume of air–fuel mixture drawn into the engine to the theoretical maximum volume that can be drawn into the cylinder.
2. A supercharger operates from the engine by a drive belt and, while it does consume some engine power, it forces a greater amount of air into the cylinders for even more power.
3. A turbocharger uses the normally wasted heat energy of the exhaust to turn an impeller at high speed. The impeller is linked to a turbine wheel on the same shaft and is used to force air into the engine.
4. There are two types of superchargers: roots-type and centrifugal.
5. A bypass valve is used to control the boost pressure on most factory-installed superchargers.
6. An intercooler is used on many turbocharged and some supercharged engines to reduce the temperature of air entering the engine for increased power.
7. A wastegate is used on most turbocharger systems to limit and control boost pressures, as well as a relief valve, to keep the speed of the turbine wheel from slowing down during engine deceleration.

REVIEW QUESTIONS

1. What are the reasons why supercharging increases engine power?
2. How does the bypass valve work on a supercharged engine?
3. What are the advantages and disadvantages of supercharging?
4. What are the advantages and disadvantages of turbocharging?
5. What turbocharger control valves are needed for proper engine operation?

CHAPTER QUIZ

1. Boost pressure is generally measured in _________.
   a. in. Hg
   b. PSI
   c. in. H_2O
   d. in. lb
2. Two types of superchargers include _________.
   a. Rotary and reciprocating
   b. Roots-type and centrifugal
   c. Double and single acting
   d. Turbine and piston
3. Which valve is used on a factory supercharger to limit boost?
   a. A bypass valve
   b. A wastegate
   c. A blow-off valve
   d. An air valve
4. How are most superchargers lubricated?
   a. By engine oil under pressure through lines from the engine
   b. By an internal oil reservoir
   c. By greased bearings
   d. No lubrication is needed because the incoming air cools the supercharger
5. How are most turbochargers lubricated?
   a. By engine oil under pressure through lines from the engine
   b. By an internal oil reservoir
   c. By greased bearings
   d. No lubrication is needed because the incoming air cools the turbocharger
6. Two technicians are discussing the term "turbo lag." Technician A says that it refers to the delay between when the exhaust leaves the cylinder and when it contacts the turbine blades of the turbocharger. Technician B says that it refers to the delay in boost pressure that occurs when the throttle is first opened. Which technician is correct?
   a. Technician A only
   b. Technician B only
   c. Both Technicians A and B
   d. Neither Technician A nor B
7. What is the purpose of an intercooler?
   a. To reduce the temperature of the air entering the engine
   b. To cool the turbocharger
   c. To cool the engine oil on a turbocharged engine
   d. To cool the exhaust before it enters the turbocharger
8. Which type of relief valve used on a turbocharged engine is noisy?
   a. A bypass valve
   b. A BOV
   c. A dump valve
   d. Both b and c
9. Technician A says that a stuck-open wastegate can cause the engine to burn oil. Technician B says that a clogged PCV system can cause the engine to burn oil. Which technician is correct?
   a. Technician A only
   b. Technician B only
   c. Both Technicians A and B
   d. Neither Technician A nor B
10. What service operation is *most* important on engines equipped with a turbocharger?
    a. Replacing the air filter regularly
    b. Replacing the fuel filter regularly
    c. Regular oil changes
    d. Regular exhaust system maintenance

chapter 13

ENGINE CONDITION DIAGNOSIS

LEARNING OBJECTIVES

After studying this chapter, the reader will be able to:

1. Diagnose typical engine-related complaints.
2. List the visual checks to determine engine condition.
3. Discuss engine noise and its relation to engine condition.
4. Describe how to perform a dry and wet compression test.
5. Explain how to perform a cylinder leakage test.
6. Perform a cylinder balance test, and determine necessary action.
7. Perform an exhaust system back-pressure test, and determine necessary action.
8. Perform engine absolute (vacuum/boost) manifold pressure tests, and determine necessary action.

This chapter will help you prepare for Engine Performance (A8) ASE certification test content area "A" (General Engine Diagnosis).

KEY TERMS

Back pressure 214
Compression test 208
Cranking vacuum test 212
Cylinder leakage test 210
Dynamic compression test 210
Idle vacuum test 212
Inches of mercury (inches Hg) 212
Paper test 207
Power balance test 211
Restricted exhaust 214
Running compression test 210
Vacuum tests 212
Wet compression test 209

If there is an engine operation problem, then the cause could be any one of many components, including the engine itself. The condition of the engine should be tested anytime the operation of the engine is not satisfactory.

TYPICAL ENGINE-RELATED COMPLAINTS

Many driveability problems are *not* caused by engine mechanical problems. A thorough inspection and testing of the ignition and fuel systems should be performed before testing for mechanical engine problems.

Typical engine mechanical-related complaints include the following:

- Excessive oil consumption
- Engine misfiring
- Loss of power
- Smoke from the engine or exhaust
- Engine noise
- Excessive blowby ● **SEE FIGURE 13–1**.

FIGURE 13–1 Blowby gases coming out of the crankcase vent hose. Excessive amounts of combustion gases flow past the piston rings and into the crankcase.

ENGINE SMOKE DIAGNOSIS

The color of engine exhaust smoke can indicate what engine problem might exist. ● **SEE CHART 13–1**.

FIGURE 13–2 White steam is usually an indication of a blown (defective) cylinder head gasket that allows engine coolant to flow into the combustion chamber where it is turned to steam.

TYPICAL EXHAUST SMOKE COLOR	POSSIBLE CAUSES
Blue	Blue exhaust indicates that the engine is burning oil. Oil is getting into the combustion chamber either past the piston rings or past the valve stem seals. Blue smoke only after start-up is usually due to defective valve stem seals.
Black	Black exhaust smoke is due to excessive fuel being burned in the combustion chamber. Typical causes include a defective or misadjusted throttle body, leaking fuel injector, or excessive fuel-pump pressure.
White (steam)	White smoke or steam from the exhaust is normal during cold weather and represents condensed steam. Every engine creates about 1 gallon of water for each gallon of gasoline burned. If the steam from the exhaust is excessive, then water (coolant) is getting into the combustion chamber. Typical causes include a defective cylinder head gasket, a cracked cylinder head, or in severe cases a cracked block. ● **SEE FIGURE 13–2.**
Note: White smoke can also be created when automatic transmission fluid (ATF) is burned. A common source of ATF getting into the engine is through a defective vacuum modulator valve used on older automatic transmissions.	

CHART 13–1

Exhaust smoke colors and possible causes.

THE DRIVER IS YOUR BEST RESOURCE

The driver of the vehicle knows a lot about the vehicle and how it is driven. *Before* diagnosis is started, always ask the following questions:

- When did the problem first occur?
- Under what conditions does it occur?
  1. When cold or when hot?
  2. During acceleration, cruise, or deceleration?
  3. How far was it driven?
  4. What recent repairs have been performed?

After the nature and scope of the problem are determined, the complaint should be verified before further diagnostic tests are performed.

TECH TIP

Your Nose Knows

Whenever diagnosing any vehicle try to use all senses including the smell. Some smells and their cause include:

- **Gasoline:** If the exhaust smells like gasoline or unburned fuel, then a fault with the ignition system is a likely cause. Unburned fuel due to lean air–fuel mixture causing a lean misfire is also possible.
- **Sweet smell:** A coolant leak often gives off a sweet smell especially if the leaking coolant flows onto the hot exhaust.
- **Exhaust smell:** Check for an exhaust leak including a possible cracked exhaust manifold, which can be difficult to find because it often does not make noise.

VISUAL CHECKS

The first and most important "test" that can be performed is a careful visual inspection.

OIL LEVEL AND CONDITION The first area for visual inspection is oil level and condition.

1. Oil level—oil should be to the proper level
2. Oil condition
   a. Using a match or lighter, try to light the oil on the dipstick; if the oil flames up, gasoline is present in the engine oil.
   b. Drip some of the engine oil from the dipstick onto the hot exhaust manifold. If the oil bubbles or boils, there is coolant (water) in the oil.
   c. Check for grittiness by rubbing the oil between your fingers.

COOLANT LEVEL AND CONDITION Most mechanical engine problems are caused by overheating. The proper operation of the cooling system is critical to the life of any engine.

NOTE: Check the coolant level in the radiator only if the radiator is cool. If the radiator is hot and the radiator cap is removed, the drop in pressure above the coolant will cause the coolant to boil immediately and can cause severe burns when the coolant explosively expands upward and outward from the radiator opening.

The cooling system should be inspected for the following conditions:

1. The coolant level in the coolant recovery container should be within the limits indicated on the overflow bottle. If this level is too low or the coolant recovery container is empty, then check the level of coolant in the radiator (only when cool) and also check the operation of the pressure cap.
2. The coolant should be checked with a hydrometer for boiling and freezing temperature. This test indicates if the concentration of the antifreeze is sufficient for proper protection.
3. Pressure-test the cooling system and look for leakage. Coolant leakage can often be seen around hoses or cooling system components because it will often cause:
   a. A grayish white stain
   b. A rusty color stain
   c. Dye stains from antifreeze (greenish or yellowish depending on the type of coolant)
4. Check for cool areas of the radiator indicating clogged sections.
5. Check operation and condition of the fan clutch, fan, and coolant pump drive belt.

OIL LEAKS Oil leaks can lead to severe engine damage if the resulting low oil level is not corrected. Besides causing an oily mess where the vehicle is parked, the oil leak can cause

FIGURE 13–3 What looks like an oil pan gasket leak can be a rocker cover gasket leak. Always look up and look for the highest place you see oil leaking; that should be repaired first.

FIGURE 13–4 The transmission and flexplate (flywheel) were removed to check the exact location of this oil leak. The rear main seal and/or the oil pan gasket could be the cause of this leak.

TECH TIP

What's Leaking?

The color of the leaks observed under a vehicle can help the technician determine and correct the cause. Some leaks, such as condensate (water) from the air-conditioning system, are normal, whereas a brake fluid leak is very dangerous. The following are colors of common leaks:

Sooty Black	Engine Oil
Yellow, green, blue, or orange	Antifreeze (coolant)
Red	Automatic transmission fluid
Murky brown	Brake or power steering fluid or very neglected antifreeze (coolant)
Clear	Air-conditioning condensate (water) (normal)

blue smoke to occur under the hood as leaking oil drips on the exhaust system. *Finding* the location of the oil leak can often be difficult. ● **SEE FIGURES 13–3 AND 13–4**. To help find the source of oil leaks follow these steps:

STEP 1 Clean the engine or area around the suspected oil leak. Use a high-powered hot-water spray to wash the engine. While the engine is running, spray the entire engine and the engine compartment. Avoid letting the water come into direct contact with the air inlet and ignition distributor or ignition coil(s).

NOTE: If the engine starts to run rough or stalls when the engine gets wet, then the secondary ignition wires (spark plug wires) or distributor cap may be defective or have weak insulation. Be certain to wipe all wires and the distributor cap dry with a soft, dry cloth if the engine stalls.

An alternative method is to spray a degreaser on the engine, then start and run the engine until warm. Engine heat helps the degreaser penetrate the grease and dirt. Use a water hose to rinse off the engine and engine compartment.

STEP 2 If the oil leak is not visible or oil seems to be coming from "everywhere," use a white talcum powder. The leaking oil will show as a dark area on the white powder. See the Tech Tip, "The Foot Powder Spray Trick."

STEP 3 Fluorescent dye can be added to the engine oil. Add about 1/2 oz (15 cc) of dye per 5 quarts of engine oil. Start the engine and allow it to run about 10 minutes to thoroughly mix the dye throughout the engine. A black

FIGURE 13–5 Using a black light to spot leaks after adding dye to the oil.

light can then be shown around every suspected oil leak location. The black light will easily show all oil leak locations because the dye will show as a bright yellow/green area. ● **SEE FIGURE 13–5.**

NOTE: Fluorescent dye works best with clean oil.

TECH TIP

The Foot Powder Spray Trick

The source of an oil or other fluid leak is often difficult to determine. A quick and easy method that works is the following. First, clean the entire area. This can best be done by using a commercially available degreaser to spray the entire area. Let it soak to loosen all accumulated oil and greasy dirt. Clean off the degreaser with a water hose. Let the area dry. Start the engine, and using spray foot powder or other aerosol powder product, spray the entire area. The leak will turn the white powder dark. The exact location of any leak can be quickly located.

NOTE: Most oil leaks appear at the bottom of the engine due to gravity. Look for the highest, most forward location for the source of the leak.

FIGURE 13–6 An accessory belt tensioner. Most tensioners have a mark that indicates normal operating location. If the belt has stretched, this indicator mark will be outside of the normal range. Anything wrong with the belt or tensioner can cause noise.

ENGINE NOISE DIAGNOSIS

An engine knocking noise is often difficult to diagnose. ● **SEE CHART 13–2**. Several items that can cause a deep engine knock include:

- **Valves clicking.** This can happen because of lack of oil to the lifters. This noise is most noticeable at idle when the oil pressure is the lowest.
- **Torque converter.** The attaching bolts or nuts may be loose on the flex plate. This noise is most noticeable at idle or when there is no load on the engine.
- **Cracked flex plate.** The noise of a cracked flex plate is often mistaken for a rod- or main-bearing noise.
- **Loose or defective drive belts or tensioners.** If an accessory drive belt is loose or defective, the flopping noise often sounds similar to a bearing knock. ● **SEE FIGURE 13–6.**
- **Piston pin knock.** This knocking noise is usually not affected by load on the cylinder. If the clearance is too great, a double knock noise is heard when the engine idles. If all cylinders are grounded out one at a time and the noise does not change, a defective piston pin could be the cause.
- **Piston slap.** A piston slap is usually caused by an undersized or improperly shaped piston or oversized cylinder bore. A piston slap is most noticeable when the engine is cold and tends to decrease or stop making noise as the piston expands during engine operation.

TYPICAL NOISES	POSSIBLE CAUSES
Clicking noise—like the clicking of a ballpoint pen	1. Loose spark plug 2. Loose accessory mount (for air-conditioning compressor, alternator, power steering pump, etc.) 3. Loose rocker arm 4. Worn rocker arm pedestal 5. Fuel pump (broken mechanical fuel pump return spring) 6. Worn camshaft 7. Exhaust leak. ● **SEE FIGURE 13–7.**
Clacking noise—like tapping on metal	1. Worn piston pin 2. Broken piston 3. Excessive valve clearance 4. Timing chain hitting cover
Knock—like knocking on a door	1. Rod bearing(s) 2. Main bearing(s) 3. Thrust bearing(s) 4. Loose torque converter 5. Cracked flex plate (drive plate)
Rattle—like a baby rattle	1. Manifold heat control valve 2. Broken harmonic balancer 3. Loose accessory mounts 4. Loose accessory drive belt or tensioner
Clatter—like rolling marbles	1. Rod bearings 2. Piston pin 3. Loose timing chain
Whine—like an electric motor running	1. Alternator bearing 2. Drive belt 3. Power steering 4. Belt noise (accessory or timing)
Clunk—like a door closing	1. Engine mount 2. Drive axle shaft U-joint or constant velocity (CV) joint

CHART 13–2

Engine-related noises and possible causes.

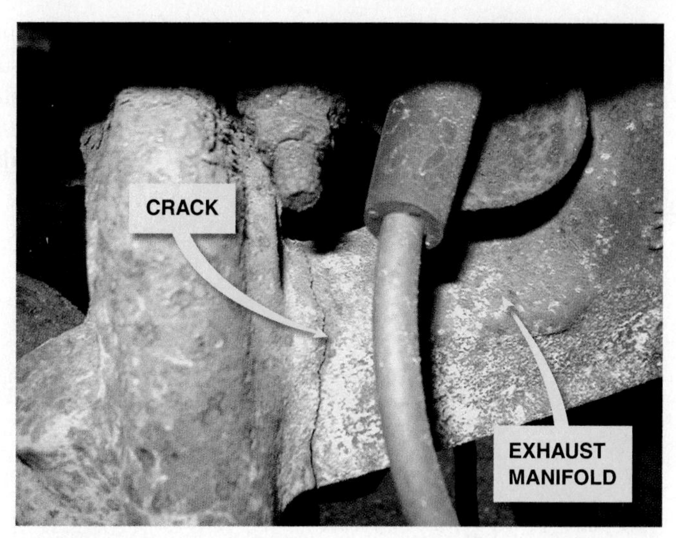

FIGURE 13–7 A cracked exhaust manifold on a Ford V-8.

- **Timing chain noise.** An excessively loose timing chain can cause a severe knocking noise when the chain hits the timing chain cover. This noise can often sound like a rod-bearing knock.
- **Rod-bearing noise.** The noise from a defective rod bearing is usually load sensitive and changes in intensity as the load on the engine increases and decreases. A rod-bearing failure can often be detected by grounding out the spark plugs one cylinder at a time. If the knocking noise decreases or is eliminated when a particular cylinder is grounded (disabled), then the grounded cylinder is the one from which the noise is originating.
- **Main-bearing knock.** A main-bearing knock often cannot be isolated to a particular cylinder. The sound can vary in intensity and may disappear at times depending on engine load.

Regardless of the type of loud knocking noise, after the external causes of the knocking noise have been eliminated, the engine should be disassembled and carefully inspected to determine the exact cause.

TECH TIP

Engine Noise and Cost

A light ticking noise often heard at one-half engine speed and associated with valve train noise is a less serious problem than many deep-sounding knocking noises. Generally, the deeper the sound of the engine noise, the more the owner will have to pay for repairs. A light "tick tick tick," though often not cheap, is usually far less expensive than a deep "knock knock knock" from the engine.

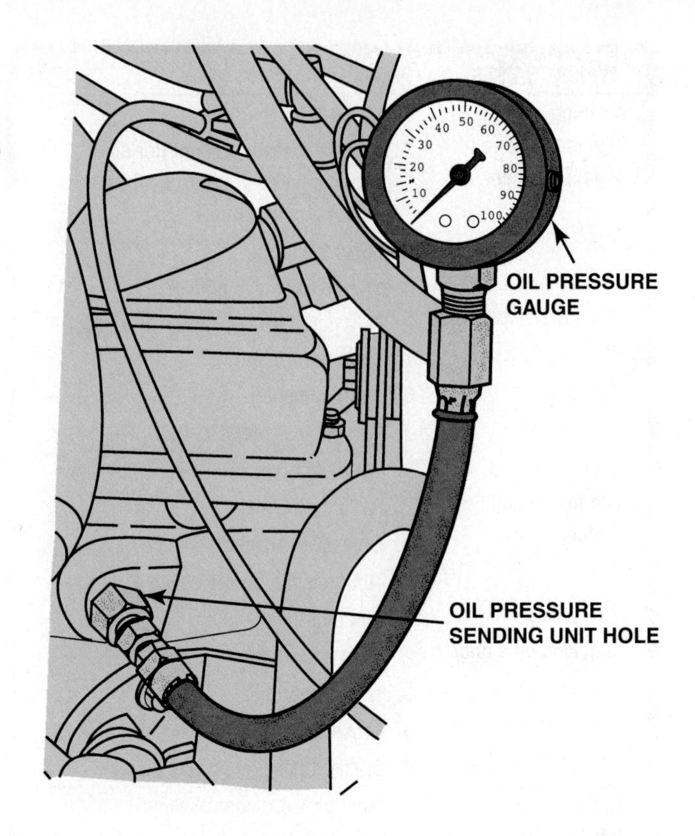

FIGURE 13–8 To measure engine oil pressure, remove the oil pressure sending (sender) unit usually located near the oil filter. Screw the pressure gauge into the oil pressure sending unit hole.

OIL PRESSURE TESTING

Proper oil pressure is very important for the operation of any engine. *Low oil pressure can cause engine wear, and engine wear can cause low oil pressure*.

If main thrust or rod bearings are worn, oil pressure is reduced because of leakage of the oil around the bearings. Oil pressure testing is usually performed with the following steps:

STEP 1. Operate the engine until normal operating temperature is achieved.

STEP 2 With the engine off, remove the oil pressure sending unit or sender, usually located near the oil filter. Thread an oil pressure gauge into the threaded hole. ● **SEE FIGURE 13–8**.

NOTE: An oil pressure gauge can be made from another gauge, such as an old air-conditioning gauge and a flexible brake hose. The threads are often the same as those used for the oil pressure sending unit.

STEP 3 Start the engine and observe the gauge. Record the oil pressure at idle and at 2500 RPM. Most vehicle manufacturers recommend a minimum oil pressure of 10 PSI per 1000 RPM. Therefore, at 2500 RPM, the oil pressure should be at least 25 PSI. Always compare your test results with the manufacturer's recommended oil pressure.

Besides engine bearing wear, other possible causes for low oil pressure include:

- Low oil level
- Diluted oil
- Stuck oil pressure relief valve

OIL PRESSURE WARNING LAMP

The red oil pressure warning lamp in the dash usually lights when the oil pressure is less than 4 to 7 PSI, depending on vehicle and engine. The oil light should not be on during driving. If the oil warning lamp is on, stop the engine immediately. Always confirm oil pressure with a reliable mechanical gauge before performing engine repairs. The sending unit or circuit may be defective.

TECH TIP

Use the KISS Test Method

Engine testing is done to find the cause of an engine problem. All the simple things should be tested first. Just remember KISS—"keep it simple, stupid." A loose alternator belt or loose bolts on a torque converter can sound just like a lifter or rod bearing. A loose spark plug can make the engine perform as if it had a burned valve. Some simple items that can cause serious problems include the following:

Oil Burning

- Low oil level
- Clogged PCV valve or system, causing blowby and oil to be blown into the air cleaner
- Clogged drainback passages in the cylinder head
- Dirty oil that has not been changed for a long time (Change the oil and drive for about 1,000 miles (1,600 kms) and change the oil and filter again.)

Noises

- Carbon on top of the piston(s) can sound like a bad rod bearing (often called a carbon knock)
- Loose torque-to-flex plate bolts (or nuts), causing a loud knocking noise

NOTE: Often this problem will cause noise only at idle; the noise tends to disappear during driving or when the engine is under load.

- A loose and/or defective drive belt, which may cause a rod- or main-bearing knocking noise (A loose or broken mount for the generator [alternator], power steering pump, or air-conditioning compressor can also cause a knocking noise.)

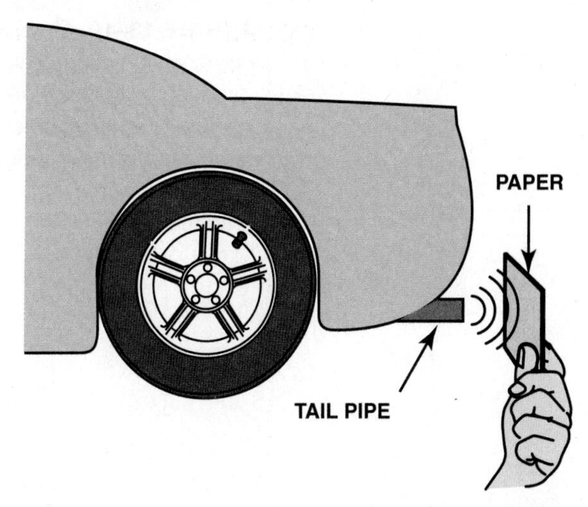

FIGURE 13–9 The paper test involves holding a piece of paper near the tailpipe of an idling engine. A good engine should produce even, outward puffs of exhaust. If the paper is sucked in toward the tailpipe, a burned valve is a possibility.

TECH TIP

The Paper Test

A soundly running engine should produce even and steady exhaust at the tailpipe. You can test this with the **paper test**. Hold a piece of paper or a 3″ × 5″ index card (even a dollar bill works) within 1 inch (25 mm) of the tailpipe with the engine running at idle. **SEE FIGURE 13–9.**

The paper should blow out evenly without "puffing." If the paper is drawn toward the tailpipe at times, the exhaust valves in one or more cylinders could be burned. Other reasons why the paper might be sucked toward the tailpipe include the following:

1. The engine could be misfiring because of a lean condition that could occur normally when the engine is cold.
2. Pulsing of the paper toward the tailpipe could also be caused by a hole in the exhaust system. If exhaust escapes through a hole in the exhaust system, air could be drawn in during the intervals between the exhaust puffs from the tailpipe to the hole in the exhaust, causing the paper to be drawn toward the tailpipe.
3. Ignition fault causing misfire.

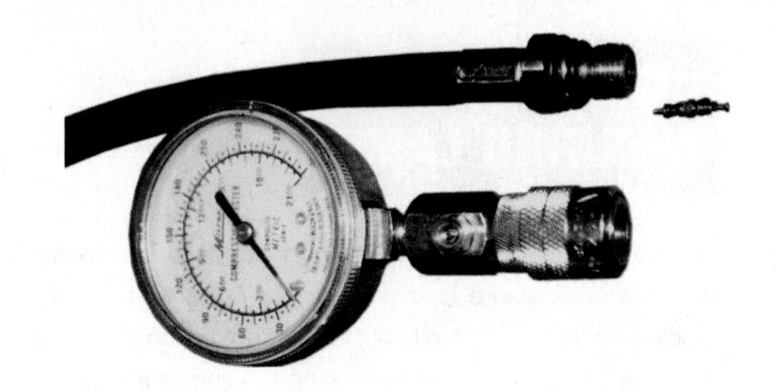

FIGURE 13–10 A two-piece compression gauge set. The threaded hose is screwed into the spark plug hole after removing the spark plug. The gauge part is then snapped onto the end of the hose.

COMPRESSION TEST

An engine **compression test** is one of the fundamental engine diagnostic tests that can be performed. For smooth engine operation, all cylinders must have equal compression. An engine can lose compression by leakage of air through one or more of only three routes:

- Intake or exhaust valve
- Piston rings (or piston, if there is a hole)
- Cylinder head gasket

For best results, the engine should be warmed to normal operating temperature before testing. An accurate compression test should be performed as follows:

STEP 1 Remove all spark plugs. This allows the engine to be cranked to an even speed. Be sure to label all spark plug wires.

CAUTION: Disable the ignition system by disconnecting the primary leads from the ignition coil or module or by grounding the coil wire after removing it from the center of the distributor cap. Also disable the fuel-injection system to prevent the squirting of fuel into the cylinder.

STEP 2 Block open the throttle. This permits the maximum amount of air to be drawn into the engine. This step also ensures consistent compression test results.

STEP 3 Thread a compression gauge into one spark plug hole and crank the engine. ● **SEE FIGURE 13–10.**

Continue cranking the engine through *four* compression strokes. Each compression stroke makes a puffing sound.

NOTE: Note the reading on the compression gauge after the first puff. This reading should be at least one-half the final reading. For example, if the final, highest reading is 150 PSI, then the reading after the first puff should be higher than 75 PSI. A low first-puff reading indicates possible weak piston rings. Release the pressure on the gauge and repeat for the other cylinders.

STEP 4 Record the highest readings and compare the results. Most vehicle manufacturers specify the minimum compression reading and the maximum allowable variation among cylinders. Most manufacturers specify a maximum difference of 20% between the highest reading and the lowest reading. For example:

If the high reading is	**150 PSI**
Subtract 20%	**−30 PSI**
Lowest allowable compression is	**120 PSI**

NOTE: To make the math quick and easy, think of 10% of 150, which is 15 (move the decimal point to the left one place). Now double it: $15 \times 2 = 30$. This represents 20%.

NOTE: During cranking, the oil pump cannot maintain normal oil pressure. Extended engine cranking, such as that which occurs during a compression test, can cause hydraulic lifters to collapse. When the engine starts, loud valve clicking noises may be heard. This should be considered normal after performing a compression test, and the noise should stop after the vehicle has been driven a short distance.

TECH TIP

The Hose Trick

Installing spark plugs can be made easier by using a rubber hose on the end of the spark plug. The hose can be a vacuum hose, fuel line, or even an old spark plug wire end. ● **SEE FIGURE 13–11**.

The hose makes it easy to start the threads of the spark plug into the cylinder head. After starting the threads, continue to thread the spark plug for several turns. Using the hose eliminates the chance of cross-threading the plug. This is especially important when installing spark plugs in aluminum cylinder heads.

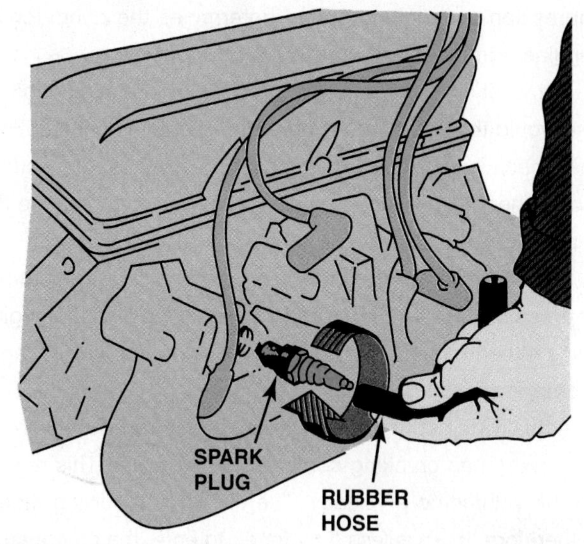

FIGURE 13–11 Use a vacuum or fuel line hose over the spark plug to install it without danger of cross-threading the cylinder head.

WET COMPRESSION TEST

If the compression test reading indicates low compression on one or more cylinders, add three squirts of oil to the cylinder and retest. This is called a **wet compression test**, when oil is used to help seal around the piston rings.

CAUTION: Do not use more oil than three squirts from a hand-operated oil squirt can. Too much oil can cause a hydrostatic lock, which can damage or break pistons or connecting rods or even crack a cylinder head.

Perform the compression test again and observe the results. If the first-puff readings greatly improve and the readings are much higher than without the oil, the cause of the low compression is worn or defective piston rings. If the compression readings increase only slightly (or not at all), then the cause of the low compression is usually defective valves. ● **SEE FIGURE 13–12**.

NOTE: During both the dry and wet compression tests, be sure that the battery and starting system are capable of cranking the engine at normal cranking speed.

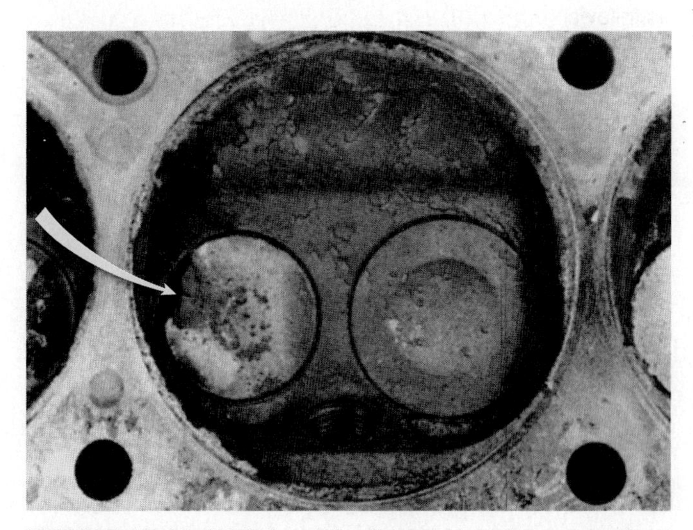

FIGURE 13–12 Badly burned exhaust valve. A compression test could have detected a problem, and a cylinder leakage test (leak-down test) could have been used to determine the exact problem.

RUNNING (DYNAMIC) COMPRESSION TEST

A compression test is commonly used to help determine engine condition and is usually performed with the engine cranking.

What is the RPM of a cranking engine? An engine idles at about 600 to 900 RPM, and the starter motor obviously cannot crank the engine as fast as the engine idles. Most manufacturers' specifications require the engine to crank at 80 to 250 cranking RPM. Therefore, a check of the engine's compression at cranking speed determines the condition of an engine that does not run at such low speeds.

But what should be the compression of a running engine? Some would think that the compression would be substantially higher, because the valve overlap of the cam is more effective at higher engine speeds, which would tend to increase the compression.

A **running compression test**, also called a **dynamic compression test**, is a compression test done with the engine running rather than during engine cranking as is done in a regular compression test.

Actually, the compression pressure of a running engine is much *lower* than cranking compression pressure. This results from the volumetric efficiency. The engine is revolving faster, and therefore, there is less *time* for air to enter the combustion chamber. With less air to compress, the compression pressure is lower. Tycompression. For most engines, the valuepically, the higher the engine RPM, the lower the running ranges are as follows:

- Compression during cranking: 125 to 160 PSI
- Compression at idle: 60 to 90 PSI
- Compression at 2000 RPM: 30 to 60 PSI

As with cranking compression, the running compression of all cylinders should be equal. Therefore, a problem is not likely to be detected by single compression values, but by *variations* in running compression values among the cylinders. Broken valve springs, worn valve guides, bent pushrods, and worn cam lobes are some items that would be indicated by a low running compression test reading on one or more cylinders.

PERFORMING A RUNNING COMPRESSION TEST To perform a running compression test, remove just one spark plug at a time. With one spark plug removed from the engine, use a jumper wire to *ground* the spark plug wire to a good engine ground. This prevents possible ignition coil damage. Start the engine, push the pressure release on the gauge, and read the compression. Increase the engine speed to about 2000 RPM and push the pressure release on the gauge again. Read the gauge. Stop the engine, reinstall the spark plug, and attach the spark plug wire, and repeat the test for each of the remaining cylinders. Just like the cranking compression test, the running compression test can inform a technician of the *relative* compression of all the cylinders.

FIGURE 13–13 A typical handheld cylinder leakage tester.

CYLINDER LEAKAGE TEST

One of the best tests that can be used to determine engine condition is the **cylinder leakage test**. This test involves injecting air under pressure into the cylinders one at a time. The amount and location of any escaping air helps the technician determine the condition of the engine. The air is injected into the cylinder through a cylinder leakage gauge into the spark plug hole. ● **SEE FIGURE 13–13**. To perform the cylinder leakage test, take the following steps:

STEP 1 For best results, the engine should be at normal operating temperature (upper radiator hose hot and pressurized).

STEP 2 The cylinder being tested must be at top dead center (TDC) of the compression stroke. ● **SEE FIGURE 13–14**.

NOTE: The greatest amount of wear occurs at the top of the cylinder because of the heat generated near the top of the cylinder. The piston ring flex also adds to the wear at the top of the cylinder.

STEP 3 Calibrate the cylinder leakage unit as per manufacturer's instructions.

STEP 4 Inject air into the cylinders one at a time, rotating the engine as necessitated by firing order to test each cylinder at TDC on the compression stroke.

FIGURE 13–14 A whistle stop used to find top dead center. Remove the spark plug and install the whistle stop, then rotate the engine by hand. When the whistle stops making a sound, the piston is at the top.

STEP 5 Evaluate the results:

Less than 10% leakage: good
Less than 20% leakage: acceptable
Less than 30% leakage: poor
More than 30% leakage: definite problem

NOTE: If leakage seems unacceptably high, repeat the test, being certain that it is being performed correctly and that the cylinder being tested is at TDC on the compression stroke.

STEP 6 Check the source of air leakage.

a. If air is heard escaping from the oil filler cap, the *piston rings* are worn or broken.

b. If air is observed bubbling out of the radiator, there is a possible blown *head gasket* or cracked *cylinder head*.

c. If air is heard coming from the throttle body or air inlet on fuel injection-equipped engines, there is a defective *intake valve(s)*.

d. If air is heard coming from the tailpipe, there is a defective *exhaust valve(s)*.

CYLINDER POWER BALANCE TEST

Most large engine analyzers and scan tools have a cylinder power balance feature. The purpose of a cylinder **power balance test** is to determine if all cylinders are contributing power equally. It determines this by shorting out one cylinder at a time. If the engine speed (RPM) does not drop as much for one cylinder as for other cylinders of the same engine, then the shorted cylinder must be weaker than the other cylinders. For example:

Cylinder Number	RPM Drop When Ignition Is Shorted
1	75
2	70
3	15
4	65
5	75
6	70

Cylinder #3 is the weak cylinder.

NOTE: Most automotive test equipment use automatic means for testing cylinder balance. Be certain to correctly identify the offending cylinder. Cylinder #3 as identified by the equipment may be the third cylinder in the firing order instead of the actual cylinder #3.

POWER BALANCE TEST PROCEDURE

When point-type ignition was used on all vehicles, the common method for determining which, if any, cylinder was weak was to remove the spark plug wire of the spark plugs one at a time while observing the readings on a tachometer and a vacuum gauge. This method is not recommended on any vehicle with any type of electronic ignition. If any of the spark plug wires are removed from a spark plug with the engine running, the ignition coil tries to supply increasing levels of voltage attempting to jump the increasing gap as the plug wires are removed. This high voltage could easily track the ignition coil, damage the ignition module, or both.

The acceptable method of canceling cylinders, which will work on all types of ignition systems, including distributorless, is to *ground* the secondary current for each cylinder. ● **SEE FIGURE 13–15**. The cylinder with the least RPM drop is the cylinder not producing its share of power.

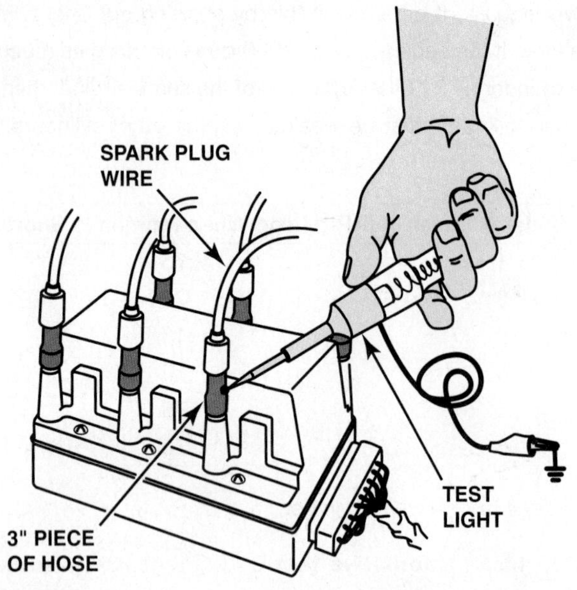

FIGURE 13–15 Using a vacuum hose and a test light to ground one cylinder at a time on a distributorless ignition system. This works on all types of ignition systems and provides a method for grounding out one cylinder at a time without fear of damaging any component. To avoid possible damage to the catalytic converter, do not short out a cylinder for longer than five seconds.

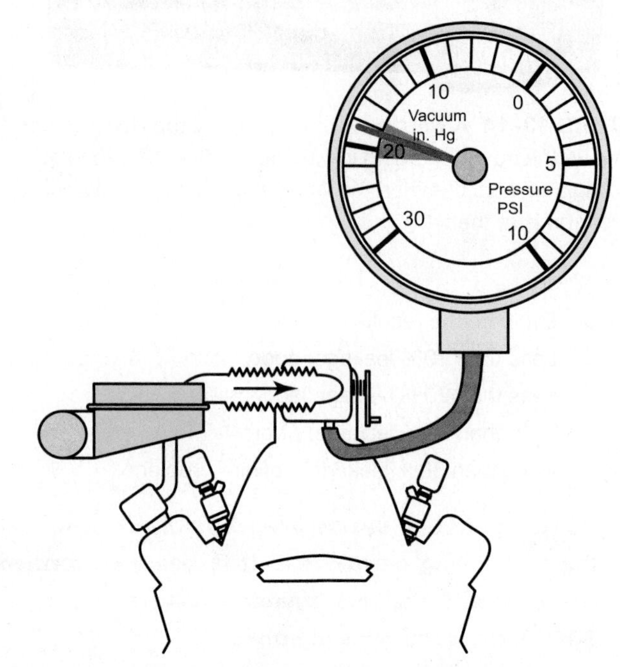

FIGURE 13–16 An engine in good mechanical condition should produce 17 to 21 inches Hg of vacuum at idle at sea level.

VACUUM TESTS

Vacuum is pressure below atmospheric pressure and is measured in **inches** (or millimeters) **of mercury (Hg)**. An engine in good mechanical condition will run with high manifold vacuum. Manifold vacuum is developed by the pistons as they move down on the intake stroke to draw the charge from the throttle body and intake manifold. Air to refill the manifold comes past the throttle plate into the manifold. Vacuum will increase anytime the engine turns faster or has better cylinder sealing while the throttle plate remains in a fixed position. Manifold vacuum will decrease when the engine turns more slowly or when the cylinders no longer do an efficient job of pumping. **Vacuum tests** include testing the engine for **cranking vacuum**, **idle vacuum**, and vacuum at 2500 RPM.

CRANKING VACUUM TEST Measuring the amount of manifold vacuum during cranking is a quick and easy test to determine if the piston rings and valves are properly sealing. (For accurate results, the engine should be warm and the throttle closed.) To perform the cranking vacuum test, take the following steps:

STEP 1 Disable the ignition or fuel injection.

STEP 2 Connect the vacuum gauge to a manifold vacuum source.

STEP 3 Crank the engine while observing the vacuum gauge.

Cranking vacuum should be higher than 2.5 inches of mercury. (Normal cranking vacuum is 3 to 6 inches Hg.) If it is lower than 2.5 inches Hg, then the following could be the cause:

- Too slow a cranking speed
- Worn piston rings
- Leaking valves
- Excessive amounts of air bypassing the throttle plate (This could give a false low vacuum reading. Common sources include a throttle plate partially open or a high-performance camshaft with excessive overlap.)

IDLE VACUUM TEST An engine in proper condition should idle with a steady vacuum between 17 and 21 inches Hg. ● **SEE FIGURE 13–16.**

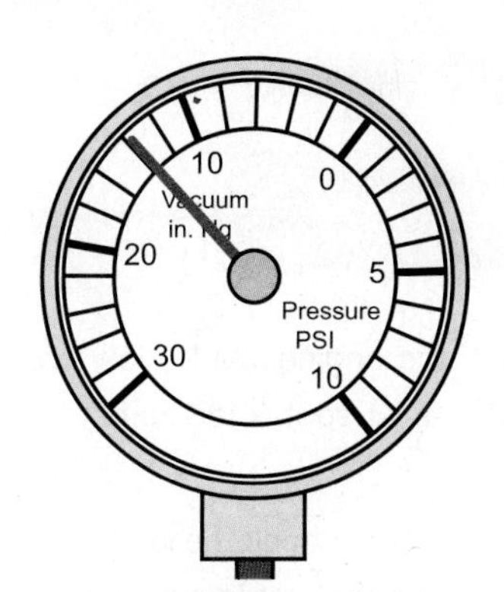

FIGURE 13–17 A steady but low reading could indicate retarded valve or ignition timing.

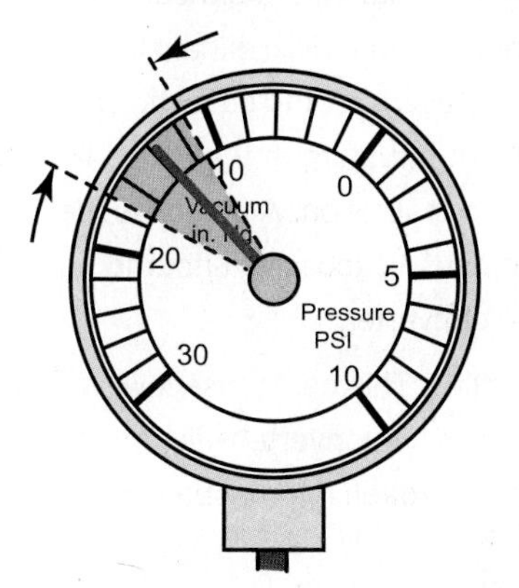

FIGURE 13–18 A gauge reading with the needle fluctuating 3 to 9 inches Hg below normal often indicates a vacuum leak in the intake system.

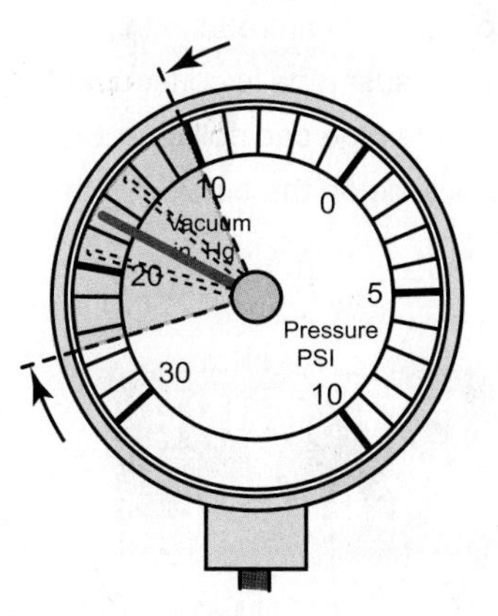

FIGURE 13–19 A leaking head gasket can cause the needle to vibrate as it moves through a range from below to above normal.

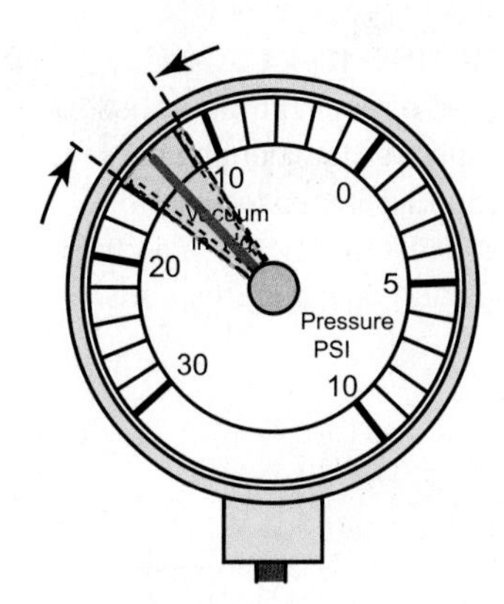

FIGURE 13–20 An oscillating needle 1 or 2 inches Hg below normal could indicate an incorrect air–fuel mixture (either too rich or too lean).

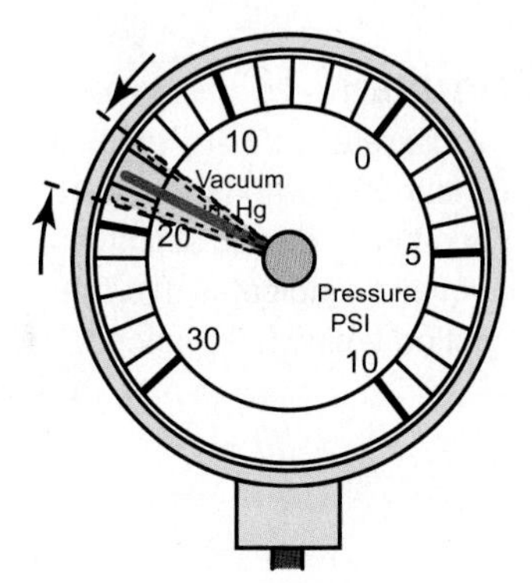

FIGURE 13–21 A rapidly vibrating needle at idle that becomes steady as engine speed is increased indicates worn valve guides.

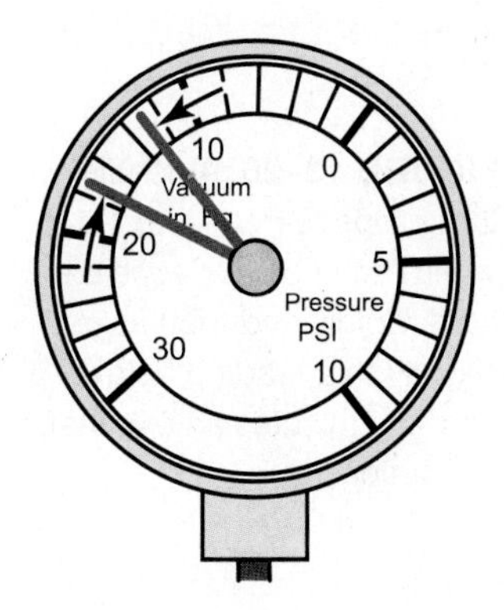

FIGURE 13–22 If the needle drops 1 or 2 inches Hg from the normal reading, one of the engine valves is burned or not seating properly.

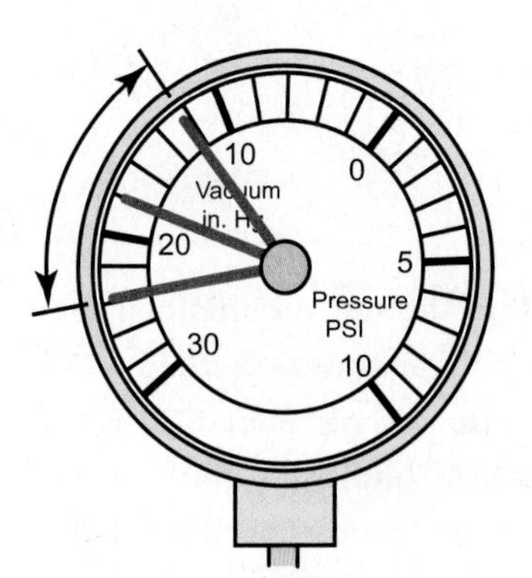

FIGURE 13–23 Weak valve springs will produce a normal reading at idle, but as engine speed increases, the needle will fluctuate rapidly between 12 and 24 inches Hg.

NOTE: Engine vacuum readings vary with altitude. A reduction of 1 inch Hg per 1,000 feet (300 meters) of altitude should be subtracted from the expected values if testing a vehicle above 1,000 feet (300 meters).

LOW AND STEADY VACUUM If the vacuum is lower than normal, yet the gauge reading is steady, the most common causes include:

- Retarded ignition timing
- Retarded cam timing (check timing chain for excessive slack or timing belt for proper installation)

● **SEE FIGURE 13–17.**

FLUCTUATING VACUUM If the needle drops, then returns to a normal reading, then drops again, and again returns, this indicates a sticking valve. A common cause of sticking valves is lack of lubrication of the valve stems. ● **SEE FIGURES 13–18 THROUGH 13–26.** If the vacuum gauge fluctuates above and below a center point, burned valves or weak valve springs may be indicated. If the fluctuation is slow and steady, unequal fuel mixture could be the cause.

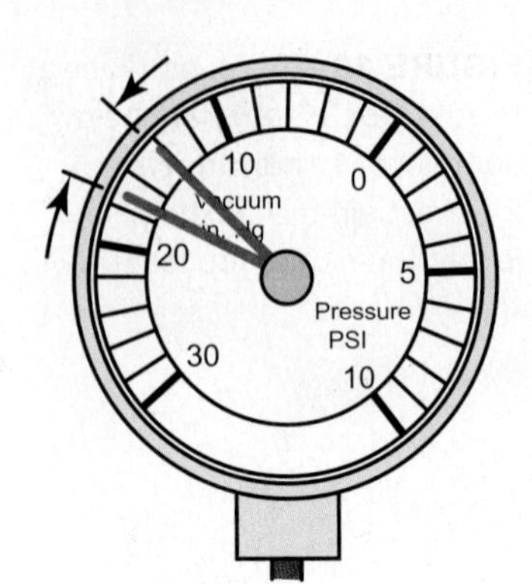

FIGURE 13–24 A steady needle reading that drops 2 or 3 inches Hg when the engine speed is increased slightly above idle indicates that the ignition timing is retarded.

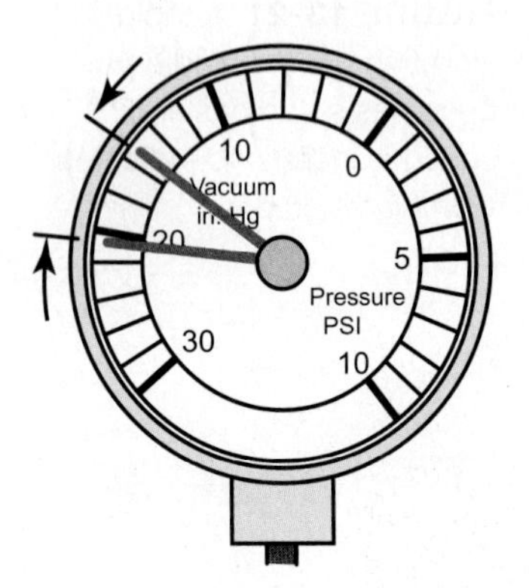

FIGURE 13–25 A steady needle reading that rises 2 or 3 inches Hg when the engine speed is increased slightly above idle indicates that the ignition timing is advanced.

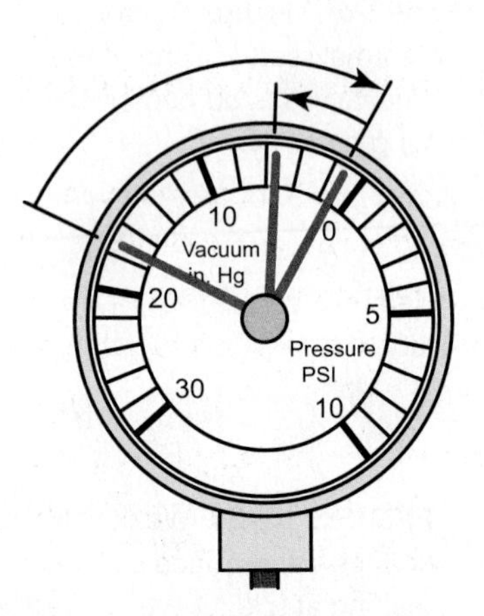

FIGURE 13–26 A needle that drops to near zero when the engine is accelerated rapidly and then rises slightly to a reading below normal indicates an exhaust restriction.

NOTE: A common trick that some technicians use is to squirt some automatic transmission fluid (ATF) down the throttle body or into the air inlet of a warm engine. Often the idle quality improves and normal vacuum gauge readings are restored. The use of ATF does create excessive exhaust smoke for a short time, but it should not harm oxygen sensors or catalytic converters.

EXHAUST RESTRICTION TEST

If the exhaust system is restricted, the engine will be low on power, yet smooth. Common causes of **restricted exhaust** include the following:

- **Clogged catalytic converter.** Always check the ignition and fuel-injection systems for faults that could cause excessive amounts of unburned fuel to be exhausted. Excessive unburned fuel can overheat the catalytic converter and cause the beads or structure of the converter to fuse together, creating the restriction. A defective fuel delivery system could also cause excessive unburned fuel to be dumped into the converter.
- **Clogged or restricted muffler.** This can cause low power. Often a defective catalytic converter will shed particles that can clog a muffler. Broken internal baffles can also restrict exhaust flow.
- **Damaged or defective piping.** This can reduce the power of any engine. Some exhaust pipe is constructed with double walls, and the inside pipe can collapse and form a restriction that is not visible on the outside of the exhaust pipe.

TESTING BACK PRESSURE WITH A VACUUM GAUGE

A vacuum gauge can be used to measure manifold vacuum at a high idle (2000 to 2500 RPM). If the exhaust system is restricted, pressure increases in the exhaust system. This pressure is called **back pressure**. Manifold vacuum will drop gradually if the engine is kept at a constant speed if the exhaust is restricted.

The reason the vacuum will drop is that all exhaust leaving the engine at the higher engine speed cannot get through the restriction. After a short time (within one minute), the exhaust tends to "pile up" above the restriction and eventually remains in the cylinder of the engine at the end of the exhaust stroke. Therefore, at the beginning of the intake stroke, when the piston traveling downward should be lowering the pressure (raising the vacuum) in the intake manifold, the extra exhaust in the cylinder *lowers* the normal vacuum. If the exhaust restriction is severe enough, the vehicle can become undriveable because cylinder filling cannot occur except at idle.

FIGURE 13–27 A technician-made adapter used to test exhaust system back pressure.

FIGURE 13–28 A tester that uses a blue liquid to check for exhaust gases in the exhaust, which would indicate a head gasket leak problem.

TESTING BACK PRESSURE WITH A PRESSURE GAUGE

Exhaust system back pressure can be measured directly by installing a pressure gauge into an exhaust opening. This can be accomplished in one of the following ways:

- **With an oxygen sensor.** Use a back pressure gauge and adapter or remove the inside of an old, discarded oxygen sensor and thread in an adapter to convert to a vacuum or pressure gauge.

  NOTE: An adapter can easily be made by inserting a metal tube or pipe. A short section of brake line works great. The pipe can be brazed to the oxygen sensor housing or it can be glued in with epoxy. An 18-millimeter compression gauge adapter can also be adapted to fit into the oxygen sensor opening. ● SEE FIGURE 13–27.

- **With the exhaust gas recirculation (EGR) valve.** Remove the EGR valve and fabricate a plate to connect to a pressure gauge.
- **With the air-injection reaction (AIR) check valve.** Remove the check valve from the exhaust tubes leading down to the exhaust manifold. Use a rubber cone with a tube inside to seal against the exhaust tube. Connect the tube to a pressure gauge.

At idle, the maximum back pressure should be less than 1.5 PSI (10 kPa), and it should be less than 2.5 PSI (15 kPa) at 2500 RPM.

DIAGNOSING HEAD GASKET FAILURE

Several items can be used to help diagnose a head gasket failure:

- **Exhaust gas analyzer.** With the radiator cap removed, place the probe from the exhaust analyzer above the radiator filler neck. If the HC reading increases, the exhaust (unburned hydrocarbons) is getting into the coolant from the combustion chamber.
- **Chemical test.** A chemical tester using blue liquid is also available. The liquid turns yellow if combustion gases are present in the coolant. ● **SEE FIGURE 13–28.**
- **Bubbles in the coolant.** Remove the coolant pump belt to prevent pump operation. Remove the radiator cap and start the engine. If bubbles appear in the coolant before it begins to boil, a defective head gasket or cracked cylinder head is indicated.
- **Excessive exhaust steam.** If excessive water or steam is observed coming from the tailpipe, this means that coolant is getting into the combustion chamber from a defective head gasket or a cracked head. If there is leakage between cylinders, the engine usually misfires and a power balancer test and/or compression test can be used to confirm the problem.

If any of the preceding indicators of head gasket failure occur, remove the cylinder head(s) and check all of the following:

1. Head gasket
2. Sealing surfaces—for warpage
3. Castings—for cracks

NOTE: A leaking thermal vacuum valve can cause symptoms similar to those of a defective head gasket. Most thermal vacuum valves thread into a coolant passage, and they often leak only after they get hot.

DASH WARNING LIGHTS

Most vehicles are equipped with several dash warning lights often called "telltale" or "idiot" lights. These lights are often the only warning a driver receives that there may be engine problems. A summary of typical dash warning lights and their meanings follows.

OIL (ENGINE) LIGHT The red oil light indicates that the engine oil pressure is too low (usually lights when oil pressure is 4 to 7 PSI [20 to 50 kPa]). Normal oil pressure should be 10 to 60 PSI (70 to 400 kPa) or 10 PSI per 1000 engine RPM.

When this light comes on, the driver should shut off the engine immediately and check the oil level and condition for possible dilution with gasoline caused by a fuel system fault. If the oil level is okay, then there is a possible serious engine problem or a possible defective oil pressure sending (sender) unit. The automotive technician should always check the oil pressure using a reliable mechanical oil pressure gauge if low oil pressure is suspected.

NOTE: Some automobile manufacturers combine the dash warning lights for oil pressure and coolant temperature into one light, usually labeled "engine." Therefore, when the engine light comes on, the technician should check for possible coolant temperature and/or oil pressure problems.

COOLANT TEMPERATURE LIGHT Most vehicles are equipped with a coolant temperature gauge or dash warning light. The warning light may be labeled "coolant," "hot," or "temperature." If the coolant temperature warning light comes on during driving, this usually indicates that the coolant temperature is above a safe level, or above about 250°F (120°C). Normal coolant temperature should be about 200°F to 220°F (90°C to 105°C).

If the coolant temperature light comes on during driving, the following steps should be followed to prevent possible engine damage:

1. Turn off the air conditioning and turn on the heater. The heater will help get rid of some of the heat in the cooling system.
2. Raise the engine speed in neutral or park to increase the circulation of coolant through the radiator.
3. If possible, turn the engine off and allow it to cool (this may take over an hour).
4. Do not continue driving with the coolant temperature light on (or the gauge reading in the red warning section or above 260°F) or serious engine damage may result.

NOTE: If the engine does not feel or smell hot, it is possible that the problem is a faulty coolant temperature sensor or gauge.

TECH TIP

Misfire Diagnosis

If a misfire goes away with propane added to the air inlet, suspect a lean injector.

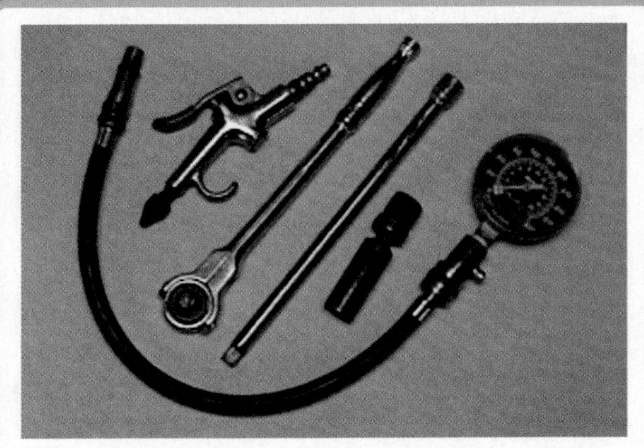

1 The tools and equipment needed to perform a compression test include a compression gauge, an air nozzle, and the socket ratchets and extensions that may be necessary to remove the spark plugs from the engine.

2 To prevent ignition and fuel-injection operation while the engine is being cranked, remove both the fuel-injection fuse and the ignition fuse. If the fuses cannot be removed, disconnect the wiring connectors for the injectors and the ignition system.

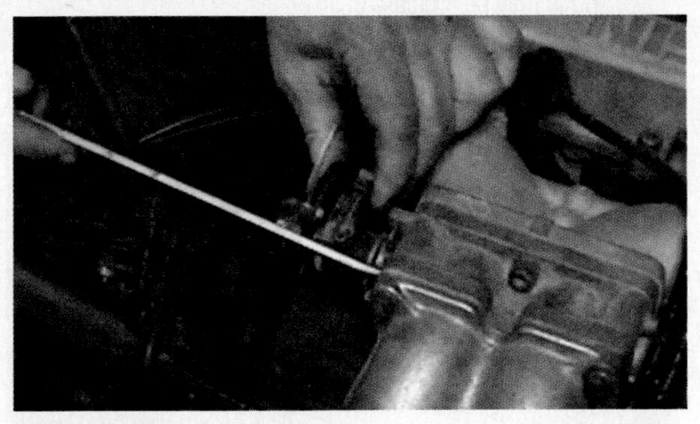

3 Block open the throttle (and choke, if the engine is equipped with a carburetor). Here a screwdriver is being used to wedge the throttle linkage open. Keeping the throttle open ensures that enough air will be drawn into the engine so that the compression test results will be accurate.

4 Before removing the spark plugs, use an air nozzle to blow away any dirt that may be around the spark plug. This step helps prevent debris from getting into the engine when the spark plugs are removed.

5 Remove all of the spark plugs. Be sure to mark the spark plug wires so that they can be reinstalled onto the correct spark plugs after the compression test has been performed.

6 Select the proper adapter for the compression gauge. The threads on the adapter should match those on the spark plug.

CONTINUED ▶

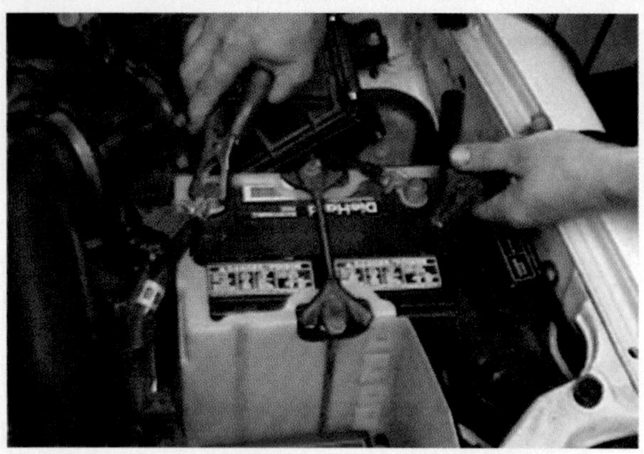

7 If necessary, connect a battery charger to the battery before starting the compression test. It is important that consistent cranking speed be available for each cylinder being tested.

8 Make a note of the reading on the gauge after the first "puff," which indicates the first compression stroke that occurred on that cylinder as the engine was being rotated. If the first puff reading is low and the reading gradually increases with each puff, weak or worn piston rings may be indicated.

9 After the engine has been cranked for four "puffs," stop cranking the engine and observe the compression gauge.

10 Record the first puff and this final reading for each cylinder. The final readings should all be within 20% of each other.

11 If a cylinder(s) is lower than most of the others, use an oil can and use two squirts of engine oil into the cylinder and repeat the compression test. This is called performing a wet compression test.

12 If the gauge reading is now much higher than the first test results, then the cause of the low compression is due to worn or defective piston rings. The oil in the cylinder temporarily seals the rings, which causes the higher reading.

SUMMARY

1. The first step in diagnosing engine condition is to perform a thorough visual inspection, including a check of oil and coolant levels and condition.
2. Oil leaks can be found by using a white powder or a fluorescent dye and a black light.
3. Many engine-related problems make a characteristic noise.
4. Oil analysis by an engineering laboratory can reveal engine problems by measuring the amount of dissolved metals in the oil.
5. A compression test can be used to test the condition of valves and piston rings.
6. A cylinder leakage test fills the cylinder with compressed air, and the gauge indicates the percentage of leakage.
7. A cylinder balance test indicates whether all cylinders are working okay.
8. Testing engine vacuum is another procedure that can help the service technician determine engine condition.

REVIEW QUESTIONS

1. Describe the visual checks that should be performed on an engine if a mechanical malfunction is suspected.
2. List three simple items that could cause excessive oil consumption.
3. List three simple items that could cause engine noises.
4. Describe how to perform a compression test and how to determine what is wrong with an engine based on a compression test result.
5. Describe the cylinder leakage test.
6. Describe how a vacuum gauge would indicate if the valves were sticking in their guides.
7. Describe the test procedure for determining if the exhaust system is restricted (clogged) using a vacuum gauge.

CHAPTER QUIZ

1. Technician A says that the paper test could detect a burned valve. Technician B says that a grayish white stain on the engine could be a coolant leak. Which technician is correct?
   - **a.** Technician A only
   - **b.** Technician B only
   - **c.** Both Technicians A and B
   - **d.** Neither Technician A nor B
2. Two technicians are discussing oil leaks. Technician A says that an oil leak can be found using a fluorescent dye in the oil with a black light to check for leaks. Technician B says that a white spray powder can be used to locate oil leaks. Which technician is correct?
   - **a.** Technician A only
   - **b.** Technician B only
   - **c.** Both Technicians A and B
   - **d.** Neither Technician A nor B
3. Which of the following is the *least likely* to cause an engine noise?
   - **a.** Carbon on the pistons
   - **b.** Cracked exhaust manifold
   - **c.** Loose accessory drive belt
   - **d.** Vacuum leak
4. A good engine should produce how much compression during a running (dynamic) compression test at idle?
   - **a.** 150 to 200 PSI
   - **b.** 100 to 150 PSI
   - **c.** 60 to 90 PSI
   - **d.** 30 to 60 PSI
5. A smoothly operating engine depends on ______________.
   - **a.** High compression on most cylinders
   - **b.** Equal compression between cylinders
   - **c.** Cylinder compression levels above 100 PSI (700 kPa) and within 70 psi (500 kPa) of each other
   - **d.** Compression levels below 100 PSI (700 kPa) on most cylinders
6. A good reading for a cylinder leakage test would be ______________.
   - **a.** Within 20% between cylinders
   - **b.** All cylinders below 20% leakage
   - **c.** All cylinders above 20% leakage
   - **d.** All cylinders above 70% leakage and within 7% of each other

7. Technician A says that during a power balance test, the cylinder that causes the biggest RPM drop is the weak cylinder. Technician B says that if one spark plug wire is grounded out and the engine speed does not drop, a weak or dead cylinder is indicated. Which technician is correct?
   a. Technician A only
   b. Technician B only
   c. Both Technicians A and B
   d. Neither Technician A nor B
8. *Cranking* vacuum should be ______________.
   a. 2.5 inches Hg or higher
   b. Over 25 inches Hg
   c. 17 to 21 inches Hg
   d. 6 to 16 inches Hg
9. Technician A says that a leaking head gasket can be tested for using a chemical tester. Technician B says that leaking head gasket can be found using an exhaust gas analyzer. Which technician is correct?
   a. Technician A only
   b. Technician B only
   c. Both Technicians A and B
   d. Neither Technician A nor B
10. The low oil pressure warning light usually comes on ______________.
   a. Whenever an oil change is required
   b. Whenever oil pressure drops dangerously low (4 to 7 PSI)
   c. Whenever the oil filter bypass valve opens
   d. Whenever the oil filter antidrainback valve opens

chapter 14

IN-VEHICLE ENGINE SERVICE

LEARNING OBJECTIVES

After studying this chapter, the reader will be able to:

1. Diagnose and replace the thermostat.
2. Diagnose and replace the water pump.
3. Diagnose and replace an intake manifold gasket.
4. Determine and verify correct cam timing.
5. Remove and replace a timing belt.
6. Describe how to adjust valves.
7. Explain hybrid engine precautions.

This chapter will help you prepare for Engine Repair (A8) ASE certification test content area "A" (General Engine Diagnosis).

KEY TERMS

EREVs 226
Fretting 223
HEVs 226
Idle stop 226
Skewed 222

FIGURE 14–1 If the thermostat has a jiggle valve, it should be placed toward the top to allow air to escape. If a thermostat were to become stuck open or open too soon, this can set a diagnostic trouble code P0128 (coolant temperature below thermostat regulating temperature).

THERMOSTAT REPLACEMENT

FAILURE PATTERNS All thermostat valves move during operation to maintain the desired coolant temperature. Thermostats can fail in the following ways:

- **Stuck Open**—If a thermostat fails open or partially open, the operating temperature of the engine will be less than normal. ● **SEE FIGURE 14–1**.
- **Stuck Closed**—If the thermostat fails closed or almost closed, the engine will likely overheat.
- **Stuck Partially Open**—This will cause the engine to warm up slowly if at all. This condition can cause the Powertrain Control Module (PCM) to set a P0128 diagnostic trouble code (DTC), which means that the engine coolant temperature does not reach the specified temperature.
- **Skewed**—A **skewed** thermostat works, but not within the correct temperature range. Therefore, the engine could overheat or operate cooler than normal or even do both.

REPLACEMENT PROCEDURE Before replacing the thermostat, double-check that the cooling system problem is not due to another fault, such as being low on coolant or an inoperative cooling fan. Check service information for the specified procedure to follow to replace the thermostat. Most recommended procedures include the following steps:

STEP 1 Allow the engine to cool for several hours so the engine and the coolant should be at room temperature.

STEP 2 Drain the coolant into a suitable container. Most vehicle manufacturers recommend that new coolant be used and the old coolant disposed of properly or recycled.

STEP 3 Remove any necessary components to get access to the thermostat.

STEP 4 Remove the thermostat housing and thermostat.

STEP 5 Replace the thermostat housing gasket and thermostat. Torque all fasteners to specifications.

STEP 6 Refill the cooling system with the specified coolant and bleed any trapped air from the system.

STEP 7 Pressurize the cooling system to verify that there are no leaks around the thermostat housing.

STEP 8 Run the engine until it reaches normal operating temperature and check for leaks.

STEP 9 Verify that the engine is reaching correct operating temperature.

FIGURE 14–2 Use caution if using a steel scraper to remove a gasket from aluminum parts. It is best to use a wood or plastic scraper.

WATER PUMP REPLACEMENT

NEED FOR REPLACEMENT A water pump will require replacement if any of the following conditions are present:

- Leaking coolant from the weep hole
- Bearing noisy or loose
- Lack of proper coolant flow caused by worn or slipping impeller blades

REPLACEMENT GUIDELINES After diagnosis has confirmed that the water pump requires replacement, check service information for the exact procedure to follow. The steps usually include the following:

STEP 1 Allow the engine to cool to room temperature.

STEP 2 Drain the coolant and dispose of properly or recycle.

STEP 3 Remove engine components to gain access to the water pump as specified in service information.

STEP 4 Remove the water pump assembly.

STEP 5 Clean the gasket surfaces and install the new water pump using a new gasket or seal as needed. ● **SEE FIGURE 14–2**. Torque all fasteners to factory specifications.

STEP 6 Install removed engine components.

STEP 7 Fill the cooling system with the specified coolant.

STEP 8 Run the engine, check for leaks, and verify proper operation.

FIGURE 14–3 An intake manifold gasket that failed and allowed coolant to be drawn into the cylinder(s).

INTAKE MANIFOLD GASKET INSPECTION

CAUSES OF FAILURE Many V-type engines leak oil and coolant, or experience an air (vacuum) leak caused by a leaking intake manifold gasket. This failure can be contributed to one or more of the following:

1. Expansion/contraction rate difference between the cast-iron head and the aluminum intake manifold can cause the intake manifold gasket to be damaged by the relative motion of the head and intake manifold. This type of failure is called **fretting.**
2. Plastic (Nylon 6.6) gasket deterioration caused by the coolant. ● **SEE FIGURE 14–3.**

DIAGNOSIS OF LEAKING INTAKE MANIFOLD GASKET

Because intake manifold gaskets are used to seal oil, air, and coolant in most cases, determining that the intake manifold gasket is the root cause can be a challenge. To diagnose a possible leaking intake manifold gasket, perform the following tests:

Visual inspection—Check for evidence of oil or coolant between the intake manifold and the cylinder heads.

Coolant level—Check the coolant level and determine if the level has been dropping. A leaking intake manifold gasket can cause coolant to leak and then evaporate, leaving no evidence of the leak.

Air (vacuum) leak—If there is a stored diagnostic trouble code (DTC) for a lean exhaust (P0171, P0172, or P0174), a leaking intake manifold gasket could be the cause. Use propane to check if the engine changes when dispensed around the intake manifold gasket. If the engine changes in speed or sound, then this test verifies that an air leak is present.

FIGURE 14–4 The lower intake manifold attaches to the cylinder heads.

FIGURE 14–5 The upper intake manifold, often called a plenum, attaches to the lower intake manifold.

INTAKE MANIFOLD GASKET REPLACEMENT

When replacing the intake manifold gasket, always check service information for the exact procedure to follow. The steps usually include the following:

STEP 1 Be sure the engine has been off for about an hour and then drain the coolant into a suitable container.

STEP 2 Remove covers and other specified parts needed to get access to the retaining bolts.

STEP 3 To help ensure that the manifold does not warp when removed, loosen all fasteners in the reverse order of the tightening sequence. This means that the bolts should be loosened starting at the ends and working toward the center.

STEP 4 Remove the upper intake manifold (plenum), if equipped, and inspect for faults. ● **SEE FIGURES 14–4 AND 14–5.**

STEP 5 Remove the lower intake manifold using the same bolt removal procedure of starting at the ends and working toward the center.

STEP 6 Thoroughly clean the area and replace the intake manifold if needed. Check that the correct replacement manifold is being used, and even the current part could look different from the original. ● **SEE FIGURE 14–6.**

STEP 7 Install the intake manifold using new gaskets as specified. Some designs use gaskets that are reusable. Replace as needed.

STEP 8 Torque all fasteners to factory specifications and in the proper sequences. The tightening sequences usually start at the center and work outward to the ends.

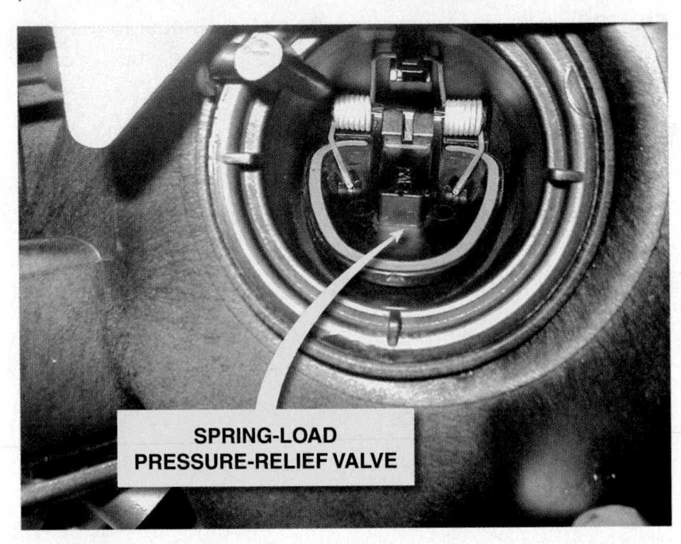

FIGURE 14–6 Some plastic intake manifolds are equipped with a pressure relief valve that would open in the event of a backfire condition to prevent the higher internal pressures from causing damage to the manifold.

CAUTION: Double-check the torque specifications and be sure to use the correct values. Many intake manifolds use fasteners that are torqued to values expressed in pound-inches and not pound-feet.

STEP 9 Reinstall all parts needed to allow the engine to start and run, including refilling the coolant if needed.

STEP 10 Start the engine and check for leaks and proper engine operation.

STEP 11 Reset or relearn the idle if specified, using a scan tool.

STEP 12 Install all of the remaining parts and perform a test drive to verify proper operation and no leaks.

FIGURE 14–7 A single overhead camshaft engine with a timing belt that also rotates the water pump.

TIMING BELT REPLACEMENT

NEED FOR REPLACEMENT Timing belts have a limited service and a specified replacement interval ranging from 60,000 miles (97,000 km) to about 100,000 miles (161,000 km). Timing belts are required to be replaced if any of the following conditions occurs:

- Meets or exceeds the vehicle manufacturer's recommended timing belt replacement interval.
- The timing belt has been contaminated with coolant or engine oil.
- The timing belt has failed (missing belt teeth or broken).

TIMING BELT REPLACEMENT GUIDELINES Before replacing the timing belt, check service information for the recommended procedure to follow. Most timing belt replacement procedures include the following steps:

STEP 1 Allow the engine to cool before starting to remove components to help eliminate the possibility of personal injury or warpage of the parts.

STEP 2 Remove all necessary components to gain access to the timing belt and timing marks.

STEP 3 If the timing belt is not broken, rotate the engine until the camshaft and crankshaft timing marks are aligned according to the specified marks. ● **SEE FIGURE 14–7**.

STEP 4 Loosen or remove the tensioner as needed to remove the timing belt.

STEP 5 Replace the timing belt and any other recommended items. Components that some vehicle manufacturers recommend replacing in addition to the timing belt include:

- Tensioner assembly
- Water pump
- Camshaft oil seal(s)
- Front crankshaft seal

STEP 6 Check (verify) that the camshaft timing is correct by rotating the engine several revolutions.

STEP 7 Install enough components to allow the engine to start to verify proper operation. Check for any leaks, especially if seals have been replaced.

STEP 8 Complete the reassembly of the engine and perform a test drive before returning the vehicle to the customer.

FIGURE 14–8 A Toyota/Lexus hybrid electric vehicle has a ready light. If the ready light is on, the engine can start at anytime without warning.

FIGURE 14–9 Always use the viscosity of oil as specified on the oil fill cap.

HYBRID ENGINE PRECAUTIONS

HYBRID VEHICLE ENGINE OPERATION Gasoline engines used in **hybrid electric vehicles (HEVs)** and in **extended range electric vehicles (EREVs)** can be a hazard to be around under some conditions. These vehicles are designed to stop the gasoline engines unless needed. This feature is called **idle stop**. This means that the engine is not running, but could start at any time if the computer detects the need to charge the hybrid batteries or other issue that requires the gasoline engine to start and run.

PRECAUTIONS Always check service information for the exact procedures to follow when working around or under the hood of a hybrid electric vehicle. These precautions could include:

- Before working under the hood or around the engine, be sure that the ignition is off and the key is out of the ignition.
- Check that the "Ready" light is off. ● **SEE FIGURE 14–8.**
- Do not touch any circuits that have orange electrical wires or conduit. The orange color indicates dangerous high-voltage wires, which could cause serious injury or death if touched.
- Always use high-voltage linesman's gloves whenever depowering the high-voltage system.

HYBRID ENGINE SERVICE The gasoline engine in most hybrid electric vehicles specifies low-viscosity engine oil as a way to achieve maximum fuel economy. ● **SEE FIGURE 14–9.** The viscosity required is often:

- SAE 0W-20
- SAE 5W-20

Many shops do not keep this viscosity in stock so preparations need to be made to get and use the specified engine oil.

In addition to engine oil, some hybrid electric vehicles such as the Honda Insight (1999–2004) require special spark plugs. Check service information for the specified service procedures and parts needed if a hybrid electric vehicle is being serviced.

1 Before starting the process of adjusting the valves, look up the specifications and exact procedures. The technician is checking this information from a computer CD-ROM-based information system.

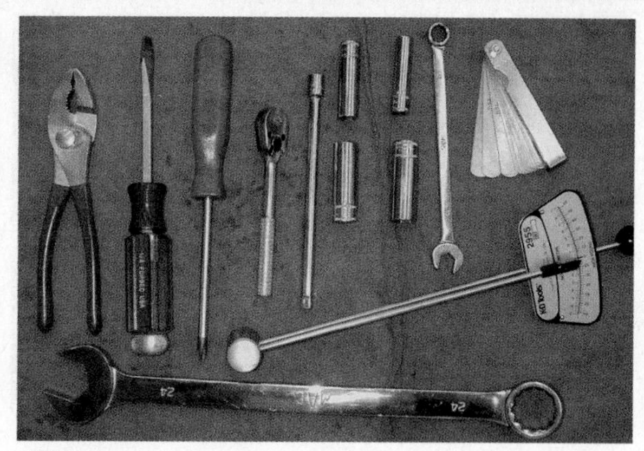

2 The tools necessary to adjust the valves on an engine with adjustable rocker arms include basic hand tools, feeler gauge, and a torque wrench.

3 An overall view of the four-cylinder engine that is due for a scheduled valve adjustment according to the vehicle manufacturer's recommendations.

4 Start the valve adjustment procedure by first disconnecting and labeling, if necessary, all vacuum lines that need to be removed to gain access to the valve cover.

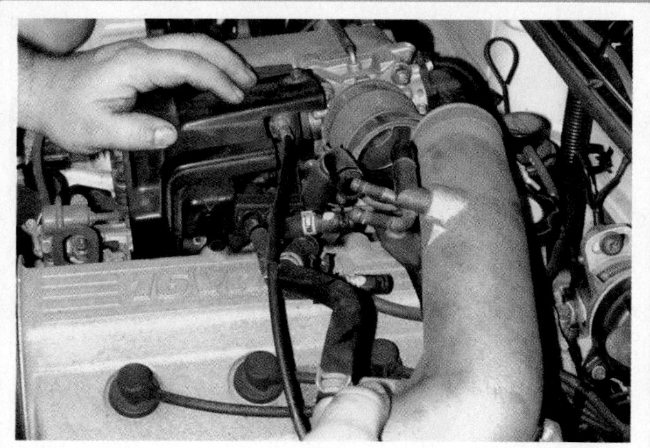

5 The air intake tube is being removed from the throttle body.

6 With all vacuum lines and the intake tube removed, the valve cover can be removed after removing all retaining bolts.

CONTINUED ▶

7 Notice how clean the engine appears. This is a testament of proper maintenance and regular oil changes by the owner.

8 To help locate how far the engine is being rotated, the technician is removing the distributor cap to be able to observe the position of the rotor.

9 The engine is rotated until the timing marks on the front of the crankshaft line up with zero degrees—top dead center (TDC)—with both valves closed on #1 cylinder.

10 With the rocker arms contacting the base circle of the cam, insert a feeler gauge of the specified thickness between the camshaft and the rocker arm. There should be a slight drag on the feeler gauge.

11 If the valve clearance (lash) is not correct, loosen the retaining nut and turn the valve adjusting screw with a screwdriver to achieve the proper clearance.

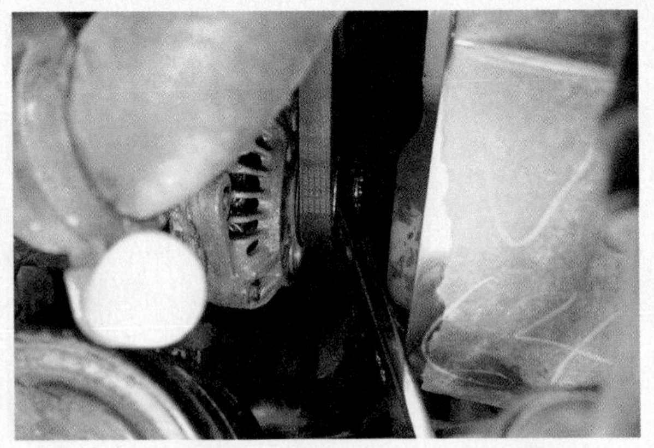

12 After adjusting the valves that are closed, rotate the engine one full rotation until the engine timing marks again align.

13 The engine is rotated until the timing marks again align indicating that the companion cylinder will now be in position for valve clearance measurement.

14 On some engines, it is necessary to watch the direction the rotor is pointing to help determine how far to rotate the engine. Always follow the vehicle manufacturer's recommended procedure.

15 The technician is using a feeler gauge that is one-thousandth of an inch thinner and another one-thousandth of an inch thicker than the specified clearance as a double check that the clearance is correct.

16 Adjusting a valve takes both hands—one to hold the wrench to loosen and tighten the lock nut and one to turn the adjusting screw. Always double-check the clearance after an adjustment is made.

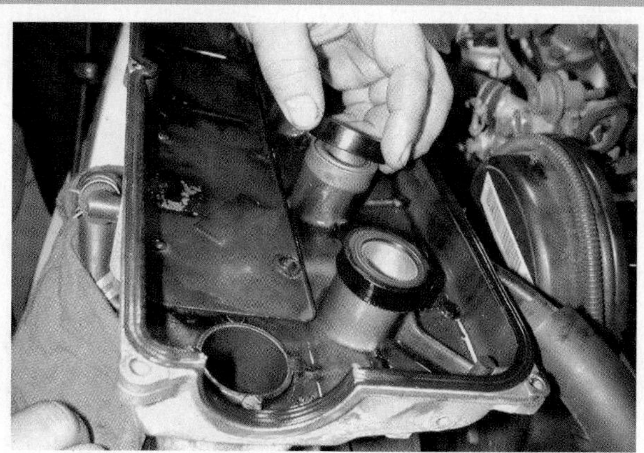

17 After all valves have been properly measured and adjusted as necessary, start the reassembly process by replacing all gaskets and seals as specified by the vehicle manufacturer.

18 Reinstall the valve cover being careful to not pinch a wire or vacuum hose between the cover and the cylinder head.

CONTINUED ▶

19 Use a torque wrench and torque the valve cover retaining bolts to factory specifications.

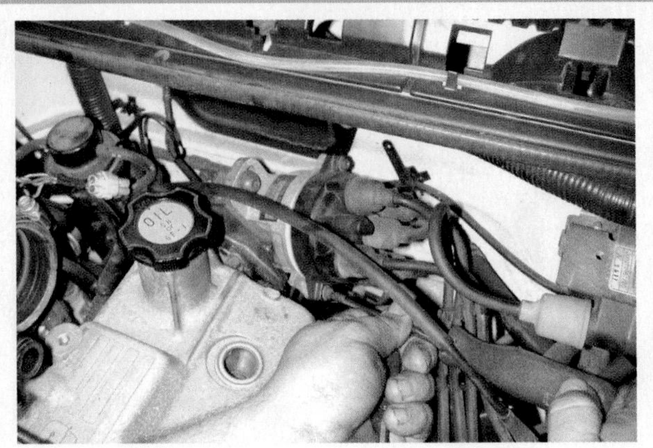

20 Reinstall the distributor cap.

21 Reinstall the spark plug wires and all brackets that were removed to gain access to the valve cover.

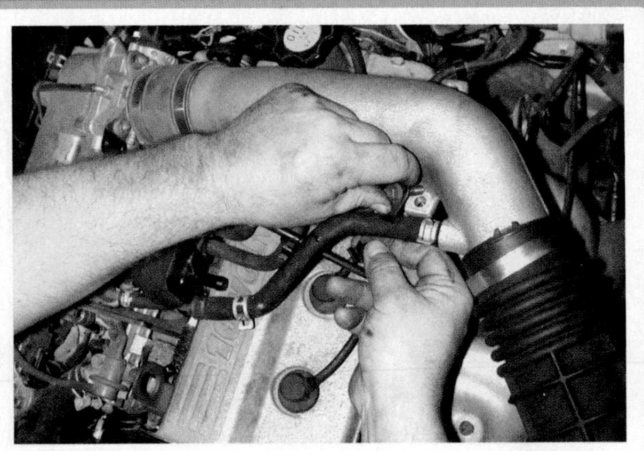

22 Reconnect all vacuum and air hoses and tubes. Replace any vacuum hoses that are brittle or swollen with new ones.

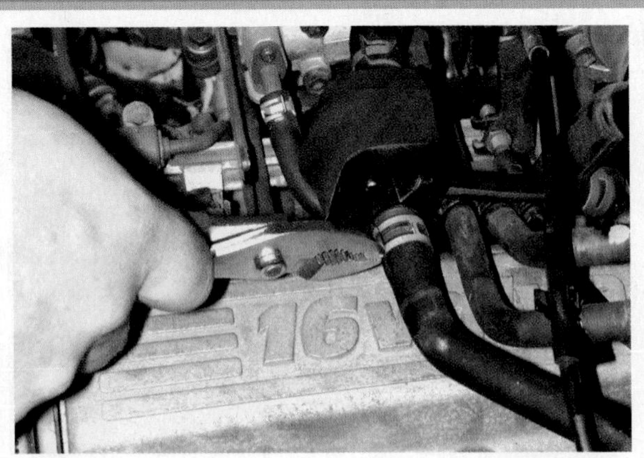

23 Be sure that the clips are properly installed. Start the engine and check for proper operation.

24 Double-check for any oil or vacuum leaks after starting the engine.

SUMMARY

1. Thermostats can fail in the following ways:
   - Stuck open
   - Stuck closed
   - Stuck partially open
   - Skewed
2. A water pump should be replaced if any of the following conditions are present:
   - Leaking from the weep hole
   - Noisy bearing
   - Loose bearing
   - Lack of normal circulation due to worn impeller blades
3. A leaking intake manifold gasket can cause coolant to get into the oil or oil into the coolant, as well as other faults, such as a poor running engine.
4. When a timing belt is replaced, most vehicle manufacturers also recommend that the following items be replaced:
   - Tensioner assembly
   - Water pump
   - Camshaft seal(s)
   - Front crankshaft seal
5. When working on a Toyota/Lexus hybrid electric vehicle (HEV), be sure that the key is off and out of the ignition and the READY light is off.

REVIEW QUESTIONS

1. How can a thermostat fail?
2. How can a water pump fail requiring replacement?
3. What will happen to the engine if the intake manifold gasket fails?
4. Why must timing belts be replaced?
5. Why is it important that the READY light be out on the dash before working under the hood of a hybrid electric vehicle?

CHAPTER QUIZ

1. A thermostat can fail in which way?
   a. Stuck open
   b. Stuck closed
   c. Stuck partially open
   d. Any of the above
2. A skewed thermostat means it is __________.
   a. Working, but not at the correct temperature
   b. Not working
   c. Missing the thermo wax in the heat sensor
   d. Contaminated with coolant
3. Coolant drained from the cooling system when replacing a thermostat or water pump should be __________.
   a. Reused
   b. Disposed of properly or recycled
   c. Filtered and reinstalled after the repair
   d. Poured down a toilet
4. A water pump can fail to provide the proper amount of flow of coolant through the cooling system if what has happened?
   a. The coolant is leaking from the weep hole
   b. The bearing is noisy
   c. The impeller blades are worn or slipping on the shaft
   d. A bearing failure has caused the shaft to become loose
5. What factor causes intake manifold gaskets on a V-type engine to fail?
   a. Fretting
   b. Coolant damage
   c. Relative movement between the intake manifold and the cylinder head
   d. All of the above
6. A defective thermostat can cause the Powertrain Control Module to set what diagnostic trouble code (DTC)?
   a. P0171
   b. P0172
   c. P0128
   d. P0300
7. A replacement plastic intake manifold may have a different design or appearance from the original factory-installed part.
   a. True
   b. False
8. The torque specifications for many plastic intake manifolds are in what unit?
   a. Pound-inches
   b. Pound-feet
   c. Ft-lbs per minute
   d. Lb-ft per second
9. When replacing a timing belt, what other part(s) many experts and vehicle manufacturers recommend to be replaced?
   a. Tensioner assembly
   b. Water pump
   c. Camshaft oil seal(s)
   d. All of the above
10. Hybrid electric vehicles usually require special engine oil of what viscosity?
    a. SAE 5W-30
    b. SAE 10W-30
    c. SAE 0W-20
    d. SAE 5W-40

chapter 15

ADVANCED STARTING AND CHARGING SYSTEMS DIAGNOSIS

LEARNING OBJECTIVES

After studying this chapter, the reader will be able to:

1. Describe the purpose and function of a battery.
2. Discuss methods that can be used to check the condition of a battery.
3. Describe how to perform a battery drain test and how to isolate the cause.
4. Explain how to test the condition of the starter.
5. List the steps necessary to perform a voltage-drop test.
6. Explain how to test the alternator.

This chapter will help you prepare for Engine Repair (A8) ASE certification test content area "A" (General Diagnosis).

KEY TERMS

AC ripple voltage 253
Alternator 233
Ampere-hour 233
Battery 233
Battery electrical drain test 240
Battery voltage correction factor 234
CA 233
Capacity test 237
CCA 233
Charging circuit 233
Conductance testing 238
Cranking circuit 233
DE 251
ELD 256
IOD 240
Load test 237
LRC 256
MCA 233
Neutral safety switch 244
Open-circuit battery voltage test 235
Parasitic load 240
Reserve capacity 233
Ripple current 254
SRE 251
State of charge 239
Surface charge 235
Voltage-drop test 247

FIGURE 15–1 This battery shows a large "1000" on the front panel but this is the CA rating and not the more important CCA rating. Always compare batteries with the same rating.

Just as in the old saying "If Mother isn't happy—no one is happy," the battery, the starter, and the charging system have to function correctly for the engine performance to be satisfactory.

PURPOSE AND FUNCTION OF A BATTERY

The primary purpose of an automotive **battery** is to provide a source of electrical power for starting and for electrical demands that exceed alternator output. The battery also acts as a voltage stabilizer for the entire electrical system. The battery is a voltage stabilizer because it acts as a reservoir where large amounts of current (amperes) can be removed quickly during starting and replaced gradually by the **alternator** during charging. The battery *must* be in good (serviceable) condition before the charging system and the cranking system can be tested. For example, if a battery is discharged, the **cranking circuit** (starter motor) could test as being defective because the battery voltage might drop below specifications. The **charging circuit** could also test as being defective because of a weak or discharged battery. It is important to test the vehicle battery before further testing of the cranking or charging system.

BATTERY RATINGS

Batteries are rated according to the amount of current they can produce under specific conditions.

COLD-CRANKING AMPERES Every automotive battery must be able to supply electrical power to crank the engine in cold weather and still provide voltage high enough to operate the ignition system for starting. The cold-cranking power of a battery is the number of amperes that can be supplied at 0°F (−18°C) for 30 seconds while the battery still maintains a voltage of 1.2 volts per cell or higher. This means that the battery voltage would be 7.2 volts for a 12 volt battery and 3.6 volts for a 6 volt battery. The cold-cranking performance rating is called **cold-cranking amperes (CCA)**. Try to purchase a battery that offers the highest CCA for the money. ● **SEE FIGURE 15–1**.

CRANKING AMPERES **Cranking amperes (CA)** are not the same as CCA, but are often advertised and labeled on batteries. The designation CA refers to the number of amperes that can be supplied by the battery at 32°F (0°C). This rating results in a higher number than the more stringent rating of CCA.

MARINE CRANKING AMPERES **Marine cranking amperes (MCA)** rating is similar to the cranking amperes (CA) rating and is tested at 32°F (0°C).

AMPERE-HOUR RATING The **Ampere-hour (Ah)** is how many amperes can be discharged from the battery before dropping to 10.5 volts over a 20-hour period. A battery that is able to supply 3.75 amperes for 20 hours has a rating of 75 ampere-hours ($3.75 \times 20 = 75$).

RESERVE CAPACITY The **reserve capacity** rating for batteries is *the number of minutes* for which the battery can produce 25 amperes and still have a battery voltage of 1.75 volts per cell (10.5 volts for a 12 volt battery). This rating is actually a measurement of the time for which a vehicle can be driven in the event of a charging system failure.

? FREQUENTLY ASKED QUESTION

How Can a Defective Battery Affect Engine Performance?

A weak or discharged battery should be replaced as soon as possible. A weak battery causes a constant load on the alternator that can cause the stator windings to overheat and fail. Low battery voltage also affects the electronic fuel-injection system. The computer senses low battery voltage and increases the fuel injector on-time to help compensate for the lower voltage to the fuel pump and fuel injectors. This increase in injector pulse time is added to the calculated pulse time and is sometimes called the **battery voltage correction factor**. Reduced fuel economy could therefore be the result of a weak or defective battery.

? FREQUENTLY ASKED QUESTION

Should Batteries Be Kept Off of Concrete Floors?

All batteries should be stored in a cool, dry place when not in use. Many technicians have been warned not to store or place a battery on concrete. According to battery experts, it is the temperature difference between the top and the bottom of the battery that causes a difference in the voltage potential between the top (warmer section) and the bottom (colder section). It is this difference in temperature that causes self-discharge to occur.

In fact, submarines cycle seawater around their batteries to keep all sections of the battery at the same temperature to help prevent self-discharge.

Therefore, always store or place batteries up off the floor and in a location where the entire battery can be kept at the same temperature, avoiding extreme heat and freezing temperatures. Concrete cannot drain the battery directly, because the case of the battery is a very good electrical insulator.

BATTERY SERVICE SAFETY CONSIDERATIONS

Batteries contain acid and release explosive gases (hydrogen and oxygen) during normal charging and discharging cycles. To help prevent physical injury or damage to the vehicle, always adhere to the following safety procedures:

1. Whenever working on any electrical component on a vehicle, disconnect the negative battery cable from the battery. When the negative cable is disconnected, all electrical circuits in the vehicle will be open, which will prevent accidental electrical contact between an electrical component and ground. Any electrical spark has the potential to cause explosion and personal injury.
2. Wear eye protection whenever working around any battery.
3. Wear protective clothing to avoid skin contact with battery acid.
4. Always adhere to all safety precautions as stated in the service procedures for the equipment used for battery service and testing.
5. Never smoke or use an open flame around any battery.

FREQUENTLY ASKED QUESTION

What Can Cause a Battery to Explode?

Batteries discharge hydrogen and oxygen gases when being charged. If there happens to be a flame or spark, the hydrogen will burn. The oxygen can also help contribute to an explosion of a small pocket of hydrogen.

BATTERY VISUAL INSPECTION

The battery and battery cables should be included in the list of items checked during a thorough visual inspection. Check the battery cables for corrosion and tightness. ● **SEE FIGURE 15–2.**

NOTE: On side-post batteries, grasp the battery cable near the battery and attempt to move the cable in a clockwise direction in an attempt to tighten the battery connection.

If possible, remove the covers and observe the level of the electrolyte. ● **SEE FIGURE 15–3.**

FIGURE 15–2 Corrosion on a battery cable could be an indication that the battery is either being overcharged or is sulfated, creating a lot of gassing of the electrolyte.

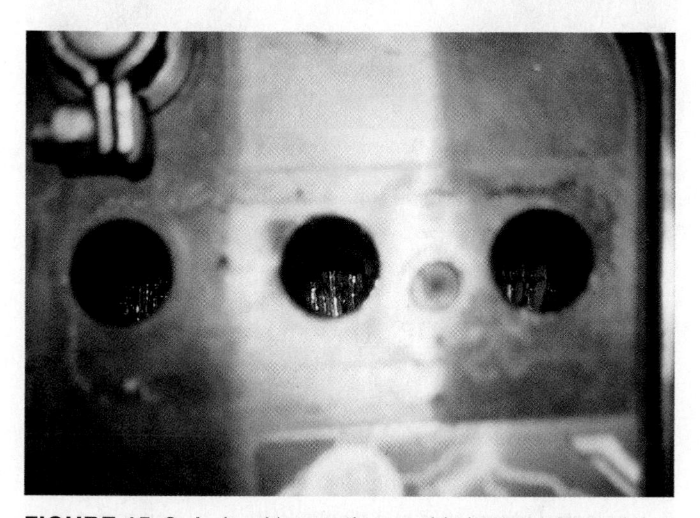

FIGURE 15–3 A visual inspection on this battery showed that the electrolyte level was below the plates in all cells.

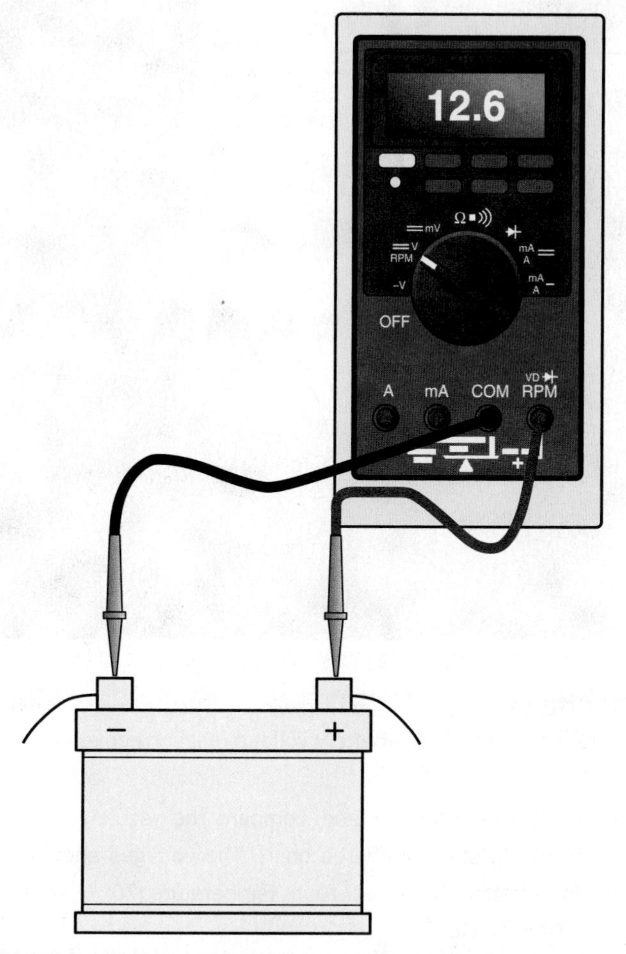

FIGURE 15–4 Using a DMM to measure the open-circuit voltage of a battery.

BATTERY VOLTAGE TEST

Testing the battery voltage with a voltmeter is a simple method for determining the state of charge of any battery. ● **SEE FIGURE 15–4**. The voltage of a battery does not necessarily indicate whether the battery can perform satisfactorily, but it does indicate to the technician more about the battery's condition than a simple visual inspection. A battery that *looks* good may not be good. This test is commonly called an **open-circuit battery voltage test** because it is conducted with an open circuit—with no current flowing and no load applied to the battery.

To test battery voltage, perform the following steps:

1. Connect a voltmeter to the positive (+) and negative (–) terminals of the battery. Set the voltmeter to read DC volts.
2. If the battery has just been charged or the vehicle has recently been driven, it is necessary to remove the surface charge from the battery before testing. A **surface charge** is a charge of higher-than-normal voltage that is only on the surface of the battery plates. The surface charge is quickly removed whenever the battery is loaded and therefore does not accurately represent the true state of charge of the battery.
3. To remove the surface charge, turn the headlights on high beam (brights) for one minute, then turn the headlights off and wait two minutes.

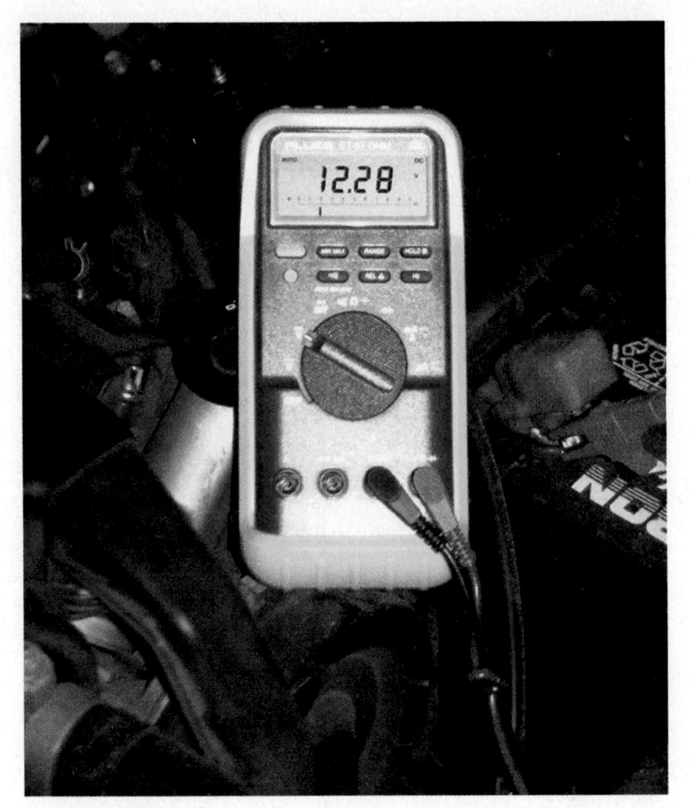

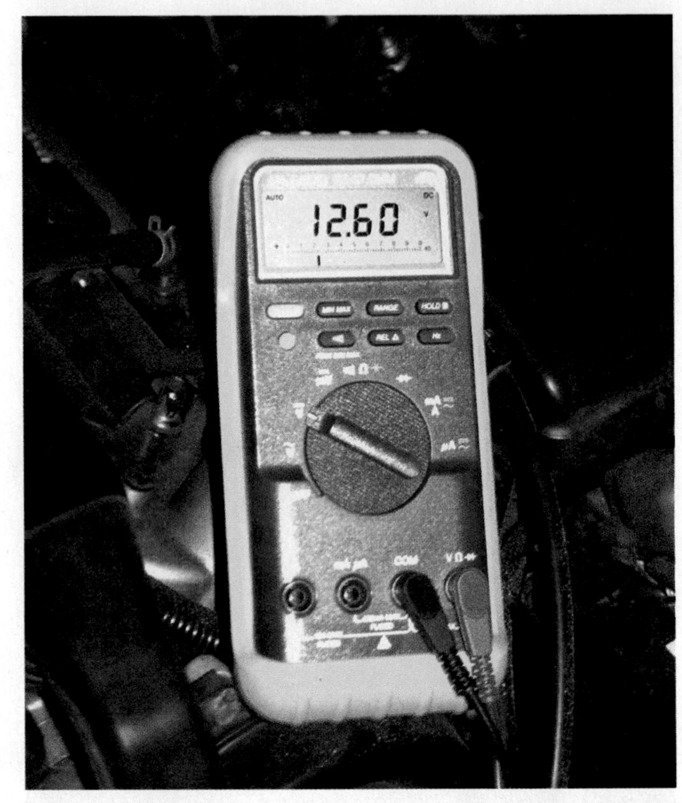

(a) (b)

FIGURE 15–5 (a) Voltmeter showing the battery voltage when the headlights were on (engine off) for one minute. (b) Headlights were turned off and the battery voltage quickly recovered to indicate 12.6 volts.

4. Read the voltmeter and compare the results with the following state-of-charge chart. The voltages shown are for a battery at or near room temperature (70° to 80°F or 21° to 27°C).

NOTE: Watch the voltmeter when the headlights are turned on. A new good battery will indicate a gradual drop in voltage, whereas a weak battery will indicate a more rapid drop in voltage. Soon the voltage will stop dropping and will stabilize. A good new battery will likely stabilize above 12 volts. A weak older battery may drop below 11 volts. After turning off the headlights, the faster the recovery, generally, the better the battery.

SEE FIGURE 15–5 AND CHART 15–1.

BATTERY VOLTAGE (V)	STATE OF CHARGE
12.6 or higher	100% charged
12.4	75% charged
12.2	50% charged
12.0	25% charged
11.9 or lower	Discharged

CHART 15–1

Battery voltage can be one indicator of the state-of-charge (SOC) of a battery.

TECH TIP

Use a Scan Tool to Check the Battery, Starter, and Alternator!

General Motors and Chrysler vehicles as well as selected others that can display data to a scan tool can be easily checked for proper operating voltage. Most scan tools can display battery or system voltage and engine speed in RPM (revolutions per minute). Connect a scan tool to the data link connector (DLC) and perform the following while watching the scan tool display (**SEE FIGURE 15–6**).

Many scan tools are also capable of recording or graphing engine data while cranking including:

- RPM during cranking should be 80 to 250 RPMs.
- Battery voltage during cranking should be above 9.6 volts.

NOTE: Normal readings for a good battery and starter would be 10.5 to 11.5 volts.

- Battery voltage after engine starts should be 13.5 to 15.0 volts.

FIGURE 15–7 A Bear Automotive starting and charging tester. This tester automatically loads the battery for 15 seconds to remove the surface charge, waits 30 seconds to allow the battery to recover, and then loads the battery again. The LCD indicates the status of the battery.

FIGURE 15–6 Using a scan tool to check battery voltage.

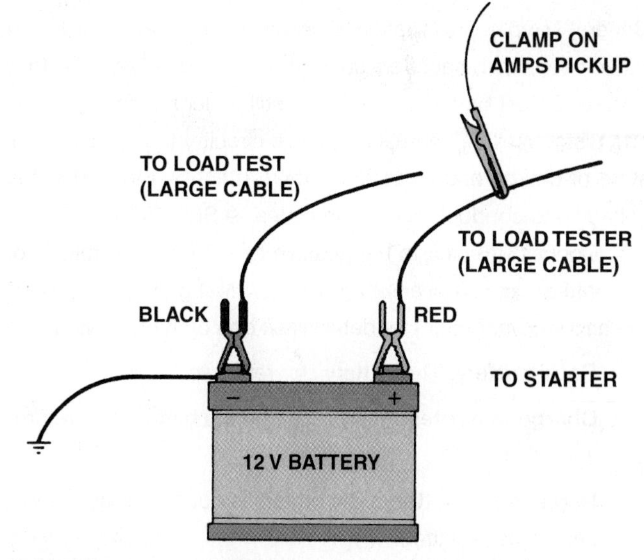

FIGURE 15–8 A typical battery load tester hookup.

BATTERY LOAD TESTING

One method to determine the condition of any battery is the **load test**, also known as a **capacity test**. Most automotive starting and charging testers use a carbon pile to create an electrical load on the battery. The amount of the load is determined by the original capacity of the battery being tested. The capacity is measured in cold-cranking amperes, which is the number of amperes that a battery can supply at 0°F (−18°C) for 30 seconds. An older type of battery rating is called the ampere-hour rating. The proper electrical load to be used to test a battery is one-half of the CCA rating or three times the ampere-hour rating, with a minimum of a 150 ampere load. Apply the load for a full 15 seconds and observe the voltmeter at the end of the 15-second period while the battery is still under load. A good battery should indicate above 9.6 volt.

NOTE: This test is sometimes called the *one minute test*, because many battery manufacturers recommend performing the load test twice, using the first load period (15 seconds) to remove the surface charge on the battery, then waiting for 30 seconds to allow time for the battery to recover, and then loading the battery again for 15 seconds. Total time required is 60 seconds (15 + 30 + 15 = 60 seconds, or 1 minute). This method provides a true indication of the condition of the battery. ● **SEE FIGURES 15–7 AND 15–8.**

If the battery fails the load test, recharge the battery and retest. If the battery fails the load test again, replace the battery.

FIGURE 15–9 An electronic battery tester.

CONDUCTANCE TESTING

General Motors Corporation, Chrysler Corporation, Ford, and other vehicle manufacturers specify that a **conductance testing** be used to test batteries in vehicles still under factory warranty. The tester uses its internal electronic circuitry to determine the state of charge and capacity of the battery by measuring the voltage and conductance of the plates. ● **SEE FIGURE 15–9.**

Connect the unit to the positive and negative terminals of the battery, and after entering the CCA rating (if known), push the arrow keys. The tester determines one of the following:

- **Good battery.** The battery can return to service.
- **Charge and retest.** Fully recharge the battery and return it to service.
- **Replace the battery.** The battery is not serviceable and should be replaced.
- **Bad cell—replace.** The battery is not serviceable and should be replaced.

CAUTION: Test results can be incorrectly reported on the display if proper, clean connections to the battery are not made. Also be sure that all accessories and the ignition switch are in the off position.

? FREQUENTLY ASKED QUESTION

What Are Some Symptoms of a Weak or Defective Battery?

There are several warning signs that may indicate that a battery is near the end of its useful life, including:

- **Uses water in one or more cells.** This indicates that the plates are sulfated and that, during the charging process, the water in the electrolyte is being turned into separate hydrogen and oxygen gases.
- **Excessive corrosion on battery cables or connections.** Corrosion is more likely to occur if the battery is sulfated, creating hot spots on the plates. When the battery is being charged, the acid fumes are forced out of the vent holes and get onto the battery cables, connections, and even on the tray underneath the battery.
- **Slower-than-normal engine cranking.** When the capacity of the battery is reduced due to damage or age, it is less likely to supply the necessary current for starting the engine, especially during cold weather.

JUMP STARTING

To safely jump-start a vehicle without doing any harm, use the following procedure:

1. Be certain the ignition switch is off on both vehicles.
2. Connect good-quality copper jumper cables as indicated in the guide in ● **FIGURE 15–10.**
3. Start the vehicle with the good battery and allow it to run for 5 to 10 minutes. This allows the alternator of the good vehicle to charge the battery on the disabled vehicle.
4. Start the disabled vehicle and, after the engine is operating smoothly, disconnect the jumper cables in the reverse order of step 2.

   NOTE: To help prevent accidental touching of the jumper cables, simply separate them into two cables and attach using wire (cable) ties or tape so that the clamps are offset from each other, making it impossible for them to touch.

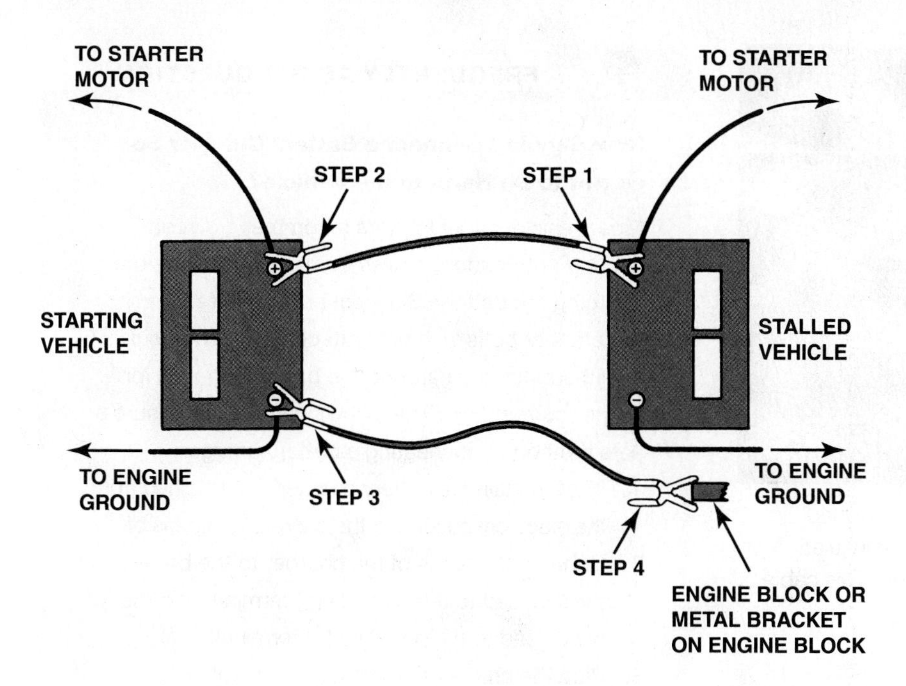

FIGURE 15–10 Jumper cable usage guide.

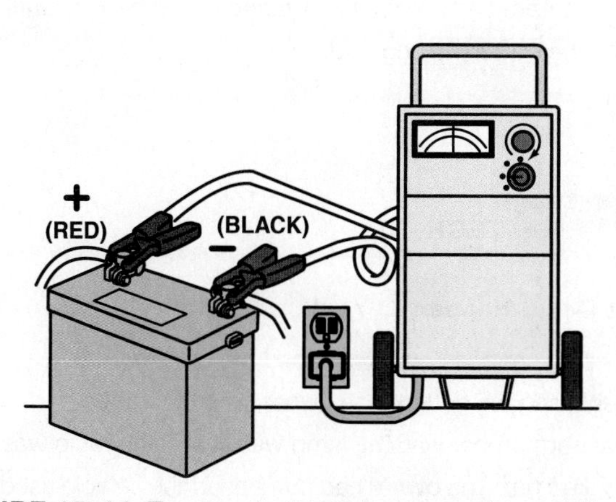

FIGURE 15–11 To use a battery charger, make sure the charger is connected to the battery before plugging in the charger.

OPEN CIRCUIT VOLTAGE (V)	STATE OF CHARGE (%)	CHARGING TIME (MIN) TO FULL CHARGE AT 80°F (27°C)* AT 60 A					
		at 60 A	at 50 A	at 40 A	at 30 A	at 20 A	at 10 A
12.6	100	Full Charge					
12.4	75	15	20	27	35	48	90
12.2	50	35	45	55	75	95	180
12.0	25	50	65	85	115	145	280
11.8	0	65	85	110	150	195	370

CHART 15–2

A chart that can be used to estimate the charging time based on battery voltage and charging rate.

* If colder, allow additional time.

BATTERY CHARGING

If the **state of charge** of a battery is low, it must be recharged. It is best to slow-charge any battery to prevent possible overheating damage to the battery. Remember, it may take 8 hours or more to charge a fully discharged battery. The initial charge rate should be about 35 amperes for 30 minutes to help start the charging process. Fast-charging a battery increases the temperature of the battery and can cause warping of the plates inside the battery. Fast-charging also increases the amount of gassing (release of hydrogen and oxygen), which can create a health and fire hazard. The battery temperature should not exceed 125°F (hot to the touch). Most batteries should be charged at a rate equal to 1% of the battery's CCA rating. ● **SEE FIGURE 15–11**.

Fast charge: 15 amperes maximum

Slow charge: 5 amperes maximum

● **SEE CHART 15–2** for battery charging times at various battery voltages and charging rates.

FIGURE 15–12 This battery cable was found corroded underneath. The corrosion had eaten through the insulation yet was not noticeable without careful inspection. This cable should be replaced.

BATTERY SERVICE

Before returning the vehicle to the customer, check and service the following items as necessary:

1. Neutralize and clean any corrosion from the battery terminals using a solution of baking soda and water.
2. Carefully inspect the battery cables by visual inspection. ● **SEE FIGURE 15–12.**
3. Check the tightness and cleanliness of all ground connections.

BATTERY ELECTRICAL DRAIN TEST

The **battery electrical drain test** determines if some component or circuit in a vehicle or truck is causing a drain on the battery when everything is off. This test is also called the **ignition off-draw (IOD)** or **parasitic load** test. This test should be performed whenever one of the following conditions exists:

1. Whenever a battery is being charged or replaced (a battery drain could have been the cause for charging or replacing the battery).
2. Whenever the battery is suspected of being drained.

Normal battery drain on a vehicle equipped with electronic radio, climate control, computerized fuel injection, and so forth, is usually about 20 to 30 mA (0.02 to 0.03 A). Most vehicle manufacturers recommend repairing the cause of any drain that exceeds 50 mA (0.05 A).

? FREQUENTLY ASKED QUESTION

How Should I Connect a Battery Charger So as Not to Do Harm to the Vehicle?

Most vehicle manufacturers recommend disconnecting both battery cables from the battery before charging the battery. Side-post batteries require adapters or bolts with nuts attached to permit sufficient surface area around the battery terminal for proper current flow. The following steps will ensure a safe method of connecting a battery charger:

1. Make certain the battery charger is unplugged from the electrical outlet and the charger control is off.
2. Connect the leads of the charger to the battery—the red lead to the positive (+) terminal and the black lead to the negative (−) terminal.
3. Plug the charger into the electrical outlet.
4. Set the controls to the fast (high) rate for about 30 minutes or until the battery starts to take a charge. After 30 minutes, reduce the charge rate to about 1% of the CCA rating of the battery until the battery is charged.

TECH TIP

It Could Happen to You!

The owner of a Toyota replaced the battery. After replacing the battery, the owner noted that the "airbag" amber warning lamp was lit and the radio was locked out. The owner had purchased the vehicle used from a dealer and did not know the 4-digit security code needed to unlock the radio. Determined to fix it, the owner tried three 4-digit numbers, hoping that one of them would work. However, after three tries, the radio became permanently disabled.

Frustrated, the owner went to a dealer. It cost over $300 to fix the problem. A special tool was required to easily reset the airbag lamp. The radio had to be removed, sent to an out-of-state authorized radio service center, and then reinstalled into the vehicle.

Therefore, before disconnecting the battery, please be certain that the owner has the security code for a security-type radio. A "memory saver" may be needed to keep the radio powered when the battery is being disconnected. ● **SEE FIGURE 15–13.**

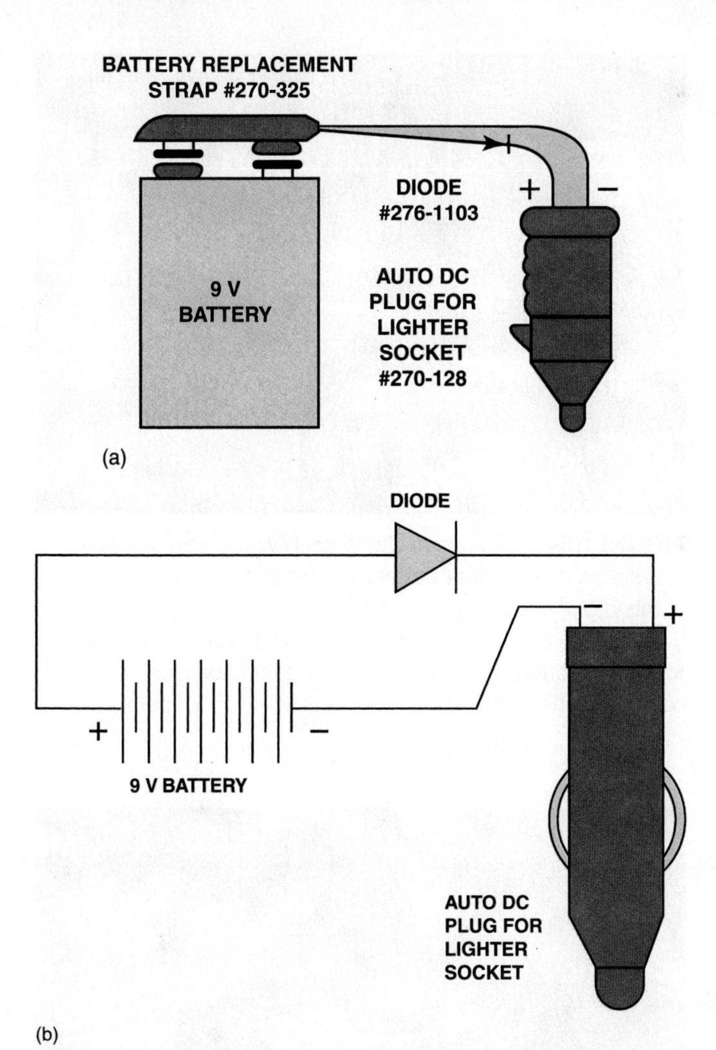

FIGURE 15–13 (a) Memory saver. The part numbers represent components from Radio Shack®. (b) A schematic drawing of the same memory saver.

Use a MIN/MAX Feature to Check for Battery Electrical Drain

Most digital multimeters that feature a "data hold," MIN/MAX, or recording feature can be used when the meter is set up to read DC amperes. This is especially helpful if the battery drain is not found during routine tests in the shop. The cause or source of this drain may only occur when the vehicle cools down at night or after it sits for several hours. Connect the ammeter in series with the disconnected negative battery cable and set the meter to record. Refer to the meter instruction booklet if necessary to be assured of a proper setup. The next morning, check the meter for the maximum, minimum, and average readings. For example,

MAX = 0.89 A (over specifications of 0.05 A)
MIN = 0.02 A (typical normal reading)
Average = 0.76 A

Because the average is close to the maximum, the battery electrical drain was taking place during most of the duration of the test.

NOTE: Some manufacturers relate maximum allowable parasitic load to the size of the battery. The higher the battery capacity, the greater the allowable load. The maximum allowable drain on a battery can be calculated by dividing the reserve capacity of the battery in minutes by four to get the maximum allowable drain in milliamps. For example, if a battery had a reserve capacity of 100 minutes, it would have a maximum allowable parasitic load of 25 mA (100 ÷ 4 = 25 mA).

NOTE: Many electronic components do draw a slight amount of current from the battery all the time with the ignition off. These components include:

1. **Digital clocks**
2. **Electronically tuned radios for station memory and clock circuits (if the vehicle is so equipped)**
3. **The engine control computer (if the vehicle is so equipped), through slight diode leakage**
4. **The alternator, through slight diode leakage**

These components may cause a voltmeter to read full battery voltage if it is connected between the negative battery terminal and the removed end of the negative battery cable. Using a voltmeter to measure battery drain is *not* recommended by most vehicle manufacturers. The high internal resistance of the voltmeter results in an irrelevant reading that does not tell the technician if there is a problem.

BATTERY ELECTRICAL DRAIN TESTING USING AN AMMETER

The ammeter method is the most accurate way to test for a possible battery drain. Connect an ammeter in series between the battery terminal (post) and the disconnected cable. (Normal battery drain is 0.020 to 0.030 A and any drain greater than 0.050 A should be found and corrected.) Many digital multimeters have an ammeter scale that can be used to safely and accurately test for an abnormal parasitic drain.

CAUTION: Some vehicle manufacturers recommend that a test light be used before connecting an ammeter when checking for a battery drain. If the drain is large enough to light a test light, the ammeter may be damaged. Be certain to use an ammeter that is rated to read the anticipated amperage.

FIGURE 15–14 This mini clamp-on DMM is being used to measure the amount of battery electrical drain that is present. In this case, a reading of 20 mA (displayed on the meter as 00.02 A) is within the normal range of 20 to 30 mA. Be sure to clamp around all of the positive battery cables or all of the negative battery cables, whichever is easiest to clamp.

PROCEDURE FOR BATTERY ELECTRICAL DRAIN TEST

The fastest and easiest method to measure battery electrical drain is to connect an inductive DC ammeter that is capable of measuring low current (10 mA). ● **SEE FIGURE 15–14** for an example of a clamp-on digital multimeter being used to measure battery drain.

Following is the procedure for performing the battery electrical drain test using an ammeter:

1. Make certain that all lights, accessories, and the ignition are off.
2. Check all vehicle doors to be certain that the interior courtesy (dome) lights are off.
3. Disconnect the *negative* (–) battery cable and install a parasitic load tool as shown in ● **FIGURE 15–15**.
4. Start the engine and drive the vehicle about 10 minutes, being sure to turn on all the lights and accessories, including the radio.
5. Turn the engine and all accessories off, including the underhood light.
6. Connect an ammeter across the parasitic load tool switch and wait 10 minutes or longer for all computers to go to sleep and circuits to shut down.

FIGURE 15–15 After connecting the shutoff tool, start the engine and operate all accessories. Stop the engine and turn off everything. Connect the ammeter across the shutoff switch in parallel. Wait 20 minutes. This time allows all electronic circuits to "time out" or shut down. Open the switch—all current now will flow through the ammeter. A reading greater than specified, usually greater than 50 mA (0.05 A), indicates a problem that should be corrected.

7. Open the switch on the load tool and read the battery electrical drain on the meter display.

   Results: Normal = 10 to 30 mA (0.02 to 0.03 A)

   Maximum allowable = 50 mA (0.05 A) (Industry standards—some vehicle manufacturers' specifications can vary)

   Be sure to reset the clock and anti-theft radio, if equipped. ● **SEE FIGURE 15–16.**

FIGURE 15–16 The battery was replaced in this Acura and the radio displayed "code" when the replacement battery was installed. Thankfully, the owner had the five-digit code required to unlock the radio.

FINDING THE SOURCE OF THE DRAIN

If there is a drain, check and temporarily disconnect the following components:

1. Cell phone or MP3 player still connected to the vehicle
2. Glove compartment light
3. Trunk light

If after disconnecting these components the battery drain can still light the test light or draw more than 50 mA (0.05 A), disconnect one fuse at a time from the fuse box until the test light goes out or the ammeter reading drops. If the drain drops to normal after one fuse is disconnected, the source of the drain is located in that particular circuit, as labeled on the fuse box. As fuses are pulled, they should not be reinstalled until the end of the test. Reinstalling a fuse can reset a module and foul up the test. Start at the fuses farthest from the battery and work toward the battery until the faulty circuit is found. Note that many vehicles have multiple fuse boxes. Continue to disconnect the *power-side* wire connectors from each component included in that particular circuit until the ammeter reads a normal amount of draw. The source of the battery drain can then be traced to an individual component or part of one circuit. If none of the fuses causes the drain to stop, disconnect the alternator output lead. A shorted diode in the alternator could be the cause.

WHAT TO DO IF A BATTERY DRAIN STILL EXISTS AFTER ALL THE FUSES ARE DISCONNECTED

If all the fuses have been disconnected and the drain still exists, the source of the drain has to be between the battery and the fuse box. The most common sources of drain under the hood include the following:

1. **The alternator.** Disconnect the alternator wires and retest. If the draw is now within acceptable limits, the problem is a defective diode(s) in the alternator.
2. **The starter solenoid (relay) or wiring near its components.** These are also a common source of battery drain, due to high current flows and heat, which can damage the wire or insulation.

CRANKING CIRCUIT

The cranking circuit includes those mechanical and electrical components required to crank the engine for starting. The cranking force in the early 1900s was the driver's arm. Modern cranking circuits include the following:

1. **Starter motor.** The starter is normally a 0.5 to 2.6 horsepower (0.4 to 2.0 kilowatts) electric motor that can develop nearly 8 horsepower (6 kilowatts) for a very short time when first cranking a cold engine.
2. **Battery.** The battery must be of the correct capacity and be at least 75% charged to provide the necessary current and voltage for correct operation of the starter.
3. **Starter solenoid or relay.** The high current required by the starter must be able to be turned on and off. A large switch would be required if the current were controlled by the driver directly. Instead, a small current switch (ignition switch) operates a solenoid or relay that controls the high starter current.
4. **Starter drive.** The starter drive uses a small gear that contacts the engine flywheel gear and transmits starter motor power to rotate the engine.
5. **Ignition switch.** The ignition switch and safety control switches control the starter motor operation. ● **SEE FIGURES 15–17 AND 15–18.**

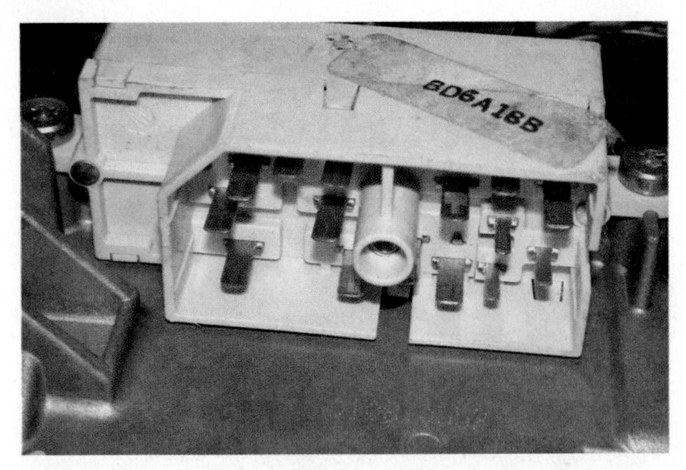

FIGURE 15–17 A typical ignition switch showing all of the electrical terminals after the connector has been removed.

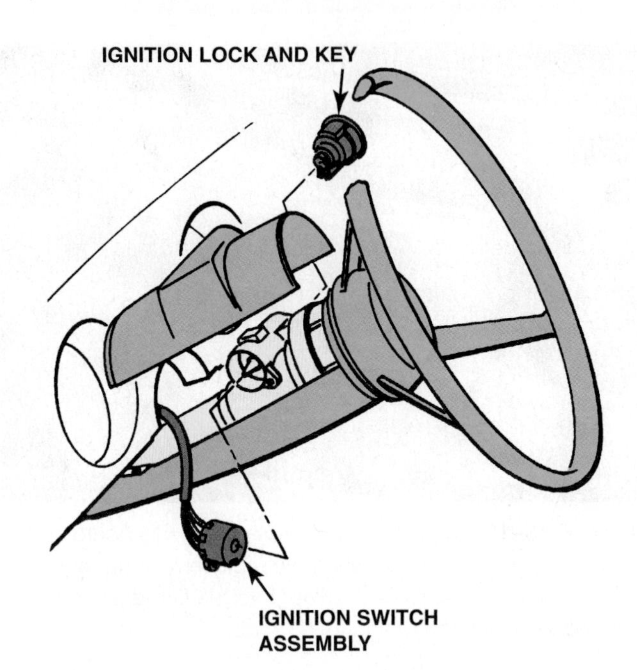

FIGURE 15–18 Some column-mounted ignition switches act directly on the contact points, whereas others use a link from the lock cylinder to the ignition switch.

The engine is cranked by an electric motor that is controlled by a key-operated ignition switch or the PCM on vehicles equipped with electronic starting. The ignition switch will not operate the starter unless the automatic transmission is in neutral or park. This is to prevent an accident that might result from the vehicle moving forward or backward when the engine is started. Many automobile manufacturers use a **neutral safety switch** or a clutch switch that opens the circuit between the ignition switch and the starter to prevent starter motor operation unless the gear selector is in neutral or park. The safety switch can either be attached to the steering column inside the vehicle near the floor or on the side of the transmission/transaxle. According to vehicle manufacturing engineers, starters can be expected to start an engine 25,000 times during normal life of the vehicle. ● **SEE FIGURE 15–19.**

FIGURE 15–19 A typical solenoid-operated starter.

DIAGNOSING STARTER PROBLEMS USING VISUAL INSPECTION

For proper operation, all starters require that the vehicle battery be at least 75% charged and that both power-side and ground-side battery cables be free from excessive voltage drops. The following should be carefully checked as part of a thorough visual inspection:

- Carefully check the battery cables for tightness both at the battery and at the starter, and block connections. ● **SEE FIGURE 15–20.**
- Check to see if the heat shield (if equipped) is in place.

FIGURE 15–20 Carefully inspect all battery terminals for corrosion.

FIGURE 15–21 Always check the battery, using a conductance or load tester. A battery showing a green charge indicator does not mean that the battery is good.

FIGURE 15–22 When connecting a starter tester such as a Sun VAT 45 to the vehicle, make certain that the inductive probe is placed over all of the cables or wires from the battery.

- Check for any nonstock add-on accessories or equipment that may drain the battery such as a sound system, extra lighting, and so on.
- Crank the engine. Feel the battery cables and connections. If any cables or connections are hot to the touch, then an excessive voltage drop is present or the starter is drawing too much current. The engine itself could be binding. Repair or replace the components or connections as needed.

STARTER TESTING ON THE VEHICLE

CHECK BATTERY Before performing a starter amperage test, be certain that the battery is sufficiently charged (75% or more) and capable of supplying adequate starting current. ● **SEE FIGURE 15–21.**

STARTER AMPERAGE TEST A starter amperage test should be performed whenever the starter fails to operate normally (is slow in cranking) or as part of a routine electrical system inspection. ● **SEE FIGURE 15–22.** Some service manuals specify normal starter amperage for starter motors being tested on the vehicle; however, most service manuals only give the specifications for bench-testing a starter without a load applied. These specifications are helpful in making certain that a repaired starter meets exact specifications, but they do not apply to starter testing on the vehicle. If exact specifications are not available, the following can be used as general maximum specifications for testing a starter on the vehicle. Any ampere reading lower than these are acceptable:

- Four-cylinder engines = 150 to 185 amperes (normally less than 100 A.)
- Six-cylinder engines = 160 to 200 amperes (normally less than 125 A.)
- Eight-cylinder engines = 185 to 250 amperes (normally less than 150 A.)

Excessive current draw may indicate one or more of the following:

1. Low battery voltage (discharged or defective battery).
2. Binding of starter armature as a result of worn bushings
3. Oil too thick (viscosity too high) for weather conditions
4. Shorted or grounded starter windings or cables
5. Tight or seized engine

CASE STUDY

The Case of the No Crank

A four-cylinder engine would not crank. Previously the customer said that once in a while, the starter seemed to lock up when the vehicle sat overnight but would then finally crank. The problem only occurred in the morning and the engine would crank and start normally the rest of the day.

The vehicle finally would not start and was towed to the shop. The service technician checked the current draw of the starter and it read higher than the scale on the ammeter. The technician then attempted to rotate the engine by hand and found that the engine would not rotate. Based on this history of not cranking normally in the morning, the technician removed the spark plugs and attempted to crank the engine. This time the engine cranked and coolant was seen shooting from cylinders number two and three. Apparently coolant leaked into the cylinders, due to a fault with the head gasket, causing the engine to hydro-lock or not rotate due to liquid being trapped on top of the piston. Replacing the bad gasket solved the cranking problems in the morning.

Summary:

- **Complaint—**Customer stated that the engine would not crank in the morning.
- **Cause—**Tests confirmed that coolant was leaking into the cylinders causing a hydrostatic lock.
- **Correction—**Replacing the head gasket solved the engine starting problem.

TECH TIP

Watch the Dome Light

When diagnosing any starter-related problem, open the door of the vehicle and observe the brightness of the dome or interior light(s). The brightness of any electrical lamp is proportional to the voltage.

Normal operation of the starter results in a slight dimming of the dome light.

- *If the light remains bright*, the problem is usually an open circuit in the control circuit.
- *If the light goes out or almost goes out*, the problem is usually a shorted or grounded armature or field coils inside the starter.

A poor electrical connection that opens under load could also be the cause.

TECH TIP

Don't Hit That Starter!

In the past, it was common to see service technicians hitting a starter in their effort to diagnose a no-crank condition. Often the shock of the blow to the starter aligned or moved the brushes, armature, and bushings. Many times, the starter functioned after being hit—even if only for a short time.

However, most of today's starters use permanent-magnet fields, and the magnets can be easily broken if hit. A magnet that is broken becomes two weaker magnets. Some early permanent-magnet (PM) starters used magnets that were glued or bonded to the field housing. If struck with a heavy tool, the magnets could be broken, with parts of the magnet falling onto the armature and into the bearing pockets, making the starter impossible to repair or rebuild.

TESTING A STARTER USING A SCAN TOOL

A scan tool can be used on most vehicles to check the cranking system. Follow these steps:

1. Connect the scan tool according to the manufacturer's instructions.
2. Select battery voltage and engine RPM on the scan tool.
3. Select "snapshot" and start recording or graphing if the scan tool is capable.
4. Crank the engine. Stop the scan tool recording.
5. Retrieve the scan data and record cranking RPM and battery voltage during cranking. Cranking RPM should be between 80 and 250 RPM. Battery voltage during cranking should be higher than 9.6 volts.

? FREQUENTLY ASKED QUESTION

Is Voltage Drop the Same as Resistance?

Many technicians have asked the question, Why measure voltage drop when the resistance can be easily measured using an ohmmeter? Think of a battery cable with all the strands of the cable broken, except for one. If an ohmmeter is used to measure the resistance of the cable, the reading would be very low, probably less than 1 ohm. However, the cable is not capable of conducting the amount of current necessary to crank the engine. In less severe cases, several strands can be broken and affect the operation of the starter motor. Although the resistance of the battery cable will not indicate any increased resistance, the restriction to current flow will cause heat and a drop in the voltage available at the starter. Because resistance is not effective until current flows, measuring the voltage drop (differences in voltage between two points) is the most accurate method of determining the true resistance in a circuit.

How much is too much? According to Bosch Corporation, all electrical circuits should have a maximum of 3% loss of the voltage of the circuit to resistance. Therefore, in a 12 volt circuit, the maximum loss of voltage in cables and connections should be 0.36 volt ($12 \times 0.03 = 0.36$ volt). The remaining 97% of the circuit voltage (11.64 volt) is available to operate the electrical device (load). Just remember:

- **Low voltage drop = low resistance**
- **High voltage drop = high resistance**

VOLTAGE-DROP TESTING

PURPOSE OF VOLTAGE-DROP TESTING **Voltage drop** is the drop in voltage that occurs when current is flowing through a resistance. For example, a voltage drop is the difference between voltage at the source and voltage at the electrical device to which it is flowing. The higher the voltage drop, the greater the resistance in the circuit. Even though voltage-drop testing can be performed on any electrical circuit, the most common areas of testing include the cranking circuit and the charging circuit wiring and connections.

RESULTS OF EXCESSIVE VOLTAGE DROP A high voltage drop (high resistance) in the cranking circuit wiring can cause slow engine cranking with less-than-normal starter amperage drain as a result of the excessive circuit resistance. If the voltage drop is high enough, such as could be caused by dirty battery terminals, the starter may not operate. A typical symptom of low battery voltage or high resistance in the cranking circuit is a "clicking" of the starter solenoid.

PERFORMING A VOLTAGE-DROP TEST Voltage-drop testing of the wire involves connecting any voltmeter (on the low scale) to the suspected high-resistance cable ends and cranking the engine. ● **SEE FIGURES 15–23, 15–24, AND 15–25.**

NOTE: Before a difference in voltage (voltage drop) can be measured between the ends of a battery cable, current must be flowing through the cable. *Resistance is not effective unless current is flowing.* If the engine is not being cranked, current is not flowing through the battery cables and the voltage drop cannot be measured.

Crank the engine with a voltmeter connected to the battery and record the reading. Crank the engine with the voltmeter connected across the starter and record the reading. If the difference in the two readings exceeds 0.5 volt, perform the following steps to determine the exact location of the voltage drop:

1. Connect the positive voltmeter test lead to the most positive end of the cable being tested. The most positive end of a cable is the end closest to the positive terminal of the battery.

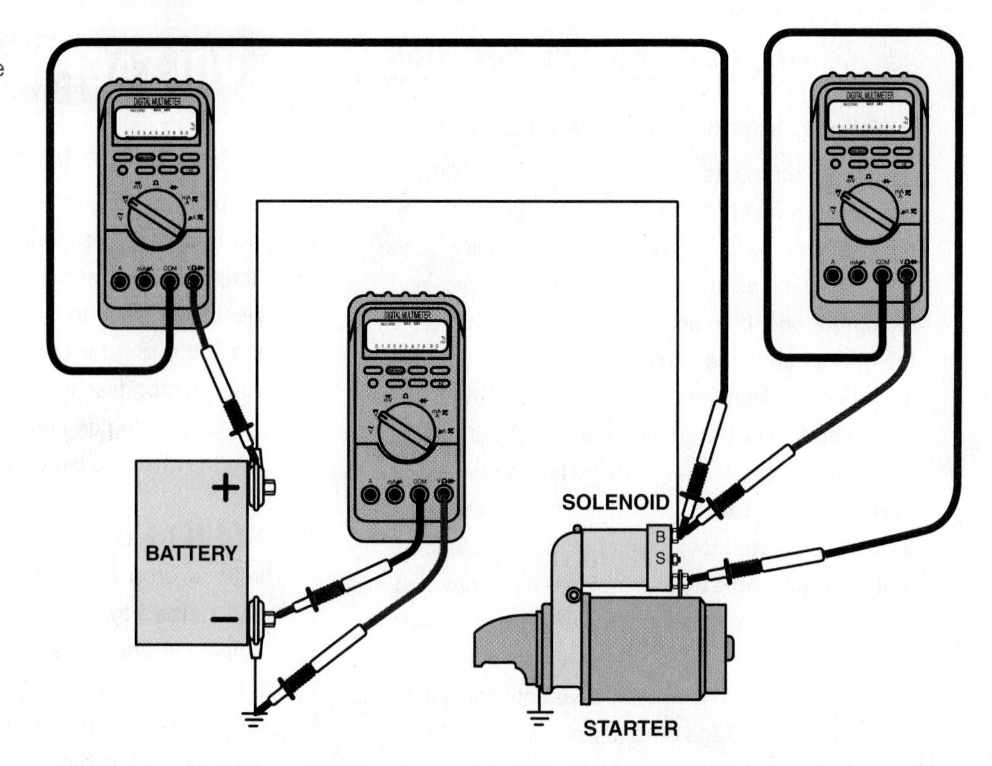

FIGURE 15–23 Voltmeter hookups for voltage-drop testing of a GM-type cranking circuit.

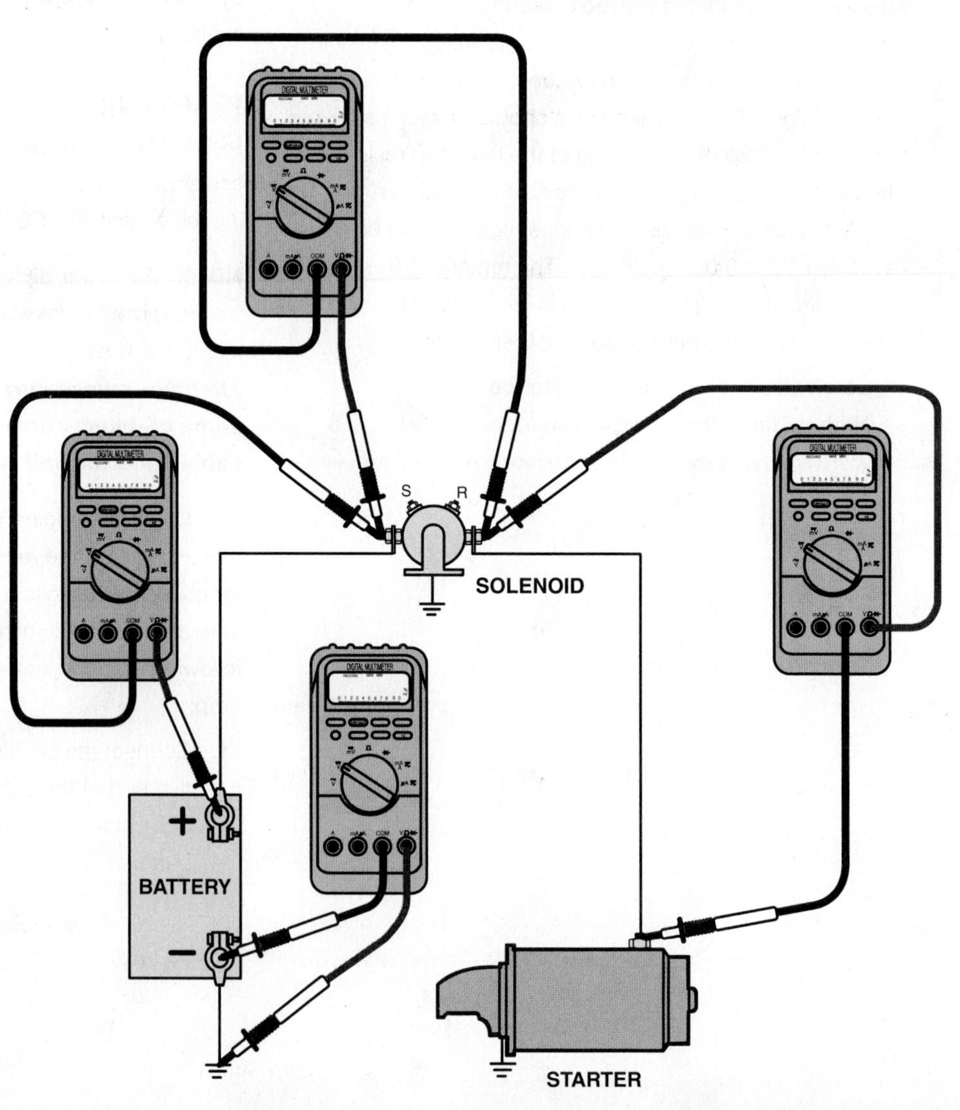

FIGURE 15–24 Voltmeter hookups for voltage-drop testing of a Ford-type cranking circuit.

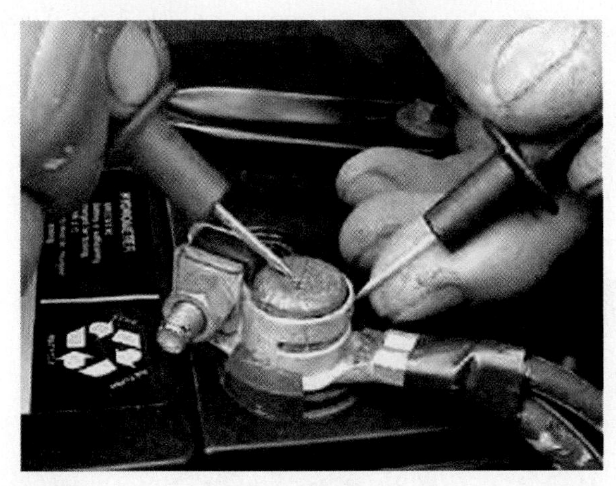

FIGURE 15–25 Using the voltmeter leads from a starting and charging test unit to measure the voltage drop between the battery terminal (red lead) and the cable end (black lead). The engine must be cranked to cause current to flow through this connection.

2. Connect the negative voltmeter test lead to the other end of the cable being tested. With no current flowing through the cable, the voltmeter should read zero because there is the same voltage at both ends of the cable.
3. Crank the engine. The voltmeter should read less than 0.2 volt.
4. Evaluate the results. If the voltmeter reads zero, the cable being tested has no resistance and is good. If the voltmeter reads higher than 0.2 volt, the cable has excessive resistance and should be replaced. However, before replacing the cable, make certain that the connections at both ends of the cable being tested are clean and tight.

● **SEE FIGURE 15–26.**

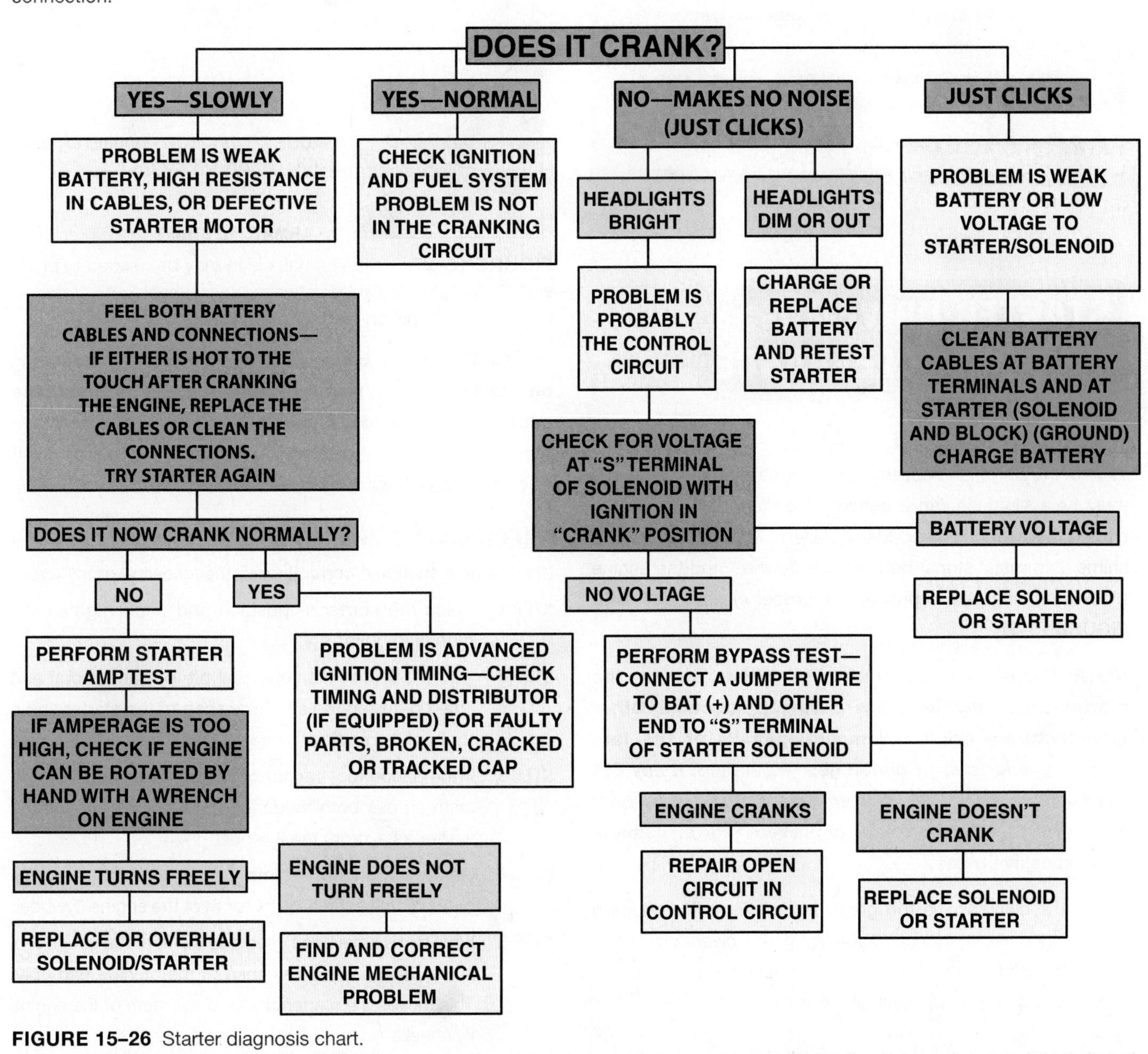

FIGURE 15–26 Starter diagnosis chart.

TECH TIP

The Touch Test

If a cable or connection is hot to the touch, there is electrical resistance in the cable or connection. The resistance changes electrical energy into heat energy. Therefore, if a voltmeter is not available, touch the battery cables and connections while cranking the engine. If any cable or connection is warm or hot to the touch, it should be cleaned or replaced.

NOTE: Some experts recommend replacing the entire battery cable if the cable ends become corroded or otherwise unusable. Many "temporary" cable ends do not provide adequate contact areas for the cable and allow the end of the cable strands to be exposed to battery acid corrosion. Also, never pound or hammer a battery cable onto a battery post. Always use a spreader tool to open the clamp wide enough to fit the battery posts.

STARTER DRIVE-TO-FLYWHEEL CLEARANCE

NEED FOR PROPER CLEARANCE For the proper operation of the starter and absence of abnormal starter noise, there must be a slight clearance between the starter pinion and the engine flywheel ring gear. Many General Motors starters use shims (thin metal strips) between the flywheel and the engine block mounting pad to provide the proper clearance. ● **SEE FIGURE 15–27.**

NOTE: Some manufacturers use shims under the starter drive end housings during production. Other manufacturers *grind* the mounting pads at the factory for proper starter pinion gear clearance. If *any* GM starter is replaced, the starter pinion *must* be checked and corrected as necessary to prevent starter damage and excessive noise.

If the clearance is too great, the starter will produce a high-pitched whine *during* cranking. If the *clearance is too small*, the starter will produce a high-pitched whine *after* the engine starts, just as the ignition key is released.

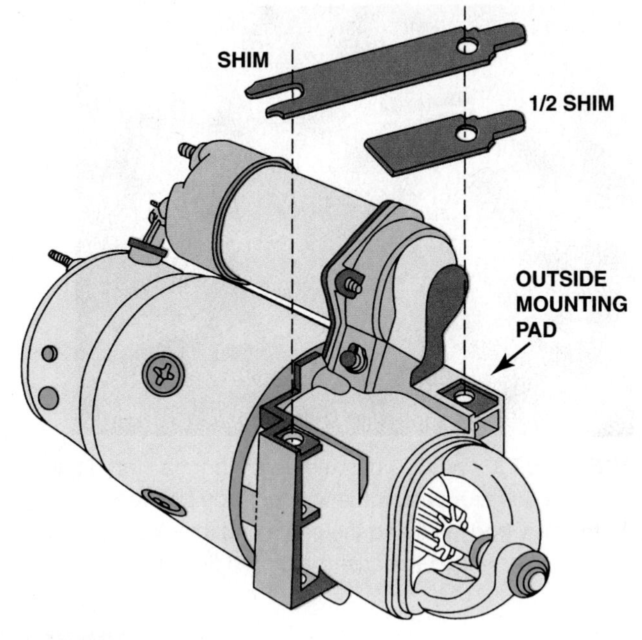

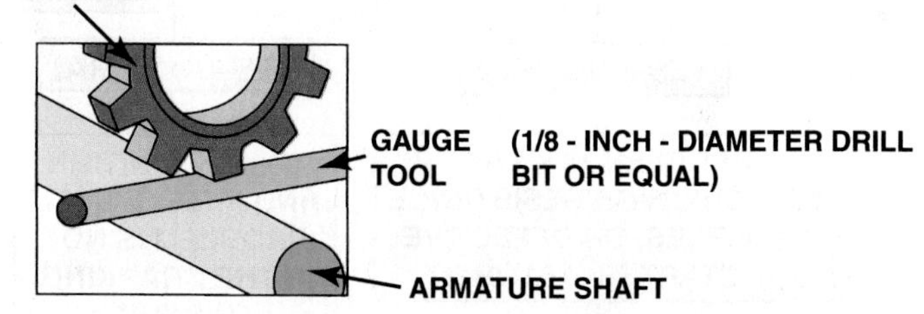

FIGURE 15–27 A shim (or half shim) may be needed to provide the proper clearance between the flywheel teeth of the engine and the pinion teeth of the starter.

NOTE: The major cause of broken drive-end housings on starters is too small a clearance. If the clearance cannot be measured, it is better to put a shim between the engine block and the starter than to leave one out and chance breaking a drive-end housing.

CHECKING FOR PROPER CLEARANCE To be sure that the starter is shimmed correctly, use the following procedure:

STEP 1 Place the starter in position and finger-tighten the mounting bolts.

STEP 2 Use a 1/8-inch-diameter drill bit (or gauge tool) and insert between the armature shaft of the starter and a tooth of the engine flywheel.

STEP 3 If the gauge tool cannot be inserted, use a full-length shim across both mounting holes, which moves the starter away from the flywheel.

STEP 4 Remove a shim or shims if the gauge tool is loose between the shaft and the tooth of the engine flywheel.

STEP 5 If no shims have been used and the fit of the gauge tool is too loose, add a half shim to the outside pad only. This moves the starter closer to the teeth of the engine flywheel.

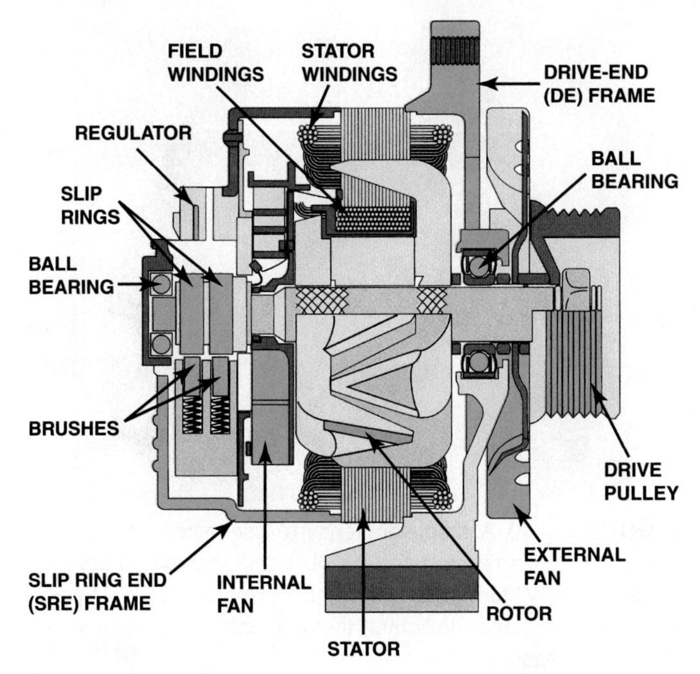

FIGURE 15–28 Cutaway view of a typical alternator.

CHARGING CIRCUIT

ALTERNATOR CONSTRUCTION An alternator is constructed of a two-piece cast-aluminum housing. Aluminum is used because of its lightweight, nonmagnetic properties and heat transfer properties, which are needed to help keep the alternator cool. A front ball bearing is pressed into the front housing (called the **drive-end [DE]** housing) to provide the support and friction reduction necessary for the belt-driven rotor assembly. The rear housing (called the **slip ring end [SRE]**) usually contains a roller-bearing support for the rotor and mounting for the brushes, diodes, and internal voltage regulator (if the alternator is so equipped). ● **SEE FIGURE 15–28.**

CHARGING CIRCUIT TESTING The charge indicator light on the dash should be on with the key on, engine off (KOEO), but should be off when the engine is running (KOER). If the charge light remains on with the engine running, check the charging system voltage. To measure charging system voltage, connect the test leads of a digital multimeter to the positive (+) and negative (–) terminals of the battery. Set the multimeter to read DC volts.

? FREQUENTLY ASKED QUESTION

How Many Horsepower Does an alternator Require to Operate?

Many technicians are asked how much power certain accessories require. A 100 A alternator requires about 2 horsepower (hp) from the engine. One horsepower is equal to 746 watts (W). Watts are calculated by multiplying amperes times volts.

$$\text{Power in W} = 100\text{ A} \times 14.5\text{ V} = 1450\text{ W}$$

$$1\text{ hp} = 746\text{ W}$$

Therefore, 1450 W is about 2 hp.

Allowing about 20% for mechanical and electrical losses adds another 0.4 hp. Therefore, when anyone asks how much power it takes to produce 100 A from an alternator, the answer is about 2.4 hp.

TECH TIP

The Dead Rat Smell Test

When checking for the root cause of an alternator failure, the wise technician should sniff (smell) the alternator! If the alternator smells like a dead rat (rancid), the stator windings have been overheated by trying to charge a discharged or defective battery. If the battery voltage is continuously low, the voltage regulator will continue supplying full-field current to the alternator. The voltage regulator is designed to cycle on and off to maintain a narrow charging system voltage range.

If the battery voltage is continually below the cutoff point of the voltage regulator, the alternator is continually producing current in the stator windings. This constant charging can often overheat the stator and burn the insulating varnish covering the stator windings. If the alternator fails the sniff test, the technician should replace the alternator *and* replace or recharge and test the battery.

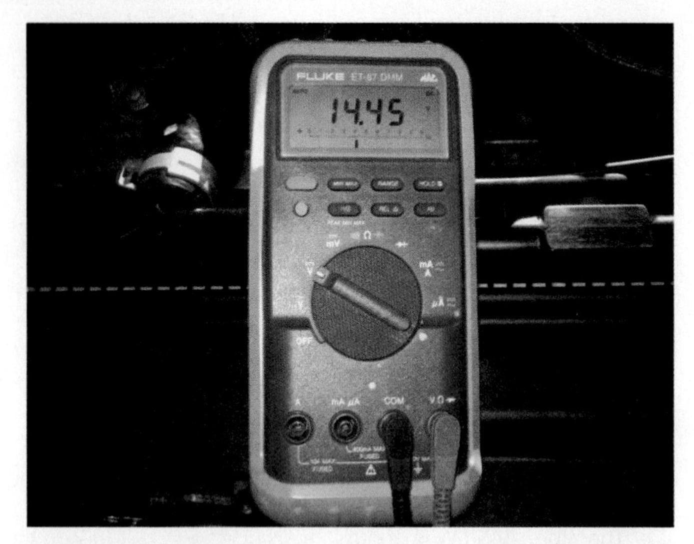

FIGURE 15–29 The digital multimeter should be set to read DC volts and the red lead connected to the battery positive (+) terminal and the black meter lead connected to the negative (−) battery terminal.

CHARGING SYSTEM VOLTAGE SPECIFICATIONS

Most alternators are designed to supply between 13.5 and 15.0 volts at 2000 engine RPM. Be sure to check the vehicle manufacturer's specifications. For example, most General Motors Corporation vehicles specify a charging voltage of 14.7 volts ± 0.5 (or between 14.2 and 15.2 volts) at 2000 RPM and no load.

CHARGING SYSTEM VOLTAGE TEST PROCEDURE Charging system voltage tests should be performed on a vehicle with a battery at least 75% charged. If the battery is discharged (or defective), the charging voltage may be below specifications. To measure charging system voltage, follow these steps:

1. Connect the voltmeter as shown in ● **FIGURE 15–29.**
2. Set the meter to read DC volts.
3. Start the engine and raise to a fast idle (about 2000 RPM).
4. Read the voltmeter and compare with specifications. If lower than specifications, charge the battery and test for excessive charging circuit voltage drop and for a possible open in the sensing wire before replacing the alternator.

NOTE: If the voltmeter reading rises, then becomes lower as the engine speed is increased, the alternator drive (accessory drive) belt is loose or slipping.

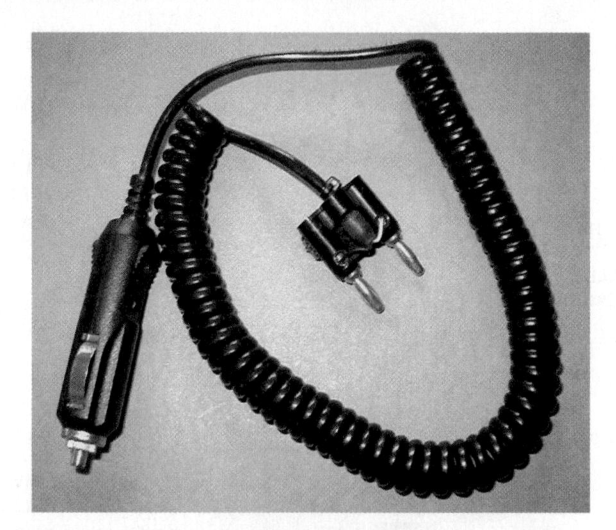

FIGURE 15–30 A simple and easy-to-use tester can be made from a lighter plug and double banana plug that fits the "COM" and "V" terminals of most digital meters. By plugging the lighter plug into the lighter, the charging circuit voltage can be easily measured.

TECH TIP

The Lighter Plug Trick

Battery voltage measurements can be read through the lighter socket. ● **SEE FIGURE 15–30.** Simply construct a test tool using a lighter plug at one end of a length of two-conductor wire and the other end connected to a double banana plug. The double banana plug will fit most meters in the common (COM) terminal and the volt terminal of the meter.

TESTING AN ALTERNATOR USING A SCAN TOOL

A scan tool can be used on most General Motors and Chrysler Corporation vehicles and others that have datastream information. Follow these steps:

1. Connect the scan tool according to the manufacturers' instructions.
2. Select battery voltage and engine RPM on the scan tool.
3. Start the engine and operate at 2000 RPM.
4. Observe the battery voltage. This voltage should be between 13.5 and 15.0 volts (or within manufacturers' specifications).

TECH TIP

The Hand Cleaner Trick

Lower-than-normal alternator output could be the result of a loose or slipping drive belt. All belts (V and serpentine multigroove) use an interference angle between the angle of the V's of the belt and the angle of the V's on the pulley. A belt wears this interference angle off the edges of the V of the belt. As a result, the belt may start to slip and make a squealing sound even if tensioned properly.

A common trick used to determine if the noise is belt related is to use grit-type hand cleaner or scouring powder. With the engine off, sprinkle some powder onto the pulley side of the belt. Start the engine. The excess powder will fly into the air, so get away from under the hood when the engine starts. If the belts are now quieter, you know that it was the glazed belt that made the noise.

NOTE: Often, the noise sounds exactly like a noisy bearing. Therefore, before you start removing and replacing parts, try the hand cleaner trick.

The grit from the hand cleaner will often remove the glaze from the belt and the noise will not return. If the belt is worn or loose, however, the noise will return and the belt should be replaced. A fast, alternative method to check for belt noise is to spray water from a squirt bottle at the belt with the engine running. If the noise stops, the belt is the cause of the noise. The water quickly evaporates and therefore, unlike the gritty hand cleaner, water simply finds the problem—it does not provide a short-term fix.

NOTE: The scan tool voltage should be within 0.5 volt of the charging voltage as tested at the battery. If the scan tool indicates a voltage lower than actual battery voltage by more than 0.5 volt, check all power and ground connections at the computer for corrosion or defects.

MEASURING THE AC RIPPLE FROM THE ALTERNATOR TELLS A LOT ABOUT ITS CONDITION. IF THE AC RIPPLE IS ABOVE 500 MILLIVOLTS, OR 0.5 VOLTS, LOOK FOR A PROBLEM IN THE DIODES OR STATOR. IF THE RIPPLE IS BELOW 500 MILLIVOLTS, CHECK THE ALTERNATOR OUTPUT TO DETERMINE ITS CONDITION.

FIGURE 15–31 AC ripple at the output terminal of the battery is more accurate than testing at the battery due to the resistance of the wiring between the alternator and the battery. The reading shown on the meter is only 78 mV (0.078 V), far below what the reading would be if a diode were defective.

AC RIPPLE VOLTAGE CHECK

A good alternator should produce only a small amount of AC voltage. It is the purpose of the diodes in the alternator to rectify AC voltage into DC voltage. **AC ripple voltage** is the AC part of the DC charging voltage produced by the alternator. If the AC ripple voltage is higher than 0.5 volt this can cause engine performance problems because the AC voltage can interfere with sensor signals. The procedure to check for AC voltage includes the following steps:

1. Set the digital meter to read AC volts.
2. Start the engine and operate it at 2000 RPM (fast idle).
3. Connect the voltmeter leads to the positive and negative battery terminals.
4. Turn on the headlights to provide an electrical load on the alternator.

NOTE: A higher, more accurate reading can be obtained by touching the meter lead to the output terminal of the alternator as shown in ● FIGURE 15–31.

FIGURE 15–32 A mini clamp-on digital multimeter can be used to measure alternator output and unwanted AC current by switching the meter to read DC amperes.

The results should be interpreted as follows: If the diodes are good, the voltmeter should read *less* than 0.4 volt AC. If the reading is *over* 0.5 volt AC, the rectifier diodes or stator are defective indicating that the alternator should be replaced.

NOTE: This test will *not* test for a defective diode trio, which is used in some alternators to power the field circuit internally and to turn off the dash charge light.

AC CURRENT CHECK

The amount of AC current (also called **ripple current**) in amperes flowing from the alternator to the battery can be measured using a clamp-on digital multimeter set to read AC amperes. Attach the clamp of the meter around the alternator output wire or all of the positive or negative battery cables if the output wire is not accessible. Start the engine and turn on all lights and accessories to load the alternator and read the meter display. The maximum allowable AC current (amperes) from the alternator is less than 10% of the rated output of the alternator. Because most newer alternators produce about 100 amperes DC, the maximum allowable AC amperes would be 10 amperes. If the reading is above 10 A (or 10%), this indicates that the rectifier diodes or a fault with the stator windings is present. ● **SEE FIGURE 15–32.**

CHARGING SYSTEM VOLTAGE-DROP TESTING

PURPOSE OF CHARGING SYSTEM VOLTAGE-DROP TESTING For the proper operation of any charging system, there must be good electrical connections between the battery positive terminal and the alternator output terminal. The alternator must also be properly grounded to the engine block.

Many vehicle manufacturers run the lead from the output terminal of the alternator to other connectors or junction blocks that are electrically connected to the positive terminal of the battery. If there is high resistance (a high voltage drop) in these connections or in the wiring itself, the battery will not be properly charged.

CHARGING SYSTEM VOLTAGE-DROP TESTING PROCEDURE When there is a suspected charging system problem (with or without a charge indicator light on), simply follow these steps to measure the voltage drop of the insulated (power-side) charging circuit:

1. Start the engine and run it at a fast idle (about 2000 engine RPM).
2. Turn on the headlights to ensure an electrical load on the charging system.
3. Using any voltmeter, connect the positive test lead (usually red) to the output terminal of the alternator. Attach the negative test lead (usually black) to the positive post of the battery.

The results should be interpreted as follows:

1. If there is less than a 0.4 volt reading, then all wiring and connections are satisfactory.
2. If the voltmeter reads higher than 0.4 volt, there is excessive resistance (voltage drop) between the alternator output terminal and the positive terminal of the battery.
3. If the voltmeter reads battery voltage (or close to battery voltage), there is an open circuit between the battery and the alternator output terminal (look for a positive open fusible link).

To determine whether the alternator is correctly grounded, maintain the engine speed at 2000 RPM with the headlights on. Connect the positive voltmeter lead to the case of the alternator and the negative voltmeter lead to the negative terminal of the battery. The voltmeter should read less than 0.2 volt if the alternator is properly grounded. If the reading is over 0.2 volt, connect one end of an auxiliary ground wire to the case of the alternator and the other end to a good engine ground. ● **SEE FIGURE 15–33.**

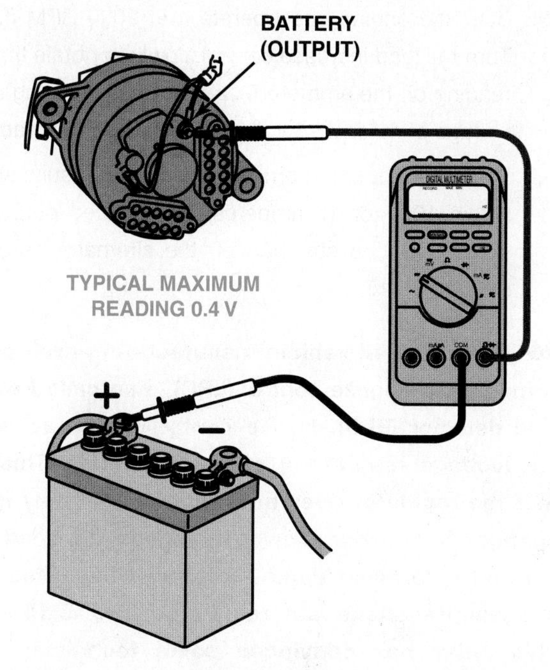

VOLTAGE DROP - INSULATED CHARGING CIRCUIT

ENGINE AT 2,000 RPM.
CHARGING SYSTEM
LOADED TO 20 A

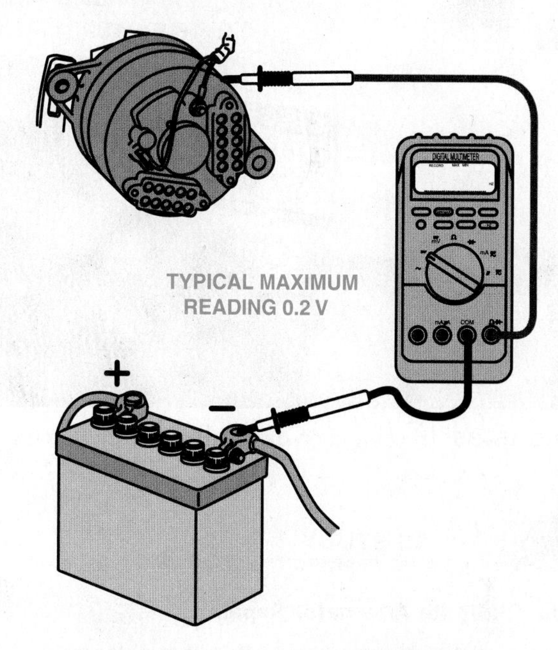

VOLTAGE DROP - CHARGING GROUND CIRCUIT

FIGURE 15–33 Voltmeter hookup to test the voltage drop of the charging circuit.

TECH TIP

"2 to 4"

Most voltage-drop specifications range between 0.2 and 0.4 volt. Generally, if the voltage loss (voltage drop) in a circuit exceeds 0.5 volt (1/2 volt), the wiring in that circuit should be repaired or replaced. During automotive testing, it is sometimes difficult to remember the exact specification for each test; therefore, the technician can simply remember "2 to 4" and that any voltage drop over that may indicate a problem.

LARGE CABLES FROM TESTER
NEGATIVE LOAD LEAD
POSITIVE LOAD LEAD
BLACK
RED
GREEN CLAMP-ON AMPS PICKUP
TO STARTER MOTOR
12 V BATTERY

TEST LEAD CONNECTIONS FOR TESTING THE STARTING SYSTEM, CHARGING SYSTEM, VOLTAGE REGULATOR, AND DIODE STATOR.

FIGURE 15–34 Typical hookup of a starting and charging tester.

ALTERNATOR OUTPUT TEST

A charging circuit may be able to produce correct charging circuit voltage, but not be able to produce adequate amperage output. If in doubt about charging system output, first check the condition of the alternator drive belt. With the engine off, attempt to rotate the fan of the alternator by hand. Replace tensioner or tighten drive belt if the alternator fan can be rotated by hand. ● **SEE FIGURE 15–34** for typical test equipment hookup.

The testing procedure for alternator output is as follows:

1. Connect the starting and charging test leads according to the manufacturers' instructions.
2. Turn the ignition switch on (engine off) and observe the ammeter. This is the ignition circuit current, and it should be about 2 to 8 amperes.

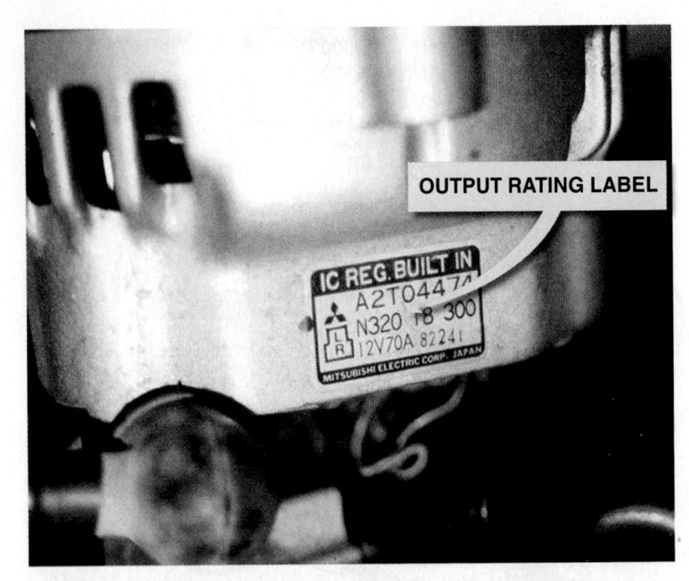

FIGURE 15–35 The output on this alternator is printed on a label.

CASE STUDY

The 2-Minute Alternator Repair

A Chevrolet pickup truck was brought to a dealer for routine service. The owner stated that the battery required a jump start after a weekend of sitting. The battery voltage was 12.4 volts (about 75% charged), but the charging voltage was also 12.4 volts at 2000 RPM. Because normal charging voltage should be 13.5 to 15.0 volts, it was obvious that the charging system was not operating correctly.

The technician checked the dash and found that the "charge" light was not on even though the rear bearing was not magnetized, indicating that the voltage regulator was not working. Before removing the alternator for service, the technician checked the wiring connection on the alternator. When the two-lead regulator connector was removed, the connector was discovered to be rusty. After the contacts were cleaned, the charging system was restored to normal operation. The technician had learned that the simple things should always be checked first before tearing into a big (or expensive) repair.

Summary:

- **Complaint**—Customer stated that the truck battery had to be jump-started after sitting for a weekend.
- **Cause**—Tests confirmed that the alternator was not charging and a rusty connection at the alternator was found during a visual inspection.
- **Correction**—Cleaning the electrical terminals at the alternator restored proper operation of the charging system.

3. Start the engine and operate it at 2000 RPM (fast idle). Turn the load increase control slowly to obtain the highest reading on the ammeter scale while maintaining a battery voltage of at least 13 volts. Note the ampere reading.
4. Total the amperes from steps 2 and 3. Results should be within 10% (or 15 amperes) of the rated output. Rated output may be stamped on the alternator as shown in ● **FIGURE 15–35**.

NOTE: Almost all vehicle manufacturers are now using some load response control (LRC), also called electronic load detector (ELD), in the control of the voltage output (voltage regulators) of the alternator. This means that the regulator does not react immediately to a load change, but rather slowly increases the load on the alternator to avoid engine idle problems. This gradual increase of voltage may require as long as 15 seconds. This delay has convinced some technicians that a problem exists in the alternator/regulator or computer control of the alternator.

NOTE: When applying a load to the battery with a carbon pile tester during an alternator output test, do not permit the battery voltage to drop below 13 volts. Most alternators will produce their maximum output (in amperes) above 13 volts.

TESTING AN ALTERNATOR USING A SCOPE

Defective diodes and open or shorted stators can be detected on an ignition scope. Connect the scope leads as usual, *except* for the coil negative connection, which attaches to the alternator output ("BAT") terminal. With the pattern selection set to "raster" (stacked), start the engine and run to approximately 1000 RPM (slightly higher-than-normal idle speed). The scope should show an even ripple pattern reflecting the slight alternating up-and-down level of the alternator output voltage.

If the alternator is controlled by an electronic voltage regulator, the rapid on-and-off cycling of the field current can create vertical spikes evenly throughout the pattern. These spikes are normal. If the ripple pattern is jagged or uneven, a defective diode (open or shorted) or a defective stator is indicated. ● **SEE FIGURES 15–36 THROUGH 15–38**. If the alternator scope pattern does not show even ripples, the alternator should be replaced.

FIGURE 15–36 Normal alternator scope pattern. This AC ripple is on top of a DC voltage line. The ripple should be less than 0.50 volt high.

FIGURE 15–37 Alternator pattern indicating a shorted diode.

FIGURE 15–38 Alternator pattern indicating an open diode.

CASE STUDY

The Start/Stall/Start/Stall Problem

A Chevrolet four-cylinder engine would stall every time it was started. The engine cranked normally and the engine started quickly. It would just stall once it had run for about 1 second. After hours of troubleshooting, it was discovered that if the "gages" fuse was removed, the engine would start and run normally. Because the alternator was powered by the "gages" fuse, the charging voltage was checked and found to be over 16 volts just before the engine stalled. Replacing the alternator fixed the problem. The powertain control module (PCM) shut down to prevent damage when the voltage exceeded 16 volts.

Summary:

- **Complaint—**Customer stated that the engine would always start but would stall.
- **Cause—**Tests confirmed that the alternator charging over 16 volts which caused the PCM to shut down.
- **Correction–**The alternator was replaced which fixed the overcharging condition and the engine continued to run after it was started.

SUMMARY

1. Batteries can be tested with a voltmeter to determine the state of charge. A battery load test loads the battery to one-half of its CCA rating. A good battery should be able to maintain above 9.6 volts for the entire 15-second test period.
2. A battery drain test should be performed if the battery runs down.
3. Proper operation of the starter motor depends on the battery being at least 75% charged and the battery cables being of the correct size (gauge) and having no more than a 0.2 volt drop.
4. Cranking system voltage-drop testing includes cranking the engine, measuring the drop in voltage from the battery to the starter, and measuring the drop in voltage from the negative terminal of the battery to the engine block.
5. The cranking circuit should be tested for proper amperage draw.
6. An open in the control circuit can prevent starter motor operation.
7. Charging system testing requires that the battery be at least 75% charged to be assured of accurate test results. The charge indicator light should be on with the ignition switch on, but should go out whenever the engine is running. Normal charging voltage (at 2000 engine RPM) is 13.5 to 15.0 volts.
8. To check for excessive resistance in the wiring between the alternator and the battery, perform a voltage-drop test.

REVIEW QUESTIONS

1. Describe the results of a voltmeter battery state-of-charge test.
2. List the steps for performing a battery load test.
3. Explain how to perform a battery drain test.
4. Explain how to perform a voltage-drop test of the cranking circuit.
5. Describe how to test the voltage drop of the charging circuit.
6. Discuss how to measure the maximum amperage output of an alternator.

CHAPTER QUIZ

1. A battery high-rate discharge (load capacity) test is being performed on a 12 volt battery. Technician A says that a good battery should have a voltage reading of higher than 9.6 volts while under load at the end of the 15-second test. Technician B says that the battery should be discharged (loaded to two times its CCA rating). Which technician is correct?
   a. Technician A only
   b. Technician B only
   c. Both Technicians A and B
   d. Neither Technician A nor B
2. Normal battery drain (parasitic drain) with a vehicle with many computer and electronic circuits is _________.
   a. 20 to 30 milliamperes
   b. 2 to 3 amperes
   c. 150 to 300 milliamperes
   d. None of the above
3. When jump-starting, _________.
   a. The last connection should be the positive post of the dead battery
   b. The last connection should be the engine block of the dead vehicle
   c. The alternator must be disconnected on both vehicles
   d. Both a and c
4. Technician A says that a discharged battery (lower-than-normal battery voltage) can cause solenoid clicking. Technician B says that a discharged battery or dirty (corroded) battery cables can cause solenoid clicking. Which technician is correct?
   a. Technician A only
   b. Technician B only
   c. Both Technicians A and B
   d. Neither Technician A nor B
5. Slow cranking can be caused by all of the following *except* _________.
   a. A low or discharged battery
   b. Corroded or dirty battery cables
   c. Engine mechanical problems
   d. An open neutral safety switch
6. If the starter turns slowly when engaged, a possible cause is _________.
   a. A worn or defective starter
   b. A defective solenoid
   c. A disconnected battery cable
   d. An open ignition switch
7. An acceptable charging circuit voltage on a 12 volt system is _________.
   a. 13.5 to 15.0 volts
   b. 12.6 to 15.6 volts
   c. 12.0 to 14.0 volts
   d. 14.9 to 16.1 volts
8. Technician A says that a voltage-drop test of the charging circuit should only be performed when current is flowing through the circuit. Technician B says that to measure the voltage drop of the charging system, connect the leads of a voltmeter to the positive and negative terminals of the battery. Which technician is correct?
   a. Technician A only
   b. Technician B only
   c. Both Technicians A and B
   d. Neither Technician A nor B
9. Testing the electrical system through the lighter plug using a digital meter can test _________.
   a. Charging system current
   b. Charging system voltage
   c. Cranking system current
   d. All of the above
10. The maximum acceptable AC ripple voltage is _________.
    a. 0.010 volt (10 mV)
    b. 0.050 volt (50 mV)
    c. 0.100 volt (100 mV)
    d. 0.400 volt (400 mV)

IGNITION SYSTEM COMPONENTS AND OPERATION

LEARNING OBJECTIVES

After studying this chapter, the reader will be able to:

1. Explain how the ignition system and ignition coils work.
2. Discuss crankshaft position sensor and pickup coil operation.
3. Discuss knock sensors and ignition control circuits.
4. Describe the operation of distributor ignition.
5. Describe the operation of waste-spark and coil-on-plug ignition systems.

This chapter will help you prepare for Engine Repair (A8) ASE certification test content area "B" (Ignition System Diagnosis and Repair).

KEY TERMS

Bypass ignition 273
Coil-on-plug (COP) ignition (also coil-by-plug, coil-near-plug, coil-over-plug) 274
Companion cylinder 271
Compression-sensing ignition 274
Detonation 277
Distributorless ignition system (DIS) 271
Distributor ignition (DI) 261
Dwell 273
Electromagnetic interference (EMI) 274
Electronic control unit (ECU) 269
Electronic ignition (EI) 261
Electronic ignition system (EIS) 269
Electronic spark timing (EST) 273
Hall-effect switch 263
Igniter 261
Ignition coil 261
Ignition control module (ICM) 261
Ignition timing 276
Inductive reactance 261
Initial timing 276
Ion-sensing ignition 276
Iridium spark plugs 278
Knock sensor (KS) 277
Magnetic pulse generator 263
Magnetic sensor 264
Mutual induction 262
Optical sensors 266
Paired cylinder 271
Pickup coil (pulse generator) 263
Ping 277
Platinum spark plugs 278
Polarity 261
Primary ignition circuit 262
Saturation 262
Schmitt trigger 264
Secondary ignition circuit 262
Self-induction 261
Spark knock 277
Spark output (SPOUT) 273
Switching 263
Transistor 263
Turns ratio 261
Up-integrated ignition 274
Waste-spark ignition 271

FIGURE 16–1 A point-type ignition system showing the distributor cam which opens the points.

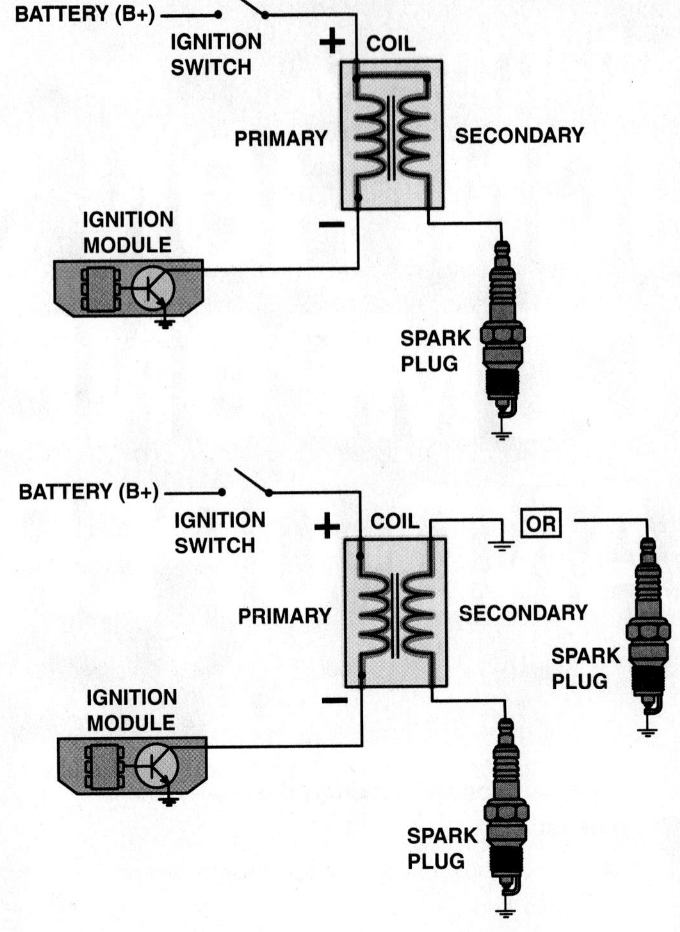

FIGURE 16–2 Some ignition coils are electrically connected, called married (top figure), whereas others use separate primary and secondary windings, called divorced (lower figure). The polarity (positive or negative) of a coil is determined by the direction in which the coil is wound.

IGNITION SYSTEM

PURPOSE AND FUNCTION The ignition system includes components and wiring necessary to create and distribute a high voltage (up to 40,000 volts or more) and send to the spark plug. A high-voltage arc occurs across the gap of a spark plug inside the combustion chamber. The spark raises the temperature of the air–fuel mixture and starts the combustion process inside the cylinder.

BACKGROUND All ignition systems apply battery voltage (close to 12 volts) to the positive side of the ignition coil(s) and pulse the negative side to ground on and off.

EARLY IGNITION SYSTEMS. Before the mid-1970s, ignition systems used a mechanically opened set of contact points to make and break the electrical connection to ground. A cam lobe, located and driven by the distributor, opened the points. There was one lobe for each cylinder. The points used a rubbing block that was lubricated by applying a thin layer of grease on the cam lobe at each service interval. Each time the points opened, a spark was created in the ignition coil. The high-voltage spark then traveled to each spark plug through the distributor cap and rotor and the spark plug wires. The distributor was used twice in the creation of the spark.

1. It was connected to the camshaft that rotated the distributor cam causing the points to open and close.
2. It used a rotor to direct the high-voltage spark from the coil entering the center of the distributor cap to inserts connected to spark plug wires to each cylinder. ● **SEE FIGURE 16–1**.

ELECTRONIC IGNITION. Since the mid-1970s, ignition systems have used sensors, such as a pickup coil and reluctor (trigger wheel), to trigger or signal an electronic module that switches the primary ground circuit of the ignition coil.

DISTRIBUTOR IGNITION (DI) is the term specified by the Society of Automotive Engineers (SAE) for an ignition system that uses a distributor.

ELECTRONIC IGNITION (EI) is the term specified by the SAE for an ignition system that does not use a distributor. Types of EI systems include:

1. **Waste-spark system.** This type of system uses one ignition coil to fire the spark plugs for two cylinders at the same time. ● **SEE FIGURE 16–2**.
2. **Coil-on-plug system.** This type of system uses a single ignition coil for each cylinder with the coil placed above or near the spark plug.

IGNITION SYSTEM OPERATION

The ignition system includes components and wiring necessary to create and distribute a high voltage (up to 40,000 volts or more). All ignition systems apply voltage close to battery voltage (12 volts) to the positive side of the ignition coil and pulse the negative side to ground. When the coil negative lead is grounded, the primary (low-voltage) circuit of the coil is complete and a magnetic field is created around the coil windings. When the circuit is opened, the magnetic field collapses and induces a high-voltage spark in the secondary winding of the ignition coil. Early ignition systems used a mechanically opened set of contact points to make and break the electrical connection to ground. Electronic ignition uses a sensor, such as a pickup coil and reluctor (trigger wheel), or trigger to signal an electronic module that makes and breaks the primary connection of the ignition coil.

NOTE: Distributor ignition (DI) is the term specified by the Society of Automotive Engineers (SAE) for an ignition system that uses a distributor. Electronic ignition (EI) is the term specified by the SAE for an ignition system that does not use a distributor.

IGNITION COILS

PURPOSE AND FUNCTION The heart of any ignition system is the **ignition coil**. The coil creates a high-voltage spark by electromagnetic induction. Many ignition coils contain two separate but electrically connected windings of copper wire. Other coils are true transformers in which the primary and secondary windings are not electrically connected.

COIL CONSTRUCTION The center of an ignition coil contains a core of laminated soft iron (thin strips of soft iron). This core increases the magnetic strength of the coil. Surrounding the laminated core are approximately 20,000 turns of fine wire (approximately 42 gauge). These windings are called the secondary coil windings. Surrounding the secondary windings are approximately 150 turns of heavy wire (approximately 21 gauge). These windings are called the primary coil windings. The secondary winding has about 100 times the number of turns of the primary winding, referred to as the **turns ratio** (approximately 100:1). The E coil is so named because the laminated, soft-iron core is E-shaped, with the coil wire turns wrapped around the center "finger" of the E and the primary winding wrapped inside the secondary winding. ● **SEE FIGURES 16–3 AND 16–4.**

FIGURE 16–3 The steel lamination used in an E coil helps increase the magnetic field strength, which helps the coil produce higher energy output for a more complete combustion in the cylinders.

The primary windings of the coil extend through the case of the coil and are labeled as "positive" and "negative." The positive terminal of the coil attaches to the ignition switch, which supplies current from the positive battery terminal. The negative terminal is attached to an **ignition control module (ICM** or **igniter**), which opens and closes the primary ignition circuit by opening or closing the ground return path of the circuit. When the ignition switch is on, voltage should be available at *both* the positive terminal and the negative terminal of the coil if the primary windings of the coil have continuity. The labeling of positive (+) and negative (–) of the coil indicates that the positive terminal is *more* positive (closer to the positive terminal of the battery) than the negative terminal of the coil. This condition is called the coil **polarity**. The polarity of the coil must be correct to ensure that electrons will flow from the hot center electrode of the spark plug on DI systems. *The polarity of an ignition coil is determined by the direction of rotation of the coil windings.* The correct polarity is then indicated on the primary terminals of the coil. If the coil primary leads are reversed, the voltage required to fire the spark plugs is increased by 40%. The coil output voltage is directly proportional to the ratio of primary to secondary turns of wire used in the coil.

SELF-INDUCTION When current starts to flow into a coil, an opposing current is created in the windings of the coil. This opposing current generation is caused by **self-induction** and is called **inductive reactance**. Inductive reactance is similar to resistance because it opposes any changes (increase or decrease) in current flow in a coil. Therefore, when an ignition coil is first energized, there is a slight delay of approximately 0.01 second before the ignition coil reaches its maximum

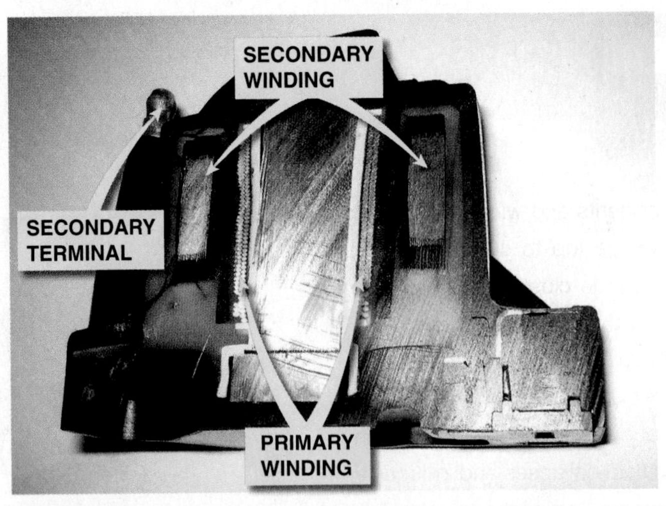

FIGURE 16–4 The primary windings are inside the secondary windings on this General Motors coil.

magnetic field strength. The point at which a coil's maximum magnetic field strength is reached is called **saturation**.

MUTUAL INDUCTION In an ignition coil there are two windings, a primary and a secondary winding. When a *change* occurs in the magnetic field of one coil winding, a change also occurs in the other coil winding. Therefore, if the current is stopped from flowing (circuit is opened), the collapsing magnetic field cuts across the turns of the secondary winding and creates a high voltage in the secondary winding. This generation of an electric current in both coil windings is called **mutual induction**. The collapsing magnetic field also creates a voltage of up to 250 volts in the primary winding.

HOW IGNITION COILS CREATE 40,000 VOLTS All ignition systems use electromagnetic induction to produce a high-voltage spark from the ignition coil. Electromagnetic induction means that a current can be created in a conductor (coil winding) by a moving magnetic field. The magnetic field in an ignition coil is produced by current flowing through the primary windings of the coil. The current for the primary winding is supplied through the ignition switch to the positive terminal of the ignition coil. The negative terminal is connected to the ground return through an electronic ignition module (igniter).

If the primary circuit is completed, current (approximately 2 to 6 A) can flow through the primary coil windings. This flow creates a strong magnetic field inside the coil. When the primary coil winding ground return path connection is opened, the magnetic field collapses and induces a voltage of 250 to 400 volts in the primary winding of the coil and a high-voltage (20,000 to 40,000 volts) low-amperage (20 to 80 mA) current in the secondary coil windings. This high-voltage pulse flows through the coil wire (if the vehicle is so equipped), distributor cap, rotor, and spark plug wires to the spark plugs. For each spark that occurs, the coil must be charged with a magnetic field and then discharged. The ignition components that regulate the current in the coil primary winding by turning it on and off are known collectively as the **primary ignition circuit**. The components necessary to create and distribute the high voltage produced in the secondary windings of the coil are called the **secondary ignition circuit**. ● **SEE FIGURE 16–4**. These circuits include the following components:

PRIMARY IGNITION CIRCUIT

1. Battery
2. Ignition switch
3. Primary windings of coil
4. Pickup coil (crank sensor)
5. Ignition module (igniter)
6. PCM

SECONDARY IGNITION CIRCUIT

1. Secondary windings of coil
2. Distributor cap and rotor (if the vehicle is so equipped)
3. Spark plug wires
4. Spark plugs

● **SEE FIGURE 16–5.**

FREQUENTLY ASKED QUESTION

How Does the Computer Control the Ignition?

The Powertrain Control Module (PCM) plays a key role in the final functioning of the ignition circuits. The PCM receives signals from all of the engine sensors and based on this information, uses an algorithm (computer program) to determine the best time to fire the spark plugs. For example, the sensors and how they could affect when the spark occurs include:

Engine Coolant Temperature (ECT)—The colder the engine, the more spark advance may be needed to achieve the highest possible engine output torque with lowest exhaust emissions.

Throttle Position (TP) sensor—The PCM uses the TP sensor information to not only determine where the accelerator position is located but also at what rate it is changing. If the accelerator pedal is rapidly being depressed, then spark timing may be delayed (retarded slightly to help prevent spark knock.

Manifold Absolute Pressure (MAP) sensor—The MAP sensor is used to detect engine load. During a heavy load, less spark advance is needed to help prevent spark knock, whereas more spark advance is needed under light load conditions for the engine to achieve maximum fuel economy and the lowest possible exhaust emissions.

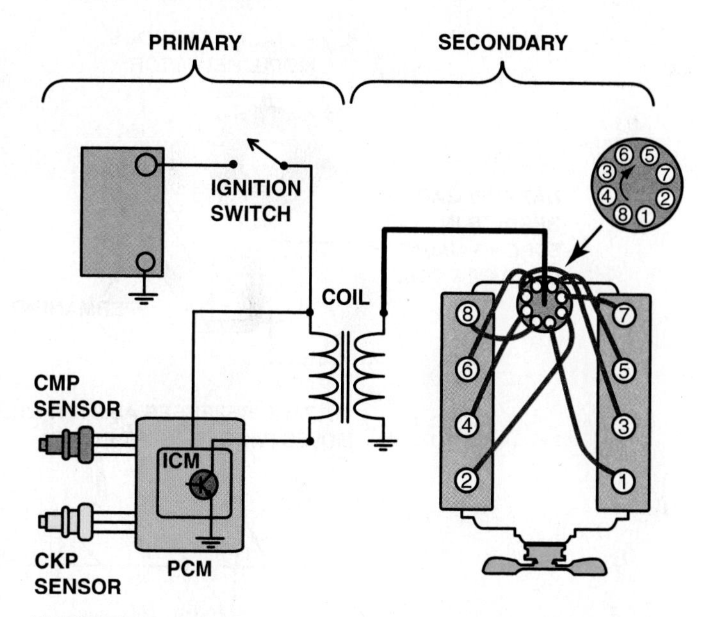

FIGURE 16–5 The primary ignition system is used to trigger and therefore create the secondary (high-voltage) spark from the ignition coil.

IGNITION SWITCHING AND TRIGGERING

For any ignition system to function, the primary current must be turned on to charge the coil and off to allow the coil to discharge, creating a high-voltage spark. This turning on and off of the primary circuit is called **switching**. The unit that does the switching is an electronic switch, such as a power transistor. This power transistor can be located in any of the following locations:

- In the ignition control module (ICM)
- In the PCM (computer)

NOTE: On some coil-on-plug systems, the ICM is part of the ignition coil itself and is serviced as an assembly.

The device that signals the switching of the coil on and off or just on in most instances, is called the trigger. A trigger is typically a pickup coil in some distributor-type ignitions and a crankshaft position sensor (CKP) on electronic (waste-spark and coil-on-plug) and many distributor-type ignitions. There are three types of devices used for triggering, including the magnetic sensor, Hall-effect switch, and optical sensor.

PRIMARY CIRCUIT OPERATION

To get a spark out of an ignition coil, the primary coil circuit must be turned on and off. This primary circuit current is controlled by a **transistor** (electronic switch) inside the ignition module or (igniter) that in turn is controlled by one of several devices, including:

- **Pickup coil (pulse generator)**—A simple and common ignition electronic switching device is the magnetic pulse generator system. Most manufacturers use the rotation of the distributor shaft to time the voltage pulses. The **magnetic pulse generator** is installed in the distributor housing. The pulse generator consists of a trigger wheel (reluctor) and a pickup coil. The pickup coil consists of an iron core wrapped with fine wire, in a coil at one end and attached to a permanent magnet at the other end. The center of the coil is called the pole piece. The pickup coil signal triggers the transistor inside the module and is also used by the computer for piston position information and engine speed (RPM). ● **SEE FIGURES 16–6 AND 16–7.**
- **Hall-effect switch** —A **Hall-effect switch** also uses a stationary sensor and rotating trigger wheel (shutter). ● **SEE FIGURE 16–8.** Unlike the magnetic pulse

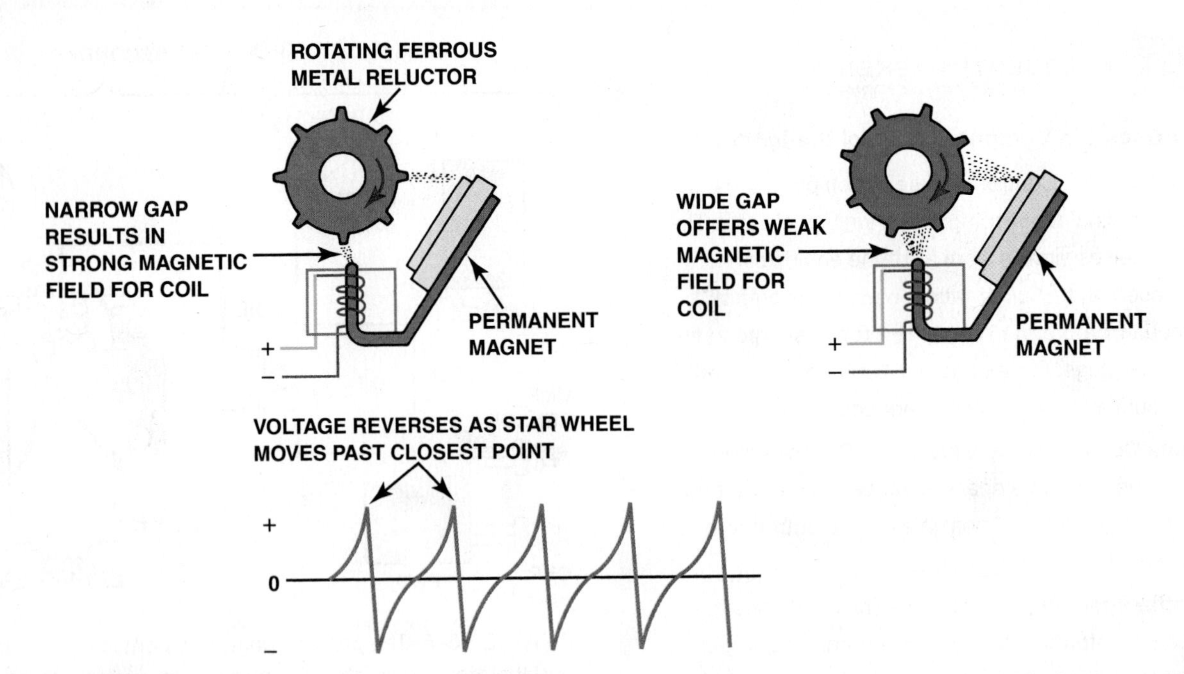

FIGURE 16–6 Operation of a typical pulse generator (pickup coil). At the bottom is a line drawing of a typical scope pattern of the output voltage of a pickup coil. The module receives this voltage from the pickup coil and opens the ground circuit to the ignition coil when the voltage starts down from its peak (just as the reluctor teeth start moving away from the pickup coil).

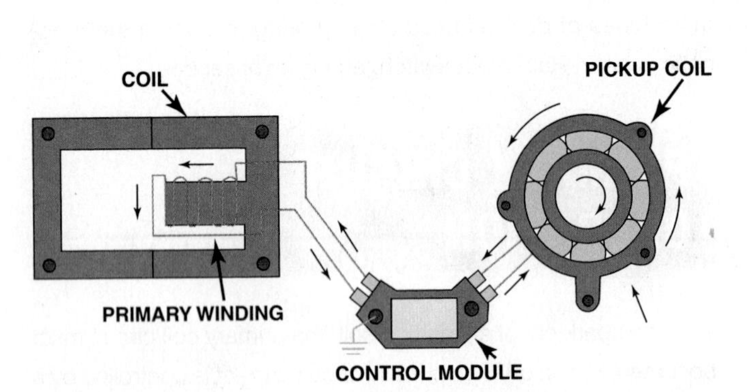

FIGURE 16–7 The varying voltage signal from the pickup coil triggers the ignition module. The ignition module grounds and ungrounds the primary winding of the ignition coil, creating a high-voltage spark.

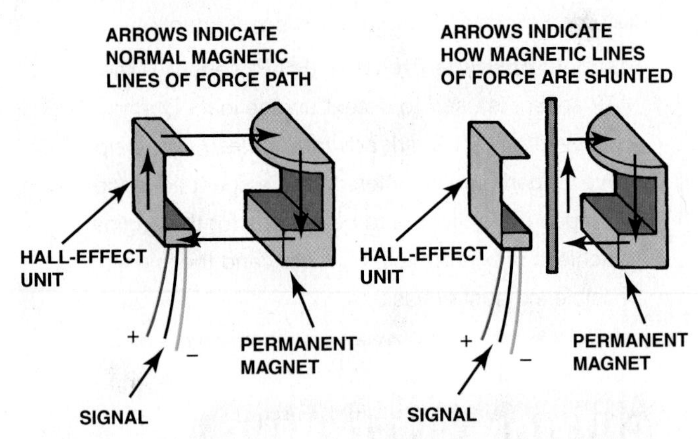

FIGURE 16–8 Hall-effect switches use metallic shutters to shunt magnetic lines of force away from a silicon chip and related circuits. All Hall-effect switches produce a square wave output for every accurate triggering.

generator, the Hall-effect switch requires a small input voltage to generate an output or signal voltage. Hall effect is the ability to generate a voltage signal in semiconductor material (gallium arsenate crystal) by passing current through it in one direction and applying a magnetic field to it at a right angle to its surface. If the input current is held steady and the magnetic field fluctuates, an output voltage is produced that changes in proportion to field strength. Most Hall-effect switches in distributors have a Hall element or device, a permanent magnet, and a rotating ring of metal blades (shutters) similar to a trigger wheel (another method uses a stationary sensor with a rotating magnet). Some blades are designed to hang down, typically found in Bosch and Chrysler systems; others may be on a separate ring on the distributor shaft, typically found in GM and Ford Hall-effect distributors. When the shutter blade enters the gap between the magnet and the Hall element, it creates a magnetic shunt that changes the field strength through the Hall element. This analog signal is sent to a **Schmitt trigger** inside the sensor itself, which converts the analog signal into a digital signal. A digital (on or off) voltage signal is created at a varying frequency to the ignition module or onboard computer. ● **SEE FIGURES 16–9 AND 16–10.**

- **Magnetic crankshaft position sensor**—A **magnetic sensor** uses the changing strength of the magnetic field surrounding a coil of wire to signal the module and

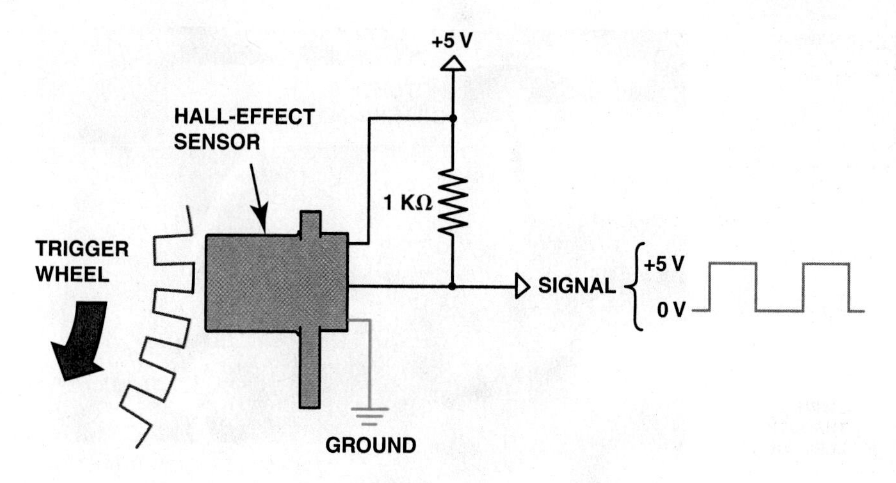

FIGURE 16–9 A Hall-effect sensor produces a digital on-off voltage signal whether it is used with a blade or a notched wheel.

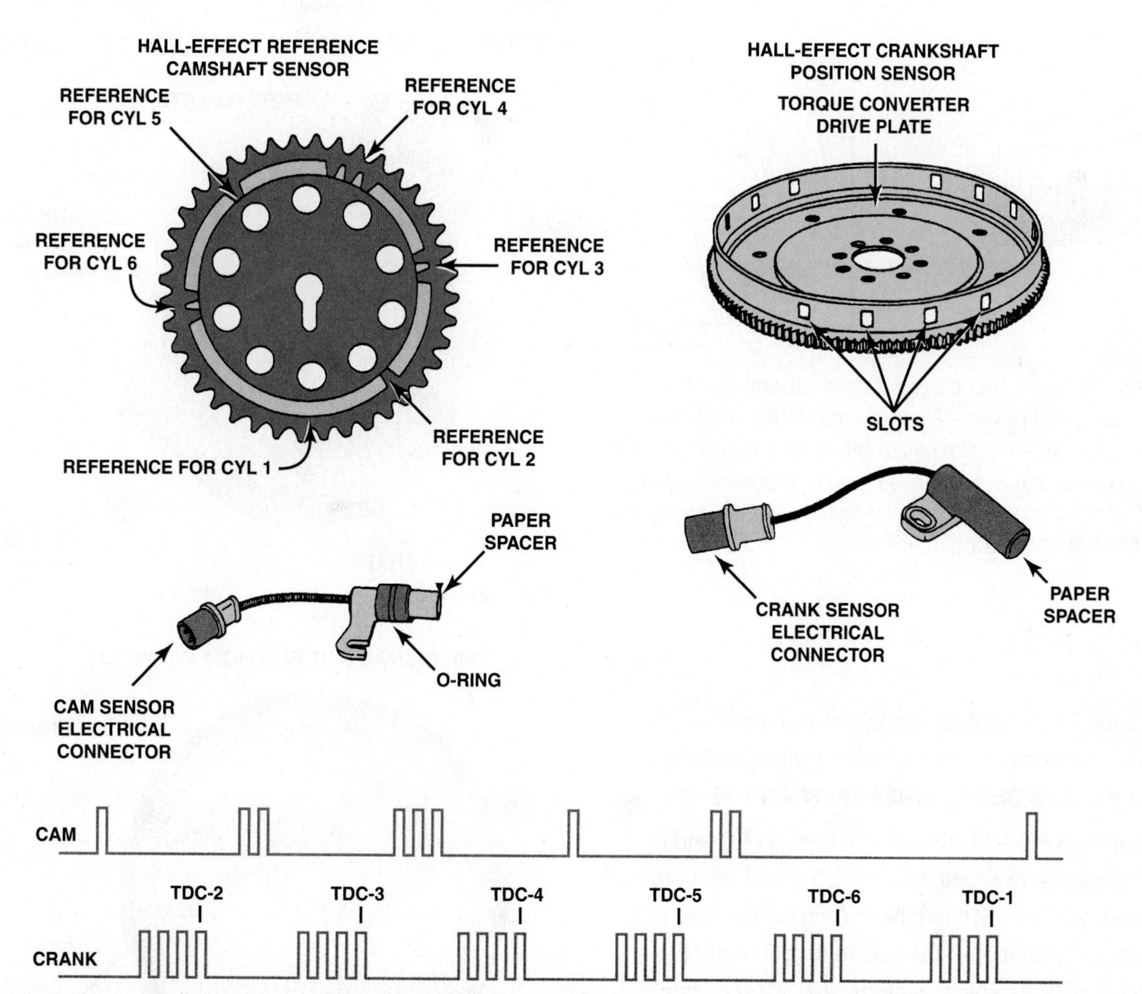

FIGURE 16–10 Some Hall-effect sensors look like magnetic sensors. This Hall-effect camshaft reference sensor and crankshaft position sensor have an electronic circuit built in that creates a 0 to 5 volt signal as shown at the bottom. These Hall-effect sensors have three wires: a power supply (8 volts) from the computer (controller); a signal (0 to 5 volts); and a signal ground.

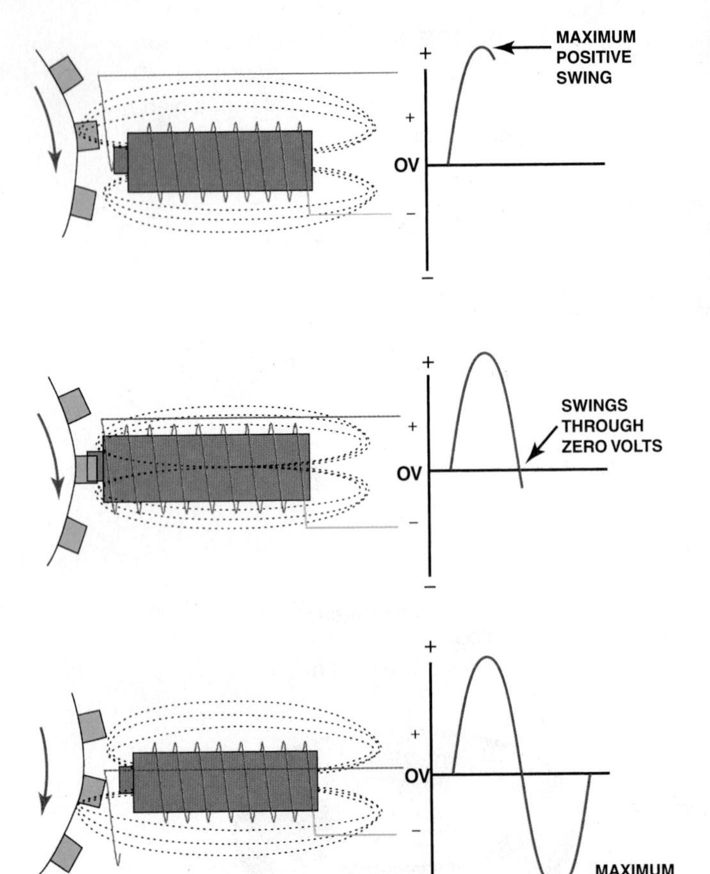

FIGURE 16–11 A magnetic sensor uses a permanent magnet surrounded by a coil of wire. The notches of the crankshaft (or camshaft) create a variable magnetic field strength around the coil. When a metallic section is close to the sensor, the magnetic field is stronger because metal is a better conductor of magnetic lines of force than air.

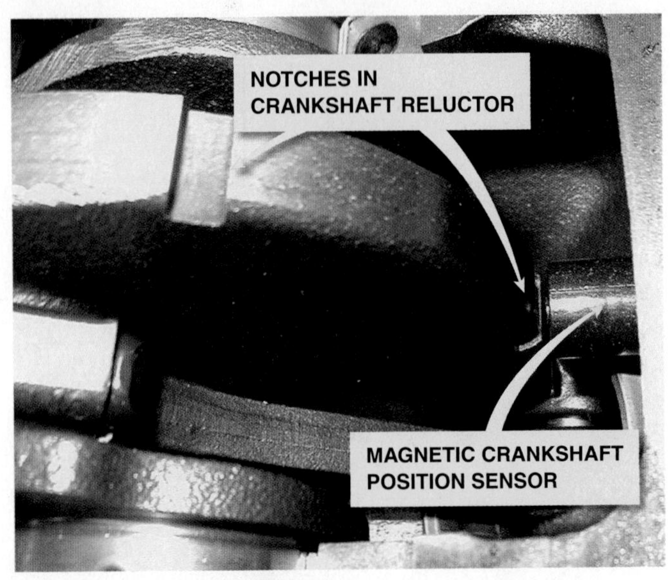

FIGURE 16–12 A typical magnetic crankshaft position sensor.

computer. This signal is used by the electronics in the module and computer as to piston position and engine speed (RPM). ● **SEE FIGURES 16–11 AND 16–12.**

- **Optical sensors**—These use light from a LED and a phototransistor to signal the computer. An interrupter disc between the LED and the phototransistor has slits that allow the light from the LED to trigger the phototransistor on the other side of the disc. Most **optical sensors** (usually located inside the distributor) use two rows of slits to provide individual cylinder recognition (low-resolution) and precise distributor angle recognition (high-resolution) signals. ● **SEE FIGURE 16–13.**

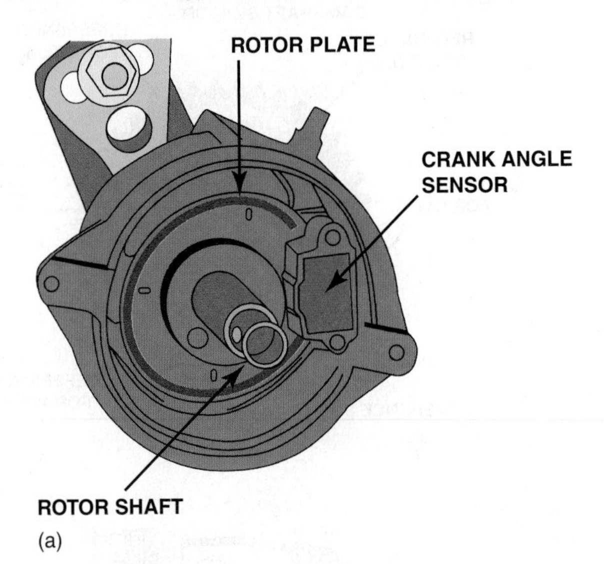

(a)

180° SIGNAL SLIT FOR NO. 1 CYLINDER

1° SIGNAL SLIT

180°
SIGNAL SLIT

ROTOR PLATE

(b)

FIGURE 16–13 (a) Typical optical distributor. (b) Cylinder 1 slit signals the computer the piston position for cylinder 1. The 1-degree slits provide accurate engine speed information to the computer.

FIGURE 16–14 A light shield over an optical sensor being installed before the rotor is attached.

TECH TIP

Optical Distributors Do Not Like Light

Optical distributors use the light emitted from LEDs to trigger phototransistors. Most optical distributors use a shield between the distributor rotor and the optical interrupter ring. Sparks jump the gap from the rotor tip to the distributor cap inserts. This shield blocks the light from the electrical arc from interfering with the detection of the light from the LEDs.

If this shield is not replaced during service, the light signals are reduced and the engine may not operate correctly. ● **SEE FIGURE 16–14.** This can be difficult to detect because nothing looks wrong during a visual inspection. Remember that all optical distributors must be shielded between the rotor and the interrupter ring.

TECH TIP

The Tachometer Trick

When diagnosing a no-start or intermittent misfire condition, check the operation of the tachometer. If the tachometer does not indicate engine speed (no-start condition) or drops toward zero (engine misfire), then the problem is due to a defect in the *primary* ignition circuit. The tachometer gets its signal from the pulsing of the primary winding of the ignition coil. The following components in the primary circuit could cause the tachometer to not work when the engine is cranking:

- Pickup coil
- Crankshaft position sensor
- Ignition module (igniter)
- Coil primary wiring

If the vehicle is not equipped with a tachometer, connect a scan tool and monitor engine RPM. Remember the following:

- No tachometer reading means the problem is in the primary ignition circuit.
- Tachometer reading okay means the problem is in the secondary ignition circuit or is a fuel-related problem.

DISTRIBUTOR IGNITION

FIRING ORDER Firing order means the order that the spark is distributed to the correct spark plug at the right time. The firing order of an engine is determined by crankshaft and camshaft design. The firing order is determined by the location of the spark plug wires in the distributor cap of an engine equipped with a distributor. The firing order is often cast into the intake manifold for easy reference. ● **SEE FIGURES 16–15.**

GENERAL MOTORS HEI ELECTRONIC IGNITION General Motor's HEI distributors use 8-mm-diameter spark plug wires that use female connections to the distributor cap towers. HEI coils must be replaced (if defective) with the exact replacement style. HEI coils differ and can be identified by the colors of the primary leads. The primary coil leads can be either white and red or yellow and red. The correct color of lead coil must be used for replacement. The colors of the leads indicate the direction in which the coil is wound, and therefore its polarity. ● **SEE FIGURES 16–16 AND 16–17.**

FIGURE 16–15 The firing order is cast or stamped on the intake manifold on most engines that have a distributor ignition.

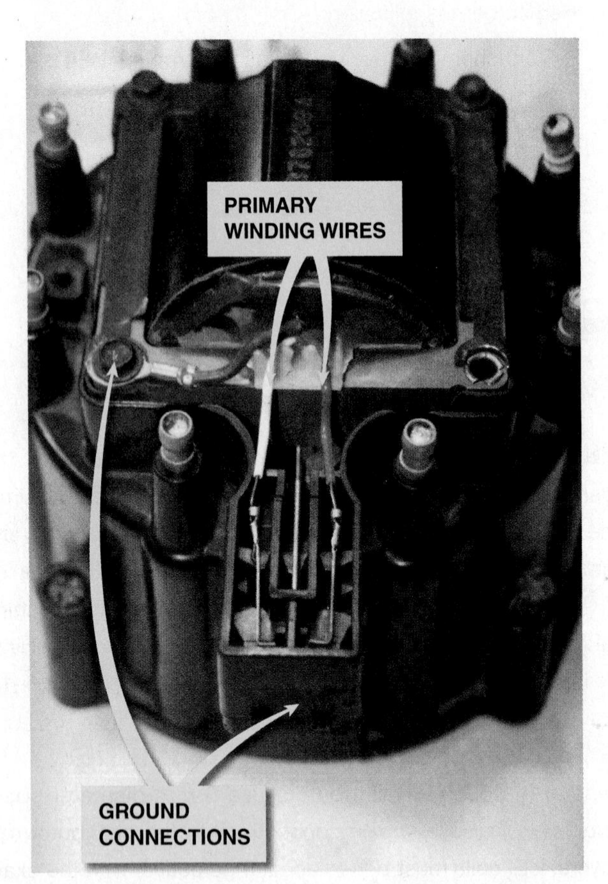

FIGURE 16–16 A typical General Motors HEI coil installed in the distributor cap. When the coil or distributor cap is replaced, check that the ground clip is transferred from the old distributor cap to the new. Without proper grounding, coil damage is likely. There are two designs of HEI coils. One uses red and white wire as shown, and the other design, which has reversed polarity, uses red and yellow wire for the coil primary.

FIGURE 16–17 This distributor ignition system uses a remotely mounted ignition coil.

FORD ELECTRONIC IGNITION Ford electronic ignition systems all function similarly, even though over the years the system has been called by various names.

The EEC IV system uses the thick-film-integration (TFI) ignition system. This system uses a smaller control module attached to the distributor and uses an air-cooled epoxy E coil. ● **SEE FIGURE 16–18**. Thick-film integration means that all electronics are manufactured on small layers built up to form a thick film. Construction includes using pastes of different electrical resistances that are deposited on a thin, flat ceramic material by a process similar to silk-screen printing. These resistors are connected by tracks of palladium silver paste. Then the chips that form the capacitors, diodes, and integrated circuits are soldered directly to the palladium silver tracks. The thick-film manufacturing process is highly automated.

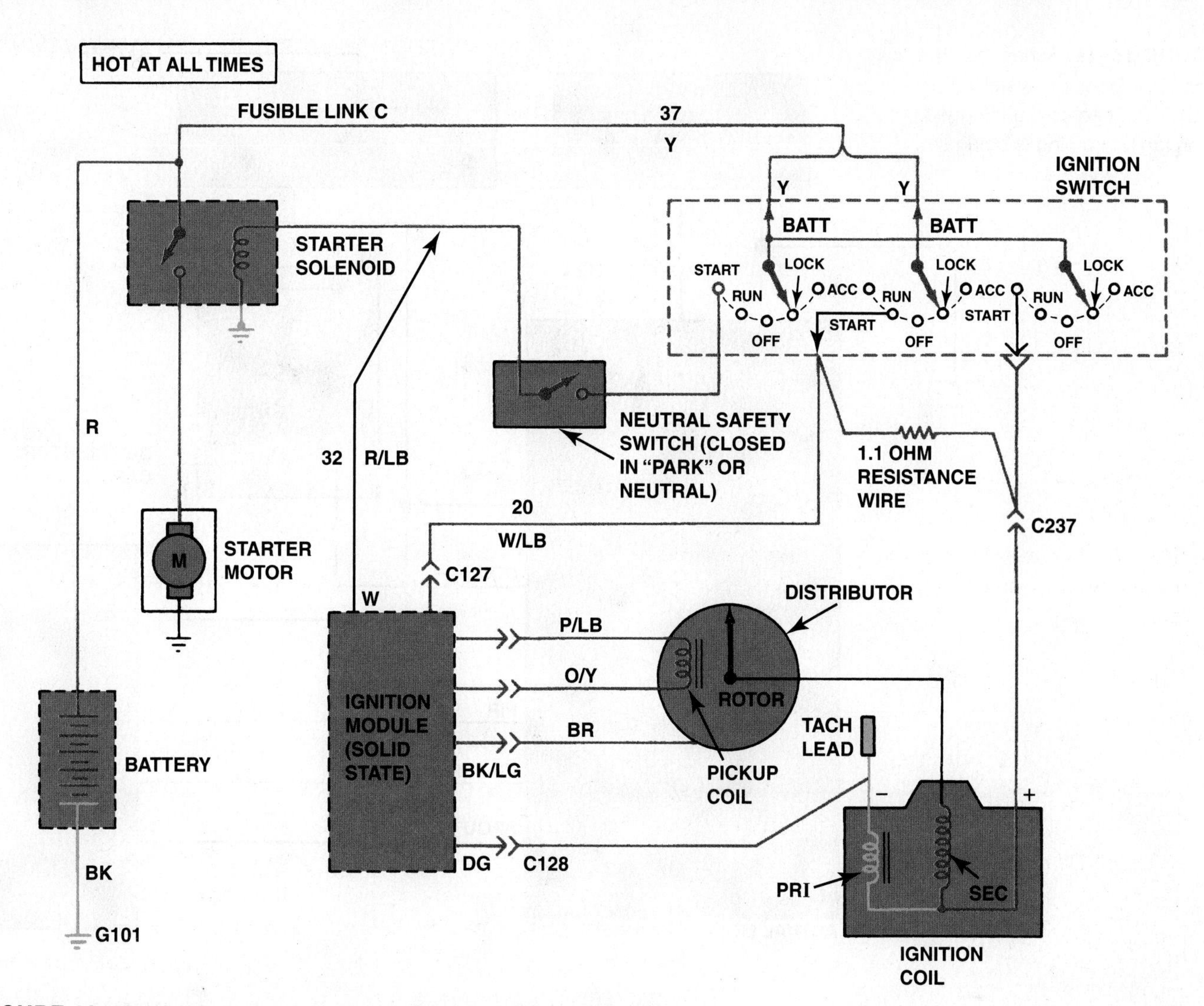

FIGURE 16–18 Wiring diagram of a typical Ford electronic ignition.

OPERATION OF FORD DISTRIBUTOR IGNITION Ford DI systems function in basically the same way regardless of year and name. Under the distributor cap and rotor is a magnetic pickup assembly. This assembly produces a small alternating electrical pulse (approximately 1.5 volts) when the distributor armature rotates past the pickup assembly (stator). This low-voltage pulse is sent to the ignition module. The ignition module then switches (through transistors) off the primary ignition coil current. When the ignition coil primary current is stopped quickly, a high-voltage "spike" discharges from the coil secondary winding. The coil current is controlled in the module circuits by decreasing dwell (coil-charging time), depending on various factors determined by operating conditions. ● **SEE FIGURE 16–19.**

CHRYSLER DISTRIBUTOR IGNITION Chrysler was the first domestic manufacturer to produce electronic ignition as standard equipment. The Chrysler system consists of a pulse generator unit in the distributor (pickup coil and reluctor). Chrysler's name for their electronic ignition is **electronic ignition system (EIS)**, and the control unit (module) is called the **electronic control unit (ECU)**.

The pickup coil in the distributor (pulse generator) generates the signal to open and close the primary coil circuit. ● **SEE FIGURE 16–20.**

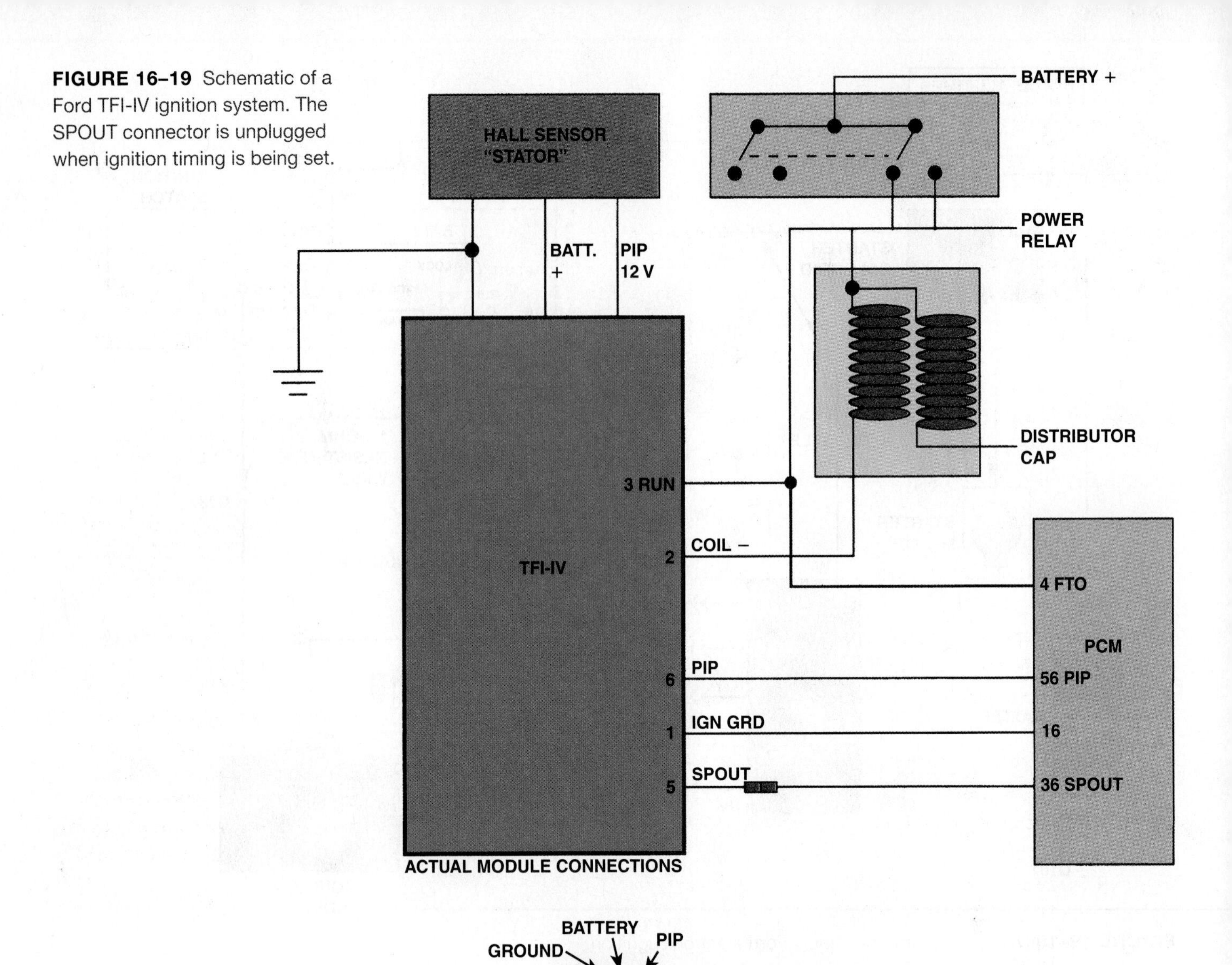

FIGURE 16–19 Schematic of a Ford TFI-IV ignition system. The SPOUT connector is unplugged when ignition timing is being set.

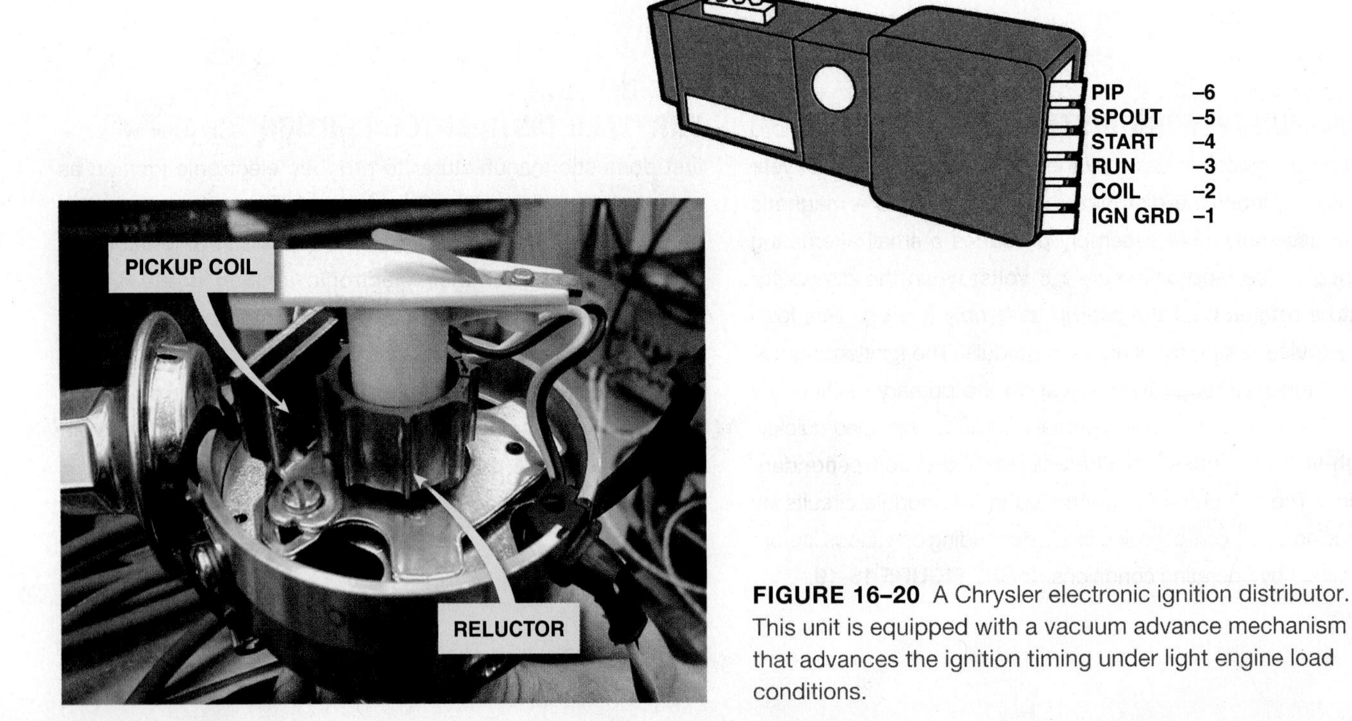

FIGURE 16–20 A Chrysler electronic ignition distributor. This unit is equipped with a vacuum advance mechanism that advances the ignition timing under light engine load conditions.

WASTE-SPARK IGNITION SYSTEMS

Waste-spark ignition is another name for **distributorless ignition system (DIS)** or electronic ignition (EI). Waste-spark ignition was introduced in the mid-1980s and uses the onboard computer to fire the ignition coils. These systems were first used on some Saabs and General Motors engines. A four-cylinder engine uses two ignition coils and a six-cylinder engine uses three ignition coils. Each coil is a true transformer in which the primary winding and secondary winding are not electrically connected. Each end of the secondary winding is connected to a cylinder exactly opposite the other in the firing order, which is called a **companion (paired) cylinder.** ● **SEE FIGURE 16–21.** This means that *both* spark plugs fire at the same time (within nanoseconds of each other). When one cylinder (e.g., 6) is on the compression stroke, the other cylinder (3) is on the exhaust stroke. This spark that occurs on the exhaust stroke is called the waste spark, because it does no useful work and is only used as a ground path for the secondary winding of the ignition coil. The voltage required to jump the spark plug gap on cylinder 3 (the exhaust stroke) is only 2 to 3 kV and provides the *ground circuit* for the secondary coil circuit. The remaining coil energy is used by the cylinder on the compression stroke. One spark plug of each pair always fires straight polarity and the other cylinder always fires reverse polarity. Spark plug life is not greatly affected by the reverse polarity. If there is only one defective spark plug wire or spark plug, two cylinders may be affected.

The coil polarity is determined by the direction the coil is wound (left-hand rule for conventional current flow) and cannot be changed. ● **SEE FIGURE 16–22.** Each spark plug for a particular cylinder always will be fired either with straight or reversed polarity, depending on its location in the engine and how the coils are wired. However, the compression and waste-spark condition flip-flops. When one cylinder is on compression, such as cylinder number 1, then the paired cylinder (number 4) is on the exhaust stroke. During the next rotation of the crankshaft, cylinder number 4 is on the compression stroke and cylinder number 1 is on the exhaust stroke.

Cylinder 1	Always fires straight polarity, one time, requiring 10 to 12 kV and one time, requiring 3 to 4 kV.
Cylinder 4	Always fires reverse polarity, one time, requiring 10 to 12 kV and one time, requiring 3 to 4 kV.

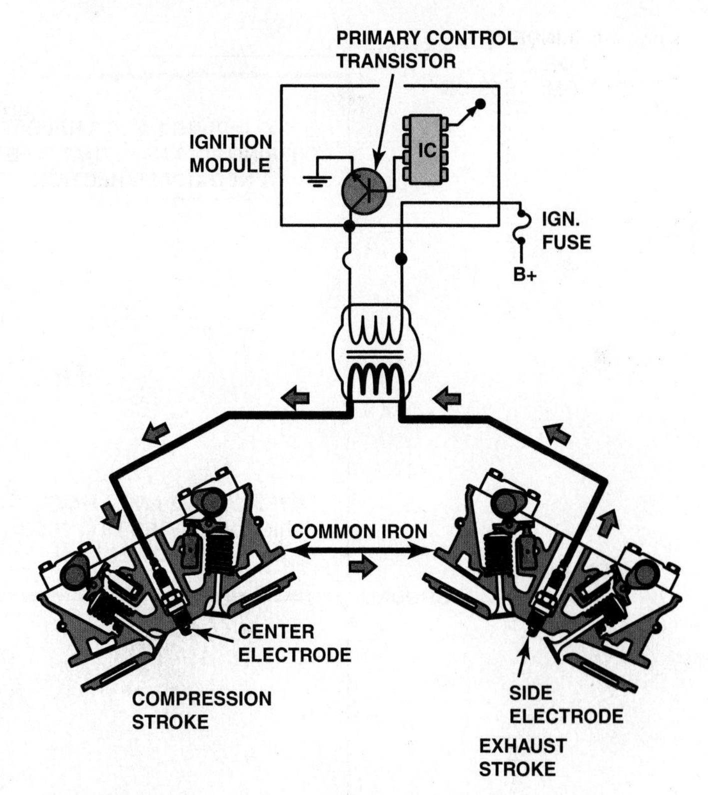

FIGURE 16–21 A waste-spark system fires one cylinder while its piston is on the compression stroke and into paired or companion cylinders while it is on the exhaust stroke. In a typical engine, it requires only about 2 to 3 kV to fire the cylinder on the exhaust strokes. The remaining coil energy is available to fire the spark plug under compression (typically about 8 to 12 kV).

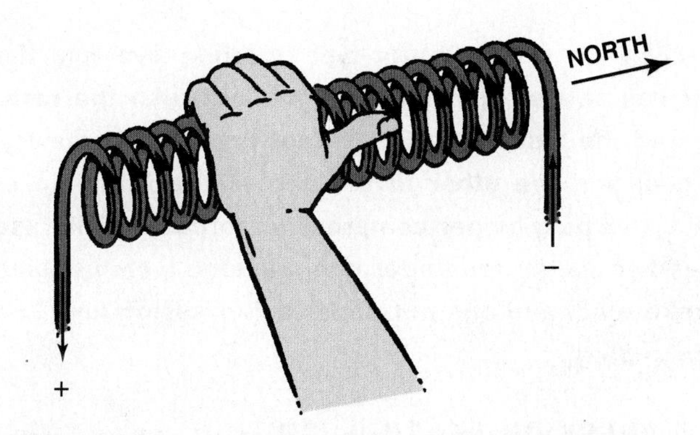

FIGURE 16–22 The left-hand rule states that if a coil is grasped with the left hand, the fingers will point in the direction of current flow and the thumb will point toward the north pole.

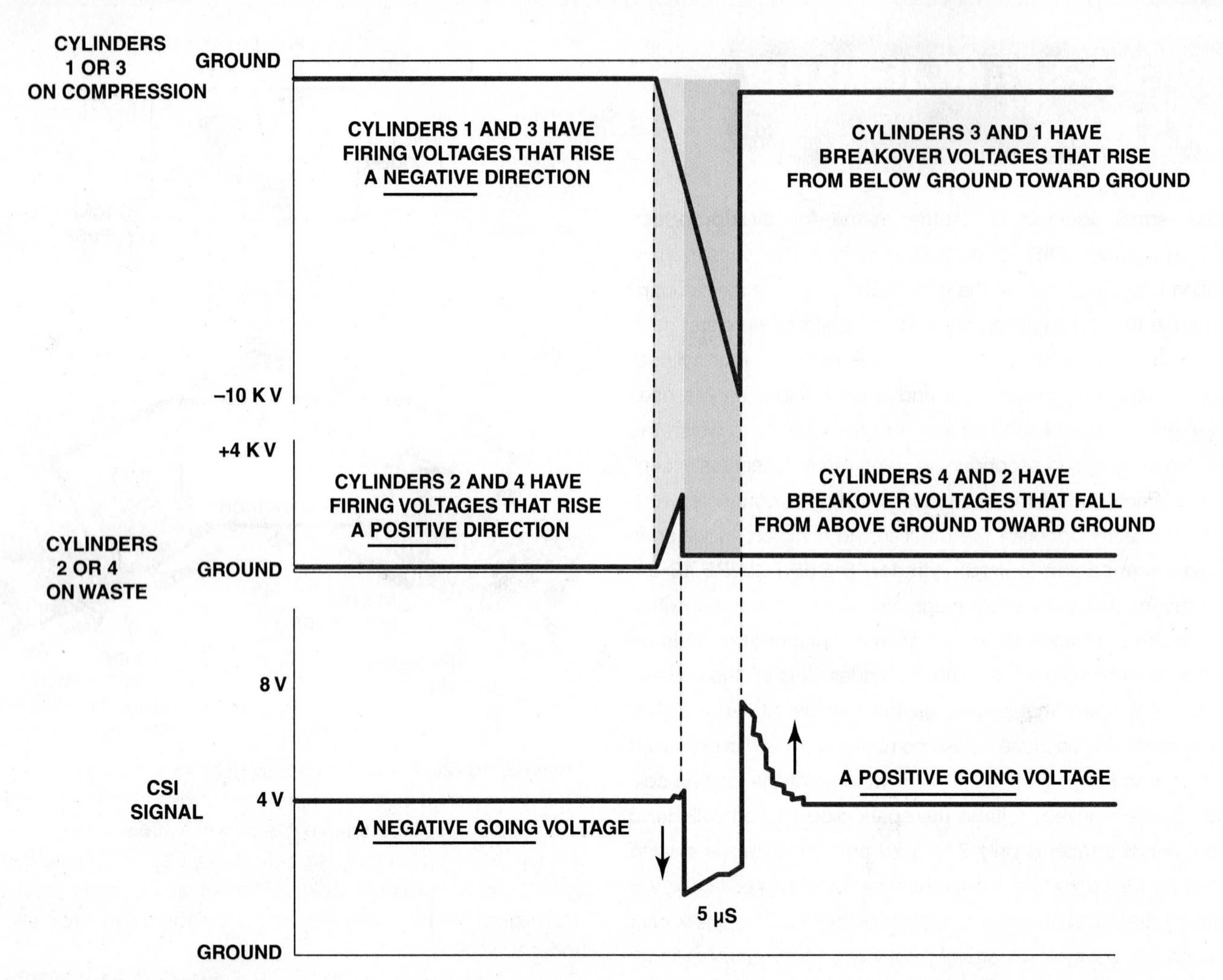

FIGURE 16–23 The slight (5 microsecond) difference in the firing of the companion cylinders is enough time to allow the PCM to determine which cylinder is firing on the compression stroke. The compression sensing ignition (CSI) signal is then processed by the PCM, which then determines which cylinder is on the compression stroke.

NOTE: With a distributor-type ignition system, the coil has two air gaps to fire: one between the rotor tip and the distributor insert (not under compression forces) and the other in the gap at the firing tip of the spark plug (under compression forces). A DIS also fires two gaps: one under compression (compression stroke plug) and one not under compression (exhaust stroke plug).

COMPRESSION-SENSING IGNITION Some waste-spark ignition systems, such as those used on Saturns, use the voltage required to fire the cylinders to determine cylinder position. It requires a higher voltage to fire a spark plug under compression than it does when the spark plug is being fired on the exhaust stroke. The electronics in the coil and the PCM can detect which of the two companion (paired) cylinders that are fired at the same time requires the higher voltage, and therefore indicates the cylinder that is on the compression stroke. For example, a typical four-cylinder engine equipped with a waste-spark ignition system will fire both cylinders 1 and 4. If cylinder 4 requires a higher voltage to fire, as determined by the electronics connected to the coil, then the PCM assumes that cylinder 4 is on the compression stroke. Engines equipped with compression-sensing ignition systems do not require the use of a camshaft position sensor to determine specific cylinder numbers. ● **SEE FIGURE 16–23.**

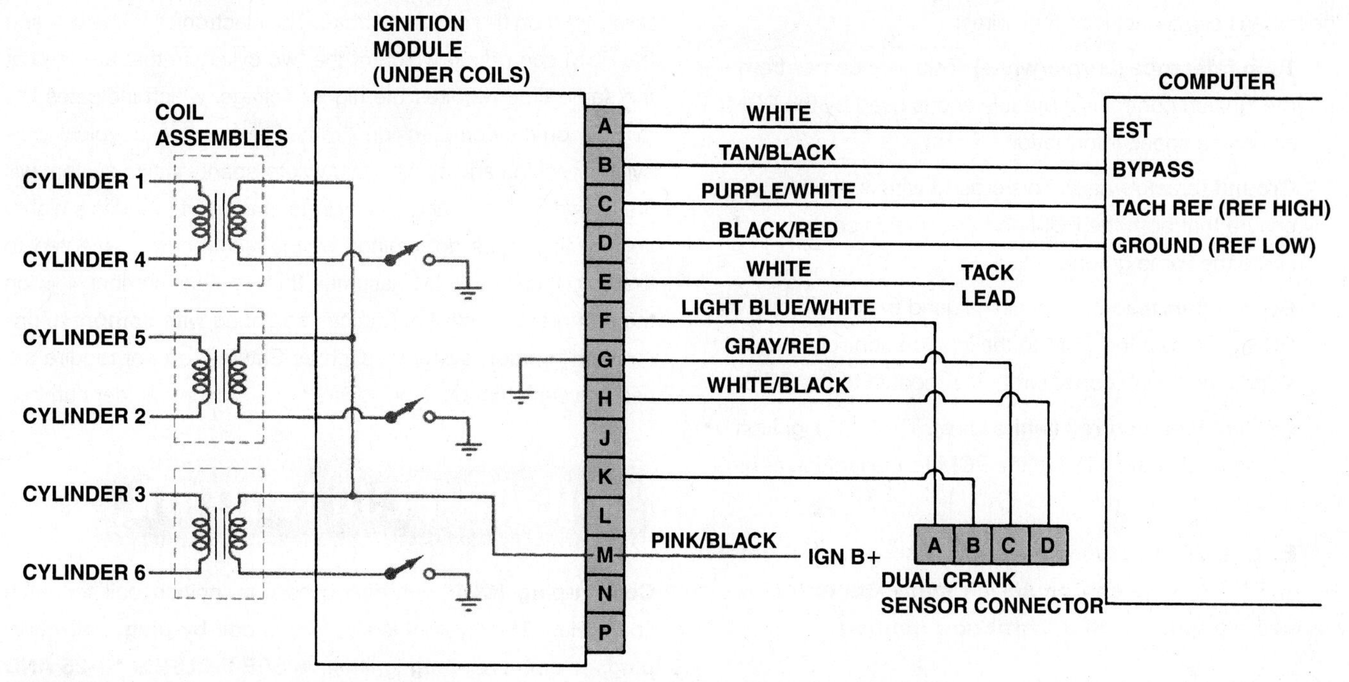

FIGURE 16–24 Typical wiring diagram of a V-6 distributorless (direct fire) ignition system.

TECH TIP

Odds Fire Straight

Waste-spark ignition systems fire two spark plugs at the same time. Most vehicle manufacturers use a waste-spark system that fires the odd-numbered cylinders (1, 3, and 5) by straight polarity (current flow from the top of the spark plug through the gap and to the ground electrode). The even-numbered cylinders (2, 4, and 6) are fired reverse polarity, meaning that the spark jumps from the side electrode to the center electrode. Some vehicle manufacturers equip their vehicles with platinum plugs with the expansive platinum alloy only on one electrode as follows:

- On odd-numbered cylinders (1, 3, 5), the platinum is on the center electrode.
- On even-numbered cylinders (2, 4, 6), the platinum is on the ground electrode.

Replacement spark plugs use platinum on both electrodes (double platinum) and can, therefore, be placed in any cylinder location.

IGNITION CONTROL CIRCUITS

Ignition control (IC) is the OBD-II terminology for the output signal from the PCM to the ignition system that controls engine timing. Previously, each manufacturer used a different term to describe this signal. For instance, Ford referred to this signal as **spark output (SPOUT)** and General Motors referred to this signal as **electronic spark timing (EST)**. This signal is now referred to as the ignition control (IC) signal. The ignition control signal is usually a digital output that is sent to the ignition system as a timing signal. If the ignition system is equipped with an ignition module, then this signal is used by the ignition module to vary the timing as engine speed and load changes. If the PCM directly controls the coils, such as most coil-on-plug ignition systems, then this IC signal directly controls the coil primary and there is a separate IC signal for each ignition coil. The IC signal controls the time that the coil fires; it either advances or retards the timing. On many systems, this signal controls the duration of the primary current flow in the coil, which is referred to as the **dwell**.

BYPASS IGNITION CONTROL A bypass-type ignition control means that the engine starts using the ignition module for timing control and then switches to the PCM for timing control after the engine starts. A **bypass ignition** is commonly used on General Motors engines equipped with distributor ignition (DI), as well as those equipped with waste-spark ignition. ● **SEE FIGURE 16–24**.

The bypass circuit includes four wires:

- **Tach reference (purple/white).** This wire comes from the ignition control (IC) module and is used by the PCM as engine speed information.
- **Ground (black/white).** This ground wire is used to ensure that both the PCM and the ignition control module share the same ground.
- **Bypass (tan/black).** This wire is used to conduct a 5 volt DC signal from the PCM to the ignition control module to switch the timing control from the module to the PCM.
- **EST (ignition control) (white wire).** This is the ignition timing control signal from the PCM to the ignition control module.

NOTE: It is this bypass wire that is disconnected before the ignition timing can be set on many General Motors engines equipped with a distributor ignition.

DIAGNOSING A BYPASS IGNITION SYSTEM One advantage of a bypass-type ignition is that the engine will run without the computer because the module can do the coil switching and can, through electronic circuits inside the module, provide for some spark advance as the engine speed increases. This is a safety feature that helps protect the catalytic converter if the ignition control from the PCM is lost. Therefore, if there is a problem, use a digital meter and check for the presence of 5 volts on the tan bypass wire. If there is not 5 volts present with the engine running, then the PCM or the wiring is at fault.

UP-INTEGRATED IGNITION CONTROL Most coil-on-plug and many waste-spark-type ignition systems use the PCM for ignition timing control. This type of ignition control is called **up-integrated** because all timing functions are interpreted in the PCM, rather than being split between the ignition control module and the PCM. The ignition module, if even used, contains the power transistor for coil switching. The signal as to when the coil fires, is determined and controlled from the PCM.

Unlike a bypass ignition control circuit, it is not possible to separate the PCM from the ignition coil control to help isolate a fault.

COMPRESSION-SENSING IGNITION

Some waste-spark ignition systems, such as those used on Saturns, use the voltage required to fire the cylinders to determine cylinder position. It requires a higher voltage to fire a spark plug under compression than it does when the spark plug is being fired on the exhaust stroke. The electronics in the coil and the PCM can detect which of the two cylinders that are fired at the same time requires the higher voltage, which indicates the cylinder on the compression stroke. For example, a typical four-cylinder engine equipped with a waste-spark ignition system will fire both cylinders 1 and 4. If cylinder number 4 requires a higher voltage to fire, as determined by the electronics connected to the coil, then the PCM assumes that cylinder number 4 is on the compression stroke. Engines equipped with **compression-sensing ignition** systems, such as Saturns, do not require the use of a camshaft position sensor to determine cylinder number.

COIL-ON-PLUG IGNITION

Coil-on-plug (COP) ignition uses one ignition coil for each spark plug. This system is also called **coil-by-plug, coil-near-plug,** or **coil-over-plug** ignition. ● **SEE FIGURES 16–25 AND 16–26.** The coil-on-plug system eliminates the spark plug wires, which are often sources of **electromagnetic interference (EMI)** that can cause problems to some computer signals. The vehicle computer controls the timing of the spark. Ignition timing also can be changed (retarded or advanced) on a cylinder-by-cylinder basis for maximum performance and to respond to knock sensor signals. ● **SEE FIGURE 16–27.**

There are two basic types of coil-on-plug ignition:

- **Two-wire** —This design uses the vehicle computer to control the firing of the ignition coil. The two wires include ignition voltage feed and the pulse ground wire, which is controlled by the computer. All ignition timing and dwell control are handled by the computer.
- **Three-wire** —This design includes an ignition module at each coil. The three wires are:
  - Ignition voltage
  - Ground
  - Pulse from the computer to the built-in module

SAFETY TIP

Never Disconnect a Spark Plug Wire When the Engine Is Running!

Ignition systems produce a high-voltage pulse necessary to ignite a lean air–fuel mixture. If you disconnect a spark plug wire when the engine is running, this high-voltage spark could cause personal injury or damage to the ignition coil and/or ignition module.

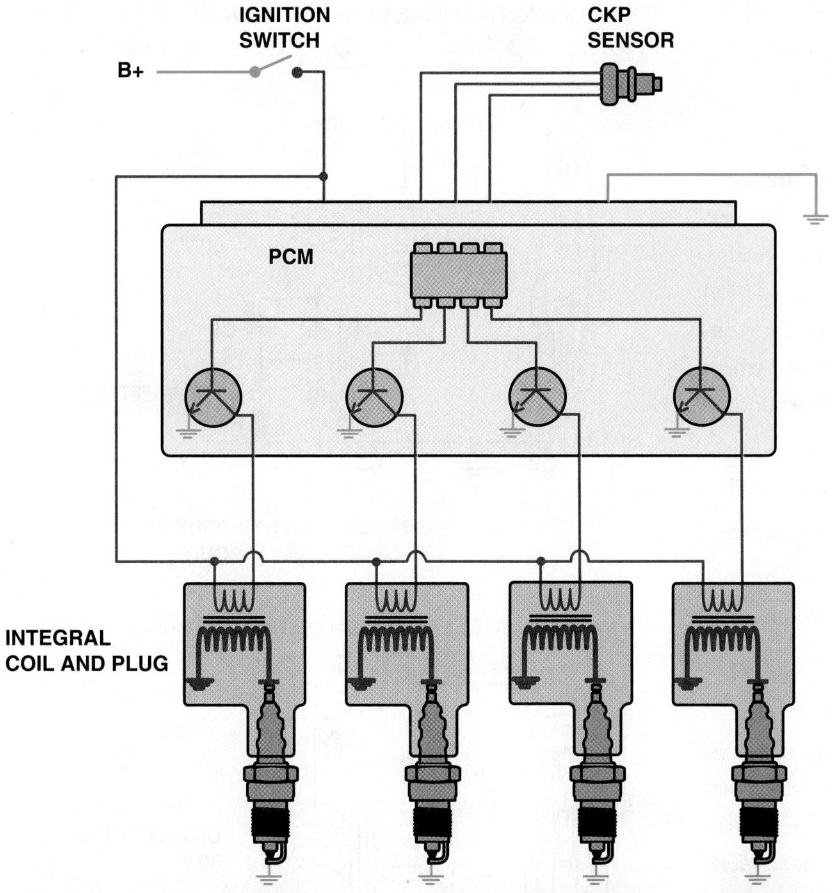

FIGURE 16–25 A typical two-wire coil-on-plug ignition system showing the triggering and the switching being performed by the PCM from input from the crankshaft position sensor.

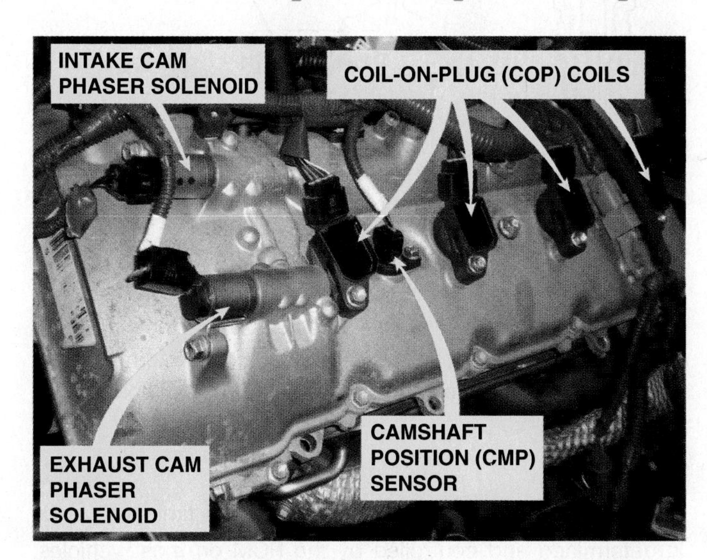

FIGURE 16–26 An overhead camshaft engine equipped with variable valve timing on both the intake and exhaust camshafts and coil-on-plug ignition.

FIGURE 16–27 Chrysler Hemi V-8 that has two spark plugs per cylinder. The coil on top of one spark fires that plug plus, through a spark plug wire, fires a plug in the companion cylinder.

General Motors vehicles use a variety of coil-on-plug-type ignition systems. Many V-8 engines use a coil-near-plug system with individual coils and modules for each individual cylinder that are placed on the valve covers. Short secondary ignition spark plug wires are used to connect the output terminal of the ignition coil to the spark plug.

Most newer Chrysler engines use coil-over-plug-type ignition systems. Each coil is controlled by the PCM, which can vary the ignition timing separately for each cylinder based on signals the PCM receives from the knock sensor(s). For example, if the knock sensor detects that a spark knock has occurred after firing cylinder 3, then the PCM will continue to monitor cylinder 3 and retard timing on just this one cylinder if necessary to prevent engine-damaging detonation.

ION-SENSING IGNITION

In an **ion-sensing ignition** system, the spark plug itself becomes a sensor. The ignition control (IC) module applies a voltage of about 100 to 400 volts DC across the spark plug gap after the ignition event to sense the plasma inside the cylinder. ● **SEE FIGURE 16–28**. The coil discharge voltage (10 to 15 kV) is electrically isolated from the ion-sensing circuit. The combustion flame is ionized and will conduct some electricity, which can be accurately measured at the spark plug gap. The purpose of this circuit includes:

- Misfire detection (required by OBD-II regulations)
- Knock detection (eliminates the need for a knock sensor)
- Ignition timing control (to achieve the best spark timing for maximum power with lowest exhaust emissions)
- Exhaust gas recirculation (EGR) control
- Air–fuel ratio control on an individual cylinder basis

Ion-sensing ignition systems still function the same as conventional coil-on-plug designs, but the engine does not need to be equipped with a camshaft position sensor for misfire detection, or a knock sensor because both of these faults are achieved using the electronics inside the ignition control circuits. Ion-sensing ignition is used in the Saab four- and six-cylinder engines and on many Harley-Davidson motorcycles.

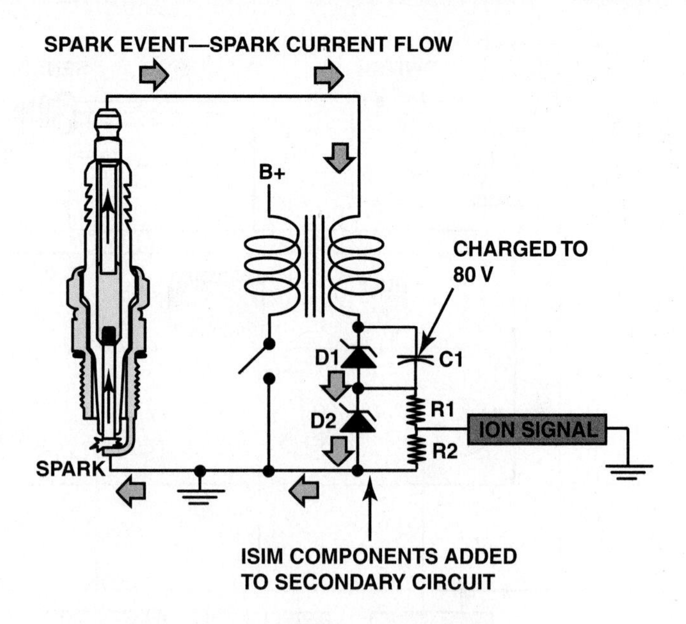

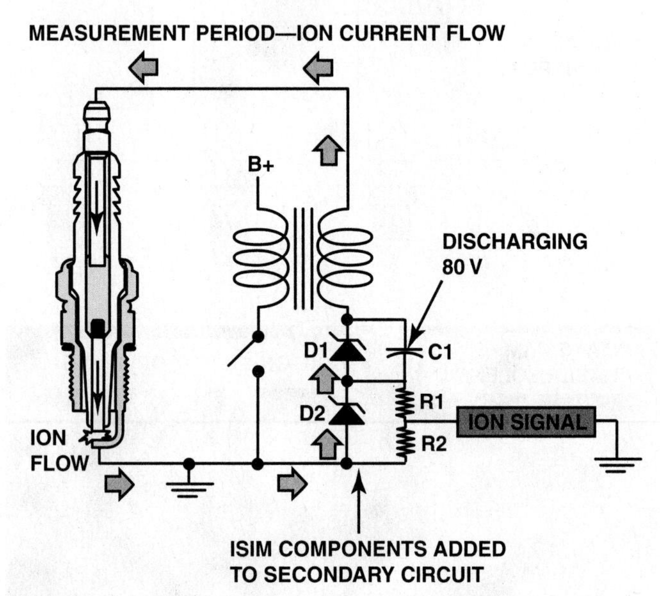

FIGURE 16–28 A DC voltage is applied across the spark plug gap after the plug fires and the circuit can determine if the correct air–fuel ratio was present in the cylinder and if knock occurred.

IGNITION TIMING

THE NEED FOR SPARK ADVANCE **Ignition timing** refers to when the spark plug fires in relation to piston position. The time when the spark occurs depends on engine speed, and therefore, must be advanced (spark plugs fire some) as the engine rotates faster. The ignition in the cylinder takes a certain amount of time, usually 30 milliseconds (30/1000 of a second). This burning time is relatively constant throughout the entire engine speed range. For maximum efficiency from the expanding gases inside the combustion chamber, the burning of the air–fuel mixture should end by about 10° after top dead center (ATDC). If the burning of the mixture is still occurring after that point, the expanding gases do not exert much force on the piston because it is moving away from the gases (the gases are "chasing" the piston).

Therefore, to achieve the goal of having the air–fuel mixture be completely burned by the time the piston reaches 10° after top dead center, the spark must be advanced (occur sooner) as the engine speed increases. This timing advance is determined and controlled by the PCM on most vehicles. ● **SEE FIGURE 16–29**.

INITIAL TIMING If the engine is equipped with a distributor, it may be possible to adjust the base or the **initial timing**. The initial timing is usually set to fire the spark plug between zero degrees (top dead center or TDC) or slightly before TDC (BTDC). Ignition timing does change as the timing chain or gear wears and readjustment is often necessary on high-mileage engines. ● **SEE FIGURE 16–30**. Waste-spark and coil-on-plug ignitions cannot be adjusted.

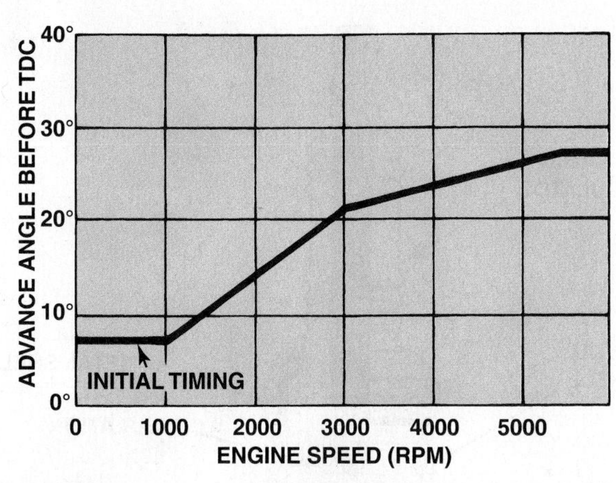

FIGURE 16–29 The initial timing is where the spark plug fires at idle speed. The computer then advances the timing based on engine speed and other factors.

FIGURE 16–30 Ignition timing marks are found on the harmonic balancers that are equipped with distributor ignition.

KNOCK SENSORS

Knock sensors are used to detect abnormal combustion, often called **ping**, **spark knock**, or **detonation**. Whenever abnormal combustion occurs, a rapid pressure increase occurs in the cylinder, creating a vibration in the engine block. It is this vibration that is detected by the knock sensor. The signal from the knock sensor is used by the PCM to retard the ignition timing until the knock is eliminated, thereby reducing the damaging effects of the abnormal combustion on pistons and other engine parts.

Inside the knock sensor is a piezoelectric element that generates a voltage when pressure or a vibration is applied to the unit. The knock sensor is tuned to the engine knock frequency, which ranges from 5 to 10 kHz, depending on the engine design. The voltage signal from the **knock sensor (KS)** is sent to the PCM. The PCM retards the ignition timing until the knocking stops.

DIAGNOSING THE KNOCK SENSOR If a knock sensor diagnostic trouble code (DTC) is present, follow the specified testing procedure in the service information. A scan tool can be used to check the operation of the knock sensor, using the following procedure:

STEP 1 Start the engine and connect a scan tool to monitor ignition timing and/or knock sensor activity.

STEP 2 Create a simulated engine knocking sound by tapping on the engine block or cylinder head with a soft-faced mallet.

STEP 3 Observe the scan tool display. The vibration from the tapping should have been interpreted by the knock sensor as a knock, resulting in a knock sensor signal and a reduction in the spark advance.

A knock sensor also can be tested using a digital storage oscilloscope. ● **SEE FIGURE 16–31**.

NOTE: Some engine computers are programmed to ignore knock sensor signals when the engine is at idle speed to avoid having the noise from a loose accessory drive belt, or other accessory, interpreted as engine knock. Always follow the vehicle manufacturer's recommended testing procedure.

REPLACING A KNOCK SENSOR If replacing a knock sensor, be sure to purchase the exact replacement needed, because they often look the same, but the frequency range can vary according to engine design, as well as where it is located on the engine. Always tighten the knock sensor using a torque wrench and tighten to the specified torque to avoid causing damage to the piezoelectric element inside the sensor.

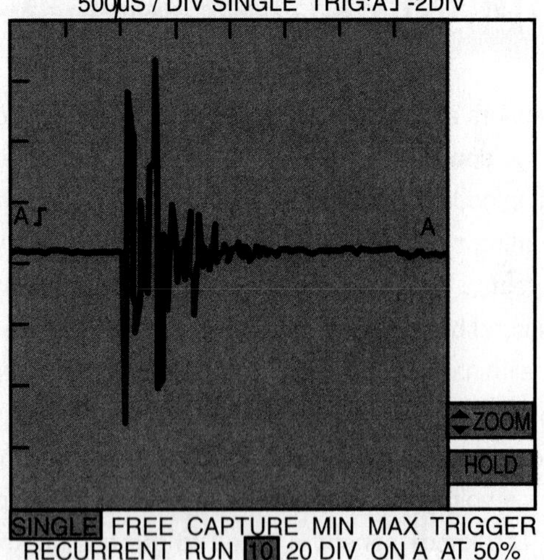

FIGURE 16–31 A typical waveform from a knock sensor during a spark knock event. This signal is sent to the computer which in turn retards the ignition timing. This timing retard is accomplished by an output command from the computer to either a spark advance control unit or directly to the ignition module.

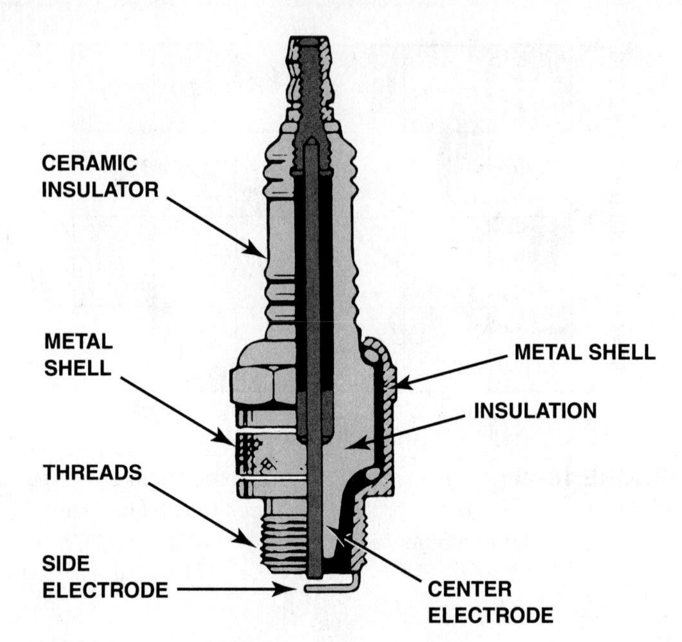

FIGURE 16–32 Parts of a typical spark plug.

SPARK PLUGS

Spark plugs are manufactured from ceramic insulators inside a steel shell. The threads of the shell are rolled and a seat is formed to create a gastight seal with the cylinder head. ● SEE FIGURE 16–32. The physical difference in spark plugs includes:

- **Reach.** This is the length of the threaded part of the plug.
- **Heat range.** The heat range of the spark plug refers to how rapidly the heat created at the tip is transferred to the cylinder head. A plug with a long ceramic insulator path will run hotter at the tip than a spark plug that has a shorter path because the heat must travel farther. ● SEE FIGURE 16–33.
- **Type of seat.** Some spark plugs use a gasket and others rely on a tapered seat to seal.

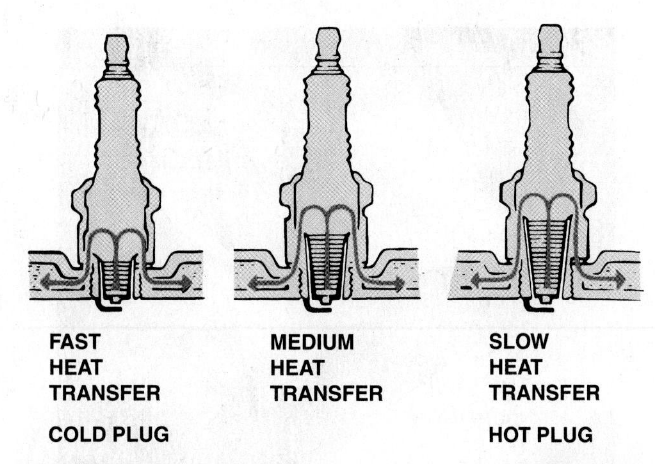

FIGURE 16–33 The heat range of a spark plug is determined by the distance the heat has to flow from the tip to the cylinder head.

RESISTOR SPARK PLUGS Most spark plugs include a resistor in the center electrode, which helps to reduce electromagnetic noise or radiation from the ignition system. The closer the resistor is to the actual spark or arc, the more effective it becomes. The value of the resistor is usually between 2,500 and 7,500 ohms.

PLATINUM SPARK PLUGS **Platinum spark plugs** have a small amount of the precious metal platinum welded onto the end of the center electrode, as well as on the ground or side electrode. Platinum is a grayish-white metal that does not react with oxygen and therefore, will not erode away as can occur with conventional nickel alloy spark plug electrodes. Platinum is also used as a catalyst in catalytic converters where it is able to start a chemical reaction without itself being consumed.

IRIDIUM SPARK PLUGS Iridium is a white precious metal and is the most corrosion-resistant metal known. Most **iridium spark plugs** use a small amount of iridium welded onto the tip of a small center electrode 0.0015 to 0.002 inches (0.4 to 0.6 mm) in diameter. The small diameter reduces the voltage required to jump the gap between the center and the side electrode, thereby reducing possible misfires. The ground or side electrode is usually tipped with platinum to help reduce electrode gap wear.

SUMMARY

1. All inductive ignition systems supply battery voltage to the positive side of the ignition coil and pulse the negative side of the coil on and off to ground to create a high-voltage spark.
2. If an ignition system uses a distributor, it is a distributor ignition (DI) system.
3. If an ignition system does not use a distributor, it is called an electronic ignition (EI) system.
4. A waste-spark ignition system fires two spark plugs at the same time.
5. A coil-on-plug ignition system uses an ignition coil for each spark plug.

REVIEW QUESTIONS

1. How can 12 volts from a battery be changed to 40,000 volts for ignition?
2. How does a magnetic sensor work?
3. How does a Hall-effect sensor work?
4. How does a waste-spark ignition system work?

CHAPTER QUIZ

1. The primary (low-voltage) ignition system must be working correctly before any spark occurs from a coil. Which component is *not* in the primary ignition circuit?
   a. Spark plug wiring
   b. Ignition module (igniter)
   c. Pickup coil (pulse generator)
   d. Ignition switch
2. The ignition module has direct control over the firing of the coil(s) of an EI system. Which component(s) triggers (controls) the module?
   a. Pickup coil
   b. Computer
   c. Crankshaft sensor
   d. All of the above
3. A reluctor is a _______.
   a. Type of sensor used in the secondary circuit.
   b. Notched ring or pointed wheel
   c. Type of optical sensor
   d. Type of Hall effect sensor
4. HEI and EIS are examples of _______.
   a. Waste-spark systems
   b. Coil-on-plug ignition systems
   c. Distributor ignition systems
   d. Pickup coil types
5. Coil polarity is determined by the _______.
   a. Direction of rotation of the coil windings
   b. Turns ratio
   c. Direction of laminations
   d. Saturation direction
6. Because of _______, an ignition coil cannot be fully charged (reach magnetic saturation) until after a delay of about 10 milliseconds
   a. Voltage drop across the ignition switch and related wiring
   b. Resistance in the coil windings
   c. Inductive reactance
   d. Saturation
7. The pulse generator _______.
   a. Fires the spark plug directly
   b. Signals the electronic control unit (module)
   c. Signals the computer that fires the spark plug directly
   d. Is used as a tachometer reference signal by the computer and has no other function
8. Two technicians are discussing distributor ignition. Technician A says that the pickup coil or optical sensor in the distributor is used to pulse the ignition module (igniter). Technician B says that some distributor ignition systems have the ignition coil inside the distributor cap. Which technician is correct?
   a. Technician A only
   b. Technician B only
   c. Both Technicians A and B
   d. Neither Technician A nor B
9. A waste-spark-type ignition system _______.
   a. Fires two spark plugs at the same time
   b. Fires one spark plug with reverse polarity
   c. Fires one spark plug with straight polarity
   d. All of the above
10. An ion-sensing ignition system allows the ignition system itself to be able to _______.
    a. Detect misfire
    b. Detect spark knock
    c. Detect rich or lean air–fuel mixture
    d. All of the above

chapter 17

IGNITION SYSTEM DIAGNOSIS AND SERVICE

LEARNING OBJECTIVES

After studying this chapter, the reader will be able to:

1. Describe the procedure used to check for spark.
2. List the steps necessary to check and/or adjust ignition timing on engines equipped with a distributor.
3. Explain the construction and operation of different types of spark plugs and discuss how to inspect the spark plug wire.
4. Describe how to test the ignition system using an oscilloscope.
5. Inspect and test ignition system pickup sensor or triggering devices.
6. Discuss what to inspect and look for during a visual inspection of the ignition system.
7. Diagnose ignition system related problems.
8. Inspect and test ignition system secondary circuit wiring and components.

This chapter will help you prepare for Engine Repair (A8) ASE certification test content area "B" (Ignition System Diagnosis and Repair).

KEY TERMS

Automatic shutdown (ASD) relay 282
Base timing 295
Burn kV 301
Distributor cap 288
Dwell section 300
Firing line 298
Firing order 297
Intermediate oscillations 299
Millisecond (ms) sweep 301
Raster 300
Remove and replace (R & R) 291
Rotor gap 303
Spark line 299
Spark tester 281
Superimposed 300

FIGURE 17–1 A spark tester looks like a regular spark plug with an alligator clip attached to the shell. This tester has a specified gap that requires at least 25,000 volts (25 kV) to fire.

FIGURE 17–2 A close-up showing the recessed center electrode on a spark tester. It is recessed 3/8 inch into the shell and the spark must then jump another 3/8 inch to the shell for a total gap of 3/4 inch.

CHECKING FOR SPARK

In the event of a no-start condition, the first step should be to check for secondary voltage out of the ignition coil or to the spark plugs. If the engine is equipped with a separate ignition coil, remove the coil wire from the center of the distributor cap, install a **spark tester,** and crank the engine. See the Tech Tip "Always Use a Spark Tester." A good coil and ignition system should produce a blue spark at the spark tester. ● **SEE FIGURES 17–1 AND 17–2.**

If the ignition system being tested does not have a separate ignition coil, disconnect any spark plug wire from a spark plug and, while cranking the engine, test for spark available at the spark plug wire, again using a spark tester.

NOTE: An intermittent spark should be considered a no-spark condition.

Typical causes of a no-spark (intermittent spark) condition include the following:

1. Weak ignition coil
2. Low or no voltage to the primary (positive) side of the coil
3. High resistance or open coil wire, or spark plug wire
4. Negative side of the coil not being pulsed by the ignition module, also called an ignition control module (ICM)
5. Defective pickup coil
6. Defective module

TECH TIP

Always Use a Spark Tester

A spark tester looks like a spark plug except it has a recessed center electrode and no side electrode. The tester commonly has an alligator clip attached to the shell so that it can be clamped on a good ground connection on the engine. A good ignition system should be able to cause a spark to jump this wide gap at atmospheric pressure. Without a spark tester, a technician might assume that the ignition system is okay, because it can spark across a normal, grounded spark plug. The voltage required to fire a standard spark plug when it is out of the engine and not under pressure is about 3,000 volts or less. An electronic ignition spark tester requires a minimum of 25,000 volts to jump the 3/4 inch gap. Therefore, never assume that the ignition system is okay because it fires a spark plug—always use a spark tester. *Remember that an intermittent spark across a spark tester should be interpreted as a no-spark condition.*

ELECTRONIC IGNITION TROUBLESHOOTING PROCEDURE

When troubleshooting any electronic ignition system for no spark, follow these steps to help pinpoint the exact cause of the problem:

STEP 1 Turn the ignition on (engine off) and, using either a voltmeter or a test light, test for battery voltage available at the positive terminal of the ignition coil. If the voltage is not available, check for an open circuit at the ignition switch or wiring. Also check the condition of the ignition fuse (if used).

NOTE: Many Chrysler group products use an automatic shutdown (ASD) relay to power the ignition coil. The ASD relay will not supply voltage to the coil unless the engine is cranking and the computer senses a crankshaft sensor signal. This little known fact has fooled many technicians.

STEP 2 Connect the voltmeter or test light to the negative side of the coil and crank the engine. The voltmeter should fluctuate or the test light should blink, indicating that the primary coil current is being turned on and off. If there is no pulsing of the negative side of the coil, then the problem is a defective pickup, electronic control module, or wiring.

CASE STUDY

The Weird Running Chevrolet Truck

An older Chevrolet pickup truck equipped with a V-8 engine was towed into a shop because it would not start. A quick check of the ignition system showed that the pickup coil had a broken wire below it and the ignition control module. The distributor was removed from the engine and the distributor shaft was removed, cleaned, and a replacement pickup coil was installed. The engine started but ran rough and hesitated when the accelerator pedal was depressed. After an hour of troubleshooting, a careful inspection of the new pickup coil showed that the time core had six instead of eight points, meaning that the new pickup coil was meant for a V-6 instead of a V-8 engine. Replacing the pickup coil again solved the problem.

Summary:

- **Complaint**—Customer stated that the truck would not start.
- **Cause**—A visual inspection was used to determine a that a pickup coil wire was broken and was replaced but with the wrong part that was in the correct box.
- **Correction**—Replacing the pickup coil with the right part fixed the truck.

IGNITION COIL TESTING USING AN OHMMETER

If an ignition coil is suspected of being defective, a simple ohmmeter check can be performed to test the resistance of the primary and secondary winding inside the coil. For accurate resistance measurements, the wiring to the coil should be removed before testing. To test the primary coil winding resistance, take the following steps (● **SEE FIGURE 17–3**):

STEP 1 Set the meter to read low ohms.

STEP 2 Measure the resistance between the positive terminal and the negative terminal of the ignition coil. Most coils will give a reading between 1 and 3 ohms; however, some coils should indicate less than 1 ohm. Check the manufacturer's specifications for the exact resistance values.

To test the secondary coil winding resistance, follow these steps:

STEP 1 Set the meter to read kilohms (kΩ).

STEP 2 Measure the resistance between either primary terminal and the secondary coil tower. The normal resistance of most coils ranges between 6,000 and 30,000 ohms. Check the manufacturer's specifications for the exact resistance values.

NOTE: Many ignition coils use a screw that is inside the secondary tower of the ignition coil. If this screw is loose, an intermittent engine misfire could occur. The secondary coil would also indicate high resistance if this screw was loose.

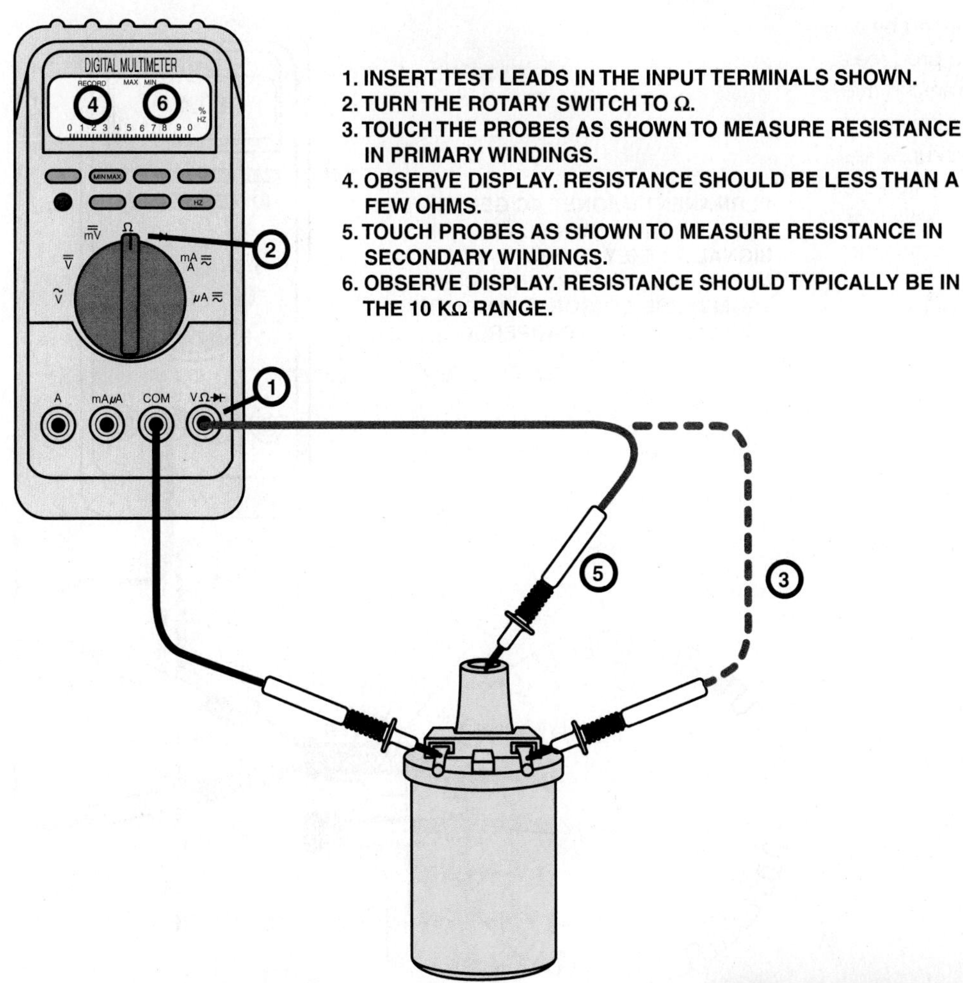

FIGURE 17–3 Checking an ignition coil using a multimeter set to read ohms.

PICKUP COIL TESTING

The pickup coil, located under the distributor cap on many electronic ignition engines, can cause a no-spark condition if defective. The pickup coil must generate an AC voltage pulse to the ignition module so that the module can pulse the ignition coil.

A pickup coil contains a coil of wire, and the resistance of this coil should be within the range specified by the manufacturer. ● **SEE FIGURE 17–4**. Some common specifications include the following:

Manufacturer	Pickup Coil Resistance (Ohms)
General Motors	500–1,500 (white and green leads)
Ford	400–1,000 (orange and purple leads)
Chrysler	150–900 (orange and black leads)

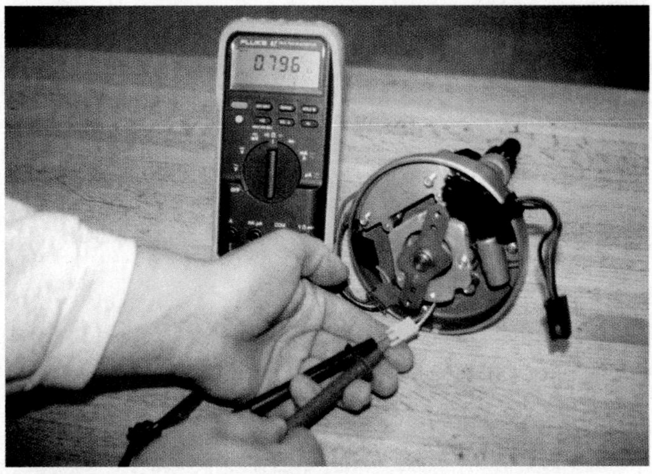

FIGURE 17–4 Measuring the resistance of an HEI pickup coil using a digital multimeter set to the ohms position. The reading on the face of the meter is 0.796 kO or 796 ohms in the middle of the 500 to 1,500 ohm specifications.

Also check that the pickup coil windings are insulated from ground by checking for continuity using an ohmmeter. With one ohmmeter lead attached to ground, touch the other lead of the ohmmeter to the pickup coil terminal. The ohmmeter should read OL (over limit) with the ohmmeter set on the high scale. If the pickup coil resistance is not within the specified range, or if it has continuity to ground, replace the pickup coil assembly.

The pickup coil also can be tested for proper voltage output. During cranking, most pickup coils should produce a minimum of 0.25 volt AC. This can be tested with the distributor out of the vehicle by rotating the distributor drive gear by hand.

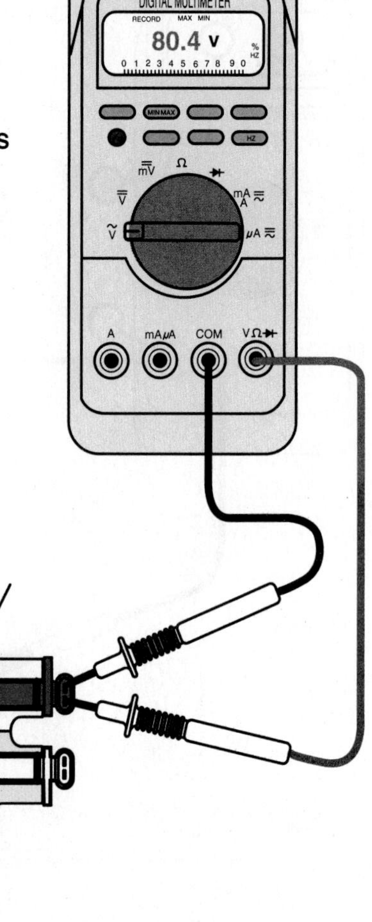

FIGURE 17–5 An AC voltage is produced by a magnetic sensor. Most sensors should produce at least 0.1 volt AC while the engine is cranking and can exceed 100 volts with the engine running if the pickup wheel has many teeth. If the pickup wheel has only a few teeth, you may need to switch the meter to read DC volts and watch the display for a jump in voltage as the teeth pass the magnetic sensor.

TESTING MAGNETIC SENSORS

First of all, magnetic sensors must be tested to see if they will stick to iron or steel, indicating that the magnetic strength of the sensors is okay. If the permanent magnet inside the sensor has cracked, the result is two weak magnets.

If the sensor is removed from the engine, hold a metal (steel) object against the end of the sensor. It should exert a strong magnetic pull on the steel object. If not, replace the sensor. Second, the sensor can be tested using a digital meter set to read AC volts. ● **SEE FIGURE 17–5.**

TESTING HALL-EFFECT SENSORS

As with any other sensor, the output of the Hall-effect sensor should be tested first. Using a digital voltmeter, check for the presence of changing voltage (pulsed on and off or digital DC) when the engine is being cranked. The best test is to use an oscilloscope and observe the waveform. ● **SEE FIGURE 17–6.**

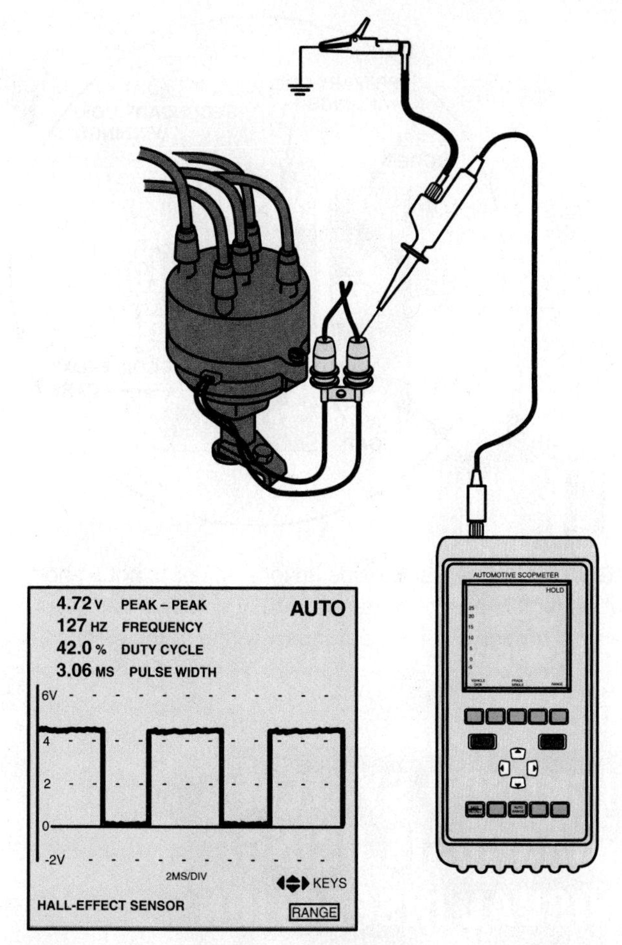

FIGURE 17–6 (a) The connection required to test a Hall-effect sensor. (b) A typical waveform from a Hall-effect sensor.

TESTING OPTICAL SENSORS

Optical sensors will not operate if dirty or covered in oil. Perform a thorough visual inspection and look for an oil leak that could cause dirty oil to get on the LED or phototransistor. Also be sure that the light shield is securely fastened and that the seal is lightproof. An optical sensor also can be checked using an oscilloscope. ● **SEE FIGURE 17–7**. Because of the speed of the engine and the number of slits in the optical sensor disk, a scope is one of the only tools that can capture useful information. For example, a Nissan has 360 slits and if it is running at 2000 RPM, a signal is generated 720,000 times per minute or 12,000 times per second.

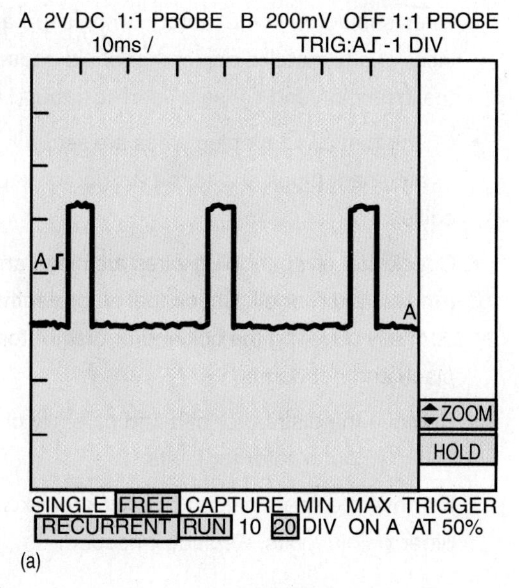

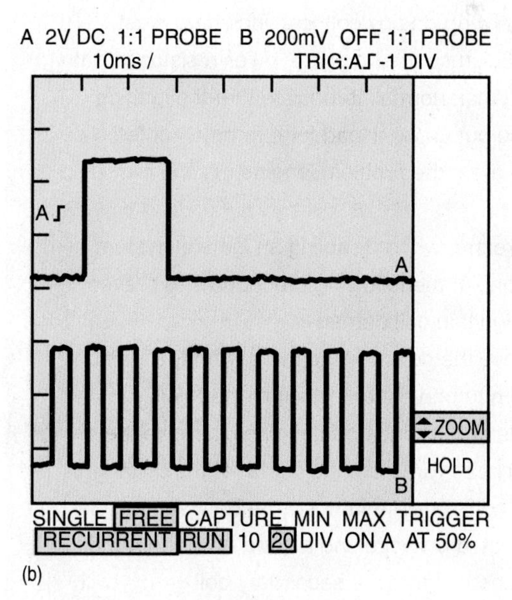

FIGURE 17–7 (a) The low-resolution signal has the same number of pulses as the engine has cylinders. (b) A dual-trace pattern showing both the low-resolution signal and the high-resolution signals that usually represent 1 degree of rotation.

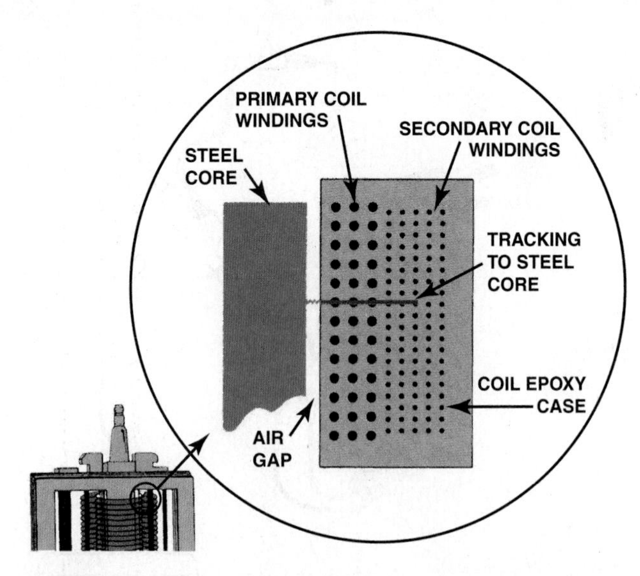

FIGURE 17–8 A track inside an ignition coil is not a short, but rather it is a low-resistance path or hole that has been burned through from the secondary wiring to the steel core.

TECH TIP

Bad Wire? Replace the Coil!

When performing engine testing (such as a compression test), always ground the coil wire. Never allow the coil to discharge without a path to ground for the spark. High-energy electronic ignition systems can produce 40,000 volts or more of electrical pressure. If the spark cannot arc to ground, the coil energy can (and usually does) arc inside the coil itself, creating a low-resistance path to the primary windings or the steel laminations of the coil. ● **SEE FIGURE 17–8.** This low-resistance path is called a track and could cause an engine misfire under load even though all of the remaining component parts of the ignition system are functioning correctly. Often these tracks do not show up on any coil test, including most scopes. Because the track is a lower-resistance path to ground than normal, it requires that the ignition system be put under a load for it to be detected, and even then, the problem (engine misfire) may be intermittent.

Therefore, when disabling an ignition system, perform one of the following procedures to prevent possible ignition coil damage:

1. Remove the power source wire from the ignition system to prevent any ignition operation.
2. On distributor-equipped engines, remove the secondary coil wire from the center of the distributor cap and connect a jumper wire between the disconnected coil wire and a good engine ground. This ensures that the secondary coil energy will be safely grounded and prevents high-voltage coil damage.

IGNITION SYSTEM DIAGNOSIS USING VISUAL INSPECTION

One of the first steps in the diagnosis process is to perform a thorough visual inspection of the ignition system, including the following:

- Check all spark plug wires for proper routing. All plug wires should be in the factory wiring separator and be clear of any metallic object that could cause damage to the insulation and cause a short-to-ground fault.
- Check that all spark plug wires are securely attached to the spark plugs and to the distributor cap or ignition coil(s).
- Check that all spark plug wires are clean and free from excessive dirt or oil. Check that all protective covers normally covering the coil and/or distributor cap are in place and not damaged.
- Remove the distributor cap and carefully check the cap and distributor rotor for faults.
- Remove the spark plugs and check for excessive wear or other visible faults. Replace if needed.

NOTE: According to research conducted by General Motors, about one-fifth (20%) of all faults are detected during a *thorough visual inspection!*

FIGURE 17–9 If the coil is working, the end of the magnetic pickup tool will move with the changes in the magnetic field around the coil.

TECH TIP

The Magnetic Pickup Tool Test

All ignition coils are pulsed on and off by the ignition control module or PCM. When the coil charges and discharges, the magnetic field around the coil changes. This pulsing of the coil can be observed by holding the magnetic end of a pickup tool near an operating ignition coil. The magnet at the end of the pickup tool will move as the magnetic field around the coil changes. ● **SEE FIGURE 17–9.**

TESTING FOR POOR PERFORMANCE

Many diagnostic equipment manufacturers offer methods for testing distributorless ignition systems on an oscilloscope. If using this type of equipment, follow the manufacturer's recommended procedures and interpretation of the specific test results.

A simple method of testing distributorless (waste-spark systems) ignition with the engine off involves removing the spark plug wires (or connectors) from the spark plugs (or coils or distributor cap) and installing short lengths (2 inches) of rubber vacuum hose in series.

NOTE: For best results, use rubber hose that is electrically conductive. Measure the vacuum hose with an ohmmeter. Suitable vacuum hose should give a reading of less than 10,000 ohms (10 kΩ) for a length of about 2 inches. ● SEE FIGURES 17–9 AND 17–10.

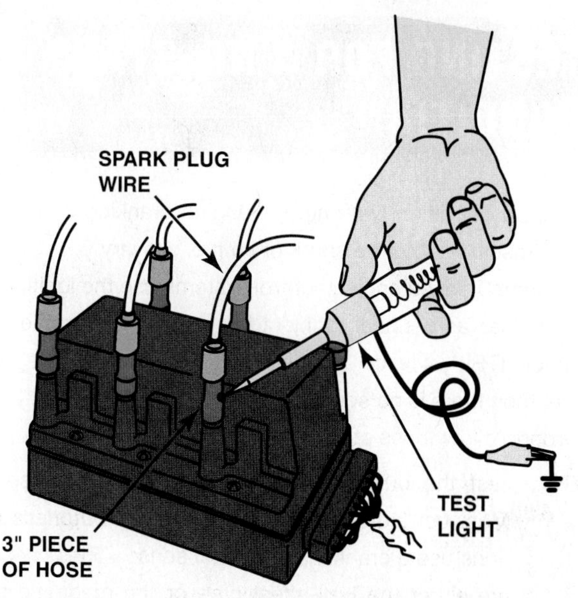

FIGURE 17–10 Using a vacuum hose and a grounded test light to ground one cylinder at a time on a DIS. This works on all types of ignition systems and provides a method for grounding out one cylinder at a time without fear of damaging any component.

STEP 1 Start the engine and ground out each cylinder one at a time by touching the tip of a grounded test light to the rubber vacuum hose. Even though the computer will increase idle speed and fuel delivery to compensate for the grounded spark plug wire, a technician should watch for a change in the operation of the engine. If no change is observed or heard, the cylinder being grounded is obviously weak or defective. Check the spark plug wire or connector with an ohmmeter to be certain of continuity.

STEP 2 Check all cylinders by grounding them out one at a time. If one weak cylinder is found (very little RPM drop), check the other cylinder using the same ignition coil (except on engines that use an individual coil for each cylinder). If both cylinders are affected, the problem could be an open spark plug wire, defective spark plug, or defective ignition coil.

STEP 3 To help eliminate other possible problems and determine exactly what is wrong, switch the suspected ignition coil to another position (if possible).

- If the problem now affects the other cylinders, the ignition coil is defective and must be replaced.
- If the problem does not "change positions" with changing the position of the ignition coil, the control module affecting the suspected coil or either cylinder's spark plug or spark plug wire could be defective.

TESTING FOR A NO-START CONDITION

A no-start condition (with normal engine cranking speed) can be the result of either no spark or no fuel delivery.

Computerized engine control systems use the ignition primary pulses as a signal to inject fuel—a port or throttle-body injection (TBI) style of fuel—injection system. If there is no pulse, then there is no squirt of fuel. To determine exactly what is wrong, follow these steps:

STEP 1 Test the output signal from the crankshaft sensor. Most computerized engines with distributorless ignitions use a crankshaft position sensor. These sensors are either the Hall-effect type or the magnetic type. The sensors must be able to produce either a sine or a digital signal. A meter set on AC volts should read a voltage across the sensor leads when the engine is being cranked. If there is no AC voltage output, replace the sensor.

STEP 2 If the sensor tests okay in step 1, check for a changing AC voltage signal at the ignition module.

NOTE: Step 2 checks the wiring between the crankshaft position sensor and the ignition control module.

STEP 3 If the ignition control module is receiving a changing signal from the crankshaft position sensor, it must be capable of switching the power to the ignition coils on and off. Remove a coil or coil package, and with the ignition switched to on (run), check for voltage at the positive terminal of the coil(s).

NOTE: Several manufacturers program the current to the coils to be turned off within several seconds of the ignition being switched to on if no pulse is received by the computer. This circuit design helps prevent ignition coil damage in the event of a failure in the control circuit or driver error, by keeping the ignition switch on (run) without operating the starter (start position). Some Chrysler engines do not supply power to the positive (+) side of the coil until a crank pulse is received by the computer which then energizes the coil(s) and injectors through the automatic shutdown (ASD) relay.

STEP 4 If the module is not pulsing the negative side of the coil or not supplying battery voltage to the positive side of the coil, replace the ignition control module.

NOTE: Before replacing the ignition control module, be certain that it is properly grounded (where applicable) and that the module is receiving ignition power from the ignition circuit.

CAUTION: Most distributorless (waste-spark) ignition systems can produce 40,000 volts or more, with energy levels high enough to cause personal injury. Do not open the circuit of an electronic ignition secondary wire, because damage to the system (or to you) can occur.

DISTRIBUTOR INDEXING

A few engines using a distributor also use it to house a camshaft position (CMP) sensor. One purpose of this sensor is to properly initiate the fuel-injection sequence. Some of these engines use a positive distributor position notch or clamp that allows the distributor to be placed in only one position, while others use a method of indexing to verify the distributor position. If a distributor is not indexed correctly, the following symptoms may occur:

- Surging (especially at idle speed)
- Light bucking
- Intermittent engine misfiring

This will most likely occur when the vehicle is at operating temperature, and under a light load at approximately 2000 RPM.

A misindexed distributor may cause these conditions.

NOTE: This is *not* the same as setting the ignition timing. Indexing the distributor does not affect the ignition timing.

Always use the factory procedure as stated in service information.

Some of the methods may require a scan tool, while others require the use of a voltmeter to verify position. Jeep, late model Chrysler V-6 and V-8 engines, and some GM trucks require indexing. ● **SEE FIGURE 17–11.**

SECONDARY IGNITION INSPECTION

DISTRIBUTOR CAP AND ROTOR Inspect a **distributor cap** for a worn or cracked center carbon insert, excessive side insert wear or corrosion, cracks, or carbon tracks, and check the towers for burning or corrosion by removing spark

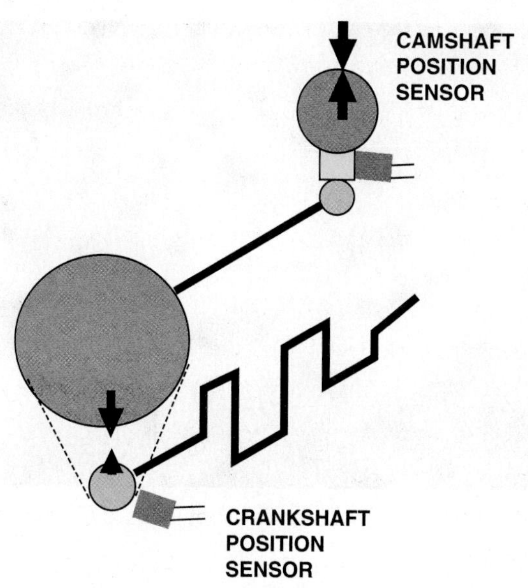

FIGURE 17–11 The relationship between the crankshaft position (CKP) sensor and the camshaft position (CMP) sensor is affected by wear in the timing gear and/or chain.

FIGURE 17–13 When checking a waste spark-type ignition system, check that the secondary wires are attached to the correct coil terminal and that the wiring is correctly routed to help avoid cross-fire.

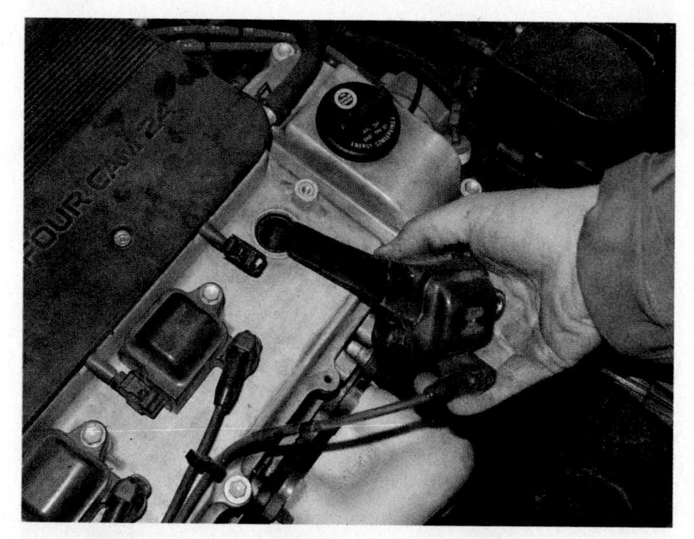

FIGURE 17–12 When checking a coil-on-plug (COP) assembly, check that the primary and secondary wiring looks normal and that the coil is not discolored from arcing or corrosion.

FIGURE 17–14 Corroded terminals on a waste-spark coil can cause misfire diagnostic trouble codes to be set.

plug wires from the distributor cap one at a time. Remember, a defective distributor cap affects starting and engine performance, especially in high-moisture conditions. If a carbon track is detected, it is most likely the result of a high-resistance or open spark plug wire. Replacement of a distributor cap because of a carbon track without checking and replacing the defective spark plug wire(s) often will result in the new distributor cap failing in a short time. It is recommended that the distributor cap and rotor be inspected every year and replaced if defective. The rotor should be replaced every time the spark plugs are replaced, because all ignition current flows through the rotor. Generally, distributor caps should only need replacement after every three or four years of normal service.

COP AND WASTE SPARK INSPECTION If a misfire is being diagnosed, perform a thorough visual inspection of the coil-on-lug (COP) assembly and look for evidence of carbon track or heat-related faults to the plug boot. ● **SEE FIGURE 17–12.**

Check ignition coils of waste spark and coil-on-plug (COP) systems for signs of carbon tracks (black lines) or corrosion.

When checking a waste spark-type ignition system, check that the coils are clean and that the spark plug wires are attached to the specified coil and coil terminal. ● **SEE FIGURES 17–13 AND 17–14.**

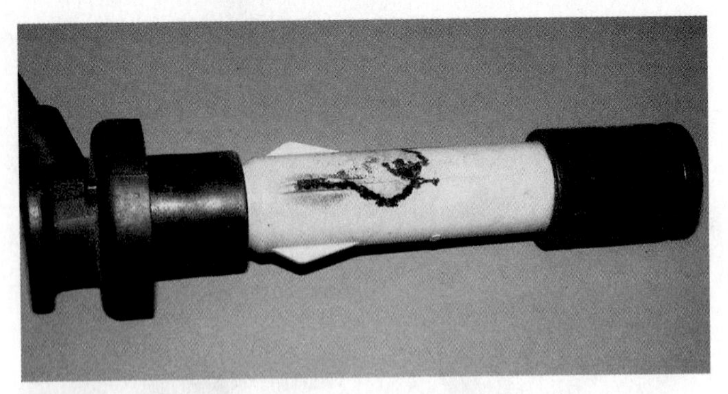

FIGURE 17–15 This spark plug boot on an overhead camshaft engine has been arcing to the valve cover causing a misfire to occur.

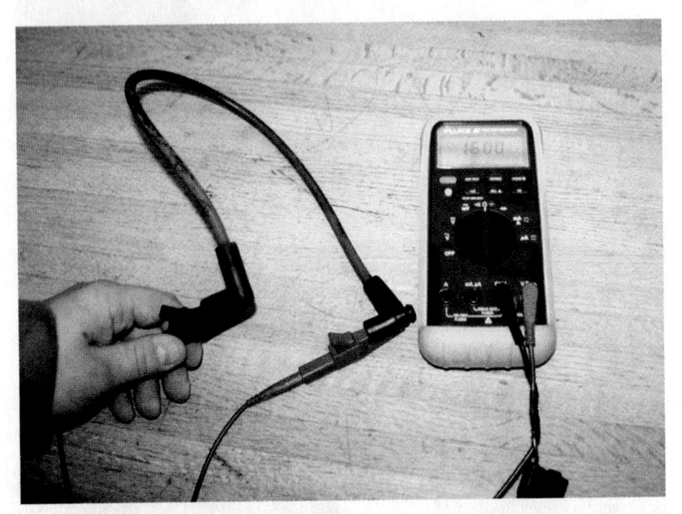

FIGURE 17–16 Measuring the resistance of a spark plug wire with a multimeter set to the ohms position. The reading of 16.03 kO (16.030 ohms) is okay because the wire is about 2 feet long. Maximum allowable resistance for a spark plug wire this long would be 20 kO (20,000 ohms). High resistance spark plug wires can cause an engine misfire especially during acceleration.

SPARK PLUG WIRE INSPECTION Spark plug wires should be visually inspected for cuts or defective insulation and checked for resistance with an ohmmeter. Good spark plug wires should measure less than 10,000 ohms per foot of length. ● **SEE FIGURES 17–15 AND 17–16.** Faulty spark plug wire insulation can cause hard starting or no starting in damp weather conditions.

TECH TIP

Spark Plug Wire Pliers Are a Good Investment

Spark plug wires are often difficult to remove. Using a good-quality spark plug wire plier, such as shown in ● **FIGURE 17–17**, saves time and reduces the chance of harming the wire during removal.

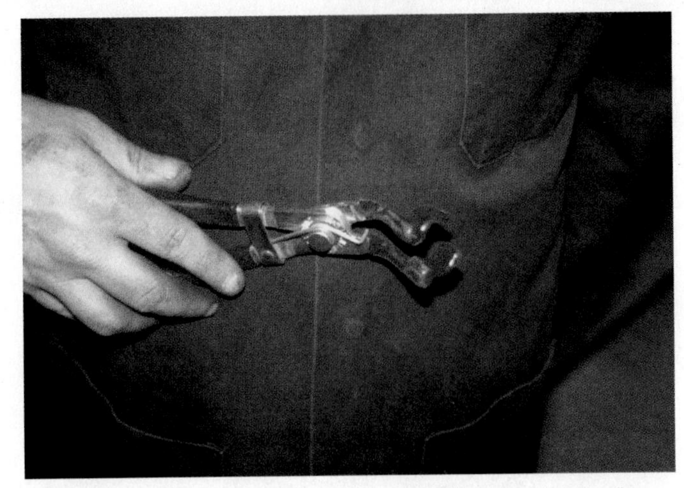

FIGURE 17–17 Spark plug wire boot pliers is a handy addition to any tool box.

TECH TIP

Route the Wires Right!

High voltage is present through spark plug wires when the engine is running. Surrounding the spark plugs is a magnetic field that can affect other circuits or components of the vehicle. For example, if a spark plug wire is routed too closely to the signal wire from a mass airflow (MAF) sensor, the induced signal from the ignition wire could create a false MAF signal to the computer. The computer, not able to detect that the signal was false, would act on the MAF signal and command the appropriate amount of fuel based on the false MAF signal.

To prevent any problems associated with high-voltage spark plug wires, be sure to route them as manufactured using all the factory holding brackets and wiring combs. ● **SEE FIGURE 17–18.** If the factory method is unknown, most factory service information shows the correct routing.

SPARK PLUG SERVICE

THE NEED FOR SERVICE Spark plugs should be inspected when an engine performance problem occurs and should be replaced at specified intervals to ensure proper ignition system performance. Most spark plugs have a service life of over 20,000 miles (32,000 kilometers). Platinum-tipped original equipment spark plugs have a typical service life of 60,000 to 100,000 miles (100,000 to 160,000 kilometers). Used spark plugs should *not* be cleaned and reused unless absolutely

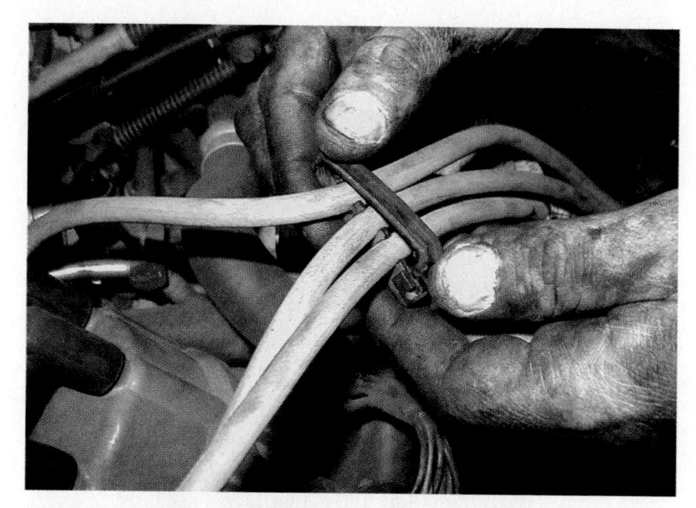

FIGURE 17–18 Always take the time to install spark plug wires back into the original holding brackets (wiring combs).

FIGURE 17–20 A spark plug thread chaser is a low-cost tool that hopefully will not be used often, but is necessary to use to clean the threads before new spark plugs are installed.

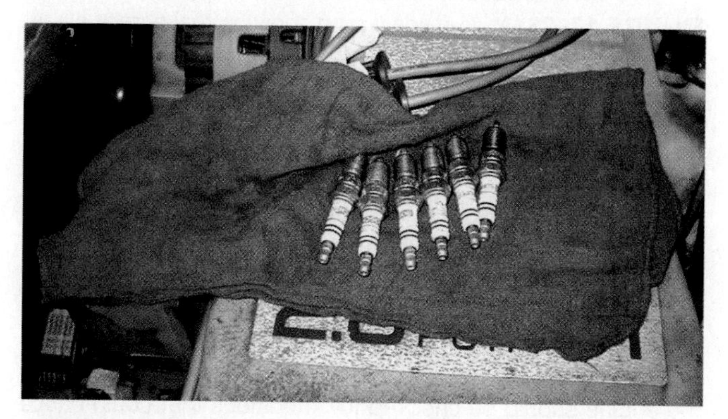

FIGURE 17–19 When removing spark plugs, it is wise to arrange them so that they can be compared and any problem can be identified with a particular cylinder.

FIGURE 17–21 Since 1991, General Motors engines have been equipped with slightly (1/8 inch or 3 mm) longer spark plugs. This requires that a longer spark plug socket should be used to prevent the possibility of cracking a spark plug during installation. The longer socket is shown next to a normal 5/8 inch spark plug socket.

necessary. The labor required to **remove and replace (R & R)** spark plugs is the same whether the spark plugs are replaced or cleaned. Although cleaning spark plugs often restores proper engine operation, the service life of cleaned spark plugs is definitely shorter than that of new spark plugs. *Platinum-tipped spark plugs should not be regapped!* Using a gapping tool can break the platinum after it has been used in an engine.

Be certain that the engine is cool before removing spark plugs, especially on engines with aluminum cylinder heads. To help prevent dirt from getting into the cylinder of an engine while removing a spark plug, use compressed air or a brush to remove dirt from around the spark plug before removal. ● **SEE FIGURES 17–19 THROUGH 17–21.**

SPARK PLUG INSPECTION Spark plugs are the windows to the inside of the combustion chamber. A thorough visual inspection of the spark plugs often can lead to the root cause of an engine performance problem. Two indications on

FIGURE 17–22 A normally worn spark plug that has a tapered platinum-tipped center electrode.

FIGURE 17–23 A worn spark plug showing fuel and/or oil deposits.

spark plugs and their possible root causes in engine performance include the following:

1. **Carbon fouling.** If the spark plug(s) has *dry black carbon* (soot), the usual causes include:
   - Excessive idling
   - Slow-speed driving under light loads that keeps the spark plug temperatures too low to burn off the deposits
   - Overrich air–fuel mixture
   - Weak ignition system output
2. **Oil fouling.** If the spark plug has *wet, oily* deposits with little electrode wear, oil may be getting into the combustion chamber from the following:
   - Worn or broken piston rings
   - Defective or missing valve stem seals

NOTE: If the deposits are heavier on the side of the plug facing the intake valve, the cause is usually due to excessive valve stem clearance or defective intake valve stem seals.

When removing spark plugs, place them in order so that they can be inspected to check for engine problems that might affect one or more cylinders. All spark plugs should be in the same condition, and the color of the center insulator should be light tan or gray. If all the spark plugs are black or dark, the engine should be checked for conditions that could cause an overly rich air–fuel mixture or possible oil burning. If only one or a few spark plugs are black, check those cylinders for proper firing (possible defective spark plug wire) or an engine condition affecting only those particular cylinders. ● **SEE FIGURES 17–22 THROUGH 17–25.**

If all spark plugs are white, check for possible overadvanced ignition timing or a vacuum leak causing a lean air–fuel mixture. If only one or a few spark plugs are white, check for a vacuum leak affecting the fuel mixture only to those particular cylinders.

NOTE: The engine computer "senses" rich or lean air–fuel ratios by means of input from the oxygen sensor. If one cylinder is lean, the computer may make all other cylinders richer to compensate.

FIGURE 17–24 A spark plug from an engine that had a blown head gasket. The white deposits could be from the additives in the coolant.

FIGURE 17–25 A platinum tipped spark plug that is fuel soaked indicating a fault with the fuel system or the ignition system causing the spark plug to not fire.

	TORQUE WITH TORQUE WRENCH (LB-FT)		TORQUE WITHOUT TORQUE WRENCH (TURNS)	
SPARK PLUG	CAST-IRON HEAD	ALUMINUM HEAD	CAST-IRON HEAD	ALUMINUM HEAD
Gasket				
14 mm	26–30	18–22	1/4	1/4
18 mm	32–38	28–34	1/4	1/4
Tapered seat				
14 mm	7–15	7–15	1/16 (snug)	1/16 (snug)
18 mm	15–20	15–20	1/16 (snug)	1/16 (snug)

CHART 17–1

Typical spark plug tightening torque based on type of spark plug and the cylinder head material.

Inspect all spark plugs for wear by first checking the condition of the center electrode. As a spark plug wears, the center electrode becomes rounded. If the center electrode is rounded, higher ignition system voltage is required to fire the spark plug. When installing spark plugs, always use the correct tightening torque to ensure proper heat transfer from the spark plug shell to the cylinder head. ● **SEE CHART 17–1**.

NOTE: General Motors does not recommend the use of antiseize compound on the threads of spark plugs being installed in an aluminum cylinder head, because the spark plug will be overtightened. This excessive tightening torque places the threaded portion of the spark plug too far into the combustion chamber where carbon can accumulate and result in the spark plugs being difficult to remove. If antiseize compound is used on spark plug threads, reduce the tightening torque by 40%. Always follow the vehicle manufacturer's recommendations.

TECH TIP

Two-Finger Trick

To help prevent overtightening a spark plug when a torque wrench is not available, simply use two fingers on the ratchet handle. Even the strongest service technician cannot overtighten a spark plug by using two fingers.

TECH TIP

Use Original Equipment Manufacturer's Spark Plugs

A technician at an independent service center replaced the spark plugs in a Pontiac with new Champion brand spark plugs of the correct size, reach, and heat range. When the customer returned to pay the bill, he inquired as to the brand name of the replacement parts used for the tune-up. When told that Champion spark plugs were used, he stopped signing his name on the check he was writing. He said that he owned 1,000 shares of General Motors stock and he owned two General Motors vehicles and he expected to have General Motors parts used in his General Motors vehicles. The service manager had the technician replace the spark plugs with AC brand spark plugs because this brand was used in the engine when the vehicle was new. Even though most spark plug manufacturers produce spark plugs that are correct for use, many customers prefer that original equipment manufacturer (OEM) spark plugs be used in their engines.

FIGURE 17–26 A water spray bottle is an excellent diagnostic tool to help find an intermittent engine misfire caused by a break in a secondary ignition circuit component.

QUICK AND EASY SECONDARY IGNITION TESTS

Engine running problems often are caused by defective or out-of-adjustment ignition components. Many ignition problems involve the high-voltage secondary ignition circuit. Following are some quick and easy secondary ignition tests:

TEST 1 If there is a crack in a distributor cap, coil, or spark plug, or if there is a defective spark plug wire, a spark may be visible at night. Because the highest voltage is required during partial throttle acceleration, the technician's assistant should accelerate the engine slightly with the gear selector in drive or second gear (if manual transmission) and the brake firmly applied. If any spark is visible or a "snapping" sound is heard, the location should be closely inspected and the defective parts replaced. A blue glow or "corona" around the shell of the spark plug is normal and not an indication of a defective spark plug.

TEST 2 For intermittent problems, use a spray bottle to apply a water mist to the spark plugs, distributor cap, and spark plug wires. ● **SEE FIGURE 17–26**. With the engine running, the water may cause an arc through any weak insulating materials and cause the engine to misfire or stall.

NOTE: Adding a little salt or liquid soap to the water makes the water more conductive and also makes it easier to find those hard-to-diagnose intermittent ignition faults.

TEST 3 To determine if the rough engine operation is due to secondary ignition problems, connect a 12 volt test light to the negative side (sometimes labeled "tach") of the coil. Connect the other lead of the test light to the positive lead of the coil. With the engine running, the test light should be dim and steady in brightness. If there is high resistance in the secondary circuit (such as that caused by a defective spark plug wire), the test light will pulse brightly at times. If the test light varies noticeably, this indicates that the secondary voltage cannot find ground easily and is feeding back through the primary windings of the coil. This feedback causes the test light to become brighter.

FIGURE 17–27 Typical timing marks. The numbers of the degrees are on the stationary plate and the notch is on the harmonic balancer.

TECH TIP

"Turn the Key" Test

If the ignition timing is correct, a warm engine should start immediately when the ignition key is turned to the start position. If the engine cranks a long time before starting, the ignition timing may be retarded. If the engine cranks slowly, the ignition timing may be too far advanced. However, if the engine starts immediately, the ignition timing, although it may not be exactly set according to specification, is usually adjusted fairly close to specifications. When a starting problem is experienced, check the ignition timing first, before checking the fuel system or the cranking system for a possible problem. This procedure can be used to help diagnose a possible ignition timing problem quickly without tools or equipment.

IGNITION TIMING

Ignition timing should be checked and adjusted according to the manufacturer's specifications and procedures on some vehicles equipped with distributor-type ignition systems. Generally, for testing, engines must be at idle with computer engine controls put into **base timing,** the timing of the spark before the computer advances the timing. To be assured of the proper ignition timing, follow exactly the timing procedure indicated on the underhood emission (VECI) decal. ● **SEE FIGURE 17–27** for a typical ignition timing plate and mark.

If the ignition timing is too far *advanced*, for example, if it is set at 12 degrees before top dead center (BTDC) instead of 8 degrees BTDC, the following symptoms may occur:

1. Engine ping or spark knock may be heard, especially while driving up a hill or during acceleration.
2. Cranking (starting) may be slow and jerky, especially when the engine is warm.
3. The engine may overheat.

If the ignition timing is too far *retarded*, for example, if it is set at 4 degrees BTDC instead of 8 degrees BTDC, the following symptoms may occur:

1. The engine may lack power and performance.
2. The engine may require a long period of starter cranking before starting.
3. Poor fuel economy may result.
4. The engine may overheat.

PRETIMING CHECKS Before the ignition timing is checked or adjusted, the following items should be checked to ensure accurate timing results:

1. The engine should be at normal operating temperature (the upper radiator hose should be hot and pressurized).
2. The engine should be at the correct timing RPM (check the specifications).
3. The vacuum hoses should be removed, and the hose from the vacuum advance unit on the distributor (if the vehicle is so equipped) should be plugged unless otherwise specified.
4. If the engine is computer equipped, check the timing procedure specified by the manufacturer. This may include disconnecting a "set timing" connector wire, grounding a diagnostic terminal, disconnecting a four-wire connector, or similar procedure.

NOTE: General Motors specifies many different timing procedures depending on the engine, type of fuel system, and type of ignition system. Always consult the emission decal under the hood or service information for the exact procedure to follow.

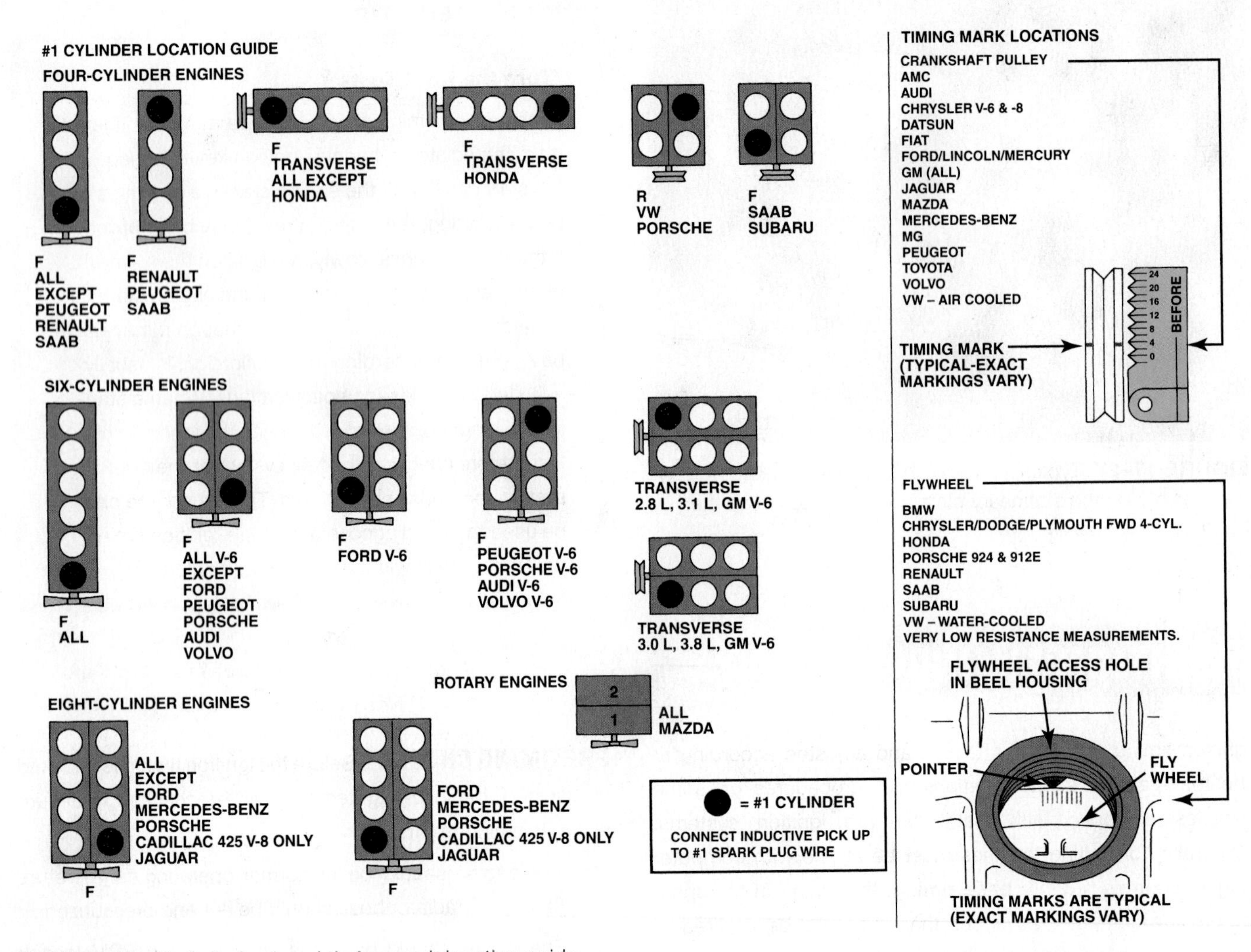

FIGURE 17–28 Cylinder 1 and timing mark location guide.

TIMING LIGHT CONNECTIONS For checking or adjusting ignition timing, make the timing light connections as follows:

1. Connect the timing light battery leads to the vehicle battery: the red to the positive terminal and the black to the negative terminal.
2. Connect the timing light high-tension lead to spark plug cable 1.

DETERMINING CYLINDER 1 The following will help in determining cylinder 1:

1. **Four- or six-cylinder engines.** On all inline four- and six-cylinder engines, cylinder 1 is the *most forward* cylinder.
2. **V-6 or V-8 engines.** Most V-type engines use the left front (driver's side) cylinder as cylinder 1, except for Ford engines and some Cadillacs, which use the right front (passenger's side) cylinder.
3. **Sideways (transverse) engines.** Most front-wheel-drive vehicles with engines installed sideways use the cylinder to the far right (passenger's side) as cylinder 1 (plug wire closest to the drive belt[s]).

Follow this rule of thumb: If cylinder 1 is unknown for a given type of engine, it is the *most forward* cylinder as viewed from above (except in Pontiac V-8 engines). ● **SEE FIGURE 17–28** for typical cylinder 1 locations.

NOTE: Some engines are not timed off of cylinder 1. For example, Jaguar inline six-cylinder engines before 1988 used cylinder 6, but the cylinders were numbered from the firewall (bulkhead) forward. Therefore, cylinder 6 was the most forward cylinder. Always check for the specifications and procedures for the vehicle being tested.

(a)

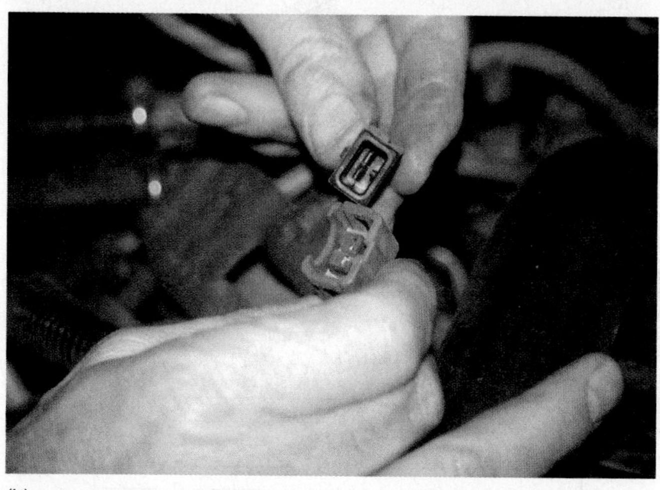

(b)

FIGURE 17–29 (a) Typical SPOUT connector as used on many Ford engines equipped with distributor ignition (DI). (b) The connector must be opened (disconnected) to check and/or adjust the ignition timing. On DIS/EDIS systems, the connector is called SPOUT/SAW (spark output/spark angle word).

TECH TIP

Two Marks Are the Key to Success

When a distributor is removed from an engine, always mark the direction the rotor is pointing to ensure that the distributor is reinstalled in the correct position. Because of the helical cut on the distributor drive gear, the rotor rotates as the distributor is being removed from the engine. To help reinstall a distributor without any problems, simply make another mark where the rotor is pointing just as the distributor is lifted out of the engine. Then to reinstall, simply line up the rotor to the second mark and lower the distributor into the engine. The rotor should then line up with the original mark as a double check.

NOTE: If cylinder 1 is difficult to reach, such as up against the bulkhead (firewall) or close to an exhaust manifold, simply use the opposite cylinder in the firing order (paired cylinder). The timing light will not detect the difference and will indicate the correct position of the timing mark in relation to the pointer or degree mark.

CHECKING OR ADJUSTING IGNITION TIMING Use the following steps for checking or adjusting ignition timing:

1. Start the engine and adjust the speed to that specified for ignition timing.
2. With the timing light aimed at the stationary timing pointer, observe the position of the timing mark on the vibration damper with the light flashing. Refer to the manufacturer's specifications on underhood decal for the correct setting. ● **SEE FIGURE 17–29.**

NOTE: If the timing mark appears ahead of the pointer in relation to the direction of crankshaft rotation, the timing is advanced. If the timing mark appears behind the pointer in relation to the direction of crankshaft rotation, the timing is retarded.

3. To adjust timing, loosen the distributor locking bolt or nut and turn the distributor housing until the timing mark is in correct alignment. Turn the distributor housing in the direction of rotor rotation to retard the timing and against rotor rotation to advance the timing.
4. After adjusting the timing to specifications, carefully tighten the distributor locking bolt. Sometimes it is necessary to readjust the timing after the initial setting because the distributor may rotate slightly when the hold-down bolt is tightened.

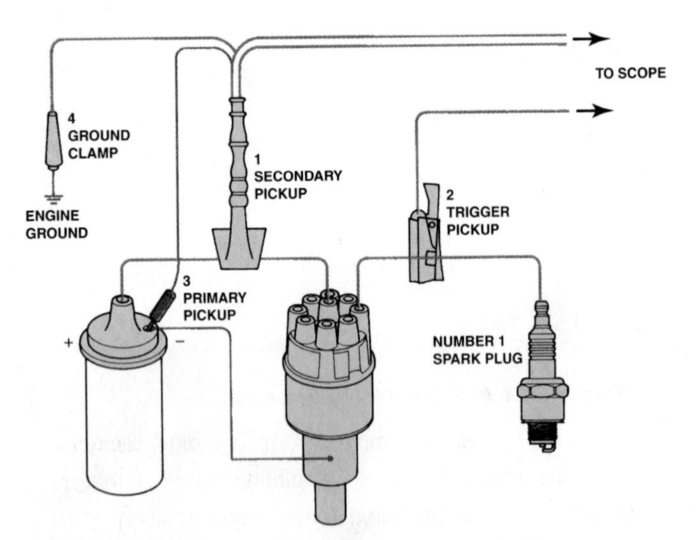

FIGURE 17–30 Typical engine analyzer hookup that includes a scope display. (1) Coil wire on top of the distributor cap if integral type of coil; (2) number 1 spark plug connection; (3) negative side of the ignition coil; (4) ground (negative) connection of the battery.

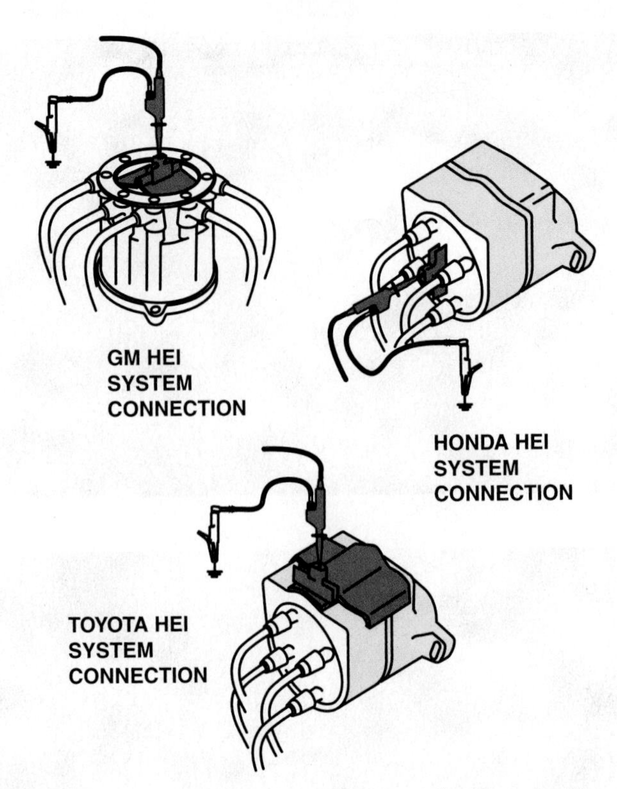

FIGURE 17–31 Clip-on adapters are used with an ignition system that uses an integral ignition coil.

SCOPE-TESTING THE IGNITION SYSTEM

Any automotive scope with the correct probes or adapters will show an ignition system pattern. All ignition systems must charge and discharge an ignition coil. With the engine off, most scopes will display a horizontal line. With the engine running, this horizontal (zero) line is changed to a pattern that will have sections both above and below the zero line. Sections of this pattern that are above the zero line indicate that the ignition coil is discharging. Sections of the scope pattern below the zero line indicate charging of the ignition coil. The height of the scope pattern indicates voltage. The length (from left to right) of the scope pattern indicates time. ● **SEE FIGURES 17–30 AND 17–31** for typical scope hookups.

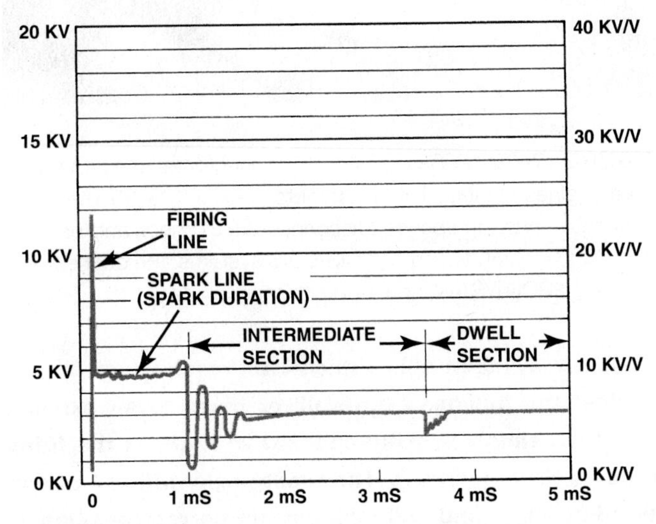

FIGURE 17–32 Typical secondary ignition oscilloscope pattern.

FIRING LINE The leftmost vertical (upward) line is called the **firing line.** The height of the firing line should be between 5,000 and 15,000 volts (5 and 15 kV) with not more than a 3 kV difference between the highest and the lowest cylinder's firing line. ● **SEE FIGURES 17–32 AND 17–33.**

The height of the firing line indicates the *voltage* required to fire the spark plug. It requires a high voltage to make the air inside the cylinder electrically conductive (to ionize the air). A higher than normal height (or height higher than that of other cylinders) can be caused by one or more of the following:

1. Spark plug gapped too wide
2. Lean fuel mixture
3. Defective spark plug wire

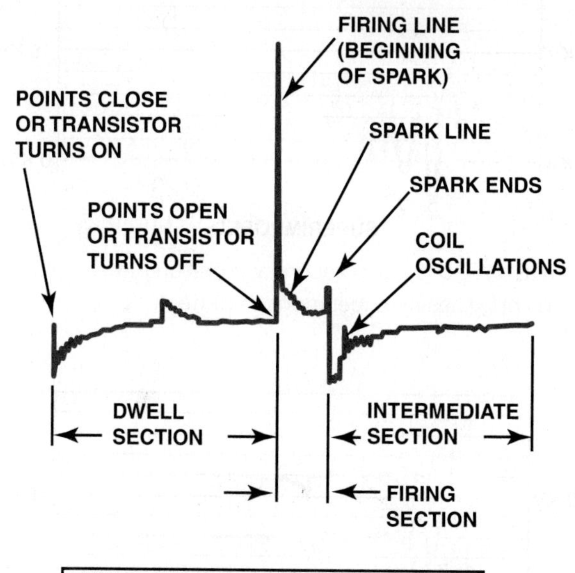

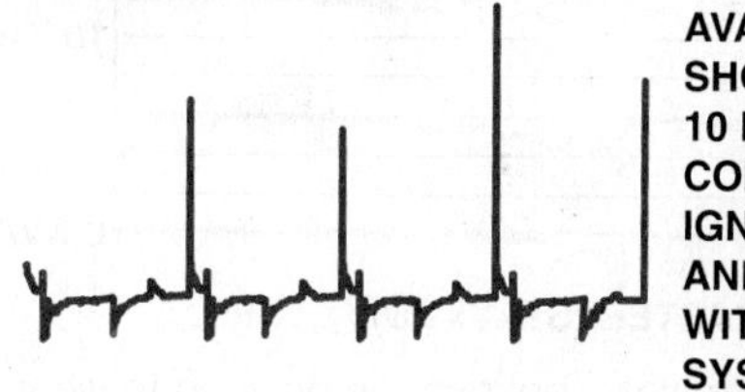

FIGURE 17–33 A single cylinder is shown at the top and a four-cylinder engine at the bottom.

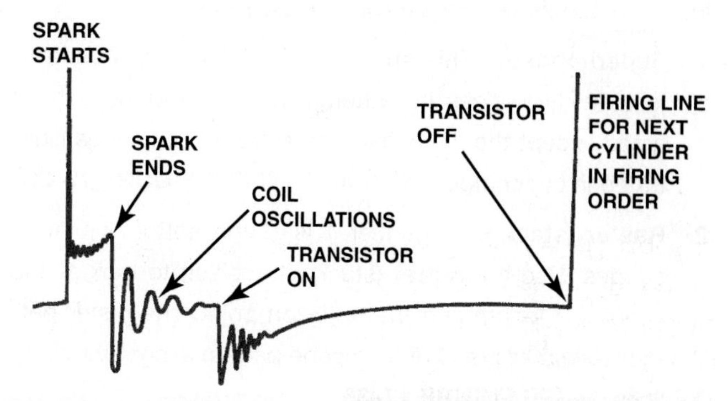

FIGURE 17–34 Drawing shows what is occurring electrically at each part of the scope pattern.

If the firing lines are higher than normal for *all* cylinders, then possible causes include one or more of the following:

1. Worn distributor cap and/or rotor (if the vehicle is so equipped)
2. Excessive wearing of all spark plugs
3. Defective coil wire (the high voltage could still jump across the open section of the wire to fire the spark plugs)

SPARK LINE The **spark line** is a short horizontal line immediately after the firing line. The height of the spark line represents the voltage required to maintain the spark across the spark plug after the spark has started. The height of the spark line should be one-fourth of the height of the firing line (between 1.5 and 2.5 kV). The length (from left to right) of the line represents the length of time for which the spark lasts (duration or burn time). The spark duration should be between 0.8 and 2.2 milliseconds (usually between 1.0 and 2.0 ms). The spark stops at the end (right side) of the spark line, as shown in ● **FIGURE 17–34**.

INTERMEDIATE OSCILLATIONS After the spark has stopped, some energy remains in the coil. This remaining energy dissipates in the coil windings and the entire secondary circuit. The **intermediate oscillations** are also called the "ringing" of the coil as it is pulsed.

The secondary pattern amplifies any voltage variation occurring in the primary circuit because of the turns ratio between the primary and secondary windings of the ignition coil. A correctly operating ignition system should display five or more "bumps" (oscillations) (three or more for a GM HEI system).

TRANSISTOR-ON POINT After the intermediate oscillations, the coil is empty (not charged), as indicated by the scope pattern being on the zero line for a short period. When the transistor turns on an electronic system, the coil is being charged.

Note that the charging of the coil occurs slowly (coil-charging oscillations) because of the inductive reactance of the coil.

DWELL SECTION Dwell is the amount of time that the current is charging the coil from the transistor-on point to the transistor-off point. At the end of the **dwell section** is the beginning of the next firing line. This point is called "transistor off" and indicates that the primary current of the coil is stopped, resulting in a high-voltage spark out of the coil.

PATTERN SELECTION Ignition oscilloscopes use three positions to view certain sections of the basic pattern more closely. These three positions are as follows:

1. **Superimposed.** This **superimposed** position is used to look at differences in patterns between cylinders in all areas except the firing line. There are no firing lines illustrated in superimposed positions. ● **SEE FIGURE 17–35.**
2. **Raster (stacked).** Cylinder 1 is at the bottom on most scopes. Use the **raster** (stacked) position to look at the spark line length and transistor-on point. The raster pattern shows all areas of the scope pattern except the firing lines. ● **SEE FIGURE 17–36.**
3. **Display (parade).** Display (parade) is the only position in which firing lines are visible. The firing line section for cylinder 1 is on the far right side of the screen, with the remaining portions of the pattern on the left side. This selection is used to compare the height of firing lines among all cylinders. ● **SEE FIGURE 17–37.**

READING THE SCOPE ON DISPLAY (PARADE) Start the engine and operate at approximately 1000 RPM to ensure a smooth and accurate scope pattern. Firing lines are visible only on the display (parade) position. The firing lines should all be 5 to 15 kV in height and be within 3 kV of each other. If one or more cylinders have high firing lines, this could indicate a defective (open) spark plug wire, a spark plug gapped too far, or a lean fuel mixture affecting only those cylinders.

A lean mixture (not enough fuel) requires a higher voltage to ignite because there are fewer droplets of fuel in the cylinder for the spark to use as "stepping stones" for the voltage to jump across. Therefore, a lean mixture is less conductive than a rich mixture.

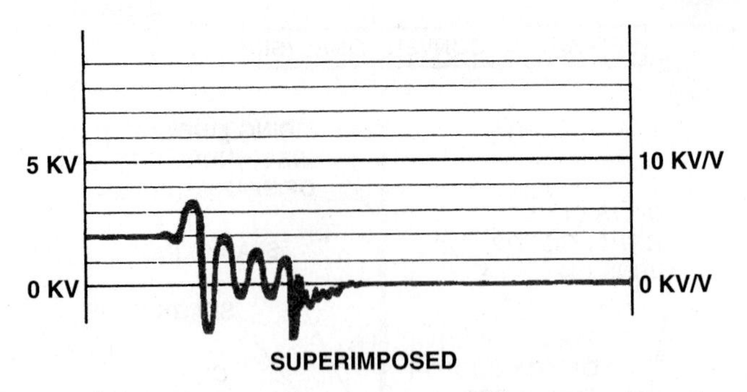

FIGURE 17–35 Typical secondary ignition pattern. Note the lack of firing lines on superimposed pattern.

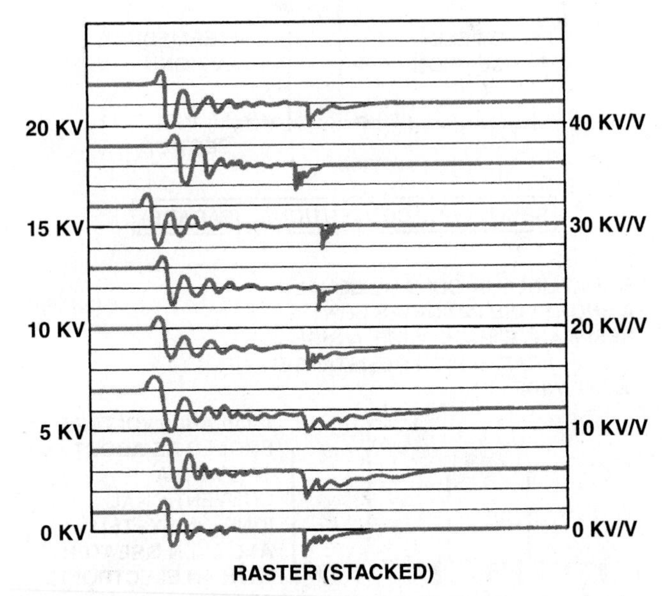

FIGURE 17–36 Raster is the best scope position to view the spark lines of all the cylinders to check for differences. Most scopes display the cylinder 1 at the bottom. The other cylinders are positioned by firing order above cylinder 1.

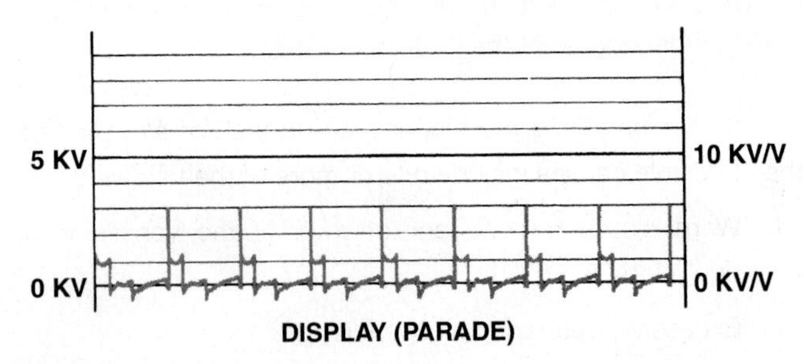

FIGURE 17–37 Display is the only position to view the firing lines of all cylinders. Cylinder 1 is displayed on the left (except for its firing line, which is shown on the right). The cylinders are displayed from left to right by firing order.

CASE STUDY

A Technician's Toughie

The owner of a Honda Civic complained that engine did not run smoothly and the "Check Engine" light was on. The service technician retrieved a P0300 (random misfire detected) as well as a P0303 (cylinder number three misfire detected). A scope was connected to each of the coils one at a time and the secondary pattern looked perfect on all four cylinders. All four coils and spark plugs were removed yet they all looked normal. The spark plug for cylinder #3 was moved to cylinder #1 and then the coils were re-installed and the vehicle driven on a test drive. A P0301 was then retrieved which indicated that the problem was due to the spark plug itself. Replacing the spark plug with a new one solved the misfire problem. The plug was apparently cracked yet not seen. The scope showed a normal secondary pattern because the voltage needed to jump to ground through the crack in the plug was about the same as would be required to jump the gap inside the combustion chamber.

Summary:

- **Complaint**—Customer stated that the engine ran poorly.
- **Cause**—Tests confirmed that one spark plug was found to be cracked.
- **Correction**—Replacing the spark plug solved the engine misfire problem.

READING THE SPARK LINES Spark lines can easily be seen on either superimposed or raster (stacked) position. On the raster position, each individual spark line can be viewed.

The spark lines should be level and one-fourth as high as the firing lines (1.5 to 2.5 kV, but usually less than 2 kV). The spark line voltage is called the **burn kV.** The *length* of the spark line is the critical factor for determining proper operation of the engine because it represents the spark duration or burn time. There is only a limited amount of energy in an ignition coil. If most of the energy is used to ionize the air gaps of the rotor and the spark plug, there may not be enough energy remaining to create a spark duration long enough to completely burn the air–fuel mixture. Many scopes are equipped with a **millisecond (ms) sweep.** This means that the scope will sweep only that portion of the pattern that can be shown during a 5- or 25-ms setting. Following are guidelines for spark line length:

- 0.8 ms—too short
- 1.5 ms—average
- 2.2 ms—too long

If the spark line is too short, possible causes include the following:

1. Spark plug(s) gap is too wide
2. Rotor tip to distributor cap insert distance gap is too wide (worn cap or rotor)
3. High-resistance spark plug wire
4. Air–fuel mixture too lean (vacuum leak, broken valve spring, etc.)

If the spark line is too long, possible causes include the following:

1. Fouled spark plug(s)
2. Spark plug(s) gap is too narrow
3. Shorted spark plug or spark plug wire

Many scopes do not have a millisecond scale. Some scopes are labeled in degrees and/or percentage (%) of dwell. The following chart can be used to determine acceptable spark line length. ● **SEE CHART 17–2.**

Normal Spark Line Length (at 700 to 1200 RPM)

NUMBER OF CYLINDERS	MILLISECONDS	PERCENTAGE (%) OF DWELL SCALE	DEGREES (°)
4	1.0–2.0	3–6	3–5
6	1.0–2.0	4–9	2–5
8	1.0–2.0	6–13	3–6

CHART 17–2

Spark line length depends on the number of cylinders and the engine speed.

SPARK LINE SLOPE Downward-sloping spark lines indicate that the voltage required to maintain the spark duration is decreasing during the firing of the spark plug. This downward slope usually indicates that the spark energy is finding ground through spark plug deposits (the plug is fouled) or other ignition problems. ● **SEE FIGURE 17–38.**

An upward-sloping spark line usually indicates a mechanical engine problem. A defective piston ring or valve would tend to seal better in the increasing pressures of combustion. As the spark plug fires, the effective increase in pressures increases

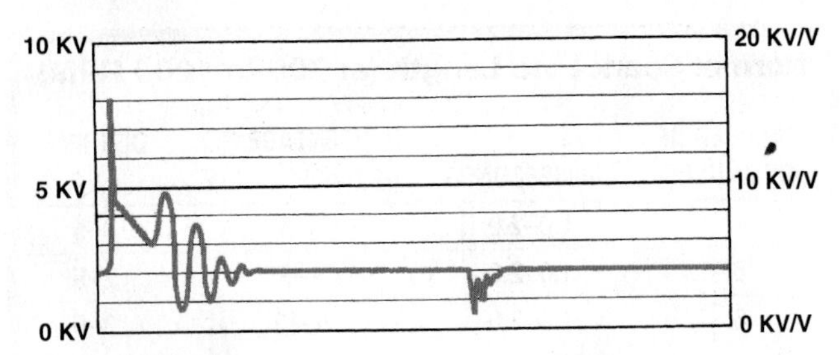

FIGURE 17–38 A downward-sloping spark line usually indicates high secondary ignition system resistance or an excessively rich air–fuel mixture.

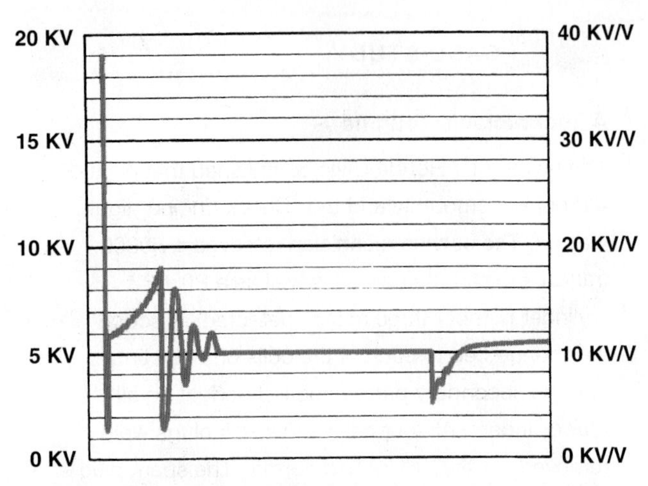

FIGURE 17–39 An upward-sloping spark line usually indicates a mechanical engine problem or a lean air–fuel mixture.

the voltage required to maintain the spark, and the height of the spark line rises during the duration of the spark. ● **SEE FIGURE 17–39.**

An upward-sloping spark line can also indicate a lean air–fuel mixture. Typical causes include:

1. Clogged injector(s)
2. Vacuum leak
3. Sticking intake valve

● **SEE FIGURE 17–40** for an example showing the relationship between the firing line and the spark line.

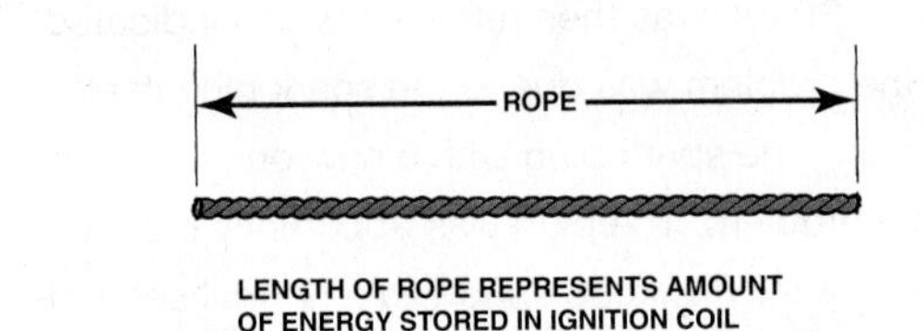

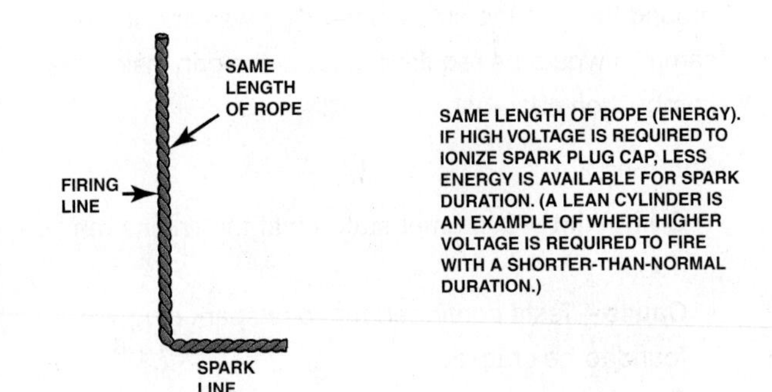

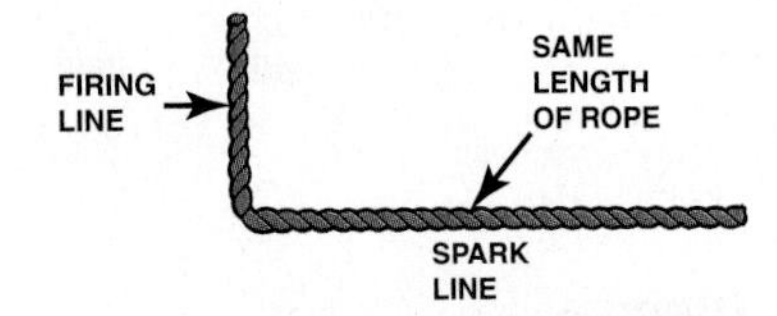

FIGURE 17–40 The relationship between the height of the firing line and length of the spark line can be illustrated using a rope. Because energy cannot be destroyed, the stored energy in an ignition coil must dissipate totally, regardless of engine operating conditions.

READING THE INTERMEDIATE SECTION The intermediate section should have three or more oscillations (bumps) for a correctly operating ignition system. Because approximately 250 volts are in the primary ignition circuit when the spark stops flowing across the spark plugs, this voltage is reduced by about 75 volts per oscillation. Additional resistances in the primary circuit would decrease the number of oscillations. If there are fewer than three oscillations, possible problems include the following:

1. Shorted ignition coil
2. Loose or high-resistance primary connections on the ignition coil or primary ignition wiring

ELECTRONIC IGNITION AND THE DWELL SECTION Electronic ignitions also use a dwell period to charge the coil. Dwell is not adjustable with electronic ignition, but it does change with increasing RPM with many electronic ignition systems. This change in dwell with RPM should be considered normal.

Many EI systems also produce a "hump" in the dwell section, which reflects a current-limiting circuit in the control module. These current-limiting humps may have slightly different shapes depending on the exact module used. For example, the humps produced by various GM HEI modules differ slightly.

DWELL VARIATION (ELECTRONIC IGNITION) A worn distributor gear, worn camshaft gear, or other distributor problem may cause engine performance problems, because the signal created in the distributor will be affected by the inaccurate distributor operation. However, many electronic ignitions vary the dwell electronically in the module to maintain acceptable current flow levels through the ignition coil and module without the use of a ballast resistor.

NOTE: Distributorless ignition systems also vary dwell time electronically within the engine computer or ignition module.

COIL POLARITY With the scope connected and the engine running, observe the scope pattern in the superimposed mode. If the pattern is upside down, the primary wires on the coil may be reversed, causing the coil polarity to be reversed.

NOTE: Check the scope hookup and controls before deciding that the coil polarity is reversed.

ACCELERATION CHECK With the scope selector set on the display (parade) position, rapidly accelerate the engine (gear selector in park or neutral with the parking brake on). The results should be interpreted as follows:

1. All firing lines should rise evenly (not to exceed 75% of maximum coil output) for properly operating spark plugs.
2. If the firing lines on one or more cylinders fail to rise, this indicates fouled spark plugs.

ROTOR GAP VOLTAGE The **rotor gap** voltage test measures the voltage required to jump the gap (0.030 to 0.050 inch or 0.8 to 1.3 mm) between the rotor and the inserts (segments) of the distributor cap. Select the display (parade) scope pattern and remove a spark plug wire using a jumper wire to provide a good ground connection. Start the engine and observe the height of the firing line for the cylinder being tested. Because the spark plug wire is connected directly to ground, the firing line height on the scope will indicate the voltage required to jump the air gap between the rotor and the distributor cap insert. The normal rotor gap voltage is 3 to 7 kV, and the voltage should not exceed 8 kV. If the rotor gap voltage indicated is near or above 8 kV, inspect and replace the distributor cap and/or rotor as required.

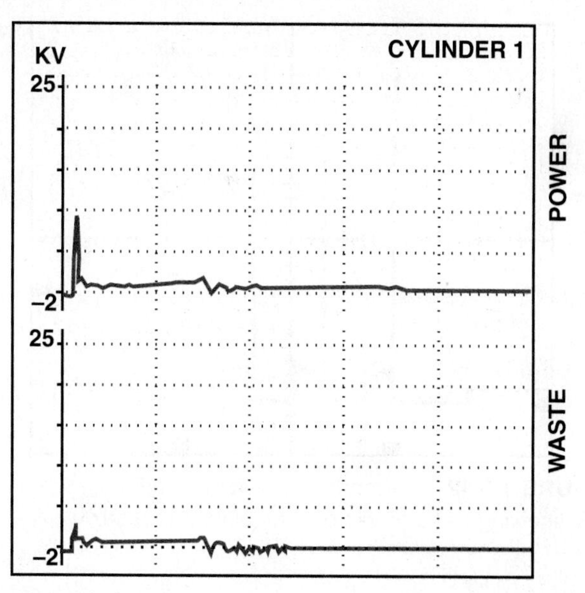

FIGURE 17–41 A dual-trace scope pattern showing both the power and the waste spark from the same coil (cylinders 1 and 6). Note that the firing line is higher on the cylinder that is under compression (power); otherwise, both patterns are almost identical.

SCOPE-TESTING A WASTE-SPARK IGNITION SYSTEM

A handheld digital storage oscilloscope can be used to check the pattern of each individual cylinder. Some larger scopes can be connected to all spark plug wires and therefore are able to display both power and waste-spark waveforms. ● **SEE FIGURE 17–41**. Because the waste spark does not require as high a voltage level as the cylinder on the power stroke, the waste form will be normally lower.

SCOPE-TESTING A COIL-ON-PLUG IGNITION SYSTEM

On a coil-on-plug type of ignition system, each individual coil can be shown on a scope and using the proper cables and adapters, the waveform for all of the cylinders can be viewed at the same time. Always follow the scope equipment manufacturer's instructions. Many Ford coil-on-plug systems use a

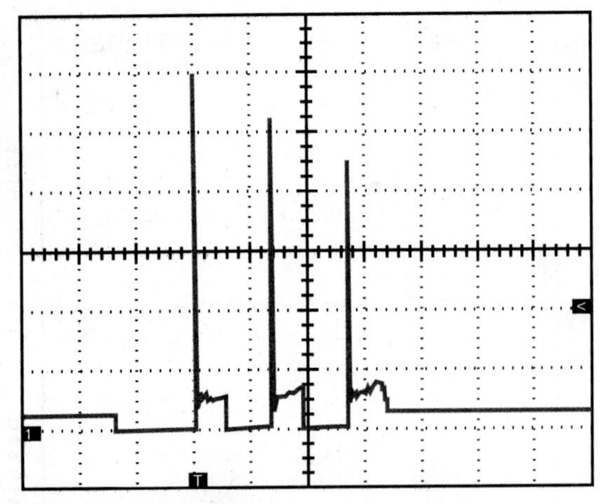

FIGURE 17–42 A secondary waveform of a Ford 4.6 liter V-8, showing three sparks occurring at idle speed.

triple-strike secondary spark event. The spark plugs are fired three times when the engine is at idle speed to improve idle quality and to reduce exhaust emissions. Above certain engine speeds, the ignition system switches to a single-fire event. **SEE FIGURE 17–42.**

IGNITION SYSTEM SYMPTOM GUIDE

Problem	Possible Causes and/or Solutions
No spark out of the coil	• Possible open in the ignition switch circuit • Possible defective ignition module (if electronic ignition coil) • Possible defective pickup coil or Hall-effect switch (if electronic ignition) • Possible shorted condenser
Weak spark out of the coil	• Possible high-resistance coil wire or spark plug wire • Possible poor ground between the distributor or module and the engine block
Engine misfire	• Possible defective (open) spark plug wire • Possible worn or fouled spark plugs • Possible defective pickup coil • Possible defective module • Possible poor electrical connections at the pickup coil and/or module

SUMMARY

1. A thorough visual inspection should be performed on all ignition components when diagnosing an engine performance problem.
2. Platinum spark plugs should not be regapped after use in an engine.
3. A secondary ignition scope pattern includes a firing line, spark line, intermediate oscillations, and transistor-on and transistor-off points.
4. The slope of the spark line can indicate incorrect air–fuel ratio or other engine problems.

REVIEW QUESTIONS

1. Why should a spark tester be used to check for spark rather than a standard spark plug?
2. How do you test a pickup coil for resistance and AC voltage output?
3. What harm can occur if the engine is cranked or run with an open (defective) spark plug wire?
4. What are the sections of a secondary ignition scope pattern?
5. What can the slope of the spark line indicate about the engine?

CHAPTER QUIZ

1. Technician A says that the firing line shows the voltage that is required to fire the spark plug. Technician B says that spark line shows the duration of the spark inside the cylinder. Which technician is correct?
   - **a.** Technician A only
   - **b.** Technician B only
   - **c.** Both Technicians A and B
   - **d.** Neither Technician A nor B
2. Technician A says that a defective spark plug wire or boot can cause an engine misfire. Technician B says that a tracked ignition coil can cause an engine misfire. Which technician is correct?
   - **a.** Technician A only
   - **b.** Technician B only
   - **c.** Both Technicians A and B
   - **d.** Neither Technician A nor B
3. The ________ sends a pulse signal to an electronic ignition module.
   - **a.** Ballast resistor
   - **b.** Pickup coil
   - **c.** Ignition coil
   - **d.** Condenser
4. Typical primary coil resistance specifications usually range from ________ ohms.
   - **a.** 100 to 450
   - **b.** 500 to 1,500
   - **c.** Less than 1 to 3
   - **d.** 6,000 to 30,000
5. Typical secondary coil resistance specifications usually range from ________ ohms.
   - **a.** 100 to 450
   - **b.** 500 to 1,500
   - **c.** 1 to 3
   - **d.** 6,000 to 30,000
6. Technician A says that an engine will not start and run if the ignition coil is tracked. Technician B says the engine will not start if the crankshaft position sensor fails. Which technician is correct?
   - **a.** Technician A only
   - **b.** Technician B only
   - **c.** Both Technicians A and B
   - **d.** Neither Technician A nor B
7. Technician A says that a distributor rotor can burn through and cause an engine misfire during acceleration. Technician B says that a defective spark plug wire can cause an engine misfire during acceleration. Which technician is correct?
   - **a.** Technician A only
   - **b.** Technician B only
   - **c.** Both Technicians A and B
   - **d.** Neither Technician A nor B
8. The secondary ignition circuit can be tested using ________.
   - **a.** An ohmmeter
   - **b.** Visual inspection
   - **c.** An ammeter
   - **d.** Both a and b
9. Two technicians are discussing a no-start (no-spark) condition. Technician A says that an open pickup coil could be the cause. Technician B says that a defective ignition control module (ICM) could be the cause. Which technician is correct?
   - **a.** Technician A only
   - **b.** Technician B only
   - **c.** Both Technicians A and B
   - **d.** Neither Technician A nor B
10. Which sensor produces a square wave signal?
    - **a.** Magnetic sensor
    - **b.** Hall-effect sensor
    - **c.** Optical sensor
    - **d.** Both b and c

chapter 18

COMPUTER AND NETWORK FUNDAMENTALS

LEARNING OBJECTIVES

After studying this chapter, the reader will be able to:

1. Explain the purpose and function of onboard computers.
2. List the various parts of an automotive computer.
3. List input sensors and output device controlled by the computer.
4. Explain the parts and characteristics of digital computers.

This chapter will help you prepare for Engine Repair (A6) ASE certification test content area "A" (General Electrical/Electronic Systems Diagnosis).

KEY TERMS

Actuator 308
Analog-to-digital (AD) converter 310
Central processing unit (CPU) 310
Clock generator 311
Controller 312
Controller area network (CAN) 316
Digital computer 310
Duty cycle 310
EEPROM 308
E^2PROM 308
Electronic control assembly (ECA) 312
Electronic control module (ECM) 312
Electronic control unit (ECU) 312
Engine mapping 311
Input conditioning 307
Keep-alive memory (KAM) 308
Multiplexing 314
Network 314
Nonvolatile RAM 308
Powertrain Control Module (PCM) 307
Programmable read-only memory (PROM) 307
Random-access memory (RAM) 308
Read-only memory (ROM) 307
Serial data 314
Splice pack 314
Terminating resistors 317

COMPUTER CONTROL

Modern automotive control systems consist of a network of electronic sensors, actuators, and computer modules designed to regulate the powertrain and vehicle support systems. The **Powertrain Control Module (PCM)** is the heart of this system. It coordinates engine and transmission operation, processes data, maintains communications, and makes the control decisions needed to keep the vehicle operating.

Automotive computers use voltage to send and receive information. Voltage is electrical pressure and does not flow through circuits, but voltage can be used as a signal. A computer converts input information or data into voltage signal combinations that represent number combinations. The number combinations can represent a variety of information—temperature, speed, or even words and letters. A computer processes the input voltage signals it receives by computing what they represent, and then delivering the data in computed or processed form.

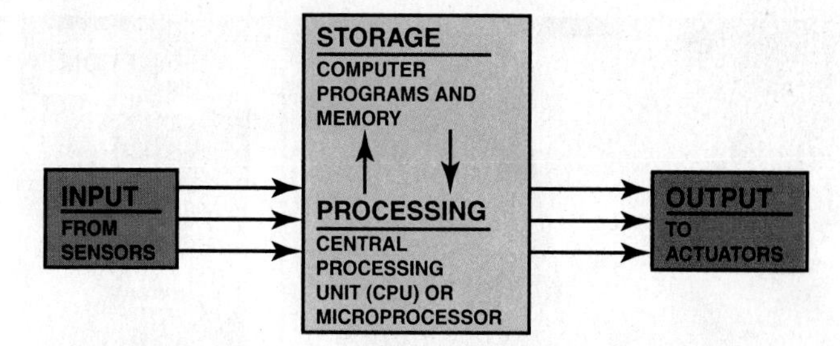

FIGURE 18–1 All computer systems perform four basic functions: input, processing, storage, and output.

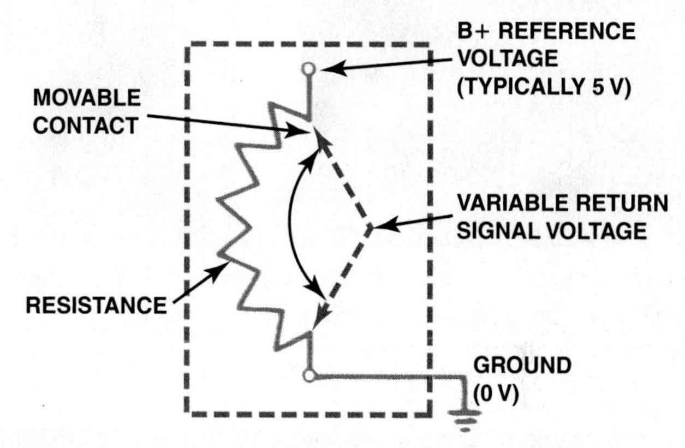

FIGURE 18–2 A potentiometer uses a movable contact to vary resistance and send an analog voltage to the PCM.

THE FOUR BASIC COMPUTER FUNCTIONS

The operation of every computer can be divided into the following four basic functions: ● **SEE FIGURE 18–1**.

- Input
- Processing
- Storage
- Output

These basic functions are not unique to computers; they can be found in many noncomputer systems. However, we need to know how the computer handles these functions.

INPUT First, the computer receives a voltage signal (input) from an input device. The device can be as simple as a button or a switch on an instrument panel, or a sensor on an automotive engine. ● **SEE FIGURE 18–2** for a typical type of automotive sensor.

Vehicles use various mechanical, electrical, and magnetic sensors to measure factors such as vehicle speed, engine RPM, air pressure, oxygen content of exhaust gas, airflow, and engine coolant temperature. Each sensor transmits its information in the form of voltage signals. The computer receives these voltage signals, but before it can use them, the signals must undergo a process called **input conditioning**. This process includes amplifying voltage signals that are too small for the computer circuitry to handle. Input conditioners generally are located inside the computer, but a few sensors have their own input-conditioning circuitry.

PROCESSING Input voltage signals received by a computer are processed through a series of electronic logic circuits maintained in its programmed instructions. These logic circuits change the input voltage signals, or data, into output voltage signals or commands.

STORAGE The program instructions for a computer are stored in electronic memory. Some programs may require that certain input data be stored for later reference or future processing. In others, output commands may be delayed or stored before they are transmitted to devices elsewhere in the system.

Computers have two types of memory: permanent and temporary. Permanent memory is called **read-only memory (ROM)** because the computer can only read the contents; it cannot change the data stored in it. This data is retained even when power to the computer is shut off. Part of the ROM is built into the computer, and the rest is located in an IC chip called a **programmable read-only memory (PROM)** or calibration assembly. ● **SEE FIGURE 18–3**. Many chips are erasable, meaning that the program can be changed. These chips are

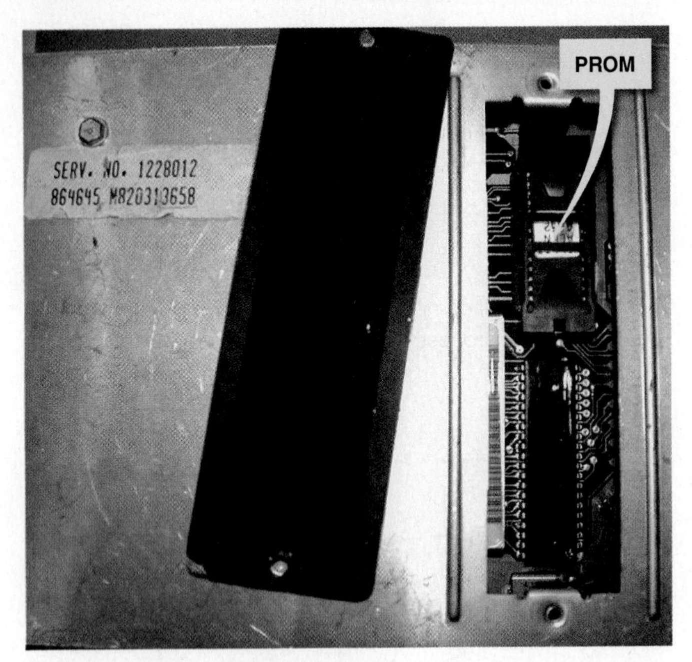

FIGURE 18–3 A replaceable PROM used in an older General Motors computer. Notice that the sealed access panel has been removed to gain access.

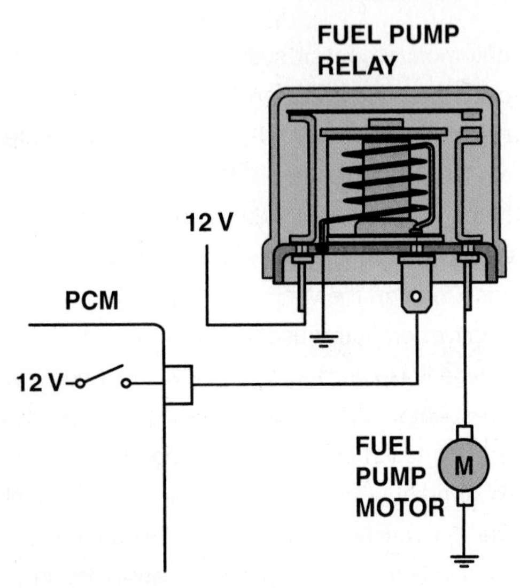

FIGURE 18–4 A typical output driver. In this case, the PCM applies voltage to the fuel pump relay coil to energize the fuel pump.

called erasable programmable read-only memory or EPROM. Since the early 1990s most programmable memory has been electronically erasable, meaning that the program in the chip can be reprogrammed by using a scan tool and the proper software. This computer reprogramming is usually called *reflashing*. These chips are electrically erasable programmable read-only memory, abbreviated **EEPROM** or **E^{2}PROM**. All vehicles equipped with onboard diagnosis second generation, called OBD II, are equipped with EEPROMs.

Temporary memory is called **random-access memory (RAM)** because the microprocessor can write or store new data into it as directed by the computer program, as well as read the data already in it. Automotive computers use two types of RAM memory: **volatile** and **nonvolatile**. Volatile RAM memory is lost whenever the ignition is turned off. However, a type of volatile RAM called **keep-alive memory (KAM)** can be wired directly to battery power. This prevents its data from being erased when the ignition is turned off. Both RAM and KAM have the disadvantage of losing their memory when disconnected from their power source. One example of RAM and KAM is the loss of station settings in a programmable radio when the battery is disconnected. Since all the settings are stored in RAM, they have to be reset when the battery is reconnected. System trouble codes are commonly stored in RAM and can be erased by disconnecting the battery.

Nonvolatile RAM memory can retain its information even when the battery is disconnected. One use for this type of RAM is the storage of odometer information in an electronic speedometer. The memory chip retains the mileage accumulated by the vehicle. When speedometer replacement is necessary, the odometer chip is removed and installed in the new speedometer unit. KAM is used primarily in conjunction with adaptive strategies.

OUTPUT After the computer has processed the input signals, it sends voltage signals or commands to other devices in the system, such as system actuators. An **actuator** is an electrical or mechanical device that converts electrical energy into heat, light, or motion, such as adjusting engine idle speed, altering suspension height, or regulating fuel metering.

Computers also can communicate with, and control, each other through their output and input functions. This means that the output signal from one computer system can be the input signal for another computer system through a network.

Most outputs work electrically in one of three ways:

- Switched
- Pulse-width modulated
- Digital

A switched output is an output that is either on or off. In many circuits, the PCM uses a relay to switch a device on or off. This is because the relay is a low-current device that can switch a higher-current device. Most computer circuits cannot handle a lot of current. By using a relay circuit as shown in ● **FIGURE 18–4**, the PCM provides the output control to the relay, which in turn

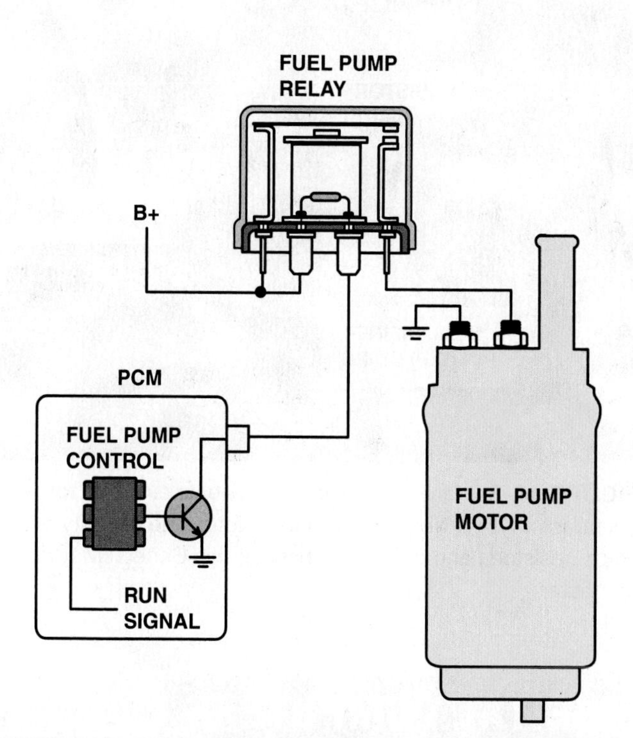

FIGURE 18–5 A typical low-side driver (LSD) which uses a control module to control the ground side of the relay coil.

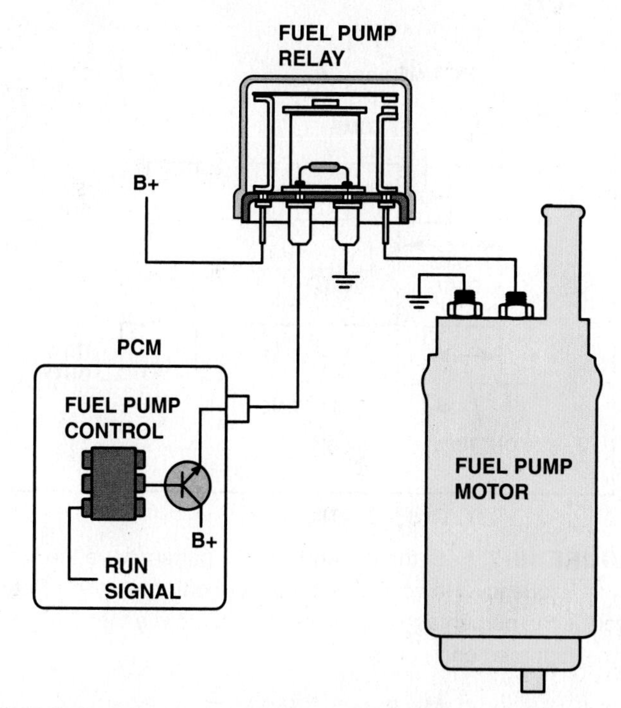

FIGURE 18–6 A typical module-controlled high-side driver (HSD) where the module itself supplies the electrical power to the device. The logic circuit inside the module can detect circuit faults including continuity of the circuit and if there is a short-to-ground in the circuit being controlled.

provides the output control to the device. The relay coil, which the PCM controls, typically draws less than 0.5 amperes. The device that the relay controls may draw 30 amperes or more. These switches are actually transistors, often called output drivers.

LOW-SIDE DRIVERS Low-side drivers, often abbreviated LSD, are transistors that complete the ground path in the circuit. Ignition voltage is supplied to the relay as well as battery voltage. The computer output is connected to the ground side of the relay coil. The computer energizes the fuel pump relay by turning the transistor on and completing the ground path for the relay coil. A relatively low current flows through the relay coil and transistor that is inside the computer. This causes the relay to switch and provides the fuel pump with battery voltage. The majority of switched outputs have typically been low-side drivers. ● **SEE FIGURE 18–5**. Low-side drivers can often perform a diagnostic circuit check by monitoring the voltage from the relay to check that the control circuit for the relay is complete. A low-side driver, however, cannot detect a short-to-ground.

HIGH-SIDE DRIVERS High-side drivers, often abbreviated HSD, control the power side of the circuit. In these applications when the transistor is switched on, voltage is applied to the device. A ground has been provided to the device so when the high-side driver switches the device will be energized. In some applications, high-side drivers are used instead of low-side drivers to provide better circuit protection. General Motors vehicles have used a high-side driver to control the fuel pump relay instead of a low-side driver. In the event of an accident, should the circuit to the fuel pump relay become grounded, a high-side driver would cause a short circuit, which would cause the fuel pump relay to de-energize. High-side drivers inside modules can detect electrical faults such as a lack of continuity when the circuit is not energized. ● **SEE FIGURE 18–6**.

PULSE WIDTH MODULATION Pulse width modulation (PWM) is a method of controlling an output using a digital signal. Instead of just turning devices on or off, the computer can control output devices more precisely by using pulse width modulation. For example, a vacuum solenoid could be a pulse-width modulated device. If the vacuum solenoid is controlled by a switched driver, switching either on or off would mean that either full vacuum would flow through the solenoid or no vacuum would flow through the solenoid. However, to control the amount of vacuum that flows through the solenoid, pulse width modulation could be used. A PWM signal is a digital signal, usually 0 and 12 volts, that is cycling at a fixed frequency. Varying the length of time that the signal is on, provides

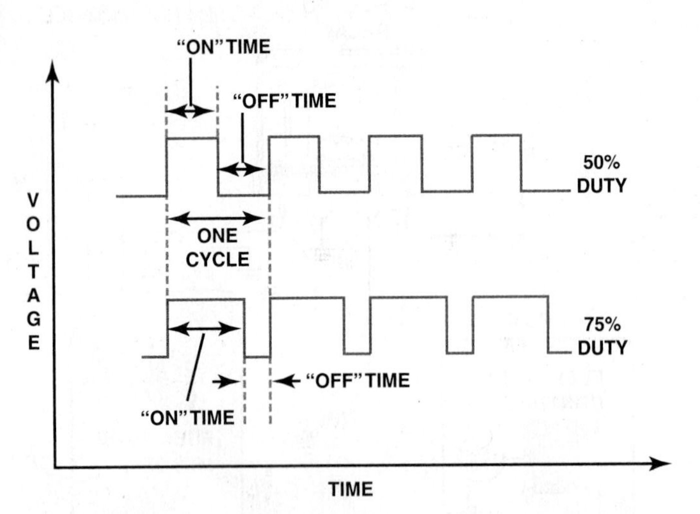

FIGURE 18–7 Both the top and bottom pattern have the same frequency. However, the amount of on-time varies. Duty cycle is the percentage of the time during a cycle that the signal is turned on.

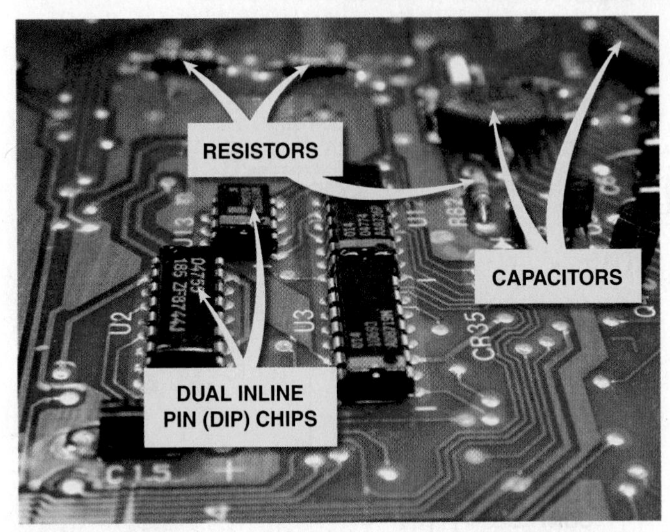

FIGURE 18–8 Many electronic components are used to construct a typical vehicle computer. Notice the quantity of chips, resistors, and capacitors used in this General Motors computer.

a signal that can vary the on and off time of an output. The ratio of on-time relative to the period of the cycle is referred to as **duty cycle**. ● **SEE FIGURE 18–7**. Depending on the frequency of the signal, which is usually fixed, this signal would turn the device on and off a fixed number of times per second. When, for example, the voltage is high (12 volts) 90% of the time and low (0 volts) the other 10% of the time, the signal has a 90% duty cycle. In other words, if this signal were applied to the vacuum solenoid, the solenoid would be on 90% of the time. This would allow more vacuum to flow through the solenoid. The computer has the ability to vary this on and off time or pulse width modulation at any rate between 0 and 100%.

A good example of pulse width modulation is the cooling fan speed control. The speed of the cooling fan is controlled by varying the amount of on-time that the battery voltage is applied to the cooling fan motor.

100% duty cycle—the fan runs at full speed
75% duty cycle—the fan runs at 3/4 speed
50% duty cycle—the fan runs at 1/2 speed
25% duty cycle—the fan runs at 1/4 speed

The use of PWM, therefore, results in very precise control of an output device to achieve the amount of cooling needed and conserve electrical energy compared to simply timing the cooling fan on high when needed. PWM may be used to control vacuum through a solenoid, the amount of purge of the evaporative purge solenoid, the speed of a fuel pump motor, control of a linear motor, or even the intensity of a lightbulb.

DIGITAL COMPUTERS

In a **digital computer**, the voltage signal or processing function is a simple high/low, yes/no, on/off signal. The digital signal voltage is limited to two voltage levels: high voltage and low voltage. Since there is no stepped range of voltage or current in between, a digital binary signal is a "square wave."

The signal is called "digital" because the on and off signals are processed by the computer as the digits or numbers 0 and 1. The number system containing only these two digits is called the binary system. Any number or letter from any number system or language alphabet can be translated into a combination of binary 0s and 1s for the digital computer.

A digital computer changes the analog input signals (voltage) to digital bits (*bi*nary digi*ts*) of information through an **analog-to-digital (AD) converter** circuit. The binary digital number is used by the computer in its calculations or logic networks. Output signals usually are digital signals that turn system actuators on and off.

The digital computer can process thousands of digital signals per second because its circuits are able to switch voltage signals on and off in billionths of a second. ● **SEE FIGURE 18–8.**

PARTS OF A COMPUTER

The software consists of the programs and logic functions stored in the computer's circuitry. The hardware is the mechanical and electronic parts of a computer.

CENTRAL PROCESSING UNIT (CPU). The microprocessor is the **central processing unit** of a computer. Since it performs the

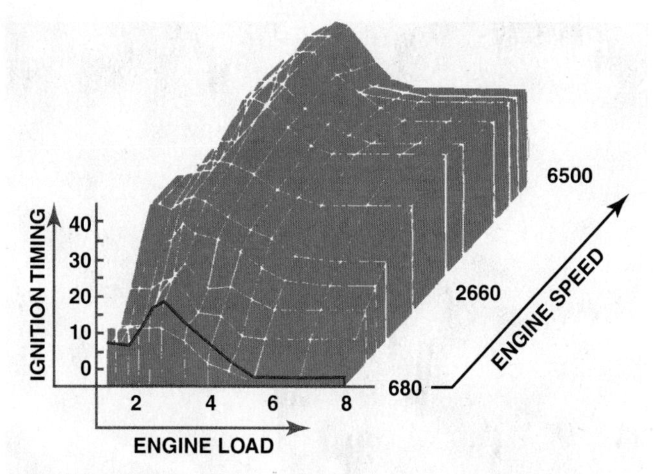

FIGURE 18–9 Typical ignition timing map developed from testing and used by the vehicle computer to provide the optimum ignition timing for all engine speeds and load combinations.

essential mathematical operations and logic decisions that make up its processing function, the CPU can be considered to be the brain of a computer. Some computers use more than one microprocessor, called a coprocessor.

COMPUTER MEMORY. Other IC devices store the computer operating program, system sensor input data, and system actuator output data, information that is necessary for CPU operation.

COMPUTER PROGRAMS

By operating a vehicle on a dynamometer and manually adjusting the variable factors such as speed, load, and spark timing, it is possible to determine the optimum output settings for the best driveability, economy, and emission control. This is called **engine mapping.** ● **SEE FIGURE 18–9.**

Engine mapping creates a three-dimensional performance graph that applies to a given vehicle and powertrain combination. Each combination is mapped in this manner to produce a PROM. This allows an automaker to use one basic computer for all models; a unique PROM individualizes the computer for a particular model. Also, if a driveability problem can be resolved by a change in the program, the manufacturers can release a revised PROM to supersede the earlier part.

Many older vehicle computers used a single PROM that plugged into the computer. ● **SEE FIGURE 18–10.** Some Ford computers used a larger "calibration module" that contained the system PROM.

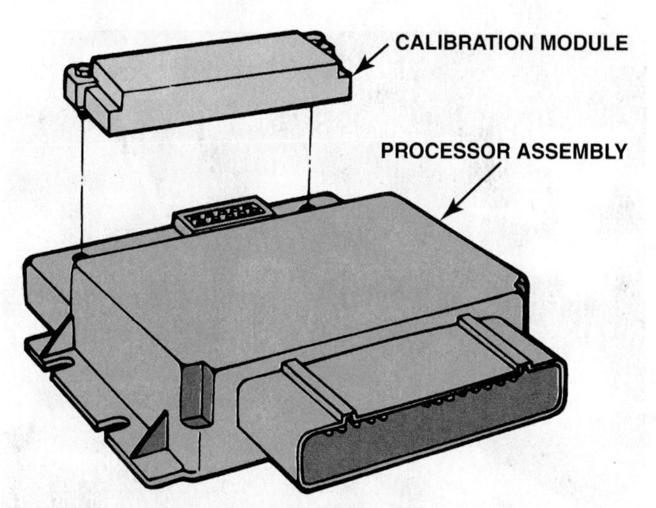

FIGURE 18–10 The calibration module on many Ford computers contains a system PROM.

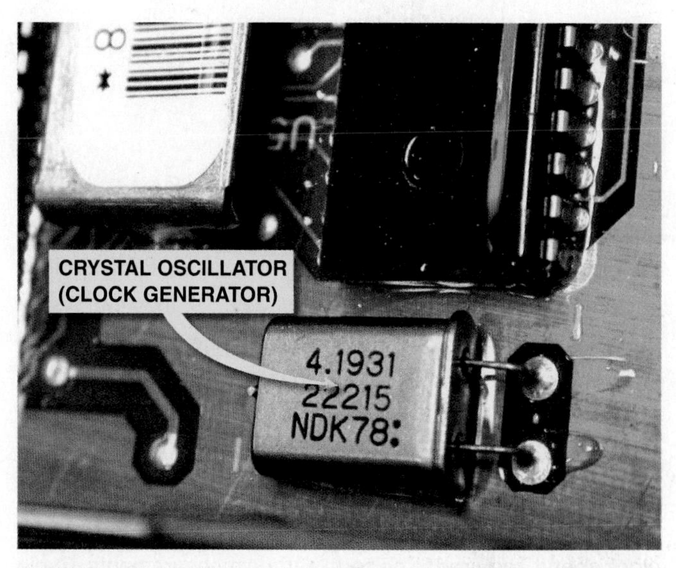

FIGURE 18–11 The clock generator produces a series of pulses that are used by the microprocessor and other components to stay in step with each other at a steady rate.

NOTE: If the onboard computer needs to be replaced, the PROM or calibration module must be removed from the defective unit and installed in the replacement computer. Since the mid-1990s, computers must be programmed or *flashed* before being put into service.

CLOCK RATES AND TIMING

The microprocessor receives sensor input voltage signals, processes them by using information from other memory units, and then sends voltage signals to the appropriate actuators. The microprocessor communicates by transmitting long strings of 0s and 1s in a language called binary code. But the microprocessor must have some way of knowing when one signal ends and another begins. That is the job of a crystal oscillator called a **clock generator.** ● **SEE FIGURE 18–11.** The computer's crystal

FIGURE 18–12 This Powertrain Control Module (PCM) is located under the hood on this Chevrolet pickup truck.

FIGURE 18–13 This PCM on a Chrysler vehicle can only be seen by hoisting the vehicle because it is located next to the radiator, and in the airflow to help keep it cool.

oscillator generates a steady stream of one-bit-long voltage pulses. Both the microprocessor and the memories monitor the clock pulses while they are communicating. Because they know how long each voltage pulse should be, they can distinguish between a 01 and a 0011. To complete the process, the input and output circuits also watch the clock pulses.

COMPUTER SPEEDS Not all computers operate at the same speed; some are faster than others. The speed at which a computer operates is specified by the cycle time, or clock speed, required to perform certain measurements. Cycle time or clock speed is measured in megahertz (4.7, 8.0, 15, 18 MHz, etc.).

BAUD RATE The computer transmits bits of a serial data stream at precise intervals. The computer's processing speed is called the baud rate, or bits per second. Just as mph helps in estimating the length of time required to travel a certain distance, the baud rate is useful in estimating how long a given computer will need to transmit a specified amount of data to another computer. Storage of a single character requires eight bits per byte, plus an additional two bits to indicate stop and start. This means that transmission of one character requires 10 bits. Dividing the baud rate by 10 tells us the maximum number of words per second that can be transmitted. For example, if the computer has a baud rate of 600, approximately 60 words can be received or sent per minute.

Automotive computers have evolved from a baud rate of 160 used in the early 1980s to a baud rate as high as 500,000 for some networks. The speed of data transmission is an important factor both in system operation and in system troubleshooting.

CONTROL MODULE LOCATIONS The onboard automotive computer has many names. It may be called an **electronic control unit (ECU)**, **electronic control module (ECM)**, **electronic control assembly (ECA)**, or a **controller**, depending on the manufacturer and the computer application. The Society of Automotive Engineers (SAE) bulletin, J-1930, standardizes the name as a Powertrain Control Module (PCM). The computer hardware is all mounted on one or more circuit boards and installed in a metal case to help shield it from electromagnetic interference (EMI). The wiring harnesses that link the computer to sensors and actuators connect to multipin connectors or edge connectors on the circuit boards.

Onboard computers range from single-function units that control a single operation to multifunction units that manage all of the separate (but linked) electronic systems in the vehicle. They vary in size from a small module to a notebook-sized box. Most other engine computers are installed in the passenger compartment either under the instrument panel or in a side kick panel where they can be shielded from physical damage caused by temperature extremes, dirt, and vibration, or interference by the high currents and voltages of various underhood systems. ● **SEE FIGURES 18–12 AND 18–13.**

COMPUTER INPUT SENSORS

The vehicle computer uses the signals (voltage levels) from the following engine sensors:

- **Engine speed (RPM or revolutions per minute) sensor.** This signal comes from the primary signal in the ignition module.
- **MAP (manifold absolute pressure) sensor.** This sensor detects engine load. The computer uses this information for fuel delivery and for onboard diagnosis of other sensors and systems such as the exhaust gas recirculation (EGR) system.
- **MAF (mass airflow) sensor.** This sensor measures the mass (weight and density) of the air entering the engine. The computer uses this information to determine the amount of fuel needed by the engine.
- **ECT (engine coolant temperature) sensor.** This sensor measures the temperature of the engine coolant needed by the computer to determine the amount of fuel and spark advance. This is a major sensor, especially when the engine is cold and when the engine is first started.
- **O2S (oxygen sensor).** This sensor measures the oxygen in the exhaust stream. These sensors are used for fuel control and to check other sensors and systems.
- **TP (throttle position) sensor.** This sensor measures the throttle opening and is used by the computer to control fuel delivery as well as spark advance and the shift points of the automotive transmission/transaxle.
- **VS (vehicle speed) sensor.** This sensor measures the vehicle speed using a sensor located at the output of the transmission/transaxle or by monitoring sensors at the wheel speed sensors.
- **Knock sensor.** The voltage signal from the knock sensor **(KS)** is sent to the PCM. The PCM retards the ignition timing until the knocking stops.

COMPUTER OUTPUTS

A vehicle computer can do just two things.

- Turn a device on.
- Turn a device off.

The computer can turn devices such as fuel injectors on and off rapidly or keep them on for a certain amount of time. Typical output devices include the following:

- **Fuel injectors.** The computer can vary the amount of time the injectors are held open, thereby controlling the amount of fuel supplied to the engine.
- **Ignition timing.** The computer can trigger the signal to the ignition module to fire the spark plugs based on information from the sensors. The spark is advanced when the engine is cold and/or when the engine is operating under light load conditions.
- **Transmission shifting.** The computer provides a ground to the shift solenoids and torque converter clutch solenoid. The operation of the automatic transmission/transaxle is optimized based on vehicle sensor information.
- **Idle speed control.** The computer can pulse the idle speed control (ISC) or idle air control (IAC) device to maintain engine idle speed and to provide an increased idle speed when needed, such as when the air-conditioning system is operating.
- **Evaporative emission control solenoids.** The computer can control the flow of gasoline fumes from the charcoal canister to the engine and seal off the system to perform a fuel system leak detection test as part of the OBD II onboard diagnosis.

MODULE COMMUNICATION AND NETWORKS

Since the 1990s, vehicles use modules to control most of the electrical component operation. A typical vehicle will have 10 or more modules and they communicate with each other over data lines or hard wiring, depending on the application.

SERIAL DATA **Serial data** is data that is transmitted by a series of rapidly changing voltage signals pulsed from low to high or from high to low. Most modules are connected together in a network because of the following advantages:

- A decreased number of wires is needed, thereby saving weight, cost, as well as helping with installation at the factory, and decreased complexity, making servicing easier.
- Common sensor data can be shared with those modules that may need the information, such as vehicle speed, outside air temperature, and engine coolant temperature.

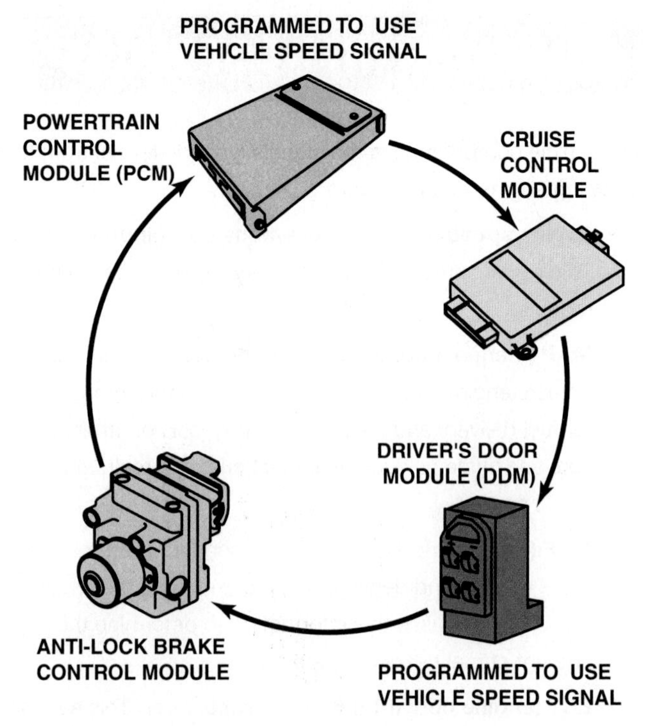

FIGURE 18–14 A network allows all modules to communicate with other modules.

MULTIPLEXING **Multiplexing** is the process of sending multiple signals of information at the same time over a signal wire and then separating the signals at the receiving end. This system of intercommunication of computers or processors is referred to as a **network**. ● **SEE FIGURE 18–14**. By connecting the computers together on a communications network, they can easily share information back and forth. This multiplexing has a number of advantages, including:

- The elimination of redundant sensors and dedicated wiring for these multiple sensors.
- The reduction of the number of wires, connectors, and circuits.
- Addition of more features and option content to new vehicles.
- Weight reduction, increasing fuel economy.
- Allows features to be changed with software upgrades instead of component replacement.

The three most common types of networks used on General Motors vehicles include:

1. **Ring link networks.** In a ring-type network, all modules are connected to each other by a serial data line in a line until all are connected in a ring. ● **SEE FIGURE 18–15.**
2. **Star link.** In a star link network, a serial data line attaches to each module and then each is connected to a central point. This central point is called a **splice pack**, abbreviated SP such as in "SP 306." The splice pack uses a bar to splice all of the serial lines together. Some GM vehicles use two or more splice packs to tie the modules together. When more than one splice pack is used, a serial data line connects one splice pack to the others. In most applications the bus bar used in each splice pack can be removed. When the bus bar is removed a special tool (J 42236) can be installed in place of the removed bus bar. Using this tool, the serial data line for each module can be isolated and tested for a possible problem. Using the special tool at the splice pack makes diagnosing this type of network easier than many others. ● **SEE FIGURE 18–16** for an example of a star link network system.
3. **Ring/star hybrid.** In a ring/star network, the modules are connected using both types of network configuration. Check service information (SI) for details on how this network is connected on the vehicle being diagnosed and always follow the recommended diagnostic steps.

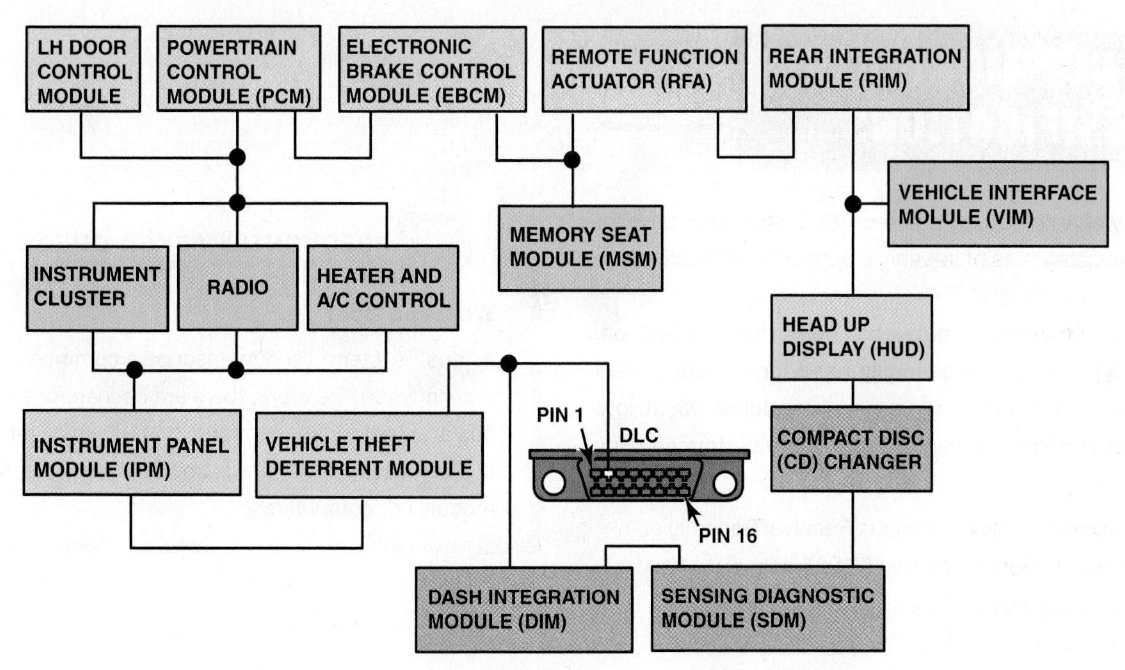

FIGURE 18–15 A ring link network reduces the number of wires it takes to interconnect all of the modules.

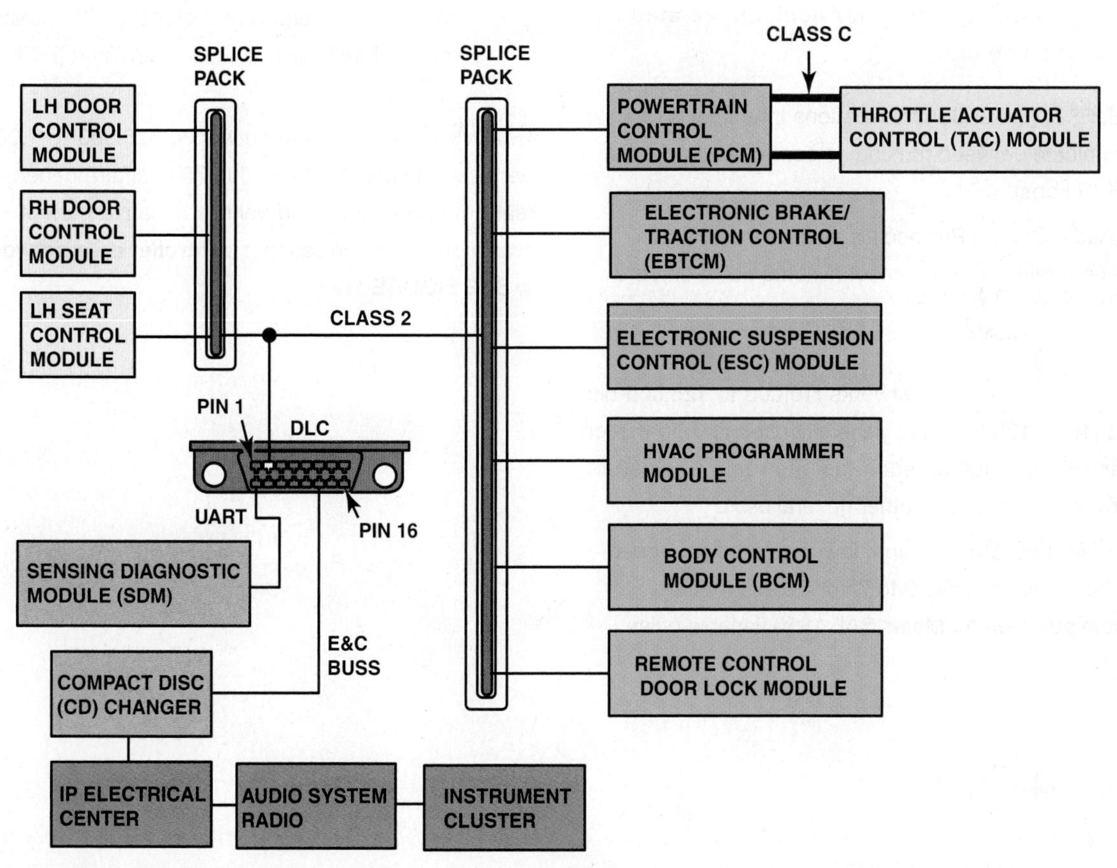

FIGURE 18–16 A star link-type network where all of the modules are connected together using splice packs.

SAE COMMUNICATION CLASSIFICATIONS

The Society of Automotive Engineers (SAE) standards have following three categories of in-vehicle network communications:

CLASS A Low-speed networks (less than 10,000 bits per second [10 kbs]) are generally used for trip computers, entertainment, and other convenience features. Most low-speed Class A communication functions are performed using the following:

- UART (Universal Asynchronous Receive/Transmit) standard used by General Motors (8192 bps).
- CCD (Chrysler Collision Detection) used by Chrysler (7812.5 bps).

  NOTE: The "collision" in CCD-type bus communication refers to the program that avoids conflicts of information exchange within the bus, and does not refer to airbags or other accident-related circuits of the vehicle.

- Chrysler SCI (Serial Communications Interface) is used to communicate between the engine controller and a scan tool (62.5 kbps).
- ACP (Audio Control Protocol) is used for remote control of entertainment equipment (twisted pairs) on Ford vehicles.

CLASS B Medium-speed networks (10,000 to 125,000 bits per second [10 to 125 kbs]) are generally used for information transfer among modules, such as instrument clusters, temperature sensor data, and other general uses.

- General Motors GMLAN, both low- and medium-speed and Class 2, which uses 0 to 7 volt pulses with an available pulse width. Meets SAE 1850 variable pulse width (VPW).
- Chrysler Programmable Communication Interface (PCI). Meets SAE standard J-1850 pulse-width modulated (PWM).
- Ford Standard Corporate Protocol (SCP). Meets SAE standard J-1850 pulse-width modulated (PWM).

CLASS C High-speed networks (125,000 to 1,000,000 bits per second [125,000 to 1,000,000 kbs]) are generally used for real-time powertrain and vehicle dynamic control. Most high-speed bus communication is **controller area network** or **CAN**. ● **SEE FIGURE 18–17.**

? FREQUENTLY ASKED QUESTION

What Is a Bus?

A "bus" is a term used to describe a communication network. Therefore, there are *connections to the bus* and *bus communications*, both of which refer to digital messages being transmitted among electronic modules or computers.

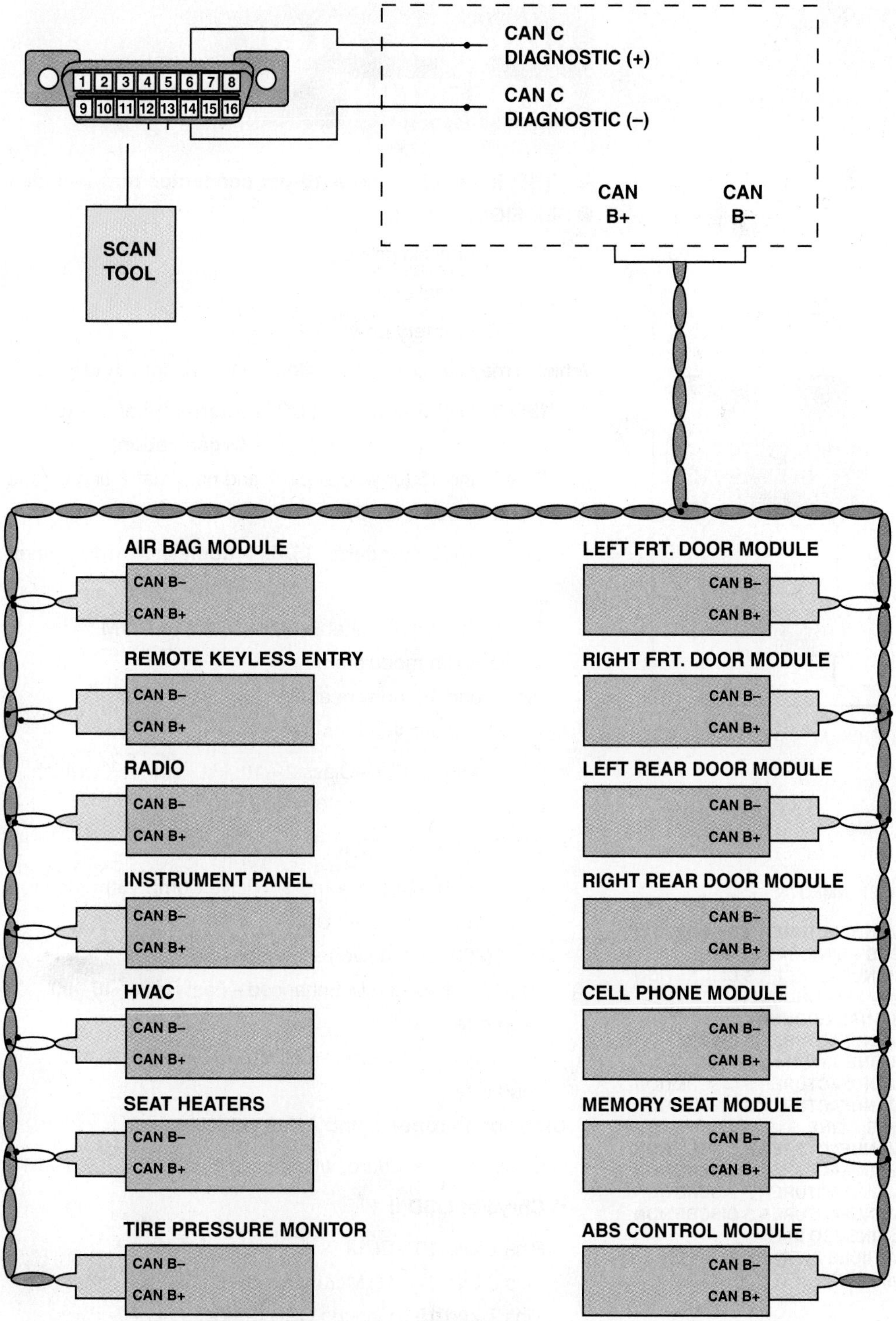

FIGURE 18–17 A typical bus system showing module CAN communications and twisted pairs of wire.

MODULE COMMUNICATION DIAGNOSIS

Most vehicle manufacturers specify that a scan tool be used to diagnose modules and module communications. Always follow the recommended testing procedures, which usually require the use of a factory scan tool.

Some tests of the communication bus (network) and some of the service procedures require the service technician to attach a DMM, set to DC volts, to monitor communications. A variable voltage usually indicates that messages are being sent and received.

Most high-speed bus systems use resistors at each end called **terminating resistors**. These resistors are used to help reduce interference into other systems in the vehicle.

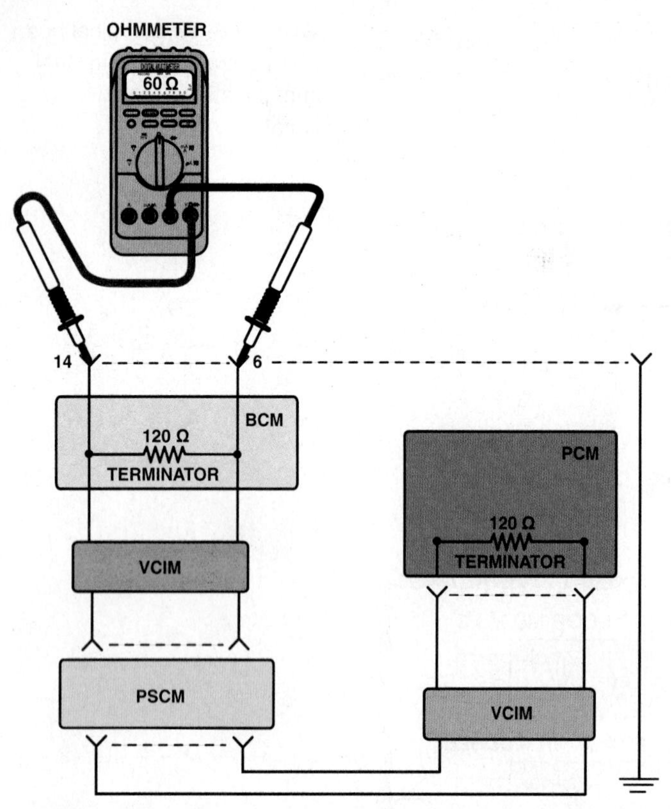

FIGURE 18–18 Checking the terminating resistors using an ohmmeter at the DLC.

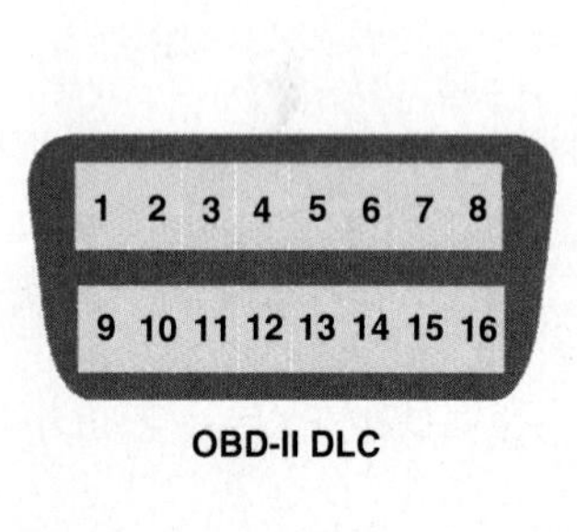

PIN NO.	ASSIGNMENTS
1.	MANUFACTURER'S DISCRETION
2.	BUS + LINE, SAE J1850
3.	MANUFACTURER'S DISCRETION
4.	CHASSIS GROUND
5.	SIGNAL GROUND
6.	MANUFACTURER'S DISCRETION
7.	K LINE, ISO 9141
8.	MANUFACTURER'S DISCRETION
9.	MANUFACTURER'S DISCRETION
10.	BUS – LINE, SAE J1850
11.	MANUFACTURER'S DISCRETION
12.	MANUFACTURER'S DISCRETION
13.	MANUFACTURER'S DISCRETION
14.	MANUFACTURER'S DISCRETION
15.	L LINE, ISO 9141
16.	VEHICLE BATTERY POSITIVE (4A MAX)

FIGURE 18–19 Sixteen-pin OBD II DLC with terminals identified. Scan tools use the power pin (16) ground pin (4) for power so that a separate cigarette lighter plug is not necessary on OBD II vehicles.

Usually two 120-ohm resistors are installed at each end and are therefore connected electrically in parallel. Two 120-ohm resistors connected in parallel would measure 60 ohms if being tested using an ohmmeter. ● **SEE FIGURE 18–18.**

OBD II DATA LINK CONNECTOR

All OBD II vehicles use a 16-pin connector that includes: ● SEE FIGURE 18–19.

Pin 4 = chassis ground

Pin 5 = signal ground

Pin 16 = battery power (4A max)

Vehicles may use one of the following two major standards:

- **ISO 9141-2 Standard (ISO = International Standards Organization)**

  Pins 7 and 15 (or wire at pin 7 and no pin at 2 or a wire at 7 and at 2 and/or 10)

- **SAE J-1850 Standard (SAE = Society of Automotive Engineers)**

  Two types: **VPW** (variable pulse width) or **PWM** (pulse-width modulated)
  Pins 2 and 10 (no wire at pin 7)

General Motors vehicles use:

- SAE J-1850 (VPW—Class 2—10.4 kb) standard, which uses pins 2, 4, 5, and 16 and not 10
- **GM Domestic OBD II**

  Pin 1 and 9—CCM (Comprehensive Component Monitor) slow baud rate—8192 UART
  Pin 1 (2006 and newer)—low-speed GMLAN
  Pins 2 and 10—OEM Enhanced—Fast Rate—40,500 baud rate
  Pins 7 and 15—Generic OBD II—ISO 9141—10,400 baud rate

Chrysler, European, and Asian vehicles use:

- ISO 9141-2 standard, which uses pins 4, 5, 7, 15, and 16
- **Chrysler OBD II**

  Pins 2 and 10—CCM
  Pins 3 and 14—OEM Enhanced—60,500 baud rate
  Pins 7 and 15—Generic OBD II—ISO 9141—10,400 baud rate

Ford vehicles use:

- SAE J-1850 (PWM—41.6 kb) standard, which uses pins 2, 4, 5, 10, and 16
- **Ford Domestic OBD II**

  Pins 2 and 10—CCM
  Pins 6 and 14—OEM Enhanced—Class C—40,500 baud rate
  Pins 7 and 15—Generic OBD II—ISO 9141—10,400 baud rate

SUMMARY

1. The Society of Automotive Engineers (SAE) standard J-1930 specifies that the term "Powertrain Control Module" (PCM) be used for the computer that controls the engine and transmission in a vehicle.
2. The four basic computer functions are input, processing, storage, and output.
3. Read-only memory (ROM) can be programmable (PROM), erasable (EPROM), or electrically erasable (EEPROM).
4. Computer input sensors include engine speed (RPM), MAP, MAF, ECT, O2S, TP, and VS.
5. A computer can only turn a device on or turn a device off, but it can do the operation rapidly.

REVIEW QUESTIONS

1. What part of the vehicle computer is considered to be the brain?
2. What is the difference between volatile and nonvolatile RAM?
3. List four input sensors.
4. List four output devices.

CHAPTER QUIZ

1. What unit of electricity is used as a signal for a computer?
   a. Volt
   b. Ohm
   c. Ampere
   d. Watt
2. The four basic computer functions are _________.
   a. Writing, processing, printing, and remembering
   b. Input, processing, storage, and output
   c. Data gathering, processing, output, and evaluation
   d. Sensing, calculating, actuating, and processing
3. All OBD II vehicles use what type of read-only memory?
   a. ROM
   b. PROM
   c. EPROM
   d. EEPROM
4. The "brain" of the computer is the _________.
   a. PROM
   b. RAM
   c. CPU
   d. AD converter
5. Computer processing speed is measured in _________.
   a. Baud rate
   b. Clock speed (Hz)
   c. Voltage
   d. Bytes
6. Which item is a computer input sensor?
   a. RPM
   b. Throttle position angle
   c. Engine coolant temperature
   d. All of the above
7. Which item is a computer output device?
   a. Fuel injector
   b. Transmission shift solenoid
   c. Evaporative emission control solenoid
   d. All of the above
8. The SAE term for the vehicle computer is _________.
   a. PCM
   b. ECM
   c. ECA
   d. Controller
9. What two things can a vehicle computer actually perform (output)?
   a. Store and process information
   b. Turn something on or turn something off
   c. Calculate and vary temperature
   d. Control fuel and timing only
10. Through which type of circuit analog signals from sensors are changed to digital signals for processing by the computer?
    a. Digital
    b. Analog
    c. AD converter
    d. PROM

chapter 19

ONBOARD DIAGNOSIS

LEARNING OBJECTIVES

After studying this chapter, the reader will be able to:

1. Explain the purpose and function of onboard diagnosis.
2. Explain the information that can be obtained from an on-board diagnostics monitor and the criteria to enable an OBD monitor.
3. List continuous and noncontinuous monitors.
4. Explain the numbering designation of OBD-II diagnostic trouble codes.

This chapter will help you prepare for Engine Performance (A8) ASE certification test content area "E" (Computerized Engine Controls Diagnosis and Repair)

KEY TERMS

California Air Resources Board (CARB) 321
Comprehensive component monitor (CCM) 322
Diagnostic executive 322
Enable criteria 325
Federal Test Procedure (FTP) 321
Freeze-frame 322
Functionality 323
Malfunction indicator lamp (MIL) 321
On-board diagnostics (OBDs) 321
Parameter identification display (PID) 329
Rationality 323
Society of Automotive Engineers (SAE) 326
Task manager 322

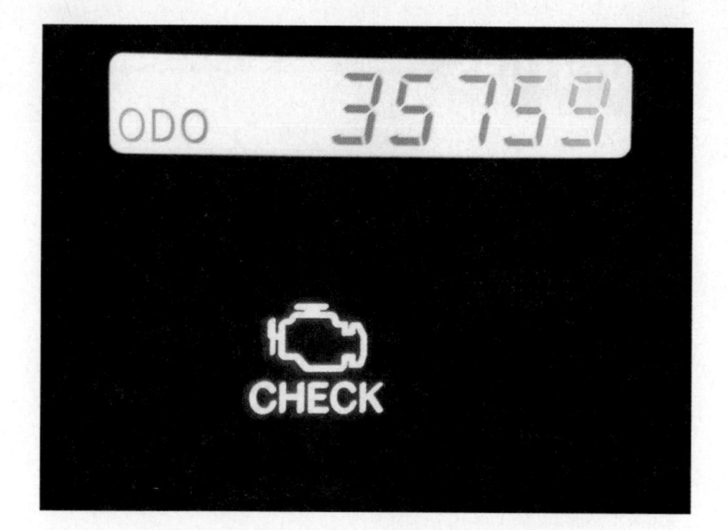

FIGURE 19–1 A typical malfunction indicator lamp (MIL) often labeled "check engine."

ON-BOARD DIAGNOSTICS GENERATION-II (OBD-II) SYSTEMS

PURPOSE AND FUNCTION OF OBD II During the 1980s, most manufacturers began equipping their vehicles with full-function control systems capable of alerting the driver of a malfunction and of allowing the technician to retrieve codes that identify circuit faults. These early diagnostic systems were meant to reduce emissions and speed up vehicle repair.

The automotive industry calls these systems **on-board diagnostics (OBDs)**. The **California Air Resources Board (CARB)** developed the first regulation requiring manufacturers selling vehicles in that state to install OBD. OBD Generation I (OBD I) applies to all vehicles sold in California beginning with the 1988 model year. It specifies the following requirements:

1. An instrument panel warning lamp able to alert the driver of certain control system failures, now called a **malfunction indicator lamp (MIL)**. ● **SEE FIGURES 19–1 AND 19–2**.
2. The system's ability to record and transmit DTCs for emission-related failures.
3. Electronic system monitoring of the HO2S, EGR valve, and evaporative purge solenoid. Although not U.S. EPA-required, during this time most manufacturers also equipped vehicles sold outside of California with OBD I.

By failing to monitor the catalytic converter, the evaporative system for leaks, and the presence of engine misfire, OBD I did not do enough to lower automotive emissions. This led the CARB and the EPA to develop OBD Generation II (OBD II).

OBD-II OBJECTIVES Generally, the CARB defines an OBD-II-equipped vehicle by its ability to do the following:

1. Detect component degradation or a faulty emission-related system that prevents compliance with federal emission standards.
2. Alert the driver of needed emission-related repair or maintenance.
3. Use standardized DTCs and accept a generic scan tool.

These requirements apply to all 1996 and later model light-duty vehicles. The Clean Air Act of 1990 directed the EPA to develop new regulations for OBD. The primary purpose of OBD II is emission-related, whereas the primary purpose of OBD I (1988) was to detect faults in sensors or sensor circuits. OBD-II regulations require that not only sensors be tested but also all exhaust emission control devices, and that they be verified for proper operation.

All new vehicles must pass the **Federal Test Procedure (FTP)** for exhaust emissions while being tested for 505 seconds on rollers that simulate the urban drive cycle around downtown Los Angeles.

NOTE: IM 240 is simply a shorter 240-second version of the 505-second federal test procedure.

FIGURE 19–2 The malfunction indicator lamp might be labeled "service engine soon," which can cause some customer concern if the vehicle had just recently been serviced such as an oil change.

The regulations for OBD-II vehicles state that the vehicle computer must be capable of testing for, and determining, if the exhaust emissions are within 1.5 times the FTP limits. To achieve this goal, the computer must do the following:

1. Test all exhaust emission system components for correct operation.
2. Actively operate the system and measure the results.
3. Continuously monitor all aspects of the engine operation to be certain that the exhaust emissions do not exceed 1.5 times the FTP.
4. Check engine operation for misfire.
5. Turn on the MIL (check engine) if the computer senses a fault in a circuit or system.
6. Record a **freeze-frame**, which is a snapshot of all of the engine data at the time the DTC was set.
7. Flash the MIL if an engine misfire occurs that could damage the catalytic converter.

DIAGNOSTIC EXECUTIVE AND TASK MANAGER

On OBD-II systems, the PCM incorporates a special segment of software. On Ford and GM systems, this software is called the **diagnostic executive**. On Chrysler systems, it is called the **task manager**. This software program is designed to manage the operation of all OBD-II monitors by controlling the sequence of steps necessary to execute the diagnostic tests and monitors.

MONITORS

A monitor is an organized method of testing a specific part of the system. Monitors are simply tests that the computer performs to evaluate components and systems. If a component or system failure is detected while a monitor is running, a DTC will be stored and the MIL illuminated by the second trip. The two types of monitors are continuous and noncontinuous.

CONTINUOUS MONITORS As required conditions are met, continuous monitors begin to run. These continuous monitors will run for the remainder of the vehicle drive cycle. The three continuous monitors are as follows:

- **Comprehensive component monitor (CCM).** This monitor watches the sensors and actuators in the OBD-II system. Sensor values are constantly compared with known-good values stored in the PCM's memory.

  The CCM is an internal program in the PCM designed to monitor a failure in any electronic component or circuit (including emission-related and nonemission-related circuits) that provide input or output signals to the PCM. The PCM considers that an input or output signal is inoperative when a failure exists due to an open circuit, out-of-range value, or if an onboard rationality check fails. If an emission-related fault is detected, the PCM will set a code and activate the MIL (requires two consecutive trips).

  Many PCM sensors and output devices are tested at key-on or immediately after engine start-up. However, some devices, such as the IAC, are only tested by the CCM after the engine meets certain engine conditions. The number of times the CCM must detect a fault before it will activate the MIL depends upon the manufacturer, but most require two consecutive trips to activate the MIL. The components tested by the CCM include:

  Four-wheel-drive low switch

  Brake switch

  Camshaft (CMP) and crankshaft (CKP) sensors

  Clutch switch (manual transmissions/transaxles only)

  Cruise servo switch

  Engine coolant temperature (ECT) sensor

  EVAP purge sensor or switch

  Fuel composition sensor

  Intake air temperature (IAT) sensor

  Knock sensor (KS)

  Manifold absolute pressure (MAP) sensor

Mass airflow (MAF) sensor

Throttle-position (TP) sensor

Transmission temperature sensor

Transmission turbine speed sensor

Vacuum sensor

Vehicle speed (VS) sensor

EVAP canister purge and EVAP purge vent solenoid

Idle air control (IAC) solenoid

Ignition control system

Transmission torque converter clutch solenoid

Transmission shift solenoids

- **Misfire monitor.** This monitor looks at engine misfire. The PCM uses the information received from the crankshaft position sensor (CKP) to calculate the time between the edges of the reluctor, as well as the rotational speed and acceleration. By comparing the acceleration of each firing event, the PCM can determine if a cylinder is not firing correctly.

  Misfire type A. Upon detection of a misfire type A (200 revolutions), which would cause catalyst damage, the MIL will blink once per second during the actual misfire, and a DTC will be stored.

  Misfire type B. Upon detection of a misfire type B (1,000 revolutions), which will exceed 1.5 times the EPA federal test procedure (FTP) standard or cause a vehicle to fail an inspection and maintenance tailpipe emissions test, the MIL will illuminate and a DTC will be stored.

The DTC associated with multiple cylinder misfire for a type A or type B misfire is DTC P0300. The DTCs associated with an individual cylinder misfire for a type A or type B misfire are DTCs P0301, P0302, P0303, P0304, P0305, P0306, P0307, P0308, P0309, and P0310.

- **Fuel trim monitor.** The PCM continuously monitors short- and long-term fuel trim. Constantly updated adaptive fuel tables are stored in long-term memory (KAM), and used by the PCM for compensation due to wear and aging of the fuel system components. The MIL will illuminate when the PCM determines the fuel trim values have reached and stayed at their limits for too long a period of time.

NONCONTINUOUS MONITORS Noncontinuous monitors run (at most) once per vehicle drive cycle. The noncontinuous monitors are as follows:

O2S monitor

O2S heater monitor

Catalyst monitor

EGR monitor

EVAP monitor

Secondary AIR monitor

Transmission monitor

PCV system monitor

Thermostat monitor

Once a noncontinuous monitor has run to completion, it will not be run again until the conditions are met during the next vehicle drive cycle. Also after a noncontinuous monitor has run to completion, the readiness status on your scan tool will show "complete" or "done" for that monitor. Monitors that have not run to completion will show up on your scanner as "incomplete."

OBD-II MONITOR INFORMATION

COMPREHENSIVE COMPONENT MONITOR The circuits and components covered by the comprehensive component monitor (CCM) do not include those directly covered by any other monitors.

However, OBD II also requires that inputs from powertrain components to the PCM be tested for **rationality**, and that outputs to powertrain components from the PCM be tested for **functionality**. Both inputs and outputs are to be checked *electrically*. Rationality means that the PCM checks that the input values are plausible and rational based on the readings from all of the other sensors.

Example:

TP	3 Volt
MAP	18 inch Hg
RPM	700 RPM
PRNDL	Park

NOTE: Comprehensive component monitors are continuous. Therefore enabling conditions do not apply.

The comprehensive component monitor (CCM) performs the following:

- Monitor runs continuously
- Monitor includes sensors, switches, relays, solenoids, and PCM hardware
- All are checked for opens, shorts-to-ground, and shorts-to-voltage
- Inputs are checked for rationality
- Outputs are checked for functionality
- Most are one-trip DTCs
- Freeze-frame is priority 3
- Three consecutive good trips are used to extinguish the MIL
- Forty warm-up cycles are used to erase DTC and freeze-frame
- Two minutes run time without reoccurrence of the fault constitutes a "good trip"

CONTINUOUS RUNNING MONITORS

- Monitors run continuously, only stop if they fail
- Fuel system: rich/lean
- Misfire: catalyst damaging/FTP (emissions)
- Two-trip faults (except early generation catalyst damaging misfire)
- MIL, DTC, freeze-frame after two consecutive faults
- Freeze-frame is priority 2 on first trip
- Freeze-frame is priority 4 on maturing trip
- Three consecutive good trips in a similar condition window are used to extinguish the MIL
- Forty warm-up cycles are used to erase DTC and freeze-frame (80 to erase one-trip failure if similar conditions cannot be met)

ONCE PER TRIP MONITORS

- Monitor runs once per trip, pass or fail
- O_2 response, O_2 heaters, EGR, purge flow EVAP leak, secondary air, catalyst
- Two-trip DTCs
- MIL, DTC, freeze-frame after two consecutive faults
- Freeze-frame is priority 1 on first trip
- Freeze-frame is priority 3 on maturing trip
- Three consecutive good trips are used to extinguish the MIL
- Forty warm-up cycles are used to erase DTC and freeze-frame

EXPONENTIALLY WEIGHTED MOVING AVERAGE MONITORS The exponentially weighted moving average (EWMA) monitor is a mathematical method used to determine performance.

- Catalyst monitor
- EGR monitor
- PCM runs six consecutive failed tests; fails in one trip
- Three consecutive failed tests on next trip, then fails
- Freeze-frame is priority 3
- Three consecutive good trips are used to extinguish the MIL
- Forty warm-up cycles are used to erase DTC and freeze-frame

NONCONTINUOUS MONITORS Noncontinuous monitors run (at most) once per vehicle drive cycle. The noncontinuous monitors are as follows:

O2S monitor
O2S heater monitor
Catalyst monitor
EGR monitor
EVAP monitor
Secondary AIR monitor
Transmission monitor
PCV system monitor
Thermostat monitor

Once a noncontinuous monitor has run to completion, it will not be run again until the conditions are met during the next vehicle drive cycle. Also after a noncontinuous monitor has run to completion, the readiness status on your scan tool will show "complete" or "done" for that monitor. Monitors that have not run to completion will show up on your scanner as "incomplete."

ENABLING CRITERIA

With so many different tests (monitors) to run, the PCM needs an internal director to keep track of when each monitor should run. As mentioned, different manufacturers have different names for this director, such as the diagnostic executive or the task manager. Each monitor has enabling criteria. These criteria are a set of conditions that must be met before the task manager will give the go-ahead for each monitor to run. Most enabling criteria follow simple logic, for example:

- The task manager will not authorize the start of the O2S monitor until the engine has reached operating temperature and the system has entered closed loop.
- The task manager will not authorize the start of the EGR monitor when the engine is at idle, because the EGR is always closed at this time.

Because each monitor is responsible for testing a different part of the system, the enabling criteria can differ greatly from one monitor to the next. The task manager must decide when each monitor should run, and in what order, to avoid confusion.

There may be a conflict if two monitors were to run at the same time. The results of one monitor might also be tainted if a second monitor were to run simultaneously. In such cases, the task manager decides which monitor has a higher priority. Some monitors also depend on the results of other monitors before they can run.

A monitor may be classified as pending if a failed sensor or other system fault is keeping it from running on schedule.

The task manager may suspend a monitor if the conditions are not correct to continue. For example, if the catalyst monitor is running during a road test and the PCM detects a misfire, the catalyst monitor will be suspended for the duration of the misfire.

TRIP A trip is defined as a key-on condition that contains the necessary conditions for a particular test to be performed followed by a key-off. These conditions are called the **enable criteria**. For example, for the EGR test to be performed, the engine must be at normal operating temperature and decelerating for a minimum amount of time. Some tests are performed when the engine is cold, whereas others require that the vehicle be cruising at a steady highway speed.

WARM-UP CYCLE Once a MIL is deactivated, the original code will remain in memory until 40 warm-up cycles are completed without the fault reappearing. A warm-up cycle is defined as a trip with an engine temperature increase of at least 40°F and where engine temperature reaches at least 160°F (71°C).

MIL CONDITION: OFF This condition indicates that the PCM has not detected any faults in an emissions-related component or system, or that the MIL circuit is not working.

MIL CONDITION: ON STEADY This condition indicates a fault in an emissions-related component or system that could affect the vehicle emission levels.

MIL CONDITION: FLASHING This condition indicates a misfire or fuel control system fault that could damage the catalytic converter.

NOTE: In a misfire condition with the MIL on steady, if the driver reaches a vehicle speed and load condition with the engine misfiring at a level that could cause catalyst damage, the MIL would start flashing. It would continue to flash until engine speed and load conditions caused the level of misfire to subside. Then the MIL would go back to the on-steady condition. This situation might result in a customer complaint of a MIL with an intermittent flashing condition.

MIL: OFF The PCM will turn off the MIL if any of the following actions or conditions occurs:

- The codes are cleared with a scan tool.
- Power to the PCM is removed at the battery or with the PCM power fuse for an extended period of time (may be up to several hours or longer).
- A vehicle is driven on three consecutive trips with a warm-up cycle and meets all code set conditions without the PCM detecting any faults.

The PCM will set a code if a fault is detected that could cause tailpipe emissions to exceed 1.5 times the FTP standard; however, the PCM will not deactivate the MIL until the vehicle has been driven on three consecutive trips with vehicle conditions similar to actual conditions present when the fault was detected. This is not merely three vehicle start-ups and trips. It means three trips during which certain engine operating conditions are met so that the OBD-II monitor that found the fault can run again and pass the diagnostic test.

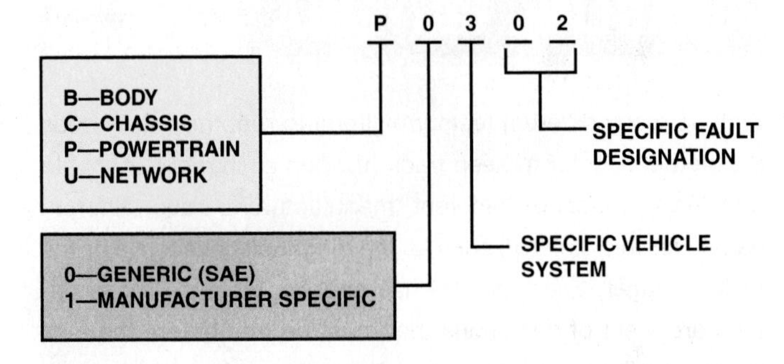

FIGURE 19–3 OBD-II DTC identification format.

OBD-II DTC NUMBERING DESIGNATION

A scan tool is required to retrieve DTCs from an OBD-II vehicle. Every OBD-II scan tool will be able to read all generic **Society of Automotive Engineers (SAE)** DTCs from any vehicle. ● **SEE FIGURE 19–3** for definitions and explanations of OBD alphanumeric DTCs. The diagnostic trouble codes (DTCs) are grouped into major categories, depending on the location of the fault on the system involved.

Pxxx codes—powertrain DTCs (engine, transmission-related faults)

Bxxx codes—body DTCs (accessories, interior-related faults)

Cxxx codes—chassis DTCs (suspension and steering-related faults)

Uxxx codes—network DTCs (module communication-related faults)

DTC NUMBERING EXPLANATION The number in the hundredth position indicates the specific vehicle system or subgroup that failed. This position should be consistent for P0xxx and P1xxx type codes. The following numbers and systems were established by SAE:

- P0100—Air metering and fuel system fault
- P0200—Fuel system (fuel injector only) fault
- P0300—Ignition system or misfire fault
- P0400—Emission control system fault
- P0500—Idle speed control, vehicle speed (VS) sensor fault
- P0600—Computer output circuit (relay, solenoid, etc.) fault
- P0700—Transaxle, transmission faults

NOTE: The tens and ones numbers indicate the part of the system at fault.

TYPES OF DTCS Not all OBD-II DTCs are of the same importance for exhaust emissions. Each type of DTC has different requirements for it to set, and the computer will only turn on the MIL for emissions-related DTCs.

TYPE A CODES A type A DTC is emission-related and will cause the MIL to be turned on the *first trip* if the computer has detected a problem. Engine misfire or a very rich or lean air–fuel ratio, for example, would cause a type A DTC. These codes alert the driver to an emission problem that may cause damage to the catalytic converter.

TYPE B CODES A type B code will be stored and the MIL will be turned on during the *second consecutive trip*, alerting the driver to the fact that a diagnostic test was performed and failed.

NOTE: Type A and B codes are emission-related codes that will cause the lighting of the malfunction indicator lamp (MIL), usually labeled "check engine" or "service engine soon."

TYPE C AND D CODES. Type C and D codes are for use with non-emission-related diagnostic tests; they will cause the lighting of a "service" lamp (if the vehicle is so equipped). Type C codes are also called type C1 codes and D codes are also called type C0 codes.

DIAGNOSTIC TROUBLE CODE PRIORITY CARB has also mandated that all diagnostic trouble codes (DTCs) be stored according to individual priority. DTCs with a higher priority overwrite those with a lower priority. The OBD-II System DTC Priority is listed below.

Priority 0—Non-emission-related codes

Priority 1—One-trip failure of two-trip fault for non-fuel, non-misfire codes

Priority 2—One-trip failure of two-trip fault for fuel or misfire codes

Priority 3—Two-trip failure or matured fault of non-fuel, non-misfire codes

Priority 4—Two-trip failure or matured fault for fuel or misfire codes

Monitor Name	Monitor Type (How Often It Completes)	Number of Faults on Separate Trips to Set a Pending DTC	Number of Separate Consecutive Trips to Light MIL, Store a DTC	Number of Trips with No Faults to Erase a Maturing DTC	Number of Trips with No Fault to Turn the MIL Off	Number of Warm-Up Cycles to Erase DTC after MIL Is Turned Off
CCM	Continuous (when trip conditions allow it)	1	2	1–Trip	3–Trips	40
Catalyst	Once per drive cycle	1	3	1–Trip	3–OBD-II drive cycle	40
Misfire Type A	Continuous		1	1–Trip	3–Similar conditions	80
Misfire Type B	Continuous	1	2	1–Trip	3–Similar conditions	80
Fuel System	Continuous	1	2	1–Trip	3–Similar conditions	80
Oxygen Sensor	Once per trip	1	2	1–Trip	3–Trips	40
EGR	Once per trip	1	2	1–Trip	3–Trips	40
EVAP	Once per trip	1	1	1–Trip	3–Trips	40
AIR	Once per trip	1	2	1–Trip	3–Trips	40

CHART 19–1

PCM Determination of Faults Chart

OBD-II FREEZE-FRAME

To assist the service technician, OBD II requires the computer to take a "snapshot" or freeze-frame of all data at the instant an emission-related DTC is set. A scan tool is required to retrieve this data.

NOTE: Although OBD II requires that just one freeze-frame of data be stored, the instant an emission-related DTC is set, vehicle manufacturers usually provide expanded data about the DTC beyond that required. However, retrieving this enhanced data usually requires the use of the vehicle-specific scan tool.

Freeze-frame items include:

- Calculated load value
- Engine speed (RPM)
- Short-term and long-term fuel trim percent
- Fuel system pressure (on some vehicles)
- Vehicle speed (mph)
- Engine coolant temperature
- Intake manifold pressure
- Closed-open-loop status
- Fault code that triggered the freeze-frame
- If a misfire code is set, identify which cylinder is misfiring

A DTC should not be cleared from the vehicle computer memory unless the fault has been corrected and the technician is so directed by the diagnostic procedure. If the problem that caused the DTC to be set has been corrected, the computer will automatically clear the DTC after 40 consecutive warm-up cycles with no further faults detected (misfire and excessively rich or lean condition codes require 80 warm-up cycles). The codes can also be erased by using a scan tool. ● **SEE CHART 19–1**.

NOTE: Disconnecting the battery may not erase OBD-II DTCs or freeze-frame data. Most vehicle manufacturers recommend using a scan tool to erase DTCs rather than disconnecting the battery, because the memory for the radio, seats, and learned engine operating parameters is lost if the battery is disconnected.

FREQUENTLY ASKED QUESTION

What Are Pending Codes?

Pending codes are set when operating conditions are met and the component or circuit is not within the normal range, yet the conditions have not yet been met to set a DTC. For example, a sensor may require two consecutive faults before a DTC is set. If a scan tool displays a pending code or a failure, a driveability concern could also be present. The pending code can help the technician to determine the root cause before the customer complains of a check engine light indication.

ENABLING CONDITIONS OR CRITERIA

These are the exact engine operating conditions required for a diagnostic monitor to run.

Example:

Specific RPM
Specific ECT, MAP, run time, VSS, etc.

PENDING Under some situations the PCM will not run a monitor if the MIL is illuminated and a fault is stored from another monitor. In these situations, the PCM postpones monitors pending a resolution of the original fault. The PCM does not run the test until the problem is remedied.

For example, when the MIL is illuminated for an oxygen sensor fault, the PCM does not run the catalyst monitor until the oxygen sensor fault is remedied. Since the catalyst monitor is based on signals from the oxygen sensor, running the test would produce inaccurate results.

CONFLICT There are also situations when the PCM does not run a monitor if another monitor is in progress. In these situations, the effects of another monitor running could result in an erroneous failure. If this conflict is present, the monitor is not run until the conflicting condition passes. Most likely, the monitor will run later after the conflicting monitor has passed.

For example, if the fuel system monitor is in progress, the PCM does not run the EGR monitor. Since both tests monitor changes in air–fuel ratio and adaptive fuel compensation, the monitors conflict with each other.

SUSPEND Occasionally, the PCM may not allow a two-trip fault to mature. The PCM will suspend the maturing fault if a condition exists that may induce erroneous failure. This prevents illuminating the MIL for the wrong fault and allows more precise diagnosis.

For example, if the PCM is storing a one-trip fault for the oxygen sensor and the EGR monitor, the PCM may still run the EGR monitor but will suspend the results until the oxygen sensor monitor either passes or fails. At that point, the PCM can determine if the EGR system is actually failing or if an oxygen sensor is failing.

PCM TESTS

RATIONALITY TEST While input signals to the PCM are constantly being monitored for electrical opens and shorts, they are also tested for rationality. This means that the input signal is compared against other inputs and information to see if it makes sense under the current conditions.

PCM sensor inputs that are checked for rationality include:

- MAP sensor
- O_2 sensor
- ECT
- Camshaft position sensor (CMP)
- VS sensor
- Crankshaft position sensor (CKP)
- IAT sensor
- TP sensor
- Ambient air temperature sensor
- Power steering switch
- O_2 sensor heater
- Engine controller
- Brake switch
- P/N switch
- Transmission controls

FUNCTIONALITY TEST A functionality test refers to PCM inputs checking the operation of the outputs.

Example:

PCM commands the IAC open; expected change in engine RPM is not seen
IAC 60 counts
RPM 700 RPM

PCM outputs that are checked for functionality include:

- EVAP canister purge solenoid
- EVAP purge vent solenoid
- Cooling fan
- Idle air control solenoid
- Ignition control system
- Transmission torque converter clutch solenoid
- Transmission shift solenoids (A,B,1–2, etc.)

ELECTRICAL TEST Refers to the PCM check of both input and outputs for the following:

- Open
- Shorts
- Ground

Example:

ECT
Shorted high (input to PCM) above capable voltage, that is, 5 volt sensor with 12 volt input to PCM would indicate a short to voltage or a short high.

	Monitor Type	Conditions to Set DTC and Illuminate MIL	Extinguish MIL	Clear DTC Criteria	Applicable DTC
Comprehensive Monitor	Continuous 1-trip monitor	(See note below) Input and output failure— rationally, functionally, electrically	3 consecutive pass trips	40 warm-up cycles	P0123

NOTE: The number of times the comprehensive component monitor must detect a fault depends on the vehicle manufacturer. On some vehicles, the comprehensive component monitor will activate the MIL as soon as it detects a fault. On other vehicles, the comprehensive component monitor must fail two times in a row.

- Freeze-frame captured on first-trip failure.
- Enabling conditions: Many PCM sensors and output devices are tested at key-on or immediately after engine start-up. However, some devices (ECT, idle speed control) are only tested by the comprehensive component monitor after the engine meets particular engine conditions.
- Pending: No pending condition
- Conflict: No conflict conditions
- Suspend: No suspend conditions

GLOBAL OBD-II

All OBD-II vehicles must be able to display data on a global (also called *generic*) scan tool under nine different modes of operation. These modes include:

Mode One Current power train data (**parameter identification** or **PID**)

Mode Two Freeze-frame data

Mode Three Diagnostic trouble codes

Mode Four Clear and reset diagnostic trouble codes (DTCs), freeze-frame data, and readiness status monitors for noncontinuous monitors only

Mode Five Oxygen sensor monitor test results

Mode Six Onboard monitoring of test results for noncontinuously monitored systems

Mode Seven Onboard monitoring of test results for continuously monitored systems

Mode Eight Bidirectional control of onboard systems

Mode Nine Module identification

Mode Ten Permanent diagnostic trouble codes (DTCs). These codes cannot be erased using a scan tool or a battery disconnect but instead can only be cleared by the PCM after it has performed systems checks to verify that the problem has been corrected.

The global (generic) data is used by most state emission programs.

ACCESSING GLOBAL OBD II

Global (generic) OBD II is used by inspectors where emission testing is performed. Aftermarket scan tools are designed to retrieve global OBD II; however, some original equipment scan tools, such as the Tech 2 used on General Motors vehicles, are not able to retrieve the information without special software.

Global OBD II is accessible using ISO-9141-2, KWP 2000, J1850 PWM, J1850 VPW, and CAN.

SNAP-ON SOLUS From the main menu select "Generic OBD II/EOBD" and then follow the on-screen instructions to select the desired test.

SNAP-ON MODIS Select the scanner using the down arrow key and then select "Global OBD II." Follow on-screen instructions to get to "start communication" and then to the list of options to view.

OTC GENISYS From the main menu select "Global OBD II" and then follow the on-screen instructions. Select "special tests" to get access to mode $06 information and parameters.

MASTER TECH From the main menu, select "Global OBD II." At the next screen, select "OBD II functions," then "system tests," and then "other results" to obtain mode $06 data. ● SEE FIGURE 19–4.

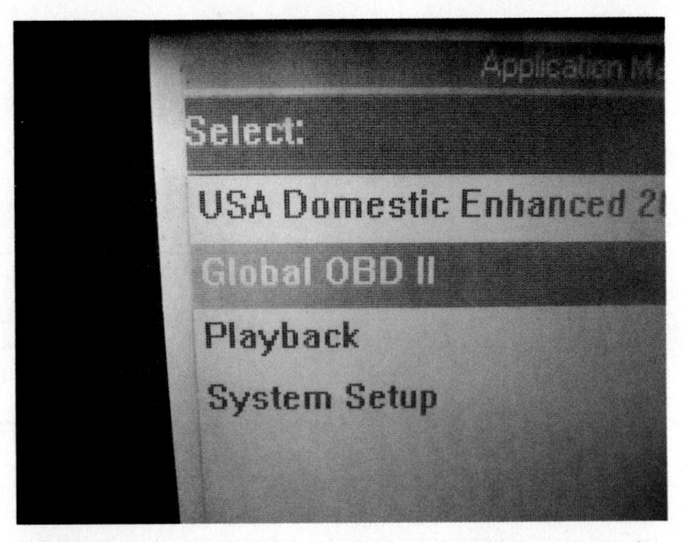

FIGURE 19–4 Global OBD II can be accessed from the main menu on all aftermarket and some original equipment scan tools.

DIAGNOSING PROBLEMS USING MODE SIX

Mode six information can be used to diagnose faults by following three steps:

1. Check the monitor status before starting repairs. This step will show how the system failed.
2. Look at the component or parameter that triggered the fault. This step will help pin down the root cause of the failure.
3. Look to the monitor enable criteria, which will show what it takes to fail or pass the monitor.

? FREQUENTLY ASKED QUESTION

How Can You Tell Generic from Factory?

When using a scan tool on an OBD II–equipped vehicle, if the display asks for make, model, and year, then the factory or enhanced part of the PCM is being accessed. If the generic or global part of the PCM is being scanned, then there is no need to know the vehicle details.

SUMMARY

1. If the MIL is on, retrieve the DTC and follow the manufacturer's recommended procedure to find the root cause of the problem.
2. All monitors must have the enable criteria achieved before a test is performed.
3. OBD-II vehicles use common generic DTCs.
4. OBD II includes generic (SAE), as well as vehicle manufacturer-specific DTCs, and data display.

REVIEW QUESTIONS

1. What does the PCM do during a trip to test emission-related components?
2. What is the difference between a type A and type B OBD-II DTC?
3. What is the difference between a trip and a warm-up cycle?
4. What could cause the MIL to flash?

CHAPTER QUIZ

1. A freeze-frame is generated on an OBD-II vehicle ________.
   a. When a type C or D diagnostic trouble code is set
   b. When a type A or B diagnostic trouble code is set
   c. Every other trip
   d. When the PCM detects a problem with the O2S
2. An ignition misfire or fuel mixture problem is an example of what type of DTC?
   a. Type A
   b. Type B
   c. Type C
   d. Type D
3. The comprehensive component monitor checks computer-controlled devices for ________.
   a. opens
   b. shorts-to-ground
   c. rationality
   d. All of the above
4. OBD II has been on all passenger vehicles in the United States since ________.
   a. 1986
   b. 1991
   c. 1996
   d. 2000
5. Which is a continuous monitor?
   a. Fuel system monitor
   b. EGR monitor
   c. Oxygen sensor monitor
   d. Catalyst monitor
6. DTC P0302 is a ________.
   a. Generic DTC
   b. Vehicle manufacturer-specific DTC
   c. Idle speed-related DTC
   d. Transmission/transaxle-related DTC
7. Global (generic) OBD II contains some data in what format?
   a. Plain English
   b. Hexadecimal
   c. Roman numerals
   d. All of the above
8. By looking at the way diagnostic trouble codes are formatted, which DTC could indicate that the gas cap is loose or defective?
   a. P0221
   b. P1301
   c. P0442
   d. P1603
9. The computer will automatically clear a DTC if there are no additional detected faults after ________.
   a. Forty consecutive warm-up cycles
   b. Eighty warm-up cycles
   c. Two consecutive trips
   d. Four key-on/key-off cycles
10. A pending code is set when a fault is detected on ______.
    a. A one-trip fault item
    b. The first fault of a two-trip failure
    c. The catalytic converter efficiency
    d. Thermostat problem (too long to closed-loop status)

chapter 20

TEMPERATURE SENSORS

LEARNING OBJECTIVES

After studying this chapter, the reader will be able to:

1. Describe the purpose and function of engine coolant temperature sensors.
2. Describe how to inspect and test temperature sensors.
3. Diagnose emissions and drivability problems resulting from malfunctions in the intake air temperature control systems.
4. Discuss how automatic transmission fluid temperature sensor values transmission operation.

This chapter will help you prepare for Engine Repair (A8) ASE certification test content area "E" (Computerized Engine Controls Diagnosis and Repair).

KEY TERMS

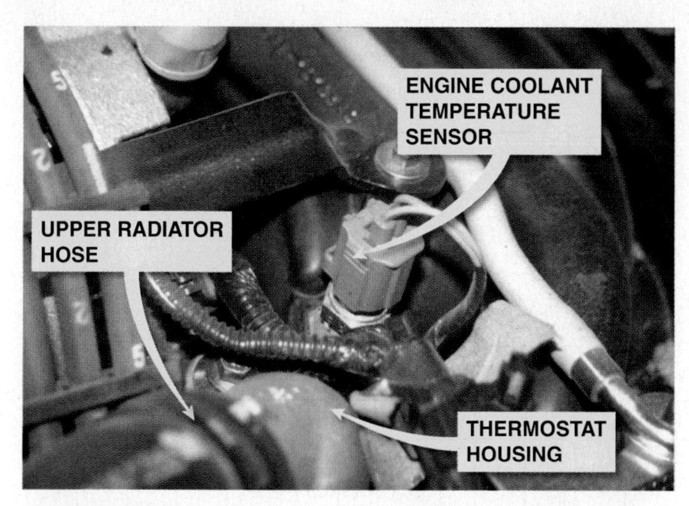

FIGURE 20–1 A typical engine coolant temperature (ECT) sensor. ECT sensors are located near the thermostat housing on most engines.

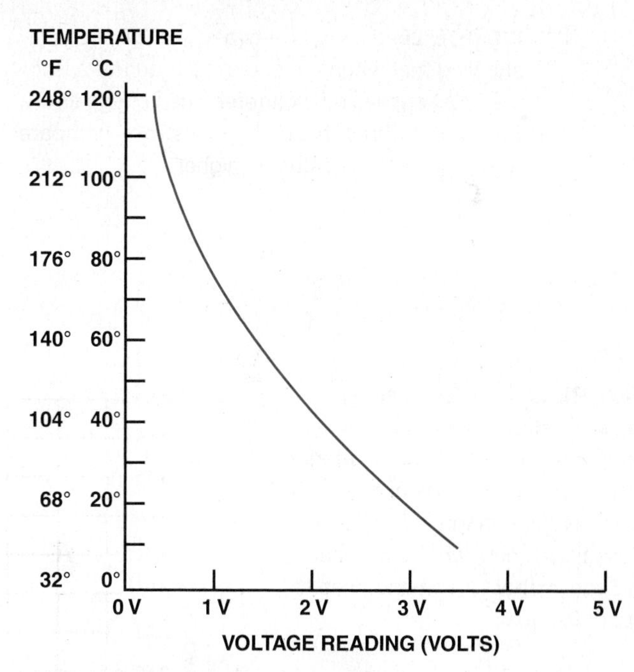

FIGURE 20–2 A typical ECT sensor temperature versus voltage curve.

ENGINE COOLANT TEMPERATURE SENSORS

PURPOSE AND FUNCTION Computer-equipped vehicles use an **engine coolant temperature (ECT)** sensor. When the engine is cold, the fuel mixture must be richer to prevent stalling and engine stumble. When the engine is warm, the fuel mixture can be leaner to provide maximum fuel economy with the lowest possible exhaust emissions. Because the computer controls spark timing and fuel mixture, it will need to know the engine temperature. An engine coolant temperature sensor screwed into the engine coolant passage will provide the computer with this information. ● **SEE FIGURE 20–1**. This will be the most important (high-authority) sensor while the engine is cold. The ignition timing can also be tailored to engine (coolant) temperature. A hot engine cannot have the spark timing as far advanced as can a cold engine. The ECT sensor is also used as an important input for the following:

- Idle air control (IAC) position
- Oxygen sensor closed-loop status
- Canister purge on/off times
- Idle speed

ECT SENSOR CONSTRUCTION Engine coolant temperature sensors are constructed of a semiconductor material that decreases in resistance as the temperature of the sensor increases. Coolant sensors have very high resistance when the coolant is cold and low resistance when the coolant is hot. This is referred to as having a **negative temperature coefficient (NTC)**, which is opposite to the situation with most other electrical components. ● **SEE FIGURE 20–2**. Therefore, if the coolant sensor has a poor connection (high resistance) at the wiring connector, the computer will supply a richer-than-normal fuel mixture based on the resistance of the coolant sensor. Poor fuel economy and a possible-rich code can be caused by a defective sensor or high resistance in the sensor wiring. If the sensor was shorted or defective and had too low a resistance, a leaner-than-normal fuel mixture would be supplied to the engine. A too-lean fuel mixture can cause driveability problems and a possible-lean computer code.

STEPPED ECT CIRCUITS Some vehicle manufacturers use a step-up resistor to effectively broaden the range of the ECT sensor. Chrysler and General Motors vehicles use the same sensor as a non-stepped ECT circuit, but instead apply the sensor voltage through two different resistors.

- When the temperature is low, usually below 120°F (50°C), the ECT sensor voltage is applied through a high-value resistor inside the PCM.
- When the temperature is high, usually above 120°F (50°C), the ECT sensor voltage is applied through a much lower resistance value inside the PCM. ● **SEE FIGURE 20–3**.

The purpose of this extra circuit is to give the PCM a more accurate reading of the engine coolant temperature compared to the same sensor with only one circuit. ● **SEE FIGURE 20–4**.

FIGURE 20–3 A typical two-step ECT circuit showing that when the coolant temperature is low, the PCM applies a 5 volt reference voltage to the ECT sensor through a higher resistance compared to when the temperature is higher.

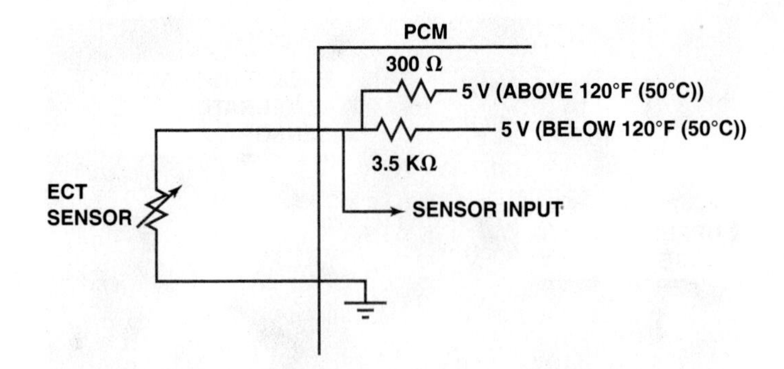

FIGURE 20–4 The transition between steps usually occurs at a temperature that would not interfere with cold engine starts or the cooling fan operation. In this example, the transition occurs when the sensor voltage is about 1 volt and rises to about 3.6 volts.

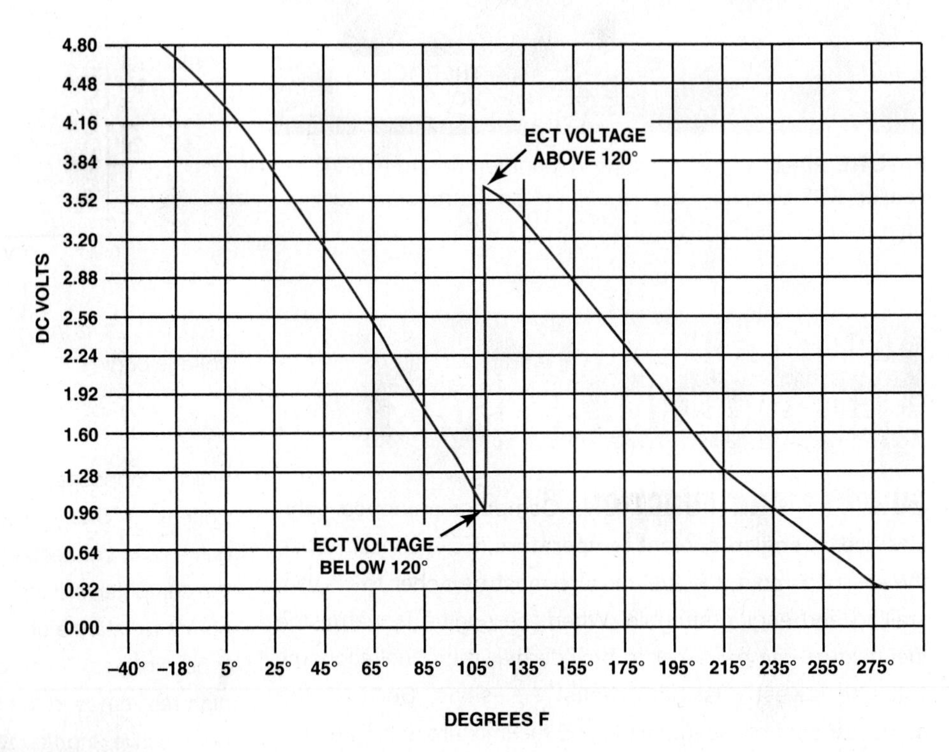

TESTING THE ENGINE COOLANT TEMPERATURE SENSOR

TESTING THE ENGINE COOLANT TEMPERATURE BY VISUAL INSPECTION The correct functioning of the engine coolant temperature sensor depends on the following items that should be checked or inspected:

- **Properly filled cooling system.** Check that the radiator reservoir bottle is full and that the radiator itself is filled to the top.

  CAUTION: Be sure that the radiator is cool before removing the radiator cap to avoid being scalded by hot coolant.

  The ECT sensor must be submerged in coolant to be able to indicate the proper coolant temperature.

- **Proper pressure maintained by the radiator cap.** If the radiator cap is defective and cannot allow the cooling system to become pressurized, air pockets could develop. These air pockets could cause the engine to operate at a hotter-than-normal temperature and prevent proper temperature measurement, especially if the air pockets occur around the sensor.
- **Proper antifreeze-water mixture.** Most vehicle manufacturers recommend a 50/50 mixture of antifreeze and water as the best compromise between freezing protection and heat transfer ability.
- **Proper operation of the cooling fan.** If the cooling fan does not operate correctly, the engine may overheat.

TESTING THE ECT USING A MULTIMETER Both the resistance (in ohms) and the voltage drop across the sensor can be measured and compared with specifications.

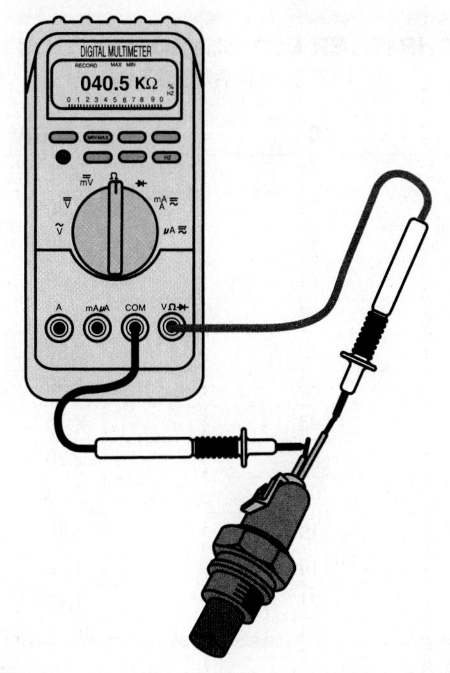

FIGURE 20–5 Measuring the resistance of the ECT sensor. The resistance measurement can then be compared with specifications. (Courtesy of Fluke Corporation)

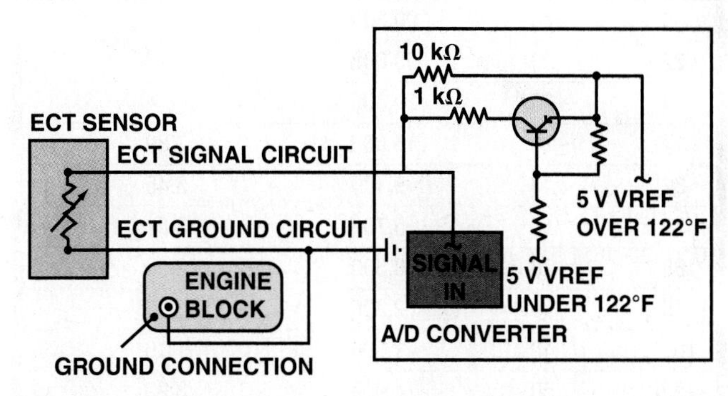

FIGURE 20–6 When the voltage drop reaches approximately 1.20 volts, the PCM turns on a transistor. The transistor connects a 1 kΩ resistor in parallel with the 10 kΩ resistor. Total circuit resistance now drops to around 909 ohms. This function allows the PCM to have full binary control at cold temperatures up to approximately 122°F, and a second full binary control at temperatures greater than 122°F.

● **SEE FIGURE 20–5**. See the following charts showing examples of typical engine coolant temperature sensor specifications. Some vehicles use the PCM to attach another resistor in the ECT circuit to provide a more accurate measure of the engine temperature. ● **SEE FIGURE 20–6**.

If resistance values match the approximate coolant temperature and there is still a coolant sensor trouble code, the problem is generally in the wiring between the sensor and the computer. Always consult the manufacturers' recommended procedures for checking this wiring. If the resistance values do not match, the sensor may need to be replaced.

GENERAL MOTORS ECT SENSOR WITHOUT PULL-UP RESISTOR

°F	°C	OHMS	VOLTAGE DROP ACROSS SENSOR
–40	–40	100,000 +	4.95
18	–8	14,628	4.68
32	0	9,420	4.52
50	10	5,670	4.25
68	20	3,520	3.89
86	30	2,238	3.46
104	40	1,459	2.97
122	50	973	2.47
140	60	667	2.00
158	70	467	1.59
176	80	332	1.25
194	90	241	0.97
212	100	177	0.75

GENERAL MOTORS ECT SENSOR WITH PULL-UP RESISTOR

°F	°C	OHMS	VOLTAGE DROP ACROSS SENSOR
−40	−40	100,000	5
−22	−30	53,000	4.78
−4	−20	29,000	4.34
14	−10	16,000	3.89
32	0	9,400	3.45
50	10	5,700	3.01
68	20	3,500	2.56
86	30	2,200	1.80
104	40	1,500	1.10
122	50	970	3.25
140	60	670	2.88
158	70	470	2.56
176	80	330	2.24
194	90	240	1.70
212	100	177	1.42
230	110	132	1.15
248	120	100	0.87

FORD ECT SENSOR

°F	°C	RESISTANCE (Ω)	VOLTAGE (V)
50	10	58,750	3.52
68	20	37,300	3.06
86	30	24,270	2.26
104	40	16,150	2.16
122	50	10,970	1.72
140	60	7,600	1.35
158	70	5,370	1.04
176	80	3,840	0.80
194	90	2,800	0.61
212	100	2,070	0.47
230	110	1,550	0.36
248	120	1,180	0.28

CHRYSLER ECT SENSOR WITHOUT PULL-UP RESISTOR

°F	°C	VOLTAGE (V)
130	54	3.77
140	60	3.60
150	66	3.40
160	71	3.20
170	77	3.02
180	82	2.80
190	88	2.60
200	93	2.40
210	99	2.20
220	104	2.00
230	110	1.80
240	116	1.62
250	121	1.45

CHRYSLER ECT SENSOR WITH PULL-UP RESISTOR

°F	°C	VOLTS
−20	−29	4.70
−10	−23	4.57
0	−18	4.45
10	−12	4.30
20	−7	4.10
30	−1	3.90
40	4	3.60
50	10	3.30
60	16	3.00
70	21	2.75
80	27	2.44
90	32	2.15
100	38	1.83
		PULL-UP RESISTOR SWITCHED BY PCM
110	43	4.20
120	49	4.10
130	54	4.00
140	60	3.60
150	66	3.40
160	71	3.20
170	77	3.02
180	82	2.80
190	88	2.60
200	93	2.40
210	99	2.20
220	104	2.00
230	110	1.80
240	116	1.62
250	121	1.45

NISSAN ECT SENSOR

°F	°C	RESISTANCE (Ω)
14	–10	7,000–11,400
68	20	2,100–2,900
122	50	680–1,000
176	80	260–390
212	100	180–200

MERCEDES ECT

°F	°C	VOLTAGE (DCV)
60	20	3.5
86	30	3.1
104	40	2.7
122	50	2.3
140	60	1.9
158	70	1.5
176	80	1.2
194	90	1.0
212	100	0.8

EUROPEAN BOSCH ECT SENSOR

°F	°C	RESISTANCE (Ω)
32	0	6,500
50	10	4,000
68	20	3,000
86	30	2,000
104	40	1,500
122	50	900
140	60	650
158	70	500
176	80	375
194	90	295
212	100	230

HONDA ECT SENSOR (RESISTANCE CHART)

°F	°C	RESISTANCE (Ω)
0	–18	15,000
32	0	5,000
68	20	3,000
104	40	1,000
140	60	500
176	80	400
212	100	250

HONDA ECT SENSOR (VOLTAGE CHART)

°F	°C	VOLTAGE (V)
0	–18	4.70
10	–12	4.50
20	–7	4.29
30	–1	4.10
40	4	3.86
50	10	3.61
60	16	3.35
70	21	3.08
80	27	2.81
90	32	2.50
100	38	2.26
110	43	2.00
120	49	1.74
130	54	1.52
140	60	1.33
150	66	1.15
160	71	1.00
170	77	0.88
180	82	0.74
190	88	0.64
200	93	0.55
210	99	0.47

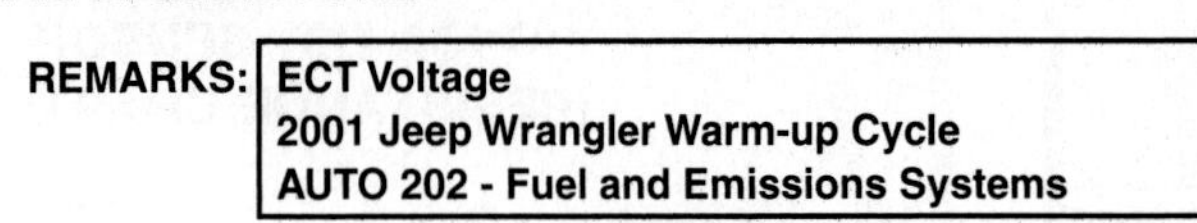

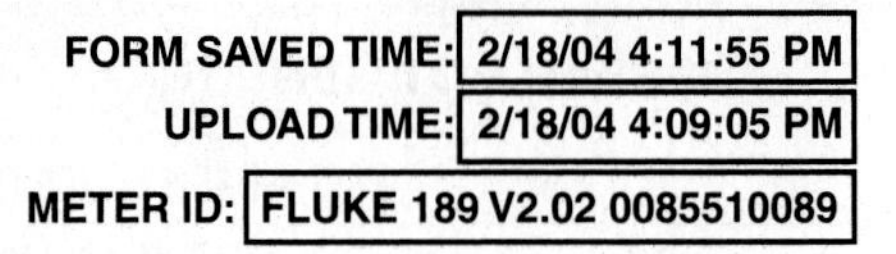

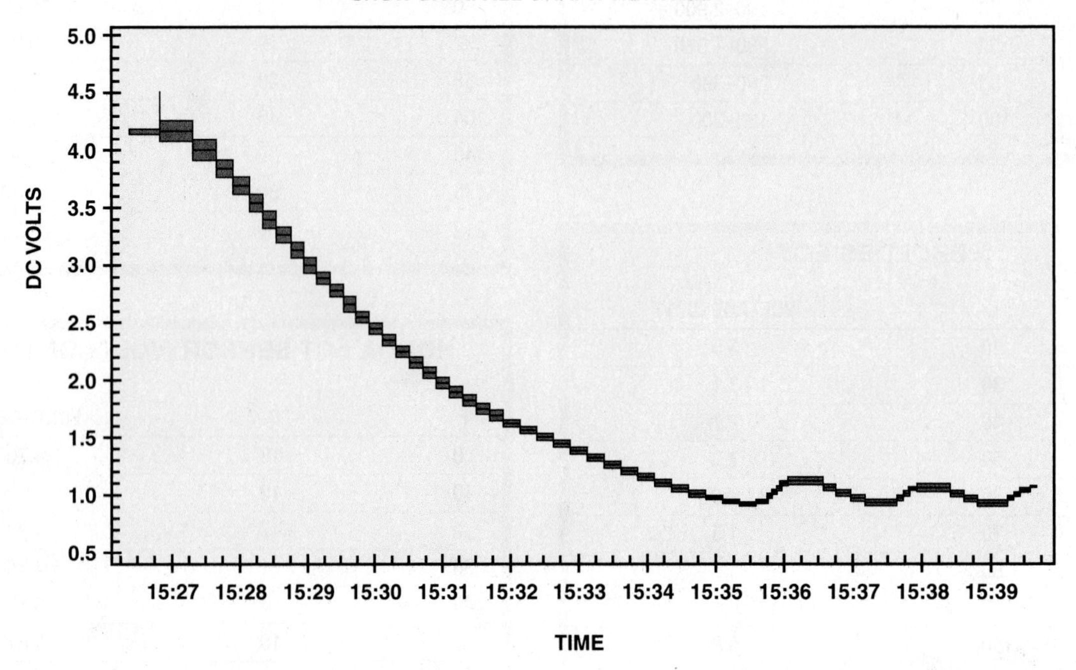

FIGURE 20–7 An ECT sensor being tested using a digital meter set to DC volts. A chart showing the voltage decrease of the ECT sensor as the temperature increases from a cold start. The bumps at the bottom of the waveform represent temperature decreases when the thermostat opens and is controlling coolant temperature.

Normal operating temperature varies with vehicle make and model. Some vehicles are equipped with a thermostat with an opening temperature of 180°F (82°C), whereas other vehicles use a thermostat that is 195°F (90°C) or higher. Before replacing the ECT sensor, be sure that the engine is operating at the temperature specified by the manufacturer. Most manufacturers recommend checking the ECT sensor after the cooling fan has cycled twice, indicating a fully warmed engine. To test for voltage at the ECT sensor, select DC volts on a digital meter and carefully back probe the sensor wire and read the voltage. ● **SEE FIGURE 20–7.**

NOTE: Many manufacturers install another resistor in parallel inside the computer to change the voltage drop across the ECT sensor. This is done to expand the scale of the ECT sensor and to make the sensor more sensitive. Therefore, if measuring *voltage* at the ECT sensor, check with the service manual for the proper voltage at each temperature.

TESTING THE ECT SENSOR USING A SCAN TOOL Follow the scan tool manufacturer's instructions and connect a scan tool to the data link connector (DLC) of the vehicle. Comparing the temperature of the engine coolant as displayed on a scan tool with the actual temperature of the engine is an excellent method to test an engine coolant temperature sensor.

1. Record the scan tool temperature of the coolant (ECT).
2. Measure the actual temperature of the coolant using an infrared pyrometer or contact-type temperature probe.

NOTE: Often the coolant temperature gauge in the dash of the vehicle can be used to compare with the scan tool temperature. Although not necessarily accurate, it may help to diagnose a faulty sensor, especially if the temperature shown on the scan tool varies greatly from the temperature indicated on the dash gauge.

The maximum difference between the two readings should be 10°F (5°C). If the actual temperature varies by more than 10°F from the temperature indicated on the scan tool, check the ECT sensor wiring and connector for damage or corrosion. If the connector and wiring are okay, check the sensor with a DVOM for resistance and compare with the actual engine temperature chart. If that checks out okay, check the computer.

NOTE: Some manufacturers use two coolant sensors, one for the dash gauge and one for the computer.

FIGURE 20–8 The IAT sensor on this General Motors 3800 V-6 engine is in the air passage duct between the air cleaner housing and the throttle body.

TECH TIP

Quick and Easy ECT Test

To check that the wiring and the computer are functioning, regarding the ECT sensor, connect a scan tool and look at the ECT temperature display.

STEP 1 Unplug the connector from the ECT sensor. The temperature displayed on the scan tool should read about –40°C.

NOTE: –40° Celsius is also –40° Fahrenheit. This is the point where both temperature scales meet.

STEP 2 With the connector still removed from the ECT sensor, use a fused jumper lead and connect the two terminals of the connector together. The scan tool should display about 285°F (140°C).

This same test procedure will work for the IAT and most other temperature sensors.

INTAKE AIR TEMPERATURE SENSOR

PURPOSE AND FUNCTION The intake air temperature (IAT) sensor is a negative temperature coefficient (NTC) thermistor that decreases in resistance as the temperature of the sensor increases. The IAT sensor can be located in one of the following locations:

- In the air cleaner housing
- In the air duct between the air filter and the throttle body, as shown in ● **FIGURE 20–8**
- Built into the mass air flow (MAF) or airflow sensor
- Screwed into the intake manifold where it senses the temperature of the air entering the cylinders

NOTE: An IAT installed in the intake manifold is the most likely to suffer damage due to an engine backfire, which can often destroy the sensor.

The purpose and function of the intake air temperature sensor is to provide the engine computer (PCM) the temperature of the air entering the engine. The IAT sensor information is used for fuel control (adding or subtracting fuel) and spark timing, depending on the temperature of incoming air.

The IAT sensor is used as an input sensor by the PCM to control many functions including:

- If the air temperature is low, the PCM will modify the amount of fuel delivery and add fuel.
- If the air temperature is high, the PCM will subtract the calculated amount of fuel.
- Spark timing is also changed, depending on the temperature of the air entering the engine. The timing is advanced if the temperature is cold and retarded from the base-programmed timing if the temperature is hot.
- Cold air is more dense, contains more oxygen, and therefore requires a richer mixture to achieve the proper air-fuel mixture. Air at 32°F (0°C) is 14% denser than air at 100°F (38°C).
- Hot air is less dense, contains less oxygen, and therefore requires less fuel to achieve the proper air-fuel mixture.

The IAT sensor is a low-authority sensor and is used by the computer to modify the amount of fuel and ignition timing as determined by the engine coolant temperature sensor.

The IAT sensor is used by the PCM as a backup in the event that the ECT sensor is determined to be inoperative.

TECH TIP

Poor Fuel Economy? Black Exhaust Smoke? Look at the IAT

If the intake air temperature sensor is defective, it may be signaling the computer that the intake air temperature is extremely cold when in fact it is warm. In such a case, the computer will supply a mixture that is much richer than normal.

If a sensor is physically damaged or electrically open, the computer will often set a diagnostic trouble code (DTC). This DTC is based on the fact that the sensor temperature did not change for a certain amount of time, usually about nine minutes. If, however, the wiring or the sensor itself has excessive resistance, a DTC will not be set and the result will be lower-than-normal fuel economy, and in serious cases, black exhaust smoke from the tailpipe during acceleration.

NOTE: Some engines use a throttle-body temperature (TBT) sensor to sense the temperature of the air entering the engine, instead of an intake air temperature sensor.

Engine temperature is most accurately determined by looking at the engine coolant temperature sensor. In certain conditions, the IAT has an effect on performance and driveability. One such condition is a warm engine being stopped in very cold weather. In this case, when the engine is restarted, the ECT may be near normal operating temperature such as 200°F (93°C) yet the air temperature could be -20°F (-30°C). In this case, the engine requires a richer mixture due to the cold air than the ECT would seem to indicate.

TESTING THE INTAKE AIR TEMPERATURE SENSOR

If the intake air temperature sensor circuit is damaged or faulty, a diagnostic trouble code is set and the malfunction indicator lamp (MIL) may or may not turn on depending on the condition and the type and model of the vehicle. To diagnose the IAT sensor follow these steps:

STEP 1 After the vehicle has been allowed to cool for several hours, use a scan tool, observe the IAT, and compare it to the engine coolant temperature. The two temperatures should be within 5°F of each other.

STEP 2 Perform a thorough visual inspection of the sensor and the wiring. If the IAT is screwed into the intake manifold, remove the sensor and check for damage.

STEP 3 Check the voltage and compare to the following chart.

INTAKE AIR TEMPERATURE SENSOR TEMPERATURE VS. RESISTANCE AND VOLTAGE DROP (APPROXIMATE)

°F	°C	OHMS	VOLTAGE DROP ACROSS THE SENSOR
–40	–40	100,000	4.95
+18	–8	15,000	4.68
32	0	9,400	4.52
50	10	5,700	4.25
68	20	3,500	3.89
86	30	2,200	3.46
104	40	1,500	2.97
122	50	1,000	2.47
140	60	700	2.00
158	70	500	1.59
176	80	300	1.25
194	90	250	0.97
212	100	200	0.75

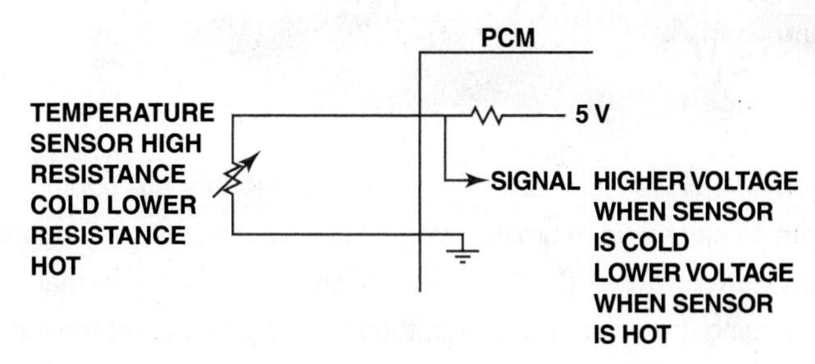

FIGURE 20–9 A typical temperature sensor circuit.

? FREQUENTLY ASKED QUESTION

What Exactly Is an NTC Sensor?

A negative temperature coefficient (NTC) thermistor is a semiconductor whose resistance decreases as the temperature increases. In other words, the sensor becomes more electrically conductive as the temperature increases. Therefore, when a voltage is applied, typically 5 volts, the signal voltage is high when the sensor is cold because the sensor has a high resistance and little current flows through to ground. ● **SEE FIGURE 20–9.**

However, when the temperature increases, the sensor becomes more electrically conductive and takes more of the 5 volts to ground, resulting in a lower signal voltage as the sensor warms.

TRANSMISSION FLUID TEMPERATURE SENSOR

The **transmission fluid temperature (TFT)**, also called *transmission oil temperature (TOT),* sensor is an important sensor for the proper operation of the automatic transmission. A TFT sensor is a NTC thermistor that decreases in resistance as the temperature of the sensor increases.

GENERAL MOTORS
Transaxle Sensor—Temperature to Resistance (approximate)

°F	°C	RESISTANCE (Ω)
32	0	7,987–10,859
50	10	4,934–6,407
68	20	3,106–3,923
86	30	1,991–2,483
104	40	1,307–1,611
122	50	878–1,067
140	60	605–728
158	70	425–507
176	80	304–359
194	90	221–259
212	100	163–190

CHRYSLER
Sensor Resistance (Ohms)—Transmission Temperature Sensor

°F	°C	RESISTANCE (Ω)
−40	−40	291,490–381,710
−4	−20	85,850–108,390
14	−10	49,250–61,430
32	0	29,330–35,990
50	10	17,990–21,810
68	20	11,370–13,610
77	25	9,120–10,880
86	30	7,370–8,750
104	40	4,900–5,750
122	50	3,330–3,880
140	60	2,310–2,670
158	70	1,630–1,870
176	80	1,170–1,340
194	90	860–970
212	100	640–720
230	110	480–540
248	120	370–410

FORD Transmission Fluid Temperature		
°F	°C	RESISTANCE (Ω)
−40 to −4	−40 to −20	967–284 K
−3 to 31	−19 to −1	284–100 K
32 to 68	0 to 20	100–37 K
69 to 104	21 to 40	37–16 K
105 to 158	41 to 70	16–5 K
159 to 194	71 to 90	5–2.7 K
195 to 230	91 to 110	2.7–1.5 K
231 to 266	111 to 130	1.5–0.8 K
267 to 302	131 to 150	0.8–0.54 K

The transmission fluid temperature signal is used by the Powertrain Control Module (PCM) to perform certain strategies based on the temperature of the automatic transmission fluid. For example:

- If the temperature of the automatic transmission fluid is low (typically below 32°F [0°C]), the shift points may be delayed and overdrive disabled. The torque converter clutch also may not be applied to assist in the heating of the fluid.
- If the temperature of the automatic transmission fluid is high (typically above 260°F [130°C]), the overdrive is disabled and the torque converter clutch is applied to help reduce the temperature of the fluid.

NOTE: Check service information for the exact shift strategy based on high and low transmission fluid temperatures for the vehicle being serviced.

CYLINDER HEAD TEMPERATURE SENSOR

Some vehicles are equipped with **cylinder head temperature (CHT)** sensors.

VW Golf

14°F (−10°C) = 11,600 Ω

68°F (20°C) = 2,900 Ω

176°F (80°C) = 390 Ω

ENGINE FUEL TEMPERATURE (EFT) SENSOR

Some vehicles, such as many Ford vehicles that are equipped with an electronic returnless type of fuel injection, use an **engine fuel temperature (EFT)** sensor to give the PCM information regarding the temperature and, therefore, the density of the fuel.

EXHAUST GAS RECIRCULATION (EGR) TEMPERATURE SENSOR

Some engines, such as Toyota, are equipped with exhaust gas recirculation (EGR) temperature sensors. EGR is a well-established method for reduction of NO_X emissions in internal combustion engines. The exhaust gas contains unburned hydrocarbons, which are recirculated in the combustion process. Recirculation is controlled by valves, which operate as a function of exhaust gas speed, load, and temperature. The gas reaches a temperature of about 850°F (450°C) for which a special heavy-duty glass-encapsulated NTC sensor is available.

The PCM monitors the temperature in the exhaust passage between the EGR valve and the intake manifold. If the temperature increases when the EGR is commanded on, the PCM can determine that the valve or related components are functioning.

ENGINE OIL TEMPERATURE SENSOR

Engine oil temperature sensors are used on many General Motors vehicles and are used as an input to the oil life monitoring system. The computer program inside the PCM calculates engine oil life based on run time, engine RPM, and oil temperature.

TEMPERATURE SENSOR DIAGNOSTIC TROUBLE CODES

The OBD-II diagnostic trouble codes that relate to temperature sensors include both high- and low-voltage codes, as well as intermittent codes. ● **SEE CHART 20–1**.

DIAGNOSTIC TROUBLE CODE	DESCRIPTION	POSSIBLE CAUSES
P0112	IAT sensor low voltage	▪ IAT sensor internally shorted-to-ground ▪ IAT sensor wiring shorted-to-ground ▪ IAT sensor damaged by backfire (usually associated with IAT sensors that are mounted in the intake manifold) ▪ Possible defective PCM
P0113	IAT sensor high voltage	▪ IAT sensor internally (electrically) open ▪ IAT sensor signal, circuit, or ground circuit open ▪ Possible defective PCM
P0117	ECT sensor low voltage	▪ ECT sensor internally shorted-to-ground ▪ The ECT sensor circuit wiring shorted-to-ground ▪ Possible defective PCM
P0118	ECT sensor high voltage	▪ ECT sensor internally (electrically) open ▪ ECT sensor signal, circuit, or ground circuit open ▪ Engine operating in an overheated condition ▪ Possible defective PCM

CHART 20–1

Selected temperature sensor-related diagnostic trouble codes.

SUMMARY

1. The ECT sensor is a high-authority sensor at engine start-up and is used for closed-loop control, as well as idle speed.
2. All temperature sensors decrease in resistance as the temperature increases. This is called negative temperature coefficient (NTC).
3. The ECT and IAT sensors can be tested visually, as well as by using a digital multimeter or a scan tool.
4. Some vehicle manufacturers use a stepped ECT circuit inside the PCM to broaden the accuracy of the sensor.
5. Other temperature sensors include transmission fluid temperature (TFT), engine fuel temperature (EFT), exhaust gas recirculation (EGR) temperature, and engine oil temperature.

REVIEW QUESTIONS

1. How does a typical NTC temperature sensor work?
2. What is the difference between a stepped and a non-stepped ECT circuit?
3. What temperature should be displayed on a scan tool if the ECT sensor is unplugged with the key on, engine off?
4. What are the three ways that temperature sensors can be tested?
5. If the transmission fluid temperature (TFT) sensor were to fail open (as if it were unplugged), what would the PCM do to the transmission shifting points?

CHAPTER QUIZ

1. The sensor that most determines fuel delivery when a fuel-injected engine is first started is the ________.
   - **a.** O2S
   - **b.** ECT sensor
   - **c.** Engine MAP sensor
   - **d.** IAT sensor
2. What happens to the voltage measured at the ECT sensor when the thermostat opens?
   - **a.** Increases slightly
   - **b.** Increases about 1 volt
   - **c.** Decreases slightly
   - **d.** Decreases about 1 volt
3. Two technicians are discussing a stepped ECT circuit. Technician A says that the sensor used for a stepped circuit is different than one used in a non-stepped circuit. Technician B says that a stepped ECT circuit uses different internal resistance inside the PCM. Which technician is correct?
   - **a.** Technician A only
   - **b.** Technician B only
   - **c.** Both Technicians A and B
   - **d.** Neither Technician A nor B
4. When testing an ECT sensor on a vehicle, a digital multimeter can be used and the signal wires back probed. What setting should the technician use to test the sensor?
   - **a.** AC volts
   - **b.** DC volts
   - **c.** Ohms
   - **d.** Hz (hertz)
5. When testing the ECT sensor with the connector disconnected, the technician should select what position on the DMM?
   - **a.** AC volts
   - **b.** DC volts
   - **c.** Ohms
   - **d.** Hz (hertz)
6. When checking the ECT sensor with a scan tool, about what temperature should be displayed if the connector is removed from the sensor with the key on, engine off?
   - **a.** 284°F (140°C)
   - **b.** 230°F (110°C)
   - **c.** 120°F (50°C)
   - **d.** −40°F (−40°C)
7. Two technicians are discussing the IAT sensor. Technician A says that the IAT sensor is more important to the operation of the engine (higher authority) than the ECT sensor. Technician B says that the PCM will add fuel if the IAT indicates that the incoming air temperature is cold. Which technician is correct?
   - **a.** Technician A only
   - **b.** Technician B only
   - **c.** Both Technicians A and B
   - **d.** Neither Technician A nor B
8. A typical IAT or ECT sensor reads about 3,000 ohms when tested using a DMM. This resistance represents a temperature of about ________.
   - **a.** −40°F (−40°C)
   - **b.** 70°F (20°C)
   - **c.** 120°F (50°C)
   - **d.** 284°F (140°C)
9. If the transmission fluid temperature (TFT) sensor indicates cold automatic transmission fluid temperature, what would the PCM do to the shifts?
   - **a.** Normal shifts and normal operation of the torque converter clutch
   - **b.** Disable torque converter clutch; normal shift points
   - **c.** Delayed shift points and torque converter clutch disabled
   - **d.** Normal shifts but overdrive will be disabled
10. A P0118 DTC is being discussed. Technician A says that the ECT sensor could be shorted internally. Technician B says that the signal wire could be open. Which technician is correct?
    - **a.** Technician A only
    - **b.** Technician B only
    - **c.** Both Technicians A and B
    - **d.** Neither Technician A nor B

chapter 21

THROTTLE POSITION (TP) SENSORS

LEARNING OBJECTIVES

After studying this chapter, the reader will be able to:

1. Discuss how throttle position sensors work.
2. Describe how to test the TP sensor using a scan tool.
3. Describe how the operation of the TP sensor affects vehicle operation.
4. Discuss the PCM uses for the TP sensor.

This chapter will help you prepare for Engine Repair (A8) ASE certification test content area "E" (Computerized Engine Controls Diagnosis and Repair).

KEY TERMS

FIGURE 21–1 A typical TP sensor mounted on the throttle plate of this port-injected engine.

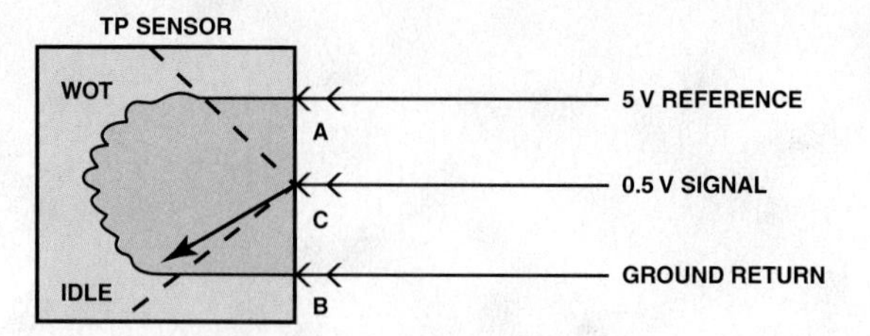

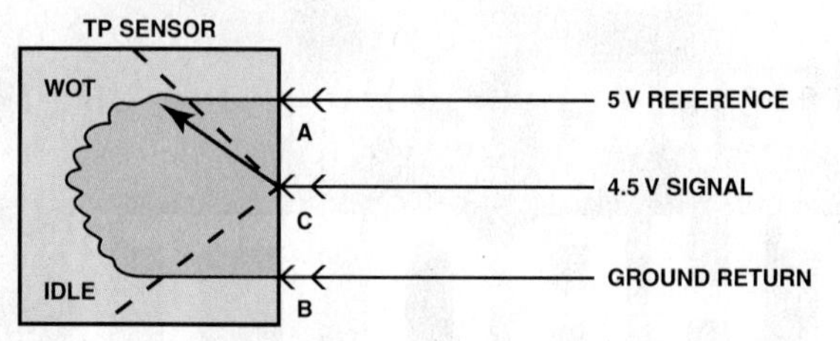

FIGURE 21–2 The signal voltage from a throttle position increases as the throttle is opened because the wiper arm is closer to the 5 volt reference. At idle, the resistance of the sensor winding effectively reduces the signal voltage output to the computer.

THROTTLE POSITION SENSOR CONSTRUCTION

Most computer-equipped engines use a **throttle position (TP) sensor** to signal to the computer the position of the throttle. ● **SEE FIGURE 21–1**. The TP sensor consists of a **potentiometer**, a type of variable resistor.

POTENTIOMETERS A potentiometer is a variable-resistance sensor with three terminals. One end of the resistor receives reference voltage, while the other end is grounded. The third terminal is attached to a movable contact that slides across the resistor to vary its resistance. Depending on whether the contact is near the supply end or the ground end of the resistor, return voltage is high or low. ● **SEE FIGURE 21–2**.

Throttle position (TP) sensors are among the most common potentiometer-type sensors. The computer uses their input to determine the amount of throttle opening and the rate of change.

A typical sensor has three wires:

- A 5 volt reference feed wire from the computer
- Signal return (a ground wire back to the computer)
- A voltage signal wire back to the computer; as the throttle is opened, the voltage to the computer changes

Normal throttle position voltage on most vehicles is about 0.5 volt at idle (closed throttle) and 4.5 volts at wide-open throttle (WOT).

NOTE: The TP sensor voltage at idle is usually about 10% of the TP sensor voltage when the throttle is wide open, but can vary from as low as 0.3 to 1.2 volts, depending on the make and model of vehicle.

TP SENSOR COMPUTER INPUT FUNCTIONS

- The computer senses any change in throttle position and changes the fuel mixture and ignition timing. The actual change in fuel mixture and ignition timing is also partly determined by the other sensors, such as the manifold pressure (engine vacuum), engine RPM, the coolant temperature, and oxygen sensor(s). Some throttle position sensors are adjustable and should be set according to the exact engine manufacturer's specifications.
- The throttle position sensor used on fuel-injected vehicles acts as an "electronic accelerator pump." This means that the computer will pulse additional fuel from the injectors when the throttle is depressed. Because the air can quickly flow into the engine when the throttle is opened, additional fuel must be supplied to prevent the air–fuel mixture from going lean, causing the engine to hesitate when the throttle is depressed. If the TP sensor is unplugged or defective, the engine may still operate satisfactorily, but hesitate upon acceleration.
- The PCM supplies the TP sensor with a regulated voltage that ranges from 4.8 to 5.1 volts. This reference voltage is usually referred to as a 5 volt reference or "Vref." The TP output signal is an input to the PCM, and the TP sensor ground also flows through the PCM.

See the Ford throttle position sensor chart for an example of how sensor voltage changes with throttle angle.

Ford Throttle Position (TP) Sensor Chart

THROTTLE ANGLE (DEGREES)	VOLTAGE (V)
0	0.50
10	0.97
20	1.44
30	1.90
40	2.37
50	2.84
60	3.31
70	3.78
80	4.24

NOTE: Generally, any reading higher than 80% represents wide-open throttle to the computer.

PCM USES FOR THE TP SENSOR

The TP sensor is used by the Powertrain Control Module (PCM) for the following reasons.

- **Clear Flood Mode** If the throttle is depressed to the floor during engine cranking, the PCM will either greatly reduce or entirely eliminate any fuel-injector pulses to aid in cleaning a flooded engine. If the throttle is depressed to the floor and the engine is not flooded with excessive fuel, the engine may not start.
- **Torque Converter Clutch Engagement and Release** The torque converter clutch will be released if the PCM detects rapid acceleration to help the transmission deliver maximum torque to the drive wheels. The torque converter clutch is applied when the vehicle is lightly accelerating and during cruise conditions to improve fuel economy.
- **Rationality Testing for MAP and MAF Sensors** As part of the rationality tests for the MAP and/or MAF sensor, the TP sensor signal is compared to the reading from other sensors to determine if they match. For example, if the throttle position sensor is showing wide-open throttle, the MAP and/or MAF reading should also indicate that this engine is under a heavy load. If not, a diagnostic trouble code could be set for the TP, as well as the MAP and/or MAF sensors.
- **Automatic Transmission Shift Points** The shift points are delayed if the throttle is opened wide to allow the engine speed to increase, thereby producing more power and aiding in the acceleration of the vehicle. If the throttle is barely open, the shift point occurs at the minimum speed designed for the vehicle.
- **Target Idle Speed (Idle Control Strategy)** When the TP sensor voltage is at idle, the PCM then controls idle speed using the idle air control (IAC) and/or spark timing variation to maintain the commanded idle speed. If the TP sensor indicates that the throttle has moved off idle, fuel delivery and spark timing are programmed for acceleration. Therefore, if the throttle linkage is stuck or binding, the idle speed may not be correct.
- **Air-Conditioning Compressor Operation** The TP sensor is also used as an input sensor for air-conditioning compressor operation. If the PCM detects that the

throttle is at or close to wide open, the air-conditioning compressor is disengaged.

- **Backs Up Other Sensors** The TP sensor is used as a backup to the MAP sensor and/or MAF in the event the PCM detects that one or both are not functioning correctly. The PCM then calculates fuel needs and spark timing based on the engine speed (RPM) and throttle position.

TESTING THE THROTTLE POSITION SENSOR

A TP sensor can be tested using one or more of the following tools:

- A digital voltmeter with three test leads connected in series between the sensor and the wiring harness connector or back probing using T-pins or other recommended tool that will not cause harm to the connector or wiring.
- A scan tool or a specific tool recommended by the vehicle manufacturer.
- A breakout box that is connected in series between the computer and the wiring harness connector(s). A typical breakout box includes test points at which TP voltages can be measured with a digital voltmeter.
- An oscilloscope.

Use jumper wires, T-pins to back-probe the wires, or a breakout box to gain electrical access to the wiring to the TP sensor. ● **SEE FIGURE 21–3.**

FIGURE 21–3 A meter lead connected to a T-pin that was gently pushed along the signal wire of the TP sensor until the point of the pin touched the metal terminal inside the plastic connector.

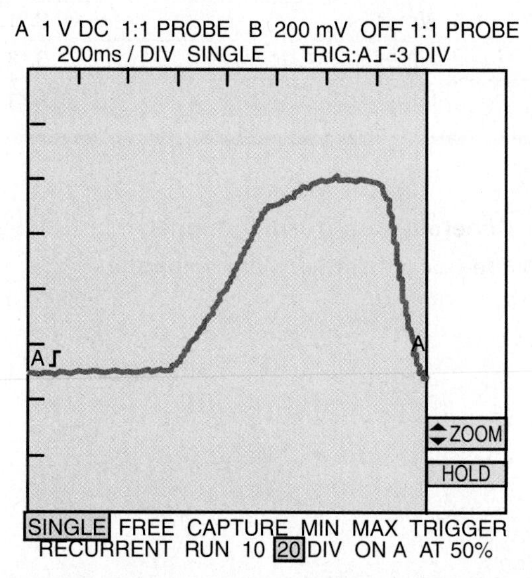

FIGURE 21–4 A typical waveform of a TP sensor signal as recorded on a DSO when the accelerator pedal was depressed with the ignition switch on (engine off). Clean transitions and the lack of any glitches in this waveform indicate a good sensor.

NOTE: The procedure that follows is the usual method used by many manufacturers. Always refer to service information for the exact recommended procedure and specifications for the vehicle being tested.

The procedure for testing the sensor using a digital multimeter is as follows:

1. Turn the ignition switch on (engine off).
2. Set the digital meter to read to DC volts and measure the voltage between the signal wire and ground (reference low) wire. The voltage should be about 0.5 volt.

   NOTE: Consult the service information for exact wire colors or locations.

3. With the engine still not running (but with the ignition still on), slowly increase the throttle opening. The voltage signal from the TP sensor should also increase. Look for any "dead spots" or open circuit readings as the throttle is increased to the wide-open position. ● **SEE FIGURE 21–4** for an example of how a good TP sensor would look when tested with a digital storage oscilloscope (DSO).

NOTE: Use the accelerator pedal to depress the throttle because this applies the same forces on the TP sensor as the driver does during normal driving. Moving the throttle by hand under the hood may not accurately test the TP sensor.

FIGURE 21–5 Checking the 5 volt reference from the computer being applied to the TP sensor with the ignition switch on (engine off).

4. With the voltmeter still connected, slowly return the throttle down to the idle position. The voltage from the TP sensor should also decrease evenly on the return to idle.

The TP sensor voltage at idle should be within the acceptable range as specified by the manufacturer. Some TP sensors can be adjusted by loosening their retaining screws and moving the sensor in relation to the throttle opening. This movement changes the output voltage of the sensor.

TECH TIP

Check Power and Ground Before Condemning a Bad Sensor

Most engine sensors use a 5 volt reference and a ground. If the 5 volts to the sensor is too high (shorted to voltage) or too low (high resistance), then the sensor output will be **skewed** or out of range. Before replacing the sensor that did not read correctly, measure both the 5 volt reference and ground. To measure the ground, simply turn the ignition on (engine off) and touch one test lead of a DMM set to read DC volts to the sensor ground and the other to the negative terminal of the battery. Any reading higher than 0.2 volt (200 mV) represents a poor ground. ● **SEE FIGURES 21–5 AND 21–6.**

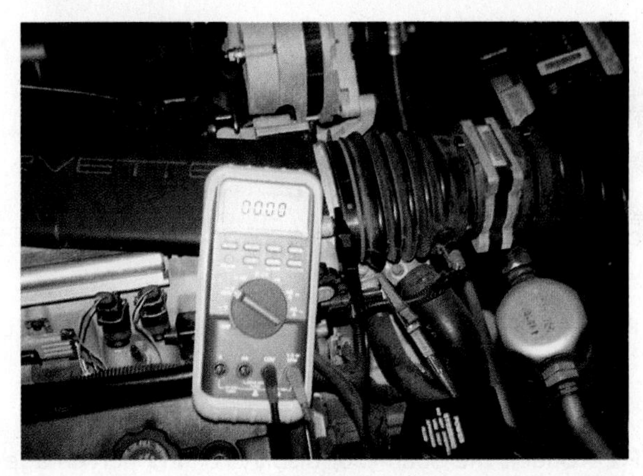

FIGURE 21–6 Checking the voltage drop between the TP sensor ground and a good engine ground with the ignition on (engine off). A reading of greater than 0.2 volt (200 mV) represents a bad computer ground.

All TP sensors should also provide a smooth transition voltage reading from idle to WOT and back to idle. Replace the TP sensor if erratic voltage readings are obtained or if the correct setting at idle cannot be obtained.

TESTING A TP SENSOR USING THE MIN/MAX FUNCTION

Many digital multimeters are capable of recording voltage readings over time and then displaying the minimum, maximum, and average readings. To perform a MIN/MAX test of the TP sensor, manually set the meter to read higher than 4 volts.

STEP 1 Connect the red meter lead to the signal wire and the black meter lead to a good ground on the ground return wire at the TP sensor.

STEP 2 With the ignition on, engine off, slowly depress and release the accelerator pedal from inside the vehicle.

STEP 3 Check the minimum and maximum voltage reading on the meter display. Any 0- or 5 volt reading would indicate a fault or short in the TP sensor.

TESTING THE TP SENSOR USING A SCAN TOOL

A scan tool can be used to check for proper operation of the throttle position sensor using the following steps:

STEP 1 With the key on, engine off, the TP sensor voltage display should be about 0.5 volt, but can vary from as low as 0.3 volt to as high as 1.2 volts.

STEP 2 Check the scan tool display for the percentage of throttle opening. The reading should be zero and gradually increase in percentage as the throttle is depressed.

STEP 3 The idle air control (IAC) counts should increase as the throttle is opened and decrease as the throttle is closed. Start the engine and observe the IAC counts as the throttle is depressed.

STEP 4 Start the engine and observe the TP sensor reading. Use a wedge or thin object to increase the throttle opening slightly. The throttle percentage reading should increase. Shut off and restart the engine. If the percentage of throttle opening returns to 0%, the PCM determines that the increased throttle opening is now the new minimum and resets the idle position of the TP sensor. Remove the wedge and cycle the ignition key. The throttle position sensor should again read zero percentage.

NOTE: Some engine computers are not capable of resetting the throttle position sensor.

TP SENSOR DIAGNOSTIC TROUBLE CODES

The diagnostic trouble codes (DTCs) associated with the throttle position sensor include the following:

DIAGNOSTIC TROUBLE CODE	DESCRIPTION	POSSIBLE CAUSES
P0122	TP sensor low voltage	▪ TP sensor internally shorted-to-ground ▪ TP sensor wiring shorted-to-ground ▪ TP sensor or wiring open
P0123	TP sensor high voltage	▪ TP sensor internally shorted to 5 volt reference ▪ TP sensor ground open ▪ TP sensor wiring shorted-to-voltage
P0121	TP sensor signal does not agree with MAP	▪ Defective TP sensor ▪ Incorrect vehicle-speed (VS) sensor signal ▪ MAP sensor out-of-calibration or defective

SUMMARY

1. A throttle position (TP) sensor is a three-wire variable resistor called a potentiometer.
2. The three wires on the TP sensor include a 5 volt reference voltage from the PCM, plus the signal wire to the PCM, and a ground, which also goes to the PCM.
3. The TP sensor is used by the PCM for clear flood mode, torque converter engagement and release, and automotive transmission shift points, as well as for rationality testing for the MAP and MAF sensors.
4. The TP sensor signal voltage should be about 0.5 volt at idle and increase to about 4.5 volts at wide-open throttle (WOT).
5. A TP sensor can be tested using a digital multimeter, a digital storage oscilloscope (DSO), or a scan tool.

REVIEW QUESTIONS

1. What is the purpose of each of the three wires on a typical TP sensor?
2. What all does the PCM do with the TP sensor signal voltage?
3. What is the procedure to follow when checking the 5 volt reference and TP sensor ground?
4. How can a TP sensor be diagnosed using a scan tool?

CHAPTER QUIZ

1. Which sensor is generally considered to be the electronic accelerator pump of a fuel-injected engine?
   a. O2S
   b. ECT sensor
   c. Engine MAP sensor
   d. TP sensor
2. Typical TP sensor voltage at idle is about ________.
   a. 2.50 to 2.80 volts
   b. 0.5 volt or 10% of WOT TP sensor voltage
   c. 1.5 to 2.8 volts
   d. 13.5 to 15.0 volts
3. A TP sensor is what type of sensor?
   a. Rheostat
   b. Voltage generating
   c. Potentiometer
   d. Piezoelectric
4. Most TP sensors have how many wires?
   a. 1
   b. 2
   c. 3
   d. 4
5. Which sensor does the TP sensor back up if the PCM determines that a failure has occurred?
   a. Oxygen sensor
   b. MAF sensor
   c. MAP sensor
   d. Either b or c
6. Which wire on a TP sensor should be back-probed to check the voltage signal to the PCM?
   a. 5 volt reference (Vref)
   b. Signal
   c. Ground
   d. Meter should be connected between the 5 volt reference and the ground
7. After a TP sensor has been tested using the MIN/MAX function on a DMM, a reading of zero volts is displayed. What does this reading indicate?
   a. The TP sensor is open at one point during the test.
   b. The TP sensor is shorted.
   c. The TP sensor signal is shorted to 5 volt reference.
   d. Both b and c are possible.
8. After a TP sensor has been tested using the MIN/MAX function on a DMM, a reading of 5 volts is displayed. What does this reading indicate?
   a. The TP sensor is open at one point during the test.
   b. The TP sensor is shorted.
   c. The TP sensor signal is shorted to 5 volt reference.
   d. Both b and c are possible.
9. A technician attaches one lead of a digital voltmeter to the ground terminal of the TP sensor and the other meter lead to the negative terminal of the battery. The ignition is switched to on, engine off and the meter displays 37.3 mV. Technician A says that this is the signal voltage and is a little low. Technician B says that the TP sensor ground circuit has excessive resistance. Which technician is correct?
   a. Technician A only
   b. Technician B only
   c. Both Technicians A and B
   d. Neither Technician A nor B
10. A P0122 DTC is retrieved using a scan tool. This DTC means ________.
    a. The TP sensor voltage is low
    b. The TP sensor could be shorted-to-ground
    c. The TP sensor signal circuit could be shorted-to-ground
    d. All of the above.

chapter 22

MAP/BARO SENSORS

LEARNING OBJECTIVES

After studying this chapter, the reader will be able to:

1. Discuss the variations in pressure that can occur within an engine.
2. Discuss how MAP sensors work.
3. List the methods that can be used to test MAP sensors.
4. Describe the symptoms of a failed MAP sensor.
5. Describe how the BARO sensor is used to test altitude.

This chapter will help you prepare for Engine Repair (A8) ASE certification test content area "E" (Computerized Engine Controls Diagnosis and Repair).

KEY TERMS

Barometric manifold absolute pressure (BMAP) sensor 358
Barometric pressure (BARO) sensor 358
Manifold absolute pressure (MAP) sensor 354
Piezoresistivity 354
Pressure differential 353
Speed density 357
Vacuum 353

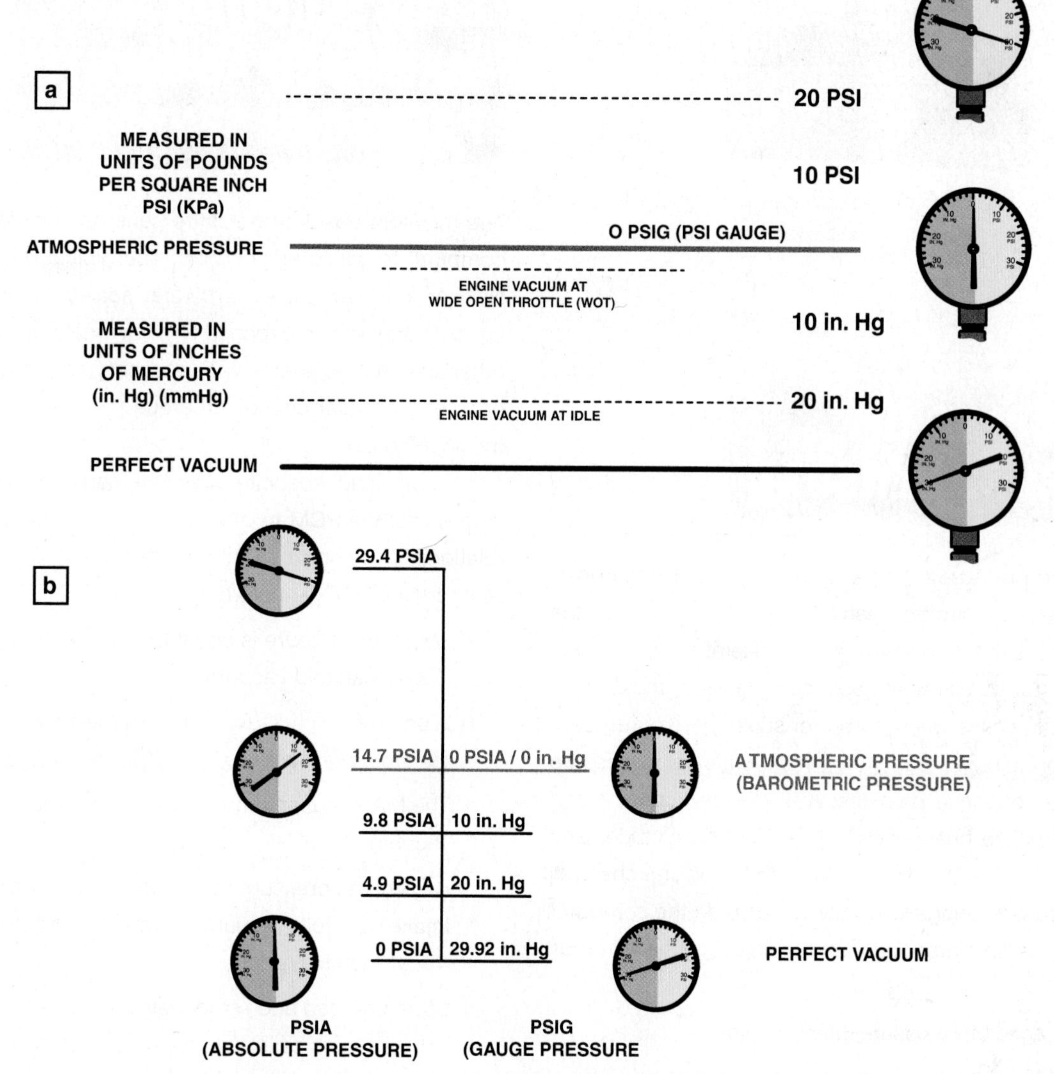

FIGURE 22–1 (a) As an engine is accelerated under a load, the engine vacuum drops. This drop in vacuum is actually an increase in absolute pressure in the intake manifold. A MAP sensor senses all pressures greater than that of a perfect vacuum. (b) The relationship between absolute pressure, vacuum, and gauge pressure.

AIR PRESSURE—HIGH AND LOW

Think of an internal combustion engine as a big air pump. As the pistons move up and down in the cylinders, they pump in air and fuel for combustion and pump out exhaust gases. They do this by creating a difference in air pressure. The air outside an engine has weight and exerts pressure, as does the air inside an engine.

As a piston moves down on an intake stroke with the intake valve open, it creates a larger area inside the cylinder for the air to fill. This lowers the air pressure within the engine. Because the pressure inside the engine is lower than the pressure outside, air flows into the engine to fill the low-pressure area and equalize the pressure.

The low pressure within the engine is called **vacuum.** Vacuum causes the higher-pressure air on the outside to flow into the low-pressure area inside the cylinder. The difference in pressure between the two areas is called a **pressure differential.** ● **SEE FIGURE 22–1.**

FIGURE 22–2 A plastic MAP sensor used for training purposes showing the electronic circuit board and electrical connections.

PRINCIPLES OF PRESSURE SENSORS

Intake manifold pressure changes with changing throttle positions. At wide-open throttle, manifold pressure is almost the same as atmospheric pressure. On deceleration or at idle, manifold pressure is below atmospheric pressure, thus creating a vacuum. In cases where turbo- or supercharging is used, under part- or full-load condition, intake manifold pressure rises above atmospheric pressure. Also, oxygen content and barometric pressure change with differences in altitude, and the computer must be able to compensate by making changes in the flow of fuel entering the engine. To provide the computer with changing airflow information, a fuel-injection system may use the following:

- Manifold absolute pressure (MAP) sensor
- Manifold absolute pressure sensor plus barometric absolute pressure (BARO) sensor
- Barometric and manifold absolute pressure sensors combined (BMAP)

The **manifold absolute pressure (MAP) sensor** may be a ceramic capacitor diaphragm, an aneroid bellows, or a piezoresistive crystal. It has a sealed vacuum reference input on one side; the other side is connected (vented) to the intake manifold. This sensor housing also contains signal conditioning circuitry. ● **SEE FIGURE 22–2**. Pressure changes in the manifold cause the sensor to deflect, varying its analog or digital return signal to the computer. As the air pressure increases, the MAP sensor generates a higher voltage or frequency return signal to the computer.

CONSTRUCTION OF MANIFOLD ABSOLUTE PRESSURE SENSORS

The manifold absolute pressure sensor is used by the engine computer to sense engine load. The typical MAP sensor consists of a ceramic or silicon wafer sealed on one side with a perfect vacuum and exposed to intake manifold vacuum on the other side. As the engine vacuum changes, the pressure difference on the wafer changes the output voltage or frequency of the MAP sensor.

A manifold absolute pressure sensor is used on many engines for the PCM to determine the load on the engine. The relationship among barometer pressure, engine vacuum, and MAP sensor voltage includes:

- Absolute pressure is equal to barometric pressure minus intake manifold vacuum.
- A decrease in manifold vacuum means an increase in manifold pressure.
- The MAP sensor compares manifold vacuum to a perfect vacuum.
- Barometric pressure minus MAP sensor reading equals intake manifold vacuum. Normal engine vacuum is 17–21 inch Hg.
- Supercharged and turbocharged engines require a MAP sensor that is calibrated for pressures above atmospheric, as well as for vacuum.

SILICON-DIAPHRAGM STRAIN GAUGE MAP SENSOR This is the most commonly used design for a MAP sensor and the output is a DC analog (variable) voltage. One side of a silicon wafer is exposed to engine vacuum and the other side is exposed to a perfect vacuum.

There are four resistors attached to the silicon wafer, which changes in resistance when strain is applied to the wafer. This change in resistance due to strain is called **piezoresistivity**. The resistors are electrically connected to a Wheatstone bridge circuit and then to a differential amplifier, which creates a voltage in proportion to the vacuum applied.

A typical General Motors MAP sensor voltage varies from 0.88 to 1.62 at engine idle.

- 17 inch Hg is equal to about 1.62 volts
- 21 inch Hg is equal to about 0.88 volts

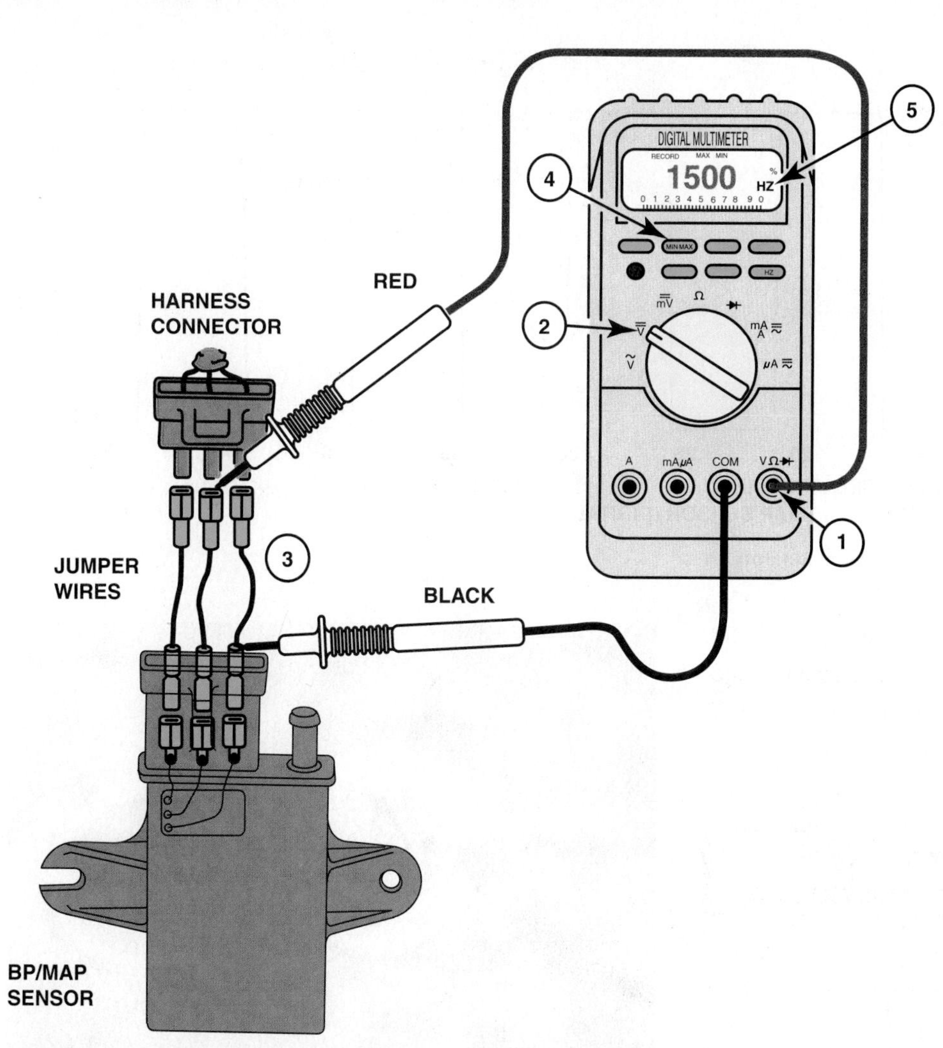

FIGURE 22–3 MAP sensors use three wires: 1. 5 volt reference from the PCM 2. Sensor signal (output signal) 3. Ground. A DMM set to test a MAP sensor. (1) Connect the red meter lead to the V meter terminal and the black meter lead to the COM meter terminal. (2) Select DC volts. (3) Connect the test leads to the sensor signal wire and the ground wire. (4) Select hertz (Hz) if testing a MAP sensor whose output is a varying frequency; otherwise keep it on DC volts. (5) Read the change of voltage (frequency) as the vacuum is applied to the sensor. Compare the vacuum reading and the frequency (or voltage) reading to the specifications. (Courtesy of Fluke Corporation)

ENGINE LOAD	MANIFOLD VACUUM	MANIFOLD ABSOLUTE PRESSURE	MAP SENSOR VOLT SIGNAL
Heavy (WOT)	Low (almost 0 inch Hg)	High (almost atmospheric)	High (4.6–4.8 V)
Light (idle)	High (17–21 inch Hg)	Low (lower than atmospheric)	Low (0.8–1.6 V)

CHART 22–1

Engine load and how it is related to engine vacuum, and MAP sensor reading.

Ford MAP Sensor Chart

MAP SENSOR OUTPUT (HZ)	ENGINE OPERATING CONDITIONS	INTAKE MANIFOLD VACUUM (INCH HG)
156–159 Hz	Key on, engine off	0 inch Hg
102–109 Hz	Engine at idle (sea level)	17–21 inch Hg
156–159 Hz	Engine at wide-open throttle (WOT)	About 0 inch Hg

CHART 22–2

Ford MAP sensor values showing relation to engine vacuum.

Therefore, a good reading should be about 1.0 volt from the MAP sensor on a sound engine at idle speed. See the following chart that shows engine load, engine vacuum, and MAP. ● **SEE CHART 22–1.**

CAPACITOR-CAPSULE MAP SENSOR A capacitor-capsule is a type of MAP sensor used by Ford, which uses two ceramic (alumina) plates with an insulating washer spacer in the center to create a capacitor. Changes in engine vacuum cause the plates to deflect, which changes the capacitance. The electronics in the sensor then generate a varying digital frequency output signal, which is proportional to the engine vacuum. ● **SEE FIGURE 22–3.** ● **SEE FIGURE 22–4** for a scope waveform of a digital MAP sensor. Also ● **SEE CHART 22–2.**

CERAMIC DISC MAP SENSOR The ceramic disc MAP sensor is used by Chrysler and it converts manifold pressure into a capacitance discharge. The discharge controls

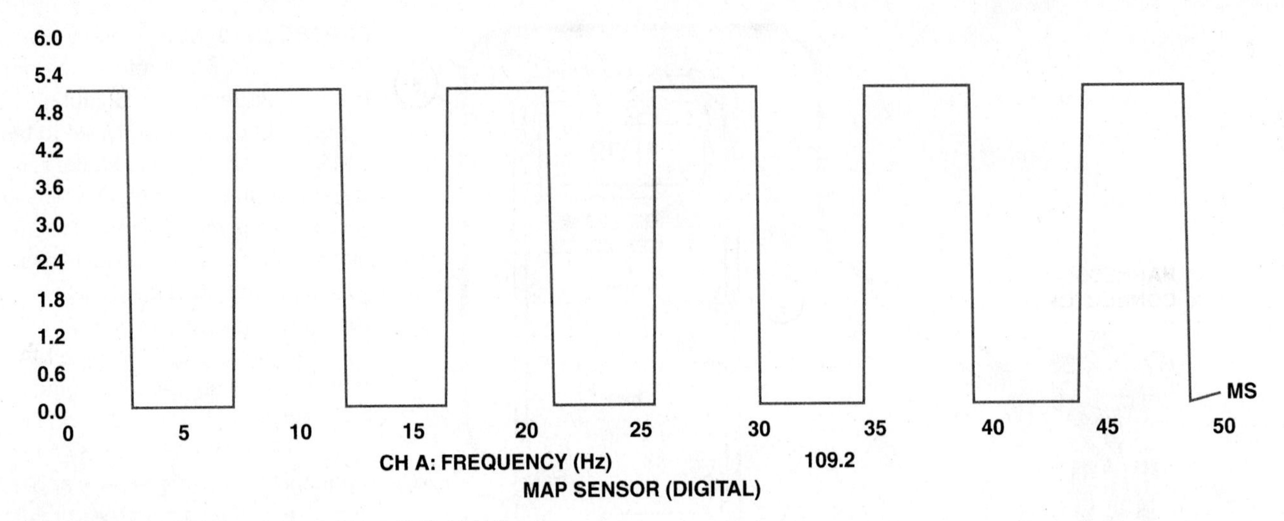

FIGURE 22–4 A waveform of a typical digital MAP sensor.

the amount of voltage delivered by the sensor to the PCM. The output is the same as the previously used strain gauge/Wheatstone bridge design and is interchangeable. ● **SEE FIGURE 22–5.** Also ● **SEE CHART 22–3.**

Chrysler MAP Sensor Chart

VACUUM (IN. HG)	MAP SENSOR SIGNAL VOLTAGE (V)
0.5	4.8
1.0	4.6
3.0	4.1
5.0	3.8
7.0	3.5
10.0	2.9
15.0	2.1
20.0	1.2
25.0	0.5

CHART 22–3

Chyrsler MAP sensor values compared to engine vacuum.

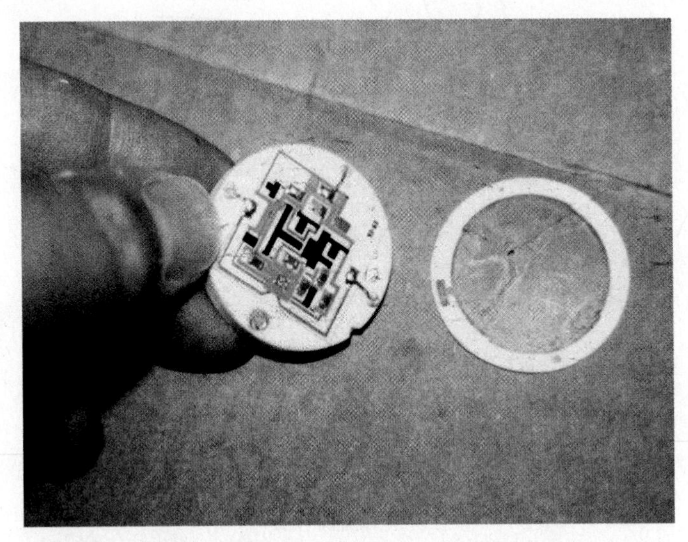

FIGURE 22–5 Shown is the electronic circuit inside a ceramic disc MAP sensor used on many Chrysler engines. The black areas are carbon resistors that are applied to the ceramic, and lasers are used to cut lines into these resistors during testing to achieve the proper operating calibration.

TECH TIP

If It's Green, It's a Signal Wire

Ford-built vehicles usually use a green wire as the signal wire back to the computer from the sensors. It may not be a solid green, but if there is green somewhere on the wire, then it is the signal wire. The other wires are the power and ground wires to the sensor.

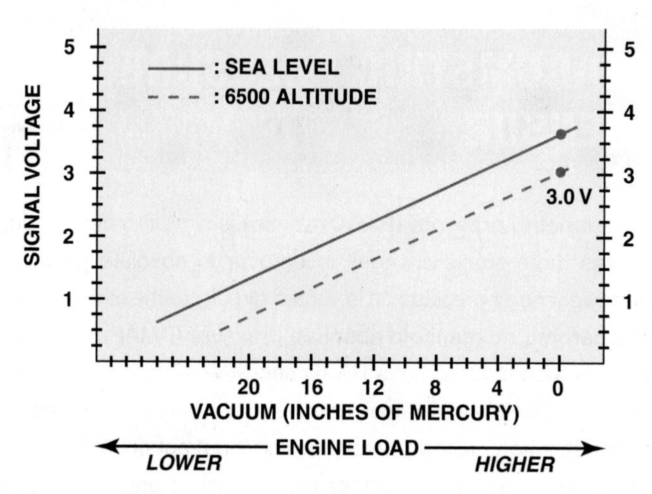

FIGURE 22–6 Altitude affects the MAP sensor voltage.

PCM USES OF THE MAP SENSOR

The PCM uses the MAP sensor to determine the following:

- **The load on the engine.** The MAP sensor is used on a **speed density**-type fuel-injection system to determine the load on the engine, and therefore the amount of fuel needed. On engines equipped with a mass air flow (MAF) sensor, the MAP is used as a backup to the MAF, for diagnosis of other sensors, and systems such as the EGR system.
- **Altitude, fuel, and spark control calculations.** At key on, the MAP sensor determines the altitude (acts as a BARO sensor) and adjusts the fuel delivery and spark timing accordingly.
  - If the altitude is high, generally over 5,000 feet (1,500 meters), the PCM will reduce fuel delivery and advance the ignition timing.
  - The altitude is also reset when the engine is accelerated to wide-open throttle and the MAP sensor is used to reset the altitude reading. ● SEE FIGURE 22–6.
- **EGR system operation.** As part of the OBD-II standards, the exhaust gas recirculation (EGR) system must be checked for proper operation. One method used by many vehicle manufacturers is to command the EGR valve on and then watch the MAP sensor signal. The opening of the EGR pintle should decrease engine vacuum. If the MAP sensor does not react with the specified drop in manifold vacuum (increase in manifold pressure), an EGR flow rate problem diagnostic trouble code is set.

TECH TIP

Use the MAP Sensor as a Vacuum Gauge

A MAP sensor measures the pressure inside the intake manifold compared with absolute zero (perfect vacuum). For example, an idling engine that has 20 inches of mercury (inch Hg) of vacuum has a lower pressure inside the intake manifold than when the engine is under a load and the vacuum is at 10 inch Hg. A decrease in engine vacuum results in an increase in manifold pressure. A normal engine should produce between 17 and 21 inch Hg at idle. Comparing the vacuum reading with the voltage reading output of the MAP sensor indicates that the reading should be between 1.62 and 0.88 volt or 109 to 102 Hz or lower on Ford MAP sensors. Therefore, a digital multimeter (DMM), scan tool, or scope can be used to measure the MAP sensor voltage and be used instead of a vacuum gauge.

NOTE: This chart was developed by testing a MAP sensor at a location about 600 feet above sea level. For best results, a chart based on your altitude should be made by applying a known vacuum, and reading the voltage of a known-good MAP sensor. Vacuum usually drops about 1 inch per 1,000 feet of altitude.

Vacuum (in. Hg)	GM (DC volts)	Ford (Hz)
0	4.80	156–159
1	4.52	
2	4.46	
3	4.26	
4	4.06	
5	3.88	141–143
6	3.66	
7	3.50	
8	3.30	
9	3.10	
10	2.94	127–130
11	2.76	
12	2.54	
13	2.36	
14	2.20	
15	2.00	114–117
16	1.80	
17	1.62	
18	1.42	108–109
19	1.20	
20	1.10	102–104
21	0.88	
22	0.66	

- **Detect deceleration (vacuum increases).** The engine vacuum rises when the accelerator is released, which changes the MAP sensor voltage. When deceleration is detected by the PCM, fuel is either stopped or greatly reduced to improve exhaust emissions.
- **Monitor engine condition.** As an engine wears, the intake manifold vacuum usually decreases. The PCM is programmed to detect the gradual change in vacuum and is able to keep the air-fuel mixture in the correct range. If the PCM were not capable of making adjustments for engine wear, the lower vacuum could be interpreted as increased load on the engine, resulting in too much fuel being injected, thereby reducing fuel economy and increasing exhaust emissions.
- **Load detection for returnless-type fuel injection.** On fuel delivery systems that do not use a return line back to the fuel tank, the engine load calculation for the fuel needed is determined by the signals from the MAP sensor.
- **Altitude and MAP sensor values.** On an engine equipped with a speed density–type fuel injection, the MAP sensor is the most important sensor needed to determine injection pulse width. Changes in altitude change the air density as well as weather conditions. Barometric pressure and altitude are inversely related:
  - As altitude increases—barometric pressure decreases
  - As altitude decreases—barometric pressure increases

  As the ignition switch is turned from off to the start position, the PCM reads the MAP sensor value to determine atmospheric and air pressure conditions. This barometric pressure reading is updated every time the engine is started and whenever wide-open throttle is detected. The barometric pressure reading at that time is updated. ● SEE CHART 22–4.

Altitude and MAP Sensor Voltage	
ALTITUDE	MAP SENSOR VOLTAGE (KEY ON, ENGINE OFF)
Sea level	4.6–4.8 V
2,500 (760 m)	4.0 V
5,000 (1,520 m)	3.7 V
7,500 (2,300 m)	3.35 V
10,000 (3,050 m)	3.05 V
12,500 (3,800 m)	2.80 V
15,000 (4,600 m)	2.45 V

CHART 22–4

Comparison between MAP sensor voltage and altitude.

BAROMETRIC PRESSURE SENSOR

A **barometric pressure (BARO) sensor** is similar in design, but senses more subtle changes in barometric absolute pressure (atmospheric air pressure). It is vented directly to the atmosphere. The **barometric manifold absolute pressure (BMAP) sensor** is actually a combination of a BARO and MAP sensor in the same housing. The BMAP sensor has individual circuits to measure barometric and manifold pressure. This input not only allows the computer to adjust for changes in atmospheric pressure due to weather, but also is the primary sensor used to determine altitude.

NOTE: A MAP sensor and a BARO sensor are usually the same sensor, but the MAP sensor is connected to the manifold and a BARO sensor is open to the atmosphere. The MAP sensor is capable of reading barometric pressure just as the ignition switch is turned to the on position before the engine starts. Therefore, altitude and weather changes are available to the computer. During mountainous driving, it may be an advantage to stop and then restart the engine so that the engine computer can take another barometric pressure reading and recalibrate fuel delivery based on the new altitude. See the Ford/BARO altitude chart for an example of how altitude affects intake manifold pressure. The computer on some vehicles will monitor the throttle position sensor and use the MAP sensor reading at wide-open throttle (WOT) to update the BARO sensor if it has changed during driving. ● SEE CHART 22–5.

NOTE: Some older Chrysler brand vehicles were equipped with a combination BARO and IAT sensor. The sensor was mounted on the bulkhead (firewall) and sensed the underhood air temperature.

Ford MAP/BARO Altitude Chart	
ALTITUDE (FEET)	VOLTS (V)
0	1.59
1,000	1.56
2,000	1.53
3,000	1.50
4,000	1.47
5,000	1.44
6,000	1.41
7,000	1.39

CHART 22–5

Comparison between Ford MAP/BARO sensor voltage and altitude.

CASE STUDY

The Cavalier Convertible Story

The owner of a Cavalier convertible stated to a service technician that the "check engine" (MIL) was on. The technician found a diagnostic trouble code (DTC) for a MAP sensor. The technician removed the hose at the MAP sensor and discovered that gasoline had accumulated in the sensor and dripped out of the hose as it was being removed. The technician replaced the MAP sensor and test drove the vehicle to confirm the repair. Almost at once the check engine light came on with the same MAP sensor code. After several hours of troubleshooting without success in determining the cause, the technician decided to start over again. Almost at once, the technician discovered that no vacuum was getting to the MAP sensor where a vacuum gauge was connected with a T-fitting in the vacuum line to the MAP sensor. The vacuum port in the base of the throttle body was clogged with carbon. After a thorough cleaning, and clearing the DTC, the Cavalier again performed properly and the check engine light did not come on again. The technician had assumed that if gasoline was able to reach the sensor through the vacuum hose, surely vacuum could reach the sensor. The technician learned to stop assuming when diagnosing a vehicle and concentrate more on testing the simple things first.

Summary:

- **Complaint—**Customer stated that the "Check Engine" warning light was on.
- **Cause—**The vacuum post to the MAP sensor was found to be clogged.
- **Correction—**Cleaning the vacuum port at the throttle body restored proper operation of the MAP sensor.

TESTING THE MAP SENSOR

Most pressure sensors operate on 5 volts from the computer and return a signal (voltage or frequency) based on the pressure (vacuum) applied to the sensor. If a MAP sensor is being tested, make certain that the vacuum hose and hose fittings are sound and making a good, tight connection to a manifold vacuum source on the engine.

Four different types of test instruments can be used to test a pressure sensor:

1. A digital voltmeter with three test leads connected in series between the sensor and the wiring harness connector or back-probe the terminals.
2. A scope connected to the sensor output, power, and ground.
3. A scan tool or a specific tool recommended by the vehicle manufacturer.
4. A breakout box connected in series between the computer and the wiring harness connection(s). A typical breakout box includes test points at which pressure sensor values can be measured with a digital voltmeter set on DC volts (or frequency counter, if a frequency-type MAP sensor is being tested).

NOTE: Always check service information for the exact testing procedures and specifications for the vehicle being tested.

TESTING THE MAP SENSOR USING A DMM OR SCOPE

Use jumper wires, T-pins to back-probe the connector, or a breakout box to gain electrical access to the wiring to the pressure sensor. Most pressure sensors use three wires:

1. A 5 volt wire from the computer
2. A variable-signal wire back to the computer
3. A ground or reference low wire

The procedure for testing the sensor is as follows:

1. Turn the ignition on (engine off)
2. Measure the voltage (or frequency) of the sensor output
3. Using a hand-operated vacuum pump (or other variable vacuum source), apply vacuum to the sensor

A good pressure sensor should change voltage (or frequency) in relation to the applied vacuum. If the signal does not change or the values are out of range according to the manufacturers' specifications, the sensor must be replaced.

TECH TIP

Visual Check of the MAP Sensor

A defective vacuum hose to a MAP sensor can cause a variety of driveability problems including poor fuel economy, hesitation, stalling, and rough idle. A small air leak (vacuum leak) around the hose can cause these symptoms and often set a trouble code in the vehicle computer. When working on a vehicle that uses a MAP sensor, make certain that the vacuum hose travels consistently *downward* on its route from the sensor to the source of manifold vacuum. Inspect the hose, especially if another technician has previously replaced the factory-original hose. It should not be so long that it sags down at any point. Condensed fuel and/or moisture can become trapped in this low spot in the hose and cause all types of driveability problems and MAP sensor codes.

When checking the MAP sensor, if anything comes out of the sensor itself, it should be replaced. This includes water, gasoline, or any other substance.

TESTING THE MAP SENSOR USING A SCAN TOOL

A scan tool can be used to test a MAP sensor by monitoring the injector pulse width (in milliseconds) when vacuum is being applied to the MAP sensor using a hand-operated vacuum pump. ● **SEE FIGURE 22–7.**

STEP 1 Apply about 20 inch Hg of vacuum to the MAP sensor and start the engine.

STEP 2 Observe the injector pulse width. On a warm engine, the injector pulse width will normally be 1.5 to 3.5 milliseconds.

STEP 3 Slowly reduce the vacuum to the MAP sensor and observe the pulse width. A lower vacuum to the MAP sensor indicates a heavier load on the engine and the injector pulse width should increase.

NOTE: If 23 inch Hg or more vacuum is applied to the MAP sensor with the engine running, this high vacuum will often stall the engine. The engine stalls because the high vacuum is interpreted by the PCM to indicate that the engine is being decelerated, which shuts off the fuel. During engine deceleration, the PCM shuts off the fuel injectors to reduce exhaust emissions and increase fuel economy.

FIGURE 22–7 A typical hand-operated vacuum pump.

MAP/BARO DIAGNOSTIC TROUBLE CODES

The diagnostic trouble codes (DTCs) associated with the MAP and BARO sensors include:

DIAGNOSTIC TROUBLE CODE	DESCRIPTION	POSSIBLE CAUSES
P0106	BARO sensor out-of-range at key on	▪ MAP sensor fault ▪ MAP sensor O-ring damaged or missing
P0107	MAP sensor low voltage	▪ MAP sensor fault ▪ MAP sensor signal circuit shorted-to-ground ▪ MAP sensor 5 volt supply circuit open
P0108	Map sensor high voltage	▪ MAP sensor fault ▪ MAP sensor O-ring damaged or missing ▪ MAP sensor signal circuit shorted-to-voltage

SUMMARY

1. Pressure below atmospheric pressure is called vacuum and is measured in inches of mercury.
2. A manifold absolute pressure sensor uses a perfect vacuum (zero absolute pressure) in the sensor to determine the pressure.
3. Three types of MAP sensors include:
   - Silicon-diaphragm strain gauge
   - Capacitor-capsule design
   - Ceramic disc design
4. A heavy engine load results in low intake manifold vacuum and a high MAP sensor signal voltage.
5. A light engine load results in high intake manifold vacuum and a low MAP sensor signal voltage.
6. A MAP sensor is used to detect changes in altitude, as well as check other sensors and engine systems.
7. A MAP sensor can be tested by visual inspection, testing the output using a digital meter or scan tool.

REVIEW QUESTIONS

1. What is the relationship among atmospheric pressure, vacuum, and boost pressure in PSI?
2. What are two types (construction) of MAP sensors?
3. What is the MAP sensor signal voltage or frequency at idle on a typical General Motors, Chrysler, and Ford engine?
4. What are three uses of a MAP sensor by the PCM?

CHAPTER QUIZ

1. As the load on an engine increases, the manifold vacuum decreases and the manifold absolute pressure ________.
   a. Increases
   b. Decreases
   c. Changes with barometric pressure only (altitude or weather)
   d. Remains constant (absolute)
2. A typical MAP sensor compares the vacuum in the intake manifold to ________.
   a. Atmospheric pressure
   b. A perfect vacuum
   c. Barometric pressure
   d. The value of the IAT sensor
3. Which statement is *false?*
   a. Absolute pressure is equal to barometric pressure plus intake manifold vacuum.
   b. A decrease in manifold vacuum means an increase in manifold pressure.
   c. The MAP sensor compares manifold vacuum to a perfect vacuum.
   d. Barometric pressure minus the MAP sensor reading equals intake manifold vacuum.
4. Which design of MAP sensor produces a frequency (digital) output signal?
   a. Silicon-diaphragm strain gauge
   b. Piezoresistivity design
   c. Capacitor-capsule
   d. Ceramic disc
5. The frequency output of a digital MAP sensor is reading 114 Hz. What is the approximate engine vacuum?
   a. Zero
   b. 5 inch Hg
   c. 10 inch Hg
   d. 15 inch Hg
6. Which is *not* a purpose or function of the MAP sensor?
   a. Measures the load on the engine
   b. Measures engine speed
   c. Calculates fuel delivery based on altitude
   d. Helps diagnose the EGR system
7. When measuring the output signal of a MAP sensor on a General Motors vehicle, the digital multimeter should be set to read ________.
   a. DC V
   b. AC V
   c. Hz
   d. DC A
8. Two technicians are discussing testing MAP sensors. Technician A says that the MAP sensor voltage on a General Motors vehicle at idle should be about 1.0 volt. Technician B says that the MAP sensor frequency on a Ford vehicle at idle should be about 105 to 108 Hz. Which technician is correct?
   a. Technician A only
   b. Technician B only
   c. Both Technicians A and B
   d. Neither Technician A nor B
9. Technician A says that MAP sensors use a 5 volt reference voltage from the PCM. Technician B says that the MAP sensor voltage will be higher at idle at high altitudes compared to when the engine is operating at near sea level. Which technician is correct?
   a. Technician A only
   b. Technician B only
   c. Both Technicians A and B
   d. Neither Technician A nor B
10. A P0107 DTC is being discussed. Technician A says that a defective MAP sensor could be the cause. Technician B says that a MAP sensor signal wire shorted-to-ground could be the cause. Which technician is correct?
    a. Technician A only
    b. Technician B only
    c. Both Technicians A and B
    d. Neither Technician A nor B

chapter 23

MASS AIR FLOW SENSORS

LEARNING OBJECTIVES

After studying this chapter, the reader will be able to:

1. Discuss how MAF sensors work.
2. List the methods that can be used to test MAF sensors.
3. Describe the symptoms of a failed MAF sensor.

This chapter will help you prepare for Engine Repair (A8) ASE certification test content area "E" (Computerized Engine Controls Diagnosis and Repair).

KEY TERMS

False air 367
Mass airflow (MAF) sensor 364
Speed density 363
Tap test 367
Vane airflow (VAF) sensor 363

SPEED DENSITY SYSTEMS

Engines that do not use an airflow meter or sensor rely on calculating the amount of air entering the engine by using the MAP sensor and engine speed as the major factors. The method of calculating the amount of fuel needed by the engine is called **speed density**.

AIRFLOW SENSORS

Older electronic fuel-injection systems that use airflow volume for fuel calculation usually have a movable vane in the intake stream. The vane is part of the **vane airflow (VAF) sensor**. The vane is deflected by intake airflow. ● **SEE FIGURE 23–1.**

The vane airflow sensor used in Bosch L-Jetronic, Ford, and most Japanese electronic port fuel-injection systems is a movable vane connected to a laser-calibrated potentiometer. The vane is mounted on a pivot pin and is deflected by intake airflow proportionate to air velocity. As the vane moves, it also moves the potentiometer. This causes a change in the signal voltage supplied to the computer. ● **SEE FIGURE 23–2.** For example, if the reference voltage is 5 volts, the potentiometer's signal to the computer will vary from a 0 voltage signal (no airflow) to almost a 5 volt signal (maximum airflow). In this way, the potentiometer provides the information the computer needs to vary the injector pulse width proportionate to airflow. There is a special "dampening chamber" built into the VAF to smooth out vane pulsations which would be created by intake manifold air-pressure fluctuations caused by the valve opening and closing. Many vane airflow sensors include a switch to energize the electric fuel pump. This is a safety feature that prevents the operation of the fuel pump if the engine stalls.

FREQUENTLY ASKED QUESTION

What Is the Difference Between an Analog and a Digital MAF Sensor?

Some MAF sensors produce a digital DC voltage signal whose frequency changes with the amount of airflow through the sensor. The frequency range also varies with the make of sensor and can range from 0–300 Hz for older General Motors MAF sensors to 1,000–9,000 Hz for most newer designs.

Some MAF sensors, such as those used by Ford and others, produce a changing DC voltage, rather than frequency, and range from 0 to 5 volts DC.

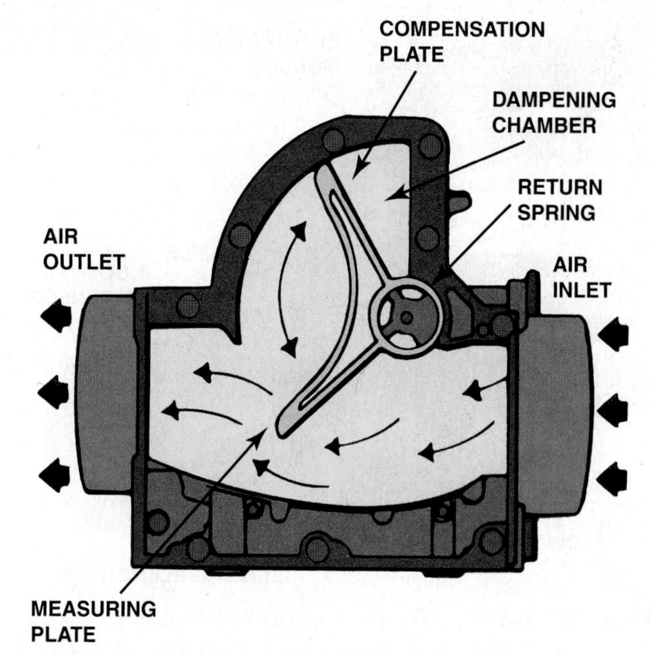

FIGURE 23–1 A vane air flow (VAF) sensor.

FIGURE 23–2 A typical air vane sensor with the cover removed. The movable arm contacts a carbon resistance path as the vane opens. Many air vane sensors have electrical contacts which are used to send current to the electric fuel pump when air flows through the sensor during engine cranking and running.

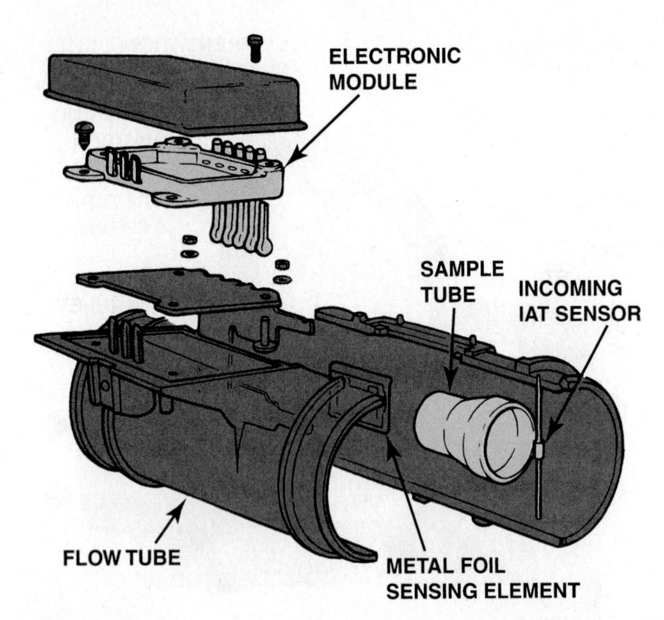

FIGURE 23–3 This five-wire mass air flow sensor consists of a metal foil sensing unit, an intake air temperature (IAT) sensor, and the electronic module.

FIGURE 23–4 The sensing wire in a typical hot wire mass air flow sensor.

MASS AIRFLOW SENSOR TYPES

PURPOSE AND FUNCTION The purpose and function of mass air flow sensors is to measure the amount of air entering the engine.

There are several types of mass airflow sensors.

HOT FILM SENSOR The hot film sensor uses a temperature-sensing resistor (thermistor) to measure the temperature of the incoming air. Through the electronics within the sensor, a conductive film is kept at a temperature 70°C above the temperature of the incoming air. ● **SEE FIGURE 23–3.**

Because the amount and density of the air both tend to contribute to the cooling effect as the air passes through the sensor, this type of sensor can actually produce an output based on the *mass* of the airflow. *Mass equals volume times density*. For example, cold air is denser than warm air so a small amount of cold air may have the same mass as a larger amount of warm air. Therefore, a mass airflow sensor is designed to measure the mass, not the volume, of the air entering the engine.

The output of this type of sensor is usually a frequency based on the amount of air entering the sensor. The more air that enters the sensor, the more the hot film is cooled. The electronics inside the sensor, therefore, increase the current flow through the hot film to maintain the 70°C temperature differential between the air temperature and the temperature of the hot film. This change in current flow is converted to a frequency output that the computer can use as a measurement of airflow. Most of these types of sensors are referred to as **mass airflow (MAF) sensors** because, unlike the air vane sensor, the MAF sensor takes into account relative humidity, altitude, and temperature of the air. The denser the air, the greater the cooling effect on the hot film sensor and the greater the amount of fuel required for proper combustion.

HOT WIRE SENSOR The hot wire sensor is similar to the hot film type, but uses a hot wire to sense the mass airflow instead of the hot film. Like the hot film sensor, the hot wire sensor uses a temperature-sensing resistor (thermistor) to measure the temperature of the air entering the sensor. ● **SEE FIGURE 23–4**. The electronic circuitry within the sensor keeps the temperature of the wire at 70°C above the temperature of the incoming air.

Both designs operate in essentially the same way. A resistor wire or screen installed in the path of intake airflow is heated to a constant temperature by electric current provided by the computer. Air flowing past the screen or wire cools it. The degree of cooling varies with air velocity, temperature, density, and humidity. These factors combine to indicate the mass of air entering the engine. As the screen or wire cools, more current is required to maintain the specified temperature. As the screen or wire heats up, less current is required. The operating principle can be summarized as follows:

- More intake air volume = cooler sensor, more current.
- Less intake air volume = warmer sensor, less current.

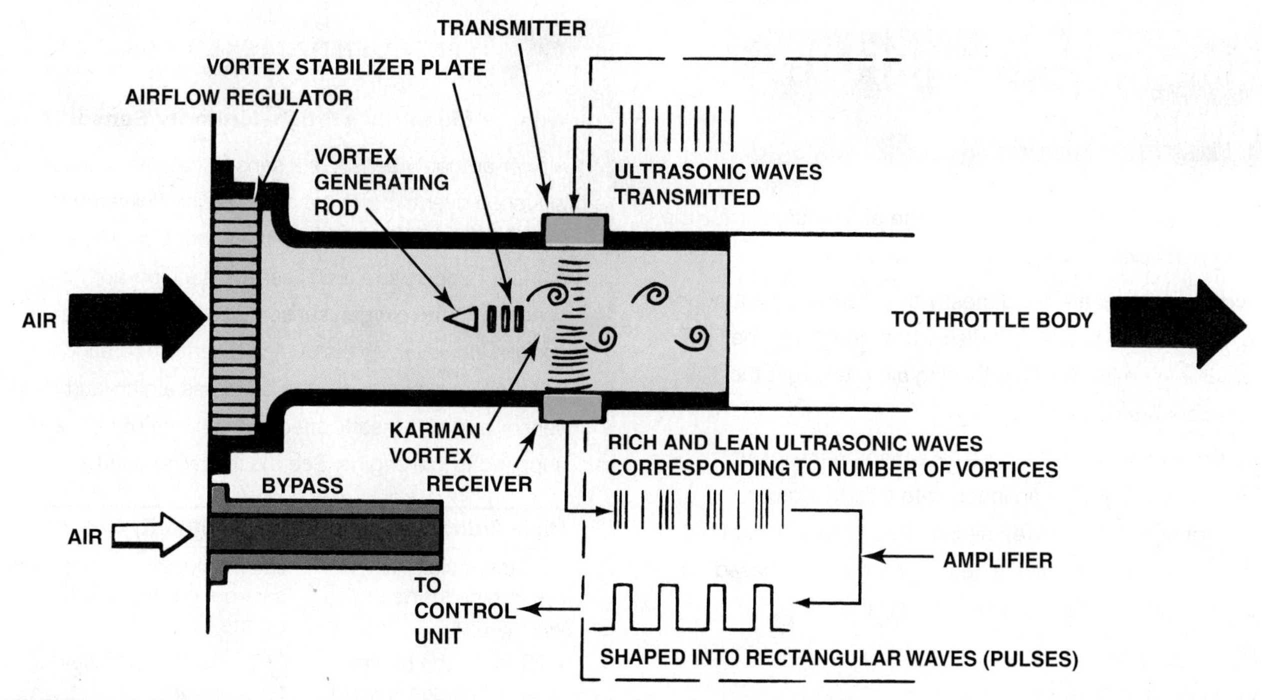

FIGURE 23–5 A Karman Vortex air flow sensor uses a triangle-shaped rod to create vortexes as the air flows through the sensor. The electronics in the sensor itself converts these vortexes to a digital square wave signal.

The computer constantly monitors the change in current and translates it into a voltage signal that is used to determine injector pulse width.

BURN-OFF CIRCUIT. Some MAF sensors use a burn-off circuit to keep the sensing wire clean of dust and dirt. A high current is passed through the sensing wire for a short time, but long enough to cause the wire to glow due to the heat. The burn-off circuit is turned on when the ignition switch is switched off after the engine has been operating long enough to achieve normal operating temperature.

KARMAN VORTEX SENSORS

In 1912, a Hungarian scientist named Theodore Van Karman observed that vortexes were created when air passed over a pointed surface. This type of sensor sends a sound wave through the turbulence created by incoming air passing through the sensor. Air mass is calculated based on the time required for the sound waves to cross the turbulent air passage.

There are two basic designs of Karman Vortex air flow sensors. The two types include:

- **Ultrasonic.** This type of sensor uses ultrasonic waves to detect the vortexes that are produced, and produce a digital (on-and-off) signal where frequency is proportional to the amount of air passing through the sensor. ● **SEE FIGURE 23–5.**
- **Pressure-type.** Chrysler uses a pressure-type Karman Vortex sensor that uses a pressure sensor to detect the vortexes. As the airflow through the sensor increases, so do the number of pressure variations. The electronics in the sensor convert these pressure variations to a square wave (digital DC voltage) signal, whose frequency is in proportion to the airflow through the sensor.

PCM USES FOR AIRFLOW SENSORS

The PCM uses the information from the airflow sensor for the following purposes:

- Airflow sensors are used mostly to determine the amount of fuel needed and base pulse-width numbers. The greater the mass of the incoming air, the longer the injectors are pulsed on.
- Airflow sensors back up the TP sensor in the event of a loss of signal or an inaccurate throttle position sensor signal. If the MAF sensor fails, then the PCM will calculate the fuel delivery needs of the engine based on throttle position and engine speed (RPM).

CASE STUDY

The Dirty MAF Sensor Story

The owner of a Buick Park Avenue equipped with a 3,800 V-6 engine complained that the engine would hesitate during acceleration, showed lack of power, and seemed to surge or miss at times. A visual inspection found everything to be like new, including a new air filter. There were no stored diagnostic trouble codes (DTCs). A look at the scan data showed airflow to be within the recommended 3 to 7 grams per second. A check of the frequency output showed the problem.

Idle frequency = 2.177 kHz (2,177 Hz)

Normal frequency at idle speed should be 2.37 to 2.52 kHz. Cleaning the hot wire of the MAF sensor restored proper operation. The sensor wire was covered with what looked like fine fibers, possibly from the replacement air filter.

Summary:

- **Complaint—**Customer stated that the engine hesitated when accelerating.
- **Cause—**Tests confirmed that the MAF sensor was was operating correctly but the frequency output at idle was not within the normal range.
- **Correction—**Cleaning the MAF sensor restored proper operation of the sensor and the engine now accelerates normally.

FREQUENTLY ASKED QUESTION

What Is Meant by a "High-Authority Sensor"?

A high-authority sensor is a sensor that has a major influence over the amount of fuel being delivered to the engine. For example, at engine start-up, the engine coolant temperature (ECT) sensor is a high-authority sensor and the oxygen sensor (O2S) is a low-authority sensor. However, as the engine reaches operating temperature, the oxygen sensor becomes a high-authority sensor and can greatly affect the amount of fuel being supplied to the engine. See the following chart.

High-Authority Sensors	Low-Authority Sensors
ECT (especially when the engine starts and is warming up)	IAT (intake air temperature) sensors modify and back up the ECT
O2S (after the engine reaches closed-loop operation)	TFT (transmission fluid temperature)
MAP	PRNDL (shift position sensor)
MAF	KS (knock sensor)
TP (high authority during acceleration and deceleration)	EFT (engine fuel temperature)

TESTING MASS AIRFLOW SENSORS

VISUAL INSPECTION Start the testing of a MAF sensor by performing a thorough visual inspection. Look at all the hoses that direct and send air, especially between the MAF sensor and the throttle body. Also check the electrical connector for:

- Corrosion
- Terminals that are bent or pushed out of the plastic connector
- Frayed wiring

MAF SENSOR OUTPUT TEST A digital multimeter, set to read DC volts, can be used to check the MAF sensor. See the chart that shows the voltage output compared with the grams per second of airflow through the sensor. Normal airflow is 3 to 7 grams per second. ● **SEE CHART 23–1.**

FIGURE 23–6 Carefully check the hose between the MAF sensor and the throttle plate for cracks or splits that could create extra (false) air into the engine that is not measured by the MAF sensor.

Analog MAF Sensor Grams per Second/ Voltage Chart

GRAMS PER SECOND	SENSOR VOLTAGE
0	0.2
2	0.7
4	1.0 (typical idle value)
8	1.5
15	2.0
30	2.5
50	3.0
80	3.5
110	4.0
150	4.5
175	4.8

CHART 23–1

Chart showing the amount of air entering the engine in grams per second compared to the sensor output voltage.

? FREQUENTLY ASKED QUESTION

What Is False Air?

Airflow sensors and mass airflow (MAF) sensors are designed to measure *all* the air entering the engine. If an air inlet hose was loose or had a hole, extra air could enter the engine without being measured. This extra air is often called **false air.** ● **SEE FIGURE 23–6.**

NOTE: If the engine runs well in reverse, yet runs terrible in any forward gear, carefully look at the inlet hose for air leaks that would open when the engine torque moves the engine slightly on its mounts.

TAP TEST With the engine running at idle speed, *gently* tap the MAF sensor with the fingers of an open hand. If the engine stumbles or stalls, the MAF sensor is defective. This test is commonly called the **tap test**.

DIGITAL METER TEST OF A MAF SENSOR A digital multimeter can be used to measure the frequency (Hz) output of the sensor and compare the reading with specifications.

The frequency output and engine speed in RPM can also be plotted on a graph to check to see if the frequency and RPM are proportional, resulting in a straight line on the graph.

MAF SENSOR CONTAMINATION

Dirt, oil, silicon, or even spider webs can coat the sensing wire. Because it tends to insulate the sensing wire at low airflow rates, a contaminated sensor often overestimates the amount of air entering the engine at idle, and therefore causes the fuel system to go rich. At higher engine speeds near wide-open throttle (WOT), the contamination can cause the sensor to underestimate

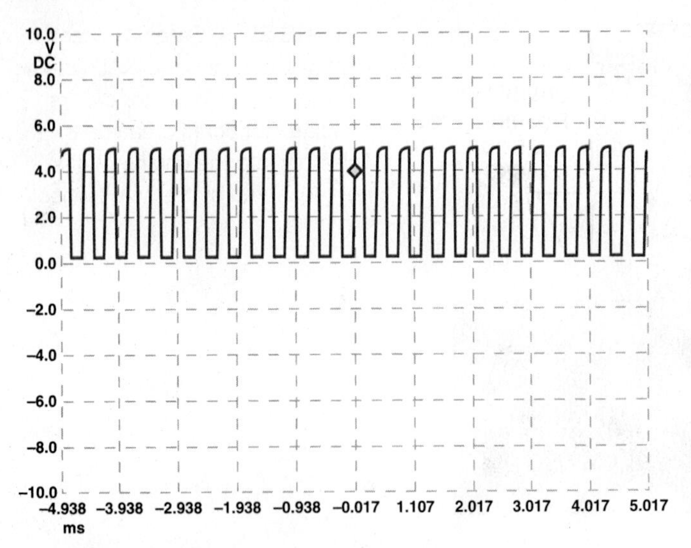

FIGURE 23–7 A scope display showing a normal Chevrolet Equinox MAF sensor at idle speed. the frequency is 2,600 Hertz (2.6kHz).

the amount of air entering the engine. As a result, the fuel system will go lean, causing spark knock and lack of power concerns. To check for contamination, check the fuel trim numbers.

If the fuel trim is negative (removing fuel) at idle, yet is positive (adding fuel) at higher engine speeds, a contaminated MAF sensor is a likely cause. Other tests for a contaminated MAF sensor are:

- At WOT, the grams per second, as read on a scan tool, should exceed 100.
- At WOT, the voltage, as read on a digital voltmeter, should exceed 4 volts for an analog sensor.
- At WOT, the frequency, as read on a meter or scan tool, should exceed 7 kHz for a digital sensor.

If the readings do not exceed these values, then the MAF sensor is contaminated.

MAF SENSOR SCOPE TESTING

A digital storage oscilloscope (DSO) can be used to monitor the operation of a MAF sensor. Connect the test leads to the signal wire and the sensor ground and following the scope instructions look for a consistent pattern that changes in frequency as the engine speed is increased. ● SEE FIGURE 23–7.

TECH TIP

The Unplug It Test

If a sensor is defective yet produces a signal to the computer, the computer will often accept the reading and make the required changes in fuel delivery and spark advance. If, however, the sensor is not reading correctly, the computer will process this wrong information and perform an action assuming that information being supplied is accurate. "If in doubt, take it out."

If the engine operates better with a sensor unplugged, then suspect that the sensor is defective. A sensor that is not supplying the correct information is said to be *skewed*. The computer will not set a diagnostic trouble code for this condition because the computer can often not detect that the sensor is supplying wrong information.

MAF-RELATED DIAGNOSTIC TROUBLE CODES

The diagnostic trouble codes (DTCs) associated with the mass airflow and air vane sensors are as follows:

DIAGNOSTIC TROUBLE CODE	DESCRIPTION	POSSIBLE CAUSES
P0100	Mass or volume airflow circuit problems	▪ Open or short in mass airflow circuit ▪ Defective MAF sensor
P0101	Mass airflow circuit range problems	▪ Defective MAF sensor (check for false air)
P0102	Mass airflow circuit low output	▪ Defective MAF sensor ▪ MAF sensor circuit open or shorted-to-ground ▪ Open 12 volt supply voltage circuit
P0103	Mass airflow circuit high output	▪ Defective MAF sensor ▪ MAF sensor circuit shorted-to-voltage

SUMMARY

1. A mass airflow sensor actually measures the density and amount of air flowing into the engine, which results in accurate engine control.
2. An air vane sensor measures the volume of the air, and the intake air temperature sensor is used by the PCM to calculate the mass of the air entering the engine.
3. A hot wire MAF sensor uses the electronics in the sensor itself to heat a wire 70°C above the temperature of the air entering the engine.

REVIEW QUESTIONS

1. How does a hot film MAF sensor work?
2. What type of voltage signal is produced by a MAF?
3. What change in the signal will occur if engine speed is increased?
4. How is a MAF sensor tested?
5. What is the purpose of a MAF sensor?
6. What are the types of airflow sensors?

CHAPTER QUIZ

1. A fuel-injection system that does not use a sensor to measure the amount (or mass) of air entering the engine is usually called a(n) ________ type of system.
   a. Air vane-controlled
   b. Speed density
   c. Mass airflow
   d. Hot wire
2. Which type of sensor uses a burn-off circuit?
   a. Hot wire MAF sensor
   b. Hot film MAF sensor
   c. Vane-type airflow sensor
   d. Both a and b
3. Which sensor has a switch that controls the electric fuel pump?
   a. VAF
   b. Hot wire MAF
   c. Hot filter MAF
   d. Karman Vortex sensor
4. Two technicians are discussing Karman Vortex sensors. Technician A says that they contain a burn-off circuit to keep them clean. Technician B says that they contain a movable vane. Which technician is correct?
   a. Technician A only
   b. Technician B only
   c. Both Technicians A and B
   d. Neither Technician A nor B
5. The typical MAF reading on a scan tool with the engine at idle speed and normal operating temperature is ________.
   a. 1 to 3 grams per second
   b. 3 to 7 grams per second
   c. 8 to 12 grams per second
   d. 14 to 24 grams per second
6. Two technicians are diagnosing a poorly running engine. There are no diagnostic trouble codes. When the MAF sensor is unplugged, the engine runs better. Technician A says that this means that the MAF is supplying incorrect airflow information to the PCM. Technician B says that this indicates that the PCM is defective. Which technician is correct?
   a. Technician A only
   b. Technician B only
   c. Both Technicians A and B
   d. Neither Technician A nor B
7. A MAF sensor on a General Motors 3,800 V-6 is being tested for contamination. Technician A says that the sensor should show over 100 grams per second on a scan tool display when the accelerator is depressed to WOT on a running engine. Technician B says that the output frequency should exceed 7,000 Hz when the accelerator pedal is depressed to WOT on a running engine. Which technician is correct?
   a. Technician A only
   b. Technician B only
   c. Both Technicians A and B
   d. Neither Technician A nor B
8. Which airflow sensor has a dampening chamber?
   a. A vane airflow
   b. A hot film MAF
   c. A hot wire MAF
   d. A Karman Vortex
9. Air that enters the engine without passing through the airflow sensor is called ____________.
   a. Bypass air
   b. Dirty air
   c. False air
   d. Measured air
10. A P0102 DTC is being discussed. Technician A says that a sensor circuit shorted-to-ground can be the cause. Technician B says that an open sensor voltage supply circuit could be the cause. Which technician is correct?
    a. Technician A only
    b. Technician B only
    c. Both Technicians A and B
    d. Neither Technician A nor B

chapter 24

OXYGEN SENSORS

LEARNING OBJECTIVES

After studying this chapter, the reader will be able to:

1. Discuss how O2 sensors work.
2. Describe the symptoms of a failed O2 sensor.
3. Interpret oxygen sensor-related diagnostic trouble codes.

This chapter will help you prepare for Engine Performance (A8) ASE certification test content area "E" (Computerized Engine Controls Diagnosis and Repair).

KEY TERMS

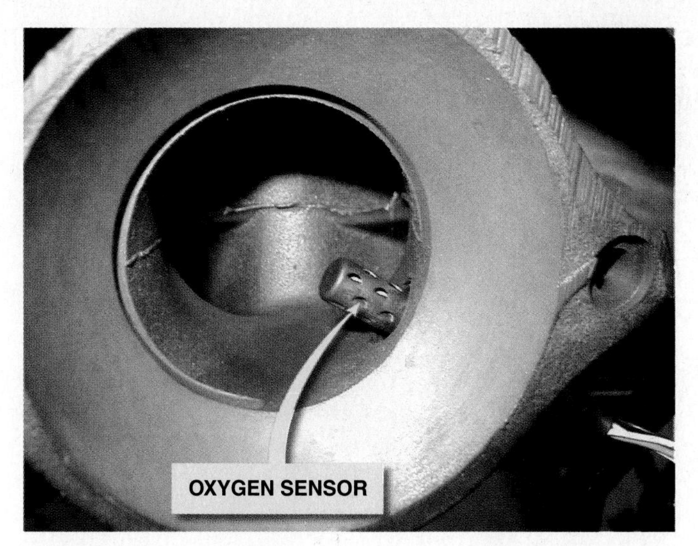

FIGURE 24–1 Many fuel-control oxygen sensors are located in the exhaust manifold near its outlet so that the sensor can detect the air–fuel mixture in the exhaust stream for all cylinders that feed into the manifold.

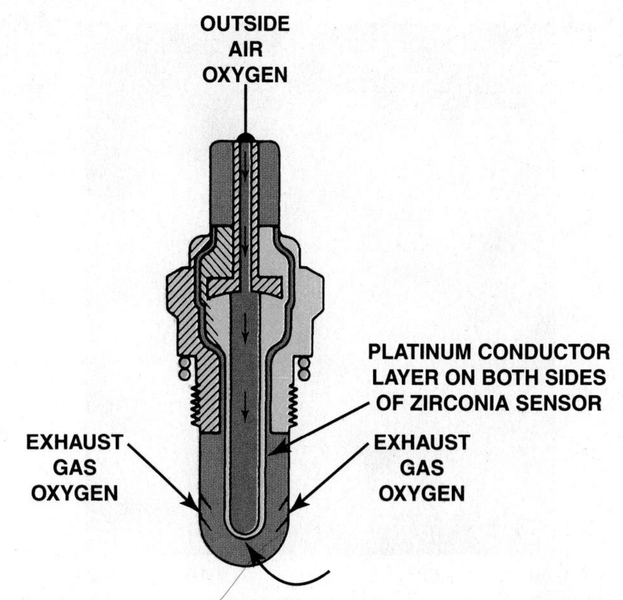

FIGURE 24–2 A cross-sectional view of a typical zirconia oxygen sensor.

OXYGEN SENSORS

PURPOSE AND FUNCTION Automotive computer systems use a sensor in the exhaust system to measure the oxygen content of the exhaust. These sensors are called **oxygen sensors (O2S)**. The oxygen sensor is installed in the exhaust manifold or located downstream from the manifold in the exhaust pipe. ● **SEE FIGURE 24–1**. The oxygen sensor is directly placed in the path of the exhaust gas stream where it monitors oxygen level in both the exhaust stream and the ambient air. In a zirconia oxygen sensor, the tip contains a thimble made of zirconium dioxide (ZrO_2), an electrically conductive material capable for generating a small voltage in the presence of oxygen. The oxygen sensor is used by the PCM to control fuel delivery.

CONSTRUCTION AND OPERATION Exhaust from the engine passes through the end of the sensor where the gases come in contact with the outer side of the thimble. Atmospheric air enters through the other end of the sensor or through the wire of the sensor and comes in contact with the inner side of the thimble. The inner and outer surfaces of the thimble are plated with platinum. The inner surface becomes a negative electrode; the outer surface is a positive electrode. The atmosphere contains a relatively constant 21% of oxygen. Rich exhaust gases contain little oxygen. Exhaust from a lean mixture contains more oxygen.

Negatively charged oxygen ions are drawn to the thimble where they collect on both the inner and outer surfaces. ● **SEE FIGURE 24–2**. Because the percentage of oxygen present in the atmosphere exceeds that in the exhaust gases, the atmosphere side of the thimble draws more negative oxygen ions than the exhaust side. The difference between the two sides creates an electrical potential, or voltage. When the concentration of oxygen on the exhaust side of the thimble is low (rich exhaust), a high voltage (0.60 to 1.0 volts) is generated between the electrodes. As the oxygen concentration on the exhaust side increases (lean exhaust), the voltage generated drops (0.00 to 0.3 volts). ● **SEE FIGURE 24–3**.

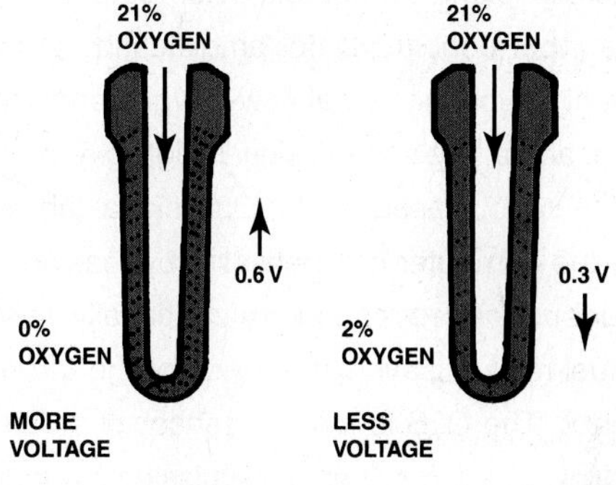

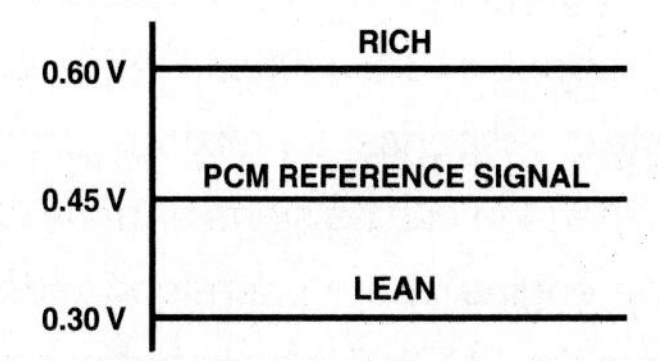

FIGURE 24–3 A difference in oxygen content between the atmosphere and the exhaust gases enables an O2S to generate voltage.

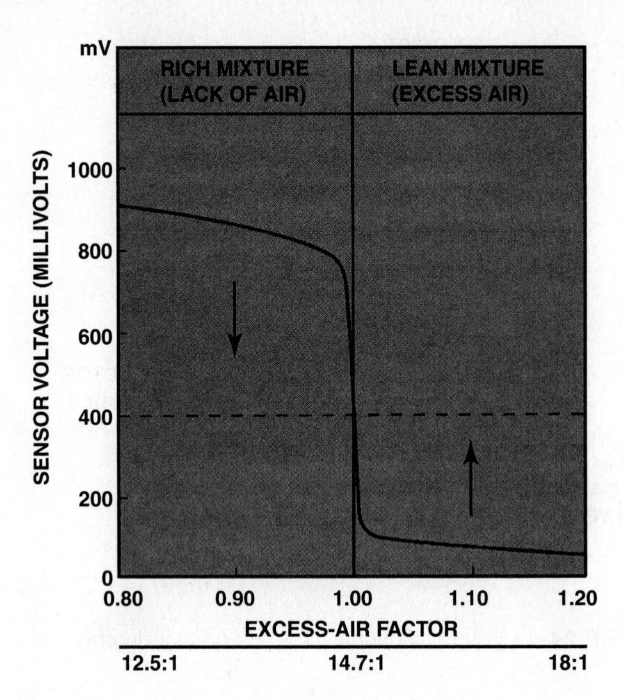

FIGURE 24–4 The oxygen sensor provides a quick response at the stoichiometric air–fuel ratio of 14.7:1.

This voltage signal is sent to the computer where it passes through the input conditioner for amplification. The computer interprets a high-voltage signal (low-oxygen content) as a rich air–fuel ratio, and a low-voltage signal (high-oxygen content) as a lean air–fuel ratio. Based on the O2S signal (above or below 0.45 volts), the computer compensates by making the mixture either leaner or richer as required to continually vary close to a 14.7:1 air–fuel ratio to satisfy the needs of the three-way catalytic converter. The O2S is the key sensor of an electronically controlled fuel metering system for emission control.

An O2S does not send a voltage signal until its tip reaches a temperature of about 572°F (300°C). Also, oxygen sensors provide their fastest response to mixture changes at about 1472°F (800°C). When the engine starts and the O2S is cold, the computer runs the engine in the open-loop mode, drawing on prerecorded data in the PROM for fuel control on a cold engine, or when O2S output is not within certain limits.

If the exhaust contains very little oxygen (O2S), the computer assumes that the intake charge is rich (too much fuel) and reduces fuel delivery. ● **SEE FIGURE 24–4**. However, when the oxygen level is high, the computer assumes that the intake charge is lean (not enough fuel) and increases fuel delivery.

There are several different types of oxygen sensors, including:

- **One-wire oxygen sensor.** The one wire of the one-wire oxygen sensor is the O2S signal wire. The ground for the O2S is through the shell and threads of the sensor and through the exhaust manifold.
- **Two-wire oxygen sensor.** The two-wire sensor has a signal wire and a ground wire for the O2S.
- **Three-wire oxygen sensor.** The three-wire sensor type uses an electric resistance heater to help get the O2S up to temperature more quickly and to help keep the sensor at operating temperature even at idle speeds. The three wires include the O2S signal, the power, and ground for the heater.
- **Four-wire oxygen sensor.** The four-wire sensor is a heated O2S (HO2S) that uses an O2S signal wire and signal ground. The other two wires are the power and ground for the heater.

FIGURE 24–5 A typical zirconia oxygen sensor.

ZIRCONIA OXYGEN SENSORS

The most common type of oxygen sensor is made from zirconia (zirconium dioxide). It is usually constructed using powder that is pressed into a thimble shape and coated with porous platinum material that acts as electrodes. All zirconia sensors use 18-mm-diameter threads with a washer. ● **SEE FIGURE 24–5.**

Zirconia oxygen sensors are constructed so that oxygen ions flow through the sensor when there is a difference between the oxygen content inside and outside of the sensor. An ion is an electrically charged particle. The greater the differences between the oxygen content between the inside and outside of the sensor, the higher the voltage created.

- **Rich mixture.** A rich mixture results in little oxygen in the exhaust stream. Compared to the outside air, this represents a large difference and the sensors create a relatively high voltage of about 1.0 volt (1,000 mV).
- **Lean mixture.** A lean mixture leaves some oxygen in the exhaust stream that did not combine with the fuel. This leftover oxygen reduces the difference between the oxygen content of the exhaust compared to the oxygen content of the outside air. As a result, the sensor voltage is low or almost 0 volt.
- **O2S voltage above 450 mV.** This is produced by the sensor when the oxygen content in the exhaust is low. This is interpreted by the engine computer (PCM) as being a rich exhaust.
- **O2S voltage below 450 mV.** This is produced by the sensor when the oxygen content is high. This is interpreted by the engine computer (PCM) as being a lean exhaust.

TITANIA OXYGEN SENSOR

The titania (titanium dioxide) oxygen sensor does not produce a voltage but rather changes in resistance with the presence of oxygen in the exhaust. All titania oxygen sensors use a four-terminal variable resistance unit with a heating element. A titania sensor samples exhaust air only and uses a reference voltage from the PCM. Titania oxide oxygen sensors use a 14 mm thread and are not interchangeable with zirconia oxygen sensors. One volt is applied to the sensor and the changing resistance of the titania oxygen sensor changes the voltage of the sensor circuit. As with a zirconia oxygen sensor, the voltage signal is above 450 mV when the exhaust is rich, and low (below 450 mV) when the exhaust is lean.

CASE STUDY

The Chevrolet Pickup Truck Story

The owner of a 1996 Chevrolet pickup truck complained that the engine ran terribly. It would hesitate and surge, yet there were no diagnostic trouble codes (DTCs). After hours of troubleshooting, the technician discovered while talking to the owner that the problem started after the transmission had been repaired, yet the transmission shop said that it was an engine problem and not related to the transmission.

A thorough visual inspection revealed that the front and rear oxygen sensor connectors had been switched. The computer was trying to compensate for an air–fuel mixture condition that did not exist. Reversing the O2S connectors restored proper operation of the truck.

Summary:

- **Complaint—**Customer stated that the truck ran terribly after the transmission was replaced.
- **Cause—**A visual inspection was used to determine a that the the upstream and downstream oxygen sensor connector had been switched.
- **Correction—**Connecting the oxygen senors to the correct harness connector restored proper engine operation.

FREQUENTLY ASKED QUESTION

Where Is HO2S1?

Oxygen sensors are numbered according to their location in the engine. On a V-type engine, heated oxygen sensor number 1 (HO2S1) is located in the exhaust system upstream of the catalytic converter on the side of the engine where the number 1 cylinder is located. ● **SEE FIGURE 24–6.**

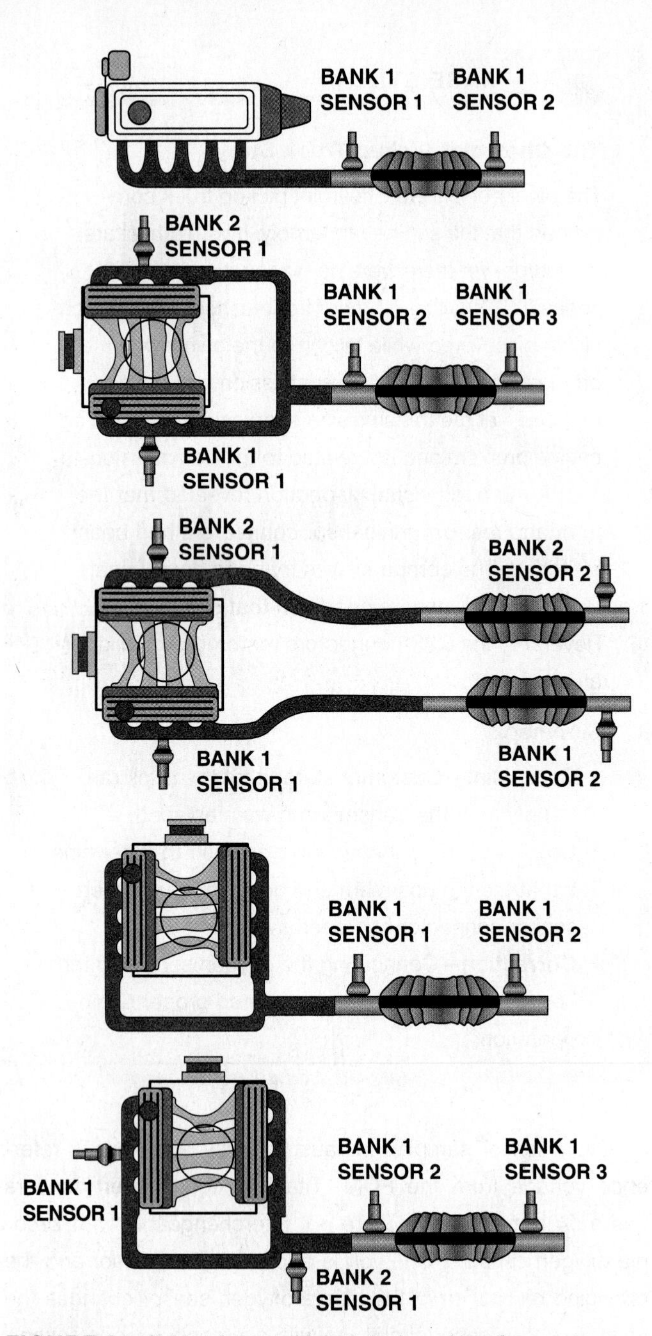

FIGURE 24–6 Number and label designations for oxygen sensors. Bank 1 is the bank where cylinder number 1 is located as shown by the red dot. Number 1 cylinder location varies by vehicle make and engine. Check service information for the location of the number 1 cylinder on the engine being checked.

CLOSED LOOP AND OPEN LOOP

The amount of fuel delivered to an engine is determined by the Powertrain Control Module (PCM) based on inputs from the engine coolant temperature (ECT), throttle position (TP) sensor, and others until the oxygen sensor is capable of supplying a usable signal. When the PCM alone (without feedback) is determining the amount of fuel needed, it is called **open-loop operation**. As soon as the oxygen sensor is capable of supplying rich and lean signals, adjustments by the computer can be made to fine-tune the correct air–fuel mixture. This checking and adjusting by the computer is called **closed-loop operation**.

PCM USES OF THE OXYGEN SENSOR

FUEL CONTROL The upstream oxygen sensors are among the main sensor(s) used for fuel control while operating in closed loop. Before the oxygen sensors are hot enough to give accurate exhaust oxygen information to the computer, fuel control is determined by other sensors and the anticipated injector pulse width determined by those sensors. After the control system achieves closed-loop status, the oxygen sensor provides feedback with actual exhaust gas oxygen content.

FUEL TRIM Fuel trim is a computer program that is used to compensate for a too rich or a too lean air–fuel exhaust as detected by the oxygen sensor(s). Fuel trim is necessary to keep the air–fuel mixture within limits to allow the catalytic converter to operate efficiently. If the exhaust is too lean or too rich for a long time, the catalytic converter can be damaged. The fuel trim numbers are determined from the signals from the oxygen sensor(s). If the engine has been operating too lean, short-term and long-term fuel time programming inside the PCM can cause an increase in the commanded injector pulse width to bring the air–fuel mixture back into the proper range. Fuel trim can be negative (subtracting fuel) or positive (adding fuel).

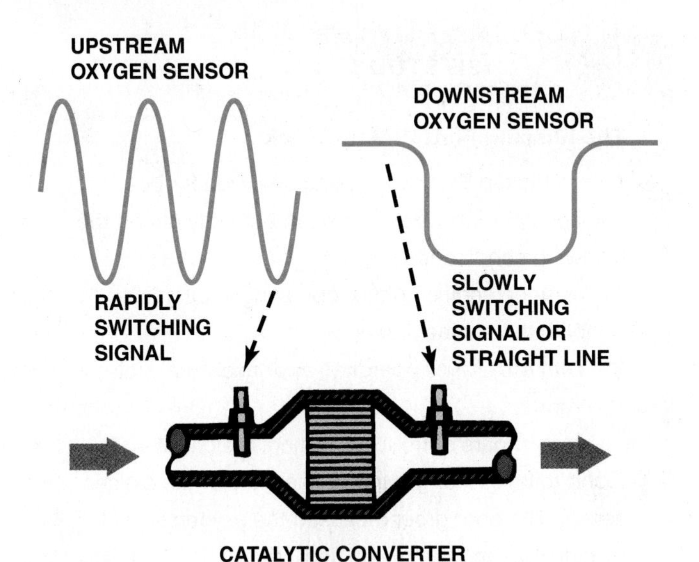

FIGURE 24–7 The OBD-II catalytic converter monitor compares the signals of the upstream and downstream oxygen sensor to determine converter efficiency.

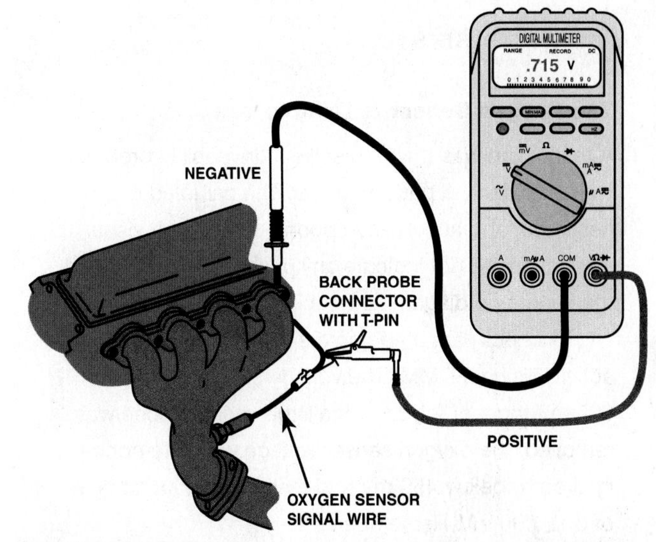

FIGURE 24–8 Testing an oxygen sensor using a DMM set on DC volts. With the engine operating in closed loop, the oxygen voltage should read over 800 mV and lower than 200 mV and be constantly fluctuating.

OXYGEN SENSOR DIAGNOSIS

The oxygen sensors are used for diagnosis of other systems and components. For example, the exhaust gas recirculation (EGR) system is tested by the PCM by commanding the valve to open during the test. Some PCMs determine whether enough exhaust gas flows into the engine by looking at the oxygen sensor response (fuel trim numbers). The upstream and downstream oxygen sensors are also used to determine the efficiency of the catalytic converter. ● **SEE FIGURE 24–7**.

TESTING AN OXYGEN SENSOR USING A DIGITAL VOLTMETER The oxygen sensor can be checked for proper operation using a digital high-impedance voltmeter.

1. With the engine off, connect the red lead of the meter to the oxygen sensor signal wire. ● **SEE FIGURE 24–8**.
2. Start the engine and allow it to reach closed-loop operation.
3. In closed-loop operation, the oxygen sensor voltage should be constantly changing as the fuel mixture is being controlled.

   The results should be interpreted as follows:

- If the oxygen sensor fails to respond, and its voltage remains at about 450 millivolts, the sensor may be defective and require replacement. Before replacing the oxygen sensor, check the manufacturers' recommended procedures.
- If the oxygen sensor reads high all the time (above 550 millivolts), the fuel system could be supplying too rich a fuel mixture or the oxygen sensor may be contaminated.

? FREQUENTLY ASKED QUESTION

What Happens to the Bias Voltage?

Some vehicle manufacturers such as General Motors Corporation have the computer apply 450 mV (0.450 V) to the O2S signal wire. This voltage is called the **bias voltage** and represents the threshold voltage for the transition from rich to lean.

This bias voltage is displayed on a scan tool when the ignition switch is turned on with the engine off. When the engine is started, the O2S becomes warm enough to produce a usable voltage and bias voltage "disappears" as the O2S responds to a rich and lean mixture. What happened to the bias voltage that the computer applied to the O2S? The voltage from the O2S simply overcame the very weak voltage signal from the computer. This bias voltage is so weak that even a 20-megohm impedance DMM will affect the strength enough to cause the voltage to drop to 426 mV. Other meters with only 10 megohms of impedance will cause the bias voltage to read less than 400 mV.

Therefore, even though the O2S voltage is relatively low powered, it is more than strong enough to override the very weak bias voltage the computer sends to the O2S.

CASE STUDY

The Oxygen Sensor Is Lying to You

A technician was trying to solve a driveability problem with an older V-6 passenger car. The car idled roughly, hesitated, and accelerated poorly. A thorough visual inspection did not indicate any possible problems and there were no diagnostic trouble codes stored.

A check was made on the oxygen sensor activity using a DMM. The voltage stayed above 600 mV most of the time. If a large vacuum hose was removed, the oxygen sensor voltage would temporarily drop to below 450 mV and then return to a reading of over 600 mV. Remember:

- High O2S readings = rich exhaust (low O_2 content in the exhaust)
- Low O2S readings = lean exhaust (high O_2 content in the exhaust)

As part of a thorough visual inspection, the technician removed and inspected the spark plugs. All the spark plugs were white, indicating a lean mixture, not the rich mixture that the oxygen sensor was indicating. The high O2S reading signaled the computer to reduce the amount of fuel, resulting in an excessively lean operation.

After replacing the oxygen sensor, the engine ran great. But what killed the oxygen sensor? The technician finally learned from the owner that the head gasket had been replaced over a year ago. The phosphate and silicate additives in the antifreeze coolant had coated the oxygen sensor.

Summary:

- **Complaint—**Customer stated that the engine idled roughly and hesitated while accelerating.
- **Cause—**Tests indicated that the oxygen sensor was skewed high.
- **Correction—**Replacing the oxygen sensor was restored to proper engine operation.

- If the oxygen sensor voltage remains low (below 350 millivolts), the fuel system could be supplying too lean a fuel mixture. Check for a vacuum leak or partially clogged fuel injector(s). Before replacing the oxygen sensor, check the manufacturer's recommended procedures.

TESTING THE OXYGEN SENSOR USING THE MIN/MAX METHOD A digital meter set on DC volts can be used to record the minimum and maximum voltage with the engine running. A good oxygen sensor should be able to produce a

CASE STUDY

The Missing Ford Pickup Truck

A Ford Pickup Truck was being analyzed for poor engine operation. The engine ran perfectly during the following conditions:

1. With the engine cold or operating in open loop
2. With the engine at idle
3. With the engine operating at or near wide-open throttle

After hours of troubleshooting, the cause was found to be a poor ground connection for the oxygen sensor. The poor ground caused the oxygen sensor voltage level to be too high, indicating to the computer that the mixture was too rich. The computer then subtracted fuel, which caused the engine to miss and run rough as the result of the now-too-lean air–fuel mixture.

Summary:

- **Complaint—**Customer stated that the engine ran poorly except when cold or at wide open throttle and at idle speed.
- **Cause—**Tests discovered a poor ground for the oxygen sensor which cause the signal to be higher than it should.
- **Correction—**Cleaning the electrical terminals at the ground connection on the engine block restored proper oxygen sensor operation.

FREQUENTLY ASKED QUESTION

Why Does the Oxygen Sensor Voltage Read 5 Volts on Many Chrysler Vehicles?

Many Chrysler vehicles apply a 5 volt reference to the signal wire of the oxygen sensor. The purpose of this voltage is to allow the computer to detect if the oxygen sensor signal circuit is opened or grounded.

- If the voltage on the signal wire is 4.5 volts or more, the computer assumes that the sensor is open.
- If the voltage on the signal wire is zero, the computer assumes that the sensor is shorted-to-ground.

If either condition exists, the computer can set a diagnostic trouble code (DTC).

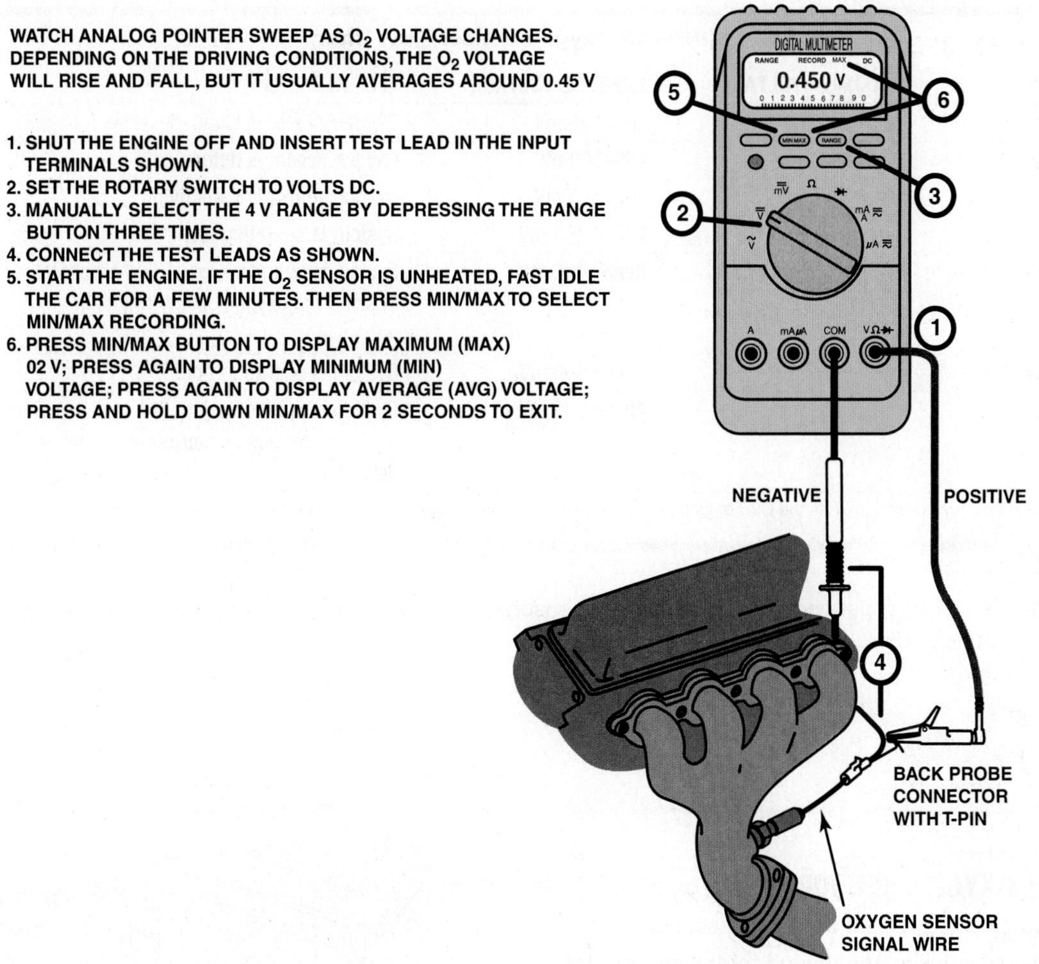

WATCH ANALOG POINTER SWEEP AS O_2 VOLTAGE CHANGES. DEPENDING ON THE DRIVING CONDITIONS, THE O_2 VOLTAGE WILL RISE AND FALL, BUT IT USUALLY AVERAGES AROUND 0.45 V

1. SHUT THE ENGINE OFF AND INSERT TEST LEAD IN THE INPUT TERMINALS SHOWN.
2. SET THE ROTARY SWITCH TO VOLTS DC.
3. MANUALLY SELECT THE 4 V RANGE BY DEPRESSING THE RANGE BUTTON THREE TIMES.
4. CONNECT THE TEST LEADS AS SHOWN.
5. START THE ENGINE. IF THE O_2 SENSOR IS UNHEATED, FAST IDLE THE CAR FOR A FEW MINUTES. THEN PRESS MIN/MAX TO SELECT MIN/MAX RECORDING.
6. PRESS MIN/MAX BUTTON TO DISPLAY MAXIMUM (MAX) 02 V; PRESS AGAIN TO DISPLAY MINIMUM (MIN) VOLTAGE; PRESS AGAIN TO DISPLAY AVERAGE (AVG) VOLTAGE; PRESS AND HOLD DOWN MIN/MAX FOR 2 SECONDS TO EXIT.

DIGITAL MULTIMETER
0.450 V
NEGATIVE
POSITIVE
BACK PROBE CONNECTOR WITH T-PIN
OXYGEN SENSOR SIGNAL WIRE

FIGURE 24–9 Using a digital multimeter to test an oxygen sensor using the MIN/MAX record function of the meter.

value of less than 300 millivolts and a maximum voltage above 800 millivolts. Replace any oxygen sensor that fails to go above 700 millivolts or lower than 300 millivolts. ● **SEE FIGURE 24–9.** See the MIN/MAX oxygen sensor test chart.

TESTING AN OXYGEN SENSOR USING A SCAN TOOL

A good oxygen sensor should be able to sense the oxygen content and change voltage outputs rapidly. How fast an oxygen sensor switches from high (above 450 millivolts) to low (below 350 millivolts) is measured in oxygen sensor **cross counts.** Cross counts are the number of times an oxygen sensor changes voltage from high to low (from low to high voltage is not counted) in 1 second (or 1.25 second, depending on scan tool and computer speed).

NOTE: On a fuel-injected engine at 2000 engine RPM, 8 to 10 cross counts is normal.

Oxygen sensor cross counts can only be determined using a scan tool or other suitable tester that reads computer data. ● **SEE CHART 24–1.**

If the cross counts are low (or zero), the oxygen sensor may be contaminated, or the fuel delivery system is delivering a constant rich or lean air–fuel mixture. To test an engine using a scan tool, follow these steps:

1. Connect the scan tool to the DLC and start the engine.
2. Operate the engine at a fast idle (2500 RPM) for two minutes to allow time for the oxygen sensor to warm to operating temperature.
3. Observe the oxygen sensor activity on the scan tool to verify closed-loop operation. Select "snapshot" mode and hold the engine speed steady and start recording.
4. Play back snapshot and place a mark beside each range of oxygen sensor voltage for each frame of the snapshot.

A good oxygen sensor and computer system should result in most snapshot values at both ends (0 to 300 and 600 to 1,000 mV). If most of the readings are in the middle, the oxygen sensor is not working correctly.

MIN/MAX Oxygen Sensor Test Chart

MINIMUM VOLTAGE	MAXIMUM VOLTAGE	AVERAGE VOLTAGE	TEST RESULTS
Below 200 mV	Above 800 mV	400–500 mV	Oxygen sensor is okay.
Above 200 mV	Any reading	400–500 mV	Oxygen sensor is defective.
Any reading	Below 800 mV	400–500 mV	Oxygen sensor is defective.
Below 200 mV	Above 800 mV	Below 400 mV	System is operating lean.*
Below 200 mV	Below 800 mV	Below 400 mV	System is operating lean. (Add propane to the intake air to see if the oxygen sensor reacts. If not, the sensor is defective.)
Below 200 mV	Above 800 mV	Above 500 mV	System is operating rich.
Above 200 mV	Above 800 mV	Above 500 mV	System is operating rich. (Remove a vacuum hose to see if the oxygen sensor reacts. If not, the sensor is defective.)

* Check for an exhaust leak upstream from the O2S or ignition misfire that can cause a false lean indication before further diagnosis.

CHART 24–1

Use this chart to check for proper operation of the oxygen sensors and fuel system after checking them using a multimeter set to read Min/Max.

TESTING AN OXYGEN SENSOR USING A SCOPE A scope can also be used to test an oxygen sensor. Connect the scope to the signal wire and ground for the sensor (if it is so equipped). ● **SEE FIGURE 24–10**. With the engine operating in closed loop, the voltage signal of the sensor should be constantly changing. ● **SEE FIGURE 24–11**. Check for rapid switching from rich to lean and lean to rich and change between once every two seconds and five times per second (0.5 to 5.0 Hz). ● **SEE FIGURES 24–12, 24–13, AND 24–14**.

NOTE: General Motors warns not to base the diagnosis of an oxygen sensor problem solely on its scope pattern. The varying voltage output of an oxygen sensor can easily be mistaken for a fault in the sensor itself, rather than a fault in the fuel delivery system.

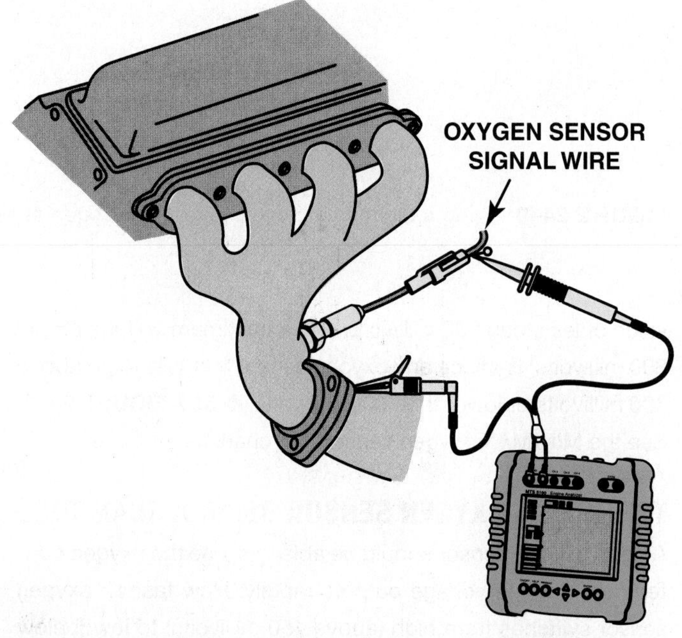

FIGURE 24–10 Connecting a handheld digital storage oscilloscope to an oxygen sensor signal wire. The use of the low-pass filter helps eliminate any low-frequency interference from affecting the scope display.

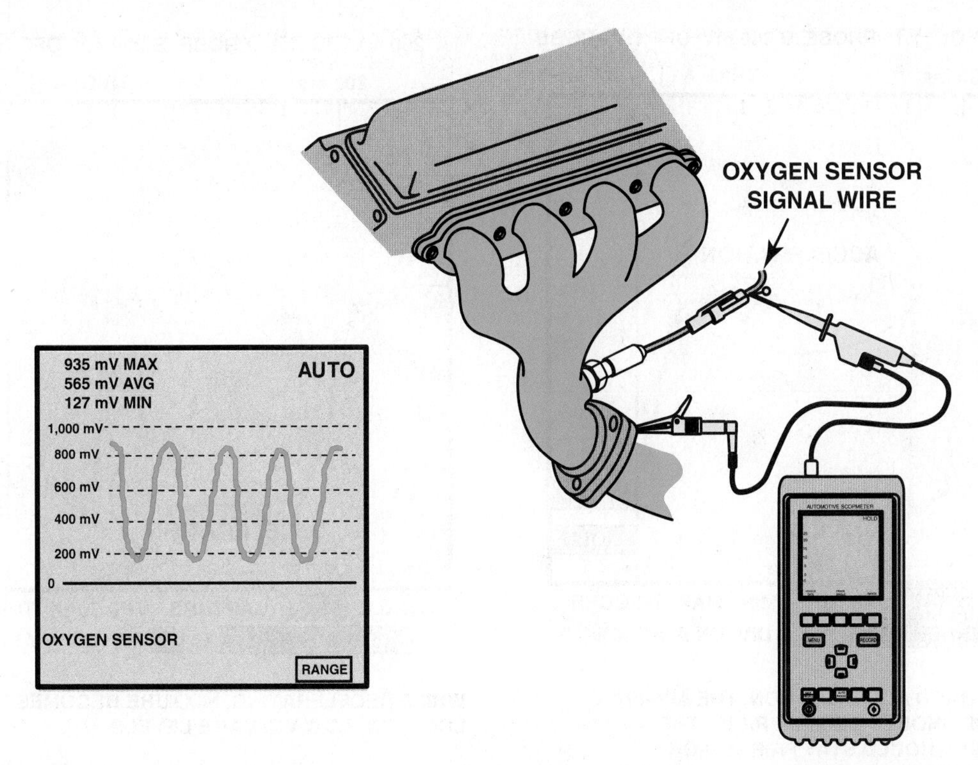

FIGURE 24–11 The waveform of a good oxygen sensor as displayed on a digital storage oscilloscope (DSO). Note that the maximum reading is above 800 mV and the minimum reading is less than 200 mV.

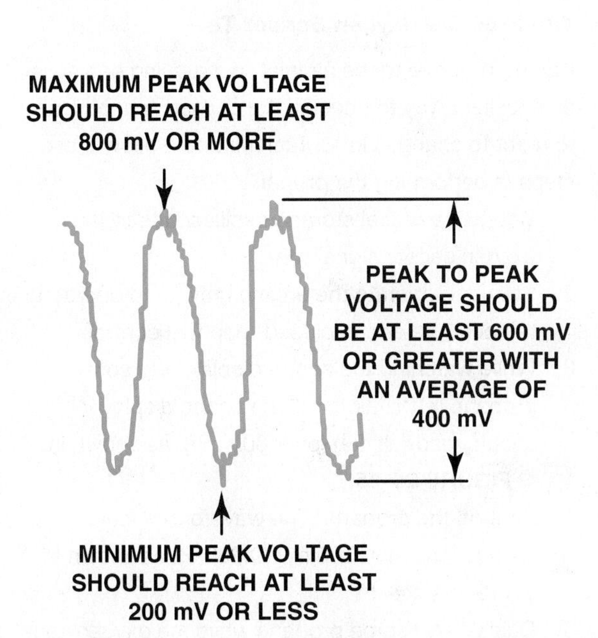

FIGURE 24–12 A typical good oxygen sensor waveform as displayed on a digital storage oscilloscope. Look for transitions that occur between once every two seconds at idle and five times per second at higher engine speeds (0.5 and 5 Hz).

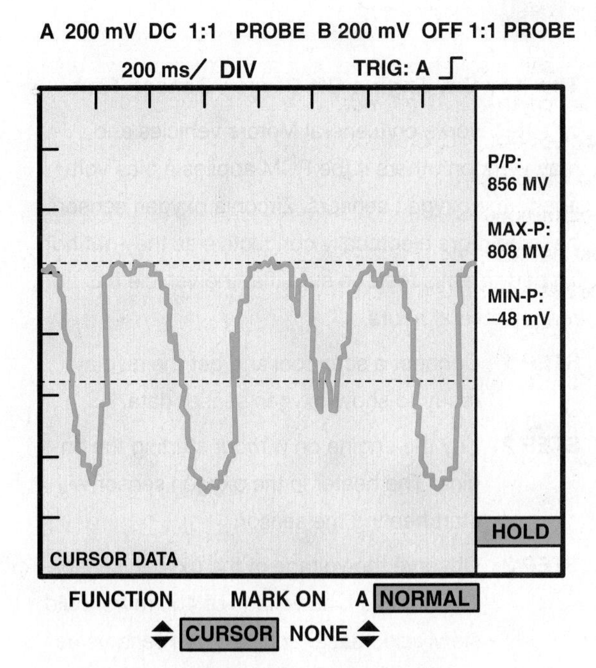

FIGURE 24–13 Using the cursors on the oscilloscope, the high- and low-oxygen sensor values can be displayed on the screen.

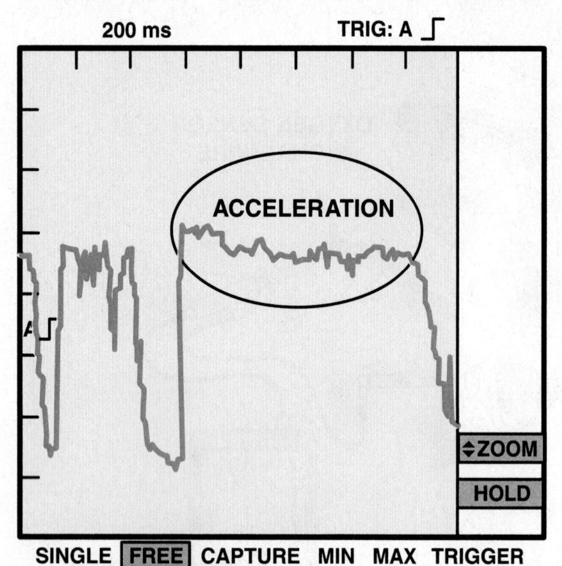

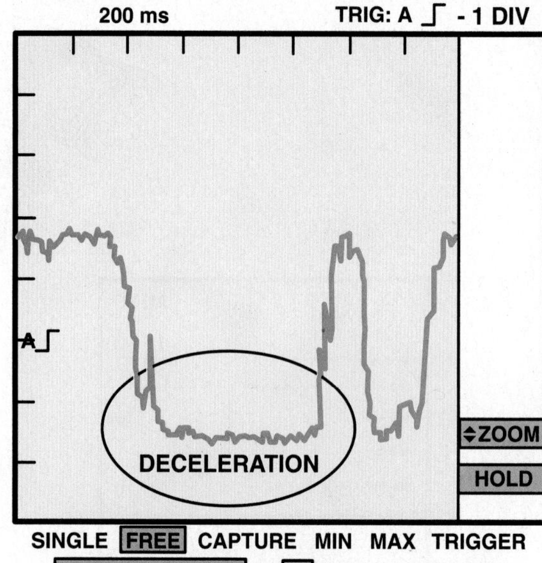

FIGURE 24–14 When the air–fuel mixture rapidly changes such as during a rapid acceleration, look for a rapid response. The transition from low to high should be less than 100 ms.

TECH TIP

The Key On, Engine Off Oxygen Sensor Test

This test works on General Motors vehicles and may work on others if the PCM applies a bias voltage to the oxygen sensors. Zirconia oxygen sensors become more electrically conductive as they get hot. To perform this test, be sure that the vehicle has not run for several hours.

STEP 1 Connect a scan tool and get the display ready to show oxygen sensor data.

STEP 2 Key the engine on *without* starting the engine. The heater in the oxygen sensor will start heating the sensor.

STEP 3 Observe the voltage of the oxygen sensor. The applied bias voltage of 450 mV should slowly decrease for all oxygen sensors as they become more electrically conductive and other bias voltage is flowing to ground.

STEP 4 A good oxygen sensor should indicate a voltage of less than 100 mV after three minutes. Any sensor that displays a higher-than-usual voltage or seems to stay higher longer than the others could be defective or skewed high.

TECH TIP

The Propane Oxygen Sensor Test

Adding propane to the air inlet of a running engine is an excellent way to check if the oxygen sensor is able to react to changes in air–fuel mixture. Follow these steps in performing the propane trick:

1. Connect a digital storage oscilloscope to the oxygen sensor signal wire.
2. Start and operate the engine until up to operating temperature and in closed-loop fuel control.
3. While watching the scope display, add some propane to the air inlet. The scope display should read full rich (over 800 mV), as shown in ● **FIGURE 24–15.**
4. Shut off the propane. The waveform should drop to less than 200 mV (0.200 V), as shown in ● **FIGURE 24–16.**
5. Quickly add some propane while the oxygen sensor is reading low and watch for a rapid transition to rich. The transition should occur in less than 100 milliseconds (ms).

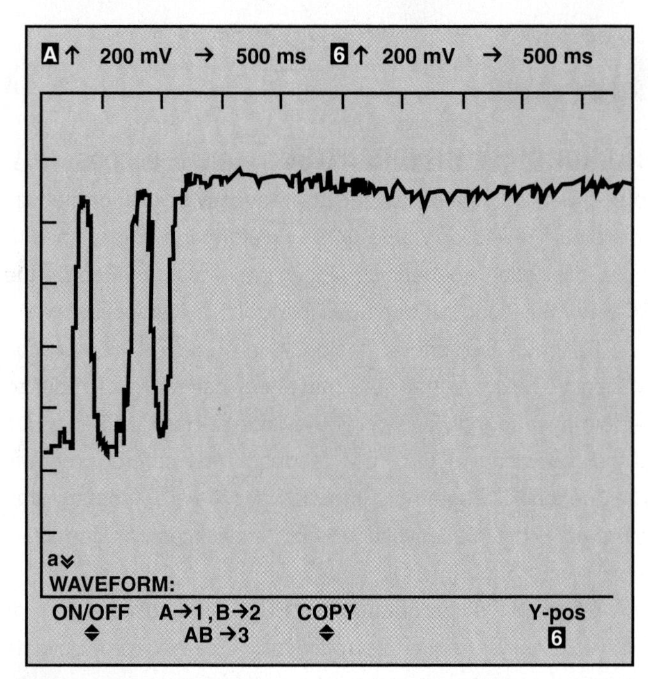

FIGURE 24–15 Adding propane to the air inlet of an engine operating in closed loop with a working oxygen sensor causes the oxygen sensor voltage to read high.

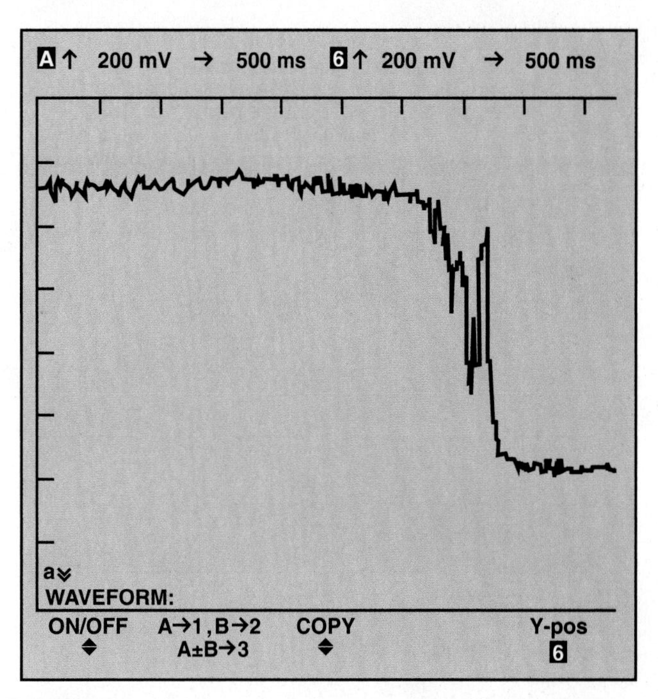

FIGURE 24–16 When the propane is shut off, the oxygen sensor should read below 200 mV.

OXYGEN SENSOR WAVEFORM ANALYSIS

As the O2S warms up, the sensor voltage begins to rise. When the sensor voltage rises above 450 mV, the PCM determines that the sensor is up to operating temperature, takes control of the fuel mixture, and begins to cycle rich and lean. At this point, the system is considered to be in closed loop. ● **SEE FIGURE 24–17**.

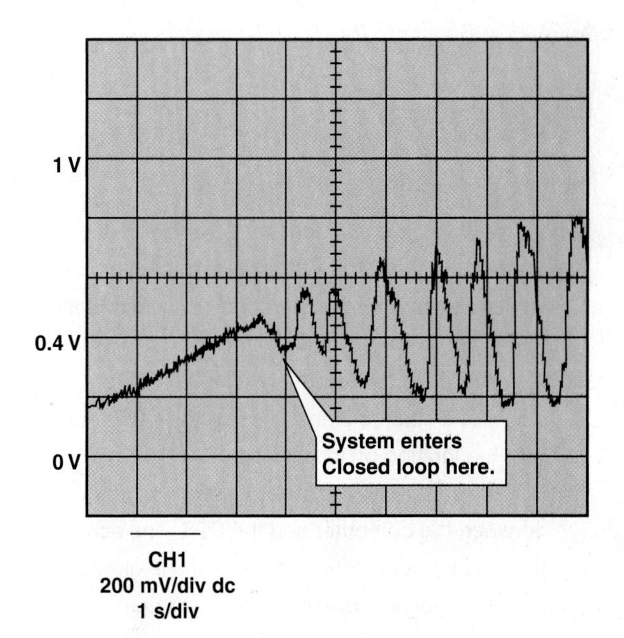

FIGURE 24–17 When the O2S voltage rises above 450 mV, the PCM starts to control the fuel mixture based on oxygen sensor activity.

FREQUENCY The frequency of the O2S is important in determining the condition of the fuel control system. The higher the frequency, the better, but the frequency must not exceed 6 Hz. For its OBD-II standards, the government has stated that a frequency greater than 6 Hz represents a misfire.

THROTTLE-BODY FUEL-INJECTION SYSTEMS. Normal TBI system rich/lean switching frequencies are from about 0.5 Hz at idle to about 3 Hz at 2500 RPM. Additionally, due to the TBI design limitations, fuel distribution to individual cylinders may not always be equal (due to unequal intake runner length, etc.). This may be normal unless certain other conditions are present at the same time.

PORT FUEL-INJECTION SYSTEMS. Specification for port fuel-injection systems is 0.5 Hz at idle to 5 Hz at 2500 RPM. ● **SEE FIGURE 24–18**. Port fuel-injection systems have more rich–lean O2S voltage transitions (cross counts) for a given amount of time than any other type of system, due to the greatly improved system design compared to TBI units.

Port fuel-injection systems take the least amount of time to react to the fuel adaptive command (e.g., changing injector pulse width).

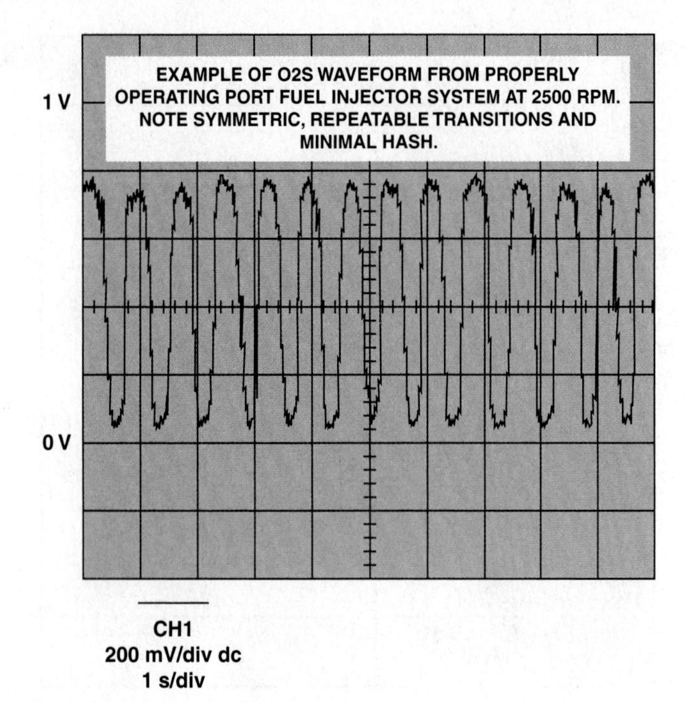

FIGURE 24–18 Normal oxygen sensor frequency is from about one to five times per second.

TECH TIP

Sensor or Wiring?

When troubleshooting a diagnostic trouble code, it is sometimes difficult to determine if the sensor itself is defective or its wiring and ground connections are defective. For example, when diagnosing an O2S code, perform the following to check the wiring:

1. Connect a scan tool and observe the O2S voltage with the ignition on (engine off).
2. Disconnect the O2S pigtail to open the circuit between the computer and the O2S. The scan tool should read 450 mV if the wiring is okay and the scan tool is showing the bias voltage.

   NOTE: Some vehicle manufacturers do not apply a bias voltage to the O2S and the reading on the scan tool may indicate zero and be okay.

3. Ground the O2S wire from the computer. The scan tool should read 0 volts if the wiring is okay.

HASH

BACKGROUND INFORMATION Hash on the O2S waveform is defined as a series of high-frequency spikes, or the fuzz (or noise) viewed on some O2S waveforms, or more specifically, oscillation frequencies higher than those created by the PCM normal feedback operation (normal rich–lean oscillations).

Hash is the critical indicator of reduced combustion efficiency. Hash on the O2S waveform can warn of reduced performance in individual engine cylinders. Hash also impedes proper operation of the PCM feedback fuel control program. The feedback program is the active software program that interprets the O2S voltage and calculates a corrective mixture control command.

Generally, the program for the PCM is not designed to process O2S signal frequencies efficiently that result from events other than normal system operation and fuel control commands. The high-frequency oscillations of the hash can cause the PCM to lose control. This, in turn, has several effects. When the operating strategy of the PCM is adversely affected, the air–fuel ratio drifts out of the catalyst window, which affects converter operating efficiency, exhaust emissions, and engine performance.

Hash on the O2S waveform indicates an exhaust charge imbalance from one cylinder to another, or more specifically, a higher oxygen content sensed from an individual combustion event. Most oxygen sensors, when working properly, can react fast enough to generate voltage deflections corresponding to a single combustion event. The bigger the amplitude of the deflection (hash), the greater the differential in oxygen content sensed from a particular combustion event.

There are vehicles that will have hash on their O2S waveforms and are operating perfectly normal. Small amounts of hash may not be of concern and larger amounts of hash may be all important. A good rule concerning hash is, if engine performance is good, there are no vacuum leaks, and if exhaust HC (hydrocarbon) and oxygen levels are okay while hash is present on the O2S waveform, then the hash is nothing to worry about.

CAUSES OF HASH Hash on the O2S signal can be caused by the following:

1. Misfiring cylinders
   - Ignition misfire
   - Lean misfire
   - Rich misfire
   - Compression-related misfire
   - Vacuum leaks
   - Injector imbalance

2. System design, such as different intake runner length
3. System design amplified by engine and component degradation caused by aging and wear
4. System manufacturing variances, such as intake tract blockage and valve stem mismachining

The spikes and hash on the waveform during a misfire event are created by incomplete combustion, which results in only partial use of the available oxygen in the cylinder. The leftover oxygen goes out the exhaust port and travels past the oxygen sensor. When the oxygen sensor "sees" the oxygen-filled exhaust charge, it quickly generates a low voltage, or spike. A series of these high-frequency spikes make up what we are calling "hash."

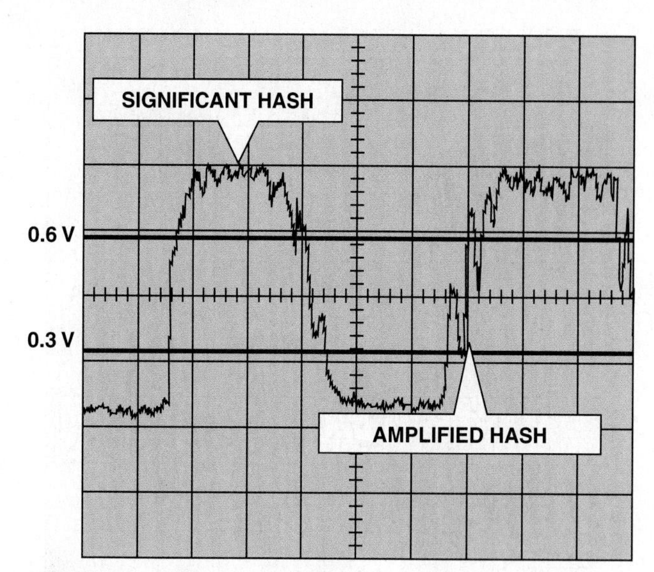

FIGURE 24–19 Significant hash can be caused by faults in one or more cylinders, whereas amplified hash is not as important for diagnosis.

CLASSIFICATIONS OF HASH

CLASS 1: AMPLIFIED AND SIGNIFICANT HASH. Amplified hash is the somewhat unimportant hash that is often present between 300 and 600 millivolts on the O2S waveform. This type of hash is usually not important for diagnosis. That is because amplified hash is created largely as a result of the electrochemical properties of the O2S itself and many times not an engine or other unrelated problem. Hash between 300 and 600 mV is not particularly conclusive, so for all practical purposes it is insignificant. ● SEE FIGURE 24–19.

Significant hash is defined as the hash that occurs above 600 mV and below 300 mV on the O2S waveform. This is the area of the waveform that the PCM is watching to determine the fuel mixture. Significant hash is important for diagnosis because it is caused by a combustion event. If the waveform exhibits class 1 hash, the combustion event problem is probably occurring in only one of the cylinders. If the event happens in a greater number of the cylinders the waveform will become class 3 or be fixed lean or rich the majority of the time.

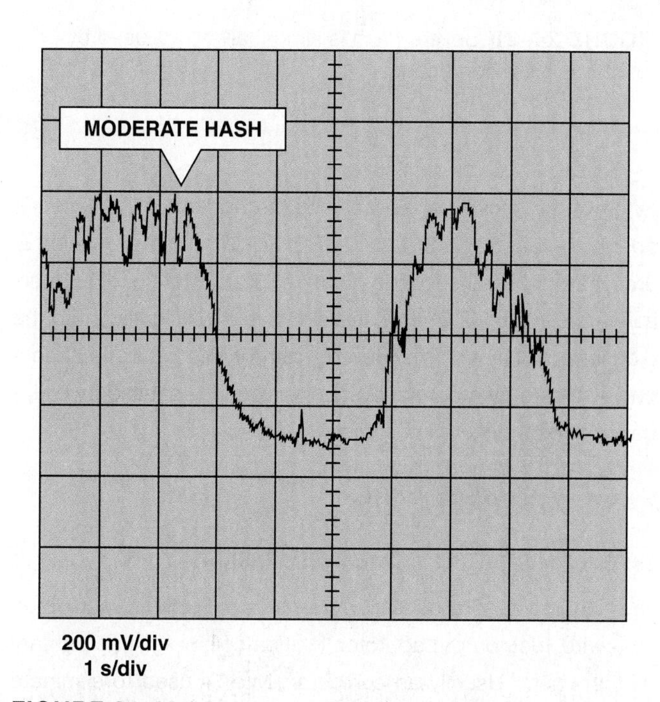

FIGURE 24–20 Moderate hash may or may not be significant for diagnosis.

CLASS 2: MODERATE HASH. Moderate hash is defined as spikes shooting downward from the top arc of the waveform as the waveform carves its arc through the rich phase. Moderate hash spikes are not greater than 150 mV in amplitude. They may get as large as 200 mV in amplitude as the O2S waveform goes through 450 mV. Moderate hash may or may not be significant to a particular diagnosis. ● SEE FIGURE 24–20. For instance, most vehicles will exhibit more hash on the O2S waveform at idle. Additionally, the engine family or type of O2S could be important factors when considering the significance of moderate hash on the O2S waveform.

CLASS 3: SEVERE HASH. Severe hash is defined as hash whose amplitude is greater than 200 mV. Severe hash may even cover the entire voltage range of the sensor for an extended period of operation. Severe hash on the DSO display appears as spikes that shoot downward, over 200 mV from the top of the operating

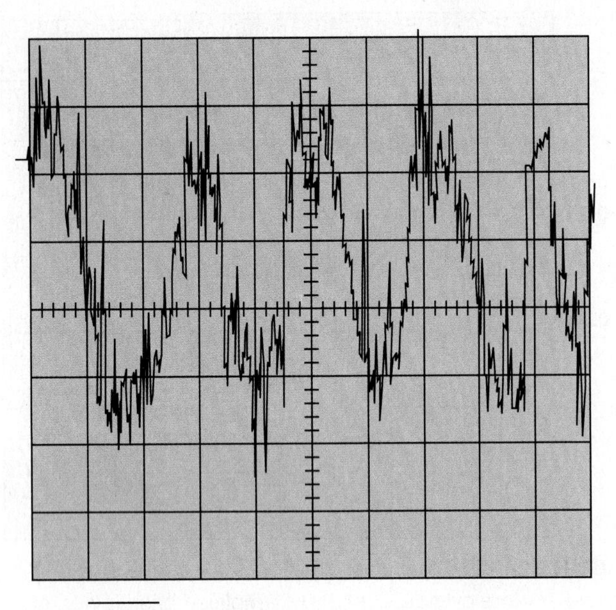

FIGURE 24–21 Severe hash is almost always caused by cylinder misfire conditions.

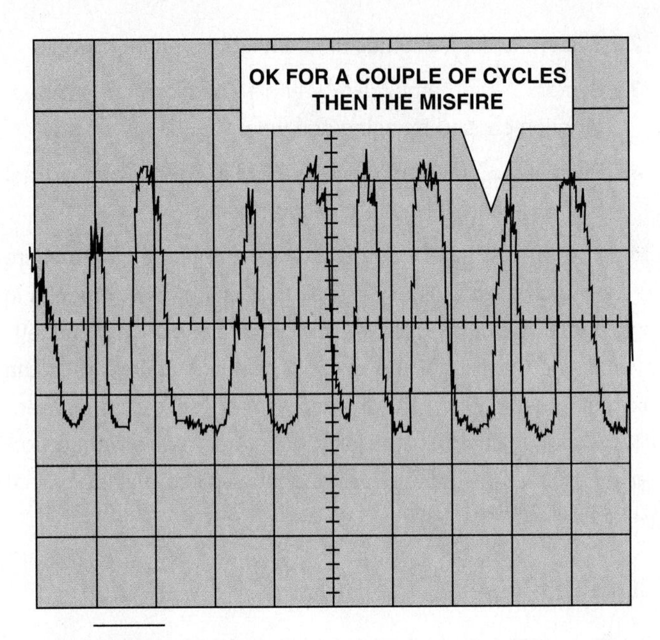

FIGURE 24–22 An ignition- or mixture-related misfire can cause hash on the oxygen sensor waveform.

range of the sensor, or as far as to the bottom of the sensor's operating range. ● **SEE FIGURE 24–21.** If severe hash is present for several seconds during a steady state engine operating mode, say 2500 RPM, it is almost always significant to the diagnosis of any vehicle. Severe hash of this nature is almost never caused by a normal system design. It is caused by cylinder misfire or mixture imbalance.

HASH INTERPRETATION

TYPES OF MISFIRES THAT CAN CAUSE HASH.

1. Ignition misfire caused by a bad spark plug, spark plug wire, distributor cap, rotor, ignition coil, or ignition primary problem. Usually an engine analyzer is used to eliminate these possibilities or confirm these problems. ● **SEE FIGURE 24–22.**
2. Rich misfire from an excessively rich fuel delivery to an individual cylinder (various potential root causes). Air–fuel ratio in a given cylinder ventured below approximately 13:1.
3. Lean misfire from an excessively lean fuel delivery to an individual cylinder (various potential root causes). Air–fuel ratio in a given cylinder ventured above approximately 17:1.
4. Compression-related misfire from a mechanical problem that reduces compression to the point that not enough heat is generated from compressing the air–fuel mixture prior to ignition, preventing combustion. This raises O2S content in the exhaust (for example, a burned valve, broken or worn ring, flat cam lobe, or sticking valve).
5. Vacuum leak misfire unique to one or two individual cylinders. This possibility is eliminated or confirmed by inducing propane around any potential vacuum leak area (intake runners, intake manifold gaskets, vacuum hoses, etc.) while watching the DSO to see when the signal goes rich and the hash changes from ingesting the propane. Vacuum leak misfires are caused when a vacuum leak unique to one cylinder or a few individual cylinders causes the air–fuel ratio in the affected cylinder(s) to venture above approximately 17:1, causing a lean misfire.
6. Injector imbalance misfire (on port fuel-injected engines only); one cylinder has a rich or lean misfire due to an individual injector(s) delivering the wrong quantity of fuel. Injector imbalance misfires are caused when an injector on one cylinder or a few individual cylinders causes the air–fuel ratio in its cylinder(s) to venture above approximately 17:1, causing a lean misfire, or below approximately 13.7:1, causing a rich misfire. ● **SEE FIGURE 24–23.**

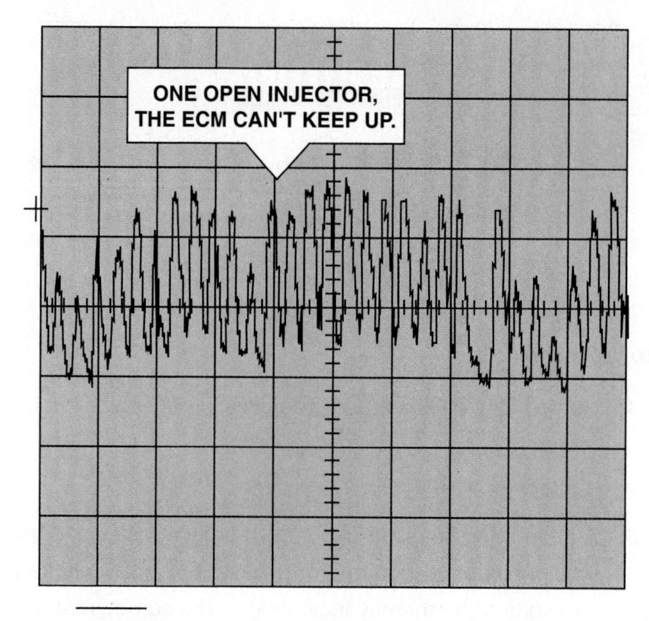

FIGURE 24–23 An injector imbalance can cause a lean or a rich misfire.

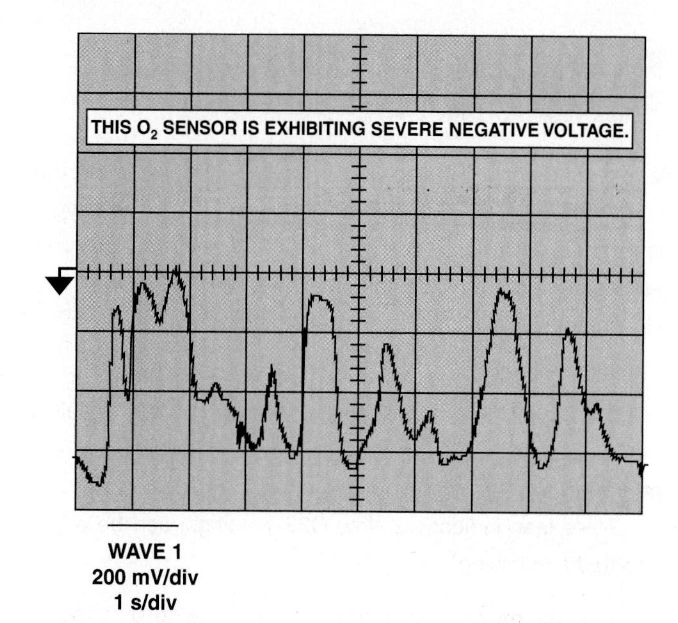

FIGURE 24–24 Negative reading oxygen sensor voltage can be caused by several problems.

OTHER RULES CONCERNING HASH ON THE O2S WAVEFORM If there is significant hash on the O2S signal that is not normal for that type of system, it will usually be accompanied by a repeatable and generally detectable engine miss at idle (e.g., a thump, thump, thump every time the cylinder fires). Generally, if the hash is significant, the engine miss will correlate in time with individual spikes seen on the waveform.

Hash that may be difficult to get rid of (and is normal in some cases) will not be accompanied by a significant engine miss that corresponds with the hash. When the individual spikes that make up the hash on the waveform do not correlate in time with an engine miss, less success can usually be found in getting rid of them by performing repairs.

A fair rule of thumb is if you are sure there are no intake vacuum leaks, and the exhaust gas HC (hydrocarbon) and oxygen levels are normal, and the engine does not run or idle rough, the hash is probably acceptable or normal.

NEGATIVE O2S VOLTAGE

When testing O2S waveforms, some oxygen sensors will exhibit some negative voltage. The acceptable amount of negative O2S voltage is −0.75 mV, provided that the maximum voltage peak exceeds 850 mV. ● **SEE FIGURE 24–24.** Testing has shown that negative voltage signals from an oxygen sensor have usually been caused due to the following:

1. Chemical poisoning of sensing element (silicon, oil, etc.).
2. Overheated engines.
3. Mishandling of new oxygen sensors (dropped and banged around, resulting in a cracked insulator).
4. Poor oxygen sensor ground.

TECH TIP

Look for Missing Shield

In rare (very rare) instances, the metal shield on the exhaust side of the oxygen sensor (the shield over the zirconia thimble) may be damaged (or missing) and may create hash on the O2S waveform that could be mistaken for bad injectors or other misfires, vacuum leaks, or compression problems. After you have checked everything, and possibly replaced the injectors, pull the O2S to check for rare situations.

LOW O2S READINGS

An oxygen sensor reading that is low could be due to other things besides a lean air–fuel mixture. Remember, an oxygen sensor senses oxygen, not unburned gas, even though a high reading generally indicates a rich exhaust (lack of oxygen) and a low reading indicates a lean mixture (excess oxygen).

FALSE LEAN If an oxygen sensor reads low as a result of a factor besides a lean mixture, it is often called a **false lean indication**.

False lean indications (low O2S readings) can be attributed to the following:

1. **Ignition misfire.** An ignition misfire due to a defective spark plug wire, fouled spark plug, and so forth, causes no burned air and fuel to be exhausted past the O2S. The O2S "sees" the oxygen (not the unburned gasoline) and the O2S voltage is low.
2. **Exhaust leak in front of the O2S.** An exhaust leak between the engine and the oxygen sensor causes outside oxygen to be drawn into the exhaust and past the O2S. This oxygen is "read" by the O2S and produces a lower-than-normal voltage. The computer interrupts the lower-than-normal voltage signal from the O2S as meaning that the air–fuel mixture is lean. The computer will cause the fuel system to deliver a richer air–fuel mixture.
3. **A spark plug misfire represents a false lean signal to the oxygen sensor.** The computer does not know that the extra oxygen going past the oxygen sensor is not due to a lean air–fuel mixture. The computer commands a richer mixture, which could cause the spark plugs to foul, increasing the rate of misfirings.

HIGH O2S READINGS

An oxygen sensor reading that is high could be due to other things beside a rich air–fuel mixture. When the O2S reads high as a result of other factors besides a rich mixture, it is often called a **false rich indication**.

False rich indication (high O2S readings) can be attributed to the following:

1. Contaminated O2S due to additives in the engine coolant or due to silicon poisoning
2. A stuck-open EGR valve (especially at idle)
3. A spark plug wire too close to the oxygen sensor signal wire, which can induce a higher-than-normal voltage in the signal wire, thereby indicating to the computer a false rich condition
4. A loose oxygen sensor ground connection, which can cause a higher-than-normal voltage and a false rich signal
5. A break or contamination of the wiring and its connectors, which could prevent reference oxygen from reaching the oxygen sensor, resulting in a false rich indication. (All oxygen sensors require an oxygen supply inside the sensor itself for reference to be able to sense exhaust gas oxygen.)

POST-CATALYTIC CONVERTER OXYGEN SENSOR TESTING

The oxygen sensor located behind the catalytic converter is used on OBD-II vehicles to monitor converter efficiency. A changing air–fuel mixture is required for the most efficient operation of the converter. If the converter is working correctly, the oxygen content after the converter should be fairly constant. ● **SEE FIGURE 24–25.**

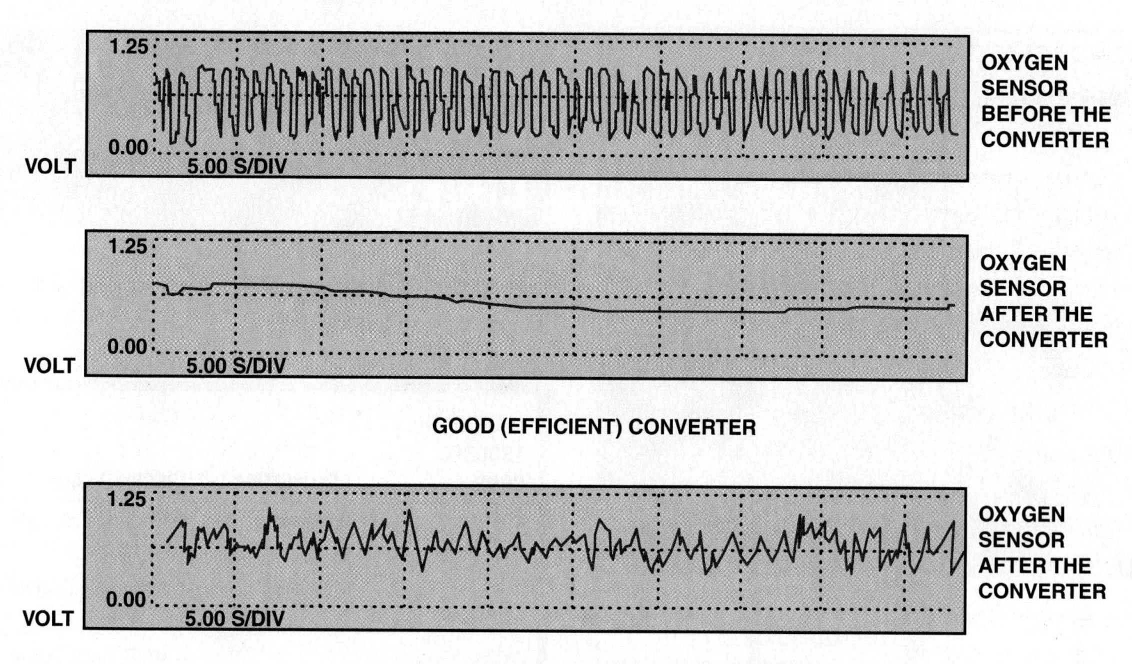

FIGURE 24–25 The post-catalytic converter oxygen sensor should display very little activity if the catalytic converter is efficient.

OXYGEN SENSOR VISUAL INSPECTION

Whenever an oxygen sensor is replaced, the old sensor should be carefully inspected to help determine the cause of the failure. This is an important step because if the cause of the failure is not discovered, it could lead to another sensor failure.

Inspection may reveal the following:

1. **Black sooty deposits** usually indicate a rich air–fuel mixture.
2. **White chalky deposits** are characteristic of silica contamination. Usual causes for this type of sensor failure include silica deposits in the fuel or a technician having used the wrong type of silicone sealant during the servicing of the engine.
3. **White sandy or gritty deposits** are characteristic of antifreeze (ethylene glycol) contamination. A defective cylinder head or intake manifold gasket, or a cracked cylinder head or engine block could be the cause. Antifreeze may also cause the oxygen sensor to become green as a result of the dye used in antifreeze.
4. **Dark brown deposits** are an indication of excessive oil consumption. Possible causes include a defective positive crankcase ventilation (PCV) system or a mechanical engine problem such as defective valve stem seals or piston rings.

CAUTION: Do not spray any silicone spray near the engine where the engine vacuum could draw the fumes into the engine. This can also cause silica damage to the oxygen sensor. Also be sure that the silicone sealer used for gaskets is rated oxygen-sensor safe.

H02S Data

Loop Status	Closed
Desired H02S Bank 1 Sen	0.998 Lambda
H02S Bank 1 Sensor 1	0.975 Lambda
Desired H02S Bank 2 Sen	0.998 Lambda
H02S Bank 2 Sensor 1	1.006 Lambda
H02S Bank 1 Sensor 1	1.43 Volts
H02S Bank 1 Sensor 2	611 mV
H02S Bank 2 Sensor 1	1.49 Volts
H02S Bank 2 Sensor 2	526 mV

1 / 40

Loop Status

Select Items | DTC | Quick Snapshot | More

FIGURE 24–26 The target lambda on this vehicle is slightly lower than 1.0 indicating that the PCM is attempting to supply the engine with an air–fuel mixture that is slightly richer than stoichiometric. Multiply the lambda number by 14.7 to find the actual air–fuel ratio.

FREQUENTLY ASKED QUESTION

What Is Lambda?

An oxygen sensor is also called a lambda sensor because the voltage changes at the air–fuel ratio of 14.7:1, which is the stoichiometric rate for gasoline. If this mixture of gasoline and air is burned, all of the gasoline is burned and uses all of the oxygen in the mixture. This exact ratio represents a lambda of 1.0. If the mixture is richer (more fuel or less air), the number is less than 1.0, such as 0.850. If the mixture is leaner than 14.7:1 (less fuel or more air), the lambda number is higher than 1.0, such as 1.130. Often, the target lambda is displayed on a scan tool. ● **SEE FIGURE 24–26.**

OXYGEN SENSOR–RELATED DIAGNOSTIC TROUBLE CODES

Diagnostic trouble codes (DTCs) associated with the oxygen sensor include the following:

DIAGNOSTIC TROUBLE CODE	DESCRIPTION	POSSIBLE CAUSES
P0131	Upstream H02S grounded	▪ Exhaust leak upstream of H02S (bank 1) ▪ Extremely lean air–fuel mixture ▪ H02S defective or contaminated ▪ H02S signal wire shorted-to-ground
P0132	Upstream H02S shorted	▪ Upstream H02S (bank 1) shorted ▪ Defective H02S ▪ Fuel-contaminated H02S
P0133	Upstream H02S slow response	▪ Open or short in heater circuit ▪ Defective or fuel-contaminated H02S ▪ EGR or fuel-system fault

SUMMARY

1. An oxygen sensor produces a voltage output signal based on the oxygen content of the exhaust stream.
2. If the exhaust has little oxygen, the voltage of the oxygen sensor will be close to 1 volt (1,000 mV) and close to zero if there is high oxygen content in the exhaust.
3. Oxygen sensors can have one, two, three, four, or more wires, depending on the style and design.
4. A wide-band oxygen sensor, also called a lean air–fuel (LAF) or linear air–fuel ratio sensor, can detect air–fuel ratios from as rich as 12:1 to as lean as 18:1.
5. The oxygen sensor signal determines fuel trim, which is used to tailor the air–fuel mixture for the catalytic converter.
6. Conditions can occur that cause the oxygen sensor to be fooled and give a false lean or false rich signal to the PCM.
7. Oxygen sensors can be tested using a digital meter, a scope, or a scan tool.

REVIEW QUESTIONS

1. How does an oxygen sensor detect oxygen levels in the exhaust?
2. What are three basic designs of oxygen sensors and how many wires may be used for each?
3. What is the difference between open-loop and closed-loop engine operation?
4. What are three ways oxygen sensors can be tested?
5. How can the oxygen sensor be fooled and provide the wrong information to the PCM?

CHAPTER QUIZ

1. The sensor that must be warmed and functioning before the engine management computer will go to closed loop is the ________.
   - **a.** O2S
   - **b.** ECT sensor
   - **c.** Engine MAP sensor
   - **d.** BARO sensor
2. The voltage output of a zirconia oxygen sensor when the exhaust stream is lean (excess oxygen) is ________.
   - **a.** Relatively high (close to 1 volt)
   - **b.** About in the middle of the voltage range
   - **c.** Relatively low (close to 0 volt)
   - **d.** Either a or b, depending on atmospheric pressure
3. Where is sensor 1, bank 1 located?
   - **a.** On the same bank where number 1 cylinder is located
   - **b.** In the exhaust system upstream of the catalytic converter
   - **c.** On the bank opposite cylinder number 1
   - **d.** Both a and b
4. A heated zirconia oxygen sensor will have how many wires?
   - **a.** 2
   - **b.** 3
   - **c.** 4
   - **d.** Either b or c
5. A high O2S voltage could be due to ________.
   - **a.** A rich exhaust
   - **b.** A lean exhaust
   - **c.** A defective spark plug wire
   - **d.** Both a and c
6. A low O2S voltage could be due to ________
   - **a.** A rich exhaust
   - **b.** A lean exhaust
   - **c.** A defective spark plug wire
   - **d.** Both b and c
7. An oxygen sensor is being tested with digital multimeter (DMM), using the MIN/MAX function. The readings are: minimum = 78 mV; maximum = 932 mV; average = 442 mV. Technician A says that the engine is operating correctly. Technician B says that the oxygen sensor is skewed too rich. Which technician is correct?
   - **a.** Technician A only
   - **b.** Technician B only
   - **c.** Both Technicians A and B
   - **d.** Neither Technician A nor B
8. An oxygen sensor is being tested using a digital storage oscilloscope (DSO). A good oxygen sensor should display how many switches per second?
   - **a.** 1 to 5
   - **b.** 5 to 10
   - **c.** 10 to 15
   - **d.** 15 to 20
9. When testing an oxygen sensor using a digital storage oscilloscope (DSO), how quickly should the voltage change when either propane is added to the intake stream or when a vacuum leak is created?
   - **a.** Less than 50 ms
   - **b.** 1 to 3 seconds
   - **c.** Less than 100 ms
   - **d.** 100 to 200 ms
10. A P0133 DTC is being discussed. Technician A says that a defective heater circuit could be the cause. Technician B says that a contaminated sensor could be the cause. Which technician is correct?
    - **a.** Technician A only
    - **b.** Technician B only
    - **c.** Both Technicians A and B
    - **d.** Neither Technician A nor B

chapter 25

WIDE-BAND OXYGEN SENSORS

LEARNING OBJECTIVES

After studying this chapter, the reader will be able to:

1. Compare dual cell wide-band sensors to single cell wide-band sensors.
2. Discuss the operation of a wide-band oxygen sensor.
3. List the methods that can be used to test oxygen sensor.
4. Discuss wide-band oxygen pattern failures.

This chapter will help you prepare for Engine Repair (A8) ASE certification test content area “C” (Fuel, Air Induction, and Exhaust Systems Diagnosis and Repair).

KEY TERMS

Air–fuel ratio sensor 398
Air reference chamber 394
Ambient air electrode 393
Ambient side electrode 393
Cup design 392
Diffusion chamber 394
Dual cell 394
Exhaust side electrode 393
Finger design 392
Light-off time (LOT) 394
Nernst cell 394
Planar design 394
Pump cell 394
Reference electrode 393
Reference voltage 394
Signal electrode 393
Single cell 398
Thimble design 392

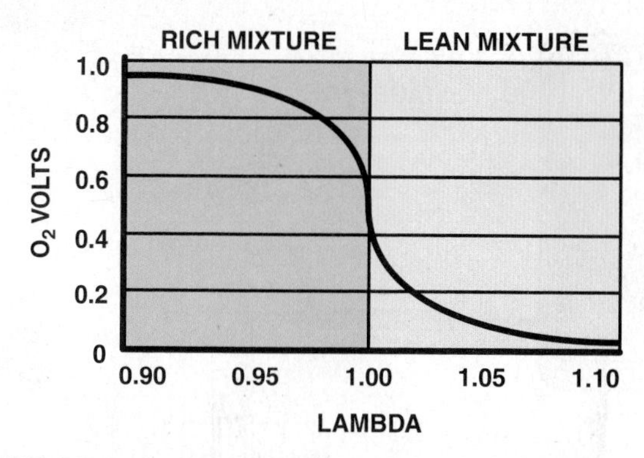

FIGURE 25–1 A conventional zirconia oxygen sensor can only reset to exhaust mixtures that are richer or leaner than 14.7:1 (lambda 1.00).

TERMINOLOGY

Wide-band oxygen sensors have been used since 1992 on some Hondas. Wide-band oxygen sensors are used by most vehicle manufacturers to ensure that the exhaust emissions can meet the current standard. Wide-band oxygen sensors are also called by various names, depending on the vehicle and/or oxygen sensor manufacturer. The terms used include:

- **Wide-band oxygen sensor**
- **Broadband oxygen sensor**
- **Wide-range oxygen sensor**
- **Air–fuel ratio (AFR) sensor**
- **Wide-range air–fuel (WRAF) sensor**
- **Lean air–fuel (LAF) sensor**
- **Air–fuel (AF) sensor**

Wide-band oxygen sensors are also manufactured in dual cell and single cell designs.

NEED FOR WIDE-BAND SENSORS

INTRODUCTION A conventional zirconia oxygen sensor reacts to an air–fuel mixture either richer or leaner than 14.7:1. This means that the sensor cannot be used to detect the exact air–fuel mixture. ● **SEE FIGURE 25–1**.

The need for more stringent exhaust emission standards such as the national-low-emission vehicle (NLEV), plus the ultra-low-emission vehicle (ULEV), and the super-ultra-low-emission vehicle (SULEV) require more accurate fuel control than can be provided by a traditional oxygen sensor.

PURPOSE AND FUNCTION A wide-band oxygen sensor is capable of supplying air–fuel ratio information to the PCM over a much broader range. The use of a wide-band oxygen sensor compared with a conventional zirconia oxygen sensor differs as follows:

1. Able to detect exhaust air–fuel ratio from as rich as 10:1 and as lean as 23:1 in some cases.
2. Cold-start activity within as little as 10 seconds.

FREQUENTLY ASKED QUESTION

How Quickly Can a Wide-Band Oxygen Sensor Achieve Closed Loop?

In a Toyota Highlander hybrid electric vehicle, the gasoline engine start is delayed for a short time when first started. It is capable of being driven immediately using electric power alone and the oxygen sensor heaters are turned on at first start. The gasoline engine often achieves closed-loop operation during *cranking* because the oxygen sensors are fully warm and ready to go at the same time the engine is started. Having the gasoline engine achieve closed loop quickly allows it to meet the stringent SULEV standards.

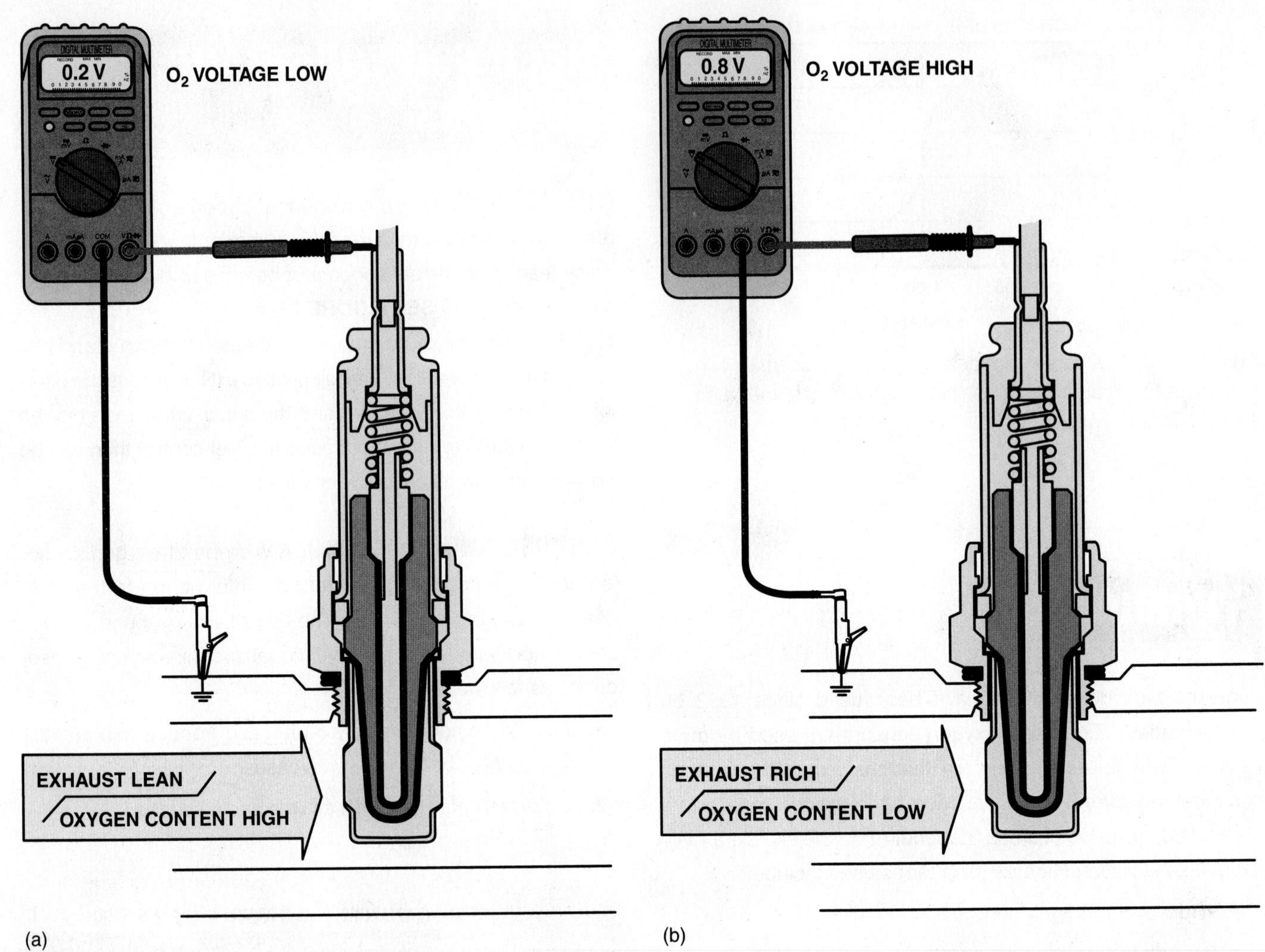

FIGURE 25–2 (a) When the exhaust is lean, the output of a zirconia oxygen sensor is below 450 mV. (b) When the exhaust is rich, the output of a zirconia oxygen sensor is above 450 mV.

CONVENTIONAL O2S REVIEW

NARROW BAND A conventional zirconia oxygen sensor (O2S) is only able to detect if the exhaust is richer or leaner than 14.7:1. A conventional oxygen sensor is therefore referred to as:

- **Two-step sensor**—either rich or lean
- **Narrow band sensor**—only informs the PCM whether the exhaust is rich or lean

The voltage at which a zirconia oxygen sensor switches from rich to lean or from lean to rich is 0.450 V (450 mV).

- Above 0.450 V = rich
- Below 0.450 V = lean

● **SEE FIGURE 25–2.**

CONSTRUCTION A typical zirconia oxygen sensor has the sensing element in the shape of a thimble and is often referred to as:

- **Thimble design**
- **Cup design**
- **Finger design**

● **SEE FIGURE 25–3.**

A typical zirconia oxygen sensor has a heater inside the thimble and does not touch the inside of the sensor. The sensor is similar to a battery that has two electrodes and an electrolyte. The electrolyte is solid and is the zirconia (zirconium dioxide).

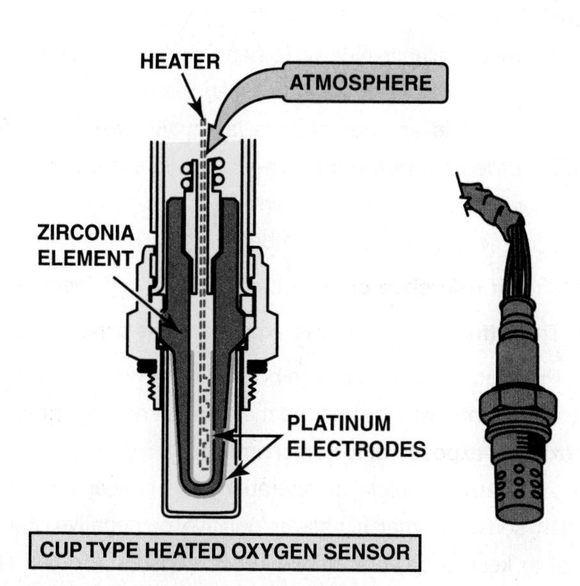

FIGURE 25–3 Most conventional zirconia oxygen sensors and some wide-band oxygen sensors use the cup-type design.

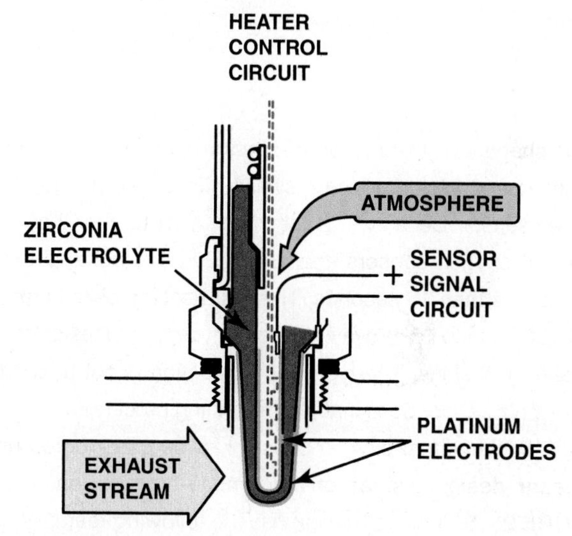

FIGURE 25–4 A typical heated zirconia oxygen sensor, showing the sensor signal circuit that uses the outer (exhaust) electrode as the negative and the ambient air side electrode as the positive.

There are also two porous platinum electrodes, which have the following functions:

- **Exhaust side electrode**—This electrode is exposed to the exhaust stream.
- **Ambient side electrode**—This electrode is exposed to outside (ambient) air and is the **signal electrode**, also called the **reference electrode** or **ambient air electrode**.

  ● **SEE FIGURE 25–4.**

The electrolyte (zirconia) is able to conduct electrons as follows:

- If the exhaust is rich, O_2 from the reference (inner) electrode tends to flow to the exhaust side electrode, which results in the generation of a voltage.
- If the exhaust is lean, O_2 flow is not needed and as a result, there is little, if any, electron movement and, therefore, no voltage is being produced.

HEATER CIRCUITS The heater circuit on conventional oxygen sensors requires 0.8 to 2.0 amperes and it keeps the sensor at about 600°F (315°C).

A wide-band oxygen sensor operates at a higher temperature than a conventional HO2S from 1,200°F to 1,400°F (650°C to 760°C). The amount of electrical current needed for a wide-band oxygen sensor is about 8 to 10 amperes.

PLANAR DESIGN In 1998, Bosch introduced a wide-band oxygen sensor that is flat and thin (1.5 mm or 0.006 inch) and not

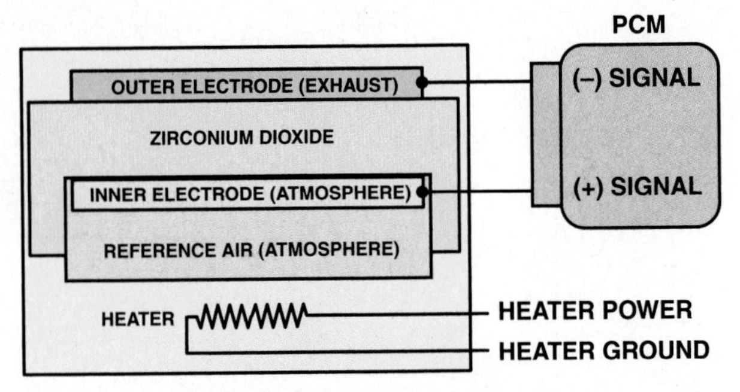

FIGURE 25–5 A planar design zirconia oxygen sensor places all of the elements together, which allows the sensor to reach operating temperature quickly.

in the shape of a thimble, as previously constructed. Now several manufacturers produce a similar planar design wide-band oxygen sensor. Because it is thin, it is easier to heat than older styles of oxygen sensors and as a result can achieve closed loop in less than 10 seconds. This fast heating, called **light-off time (LOT)**, helps improve fuel economy and reduces cold-start exhaust emissions. The type of construction is not noticed by the technician, nor does it affect the testing procedures.

A conventional oxygen sensor can be constructed using a **planar design** instead of the thimble-type design. ● **SEE FIGURE 25–5**. A planar design has the following features:

- The elements including the zirconia electrolyte and the two electrodes and heater are stacked together in a flat-type design.
- The planar design allows faster warm-up because the heater is in direct contact with the other elements.
- Planar oxygen sensors are the most commonly used. Some planar designs are used as a conventional narrow-band oxygen sensor.

The sandwich-type design of the planar style of oxygen sensor has the same elements and operates the same, but is stacked in the following way from the exhaust side to the ambient air side:

Exhaust stream

Outer electrode

Zirconia (ZiO_2) (electrolyte)

Inner electrode (reference or signal)

Outside (ambient) air

Heater

Another name for a conventional oxygen sensor is a **Nernst cell**. The Nernst cell is named for Walther Nernst, 1864–1941, a German physicist known for his work in electrochemistry.

DUAL CELL PLANAR WIDE-BAND SENSOR OPERATION

In a conventional zirconia oxygen sensor, a bias or **reference voltage** can be applied to the two platinum electrodes, and then oxygen ions can be forced (pumped) from the ambient reference air side to the exhaust side of the sensor. If the polarity is reversed, the oxygen ion can be forced to travel in the opposite direction.

A **dual cell** planar-type wide-band oxygen sensor is made like a conventional planar O2S and is labeled Nernst cell. Above the Nernst cell is another zirconia layer with two electrodes, which is called the **pump cell**. The two cells share a common ground, which is called the reference.

There are two internal chambers:

- The **air reference chamber** is exposed to ambient air.
- The **diffusion chamber** is exposed to the exhaust gases.

Platinum electrodes are on both sides of the zirconia electrolyte elements, which separate the air reference chamber and the exhaust-exposed diffusion chamber.

The basic principle of operation of a typical wide-band oxygen sensor is that it uses a positive or negative voltage signal to keep a balance between two sensors. Oxygen sensors do not measure the quantity of free oxygen in the exhaust. Instead, oxygen sensors produce a voltage that is based on the ion flow between the platinum electrodes of the sensor to maintain a stoichiometric balance.

For example:

- If there is a lean exhaust, there is oxygen in the exhaust and the ion flow from the ambient side to the exhaust side is low.
- If there were rich exhaust, the ion flow is increased to help maintain balance between the ambient air side and the exhaust side of the sensor.

The PCM can apply a small current to the pump cell electrodes, which causes oxygen ions through the zirconia into or out of the diffusion chamber. The PCM pumps O_2 ions in and out of the diffusion chamber to bring the voltage back to 0.450 volt, using the pump cell.

The operation of a wide-band oxygen sensor is best described by looking at what occurs when the exhaust is stoichiometric, rich, and lean.

STOICHIOMETRIC

- When the exhaust is at stoichiometric (14.7:1 air–fuel ratio), the voltage of the Nernst cell is 450 mV (0.450 V).

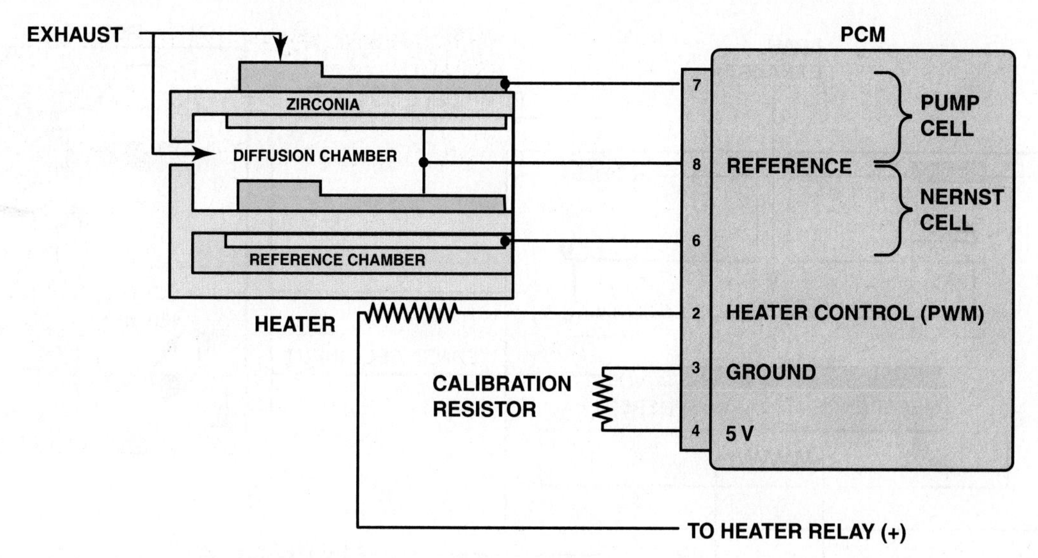

FIGURE 25–6 The reference electrodes are shared by the Nernst cell and the pump cell.

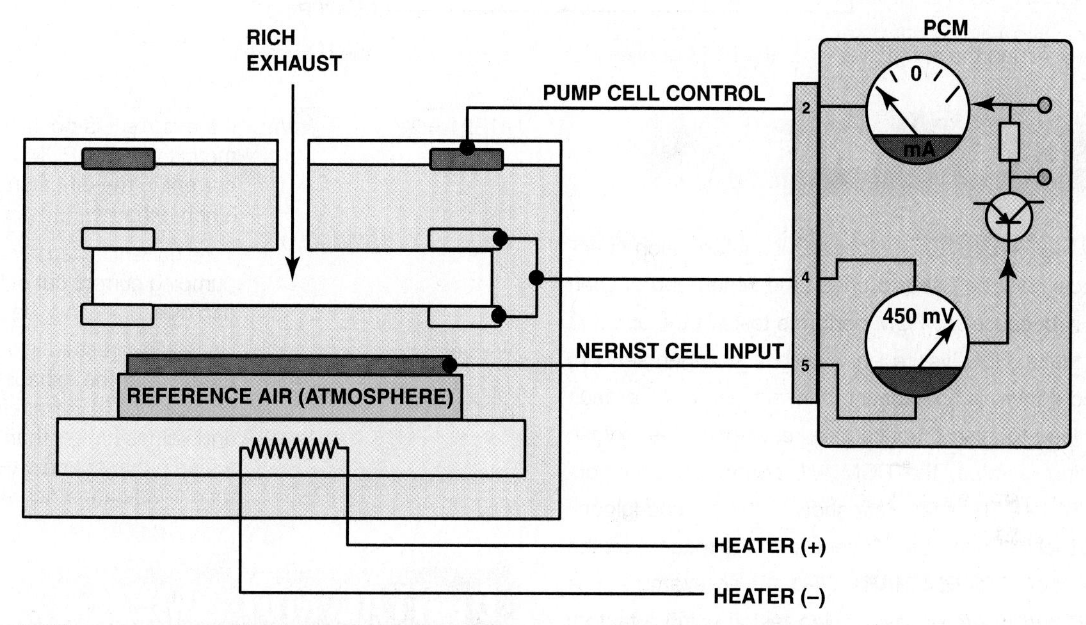

FIGURE 25–7 When the exhaust is rich, the PCM applies a negative current into the pump cell.

- The voltage between the diffusion chamber and the air reference chamber changes from 0.450 volt. This voltage will be:
  - Higher if the exhaust is rich
  - Lower if the exhaust is lean

The reference voltage remains constant, usually at 2.5 volts, but can vary depending on the year, make, and model of vehicle and the type of sensor. Typical reference voltages include:

- 2.2
- 2.5
- 2.7
- 3.3
- 3.6

● **SEE FIGURE 25–6.**

RICH EXHAUST. When the exhaust is rich, the voltage between the common (reference) electrode and the Nernst cell electrode that is exposed to ambient air is higher than 0.450 volt. The PCM applies a negative current in milliamperes to the pump cell electrode to bring the circuit back into balance. ● **SEE FIGURE 25–7.**

LEAN EXHAUST. When the exhaust is lean, the voltage between the common (reference) electrode and the Nernst cell electrode is lower than 0.450 volt. The PCM applies a positive current in milliamperes to the pump cell to bring the circuit back into balance. ● **SEE FIGURE 25–8.**

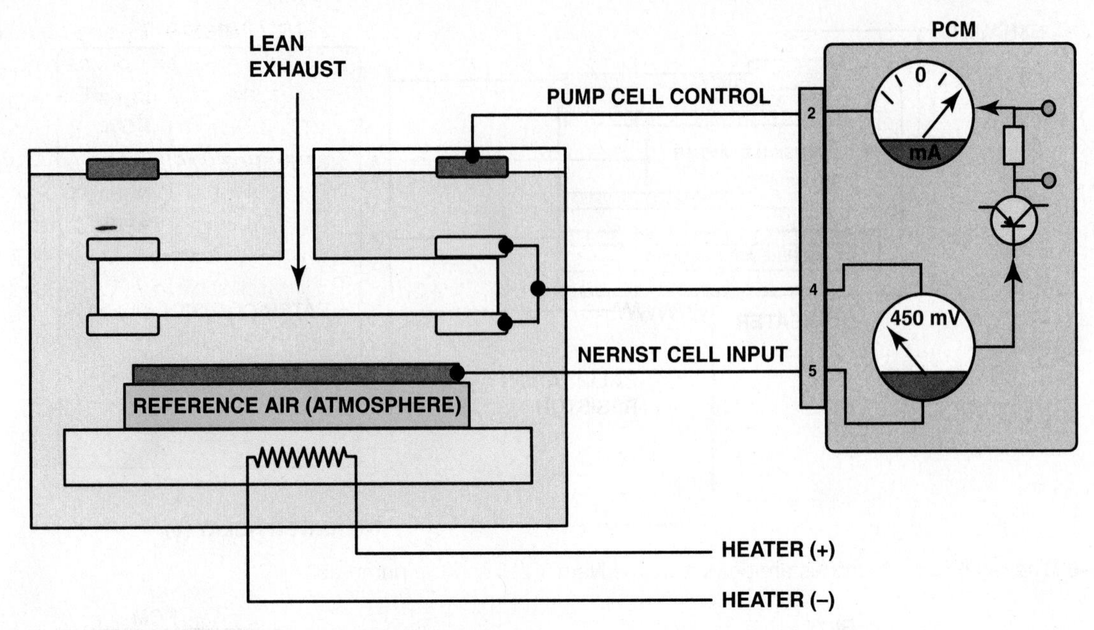

FIGURE 25–8 When the exhaust is lean, the PCM applies a positive current into the pump cell.

DUAL CELL DIAGNOSIS

SCAN TOOL DIAGNOSIS Most service information specifies that a scan tool be used to check the wide-band oxygen sensor. This is because the PCM performs tests of the unit and can identify faults. However, even wide-band oxygen sensors can be fooled if there is an exhaust manifold leak or other fault which could lead to false or inaccurate readings. If the oxygen sensor reading is false, the PCM will command an incorrect amount of fuel. The scan data shown on a generic (global) OBD-II scan tool will often be different than the reading on the factory scan tool. ● **SEE CHART 25–1** for an example of a Toyota wide-band oxygen sensor being tested using a factory scan tool and a generic OBD-II scan tool.

MASTER TECH TOYOTA (FACTORY SCAN TOOL)	OBD-II SCAN TOOL	AIR–FUEL RATIO
2.50 V	0.50 V	12.5:1
3.00 V	0.60 V	14.0:1
3.30 V	0.66 V	14.7:1
3.50 V	0.70 V	15.5:1
4.00 V	0.80 V	18.5:1

CHART 25–1

A comparison showing what a factory scan tool and a generic OBD-II scan tool might display at various air–fuel ratios.

SCAN TOOL DATA (PID) The following information will be displayed on a scan tool when looking at data for a wide-band oxygen sensor:

HOS21 = ________ mA	If the current is positive, this means that the PCM is pumping current in the diffusion gap due to a rich exhaust. If the current is negative, the PCM is pumping current out of the diffusion gap due to a lean exhaust.
Air–fuel ratio =	Usually expressed in lambda. One means that the exhaust is at stoichiometric (14.7:1 air–fuel ratio) and values higher than 1 indicate a lean exhaust and values lower than 1 indicate a rich exhaust.

DIGITAL MULTIMETER TESTING

When testing a wide-band oxygen sensor for proper operation, perform the following steps:

STEP 1 Check service information and determine the circuit and connector terminal identification.

STEP 2 Measure the calibration resistor. While the value of this resistor can vary widely, depending on the type of sensor, the calibrating resistor should still be checked for opens and shorts.

NOTE: The calibration resistor is usually located within the connector itself.

- If open, the ohmmeter will read OL (infinity ohms).
- If shorted, the ohmmeter will read zero or close to zero.

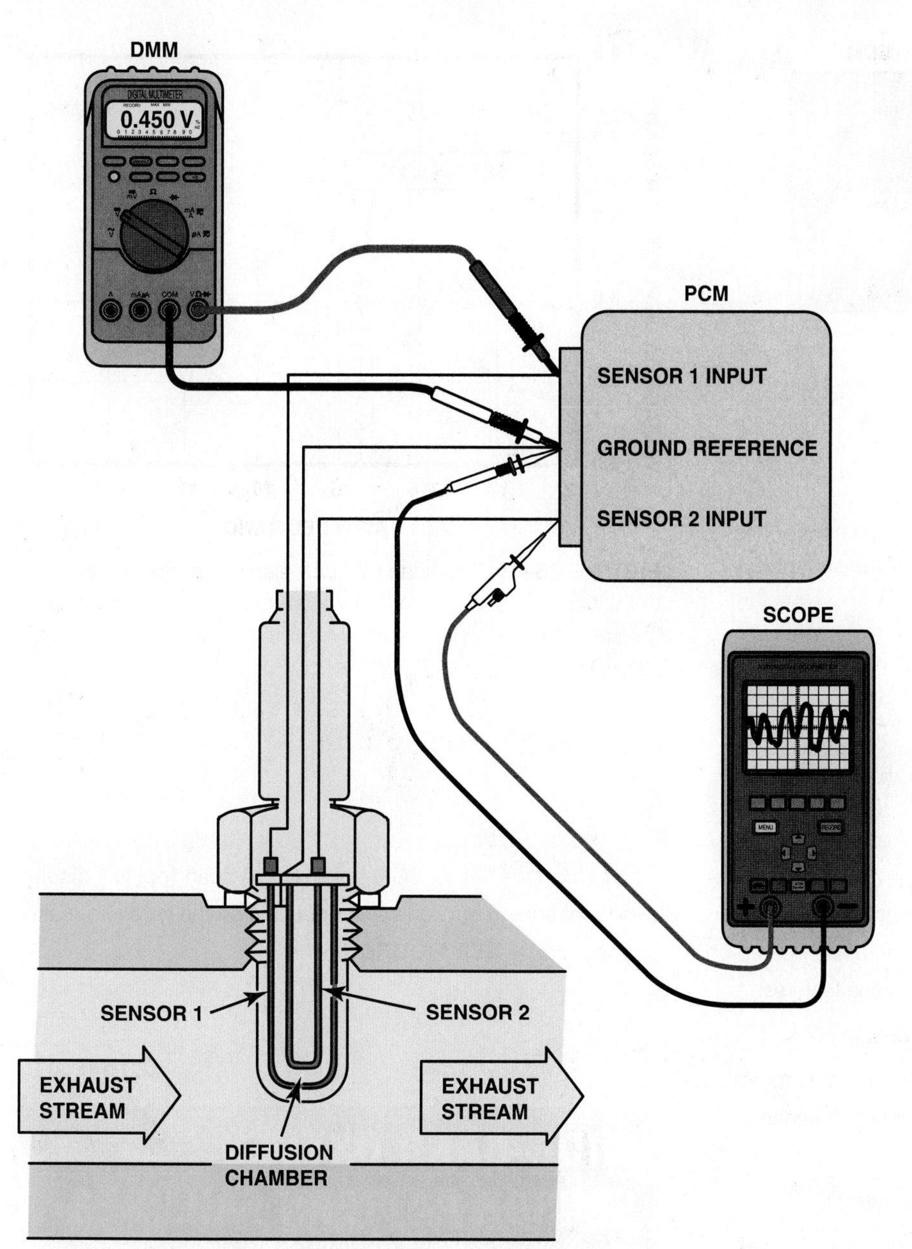

FIGURE 25–9 Testing a dual cell wide-band oxygen sensor can be done using a voltmeter or a scope. The meter reading is attached to the Nernst cell (sensor 1) and should read stoichiometric (450 mV) at all times. The scope is showing activity to the pump cell (sensor 2) with commands from the PCM to keep the Nernst cell at 14.7:1 air–fuel ratio.

STEP 3 Measure the heater circuit for proper resistance or current flow.

STEP 4 Measure the reference voltage relative to ground. This can vary but is generally 2.4 to 2.6 volts.

STEP 5 Using jumper wires, connect an ammeter and measure the current in the pump cell control wire.

RICH EXHAUST (LAMBDA LESS THAN 1.00) When the exhaust is rich, the Nernst cell voltage will move higher than 0.45 volt. The PCM will pump oxygen from the exhaust into the diffusion gap by applying a negative voltage to the pump cell.

LEAN EXHAUST (LAMBDA HIGHER THAN 1.00)

When the exhaust is lean, the Nernst cell voltage will move lower than 0.45 volt. The PCM will pump oxygen out of the diffusion gap by applying a positive voltage to the pump cell. ● **SEE FIGURE 25–9.**

The pump cell is used to pump oxygen into the diffusion gap when the exhaust is rich. The pump cell applies a negative voltage to do this.

- Positive current = lean exhaust
- Negative current = rich exhaust

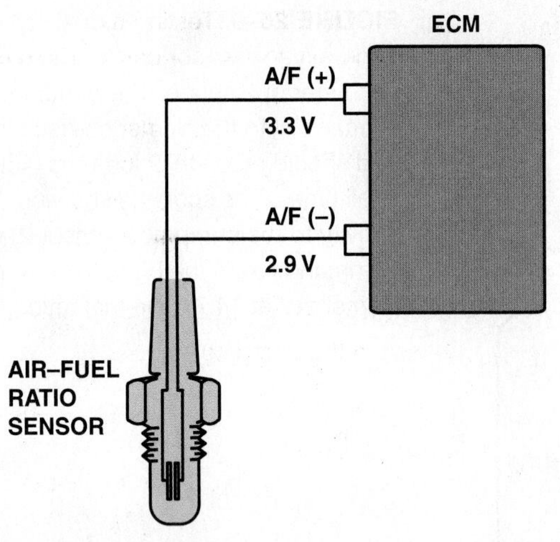

FIGURE 25–10 A single cell wide-band oxygen sensor has four wires with two for the heater and two for the sensor itself. The voltage applied to the sensor is 0.4 volt (3.3 − 2.9 = 0.4) across the two leads of the sensor.

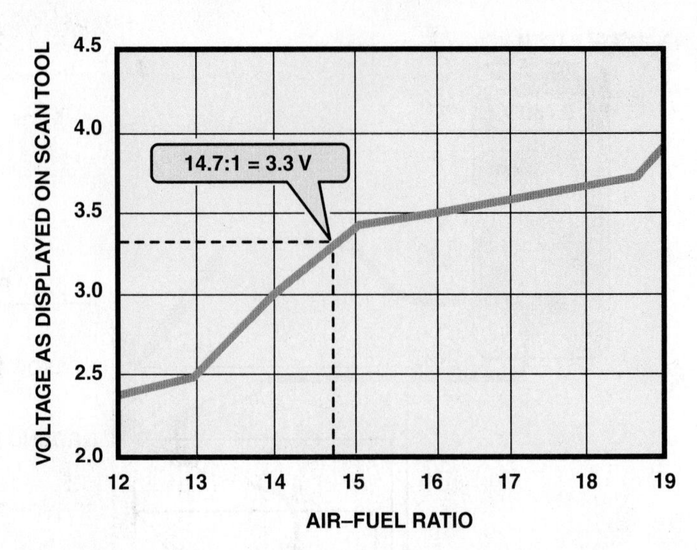

FIGURE 25–11 The scan tool can display various voltage but will often show 3.3 volts because the PCM is controlling the sensor by applying a low current to the sensor to achieve balance.

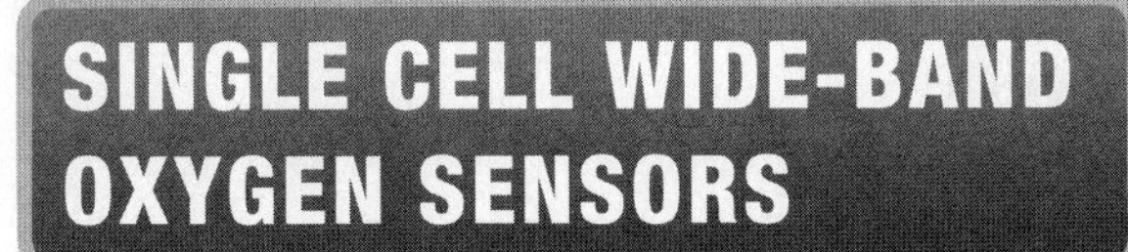

SINGLE CELL WIDE-BAND OXYGEN SENSORS

CONSTRUCTION A typical **single cell** wide-band oxygen sensor looks similar to a conventional four-wire zirconia oxygen sensor. The typical single cell wide-band oxygen sensor, usually called an **air–fuel ratio sensor**, has the following construction features:

- Can be made using the cup or planar design
- Oxygen is pumped into the diffusion layer similar to the operation of a dual cell wide-band oxygen sensor. ● **SEE FIGURE 25–10.**
- Current flow reverses positive and negative
- Consists of two cell wires and two heater wires (power and ground)
- The heater usually requires 6 amperes and the ground side is pulse-width modulated.

TESTING WITH A MILLIAMMETER The PCM controls the single cell wide-band oxygen sensor by maintaining a voltage difference of 300 mV (0.3 V) between the two sensor leads. The PCM keeps the voltage difference constant under all operating conditions by increasing or decreasing current between the element of the cell.

- Zero (0 mA) represents lambda or stoichiometric air–fuel ratio of 14.7:1
- +10 mA indicates a lean condition
- −10 mA indicates a rich condition

TESTING USING A SCAN TOOL A scan tool will display a voltage reading but can vary depending on the type and maker of scan tool. ● **SEE FIGURE 25–11.**

WIDE-BAND OXYGEN SENSOR PATTERN FAILURES

Wide-band oxygen sensors have a long life but can fail. Most of the failures will cause a diagnostic trouble code (DTC) to set, usually causing the malfunction indicator (check engine) lamp to light.

However, one type of failure may not set a DTC when the following occurs:

1. Voltage from the heater circuit bleeds into the Nernst cell.
2. This voltage will cause the engine to operate extremely lean and may or may not set a diagnostic trouble code.
3. When testing indicates an extremely lean condition, unplug the connector to the oxygen sensor. If the engine starts to operate correctly with the sensor unplugged, it confirms that the wide-band oxygen sensor has failed and requires replacement.

SUMMARY

1. Wide-band oxygen sensors are known by many different terms, including:
   - Broadband oxygen sensor
   - Wide-range oxygen sensor
   - Air–fuel ratio (AFR) sensor
   - Wide-range air–fuel (WRAF) sensor
   - Lean air–fuel (LAF) sensor
   - Air–fuel (AF) sensor
2. Wide-band oxygen sensors are manufactured using a cup or planar design and are dual cell or single cell design.
3. A wide-band oxygen sensor is capable of furnishing the PCM with exhaust air–fuel ratios as rich as 10:1 and as lean as 23:1.
4. The use of a wide-band oxygen sensor allows the engine to achieve more stringent exhaust emission standards.
5. A conventional zirconia oxygen sensor can be made in a cup shape or planar design and is sometimes called a narrow band or 2-step sensor.
6. The heater used on a conventional zirconia oxygen sensor uses up to 2 amperes and heats the sensor to about 600°F (315°C). A broadband sensor heater has to heat the sensor to 1,200°F to 1,400°F (650°C to 760°C) and requires up to 8 to 10 amperes.
7. A typical dual cell wide-band oxygen sensor uses the PCM to apply a current to the pump cell to keep the Nernst cell at 14.7:1.
   - When the exhaust is rich, the PCM applies a negative current to the pump cell.
   - When the exhaust is lean, the PCM applies a positive current to the pump cell.
8. Wide-band oxygen sensors can also be made using a single cell design.
9. Wide-band oxygen sensors can be best tested using a scan tool, but dual cell sensors can be checked with a voltmeter or scope. Single cell sensors can be checked using a milliammeter.

REVIEW QUESTIONS

1. What type of construction is used to make wide-band oxygen sensors?
2. Why are wide-band oxygen sensors used instead of conventional zirconia sensors?
3. How is the heater different for a wide-band oxygen sensor compared with a conventional zirconia oxygen sensor?
4. How does a wide-range oxygen sensor work?
5. How can a wide-band oxygen sensor be tested?

CHAPTER QUIZ

1. A wide-band oxygen sensor was first used on a Honda in what model year?
   a. 1992
   b. 1996
   c. 2000
   d. 2006
2. A wide-band oxygen sensor is capable of detecting the air–fuel mixture in the exhaust from ____________ (rich) to ____________ (lean).
   a. 12:1 to 15:1
   b. 13:1 to 16.7:1
   c. 10:1 to 23:1
   d. 8:1 to 18:1
3. A conventional zirconia oxygen sensor can be made with what designs?
   a. Cup and thimble
   b. Cup and planar
   c. Finger and thimble
   d. Dual cell and single cell
4. A wide-band oxygen sensor can be made using what design?
   a. Cup and thimble
   b. Cup and planar
   c. Finger and thimble
   d. Dual cell and single cell
5. A wide-band oxygen sensor heater could draw how much current?
   a. 0.8 to 2.0 A
   b. 2 to 4 A
   c. 6 to 8 A
   d. 8 to 10 A
6. A wide-band oxygen sensor needs to be heated to what operating temperature?
   a. 600°F (315°C)
   b. 800°F (427°C)
   c. 1,400°F (760°C)
   d. 2,000°F (1,093°C)
7. The two internal chambers of a dual cell wide-band oxygen sensor include ____________.
   a. Single and dual
   b. Nernst and pump
   c. Air reference and diffusion
   d. Inside and outside
8. When the exhaust is rich, the PCM applies a ____________ current into the pump cell.
   a. Positive
   b. Negative
9. When the exhaust is lean, the PCM applies a ____________ current into the pump cell.
   a. Positive
   b. Negative
10. A dual cell wide-band oxygen sensor can be tested using a ____________.
    a. Scan tool
    b. Voltmeter
    c. Scope
    d. All of the above

chapter 26

FUEL PUMPS, LINES, AND FILTERS

LEARNING OBJECTIVES

After studying this chapter, the reader will be able to:

1. Explain the role of fuel tanks in the fuel delivery system.
2. Discuss the different types of fuel lines.
3. Explain the different types of electric fuel pumps.
4. Describe how to test and replace fuel pumps.
5. Describe how to test and replace fuel filters.

This chapter will help you prepare for Engine Repair (A8) ASE certification test content area "C" (Fuel, Air Induction, and Exhaust Systems Diagnosis and Repair).

KEY TERMS

FUEL DELIVERY SYSTEM

Creating and maintaining a correct air–fuel mixture requires a properly functioning fuel and air delivery system. Fuel delivery (and return) systems use many if not all of the following components to make certain that fuel is available under the right conditions to the fuel-injection system:

- Fuel storage tank, filler neck, and gas cap
- Fuel tank pressure sensor
- Fuel pump
- Fuel filter(s)
- Fuel delivery lines and fuel rail
- Fuel-pressure regulator
- Fuel return line (if equipped with a return-type fuel delivery system)

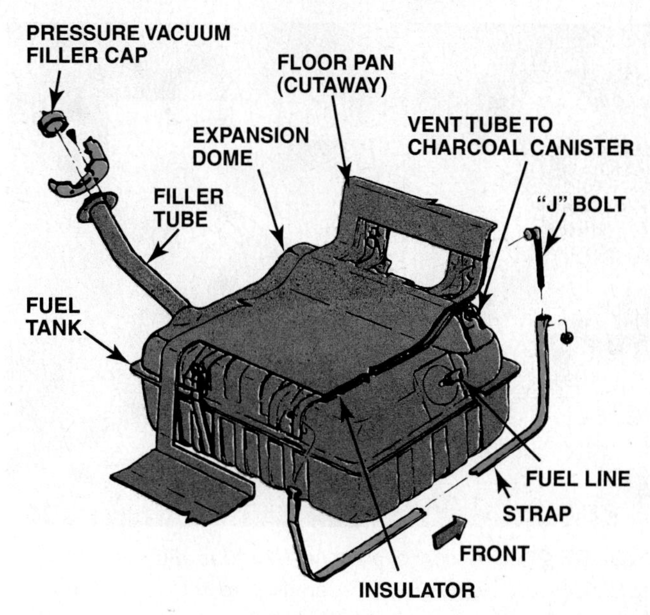

FIGURE 26–1 A typical fuel tank installation.

FUEL TANKS

A vehicle fuel tank is made of corrosion-resistant steel or polyethylene plastic. Some models, such as sport utility vehicles (SUVs) and light trucks, may have an auxiliary fuel tank.

Tank design and capacity are a compromise between available space, filler location, fuel expansion room, and fuel movement. Some later-model tanks deliberately limit tank capacity by extending the filler tube neck into the tank low enough to prevent complete filling, or by providing for expansion room. ● **SEE FIGURE 26–1**. A vertical **baffle** in this same tank limits fuel sloshing as the vehicle moves.

Regardless of size and shape, all fuel tanks incorporate most if not all of the following features:

- Inlet or filler tube through which fuel enters the tank
- Filler cap with pressure holding and relief features
- An outlet to the fuel line leading to the fuel pump or fuel injector
- Fuel pump mounted within the tank
- Tank vent system
- Fuel pickup tube and fuel level sending unit

TANK LOCATION AND MOUNTING Most vehicles use a horizontally suspended fuel tank, usually mounted below the rear of the floor pan, just ahead of or behind the rear axle. Fuel tanks are located there so that frame rails and body components protect the tank in the event of a crash. To prevent squeaks, some models have insulated strips cemented on the top or sides of the tank wherever it contacts the underbody.

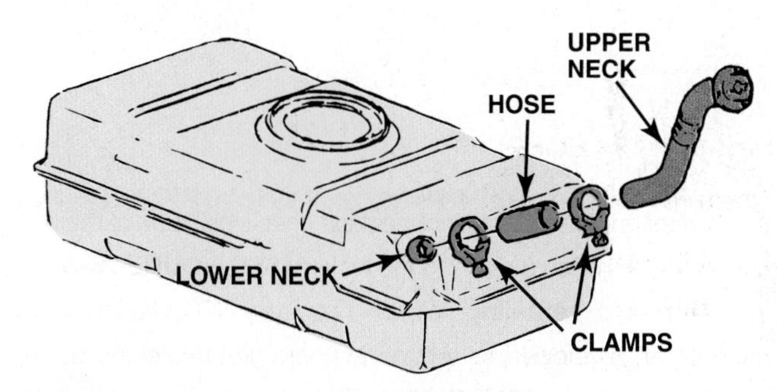

FIGURE 26–2 A three-piece filler tube assembly.

Fuel inlet location depends on the tank design and filler tube placement. It is located behind a filler cap and is often a hinged door in the outer side of either rear fender panel.

Generally, a pair of metal retaining straps holds a fuel tank in place. Underbody brackets or support panels hold the strap ends using bolts. The free ends are drawn underneath the tank to hold it in place, then bolted to other support brackets or to a frame member on the opposite side of the tank.

FILLER TUBES Fuel enters the tank through a large tube extending from the tank to an opening on the outside of the vehicle. ● **SEE FIGURE 26–2**.

Effective in 1993, federal regulations require manufacturers to install a device to prevent fuel from being siphoned through the filler neck. Federal authorities recognized methanol as a poison, and methanol used in gasoline is a definite health hazard. Additionally, gasoline is a suspected carcinogen (cancer-causing agent). To prevent siphoning, manufacturers welded a filler-neck check-ball tube in fuel tanks. To drain check ball–equipped fuel

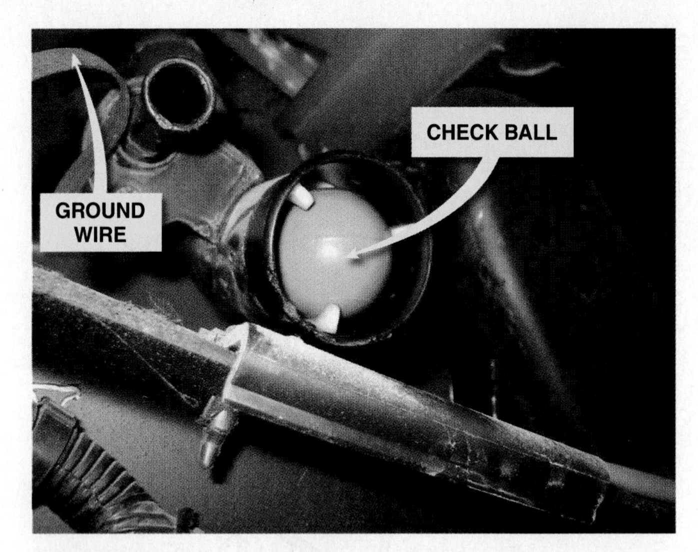

FIGURE 26–3 A view of a typical filler tube with the fuel tank removed. Notice the ground strap used to help prevent the buildup of static electricity as the fuel flows into the plastic tank. The check ball looks exactly like a ping-pong ball.

FIGURE 26–4 Vehicles equipped with onboard refueling vapor recovery usually have a reduced-size fill tube.

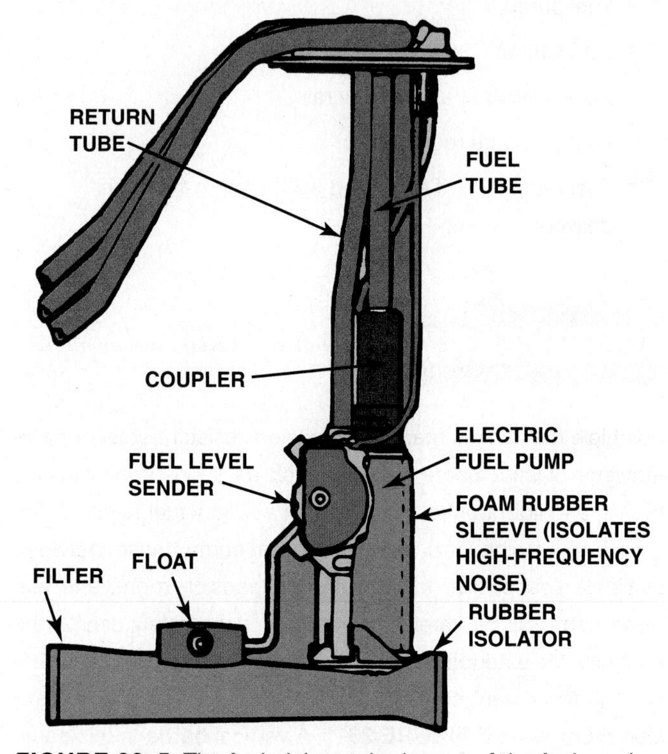

FIGURE 26–5 The fuel pickup tube is part of the fuel sender and pump assembly.

tanks, a technician must disconnect the check-ball tube at the tank and attach a siphon directly to the tank. ● **SEE FIGURE 26–3.**

Onboard refueling vapor recovery (ORVR) systems have been developed to reduce evaporative emissions during refueling. ● **SEE FIGURE 26–4.** These systems add components to the filler neck and the tank. One ORVR system utilizes a tapered filler neck with a smaller diameter tube and a check valve. When fuel flows down the neck, it opens the normally closed check valve. The vapor passage to the charcoal canister is opened. The decreased size neck and the opened air passage allow fuel and vapor to flow rapidly into the tank and the canister, respectively. When the fuel has reached a predetermined level, the check valve closes, and the fuel tank pressure increases. This forces the nozzle to shut off, thereby preventing the tank from being overfilled.

PRESSURE-VACUUM FILLER CAP Fuel and vapors are sealed in the tank by the safety filler cap. The safety cap must release excess pressure or excess vacuum. Either condition could cause fuel tank damage, fuel spills, and vapor escape. Typically, the cap will release if the pressure is over 1.5 to 2.0 PSI (10 to 14 kPa) or if the vacuum is 0.15 to 0.30 PSI (1 to 2 kPa).

FUEL PICKUP TUBE The fuel pickup tube is usually a part of the fuel sender assembly or the electric fuel pump assembly. Since dirt and sediment eventually gather on the bottom of a fuel tank, the fuel pickup tube is fitted with a filter sock or strainer to prevent contamination from entering the fuel lines. The woven plastic strainer also acts as a water separator by preventing water from being drawn up with the fuel. The filter sock usually is designed to filter out particles that are larger than 70 to 100 microns, or 30 microns if a gerotor-type fuel pump is used. One micron is 0.000039 inch ● **SEE FIGURE 26–5.**

NOTE: The human eye cannot see anything smaller than about 40 microns.

The filter is made from woven Saran resin (copolymer of vinylidene chloride and vinyl chloride). The filter blocks any water that may be in the fuel tank, unless it is completely submerged in water. In that case, it will allow water through the filter. This filter should be replaced whenever the fuel pump is replaced.

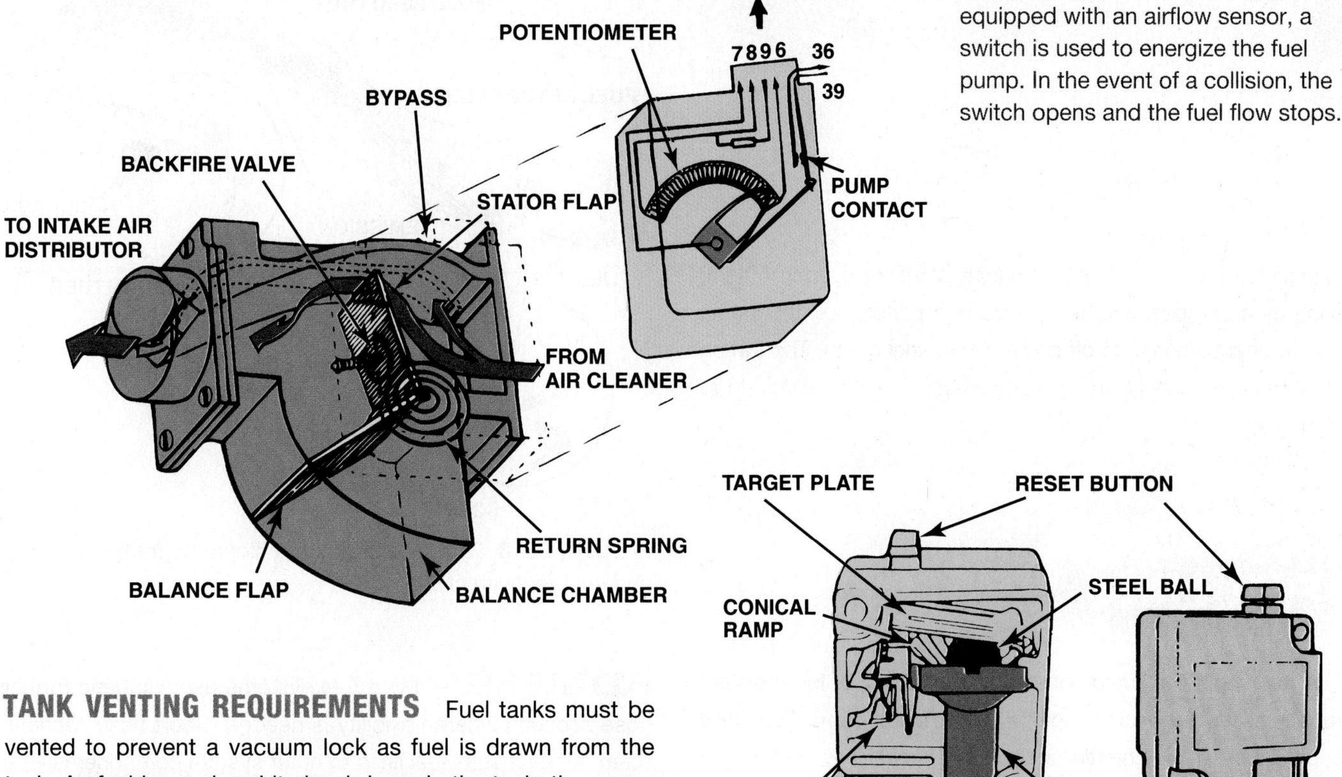

FIGURE 26–6 On some vehicles equipped with an airflow sensor, a switch is used to energize the fuel pump. In the event of a collision, the switch opens and the fuel flow stops.

TANK VENTING REQUIREMENTS Fuel tanks must be vented to prevent a vacuum lock as fuel is drawn from the tank. As fuel is used and its level drops in the tank, the space above the fuel increases. As the air in the tank expands to fill this greater space, its pressure drops. Without a vent, the air pressure inside the tank would drop below atmospheric pressure, developing a vacuum, which prevents the flow of fuel. Under extreme pressure variance, the tank could collapse. Venting the tank allows outside air to enter as the fuel level drops, preventing a vacuum from developing.

An EVAP system vents gasoline vapors from the fuel tank directly to a charcoal-filled vapor storage canister, and uses an unvented filler cap. Many filler caps contain valves that open to relieve pressure or vacuum above specified safety levels. Systems that use completely sealed caps have separate pressure and vacuum relief valves for venting.

Because fuel tanks are not vented directly to the atmosphere, the tank must allow for fuel expansion, contraction, and overflow that can result from changes in temperature or overfilling. One way is to use a dome in the top of the tank. Many General Motors vehicles use a design that includes a vertical slosh baffle which reserves up to 12% of the total tank capacity for fuel expansion.

FIGURE 26–7 Ford uses an inertia switch to turn off the electric fuel pump in case of an accident.

ROLLOVER LEAKAGE PROTECTION

All vehicles have one or more devices to prevent fuel leaks in case of vehicle rollover or a collision in which fuel may spill.

Variations of the basic one-way check valve may be installed in any number of places between the fuel tank and the engine. The valve may be installed in the fuel return line, vapor vent line, or fuel tank filler cap.

In addition to the rollover protection devices, some vehicles use devices to ensure that the fuel pump shuts off when an accident occurs. Some pumps depend upon an oil pressure or an engine speed signal to continue operating; these pumps turn off whenever the engine dies. On some air vane sensors, a microswitch is built into the sensor to switch on the fuel pump as soon as intake airflow causes the vane to lift from its rest position. ● **SEE FIGURE 26–6.**

Ford vehicles use an **inertia switch.** ● **SEE FIGURE 26–7.** The inertia switch is installed in the rear of the vehicle between the electric fuel pump and its power supply. With any sudden impact, such as a jolt from another vehicle in a parking lot, the inertia switch opens and shuts off power to the fuel pump. The switch must be reset manually by pushing a button to restore current to the pump.

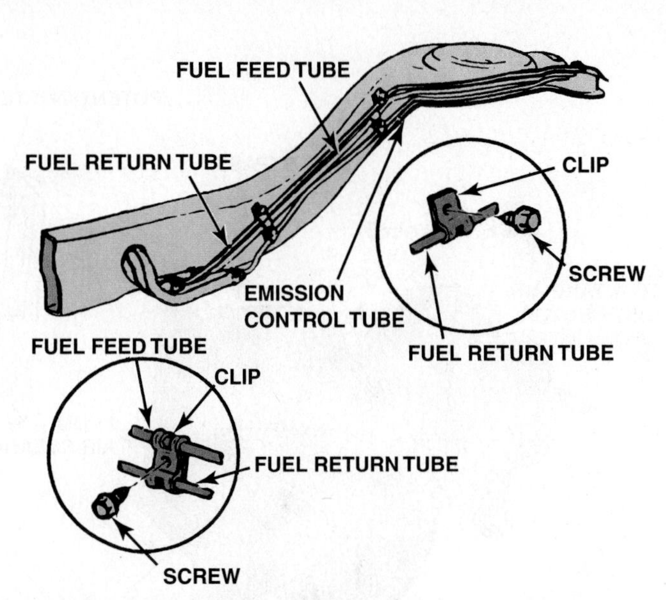

FIGURE 26–8 Fuel lines are routed along the frame or body and secured with clips.

FUEL LINES

Fuel and vapor lines made of steel, nylon tubing, or fuel-resistant rubber hoses connect the parts of the fuel system. Fuel lines supply fuel to the throttle body or fuel rail. They also return excess fuel and vapors to the tank. Depending on their function, fuel and vapor lines may be either rigid or flexible.

Fuel lines must remain as cool as possible. If any part of the line is located near too much heat, the gasoline passing through it vaporizes and **vapor lock** occurs. When this happens, the fuel pump supplies only vapor that passes into the injectors. Without liquid gasoline, the engine stalls and a hot restart problem develops.

The fuel delivery system supplies 10 to 15 PSI (69 to 103 kPa) or up to 35 PSI (241 kPa) to many throttle-body injection units and up to 50 PSI (345 kPa) for multiport fuel-injection systems. Fuel-injection systems retain residual or rest pressure in the lines for a half hour or longer when the engine is turned off to prevent hot engine restart problems. Higher-pressure systems such as these require special fuel lines.

RIGID LINES All fuel lines fastened to the body, frame, or engine are made of seamless steel tubing. Steel springs may be wound around the tubing at certain points to protect against impact damage.

Only steel tubing, or that recommended by the manufacturer, should be used when replacing rigid fuel lines. *Never substitute copper or aluminum tubing for steel tubing*. These materials do not withstand normal vehicle vibration and could combine with the fuel to cause a chemical reaction.

FLEXIBLE LINES Most fuel systems use synthetic rubber hose sections where flexibility is needed. Short hose sections often connect steel fuel lines to other system components. The fuel delivery hose inside diameter (ID) is generally larger (3/16 to 3/8 inches or 8 to 10 millimeters) than the fuel return hose ID (1/4 inches or 6 millimeters).

Fuel-injection systems require special-composition reinforced hoses specifically made for these higher-pressure systems. Similarly, vapor vent lines must be made of materials that resist fuel vapors. Replacement vent hoses are usually marked with the designation "EVAP" to indicate their intended use.

FUEL LINE MOUNTING Fuel supply lines from the tank to a throttle body or fuel rail are routed to follow the frame along the underbody of the vehicle. Vapor and return lines may be routed with the fuel supply line. All rigid lines are fastened to the frame rail or underbody with screws and clamps, or clips. ● **SEE FIGURE 26–8**.

FUEL-INJECTION LINES Hoses used for fuel-injection systems are made of materials with high resistance to oxidation and deterioration. Replacement hoses for injection systems should always be equivalent to original equipment manufacturer (OEM) hoses.

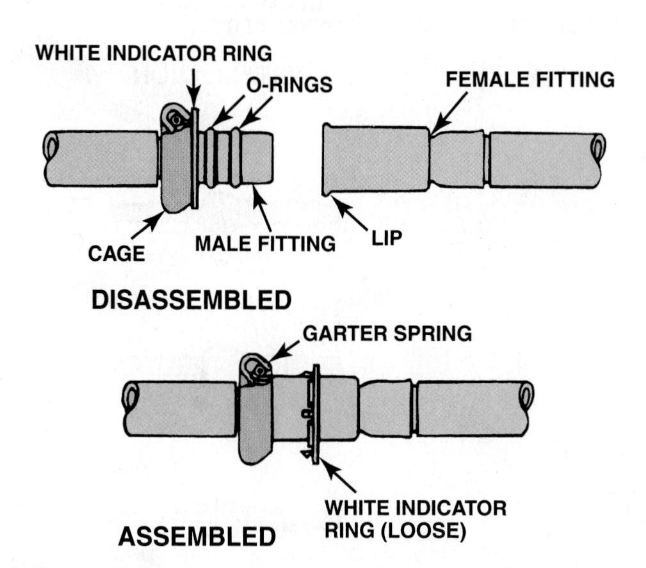

FIGURE 26–9 Some Ford metal line connections use spring-locks and O-rings.

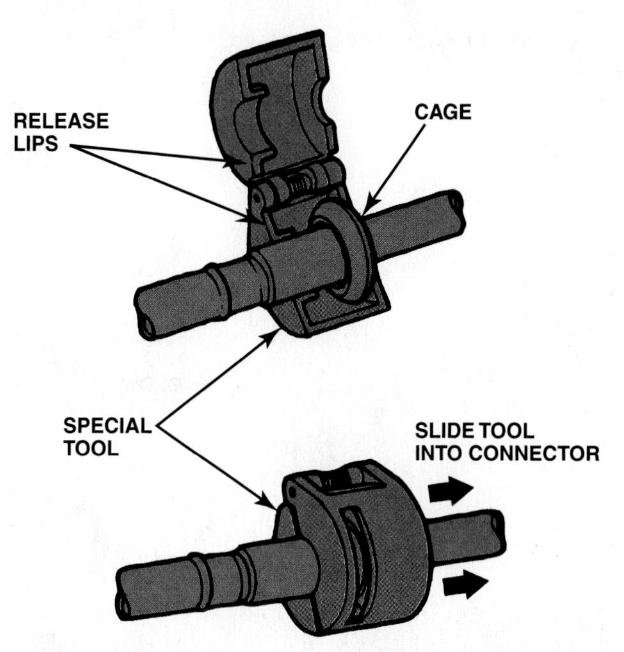

FIGURE 26–10 Ford spring-lock connectors require a special tool for disassembly.

CAUTION: ***Do not use spring-type clamps on fuel-injected engines*****—they cannot withstand the fuel pressures involved.**

FUEL-INJECTION FITTINGS AND NYLON LINES

Because of their operating pressures, fuel-injection systems often use special kinds of fittings to ensure leakproof connections. Some high-pressure fittings on GM vehicles with port fuel-injection systems use O-ring seals instead of the traditional flare connections. When disconnecting such a fitting, inspect the O-ring for damage and replace it if necessary. *Always* tighten O-ring fittings to the specified torque value to prevent damage.

Other manufacturers also use O-ring seals on fuel line connections. In all cases, the O-rings are made of special materials that withstand contact with gasoline and oxygenated fuel blends. Some manufacturers specify that the O-rings be replaced every time the fuel system connection is opened. When replacing one of these O-rings, a new part specifically designed for fuel system service must be used.

Ford also uses spring-lock connectors to join male and female ends of steel tubing. ● **SEE FIGURE 26–9**. The coupling is held together by a garter spring inside a circular cage. The flared end of the female fitting slips behind the spring to lock the coupling together.

General Motors has used nylon fuel lines with quick-connect fittings at the fuel tank and fuel filter since the early 1990s. Like the GM threaded couplings used with steel lines, nylon line couplings use internal O-ring seals. Unlocking the metal connectors requires a special quick-connector separator tool; plastic connectors can be released without the tool. ● **SEE FIGURES 26–10 AND 26–11**.

FUEL LINE LAYOUT Fuel pressures have tended to become higher to prevent vapor lock, and a major portion of the fuel routed to the fuel-injection system returns to the tank by way of a fuel return line or return-type systems. This allows better control, within limits, of heat absorbed by the gasoline as it is routed through the engine compartment. Throttle-body and multiport injection systems have typically used a pressure regulator to control fuel pressure in the throttle body or fuel rail, and also allow excess fuel not used by the injectors to return to the tank. However, the warmer fuel in the tank may create problems, such as an excessive rise in fuel vapor pressures in the tank.

FREQUENTLY ASKED QUESTION

Just How Much Fuel Is Recirculated?

Approximately 80% of the available fuel-pump volume is released to the fuel tank through the fuel-pressure regulator at idle speed. For example, a passenger vehicle cruising down the road at 60 mph gets 30 mpg. With a typical return-style fuel system pumping about 30 gallons per hour from the tank, it would therefore burn 2 gallons per hour, and return about 28 gallons per hour to the tank!

FIGURE 26–11 Typical quick-connect steps.

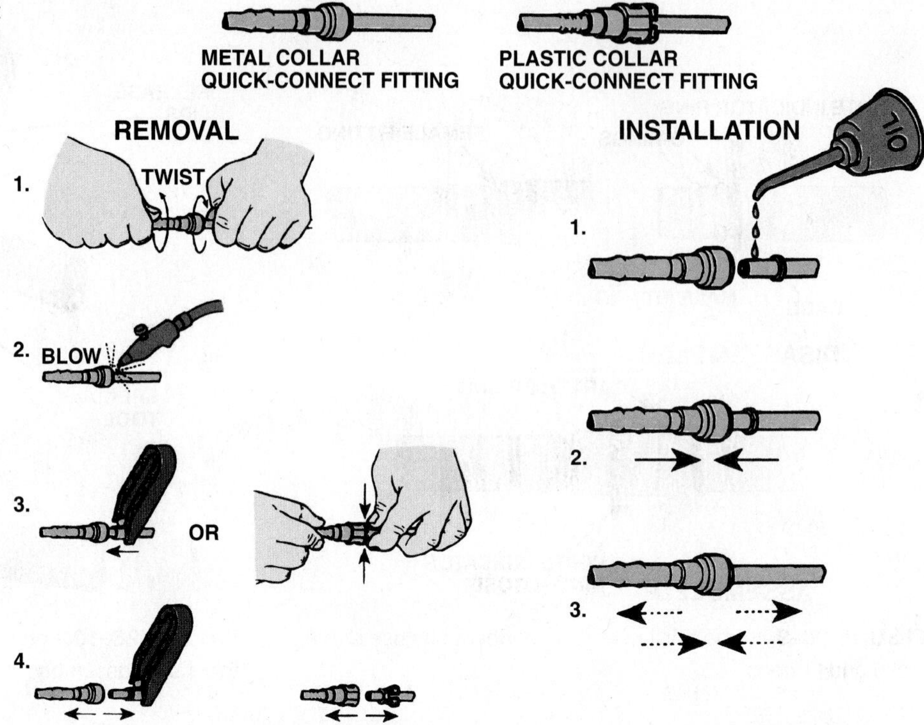

With late-model vehicles, there has been some concern about too much heat being sent back to the fuel tank, causing rising in-tank temperatures and increases in fuel vaporization and **volatile organic compound (VOC)** (hydrocarbon) emissions. To combat this problem, manufacturers have placed the pressure regulator back by the tank instead of under the hood on mechanical returnless systems. In this way, returned fuel is not subjected to the heat generated by the engine and the underhood environment. To prevent vapor lock in these systems, pressures have been raised in the fuel rail, and injectors tend to have smaller openings to maintain control of the fuel spray under pressure.

Not only must the fuel be filtered and supplied under adequate pressure, but there must also be a consistent *volume* of fuel to assure smooth engine performance even under the heaviest of loads.

? FREQUENTLY ASKED QUESTION

How Can an Electric Pump Work Inside a Gas Tank and Not Cause a Fire?

Even though fuel fills the entire pump, no burnable mixture exists inside the pump because there is no air and no danger of commutator brush arcing, igniting the fuel.

ELECTRIC FUEL PUMPS

The electric fuel pump is a pusher unit. When the pump is mounted in the tank, the entire fuel supply line to the engine can be pressurized. Because the fuel, when pressurized, has a higher boiling point, it is unlikely that vapor will form to interfere with fuel flow.

Most vehicles use the impeller or turbine pumps. ● **SEE FIGURE 26–12**. All electrical pumps are driven by a small electric motor, but the turbine pump turns at higher speeds and is quieter than the others.

POSITIVE DISPLACEMENT PUMP A positive displacement pump is a design that forces everything that enters the pump to leave the pump.

In the **roller cell** or vane pump, the impeller draws fuel into the pump, and then pushes it out through the fuel line to the injection system. All designs of pumps use a variable-sized chamber to draw in fuel. When the maximum volume has been reached, the supply port closes and the discharge opens. Fuel is then forced out the discharge as this volume decreases. The chambers are formed by rollers or gears in a rotor plate. Since this type of pump uses no valves to move the fuel, the fuel flows steadily through the pump housing. Since fuel flows steadily through the entire pump, including the electrical portion, the pump stays cool. Usually, only when a vehicle runs out of fuel is there a risk of pump damage.

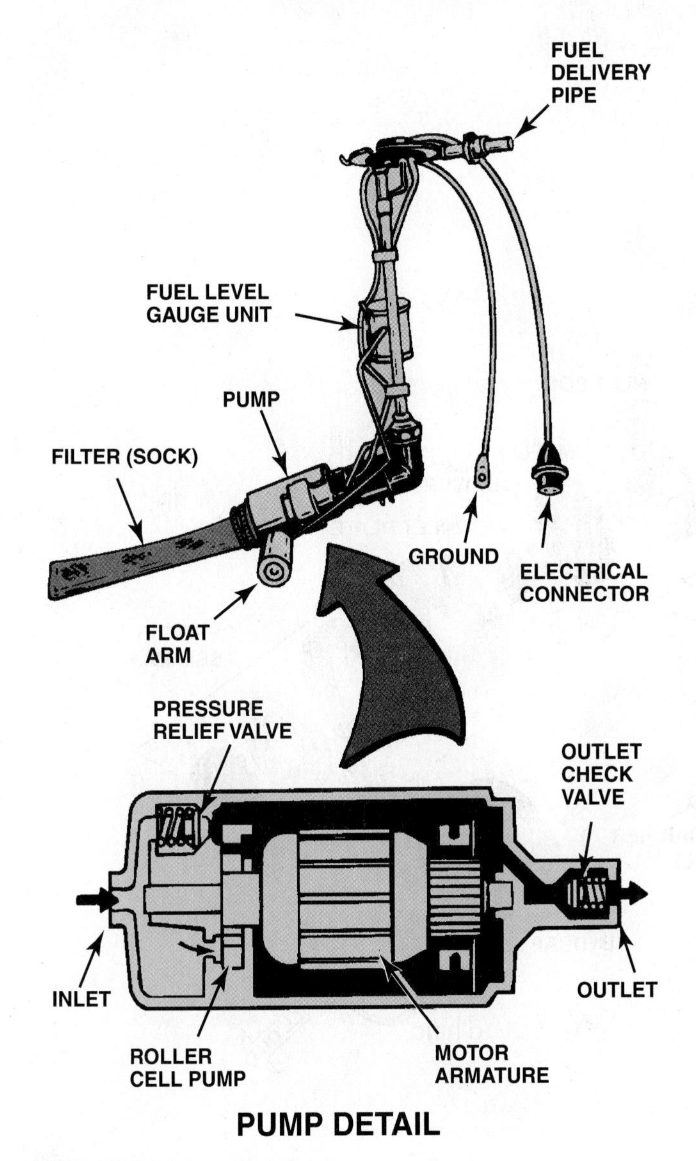

FIGURE 26–12 A roller cell-type electric fuel pump.

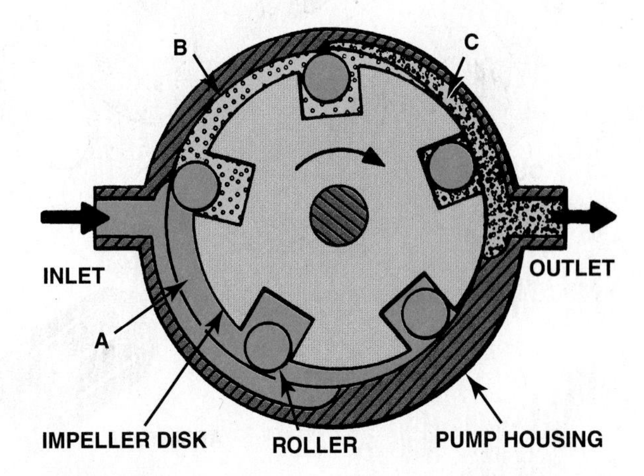

FIGURE 26–13 The pumping action of an impeller or rotary vane pump.

Most electric fuel pumps are equipped with a fuel outlet check valve that closes to maintain fuel pressure when the pump shuts off. **Residual or rest pressure** prevents vapor lock and hot-start problems on these systems.

● **FIGURE 26–13** shows the pumping action of a **rotary vane pump.** The pump consists of a central impeller disk, several rollers or vanes that ride in notches in the impeller, and a pump housing that is offset from the impeller centerline. The impeller is mounted on the end of the motor armature and spins whenever the motor is running. The rollers are free to slide in and out within the notches in the impeller to maintain sealing contact. Unpressurized fuel enters the pump, fills the spaces between the rollers, and is trapped between the impeller, the housing, and two rollers. An internal gear pump, called a **gerotor**, is another type of positive displacement pump that is often used in engine oil pumps. It uses the meshing of internal and external gear teeth to pressurize the fuel. ● **SEE FIGURE 26–14** for an example of a gerotor-type fuel pump that uses an impeller as the first stage and is used to move the fuel gerotor section where it is pressurized.

HYDROKINETIC FLOW PUMP DESIGN The word *hydro* means liquid and the term *kinetic* refers to motion, so the term **hydrokinetic pump** means that this design of pump rapidly moves the fuel to create pressure. This design of pump is a nonpositive displacement pump design.

A **turbine pump** is the most common because it tends to be less noisy. Sometimes called **turbine**, **peripheral**, and **side-channel**, these units use an impeller that accelerates the fuel particles before actually discharging them into a tract where they generate pressure via pulse exchange. Actual pump volume is controlled by using a different number of impeller blades, and in some cases a higher number of impellers, or different shapes along the side discharge channels. These units are fitted more toward lower operating pressures of less than 60 PSI. ● **SEE FIGURE 26–15** for an example of a two-stage turbine pump. The turbine impeller has a staggered blade design to minimize pump harmonic noise and to separate vapor from the liquid fuel. The end cap assembly contains a pressure relief valve and a radio-frequency interference (RFI) suppression module. The check valve is usually located in the upper fuel pipe connector assembly.

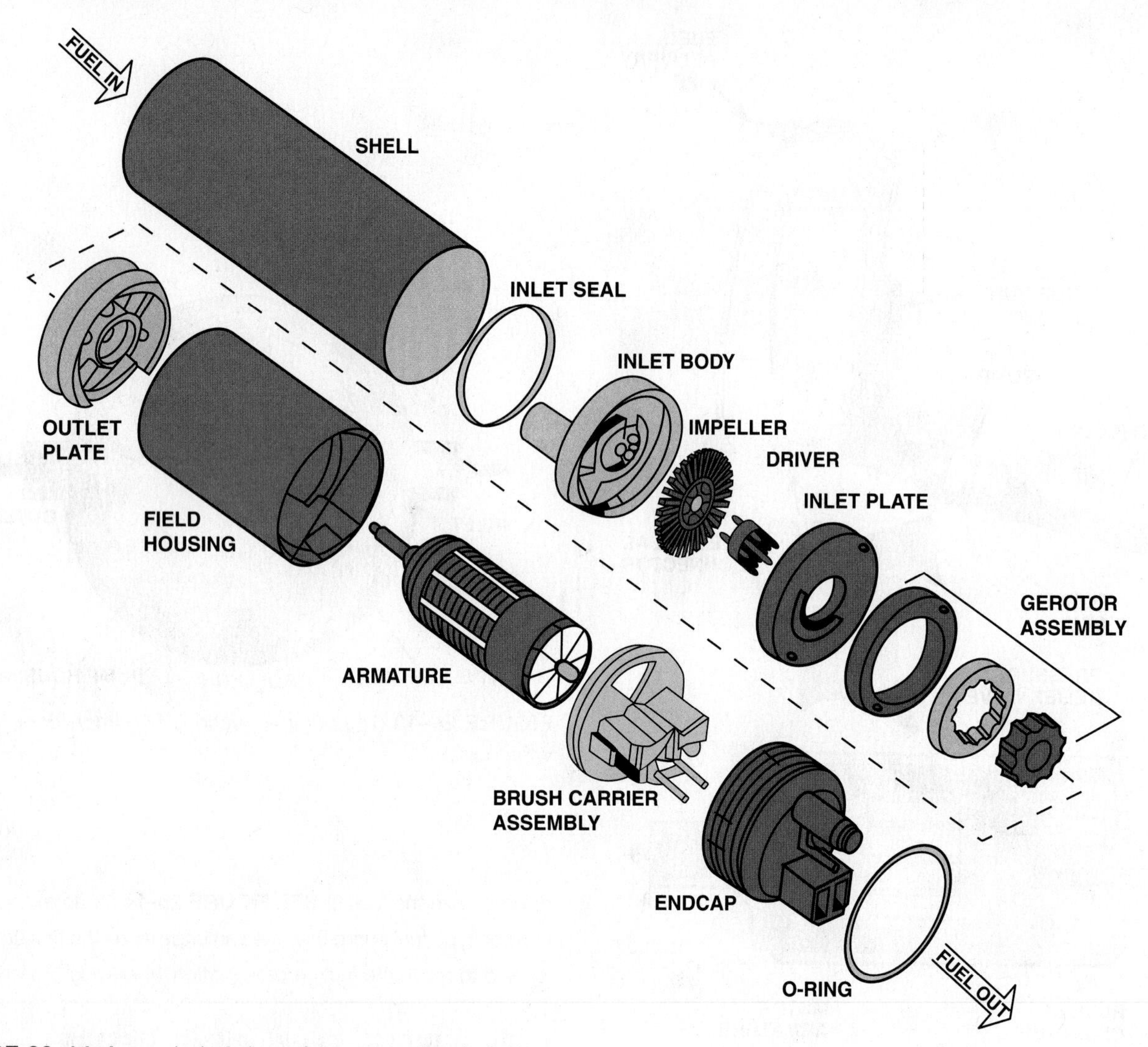

FIGURE 26–14 An exploded view of a gerotor electric fuel pump.

After it passes through the strainer, fuel is drawn into the lower housing inlet port by the impellers. It is pressurized and delivered to the convoluted fuel tube for transfer through a check valve into the fuel feed pipe. A typical electric fuel pump used on a fuel-injection system delivers about 40 to 50 gallons per hour or 0.6 to 0.8 gallons per minute at a pressure of 70 to 90 PSI.

MODULAR FUEL SENDER ASSEMBLY The modular fuel sender consists of a fuel level sensor, a turbine pump, and a jet pump. The reservoir housing is attached to the cover containing fuel pipes and the electrical connector. Fuel is transferred from the pump to the fuel pipe through a convoluted (flexible) fuel pipe. The convoluted fuel pipe eliminates the need for rubber hoses, nylon pipes, and clamps. The reservoir dampens fuel slosh to maintain a constant fuel level available to the roller vane pump; it also reduces noise.

Some of the flow, however, is returned to the jet pump for recirculation. Excess fuel is returned to the reservoir through one of the three hollow support pipes. The hot fuel quickly mixes with the cooler fuel in the reservoir; this minimizes the possibility of vapor lock. In these modules, the reservoir is filled by the jet pump. Some of the fuel from the pump is sent through the jet pump to lift fuel from the tank into the reservoir.

ELECTRIC PUMP CONTROL CIRCUITS Fuel-pump circuits are controlled by the fuel-pump relay. Fuel-pump relays are activated initially by turning the ignition key to on, which allows the pump to pressurize the fuel system. As a safety precaution, the relay de-energizes after a few seconds until the key is moved to the crank position. On some systems, once an ignition coil signal, or "tach" signal, is received by the engine control computer, indicating the engine is rotating, the relay remains energized even with the key released to the run position.

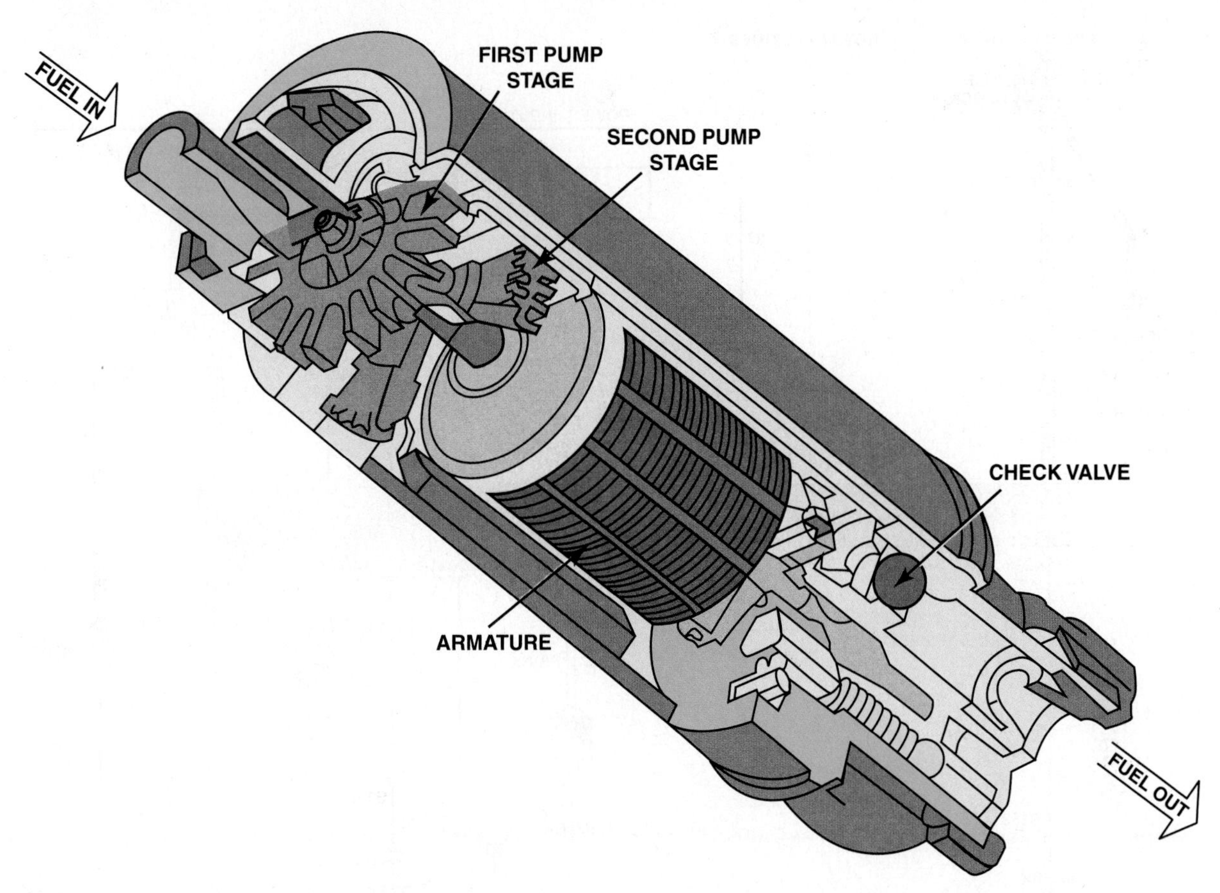

FIGURE 26–15 A cutaway view of a typical two-stage turbine electric fuel pump.

FREQUENTLY ASKED QUESTION

Why Are Many Fuel-Pump Modules Spring-Loaded?

Fuel modules that contain the fuel pickup sock, fuel pump, and fuel level sensor are often spring-loaded when fitted to a plastic fuel tank. The plastic material shrinks when cold and expands when hot, so having the fuel module spring-loaded ensures that the fuel pickup sock will always be the same distance from the bottom of the tank. ● **SEE FIGURE 26–16.**

CHRYSLER. On older Chrysler vehicles, the PCM must receive an engine speed (RPM) signal during cranking before it can energize a circuit driver inside the power module to activate an automatic shutdown (ASD) relay to power the fuel pump, ignition coil, and injectors. As a safety precaution, if the RPM signal to the logic module is interrupted, the logic module signals the power module to deactivate the ASD, turning off the pump, coil, and injectors. In some vehicles, the oil pressure switch circuit may be used as a safety circuit to activate the pump in the ignition switch run position.

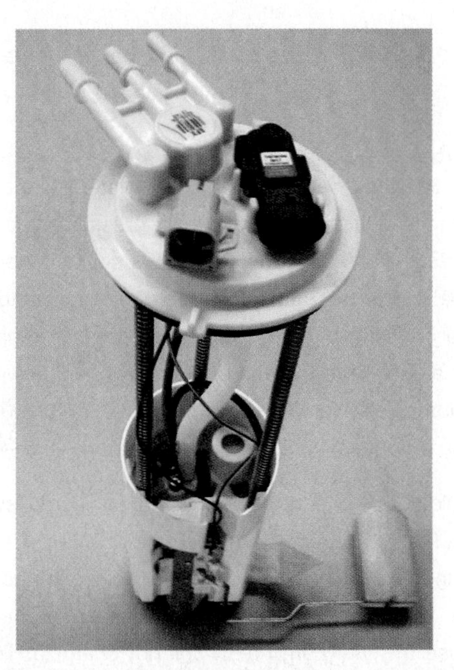

FIGURE 26–16 A typical fuel-pump module assembly, which includes the pickup strainer and fuel pump, as well as the fuel-pressure sensor and fuel level sensing unit.

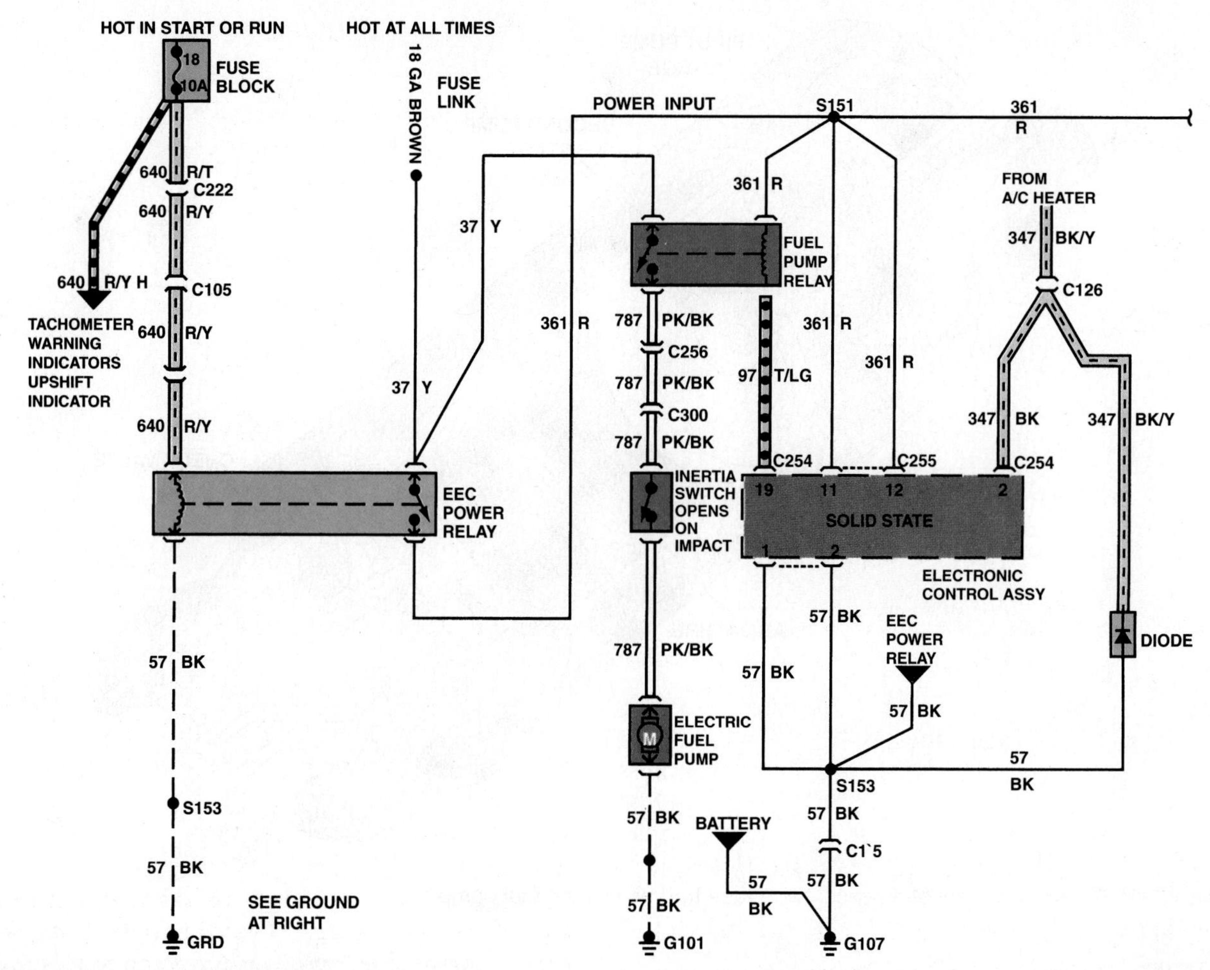

FIGURE 26–17 A schematic showing that an inertia switch is connected in series between the fuel-pump relay and the fuel pump.

GENERAL MOTORS. General Motors systems energize the pump with the ignition switch to initially pressurize the fuel lines, but then deactivate the pump if an RPM signal is not received within one or two seconds. The pump is reactivated as soon as engine cranking is detected. The oil pressure sending unit serves as a backup to the fuel-pump relay on some vehicles. In case of pump relay failure, the oil pressure switch will operate the fuel pump once oil pressure reaches about 4 PSI (28 kPa).

FORD. Older fuel-injected Fords used an inertia switch between the fuel pump relay and fuel pump.

The inertia switch opens under a specified impact, such as a collision. When the switch opens, current to the pump shuts off because the fuel-pump relay will not energize. The switch must be reset manually by opening the trunk and depressing the reset button before current flow to the pump can be restored. ● **SEE FIGURE 26–17** for a schematic of a typical fuel system that uses an inertia switch in the power feed circuit to the electric fuel pump.

Since about 2008, the inertial switch has been replaced with a signal input from the airbag module. If the airbag is deployed, the circuit to the fuel pump is opened and the fuel pump stops.

PUMP PULSATION DAMPENING Some manufacturers use an **accumulator** in the system to reduce pressure pulses and noise. Others use a pulsator located at the outlet of the fuel pump to absorb pressure pulsations that are created by the pump. These pulsators are usually used on roller vane pumps and are a source of many internal fuel leaks. ● **SEE FIGURE 26–18.**

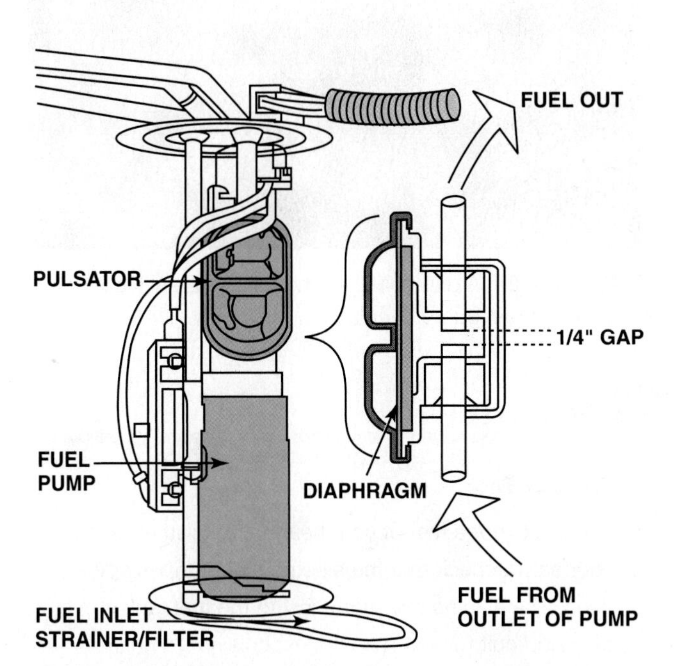

FIGURE 26–18 A typical fuel pulsator used mostly with roller vane-type pumps to help even out the pulsation in pressure that can cause noise.

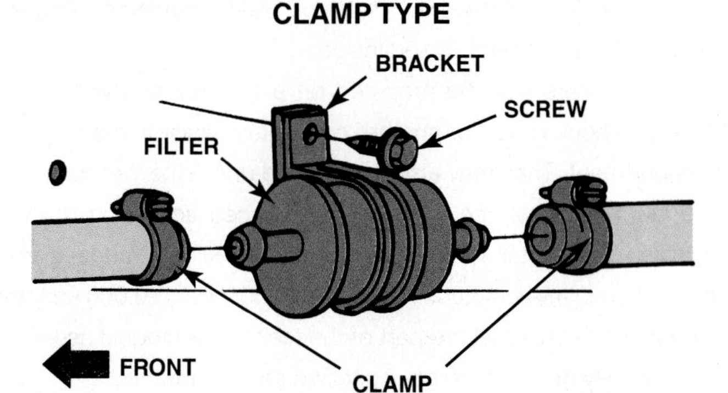

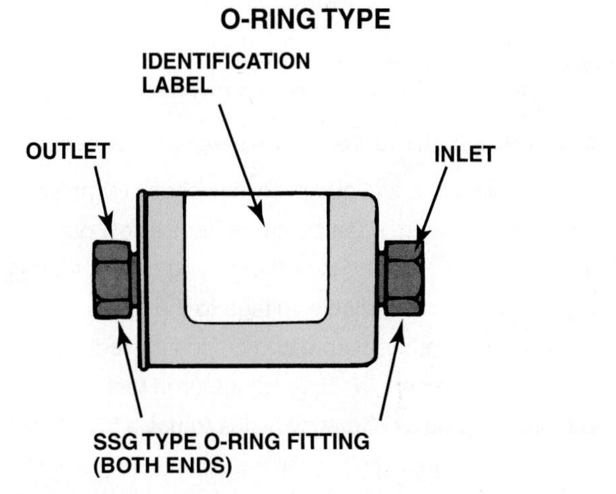

FIGURE 26–19 Inline fuel filters are usually attached to the fuel line with screw clamps or threaded connections. The fuel filter must be installed in the proper direction or a restricted fuel flow can result.

NOTE: Some experts suggest that the pulsator be removed and replaced with a standard section of fuel line to prevent the loss of fuel pressure that results when the connections on the pulsator loosen and leak fuel back into the tank.

VARIABLE SPEED PUMPS Another way to help reduce noise, current draw, and pump wear is to reduce the speed of the pump when less than maximum output is required. Pump speed and pressure can be regulated by controlling the voltage supplied to the pump with a resistor switched into the circuit, or by using a separate fuel pump driver module to supply a pulse-width modulated (PWM) voltage to the pump. With slower pump speed and pressure, less noise is produced.

FUEL FILTERS

Despite the care generally taken in refining, storing, and delivering gasoline, some impurities get into the automotive fuel system. Fuel filters remove dirt, rust, water, and other contamination from the gasoline before it can reach the fuel injectors. Most fuel filters are designed to filter particles that are 10 to 20 microns or larger in size.

The useful life of many filters is limited, but vehicles that use a returnless-type fuel-injection system usually use filters that are part of the fuel pump assembly and do not have any specified interval. This means that they should last the life of the vehicle. If fuel filters are not replaced according to the manufacturer's recommendations, they can become clogged and restrict fuel flow.

In addition to using several different types of fuel filters, a single fuel system may contain two or more filters. The inline filter is located in the line between the fuel pump and the throttle body or fuel rail. ● **SEE FIGURE 26–19**. This filter protects the system from contamination, but does not protect the fuel pump.

The inline filter usually is a metal or plastic container with a pleated paper element sealed inside.

Fuel filters may be mounted on a bracket on the fender panel, a shock tower, or another convenient place in the engine compartment. They may also be installed under the vehicle near the fuel tank. Fuel filters should be replaced according to the vehicle manufacturer's recommendations, which range from every 30,000 miles (48,000 km) to 100,000 miles (160,000 km) or longer. Fuel filters that are part of the fuel-pump module assemblies usually do not have any specified service interval.

TECH TIP

Use a Headlight to Test for Power and Ground

When replacing a fuel pump, always check for proper power and ground. If the supply voltage is low due to resistance in the circuit or the ground connection is poor, the lower available voltage to the pump will result in lower pump output and could also reduce the life of the pump. While a voltage drop test can be preformed, a quick and easy test is to use a headlight connected to the circuit. If the headlight is bright, then both the power side and the ground side of the pump circuit are normal. If the headlight is dim, then more testing will be needed to find the source of the resistance in the circuit(s) ● **SEE FIGURE 26–20.**

FIGURE 26–20 A dim headlight indicates excessive resistance in fuel pump circuit.

TECH TIP

The Ear Test

No, this is not a test of your hearing, but rather using your ear to check that the electric fuel pump is operating. The electric fuel pump inside the fuel tank is often difficult to hear running, especially in a noisy shop environment. A commonly used trick to better hear the pump is to use a funnel in the fuel filter neck. ● **SEE FIGURE 26–21.**

FUEL-PUMP TESTING

Fuel-pump testing includes many different tests and procedures. Even though a fuel pump can pass one test, it does not mean that there is not a fuel-pump problem. For example, if the pump motor is rotating slower than normal, it may be able to produce the specified pressure, but not enough volume to meet the needs of the engine while operating under a heavy load.

TESTING FUEL-PUMP PRESSURE Fuel pump-regulated pressure has become more important than ever with a more exact fuel control. Although an increase in fuel pressure does increase fuel volume to the engine, this is *not* the preferred method to add additional fuel as some units will not open correctly at the increased fuel pressure. On the other side of the discussion, many newer engines will not start when fuel pressure is just a few PSI low. Correct fuel pressure is very important for proper engine operation. Most fuel-injection systems operate at either

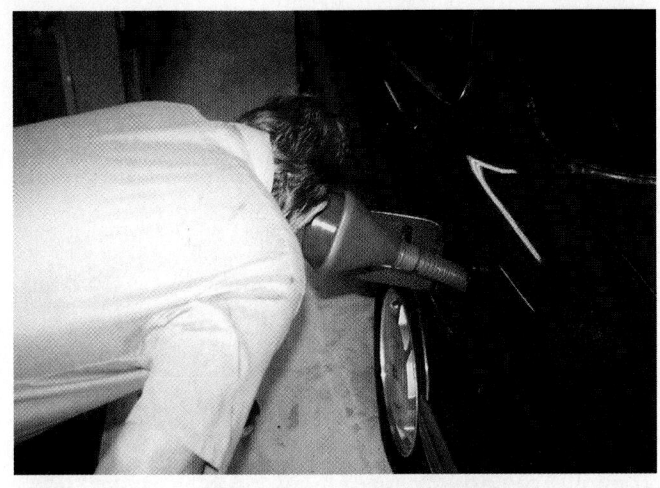

(a)

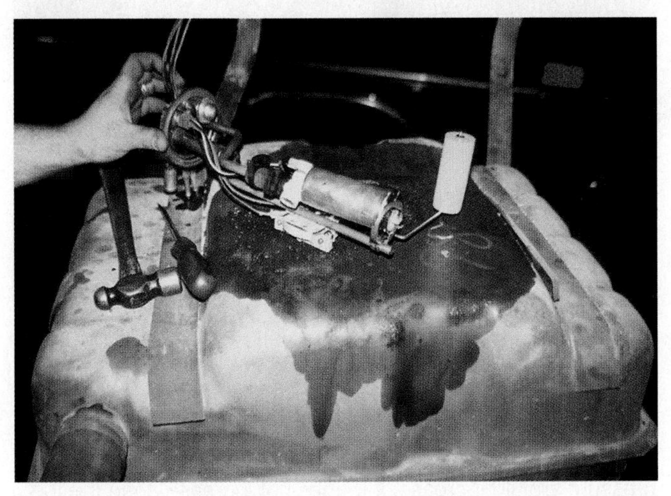

(b)

FIGURE 26–21 (a) A funnel helps in hearing if the electric fuel pump inside the gas tank is working. (b) If the pump is not running, check the wiring and current flow before going through the process of dropping the fuel tank to remove the pump.

a low pressure of about 10 PSI or a high pressure of between 35 and 45 PSI.

Normal Operating Pressure	(PSI)	Maximum Pump Pressure (PSI)
Low-pressure TBI units	9–13	18–20
High-pressure TBI units	25–35	50–70
Port fuel-injection systems	35–45	70–90
Central port fuel injection (GM)	55–64	90–110

FIGURE 26–22 The Schrader valve on this General Motors 3800 V-6 is located next to the fuel-pressure regulator.

In both types of systems, maximum fuel-pump pressure is about double the normal operating pressure to ensure that a continuous flow of cool fuel is being supplied to the injector(s) to help prevent vapor from forming in the fuel system. Although vapor or foaming in a fuel system can greatly affect engine operation, the cooling and lubricating flow of the fuel must be maintained to ensure the durability of injector nozzles.

To measure fuel-pump pressure, locate the Schrader valve and attach a fuel-pressure gauge. ● **SEE FIGURE 26–22.**

TECH TIP

The Rubber Mallet Trick

Often a no-start condition is due to an inoperative electric fuel pump. A common trick is to tap on the bottom of the fuel tank with a rubber mallet in an attempt to jar the pump motor enough to work. Instead of pushing a vehicle into the shop, simply tap on the fuel tank and attempt to start the engine. This is not a repair, but rather a confirmation that the fuel pump does indeed require replacement.

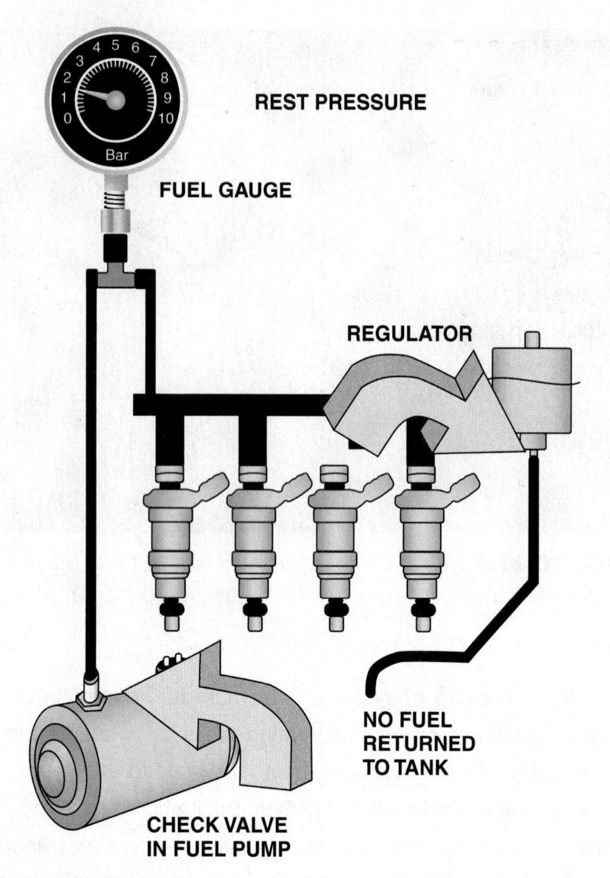

FIGURE 26–23 The fuel system should hold pressure if the system is leak free.

NOTE: Some vehicles, such as those with General Motors TBI fuel-injection systems, require a specific fuel-pressure gauge that connects to the fuel system. Always follow the manufacturers' recommendations and procedures.

REST PRESSURE TEST If the fuel pressure is acceptable, then check the system for leakdown. Observe the pressure gauge after five minutes. ● **SEE FIGURE 26–23**. The pressure should be the same as the initial reading. If not, then the pressure regulator, fuel-pump check valve, or the injectors are leaking.

DYNAMIC PRESSURE TEST To test the pressure dynamically, start the engine. If the pressure is vacuum referenced, then the pressure should change when the throttle is cycled. If it does not, then check the vacuum supply circuit. Remove the vacuum line from the regulator and inspect for any presence of fuel. ● **SEE FIGURE 26–24**. There should never be any fuel present on the vacuum side of the regulator diaphragm. When the engine speed is increased, the pressure reading should remain within the specifications.

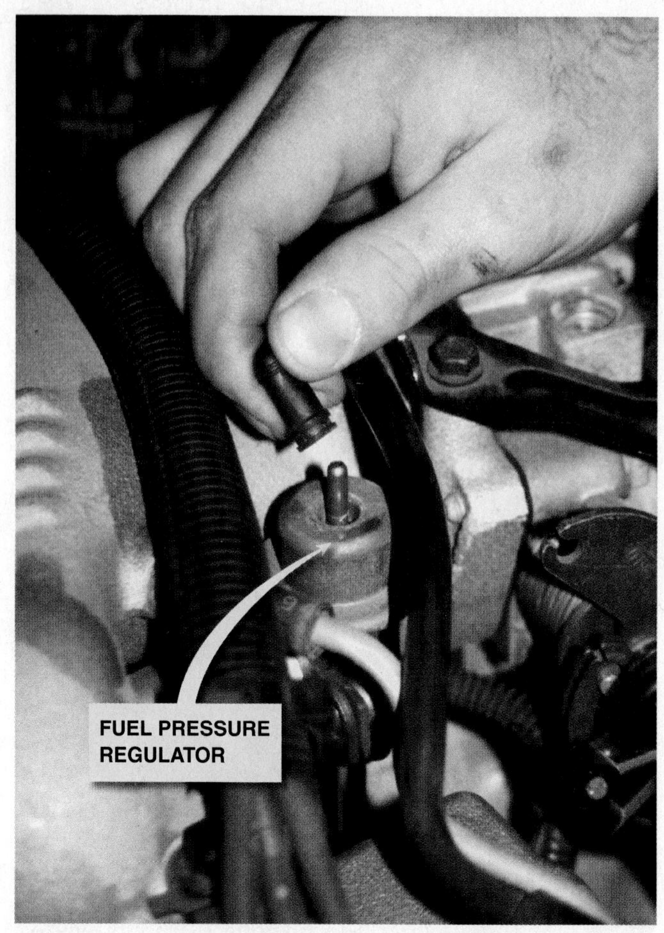

FIGURE 26–24 If the vacuum hose is removed from the fuel-pressure regulator when the engine is running, the fuel pressure should increase. If it does not increase, then the fuel pump is not capable of supplying adequate pressure or the fuel-pressure regulator is defective. If gasoline is visible in the vacuum hose, the regulator is leaking and should be replaced.

TECH TIP

The Fuel-Pressure Stethoscope Test

When the fuel pump is energized and the engine is not running, fuel should be heard flowing back to the fuel tank at the outlet of the fuel-pressure regulator. ● **SEE FIGURE 26–25**. If fuel is heard flowing through the return line, the fuel-pump pressure is higher than the regulator pressure. If no sound of fuel is heard, either the fuel pump or the fuel-pressure regulator is at fault.

Some engines do not use a vacuum-referenced regulator. The running pressure remains constant, which is typical for a mechanical returnless-type fuel system. On these systems, the pressure is higher than on return-type systems to help reduce the formation of fuel vapors in the system.

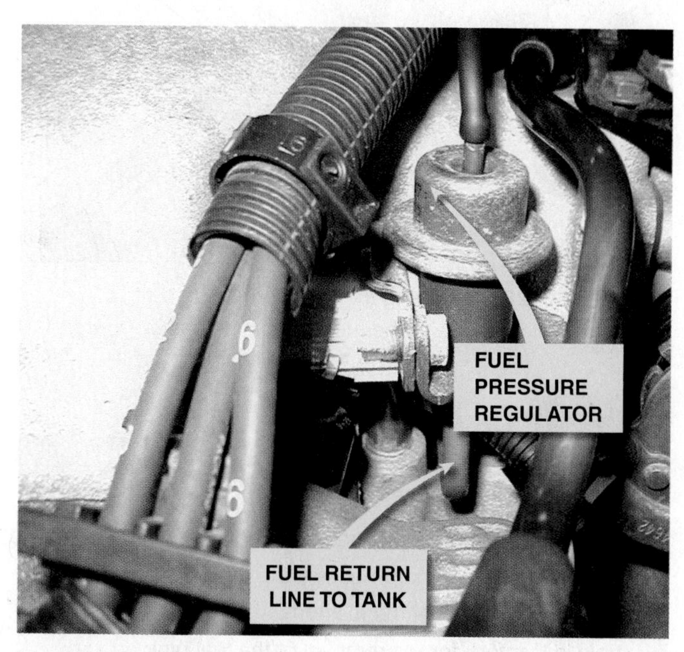

FIGURE 26–25 Fuel should be heard returning to the fuel tank at the fuel return line if the fuel pump and fuel-pressure regulator are functioning correctly.

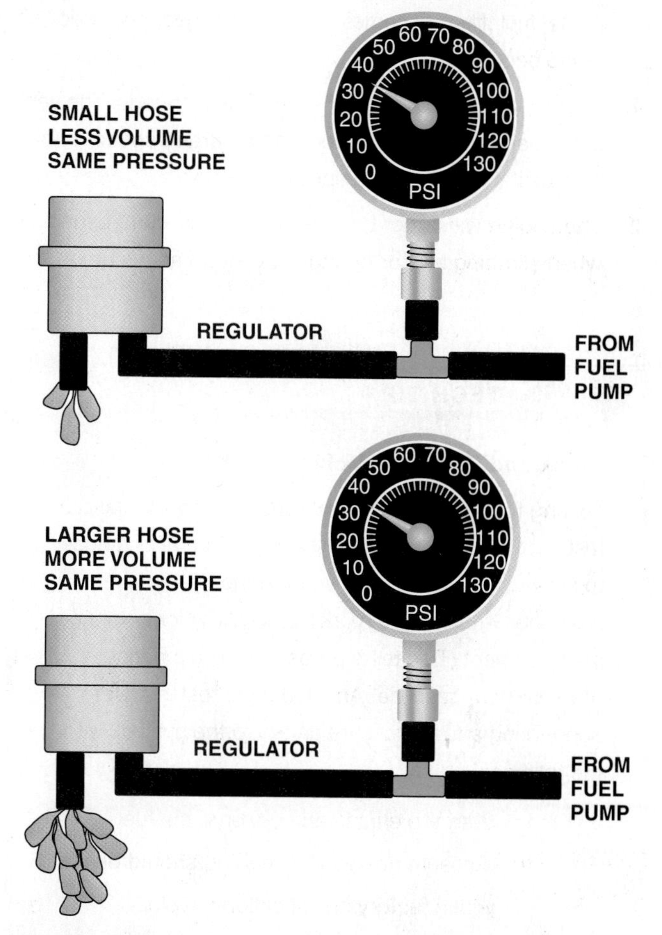

FIGURE 26–26 A fuel-pressure reading does not confirm that there is enough fuel volume for the engine to operate correctly.

TESTING FUEL-PUMP VOLUME Fuel pressure alone is not enough for proper engine operation. ● **SEE FIGURE 26–26.** Sufficient fuel capacity (flow) should be at least 2 pints (1 liter) every 30 seconds or 1 pint in 15 seconds. Fuel flow specifications are usually expressed in gallons per minute. A typical specification would be 0.5 gallons per minute or more. Volume testing is shown in ● **FIGURE 26–27.**

All fuel must be filtered to prevent dirt and impurities from damaging the fuel system components and/or engine. The first filter is inside the gas tank and is usually not replaceable separately but is attached to the fuel pump (if the pump is electric) and/or fuel gauge sending unit. The replaceable fuel filter is usually located between the fuel tank and the fuel rail or inlet to the fuel-injection system. Most vehicle manufacturers state in service information when to replace the fuel filter. Most newer vehicles, that use returnless-type fuel-injection systems, do not have replaceable filters as they are built into the fuel pump module assembly. (Check the vehicle manufacturers' recommendations for exact time and mileage intervals.)

FIGURE 26–27 A fuel system tester connected in series in the fuel system so all of the fuel used flows through the meter, which displays the rate of flow and the fuel pressure.

If the fuel filter becomes partially clogged, the following are likely to occur:

1. There will be low power at higher engine speeds. The vehicle usually will not go faster than a certain speed (engine acts as if it has a built-in speed governor).
2. The engine will cut out or miss on acceleration, especially when climbing hills or during heavy-load acceleration.

TECH TIP

Quick and Easy Fuel Volume Test

Testing for pump volume involves using a specialized tester or a fuel-pressure gauge equipped with a hose to allow the fuel to be drawn from the system into a container with volume markings to allow for a volume measurement. This test can be hazardous because of expanding gasoline. An alternative test involves connecting a fuel-pressure gauge to the system with the following steps:

STEP 1 Start the engine and observe the fuel-pressure gauge. The reading should be within factory specifications (typically between 35 and 45 PSI).

STEP 2 Remove the hose from the fuel-pressure regulator. The pressure should increase if the system uses a demand-type regulator.

STEP 3 Rapidly accelerate the engine while watching the fuel-pressure gauge. If the fuel volume is okay, the fuel pressure should not drop more than 2 PSI. If the fuel pressure drops more than 2 PSI, replace the fuel filter and retest.

STEP 4 After replacing the fuel filter, accelerate the engine and observe the pressure gauge. If the pressure drops more than 2 PSI, replace the fuel pump.

NOTE: The fuel pump could still be delivering less than the specified volume of fuel, but as long as the volume needed by the engine is met, the pressure will not drop. If, however, the vehicle is pulling a heavy load, the demand for fuel volume may exceed the capacity of the pump.

FIGURE 26–28 Removing the bed from a pickup truck makes gaining access to the fuel pump a lot easier.

TECH TIP

Remove the Bed to Save Time?

The electric fuel pump is easier to replace on many General Motors pickup trucks if the bed is removed. Access to the top of the fuel tank, where the access hole is located, for the removal of the fuel tank sender unit and pump is restricted by the bottom of the pickup truck bed. It would take several people (usually other technicians in the shop) to lift the truck bed from the frame after removing only a few fasteners. ● **SEE FIGURE 26–28.**

CAUTION: Be sure to clean around the fuel pump opening so that dirt or debris does not enter the tank when the fuel pump is removed.

A weak or defective fuel pump can also be the cause of the symptoms just listed. If an electric fuel pump for a fuel-injected engine becomes weak, additional problems include the following:

1. The engine may be hard to start.
2. There may be a rough idle and stalling.
3. There may be erratic shifting of the automatic transmission as a result of engine missing due to lack of fuel-pump pressure and/or volume.

CAUTION: Be certain to consult the vehicle manufacturers' recommended service and testing procedures before attempting to test or replace any component of a high-pressure electronic fuel-injection system.

FUEL-PUMP CURRENT DRAW TEST

Another test that can and should be performed on a fuel pump is to measure the current draw in amperes. This test is most often performed by connecting a digital multimeter set to read DC amperes and test the current draw. ● **SEE FIGURE 26–29** for the hookup for vehicles equipped with a fuel-pump relay. Compare the reading to factory specifications. ● **SEE CHART 26–1** for an example of typical fuel-pump current draw readings.

NOTE: Testing the current draw of an electric fuel pump may not indicate whether the pump is good. A pump that is not rotating may draw normal current.

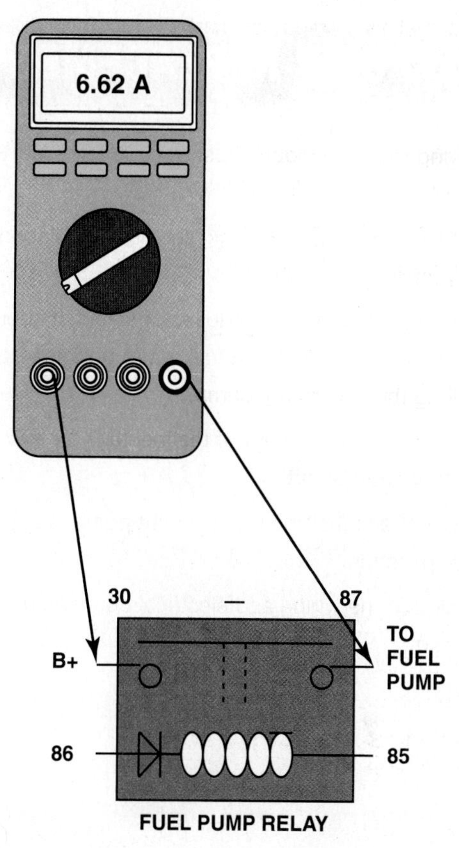

FIGURE 26–29 Hookup for testing fuel-pump current draw on any vehicle equipped with a fuel-pump relay.

Fuel-Pump Current Draw Table

AMPERAGE READING	EXPECTED VALUE	AMPERAGE TOO HIGH	AMPERAGE TOO LOW
Throttle-Body Fuel-Injection Engines	2–5 A	• Check the fuel filter. • Check for restrictions in other fuel line areas. • Replace the fuel pump.	• Check for a high-resistance connection. • Check for a high-resistance ground fault. • Replace the fuel pump.
Port Fuel-Injection Engines	4–8 A	• Check the fuel filter. • Check for restrictions in other fuel line areas. • Replace the fuel pump.	• Check for a high-resistance connection. • Check for a high-resistance ground fault. • Replace the fuel pump.
Turbo Engines	6–10 A	• Check the fuel filter. • Check for restrictions in other fuel line areas. • Replace the fuel pump.	• Check for a high-resistance connection. • Check for a high-resistance ground fault. • Replace the fuel pump.
GM CPI Truck Engines	8–12 A	• Check the fuel filter. • Check for restrictions in other fuel line areas. • Replace the fuel pump.	• Check for a high-resistance connection. • Check for a high-resistance ground fault. • Replace the fuel pump.

CHART 26–1

Fuel-pump draw and possible faults that could cause either too high or too low an amperage reading.

FUEL-PUMP REPLACEMENT

The following recommendations should be followed whenever replacing an electric fuel pump:

- The fuel-pump strainer (sock) should be replaced with the new pump.
- If the original pump had a deflector shield, it should always be used to prevent fuel return bubbles from blocking the inlet to the pump.
- Always check the interior of the fuel tank for evidence of contamination or dirt.
- Double-check that the replacement pump is correct for the application.
- Check that the wiring and electrical connectors are clean and tight.

Fuel Supply–Related Symptom Guide

PROBLEM	POSSIBLE CAUSES
Pressure too high after engine start-up.	1. Defective fuel-pressure regulator 2. Restricted fuel return line 3. Excessive system voltage 4. Restricted return line 5. Wrong fuel pump
Pressure too low after engine start-up.	1. Stuck-open pressure regulator 2. Low voltage 3. Poor ground 4. Plugged fuel filter 5. Faulty inline fuel pump 6. Faulty in-tank fuel pump 7. Partially clogged filter sock 8. Faulty hose coupling 9. Leaking fuel line 10. Wrong fuel pump 11. Leaking pulsator 12. Restricted accumulator 13. Faulty pump check valves 14. Faulty pump installation
Pressure drops off with key on/ engine off. With key off, the pressure does not hold.	1. Leaky pulsator 2. Leaking fuel-pump coupling hose 3. Faulty fuel pump (check valves) 4. Faulty pressure regulator 5. Leaking fuel injector 6. Leaking cold-start fuel injector 7. Faulty installation 8. Lines leaking

SUMMARY

1. The fuel delivery system includes the following items:
   - Fuel tank
   - Fuel pump
   - Fuel filter(s)
   - Fuel lines
2. A fuel tank is either constructed of steel with a tin plating for corrosion resistance or polyethylene plastic.
3. Fuel tank filler tubes contain an anti-siphoning device.
4. Accident and rollover protection devices include check valves and inertia switches.
5. Most fuel lines are made of nylon plastic.
6. Electric fuel-pump types include roller cell, gerotor, and turbine.
7. Fuel filters remove particles that are 10 to 20 microns or larger in size and should be replaced regularly.
8. Fuel pumps can be tested by checking:
   - Pressure
   - Volume
   - Specified current draw

REVIEW QUESTIONS

1. What are the two materials used to construct fuel tanks?
2. What are the three most commonly used pump designs?
3. What is the proper way to disconnect and connect plastic fuel line connections?
4. Where are the fuel filters located in the fuel system?
5. What accident and rollover devices are installed in a fuel delivery system?
6. What three methods can be used to test a fuel pump?

CHAPTER QUIZ

1. The first fuel filter in the sock inside the fuel tank normally filters particles larger than _________.
   a. 0.001 to 0.003 inch
   b. 0.010 to 0.030 inch
   c. 10 to 20 microns
   d. 70 to 100 microns
2. If it is tripped, which type of safety device will keep the electric fuel pump from operating?
   a. Rollover valve
   b. Inertia switch
   c. Anti-siphoning valve
   d. Check valve
3. Fuel lines are constructed from _________.
   a. Seamless steel tubing
   b. Nylon plastic
   c. Copper and/or aluminum tubing
   d. Both a and b
4. What prevents the fuel pump inside the fuel tank from catching the gasoline on fire?
   a. Electricity is not used to power the pump
   b. No air is around the motor brushes
   c. Gasoline is hard to ignite in a closed space
   d. All of the above
5. A good fuel pump should be able to supply how much fuel per minute?
   a. 1/4 pint
   b. 1/2 pint
   c. 1 pint
   d. 0.6 to 0.8 gallons
6. Technician A says that fuel pump modules are spring-loaded so that they can be compressed to fit into the opening. Technician B says that they are spring-loaded to allow for expansion and contraction of plastic fuel tanks. Which technician is correct?
   a. Technician A only
   b. Technician B only
   c. Both Technicians A and B
   d. Neither Technician A nor B
7. Most fuel filters are designed to remove particles larger than ________.
   a. 10 microns
   b. 20 microns
   c. 70 microns
   d. 100 microns
8. The amperage draw of an electric fuel pump is higher than specified. All of the following are possible causes *except:*
   a. Corroded electrical connections at the pump motor
   b. Clogged fuel filter
   c. Restriction in the fuel line
   d. Defective fuel pump
9. A fuel pump is being replaced for the third time. Technician A says that the gasoline could be contaminated. Technician B says that wiring to the pump could be corroded. Which technician is correct?
   a. Technician A only
   b. Technician B only
   c. Both Technicians A and B
   d. Neither Technician A nor B
10. The amperage draw of an electric fuel pump is lower than specified. What is the most likely cause?
    a. Corroded electrical connections at the pump motor
    b. Clogged fuel filter
    c. Restriction in the fuel line
    d. Stuck fuel pump impeller blades

chapter 27

FUEL-INJECTION COMPONENTS AND OPERATION

LEARNING OBJECTIVES

After studying this chapter, the reader will be able to:

1. Describe how throttle-body injection and port fuel-injection systems work.
2. Discuss the function of the fuel-pressure regulator and describe a vacuum-biased fuel-pressure regulator.
3. List the types of fuel-injection systems and explain their modes of operation.

This chapter will help you prepare for Engine Repair (A8) ASE certification test content area “C” (Fuel, Air Induction, and Exhaust Systems Diagnosis and Repair).

KEY TERMS

Demand delivery system (DDS) 428
Electronic air control (EAC) 432
Electronic returnless fuel system (ERFS) 427
Flare 432
Fuel rail 429
Gang fired 424
Idle speed control (ISC) motor 433
Mechanical returnless fuel system (MRFS) 427
Port fuel-injection 422
Pressure control valve (PCV) 429
Pressure vent valve (PVV) 427
Sequential fuel injection (SFI) 425
Throttle-body-injection (TBI) 422

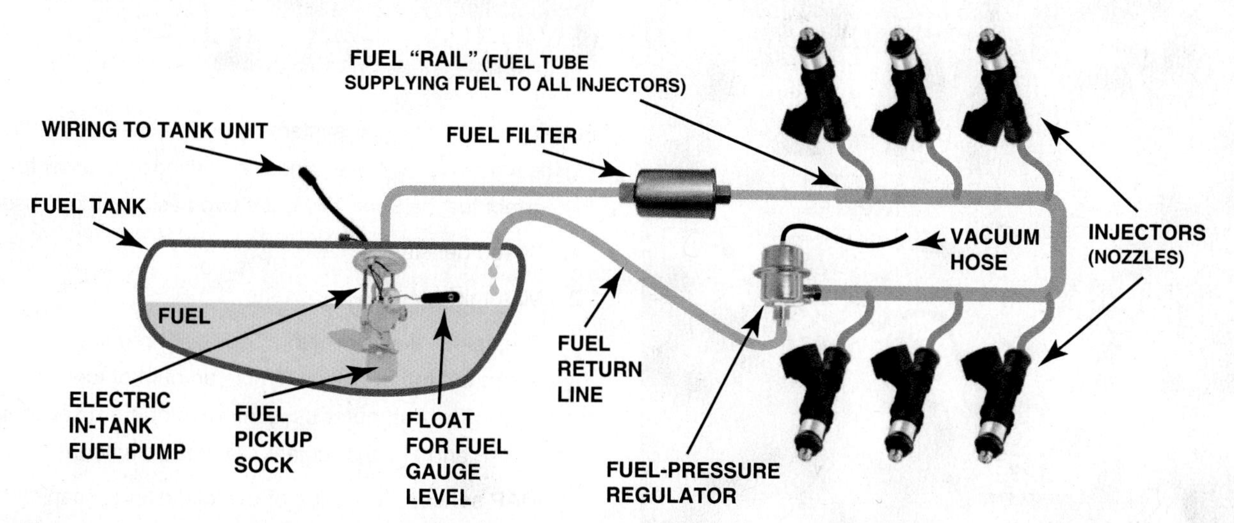

FIGURE 27–1 Typical port fuel-injection system, indicating the location of various components. Notice that the fuel-pressure regulator is located on the fuel return side of the system. The computer does not control fuel pressure, but does control the operation of the electric fuel pump (on most systems) and the pulsing on and off of the injectors.

ELECTRONIC FUEL-INJECTION OPERATION

Electronic fuel-injection systems use the computer to control the following operation of fuel injectors and other functions based on information sent to the computer from the various sensors. Most electronic fuel-injection systems share the following:

1. Electric fuel pump (usually located inside the fuel tank)
2. Fuel-pump relay (usually controlled by the computer)
3. Fuel-pressure regulator (mechanically operated spring-loaded rubber diaphragm maintains proper fuel pressure)
4. Fuel-injector nozzle or nozzles

● **SEE FIGURE 27–1**. Most electronic fuel-injection systems use the computer to control the following aspects of their operation:

1. **Pulsing the fuel injectors on and off.** The longer the injectors are held open, the greater the amount of fuel injected into the cylinder.
2. **Operating the fuel-pump relay circuit.** The computer usually controls the operation of the electric fuel pump located inside (or near) the fuel tank. The computer uses signals from the ignition switch and RPM signals from the ignition module or system to energize the fuel-pump relay circuit.

NOTE: This is a safety feature, because if the engine stalls and the tachometer (engine speed) signal is lost, the computer will shut off (de-energize) the fuel-pump relay and stop the fuel pump.

Computer-controlled fuel-injection systems are normally reliable systems if the proper service procedures are followed. Fuel-injection systems use the gasoline flowing through the injectors to lubricate and cool the injector electrical windings and pintle valves.

NOTE: The fuel does not actually make contact with the electrical windings because the injectors have O-rings at the top and bottom of the winding spool to keep fuel out.

"Two Must-Do's"

For long service life of the fuel system, always do the following:

1. Avoid operating the vehicle on a near-empty tank of fuel. The water or alcohol becomes more concentrated when the fuel level is low. Dirt that settles near the bottom of the fuel tank can be drawn through the fuel system and cause damage to the pump and injector nozzles.
2. Replace the fuel filter at regular service intervals.

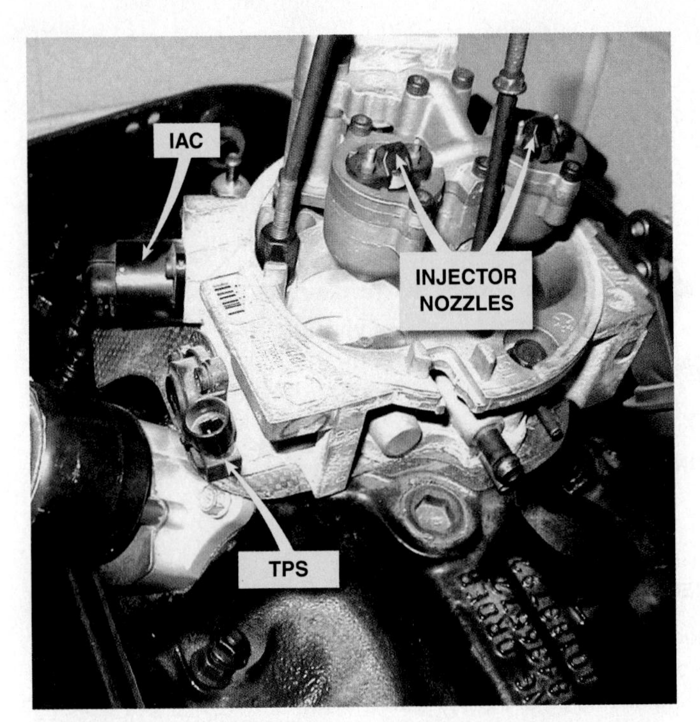

FIGURE 27–2 A dual-nozzle TBI unit on a Chevrolet 5.0 L V-8 engine. The fuel is squirted above the throttle plate where the fuel mixes with air before entering the intake manifold.

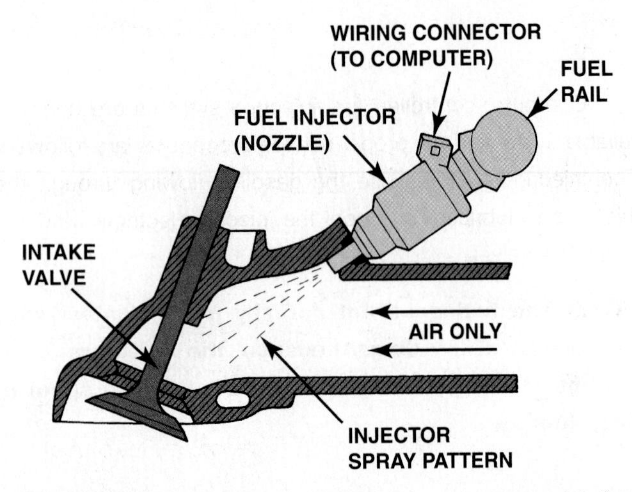

FIGURE 27–3 A typical port fuel-injection system squirts fuel into the low pressure (vacuum) of the intake manifold, about 3 inch (70 to 100 mm) from the intake valve.

There are two types of electronic fuel-injection systems:

- **Throttle-body-injection (TBI)** type. A TBI system delivers fuel from a nozzle(s) into the air above the throttle plate. ● **SEE FIGURE 27–2.**
- **Port fuel-injection** type. A port fuel-injection design uses a nozzle for each cylinder and the fuel is squirted into the intake manifold about 2 to 3 inches (70 to 100 mm) from the intake valve. ● **SEE FIGURE 27–3.**

SPEED-DENSITY FUEL-INJECTION SYSTEMS

Fuel-injection computer systems require a method for measuring the amount of air the engine is breathing in, in order to match the correct fuel delivery. There are two basic methods used:

1. Speed density
2. Mass airflow

The speed-density method does not require an air quantity sensor, but rather calculates the amount of fuel required by the engine. The computer uses information from sensors such as the MAP and TP to calculate the needed amount of fuel.

- **MAP sensor.** The value of the intake (inlet) manifold pressure (vacuum) is a direct indication of engine load.
- **TP sensor.** The position of the throttle plate and its rate of change are used as part of the equation to calculate the proper amount of fuel to inject.
- **Temperature sensors.** Both engine coolant temperature (ECT) and intake air temperature (IAT) are used to calculate the density of the air and the need of the engine for fuel. A cold engine (low-coolant temperature) requires a richer air–fuel mixture than a warm engine.

On speed-density systems, the computer calculates the amount of air in each cylinder by using manifold pressure and engine RPM. The amount of air in each cylinder is the major factor in determining the amount of fuel needed. Other sensors provide information to modify the fuel requirements. The formula used to determine the injector pulse width (PW) in milliseconds (ms) is:

Injector pulse width = MAP/BARO × RPM/maximum RPM

The formula is modified by values from other sensors, including:

- Throttle position (TP)
- Engine coolant temperature (ECT)
- Intake air temperature (IAT)
- Oxygen sensor voltage (O2S)
- Adaptive memory

A fuel injector delivers atomized fuel into the airstream where it is instantly vaporized. All throttle-body (TB) fuel-injection systems and many multipoint (port) injection systems use the speed-density method of fuel calculation.

MASS AIRFLOW FUEL-INJECTION SYSTEMS

The formula used by fuel-injection systems that use a mass airflow (MAF) sensor to calculate the injection base pulse width is:

Injector pulse width = airflow/RPM

The formula is modified by other sensor values such as:

- Throttle position
- Engine coolant temperature
- Barometric pressure
- Adaptive memory

NOTE: Many four-cylinder engines do not use a MAF sensor because, due to the time interval between intake events, some reverse airflow can occur in the intake manifold. The MAF sensor would "read" this flow of air as being additional air entering the engine, giving the PCM incorrect airflow information. Therefore, most four-cylinder engines use the speed-density method of fuel control.

THROTTLE-BODY INJECTION

The computer controls injector pulses in one of two ways:

- Synchronized
- Nonsynchronized

If the system uses a synchronized mode, the injector pulses once for each distributor reference pulse. In some vehicles, when dual injectors are used in a synchronized system, the injectors pulse alternately. In a nonsynchronized system, the injectors are pulsed once during a given period (which varies according to calibration) completely independent of distributor reference pulses.

The injector always opens the same distance, and the fuel pressure is maintained at a controlled value by the pressure regulator. The regulators used on throttle-body injection systems are not connected to a vacuum like many port fuel-injection systems. The strength of the spring inside the regulator determines at what pressure the valve is unseated, sending the fuel back to the tank and lowering the pressure. ● **SEE FIGURE 27–4.** The amount of fuel delivered by the injector depends on the amount of time (on-time) that the nozzle is open. This is the injector pulse width—the on-time in milliseconds that the nozzle is open.

The PCM commands a variety of pulse widths to supply the amount of fuel that an engine needs at any specific moment.

- A long pulse width delivers more fuel.
- A short pulse width delivers less fuel.

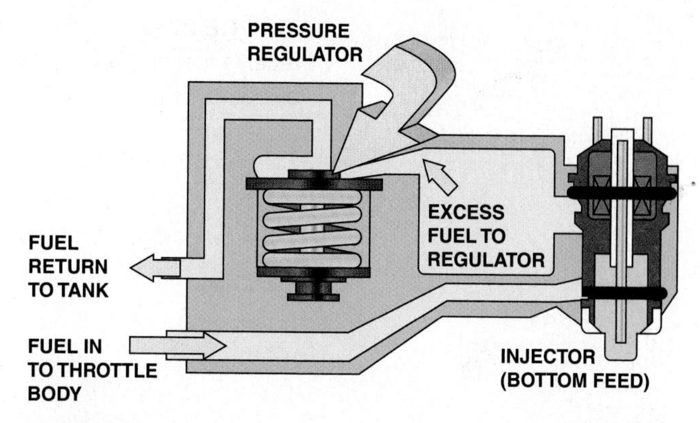

FIGURE 27–4 The tension of the spring in the fuel-pressure regulator determines the operating pressure on a throttle-body fuel-injection unit.

? FREQUENTLY ASKED QUESTION

How Do the Sensors Affect the Pulse Width?

The base pulse width of a fuel-injection system is primarily determined by the value of the MAF or MAP sensor and engine speed (RPM). However, the PCM relies on the input from many other sensors, such as the following, to modify the base pulse width as needed:

- **TP Sensor.** This sensor causes the PCM to command up to 500% (five times) the base pulse width if the accelerator pedal is depressed rapidly to the floor. It can also reduce the pulse width by about 70% if the throttle is rapidly closed.
- **ECT.** The value of this sensor determines the temperature of the engine coolant, helps determine the base pulse width, and can account for up to 60% of the determining factors.
- **BARO.** The BARO sensor compensates for altitude and adds up to about 10% under high-pressure conditions and subtracts as much as 50% from the base pulse width at high altitudes.
- **IAT.** The intake air temperature is used to modify the base pulse width based on the temperature of the air entering the engine. It is usually capable of adding as much as 20% if very cold air is entering the engine or reducing the pulse width by up to 20% if very hot air is entering the engine.
- **O2S.** This is one of the main modifiers to the base pulse width and can add or subtract up to about 20% to 25% or more, depending on the oxygen sensor activity.

FIGURE 27–5 The injectors receive fuel and are supported by the fuel rail.

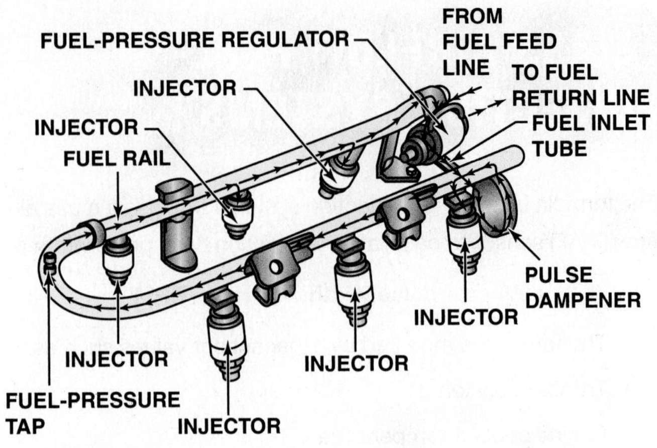

FIGURE 27–6 Cross section of a typical port fuel-injection nozzle assembly. These injectors are serviced as an assembly only; no part replacement or service is possible except for replacement of external O-ring seals.

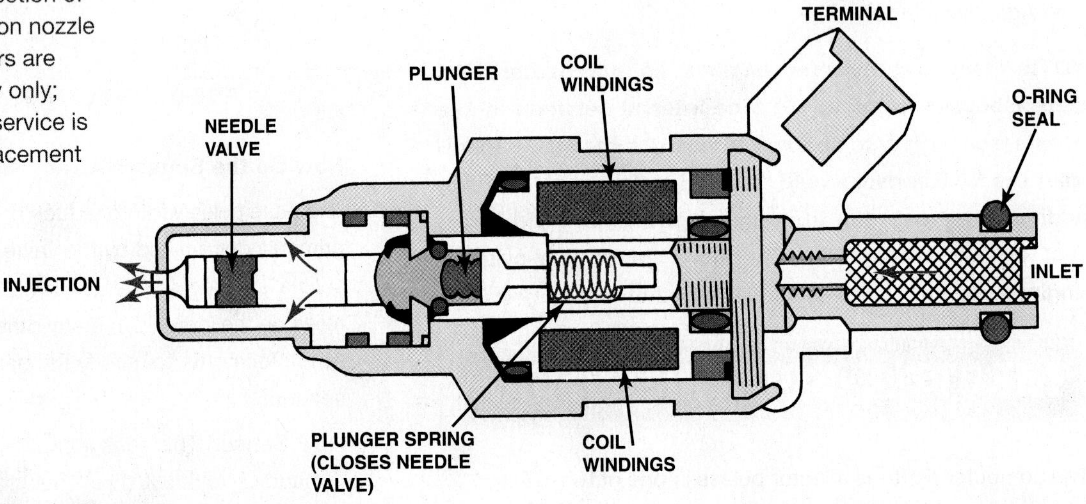

PORT-FUEL INJECTION

The advantages of port fuel-injection design also are related to characteristics of intake manifolds:

- Fuel distribution is equal to all cylinders because each cylinder has its own injector. ● **SEE FIGURE 27–5.**
- The fuel is injected almost directly into the combustion chamber, so there is no chance for it to condense on the walls of a cold intake manifold.
- Because the manifold does not have to carry fuel to properly position a TBI unit, it can be shaped and sized to tune the intake airflow to achieve specific engine performance characteristics.

An EFI injector is simply a specialized solenoid. ● **SEE FIGURE 27–6.** It has an armature winding to create a magnetic field, and a needle (pintle), a disc, or a ball valve. A spring holds the needle, disc, or ball closed against the valve seat, and when energized, the armature winding pulls open the valve when it receives a current pulse from the Powertrain Control Module (PCM). When the solenoid is energized, it unseats the valve to inject fuel.

Electronic fuel-injection systems use a solenoid-operated injector to spray atomized fuel in timed pulses into the manifold or near the intake valve. ● **SEE FIGURE 27–7.** Injectors may be sequenced and fired in one of several ways, but their pulse width is determined and controlled by the engine computer.

Port systems have an injector for each cylinder, but they do not all fire the injectors in the same way. Domestic systems use one of three ways to trigger the injectors:

- Grouped double-fire
- Simultaneous double-fire
- Sequential

GROUPED DOUBLE-FIRE This system divides the injectors into two equalized groups. The groups fire alternately; each group fires once each crankshaft revolution, or twice per four-stroke cycle. The fuel injected remains near the intake valve and enters the engine when the valve opens. This method of pulsing injectors in groups is sometimes called **gang fired**.

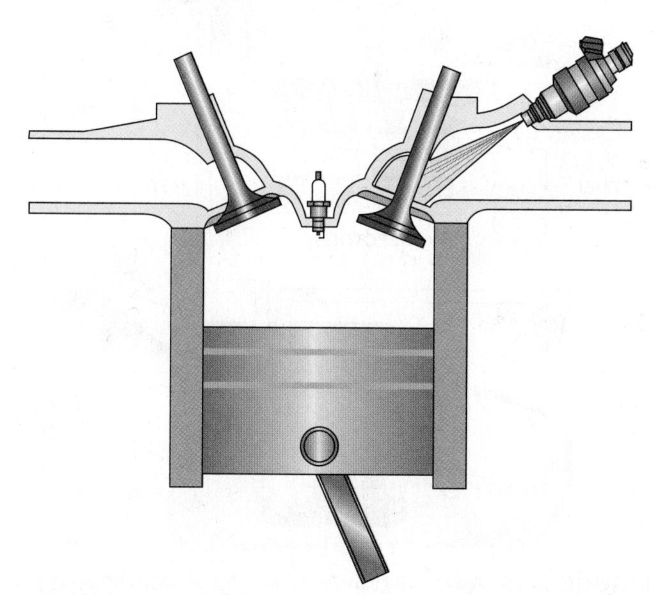

FIGURE 27–7 Port fuel injectors spray atomized fuel into the intake manifold about 3 inches (75 mm) from the intake valve.

FIGURE 27–8 A port fuel-injected engine that is equipped with long, tuned intake-manifold runners.

SIMULTANEOUS DOUBLE-FIRE This design fires all of the injectors at the same time once every engine revolution: two pulses per four-stroke cycle. Many port fuel-injection systems on four-cylinder engines use this pattern of injector firing. It is easier for engineers to program this system and it can make relatively quick adjustments in the air–fuel ratio, but it still requires the intake charge to wait in the manifold for varying lengths of time.

SEQUENTIAL Sequential firing of the injectors according to engine firing order is the most accurate and desirable method of regulating port fuel injection. However, it is also the most complex and expensive to design and manufacture. In this system, the injectors are timed and pulsed individually, much like the spark plugs are sequentially operated in firing order of the engine. This system is often called **sequential fuel injection** or **SFI**. Each cylinder receives one charge every two crankshaft revolutions, just before the intake valve opens. This means that the mixture is never static in the intake manifold and mixture adjustments can be made almost instantaneously between the firing of one injector and the next. A camshaft position sensor (CMP) signal or a special distributor reference pulse informs the PCM when the No. 1 cylinder is on its compression stroke. If the sensor fails or the reference pulse is interrupted, some injection systems shut down, while others revert to pulsing the injectors simultaneously.

The major advantage of using port injection instead of the simpler throttle-body injection is that the intake manifolds on port fuel-injected engines only contain air, not a mixture of air and fuel. This allows the engine design engineer the opportunity to design long, "tuned" intake-manifold runners that help the engine produce increased torque at low engine speeds. ● **SEE FIGURE 27–8.**

? FREQUENTLY ASKED QUESTION

How Can It Be Determined If the Injection System Is Sequential?

Look at the color of the wires at the injectors. If a sequentially fired injector is used, then one wire color (the pulse wire) will be a different color for each injector. The other wire is usually the same color because all injectors receive voltage from some source. If a group- or batch-fired injection system is being used, then the wire colors will be the same for the injectors that are group fired. For example, a V-6 group-fired engine will have three injectors with a pink and blue wire (power and pulse) and the other three will have pink and green wires.

NOTE: Some port fuel-injection systems used on engines with four or more valves per cylinder may use two injectors per cylinder. One injector is used all the time, and the second injector is operated by the computer when high-engine speed and high-load conditions are detected by the computer. Typically, the second injector injects fuel into the high-speed intake ports of the manifold. This system permits good low-speed power and throttle responses as well as superior high-speed power.

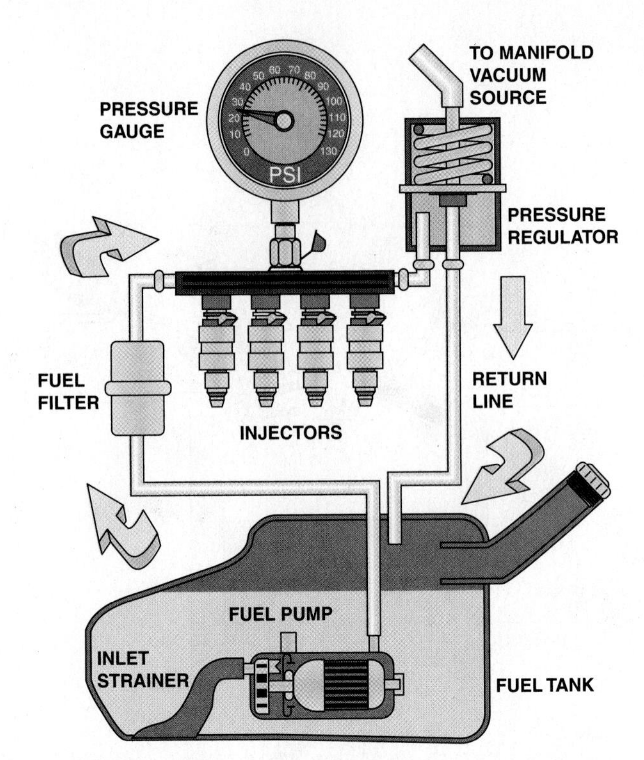

FIGURE 27–9 A typical port fuel-injected system showing a vacuum-controlled fuel-pressure regulator.

FUEL-PRESSURE REGULATOR

PURPOSE AND FUNCTION The pressure regulator and fuel pump work together to maintain the required pressure drop at the injector tips. The fuel-pressure regulator typically consists of a spring-loaded, diaphragm-operated valve in a metal housing.

Fuel-pressure regulators on fuel-return-type fuel-injection systems are installed on the return (downstream) side of the injectors at the end of the fuel rail, or are built into or mounted upon the throttle-body housing. Downstream regulation minimizes fuel-pressure pulsations caused by pressure drop across the injectors as the nozzles open. It also ensures positive fuel pressure at the injectors at all times and holds residual pressure in the lines when the engine is off. On mechanical returnless systems, the regulator is located back at the tank with the fuel filter.

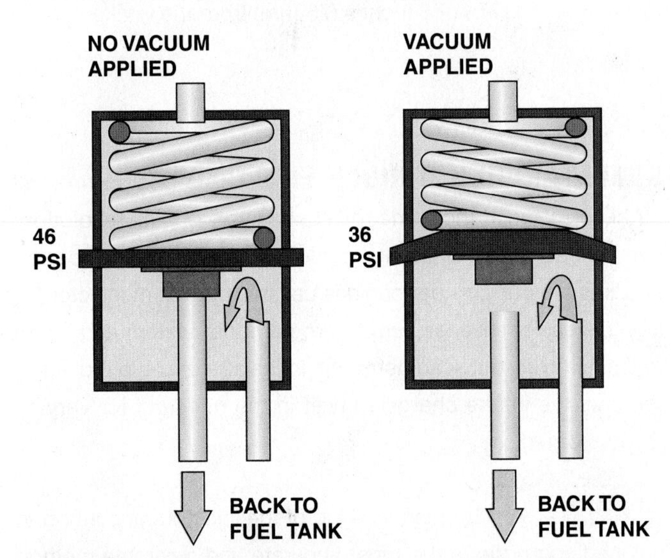

FIGURE 27–10 A typical fuel-pressure regulator that has a spring that exerts 46 pounds of force against the fuel. If 20 inches of vacuum are applied above the spring, the vacuum reduces the force exerted by the spring on the fuel, allowing the fuel to return to the tank at a lower pressure.

OPERATION In order for excess fuel (about 80% to 90% of the fuel delivered) to return to the tank, fuel pressure must overcome spring pressure on the spring-loaded diaphragm to uncover the return line to the tank. This happens when system pressure exceeds operating requirements. With TBI, the regulator is close to the injector tip, so the regulator senses essentially the same air pressure as the injector.

The pressure regulator used in a port fuel-injection system has an intake-manifold vacuum line connection on the regulator vacuum chamber. This allows fuel pressure to be modulated by a combination of spring pressure and manifold vacuum acting on the diaphragm. ● **SEE FIGURES 27–9 AND 27–10.**

In both TBI and port fuel-injection systems, the regulator shuts off the return line when the fuel pump is not running. This maintains pressure at the injectors for easy restarting after hot soak as well as reducing vapor lock.

Port fuel-injection systems generally operate with pressures at the injector of about 30 to 55 PSI (207 to 379 kPa), while TBI systems work with injector pressures of about 10 to 20 PSI (69 to 138 kPa). The difference in system pressures results from the difference in how the systems operate. Since injectors in a TBI system inject the fuel into the airflow at the manifold inlet (above the throttle), there is more time for atomization in the manifold before the air–fuel charge reaches the intake valve. This allows TBI injectors to work at lower pressures than injectors used in a port system.

FIGURE 27–11 A lack of fuel flow could be due to a restricted fuel-pressure regulator. Notice the fine screen filter. If this filter were to become clogged, higher than normal fuel pressure would occur.

TECH TIP

Don't Forget the Regulator

Some fuel-pressure regulators contain a 10 micron filter. If this filter becomes clogged, a lack of fuel flow would result. ● **SEE FIGURE 27–11.**

VACUUM-BIASED FUEL-PRESSURE REGULATOR

The primary reason why many port fuel-injected systems use a vacuum-controlled fuel-pressure regulator is to ensure that there is a constant pressure drop across the injectors. In a throttle-body fuel-injection system, the injector squirts into the atmospheric pressure regardless of the load on the engine. In a port fuel-injected engine, however, the pressure inside the intake manifold changes as the load on the engine increases.

ENGINE OPERATING CONDITION	INTAKE-MANIFOLD VACUUM	FUEL PRESSURE
Idle or cruise	High	Lower
Heavy load	Low	Higher

The computer can best calculate injector pulse width based on all sensors if the pressure drop across the injector is the same under all operating conditions. A vacuum-controlled fuel-pressure regulator allows the equal pressure drop by reducing the force exerted by the regulator spring at high vacuum (low-load condition), yet allowing the full force of the regulator spring to be exerted when the vacuum is low (high-engine-load condition).

ELECTRONIC RETURNLESS FUEL SYSTEM

This system is unique because it does not use a mechanical valve to regulate rail pressure. Fuel pressure at the rail is sensed by a pressure transducer, which sends a low-level signal to a controller. The controller contains logic to calculate a signal to the pump power driver. The power driver contains a high-current transistor that controls the pump speed using pulse width modulation (PWM). This system is called the **electronic returnless fuel system (ERFS).** ● **SEE FIGURE 27–12.** This transducer can be differentially referenced to manifold pressure for closed-loop feedback, correcting and maintaining the output of the pump to a desired rail setting. This system is capable of continuously varying rail pressure as a result of engine vacuum, engine fuel demand, and fuel temperature (as sensed by an external temperature transducer, if necessary). A **pressure vent valve (PVV)** is employed at the tank to relieve overpressure due to thermal expansion of fuel. In addition, a supply-side bleed, by means of an in-tank reservoir using a supply-side jet pump, is necessary for proper pump operation.

MECHANICAL RETURNLESS FUEL SYSTEM

The first production returnless systems employed the **mechanical returnless fuel system (MRFS)** approach. This system has a bypass regulator to control rail pressure that is located in close proximity to the fuel tank. Fuel is sent by the in-tank pump to a chassis-mounted inline filter with excess fuel returning to the tank through a short return line. ● **SEE FIGURE 27–13.** The inline filter may be mounted directly to the tank, thereby eliminating the shortened return line. Supply pressure is regulated on the downstream side of the inline filter to accommodate changing restrictions throughout the filter's service life. This system is limited to constant rail pressure (*CRP) system calibrations, whereas with ERFS, the pressure transducer can be referenced to atmospheric pressure for CRP systems or differentially referenced to intake-manifold pressure for constant differential injector pressure (**CIP) systems.

NOTE: *CRP is referenced to atmospheric pressure, has lower operating pressure, and is desirable for calibrations using speed/air density sensing. **CIP is referenced to manifold pressure, varies rail pressure, and is desirable in engines that use mass airflow sensing.

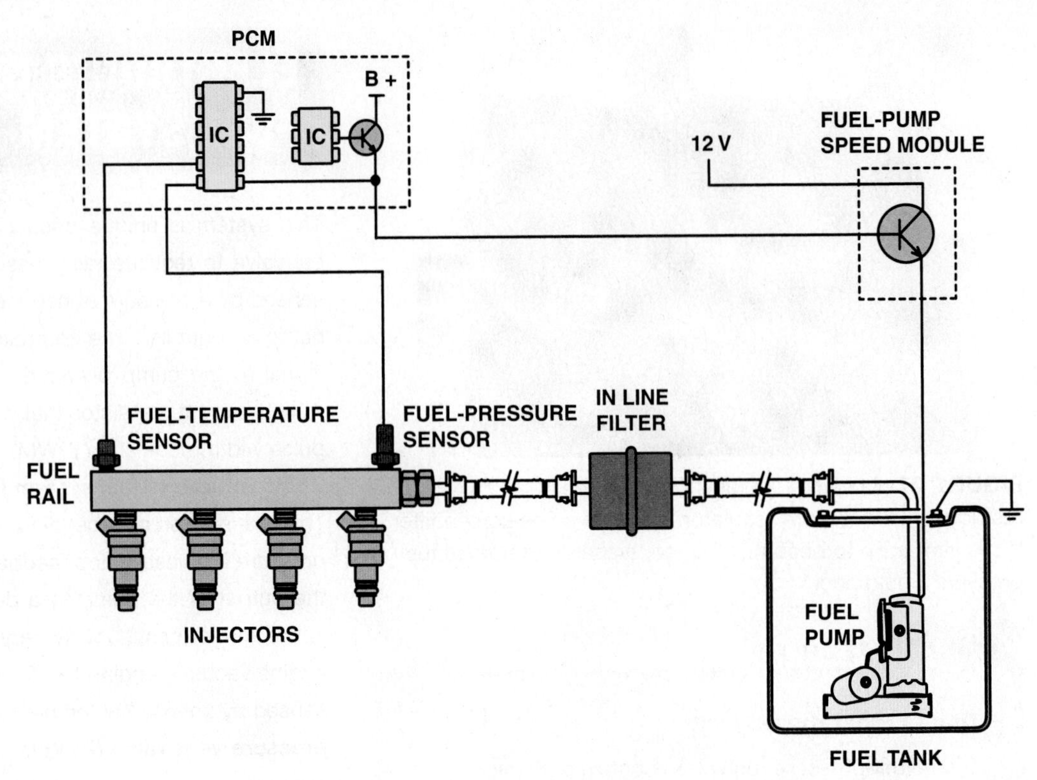

FIGURE 27–12 The fuel-pressure sensor and fuel-temperature sensor are often constructed together in one assembly to help give the PCM the needed data to control the fuel-pump speed.

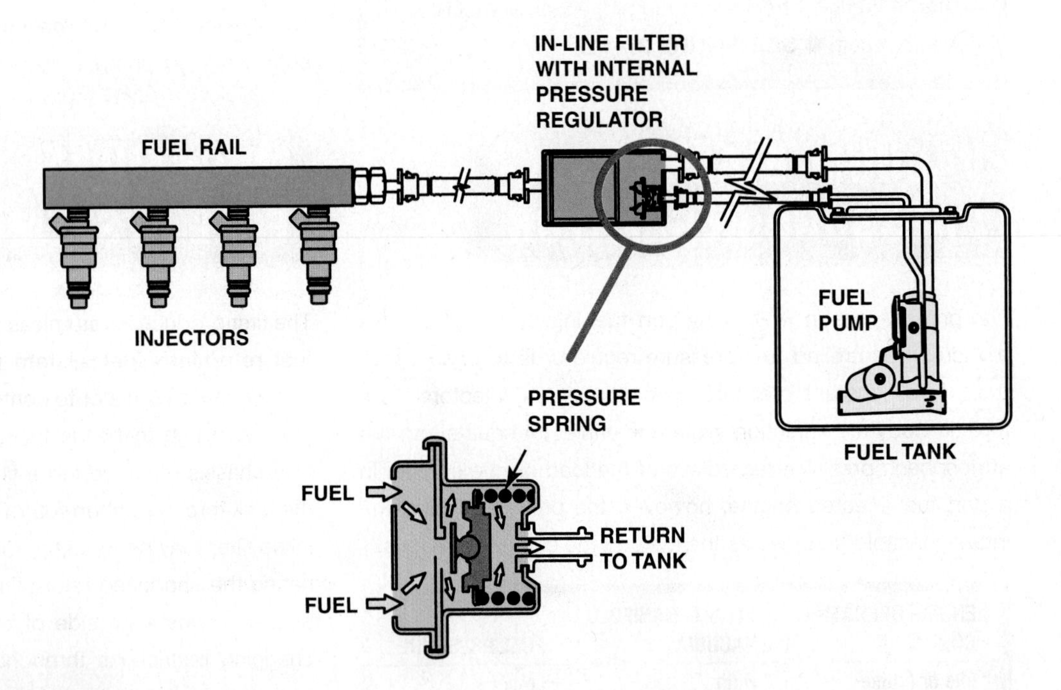

FIGURE 27–13 A mechanical returnless fuel system. The bypass regulator in the fuel tank controls fuel line pressure.

DEMAND DELIVERY SYSTEM (DDS)

Given the experience with both ERFS and MRFS, a need was recognized to develop new returnless technologies that could combine the speed control and constant injector pressure attributes of ERFS together with the cost savings, simplicity, and reliability of MRFS. This new technology also needed to address pulsation dampening/hammering and fuel transient response. Therefore, the **demand delivery system (DDS)** technology was developed. A different form of demand pressure regulator has been applied to the fuel rail. It mounts at the head or port entry and regulates the pressure downstream at the injectors by admitting the precise quantity of fuel into the rail as consumed by the engine. Having demand regulation at the rail improves pressure response to flow transients and provides rail pulsation dampening. A fuel pump and a low-cost, high-performance bypass regulator are used

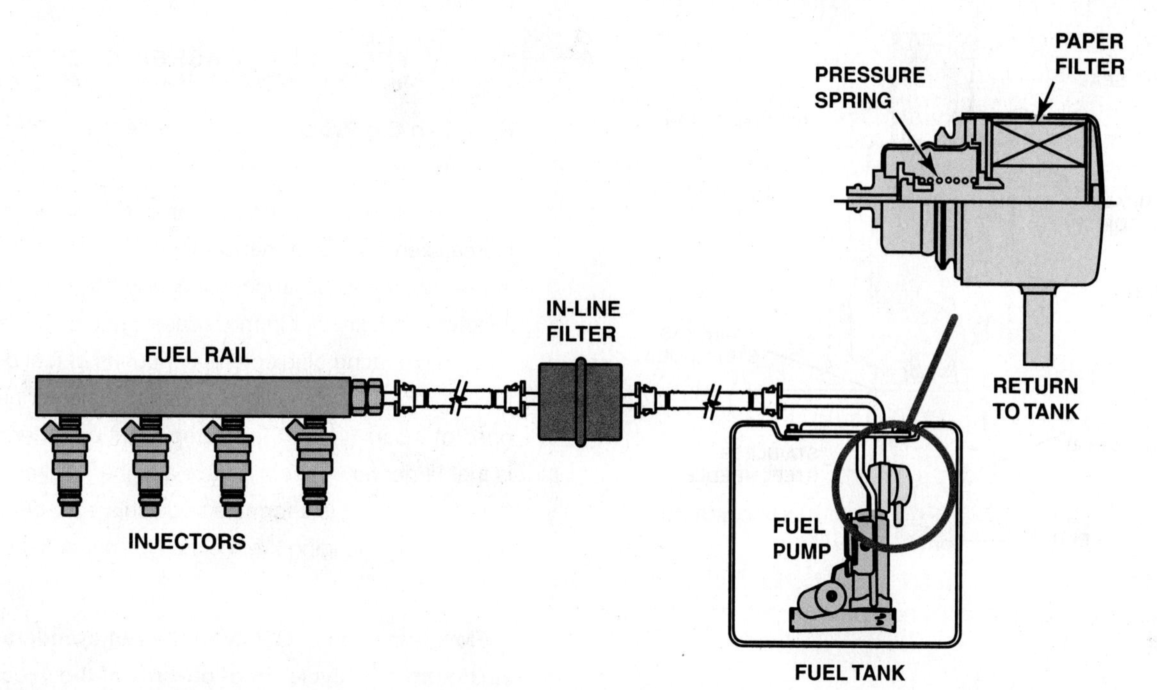

FIGURE 27–14 A demand delivery system uses a fuel-pressure regulator attached to the fuel pump assembly inside the fuel tank.

FIGURE 27–15 A rectangular-shaped fuel rail is used to help dampen fuel system pulsations and noise caused by the injectors opening and closing.

? FREQUENTLY ASKED QUESTION

Why Are Some Fuel Rails Rectangular Shaped?

A port fuel-injection system uses a pipe or tubes to deliver fuel from the fuel line to the intended fuel injectors. This pipe or tube is called the **fuel rail.** Some vehicle manufacturers construct the fuel rail in a rectangular cross section. ● **SEE FIGURE 27–15.** The sides of the fuel rail are able to move in and out slightly, thereby acting as a fuel pulsator evening out the pressure pulses created by the opening and closing of the injectors to reduce underhood noise. A round cross-sectional fuel rail is not able to deform and, as a result, some manufacturers have had to use a separate dampener.

within the appropriate fuel sender. ● **SEE FIGURE 27–14.** They supply a pressure somewhat higher than the required rail set pressure to accommodate dynamic line and filter pressure losses. Electronic pump speed control is accomplished using a smart regulator as an integral flow sensor. A **pressure control valve (PCV)** may also be used and can readily reconfigure an existing design fuel sender into a returnless sender.

FUEL INJECTORS

EFI systems use solenoid-operated injectors. ● **SEE FIGURE 27–16.** This electromagnetic device contains an armature and a spring-loaded needle valve or ball valve assembly. When the computer energizes the solenoid, voltage is applied to the solenoid coil until the current reaches a specified level. This permits a quick pull-in of the armature during turn-on. The armature is pulled off of its seat against spring force, allowing

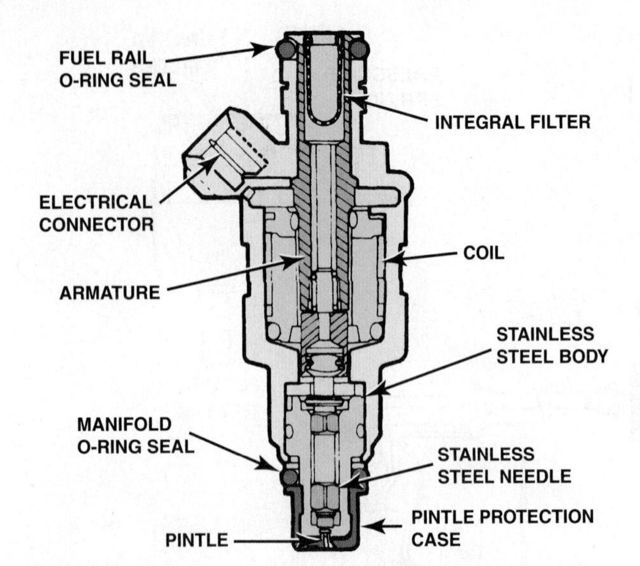

FIGURE 27–16 A multiport fuel injector. Notice that the fuel flows straight through and does not come in contact with the coil windings.

FIGURE 27–17 Each of the eight injectors shown are producing a correct spray pattern for the applications. While all throttle-body injectors spray a conical pattern, most port fuel injections do not.

How Can the Proper Injector Size Be Determined?

Most people want to increase the output of fuel to increase engine performance. Injector sizing can sometimes be a challenge, especially if the size of injector is not known. In most cases, manufacturers publish the rating of injectors, in pounds of fuel per hour (lb/hr). The rate is figured with the injector held open at 3 bars (43.5 PSI). An important consideration is that larger flow injectors have a higher minimum flow rating. Here is a formula to calculate injector sizing when changing the mechanical characteristics of an engine.

Flow rate = hp × BSFC/number of cylinders × maximum duty cycle (% of on-time of the injectors)

- **hp** is the projected horsepower. Be realistic!
- **BSFC** is brake-specific fuel consumption in pounds per horsepower-hour. Calculated values are used for this, 0.4 to 0.8 lb. In most cases, start on the low side for naturally aspirated engines and the high side for engines with forced induction.
- **Number of cylinders** is actually the number of injectors being used.
- **Maximum duty cycle** is considered at 0.8 (80%). Above this, the injector may overheat, lose consistency, or not work at all.

For example:

5.7 liter V-8 = 240 hp × 0.65/8 cylinders × 8
= 24.37 lb/hr injectors required

fuel to flow through the inlet filter screen to the spray nozzle, where it is sprayed in a pattern that varies with application. ● **SEE FIGURE 27–17**. The injector opens the same amount each time it is energized, so the amount of fuel injected depends on the length of time the injector remains open. By angling the director hole plates, the injector sprays fuel more directly at the intake valves, which further atomizes and vaporizes the fuel before it enters the combustion chamber. PFI injectors typically are a top-feed design in which fuel enters the top of the injector and passes through its entire length to keep it cool before being injected.

Ford introduced two basic designs of deposit-resistant injectors on some engines. The design, manufactured by Bosch, uses a four-hole director/metering plate similar to that used by the Rochester Multec injectors. The design manufactured by Nippondenso uses an internal upstream orifice in the adjusting tube. It also has a redesigned pintle/seat containing a wider tip opening that tolerates deposit buildup without affecting injector performance.

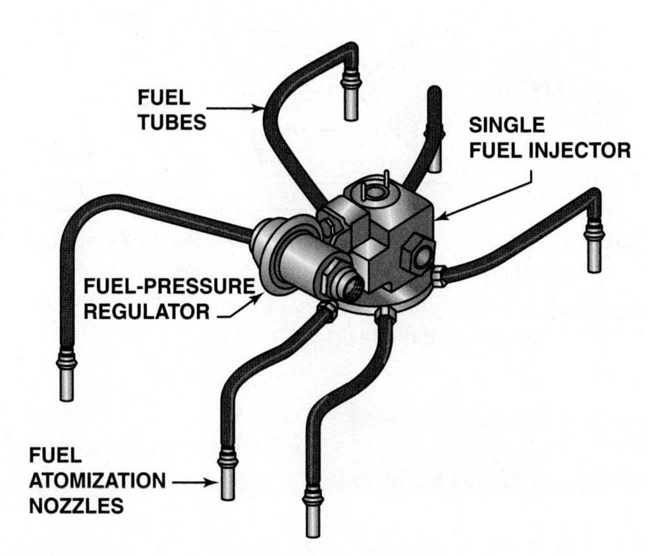

FIGURE 27–18 A central port fuel-injection system.

CENTRAL PORT INJECTION

A cross between port fuel injection and throttle-body injection, CPI was introduced in the early 1990s by General Motors. The CPI assembly consists of a single fuel injector, a pressure regulator, and six poppet nozzle assemblies with nozzle tubes. ● **SEE FIGURE 27–18**. The central sequential fuel injection (CSFI) system has six injectors in place of just one used on the CPI unit.

When the injector is energized, its armature lifts off of the six fuel tube seats and pressurized fuel flows through the nozzle tubes to each poppet nozzle. The increased pressure causes each poppet nozzle ball to also lift from its seat, allowing fuel to flow from the nozzle. This hybrid injection system combines the single injector of a TBI system with the equalized fuel distribution of a PFI system. It eliminates the individual fuel rail while allowing more efficient manifold tuning than is otherwise possible with a TBI system. Newer versions use six individual solenoids to fire one for each cylinder. ● **SEE FIGURE 27–19**.

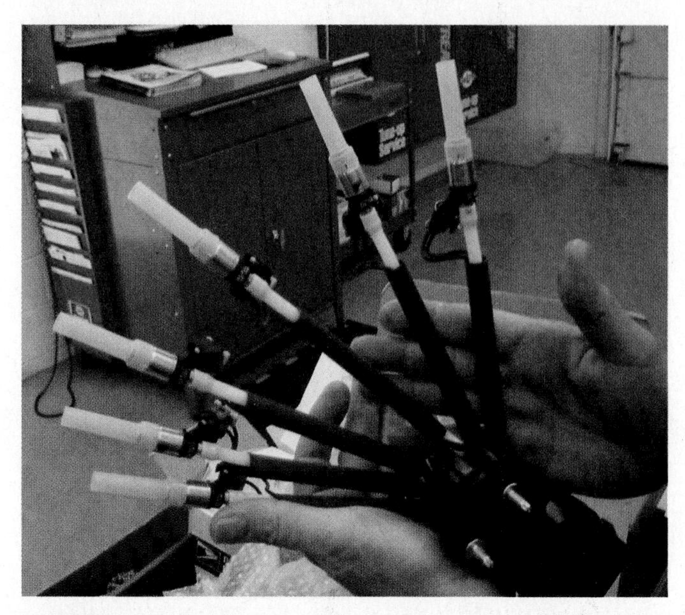

FIGURE 27–19 A factory replacement unit for a CSFI unit that has individual injectors at the ends that go into the intake manifold instead of poppet valves.

FUEL-INJECTION MODES OF OPERATION

All fuel-injection systems are designed to supply the correct amount of fuel under a wide range of engine operating conditions. These modes of operation include:

Starting (cranking)	Acceleration enrichment
Clear flood	Deceleration enleanment
Idle (run)	Fuel shutoff

STARTING MODE When the ignition is turned to the start (on) position, the engine cranks and the PCM energizes the fuel-pump relay. The PCM also pulses the injectors on, basing the pulse width on engine speed and engine coolant temperature. The colder the engine is, the greater the pulse width. Cranking mode air–fuel ratio varies from about 1.5:1 at −40°F (−40°C) to 14.7:1 at 200°F (93°C).

CLEAR FLOOD MODE If the engine becomes flooded with too much fuel, the driver can depress the accelerator pedal to greater than 80% to enter the clear flood mode. When the PCM detects that the engine speed is low (usually below 600 RPM) and the throttle-position (TP) sensor voltage is high (WOT), the injector pulse width is greatly reduced or even shut off entirely, depending on the vehicle.

OPEN-LOOP MODE Open-loop operation occurs during warm-up before the oxygen sensor can supply accurate information to the PCM. The PCM determines injector pulse width based on values from the MAF, MAP, TP, ECT, and IAT sensors.

CLOSED-LOOP MODE Closed-loop operation is used to modify the base injector pulse width as determined by feedback from the oxygen sensor to achieve proper fuel control.

ACCELERATION ENRICHMENT MODE During acceleration, the throttle-position (TP) voltage increases, indicating that a richer air–fuel mixture is required. The PCM then supplies a longer injector pulse width and may even supply extra pulses to supply the needed fuel for acceleration.

DECELERATION ENLEANMENT MODE When the engine decelerates, a leaner air–fuel mixture is required to help reduce emissions and to prevent deceleration backfire. If the deceleration is rapid, the injector may be shut off entirely for a short time and then pulsed on enough to keep the engine running.

FUEL SHUTOFF MODE Besides shutting off fuel entirely during periods of rapid deceleration, PCM also shuts off the injector when the ignition is turned off to prevent the engine from continuing to run.

FREQUENTLY ASKED QUESTION

What Is Battery Voltage Correction?

Battery voltage correction is a program built into the PCM that causes the injector pulse width to increase if there is a drop in electrical system voltage. Lower battery voltage would cause the fuel injectors to open slower than normal and the fuel pump to run slower. Both of these conditions can cause the engine to run leaner than normal if the battery voltage is low. Because a lean air–fuel mixture can cause the engine to overheat, the PCM compensates for the lower voltage by adding a percentage to the injector pulse width. This richer condition will help prevent serious engine damage. The idle speed is also increased to turn the generator (alternator) faster if low battery voltage is detected.

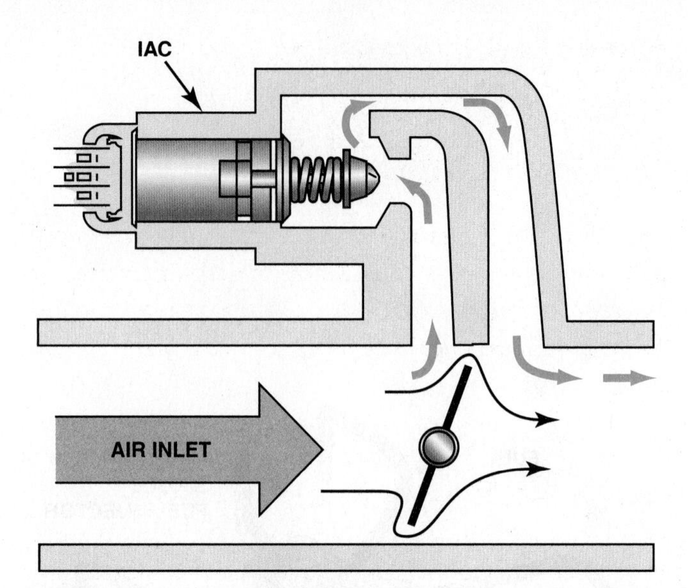

FIGURE 27–20 The small arrows indicate the air bypassing the throttle plate in the closed throttle position. This air is called minimum air. The air flowing through the IAC is the airflow that determines the idle speed.

IDLE CONTROL

Port fuel-injection systems generally use an auxiliary air bypass. ● **SEE FIGURE 27–20**. This air bypass or regulator provides needed additional airflow, and thus more fuel. The engine needs more power when cold to maintain its normal idle speed to overcome the increased friction from cold lubricating oil. It does this by opening an intake air passage to let more air into the engine just as depressing the accelerator pedal would open the throttle valve, allowing more air into the engine. The system is calibrated to maintain engine idle speed at a specified value regardless of engine temperature.

Most PFI systems use an idle air control (IAC) motor to regulate idle bypass air. The IAC is computer-controlled, and is either a solenoid-operated valve or a stepper motor that regulates the airflow around the throttle. The idle air control valve is also called an **electronic air control (EAC)** valve.

When the engine stops, most IAC units will retract outward to get ready for the next engine start. When the engine starts, the engine speed is high to provide for proper operation when the engine is cold. Then, as the engine gets warmer, the computer reduces engine idle speed gradually by reducing the number of counts or steps commanded by the IAC.

When the engine is warm and restarted, the idle speed should momentarily increase, then decrease to normal idle speed. This increase and then decrease in engine speed is often called an engine **flare**. If the engine speed does not flare, then the IAC may not be working (it may be stuck in one position).

STEPPER MOTOR OPERATION

A digital output is used to control stepper motors. Stepper motors are direct-current motors that move in fixed steps or increments from de-energized (no voltage) to fully energized (full voltage). A stepper motor often has as many as 120 steps of motion.

A common use for stepper motors is as an idle air control (IAC) valve, which controls engine idle speeds and prevents stalls due to changes in engine load. When used as an IAC, the stepper motor is usually a reversible DC motor that moves in increments, or steps. The motor moves a shaft back and forth to operate a conical valve. When the conical valve is moved back, more air bypasses the throttle plates and enters the engine, increasing idle speed. As the conical valve moves inward, the idle speed decreases.

When using a stepper motor that is controlled by the PCM, it is very easy for the PCM to keep track of the position of the stepper motor. By counting the number of steps that have been sent to the stepper motor, the PCM can determine the relative position of the stepper motor. While the PCM does not actually receive a feedback signal from the stepper motor, it does know how many steps forward or backward the motor should have moved.

A typical stepper motor uses a permanent magnet and two electromagnets. Each of the two electromagnetic windings is controlled by the computer. The computer pulses the windings and changes the polarity of the windings to cause the armature of the stepper motor to rotate 90 degrees at a time. Each 90-degree pulse is recorded by the computer as a "count" or "step"; therefore, the name given to this type of motor. ● **SEE FIGURE 27–21**.

Idle airflow in a TBI system travels through a passage around the throttle and is controlled by a stepper motor. In some applications, an externally mounted permanent magnet motor called the **idle speed control (ISC) motor** mechanically advances the throttle linkage to advance the throttle opening.

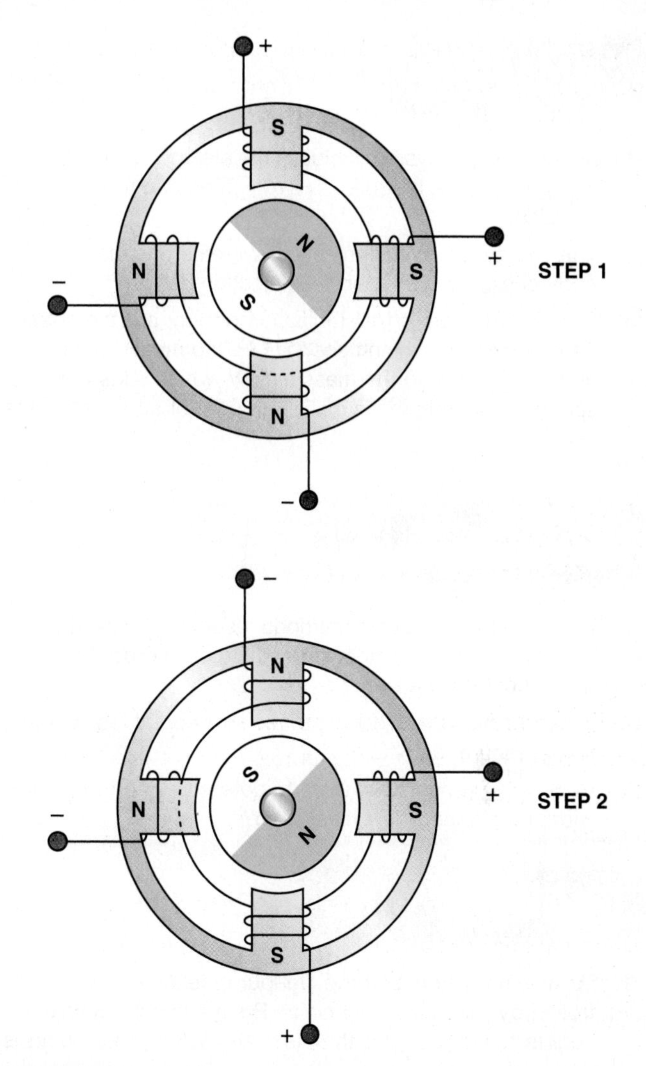

FIGURE 27–21 Most stepper motors use four wires, which are pulsed by the computer to rotate the armature in steps.

? FREQUENTLY ASKED QUESTION

Why Does the Idle Air Control Valve Use Milliamperes?

Some Chrysler vehicles, such as the Dodge minivan, use linear solenoid idle air control valves (LSIAC). The PCM uses regulated current flow through the solenoid to control idle speed and the scan tool display is in milliamperes (mA).

Closed position = 180–200 mA
Idle = 300–450 mA
Light cruise = 500–700 mA
Fully open = 900–950 mA

SUMMARY

1. A fuel-injection system includes the electric fuel pump and fuel-pump relay, fuel-pressure regulator, and fuel injectors (nozzles).
2. The two types of fuel-injection systems are the throttle-body design and the port fuel-injection design.
3. The two methods of fuel-injection control are the speed-density system, which uses the MAP to measure the load on the engine, and the mass airflow, which uses the MAF sensor to directly measure the amount of air entering the engine.
4. The amount of fuel supplied by fuel injectors is determined by how long they are kept open. This opening time is called the pulse width and is measured in milliseconds.
5. The fuel-pressure regulator is usually located on the fuel return on return-type fuel-injection systems.
6. TBI-type fuel-injection systems do not use a vacuum-controlled fuel-pressure regulator, whereas many port fuel-injection systems use a vacuum-controlled regulator to monitor equal pressure drop across the injectors.
7. Other fuel designs include the electronic returnless, the mechanical returnless, and the demand delivery systems.

REVIEW QUESTIONS

1. What are the two basic methods used for measuring the amount of air the engine is breathing in, in order to match the correct fuel delivery?
2. What is the purpose of the vacuum-controlled (biased) fuel-pressure regulator?
3. How many sensors are used to determine the base pulse width on a speed-density system?
4. How many sensors are used to determine the base pulse width on a mass airflow system?
5. What are the three types of returnless fuel-injection systems?

CHAPTER QUIZ

1. Technician A says that the fuel-pump relay is usually controlled by the PCM. Technician B says that a TBI injector squirts fuel above the throttle plate. Which technician is correct?
   a. Technician A only
   b. Technician B only
   c. Both Technicians A and B
   d. Neither Technician A nor B
2. Why are some fuel rails rectangular in shape?
   a. Increases fuel pressure
   b. Helps keep air out of the injectors
   c. Reduces noise
   d. Increases the speed of the fuel through the fuel rail
3. Which fuel-injection system uses the MAP sensor as the primary sensor to determine the base pulse width?
   a. Speed density
   b. Mass airflow
   c. Demand delivery
   d. Mechanical returnless
4. Why is a vacuum line attached to a fuel-pressure regulator on many port fuel-injected engines?
   a. To draw fuel back into the intake manifold through the vacuum hose
   b. To create an equal pressure drop across the injectors
   c. To raise the fuel pressure at idle
   d. To lower the fuel pressure under heavy engine load conditions to help improve fuel economy
5. Which sensor has the greatest influence on injector pulse width besides the MAF sensor?
   a. IAT
   b. BARO
   c. ECT
   d. TP
6. Technician A says that the port fuel-injection injectors operate using 5 volts from the computer. Technician B says that sequential fuel injectors all use a different wire color on the injectors. Which technician is correct?
   a. Technician A only
   b. Technician B only
   c. Both Technicians A and B
   d. Neither Technician A nor B
7. Which type of port fuel-injection system uses a fuel-temperature and/or fuel-pressure sensor?
   a. All port fuel-injected engines
   b. TBI units only
   c. Electronic returnless systems
   d. Demand delivery systems
8. Dampeners are used on some fuel rails to __________.
   a. Increase the fuel pressure in the rail
   b. Reduce (decrease) the fuel pressure in the rail
   c. Reduce noise
   d. Trap dirt and keep it away from the injectors
9. Where is the fuel-pressure regulator located on a vacuum-biased port fuel-injection system?
   a. In the tank
   b. At the inlet of the fuel rail
   c. At the outlet of the fuel rail
   d. Near or on the fuel filter
10. What type of device is used in a typical idle air control?
   a. DC motor
   b. Stepper motor
   c. Pulsator-type actuator
   d. Solenoid

SYSTEMS

LEARNING OBJECTIVES

After studying this chapter, the reader will be able to:

1. Discuss how to troubleshoot a gasoline direct-injection system.
2. Explain how a gasoline direct-injection system works.
3. Describe the differences between port fuel-injection and gasoline direct-injection systems.
4. List the various modes of operation of a gasoline direct-injection system.

This chapter will help you prepare for Engine Repair (A8) ASE certification test content area "C" (Fuel, Air Induction, and Exhaust Systems Diagnosis and Repair).

KEY TERMS

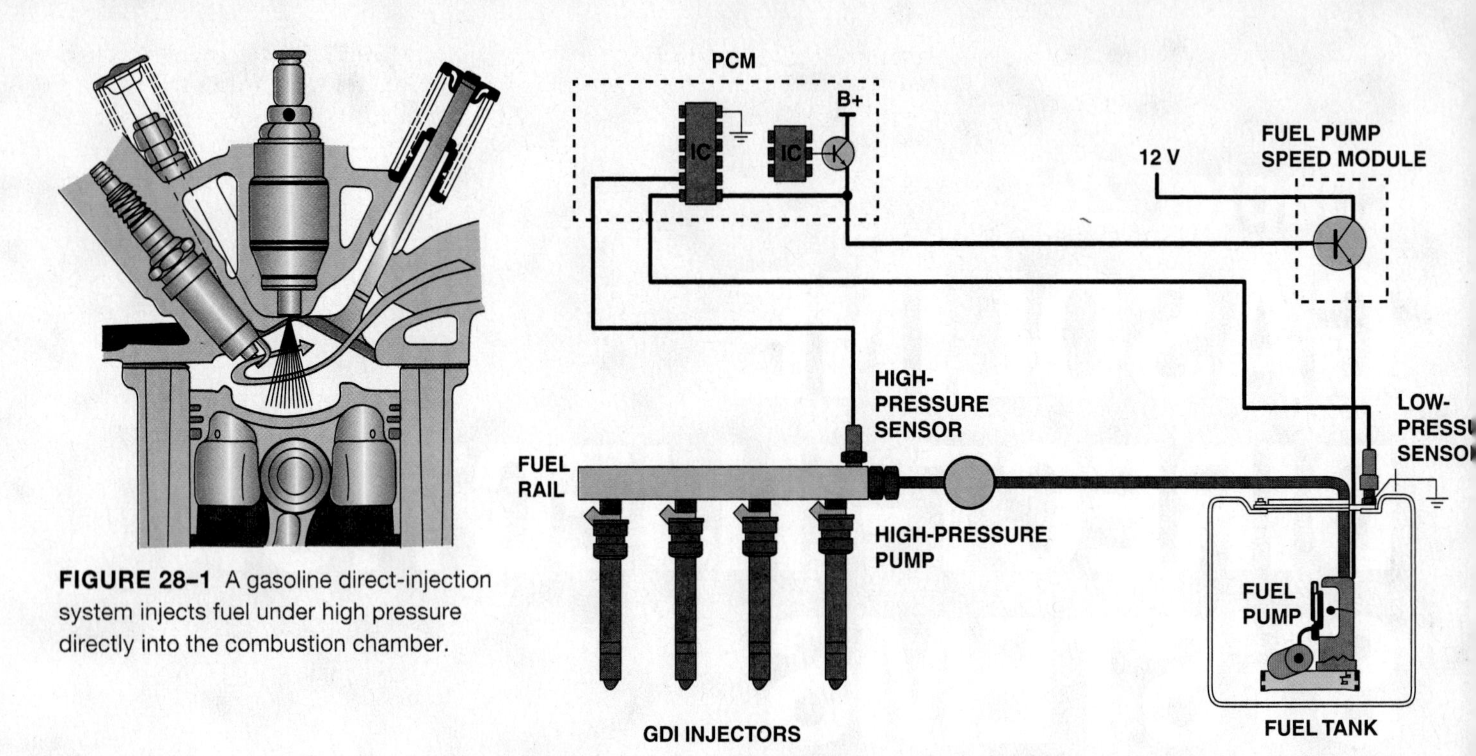

FIGURE 28–1 A gasoline direct-injection system injects fuel under high pressure directly into the combustion chamber.

FIGURE 28–2 A GDI system uses a low-pressure pump in the gas tank similar to other types of fuel-injection systems. The PCM controls the pressure of the high-pressure pump using sensor inputs.

DIRECT FUEL INJECTION

Several vehicle manufacturers such as Audi, Mitsubishi, Mercedes, BMW, Toyota/Lexus, Mazda, Ford, and General Motors are using **gasoline direct injection (GDI)** systems, which General Motors refers to as a **spark ignition direct injection (SIDI)** system. A direct-injection system sprays high-pressure fuel, up to 2,900 PSI, into the combustion chamber as the piston approaches the top of the compression stroke. With the combination of high-pressure swirl injectors and modified combustion chamber, almost instantaneous vaporization occurs. This combined with a higher compression ratio allows a direct-injected engine to operate using a leaner-than-normal air–fuel ratio, which results in improved fuel economy with higher power output and reduced exhaust emissions. ● **SEE FIGURE 28–1.**

ADVANTAGES OF GASOLINE DIRECT INJECTION The use of direct injection compared with port fuel injection has many advantages including:

- Improved fuel economy due to reduced pumping losses and heat loss
- Allows a higher compression ratio for higher engine efficiency
- Allows the use of lower-octane gasoline
- The volumetric efficiency is higher
- Less need for extra fuel for acceleration
- Improved cold starting and throttle response
- Allows the use of greater percentage of EGR to reduce exhaust emissions
- Up to 25% improvement in fuel economy
- 12% to 15% reduction in exhaust emissions

DISADVANTAGES OF GASOLINE DIRECT INJECTION

- Higher cost due to high-pressure pump and injectors
- More components compared with port fuel injection
- Due to the high compression, a NO_x storage catalyst is sometimes required to meet emission standards, especially in Europe. (● **SEE FIGURE 28–2.**)
- Uses up to six operating modes depending on engine load and speed, which requires more calculations to be performed by the Powertrain Control Module (PCM).

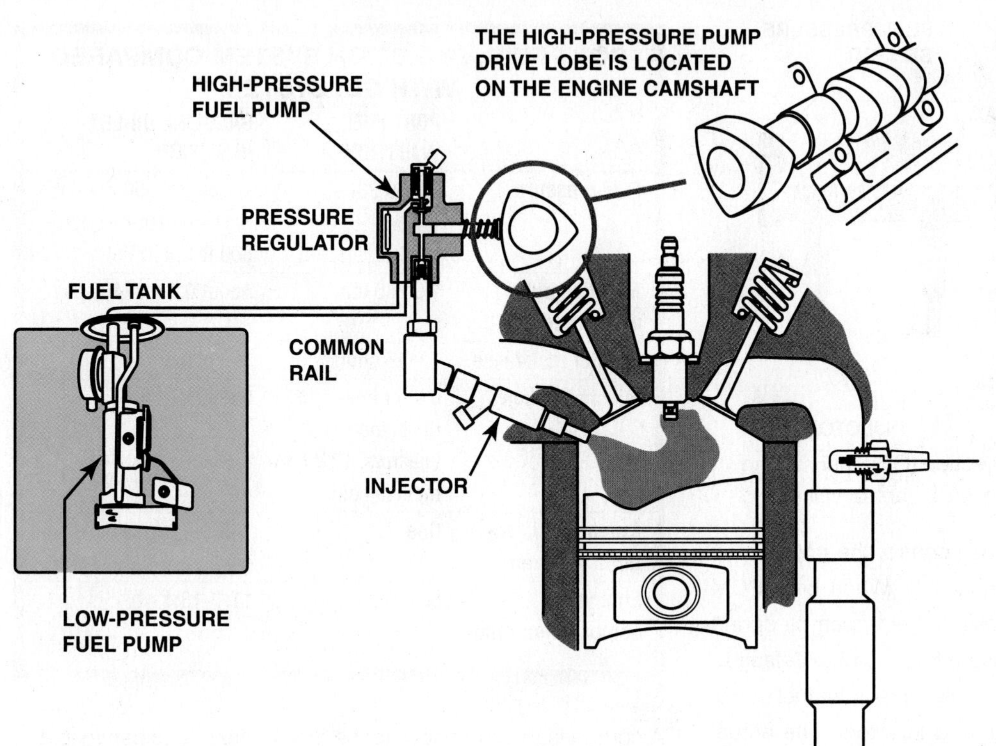

FIGURE 28–3 A typical direct-injection system uses two pumps—one low-pressure electric pump in the fuel tank and the other a high-pressure pump driven by the camshaft. The high pressure fuel system operates at a pressure as low as 500 PSI during light load conditions and as high as 2,900 PSI under heavy loads.

DIRECT-INJECTION FUEL DELIVERY SYSTEM

LOW-PRESSURE SUPPLY PUMP The fuel pump in the fuel tank supplies fuel to the high-pressure fuel pump at a pressure of approximately 60 PSI. The fuel filter is located in the fuel tank and is part of the fuel pump assembly. It is not usually serviceable as a separate component; the engine control module (ECM) controls the output of the high-pressure pump, which has a range between 500 PSI (3,440 kPa) and 2,900 PSI (15,200 kPa) during engine operation. ● **SEE FIGURES 28–2 AND 28–3.**

HIGH-PRESSURE PUMP In a General Motors system, the engine control module (ECM) controls the output of the high-pressure pump, which has a range between 500 PSI (3,440 kPa) and 2,900 PSI (15,200 kPa) during engine operation. The high-pressure fuel pump connects to the pump in the fuel tank through the low-pressure fuel line. The pump consists of a single-barrel piston pump, which is driven by the engine camshaft. The pump plunger rides on a three-lobed cam on the camshaft. The high-pressure pump is cooled and lubricated by the fuel itself. ● **SEE FIGURE 28–4.**

FUEL RAIL The fuel rail stores the fuel from the high-pressure pump and stores high-pressure fuel for use to each injector. All injectors get the same pressure fuel from the fuel rail.

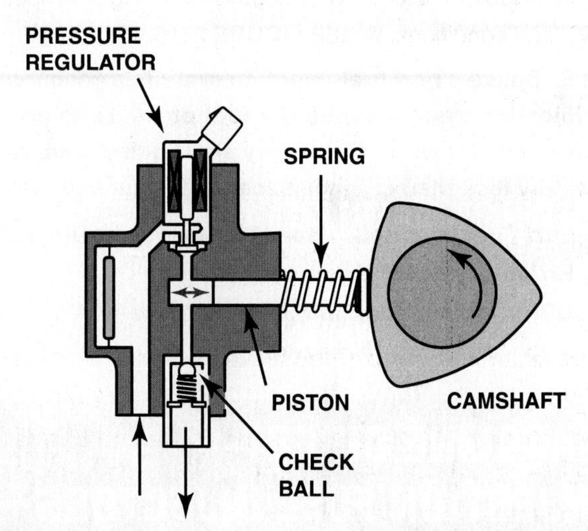

FIGURE 28–4 A typical camshaft-driven high-pressure pump used to increase fuel pressure to 2,000 PSI or higher.

FUEL PRESSURE REGULATOR An electric pressure-control valve is installed between the pump inlet and outlet valves. The fuel rail pressure sensor connects to the PCM with three wires:

- 5 volt reference
- ground
- signal

The sensor signal provides an analog signal to the PCM that varies in voltage as fuel rail pressure changes. Low pressure results in a low-voltage signal and high pressure results in a high-voltage signal.

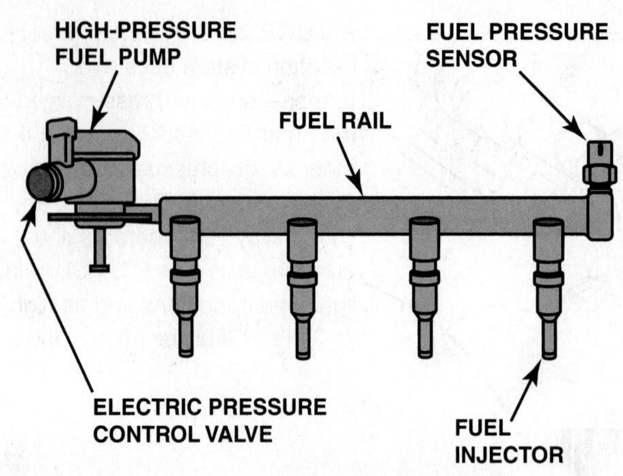

FIGURE 28–5 A gasoline direct-injection (GDI) fuel rail and pump assembly with the electric pressure control valve.

The PCM uses internal drivers to control the power feed and ground for the pressure control valve. When both PCM drivers are deactivated, the inlet valve is held open by spring pressure. This causes the high pressure fuel pump to default to low-pressure mode. The fuel from the high-pressure fuel pump flows through a line to the fuel rail and injectors. The actual operating pressure can vary from as low as 500 PSI (3,440 kPa) at idle to over 2,000 PSI (13,800 kPa) during high speed or heavy load conditions. ● **SEE FIGURE 28–5.**

NOTE: Unlike a port fuel-injection system, a gasoline direct injection system varies the fuel pressure to achieve greater fuel delivery using a very short pulse time, which is usually less than one millisecond. To summarize:

- **Port Fuel Injection = constant fuel pressure but variable injector pulse-width.**
- **GDI = almost constant injector pulse-width with varying fuel pressure.**

GASOLINE DIRECT-INJECTION FUEL INJECTORS

Each high-pressure fuel injector assembly is an electrically magnetic injector mounted in the cylinder head. In the GDI system, the PCM controls each fuel injector with 50 to 90 volts (usually 60 to 70 volts), depending on the system, which is created by a boost capacitor in the PCM. During the high-voltage boost phase, the capacitor is discharged through an injector, allowing for initial injector opening. The injector is then held open with 12 volts. The high-pressure fuel injector has a small slit or six precision-machined holes that generate the desired spray pattern. The injector also has an extended tip to allow for cooling from a water jacket in the cylinder head.

● **SEE CHART 28–1** for an overview of the differences between a port fuel-injection system and a gasoline direct-injection system.

PORT FUEL-INJECTION SYSTEM COMPARED WITH GDI SYSTEM

	PORT FUEL INJECTION	GASOLINE DIRECT INJECTION
Fuel pressure	35–60 PSI	Lift pump—50 to 60 PSI High-pressure pump—500 to 2,900 PSI
Injection pulse width at idle	1.5–3.5 ms	About 0.4 ms (400 μs)
Injector resistance	12–16 ohms	1–3 ohms
Injector voltage	6 V for low-resistance injectors, 12 V for most injectors	50–90 V
Number of injections per event	One	1–3
Engine compression ratio	8:1–11:1	11:1–13:1

CHART 28–1

A comparison chart showing the major differences between a port fuel-injection system and a gasoline direct-injection system.

MODES OF OPERATION

The two basic modes of operation include:

1. **Stratified mode.** In this mode of operation, the air–fuel mixture is richer around the spark plug than it is in the rest of the cylinder.
2. **Homogeneous mode.** In this mode of operation, the air–fuel mixture is the same throughout the cylinder.

There are variations of these modes that can be used to fine-tune the air–fuel mixture inside the cylinder. For example, Bosch, a supplier to many vehicle manufacturers, uses the following six modes of operation:

- **Homogeneous mode.** In this mode, the injector is pulsed one time to create an even air–fuel mixture in the cylinder. The injection occurs during the intake stroke. This mode is used during high-speed and/or high-torque conditions.
- **Homogeneous lean mode.** Similar to the homogeneous mode except that the overall air–fuel mixture is slightly lean for better fuel economy. The injection occurs during the intake stroke. This mode is used under steady, light-load conditions.
- **Stratified mode.** In this mode of operation, the injection occurs just before the spark occurs resulting in lean combustion, reducing fuel consumption.
- **Homogeneous stratified mode.** In this mode, there are two injections of fuel:

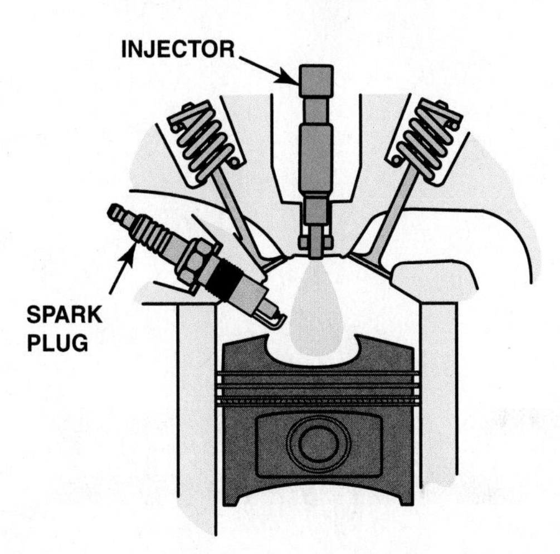

FIGURE 28–6 In this design, the fuel injector is at the top of the cylinder and sprays fuel into the cavity of the piston.

- The first injection is during the intake stroke.
- The second injection is during the compression stroke.

As a result of these double injections, the rich air–fuel mixture around the spark plug is ignited first. Then, the rich mixture ignites the leaner mixture. The advantages of this mode include lower exhaust emissions than the stratified mode and less fuel consumption than the homogeneous lean mode.

- **Homogeneous knock protection mode.** The purpose of this mode is to reduce the possibility of spark knock from occurring under heavy loads at low engine speeds. There are two injections of fuel:
  - The first injection occurs on the intake stroke.
  - The second injection occurs during the compression stroke with the overall mixture being stoichiometric.

  As a result of this mode, the PCM does not need to retard ignition timing as much to operate knock-free.
- **Stratified catalyst heating mode.** In this mode, there are two injections:
  - The first injection is on the compression stroke just before combustion.
  - The second injection is after combustion occurs to heat the exhaust. This mode is used to quickly warm the catalytic converter and to burn the sulfur from the NO_x catalyst.

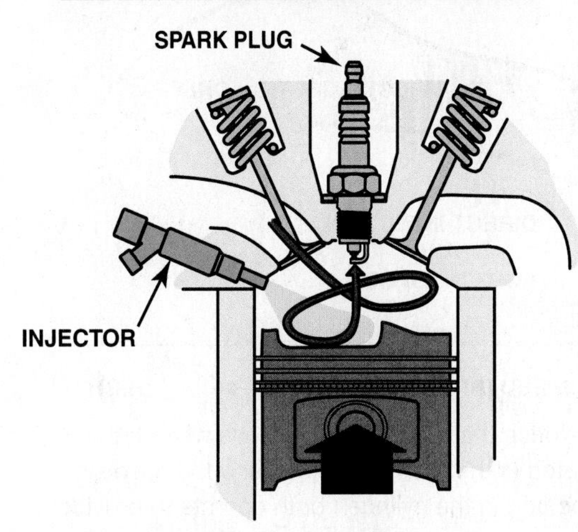

FIGURE 28–7 The side injector combines with the shape of the piston to create a swirl as the piston moves up on the compression stroke.

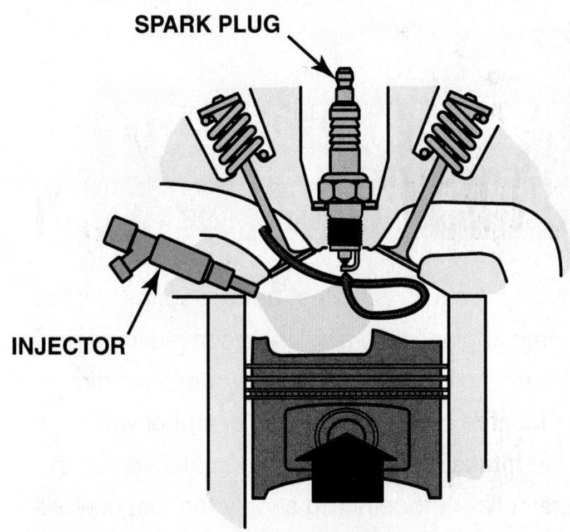

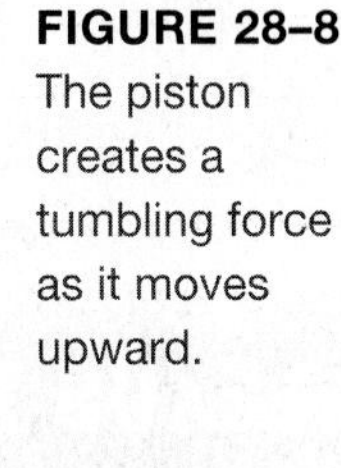

FIGURE 28–8 The piston creates a tumbling force as it moves upward.

PISTON TOP DESIGNS

Gasoline direct-injection (GDI) systems use a variety of shapes of piston and injector locations depending on make and model of engine. Three of the most commonly used designs include:

- **Spray-guided combustion.** In this design, the injector is placed in the center of the combustion chamber and injects fuel into the dished-out portion of the piston. The shape of the piston helps guide and direct the mist of fuel in the combustion chamber. ● **SEE FIGURE 28–6.**
- **Swirl combustion.** This design uses the shape of the piston and the position of the injector at the side of the combustion chamber to create turbulence and swirl of the air–fuel mixture. ● **SEE FIGURE 28–7.**
- **Tumble combustion.** Depending on when the fuel is injected into the combustion chamber helps determine how the air–fuel mixture is moved or tumbled. ● **SEE FIGURE 28–8.**

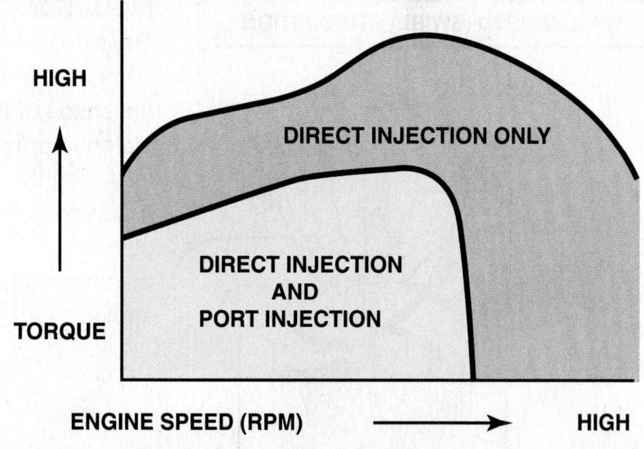

FIGURE 28–9 Notice that there are conditions when the port fuel-injector, located in the intake manifold, and the gasoline direct injector, located in the cylinder, both operate to provide the proper air–fuel mixture.

LEXUS PORT- AND DIRECT-INJECTION SYSTEMS

OVERVIEW Many Lexus vehicles use gasoline direct injection and in some engines they also use a conventional port fuel-injection system. The Lexus D-4S system combines direct-injection injectors located in the combustion chamber with port fuel-injectors in the intake manifold near the intake valve. The two injection systems work together to supply the fuel needed by the engine. ● **SEE FIGURE 28–9** for how the two systems are used throughout the various stages of engine operation.

COLD-START WARM-UP To help reduce exhaust emissions after a cold start, the fuel system uses a stratified change mode. This results in a richer air–fuel mixture near the spark plug and allows for the spark to be retarded to increase the temperature of the exhaust. As a result of the increased exhaust temperature, the catalytic converter rapidly reaches operating temperature, which reduces exhaust emissions.

ENGINE START SYSTEM

An engine equipped with gasoline direct injection could use the system to start the engine. This is most useful during idle stop mode when the engine is stopped while the vehicle is at a traffic light to save fuel. The steps used in the Mazda start-stop system, called the *smart idle stop system (SISS)*, allow the engine to be started without a starter motor and include the following steps:

STEP 1 The engine is stopped. The normal stopping position of an engine when it stops is 70 degrees before top dead center, plus or minus 20 degrees. This is because the engine stops with one cylinder on the compression stroke and the PCM can determine the cylinder position, using the crankshaft and camshaft position sensors.

STEP 2 When a command is made to start the engine by the PCM, fuel is injected into the cylinder that is on the compression stroke and ignited by the spark plug.

STEP 3 The piston on the compression stroke is forced downward forcing the crankshaft to rotate counterclockwise or in the opposite direction to normal operation.

STEP 4 The rotation of the crankshaft then forces the companion cylinder toward the top of the cylinder.

STEP 5 Fuel is injected and the spark plug is fired, forcing the piston down, causing the crankshaft to rotate in the normal (clockwise) direction. Normal combustion events continue allowing the engine to keep running.

GASOLINE DIRECT-INJECTION SERVICE

NOISE ISSUES Gasoline direct-injection systems operate at high pressure and the injectors can often be heard with the engine running and the hood open. This noise can be a customer concern because the clicking sound is similar to noisy valves. If a noise issue is the customer concern, check the following:

- Check a similar vehicle to determine if the sound is louder or more noticeable than normal.
- Check that nothing under the hood is touching the fuel rail. If another line or hose is in contact with the fuel rail, the sound of the injectors clicking can be transmitted throughout the engine, making the sound more noticeable.
- Check for any technical service bulletins (TSBs) that may include new clips or sound insulators to help reduce the noise.

CARBON ISSUES Carbon is often an issue in engines equipped with gasoline direct-injection systems. Carbon can affect engine operation by accumulating in two places:

- **On the injector itself.** Because the injector tip is in the combustion chamber, fuel residue can accumulate on the injector, reducing its ability to provide the proper spray pattern and amount of fuel. Some injector designs are more likely to be affected by carbon than others. For example, if the injector uses small holes, these tend to become clogged more than an injector that uses a single slit opening where the fuel being sprayed out tends to blast away any carbon. ● **SEE FIGURE 28–10.**
- **The backside of the intake valve.** This is a common place for fuel residue and carbon to accumulate on engines equipped with gasoline direct injection. The accumulation of carbon on the intake valve can become so severe that the engine will start and idle, but lack power to accelerate the vehicle. The carbon deposits restrict the airflow into the cylinder enough to decrease engine power.

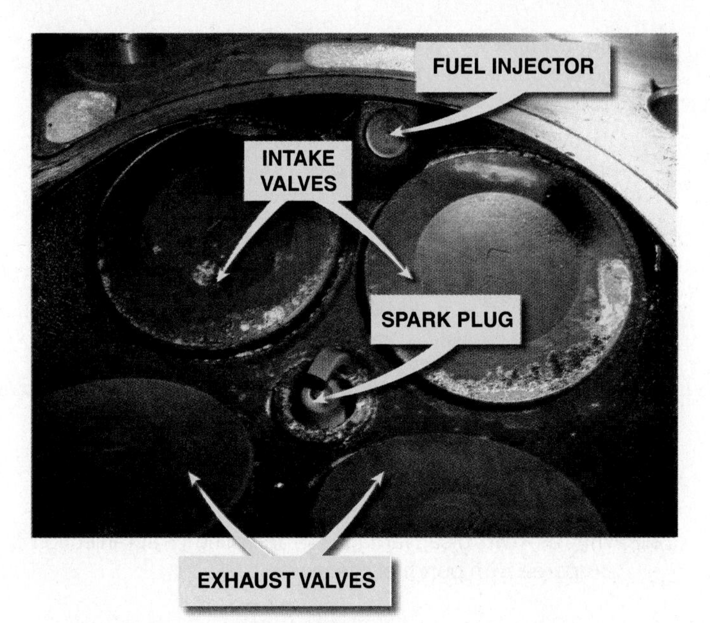

FIGURE 28–10 There may become a driveability issue because the gasoline direct-injection injector is exposed to combustion carbon and fuel residue.

NOTE: Lexus engines that use both port and gasoline direct-injection injectors do not show intake valve deposits. It is thought that the fuel being sprayed onto the intake valve from the port injector helps keep the intake valve clean.

CARBON CLEANING. Most experts recommend the use of Techron®, a fuel system dispersant, to help keep carbon from accumulating. The use of a dispersant every six months or every 6,000 miles has proven to help prevent injector and intake valve deposits.

If the lack of power is discovered and there are no stored diagnostic trouble codes, a conventional carbon cleaning procedure will likely restore power if the intake valves are coated.

SUMMARY

1. A gasoline direct-injection system uses a fuel injector that delivers a short squirt of fuel directly into the combustion chamber rather than in the intake manifold, near the intake valve on a port fuel-injection system.
2. The advantages of using gasoline direct injection instead of port fuel injection include:
   - Improved fuel economy
   - Reduced exhaust emissions
   - Greater engine power
3. Some of the disadvantages of gasoline direct-injection systems compared with a port fuel-injection system include:
   - Higher cost
   - The need for NO_x storage catalyst in some applications
   - More components
4. The operating pressure can vary from as low as 500 PSI during some low-demand conditions to as high as 2,900 PSI.

5. The fuel injectors are open for a very short period of time and are pulsed using a 50 to 90 volt pulse from a capacitor circuit.
6. GDI systems can operate in many modes, which are separated into the two basic modes:
   - Stratified mode
   - Homogeneous mode
7. GDI can be used to start an engine without the use of a starter motor for idle-stop functions.
8. GDI does create a louder clicking noise from the fuel injectors than port fuel-injection injectors.
9. Carbon deposits on the injector and the backside of the intake valve are a common problem with engines equipped with gasoline direct-injection systems.

REVIEW QUESTIONS

1. What are two advantages of gasoline direct injection compared with port fuel injection?
2. What are two disadvantages of gasoline direct injection compared with port fuel injection?
3. How is the fuel delivery system different from a port fuel-injection system?
4. What are the basic modes of operation of a GDI system?

CHAPTER QUIZ

1. Where is the fuel injected in an engine equipped with gasoline direct injection?
   - **a.** Into the intake manifold near the intake valve
   - **b.** Directly into the combustion chamber
   - **c.** Above the intake port
   - **d.** In the exhaust port
2. The fuel pump inside the fuel tank on a vehicle equipped with gasoline direct injection produces about what fuel pressure?
   - **a.** 5 to 10 PSI
   - **b.** 10 to 20 PSI
   - **c.** 20 to 40 PSI
   - **d.** 50 to 60 PSI
3. The high-pressure fuel pumps used in gasoline direct-injection (GDI) systems are powered by ________.
   - **a.** Electricity (DC motor)
   - **b.** Electricity (AC motor)
   - **c.** The camshaft
   - **d.** The crankshaft
4. The high-pressure fuel pump pressure is regulated by using ________.
   - **a.** An electric pressure-control valve
   - **b.** A vacuum-biased regulator
   - **c.** A mechanical regulator at the inlet to the fuel rail
   - **d.** A non-vacuum biased regulator
5. The fuel injectors operate under a fuel pressure of about ________.
   - **a.** 35 to 45 PSI
   - **b.** 90 to 150 PSI
   - **c.** 500 to 2,900 PSI
   - **d.** 2,000 to 5,000 PSI
6. The fuel injectors used on a gasoline direct-injection system are pulsed on using what voltage?
   - **a.** 12 to 14 volt
   - **b.** 50 to 90 volt
   - **c.** 100 to 110 volt
   - **d.** 200 to 220 volt
7. Which mode of operation results in a richer air–fuel mixture near the spark plug?
   - **a.** Stoichiometric
   - **b.** Homogeneous
   - **c.** Stratified
   - **d.** Knock protection
8. Some engines that use a gasoline direct-injection system also have port injection.
   - **a.** True
   - **b.** False
9. A gasoline direct-injection system can be used to start an engine without the need for a starter.
   - **a.** True
   - **b.** False
10. A lack of power from an engine equipped with gasoline direct injection could be due to ________.
    - **a.** Noisy injectors
    - **b.** Carbon on the injectors
    - **c.** Carbon on the intake valves
    - **d.** Both b and c

chapter 29

ELECTRONIC THROTTLE CONTROL SYSTEM

LEARNING OBJECTIVES

After studying this chapter, the reader will be able to:

1. Describe electronic throttle control systems and explain how the position of the accelerator pedal is detected.
2. Explain how an electronic throttle control system works.
3. List the parts of a typical electronic throttle control system.
4. Describe how to diagnose faults in an electronic throttle control system.
5. Explain how to service an electronic throttle system.

This chapter will help you prepare for ASE content area "E" (Computerized Engine Controls Diagnosis and Repair).

KEY TERMS

Accelerator pedal position (APP) sensor 444
Coast-down stall 451
Default position 445
Drive-by-wire 444
Electronic throttle control (ETC) 444
Fail safe position 445
Neutral position 445
Servomotor 445
Throttle position (TP) sensor 444

ELECTRONIC THROTTLE CONTROL (ETC) SYSTEM

ADVANTAGES OF ETC The absence of any mechanical linkage between the throttle pedal and the throttle body requires the use of an electric actuator motor. The electronic throttle system has the following advantages over the conventional cable:

- Eliminates the mechanical throttle cable, thereby reducing the number of moving parts.
- Eliminates the need for cruise control actuators and controllers.
- Helps reduce engine power for traction control (TC) and electronic stability control (ESC) systems.
- Used to delay rapid applications of torque to the transmission/transaxle to help improve driveability and to smooth shifts.
- Helps reduce pumping losses by using the electronic throttle to open at highway speeds with greater fuel economy. The electronic throttle control (ETC) opens the throttle to maintain engine and vehicle speed as the Powertrain Control Module leans the air–fuel ratio, retards ignition timing, and introduces additional exhaust gas recirculation (EGR) to reducing pumping losses.
- Used to provide smooth engine operation, especially during rapid acceleration.
- Eliminates the need for an idle air control valve.

The electronic throttle can be called **drive-by-wire**, but most vehicle manufacturers use the term **electronic throttle control (ETC)** to describe the system that opens the throttle valve electrically.

PARTS INVOLVED The typical ETC system includes the following components:

1. **Accelerator pedal position (APP)** sensor, also called *accelerator pedal sensor (APS)*
2. The electronic throttle actuator (servomotor), which is part of the electronic throttle body
3. A **throttle position (TP) sensor**
4. An electronic control unit, which is usually the Powertrain Control Module

   ● **SEE FIGURE 29–1.**

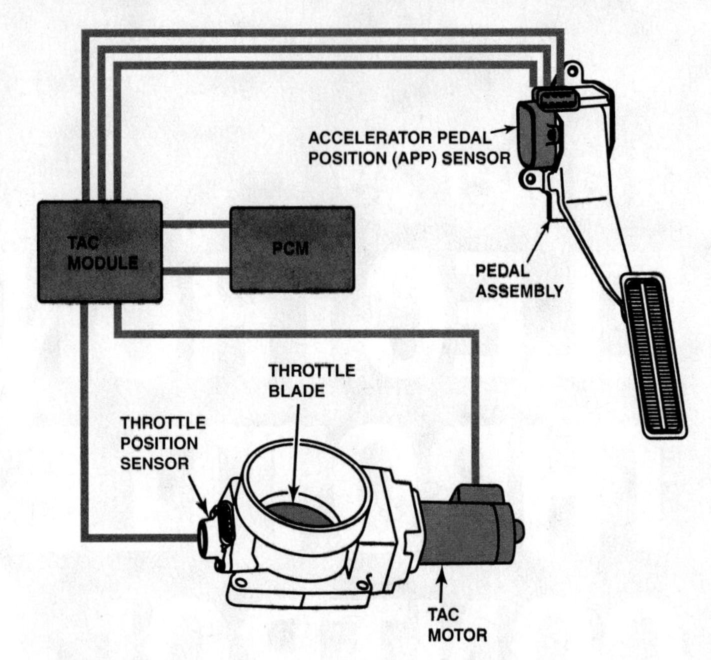

FIGURE 29–1 The throttle pedal is connected to the accelerator pedal position (APP) sensor. The electronic throttle body includes a throttle position sensor to provide throttle angle feedback to the vehicle computer. Some systems use a Throttle Actuator Control (TAC) module to operate the throttle blade (plate).

NORMAL OPERATION OF THE ETC SYSTEM

Driving a vehicle equipped with an electronic throttle control system is about the same as driving a vehicle with a conventional mechanical throttle cable and throttle valve. However, the driver may notice some differences, which are to be considered normal. These normal conditions include:

- The engine may not increase above idle speed when depressing the accelerator pedal when the gear selector is in PARK.
- If the engine speed does increase when the accelerator is depressed with the transmission in PARK or NEUTRAL, the engine speed will likely be limited to less than 2000 RPM.
- While accelerating rapidly, there is often a slight delay before the engine responds. ● **SEE FIGURE 29–2.**
- While at cruise speed, the accelerator pedal may or may not cause the engine speed to increase if the accelerator pedal is moved slightly.

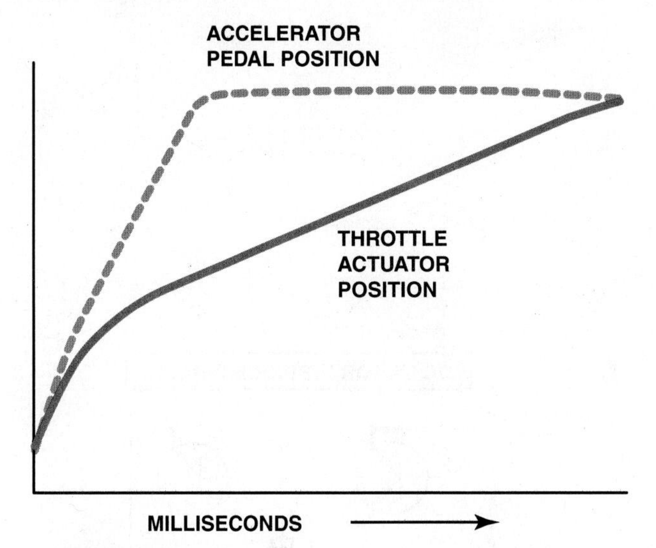

FIGURE 29–2 The opening of the throttle plate can be delayed as long as 30 milliseconds (0.030 sec) to allow time for the amount of fuel needed to catch up to the opening of the throttle plate.

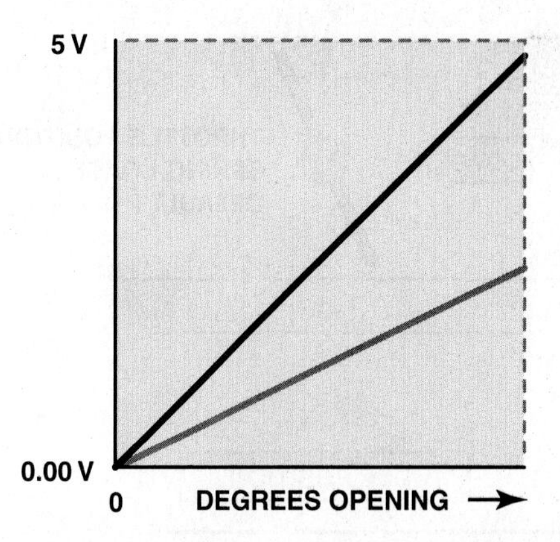

FIGURE 29–3 A typical accelerator pedal position (APP) sensor, showing two different output voltage signals that are used by the PCM to determine accelerator pedal position. Two (or three in some applications) are used as a double check because this is a safety-related sensor.

ACCELERATOR PEDAL POSITION SENSOR

CABLE-OPERATED SYSTEM Honda Accords until 2008 model year used a cable attached to the accelerator pedal to operate the APP sensor located under the hood. A similar arrangement was used in Dodge RAM trucks in 2003. In both of these applications, the throttle cable was simply moving the APP sensor and not moving the throttle plate. The throttle plate is controlled by the PCM and moved by the electronic throttle control motor.

TWO SENSORS The accelerator pedal position sensor uses two and sometimes three separate sensors, which act together to give accurate accelerator pedal position information to the controller, but also are used to check that the sensor is working properly. They function just like a throttle position sensor, and two are needed for proper system function. One APP sensor output signal increases as the pedal is depressed and the other signal decreases. The controller compares the signals with a look-up table to determine the pedal position. Using two or three signals improves redundancy should one sensor fail, and allows the PCM to quickly detect a malfunction. When three sensors are used, the third signal can either decrease or increase with pedal position, but its voltage range will still be different from the other two. ● **SEE FIGURE 29–3**.

THROTTLE BODY ASSEMBLY

The throttle body assembly contains the following components:

- Throttle plate
- Electric actuator DC motor
- Dual throttle position (TP) sensors
- Gears used to multiply the torque of the DC motor
- Springs used to hold the throttle plate in the default location

THROTTLE PLATE AND SPRING The throttle plate is held slightly open by a concentric clock spring. The spring applies a force that will close the throttle plate if power is lost to the actuator motor. The spring is also used to open the throttle plate slightly from the fully closed position.

ELECTRONIC THROTTLE BODY MOTOR The actuator is a DC electric motor and is often called a **servomotor**. The throttle plate is held in a **default position** by a spring inside the throttle body assembly. This partially open position, also called the **neutral position** or the **fail safe position**, is about 16% to 20% open. This default position varies depending on the vehicle and usually results in an engine speed of 1200 to 1500 RPM.

- The throttle plate is driven closed to achieve speeds lower than the default position, such as idle speed.

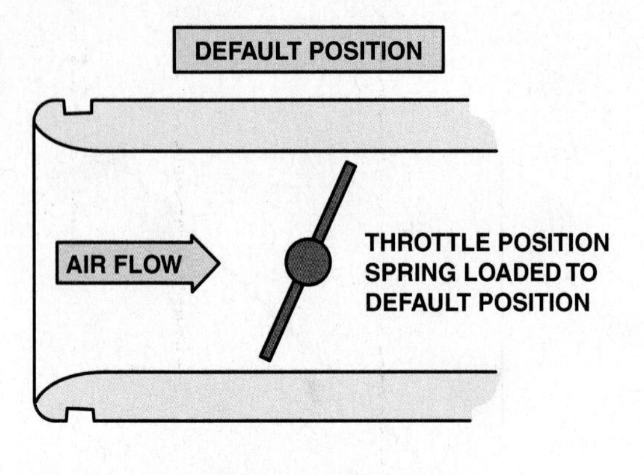

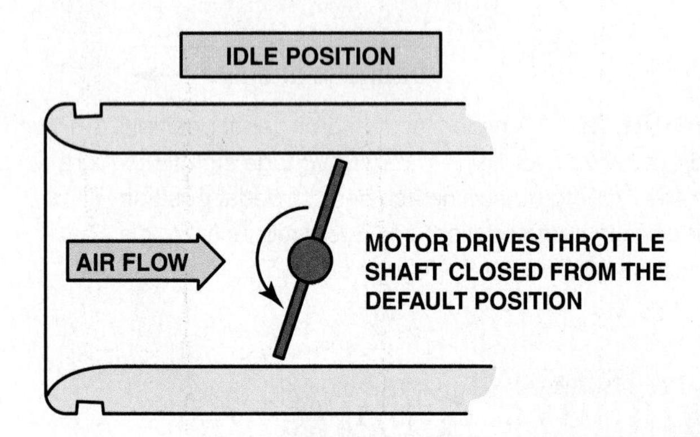

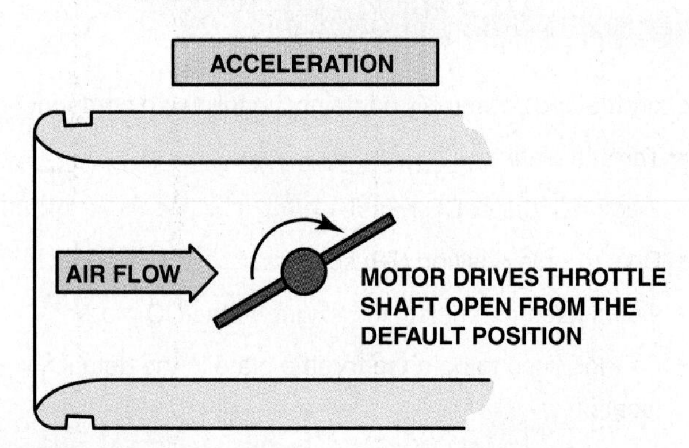

FIGURE 29–4 The default position for the throttle plate is in slightly open position. The servomotor then is used to close it for idle and open it during acceleration.

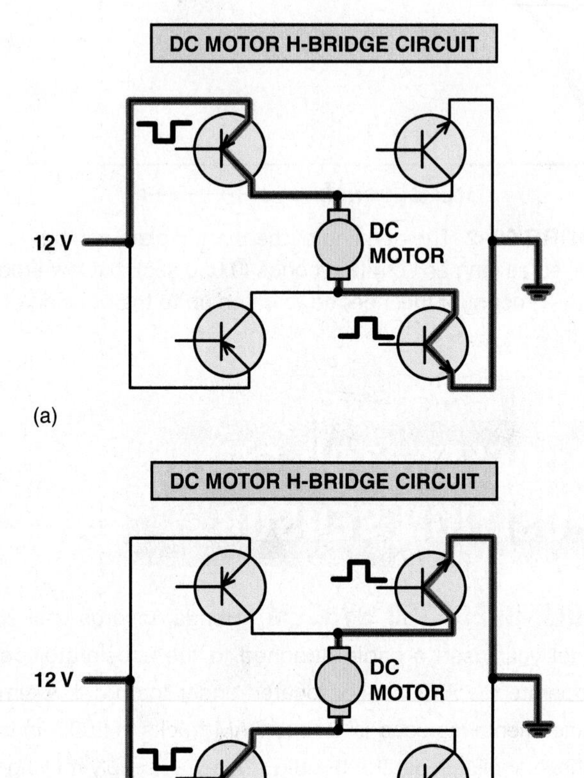

FIGURE 29–5 (a) An H-bridge circuit is used to control the direction of the DC electric motor of the electronic throttle control unit. (b) To reverse the direction of operation, the polarity of the current through the motor is reversed.

- The throttle plate is driven open to achieve speeds higher than the default position, such as during acceleration. ● **SEE FIGURE 29–4**.

The throttle plate motor is driven by a bidirectional pulse-width modulated (PWM) signal from the PCM or electronic throttle control module using an H-bridge circuit. ● **SEE FIGURE 29–5**.

The H-bridge circuit is controlled by the Powertrain Control Module by:

- Reversing the polarity of power and ground brushes to the DC motor
- Pulse-width modulating (PWM) the current through the motor

The PCM monitors the position of the throttle from the two throttle position (TP) sensors. The PCM then commands the throttle plate to the desired position. ● **SEE FIGURE 29–6**.

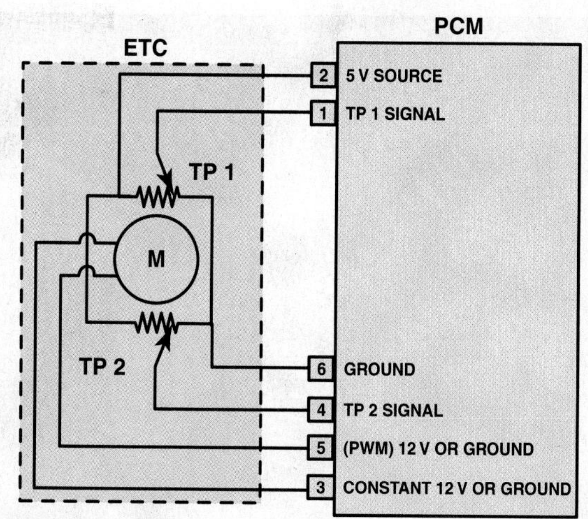

FIGURE 29–6 Schematic of a typical electronic throttle control (ETC) system. Note that terminal #5 is always pulse-width modulated and that terminal #3 is always constant, but both power and ground are switched to change the direction of the motor.

FREQUENTLY ASKED QUESTION

What Is the "Spring Test"?

The spring test is a self-test performed by the PCM whenever the engine is started. The PCM operates the throttle to check if it can react to the command and return to the default (home) position. This self-test is used by the PCM to determine that the spring and motor are working correctly and may be noticed by some vehicle owners by the following factors:

- A slight delay in the operation of the starter motor. It is when the ignition is turned to the on position that the PCM performs the test. While it takes just a short time to perform the test, it can be sensed by the driver that there could be a fault in the ignition switch or starter motor circuits.
- A slight "clicking" sound may also be heard coming from under the hood when the ignition is turned on. This is normal and is related to the self-test on the throttle as it opens and closes.

FREQUENTLY ASKED QUESTION

Why Not Use a Stepper Motor for ETC?

A stepper motor is a type of motor that has multiple windings and is pulsed by a computer to rotate a certain number of degrees when pulsed. The disadvantage is that a stepper motor is too slow to react compared with a conventional DC electric motor and is the reason a stepper motor is not used in electronic throttle control systems.

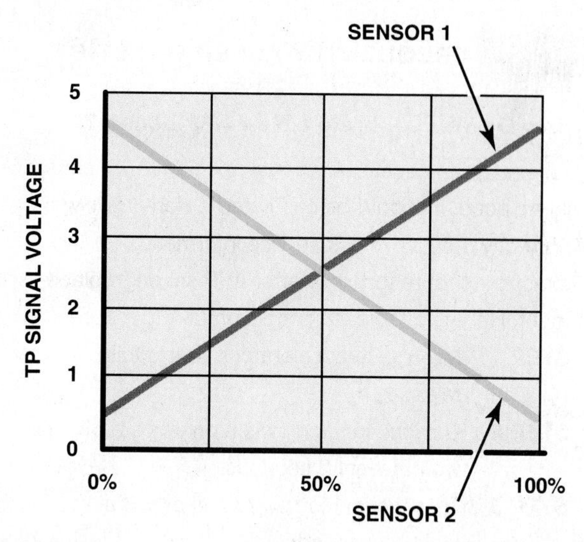

FIGURE 29–7 The two TP sensors used on the throttle body of an electronic throttle body assembly produce opposite voltage signals as the throttle is opened. The total voltage of both combined at any throttle plate position is 5 volts.

THROTTLE POSITION (TP) SENSOR

Two throttle position sensors are used in the throttle body assembly to provide throttle position signals to the PCM. Two sensors are used as a fail-safe measure and for diagnosis. There are two types of TP sensors used in electronic throttle control systems: potentiometers and Hall-effect.

THREE-WIRE POTENTIOMETER SENSORS These sensors use a 5 volt reference from the PCM and produce an analog (variable) voltage signal that is proportional to the throttle plate position. The two sensors produce opposite signals as the throttle plate opens:

- One sensor starts at low voltage (about 0.5 volt) and increases as the throttle plate is opened.
- The second sensor starts at a higher voltage (about 4.5 volt) and produces a lower voltage as the throttle plate is opened. ● SEE FIGURE 29–7.

HALL-EFFECT TP SENSORS Some vehicle manufacturers such as Honda use a noncontact Hall-effect throttle position sensor. Because there is no physical contact, this type of sensor is less likely to fail due to wear.

? FREQUENTLY ASKED QUESTION

How Do You Calibrate a New APP Sensor?

Whenever an accelerator pedal position (APP) sensor is replaced, it should be calibrated before it will work correctly. Always check service information for the exact procedure to follow after APP sensor replacement. Here is a typical example of the procedure:

STEP 1 Make sure accelerator pedal is fully released.

STEP 2 Turn the ignition switch on (engine off) and wait at least 2 seconds.

STEP 3 Turn the ignition switch off and wait at least 10 seconds.

STEP 4 Turn the ignition switch on (engine on) and wait at least 2 seconds.

STEP 5 Turn the ignition switch off and wait at least 10 seconds.

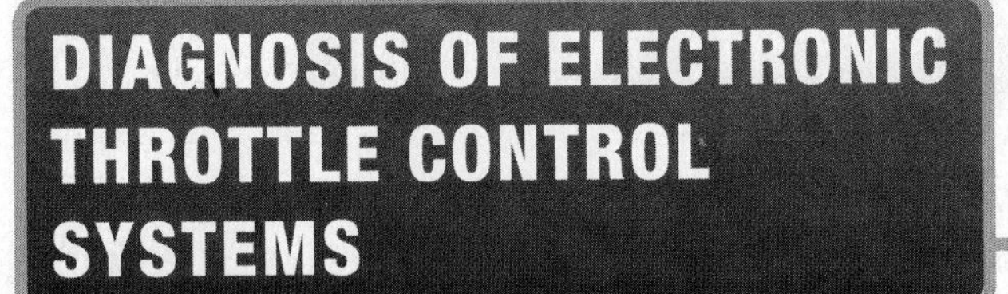

DIAGNOSIS OF ELECTRONIC THROTTLE CONTROL SYSTEMS

FAULT MODE Electronic throttle control systems can have faults like any other automatic system. Due to the redundant sensors in accelerator pedal position sensors and throttle position sensor, many faults result in a "limp home" situation instead of a total failure. The limp home mode is also called the "fail-safe mode" and indicates the following actions performed by the Powertrain Control Module:

- Engine speed is limited to the default speed (about 1200 to 1600 RPM).
- There is slow or no response when the accelerator pedal is depressed.
- The cruise control system is disabled.
- A diagnostic trouble code (DTC) is set.
- An ETC warning lamp on the dash will light. The warning lamp may be labeled differently, depending on the vehicle manufacturer. For example:
    - **General Motors vehicle**—Reduced power lamp (● **SEE FIGURE 29–8)**

(a)

(b)

FIGURE 29–8 (a) A "reduced power" warning light indicates a fault with the electronic throttle control system on some General Motors vehicles. (b) A symbol showing an engine with an arrow pointing down is used on some General Motors vehicles to indicate a fault with the electronic throttle control system.

    - **Ford**—Wrench symbol (amber or green) (● **SEE FIGURE 29–9)**
    - **Chrysler**—Red lightning bolt symbol (● **SEE FIGURE 29–10)**
- The engine will run and can be driven slowly. This limp-in mode operation allows the vehicle to be driven off of the road and to a safe location.

The ETC may enter the limp-in mode if any of the following has occurred:

- Low battery voltage has been detected
- PCM failure

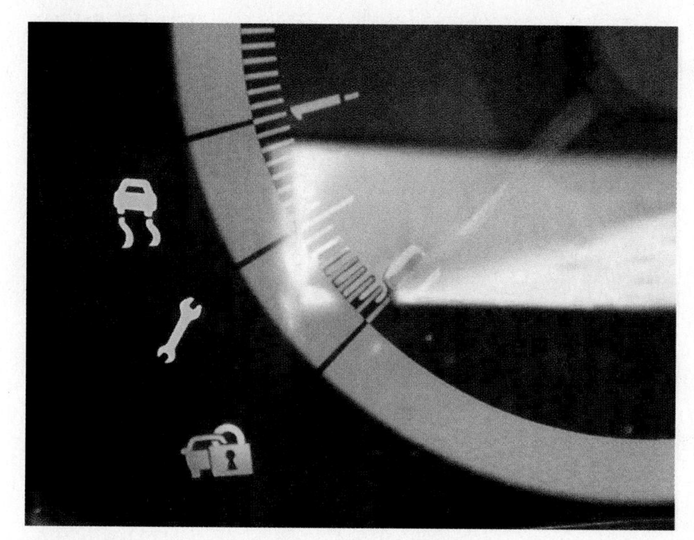

FIGURE 29–9 A wrench symbol warning lamp on a Ford vehicle. The symbol can also be green.

FIGURE 29–10 A symbol used on a Chrysler vehicle indicating a fault with the electronic throttle control.

- One TP and the MAP sensor have failed
- Both TP sensors have failed
- The ETC actuator motor has failed
- The ETC throttle spring has failed

VACUUM LEAKS The electronic throttle control system is able to compensate for many vacuum leaks. A vacuum leak at the intake manifold, for example, will allow air into the engine that is not measured by the mass airflow sensor. The ETC system will simply move the throttle as needed to achieve the proper idle speed to compensate for the leak.

DIAGNOSTIC PROCEDURE If a fault occurs in the ETC system, check service information for the specified procedure to follow for the vehicle being checked. Most vehicle service information includes the following steps:

STEP 1 Verify the customer concern.

STEP 2 Use a factory scan tool or an aftermarket scan tool with original equipment capability and check for diagnostic trouble codes (DTCs).

STEP 3 If there are stored diagnostic trouble codes, follow service information instructions for diagnosing the system.

STEP 4 If there are no stored diagnostic trouble codes, check scan tool data for possible fault areas in the system.

SCAN TOOL DATA Scan data related to the electronic throttle control system can be confusing. Typical data and the meaning include:

- **APP indicated angle.** The scan tool will display a percentage ranging from 0% to 100%. When the throttle is released, the indicated angle should be 0%. When the throttle is depressed to wide open, the reading should indicate 100%.
- **TP desired angle.** The scan tool will display a percentage ranging from 0% to 100%. This represents the desired throttle angle as commanded by the driver of the vehicle.

CASE STUDY

The High Idle Toyota

The owner of a Toyota Camry complained that the engine would idle at over 1200 RPM compared with a normal 600 to 700 RPM. The vehicle would also not accelerate. Using a scan tool, a check for diagnostic trouble codes showed one code: P2101—"TAC motor circuit low."

Checking service information led to the inspection of the electronic throttle control throttle body assembly. With the ignition key out of the ignition and the inlet air duct off the throttle body, the technician used a screwdriver to see if the throttle plate worked.

Normal operation—The throttle plate should move and then spring back quickly to the default position.

Abnormal operation—If the throttle plate stays where it is moved or does not return to the default position, there is a fault with the throttle body assembly. ● **SEE FIGURE 29–11.**

The technician replaced the throttle body assembly with an updated version and proper engine operation was restored. The technician disassembled the old throttle body and found it was corroded inside due to moisture entering the unit through the vent hose. ● **SEE FIGURE 29–12.**

Summary:

- **Complaint**—Customer stated that the engine would idle at over 2,000 RPM.
- **Cause**—A stored P2101 DTC was stored indicating a fault with the throttle body assembly.
- **Correction**—The throttle body was replaced with an improved version that placed the vent tube in a different position to help avoid water getting into the assembly.

FIGURE 29–11 The throttle plate stayed where it was moved, which indicates that there is a problem with the electronic throttle body control assembly.

FIGURE 29–12 A corroded electronic throttle control assembly shown with the cover removed.

- **TP indicated angle.** The TP indicated angle is the angle of the measured throttle opening and it should agree with the TP desired angle.
- **TP sensors 1 and 2.** The scan tool will display "agree" or "disagree." If the PCM or throttle actuator control (TAC) module receives a voltage signal from one of the TP sensors that is not in the proper relationship to the other TP sensor, the scan tool will display *disagree*.

ETC THROTTLE FOLLOWER TEST

On some vehicles, such as many Chrysler vehicles, the operation of the electronic throttle control can be tested using a factory or factory-level scan tool. To perform this test, use the "throttle follower test" procedure as shown on the scan tool. An assistant is needed to check that the throttle plate is moving as the accelerator pedal is depressed. This test cannot be done normally because the PCM does not normally allow the throttle plate to be moved unless the engine is running.

SERVICING ELECTRONIC THROTTLE SYSTEMS

ETC-RELATED PERFORMANCE ISSUES The only service that an electronic throttle control system may require is a cleaning of the throttle body. Throttle body cleaning is a routine service procedure on port fuel-injected engines and is still needed when the throttle is being opened by an electric motor rather than a throttle cable tied to a mechanical accelerator pedal. The throttle body may need cleaning if one or more of the following symptoms are present:

- Lower than normal idle speed
- Rough idle
- Engine stalls when coming to a stop (called a **coast-down stall**)

If any of the above conditions exists, a throttle body cleaning will often correct these faults.

CAUTION: Some vehicle manufacturers add a non-stick coating to the throttle assembly and warn that cleaning could remove this protective coating. Always follow the vehicle manufacturer's recommended procedures.

THROTTLE BODY CLEANING PROCEDURE Before attempting to clean a throttle body on an engine equipped with an electronic throttle control system, be sure that the ignition key is out of the vehicle and the ready light is off if working on a Toyota/Lexus hybrid electric vehicle to avoid the possibility of personal injury.

WARNING

The electric motor that operates the throttle plate is strong enough to cut off a finger. ● **SEE FIGURE 29–13.**

To clean the throttle, perform the following steps:

STEP 1 With the ignition off and the key removed from the ignition, remove the air inlet hose from the throttle body.

STEP 2 Spray throttle body cleaner onto a shop cloth.

STEP 3 Open the throttle body and use the shop cloth to remove the varnish and carbon deposits from the throttle body housing and throttle plate.

FIGURE 29–13 Notice the small motor gear on the left drives a larger plastic gear (black), which then drives the small gear in mesh with the section of a gear attached to the throttle plate. This results in a huge torque increase from the small motor and helps explain why it could be dangerous to insert a finger into the throttle body assembly.

CAUTION: Do not spray cleaner into the throttle body assembly. The liquid cleaner could flow into and damage the throttle position (TP) sensors.

STEP 4 Reinstall the inlet hose being sure that there are no air leaks between the hose and the throttle body assembly.

STEP 5 Start the engine and allow the PCM to learn the correct idle. If the idle is not correct, check service information for the specified procedures to follow to perform a throttle relearn.

THROTTLE BODY RELEARN PROCEDURE When installing a new throttle body or Powertrain Control Module or sometimes after cleaning the throttle body, the throttle position has to be learned by the PCM. After the following conditions have been met, a typical throttle body relearn procedure for a General Motors vehicle includes:

- Accelerator pedal released
- Battery voltage higher than 8 volts
- Vehicle speed must be zero
- Engine coolant temperature (ECT) higher than 40°F (5°C) and lower than 212°F (100°C)
- Intake air temperature (IAT) higher than 40°F (5°C)
- No throttle diagnostic trouble codes set

If all of the above conditions are met, perform the following steps:

STEP 1 Turn the ignition on (engine off) for 30 seconds.

STEP 2 Turn the ignition off and wait 30 seconds.

Start the engine and the idle learn procedure should cause the engine to idle at the correct speed.

SUMMARY

1. Using an electronic throttle control (ETC) system on an engine has many advantages over a conventional method that uses a mechanical cable between the accelerator pedal and the throttle valve.
2. The major components of an electronic throttle control system include:
   - Accelerator pedal position (APP) sensor
   - Electronic throttle control actuator motor and spring
   - Throttle position (TP) sensor
   - Electronic control unit
3. The throttle position (TP) sensor is actually two sensors that share the 5 volt reference from the PCM and produce opposite signals as a redundant check.
4. Limp-in mode is commanded if there is a major fault in the system, which can allow the vehicle to be driven enough to be pulled off the road to safety.
5. The diagnostic procedure for the ETC system includes verifying the customer concern, using a scan tool to check for diagnostic trouble codes, and checking the value of the TP and APP sensors.
6. Servicing the ETC system includes cleaning the throttle body and throttle plate.

REVIEW QUESTIONS

1. What parts can be deleted if an engine uses an electronic throttle control (ETC) system instead of a conventional accelerator pedal and cable to operate the throttle valve?
2. How can the use of an ETC system improve fuel economy?
3. How is the operation of the throttle different on a system that uses an ETC system compared with a conventional mechanical system?
4. What component parts are included in an ETC system?
5. What is the default or limp-in position of the throttle plate?
6. What dash warning light indicates a fault with the ETC system?

CHAPTER QUIZ

1. The use of an ETC system allows the elimination of all *except* ____________.
   a. Accelerator pedal
   b. Mechanical throttle cable (most systems)
   c. Cruise control actuator
   d. Idle air control
2. To what extent is the throttle plate spring loaded to hold the throttle slightly open?
   a. 3% to 5%
   b. 8% to 10%
   c. 16% to 20%
   d. 22% to 28%
3. What type of electric motor is the throttle plate actuator motor?
   a. Stepper motor
   b. DC motor
   c. AC motor
   d. Brushless motor
4. The actuator motor is controlled by the PCM through what type of circuit?
   a. Series
   b. Parallel
   c. H-bridge
   d. Series-parallel
5. When does the PCM perform a self-test of the ETC system?
   a. During cruise speed when the throttle is steady
   b. During deceleration
   c. During acceleration
   d. When the ignition switch is first rotated to the on position before the engine starts
6. What type is the throttle position sensor used in the throttle body assembly of an ETC system?
   a. A single potentiometer
   b. Two potentiometers that read in the opposite direction
   c. A Hall-effect sensor
   d. Either b or c
7. A green wrench symbol is displayed on the dash. What does this mean?
   a. A fault in the ETC in a Ford vehicle has been detected
   b. A fault in the ETC in a Honda vehicle has been detected
   c. A fault in the ETC in a Chrysler vehicle has been detected
   d. A fault in the ETC in a General Motors vehicle has been detected
8. A technician is checking the operation of the electronic throttle control system by depressing the accelerator pedal with the ignition in the on (run) position (engine off). What is the most likely result if the system is functioning correctly?
   a. The throttle goes to wide open when the accelerator pedal is depressed all the way
   b. No throttle movement
   c. The throttle will open partially but not all of the way
   d. The throttle will perform a self-test by closing and then opening to the default position

9. With the ignition off and the key out of the ignition, what should happen if a technician uses a screwdriver and pushes on the throttle plate in an attempt to open the valve?
   a. Nothing. The throttle should be kept from moving by the motor, which is not energized with the key off.
   b. The throttle should move and stay where it is moved and not go back unless moved back.
   c. The throttle should move, and then spring back to the home position when released.
   d. The throttle should move closed, but not open further than the default position.

10. The throttle body may be cleaned (if recommended by the vehicle manufacturer) if what conditions are occurring?
   a. Coast-down stall
   b. Rough idle
   c. Lower-than-normal idle speed
   d. Any of the above

chapter 30

FUEL-INJECTION SYSTEM DIAGNOSIS AND SERVICE

LEARNING OBJECTIVES

After studying this chapter, the reader will be able to:

1. Explain how to check a fuel-pressure regulator.
2. Describe how to test fuel injectors.
3. Explain how to diagnose electronic fuel-injection problems.
4. Describe how to service the fuel-injection system.

This chapter will help you prepare for Engine Repair (A8) ASE certification test content area "C" (Fuel, Air Induction, and Exhaust Systems Diagnosis and Repair).

KEY TERMS

PORT FUEL-INJECTION PRESSURE REGULATOR DIAGNOSIS

Most port fuel-injected engines use a vacuum hose connected to the fuel-pressure regulator. At idle, the pressure inside the intake manifold is low (high vacuum). Manifold vacuum is applied above the diaphragm inside the fuel-pressure regulator. This reduces the pressure exerted on the diaphragm and results in a lower, about 10 PSI (69 kPa), fuel pressure applied to the injectors. To test a vacuum-controlled fuel-pressure regulator, follow these steps:

1. Connect a fuel-pressure gauge to monitor the fuel pressure.
2. Locate the fuel-pressure regulator and disconnect the vacuum hose from the regulator.

   NOTE: If gasoline drips out of the vacuum hose when removed from the fuel-pressure regulator, the regulator is defective and will require replacement.

3. With the engine running at idle speed, reconnect the vacuum hose to the fuel-pressure regulator while watching the fuel-pressure gauge. The fuel pressure should drop (about 10 PSI or 69 kPa) when the hose is reattached to the regulator.
4. Using a hand-operated vacuum pump, apply vacuum (20 inches Hg) to the regulator. The regulator should hold vacuum. If the vacuum drops, replace the fuel-pressure regulator. ● **SEE FIGURE 30–1.**

NOTE: Some vehicles do not use a vacuum-regulated fuel-pressure regulator. Many of these vehicles use a regulator located inside the fuel tank that supplies a constant fuel pressure to the fuel injectors.

FIGURE 30–1 If the vacuum hose is removed from the fuel-pressure regulator when the engine is running, the fuel pressure should increase. If it does not increase, then the fuel pump is not capable of supplying adequate pressure or the fuel-pressure regulator is defective. If gasoline is visible in the vacuum hose, the regulator is leaking and should be replaced.

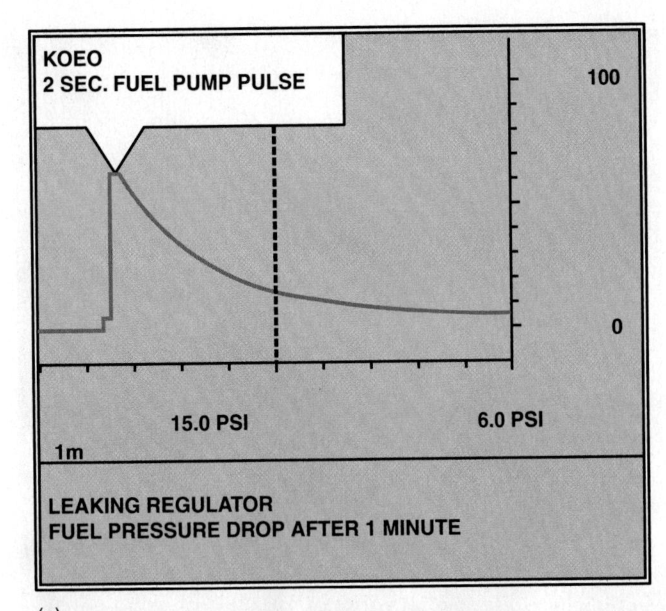

(a)

100

0

32.6 PSI

28.0 PSI

10m

(b)

FIGURE 30–2 (a) A fuel-pressure graph after key on, engine off (KOEO) on a TBI system. (b) Pressure drop after 10 minutes on a normal port fuel-injection system.

TECH TIP

Pressure Transducer Fuel Pressure Test

Using a pressure transducer and a **graphing multimeter (GMM)** or digital storage oscilloscope (DSO) allows the service technician to view the fuel pressure over time. ● **SEE FIGURE 30–2(a)**. Note that the fuel pressure dropped from 15 PSI down to 6 PSI on a TBI-equipped vehicle after just one minute. A normal pressure holding capability is shown in ● **FIGURE 30–2(b)**. when the pressure dropped only about 10% after 10 minutes on a port–fuel–injection system.

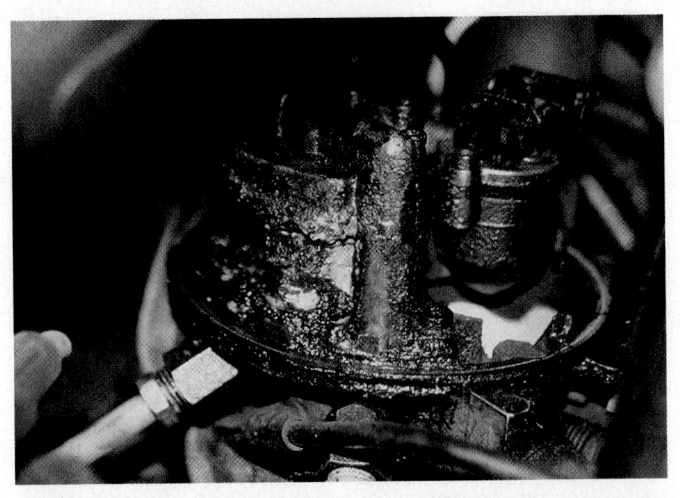

FIGURE 30–3 A clogged PCV system caused the engine oil fumes to be drawn into the air cleaner assembly. This is what the technician discovered during a visual inspection.

DIAGNOSING ELECTRONIC FUEL-INJECTION PROBLEMS USING VISUAL INSPECTION

All fuel-injection systems require the proper amount of clean fuel delivered to the system at the proper pressure and the correct amount of filtered air. The following items should be carefully inspected before proceeding to more detailed tests:

- Check the air filter and replace as needed.
- Check the air induction system for obstructions.
- Check the conditions of all vacuum hoses. Replace any hose that is split, soft (mushy), or brittle.
- Check the positive crankcase ventilation (PCV) valve for proper operation or replacement as needed. ● **SEE FIGURE 30–3**.

NOTE: The use of an incorrect PCV valve can cause a rough idle or stalling.

- Check all fuel-injection electrical connections for corrosion or damage.
- Check for gasoline at the vacuum port of the fuel-pressure regulator if the vehicle is so equipped. Gasoline in the vacuum hose at the fuel-pressure regulator indicates that the regulator is defective and requires replacement.

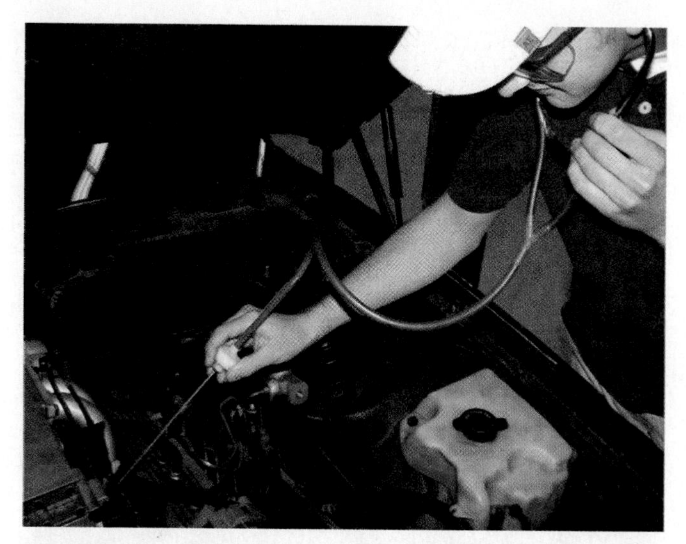

FIGURE 30–4 All fuel injectors should make the same sound with the engine running at idle speed. A lack of sound indicates a possible electrically open injector or a break in the wiring. A defective computer could also be the cause of a lack of clicking (pulsing) of the injectors.

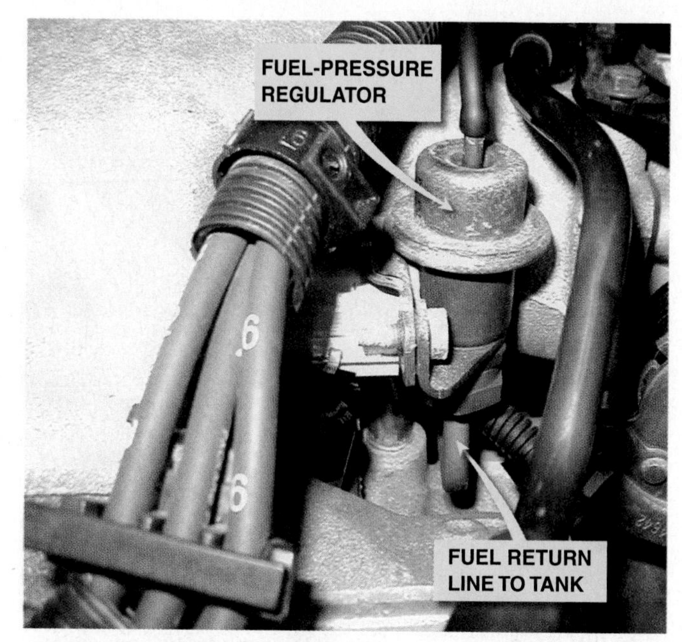

FIGURE 30–5 Fuel should be heard returning to the fuel tank at the fuel return line if the fuel-pump and fuel-pressure regulator are functioning correctly.

TECH TIP

Stethoscope Fuel-Injection Test

A commonly used test for injector operation is to listen to the injector using a stethoscope with the engine operating at idle speed. ● **SEE FIGURE 30–4.** All injectors should produce the same clicking sound. If any injector makes a clunking or rattling sound, it should be tested further or replaced. With the engine still running, place the end of the stethoscope probe to the return line from the fuel-pressure regulator. ● **SEE FIGURE 30–5.** Fuel should be heard flowing back to the fuel tank if the fuel-pump pressure is higher than the fuel-regulator pressure. If no sound of fuel is heard, then either the fuel pump or the fuel-pressure regulator is at fault.

TECH TIP

Quick and Easy Leaking Injector Test

Leaking injectors may be found by disabling the ignition, unhooking all injectors, and checking exhaust for hydrocarbons (HC) using a gas analyzer while cranking the engine (maximum HC = 300 PPM).

SCAN TOOL VACUUM LEAK DIAGNOSIS

If a vacuum (air) leak occurs on an engine equipped with a speed-density-type of fuel injection, the extra air would cause the following to occur:

- The idle speed increases due to the extra air just as if the throttle pedal was depressed.
- The MAP sensor reacts to the increased air from the vacuum leak as an additional load on the engine.
- The computer increases the injector pulse width slightly longer due to the signal from the MAP sensor.
- The air–fuel mixture remains unchanged.
- The idle air control (IAC) counts will decrease, thereby attempting to reduce the engine speed to the target idle speed stored in the computer memory. ● **SEE FIGURE 30-6.**

Therefore, one of the best indicators of a vacuum leak on a speed-density fuel-injection system is to look at the IAC counts or percentage. Normal **IAC counts** or percentage is usually 15 to 25. A reading of less than 5 indicates a vacuum leak.

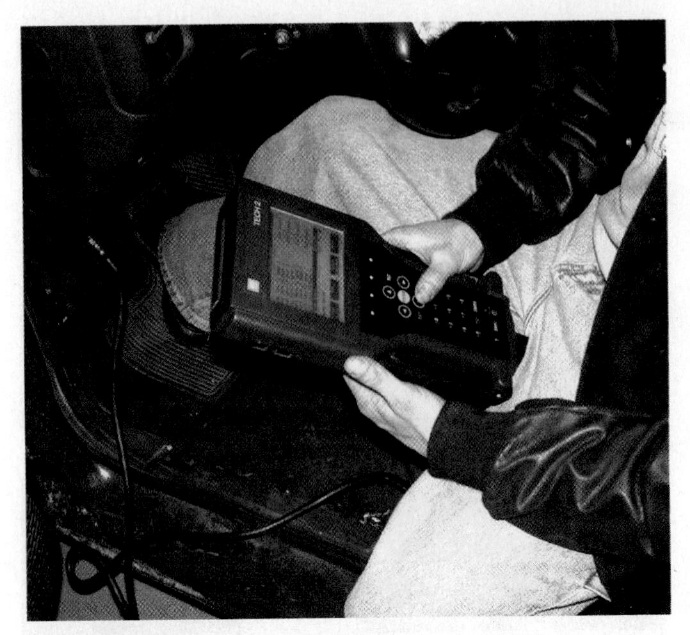

FIGURE 30–6 Using a scan tool to check for IAC counts or percentage as part of a diagnostic routine.

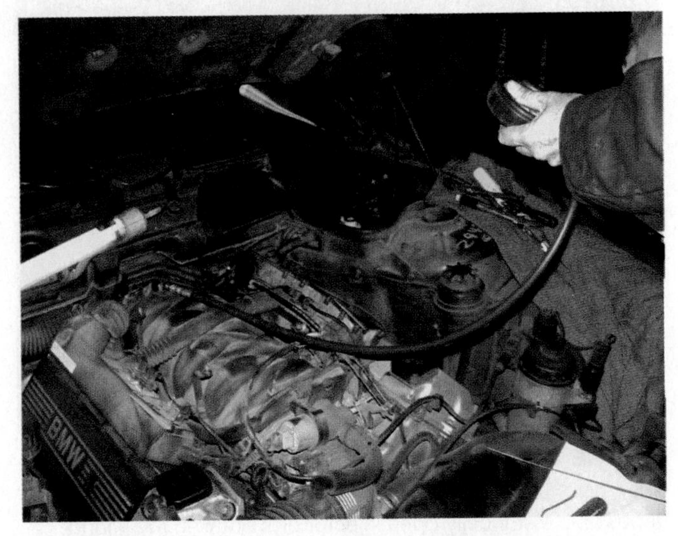

FIGURE 30–7 Checking the fuel pressure using a fuel-pressure gauge connected to the Schrader valve.

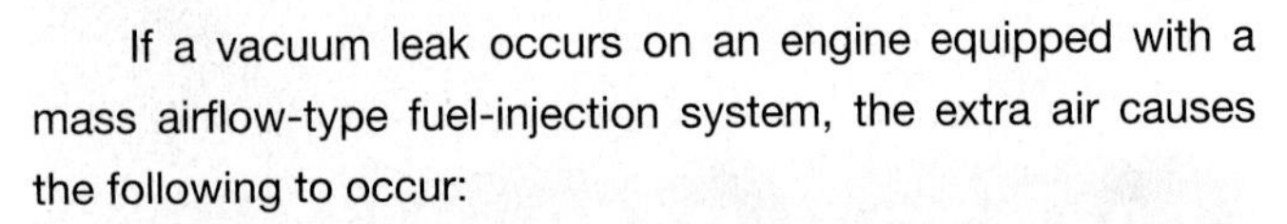

If a vacuum leak occurs on an engine equipped with a mass airflow-type fuel-injection system, the extra air causes the following to occur:

- The engine will operate leaner than normal because the extra air has not been measured by the MAF sensor.
- The idle speed will likely be lower due to the leaner-than-normal air–fuel mixture.
- The idle air control (IAC) counts or percentage will often increase in an attempt to return the engine speed to the target speed stored in the computer.

TECH TIP

No Spark, No Squirt

Most electronic fuel-injection computer systems use the ignition primary (pickup coil or crank sensor) pulse as the trigger for when to inject (squirt) fuel from the injectors (nozzles). If this signal were not present, no fuel would be injected. Because this pulse is also necessary to trigger the module to create a spark from the coil, it can be said that "no spark" could also mean "no squirt." Therefore, if the cause of a no-start condition is observed to be a lack of fuel injection, do not start testing or replacing fuel-system components until the ignition system is checked for proper operation.

PORT FUEL-INJECTION SYSTEM DIAGNOSIS

To determine if a port fuel-injection system—including the fuel pump, injectors, and fuel-pressure regulator—is operating correctly, take the following steps:

1. Attach a fuel-pressure gauge to the Schrader valve on the fuel rail. ● **SEE FIGURE 30–7.**
2. Turn the ignition key on or start the engine to build up the fuel-pump pressure (to about 35 to 45 PSI).
3. Wait 20 minutes and observe the fuel pressure retained in the fuel rail and note the PSI reading. The fuel pressure should not drop more than 20 PSI (140 kPa) in 20 minutes. If the drop is less than 20 PSI in 20 minutes, everything is okay; if the drop is *greater,* then there is a possible problem with:
   - The check valve in the fuel pump
   - Injectors, lines, or fittings
   - A fuel-pressure regulator

   To determine which unit is defective, perform the following:

- Reenergize the electric fuel pump.
- Clamp the fuel *supply* line, and wait 10 minutes (see CAUTION). If the pressure drop does not occur, replace the fuel pump. If the pressure drop still occurs, continue with the next step.

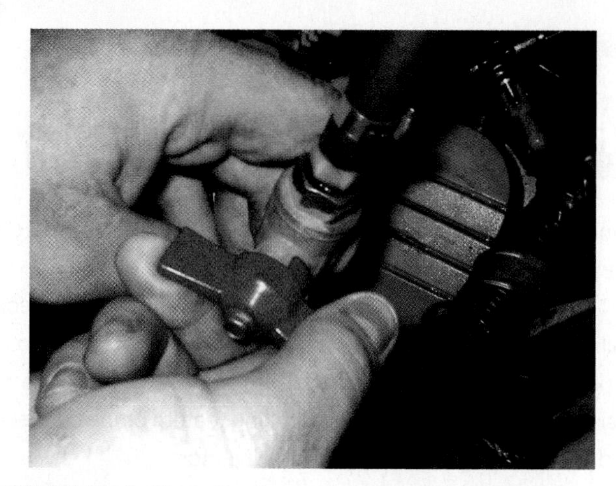

FIGURE 30–8 Shutoff valves must be used on vehicles equipped with plastic fuel lines to isolate the cause of a pressure drop in the fuel system.

- Repeat the pressure buildup of the electric pump and clamp the fuel return line. If the pressure drop time is now okay, replace the fuel-pressure regulator.
- If the pressure drop still occurs, one or more of the injectors is leaking. Remove the injectors with the fuel rail and hold over paper. Replace those injectors that drip one or more drops after 10 minutes with pressurized fuel.

CAUTION: Do not clamp plastic fuel lines. Connect shutoff valves to the fuel system to shut off supply and return lines. ● SEE FIGURE 30–8.

TESTING FOR AN INJECTOR PULSE

One of the first checks that should be performed when diagnosing a no-start condition is whether the fuel injectors are being pulsed by the computer. Checking for proper pulsing of the injector is also important in diagnosing a weak or dead cylinder.

A **noid light** is designed to electrically replace the injector in the circuit and to flash if the injector circuit is working correctly. ● **SEE FIGURE 30–9**. To use a noid light, disconnect the electrical connector at the fuel injector and plug the noid light into the injector harness connections. Crank or start the engine. The noid light should flash regularly.

(a)

(b)

FIGURE 30–9 (a) Noid lights are usually purchased as an assortment so that one is available for any type or size of injector wiring connector. (b) The connector is unplugged from the injector and a noid light is plugged into the injector connector. The noid light should flash when the engine is being cranked if the power circuit and the pulsing to ground by the computer are functioning okay.

NOTE: The term *noid* is simply an abbreviation of the word sole*noid*. Injectors use a movable iron core and are therefore solenoids. Therefore, a noid light is a replacement for the solenoid (injector).

Possible noid light problems and causes include the following:

1. **The light is off and does not flash.** The problem is an open in either the power side or ground side (or both) of the injector circuit.

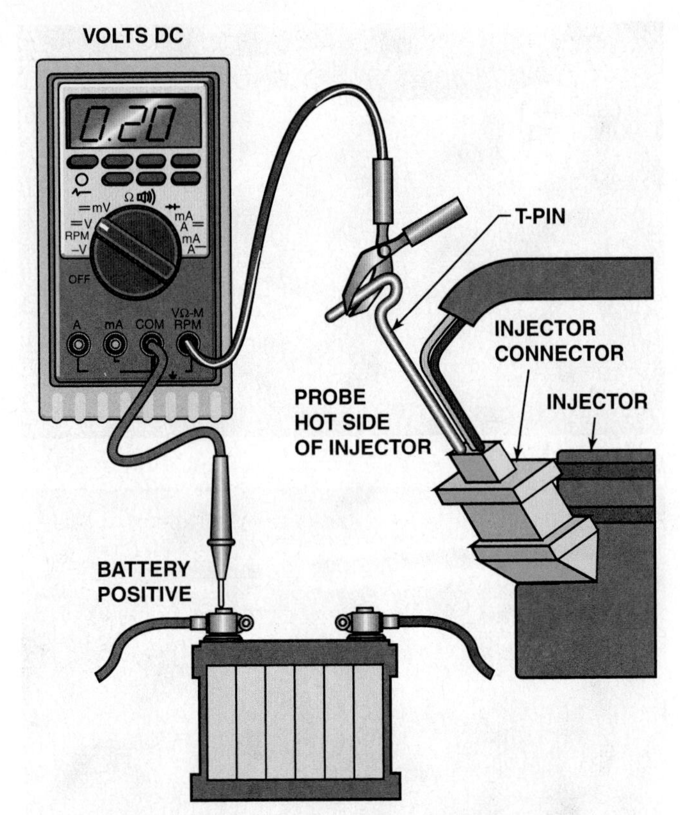

FIGURE 30–10 Use a DMM set to read DC volts to check the voltage drop of the positive circuit to the fuel injector. A reading of 0.5 volt or less is generally considered to be acceptable.

2. **The noid light flashes dimly.** A dim noid light indicates excessive resistance or low voltage available to the injector. Both the power and ground side must be checked.
3. **The noid light is on and does not flash.** If the noid light is on, then both a power and a ground are present. Because the light does not flash (blink) when the engine is being cranked or started, then a short-to-ground fault exists either in the computer itself or in the wiring between the injector and the computer.

CAUTION: A noid lamp must be used with caution. The computer may show a good noid light operation and have low supply voltage. ● SEE FIGURE 30–10.

CHECKING FUEL-INJECTOR RESISTANCE

Each port fuel injector must deliver an equal amount of fuel or the engine will idle roughly or perform poorly.

The electrical balance test involves measuring the injector coil-winding resistance. For best engine operation, all injectors should have the same electrical resistance. To measure the resistance, carefully release the locking feature of the connector and remove the connector from the injector.

NOTE: Some engines require specific procedures to gain access to the injectors. Always follow the manufacturers' recommended procedures.

With an ohmmeter, measure the resistance across the injector terminals. Be sure to use the low-ohms feature of the digital ohmmeter to read in tenths (0.1) of an ohm. ● **SEE FIGURES 30–11 AND 30–12.** Check service information for the resistance specification of the injectors. Measure the resistance of all of the injectors. Replace any injector that does not fall within the resistance range of the specification. The resistance of the injectors should be measured twice—once when the engine (and injectors) are cold and once after the engine has reached normal operating temperature. If any injector measures close to specification, make certain that the terminals of the injector are electrically sound, and perform other tests to confirm an injector problem before replacement.

TYPICAL RESISTANCE VALUES There are two basic types of injectors, which have an effect on their resistance including:

1. **Low-resistance injectors.** The features of a low-resistance injectors include:
   - Uses a "peak and hold" type firing where a high current, usually about 4 amperes, is used to open the injector, then it is held open by using a lower current, which is usually about 1 ampere.

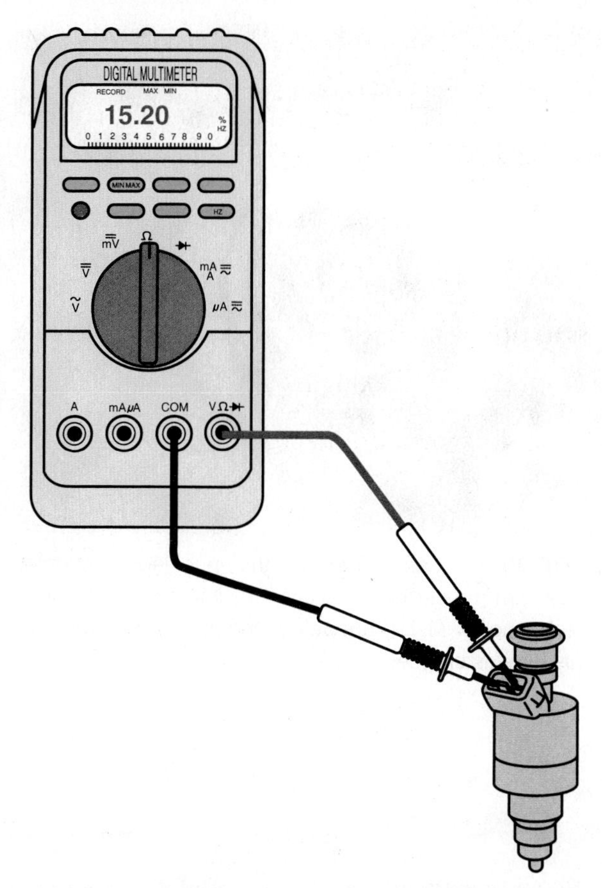

FIGURE 30–11 Connections and settings necessary to measure fuel-injector resistance.

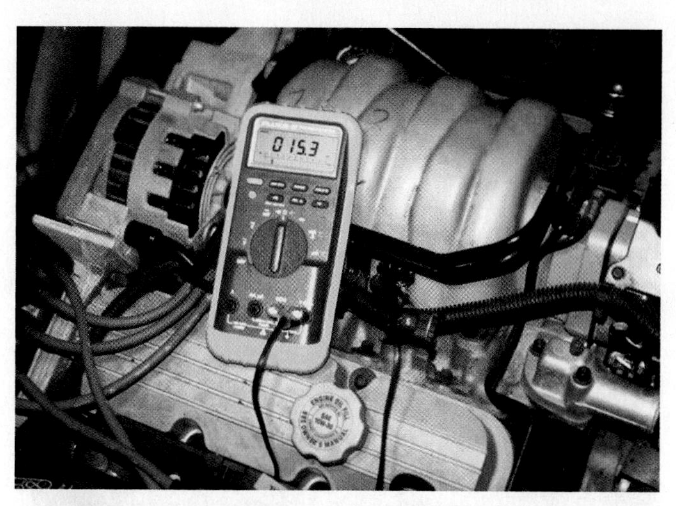

FIGURE 30–12 To measure fuel-injector resistance, a technician constructed a short wiring harness with a double banana plug that fits into the V and COM terminals of the meter and an injector connector at the other end. This setup makes checking resistance of fuel injectors quick and easy.

- All throttle body injection (TBI) injectors and some port fuel injectors are low-resistance injectors and are fired using a peak-and-hold circuit by the PCM.
- The resistance value of a peak-and-hold-type injector is usually 1.5 to 4.0 ohms.

2. **Higher-resistance injectors.** The features of a higher-resistance injectors include:
   - Uses a constant low current, usually 1 ampere, to open the injector.
   - Is called a "saturated" type of injector because the current flows until the magnetic field is strong enough to open the injector.
   - Most port fuel injectors are of the saturated type.
   - The resistance value of a saturated injector is usually 12 to 16 ohms.

? FREQUENTLY ASKED QUESTION

How Does the Fiat Chrysler Multiair System Work?

Some Chrysler and Fiat brand vehicles use a type of fuel-injection system that includes the following unique features:

- The engine has one overhead camshaft but only the exhaust cam lobes actually open the exhaust valves.
- The intake camshaft lobes are used to pressurize engine oil, which is then directed to a solenoid that is pulse-width modulated.
- The oil from the solenoid is sent to a piston on top of the intake valves, which are opened by the piston.
- The timing and valve lift are determined by the PCM that pulses the control solenoid to allow oil to open the valve. ● **SEE FIGURE 30–13.**

Because the intake values are opened using pressured engine oil, it is critical that the specified oil be used and changed at the specified interval. Some customers complain of a "clatter" from the engine, especially at idle, which is normal for this engine and is due to the operation of the control solenoids.

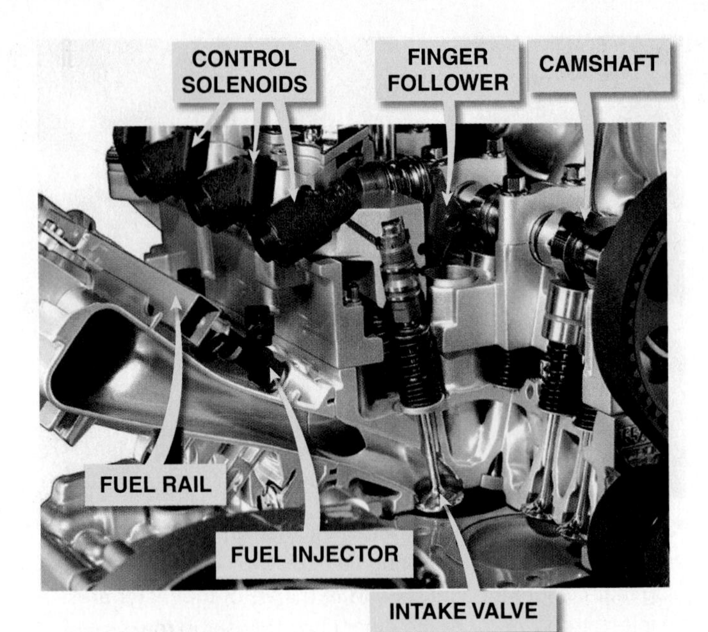

FIGURE 30–13 In a multiair engine design, the exhaust valves are opened by the exhaust camshaft lobes. Intake valves are opened by the high-pressure engine oil, the high pressure being produced by a lobe-actuated piston and controlled by a PCM-controlled solenoids.

FIGURE 30–14 If an injector has the specified resistance, this does not mean that it is okay. This injector had the specified resistance yet it did not deliver the correct amount of fuel because it was clogged.

MEASURING RESISTANCE OF INDIVIDUAL INJECTORS

While there are many ways to check injectors, the first test is to measure the resistance of the coil inside and compare it to factory specifications. ● **SEE FIGURE 30–14**. If the injectors are not accessible, check service information for the location of the electrical connector for the injectors. Unplug the connector and measure the resistance of each injector at the injector side of the connector. Use service information to determine the wire colors for the power side and the pulse side of each injector.

TECH TIP

Equal Resistance Test

All fuel injectors should measure the specified resistance. However the specification often indicates the temperature of the injectors be at room temperature and of course will vary according to the temperature. Rather than waiting for all of the injectors to achieve room temperature, measure the resistance and check that they are all within 0.4 ohm of each other. To determine the difference, record the resistance of each injector and then subtract the lowest resistance reading from the highest resistance reading to get the difference. If more than 0.4 ohm then further testing will be needed to verify defective injector(s).

FIGURE 30–15 Connect a fuel-pressure gauge to the fuel rail at the Schrader valve.

PRESSURE-DROP BALANCE TEST

The pressure balance test involves using an electrical timing device to pulse the fuel injectors on for a given amount of time, usually 500 milliseconds or 0.5 seconds, and observing the drop in pressure that accompanies the pulse. If the *fuel flow* through each injector is equal, the drop in pressure in the system will be equal. Most manufacturers recommend that the pressures be within about 1.5 PSI (10 kPa) of each other for satisfactory engine performance. This test method not only tests the electrical functioning of the injector (for definite time and current pulse), but also tests for mechanical defects that could affect fuel flow amounts.

The purpose of running this injector balance test is to determine which injector is restricted, inoperative, or delivering fuel differently than the other injectors. Replacing a complete set of injectors can be expensive. The basic tools needed are:

- Accurate pressure gauge with pressure relief
- Injector pulser with time control
- Necessary injector connection adapters
- Safe receptacle for catching and disposing of any fuel released

STEP 1 Attach the pressure gauge to the fuel delivery rail on the supply side. Make sure the connections are safe and leakproof.

STEP 2 Attach the injector pulser to the first injector to be tested.

STEP 3 Turn the ignition key to the on position to prime the fuel rail. Note the static fuel-pressure reading. ● **SEE FIGURE 30–15.**

STEP 4 Activate the pulser for the timed firing pulses.

STEP 5 Note and record the new static rail pressure after the injector has been pulsed.

STEP 6 Reenergize the fuel pump and repeat this procedure for all of the engine injectors.

STEP 7 Compare the two pressure readings and compute the pressure drop for each injector. Compare the pressure drops of the injectors to each other. Any variation in pressure drops will indicate an uneven fuel delivery rate between the injectors.

For example:

Injector	1	2	3	4	5	6
Initial pressure	40	40	40	40	40	40
Second pressure	30	30	35	30	20	30
Pressure drop	10	10	5	10	20	10
Possible problem	OK	OK	Restriction	OK	Leak	OK

FIGURE 30–16 An injector tester being used to check the voltage drop through the injector while the tester is sending current through the injectors. This test is used to check the coil inside the injector. This same tester can be used to check for equal pressure drop of each injector by pulsing the injector on for 500 ms.

FIGURE 30–17 A digital storage oscilloscope can be easily connected to an injector by carefully back-probing the electrical connector.

INJECTOR VOLTAGE-DROP TESTS

Another test of injectors involves pulsing the injector and measuring the voltage drop across the windings as current is flowing. A typical voltage-drop tester is shown in ● **FIGURE 30–16.** The tester, which is recommended for use by General Motors Corporation, pulses the injector while a digital multimeter is connected to the unit, which will display the voltage drop as the current flows through the winding.

CAUTION: Do not test an injector using a pulse-type tester more than one time without starting the engine to help avoid a hydrostatic lock caused by the flow of fuel into the cylinder during the pulse test.

Record the highest voltage drop observed on the meter display during the test. Repeat the voltage-drop test for all of the injectors. The voltage drop across each injector should be within 0.1 volt of each other. If an injector has a higher-than-normal voltage drop, the injector windings have higher-than-normal resistance.

SCOPE-TESTING FUEL INJECTORS

A scope (analog or digital storage) can be connected into each injector circuit. There are three types of injector drive circuits and each type of circuit has its own characteristic pattern. ● **SEE FIGURE 30–17** for an example of how to connect a scope to read a fuel-injector waveform.

SATURATED SWITCH TYPE In a saturated switch-type injector-driven circuit, voltage (usually a full 12 volts) is applied to the injector. The ground for the injector is provided by the vehicle computer. When the ground connection is completed, current flows through the injector windings. Due to the resistance and inductive reactance of the coil itself, it requires a fraction of a second (about 3 milliseconds or 0.003 seconds) for the coil to reach **saturation** or maximum current flow. Most saturated switch-type fuel injectors have 12 to 16 ohms of resistance. This resistance, as well as the computer switching circuit, control and limit the current flow through the injector. A voltage spike occurs when the

PEAK VOLTAGE CAUSED BY COLLAPSE OF THE INJECTOR COIL
DRIVER TRANSISTOR TURNS ON, PULLING THE INJECTOR PINTLE AWAY FROM ITS SEAT, FUEL FLOW BEGINS
BATTERY / SOURCE VOLTAGE SUPPLIED TO INJECTOR
DRIVER TRANSISTOR TURNS OFF, STOPPING FUEL FLOW
INJECTOR ON-TIME

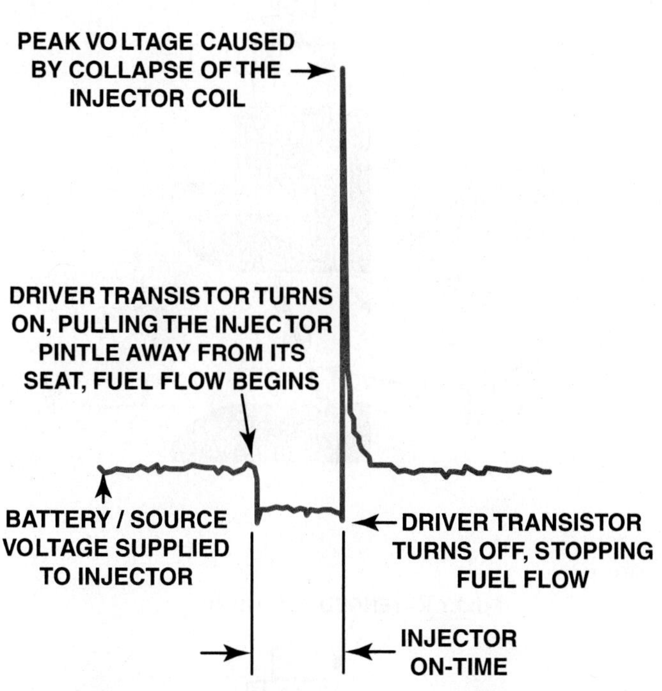

FIGURE 30–18 The injector on-time is called the pulse width.

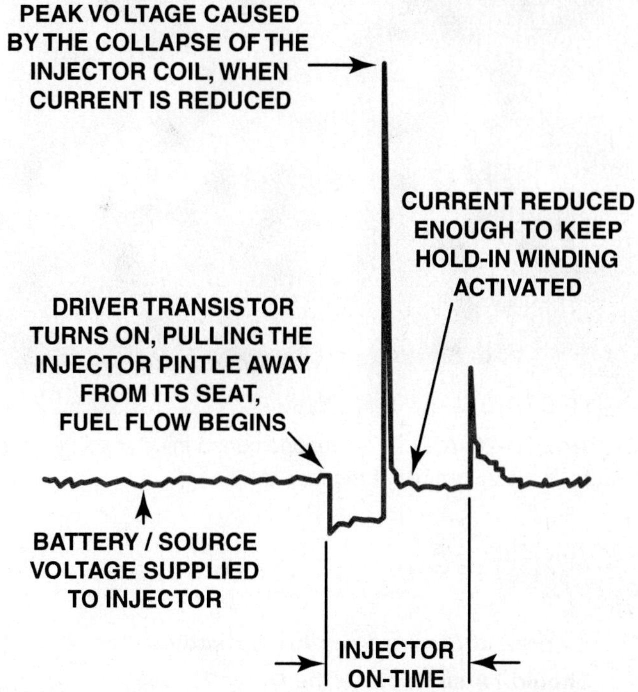

PEAK VOLTAGE CAUSED BY THE COLLAPSE OF THE INJECTOR COIL, WHEN CURRENT IS REDUCED
CURRENT REDUCED ENOUGH TO KEEP HOLD-IN WINDING ACTIVATED
DRIVER TRANSISTOR TURNS ON, PULLING THE INJECTOR PINTLE AWAY FROM ITS SEAT, FUEL FLOW BEGINS
BATTERY / SOURCE VOLTAGE SUPPLIED TO INJECTOR
INJECTOR ON-TIME

FIGURE 30–19 A typical peak-and-hold fuel-injector waveform. Most fuel injectors that measure less than 6 ohms will usually display a similar waveform.

computer shuts off (opens the injector ground-side circuit) the injectors. ● **SEE FIGURE 30–18.**

PEAK-AND-HOLD TYPE A **peak-and-hold** type is typically used for TBI and some port low-resistance injectors. Full battery voltage is applied to the injector and the ground side is controlled through the computer. The computer provides a high initial current flow (about 4 amperes) to flow through the injector windings to open the injector core. Then the computer reduces the current to a lower level (about 1 ampere). The hold current is enough to keep the injector open, yet conserves energy and reduces the heat buildup that would occur if the full current flow remains on as long as the injector is commanded on. Typical peak-and-hold-type injector resistance ranges from 2 to 4 ohms.

The scope pattern of a typical peak-and-hold-type injector shows the initial closing of the ground circuit, then a voltage spike as the current flow is reduced. Another voltage spike occurs when the lower level current is turned off (opened) by the computer. ● **SEE FIGURE 30–19.**

PULSE-WIDTH MODULATED TYPE A pulse-width modulated type of injector drive circuit uses lower-resistance coil injectors. Battery voltage is available at the positive terminal of the injector and the computer provides a variable-duration connection to ground on the negative side of the injector. The computer can vary the time intervals that the injector is grounded for very precise fuel control.

Each time the injector circuit is turned off (ground circuit opened), a small voltage spike occurs. It is normal to see multiple voltage spikes on a scope connected to a pulse-width modulated type of fuel injector.

FIGURE 30–20 A set of six reconditioned injectors. The sixth injector is barely visible at the far right.

FREQUENTLY ASKED QUESTION

If Three of Six Injectors Are Defective, Should I Also Replace the Other Three?

This is a good question. Many service technicians "recommend" that the three good injectors be replaced along with the other three that tested as being defective. The reasons given by these technicians include:

- All six injectors have been operating under the same fuel, engine, and weather conditions.
- The labor required to replace all six is just about the same as replacing only the three defective injectors.
- Replacing all six at the same time helps ensure that all of the injectors are flowing the same amount of fuel so that the engine is operating most efficiently.

With these ideas in mind, the customer should be informed and offered the choice. Complete sets of injectors such as those in ● **FIGURE 30–20** can be purchased at a reasonable cost.

IDLE AIR SPEED CONTROL DIAGNOSIS

On an engine equipped with fuel injection (TBI or port injection), the idle speed is controlled by increasing or decreasing the amount of air bypassing the throttle plate. Again, an electronic stepper motor or pulse-width modulated solenoid is used to maintain the correct idle speed. This control is often called the **idle air control (IAC).** ● **SEE FIGURES 30–21 THROUGH 30–23.**

When the engine stops, most IAC units will retract outward to get ready for the next engine start. When the engine starts, the engine speed is high to provide for proper operation when the engine is cold. Then, as the engine gets warmer, the computer reduces engine idle speed gradually by reducing the number of counts or steps commanded by the IAC.

When the engine is warm and restarted, the idle speed should momentarily increase, then decrease to normal idle speed. This increase and then decrease in engine speed is often called an engine-flare. If the engine speed does not flare, then the IAC may not be working (it may be stuck in one position).

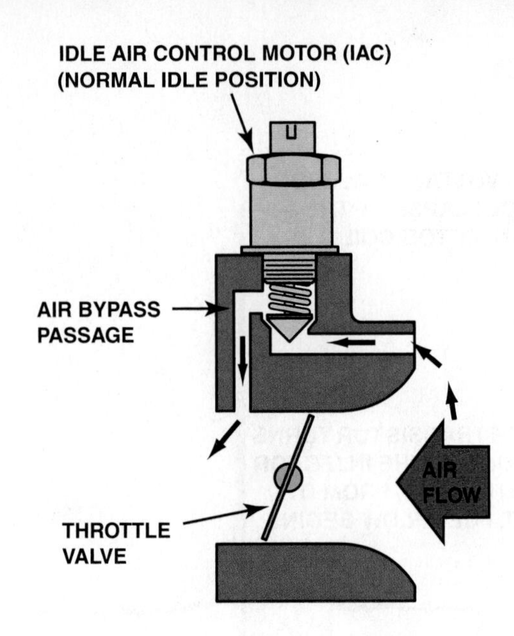

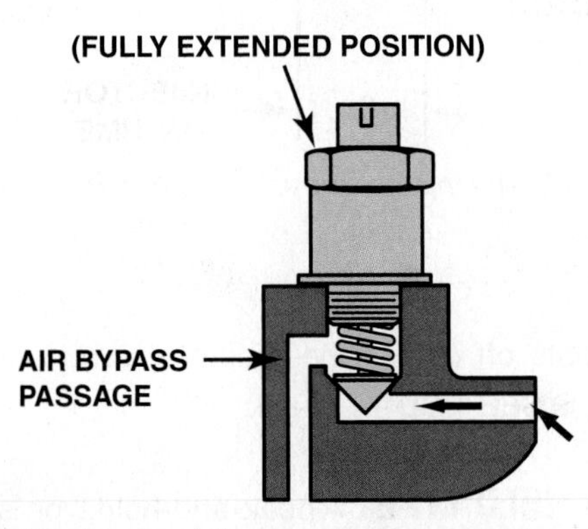

FIGURE 30–21 An IAC controls idle speed by controlling the amount of air that passes around the throttle plate. More airflow results in a higher idle speed.

FIGURE 30–22 A typical IAC.

FIGURE 30–23 Some IAC units are purchased with the housing as shown. Carbon buildup in these passages can cause a rough or unstable idling or stalling.

FIGURE 30–24 When the cover is removed from the top of the engine, a mouse or some other animal nest is visible. The animal had already eaten through a couple of injector wires. At least the cause of the intermittent misfire was discovered.

CASE STUDY

There Is No Substitute for a Thorough Visual Inspection

An intermittent "check engine" light and a random-misfire diagnostic trouble code (DTC) P0300 was being diagnosed. A scan tool did not provide any help because all systems seemed to be functioning normally. Finally, the technician removed the engine cover and discovered a mouse nest. ● **SEE FIGURE 30–24.**

Summary:

- **Complaint—**Customer stated that the "Check Engine" light was on.
- **Cause—**A stored P0300 DTC was stored indicating a random misfire had been detected caused by an animal that had partially eaten some fuel injector wires.
- **Correction—**The mouse nest was removed and the wiring was repaired.

FUEL-INJECTION SERVICE

After many years of fuel-injection service, some service technicians still misunderstand the process of proper fuel-system handling. Much has been said over the years with regard to when and how to perform injector cleaning. Some manufacturers have suggested methods of cleaning while others have issued bulletins to prohibit any cleaning at all.

All engines using fuel injection do require some type of fuel-system maintenance. Normal wear and tear with today's underhood temperatures and changes in gasoline quality contribute to the buildup of olefin wax, dirt, water, and many other additives. Unique to each engine is an air-control design that also may contribute different levels of carbon deposits, such as oil control.

Fuel-injection system service should include the following operations:

1. **Check fuel-pump operating pressure and volume.** The missing link here is volume. Most working technicians assume that if the pressure is correct, the volume is also okay. Hook up a fuel-pressure tester to the fuel rail inlet to quickly test the fuel pressure with the engine running. At the same time, test the volume of the pump by sending fuel into the holding tank. (One ounce per second is the usual specification.) ● **SEE FIGURE 30–25.** A two-line system tester is the recommended procedure to use and is attached to the fuel inlet and the return on the fuel rail. The vehicle onboard system is looped and returns fuel to the tank.
2. **Test the fuel-pressure regulator for operation and leakage.** At this time, the fuel-pressure regulator would be tested for operational pressure and proper regulation, including leakage. (This works well as the operator has total control of

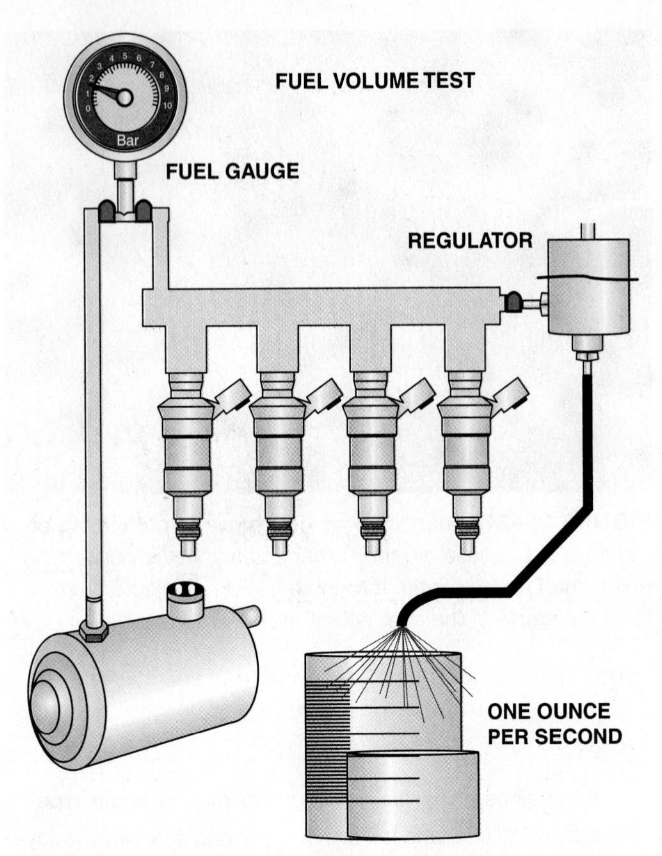

FIGURE 30–25 Checking fuel-pump volume using a hose from the outlet of the fuel-pressure regulator into a calibrated container.

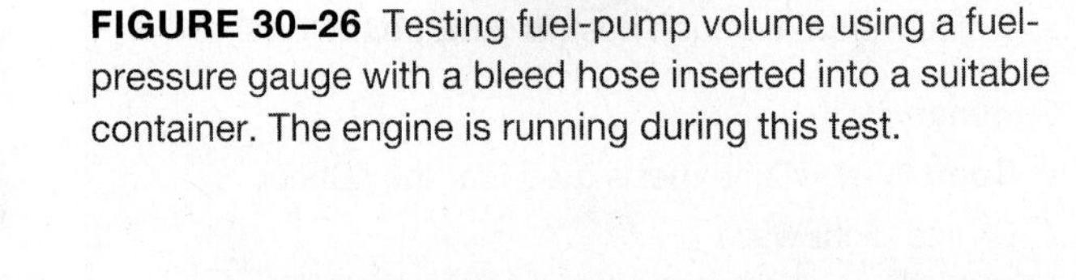

FIGURE 30–26 Testing fuel-pump volume using a fuel-pressure gauge with a bleed hose inserted into a suitable container. The engine is running during this test.

rail pressure with a unit control valve.) Below are some points to consider:

- Good pressure does not mean proper volume. For example, a clogged filter may test okay on pressure but the restriction may not allow proper volume under load. ● **SEE FIGURE 30–26.**
- It is a good idea to use the vehicle's own gasoline to service the system versus a can of shop gasoline that has been sitting around for some time.
- Pressure regulators do fail and a lot more do not properly shut off fuel, causing higher-than-normal pump wear and shorter service life.

3. **Flush the entire fuel rail and upper fuel-injector screens including the fuel-pressure regulator.** Raise the input pressure to a point above regulator setting to allow a constant flow of fuel through the inlet pressure side of the system, through the fuel rail, and out the open fuel-pressure regulator. In most cases the applied pressure is 75 to 90 PSI (517 to 620 kPa), but will be maintained by the presence of a regulator. At this point, cleaning chemical is added to the fuel at a 5:1 mixture and allowed to flow through the system for 15 to 30 minutes. ● **SEE FIGURE 30–27.** Results are best on a hot engine with the fuel supply looped and the engine not running. Below are some points to consider:

- This flush is the fix most vehicles need first. The difference is that the deposits are removed to a remote tank and filter versus attempting to soften the deposits and blow them through the upper screens.
- Most injectors use a 10-micron final filter screen. A 25% restriction in the upper screen would increase the injector on-time approximately 25%.
- **Clean the fuel injectors.** Start the engine and adjust the output pressure closer to regulator pressure or lower than in the previous steps. Lower pressure will cause the pulse width to open up somewhat longer and allow the injectors to be cleaned. Slow speed (idle) position will take a longer time frame and operating temperature will be reached. Clean injectors are the objective, but the chemical should also decarbon the engine valves, pistons, and oxygen sensor.

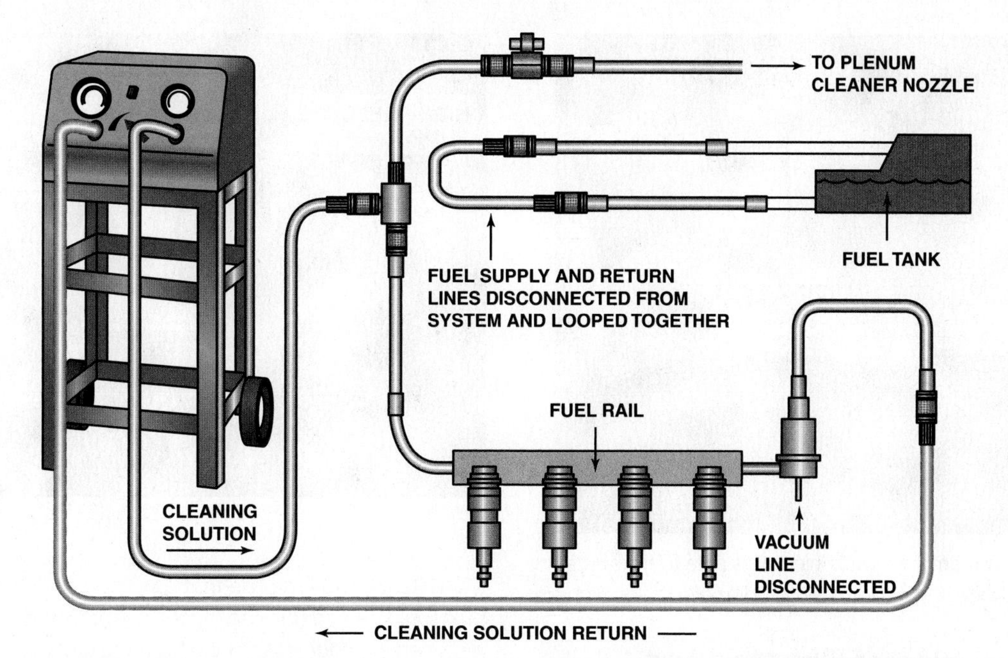

FIGURE 30–27 A typical two-line cleaning machine hookup, showing an extension hose that can be used to squirt a cleaning solution into the throttle body while the engine is running on the cleaning solution and gasoline mixture.

TECH TIP

Check the Injectors at the "Bends and the Ends"

Injectors that are most likely to become restricted due to clogging of the filter basket screen are the injectors at the ends of the rail especially on returnless systems where dirt can accumulate. Also the injectors that are located at the bends of the fuel rail are also subject to possible clogging due to the dirt being deposited where the fuel makes a turn in the rail.

4. **Decarbon the engine assembly.** On most vehicles, the injector spray will help the decarboning process. On others, you may need to enhance the operation with external addition of a mixture through the PCV hose, throttle plates, or idle air controls.
5. **Clean the throttle plate and idle air control passages.** Doing this service alone on most late-model engines will show a manifold vacuum increase of up to 2 inches Hg. ● **SEE FIGURE 30–28**. This works well as air is drawn into IAC passages on a running engine and will clean the passages without IAC removal.
6. **Relearn the onboard computer.** Some vehicles may have been running in such a poor state of operation that the onboard computer may need to be relearned. Consult service information for the suggested relearn procedures for each particular vehicle.

FIGURE 30–28 To thoroughly clean a throttle body, it is sometimes best to remove it from the vehicle.

TECH TIP

Use an Injector Tester

The best way to check injectors is to remove them all from the engine and test them using an injector tester. A typical injector tester uses a special nonflammable test fluid that has the same viscosity as gasoline. The tester pulses the injectors, and the amount of fuel delivered as well as the spray pattern can be seen. Many testers are capable of varying the frequency of the pulse as well as the duration that helps find intermittent injector faults. ● **SEE FIGURE 30–29.**

All of the previously listed steps may be performed using a *two-line* fuel-injector service unit such as Carbon Clean, Auto Care, Injector Test, DeCarbon, or Motor-Vac.

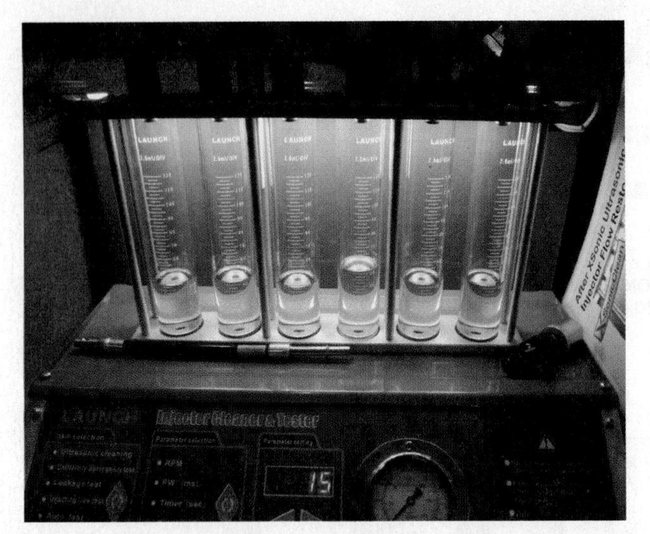

FIGURE 30–29 The amount each injector is able to flow is displayed in glass cylinders are each injector for a quick visual check.

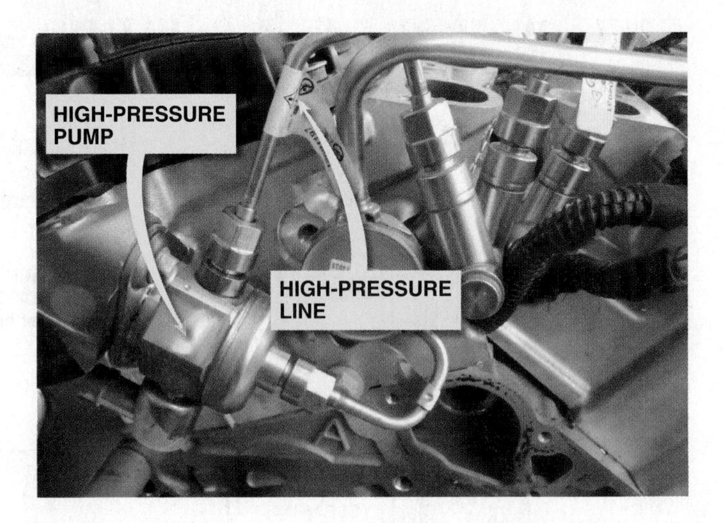

FIGURE 30–30 The line that has the yellow tag is a high-pressure line and this line must be replaced with a new part if removed even for a few minutes to gain access to another part.

Fuel-Injection Symptom Chart

Symptom	Possible Causes
Hard cold starts	• Low fuel pressure • Leaking fuel injectors • Contaminated fuel • Low-volatility fuel • Dirty throttle plate
Garage stalls	• Low fuel pressure • Insufficient fuel volume • Restricted fuel injector • Contaminated fuel • Low-volatility fuel
Poor cold performance	• Low fuel pressure • Insufficient fuel volume • Contaminated fuel • Low-volatility fuel
Tip-in hesitation (hesitation just as the accelerator pedal is depressed)	• Low fuel pressure • Insufficient fuel volume • Intake valve deposits • Contaminated fuel • Low-volatility fuel

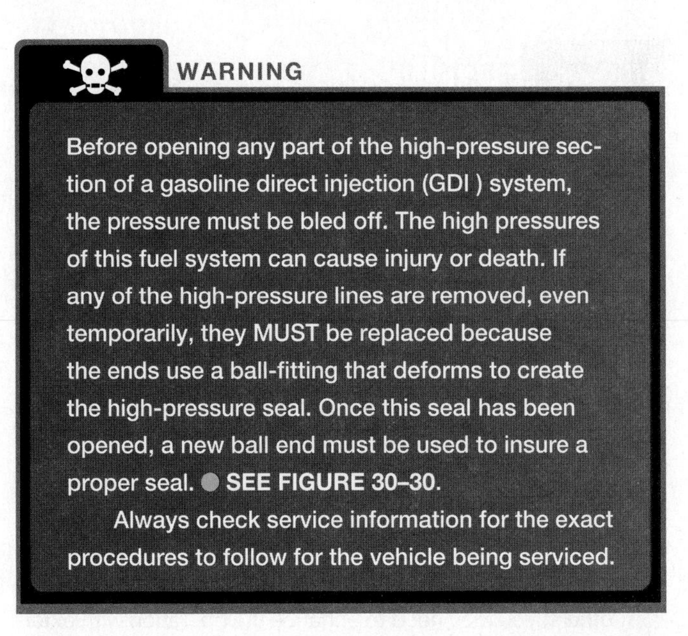

WARNING

Before opening any part of the high-pressure section of a gasoline direct injection (GDI) system, the pressure must be bled off. The high pressures of this fuel system can cause injury or death. If any of the high-pressure lines are removed, even temporarily, they MUST be replaced because the ends use a ball-fitting that deforms to create the high-pressure seal. Once this seal has been opened, a new ball end must be used to insure a proper seal. ● **SEE FIGURE 30–30.**

Always check service information for the exact procedures to follow for the vehicle being serviced.

FUEL-SYSTEM SCAN TOOL DIAGNOSTICS

Diagnosing a faulty fuel system can be a difficult task. However, it can be made easier by utilizing the information available via the serial data stream. By observing the long-term fuel trim and the short-term fuel trim, we can determine how the fuel system is performing. Short-term fuel trim and long-term fuel trim can help zero in on specific areas of trouble. Readings should be taken at idle and at 3000 RPM.

Condition	Long-Term Fuel Trim at Idle	Long-Term Fuel Trim at 3000 RPM
System normal	0% ± 10%	0% ± 10%
Vacuum leak	HIGH	OK
Fuel flow problem	OK	HIGH
Low fuel pressure	HIGH	HIGH
High fuel pressure	*OK or LOW	*OK or LOW

*High fuel pressure will affect trim at idle, at 3000 RPM, or both.

CHART 30–1

Fuel trim levels and possible causes if not within 10%.

FUEL-PUMP RELAY CIRCUIT DIAGNOSIS

1 The tools needed to diagnose a circuit containing a relay include a digital multimeter (DMM), a fused jumper wire, and an assortment of wiring terminals.

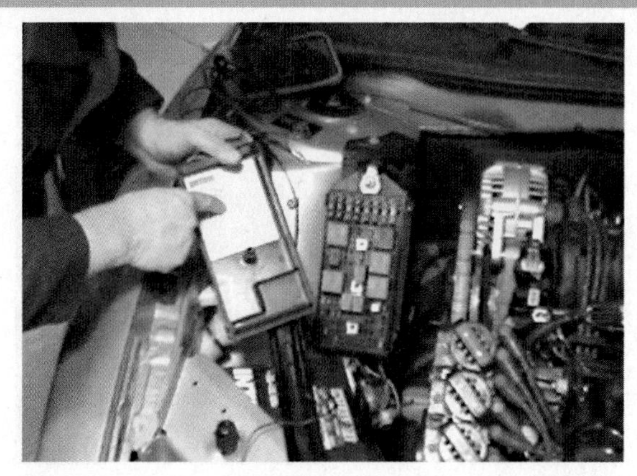

2 Start the diagnosis by locating the relay center. It is under the hood on this General Motors vehicle, so access is easy. Not all vehicles are this easy.

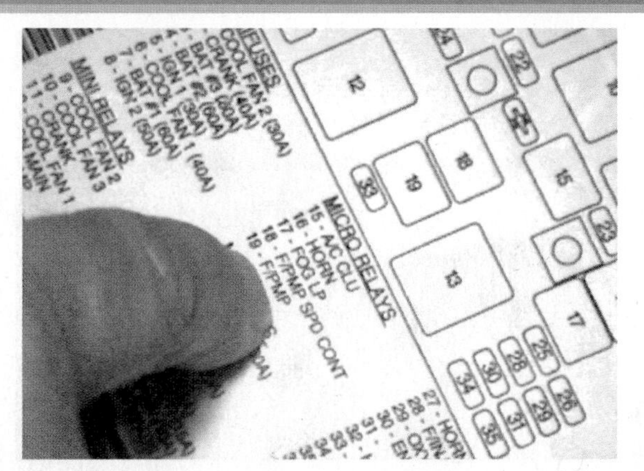

3 The chart under the cover for the relay center indicates the location of the relay that controls the electric fuel pump.

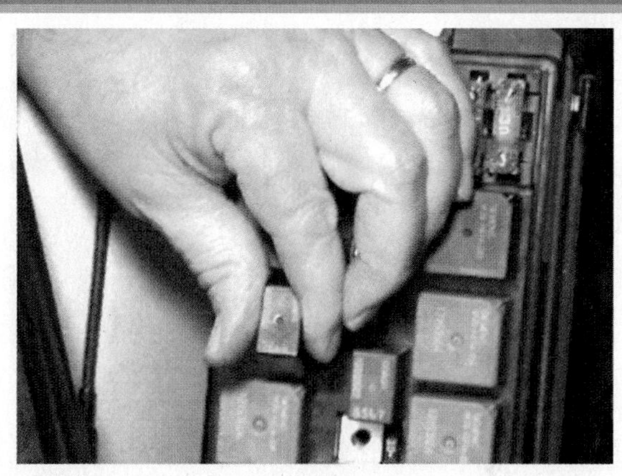

4 Locate the fuel-pump relay and remove by using a puller if necessary. Try to avoid rocking or twisting the relay to prevent causing damage to the relay terminals or the relay itself.

5 Terminals 85 and 86 represent the coil inside the relay. Terminal 30 is the power terminal, 87a is the normally closed contact, and 87 is the normally open contact.

6 The terminals are also labeled on most relays.

CONTINUED ▶

FUEL-PUMP RELAY CIRCUIT DIAGNOSIS (CONTINUED)

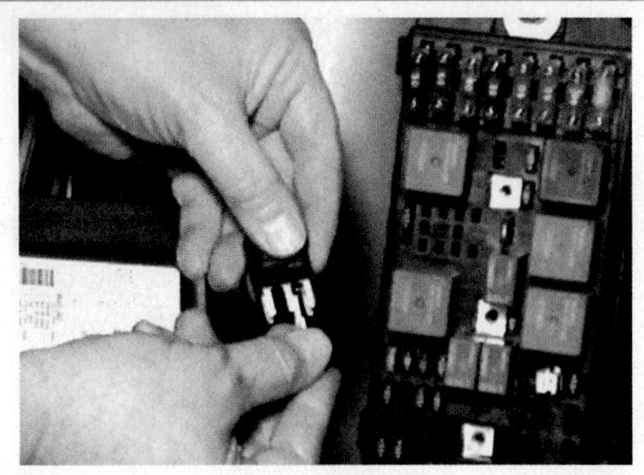

7 To help make good electrical contact with the terminals without doing any harm, select the proper-size terminal from the terminal assortment.

8 Insert the terminals into the relay socket in 30 and 87.

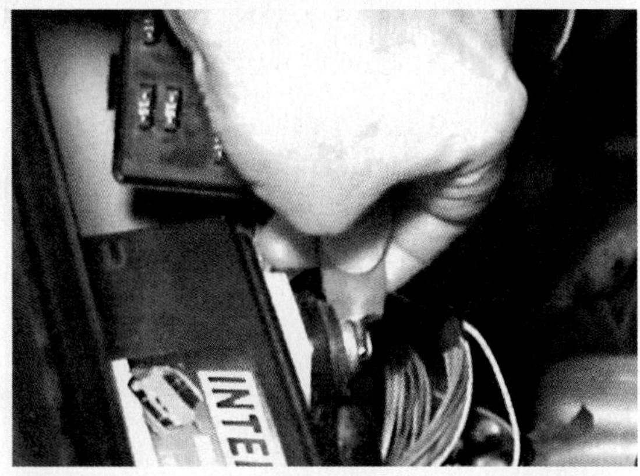

9 To check for voltage at terminal 30, use a test light or a voltmeter. Start by connecting the alligator clip of the test light to the positive (+) terminal of the battery.

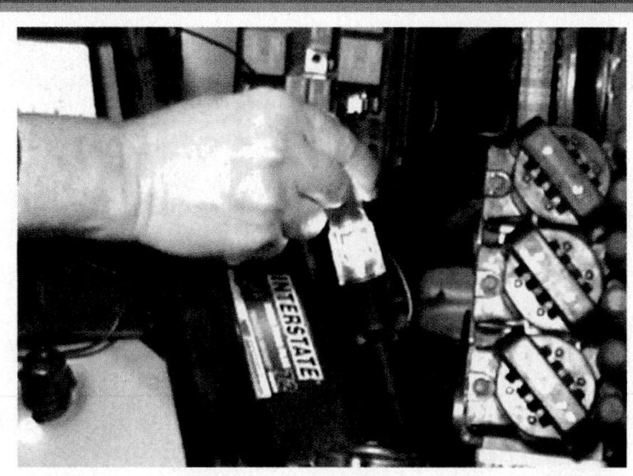

10 Touch the test light to the negative (−) terminal of the battery or a good engine ground to check the test light.

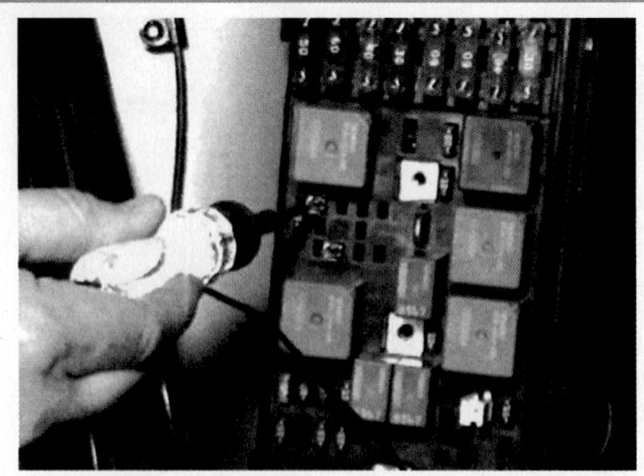

11 Use the test light to check for voltage at terminal 30 of the relay. The ignition may have to be in the on (run) position.

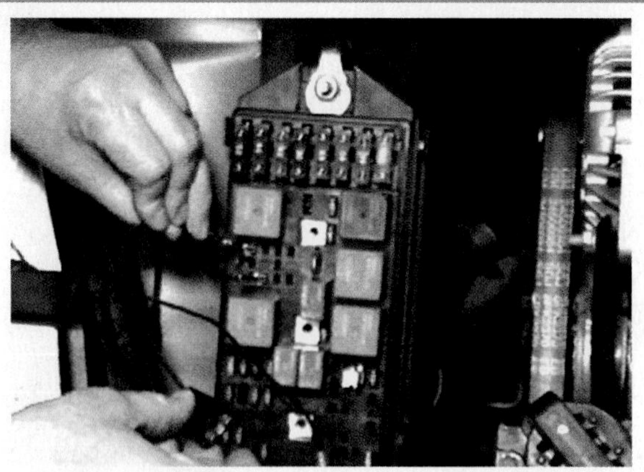

12 To check to see if the electric fuel pump can be operated from the relay contacts, use a fused jumper wire and touch the relay contacts that correspond to terminals 30 and 87 of the relay.

13 Connect the leads of the meter to contacts 30 and 87 of the relay socket. The reading of 4.7 amperes is okay because the specification is 4 to 8 amperes.

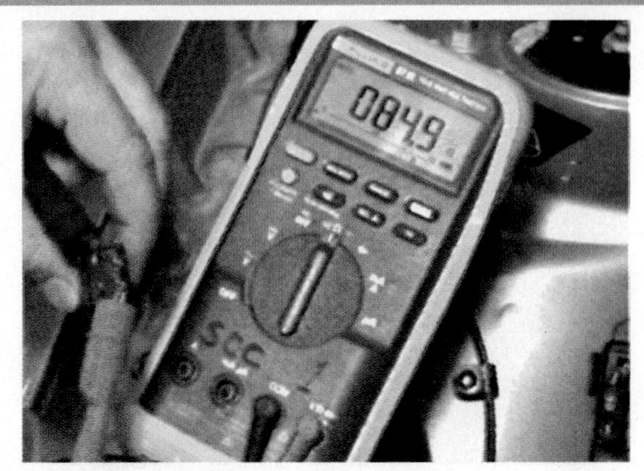

14 Set the meter to read ohms (Ω) and measure the resistance of the relay coil. The usual reading for most relays is between 60 and 100 ohms.

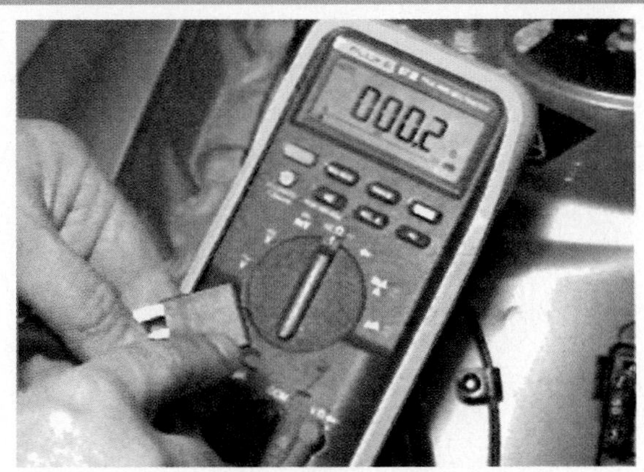

15 Measure between terminal 30 and 87a. Terminal 87a is the normally closed contact, and there should be little, if any, resistance between these two terminals, as shown.

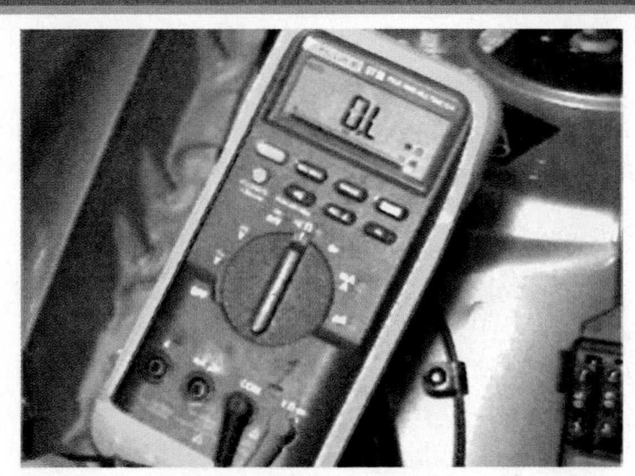

16 To test the normally open contacts, connect one meter lead to terminal 30 and the other lead to terminal 87. The ohmmeter should show an open circuit by displaying OL.

17 Connect a fused jumper wire to supply 12 volts to terminal 86 and a ground to terminal 85 of the relay. If the relay clicks, then the relay coil is able to move the armature (movable arm) of the relay.

18 After testing, be sure to reinstall the relay and the relay cover.

FUEL INJECTOR CLEANING

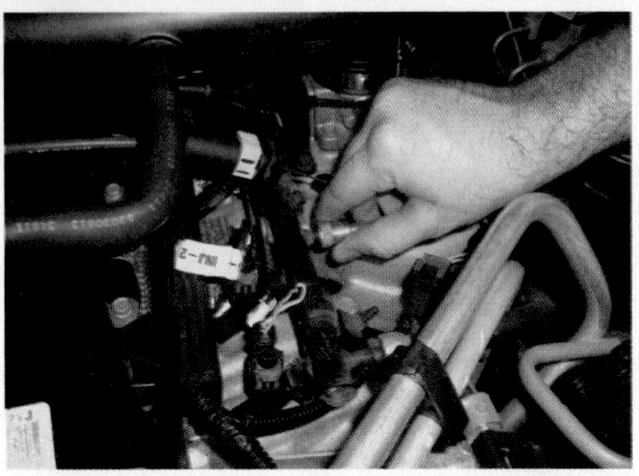

1 Start the fuel injector cleaning process by bringing the vehicle's engine up to operating temperature. Shut off the engine, remove the cap from the fuel rail test port, and install the appropriate adapter.

2 The vehicle's fuel pump is disabled by removing its relay or fuse. In some cases, it may be necessary to disconnect the fuel pump at the tank if the relay or fuse powers more than just the pump.

3 Turn the outlet valve of the canister to the OFF or CLOSED position.

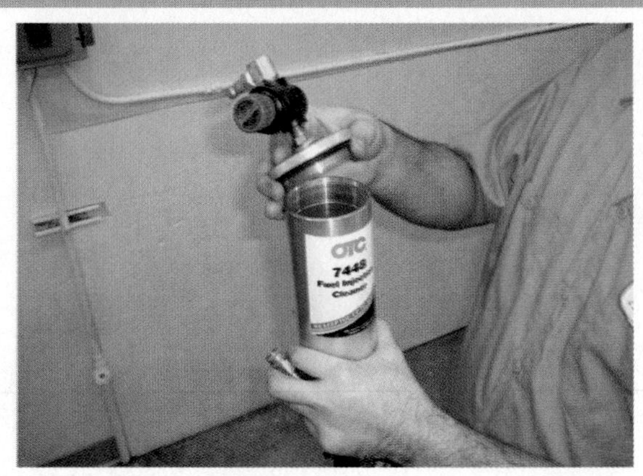

4 Remove the fuel injector cleaning canister's top and regulator assembly. Note that there is an O-ring seal located here that must be in place for the canister's top to seal properly.

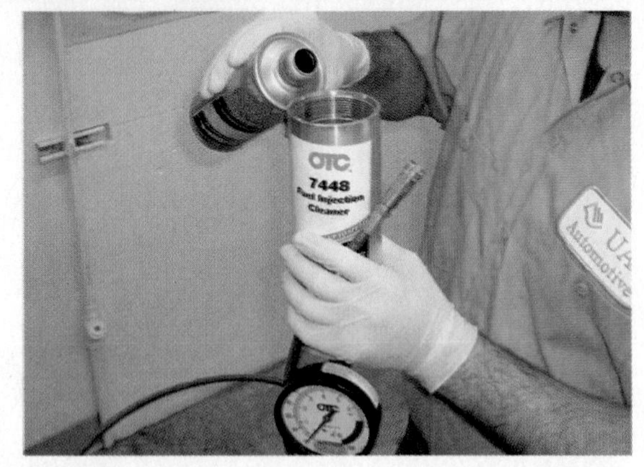

5 Pour the injection system cleaning fluid into the open canister. Rubber gloves are highly recommended for this step as the fluid is toxic.

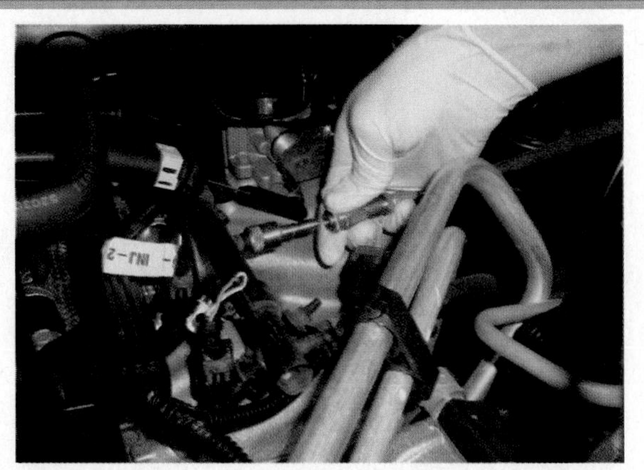

6 Replace the canister's top (making sure it is tight) and connect its hose to the fuel rail adapter. Be sure that the hose is routed away from exhaust manifolds and other hazards.

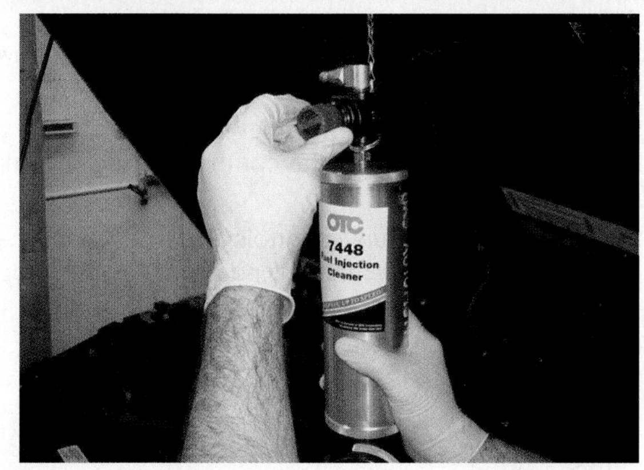

7 Hang the canister from the vehicle's hood and adjust the air pressure regulator to full OPEN position (CCW).

8 Connect shop air to the canister and adjust the air pressure regulator to the desired setting. Canister pressure can be read directly from the gauge.

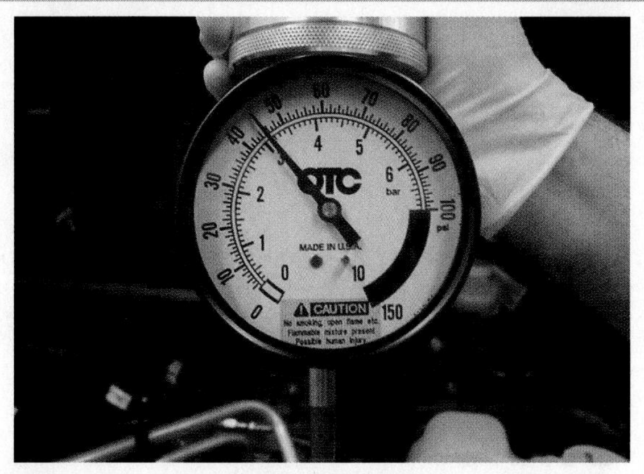

9 Canister pressure should be adjusted to 5 PSI below system fuel pressure. An alternative for return-type systems is to block the fuel return line to the tank.

10 Open the outlet valve on the canister.

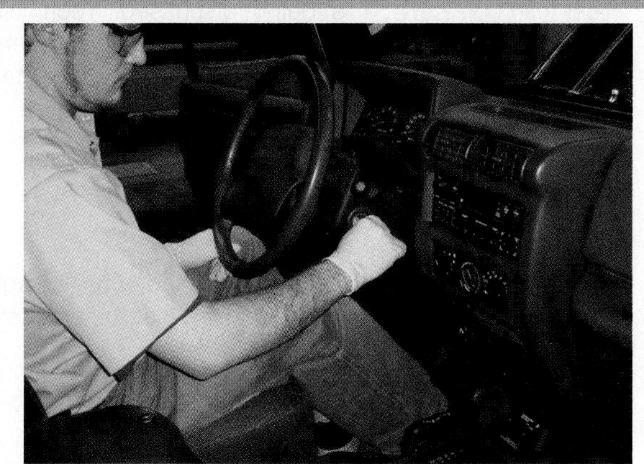

11 Start the vehicle's engine and let run at 1000–1500 RPM. The engine is now running on fuel injector cleaning fluid provided by the canister.

12 Continue the process until the canister is empty and the engine stalls. Remove the cleaning equipment, enable the vehicle's fuel pump, and run the engine to check for leaks.

SUMMARY

1. A typical port fuel-injection system uses an individual fuel injector for each cylinder and squirts fuel directly into the intake manifold about 3 inches (80 mm) from the intake valve.
2. A typical fuel-injection system fuel pressure should not drop more than 20 PSI in 20 minutes.
3. A noid light can be used to check for the presence of an injector pulse.
4. Injectors can be tested for resistance and should be within 0.3 to 0.4 ohms of each other.
5. Different designs of injectors have a different scope waveform depending on how the computer pulses the injector on and off.
6. An idle air control unit controls idle speed and can be tested for proper operation using a scan tool or scope.

REVIEW QUESTIONS

1. List the ways fuel injectors can be tested.
2. List the steps necessary to test a fuel-pressure regulator.
3. Describe why it may be necessary to clean the throttle plate of a port fuel-injected engine.

CHAPTER QUIZ

1. Most port fuel-injected engines operate on how much fuel pressure?
   a. 3 to 5 PSI (21 to 35 kPa)
   b. 9 to 13 PSI (62 to 90 kPa)
   c. 35 to 45 PSI (240 to 310 kPa)
   d. 55 to 65 PSI (380 to 450 kPa)
2. Fuel injectors can be tested using ________.
   a. An ohmmeter
   b. A stethoscope
   c. A scope
   d. All of the above
3. Throttle-body fuel-injection systems use what type of injector driver?
   a. Peak and hold
   b. Saturated switch
   c. Pulse-width modulated
   d. Pulsed
4. Port fuel-injection systems generally use what type of injector driver?
   a. Peak and hold
   b. Saturated switch
   c. Pulse-width modulated
   d. Pulsed
5. The vacuum hose from the fuel-pressure regulator was removed from the regulator and gasoline dripped out of the hose. Technician A says that is normal and that everything is okay. Technician B says that one or more of the injectors may be defective, causing the fuel to get into the hose. Which technician is correct?
   a. Technician A only
   b. Technician B only
   c. Both Technicians A and B
   d. Neither Technician A nor B
6. The fuel pressure drops rapidly when the engine is turned off. Technician A says that one or more injectors could be leaking. Technician B says that a defective check valve in the fuel pump could be the cause. Which technician is correct?
   a. Technician A only
   b. Technician B only
   c. Both Technicians A and B
   d. Neither Technician A nor B
7. In a typical port fuel-injection system, which injectors are most subject to becoming restricted?
   a. Any of them equally
   b. The injectors at the end of the rail on a returnless system
   c. The injectors at the bends in the rail
   d. Either b or c
8. What component pulses the fuel injector on most vehicles?
   a. Electronic control unit (computer)
   b. Ignition module
   c. Crankshaft sensor
   d. Both b and c
9. Fuel-injection service is being discussed. Technician A says that the throttle plate(s) should be cleaned. Technician B says that the fuel rail should be cleaned. Which technician is correct?
   a. Technician A only
   b. Technician B only
   c. Both Technicians A and B
   d. Neither Technician A nor B
10. If the throttle plate needs to be cleaned, what symptoms will be present regarding the operation of the engine?
   a. Stalls
   b. Rough idle
   c. Hesitation on acceleration
   d. All of the above

chapter 31

VEHICLE EMISSIONS STANDARDS AND TESTING

LEARNING OBJECTIVES

After studying this chapter, the reader will be able to:

1. Discuss emissions standards.
2. Identify the reasons why excessive amounts of HC, CO, and NO_x exhaust emissions are created.
3. Diagnose driveability and emissions problems resulting from malfunctions of interrelated systems.
4. Describe how to test for various emissions products.

This chapter will help you prepare for ASE A8 certification test content area "D" (Emissions Control Systems Diagnosis and Repair) and ASE L1 certification test content area "F" (I/M Failure Diagnosis).

KEY TERMS

EMISSIONS STANDARDS IN THE UNITED STATES

In the United States, emissions standards are managed by the Environmental Protection Agency (EPA) as well as some U.S. state governments. Some of the strictest standards in the world are formulated in California by the California Air Resources Board (CARB).

TIER 1 AND TIER 2 Federal emissions standards are set by the Clean Air Act Amendments (CAAA) of 1990 grouped by tier. All vehicles sold in the United States must meet Tier 1 standards that went into effect in 1994 and are the least stringent. Additional Tier 2 standards have been optional since 2001, and fully adopted by 2009. The current Tier 1 standards are different between automobiles and light trucks (SUVs, pickup trucks, and minivans), but Tier 2 standards will be the same for both types.

There are several ratings that can be given to vehicles, and a certain percentage of a manufacturer's vehicles must meet different levels in order for the company to sell its products in affected regions. Beyond Tier 1, and in order by stringency, are the following levels:

- **TLEV (Transitional Low-Emission Vehicle).** More stringent for HC than Tier 1.
- **LEV** (also known as **LEV I**) **(Low-Emission Vehicle).** An intermediate California standard about twice as stringent as Tier 1 for HC and NO_X.
- **ULEV** (also known as **ULEV I**) **(Ultra-Low-Emission Vehicle).** A stronger California standard emphasizing very low HC emissions.
- **ULEV II (Ultra-Low-Emission Vehicle).** A cleaner-than-average vehicle certified under the Phase II LEV standard. Hydrocarbon and carbon monoxide emissions levels are nearly 50% lower than those of a LEV II-certified vehicle. ● **SEE FIGURE 31–1.**
- **SULEV (Super-Ultra-Low-Emission Vehicle).** A California standard even tighter than ULEV, including much lower HC and NO_X emissions; roughly equivalent to Tier 2 Bin 2 vehicles.
- **ZEV (Zero-Emission Vehicle).** A California standard prohibiting any tailpipe emissions. The ZEV category is largely restricted to electric vehicles and hydrogen-fueled vehicles. In these cases, any emissions that are created are produced at another site, such as a power plant or hydrogen reforming center, unless such sites run on renewable energy.

FIGURE 31–1 The underhood decal showing that this Lexus RX-330 meets both national (Tier 2; BIN 5) and California LEV-II (ULEV) regulation standards.

NOTE: A battery-powered electric vehicle charged from the power grid will still be up to 10 times cleaner than even the cleanest gasoline vehicles over their respective lifetimes.

- **PZEV (Partial Zero-Emission Vehicle).** Compliant with the SULEV standard; additionally has near-zero evaporative emissions and a 15-year/150,000-mile warranty on its emission control equipment.

Tier 2 standards are even more stringent. Tier 2 variations are appended with "II," such as LEV II or SULEV II. Other categories have also been created:

- **ILEV (Inherently Low-Emission Vehicle)**
- **AT-PZEV (Advanced Technology Partial Zero-Emission Vehicle).** If a vehicle meets the PZEV standards and is using high-technology features, such as an electric motor or high-pressure gaseous fuel tanks for compressed natural gas, it qualifies as an AT-PZEV. Hybrid electric vehicles such as the Toyota Prius can qualify, as can internal combustion engine vehicles that run on natural gas (CNG), such as the Honda Civic GX. These vehicles are classified as "partial" ZEV because they receive partial credit for the number of ZEV vehicles that automakers would otherwise be required to sell in California.
- **NLEV (National Low-Emission Vehicle).** All vehicles nationwide must meet this standard, which started in 2001.

FEDERAL EPA BIN NUMBER The higher the tier number, the newer the regulation; the lower the bin number, the cleaner the vehicle. The Toyota Prius is a very clean Bin 3, while the Hummer H2 is a dirty Bin 11. ● **SEE CHARTS 31–1, 31–2, AND 31–3.**

CERTIFICATION LEVEL	NMOG (G/MI)	CO (G/MI)	NO_X (G/MI)
Bin 1	0.0	0.0	0.0
Bin 2	0.010	2.1	0.02
Bin 3	0.055	2.1	0.03
Bin 4	0.070	2.1	0.04
Bin 5	0.090	4.2	0.07
Bin 6	0.090	4.2	0.10
Bin 7	0.090	4.2	0.15
Bin 8a	0.125	4.2	0.20
Bin 8b	0.156	4.2	0.20
Bin 9a	0.090	4.2	0.30
Bin 9b	0.130	4.2	0.30
Bin 9c	0.180	4.2	0.30
Bin 10a	0.156	4.2	0.60
Bin 10b	0.230	6.4	0.60
Bin 10c	0.230	6.4	0.60
Bin 11	0.230	7.3	0.90

CHART 31-1

EPA Tier 2—120,000 mile tailpipe emission limits. After January 2007, the highest allowable Bin is 8.

Source: Data compiled from the Environmental Protection Agency (EPA).

Note: The bin number is determined by the type and weight of the vehicle.

U.S. EPA VEHICLE INFORMATION PROGRAM (THE HIGHER THE SCORE, THE LOWER THE EMISSIONS)	
SELECTED EMISSIONS STANDARDS	SCORE
Bin 1 and ZEV	10
PZEV	9.5
Bin 2	9
Bin 3	8
Bin 4	7
Bin 5 and LEV II cars	6
Bin 6	5
Bin 7	4
Bin 8	3
Bin 9a and LEV I cars	2
Bin 9b	2
Bin 10a	1
Bin 10b and Tier 1 cars	1
Bin 11	0

CHART 31-2

Air pollution score.

Source: Courtesy of the Environmental Protection Agency (EPA).

MINIMUM FUEL ECONOMY (MPG) COMBINED CITY-HIGHWAY LABEL VALUE					
SCORE	GASOLINE	DIESEL	E-85	LPG	CNG*
10	44	50	31	28	33
9	36	41	26	23	27
8	30	35	22	20	23
7	26	30	19	17	20
6	23	27	17	15	18
5	21	24	15	14	16
4	19	22	14	12	14
3	17	20	12	11	13
2	16	18	—	—	12
1	15	17	11	10	11
0	14	16	10	9	10

CHART 31-3

Greenhouse gas score.

Source: Courtesy of the Environmental Protection Agency (EPA).

*CNG assumes a gallon equivalent of 121.5 cubic feet.

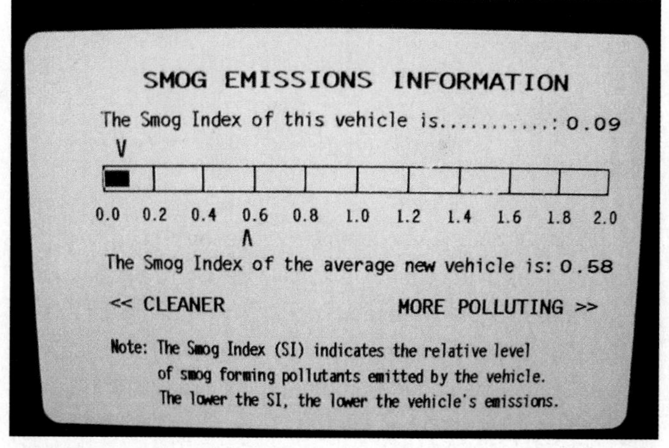

FIGURE 31-2 This label on a Toyota Camry hybrid shows the relative smog-producing emissions, but this does not include carbon dioxide (CO_2), which may increase global warming.

SMOG EMISSION INFORMATION New vehicles are equipped with a sticker that shows the relative level of smog-causing emissions created by the vehicle compared to others on the market. Smog-causing emissions include unburned hydrocarbons (HC) and oxides of nitrogen (NO_X). ● **SEE FIGURE 31-2.**

CALIFORNIA STANDARDS The pre-2004 California Air Resources Board (CARB) standards as a whole were known as LEV I. Within that, there were four possible ratings: Tier 1, TLEV, LEV, and ULEV. The newest CARB rating system (since January 1, 2004) is known as LEV II. Within that rating system there are three primary ratings: LEV, ULEV, and SULEV. States other than California are given the option to use the federal EPA standards, or they can adopt California's standards.

TIER 3 STANDARDS

Starting in 2017, the Tier 3 sets new vehicle emissions standards and lowers the sulfur content of gasoline from the current level of 30 ppm to 10 ppm because the vehicle and its fuel are an integrated system. Tier 3 is designed to match with the California Air Resources Board (CARB) Low Emission Vehicle (LEV III) program so vehicle manufacturers can sell the same vehicles in all 50 states.

EUROPEAN STANDARDS

Europe has its own set of standards that vehicles must meet, which includes the following tiers:

- Euro I (1992–1995)
- Euro II (1995–1999)
- Euro III (1999–2005)
- Euro IV (2005–2008)
- Euro V (2008+)

EXHAUST ANALYSIS TESTING

The Clean Air Act Amendments require enhanced I/M programs in areas of the country that have the worst air quality and the Northeast Ozone Transport region. The states must submit to the EPA a **State Implementation Plan (SIP)** for their programs. Each enhanced I/M program is required to include as a minimum the following items:

- Computerized emission analyzers
- Visual inspection of emission control items
- Minimum waiver limit (to be increased based on the inflation index)
- Remote on-road testing of one-half of 1% of the vehicle population
- Registration denial for vehicles not passing an I/M test
- Denial of waiver for vehicles that are under warranty or that have been tampered with
- Annual inspections
- OBD-II systems check for 1996 and newer vehicles

FEDERAL TEST PROCEDURE (FTP) The **Federal Test Procedure (FTP)** is the test used to certify all new vehicles before they can be sold. Once a vehicle meets these standards, it is certified by the EPA for sale in the United States. The FTP test procedure is a loaded-mode test lasting for a total duration of 505 seconds and is designed to simulate an urban driving trip. A cold start-up representing a morning start and a hot start after a soak period is part of the test. In addition to this drive cycle, a vehicle must undergo evaporative testing. Evaporative emissions are determined using the Sealed Housing for Evaporative Determination (SHED) test, which measures the evaporative emissions from the vehicle after a heat-up period representing a vehicle sitting in the sun. In addition, the vehicle is driven and then tested during the hot soak period.

FIGURE 31–3 Photo of a sign taken at an emissions test facility.

NOTE: A SHED is constructed entirely of stainless steel. The walls, floors, and ceiling, plus the door, are all constructed of stainless steel because it does not absorb hydrocarbons, which could offset test results.

The FTP is a much more stringent test of vehicle emissions than is any test type that uses equipment that measures percentages of exhaust gases. The federal emissions standards for each model year vehicle are the same for that model regardless of what size engine the vehicle is equipped with. This is why larger V-8 engines often are equipped with more emission control devices than smaller four- and six-cylinder engines.

I/M TEST PROGRAMS There are a variety of I/M testing programs that have been implemented by the various states. These programs may be centralized testing programs or decentralized testing programs. Each state is free to develop a testing program suitable to their needs as long as they can demonstrate to the EPA that their plan will achieve the attainment levels set by the EPA. This approach has led to a variety of different testing programs. ● **SEE FIGURE 31–3.**

FIGURE 31–4 A vehicle being tested during an enhanced emissions test.

VISUAL TAMPERING CHECKS Visual tampering checks may be part of an I/M testing program and usually include checking for the following items:

- Catalytic converter
- Fuel tank inlet restrictor
- Exhaust gas recirculation (EGR)
- Evaporative emission system
- Air-injection reaction system (AIR)
- Positive crankcase ventilation (PCV)

If any of these systems are missing, not connected, or tampered with, the vehicle will fail the emissions test and will have to be repaired/replaced by the vehicle owner. Any cost associated with repairing or replacing these components may not be used toward the waiver amount required for the vehicle.

ONE-SPEED AND TWO-SPEED IDLE TEST The one-speed and two-speed idle test measures the exhaust emissions from the tailpipe of the vehicle at idle and/or at 2500 RPM. This uses stand-alone exhaust gas sampling equipment that measures the emissions in percentages. Each state chooses the standards that the vehicle has to meet in order to pass the test. The advantage to using this type of testing is that the equipment is relatively cheap and allows states to have decentralized testing programs because many facilities can afford the necessary equipment required to perform this test.

LOADED MODE TEST The loaded mode test uses a dynamometer that places a "single weight" load on the vehicle. The load applied to the vehicle varies with the speed of the vehicle. Typically, a four-cylinder vehicle speed would be 24 mph, a six-cylinder vehicle speed would be 30 mph, and an eight-cylinder vehicle speed would be 34 mph. Conventional stand-alone sampling equipment is used to measure HC and CO emissions. This type of test is classified as a Basic I/M test by the EPA. ● **SEE FIGURE 31–4.**

ACCELERATION SIMULATION MODE (ASM) The **ASM-type** of test uses a dynamometer that applies a heavy load on the vehicle at a steady-state speed. The load applied to the vehicle is based on the acceleration rate on the second simulated hill of the FTP. This acceleration rate is 3.3 mph/sec/sec (read as 3.3 mph per second per second, which is the unit of acceleration). There are different ASM tests used by different states.

The **ASM 50/15 test** places a load of 50% on the vehicle at a steady 15 mph. This load represents 50% of the horsepower required to simulate the FTP acceleration rate of 3.3 mph/sec. This type of test produces relatively high levels of NO_X emissions; therefore, it is useful in detecting vehicles that are emitting excessive NO_X.

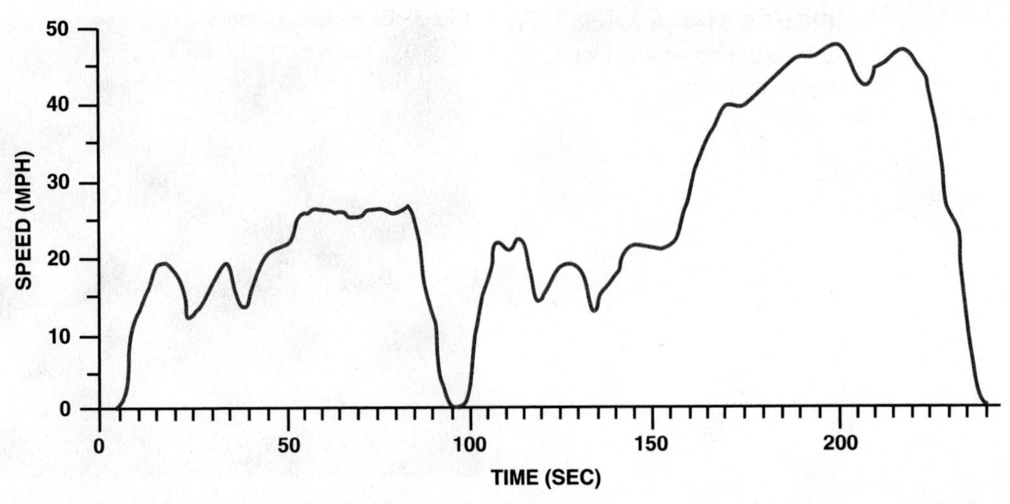

FIGURE 31–5 Trace showing the Inspection/Maintenance 240 test. The test duplicates an urban test loop around Los Angeles, California. The first "hump" in the curve represents the vehicle being accelerated to about 20 mph, then driving up a small hill to about 30 mph and coming to a stop. At about 94 seconds, the vehicle stops and again accelerates while climbing a hill and speeding up to about 50 mph during this second phase of the test.

The **ASM 25/25 test** places a 25% load on the vehicle while it is driven at a steady 25 mph. This represents 25% of the load required to simulate the FTP acceleration rate of 3.3 mph/sec. Because this applies a smaller load on the vehicle at a higher speed, it will produce a higher level of HC and CO emissions than the ASM 50/15. NO_x emissions will tend to be lower with this type of test.

I/M 240 TEST The **I/M 240 test** is the EPA's enhanced test. It is actually a portion of the 505-second FTP test used by the manufacturers to certify their new vehicles. The "240" stands for 240 seconds of drive time on a dynamometer. This is a loaded-mode transient test that uses constant volume sampling equipment to measure the exhaust emissions in mass just as is done during the FTP. The I/M 240 test simulates the first two hills of the FTP drive cycle. ● **FIGURE 31–5** shows the I/M 240 drive trace.

OBD-II TESTING In 1999, the EPA requested that states adopt OBD-II systems testing for 1996 and newer vehicles. The OBD-II system is designed to illuminate the MIL light and store trouble codes any time a malfunction exists that would cause the vehicle emissions to exceed 1 1/2 times the FTP limits. If the OBD-II system is working correctly, the system should be able to detect a vehicle failure that would cause emissions to increase to an unacceptable level. The EPA has determined that the OBD-II system should detect emission failures of a vehicle even before that vehicle would fail an emissions test of the type that most states are employing. Furthermore, the EPA has determined that, as the population of OBD-II-equipped vehicles increases and the population of older non-OBD-II-equipped vehicles decreases, tailpipe testing will no longer be necessary.

The OBD-II testing program consists of a computer that can scan the vehicle OBD-II system using the DLC connector. The technician first performs a visual check of the vehicle MIL light to determine if it is working correctly. Next, the computer is connected to the vehicle's DLC connector. The computer will scan the vehicle OBD-II system and determine if there are any codes stored that are commanding the MIL light on. In addition, it will scan the status of the readiness monitors and determine if they have all run and passed. If the readiness monitors have all run and passed, it indicates that the OBD-II system has tested all the components of the emission control system. An OBD-II vehicle would fail this OBD-II test if:

- The MIL light does not come on with the key on, engine off
- The MIL is commanded on
- A number (varies by state) of the readiness monitors has not been run

If none of these conditions are present, the vehicle will pass the emissions test.

FIGURE 31–6 A partial stream sampling exhaust probe being used to measure exhaust gases in parts per million (PPM) or percent (%).

REMOTE SENSING The EPA requires that, in high-enhanced areas, states perform on-the-road testing of vehicle emissions. The state must sample 0.5% of the vehicle population base in high-enhanced areas. This may be accomplished by using a remote sensing device. This type of sensing may be done through equipment that projects an infrared light through the exhaust stream of a passing vehicle. The reflected beam can then be analyzed to determine the pollutant levels coming from the vehicle. If a vehicle fails this type of test, the vehicle owner will receive notification in the mail that he or she must take the vehicle to a test facility to have the emissions tested.

RANDOM ROADSIDE TESTING Some states may implement random roadside testing that would usually involve visual checks of the emission control devices to detect tampering. Obviously, this method is not very popular as it can lead to traffic tie-ups and delays on the part of commuters.

Exhaust analysis is an excellent tool to use for the diagnosis of engine performance concerns. In areas of the country that require exhaust testing to be able to get license plates, exhaust analysis must be able to:

- Establish a baseline for failure diagnosis and service.
- Identify areas of engine performance that are and are not functioning correctly.
- Determine that the service and repair of the vehicle have been accomplished and are complete.

EXHAUST ANALYSIS AND COMBUSTION EFFICIENCY

A popular method of engine analysis, as well as emissions testing, involves the use of five-gas exhaust analysis equipment. ● **SEE FIGURE 31–6**. The five gases analyzed and their significance include:

- **Hydrocarbons** Hydrocarbons (HC) are unburned gasoline and are measured in parts per million (PPM). A correctly operating engine should burn (oxidize) almost all the gasoline; therefore, very little unburned gasoline should be present in the exhaust. Acceptable levels of HC are 50 PPM or less. High levels of HC could be due to excessive oil consumption caused by weak piston rings or worn valve guides. The most common cause of excessive HC emissions is a fault in the ignition system. Items that should be checked include:
  - Spark plugs
  - Spark plug wires
  - Distributor cap and rotor (if the vehicle is so equipped)
  - Ignition timing (if possible)
  - Ignition coil
- **Carbon Monoxide** Carbon monoxide (CO) is unstable and will easily combine with any oxygen to form stable carbon dioxide (CO_2). The fact that CO combines with oxygen is the reason that CO is a poisonous gas (in the lungs, it combines with oxygen to form CO_2 and deprives the brain of oxygen). CO levels of a properly operating engine should be less than 0.5%. High levels of CO can be caused by clogged or restricted crankcase ventilation

FREQUENTLY ASKED QUESTION

What Does NMHC Mean?

NMHC means **non-methane hydrocarbon** and it is the standard by which exhaust emissions testing for hydrocarbons is evaluated. Methane is natural gas and can come from animals, animal waste, and other natural sources. By not measuring methane gas, all background sources are eliminated, giving better results as to the true amount of unburned hydrocarbons that are present in the exhaust stream.

devices such as the PCV valve, hose(s), and tubes. Other items that might cause excessive CO include:

- Clogged air filter
- Incorrect idle speed
- Too-high fuel-pump pressure
- Any other items that can cause a rich condition

- **Carbon Dioxide (CO_2)** Carbon dioxide (CO_2) is the result of oxygen in the engine combining with the carbon of the gasoline. An acceptable level of CO_2 is between 12% and 15%. A high reading indicates an efficiently operating engine. If the CO_2 level is low, the mixture may be either too rich or too lean.
- **Oxygen** The next gas is oxygen (O_2). There is about 21% oxygen in the atmosphere, and most of this oxygen should be "used up" during the combustion process to oxidize all the hydrogen and carbon (hydrocarbons) in the gasoline. Levels of O_2 should be very low (about 0.5%). High levels of O_2, especially at idle, could be due to an exhaust system leak.

NOTE: Adding 10% alcohol to gasoline provides additional oxygen to the fuel and will result in lower levels of CO and higher levels of O_2 in the exhaust.

- **Oxides of Nitrogen (NO_x)** An oxide of nitrogen (NO) is a colorless, tasteless, and odorless gas when it leaves the engine, but as soon as it reaches the atmosphere and mixes with more oxygen, nitrogen oxides (NO_2) are formed. NO_2 is reddish-brown and has an acid and pungent smell. NO and NO_2 are grouped together and referred to as NO_x, where *x* represents any number of oxygen atoms. NO_x, the symbol used to represent all oxides of nitrogen, is the fifth gas commonly tested using a five-gas analyzer. The exhaust gas recirculation (EGR) system is the major controlling device limiting the formation of NO_x.

Acceptable exhaust emissions are given in **CHART 31–4**.

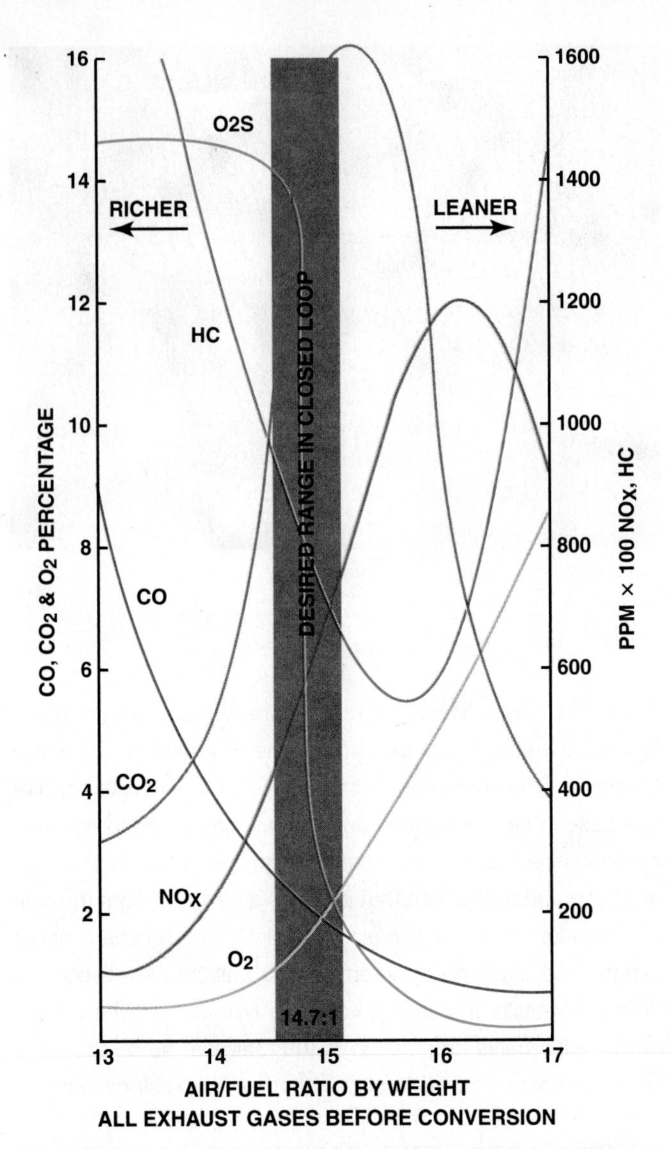

FIGURE 31–7 Exhaust emissions are very complex. When the air–fuel mixture becomes richer, some exhaust emissions are reduced, while others increase.

	WITHOUT CATALYTIC CONVERTER	WITH CATALYTIC CONVERTER
HC	300 PPM or less	30–50 PPM or less
CO	3% or less	0.3%–0.5% or less
O_2	0%–2%	0%–2%
CO_2	12%–15% or higher	12%–15% or higher
NO_x	Less than 100 PPM at idle and less than 1000 PPM at WOT	Less than 100 PPM at idle and less than 1000 PPM at WOT

CHART 31–4

Typical specifications for gases with and without a catalytic converter. If the readings are about right for a vehicle that does not have a converter yet it is equipped with one, then the catalytic converter is likely not functioning.

● SEE FIGURE 31–7.

? FREQUENTLY ASKED QUESTION

How Can My Worn-Out, Old, High-Mileage Vehicle Pass an Exhaust Emissions Test?

Age and mileage of a vehicle are generally not factors when it comes to passing an exhaust emissions test. Regular maintenance is the most important factor for passing an enhanced Inspection and Maintenance (I/M) exhaust analysis test. Failure of the vehicle owner to replace broken accessory drive belts, leaking air pump tubes, defective spark plug wires, or a cracked exhaust manifold can lead to failure of other components such as the catalytic converter. Tests have shown that if the vehicle is properly cared for, even an engine that has run 300,000 miles (483,000 km) can pass an exhaust emissions test.

HC TOO HIGH

High hydrocarbon exhaust emissions are usually caused by an engine misfire. What burns the fuel in an engine? The ignition system ignites a spark at the spark plug to ignite the *proper* mixture inside the combustion chamber. If a spark plug does not ignite the mixture, the resulting unburned fuel is pushed out of the cylinder on the exhaust stroke by the piston through the exhaust valves and into the exhaust system. Therefore, if any of the following ignition components or adjustments are not correct, excessive HC emission is likely:

1. Defective or worn spark plugs
2. Defective or loose spark plug wires
3. Defective distributor cap and/or rotor
4. Incorrect ignition timing (either too far advanced or too far retarded)
5. A lean air–fuel mixture can also cause a misfire. This condition is referred to as a lean misfire. A lean air–fuel mixture can be caused by low fuel pump pressure, a clogged fuel filter, or a restricted fuel injector.

NOTE: To make discussion easier in future reference to these items, this list of ignition components and check can be referred to simply as "spark stuff."

TECH TIP

CO Equals O_2

If the exhaust is rich, CO emissions will be higher than normal. If the exhaust is lean, O_2 emissions will be higher than normal. Therefore, if the CO reading is the same as the O_2 reading, then the engine is operating correctly. For example, if both CO and O_2 are 0.5% and the engine develops a vacuum leak, the O_2 will rise. If a fuel-pressure regulator were to malfunction, the resulting richer air–fuel mixture would increase CO emissions. Therefore, if both the rich indicator (CO) and the lean indicator (O_2) are equal, the engine is operating correctly.

CO TOO HIGH

Excessive carbon monoxide is an indication of too rich an air–fuel mixture. CO is the rich indicator. The higher the CO reading, the richer the air–fuel mixture. High concentrations of CO indicate that not enough oxygen was available for the amount of fuel. Common causes of high CO include:

- Too-high fuel-pump pressure
- Defective fuel-pressure regulator
- Clogged air filter or PCV valve

NOTE: One technician remembers "CO" as meaning "clogged oxygen" and always looks for restricted airflow into the engine whenever high CO levels are detected.

- Defective injectors

MEASURING OXYGEN (O_2) AND CARBON DIOXIDE (CO_2)

Two gas exhaust analyzers (HC and CO) work well, but both HC and CO are consumed (converted) inside the catalytic converter. The amount of leftover oxygen coming out of the tailpipe is an indication of leanness. The higher the O_2 level, the leaner the exhaust. Oxygen therefore is the **lean indicator**. Acceptable levels of O_2 are 0% to 2%.

TECH TIP

How to Find a Leak in the Exhaust System

A hole in the exhaust system can dilute the exhaust gases with additional oxygen (O_2). ● **SEE FIGURE 31–8.**

This additional O_2 in the exhaust can lead the service technician to believe that the air–fuel mixture is too lean. To help identify an exhaust leak, perform an exhaust analysis at idle and at 2500 RPM (fast idle) and compare with the following:

- If the O_2 is high at idle and at 2500 RPM, the mixture is lean at both idle and at 2500 RPM.
- If the O_2 is low at idle and high at 2500 RPM, this usually means that the vehicle is equipped with a working AIR pump.
- If the O_2 is high at idle, but okay at 2500 RPM, a hole in the exhaust or a small vacuum leak that is "covered up" at higher speed is indicated.

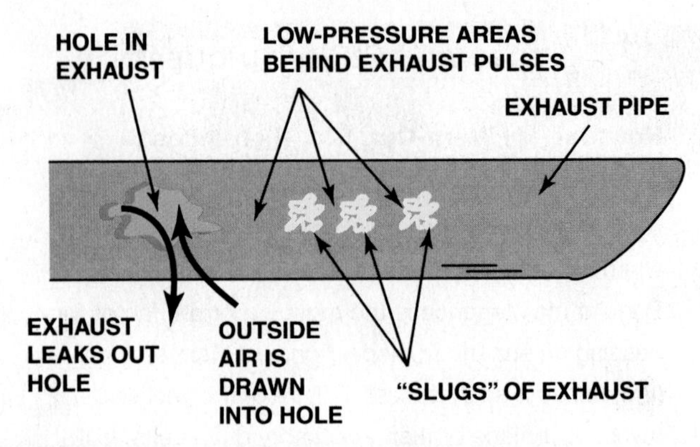

FIGURE 31–8 A hole in the exhaust system can cause outside air (containing oxygen) to be drawn into the exhaust system. This extra oxygen can be confusing to a service technician because the extra O_2 in the exhaust stream could be misinterpreted as a too-lean air–fuel mixture.

NOTE: A hole in the exhaust system can draw outside air (oxygen) into the exhaust system. Therefore, to be assured of an accurate reading, carefully check the exhaust system for leaks. Using a smoke machine is an easy method to locate leaks in the exhaust system.

Carbon dioxide (CO_2) is a measure of efficiency. The higher the level of CO_2 in the exhaust stream, the more efficiently the engine is operating. Levels of 12% to 17% are considered to be acceptable. Because CO_2 levels peak at an air–fuel mixture of 14.7:1, a lower level of CO_2 indicates either a too-rich or a too-lean condition. The CO_2 measurement by itself does not indicate which condition is present. For example:

$CO_2 = 8\%$ (This means efficiency is low and the air–fuel mixture is not correct.)

Look at O_2 and CO levels.

A high O_2 indicates lean and a high CO indicates rich.

TECH TIP

Your Nose Knows

Using the nose, a technician can often identify a major problem without having to connect the vehicle to an exhaust analyzer. For example:

- The strong smell of exhaust is due to excessive unburned hydrocarbon (HC) emissions. Look for an ignition system fault that could prevent the proper burning of the fuel. A vacuum leak could also cause a lean misfire and cause excessive HC exhaust emissions.
- If your eyes start to burn or water, suspect excessive oxides of nitrogen (NO_x) emissions. The oxides of nitrogen combine with the moisture in the eyes to form a mild solution of nitric acid. The acid formation causes the eyes to burn and water. Excessive NO_x exhaust emissions can be caused by:
  - A vacuum leak causing higher-than-normal combustion chamber temperature
  - Overadvanced ignition timing causing higher-than-normal combustion chamber temperature
  - Lack of proper amount of exhaust gas recirculation (EGR) (This is usually noticed above idle on most vehicles.)
- Dizzy feeling or headache. This is commonly caused by excessive carbon monoxide (CO) exhaust emissions. Get into fresh air as soon as possible. A probable cause of high levels of CO is an excessively rich air–fuel mixture.

PHOTOCHEMICAL SMOG FORMATION

Oxides of nitrogen are formed by high temperature—over 2,500°F (1,370°C)—and/or pressures inside the combustion chamber. Oxides of nitrogen contribute to the formation of photochemical **smog** when sunlight reacts chemically with NO_X and unburned hydrocarbons (HC). "Smog" is a term derived by combining the words *smoke* and *fog*. Ground-level ozone is a constituent of smog. **Ozone** is an enriched oxygen molecule with three atoms of oxygen (O_3) instead of the normal two atoms of oxygen (O_2).

Ozone in the upper atmosphere is beneficial because it blocks out harmful ultraviolet rays that contribute to skin cancer. However, at ground level, this ozone (smog) is an irritant to the respiratory system.

TESTING FOR OXIDES OF NITROGEN

Because the formation of NO_X occurs mostly under load, the most efficient method to test for NO_X is to use a portable exhaust analyzer that can be carried in the vehicle while the vehicle is being driven under a variety of conditions.

SPECIFICATIONS FOR NO_X From experience, a maximum reading of 1,000 parts per million (PPM) of NO_X under loaded driving conditions will generally mean that the vehicle will pass an enhanced I/M roller test. A reading of over 100 PPM at idle should be considered excessive.

TECH TIP

Check for Dog Food?

A commonly experienced problem in many parts of the country involves squirrels or other animals placing dog food into the air intake ducts of vehicles. Dog food is often found packed tight in the ducts against the air filter. An air intake restriction occurs and drives the fuel mixture richer than normal and reduces engine power and vehicle performance as well as creating high CO exhaust emissions.

CASE STUDY

The Case of the Retarded Exhaust Camshaft

A Toyota equipped with a double overhead camshaft (DOHC) inline six-cylinder engine failed the state-mandated enhanced exhaust emissions test for NO_X. The engine ran perfectly without spark knocking (ping), which is usually a major reason for excessive NO_X emissions. The technician checked the following:

- The ignition timing, which was found to be set to specifications (if too far advanced, can cause excessive NO_X)
- The cylinders, which were decarbonized using top engine cleaner
- The EGR valve, which was inspected and the EGR passages cleaned

After all the items were completed, the vehicle was returned to the inspection station where the vehicle again failed for excessive NO_X emissions (better, but still over the maximum allowable limit).

After additional hours of troubleshooting, the technician decided to go back to basics and start over again. A check of the vehicle history with the owner indicated that the only previous work performed on the engine was a replacement timing belt over a year before. The technician discovered that the exhaust cam timing was retarded two teeth, resulting in late closing of the exhaust valve. The proper exhaust valve timing resulted in a slight amount of exhaust being retained in the cylinder. This extra exhaust was added to the amount supplied by the EGR valve and helped reduce NO_X emissions. After repositioning the timing belt, the vehicle passed the emissions test well within the limits.

Summary:

- **Complaint—**Customer stated that the vehicle failed an emission test due to excessive NOx exhaust emissions.
- **Cause—**The exhaust cam was discovered to be retarded by two teeth as a result of the timing belt being incorrectly installed during a previous repair.
- **Correction—**The timing belt was properly aligned and the vehicle passed the emission test.

SEE CHART 31–5 for a summary of the exhaust gases with some possible causes.

GAS	CAUSE AND CORRECTION
High HC	Engine misfire or incomplete burning of fuel caused by: 1. Ignition system fault 2. Lean misfire 3. Too low an engine temperature (thermostat)
High CO	Rich condition caused by: 1. Leaking fuel injectors or fuel-pressure regulator 2. Clogged air filter or PCV system 3. Excessive fuel pressure
High HC and CO	Excessively rich condition caused by: 1. All items included under high CO 2. Fouled spark plugs causing a misfire to occur 3. Possible nonoperating catalytic converter
High NO_x	Excessive combustion chamber temperature caused by: 1. Nonoperating EGR valve 2. Clogged EGR passages 3. Engine operating temperature too high due to cooling system restriction, worn water pump impeller, or other faults in the cooling system 4. Lean air–fuel mixture 5. High compression due to excessive carbon buildup in the cylinders

CHART 31–5

Exhaust gases with some possible causes.

CASE STUDY

O2S Shows Rich, But Pulse Width Is Low

A service technician was attempting to solve a driveability problem. The computer did not indicate any diagnostic trouble codes (DTCs). A check of the oxygen sensor voltage indicated a higher-than-normal reading almost all the time. The pulse width to the port injectors was lower than normal. The lower-than-normal pulse width indicates that the computer is attempting to reduce fuel flow into the engine by decreasing the amount of on-time for all the injectors.

What could cause a rich mixture if the injectors were being commanded to deliver a lean mixture? Finally the technician shut off the engine and took a careful look at the entire fuel-injection system. When the vacuum hose was removed from the fuel-pressure regulator, fuel was found dripping from the vacuum hose. The problem was a defective fuel-pressure regulator that allowed an uncontrolled amount of fuel to be drawn by the intake manifold vacuum into the cylinders. While the computer tried to reduce fuel by reducing the pulse width signal to the injectors, the extra fuel being drawn directly from the fuel rail caused the engine to operate with too rich an air–fuel mixture.

Summary:

- **Complaint**—Customer stated that the engine did not perform correctly.
- **Cause**—No stored diagnostic trouble codes (DTCs) were found but the oxygen sensor reading was higher than normal indicating that the exhaust air-fuel mixture was too rich.
- **Correction**—The fuel pressure regulator was found to be leaking causing fuel to be drawn into the intake causing the richer- then-normal air-fuel mixture. Replacing the fuel pressure regulator solved the driveability complaint.

SUMMARY

1. Excessive hydrocarbon (HC) exhaust emissions are created by a lack of proper combustion such as a fault in the ignition system, too lean an air–fuel mixture, or too-cold engine operation.
2. Excessive carbon monoxide (CO) exhaust emissions are usually created by a rich air–fuel mixture.
3. Excessive oxides of nitrogen (NO_x) exhaust emissions are usually created by excessive heat or pressure in the combustion chamber or a lack of the proper amount of exhaust gas recirculation (EGR).
4. Carbon dioxide (CO_2) levels indicate efficiency. The higher the CO_2, the more efficient the engine operation.
5. Oxygen (O_2) indicates leanness. The higher the O_2, the leaner the air–fuel mixture.
6. A vehicle should be driven about 20 miles, especially during cold weather, to allow the engine to be fully warm before an enhanced emissions test.

REVIEW QUESTIONS

1. List the five exhaust gases and their maximum allowable readings for a fuel-injected vehicle equipped with a catalytic converter.
2. List two causes of a rich exhaust.
3. List two causes of a lean exhaust.
4. List those items that should be checked if a vehicle fails an exhaust test for excessive NO_x emissions.

CHAPTER QUIZ

1. Technician A says that high HC emission levels are often caused by a fault in the ignition system. Technician B says that high CO_2 emissions are usually caused by a richer-than-normal air–fuel mixture. Which technician is correct?
   - **a.** Technician A only
   - **b.** Technician B only
   - **c.** Both Technicians A and B
   - **d.** Neither Technician A nor B
2. HC and CO are high and CO_2 and O_2 are low. This could be caused by a ________.
   - **a.** Rich mixture
   - **b.** Lean mixture
   - **c.** Defective ignition component
   - **d.** Clogged EGR passage
3. Which gas is generally considered to be the rich indicator? (The higher the level of this gas, the richer the air–fuel mixture.)
   - **a.** HC
   - **b.** CO
   - **c.** CO_2
   - **d.** O_2
4. Which gas is generally considered to be the lean indicator? (The higher the level of this gas, the leaner the air–fuel mixture.)
   - **a.** HC
   - **b.** CO
   - **c.** CO_2
   - **d.** O_2
5. Which exhaust gas indicates efficiency? (The higher the level of this gas, the more efficient the engine operates.)
   - **a.** HC
   - **b.** CO
   - **c.** CO_2
   - **d.** O_2
6. All of the gases are measured in percentages *except* ________.
   - **a.** HC
   - **b.** CO
   - **c.** CO_2
   - **d.** O_2
7. After the following exhaust emissions were measured, how was the engine operating?

   HC = 766 PPM CO_2 = 8.2% CO = 4.6% O_2 = 0.1%
   - **a.** Too rich
   - **b.** Too lean
8. Technician A says that carbon inside the engine can cause excessive NO_x to form. Technician B says that excessive NO_x could be caused by a cooling system fault causing the engine to operate too hot. Which technician is correct?
   - **a.** Technician A only
   - **b.** Technician B only
   - **c.** Both Technicians A and B
   - **d.** Neither Technician A nor B
9. A clogged EGR passage could cause excessive ________ exhaust emissions.
   - **a.** HC
   - **b.** CO
   - **c.** NO_x
   - **d.** CO_2
10. An ignition fault could cause excessive ________ exhaust emissions.
    - **a.** HC
    - **b.** CO
    - **c.** NO_x
    - **d.** CO_2

chapter 32

EMISSION CONTROL DEVICES OPERATION AND DIAGNOSIS

LEARNING OBJECTIVES

After studying this chapter, the reader will be able to:

1. Explain what causes smog.
2. Describe the purpose and function of the exhaust gas recirculation system.
3. Explain methods for diagnosing and testing for faults in the exhaust gas recirculation system.
4. Explain methods for diagnosing and testing faults in the PCV.
5. Describe the purpose and function of the positive crankcase ventilation and the secondary air injection (SAI) system.
6. Describe the purpose and function of the evaporative emission control system.
7. Describe how to inspect, test, and service components of the evaporative emission control system.
8. Discuss how the evaporative emission control system is tested under OBC-II regulations.
9. Describe how to inspect and test components of secondary air injection systems.
10. Describe how to diagnose emissions and driveability problems resulting from malfunctions in the secondary air injection system.
11. Describe the purpose and function of the catalytic converter.

This chapter will help you prepare for the ASE Engine Performance (A8) certification test content area "D" (Emission Control Systems).

KEY TERMS

Adsorption 512
Backpressure 508
Catalyst 505
Catalytic converter 505
Cerium 506
Delta pressure feedback EGR (DPFE) 494
EGR valve position (EVP) 494
Exhaust gas recirculation (EGR) 491
Inert 491
Leak detection pump (LDP) 515
Linear EGR 494
Negative backpressure 493
Oxygen storage capacity (OSC) 507
Palladium 505
Platinum 505
Positive backpressure 493
Preconverter 506
Pressure feedback EGR (PFE) 494
Pup converter 506
Rhodium 505
Three-way catalytic converter (TWC) 505
Washcoat 505

FIGURE 32–1 Notice the reddish-brown haze caused by nitrogen oxides that is often over many major cities.

SMOG

The common term used to describe air pollution is *smog*, a word that combines the two words *sm*oke and f*og*. Smog is formed in the atmosphere when sunlight combines with unburned fuei (hydrocarbon, or HC) and oxides of nitrogen (NO_x) produced during the combustion process inside the cylinders of an engine. Smog is ground-level ozone (O_3), a strong irritant to the lungs and eyes.

NOTE: Although upper-atmospheric ozone is desirable because it blocks out harmful ultraviolet rays from the sun, ground-level ozone is considered to be unhealthy smog.

- **HC (unburned hydrocarbons).** Excessive HC emissions (unburned fuel) are controlled by the evaporative system (charcoal canister), the positive crankcase ventilation (PCV) system, the air-pump system, and the catalytic converter.
- **CO (carbon monoxide).** Excessive CO emissions are controlled by the PCV system, the air-pump system, and the catalytic converter.
- **NO_x (oxides of nitrogen).** Excessive NO_x emissions are controlled by the exhaust gas recirculation (EGR) system and the catalytic converter. An oxide of nitrogen (NO) is a colorless, tasteless, and odorless gas when it leaves the engine, but as soon as it reaches the atmosphere and mixes with more oxygen, nitrogen oxides (NO_2) are formed, which appear as reddish-brown. ● **SEE FIGURE 32–1.**

EXHAUST GAS RECIRCULATION SYSTEMS

Exhaust gas recirculation (EGR) is an emission control that lowers the amount of nitrogen oxides (NO_x) formed during combustion. In the presence of sunlight, NO_x reacts with hydrocarbons in the atmosphere to form ozone (O_3) or photochemical smog, an air pollutant.

NO_x FORMATION Nitrogen (N_2) and oxygen (O_2) molecules are separated into individual atoms of nitrogen and oxygen during the combustion process. These then bond to form NO_x (NO, NO_2). When combustion flame-front temperatures exceed 2,500°F (1,370°C), NO_x formation increases dramatically.

CONTROLLING NO_x To handle the NO_x generated above 2,500°F (1,370°C), the most efficient method to meet NO_x emissions without significantly affecting engine performance, fuel economy, and other exhaust emissions is to use exhaust gas recirculation. The EGR system routes small quantities, usually between 6% and 10%, of exhaust gas into the intake manifold. Here, the exhaust gas mixes with and takes the place of some of the intake charge. This leaves less room for the intake charge to enter the combustion chamber. The recirculated exhaust gas is **inert** (chemically inactive) and does not enter into the combustion process. The result is a lower peak combustion temperature. As the combustion temperature is lowered, the production of oxides of nitrogen is also reduced.

The EGR system has some means of interconnecting the exhaust and intake manifolds. ● **SEE FIGURES 32–2 AND 32–3.** The interconnecting passage is controlled by the EGR valve. On V-type engines, the intake manifold crossover is used as a source of exhaust gas for the EGR system. A cast passage connects the exhaust crossover to the EGR valve. The gas is sent from the EGR valve to openings in the manifold. On inline-type engines, an external tube is generally used to carry exhaust gas to the EGR valve. This tube is often designed to be long so that the exhaust gas is cooled before it enters the EGR valve.

NOTE: The amount of EGR is subtracted from the mass airflow calculations. While the EGR gases do occupy space, they do not affect the air–fuel mixture.

EGR SYSTEM OPERATION Since small amounts of exhaust are all that is needed to lower peak combustion temperatures, the orifice that the exhaust passes through is

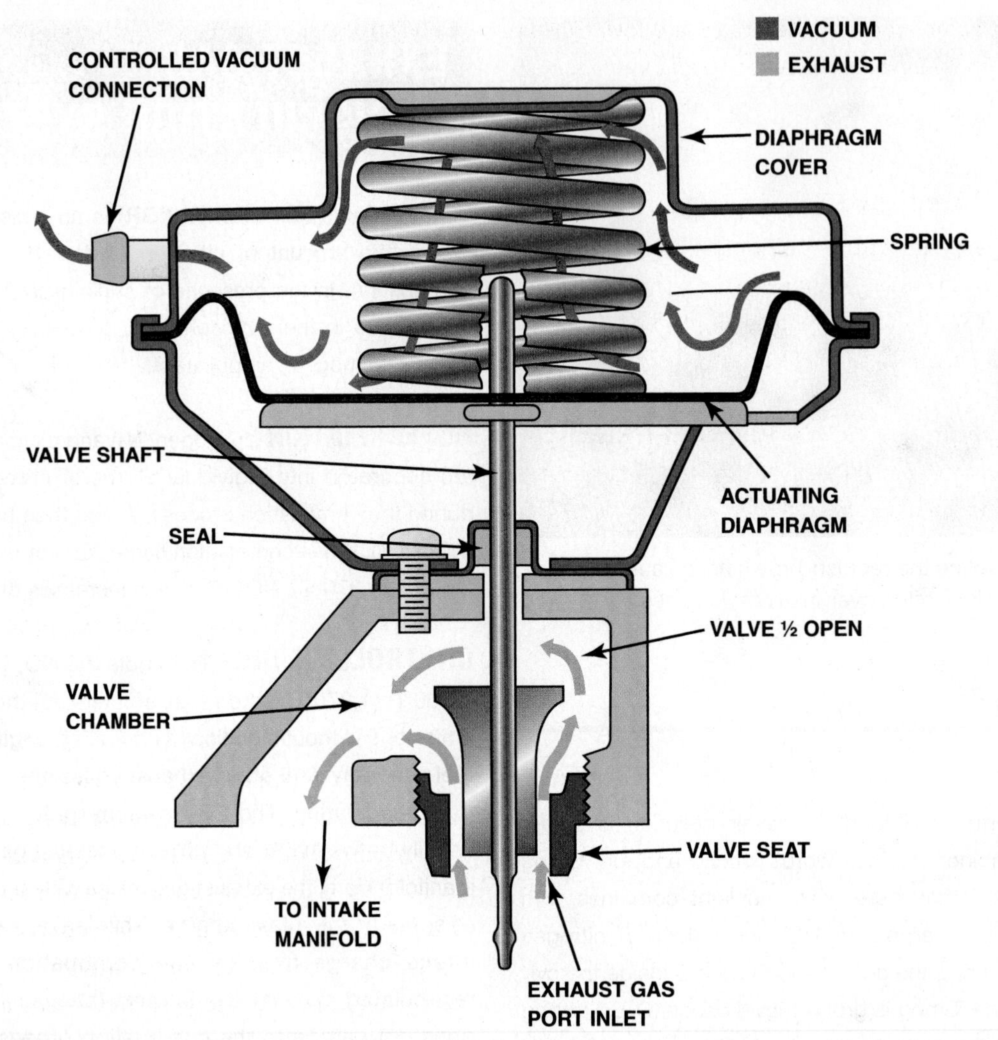

FIGURE 32–2 When the EGR valve opens, exhaust gases flow through the valve and into passages in the intake manifold.

FIGURE 32–3 A vacuum-operated EGR valve. The vacuum to the EGR valve is computer controlled by the EGR valve control solenoid.

small. Because combustion temperatures are low, EGR is usually *not* required during the following conditions:

- Idle speed
- When the engine is cold
- At wide-open throttle (WOT)

The level of NO_x emission changes according to engine speed, temperature, and load. EGR is not used at wide-open throttle (WOT) because it would reduce engine performance and the engine does not operate under these conditions for a long period of time.

In addition to lowering NO_x levels, the EGR system also helps control detonation. Detonation, or ping, occurs when high pressure and heat cause the air–fuel mixture to ignite. This uncontrolled combustion can severely damage the engine.

Using the EGR system allows for greater ignition timing advance and for the advance to occur sooner without detonation problems, which increases power and efficiency.

POSITIVE AND NEGATIVE BACKPRESSURE EGR VALVES Some EGR valves used on older engines are designed with a small valve inside that bleeds off any applied vacuum and prevents the valve from opening. These types of EGR valves require a positive backpressure in the exhaust system. This is called a **positive backpressure** EGR valve. At low engine speeds and light engine loads, the EGR system is not needed, and the backpressure in it is also low. Without sufficient backpressure, the EGR valve does not open even though vacuum may be present at the EGR valve.

On each exhaust stroke, the engine emits an exhaust "pulse." Each pulse represents a positive pressure. Behind each pulse is a small area of low pressure. Some EGR valves react to this low-pressure area by closing a small internal valve, which allows the EGR valve to be opened by vacuum. This type of EGR valve is called a **negative backpressure** EGR valve. The following conditions must occur:

1. Vacuum must be applied to the EGR valve itself. This is usually ported vacuum on some TBI fuel-injected systems. The vacuum source is often manifold vacuum and is controlled by the computer through a solenoid valve.
2. Exhaust backpressure must be present to close an internal valve inside the EGR to allow the vacuum to move the diaphragm.

NOTE: The installation of a low-restriction exhaust system could prevent the proper operation of the backpressure-controlled EGR valve.

FIGURE 32–4 An EGR valve position sensor on top of an EGR valve.

COMPUTER-CONTROLLED EGR SYSTEMS Many computer-controlled EGR systems have one or more solenoids controlling the EGR vacuum. The computer controls a solenoid to shut off vacuum to the EGR valve at cold engine temperatures, idle speed, and wide-open throttle operation. If two solenoids are used, one acts as an off/on control of supply vacuum, while the second solenoid vents vacuum when EGR flow is not desired or needs to be reduced. The second solenoid is used to control a vacuum air bleed, allowing atmospheric pressure in to modulate EGR flow according to vehicle operating conditions.

TECH TIP

Find the Root Cause

Excessive backpressure, such as that caused by a partially clogged exhaust system, could cause the plastic sensors on the EGR valve to melt. Always check for a restricted exhaust whenever replacing a failed EGR valve sensor.

The top of the valve contains a vacuum regulator and EGR pintle-position sensor in one assembly sealed inside a nonremovable plastic cover. The pintle-position sensor provides a voltage output to the PCM, which increases as the duty cycle increases, allowing the PCM to monitor valve operation. ● **SEE FIGURE 32–4.**

FIGURE 32–5 A General Motors linear EGR Valve.

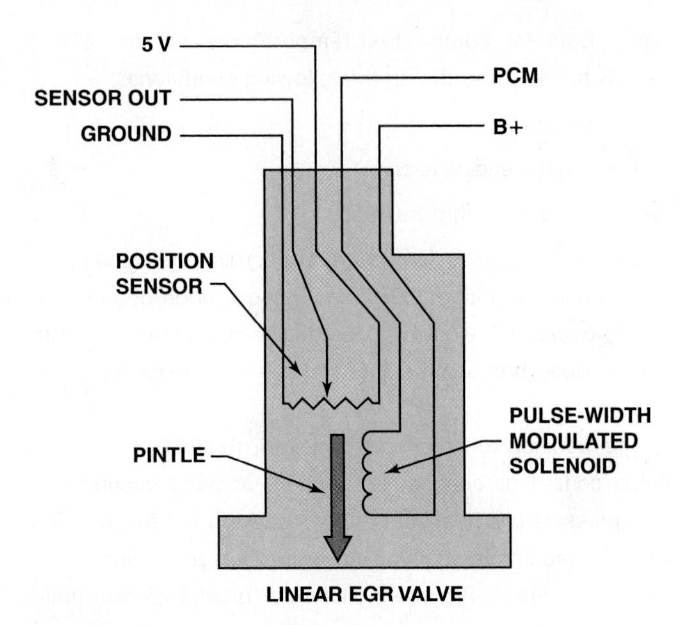

FIGURE 32–6 The EGR valve pintle is pulse-width modulated and a three-wire potentiometer provides pintle-position information back to the PCM.

EGR VALVE POSITION SENSORS Late-model, computer-controlled EGR systems use a sensor to indicate EGR operation. On-board diagnostics generation-II (OBD-II) EGR system monitors require an EGR sensor to do their job. A linear potentiometer on the top of the EGR valve stem indicates valve position for the computer. This is called an **EGR valve position (EVP)** sensor. Some later-model Ford EGR systems, however, use a feedback signal provided by an EGR exhaust backpressure sensor, which converts the exhaust backpressure to a voltage signal. This sensor is called a **pressure feedback EGR (PFE)** sensor.

DIGITAL EGR VALVES GM introduced a digital EGR valve design on some engines. Unlike vacuum-operated EGR valves, the digital EGR valve consists of three solenoids controlled by the PCM. Each solenoid controls a different size orifice in the base—small, medium, and large. The PCM controls each solenoid ground individually. It can produce any of seven different flow rates, using the solenoids to open the three valves in different combinations. The digital EGR valve offers precise control, and using a swivel pintle design helps prevent carbon deposit problems.

LINEAR EGR Most General Motors and many other vehicles use a **linear EGR** that contains a pulse-width modulated solenoid to precisely regulate exhaust gas flow and a feedback potentiometer that signals the computer regarding the actual position of the valve. ● SEE FIGURES 32–5 AND 32–6.

OBD-II EGR MONITORING STRATEGIES

In 1996, the U.S. EPA began requiring OBD-II systems in all passenger cars and most light-duty trucks. These systems include emissions system monitors that alert the driver and the technician if an emissions system is malfunctioning. To be certain the EGR system is operating, the PCM runs a functional test of the system, when specific operating conditions exist. The OBD-II system tests by opening and closing the EGR valve. The PCM monitors an EGR function sensor for a change in signal voltage. If the EGR system fails, a diagnostic trouble code (DTC) is set. If the system fails two consecutive times, the malfunction indicator light (MIL) is lit.

Chrysler monitors the difference in the exhaust oxygen sensor's voltage activity as the EGR valve opens and closes. Oxygen in the exhaust decreases when the EGR valve is open and increases when the EGR valve is closed. The PCM sets a DTC if the sensor signal does not change.

Most Fords use an EGR monitor test sensor called a **delta pressure feedback EGR (DPFE) sensor**. This sensor measures the pressure differential between two sides of a metered orifice positioned just below the EGR valve's exhaust side. Pressure between the orifice and the EGR valve decreases when the EGR opens because it becomes exposed to the lower pressure in the intake. The DPFE sensor recognizes this pressure drop, compares it to the relatively higher pressure on the exhaust side of the orifice, and signals the value of the pressure difference to the

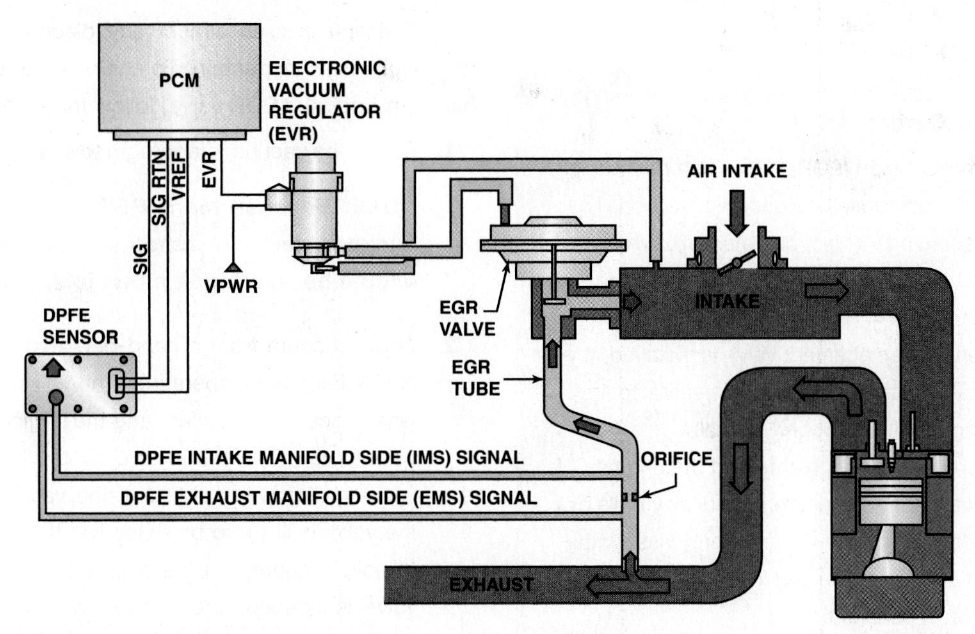

FIGURE 32–7 A DPFE sensor and related components.

DPFE EGR Sensor Chart

Pressure			Voltage
PSI	**Inch Hg**	**KPa**	**Volts**
4.34	8.83	29.81	4.56
3.25	6.62	22.36	3.54
2.17	4.41	14.90	2.51
1.08	2.21	7.46	1.48
0	0	0	0.45

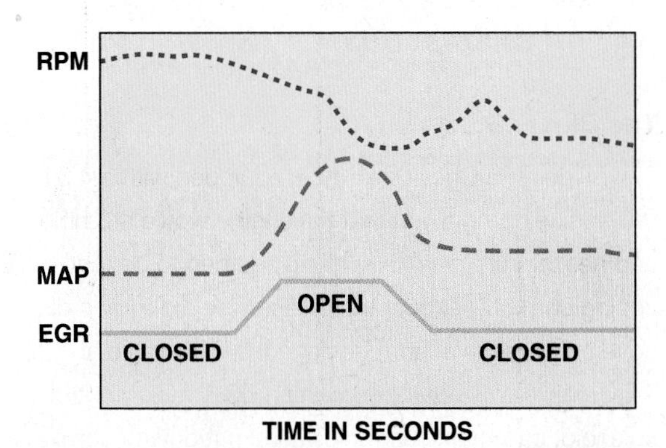

FIGURE 32–8 An OBD-II active test. The PCM opens the EGR valve and then monitors the MAP sensor and/or engine speed (RPM) to verify that it meets acceptable values.

PCM. ● **SEE FIGURE 32–7**. When the EGR valve is closed, the exhaust-gas pressure on both sides of the orifice is equal.

The OBD-II EGR monitor for this second system runs when programmed operating conditions (enable criteria) have been met. The monitor evaluates the pressure differential while the PCM commands the EGR valve to open. Like other systems, the monitor compares the measured value with the look-up table value. If the pressure differential falls outside the acceptable value, a DTC sets.

Many vehicle manufacturers use the manifold absolute pressure (MAP) sensor as the EGR monitor on some applications. After meeting the enable criteria (operating condition requirements), the EGR monitor is run. The PCM monitors the MAP sensor while it commands the EGR valve to open. The MAP sensor signal should change in response to the sudden change in manifold pressure or the fuel trim changes created by a change in the oxygen sensor voltage. If the signal value falls outside the acceptable value in the look-up table, a DTC sets. ● **SEE FIGURE 32–8**. If the EGR fails on two consecutive trips the PCM lights the MIL.

DIAGNOSING A DEFECTIVE EGR SYSTEM

If the EGR valve is not opening or the flow of the exhaust gas is restricted, then the following symptoms are likely:

- Ping (spark knock or detonation) during acceleration or during cruise (steady-speed driving)
- Excessive oxides of nitrogen (NO_X) exhaust emissions

  If the EGR valve is stuck open or partially open, then the following symptoms are likely:

- Rough idle or frequent stalling
- Poor performance/low power, especially at low engine speed

TECH TIP

Watch Out for Carbon Balls!

Exhaust gas recirculation (EGR) valves can get stuck and open partially because of a chunk of carbon. The EGR valve or solenoid will test as defective. When the valve (or solenoid) is removed, small chunks or balls of carbon often fall into the exhaust manifold passage. When the replacement valve is installed, the carbon balls can be drawn into the new valve again, causing the engine to idle roughly or stall.

To help prevent this problem, start the engine with the EGR valve or solenoid removed. Any balls or chunks of carbon will be blown out of the passage by the exhaust. Stop the engine and install the replacement EGR valve or solenoid.

CASE STUDY

The Blazer Story

The owner of a Chevrolet Blazer equipped with a 4.3 L, V-6 engine complained that the engine would stumble and hesitate at times. Everything seemed to be functioning correctly, except that the service technician discovered a weak vacuum going to the EGR valve at idle. This vehicle was equipped with an EGR valve-control solenoid, called an electronic vacuum regulator valve or EVRV by General Motors Corporation. The computer pulses the solenoid to control the vacuum that regulates the operation of the EGR valve. The technician checked service information for details on how the system worked. The technician discovered that vacuum should be present at the EGR valve only when the gear selector indicates a drive gear (drive, low, reverse). Because the technician discovered the vacuum at the solenoid to be leaking, the solenoid was obviously defective and required replacement. After replacement of the solenoid (EVRV), the hesitation problem was solved.

Summary:

- **Complaint—**Customer stated that the engine would stumble and hesitate.
- **Cause—**No stored diagnostic trouble codes were found but the service technician discovered a leaking EGR control solenoid.
- **Correction—**The EGR control solenoid was replaced and the engine was restored to normal operation.

The first step in almost any diagnosis is to perform a thorough visual inspection. To check for proper operation of a vacuum-operated EGR valve, follow these steps:

1. Check the vacuum diaphragm to see if it can hold vacuum.

   NOTE: Because many EGR valves require exhaust backpressure to function correctly, the engine should be running at a fast idle.

2. Apply vacuum from a hand-operated vacuum pump and check for proper operation. The valve itself should move when vacuum is applied, and the engine operation should be affected. The EGR valve should be able to hold the vacuum that was applied. If the vacuum drops off, then the valve is likely to be defective. If the EGR valve is able to hold vacuum, but the engine is not affected when the valve is opened, then the exhaust passage(s) must be checked for restriction. See the Tech Tip "The Snake Trick." If the EGR valve will not hold vacuum, the valve itself is likely to be defective and require replacement.
3. Connect a vacuum gauge to an intake manifold vacuum source and monitor the engine vacuum at idle (should be 17 to 21 inch Hg at sea level). Raise the speed of the engine to 2500 RPM and note the vacuum reading (should be 17 to 21 inch Hg or higher). Activate the EGR valve using a scan tool or vacuum pump, if vacuum controlled, and observe the vacuum gauge. The results are as follows:
   - The vacuum should drop 6 to 8 inch Hg.
   - If the vacuum drops less than 6 to 8 inch Hg, the valve or the EGR passages are clogged.

EGR-Related OBD-II Diagnostic Trouble Codes

Diagnostic Trouble Code	Description	Possible Causes
P0400	Exhaust gas recirculation flow problems	• EGR valve • EGR valve hose or electrical connection • Defective PCM
P0401	Exhaust gas recirculation flow insufficient	• EGR valve • Clogged EGR ports or passages
P0402	Exhaust gas recirculation flow excessive	• Stuck-open EGR valve • Vacuum hose(s) misrouted • Electrical wiring shorted

FIGURE 32–9 Removing the EGR passage plugs from the intake manifold on a Honda.

TECH TIP

The Snake Trick

The EGR passages on many intake manifolds become clogged with carbon, which reduces the flow of exhaust and the amount of exhaust gases in the cylinders. This reduction can cause spark knock (detonation) and increased emissions of oxides of nitrogen (especially important in areas with enhanced exhaust emissions testing).

To quickly and easily remove carbon from exhaust passages, cut an approximately 1 foot (30 cm) length from stranded wire, such as garage door guide wire or an old speedometer cable. Flare the end and place the end of the wire into the passage. Set your drill on reverse, turn it on, and the wire will pull its way through the passage, cleaning the carbon as it goes, just like a snake in a drain pipe. Some vehicles, such as Hondas, require that plugs be drilled out to gain access to the EGR passages, as shown in ● **FIGURE 32–9.**

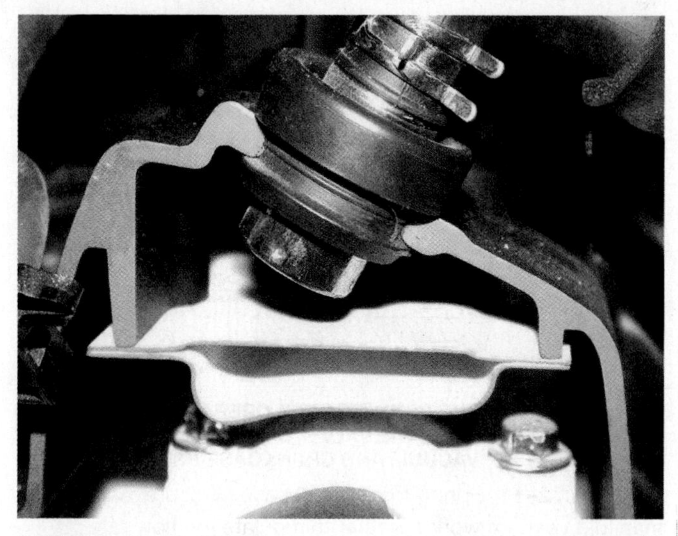

FIGURE 32–10 A PCV valve shown in a cutaway valve cover showing the baffles that prevent liquid oil from being drawn into the intake manifold.

CRANKCASE VENTILATION

The problem of crankcase ventilation has existed since the beginning of the automobile, because no piston ring, new or old, can provide a perfect seal between the piston and the cylinder wall. When an engine is running, the pressure of combustion forces the piston downward. This same pressure also forces gases and unburned fuel from the combustion chamber, past the piston rings, and into the crankcase. This process of gases leaking past the rings is called blowby, and the gases form crankcase vapors.

These combustion by-products, particularly unburned hydrocarbons caused by blowby, must be ventilated from the crankcase. However, the crankcase cannot be vented directly to the atmosphere, because the hydrocarbon vapors add to air pollution. Positive crankcase ventilation (PCV) systems were developed to ventilate the crankcase and recirculate the vapors to the engine's induction system so they can be burned in the cylinders. All systems use a PCV valve, calibrated orifice or separator, an air inlet filter, and connecting hoses. ● **SEE FIGURE 32–10**. An oil/vapor or oil/water separator is used in some systems instead of a valve or orifice, particularly with turbocharged and fuel-injected engines. The oil/vapor separator lets oil condense and drain back into the crankcase. The oil/water separator accumulates moisture and prevents it from freezing during cold engine starts.

The air for the PCV system is drawn after the air cleaner filter, which acts as a PCV filter.

NOTE: Some older designs drew from the dirty side of the air cleaner, where a separate crankcase ventilation filter was used.

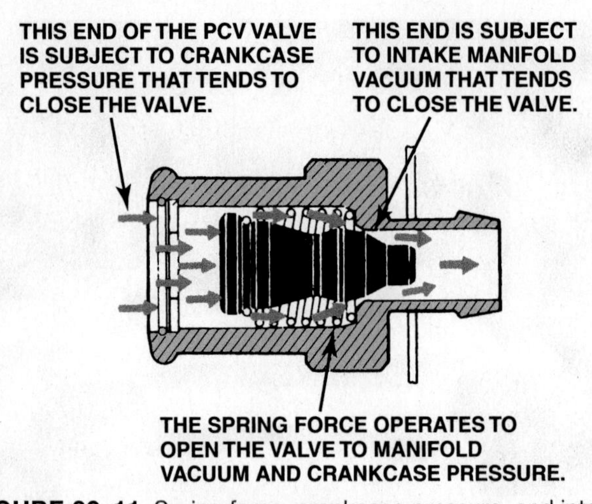

FIGURE 32–11 Spring force, crankcase pressure, and intake manifold vacuum work together to regulate the flow rate through the PCV valve.

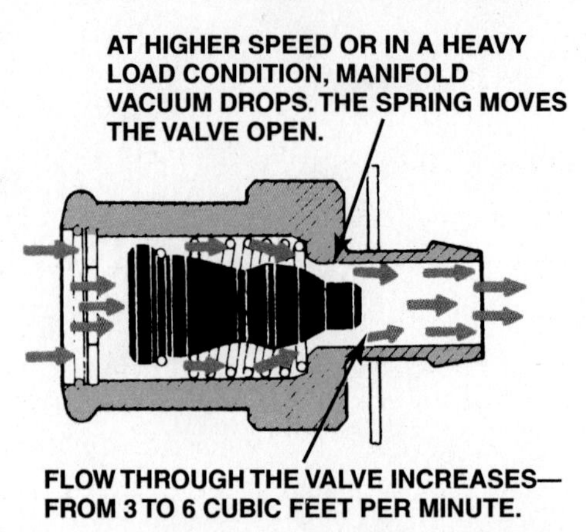

FIGURE 32–13 Air flows through the PCV valve during acceleration and when the engine is under a heavy load.

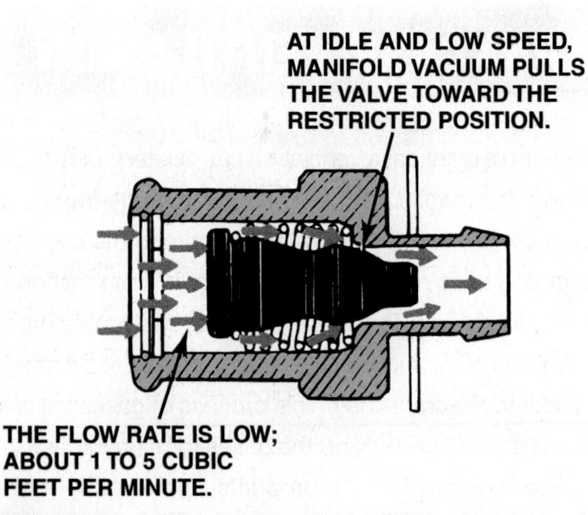

FIGURE 32–12 Air flows through the PCV valve during idle, cruising, and light-load conditions.

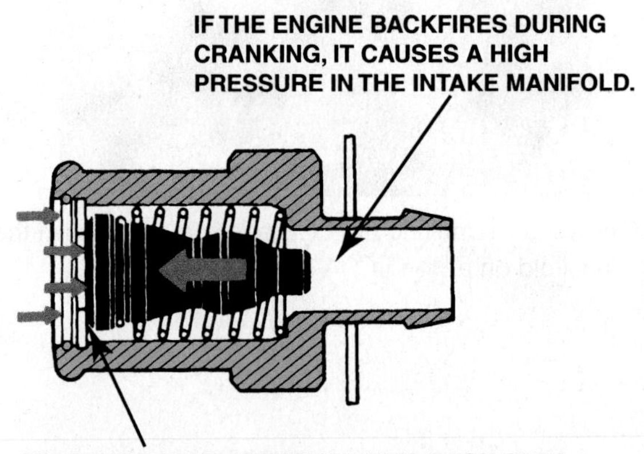

FIGURE 32–14 PCV valve operation in the event of a backfire.

PCV VALVES

The PCV valve in most systems is a one-way valve containing a spring-operated plunger that controls valve flow rate. ● **SEE FIGURE 32–11.** Flow rate is established for each engine and a valve for a different engine should not be substituted. The flow rate is determined by the size of the plunger and the holes inside the valve. PCV valves usually are located in the valve cover or intake manifold.

The PCV valve regulates airflow through the crankcase under all driving conditions and speeds. When manifold vacuum is high (at idle, cruising, and light-load operation), the PCV valve restricts the airflow to maintain a balanced air–fuel ratio. ● **SEE FIGURE 32–12.** It also prevents high intake manifold vacuum from pulling oil out of the crankcase and into the intake manifold. Under high speed or heavy loads, the valve opens and allows maximum air flow. ● **SEE FIGURE 32–13.** If the engine backfires, the valve will close instantly to prevent a crankcase explosion. ● **SEE FIGURE 32–14.**

CASE STUDY

The Whistling Engine

An older vehicle was being diagnosed for a whistling sound whenever the engine was running, especially at idle. It was finally discovered that the breather in the valve cover was plugged and caused high vacuum in the crankcase. The engine was sucking air from what was likely the rear main seal lip, making the "whistle" noise. After replacing the breather and PCV, the noise stopped.

Summary:

- **Complaint—**Customer stated that the engine made a whistling sound.
- **Cause—**The crankcase vent breather was found to be clogged causing air to be drawn into the engine through the rear main lip seal creating the whistling sound.
- **Correction—**The crankcase breather was replaced and the noise stopped.

ORIFICE-CONTROLLED SYSTEMS

The closed PCV system used on some four-cylinder engines contains a calibrated orifice instead of a PCV valve. The orifice may be located in the valve cover or intake manifold, or in a hose connected between the valve cover, air cleaner, and intake manifold.

While most orifice flow control systems work the same as a PCV valve system, they may not use fresh air scavenging of the crankcase. Crankcase vapors are drawn into the intake manifold in calibrated amounts depending on manifold pressure and the orifice size. If vapor availability is low, as during idle, air is drawn in with the vapors. During off-idle operation, excess vapors are sent to the air cleaner.

At idle, PCV flow is controlled by a 0.050 inch (1.3 mm) orifice. As the engine moves off-idle, ported vacuum pulls a spring-loaded valve off of its seat, allowing PCV flow to pass through a 0.090 inch (2.3 mm) orifice.

SEPARATOR SYSTEMS Turbocharged and many fuel-injected engines use an oil/vapor or oil/water separator and a calibrated orifice instead of a PCV valve. In the most common applications, the air intake throttle body acts as the source for crankcase ventilation vacuum and a calibrated orifice acts as the metering device.

TECH TIP

Check for Oil Leaks with the Engine Off

The owner of an older vehicle equipped with a V-6 engine complained to his technician that he smelled burning oil, but only *after* shutting off the engine. The technician found that the rocker cover gaskets were leaking. But why did the owner notice the smell of hot oil only when the engine was shut off? Because of the positive crankcase ventilation (PCV) system, engine vacuum tends to draw oil away from gasket surfaces. But when the engine stops, engine vacuum disappears and the oil remaining in the upper regions of the engine will tend to flow down and out through any opening. Therefore, a good technician should check an engine for oil leaks not only with the engine running but also shortly after shutdown.

PCV SYSTEM DIAGNOSIS

When intake air flows freely, the PCV system functions properly, as long as the PCV valve or orifice is not clogged. Modern engine design includes the air and vapor flow as a calibrated part of the air–fuel mixture. In fact, some engines receive as much as 30% of their idle air through the PCV system. For this reason, a flow problem in the PCV system results in driveability problems.

A blocked or plugged PCV system is a major cause of high oil consumption, and contributes to many oil leaks. Before expensive engine repairs are attempted, check the condition of the PCV system.

PCV SYSTEM PERFORMANCE CHECK A properly operating positive crankcase ventilation system should be able to draw vapors from the crankcase and into the intake manifold. If the pipes, hoses, and PCV valve itself are not restricted, vacuum is applied to the crankcase. A slight vacuum is created in the crankcase (usually less than 1 inch Hg if measured at the dipstick) and is also applied to other areas of the engine. Oil drain-back holes provide a path for oil to drain back into the oil pan. These holes also allow crankcase vacuum to be applied under the rocker covers and in the valley area of most V-type engines. There are several methods that can be used to test a PCV system.

THE RATTLE TEST The rattle test is performed by simply removing the PCV valve and shaking it in your hand.

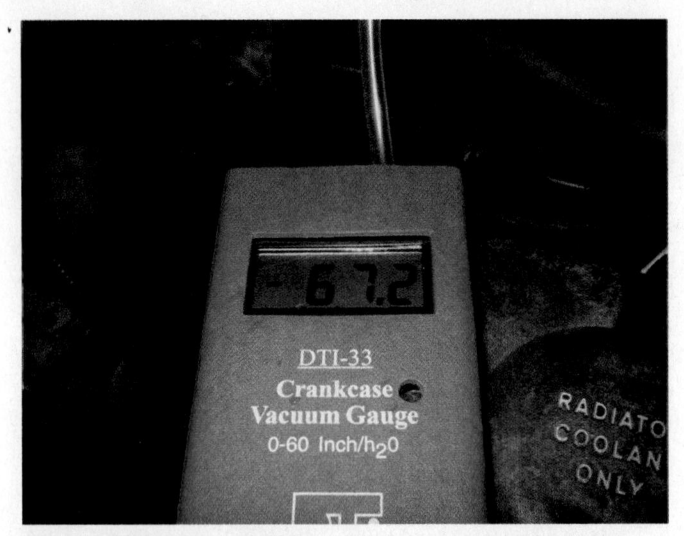

FIGURE 32–15 Using a gauge that measures vacuum in units of inches of water to test the vacuum at the dipstick tube, checking that the PCV system is capable of drawing a vacuum on the crankcase (28 inches of water equals 1 PSI or about 2 inch Hg of vacuum).

- If the PCV valve does *not* rattle, it is definitely defective and must be replaced.
- If the PCV valve *does* rattle, it does not necessarily mean that the valve is good. All PCV valves contain springs that can become weaker with age and heating and cooling cycles. Replace any PCV valve with the *exact* replacement according to vehicle manufacturers' recommended intervals (usually every three years or 36,000 miles, or 60,000 km).

THE 3 × 5 CARD TEST Remove the oil-fill cap (where oil is added to the engine) and start the engine.

NOTE: Use care on some overhead camshaft engines. With the engine running, oil may be sprayed from the open oil-fill opening.

Hold a 3 × 5 card over the opening (a dollar bill or any other piece of paper can be used for this test).

- If the PCV system, including the valve and hoses, is functioning correctly, the card should be held down on the oil-fill opening by the slight vacuum inside the crankcase.
- If the card will not stay, carefully inspect the PCV valve, hose(s), and manifold vacuum port for carbon buildup (restriction). Clean or replace as necessary.

? FREQUENTLY ASKED QUESTION

What Are the Wires for at the PCV Valve?

Ford uses an electric heater to prevent ice from forming inside the PCV valve causing blockage. Water is a by-product of combustion and resulting moisture can freeze when the outside air temperature is low.

General Motors and others clip a heater hose to the PCV hose to provide the heat needed to prevent an ice blockage.

NOTE: On some four-cylinder engines, the 3 × 5 card may vibrate on the oil-fill opening when the engine is running at idle speed. This is normal because of the time intervals between intake strokes on a four-cylinder engine.

THE SNAP-BACK TEST The proper operation of the PCV valve can be checked by placing a finger over the inlet hole in the valve when the engine is running and removing the finger rapidly. Repeat several times. The valve should "snap back." If the valve does not snap back, replace the valve.

CRANKCASE VACUUM TEST Sometimes the PCV system can be checked by testing for a weak vacuum at the oil dipstick tube using an inches-of-water manometer or gauge as follows:

STEP 1 Remove the oil-filler cap and cover the opening.

STEP 2 Remove the oil-level indicator (dipstick).

STEP 3 Connect a water manometer or gauge to the dipstick tube.

STEP 4 Start the engine and observe the gauge at idle and at 2500 RPM. ● **SEE FIGURE 32–15.**

The gauge should show some vacuum, especially at 2500 RPM. If not, carefully inspect the PCV system for blockages or other faults.

FIGURE 32–16 Most PCV valves used on newer vehicles are secured with fasteners, which makes it more difficult to disconnect and thereby less likely to increase emissions.

PCV MONITOR

Starting with 2004 and newer vehicles, all vehicles must be checked for proper operation of the PCV system. The PCV monitor will fail if the PCM detects an opening between the crankcase and the PCV valve or between the PCV valve and the intake manifold. ● **SEE FIGURE 3–16.**

PCV-Related Diagnostic Trouble Code

Diagnostic Trouble Code	Description	Possible Causes
P1480	PCV solenoid circuit fault	• Defective PCV solenoid • Loose or corroded electrical connection • Loose defective vacuum hoses/connections

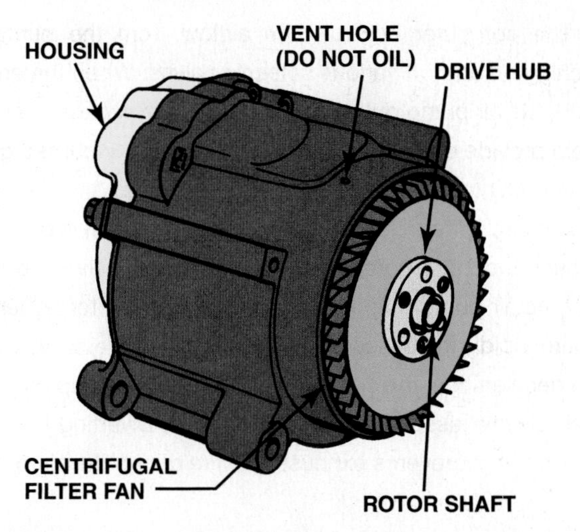

FIGURE 32–17 A typical belt-driven AIR pump. Air enters through the revolving fins behind the drive pulley. The fins act as an air filter because dirt is heavier than air and therefore the dirt is deflected off of the fins at the same time air is being drawn into the pump.

SECONDARY AIR INJECTION SYSTEM

An air pump provides the air necessary for the oxidizing process inside the catalytic converter. The system of adding air to the exhaust is commonly called secondary air injection (SAI).

NOTE: This system is commonly called AIR, meaning air injection reaction. Therefore, an AIR pump does pump air.

The AIR pump, sometimes referred to as a smog pump *or* thermactor pump, is mounted at the front of the engine and driven by a belt from the crankshaft pulley. It pulls fresh air in through an external filter and pumps the air under slight pressure to each exhaust port through connecting hoses or a manifold.

A secondary air injection system includes the following components:

- A belt-driven pump with inlet air filter (older models). ● **SEE FIGURE 32–17.**
- An electronic air pump (newer models)
- One or more air distribution manifolds and nozzles
- One or more exhaust check valves
- Connecting hoses for air distribution
- Air management valves and solenoids on all newer applications

With the introduction of NO_X reduction converters (also called dual-bed, three-way converters, or TWC), the output of the AIR pump is sent to the center of the converter where the extra air can help oxidize HC and CO into H_2O and CO_2.

The computer controls the airflow from the pump by switching on and off various solenoid valves. When the engine is cold, the air pump output is directed to the exhaust manifold to help provide enough oxygen to convert HC (unburned gasoline) and CO (carbon monoxide) to H_2O (water) and CO_2 (carbon dioxide). When the engine becomes warm and is operating in closed loop, the computer operates the air valves so as to direct the air pump output to the catalytic converter. When the vacuum rapidly increases above the normal idle level, as during rapid deceleration, the computer diverts the air pump output to the air cleaner assembly to silence the air. Diverting the air to the air cleaner prevents exhaust backfire during deceleration.

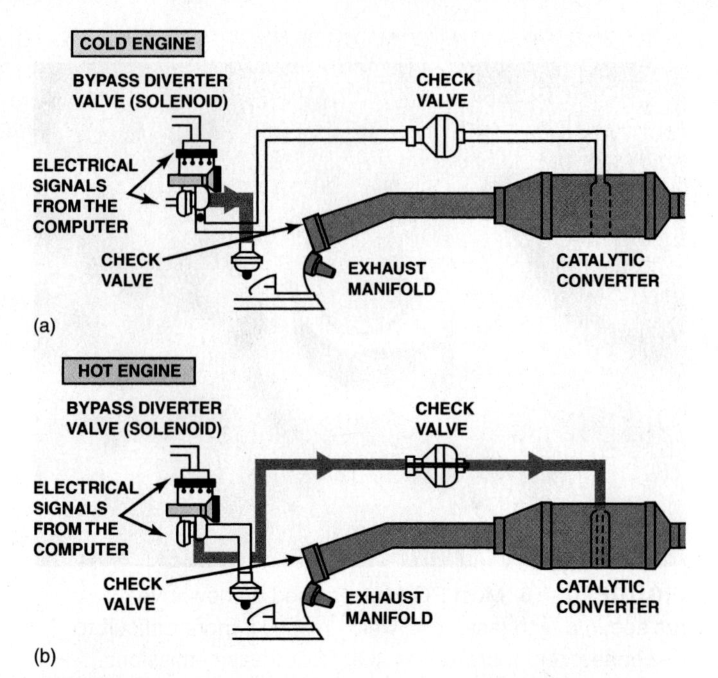

FIGURE 32–18 (a) When the engine is cold and before the oxygen sensor is hot enough to achieve closed-loop, the airflow from the air pump is directed to the exhaust manifold(s) through the one-way check valves that keep the exhaust gases from entering the switching solenoids and the pump itself. (b) When the engine achieves closed-loop, the air is directed to the catalytic converter.

AIR DISTRIBUTION MANIFOLDS AND NOZZLES

The air-injection system sends air from the pump to a nozzle installed near each exhaust port in the cylinder head. This provides equal air injection for the exhaust from each cylinder and makes it available at a point in the system where exhaust gases are the hottest.

Air is delivered to the exhaust system in one of two ways:

- An external air manifold, or manifolds, distributes the air through injection tubes with stainless steel nozzles. The nozzles are threaded into the cylinder heads or exhaust manifolds close to each exhaust valve. This method is used primarily with smaller engines.
- An internal air manifold distributes the air to the exhaust ports near each exhaust valve through passages cast in the cylinder head or the exhaust manifold. This method is used mainly with larger engines.

EXHAUST CHECK VALVES All air-injection systems use one or more one-way check valves to protect the air pump and other components from reverse exhaust flow. A check valve contains a spring-type metallic disc or reed that closes the air line under exhaust backpressure. Check valves are located between the air manifold and the diverter valve. If exhaust pressure exceeds injection pressure, or if the air pump fails, the check valve spring closes the valve to prevent reverse exhaust flow.

All air pump systems use one-way check valves to allow air to flow into the exhaust manifold and to prevent the hot exhaust from flowing into the valves on the air pump itself. **SEE FIGURE 32–18.**

NOTE: These check valves commonly fail, resulting in excessive exhaust emissions (CO especially). When the check valve fails, hot exhaust can travel up to and destroy the switching valve(s) and air pump itself.

BELT-DRIVEN AIR PUMPS The belt-driven air pump uses a centrifugal filter just behind the drive pulley. As the pump rotates, underhood air is drawn into the pump and slightly compressed. The air is then directed to:

- The exhaust manifold when the engine is cold to help oxidize CO and HC into CO_2 and H_2O
- The catalytic converter on many models to help provide the extra oxygen needed for the efficient conversion of CO and HC into CO_2 and H_2O
- The air cleaner during deceleration or wide-open throttle (WOT) engine operation

ELECTRIC MOTOR-DRIVEN AIR PUMPS This type of pump is generally used only during cold engine operation and is computer controlled. The air injection reaction (AIR) system helps reduce hydrocarbon (HC) and carbon monoxide (CO). It also helps to warm up the three-way catalytic converters

FIGURE 32–19 A typical electric motor-driven AIR pump. This unit is on a Chevrolet Corvette and only works when the engine is cold.

quickly on engine start-up so conversion of exhaust gases may occur sooner.

The AIR pump and solenoid are controlled by the PCM. The PCM turns on the AIR pump by providing the ground to complete the circuit which energizes the AIR pump solenoid relay. When air to the exhaust ports is desired, the PCM energizes the relay in order to turn on the solenoid and the AIR pump. ● **SEE FIGURE 32–19.**

The PCM turns on the AIR pump during start-up any time the engine coolant temperature is above 32°F (0°C). A typical electric AIR pump operates for a maximum of 240 seconds, or until the system enters closed-loop operation. The AIR system is disabled under the following conditions:

- The PCM recognizes a problem and sets a diagnostic trouble code.
- The AIR pump has been on for 240 seconds.
- The engine speed is more than 2825 RPM.
- The manifold absolute pressure (MAP) is less than 6 inch Hg (20 kPa).
- Increased temperature detected in three-way catalytic converter during warm-up.
- The short- and long-term fuel trim are not in their normal ranges.
- Power enrichment is detected.

If no air (oxygen) enters the exhaust stream at the exhaust ports, the HC and CO emission levels will be higher than normal. Air flowing to the exhaust ports at all times could increase temperature of the three-way catalytic converter (TWC).

The diagnostic trouble codes P0410 and/or P0418 set if there is a malfunction in the following components:

- The AIR pump
- The AIR solenoid
- The AIR pump solenoid relay
- Leaking hoses or pipes
- Leaking check valves
- The circuits going to the AIR pump and the AIR pump solenoid relay

The AIR pump is an electric-type pump that requires no periodic maintenance. To check the operation of the AIR pump, the engine should be at normal operating temperature in neutral at idle. Using a scan tool, enable the AIR pump system and watch the heated oxygen sensor (HO2S) voltages for both bank 1 and bank 2 HO2S. The HO2S voltages for both sensors should remain under 350 mV because air is being directed to the exhaust ports. If the HO2S voltages remain low during this test, the AIR pump, solenoid, and shut-off valve are operating satisfactorily. If the HO2S voltage does not remain low when the AIR pump is enabled, inspect for the following:

- Voltage at the AIR pump when energized
- A seized AIR pump
- The hoses, vacuum lines, pipes, and all connections for leaks and proper routing
- Airflow going to the exhaust ports

- AIR pump for proper mounting
- Hoses and pipes for deterioration or holes

If a leak is suspected on the pressure side of the system, or if a hose or pipe has been disconnected on the pressure side, the connections should be checked for leaks with a soapy water solution. With the AIR pump running, bubbles form if a leak exists.

The check valves should be inspected whenever the hose is disconnected or whenever check valve failure is suspected. An AIR pump that had become inoperative and had shown indications of having exhaust gases in the outlet port would indicate check valve failure.

SECONDARY AIR INJECTION SYSTEM DIAGNOSIS

The air pump system should be inspected if an exhaust emissions test failure occurs. In severe cases, the exhaust will enter the air cleaner assembly, resulting in a horribly running engine because the extra exhaust displaces the oxygen needed for proper combustion. With the engine running, check for normal operation:

Engine Operation	Normal Operation of a Typical Secondary Air Injection (SAI) System
Cold engine (open-loop operation)	Air is diverted to the exhaust manifold(s) or cylinder head
Warm engine (closed-loop operation)	Air is diverted to the catalytic converter
Deceleration	Air is diverted to the air cleaner assembly
Wide-open throttle	Air is diverted to the air cleaner assembly

VISUAL INSPECTION Carefully inspect all air injection reaction (AIR) system hoses and pipes. Any pipes that have holes and leak air or exhaust require replacement. The check valve(s) should be checked when a pump has become inoperative. Exhaust gases could have gotten past the check valve and damaged the pump. Check the drive belt on an engine-driven pump for wear and proper tension.

FOUR-GAS EXHAUST ANALYSIS An AIR system can be easily tested using an exhaust gas analyzer. Follow these steps:

1. Start the engine and allow it to run until normal operating temperature is achieved.
2. Connect the analyzer probe to the tailpipe and observe the exhaust readings for hydrocarbons (HC) and carbon monoxide (CO).
3. Using the appropriate pinch-off pliers, shut off the airflow from the AIR system. Observe the HC and CO readings. If the AIR system is working correctly, the HC and CO should increase when the AIR system is shut off.
4. Record the O_2 reading with the AIR system still inoperative. Unclamp the pliers and watch the O_2 readings. If the system is functioning correctly, the O_2 level should increase by 1% to 4%.

SAI-Related Diagnostic Trouble Codes

Diagnostic Trouble Code	Description	Possible Causes
P0411	SAI system problem	• Defective AIR solenoid • Loose or corroded electrical connections • Loose, missing, or defective rubber hose(s)

FIGURE 32–20 Most catalytic converters are located as close to the exhaust manifold as possible as seen in this display of a Chevrolet Corvette.

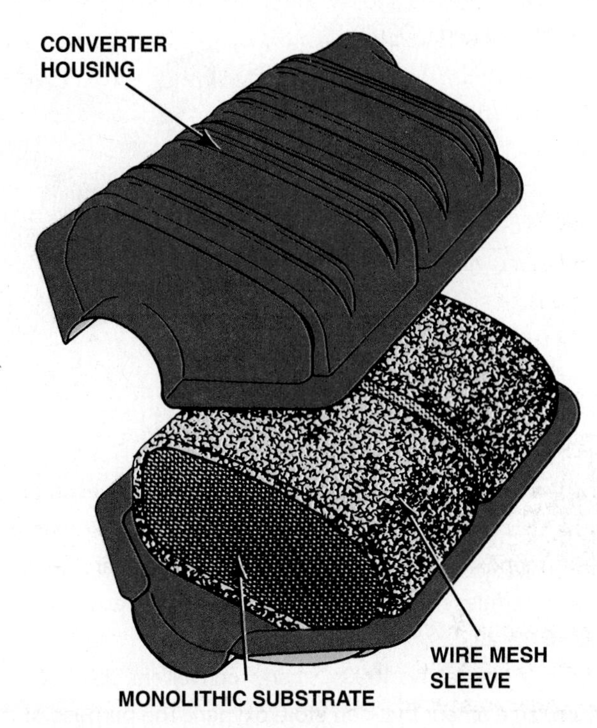

FIGURE 32–21 A typical catalytic converter with a monolithic substrate.

CATALYTIC CONVERTERS

A **catalytic converter** is an after-treatment device used to reduce exhaust emissions outside of the engine. This device is installed in the exhaust system between the exhaust manifold and the muffler, and usually is positioned beneath the passenger compartment. The location of the converter is important, since as much of the exhaust heat as possible must be retained for effective operation. The nearer it is to the engine, the better. ● **SEE FIGURE 32–20**.

CERAMIC MONOLITH CATALYTIC CONVERTER

Most catalytic converters are constructed of a ceramic material in a honeycomb shape with square openings for the exhaust gases. There are approximately 400 openings per square inch (62 per sq cm) and the wall thickness is about 0.006 inch (1.5 mm). The substrate is then coated with a porous alumina material called the **washcoat**, which makes the surface rough. The catalytic materials are then applied on top of the washcoat. The substrate is contained within a round or oval shell made by welding together two stamped pieces of aluminum or stainless steel. ● **SEE FIGURE 32–21**.

The ceramic substrate in monolithic converters is not restrictive, but the converter breaks more easily when subject to shock or severe jolts and is more expensive to manufacture. Monolithic converters can be serviced only as a unit.

An exhaust pipe is connected to the manifold or header to carry gases through a catalytic converter and then to the muffler or silencer. V-type engines usually route the exhaust into one catalytic converter.

CATALYTIC CONVERTER OPERATION The converter contains small amounts of **rhodium**, **palladium**, and **platinum**. These elements act as catalysts. A **catalyst** is an element that starts a chemical reaction without becoming a part of, or being consumed in, the process. In a **three-way catalytic converter (TWC)** all three exhaust emissions (NO_X, HC, and CO) are converted to carbon dioxide (CO_2) and water (H_2O). As the exhaust gas passes through the catalyst, oxides of nitrogen are chemically reduced (i.e., nitrogen and oxygen are separated)

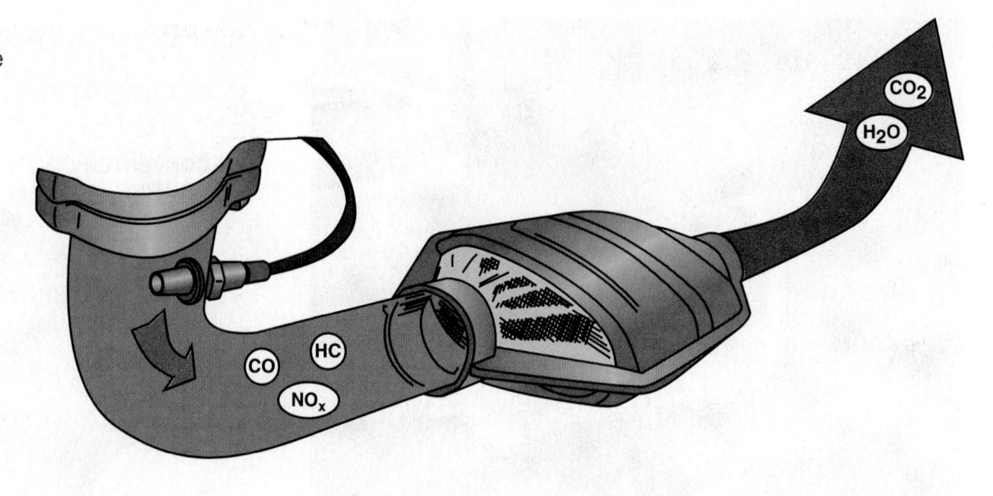

FIGURE 32–22 The three-way catalytic converter first separates the NO_X into nitrogen and oxygen and then converts the HC and CO into harmless H_2O and CO_2.

in the first section of the catalytic converter. In the second section of the catalytic converter, most of the hydrocarbons and carbon monoxide remaining in the exhaust gas are oxidized to form harmless carbon dioxide and water vapor. ● **SEE FIGURE 32–22.**

Since the early 1990s, many converters also contain **cerium**, an element that can store oxygen. The purpose of the cerium is to provide oxygen to the oxidation bed of the converter when the exhaust is rich and lacks enough oxygen for proper oxidation. When the exhaust is lean, the cerium absorbs the extra oxygen. The converter must have a varying rich-to-lean exhaust for proper operation:

- A rich exhaust is required for reduction—stripping the oxygen from the nitrogen in NO_X
- A lean exhaust is required to provide the oxygen necessary to oxidize HC and CO (combining oxygen with HC and CO to form H_2O and CO_2)

If the catalytic converter is not functioning correctly, check to see that the air–fuel mixture being supplied to the engine is correct and that the ignition system is free of defects.

CONVERTER LIGHT-OFF The catalytic converter does not work when cold and it must be heated to its light-off temperature of close to 500°F (260°C) before it starts working at 50% effectiveness. When fully effective, the converter reaches a temperature range of 900° to 1,600°F (482° to 871°C). In spite of the intense heat, however, catalytic reactions do not generate a flame associated with a simple burning reaction. Because of the extreme heat (almost as hot as combustion chamber temperatures), a converter remains hot long after the engine is shut off. Most vehicles use a series of heat shields to protect the passenger compartment and other parts of the chassis from excessive heat. Vehicles have been known to start fires because of the hot converter causing tall grass or dry leaves beneath the just-parked vehicle to ignite, especially if the engine is idling.

CONVERTER USAGE A catalytic converter must be located as close as possible to the exhaust manifold to work effectively. The farther back the converter is positioned in the exhaust system, the more gases cool before they reach the converter. Since positioning in the exhaust system affects the oxidation process, cars that use only an oxidation converter generally locate it underneath the front of the passenger compartment.

Some vehicles have used a small, quick-heating oxidation converter called a **preconverter**, **pup**, or mini-converter that connects directly to the exhaust manifold outlet. These have a small catalyst surface area close to the engine that heats up rapidly to start the oxidation process more quickly during cold engine warm-up. For this reason, they were often called light-off converters, or LOC. The oxidation reaction started in the LOC is completed by the larger main converter under the passenger compartment.

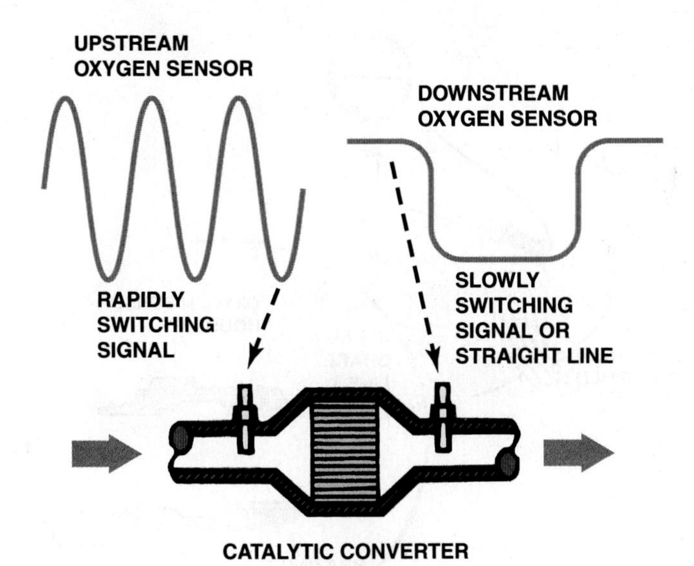

FIGURE 32–23 The OBD-II catalytic converter monitor compares the signals of the upstream and downstream O2Ss to determine converter efficiency.

OBD-II CATALYTIC CONVERTER PERFORMANCE

With OBD-II-equipped vehicles, catalytic converter performance is monitored by heated oxygen sensors (HO2Ss), both before and after the converter. The converters used on these vehicles have what is known as **OSC** or **oxygen storage capacity**. OSC is due mostly to the cerium coating in the catalyst rather than the precious metals used. When the TWC is operating as it should, the postconverter HO2S is far less active than the preconverter sensor. The converter stores, then releases the oxygen during normal reduction and oxidation of the exhaust gases, smoothing out the variations in O_2 being released.

Where a cycling sensor voltage output is expected before the converter, because of the converter action, the postconverter HO2S should read a steady signal without much fluctuation. ● **SEE FIGURE 32–23.**

CONVERTER-DAMAGING CONDITIONS

Since converters have no moving parts, they require no periodic service. Under federal law, catalyst effectiveness is warranted for 80,000 miles or eight years.

The three main causes of premature converter failure are:

- **Contamination.** Substances that can destroy the converter include exhaust that contains excess engine oil, antifreeze, sulfur (from poor fuel), and various other chemical substances.
- **Excessive temperatures.** Although a converter operates at high temperature, it can be destroyed by excessive temperatures. This most often occurs either when too much unburned fuel enters the converter, or with excessively lean mixtures. Excessive temperatures may be caused by long idling periods on some vehicles, since more heat develops at those times than when driving at normal highway speeds. Severe high temperatures can cause the converter to melt down, leading to the internal parts breaking apart and either clogging the converter or moving downstream to plug the muffler. In either case, the restricted exhaust flow severely reduces engine power.
- **Improper air–fuel mixtures.** Rich mixtures or raw fuel in the exhaust can be caused by engine misfiring, or an excessively rich air–fuel mixture resulting from a defective coolant temp sensor or defective fuel injectors. Lean mixtures are commonly caused by intake manifold leaks. When either of these circumstances occurs, the converter can become a catalytic furnace, causing the previously described damage.

To avoid excessive catalyst temperatures and the possibility of fuel vapors reaching the converter, follow these rules:

1. Do not try to start the engine by pushing the vehicle. Use jumper cables or a jump box to start the engine.
2. Do not crank an engine for more than 40 seconds when it is flooded or firing intermittently.
3. Do not turn off the ignition switch when the vehicle is in motion.
4. Do not disconnect a spark plug wire for more than 30 seconds.
5. Repair engine problems such as dieseling, misfiring, or stumbling as soon as possible.

?

FREQUENTLY ASKED QUESTION

Can a Catalytic Converter Be Defective Without Being Clogged?

Yes. Catalytic converters can fail by being chemically damaged or poisoned without being mechanically clogged. Therefore, the catalytic converter should not only be tested for physical damage (clogging) by performing a backpressure or vacuum test and a rattle test but also a test for temperature rise, usually with a pyrometer, or propane test, to check the efficiency of the converter.

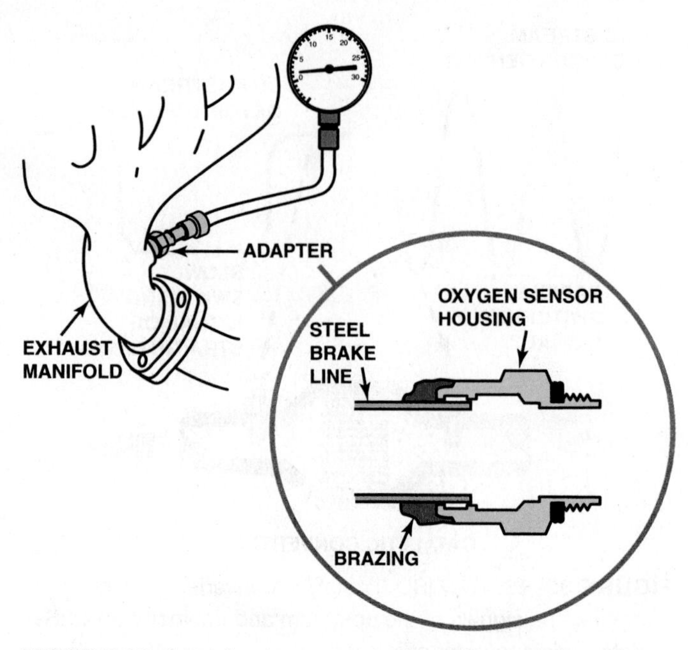

FIGURE 32–24 A backpressure tool can be made by using an oxygen sensor housing and using epoxy or braze to hold the tube to the housing.

DIAGNOSING CATALYTIC CONVERTERS

THE TAP TEST The simple tap test involves tapping (not pounding) on the catalytic converter using a rubber mallet. If the substrate inside the converter is broken, the converter will rattle when hit. If the converter rattles, a replacement converter is required.

TESTING BACKPRESSURE WITH A VACUUM GAUGE A vacuum gauge can be used to measure manifold vacuum at a high idle (2000 to 2500 RPM). If the exhaust system is restricted, pressure increases in the exhaust system. This pressure is called **backpressure**. Manifold vacuum will drop gradually if the engine is kept at a constant speed if the exhaust is restricted.

The reason the vacuum will drop is that all the exhaust leaving the engine at the higher engine speed cannot get through the restriction. After a short time (within one minute), the exhaust tends to "pile up" above the restriction and eventually remains in the cylinder of the engine at the end of the exhaust stroke. Therefore, at the beginning of the intake stroke, when the piston traveling downward should be lowering the pressure (raising the vacuum) in the intake manifold, the extra exhaust in the cylinder *lowers* the normal vacuum. If the exhaust restriction is severe enough, the vehicle can become undriveable because cylinder filling cannot occur except at idle.

TESTING BACKPRESSURE WITH A PRESSURE GAUGE Exhaust system backpressure can be measured directly by installing a pressure gauge in an exhaust opening. This can be accomplished in one of the following ways:

1. To test an oxygen sensor, remove the inside of an old, discarded oxygen sensor and thread in an adapter to convert it to a vacuum or pressure gauge.

   NOTE: An adapter can be easily made by inserting a metal tube or pipe. A short section of brake line works great. The pipe can be brazed to the oxygen sensor housing or it can be glued with epoxy. An 18-millimeter compression gauge adapter can also be adapted to fit into the oxygen sensor opening. ● SEE FIGURE 32–24.

2. To test an exhaust gas recirculation (EGR) valve, remove the EGR valve and fabricate a plate.
3. To test an air injection reaction (AIR) check valve, remove the check valve from the exhaust tubes leading to the exhaust manifold. Use a rubber cone with a tube inside to seal against the exhaust tube. Connect the tube to a pressure gauge.

At idle the maximum backpressure should be less than 1.5 PSI (10 kPa), and it should be less than 2.5 PSI (15 kPa) at 2500 RPM.

TESTING A CATALYTIC CONVERTER FOR TEMPERATURE RISE A properly working catalytic converter should be able to reduce NO_X exhaust emissions into nitrogen (N) and

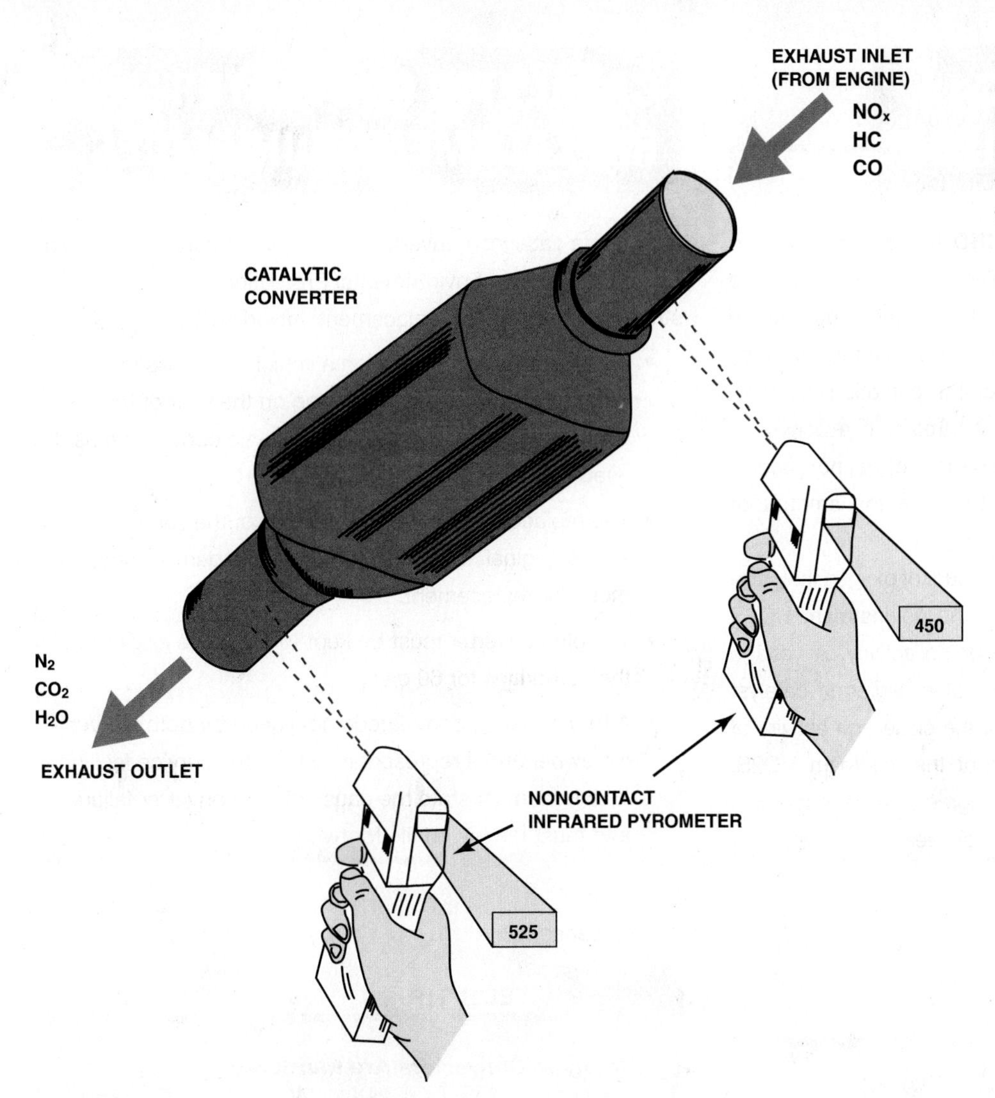

FIGURE 32–25 The temperature of the outlet should be at least 10% hotter than the temperature of the inlet. if a converter is not working, the inlet temperature will be hotter than the outlet temperature.

oxygen and oxidize unburned hydrocarbon and carbon monoxide into harmless carbon dioxide and water vapor. During these chemical processes, the catalytic converter should increase in temperature at least 10% if the converter is working properly. To test the converter, operate the engine at 2500 RPM for at least 2 minutes to fully warm up the converter. Measure the inlet and the outlet temperatures using an infrared pyrometer as shown in ● **FIGURE 32–25.**

NOTE: If the engine is extremely efficient, the converter may not have any excessive unburned hydrocarbons or carbon monoxide to convert! In this case, a spark plug wire could be grounded out using a vacuum hose and a test light to create some unburned hydrocarbon in the exhaust. Do not ground out a cylinder for longer than 10 seconds or the excessive amount of unburned hydrocarbon could overheat and damage the converter.

CATALYTIC CONVERTER EFFICIENCY TESTS

The efficiency of a catalytic converter can be determined using an exhaust gas analyzer.

OXYGEN LEVEL TEST. With the engine warm and in closed-loop, check the oxygen and carbon monoxide levels.

- If O_2 is zero, go to the snap-throttle test.
- If O_2 is greater than zero, check the CO level.
- If CO is greater than zero, the converter is *not* functioning correctly.

SNAP-THROTTLE TEST. With the engine warm and in closed-loop, snap the throttle to wide open (WOT) in park or neutral and observe the oxygen reading.

- O_2 reading should not exceed 1.2%; if it does, the converter is *not* working.
- If the O_2 rises to 1.2%, the converter may have low efficiency.
- If the O_2 remains below 1.2%, then the converter is okay.

OBD-II CATALYTIC CONVERTER MONITOR

The catalytic converter monitor of OBD II uses an upstream and downstream HO2S to test catalyst efficiency. When the engine combusts a lean air–fuel mixture, higher amounts of oxygen flow through the exhaust into the converter. The catalyst materials absorb this oxygen for the oxidation process, thereby removing it from the exhaust stream. If a converter cannot absorb enough oxygen, oxidation does not occur. Engineers established a correlation between the amount of oxygen absorbed and converter efficiency.

The OBD-II system monitors the amount of oxygen the catalyst retains. A voltage waveform from the downstream HO2S of a good catalyst should have little or no activity. A voltage waveform from the downstream HO2S of a degraded catalyst shows a lot of activity. In other words, the closer the activity of the downstream HO2S matches that of the upstream HO2S, the greater the degree of converter degradation. In operation, the OBD-II monitor compares activity between the two exhaust oxygen sensors.

TECH TIP

Aftermarket Catalytic Converters

Some replacement aftermarket (nonfactory) catalytic converters do not contain the same amount of cerium as the original part. Cerium is the element that is used in catalytic converters to store oxygen. As a result of the lack of cerium, the correlation between the oxygen storage and the conversion efficiency may be affected enough to set a false diagnostic trouble code (P0422).

CATALYTIC CONVERTER REPLACEMENT GUIDELINES

Because a catalytic converter is a major exhaust gas emission control device, the Environmental Protection Agency (EPA) has strict guidelines for its replacement, including:

- If a converter is replaced on a vehicle with less than 80,000 miles/8 years, depending on the year of the vehicle, an original equipment catalytic converter *must* be used as a replacement.
- The replacement converter must be of the same design as the original. If the original had an air pump fitting, so must the replacement.
- The old converter must be kept for possible inspection by the authorities for 60 days.
- A form must be completed and signed by both the vehicle owner and a representative from the service facility. This form must state the cause of the converter failure and must remain on file for two years.

TECH TIP

Catalytic Converters Are Murdered

Catalytic converters start a chemical reaction but do not enter into the chemical reaction. Therefore, catalytic converters do not wear out and they do not die of old age. If a catalytic converter is found to be defective (nonfunctioning or clogged), look for the *root* cause. Remember this:

> "Catalytic converters do not commit suicide—they're murdered."

Items that should be checked when a defective catalytic converter is discovered include all components of the ignition and fuel systems. Excessive unburned fuel can cause the catalytic converter to overheat and fail. The oxygen sensor must be working and fluctuating from 0.5 to 5 Hz (times per second) to provide the necessary air–fuel mixture variations for maximum catalytic converter efficiency.

Catalytic Converter-Related Diagnostic Trouble Code		
Diagnostic Trouble Code	**Description**	**Possible Causes**
P0420	Catalytic converter efficiency failure	• Engine mechanical fault • Exhaust leaks • Fuel contaminants, such as engine oil, coolant, or sulfur

FIGURE 32–26 A capless system from a Ford Flex does not use a replaceable cap and instead is spring-loaded closed.

EVAPORATIVE EMISSION CONTROL SYSTEM

The purpose of the evaporative (EVAP) emission control system is to trap and hold gasoline vapors also called volatile organic compounds, or VOCs. The charcoal canister is part of an entire system of hoses and valves called the evaporative control system. These vapors are instead routed into a charcoal canister, from where they go to the intake airflow so they are burned in the engine.

COMMON COMPONENTS The fuel tank filler caps used on vehicles with modern EVAP systems are a special design. Most EVAP fuel tank filler caps have pressure-vacuum relief built into them. When pressure or vacuum exceeds a calibrated value, the valve opens. Once the pressure or vacuum has been relieved, the valve closes. If a sealed cap is used on an EVAP system that requires a pressure-vacuum relief design, a vacuum lock may develop in the fuel system, or the fuel tank may be damaged by fuel expansion or contraction. ● **SEE FIGURE 32–26.**

? FREQUENTLY ASKED QUESTION

When Filling My Fuel Tank, Why Should I Stop When the Pump Clicks Off?

Every fuel tank has an upper volume chamber that allows for expansion of the fuel when hot. The volume of the chamber is between 10% and 20% of the volume of the tank. For example, if a fuel tank had a capacity of 20 gallons, the expansion chamber volume would be from 2 to 4 gallons. A hose is attached at the top of the chamber and vented to the charcoal canister. If extra fuel is forced into this expansion volume, liquid gasoline can be drawn into the charcoal canister. This liquid fuel can saturate the canister and create an overly rich air–fuel mixture when the canister purge valve is opened during normal vehicle operation. This extra-rich air–fuel mixture can cause the vehicle to fail an exhaust emissions test, reduce fuel economy, and possibly damage the catalytic converter. To avoid problems, simply add fuel to the next dime's worth after the nozzle clicks off. This will ensure that the tank is full, yet not overfilled.

FIGURE 32–27 A charcoal canister can be located under the hood or underneath the vehicle.

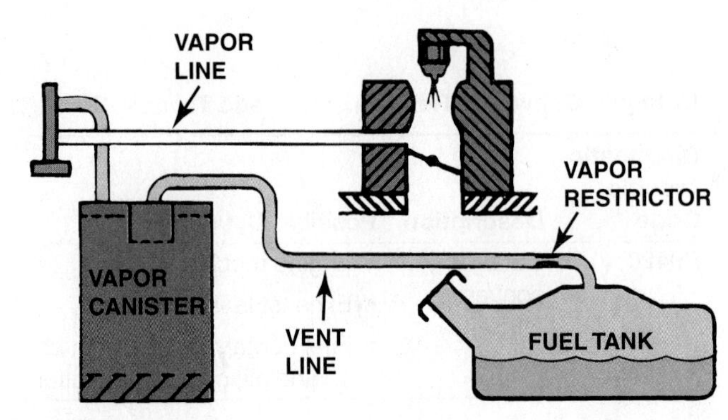

FIGURE 32–28 The evaporative emission control system includes all of the lines, hoses, and valves, plus the charcoal canister.

HOW THE EVAPORATIVE CONTROL SYSTEM WORKS

The canister is located under the hood or underneath the vehicle, and is filled with activated charcoal granules that can hold up to one-third of their own weight in fuel vapors. A vent line connects the canister to the fuel tank. ● **SEE FIGURE 32–27.**

NOTE: Some vehicles with large or dual fuel tanks may have dual canisters.

Activated charcoal is an effective vapor trap because of its great surface area. Each gram of activated charcoal has a surface area of 1,100 square meters, or more than a quarter acre. Typical canisters hold either 300 or 625 grams of charcoal *with a surface area equivalent to 80 or 165 football fields.* **Adsorption** attaches the fuel vapor molecules to the carbon surface. This attaching force is not strong, so the system purges the vapor molecules quite simply by sending a fresh airflow through the charcoal. ● **SEE FIGURE 32–28.**

VAPOR PURGING During engine operation, stored vapors are drawn from the canister into the engine through a hose connected to the throttle body or the air cleaner. This "purging" process mixes HC vapors from the canister with the existing air–fuel charge.

COMPUTER-CONTROLLED PURGE Canister purging on engines with electronic fuel management systems is regulated by the Powertrain Control Module (PCM). This is done by a microprocessor-controlled vacuum solenoid, and one or more purge valves. ● **SEE FIGURE 32–29.** Under normal conditions,

Pressure Conversions

PSI	Inch Hg	Inch H_2O
14.7	29.93	407.19
1.0	2.036	27.7
0.9	1.8	24.93
0.8	1.63	22.16
0.7	1.43	19.39
0.6	1.22	16.62
0.5	1.018	13.85
0.4	0.814	11.08
0.3	0.611	8.31
0.2	0.407	5.54
0.1	0.204	2.77
0.09	0.183	2.49
0.08	0.163	2.22
0.07	0.143	1.94
0.06	0.122	1.66
0.05	0.102	1.385

1 PSI = 28 inches of water; 1/4 PSI = 7 inches of water.

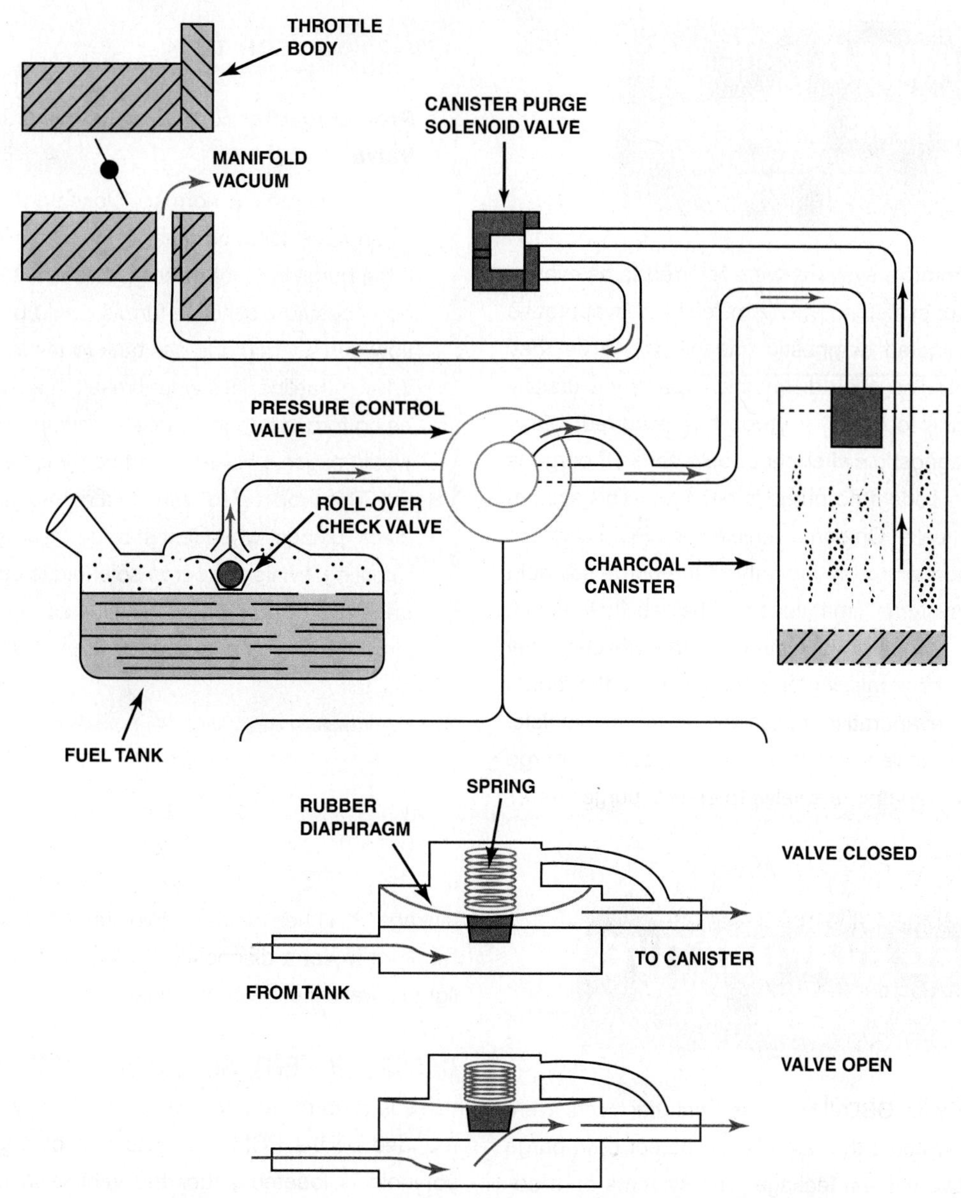

FIGURE 32–29 A typical evaporative emission control system. Note that when the computer turns on the canister purge solenoid valve, manifold vacuum draws any stored vapors from the canister into the engine. Manifold vacuum also is applied to the pressure control valve. When this valve opens, fumes from the fuel tank are drawn into the charcoal canister and eventually into the engine. When the solenoid valve is turned off (or the engine stops and there is no manifold vacuum), the pressure control valve is spring-loaded shut to keep vapors inside the fuel tank from escaping to the atmosphere.

most engine control systems permit purging only during closed-loop operation at cruising speeds. During other engine operation conditions, such as open-loop mode, idle, deceleration, or wide-open throttle, the PCM prevents canister purging.

Pressures can build inside the fuel system and are usually measured in units of inches of water, abbreviated inch H_2O (28 inches of water equals 1 PSI). Pressure buildup is a function of:

- Fuel evaporation rates (volatility)
- Gas tank size (fuel surface area and volume)
- Fuel level (liquid versus vapor)
- Fuel slosh (driving conditions)
- Temperature (ambient, in-tank, close to the tank)
- Returned fuel from the rail

NONENHANCED EVAPORATIVE CONTROL SYSTEMS

Prior to 1996, evaporative systems were referred to as evaporative (EVAP) control systems. This term refers to evaporative systems that had limited diagnostic capabilities. While they are often PCM controlled, their diagnostic capability is usually limited to their ability to detect if purge has occurred. Many systems have a diagnostic switch that could sense if purge is occurring and set a code if no purge is detected. This system does not check for leaks. On some vehicles, the PCM also has the capability of monitoring the integrity of the purge solenoid and circuit. These systems' limitations are their ability to check the integrity of the evaporative system on the vehicle. They could not detect leaks or missing or loose gas caps that could lead to excessive evaporative emissions from the vehicle. Nonenhanced evaporative systems use either a canister purge solenoid or a vapor management valve to control purge vapor.

TECH TIP

Problems After Refueling? Check the Purge Valve

The purge valve is normally closed and open only when the PCM is commanding the system to purge. If the purge solenoid were to become stuck in the open position, gasoline fumes would be allowed to flow directly from the gas tank to the intake manifold. When refueling, this would result in a lot of fumes being forced into the intake manifold and as a result would cause a hard-to-start condition after refueling. This would also result in a rich exhaust and likely black exhaust when first starting the engine after refueling. While the purge solenoid is usually located under the hood of most vehicles and is less subject to rust and corrosion than the vent valve, it can still fail.

ENHANCED EVAPORATIVE CONTROL SYSTEM

Beginning in 1996 with OBD-II vehicles, manufacturers were required to install systems that are able to detect both purge flow and evaporative system leakage. The systems on models produced between 1996 and 2000 have to be able to detect a leak as small as 0.040 inch diameter. Beginning in the model year 2000, the enhanced systems started a phase-in of 0.020 inch-diameter leak detection.

NOTE: Chrysler says that a 0.020 inch-diameter leak in an EVAP system can yield an average of about 1.35 grams of HC per mile driven.

All vehicles built after 1995 have enhanced evaporative systems that have the ability to detect purge flow and system leakage. If either of these two functions fails, the system is required to set a diagnostic trouble code and turn on the MIL light to warn the driver of the failure.

CANISTER VENT SOLENOID VALVE The canister vent valve is a normally open valve and is only closed when commanded by the PCM during testing of the system. The vent solenoid is located under the vehicle in most cases and is exposed to the environment making this valve subject to rust and corrosion.

CANISTER PURGE SOLENOID (CPS) VALVE The purge solenoid, also called the canister purge (CANP) solenoid is normally closed and is pulsed open by the PCM during purging.

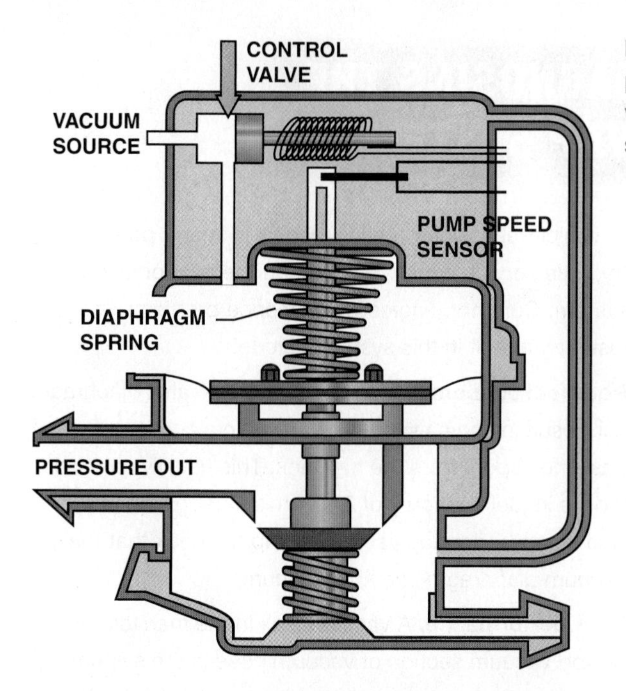

FIGURE 32–30 A leak detection pump (LDP) used on some Chrysler vehicles to pressurize (slightly) the fuel system to check for leaks.

LEAK DETECTION PUMP SYSTEM

Many Chrysler vehicles use a **leak detection pump (LDP)** as part of the evaporative control system diagnosis equipment. ● **SEE FIGURE 32–30**. The system works as follows:

- The purge solenoid is normally closed. The conventional purge solenoid is ground-side controlled by the PCM. The proportional purge solenoid is feed-side controlled by the PCM. The PCM will energize the solenoid to purge fuel vapors from the canister and to lower tank pressure.
- The vent valve in the LDP is normally open. Filtered fresh air is drawn through the LDP to the canister.
- The solenoid on the LDP normally blocks manifold vacuum from the engine. When grounded by the PCM, manifold vacuum is allowed to pass through the solenoid and into the upper diaphragm chamber. Vacuum will pull the diaphragm back against spring pressure (rated at 7.5 inch H_2O). As the diaphragm is pulled up, fresh air is drawn into the pressure side of the LDP diaphragm through the inlet reed valve. This is the LDP intake stroke. When the diaphragm is pulled up against spring pressure, the normally closed contacts of the LDP reed switch open. The LDP reed switch is the only input regarding EVAP system pressure for the PCM. With the switch contacts open, the PCM knows the diaphragm has been drawn upwards. When the LDP solenoid is de-energized by the PCM, fresh air is allowed to enter the top side of the diaphragm chamber displacing the vacuum that was there. Spring pressure (7.5 inch H_2O) forces the air in the lower diaphragm chamber out through the outlet (exhaust) reed valve. This is the LDP exhaust stroke. The LDP switch contacts close once the diaphragm returns to its original at-rest position.
- The PCM checks for EVAP leaks by first de-energizing the purge solenoid (normally closed), and then rapidly cycling the LDP solenoid and watching the LDP switch. Once pressure (7.5 inch H_2O) is built up in the system, the diaphragm will be seated upward against spring pressure. The PCM knows this since it is monitoring the LDP switch. So, the PCM compares LDP switch position against LDP solenoid cycling time to determine if leakage is present.
- When manually checking for leaks (for example, smoke machine), the vent valve must be closed. Closing of the vent valve requires that the LDP solenoid be energized and that a vacuum source be applied to the LDP solenoid. This will enable the LDP diaphragm to stroke upward, thereby allowing the vent valve spring to close the vent valve.

PUMP PERIOD The time between LDP solenoid off and LDP switch close is called the pump period. This time period is inversely proportional to the size of the leak. The shorter the pump period, the larger the leak. The longer the pump period, the smaller the leak.

EVAP large leak (>0.080): less than 0.9 seconds

EVAP medium leak (0.040 to 0.080): 0.9 to 1.2 seconds

EVAP small leak (0.020 to 0.040): 1.2 to 6 seconds

ONBOARD REFUELING VAPOR RECOVERY

The onboard refueling vapor recovery (ORVR) system was first introduced on some 1998 vehicles. Previously designed EVAP systems allowed fuel vapor to escape to the atmosphere during refueling.

The primary feature of most ORVR systems is the restricted tank filler tube, which is about 1 inch (25 mm) in diameter. This reduced-size filler tube creates an aspiration effect, which tends to draw outside air into the filler tube. During refueling, the fuel tank is vented to the charcoal canister, which captures the gas fumes and with air flowing into the filler tube, no vapors can escape to the atmosphere.

STATE INSPECTION EVAP TESTS

In some states, a periodic inspection and test of the fuel system are mandated along with a dynamometer test. The emissions inspection includes tests on the vehicle before and during the dynamometer test. Before the running test, the fuel tank and cap, fuel lines, canister, and other fuel system components must be inspected and tested to ensure that they are not leaking gasoline vapors into the atmosphere.

First, the fuel tank cap is tested to ensure that it is sealing properly and holds pressure within specs. Next, the cap is installed on the vehicle, and using a special adapter, the EVAP system is pressurized to approximately 0.5 PSI and monitored for 2 minutes. Pressure in the tank and lines should not drop below approximately 0.3 PSI.

If the cap or system leaks, hydrocarbon emissions are likely being released, and the vehicle fails the test. If the system leaks, an ultrasonic leak detector may be used to find the leak.

Finally, with the engine warmed up and running at a moderate speed, the canister purge line is tested for adequate flow using a special flow meter inserted into the system. In one example, if the flow from the canister to the intake system when the system is activated is at least one liter per minute, then the vehicle passes the canister purge test.

DIAGNOSING THE EVAP SYSTEM

Before vehicle emissions testing began in many parts of the country, little service work was done on the evaporative emission system. Common engine-performance problems that can be caused by a fault in this system include:

- **Poor fuel economy.** A leak in a vacuum-valve diaphragm can result in engine vacuum drawing in a constant flow of gasoline vapors from the fuel tank. This usually results in a drop in fuel economy of 2 to 4 miles per gallon (mpg). Use a hand-operated vacuum pump to check that the vacuum diaphragm can hold vacuum.
- **Poor performance.** A vacuum leak in the manifold or ported vacuum section of vacuum hose in the system can cause the engine to run rough. Age, heat, and time all contribute to the deterioration of rubber hoses.

Enhanced exhaust emissions (I/M-240) testing tests the evaporative emission system. A leak in the system is tested by pressurizing the entire fuel system to a level below 1 pound per square inch or 1 PSI (about 14 inches of water). The system is typically pressurized with nitrogen, a nonflammable gas that makes up 78% of our atmosphere. The pressure in the system is then shut off and the pressure monitored. If the pressure drops below a set standard, then the vehicle fails the test. This test determines if there is a leak in the system.

NOTE: To help pass the evaporative section of an enhanced emissions test, arrive at the test site with less than a half-tank of fuel. This means that the rest of the volume of the fuel tank is filled with air. It takes longer for the pressure to drop from a small leak when the volume of the air is greater compared to when the tank is full and the volume of air remaining in the tank is small.

LOCATING LEAKS IN THE SYSTEM

Leaks in the evaporative emission control system will cause the "check engine" (malfunction indictor light (MIL)) to light on most vehicles. ● **SEE FIGURE 32–31**. A leak will also cause a gas smell, which would be most noticeable if the vehicle were parked in an enclosed garage. The first step is to determine if there is leak in the system by setting the EVAP tester to either a

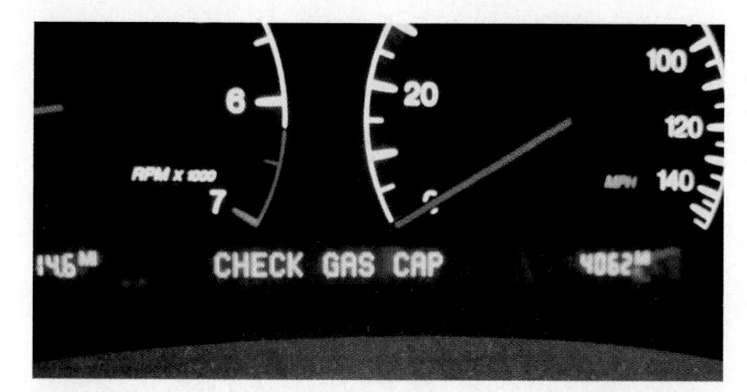

FIGURE 32–31 Some vehicles will display a message if an evaporative control system leak is detected that could be the result of a loose gas cap.

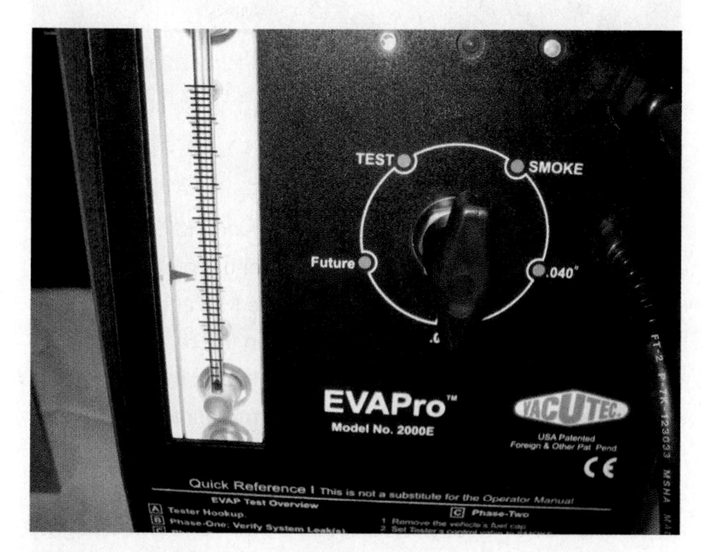

FIGURE 32–32 To test for a leak, this tester was set to the 0.020 inch hole and turned on. The ball rose in the scale on the left and the red arrow was moved to that location. If when testing the system for leaks, the ball rises higher than the arrow, then the leak is larger than 0.020 inch. If the ball does not rise to the level of the arrow, the leak is smaller than 0.020 inch.

FIGURE 32–33 This unit is applying smoke to the fuel tank through an adapter and the leak was easily found to be the gas cap seal.

FIGURE 32–34 An emission tester that uses nitrogen to pressurize the fuel system.

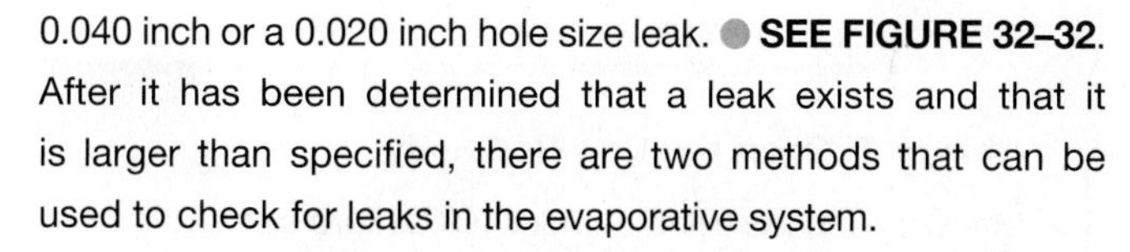

0.040 inch or a 0.020 inch hole size leak. ● **SEE FIGURE 32–32.** After it has been determined that a leak exists and that it is larger than specified, there are two methods that can be used to check for leaks in the evaporative system.

- **Smoke machine testing.** The most efficient method of leak detection is to introduce smoke under low pressure from a machine specifically designed for this purpose. ● **SEE FIGURE 32–33.**
- **Nitrogen gas pressurization.** This method uses nitrogen gas under a very low pressure (lower than 1 PSI) in the fuel system. The service technician then listens for the escaping air, using amplified headphones. ● **SEE FIGURE 32–34.**

EVAPORATIVE SYSTEM MONITOR

OBD-II computer programs not only detect faults, but also *periodically test various systems* and alert the driver before emissions-related components are harmed by system faults. Serious faults cause a blinking malfunction indicator lamp (MIL) or even an engine shutdown; less serious faults may simply store a code but not illuminate the MIL.

The OBD-II requirements did not radically affect fuel system design. However, one new component, a fuel evaporative canister purge line pressure sensor, was added for monitoring purge line pressure during tests. The OBD-II requirements state that vehicle fuel systems are to be routinely tested *while underway* by the PCM management system.

All OBD-II vehicles—during normal driving cycles and under specific conditions—experience a canister purge system pressure test, as commanded by the PCM. While the vehicle is being driven, the vapor line between the canister and the purge valve is monitored for pressure changes. When the canister purge solenoid is open, the line should be under a vacuum since vapors must be drawn from the canister into the intake system. However, when the purge solenoid is closed, there should be no vacuum in the line. The pressure sensor detects if a vacuum is present or not, and the information is compared to the command given to the solenoid. If, during the canister purge cycle, no vacuum exists in the canister purge line, a code is set indicating a possible fault, which could be caused by an inoperative or clogged solenoid or a blocked or leaking canister purge fuel line. Likewise, if vacuum exists when no command for purge is given, a stuck solenoid is evident, and a code is set.

The EVAP system monitor tests for purge volume and leaks. Most applications purge the charcoal canister by venting the vapors into the intake manifold during cruise. To do this, the PCM typically opens a solenoid-operated purge valve installed in the purge line leading to the intake manifold.

A typical EVAP monitor first closes off the system to atmospheric pressure and opens the purge valve during cruise operation. A fuel tank pressure (FTP) sensor then monitors the rate with which vacuum increases in the system. The monitor uses this information to determine the purge volume flow rate. To test for leaks, the EVAP monitor closes the purge valve, creating a completely closed system. The fuel tank pressure sensor then monitors the leak down rate. If the rate exceeds PCM-stored values, a leak greater than or equal to the OBD-II standard of 0.040 inch (1.0 mm) or 0.020 inch (0.5 mm) exists. After two consecutive failed trips testing either purge volume or the presence of a leak, the PCM lights the MIL and sets a DTC.

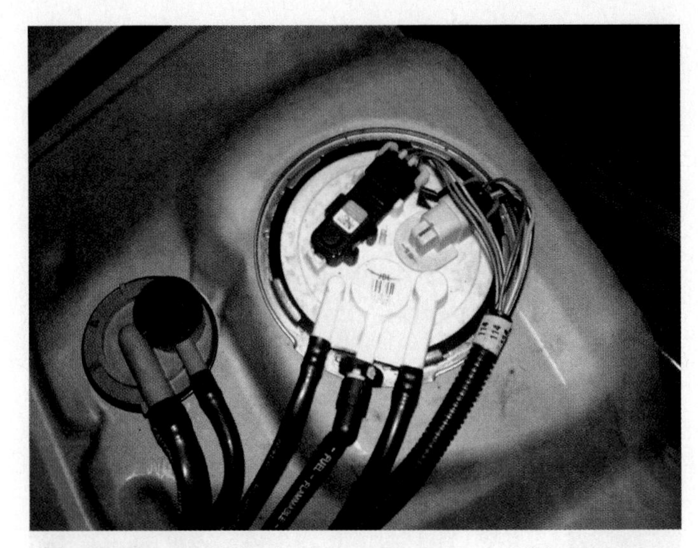

FIGURE 32–35 The fuel tank pressure sensor (black unit with three wires) looks like a MAP sensor and is usually located on top of the fuel pump module (white unit).

The fuel tank pressure sensor is often the same part as the MAP sensor, and instead of monitoring intake manifold absolute pressure, it is used to monitor fuel tank pressure. ● **SEE FIGURE 32–35.**

ENGINE OFF NATURAL VACUUM System integrity (leakage) can also be checked after the engine is shut off. The premise is that a warm evaporative system will cool down after the engine is shut off and the vehicle is stable. A slight vacuum will be created in the gas tank during this cool down period. If a specific level of vacuum is reached and maintained, the system is said to have integrity (no leakage).

TECH TIP

Always Tighten the Cap Correctly

Many diagnostic trouble codes (DTCs) are set because the gas cap has not been properly installed. To be sure that a screw-type gas cap is properly sealed, tighten the cap until it clicks three times. The clicking is a ratchet device and the clicking does not harm the cap. Therefore, if a P0440 or similar DTC is set, check the cap. ● **SEE FIGURE 32–36.**

FIGURE 32–36 This Toyota cap has a warning—the check engine light will come on if not tightened until one click.

FIGURE 32–37 The fuel level must be above 15% and below 85% before the EVAP monitor will run on most vehicles.

TYPICAL EVAP MONITOR

The PCM will run the EVAP monitor when the following enable criteria are met. Typical enable criteria include:

- Cold start
- BARO greater than 70 kPa (20.7 inch Hg or 10.2 PSI)
- IAT between 39° and 86°F at engine start-up
- ECT between 39° and 86°F at engine start-up
- ECT and IAT within 39°F of each other at engine start-up
- Fuel level within 15% to 85%
- TP sensor between 9% and 35%

RUNNING THE EVAP MONITOR The following four tests are performed during a typical GM EVAP monitor. A DTC is assigned to each test.

1. **Weak Vacuum Test (P0440—large leak).** This test identifies gross leaks. During the monitor, the vent solenoid is closed and the purge solenoid is duty cycled. The FTP should indicate a vacuum of approximately 6 to 10 inch H_2O.
2. **Small Leak Test (P0442—small leak).** After the large leak test passes, the PCM checks for a small leak by keeping the vent solenoid closed and closing the purge solenoid. The system is now sealed. The PCM measures the change in FTP voltage over time.
3. **Excess Vacuum Test (P0446).** This test checks for vent path restrictions. With the vent solenoid open and purge commanded, the PCM should not see excessive vacuum in the EVAP system. Typical EVAP system vacuum with the vent solenoid open is about 5 to 6 inch H_2O.

TECH TIP

Keep the Fuel Tank Properly Filled

Most evaporative system monitors will not run unless the fuel level is between 15% and 85%. In other words, if a driver always runs with close to an empty tank or always tries to keep the tank full, the EVAP monitor may not run. ● **SEE FIGURE 32–37.**

4. **Purge Solenoid Leak Test (P1442).** With the purge solenoid closed and vent solenoid closed, no vacuum should be present in the system. If there is vacuum present, the purge solenoid may be leaking.

EVAP System-Related Diagnostic Trouble Codes

Diagnostic Trouble Code	Description	Possible Causes
P0440	Evaporative system fault	• Loose gas cap • Defective EVAP vent • Cracked charcoal canister • EVAP vent or purge vapor line problems
P0442	Small leak detected	• Loose gas cap • Defective EVAP vent or purge solenoid • EVAP vent or purge line problems
P0446	EVAP canister vent blocked	• EVAP vent or purge solenoid electrical problems • Restricted EVAP canister vent line

SUMMARY

1. Recirculating 6% to 10% inert exhaust gases back into the intake system reduces peak temperature inside the combustion chamber and reduces NO_X exhaust emissions.
2. EGR is usually not needed at idle, at wide-open throttle, or when the engine is cold.
3. Many EGR systems use a feedback potentiometer to signal the PCM the position of the EGR valve pintle.
4. OBD-II regulation requires that the flow rate be tested and then is achieved by opening the EGR valve and observing the reaction of the MAP sensor.
5. Positive crankcase ventilation (PCV) systems use a valve or a fixed orifice to transfer and control the fumes from the crankcase back into the intake system.
6. A PCV valve regulates the flow of fumes depending on engine vacuum and seals the crankcase vent in the event of a backfire.
7. As much as 30% of the air needed by the engine at idle speed flows through the PCV system.
8. The AIR system forces air at low pressure into the exhaust to reduce CO and HC exhaust emissions.
9. A catalytic converter is an after-treatment device that reduces exhaust emissions outside of the engine. A catalyst is an element that starts a chemical reaction but is not consumed in the process.
10. The catalyst material used in a catalytic converter includes rhodium, palladium, and platinum.
11. The OBD-II system monitor compares the relative activity of a rear oxygen sensor to the precatalytic oxygen sensor to determine catalytic converter efficiency.
12. The purpose of the evaporative emission (EVAP) control system is to reduce the release of volatile organic compounds (VOCs) into the atmosphere.
13. A carbon (charcoal) canister is used to trap and hold gasoline vapors until they can be purged and run into the engine to be burned.
14. OBD-II regulation requires that the evaporative emission control system be checked for leakage and proper purge flow rates.
15. External leaks can best be located by pressurizing the fuel system with low-pressure smoke.

REVIEW QUESTIONS

1. How does the use of exhaust gas reduce NO_X exhaust emission?
2. How does the DPFE sensor work?
3. What exhaust emissions does the SAI system control?
4. How does a catalytic converter reduce NO_X to nitrogen and oxygen?
5. How does the computer monitor catalytic converter performance?
6. What components are used in a typical evaporative emission control system?
7. How does the computer control the purging of the vapor canister?

CHAPTER QUIZ

1. Two technicians are discussing clogged EGR passages. Technician A says clogged EGR passages can cause excessive NO_X exhaust emission. Technician B says that clogged EGR passages can cause the engine to ping (spark knock or detonation). Which technician is correct?
   a. Technician A only
   b. Technician B only
   c. Both Technicians A and B
   d. Neither Technician A nor B
2. An EGR valve that is partially stuck open would *most likely* cause what condition?
   a. Rough idle/stalling
   b. Excessive NO_X exhaust emissions
   c. Ping (spark knock or detonation)
   d. Missing at highway speed
3. How much air flows through the PCV system when the engine is at idle speed?
   a. 1% to 3%
   b. 5% to 10%
   c. 10% to 20%
   d. Up to 30%
4. Technician A says that if a PCV valve rattles, then it is okay and does not need to be replaced. Technician B says that if a PCV valve does not rattle, it should be replaced. Which technician is correct?
   a. Technician A only
   b. Technician B only
   c. Both Technicians A and B
   d. Neither Technician A nor B

5. The switching valves on the AIR pump have failed several times. Technician A says that a defective exhaust check valve could be the cause. Technician B says that a restricted exhaust system could be the cause. Which technician is correct?
   a. Technician A only
   b. Technician B only
   c. Both Technicians A and B
   d. Neither Technician A nor B
6. Two technicians are discussing testing a catalytic converter. Technician A says that a vacuum gauge can be used and observed to see if the vacuum drops with the engine at idle for 30 seconds. Technician B says that a pressure gauge can be used to check for backpressure. Which technician is correct?
   a. Technician A only
   b. Technician B only
   c. Both Technicians A and B
   d. Neither Technician A nor B
7. At about what temperature does oxygen combine with the nitrogen in the air to form NO_x?
   a. 500°F (260°C)
   b. 750°F (400°C)
   c. 1,500°F (815°C)
   d. 2,500°F (1,370°C)
8. A P0401 is being discussed. Technician A says that a stuck-closed EGR valve could be the cause. Technician B says that clogged EGR ports could be the cause. Which technician is correct?
   a. Technician A only
   b. Technician B only
   c. Both Technicians A and B
   d. Neither Technician A nor B
9. Which EVAP valve(s) is (are) normally closed?
   a. Canister purge valve
   b. Canister vent valve
   c. Both canister purge and canister vent valve
   d. Neither canister purge nor canister vent valve
10. What must be the fuel level before an evaporative emission monitor will run?
    a. At least 75% full
    b. Over 25%
    c. Between 15% and 85%
    d. The level of the fuel in the tank is not needed to run the monitor test

chapter 33

SCAN TOOLS AND ENGINE PERFORMANCE DIAGNOSIS

LEARNING OBJECTIVES

After studying this chapter, the reader will be able to:

1. List the steps of the diagnostic process.
2. Discuss the types of scan tools that are used to assess vehicle components.
3. Explain the troubleshooting procedures to follow if no diagnostic trouble code has been set.
4. Explain the troubleshooting procedures to follow if a diagnostic trouble code has been set.
5. Describe the methods that can be used to reprogram (reflash) a vehicle computer.

This chapter will help you prepare for the ASE computerized engine controls diagnosis (A8) certification test content area "E."

KEY TERMS

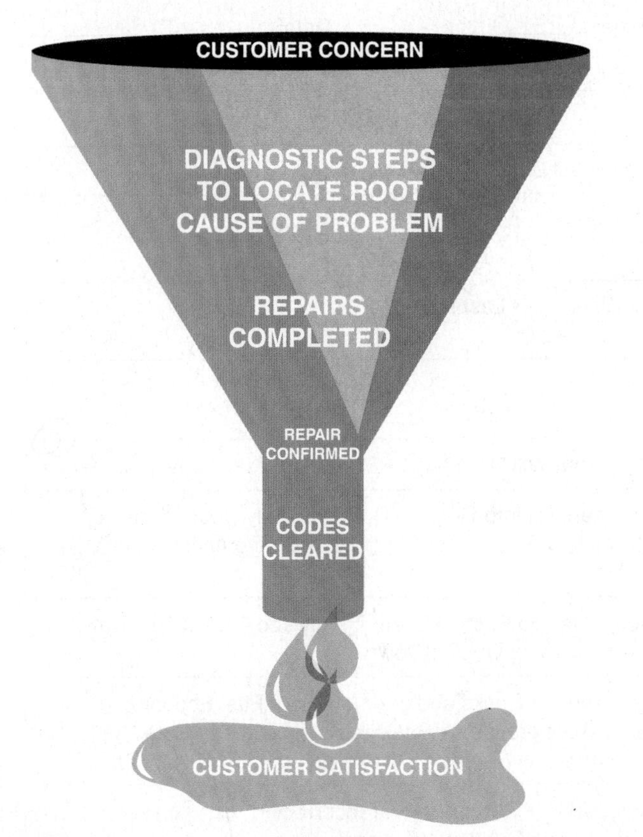

FIGURE 33–1 A funnel is one way to visualize the diagnostic process. The purpose is to narrow the possible causes of a concern until the root cause is determined and corrected.

FIGURE 33–2 Step 1 is to verify the customer concern or problem. If the problem cannot be verified, then the repair cannot be performed.

THE EIGHT-STEP DIAGNOSTIC PROCEDURE

It is important that all automotive service technicians know how to diagnose and troubleshoot engine computer systems. The diagnostic process is a strategy that eliminates known good components or systems in order to find the root cause of automotive engine performance problems. All vehicle manufacturers recommend a diagnostic procedure, and the plan suggested in this chapter combines most of the features of these plans plus additional steps developed over years of real-world problem solving.

Many different things can cause an engine performance problem or concern. The service technician has to narrow the possibilities to find the cause of the problem and correct it. A funnel is a way of visualizing a diagnostic procedure. ● **SEE FIGURE 33–1**. At the wide top are the symptoms of the problem; the funnel narrows as possible causes are eliminated until the root cause is found and corrected at the bottom of the funnel.

All problem diagnosis deals with symptoms that could be the result of many different causes. The wide range of possible solutions must be narrowed to the most likely and these must eventually be further narrowed to the actual cause. The following section describes eight steps the service technician can take to narrow the possibilities to one cause.

STEP 1 VERIFY THE PROBLEM (CONCERN) Before a minute is spent on diagnosis, be certain that a problem exists. If the problem cannot be verified, it cannot be solved or tested to verify that the repair was complete. ● **SEE FIGURE 33–2**.

The driver of the vehicle knows much about the vehicle and how it is driven. *Before* diagnosis, always ask the following questions:

- Is the malfunction indicator light (check engine) on?
- What was the temperature outside?
- Was the engine warm or cold?

ENGINE PERFORMANCE DIAGNOSIS WORKSHEET

(To Be Filled Out By the Vehicle Owner)

Name: ______________________ Mileage: ______________ Date: ______________

Make: ______________ Model: ______________ Year: ______________ Engine: ______________

(Please Circle All That Apply in All Categories)	
Describe Problem:	
When Did the Problem First Occur?	• Just Started • Last Week • Last Month • Other ______________
List Previous Repairs in the Last 6 Months:	
Starting Problems	• Will Not Crank • Cranks, but Will Not Start • Starts, but Takes a Long Time
Engine Quits or Stalls	• Right after Starting • When Put into Gear • During Steady Speed Driving • Right after Vehicle Comes to a Stop • While Idling • During Acceleration • When Parking
Poor Idling Conditions	• Is Too Slow at All Times • Is Too Fast • Intermittently Too Fast or Too Slow • Is Rough or Uneven • Fluctuates Up and Down
Poor Running Conditions	• Runs Rough • Lacks Power • Bucks and Jerks • Poor Fuel Economy • Hesitates or Stumbles on Acceleration • Backfires • Misfires or Cuts Out • Engine Knocks, Pings, Rattles • Surges • Dieseling or Run-On
Auto. Transmission Problems	• Improper Shifting (Early/Late) • Changes Gear Incorrectly • Vehicle Does Not Move when in Gear • Jerks or Bucks
Usually Occurs	• Morning • Afternoon • Anytime
Engine Temperature	• Cold • Warm • Hot
Driving Conditions During Occurrence	• Short—Less Than 2 Miles • 2–10 Miles • Long—More Than 10 Miles • Stop and Go • While Turning • While Braking • At Gear Engagement • With A/C Operating • With Headlights On • During Acceleration • During Deceleration • Mostly Downhill • Mostly Uphill • Mostly Level • Mostly Curvy • Rough Road
Driving Habits	• Mostly City Driving • Highway • Park Vehicle Inside • Park Vehicle Outside **Drive Per Day:** • Less Than 10 Miles • 10–50 • More Than 50
Gasoline Used	**Fuel Octane:** • 87 • 89 • 91 • More Than 91 **Brand:** ______________
Temperature when Problem Occurs	• 32–55° F • Below Freezing (32° F) • Above 55° F
Check Engine Light/ Dash Warning Light	• Light on Sometimes • Light on Always • Light Never On
Smells	• Hot • Gasoline • Oil Burning • Electrical
Noises	• Rattle • Knock • Squeak • Other

FIGURE 33–3 A form that the customer should fill out if there is a driveablilty concern to help the service technician more quickly find the root cause.

- Was the problem during starting, acceleration, cruise, or some other condition?
- How far had the vehicle been driven?
- Were any dash warning lights on? If so, which one(s)?
- Has there been any service or repair work performed on the vehicle lately?

NOTE: This last question is very important. Many engine performance faults are often the result of something being knocked loose or a hose falling off during repair work. Knowing that the vehicle was just serviced before the problem began may be an indicator as to where to look for the solution to a problem.

After the nature and scope of the problem are determined, the complaint should be verified before further diagnostic tests are performed. A sample form that customers could fill out with details of the problem is shown in ● **FIGURE 33–3.**

NOTE: Because drivers differ, it is sometimes the best policy to take the customer on the test-drive to verify the concern.

STEP 2 PERFORM A THOROUGH VISUAL INSPECTION AND BASIC TESTS The visual inspection is the most important aspect of diagnosis! Most experts agree that between 10% and 30% of all engine performance, problems can be found simply by performing a *thorough* visual inspection. The inspection should include the following:

FIGURE 33–4 This is what was found when removing an air filter from a vehicle that had a lack-of-power concern. Obviously, the nuts were deposited by squirrels or some other animal, blocking a lot of the airflow into the engine.

- **Check for obvious problems (basics, basics, basics).**

  Fuel leaks

  Vacuum hoses that are disconnected or split

  Corroded connectors

  Unusual noises, smoke, or smell

  Check the air cleaner and air duct (squirrels and other small animals can build nests or store dog food in them). ● **SEE FIGURE 33–4.**
- **Check everything that does and does not work.** This step involves turning things on and observing that everything is working properly.
- **Look for evidence of previous repairs.** Any time work is performed on a vehicle, there is always a risk that something will be disturbed, knocked off, or left disconnected.
- **Check oil level and condition.** Another area for visual inspection is oil level and condition.

  Oil level. Oil should be to the proper level.

  Oil condition. Using a match or lighter, try to light the oil on the dipstick; if the oil flames up, gasoline is present in the engine oil. Drip some engine oil from the dipstick onto the hot exhaust manifold. If the oil bubbles or boils, coolant (water) is present in the oil. Check for grittiness by rubbing the oil between your fingers.

NOTE: Gasoline in the oil will cause the engine to run rich by drawing fuel through the positive crankcase ventilation (PCV) system.

TECH TIP

"Original Equipment" Is Not a Four-Letter Word

To many service technicians, an original-equipment part is considered to be only marginal and to get the really "good stuff" an aftermarket (renewal market) part has to be purchased. However, many problems can be traced to the use of an aftermarket part that has failed early in its service life. Technicians who work at dealerships usually go immediately to an aftermarket part that is observed during a visual inspection. It has been their experience that simply replacing the aftermarket part with the factory original-equipment (OE) part often solves the problem.

Original equipment parts are *required* to pass quality and durability standards and tests at a level not required of aftermarket parts. The technician should be aware that the presence of a new part does not necessarily mean that the part is good.

TECH TIP

Smoke Machine Testing

Vacuum (air) leaks can cause a variety of driveability problems and are often difficult to locate. One good method is to use a machine that generates a stream of smoke. Connecting the outlet of the **smoke machine** to the hose that was removed from the vacuum brake booster allows smoke to enter the intake manifold. Any vacuum leaks will be spotted by observing smoke coming out of the leak.
● **SEE FIGURE 33–5.**

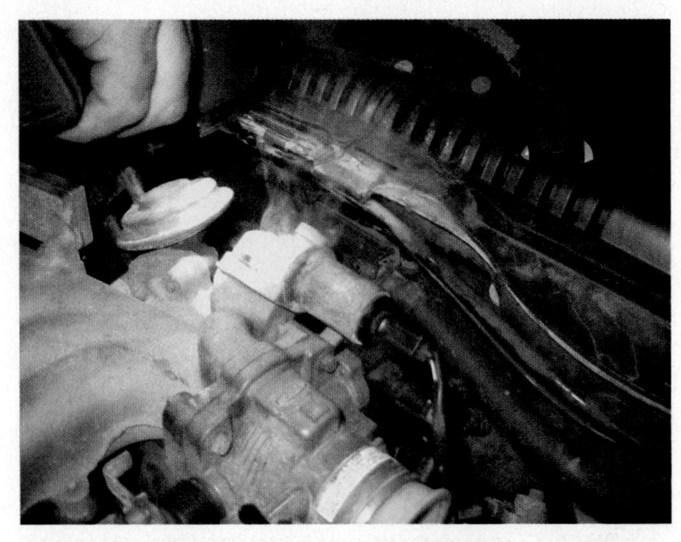

FIGURE 33–5 Using a bright light makes seeing where the smoke is coming from easier. In this case, smoke was added to the intake manifold with the inlet blocked with a yellow plastic cap and smoke was seen escaping past a gasket at the idle air control.

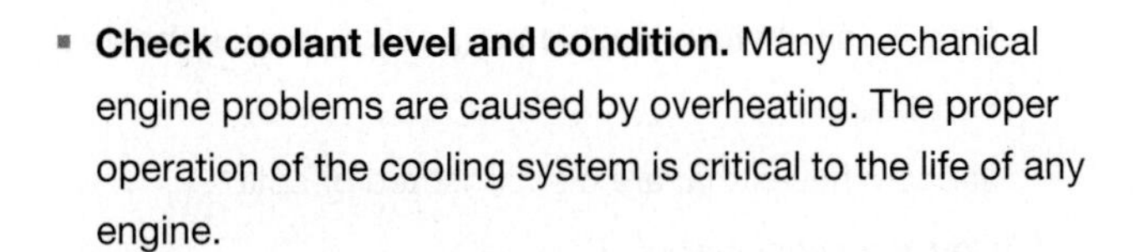

- **Check coolant level and condition.** Many mechanical engine problems are caused by overheating. The proper operation of the cooling system is critical to the life of any engine.

  NOTE: Check the coolant level in the radiator only if the radiator is cool. If the radiator is hot and the radiator cap is removed, the drop in pressure above the coolant will cause the coolant to boil immediately, which can cause severe burns because the coolant expands explosively upward and outward from the radiator opening.

- **Use the paper test.** A soundly running engine should produce even and steady exhaust at the tailpipe. For the **paper test**, hold a piece of paper (even a dollar bill works) or a 3-by-5 inch card within 1 inch (2.5 cm) of the tailpipe with the engine running at idle. The paper should blow evenly away from the end of the tailpipe without "puffing" or being drawn inward toward the end of the tailpipe. If the paper is at times drawn *toward* the tailpipe, the valves in one or more cylinders could be burned. Other reasons why the paper might be drawn toward the tailpipe include the following:

  1. The engine could be misfiring because of a lean condition that could occur normally when the engine is cold.
  2. Pulsing of the paper toward the tailpipe could also be caused by a hole in the exhaust system. If exhaust escapes through a hole in the exhaust system, air could be drawn—in the intervals between the exhaust puffs—from the tailpipe to the hole in the exhaust, causing the paper to be drawn toward the tailpipe.

- **Ensure adequate fuel level.** Make certain that the fuel tank is at least one-fourth to one-half full; if the fuel level is low it, is possible that any water or alcohol at the bottom of the fuel tank is more concentrated and can be drawn into the fuel system.
- **Check the battery voltage.** The voltage of the battery should be at least 12.4 volts and the charging voltage (engine running) should be 13.5 to 15.0 volts at 2000 RPM. Low battery voltage can cause a variety of problems including reduced fuel economy and incorrect (usually too high) idle speed. Higher-than-normal battery voltage can also cause the PCM problems and could cause damage to electronic modules.
- **Check the spark using a spark tester.** Remove one spark plug wire and attach the removed plug wire to the spark tester. Attach the grounding clip of the spark tester to a good clean engine ground, start or crank the engine, and observe the spark tester. ● **SEE FIGURE 33–6.** The spark at the spark tester should be steady and consistent. If an

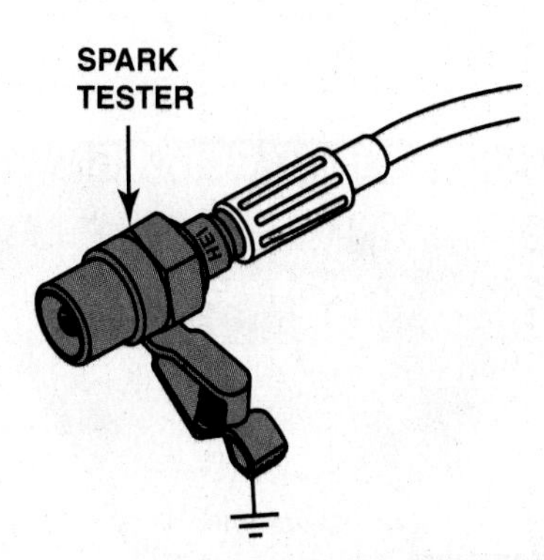

FIGURE 33–6 A spark tester connected to a spark plug wire or coil output. A typical spark tester will fire only if at least 25,000 volts is available from the coil, making a spark tester a useful tool. Do not use one that just lights when a spark is present, because it does not require more than about 2,000 volts to light.

intermittent spark occurs, then this condition should be treated as a no-spark condition. If this test does not show satisfactory spark, carefully inspect and test all components of the primary and secondary ignition systems.

NOTE: Do not use a standard spark plug to check for proper ignition system voltage. An electronic ignition spark tester is designed to force the spark to jump about 0.75 inch (19 mm). This amount of gap requires between 25,000 and 30,000 volts (25 to 30 kV) at atmospheric pressure, which is enough voltage to ensure that a spark can occur under compression inside an engine.

- **Check the fuel-pump pressure.** Checking the fuel-pump pressure is relatively easy on many port fuel-injected engines. Often the cause of intermittent engine performance is due to a weak electric fuel pump or clogged fuel filter. Checking fuel pump pressure early in the diagnostic process eliminates low fuel pressure as a possibility.

STEP 3 RETRIEVE THE DIAGNOSTIC TROUBLE CODES (DTCs) If a diagnostic trouble code (DTC) is present in the computer memory, it may be signaled by illuminating a malfunction indicator lamp (MIL), commonly labeled "check engine" or "service engine soon." ● **SEE FIGURE 33–7**. Any

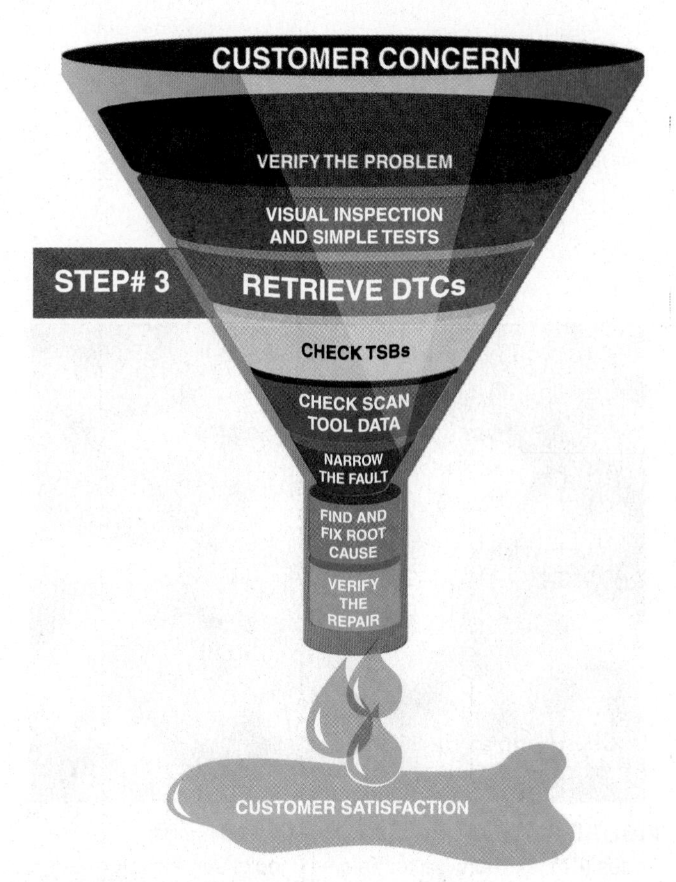

FIGURE 33–7 Step 3 in the diagnostic process is to retrieve any stored diagnostic trouble codes.

code(s) that is displayed when the MIL is *not* on is called a **pending code**. Because the MIL is not on, this indicates that the fault has not repeated to cause the PCM to turn on the MIL. Although this pending code is helpful to the technician to know that a fault has, in the past, been detected, further testing will be needed to find the root cause of the problem.

STEP 4 CHECK FOR TECHNICAL SERVICE BULLETINS (TSBs) Check for corrections in **technical service bulletins (TSBs)** that match the symptoms. ● **SEE FIGURE 33–8**. According to studies performed by automobile manufacturers, as many as 30% of vehicles can be repaired following the information, suggestions, or replacement parts found in a service bulletin. DTCs must be known before searching for service bulletins, because bulletins often include information on solving problems that involve a stored diagnostic trouble code.

STEP 5 LOOK CAREFULLY AT SCAN TOOL DATA Vehicle manufacturers have been giving the technician more and more data on a scan tool connected to the **data link**

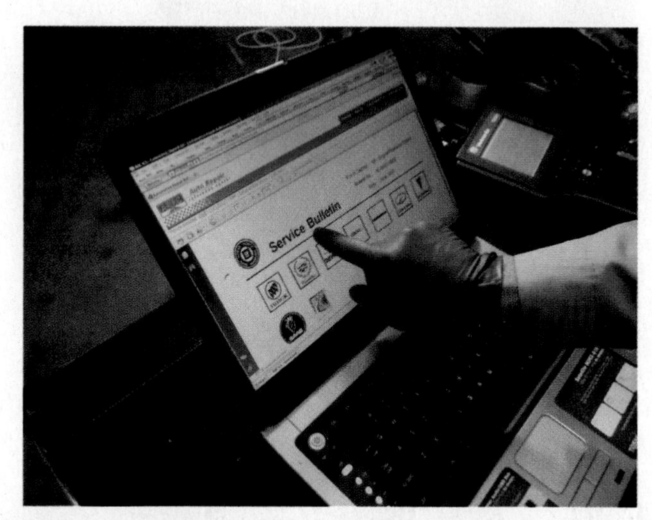

FIGURE 33–8 After checking for stored diagnostic trouble codes (DTCs), the wise technician checks service information for any technical service bulletins that may relate to the vehicle being serviced.

connector (DLC). ● **SEE FIGURE 33–9.** Beginning technicians are often observed scrolling through scan data without a real clue about what they are looking for. When asked, they usually reply that they are looking for something unusual, as if the screen will flash a big message "LOOK HERE—THIS IS NOT CORRECT." That statement does not appear on scan tool displays. The best way to look at scan data is in a definite sequence and with specific, selected bits of data that can tell the most about the operation of the engine, such as the following:

- Engine coolant temperature (ECT) is the same as intake air temperature (IAT) after the vehicle sits for several hours.
- Idle air control (IAC) valve is being commanded to an acceptable range.
- Oxygen sensor (O2S) is operating properly:

  1. Readings below 200 mV at times
  2. Readings above 800 mV at times
  3. Rapid transitions between rich and lean

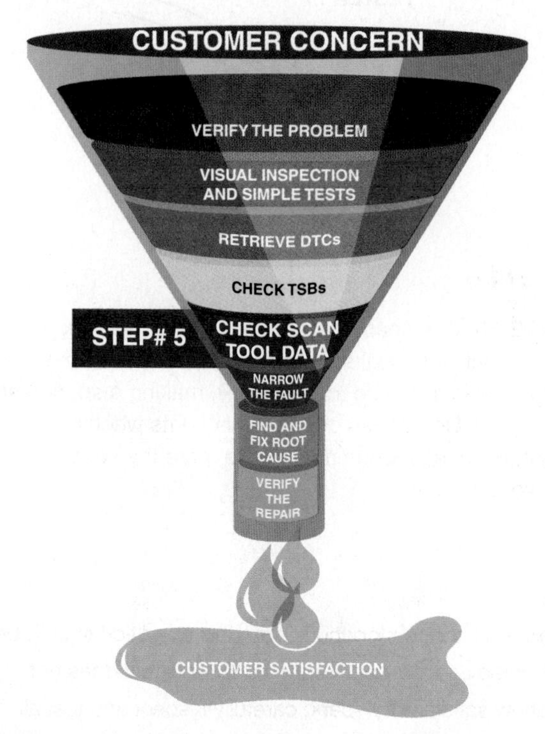

FIGURE 33–9 Looking carefully at the scan tool data is very helpful in locating the source of a problem.

STEP 6 NARROW THE PROBLEM TO A SYSTEM OR CYLINDER Narrowing the focus to a system or individual cylinder is the hardest part of the entire diagnostic process.

- Perform a cylinder power balance test.
- If a weak cylinder is detected, perform a compression and a cylinder leakage test to determine the probable cause.

STEP 7 REPAIR THE PROBLEM AND DETERMINE THE ROOT CAUSE The repair or part replacement must be performed following vehicle manufacturer's recommendations and be certain that the root cause of the problem has been found. Also follow the manufacturer's recommended repair procedures and methods.

STEP 8 VERIFY THE REPAIR AND CLEAR ANY STORED DTCS ● **SEE FIGURE 33–10.**

- Test-drive to verify that the original problem (concern) is fixed.

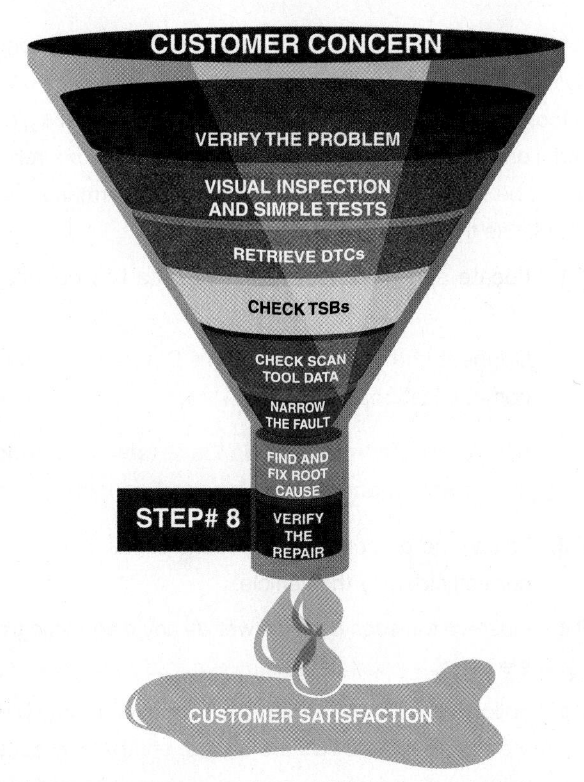

FIGURE 33–10 Step 8 is very important. Be sure that the customer's concern has been corrected.

- Verify that no additional problems have occurred during the repair process.
- Check for and then clear all diagnostic trouble codes. (This step ensures that the computer will not make any changes based on a stored DTC, but should not be performed if the vehicle is going to be tested for emissions because all of the monitors will need to be run and pass.)
- Return the vehicle to the customer and double-check the following:
  1. The vehicle is clean.
  2. The radio is turned off.
  3. The clock is set to the right time and the radio stations have been restored if the battery was disconnected during the repair procedure.

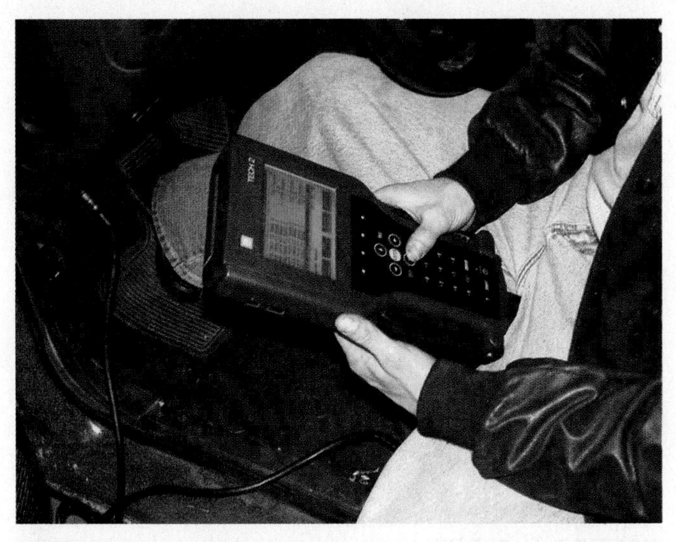

FIGURE 33–11 A TECH 2 scan tool is the factory scan tool used on General Motors vehicles.

TECH TIP

One Test Is Worth 1,000 "Expert" Opinions

Whenever any vehicle has an engine performance or driveability concern, certain people always say:

"Sounds like it's a bad injector."

"I'll bet you it's a bad computer."

"I had a problem just like yours yesterday and it was a bad EGR valve."

Regardless of the skills and talents of those people, it is still more accurate to perform tests on the vehicle than to rely on feelings or opinions of others who have not even seen the vehicle. Even your own opinion should not sway your thinking. Follow a plan, perform tests, and the test results will lead to the root cause.

SCAN TOOLS

Scan tools are the workhorse for any diagnostic work on all vehicles. Scan tools can be divided into two basic groups:

1. **Factory scan tools.** These are the scan tools required by all dealers that sell and service the brand of vehicle. Examples of factory scan tools include:
   - **General Motors**—Tech 2 or MDI (Multiple Diagnostic Interface). ● **SEE FIGURE 33–11**.
   - **Ford**—New Generation Star (NGS) or IDS (Integrated Diagnostic System)
   - **Chrysler**—DRB-III, Star Scan, or wiTECH.

FIGURE 33–12 A Bluetooth adapter that plugs into the DLC and transmits global OBD II information to a smart phone that has a scan tool app installed.

- **Honda**—HDS or Master Tech
- **Toyota**—Master Tech

All factory scan tools are designed to provide bidirectional capability that allows the service technician the opportunity to operate components using the scan tool thereby confirming that the component is able to work when commanded. Also all factory scan tools are capable of displaying all factory parameters.

2. **Aftermarket scan tools.** These scan tools are designed to function on more than one brand of vehicle. Examples of aftermarket scan tools include:
   - **Snap-on** (various models including the MT2500 and Modis)
   - **OTC** (various models including Genisys and Task Master)
   - **AutoEnginuity** and other programs that use a laptop or handheld computer for the display

While many aftermarket scan tools can display most if not all of the parameters of the factory scan tool, there can be a difference when trying to troubleshoot some faults. ● **SEE FIGURE 33–12**.

RETRIEVAL OF DIAGNOSTIC INFORMATION

To retrieve diagnostic information from the Powertrain Control Module (PCM), a scan tool is needed. If a factory or factory-level scan tool is used, then all of the data can be retrieved. If a global (generic) only type scan tool is used, only the emissions-related data can be retrieved. To retrieve diagnostic information from the PCM, use the following steps:

STEP 1 Locate and gain access to the data link connector (DLC).

STEP 2 Connect the scan tool to the DLC and establish communication.

NOTE: If no communication is established, follow the vehicle manufacturer's specified instructions.

STEP 3 Follow the on-screen instructions of the scan tool to correctly identify the vehicle.

STEP 4 Observe the scan data, as well as any diagnostic trouble codes.

STEP 5 Follow vehicle manufacturer's instructions if any DTCs are stored. If no DTCs are stored, compare all sensor values with a factory acceptable range chart to see if any sensor values are out-of-range.

Parameter Identification (PID)

Scan Tool Parameter	Units Displayed	Typical Data Value
Engine Idling/Radiator Hose Hot/Closed Throttle/ Park or Neutral/Closed Loop/Accessories Off/ Brake Pedal Released		
3X Crank Sensor	RPM	Varies
24X Crank Sensor	RPM	Varies
Actual EGR Position	Percent	0
BARO	kPa/Volts	65–110 kPa/ 3.5–4.5 volts
CMP Sensor Signal Present	Yes/No	Yes
Commanded Fuel Pump	On/Off	On
Cycles of Misfire Data	Counts	0–99
Desired EGR Position	Percent	0

(CONTINUED)

Scan Tool Parameter	Units Displayed	Typical Data Value
ECT	°C/°F	Varies
EGR Duty Cycle	Percent	0
Engine Run Time	Hr: Min: Sec	Varies
EVAP Canister Purge	Percent	Low and Varying
EVAP Fault History	No Fault/ Excess Vacuum/ Purge Valve Leak/ Small Leak/ Weak Vacuum	No Fault
Fuel Tank Pressure	Inches of H_2O/ Volts	Varies
HO2S Sensor 1	Ready/ Not Ready	Ready
HO2S Sensor 1	Millivolts	0–1,000 and Varying
HO2S Sensor 2	Millivolts	0–1,000 and Varying
HO2S X Counts	Counts	Varies
IAC Position	Counts	15–25 preferred
IAT	°C/°F	Varies
Knock Retard	Degrees	0
Long-Term FT	Percent	0–10
MAF	Grams per second	3–7
MAF Frequency	Hz	1,200–3,000 (depends on altitude and engine load)
MAP	kPa/Volts	20–48 kPa/0.75–2 Volts (depends on altitude)
Misfire Current Cyl. 1–10	Counts	0
Misfire History Cyl. 1–10	Counts	0
Short-Term FT	Percent	0-10
Start-Up ECT	°C/°F	Varies
Start-Up IAT	°C/°F	Varies
Total Misfire Current Count	Counts	0
Total Misfire Failures	Counts	0
Total Misfire Passes	Counts	0
TP Angle	Percent	0
TP Sensor	Volts	0.20–0.74
Vehicle Speed	Mph/Km/h	0

Note: Viewing the PID screen on the scanner is useful in determining if a problem is occurring at the present time.

TROUBLESHOOTING USING DIAGNOSTIC TROUBLE CODES

Pinning down causes of the actual problem can be accomplished by trying to set the opposite code. For example, if a code indicates an open throttle position (TP) sensor (high resistance), clear the code and create a shorted (low-resistance) condition. This can be accomplished by using a jumper wire and connecting the signal terminal to the 5-volt reference terminal. This should set a diagnostic trouble code.

- **If the opposite code sets,** this indicates that the wiring and connector for the sensor is okay and the sensor itself is defective (open).
- **If the same code sets,** this indicates that the wiring or electrical connection is open (has high resistance) and is the cause of the setting of the DTC.

METHODS FOR CLEARING DIAGNOSTIC TROUBLE CODES Clearing diagnostic trouble codes from a vehicle computer sometimes needs to be performed. There are three methods that can be used to clear stored diagnostic trouble codes.

CAUTION: Clearing diagnostic trouble codes (DTCs) also will clear all of the noncontinuous monitors.

- **Clearing codes—Method 1.** The preferred method of clearing codes is by using a scan tool. This is the method recommended by most vehicle manufacturers if the procedure can be performed on the vehicle. The computer of some vehicles cannot be cleared with a scan tool.
- **Clearing codes—Method 2.** If a scan tool is not available or a scan tool cannot be used on the vehicle being serviced, the power to the computer can be disconnected.
  1. Disconnect the fusible link (if so equipped) that feeds the computer.
  2. Disconnect the fuse or fuses that feed the PCM.

  NOTE: The fuse may not be labeled as a PCM fuse. For example, many Toyotas can be cleared by disconnecting the fuel-injection fuse. Some vehicles require that two fuses be disconnected to clear any stored codes.
- **Clearing codes—Method 3.** If the other two methods cannot be used, the negative (−) battery cable can be disconnected to clear stored diagnostic trouble codes.

NOTE: Because of the adaptive learning capacity of the PCM, a vehicle may fail an exhaust emissions test if the vehicle is not driven enough to allow the computer to run all of the monitors.

CAUTION: By disconnecting the battery, the radio presets and clock information will be lost. They should be reset before returning the vehicle to the customer. If the radio has a security code, the code must be entered before the radio will function. Before disconnecting the battery, always check with the vehicle owner to be sure that the code is available.

RETRIEVING CODES PRIOR TO 1996

FLASH CODES Most vehicles from the early 1980s through 1995 used some method to retrieve diagnostic trouble codes. For example, General Motors diagnostic trouble codes could be retrieved by using a metal tool and contacting terminals A and B of the 12-pin DLC. ● **SEE FIGURE 33–13.**

This method is called flash code retrieval because the MIL will flash to indicate diagnostic trouble codes. The steps of the method are as follows:

1. Turn the ignition switch to on (engine off). The "check engine" light or "service engine soon" light should be on. If the amber malfunction indicator light (MIL) is not on, a problem exists within the light circuit.
2. Connect terminals A and B at the DLC.
3. Observe the MIL. A code 12 (one flash, then a pause, then two flashes) reveals that there is no engine speed indication to the computer. Because the engine is not running, this simply indicates that the computer diagnostic system is working correctly.

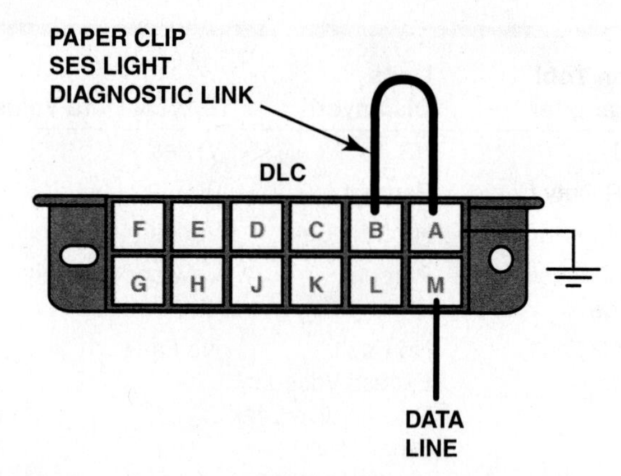

FIGURE 33–13 To retrieve flash codes from an OBD-I General Motors vehicle, connect terminals A and B with the ignition on–engine off. The M terminal is used to retrieve data from the sensors to a scan tool.

Quick and Easy Chrysler Code Retrieval

Most Chrysler-made vehicles (Dodge, Ram, and Chrysler) can display the diagnostic trouble code on the dash by turning the ignition switch on and then off and then on three times with the last time being on. This makes it easy for anyone to see if there are any stored trouble codes without having to use a scan tool. This works on vehicles built after 1996 too. ● **SEE FIGURE 33–14.**

Do Not Lie to a Scan Tool!

Because computer calibration may vary from year to year, using the incorrect year for the vehicle while using a scan tool can cause the data retrieved to be incorrect or inaccurate.

FIGURE 33–14 Diagnostic trouble codes (DTCs) from Chrysler and Dodge vehicles can be retrieved by turning the ignition switch to on and then off three times.

RETRIEVAL METHODS Check service information for the exact procedure to follow to retrieve diagnostic trouble codes. Depending on the exact make, model, and year of manufacture, the procedure can include the use of one or more of the following:

- Scan tool
- Special tester
- Fused jumper wire
- Test light

DLC LOCATIONS

The data link connector (DLC) is a standardized 16-cavity connector to which a scan tool can be connected to retrieve diagnostic information from the vehicle's computers. The normal location is under the dash on the driver's side of the vehicle. It can be covered, and if it is, then it should be easy to remove the cover without the use of any tool, such as when located underneath the ash tray. ● **SEE FIGURE 33–15.**

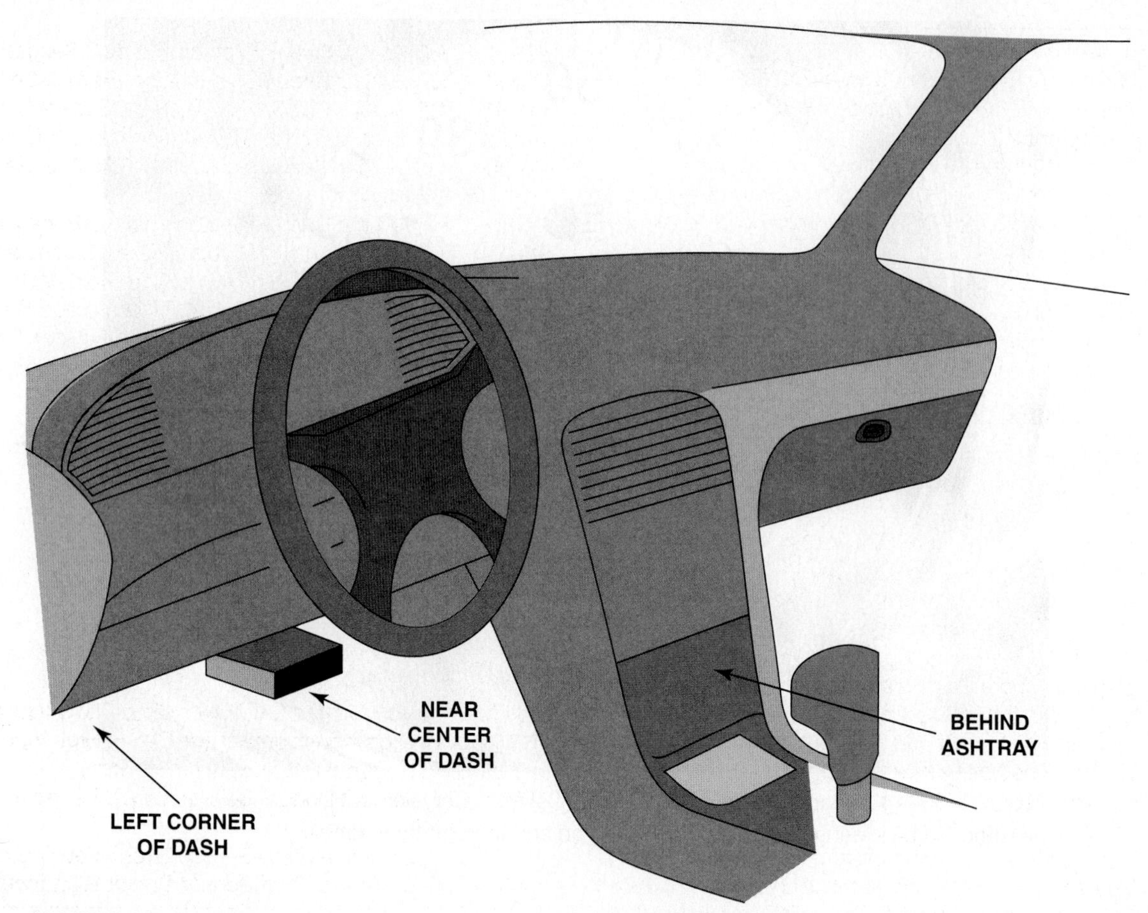

FIGURE 33–15 The data link connector (DLC) can be located in various locations.

OBD-II DIAGNOSIS

Starting with the 1996 model year, all vehicles sold in the United States must use the same type of 16-pin data link connector (DLC) and must monitor emissions-related components. ● **SEE FIGURE 33–16.**

RETRIEVING OBD-II CODES A scan tool is required to retrieve diagnostic trouble codes from most OBD-II vehicles. Every OBD-II scan tool will be able to read all generic Society of Automotive Engineers (SAE) DTCs from any vehicle.

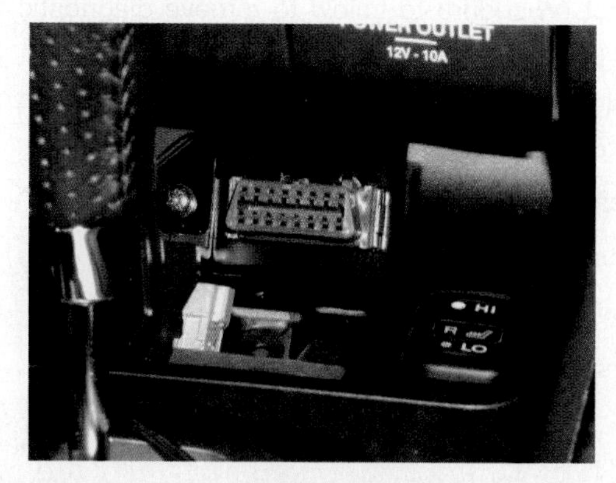

FIGURE 33–16 A typical OBD-II data link connector (DLC). The location varies with make and model and may even be covered, but a tool is not needed to gain access. Check service information for the exact location if needed.

Fuel and Air Metering System

P0100 Mass or Volume Airflow Circuit Problem
P0101 Mass or Volume Airflow Circuit Range or Performance Problem
P0102 Mass or Volume Airflow Circuit Low Input
P0103 Mass or Volume Airflow Circuit High Input
P0105 Manifold Absolute Pressure or Barometric Pressure Circuit Problem
P0106 Manifold Absolute Pressure or Barometric Pressure Circuit Range or Performance Problem
P0107 Manifold Absolute Pressure or Barometric Pressure Circuit Low Input
P0108 Manifold Absolute Pressure or Barometric Pressure Circuit High Input
P0110 Intake Air Temperature Circuit Problem
P0111 Intake Air Temperature Circuit Range or Performance Problem
P0112 Intake Air Temperature Circuit Low Input
P0113 Intake Air Temperature Circuit High Input
P0115 Engine Coolant Temperature Circuit Problem
P0116 Engine Coolant Temperature Circuit Range or Performance Problem
P0117 Engine Coolant Temperature Circuit Low Input
P0118 Engine Coolant Temperature Circuit High Input
P0120 Throttle Position Circuit Problem
P0121 Throttle Position Circuit Range or Performance Problem
P0122 Throttle Position Circuit Low Input
P0123 Throttle Position Circuit High Input
P0125 Excessive Time to Enter Closed-Loop Fuel Control
P0128 Coolant Temperature Below Thermostat Regulating Temperature
P0130 O2 Sensor Circuit Problem (Bank 1* Sensor 1)
P0131 O2 Sensor Circuit Low Voltage (Bank 1* Sensor 1)
P0132 O2 Sensor Circuit High Voltage (Bank 1* Sensor 1)
P0133 O2 Sensor Circuit Slow Response (Bank 1* Sensor 1)
P0134 O2 Sensor Circuit No Activity Detected (Bank 1* Sensor 1)
P0135 O2 Sensor Heater Circuit Problem (Bank 1* Sensor 1)
P0136 O2 Sensor Circuit Problem (Bank 1* Sensor 2)
P0137 O2 Sensor Circuit Low Voltage (Bank 1* Sensor 2)
P0138 O2 Sensor Circuit High Voltage (Bank 1* Sensor 2)
P0139 O2 Sensor Circuit Slow Response (Bank 1* Sensor 2)
P0140 O2 Sensor Circuit No Activity Detected (Bank 1* Sensor 2)
P0141 O2 Sensor Heater Circuit Problem (Bank 1* Sensor 2)
P0142 O2 Sensor Circuit Problem (Bank 1* Sensor 3)
P0143 O2 Sensor Circuit Low Voltage (Bank 1* Sensor 3)
P0144 O2 Sensor Circuit High Voltage (Bank 1* Sensor 3)
P0145 O2 Sensor Circuit Slow Response (Bank 1* Sensor 3)
P0146 O2 Sensor Circuit No Activity Detected (Bank 1* Sensor 3)
P0147 O2 Sensor Heater Circuit Problem (Bank 1* Sensor 3)
P0150 O2 Sensor Circuit Problem (Bank 2 Sensor 1)
P0151 O2 Sensor Circuit Low Voltage (Bank 2 Sensor 1)
P0152 O2 Sensor Circuit High Voltage (Bank 2 Sensor 1)
P0153 O2 Sensor Circuit Slow Response (Bank 2 Sensor 1)
P0154 O2 Sensor Circuit No Activity Detected (Bank 2 Sensor 1)
P0155 O2 Sensor Heater Circuit Problem (Bank 2 Sensor 1)
P0156 O2 Sensor Circuit Problem (Bank 2 Sensor 2)
P0157 O2 Sensor Circuit Low Voltage (Bank 2 Sensor 2)
P0158 O2 Sensor Circuit High Voltage (Bank 2 Sensor 2)
P0159 O2 Sensor Circuit Slow Response (Bank 2 Sensor 2)
P0160 O2 Sensor Circuit No Activity Detected (Bank 2 Sensor 2)
P0161 O2 Sensor Heater Circuit Problem (Bank 2 Sensor 2)
P0162 O2 Sensor Circuit Problem (Bank 2 Sensor 3)
P0163 O2 Sensor Circuit Low Voltage (Bank 2 Sensor 3)
P0164 O2 Sensor Circuit High Voltage (Bank 2 Sensor 3)
P0165 O2 Sensor Circuit Slow Response (Bank 2 Sensor 3)
P0166 O2 Sensor Circuit No Activity Detected (Bank 2 Sensor 3)
P0167 O2 Sensor Heater Circuit Problem (Bank 2 Sensor 3)
P0170 Fuel Trim Problem (Bank 1*)
P0171 System Too Lean (Bank 1*)
P0172 System Too Rich (Bank 1*)
P0173 Fuel Trim Problem (Bank 2)
P0174 System Too Lean (Bank 2)
P0175 System Too Rich (Bank 2)
P0176 Fuel Composition Sensor Circuit Problem
P0177 Fuel Composition Sensor Circuit Range or Performance
P0178 Fuel Composition Sensor Circuit Low Input
P0179 Fuel Composition Sensor Circuit High Input
P0180 Fuel Temperature Sensor Problem
P0181 Fuel Temperature Sensor Circuit Range or Performance
P0182 Fuel Temperature Sensor Circuit Low Input
P0183 Fuel Temperature Sensor Circuit High Input

Fuel and Air Metering (Injector Circuit)

P0201 Injector Circuit Problem—Cylinder 1
P0202 Injector Circuit Problem—Cylinder 2
P0203 Injector Circuit Problem—Cylinder 3
P0204 Injector Circuit Problem—Cylinder 4
P0205 Injector Circuit Problem—Cylinder 5
P0206 Injector Circuit Problem—Cylinder 6
P0207 Injector Circuit Problem—Cylinder 7
P0208 Injector Circuit Problem—Cylinder 8
P0209 Injector Circuit Problem—Cylinder 9
P0210 Injector Circuit Problem—Cylinder 10
P0211 Injector Circuit Problem—Cylinder 11
P0212 Injector Circuit Problem—Cylinder 12
P0213 Cold Start Injector 1 Problem
P0214 Cold Start Injector 2 Problem

Ignition System or Misfire

P0300 Random Misfire Detected
P0301 Cylinder 1 Misfire Detected
P0302 Cylinder 2 Misfire Detected
P0303 Cylinder 3 Misfire Detected
P0304 Cylinder 4 Misfire Detected
P0305 Cylinder 5 Misfire Detected
P0306 Cylinder 6 Misfire Detected
P0307 Cylinder 7 Misfire Detected
P0308 Cylinder 8 Misfire Detected
P0309 Cylinder 9 Misfire Detected
P0310 Cylinder 10 Misfire Detected
P0311 Cylinder 11 Misfire Detected
P0312 Cylinder 12 Misfire Detected

Ignition System or Misfire—Continued

P0320 Ignition or Distributor Engine Speed Input Circuit Problem
P0321 Ignition or Distributor Engine Speed Input Circuit Range or Performance
P0322 Ignition or Distributor Engine Speed Input Circuit No Signal
P0325 Knock Sensor 1 Circuit Problem
P0326 Knock Sensor 1 Circuit Range or Performance
P0327 Knock Sensor 1 Circuit Low Input
P0328 Knock Sensor 1 Circuit High Input
P0330 Knock Sensor 2 Circuit Problem
P0331 Knock Sensor 2 Circuit Range or Performance
P0332 Knock Sensor 2 Circuit Low Input
P0333 Knock Sensor 2 Circuit High Input
P0335 Crankshaft Position Sensor Circuit Problem
P0336 Crankshaft Position Sensor Circuit Range or Performance
P0337 Crankshaft Position Sensor Circuit Low Input
P0338 Crankshaft Position Sensor Circuit High Input

Auxiliary Emissions Control

P0400 Exhaust Gas Recirculation Flow Problem
P0401 Exhaust Gas Recirculation Flow Insufficient Detected
P0402 Exhaust Gas Recirculation Flow Excessive Detected
P0405 Air Conditioner Refrigerant Charge Loss
P0410 Secondary Air Injection System Problem
P0411 Secondary Air Injection System Insufficient Flow Detected
P0412 Secondary Air Injection System Switching Valve or Circuit Problem
P0413 Secondary Air Injection System Switching Valve or Circuit Open
P0414 Secondary Air Injection System Switching Valve or Circuit Shorted
P0420 Catalyst System Efficiency below Threshold (Bank 1*)
P0421 Warm Up Catalyst Efficiency below Threshold (Bank 1*)
P0422 Main Catalyst Efficiency below Threshold (Bank 1*)
P0423 Heated Catalyst Efficiency below Threshold (Bank 1*)
P0424 Heated Catalyst Temperature below Threshold (Bank 1*)
P0430 Catalyst System Efficiency below Threshold (Bank 2)
P0431 Warm Up Catalyst Efficiency below Threshold (Bank 2)
P0432 Main Catalyst Efficiency below Threshold (Bank 2)
P0433 Heated Catalyst Efficiency below Threshold (Bank 2)
P0434 Heated Catalyst Temperature below Threshold (Bank 2)
P0440 Evaporative Emission Control System Problem
P0441 Evaporative Emission Control System Insufficient Purge Flow
P0442 Evaporative Emission Control System Leak Detected
P0443 Evaporative Emission Control System Purge Control Valve Circuit Problem
P0444 Evaporative Emission Control System Purge Control Valve Circuit Open
P0445 Evaporative Emission Control System Purge Control Valve Circuit Shorted
P0446 Evaporative Emission Control System Vent Control Problem
P0447 Evaporative Emission Control System Vent Control Open
P0448 Evaporative Emission Control System Vent Control Shorted
P0450 Evaporative Emission Control System Pressure Sensor Problem
P0451 Evaporative Emission Control System Pressure Sensor Range or Performance
P0452 Evaporative Emission Control System Pressure Sensor Low Input
P0453 Evaporative Emission Control System Pressure Sensor High Input

Vehicle Speed Control and Idle Control

P0500 Vehicle Speed Sensor Problem
P0501 Vehicle Speed Sensor Range or Performance
P0502 Vehicle Speed Sensor Low Input
P0505 Idle Control System Problem
P0506 Idle Control System RPM Lower Than Expected
P0507 Idle Control System RPM Higher Than Expected
P0510 Closed Throttle Position Switch Problem

Computer Output Circuit

P0600 Serial Communication Link Problem
P0605 Internal Control Module (Module Identification Defined by J1979)

Transmission

P0703 Brake Switch Input Problem
P0705 Transmission Range Sensor Circuit Problem (PRNDL Input)
P0706 Transmission Range Sensor Circuit Range or Performance
P0707 Transmission Range Sensor Circuit Low Input
P0708 Transmission Range Sensor Circuit High Input
P0710 Transmission Fluid Temperature Sensor Problem
P0711 Transmission Fluid Temperature Sensor Range or Performance
P0712 Transmission Fluid Temperature Sensor Low Input
P0713 Transmission Fluid Temperature Sensor High Input
P0715 Input or Turbine Speed Sensor Circuit Problem
P0716 Input or Turbine Speed Sensor Circuit Range or Performance
P0717 Input or Turbine Speed Sensor Circuit No Signal
P0720 Output Speed Sensor Circuit Problem
P0721 Output Speed Sensor Circuit Range or Performance
P0722 Output Speed Sensor Circuit No Signal
P0725 Engine Speed Input Circuit Problem
P0726 Engine Speed Input Circuit Range or Performance
P0727 Engine Speed Input Circuit No Signal
P0730 Incorrect Gear Ratio
P0731 Gear 1 Incorrect Ratio
P0732 Gear 2 Incorrect Ratio
P0733 Gear 3 Incorrect Ratio
P0734 Gear 4 Incorrect Ratio
P0735 Gear 5 Incorrect Ratio

Transmission—Continued

P0736	Reverse Incorrect Ratio
P0740	Torque Converter Clutch System Problem
P0741	Torque Converter Clutch System Performance or Stuck Off
P0742	Torque Converter Clutch System Stuck On
P0743	Torque Converter Clutch System Electrical
P0745	Pressure Control Solenoid Problem
P0746	Pressure Control Solenoid Performance or Stuck Off
P0747	Pressure Control Solenoid Stuck On
P0748	Pressure Control Solenoid Electrical
P0750	Shift Solenoid A Problem
P0751	Shift Solenoid A Performance or Stuck Off
P0752	Shift Solenoid A Stuck On
P0753	Shift Solenoid A Electrical
P0755	Shift Solenoid B Problem
P0756	Shift Solenoid B Performance or Stuck Off
P0757	Shift Solenoid B Stuck On
P0758	Shift Solenoid B Electrical
P0760	Shift Solenoid C Problem
P0761	Shift Solenoid C Performance or Stuck Off
P0762	Shift Solenoid C Stuck On
P0763	Shift Solenoid C Electrical
P0765	Shift Solenoid D Problem
P0766	Shift Solenoid D Performance or Stuck Off
P0767	Shift Solenoid D Stuck On
P0768	Shift Solenoid D Electrical
P0770	Shift Solenoid E Problem
P0771	Shift Solenoid E Performance or Stuck Off
P0772	Shift Solenoid E Stuck On
P0773	Shift Solenoid E Electrical

* The side of the engine where number one cylinder is located.

OBD-II ACTIVE TESTS

The vehicle computer must run tests on the various emission-related components and turn on the malfunction indicator lamp (MIL) if faults are detected. OBD II is an *active* computer analysis system because it actually tests the operation of the oxygen sensors, exhaust gas recirculation system, and so forth whenever conditions permit. It is the purpose and function of the Powertrain Control Module (PCM) to monitor these components and perform these active tests.

For example, the PCM may open the EGR valve momentarily to check its operation while the vehicle is decelerating. A change in the manifold absolute pressure (MAP) sensor signal will indicate to the computer that the exhaust gas is, in fact, being introduced into the engine. Because these tests are active and certain conditions must be present before these tests can be run, the computer uses its internal diagnostic program to keep track of all the various conditions and to schedule active tests so that they will not interfere with each other.

OBD-II DRIVE CYCLE The vehicle must be driven under a variety of operating conditions for all active tests to be performed. A **trip** is defined as an engine-operating drive cycle that contains the necessary conditions for a particular test to be performed. For example, for the EGR test to be performed, the engine has to be at normal operating temperature and decelerating for a minimum amount of time. Some tests are performed when the engine is cold, whereas others require that the vehicle be cruising at a steady highway speed.

TYPES OF OBD-II CODES Not all OBD-II diagnostic trouble codes are of the same importance for exhaust emissions. Each type of DTC has different requirements for it to set, and the computer will only turn on the MIL for emissions-related DTCs.

TYPE A CODES. A type A diagnostic trouble code is emissions-related and will cause the MIL to be turned on at the *first trip* if the computer has detected a problem. Engine misfire or a very rich or lean air–fuel ratio, for example, would cause a type A diagnostic trouble code. These codes alert the driver to an emissions problem that may cause damage to the catalytic converter.

TYPE B CODES. A type B code will be stored and the MIL will be turned on during the *second consecutive trip*, alerting the driver to the fact that a diagnostic test was performed and failed.

NOTE: Type A and Type B codes are emissions-related and will cause the lighting of the malfunction indicator lamp, usually labeled "check engine" or "service engine soon."

TYPE C AND D CODES. Type C and type D codes are for use with nonemissions-related diagnostic tests. They will cause the lighting of a "service" lamp (if the vehicle is so equipped).

OBD-II FREEZE-FRAME To assist the service technician, OBD II requires the computer to take a "snapshot" or freeze-frame of all data at the instant an emissions-related DTC is set. A scan tool is required to retrieve this data. CARB and EPA regulations require that the controller store specific freeze-frame (engine-related) data when the first emissions-related fault is detected. The data stored in freeze-frame can only be replaced by data from a trouble code with a higher priority such as a trouble related to a fuel system or misfire monitor fault.

NOTE: Although OBD II requires that just one freeze-frame of data be stored, the instant an emissions-related DTC is set, vehicle manufacturers usually provide expanded data about the DTC beyond that required. However, retrieving enhanced data usually requires the use of the vehicle-specific scan tool.

The freeze-frame has to contain data values that occurred at the time the code was set (these values are provided in standard units of measurement). Freeze-frame data is recorded during the first trip on a two-trip fault. As a result, OBD-II systems record the data present at the time an emissions-related code is recorded and the MIL activated. This data can be accessed and displayed on a scan tool. Freeze-frame data is one frame or one instant in time. Freeze-frame data is not updated (refreshed) if the same monitor test fails a second time.

REQUIRED FREEZE-FRAME DATA ITEMS.

- Code that triggered the freeze-frame
- A/F ratio, airflow rate, and calculated engine load
- Base fuel-injector pulse width
- ECT, IAT, MAF, MAP, TP, and VS sensor data
- Engine speed and amount of ignition spark advance
- Open- or closed-loop status
- Short-term and long-term fuel trim values
- For misfire codes—identify the cylinder that misfired

NOTE: All freeze-frame data will be lost if the battery is disconnected, power to the PCM is removed, or the scan tool is used to erase or clear trouble codes.

DIAGNOSING INTERMITTENT MALFUNCTIONS Of all the different types of conditions that you will see, the hardest to accurately diagnose and repair are intermittent malfunctions. These conditions may be temperature related (only occur when the vehicle is hot or cold), or humidity related (only occur when it is raining). Regardless of the conditions that will cause the malfunction to occur, you must diagnose and correct the condition.

When dealing with an intermittent concern, you should determine the conditions when the malfunction occurs, and then try to duplicate those conditions. If a cause is not readily apparent to you, ask the customer when the symptom occurs. Ask if there are any conditions that seem to be related to, or cause the concern.

Another consideration when working on an OBD-II-equipped vehicle is whether a concern is intermittent, or if it only occurs when a specific diagnostic test is performed by the PCM. Since OBD-II systems conduct diagnostic tests only under very precise conditions, some tests may only be run once during an ignition cycle. Additionally, if the requirements needed to perform the test are not met, the test will not run during an ignition cycle. This type of onboard diagnostics could be mistaken as "intermittent" when, in fact, the tests are only infrequent (depending on how the vehicle is driven). Examples of this type of diagnostic test are HO2S heaters, evaporative canister purge, catalyst efficiency, and EGR flow. When diagnosing intermittent concerns on an OBD-II-equipped vehicle, a logical diagnostic strategy is essential. The use of stored freeze-frame information can also be very useful when diagnosing an intermittent malfunction if a code has been stored.

SERVICE/FLASH PROGRAMMING

Designing a program that allows an engine to meet strict air quality and fuel economy standards while providing excellent performance is no small feat. However, this is only part of the challenge facing engineers assigned with the task of developing OBD-II software. The reason for this is the countless variables involved with running the diagnostic monitors. Although programmers do their best to factor in any and all operating conditions when writing this complex code, periodic revisions are often required.

Reprogramming consists of downloading new calibration files from a scan tool, personal computer, or modem into the PCM's electronically erasable programmable read-only

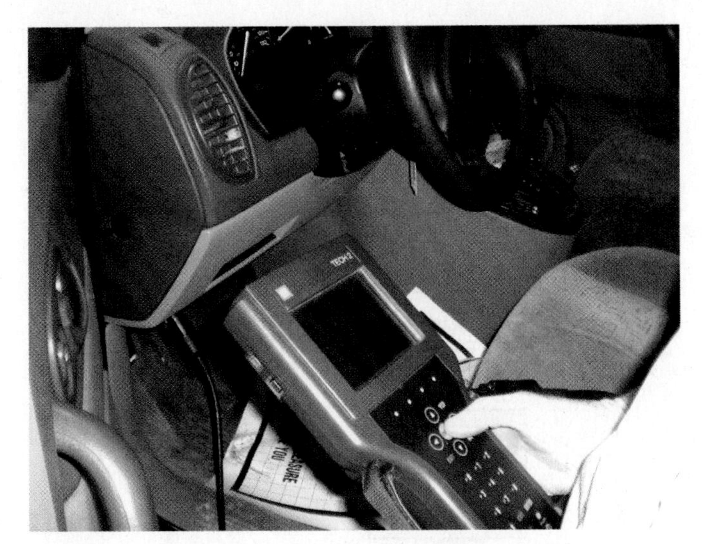

FIGURE 33–17 The first step in the reprogramming procedure is to determine the current software installed using a scan tool. Not all scan tools can be used. In most cases using the factory scan tool is needed for reprogramming unless the scan tool is equipped to handle reprogramming.

FIGURE 33–18 Follow the on-screen instructions.

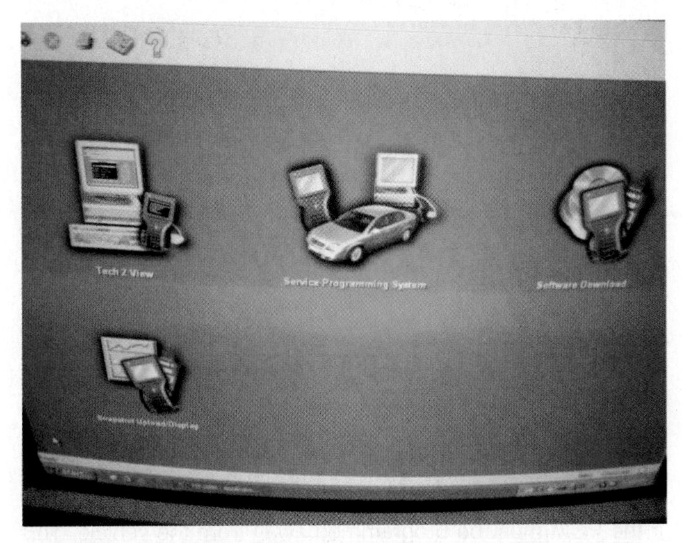

FIGURE 33–19 An Internet connection is usually needed to perform updates although some vehicle manufacturers use CDs which are updated regularly at a cost to the shop.

memory (EEPROM). This can be done on or off the vehicle using the appropriate equipment. Since reprogramming is not an OBD-II requirement however, many vehicles will need a new PCM in the event software changes become necessary. Physically removing and replacing the PROM chip is no longer possible.

The following are three industry-standard methods used to reprogram the EEPROM:

- Remote programming
- Direct programming
- Off-board programming

REMOTE PROGRAMMING. Remote programming uses the scan tool to transfer data from the manufacturer's shop PC to the vehicle's PCM. This is accomplished by performing the following steps:

- Connect the scan tool to the vehicle's DLC. ● **SEE FIGURE 33–17.**
- Enter the vehicle information into the scan tool through the programming application software incorporated in the scan tool. ● **SEE FIGURE 33–18.**
- Download VIN and current EEPROM calibration using a scan tool.
- Disconnect the scan tool from the DLC and connect the tool to the shop PC.
- Download the new calibration from the PC to the scan tool. ● **SEE FIGURE 33–19.**
- Reconnect the scan tool to the vehicle's DLC and download the new calibration into the PCM.

FIGURE 33–20 Connecting cables and a computer to perform off-board programming.

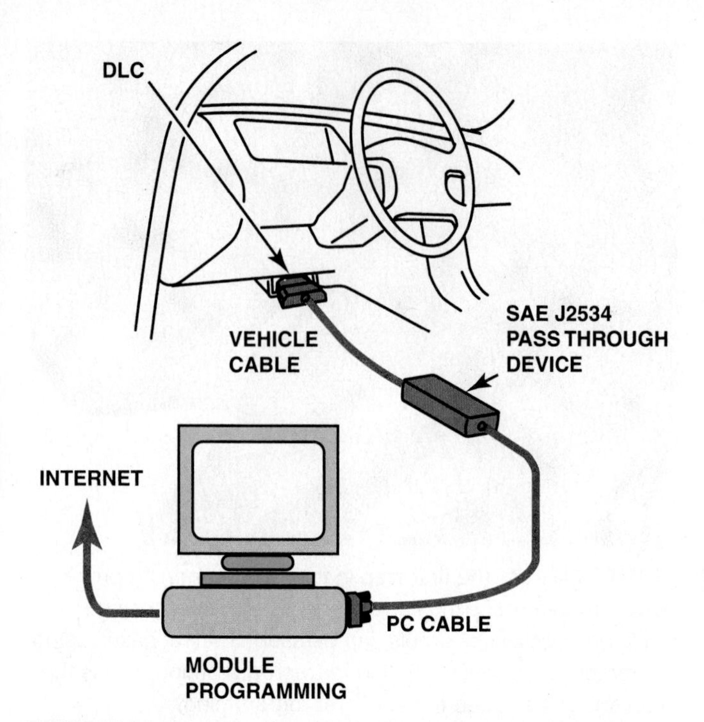

FIGURE 33–21 The J2534 pass-through reprogramming system does not need a scan tool to reflash the PCM on most 2004 and newer vehicles.

CAUTION: Before programming, the vehicle's battery must be between 11 and 14 volts. Do not attempt to program while charging the battery unless using a special battery charger that does not produce excessive ripple voltage such as the Midtronics PSC-300 (30 amp) or PSC-550 (55 amp), or similar as specified by the vehicle manufacturer.

DIRECT PROGRAMMING. Direct programming does utilize a connection between the shop PC and the vehicle DLC.

OFF-BOARD PROGRAMMING. Off-board programming is used if the PCM must be programmed away from the vehicle. This is performed using the off-board programming adapter. ● **SEE FIGURE 33–20.**

J2534 REPROGRAMMING Legislation has mandated that vehicle manufacturers meet the SAE J2534 standards for all emissions-related systems on all new vehicles starting with model year 2004. This standard enables independent service repair operators to program or reprogram emissions-related ECMs from a wide variety of vehicle manufacturers with a single tool. ● **SEE FIGURE 33–21.** A J2534 compliant pass-through system is a standardized programming and diagnostic system. It uses a personal computer (PC) plus a standard interface to a software device driver, and a hardware vehicle communication interface. The interface connects to a PC, and to a programmable ECM on a vehicle through the J1962 data link connector (DLC). This system allows programming of all vehicle manufacturer ECMs using a single set of programming hardware. Programming software made available by the vehicle manufacturer must be functional with a J2534 compliant pass-through system.

The software for a typical pass-through application consists of two major components:

- The part delivered by the company that furnishes the hardware for J2534 enables the pass-through vehicle communication interface to communicate with the PC and provides for all Vehicle Communication Protocols as required by SAE J2534. It also provides for the software interface to work with the software applications as provided for by the vehicle manufacturers. ● **SEE FIGURE 33–22.**
- The second part of the pass-through enabling software is provided for by the vehicle manufacturers. This is normally a subset of the software used with their original equipment manufacturer (OEM) tools and their website will indicate how to obtain this software and under what conditions it can be used. Refer to the National Automotive Service Task Force (NASTF) website for the addresses for all vehicle manufacturers' service information and cost, *www.NASTF.org*.

Since the majority of vehicle manufacturers make this software available in downloadable form, having an Internet browser (Explorer/Netscape) and connection is a must.

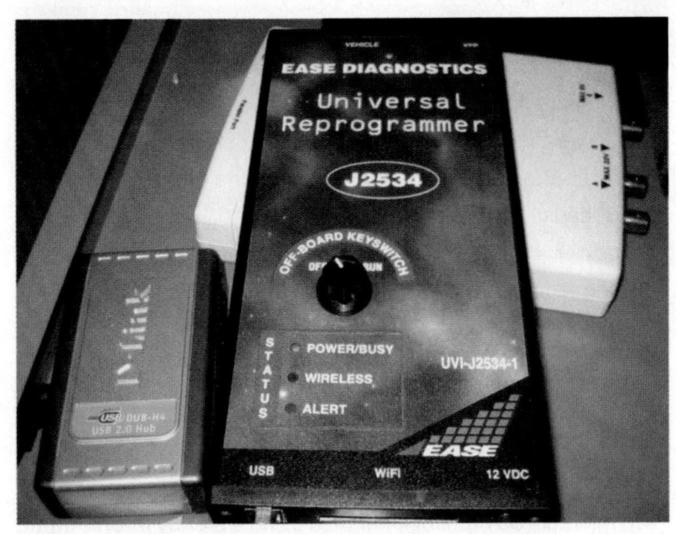

FIGURE 33–22 A typical J2534 universal reprogrammer that uses the J2534 standards.

MANUFACTURER'S DIAGNOSTIC ROUTINES

Each vehicle manufacturer has established their own diagnostic routines and they should be followed. Most include the following steps:

STEP 1 Retrieve diagnostic trouble codes.

STEP 2 Check for all technical service bulletins that could be related to the stored DTC.

STEP 3 If there are multiple DTCs, the diagnostic routine may include checking different components or systems instead of when only one DTC was stored.

STEP 4 Perform system checks.

STEP 5 Perform a road test matching the parameters recorded in the freeze-frame to check that the repair has corrected the malfunction.

STEP 6 Repeat the road test to cause the MIL to be extinguished.

NOTE: Do not clear codes (DTCs) unless instructed by the service information.

Following the vehicle manufacturer's specific diagnostic routines will ensure that the root cause is found and the repair verified. This is important for customer satisfaction.

COMPLETING SYSTEM REPAIRS

After the repair has been successfully completed, the vehicle should be driven under similar conditions that caused the original concern. Verify that the problem has been corrected. To perform this test-drive, it is helpful to have a copy of the freeze-frame parameters that were present when the DTC was set. By driving under similar conditions, the PCM may perform a test of the system and automatically extinguish the MIL. This is the method preferred by most vehicle manufacturers. The DTC can be cleared using a scan tool, but then that means that monitors will have to be run and the vehicle may fail an emissions inspection if driven directly to the testing station.

PROCEDURES FOR RESETTING THE PCM

The PCM can be reset or cleared of previously set DTCs and freeze-frame data in the following ways:

1. **Driving the Vehicle.** Drive the vehicle under similar conditions that were present when the fault occurred. If the conditions are similar and the PCM performed the noncontinuous monitor test and it passed three times, then the PCM will extinguish the MIL. This is the method preferred by most vehicle manufacturers; however, this method could be time consuming. If three passes cannot be achieved, the owner of the vehicle will have to be told that even though the check engine light (MIL) is on, the problem has been corrected and the MIL should go out in a few days of normal driving.
2. **Clear DTCs Using a Scan Tool.** A scan tool can be used to clear the diagnostic trouble code (DTC), which will also delete all of the freeze-frame data. The advantage of using a scan tool is that the check engine (MIL) will be out and the customer will be happy that the problem (MIL on) has been corrected. Do not use a scan tool to clear a DTC if the vehicle is going to be checked soon at a test station for state-mandated emissions tests.

3. **Battery Disconnect.** Disconnecting the negative battery cable will clear the DTCs and freeze-frame on many vehicles but not all. Besides clearing the DTCs, disconnecting the battery for about 20 minutes will also erase radio station presets and other memory items in many cases. Most vehicle manufacturers do not recommend that the battery be disconnected to clear DTCs and it may not work on some vehicles.

The Brake Pedal Trick

If the vehicle manufacturer recommends that battery power be disconnected, first disconnect the negative battery cable and then depress the brake pedal. Because the brake lights are connected to battery power, depressing the brake pedal causes all of the capacitors in the electrical system and computer(s) to discharge through the brake lights.

ROAD TEST (DRIVE CYCLE)

Use the freeze-frame data and test-drive the vehicle so that the vehicle is driven to match the conditions displayed on the freeze-frame. If the battery has been disconnected, then the vehicle may have to be driven under conditions that allow the PCM to conduct monitor tests. This drive pattern is called a **drive cycle**. The drive cycle is different for each vehicle manufacturer but a universal drive cycle may work in many cases. In many cases, performing a universal drive cycle will reset most monitors in most vehicles.

UNIVERSAL DRIVE CYCLE

PRECONDITIONING: PHASE 1.

MIL must be off.

No DTCs present.

Fuel fill between 15% and 85%.

Cold start—Preferred = eight hour soak at 68°F to 86°F.

Alternative = ECT below 86°F.

1. With the ignition off, connect scan tool.
2. Start engine and drive between 20 and 30 mph for 22 minutes, allowing speed to vary.
3. Stop and idle for 40 seconds, gradually accelerate to 55 mph.
4. Maintain 55 mph for 4 minutes using a steady throttle input.
5. Stop and idle for 30 seconds, then accelerate to 30 mph.
6. Maintain 30 mph for 12 minutes.
7. Repeat steps 4 and 5 four times.

   Using scan tool, check readiness. If insufficient readiness set, continue to universal drive trace phase II.

   Important: (Do not shut off engine between phases).

Phase II:

1. Bring vehicle to a stop and idle for 45 seconds, then accelerate to 30 mph.
2. Maintain 30 mph for 22 minutes.
3. Repeat steps 1 and 2 three times.
4. Bring vehicle to a stop and idle for 45 seconds, then accelerate to 35 mph.
5. Maintain speed between 30 and 35 mph for 4 minutes.
6. Bring vehicle to a stop and idle for 45 seconds, then accelerate to 30 mph.
7. Maintain 30 mph for 22 minutes.
8. Repeat steps 6 and 7 five times.
9. Using scan tool, check readiness.

TECH TIP

Drive the Light Out

If working on a vehicle that is subject to state emissions testing, it is best to not clear codes. When diagnostic trouble codes are cleared, all of the monitors have to be rerun and this can be a time-consuming job. Instead of clearing the code, simply drive the vehicle until the PCM clears the code. This will likely take less time compared to trying to drive the vehicle under varying conditions to run all of the monitors.

SUMMARY

1. Funnel diagnostics is a visual approach to a diagnostic procedure and involves the following steps:
   - **STEP 1** Verify the problem (concern)
   - **STEP 2** Perform a thorough visual inspection and basic tests
   - **STEP 3** Retrieve the diagnostic trouble codes (DTCs)
   - **STEP 4** Check for technical service bulletins (TSBs)
   - **STEP 5** Look carefully at scan tool data
   - **STEP 6** Narrow the problem to a system or cylinder
   - **STEP 7** Repair the problem and determine the root cause
   - **STEP 8** Verify the repair and check for any stored DTCs
2. Care should be taken to not induce high voltage or current around any computer or computer-controlled circuit or sensor.
3. A thorough visual inspection is important during the diagnosis and troubleshooting of any engine performance problem or electrical malfunction.
4. If the MIL is on, retrieve the DTC and follow the manufacturer's recommended procedure to find the root cause of the problem.
5. OBD-II vehicles use a 16-pin DLC and common DTCs.

REVIEW QUESTIONS

1. Explain the procedure to follow when diagnosing a vehicle with stored DTCs using a scan tool.
2. Discuss what the PCM does during a drive cycle to test emissions-related components.
3. Explain the difference between a type A and type B OBD-II diagnostic trouble code.
4. List three things that should be checked as part of a thorough visual inspection.
5. List the eight-step funnel diagnostic procedure.
6. Explain why a bulletin search should be performed after stored DTCs are retrieved.
7. List the three methods that can be used to reprogram a PCM.

CHAPTER QUIZ

1. Technician A says that the first step in the diagnostic process is to verify the problem (concern). Technician B says the second step is to perform a thorough visual inspection. Which technician is correct?
   a. Technician A only
   b. Technician B only
   c. Both Technicians A and B
   d. Neither Technician A nor B
2. Which item is *not* important to know before starting the diagnosis of an engine performance problem?
   a. List of previous repairs
   b. The brand of engine oil used
   c. The type of gasoline used
   d. The temperature of the engine when the problem occurs
3. A paper test can be used to check for a possible problem with __________.
   a. The ignition system (bad spark plug wire)
   b. A faulty injector on a multiport engine
   c. A burned valve
   d. All of the above
4. Which step should be performed *last* when diagnosing an engine performance problem?
   a. Checking for any stored diagnostic trouble codes
   b. Checking for any technical service bulletins (TSBs)
   c. Performing a thorough visual inspection
   d. Verifying the repair
5. Technician A says that if the opposite DTC can be set, the problem is the component itself. Technician B says if the opposite DTC cannot be set, the problem is with the wiring or grounds. Which technician is correct?
   a. Technician A only
   b. Technician B only
   c. Both Technicians A and B
   d. Neither Technician A nor B
6. The preferred method to clear diagnostic trouble codes (DTCs) is to __________.
   a. Disconnect the negative battery cable for 10 seconds
   b. Use a scan tool
   c. Remove the computer (PCM) power feed fuse
   d. Cycle the ignition key on and off 40 times
7. Which is the factory scan tool for Chrysler brand vehicles equipped with CAN?
   a. Star Scan
   b. Tech 2
   c. NGS
   d. Master Tech
8. Technician A says that reprogramming a PCM using the J2534 system requires a factory scan tool, while Technician B says it requires Internet access. Which technician is correct?
   a. Technician A only
   b. Technician B only
   c. Both Technicians A and B
   d. Neither Technician A nor B
9. Technician A says that knowing if there are any stored diagnostic trouble codes (DTCs) may be helpful when checking for related technical service bulletins (TSBs). Technician B says that only a factory scan tool should be used to retrieve DTCs. Which technician is correct?
   a. Technician A only
   b. Technician B only
   c. Both Technicians A and B
   d. Neither Technician A nor B
10. Which method can be used to reprogram a PCM?
    a. Remote
    b. Direct
    c. Off-board
    d. All of the above

appendix 1

SAMPLE ENGINE PERFORMANCE (A8) ASE-TYPE CERTIFICATION TEST WITH ANSWERS

1. An injector pulse can be tested for by using a __________.
   a. Spark tester
   b. Vacuum hose and a test light
   c. NOID light
   d. DVOM on "pulse ck"
2. A starter motor is drawing too many amperes (current). Technician A says that this could be due to low battery voltage. Technician B says that it could be due to a defective starter motor. Which technician is correct?
   a. Technician A only
   b. Technician B only
   c. Both Technicians A and B
   d. Neither Technician A nor B
3. All of the following could be a cause of excessive starter ampere draw *except* __________.
   a. A misadjusted starter pinion gear
   b. A loose starter housing
   c. Armature wires separated from the commutator
   d. A bent armature
4. The starter motor armature has been rubbing on the pole shoes. The probable cause is __________.
   a. A bent starter shaft
   b. A worn commutator on the armature
   c. Worn starter bushing(s)
   d. Both a and c
5. A starter cranks for a while, then whines. Technician A says that the starter solenoid may be bad. Technician B says that the starter drive may be bad. Which technician is correct?
   a. Technician A only
   b. Technician B only
   c. Both Technicians A and B
   d. Neither Technician A nor B
6. On a negative ground battery system __________.
   a. Disconnect the ground cable first and reconnect the positive cable first
   b. Disconnect the ground cable first and reconnect the positive cable last
   c. Disconnect the positive cable first and reconnect the ground cable first
   d. Disconnect the positive cable first and reconnect the ground cable last
7. A technician is checking the charging system for low output. A voltage drop of 1.67 volts is found between the generator (alternator) output terminal and the battery positive terminal. Technician A says that a corroded connector could be the cause. Technician B says that a defective rectifier diode could be the cause of the voltage drop. Which technician is correct?
   a. Technician A only
   b. Technician B only
   c. Both Technicians A and B
   d. Neither Technician A nor B
8. A technician places a hand on top of a carburetor with the engine running. The engine runs better. Technician A says the choke is defective. Technician B says the engine has a vacuum leak. Which technician is correct?
   a. Technician A only
   b. Technician B only
   c. Both Technicians A and B
   d. Neither Technician A nor B
9. A driver turns the ignition switch to "start" and nothing happens (the dome light remains bright). Technician A says that dirty battery connections or a defective or discharged battery could be the cause. Technician B says that an *open* control circuit such as a defective neutral safety switch could be the cause. Which technician is correct?
   a. Technician A only
   b. Technician B only
   c. Both Technicians A and B
   d. Neither Technician A nor B
10. Engine ping during acceleration can be caused if the ignition timing is __________.
    a. Advanced
    b. Retarded
11. Normal battery drain (parasitic drain) on a vehicle with many computer and electronic circuits is __________.
    a. 20 to 30 mA
    b. 2 to 3 A
    c. 150 to 300 mA
    d. 0.3 to 0.4 A
12. When jump starting __________.
    a. The last connection should be the positive post of the dead battery
    b. The last connection should be the engine block of the dead vehicle
    c. The generator (alternator) must be disconnected on both vehicles
    d. The bumpers should touch to provide a good ground between the vehicles
13. Technician A says that a voltage-drop test of the charging circuit should only be performed when current is flowing through the circuit. Technician B says to connect the lead of a voltmeter to the positive and negative terminals of the battery to measure the voltage drop of the charging system. Which technician is correct?
    a. Technician A only
    b. Technician B only
    c. Both Technicians A and B
    d. Neither Technician A nor B

14. A pickup coil is being measured with a digital multimeter set to the kilo ohm position. The specification for the resistance is 500 to 1,500 ohms. The digital face reads 0.826. Technician A says that the coil is okay. Technician B says that the resistance is below specifications. Which technician is correct?
   a. Technician A only
   b. Technician B only
   c. Both Technicians A and B
   d. Neither Technician A nor B
15. A blown head gasket is suspected on a 5-year-old vehicle. The service technician should perform which of the following tests to confirm the problem?
   a. Running compression test, vacuum test
   b. Leakdown test, compression test
   c. PCV test, oil pressure test
   d. Dipstick test, timing chain slack test
16. Technician A says to check for spark by connecting a spark tester to the end of a spark plug wire. Technician B says that a regular spark plug should be connected to the end of a spark plug wire to check for spark. Which technician is correct?
   a. Technician A only
   b. Technician B only
   c. Both Technicians A and B
   d. Neither Technician A nor B
17. A defective (open) spark plug wire was found in an engine that was misfiring during acceleration. Technician A says that the distributor and rotor should be carefully inspected and replaced if necessary because the defective wire could have caused a carbon track. Technician B says that the ignition coil should be replaced because the bad wire could have caused the coil to become tracked internally. Which technician is correct?
   a. Technician A only
   b. Technician B only
   c. Both Technicians A and B
   d. Neither Technician A nor B
18. Technician A says that all spark plugs should be gapped before being installed in the engine. Technician B says that platinum spark plugs should *not* be regapped after having been used in an engine. Which technician is correct?
   a. Technician A only
   b. Technician B only
   c. Both Technicians A and B
   d. Neither Technician A nor B
19. The vacuum hose to a MAP sensor became disconnected. Technician A says that the lack of vacuum to the sensor will cause the computer to provide a rich mixture to the engine. Technician B says that the computer will supply a lean mixture to the engine. Which technician is correct?
   a. Technician A only
   b. Technician B only
   c. Both Technicians A and B
   d. Neither Technician A nor B
20. The connector to the throttle position (TP) sensor became disconnected. Technician A says that the engine may not idle correctly unless it is reconnected. Technician B says that the engine may hesitate on acceleration. Which technician is correct?
   a. Technician A only
   b. Technician B only
   c. Both Technicians A and B
   d. Neither Technician A nor B
21. An oxygen sensor (O2S) is being tested. Technician A says that the O2S voltage should fluctuate from above 800 mV to below 200 mV. Technician B says the O2S has to be above 600°F (315°C) before testing can begin. Which technician is correct?
   a. Technician A only
   b. Technician B only
   c. Both Technicians A and B
   d. Neither Technician A nor B
22. Technician A says that OBD-II generic codes are the same for all OBD-II vehicles. Technician B says that the DLC is located under the hood on all OBD-II vehicles. Which technician is correct?
   a. Technician A only
   b. Technician B only
   c. Both Technicians A and B
   d. Neither Technician A nor B
23. Ignition timing on an engine equipped with computer-controlled distributor ignition (DI) is being discussed. Technician A says that the computer can only advance the timing. Technician B says that the signal from the knock sensor (KS) can cause the timing to retard. Which technician is correct?
   a. Technician A only
   b. Technician B only
   c. Both Technicians A and B
   d. Neither Technician A nor B
24. Two technicians are discussing jump starting a computer-equipped vehicle with another computer-equipped vehicle. Technician A says that the ignition of both vehicles should be in the off position while making the jumper cable connections. Technician B says that the computer-equipped vehicles should not be jump started. Which technician is correct?
   a. Technician A only
   b. Technician B only
   c. Both Technicians A and B
   d. Neither Technician A nor B
25. A fully charged 12 volt battery should measure ___________.
   a. 12.6 volts
   b. 12.4 volts
   c. 12.2 volts
   d. 12.0 volts
26. Two technicians are discussing an OBD-II diagnostic trouble code (DTC) P0301. Technician A says that this is a generic code. Technician B says that the code is a manufacturer's specific code. Which technician is correct?
   a. Technician A only
   b. Technician B only
   c. Both Technicians A and B
   d. Neither Technician A nor B

27. Battery voltage during cranking is below specifications. Technician A says that a fault in the engine may be the cause. Technician B says that the starter motor may be defective. Which technician is correct?
    a. Technician A only
    b. Technician B only
    c. Both Technicians A and B
    d. Neither Technician A nor B

28. An engine cranks but will not start. No spark is available at the end of a spark plug wire with a spark tester connected and the engine cranked. Technician A says that a defective pickup coil could be the cause. Technician B says that a defective ignition module could be the cause. Which technician is correct?
    a. Technician A only
    b. Technician B only
    c. Both Technicians A and B
    d. Neither Technician A nor B

29. An engine misfire is being diagnosed. One spark plug wire measured "OL" on a digital ohmmeter set to the K ohm scale. Technician A says that the spark plug should be replaced. Technician B says that the ignition coil may also need to be replaced because it may be *tracked* due to the high voltage created by the defective spark plug wire. Which technician is correct?
    a. Technician A only
    b. Technician B only
    c. Both Technicians A and B
    d. Neither Technician A nor B

30. An engine equipped with a turbocharger is burning oil (blue exhaust smoke all the time). Technician A says that a defective wastegate could be the cause. Technician B says that a plugged PCV system could be the cause. Which technician is correct?
    a. Technician A only
    b. Technician B only
    c. Both Technicians A and B
    d. Neither Technician A nor B

31. An engine idles roughly and stalls occasionally. Technician A says that using fuel with too high an RVP level could be the cause. Technician B says that using winter-blend gasoline during warm weather could be the cause. Which technician is correct?
    a. Technician A only
    b. Technician B only
    c. Both Technicians A and B
    d. Neither Technician A nor B

32. A compression test gave the following results:

    Cylinder 1: 155
    Cylinder 2: 140
    Cylinder 3: 110
    Cylinder 4: 105

    Technician A says that a defective (burned) valve is the most likely cause. Technician B says that a leaking intake manifold gasket could be the cause. Which technician is correct?
    a. Technician A only
    b. Technician B only
    c. Both Technicians A and B
    d. Neither Technician A nor B

33. Two technicians are discussing a compression test. Technician A says that the engine should be turned over with the pressure gauge installed for "four puffs." Technician B says that the maximum difference between the highest-reading cylinder and the lowest-reading cylinder should be 20%. Which technician is correct?
    a. Technician A only
    b. Technician B only
    c. Both Technicians A and B
    d. Neither Technician A nor B

34. Technician A says that oil should be squirted into all of the cylinders before taking a compression test. Technician B says that if the compression greatly increases when some oil is squirted into the cylinders, it indicates defective or worn piston rings. Which technician is correct?
    a. Technician A only
    b. Technician B only
    c. Both Technicians A and B
    d. Neither Technician A nor B

35. During a cylinder leakage (leak-down) test, air is noticed coming out of the oil-fill opening. Technician A says that the oil filter may be clogged. Technician B says that the piston rings may be worn or defective. Which technician is correct?
    a. Technician A only
    b. Technician B only
    c. Both Technicians A and B
    d. Neither Technician A nor B

36. A cylinder leakage (leak-down) test indicates 30% leakage, and air is heard coming out of the air inlet. Technician A says that this is a normal reading for a slightly worn engine. Technician B says that one or more intake valves are defective. Which technician is correct?
    a. Technician A only
    b. Technician B only
    c. Both Technicians A and B
    d. Neither Technician A nor B

37. Two technicians are discussing a cylinder power balance test. Technician A says the more the engine RPM drops, the weaker the cylinder. Technician B says that all cylinder RPM drops should be within 50 RPM of each other. Which technician is correct?
    a. Technician A only
    b. Technician B only
    c. Both Technicians A and B
    d. Neither Technician A nor B

38. Technician A says that cranking vacuum should be the same as idle vacuum. Technician B says that a sticking valve is indicated by a floating vacuum gauge needle reading. Which technician is correct?
    a. Technician A only
    b. Technician B only
    c. Both Technicians A and B
    d. Neither Technician A nor B

39. Technician A says that black exhaust smoke is an indication of too rich an air–fuel mixture. Technician B says that white smoke (steam) is an indication of coolant being burned in the engine. Which technician is correct?
    a. Technician A only
    b. Technician B only
    c. Both Technicians A and B
    d. Neither Technician A nor B

40. Excessive exhaust system back pressure has been measured. Technician A says that the catalytic converter may be clogged. Technician B says that the muffler may be clogged. Which technician is correct?
   a. Technician A only
   b. Technician B only
   c. Both Technicians A and B
   d. Neither Technician A nor B

41. A head gasket failure is being diagnosed. Technician A says that an exhaust analyzer can be used to check for HC when the tester probe is held above the radiator coolant. Technician B says that a chemical-coated paper changes color in the presence of combustion gases. Which technician is correct?
   a. Technician A only
   b. Technician B only
   c. Both Technicians A and B
   d. Neither Technician A nor B

42. Technician A says that catalytic converters should last the life of the vehicle unless damaged or contaminated. Technician B says that catalytic converters wear out and should be replaced every 50,000 miles (80,000 km). Which technician is correct?
   a. Technician A only
   b. Technician B only
   c. Both Technicians A and B
   d. Neither Technician A nor B

43. A technician is measuring the battery voltage while cranking the engine and observes 11.2 volts on the voltmeter. Technician A says that the starter may be defective. Technician B says that the battery or cables may be defective. Which technician is correct?
   a. Technician A only
   b. Technician B only
   c. Both Technicians A and B
   d. Neither Technician A nor B

44. The charging system voltage is found to be lower than specified by the vehicle manufacturer. Technician A says that a loose or defective drive belt could be the cause. Technician B says that a defective generator (alternator) could be the cause. Which technician is correct?
   a. Technician A only
   b. Technician B only
   c. Both Technicians A and B
   d. Neither Technician A nor B

45. The fuel pressure on a port-injected engine drops to zero in less than 20 minutes after the engine is turned off. Technician A says to pinch off the fuel return line and if the pressure stays high, the regulator is defective. Technician B says to pinch off the fuel supply line and if the pressure drops off, the problem could be a leaking fuel injector(s). Which technician is correct?
   a. Technician A only
   b. Technician B only
   c. Both Technicians A and B
   d. Neither Technician A nor B

46. Technician A says that if the thermostat is removed from the engine, the engine may overheat. Technician B says that coolant bypasses the thermostat when the thermostat is closed. Which technician is correct?
   a. Technician A only
   b. Technician B only
   c. Both Technicians A and B
   d. Neither Technician A nor B

47. Two technicians are discussing pressure testing the cooling system to check for leaks. Technician A says to pump up the radiator to 20 to 25 PSI and watch for the pressure to drop. Technician B says that the entire cooling system is being pressure checked except the cap by pressurizing the radiator. Which technician is correct?
   a. Technician A only
   b. Technician B only
   c. Both Technicians A and B
   d. Neither Technician A nor B

48. A 4-year-old pickup truck has a high CO reading. Technician A says that a hole in the exhaust downstream from the oxygen sensor could be the cause. Technician B says a defective (electrically open) injector could be the cause. Which technician is correct?
   a. Technician A only
   b. Technician B only
   c. Both Technicians A and B
   d. Neither Technician A nor B

49. A vehicle that has a higher than normal HC and CO reading is most likely running ____________.
   a. Rich
   b. Lean

50. A short-term fuel trim of 0% and a long-term fuel trim of +20% means ____________.
   a. The engine is running lean now
   b. The engine has a history of running rich
   c. The engine has a history of running lean
   d. The engine is running rich now

51. The catalytic converter can be tested using ____________.
   a. A vacuum gauge connected to the intake manifold vacuum
   b. A temperature measuring tool
   c. A pressure gauge attached in the place of the O2S
   d. All of the above

52. A four-cylinder TBI engine has a rich DTC. Technician A says that this could be caused by a faulty injector. Technician B says a faulty MAP sensor could be the cause. Which technician is correct?
   a. Technician A only
   b. Technician B only
   c. Both Technicians A and B
   d. Neither Technician A nor B

53. A pickup coil is being measured with a digital multimeter set to the ohm scale. The meter reads 0.01 ohm between the pickup coil lead and a good engine ground. Technician A says that the pickup coil is electronically open. Technician B says that the pickup coil is shorted to ground. Which technician is correct?
   a. Technician A only
   b. Technician B only
   c. Both Technicians A and B
   d. Neither Technician A nor B

54. An engine has a rough idle but runs okay above idle speed. Technician A says that the EGR valve could be partially stuck open. Technician B says that the thermostat may be stuck open. Which technician is correct?
   a. Technician A only
   b. Technician B only
   c. Both Technicians A and B
   d. Neither Technician A nor B

55. An oxygen sensor reads lower than normal. Technician A says that the engine may have a vacuum leak (intake manifold air leak). Technician B says that the exhaust manifold may be cracked. Which technician is correct?
   a. Technician A only
   b. Technician B only
   c. Both Technicians A and B
   d. Neither Technician A nor B

56. An engine has a defective spark plug. Technician A says that the O2S will read lower than normal due to the misfire. Technician B says that the O2S will read higher than normal due to the misfire. Which technician is correct?
   a. Technician A only
   b. Technician B only
   c. Both Technicians A and B
   d. Neither Technician A nor B

57. An engine with a cracked spark plug is being analyzed using an ignition scope. Technician A says that the cylinder with the defective spark plug will have a shorter than normal firing line. Technician B says that the spark line will be longer than normal for the cylinder with the cracked spark plug. Which technician is correct?
   a. Technician A only
   b. Technician B only
   c. Both Technicians A and B
   d. Neither Technician A nor B

58. The idle speed of a port fuel-injected engine is too fast. Technician A says that the PCV valve may be defective. Technician B says the EGR valve may be stuck closed. Which technician is correct?
   a. Technician A only
   b. Technician B only
   c. Both Technicians A and B
   d. Neither Technician A nor B

59. An engine starts and idles okay but it misfires above idle. Technician A says that the fuel pump may be weak. Technician B says that the secondary ignition system may have a fault. Which technician is correct?
   a. Technician A only
   b. Technician B only
   c. Both Technicians A and B
   d. Neither Technician A nor B

60. A customer comments that the engine misfires when going up a hill or on hard acceleration. When viewing the scope, the technician sees that the #5 firing line is about 5 to 6 kV higher than the rest, and the spark line slants down from the firing line to the coil oscillations. What is the most likely cause?
   a. A fuel-fouled plug on #5
   b. A plug with a worn electrode on #5
   c. A high-resistance plug wire on #5
   d. This is a normal pattern

61. A high O2S voltage could be due to ____________.
   a. A rich exhaust
   b. A lean exhaust
   c. A defective spark plug wire
   d. Both a and c

62. A low O2S voltage could be due to ____________.
   a. A rich exhaust
   b. A lean exhaust
   c. A defective spark plug wire
   d. Both b and c

63. A fuel-injected vehicle is tested on a four-gas exhaust analyzer.

   HC = 102 PPM   CO = 0.3% O_2 = 6.3% CO_2 = 6.1%

   Technician A says that everything is okay including the TP setting because the O_2 and CO_2 are about equal. Technician B says that the engine is running lean. Which technician is correct?
   a. Technician A only
   b. Technician B only
   c. Both Technicians A and B
   d. Neither Technician A nor B

64. A technician is working on a vehicle equipped with a port-injected engine. After connecting the vehicle to a scan tool, the technician finds it has a long-term fuel trim of +20%. Technician A says that an exhaust leak in front of the oxygen sensor could cause this. Technician B says that a defective plug wire could cause this. Which technician is correct?
   a. Technician A only
   b. Technician B only
   c. Both Technicians A and B
   d. Neither Technician A nor B

65. The oil pressure warning light comes on when the oil pressure reaches about ____________.
   a. 1 PSI
   b. 4 PSI
   c. 8 PSI
   d. 12 PSI

66. The spark line is short in duration (0.60 ms maximum), and the firing line is low on all cylinders. What is the most likely cause?
   a. The cap to rotor button has excessive resistance.
   b. The rotor to distributor air gap is too close.
   c. The secondary is shorted between its windings.
   d. The primary coil resistance is too low.

67. Two technicians are discussing catalytic converters. Technician A says that a nonworking (chemically inert) catalytic converter will test as being clogged during a vacuum or back pressure test. Technician B says that the temperature of the inlet and outlet of the converter can detect if it is working okay. Which technician is correct?
   a. Technician A only
   b. Technician B only
   c. Both Technicians A and B
   d. Neither Technician A nor B

68. HC and CO_2 are high and CO and O_2 are low. The most likely cause is ____________.
   a. Too-rich conditions
   b. Fault in the secondary ignition system
   c. Stuck-open EGR
   d. Lean misfire

69. A typical O2S sensor output can be measured for changing ______________.

a. DC volts
b. AC volts
c. Ohms (resistance)
d. Frequency (hertz)

70. Two technicians are discussing excessive HC exhaust emissions. Technician A says that a stuck-open thermostat could be the cause. Technician B says that a lean misfire could be the cause. Which technician is correct?

a. Technician A only
b. Technician B only
c. Both Technicians A and B
d. Neither Technician A nor B

ANSWERS FOR SAMPLE ENGINE PERFORMANCE (A8) ASE-TYPE CERTIFICATION TEST

1. c
2. c
3. c
4. d
5. b
6. a
7. a
8. b
9. b
10. a
11. a
12. b
13. a
14. a
15. b
16. a
17. c
18. c
19. a
20. c
21. c
22. a
23. c
24. a
25. a
26. a
27. c
28. c
29. c
30. b
31. c
32. d
33. c
34. b
35. b
36. b
37. b
38. d
39. c
40. c
41. c
42. a
43. d
44. c
45. c
46. c
47. b
48. d
49. a
50. c
51. d
52. c
53. b
54. a
55. c
56. a
57. c
58. a
59. c
60. c
61. a
62. d
63. b
64. c
65. b
66. c
67. b
68. b
69. a
70. c

appendix 2

2013 NATEF CORRELATION CHART (A8)

MLR—Maintenance & Light Repair
AST—Auto Service Technology (Includes MLR)
MAST—Master Auto Service Technology (Includes MLR and AST)

ENGINE PERFORMANCE (A8)

TASK	PRIORITY	MLR	AST	MAST	TEXT PAGE #	TASK PAGE #
A. GENERAL: ENGINE DIAGNOSIS						
1. Identify and interpret engine performance concerns; determine necessary action.	P-1		✔	✔	523	44, 55, 154, 155, 177
2. Research applicable vehicle and service information, vehicle service history, service precautions, and technical service bulletins.	P-1	✔	✔	✔	2–4	4–9, 12, 13, 16, 24, 29,45, 68, 69, 87,88, 135, 136, 153, 178–180
3. Diagnose abnormal engine noises or vibration concerns; determine necessary action.	P-3		✔	✔	204–205	38,46
4. Diagnose the cause of excessive oil consumption, coolant consumption, unusual exhaust color, odor, and sound; determine necessary action.	P-2		✔	✔	201	25, 41, 42, 54, 92
5. Perform engine absolute (vacuum/boost) manifold pressure tests; determine necessary action.	P-1	✔	✔	✔	212–214	47
6. Perform cylinder power balance test; determine necessary action.	P-1	✔	✔	✔	211	48, 49
7. Perform cylinder cranking and running compression tests; determine necessary action.	P-1	✔	✔	✔	208–210	50–52
8. Perform cylinder leakage test; determine necessary action.	P-1	✔	✔	✔	210–211	53
9. Diagnose engine mechanical, electrical, electronic, fuel, and ignition concerns; determine necessary action.	P-2		✔	✔	523–531	55
10. Verify engine operating temperature; determine necessary action.	P-1	✔	✔	✔	134	26–28, 56, 59
11. Verify correct camshaft timing.	P-1		✔	✔	225	62, 139
B. COMPUTERIZED CONTROLS DIAGNOSIS AND REPAIR						
1. Retrieve and record diagnostic trouble codes, OBD monitor status, and freeze frame data; clear codes when applicable.	P-1		✔	✔	527–531	181, 184, 185
2. Access and use service information to perform step-by-step (troubleshooting) diagnosis.	P-1		✔	✔	523	57, 182, 183, 186

TASK	PRIORITY	MLR	AST	MAST	TEXT PAGE #	TASK PAGE #
3. Perform active tests of actuators using a scan tool; determine necessary action.	P-2		✔	✔	529–530	58, 141
4. Describe the importance of running all OBD-II monitors for repair verification.	P-1	✔	✔	✔	322	93
5. Diagnose the causes of emissions or driveability concerns with stored or active diagnostic trouble codes; obtain, graph, and interpret scan tool data.	P-1			✔	326–329	89, 90
6. Diagnose emissions or driveability concerns without stored diagnostic trouble codes; determine necessary action.	P-1			✔	330	20, 91, 137, 138
7. Inspect and test computerized engine control system sensors, powertrain/engine control module (PCM/ECM), actuators, and circuits using a graphing multimeter (GMM)/digital storage oscilloscope (DSO); perform necessary action.	P-2			✔	285; 299; 375	94–124, 132
8. Diagnose driveability and emissions problems resulting from malfunctions of interrelated systems (cruise control, security alarms, suspension controls, traction controls, A/C, automatic transmissions, non-OEM installed accessories, or similar systems); determine necessary action.	P-3			✔	523–529	132, 140, 142, 187, 188
C. IGNITION SYSTEM DIAGNOSIS AND REPAIR						
1. Diagnose (troubleshoot) ignition system related problems such as no-starting, hard starting, engine misfire, poor driveability, spark knock, power loss, poor mileage, and emissions concerns; determine necessary action.	P-2		✔	✔	282	70, 71, 75, 80-83
2. Inspect and test crankshaft and camshaft position sensor(s); perform necessary action.	P-1		✔	✔	284–285	72-74, 76–78, 84
3. Inspect, test, and/or replace ignition control module, powertrain/engine control module; reprogram as necessary.	P-3		✔	✔	286	85
4. Remove and replace spark plugs; inspect secondary ignition components for wear and damage.	P-1	✔	✔	✔	289–294	79, 86
D. FUEL, AIR INDUCTION, AND EXHAUST SYSTEMS DIAGNOSIS AND REPAIR						
1. Diagnose (troubleshoot) hot or cold no-starting, hard starting, poor drivability, incorrect idle speed, poor idle, flooding, hesitation, surging, engine misfire, power loss, stalling, poor mileage, dieseling, and emissions problems; determine necessary action.	P-2			✔	455–457	143
2. Check fuel for contaminants; determine necessary action.	P-2		✔	✔	102; 105	17–19
3. Inspect and test fuel pumps and pump control systems for pressure, regulation, and volume; perform necessary action.	P-1		✔	✔	412–417	126–130
4. Replace fuel filter(s).	P-1	✔	✔	✔	411	131
5. Inspect, service, or replace air filters, filter housings, and intake duct work.	P-1	✔	✔	✔	456	33
6. Inspect throttle body, air induction system, intake manifold and gaskets for vacuum leaks and/or unmetered air.	P-2		✔	✔	367; 456; 525	61, 134, 152
7. Inspect and test fuel injectors.	P-2		✔	✔	458–465	144–151

TASK	PRIORITY	MLR	AST	MAST	TEXT PAGE #	TASK PAGE #
8. Verify idle control operation.	P-1		✔	✔	466	188
9. Inspect integrity of the exhaust manifold, exhaust pipes, muffler(s), catalytic converter(s), resonator(s), tail pipe(s), and heat shields; perform necessary action.	P-1	✔	✔	✔	170–173	34, 43
10. Inspect condition of exhaust system hangers, brackets, clamps, and heat shields; repair or replace as needed.	P-1	✔	✔	✔	171–173	35
11. Perform exhaust system back-pressure test; determine necessary action.	P-2		✔	✔	214; 508	170
12. Check and refill diesel exhaust fluid (DEF).	P-3	✔	✔	✔	–	14
13. Test the operation of turbocharger/supercharger systems; determine necessary action.	P-3			✔	188–198	39
E. EMISSIONS CONTROL SYSTEMS DIAGNOSIS AND REPAIR						
1. Diagnose oil leaks, emissions, and drivability concerns caused by the positive crankcase ventilation (PCV) system; determine necessary action.	P-3		✔	✔	499	161
2. Inspect, test, and service positive crankcase ventilation (PCV) filter/breather cap, valve, tubes, orifices, and hoses; perform necessary action.	P-2	✔	✔	✔	500	162, 163
3. Diagnose emissions and drivability concerns caused by the exhaust gas recirculation (EGR) system; determine necessary action.	P-3		✔	✔	495	156
4. Diagnose emissions and drivability concerns caused by the secondary air injection and catalytic converter systems; determine necessary action.	P-2			✔	504; 508	164
5. Diagnose emissions and drivability concerns caused by the evaporative emissions control system; determine necessary action.	P-2			✔	516	172
6. Inspect and test electrical/electronic sensors, controls, and wiring of exhaust gas recirculation (EGR) systems; perform necessary action.	P-2		✔	✔	494–495	158
7. Inspect, test, service, and replace components of the EGR system including tubing, exhaust passages, vacuum/pressure controls, filters, and hoses; perform necessary action.	P-2		✔	✔	496–497	157, 159, 160
8. Inspect and test electrical/electronically operated components and circuits of air injection systems; perform necessary action.	P-3		✔	✔	503	165, 166
9. Inspect and test catalytic converter efficiency.	P-2		✔	✔	509	167–169
10. Inspect and test components and hoses of the evaporative emissions control system; perform necessary action.	P-1		✔	✔	516	172, 173
11. Interpret diagnostic trouble codes (DTCs) and scan tool data related to the emissions control systems; determine necessary action.	P-1		✔	✔	496; 501; 504; 511; 519	174–176

GLOSSARY

Aboveground storage tank (AGST) A storage tank that stores used oil and is located aboveground.

AC ripple voltage An alternating current voltage that rides on top of a DC charging current output from an AC generator (alternator).

Acceleration simulation mode (ASM) Uses a dynamometer that applies a heavy load on the vehicle at a steady-state speed.

Accelerator pedal position (APP) sensor A sensor that is used to monitor the position and rate of change of the accelerator pedal.

Accumulator A temporary location for fluid under pressure.

Active fuel management (AFM) A term used by General Motors to describe their variable displacement system. Previously called displacement on demand.

Actuator An electrical or mechanical device that converts electrical energy into a mechanical action, such as adjusting engine idle speed, altering suspension height, or regulating fuel metering.

Adjustable wrench A wrench that has a movable jaw to allow it to fit many sizes of fasteners.

Adsorption Attaches the fuel vapor molecules to the carbon surface.

AFV Alternative-fuel vehicle.

AGST Aboveground storage tank, used to store used oil.

Air–fuel ratio The ratio of air to fuel in an intake charge as measured by weight.

Air–fuel ratio sensor A term used to describe a wide-band oxygen sensor.

Air reference chamber This electrode is exposed to outside (ambient) air and is the signal electrode, also called the reference electrode or ambient air electrode.

Alternator An electric generator that produces alternating current; also called an AC generator.

Ambient air electrode This electrode is exposed to outside (ambient) air and is the signal electrode, also called the reference electrode or ambient air electrode.

Ambient side electrode This electrode is exposed to outside (ambient) air and is the signal electrode, also called the reference electrode or ambient air electrode.

Ampere-hour A battery rating that combines the amperage output times the amount of time in hours that a battery is able to supply.

Analog-to-digital (AD) converter An electronic circuit that converts analog signals into digital signals that can then be used by a computer.

Anhydrous ethanol A type of ethanol that has almost zero absorbed water.

Annealing A heat-treating process that removes the brittleness at the outer surface of the material to reduce the chance of cracking from the temperature changes.

Antiknock Index (AKI) The pump octane.

API gravity An arbitrary scale expressing the gravity or density of liquid petroleum products devised jointly by the American Petroleum Institute and the National Bureau of Standards.

Asbestosis A health condition where asbestos causes scar tissue to form in the lungs causing shortness of breath.

ASD Automatic Shutdown Relay.

ASM 50/15 test Places a load of 50% on the vehicle at a steady 15 mph. This load represents 50% of the horsepower required to simulate the FTP acceleration rate of 3.3 mph/sec.

ASM 25/25 test Places a 25% load on the vehicle while it is driven at a steady speed of 25 mph. This represents 25% of the load required to simulate the FTP acceleration rate of 3.3 mph/sec.

ASTM American Society for Testing Materials.

B20 A blend of 20% biodiesel with 80% petroleum diesel.

Backpressure The exhaust system's resistance to flow. Measured in pounds per square inch (PSI).

Baffle A plate or shield used to direct the flow of a liquid or gas.

Bar When air is pumped into the cylinder, the combustion chamber receives an increase of air pressure known as boost and is measured in pounds per square inch (PSI), atmospheres (ATM), or bar.

BARO sensor A sensor used to measure barometric pressure.

Barometric manifold absolute pressure (BMAP) sensor A sensor that measures both the barometric pressure and the absolute pressure in the intake manifold.

Base timing The timing of the spark before the computer advances the timing.

Battery A battery stores electrical energy in the form of a chemical reaction that can be reversed.

Battery electrical drain test A test used to determines if some component or circuit in a vehicle or truck is causing a drain on the battery when everything is off. This test is also called the ignition off-draw (IOD) or parasitic load test.

Battery voltage correction factor The PCM senses low battery voltage and increases the fuel injector on-time to help compensate for the lower voltage to the fuel pump and fuel injectors. This increase in injector pulse time is added to the calculated pulse time.

BCI Battery Council International.

Bench grinder An electric-powered grinding stone usually combined with a wire wheel and mounted to a bench.

Bias voltage A weak signal voltage applied to an oxygen sensor by the PCM. This weak signal voltage is used by the PCM to detect when the oxygen sensor has created a changing voltage and for diagnosis of the oxygen sensor circuit.

Biodiesel A renewable fuel manufactured from vegetable oils, animal fats, or recycled restaurant grease.

Biomass Nonedible farm products, such as cornstalks, cereal straws, and plant wastes from industrial processes, such as sawdust and paper pulp, used in making ethanol.

Block The foundation of any engine. All other parts are either directly or indirectly attached to the block of an engine.

BMAP sensor A sensor that has individual circuits to measure barometric and manifold pressure. This input not only allows the computer to adjust for changes in atmospheric pressure due to weather, but also is the primary sensor used to determine altitude.

Bolts A threaded fastener use to attach two parts. The threaded end can be installed into a casting such as an engine block or a nut used to join two parts.

Boost An increase in air pressure above atmospheric. Measured in pounds per square inch (PSI).

Bore The inside diameter of the cylinder in an engine.

Boundary lubrication An oil film that is thick enough to keep the surfaces from seizing, but can allow some contact to occur.

Boxer A type of engine design that is flat and has opposing cylinders. Called a boxer because the pistons on one side resemble a boxer during engine operation. Also called a pancake engine.

Breaker bar A handle used to rotate a socket; also called a flex handle.

British thermal unit (BTU) A unit of heat measurement.

Bump cap A hat that is made from plastic and is hard enough to protect the head from bumps.

Burn kV Spark line voltage.

Bypass A passage that allows coolant to flow around a closed thermostat in an engine cooling system.

Bypass ignition Commonly used on General Motors engines equipped with distributor ignition (DI), as well as those equipped with waste-spark ignition.

Bypass valve Allows intake air to flow directly into the intake manifold bypassing the supercharger.

CA Cranking amperes. A rating for batteries.

CAA Clean Air Act. Federal legislation passed in 1970 that established national air quality standards.

Calibration codes Codes used on many Powertrain Control Modules (PCM).

California Air Resources Board (CARB) A state of California agency that regulates the air quality standards for the state.

Cam-in-block design An engine where the camshaft is located in the block rather than in the cylinder head.

Camshaft A shaft in an engine that is rotated by the crankshaft by a belt or chain and used to open valves.

Campaign A recall where vehicle owners are contacted to return a vehicle to a dealer for corrective action.

Controller area network (CAN) A type of serial data transmission.

Capacity test A battery test that tests a battery by applying an electric load.

Casting number An identification code cast into an engine block or other large cast part of a vehicle.

Catalysts Platinum and palladium used in the catalytic converter to combine oxygen (O_2) with hydrocarbons (HC) and carbon monoxide (CO) to form nonharmful tailpipe emissions of water (H_2O) and carbon dioxide (CO_2).

Catalytic converter An emission control device located in the exhaust system that changes HC and CO into harmless H_2O and CO_2. In a three-way catalyst, NO_x is also separated into harmless, separate N and O.

Catalytic cracking Breaking hydrocarbon chains using heat in the presence of a catalyst.

Cavitation A condition that can occur in a cooling system where the inlet pressure is lowered too much, the coolant at the pump inlet can boil, producing vapor. The pump will then spin the coolant vapors and not pump coolant.

CCA Cold Cranking Amps. A rating of a battery tested at 0° F.

CCM Comprehensive component monitor.

Cellulose ethanol Ethanol produced from biomass feedstock such as agricultural and industrial plant wastes.

Cellulosic biomass Composed of cellulose and lignin, with smaller amounts of proteins, lipids (fats, waxes, and oils), and ash.

Centrifugal pump A type of pump used for water pumps where a large volume of liquid can be moved using a rotating impeller without building pressure.

Cerium An element that can store oxygen.

CFR Code of Federal Regulations.

Charging circuit The components, wiring, and connectors needed to keep the battery charged.

Cheater bar A pipe or other object used to lengthen the handle of a ratchet or breaker bar. Not recommended to be used as the extra force can cause the socket or ratchet to break.

Chisels A type of hand tool used to mark or cut strong material such as steel.

Clock generator A crystal that determines the speed of computer circuits.

Closed-end wrench A type of hand tool that has an end that surrounds the head of a bolt or nut.

Closed-loop operation A phase of computer-controlled engine operation in which oxygen sensor feedback is used to calculate air–fuel mixture.

Cloud point The low-temperature point at which the waxes present in most diesel fuel tend to form wax crystals that clog the fuel filter.

CNG Compressed natural gas.

Coal-to-liquid (CTL) A refining process in which coal is converted to liquid fuel.

Coast-down stall A condition that results in the engine stalling when coasting to a stop.

Coil-on-plug (COP) ignition An ignition system without a distributor, where each spark plug has an ignition coil.

Combustion The rapid burning of the air–fuel mixture in the engine cylinders, creating heat and pressure.

Combustion chamber The space left within the cylinder when the piston is at the top of its combustion chamber.

Companion cylinder Two cylinders that share an ignition coil on a waste-spark-type ignition system.

Compressed natural gas (CNG) An alternative fuel that uses natural gas compressed at high pressures and used as a vehicle fuel.

Compression ratio The ratio of the volume in the engine cylinder with the piston at bottom dead center (BDC) to the volume at top dead center (TDC).

Compression-sensing ignition A type of waste-spark ignition system that does not require the use of a camshaft position sensor to determine cylinder number.

Compression test An engine test that helps determine the condition of an engine based on how well each cylinder is able to compress the air on the compression stroke.

Compressor bypass valve (CBV) This type of relief valve routes the pressurized air to the inlet side of the turbocharger for reuse and is quiet during operation.

Conductance testing A type of electronic battery tester that determines the condition and capacity of a battery by measuring the conductance of the cells.

Connecting rod A metal rod that connects the piston to the crankshaft.

Controller A term that is usually used to refer to a computer or an electronic control unit (ECU).

Coolant recovery system A type of cooling system where the coolant is drawn back into a plastic container when the coolant cools.

Core tubes Oval-shaped hollow tubes where coolant flows through in a radiator.

CPU Central processor unit.

Cracking A refinery process in which hydrocarbons with high boiling points are broken into hydrocarbons with low boiling points.

Cranking circuit The components, wiring and connectors needed to crank the engine.

Cranking vacuum test Measuring the amount of manifold vacuum during cranking.

Crankshaft The part of an engine that transfers the up and down motion of the pistons to rotary motion.

Cross counts The number of times an oxygen sensor changes voltage from high to low (from low to high voltage is not counted) in 1 second (or 1.25 second, depending on scan tool and computer speed).

Cup design A design of an oxygen sensor that uses a shape like a cup or thimble.

Cycle A series of events such as the operation of the four strokes of an engine that repeats.

Cylinder The part of an engine that is round and houses the piston.

Cylinder cutoff system A term used to describe a system where some cylinders are deactivated to reduce fuel consumption.

Cylinder head temperature (CHT) sensor A temperature sensor mounted on the cylinder head and used by the PCM to determine fuel delivery.

Cylinder leakage test A test that involves injecting air under pressure into the cylinders one at a time. The amount and location of any escaping air helps the technician determine the condition of the engine.

Data link connector (DLC) The electrical connector where a scan tool is connected to access the computer of the vehicle.

DE Abbreviation for the drive end housing of a starter or generator (alternator).

Default position The position of the throttle plate in an electronic throttle control without any signals from the controller.

Delta Pressure Feedback EGR (DPFE) sensor This sensor measures the pressure differential between two sides of a metered orifice positioned just below the EGR valve's exhaust side.

emand delivery system (DDS) A type of electronic fuel injection system.

Detonation A violent explosion in the combustion chamber created by uncontrolled burning of the air–fuel mixture; often causes a loud, audible knock. Also known as spark knock or ping.

Diagnostic executive Software program designed to manage the operation of all OBD-II monitors by controlling the sequence of steps necessary to execute the diagnostic tests and monitors.

Diesel exhaust fluid (DEF) A colorless, odorless, and nontoxic liquid used to reduce NO_x emissions produced in a diesel engine by injecting urea into the exhaust stream.

Diffusion chamber A section or part of a wide-band oxygen sensor that is exposed to exhaust gases.

Diesohol Standard #2 diesel fuel combined with up to 15% ethanol.

Digital computer A computer that uses on and off signals only. Uses an A to D converter to change analog signals to digital before processing.

Direct injection A fuel-injection system design in which gasoline is injected directly into the combustion chamber.

DIS Distributorless ignition system. Also called direct-fire ignition system.

Displacement The total volume displaced or swept by the cylinders in an internal combustion engine.

Displacement on demand (DOD) A term used by General Motors to describe their variable displacement system.

Distillation The process of purification through evaporation and then condensation of the desired liquid.

Distillation curve A graph that plots the temperatures at which the various fractions of a fuel evaporate.

Distributor cap Provides additional space between the spark plug connections to help prevent crossfire.

Double overhead camshaft (DOHC) An engine design that has two overhead camshafts. One camshaft operates the intake valves and the other is used for the exhaust valves.

Drive-by-wire A term used to describe an engine equipped with an electronic throttle control (ETC) system.

Drive cycle Driving the vehicle under conditions that allow the PCM to conduct monitor tests.

Drive size The size in fractions of an inch of the square drive for sockets.

Driveability index (DI) A calculation of the various boiling temperatures of gasoline that once compiled can indicate the fuel's ability to perform well at low temperatures.

Dual cell A design of a wide-band oxygen sensor that uses two cells.

Dump valve Features an adjustable spring design that keeps the valve closed until a sudden release of the throttle. The resulting pressure increase opens the valve and vents the pressurized air directly into the atmosphere.

Duty cycle Refers to the percentage of on-time of the signal during one complete cycle.

Dwell The number of degrees of distributor cam rotation that the points are closed.

Dwell section The amount of time that the current is charging the coil from the transistor-on point to the transistor-off point.

Dynamic compression test A compression test done with the engine running rather than during engine cranking as is done in a regular compression test.

E10 A fuel blend of 10% ethanol and 90% gasoline.

E^{2}PROM Electrically erasable programmable read-only memory.

E85 A fuel blend of 85% ethanol and 15% gasoline.

ECA Electronic Control Module. The name used by Ford to describe the computer used to control spark and fuel on older-model vehicles.

ECM Electronic control module on a vehicle.

ECT Engine coolant temperature.

ECU Electronic control unit on a vehicle.

E-diesel Standard #2 diesel fuel combined with up to 15% ethanol. Also known as diesohol.

EEPROM Electronically erasable programmable read-only memory.

EGR valve position (EVP) A linear potentiometer on the top of the EGR valve stem indicates valve position for the computer.

ELD Abbreviation for electrical load detection, a circuit used in the charging system to allow the system to work only when needed thereby improving fuel economy.

Electromagnetic interference (EMI) An undesirable electronic signal. It is caused by a magnetic field building up and collapsing, creating unwanted electrical interference on a nearby circuit.

Electronic air control (EAC) The idle air control valve.

Electronic ignition (EI) Electronic ignition (EI) is the term specified by the SAE for an ignition system that does not use a distributor.

Electronic ignition system (EIS) A term used to describe the type of Chrysler ignition system.

Electronic returnless fuel system (ERFS) A fuel delivery system that does not return fuel to the tank.

Electronic spark timing (EST) The computer controls spark timing advance.

Electronic throttle control (ETC) A system that moves the throttle plate using an electric motor instead of a mechanical linkage from the accelerator pedal.

Enable criteria Operating condition requirements.

Engine fuel temperature (EFT) sensor A temperature sensor located on the fuel rail that measures the temperature of the fuel entering the engine.

Engine mapping A computer program that uses engine test data to determine the best fuel–air ratio and spark advance to use at each speed of the engine for best performance.

EPA Environmental Protection Agency.

EREV Abbreviation for extended range electric vehicles.

Ethanol Grain alcohol that is blended with gasoline to produce motor fuel. Also known as ethyl alcohol.

Ethyl alcohol See *ethanol*.

Ethylene glycol Used with water for use as a coolant.

EVCP Exhaust valve cam phaser

Exhaust gas recirculation (EGR) An emission control device to reduce NO_X (oxides of nitrogen).

Exhaust side electrode The electrode of a wide-band oxygen sensor that is exposed to the exhaust stream.

Exhaust valve The valve in an engine that opens to allow the exhaust to escape into the exhaust manifold.

Extension A socket wrench tool used between a ratchet or breaker bar and a socket.

External combustion engine A type of engine that burns fuel from outside the engine itself such as a steam engine.

Eye wash station A water fountain designed to rinse the eyes with a large volume of water.

Fail safe position A term used to describe the default position for the throttle plate in an electronic throttle control (ETC) system.

False air A term used to describe air that enters the engine without being measured by the mass air flow sensor.

False lean indication Occurs when an oxygen sensor reads low as a result of a factor besides a lean mixture.

False rich indication A high oxygen sensor voltage reading that is not the result of a rich exhaust. Some common causes for this false rich indication include a contaminated oxygen sensor and having the signal wire close to a high voltage source such as a spark plug wire.

FFV Flexible fuel vehicle.

Files A hand tool that is used to smooth rough or sharp edges from metal.

Finger design A design of an oxygen sensor that uses a shape like a cup or thimble.

Fins Thin metal pieces attached to coolant tubes to help transfer heat from the coolant to the outside air.

Fire blanket A fireproof wool blanket used to cover a person who is on fire and smother the fire.

Fire extinguisher classes The types of fires that a fire extinguisher is designed to handle are referred to as fire classes.

Firing line The leftmost vertical (upward) line.

Firing order The order that the spark is distributed to the correct spark plug at the right time.

Fisher-Tropsch A method to create synthetic liquid fuel from coal.

Flare An increase and then decrease in engine speed.

Flash point The temperature at which the vapors on the surface of the fuel will ignite if exposed to an open flame.

Flex fuel Flex-fuel vehicles are capable of running on straight gasoline or gasoline/ethanol blends.

Four-stroke cycle An engine design that requires four strokes to complete one cycle with each stroke requiring 180 degrees of crankshaft rotation.

Freeze-frame A snapshot of all of the engine data at the time the DTC was set.

Fretting A term used to describe the shedding of a gasket caused by the expansion and contraction of the two surfaces on the sides of the gasket.

FTD Fischer-Tropsch diesel.

FTP Federal Test Procedure.

Fuel rail A term used to describe the tube that delivers the fuel from the fuel line to the individual fuel injectors.

Fuel compensation sensor A sensor used in flex-fuel vehicles that provides information to the PCM on the ethanol content and temperature of the fuel as it is flowing through the fuel delivery system.

Functionality Refers to PCM inputs checking the operation of the outputs.

Fungible A term used to describe a product that has the same grade or meets the same specifications, such as oil, that can be interchanged with another product without any affect.

Gang fired Pulsing injectors in groups.

Gasoline Refined petroleum product that is used primarily as a motor fuel. Gasoline is made up of many different hydrocarbons and also contains additives for enhancing its performance in an ICE.

Gasoline direct injection (GDI) A fuel-injection system design in which gasoline is injected directly into the combustion chamber.

Gas-to-liquid (GTL) A refining process in which natural gas is converted into liquid fuel.

GAWR Gross axle weight rating. A rating of the load capacity of a vehicle and included on placards on the vehicle and in the owner's manual.

Gerotor A type of positive displacement pump that is often used in engine oil pumps. It uses the meshing of internal and external gear teeth to pressurize the fuel.

Glow plug A heating element that uses 12 volts from the battery and aids in the starting of a cold engine.

GMM Graphing multimeter. A cross between a digital meter and a digital storage oscilloscope.

Grade The strength rating of a bolt.

Grain alcohol See *ethanol*.

Ground side switching A type of computer control of a circuit where the PCM controls the ground side of the circuit. Commonly used to control the oil control valve in a variable valve timing system.

GVWR Gross vehicle weight rating. The total weight of the vehicle including the maximum cargo.

Hacksaws A type of hand tool that is used to cut metal or other hard materials.

Hall-effect switch A semiconductor moving relative to a magnetic field, creating a variable voltage output. A type of electromagnetic sensor used in electronic ignition and other systems and used to determine position. Named for Edwin H. Hall, who discovered the Hall effect in 1879.

Hammer A hand tool that is used to apply force by swinging.

Hangers Made of rubberized fabric with metal ends that hold the muffler and tailpipe in position so that they do not touch any metal part. This helps to isolate the exhaust noise from the rest of the vehicle.

Hazardous waste materials A classification of materials that can cause harm to people or the environment.

Heat of compression Air is compressed until its temperature reaches about 1,000°F.

Helmholtz resonator Used on the intake duct between the air filter and the throttle body to reduce air intake noise during engine acceleration.

HEPA vacuum High-efficiency particulate air filter vacuum used to clean brake dust.

HEUI Hydraulic Electronic Unit Injection, a type of diesel injector system.

HEV Abbreviation for hybrid electric vehicle.

High-pressure common rail (HPCR) Diesel fuel under high pressure, over 20,000 PSI (138,000 kPa), is applied to the injectors, which are opened by a solenoid controlled by the computer. Because the injectors are computer controlled, the combustion process can be precisely controlled to provide maximum engine efficiency with the lowest possible noise and exhaust emissions.

Homogeneous mode A mode of operation in a gasoline direct injection system where the air–fuel mixture is the same throughout the cylinder.

Hydraulic electronic unit injection (HEUI) A type of diesel fuel injector that uses high-pressure engine oil to operate the injector.

Hydrocracking A refinery process that converts hydrocarbons with a high boiling point into ones with low boiling points.

Hydrodynamic lubrication An increase in oil pressure that occurs when oil is forced between two moving objects such as between a crankshaft journals and the bearing.

Hydrokinetic pump This design of pump rapidly moves the fuel to create pressure.

I/M 240 test It is a portion of the 505-second FTP test used by the manufacturers to certify their new vehicles. The "240" stands for 240 seconds of drive time on a dynamometer.

IAC Idle air control.

IAC counts The commanded position of a typical idle air control valve.

Idle speed control (ISC) motor A motor, usually a stepper motor, used to move a pintle that allows more or less air past the throttle plate thereby controlling idle speed.

Idle stop A block or screw that stops the throttle plate from closing all of the way which could cause it to stick in the housing.

Idle vacuum test A test performed using a vacuum gauge attached to the intake manifold of an engine running at idle speed.

Igniter Ignition Control Module.

Ignition coil An electrical device that consists of two separate coils of wire: a primary and a secondary winding. The purpose of an ignition coil is to produce a high-voltage (20,000 to 40,000 volts), low-amperage (about 80 mA) current necessary for spark ignition.

Ignition control module (ICM) Controls (turns on and off) the primary ignition current of an electronic ignition system.

Ignition timing The exact point of ignition in relation to piston position.

Impeller The rotating part inside a water pump that is rotated by the engine and is used to move coolant.

Inches of mercury (in. Hg) A measurement of vacuum; pressure below atmospheric pressure.

Indirect injection (IDI) Fuel is injected into a small prechamber, which is connected to the cylinder by a narrow opening. The initial combustion takes place in this prechamber. This has the effect of slowing the rate of combustion, which tends to reduce noise.

Inductive reactance An opposing current created in a conductor whenever there is a charging current flow in a conductor.

Inert Chemically inactive.

Inertia switch Turns off the electric fuel pump in an accident.

Initial timing When the spark plug fires at idle speed. The computer then advances the timing based on engine speed and other factors.

Injection pump Delivers fuel to the injectors at a high pressure and at timed intervals. Each injector sprays fuel into the combustion chamber at the precise moment required for efficient combustion.

Input conditioning What the computer does to the input signals to make them useful; usually includes an analog to digital converter and other electronic circuits that eliminate electrical noise.

Intake valve A valve in an engine used to allow air or air and fuel into the combustion chamber.

IOD Ignition off draw. Another name used to describe battery electrical drain.

Intercooler Similar to a radiator, wherein outside air can pass through, which cools the pressurized heated air.

Intermediate oscillations Also called the "ringing" of the coil as it is pulsed as viewed on a scope.

Internal combustion engine A term used to describe a normal engine used in vehicles where the fuel is consumed inside the engine itself.

Ion-sensing ignition A type of coil-on-plug ignition that uses a signal voltage across the spark plug gap after the plug has fired to determine the air–fuel mixture and if spark knock occurred.

Iridium spark plugs Use a small amount of iridium welded onto the tip of a small center electrode 0.0015 to 0.002 inch (0.4 to 0.6 mm) in diameter. The small diameter reduces the voltage required to jump the gap between the center and the side electrode, thereby reducing possible misfires. The ground or side electrode is usually tipped with platinum to help reduce electrode sap wear.

KAM Keep-alive memory.

Knock sensor (KS) A sensor that can detect engine spark knock.

Leak detection pump (LDP) Chrysler uses an electric pump that pressurizes the fuel system to check for leaks by having the PCM monitor the fuel tank pressure sensor.

Lean indicator The higher the oxygen (O_2) levels in the exhaust the leaner the air–fuel mixture.

LED Abbreviation for light emitting diode.

Lift pump The diesel fuel is drawn from the fuel tank by the lift pump and delivers the fuel to the injection pump.

Light-off time (LOT) The time it takes for an oxygen sensor to be start to work.

Linear EGR Contains a stepper motor to precisely regulate exhaust gas flow and a feedback potentiometer that signals to the computer the actual position of the valve.

Liquefied petroleum gas (LPG) Sold as compressed liquid propane that is often mixed with about 10% of other gases such as butane, propylene, butylenes, and mercaptan to give the colorless and odorless propane a smell.

Load test A type of battery test where the battery is placed under an electrical load.

Longitudinal header A passage in an engine block where oil can flow. Also called an oil gallery.

LP-gas another name for liquefied petroleum gas (LPG) which is also sold as propane.

LPG See *liquefied petroleum gas*.

LRC Abbreviation for load response control; also known as electronic load detection. A system used in many charging systems that activate the generator (alternator) when an electrical load is detected and not all of the time to help improve fuel economy.

M85 Internal combustion engine fuel containing 85% methanol and 15% gasoline.

Magnetic pulse generator The pulse generator consists of a trigger wheel (reluctor) and a pickup coil. The pickup coil consists of an iron core wrapped with fine wire, in a coil at one end and attached to a permanent magnet at the other end. The center of the coil is called the pole piece.

Magnetic sensor Uses a permanent magnet surrounded by a coil of wire. The notches of the crankshaft (or camshaft) create a variable magnetic field strength around the coil. When a metallic section is close to the sensor, the magnetic field is stronger because metal is a better conductor of magnetic lines of force than air.

Malfunction indicator lamp (MIL) *This amber, dashboard* warning light may be labeled check engine or service engine soon.

MAP sensor A sensor used to measure the pressure inside the intake manifold compared to a perfect vacuum.

Mass airflow (MAF) sensor Measures the density and amount of air flowing into the engine, which results in accurate engine control.

MCA Abbreviation for marine cranking amperes battery rating.

MDS Multiple Displacement System. A Chrysler term used to describe their variable displacement system.

Mechanical force A force applied to an object.

Mechanical power A force applied to an object which results in movement or motion.

Mechanical returnless fuel system (MRFS) A returnless fuel delivery system design that uses a mechanical pressure regulator located in the fuel tank.

Mercury A heavy metal.

Methanol Typically manufactured from natural gas. Methanol content, including co-solvents, in unleaded gasoline is limited by law to 5%.

Methanol-to-gasoline (MTG) A refining process in which methanol is converted into liquid gasoline.

Metric bolts Bolts manufactured and sized in the metric system of measurement.

Micron One micron is one-millionth of a meter and is equal to 0.000039 inch.

Millisecond (ms) sweep The scope will sweep only that portion of the pattern that can be shown during a 5- or 25-ms setting.

MSDS Material safety data sheet.

Multiplexing A process of sending multiple signals of information at the same time over a signal wire.

Mutual induction The generation of an electric current due to a changing magnetic field of an adjacent coil.

Naturally (normally) aspirated An engine that uses atmospheric pressure for intake.

Negative backpressure An EGR valve that reacts to a low-pressure area by closing a small internal valve, which allows the EGR valve to be opened by vacuum.

Negative temperature coefficient (NTC) Usually used in reference to a temperature sensor (coolant or air temperature). As the temperature increases, the resistance of the sensor decreases.

Nernst cell Another name for a conventional oxygen sensor. The Nernst cell is named for Walther Nernst, 1864–1941, a German physicist known for his work in electrochemistry.

Network A communications system used to link multiple computers or modules.

Neutral position A term used to describe the home or the default position of the throttle plate in an electronic throttle control system.

Neutral safety switch An electrical switch that allows the starter to be energized only if the gear selector is in neutral or park.

NGV Natural gas vehicle.

Noid light Designed to electrically replace the injector in the circuit and to flash if the injector circuit is working correctly.

Non-methane hydrocarbon (NMHC) The standard by which exhaust emission testing for hydrocarbons is evaluated.

Nonprincipal end Opposite the principal end and is generally referred to as the front of the engine, where the accessory belts are used.

Nonvolatile RAM Computer memory capability that is not lost when power is removed. See also *read-only memory (ROM)*.

Nuts A female threaded fastener that is used with a bolt to hold two parts together.

OBDs On-board diagnostics.

Octane rating The measurement of a gasoline's ability to resist engine knock. The higher the octane rating, the less prone the gasoline is to cause engine knock (detonation).

Oil control valve (OCV) The camshaft position actuator oil control valve (OCV) directs oil from the oil feed in the head to the appropriate camshaft position actuator oil passages.

Oil gallery A passage in an engine block where oil can flow.

Opacity The percentage of light that is blocked by the exhaust smoke.

Open-circuit battery voltage test A test of battery condition that is performed without a load on the battery.

Open-end wrench A type of wrench that allows access to the flats of a bolt or nut from the side.

Open-loop operation A phase of computer-controlled engine operation where air–fuel mixture is calculated in the absence of oxygen sensor signals. During open loop, calculations are based primarily on throttle position, engine RPM, and engine coolant temperature.

Optical sensors Use light from a LED and a phototransistor to signal the computer.

ORVR Onboard refueling vapor recovery.

OSC Oxygen storage capacity.

OSHA Occupational Safety and Health Administration.

Oxygen sensor (O2S) A sensor in the exhaust system to measure the oxygen content of the exhaust.

Oxygenated fuels Fuels such as ETBE or MTBE that contain extra oxygen molecules to promote cleaner burning. Oxygenated fuels are used as gasoline additives to reduce CO emissions.

Ozone Oxygen-rich (O_3) gas created by sunlight reaction with unburned hydrocarbons (HC) and oxides of nitrogen (NO_x); also called smog.

Palladium A catalyst that starts a chemical reaction without becoming a part of, or being consumed in, the process.

Pancake engine See *boxer*.

Paper test Hold a piece of paper or a 3 × 5 index card (even a dollar bill works) within 1 inch (2.5 cm) of the tailpipe with the engine running at idle. The paper should blow out evenly without "puffing." If the paper is drawn toward the tailpipe at times, the exhaust valves in one or more cylinders could be burned.

Paired cylinders Another name used to describe companion cylinders in a waste-spark-type ignition system.

Parasitic load A term used to describe a battery drain with all circuits off. Also called battery electrical drain or ignition off draw.

Parameter identification display (PID) The information found in the vehicle data stream as viewed on a scan tool.

Powertrain Control Module (PCM) The onboard computer that controls both the engine management and transmission functions of the vehicle.

PCV Pressure control valve.

Peak-and-hold injector A type of injector driver that uses full battery voltage to the injector and the ground side is controlled through the computer. The computer provides a high initial current flow (about 4 amperes) through the injector windings to open the injector core. Then the computer reduces the current to a lower level (about 1 ampere).

Pending code A code(s) that is displayed on a scan tool when the MIL is not on. Because the MIL is not on, this indicates that the fault has not repeated to cause the PCM to turn on the MIL.

Peripheral pump Turbine pump.

Petrodiesel Another term for petroleum diesel, which is ordinary diesel fuel refined from crude oil.

Petroleum Another term for crude oil. The literal meaning of petroleum is "rock oil."

PFE Pressure feedback EGR sensor.

Pickup coil An ignition electronic triggering device in the magnetic pulse generator system.

Piezoresistivity Change in resistance due to strain.

Pinch weld seam A strong section under a vehicle where two body panels are welded together.

Ping Secondary rapid burning of the last 3% to 5% of the air–fuel mixture in the combustion chamber causes a second flame front that collides with the first flame front causing a knock noise. Also called detonation or spark knock.

Piston stroke A one-way piston movement between the top and bottom of the cylinder.

Pitch The pitch of a threaded fastener refers to the number of threads per inch.

Planar design A design of oxygen sensor where the elements including the zirconia electrolyte and the two electrodes and heater are stacked together in a flat-type design.

Platinum spark plug A spark plug that has a small amount of the precious metal platinum welded onto the end of the center electrode, as well as on the ground or side electrode. Platinum is a grayish white metal that does not react with oxygen and, therefore, will not erode away as can occur with conventional nickel alloy spark plug electrodes.

Platinum A catalyst that starts a chemical reaction without becoming a part of, or being consumed in, the process.

Plenum A chamber, located between the throttle body and the runners of the intake manifold, used to distribute the intake charge more evenly and efficiently.

Pliers A hand tool with two jaws used to grasp or turn a part.

Polarity The condition of being positive or negative in relation to a magnetic pole.

Pop tester A device used for checking a diesel injector nozzle for proper spray pattern. The handle is depressed and pop off pressure is displayed on the gauge.

Port fuel-injection Uses a nozzle for each cylinder and the fuel is squirted into the intake manifold about 2 to 3 inches (70 to 100 mm) from the intake valve.

Positive backpressure An EGR valve that is designed with a small valve inside that bleeds off any applied vacuum and prevents the valve from opening unless there is exhaust backpressure applied to the valve.

Positive displacement All of the air that enters is forced through the roots-type supercharger.

Positive displacement pump A type of pump where *each rotation of the pump delivers the same volume of oil and everything that enters must exit.*

Potentiometer A 3-terminal variable resistor that varies the voltage drop in a circuit.

Power balance test Determines if all cylinders are contributing power equally. It determines this by shorting out one cylinder at a time.

Power side switching A type of computer control of a circuit where the PCM controls the power side of the circuit. Commonly used to control the oil control valve in a variable valve timing system.

PPE Abbreviation for personal protective equipment.

PPO Pure plant oil.

Preconverter A small, quick heating oxidation converter.

Pressure control valve (PCV) A valve used to control the fuel system pressure on a demand delivery-type fuel system.

Pressure differential A difference in pressure from one brake circuit to another.

Pressure regulating valve A valve that is used to limit the upper pressure of an oil pump. Also called a relief valve.

Pressure vent valve (PVV) A valve located in the fuel tank to prevent overpressure due to the thermal expansion of the fuel.

Primary ignition circuit The ignition components that regulate the current in the coil primary winding by turning it on and off.

Principal end The end of the engine that the flywheel is attached to.

PROM Programmable read-only memory.

Propane See *liquified petroleum gas*.

Pump cell A pump cell is the area above the Nernst cell which is another zirconia layer with two electrodes. The two cells share a common ground, which is called the reference.

Punches A type of hand tool used to drive roll pins.

Pup converter See *preconverter*.

Pushrod engine When the camshaft is located in the block, the valves are operated by lifters, pushrods, and rocker arms.

PWM Pulse-width modulation.

R & R Remove and replace.

RAM Random-access memory.

Raster Stacked.

Rationality Refers to a PCM comparison of input value to values.

Ratchet A type of hand tool that is used to rotate a socket and is reversible.

RCRA Resource Conservation and Recovery Act.

Recall A notification to the owner of a vehicle that a safety issue needs to be corrected.

Reference voltage In a conventional zirconia oxygen sensor, a bias or reference voltage can be applied to the two platinum electrodes, and then oxygen ions can be forced (pumped) from the ambient reference air side to the exhaust side of the sensor.

Reformulated gasoline (RFG) Has oxygenated additives and is refined to reduce both the lightest and heaviest hydrocarbon content from gasoline in order to promote cleaner burning.

Regeneration A process of taking the kinetic energy of a moving vehicle and converting it to electrical energy and storing it in a battery.

Reid vapor pressure (RVP) A method of determining vapor pressure of gasoline and other petroleum products. Widely used in the petroleum industry as an indicator of the volatility of gasoline.

Reserve capacity The number of minutes a battery can produce 25 amperes and still maintain a battery voltage of 1.75 volts per cell (10.5 volts for a 12 volts battery).

Residual or rest pressure Prevents vapor lock and hot-start problems on these systems.

Restricted exhaust The engine will be low on power, yet smooth.

Reverse cooling A type of cooling system where the coolant flows from the radiator to the cylinder head(s) before flowing to the engine block.

Rhodium A catalyst that starts a chemical reaction without becoming a part of, or being consumed in, the process.

Right-to-know laws Laws that state that employees have a right to know when the materials they use at work are hazardous.

Ripple current The unwanted AC current from a generator (alternator).

Roller cell Vane pump.

ROM Read-only memory.

Roots-type Subcharger Named for Philander and Francis Roots, two brothers from Connersville, Indiana, who patented the design in 1860 as a type of water pump to be used in mines. Later it was used to move air and is used today on two-stroke cycle Detroit diesel engines and other supercharged engines. The roots-type supercharger is called a positive displacement design because all of the air that enters is forced through the unit.

Rotary engine Operates on the four-stroke cycle but uses a rotor instead of a piston and crankshaft to achieve intake, compression, power, and exhaust stroke.

Rotary vane pump The pump consists of a central impeller disk, several rollers or vanes that ride in notches in the impeller, and a pump housing that is offset from the impeller centerline.

Rotor gap Measures the voltage required to jump the gap (0.030 to 0.050 in. or 0.8 to 1.3 mm) between the rotor and the inserts (segments) of the distributor cap.

Running compression test A test that can inform a technician of the relative compression of all the cylinders.

SAE Society of Automotive Engineers.

Saturation The point of maximum magnetic field strength of a coil.

Schmitt trigger Converts the analog signal into a digital signal.

Screwdriver A type of hand tool used to install or remove screws.

Scroll The scroll is a smoothly curved passage that changes the fluid flow direction with minimum loss in velocity.

Secondary ignition circuit The components necessary to create and distribute the high voltage produced in the secondary windings of the coil.

Selective catalytic reduction (SCR) A method used to reduce NO_x emissions produced in a diesel engine by injecting urea into the exhaust stream.

Self-induction The generation of an electric current in the wires of a coil created when the current is first connected or disconnected.

Sequential fuel injection (SFI) A fuel-injection system in which injectors are pulsed individually in sequence with the firing order.

Serial data Data that are transmitted by a series of rapidly changing voltage signals.

Servomotor An electric motor that moves an actuator such as the throttle plate in an electronic throttle control system.

Side-channel pump Turbine pump.

Signal electrode This electrode is exposed to outside (ambient) air and is the signal electrode, also called the reference electrode or ambient air electrode.

Silicone coupling A type of thermal fan that is used to disengage the fan so that it uses little power at high engine speeds and minimizes noise.

Single cell A type of wide-band oxygen sensor that uses one cell using four wires; two for the heater and two cell wires.

Single overhead camshaft (SOHC) A type of engine design that uses one overhead camshaft to operate both the intake and the exhaust valves on each bank of cylinders.

SIP State Implementation Plan.

Skewed An output from a sensor that moves in the correct direction but does not accurately measure the condition it is designed to measure.

Smog The term used to describe a combination of smoke and fog. Formed by NO_x and HC with sunlight.

Smoke machine A machine that generates smoke under a slight pressure which can be used to find leaks. Usually used to find leaks in the EVAP system.

Snips A type of hand tool used to cut sheet metal and other thin materials.

Socket A tool that fits over the head of a bolt or nut and is rotated by a ratchet or breaker bar.

Socket adapter An adapter that allows the use of one size of driver (ratchet or breaker bar) to rotate another drive size of socket.

Solvent Usually colorless liquids that are used to remove grease and oil.

Spark ignition direct injection (SIDI) GM's name for gasoline direct injection system.

Spark knock See *Detonation*.

Spark line A short horizontal line immediately after the firing line.

Spark output The term that Ford used to describe the OBD-II terminology for the output signal from the PCM to the ignition system that controls engine timing.

Spark tester Looks like a spark plug except it has a recessed center electrode and no side electrode. The tester commonly has an alligator clip attached to the shell so that it can be clamped on a good ground connection on the engine.

Speed density The method of calculating the amount of fuel needed by the engine.

Splice pack A central point where many serial data lines jam together, often abbreviated SP.

Spline phaser A type of variable valve timing actuator that uses a piston inside the cam phaser and rides along the helical splines, which compresses a coil spring. This movement causes the cam phaser gear and the camshaft to move in an opposite direction.

Spontaneous combustion A condition that can cause some materials, such as oily rags, to catch fire without a source of ignition.

SST Abbreviation for special service tools.

SPOUT The term used by Ford to describe the "spark out" signal from the ICM to the PCM.

SRE Abbreviation for slip ring end part of a generator (alternator).

State of charge The amount expressed in a percentage that a battery is changed.

Stratified mode A mode of engine operation where the air–fuel mixture is richer around the spark plug than it is in the rest of the cylinder.

SOHC Abbreviation for single overhead camshaft; a type of engine design.

Stoichiometric ratio The ideal mixture or ratio at which all of the fuel combines with all of the oxygen in the air and burns completely.

Straight vegetable oil (SVO) Vegetable oil, a triglyceride with a glycerin component joining three hydrocarbon chains of 16 to 18 carbon atoms each.

Stroke The distance the piston travels in the cylinder of an engine.

Stud A short rod with threads on both ends.

Superimposed A position used to look at differences in patterns between cylinders in all areas except the firing line.

Supercharger An engine driven device that forces air under pressure into the intake manifold to increase engine power.

Supercharging A method uses a pump to pack a denser air–fuel charge into the cylinders.

Surface charge A charge on a battery that just on the surface of the plates and does not indicate the true state of charge of the battery.

Surge tank A reservoir mounted at the highest point in the cooling system.

Switchgrass A feedstock for ethanol production that requires very little energy or fertilizer to cultivate.

Switching Turning on and off of the primary circuit.

Syncrude A product of a process where coal is broken down to create liquid products. First the coal is reacted with hydrogen (H2) at high temperatures and pressure with a catalyst.

Syn-gas Synthesis gas generated by a reaction between coal and steam. Syn-gas is made up of mostly hydrogen and carbon monoxide and is used to make methanol. Syn-gas is also known as town gas.

Synthetic fuel Fuels generated through synthetic processes such as Fischer-Tropsch.

Tap test Involves tapping (not pounding) on the catalytic converter using a rubber mallet.

Task manager A term Chrysler uses to describe the software program that is designed to manage the operation of all OBD-II monitors by controlling the sequence of steps necessary to execute the diagnostic tests and monitors.

TBI Throttle-body injection.

Tensile strength The maximum stress used under tension (lengthwise force) without causing failure.

Terminating resistors Resistors placed at the end of a high-speed serial data circuit to help reduce electromagnetic interference.

Tetraethyl lead (TEL) A liquid added to gasoline in the early 1920s to reduce the tendency to knock.

Thermostat A device that controls the flow in a system such as the engine cooling system based on temperature.

Thermostatic spring The thermostatic spring operates a valve that allows the fan to freewheel when the radiator is cold.

Thimble design A design of an oxygen sensor that uses a shape like a cup or thimble.

Throttle-body temperature (TBT) A temperature sensor that is mounted on the throttle body and measures the temperature of the air entering the engine.

Throttle position (TP) sensor The sensor that provides feedback concerning the position of the throttle plate.

Top dead center (TDC) The highest point in the cylinder that the piston can travel. The measurement from bottom dead center (BDC) to top dead center (TDC) determines the stroke length of the crankshaft.

Transistor A semiconductor device that can operate as an amplifier or an electrical switch.

Transmission fluid temperature (TFT) sensor A sensor located inside an automatic transmission/transaxle that measures the temperature of the fluid.

Trip The vehicle must be driven under a variety of operating conditions for all active tests to be performed. A trip is defined as an engine-operating drive cycle that contains the necessary conditions for a particular test to be performed.

Trouble light A light designed to help service technicians see while working on a vehicle.

TSB Technical service bulletin.

Turbine pump Turns at higher speeds and is quieter than the other electric pumps.

Turbo lag The delay between acceleration and turbo boost.

Turbocharger An exhaust-powered supercharger.

Turn ratio The number of times that the secondary windings in an ignition coil exceed that of the primary winding.

TWC Three-way converter.

UCG Underground coal gasification.

UNC Unified national coarse. A type of thread used on fasteners.

Underground coal gasification (UCG) A process performed underground where coal is turned into a liquid fuel.

Underground storage tank (UST) A type of oil tank that is located underground.

UNF Unified national fine. A type of thread used on fasteners.

Universal joint A joint in a steering or drive shaft that allows torque to be transmitted at an angle.

Up-integrated ignition Ignition control where all timing functions are performed in the PCM, rather than being split between the ignition control module and the PCM.

Urea Urea is a colorless, odorless, and nontoxic liquid. Urea is called diesel exhaust fluid (DEF) in North America and AdBlue in Europe. The urea is used to reduce NO_x emissions produced in a diesel engine by injecting urea into the exhaust stream.

Used cooking oil (UCO) A term used when the oil may or may not be pure vegetable oil.

Used oil Any petroleum-based or synthetic oil that has been used.

Vacuum Any pressure less than atmospheric pressure (14.7 PSI).

Vacuum test Testing the engine for cranking vacuum, idle vacuum, and vacuum at 2500 RPM.

VAF Vane air flow.

Vane phaser A vane phaser is an actuator used in most variable valve timing systems and uses a rotor with four vanes, which is connected to the end of the camshaft. The rotor is located inside the stator, which is bolted to the cam sprocket. The stator and rotor are not connected. Oil pressure is controlled on both sides of the vanes of the rotor, which creates a hydraulic link between the two parts. The oil control valve varies the balance of pressure on either side of the vanes and thereby controls the position of the camshaft. A return spring is used under the reluctor of the phaser to help return it to the home or zero degrees position.

Vapor lock A lean condition caused by vaporized fuel in the fuel system.

Variable displacement system A term used to describe a system where some cylinders are deactivated to reduce fuel consumption.

Variable fuel sensor See *fuel compensation sensor*.

VECI Vehicle emission control information. This sticker is located under the hood on all vehicles and includes emissions-related information that is important to the service technician.

V-FFV Abbreviation for a virtual flexible fuel vehicle. This type of vehicle does not use a fuel sensor and instead uses the oxygen sensor(s) to detect the amount of alcohol being used in the system.

VIN Vehicle identification number.

Viscosity The resistance to flow or thickness of a liquid such as oil.

VOC Volatile organic compound.

Volatility A measurement of the tendency of a liquid to change to vapor. Volatility is measured using RVP, or Reid vapor pressure.

Voltage-drop test An electrical test performed with current flowing through the circuit to determine what voltage is lost to resistance.

Volumetric efficiency The ratio between the amount of air–fuel mixture that actually enters the cylinder and the amount that could enter under ideal conditions expressed in percent.

VTEC Variable Valve Timing and Lift Electronic. A term used by Honda/Acura to describe their variable timing and lift system.

VVT Variable valve timing

Wankel engine Rotary engine.

Washcoat A porous alumina material that makes the surface rough.

Washers Flat metal discs with a hole in the center used under threaded fasteners to help spread the clamping force over a wider area.

Waste vegetable oil (WVO) This oil could include animal or fish oils from cooking.

Wastegate A valve similar to a door that can open and close. The wastegate is a bypass valve at the exhaust inlet to the turbine. It allows all of the exhaust into the turbine, or it can route part of the exhaust past the turbine to the exhaust system.

Waste-spark ignition Introduced in the mid-1980s, it uses the onboard computer to fire the ignition coils.

Water–fuel separator Separates water and fuel in a diesel engine.

Wet compression test A test that uses oil to help seal around the piston rings.

WHMIS Workplace Hazardous Materials Information Systems.

Windage tray A baffle in an oil pan used to keep oil from getting in contact with the crankshaft during engine operation.

Wrench A hand tool used to grasp and rotate a threaded fastener.

WWFC Abbreviation for world wide fuel charter.

INDEX

D

E

F

G

H

I

J

K

L

P

R

T

U

V

W

Z

AUTOMOTIVE ELECTRICAL, HEATING, AND AIR CONDITIONING SYSTEMS

James D. Halderman

Edited by Jeffrey Rehkopf

Taken from

Automotive Electricity and Electronics, Fourth Edition
By James D. Halderman

Automotive Heating and Air Conditioning, Seventh Edition
By James D. Halderman

Hybrid and Alternative Fuel Vehicles, Third Edition
By James D. Halderman

PEARSON

Boston Columbus Hoboken Indianapolis New York San Francisco
Amsterdam Cape Town Dubai London Madrid Milan Munich Paris Montréal
Toronto Delhi Mexico City São Paulo Sydney Hong Kong Seoul Singapore Taipei Tokyo

Editorial Director: Andrew Gilfillan
Editorial Assistant: Nancy Kesterson
Director of Marketing: David Gesell
Program Manager: Holly Shufeldt
Project Manager: Janet Portisch
Senior Art Director: Diane Ernsberger
Cover Designer: Cenveo Publishing Services
Media Project Manager: April Cleland
Full-Service Project Management and Composition: Integra Software Services, Ltd.
Printer/Binder: LSC Communications, Inc.
Cover Printer: LSC Communications, Inc.

Taken from
Automotive Electricity and Electronics, Fourth Edition
By James D. Halderman
Copyright © 2014 by Pearson Education

Automotive Heating and Air Conditioning, Seventh Edition
By James D. Halderman
Copyright © 2015 by Pearson Education

Hybrid and Alternative Fuel Vehicles, Third Edition
By James D. Halderman
Copyright © 2016 by Pearson Education

ISBN 10: 0-13-351655-5
ISBN 13: 978-0-13-351655-5

BRIEF CONTENTS

Technical and Content Reviewers

TECHNICAL AND CONTENT REVIEWERS The following people helped in checking this text for technical accuracy and clarity of presentation. Their suggestions and recommendations were included in the final version of the page proofs. Their valuable input helped make this textbook clear and technically accurate while maintaining the easy-to-read style that has made other books in this series so popular. While too many to name individually, thanks are also due to the many members of IAGMASP who twisted your editor's arm and ear at the annual conferences.

Alan Crouch
J. Sargeant Reynolds Community College

Josh Gilbert
Guilford Technical College

Joe Jackson
Ranken Technical College

Marty Kamimoto
Fresno City College

Darrin Marshall
University of Alaska, Anchorage

Kelly Smith
University of Alaska, Anchorage

Martin Smith
British Columbia Institute of Technology, Canada

Chuck Taylor
Sinclair Community College, Dayton OH

Timothy A. Wawerczyk
Broward College

Raytheon
Professional Services, LLC

Michael Parker
Lakes Region Community College,
Laconia, NH

Christopher Gallo
Camden County College

chapter 1

ELECTRICAL FUNDAMENTALS

LEARNING OBJECTIVES

After studying this chapter, the reader will be able to:

1. Define electricity.
2. Explain the units of electrical measurement.
3. Discuss the relationship among volts, amperes, and ohms.
4. Discuss the sources of electricity as used in automotive applications.
5. Discuss the characteristics of resistance in conductors.

This chapter will help you prepare for the ASE Electrical/Electronic Systems (A6) certification test content area "A" (General Electrical/Electronic System Diagnosis).

KEY TERMS

Ammeter 6
Ampere 6
Bound electrons 4
Conductors 4
Conventional theory 5
Coulomb 6
Electrical potential 6
Electricity 2
Electrochemistry 8
Electromotive force (EMF) 7
Electron theory 5
Free electrons 4
Insulators 5
Ion 3
Neutral charge 2
Ohmmeter 7
Ohms 7
Peltier effect 8
Photoelectricity 8
Piezoelectricity 8
Potentiometer 10
Resistance 7
Rheostat 11
Semiconductor 5
Static electricity 7
Thermocouple 8
Thermoelectricity 8
Valence ring 3
Voltmeter 7
Watt 7

 STC OBJECTIVES

GM Service Technical College topics covered in this chapter are as follows:

1. Define electrical potential and identify sources of electrical potential. (08510.05W)
2. Identify the properties of electricity.
3. Identify current flow through a basic electrical circuit.
4. Identify types of conductors.
5. Identify types of insulators.

INTRODUCTION

The electrical system is one of the most important systems in a vehicle today. Every year more and more components and systems use electricity. Those technicians who really know and understand automotive electrical and electronic systems will be in great demand.

Electricity may be difficult for some people to learn for the following reasons.

- It cannot be seen.
- Only the results of electricity can be seen.
- It has to be detected and measured.
- The test results have to be interpreted.

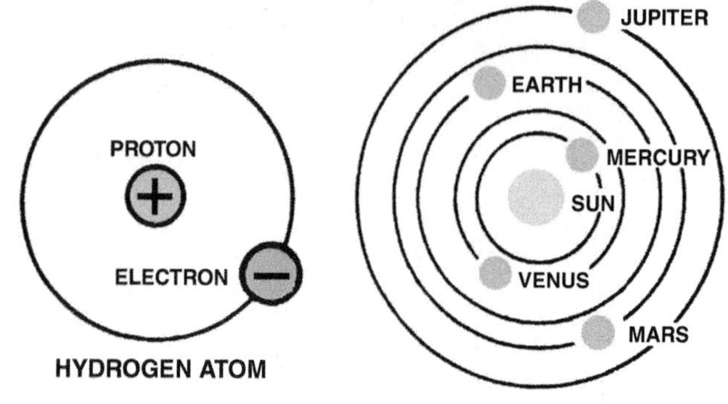

FIGURE 1–1 In an atom (left), electrons orbit protons in the nucleus just as planets orbit the sun in our solar system (right).

ELECTRICITY

BACKGROUND Our universe is composed of matter, which is *anything* that has mass and occupies space. All matter is made from slightly over 100 individual components called *elements*. The smallest particle that an element can be broken into and still retain the properties of that element is known as an atom. **SEE FIGURE 1–1**.

DEFINITION **Electricity** is the movement of electrons from one atom to another. The dense center of each atom is called the nucleus. The nucleus contains:

- *Protons*, which have a positive charge
- *Neutrons*, which are electrically neutral (have no charge)

Electrons, which have a negative charge, surround the nucleus in orbits. Each atom contains an equal number of electrons and protons. The physical aspect of all protons, electrons, and neutrons is the same for all atoms. It is the *number* of electrons and protons in the atom that determines the material and how electricity is conducted. Because the number of negative-charged electrons is balanced with the same number of positive-charged protons, an atom has a **neutral charge** (no charge).

NOTE: As an example of the relative sizes of the parts of an atom, consider that if an atom were magnified so that the nucleus were the size of the period at the end of this sentence, the whole atom would be bigger than a house.

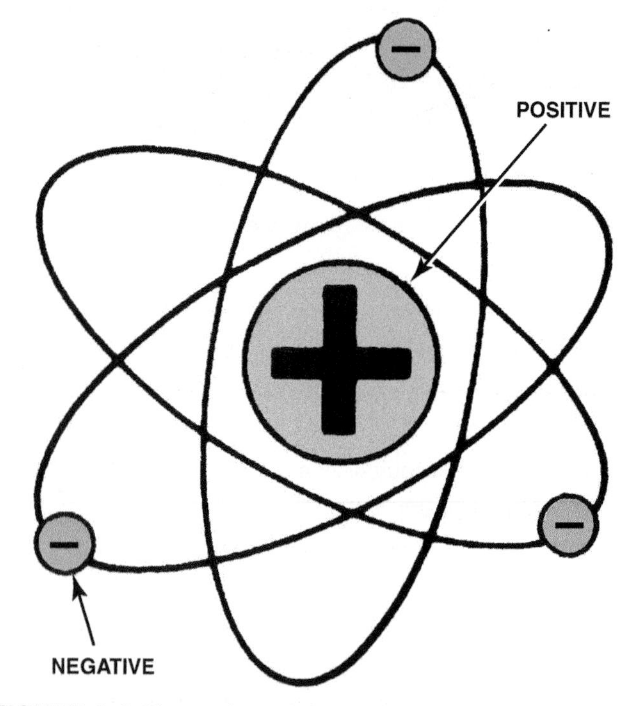

FIGURE 1–2 The nucleus of an atom has a positive (+) charge and the surrounding electrons have a negative (–) charge.

POSITIVE AND NEGATIVE CHARGES The parts of the atom have different charges. The orbiting electrons are negatively charged, while the protons are positively charged. Positive charges are indicated by the "plus" sign (+), and negative charges by the "minus" sign (–), as shown in **FIGURE 1–2**.

These same + and – signs are used to identify parts of an electrical circuit. Neutrons have no charge at all. They are neutral. In a normal, or balanced, atom, the number of negative particles equals the number of positive particles. That is, there are as many electrons as there are protons. **SEE FIGURE 1–3**.

MAGNETS AND ELECTRICAL CHARGES An ordinary magnet has two ends, or poles. One end is called the south pole, and the other is called the north pole. If two magnets are brought

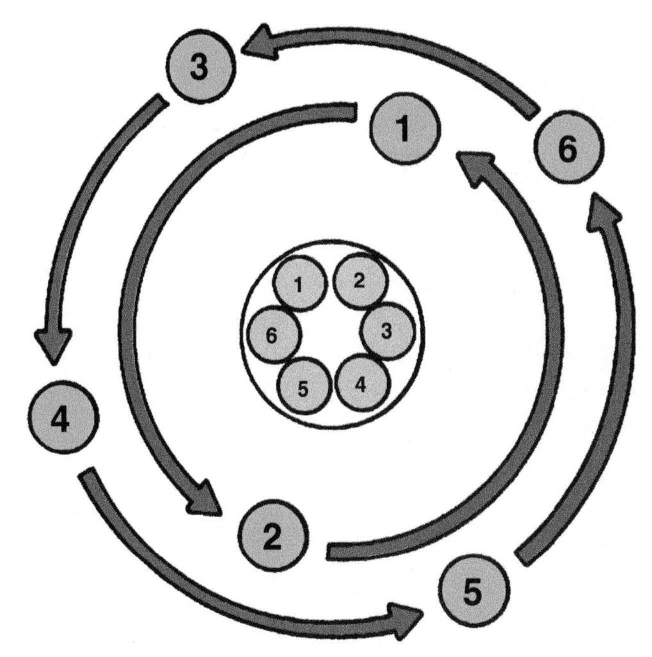

FIGURE 1–3 This figure shows a balanced atom. The number of electrons is the same as the number of protons in the nucleus.

FIGURE 1–4 Unlike charges attract and like charges repel.

close to each other with like poles together (south to south or north to north), the magnets will push each other apart, because like poles repel each other. If the opposite poles of the magnets are brought close to each other, south to north, the magnets will snap together, because unlike poles attract each other.

The positive and negative charges within an atom are like the north and south poles of a magnet. Charges that are alike will repel each other, similar to the poles of a magnet. ● **SEE FIGURE 1–4.**

That is why the negative electrons continue to orbit around the positive protons. They are attracted and held by the opposite charge of the protons. The electrons keep moving in orbit because they repel each other.

IONS When an atom loses any electrons, it becomes unbalanced. It will have more protons than electrons, and therefore will have a positive charge. If it gains more electrons than protons, the atom will be negatively charged. When an atom is not balanced, it becomes a charged particle called an **ion**. Ions try to regain their balance of equal protons and electrons by exchanging electrons with neighboring atoms. The flow of electrons during the "equalization" process is defined as the flow of electricity. ● **SEE FIGURE 1–5.**

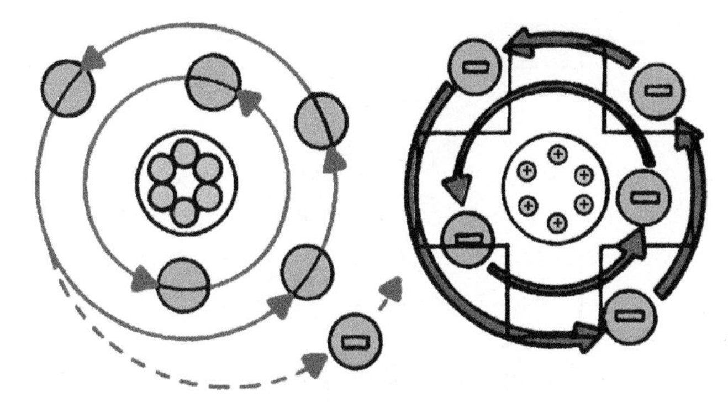
FIGURE 1–5 An unbalanced, positively charged atom (ion) will attract electrons from neighboring atoms.

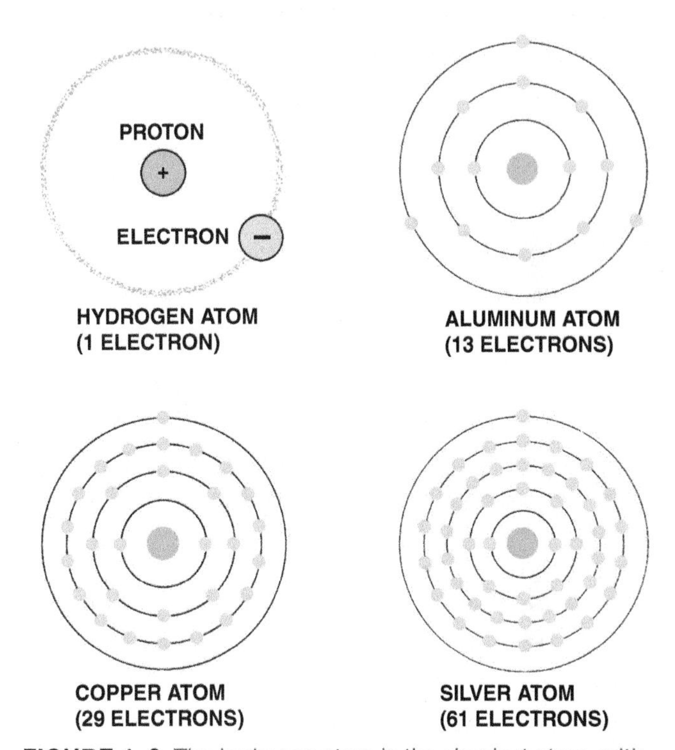

FIGURE 1–6 The hydrogen atom is the simplest atom, with only one proton, one neutron, and one electron. More complex elements contain higher numbers of protons, neutrons, and electrons.

ELECTRON SHELLS Electrons orbit around the nucleus in definite paths. These paths form shells, like concentric rings, around the nucleus. Only a specific number of electrons can orbit within each shell. If there are too many electrons for the first and closest shell to the nucleus, the others will orbit in additional shells until all electrons have an orbit within a shell. There can be as many as seven shells around a single nucleus. ● **SEE FIGURE 1–6.**

FREE AND BOUND ELECTRONS The outermost electron shell or ring, called the **valence ring**, is the most important part of understanding electricity. The number of electrons in this

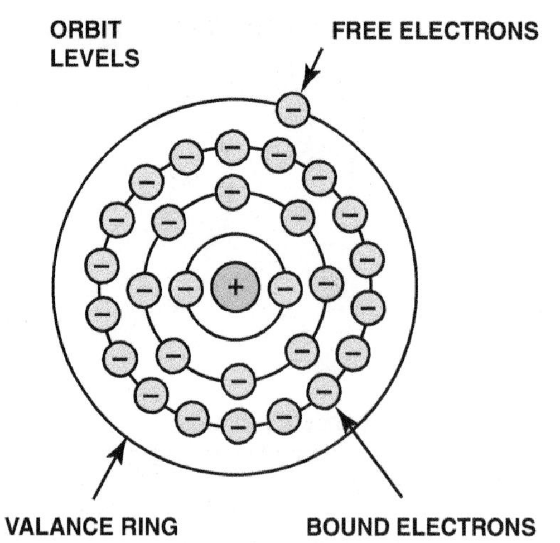

FIGURE 1–7 As the number of electrons increases, they occupy increasing energy levels that are farther from the center of the atom.

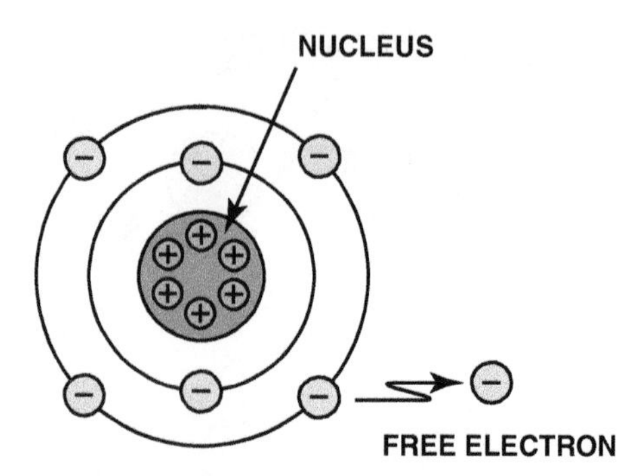

FIGURE 1–8 Electrons in the outer orbit, or shell, can often be drawn away from the atom and become free electrons.

FREQUENTLY ASKED QUESTION

Is Water a Conductor?

Pure water is an insulator; however, if anything is in the water, such as salt or dirt, then the water becomes conductive. Because it is difficult to keep it from becoming contaminated, water is usually thought of as being capable of conducting electricity, especially high-voltage household 110 or 220 volt outlets.

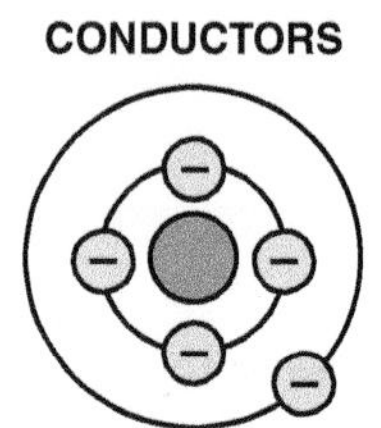

FIGURE 1–9 A conductor is any element that has one to three electrons in its outer orbit.

outer ring determines the valence of the atom and indicates its capacity to combine with other atoms.

If the valence ring of an atom has three or fewer electrons in it, the ring has room for more. The electrons there are held very loosely, and it is easy for a drifting electron to join the valence ring and push another electron away. These loosely held electrons are called **free electrons**. When the valence ring has five or more electrons in it, it is fairly full. The electrons are held tightly, and it is hard for a drifting electron to push its way into the valence ring. These tightly held electrons are called **bound electrons.** ● **SEE FIGURES 1–7 AND 1–8.**

The movement of these drifting electrons is called current. Current can be small, with only a few electrons moving, or it can be large, with a tremendous number of electrons moving. Electric current is the controlled, directed movement of electrons from atom to atom within a conductor.

CONDUCTORS **Conductors** are materials with fewer than four electrons in their atom's outer orbit. ● **SEE FIGURE 1–9.**

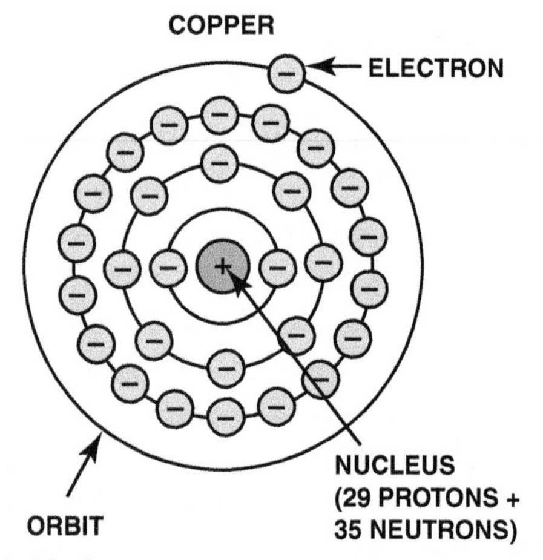

FIGURE 1–10 Copper is an excellent conductor of electricity because it has just one electron in its outer orbit, making it easy to be knocked out of its orbit and flow to other nearby atoms. This causes electron flow, which is the definition of electricity.

Copper is an excellent conductor because it has only one electron in its outer orbit. This orbit is far enough away from the nucleus of the copper atom that the pull or force holding the outermost electron in orbit is relatively weak. ● **SEE FIGURE 1–10.**

Copper is the conductor most used in vehicles because the price of copper is reasonable compared to the relative cost

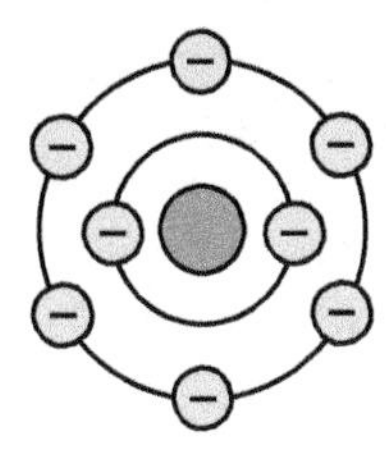

FIGURE 1–11 Insulators are elements with five to eight electrons in the outer orbit.

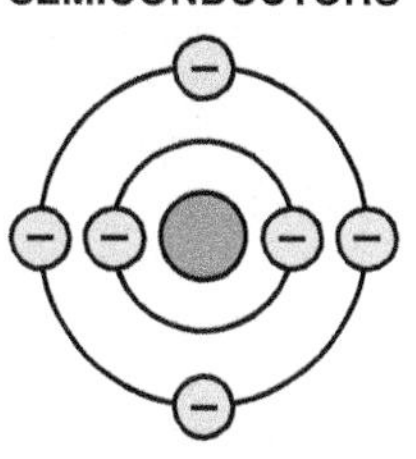

FIGURE 1–12 Semiconductor elements contain exactly four electrons in the outer orbit.

of other conductors with similar properties. Examples of other commonly used conductors include:

- Silver
- Gold
- Aluminum
- Steel
- Cast iron

INSULATORS Some materials hold their electrons very tightly; therefore, electrons do not move through them very well. These materials are called insulators. **Insulators** are materials with more than four electrons in their atom's outer orbit. Because they have more than four electrons in their outer orbit, it becomes easier for these materials to acquire (gain) electrons than to release electrons. ● **SEE FIGURE 1–11**.

Examples of insulators include:

- Vinyl, rubber, and plastics
- Paper
- Wood
- Glass
- Ceramics (spark plugs)
- Varnish and enamels (copper windings in generators and starters)

SEMICONDUCTORS Materials with exactly four electrons in their outer orbit are neither conductors nor insulators, but are called **semiconductors**. Semiconductors can be either an insulator or a conductor in different design applications. ● **SEE FIGURE 1–12**.

Examples of semiconductors include:

- Silicon
- Germanium
- Carbon

Semiconductors are used in diodes, transistors, computers, and other electronic devices.

FIGURE 1–13 Current electricity is the movement of electrons through a conductor.

HOW ELECTRONS MOVE THROUGH A CONDUCTOR

CURRENT FLOW The following events occur if a source of power, such as a battery, is connected to the ends of a conductor—a positive charge (lack of electrons) is placed on one end of the conductor and a negative charge (excess of electrons) is placed on the opposite end of the conductor. For current to flow, there *must* be an imbalance of excess electrons at one end of the circuit and a deficiency of electrons at the opposite end of the circuit.

- The negative charge will repel the free electrons from the atoms of the conductor, whereas the positive charge on the opposite end of the conductor will attract electrons.
- As a result of this attraction of opposite charges and repulsion of like charges, electrons will flow through the conductor. ● **SEE FIGURE 1–13**.

CONVENTIONAL THEORY VERSUS ELECTRON THEORY

- **Conventional theory.** It was once thought that electricity had only one charge and moved from positive to negative. This theory of the flow of electricity through a conductor is called the **conventional theory** of current flow. ● **SEE FIGURE 1–14**.
- **Electron theory.** The discovery of the electron and its negative charge led to the **electron theory**, which states that there is electron flow from negative to positive. Most automotive applications use the conventional theory. We will use the conventional theory (positive to negative) unless stated otherwise.

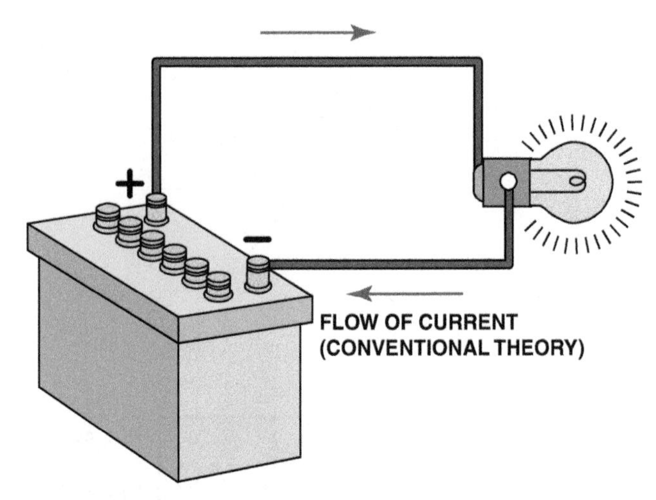

FIGURE 1–14 Conventional theory states that current flows through a circuit from positive (+) to negative (–). Automotive electricity uses the conventional theory in all electrical diagrams and schematics.

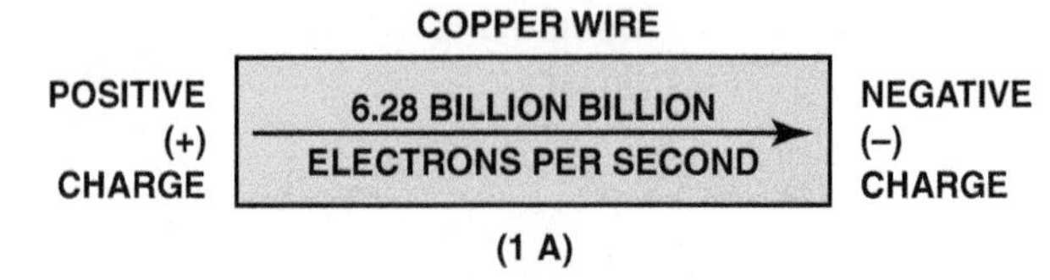

FIGURE 1–15 One ampere is the movement of 1 coulomb (6.28 billion billion electrons) past a point in 1 second.

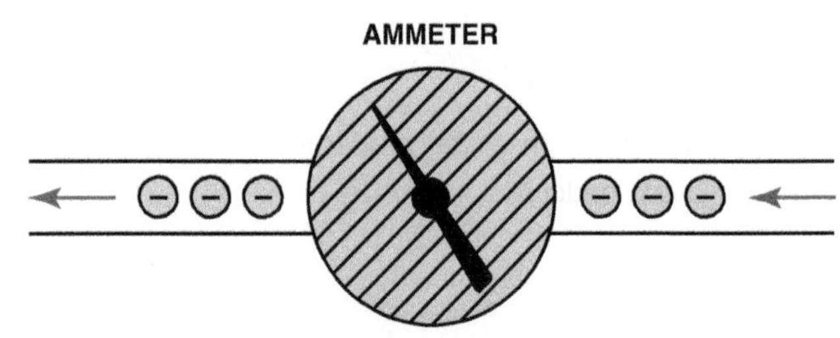

FIGURE 1–16 An ammeter is installed in the path of the electrons similar to a water meter used to measure the flow of water in gallons per minute. The ammeter displays current flow in amperes.

UNITS OF ELECTRICITY

Electricity is measured using meters or other test equipment. The three fundamentals of electricity-related units include the ampere, volt, and ohm.

AMPERES The **ampere** is the unit used throughout the world to measure current flow. When 6.28 billion billion electrons (the name for this large number of electrons is a **coulomb**) move past a certain point in 1 second, this represents 1 ampere of current. ● **SEE FIGURE 1–15.**

The ampere is the electrical unit for the amount of electron flow, just as "gallons per minute" is the unit that can be used to measure the quantity of water flow. It is named for the French electrician, André Marie Ampére (1775–1836). The conventional abbreviations and measurement for amperes are as follows:

1. The ampere is the unit of measurement for the amount of current flow.
2. *A* and *amps* are acceptable abbreviations for *amperes.*
3. The capital letter *I*, for *intensity*, is used in mathematical calculations to represent amperes.
4. Amperes do the actual work in the circuit. It is the actual movement of the electrons through a light bulb or motor that actually makes the electrical device work. Without amperage through a device it will not work at all.
5. Amperes are measured by an **ammeter** (not ampmeter). ● **SEE FIGURE 1–16.**

VOLTS The volt is the unit of measurement for electrical pressure. It is named for an Italian physicist, Alessandro Volta (1745–1827). The comparable unit using water pressure as an example would be pounds per square inch (psi). It is possible to have very high pressures (volts) and low water flow (amperes). It is also possible to have high water flow (amperes) and low pressures (volts). Voltage is also called **electrical potential**, because if there is voltage present in a conductor, there is a potential (possibility) for current flow. This electrical pressure is a result of the following:

- Excess electrons remain at one end of the wire or circuit.
- There is a lack of electrons at the other end of the wire or circuit.
- The natural effect is to equalize this imbalance, creating a pressure to allow the movement of electrons through a conductor.
- It is possible to have pressure (volts) without any flow (amperes). For example, a fully charged 12 volt battery placed on a workbench has 12 volts of pressure potential, but because there is no conductor (circuit) connected between the positive and negative terminals of the battery, there is no flow (amperes). Current will only flow when there is pressure and a circuit for the electrons to flow in order to "equalize" to a balanced state.

Voltage does *not* flow through conductors, but voltage does cause current (in amperes) to flow through conductors. ● **SEE FIGURE 1–17.**

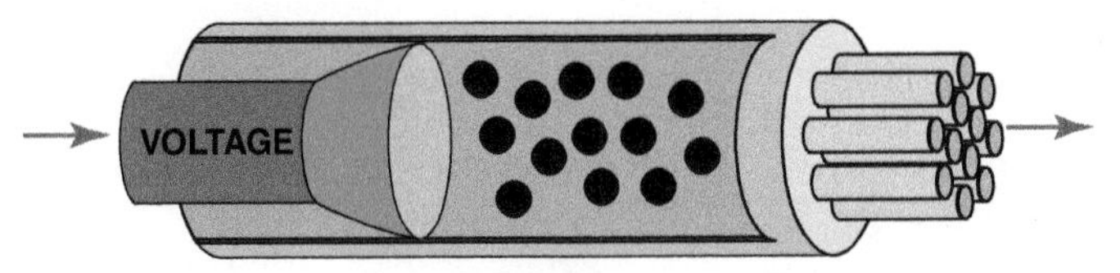

FIGURE 1–17 Voltage is the electrical pressure that causes the electrons to flow through a conductor.

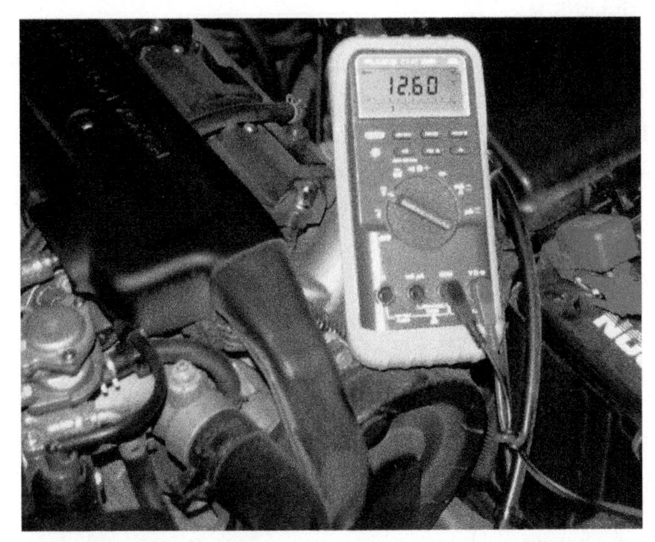

FIGURE 1–18 This digital multimeter set to read DC volts is being used to test the voltage of a vehicle battery. Most multimeters can also measure resistance (ohms) and current flow (amperes).

The conventional abbreviations and measurement for voltage are as follows:

1. The volt is the unit of measurement for the amount of electrical pressure.
2. **Electromotive force**, abbreviated **EMF**, is another way of indicating voltage.
3. *V* is the generally accepted abbreviation for *volts*.
4. The symbol used in calculations is *E*, for *electromotive force*.
5. Volts are measured by a **voltmeter**. ● **SEE FIGURE 1–18.**

OHMS **Resistance** to the flow of current through a conductor is measured in units called **ohms**, named after the German physicist George Simon Ohm (1787–1854). The resistance to the flow of free electrons through a conductor results from the countless collisions the electrons cause within the atoms of the conductor. ● **SEE FIGURE 1–19.**

The conventional abbreviations and measurement for resistance are as follows:

1. The ohm is the unit of measurement for electrical resistance.
2. The symbol for ohms is Ω (Greek capital letter omega), the last letter of the Greek alphabet.

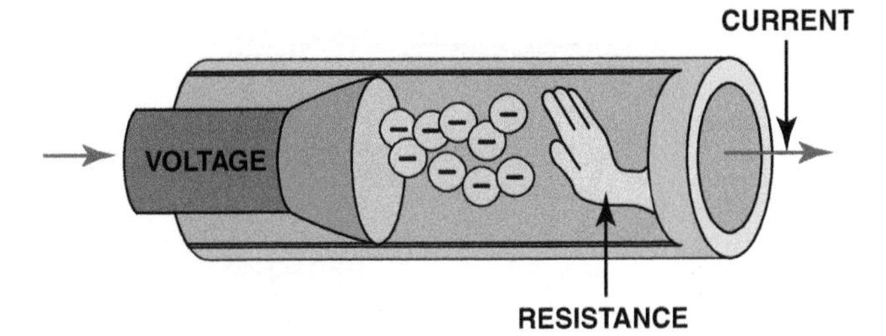

FIGURE 1–19 Resistance to the flow of electrons through a conductor is measured in ohms.

3. The symbol used in calculations is *R*, for *resistance*.
4. Ohms are measured by an **ohmmeter**.
5. Resistance to electron flow depends on the material used as a conductor.

WATTS A **watt** is the electrical unit for *power,* the capacity to do work. It is named after a Scottish inventor, James Watt (1736–1819). The symbol for power is *P*. Electrical power is calculated as amperes times volts:

$$P \text{ (power)} = I \text{ (amperes)} \times E \text{ (volts)}$$

The formula can also be used to calculate the amperage if the wattage and the voltage are known. For example, a 100 watt light bulb powered by 120 volts AC in the shop requires how many amperes?

A (amperes) $= P$ (watts) $\div E$ (volts)

$A = 0.83$ amperes

SOURCES OF ELECTRICITY

FRICTION When certain different materials are rubbed together, the friction causes electrons to be transformed from one to the other. Both materials become electrically charged. These charges are not in motion, but stay on the surface where they were deposited. Because the charges are stationary, or static, this type of voltage is called **static electricity**. Walking across a carpeted floor creates a buildup of a static charge in your body which is an insulator and then the charge is discharged when you touch a metal conductor. Vehicle tires rolling on pavement often create static electricity that interferes with radio reception.

HEAT When pieces of two different metals are joined together at both ends and one junction is heated, current passes through the metals. The current is very small, only millionths of an

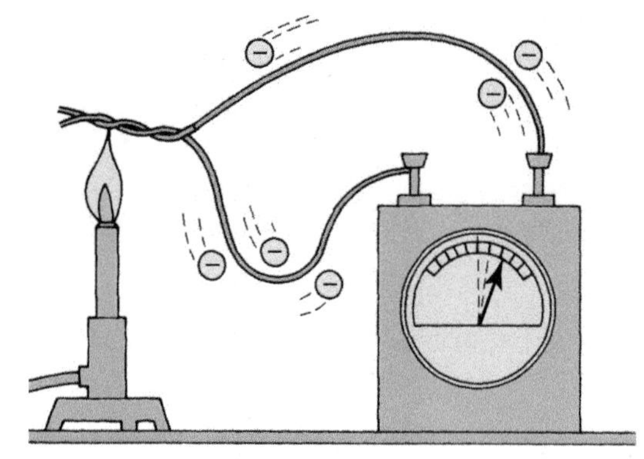

FIGURE 1–20 Electron flow is produced by heating the connection of two different metals.

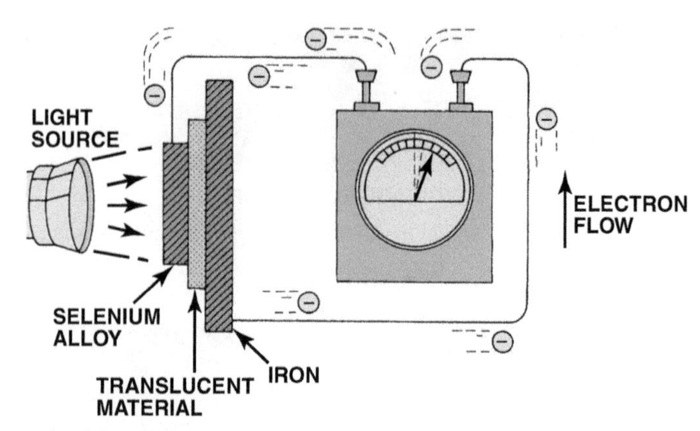

FIGURE 1–21 Electron flow is produced by light striking a light-sensitive material.

ampere, but this is enough to use in a temperature-measuring device called a **thermocouple**. This form of voltage is called **thermoelectricity**. ● **SEE FIGURE 1–20**.

Thermoelectricity has been known for over a century. In 1823, a German physicist, Thomas Johann Seebeck, discovered that a voltage was developed in a loop containing two dissimilar metals, provided the two junctions were maintained at different temperatures. A decade later, a French scientist, Jean Charles Athanase Peltier, found that electrons moving through a solid can carry heat from one side of the material to the other side. This effect is called the **Peltier effect**. A Peltier effect device can be used in heated and/or cooled seats, with the current flowing in one direction for cooling and reversing the current for heating.

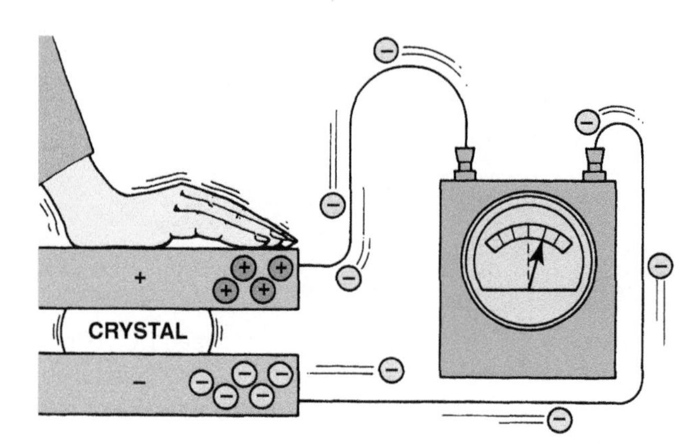

FIGURE 1–22 Electron flow is produced by pressure on certain crystals.

LIGHT In 1839, Edmond Becquerel noticed that by shining a beam of sunlight over two different liquids, he could develop an electric current. When certain metals are exposed to light, some of the light energy is transferred to the free electrons of the metal. This excess energy breaks the electrons loose from the surface of the metal. They can then be collected and made to flow in a conductor. ● **SEE FIGURE 1–21**.

This **photoelectricity** is widely used in light-measuring devices such as photographic exposure meters and automatic headlamp dimmers.

PRESSURE The first experimental demonstration of a connection between the generation of a voltage due to pressure applied to a crystal was published in 1880 by Pierre and Jacques Curie. Their experiment consisted of voltage being produced when prepared crystals, such as quartz, topaz, and Rochelle salt, had a force applied. ● **SEE FIGURE 1–22**.

This current is used in crystal microphones, underwater hydrophones, and certain stethoscopes. The voltage created is called **piezoelectricity**. A gas grille igniter uses the principle of piezoelectricity to produce a spark, and engine knock sensors (KS) use piezoelectricity to create a voltage signal for use as an input for an engine computer input signal. ● **SEE FIGURE 1–23**.

CHEMICAL Two different materials (usually metals) placed in a conducting and reactive chemical solution create a difference in potential, or voltage, between them. This principle is called **electrochemistry** and is the basis of the automotive battery.

MAGNETISM Electricity can be produced if a conductor is moved through a magnetic field or a moving magnetic field is moved near a conductor. This is the principle on which many automotive devices work, including the following:

- Starter motor
- Alternator
- Ignition coils
- Solenoids and relays

FIGURE 1–23 Engine knock sensors (KS) detect spark knock in the combustion chamber. The knock (spark ping) causes the sensor to output a voltage signal. (Courtesy of General Motors)

Diameter:		Higher Resistance
		Lower Resistance
Length:		Higher Resistance
		Lower Resistance
Condition:		Higher Resistance
		Lower Resistance
Temperature:	Hotter	Higher Resistance
	Cooler	Lower Resistance

FIGURE 1–24 Factors affecting resistance in a conductor. (Courtesy of General Motors)

RESISTANCE IN CONDUCTORS

All conductors have some resistance to current flow. The following are principles of conductors and their resistance. ● **SEE FIGURE 1–24.**

- **If the conductor length is doubled, its resistance doubles.** This is the reason why battery cables are designed to be as short as possible.
- **If the conductor diameter is increased, its resistance is reduced.** This is the reason starter motor cables are larger in diameter than other wiring in the vehicle.
- **As the temperature increases, the resistance of the conductor also increases.** This is the reason for installing heat shields on some starter motors. The heat shield helps to protect the copper wiring inside the starter from excessive engine heat, reducing resistance in the starter windings.
- **If the conductor is damaged its resistance is increased.**
- **Materials used in the conductor have an impact on its resistance.** Silver has the lowest resistance among all materials used as conductors, but is expensive. Copper is the next lowest in resistance and is reasonably priced. ● **SEE CHART 1–1** for a comparison of materials.

1	Silver
2	Copper
3	Gold
4	Aluminum
5	Tungsten
6	Zinc
7	Brass (copper and zinc)
8	Platinum
9	Iron
10	Nickel
11	Tin
12	Steel
13	Lead

CHART 1–1

Conductor ratings (starting with the best).

? FREQUENTLY ASKED QUESTION

Why Is Gold Used if Copper Has Lower Resistance?

Copper is used for most automotive electrical components and wiring because it has low resistance and is reasonably priced. Gold is used in airbag connections and sensors because it does not corrode. Gold can be buried for hundreds of years and when dug up it is just as shiny as ever.

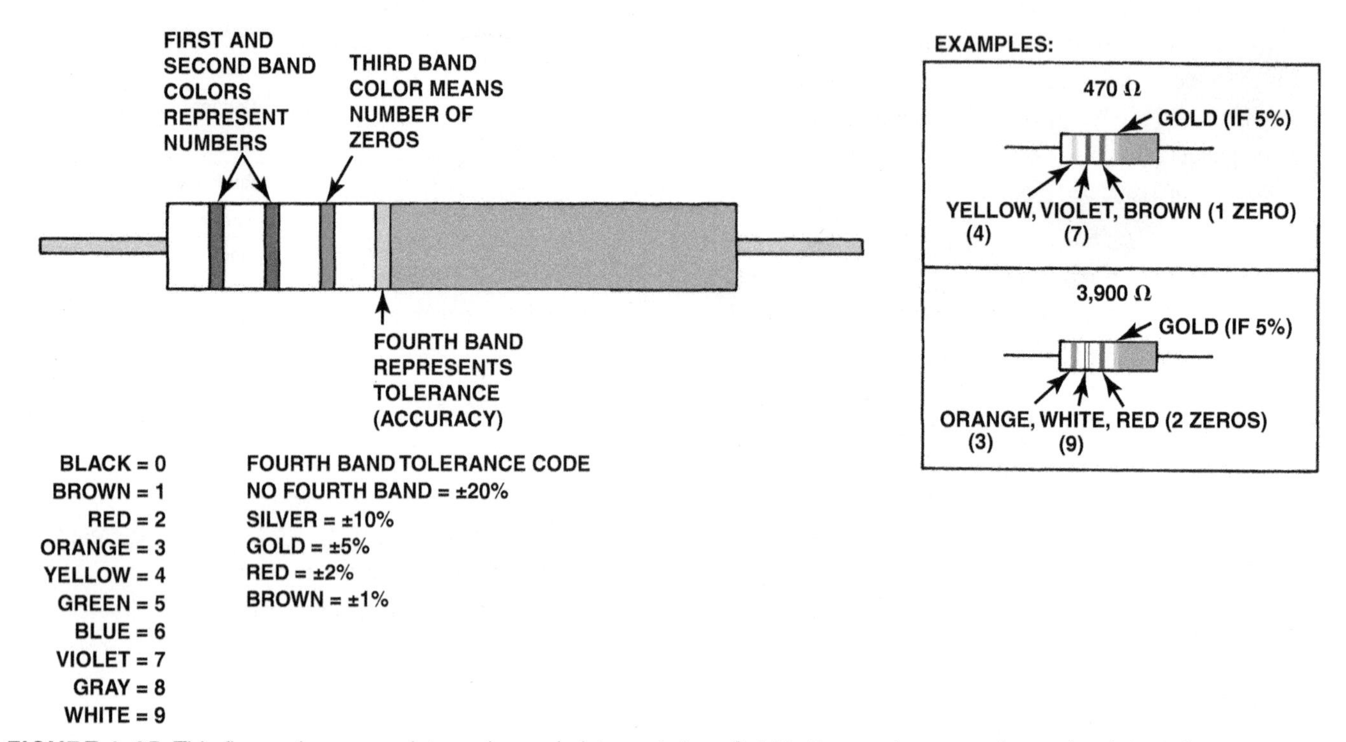

FIGURE 1–25 This figure shows a resistor color-code interpretation. Gold is the most commonly used resistor tolerance.

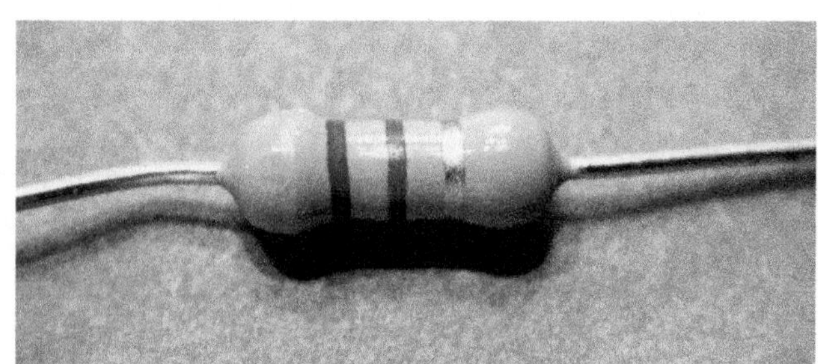

FIGURE 1–26 A typical carbon resistor.

THROTTLE POSITION (TP) SENSOR

5 v REFERENCE VOLTAGE

SIGNAL VOLTAGE (VARIABLE WITH POSITION OF MOVABLE CONTACT)

GROUND (O VOLTS)

MOVABLE CONTACT

FIGURE 1–27 A three-wire variable resistor is called a potentiometer.

RESISTORS

FIXED RESISTORS Resistance is the opposition to current flow. Resistors represent an electrical load, or resistance to current flow. Most electrical and electronic devices use resistors of specific values to limit and control the flow of current. Resistors can be made from carbon or from other materials that restrict the flow of electricity and are available in various sizes and resistance values. Most resistors have a series of painted color bands around them. These color bands are coded to indicate the degree of resistance. ● **SEE FIGURES 1–25 AND 1–26.**

VARIABLE RESISTORS Two basic types of mechanically operated variable resistors are used in automotive applications. The difference between the two is that the potentiometer produces a variable voltage, whereas the rheostat produces a variable current.

- A **potentiometer** is a three-terminal variable resistor where a movable contact provides a variable voltage output. ● **SEE FIGURE 1–27**. Potentiometers are used as an input to the various vehicle computer systems. These include throttle position (TP) sensors, accelerator pedal position (APP) sensors, audio controls, and many other systems that need to sense movement.

- Another type of mechanically operated variable resistor is the **rheostat**. A rheostat is a *two*-terminal unit in which all of the current flows through the movable arm. ● **SEE FIGURE 1–28**. A rheostat was commonly used as a dash light dimmer control on older vehicles.

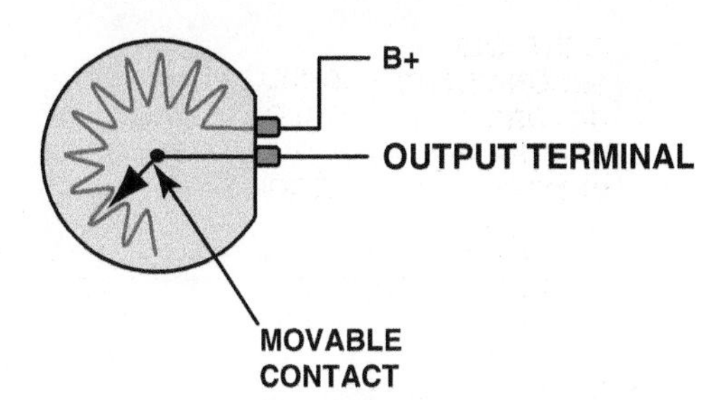

FIGURE 1–28 A two-wire variable resistor is called a rheostat.

SUMMARY

1. Electricity is the movement of electrons from one atom to another.
2. In order for current to flow in a circuit or wire, there must be an excess of electrons at one end and a deficiency of electrons at the other end.
3. Automotive electricity uses the conventional theory that electricity flows from positive to negative.
4. The ampere is the measure of the amount of current flow.
5. Voltage is the unit of electrical pressure.
6. The ohm is the unit of electrical resistance.
7. Sources of electricity include friction, heat, light, pressure, chemical and magnetism.

REVIEW QUESTIONS

1. What is electricity?
2. What are ampere, volt, and ohm?
3. Give three examples of conductors and three examples of insulators?
4. What are some sources of electricity?

CHAPTER QUIZ

1. An electrical conductor is an element with ________ electrons in its outer orbit.
   a. Less than 2
   b. Less than 4
   c. Exactly 4
   d. More than 4
2. Like charges ________.
   a. Attract
   b. Repel
   c. Neutralize each other
   d. Add
3. Carbon and silicon are examples of ________.
   a. Semiconductors
   b. Insulators
   c. Conductors
   d. Photoelectric materials
4. Which unit of electricity does the work in a circuit?
   a. Volt
   b. Ampere
   c. Ohm
   d. Coulomb
5. As temperature increases, ________.
   a. The resistance of a conductor decreases
   b. The resistance of a conductor increases
   c. The resistance of a conductor remains the same
   d. The voltage of the conductor decreases
6. The ________ is a unit of electrical pressure.
   a. Coulomb
   b. Volt
   c. Ampere
   d. Ohm
7. Technician A says that a two-wire variable resistor is called a rheostat. Technician B says that a three-wire variable resistor is called a potentiometer. Which technician is correct?
   a. Technician A only
   b. Technician B only
   c. Both Technicians A and B
   d. Neither Technician A nor B
8. The principle used in an automotive battery is called which of these?
   a. Electrochemistry
   b. Piezoelectricity
   c. Thermoelectricity
   d. Photoelectricity
9. The fact that a voltage can be created by exerting force on a crystal is used in which type of sensor?
   a. Throttle position (TP)
   b. Manifold absolute pressure (MAP)
   c. Barometric pressure (BARO)
   d. Knock sensor (KS)
10. A potentiometer, a three-wire variable resistance, is used in which type of sensor?
    a. Throttle position
    b. Manifold absolute pressure
    c. Barometric pressure
    d. Knock sensor

chapter 2

ELECTRICAL CIRCUITS AND OHM'S LAW

LEARNING OBJECTIVES

After studying this chapter, the reader will be able to:

1. Identify the parts of a complete circuit.
2. Describe the characteristics of different types of circuit.
3. Explain Ohm's law as it applies to automotive circuits.
4. Explain Watt's law as it applies to automotive circuits.

This chapter will help you prepare for the ASE Electrical/Electronic Systems (A6) certification test content area "A" (General Electrical/Electronic System Diagnosis).

KEY TERMS

GM STC OBJECTIVES

GM Service Technical College topics covered in this chapter are as follows:

1. Use a schematic or diagram to identify loads, power sources, and grounds.
2. Using the proper tools, apply Ohm's law to circuit diagnosis.

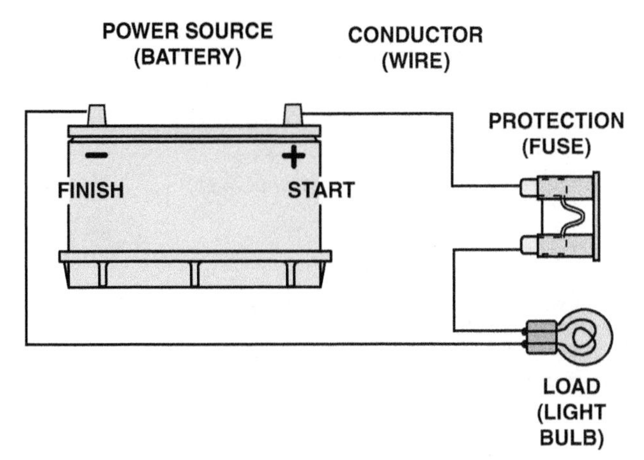

FIGURE 2–1 All complete circuits must have a power source, a power path, protection (fuse), an electrical load (light bulb in this case), and a return path back to the power source.

CIRCUITS

DEFINITION A **circuit** is a complete path that electrons travel from a power source (such as a battery) through a **load** such as a light bulb and back to the power source. It is called a *circuit* because the current must start and finish at the same place (power source).

For *any* electrical circuit to work at all, it must be continuous from the battery (power), through all the wires and components, and back to the battery (ground). A circuit that is continuous throughout is said to have **continuity**.

PARTS OF A COMPLETE CIRCUIT Every **complete circuit** contains the following parts. ● **SEE FIGURE 2–1.**

1. A **power source**, such as a vehicle's battery.
2. Protection from harmful overloads (excessive current flow). (Fuses, circuit breakers, and fusible links are examples of electrical circuit protection devices.)
3. The power path for the current to flow through from the power source to the resistance. (This path from a power source to the load—a light bulb in this example—is usually an insulated copper wire.)
4. The **electrical load** or resistance which converts electrical energy into heat, light, or motion.
5. A **return path (ground)** for the electrical current from the load back to the power source so that there is a *complete* circuit. (This return, or ground, path is usually the metal body, frame, ground wires, and engine block of the vehicle. ● **SEE FIGURE 2–2.**)
6. Switches and controls that turn the circuit on and off. (● **SEE FIGURE 2–3.**)

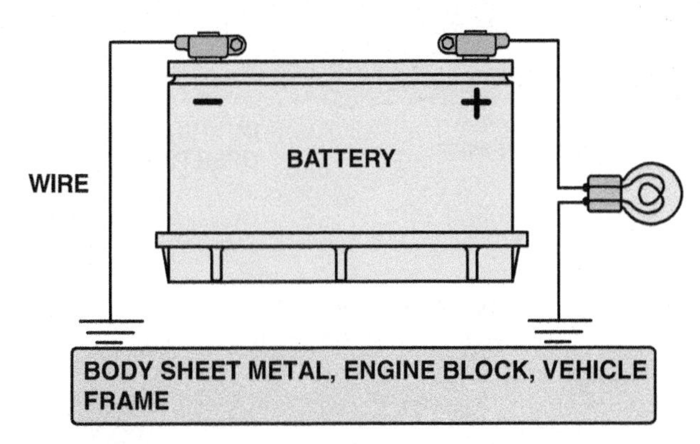

FIGURE 2–2 The return path back to the battery can be any electrical conductor, such as a copper wire or the metal frame or body of the vehicle.

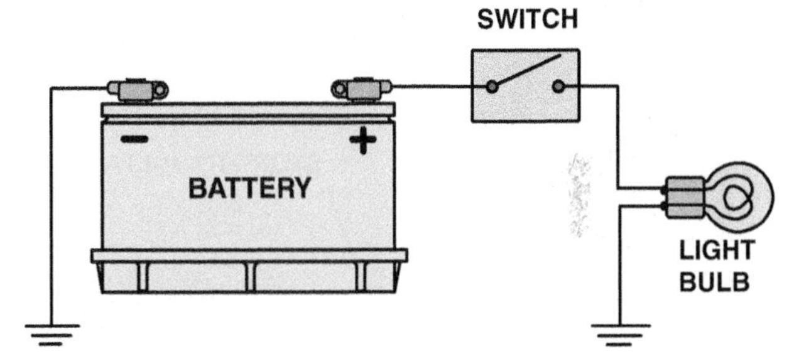

FIGURE 2–3 An electrical switch opens the circuit and no current flows. The switch can be located in the power path (shown) or in the return (ground) path wire.

CIRCUIT FAULT TYPES

OPEN CIRCUITS An **open circuit** is any circuit that is *not* complete, or that lacks continuity, such as a broken wire. ● **SEE FIGURE 2–4.**

Open circuits have the following features.

1. *No current at all* will flow through an open circuit.
2. An open circuit may be created by a break in the circuit or by a switch that opens (turns off) the circuit and prevents the flow of current.
3. In any circuit containing a power load and ground, an opening anywhere in the circuit will cause the circuit not to work.
4. A light switch in a home and the headlight switch in a vehicle are examples of devices that open a circuit to control its operation.
5. A fuse will blow (open) when the current in the circuit exceeds the fuse rating. This stops the current flow to prevent any harm to the components or wiring as a result of the fault.

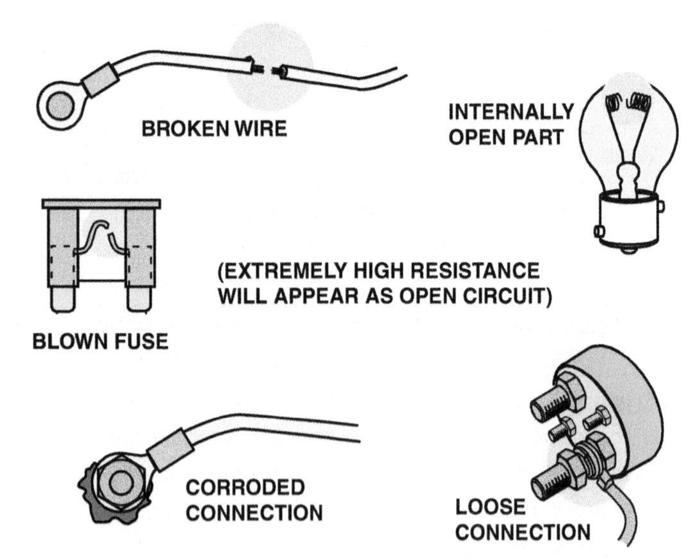

FIGURE 2–4 Examples of common causes of open circuits.

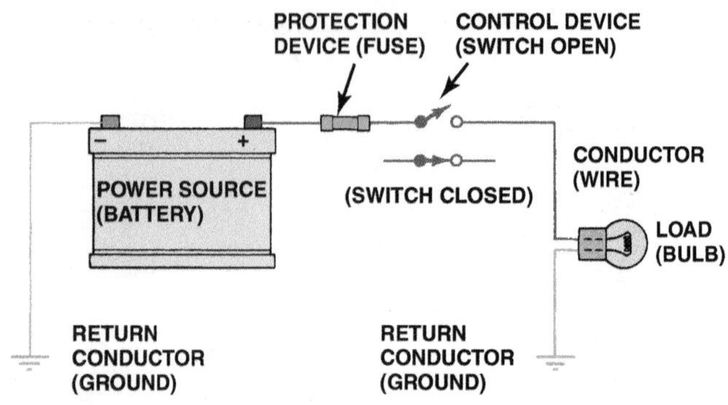

FIGURE 2–6 A fuse or circuit breaker opens the circuit to prevent possible overheating damage in the event of a short circuit.

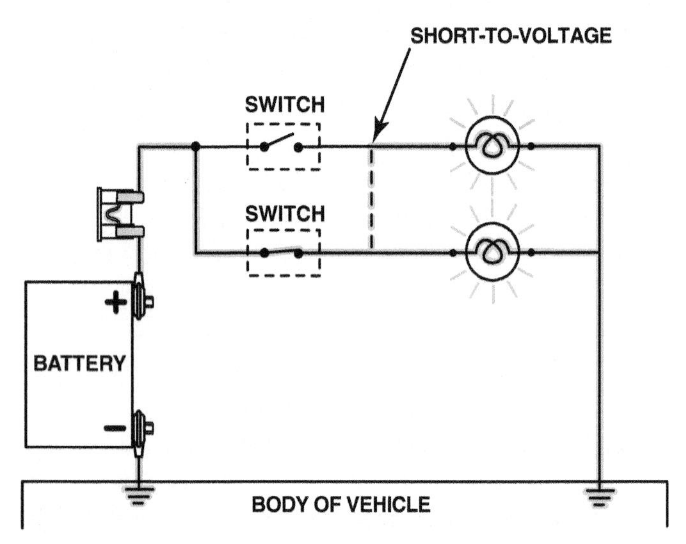

FIGURE 2–5 A short circuit permits electrical current to bypass some or all of the resistance in the circuit.

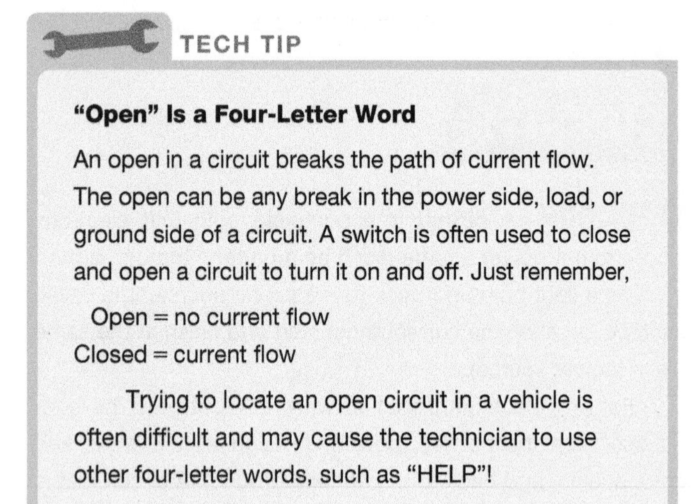

TECH TIP

"Open" Is a Four-Letter Word

An open in a circuit breaks the path of current flow. The open can be any break in the power side, load, or ground side of a circuit. A switch is often used to close and open a circuit to turn it on and off. Just remember,

Open = no current flow
Closed = current flow

Trying to locate an open circuit in a vehicle is often difficult and may cause the technician to use other four-letter words, such as "HELP"!

SHORT-TO-VOLTAGE *If a wire (conductor) or component is shorted to voltage*, it is commonly referred to as being **shorted**. A **short-to-voltage** occurs when the power side of one circuit is electrically connected to the power side of another circuit. ● **SEE FIGURE 2–5.**

A short circuit has the following features.

1. It is a complete circuit in which the current usually bypasses *some* or *all* of the resistance in the circuit.
2. It involves the power side of the circuit.
3. It involves a copper-to-copper connection (two power-side wires touching together).
4. It is also called a *short-to-voltage*.
5. It usually affects more than one circuit. In this case, if one circuit is electrically connected to another circuit, one of the circuits may operate when it is not supposed to because it is being supplied power from another circuit.
6. It *may* or *may not* blow a fuse. ● **SEE FIGURE 2–6.**

SHORT-TO-GROUND A **short-to-ground** is a type of short circuit that occurs when the current bypasses part of the normal circuit and flows directly to ground. A short-to-ground has the following features.

1. Because the ground return circuit is metal (vehicle frame, engine, or body), it is often identified as having current flowing from copper to steel.
2. It occurs at any place where a power path wire accidentally touches a return path wire or conductor. ● **SEE FIGURE 2–7.**

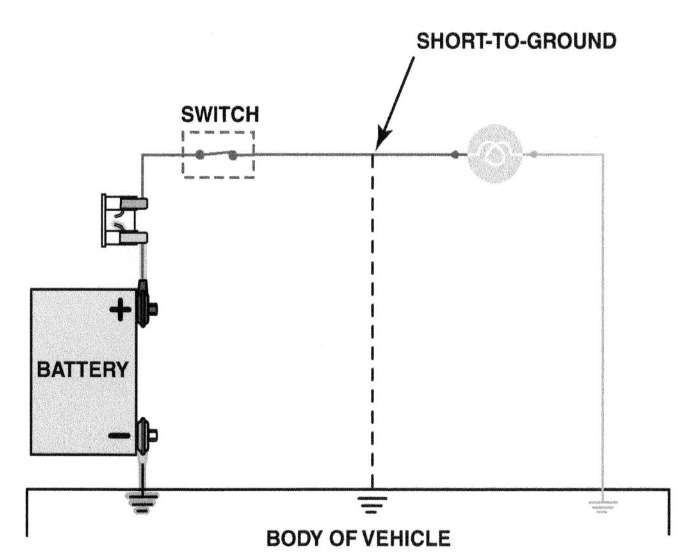

FIGURE 2–7 A short-to-ground affects the power side of the circuit. Current flows directly to the ground return, bypassing some or all of the electrical loads in the circuit. There is no current in the circuit past the short. A short-to-ground will also cause the fuse to blow.

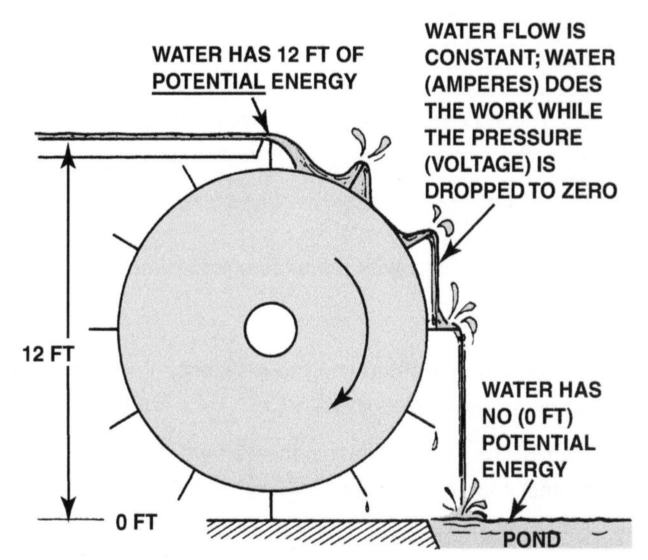

FIGURE 2–8 Electrical flow through a circuit is similar to water flowing over a waterwheel. The more the water (amperes in electricity), the greater the amount of work (waterwheel). The amount of water remains constant, yet the pressure (voltage in electricity) drops as the current flows through the circuit.

3. A defective component or circuit that is shorted to ground is commonly called **grounded**.
4. A short-to-ground almost always results in a blown fuse, damaged connectors, or melted wires.

TECH TIP

Think of a Waterwheel

A beginner technician cleaned the positive terminal of the battery on a vehicle with a slow cranking engine. When questioned by the shop foreman as to why only the positive post had been cleaned, the technician responded that the negative terminal was "only a ground." The foreman reminded the technician that the current, in amperes, is constant throughout a series circuit (such as the cranking motor circuit). If 200 amperes leave the positive post of the battery, then 200 amperes must return to the battery through the negative post.

The technician could not understand how electricity can do work (crank an engine), yet return the same amount of current, in amperes, as left the battery. The shop foreman explained that even though the current is constant throughout the circuit, the voltage (electrical pressure or potential) drops to zero in the circuit. To explain further, the shop foreman drew a waterwheel. ● **SEE FIGURE 2–8**.

As water drops from a higher level to a lower level, high potential energy (or voltage) is used to turn the waterwheel and results in low potential energy (or lower voltage). The same amount of water (or amperes) reaches the pond under the waterwheel as started the fall above the waterwheel. As current (amperes) flows through a conductor, it performs work in the circuit (turns the waterwheel) while its voltage (potential) drops.

HIGH RESISTANCE High resistance can be caused by any of the following:

- Corroded connections or sockets
- Loose terminals in a connector
- Loose power or ground connections

If there is high resistance anywhere in a circuit, it may cause the following problems.

1. Slow operation of a motor-driven unit, such as the windshield wipers or blower motor
2. Dim lights
3. "Clicking" of relays or solenoids
4. No operation of a circuit or electrical component

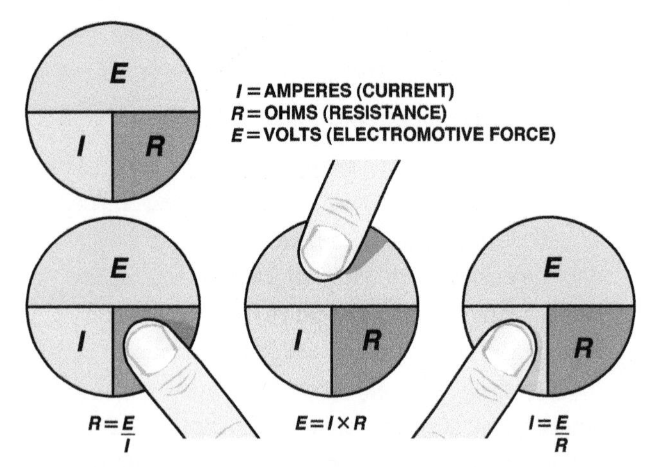

FIGURE 2–9 To calculate one unit of electricity when the other two are known, simply use your finger and cover the unit you do not know. For example, if both voltage (*E*) and resistance (*R*) are known, cover the letter *I* (amperes). Notice that the letter *E* is above the letter *R*, so divide the resistor's value into the voltage to determine the current in the circuit.

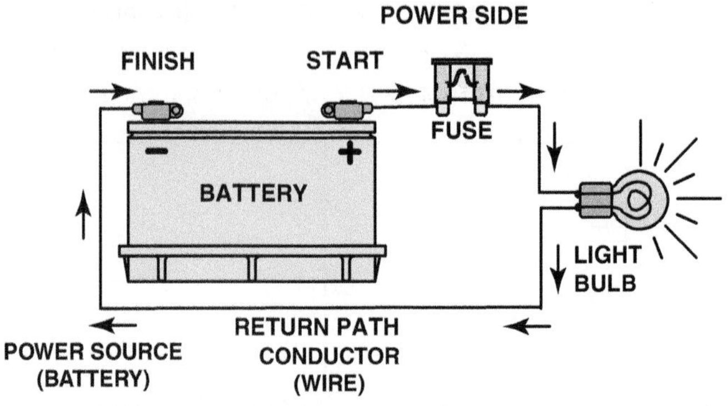

FIGURE 2–10 This closed circuit includes a power source, power-side wire, circuit protection (fuse), resistance (bulb), and return path wire. In this circuit, if the battery has 12 volts and the electrical load has 4 ohms, then the current through the circuit is 3 amperes.

OHM'S LAW

DEFINITION The German physicist George Simon Ohm established that electric pressure (EMF) in volts, electrical resistance in ohms, and the amount of current in amperes flowing through any circuit are all related. **Ohm's law** states:

It requires 1 volt to push 1 ampere through 1 ohm of resistance.

This means that if the voltage is doubled, then the number of amperes of current flowing through a circuit will also double if the resistance of the circuit remains the same.

FORMULAS Ohm's law can also be stated as a simple formula used to calculate one value of an electrical circuit if the other two are known. If, for example, the current (*I*) is unknown but the voltage (*E*) and resistance (*R*) are known, then Ohm's law can be used to find the answer. ● **SEE FIGURE 2–9.**

$$I = \frac{E}{R}$$

where

I = Current in amperes (A)

E = Electromotive force (EMF) in volts (V)

R = Resistance in ohms (Ω)

VOLTAGE	RESISTANCE	AMPERAGE
Up	Down	Up
Up	Same	Up
Up	Up	Same
Same	Down	Up
Same	Same	Same
Same	Up	Down
Down	Up	Down
Down	Same	Down

CHART 2–1

Ohm's law relationship with the three units of electricity.

1. Ohm's law can determine the resistance if the volts and amperes are known: $R = \frac{E}{I}$
2. Ohm's law can determine the *voltage* if the resistance (ohms) and amperes are known: $E = I \times R$
3. Ohm's law can determine the amperes if the resistance and voltage are known: $I = \frac{E}{R}$

NOTE: Before applying Ohm's law, be sure that each unit of electricity is converted into base units. For example, 10 KΩ should be converted to 10,000 Ω and 10 mA should be converted into 0.010 A. See Chapter 4 for more information on prefixes and conversions.

● **SEE CHART 2–1.**

OHM'S LAW APPLIED TO SIMPLE CIRCUITS If a battery with 12 volts is connected to a resistor of 4 ohms, as shown in ● **FIGURE 2–10**, how many amperes will flow through the circuit?

Using Ohm's law, we can calculate the number of amperes that will flow through the wires and the resistor. Remember, if two factors are known (volts and ohms in this example), the remaining factor (amperes) can be calculated using Ohm's law.

$$I = \frac{E}{R} = \frac{12V}{4\Omega} = A$$

The values for the voltage (12) and the resistance (4) were substituted for the variables E and R, and I is thus 3 amperes

$$\left(\frac{12}{4} = 3\right)$$

If we want to connect a resistor to a 12 volt battery, we now know that this simple circuit requires 3 amperes to operate. This may help us for two reasons.

1. We can now determine the wire diameter that we will need based on the number of amperes flowing through the circuit.
2. The correct fuse rating can be selected to protect the circuit.

WATT'S LAW

BACKGROUND James Watt (1736–1819), a Scottish inventor, first determined the power of a typical horse while measuring the amount of coal being lifted out of a mine. The power of one horse was determined to be 33,000 foot-pounds per minute. Electricity can also be expressed in a unit of power called a watt and the relationship is known as **Watt's law**, which states:

A watt is a unit of electrical power represented by a current of 1 ampere through a circuit with a potential difference of 1 volt.

FORMULAS A **watt** is a unit of electrical power represented by a current of 1 ampere through a circuit with a potential difference of 1 volt.

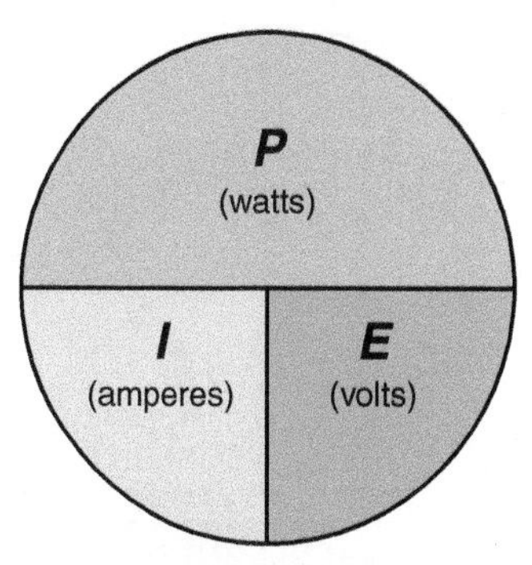

FIGURE 2–11 To calculate one unit when the other two are known, simply cover the unknown unit to see what unit needs to be divided or multiplied to arrive at the solution.

The symbol for a watt is the capital letter *W*. The formula for watts is:

$$W = I \times E$$

Another way to express this formula is to use the letter *P* to represent the unit of power. The formula then becomes:

$$P = I \times E$$

NOTE: An easy way to remember this equation is that it spells "pie."

Engine power is commonly rated in watts or kilowatts (1,000 watts equal 1 kilowatt), because 1 horsepower is equal to 746 watts. For example, a 200 horsepower engine can be rated as having the power equal to 149,200 watts or 149.2 kilowatts (kW).

To calculate watts, both the current in amperes and the voltage in the circuit must be known. If any two of these factors are known, then the other remaining factor can be determined by the following equations:

$P = I \times E$ **(watts equal amperes times voltage)**

$I = \frac{P}{E}$ **(amperes equal watts divided by voltage)**

$E = \frac{P}{I}$ **(voltage equals watts divided by amperes)**

A Watt's circle can be drawn and used like the Ohm's law circle diagram. ● **SEE FIGURE 2–11.**

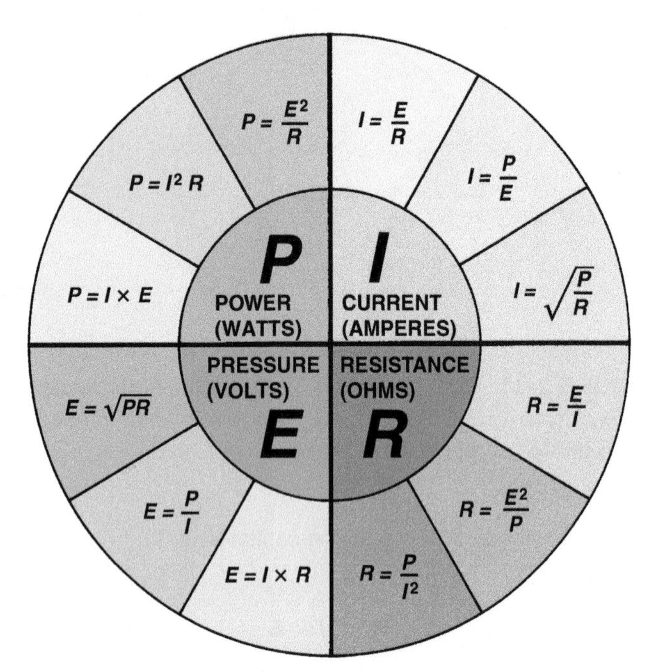

FIGURE 2–12 "Magic circle" of most formulas for problems involving electrical calculations. Each quarter of the "pie" has formulas used to solve for a particular unknown value: current (amperes), in the upper right segment; resistance (ohms), in the lower right; voltage (E), in the lower left; and power (watts), in the upper left.

MAGIC CIRCLE The formulas for calculating any combination of electrical units are shown in **FIGURE 2–12**.

It is almost impossible to remember all of these formulas, so this one circle showing all of the formulas is nice to have available if needed.

TECH TIP

Wattage Increases by the Square of the Voltage

The brightness of a light bulb, such as an automotive headlight or courtesy light, depends on the number of watts available. The watt is the unit by which electrical power is measured. If the battery voltage drops, even slightly, the light becomes noticeably dimmer. The formula for calculating power (P) in watts is $P = I \times E$. This can also be expressed as Watts = Amps × Volts.

According to Ohm's law, $I = \frac{E}{R}$. Therefore, $\frac{E}{R}$ can be substituted for I in the previous formula resulting in $P = \frac{E}{R} \times E$ or $P = \frac{E^2}{R}$.

E^2 means E multiplied by itself. A small change in the voltage (E) has a big effect on the total brightness of the bulb. (Remember, household light bulbs are sold according to their wattage.) Therefore, if the voltage to an automotive bulb is reduced, such as by a poor electrical connection, the brightness of the bulb is *greatly* affected. A poor electrical ground causes a voltage drop. The voltage at the bulb is reduced and the bulb's brightness is reduced.

SUMMARY

1. All complete electrical circuits have a power source (such as a battery), a circuit protection device (such as a fuse), a power-side wire or path, an electrical load, a ground return path, and a switch or a control device.
2. A short-to-voltage involves a copper-to-copper connection and usually affects more than one circuit.
3. A short-to-ground usually involves a power path conductor coming in contact with a return (ground) path conductor and usually causes the fuse to blow.
4. An open is a break in the circuit resulting in absolutely no current flow through the circuit.

REVIEW QUESTIONS

1. What is included in a complete electrical circuit?
2. What is the difference between a short-to-voltage and a short-to-ground?
3. What is the difference between an electrical open and a short?
4. What is Ohm's law?
5. What happens to current flow (amperes) and wattage if the resistance of a circuit is increased because of a corroded connection?
6. State Watt's law.

CHAPTER QUIZ

1. If an insulated wire gets rubbed through a part of the insulation and the wire conductor touches the steel body of a vehicle, the type of failure would be called a(n) _______.
   a. Short-to-voltage
   b. Short-to-ground
   c. Open
   d. Chassis ground
2. If two insulated power wires were to melt together at the point where the copper conductors touched each other, the type of failure would be called a(n) _______.
   a. Short-to-voltage
   b. Short-to-ground
   c. Open
   d. Floating ground
3. If 12 volts are being applied to a resistance of 3 ohms, _______ amperes will flow.
   a. 12
   b. 3
   c. 4
   d. 36
4. How many watts are consumed by a light bulb if 1.2 amperes are measured when 12 volts are applied?
   a. 14.4 watts
   b. 144 watts
   c. 10 watts
   d. 0.10 watt
5. How many watts are consumed by a starter motor if it draws 150 amperes at 10 volts?
   a. 15 watts
   b. 150 watts
   c. 1,500 watts
   d. 15,000 watts
6. High resistance in an electrical circuit can cause _______.
   a. Dim lights
   b. Slow motor operation
   c. Clicking of relays or solenoids
   d. All of the above
7. If the voltage increases in a circuit, what happens to the current (amperes) if the resistance remains the same?
   a. Increases
   b. Decreases
   c. Remains the same
   d. Cannot be determined
8. If 200 amperes flow from the positive terminal of a battery and operate the starter motor, how many amperes will flow back to the negative terminal of the battery?
   a. Cannot be determined
   b. Zero
   c. One half (about 100 amperes)
   d. 200 amperes
9. What is the symbol for voltage used in calculations?
   a. R
   b. E
   c. EMF
   d. I
10. Which circuit failure is most likely to cause the fuse to blow?
    a. Open
    b. Short-to-ground
    c. Short-to-voltage
    d. High resistance

chapter 3

SERIES, PARALLEL, AND SERIES–PARALLEL CIRCUITS

LEARNING OBJECTIVES

After studying this chapter, the reader will be able to:

1. Identify a series circuit.
2. Identify a parallel circuit.
3. Identify a series–parallel circuit.
4. Calculate the total resistance in a parallel circuit.
5. State Kirchhoff's voltage law.
6. Calculate voltage drops in a series circuit.
7. Explain series and parallel circuit laws.
8. State Kirchhoff's current law.
9. Identify where faults in a series–parallel circuit can be detected or determined.

This chapter will help you prepare for the ASE Electrical/Electronic Systems (A6) certification test content area "A" (General Electrical/Electronic System Diagnosis).

KEY TERMS

GM STC OBJECTIVES

GM Service Technical College topics covered in this chapter are as follows:

1. Using the proper tools, measure total circuit voltage in a series circuit.
2. Using the proper tools, measure total circuit current in a series circuit.
3. Using the proper tools, measure total circuit resistance in a series circuit.
4. Using the proper tools, measure voltage drop at each load in a series circuit.
5. Using the proper tools, measure total circuit voltage in a parallel circuit.
6. Using the proper tools, measure total circuit current in a parallel circuit.
7. Using the proper tools, measure total circuit resistance in a parallel circuit.
8. Using the proper tools, measure parallel circuit current through each load on each branch.
9. Using the proper tools, measure total circuit voltage in a series–parallel circuit.
10. Using the proper tools, measure total circuit current in a series–parallel circuit.
11. Using the proper tools, measure total circuit resistance in a series–parallel circuit.
12. Using the proper tools, measure circuit voltage drop at each load in a series–parallel circuit.
13. Using the proper tools, measure series–parallel circuit current through each load.
14. Basic circuit elements (08510.08W).
15. How to use a basic electrical circuit to calculate current, voltage, and resistance using Ohm's law.

SERIES CIRCUITS

DEFINITION A **series circuit** is a complete circuit that has only one path for current to flow through all of the electrical loads. Electrical components such as fuses and switches are generally not considered to be included in the determination of a series circuit.

CONTINUITY The circuit must be continuous without any breaks. This is called **continuity**. Every circuit must have continuity in order for current to flow through the circuit. Because there is only one path for current to flow, the current is the same everywhere in a complete series circuit.

NOTE: Because an electrical load needs both a power and a ground to operate, a break (open) anywhere in a series circuit will cause the current in the circuit to stop.

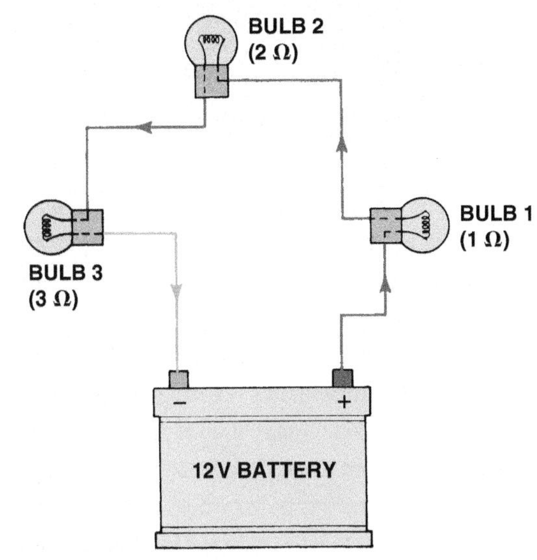

FIGURE 3–1 A series circuit with three bulbs. All current flows through all resistances (bulbs). The total resistance of the circuit is the sum of the individual resistances of each bulb, and the bulbs will light dimly because of the increased resistance and the reduction of current flow (amperes) through the circuit.

OHM'S LAW AND SERIES CIRCUITS

SERIES CIRCUIT TOTAL RESISTANCE A series circuit is a circuit containing more than one resistance in which all current must flow through all resistances in the circuit. Ohm's law can be used to calculate the value of one unknown (voltage, resistance, or amperes) if the other two values are known.

Because *all* current flows through all resistances, the total resistance is the sum (addition) of all resistances. ● **SEE FIGURE 3–1**.

The total resistance of the circuit shown here is 6 ohms (1 Ω + 2 Ω + 3 Ω). The formula for total resistance (R_T) for a series circuit is:

$$R_T = R_1 + R_2 + R_3 + \cdots$$

Ohm's law can be used to calculate the current (in amps) that will flow in the circuit. Ohm's law states that amperage (I) is equal to the voltage (E) divided by the resistance (R). The formula is:

$$I = \frac{E}{R} = \frac{12\text{ V}}{6\ \Omega} = 2\text{ A}$$

Therefore, with a total resistance of 6 ohms using a 12-volt battery in the series circuit shown, 2 amperes of current will flow through the entire circuit. If the amount of resistance in a series circuit is reduced, more current will flow.

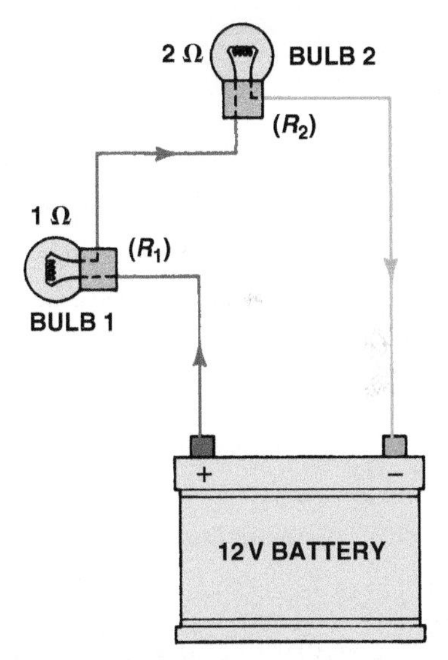

FIGURE 3–2 A series circuit with two bulbs.

For example, in ● **FIGURE 3–2**, one resistance (3-ohm bulb) has been eliminated compared to Figure 3–1, and now the total resistance is 3 ohms (1 Ω + 2 Ω).

Using Ohm's law to calculate current flow yields 4 amperes.

$$I = \frac{E}{R} = \frac{12\text{ V}}{3\ \Omega} = 4\text{ A}$$

Notice that the current flow was doubled (4 amperes instead of 2 amperes) when the resistance was cut in half (from 6 to 3 ohms). The current flow would also double if the applied voltage was doubled.

Farsighted Quality of Electricity

Electricity almost seems to act as if it knows what resistances are ahead on the long trip through a circuit. If the trip through the circuit has many high-resistance components, very few electrons (amperes) will choose to attempt to make the trip. If a circuit has little or no resistance (e.g., a short circuit), then as many electrons (amperes) as possible attempt to flow through the complete circuit. If another load, such as a light bulb, were added in series, the current flow would decrease and the bulbs would be dimmer than before the other bulb was added. If the flow exceeds the capacity of the fuse or the circuit breaker, then the circuit is opened and all current flow stops.

KIRCHHOFF'S VOLTAGE LAW

DEFINITION A German physicist, Gustav Robert Kirchhoff (1824–1887), developed laws about electrical circuits. His second law, **Kirchhoff's voltage law**, concerns voltage drops. It states:

> The voltage around any closed circuit is equal to the sum (total) of the voltage drops across the resistances.

For example, the voltage that flows through a series circuit drops with each resistor in a manner similar to that in which the strength of an athlete drops each time a strenuous physical feat is performed. The greater the resistance is, the greater the drop in voltage.

APPLYING KIRCHHOFF'S VOLTAGE LAW Kirchhoff states in his second law that the voltage will drop in proportion to the resistance and that the total of all voltage drops will equal the applied voltage. ● **SEE FIGURE 3–3.**

Using ● **FIGURE 3–4**, the total resistance of the circuit can be determined by adding the individual resistances $2\,\Omega + 4\,\Omega + 6\,\Omega = 12\,\Omega$.

The current through the circuit is determined by using Ohm's law, $I = \frac{E}{R} = \frac{12\text{ V}}{12\,\Omega} = 1\text{ A}$.

Therefore, in the circuit shown, the following values are known.

Resistance = 12 Ω

Voltage = 12 V

Current = 1 A

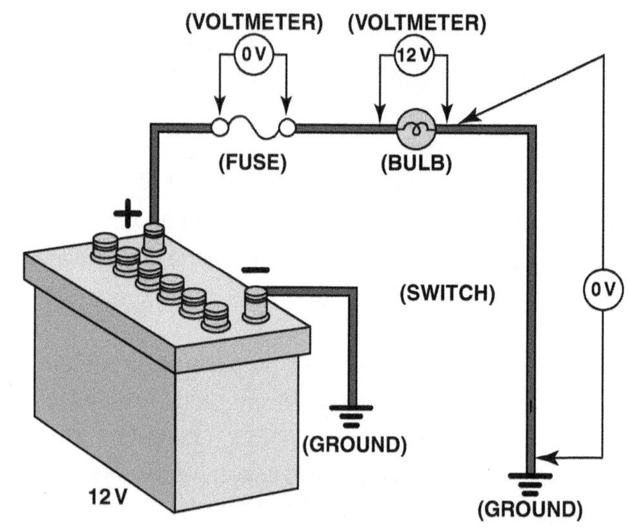

FIGURE 3–3 As current flows through a circuit, the voltage drops in proportion to the amount of resistance in the circuit. Most, if not all, of the resistance should occur across the load such as the bulb in this circuit. All of the other components and wiring should produce little, if any, voltage drop. If a wire or connection did cause a voltage drop, less voltage would be available to light the bulb and the bulb would be dimmer than normal.

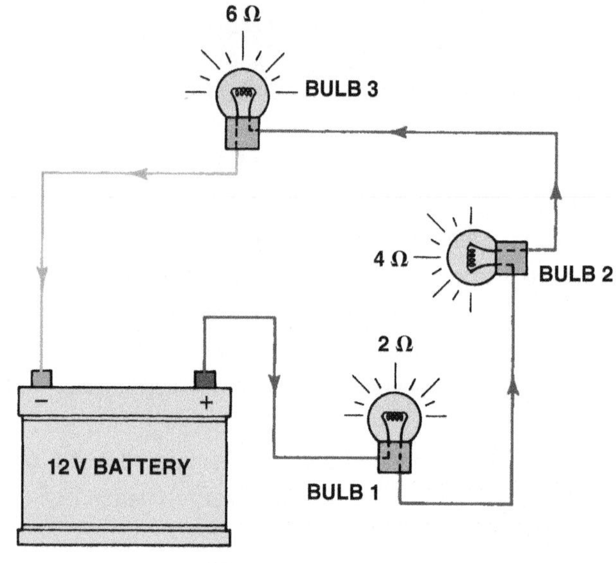

FIGURE 3–4 In a series circuit, the voltage is dropped or lowered by each resistance in the circuit. The higher the resistance is, the greater the drop in voltage.

Everything is known *except* the voltage drop caused by each resistance. A **voltage drop** is the drop in voltage across a resistance when current flows through a complete circuit. In other words, a voltage drop is the amount of voltage (electrical pressure) required to push electrons through a resistance. The voltage drop can be determined by using Ohm's law and calculating for voltage (*E*) using the value of each resistance individually, as follows:

$$E = I \times R$$

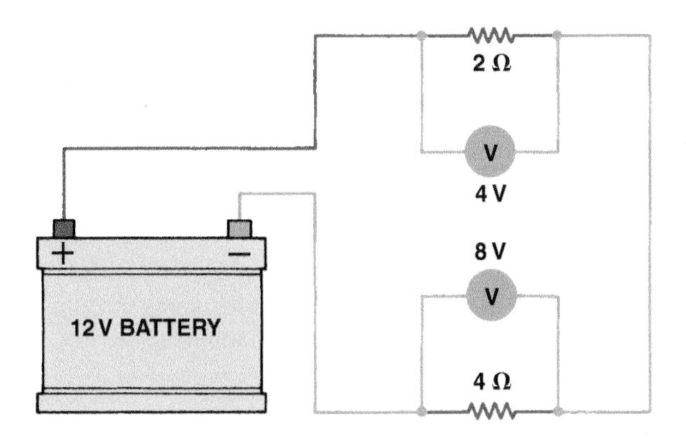

A. $I = E/R$ (TOTAL "R" = 6 Ω)
12 V/6 Ω = 2 A

B. $E = I/R$ (VOLTAGE DROP)
AT 2 Ω RESISTANCE =
$E = 2 \times 2 = 4$ V
AT 4 Ω RESISTANCE =
$E = 2 \times 4 = 8$ V

C. 4 + 8 = 12 V
SUM OF VOLTAGE DROP
EQUALS APPLIED VOLTAGE

FIGURE 3–5 A voltmeter reads the differences of voltage between the test leads. The voltage read across a resistance is the voltage drop that occurs when current flows through a resistance. A voltage drop is also called an "IR" drop because it is calculated by multiplying the current (I) through the resistance (electrical load) by the value of the resistance (R).

where

E = Voltage

I = Current in the circuit (Remember, the current is constant in a series circuit; only the voltage varies.)

R = Resistance of only one of the resistances

The voltage drops are as follows:

Voltage drop for bulb 1: $E = I \times R = 1\text{ A} \times 2\,\Omega = 2\text{ V}$

Voltage drop for bulb 2: $E = I \times R = 1\text{ A} \times 4\,\Omega = 4\text{ V}$

Voltage drop for bulb 3: $E = I \times R = 1\text{ A} \times 6\,\Omega = 6\text{ V}$

NOTE: Notice that the voltage drop is proportional to the resistance. In other words, the higher the resistance is, the greater the voltage drop. A 6-ohm resistance dropped the voltage three times as much as the voltage drop created by the 2-ohm resistance.

According to Kirchhoff, the sum (addition) of the voltage drops should equal the applied voltage (battery voltage).

Total of voltage drops = 2 V + 4 V + 6 V = 12 V = Battery voltage

This proves Kirchhoff's second (voltage) law. Another example is illustrated in **FIGURE 3–5**.

VOLTAGE DROPS

VOLTAGE DROPS USED IN CIRCUITS A voltage drop indicates resistance in the circuit. Often a voltage drop is not wanted in a circuit because it causes the electrical load to not operate correctly. Some automotive electrical systems use voltage drops in cases such as the following:

1. **Dash lights.** Older vehicles were equipped with a method of dimming the brightness of the dash lights by turning a variable resistor. The resistance value of this type of resistor can be increased or decreased and therefore varies the voltage to the dash light bulbs. A high voltage to the bulbs causes them to be bright, and a low voltage results in a dim light.
2. **Blower motor** (heater or air conditioning fan). Speeds can be controlled by a fan switch sending current through high-, medium-, or low-resistance wire resistors. The highest resistance will drop the voltage the most, causing the motor to run at the lowest speed. The highest speed of the motor will occur when *no* resistance is in the circuit and full battery voltage is switched to the blower motor.

USING VOLTAGE DROP WHEN TESTING CIRCUITS Any resistance in a circuit causes a voltage drop, whether the resistance is intentional (designed in to the circuit) or not (a loose connection). The amount of voltage dropped will be proportional to the amount of the resistance.

Because any resistance, wanted or unwanted, will cause a voltage drop, a voltmeter can be used to measure the effects of resistance in a circuit. Measuring voltage drop is the preferred method recommended by most manufacturers to locate or test a circuit for excessive resistance.

SERIES CIRCUIT LAWS

Electrical loads or resistances connected in series follow the following **series circuit laws**.

LAW 1 The total resistance in a series circuit is the sum total of the individual resistances. The resistance values of each electrical load are simply added together.

FREQUENTLY ASKED QUESTION

Why Check the Voltage Drop Instead of Measuring the Resistance?

Imagine a wire with all strands cut except for one. An ohmmeter can be used to check the resistance of this wire and the resistance would be low, indicating that the wire was okay, but this one small strand cannot properly carry the current (amperes) in the circuit. A voltage drop test is therefore a better test to determine the resistance in components for two reasons:

- An ohmmeter can only test a wire or component that has been disconnected from the circuit and is not carrying current. The resistance can, and does, change when current flows.
- A voltage drop test is a dynamic test because as the current flows through a component, the conductor increases in temperature, which in turn increases resistance. This means that a voltage drop test is testing the circuit during normal operation and is therefore the most accurate way of determining circuit conditions.

A voltage drop test is also easier to perform because the resistance does not have to be known, only that the loss of voltage in a circuit should be less than 3%, or less than about 0.36 volt for any 12-volt circuit.

LAW 2 The current is the same throughout the circuit. ● **SEE FIGURE 3–6**.

If 2 amperes of current leave the battery, 2 amperes of current return to the battery.

LAW 3 Although the current (in amperes) is constant, the voltage drops across each resistance in the circuit can vary at each resistor. The voltage drop across each load is proportional to the value of the resistance compared to the total resistance. For example, if the resistance of each resistor in a two-resistor circuit is half of the total resistance, the voltage drop across that resistance will be half of the applied voltage. The sum total of all individual voltage drops equals the applied source voltage.

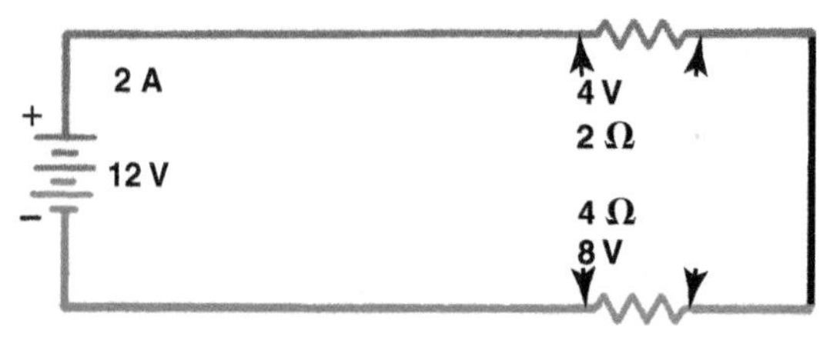

FIGURE 3–6 In this series circuit with a 2-ohm resistor and a 4-ohm resistor, current (2 amperes) is the same throughout, even though the voltage drops across each resistor are different.

TECH TIP

Light Bulbs and Ohm's Law

If the resistance of a typical automotive light bulb is measured at room temperature, the resistance will often be around 1 ohm. If 12 volts were to be applied to this bulb, a calculated current of 12 amperes would be expected $\left(I = \frac{E}{R} = 12 \div 1 = 12\text{ A}\right)$. However, as current flows through the filament of the bulb, it heats up and becomes incandescent, thereby giving off light. When the bulb is first connected to a power source and current starts to flow, a high amount of current, called surge current, flows through the filament. Then, within a few thousandths of a second, the current flow is reduced to about 10% of the surge current due to the increasing resistance of the filament, resulting in an actual current flow of about 1.2 A or about 100 ohms of resistance when the bulb is working.

As a result, using Ohm's law to calculate current flow does not take into account the differences in temperature of the components during actual operation.

SERIES CIRCUIT EXAMPLES

Each of the four examples includes solving for the following:

- Total resistance in the circuit
- Current flow (amperes) through the circuit
- Voltage drop across each resistance

Example 1:

(● **SEE FIGURE 3–7**.)

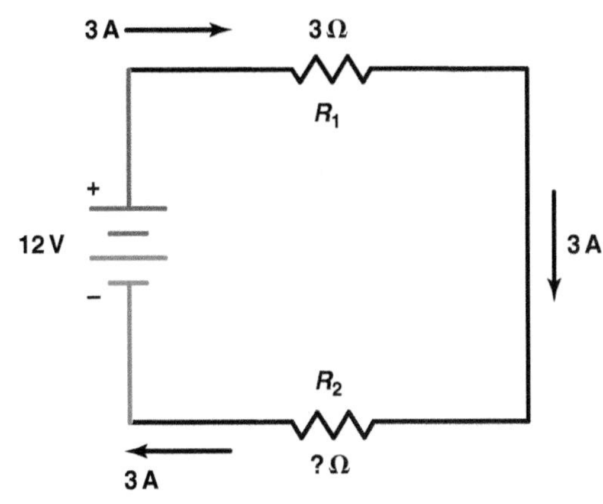

FIGURE 3–7 Example 1.

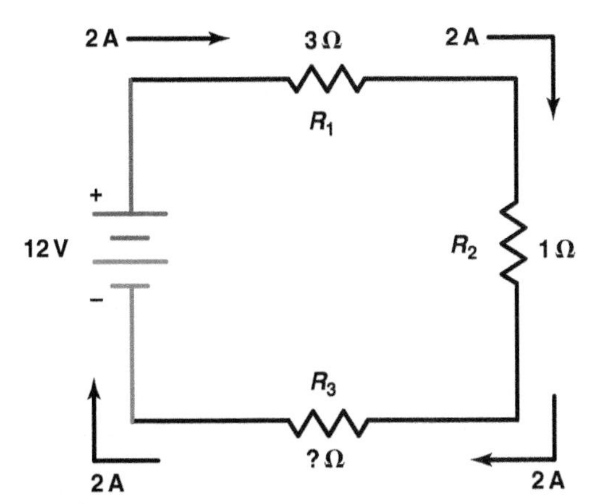

FIGURE 3–8 Example 2.

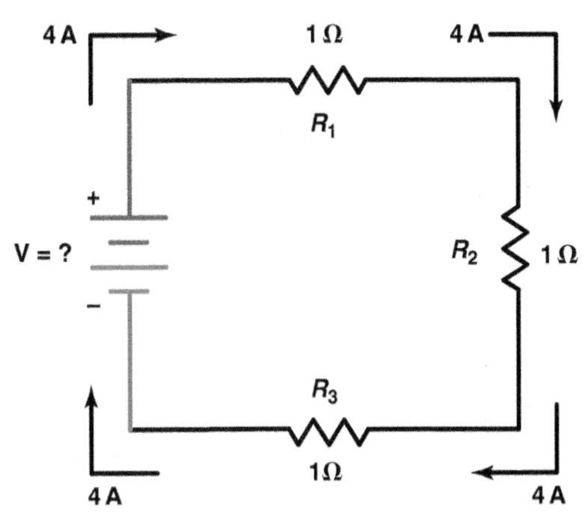

FIGURE 3–9 Example 3.

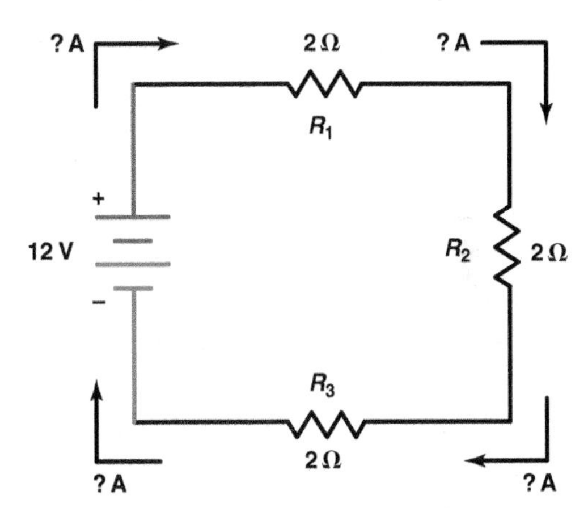

FIGURE 3–10 Example 4.

The unknown in this problem is the value of R_2. Because the source voltage and the circuit current are known, the total circuit resistance can be calculated using Ohm's law.

$$R_{Total} = \frac{E}{I} = 12\text{ v} \div 3\text{ A} = 4\ \Omega$$

Because R_1 is 3 ohms and the total resistance is 4 ohms, therefore the value of R_2 is 1 ohm.

Example 2:

(**SEE FIGURE 3–8**.)

The unknown in this problem is the value of R_3. The total resistance, however, can be calculated using Ohm's law.

$$R_{Total} = \frac{E}{I} = 12\text{ v} \div 2\text{ A} = 6\ \Omega$$

The total resistance of R_1 (3 ohms) and R_2 (1 ohm) equals 4 ohms, so that the value of R_3 is the difference between the total resistance (6 ohms) and the value of the known resistance (4 ohms).

$$6 - 4 = 2\ \Omega = R_3$$

Example 3:

(**SEE FIGURE 3–9**.)

The unknown value in this problem is the voltage of the battery. To solve for voltage, use Ohm's law ($E - I \times R$). The R in this problem refers to the total resistance (R_T). The total resistance of a series circuit is determined by adding the values of the individual resistors.

$$R_T = 1\ \Omega + 1\ \Omega + 1\ \Omega$$
$$R_T = 3\ \Omega$$

Placing the value for the total resistance (3 Ω) into the equation results in a battery voltage of 12 volts.

$$E = 4\text{ A} \times 3\ \Omega$$
$$E = 12\text{ volts}$$

Example 4:

(**SEE FIGURE 3–10**.)

The unknown in this example is the current (amperes) in the circuit. To solve for current, use Ohm's law.

$$I = \frac{E}{R} = 12\text{ v} \div 6\ \Omega = 2\text{ A}$$

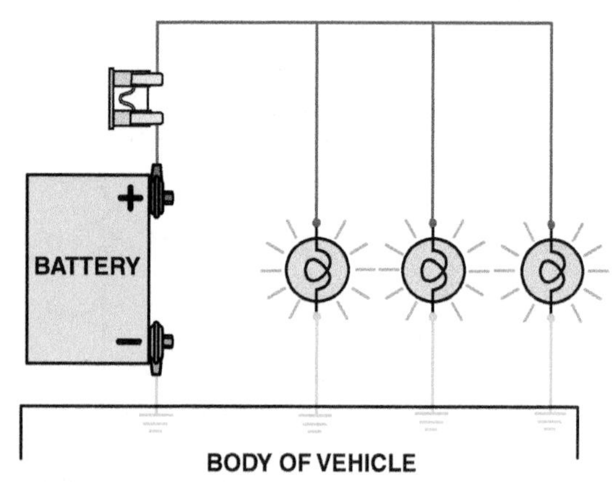

FIGURE 3–11 A typical parallel circuit used in vehicles includes many of the interior and exterior lights.

Notice that the total resistance in the circuit (6 ohms) was used in this example, which is the total of the three individual resistors (2 Ω + 2 Ω + 2 Ω = 6 Ω). The current through the circuit is 2 amperes.

PARALLEL CIRCUITS

DEFINITION A **parallel circuit** is a complete circuit that has more than one path for the current to flow. The separate paths that split and meet at junction points (splices) are called **branches, legs,** or **shunts.** The current flow through each branch or leg varies depending on the resistance in that branch. A break or open in one leg or section of a parallel circuit does not stop the current flow through the remaining legs of the parallel circuit. Most circuits in vehicles are parallel circuits and each branch is connected to the 12 volt power supply. ● **SEE FIGURE 3–11.**

KIRCHHOFF'S CURRENT LAW

DEFINITION **Kirchhoff's current law** (his first law) states:

> The current flowing into any junction of an electrical circuit is equal to the current flowing out of that junction.

Kirchhoff's law states that the amount of current flowing into junction A will equal the current flowing out of junction A.

Example:

Because the 6-ohm leg requires 2 amperes and the 3-ohm resistance leg requires 4 amperes, it is necessary that the wire from the battery to junction A be capable of handling 6 amperes. Also notice that the sum of the current flowing *out* of a junction (2 + 4 = 6 A) is equal to the current flowing *into* the junction (6 A), proving Kirchhoff's current law. ● **SEE FIGURE 3–12.**

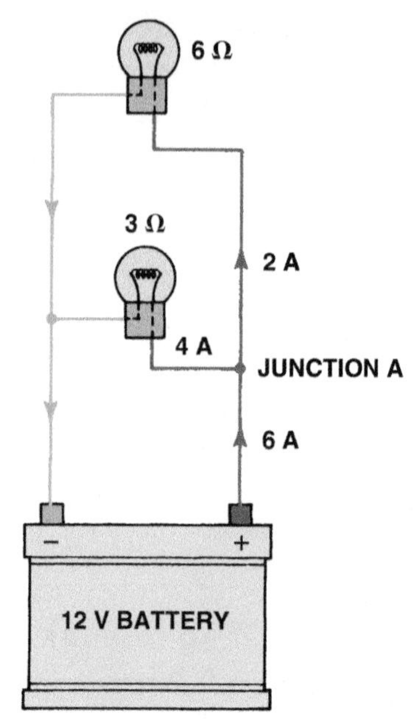

FIGURE 3–12 The amount of current flowing into junction point A equals the total amount of current flowing out of the junction.

PARALLEL CIRCUIT LAWS

LAW 1 The total resistance of a parallel circuit is always less than that of the lowest-resistance leg. This occurs because not all of the current flows through each leg or branch. With many branches, more current can flow from the battery just as more vehicles can travel on a road with five lanes compared to only one or two lanes.

LAW 2 The voltage is the same for each leg of a parallel circuit.

LAW 3 The sum of the individual currents in each leg will equal the total current. The amount of current flow through a parallel circuit may vary for each leg depending on the resistance of that leg. The current flowing through each leg results in the same voltage drop (from the power side to the ground side) as for every other leg of the circuit. ● **SEE FIGURE 3–13.**

NOTE: A parallel circuit drops the voltage from source voltage to zero (ground) across the resistance in each leg of the circuit.

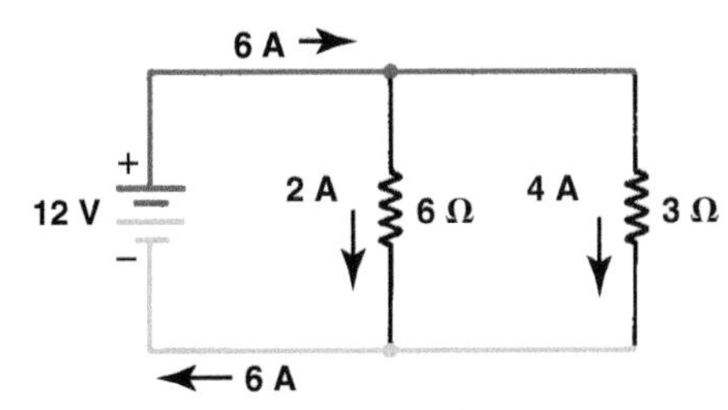

FIGURE 3–13 The current in a parallel circuit splits (divides) according to the resistance in each branch. Each branch has 12 volts applied to the resistors.

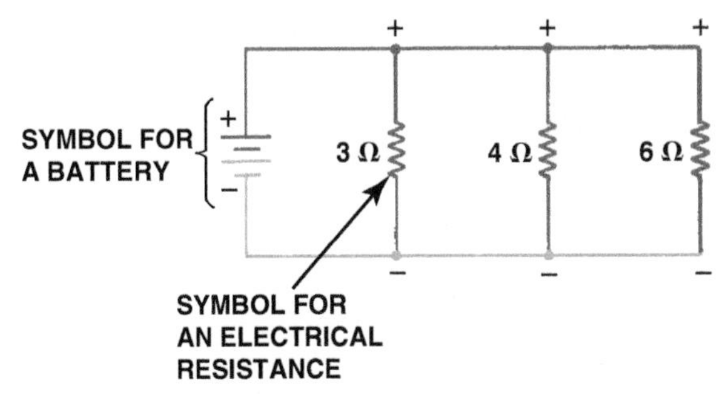

FIGURE 3–14 In a typical parallel circuit, each resistance has power and ground and each leg operates independently of the other legs of the circuit.

DETERMINING TOTAL RESISTANCE IN A PARALLEL CIRCUIT

There are five methods commonly used to determine total resistance in a parallel circuit.

Determining the total resistance of components connected in parallel can be very useful in automotive service.

An example might be when installing extra lighting. The technician must determine the total current of the lights connected in parallel in order to select the proper gauge wire and protection device.

There are five methods that can be used to determine the total resistance in a parallel circuit.

METHOD 1 The total *current* (in amperes) can be calculated first by treating each leg of the parallel circuit as a simple circuit. ● **SEE FIGURE 3–14.**

Each leg has its own power (+) and ground (–) and, therefore, the current through each leg is independent of the current through any other leg.

Current through the 3 Ω resistance can be found using the equation $I = \frac{E}{R} = \frac{12\text{ V}}{3\ \Omega} = 4\text{ A}$

Current through the 4 Ω resistance can be found using the equation $I = \frac{E}{R} = \frac{12\text{ V}}{4\ \Omega} = 3\text{ A}$

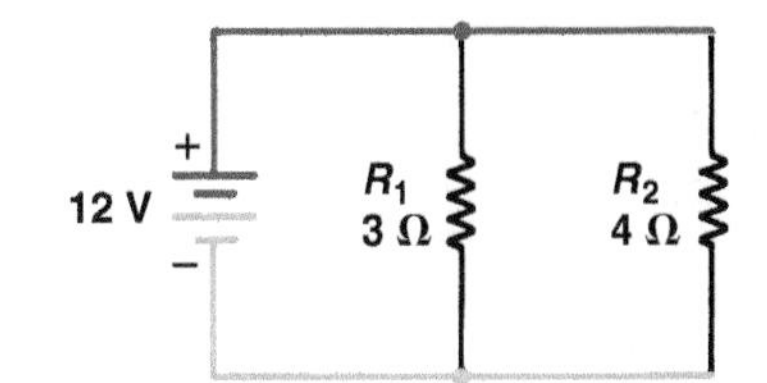

FIGURE 3–15 A schematic showing two resistors in parallel connected to a 12 volt battery.

Current through the 4 Ω resistance can be found using the equation $I = \frac{E}{R} = \frac{12\text{ V}}{6\ \Omega} = 2\text{ A}$

The total current flowing from the battery is the sum total of the individual currents for each leg. Total current from the battery is, therefore, 9 amperes (4 A + 3 A + 2 A = 9 A).

If **total circuit resistance** (R_T) is needed, Ohm's law can be used to calculate it because voltage (*E*) and current (*I*) are now known.

$$R_T = \frac{E}{I} = \frac{12\text{ V}}{9\text{ A}} = 1.33\ \Omega$$

Note that the total resistance (1.33 Ω) is smaller than that of the smallest-resistance leg of the parallel circuit. This characteristic of a parallel circuit holds true because not all of the total current flows through all resistances as in a series circuit.

Because the current has alternative paths to ground through the various legs of a parallel circuit, as additional resistances (legs) are added to a parallel circuit, the total current from the battery (power source) increases.

Additional current can flow when resistances are added in parallel, because each leg of a parallel circuit has its own power and ground and the current flowing through each leg is strictly dependent on the resistance of *that* leg.

METHOD 2 If only two resistors are connected in parallel, the total resistance (R_T) can be found using the formula $R_T = \left(\frac{R_1 \times R_2}{R_1 + R_2}\right)$. For example, using the circuit in ● **FIGURE 3–15** and substituting 3 ohms for R_1 and 4 ohms for R_2, $R_T = \frac{(3 \times 4)}{(3 + 4)} = \frac{12}{7} = 1.7\ \Omega$.

Note that the total resistance (1.7 Ω) is smaller than that of the smallest-resistance leg of the circuit.

NOTE: Which resistor is R_1 and which is R_2 is not important. The position in the formula makes no difference in the multiplication and addition of the resistor values.

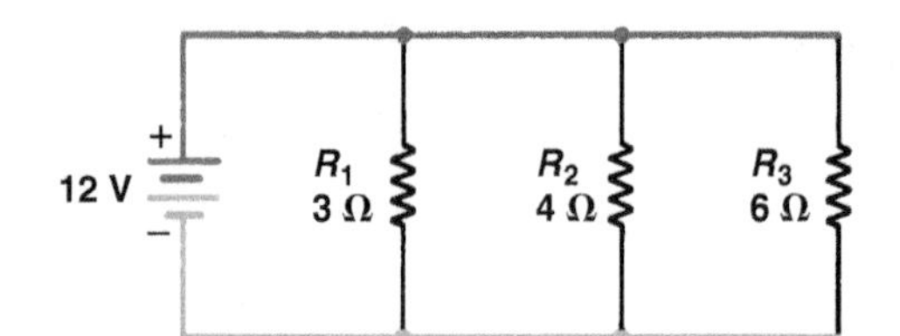

FIGURE 3–16 A parallel circuit with three resistors connected to a 12 volt battery.

This formula can be used for more than two resistances in parallel, but only two resistances can be calculated at a time. After solving for R_T for two resistors, use the value of R_T as R_1 and the additional resistance in parallel as R_2. Then solve for another R_T. Continue the process for all resistance legs of the parallel circuit. However, note that it might be easier to solve for R_T when there are more than two resistances in parallel by using either method 3 or method 4.

METHOD 3 A formula that can be used to find the total resistance for any number of resistances in parallel is $\frac{1}{R_T} = \frac{1}{R_1} + \frac{1}{R_2} + \frac{1}{R_3} + \cdots$

To solve for R_T for the three resistance legs in **FIGURE 3–16**, substitute the values of the resistances for R_1, R_2, and R_3: $\frac{1}{R_T} = \frac{1}{3} + \frac{1}{4} + \frac{1}{6}$.

The fractions cannot be added together unless they all have the same denominator. The lowest common denominator in this example is 12. Therefore, $\frac{1}{3}$ becomes $\frac{4}{12}$, $\frac{1}{4}$ becomes $\frac{3}{12}$, and $\frac{1}{6}$ becomes $\frac{2}{12}$. $\frac{1}{R_T} = \frac{4}{12} + \frac{3}{12} + \frac{2}{12}$ or $\frac{9}{12}$. Cross multiplying $R_T = \frac{12}{9} = 1.33\ \Omega$.

Note that the result (1.33 Ω) is the same regardless of the method used (see Method 1). The most difficult part of using this method (besides using fractions) is determining the lowest common denominator, especially for circuits containing a wide range of ohmic values for the various legs. For an easier method using a calculator, see Method 4.

METHOD 4 This method uses an electronic calculator, commonly available at very low cost. Instead of determining the lowest common denominator as in Method 3, one can use the electronic calculator to convert the fractions to decimal equivalents. The memory buttons on most calculators can be used to keep a running total of the fractional values.

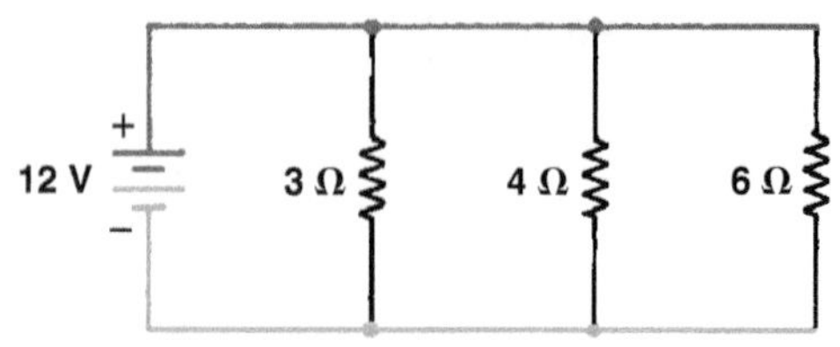

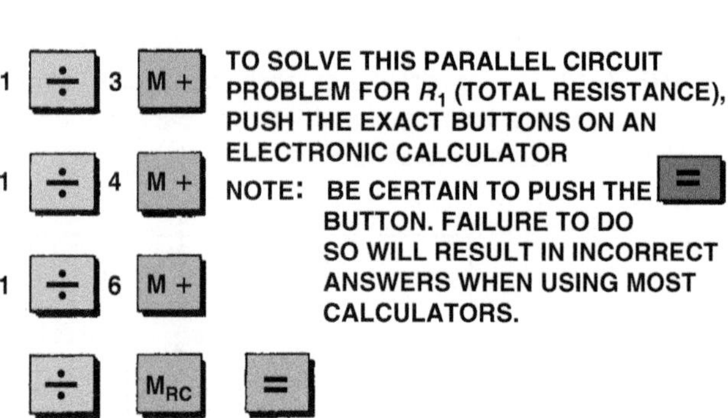

(ANSWER = 1.3333)

FIGURE 3–17 Using an electronic calculator to determine the total resistance of a parallel circuit.

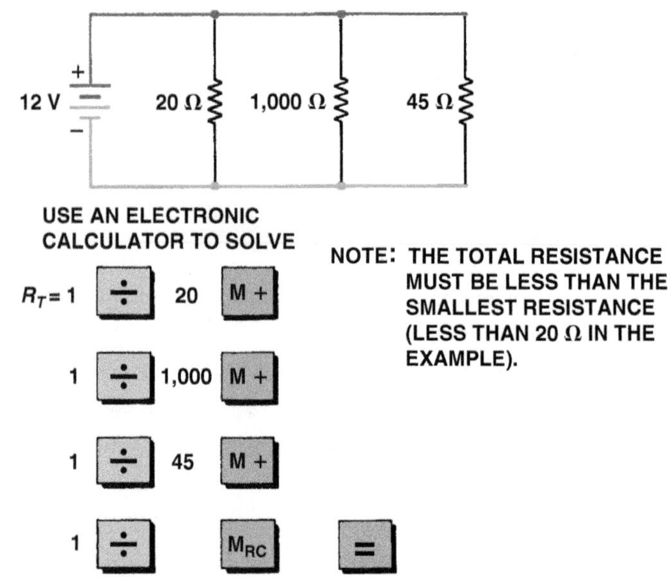

FIGURE 3–18 Another example of how to use an electronic calculator to determine the total resistance of a parallel circuit. The answer is 13.45 ohms. Notice that the effective resistance of this circuit is less than the resistance of the lowest branch (20 ohms).

Using **FIGURE 3–17**, calculate the total resistance (R_T) by pushing the indicated buttons on the calculator. **ALSO SEE FIGURE 3–18.**

NOTE: This method can be used to find the total resistance of *any number* of resistances in parallel.

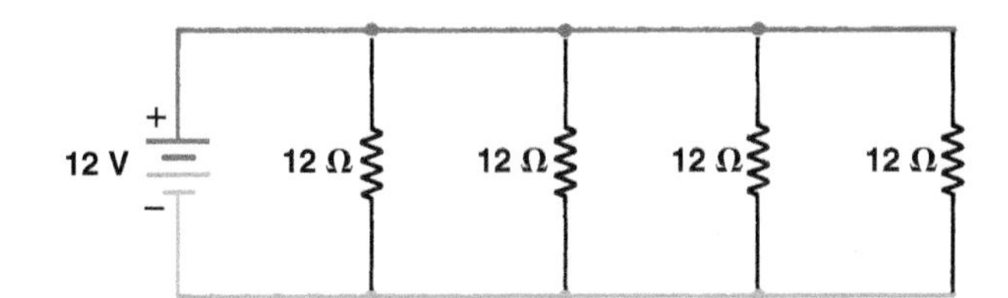

FIGURE 3–19 A parallel circuit containing four 12-ohm resistors. When a circuit has more than one resistor of equal value, the total resistance can be determined by simply dividing the value of the resistance (12 ohms in this example) by the number of equal-value resistors (4 in this example) to get 3 ohms.

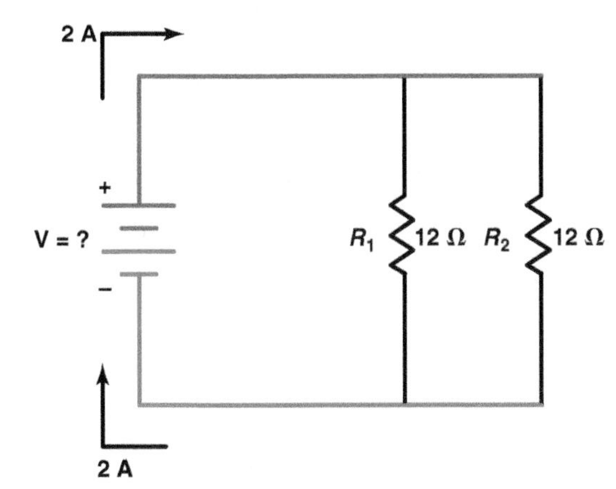

FIGURE 3–20 Example 1.

The memory recall (MRC) and equals (=) buttons invert the answer to give the correct value for total resistance (1.33 Ω). The inverse $\left(\frac{1}{X} \text{ or } X^{-1}\right)$ button can be used with the sum (SUM) button on scientific calculators without using the memory button.

METHOD 5 This method can be easily used when two or more resistances connected in parallel are all of the same value. ● **SEE FIGURE 3–19.**

To calculate the total resistance (R_T) of equal-value resistors, divide the number of equal resistors into the value of the resistance: R_T Value of one equal resistance/Number of equal resistances $= \frac{12\ \Omega}{4} = 3\ \Omega$.

NOTE: Because most automotive and light-truck electrical circuits involve multiple use of the same resistance, this method is the most useful. For example, if six additional 12-ohm lights were added to a vehicle, the additional lights would represent just 2 ohms of resistance $\left(\frac{12\ \Omega}{6} \text{ lights} = 2\right)$. Therefore, 6 amperes of additional current would be drawn by the additional lights $\left(I = \frac{E}{R} = \frac{12\text{ V}}{2\ \Omega} = 6\text{ A}\right)$.

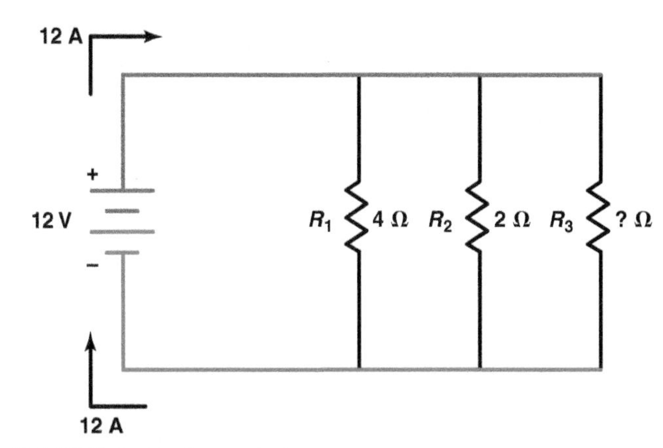

FIGURE 3–21 Example 2.

PARALLEL CIRCUIT CALCULATION EXAMPLES

Each of the four examples includes solving for the following:

- Total resistance
- Current flow (amperes) through each branch as well as total current flow
- Voltage drop across each resistance

Example 1:

(● **SEE FIGURE 3–20.**)

In this example, the voltage of the battery is unknown and the equation to be used is $E = I \times R$, where R represents the total resistance of the circuit. Using the equation for two resistors in parallel, the total resistance is 6 ohms.

$$R_T = \frac{R_1 \times R_2}{R_1 + R_2} = \frac{12 \times 12}{12 + 12} = \frac{144}{24} = 6\ \Omega$$

Placing the value of the total resistors into the equation results in a value for the battery voltage of 12 volts.

$$E = I \times R$$

$$E = 2\text{ A} \times 6\ \Omega$$

$$E = 12\text{ V}$$

Example 2:

(● **SEE FIGURE 3–21.**)

In this example, the value of R_3 is unknown. Because the voltage (12 volts) and the current (12 amperes) are known, it is easier to solve for the unknown resistance by treating each branch or leg as a separate circuit. Using Kirchhoff's law, the total current equals the total current flow through

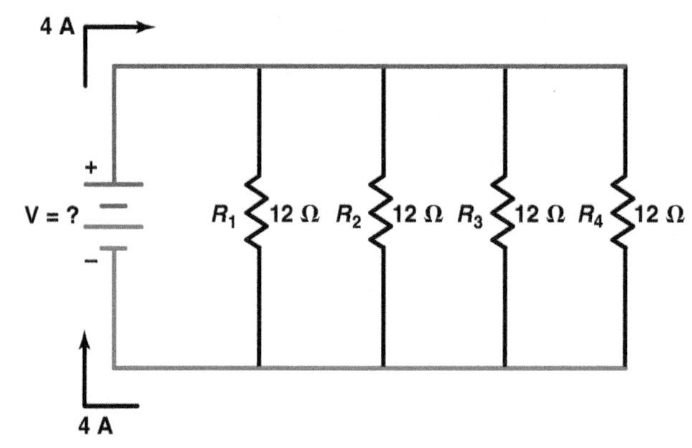

FIGURE 3–22 Example 3.

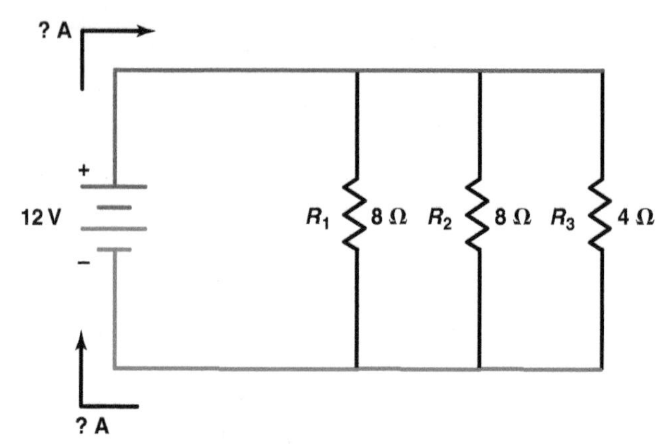

FIGURE 3–23 Example 4.

each branch. The current flow through R_1 is 3 amperes $\left(I = \frac{E}{R} = \frac{12\text{ V}}{4\ \Omega} = 3\text{ A}\right)$ and the current flow through R_2 is 6 amperes $\left(I = \frac{E}{R} = \frac{12\text{ V}}{2\ \Omega} = 6\text{ A}\right)$. Therefore, the total current through the two known branches equals 9 amperes (3 A + 6 A = 9 A). Because there are 12 amperes leaving and returning to the battery, the current flow through R_3 must be 3 amperes (12 A − 9 A = 3 A). The resistance must therefore be $4\ \Omega \left(I = \frac{E}{R} = \frac{12\text{ V}}{4\ \Omega} = 3\text{ A}\right)$.

Example 3:

(● **SEE FIGURE 3–22.**)

In this example, the voltage of the battery is unknown. The equation to solve for voltage according to Ohm's law is:

$$E = I \times R$$

The R in this equation refers to the total resistance. Because there are four resistors of equal value, the total can be determined by the following equation.

$$R_{Total} = \frac{\text{Value of resistors}}{\text{Number of equal resistors}} = \frac{12\ \Omega}{4} = 3\ \Omega$$

Inserting the value of the total resistance of the parallel circuit (3 Ω) into Ohm's law results in a battery voltage of 12 volts.

$$E = 4\text{ A} \times 3\ \Omega$$
$$E = 12\text{ V}$$

Example 4:

(● **SEE FIGURE 3–23.**)

The unknown value is the amount of current in the circuit. The Ohm's law equation for determining current is:

$$I = \frac{E}{R}$$

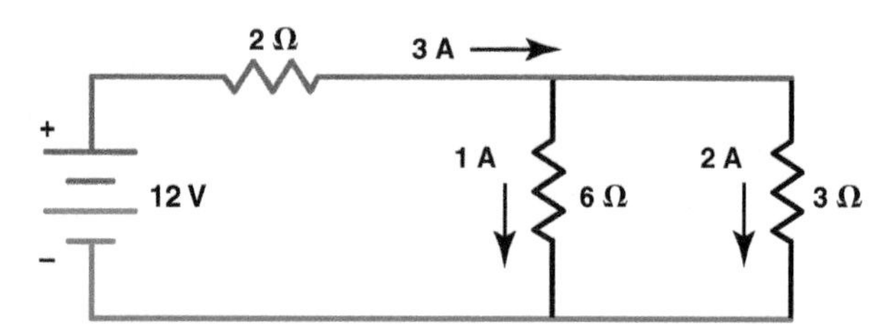

FIGURE 3–24 A series–parallel circuit.

The R represents the total resistance. Because there are two equal resistances (8 Ω), these two can be replaced by one resistance of 4 ohm $\left(R_{Total} = \frac{\text{Value}}{\text{Number}} = \frac{8\ \Omega}{2} = 4\ \Omega\right)$.

The total resistance of this parallel circuit containing two 8-ohm resistors and one 4-ohm resistor is 2 ohms (two 8-ohm resistors in parallel equals one 4 ohm. Then you have two 4-ohm resistors in parallel which equals 2 ohms). The current flow from the battery is then calculated to be 6 amperes.

$$I = \frac{E}{R} = \frac{12\text{ V}}{2\ \Omega} = 6\text{ A}$$

SERIES–PARALLEL CIRCUITS

DEFINITION **Series–parallel circuits** are a combination of series and parallel segments in one complex circuit. A series–parallel circuit is also called a **compound** or **combination circuit**. Many automotive circuits include sections that are in parallel and in series.

One example of a series–parallel circuit is a dash light dimming circuit. A variable resistor is used to send an input to the body control module (BCM), which in turn controls the brightness of various switch lamps. The pulsed voltage from the BCM acts as a variable voltage source in series with the instrument lamps, which are wired in parallel. ● **SEE FIGURE 3–25.**

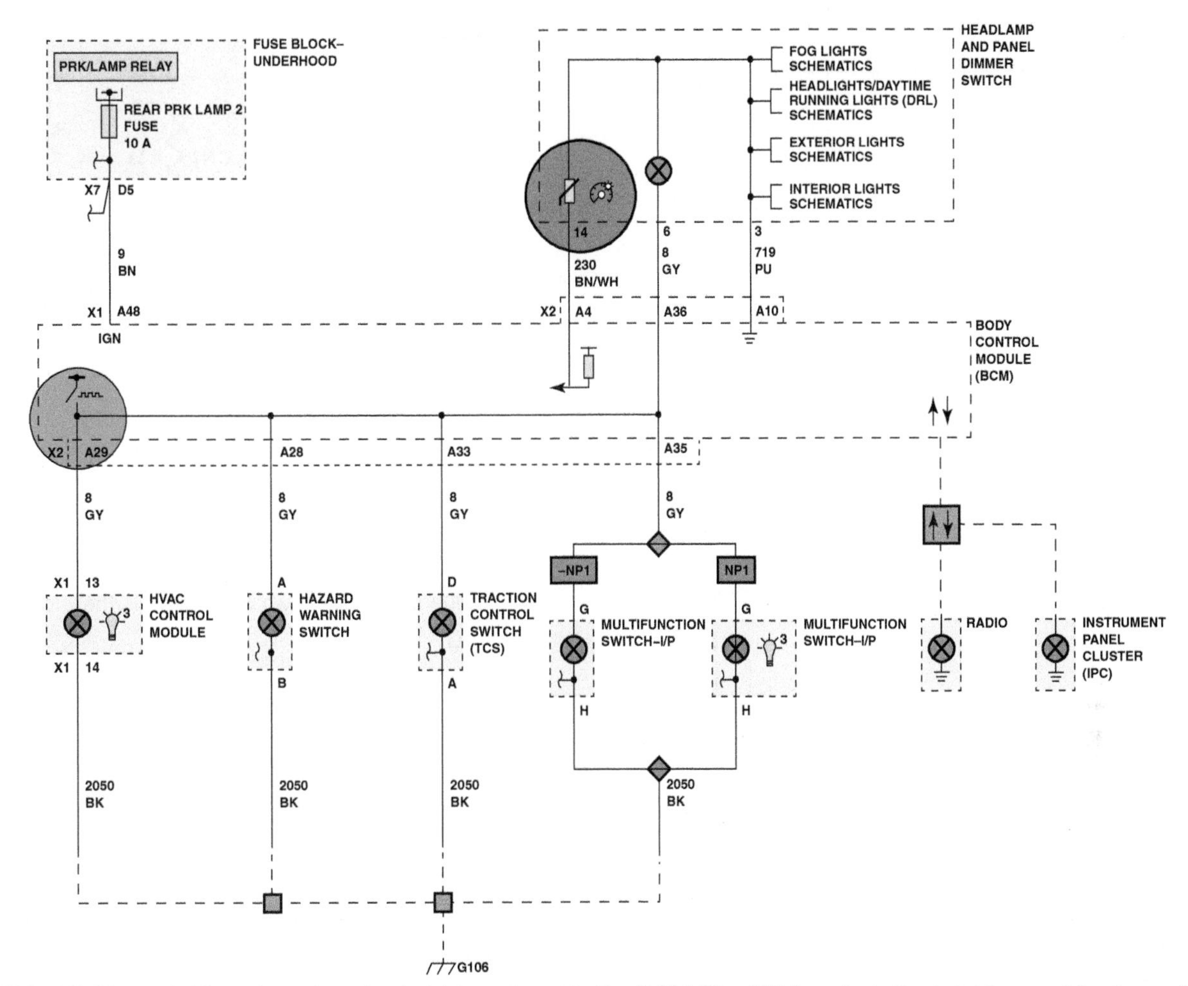

FIGURE 3–25 The variable resistor (purple circle) is an input to the BCM. The BCM controls the brightness of the lamps by pulsing the voltage rapidly (blue circle). (Courtesy of General Motors)

A series–parallel circuit includes both parallel loads or resistances, plus additional loads or resistances that are electrically connected in series.

SERIES–PARALLEL CIRCUIT FAULTS If a conventional parallel circuit, such as a taillight circuit, had an electrical fault that increased the resistance in one branch of the circuit, then the amount of current flow through that one branch will be reduced. The added resistance caused by the fault would create a voltage drop at that point. As a result of this drop in voltage, a lower voltage would be applied and the bulb in the taillight would be dimmer than normal because the brightness of the bulb depends on the voltage and current applied. If, however, the added resistance occurred in a part of the circuit that fed both taillights, then both taillights would be dimmer than normal. In this case, the added resistance created a series–parallel circuit that was originally just a simple parallel circuit.

SOLVING SERIES–PARALLEL CIRCUIT CALCULATION PROBLEMS

The key to solving series–parallel circuit problems is to combine or simplify as much as possible. For example, if there are two loads or resistances in series within a parallel branch or leg, then the circuit can be made simpler if the two are first added together before attempting to solve the parallel section. ● **SEE FIGURE 3–26.**

These examples show how to solve series–parallel calculations. The examples show how to solve for:

- Total resistance
- Current flow (amperes) through each branch, as well as total current flow
- Voltage drop across each resistance

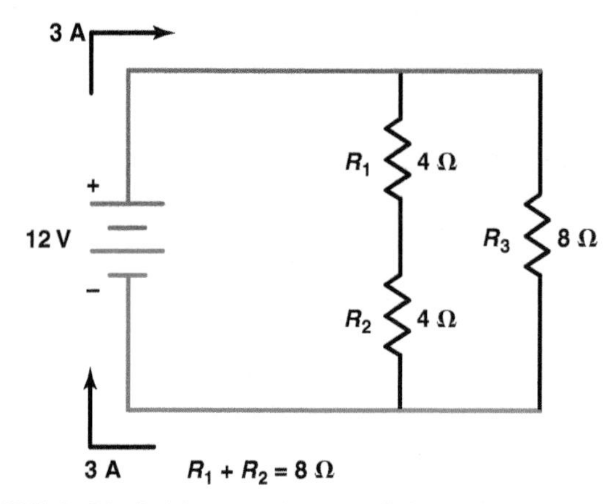

FIGURE 3–26 Solving a series–parallel circuit problem.

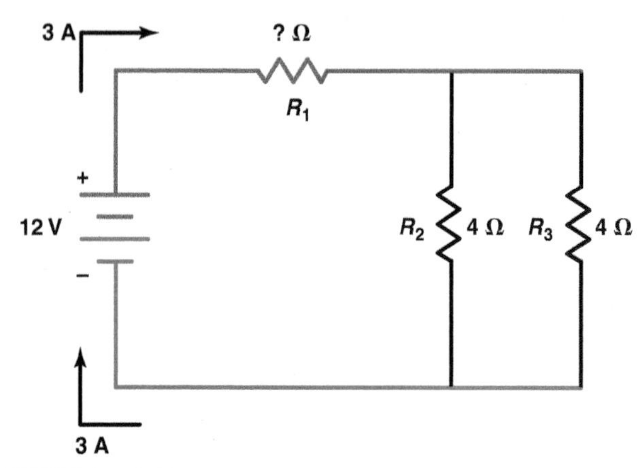

FIGURE 3–27 Example 1.

Example 1:

(SEE FIGURE 3–27.)

The unknown resistor is in series with the other two resistances, which are connected in parallel. The Ohm's law equation to determine resistance is:

$$R = \frac{E}{I} = \frac{12\text{ V}}{3} = 4\ \Omega$$

The total resistance of the circuit is therefore 4 ohms and the value of the unknown can be determined by subtracting the value of the two resistors that are connected in parallel. The parallel branch resistance is 2 ohms.

$$R_T = \frac{4 \times 4}{4 + 4} = \frac{16}{8} = 2\ \Omega$$

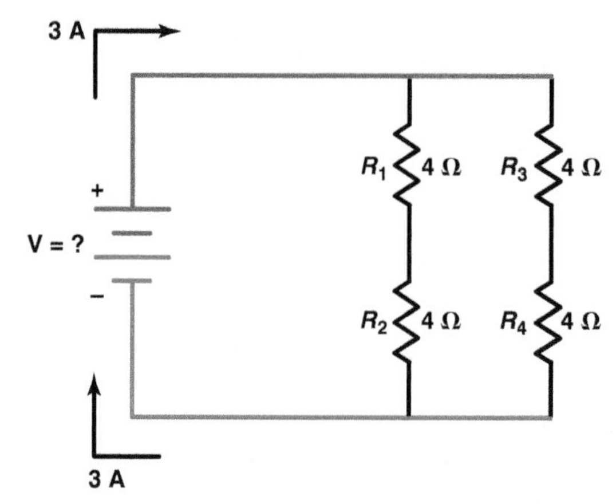

FIGURE 3–28 Example 2.

The value of the unknown resistance is therefore 2 ohms. Total $R = 4\ \Omega - 2\ \Omega = 2\ \Omega$.

Example 2:

(SEE FIGURE 3–28.)

The unknown unit in this circuit is the voltage of the battery. The Ohm's law equation is:

$$E = I \times R$$

Before solving the problem, the total resistance must be determined. Because each branch contains two 4-ohm resistors in series, the value in each branch can be added to help simplify the circuit. By adding the resistors in each branch together, the parallel circuit now consists of two 8-ohm resistors.

$$R_T = \frac{R_1 \times R_2}{R_1 + R_2} = \frac{8 \times 8}{8 + 8} = \frac{64}{16} = 4\ \Omega$$

Inserting the value for the total resistance into the Ohm's law equation results in a value of 12 volts for the battery voltage.

$$E = I \times R$$
$$E = 3\text{ A} \times 4\ \Omega$$
$$E = 12\text{ V}$$

SUMMARY

1. Series circuits:
   a. In a simple series circuit, the current remains constant throughout, but the voltage drops as current flows through the resistances of the circuit.
   b. The voltage drop across each resistance or load is directly proportional to the value of the resistance compared to the total resistance in the circuit.
   c. The sum (total) of the voltage drops equals the applied voltage (Kirchhoff's voltage law).
   d. An open or a break anywhere in a series circuit stops all current from flowing.
2. Parallel circuits:
   a. A parallel circuit, such as is used for all automotive lighting, has the same voltage available to each resistance (bulb).
   b. The total resistance of a parallel circuit is always lower than the smallest resistance.
   c. The separate paths that split and meet at junction points (splices) are called branches, legs, or shunts.
   d. Kirchhoff's current law states: "The current flowing into a junction of an electrical circuit is equal to current flowing out of that junction."
3. Series–parallel circuits:
   a. A series–parallel circuit is also called a compound circuit or a combination circuit.
   b. A series–parallel circuit is a combination of a series and a parallel circuit, which does not include fuses or switches.
   c. A fault in a series portion of the circuit would affect the operation if the series part was in the power or the ground side of the parallel portion of the circuit.
   d. A fault in one leg of a series–parallel circuit will affect just the component(s) in that one leg.

REVIEW QUESTIONS

1. What is Kirchhoff's voltage law?
2. What would current (amperes) do if the voltage were doubled in a circuit?
3. What would current (amperes) do if the resistance in the circuit were doubled?
4. What is the formula for voltage drop?
5. Why is the total resistance of a parallel circuit less than the smallest resistance?
6. Why are parallel circuits (instead of series circuits) used in most automotive applications?
7. What does Kirchhoff's current law state?
8. What would be the effect of an open circuit in one leg of a parallel portion of a series–parallel circuit?
9. What would be the effect of an open circuit in a series portion of a series–parallel circuit?

CHAPTER QUIZ

1. The amperage in a series circuit is ____________.
   a. The same anywhere in the circuit
   b. Varies in the circuit due to the different resistances
   c. High at the beginning of the circuit and decreases as the current flows through the resistance
   d. Always less returning to the battery than leaving the battery
2. The sum of the voltage drops in a series circuit equals the ____________.
   a. Amperage
   b. Resistance
   c. Source voltage
   d. Wattage
3. If the resistance and the voltage are known, what is the formula for finding the current (amperes)?
   a. $E = I \times R$
   b. $I = E \times R$
   c. $R = E \times I$
   d. $I = E/R$
4. A series circuit has three resistors of 4 ohms each. The voltage drop across each resistor is 4 volts. Technician A says that the source voltage is 12 volts. Technician B says that the total resistance is 12 ohms. Which technician is correct?
   a. Technician A only
   b. Technician B only
   c. Both Technicians A and B
   d. Neither Technician A nor B
5. If a 12-volt battery is connected to a series circuit with three resistors of 2 ohms, 4 ohms, and 6 ohms, how much current will flow through the circuit?
   a. 1 amp
   b. 2 amp
   c. 3 amp
   d. 4 amp

6. A series circuit has two 10-ohm bulbs. A third bulb is added in series. Technician A says that the three bulbs will be dimmer than when only two bulbs were in the circuit. Technician B says that the current in the circuit will increase. Which technician is correct?
   a. Technician A only
   b. Technician B only
   c. Both Technicians A and B
   d. Neither Technician A nor B

7. Technician A says that the sum of the voltage drops in a series circuit should equal the source voltage. Technician B says that the current (amperes) varies depending on the value of the resistance in a series circuit. Which technician is correct?
   a. Technician A only
   b. Technician B only
   c. Both Technicians A and B
   d. Neither Technician A nor B

8. Two bulbs are connected in parallel to a 12-volt battery. One bulb has a resistance of 6 ohms and the other bulb has a resistance of 2 ohms. Technician A says that only the 2-ohm bulb will light because all of the current will flow through the path with the least resistance and no current will flow through the 6-ohm bulb. Technician B says that the 6-ohm bulb will be dimmer than the 2-ohm bulb. Which technician is correct?
   a. Technician A only
   b. Technician B only
   c. Both Technicians A and B
   d. Neither Technician A nor B

9. Calculate the total resistance and current in a parallel circuit with three resistors of 4 Ω, 8 Ω, and 16 Ω, using any one of the five methods (calculator suggested). What is the total resistance and current?
   a. 27 ohms (0.4 ampere)
   b. 14 ohms (0.8 ampere)
   c. 4 ohms (3.0 amperes)
   d. 2.3 ohms (5.3 amperes)

10. A vehicle has four parking lights all connected in parallel and one of the bulbs burns out. Technician A says that this could cause the parking light circuit fuse to blow (open). Technician B says that it would decrease the current in the circuit. Which technician is correct?
    a. Technician A only
    b. Technician B only
    c. Both Technicians A and B
    d. Neither Technician A nor B

chapter 4

CIRCUIT TESTERS AND DIGITAL METERS

LEARNING OBJECTIVES

After studying this chapter, the reader will be able to:

1. Discuss how to safely set up and use a fused jumper wire, a test light, and a logic probe.
2. Explain how to safely and properly use a digital meter.
3. Explain how to set a digital meter to read voltage, resistance, and current.
4. Explain meter terms and ratings.
5. Interpret meter readings and compare to factory specifications.

This chapter will help you prepare for the ASE Electrical/Electronic Systems (A6) certification test content area "A" (General Electrical/Electronic System Diagnosis).

KEY TERMS

GM STC OBJECTIVES

GM Service Technical College topics covered in this chapter are as follows:

1. Interpret the DMM meter functions and screens.
2. Using the proper tools, measure duty cycle in a pulse width modulated circuit.
3. Using the proper tools, measure frequency in a pulse width modulated circuit.
4. Using the proper tools, apply voltage drop to circuit diagnosis.
5. Using the proper tools, apply amperage testing to circuit diagnosis.
6. Test voltage and ground circuits using electrical/electronic tools.

FIGURE 4–1 A technician-made fused jumper lead, which is equipped with a red 10-ampere fuse. This fused jumper wire uses terminals for testing circuits at a connector instead of alligator clips.

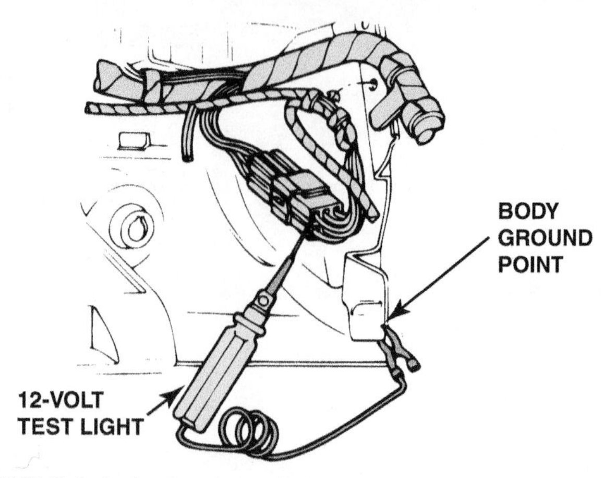

FIGURE 4–2 A 12-volt test light is attached to a good ground while probing for power.

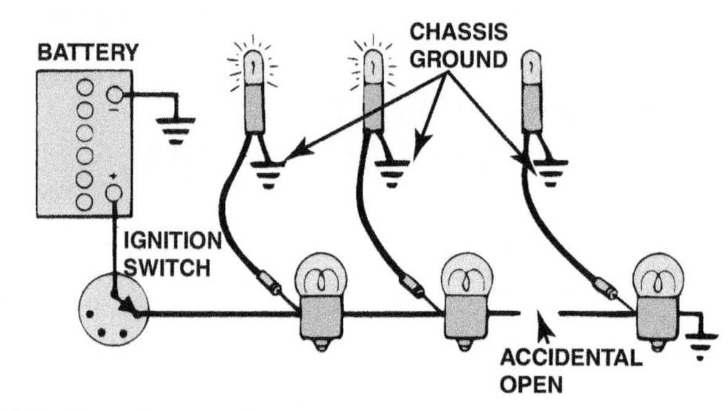

FIGURE 4–3 A test light can be used to locate an open in a circuit. Note that the test light is grounded at a different location than the circuit itself.

FUSED JUMPER WIRE

DEFINITION A fused jumper wire is used to check a circuit by bypassing the switch or to provide a power or ground to a component. A fused jumper wire, also called a test lead, can be purchased or made by the service technician. ● **SEE FIGURE 4–1.**

It should include the following features.

- **Fused.** A typical fused jumper wire has a blade-type fuse that can be easily replaced. A 10-ampere fuse (red color) is often the value used.
- **Alligator clip ends.** Alligator clips at the ends allow the fused jumper wire to be clipped to a ground or power source while the other end is attached to the power side or ground side of the unit being tested.
- **Good-quality insulated wire.** Most purchased jumper wire is about 14 gauge stranded copper wire with a flexible rubberized insulation to allow it to move easily even in cold weather.

USES OF A FUSED JUMPER WIRE A fused jumper wire can be used to help diagnose a component or circuit by performing the following procedures.

- **Supply power or ground.** If a component, such as a horn, does not work, a fused jumper wire can be used to supply a temporary power and/or ground. Start by unplugging the electrical connector from the device and connect a fused jumper lead to the power terminal. Another fused jumper wire may be needed to provide the ground. If the unit works, the problem is in the power side or ground side circuit.

CAUTION: Never use a fused jumper wire to bypass any resistance or load in the circuit. The increased current flow could damage the wiring and could blow the fuse on the jumper lead.

TEST LIGHTS

NONPOWERED TEST LIGHT A 12-volt test light is one of the simplest testers that can be used to detect electricity. A **test light** is simply a light bulb with a probe and a ground wire attached. ● **SEE FIGURE 4–2.**

It is used to detect battery voltage potential at various test points. Battery voltage cannot be seen or felt and can be detected only with test equipment.

The ground clip is connected to a clean ground on either the negative terminal of the battery or a clean metal part of the body and the probe touched to terminals or components. If the test light comes on, this indicates that voltage is available. ● **SEE FIGURE 4–3.**

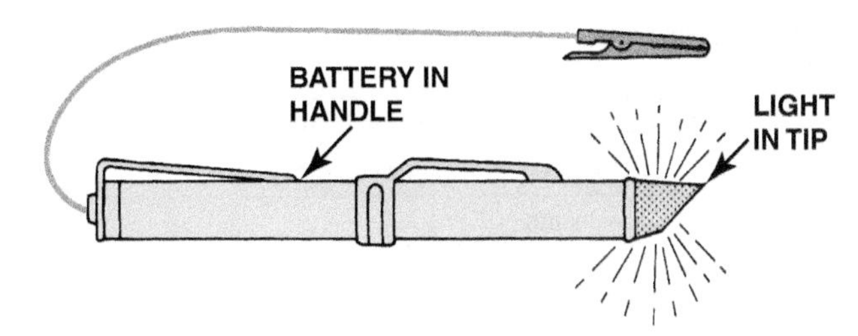

FIGURE 4–4 A continuity light should not be used on computer circuits because the applied voltage can damage delicate electronic components or circuits.

A purchased test light could be labeled a "12 volt test light." Do not purchase a test light designed for household current (110 or 220 volts), as it will not light with 12 to 14 volts.

USES OF A 12-VOLT TEST LIGHT A 12-volt test light can be used to check the following:

- **Electrical power.** If the test light comes on, then there is power available. It will not, however, indicate the voltage level or if there is enough current available to operate an electrical load. This only indicates that there is enough voltage and current to light the test light (about 0.25 A).
- **Grounds.** A test light can be used to check for grounds by attaching the clip of the test light to the positive terminal of the battery or to any 12-volt electrical terminal. The tip of the test light can then be used to touch the ground wire. If there is a ground connection, the test light will come on.

CONTINUITY TEST LIGHTS A **continuity light** is similar to a test light but includes a battery for self-power. A continuity light illuminates whenever it is connected to both ends of a wire that has continuity or is not broken. ● **SEE FIGURE 4–4**.

CAUTION: The use of a self-powered (continuity) test light is not recommended on any electronic circuit, because a continuity light contains a battery and applies voltage; therefore, it may harm delicate electronic components.

HIGH-IMPEDANCE TEST LIGHT A high-impedance test light has a high internal resistance and therefore draws very low current in order to light. High-impedance test lights are safe to use on computer circuits because they will not affect the circuit current in the same way as conventional 12-volt test lights when connected to a circuit. There are two types of high-impedance test lights.

- Some test lights use an electronic circuit to limit the current flow, to avoid causing damage to electronic devices.

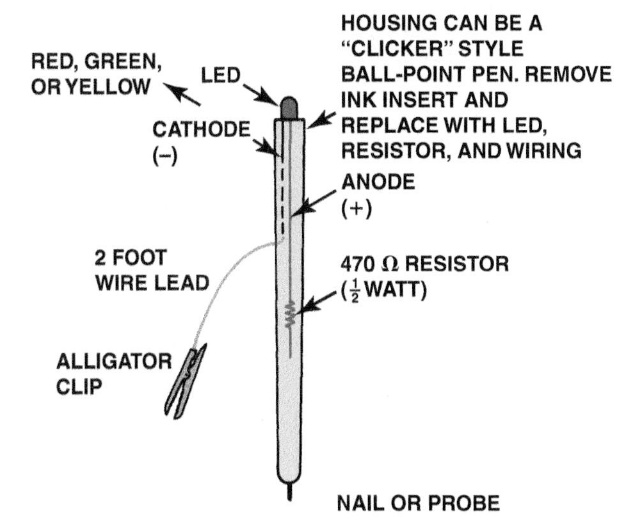

FIGURE 4–5 An LED test light can be easily made using low cost components and an old ink pen. With the 470 ohm resistor in series with the LED, this tester only draws 0.025 ampere (25 milliamperes) from the circuit being tested. This low current draw helps assure the technician that the circuit or component being tested will not be damaged by excessive current flow.

- An **LED test light** uses a light-emitting diode (LED) instead of a standard automotive bulb for a visual indication of voltage. An LED test light requires only about 25 milliamperes (0.025 ampere) to light; therefore, it can be used on electronic circuits as well as on standard circuits.

● **SEE FIGURE 4–5** for construction details for a homemade LED test light.

LOGIC PROBE

PURPOSE AND FUNCTION A **logic probe** is an electronic device that lights up a red (usually) LED if the probe is touched to battery voltage. If the probe is touched to ground, a green (usually) LED lights. ● **SEE FIGURE 4–6**.

A logic probe can "sense" the difference between high- and low-voltage levels, which explains the name *logic.*

- A typical logic probe can also light another light (often amber color) when a change in voltage occurs.
- Some logic probes will flash the red light when a pulsing voltage signal is detected.
- Some will flash the green light when a pulsing ground signal is detected.

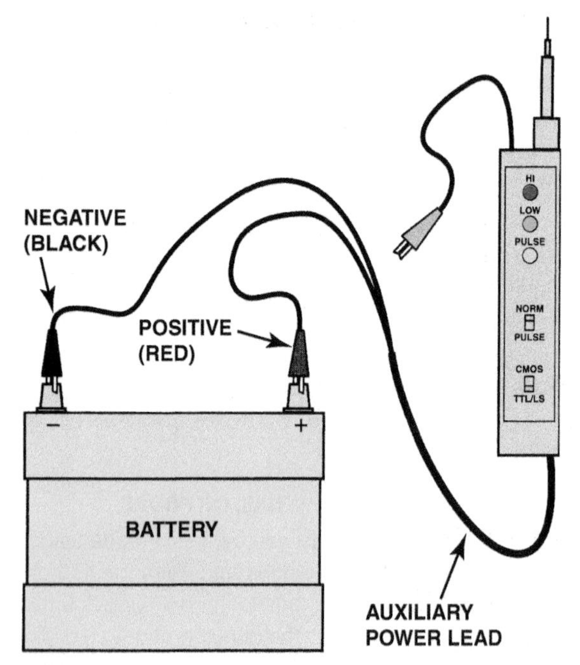

FIGURE 4–6 A logic probe connected to the vehicle battery. When the tip probe is connected to a circuit, it can check for power, ground, or a pulse.

SYMBOL	MEANING
AC	Alternating current or voltage
DC	Direct current or voltage
V	Volts
mV	Millivolts (1/1,000 volts)
A	Ampere (amps), current
mA	Milliampere (1/1,000 amps)
%	Percent (for duty cycle readings only)
Ω	Ohms, resistance
kΩ	Kilohm (1,000 ohms), resistance
MΩ	Megohm (1,000,000 ohms), resistance
Hz	Hertz (cycles per second), frequency
kHz	Kilohertz (1,000 cycles/sec.), frequency
ms	Milliseconds (1/1,000 sec.) for pulse width measurements

CHART 4–1

Common symbols and abbreviations used on digital meters.

This feature is helpful when checking for a variable voltage output from a computer or ignition sensor.

USING A LOGIC PROBE A logic probe must first be connected to a power and ground source such as the vehicle battery. This connection powers the probe and gives it a reference low (ground).

Most logic probes also make a distinctive sound for each high- and low-voltage level. This makes troubleshooting easier when probing connectors or component terminals. A sound (usually a beep) is heard when the probe tip is touched to a changing voltage source. The changing voltage also usually lights the pulse light on the logic probe. Therefore, the probe can be used to check components such as:

- Pickup coils
- Hall-effect sensors
- Magnetic sensors

DIGITAL MULTIMETERS

TERMINOLOGY **Digital multimeter (DMM)** and **digital volt-ohm-meter (DVOM)** are terms commonly used for electronic **high-impedance test meters.** *High impedance* means that the electronic internal resistance of the meter is high enough to prevent excessive current draw from any circuit being tested. Most meters today have a minimum of 10 million ohms (10 megohms) of resistance. This high internal resistance between the meter leads is present only when measuring volts. The high resistance in the meter itself reduces the amount of current flowing through the meter when it is being used to measure voltage, leading to more accurate test results because the meter does not change the load on the circuit. High-impedance meters are required for measuring computer circuits.

CAUTION: Analog (needle-type) meters are almost always lower than 10 megohms and should not be used to measure any computer or electronic circuit. Connecting an analog meter to a computer circuit could damage the computer or other electronic modules.

A high-impedance meter can be used to measure any automotive circuit within the ranges of the meter. ● **SEE FIGURE 4–7.**

The common abbreviations for the units that many meters can measure are often confusing. ● **SEE CHART 4–1** for the most commonly used symbols and their meanings.

MEASURING VOLTAGE A voltmeter measures the *pressure* or potential of electricity in units of volts. A voltmeter is connected to a circuit in parallel. Voltage can be measured by selecting either AC or DC volts.

- **DC volts (DCV).** This setting is the most common for automotive use. Use this setting to measure battery voltage and voltage to all lighting and accessory circuits.

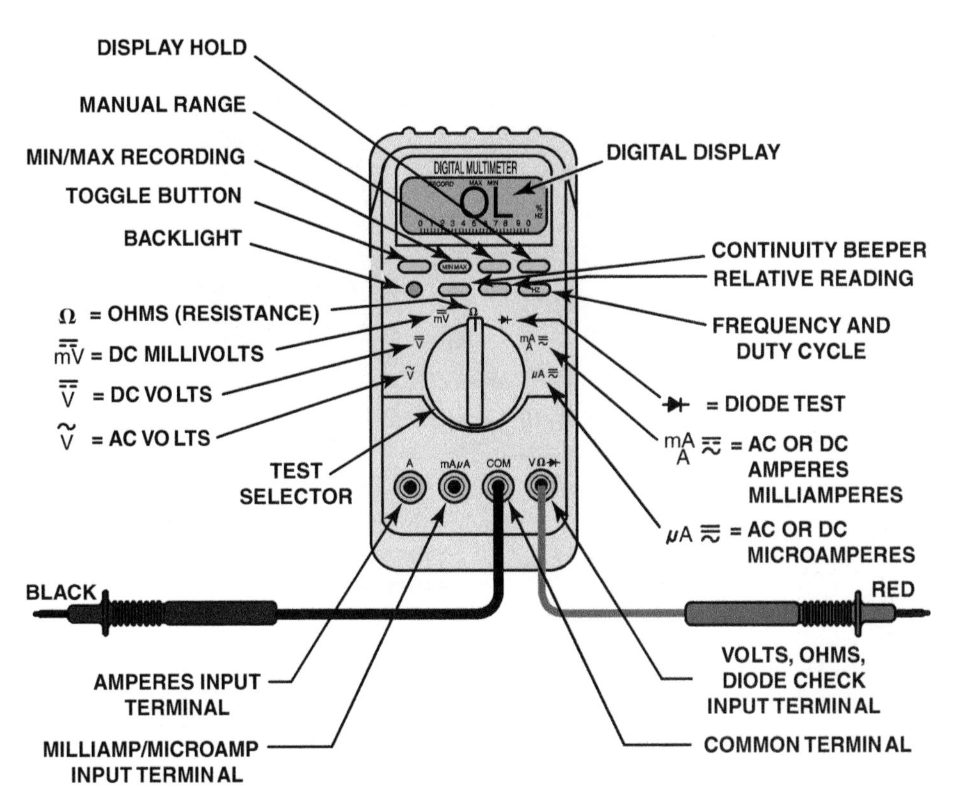

FIGURE 4–7 Typical digital multimeter. The black meter lead is always placed in the COM terminal. The red meter test lead should be in the volt-ohm terminal except when measuring current in amperes.

FIGURE 4–8 Typical digital multimeter (DMM) set to read DC volts.

- **AC volts (ACV).** This setting is used to check for unwanted AC voltage from alternators and some sensors.
- **Range.** The range is automatically set for most meters but can be manually ranged if needed.

● SEE FIGURES 4–8 AND 4–9.

MEASURING RESISTANCE An ohmmeter measures the resistance in ohms of a component or circuit section when no current is flowing through the circuit. An ohmmeter contains a battery (or other power source) and is connected in series with the component or wire being measured. When the leads are connected to a component, current flows through the test leads and the difference in voltage (voltage drop) between the leads is measured as resistance. Note the following facts about using an ohmmeter.

- Zero ohms on the scale means that there is no resistance between the test leads, thus indicating continuity or a continuous path for the current to flow in a closed circuit.
- Infinity means no connection, as in an open circuit.
- Ohmmeters have no required polarity even though red and black test leads are used for resistance measurement.

CAUTION: The circuit must be electrically open with no current flowing when using an ohmmeter. If current is flowing when an ohmmeter is connected, the reading will be incorrect and the meter can be destroyed.

Different meters have different ways of indicating infinity resistance, or a reading higher than the scale allows. Examples of an over limit display include:

- **OL**, meaning **over limit** or overload
- Flashing or solid number 1
- Flashing or solid number 3 on the left side of the display

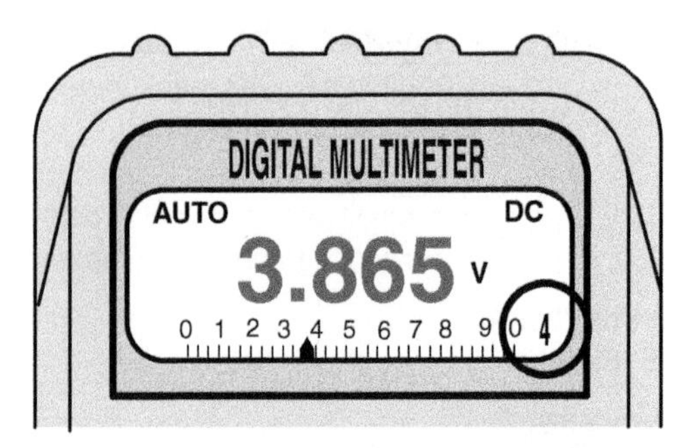

BECAUSE THE SIGNAL READING IS BELOW 4 VOLTS, THE METER AUTORANGES TO THE 4-VOLT SCALE. IN THE 4-VOLT SCALE, THIS METER PROVIDES THREE DECIMAL PLACES.

(a)

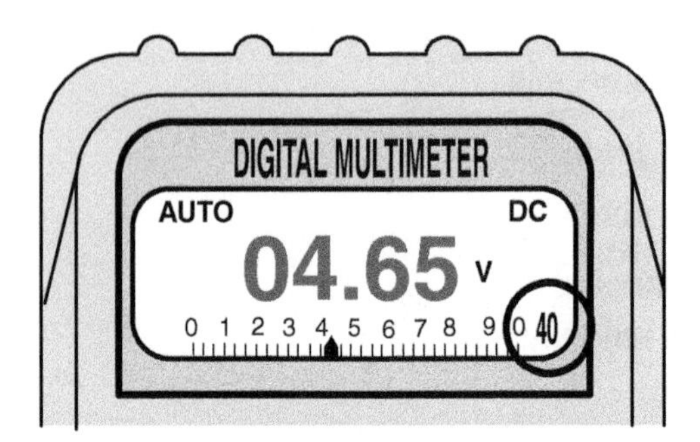

WHEN THE VOLTAGE EXCEEDED 4 VOLTS, THE METER AUTORANGES INTO THE 40-VOLT SCALE. THE DECIMAL POINT MOVES ONE PLACE TO THE RIGHT LEAVING ONLY TWO DECIMAL PLACES.

(b)

FIGURE 4–9 A typical autoranging digital multimeter automatically selects the proper scale to read the voltage being tested. The scale selected is usually displayed on the meter face. (a) Note that the display indicates "4," meaning that this range can read up to 4 volts. (b) The range is now set to the 40-volt scale, meaning that the meter can read up to 40 volts on the scale. Any reading above this level will cause the meter to reset to a higher scale. If not set on autoranging, the meter display would indicate OL if a reading exceeds the limit of the scale selected.

Check the meter instructions for the exact display used to indicate an open circuit or over range reading. ● **SEE FIGURES 4–10 AND 4–11.**

To summarize, open and zero readings are as follows:

0.00 Ω = Zero resistance (component or circuit has continuity)

OL = An open circuit or reading is higher than the scale selected (no current flows)

MEASURING AMPERES An ammeter measures the flow of *current* through a complete circuit in units of amperes. The ammeter has to be installed in the circuit (in series) so that it can measure all the current flow in that circuit, just as a water flow meter would measure the amount of water flow (cubic feet per minute, for example). ● **SEE FIGURE 4–12.**

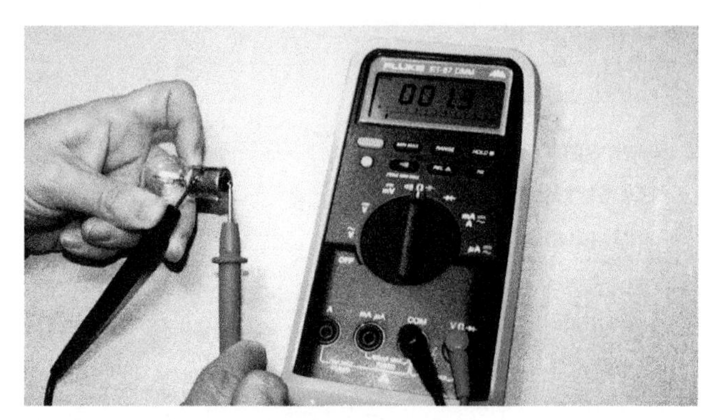

FIGURE 4–10 Using a digital multimeter set to read ohms (Ω) to test this light bulb. The meter reads the resistance of the filament.

FREQUENTLY ASKED QUESTION

How Much Voltage Does an Ohmmeter Apply?

Most digital meters that are set to measure ohms (resistance) apply 0.3 to 1 volt to the component being measured. The voltage comes from the meter itself to measure the resistance. Two things are important to remember about an ohmmeter.

1. The component or circuit must be disconnected from any electrical circuit while the resistance is being measured.
2. Because the meter itself applies a voltage (even though it is relatively low), a meter set to measure ohms can damage electronic circuits. Computer or electronic chips can be easily damaged if subjected to only a few milliamperes of current, similar to the amount an ohmmeter applies when a resistance measurement is being performed.

NOTE: The J-39200-A Multimeter Kit contains a Fluke 87-V DMM. This meter defaults to AC current when switched to the AMP settings. Be sure to switch the meter to DC AMPS when measuring DC current.

CAUTION: An ammeter must be installed in series with the circuit to measure the current flow in the circuit. If a meter set to read amperes is connected in parallel, such as across a battery, the meter or the leads may be destroyed, or the fuse will blow, by the current available across the battery. Some digital multimeters (DMMs) beep if the unit selection does not match the test lead connection on the meter. However, in a noisy shop, this beep sound may be inaudible.

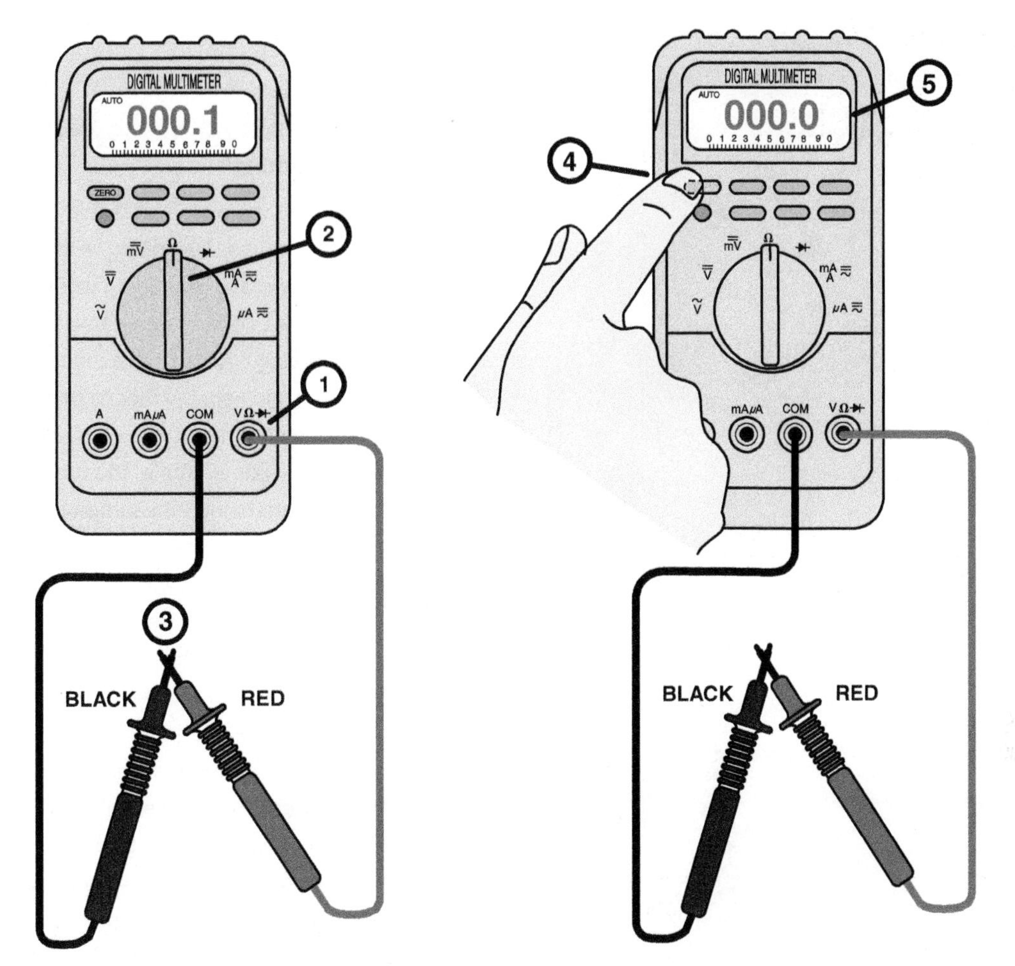

FIGURE 4–11 Many digital multimeters can have the display indicate zero to compensate for test lead resistance. (1) Connect leads in the V Ω and COM meter terminals. (2) Select the Ω scale. (3) Touch the two meter leads together. (4) Push the "zero" or "relative" button on the meter. (5) The meter display will now indicate zero ohms of resistance.

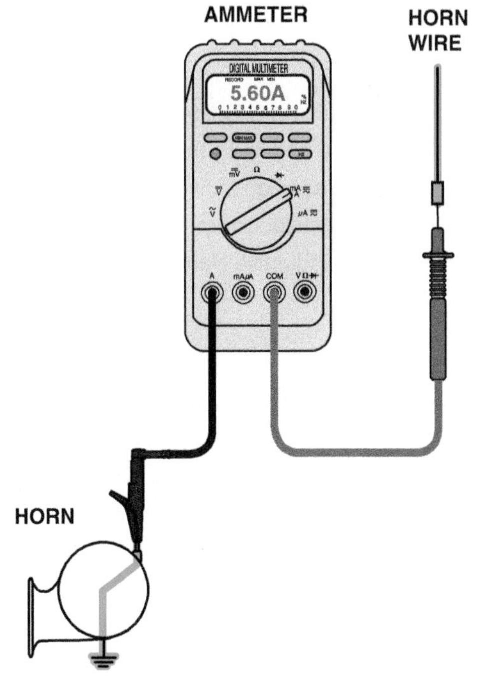

FIGURE 4–12 Measuring the current flow required by a horn requires that the ammeter be connected to the circuit in series and the horn button be depressed by an assistant.

? FREQUENTLY ASKED QUESTION

What Does "CE" Mean on Many Meters?

The "CE" means that the meter meets the newest European Standards and the letters CE stands for a French term for *Conformite' Europeenne* meaning European Conformity in French.

Digital meters require that the meter leads be moved to the ammeter terminals. Most digital meters have an ampere scale that can accommodate a maximum of 10 amperes.

INDUCTIVE AMMETERS

OPERATION **Inductive ammeters** do not make physical contact with the circuit. They measure the strength of the magnetic field surrounding the wire carrying the current and use a Hall-effect sensor to measure current. The Hall-effect sensor

FIGURE 4–13 An inductive ammeter clamp is used with all starting and charging testers to measure the current flow through the battery cables.

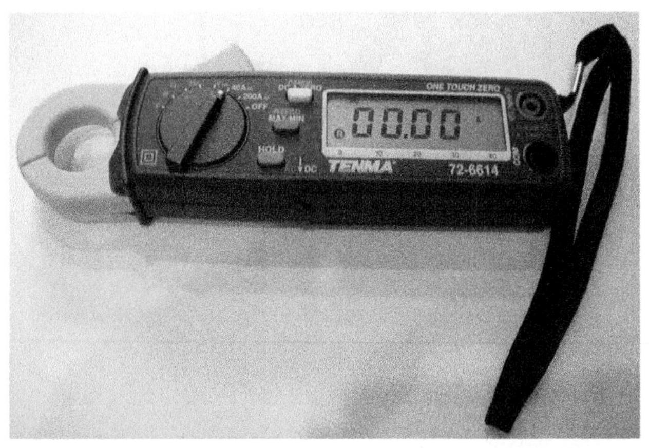

FIGURE 4–14 A typical mini clamp-on-type digital multimeter. This meter is capable of measuring alternating current (AC) and direct current (DC) without requiring that the circuit be disconnected to install the meter in series. The jaws are simply placed over the wire and current flow through the circuit is displayed.

TECH TIP

Over Limit Display Does Not Mean the Meter Is Reading "Nothing"

The meaning of the over limit display on a digital meter often confuses beginning technicians. When asked what the meter is reading when an over limit (OL) is displayed on the meter face, the response is often, "Nothing." Many meters indicate *over limit* or *over load,* which simply means that the reading is over the maximum that can be displayed for the selected range. For example, the meter will display OL if 12 volts are being measured but the meter has been set to read a maximum of 4 volts.

Autoranging meters adjust the range to match what is being measured. Here OL means a value higher than the meter can read (unlikely on the voltage scale for automobile usage), or infinity when measuring resistance (ohms). Therefore, OL means infinity when measuring resistance or an open circuit is being indicated. The meter will read 00.0 if the resistance is zero, so "nothing" in this case indicates continuity (zero resistance), whereas OL indicates infinity resistance. Therefore, when talking with another technician about a meter reading, make sure you know exactly what the reading on the face of the meter means. Also be sure that you are connecting the meter leads correctly. ● **SEE FIGURE 4–15.**

detects the strength of the magnetic field that surrounds the wire carrying an electrical current. ● **SEE FIGURE 4–13.**

This means that the meter probe surrounds the wire(s) carrying the current and measures the strength of the magnetic field that surrounds any conductor carrying a current.

AC/DC CLAMP-ON DIGITAL MULTIMETERS An **AC/DC clamp-on digital multimeter (DMM)** is a useful meter for automotive diagnostic work. ● **SEE FIGURE 4–14.**

The major advantage of the clamp-on-type meter is that there is no need to break the circuit to measure current (amperes). Simply clamp the jaws of the meter around the power lead(s) or ground lead(s) of the component being measured and read the display. Most clamp-on meters can also measure alternating current, which is helpful in the diagnosis of an alternator problem. Volts, ohms, frequency, and temperature can also be measured with the typical clamp-on DMM, but use conventional meter leads. The inductive clamp is only used to measure amperes.

DIODE CHECK, PULSE WIDTH, AND FREQUENCY

DIODE CHECK Diode check is a meter function that can be used to check diodes including light-emitting diodes (LEDs).

The meter is able to text diodes in the following way:

- The meter applies roughly a 3-volt DC signal to the text leads.

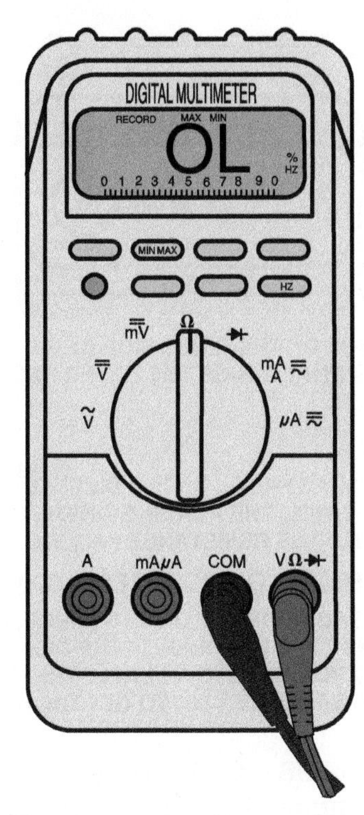

FIGURE 4–15 Typical digital multimeter showing OL (over limit) on the readout with the ohms (Ω) unit selected. This usually means that the unit being measured is open (infinity resistance) and has no continuity.

- The voltage is high enough to cause a diode to work and the meter will display:
  1. 0.4 to 0.7 volt when testing silicon diodes such as those found in alternators
  2. 1.5 to 2.3 volts when testing LEDs such as those found in some lighting applications

PULSE WIDTH Pulse width is the amount of time by percentage that a signal is on compared to being off.

- 100% pulse width indicates that a device is being commanded on all of the time.
- 50% pulse width indicates that a device is being commanded on half of the time.
- 25% pulse width indicates that a device is being commanded on just 25% of the time.

Pulse width is used to measure the on time for fuel injectors and other computer-controlled solenoid and devices.

FREQUENCY Frequency is a measure of how many times per second a signal changes. Frequency is measured in a unit called hertz, formerly termed "cycles per second."

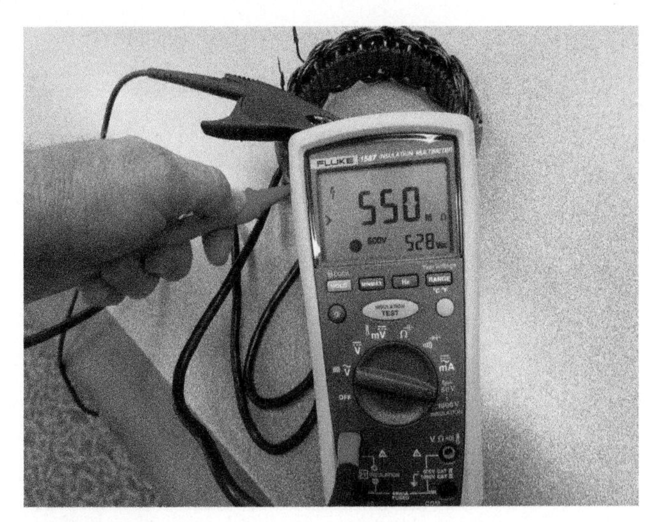

FIGURE 4–16 The Fluke 1587 Insulation Multimeter includes a megohmmeter function for measuring high resistance. Here the meter shows that the generator stator windings are insulated from the laminated frame to a level of 550 Mega ohms. (Courtesy of Jeffrey Rehkopf)

Frequency measurements are used when checking the following:

- Mass airflow (MAF) sensors for proper operation
- Ignition primary pulse signals when diagnosing a no-start condition
- Checking a wheel speed sensor

SPECIALTY METERS

There may be some service procedures that ask the technician to measure components in a way that the standard digital multimeter (Fluke 87-V, for example) cannot measure. Very specific low-resistance measurements can identify faults within a stator winding. Very high-specific Ohm readings can not only indicate that wiring is insulated from ground but also indicate the quality of that insulation.

MEGOHMMETER A megohmmeter (also called a "megger") is a special type of ohmmeter that can measure in units up to 550 megaohms. Where the Fluke 87-V indicates high resistance or an open circuit as "OL," the megohmmeter can make an actual measurement of the resistance. The meter applies a high voltage (from 100 to 1,000 volts AC) to the windings and then probes for any leakage to the housing or to other wires during the test. A high reading of 550 MΩ indicates that the wiring is indeed completely isolated from the stator housing and laminations, even under high-voltage conditions. ● **SEE FIGURE 4–16.**

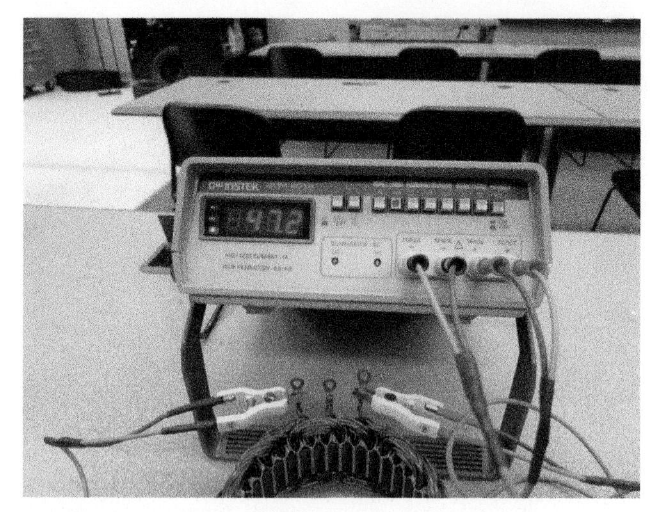

FIGURE 4–17 This milliohmmeter is set to the 200 mΩ range for measuring generator stator windings. The reading is 47.2 milliohms (47.2 mΩ) on all three windings, showing that the windings all have the same amount of resistance. (Courtesy of Jeffrey Rehkopf)

Circuit isolation is critical to the operation of hybrid and electric vehicles, where the high-voltage (HV) components have to remain completely insulated and isolated from vehicle ground. The megohmmeter may be used to locate and verify faulty circuits. It is also useful after a repair to verify that the repaired circuit is isolated from ground, as is required for the HV systems to operate correctly.

MILLIOHMMETER The milliohmmeter is an instrument that can accurately measure resistances of less than 1 ohm, which a standard multimeter cannot do. This is useful when checking the windings of electric motors, generator (alternator) stator windings, and motor/generator (hybrid/electric vehicles) stator windings. ● **SEE FIGURE 4–17.**

Where the Fluke 87-V can read that the resistance is low (0 to .2 Ω and fluctuating), the milliohmmeter will measure exactly what that low resistance is. For example, in a three-phase stator winding, a damaged stator winding will usually read at a higher resistance when compared to the other two windings (perhaps 310 mΩ compared to 145 mΩ), indicating a bad stator assembly.

ELECTRICAL UNIT PREFIXES

DEFINITIONS Electrical units are measured in numbers such as 12 volts, 150 amperes, and 470 ohms. Large units over 1,000 may be expressed in kilo units. **Kilo (k)** means 1,000. ● **SEE FIGURE 4–18.**

4,700 ohms = 4.7 kilohms (kΩ)

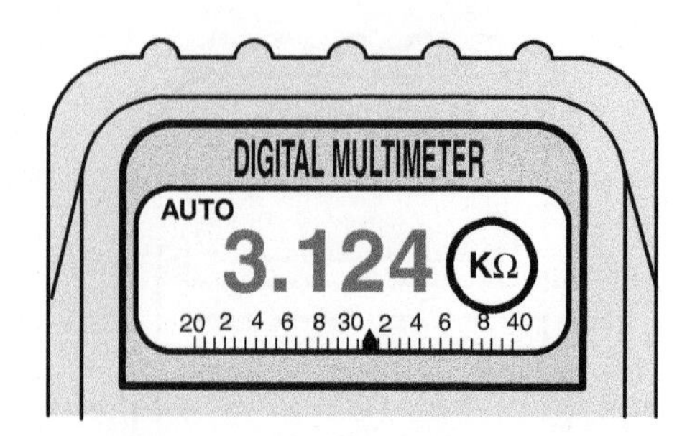

THE SYMBOL ON THE RIGHT SIDE OF THE DISPLAY INDICATES WHAT RANGE THE METER HAS BEEN SET TO READ.

Ω = OHMS

IF THE ONLY SYMBOL ON THE DISPLAY IS THE OHMS SYMBOL, THE READING ON THE DISPLAY IS EXACTLY THE RESISTANCE IN OHMS.

KΩ = KILOHMS = OHMS TIMES 1,000

A "K" IN FRONT OF THE OHMS SYMBOL MEANS "KILOHMS"; THE READING ON THE DISPLAY IS IN KILOHMS. YOU HAVE TO MULTIPLY THE READING ON THE DISPLAY BY 1,000 TO GET THE RESISTANCE IN OHMS.

MΩ = MEGOHMS = OHMS TIMES 1,000,000

A "M" IN FRONT OF THE OHMS SYMBOL MEANS "MEGOHMS"; THE READING ON THE DISPLAY IS IN MEGOHMS. YOU HAVE TO MULTIPLY THE READING ON THE DISPLAY BY 1,000,000 TO GET THE RESISTANCE IN OHMS.

FIGURE 4–18 Always look at the meter display when a measurement is being made, especially if using an autoranging meter.

If the value is over 1 million (1,000,000), then the prefix **mega (M)** is often used. For example:

1,100,000 volts = 1.1 megavolts (MV)

4,700,000 ohms = 4.7 ohms (MΩ)

Sometimes a circuit conducts so little current that a smaller unit of measure is required. Small units of measure expressed in 1/1,000 are prefixed by **milli (m)**. To summarize:

mega (M) = 1,000,000 (decimal point six places to the right = 1,000,000)

kilo (k) = 1,000 (decimal point three places to the right = 1,000)

milli (m) = 1/1,000 (decimal point three places to the left = 0.001)

NOTE: Lowercase *m* equals a small unit (milli), whereas a capital *M* represents a large unit (mega).

● **SEE CHART 4–2.**

TO/ FROM	MEGA	KILO	BASE	MILLI
Mega	0 places	3 places to the right	6 places to the right	9 places to the right
Kilo	3 places to the left	0 places	3 places to the right	6 places to the right
Base	6 places to the left	3 places to the left	0 places	3 places to the right
Milli	9 places to the left	6 places to the left	3 places to the left	0 places

CHART 4–2

A conversion chart showing the decimal point location for the various prefixes.

PREFIXES The prefixes can be confusing because most digital meters can express values in more than one unit, especially if the meter is autoranging. For example, an ammeter reading may show 36.7 mA on autoranging. When the scale is changed to amperes ("A" in the window of the display), the number displayed will be 0.037 A. Note that the resolution of the value is reduced.

NOTE: Always check the face of the meter display for the unit being measured. To best understand what is being displayed on the face of a digital meter, select a manual scale and move the selector until *whole units appear,* such as "A" for amperes instead of "mA" for milliamperes.

HOW TO READ DIGITAL METERS

STEPS TO FOLLOW Getting to know and use a digital meter takes time and practice. The first step is to read, understand, and follow all safety and operational instructions that come with the meter. Use of the meter usually involves the following steps.

STEP 1 **Select the proper unit of electricity for what is being measured.** This unit could be volts, ohms (resistance), or amperes (amount of current flow). If the meter is not autoranging, select the proper scale for the anticipated reading. For example, if a 12-volt battery is being measured, select a meter reading range that is higher than the voltage but not too high. A 20- or 30-volt range will accurately show the voltage of a 12-volt battery. If a 1,000-volt scale is selected, a 12-volt reading may not be accurate.

TECH TIP

Think of Money

Digital meter displays can often be confusing. The display for a battery measured as 12 1/2 volts would be 12.50 V, just as $12.50 is 12 dollars and 50 cents. A 1/2 volt reading on a digital meter will be displayed as 0.50 V, just as $0.50 is half of a dollar.

It is more confusing when low values are displayed. For example, if a voltage reading is 0.063 volt, an autoranging meter will display 63 millivolts (63 mV), or 63/1,000 of a volt, or $63 of $1,000. (It takes 1,000 mV to equal 1 volt.) Think of millivolts as one-tenth of a cent, with 1 volt being $1.00. Therefore, 630 millivolts are equal to $0.63 of $1.00 (630 tenths of a cent, or 63 cents).

To avoid confusion, try to manually range the meter to read base units (whole volts). If the meter is ranged to base unit volts, 63 millivolts would be displayed as 0.063 or maybe just 0.06, depending on the display capabilities of the meter.

STEP 2 **Place the meter leads into the proper input terminals.**

- The black lead is inserted into the common (COM) terminal. This meter lead usually stays in this location for all meter functions.
- The red lead is inserted into the volt, ohm, or diode check terminal usually labeled "VΩ" when voltage, resistance, or diodes are being measured.
- When current flow in amperes is being measured, most digital meters require that the red test lead be inserted in the ammeter terminal, usually labeled "A" or "mA."

CAUTION: If the meter leads are inserted into ammeter terminals, even though the selector is set to volts, the meter may be damaged or an internal fuse may blow if the test leads touch both terminals of a battery.

STEP 3 **Measure the component being tested.** Carefully note the decimal point and the unit on the face of the meter.

- **Meter lead connections.** If the meter leads are connected to a battery backward (red to the battery negative, for example), the display will still show the correct reading, but a negative sign (−) will be

	VOLTAGE BEING MEASURED					
	0.01 V (10 mV)	**0.150 V (150 mV)**	**1.5 V**	**10.0 V**	**12.0 V**	**120 V**
Scale Selected			**Voltmeter will display:**			
200 mV	10.0	150.0	OL	OL	OL	OL
2 V	0.100	0.150	1.500	OL	OL	OL
20 V	0.1	1.50	1.50	10.00	12.00	OL
200 V	00.0	01.5	01.5	10.0	12.0	120.0
2 kV	00.00	00.00	000.1	00.10	00.12	0.120
Autorange	10.0 mV	15.0 mV	1.50	10.0	12.0	120.0

	RESISTANCE BEING MEASURED					
	10 OHMS	**100 OHMS**	**470 OHMS**	**1 KILOHM**	**220 KILOHMS**	**1 MEGOHM**
Scale Selected			**Ohmmeter will display:**			
400 ohms	10.0	100.0	OL	OL	OL	OL
4 kilohms	010	100	0.470 k	1000	OL	OL
40 kilohms	00.0	0.10 k	0.47 k	1.00 k	OL	OL
400 kilohms	000.0	00.1 k	00.5 k	0.10 k	220.0 k	OL
4 megohms	00.00	0.01 M	0.05 M	00.1 M	0.22 M	1.0 M
Autorange	10.0	100.0	470.0	1.00 k	220 k	1.00 M

	CURRENT BEING MEASURED					
	50 mA	**150 mA**	**1.0 A**	**7.5 A**	**15.0 A**	**25.0 A**
Scale Selected			**Ammeter will display:**			
40 mA	OL	OL	OL	OL	OL	OL
400 mA	50.0	150	OL	OL	OL	OL
4 A	0.05	0.00	1.00	OL	OL	OL
40 A	0.00	0.000	01.0	7.5	15.0	25.0
Autorange	50.0 mA	150.0 mA	1.00	7.5	15.0	25.0

CHART 4–3

Sample meter readings using manually set and autoranging selection on the digital meter control.

displayed in front of the number. The correct polarity is not important when measuring resistance (ohms) except where indicated, such as measuring a diode.

- **Autorange.** Many meters automatically default to the autorange position and the meter will display the value in the most readable scale. The meter can be manually ranged to select other levels or to lock in a scale for a value that is constantly changing.

  If a 12-volt battery is measured with an autoranging meter, the correct reading of 12.0 is given. "AUTO" and "V" should show on the face of the meter. For example, if a meter is manually set to the 2 kilohm scale, the highest that the meter will read is 2,000 ohms. If the reading is over 2,000 ohms, the meter will display OL. ● **SEE CHART 4–3.**

STEP 4 **Interpret the reading.** This is especially difficult on autoranging meters, where the meter itself selects the proper scale. The following are two examples of different readings.

Example 1: A voltage drop is being measured. The specifications indicate a maximum voltage drop of 0.2 volt. The meter reads "AUTO" and "43.6 mV." This reading means that the voltage drop is 0.0436 volt, or 43.6 mV, which is far lower than the 0.2 volt (200 mV). Because the number showing on the meter face is much larger than the specifications, many beginner technicians are led to believe that the voltage drop is excessive.

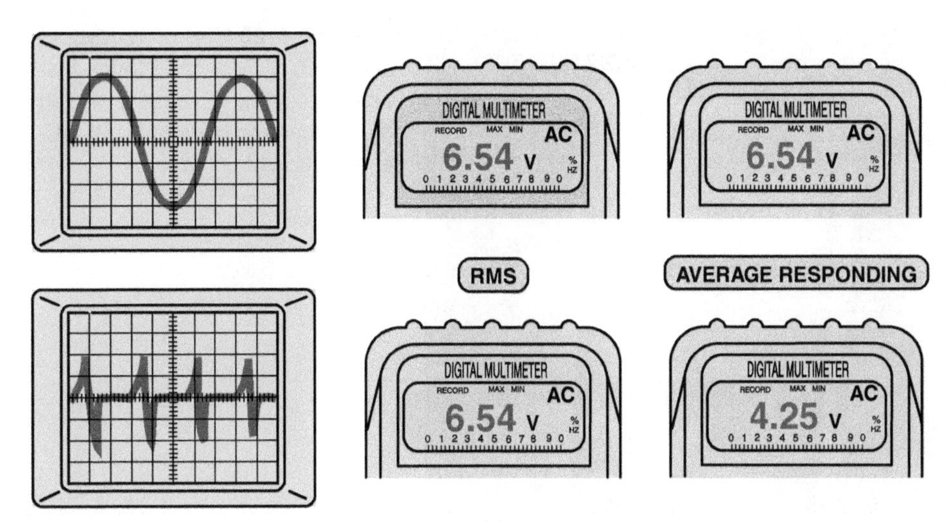

FIGURE 4–19 When reading AC voltage signals, a true RMS meter (such as a Fluke 87) provides a different reading than an average responding meter (such as a Fluke 88). The only place this difference is important is when a reading is to be compared with a specification.

TECH TIP

Purchase a Digital Meter That Will Work for Automotive Use

Try to purchase a digital meter that is capable of reading the following:

- DC volts
- AC volts
- DC amperes (up to 10 A or more is helpful)
- Ohms (Ω) up to 40 MΩ (40 million ohms)
- Diode check

Additional features for advanced automotive diagnosis include:

- Frequency (hertz, abbreviated Hz)
- Temperature probe (°F and/or °C)
- Pulse width (millisecond, abbreviated ms)
- Duty cycle (%)
- Min/max recording
- Hold feature
- Backlighting
- CAT III rating

NOTE: Pay attention to the units displayed on the meter face and convert to whole units.

Example 2: A spark plug wire is being measured. The reading should be less than 10,000 ohms for each foot in length if the wire is okay. The wire being tested is 3 ft long (maximum allowable resistance is 30,000 ohms). The meter reads "AUTO" and "14.85 kΩ." This reading is equivalent to 14,850 ohms.

NOTE: When converting from kilohms to ohms, make the decimal point a comma.

Because this reading is well below the specified maximum allowable, the spark plug wire is okay.

RMS VERSUS AVERAGE Alternating current voltage waveforms can be true sinusoidal or nonsinusoidal. A true sine wave pattern measurement will be the same for both **root-mean-square (RMS)** and average reading meters. RMS and averaging are two methods used to measure the true effective rating of a signal that is constantly changing. ● **SEE FIGURE 4–19.**

Only true RMS meters are accurate when measuring nonsinusoidal AC waveforms, which are seldom used in automotive applications.

RESOLUTION, DIGITS, AND COUNTS **Meter resolution** refers to how small or fine a measurement the meter can make. By knowing the resolution of a DMM you can determine whether the meter could measure down to only 1 volt or down to 1 millivolt (1/1,000 of a volt).

You would not buy a ruler marked in 1 inch segments (or centimeters) if you had to measure down to 1/4 in. (or 1 mm). A thermometer that only measured in whole degrees is not of much use when your normal temperature is 98.6°F. You need a thermometer with 0.1° *resolution.*

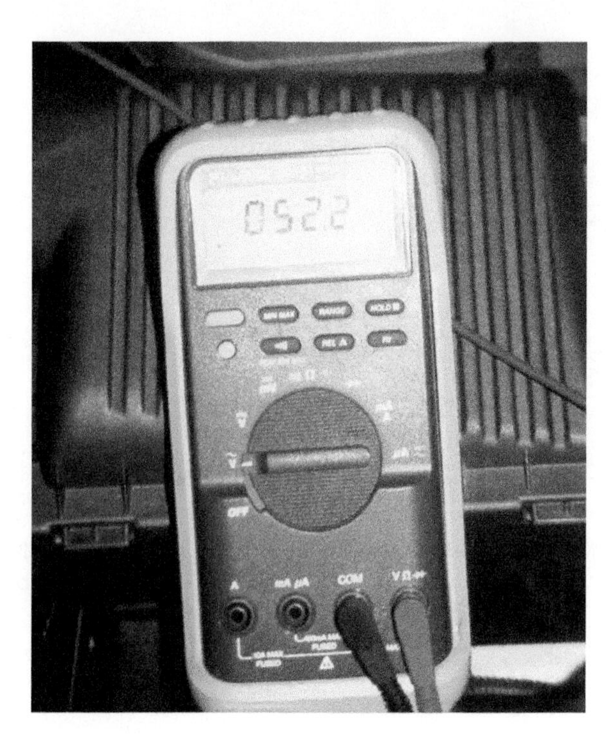

FIGURE 4–20 This meter display shows 052.2 AC volts. Notice that the zero beside the 5 indicates that the meter can read over 100 volts AC with a resolution of 0.1 volt.

The terms *digits* and *counts* are used to describe a meter's resolution. DMMs are grouped by the number of counts or digits they display.

- A 3 1/2-digit meter can display three full digits ranging from 0 to 9, and one "half" digit that displays only a 1 or is left blank. A 3 1/2-digit meter will display up to 1,999 counts of resolution.
- A 4 1/2-digit meter can display up to 19,000 counts of resolution. It is more precise to describe a meter by counts of resolution than by 3 1/2 or 4 1/2 digits. Some 3 1/2-digit meters have enhanced resolution of up to 3,200 or 4,000 counts.

Meters with more counts offer better resolution for certain measurements. For example, a 1,999-count meter will not be able to measure down to a tenth of a volt when measuring 200 volts or more. ● **SEE FIGURE 4–20.**

However, a 3,200-count meter will display a tenth of a volt up to 320 volts. Digits displayed to the far right of the display may at times flicker or constantly change. This is called *digit rattle* and represents a changing voltage being measured on the ground (COM terminal of the meter lead). High-quality meters are designed to reject this unwanted voltage.

ACCURACY **Meter accuracy** is the largest allowable error that will occur under specific operating conditions. In other words, it is an indication of how close the DMM's displayed measurement is to the actual value of the signal being measured.

Accuracy for a DMM is usually expressed as a percent of reading. An accuracy of ±1% of reading means that for a displayed reading of 100.0 V, the actual value of the voltage could be anywhere between 99.0 and 101.0 volts. Thus, the lower the percent of accuracy is, the better.

- Unacceptable = 1.00%
- Okay = 0.50% (1/2%)
- Good = 0.25% (1/4%)
- Excellent = 0.10% (1/10%)

For example, if a battery had 12.6 volts, a meter could read between the following, based on its accuracy.

± 0.1%	high = 12.61
	low = 12.59
± 0.25%	high = 12.63
	low = 12.57
± 0.50%	high = 12.66
	low = 12.54
± 1.00%	high = 12.73
	low = 12.47

Before you purchase a meter, check the accuracy. Accuracy is usually indicated on the specifications sheet for the meter.

SAFETY TIP

Meter Usage on Hybrid Electric Vehicles

Many hybrid electric vehicles use system voltage as high as 650 volts DC. Be sure to follow all vehicle manufacturer's testing procedures; and if a voltage measurement is needed, be sure to use a meter and test leads that are designed to insulate against high voltages. The **International Electrotechnical Commission (IEC)** has several categories of voltage standards for meter and meter leads. These categories are ratings for overvoltage protection and are rated CAT I, CAT II, CAT III, and CAT IV. The higher the category, the greater the protection against voltage spikes caused by high-energy circuits. Under each category there are various energy and voltage ratings.

CAT I Typically a CAT I meter is used for low-energy voltage measurements such as at wall outlets in the home. Meters with a CAT I rating are usually rated at 300 to 800 volts.

CAT II This higher-rated meter would be typically used for checking higher-energy-level voltages at the fuse panel in the home. Meters with a CAT II rating are usually rated at 300 to 600 volts.

CAT III This minimum-rated meter should be used for hybrid vehicles. The CAT III category is designed for high-energy levels and voltage measurements at the service pole at the transformer. Meters with this rating are usually rated at 600 to 1,000 volts.

CAT IV CAT IV meters are for clamp-on meters only. If a clamp-on meter also has meter leads for voltage measurements, that part of the meter will be rated as CAT III.

NOTE: Always use the highest CAT rating meter, especially when working with hybrid vehicles. A CAT III, 600-volt meter is safer than a CAT II, 1,000-volt meter because of the energy level of the CAT ratings.

Therefore, for best personal protection, use only meters and meter leads that are CAT III or CAT IV rated when measuring voltage on a hybrid vehicle. ● **SEE FIGURES 4–21 AND 4–22**.

FIGURE 4–21 Be sure to use only a meter that is CAT III rated when taking electrical voltage measurements on a hybrid vehicle.

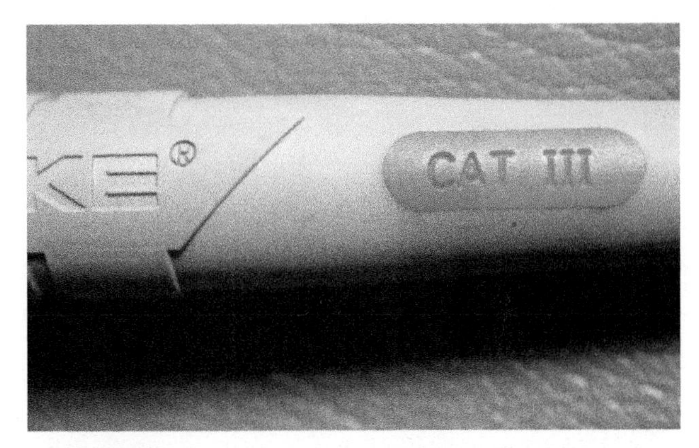

FIGURE 4–22 Always use meter leads that are CAT III rated on a meter that is also CAT III rated, to maintain the protection needed when working on hybrid vehicles.

TECH TIP

Essential Tool J-39200-A*

If the technician already has a Fluke 87 multimeter, it may be time for a new one. Note the distinction between tool numbers. The original number was J-39200 and the present number is J-39200-A.

The J-39200 has been an essential tool since 1992. The original tool was built to the IEC348 standard, not the present IEC61010 standard, which had not been established at that time. The J-39200-A presently being offered is built to the more stringent IEC61010 standard and is CAT III rated. The original J-39200 is not category rated. It does not have the overvoltage protection of a CAT III instrument, but is rated for use up to 1,000 V.

The biggest concern, however, is whether existing J-39200 multimeters, after years of hard service, are still in good repair. Many—probably most—of these instruments have not been calibrated or serviced since new.

- Do they have high-energy fuses?
- Are the test leads damaged, with exposed conductors?
- Are strain reliefs damaged, causing intermittent readings?

Working safely has two parts, and you are responsible for both parts:

1. Having safe equipment
2. Using that equipment in a safe manner

If your old J-39200 is in doubtful condition, and you decide that working safely requires you to replace it, follow the guidelines in the chapter, and obtain a new instrument that is CAT III from Dealer Equipment. The J-39200-A is such an instrument. ● **SEE FIGURE 4–23.**

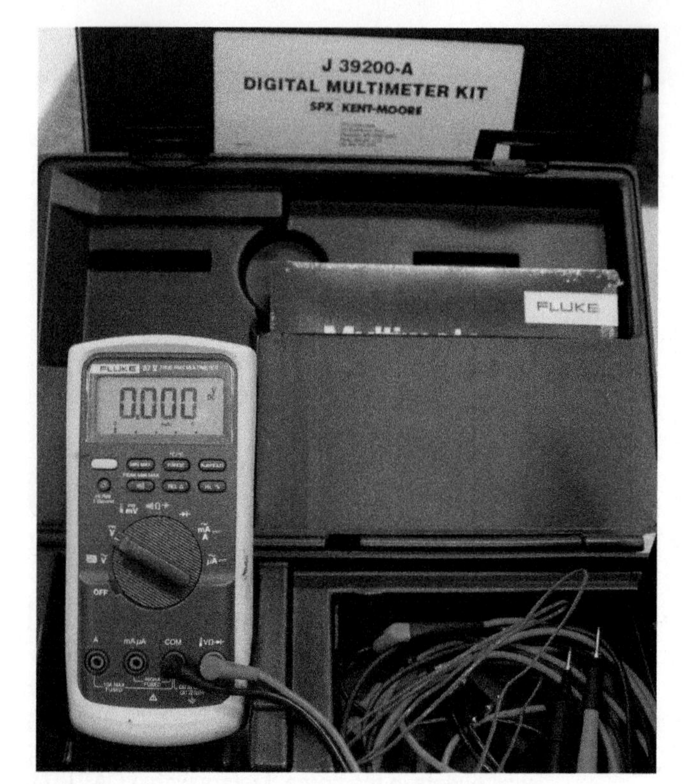

FIGURE 4–23 The J-39200-A kit is the recommended tool for General Motors technicians. It contains a Fluke 87-V meter, two sets of test leads, a thermocouple for measuring temperature, and spare fuses. (Courtesy of Jeffrey Rehkopf)

*GM Techlink Newsletter, January 2008.

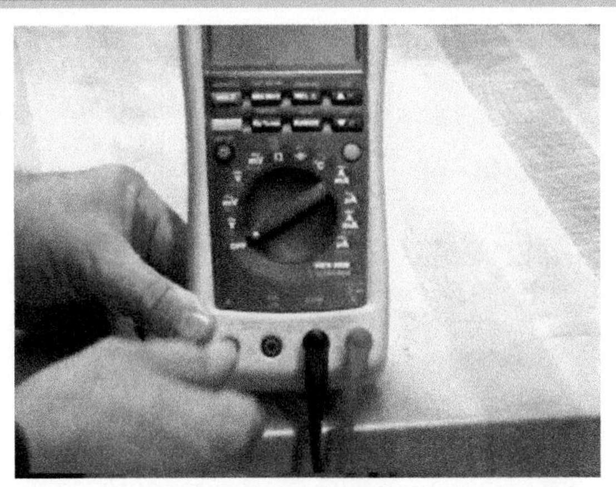

1 For most electrical measurements, the black meter lead is inserted in the terminal labeled COM and the red meter lead is inserted into the terminal labeled V.

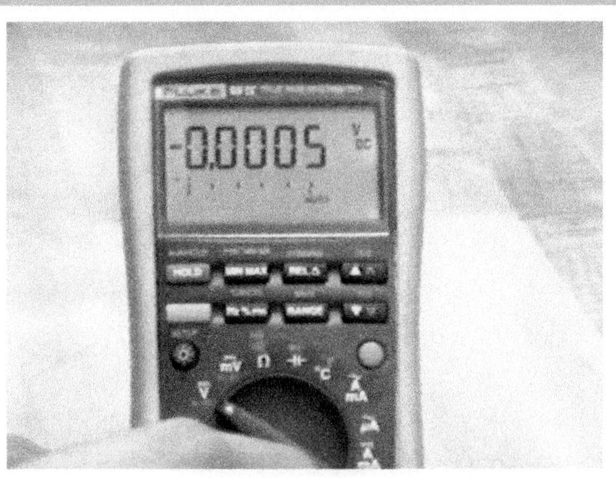

2 To use a digital meter, turn the power switch and select the unit of electricity to be measured. In this case, the rotary switch is turned to select DC volts V.

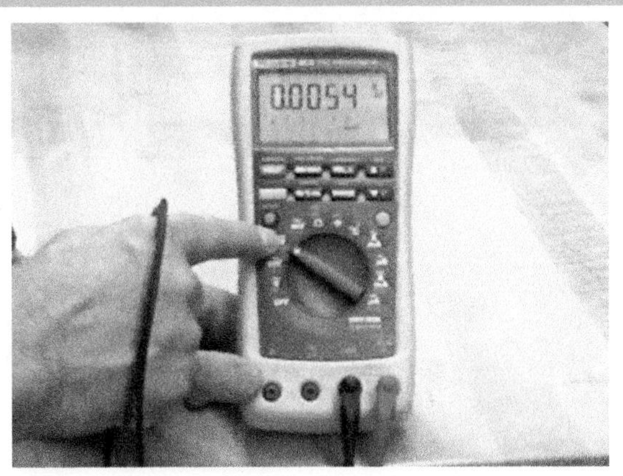

3 For most automotive electrical use, such as measuring battery voltage, select DC volts.

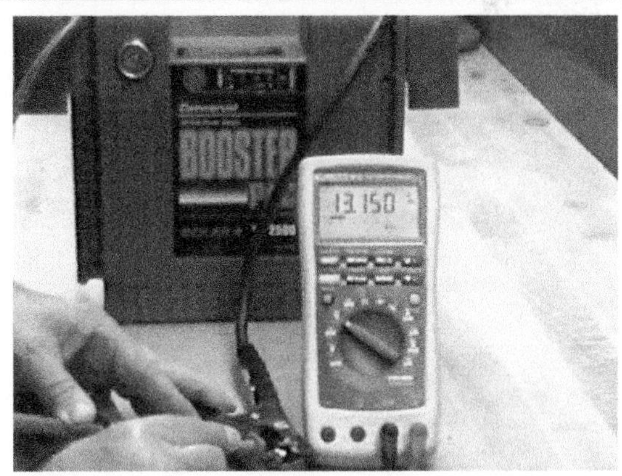

4 Connect the red meter lead to the positive (+) terminal of a battery and the black meter lead to the negative (−) terminal of a battery. The meter reads the voltage difference between the leads.

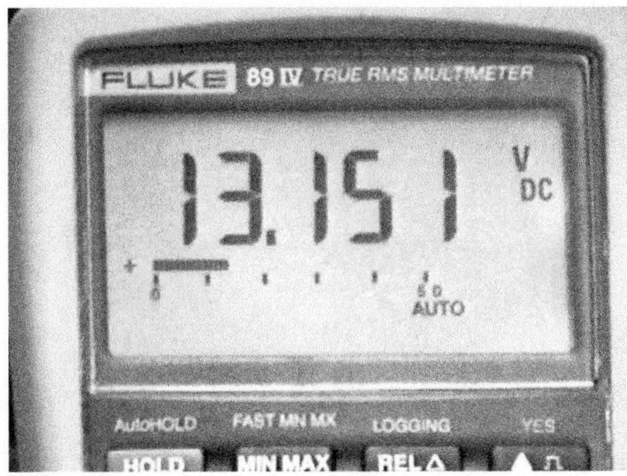

5 This jump start battery unit measures 13.151 volts with the meter set on autoranging on the DC voltage scale.

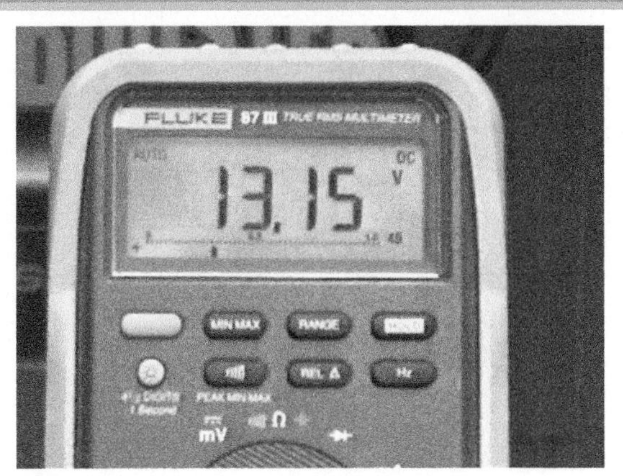

6 Another meter (Fluke 87 III) displays four digits when measuring the voltage of the battery jump start unit.

CONTINUED ▶

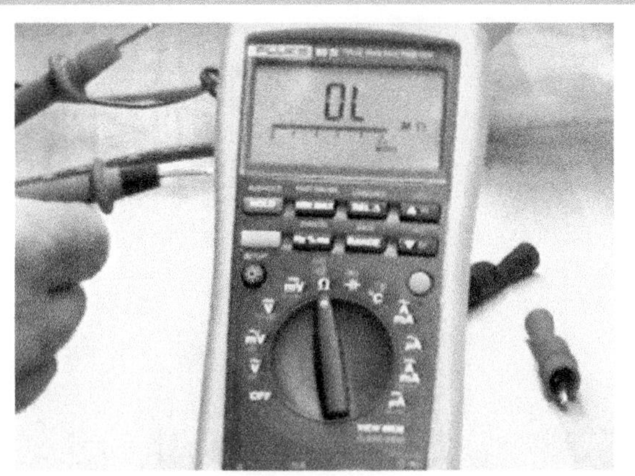

7 To measure resistance, turn the rotary dial to the ohm (Ω) symbol. With the meter leads separated, the meter display reads OL (over limit).

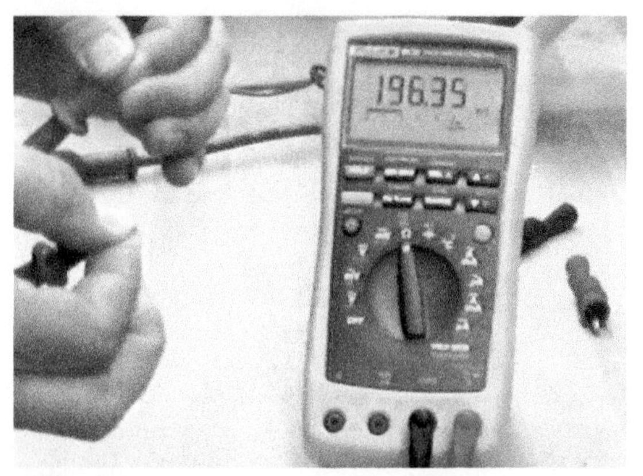

8 The meter can read your own body resistance if you grasp the meter lead terminals with your fingers. The reading on the display indicates 196.35 kΩ.

9 When measuring anything, be sure to read the symbol on the meter face. In this case, the meter is reading 291.10 kΩ.

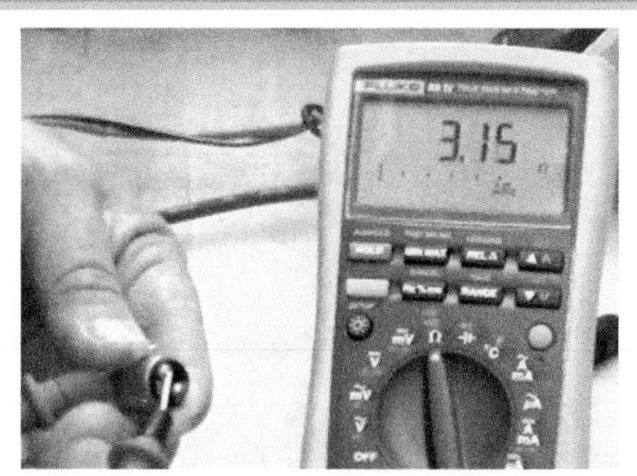

10 A meter set on ohms can be used to check the resistance of a light bulb filament. In this case, the meter reads 3.15 ohms. If the bulb were bad (filament open), the meter would display OL.

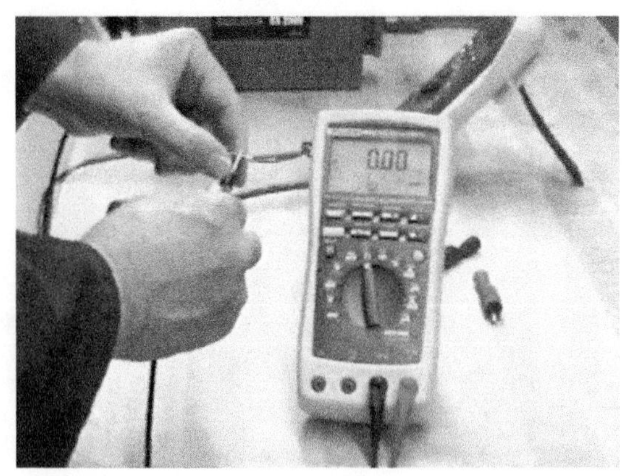

11 A digital meter set to read ohms should measure 0.00 as shown when the meter leads are touched together.

12 The large letter V means volts and the wavy symbol over the V means that the meter measures alternating current (AC) voltage if this position is selected.

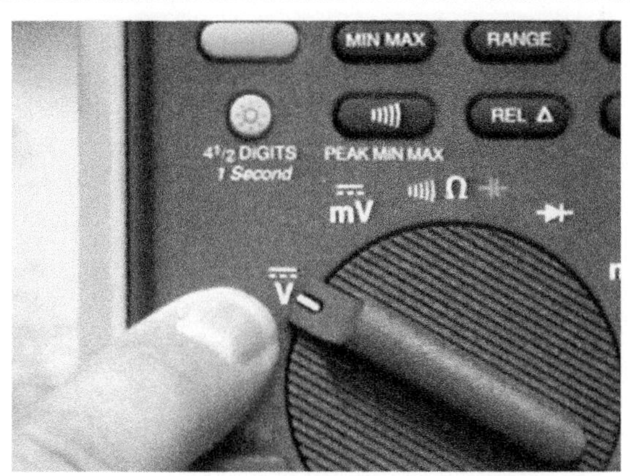

13 The next symbol is a V with a dotted and a straight line overhead. This symbol stands for direct current (DC) volts. This position is most used for automotive service.

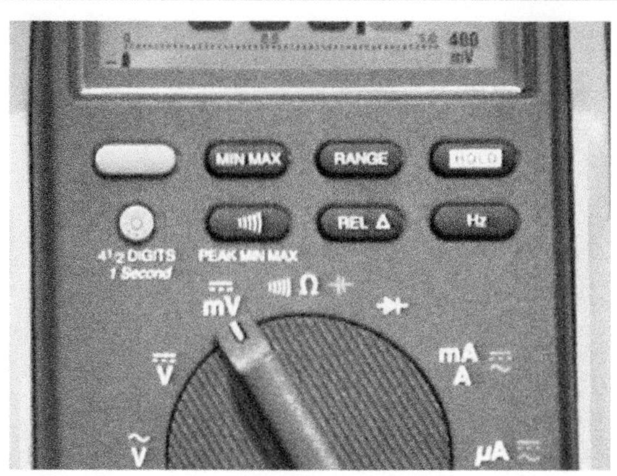

14 The symbol mV indicates millivolts or 1/1,000 of a volt (0.001). The solid and dashed line above the mV means DC mV.

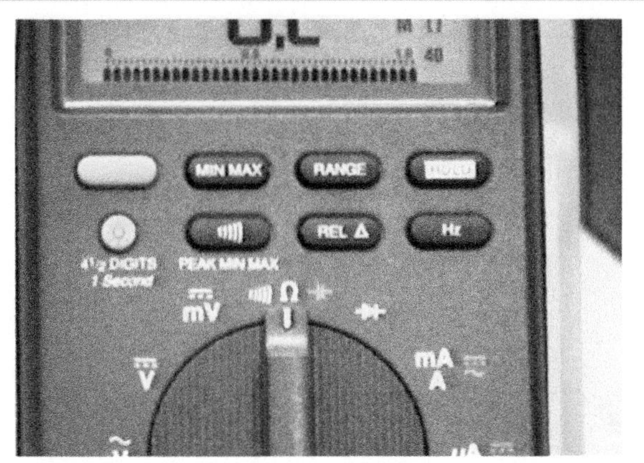

15 The rotary switch is turned to Ω (ohms) unit of resistance measure. The symbol to the left of the Ω symbol is the beeper or continuity indicator.

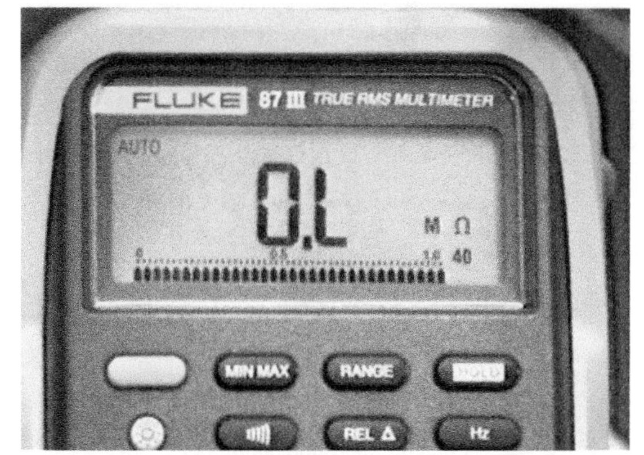

16 Notice that AUTO is in the upper left and the MΩ is in the lower right. This MΩ means megaohms or that the meter is set to read in millions of ohms.

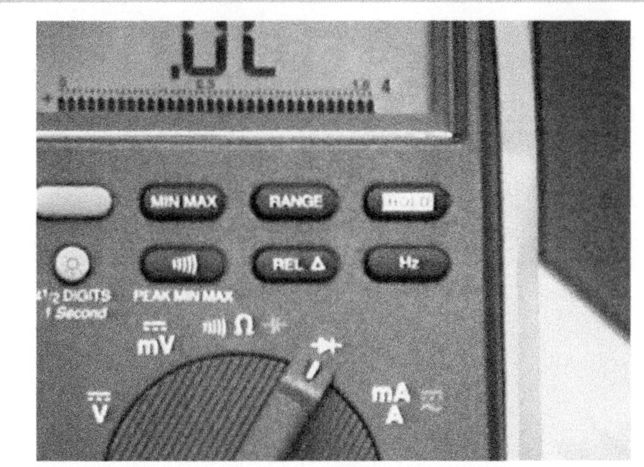

17 The symbol shown is the symbol of a diode. In this position, the meter applies a voltage to a diode and the meter reads the voltage drop across the junction of a diode.

18 One of the most useful features of this meter is the MIN/MAX feature. By pushing the MIN/MAX button, the meter will be able to display the highest (MAX) and the lowest (MIN) reading.

CONTINUED ▶

19 Pushing the MIN/MAX button puts the meter into record mode. Note the 100 ms and "rec" on the display. In this position, the meter is capturing any voltage change that lasts 100 ms (0.1 seconds) or longer.

20 To increase the range of the meter, touch the range button. Now the meter is set to read voltage up to 40 volts DC.

21 Pushing the range button one more time changes the meter scale to the 400-voltage range. Notice that the decimal point has moved to the right.

22 Pushing the range button again changes the meter to the 4,000-volt range. This range is not suitable to use in automotive applications.

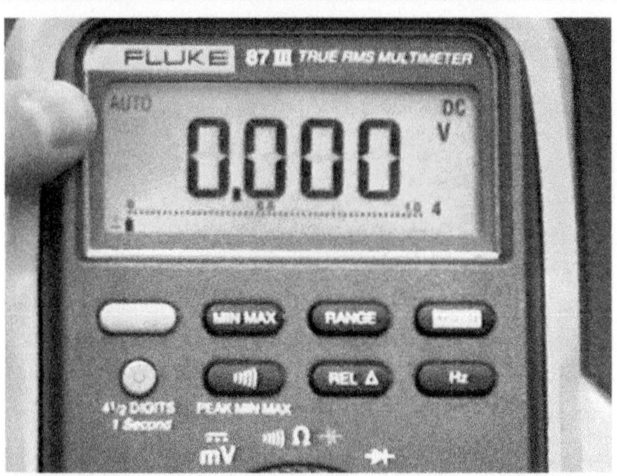

23 By pushing and holding the range button, the meter will reset to autorange. Autorange is the preferred setting for most automotive measurements except when using MIN/MAX record mode.

SUMMARY

1. Circuit testers include test lights and fused jumper leads.
2. Digital multimeter (DMM) and digital volt-ohm-meter (DVOM) are terms commonly used for electronic high-impedance test meters.
3. Use of a high-impedance digital meter is required on any computer-related circuit or component.
4. Ammeters measure current and must be connected in series in the circuit.
5. Voltmeters measure voltage and are connected in parallel.
6. Ohmmeters measure resistance of a component and must be connected in parallel, with the circuit or component disconnected from power.
7. Logic probes can indicate the presence of power, ground, or pulsed signals.

REVIEW QUESTIONS

1. Why should high-impedance meters be used when measuring voltage on computer-controlled circuits?
2. How is an ammeter connected to an electrical circuit?
3. Why must an ohmmeter be connected to a disconnected circuit or component?

CHAPTER QUIZ

1. Inductive ammeters work because of what principle?
   a. Magic
   b. Electrostatic electricity
   c. A magnetic field surrounds any wire carrying a current
   d. Voltage drop as it flows through a conductor
2. A meter used to measure amperes is called a(n) ________.
   a. Amp meter
   b. Ampmeter
   c. Ammeter
   d. Coulomb meter
3. A voltmeter should be connected to the circuit being tested ________.
   a. In series
   b. In parallel
   c. Only when no power is flowing
   d. Both a and c
4. An ohmmeter should be connected to the circuit or component being tested ________.
   a. With current flowing in the circuit or through the component
   b. When connected to the battery of the vehicle to power the meter
   c. Only when no power is flowing (electrically open circuit)
   d. Both b and c
5. A high-impedance meter ________.
   a. Measures a high amount of current flow
   b. Measures a high amount of resistance
   c. Can measure a high voltage
   d. Has a high internal resistance
6. A meter is set to read DC volts on the 4-volt scale. The meter leads are connected at a 12-volt battery. The display will read ________.
   a. 0.00
   b. OL
   c. 12 V
   d. 0.012 V
7. What could happen if the meter leads were connected to the positive and negative terminals of the battery while the meter and leads were set to read amperes?
   a. Could blow an internal fuse or damage the meter
   b. Would read volts instead of amperes
   c. Would display OL
   d. Would display 0.00
8. The highest amount of resistance that can be read by the meter set to the 2 kΩ scale is ________.
   a. 2,000 ohms
   b. 200 ohms
   c. 200 kΩ (200,000 ohms)
   d. 20,000,000 ohms
9. If a digital meter face shows 0.93 when set to read kΩ, the reading means ________.
   a. 93 ohms
   b. 930 ohms
   c. 9,300 ohms
   d. 93,000 ohms
10. A reading of 432 shows on the face of the meter set to the millivolt scale. The reading means ________.
    a. 0.432 volt
    b. 4.32 volts
    c. 43.2 volts
    d. 4,320 volts

chapter 5

OSCILLOSCOPES AND GRAPHING SCAN TOOLS

LEARNING OBJECTIVES

After studying this chapter, the reader will be able to:

1. Use a digital storage oscilloscope to measure voltage signals.
2. Interpret meter and scope readings and determine if the values are within factory specifications.
3. Explain time base and volts per division settings.

This chapter will help you prepare for the ASE Electrical/Electronic Systems (A6) certification test content area "A" (General Electrical/Electronic System Diagnosis).

KEY TERMS

AC coupling 59
BNC connector 62
Cathode ray tube (CRT) 57
Channel 60
DC coupling 59
Digital storage oscilloscope (DSO) 57
Division 57
Duty cycle 59
External trigger 62
Frequency 59
Graticule 57
Hertz 59
Oscilloscope (scope) 57
Pulse train 59
Pulse width 59
PWM 59
Time base 57
Trigger level 62
Trigger slope 62

GM STC OBJECTIVES

GM Service Technical College topics covered in this chapter are as follows:

1. Perform voltage testing at the DLC on GMLAN.
2. Test control circuit using electrical/electronic tools.
3. Test GMLAN low speed using electrical/electronic tools.
4. Test Class 2 using electrical/electronic tools.

TYPES OF OSCILLOSCOPES

TERMINOLOGY An **oscilloscope** (usually called a **scope**) is a visual voltmeter with a timer that shows when a voltage changes. Following are several types of oscilloscopes.

- An *analog scope* uses a **cathode ray tube (CRT)** similar to a television screen to display voltage patterns. The scope screen displays the electrical signal constantly.
- A *digital scope* commonly uses a liquid crystal display (LCD), but a CRT may also be used on some digital scopes. A digital scope takes samples of the signals that can be stopped or stored and is therefore called a **digital storage oscilloscope**, or **DSO**.
  - A digital scope does not capture each change in voltage but instead captures voltage levels over time and stores them as dots. Each dot is a voltage level. Then the scope displays the waveforms using the thousands of dots (each representing a voltage level) and then electrically connects the dots to create a waveform.
  - A DSO can be connected to a sensor output signal wire and can record over a long period of time the voltage signals. Then it can be replayed and a technician can see if any faults were detected. This feature makes a DSO the perfect tool to help diagnose intermittent problems.
  - A digital storage scope, however, can sometimes miss faults called glitches that may occur between samples captured by the scope. This is why a DSO with a high "sampling rate" is preferred. Sampling rate means that a scope is capable of capturing voltage changes that occur over a very short period of time. Some digital storage scopes have a capture rate of 25 million (25,000,000) samples per second. This means that the scope can capture a glitch (fault) that lasts just 40 nano (0.00000040) seconds long.
  - A scope has been called "a voltmeter with a clock."
    - The voltmeter part means that a scope can capture and display changing voltage levels.
    - The clock part means that the scope can display these changes in voltage levels within a specific time period; and with a DSO it can be replayed so that any faults can be seen and studied.

OSCILLOSCOPE DISPLAY GRID A typical scope face usually has eight or ten grids vertically (up and down) and ten grids horizontally (left to right). The transparent scale (grid), used for reference measurements, is called a **graticule**. This arrangement is commonly 8 × 10 or 10 × 10 divisions. ● **SEE FIGURE 5–1**.

NOTE: These numbers originally referred to the metric dimensions of the graticule in centimeters. Therefore, an 8 × 10 display would be 8 cm (80 mm or 3.14 inch) high and 10 cm (100 mm or 3.90 inch) wide.

- Voltage is displayed on a scope starting with zero volts at the bottom and higher voltage being displayed vertically.
- The scope illustrates time left to right. The pattern starts on the left and sweeps across the screen from left to right.

MILLISECONDS PER DIVISION (MS/DIV)	TOTAL TIME DISPLAYED
1 ms	10 ms (0.010 sec.)
10 ms	100 ms (0.100 sec.)
50 ms	500 ms (0.500 sec.)
100 ms	1 sec. (1.000 sec.)
500 ms	5 sec. (5.0 sec.)
1,000 ms	10 sec. (10.0 sec.)

CHART 5–1

The time base is milliseconds (ms) and total time of an event that can be displayed.

SCOPE SETUP AND ADJUSTMENTS

SETTING THE TIME BASE Most scopes use 10 graticules from left to right on the display. Setting the **time base** means setting how much time will be displayed in each block called a **division**. For example, if the scope is set to read 2 seconds per division (referred to as *s/div*), then the total displayed would be 20 seconds (2 × 10 divisions = 20 seconds). The time base should be set to an amount of time that allows two to four events to be displayed. Milliseconds (0.001 seconds) are commonly used in scopes when adjusting the time base. Sample time is milliseconds per division (indicated as *ms/div*) and total time. ● **SEE CHART 5–1**.

NOTE: Increasing the time base reduces the number of samples per second.

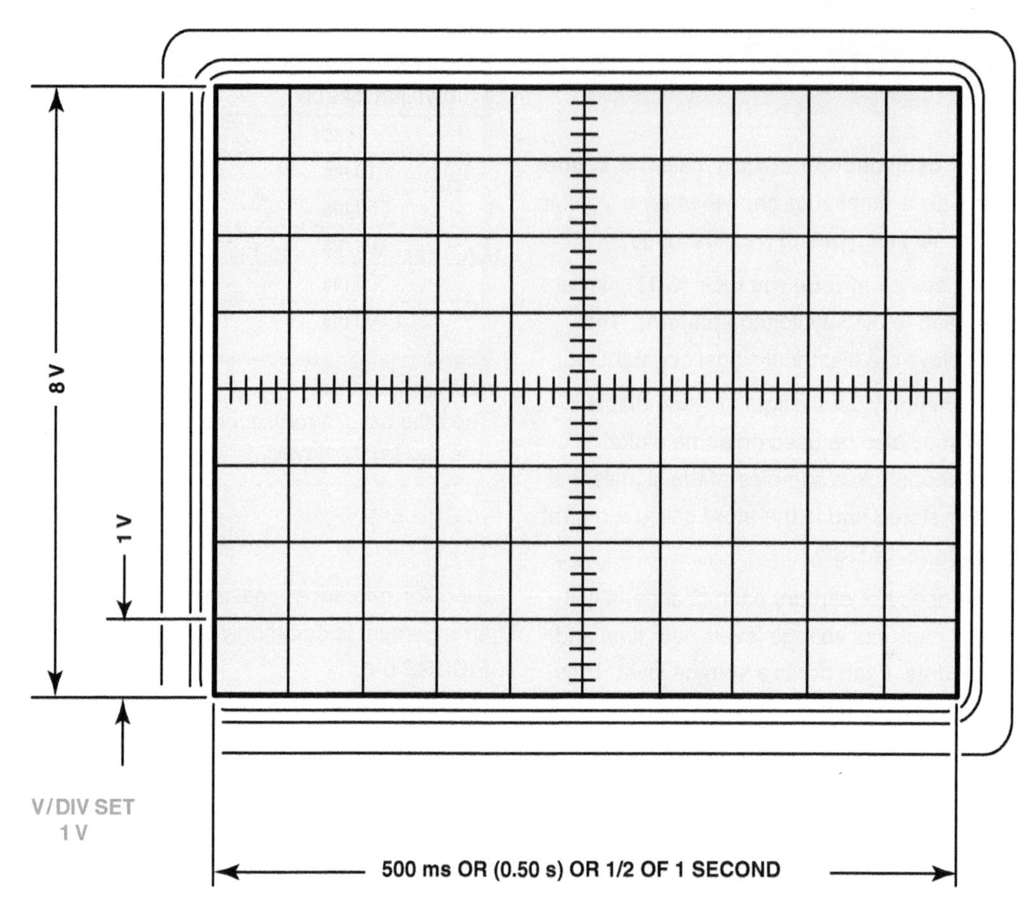

FIGURE 5–1 A scope display allows technicians to take measurements of voltage patterns. In this example, each vertical division is 1 volt and each horizontal division is set to represent 50 milliseconds.

The horizontal scale is divided into 10 divisions (sometimes called *grats*). If each division represents 1 second of time, then the total time period displayed on the screen will be 10 seconds. The time per division is selected so that several events of the waveform are displayed. Time per division settings can vary greatly in automotive use, including:

- MAP/MAF sensors: 2 ms/div (20 ms total)
- Network (CAN) communications network: 2 ms/div (20 ms total)
- Throttle position (TP) sensor: 100 ms per division (1 sec. total)
- Fuel injector: 2 ms/div (20 ms total)
- Oxygen sensor: 1 seconds per division (10 seconds total)
- Primary ignition: 10 ms/div (100 ms total)
- Secondary ignition: 10 ms/div (100 ms total)
- Voltage measurements: 5 ms/div (50 ms total)

The total time displayed on the screen allows comparisons to see if the waveform is consistent or is changing. Multiple waveforms shown on the display at the same time also allow for measurements to be seen more easily. ● **SEE FIGURE 5–2** for an example of a throttle position sensor waveform created by measuring the voltage output as the throttle was depressed and then released.

VOLTS PER DIVISION The volts per division, abbreviated *V/div*, should be set so that the entire anticipated waveform can be viewed. Examples include:

Throttle position (TP) sensor: 1 V/div (8 V total)

Battery, starting and charging: 2 V/div (16 V total)

Oxygen sensor: 200 mV/div (1.6 V total)

Notice from the examples that the total voltage to be displayed exceeds the voltage range of the component being tested. This ensures that all the waveform will be displayed. It also allows for some unexpected voltage readings. For example, an oxygen sensor should read between 0 and 1 Volt (1,000 mV). By setting the V/div to 200 mV, up to 1.6 V (1,600 mV) will be displayed.

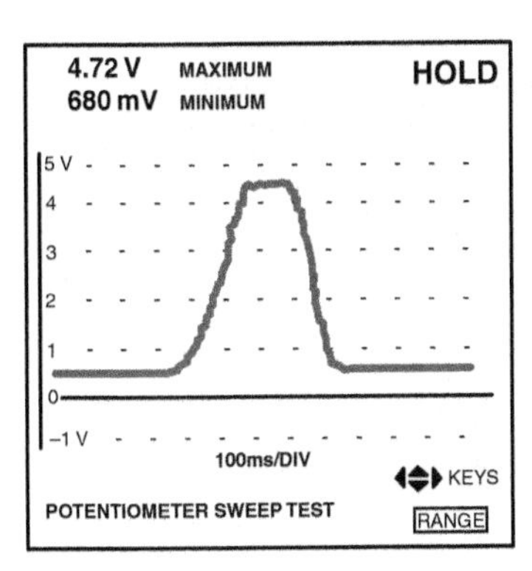

FIGURE 5–2 The display on a digital storage oscilloscope (DSO) displays the entire waveform of a throttle position (TP) sensor from idle to wide-open throttle and then returns to idle. The display also indicates the maximum (4.72 V) and minimum (680 mV or 0.68 V) readings. The display does not show anything until the throttle is opened, because the scope has been set up to start displaying a waveform only after a certain voltage level has been reached. This voltage is called the trigger or trigger point.

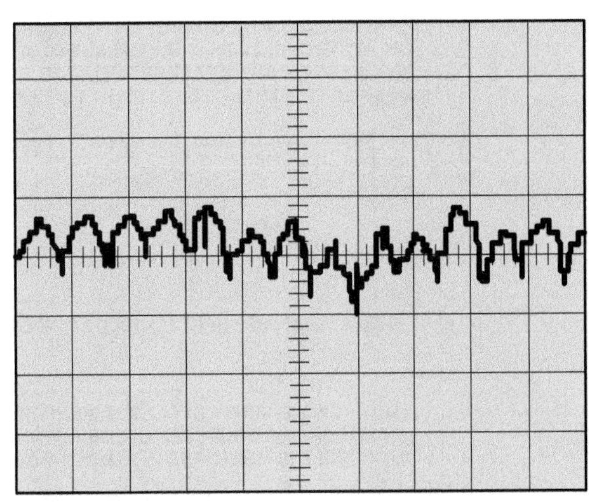

FIGURE 5–3 Ripple voltage is created from the AC voltage from an alternator. Some AC ripple voltage is normal but if the AC portion exceeds 0.5 volt, then a bad diode is the most likely cause. Excessive AC ripple can cause many electrical and electronic devices to work incorrectly.

DC AND AC COUPLING

DC COUPLING **DC coupling** is the most used position on a scope because it allows the scope to display both alternating current (AC) voltage signals and direct current (DC) voltage signals present in the circuit. The AC part of the signal will ride on top of the DC component. For example, if the engine is running and the charging voltage is 14.4 volts DC, this will be displayed as a horizontal line on the screen. Any AC ripple voltage leaking past the alternator diodes will be displayed as an AC signal on top of the horizontal DC voltage line. Therefore, both components of the signal can be observed at the same time.

AC COUPLING When the **AC coupling** position is selected, a capacitor is placed into the meter lead circuit, which effectively blocks all DC voltage signals but allows the AC portion of the signal to pass and be displayed. AC coupling can be used to show output signal waveforms from sensors such as:

- Distributor pickup coils
- Magnetic wheel speed sensors
- Magnetic crankshaft position sensors
- Magnetic camshaft position sensors
- The AC ripple from an alternator. **SEE FIGURE 5–3.**
- Magnetic vehicle speed sensors

NOTE: Check the instructions from the scope manufacturer for the recommended settings to use. Sometimes it is necessary to switch from DC coupling to AC coupling or from AC coupling to DC coupling to properly see some waveforms.

PULSE TRAINS

DEFINITION Scopes can show all voltage signals. Among the most commonly found in automotive applications is a DC voltage that varies up and down and does not go below zero like an AC voltage. A DC voltage that turns on and off in a series of pulses is called a **pulse train**. Pulse trains differ from an AC signal in that they do not go below zero. An alternating voltage goes above and below zero voltage. Pulse train signals can vary in several ways. **SEE FIGURE 5–4.**

FREQUENCY **Frequency** is the number of cycles per second measured in **hertz**. The engine revolutions per minute (RPM) signal is an example of a signal that can occur at various frequencies. At low engine speed, the ignition pulses occur fewer times per second (lower frequency) than when the engine is operated at higher engine speeds (RPM).

DUTY CYCLE **Duty cycle** refers to the percentage of on-time of the signal during one complete cycle. As on-time increases, the amount of time the signal is off decreases and is usually measured in percentage. Duty cycle is also called **pulse-width modulation (PWM)** and can be measured in degrees. **SEE FIGURE 5–5.**

PULSE WIDTH The **pulse width** is a measure of the actual on-time measured in milliseconds. Fuel injectors are usually controlled by varying the pulse width. **SEE FIGURE 5–6.**

1. FREQUENCY—FREQUENCY IS THE NUMBER OF CYCLES THAT TAKE PLACE PER SECOND. THE MORE CYCLES THAT TAKE PLACE IN ONE SECOND, THE HIGHER THE FREQUENCY READING. FREQUENCIES ARE MEASURED IN HERTZ, WHICH IS THE NUMBER OF CYCLES PER SECOND. AN 8-HERTZ SIGNAL CYCLES EIGHT TIMES PER SECOND.

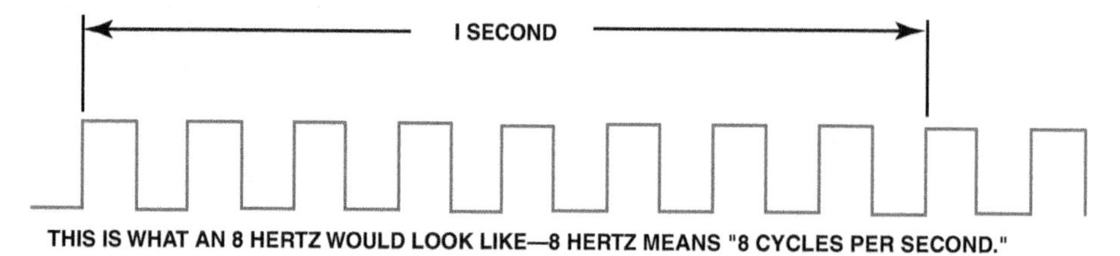

THIS IS WHAT AN 8 HERTZ WOULD LOOK LIKE—8 HERTZ MEANS "8 CYCLES PER SECOND."

2. DUTY CYCLE—DUTY CYCLE IS A MEASUREMENT COMPARING THE SIGNAL ON-TIME TO THE LENGTH OF ONE COMPLETE CYCLE. AS ON-TIME INCREASES, OFF-TIME DECREASES. DUTY CYCLE IS MEASURED IN PERCENTAGE OF ON-TIME. A 60% DUTY CYCLE IS A SIGNAL THAT IS ON 60% OF THE TIME, AND OFF 40% OF THE TIME. ANOTHER WAY TO MEASURE DUTY CYCLE IS DWELL, WHICH IS MEASURED IN DEGREES INSTEAD OF PERCENT.

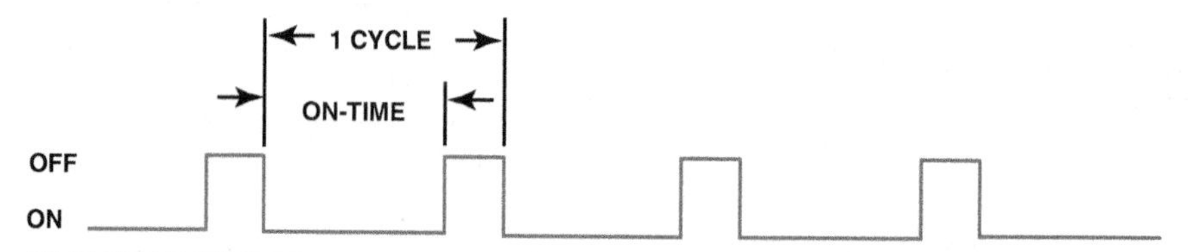

DUTY CYCLE IS THE RELATIONSHIP BETWEEN ONE COMPLETE CYCLE AND THE SIGNAL'S ON-TIME. A SIGNAL CAN VARY IN DUTY CYCLE WITHOUT AFFECTING THE FREQUENCY.

3. PULSE WIDTH—PULSE WIDTH IS THE ACTUAL ON-TIME OF A SIGNAL, MEASURED IN MILLISECONDS. WITH PULSE WIDTH MEASUREMENTS, OFF-TIME DOESN'T REALLY MATTER—THE ONLY REAL CONCERN IS HOW LONG THE SIGNAL IS ON. THIS IS A USEFUL TEST FOR MEASURING CONVENTIONAL INJECTOR ON-TIME, TO SEE THAT THE SIGNAL VARIES WITH LOAD CHANGE.

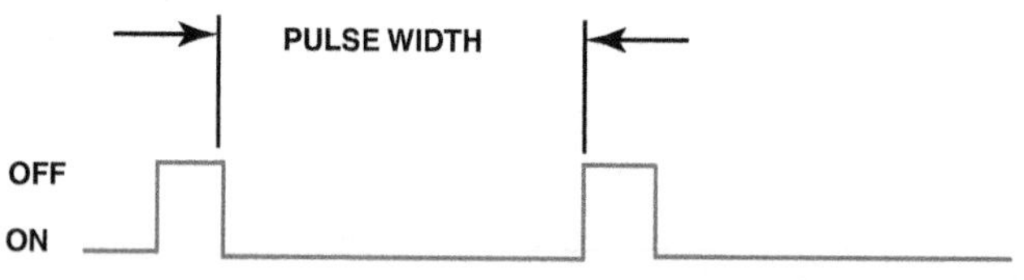

PULSE WIDTH IS THE ACTUAL TIME A SIGNAL IS ON, MEASURED IN MILLISECONDS. THE ONLY THING BEING MEASURED IS HOW LONG THE SIGNAL IS ON.

FIGURE 5–4 A pulse train is any electrical signal that turns on and off, or goes high and low in a series of pulses. Ignition module and fuel-injector pulses are examples of a pulse train signal.

NUMBER OF CHANNELS

DEFINITION Scopes are available that allow the viewing of more than one sensor or event at the same time on the display. The number of events, which require leads for each, is called a **channel**. A channel is an input to a scope. Commonly available scopes include:

- **Single channel.** A single channel scope is capable of displaying only one sensor signal waveform at a time.
- **Two channel.** A two-channel scope can display the waveform from two separate sensors or components at the same time. This feature is very helpful when testing the camshaft and crankshaft position sensors on an engine to see if they are properly timed. ● **SEE FIGURE 5–7.**
- **Four channel.** A four-channel scope allows the technician to view up to four different sensors or actuators on one display.

NOTE: Often the capture speed of the signals is slowed when using more than one channel.

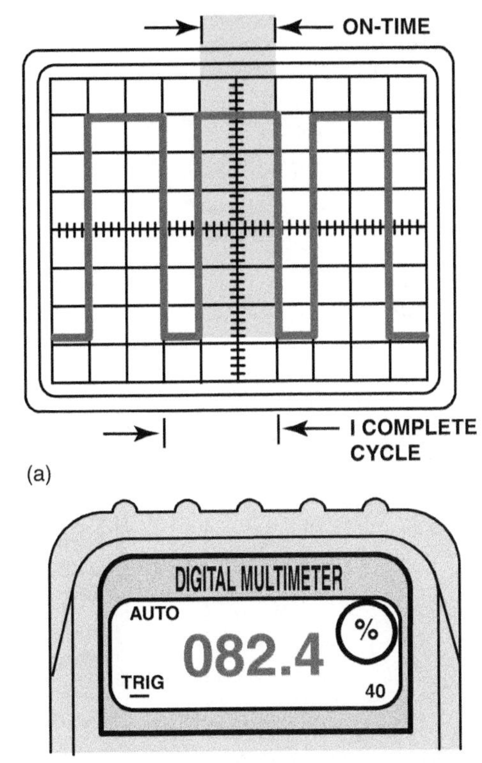

FIGURE 5–5 (a) A scope representation of a complete cycle showing both on-time and off-time. (b) A meter display indicating the on-time duty cycle in a percentage (%). Note the trigger and negative (−) symbol. This indicates that the meter started recording the percentage of on-time when the voltage dropped (start of on-time).

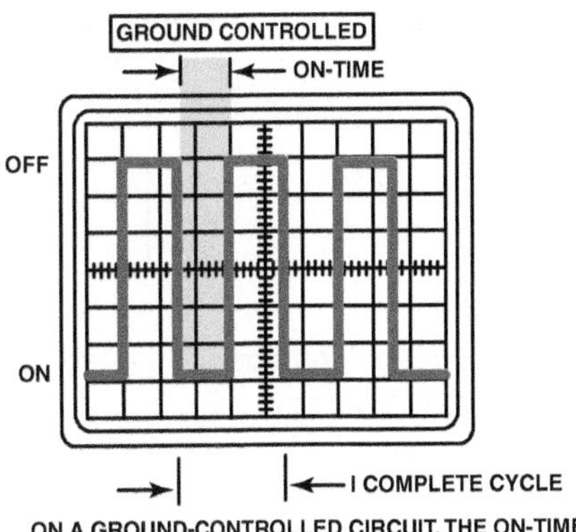

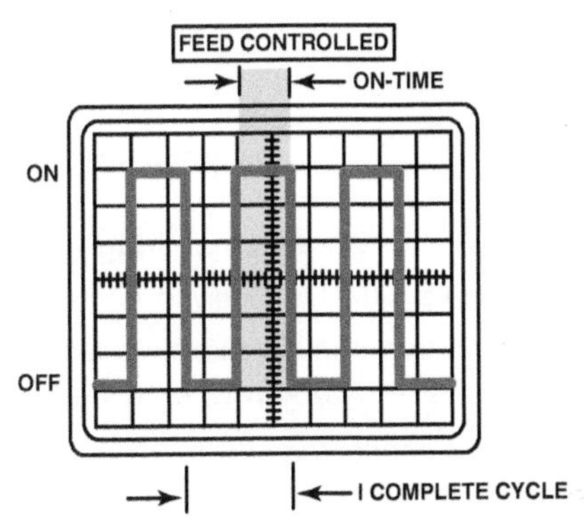

FIGURE 5–6 Most automotive computer systems control the device by opening and closing the ground to the component.

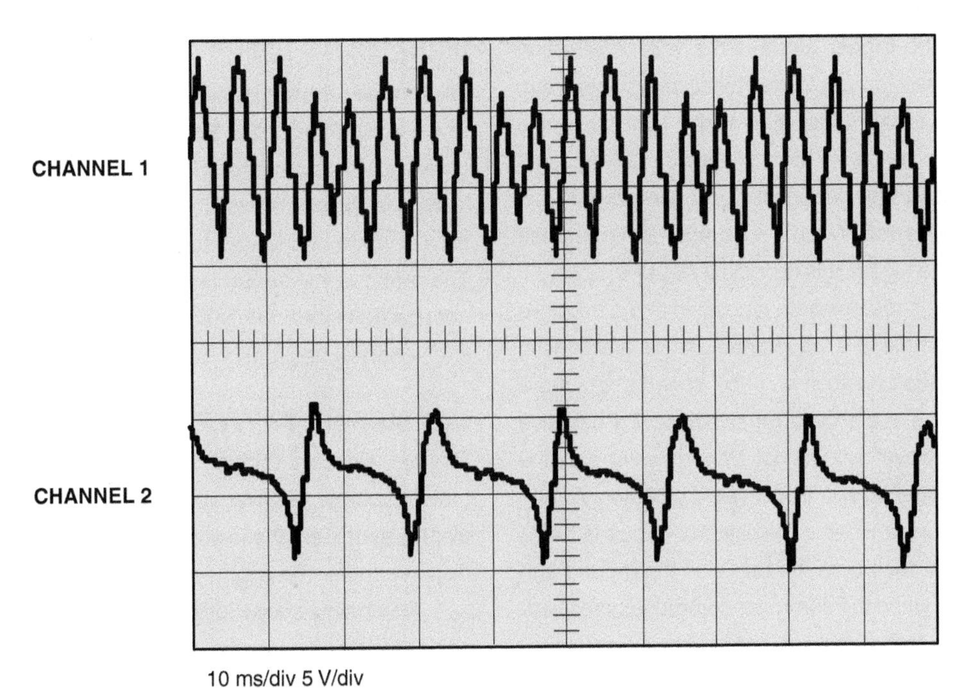

FIGURE 5–7 A two-channel scope being used to compare two signals on the same vehicle.

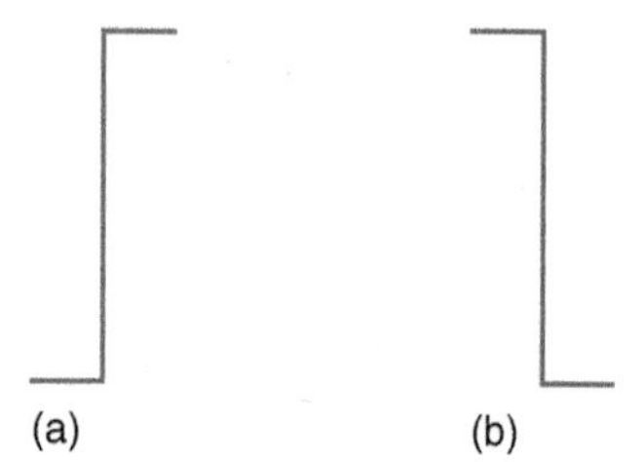

FIGURE 5–8 (a) A symbol for a positive trigger—a trigger occurs at a rising (positive) edge of the signal (waveform). (b) A symbol for a negative trigger—a trigger occurs at a falling (negative) edge of the signal (waveform).

TRIGGERS

EXTERNAL TRIGGER An **external trigger** is when the waveform starts when a signal is received from another external source rather than from the signal pickup lead. A common example of an external trigger comes from the probe clamp around the cylinder #1 spark plug wire to trigger the start of an ignition pattern.

TRIGGER LEVEL **Trigger level** is the voltage that must be detected by the scope before the pattern will be displayed. A scope will only start displaying a voltage signal when it is triggered or is told to start. The trigger level must be set to start the display. If the pattern starts at 1 volt, then the trace will begin displaying on the left side of the screen *after* the trace has reached 1 volt.

TRIGGER SLOPE The **trigger slope** is the voltage direction that a waveform must have in order to start the display. Most often, the trigger to start a waveform display is taken from the signal itself. Besides trigger voltage level, most scopes can be adjusted to trigger only when the voltage rises past the trigger-level voltage. This is called a *positive slope*. When the voltage falling past the higher level activates the trigger, this is called a *negative slope*.

The scope display indicates both a positive and a negative slope symbol. For example, if a waveform such as a magnetic sensor used for crankshaft position or wheel speed starts moving upward, a positive slope should be selected. If a negative slope is selected, the waveform will not start showing until the voltage reaches the trigger level in a downward direction. A negative slope should be used when a fuel-injector circuit is being analyzed. In this circuit, the computer provides the ground and the voltage level drops when the computer commands the injector on. Sometimes the technician needs to change from negative to positive or positive to negative trigger if a waveform is not being shown correctly. ● **SEE FIGURE 5–8.**

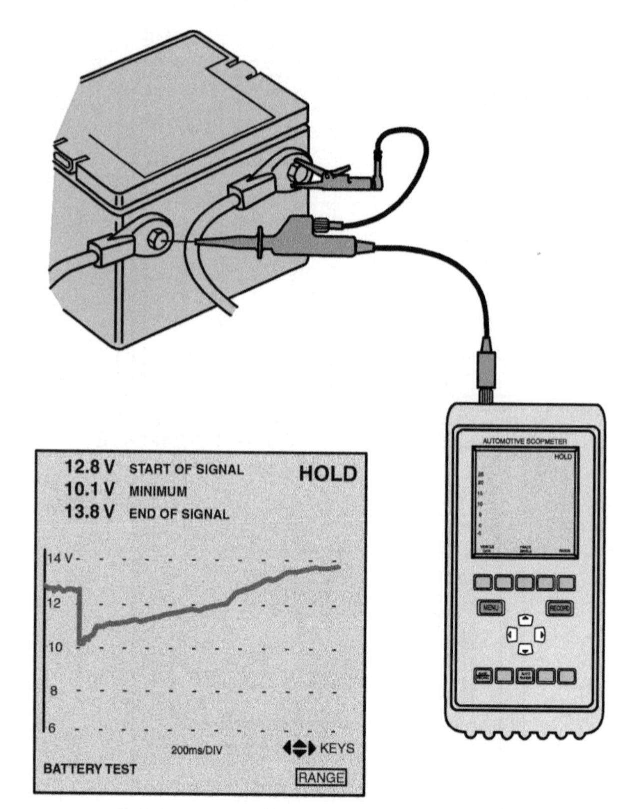

FIGURE 5–9 Constant battery voltage is represented by a flat horizontal line. In this example, the engine was started and the battery voltage dropped to about 10 volts, as shown on the left side of the scope display. When the engine started, the alternator started to charge the battery and the voltage is shown as climbing.

USING A SCOPE

USING SCOPE LEADS Most scopes, both analog and digital, normally use the same test leads. These leads usually attach to the scope through a **BNC connector**, a miniature standard coaxial cable connector. BNC is an international standard that is used in the electronics industry. If using a BNC connector, be sure to connect one lead to a good clean, metal engine ground. The probe of the scope lead attaches to the circuit or component being tested. Many scopes use one ground lead and then each channel has it own signal pickup lead.

MEASURING BATTERY VOLTAGE WITH A SCOPE One of the easiest things to measure and observe on a scope is battery voltage. A lower voltage can be observed on the scope display as the engine is started and a higher voltage should be displayed after the engine starts. ● **SEE FIGURE 5–9.**

An analog scope displays rapidly and cannot be set to show or freeze a display. Therefore, even though an analog scope shows all voltage signals, it is easy to miss a momentary glitch on an analog scope.

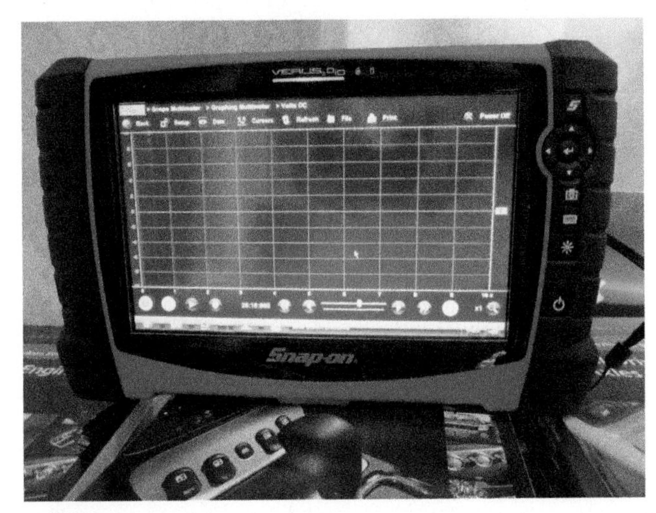

FIGURE 5–10 This scan tool includes a built-in four-channel oscilloscope. (Courtesy of Jeffrey Rehkopf)

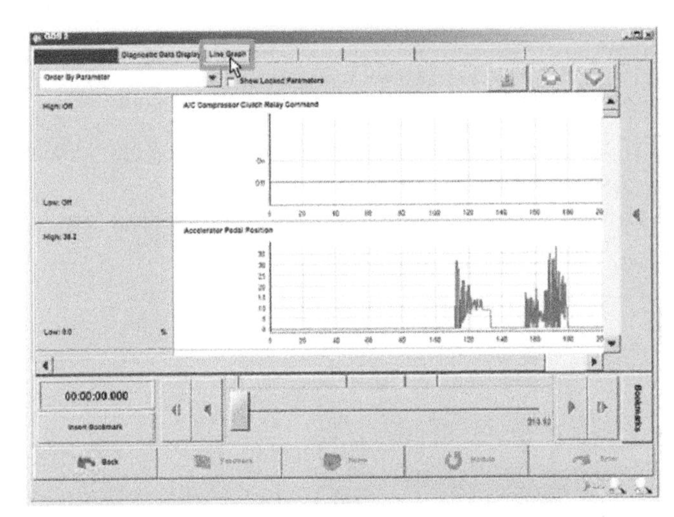

FIGURE 5–11 Although it is not an oscilloscope, GDS2 has graphing features as part of the software package. (Courtesy of General Motors)

CAUTION: Check the instructions for the scope being used before attempting to scope household AC circuits. Some scopes, such as the Snap-On MODIS, are not designed to measure high-voltage AC circuits.

GRAPHING SCAN TOOLS

Many scan tools are capable of displaying the voltage levels captured by the scan tool through the data link connector (DLC) on a screen. This feature is helpful where seeing changes in voltage levels is difficult to detect by looking at numbers that are constantly changing. Read and follow the instructions for the scan tool being used. ● **SEE FIGURE 5–10.**

General Motors Global Diagnostic System 2 (GDS2) has extensive graphing capabilities, allowing the technician to select and overlay many data parameters for a detailed look at system operation. ● **SEE FIGURE 5–11.**

SUMMARY

1. Analog oscilloscopes use a cathode ray tube to display voltage patterns.
2. The waveforms shown on an analog oscilloscope cannot be stored for later viewing.
3. A digital storage oscilloscope (DSO) creates an image or waveform on the display by connecting thousands of dots captured by the scope leads.
4. An oscilloscope display grid is called a graticule. Each of the 8 × 10 or 10 × 10 dividing boxes is called a division.
5. Setting the time base means establishing the amount of time each division represents.
6. Setting the volts per division allows the technician to view either the entire waveform or just part of it.
7. DC coupling and AC coupling are two selections that can be made to observe different types of waveforms.
8. A graphing scan tool may have a built-in oscilloscope or graphing capabilities for data parameters.
9. Oscilloscopes display voltage over time. A DSO can capture and store a waveform for viewing later.

REVIEW QUESTIONS

1. What are the differences between an analog and a digital oscilloscope?
2. What is the difference between DC coupling and AC coupling?
3. Why are DC signals that change called pulse trains?
4. What is the difference between an oscilloscope and a graphing scan tool?

CHAPTER QUIZ

1. Technician A says an analog scope can store the waveform for viewing later. Technician B says that the trigger level has to be set on most scopes to be able to view a changing waveform. Which technician is correct?
   - **a.** Technician A only
   - **b.** Technician B only
   - **c.** Both Technicians A and B
   - **d.** Neither Technician A nor B
2. An oscilloscope display is called a _______.
   - **a.** Grid
   - **b.** Graticule
   - **c.** Division
   - **d.** Box
3. A signal showing the voltage of a battery displayed on a digital storage oscilloscope (DSO) is being discussed. Technician A says that the display will show one horizontal line above the zero line. Technician B says that the display will show a line sloping upward from zero to the battery voltage level. Which technician is correct?
   - **a.** Technician A only
   - **b.** Technician B only
   - **c.** Both Technicians A and B
   - **d.** Neither Technician A nor B
4. Setting the time base to 50 ms per division will allow the technician to view a waveform how long in duration?
   - **a.** 50 ms
   - **b.** 200 ms
   - **c.** 400 ms
   - **d.** 500 ms
5. A throttle position sensor waveform is going to be observed. At what setting should the volts per division be set to see the entire waveform from 0 to 5 volts?
   - **a.** 0.5 V/div
   - **b.** 1.0 V/div
   - **c.** 2.0 V/div
   - **d.** 5.0 V/div
6. Two technicians are discussing the DC coupling setting on a DSO. Technician A says that the position allows both the DC and AC signals of the waveform to be displayed. Technician B says that this setting allows just the DC part of the waveform to be displayed. Which technician is correct?
   - **a.** Technician A only
   - **b.** Technician B only
   - **c.** Both Technicians A and B
   - **d.** Neither Technician A nor B
7. Voltage signals (waveforms) that do not go below zero are called _______.
   - **a.** AC signals
   - **b.** Pulse trains
   - **c.** Pulse width
   - **d.** DC coupled signals
8. Cycles per second are expressed in _______.
   - **a.** Hertz
   - **b.** Duty cycle
   - **c.** Pulse width
   - **d.** Slope
9. Oscilloscopes use what type of lead connector?
   - **a.** Banana plugs
   - **b.** Double banana plugs
   - **c.** Single conductor plugs
   - **d.** BNC
10. GDS2 is a type of _______.
    - **a.** Scan tool
    - **b.** Oscilloscope
    - **c.** Computer
    - **d.** Connector

chapter 6
AUTOMOTIVE WIRING AND WIRE REPAIR

LEARNING OBJECTIVES

After studying this chapter, the reader will be able to:

1. Explain the wire gauge numbering system.
2. Describe how fusible links and fuses protect circuits and wiring.
3. List the steps for performing a proper wire repair.
4. Perform solder repair of electrical wiring.
5. Discuss circuit breakers and PTC electronic circuit protection devices.
6. Explain the types of electrical conduit.

This chapter will help you prepare for the ASE Electrical/Electronic Systems (A6) certification test content area "A" (General Electrical/Electronic System Diagnosis).

GM STC OBJECTIVES

GM Service Technical College topics covered in this chapter are as follows:

1. Identify the different types of TPAs.
2. Identify the different types of CPAs.
3. Identify handling procedures for the different types of TPAs and CPAs.
4. Identify tools required to release a terminal.
5. Identify tools required for terminal replacement.
6. Demonstrate the appropriate method for performing a drag test to identify the integrity of terminals in a connection system.
7. Complete a wire repair incorporating a splice clip.
8. Correctly install a wiring splice sleeve.
9. Correctly replace a connector terminal.

KEY TERMS

Adhesive-lined heat shrink tubing 78
American wire gauge (AWG) 66
Auto link 70
Battery cables 68
Braided ground straps 67
Circuit breakers 71
Cold solder joint 78
CPA 74
Crimp-and-seal connectors 79
Drag test 76
Fuse link 70
Fuses 69
Fusible link 72
Heat shrink tubing 78
Jumper cables 68
Lock tang 75
Metric wire gauge 66
Pacific fuse element 70
Primary wire 67
PTC circuit protection 72
Rosin-core solder 77
Terminal 74
Twisted pair 68

1.	Silver
2.	Copper
3.	Gold
4.	Aluminum
5.	Tungsten
6.	Zinc
7.	Brass (copper and zinc)
8.	Platinum
9.	Iron
10.	Nickel
11.	Tin
12.	Steel
13.	Lead

CHART 6–1

The list of relative conductivity of metals, showing silver to be the best.

WIRE GAUGE DIAMETER TABLE	
AMERICAN WIRE GAUGE (AWG)	WIRE DIAMETER IN INCHES
20	0.03196118
18	0.040303
16	0.0508214
14	0.064084
12	0.08080810
10	0.10189
8	0.128496
6	0.16202
5	0.18194
4	0.20431
3	0.22942
2	0.25763
1	0.2893
0	0.32486
00	0.3648

CHART 6–2

American wire gauge (AWG) number and the actual conductor diameter in inches.

AUTOMOTIVE WIRING

DEFINITION AND TERMINOLOGY Most automotive wire is made from strands of copper covered by plastic insulation. Copper is an excellent conductor of electricity that is reasonably priced and very flexible. However, solid copper wire can break when moved repeatedly; therefore, most copper wiring is constructed of multiple small strands that allow for repeated bending and moving without breaking. Solid copper wire is generally used for components such as starter armature and alternator stator windings that do not bend or move during normal operation. Copper is the best electrical conductor besides silver, which is a great deal more expensive. The conductivity of various metals is rated in ● **CHART 6–1**.

AMERICAN WIRE GAUGE Wiring is sized and purchased according to gauge size as assigned by the **American wire gauge (AWG)** system. AWG numbers can be confusing because as the gauge number *increases,* the size of the conductor wire *decreases*. Therefore, a 14 gauge wire is smaller than a 10 gauge wire. The *greater* the amount of current (in amperes) that is flowing through a wire, the *larger the diameter (smaller gauge number)* that will be required. ● **SEE CHART 6–2**, which compares the AWG number to the actual wire diameter in inches. The diameter refers to the diameter of the metal conductor and does not include the insulation.

Following are general applications for the most commonly used wire gauge sizes. Always check the installation instructions or the manufacturer's specifications for wire gauge size before replacing any automotive wiring.

- 20 to 22 gauge: radio speaker wires
- 18 gauge: small bulbs and short leads
- 16 gauge: taillights, gas gauge, turn signals, windshield wipers
- 14 gauge: horn, radio power lead, headlights, cigarette lighter, brake lights
- 12 gauge: headlight switch-to-fuse box, rear window defogger, power windows and locks
- 10 gauge: alternator-to-battery
- 4, 2, or 0 (1/0) gauge: battery cables

METRIC WIRE GAUGE Most manufacturers indicate on the wiring diagrams the **metric wire gauge** sizes measured in square millimeters (mm^2) of cross-sectional area. The following chart gives conversions or comparisons between metric gauge and AWG sizes. Notice that the metric wire size increases with size (area), whereas the AWG size gets smaller with larger size wire. ● **SEE CHART 6–3.**

The AWG number should be decreased (wire size increased) with increased lengths of wire. ● **SEE CHART 6–4.**

For example, a trailer may require 14 gauge wire to light all the trailer lights, but if the wire required is over 25 feet long,

METRIC SIZE (MM²)	AWG SIZE
0.5	20
0.8	18
1.0	16
2.0	14
3.0	12
5.0	10
8.0	8
13.0	6
19.0	4
32.0	2
52.0	0

CHART 6–3

Metric wire size in squared millimeters (mm^2) conversion chart to American wire gauge (AWG).

12 V	RECOMMENDED WIRE GAUGE (AWG) (FOR LENGTH IN FEET)*						
AMPS	3′	5′	7′	10′	15′	20′	25′
5	18	18	18	18	18	18	18
7	18	18	18	18	18	18	16
10	18	18	18	18	16	16	16
12	18	18	18	18	16	16	14
15	18	18	18	18	14	14	12
18	18	18	16	16	14	14	12
20	18	18	16	16	14	12	10
22	18	18	16	16	12	12	10
24	18	18	16	16	12	12	10
30	18	16	16	14	10	10	10
40	18	16	14	12	10	10	8
50	16	14	12	12	10	10	8
100	12	12	10	10	6	6	4
150	10	10	8	8	4	4	2
200	10	8	8	6	4	4	2

* When mechanical strength is a factor, use the next larger wire gauge.

CHART 6–4

Recommended AWG wire size increases as the length increases because all wires have internal resistance. The longer the wire is, the greater the resistance. The larger the diameter is, the lower the resistance.

12 gauge wire should be used. Most automotive wire, except for spark plug wire, is often called **primary wire** (named for the voltage range used in the primary ignition circuit) because it is designed to operate at or near battery voltage.

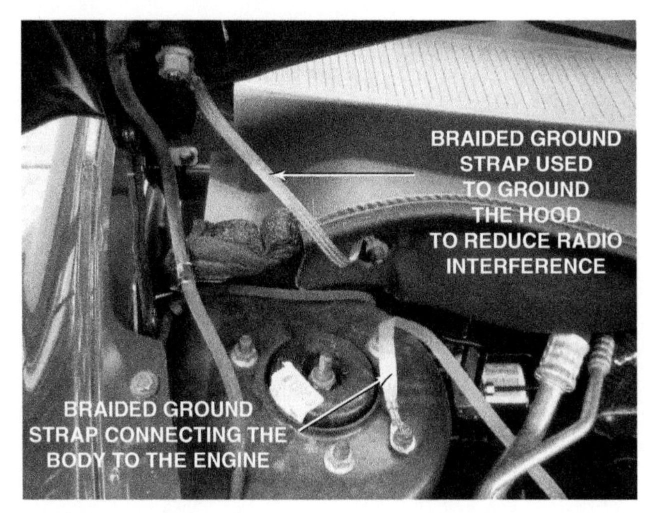

FIGURE 6–1 All lights and accessories ground to the body of the vehicle. Body ground wires such as this one are needed to conduct all of the current from these components back to the negative terminal of the battery. The body ground wire connects the body to the engine. Most battery negative cables attach to the engine.

FREQUENTLY ASKED QUESTION

Do They Make 13 Gauge Wire?

Yes. AWG sizing of wire includes all gauge numbers, including 13, even though the most commonly used sizes are even numbered, such as 12, 14, or 16.

Because the sizes are so close, wire in every size is not commonly stocked, but can be ordered for a higher price. Therefore, if a larger wire size is needed, it is common practice to select the next lower, even-numbered gauge.

GROUND WIRES

PURPOSE AND FUNCTION All vehicles use ground wires between the engine and body and/or between the body and the negative terminal of the battery. The two types of ground wires are:

- Insulated copper wire
- Braided ground straps

Braided grounds straps are uninsulated. It is not necessary to insulate a ground strap because it does not matter if it touches metal, as it already attaches to ground. Braided ground straps are more flexible than stranded wire. Because the engine will move slightly on its mounts, the braided ground strap must be able to flex without breaking. ● **SEE FIGURE 6–1**.

? FREQUENTLY ASKED QUESTION

What Is a Twisted Pair?

A **twisted pair** is used to transmit low-voltage signals using two wires that are twisted together. Electromagnetic interference can create a voltage in a wire and twisting the two signal wires cancels out the induced voltage. A twisted pair means that the two wires have at least nine turns per foot (turns per meter). A rule of thumb is a twisted pair should have one twist per inch of length.

FIGURE 6–2 Battery cables are designed to carry heavy starter current and are therefore usually 4 gauge or larger wire. Note that this battery has a thermal blanket covering to help protect the battery from high underhood temperatures. The wiring is also covered with plastic conduit called split-loom tubing.

SKIN EFFECT The braided strap also dampens out some radio-frequency interference that otherwise might be transmitted through standard stranded wiring due to the skin effect.

The skin effect is the term used to describe how high-frequency AC electricity flows through a conductor. Direct current flows through a conductor, but alternating current tends to travel through the outside (skin) of the conductor. Because of the skin effect, most audio (speaker) cable is constructed of many small-diameter copper wires instead of fewer larger strands, because the smaller wire has a greater surface area and therefore results in less resistance to the flow of AC voltage.

NOTE: Body ground wires are necessary to provide a circuit path for the lights and accessories that ground to the body and flow to the negative battery terminal.

BATTERY CABLES

Battery cables are the largest wires used in the automotive electrical system. The cables are usually 4 gauge, 2 gauge, or 1 gauge wires (19 mm^2 or larger). ● **SEE FIGURE 6–2**.

Wires larger than 1 gauge are called 0 gauge (pronounced "ought"). Larger cables are labeled 2/0 or 00 (2 ought) and 3/0 or 000 (3 ought). Electrical systems that are 6 volts require battery cables two sizes larger than those used for 12-volt electrical systems, because the lower voltage used in antique vehicles resulted in twice the amount of current (amperes) to supply the same electrical power.

JUMPER CABLES

Jumper cables are 4 to 2/0 gauge electrical cables with large clamps attached and are used to connect the discharged battery of one vehicle to the good battery of another vehicle. Good-quality jumper cables are necessary to prevent excessive voltage drops caused by cable resistance. Aluminum wire jumper cables should not be used, because even though aluminum is a good electrical conductor (although not as good as copper), it is less flexible and can crack and break when bent or moved repeatedly. The size should be 6 gauge or larger.

1/0 AWG welding cable can be used to construct an excellent set of jumper cables using welding clamps on both ends. Welding cable is usually constructed of many very fine strands of wire, which allow for easier bending of the cable as the strands of fine wire slide against each other inside the cable.

NOTE: Always check the wire gauge of any battery cables or jumper cables and do not rely on the outside diameter of the wire. Many lower-cost jumper cables use smaller gauge wire, but may use thick insulation to make the cable look as if it is the correct size wire.

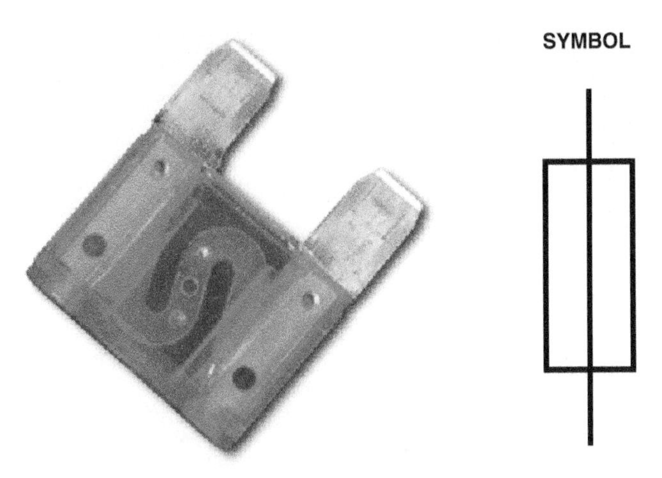

FIGURE 6–3 A fuse, showing the schematic symbol. (Courtesy of General Motors)

FIGURE 6–4 This underhood bussed electrical center (UBEC) contains fuses and relays. (Courtesy of Jeffrey Rehkopf)

FUSES AND CIRCUIT PROTECTION DEVICES

CONSTRUCTION **Fuses** should be used in every circuit to protect the wiring from overheating and damage caused by excessive current flow as a result of a short circuit or other malfunction. The schematic symbol for a fuse is shown in ● **FIGURE 6–3**.

A fuse is constructed of a fine tin conductor inside a glass, plastic, or ceramic housing. The tin is designed to melt and open the circuit if excessive current flows through the fuse. Each fuse is rated according to its maximum current-carrying capacity.

Many fuses are used to protect more than one circuit of the automobile. Fuses are located in a fuse block or electrical center (also called a bussed electrical center) along with other components such as circuit relays. The locations of the fuse blocks or electrical centers on the vehicle are typically under the hood, within the instrument panel (I/P), in the passenger compartment (center console), or the body rear (under the seat or in the trunk area). ● **SEE FIGURE 6–4**.

A typical example is the fuse for the cigarette lighter that also protects many other circuits, such as those for the courtesy lights, clock, and other circuits. A fault in one of these circuits can cause this fuse to melt, which will prevent the operation of all other circuits that are protected by the fuse.

NOTE: The SAE term for a cigarette lighter is *cigar lighter* because the diameter of the heating element is large enough for a cigar. The term *cigarette lighter* will be used throughout this book because it is the most common usage.

NORMAL CURRENT IN THE CIRCUIT (AMPERES)	FUSE RATING (AMPERES)
7.5	10
16	20
24	30

CHART 6–5

The fuse rating should be 20% higher than the maximum current in the circuit to provide the best protection for the wiring and the component being protected.

FUSE RATINGS Fuses are used to protect the wiring and components in the circuit from damage if an excessive amount of current flows. The fuse rating is normally about 20% higher than the normal current in the circuit. See ● **CHART 6–5** for a typical fuse rating based on the normal current in the circuit. In other words, the normal current flow should be about 80% of the fuse rating.

BLADE FUSES Colored blade-type fuses are also referred to as ATO fuses and have been used since 1977. The color of the plastic of blade fuses indicates the maximum current flow, measured in amperes.

See ● **CHART 6–6** for the color and the amperage rating of blade fuses.

Each fuse has an opening in the top of its plastic portion to allow access to its metal contacts for testing purposes. ● **SEE FIGURE 6–5**.

MINI FUSES To save space, many vehicles use mini (small) blade fuses. ● **SEE FIGURE 6–6**. Not only do they save space but they also allow the vehicle design engineer to fuse individual circuits instead of grouping many different components

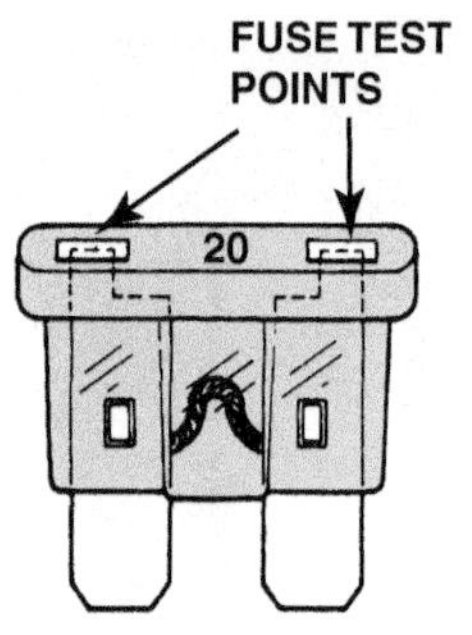

FIGURE 6–5 Blade-type fuses can be tested through openings in the plastic at the top of the fuse.

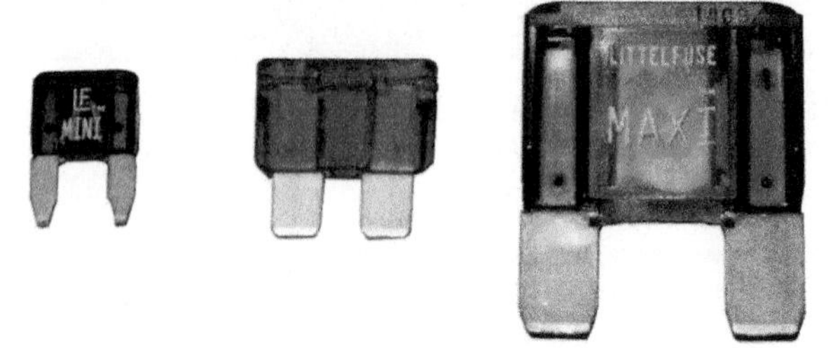

FIGURE 6–6 Three sizes of blade-type fuses: mini on the left, standard or ATO type in the center, and maxi on the right.

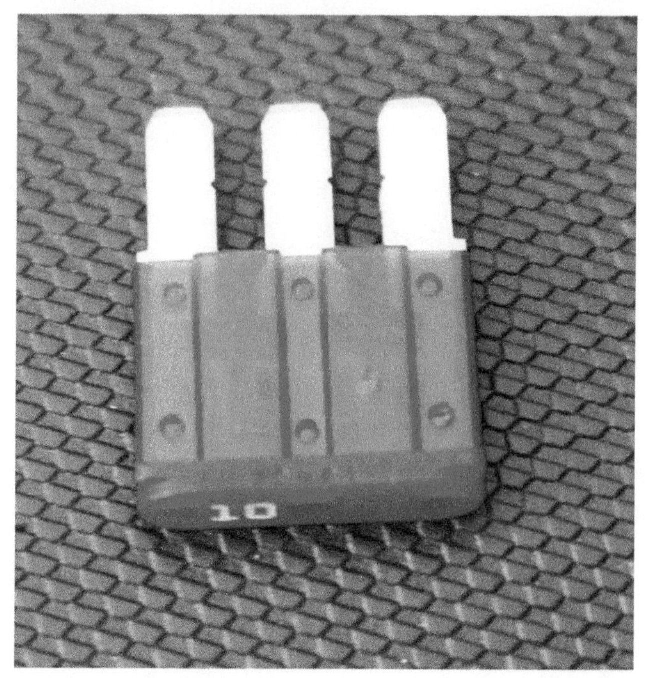

FIGURE 6–7 The micro3 fuse is used to power two circuits, with the power feeding from the center leg to each of the outer legs. (Courtesy of Jeffrey Rehkopf and Nimnicht Chevrolet)

AMPERAGE RATING	COLOR
1	Dark green
2	Gray
2.5	Purple
3	Violet
4	Pink
5	Tan
6	Gold
7.5	Brown
9	Orange
10	Red
14	Black
15	Blue
20	Yellow
25	White
30	Green

CHART 6–6

The amperage rating and the color of the blade fuse are standardized.

on one fuse. This improves customer satisfaction because if one component fails, it only affects that one circuit without stopping electrical power to several other circuits as well. This makes troubleshooting a lot easier too, because each circuit is separate. ● **SEE CHART 6–7** for the amperage rating and corresponding fuse color for mini fuses.

AMPERAGE RATING	COLOR
5	Tan
7.5	Brown
10	Red
15	Blue
20	Yellow
25	Natural
30	Green

CHART 6–7

Mini fuse amperage rating and colors.

MICRO 3 FUSES Some vehicles may use a micro3 size fuse on certain circuits. This fuse is similar to a mini fuse except that it has three blades instead of two. The center blade is the voltage feed (12 volts) and each outside leg feeds a different circuit. ● **SEE FIGURE 6–7**.

MAXI FUSES Maxi fuses are a large version of blade fuses and are used to replace fusible links in many vehicles. Maxi fuses are rated up to 80 amperes or more. ● **SEE CHART 6–8** for the amperage rating and corresponding color for maxi fuses.

PACIFIC FUSE ELEMENT First used in the late 1980s, **Pacific fuse elements** (also called a **fuse link** or **auto link**) are used to protect wiring from a direct short-to-ground. The

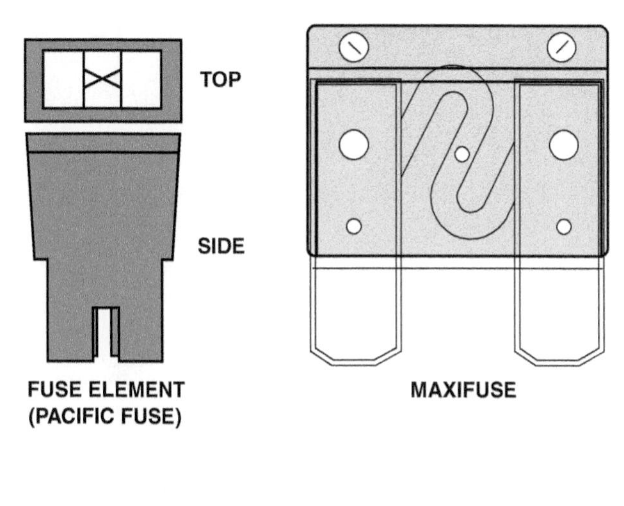

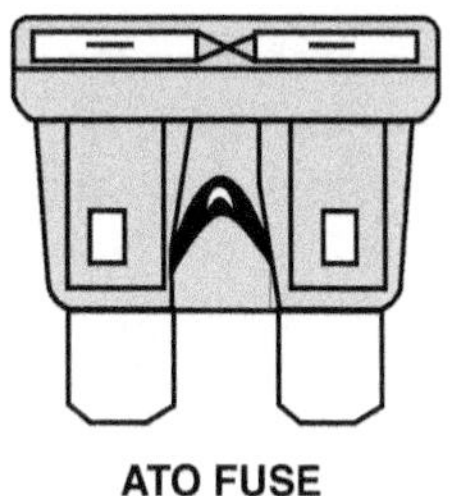

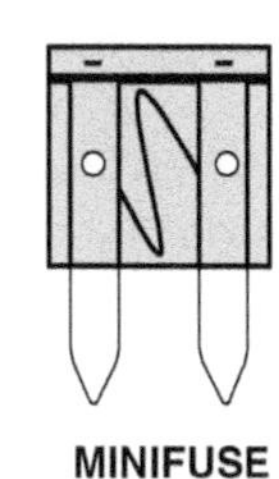

FIGURE 6–8 A comparison of the various types of protective devices used in most vehicles.

AMPERAGE RATING	COLOR
20	Yellow
30	Green
40	Amber
50	Red
60	Blue
70	Brown
80	Natural

CHART 6–8

Maxi fuse amperage rating and colors.

housing contains a short link of wire sized for the rated current load. The transparent top allows inspection of the link inside. ● **SEE FIGURE 6–8**.

TESTING FUSES It is important to test the condition of a fuse if the circuit being protected by the fuse does not operate. Most blown fuses can be detected quickly because the center conductor is melted. Fuses can also fail and open the circuit because of a poor connection in the fuse itself or in the fuse holder. Therefore, just because a fuse "looks okay" does not mean that it *is* okay. All fuses should be tested with a test light. The test light should be connected to first one side of the fuse and then the other. A test light should light on both sides. If the test light only lights on one side, the fuse is blown or open. If the test light does not light on either side of the fuse, then that circuit is not being supplied power. ● **SEE FIGURE 6–9**. An ohmmeter can also be used to test fuses.

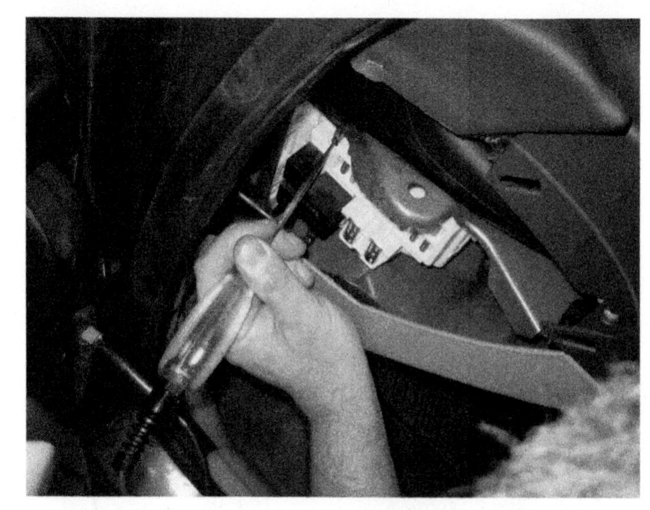

FIGURE 6–9 To test a fuse, use a test light to check for power at the power side of the fuse. The ignition switch and lights may have to be on before some fuses receive power. If the fuse is good, the test light should light on both sides (power side and load side) of the fuse.

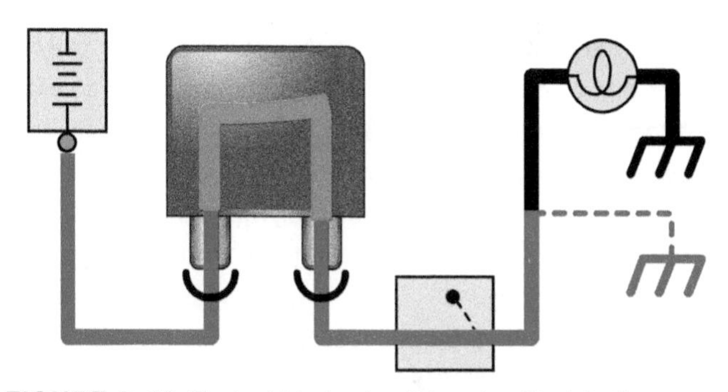

FIGURE 6–10 Typical blade circuit breaker fits into the same space as a blade fuse. If excessive current flows through the bimetallic strip, the strip bends and opens the contacts and stops current flow. When the circuit breaker cools, the contacts close again, completing the electrical circuit. (Courtesy of General Motors)

CIRCUIT BREAKERS Circuit breakers are used to prevent harmful overload (excessive current flow) in a circuit by opening the circuit and stopping the current flow to prevent overheating and possible fire caused by hot wires or electrical components. **Circuit breakers** are mechanical units made of two different metals (bimetallic) that deform when heated and open a set of contact points that work in the same manner as an "off" switch. ● **SEE FIGURE 6–10**.

Cycling-type circuit breakers, therefore, are reset when the current stops flowing, which causes the bimetallic strip to

FIGURE 6–11 Older electrical symbols for a circuit breaker.

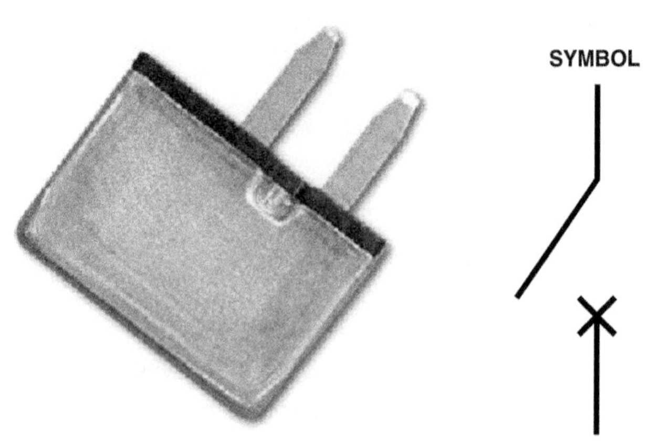

FIGURE 6–12 A circuit breaker and its schematic symbol. (Courtesy of General Motors)

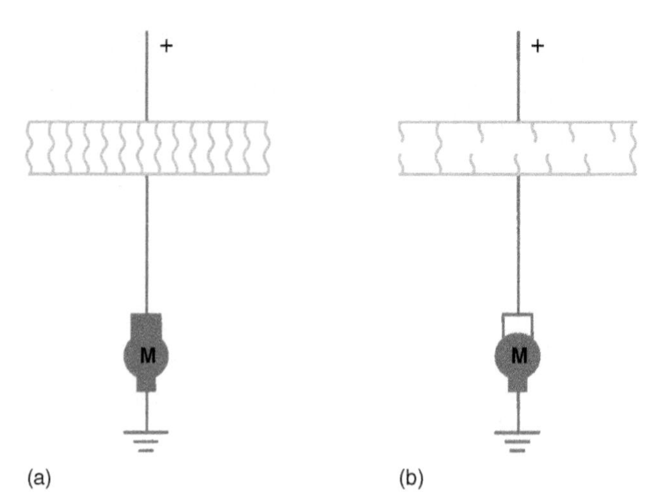

FIGURE 6–13 (a) The normal operation of a PTC circuit protector such as in a power window motor circuit showing the many conducting paths. With normal current flow, the temperature of the PTC circuit protector remains normal. (b) When current exceeds the amperage rating of the PTC circuit protector, the polymer material that makes up the electronic circuit protector increases in resistance. As shown, a high-resistance electrical path still exists even though the motor will stop operating as a result of the very low current flow through the very high resistance. The circuit protector will not reset or cool down until voltage is removed from the circuit.

cool and the circuit to close again. A circuit breaker is used in circuits that could affect the safety of passengers if a conventional nonresetting fuse were used. The headlight circuit is an excellent example of the use of a circuit breaker rather than a fuse. A short or grounded circuit anywhere in the headlight circuit could cause excessive current flow and, therefore, the opening of the circuit. Obviously, a sudden loss of headlights at night could have disastrous results. A circuit breaker opens and closes the circuit rapidly, thereby protecting the circuit from overheating and also providing sufficient current flow to maintain at least partial headlight operation.

Circuit breakers are also used in other circuits where conventional fuses could not provide for the surges of high current commonly found in those circuits. ● **SEE FIGURES 6–11 AND 6–12** for the electrical symbols used to represent a circuit breaker.

Examples are the circuits for the following accessories.

1. Power seats
2. Power door locks
3. Power windows

PTC CIRCUIT PROTECTORS **Positive temperature coefficient (PTC) circuit protectors** are solid state (without moving parts). Like all other circuit protection devices, PTCs are installed in series in the circuit being protected. If excessive current flows, the temperature and resistance of the PTC increase.

This increased resistance reduces current flow (amperes) in the circuit and may cause the electrical component in the circuit not to function correctly. For example, when a PTC circuit protector is used in a power window circuit, the increased resistance causes the operation of the power window to be much slower than normal.

Unlike circuit breakers or fuses, PTC circuit protection devices do *not* open the circuit, but rather provide a very high resistance between the protector and the component. ● **SEE FIGURE 6–13.**

The electronic control units (computers) used in most vehicles today incorporate overload protection devices. Therefore, when a component fails to operate, do not blame the computer. The current control device is controlling current flow to protect the computer. Components that do not operate correctly should be checked for proper resistance and current draw.

FUSIBLE LINKS A **fusible link** is a type of fuse that consists of a short length (6 to 9 inch long) of standard copper-strand wire covered with a special nonflammable insulation. This wire is usually four wire numbers smaller than the wire of the circuits it protects. For example, a 12 gauge circuit is protected by a 16 gauge fusible link. The special thick insulation over the wire may make it look larger than other wires of the same gauge number. ● **SEE FIGURE 6–14.**

If excessive current flow (caused by a short-to-ground or a defective component) occurs, the fusible link will melt in half and open the circuit to prevent a fire hazard. Some fusible links

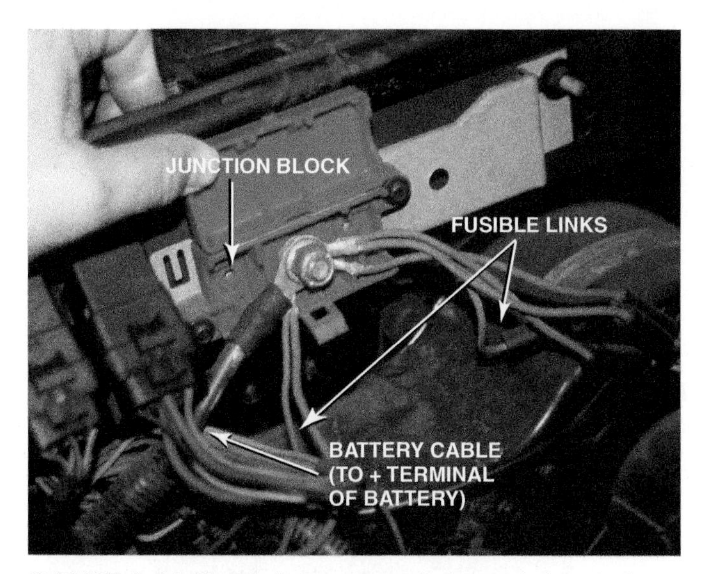

FIGURE 6–14 Fusible links are usually located close to the battery and are usually attached to a junction block. Notice that they are only 6 to 9 inch long and feed more than one fuse from each fusible link.

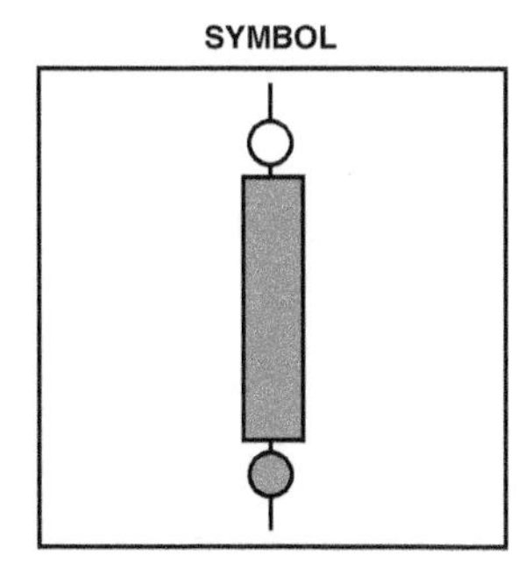

FIGURE 6–15 The electrical symbol for a fuse link. (Courtesy of General Motors)

are identified with "fusible link" tags at the junction between the fusible link and the standard chassis wiring, which represent only the junction. Fusible links are the backup system for circuit protection. All current except the current used by the starter motor flows through fusible links and then through individual circuit fuses. It is possible that a fusible link will melt and not blow a fuse. Fusible links are installed as close to the battery as possible so that they can protect the wiring and circuits coming directly from the battery. ● **SEE FIGURE 6–15.**

MEGA FUSES Many newer vehicles are equipped with mega fuses instead of fusible links to protect high-amperage circuits. Circuits often controlled by mega fuses include:

- Charging circuit
- HID headlights
- Heated front or rear glass
- Multiple circuits usually protected by mega fuses

FIGURE 6–16 A 125 ampere rated mega fuse used to protect the circuit from the alternator.

- Mega fuse rating for vehicles, including 80, 100, 125, 150, 175, 200, 225, and 250 amperes

● **SEE FIGURE 6–16.**

CHECKING FUSIBLE LINKS AND MEGA FUSES Fusible links and mega fuses are usually located near where electrical power is sent to other fuses or circuits, such as:

- Starter solenoid battery terminals
- Power distribution centers
- Output terminals of alternators
- Positive terminals of the battery

Fusible links can melt and not show any external evidence of damage. To check a fusible link, gently pull on each end to see if it stretches. If the insulation stretches, then the wire inside has melted and the fusible link must be replaced after determining what caused the link to fail.

Another way to check a fusible link is to use a test light or a voltmeter and check for available voltage at both ends of the fusible link. If voltage is available at only one end, then the link is electrically open and should be replaced.

REPLACING A FUSIBLE LINK If a fusible link is found to be melted, perform the following steps.

STEP 1 Determine why the fusible link failed and repair the fault.

STEP 2 Check service information for the exact length, gauge, and type of fusible link required.

STEP 3 Replace the fusible link with the specified fusible link wire and according to the instructions found in the service information.

CAUTION: Always use the *exact* length of fusible link wire required because if it is too short, it will not have enough resistance to generate the heat needed to melt the wire and protect the circuits or components. If the wire is too long, it could melt during normal operation of the circuits it is protecting. Fusible link wires are usually longer than 6 inch and shorter than 9 inch.

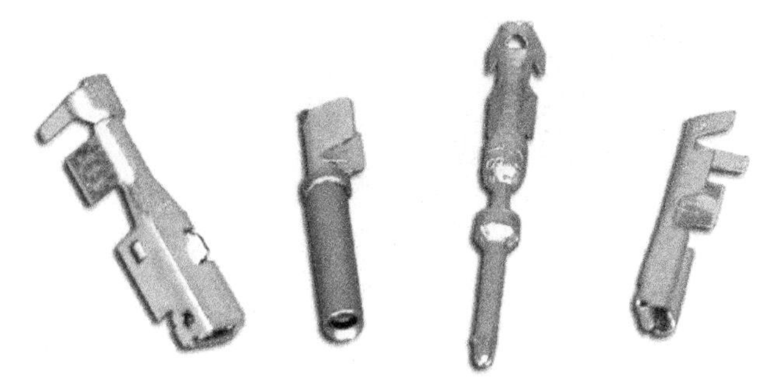

FIGURE 6–17 A wire terminal is a metal component that makes the electrical connection. (Courtesy of General Motors)

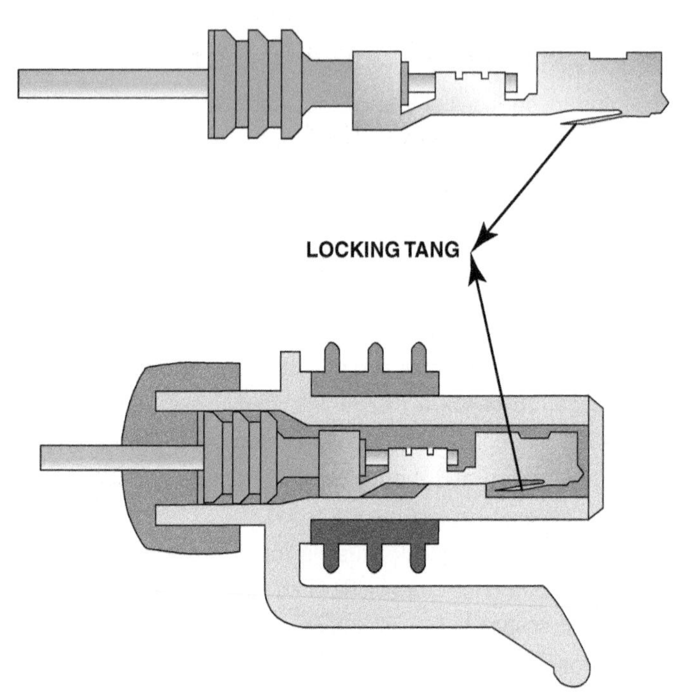

FIGURE 6–18 The terminal snaps into the plastic connector body. (Courtesy of General Motors)

TERMINALS AND CONNECTORS

A **terminal** is a metal fastener attached to the end of a wire, which makes the electrical connection. The term "connector" usually refers to the plastic portion that snaps or connects together, thereby making the mechanical connection. Wire terminal ends usually snap into and are held by a connector. ● **SEE FIGURES 6–17 AND 6–18**. Male and female connectors can then be snapped together, thereby completing an electrical connection. Connectors exposed to the environment are also equipped with a weather-tight seal. ● **SEE FIGURE 6–19**.

There are many different designs of terminals and connectors. Always refer to service information (SI) for details.

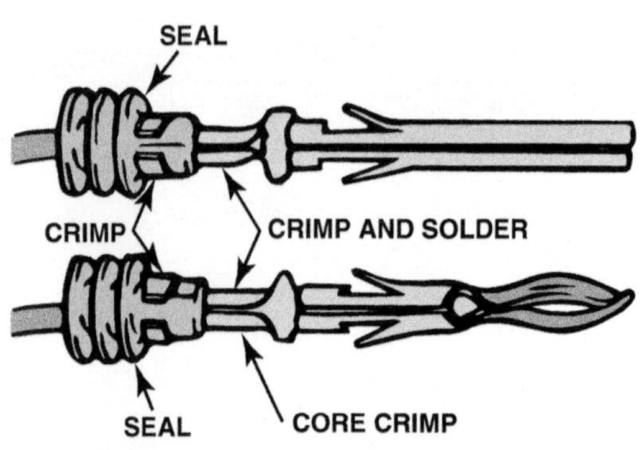

FIGURE 6–19 Some terminals have seals attached to help seal the electrical connections.

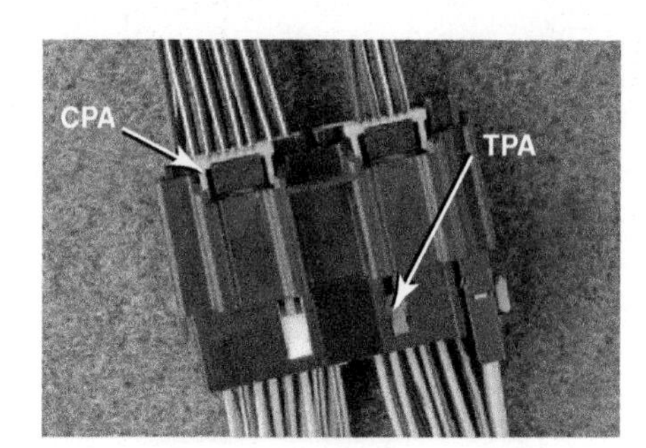

FIGURE 6–20 The CPA must be removed before the release tab can be pressed. The TPA secures the wire terminals into the connector body. (Courtesy of General Motors)

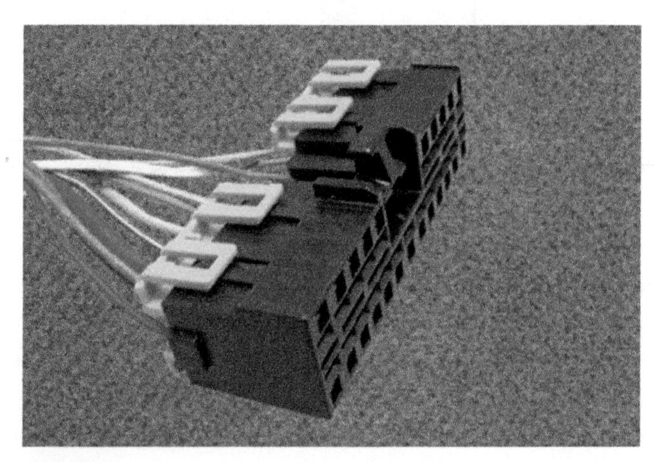

FIGURE 6–21 The TPA clips onto the back of the connector to help retain the terminals into the connector body. (Courtesy of General Motors)

Important connections are secured by using a secondary retaining device, called a **connector position assurance (CPA)** clip or plug. The CPA keeps the latch of the connector from releasing accidentally. ● **SEE FIGURE 6–20**.

Most connectors also include a terminal position assurance (TPA) clip or comb that retains the wire terminals into the connector. The TPA must be removed before the terminals can be removed from the connector for repair. ● **SEE FIGURE 6–21**.

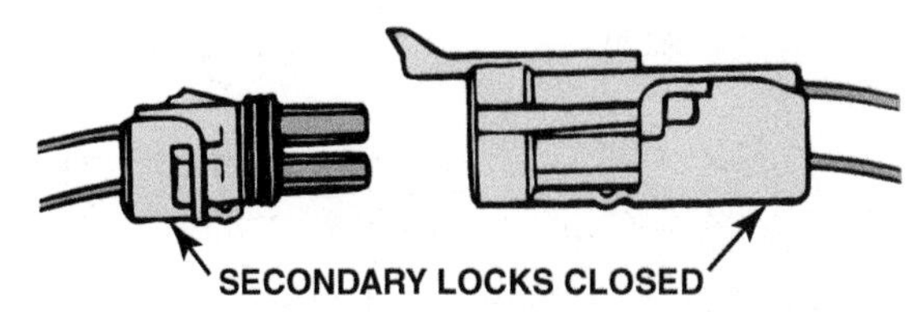

FIGURE 6–22 Separate a connector by opening the lock and pulling the two apart.

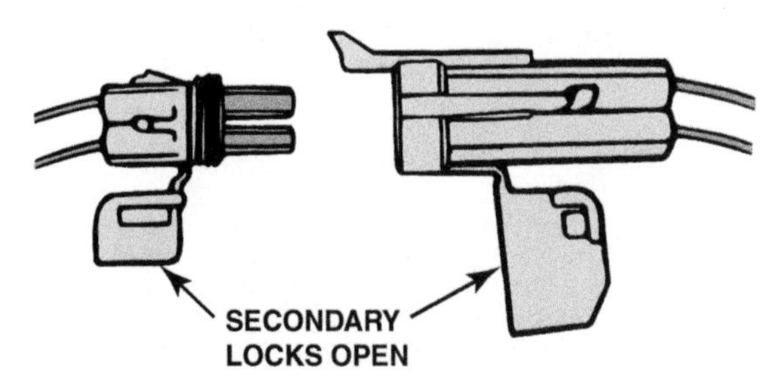

FIGURE 6–23 The secondary locks help retain the terminals in the connector.

TECH TIP

Find the Root Cause

If a mega fuse or fusible link fails, find the root cause before replacing it. A mega fuse can fail due to vibration or physical damage as a result of a collision or corrosion. Check to see if the fuse itself is loose and can be moved by hand. If loose, then simply replace the mega fuse. If a fusible link or mega fuse has failed due to excessive current, check for evidence of a collision or any other reason that could cause an excessive amount of current to flow. This inspection should include each electrical component being supplied current from the fusible link. After being sure that the root cause has been found and corrected, replace the fusible link or mega fuse.

Terminals are retained in connectors by the use of a **lock tang**. Removing a terminal from a connector includes the following steps.

STEP 1 Release the connector position assurance (CPA)

STEP 2 Separate the male and female connector by opening the lock. ● **SEE FIGURE 6–22**.

STEP 3 Release the secondary lock, if equipped. ● **SEE FIGURE 6–23**.

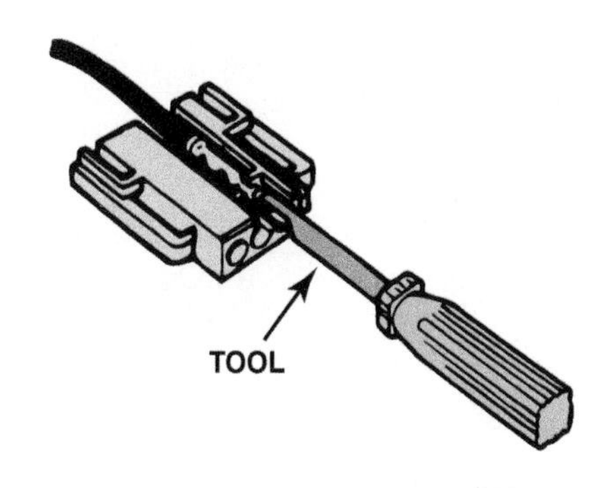

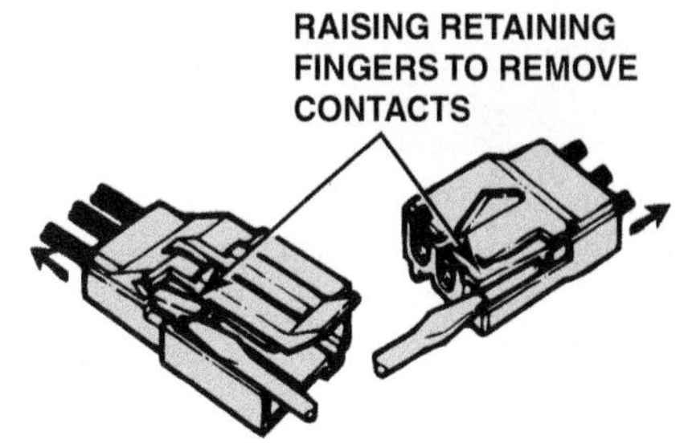

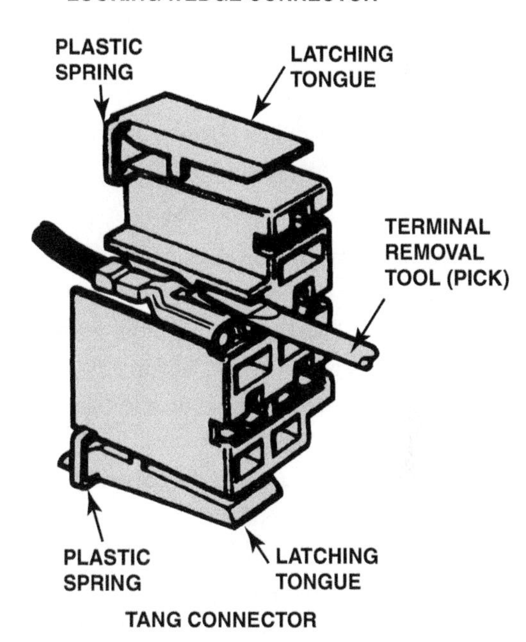

FIGURE 6–24 Use a small removal tool, sometimes called a pick, to release terminals from the connector.

STEP 4 Using a pick, look for the slot in the plastic connector where the lock tang is located, depress the lock tang, and gently remove the terminal from the connector. ● **SEE FIGURE 6–24**.

MICRO .64 TERMINALS Many GM applications use a micro 64 terminal in high density connectors. The "64" refers to the diameter of the terminal pin which is .64 mm. When working with these terminals always use the proper tools and test equipment to avoid spreading the terminals, resulting in intermittent

FIGURE 6–25 Remove the TPA (pink) and then insert the tool into the release canals; gently push and pull on the terminal until it releases. (Courtesy of General Motors)

TECH TIP

Look for the "Green Crud"

Corroded connections are a major cause of intermittent electrical problems and open circuits. The usual sequence of conditions is as follows:

1. **Heat causes expansion.** This heat can be from external sources such as connectors being too close to the exhaust system. Another possible source of heat is a poor connection at the terminal, causing a voltage drop and heat due to the electrical resistance.
2. **Condensation occurs when a connector cools.** The moisture in the condensation causes rust and corrosion.
3. **Water gets into the connector.** The solution is, if corroded connectors are noticed, the terminal should be cleaned and the condition of the electrical connection to the wire terminal end(s) confirmed. Many vehicle manufacturers recommend using a dielectric silicone or lithium-based grease inside connectors to prevent moisture from getting into and attacking the connector.

connections. To remove a terminal first remove the TPA and then use the special pick tool to release the terminal from the connector. ● **SEE FIGURE 6–25.**

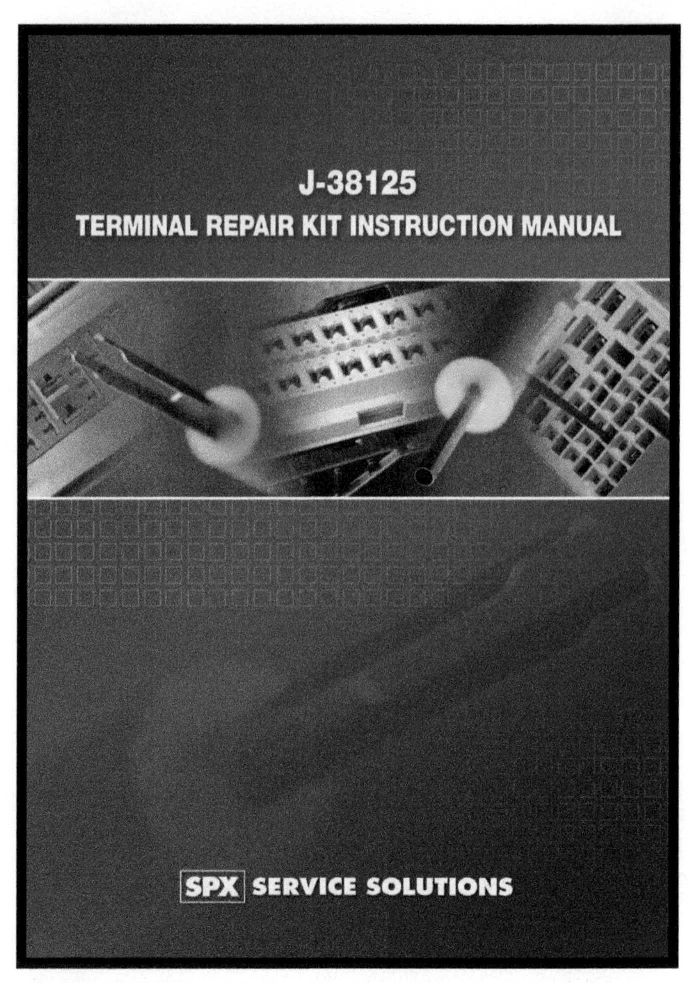

FIGURE 6–26 This manual is available from the General Motors Centre of Learning website. (Courtesy of General Motors)

TECH TIP

Read the Instructions

When servicing or repairing any GM wiring system always refer to the instructions contained in the J-38125 Terminal Repair Kit Instruction Manual. This manual shows all the connectors and terminals used by General Motors and shows how they come apart and what tools should be used when replacing damaged terminals. ● **SEE FIGURE 6–26.**

DRAG TEST A **drag test** is used to check whether the terminal has the proper fit and tension to produce a solid connection. The test uses a matching male terminal to test the fit in the female terminal. When the terminal is pulled out there should be a slight drag; also, the test terminal should not fall out if held hanging down. ● **SEE FIGURES 6–27 AND 6–28.**

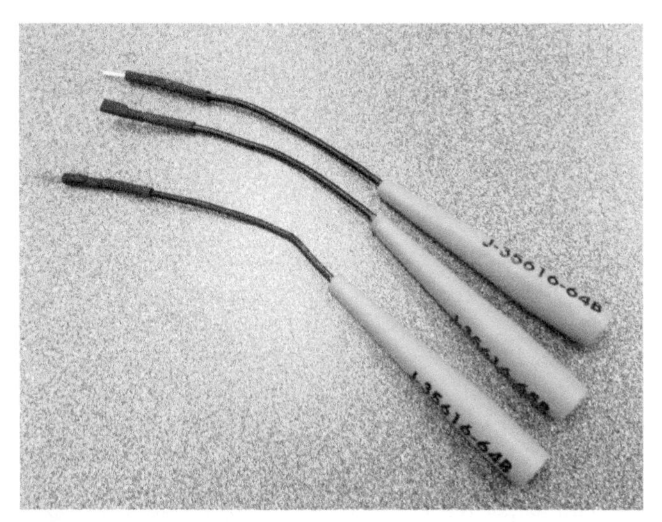

FIGURE 6–27 Micro 64 test leads. (Courtesy of Jeffrey Rehkopf)

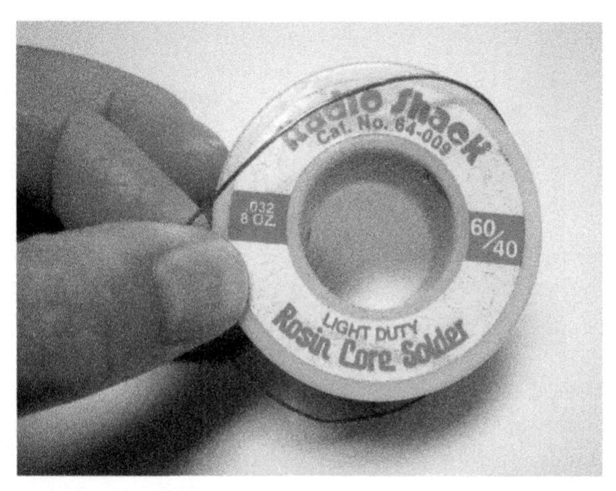

FIGURE 6–29 Always use rosin-core solder for electrical or electronic soldering. Also, use small-diameter solder for small soldering irons. Use large-diameter solder only for large-diameter (large-gauge) wire and higher-wattage soldering irons (guns).

FIGURE 6–28 When held hanging down the lead should not fall out. If it does the terminal will have to be replaced. (Courtesy of Jeffrey Rehkopf)

WIRE REPAIR

SOLDER Many manufacturers recommend that all wiring repairs be soldered. Solder is an alloy of tin and lead used to make a good electrical contact between two wires or connections in an electrical circuit. However, a flux must be used to help clean the area and to help make the solder flow. Therefore, solder is made with a resin (rosin) contained in the center, called **rosin-core solder**.

CAUTION: Never use acid-core solder to repair electrical wiring as the acid will cause corrosion.

● **SEE FIGURE 6–29.**

An acid-core solder is also available but should only be used for soldering sheet metal. Solder is available with various percentages of tin and lead in the alloy. Ratios are used to identify these various types of solder, with the first number denoting the percentage of tin in the alloy and the second number giving the percentage of lead. The most commonly used solder is 50/50, which means that 50% of the solder is tin and the other 50% is lead. The percentages of each alloy primarily determine the melting point of the solder.

- 60/40 solder (60% tin/40% lead) melts at 361°F (183°C).
- 50/50 solder (50% tin/50% lead) melts at 421°F (216°C).
- 40/60 solder (40% tin/60% lead) melts at 460°F (238°C).

NOTE: The melting points stated here can vary depending on the purity of the metals used.

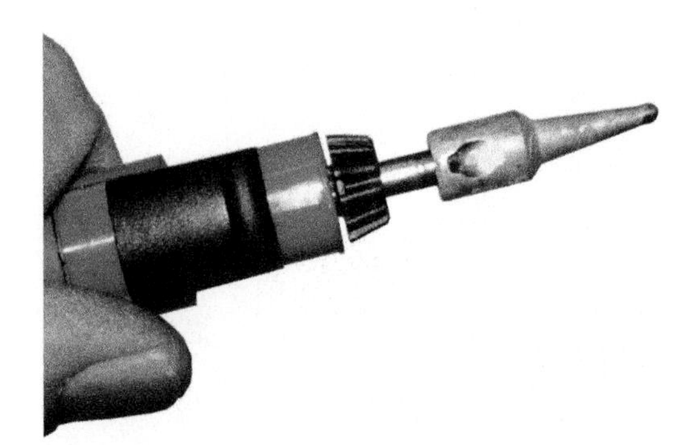

FIGURE 6–30 A butane-powered soldering tool. The cap has a built-in striker to light a converter in the tip of the tool. This handy soldering tool produces the equivalent of 60 watts of heat. It operates for about 1/2 hour on one charge from a commonly available butane refill dispenser.

Because of the lower melting point, 60/40 solder is the most highly recommended solder to use, followed by 50/50.

SOLDERING GUNS When soldering wires, be sure to heat the wires (not the solder) using:

- An electric soldering gun or soldering pencil (60 to 150 watt rating)
- Butane-powered tool that uses a flame to heat the tip (about 60 watt rating) ● **SEE FIGURE 6–30.**

SOLDERING PROCEDURE Soldering a wiring splice includes the following steps.

STEP 1 While touching the soldering gun to the splice, apply solder to the junction of the gun and the wire.

STEP 2 The solder will start to flow. Do not move the soldering gun.

STEP 3 Just keep feeding more solder into the splice as it flows into and around the strands of the wire.

STEP 4 After the solder has flowed throughout the splice, remove the soldering gun and the solder from the splice and allow the solder to cool slowly.

The solder should have a shiny appearance. Dull-looking solder may be caused by not reaching a high enough temperature, which results in a **cold solder joint**. Reheating the splice and allowing it to cool often restores the shiny appearance.

CRIMPING TERMINALS Terminals can be crimped to create a good electrical connection if the proper type of crimping tool is used. Most vehicle manufacturers recommend that a W-shaped crimp be used to force the strands of the wire into a tight space. ● **SEE FIGURE 6–31.**

Most vehicle manufacturers also specify that all hand-crimped terminals or splices be soldered. ● **SEE FIGURE 6–32.**

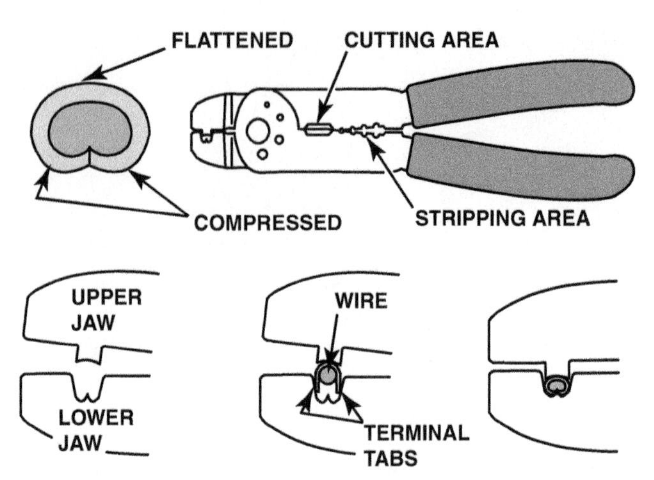

FIGURE 6–31 Notice that to create a good crimp the open part of the terminal is placed in the jaws of the crimping tool toward the anvil or the W-shape part.

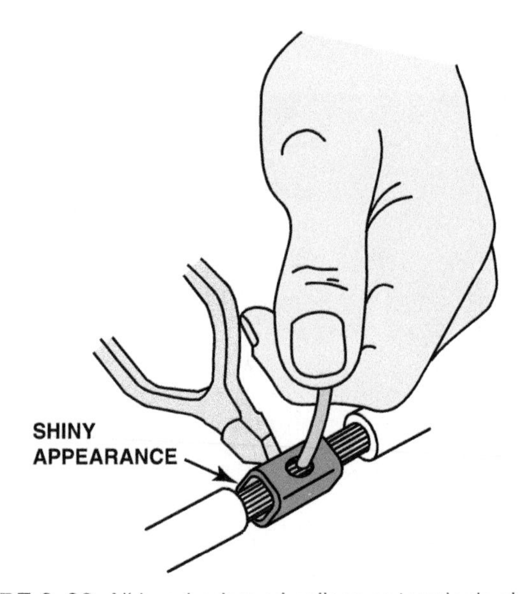

FIGURE 6–32 All hand-crimped splices or terminals should be soldered to be assured of a good electrical connection.

HEAT SHRINK TUBING **Heat shrink tubing** is usually made from polyvinyl chloride (PVC) or polyolefin and shrinks to about half of its original diameter when heated; this is usually called a 2:1 shrink ratio. Heat shrink by itself does not provide protection against corrosion, because the ends of the tubing are not sealed against moisture. DaimlerChrysler Corporation recommends that all wire repairs that may be exposed to the elements be repaired and sealed using **adhesive-lined heat shrink tubing**. The tubing is usually made from flame-retardant flexible polyolefin with an internal layer of special thermoplastic adhesive. When heated, this tubing shrinks to one-third of its original diameter (3:1 shrink ratio) and the adhesive melts and seals the ends of the tubing. ● **SEE FIGURE 6–33.**

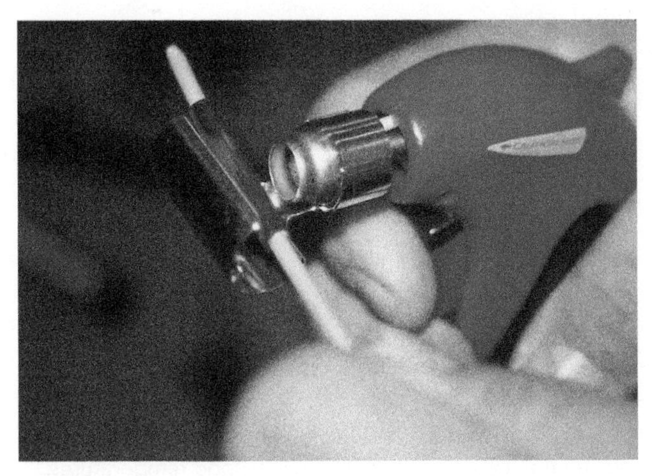

FIGURE 6–33 A butane torch especially designed for use on heat shrink applies heat without an open flame, which could cause damage.

FIGURE 6–34 A typical crimp-and-seal connector. This type of connector is first lightly crimped to retain the ends of the wires and then it is heated. The tubing shrinks around the wire splice, and thermoplastic glue melts on the inside to provide an effective weather-resistant seal.

CRIMP-AND-SEAL CONNECTORS General Motors Corporation recommends the use of crimp-and-seal connectors as the method for wire repair. **Crimp-and-seal connectors** contain a sealant and shrink tubing in one piece and are *not* simply butt connectors. ● **SEE FIGURE 6–34**.

The usual procedure specified for making a wire repair using a crimp-and-seal connector is as follows:

STEP 1 Strip the insulation from the ends of the wire (about 5/16 inch or 8 mm).

STEP 2 Select the proper size of crimp-and-seal connector for the gauge of wire being repaired. Insert the wires into the splice sleeve and crimp.

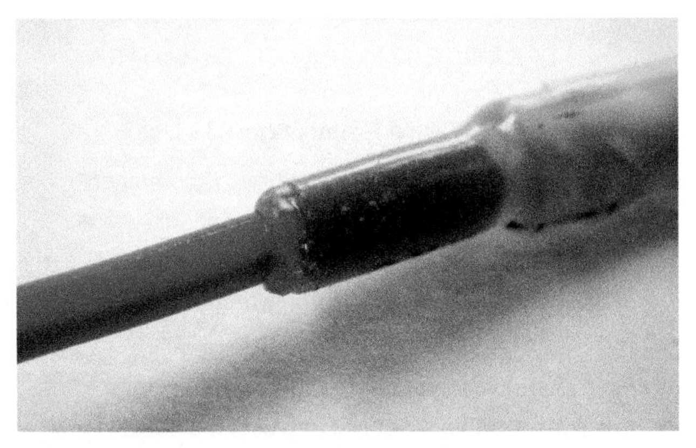

FIGURE 6–35 Heating the crimp-and-seal connector melts the glue and forms an effective seal against moisture.

NOTE: Use only the specified crimping tool to help prevent the pliers from creating a hole in the cover.

STEP 3 Apply heat to the connector until the sleeve shrinks down around the wire and a small amount of sealant is observed around the ends of the sleeve, as shown in ● **FIGURE 6–35**.

ALUMINUM WIRE REPAIR Some vehicle manufacturers used plastic-coated solid aluminum wire for some body wiring. Because aluminum wire is brittle and can break as a result of vibration, it is used only where there is no possible movement of the wire, such as along the floor or sill area. This section of wire is stationary, and the wire changes back to copper at a junction terminal after the trunk or rear section of the vehicle, where movement of the wiring may be possible.

If any aluminum wire must be repaired or replaced, the following procedure should be used to be assured of a proper repair. The aluminum wire is usually found protected in a plastic conduit. This conduit is then normally slit, after which the wires can easily be removed for repair.

STEP 1 Carefully strip only about 1/4 inch (6 mm) of insulation from the aluminum wire, being careful not to nick or damage the aluminum wire case.

STEP 2 Use a crimp connector to join two wires together. Do *not* solder an aluminum wire repair. Solder will not readily adhere to aluminum because the heat causes an oxide coating on the surface of the aluminum.

STEP 3 The spliced, crimped connection must be coated with petroleum jelly to prevent corrosion.

STEP 4 The coated connection should be covered with shrinkable plastic tubing or wrapped with electrical tape to seal out moisture.

FREQUENTLY ASKED QUESTION

What Method of Wire Repair Should I Use?

Good question. Vehicle manufacturers recommend all wire repairs performed under the hood, or where the repair could be exposed to the elements, be weatherproof. The most commonly recommended methods include the following:

- **Crimp and seal connector.** These connectors are special and are not like low-cost insulated-type crimp connectors. This type of connector is recommended by General Motors and others and is sealed using heat after the mechanical crimp has secured the wire ends together.
- **Solder and adhesive-lined heat shrink tubing.** This method is recommended by Chrysler and it uses the special heat shrink that has glue inside that melts when heated to form a sealed connection. Regular heat shrink tubing can be used inside a vehicle, but should not be used where it can be exposed to the elements.
- **Solder and electrical tape.** This is acceptable to use inside the vehicle where the splice will not be exposed to the outside elements. It is best to use a crimp and seal even on the inside of the vehicle for best results.

FREQUENTLY ASKED QUESTION

What Is in Lead-Free Solder?

Lead is an environmental and a health concern and all vehicle manufacturers are switching to lead-free solder. Lead-free solder does not contain lead but usually a very high percentage of tin. Several formulations of lead-free solder include:

- 95% Tin; 5% Antimony (melting temperature 450°F (245°C))
- 97% Tin; 3% Copper (melting temperature 441°F (227°C))
- 96% Tin; 4% Silver (melting temperature 443°F (228°C))

FIGURE 6–36 Conduit that has a paint strip is constructed of plastic that can withstand high underhood temperatures.

ELECTRICAL CONDUIT

Electrical conduit covers and protects wiring. The color used on electrical convoluted conduit tells the technician a lot if some information is known, such as the following:

- **Black conduit with a green or blue stripe.** This conduit is designed for high temperatures and is used under the hood and near hot engine parts. Do not replace high-temperature conduit with low-temperature conduit that does not have a stripe when performing wire repairs. ● **SEE FIGURE 6–36.**
- **Blue conduit.** This color conduit is used to cover wires that have voltages ranging from 12 to 42 volts. Circuits that use this high voltage usually are for the electric power steering. While 42 volts does not represent a shock hazard, an arc will be maintained if a line circuit is disconnected. Use caution around these circuits. ● **SEE FIGURE 6–37.**
- **Orange conduit.** This color conduit is used to cover wiring that carries high-voltage current from 144 to 650 volts. These circuits are found in hybrid electric vehicles (HEVs). An electric shock from these wires can be fatal, so extreme caution has to be taken when working on or near the components that have orange conduit. Follow the vehicle manufacturer's instruction for de-powering the high-voltage circuits before work begins on any of the high-voltage components. ● **SEE FIGURE 6–38.**

FIGURE 6–37 Blue conduit is used to cover circuits that carry up to 42 volts.

FIGURE 6–38 Always follow the vehicle manufacturer's instructions which include the use of linesman's (high-voltage) gloves if working on circuits that are covered in orange conduit.

SUMMARY

1. The higher the AWG size number, the smaller the wire diameter.
2. Metric wire is sized in square millimeters (mm^2) and the higher the number, the larger the wire.
3. All circuits should be protected by a fuse, fusible link, or circuit breaker. The current in the circuit should be about 80% of the fuse rating.
4. A terminal is the metal end of a wire, whereas a connector is the plastic housing for the terminal.
5. All wire repair should use either soldering or a crimp-and-seal connector.

REVIEW QUESTIONS

1. What is the difference between the American wire gauge (AWG) system and the metric system?
2. What is the difference between a wire and a cable?
3. What is the difference between a terminal and a connector?
4. How do fuses, PTC circuit protectors, circuit breakers, and fusible links protect a circuit?
5. How should a wire repair be done if the repair is under the hood where it is exposed to the outside?

CHAPTER QUIZ

1. The higher the AWG number, ________.
   - **a.** The smaller the wire diameter
   - **b.** The larger the wire diameter
   - **c.** The thicker the insulation
   - **d.** The more strands in the conductor core
2. Metric wire size is measured in units of ________.
   - **a.** Meters
   - **b.** Cubic centimeters
   - **c.** Square millimeters
   - **d.** Cubic millimeters
3. Which statement is true about fuse ratings?
   - **a.** The fuse rating should be less than the maximum current for the circuit.
   - **b.** The fuse rating should be higher than the normal current for the circuit.
   - **c.** Of the fuse rating, 80% should equal the current in the circuit.
   - **d.** Both b and c

4. Which statements are true about wire, terminals, and connectors?
   - **a.** Wire is called a lead, and the metal end is a connector.
   - **b.** A connector is usually a plastic piece where terminals lock in.
   - **c.** A lead and a terminal are the same thing.
   - **d.** Both a and c
5. The type of solder that should be used for electrical work is _______.
   - **a.** Rosin core
   - **b.** Acid core
   - **c.** 60/40 with no flux
   - **d.** 50/50 with acid paste flux
6. A technician is performing a wire repair on a circuit under the hood of the vehicle. Technician A says to use solder and adhesive-lined heat shrink tubing or a crimp and seal connector. Technician B says to solder and use electrical tape. Which technician is correct?
   - **a.** Technician A only
   - **b.** Technician B only
   - **c.** Both Technicians A and B
   - **d.** Neither Technician A nor B
7. Two technicians are discussing fuse testing. Technician A says that a test light should light on both test points of the fuse if it is okay. Technician B says the fuse is defective if a test light only lights on one side of the fuse. Which technician is correct?
   - **a.** Technician A only
   - **b.** Technician B only
   - **c.** Both Technicians A and B
   - **d.** Neither Technician A nor B
8. If a wire repair, such as that made under the hood or under the vehicle, is exposed to the elements, which type of repair should be used?
   - **a.** Wire nuts and electrical tape
   - **b.** Solder and adhesive-lined heat shrink or crimp-and-seal connectors
   - **c.** Butt connectors
   - **d.** Rosin-core solder and electrical tape
9. Many ground straps are uninsulated and braided because _______.
   - **a.** They are more flexible to allow movement of the engine without breaking the wire.
   - **b.** They are less expensive than conventional wire.
   - **c.** They help dampen radio-frequency interference (RFI).
   - **d.** Both a and c
10. What causes a fuse to blow?
    - **a.** A decrease in circuit resistance
    - **b.** An increase in the current flow through the circuit
    - **c.** A sudden decrease in current flow through the circuit
    - **d.** Both a and b

WIRING SCHEMATICS AND CIRCUIT TESTING

LEARNING OBJECTIVES

After studying this chapter, the reader will be able to:

1. Interpret wiring schematics.
2. Locate shorts, grounds, opens, and resistance problems in electrical circuits, and determine necessary action.
3. Inspect and test switches, connectors, relays, solid state devices, and wires of electrical circuits, and perform necessary action.

This chapter will help you prepare for the ASE Electrical/Electronic Systems (A6) certification test content area "A" (General Electrical/Electronic System Diagnosis).

KEY TERMS

Coil 93
DPDT 91
DPST 91
Gauss gauge 99
Momentary switch 91
N.C. 91
N.O. 91
Poles 91
Relay 92
Short circuit 98
SPDT 91
SPST 91
Terminal 88
Throws 91
Tone generator tester 100
Wiring schematic 84

GM STC OBJECTIVES

GM Service Technical College topics covered in this chapter are as follows:

1. Electrical component symbols (08510.18W).
2. Schematic details.
3. The path that current flows.

FIGURE 7–1 The center wire is a solid color wire, meaning that the wire has no other identifying tracer or stripe color. The two end wires could be labeled "BRN/WHT," indicating a brown wire with a white tracer or stripe.

WIRING SCHEMATICS AND SYMBOLS

TERMINOLOGY The service manuals of automotive manufacturers include wiring schematics of every electrical circuit in a vehicle. A **wiring schematic**, sometimes called a *diagram*, shows electrical components and wiring using symbols and lines to represent components and wires. A typical wiring schematic may include all of the circuits combined on several large foldout sheets, or they may be broken down to show individual circuits. All circuit schematics or diagrams include:

- Power-side wiring of the circuit
- All splices
- Connectors
- Wire size
- Wire color
- Trace color (if any)
- Circuit number
- Electrical components
- Ground return paths
- Fuses and switches

CIRCUIT INFORMATION Many wiring schematics include numbers and letters near components and wires that may confuse readers of the schematic. Most letters used near or on a wire identify the color or colors of the wire.

- The first color or color abbreviation is the color of the wire insulation.
- The second color (if mentioned) is the color of the stripe or tracer on the base color. ● **SEE FIGURE 7–1.**

Wires with different color tracers are indicated by both colors with a slash (/) between them. For example, BRN/WHT means a brown wire with a white stripe or tracer. ● **SEE CHART 7–1.**

WIRE SIZE Wire size is shown on all schematics. ● **FIGURE 7–2** illustrates a rear side-marker bulb circuit diagram where "0.8" indicates the metric wire gauge size in square millimeters (mm^2) and "PPL" indicates a solid purple wire.

The wire diagram also shows that the color of the wire changes at the number C210. This stands for "connector #210" and is used for reference purposes. The symbol for the connection can vary depending on the manufacturer. The color change from purple (PPL) to purple with a white tracer (PPL/WHT) is not important except for knowing where the wire changes color in the circuit. The wire gauge has remained the same on both sides of the connection (0.8 mm^2 or 18 gauge). The ground circuit is the "0.8 BLK" wire. ● **FIGURE 7–3 AND 7–4** show many of the electrical and electronic symbols that are used in wiring and circuit diagrams.

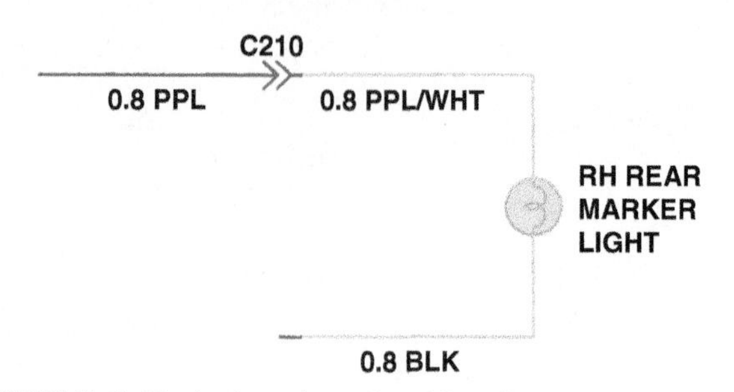

FIGURE 7–2 Typical section of a wiring diagram. Notice that the wire color changes at connection C210. The "0.8" represents the metric wire size in square millimeters.

ABBREVIATION	COLOR
BRN	Brown
BLK	Black
GRN	Green
WHT	White
PPL	Purple
PNK	Pink
TAN	Tan
BLU	Blue
YEL	Yellow
ORN	Orange
DK BLU	Dark blue
LT BLU	Light blue
DK GRN	Dark green
LT GRN	Light green
RED	Red
GRY	Gray
VIO	Violet

CHART 7–1

Typical abbreviations used on schematics to show wire color. Some vehicle manufacturers use two letters to represent a wire color. Check service information for the color abbreviations used.

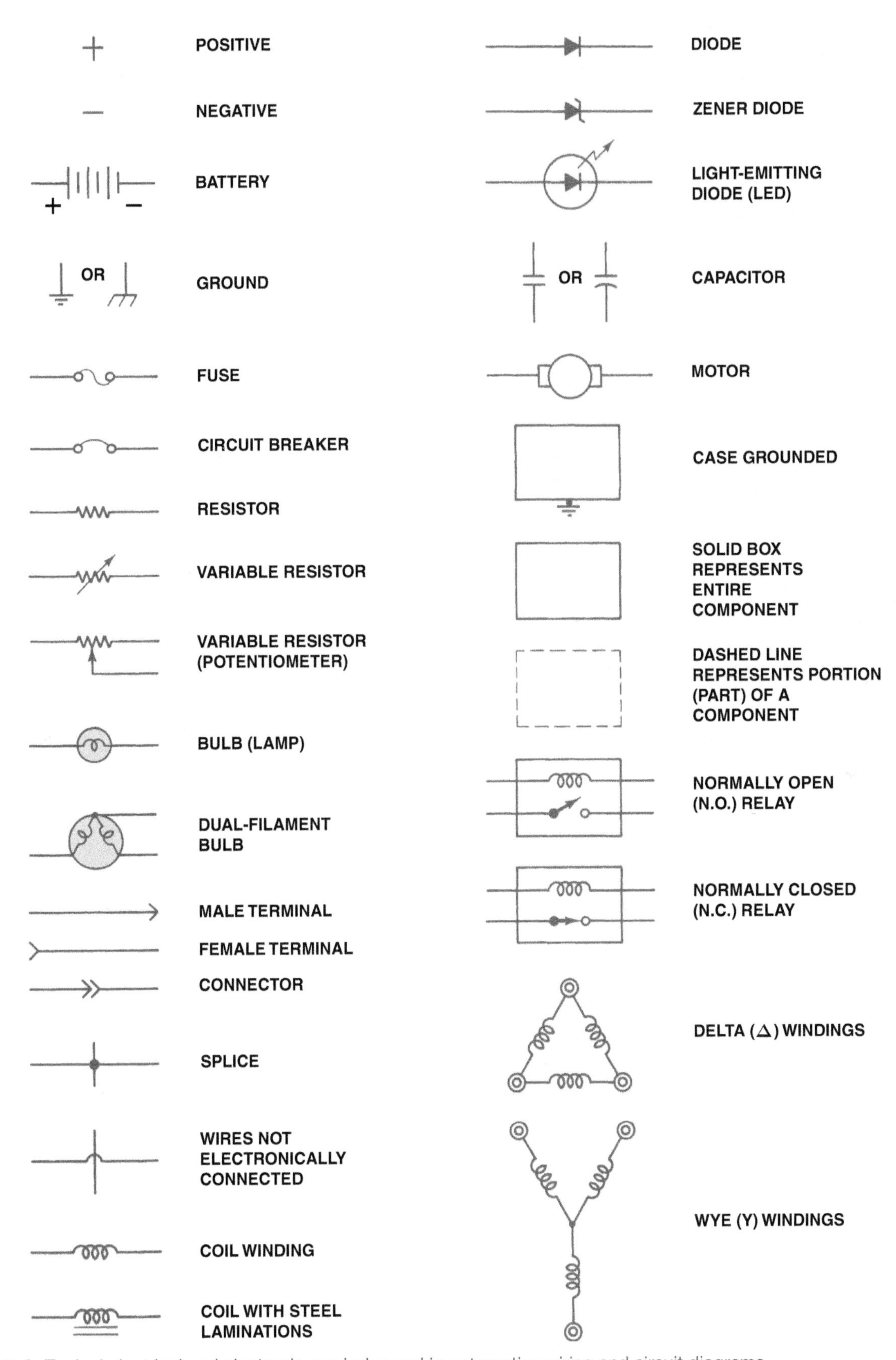

FIGURE 7–3 Typical electrical and electronic symbols used in automotive wiring and circuit diagrams.

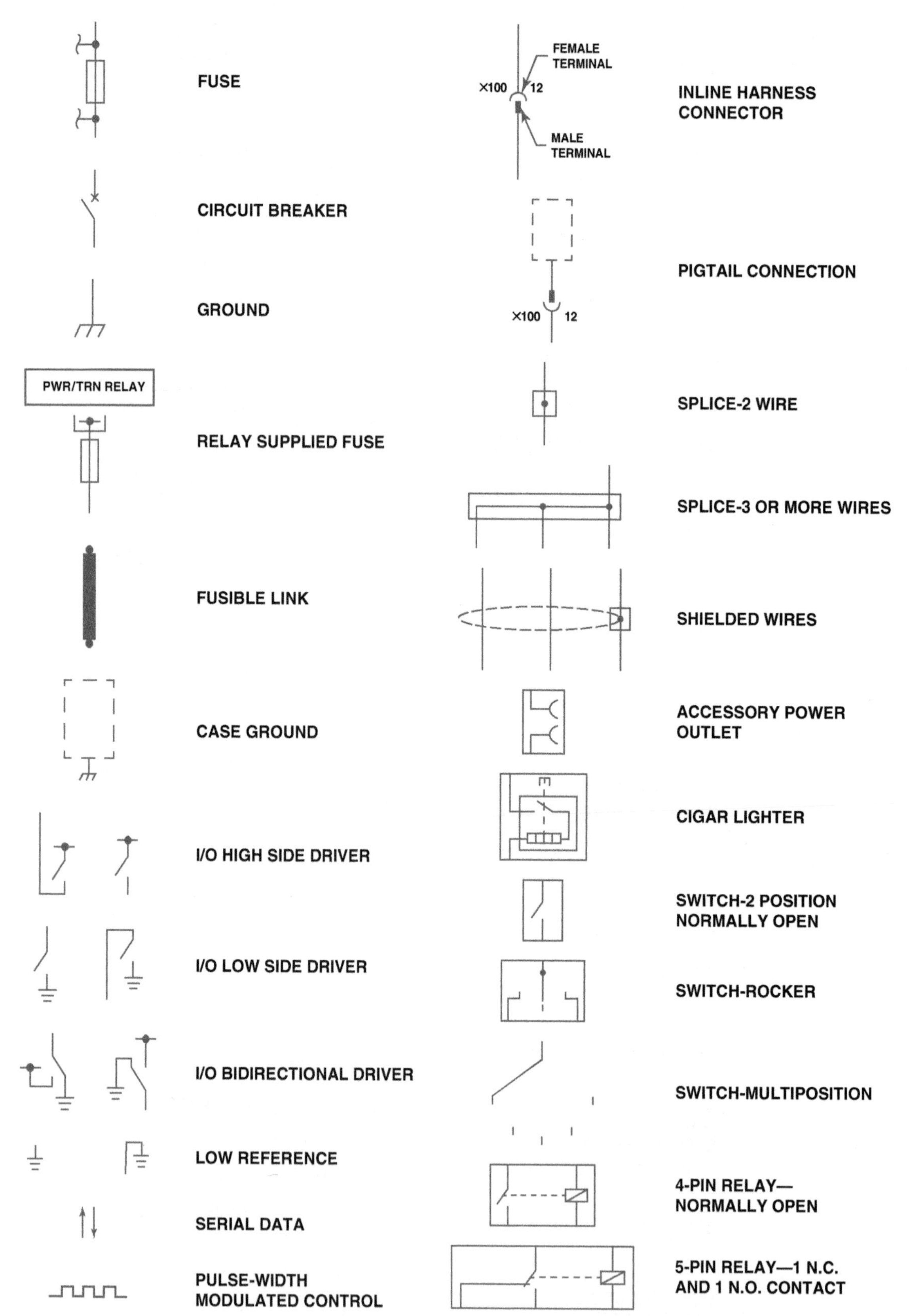

FIGURE 7–4 Starting in 2008 GM began using these updated "Global" electrical symbols on electrical schematics. (Courtesy of General Motors)

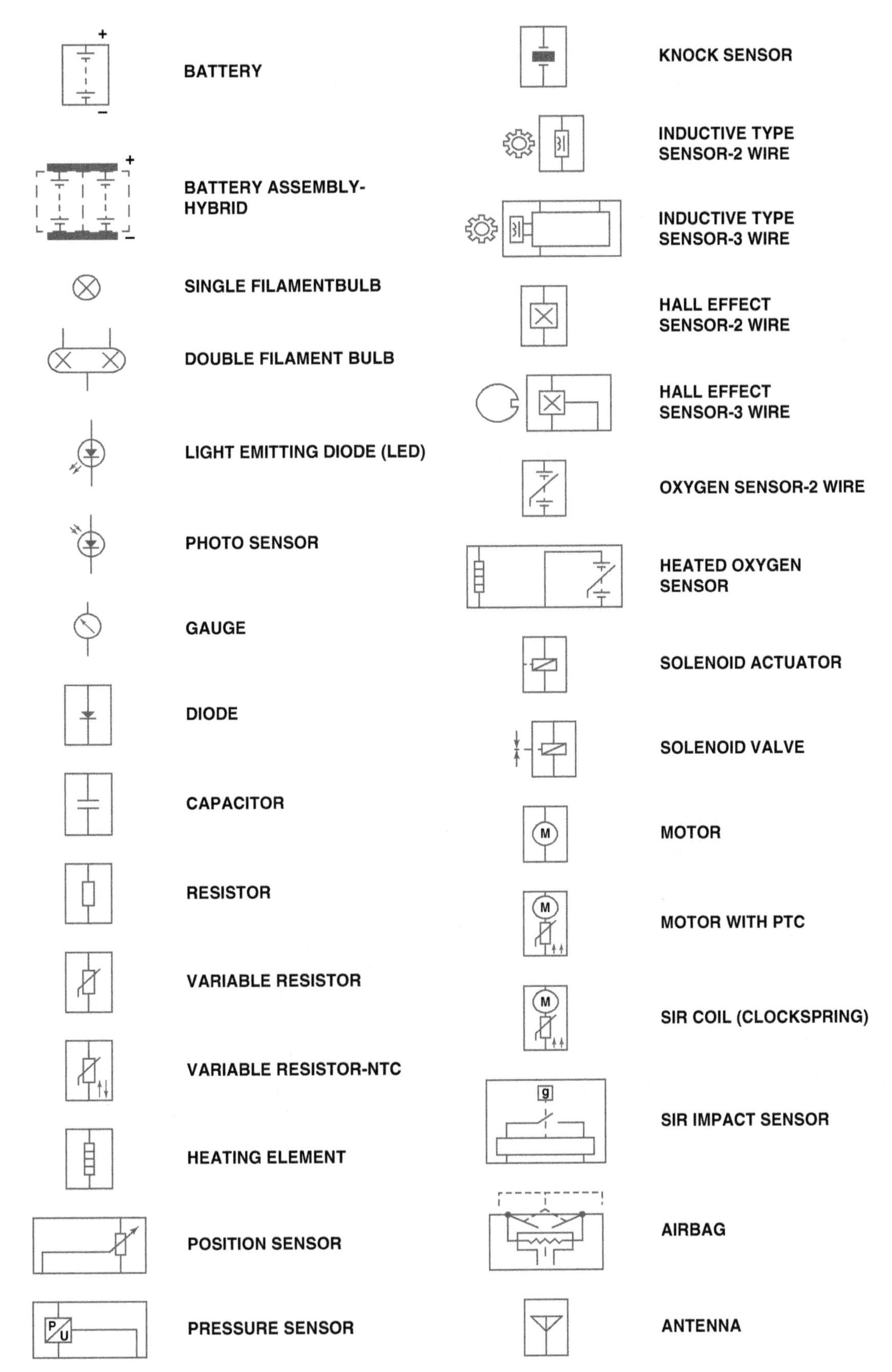

FIGURE 7–4 *(Continued)*

FIGURE 7–5 In this typical connector, note that the positive terminal is usually a female connector.

FIGURE 7–6 The symbol for a battery. The positive plate of a battery is represented by the longer line and the negative plate by the shorter line. The voltage of the battery is usually stated next to the symbol.

TECH TIP

Read the Arrows

Wiring diagrams indicate connections by symbols that look like arrows. **SEE FIGURE 7–5.**

Do *not* read these "arrows" as pointers showing the direction of current flow. Also observe that the power side (positive side) of the circuit is usually the female end of the connector. If a connector becomes disconnected, it will be difficult for the circuit to become shorted to ground or to another circuit because the wire is recessed inside the connector.

SCHEMATIC SYMBOLS

In a schematic drawing, photos or line drawings of actual components are replaced with a symbol that represents the actual component. The following discussion centers on these symbols and their meanings.

BATTERY The plates of a battery are represented by long and short lines. **SEE FIGURE 7–6.**

The longer line represents the positive plate of a battery and the shorter line represents the negative plate of the battery. Therefore, each pair of short and long lines represents one cell of a battery. Because each cell of a typical automotive lead-acid battery has 2.1 volts, a battery symbol showing a 12-volt battery should have six pairs of lines. However, most battery symbols simply use two or three pairs of long and short lines and then list the voltage of the battery next to the symbol. As a result, the battery symbols are shorter and yet clear, because the voltage is stated. The positive terminal of the battery is often indicated with a plus sign (+), representing the positive post of the battery, and is placed next to the long line of the end cell. The negative terminal of the battery is represented by a negative sign (–) and is placed next to the shorter cell line. The negative battery terminal is connected to ground. **SEE FIGURE 7–7.**

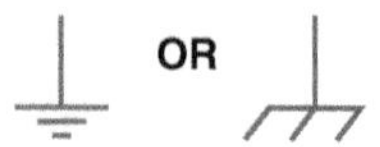

FIGURE 7–7 The ground symbol on the left represents earth ground. The ground symbol on the right represents a chassis ground.

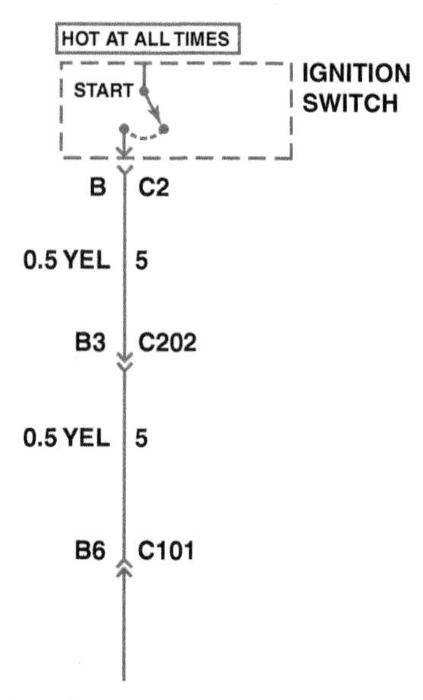

FIGURE 7–8 Starting at the top, the wire from the ignition switch is attached to terminal B of connector C2, the wire is 0.5 mm^2 (20 gauge AWG), and is yellow. The circuit number is 5. The wire enters connector C202 at terminal B3.

WIRING Electrical wiring is shown as straight lines and with a few numbers and/or letters to indicate the following:

- **Wire size.** This can be either AWG, such as 18 gauge, or in square millimeters, such as 0.8 mm^2.
- **Circuit numbers.** Each wire in part of a circuit is labeled with the circuit number to help the service technician trace the wiring and to provide an explanation of how the circuit should work.
- **Wire color.** Most schematics also indicate an abbreviation for the color of the wire and place it next to the wire. Many wires have two colors: a solid color and a stripe color. In this case, the solid color is listed, followed by a dark slash (/) and the color of the stripe. For example, Red/Wht would indicate a red wire with a white tracer. **SEE FIGURE 7–8.**
- **Terminals.** The metal part attached at the end of a wire is called a **terminal**. A symbol for a terminal is shown in **FIGURE 7–9.**
- **Splices.** When two wires are electrically connected, the junction is shown with a black dot. The identification of the splice is an "S" followed by three numbers, such as S103.

FIGURE 7–9 The electrical terminals are usually labeled with a letter or number.

FIGURE 7–10 Two wires that cross at the dot indicate that the two are electrically connected.

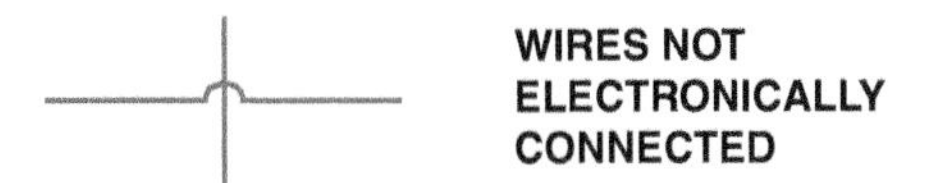

FIGURE 7–11 Wires that cross, but do not electrically contact each other, are shown with one wire bridging over the other.

● **SEE FIGURE 7–10.** When two wires cross in a schematic that are not electrically connected, one of the wires is shown as going over the other wire and does not connect. ● **SEE FIGURE 7–11.**

- **Connectors.** An electrical connector is a plastic part that contains one or more terminals. Although the terminals provide the electrical connection in a circuit, it is the plastic connector that keeps the terminals together mechanically.
- **Location.** Connections are usually labeled "C" and then three numbers. The three numbers indicate the general location of the connector. Normally, the connector number represents the general area of the vehicle, including:

100–199	Under the hood
200–299	Under the dash
300–399	Passenger compartment
400–499	Rear package or trunk area
500–599	Left-front door
600–699	Right-front door
700–799	Left-rear door
800–899	Right-rear door

Even-numbered connectors are on the right (passenger side) of the vehicle and odd-numbered connectors are on the left (driver's side) of the vehicle. For example, C102 is a connector located under the hood (between 100 and 199) on the right side of the vehicle (even number 102). ● **SEE FIGURE 7–12.**

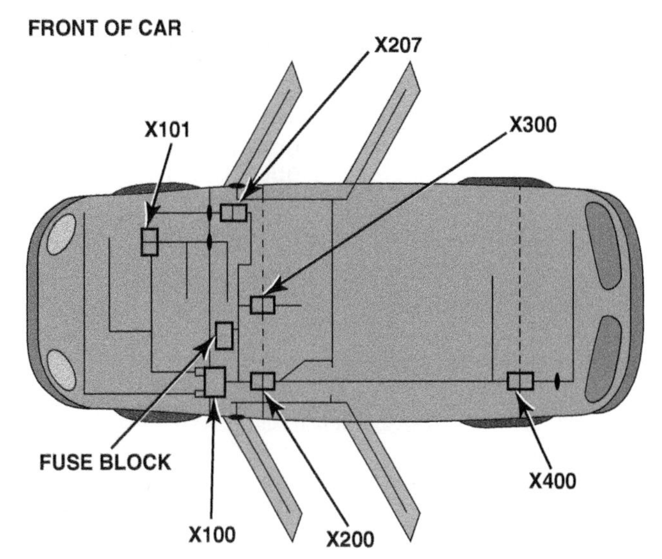

FIGURE 7–12 Connectors (C or X), grounds (G), and splices (S or JX) are followed by a number, generally indicating the location in the vehicle. For example, G209 is a ground connection located under the dash. (Courtesy of General Motors)

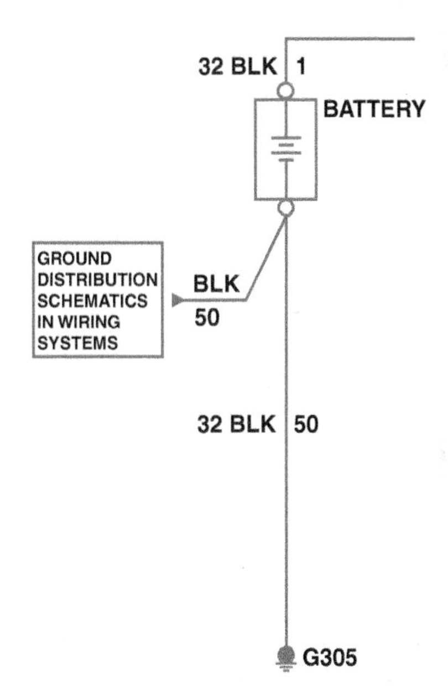

FIGURE 7–13 The ground for the battery is labeled G305, indicating the ground connector is located in the passenger compartment of the vehicle. The ground wire is black (BLK), the circuit number is 50, and the wire is 32 mm^2 (2 gauge AWG).

- **Grounds and splices.** These are also labeled using the same general format as connectors. Therefore, a ground located under the dash on the driver's side could be labeled G305 (G means "ground" and "305" means that it is located in the passenger compartment). ● **SEE FIGURE 7–13.**

NOTE: Beginning with new models in 2008 GM changed the labeling standards for many components. ● SEE CHART 7–2.

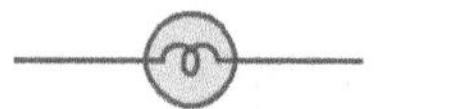

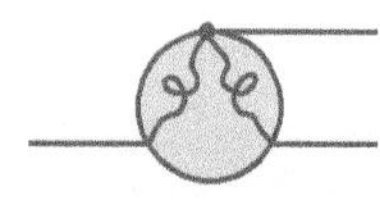

FIGURE 7–14 The symbol for light bulbs shows the filament inside a circle, which represents the glass ampoule of the bulb.

OLD	NEW
Component Connectors	
C1, C2, C3	X1, X2, X3
C100, C200, C300	X100, X200, X300
Splices	
S100, S200, S300	J100, J200, J300
Splice Packs	
SP100, SP200, SP300	JX100, JX200, JX300
Grounds	
G100, G200, G300	G100, G200, G300

CHART 7–2

The new GM global schematics use new labels for the various components. While the prefix has changed, the numbering system is the same. Some examples are shown here. Refer to SI for more information. (Courtesy of General Motors)

ELECTRICAL COMPONENTS

Most electrical components have their own unique symbol that shows the basic function or parts.

BULBS Light bulbs often use a filament, which heats and then gives off light when electrical current flows. The symbol used for a light bulb is a circle with a filament inside. A dual-filament bulb, such as is used for taillights and brake light/turn signals, is shown with two filaments. ● **SEE FIGURE 7–14.**

ELECTRIC MOTORS An electric motor symbol shows a circle with the letter *M* in the center and two electrical connections, one at the top and one at the bottom. ● **SEE FIGURE 7–15** for an example of a cooling fan motor.

RESISTORS Although resistors are usually part of another component, the symbol appears on many schematics and wiring diagrams. A resistor symbol is a jagged line representing resistance to current flow. If the resistor is variable, such as a thermistor, an arrow is shown running through the symbol of a fixed resistor. A potentiometer is a three-wire variable resistor, shown with an arrow pointing toward the resistance part of a fixed resistor. ● **SEE FIGURE 7–16.**

A two-wire rheostat is usually shown as part of another unit, such as a fuel level sensing unit. ● **SEE FIGURE 7–17.**

CAPACITORS Capacitors are usually part of an electronic component, but not a replaceable component unless the vehicle is an older model. Many older vehicles used capacitors to reduce radio interference and were installed inside alternators or were attached to wiring connectors. ● **SEE FIGURE 7–18.**

FIGURE 7–15 An electric motor symbol shows a circle with the letter *M* in the center and two black sections that represent the brushes of the motor. This symbol is used even though the motor is a brushless design.

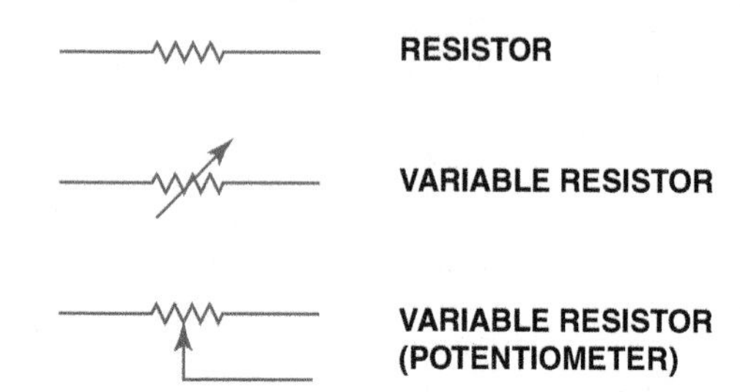

FIGURE 7–16 Resistor symbols vary depending on the type of resistor.

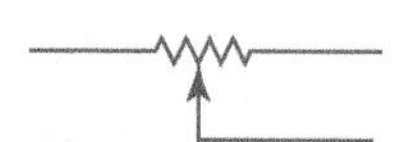

FIGURE 7–17 A rheostat uses only two wires—one is connected to a voltage source and the other is attached to the movable arm.

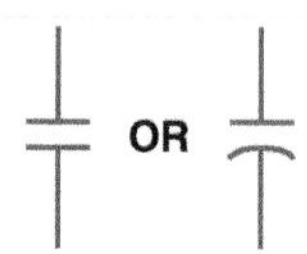

FIGURE 7–18 Symbols used to represent capacitors. If one of the lines is curved, this indicates that the capacitor being used has a polarity, while the one without a curved line can be installed in the circuit without concern about polarity.

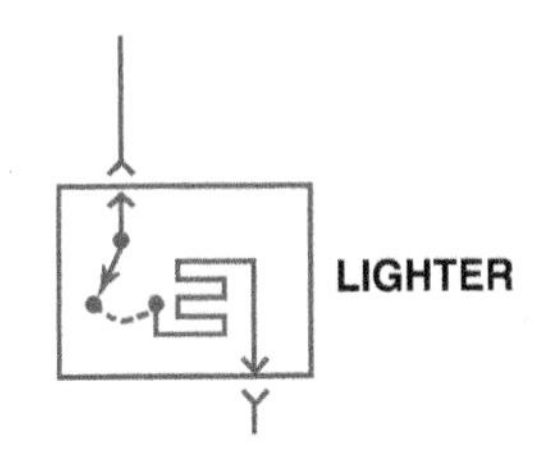

FIGURE 7–19 The grid-like symbol represents an electrically heated element.

FIGURE 7–20 A dashed outline represents a portion (part) of a component.

FIGURE 7–21 A solid box represents an entire component.

ELECTRIC HEATED UNIT Electric grid-type rear window defoggers and cigarette lighters are shown with a square box-type symbol. **SEE FIGURE 7–19.**

BOXED COMPONENTS If a component is shown in a box using a solid line, the box is the entire component. If a box uses dashed lines, it represents part of a component. A commonly used dashed-line box is a fuse panel. Often, just one or two fuses are shown in a dashed-line box. This means that a fuse panel has more fuses than shown. **SEE FIGURES 7–20 AND 7–21.**

SEPARATE REPLACEABLE PART Often components are shown on a schematic that cannot be replaced, but are part of a complete assembly. When looking at a schematic of General Motors vehicles, the following is shown.

- If a part name is underlined, it is a replaceable part.
- If a part is not underlined, it is not available as a replaceable part, but is included with other components shown and sold as an assembly.
- If the case itself is grounded, the ground symbol is attached to the component as shown in **FIGURE 7–22.**

FIGURE 7–22 This symbol represents a component that is case grounded.

SWITCHES Electrical switches are drawn on a wiring diagram in their normal position. This can be one of two possible positions.

- **Normally open.** The switch is not connected to its internal contacts and no current will flow. This type of switch is labeled **N.O.**
- **Normally closed.** The switch is electrically connected to its internal contacts and current will flow through the switch. This type of switch is labeled **N.C.**

Other switches can use more than two contacts.

The **poles** refer to the number of circuits completed by the switch and the **throws** refer to the number of output circuits. A **single-pole, single-throw (SPST)** switch has only two positions, on or off. A **single-pole, double-throw (SPDT)** switch has three terminals, one wire in and two wires out. A headlight dimmer switch is an example of a typical SPDT switch. In one position, the current flows to the low-filament headlight; in the other, the current flows to the high-filament headlight.

NOTE: A SPDT switch is not an on or off type of switch but instead directs power from the source to either the high-beam lamps or the low-beam lamps.

There are also **double-pole, single-throw (DPST)** switches and **double-pole, double-throw (DPDT)** switches. **SEE FIGURE 7–23.**

NOTE: All switches are shown on schematics in their normal position. This means that the headlight switch will be shown normally off, as are most other switches and controls.

MOMENTARY SWITCH A **momentary switch** is a switch primarily used to send a voltage signal to a module or controller to request that a device be turned on or off. The switch makes momentary contact and then returns to the open position. A horn switch is a commonly used momentary switch. The symbol that represents a momentary switch uses two

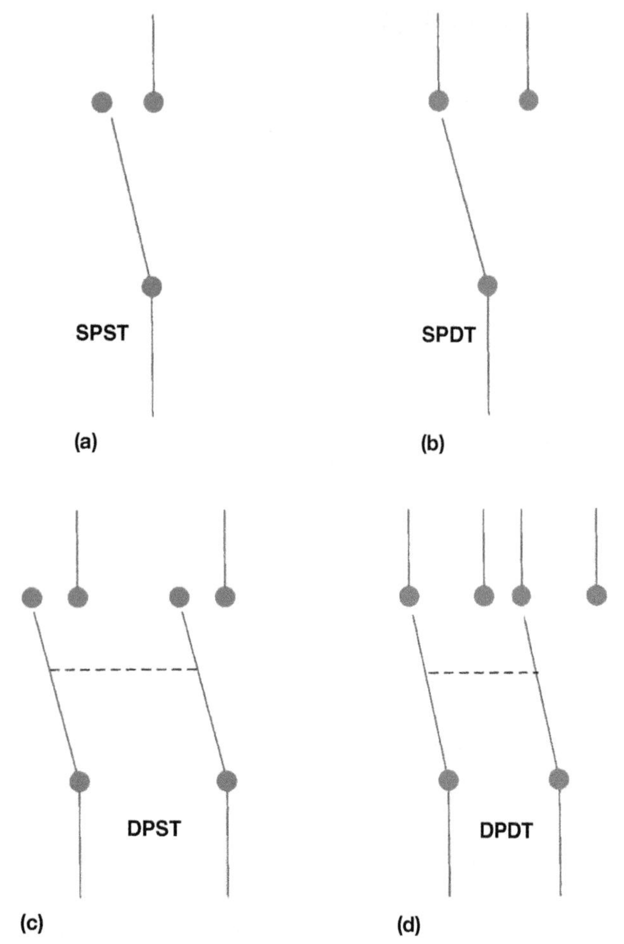

FIGURE 7–23 (a) A symbol for a single-pole, single-throw (SPST) switch. This type of switch is normally open (N.O.) because nothing is connected to the terminal that the switch is contacting in its normal position. (b) A single-pole, double-throw (SPDT) switch has three terminals. (c) A double-pole, single-throw (DPST) switch has two positions (off and on) and can control two separate circuits. (d) A double-pole, double-throw (DPDT) switch has six terminals—three for each pole. Note: Both (c) and (d) also show a dotted line between the two arms indicating that they are mechanically connected, called a "ganged switch."

dots for the contact with a switch above them. A momentary switch can be either normally open or normally closed. ● **SEE FIGURE 7–24.**

A momentary switch, for example, can be used to lock or unlock a door or to turn the air-conditioning on or off. If the device is currently operating, the signal from the momentary switch will turn it off, and if it is off, the switch will signal the module to turn it on. The major advantage of momentary switches is that they can be lightweight and small, because the switch does not carry any heavy electrical current, just a small voltage signal. Most momentary switches use a membrane constructed of foil and plastic.

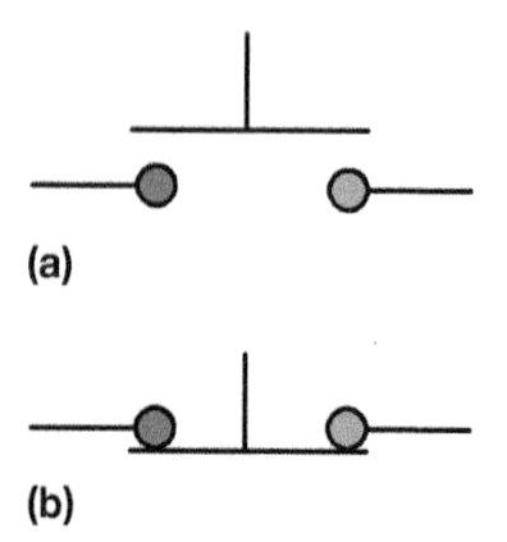

FIGURE 7–24 (a) A symbol for a normally open (N.O.) momentary switch. (b) A symbol for a normally closed (N.C.) momentary switch.

TECH TIP

Color-Coding Is Key to Understanding

Whenever diagnosing an electrical problem, it is common practice to print out the schematic of the circuit and then take it to the vehicle. A meter is then used to check for voltage at various parts of the circuit to help determine where there is a fault. The diagnosis can be made easier if the parts of the circuit are first color coded using markers or color pencils. A color-coding system that has been widely used is one developed by Jorge Menchu (www.aeswave.com).

The colors represent voltage conditions in various parts of a circuit. Once the circuit has been color coded, then the circuit can be tested using the factory wire colors as a guide. ● **SEE FIGURE 7–25.**

RELAY TERMINAL IDENTIFICATION

DEFINITION A **relay** is a magnetic switch that uses a movable armature to control a high-amperage circuit by using a low-amperage electrical switch.

ISO RELAY TERMINAL IDENTIFICATION Most automotive relays adhere to common terminal identification. The primary source for this common identification comes from the standards established by the International Standards Organization (ISO). Knowing this terminal information will help in the correct diagnosis and troubleshooting of any circuit containing a relay. ● **SEE FIGURES 7–26 AND 7–27.**

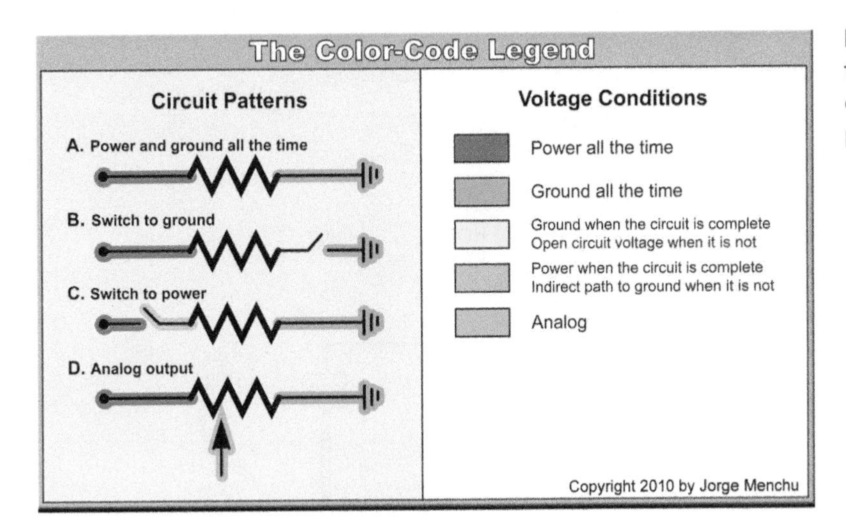

FIGURE 7–25 Using a marker and color-coding the various parts of the circuit makes the circuit easier to understand and helps diagnosing electrical problems easier. (Courtesy of Jorge Menchu)

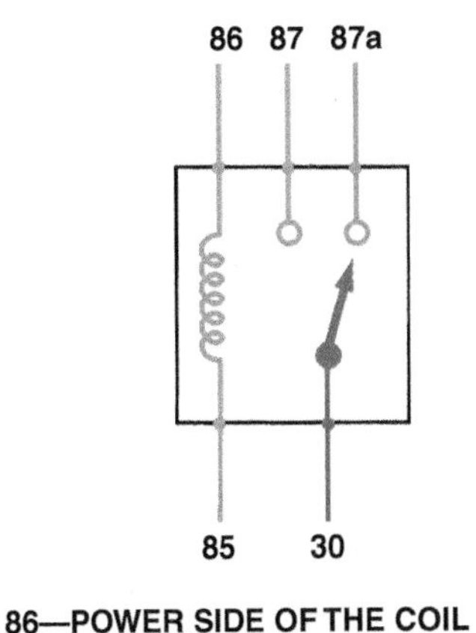

86—POWER SIDE OF THE COIL
85—GROUND SIDE OF THE COIL

(MOSTLY RELAY COILS HAVE BETWEEN 50 AND 150 Ω OF RESISTANCE)

30—COMMON POWER FOR RELAY CONTACTS
87—NORMALLY OPEN OUTPUT (N.O.)
87a—NORMALLY CLOSED OUTPUT (N.C.)

FIGURE 7–26 A relay uses a movable arm to complete a circuit whenever there is a power at terminal 86 and a ground at terminal 85. A typical relay only requires about 1/10 ampere through the relay coil. The movable arm then closes the contacts (#30 to #87) and can relay 30 amperes or more.

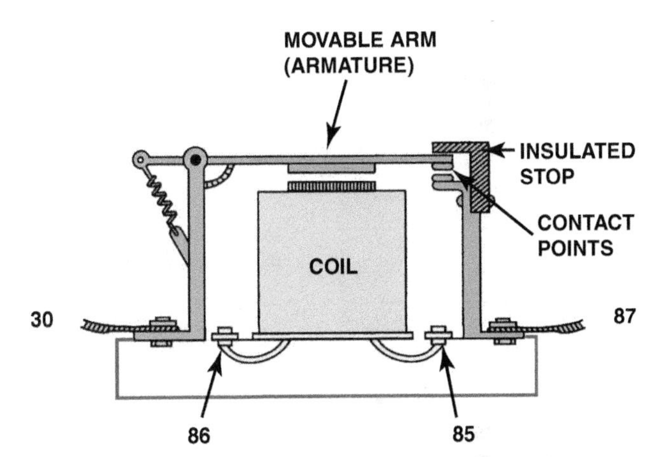

FIGURE 7–27 A cross-sectional view of a typical four-terminal relay. Current flowing through the coil (terminals 86 and 85) causes the movable arm (called the armature) to be drawn toward the coil magnet. The contact points complete the electrical circuit connected to terminals 30 and 87.

Relays are found in many circuits because they are capable of being controlled by computers, yet are able to handle enough current to power motors and accessories. Relays include the following components and terminals.

RELAY OPERATION

1. **Coil** (terminals 85 and 86)
   - A coil provides the magnetic pull to a movable armature (arm).
   - The resistance of most relay coils ranges from 50 to 150 ohms, but is usually between 60 and 100 ohms.
   - The ISO identification of the coil terminals are 86 and 85. The terminal number 86 represents the power to the relay coil and the terminal labeled 85 represents the ground side of the relay coil.
   - The relay coil can be controlled by supplying either power or ground to the relay coil winding.
   - The coil winding represents the *control circuit,* which uses low current to control the higher current through the other terminals of the relay. ● **SEE FIGURE 7–28.**

FIGURE 7–28 These types of relays can be found on General Motors vehicles. There may be a schematic of the relay wiring on the case. (Courtesy of Jeffrey Rehkopf)

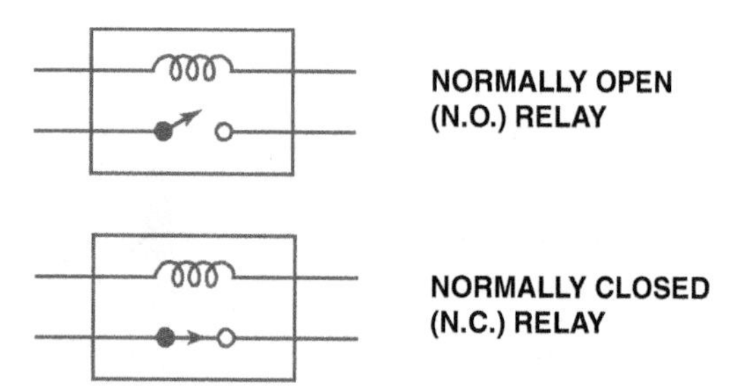

FIGURE 7–29 All schematics are shown in their normal, nonenergized position.

2. Other terminals used to control the load current
   - The higher amperage current flow through a relay flows through terminals 30 and 87, and often 87a.
   - Terminal 30 is usually where power is applied to a relay. Check service information for the exact operation of the relay being tested.
   - When the relay is at rest without power and ground to the coil, the armature inside the relay electrically connects terminals 30 and 87a if the relay has five terminals. When there is power at terminal 85 and a ground at terminal 86 of the relay, a magnetic field is created in the coil winding, which draws the armature of the relay toward the coil. The armature, when energized electrically, connects terminals 30 and 87.

The maximum current through the relay is determined by the resistance of the circuit, and relays are designed to safely handle the designed current flow. ● **SEE FIGURES 7–29 AND 7–30.**

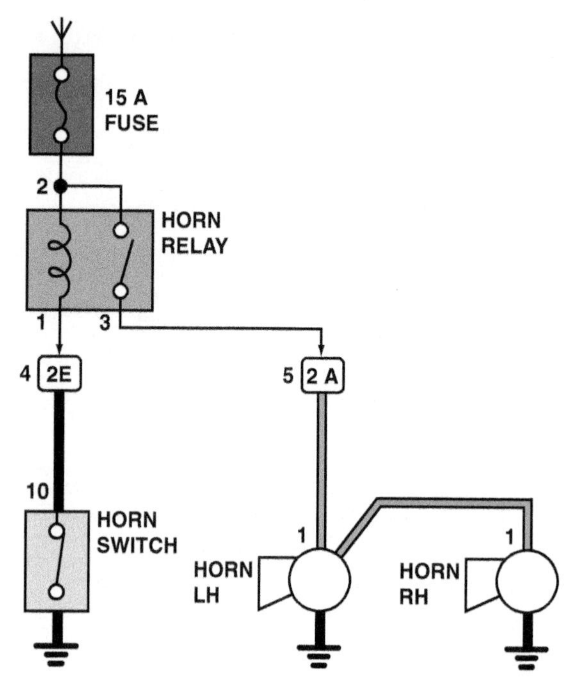

FIGURE 7–30 A typical horn circuit. Note that the relay contacts supply the heavy current to operate the horn when the horn switch simply completes a low-current circuit to ground, causing the relay contacts to close.

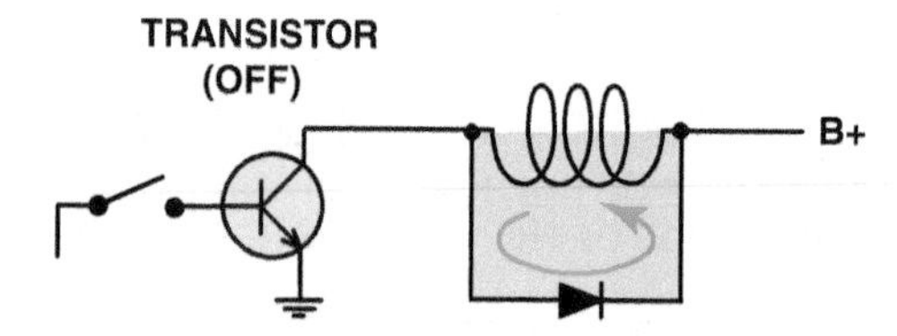

FIGURE 7–31 When the relay or solenoid coil current is turned off, the stored energy in the coil flows through the clamping diode and effectively reduces voltage spike.

RELAY VOLTAGE SPIKE CONTROL Relays contain a coil and when power is removed, the magnetic field surrounding the coil collapses, creating a voltage to be induced in the coil winding. This induced voltage can be as high as 100 volts or more and can cause problems with other electronic devices in the vehicle. For example, the short high-voltage surge can be heard as a "pop" in the radio. To reduce the induced voltage, some relays contain a diode connected across the coil. ● **SEE FIGURE 7–31.**

TECH TIP

Divide the Circuit in Half

When diagnosing any circuit that has a relay, start testing at the relay and divide the circuit in half.

- **High current portion:** Remove the relay and check that there are 12 volts at the terminal 30 socket. If there is, then the power side is okay. Use an ohmmeter and check between terminal 87 socket and ground. If the load circuit has continuity, there should be some resistance. If OL, the circuit is electrically open.
- **Control circuit (low current):** With the relay removed from the socket, check that there is 12 volts to terminal 86 with the ignition on and the control switch on. If not, check service information to see if power should be applied to terminal 86, then continue troubleshooting the switch power and related circuit.
- **Check the relay itself:** Use an ohmmeter and measure for continuity and resistance.
    - Between terminals 85 and 86 (coil), there should be 60 to 100 ohms. If not, replace the relay.
    - Between terminals 30 and 87 (high-amperage switch controls), there should be continuity (low ohms) when there is power applied to terminal 85 and a ground applied to terminal 86 that operates the relay. If OL is displayed on the meter set to read ohms, the circuit is open which requires that the reply be replaced.
    - Between terminals 30 and 87a (if equipped), with the relay turned off, there should be low resistance (less than 5 ohms).

When the current flows through the coil, the diode is not part of the circuit because it is installed to block current. However, when the voltage is removed from the coil, the resulting voltage induced in the coil windings has a reversed polarity to the applied voltage. Therefore, the voltage in the coil is applied to the coil in a forward direction through the diode, which conducts the current back into the winding. As a result, the induced voltage spike is eliminated.

Most relays use a resistor connected in parallel with the coil winding. The use of a resistor, typically about 400 to 600 ohms, reduces the voltage spike by providing a path for the voltage created in the coil to flow back through the coil windings when the coil circuit is opened. See **FIGURE 7–32**.

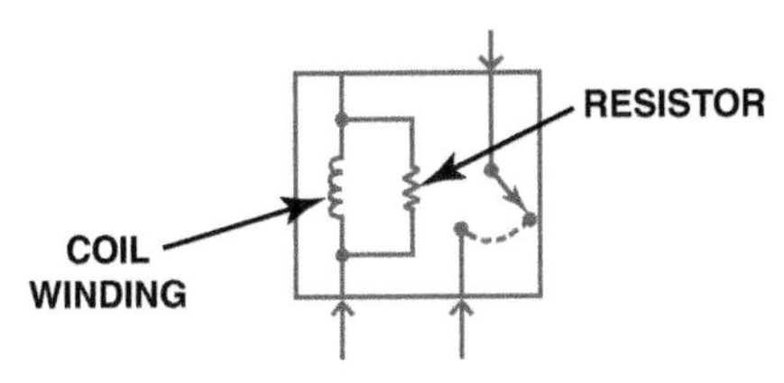

FIGURE 7–32 A resistor used in parallel with the coil windings is a common spike reduction method used in many relays.

FREQUENTLY ASKED QUESTION

What Is the Difference Between a Relay and a Solenoid?

Often, these terms are used differently among vehicle manufacturers, which can lead to some confusion.

Relay: A relay is an electromagnetic switch that uses a movable arm. Because a relay uses a movable arm, it is generally limited to current flow not exceeding 30 amperes.

Solenoid: A solenoid is an electromagnetic switch that uses a movable core. Because of this type of design, a solenoid is capable of handling 200 amperes or more and is used in the starter motor circuit and other high-amperage applications, such as in the glow plug circuit of diesel engines.

LOCATING AN OPEN CIRCUIT

TERMINOLOGY An open circuit is a break in the electrical circuit that prevents current from flowing and operating an electrical device. Examples of open circuits include:

- Blown (open) light bulbs
- Cut or broken wires
- Disconnected or partially disconnected electrical connectors
- Electrically open switches
- Loose or broken ground connections or wires
- Blown fuse

PROCEDURE TO LOCATE AN OPEN CIRCUIT The typical procedure for locating an open circuit involves the following steps.

STEP 1 **Perform a thorough visual inspection.** Check the following:

- Look for evidence of a previous repair. Often, an electrical connector or ground connection can be accidentally left disconnected.
- Look for evidence of recent body damage or body repairs. Movement due to a collision can cause metal to move, which can cut wires or damage connectors or components.

STEP 2 **Print out the schematic.** Trace the circuit and check for voltage at certain places. This will help pinpoint the location of the open circuit.

STEP 3 **Check everything that does and does not work.** Often, an open circuit will affect more than one component. Check the part of the circuit that is common to the other components that do not work.

STEP 4 **Check for voltage.** Voltage is present up to the location of the open circuit fault. For example, if there is battery voltage at the positive terminal and the negative (ground) terminal of a two-wire light bulb socket with the bulb plugged in, then the ground circuit is open.

COMMON POWER OR GROUND

When diagnosing an electrical problem that affects more than one component or system, check the electrical schematic for a common power source or a common ground. ● **SEE FIGURE 7–33** for an example of lights being powered by one fuse (power source).

- Underhood light
- Inside lighted mirrors
- Dome light
- Left-side courtesy light
- Right-side courtesy light

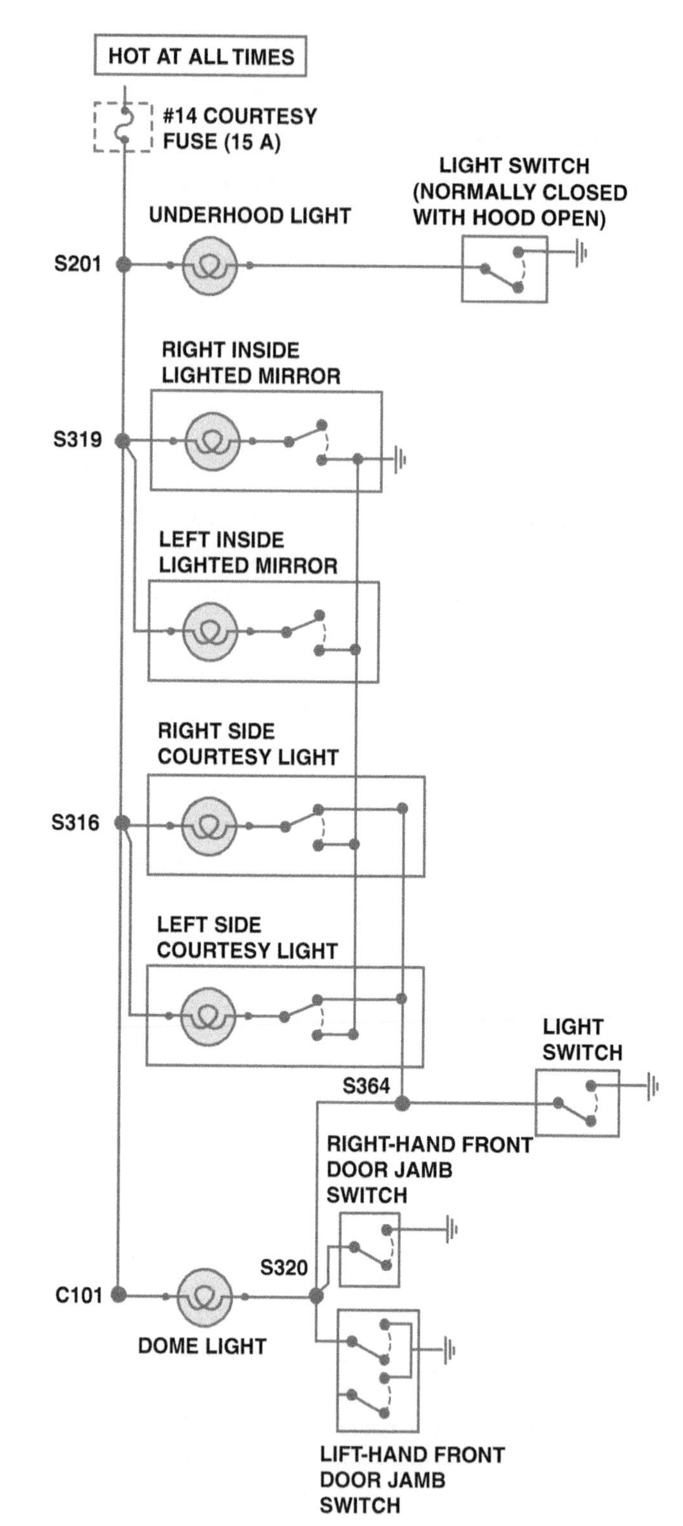

FIGURE 7–33 A typical wiring diagram showing multiple switches and bulbs powered by one fuse.

Therefore, if a customer complains about one or more of the items listed, check the fuse and the common part of the circuit that feeds all of the affected lights. Check for a common ground if several components that seem unrelated are not functioning correctly.

REAL WORLD FIX

The Electric Mirror Fault Story

Often, a customer will notice just one fault even though other lights or systems may not be working correctly. For example, a customer noticed that the electric mirrors stopped working. The service technician checked all electrical components in the vehicle and discovered that the interior lights were also not working.

The interior lights were not mentioned by the customer as being a problem most likely because the driver only used the vehicle in daylight hours.

The service technician found the interior light and power accessory fuse blown. Replacing the fuse restored the proper operation of the electric outside mirror and the interior lights. However, what caused the fuse to blow? A visual inspection of the dome light, next to the electric sunroof, showed an area where a wire was bare. Evidence showed the bare wire had touched the metal roof, which could cause the fuse to blow. The technician covered the bare wire with a section of vacuum hose and then taped the hose with electrical tape to complete the repair.

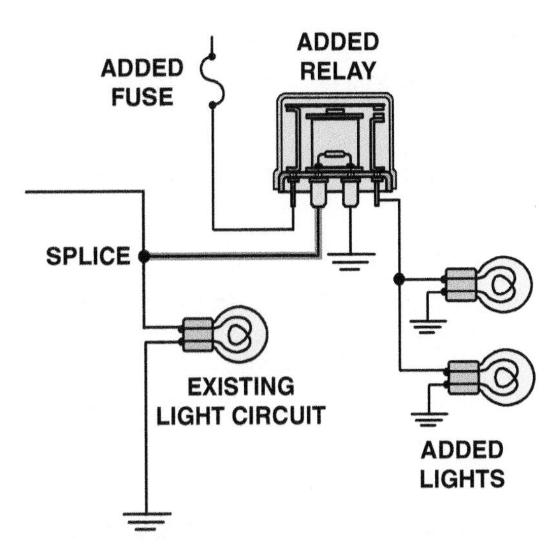

FIGURE 7–34 To add additional lighting, simply tap into an existing light wire and connect a relay. Whenever the existing light is turned on, the coil of the relay is energized. The arm of the relay then connects power from another circuit (fuse) to the auxiliary lights without overloading the existing light circuit.

TECH TIP

Do It Right—Install a Relay

Often the owners of vehicles, especially owners of pickup trucks and sport utility vehicles (SUVs), want to add additional electrical accessories or lighting. It is tempting in these cases to simply splice into an existing circuit. However, when another circuit or component is added, the current that flows through the newly added component is also added to the current for the original component. This additional current can easily overload the fuse and wiring. Do not simply install a larger amperage fuse; the wire gauge size was not engineered for the additional current and could overheat.

The solution is to install a relay, which uses a small coil to create a magnetic field that causes a movable arm to switch on a higher current circuit. The typical relay coil has from 50 to 150 ohms (usually 60 to 100 ohms) of resistance and requires just 0.24 to 0.08 ampere when connected to a 12-volt source. This small additional current will not be enough to overload the existing circuit. ● **SEE FIGURE 7–34** for an example of how additional lighting can be added.

CIRCUIT TROUBLESHOOTING PROCEDURE

Follow these steps when troubleshooting wiring problems.

STEP 1 Verify the malfunction. If, for example, the backup lights do not operate, make certain that the ignition is on (key on, engine off), with the gear selector in reverse, and check for operation of the backup lights.

STEP 2 Check everything else that does or does not operate correctly. For example, if the taillights are also not working, the problem could be a loose or broken ground connection in the trunk area that is shared by both the backup lights and the taillights.

FIGURE 7–35 Always check the simple things first. Check the fuse for the circuit you are testing. Maybe a fault in another circuit controlled by the same fuse could have caused the fuse to blow. Use a test light to check that both sides of the fuse have voltage. (Courtesy of Jeffrey Rehkopf)

STEP 3 Check the fuse for the backup lights. ● **SEE FIGURE 7–35.**

STEP 4 Check for voltage at the backup light socket. This can be done using a test light or a voltmeter.

If voltage is available at the socket, the problem is either a defective bulb or a poor ground at the socket or a ground wire connection to the body or frame. If no voltage is available at the socket, consult a wiring diagram for the type of vehicle being tested. The wiring diagram should show all of the wiring and components included in the circuit. For example, the backup light current must flow through the fuse and ignition switch to the gear selector switch before traveling to the rear backup light socket. As stated in the second step, the fuse used for the backup lights may also be used for other vehicle circuits.

The wiring diagram can be used to determine all other components that share the same fuse. If the fuse is blown (open circuit), the cause can be a short in any of the circuits sharing the same fuse. Because the backup light circuit current must be switched on and off by the gear selector switch, an open in the switch can also prevent the backup lights from functioning.

FREQUENTLY ASKED QUESTION

Where to Start?

The common question is, where does a technician start the troubleshooting when using a wiring diagram (schematic)?

HINT 1 If the circuit contains a relay, start your diagnosis at the relay. The entire circuit can be tested at the terminals of the relay.

HINT 2 The easiest first step is to locate the unit on the schematic that is not working at all or not working correctly.

a. Trace where the unit gets its ground connection.

b. Trace where the unit gets its power connection.

Often a ground is used by more than one component. Therefore, ensure that everything else is working correctly. If not, then the fault may lie at the common ground (or power) connection.

HINT 3 Divide the circuit in half by locating a connector or a part of the circuit that can be accessed easily. Then check for power and ground at this midpoint. This step could save you much time.

HINT 4 Use a fused jumper wire to substitute a ground or a power source to replace a suspected switch or section of wire.

LOCATING A SHORT CIRCUIT

TERMINOLOGY A short circuit usually blows a fuse, and a replacement fuse often also blows in the attempt to locate the source of the short circuit. A **short circuit** is an electrical connection to another wire or to ground before the current flows through some or all of the resistance in the circuit. A short-to-ground will always blow a fuse and usually involves a wire on the power side of the circuit coming in contact with metal.

Therefore, a thorough visual inspection should be performed around areas involving heat or movement, especially if there is evidence of a previous collision or previous repair that may not have been properly completed.

A short-to-voltage may or may not cause the fuse to blow and usually affects another circuit. Look for areas of heat or movement where two power wires could come in contact with each other. Several methods can be used to locate the short.

FUSE REPLACEMENT METHOD Disconnect one component at a time and then replace the fuse. If the new fuse blows, continue the process until you determine the location of the short. This method uses many fuses and is *not* a preferred method for finding a short circuit.

CIRCUIT BREAKER METHOD Another method is to connect an automotive circuit breaker to the contacts of the fuse holder with alligator clips. Circuit breakers are available that plug directly into the fuse panel, replacing a blade-type fuse. The circuit breaker will alternately open and close the circuit, protecting the wiring from possible overheating damage while still providing current flow through the circuit.

NOTE: A heavy-duty (HD) flasher can also be used in place of a circuit breaker to open and close the circuit. Wires and terminals must be made to connect the flasher unit where the fuse normally plugs in.

All components included in the defective circuit should be disconnected one at a time until the circuit breaker stops clicking. The unit that was disconnected and stopped the circuit breaker clicking is the unit causing the short circuit. If the circuit breaker continues to click with all circuit components unplugged, the problem is in the wiring *from* the fuse panel *to* any one of the units in the circuit. Visual inspection of all the wiring or further disconnecting will be necessary to locate the problem.

TEST LIGHT METHOD To use the test light method, simply remove the blown fuse and connect a test light to the terminals of the fuse holder (polarity does not matter). If there is a short circuit, current will flow from the power side of the fuse holder through the test light and on to ground through the short circuit, and the test light will then light. Unplug the connectors or components protected by the fuse until the test light goes out. The circuit that was disconnected, which caused the test light to go out, is the circuit that is shorted.

BUZZER METHOD The buzzer method is similar to the test light method, but uses a buzzer to replace a fuse and act as an electrical load. The buzzer will sound if the circuit is shorted and will stop when the part of the circuit that is grounded is unplugged.

OHMMETER METHOD The fourth method uses an ohmmeter connected to the fuse holder and ground. This is the recommended method of finding a short circuit, as an ohmmeter will indicate low ohms when connected to a short circuit. However, an ohmmeter should never be connected to an operating circuit. The correct procedure for locating a short using an ohmmeter is as follows:

1. Connect one lead of an ohmmeter (set to a low scale) to a good clean metal ground and the other lead to the circuit (load) side of the fuse holder.

   CAUTION: Connecting the lead to the power side of the fuse holder will cause current to flow through and damage the ohmmeter.

2. The ohmmeter will read zero or almost zero ohms if the circuit or a component in the circuit is shorted.
3. Disconnect one component in the circuit at a time and watch the ohmmeter. If the ohmmeter reading shoots to a high value or infinity, the component just unplugged was the source of the short circuit.
4. If all of the components have been disconnected and the ohmmeter still reads low ohms, then disconnect electrical connectors until the ohmmeter reads high ohms. The location of the short-to-ground is then between the ohmmeter and the disconnected connector.

NOTE: Some meters, such as the Fluke 87, can be set to beep (alert) when the circuit closes or when the circuit opens—a very useful feature.

GAUSS GAUGE METHOD If a short circuit blows a fuse, a special pulsing circuit breaker (similar to a flasher unit) can be installed in the circuit in place of the fuse. Current will flow through the circuit until the circuit breaker opens the circuit. As soon as the circuit breaker opens the circuit, it closes again. This on-and-off current flow creates a pulsing magnetic field around the wire carrying the current. A **Gauss gauge** is a handheld meter that responds to weak magnetic fields. It is used to observe this pulsing magnetic field, which is indicated on the gauge as needle movement. This pulsing magnetic field will register on the Gauss gauge even through the metal body of the vehicle. A needle-type compass can also be used to observe the pulsing magnetic field. ● **SEE FIGURES 7–36 AND 7–37.**

(a)

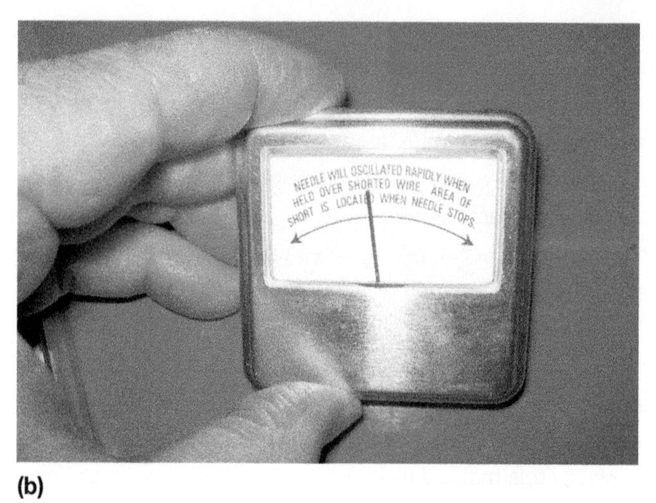

(b)

FIGURE 7–36 (a) After removing the blown fuse, a pulsing circuit breaker is connected to the terminals of the fuse. (b) The circuit breaker causes current to flow, then stop, then flow again, through the circuit up to the point of the short-to-ground. By observing the Gauss gauge, the location of the short is indicated near where the needle stops moving due to the magnetic field created by the flow of current through the wire.

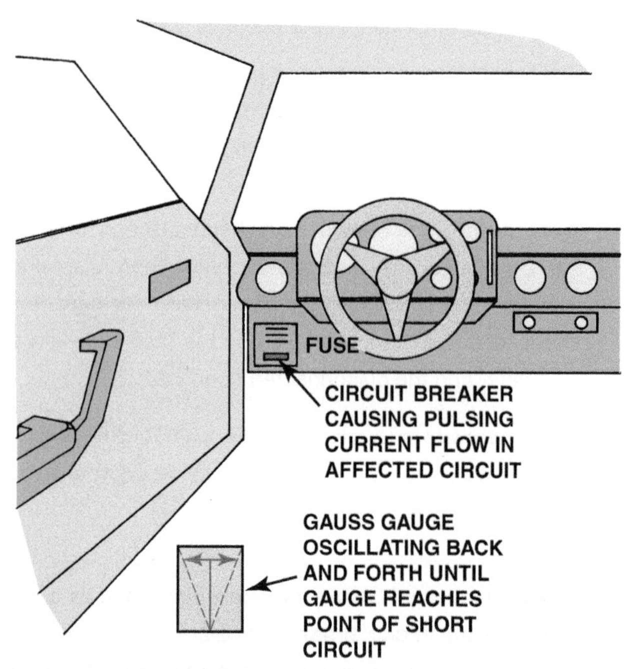

FIGURE 7–37 A Gauss gauge can be used to determine the location of a short circuit even behind a metal panel.

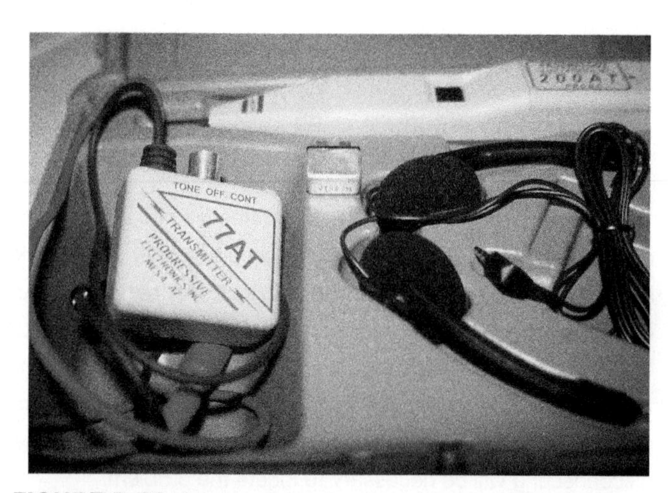

FIGURE 7–38 A tone generator-type tester used to locate open circuits and circuits that are shorted-to-ground. Included with this tester is a transmitter (tone generator), receiver probe, and headphones for use in noisy shops.

ELECTRONIC TONE GENERATOR TESTER An electronic tone generator tester can be used to locate a short-to-ground or an open circuit. Similar to test equipment used to test telephone and cable television lines, a **tone generator tester** generates a tone that can be heard through a receiver (probe). ● **SEE FIGURE 7–38.**

The tone will be generated as long as there is a continuous electrical path along the circuit. The signal will stop if there is a short-to-ground or an open in the circuit. ● **SEE FIGURE 7–39.**

The windings in the solenoids and relays will increase the strength of the signal in these locations.

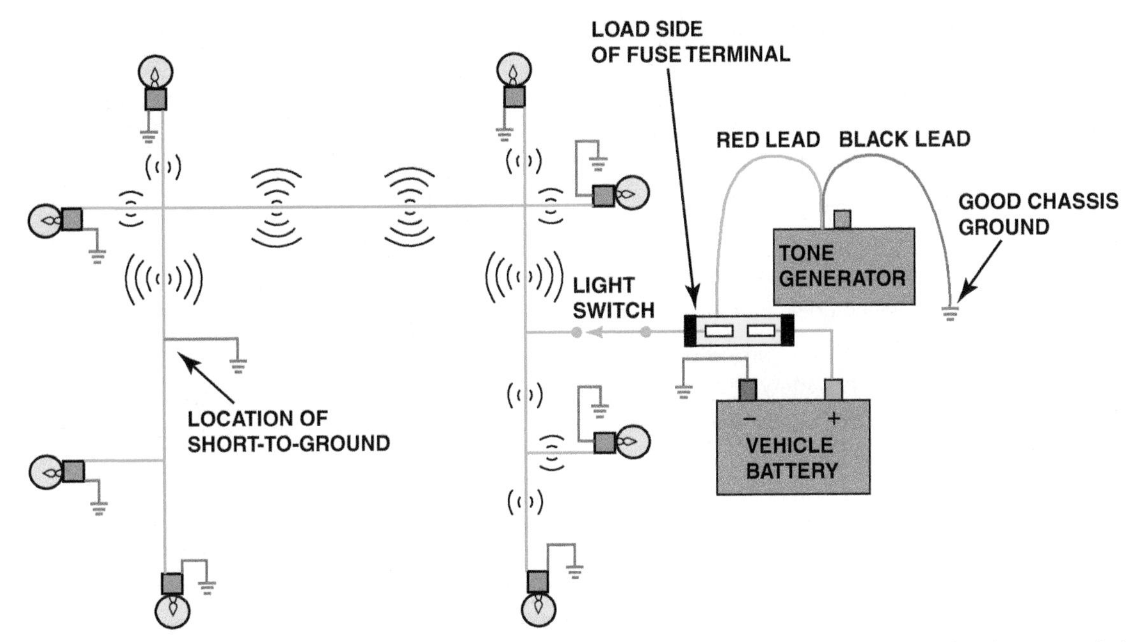

FIGURE 7–39 To check for a short-to-ground using a tone generator, connect the black transmitter lead to a good chassis ground and the red lead to the load side of the fuse terminal. Turn the transmitter on and check for tone signal with the receiver. Using a wiring diagram, follow the strongest signal to the location of the short-to-ground. There will be no signal beyond the fault, either a short-to-ground as shown or an open circuit.

ELECTRICAL TROUBLE-SHOOTING GUIDE

When troubleshooting any electrical component, remember the following hints to find the problem faster and more easily.

1. For a device to work, it must have two things: power and ground.
2. If there is no power to a device, an open power side (blown fuse, etc.) is indicated.
3. If there is power on both sides of a device, an open ground is indicated.
4. If a fuse blows immediately, a grounded power-side wire is indicated.
5. Most electrical faults result from heat or movement.
6. Most noncomputer-controlled devices operate by opening and closing the power side of the circuit (power-side switch).
7. Most computer-controlled devices operate by opening and closing the ground side of the circuit (ground-side switch).

TECH TIP

Heat or Movement

Electrical shorts are commonly caused either by movement, which causes the insulation around the wiring to be worn away, or by heat melting the insulation. When checking for a short circuit, first check the wiring that is susceptible to heat, movement, and damage.

1. **Heat.** Wiring near heat sources, such as the exhaust system, cigarette lighter, or alternator
2. **Wire movement.** Wiring that moves, such as in areas near the doors, trunk, or hood
3. **Damage.** Wiring subject to mechanical injury, such as in the trunk, where heavy objects can move around and smash or damage wiring, can also occur as a result of an accident or a previous repair

TECH TIP

Wiggle Test

Intermittent electrical problems are common yet difficult to locate. To help locate these hard-to-find problems, try operating the circuit and then start wiggling the wires and connections that control the circuit. If in doubt where the wiring goes, try moving all the wiring starting at the battery. Pay particular attention to wiring running near the battery or the windshield washer container. Corrosion can cause wiring to fail, and battery acid fumes and alcohol-based windshield washer fluid can start or contribute to the problem. If you notice any change in the operation of the device being tested while wiggling the wiring, look closer in the area you were wiggling until you locate and correct the actual problem.

STEP-BY-STEP TROUBLESHOOTING PROCEDURE

Knowing what should be done and when it should be done is a major concern for many technicians trying to repair an electrical problem. The following field-tested procedure provides a step-by-step guide for troubleshooting an electrical fault.

STEP 1 Determine the customer concern (complaint) and get as much information as possible from the customer or service advisor.

- **a.** When did the problem start?
- **b.** Under what conditions does the problem occur?
- **c.** Have there been any recent previous repairs to the vehicle which could have created the problem?

STEP 2 Verify the customer's concern by actually observing the fault.

STEP 3 Perform a thorough visual inspection and be sure to check everything that does and does not work.

STEP 4 Check for technical service bulletins (TSBs).

STEP 5 Locate the wiring schematic for the circuit being diagnosed.

FIGURE 7–40 Antistatic spray can be used by customers to prevent being shocked when they touch a metal object like the door handle.

REAL WORLD FIX

Shocking Experience

A customer complained that after driving for a while, he got a static shock whenever he grabbed the door handle when exiting the vehicle. The customer thought that there must be an electrical fault and that the shock was coming from the vehicle itself. In a way, the shock was caused by the vehicle, but it was not a fault. The service technician sprayed the cloth seats with an antistatic spray and the problem did not reoccur. Obviously, a static charge was being created by the movement of the driver's clothing on the seats and then discharged when the driver touched the metal door handle. ● **SEE FIGURE 7–40.**

STEP 6 Check the factory service information and follow the troubleshooting procedure.

- **a.** Determine how the circuit works.
- **b.** Determine which part of the circuit is good, based on what works and what does not work.
- **c.** Isolate the problem area.

NOTE: Split the circuit in half to help isolate the problem and start at the relay (if the circuit has a relay).

STEP 7 Determine the root cause and repair the vehicle.

STEP 8 Verify the repair and complete the work order by listing the three Cs (complaint, cause, and correction).

SUMMARY

1. Most wiring diagrams include the wire color, circuit number, and wire gauge.
2. The number used to identify connectors, grounds, and splices usually indicates where they are located in the vehicle.
3. All switches and relays on a schematic are shown in their normal position either normally closed (N.C.) or normally open (N.O.).
4. A typical relay uses a small current through a coil (terminals 85 and 86) to operate the higher current part (terminals 30 and 87).
5. A short-to-voltage affects the power side of the circuit and usually involves more than one circuit.
6. A short-to-ground usually causes the fuse to blow and usually affects only one circuit.
7. Most electrical faults are a result of heat or movement.

REVIEW QUESTIONS

1. List the numbers used on schematics to indicate grounds, splices, and connectors and where they are used in the vehicle.
2. List and identify the terminals of a typical ISO-type relay.
3. List three methods that can be used to help locate a short circuit.
4. How can a tone generator be used to locate a short circuit?

CHAPTER QUIZ

1. On a wiring diagram, S110 with a "0.8 BRN/BLK" means ____________.
   a. Circuit #.8, spliced under the hood
   b. A connector with 0.8 mm^2 wire
   c. A splice of a brown with black stripe, wire size being 0.8 mm^2 (18 gauge AWG)
   d. Both a and b
2. Where is connector X250 located on the vehicle?
   a. Under the hood
   b. Under the dash
   c. In the passenger compartment
   d. In the trunk
3. All switches illustrated in schematics are ____________.
   a. Shown in their normal position
   b. Always shown in their on position
   c. Always shown in their off position
   d. Shown in their on position except for lighting switches
4. When testing a relay using an ohmmeter, which two terminals should be touched to measure the coil resistance?
   a. 87 and 30
   b. 86 and 85
   c. 87a and 87
   d. 86 and 87
5. Technician A says that a good relay should measure between 60 and 100 ohms across the coil terminals. Technician B says that OL should be displayed on an ohmmeter when touching terminals 30 and 87. Which technician is correct?
   a. Technician A only
   b. Technician B only
   c. Both Technicians A and B
   d. Neither Technician A nor B
6. Which relay terminal is the normally closed (N.C.) terminal?
   a. 30
   b. 85
   c. 87
   d. 87a
7. Technician A says that there is often more than one circuit being protected by each fuse. Technician B says that more than one circuit often shares a single ground connector. Which technician is correct?
   a. Technician A only
   b. Technician B only
   c. Both Technicians A and B
   d. Neither Technician A nor B
8. Two technicians are discussing finding a short-to-ground using a test light. Technician A says that the test light, connected in place of the fuse, will light when the circuit that has the short is disconnected. Technician B says that the test light should be connected to the positive (+) and negative (–) terminals of the battery during this test. Which technician is correct?
   a. Technician A only
   b. Technician B only
   c. Both Technicians A and B
   d. Neither Technician A nor B
9. A short circuit can be located using a ____________.
   a. Test light
   b. Gauss gauge
   c. Tone generator
   d. All of the above
10. For an electrical device to operate, it must have ____________.
    a. Power and a ground
    b. A switch and a fuse
    c. A ground and fusible link
    d. A relay to transfer the current to the device

AND CAPACITORS

LEARNING OBJECTIVES

After studying this chapter, the reader will be able to:

1. Explain capacitance.
2. Explain how a capacitor can be used to filter electronic noise, can store electrical charge, and can be used as a timer circuit.

This chapter will help you prepare for the ASE Electrical/Electronic Systems (A6) certification test content area "A" (General Electrical/Electronic System Diagnosis).

KEY TERMS

 STC OBJECTIVES

GM Service Technical College topics covered in this chapter are as follows:

1. Identifying important types, characteristics, and diagnosis of various solid state electrical component (18043.03W-R3).
2. Electrical components (08510.15W).

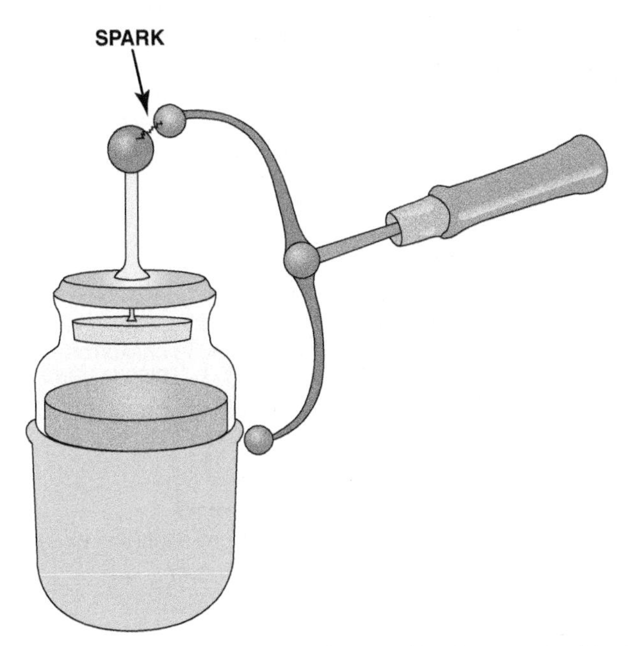

FIGURE 8–1 A Leyden jar can be used to store an electrical charge.

CAPACITANCE

DEFINITION **Capacitance** is the ability of an object or surface to store an electrical charge. Around 1745, Ewald Christian von Kliest and Pieter van Musschenbroek independently discovered capacitance in an electric circuit. While engaged in separate studies of electrostatics, they discovered that an electric charge could be stored for a period of time. They used a device, now called a **Leyden jar**, for their experimentation, which consisted of a glass jar filled with water, with a nail piercing the stopper and dipping into the water. ● **SEE FIGURE 8–1**.

The two scientists connected the nail to an electrostatic charge. After disconnecting the nail from the source of the charge, they felt a shock by touching the nail, demonstrating that the device had stored the charge.

In 1747, John Bevis lined both the inside and outside of the jar with foil. This created a capacitor with two conductors (the inside and outside metal foil layers) equally separated by the insulating glass. The Leyden jar was also used by Benjamin Franklin to store the charge from lightning as well as in other experiments. The natural phenomenon of lightning includes capacitance, because huge electrical fields develop between cloud layers or between clouds and the earth prior to a lightning strike.

NOTE: Capacitors are also called **condensers**. This term developed because electric charges collect, or condense, on the plates of a capacitor much like water vapor collects and condenses on a cold bottle or glass.

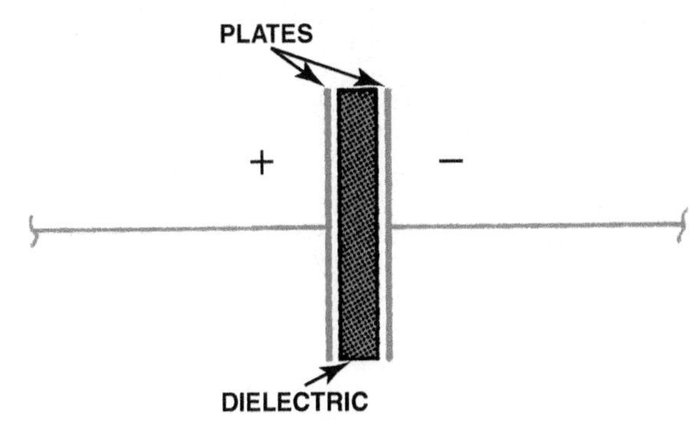

FIGURE 8–2 This simple capacitor, made of two plates separated by an insulating material, is called a dielectric.

MATERIAL	DIELECTRIC CONSTANT
Vacuum	1
Air	1.00059
Polystyrene	2.5
Paper	3.5
Mica	5.4
Flint glass	9.9
Methyl alcohol	35
Glycerin	56.2
Pure water	81

CHART 8–1

The higher the dielectric constant is, the better the insulating properties between the plates of the capacitor.

CAPACITOR CONSTRUCTION AND OPERATION

CONSTRUCTION A capacitor (also called a condenser) consists of two conductive plates with an insulating material between them. The insulating material is commonly called a **dielectric**. This substance is a poor conductor of electricity and can include air, mica, ceramic, glass, paper, plastic, or any similar nonconductive material. The dielectric constant is the relative strength of a material against the flow of electrical current. The higher the number is, the better the insulating properties. ● **SEE CHART 8–1**.

OPERATION When a capacitor is placed in a closed circuit, the voltage source (battery) forces electrons around the circuit. Because electrons cannot flow through the dielectric of the capacitor, excess electrons collect on what becomes the negatively charged plate. At the same time, the other plate loses electrons and, therefore, becomes positively charged. ● **SEE FIGURE 8–2**.

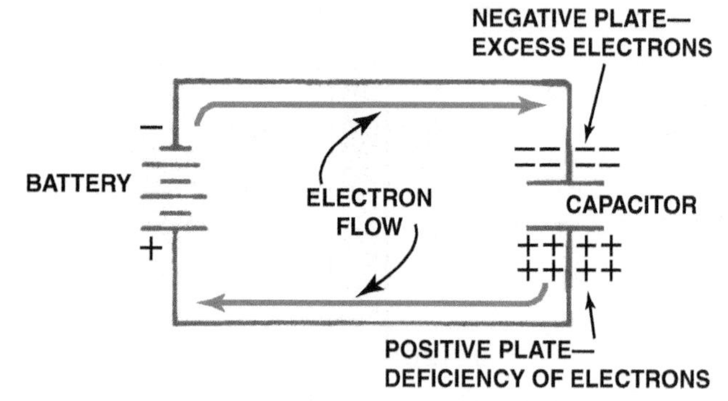

FIGURE 8–3 As the capacitor is charging, the battery forces electrons through the circuit.

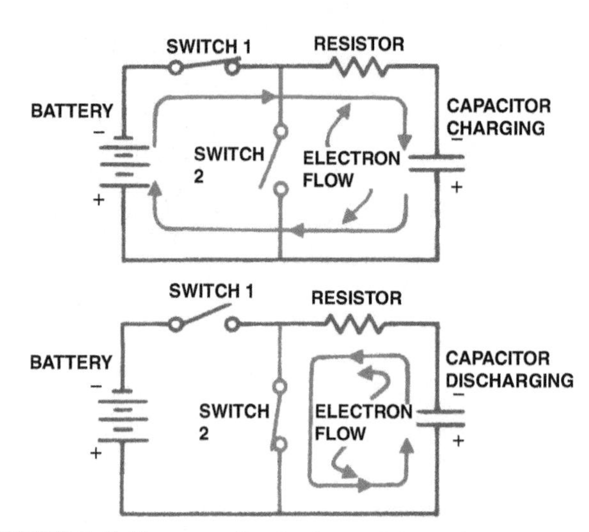

FIGURE 8–5 The capacitor is charged through one circuit (top) and discharged through another (bottom).

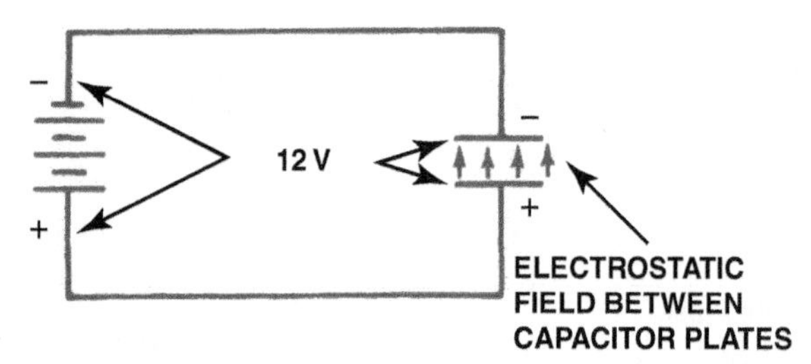

FIGURE 8–4 When the capacitor is charged, there is equal voltage across the capacitor and the battery. An electrostatic field exists between the capacitor plates. No current flows in the circuit.

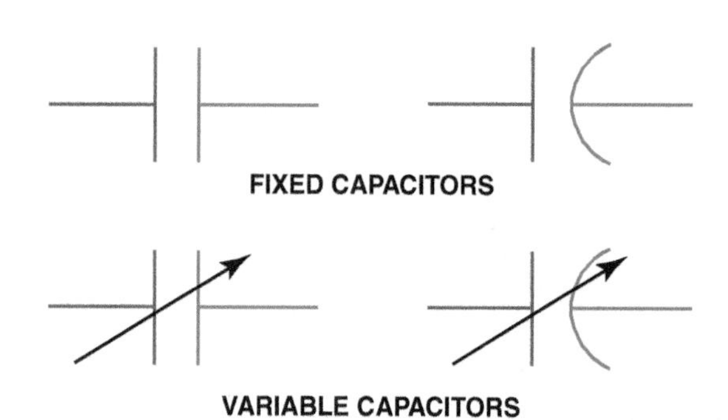

FIGURE 8–6 Capacitor symbols are shown in electrical diagrams. The negative plate is often shown curved.

Current continues until the voltage charge across the capacitor plates becomes the same as the source voltage. At that time, the negative plate of the capacitor and the negative terminal of the battery are at the same negative potential. ● **SEE FIGURE 8–3.**

The positive plate of the capacitor and the positive terminal of the battery are also at equal positive potentials. There is then a voltage charge across the battery terminals and an equal voltage charge across the capacitor plates. The circuit is in balance, and there is no current. An electrostatic field now exists between the capacitor plates because of their opposite charges. It is this field that stores energy. In other words, a charged capacitor is similar to a charged battery. ● **SEE FIGURE 8–4.**

If the circuit is opened, the capacitor will hold its charge until it is connected into an external circuit through which it can discharge. When the charged capacitor is connected to an external circuit, it discharges. After discharging, both plates of the capacitor are neutral because all the energy from a circuit stored in a capacitor is returned when it is discharged. ● **SEE FIGURE 8–5.**

Theoretically, a capacitor holds its charge indefinitely. Actually, the charge slowly leaks off the capacitor through the dielectric. The better the dielectric, the longer the capacitor holds its charge. To avoid an electrical shock, any capacitor should be treated as if it were charged until it is proven to be discharged. To safely discharge a capacitor, use a test light with the clip attached to a good ground and touch the pigtail or terminal with the point of the test light. ● **SEE FIGURE 8–6** for the symbol for capacitors as used in electrical schematics.

FACTORS OF CAPACITANCE

Capacitance is governed by three factors.

- The surface area of the plates
- The distance between the plates
- The dielectric material

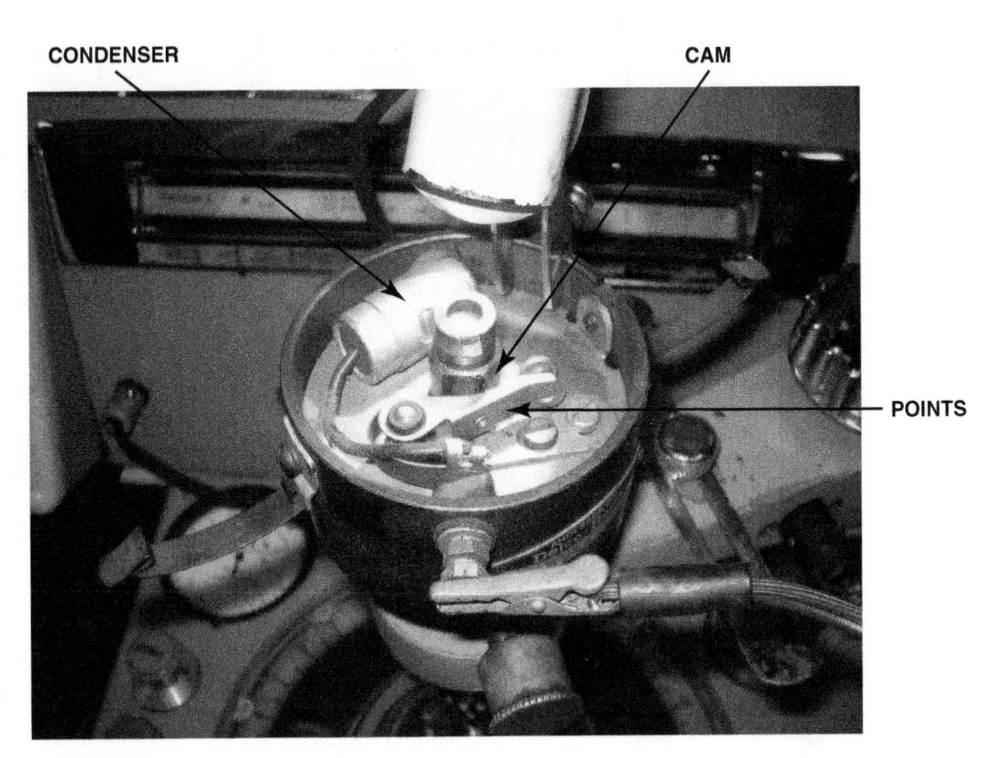

FIGURE 8–7 A point-type distributor shown with the condenser from an old vehicle being tested on a distributor machine.

FREQUENTLY ASKED QUESTION

What Are "Points and Condenser"?

Points and condenser are used in point-type ignition systems.

Points. A set of points uses one stationary contact and a movable contact that is opened by a cam lobe inside the ignition distributor. When the points are closed, current flows through the primary windings of the ignition coil and creates a strong magnetic field. As the engine rotates, the distributor can open the contact points, which opens the circuit to the coil. The stored magnetic field in the coil collapses and generates a high-voltage arc from the secondary winding of the coil. It is this spark that is sent to the spark plugs that ignite the air–fuel mixture inside the engine.

Condenser. The condenser (capacitor) is attached to the points and the case of the condenser is grounded. When the points start to open, the charge built up in the primary winding of the coil would likely start to arc across the opening points. To prevent the points from arcing and to increase how rapidly the current is turned off, the condenser stores the current temporarily.

Points and condenser were used in vehicles and small gasoline engines until the mid-1970s. ● **SEE FIGURE 8–7**.

The larger the surface area of the plates is, the greater the capacitance, because more electrons collect on a larger plate area than on a small one. The closer the plates are to each other, the greater the capacitance, because a stronger electrostatic field exists between charged bodies that are close together. The insulating qualities of the dielectric material also affect capacitance. The capacitance of a capacitor is higher if the dielectric is a very good insulator.

MEASUREMENT OF CAPACITANCE Capacitance is measured in **farads**, which is named after Michael Faraday (1791–1867), an English physicist. The symbol for farads is the letter *F*. If a charge of 1 coulomb is placed on the plates of a capacitor and the potential difference between them is 1 volt, then the capacitance is defined to be 1 farad, or 1 F. One coulomb is equal to the charge of 6.25×10^{18} electrons. One farad is an extremely large quantity of capacitance. Microfarads (0.000001 farad), or μF, are more commonly used.

The capacitance of a capacitor is proportional to the quantity of charge that can be stored in it for each volt difference in potential.

USES FOR CAPACITORS

SPIKE SUPPRESSION A capacitor can be used in parallel to a coil to reduce the resulting voltage spike that occurs when the circuit is opened. The energy stored to the magnet field of

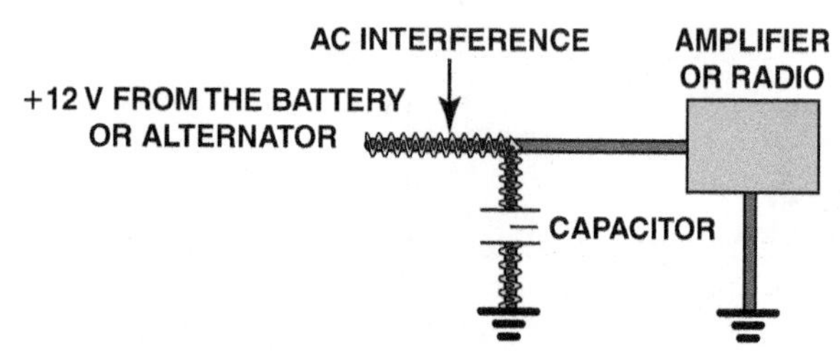

FIGURE 8–8 A capacitor blocks direct current (DC) but passes alternating current (AC). A capacitor makes a very good noise suppressor because most of the interference is AC and the capacitor will conduct this AC to ground before it can reach the radio or amplifier.

FIGURE 8–9 A 1 farad capacitor used to boost the power to large speakers.

the coil is rapidly released at this time. The capacitor acts to absorb the high voltage produced and stop it from interfering with other electronic devices, such as automotive radio and video equipment.

NOISE FILTERING Interference in a sound system or radio is usually due to alternating current (AC) voltage created somewhere in the vehicle, such as in the alternator. A capacitor does the following:

- Blocks the flow of direct current (DC)
- Allows alternating current (AC) to pass

By connecting a capacitor (condenser) to the power lead of the radio or sound system amplifier, the AC voltage passes through the capacitor to the ground where the other end of the capacitor is connected. Therefore, the capacitor provides a path for the AC without affecting the DC power circuit. ● **SEE FIGURE 8–8.**

Because a capacitor stores a voltage charge, it opposes or slows any voltage change in a circuit. Therefore, capacitors are often used as voltage "shock absorbers." On older vehicles a capacitor may be found attached to one terminal of the ignition coil. In this application, the capacitor absorbs and dampens changes in ignition voltage that interfere with AM radio reception.

SUPPLEMENTAL POWER SOURCE A capacitor can be used to supply electrical power for short bursts in an audio system to help drive the speakers. Woofers and subwoofers require a lot of electrical current that often cannot be delivered by the amplifier itself. ● **SEE FIGURE 8–9.**

Capacitors are used in the air bag sensing and diagnostic module (SDM) as a backup power source in case the battery is disconnected in a collision. ● **SEE FIGURE 8–10.**

FIGURE 8–10 The capacitors inside the module supply power to the SDM even after the battery is disconnected. (Courtesy of Jeffrey Rehkopf)

In a hybrid vehicle powertrain inverter module (PIM) large capacitors are used to store voltage for quick use by the drive motors. The capacitors can also quickly absorb the energy generated during regenerative braking, later using that stored energy to help recharge the high voltage (HV) battery. ● **SEE FIGURE 8–11.**

TIMER CIRCUITS Capacitors are used in electronic circuits as part of a timer, to control window defoggers, interior lighting, pulse wipers, and automatic headlights. The capacitors store energy and then are allowed to discharge through a resistance load. The greater the capacity of the capacitor and the higher the resistance load, the longer the time it takes for the capacitor to discharge.

FIGURE 8–11 Capacitors as used inside of a hybrid vehicle inverter module. (Courtesy of Jeffrey Rehkopf)

HIGH-VOLTAGE INJECTOR DRIVERS Capacitors are used on Duramax diesel engines equipped with Bosch piezoelectric fuel injectors. These injectors operate a high voltage, as indicated by the orange color of the injector harness. The ECM supplies high voltage to charge the capacitors. This voltage is supplied to the injectors at up to 160 v at 20 amps and can peak up to 240 v. This causes the injector to open. The capacitor discharges through the injector for initial opening and then holds the injector open with 12 v.

CAUTION: Do not make contact with the fuel injector harness, ECM, or fuel injectors while the ignition is in the "on" or "run" position. Use certified insulated gloves EL-48286. These Class 0 gloves are rated at 1000 v. Check for functionality and check the expiration date of the gloves.

On a spark ignition direct injection (SIDI) engine, the ECM contains a converter that steps up voltage from 12 volts to 65 volts and charges a capacitor. The capacitor provides 65 volts to open the injector. Then the ECM provides pulse-width modulated 12 volts to hold the injector open for the prescribed time.

CONDENSER MICROPHONES A microphone converts sound waves into an electric signal. All microphones have a diaphragm that vibrates as sound waves strike. The vibrating diaphragm in turn causes an electrical component to create an output flow of current at a frequency proportional to the sound waves. A condenser microphone uses a capacitor for this purpose.

In a condenser microphone, the diaphragm is the negatively charged plate of a charged capacitor. When a sound wave compresses the diaphragm, the diaphragm is moved closer to the positive plate. Decreasing the distance between the plates increases the electrostatic attraction between them, which results in a flow of current to the negative plate. As the diaphragm moves out in response to sound waves, it also moves further from the positive plate. Increasing the distance between the plates decreases the electrostatic attraction between them. This results in a flow of current back to the positive plate. These alternating flows of current provide weak electronic signals that travel to an amplifier and then to a loudspeaker.

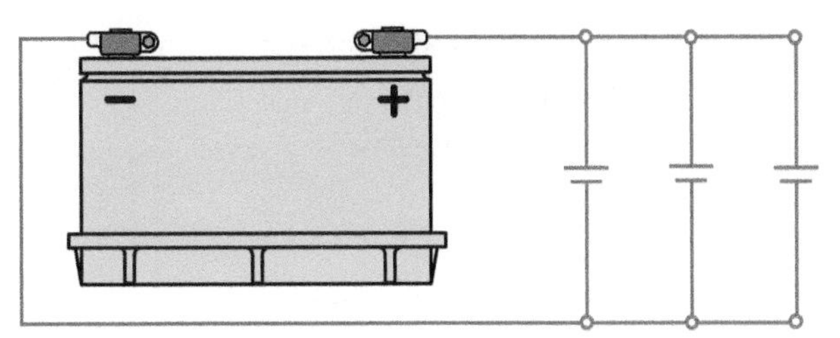

FIGURE 8–12 Capacitors in parallel effectively increase the capacitance.

CAPACITORS IN CIRCUITS

CAPACITORS IN PARALLEL CIRCUITS Capacitance can be increased in a circuit by connecting capacitors in parallel. For example, if a greater boost is needed for a sound system, then additional capacitors should be connected in parallel because their value adds together. ● **SEE FIGURE 8–12**.

We know that capacitance of a capacitor can be increased by increasing the size of its plates. Connecting two or more capacitors in parallel in effect increases plate size. Increasing plate area makes it possible to store more charge and therefore creates greater capacitance. To determine total capacitance of several parallel capacitors, simply add up their individual values. The following is the formula for calculating total capacitance in a circuit containing capacitors in parallel.

$$C_T = C_1 + C_2 + C_3 \ldots$$

For example, 220 μF + 220 μF = 400 μF when connected in parallel.

CAPACITORS IN SERIES CIRCUITS Capacitance in a circuit can be decreased by placing capacitors in series, as shown in ● **FIGURE 8–13**.

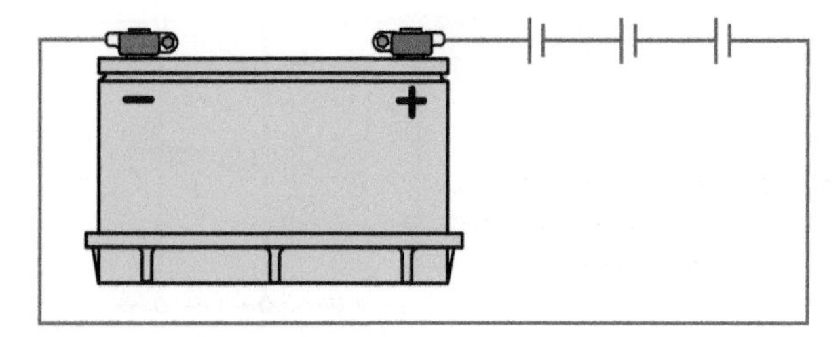

FIGURE 8–13 Capacitors in series decrease the capacitance.

We know that capacitance of a capacitor can be decreased by placing the plates further apart. Connecting two or more capacitors in series in effect increases the distance between the plates and thickness of the dielectric, thereby decreasing the amount of capacitance.

Following is the formula for calculating total capacitance in a circuit containing two capacitors in series.

$$C_T = \frac{C_1 \times C_2}{C_1 + C_2}$$

For example, $\frac{220\ \mu F \times 220\ \mu F}{220\ \mu F + 220\ \mu F} = \frac{48{,}400}{440} = 110\ \mu F$

NOTE: Capacitors are often used to reduce radio interference or to improve the performance of a high-power sound system. Additional capacitance can, therefore, be added by attaching another capacitor in parallel.

SUPPRESSION CAPACITORS Capacitors are installed across many circuits and switching points to absorb voltage fluctuations. Among other applications, they are used across the following:

- The primary circuit of some electronic ignition modules
- The output terminal of most alternators
- The armature circuit of some electric motors

Radio choke coils reduce current fluctuations resulting from self-induction. They are often combined with capacitors to act as electromagnetic interference (EMI) filter circuits for windshield wiper and electric fuel pump motors. Filters also may be incorporated in wiring connectors.

SUMMARY

1. Capacitors (condensers) are used in numerous automotive applications.
2. Capacitors can block direct current and pass alternating current.
3. Capacitors are used to control radio-frequency interference and are installed in various electronic circuits to control unwanted noise.
4. Capacitors connected in series reduce the capacitance, whereas if connected in parallel increase the capacitance.

REVIEW QUESTIONS

1. How does a capacitor store an electrical charge?
2. How should two capacitors be electrically connected if greater capacitance is needed?
3. Where can a capacitor be used as a power source?
4. How can a capacitor be used as a noise filter?

CHAPTER QUIZ

1. A capacitor ________.
   a. Stores electrons
   b. Passes AC
   c. Blocks DC
   d. All of the above
2. To increase the capacity, capacitors should be connected in ________.
   a. Series
   b. Parallel
   c. With resistors connected between the leads
   d. Series-parallel
3. Capacitors are commonly used as a ________.
   a. Voltage supply
   b. Timer
   c. Noise filter
   d. All of the above
4. A charged capacitor acts like a ________.
   a. Switch
   b. Battery
   c. Resistor
   d. Coil
5. The unit of measurement for capacitor rating is the ________.
   a. Ohm
   b. Volt
   c. Farad
   d. Ampere
6. Two technicians are discussing the operation of a capacitor. Technician A says that a capacitor can create electricity. Technician B says that a capacitor can store electricity. Which technician is correct?
   a. Technician A only
   b. Technician B only
   c. Both Technicians A and B
   d. Neither Technician A nor B
7. Capacitors block the flow of ________ current but allow ________ current to pass.
   a. Strong; weak
   b. AC; DC
   c. DC; AC
   d. Weak; strong
8. To increase the capacity, what could be done?
   a. Connect another capacitor in series
   b. Connect another capacitor in parallel
   c. Add a resistor between two capacitors
   d. Both a and b
9. A capacitor can be used in what components?
   a. Microphone
   b. Radio
   c. Speaker
   d. All of the above
10. A capacitor used for spike protection will normally be placed in ________ to the load or circuit.
    a. Series
    b. Parallel
    c. Either series or parallel
    d. Parallel with a resistor in series

chapter 9

MAGNETISM AND ELECTROMAGNETISM

LEARNING OBJECTIVES

After studying this chapter, the reader will be able to:

1. Explain how magnetism and voltage are related.
2. Explain how an electromagnet works.
3. Describe how an ignition coil works.

This chapter will help you prepare for the ASE Electrical/Electronic Systems (A6) certification test content area "A" (General Electrical/Electronic System Diagnosis).

KEY TERMS

Ampere-turns 117
Counter electromotive force (CEMF) 121
Electromagnetic interference (EMI) 123
Flux density 114
Flux lines 113
Ignition control module (ICM) 123
Left-hand rule 115
Lenz's law 120
Magnetic flux 113
Magnetic induction 114
Magnetism 113
Mutual induction 121
Permeability 114
Pole 113
Relay 118
Reluctance 115
Residual magnetism 114
Turns ratio 122

GM STC OBJECTIVES

GM Service Technical College topics covered in this chapter are as follows:

1. Properties of permanent magnets and electromagnets (18043.02W-R4).
2. The relationship between electricity and magnetism (08510.11W).
3. The use of magnetism in the operation of relays and solenoids.
4. The use of electromagnetic induction in the operation of transformers and generators.

FIGURE 9–1 A freely suspended natural magnet (lodestone) will point toward the magnetic north pole.

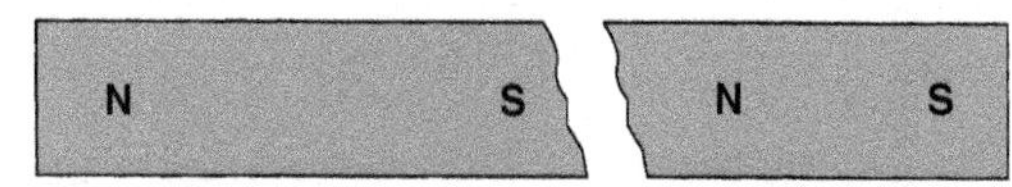

FIGURE 9–2 If a magnet breaks or is cracked, it becomes two weaker magnets.

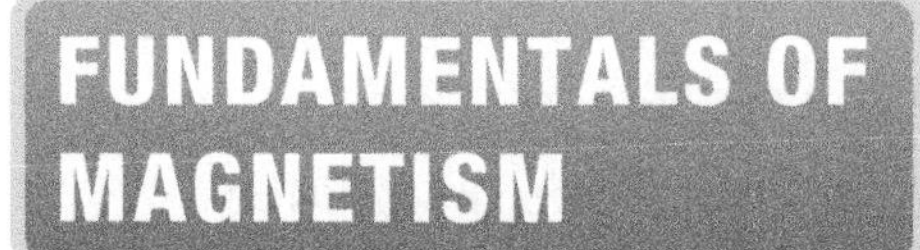

FUNDAMENTALS OF MAGNETISM

DEFINITION **Magnetism** is a form of energy that is caused by the motion of electrons in some materials. It is recognized by the attraction it exerts on other materials. Like electricity, magnetism cannot be seen. It can be explained in theory, however, because it is possible to see the results of magnetism and recognize the actions that it causes. Magnetite is the most naturally occurring magnet. Naturally magnetized pieces of magnetite, called *lodestone,* will attract and hold small pieces of iron. ● **SEE FIGURE 9–1.**

Many other materials can be artificially magnetized to some degree, depending on their atomic structure. Soft iron is very easy to magnetize, whereas some materials, such as aluminum, glass, wood, and plastic, cannot be magnetized at all.

TECH TIP

A Cracked Magnet Becomes Two Magnets

Magnets are commonly used in vehicle crankshaft, camshaft, and wheel speed sensors. If a magnet is struck and cracks or breaks, the result is two smaller-strength magnets. Because the strength of the magnetic field is reduced, the sensor output voltage is also reduced. A typical problem occurs when a magnetic crankshaft sensor becomes cracked, resulting in a no-start condition. Sometimes the cracked sensor works well enough to start an engine that is cranking at normal speeds but will not work when the engine is cold. ● **SEE FIGURE 9–2.**

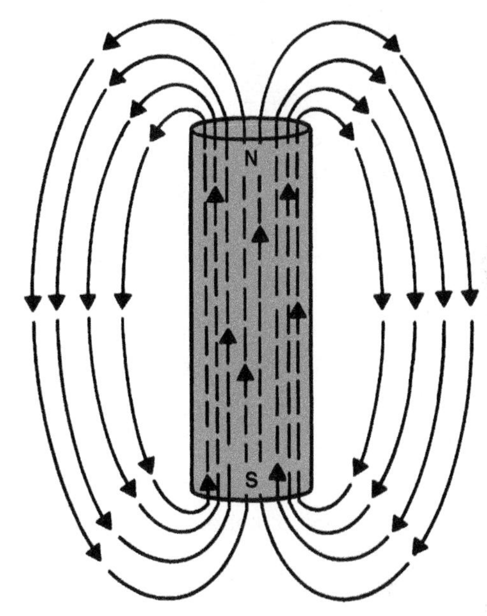

FIGURE 9–3 Magnetic lines of force leave the north pole and return to the south pole of a bar magnet.

LINES OF FORCE The lines that create a field of force around a magnet are believed to be caused by the way groups of atoms are aligned in the magnetic material. In a bar magnet, the lines are concentrated at both ends of the bar and form closed, parallel loops in three dimensions around the magnet. Force does not flow along these lines the way electrical current flows, but the lines *do* have direction. They come out of the north end, or **pole**, of the magnet and enter at the other end. ● **SEE FIGURE 9–3.**

The opposite ends of a magnet are called its north and south poles. In reality, they should be called the "north seeking" and "south seeking" poles, because they seek the earth's North Pole and South Pole, respectively.

The more lines of force that are present, the stronger the magnet becomes. The magnetic lines of force, also called **magnetic flux** or **flux lines**, form a magnetic field. The terms *magnetic field, lines of force, flux*, and *flux lines* are used interchangeably.

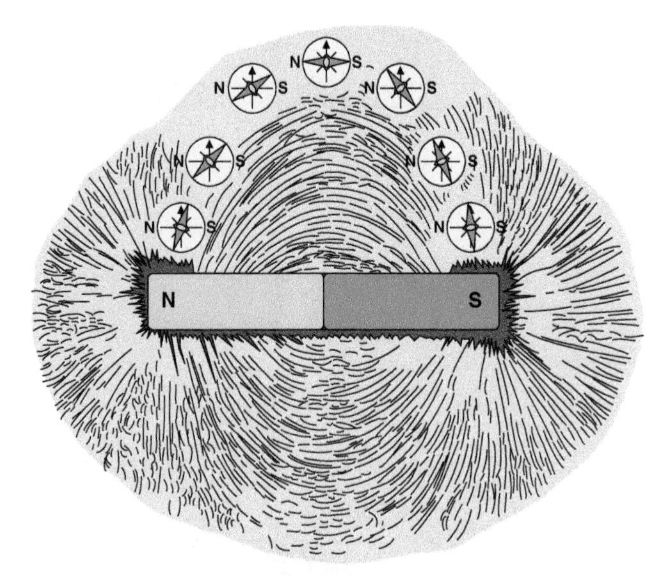

FIGURE 9–4 Iron filings and a compass can be used to observe the magnetic lines of force.

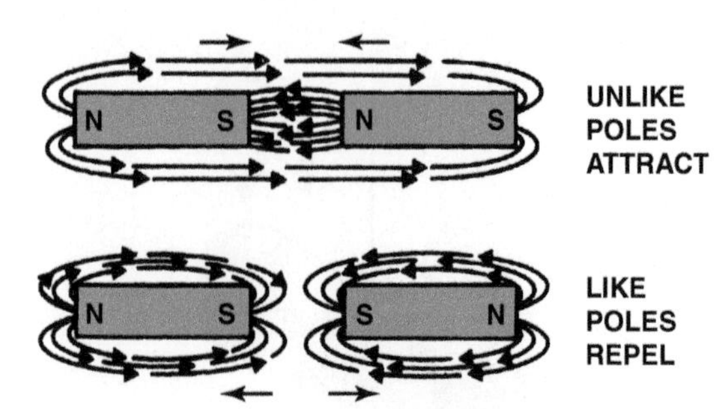

FIGURE 9–5 Magnetic poles behave like electrically charged particles—unlike poles attract and like poles repel.

TECH TIP

Magnetize a Steel Needle

A piece of steel can be magnetized by rubbing a magnet in one direction along the steel. This causes the atoms to line up in the steel, so it acts like a magnet. The steel often will not remain magnetized, whereas the true magnet is permanently magnetized.

When soft iron or steel is used, such as a paper clip, it will lose its magnetism quickly. The atoms in a magnetized needle can be disturbed by heating it or by dropping the needle on a hard object, which would cause the needle to lose its magnetism. Soft iron is used inside ignition coils because it will not keep its magnetism.

Flux density refers to the number of flux lines per unit of area. A magnetic field can be measured using a Gauss gauge, named for German scientist Johann Carl Friedrick Gauss (1777–1855).

Magnetic lines of force can be seen by spreading fine iron filings or dust on a piece of paper laid on top of a magnet. A magnetic field can also be observed by using a compass. A compass is simply a thin magnet or magnetized iron needle balanced on a pivot. The needle will rotate to point toward the opposite pole of a magnet. The needle can be very sensitive to small magnetic fields. Because it is a small magnet, a compass usually has one north end (marked N) and one south end (marked S). ● **SEE FIGURE 9–4**.

MAGNETIC INDUCTION If a piece of iron or steel is placed in a magnetic field, it will also become magnetized. This process of creating a magnet by using a magnetic field is called **magnetic induction**.

If the metal is then removed from the magnetic field, and it retains some magnetism, this is called **residual magnetism**.

ATTRACTING OR REPELLING The poles of a magnet are called north (N) and south (S) because, when a magnet is suspended freely, the poles tend to point toward the earth's North Pole and South Pole. Magnetic flux lines exit from the north pole and bend around to enter the south pole. An equal number of lines exit and enter, so magnetic force is equal at both poles of a magnet. Flux lines are concentrated at the poles, and therefore magnetic force (flux density) is stronger at the ends.

Magnetic poles behave like positively and negatively charged particles. When unlike poles are placed close together, the lines exit from one magnet and enter the other. The two magnets are pulled together by flux lines. If like poles are placed close together, the curving flux lines meet head on, forcing the magnets apart. Therefore, like poles of a magnet repel and unlike poles attract. ● **SEE FIGURE 9–5**.

PERMEABILITY Magnetic flux lines cannot be insulated. There is no known material through which magnetic force does not pass, if the force is strong enough. However, some materials allow the force to pass though more easily than others. This degree of passage is called **permeability**. Iron allows magnetic flux lines to pass through much more easily than air, so iron is highly permeable.

FIGURE 9–6 A crankshaft position sensor and reluctor (notched wheel) (Courtesy of General Motors).

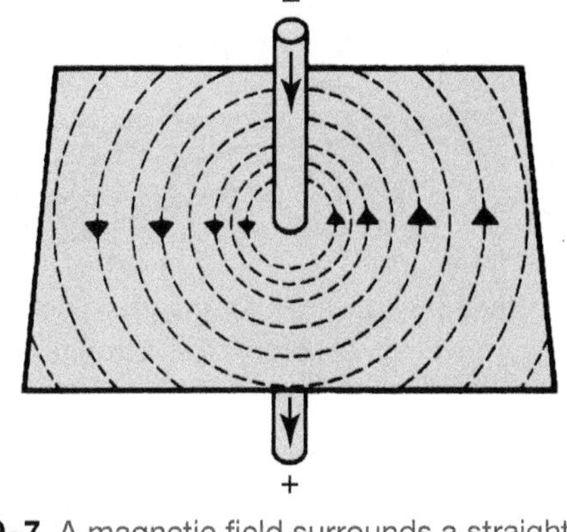

FIGURE 9–7 A magnetic field surrounds a straight, current-carrying conductor.

An example of this characteristic is the use of a reluctor wheel in magnetic-type camshaft position (CMP) and crankshaft position (CKP) sensors. The teeth on a reluctor cause the magnetic field to increase as each tooth gets closer to the sensor and decrease as the tooth moves away, thus creating an AC voltage signal. ● SEE FIGURE 9–6.

RELUCTANCE Although there is no absolute insulation for magnetism, certain materials resist the passage of magnetic force. This can be compared to resistance without an electrical circuit. Air does not allow easy passage, so air has a high **reluctance**. Magnetic flux lines tend to concentrate in permeable materials and avoid materials with high reluctance. As with electricity, magnetic force follows the path of least resistance.

ELECTROMAGNETISM

DEFINITION Scientists did not discover that current-carrying conductors also are surrounded by a magnetic field until 1820. These fields may be made many times stronger than those surrounding conventional magnets. Also, the magnetic field strength around a conductor may be controlled by changing the current.

- As current increases, more flux lines are created and the magnetic field expands.
- As current decreases, the magnetic field contracts. The magnetic field collapses when the current is shut off.
- The interaction and relationship between magnetism and electricity is known as electromagnetism.

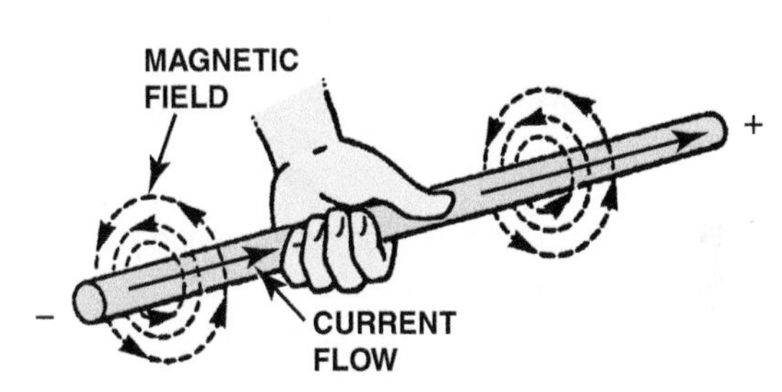

FIGURE 9–8 The left-hand rule for magnetic field direction is used with the electron flow theory.

CREATING AN ELECTROMAGNET An easy way to create an electromagnet is to wrap a nail with 20 turns of insulated wire and connect the ends to the terminals of a 1.5-volt dry cell battery. When energized, the nail will become a magnet and will be able to pick up tacks or other small steel objects.

STRAIGHT CONDUCTOR The magnetic field surrounding a straight, current-carrying conductor consists of several concentric cylinders of flux that are the length of the wire. The amount of current flow (amperes) determines how many flux lines (cylinders) there will be and how far out they extend from the surface of the wire. ● SEE FIGURE 9–7.

LEFT-HAND AND RIGHT-HAND RULES Magnetic flux cylinders have direction, just as the flux lines surrounding a bar magnet have direction. The **left-hand rule** is a simple way to determine this direction. When you grasp a conductor with your left hand so that your thumb points in the direction of electron flow (− to +) through the conductor, your fingers curl around the wire in the direction of the magnetic flux lines. ● SEE FIGURE 9–8.

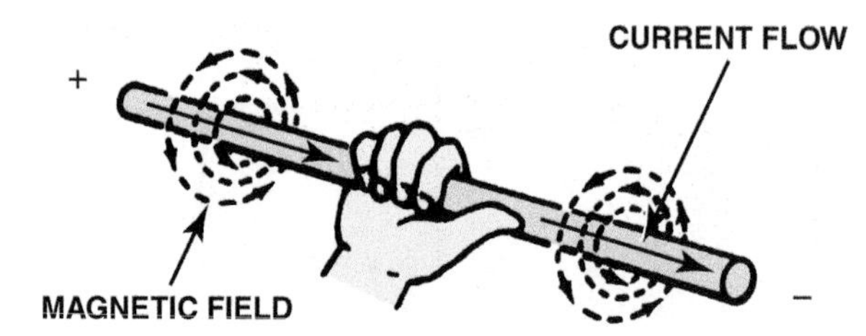

FIGURE 9–9 The right-hand rule for magnetic field direction is used with the conventional theory of electron flow.

FIGURE 9–10 Conductors with opposing magnetic fields will move apart into weaker fields.

Electricity and Magnetism

Electricity and magnetism are closely related because any electrical current flowing through a conductor creates a magnetic field. Any conductor moving through a magnetic field creates an electrical current. This relationship can be summarized as follows:

- Electricity creates magnetism.
- Magnetism creates electricity.

From a service technician's point of view, this relationship is important because wires carrying current should always be routed as the factory intended to avoid causing interference with another circuit or electronic component. This is especially important when installing or servicing spark plug wires, which carry high voltages and can cause high electromagnetic interference.

Most automotive circuits use the conventional theory of current (+ to −) and, therefore, the right-hand rule is used to determine the direction of the magnetic flux lines. ● **SEE FIGURE 9–9**.

FIELD INTERACTION The cylinders of flux surrounding current-carrying conductors interact with other magnetic fields. In the following illustrations, the cross symbol (+) indicates current moving inward or away from you. It represents the tail of an arrow. The dot symbol (•) represents an arrowhead and indicates current moving outward. If two conductors carry current in opposite directions, their magnetic fields also carry current in opposite directions (according to the left-hand rule). If they are placed side by side, then the opposing flux lines between the conductors create a strong magnetic field. Current-carrying conductors tend to move out of a strong field into a weak field, so the conductors move away from each other. ● **SEE FIGURE 9–10.**

If the two conductors carry current in the same direction, then their fields are in the same direction. The flux lines between the two conductors cancel each other out, leaving a very weak field between them. The conductors are drawn into this weak field, and they tend to move toward each other.

MOTOR PRINCIPLE Electric motors, such as vehicle starter motors, use this magnetic field interaction to convert electrical energy into mechanical energy. If two conductors carrying current in opposite directions are placed between strong north and south poles, the magnetic field of the conductor interacts with the magnetic fields of the poles. The counterclockwise field of the top conductor adds to the fields of the poles and creates a strong field beneath the conductor. The conductor then tries to move up to get out of this strong field. The clockwise field of the lower conductor adds to the field of the poles and creates a strong field above the conductor. The conductor then tries to move down to get out of this strong field. These forces cause the center of the motor, where the conductors are mounted, to turn clockwise. ● **SEE FIGURE 9–11.**

COIL CONDUCTOR If several loops of wire are made into a coil, then the magnetic flux density is strengthened. Flux

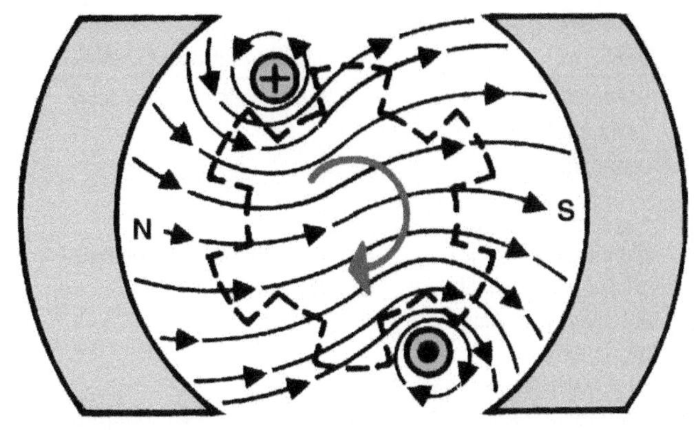

FIGURE 9–11 Electric motors use the interaction of magnetic fields to produce mechanical energy.

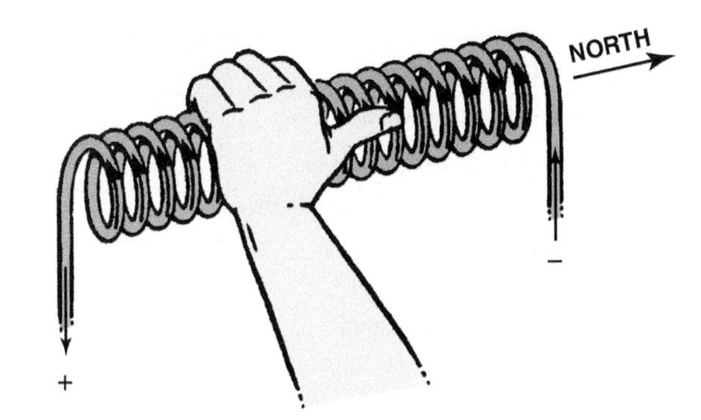

FIGURE 9–13 The left-hand rule for coils is shown.

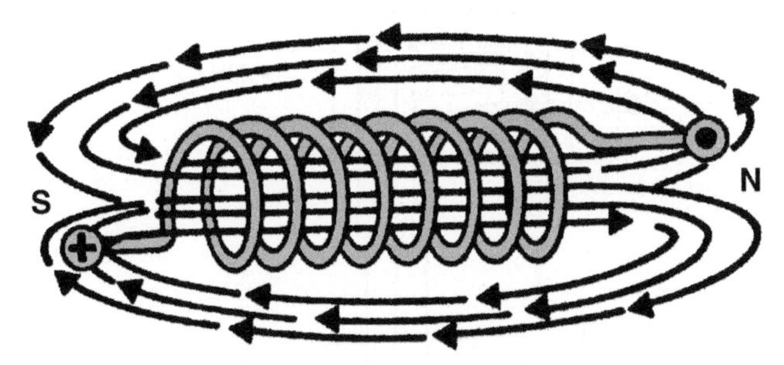

FIGURE 9–12 The magnetic lines of flux surrounding a coil look similar to those surrounding a bar magnet.

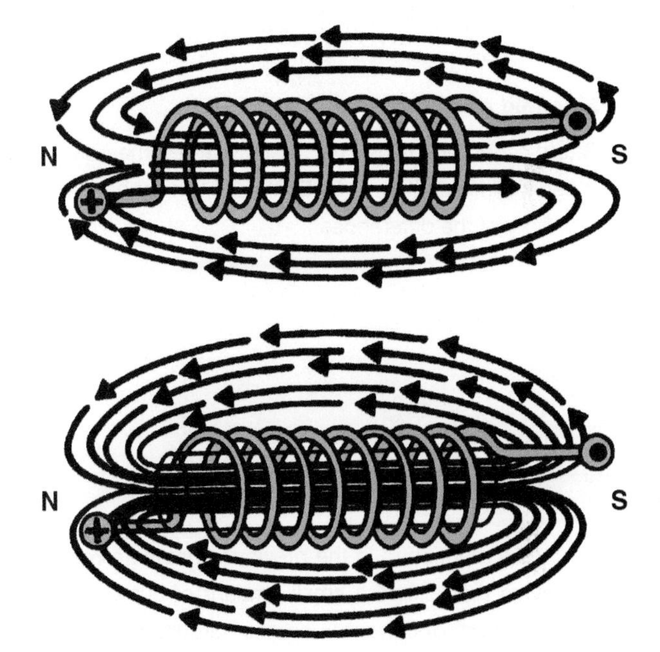

FIGURE 9–14 An iron core concentrates the magnetic lines of force surrounding a coil.

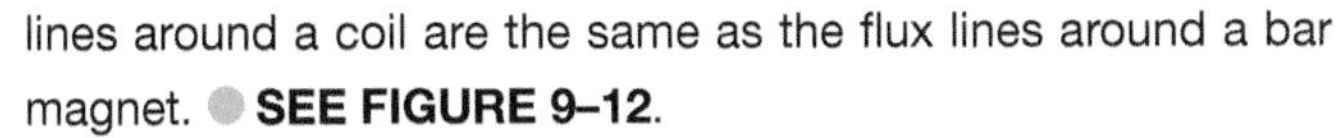

lines around a coil are the same as the flux lines around a bar magnet. SEE FIGURE 9–12.

They exit from the north pole and enter at the south pole. Use the left-hand rule to determine the north pole of a coil, as shown in FIGURE 9–13.

Grasp the coil with your left hand so that your fingers point in the direction of electron flow; your thumb will point toward the north pole of the coil.

ELECTROMAGNETIC STRENGTH The magnetic field surrounding a current-carrying conductor can be strengthened (increased) in three ways.

- Place a soft iron core in the center of the coil.
- Increase the number of turns of wire in the coil.
- Increase the current flow through the coil windings.

Because soft iron is highly permeable, magnetic flux lines pass through it easily. If a piece of soft iron is placed inside a coiled conductor, the flux lines concentrate in the iron core, rather than pass through the air, which is less permeable. The concentration of force greatly increases the strength of the magnetic field inside the coil. Increasing the number of turns in a coil and/or increasing the current flow through the coil results in greater field strength and is proportional to the number of turns. The magnetic field strength is often expressed in the units called **ampere-turns**. Coils with an iron core are called electromagnets. SEE FIGURE 9–14.

	CONSTRUCTION	AMPERAGE RATING (AMPERE)	USES	CALLED IN SERVICE INFORMATION
Relay	Uses a movable arm Coil: 60–100 ohms requiring 0.12–0.20 ampere to energize	1–30	Lower current switching, lower cost, more commonly used	Electromagnetic switch or relay
Solenoid	Uses a movable core Coil(s): 0.2–0.6 ohm requiring 20–60 amperes to energize	30–400	Higher cost, used in starter motor circuits and other high-amperage applications	Solenoid, relay, or electromagnetic switch

CHART 9–1

Comparison between a relay and a solenoid.

FREQUENTLY ASKED QUESTION

Solenoid or Relay?

Often, either term is used to describe the same part in service information. ● **SEE CHART 9–1** for a summary of the differences.

USES OF ELECTROMAGNETISM

RELAYS As mentioned in the previous chapter, a **relay** is a control device that allows a small amount of current to control a large amount of current in another circuit. A simple relay contains an electromagnetic coil in series with a battery and a switch. Near the electromagnet is a movable flat arm, called an *armature*, of some material that is attracted by a magnetic field. ● **SEE FIGURE 9–15.**

The armature pivots at one end and is held a small distance away from the electromagnet by a spring (or by the spring steel of the movable arm itself). A contact point, made of a good conductor, is attached to the free end of the armature. Another contact point is fixed a small distance away. The two contact points are wired in series with an electrical load and the battery.

When the switch is closed, the following occurs.

1. Current travels from the battery through a coil, creating an electromagnet.
2. The magnetic field created by the current attracts the armature, pulling it down until the contact points close.
3. Closing the contacts allows current in the heavy current circuit from the battery to the load.

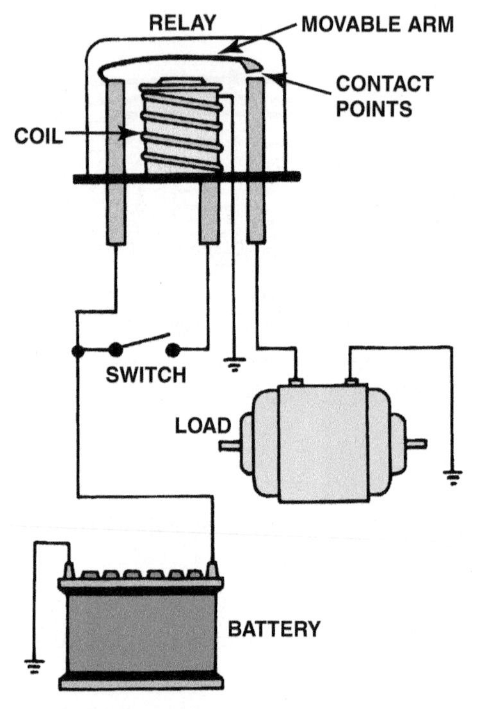

FIGURE 9–15 An electromagnetic switch that has a movable arm is referred to as a relay.

When the switch is open, the following occurs.

1. The electromagnet loses its magnetism when the current is shut off.
2. Spring pressure lifts the arm back up.
3. The heavy current circuit is broken by the opening of the contact points.

Relays may also be designed with normally closed contacts that open when current passes through the electromagnetic coil.

SOLENOID A solenoid is an example of an electromagnetic switch. A solenoid uses a movable core rather than a movable arm and is generally used in higher-amperage applications. A solenoid can be a separate unit or attached to a starter such as a starter solenoid. ● **SEE FIGURE 9–16.**

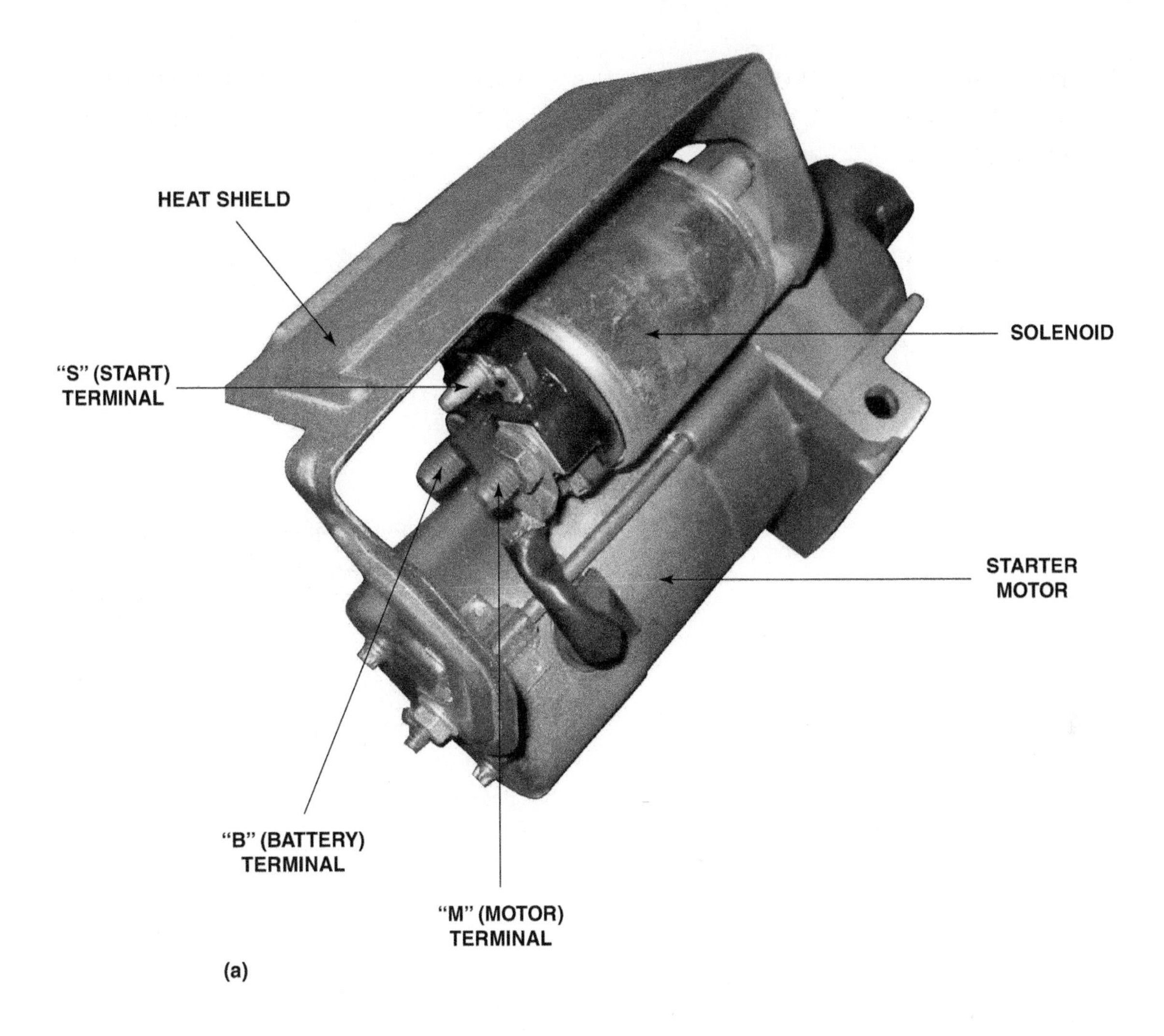

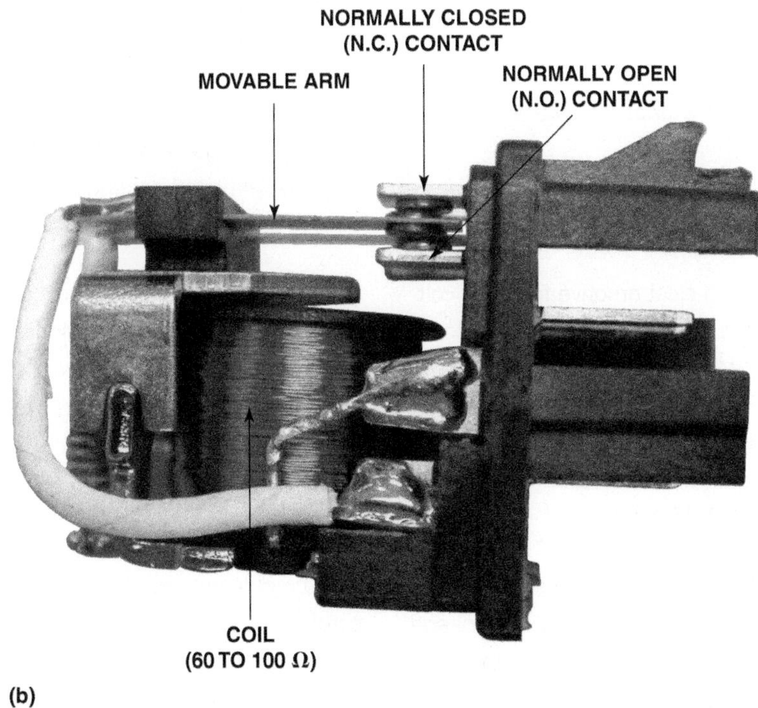

FIGURE 9–16 (a) A starter with attached solenoid. All of the current needed by the starter flows through the two large terminals of the solenoid and through the solenoid contacts inside. (b) A relay is designed to carry lower current compared to a solenoid and uses a movable arm.

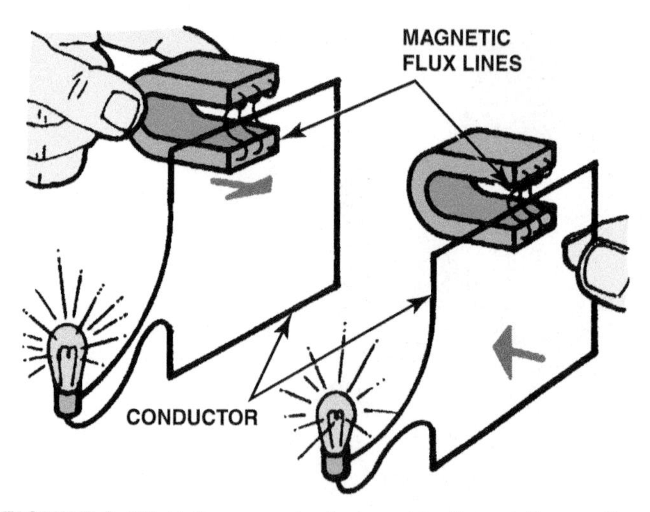

FIGURE 9–17 Voltage can be induced by the relative motion between a conductor and magnetic lines of force.

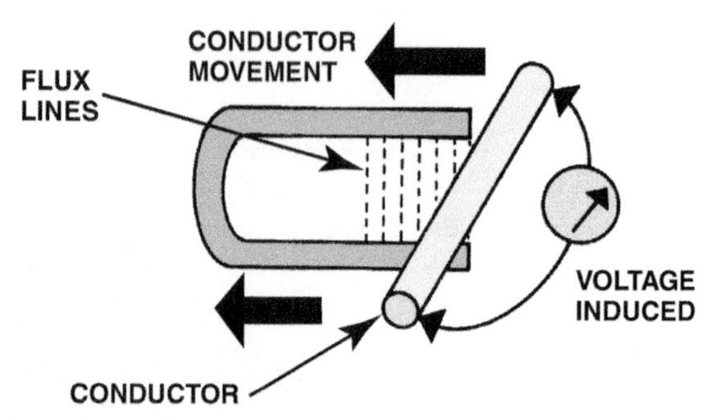

FIGURE 9–18 Maximum voltage is induced when conductors cut across the magnetic lines of force (flux lines) at a 90 degree angle.

ELECTROMAGNETIC INDUCTION

PRINCIPLES INVOLVED Electricity can be produced by using the relative movement of an electrical conductor and a magnetic field. The following three items are necessary to produce electricity (voltage) from magnetism.

1. Electrical conductor (usually a coil of wire)
2. Magnetic field
3. Movement of either the conductor or the magnetic field

Therefore:

- Electricity creates magnetism.
- Magnetism can create electricity.

Magnetic flux lines create an electromotive force, or voltage, in a conductor if either the flux lines or the conductor is moving. This movement is called *relative motion*. This process is called induction, and the resulting electromotive force is called *induced voltage*. This creation of a voltage (electricity) in a conductor by a moving magnetic field is called electromagnetic induction. ● **SEE FIGURE 9–17.**

VOLTAGE INTENSITY Voltage is induced when a conductor cuts across magnetic flux lines. The amount of the voltage depends on the rate at which the flux lines are broken. The more flux lines that are broken per unit of time, the greater the induced voltage. If a single conductor breaks 1 million flux lines per second, 1 volt is induced.

There are four ways to increase induced voltage.

- Increase the strength of the magnetic field, so there are more flux lines.
- Increase the number of conductors that are breaking the flux lines.
- Increase the speed of the relative motion between the conductor and the flux lines so that more lines are broken per time unit.
- Increase the angle between the flux lines and the conductor to a maximum of 90 degrees. There is no voltage induced if the conductors move parallel to, and do not break, any flux lines.

Maximum voltage is induced if the conductors break flux lines at 90 degrees. Induced voltage varies proportionately at angles between 0 and 90 degrees. ● **SEE FIGURE 9–18.**

Voltage can be induced electromagnetically and can be measured. Induced voltage creates current. The direction of induced voltage (and the direction in which current moves) is called *polarity* and depends upon the direction of the flux lines, as well as the direction of relative motion.

LENZ'S LAW An induced current moves so that its magnetic field opposes the motion that induced the current. This principle is called **Lenz's law**. The relative motion of a conductor and a magnetic field is opposed by the magnetic field of the current it has induced.

SELF-INDUCTION When current begins to flow in a coil, the flux lines expand as the magnetic field forms and strengthens. As current increases, the flux lines continue to expand, cutting across the wires of the coil and actually inducing another voltage within the same coil. Following Lenz's law, this self-induced voltage tends to *oppose* the current that produces it. If the current continues to increase, the second voltage opposes the increase. When the current stabilizes, the countervoltage is no

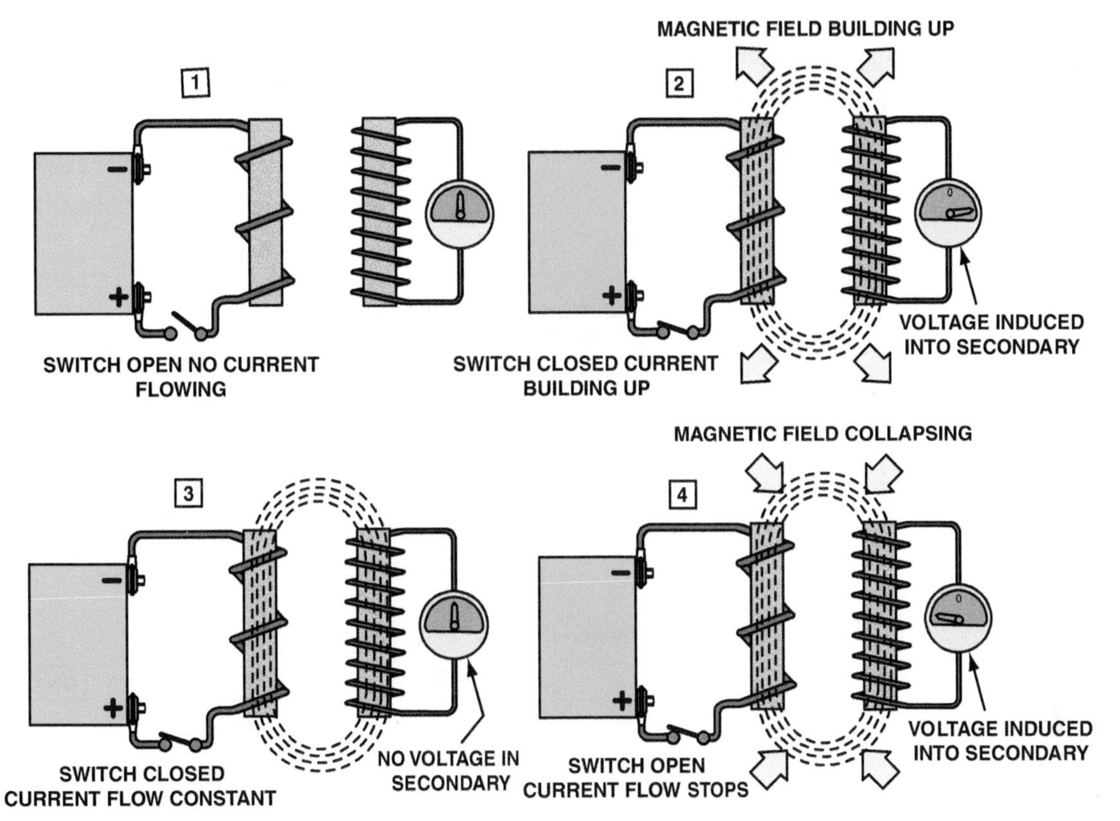

FIGURE 9–19 Mutual induction occurs when the expansion or collapse of a magnetic field around one coil induces a voltage in a second coil.

longer induced because there are no more expanding flux lines (no relative motion). When current to the coil is shut off, the collapsing magnetic flux lines self-induce a voltage in the coil that tries to maintain the original current. The self-induced voltage *opposes* and *slows* the *decrease* in the original current. The self-induced voltage that opposes changes in current flow is an inductor called **counter electromotive force (CEMF)**.

MUTUAL INDUCTION When two coils are close together, energy may be transferred from one to the other by magnetic coupling called mutual induction. **Mutual induction** means that the expansion or collapse of the magnetic field around one coil induces a voltage in the second coil.

IGNITION COILS

IGNITION COIL WINDINGS Ignition coils use two windings and are wound on the same iron core.

- One coil winding is connected to a battery through a switch and is called the *primary winding*.
- The other coil winding is connected to an external circuit and is called the *secondary winding*.

When the switch is open, there is no current in the primary winding. There is no magnetic field and, therefore, no voltage in the secondary winding. When the switch is closed, current is introduced and a magnetic field builds up around both windings. The primary winding thus changes electrical energy from the battery into magnetic energy of the expanding field. As the field expands, it cuts across the secondary winding and induces a voltage in it. A meter connected to the secondary circuit shows current. ● SEE FIGURE 9–19.

When the magnetic field has expanded to its full strength, it remains steady as long as the same amount of current exists. The flux lines have stopped their cutting action. There is no relative motion and no voltage in the secondary winding, as shown on the meter.

When the switch is opened, primary current stops and the field collapses. As it does, flux lines cut across the secondary winding but in the opposite direction. This induces a secondary voltage with current in the opposite direction, as shown on the meter.

Mutual induction is used in ignition coils. In an ignition coil, low-voltage primary current induces a very high secondary voltage because of the different number of turns in the primary and secondary windings. Because the voltage is increased, an ignition coil is also called a *step-up transformer*.

FIGURE 9–20 Some ignition coils are electrically connected, called "married" (top figure), whereas others use separated primary and secondary windings, called "divorced" (lower figure).

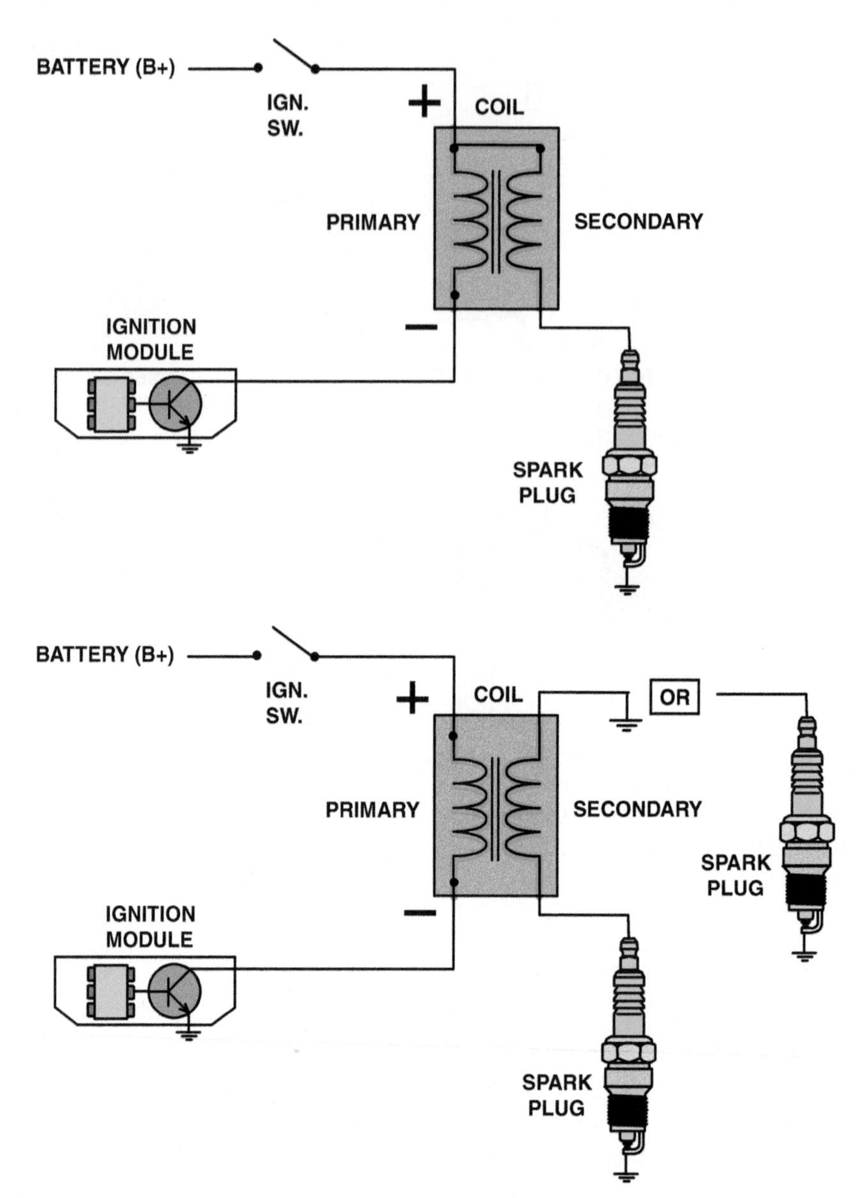

- **Electrically connected windings.** Many ignition coils contain two separate but electrically connected windings of copper wire. This type of coil is called a "married" type and is used in older distributor-type ignition systems and in many coil-on-plug (COP) designs.
- **Electrically insulated windings.** Other coils are true transformers in which the primary and secondary windings are not electrically connected. This type of coil is often called a "divorced" type and is used in all waste-spark-type ignition systems.

SEE FIGURE 9–20.

IGNITION COIL CONSTRUCTION The center of an ignition coil contains a core of laminated soft iron (thin strips of soft iron). This core increases the magnetic strength of the coil. Surrounding the laminated core are approximately 20,000 turns of fine wire (approximately 42 gauge). These windings are called the secondary coil windings. Surrounding the secondary windings are approximately 150 turns of heavy wire (approximately 21 gauge). These windings are called the primary coil windings. The secondary winding has about 100 times the number of turns of the primary winding, referred to as the **turns ratio** (approximately 100:1). In many coils, these windings are surrounded with a thin metal shield and insulating paper, and placed into a metal container. The metal container and shield help retain the magnetic field produced in the coil windings. The primary and secondary windings produce heat because of the electrical resistance in the turns of wire. Many coils contain

FIGURE 9–21 A GM waste-spark ignition coil showing the section of laminations that is shaped like the letter E. These mild steel laminations improve the efficiency of the coil.

FIGURE 9–22 The coil-on-plug (COP) design typically uses a bobbin-type coil.

oil to help cool the ignition coil. Other coil designs include the following:

- **Air-cooled, epoxy-sealed E coil.** The *E coil* is so named because the laminated, soft iron core is E shaped, with the coil wire turns wrapped around the center "finger" of the E and the primary winding wrapped inside the secondary winding. ● **SEE FIGURE 9–21.**
- **Spool design.** Used mostly for coil-on plug design, the coil windings are wrapped around a nylon or plastic spool or bobbin. ● **SEE FIGURE 9–22.**

IGNITION COIL OPERATION The negative terminal is attached to an **ignition control module (ICM, or igniter)**, which opens and closes the primary ignition circuit by opening or closing the ground return path of the circuit. When the ignition switch is on, voltage should be available at *both* the positive terminal and the negative terminal of the coil if the primary windings of the coil have continuity.

A spark is created by the following sequence of events.

- A magnetic field is created in the primary winding of the coil when there is 12 volts applied to the primary coil winding and the ignition control module grounds the other end on the coil.
- When the ignition control module (or powertrain control module) opens the ground circuit, the stored magnetic field collapses and creates a high voltage (up to 40,000 volts or more) in the secondary winding.
- The high-voltage pulse then flows to the spark plug and creates a spark at the ground electrode inside the engine that ignites the air–fuel mixture inside the cylinder.

ELECTROMAGNETIC INTERFERENCE

DEFINITION Until the advent of the onboard computer, **electromagnetic interference (EMI)** was not a source of real concern to automotive engineers. The problem was mainly one of *radio-frequency interference (RFI)*, caused primarily by the use of secondary ignition cables. Using spark plug wires that contained a high-resistance, nonmetallic core made of carbon, linen, or fiberglass strands impregnated with graphite mostly solved RFI from the secondary ignition system. RFI is a part of electromagnetic interference, which deals with interference that affects radio reception. All electronic devices used in vehicles are affected by EMI/RFI.

TECH TIP

Cell Phone Interference

A cellular phone emits a weak signal if it is turned on, even though it is not being used. This signal is picked up and tracked by cell phone towers. When the cell phone is called, it emits a stronger signal to notify the tower that it is on and capable of receiving a phone call. It is this "handshake" signal that can cause interference in the vehicle. Often this signal causes some static in the radio speakers even though the radio is off, but it can also cause a false antilock brake (ABS) trouble code to set. These signals from the cell phone create a voltage that is induced in the wires of the vehicle. Because the cell phone usually leaves with the customer, the service technician is often unable to verify the customer concern.

Remember, the interference occurs right *before* the cell phone rings. To fix the problem, connect an external antenna to the cell phone. This step will prevent the induction of a voltage in the wiring of the vehicle.

HOW EMI IS CREATED Whenever there is current in a conductor, an electromagnetic field is created. When current stops and starts, as in a spark plug cable or a switch that opens and closes, the field strength changes. Each time this happens, it creates an electromagnetic signal wave. If it happens rapidly enough, the resulting high-frequency signal waves, or EMI, interfere with radio and television transmission or with other electronic systems such as those under the hood. This is an undesirable side-effect of the phenomenon of electromagnetism.

Static electric charges caused by friction of the tires with the road, or the friction of engine drive belts contacting their pulleys, also produce EMI. Drive axles, driveshafts, and clutch or brake lining surfaces are other sources of static electric charges.

There are four ways of transmitting EMI, all of which can be found in a vehicle.

- Conductive coupling is actual physical contact through circuit conductors.
- Capacitive coupling is the transfer of energy from one circuit to another through an electrostatic field between two conductors.
- Inductive coupling is the transfer of energy from one circuit to another as the magnetic fields between two conductors form and collapse.
- Electromagnetic radiation is the transfer of energy by the use of radio waves from one circuit or component to another.

EMI SUPPRESSION DEVICES There are four general ways in which EMI is reduced.

- **Resistance suppression.** Adding resistance to a circuit to suppress RFI works only for high-voltage systems. This has been done by the use of resistance spark plug cables, resistor spark plugs, and the silicone grease used on the distributor cap and rotor of some electronic ignitions.
- **Suppression capacitors and coils.** Capacitors are installed across many circuits and switching points to absorb voltage fluctuations. Among other applications, they are used across the following:
  - The primary circuit of some electronic ignition modules
  - The output terminal of most alternators
  - The armature circuit of some electric motors

Coils reduce current fluctuations resulting from self-induction. They are often combined with capacitors to act as EMI filter circuits for windshield wiper and electric fuel pump motors. Filters also may be incorporated in wiring connectors.

- **Shielding.** The circuits of onboard computers are protected to some degree from external electromagnetic waves by their metal housings.
- **Ground wires or straps.** Ground wires or braided straps between the engine and chassis of an automobile help suppress EMI conduction and radiation by providing a low-resistance circuit ground path. Such suppression ground straps are often installed between rubber-mounted components and body parts. On some models, ground straps are installed between body parts, such as between the hood and a fender panel, where no electrical circuit exists. The strap has no other job than to suppress EMI. Without it, the sheet-metal body and hood could function as a large capacitor. The space between the fender and hood could form an electrostatic field and couple with the computer circuits in the wiring harness routed near the fender panel.

SUMMARY

1. Most automotive electrical components use magnetism, the strength of which depends on both the amount of current (amperes) and the number of turns of wire of each electromagnet.
2. The strength of electromagnets is increased by using a soft iron core.
3. Voltage can be induced from one circuit to another.
4. Electricity creates magnetism and magnetism creates electricity.
5. Radio-frequency interference (RFI) is a part of electromagnetic interference (EMI).

REVIEW QUESTIONS

1. What is the relationship between electricity and magnetism?
2. What is the difference between mutual induction and self-induction?
3. What is the result if a magnet cracks?
4. How can EMI be reduced or controlled?

CHAPTER QUIZ

1. Technician A says that magnetic lines of force can be seen by placing iron filings on a piece of paper and then holding them over a magnet. Technician B says that the effects of magnetic lines of force can be seen using a compass. Which technician is correct?
   a. Technician A only
   b. Technician B only
   c. Both Technicians A and B
   d. Neither Technician A nor B
2. Unlike magnetic poles _______, and like magnetic poles _______.
   a. Repel; attract
   b. Attract; repel
   c. Repel; repel
   d. Attract; attract
3. The conventional theory for current flow is being used to determine the direction of magnetic lines of force. Technician A says that the left-hand rule should be used. Technician B says that the right-hand rule should be used. Which technician is correct?
   a. Technician A only
   b. Technician B only
   c. Both Technicians A and B
   d. Neither Technician A nor B
4. Technician A says that a relay is an electromagnetic switch. Technician B says that a solenoid uses a movable core. Which technician is correct?
   a. Technician A only
   b. Technician B only
   c. Both Technicians A and B
   d. Neither Technician A nor B
5. Two technicians are discussing electromagnetic induction. Technician A says that the induced voltage can be increased if the speed is increased between the conductor and the magnetic lines of force. Technician B says that the induced voltage can be increased by increasing the strength of the magnetic field. Which technician is correct?
   a. Technician A only
   b. Technician B only
   c. Both Technicians A and B
   d. Neither Technician A nor B
6. An ignition coil operates using the principle(s) of _______.
   a. Electromagnetic induction
   b. Self-induction
   c. Mutual induction
   d. All of the above
7. Electromagnetic interference can be reduced by using a _______.
   a. Resistance
   b. Capacitor
   c. Coil
   d. All of the above
8. An ignition coil is an example of a _______.
   a. Solenoid
   b. Step-down transformer
   c. Step-up transformer
   d. Relay
9. Magnetic field strength is measured in _______.
   a. Ampere-turns
   b. Flux
   c. Density
   d. Coil strength
10. Two technicians are discussing ignition coils. Technician A says that some ignition coils have the primary and secondary windings electrically connected. Technician B says that some coils have totally separate primary and secondary windings that are not electrically connected. Which technician is correct?
    a. Technician A only
    b. Technician B only
    c. Both Technicians A and B
    d. Neither Technician A nor B

chapter 10

ELECTRONIC FUNDAMENTALS

LEARNING OBJECTIVES

After studying this chapter, the reader will be able to:

1. Identify semiconductor components.
2. Discuss where various electronic and semiconductor devices are used in vehicles.
3. Explain necessary precautions when working with semiconductor circuits.
4. Describe how diodes and transistors work, and how to test them.
5. Identify the causes of failure of electronic components.

This chapter will help you prepare for the ASE Electrical/Electronic Systems (A6) certification test content area "A" (General Electrical/Electronic System Diagnosis).

GM STC OBJECTIVE

GM Service Technical College topic covered in this chapter is as follows:

1. Identify important types, characteristics, and diagnosis of various solid state electrical components (18043.03W-R3).

KEY TERMS

Anode 128
Base 135
Bipolar transistor 135
Burn in 130
Cathode 128
CHMSL 134
Clamping diode 130
Darlington pair 137
Despiking diode 130
Diode 128
Doping 127
Emitter 135
ESD 145
FET 136
Forward bias 128
Germanium 127
Heat sink 137
Hole theory 128
Hybrid Powertrain Control Module (HPCM) 143
Impurities 127
Integrated circuit (IC) 137
Inverter 143
insulated gate bipolar transistors 145
Junction 128
Light emitting diode (LED) 132
MOSFET 137
motor control modules 143
NPN transistor 135
NTC 134
N-type material 127
Op-amps 139
Photodiodes 133
Photons 133
Photoresistor 133
Phototransistor 137
Peak inverse voltage (PIV) 132
Peak reverse voltage (PRV) 132
PNP transistor 135
P-type material 127
PWM 140
Rectifier bridge 135
Reverse bias 129
SCR 134
Semiconductors 127
Silicon 127
Spike protection resistor 131
Suppression diode 130
Thermistor 134
Threshold voltage 136
Transistor 135
Zener diode 129

Electronic components are the heart of computers. Knowing how electronic components work helps take the mystery out of automotive electronics.

SEMICONDUCTORS

DEFINITION Semiconductors are neither conductors nor insulators. The flow of electrical current is caused by the movement of electrons in materials, known as conductors, having *fewer* than four electrons in their atom's outer orbit. Insulators contain *more* than four electrons in their outer orbit and cannot conduct electricity because their atomic structure is stable (no free electrons).

Semiconductors are materials that contain exactly four electrons in the outer orbit of their atom structure and are, therefore, neither good conductors nor good insulators.

EXAMPLES OF SEMICONDUCTORS Two examples of semiconductor materials are **germanium** and **silicon**, which have exactly four electrons in their valance ring and no free electrons to provide current flow. However, both of these semiconductor materials can be made to conduct current if another material is added to provide the necessary conditions for electron movement.

CONSTRUCTION When another material is added to a semiconductor material in very small amounts, it is called **doping**. The doping elements are called **impurities**; therefore, after their addition, the germanium and silicon are no longer considered *pure* elements. The material added to pure silicon or germanium to make it electrically conductive represents only one atom of impurity for every *100 million* atoms of the pure semiconductor material. The resulting atoms are still electrically *neutral,* because the number of electrons still equals the number of protons of the combined materials. These combined materials are classified into two groups depending on the number of electrons in the bonding between the two materials.

- N-type materials
- P-type materials

N-TYPE MATERIAL **N-type material** is silicon or germanium that is doped with an element such as *phosphorus*, *arsenic*, or *antimony*, each having five electrons in its outer orbit. These five electrons are combined with the four electrons of the silicon or germanium to total nine electrons. There is room for only eight electrons in the bonding between the semiconductor material and the doping material. This leaves extra electrons, and even though the material is still electrically neutral, these extra electrons tend to repel other electrons outside the material. ● SEE FIGURE 10–1.

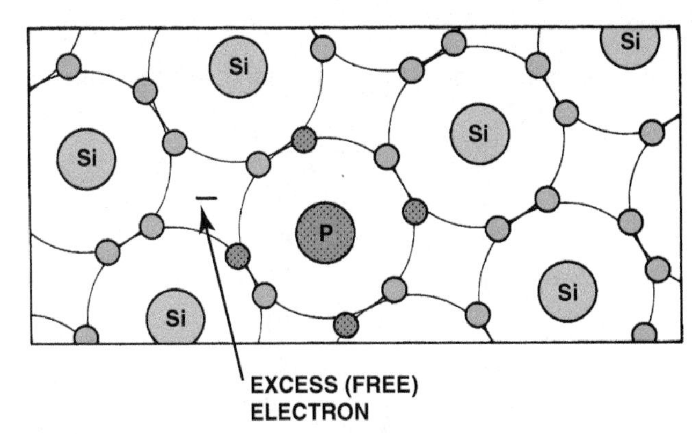

FIGURE 10–1 N-type material. Silicon (Si) doped with a material (such as phosphorus) with five electrons in the outer orbit results in an extra free electron.

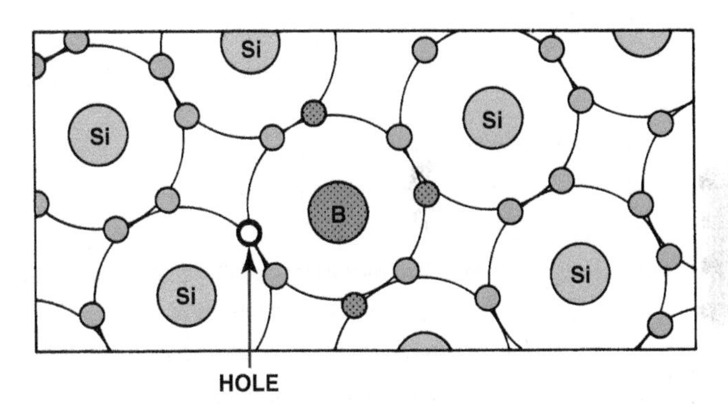

FIGURE 10–2 P-type material. Silicon (Si) doped with a material, such as boron (B), with three electrons in the outer orbit results in a hole capable of attracting an electron.

P-TYPE MATERIAL **P-type material** is produced by doping silicon or germanium with the element *boron* or the element *indium*. These impurities have only three electrons in their outer shell and, when combined with the semiconductor material, result in a material with seven electrons, one electron *less* than is required for atom bonding. This lack of one electron makes the material able to attract electrons, even though the material still has a neutral charge. This material tends to attract electrons to fill the holes for the missing eighth electron in the bonding of the materials. ● SEE FIGURE 10–2.

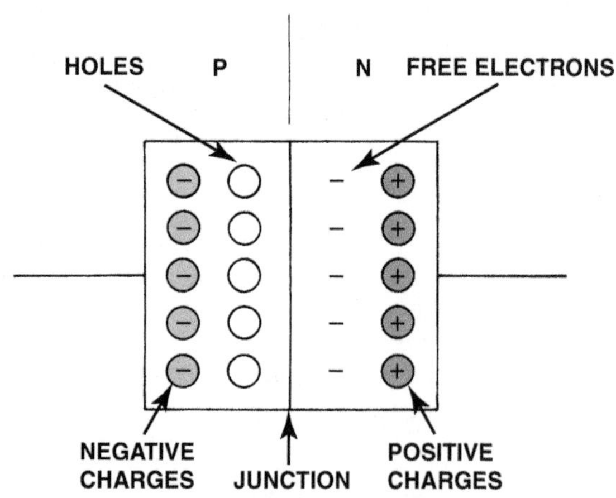

FIGURE 10–3 Unlike charges attract and the current carriers (electrons and holes) move toward the junction.

FREQUENTLY ASKED QUESTION

What Is the Hole Theory?

Current flow is expressed as the movement of electrons from one atom to another. In semiconductor and electronic terms, the movement of electrons fills the holes of the P-type material. Therefore, as the holes are filled with electrons, the unfilled holes move opposite to the flow of the electrons. This concept of hole movement is called the **hole theory** of current flow. The holes move in the direction opposite that of electron flow. For example, think of an egg carton, where if an egg is moved in one direction, the holes created move in the opposite direction. ● **SEE FIGURE 10–3.**

SUMMARY OF SEMICONDUCTORS

The following is a summary of semiconductor fundamentals.

1. The two types of semiconductor materials are P type and N type. N-type material contains extra electrons; P-type material contains holes due to missing electrons. The number of excess electrons in an N-type material must remain constant, and the number of holes in the P-type material must also remain constant. Because electrons are interchangeable, movement of electrons in or out of the material is possible to maintain a balanced material.

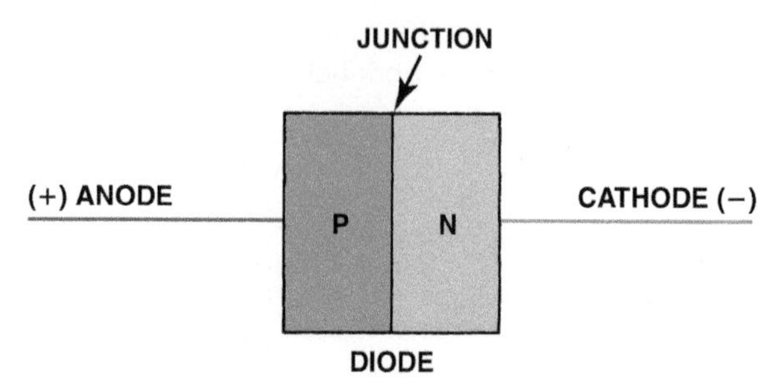

FIGURE 10–4 A diode is a component with P-type and N-type materials together. The negative electrode is called the cathode and the positive electrode is called the anode.

2. In P-type semiconductors, electrical conduction occurs mainly as a result of holes (absence of electrons). In N-type semiconductors, electrical conduction occurs mainly as a result of electrons (excess of electrons).
3. Hole movement results from the jumping of electrons into new positions.
4. Under the effect of a voltage applied to the semiconductor, electrons travel toward the positive terminal and holes move toward the negative terminal. The direction of hole current agrees with the conventional direction of current flow.

DIODES

CONSTRUCTION A **diode** is an electrical one-way check valve made by combining a P-type material and an N-type material. The word *diode* means "having two electrodes." Electrodes are electrical connections: The positive electrode is called the **anode**; the negative electrode is called the **cathode**. The point where the two types of materials join is called the **junction**. ● **SEE FIGURE 10–4.**

OPERATION The N-type material has one extra electron, which can flow into the P-type material. The P type requires electrons to fill its holes. If a battery's positive terminal (+) were connected to the diode's P-type material and negative (−) to the N-type material, then the electrons that left the N-type material and flowed into the P-type material to fill the holes would be quickly replaced by the electron flow from the battery. Current flows through a forward-bias diode for the following reasons.

- Electrons move toward the holes (P-type material).
- Holes move toward the electrons (N-type material).

● **SEE FIGURE 10–5.**

As a result, current would flow through the diode with low resistance. This condition is called **forward bias**.

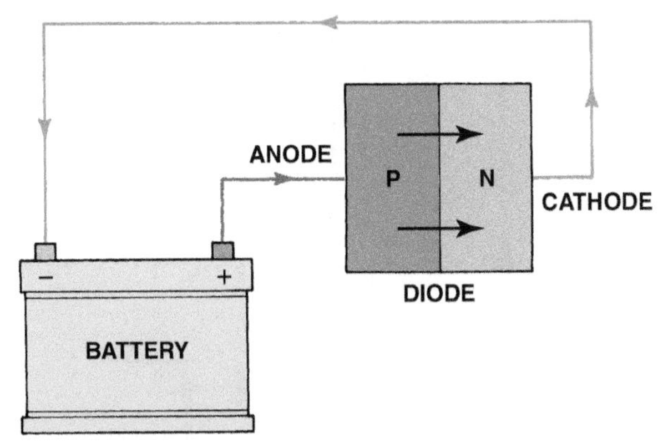

FIGURE 10–5 Diode connected to a battery with correct polarity (battery positive to P type and battery negative to N type). Current flows through the diode. This condition is called forward bias.

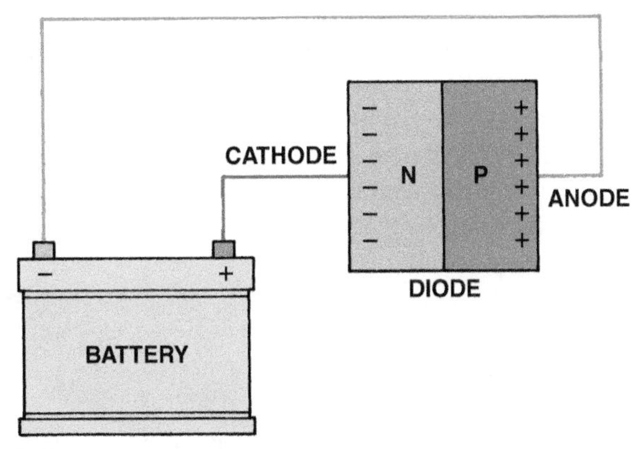

FIGURE 10–6 Diode connected with reversed polarity. No current flows across the junction between the P-type and N-type materials. This connection is called reverse bias.

If the battery connections were reversed and the positive side of the battery was connected to the N-type material, the electrons would be pulled toward the battery and away from the junction of the N-type and P-type materials. (Remember, unlike charges attract, whereas like charges repel.) Because electrical conduction requires the flow of electrons across the junction of the N-type and P-type materials and because the battery connections are actually reversed, the diode offers very high resistance to current flow. This condition is called **reverse bias.** ● **SEE FIGURE 10–6.**

Therefore, diodes allow current flow only when current of the correct polarity is connected to the circuit.

- Diodes are used in alternators to control current flow in one direction, which changes the AC voltage generated into DC voltage.
- Diodes are also used in computer controls, relays, air-conditioning circuits, and many other circuits to prevent possible damage due to reverse current flows that may be generated within the circuit. ● **SEE FIGURE 10–7.**

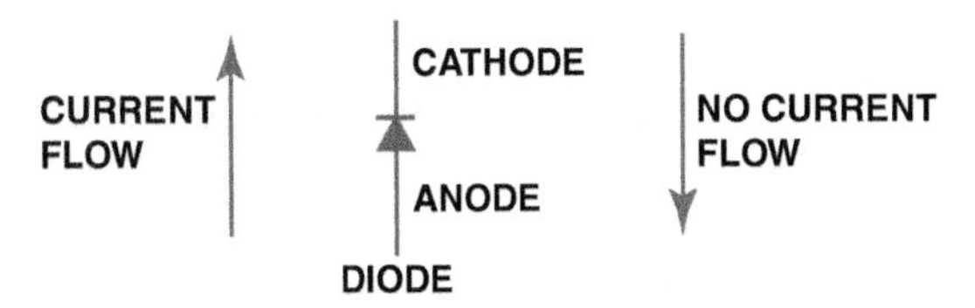

FIGURE 10–7 Diode symbol and electrode names. The stripe on one end of a diode represents the cathode end of the diode.

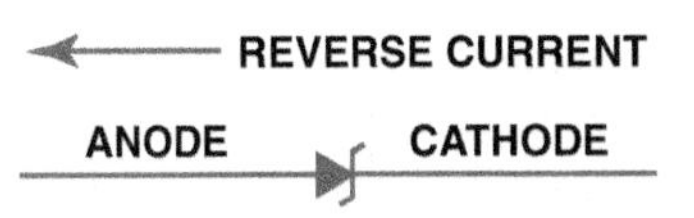

FIGURE 10–8 A zener diode blocks current flow until a certain voltage is reached, then it permits current to flow.

FREQUENTLY ASKED QUESTION

What Is the Difference Between Electricity and Electronics?

Electronics usually means that solid-state devices are used in the electrical circuits. Electricity as used in automotive applications usually means electrical current flow through resistance and loads without the use of diodes, transistors, or other electronic devices.

ZENER DIODES

CONSTRUCTION A **zener diode** is a specially constructed diode designed to operate with a reverse-bias current. Zener diodes were named in 1934 for their inventor, Clarence Melvin Zener, an American professor of physics.

OPERATION A zener diode acts as any diode in that it blocks reverse-bias current, but only up to a certain voltage. Above this certain voltage (called the *breakdown voltage* or the zener region), a zener diode will conduct current in the opposite direction without damage to the diode. A zener diode is heavily doped, and the reverse-bias voltage does not harm the material. The voltage drop across a zener diode remains practically the same before and after the breakdown voltage, and this factor makes a zener diode perfect for voltage regulation. Zener diodes can be constructed for various breakdown voltages and can be used in a variety of automotive and electronic applications, especially for electronic voltage regulators used in the charging system. ● **SEE FIGURE 10–8.**

TECH TIP

"Burn In" to Be Sure

A common term heard in the electronic and computer industry is **burn in**, which means to operate an electronic device, such as a computer, for a period from several hours to several days.

Most electronic devices fail in infancy, or during the first few hours of operation. This early failure occurs if there is a manufacturing defect, especially at the P-N junction of any semiconductor device. The junction will usually fail after only a few operating cycles.

What does this information mean to the average person? When purchasing a personal or business computer, have the computer burned in before delivery. This step helps ensure that all of the circuits have survived infancy and that the chances of chip failure are greatly reduced. Purchasing sound or television equipment that has been on display may be a good value, because during its operation as a display model, the burn-in process has been completed. The automotive service technician should be aware that if a replacement electronic device fails shortly after installation, the problem may be a case of early electronic failure.

NOTE: Whenever there is a failure of a replacement part, the technician should always check for excessive voltage or heat to and around the problem component.

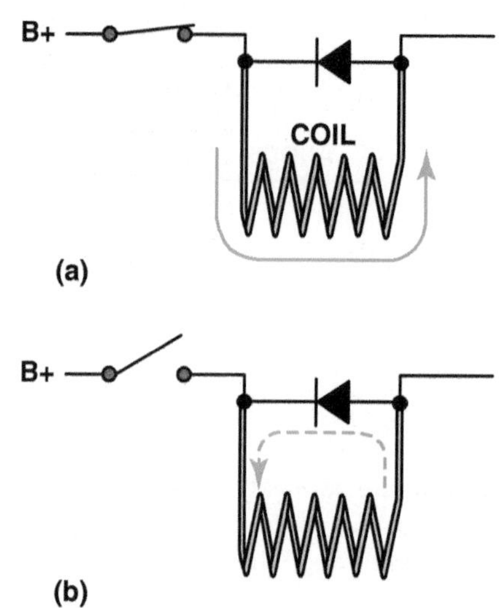

FIGURE 10–9 (a) Notice that when the coil is being energized, the diode is reverse biased and the current is blocked from passing through the diode. The current flows through the coil in the normal direction. (b) When the switch is opened, the magnetic field surrounding the coil collapses, producing a high-voltage surge in the reverse polarity of the applied voltage. This voltage surge forward biases the diode, and the surge is dissipated harmlessly back through the windings of the coil.

FIGURE 10–10 A diode connected to both terminals of the air-conditioning compressor clutch used to reduce the high-voltage spike that results when a coil (compressor clutch coil) is de-energized.

HIGH-VOLTAGE SPIKE PROTECTION

CLAMPING DIODES Diodes can be used as a high-voltage clamping device when the power (+) is connected to the cathode (–) of the diode. If a coil is pulsed on and off, a high-voltage spike is produced whenever the coil is turned off. To control and direct this possibly damaging high-voltage spike, a diode can be installed across the leads to the coil to redirect the high-voltage spike back through the coil windings to prevent possible damage to the rest of the vehicle's electrical or electronic circuits. A diode connected across the terminals of a coil to control voltage spikes is called a **clamping diode**. Clamping diodes can also be called **despiking** or **suppression diodes**. ● **SEE FIGURE 10–9.**

CLAMPING DIODE APPLICATION Diodes were first used on A/C compressor clutch coils at the same time electronic devices were first used. The diode was used to help prevent the high-voltage spike generated inside the A/C clutch coil from damaging delicate to delicate electronic circuits anywhere in the vehicle's electrical system. ● **SEE FIGURE 10–10.**

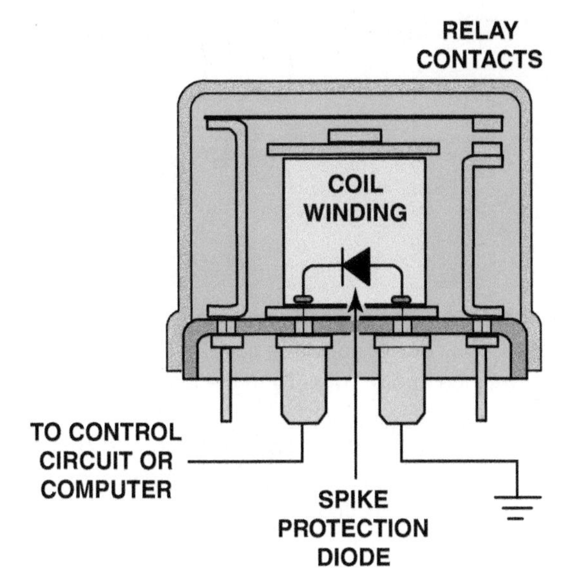

FIGURE 10–11 Spike protection diodes are commonly used in computer-controlled circuits to prevent damaging high-voltage surges that occur any time current flowing through a coil is stopped.

Because most automotive circuits eventually are electrically connected to each other in parallel, a high-voltage surge anywhere in the vehicle could damage electronic components in other circuits.

The circuits most likely to be affected by the high-voltage surge, if the diode fails, are the circuits controlling the operation of the A/C compressor clutch and any component that uses a coil, such as those of the blower motor and climate control units.

Many relays are equipped with a diode to prevent a voltage spike when the contact points open and the magnetic field in the coil winding collapses. ● **SEE FIGURE 10–11.**

DESPIKING ZENER DIODES Zener diodes can also be used to control high-voltage spikes and keep them from damaging delicate electronic circuits. Zener diodes are most commonly used in electronic fuel-injection circuits that control the firing of the injectors. If clamping diodes were used in parallel with the injection coil, the resulting clamping action would tend to delay the closing of the fuel injector nozzle. A zener diode is commonly used to clamp only the higher voltage portion of the resulting voltage spike without affecting the operation of the injector. ● **SEE FIGURE 10–12.**

DESPIKING RESISTORS All coils must use some protection against high-voltage spikes that occur when the voltage is removed from any coil. Instead of a diode installed in parallel with the coil windings, a resistor can be used, called a **spike protection resistor**. ● **SEE FIGURE 10–13.**

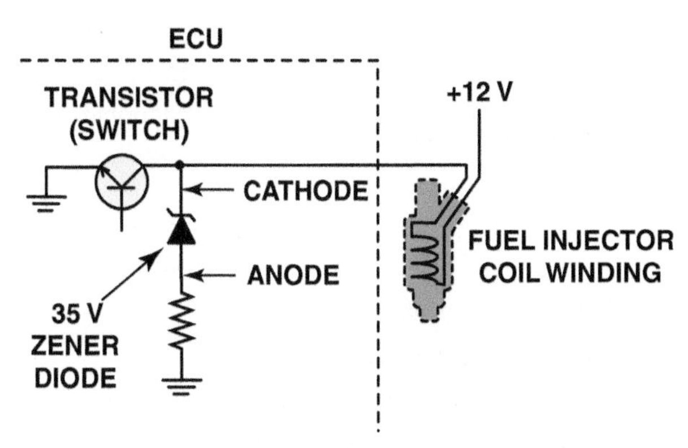

FIGURE 10–12 A zener diode is commonly used inside automotive computers to protect delicate electronic circuits from high-voltage spikes. A 35-volt zener diode will conduct any voltage spike higher than 35 volts resulting from the discharge of the fuel injector coil safely to ground through a current-limiting resistor in series with the zener diode.

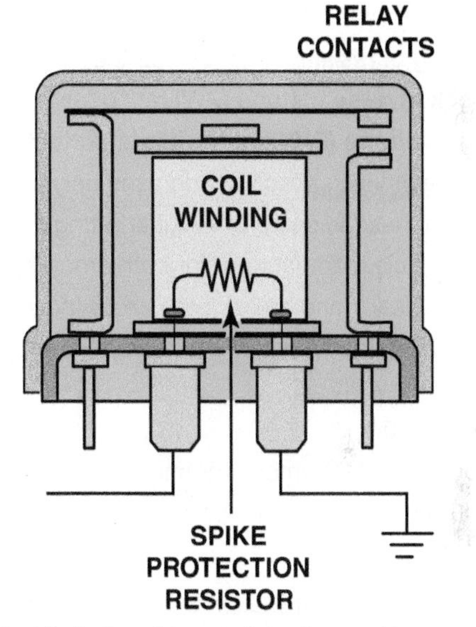

FIGURE 10–13 A despiking resistor is used in many automotive applications to help prevent harmful high-voltage surges from being created when the magnetic field surrounding a coil collapses when the coil circuit is opened.

Resistors are often preferred for two reasons.

Reason 1	Coils will usually fail when shorted rather than open, as this shorted condition results in greater current flow in the circuit. A diode installed in the reverse-bias direction cannot control this extra current, whereas a resistor in parallel can help reduce potentially damaging current flow if the coil becomes shorted.
Reason 2	The protective diode can also fail, and diodes usually fail by shorting before they blow open. If a diode becomes shorted, excessive current can flow through the coil circuit, perhaps causing damage. A resistor usually fails open and, therefore, even in failure could not in itself cause a problem.

Resistors on coils are often used in relays and in climate-control circuit solenoids to control vacuum to the various air management system doors as well as other electronically controlled appliances.

DIODE RATINGS

SPECIFICATIONS Most diodes are rated according to the following:

- Maximum current flow in the forward-bias direction. Diodes are sized and rated according to the amount of current they are designed to handle in the forward-bias direction. This rating is normally from 1 to 5 amperes for most automotive applications.
- This rating of resistance to reverse-bias voltage is called the **peak inverse voltage (PIV)** rating, or the **peak reverse voltage (PRV)** rating. It is important that the service technician specifies and uses only a replacement diode that has the same or a higher rating than specified by the vehicle manufacturer for both amperage and PIV rating. Typical 1 ampere diodes use an industry numbering code that indicates the PIV rating. For example:
  1N 4001-50 V PIV
  1N 4002-100 V PIV
  1N 4003-200 V PIV (most commonly used)
  1N 4004-400 V PIV
  1N 4005-600 V PIV
- The "1N" means that the diode has one P-N junction. A higher rating diode can be used with no problems (except for slightly higher cost, even though the highest rated diode generally costs less than $1). Never substitute a *lower*-rated diode than is specified.

DIODE VOLTAGE DROP The voltage drop across a diode is about the same voltage as that required to forward bias the diode. If the diode is made from germanium, the forward voltage is 0.3 to 0.5 volt. If the diode is made from silicon, the forward voltage is 0.5 to 0.7 volt.

NOTE: When diodes are tested using a digital multimeter, the meter will display the voltage drop across the P-N junction (about 0.5 to 0.7 volt) when the meter is set to the *diode-check* position.

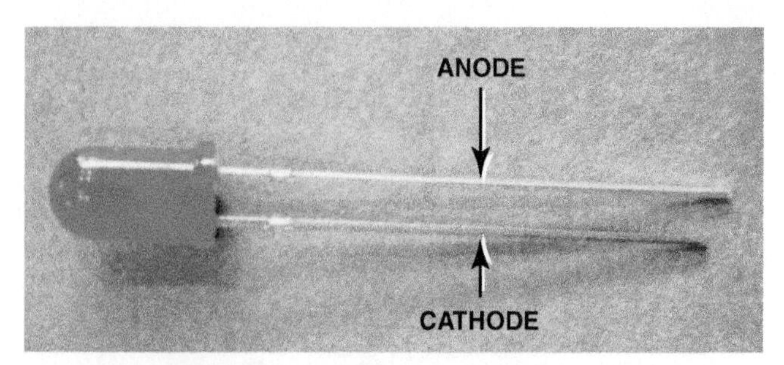

FIGURE 10–14 A typical light-emitting diode (LED). This particular LED is designed with a built-in resistor so that 12 volts DC may be applied directly to the leads without an external resistor. Normally a 300 to 500 ohm, 0.5 watt resistor is required to be attached in series with the LED, to control current flow to about 0.020 ampere (20 milliamperes) or damage to the P-N junction may occur.

LIGHT-EMITTING DIODES

OPERATION All diodes radiate some energy during normal operation. Most diodes radiate heat because of the junction barrier voltage drop (typically 0.6 volt for silicon diodes). **Light-emitting diodes (LED)** radiate light when current flows through the diode in the forward-bias direction. ● **SEE FIGURE 10–14.**

The forward-bias voltage required for an LED ranges between 1.5 and 2.2 volts.

An LED will only light if the voltage at the anode (positive electrode) is at least 1.5 to 2.2 volts higher than the voltage at the cathode (negative electrode).

NEED FOR CURRENT LIMITING If an LED were connected across a 12-volt automotive battery, the LED would light brightly, but only for a second or two. Excessive current (amperes) that flows across the P-N junction of any electronic device can destroy the junction. A resistor *must* be connected in series with every diode (including LEDs) to control current flow across the P-N junction. This protection should include the following:

1. The value of the resistor should be from 300 to 500 ohms for each P-N junction. Commonly available resistors in this range include 470, 390, and 330 ohm resistors.
2. The resistors can be connected to either the anode or the cathode end. (Polarity of the resistor does not matter.) Current flows through the LED in series with the resistor, and the resistor will control the current flow through the LED regardless of its position in the circuit.

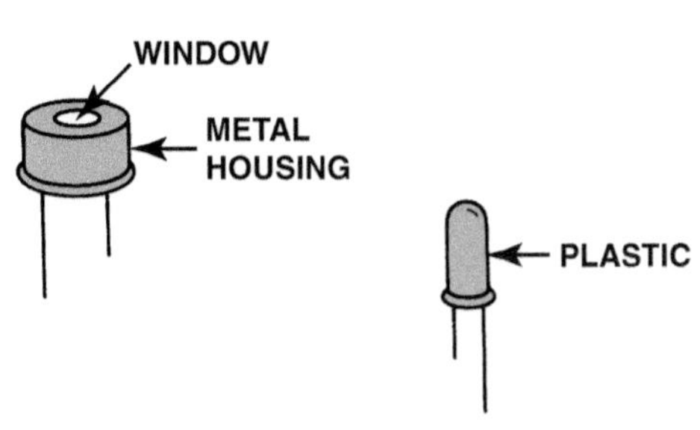

FIGURE 10–15 Typical photodiodes. They are usually built into a plastic housing so that the photodiode itself may not be visible.

FIGURE 10–16 Symbol for a photodiode. The arrows represent light striking the P-N junction of the photodiode.

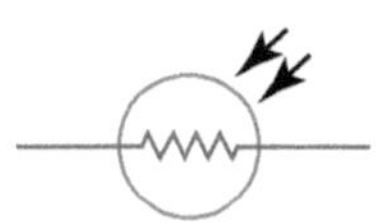

FIGURE 10–17 Either symbol may be used to represent a photoresistor.

FREQUENTLY ASKED QUESTION

How Does an LED Emit Light?

An LED contains a chip that houses P-type and N-type materials. The junction between these regions acts as a barrier to the flow of electrons between the two materials. When a voltage of 1.5 to 2.2 volts is applied to the correct polarity, current will flow across the junction. As the electrons enter the P-type material, it combines with the holes in the material and releases energy in the form of light (called **photons**). The intensity and color the light produces depends on materials used in the manufacture of the semiconductor.

LEDs are very efficient compared to conventional incandescent bulbs, which depend on heat to create light. LEDs generate very little heat, with most of the energy consumed converted directly into light. LEDs are reliable and are being used for taillights, brake lights, daytime running lights, and headlights in some vehicles.

3. Resistors protecting diodes can be actual resistors or other current-limiting loads such as lamps or coils. With the current-limiting devices to control the current, the average LED will require about 20 to 30 milliamperes (mA), or 0.020 to 0.030 ampere.

PHOTODIODES

PURPOSE AND FUNCTION All semiconductor P-N junctions emit energy, mostly in the form of heat or light such as with an LED. In fact, if an LED is exposed to bright light, a voltage potential is established between the anode and the cathode. **Photodiodes** are specially constructed to respond to various wavelengths of light with a "window" built into the housing. **SEE FIGURE 10–15.**

Photodiodes are frequently used in steering wheel controls for transmitting tuning, volume, and other information from the steering wheel to the data link and the unit being controlled. If several photodiodes are placed on the steering column end and LEDs or phototransistors are placed on the steering wheel side, then data can be transmitted between the two moving points without the interference that could be caused by physical contact types of units.

CONSTRUCTION A photodiode is sensitive to light. When light energy strikes the diode, electrons are released and the diode will conduct in the forward-bias direction. (The light energy is used to overcome the barrier voltage.)

The resistance across the photodiode decreases as the intensity of the light increases. This characteristic makes the photodiode a useful electronic device for controlling some automotive lighting systems such as automatic headlights. The symbol for a photodiode is shown in **FIGURE 10–16.**

PHOTORESISTORS

A **photoresistor** is a semiconductor material (usually cadmium sulfide) that changes resistance with the presence or absence of light.

Dark = High resistance
Light = Low resistance

Because resistance is reduced when the photoresistor is exposed to light, the photoresistor can be used to control headlight dimmer relays and automotive headlights. **SEE FIGURE 10–17.**

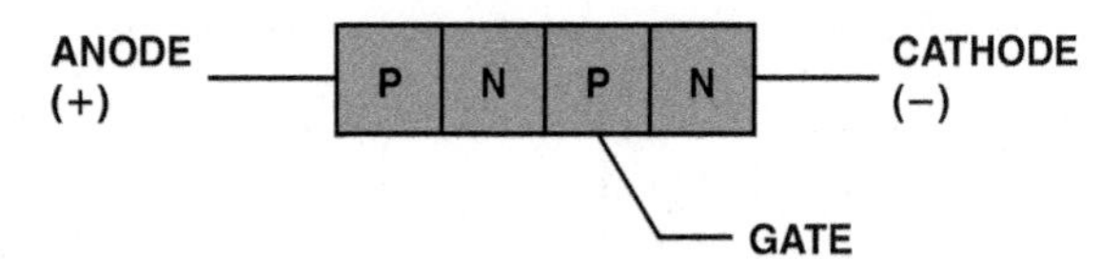

FIGURE 10–18 Symbol and terminal identification of an SCR.

SILICON-CONTROLLED RECTIFIERS

CONSTRUCTION A **silicon-controlled rectifier (SCR)** is commonly used in the electronic circuits of various automotive applications. An SCR is a semiconductor device that looks like two diodes connected end to end. ● **SEE FIGURE 10–18.**

If the anode is connected to a higher voltage source than the cathode in a circuit, no current will flow as would occur with a diode. If, however, a positive voltage source is connected to the gate of the SCR, then current can flow from anode to cathode with a typical voltage drop of 1.2 volts (double the voltage drop of a typical diode, at 0.6 volt).

Voltage applied to the gate is used to turn the SCR on. However, if the voltage source at the gate is shut off, the current will still continue to flow through the SCR until the source current is stopped.

USES OF AN SCR SCRs can be used to construct a circuit for a **center high-mounted stoplight (CHMSL)**. If this third stoplight were wired into either the left- or the right-side brake light circuit, the CHMSL would also flash whenever the turn signals were used for the side that was connected to the CHMSL. When two SCRs are used, both brake lights must be activated to supply current to the CHMSL. The current to the CHMSL is shut off when both SCRs lose their power source (when the brake pedal is released, which stops the current flow to the brake lights). ● **SEE FIGURE 10–19.**

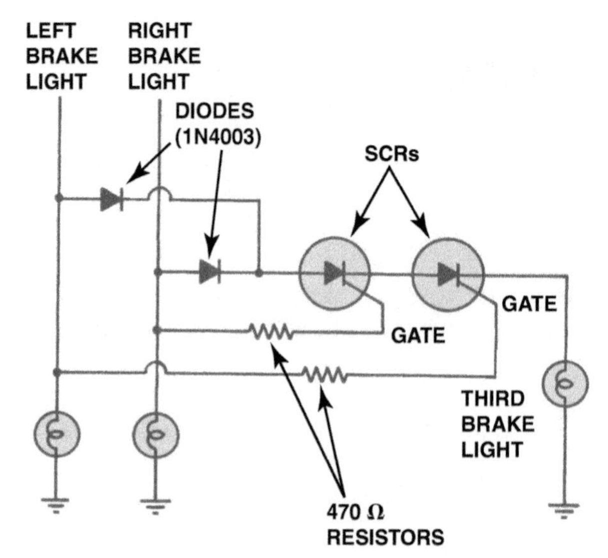

FIGURE 10–19 Wiring diagram for a center high-mounted stoplight (CHMSL) using SCRs.

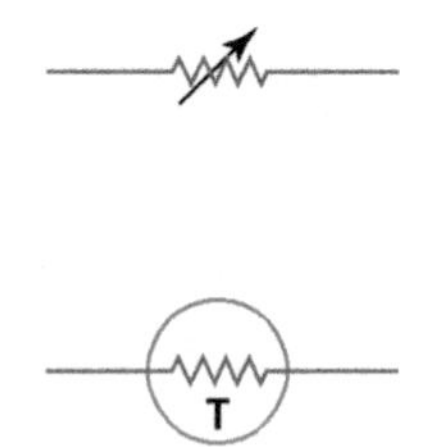

FIGURE 10–20 Symbols used to represent a thermistor.

	COPPER WIRE	NTC THERMISTOR
Cold	Lower resistance	Higher resistance
Hot	Higher resistance	Lower resistance

CHART 10–1

The resistance changes opposite that of a copper wire with changes in temperature.

THERMISTORS

CONSTRUCTION A **thermistor** is a semiconductor material such as silicon that has been doped to provide a given resistance. When the thermistor is heated, the electrons within the crystal gain energy and electrons are released. This means that a thermistor actually produces a small voltage when heated. If voltage is applied to a thermistor, its resistance decreases because the thermistor itself is acting as a current carrier rather than as a resistor at higher temperatures.

USES OF THERMISTORS A thermistor is commonly used as a temperature-sensing device for coolant temperature and intake manifold air temperature. Because thermistors operate in a manner opposite to that of a typical conductor, they are called **negative temperature coefficient (NTC)** thermistors; their resistance decreases as the temperature increases. ● **SEE CHART 10–1.**

Thermistor symbols are shown in ● **FIGURE 10–20.**

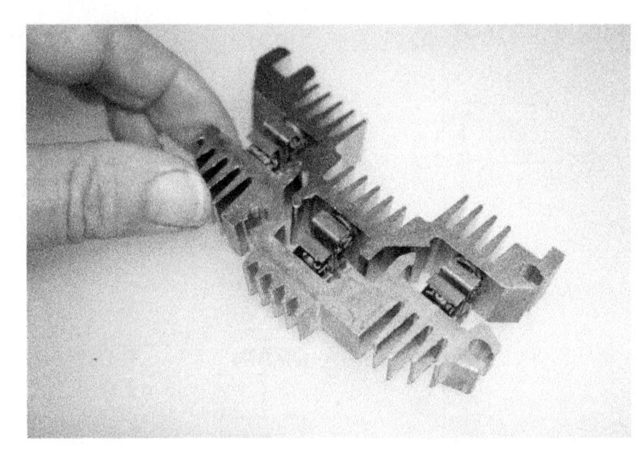

FIGURE 10–21 This rectifier bridge contains six diodes; three positive and three negative diodes. They are mounted to an aluminum-finned heat sink to keep the diodes cool during alternator operation. (Courtesy of Jeffrey Rehkopf)

RECTIFIER BRIDGES

DEFINITION The word *rectify* means "to set straight"; therefore, a rectifier is an electronic device (such as a diode) used to convert a changing voltage into a straight or constant voltage. A **rectifier bridge** is a group of diodes that is used to change alternating current (AC) into direct current (DC). A rectifier bridge is used in alternators to rectify the AC voltage produced in the stator (stationary windings) of the alternator into DC voltage. These rectifier bridges contain six diodes: one pair of diodes (one positive and one negative) for each of the three stator windings. ● **SEE FIGURE 10–21.**

TRANSISTORS

PURPOSE AND FUNCTION A **transistor** is a semiconductor device that can perform the following electrical functions.

1. Act as an electrical switch in a circuit
2. Act as an amplifier of current in a circuit
3. Regulate the current in a circuit

The word *transistor,* derived from the words *transfer* and *resistor,* is used to describe the transfer of current across a resistor. A transistor is made of three alternating sections or layers of P-type and N-type materials. This type of transistor is usually called a **bipolar transistor**.

CONSTRUCTION A transistor that has P-type material on each end, with N-type material in the center, is called a **PNP transistor**. Another type, with the exact opposite arrangement, is called an **NPN transistor**.

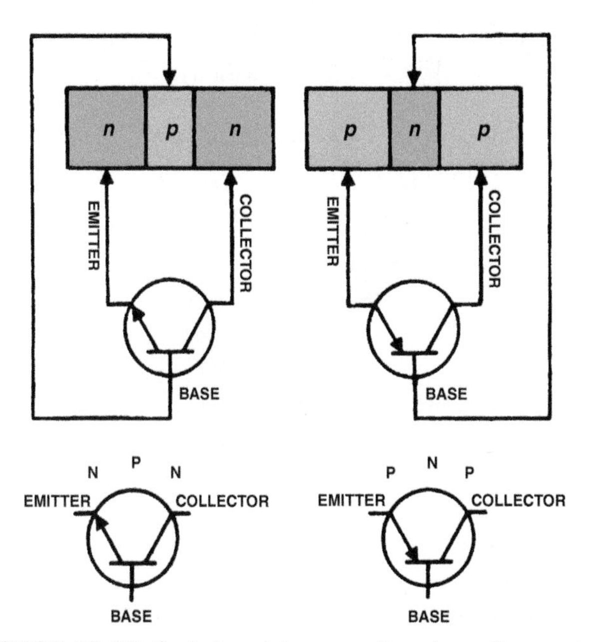

FIGURE 10–22 Basic transistor operation. A small current flowing through the base and emitter of the transistor turns on the transistor and permits a higher amperage current to flow from the collector and the emitter.

The material at one end of a transistor is called the **emitter** and the material at the other end is called the **collector**. The **base** is in the center and the voltage applied to the base is used to control current through a transistor.

TRANSISTOR SYMBOLS All transistor symbols contain an arrow indicating the emitter part of the transistor. The arrow points in the direction of current flow (conventional theory).

When an arrowhead appears in any semiconductor symbol, it stands for a P-N junction and it points from the P-type material toward the N-type material. The arrow on a transistor is always attached to the *emitter* side of the transistor. ● **SEE FIGURE 10–22.**

HOW A TRANSISTOR WORKS A transistor is similar to two back-to-back diodes that can conduct current in only one direction. As in a diode, N-type material can conduct electricity by means of its supply of free electrons, and P-type material conducts by means of its supply of positive holes.

A transistor will allow current flow if the electrical conditions allow it to switch on, in a manner similar to the working of an electromagnetic relay. The electrical conditions are determined, or switched, by means of the base, or *B*. The base will carry current only when the proper voltage and polarity are applied. The main circuit current flow travels through the other two parts of the transistor: the emitter *E* and the collector *C*. ● **SEE FIGURE 10–23.**

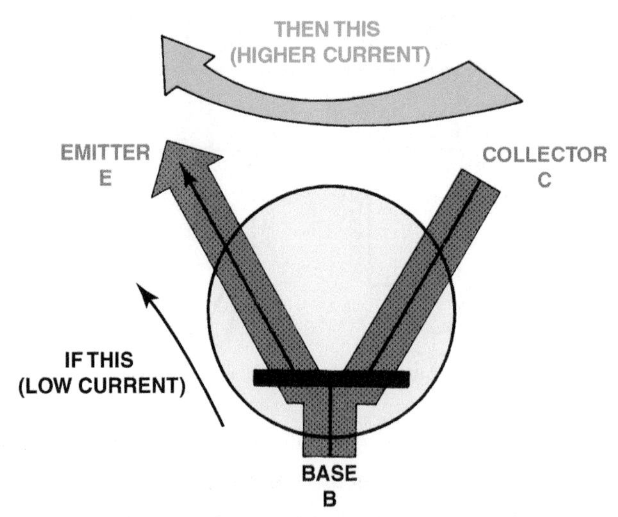

FIGURE 10–23 Basic transistor operation. A small current flowing through the base and emitter of the transistor turns on the transistor and permits a higher amperage current to flow from the collector and the emitter.

FREQUENTLY ASKED QUESTION

Is a Transistor Similar to a Relay?

Yes, in many cases a transistor is similar to a relay.

Both use a low current to control a higher current circuit. ● **SEE CHART 10–2**.

A relay can only be on or off. A transistor can provide a variable output if the base is supplied a variable current input.

	RELAY	TRANSISTOR
Low-current circuit	Coil (terminals 85 and 86)	Base and emitter
High-current circuit	Contacts terminals 30 and 87	Collector and emitter

CHART 10–2

Comparison between the control (low-current) and high-current circuits of a transistor compared to a mechanical relay.

If the base current is turned off or on, the current flow from collector to emitter is turned off or on. The current controlling the base is called the control current. The control current must be high enough to switch the transistor on or off. (This control voltage, called the **threshold voltage**, must be above approximately 0.3 volt for germanium and 0.6 volt for silicon transistors.) This control current can also "throttle" or regulate the main circuit, in a manner similar to the operation of a water faucet.

HOW A TRANSISTOR AMPLIFIES A transistor can amplify a signal if the signal is strong enough to trigger the base of a transistor on and off. The resulting on-off current flow through the transistor can be connected to a higher-powered electrical circuit. This results in a higher-powered circuit being controlled by a lower-powered circuit. This low-powered circuit's cycling is exactly duplicated in the higher-powered circuit, and therefore any transistor can be used to amplify a signal. However, because some transistors are better than others for amplification, specialized types of transistors are used for each specialized circuit function.

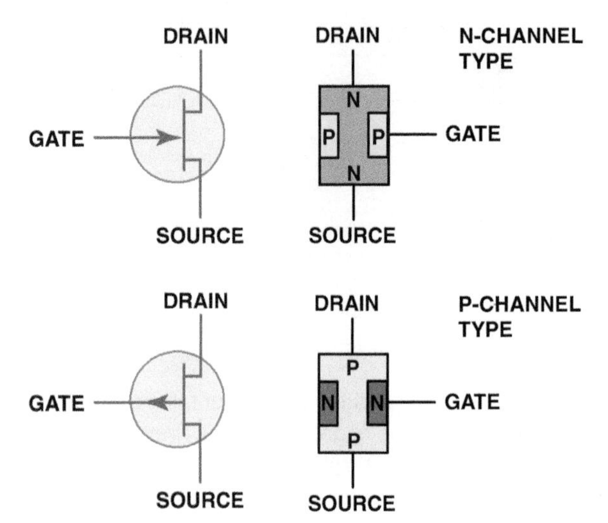

FIGURE 10–24 The three terminals of a field-effect transistor (FET) are called the source, gate, and drain.

FREQUENTLY ASKED QUESTION

What Does the Arrow Mean on a Transistor Symbol?

The arrow on a transistor symbol is always on the emitter and points toward the N-type material. The arrow on a diode also points toward the N-type material. To know which type of transistor is being shown, note which direction the arrow points.

- PNP: pointing in
- NPN: not pointing in

FIELD-EFFECT TRANSISTORS

Field-effect transistors (FETs) have been used in most automotive applications since the mid-1980s. They use less electrical current and rely mostly on the strength of a small voltage signal to control the output. The parts of a typical FET include the *source, gate,* and *drain.* ● **SEE FIGURE 10–24**.

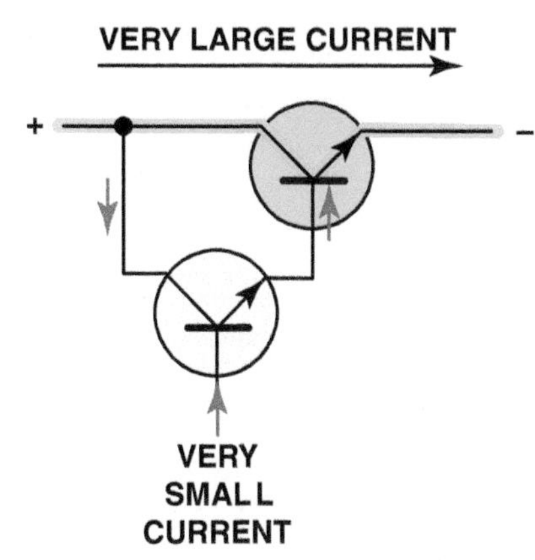

FIGURE 10–25 A Darlington pair consists of two transistors wired together, allowing for a very small current to control a larger current flow circuit.

FREQUENTLY ASKED QUESTION

What Is a Darlington Pair?

A **Darlington pair** consists of two transistors wired together. This arrangement permits a very small current flow to control a large current flow. The Darlington pair is named for Sidney Darlington, an American physicist for Bell Laboratories from 1929 to 1971. Darlington amplifier circuits are commonly used in electronic ignition systems, computer engine control circuits, and many other electronic applications. ● **SEE FIGURE 10–25.**

Many field-effect transistors are constructed of metal oxide semiconductor (MOS) materials, called **MOSFETs.** MOSFETs are highly sensitive to static electricity and can be easily damaged if exposed to excessive current or high-voltage surges (spikes). Most automotive electronic circuits use MOSFETs, which explains why it is vital for the service technician to use caution to avoid doing anything that could result in a high-voltage spike, and perhaps destroy an expensive computer module. Some vehicle manufacturers recommend that technicians wear an antistatic wristband when working with modules that contain MOSFETs. Always follow the vehicle manufacturer's instructions found in service information to avoid damaging electronic modules or circuits.

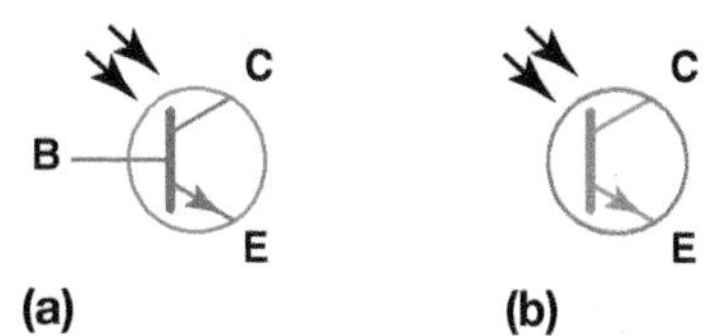

FIGURE 10–26 Symbols for a phototransistor. (a) This symbol uses the line for the base; (b) this symbol does not.

PHOTOTRANSISTORS

Similar in operation to a photodiode, a **phototransistor** uses light energy to turn on the base of a transistor. A phototransistor is an NPN transistor that has a large exposed base area to permit light to act as the control for the transistor. Therefore, a phototransistor may or may not have a base lead. If not, then it has only a collector and emitter lead. When the phototransistor is connected to a powered circuit, the light intensity is amplified by the gain of the transistor. Phototransistors, along with photo diodes, are frequently used in steering wheel controls. ● **SEE FIGURE 10–26.**

INTEGRATED CIRCUITS

PURPOSE AND FUNCTION Solid-state components are used in many electronic semiconductors and/or circuits. They are called "solid state" because they have no moving parts, just higher or lower voltage levels within the circuit. Discrete (individual) diodes, transistors, and other semiconductor devices were often used to construct early electronic ignition and electronic voltage regulators. Newer-style electronic devices use the same components, but they are now combined (integrated) into one group of circuits, and are thus called an **integrated circuit (IC).**

HEAT SINK **Heat sink** is a term used to describe any area around an electronic component that, because of its shape or design, can conduct damaging heat away from electronic parts. Examples of heat sinks include the following:

- Ribbed electronic ignition control units
- Cooling slits and cooling fan attached to an alternator

A special heat-conducting compound is used under the heat sink of the rectifier bridge inside of a generator. ● **SEE FIGURE 10–27.**

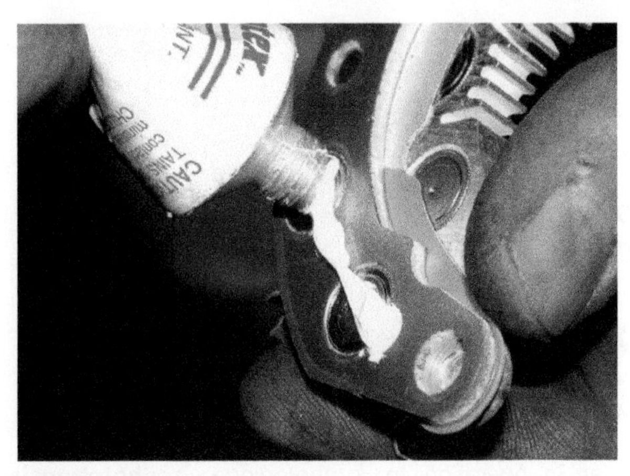

FIGURE 10–27 Silicone heat transfer compound is applied to the heat sink when it is mounted to the generator frame.

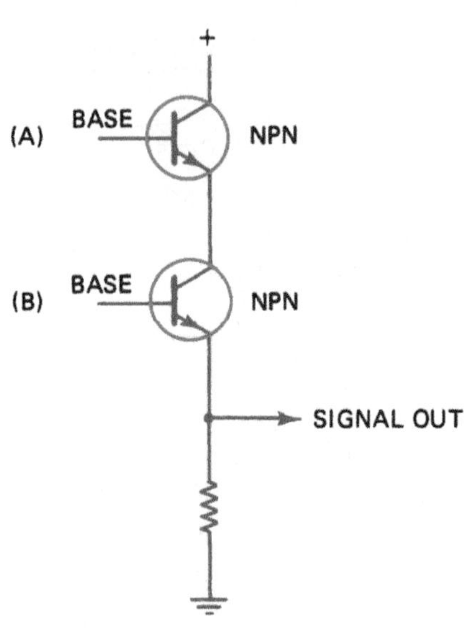

FIGURE 10–28 Typical transistor AND gate circuit using two transistors. The emitter is always the line with the arrow. Notice that both transistors must be turned on before there is voltage present at the point labeled "signal out."

FREQUENTLY ASKED QUESTION

What Causes a Transistor or Diode to Blow?

Every automotive diode and transistor is designed to operate within certain voltage and amperage ranges for individual applications. For example, transistors used for switching are designed and constructed differently from transistors used for amplifying signals.

Because each electronic component is designed to operate satisfactorily for its particular application, any severe change in operating current (amperes), voltage, or heat can destroy the *junction.* This failure can cause either an open circuit (no current flows) or a short (current flows through the component all the time when the component should be blocking the current flow).

Heat sinks are necessary to prevent damage to diodes, transistors, and other electronic components due to heat buildup. Excessive heat can damage the junction between the N-type and P-type materials used in diodes and transistors.

TRANSISTOR GATES

PURPOSE AND FUNCTION Knowledge of the basic operation of electronic gates is important in understanding how computers work. A gate is an electronic circuit whose output depends on the location and voltage of two inputs.

CONSTRUCTION Whether a transistor is on or off depends on the voltage at the base of the transistor. For the transistor to turn on, a voltage difference between the base of the transmitter and the emitter should be at least a 0.6 volt. Most electronic and computer circuits use 5 volts as a power source. If two transistors are wired together, several different outputs can be received depending on how the two transistors are wired. **SEE FIGURE 10–28.**

OPERATION If the voltage at *A* is higher than that of the emitter, the top transistor is turned on; however, the bottom transistor is off unless the voltage at *B* is also higher. If both transistors are turned on, the output signal voltage will be high. If only one of the two transistors is on, the output will be zero (off or no voltage). Because it requires both *A* and *B* to be on to result in a voltage output, this circuit is called an *AND gate*. In other words, both transistors have to be on before the gate opens and allows a voltage output. Other types of gates can be constructed using various connections to the two transistors. For example:

AND gate. Requires both transistors to be on to get an output.

OR gate. Requires either transistor to be on to get an output.

NAND (NOT-AND) gate. Output is on unless both transistors are on.

NOR (NOT-OR) gate. Output is on only when both transistors are off.

Gates represent logic circuits that can be constructed so that the output depends on the voltage (on or off; high or low) of the inputs to the bases of transistors. Their inputs can come from sensors or other circuits that monitor sensors, and their

FREQUENTLY ASKED QUESTION

What Are Logic Highs and Lows?

All computer circuits and most electronic circuits (such as gates) use various combinations of high and low voltages. High voltages are typically those above 5 volts, and low is generally considered zero (ground). However, high voltages do not *have* to begin at 5 volts. *High, or the number 1, to a computer is the presence of voltage above a certain level.* For example, a circuit could be constructed where any voltage higher than 3.8 volts would be considered high. *Low, or the number 0, to a computer is the absence of voltage or a voltage lower than a certain value.* For example, a voltage of 0.62 volt may be considered low. Various associated names and terms can be summarized.

- Logic low = Low voltage = Number 0 = Reference low
- Logic high = Higher voltage = Number 1 = Reference high

outputs can be used to operate an output device if amplified and controlled by other circuits. For example, the blower motor will be commanded on when the following events occur, to cause the control module to turn it on.

1. The ignition must be on (input).
2. The air-conditioning is commanded on.
3. The engine coolant temperature is within a predetermined limit.

If all of these conditions are met, then the control module will command the blower motor on. If any of the input signals are incorrect, the control module will not be able to perform the correct command.

OPERATIONAL AMPLIFIERS

Operational amplifiers (op-amps) are used in circuits to control and amplify digital signals. Op-amps are frequently used for motor control in climate control systems (heating and air-conditioning) airflow control door operation. Op-amps can provide the proper voltage polarity and current (amperes) to control the direction of permanent magnetic (PM) motors. The symbol for an op-amp is shown in **FIGURE 10–29**.

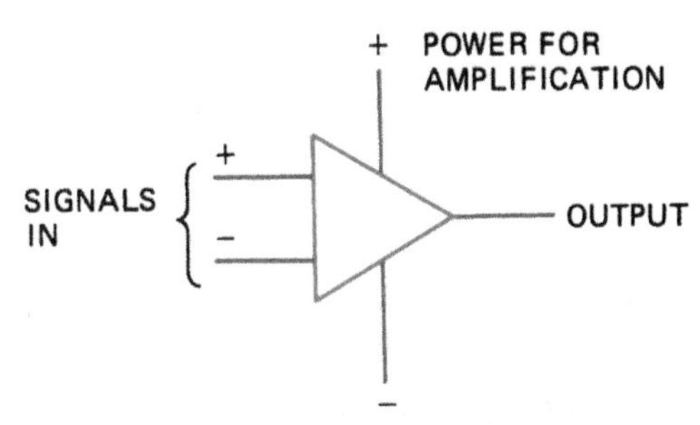

FIGURE 10–29 Symbol for an operational amplifier (op-amp).

ELECTRONIC COMPONENT FAILURE CAUSES

Electronic components such as electronic ignition modules, electronic voltage regulators, onboard computers, and any other electronic circuit are generally quite reliable; however, failure can occur. Frequent causes of premature failure include the following:

- **Poor connections.** It has been estimated that most engine computers returned as defective have simply had poor connections at the wiring harness terminal ends. These faults are often intermittent and hard to find.

  NOTE: When cleaning electronic contacts, use a pencil eraser. This cleans the contacts without harming the thin, protective coating used on most electronic terminals.

- **Heat.** The operation and resistance of electronic components and circuits are affected by heat. Electronic components should be kept as cool as possible and never hotter than 260°F (127°C).
- **Voltage spikes.** A high-voltage spike can literally burn a hole through semiconductor material. The source of these high-voltage spikes is often the discharge of a coil without proper (or with defective) despiking protection. A poor electrical connection at the battery or other major electrical connection can cause high-voltage spikes to occur, because the *entire wiring harness creates its own magnetic field*, similar to that formed around a coil. If the connection is loose and momentary loss of contact occurs, a high-voltage surge can occur through the entire electrical system. To help prevent this type of damage, ensure that all electrical connections, including grounds, are properly clean and tight.

  CAUTION: One of the major causes of electronic failure occurs during jump-starting a vehicle. Always check that the ignition switch is off on both vehicles when making the connection. Always double-check that the correct battery polarity (+ to + and − to −) is being performed.

TECH TIP

Blinking LED Theft Deterrent

A blinking (flashing) LED consumes only about 5 milliamperes (5/1,000 of 1 ampere or 0.005 A). Most alarm systems use a blinking red LED to indicate that the system is armed. A fake alarm indicator is easy to make and install.

A 470-ohm, 0.5-watt resistor limits current flow to prevent battery drain. The positive terminal (anode) of the diode is connected to a fuse that is hot at all times, such as the cigarette lighter. The negative terminal (cathode) of the LED is connected to any ignition-controlled fuse. ● **SEE FIGURE 10–30**.

When the ignition is turned off, the power flows through the LED to ground and the LED flashes. To prevent distraction during driving, the LED goes out when the ignition is on. Therefore, this fake theft deterrent is "auto setting" and no other action is required to activate it when you leave your vehicle except to turn off the ignition and remove the key as usual.

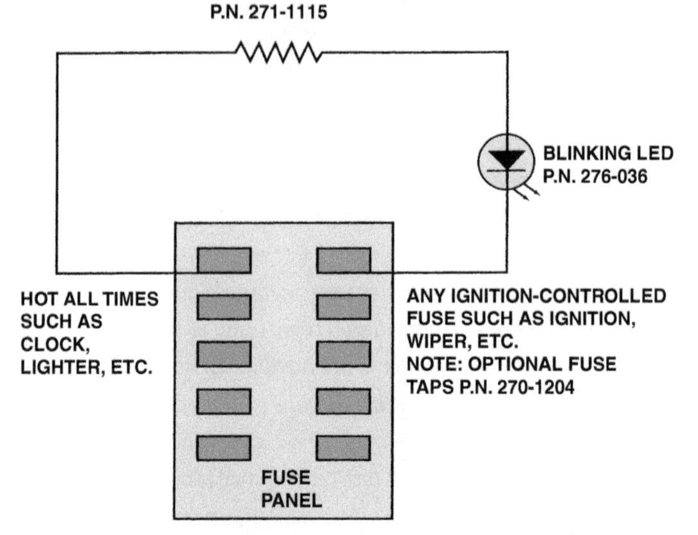

FIGURE 10–30 Schematic for a blinking LED theft deterrent.

- **Excessive current.** All electronic circuits are designed to operate within a designated range of current (amperes). If a solenoid or relay is controlled by a computer circuit, the resistance of that solenoid or relay becomes part of that control circuit. If a coil winding inside the solenoid or relay becomes shorted, the resulting lower resistance will increase the current through the circuit. Even though individual components are used with current-limiting resistors in series, the coil winding resistance is also used as a current-control component in the circuit. If a computer fails, always measure the resistance across all computer-controlled relays and solenoids. The resistance should be within specifications (generally *over* 20 ohms) for each component that is computer controlled.

NOTE: Some computer-controlled solenoids are pulsed on and off rapidly. This type of solenoid is used in many electronically shifted transmissions. Their resistance is usually about half of the resistance of a simple on-off solenoid—usually between 10 and 15 ohms. Because the computer controls the on-time of the solenoid, the solenoid and its circuit control are called **pulse-width modulated (PWM)**.

HOW TO TEST DIODES AND TRANSISTORS

TESTERS Diodes and transistors can be tested with an ohmmeter. The diode or transistor being tested must be disconnected from the circuit for the results to be meaningful.

- Use the *diode-check* position on a digital multimeter.
- In the diode-check position on a digital multimeter, the meter applies a higher voltage than when the ohms test function is selected.
- This slightly higher voltage (about 2 to 3 volts) is enough to forward bias a diode or the P-N junction of transistors.

DIODES Using the diode test position, the meter applies a voltage. The display will show the voltage drop across the diode P-N junction. A good diode should give an over limit (OL) reading with the test leads attached to each lead of the diode in one way, and a voltage reading of 0.400 to 0.600 volt when the leads are reversed. This reading is the voltage drop or the barrier voltage across the P-N junction of the diode.

1. A low-voltage reading with the meter leads attached both ways across a diode means that the diode is *shorted* and must be replaced.

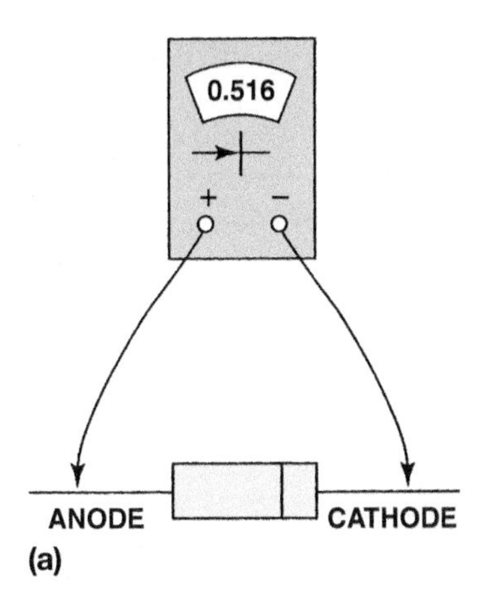

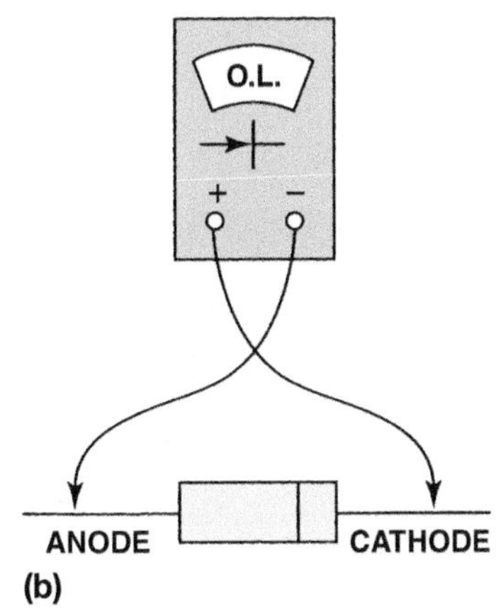

FIGURE 10–31 To check a diode, select "diode check" on a digital multimeter. The display will indicate the voltage drop (difference) between the meter leads. The meter itself applies a low-voltage signal (usually about 3 volts) and displays the difference on the display. (a) When the diode is forward biased, the meter should display a voltage between 0.500 and 0.700 volt (500 to 700 millivolts). (b) When the meter leads are reversed, the meter should read OL (over limit) because the diode is reverse biased and blocking current flow.

2. An OL reading with the meter leads attached both ways across a diode means that the diode is *open* and must be replaced.
   SEE FIGURE 10–31.

TRANSISTORS Using a digital meter set to the diode-check position, a good transistor should show a voltage drop of 0.400 to 0.600 volt between the following:

- The emitter (*E*) and the base (*B*) and between the base (*B*) and the collector (*C*) with a meter connected one way, and OL when the meter test leads are reversed.
- An OL reading (no continuity) in both directions when a transistor is tested between the emitter (*E*) and the collector (*C*). (A transistor tester can also be used if available.) **SEE FIGURE 10–32.**

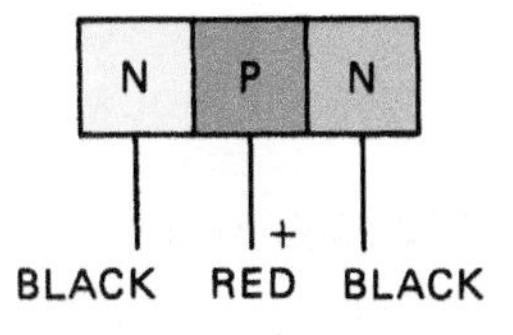

FIGURE 10–32 If the red (positive) lead of the ohmmeter (or a multimeter set to diode check) is touched to the center and the black (negative lead) touched to either end of the electrode, the meter should forward bias the P-N junction and indicate on the meter as low resistance. If the meter reads high resistance, reverse the meter leads, putting the black on the center lead and the red on either end lead. If the meter indicates low resistance, the transistor is a good PNP type. Check all P-N junctions in the same way.

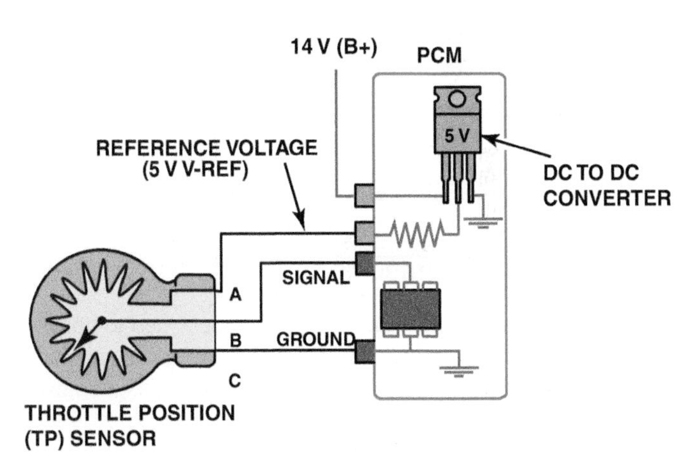

FIGURE 10–33 A DC to DC converter is built into most powertrain control modules (PCMs) and is used to supply the 5-volt reference called V-ref to many sensors used to control the internal combustion engine.

CONVERTERS AND INVERTERS

CONVERTERS DC to DC converters (usually written as DC-DC converter) are electronic devices used to transform DC voltage from one level to another higher or lower level. They are used to distribute various levels of DC voltage throughout a vehicle from a single power bus (or voltage source).

EXAMPLES OF USE One example of a DC-DC converter circuit is the circuit the PCM uses to convert 14 to 5 volts. The 5 volts is called the reference voltage, abbreviated V-ref, and is used to power many sensors in a computer-controlled engine management system. The schematic of a typical 5 volt V-ref interfacing with the TP sensor circuit is shown in **FIGURE 10–33.**

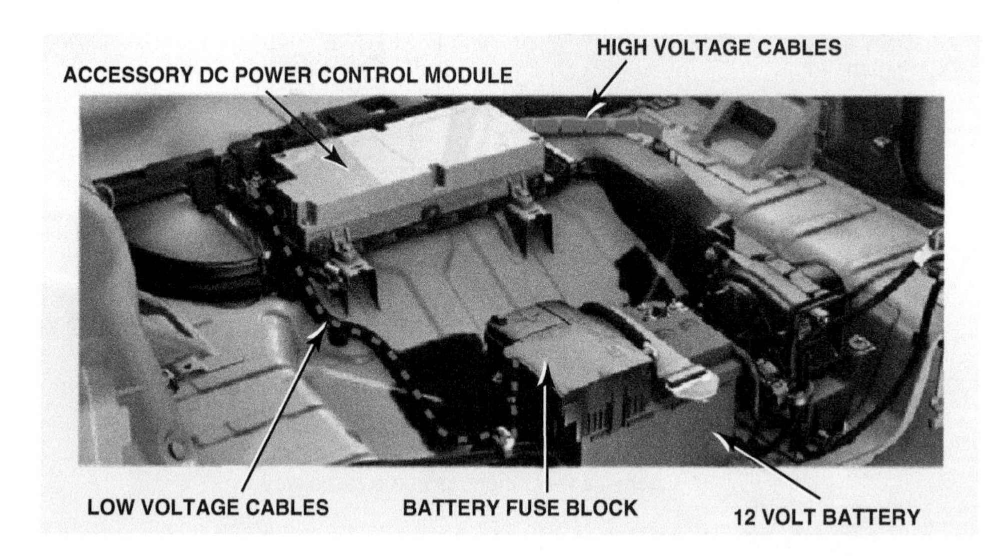

FIGURE 10–34 The accessory DC power control module is located in the rear compartment (trunk), underneath the floor panel. (Courtesy of General Motors)

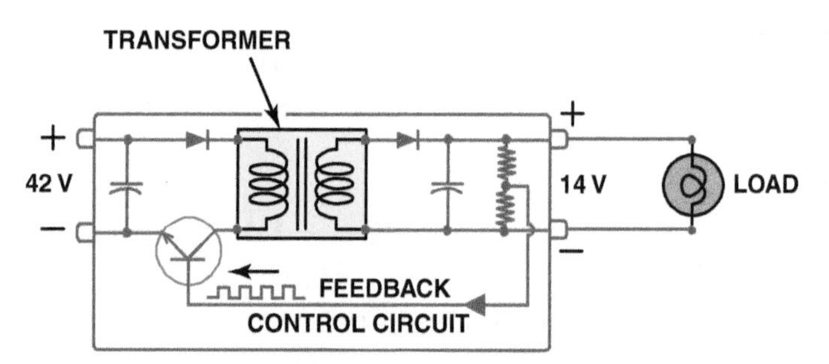

FIGURE 10–35 This DC-DC converter is designed to convert 42 volts to 14 volts, to provide 14 volts power to accessories on a hybrid electric vehicle operating with a 42-volt electrical system.

The PCM operates on 14 volts, using the principle of DC conversion to provide a constant 5 volts of sensor reference voltage to the TP sensor and others. The TP sensor demands little current, so the V-ref circuit is a low-power DC voltage converter in the range of 1 watt. The PCM uses a DC-DC converter, which is a small semiconductor device called a voltage regulator, and is designed to convert battery voltage to a constant 5 volts regardless of changes in the charging voltage.

Hybrid electric vehicles use DC-DC converters to provide higher or lower DC voltage levels and current requirements.

On the Chevrolet Volt the converter is called an accessory (DC) power control module. It converts high-voltage DC electrical energy from the drive motor battery to low-voltage DC electrical energy. The module supplies electrical energy to the vehicle's low-voltage electrical system and also maintains the 12-volt battery charge. The accessory DC power control module is capable of supplying 180 amps of 14 volts DC and takes the place of a belt-driven generator found on conventional vehicles. ● **SEE FIGURE 10–34**.

A low voltage (42 to 14 volt) DC-DC converter schematic is shown in ● **FIGURE 10–35** and represents how one type of DC-DC converter works.

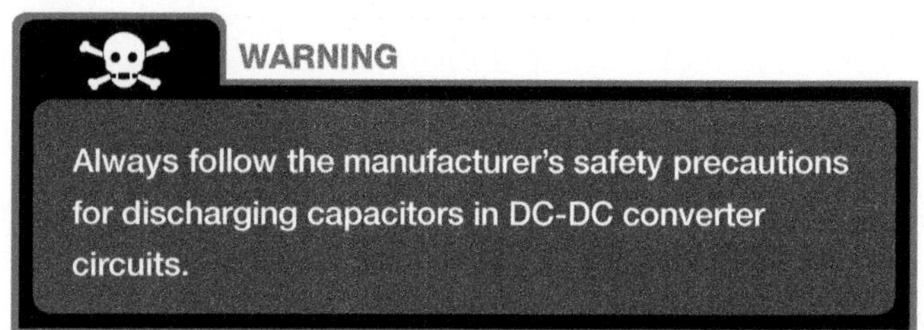

WARNING

Always follow the manufacturer's safety precautions for discharging capacitors in DC-DC converter circuits.

The central component of a converter is a transformer that physically isolates the input (42 volts) from the output (14 volts). The power transistor pulses the high-voltage coil of the transformer, and the resulting changing magnetic field induces a voltage in the coil windings of the lower voltage side of the transformer. The diodes and capacitors help control and limit the voltage and frequency of the circuit.

DC-DC CONVERTER CIRCUIT TESTING Usually a DC control voltage is used, which is supplied by a digital logic circuit to shift the voltage level to control the converter. A voltage test can indicate if the correct voltages are present when the converter is on and off.

Voltage measurements are usually specified to diagnose a DC-DC converter system. A digital multimeter (DMM) that is CAT III rated should be used.

1. Always follow the manufacturer's safety precautions when working with high-voltage circuits. These circuits are usually indicated by orange wiring.
2. Never tap into wires in a DC-DC converter circuit to access power for another circuit.
3. Never tap into wires in a DC-DC converter circuit to access a ground for another circuit.
4. Never block airflow to a DC-DC converter heat sink.
5. Never use a heat sink for a ground connection for a meter, scope, or accessory connection.

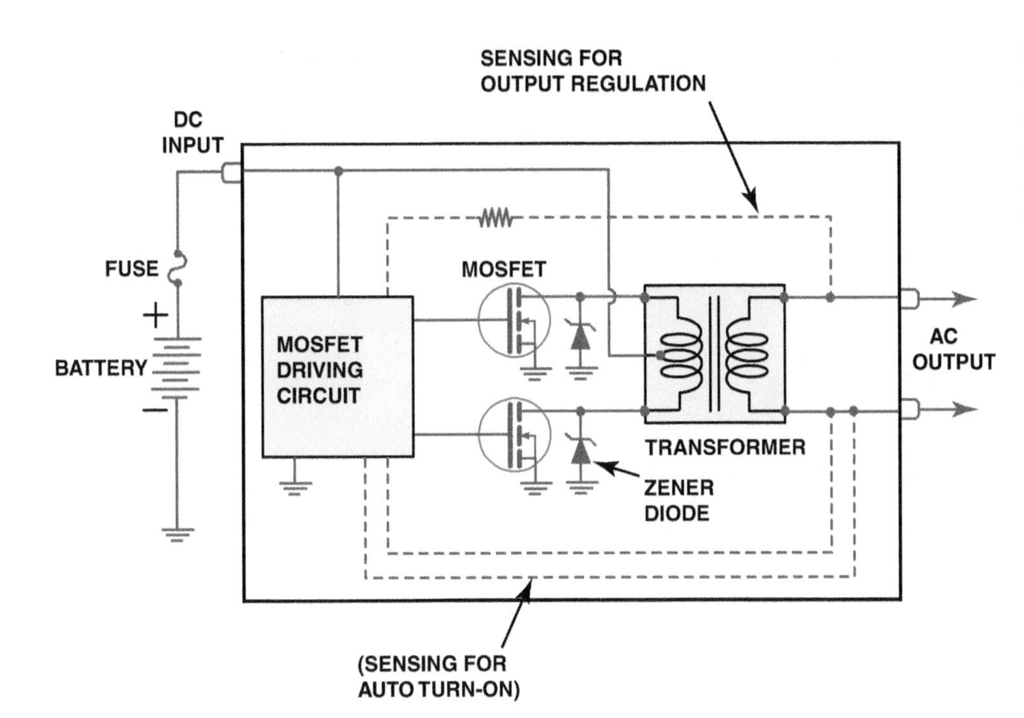

FIGURE 10–36 A typical circuit for an inverter designed to change direct current from a battery to alternating current for use by the electric motors used in a hybrid electric vehicle.

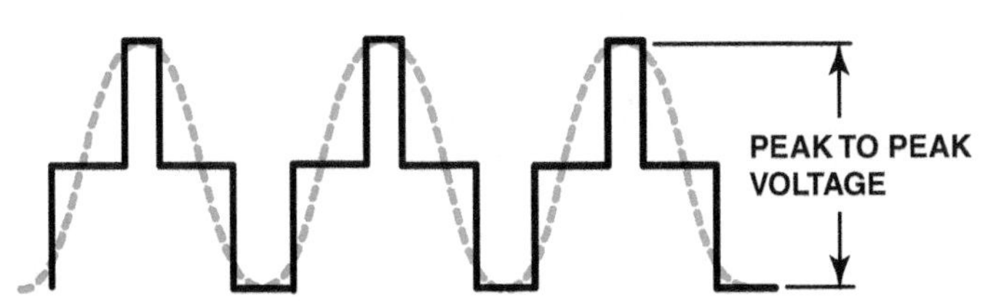

FIGURE 10–37 The switching (pulsing) MOSFETs create a waveform called a modified sine wave (solid lines) compared to a true sine wave (dotted lines).

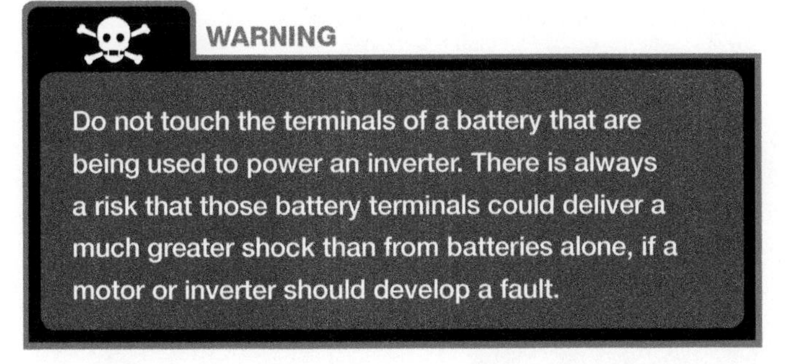

6. Never connect or disconnect a DC-DC converter while the converter is powered up.
7. Never connect a DC-DC converter to a larger voltage source than specified.

INVERTERS An **inverter** is an electronic circuit that changes direct current (DC) into alternating current (AC). In most DC-AC inverters, the switching transistors, which are usually MOSFETs, are turned on alternately for short pulses. As a result, the transformer produces a modified sine wave output, rather than a true sine wave. ● **SEE FIGURE 10–36.**

The waveform produced by an inverter is not the perfect sine wave of household AC, but is rather more like a pulsing DC that reacts similar to sine wave AC in transformers and in induction motors. ● **SEE FIGURE 10–37.**

Inverters power AC motors. An inverter converts DC power to AC power at the required frequency and amplitude. The inverter consists of three half-bridge units, and the output voltage is mostly created by a pulse-width modulation (PWM) technique. The three-phase voltage waves are shifted 120 degrees to each other, to power each of the three phases.

INVERTER APPLICATION On a hybrid vehicle the inverter function is built in as part of the drive motor/generator power inverter module. This module is usually located under the hood and manages all hybrid powertrain functions. ● **SEE FIGURE 10–38.**

The **Hybrid Powertrain Control Module (HPCM)** is located within the module. The HPCM is the main controller of vehicle powertrain operation and determines when to perform normal operating modes such as electric mode, extended range mode, and regenerative braking. ● **SEE FIGURE 10–39.**

The **motor control modules** control the speed, direction, and output torque of their respective drive motor/generators based on commands from the HPCM. This is accomplished through the sequencing actuation of high current switching transistors called insulated gate bipolar transistors. The HPCM receives feedback from the drive motor/generators, such as speed, current, direction, and temperature. The auxiliary transmission fluid pump motor is also controlled by a dedicated motor control module. Each motor control module is a

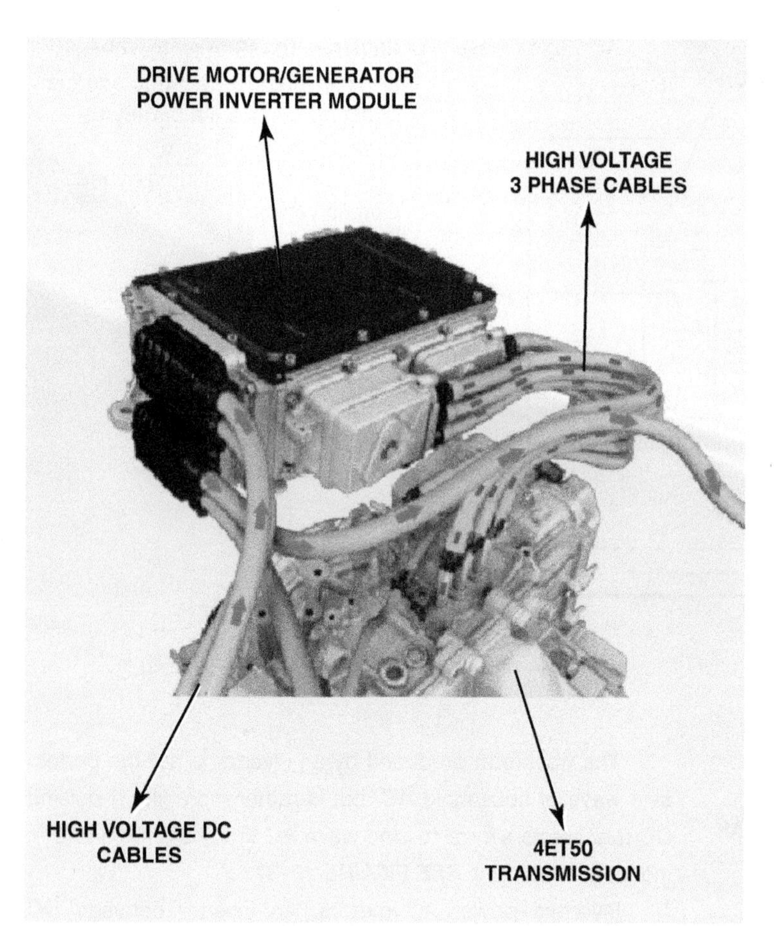

FIGURE 10–38 The drive motor/generator power inverter module (PIM) is the heart of the hybrid powertrain. (Courtesy of General Motors)

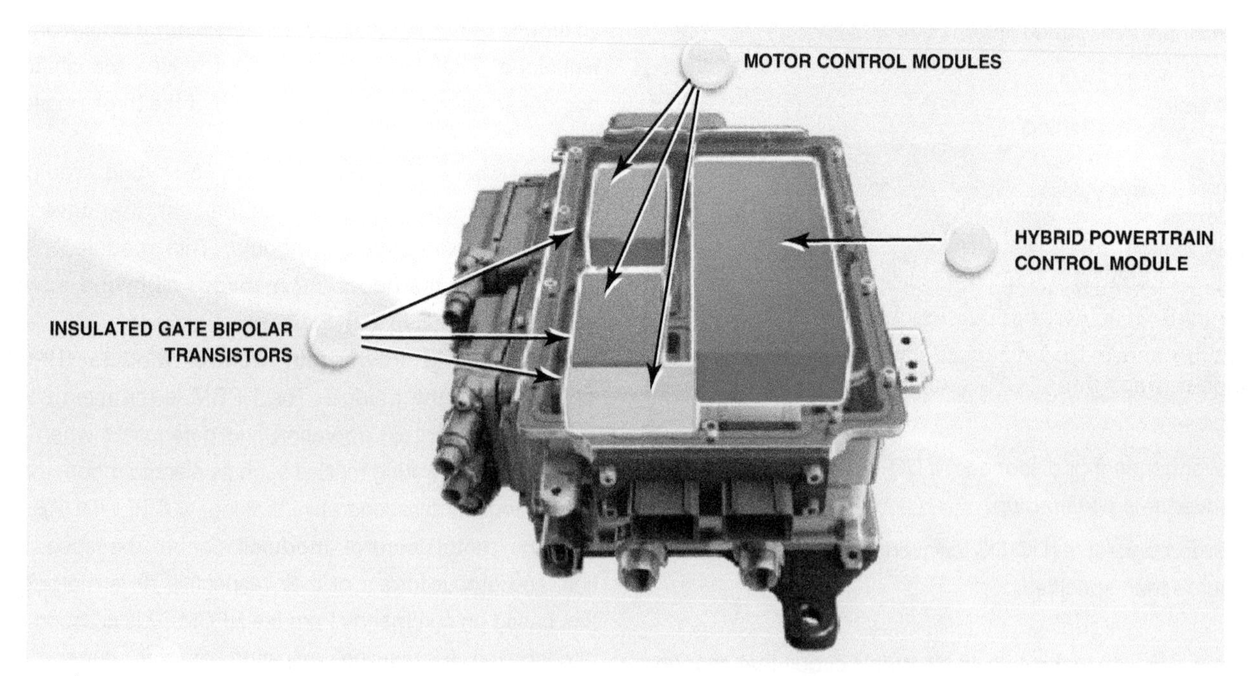

FIGURE 10–39 Internal components of the powertrain inverter module. (Courtesy of General Motors)

flash-programmable microprocessor that cannot be serviced separately from the drive motor/generator power inverter module.

The **insulated gate bipolar transistors** are electronic switches that rapidly cycle high-voltage DC electrical energy on and off. The on and off cycling of the high voltage creates three phase alternating current (AC) that is utilized by the motor/generators. This process is called electrical inversion. The insulated gate bipolar transistors also invert AC to DC. The transistors cannot be serviced separately from the drive motor generator power inverter module.*

ELECTROSTATIC DISCHARGE

DEFINITION **Electrostatic discharge (ESD)** is created when static charges build up on the human body when movement occurs. The friction of the clothing and the movement of shoes against carpet or vinyl floors cause a high voltage to build. Then when we touch a conductive material, such as a doorknob, the static charge is rapidly discharged. These charges, although just slightly painful to us, can cause severe damage to delicate electronic components. The following are typical static voltages.

- If you can feel it, it is at least 3,000 volts.
- If you can hear it, it is at least 5,000 volts.
- If you can see it, it is at least 10,000 volts.

Although these voltages seem high, the current, in amperes, is extremely low. However, sensitive electronic components such as vehicle computers, radios, and instrument panel clusters can be ruined if exposed to as little as 30 volts. This is a problem because components can be harmed at voltages lower than we can feel.

AVOIDING ESD To help prevent damage to components, follow these easy steps.

1. Keep the replacement electronic component in the protective wrapping until just before installation.
2. Before handling any electronic component, ground yourself by touching a metal surface to drain away any static charge.
3. Do not touch the terminals of electronic components.
4. If working in an area where you could come in contact with terminals, wear a static electrically grounding wrist strap available at most electronic parts stores, such as Radio Shack.

If these precautions are observed, ESD damage can be eliminated or reduced. Remember, just because the component works even after being touched does not mean that damage has not occurred. Often, a section of the electronic component may be damaged, yet will not fail until several days or weeks later.

SUMMARY

1. Semiconductors are constructed by doping semiconductor materials such as silicon.
2. N-type and P-type materials can be combined to form diodes, transistors, SCRs, and computer chips.
3. Diodes can be used to direct and control current flow in circuits and to provide despiking protection.
4. Transistors are electronic relays that can also amplify signals.
5. All semiconductors can be damaged if subjected to excessive voltage, current, or heat.
6. Never touch the terminals of a computer or electronic device; static electricity can damage electronic components.

REVIEW QUESTIONS

1. What is the difference between P-type material and N-type material?
2. How can a diode be used to suppress high-voltage surges in automotive components or circuits containing a coil?
3. How does a transistor work?
4. To what precautions should all service technicians adhere, to avoid damage to electronic and computer circuits?

* General Motors STC Course # 18420.04W-R2

CHAPTER QUIZ

1. A semiconductor is a material ________.
   - **a.** With fewer than four electrons in the outer orbit of its atoms
   - **b.** With more than four electrons in the outer orbit of its atoms
   - **c.** With exactly four electrons in the outer orbit of its atoms
   - **d.** Determined by other factors besides the number of electrons
2. The arrow in a symbol for a semiconductor device ________.
   - **a.** Points toward the negative
   - **b.** Points away from the negative
   - **c.** Is attached to the emitter on a transistor
   - **d.** Both a and c
3. A diode installed across a coil with the cathode toward the battery positive is called a(n) ________.
   - **a.** Clamping diode
   - **b.** Forward-bias diode
   - **c.** SCR
   - **d.** Transistor
4. A transistor is controlled by the polarity and current at the ________.
   - **a.** Collector
   - **b.** Emitter
   - **c.** Base
   - **d.** Both a and b
5. A transistor can ________.
   - **a.** Switch on and off
   - **b.** Amplify
   - **c.** Throttle
   - **d.** All of the above
6. Clamping diodes ________.
   - **a.** Are connected into a circuit with the positive (+) voltage source to the cathode and the negative (–) voltage to the anode
   - **b.** Are also called despiking diodes
   - **c.** Can suppress transient voltages
   - **d.** All of the above
7. A zener diode is normally used for voltage regulation. A zener diode, however, can also be used for high-voltage spike protection if connected ________.
   - **a.** Positive to anode, negative to cathode
   - **b.** Positive to cathode, ground to anode
   - **c.** Negative to anode, cathode to a resistor then to a lower voltage terminal
   - **d.** Both a and c
8. The forward-bias voltage required for an LED is ________.
   - **a.** 0.3 to 0.5 volt
   - **b.** 0.5 to 0.7 volt
   - **c.** 1.5 to 2.2 volts
   - **d.** 4.5 to 5.1 volts
9. An LED can be used in a ________.
   - **a.** Headlight
   - **b.** Taillight
   - **c.** Brake light
   - **d.** All of the above
10. Another name for a ground is ________.
    - **a.** Logic low
    - **b.** Zero
    - **c.** Reference low
    - **d.** All of the above

chapter 11

COMPUTER FUNDAMENTALS

LEARNING OBJECTIVES

After studying this chapter, the reader should be able to:

1. Explain the purpose and function of onboard computers.
2. List the various parts of an automotive computer.
3. List input sensors and output device controlled by the computer.

This chapter will help you prepare for Engine Repair (A6) ASE certification test content area "A" (General Electrical/Electronic Systems Diagnosis).

KEY TERMS

Actuator 149
Analog-to-digital (AD) converter 151
Central processing unit (CPU) 151
Clock generator 152
Controller 153
Digital computer 151
Duty cycle 151
EEPROM 149
E^2 PROM 149
Electronic control assembly (ECA) 153
Electronic control module (ECM) 153
Electronic control unit (ECU) 153
Engine mapping 152
Input conditioning 148
Keep-alive memory (KAM) 149
Powertrain Control Module (PCM) 148
Programmable read-only memory (PROM) 149
Random-access memory (RAM) 149
Read-only memory (ROM) 149

GM STC OBJECTIVES

GM Service Technical College topics covered in this chapter are as follows:

1. Identify important types, characteristics, and diagnosis of various solid state electrical components (18043.03W-R4).
2. Identify common characteristics and functions of control modules.

COMPUTER CONTROL

Modern automotive control systems consist of a network of electronic sensors, actuators, and computer modules designed to regulate the powertrain and vehicle support systems. The **powertrain control module (PCM)** is a major component of this system. It coordinates engine and transmission operation, processes data, maintains communications, and makes the control decisions needed to keep the vehicle operating.

Automotive computers use voltage to send and receive information. Voltage is electrical pressure and does not flow through circuits, but voltage can be used as a signal. A computer converts input information or data into voltage signal combinations that represent number combinations. The number combinations can represent a variety of information—temperature, speed, or even words and letters. A computer processes the input voltage signals it receives by computing what they represent, and then delivering the data in computed or processed form.

NOTE: Although the PCM is technically a computer, it is more common for manufacturers to refer to the various computers on a vehicle as "control modules."

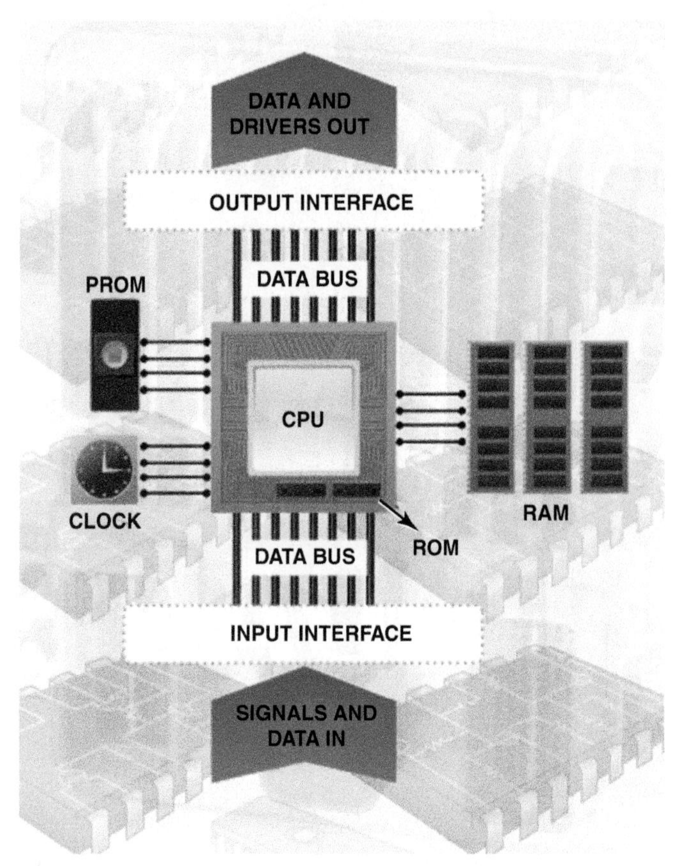

FIGURE 11–1 All computer systems perform four basic functions: input, processing, storage, and output. (Courtesy of General Motors)

THE FOUR BASIC COMPUTER FUNCTIONS

The operation of every computer can be divided into the following four basic functions: ● **SEE FIGURE 11–1**.

- Input
- Processing
- Storage
- Output

These basic functions are not unique to computers; they can be found in many noncomputer systems. However, we need to know how the computer handles these functions.

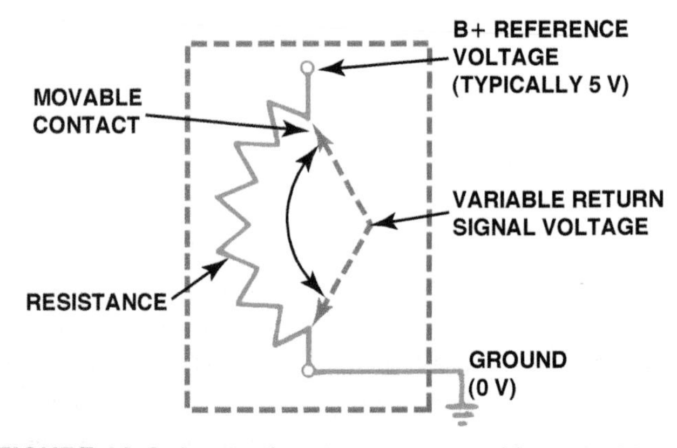

FIGURE 11–2 A potentiometer uses a movable contact to vary resistance and send an analog voltage to the PCM.

INPUT First, the computer receives a voltage signal (input) from an input device. The device can be as simple as a button or a switch on an instrument panel, or a sensor on an automotive engine. ● **SEE FIGURE 11–2** for a typical type of automotive sensor.

Vehicles use various mechanical, electrical, and magnetic sensors to measure factors such as vehicle speed, engine RPM, air pressure, oxygen content of exhaust gas, airflow, and engine coolant temperature. Each sensor transmits its information in the form of voltage signals. The computer receives these voltage signals, but before it can use them, the signals must undergo a process called **input conditioning**. This process includes amplifying voltage signals that are too small for the computer circuitry to handle. Input conditioners generally are located inside the computer, but a few sensors have their own input-conditioning circuitry.

FIGURE 11–3 A replaceable PROM used in an older General Motors computer. The access panel has been removed to gain access.

PROCESSING Input voltage signals received by a computer are processed through a series of electronic logic circuits maintained in its programmed instructions. These logic circuits change the input voltage signals, or data, into output voltage signals or commands.

STORAGE The program instructions for a computer are stored in electronic memory. Some programs may require that certain input data be stored for later reference or future processing. In others, output commands may be delayed or stored before they are transmitted to devices elsewhere in the system.

Computers have two types of memory: permanent and temporary. Permanent memory is called **read-only memory (ROM)** because the computer can only read the contents; it cannot change the data stored in it. This data is retained even when power to the computer is shut off. Part of the ROM is built into the computer, and the rest is located in an IC chip called a **programmable read-only memory (PROM)** or calibration assembly. **SEE FIGURE 11–3.** Many chips are erasable, meaning that the program can be changed. These chips are called erasable programmable read-only memory or EPROM. Since the early 1990s most programmable memory has been electronically erasable, meaning that the program in the chip can be reprogrammed by using a scan tool and the proper software. This computer reprogramming is usually called *reflashing*. These chips are electrically erasable programmable read-only memory, abbreviated **EEPROM** or **E^2 PROM**. All vehicles equipped with onboard diagnosis second generation, called OBD II, are equipped with EEPROMs.

Temporary memory is called **random-access memory (RAM)** because the microprocessor can write or store new data into it as directed by the computer program, as well as read the data already in it. Automotive computers use two types of RAM memory: volatile and nonvolatile. Volatile RAM memory is lost whenever the ignition is turned off. However, a type of volatile RAM called **keep-alive memory (KAM)** can be wired directly to battery power. This prevents its data from being erased when the ignition is turned off. Both RAM and KAM have the disadvantage of losing their memory when disconnected from their power source. One example of RAM and KAM is the loss of station settings in a programmable radio when the battery is disconnected. Since all the settings are stored in RAM, they have to be reset when the battery is reconnected. System trouble codes are commonly stored in RAM and can be erased by disconnecting the battery.

Nonvolatile RAM memory can retain its information even when the battery is disconnected. One use for this type of RAM is the storage of odometer information in an electronic speedometer. The memory chip retains the mileage accumulated by the vehicle. When speedometer replacement is necessary, the odometer chip is removed and installed in the new speedometer unit. KAM is used primarily in conjunction with adaptive strategies.

OUTPUT After the computer has processed the input signals, it sends voltage signals or commands to other devices in the system, such as system actuators. An **actuator** is an electrical or mechanical device that converts electrical energy into heat, light, or motion, such as adjusting engine idle speed, altering suspension height, or regulating fuel metering.

Computers also can communicate with, and control, each other through their output and input functions. This means that the output signal from one computer system can be the input signal for another computer system through a network.

Most outputs work electrically in one of three ways:

- Switched
- Pulse-width modulated
- Digital

A switched output is an output that is either on or off. In many circuits, the PCM uses a relay to switch a device on or off. This is because the relay is a low-current device that can switch a higher-current device. Most computer circuits cannot handle a lot of current. By using a relay circuit as shown in **FIGURE 11–4,** the PCM provides the output control to the relay, which in turn provides the output control to the device. The relay coil, which the PCM controls, typically draws less than 0.5 amperes. The device that the relay controls may draw 30 amperes or more. These switches are actually transistors, often called output drivers.

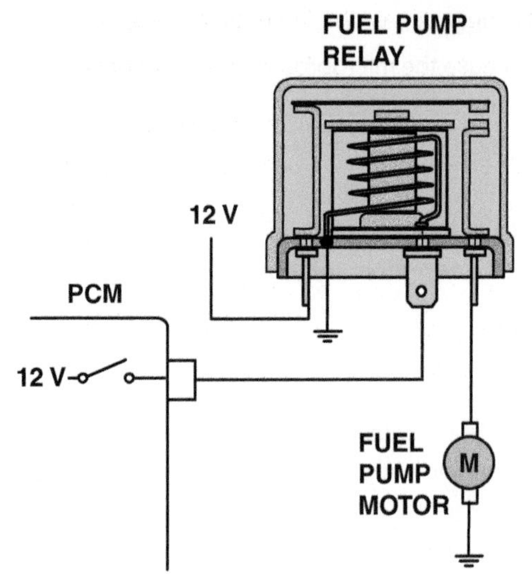

FIGURE 11–4 A typical output driver. In this case, the PCM applies voltage to the fuel pump relay coil to energize the fuel pump.

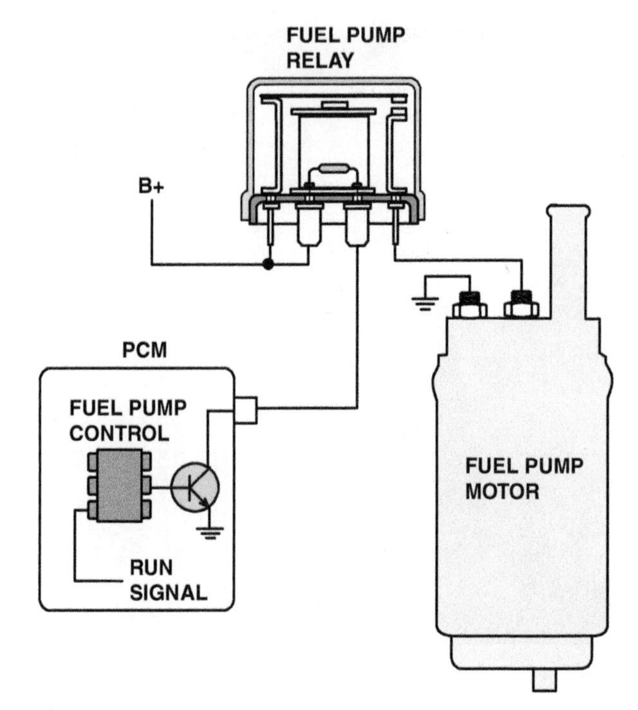

FIGURE 11–5 A typical low-side driver (LSD) which uses a control module to control the ground side of the relay coil.

LOW-SIDE DRIVERS Low-side drivers, often abbreviated LSD, are transistors that complete the ground path in the circuit. Ignition voltage is supplied to the relay as well as battery voltage. The computer output is connected to the ground side of the relay coil. The computer energizes the fuel pump relay by turning the transistor on and completing the ground path for the relay coil. A relatively low current flows through the relay coil and transistor that is inside the computer. This causes the relay to switch and provides the fuel pump with battery voltage. The majority of switched outputs have typically been low-side drivers. ● **SEE FIGURE 11–5**. Low-side drivers can often perform a diagnostic circuit check by monitoring the voltage from the relay to check that the control circuit for the relay is complete. A low-side driver, however, cannot detect a short-to-ground.

HIGH-SIDE DRIVERS High-side drivers, often abbreviated HSD, control the power side of the circuit. In these applications when the transistor is switched on, voltage is applied to the device. A ground has been provided to the device so when the high-side driver switches the device will be energized. In some applications, high-side drivers are used instead of low-side drivers to provide better circuit protection. General Motors vehicles have used a high-side driver to control the fuel pump relay instead of a low-side driver. In the event of an accident, should the circuit to the fuel pump relay become grounded, a high-side driver would cause a short circuit, which would cause the fuel pump relay to de-energize. High-side drivers inside modules can detect electrical faults such as a lack of continuity when the circuit is not energized. ● **SEE FIGURE 11–6**.

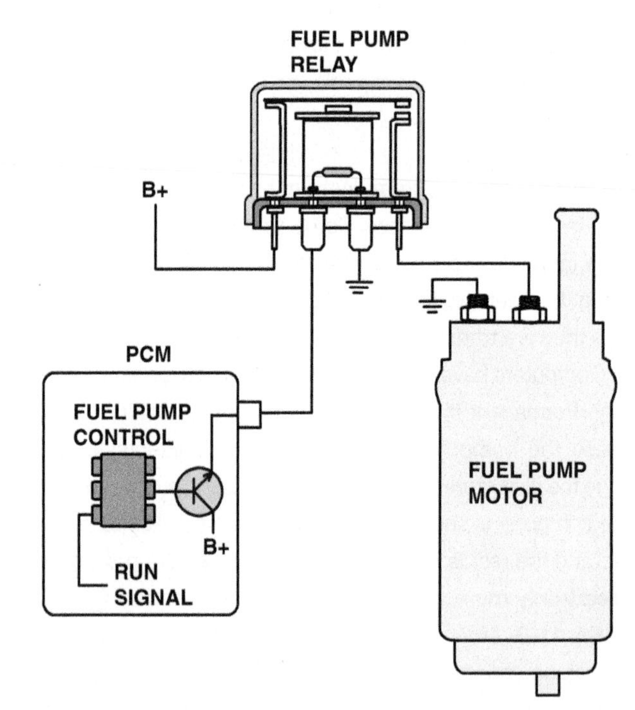

FIGURE 11–6 A typical module-controlled high-side driver (HSD) where the module itself supplies the electrical power to the device. The logic circuit inside the module can detect circuit faults including continuity of the circuit and if there is a short-to-ground in the circuit being controlled.

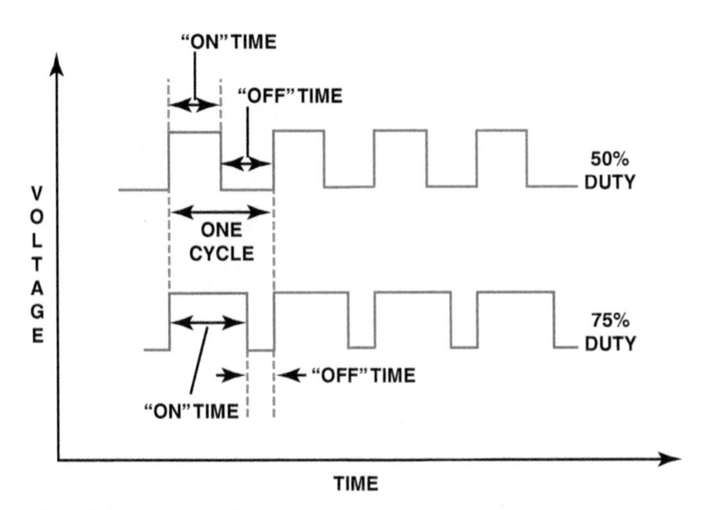

FIGURE 11–7 Both the top and bottom patterns have the same frequency. However, the amount of on-time varies. Duty cycle is the percentage of the time during a cycle that the signal is turned on.

PULSE-WIDTH MODULATION Pulse-width modulation (PWM) is a method of controlling an output using a digital signal. Instead of just turning devices on or off, the computer can control output devices more precisely by using pulse width modulation. For example, a vacuum solenoid could be a pulse-width modulated device. If the vacuum solenoid is controlled by a switched driver, switching either on or off would mean that either full vacuum would flow through the solenoid or no vacuum would flow through the solenoid. However, to control the amount of vacuum that flows through the solenoid, pulse-width modulation could be used. A PWM signal is a digital signal, usually 0 and 12 volts, that is cycling at a fixed frequency. Varying the length of time that the signal is on provides a signal that can vary the on and off time of an output. The ratio of on-time relative to the period of the cycle is referred to as **duty cycle** ● **SEE FIGURE 11–7**. Depending on the frequency of the signal, which is usually fixed, this signal would turn the device on and off a fixed number of times per second. When, for example, the voltage is high (12 volts) 90% of the time and low (0 volt) the other 10% of the time, the signal has a 90% duty cycle. In other words, if this signal were applied to the vacuum solenoid, the solenoid would be on 90% of the time. This would allow more vacuum to flow through the solenoid. The computer has the ability to vary this on and off time or pulse-width modulation at any rate between 0 and 100%.

A good example of pulse width modulation is the cooling fan speed control. The speed of the cooling fan is controlled by varying the amount of on-time that the battery voltage is applied to the cooling fan motor.

100% duty cycle—the fan runs at full speed

75% duty cycle—the fan runs at 3/4 speed

50% duty cycle—the fan runs at 1/2 speed

25% duty cycle—the fan runs at 1/4 speed

The use of PWM, therefore, results in very precise control of an output device to achieve the amount of cooling needed and conserve electrical energy compared to simply timing the cooling fan on high when needed. PWM may be used to control vacuum through a solenoid, the amount of purge of the evaporative purge solenoid, the speed of a fuel pump motor, control of a linear motor, or even the intensity of a lightbulb.

DIGITAL COMPUTERS

In a **digital computer**, the voltage signal or processing function is a simple high/low, yes/no, on/off signal. The digital signal voltage is limited to two voltage levels: high voltage and low voltage. Since there is no stepped range of voltage or current in between, a digital binary signal is a "square wave."

The signal is called "digital" because the on and off signals are processed by the computer as the digits or numbers 0 and 1. The number system containing only these two digits is called the binary system. Any number or letter from any number system or language alphabet can be translated into a combination of binary 0s and 1s for the digital computer.

A digital computer changes the analog input signals (voltage) to digital bits (*b*inary dig*its*) of information through an **analog-to-digital (AD) converter** circuit. The binary digital number is used by the computer in its calculations or logic networks. Output signals usually are digital signals that turn system actuators on and off.

The digital computer can process thousands of digital signals per second because its circuits are able to switch voltage signals on and off in billionths of a second. ● **SEE FIGURE 11–8.**

PARTS OF A COMPUTER The software consists of the programs and logic functions stored in the computer's circuitry. The hardware is the mechanical and electronic parts of a computer.

CENTRAL PROCESSING UNIT (CPU). The microprocessor is the **central processing unit** of a computer. Since it performs the essential mathematical operations and logic decisions that make up its processing function, the CPU can be considered to be the brain of a computer. Some computers use more than one microprocessor, called a coprocessor.

COMPUTER MEMORY. Other IC (integrated circuit) devices store the computer operating program, system sensor input data, and system actuator output data, information that is necessary for CPU operation.

FIGURE 11–8 Many electronic components are used to construct a typical vehicle computer. (Courtesy of General Motors)

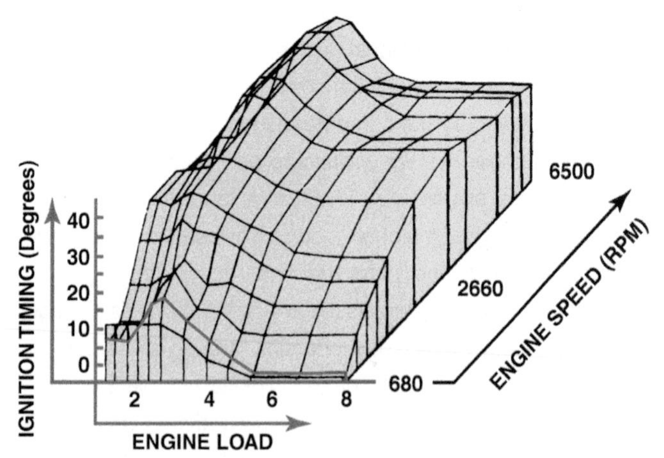

FIGURE 11–9 Typical ignition timing map developed from testing and used by the vehicle computer to provide the optimum ignition timing for all engine speeds and load combinations.

COMPUTER PROGRAMS By operating a vehicle on a dynamometer and manually adjusting the variable factors such as speed, load, and spark timing, it is possible to determine the optimum output settings for the best driveability, economy, and emission control. This is called **engine mapping.** ● **SEE FIGURE 11–9.**

Engine mapping creates a three-dimensional performance graph that applies to a given vehicle and powertrain combination. Each combination is mapped in this manner to produce a PROM. This allows an automaker to use one basic computer for all models; a unique PROM individualizes the computer for a particular model. Also, if a driveability problem can be resolved by a change in the program, the manufacturers can release a revised PROM to supersede the earlier part.

FIGURE 11–10 The clock generator produces a series of pulses that are used by the microprocessor and other components to stay in step with each other at a steady rate.

Many older vehicle computers used a single PROM that plugged into the computer. Since the mid-1990s the PROM chip is not removable. The computer must be programmed, or flashed before being put into service.

CLOCK RATES AND TIMING The microprocessor receives sensor input voltage signals, processes them by using information from other memory units, and then sends voltage signals to the appropriate actuators. The microprocessor communicates by transmitting long strings of 0s and 1s in a language called binary code. But the microprocessor must have some way of knowing when one signal ends and another begins. That is the job of a crystal oscillator called a **clock generator.** ● **SEE FIGURE 11–10.** The computer's crystal oscillator generates a steady stream of one-bit-long voltage pulses. Both the microprocessor and the memories monitor the clock pulses while they are communicating. Because they know how long each voltage pulse should be, they can distinguish between a 01 and a 0011. To complete the process, the input and output circuits also watch the clock pulses.

COMPUTER SPEEDS Not all computers operate at the same speed; some are faster than others. The speed at which a computer operates is specified by the cycle time, or clock speed, required to perform certain measurements. Cycle time or clock speed is measured in megahertz (4.7, 8.0, 15, 18 MHz, etc.).

BAUD RATE The computer transmits bits of a serial data stream at precise intervals. The computer's processing speed is called the baud rate, or bits per second. Just as mph helps

FIGURE 11–11 This powertrain control module (PCM) is located under the hood on this Chevrolet pickup truck.

FIGURE 11–12 This engine control module (ECM) measures about 6 × 6 × 1 inches. (Courtesy of General Motors)

in estimating the length of time required to travel a certain distance, the baud rate is useful in estimating how long a given computer will need to transmit a specified amount of data to another computer. Storage of a single character requires eight bits per byte, plus an additional two bits to indicate stop and start. This means that transmission of one character requires 10 bits. Dividing the baud rate by 10 tells us the maximum number of words per second that can be transmitted. For example, if the computer has a baud rate of 600, approximately 60 words can be received or sent per minute.

Automotive computers have evolved from a baud rate of 160 used in the early 1980s to a baud rate as high as 50,000,000 for some networks. The speed of data transmission is an important factor both in system operation and in system troubleshooting.

CONTROL MODULE LOCATIONS The onboard automotive computer has many names. It may be called an **electronic control unit (ECU)**, **electronic control module (ECM)**, **electronic control assembly (ECA)**, or a **controller**, depending on the manufacturer and the computer application. The Society of Automotive Engineers (SAE) bulletin, J-1930, standardizes the name as a powertrain control module (PCM). The computer hardware is all mounted on one or more circuit boards and installed in a metal case to help shield it from electromagnetic interference (EMI). The wiring harnesses that link the computer to sensors and actuators connect to multipin connectors or edge connectors on the circuit boards.

Onboard computers range from single-function units that control a single operation to multifunction units that manage all of the separate (but linked) electronic systems in the vehicle. They vary in size from a small module to a notebook-sized box. Some engine computers are installed in the passenger compartment either under the instrument panel or in a side kick panel while most GM PCMs are mounted under the hood. ● **SEE FIGURES 11–11 AND 11–12.**

CONTROL MODULE INPUTS

The powertrain control module (PCM) uses signals (voltage levels) from a variety of sensors, including the following:

ECT (Engine Coolant Temperature). The engine coolant temperature (ECT) sensor is used to measure the engine coolant temperature to help determine the amount of fuel and spark advance. This is an important sensor, especially when the engine is cold and when the engine is first started.

MAP (Manifold Absolute Pressure). The manifold absolute pressure (MAP) sensor is a strain gauge-type sensor that measures changes in the intake manifold pressure. The ECM uses this information, which indicates engine load, to calculate fuel delivery and spark timing (some vehicles) and for onboard diagnosis of other sensors and systems such as the exhaust gas recirculation (EGR) system. When the engine is not running, the manifold is at atmospheric pressure, and the MAP sensor registers barometric pressure (BARO).

MAF (Mass Airflow) Sensor. The mass airflow (MAF) sensor is positioned in the intake air duct or manifold. The MAF sensor measures the volume and density of the incoming air. This sensor uses the temperature, density, and humidity of the air to determine the mass of the incoming air. This signal tells the ECM how much airflow there is, so that the ECM can make fuel delivery and spark timing calculations.

TP (Throttle Position) Sensor. The throttle position sensor's signal is used to determine idle, wide open throttle (WOT), deceleration enleanment, and acceleration enrichment. The input is used by the computer to control fuel delivery as well as spark advance and the shift points of the transmission/transaxle.

CKP (Crankshaft Position) Sensor. The crankshaft position sensor is the most critical input to the ignition system. The crankshaft position sensor provides RPM and identifies cylinder pairs at top dead center (TDC).

CMP (Camshaft Position) Sensor. While the crankshaft position sensor identifies cylinder pairs at TDC, the camshaft position sensor identifies cylinder stroke. The camshaft position sensor sends a signal to the ECM, which uses it as a sync pulse to trigger the injectors in the proper sequence. The ECM uses the camshaft position sensor signal to determine the cylinder on the intake stroke. This enables the ECM to synchronize the ignition system and calculate true sequential fuel injection.

Vehicle Speed Sensor. The vehicle speed sensor (VSS) provides vehicle speed information to the ECM.

BARO (Barometric Sensor). Some vehicles use a separate barometric sensor, designed to measure the pressure of the atmosphere, which is affected by altitude and weather conditions.

IAT (Intake Air Temperature) Sensor. The IAT sensor is a two-wire sensor positioned in the engine air intake. Air temperature readings are of particular importance during open loop or cold engine operation. On newer vehicles, the IAT sensor has been incorporated into the MAF sensor.

Intake Air Temperature 2. The IAT 2 sensor is a variable resistor used on supercharged engines to measure the air temperature inside the engine intake manifold.

Fuel Tank Pressure. The fuel tank pressure sensor is used to detect leaks in the evaporative emissions (EVAP) system.

Knock Sensor. The knock sensor system enables the control module to control ignition timing for optimal performance while protecting the engine from potentially damaging detonation. The knock sensor is used to detect engine detonation (knock) and signal the control module to retard ignition timing.

Ceramic Resistor Card Fuel Level Sensor. Starting with 1996 and later enhanced EVAP equipped models, the fuel-level sending unit was switched to a ceramic card resistor (40 to 250 Ohm (Ω) potentiometer). This improves fuel-level sensing accuracy, which is required when the ECM performs on-board diagnostic tests.

Accelerator Pedal Position Sensor. The accelerator pedal position (APP) sensor is mounted on the accelerator pedal assembly. Newer vehicles have two individual APP sensors within one housing. Older vehicles have three individual APP sensors within one housing. The APP sensor works with the throttle position sensor and the throttle actuator control (TAC) system to provide input to the ECM regarding driver-requested APP and throttle angle at the throttle body.

Oxygen (O2) Sensors. Provide data for fuel trim and catalyst monitor.

CONTROL MODULE OUTPUTS

A vehicle computer can do just two things:

- Turn a device on.
- Turn a device off.

The computer can turn devices such as fuel injectors on and off rapidly or keep them on for a certain amount of time. Typical ECM output devices include the following:

Fuel Pump. Because the fuel pump draws high current, it is not directly controlled by the ECM. Instead, the ECM controls a fuel pump relay that provides system voltage to the fuel pump. When the ignition is turned on before the starter is engaged, the ECM energizes the fuel pump relay by providing system voltage. The ECM shuts off the fuel pump relay if an ignition reference pulse is not received within two seconds. The ECM powers the relay circuit as long as it receives ignition reference pulses.

Fuel Injector. Fuel is delivered by fuel injectors, which are controlled by the ECM. An electric fuel pump supplies a continuous supply of pressurized fuel from the fuel injector. The ECM controls fuel flow by pulse-width modulation (PWM) of the injector, referred to as on time. Engine temperature, throttle position, manifold pressure or mass airflow, engine load, intake air temperature (IAT), engine RPM, oxygen sensor (O2S), and system voltage determine the fuel injector on time.

Ignition Timing. The computer sends an ignition control (IC) signal to the ignition module or coils to fire the spark plugs based on information from the sensors. The spark is advanced when the engine is cold and/or when the engine is operating under light load conditions.

Transmission Shifting. The computer provides a ground to the shift solenoids and torque converter clutch solenoid. The operation of the automatic transmission/transaxle is optimized based on vehicle sensor information.

Malfunction Indicator Lamp. The MIL is located on the dashboard. When illuminated, the MIL indicates an engine status concern.

Idle Speed Control. The idle air control (IAC) valve is used on cable-actuated throttle bodies. The newer throttle actuator control (TAC) system does not use an idle air control valve. This idle air control valve is located in the throttle body of both throttle body injection and multi port fuel injection systems.

TAC Motor. The throttle actuator control motor controls the throttle opening according to commands from the computer. It also operates as the cruise control system if the vehicle is so equipped.

ADDITIONAL CONTROL MODULES

All vehicles are equipped with a number of control modules, depending on the model and trim level. The various modules communicate and share information over the vehicle network system. Some of the more important modules are listed here. For example, a 2013 Buick Enclave includes these modules:

Body Control Module—located under the driver's side instrument panel. Controls many vehicle functions such as interior and exterior lighting and horns.

Transmission Control Module—located inside the transmission. Controls the operation of the transmission by use of solenoids located inside the transmission.

Electronic Brake Control Module—located under the hood. Controls the antilock brake system (ABS), traction control, stability control, and other ABS functions.

K20 Engine Control Module—located under the hood, on the driver's side. In charge of all engine functions including fuel, ignition, and emission control systems. This is the master module for the OBD II diagnostic system.

Fuel Pump Flow Control Module—located under the vehicle, on the driver's side. Controls fuel delivery and volume, according to commands from the K20 ECM.

HVAC System Control Module—located in the center stack, behind the trim panel. The operator controls are mounted directly on this module, allowing complete control of the HVAC system.

Airbag Sensing and Diagnostic Module—located on the front center floor pan, under the front floor console. Maintains diagnosis and operation of the airbag system.

Theft Deterrent Module—located in the steering column, at the ignition lock cylinder. This small module wraps around the ignition lock cylinder and prevents the vehicle from running in case of a theft attempt.

MODULE PROGRAMMING

Almost all of the modules on a vehicle require a programming procedure when they are replaced. Programming may also be required when any sensor or component related to the module is replaced. For example, if a crankshaft position (CKP) sensor is replaced, the new sensor must be relearned by the ECM.

PROGRAMMING GUIDELINES

- *Do not* program a control module unless directed to by a service procedure or a service bulletin. If the control module is not properly configured with the correct calibration software, the control module will not control all of the vehicle features properly.
- Ensure the programming tool (Tech 2 or GDS2) is equipped with the latest software and is securely connected to the data link connector. If there is an interruption during programming, programming failure or control module damage may occur.
- Stable battery voltage is critical during programming. Any fluctuation, spiking, over voltage, or loss of voltage will interrupt programming. When required, install the EL-49642 SPS Programming Support Tool to maintain system voltage. If not available, connect a fully charged 12 V jumper or booster pack disconnected from the AC voltage supply. *Do not* connect a battery charger. ● **SEE FIGURE 11–13.**

FIGURE 11–13 The PSC 550 Maintainer (also called the SPS Programming Support Tool EL-49642) is a clean power supply that should be used during programming to avoid module failures. (Courtesy of General Motors)

- Turn off or disable systems that may put a load on the vehicles battery such as, interior lights, exterior lights (including daytime running lights), HVAC, radio, and so on.
- During the programming procedure, follow the SPS prompts for the correct ignition switch position.
- Clear DTCs after programming is complete. Clearing powertrain DTCs will set the Inspection/Maintenance (I/M) system status indicators to "no."

PROGRAMMING EXAMPLE Here is a typical instruction for reprogramming a transmission control module from GM service information (SI)*:

1. Install EL-49642 SPS Programming Support Tool.
2. Access the Service Programming System (SPS) and follow the on-screen instructions.
3. On the SPS-Supported Controllers screen, select Transmission Control Module—Programming and follow the on-screen instructions.
4. At the end of programming, choose the "Clear All DTCs" function on the SPS screen.
5. With a scan tool, perform the Reset Transmission Adapts.

*GM SI Document # 1556771.

SUMMARY

1. The Society of Automotive Engineers (SAE) standard J-1930 specifies that the term "powertrain control module" (PCM) be used for the computer that controls the engine and transmission in a vehicle.
2. The four basic computer functions include input, processing, storage, and output.
3. Read-only memory (ROM) can be programmable (PROM), erasable (EPROM), or electrically erasable (EEPROM).
4. Computer input sensors include engine speed (RPM), MAP, MAF, ECT, O2S, TP, and VS.
5. A computer can only turn a device on or turn a device off, but it can do the operation rapidly.
6. Control modules have to be programmed when they are replaced.

REVIEW QUESTIONS

1. What part of the vehicle computer is considered to be the brain?
2. What is the difference between volatile and nonvolatile RAM?
3. List four input sensors.
4. List four output devices.

CHAPTER QUIZ

1. What unit of electricity is used as a signal for a computer?
   - **a.** Volt
   - **b.** Ohm
   - **c.** Ampere
   - **d.** Watt
2. The four basic computer functions include _______.
   - **a.** Writing, processing, printing, and remembering
   - **b.** Input, processing, storage, and output
   - **c.** Data gathering, processing, output, and evaluation
   - **d.** Sensing, calculating, actuating, and processing
3. All OBD II vehicles use what type of read-only memory?
   - **a.** ROM
   - **b.** PROM
   - **c.** EPROM
   - **d.** EEPROM
4. The "brain" of the computer is the _______.
   - **a.** PROM
   - **b.** RAM
   - **c.** CPU
   - **d.** AD converter
5. Computer processing speed is measured in _______.
   - **a.** Baud rate
   - **b.** Clock speed (Hz)
   - **c.** Voltage
   - **d.** Bytes
6. Which item is a computer input sensor?
   - **a.** RPM
   - **b.** Throttle position angle
   - **c.** Engine coolant temperature
   - **d.** All of the above
7. Which item is a computer output device?
   - **a.** Fuel injector
   - **b.** Transmission shift solenoid
   - **c.** Evaporative emission control solenoid
   - **d.** All of the above
8. The SAE term for the vehicle computer is _______.
   - **a.** PCM
   - **b.** ECM
   - **c.** ECA
   - **d.** Controller
9. What two things can a vehicle computer actually perform (output)?
   - **a.** Store and process information
   - **b.** Turn something on or turn something off
   - **c.** Calculate and vary temperature
   - **d.** Control fuel and timing only
10. Through which type of circuit analog signals from sensors are changed to digital signals for processing by the computer?
    - **a.** Digital
    - **b.** Analog
    - **c.** AD converter
    - **d.** PROM

chapter 12

CAN AND NETWORK COMMUNICATIONS

LEARNING OBJECTIVES

After studying this chapter, the reader should be able to:

1. Explain the purpose and function of onboard network systems.
2. List the various types of vehicle networks.
3. Check for module communication errors using a scan tool.

This chapter will help you prepare for Engine Repair (A6) ASE certification test content area "A" (General Electrical/Electronic Systems Diagnosis).

KEY TERMS

Class 2 163
Controller area network (CAN) 161
Expansion bus 165
GMLAN 163
Keyword 2000 164
LIN 164
Multiplexing 159
Network 159
Serial data 159
Splice pack 159
UART 164
Terminating resistors 161

GM STC OBJECTIVES

GM Service Technical College topics covered in this chapter are as follows:

1. Diagnose GMLAN low speed using a systematic process.
2. Test GMLAN low speed using electrical/electronic tools.
3. Test Class 2 using electrical/electronic tools.
4. Test LIN network using electrical/electronic tools.
5. Diagnose GMLAN high/mid-speed using a systematic process.
6. Test GMLAN high/mid-speed using electrical/electronic tools.
7. Perform voltage testing at the DLC on GMLAN.

MODULE COMMUNICATION AND NETWORKS

Since the 1990s, vehicles have used modules to control most of the electrical component operation. A typical vehicle will have 10 or more control modules and they communicate with each other over data lines or hard wiring, depending on the application.

SERIAL DATA **Serial data** is data that is transmitted by a series of rapidly changing voltage signals pulsed from low to high or from high to low. Most modules are connected together in a network because of the following advantages:

- A decreased amount of wire is needed, thereby saving weight, cost, as well as helping with installation at the factory, and decreased complexity, making servicing easier.
- Common sensor data can be shared with those modules that may need the information, such as vehicle speed, outside air temperature, and engine coolant temperature.

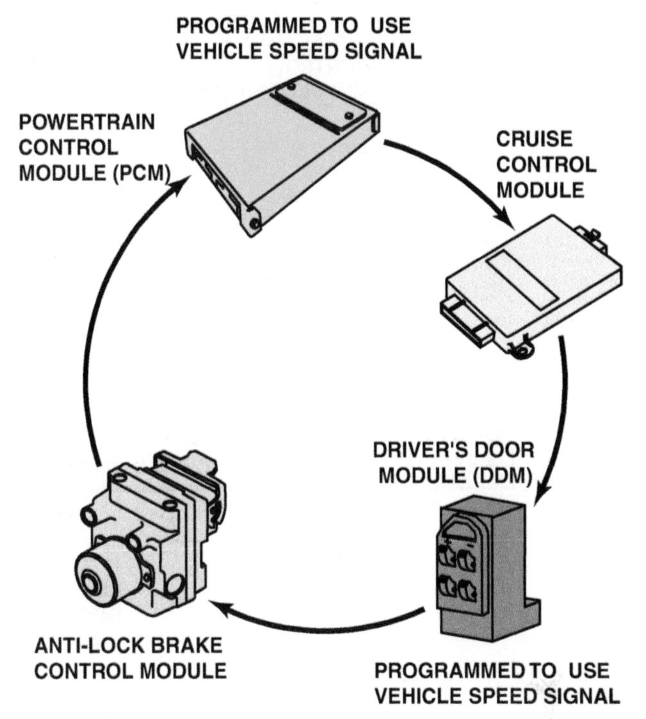

FIGURE 12–1 A network allows all modules to communicate with other modules.

MULTIPLEXING **Multiplexing** is the process of sending multiple signals of information at the same time over a signal wire and then separating the signals at the receiving end. This system of intercommunication of computers or processors is referred to as a **network**. ● **SEE FIGURE 12–1**. By connecting the computers together on a communications network, they can easily share information back and forth. This multiplexing has a number of advantages, including the following:

- The elimination of redundant sensors and dedicated wiring for these multiple sensors.
- The reduction of the number of wires, connectors, and circuits.
- Addition of more features and option content to new vehicles.
- Weight reduction, increasing fuel economy.
- Allows features to be changed with software upgrades instead of component replacement.

The most common types of networks used on General Motors vehicles include the following:

1. **Ring link networks.** In a ring-type network, all modules are connected to each other by a serial data line in a line until all are connected in a ring. ● **SEE FIGURE 12–2.**
2. **Star link.** In a star link network, a serial data line attaches to each module and then each is connected to a central point. This central point is called a **splice pack**. The splice pack uses a bar to splice all of the serial lines together. Some GM vehicles use two or more splice packs to tie the modules together. When more than one splice pack is used, a serial data line connects one splice pack to the others. In most applications the bus bar used in each splice pack can be removed. When the bus bar is removed a special tool (J 42236) can be installed in place of the removed, bus bar. Using this tool, the serial data line for each module can be isolated and tested for a possible problem. Using the special tool at the splice pack makes diagnosing this type of network easier than many others. ● **SEE FIGURE 12–3** for an example of a star link network system.
3. **Ring/Star hybrid.** In a ring/star network, the modules are connected using both types of network configuration. Check service information (SI) for details on how this network is connected on the vehicle being diagnosed and always follow the recommended diagnostic steps.

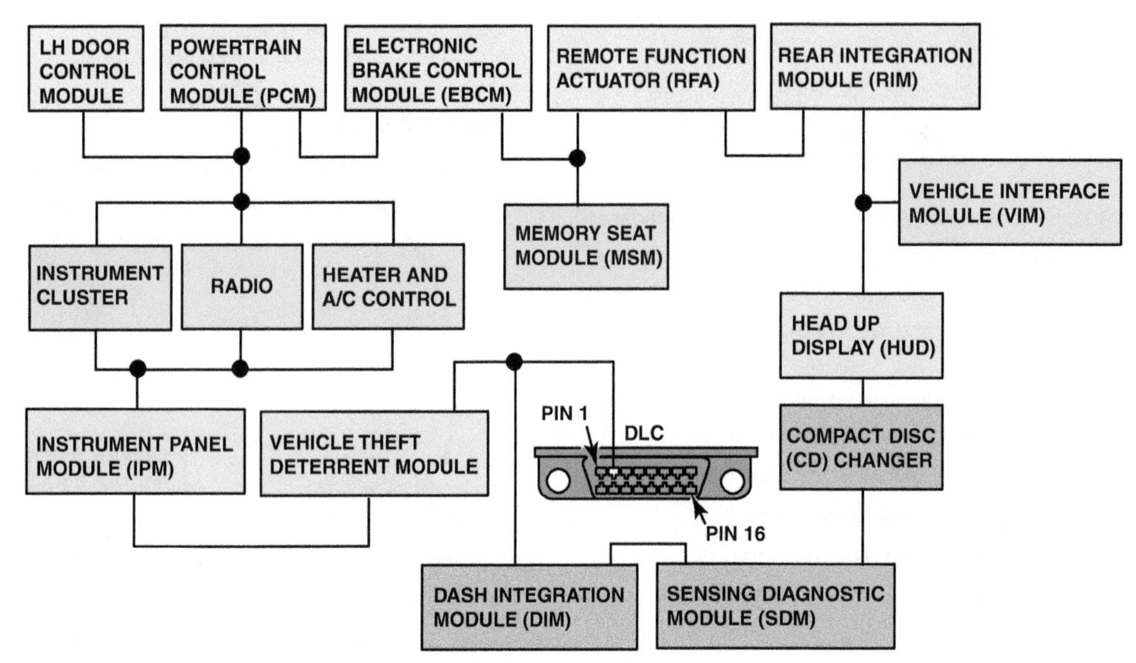

FIGURE 12–2 A ring link network reduces the number of wires it takes to interconnect all of the modules.

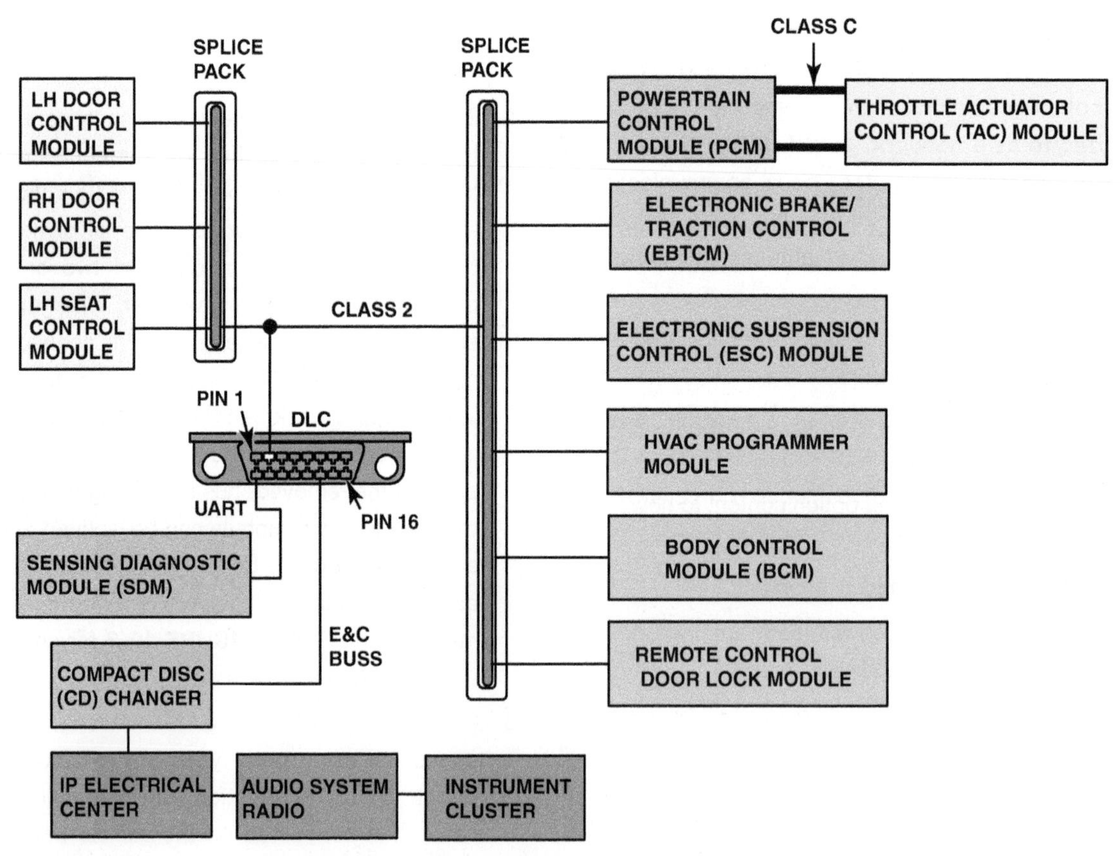

FIGURE 12–3 A star link-type network where all of the modules are connected together using splice packs.

SAE COMMUNICATION CLASSIFICATIONS

The Society of Automotive Engineers (SAE) standards include three categories of in-vehicle network communications, including the following:

CLASS A Low-speed networks (less than 10,000 bits per second [10 kbs]) are generally used for trip computers, entertainment, and other convenience features. Most low-speed Class A communication functions are performed using the following:

- UART (Universal Asynchronous Receive/Transmit) standard used by General Motors (8192 bps).
- CCD (Chrysler Collision Detection) used by Chrysler (7812.5 bps).

  NOTE: The "collision" in CCD-type bus communication refers to the program that avoids conflicts of information exchange within the bus and does not refer to airbags or other accident-related circuits of the vehicle.

- Chrysler SCI (Serial Communications Interface) is used to communicate between the engine controller and a scan tool (62.5 kbps).
- ACP (Audio Control Protocol) is used for remote control of entertainment equipment (twisted pairs) on Ford vehicles.

CLASS B Medium-speed networks (10,000 to 125,000 bits per second [10 to 125 kbs]) are generally used for information transfer among modules, such as instrument clusters, temperature sensor data, and other general uses.

- General Motors GMLAN, both low- and medium-speed and Class 2, which uses 0 to 7 volt pulses with an available pulse width. Meets SAE 1850 variable pulse width (VPW).
- Chrysler Programmable Communication Interface (PCI). Meets SAE standard J-1850 pulse-width modulated (PWM).
- Ford Standard Corporate Protocol (SCP). Meets SAE standard J-1850 pulse-width modulated (PWM).

CLASS C High-speed networks (125,000 to 1,000,000 bits per second [125,000 to 1,000,000 kbs]) are generally used for real-time powertrain and vehicle dynamic control. Most high-speed bus communication is **controller area network or CAN.** ● **SEE FIGURE 12–4.**

FREQUENTLY ASKED QUESTION

What Is a Bus?

A "bus" is a term used to describe a communication network. Therefore, there are *connections to the bus and bus communications*, both of which refer to digital messages being transmitted among electronic modules or computers.

MODULE COMMUNICATION DIAGNOSIS

Most vehicle manufacturers specify that a scan tool be used to diagnose modules and module communications. Always follow the recommended testing procedures, which usually require the use of a factory scan tool.

Some tests of the communication bus (network) and some of the service procedures require the service technician to attach a DMM, set to DC volts, to monitor communications. A variable voltage usually indicates that messages are being sent and received.

Most high-speed bus systems use resistors at each end called **terminating resistors**. These resistors are used to help reduce interference into other systems in the vehicle.

Usually two 120-ohm resistors are installed at each end and are therefore connected electrically in parallel. Two 120-ohm resistors connected in parallel would measure 60 ohms if being tested using an ohmmeter. ● **SEE FIGURE 12–5.**

GENERAL MOTORS GLOBAL COMMUNICATION ARCHITECTURE*

A Global Communication Architecture incorporates multiple networks, including single wire, dual wire, and LIN. This architecture uses smart components in multiple LIN networks to reduce wiring. It can seem confusing, but by looking at each network individually, diagnosis can be simplified.

*Adapted from GM Course # 18044.20D4.

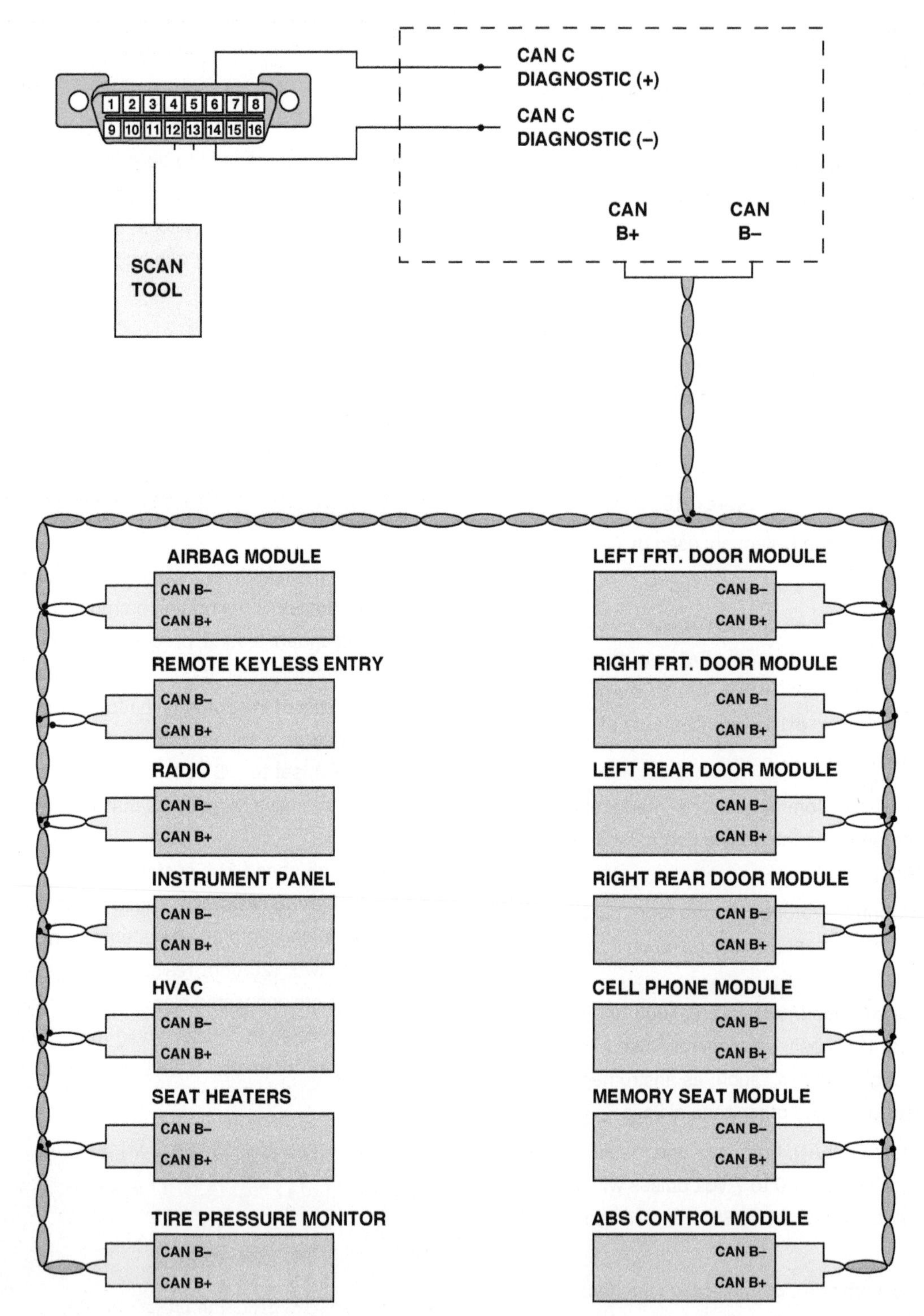

FIGURE 12–4 A typical bus system showing module CAN communications and twisted pairs of wire.

We'll be seeing more modules and smart components as GM continues to implement this system. Some advantages of this approach are given:

- Combines best features of systems
- Improves serviceability
- Allows component features and systems to be added easily
- Enables common tools and equipment to be used across vehicles

The data link connector (DLC) can be used for diagnosis using a scan tool, oscilloscope, or digital volt/ohmmeter (DVOM) using the terminals shown in **FIGURE 12–6**.

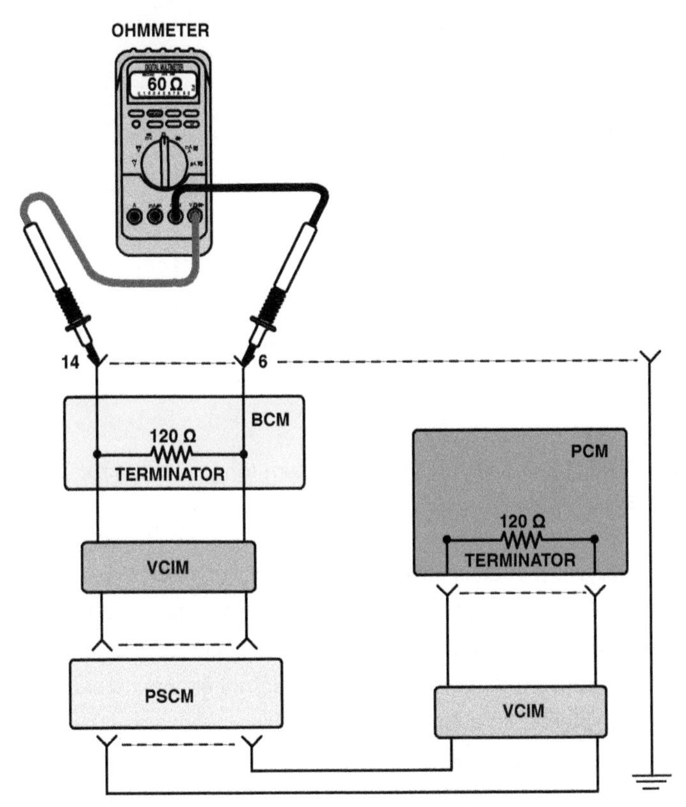

FIGURE 12–5 Checking the terminating resistors using an ohmmeter at the DLC.

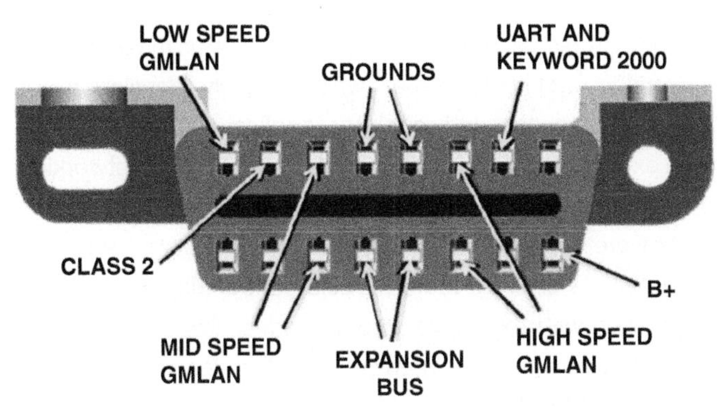

FIGURE 12–6 The data link connector (DLC) showing access points for the various networks. (Courtesy of General Motors)

SINGLE WIRE COMMUNICATION NETWORKS There are four types of single wire communication networks.

1. Class 2
2. Low-Speed GM Local Area Network (GMLAN)
3. Universal Asynchronous Receiver Transmitter (UART)
4. Keyword 2000

Faults on a single wire data communication circuit cause similar problems, regardless of protocol. When diagnosing faults, look for fluctuating voltage of not less than 1 volt or near B+ voltage.

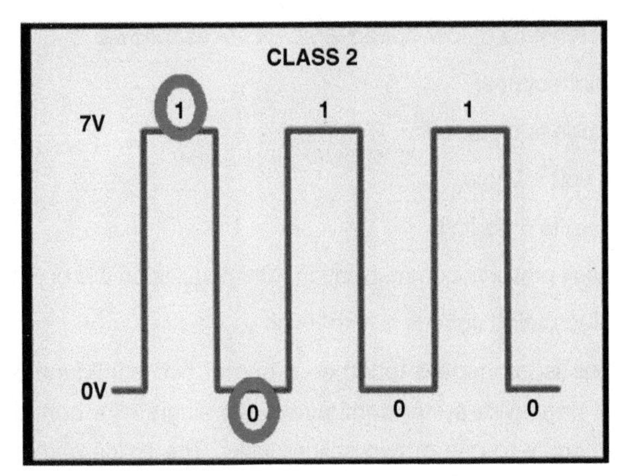

FIGURE 12–7 Class 2 networks use a 0 to 7 volt signal to communicate. (Courtesy of General Motors)

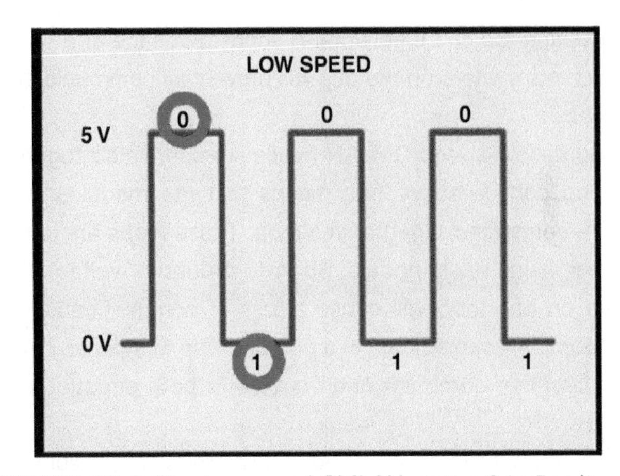

FIGURE 12–8 The low-speed GMLAN uses a 0 to 5 volt signal to communicate. (Courtesy of General Motors)

CLASS 2 **Class 2** communication networks have been used in GM vehicles since the mid-1990s. The Class 2 network can transfer data communication messages on a single wire at an average of 10.4 kilobytes per second. This network can only be found in a few GM vehicles currently in production. ● **SEE FIGURE 12–7.** Characteristics of the Class 2 system are as follows:

- Active at 7 volts (Binary code = 1)
- Inactive at ground (Binary code = 0)
- Bidirectional communication

To diagnose Class 2 communication faults:

- Probe the DLC at pin 2
- With 0 to 7 volts used to communicate, a typical voltage reading at pin 2 should be right around 1 to 3 volts and fluctuating

LOW-SPEED GMLAN Low-speed **GMLAN** communication has widely replaced Class 2 communication. This bus operates at 33.33 kilobytes per second, which is triple the speed of Class 2. ● **SEE FIGURE 12–8.**

Characteristics of low-speed GMLAN are as follows:

- Bidirectional
- Code is reversed
- 0 volt = Logic 1
- 5 volts = Logic 0
- Fault prevents communication with modules on this bus *only*
- Fluctuating voltage is 1 to 3 volts

This bus is connected together in one of two configurations, star or ring. With a star configuration, a single wire connects each module to one of two splice packs. The splice pack is a good starting point for isolating and troubleshooting the bus. Removing the splice pack comb allows measurements to be taken on each circuit leg to find which one has the fault that is corrupting the entire network. A short-to-voltage or a short-to-ground anywhere on the bus will prevent *all* communication on that bus.

Some low-speed GMLAN buses are connected together in a ring configuration. This means that the modules on a bus are connected together in a loop. These loops are joined together inside each module. Short-to-ground or voltage that occurs on one loop will cause a loss of communication on both loops. An open circuit in a ring bus won't cause any concerns because communication occurs in both directions on the loop.

To diagnose low-speed GMLAN communication faults:

- Probe the DLC at pin 1
- With 0 to 5 volts used to communicate, a typical voltage reading at pin 1 should be right around 1 to 3 volts and fluctuating
- If serial data is lost, control modules will set a no-communication code that identifies the non-communicating control module.
- A loss of serial data communications DTC does not represent a failure of the module that is reporting the code.

UART AND KEYWORD 2000 Universal Asynchronous Receive and Transmit (**UART**) and **Keyword 2000** are similar communication protocols. Diagnosing UART is similar to any other protocol. If there is a fault in the wiring, communication will be lost with one or more control modules

- Check to make sure the circuit is not shorted to ground or shorted to power
- A fault within this circuit at pin 7 at the DLC to the splice pack will only cause a communication problem with the scan tool

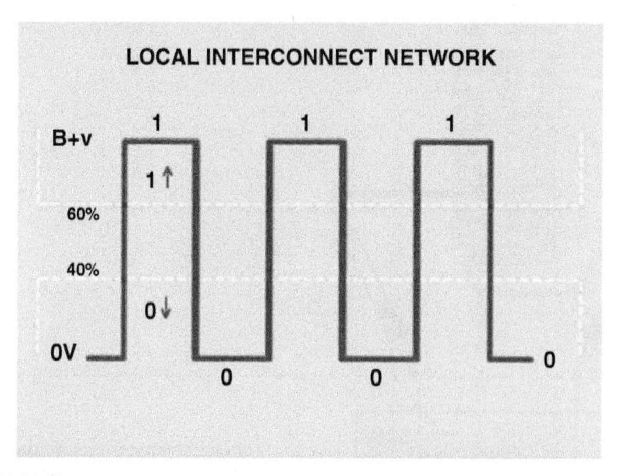

FIGURE 12–9 The LIN voltage pattern ranges from 0 to battery voltage. (Courtesy of General Motors)

Keyword 2000 is used only on a small number of vehicles. The protocol uses a single wire bidirectional data line between the module and the scan tool. Message structure is a request and response arrangement. Modules do not exchange data on these systems, so a fault within the bus will cause an interruption with scan tool diagnostics.

To diagnose UART and Keyword 2000 communication faults, probe the DLC at pin 7.

LOCAL INTERCONNECT NETWORK The **local interconnect network (LIN)** is another type of single wire communication network. It is a UART-based/single-master/multiple-slave networking architecture that was originally developed for automotive sensor and actuator applications. The master node extends communication benefits of in-vehicle networking to individual sensors and actuators. Benefits to the electrical system are reduced wiring, lighter vehicle weight, and cost effective data sharing. More module control allows enhanced diagnostics with expanded trouble codes. The LIN connects with the controller area network (CAN) at the master node. ● SEE FIGURE 12–9.

LIN has some unique characteristics:

- Does NOT connect to the DLC
- Only used for communication between a module and a component. In order for communication to take place, voltage and ground must be good at the module and at the component
- Cycles between 0 and B+ volts
- B+ volts = Logic 1
- 0 volt = Logic 0
- Under 40% of battery voltage = Logic 0
- Over 60% of battery voltage = Logic 1

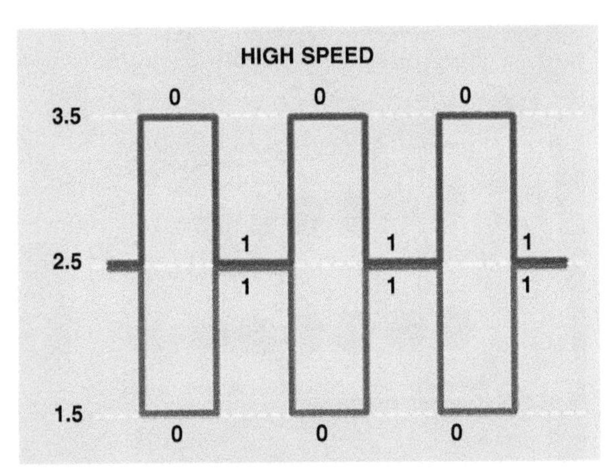

FIGURE 12–10 The voltage signature of 2-wire CAN systems are the same for high-speed GMLAN, mid-speed GMLAN, and expansion bus. One wire is driven high and the other driven low when communicating. (Courtesy of General Motors)

DUAL WIRE COMMUNICATION DATA NETWORKS

There are three types of dual wire communication data buses.

1. High-Speed GM Local Area Network (GMLAN)
2. Expansion Bus
3. Mid-Speed GM Local Area Network (GMLAN)

NOTE: These networks are diagnosed similarly, regardless of protocol.

HIGH-SPEED GMLAN High-speed GMLAN is a dual wire communication network. With dual wire networks, a short to-power or a short-to-ground has the same effect that it had on the single wire buses, meaning that the entire network will be down. However, since there are two wires, you need to check them both for faults. ● **SEE FIGURE 12–10.**

Characteristics of the high-speed GMLAN are as follows:

- More complex than single wire networks
- Short-to-ground or short-to-power will prevent *all* communication on high-speed bus
- Transmitted on two twisted wires
- Speeds up to 500 kilobytes/second
- Terminated with two 120-ohm resistors
- Positive (pin 6) and negative (pin 14) are driven to opposite extremes from rest or idle level
- 2.5 volts = Logic 1
- 2.5 +/–1 = Logic 0

Dual wire network circuits include two 120-ohm resistors. These resistors are commonly found near each end of the circuit. They help to reduce electrical noise. General diagnosis of GMLAN faults includes the following steps:

- Always refer to service information (SI) first; print out a schematic for reference.
- With the key OFF, begin testing at the DLC. For high-speed GMLAN, you would test at pins 6 and 14.
- Typical resistance at DLC should be approximately 60 ohms if no open exists.
- Higher resistance suggests an open.
- Lower resistance suggests a short between the wires.
- If there is a measurement of 120 ohms, you are not seeing both terminating resistors but, rather, an open.
- To locate the open, split the circuit and measure in each direction. There should be 120 ohms in one direction and infinite ohms in the other direction.
- Follow the circuit in the direction of the infinite resistance to help to isolate the fault location.
- With the key ON, 1.5 to 3.5 volts is used to communicate; a typical voltage reading at pins 6 and 14 should be right around 1 to 3 volts and fluctuating.
- If serial data is lost, control modules will set a no communication code that identifies the non-communicating control module.
- A loss of serial data communications DTC does not represent a failure of the module that is reporting the code.

TECH TIP

GMLAN Quick Check Using the DVOM

A multimeter can be used to verify GMLAN operation. At the DLC with the key on, the normal voltage between pin 6 and ground should be slightly higher than 2.5 volts, and fluctuating. Measuring the voltage at pin 14 should show slightly lower than 2.5 volts, and fluctuating.

NOTE: Always use the proper terminal adapters when probing at the DLC. *Do not* push the DMM probes into the connector, which can damage the terminal end.

EXPANSION BUS The **expansion bus**, otherwise known as the chassis high-speed GMLAN, is similar to high-speed GMLAN in operation. This system addresses a GMLAN bandwidth issue by creating a network for specific modules. The expansion bus is currently used in Two-mode Hybrid trucks and Volt. It is also used on global platform vehicles for both the chassis and the powertrain.

The expansion bus is unique, as it doesn't always connect to the data link connector. Sometimes you may need to test at the specific module. Characteristics of the expansion bus are as follows:

- Transmitted on two twisted wires
- Speeds up to 500 kilobytes/second
- Terminated with two 120-ohm resistors
- Positive (pin 12) and negative (pin 13) are driven to opposite extremes from rest or idle level
- 2.5 volts = Logic 1
- 2.5 +/−1 = Logic 0

To diagnose expansion bus communication faults, check to see if it is connected to the DLC.

- If connected, this bus will be at terminals 12 and 13. Otherwise, you may need to test at the specific module.
- With 1.5 to 3.5 volts used to communicate, a typical voltage reading at pins 12 and 13 should be right around 1 to 3 volts and fluctuating.

MID-SPEED GMLAN The controller area network, or CAN, protocol has been expanded into each of the dual wire networks, including the **mid-speed GMLAN**. It is very similar to the other dual wire networks, with the following characteristics:

- Starts at 2.5 volts
- Moves +/−1 volt to communicate
- Limited to 125 kilobytes per second
- Slower than other high-speed networks
- To diagnose mid-speed GMLAN communication faults, probe the DLC at terminals 3 and 11
- With 1.5 to 3.5 volts used to communicate, a typical voltage reading at pins 3 and 11 should be right around 1 to 3 volts and fluctuating

DIAGNOSTIC AIDS* When diagnosing communication DTCs or other faults, here are some points to consider. These diagnostic aids apply to the various types of GM LAN communication systems.

While diagnosing a specific customer concern or after a repair, a history U-code may be present; however, there is no associated "current" or "active" status. Loss-of-communication U-codes such as these can set for a variety of reasons. Many times, they are transparent to the vehicle operator and technician and/or have no associated symptoms. Eventually, they will erase themselves automatically after a number of fault-free

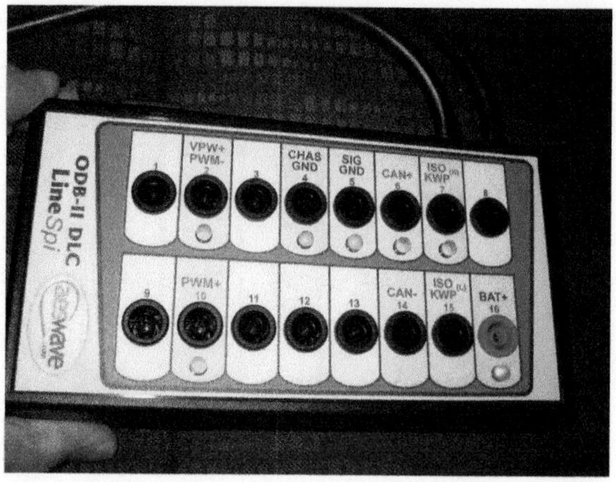

FIGURE 12–11 A breakout box (CAN test box) is a good tool to use during diagnosis of network concerns. The BOB prevents direct probing of the DLC terminals, which are easily damaged.

Let Bob Help with Diagnosis

A DLC breakout box (BOB) is a useful tool for diagnosing electrical faults and CAN bus line activity without causing possible damage to the DLC terminals. It can be used to check power and ground circuits, check active protocol lines, or connect a multimeter or oscilloscope for detailed signal analysis. SEE FIGURE 12–11.

Some useful functions of the CAN test box are as follows:

- Verify ECU activity—LEDs at pin-out display signal detection; flashing LEDs indicate ECU activity.
- Monitor OBDII data lines.
- Probe lines with scope or multimeter for detailed signal information.
- Check and monitor battery voltage—continuous numeric display of voltage; alarm warning when voltage drops below 11.6 V or rises above 15.2.

*General Motors Service Information Document # 2172453.

ignition cycles. These "false codes" or conditions would most likely be caused by one of these actions:

- A control module on the data communication circuit was disconnected while the communication circuit is awake. (Unplugging modules with the key ON.)
- Power to one or more modules was interrupted during diagnosis. (Pulling and replacing fuses with the key ON.)
- A low-battery condition was present, so some control modules stop communicating when battery voltage drops below a certain threshold. (Leaving the key ON for extended periods of time.)
- Battery power was restored to the vehicle and control modules on the communication circuit did not all reinitialize at the same time. (Reconnecting the vehicle battery with the key ON.)

If a loss-of-communication U-code appears in history for no apparent reason, it is most likely associated with one of the scenarios above. These are all temporary conditions and should never be interpreted as an intermittent fault, causing you to replace a part.

Here are some more points to consider during the diagnosis of a "no communication" concern:

- Do not replace a control module that is reporting a U-code. The U-code identifies which control module needs to be diagnosed for a communication issue.
- Communication may be available between the body control module (BCM) and the scan tool, even with the high-speed GMLAN serial data system inoperative. This condition is due to the BCM using both the high- and low-speed GMLAN systems.
- An open in the DLC ground circuit terminal 5 will allow the scan tool to operate but not communicate with the vehicle. (Check for aftermarket accessories using the back of terminal 5 for ground.)
- The engine will not start when there is a total malfunction of the high-speed GMLAN serial data bus.

Technicians may find various local area network (LAN) communication diagnostic trouble codes (DTCs) and no low-speed GMLAN communications with the scan tool. These conditions may be caused by the installation of an aftermarket navigation radio module or accessories plugged into the DLC. Some customers may comment of one or more of the following concerns:

- Vehicle will not crank
- Vehicle cranks but will not start
- Vehicle stability enhancement system warning lights and messages
- PRNDL gear indicator position errors

SUMMARY

1. There are four types of single wire communication networks.
   - Class 2
   - Low-Speed GM Local Area Network (GMLAN)
   - Universal Asynchronous Receiver Transmitter (UART)
   - Keyword 2000
2. The Society of Automotive Engineers (SAE) standards include three categories of in-vehicle network communications, including Class A, Class B, and Class C.
3. There are three types of dual wire communication data buses.
   - High-Speed GM Local Area Network (GMLAN)
   - Expansion Bus
   - Mid-Speed GM Local Area Network (GMLAN)
4. When diagnosing a communication fault, always refer to service information (SI) for a list of diagnostic aids related to the vehicle.

REVIEW QUESTIONS

1. List what could cause a module to set a "fasle" U code.
2. List the function of each pin at the data link connector.
3. Explain the difference between a GM Class 2 network and GMLAN.

CHAPTER QUIZ

1. What is serial data?
   a. A series of rapidly changing voltage signals
   b. The number that identifies the production date of a module
   c. Another name for the vehicle VIN
   d. None of these
2. Which of these SAE network categories is the slowest in speed?
   a. Class A
   b. Class B
   c. Class C
   d. They are all the same speed
3. The General Motors Class 2 communication network is a ________ system.
   a. Class A
   b. Class B
   c. Class C
   d. UART
4. A high-speed bus network normally uses two ________ ohm resistors, one at each end of the network.
   a. 60
   b. 50
   c. 120
   d. 1,200
5. Which of these is considered a single wire communication network?
   a. Class 2
   b. UART
   c. Keyword 2000
   d. All of these
6. The Class 2 communication signal is available at PIN # ________ in the data link connector.
   a. 1
   b. 2
   c. 3
   d. 14
7. Which of these single wire data systems is *not* connected directly to the DLC?
   a. UART
   b. Class 2
   c. LIN
   d. Keyword 2000
8. When probing the Class 2 network at the DLC what would be a normal reading on the multimeter?
   a. Steady 7 volts
   b. O volts except when cranking
   c. 1 to 3 volts and fluctuating
   d. A steady 4 volts
9. All of these are dual wire data networks *except* ________.
   a. High-speed GMLAN
   b. Mid-speed GMLAN
   c. Low-speed GMLAN
   d. Chassis high-speed GMLAN
10. What will happen if the high-speed GMLAN is shorted to ground?
    a. The vehicle may not start
    b. All high-speed communication will cease
    c. Some other networks will still communicate
    d. All of these are possible

chapter 13

BATTERIES

LEARNING OBJECTIVES

After studying this chapter, the reader will be able to:

1. Describe how a battery works.
2. Describe deep cycling.
3. Discuss how charge indicators work.
4. List battery ratings.

This chapter will help you prepare for the ASE Electrical/Electronic Systems (A6) certification test content area "B" (Battery Diagnosis and Service).

GM STC OBJECTIVES

GM Service Technical College topics covered in this chapter are as follows:

1. The purpose of batteries (08510.01W).
2. Battery components.
3. The types of batteries.
4. Battery properties, including voltage, capacity, and ratings.
5. Factors affecting battery life.
6. Recognize hybrid battery system components and their operation (18441.01W-R2).
7. Identify the characteristics and operation of the high-voltage Li-ion battery (18420.02W-R2).

KEY TERMS

AGM 174
Ampere hour 176
Alkaline 182
Battery Council International (BCI) 176
CA 175
CCA 175
Cells 171
Deep cycling 176
Electrolyte 172
Flooded cell battery 174
Gassing 171
Gel battery 175
Grid 170
Low-water-loss battery 170
Lithium-ion (li-ion) 180
Lithium-polymer (Li-poly) 176
Maintenance-free battery 170
MCA 176
Nickel-metal hydride (NiMH) 177
Partitions 171
Porous lead 171
Recombinant battery 175
Reserve capacity 176
Sediment chamber 170
SLA 174
SLI 170
Specific gravity 173
Sponge lead 171
SVR 174
Thermistors 180
VRLA 174

? FREQUENTLY ASKED QUESTION

What Is an SLI Battery?

Sometimes the term SLI is used to describe a type of battery. **SLI** means **starting, lighting, and ignition,** and describes the use of a typical automotive battery. Other types of batteries used in industry are usually batteries designed to be deep cycled and are usually not as suitable for automotive needs.

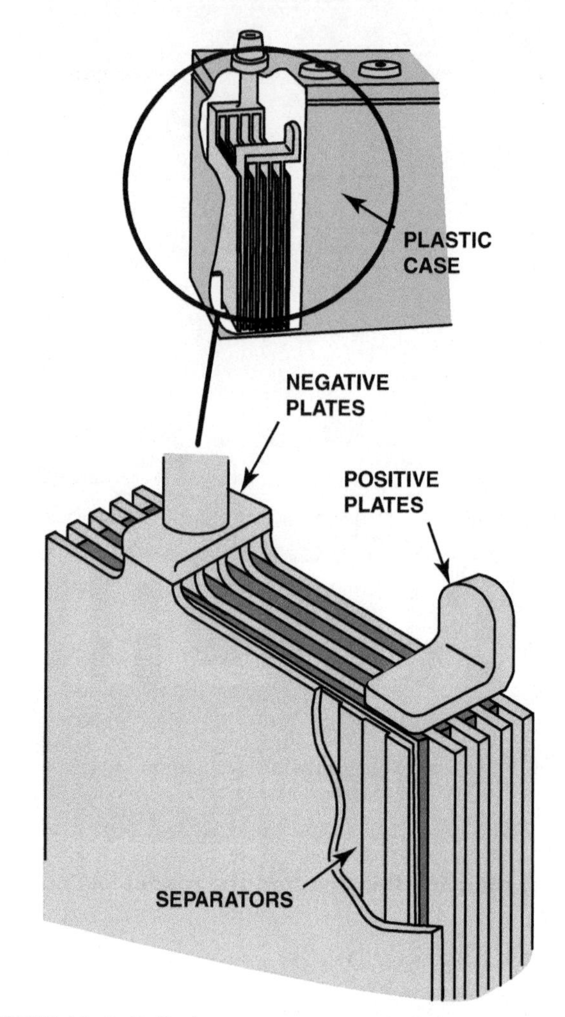

FIGURE 13–1 Batteries are constructed of plates grouped into cells and installed in a plastic case.

INTRODUCTION

PURPOSE AND FUNCTION Every electrical component in a vehicle is supplied current from the battery. The battery is one of the most important parts of a vehicle because it is the heart or foundation of the electrical system. The primary purpose of an automotive battery is to provide a source of electrical power for starting and for electrical demands that exceed alternator output.

WHY BATTERIES ARE IMPORTANT The battery also acts as a voltage stabilizer for the entire electrical system. The battery is a voltage stabilizer because it acts as a reservoir from where large amounts of current (amperes) can be used quickly during starting and replaced back gradually by the alternator during charging.

- The battery *must* be in good (serviceable) condition before the charging and cranking systems can be tested. For example, if a battery is discharged, the cranking circuit (starter motor) could test as being defective because the battery voltage might drop below specifications.
- The charging circuit could also test as being defective because of a weak or discharged battery. It is important to test the vehicle battery before further testing of the cranking or charging system.

BATTERY CONSTRUCTION

CASE Most automotive battery cases (container or covers) are constructed of polypropylene, a thin (approximately 0.08 inch or 0.02 millimeter, thick), strong, and lightweight plastic. In contrast, containers for industrial batteries and some truck batteries are constructed of a hard, thick rubber material.

Inside the case are six cells (for a 12-volt battery). Each cell has positive and negative plates. Built into the bottom of many batteries are ribs that support the lead-alloy plates and provide a space for sediment to settle, called the **sediment chamber**. This space prevents spent active material from causing a short circuit between the plates at the bottom of the battery. ● **SEE FIGURE 13–1**.

A **maintenance-free battery** uses little water during normal service because of the alloy material used to construct the battery plate grids. Maintenance-free batteries are also called **low-water-loss batteries**.

GRIDS Each positive and negative plate in a battery is constructed on a framework, or **grid**, made primarily of lead. Lead is a soft material and must be strengthened for use in an automotive battery grid. Adding antimony or calcium to the pure lead adds strength to the lead grids. ● **SEE FIGURE 13–2**.

Battery grids hold the active material and provide the electrical pathways for the current created in the plate.

FIGURE 13–2 A grid from a battery used in both positive and negative plates.

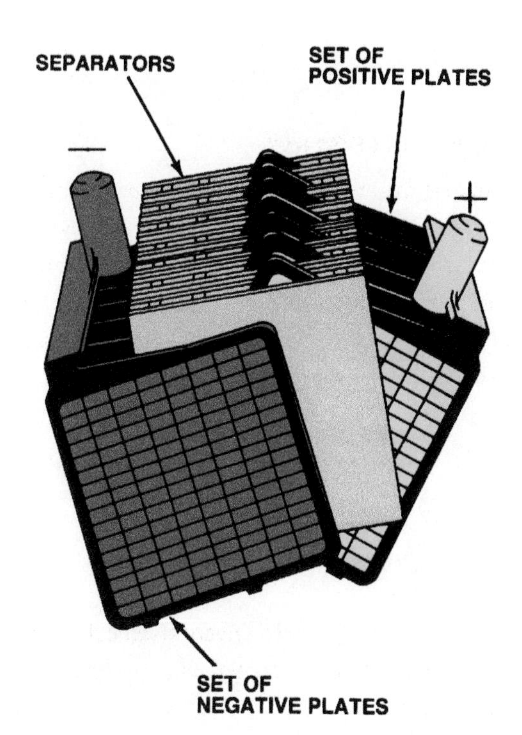

FIGURE 13–3 Two groups of plates are combined to form a battery element.

Maintenance-free batteries use calcium instead of antimony, because 0.2% calcium has the same strength as 6% antimony. A typical lead-calcium grid uses only 0.09% to 0.12% calcium. Using low amounts of calcium instead of higher amounts of antimony reduces **gassing**. Gassing is the release of hydrogen and oxygen from the battery that occurs during charging and results in water usage.

Low-maintenance batteries use a low percentage of antimony (about 2% to 3%), or use antimony only in the positive grids and calcium in the negative grids. *The percentages that make up the alloy of the plate grids constitute the major difference between standard and maintenance-free batteries.* The chemical reactions that occur inside each battery are identical regardless of the type of material used to construct the grid plates.

POSITIVE PLATES The positive plates have *lead dioxide (peroxide)* placed onto the grid framework. This process is called *pasting*. This active material can react with the sulfuric acid of the battery and is dark brown in color.

NEGATIVE PLATES The negative plates are pasted to the grid with a pure **porous lead**, called **sponge lead**, and are gray in color.

SEPARATORS The positive and the negative plates must be installed alternately next to each other without touching. Nonconducting *separators* are used, which allow room for the reaction of the acid with both plate materials, yet insulate the plates to prevent shorts. These separators are porous (with many small holes) and have ribs facing the positive plate. Separators can be made from resin-coated paper, porous rubber, fiberglass, or expanded plastic. Many batteries use envelope-type separators that encase the entire plate and help prevent any material that may shed from the plates from causing a short circuit between plates at the bottom of the battery.

CELLS **Cells** are constructed of positive and negative plates with insulating separators between each plate. Most batteries use one more negative plate than positive plate in each cell; however, many newer batteries use the same number of positive and negative plates. A cell is also called an element. Each cell is actually a 2.1-volt battery, regardless of the number of positive or negative plates used. The greater the number of plates used in each cell, the greater the amount of *current* that can be produced. Typical batteries contain four positive plates and five negative plates per cell. A 12-volt battery contains six cells connected in series, which produce 12.6 volts ($6 \times 2.1 = 12.6$) and contain 54 plates (9 plates per cell $\times$ 6 cells). If the same 12-volt battery had five positive plates and six negative plates, for a total of 11 plates per cell ($5 + 6$), or 66 plates (11 plates $\times$ 6 cells), then it would have the same voltage, but the amount of current that the battery could produce would be increased. ● **SEE FIGURE 13–3.**

The amperage capacity of a battery is determined by the amount of active plate material in the battery and the area of the plate material exposed to the electrolyte in the battery.

PARTITIONS Each cell is separated from the other cells by **partitions**, which are made of the same material as that used for the outside case of the battery. Electrical connections between cells are provided by lead connectors that loop over the top of

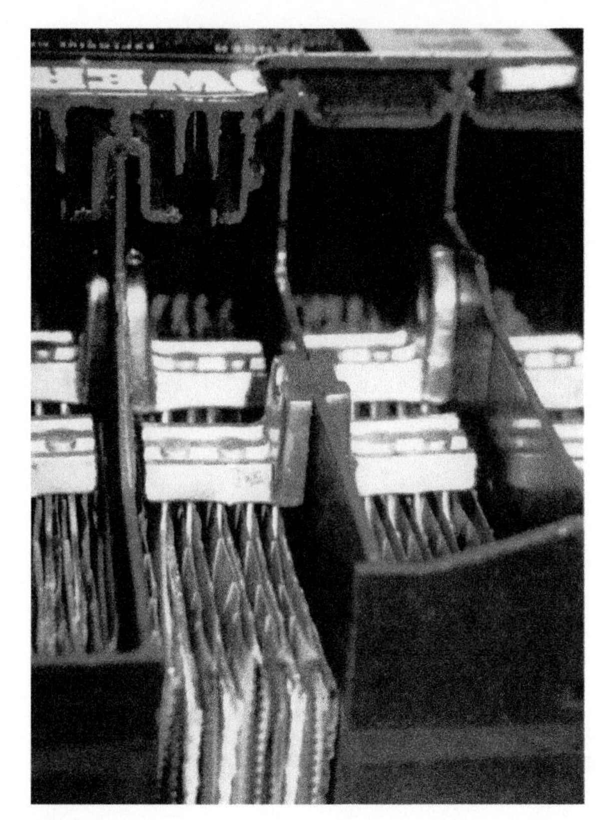

FIGURE 13–4 A cutaway battery showing the connection of the cells to each other through the partition.

the partition and connect the plates of the cells together. Many batteries connect the cells directly through the partition connectors, which provide the shortest path for the current and the lowest resistance. ● **SEE FIGURE 13–4.**

ELECTROLYTE **Electrolyte** is the term used to describe the acid solution in a battery. The electrolyte used in automotive batteries is a solution (liquid combination) of 36% sulfuric acid and 64% water. This electrolyte is used for both lead-antimony and lead-calcium (maintenance-free) batteries. The chemical symbol for this sulfuric acid solution is H_2SO_4.

H_2 = Symbol for hydrogen (the subscript 2 means that there are two atoms of hydrogen)

S = Symbol for sulfur

O_4 = Symbol for oxygen (the subscript 4 indicates that there are four atoms of oxygen)

Electrolyte is sold premixed in a proper proportion and is factory installed or added to the battery when the battery is sold. Additional electrolyte must *never* be added to any battery after the original electrolyte fill. It is normal for some water (H_2O) to escape during charging as a result of the chemical reactions. The escape of gases from a battery

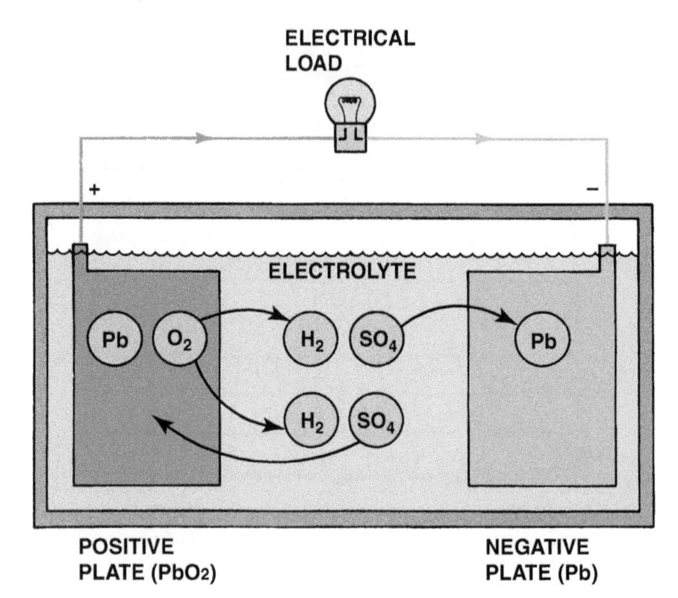

FIGURE 13–5 Chemical reaction for a lead-acid battery that is fully charged being discharged by the attached electrical load.

during charging or discharging is called gassing. Only pure distilled water should be added to a battery. If distilled water is not available, clean drinking water can be used.

HOW A BATTERY WORKS

PRINCIPLE INVOLVED The principle on which a battery works is based on a scientific principle discovered years ago which states that:

- When two dissimilar metals are placed in an acid, electrons flow between the metals if a circuit is connected between them.
- This can be demonstrated by pushing a steel nail and a piece of solid copper wire into a lemon. Connect a voltmeter to the ends of the copper wire and the nail, and voltage will be displayed.

A fully charged lead-acid battery has a positive plate of lead dioxide (peroxide) and a negative plate of lead surrounded by a sulfuric acid solution (electrolyte). The difference in potential (voltage) between lead peroxide and lead in acid is approximately 2.1 volts.

DURING DISCHARGING The positive plate lead dioxide (PbO_2) combines with the SO_4, forming $PbSO_4$ from the electrolyte and releases its O_2 into the electrolyte, forming H_2O. The negative plate also combines with the SO_4 from the electrolyte and becomes lead sulfate ($PbSO_4$). ● **SEE FIGURE 13–5.**

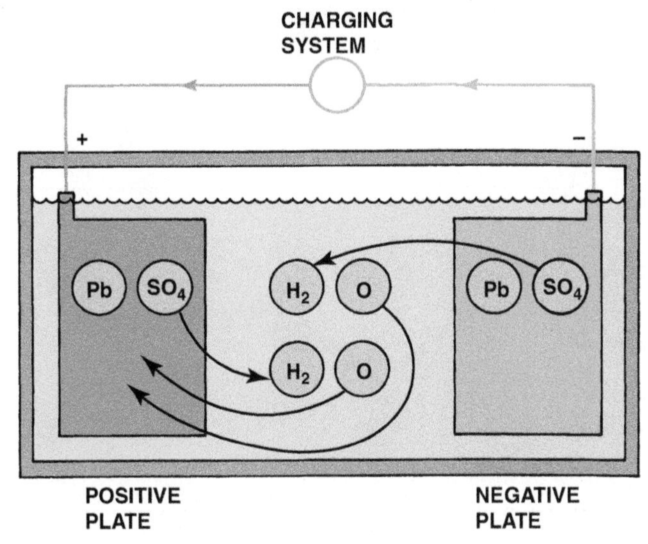

FIGURE 13–6 Chemical reaction for a lead-acid battery that is fully discharged being charged by the attached generator.

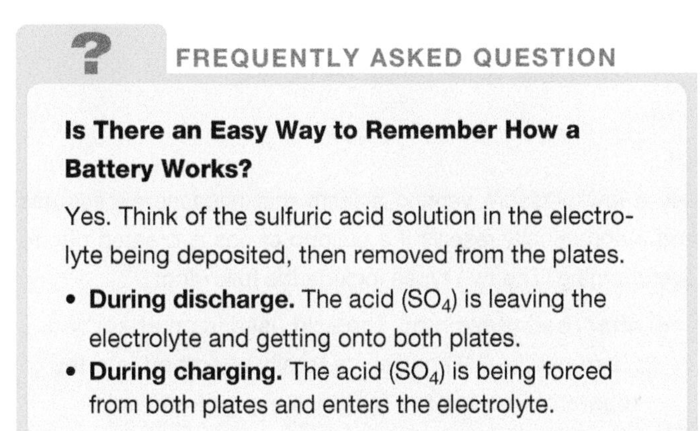

FREQUENTLY ASKED QUESTION

Is There an Easy Way to Remember How a Battery Works?

Yes. Think of the sulfuric acid solution in the electrolyte being deposited, then removed from the plates.

- **During discharge.** The acid (SO_4) is leaving the electrolyte and getting onto both plates.
- **During charging.** The acid (SO_4) is being forced from both plates and enters the electrolyte.

FULLY DISCHARGED STATE When the battery is fully discharged, both the positive and the negative plates are $PbSO_4$ (lead sulfate) and the electrolyte has become water (H_2O). As the battery is being discharged, the plates and the electrolyte approach the completely discharged state. There is also the danger of freezing when a battery is discharged, because the electrolyte is mostly water.

CAUTION: Never charge or jump-start a frozen battery because the hydrogen gas can get trapped in the ice and ignite if a spark is caused during the charging process. The result can be an explosion.

DURING CHARGING During charging, the sulfate from the acid leaves both the positive and the negative plates and returns to the electrolyte, where it becomes a normal-strength sulfuric acid solution. The positive plate returns to lead dioxide (PbO_2), the negative plate is again pure lead (Pb), and the electrolyte becomes H_2SO_4. ● **SEE FIGURE 13–6**.

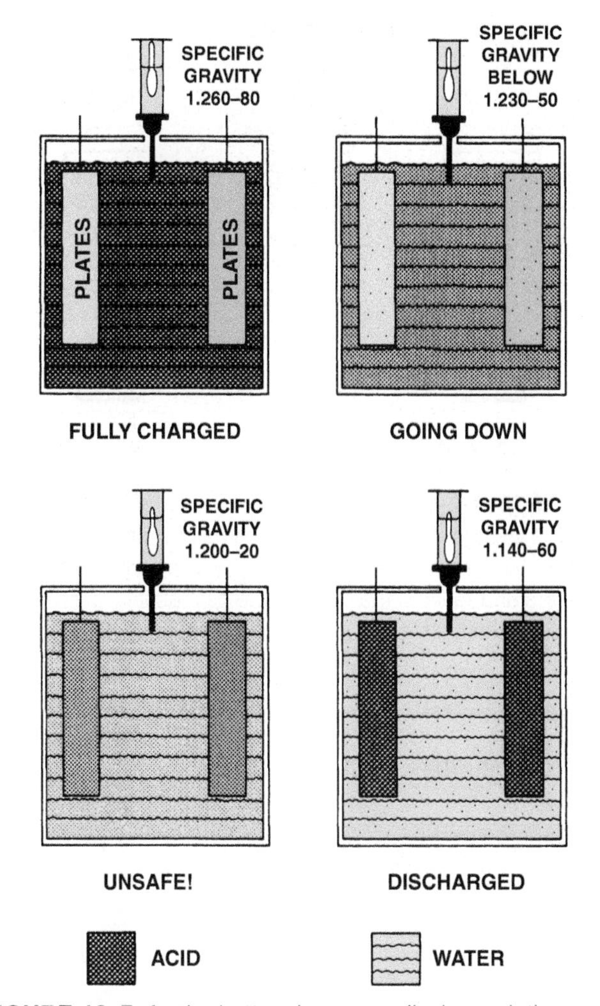

FIGURE 13–7 As the battery becomes discharged, the specific gravity of the battery acid decreases.

SPECIFIC GRAVITY

DEFINITION The amount of sulfate in the electrolyte is determined by the electrolyte's **specific gravity**, which is the ratio of the weight of a given volume of a liquid to the weight of an equal volume of water. In other words, the more dense the liquid is, the higher its specific gravity. Pure water is the basis for this measurement and is given a specific gravity of 1.000 at 80°F (27°C). Pure sulfuric acid has a specific gravity of 1.835; the *correct* concentration of water and sulfuric acid (called electrolyte—64% water, 36% acid) is 1.260 to 1.280 at 80°F. The higher the battery's specific gravity, the more fully it is charged. ● **SEE FIGURE 13–7**.

CHARGE INDICATORS Some batteries are equipped with a built-in state-of-charge indicator, commonly called *green eyes*. This indicator is simply a small, ball-type hydrometer that is installed in one cell. This hydrometer uses a plastic ball that floats

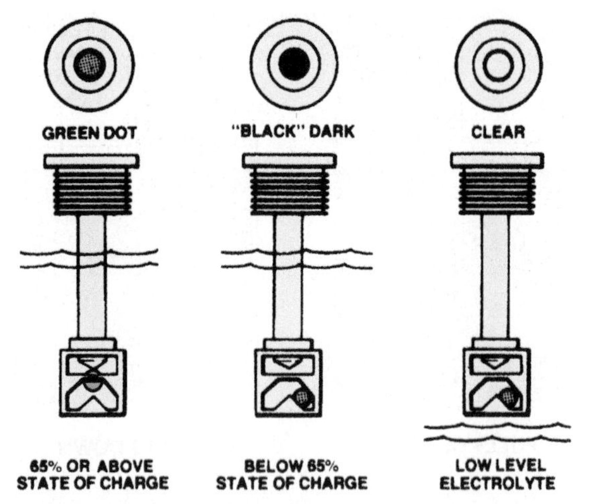

FIGURE 13–8 Typical battery charge indicator. If the specific gravity is low (battery discharged), the ball drops away from the reflective prism. When the battery is charged enough, the ball floats and reflects the color of the ball (usually green) back up through the sight glass and the sight glass is dark.

FIGURE 13–9 An absorbed glass mat battery is totally sealed and is more vibration resistant than conventional lead-acid batteries.

SPECIFIC GRAVITY	STATE OF CHARGE	BATTERY VOLTAGE (V)
1.265	Fully charged	12.6 or higher
1.225	75% charged	12.4
1.190	50% charged	12.2
1.155	25% charged	12.0
Lower than 1.120	Discharged	11.9 or lower

CHART 13–1

A comparison showing the relationship among specific gravity, battery voltage, and state of charge.

if the electrolyte density is sufficient (which it is when the battery is about 65% charged). When the ball floats, it appears in the hydrometer's sight glass, changing its color. **SEE FIGURE 13–8.**

Because the hydrometer is testing only one cell (out of six on a 12-volt battery), and because the hydrometer ball can easily stick in one position, do not trust that this is accurate information about a state of charge (SOC) of the battery.

Values of specific gravity, state of charge, and battery voltage at 80°F (27°C) are given in **CHART 13–1**.

VALVE-REGULATED LEAD-ACID BATTERIES

TERMINOLOGY There are two basic types of **valve-regulated lead-acid (VRLA)**, also called **sealed valve-regulated (SVR)** or **sealed lead-acid (SLA)**, batteries. These batteries use a low-pressure venting system that releases excess gas and automatically reseals if a buildup of gas is created due to overcharging. The two types include the following:

- **Absorbed glass mat.** The acid used in an **absorbed glass mat (AGM)** battery is totally absorbed into the separator, making the battery leakproof and spill-proof. The battery is assembled by compressing the cell about 20%, then inserting it into the container. The compressed cell helps reduce damage caused by vibration and helps keep the acid tightly against the plates. The sealed maintenance-free design uses a pressure release valve in each cell. Unlike conventional batteries that use a liquid electrolyte, called **flooded cell batteries**, most of the hydrogen and oxygen given off during charging remains inside the battery. The separator or mat is only 90% to 95% saturated with electrolyte, thereby allowing a portion of the mat to be filled with gas. The gas spaces provide channels to allow the hydrogen and oxygen gases to recombine rapidly and safely. Because the acid is totally absorbed into the glass mat separator, an AGM battery can be mounted in any direction. AGM batteries also have a longer service life, often lasting 7 to 10 years. Absorbed glass mat batteries are used as standard equipment in some vehicles such as the Chevrolet Corvette and in most Toyota hybrid electric vehicles. **SEE FIGURE 13–9.**

- **Gelled electrolyte batteries.** In a gelled electrolyte battery, silica is added to the electrolyte, which turns the electrolyte into a substance similar to gelatin. This type of battery is also called a **gel battery**.

Both types of valve-regulated lead-acid batteries are also called **recombinant battery** design. A recombinant-type battery means that the oxygen gas generated at the positive plate travels through the dense electrolyte to the negative plate. When the oxygen reaches the negative plate, it reacts with the lead, which consumes the oxygen gas and prevents the formation of hydrogen gas. It is because of this oxygen recombination that VRLA batteries do not use water.

FIGURE 13–10 A typical battery hold-down bracket. All batteries should use a bracket to prevent battery damage due to vibration and shock.

CAUSES AND TYPES OF BATTERY FAILURE

NORMAL LIFE Most automotive batteries have a useful service life of three to seven years; however, proper care can help increase the life of a battery, but abuse can shorten it. The major cause of premature battery failure is overcharging.

CHARGING VOLTAGE The automotive charging circuit, consisting of an alternator and connecting wires, must operate correctly to prevent damage to the battery.

- Charging voltages higher than 15.5 volts can damage a battery by warping the plates as a result of the heat of overcharging.
- AGM batteries can be damaged if charged at a voltage higher than 14.5 volts.

Overcharging also causes the active plate material to disintegrate and fall out of the supporting grid framework. Vibration or bumping can also cause internal damage similar to that caused by overcharging. It is important, therefore, to ensure that all automotive batteries are securely clamped with the battery hold-down bracket in the vehicle. The shorting of cell plates can occur without notice. If one of the six cells of a 12-volt battery is shorted, the resulting voltage of the battery is only 10 volts (12 − 2 = 10). With only 10 volts available, the starter *usually* will not be able to start the engine.

BATTERY HOLD-DOWNS All batteries must be attached securely to the vehicle to prevent battery damage. Normal vehicle vibrations can cause the active materials inside the battery to shed. Battery hold-down clamps or brackets help reduce vibration, which can greatly reduce the capacity and life of any battery. ● **SEE FIGURE 13–10.**

BATTERY RATINGS

Batteries are rated according to the amount of current they can produce under specific conditions.

COLD-CRANKING AMPERES Every automotive battery must be able to supply electrical power to crank the engine in cold weather and still provide battery voltage high enough to operate the ignition system for starting. The cold-cranking ampere rating of a battery is the number of amperes that can be supplied by a battery at 0°F (−18°C) for 30 seconds while the battery still maintains a voltage of 1.2 volts per cell or higher. This means that the battery voltage would be 7.2 volts for a 12-volt battery and 3.6 volts for a 6-volt battery. The cold-cranking performance rating is called **cold-cranking amperes (CCA)**. Try to purchase a battery with the highest CCA for the money. See the vehicle manufacturer's specifications for recommended battery capacity.

CRANKING AMPERES The designation **CA** refers to the number of amperes that can be supplied by a battery at 32°F (0°C). This rating results in a higher number than the more stringent CCA rating. ● **SEE FIGURE 13–11.**

FIGURE 13–11 This battery has a rating of 1,000 cranking amperes (CA) and 900 amperes using the cold-cranking amperes (CCA) rating system.

FREQUENTLY ASKED QUESTION

What Is Deep Cycling?

Deep cycling is almost fully discharging a battery and then completely recharging it. Golf cart batteries are an example of lead-acid batteries that must be designed to be deep cycled. A golf cart must be able to cover two 18-hole rounds of golf and then be fully recharged overnight. Charging is hard on batteries because the internal heat generated can cause plate warpage, so these specially designed batteries use thicker plate grids that resist warpage. Normal automotive batteries are not designed for repeated deep cycling.

FREQUENTLY ASKED QUESTION

What Determines Battery Capacity?

The capacity of any battery is determined by the amount of active plate material in the battery. A battery with a large number of thin plates can produce high current for a short period. If a few thick plates are used, the battery can produce low current for a long period. A trolling motor battery used for fishing must supply a low current for a long period of time. An automotive battery is required to produce a high current for a short period for cranking. Therefore, every battery is designed for a specific application.

MARINE CRANKING AMPERES **Marine cranking amperes (MCA)** is similar to cranking amperes and is tested at 32°F (0°C).

RESERVE CAPACITY The **reserve capacity** rating for batteries is *the number of minutes* for which the battery can produce 25 amperes and still have a battery voltage of 1.75 volts per cell (10.5 volts for a 12-volt battery). This rating is actually a measurement of the time for which a vehicle can be driven in the event of a charging system failure.

AMPERE HOUR **Ampere hour** is an older battery rating system that measures how many amperes of current the battery can produce over a period of time. For example, a battery that has a 50 amp-hour (A-H) rating can deliver 50 amperes for one hour or 1 ampere for 50 hours or any combination that equals 50 amp-hours.

BATTERY SIZES

BCI GROUP SIZES Battery sizes are standardized by the **Battery Council International (BCI)**. When selecting a replacement battery, check the specified group number in service information, battery application charts at parts stores, or the owner's manual.

TYPICAL GROUP SIZE APPLICATIONS

- **24/24F (top terminals).** Fits many Honda, Acura, Infiniti, Lexus, Nissan, and Toyota vehicles.
- **34/78 (dual terminals, both side and top posts).** Fits many General Motors pickups and SUVs, as well as midsize and larger GM sedans and large Chrysler/Dodge vehicles.
- **35 (top terminals).** Fits many Japanese brand vehicles.
- **65 (top terminals).** Fits most large Ford/Mercury passenger cars, trucks, and SUVs.
- **75 (side terminals).** Fits some General Motors small and midsize cars and some Chrysler/Dodge vehicles.
- **78 (side terminals).** Fits many General Motors pickups and SUVs, as well as midsize and larger GM sedans.

Exact dimensions can be found on the Internet by searching for BCI battery sizes.

HYBRID AND ELECTRIC VEHICLE HIGH-VOLTAGE BATTERIES

PURPOSE AND FUNCTION Most hybrid electric vehicles use dual-voltage electrical systems. The high-voltage (HV) system is used to power the electric drive (traction) motor, while a conventional 12-volt system is used to power all other aspects of vehicle operation. One advantage to using this system is that the vehicle can use any conventional electrical accessories in its design.

ELECTRIC MOTOR REQUIREMENTS HEVs use high-output electric motors to drive and assist vehicle movement. These motors are rated anywhere from 10 to 50 kW, so they consume large amounts of electrical power during operation. If a conventional 12-volt electrical system was used to power these motors, the amount of current flow required would be extremely large and the cables used to transmit this energy would also be so large as to be impractical. Also, the motors used in these systems would have large gauge windings and would be big and heavy relative to their power output.

Automotive engineers overcome this problem by increasing the voltage provided to the motors, thus decreasing the amount of current that must flow to meet the motor's wattage requirements. (See Frequently Asked Question, "Why Do Higher Voltage Motors Draw Less Current?") Smaller amounts of current flowing in the cables mean that the cables can be sized smaller, making it much more practical to place a battery in the rear of the vehicle and run cables from there to the drive motor in the engine compartment.

The motors can also be made much smaller and more powerful when they are designed to operate on higher voltages.

FREQUENTLY ASKED QUESTION

Why Do Higher Voltage Motors Draw Less Current?

Keep in mind that an electric motor is powered by wattage. Every electric motor is rated according to the amount of power (in watts) it consumes. Power is calculated using the following formula:

$$P = I \times E$$

or

$$\textit{Power (in watts)} = \textit{Current (in amperes)} \times \textit{Voltage (in volts)}$$

An electric motor rated at 144 watts will consume 12 amperes at 12 volts of applied voltage (12 volts × 12 amperes = 144 watts). If this same motor was powered with 6 volts, it would draw 24 amperes to achieve the same power output. This increase in current draw would require a much bigger cable to efficiently transmit the electric current and minimize voltage drop. The motor windings would also have to be much heavier to handle this increased current. Imagine that we power this same motor with a 144-volt battery. Now we require only 1 ampere of electrical current to operate the motor (144 volts × 1 ampere = 144 watts). The cable required to transmit this current could be sized much smaller and it will now be much easier to run the cables over the length of the car without significant power loss. Also, the electric motor can be made much smaller and more efficient when less current is needed to power it. Some hybrid systems have motors that operate at up to 650 volts in an effort to increase system efficiency.

NICKEL-METAL HYDRIDE BATTERIES

USES Most current production HEVs use **nickel-metal hydride (NiMH)** battery technology for the high-voltage battery. NiMH batteries are being used for these applications because of their performance characteristics such as specific energy, cycle life, and safety. From a manufacturing perspective, the NiMH battery is attractive because the materials used in its construction are plentiful and recyclable.

DESCRIPTION AND OPERATION Nickel-metal hydride (NiMH) uses a positive electrode made of nickel hydroxide and potassium hydroxide electrolyte. The nominal voltage of an NiMH battery cell is 1.2 volts. The negative electrode is unique, however, in that it is a hydrogen-absorbing alloy, also known as a **metal hydride**.

ELECTROLYTE NiMH batteries are known as alkaline batteries due to the alkaline (pH greater than 7) nature of the electrolyte. The electrolyte is aqueous potassium hydroxide. Potassium hydroxide works very well for this application because it does not corrode the other parts of the battery and

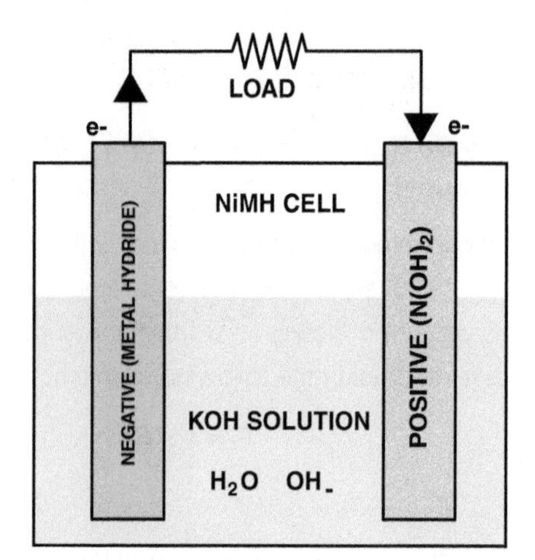

FIGURE 13–12 An NiMH cell. The unique element in a nickel-metal hydride cell is the negative electrode which is a hydrogen-absorbing alloy. The positive electrode is nickel hydroxide. The electrolyte does not enter into the chemical reaction and is able to maintain a constant conductivity regardless of the state-of-charge of the cell.

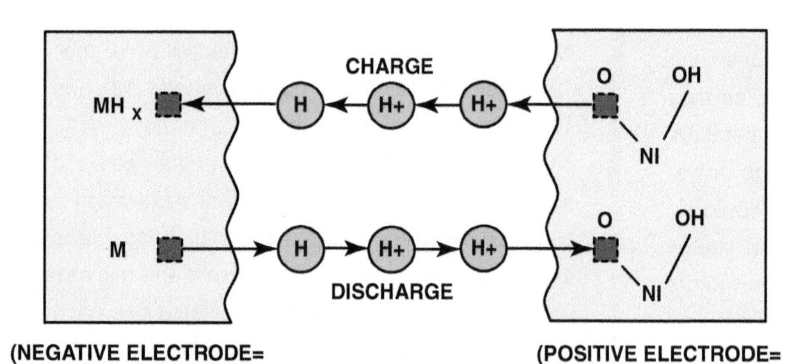

FIGURE 13–13 Chemical reactions inside an NiMH cell. Charging and discharging both involve an exchange of hydrogen ions (protons) between the two electrodes.

can be housed in a sealed steel container. Also, potassium hydroxide does not take part in the chemical reaction of the battery, so the electrolyte concentration stays constant at any given state-of-charge (SOC). These factors help the NiMH battery achieve high power performance and excellent cycle life. ● **SEE FIGURE 13–12.**

OPERATION DURING CHARGING During battery charging, hydrogen ions (protons) travel from the positive electrode to the negative electrode, where they are absorbed into the metal hydride material. The electrolyte does not participate in the reaction and acts only as a medium for the hydrogen ions to travel through.

OPERATION DURING DISCHARGING When the battery is discharged, this process reverses, with the hydrogen ions (protons) traveling from the negative electrode back to the positive electrode. The density of the electrodes changes somewhat during the charge-discharge process, but this is kept to a minimum as only protons are exchanged during battery cycling. Electrode stability due to minimal density changes is one of the reasons why the NiMH battery has very good cycle life. ● **SEE FIGURE 13–13.**

ADVANTAGES Nickel-based alkaline batteries have a number of advantages over other battery designs. These include the following:

- High specific energy.
- The nickel electrode can be manufactured with large surface areas, which increase the overall battery capacity.
- The electrolyte does not react with steel, so NiMH batteries can be housed in sealed steel containers that transfer heat reasonably well.
- The materials used in NiMH batteries are environmentally friendly and can be recycled.
- Excellent cycle life.
- Durable and safe.

DISADVANTAGES Disadvantages of the NiMH battery include the following:

- High rate of self-discharge, especially at elevated temperatures.
- Moderate levels of memory effect, although this seems to be less prominent in newer designs.
- Moderate to high cost.

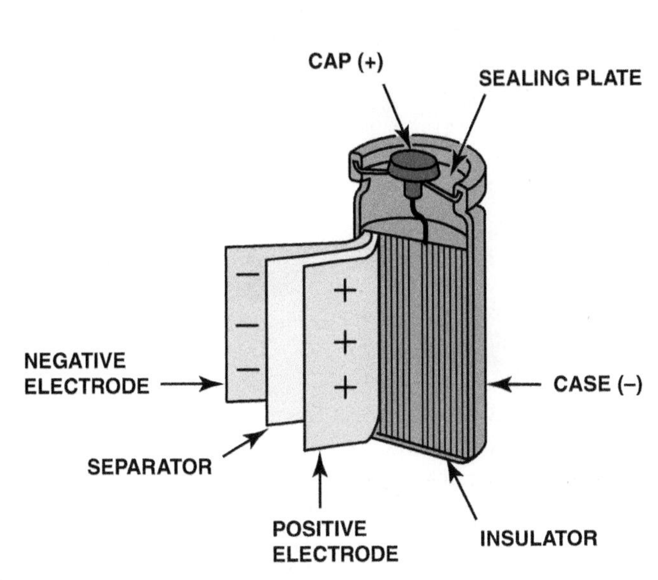

FIGURE 13–14 Cylindrical type NiMH batteries are made with stainless steel housing.

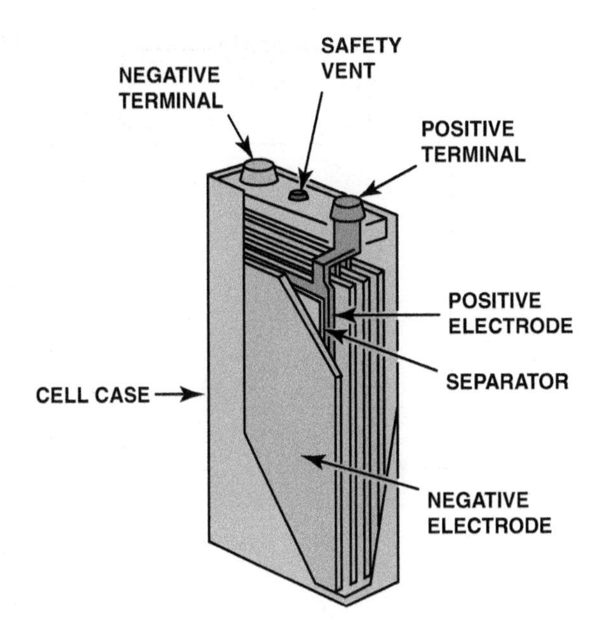

FIGURE 13–15 A prismatic NiMH cell. Prismatic cells are built with flat plates and separators similar to conventional lead-acid batteries.

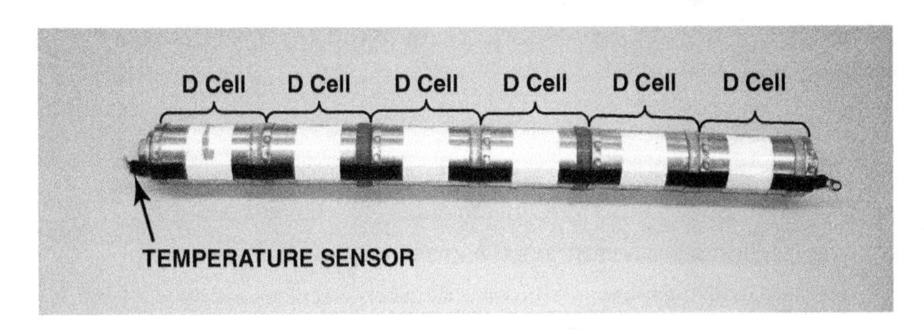

FIGURE 13–16 Each cell has 1.25 volts and a group of six as shown has 7.5 volts. These sections are then connected to other sections to create the high-voltage battery pack.

NIMH BATTERY DESIGNS There are two primary designs for an NiMH battery cell. These are as follows:

1. **Cylindrical type.** The cylindrical type has the active materials made in long ribbons and arranged in a spiral fashion inside a steel cylinder (case). The negative electrode is wound alongside the positive electrode, and the separator material holding the electrolyte is placed between them. The negative electrode is attached to the steel battery case, while the positive electrode is attached to the (−) terminal at the top of the battery. There is a self-resealing safety vent located at the top of the battery case, which will relieve internal pressure in case of overcharge, short circuiting, reverse charge, or other abuse. Cylindrical cells are often constructed very similar to a conventional "D" cell. Cylindrical cells are most often incorporated into modules with a group of six cells connected in series. This creates a single **battery module** with a 7.2-volt output. Groups of these modules can then be connected in series to create higher voltage battery packs. ● **SEE FIGURE 13–14.**
2. **Prismatic type.** The prismatic type is a rectangular or boxlike design with the active materials formed into flat plates, much like a conventional lead-acid battery. The positive and negative plates are placed alternately in the battery case, with tabs used to connect the plate groups. Separator material is placed between the plates to prevent them from touching but still allow electrolyte to circulate freely. ● **SEE FIGURE 13–15.**

Each cell of an NiMH battery produces only 1.2 volts. In order to create a battery pack that is capable of producing high voltage, many individual NiMH cells must be connected in series. ● **SEE FIGURE 13–16.**

Battery designers are limited by the nominal cell voltage of the battery technology. In the case of NiMH batteries, each cell is capable of producing only 1.2 volts. A high-voltage battery based on NiMH technology must be built using multiples

FIGURE 13–17 The GM 2 Mode hybrid battery connects the cells in series to obtain 300 volts DC. (Courtesy of Jeffrey Rehkopf)

 FREQUENTLY ASKED QUESTION

How Is SOC of an NiMH Battery Determined?

The state-of-charge (SOC) of an NiMH battery cannot be measured using cell voltage alone. Instead, SOC is determined using a complex calculation based on battery temperature, output current, and cell voltage. Accurate SOC measurements are critical for maximizing NiMH battery performance and service life.

of 1.2 volts. In order to build a 144-volt battery, 114 individual NiMH cells must be connected together in series (144 × 1.2 volts). Obviously, the higher the voltage output of the battery, the greater the number of individual battery cells that must be used to achieve the necessary voltage. ● **SEE FIGURE 13–17.**

HIGH-VOLTAGE BATTERY COOLING High operating temperatures can lower performance and cause damage to an NiMH battery pack. Most current production HEVs use air cooling to control HV battery pack temperature. Cabin air is circulated over the battery cells using an electric fan and ducting inside the vehicle. Some GM vehicles use a liquid cooling system for thermal control of the battery pack. The system not only cools the battery pack but can also warm the pack in case of extreme cold weather.

Temperature sensors (**thermistors**) are mounted in various locations in the battery pack housing to send data to the module responsible for controlling battery temperature. These inputs are used to help determine battery charge rate and cooling fan operation.

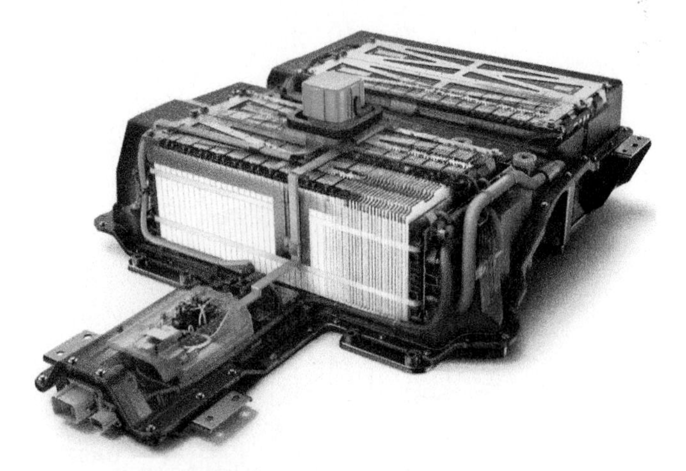

FIGURE 13–18 The Chevrolet Spark EV uses a lithium-ion HV battery pack. Note the disconnect in the top, center. (Courtesy of General Motors)

LITHIUM-ION HIGH-VOLTAGE BATTERIES

USES A battery design that shows a great deal of promise for electric vehicles (EV) and hybrid electric vehicles (HEV) applications is **lithium-ion (Li-ion)** technology. Lithium-ion batteries have been used extensively in consumer electronics since 1991 and are currently used in the following vehicles:

- Chevrolet Volt and Cadillac ELR
- Nissan Leaf
- Chevrolet Spark ● **SEE FIGURE 13–18**

DESCRIPTION A lithium-ion cell is named so because during battery cycling, lithium ions move back and forth between the positive and negative electrodes. Lithium-ion has approximately twice the specific energy of nickel-metal hydride (NiMH).

CONSTRUCTION The positive electrode in a conventional lithium-ion battery has lithium cobalt oxide as its main ingredient, with the negative electrode being made from a specialty carbon. The electrolyte is an organic solvent, and this is held in a separator layer between the two electrode plates. To prevent battery rupture and ensure safety, a pressure release valve is built into the battery's housing that will release gas if the internal pressure rises above a preset point.

OPERATION The lithium-ion cell is designed so that lithium ions can pass back and forth between the electrodes when the battery is in operation.

1. During battery discharge, lithium ions leave the anode (negative electrode) and enter the cathode (positive electrode).
2. The reverse takes place when the battery is charging.

FREQUENTLY ASKED QUESTION

How Many Types of Lithium-ion Batteries Are There?

There are numerous types of lithium-ion batteries, and the list is growing. While every component of the battery is under development, the primary difference between the various designs is the materials used for the positive electrode or cathode. The original Li-ion cell design used lithium cobalt oxide for its cathode, which has good energy storage characteristics but suffers chemical breakdown at relatively low temperatures. This failure results in the release of heat and oxygen, which often leads to a fire or explosion as the electrolyte ignites.

In order to make lithium-ion batteries safer and more durable, a number of alternative cathode materials have been formulated. One of the more promising cathode designs for automotive applications is lithium iron phosphate ($LiFePO_4$), which is stable at higher temperatures and releases less energy when it does suffer breakdown. Other lithium-ion cathode designs include the following:

- Lithium nickel cobalt oxide (LNCO)
- Lithium metal oxide (LMO)
- Nickel cobalt manganese (NCM)
- Nickel cobalt aluminum (NCA)
- Manganese oxide spinel (MnO)

Research and development continues on not only cathode design but also anodes, separator materials, and electrolyte chemistry.

ADVANTAGES Lithium-ion batteries have the following advantages:

- High specific energy
- Good high temperature performance
- Low self-discharge
- Minimal memory effect
- High nominal cell voltage. The nominal voltage of a lithium-ion cell is 3.6 volts, which is three times that of nickel-based alkaline batteries. This allows for fewer battery cells being required to produce high voltage from an HV battery.

FREQUENTLY ASKED QUESTION

What Were the Causes of Lithium-ion Battery Failure?

Three major factors are responsible for failure of lithium-ion batteries.

- Operating the cells outside their required voltage range (2 to 4 volts)
- Operating the cells outside their required temperature range of 32° to 176°F (0° to 80° C)
- Short circuits (internal or external)

For these reasons, lithium-ion battery packs in automotive applications require precise battery management using specialized cooling and safety systems.

DISADVANTAGES Disadvantages of the lithium-ion battery include:

- High cost
- Issues related to battery overheating

STATE-OF-CHARGE MANAGEMENT The HV battery in a hybrid electric vehicle is subjected to constant charging and discharging during normal operation. The battery can overheat under the following conditions:

1. The battery state-of-charge (SOC) rises above 80%.
2. The battery is placed under a load when its SOC is below 20%. In order to prevent overheating and maximize service life, the battery SOC must be carefully managed. In most hybrid electric vehicle applications, a target SOC of 60% is used, and the battery is then cycled so its SOC varies no more than 20% higher or lower than the target. **SEE FIGURE 13–19.**

NOTE: The Chevrolet Volt is designed to allow the state-of-charge to drop to 25% to 35% before the gasoline engine is started to maintain that level of state-of-charge. The cooling system and software are designed to allow this reduced SOC so that the vehicle can be driven for an extended distance without having to start the engine.

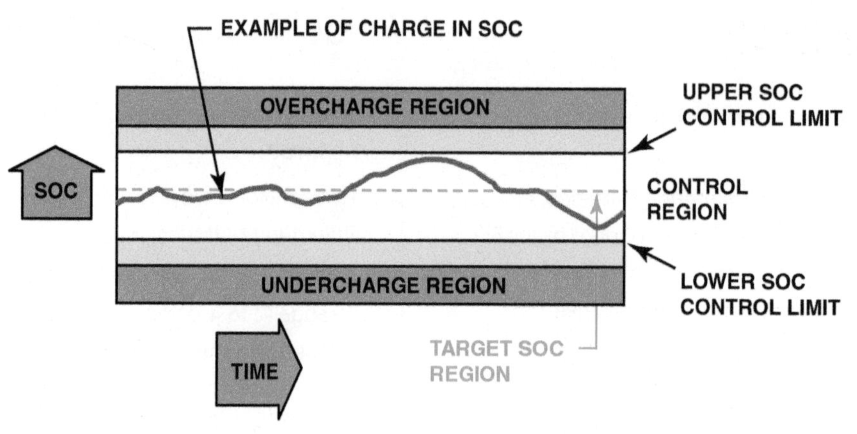

FIGURE 13–19 The HV battery pack SOC is maintained in a relatively narrow range to prevent overheating and maximize service life.

OTHER HIGH-VOLTAGE BATTERY TYPES

There are many different types of batteries that are not currently being used in electric or hybrid electric vehicles but may find other applications in future vehicles. These types of batteries include the following types:

NICKEL-CADMIUM The nickel-cadmium design is known as an **alkaline** battery, because of the alkaline nature of its electrolyte. Alkaline batteries generate electrical energy through the chemical reaction of a metal with oxygen in an alkaline electrolyte. A nickel-cadmium battery uses the following materials:

- Nickel hydroxide for the positive electrode
- Metallic cadmium for the negative electrode
- Potassium hydroxide (an alkaline solution) for the electrolyte

The nominal voltage of a Ni-Cd battery cell is 1.2 volts.

- **Advantages of Ni-Cd batteries** include the following:
  - good low-temperature performance
  - long life
  - excellent reliability
  - low maintenance requirements.
- **Disadvantages of Ni-Cd batteries** include the following:
  - Ni-Cd batteries have a specific energy that is only slightly better than lead-acid technology
  - suffers from toxicity related to its cadmium content.

LITHIUM POLYMER The **lithium-polymer (Li-poly)** battery design came out of the development of solid state electrolytes in the 1970s. Solid state electrolytes are solids that can conduct ions but do not allow electrons to move through them. Since lithium-polymer batteries use solid electrolytes, they are known as solid-state batteries. Solid polymer is much less flammable than liquid electrolytes and is able to conduct ions at temperatures above 140°F (60°C).

- **Advantages.** Li-poly batteries show good promise for EV and HEV applications for a number of reasons, including the following:
  - The lithium in the battery is in ionic form, making the battery safer because it is much less reactive than pure lithium metal.

FREQUENTLY ASKED QUESTION

How Is an Alkaline Battery Different from a Lead-Acid Battery?

Lead-acid batteries use sulfuric acid as the electrolyte, which acts as the medium between the battery's positive and negative electrodes. Acids have a pH that is below 7, and pure water has a pH of exactly 7. If electrolyte from a lead-acid battery is spilled, it can be neutralized using a solution of baking soda and water (an alkaline solution).

Alkaline batteries use an electrolyte such as potassium hydroxide, which has a pH greater than 7. This means that the electrolyte solution is basic, which is the opposite of acidic. If an alkaline battery's electrolyte is spilled, it can be neutralized using a solution of vinegar and water (vinegar is acidic). Both nickel-cadmium (Ni-Cd) and nickel-metal hydride (NiMH) batteries are alkaline battery designs.

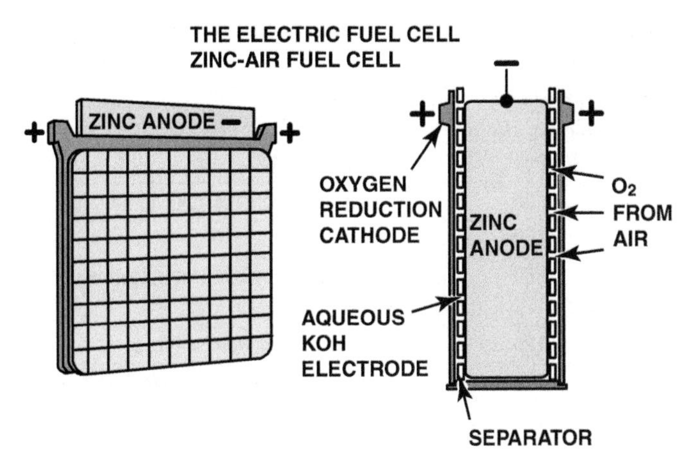

FIGURE 13–20 Zinc-air batteries are recharged by replacing the zinc anodes. These batteries are also considered to be a type of fuel cell, because the positive electrode is oxygen taken from atmospheric air.

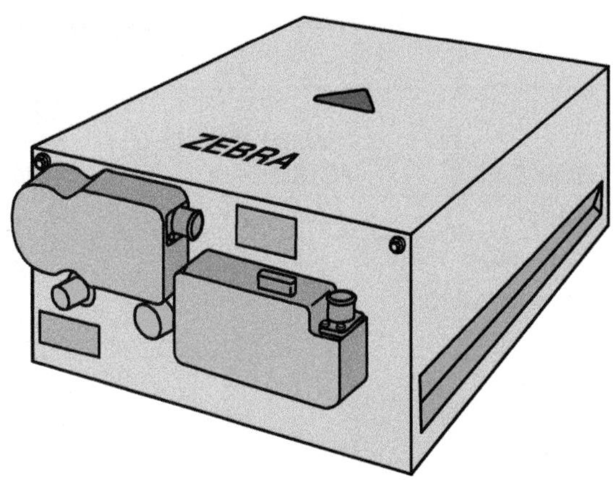

FIGURE 13–21 Sodium-metal-chloride batteries are also known as ZEBRA batteries. These batteries are lightweight (40% of the weight of lead-acid) and have a high-energy density.

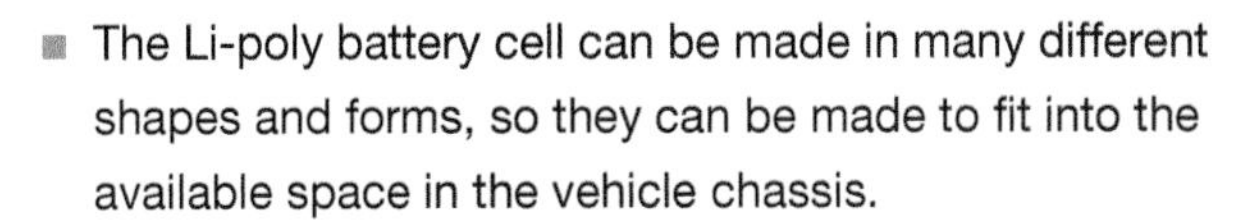

- The Li-poly battery cell can be made in many different shapes and forms, so they can be made to fit into the available space in the vehicle chassis.
- Li-poly batteries have good cycle and calendar life, and have the potential to have the highest specific energy and power of any battery technology.
- **Disadvantage.** The major disadvantage with the Li-poly battery is that it is a high-temperature design and must be operated between 176°F and 248°F (80°C and 120°C).

H2/ZINC-AIR The **zinc-air** design is a mechanically rechargeable battery. This is because it uses a positive electrode of gaseous oxygen and a sacrificial negative electrode made of zinc. The negative electrode is spent during the discharge cycle, and the battery is recharged by replacing the zinc electrodes. Zinc-air is one of several metal-air battery designs (others include aluminum-air and iron-air) that must be recharged by replacement of the negative electrode (anode). ● **SEE FIGURE 13–20.**

- **Advantage.** Zinc-air has a very high specific energy and efficiency, and the potential range of an EV vehicle equipped with a zinc-air battery is up to 600 km. Zinc-air batteries can be recharged very quickly, since a full recharge is achieved through replacement of the zinc electrodes.
- **Disadvantage.** The primary disadvantage with this design is the level of infrastructure required to make recharging practical.

ZEBRA BATTERY

- **Construction.** The ZEBRA battery is a sodium-metal-chloride battery. This battery was invented in 1985 by the *Zeolite Battery Research Africa* (ZEBRA) project. This type of battery uses two different electrolytes; first, the beta alumina similar to the sodium-sulfur design, then another layer of electrolyte between the beta alumina and the positive electrode. ● **SEE FIGURE 13–21.**
- **Advantage.** This design has been used successfully in various applications and has proven to be safe under all operating conditions. Sodium-metal-chloride technology is considered to have a very good potential for EV and HEV applications.
- **Disadvantage.** A disadvantage of the sodium-metal-chloride design is high operating temperatures.

BATTERY COMPARISON

● **CHART 13–2** shows a comparison of specific energy and nominal voltage for the various battery technologies.

BATTERY TYPE COMPARISON CHART				
BATTERY TYPE	NOMINAL VOLTAGE (V) PER CELL	THEORETICAL SPECIFIC ENERGY (WH/KG*)	PRACTICAL SPECIFIC ENERGY (WH/KG*)	MAJOR ISSUES
Lead-Acid	2.1	252	35	Heavy, low cycle life, toxic materials
Nickel-Cadmium	1.2	244	50	Toxic materials, cost
Nickel-Metal Hydride	1.2	278–800	80	Cost, high self-discharge rate,memory effect
Lithium-Ion	3.6	766	120	Safety issues, calendar life, cost
Zinc-Air	1.1	1320	110	Low power, limited cycle life, bulky
Sodium-Sulfur	2.0	792	100	High-temperature battery, safety,low power electrolyte
Sodium-Metal-Chloride (ZEBRA)	2.5	787	90	High temperature operation, low power

*Specific energy is measured in watt-hours/kilogram

CHART 13–2

Secondary-type battery comparison showing specifications and limitations.

SUMMARY

1. Maintenance-free batteries use lead-calcium grids instead of lead-antimony grids to reduce gassing.
2. When a battery is being discharged, the acid (SO_4) is leaving the electrolyte and being deposited on the plates. When the battery is being charged, the acid (SO_4) is forced off the plates and goes back into the electrolyte.
3. All batteries give off hydrogen and oxygen when being charged.
4. Batteries are rated according to CCA and reserve capacity.
5. What battery types are most used in electric and hybrid electric vehicles?

REVIEW QUESTIONS

1. Why can discharged batteries freeze?
2. What are the battery-rating methods?
3. Why can a battery explode if it is exposed to an open flame or spark?

CHAPTER QUIZ

1. When a battery becomes completely discharged, both positive and negative plates become _______ and the electrolyte becomes _______.
   a. H_2SO_4/Pb
   b. $PbSO_4/H_2O$
   c. PbO_2/H_2SO_4
   d. $PbSO_4/H_2SO_4$
2. A fully charged 12-volt battery should indicate _______.
   a. 12.6 volts or higher
   b. A specific gravity of 1.265 or higher
   c. 12 volts
   d. Both a and b

3. Deep cycling means _______.
   a. Overcharging the battery
   b. Overfilling or underfilling the battery with water
   c. The battery is almost fully discharged and then recharged
   d. The battery is overfilled with acid (H_2SO_4)

4. What makes a battery "low maintenance" or "maintenance free"?
   a. Material is used to construct the grids.
   b. The plates are constructed of different metals.
   c. The electrolyte is hydrochloric acid solution.
   d. The battery plates are smaller, making more room for additional electrolytes.

5. The positive battery plate is _______.
   a. Lead dioxide
   b. Brown in color
   c. Sometimes called lead peroxide
   d. All of the above

6. Which battery rating is tested at 0°F (−18°C)?
   a. Cold-cranking amperes (CCA)
   b. Cranking amperes (CA)
   c. Reserve capacity
   d. Battery voltage test

7. Which battery rating is expressed in minutes?
   a. Cold-cranking amperes (CCA)
   b. Cranking amperes (CA)
   c. Reserve capacity
   d. Battery voltage test

8. What battery rating is tested at 32°F (0°C)?
   a. Cold-cranking amperes (CCA)
   b. Cranking amperes (CA)
   c. Reserve capacity
   d. Battery voltage test

9. What gases are released from a battery when it is being charged?
   a. Oxygen
   b. Hydrogen
   c. Nitrogen and oxygen
   d. Hydrogen and oxygen

10. A charge indicator (eye) operates by showing green or red when the battery is charged and dark if the battery is discharged. This charge indicator detects _______.
    a. Battery voltage
    b. Specific gravity
    c. Electrolyte water pH
    d. Internal resistance of the cells

chapter 14

BATTERY TESTING AND SERVICE

LEARNING OBJECTIVES

After studying this chapter, the reader will be able to:

1. List the precautions necessary when working with batteries.
2. Describe how to inspect and clean battery cables, connectors, clamps, and hold-downs.
3. Discuss how to test batteries for open-circuit voltage and specific gravity.
4. Describe how to perform a battery load test and a conductance test.
5. Explain how to safely charge or jump-start a battery.
6. Discuss how to perform a battery drain test.

This chapter will help you prepare for the ASE Electrical/Electronic Systems (A6) certification test content area "B" (Battery Diagnosis and Service).

KEY TERMS

Battery electrical drain test 195
Dynamic voltage 188
Hydrometer 189
IOD 195
Load test 190
Open circuit voltage 188
Parasitic load test 195
Three-minute charge test 190

GM STC OBJECTIVES

GM Service Technical College topics covered in this chapter are as follows:

1. Factors affecting battery life (08510.01W).
2. Battery inspection, testing, and maintenance procedures.
3. Recall safety practices required when working on Two-mode Hybrid and EREV vehicles.

BATTERY SERVICE SAFETY CONSIDERATIONS

HAZARDS Batteries contain acid and release explosive gases (hydrogen and oxygen) during normal charging and discharging cycles.

SAFETY PROCEDURES To help prevent physical injury or damage to the vehicle, always adhere to the following safety procedures.

1. When working on any electrical component on a vehicle, disconnect the negative battery cable from the battery. When the negative cable is disconnected, all electrical circuits in the vehicle will be open, which will prevent accidental electrical contact between an electrical component and ground. Any electrical spark has the potential to cause explosion and personal injury.
2. Wear eye protection (goggles preferred) when working around any battery.
3. Wear protective clothing to avoid skin contact with battery acid.
4. Always adhere to all safety precautions as stated in the service procedures for the equipment used for battery service and testing.
5. Never smoke or use an open flame around any battery.

FIGURE 14–1 A visual inspection on this battery shows the electrolyte level was below the plates in all cells.

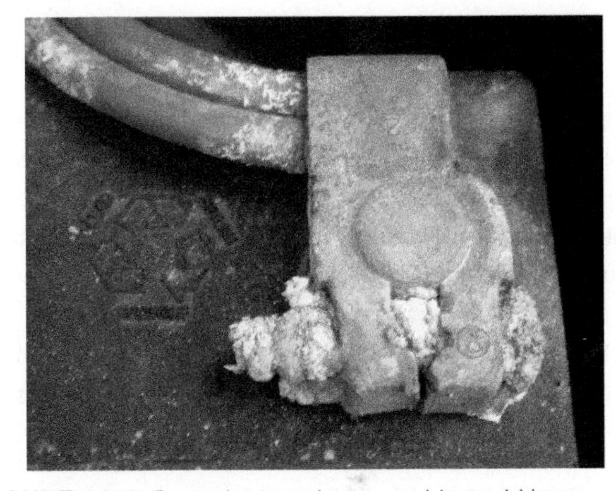

FIGURE 14–2 Corrosion on a battery cable could be an indication that the battery itself is either being overcharged or is sulfated, creating a lot of gassing of the electrolyte.

SYMPTOMS OF A WEAK OR DEFECTIVE BATTERY

The following warning signs indicate that a battery is near the end of its useful life.

- **Uses water in one or more cells.** This indicates that the plates are sulfated and that during the charging process, the water in the electrolyte is being turned into separate hydrogen and oxygen gases. ● **SEE FIGURE 14–1.**
- **Excessive corrosion on battery cables or connections.** Corrosion is more likely to occur if the battery is sulfated, creating hot spots on the plates. When the battery is being charged, the acid fumes are forced out of the vent holes and onto the battery cables, connections, and even on the battery tray underneath the battery. ● **SEE FIGURE 14–2.**
- **Slower than normal engine cranking.** When the capacity of the battery is reduced due to damage or age, it is less likely to be able to supply the necessary current for starting the engine, especially during cold weather.

BATTERY MAINTENANCE

NEED FOR MAINTENANCE Most new-style batteries are of a maintenance–free design that uses lead–calcium instead of lead–antimony plate grid construction. Because lead–calcium batteries do not release as much gas as the older-style, lead–antimony batteries, there is less consumption of water

 TECH TIP

Dynamic versus Open Circuit Voltage

Open circuit voltage is the voltage (usually of a battery) that exists *without* a load being applied. **Dynamic voltage** is the voltage of the power source (battery) with the circuit in operation. A vehicle battery, for example, may indicate that it has 12.6 volts or more, but that voltage will drop when the battery is put under a load such as cranking the engine. If the battery voltage drops too much, the starter motor will rotate more slowly and the engine may not start.

If the dynamic voltage is lower than specified, the battery may be weak or defective or the circuit may be defective.

during normal service. Also, with less gassing, less corrosion is observed on the battery terminals, wiring, and support trays. If the electrolyte level can be checked, and if it is low, add only distilled water. Distilled water is recommended by all battery manufacturers, but if distilled water is not available, clean ordinary drinking water, low in mineral content, can be used.

Battery maintenance includes making certain that the battery case is clean and checking that the battery cables and hold-down fasteners are clean and tight.

BATTERY TERMINAL CLEANING Many battery-related faults are caused by poor electrical connections at the battery. Battery cable connections should be checked and cleaned to prevent voltage drop at the connections. One common reason for an engine to not start is loose or corroded battery cable connections. Perform an inspection and check for the following conditions.

- Loose or corroded connections at the battery terminals (should not be able to be moved by hand)
- Loose or corroded connections at the ground connector on the engine block
- Wiring that has been modified to add auxiliary power for a sound system, or other electrical accessory

If the connections are loose or corroded, use 1 tablespoon of baking soda in 1 quart (liter) of water and brush this mixture onto the battery and housing to neutralize the acid. Mechanically clean the connections and wash the area with water.

BATTERY HOLD-DOWN The battery should also be secured with a hold-down bracket to prevent vibration from damaging the plates inside the battery. The hold-down bracket should be snug enough to prevent battery movement, yet not so tight as to cause the case to crack. Factory-original hold-down brackets are often available through local automobile dealers, and universal hold-down units are available through local automotive parts stores.

BATTERY VOLTAGE TEST

STATE OF CHARGE Testing the battery voltage with a voltmeter is a simple method for determining the state of charge of any battery. ● **SEE FIGURE 14–3**. The voltage of a battery does not necessarily indicate whether the battery can perform satisfactorily, but it does indicate to the technician more about the battery's condition than a simple visual inspection. A battery that "looks good" may not be good. This test is commonly called an *open circuit battery voltage test* because it is conducted with an open circuit, no current flowing, and no load applied to the battery.

1. If the battery has just been charged or the vehicle has recently been driven, it is necessary to remove the surface charge from the battery before testing. A surface charge is a charge of higher-than-normal voltage that is just on the surface of the battery plates. The surface charge is quickly removed when the battery is loaded and therefore does not accurately represent the true state of charge of the battery.
2. To remove the surface charge, turn the headlights on high beam (brights) for one minute, then turn the headlights off and wait two minutes.
3. With the engine and all electrical accessories off, and the doors shut (to turn off the interior lights), connect a voltmeter to the battery posts. Connect the red positive lead to the positive post and the black negative lead to the negative post.

   NOTE: If the meter reads negative (−), the battery has been reverse charged (has reversed polarity) and should be replaced, or the meter has been connected incorrectly.

4. Read the voltmeter and compare the results with the state of charge. The voltages shown are for a battery at or near room temperature (70°F to 80°F, or 21°C to 27°C). ● **SEE CHART 14–1**.

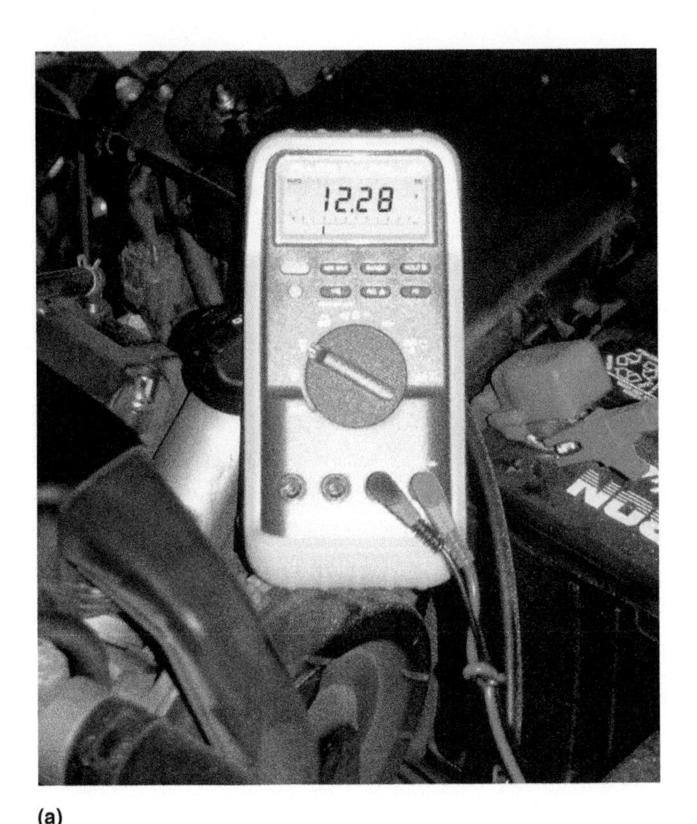

(a)

(b)

FIGURE 14–3 (a) A voltage reading of 12.28 volts indicates that the battery is not fully charged and should be charged before testing. (b) A battery that measures 12.6 volts or higher after the surface charge has been removed is 100% charged.

BATTERY VOLTAGE (V)	STATE OF CHARGE
12.6 or higher	100% charged
12.4	75% charged
12.2	50% charged
12.0	25% charged
11.9 or lower	Discharged

CHART 14–1

The estimated state of charge of a 12-volt battery after the surface charge has been removed.

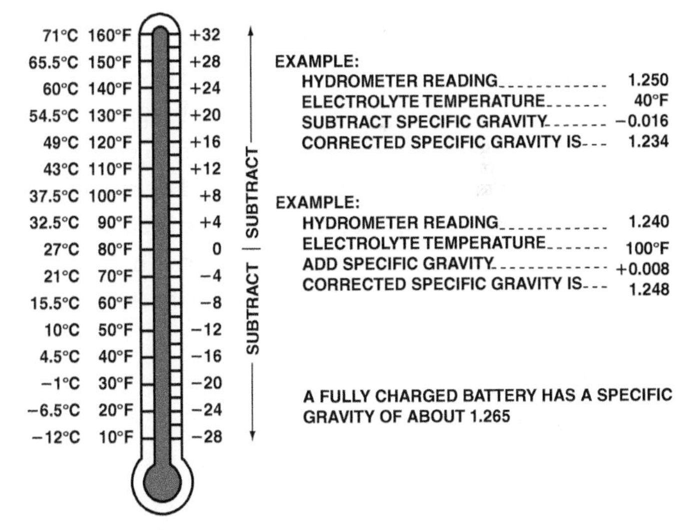

FIGURE 14–4 When testing a battery using a hydrometer, the reading must be corrected if the temperature is above or below 80°F (27°C).

HYDROMETER TESTING

If the battery has removable filler caps, the specific gravity of the electrolyte can also be checked. A **hydrometer** is a tester that measures the specific gravity. ● **SEE FIGURE 14–4.**

This test can also be performed on most maintenance-free batteries because their filler caps are removable, except for those produced by Delco (Delphi) Battery. The specific gravity test indicates the state of battery charge and can indicate a defective battery if the specific gravity of one or more cells varies by more than 0.050 from the value of the highest-reading cell. ● **SEE CHART 14–2.**

SPECIFIC GRAVITY	BATTERY VOLTAGE (V)	STATE OF CHARGE
1.265	12.6 or higher	100% charged
1.225	12.4	75% charged
1.190	12.2	50% charged
1.155	12.0	25% charged
Lower than 1.120	11.9 or lower	Discharged

CHART 14–2

Measuring the specific gravity can detect a defective battery. A battery should be at least 75% charged before being load tested.

FIGURE 14–5 This battery has cold-cranking amperes (CCA) of 550 amperes, cranking amperes (CA) of 680 amperes, and load test amperes of 270 amperes listed on the top label. Not all batteries have this complete information.

FREQUENTLY ASKED QUESTION

What Is the Three-Minute Charge Test?

A **three-minute charge test** is used to check if a battery is sulfated and is performed as follows:

- Connect a battery charger and a voltmeter to the battery terminals.
- Charge the battery at a rate of 40 amperes for three minutes.
- At the end of three minutes, read the voltmeter.

Results: If the voltage is above 15.5 volts, replace the battery. If the voltage is below 15.5 volts, the battery is not sulfated and should be charged and retested.

This is *not* a valid test for many maintenance-free batteries, such as the Delphi Freedom. Due to the high internal resistance, a discharged Delphi Freedom battery may not start to accept a charge for several hours. Always use another alternative battery test before discarding a battery based on the results of the three-minute charge test.

BATTERY LOAD TESTING

TERMINOLOGY One test to determine the condition of any battery is the **load test**. Most automotive starting and charging testers use a carbon pile to create an electrical load on the battery. The amount of the load is determined by the original CCA rating of the battery, which should be at least 75% charged before performing a load test. The capacity is measured in cold-cranking amperes, which is the number of amperes that a battery can supply at 0°F (−18°C) for 30 seconds.

TEST PROCEDURE To perform a battery load test, take the following steps.

STEP 1 **Determine the CCA rating of the battery.** The proper electrical load used to test a battery is half of the CCA rating or three times the ampere-hour rating, with a minimum 150 ampere load. ● **SEE FIGURE 14–5.**

STEP 2 **Connect the load tester to the battery.** Follow the instructions for the tester being used.

STEP 3 **Apply the load for a full 15 seconds.** Observe the voltmeter during the load testing and check the voltage at the end of the 15-second period while the battery is still under load. A good battery should indicate above 9.6 volts.

STEP 4 **Repeat the test.** Many battery manufacturers recommend performing the load test twice, using the first load period to remove the surface charge on the battery and the second test to provide a truer indication of the condition of the battery. Wait 30 seconds between tests to allow time for the battery to recover. ● **SEE FIGURE 14–6.**

Results: If the battery fails the load test, recharge the battery and retest. If the load test is failed again, replacement of the battery is required.

FIGURE 14–6 An alternator regulator battery starter tester (ARBST) automatically loads the battery with a fixed load for 15 seconds to remove the surface charge, then removes the load for 30 seconds to allow the battery to recover, and then reapplies the load for another 15 seconds. The results of the test are then displayed.

FREQUENTLY ASKED QUESTION

How Should You Test a Vehicle Equipped with Two Batteries?

Many vehicles equipped with a diesel engine use two batteries. These batteries are usually electrically connected in parallel to provide additional current (amperes) at the same voltage. ● **SEE FIGURE 14–7.**

Some heavy-duty trucks and buses connect two batteries in series to provide about the same current as one battery, but with twice the voltage, as shown in ● **FIGURE 14–8.**

To successfully test the batteries, they should be disconnected and tested separately. If just one battery is found to be defective, most experts recommend that both be replaced to help prevent future problems. Because the two batteries are electrically connected, a fault in one battery can cause the good battery to discharge into the defective battery, thereby affecting both even if just one battery is defective.

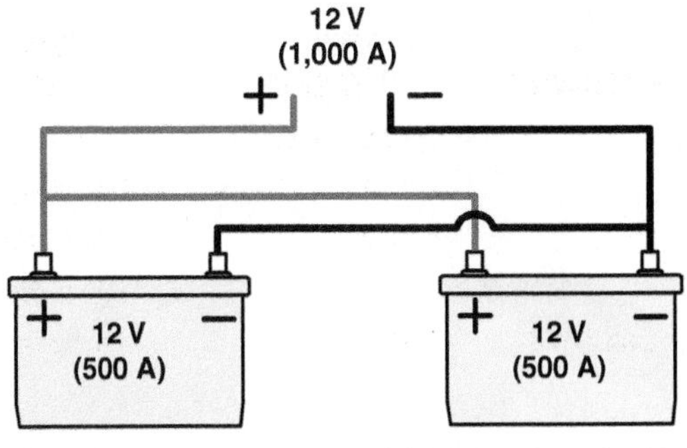

FIGURE 14–7 Most light-duty vehicles equipped with two batteries are connected in parallel as shown. Two 500 amperes, 12 volt batteries are capable of supplying 1,000 amperes at 12 volts, which is needed to start many diesel engines.

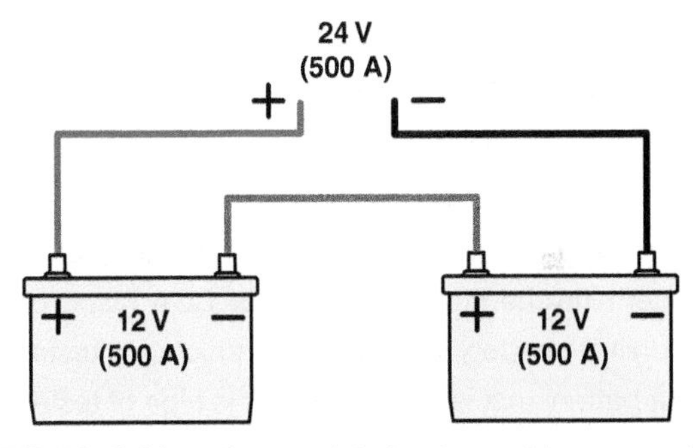

FIGURE 14–8 Many heavy-duty trucks and buses use two 12-volt batteries connected in series to provide 24 volts.

ELECTRONIC CONDUCTANCE TESTING

TERMINOLOGY General Motors Corporation, Chrysler Corporation, and Ford specify that an electronic conductance tester be used to test batteries in vehicles still under factory warranty. Conductance is a measure of how well a battery can create current. This tester sends a small signal through the battery and then measures a part of the AC response. As a battery ages, the plates can become sulfated and shed active materials from the grids, thus reducing the battery capacity. Conductance testers can be used to test flooded or absorbed glass mat-type (AGM) batteries. The unit can determine the following information about a battery.

- CCA
- State of charge
- Voltage of the battery
- Defects such as shorts and opens

FIGURE 14–9 A conductance tester is very easy to use and has proved to accurately determine battery condition if the connections are properly made. Follow the instructions on the display exactly for best results. (Courtesy of General Motors)

SAFETY TIP

Never Charge or Jump-Start a Frozen Battery

A discharged battery can freeze because the electrolyte becomes mostly water. Never attempt to charge or jump-start a vehicle that has a frozen battery. When the battery freezes, it often bulges at the sides because water expands about 9% when it freezes, forming ice crystals that occupy more space than water. The crystals can trap bubbles of hydrogen and oxygen that are created during the chemical processes in a battery. When attempting to charge or jump-start the frozen battery, these pockets of gases can explode. Because the electrolyte expands, the freezing action usually destroys the plates and can loosen the active material from the grids. It is rare for a frozen battery to be restored to useful service.

However, a conductance tester is not designed to accurately determine the state of charge or CCA rating of a new battery. Unlike a battery load test, a conductance tester can be used on a battery that is discharged. This type of tester should only be used to test batteries that have been in service. ● **SEE FIGURE 14–9**.

TEST PROCEDURE

STEP 1 Connect the unit to the positive and negative terminals of the battery. If testing a side post battery, always use the lead adapters and *never* use steel bolts, as these can cause an incorrect reading.

NOTE: Test results can be incorrectly reported on the display if proper, clean connections to the battery are not made. Also be sure that all accessories and the ignition switch are in the off position.

STEP 2 Enter the CCA rating (if known) and push the arrow keys.

STEP 3 The tester determines and displays one of the following:

- **Good battery.** The battery can return to service.
- **Charge and retest.** Fully recharge the battery and return it to service.
- **Replace the battery.** The battery is not serviceable and should be replaced.
- **Bad cell–replace.** The battery is not serviceable and should be replaced.

Some conductance testers can check the charging and cranking circuits, too.

BATTERY CHARGING

CHARGING PROCEDURE If the state of charge of a battery is low, it must be recharged. It is best to slow-charge any battery to prevent possible overheating damage to the battery. Perform the following steps.

STEP 1 **Determine the charge rate.** The charge rate is based on the current state of charge (SOC) and charging rate. ● **SEE CHART 14–3** for the recommended charging rate.

STEP 2 **Connect a battery charger to the battery.** Be sure the charger is not plugged in when connecting a charger to a battery. Always follow the battery charger's instructions for proper use.

STEP 3 **Set the charging rate.** The initial charge rate should be about 35 amperes for 30 minutes to help start the charging process. Fast-charging a battery increases the temperature of the battery and can cause warping of the plates inside the battery. Fast-charging also increases the amount of gassing (release of hydrogen and oxygen), which can create a health and fire hazard. The battery temperature should not exceed 125°F (hot to the touch).

- Fast charge: 15 amperes maximum
- Slow charge: 5 amperes maximum

● **SEE FIGURE 14–10**.

OPEN CIRCUIT VOLTAGE	BATTERY SPECIFIC GRAVITY*	STATE OF CHARGE	CHARGING TIME TO FULL CHARGE AT 80°F**					
			at 60 amps	at 50 amps	at 40 amps	at 30 amps	at 20 amps	at 10 amps
12.6	1.265	100%	FULL CHARGE					
12.4	1.225	75%	15 min.	20 min.	27 min.	35 min.	48 min.	90 min.
12.2	1.190	50%	35 min.	45 min.	55 min.	75 min.	95 min.	180 min.
12.0	1.155	25%	50 min.	65 min.	85 min.	115 min.	145 min.	260 min.
11.8	1.120	0%	65 min.	85 min.	110 min.	150 min.	195 min.	370 min.

CHART 14-3

Battery charging guideline showing the charging times that vary according to state of charge, temperature, and charging rate. It may take eight hours or more to charge a fully discharged battery.

*Correct for temperature

**If colder, it'll take longer

FIGURE 14-10 This battery charger/tester (Midtronics GR-8) features quick and efficient assessment of the condition of the battery while under a load. It also maintains the battery state of charge (SOC) during module programming and extended service work. (Courtesy of General Motors)

TECH TIP

Charge Batteries at 1% of Their CCA Rating

Many batteries are damaged due to overcharging. To help prevent damages such as warped plates and excessive release of sulfur smell gases, charge batteries at a rate equal to 1% of the battery's CCA rating. For example, a battery with a 700 CCA rating should be charged at 7 amperes ($700 \times 0.01 = 7$ amperes). No harm will occur to the battery at this charge rate even though it may take longer to achieve a full charge. This means that a battery may require eight or more hours to become fully charged depending on the battery capacity and state of charge (SOC).

CHARGING AGM BATTERIES Charging an AGM battery requires a different charger than is used to recharge a flooded-type battery. The differences include the following:

- The AGM can be charged with high current, up to 75% of the ampere-hour rating due to lower internal resistance.
- The charging voltage has to be kept at or below 14.4 volts to prevent damage.

Because most conventional battery chargers use a charging voltage of 16 volts or higher, a charger specifically designed to charge AGM batteries must be used.

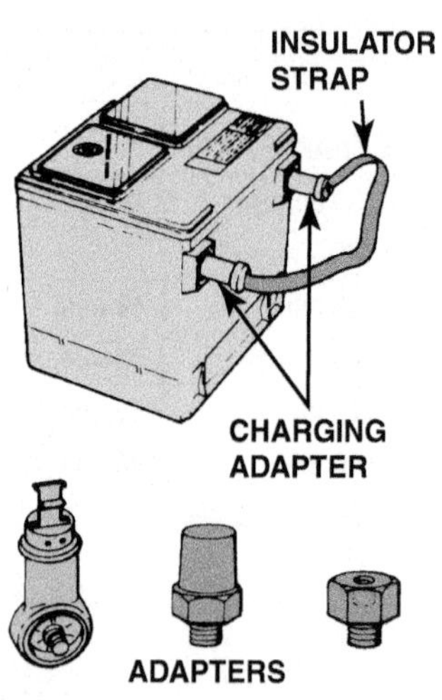

FIGURE 14–11 Adapters should be used on side terminal batteries whenever charging.

FIGURE 14–12 A typical battery jump box used to jump start vehicles. These hand-portable units have almost made jumper cables obsolete.

TECH TIP

Always Use Adapters on Side Post Batteries

Side post batteries require that an adapter be used when charging the battery, if it is removed from the vehicle. Do not use steel bolts. If a bolt is threaded into the terminal, only the parts of the threads that contact the battery terminal will be conducting all of the charging current. An adapter or a bolt with a nut attached is needed to achieve full contact with the battery terminals. **SEE FIGURE 14–11.**

Absorbed glass mat batteries are often used as auxiliary batteries in hybrid electric vehicles when the battery is located inside the vehicle.

BATTERY CHARGE TIME

The time needed to charge a completely discharged battery can be estimated by using the reserve capacity rating of the battery in minutes divided by the charging rate.

Hours needed to charge the battery = Reserve capacity ÷ Charge current

For example, if a 10 amperes charge rate is applied to a discharged battery that has a 90-minute reserve capacity, the time needed to charge the battery will be nine hours.

90 minutes ÷ 10 amperes = 9 hours

FREQUENTLY ASKED QUESTION

Should Batteries Be Kept Off of Concrete Floors?

All batteries should be stored in a cool, dry place when not in use. Many technicians have been warned not to store or place a battery on concrete. According to battery experts, it is the temperature difference between the top and the bottom of the battery that causes a difference in the voltage potential between the top (warmer section) and the bottom (colder section). It is this difference in temperature that causes self-discharge to occur.

In fact, submarines cycle seawater around their batteries to keep all sections of the battery at the same temperature to help prevent self-discharge.

Therefore, always store or place batteries up off the floor and in a location where the entire battery can be kept at the same temperature, avoiding extreme heat and freezing temperatures. Concrete cannot drain the battery directly because the case of the battery is a very good electrical insulator.

JUMP STARTING

To jump-start another vehicle with a dead battery, connect good-quality copper jumper cables or a jump box to the good battery and the dead battery, as shown in **FIGURE 14–12**.

When using jumper cables or a battery jump box, the last connection made should always be on the engine block or an

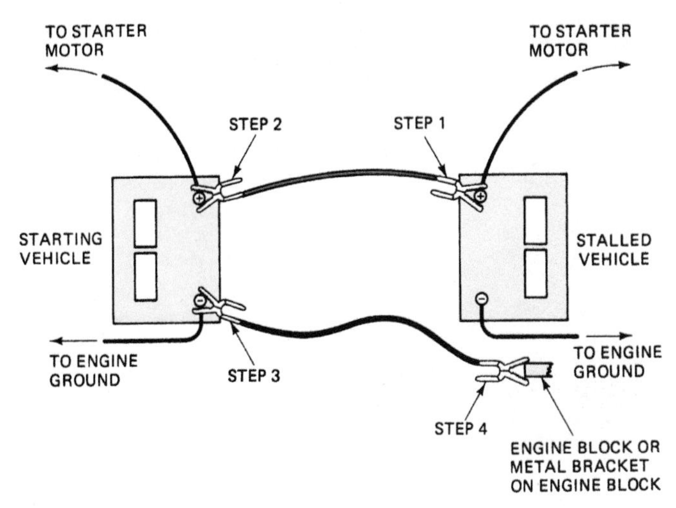

FIGURE 14–13 Jumper cable usage guide. Notice that the last connection should be the engine block of the disabled vehicle to help prevent the spark that normally occurs from igniting the gases from the battery.

FIGURE 14–14 The sticker on this battery indicates that it was built in January 2012.

engine bracket on the dead vehicle as far from the battery as possible. ● **SEE FIGURE 14–13.**

It is normal for a spark to occur when the jumper cables finally complete the jumping circuit, and this spark could cause an explosion of the gases around the battery. Many newer vehicles have special ground and/or positive power connections built away from the battery just for the purpose of jump-starting. Check the owner's manual or service information for the exact location.

TECH TIP

Look at the Battery Date Code

All major battery manufacturers stamp codes on the battery case that give the date of manufacture and other information about the battery. Most battery manufacturers use a number to indicate the year of manufacture and a letter to indicate the month of manufacture, except the letter I, because it can be confused with the number 1. For example:

A = January	G = July
B = February	H = August
C = March	J = September
D = April	K = October
E = May	L = November
F = June	M = December

The shipping date from the manufacturing plant is usually indicated by a *sticker* on the body of the battery. Almost every battery manufacturer uses just one letter and one number to indicate the month and year. ● **SEE FIGURE 14–14.**

BATTERY ELECTRICAL DRAIN TEST

TERMINOLOGY The **battery electrical drain test** determines if any component or circuit in a vehicle is causing a drain on the battery when everything is off. This test is also called the **ignition off draw (IOD)** or **parasitic load test**.

FIGURE 14–15 This mini clamp-on digital multimeter is being used to measure the amount of battery electrical drain that is present. In this case, a reading of 20 milliamperes (displayed on the meter as 00.02 ampere) is within the normal range of 20 to 30 milliamperes. Be sure to clamp around all of the positive battery cable or all of the negative battery cable, whichever is easiest to get the clamp around.

FIGURE 14–16 After connecting the shut-off tool, start the engine and operate all accessories. Stop the engine and turn off everything. Connect the ammeter across the shut-off switch in parallel. Wait 20 minutes. This time allows all electronic circuits to "time out" or shut down. Open the switch-all current now will flow through the ammeter. A reading greater than specified (usually greater than 50 milliamperes, or 0.05 ampere) indicates a problem that should be corrected.

Many electronic components draw a continuous, slight amount of current from the battery when the ignition is off. These components include the following:

1. Electronically tuned radios for station memory and clock circuits
2. Computers and controllers, through slight diode leakage
3. The alternator, through slight diode leakage

These components may cause a voltmeter to read full battery voltage if it is connected between the negative battery terminal and the removed end of the negative battery cable. Because of this fact, voltmeters should not be used for battery drain testing. This test should be performed when one of the following conditions exists.

1. When a battery is being charged or replaced (a battery drain could have been the cause for charging or replacing the battery)
2. When the battery is suspected of being drained

PROCEDURE FOR BATTERY ELECTRICAL DRAIN TEST

- **Inductive DC ammeter.** The fastest and easiest method to measure battery electrical drain is to connect an inductive DC ammeter that is capable of measuring low current (10 milliamperes). **SEE FIGURE 14–15** for an example of a clamp-on digital multimeter being used to measure battery drain.
- **DMM set to read milliamperes.** Following is the procedure for performing the battery electrical drain test using a DMM set to read DC amperes.

STEP 1 Make certain that all lights, accessories, and ignition are off.

STEP 2 Check all vehicle doors to be certain that the interior courtesy (dome) lights are off.

STEP 3 Disconnect the *negative* (−) battery cable and install a parasitic load tool, as shown in **FIGURE 14–16.**

STEP 4 Start the engine and drive the vehicle about 10 minutes, being sure to turn on all the lights and accessories including the radio.

STEP 5 Turn the engine and all accessories off including the underhood light.

STEP 6 Connect an ammeter across the parasitic load tool switch and wait 20 minutes for all computers and circuits to shut down.

STEP 7 Open the switch on the load tool and read the battery electrical drain on the meter display.

NOTE: Using a voltmeter or test light to measure battery drain is *not* recommended by most vehicle manufacturers. The high internal resistance of the voltmeter results in an irrelevant reading that does not provide the technician with adequate information about a problem.

The Chevrolet Battery Story

A 2005 Chevrolet Impala was being diagnosed for a dead battery. Testing for a battery drain (parasitic draw) showed 2.25 amperes, which was clearly over the acceptable value of 0.050 or less. At the suggestion of the shop foreman, the technician used a Tech 2 scan tool to check if all of the computers and modules went to sleep after the ignition was turned off. The scan tool display indicated that the instrument panel (IP) showed that it remained awake after all of the others had gone into sleep mode. The IP cluster was unplugged and the vehicle was tested for an electrical drain again. This time, it was only 32 milliamperes (0.032 ampere), well within the normal range. Replacing the IP cluster solved the excessive battery drain.

It Could Happen to You!

After replacing his vehicle's battery the owner noted that the "airbag" amber warning lamp was lit and the radio was locked out. The owner had purchased the vehicle used and did not know the four-digit security code needed to unlock the radio. Determined to fix the problem, the owner tried three four-digit numbers, hoping that one of them would work. However, after three tries, the radio became permanently disabled.

Frustrated, the owner went to a dealer. It cost over $300 to fix the problem. A special tool was required to easily reset the airbag lamp. The radio had to be removed and sent out of state to an authorized radio service center and then reinstalled into the vehicle.

Therefore, before disconnecting the battery, check to be certain that the owner has the security code for a security-type radio. A "memory saver" may be needed to keep the radio powered up when the battery is being disconnected. **SEE FIGURE 14–17.**

SPECIFICATIONS Results:

- Normal = 20 to 30 milliamperes (0.02 to 0.03 amperes)
- Maximum allowable = 50 milliamperes (0.05 amperes)

RESET ALL MEMORY FUNCTIONS Be sure to reset the clock, "auto up" windows, and antitheft radio if equipped.

BATTERY DRAIN AND RESERVE CAPACITY It is normal for a battery to self-discharge even if there is not an electrical load such as computer memory to drain the battery. According to General Motors, this self-discharge is about 13 milliamperes (0.013 ampere).

Some vehicle manufacturers specify a maximum allowable parasitic draw or battery drain be based on the reserve capacity of the battery. The calculation used is the reserve capacity of the battery divided by 4; this equals the maximum allowable battery drain. For example, a battery rated at 120 minutes reserve capacity should have a maximum battery drain of 30 milliamperes.

120 minutes reserve capacity ÷ 4 = 30 mA

FINDING THE SOURCE OF THE DRAIN If there is a drain, check and temporarily disconnect the following components.

1. Underhood light
2. Glove compartment light
3. Trunk light

If after disconnecting these three components the battery drain draws more than 50 milliamperes (0.05 ampere), disconnect one fuse at a time from the fuse box until the excessive drain drops to normal.

NOTE: Do not reinsert fuses after they have been removed as this action can cause modules to "wake up," leading to an inconclusive test.

If the excessive battery drain stops after one fuse is disconnected, the source of the drain is located in that particular circuit, as labeled on the fuse box. Continue to disconnect the *power-side* wire connectors from each component included in that particular circuit until the test light goes off. The source of

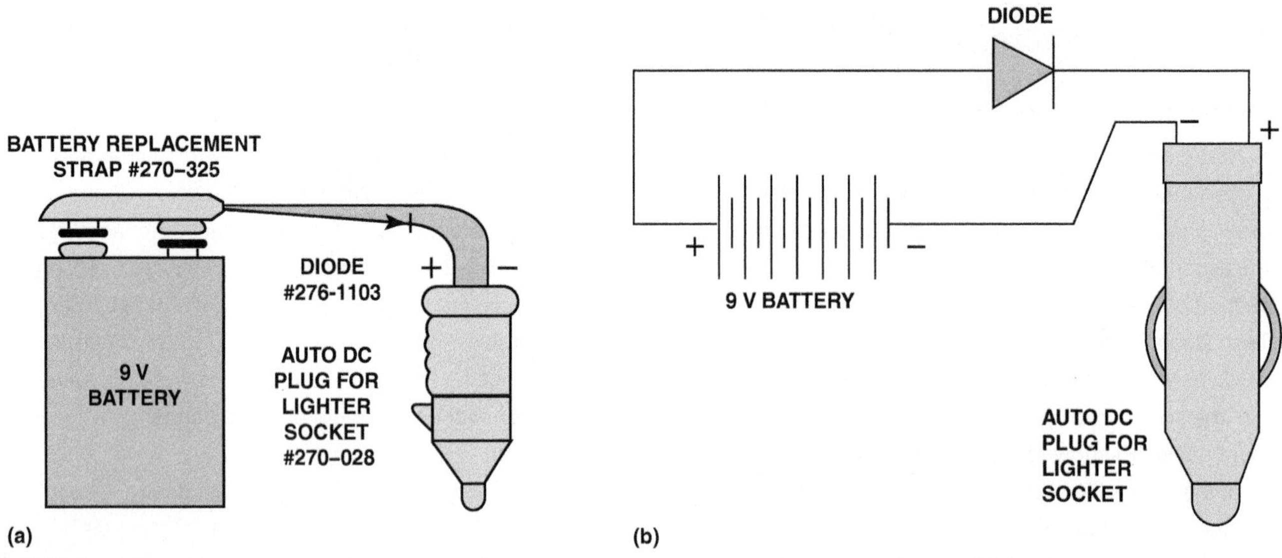

FIGURE 14–17 (a) Memory saver. The part numbers represent components from Radio Shack. (b) A schematic drawing of the same memory saver. Some experts recommend using a 12-volt lantern battery instead of a small 9-volt battery to help ensure that there will be enough voltage in the event that a door is opened while the vehicle battery is disconnected. Interior lights could quickly drain a small 9-volt battery.

the battery drain can then be traced to an individual component or part of one circuit.

WHAT TO DO IF A BATTERY DRAIN STILL EXISTS

If all the fuses have been disconnected and the drain still exists, the source of the drain has to be between the battery and the fuse box. The most common sources of drain under the hood include the following:

1. **The alternator.** Disconnect the alternator wires and retest. If the ammeter now reads a normal drain, the problem is a defective diode(s) in the alternator.
2. **The starter solenoid (relay) or wiring near its components.** These are also a common source of battery drain, due to high current flows and heat, which can damage the wire or insulation.

FIGURE 14–18 Many newer vehicles have batteries that are sometimes difficult to find. Some are located under plastic panels under the hood, under the front fender, or even under the rear seat as shown here.

HIGH-VOLTAGE BATTERY DISCONNECT

PURPOSE AND FUNCTION Most hybrid electric vehicles are equipped with a high-voltage disconnect switch or connector that is used to cut off high voltage from the rest of the system beyond the battery pack. This disconnect does not need to be removed or switched unless service work is being performed on the high-voltage circuits in the vehicle. Always

FREQUENTLY ASKED QUESTION

Where Is the Battery?

Many vehicle manufacturers today place the battery under the backseat, under the front fender, or in the trunk. SEE FIGURE 14–18.

Often, the battery is not visible even if it is located under the hood. When testing or jump-starting a vehicle, look for a battery access point.

FIGURE 14–19 The high voltage service disconnect plug is located on the battery case of this General Motors Two-Mode hybrid vehicle. (Courtesy of Jeffrey Rehkopf)

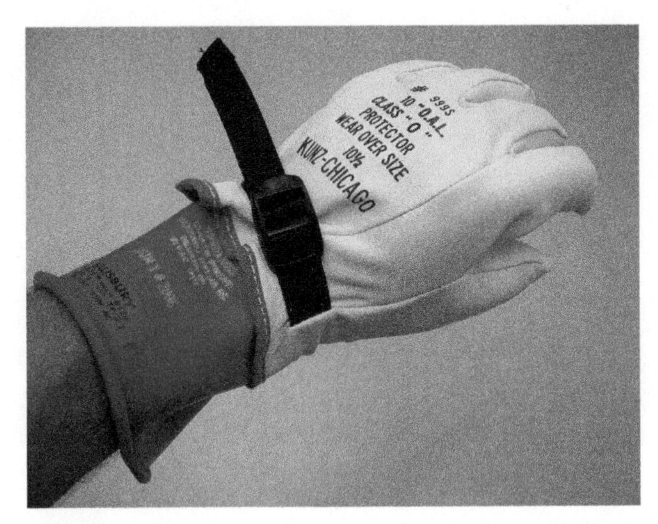

FIGURE 14–20 Appropriate personal protective equipment (PPE) must be worn whenever working on or around a hybrid vehicle high-voltage system.

TECH TIP

Check the Battery Condition First

A discharged or defective battery has lower voltage potential than a good battery that is at least 75% charged. This lower battery voltage cannot properly power the starter motor. A weak battery could also prevent the charging voltage from reaching the voltage regulator cutoff point. This lower voltage could be interpreted as indicating a defective alternator and/or voltage regulator. If the vehicle continues to operate with low system voltage, the stator winding in the alternator can be overheated, causing alternator failure.

follow the exact safety instruction as stated in service information before working around any of the high-voltage components or wiring. ● **SEE FIGURE 14–19.**

HIGH-VOLTAGE BATTERY SERVICE

During normal vehicle operation, the charge and discharge cycles of the high-voltage battery in an HEV are monitored and controlled by a separate battery module. This module monitors battery temperature, current, and voltage to calculate SOC and determine at what rate the battery should be charged. While dealerships sometimes have a special high-voltage battery charger for recharging HEV battery packs, the best charger is the vehicle itself. If the HV battery in an HEV becomes discharged, the first step ought to be getting the vehicle started to recharge the battery pack. The procedure will vary depending on the model in question. Always follow the manufacturer's specified procedures when starting a disabled hybrid vehicle.

HIGH-VOLTAGE BATTERY SAFETY PRECAUTIONS

Always keep in mind that the high-voltage batteries for an HEV can produce sufficient voltage and current to severely injure or kill. Always wear appropriate personal protective equipment (PPE) and use approved safety procedures when working around these batteries. ● **SEE FIGURE 14–20.** Precautions include the following steps:

- Do not work on the vehicle if moisture is present on the skin or anywhere on or near the vehicle.
- If service must be performed on the hybrid system, be sure to disconnect the HV battery and allow enough time for system capacitors to discharge before proceeding.
- If an electrical fire occurs, do not attempt to extinguish it using water. Use an ABC fire extinguisher or wait for firefighters to deal with it.

FIGURE 14–21 A battery service warning label from GM Two-Mode vehicle. (Courtesy of Jeffrey Rehkopf)

- ALWAYS refer to the service manual for approved safety procedures when handling the HV battery pack.
- The battery case contains liquid potassium hydroxide, a strong alkali solution. Any liquid around the battery should be checked with litmus paper to determine if it is an electrolyte spill. If an electrolyte spill has occurred, be sure to disable the HV system, and then use a mixture of vinegar and water to neutralize the solution before cleaning up with soap and water.
- Remove any clothing that has come into contact with electrolyte and flush any exposed skin with large amounts of water. If electrolyte comes in contact with the eyes, flush with large amounts of water, but do not use a neutralizing solution. Be sure to seek medical advice to prevent further injury from electrolyte contact.
- Read all warning labels and always follow the vehicle manufacturer's instructions. ● **SEE FIGURE 14–21**.

BATTERY SYMPTOM GUIDE

The following list will assist technicians in troubleshooting batteries.

Problem	Possible Causes and/or Solutions
1. Headlights are dimmer than normal.	1. Discharged battery or poor connections on the battery, engine, or body
2. Solenoid clicks.	2. Discharged battery or poor connections on the battery or an engine fault, such as coolant on top of the pistons, causing a hydrostatic lock
3. Engine is slow in cranking.	3. Discharged battery, high-resistance battery cables, or defective starter or solenoid
4. Battery will not accept a charge.	4. Possible loose battery cable connections (If the battery is a maintenance-free type, attempt to fast-charge the battery for several hours. If the battery still will not accept a charge, replace the battery.)
5. Battery is using water.	5. Check charging system for too high a voltage (If the voltage is normal, the battery is showing signs of gradual failure. Load-test and replace the battery, if necessary.)

SUMMARY

1. All batteries should be securely attached to the vehicle with hold-down brackets to prevent vibration damage.
2. Batteries can be tested with a voltmeter to determine the state of charge. A battery load test loads the battery to half of its CCA rating. A good battery should be able to maintain higher than 9.6 volts for the entire 15-second test period.
3. Batteries can be tested with a conductance tester even if discharged.

4. A battery drain test should be performed if the battery runs down.
5. Be sure that the battery charger is unplugged from power outlet when making connections to a battery.
6. Hybrid electric vehicles are equipped with a high-voltage disconnect switch or connector.
7. Appropriate personal protective equipment (PPE) must be worn whenever working on or around a hybrid vehicle high-voltage system.

REVIEW QUESTIONS

1. What are the results of a voltmeter test of a battery and its state of charge?
2. What are the steps for performing a battery load test?
3. How is a battery drain test performed?
4. Why should a battery not be fast charged?
5. What should the technician be aware of when working around hybrid vehicle batteries?

CHAPTER QUIZ

1. Technician A says that distilled or clean drinking water should be added to a battery when the electrolyte level is low. Technician B says that fresh electrolyte (solution of acid and water) should be added. Which technician is correct?
   a. Technician A only
   b. Technician B only
   c. Both Technicians A and B
   d. Neither Technician A nor B
2. All batteries should be in a secure bracket that is bolted to the vehicle to prevent physical damage to the battery.
   a. True **b.** False
3. A battery date code sticker indicates D6. What does this mean?
   a. The date it was shipped from the factory was December 2006.
   b. The date it was shipped from the factory was April 2006.
   c. The battery expires in December 2002.
   d. It was built the second day of the week (Tuesday).
4. Many vehicle manufacturers recommend that a special electrical connector be installed between the battery and the battery cable when testing for ________.
   a. Battery drain (parasitic drain)
   b. Specific gravity
   c. Battery voltage
   d. Battery charge rate
5. When load testing a battery, which battery rating is often used to determine how much load to apply to the battery?
   a. CA **c.** MCA
   b. RC **d.** CCA
6. When measuring the specific gravity of the electrolyte, the maximum allowable difference between the highest and lowest hydrometer reading is ________.
   a. 0.010 **c.** 0.050
   b. 0.020 **d.** 0.50
7. A battery high-rate discharge (load capacity) test is being performed on a 12-volt battery. Technician A says that a good battery should have a voltage reading of higher than 9.6 volts while under load at the end of the 15-second test. Technician B says that the battery should be discharged (loaded) to twice its CCA rating. Which technician is correct?
   a. Technician A only
   b. Technician B only
   c. Both Technicians A and B
   d. Neither Technician A nor B
8. When charging a lead-acid (flooded-type) battery, ________.
   a. The initial charging rate should be about 35 amperes for 30 minutes
   b. The battery may not accept a charge for several hours, yet may still be a good (serviceable) battery
   c. The battery temperature should not exceed 125°F (hot to the touch)
   d. All of the above
9. Normal battery drain (parasitic drain) in a vehicle with many computer and electronic circuits is ________.
   a. 20 to 30 milliamperes
   b. 2 to 3 amperes
   c. 150 to 300 milliamperes
   d. None of the above
10. When jump-starting, ________.
    a. The last connection should be the positive post of the dead battery
    b. The last connection should be the engine block of the dead vehicle
    c. The alternator must be disconnected on both vehicles
    d. Both a and c

chapter 15

CRANKING SYSTEM

LEARNING OBJECTIVES

After studying this chapter, the reader will be able to:

1. Describe the parts and operation of a cranking circuit.
2. Discuss how a starter motor converts electrical power into mechanical power.
3. List the different types of starters.
4. Describe the purpose and function of starter drives.

This chapter will help you prepare for the ASE Electrical/Electronic Systems (A6) certification test content area "C" (Starting System Diagnosis and Repair).

KEY TERMS

Armature 209
Brush-end housing 208
Brushes 210
CEMF 207
Commutator-end housing 208
Commutator segments 210
Compression spring 213
Drive-end housing 208
Field coils 208
Field housing 208
Field poles 208
Ground brushes 210
Hold-in winding 214
Insulated brushes 210
Mesh spring 213
Neutral safety switch 203
Overrunning clutch 213
Pole shoes 208
Pull-in winding 214
RVS 205
Starter drive 212
Through bolts 208

 STC OBJECTIVES

GM Service Technical College topics covered in this chapter are as follows:

1. GM starting systems, overview, and operation.
2. Identify starting system characteristics.

FIGURE 15–1 A typical solenoid-operated starter.

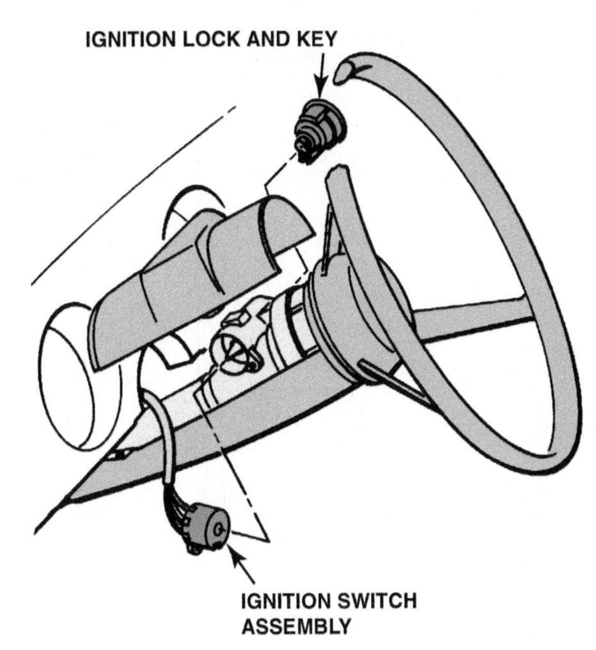

FIGURE 15–2 Some column-mounted ignition switches act directly on the electrical ignition switch itself, whereas others use a link from the lock cylinder to the ignition switch.

CRANKING CIRCUIT

PARTS INVOLVED For any engine to start, it must first be rotated using an external power source. It is the purpose and function of the cranking circuit to create the necessary power and transfer it from the battery to the starter motor, which rotates the engine.

The cranking circuit includes those mechanical and electrical components that are required to crank the engine for starting. The cranking force in the early 1900s was the driver's arm, because the driver had to physically crank the engine until it started. Modern cranking circuits include the following:

1. **Starter motor.** The starter is normally a 0.5 to 2.6 horsepower (0.4 to 2 kilowatts) electric motor that can develop nearly 8 horsepower (6 kilowatts) for a very short time when first cranking a cold engine. ● **SEE FIGURE 15–1**.
2. **Battery.** The battery must be of the correct capacity and be at least 75% charged to provide the necessary current and voltage for correct starter operation.
3. **Starter solenoid or relay.** The high current required by the starter must be able to be turned on and off. A large switch would be required if the current were controlled by the driver directly. Instead, a small current switch (ignition switch) operates a solenoid or relay that controls the high current to the starter.
4. **Starter drive.** The starter drive uses a small pinion gear that contacts the engine flywheel gear teeth and transmits starter motor power to rotate the engine.
5. **Ignition switch.** The ignition switch and safety control switches control the starter motor operation. ● **SEE FIGURE 15–2**.

CONTROL CIRCUIT PARTS AND OPERATION The engine is cranked by an electric motor that is controlled by a key-operated ignition switch. The ignition switch will not operate the starter unless the automatic transmission is in neutral or park, or the clutch pedal is depressed on manual transmission/transaxle vehicles. This is to prevent any accident that might result from the vehicle moving forward or rearward when the engine is started. The types of controls that are used to be sure that the vehicle will not move when being cranked include the following:

- Most automobile manufacturers use an electric switch called a **neutral safety switch**, which opens the circuit between the ignition switch and the starter to prevent starter motor operation, unless the gear selector is in neutral or park. The safety switch can be attached either to the steering column inside the vehicle near the floor or on the side of the transmission.
- Some manufacturers use a mechanical blocking device in the steering column to prevent the driver from turning the key switch to the start position unless the gear selector is in neutral or park.
- Manual transmission vehicles also use a safety switch to permit cranking only if the clutch is depressed. This switch is commonly called the *clutch safety switch*. ● **SEE FIGURE 15–3**.

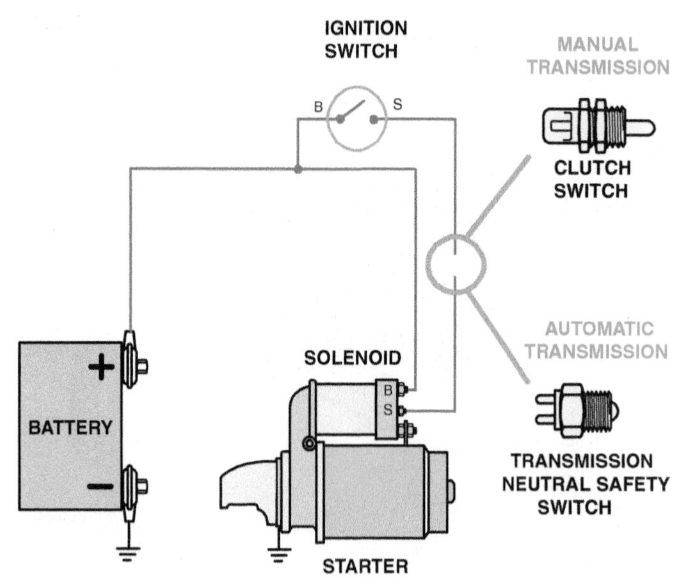

FIGURE 15–3 To prevent the engine from cranking, an electrical switch is usually installed to open the circuit between the ignition switch and the starter solenoid. (Older vehicles)

COMPUTER-CONTROLLED STARTING

OPERATION Most key-operated ignition systems and push-button-to-start systems are now using the body control module (BCM) and the engine control module (ECM) to crank the engine. The ignition switch start position or the push-to-start button is used as an input signal to the BCM. The BCM uses the vehicle data network to request starting from the ECM. **SEE FIGURE 15–4.**

Before the ECM cranks the engine, a number of conditions must be met that may include the following:

- The brake pedal is depressed.
- The gear selector is in park or neutral.
- The correct key fob (code) is present in the vehicle.

A typical vehicle start-up follows this sequence:

- The ignition key can be turned to the start position (or the START button pressed) then released, and the ECM cranks the engine (using the CRNK relay) until it senses that the engine has started.

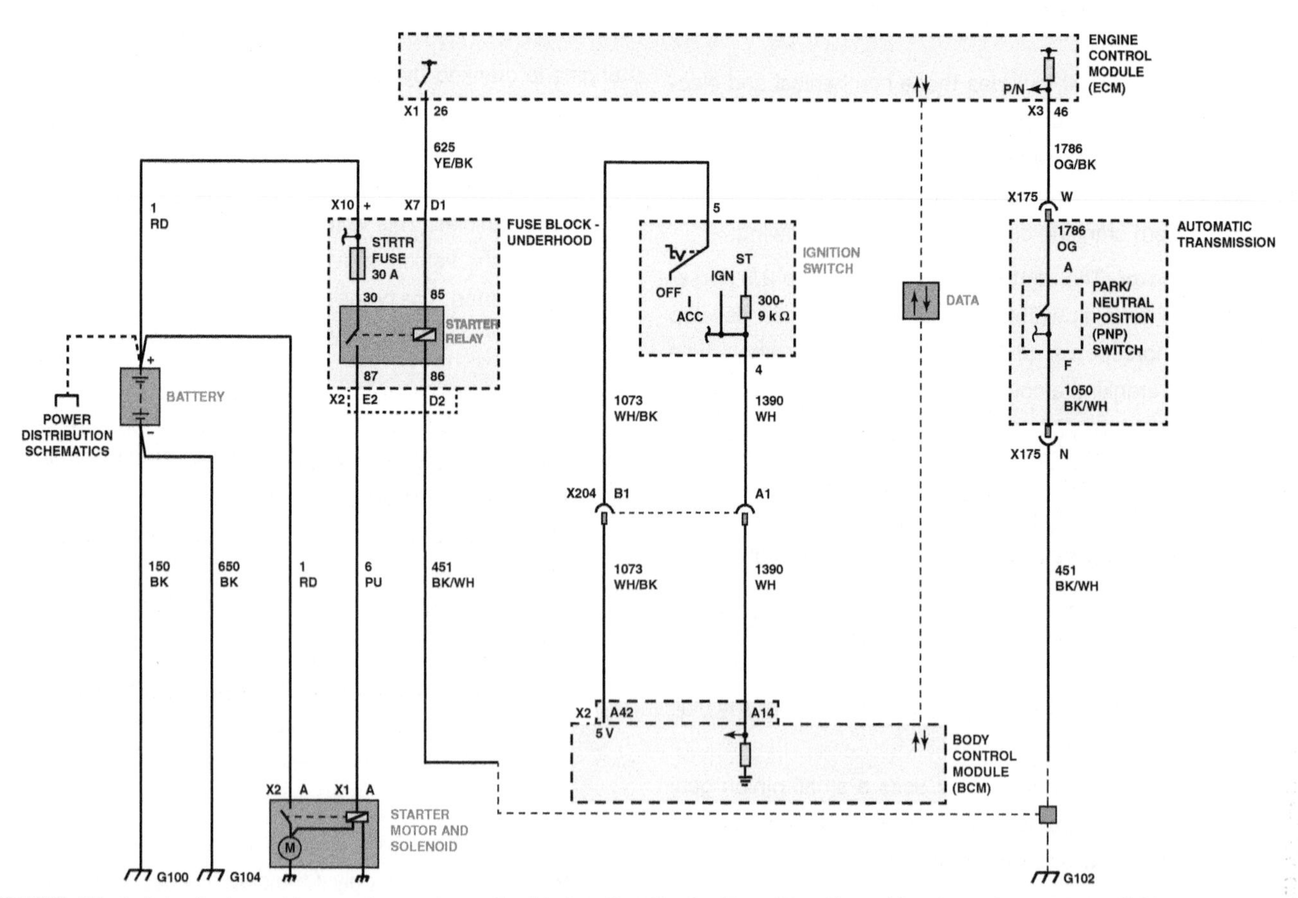

FIGURE 15–4 A typical cranking system schematic. Notice that the ignition "Start" position is an input to the BCM, which informs the ECM of the "Start" request. According to programming, the ECM then energizes the starter relay to operate the starter. (Courtesy of General Motors)

FIGURE 15–5 Instead of using an ignition key to start the engine, some vehicles are using a start button which is also used to stop the engine. (Courtesy of General Motors)

FIGURE 15–6 The top button on this key fob is the remote start button.

- The ECM can detect that the engine has started by looking at the engine speed signal (RPM).
- Normal cranking speed can vary between 100 and 250 RPM. If the engine speed exceeds 400 RPM, the ECM determines that the engine has started and opens the circuit to the relay, which in turn opens the "S" (start) terminal of the starter solenoid that stops the starter motor.

Computer-controlled starting is almost always part of the system if a push-button start is used. ● **SEE FIGURE 15–5.**

REMOTE STARTING Remote starting, sometimes called **remote vehicle start (RVS)**, is a system that allows the driver to start the engine of the vehicle from inside the house or a building at a distance of about 200 ft (65 m). The doors remain locked to reduce the possibility of theft. This feature allows the heating or air-conditioning system to start before the driver arrives. ● **SEE FIGURE 15–6.**

FIGURE 15–7 The key fob pocket on this vehicle is located in the center console. (Courtesy of General Motors)

TECH TIP

No Start Due to Key Fob Low Battery

During the normal operation or servicing of vehicles with the Passive Entry Passive Start (PEPS), or push button start, option, it may be necessary to place the remote transmitter (key fob) in the transmitter pocket in the vehicle.

Some instances in which the remote transmitter may need to be placed in the transmitter pocket include when:

- Programming new or existing remote transmitters
- The "No Remote Detected" message is displayed on the Driver Information Center
- A loss of RF communication with the remote transmitter is caused by interference
- The remote transmitter battery is low

Depending on the vehicle, the location of the transmitter pocket varies. Specific remote transmitter placement instructions must be followed to ensure communication is established between the remote transmitter and the vehicle. ● **SEE FIGURE 15–7.**

NOTE: Most remote start systems will turn off the engine after 10 minutes of run time unless reset by using the remote.

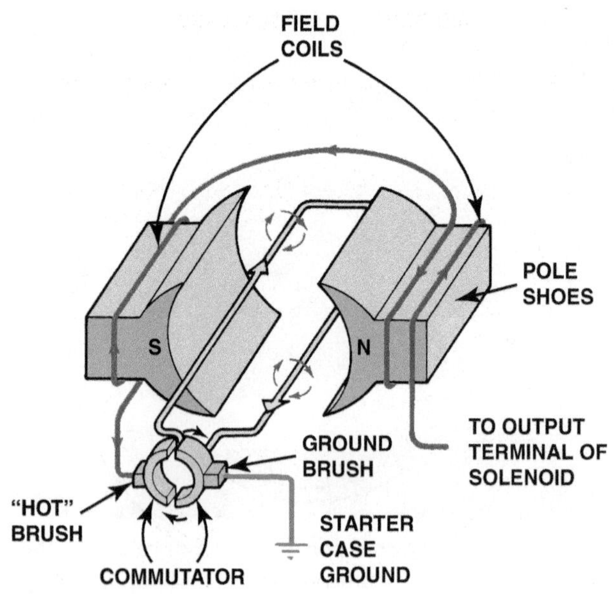

FIGURE 15–8 This series-wound electric motor shows the basic operation with only two brushes: one hot brush and one ground brush. The current flows through both field coils, then through the hot brush and the loop winding of the armature, before reaching ground through the ground brush.

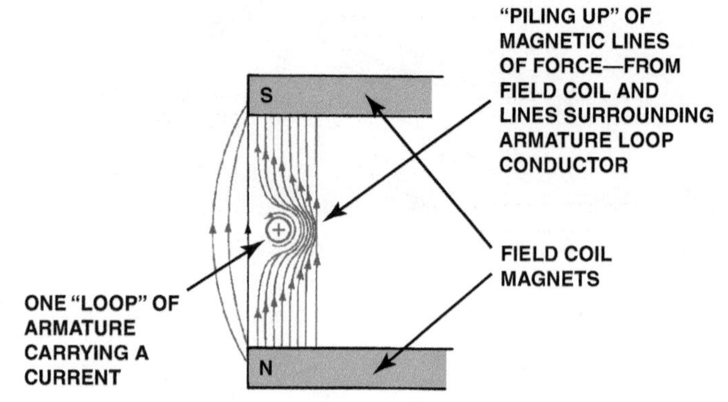

FIGURE 15–9 The interaction of the magnetic fields of the armature loops and field coils creates a stronger magnetic field on the right side of the conductor, causing the armature loop to move toward the left.

STARTER MOTOR OPERATION

PRINCIPLES A starter motor uses electromagnetic principles to convert electrical energy from the battery (up to 300 amperes) to mechanical power (up to 8 horsepower [6 kilowatts]) to crank the engine. Current for the starter motor or power circuit is controlled by a solenoid or relay, which is itself controlled by the driver-operated ignition switch.

The current travels through the brushes and into the armature windings, where other magnetic fields are created around each copper wire loop in the armature. The two strong magnetic fields created inside the starter housing create the force that rotates the armature.

Inside the starter housing is a strong magnetic field created by the field coil magnets. The armature, a conductor, is installed inside this strong magnetic field, with little clearance between the armature and the field coils.

The two magnetic fields act together, and their lines of force "bunch up" or are strong on one side of the armature loop wire and become weak on the other side of the conductor. This causes the conductor (armature) to move from the area of strong magnetic field strength toward the area of weak magnetic field strength. ● **SEE FIGURES 15–8 AND 15–9.**

The difference in magnetic field strength causes the armature to rotate. This rotation force (torque) is increased as the current flowing through the starter motor increases. The torque of a starter is determined by the strength of the magnetic fields inside the starter. Magnetic field strength is measured in ampere-turns. If the current or the number of turns of wire is increased, the magnetic field strength is increased.

The magnetic field of the starter motor is provided by two or more pole shoes and field windings. The pole shoes are made of iron and are attached to the frame with large screws or bonding material. ● **SEE FIGURE 15–10.**

● **FIGURE 15–11** shows the paths of magnetic flux lines within a four-pole motor.

The field windings are usually made of a heavy copper ribbon to increase their current-carrying capacity and electromagnetic field strength. ● **SEE FIGURE 15–12.**

Automotive starter motors usually have four pole shoes with four copper field windings to provide a strong magnetic field within the motor. Some starters use four strong permanent magnets instead of copper windings.

SERIES MOTORS A series motor develops its maximum torque at the initial start (0 RPM) and develops less torque as the speed increases.

- A series motor is commonly used for an automotive starter motor because of its high starting power characteristics.

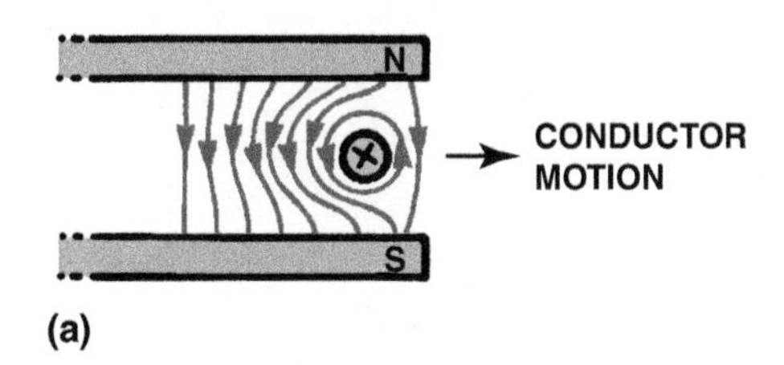

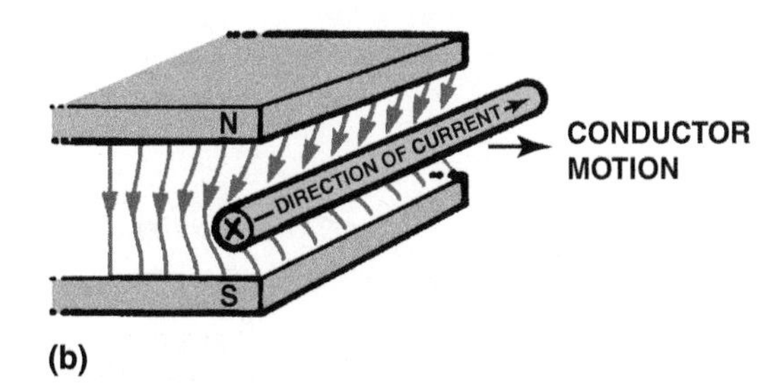

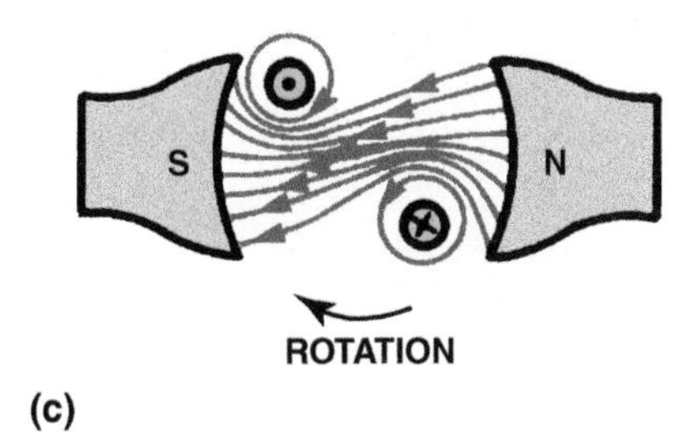

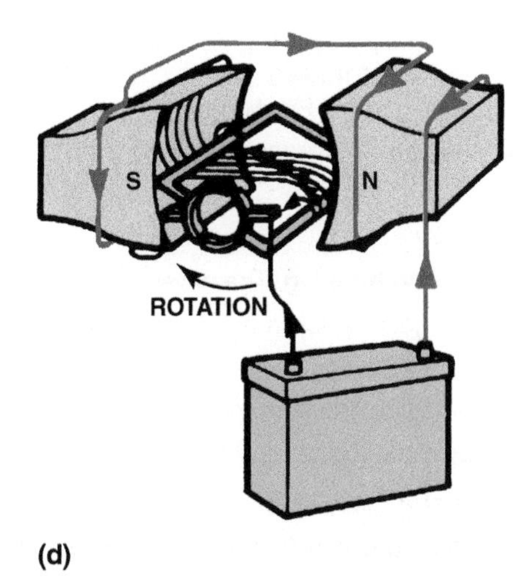

FIGURE 15–10 The armature loops rotate due to the difference in the strength of the magnetic field. The loops move from a strong magnetic field strength toward a weaker magnetic field strength.

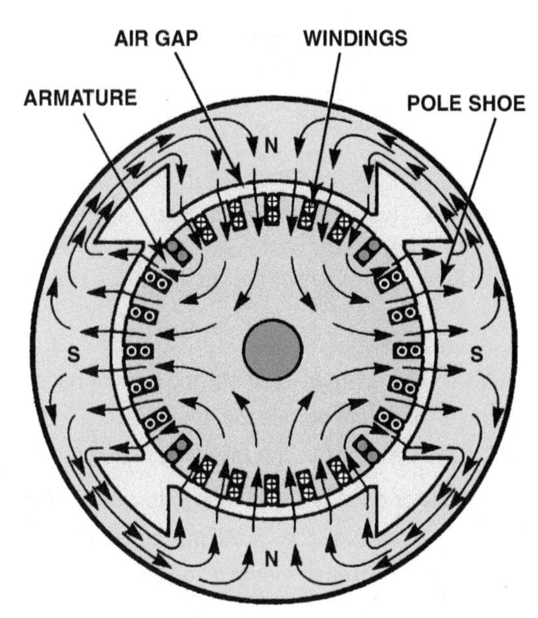

FIGURE 15–11 Magnetic lines of force in a four-pole motor.

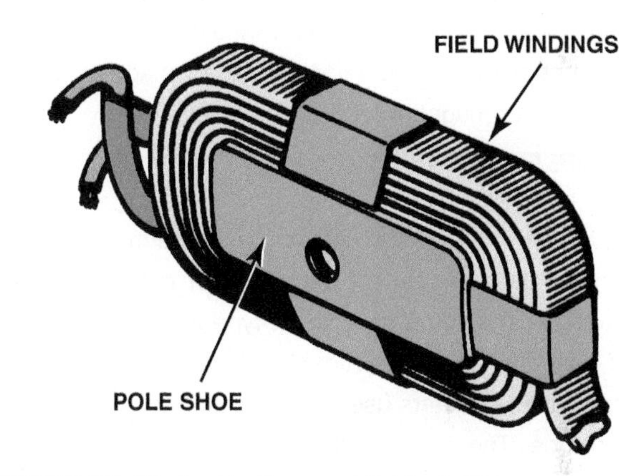

FIGURE 15–12 A pole shoe and field winding.

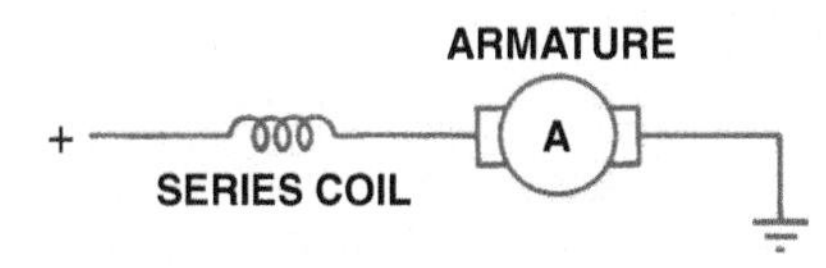

FIGURE 15–13 This wiring diagram illustrates the construction of a series-wound electric motor. Notice that all current flows through the field coils, then through the armature (in series) before reaching ground.

- A series starter motor develops less torque at high RPM, because a current is produced in the starter itself that acts against the current from the battery. Because this current works against battery voltage, it is called **counterelectromotive force**, or **CEMF**. This CEMF is produced by electromagnetic induction in the armature conductors, which are cutting across the magnetic lines of force formed by the field coils. This induced voltage operates against the applied voltage supplied by the battery, which reduces the strength of the magnetic field in the starter.
- Because the power (torque) of the starter depends on the strength of the magnetic fields, the torque of the starter decreases as the starter speed increases. A series-wound starter also draws less current at higher speeds and will keep increasing in speed under light loads. This could lead to the destruction of the starter motor unless controlled or prevented. ● **SEE FIGURE 15–13.**

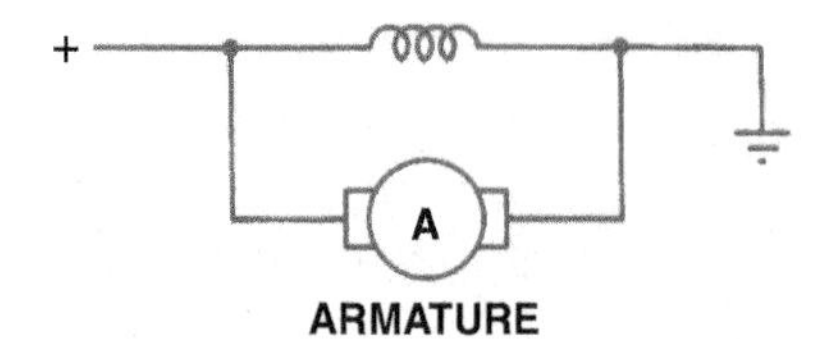

FIGURE 15–14 This wiring diagram illustrates the construction of a shunt-type electric motor, and shows the field coils in parallel (or shunt) across the armature.

SHUNT MOTORS Shunt-type electric motors have the field coils in parallel (or shunt) across the armature.

A shunt-type motor has the following features.

- A shunt motor does not decrease in torque at higher motor RPM, because the CEMF produced in the armature does not decrease the field coil strength.
- A shunt motor, however, does not produce as high a starting torque as that produced by a series-wound motor, and is not used for starters. Some small electric motors, such as those used for windshield wiper, use a shunt motor but most use permanent magnets rather than electromagnets.

● **SEE FIGURE 15–14.**

PERMANENT MAGNET MOTORS A permanent magnet (PM) starter uses permanent magnets that maintain constant field strength, the same as a shunt-type motor, so they have similar operating characteristics. To compensate for the lack of torque, all PM starters use gear reduction to multiply starter motor torque. The permanent magnets used are an alloy of neodymium, iron, and boron, and are almost 10 times more powerful than previously used permanent magnets.

COMPOUND MOTORS A compound-wound, or compound, motor has the operating characteristics of a series motor *and* a shunt-type motor, because some of the field coils are connected to the armature in series and some (usually only one) are connected directly to the battery in parallel (shunt) with the armature. The shunt-wound field coil is called a shunt coil and is used to limit the maximum speed of the starter. ● **SEE FIGURE 15–15.**

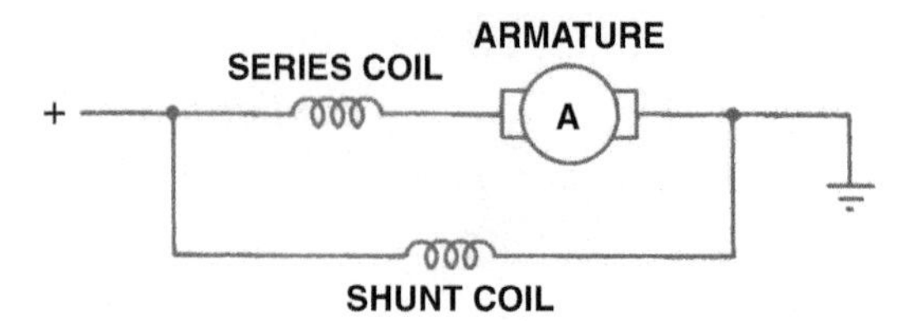

FIGURE 15–15 A compound motor is a combination of series and shunt types, using part of the field coils connected electrically in series with the armature and part in parallel (shunt).

HOW THE STARTER MOTOR WORKS

PARTS INVOLVED A starter consists of the main structural support of a starter called the main **field housing**, one end of which is called a **commutator-end** (or **brush-end) housing** and the other end a **drive-end housing**. The drive-end housing contains the drive pinion gear, which meshes with the engine flywheel gear teeth to start the engine. The commutator-end plate supports the end containing the starter brushes. **Through bolts** hold the three components together. ● **SEE FIGURE 15–16.**

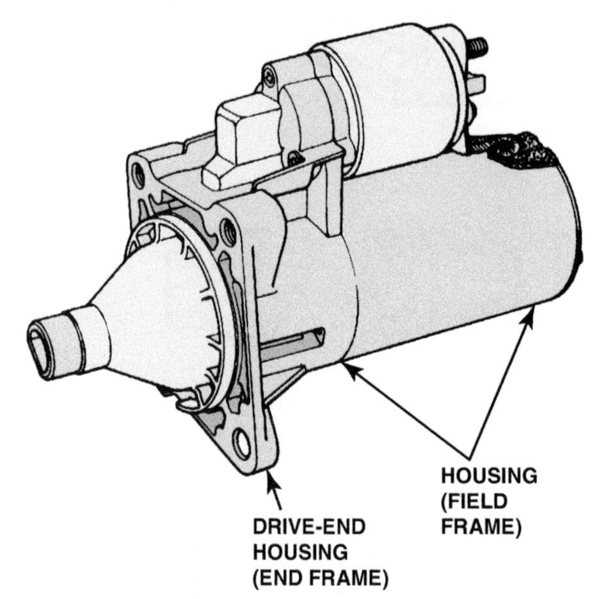

FIGURE 15–16 A typical starter motor showing the drive-end housing.

- **Field coils.** The steel housing of the starter motor contains permanent magnets or four electromagnets that are connected directly to the positive post of the battery to provide a strong magnetic field inside the starter. The four electromagnets use heavy copper or aluminum wire wrapped around a soft-iron core, which is contoured to fit against the rounded internal surface of the starter frame. The soft-iron cores are called **pole shoes**. Two of the four pole shoes are wrapped with copper wire in one direction to create a north pole magnet, and the other two pole shoes are wrapped in the opposite direction to create a south pole magnet. These magnets, when energized, create strong magnetic fields inside the starter housing and, therefore, are called **field coils**. The soft-iron cores (pole shoes) are often called **field poles.** ● **SEE FIGURE 15–17.**

FIGURE 15–17 Pole shoes and field windings installed in the housing. (Courtesy of Jeffrey Rehkopf)

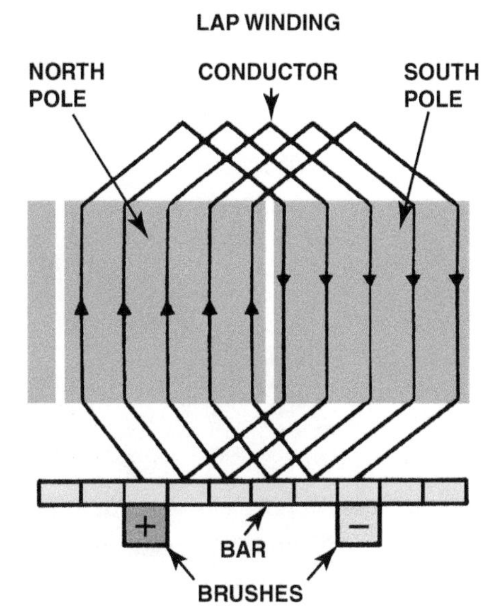

FIGURE 15–19 An armature showing how its copper wire loops are connected to the commutator.

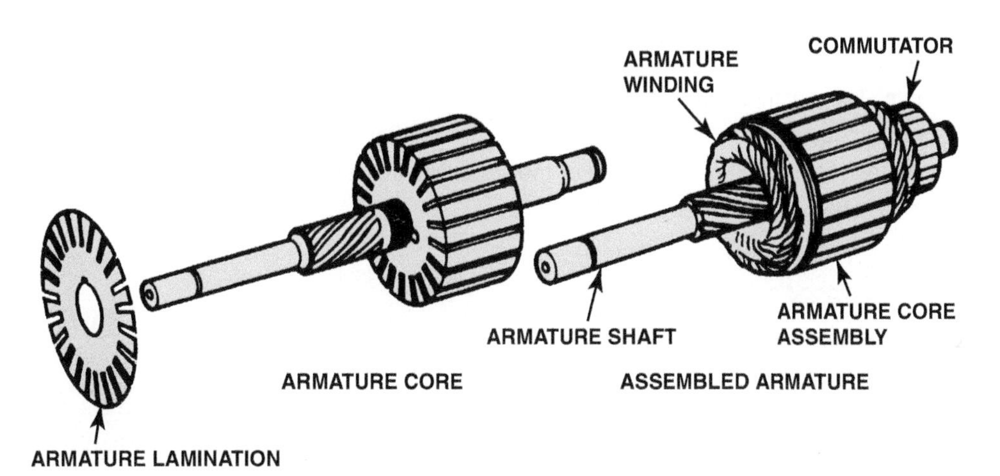

FIGURE 15–18 A typical starter motor armature. The armature core is made from thin sheet metal sections assembled on the armature shaft, which is used to increase the magnetic field strength.

- **Armature.** Inside the field coils is an **armature** that is supported with either bushings, needle bearings, or ball bearings at both ends, which permit it to rotate. The armature is constructed of thin, circular disks of steel laminated together and wound lengthwise with heavy-gauge insulated copper wire. The laminated iron core supports the copper loops of wire and helps concentrate the magnetic field produced by the coils. ● **SEE FIGURE 15–18.**

Insulation between the laminations helps to increase the magnetic efficiency in the core. For reduced resistance, the armature conductors are made of a thick copper wire. The two ends of each conductor are attached to two adjacent commutator bars.

The commutator is made of copper bars insulated from each other by mica or some other insulating material. ● **SEE FIGURE 15–19.**

The armature core, windings, and commutator are assembled on a long armature shaft. This shaft also carries the pinion gear that meshes with the engine flywheel ring gear. ● **SEE FIGURE 15–20.**

STARTER BRUSHES To supply the proper current to the armature, a four-pole motor must have four brushes riding on the commutator. Most automotive starters have two grounded and two insulated brushes, which are held against the commutator by spring force.

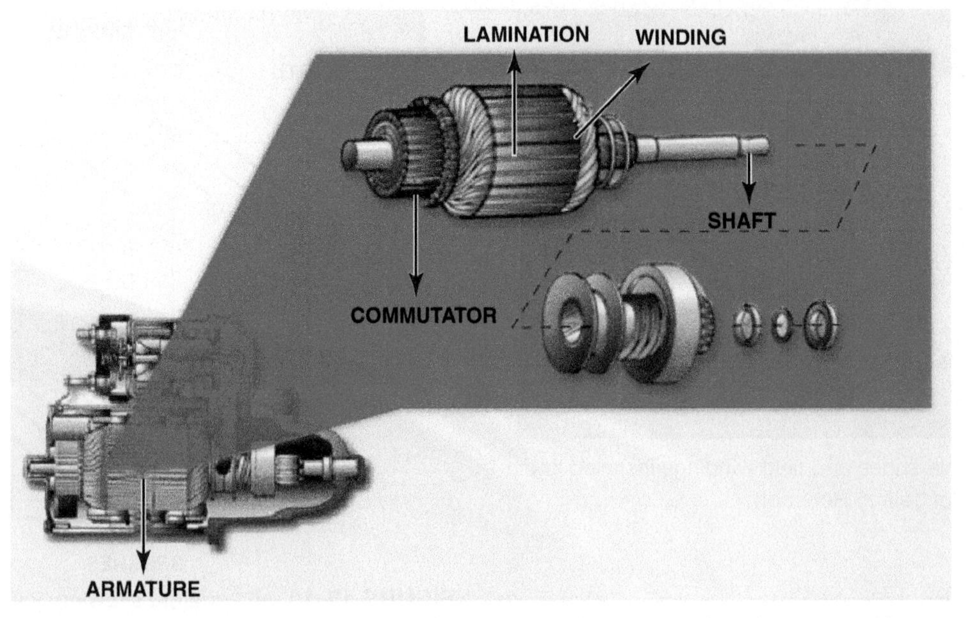

FIGURE 15–20 Starter armature, showing the starter drive (overrunning clutch) and drive pinion gear. (Courtesy of General Motors)

The ends of the copper armature windings are soldered to **commutator segments**. The electrical current that passes through the field coils is then passed to the commutator of the armature by brushes that can move over the segments of the rotating armature. These **brushes** are made of a combination of copper and carbon.

- The copper used here is a good conductor material.
- The carbon added to the starter brushes helps provide the graphite-type lubrication needed to reduce wear of the brushes and the commutator segments.

The starter uses four brushes—two brushes to transfer the current from the field coils to the armature, and two brushes to provide the ground return path for the current that flows through the armature.

The two sets of brushes include:

1. Two **insulated brushes**, which are in holders and are insulated from the housing.
2. Two **ground brushes**, which use bare, stranded copper wire connections to the brushes. The ground brush holders are not insulated and attach directly to the field housing or brush-end housing.

● **SEE FIGURE 15–21**.

PERMANENT MAGNET FIELDS Permanent magnets are used in place of the electromagnetic field coils and pole shoes in many starters today. This eliminates the motor field circuit, which in turn eliminates the potential for field coil faults and other electrical problems.

The permanent magnet starter is smaller in size and weighs less than a wound field starter. It runs at a higher RPM and is usually combined with a planetary reduction gear that drives the starter drive gear assembly. ● **SEE FIGURE 15–22**.

Don't Hit That Starter!

In the past, it was common to see service technicians hitting a starter in their effort to diagnose a no-crank condition. Often the shock of the blow to the starter aligned or moved the brushes, armature, and bushings. Many times, the starter functioned after being hit, even if only for a short time.

However, most starters today use permanent magnet fields, and the magnets can be easily broken if hit. A magnet that is broken becomes two weaker magnets. Some early permanent magnet starters used magnets that were glued or bonded to the field housing. If struck with a heavy tool, the magnets could be broken with parts of the magnet falling onto the armature and into the bearing pockets, making the starter impossible to repair or rebuild. ● **SEE FIGURE 15–23**.

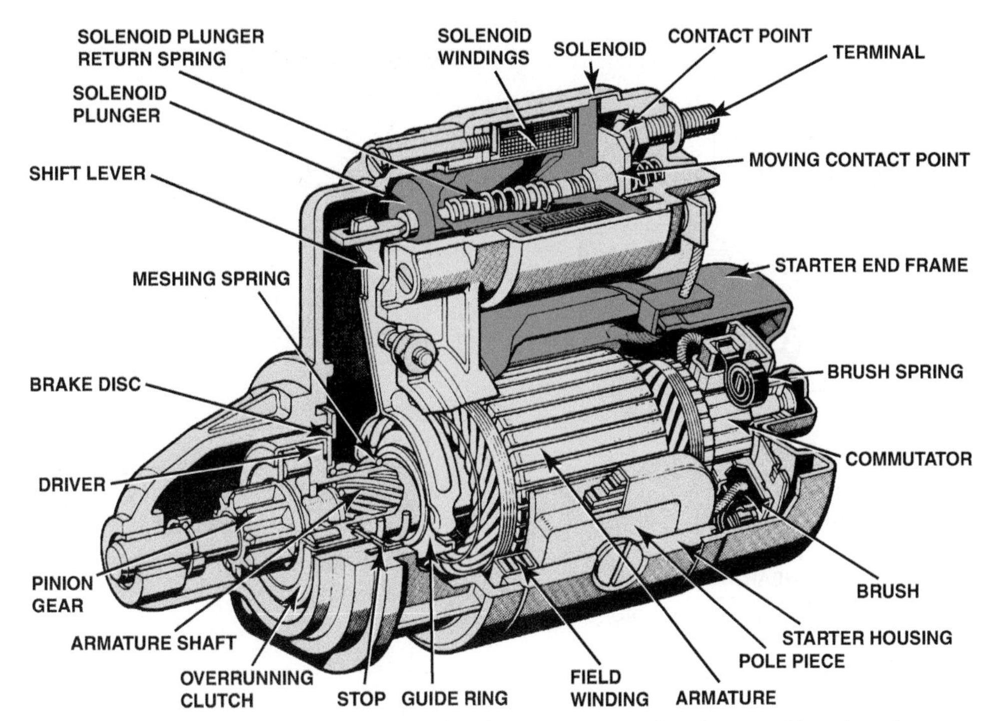

FIGURE 15–21 A cutaway of a typical starter motor showing the commutator, brushes, and brush spring.

FIGURE 15–22 The permanent magnet starter uses "Magnaquench" pole shoes instead of wire wound pole shoes. (Courtesy of General Motors)

FIGURE 15–23 This starter permanent magnet field housing was ruined when someone used a hammer on the field housing in an attempt to "fix" a starter that would not work. A total replacement is the only solution in this case.

GEAR-REDUCTION STARTERS

PURPOSE AND FUNCTION Gear-reduction starters are used by many automotive manufacturers. The purpose of the gear reduction (typically 2:1 to 4:1) is to increase starter motor speed and provide the torque multiplication necessary to crank an engine. ● **SEE FIGURE 15–24.**

As a series-wound motor increases in rotational speed, the starter produces less power, and less current is drawn from the battery because the armature generates greater CEMF as the starter speed increases. However, a starter motor's maximum torque occurs at 0 RPM and torque decreases with increasing RPM.

A smaller starter using a gear-reduction design can produce the necessary cranking power with reduced starter amperage requirements. Lower current requirements mean that smaller battery cables can be used. Many permanent magnet starters use a planetary gear set (a type of gear reduction) to provide the necessary torque for starting. ● **SEE FIGURE 15–25.**

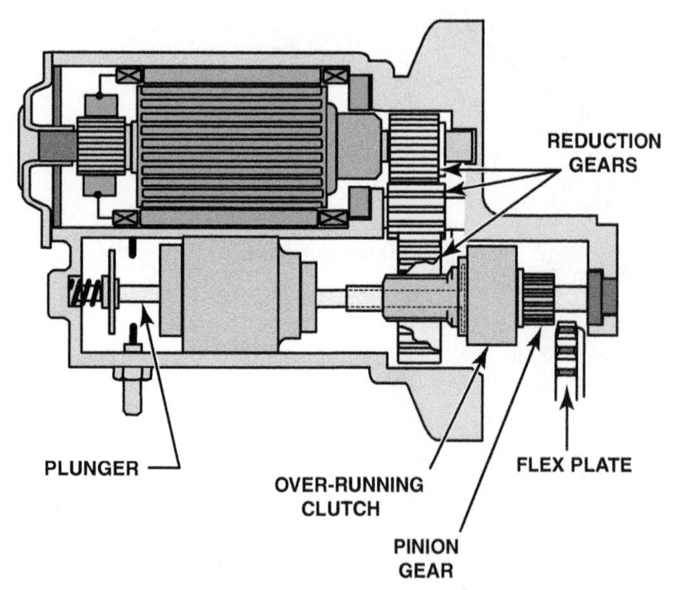

FIGURE 15–24 An off-set gear reduction starter motor.

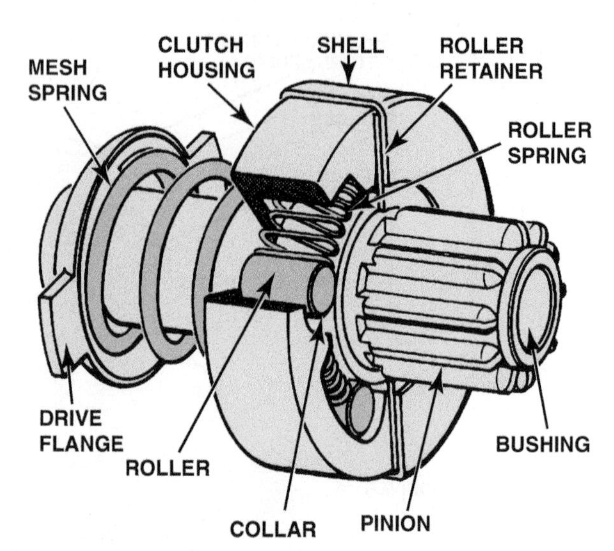

FIGURE 15–26 A cutaway of a typical starter drive showing all of the internal parts.

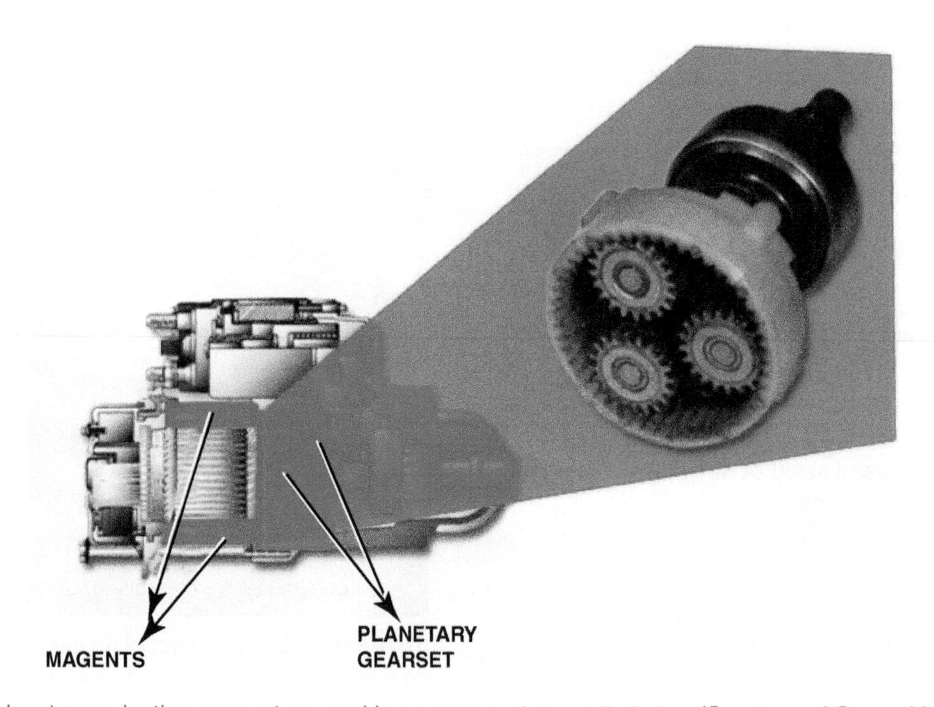

FIGURE 15–25 A planetary reduction gear set as used in a permanent magnet starter. (Courtesy of General Motors)

STARTER DRIVES

PURPOSE AND FUNCTION A **starter drive** includes a small pinion gear that meshes with and rotates the larger ring gear on the engine flywheel or flex plate for starting. The pinion gear must engage with the engine gear slightly *before* the starter motor rotates, to prevent serious damage to either the starter gear or the engine, but must be disengaged after the engine starts. The ends of the starter pinion gear are tapered to help the teeth mesh more easily without damaging the flywheel ring gear teeth. ● **SEE FIGURE 15–26.**

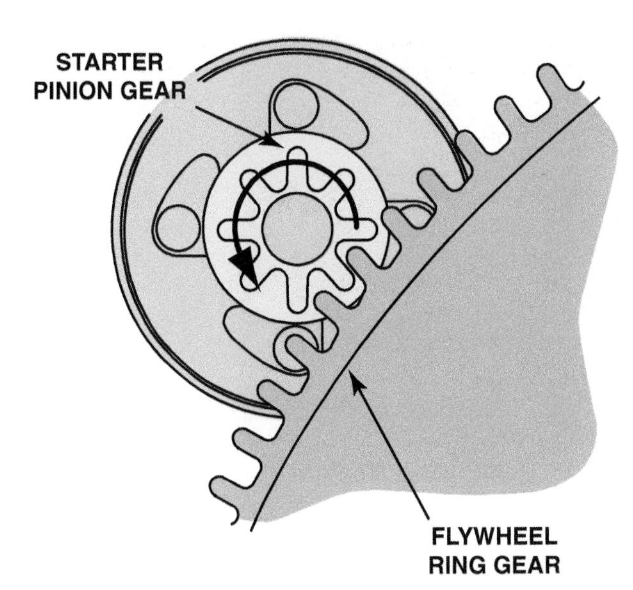

FIGURE 15–27 The ring gear to pinion gear ratio is usually 15:1 to 20:1.

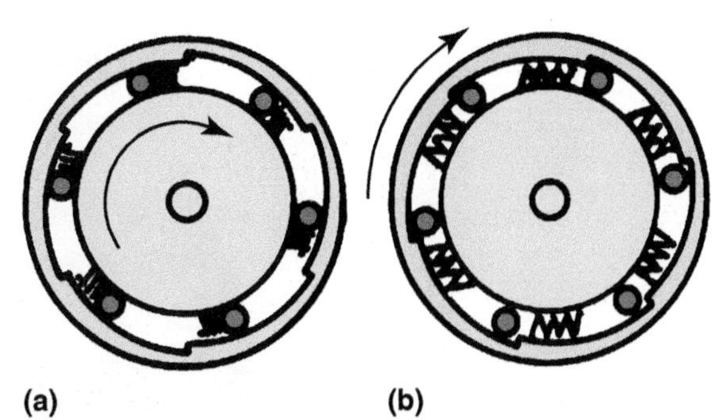

FIGURE 15–28 Operation of the overrunning clutch. (a) Starter motor is driving the starter pinion and cranking the engine. The rollers are wedged against spring force into their slots. (b) The engine has started and is rotating faster than the starter armature. Spring force pushes the rollers so they can rotate freely.

STARTER DRIVE GEAR RATIO The ratio of the number of teeth on the engine ring gear to the number on the starter pinion is between 15:1 and 20:1. A typical small starter pinion gear has 9 teeth that turn an engine ring gear with 166 teeth. This provides an 18:1 gear reduction; thus, the starter motor is rotating approximately 18 times faster than the engine. Normal cranking speed for the engine is 200 RPM (varies from 70 to 250 RPM). This means that the starter motor speed is 18 times faster, or 3600 starter RPM ($200 \times 18 = 3600$). If the engine starts and is accelerated to 2000 RPM (normal cold engine speed), the starter will be destroyed by the high speed (36,000 RPM) if the starter was not disengaged from the engine. ● **SEE FIGURE 15–27.**

STARTER DRIVE OPERATION All starter drive mechanisms use a type of one-way clutch that allows the starter to rotate the engine, but then turns freely if the engine speed is greater than the starter motor speed. This clutch, called an **overrunning clutch**, protects the starter motor from damage if the ignition switch is held in the start position after the engine starts. The overrunning clutch, which is built in as a part of the starter drive unit, uses steel balls or rollers installed in tapered notches. ● **SEE FIGURE 15–28.**

This taper forces the balls or rollers tightly into the notch, when rotating in the direction necessary to start the engine. When the engine rotates faster than the starter pinion, the balls or rollers are forced out of the narrow tapered notch, allowing the pinion gear to turn freely (overrun).

The spring between the drive tang or pulley and the overrunning clutch and pinion is called a **mesh spring**. It helps to cushion and control the engagement of the starter drive pinion with the engine flywheel gear. This spring is also called a **compression spring**, because the starter solenoid or starter yoke compresses the spring and the spring tension causes the starter pinion to engage the engine flywheel.

FREQUENTLY ASKED QUESTION

What Is a Starter Bendix?

Older-model (1930s to 1960s) starters often used a drive mechanism, made by the Bendix Company, which used inertia to engage the starter pinion with the engine flywheel gear. This type of drive is also called an inertia drive.

On these older-model starters, the small starter pinion gear was attached to a shaft with threads, and the weight of this gear caused it to be spun along the threaded shaft and mesh with the flywheel whenever the starter motor spun. If the engine speed was greater than the starter speed, the pinion gear was forced back along the threaded shaft and out of mesh with the flywheel gear. The Bendix drive mechanism has generally not been used on light-duty vehicles since the early 1960s, but some technicians still use this term when discussing a starter drive.

FAILURE MODE A starter drive is generally a dependable unit and does not require replacement unless defective or worn. The major wear occurs in the overrunning clutch section of the starter drive unit. The steel balls or rollers wear and often do not wedge tightly into the tapered notches as is necessary for engine cranking. A worn starter drive can cause the starter

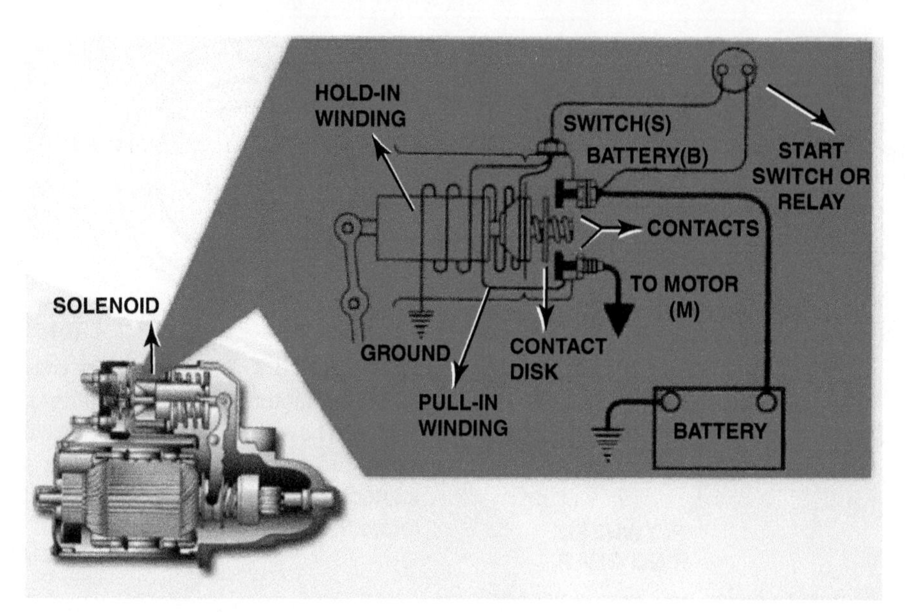

FIGURE 15–29 Wiring diagram of a typical starter solenoid. Notice that both the pull-in winding and the hold-in winding are energized when the ignition switch is first turned to the "start" position. As soon as the solenoid contact disk makes electrical contact with both the B and M terminals, the battery current is conducted to the starter motor and electrically neutralizes the pull-in winding. (Courtesy of General Motors)

motor to operate and then stop cranking the engine further creating a "whining" noise. The whine indicates that the starter motor is operating and that the starter drive is not rotating the engine flywheel. The entire starter drive is replaced as a unit. The overrunning clutch section of the starter drive cannot be serviced or repaired separately because the drive is a sealed unit. Starter drives are most likely to fail intermittently at first and then more frequently, until replacement becomes necessary to start the engine. Intermittent starter drive failure (starter whine) is often most noticeable during cold weather.

STARTER SOLENOID

SOLENOID OPERATION A starter solenoid is an electromagnetic switch containing two separate, but connected, electromagnetic windings and a high current switch. The switch may also be called a contactor or contact disk. The switch supplies battery voltage to the starter motor and at the same time a lever moves with the solenoid core and engages the starter drive.

SOLENOID WINDINGS The two internal windings contain approximately the same number of turns but are made from different-gauge wire. Both windings together produce a strong magnetic field that pulls a metal plunger into the solenoid. The plunger is attached to the starter drive through a shift fork lever. When the ignition switch is turned to the start position, the motion of the plunger into the solenoid causes the starter drive to move into mesh with the flywheel ring gear.

1. The heavier-gauge winding (called the **pull-in winding**) is needed to draw the plunger into the solenoid and is grounded through the starter motor brushes.
2. The lighter-gauge winding (called the **hold-in winding**), which is grounded through the starter frame, produces enough magnetic force to keep the plunger in position. The main purpose of using two separate windings is to permit as much current as possible to operate the starter and yet provide the strong magnetic field required to move the starter drive into engagement. ● **SEE FIGURE 15–29.**

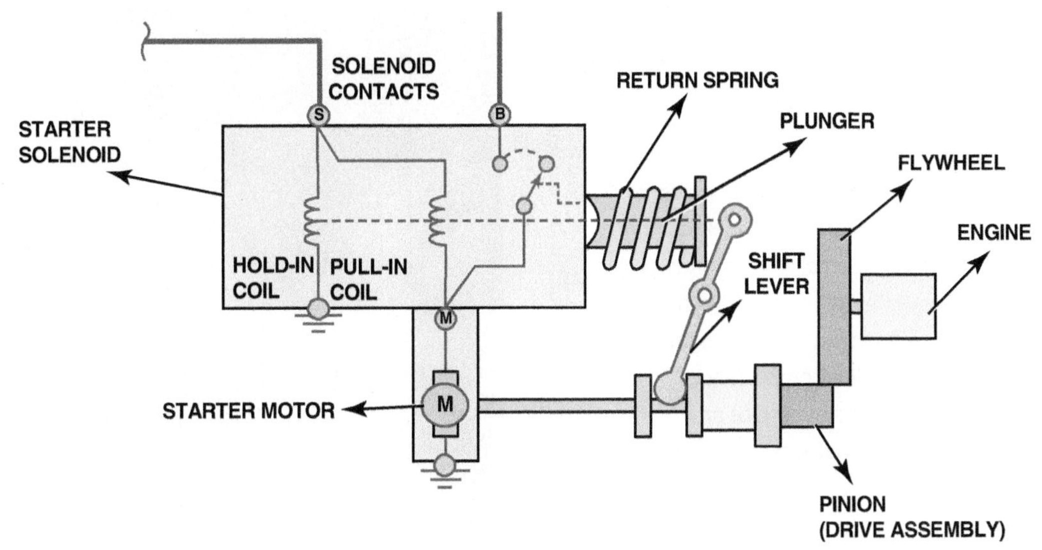

FIGURE 15–30 Power from the relay flows through the pull-in windings and the hold-in windings. The strong magnetic field begins to move the solenoid and at the same time the starter motor begins to turn, helping to mesh the pinion gear to the flywheel gear. (Courtesy of General Motors)

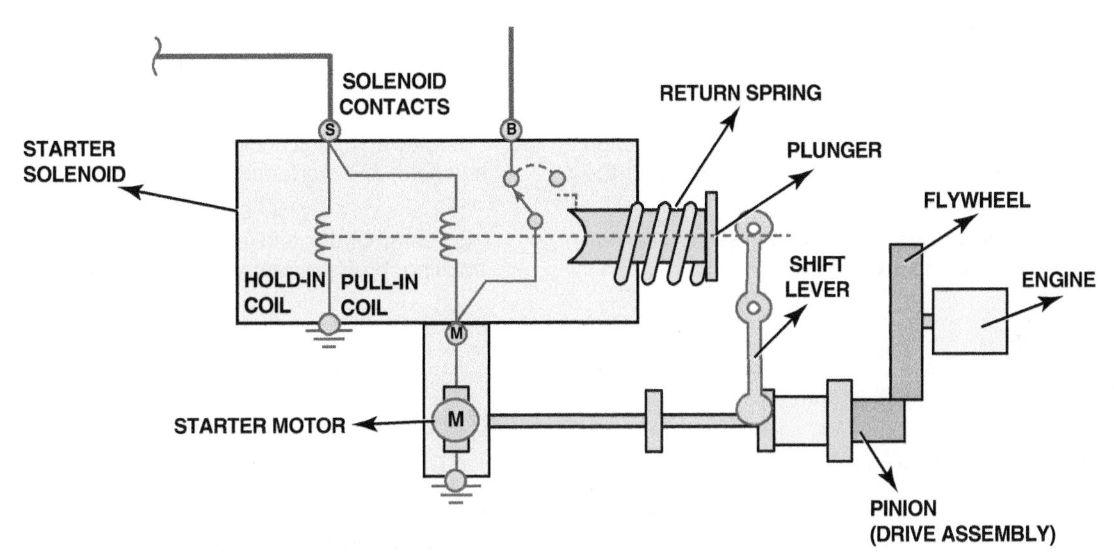

FIGURE 15–31 The contact disk bridges full battery voltage to the starter brushes and the engine cranks. The hold in windings keep the solenoid engaged. (Courtesy of General Motors)

OPERATION

1. The solenoid operates as soon as the ignition or computer-controlled relay energizes the "S" (start) terminals. At that instant, the plunger is drawn into the solenoid enough to engage the starter drive. ● **SEE FIGURE 15–30.**
2. The plunger makes contact with a metal disk that connects the battery terminal post of the solenoid to the motor terminal. This permits full battery current to flow through the solenoid to operate the starter motor. ● **SEE FIGURE 15–31.**
3. The contact disk also electrically disconnects the pull-in winding. The solenoid *has* to work to supply current to the starter. Therefore, if the starter motor operates at all, the solenoid is working, even though it may have high external resistance that could cause slow starter motor operation.

? FREQUENTLY ASKED QUESTION

How Are Starters Made So Small?

Starters and most components in a vehicle are being made as small and as light in weight as possible to help increase vehicle performance and fuel economy. A starter can be constructed smaller due to the use of gear reduction and permanent magnets to achieve the same cranking torque as a straight drive starter, but using much smaller components. ● **SEE FIGURE 15–32** for an example of an automotive starter armature that is palm size.

FIGURE 15–32 A palm-size starter armature.

SUMMARY

1. All starter motors use the principle of magnetic interaction between the field coils attached to the housing and the magnetic field of the armature.
2. The control circuit includes the ignition switch, neutral safety (clutch) switch, and solenoid.
3. The power circuit includes the battery, battery cables, solenoid, and starter motor.
4. The parts of a typical starter include the main field housing, commutator-end (or brush-end) housing, drive-end housing, brushes, armature, and starter drive.

REVIEW QUESTIONS

1. What is the difference between the control circuit and the power (motor) circuit sections of a typical cranking circuit?
2. What are the parts of a typical starter?
3. Why does a gear-reduction unit reduce the amount of current required by the starter motor?
4. What are the symptoms of a defective starter drive?

CHAPTER QUIZ

1. Starters motors operate on the principle that _______.
   - **a.** The field coils rotate in the opposite direction from the armature
   - **b.** Opposite magnetic poles repel
   - **c.** Like magnetic poles repel
   - **d.** The armature rotates from a strong magnetic field toward a weaker magnetic field
2. Series-wound electric motors _______.
   - **a.** Produce electrical power
   - **b.** Produce maximum power at 0 RPM
   - **c.** Produce maximum power at high RPM
   - **d.** Use a shunt coil

3. Technician A says that a defective solenoid can cause a starter whine. Technician B says that a defective starter drive can cause a starter whining noise. Which technician is correct?
   a. Technician A only
   b. Technician B only
   c. Both Technicians A and B
   d. Neither Technician A nor B
4. The neutral safety switch is located _______.
   a. Between the starter solenoid and the starter motor
   b. Inside the ignition switch itself
   c. Between the ignition switch and the starter solenoid
   d. In the battery cable between the battery and the starter solenoid
5. The brushes are used to transfer electrical power between _______.
   a. Field coils and the armature
   b. The commutator segments
   c. The solenoid and the field coils
   d. The armature and the solenoid
6. The faster a starter motor rotates, _______.
   a. The more current it draws from the battery
   b. The less CEMF is generated
   c. The less current it draws from the battery
   d. The greater the amount of torque produced
7. Normal cranking speed of the engine is about _______.
   a. 2000 RPM
   b. 1500 RPM
   c. 1000 RPM
   d. 200 RPM
8. A starter motor rotates about _______ times faster than the engine.
   a. 18
   b. 10
   c. 5
   d. 2
9. Permanent magnets are commonly used for what part of the starter?
   a. Armature
   b. Solenoid
   c. Field coils
   d. Commutator
10. Which unit contains a hold-in winding and a pull-in winding?
   a. Field coil
   b. Starter solenoid
   c. Armature
   d. Ignition switch

chapter 16

CRANKING SYSTEM DIAGNOSIS AND SERVICE

LEARNING OBJECTIVES

After studying this chapter, the reader will be able to:

1. Discuss how to perform a voltage drop test on the cranking circuit.
2. Perform control circuit testing and starter amperage test, and determine necessary action.
3. Explain starter motor service and bench testing.

This chapter will help you prepare for the ASE Electrical/Electronic Systems (A6) certification test content area "C" (Starting System Diagnosis and Repair).

KEY TERMS

Bench testing 224

Growler 223

Shims 225

Voltage drop 220

GM STC OBJECTIVES

GM Service Technical College topics covered in this chapter are as follows:

1. GM starting systems, overview and operation, and diagnosis.
2. Identify starting system characteristics.

STARTING SYSTEM TROUBLESHOOTING PROCEDURE

OVERVIEW The proper operation of the starting system depends on a good battery, good cables and connections, and a good starter motor. Because a starting problem can be caused by a defective component anywhere in the starting circuit, it is important to check for the proper operation of each part of the circuit to diagnose and repair the problem quickly.

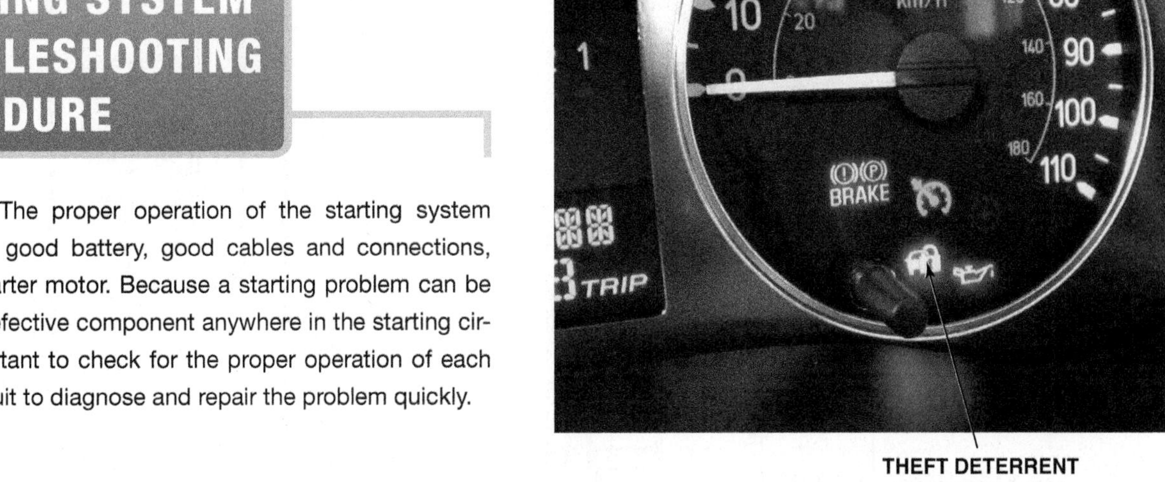

FIGURE 16–1 A theft deterrent indicator lamp of the dash. A flashing lamp usually indicates a fault in the system, and the engine may not start.

TECH TIP

Voltage Drop Is Resistance

Many technicians have asked, "Why measure voltage drop when the resistance can be easily measured using an ohmmeter?" Think of a battery cable with all the strands of the cable broken, except for one strand. If an ohmmeter were used to measure the resistance of the cable, the reading would be very low, probably less than 1 ohm. However, the cable is not capable of conducting the amount of current necessary to crank the engine. In less severe cases, several strands can be broken, thereby affecting the operation of the starter motor. Although the resistance of the battery cable will not indicate an increase, the restriction to current flow will cause heat and a drop of voltage available at the starter. Because resistance is not effective until current flows, measuring the voltage drop (differences in voltage between two points) is the most accurate method of determining the true resistance in a circuit.

How much is too much? According to Bosch Corporation, all electrical circuits should have a maximum of 3% loss of the circuit voltage to resistance. Therefore, in a 12-volt circuit, the maximum loss of voltage in cables and connections should be 0.36 volt ($12 \times 0.03 = 0.36$ volt). The remaining 97% of the circuit voltage (11.64 volts) is available to operate the electrical device (load). Just remember:

- Low-voltage drop = Low resistance
- High-voltage drop = High resistance

STEPS INVOLVED Following are the steps involved in the diagnosis of a fault in the cranking circuit:

STEP 1 **Verify the customer concern.** Sometimes the customer is not aware of how the cranking system is supposed to work, especially if it is computer controlled.

STEP 2 **Visually inspect the battery and battery connections.** The starter is the highest amperage draw device used in a vehicle and any faults, such as corrosion on battery terminals, can cause cranking system problems.

STEP 3 **Test battery condition.** Perform a battery load or conductance test on the battery to be sure that the battery is capable of supplying the necessary current for the starter.

STEP 4 **Check the control circuit.** An open or high resistance anywhere in the control circuit can cause the starter motor to not engage. Items to check include:

- "S" terminal of the starter solenoid
- Neutral safety or clutch switch
- Starter enable relay (if equipped)
- Antitheft system fault (If the engine does not crank or start and the theft indicator light is on or flashing, there is likely a fault in the theft deterrent system. Check service information for the exact procedures to follow before attempting to service the cranking circuit.) ● **SEE FIGURE 16–1.**

STEP 5 **Check voltage drop of the starter circuit.** Any high resistance in either the power side or ground side of the starter circuit will cause the starter to rotate slowly or not at all.

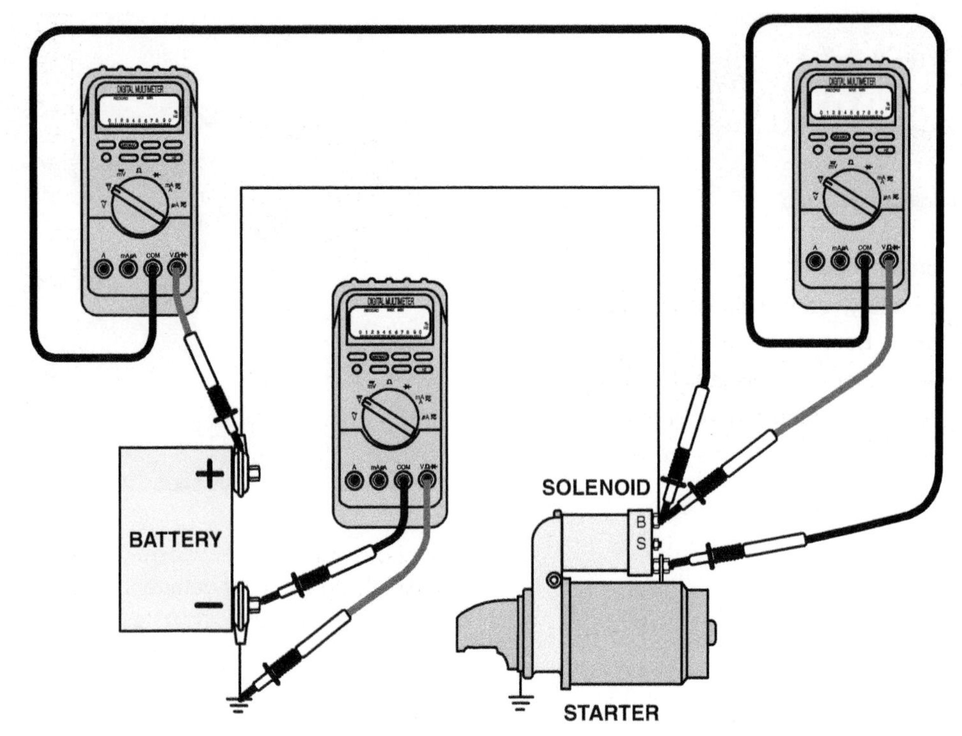

FIGURE 16–2 Voltmeter hookups for voltage drop testing of a solenoid-type cranking circuit.

VOLTAGE DROP TESTING

PURPOSE **Voltage drop** is the drop in voltage that occurs when current is flowing through a resistance. In other words, a voltage drop is the difference between voltage at the source and voltage at the electrical device to which it is flowing. The higher the voltage drop is, the greater the resistance in the circuit. Even though voltage drop testing can be performed on any electrical circuit, the most common areas of testing include the cranking circuit and the charging circuit wiring and connections. Voltage drop testing should be performed on both the power side and ground side of the circuit.

A high voltage drop (high resistance) in the cranking circuit wiring can cause slow engine cranking with less than normal starter amperage drain as a result of the excessive circuit resistance. If the voltage drop is high enough, such as that caused by dirty battery terminals, the starter may not operate. A typical symptom of high resistance in the cranking circuit is a "clicking" of the starter solenoid.

TEST PROCEDURE Voltage drop testing of the wire involves connecting a voltmeter set to read DC volts to the suspected high-resistance cable ends and cranking the engine. ● **SEE FIGURES 16–2 AND 16–3.**

NOTE: Before a difference in voltage (voltage drop) can be measured between the ends of a battery cable, current must be flowing through the cable. If the engine is not being cranked, current is not flowing through the battery cables and the voltage drop cannot be measured.

STEP 1 Disable the ignition or fuel injection as follows:

- Disconnect the primary (low-voltage) electrical connection(s) from the ignition module or ignition coils.
- Remove the fuel-injection fuse or relay, or the electrical connection leading to all of the fuel injectors.

CAUTION: Never disconnect the high-voltage ignition wires unless they are connected to ground. The high voltage that could occur when cranking can cause the ignition coil to fail (arc internally).

STEP 2 Connect one lead of the voltmeter to the starter motor battery terminal and the other end to the positive battery terminal.

STEP 3 Crank the engine and observe the reading while cranking. (Disregard the first higher reading.) The reading should be less than 0.20 volt (200 millivolts).

STEP 4 If accessible, test the voltage drop across the "B" and "M" terminals of the starter solenoid with the engine cranking. The voltage drop should be less than 0.20 volt (200 millivolts).

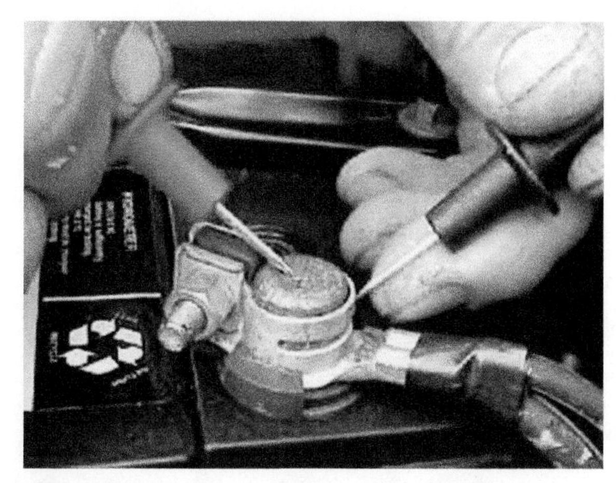

FIGURE 16–3 To test the voltage drop of the battery cable connection, place one voltmeter lead on the battery terminal and the other voltmeter lead on the cable end and crank the engine. The voltmeter will read the difference in voltage between the two leads, which should not exceed 0.20 volt (200 millivolts).

TECH TIP

A Warm Cable Equals High Resistance

If a cable or connection is warm to the touch, there is electrical resistance in the cable or connection. The resistance changes electrical energy into heat energy. Therefore, if a voltmeter is not available, touch the battery cables and connections while cranking the engine. If any cable or connection is hot to the touch, it should be cleaned or replaced.

STEP 5 Repeat the voltage drop on the ground side of the cranking circuit by connecting one voltmeter lead to the negative battery terminal and the other at the starter housing. Crank the engine and observe the voltmeter display. The voltage drop should be less than 0.2 volt (200 millivolts).

CONTROL CIRCUIT TESTING

PARTS INVOLVED The control circuit for the starting circuit includes the battery, ignition switch, neutral or clutch safety switch, theft deterrent system, a module (usually the ECM) and starter solenoid. The ignition switch provides a voltage input to a module to request starting. Current for the pull-in and hold-in windings of the solenoid do not flow through the ignition switch.

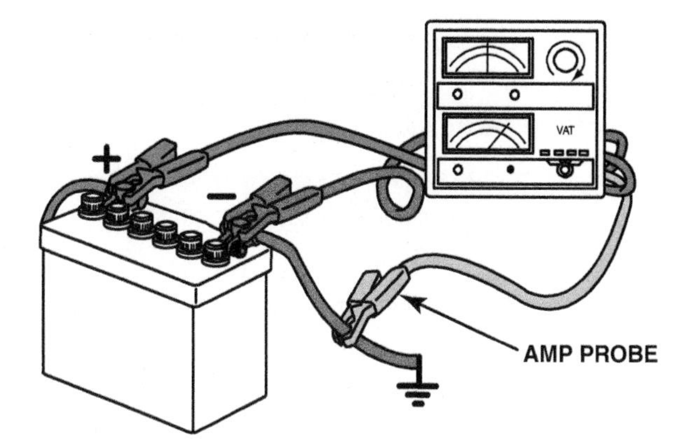

FIGURE 16–4 A starter amperage tester uses an amp probe around the positive or negative battery cables.

The module energizes a relay that provides current to the windings of the solenoid. High current then flows directly from the battery through the solenoid and to the starter motor. Therefore, an open or break anywhere in the control circuit will prevent the operation of the starter motor.

If a starter is inoperative, first check for voltage at the "S" (start) terminal of the starter solenoid. If there is no voltage while cranking check for faults with the following:

- Neutral safety or clutch switch
- Start relay
- "Start" request at the ECM using a scan tool

STARTER AMPERAGE TEST

REASON FOR A STARTER AMPERAGE TEST A starter should be tested to see if the reason for slow or no cranking is due to a fault with the starter motor or another problem. A voltage drop test is used to find out if the battery cables and connections are okay. A starter amperage draw test determines if the starter motor is the cause of a no or slow cranking concern.

TEST PREPARATION Before performing a starter amperage test, be certain that the battery is sufficiently charged (75% or more) and capable of supplying adequate starting current. Connect a starter amperage tester following the tester's instructions. ● **SEE FIGURE 16–4.**

A starter amperage test should be performed when the starter fails to operate normally (is slow in cranking) or as part of a routine electrical system inspection.

SPECIFICTIONS Some service manuals specify normal starter amperage for starter motors being tested on the vehicle;

TECH TIP

Watch the Dome Light

When diagnosing any starter-related problem, open the door of the vehicle and observe the brightness of the dome or interior light(s).

The brightness of any electrical lamp is proportional to the voltage of the battery.

Normal operation of the starter results in a slight dimming of the dome light.

If the light remains bright, the problem is usually an open in the control circuit.

If the light goes out or almost goes out, there could be a problem with the following:

- A shorted or grounded armature of field coils inside the starter
- Loose or corroded battery connections or cables
- Weak or discharged battery

however, most service manuals only give the specifications for bench testing a starter without a load applied. These specifications are helpful in making certain that a repaired starter meets exact specifications, but they do not apply to starter testing on the vehicle. If exact specifications are not available, the following can be used as general *maximum* amperage draw specifications for testing a starter on the vehicle.

- **4-cylinder engines** = 150 to 185 amperes (normally less than 100 amperes) at room temperature
- **6-cylinder engines** = 160 to 200 amperes (normally less than 125 amperes) at room temperature
- **8-cylinder engines** = 185 to 250 amperes (normally less than 150 amperes) at room temperature

Excessive current draw may indicate one or more of the following:

1. Binding of starter armature as a result of worn bushings
2. Oil too thick (viscosity too high) for weather conditions
3. Shorted or grounded starter windings or cables
4. Tight or seized engine
5. Shorted starter motor (usually caused by fault with the field coils or armature)
    - High mechanical resistance = High starter amperage draw
    - High electrical resistance = Low starter amperage draw

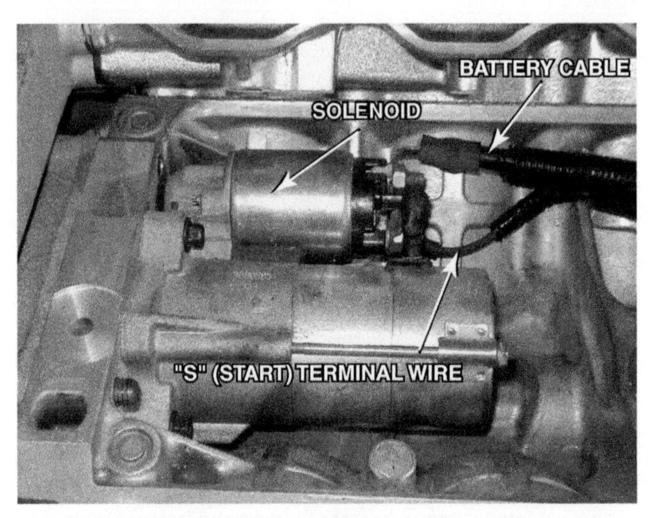

FIGURE 16–5 The starter is located under the intake manifold on this Cadillac Northstar engine.

Lower amperage draw and slow or no cranking may indicate one or more of the following:

- Dirty or corroded battery connections
- High internal resistance in the battery cable(s)
- High internal starter motor resistance
- Poor ground connection between the starter motor and the engine block

STARTER REMOVAL

PROCEDURE After testing has confirmed that a starter motor may need to be replaced, most vehicle manufacturers recommend the following general steps and procedures:

STEP 1 Disconnect the negative battery cable.

STEP 2 Hoist the vehicle safely.

NOTE: This step may not be necessary. Check service information for the specified procedure for the vehicle being serviced. Some starters are located under the intake manifold. ● SEE FIGURE 16–5.

STEP 3 Remove the starter retaining bolts and lower the starter to gain access to the wire(s) connection(s) on the starter.

STEP 4 Disconnect and label the wire(s) from the starter and remove the starter.

STEP 5 Inspect the flywheel (flexplate) for ring gear damage. Also check that the mounting holes are clean and the mounting flange is clean and smooth. Service as needed.

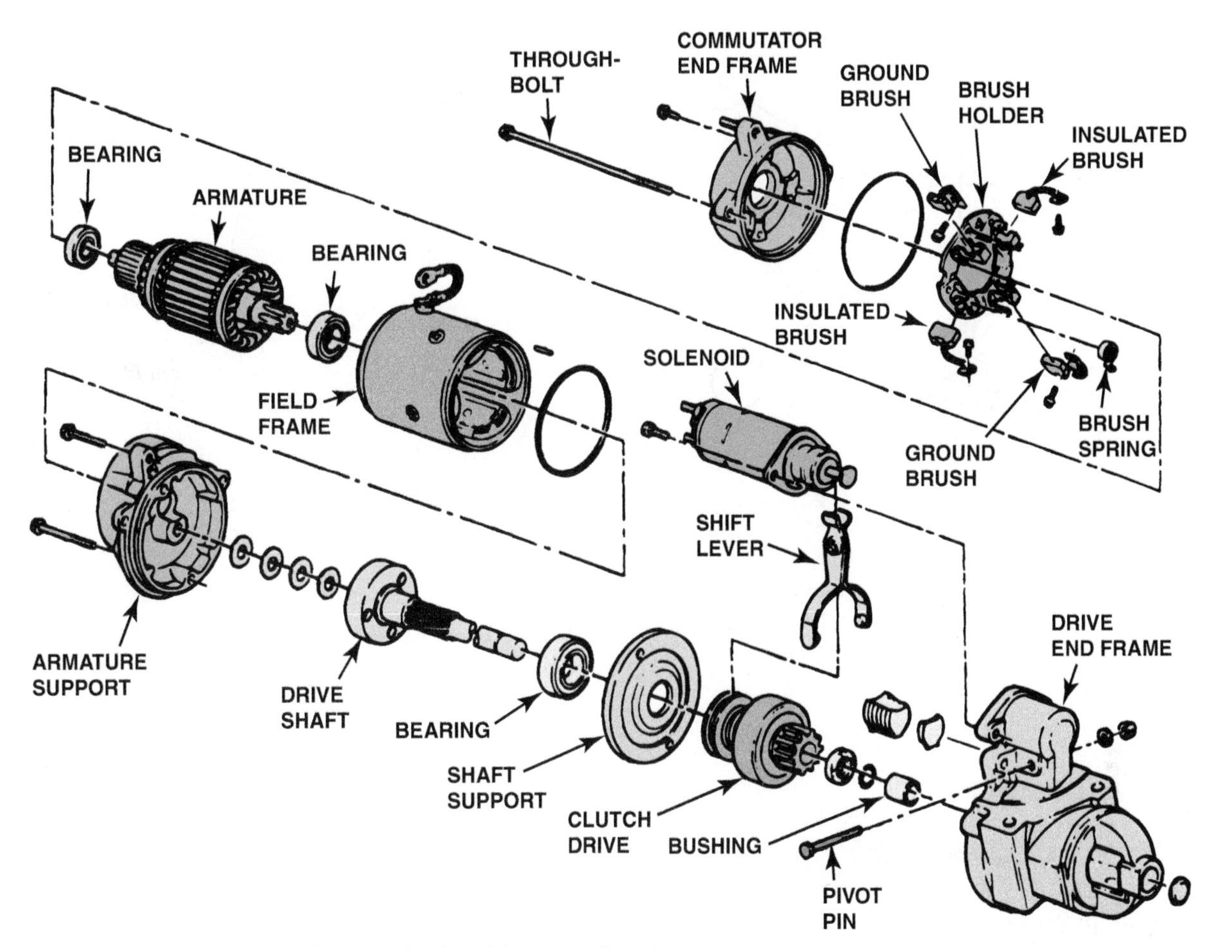

FIGURE 16–6 An exploded view of a typical solenoid-operated starter.

STARTER MOTOR SERVICE

PURPOSE Most starter motors are replaced as an assembly or not easily disassembled or serviced. However, some starters, especially on classic muscle or collector vehicles, can be serviced.

DISASSEMBLY PROCEDURE Disassembly of a starter motor usually includes the following steps:

STEP 1 Remove the starter solenoid assembly.

STEP 2 Mark the location of the through bolts on the field housing to help align them during reassembly.

STEP 3 Remove the drive-end housing and then the armature assembly.

● **SEE FIGURE 16–6.**

INSPECTION AND TESTING The various parts should be inspected and tested to see if the components can be used to restore the starter to serviceable condition.

- **Solenoid.** Check the resistance of the solenoid winding. The solenoid can be tested using an ohmmeter to check for the proper resistance in the hold-in and pull-in windings. ● **SEE FIGURE 16–7.**

Most technicians replace the solenoid whenever the starter is replaced and is usually included with a replacement starter.

- **Starter armature.** After the starter drive has been removed from the armature, it can be checked for runout using a dial indicator and V-blocks, as shown in ● **FIGURE 16–8.**
- **Growler.** Because the loops of copper wire are interconnected in the armature of a starter, an armature can be accurately tested only by use of a **growler**. A growler is a 110-volt AC test unit that generates an alternating (60 hertz) magnetic field around an armature. A starter armature is placed into the V-shaped top portion of a laminated soft-iron core surrounded by a coil of copper wire. Plug the growler into a 110-volt outlet and then follow the instructions for testing the armature.

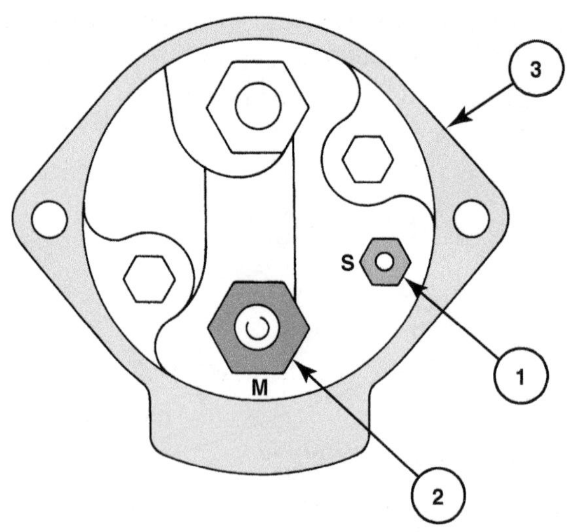

FIGURE 16–7 GM solenoid ohmmeter check. The reading between 1 and 3 (S terminal and ground) should be 0.4 to 0.6 ohm (hold-in winding). The reading between 1 and 2 (S terminal and M terminal) should be 0.2 to 0.4 ohm (pull-in winding).

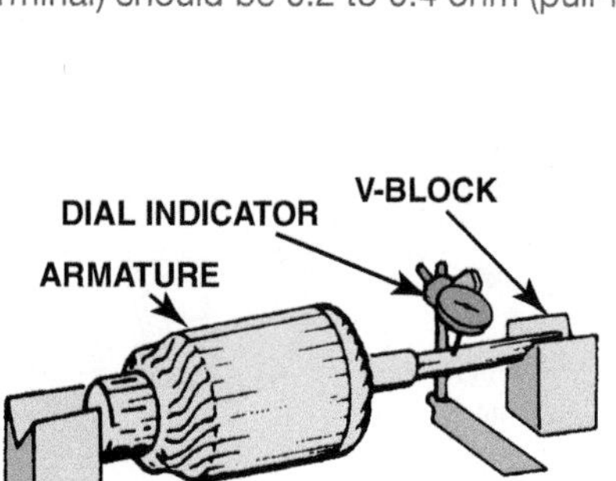

FIGURE 16–8 Measuring an armature shaft for runout using a dial indicator and V-blocks.

- **Starter motor field coils.** With the armature removed from the starter motor, the field coils should be tested for opens and grounds using a powered test light or an ohmmeter. To test for a grounded field coil, touch one lead of the tester to a field brush (insulated or hot) and the other end to the starter field housing. The ohmmeter should indicate infinity (no continuity), and the test light should *not* light. If there is continuity, replace the field coil housing assembly. The ground brushes should show continuity to the starter housing.

  NOTE: Many starters use removable field coils. These coils must be rewound using the proper equipment and insulating materials. Usually, the cost involved in replacing defective field coils exceeds the cost of a replacement starter.

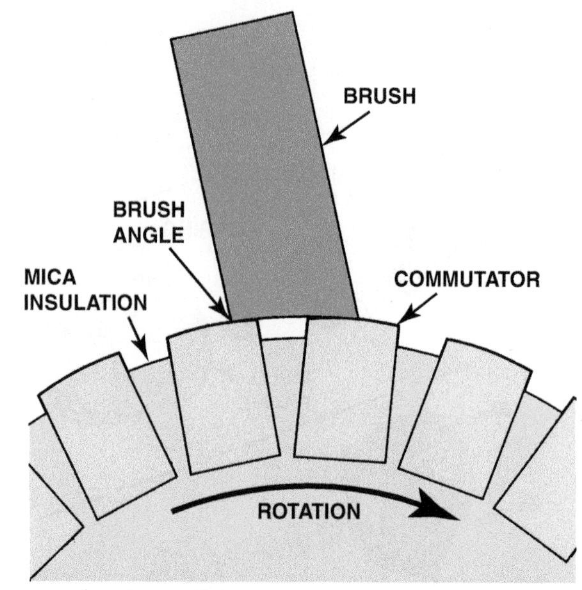

FIGURE 16–9 Replacement starter brushes should be installed so the beveled edge matches the rotation of the commutator.

- **Starter brush inspection.** Starter brushes should be replaced if the brush length is less than half of its original length (less than 0.5 inch [13 millimeters]). On some models of starter motors, the field brushes are serviced with the field coil assembly and the ground brushes with the brush holder. Many starters use brushes that are held in with screws and are easily replaced, whereas other starters may require soldering to remove and replace the brushes. ● **SEE FIGURE 16–9.**

BENCH TESTING

Every starter should be tested before installation in a vehicle. **Bench testing** is the usual method and involves clamping the starter in a vise to prevent rotation during operation and connecting heavy-gauge jumper wires (minimum 4 gauge) to both a good battery and the starter. The starter motor should rotate as fast as specifications indicate and not draw more than the free-spinning amperage permitted. A typical amperage specification for a starter being tested on a bench (not installed in a vehicle) usually ranges from 60 to 100 amperes.

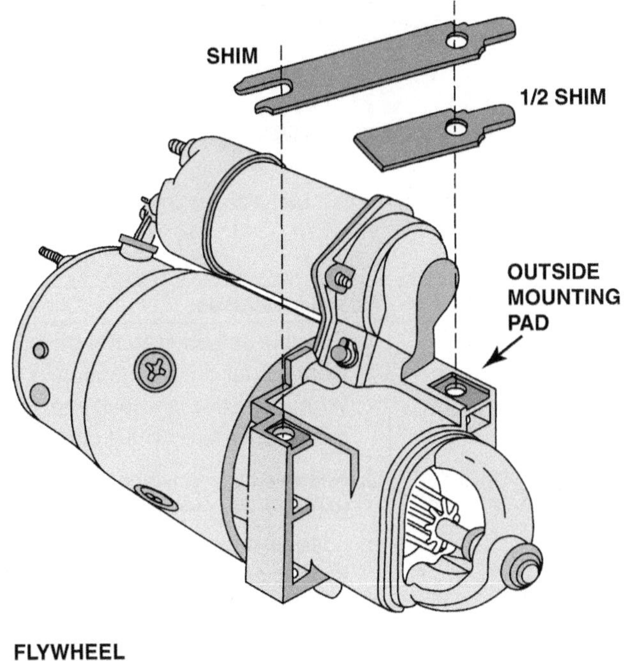

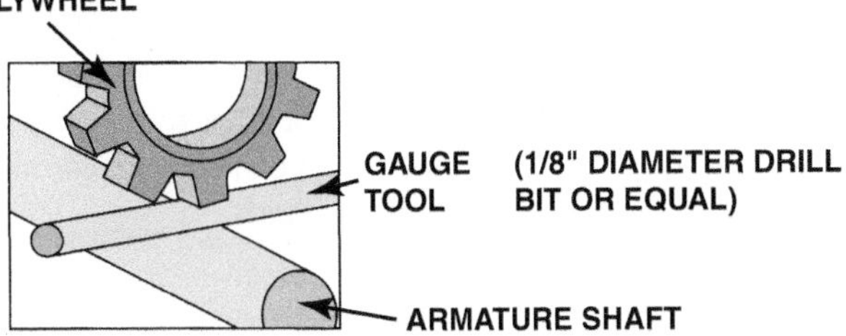

FIGURE 16–10 A shim (or half shim) may be needed to provide the proper clearance between the flywheel teeth of the engine and the pinion teeth of the starter.

STARTER INSTALLATION

After verifying that the starter assembly is functioning correctly, verify that the negative battery cable has been disconnected. Then safely hoist the vehicle, if necessary. Following are the usual steps to install a starter. Be sure to check service information for the exact procedures to follow for the vehicle being serviced.

STEP 1 Check service information for the exact wiring connections to the starter and/or the solenoid.

STEP 2 Verify that all electrical connections on the starter-motor and/or the solenoid are correct for the vehicle and that they are in good condition.

NOTE: Be sure that the locking nuts for the studs are tight. Often, the retaining nut that holds the wire to the stud will be properly tightened, but if the stud itself is loose, cranking problems can occur.

STEP 3 Attach the power and control wires.

STEP 4 Install the starter, and torque all the fasteners to factory specifications and tighten evenly.

STEP 5 Perform a starter amperage draw test and check for proper engine cranking.

CAUTION: Be sure to install all factory heat shields to help ensure problem starter operation under all weather and driving conditions.

STARTER DRIVE-TO-FLYWHEEL CLEARANCE

NEED FOR SHIMS For the proper operation of the starter and absence of abnormal starter noise, there must be a slight clearance between the starter pinion and the engine flywheel ring gear. Some starters use **shims**, which are thin metal strips between the flywheel and the engine block mounting pad to provide the proper clearance. ● **SEE FIGURE 16–10.**

Some manufacturers use shims under the starter drive-end housings during production. Other manufacturers *grind* the mounting pads at the factory for proper starter pinion gear clearance. If a GM starter is replaced, the starter pinion clearance should be checked and corrected as necessary to prevent starter damage and excessive noise.

SYMPTOMS OF CLEARANCE PROBLEMS

- If the clearance is too great, the starter will produce a high-pitched whine *during* cranking.

- If the clearance is too small, the starter may bind, crank slowly, or produce a high-pitched whine *after* the engine starts, just as the ignition key is released.

PROCEDURE FOR PROPER CLEARANCE To be sure that the starter is shimmed correctly, use the following procedure.

STEP 1 Place the starter in position and finger-tighten the mounting bolts.

STEP 2 Use a 1/8 inch diameter drill bit (or 0.020" wire gauge tool) and insert between the armature shaft and a tooth of the engine flywheel.

STEP 3 If the gauge tool cannot be inserted, use a full-length shim across both mounting holes to move the starter away from the flywheel.

STEP 4 Remove a shim (or shims) if the gauge tool is loose between the shaft and the tooth of the engine flywheel.

STEP 5 If no shims have been used and the fit of the gauge tool is too loose, add a half shim to the outside pad only. This moves the starter closer to the teeth of the engine flywheel.

STARTING SYSTEM SYMPTOM GUIDE

The following list will assist technicians in troubleshooting starting systems:

Problem	Possible Causes
1. Starter motor whines	1. Possible defective starter drive; worn starter drive engagement yoke; defective flywheel; improper starter drive to flywheel clearance
2. Starter rotates slowly	2. Possible high resistance in the battery cables or connections; possible defective or discharged battery; possible worn starter bushings, causing the starter armature to drag on the field coils; possible worn starter brushes or weak brush springs; possible defective (open or shorted) field coil; engine damage (causes high amperage draw)
3. Starter fails to rotate	3. Possible defective ignition switch or neutral safety switch, or open in the starter motor control circuit; theft deterrent system fault; possible defective starter solenoid
4. Starter produces grinding noise	4. Possible defective starter drive unit; possible defective flywheel; possible incorrect distance between the starter pinion and the flywheel; possible cracked or broken starter drive-end housing; worn or damaged flywheel or ring gear teeth
5. Starter clicks when engaged	5. Low battery voltage; loose or corroded battery connections; engine seized (causes high amperage draw)

1 This dirty and greasy starter can be restored to useful service.

2 The connecting wire between the solenoid and the starter is removed.

3 An old starter field housing is being used to support the drive-end housing of the starter as it is being disassembled. This rebuilder is using an electric impact wrench to remove the solenoid fasteners.

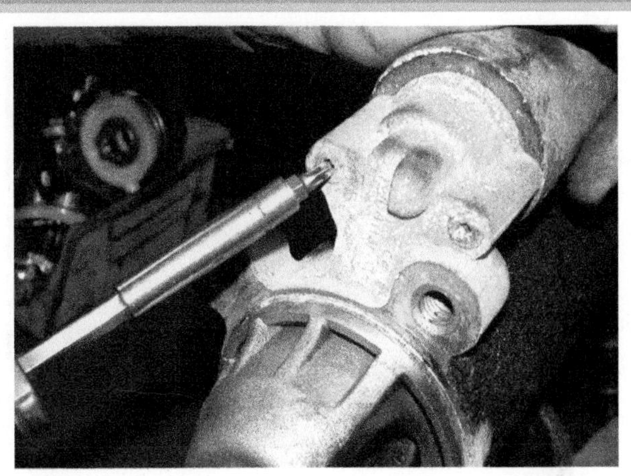

4 A Torx driver is used to remove the solenoid attaching screws.

5 After the retaining screws have been removed, the solenoid can be separated from the starter motor. This rebuilder always replaces the solenoid.

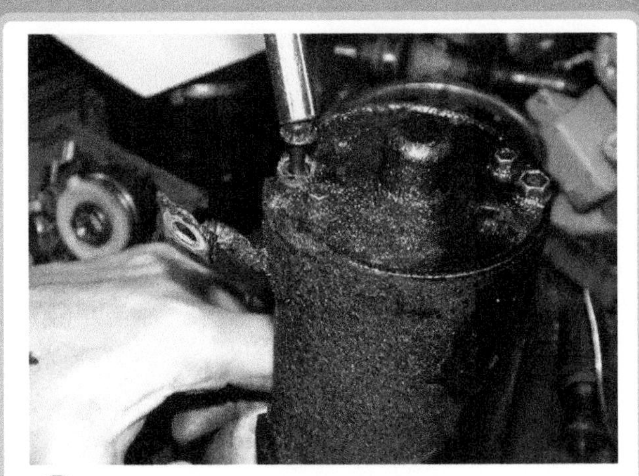

6 The through-bolts are being removed.

CONTINUED ▶

STARTER OVERHAUL (CONTINUED)

7 The brush end plate is removed.

8 The armature assembly is removed from the field frame.

9 Notice that the length of a direct-drive starter armature (top) is the same length as the overall length of a gear-reduction armature except smaller in diameter.

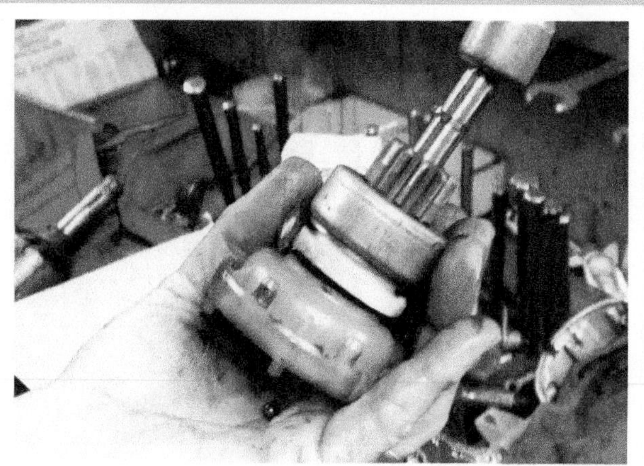

10 A light tap with a hammer dislodges the armature thrust ball (in the palm of the hand) from the center of the gear reduction assembly.

11 This figure shows the planetary ring gear and pinion gears.

12 A close-up of one of the planetary gears, which shows the small needle bearings on the inside.

13 The clip is removed from the shaft so the planetary gear assembly can be separated and inspected.

14 The shaft assembly is being separated from the stationary gear assembly.

15 The commutator on the armature is discolored and the brushes may not have been making good contact with the segments.

16 All of the starter components are placed in a tumbler with water-based cleaner. The armature is installed in a lathe and the commutator is resurfaced using emery cloth.

17 The finished commutator looks like new.

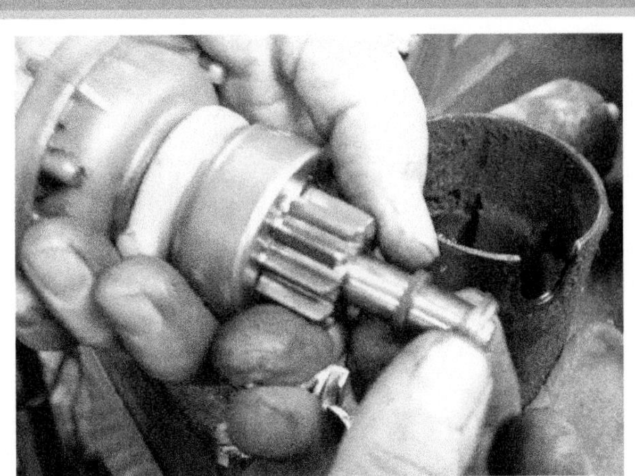

18 Starter reassembly begins by installing a new starter drive on the shaft assembly. The stop ring and stop ring retainer are then installed.

CONTINUED ▶

19 The gear-reduction assembly is positioned along with the shift fork (drive lever) into the cleaned drive-end housing.

20 After gear retainer has been installed over the gear reduction assembly, the armature is installed.

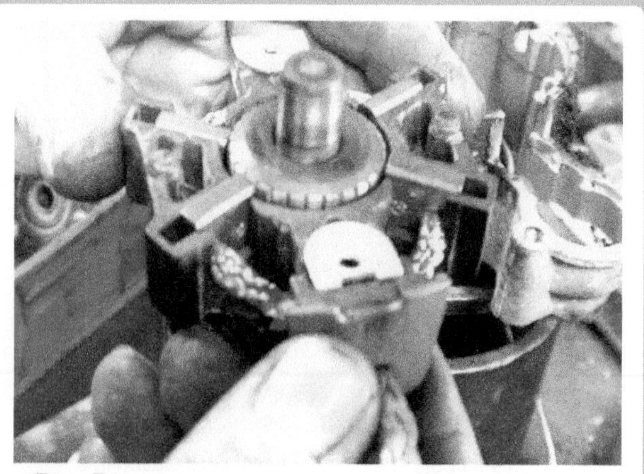

21 New brushes are being installed into the brush holder assembly.

22 The brush end plate and the through-bolts are installed, being sure that the ground connection for the brushes is clean and tight.

23 This starter was restored to useful service by replacing the solenoid, the brushes, and the starter drive assembly plus a thorough cleaning and attention to detail in the reassembly.

SUMMARY

1. Proper operation and testing of the starter motor depends on the battery being at least 75% charged and the battery cables being of the correct size (gauge) and having no more than a 0.2 volt drop.
2. Voltage drop testing includes cranking the engine, measuring the drop in voltage from the battery to the starter, and measuring the drop in voltage from the negative terminal of the battery to the engine block.
3. The cranking circuit should be tested for proper amperage draw.
4. An open in the control circuit can prevent starter motor operation.

REVIEW QUESTIONS

1. What are the parts of the cranking circuit?
2. What are the steps taken to perform a voltage drop test of the cranking circuit?
3. What are the steps necessary to replace a starter?

CHAPTER QUIZ

1. A growler is used to test what starter component?
   a. Field coils
   b. Armatures
   c. Commutator
   d. Solenoid
2. Two technicians are discussing what could be the cause of slow cranking and excessive current draw. Technician A says that an engine mechanical fault could be the cause. Technician B says that the starter motor could be binding or defective. Which technician is correct?
   a. Technician A only
   b. Technician B only
   c. Both Technicians A and B
   d. Neither Technician A nor B
3. A V-6 is being checked for starter amperage draw. The initial surge current was about 210 amperes and about 160 amperes during cranking. Technician A says the starter is defective and should be replaced because the current flow exceeds 200 amperes. Technician B says this is normal current draw for a starter motor on a V-6 engine. Which technician is correct?
   a. Technician A only
   b. Technician B only
   c. Both Technicians A and B
   d. Neither Technician A nor B
4. What component or circuit can keep the engine from cranking?
   a. Antitheft
   b. Solenoid
   c. Ignition switch
   d. All of the above
5. Technician A says that a discharged battery (lower than normal battery voltage) can cause solenoid clicking. Technician B says that a discharged battery or dirty (corroded) battery cables can cause solenoid clicking. Which technician is correct?
   a. Technician A only
   b. Technician B only
   c. Both Technicians A and B
   d. Neither Technician A nor B
6. Slow cranking by the starter can be caused by all of these *except* _______.
   a. A low or discharged battery
   b. Corroded or dirty battery cables
   c. Engine mechanical problems
   d. An open neutral safety switch
7. Bench testing of a starter should be done _______.
   a. After reassembling an old starter
   b. Before installing a new starter
   c. After removing the old starter
   d. Both a and b
8. If the clearance between the starter pinion and the engine flywheel is too great, _______.
   a. The starter will produce a high-pitched whine during cranking
   b. The starter will produce a high-pitched whine after the engine starts
   c. The starter drive will not rotate at all
   d. The solenoid will not engage the starter drive unit

9. A technician connects one lead of a digital voltmeter to the positive (+) terminal of the battery and the other meter lead to the battery terminal (B) of the starter solenoid and then cranks the engine. During cranking, the voltmeter displays a reading of 878 millivolts. Technician A says that this reading indicates that the positive battery cable has too high resistance. Technician B says that this reading indicates that the starter is defective. Which technician is correct?
   - **a.** Technician A only
   - **b.** Technician B only
   - **c.** Both Technicians A and B
   - **d.** Neither Technician A nor B

10. A vehicle equipped with a V-8 engine does not crank fast enough to start. Technician A says the battery could be discharged or defective. Technician B says that the negative cable could be loose at the battery. Which technician is correct?
   - **a.** Technician A only
   - **b.** Technician B only
   - **c.** Both Technicians A and B
   - **d.** Neither Technician A nor B

chapter 17

CHARGING SYSTEM

LEARNING OBJECTIVES

After studying this chapter, the reader will be able to:

1. Describe a generator's overrunning pulleys.
2. Describe the components and operation of an generator.
3. Discuss how a generator works.
4. Explain how the voltage produced by a generator is regulated.
5. Discuss computer-controlled generators.

This chapter will help you prepare for the ASE Electrical/Electronic Systems (A6) certification test content area "C" (Starting System Diagnosis and Repair).

KEY TERMS

Claw poles 236
Delta winding 240
Diodes 237
Drive-end (DE) housing 234
Duty cycle 245
EPM 245
OAD 235
OAP 234
Regulated Voltage Control (RVC) 245
Rotor 236
Slip-ring-end (SRE) housing 234
Stator 237
Thermistor 243

GM STC OBJECTIVES

GM Service Technical College topics covered in this chapter are as follows:

1. GM charging systems, overview, and operation.
2. Identify charging system characteristics.
3. Identify regulated voltage control systems.

FIGURE 17–1 A typical generator on a Chevrolet V-8 engine.

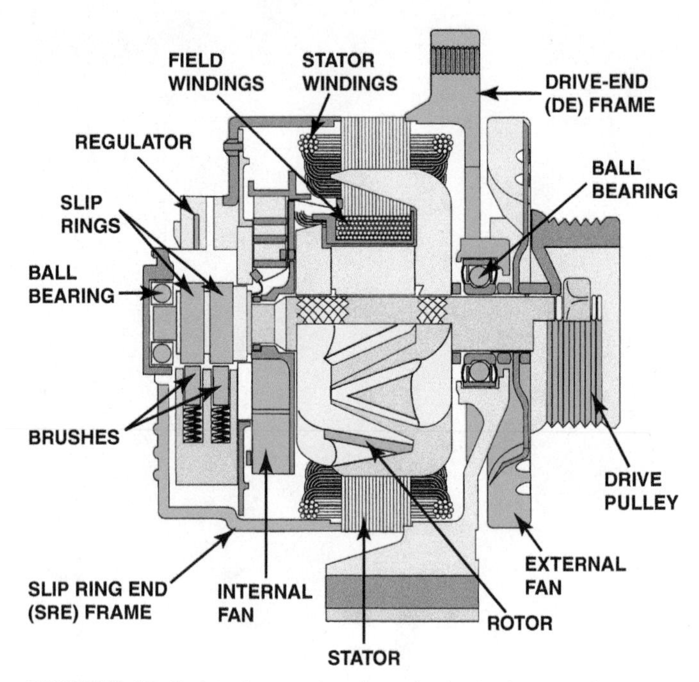

FIGURE 17–2 A cutaway drawing of a typical generator.

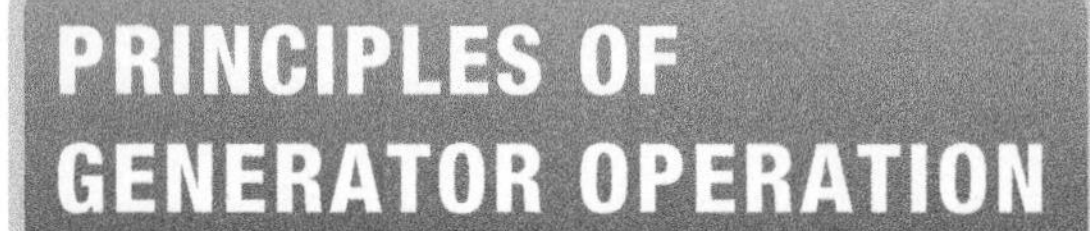

PRINCIPLES OF GENERATOR OPERATION

TERMINOLOGY It is the purpose and function of the charging system to keep the battery fully charged. The Society of Automotive Engineers (SAE) term for the unit that generates electricity is *generator.* The term **alternator** is most commonly used in the trade; General Motors uses the term *generator* in service information. Both terms are used here, depending on the subject.

PRINCIPLES All electrical generators use the principle of electromagnetic induction to generate electrical power from mechanical power. Electromagnetic induction involves the generation of electric current in a conductor when the conductor is moved through a magnetic field. The amount of current generated can be increased by the following factors:

1. Increasing the *speed* of the conductors through the magnetic field
2. Increasing the *number* of conductors passing through the magnetic field
3. Increasing the *strength* of the magnetic field

CHANGING AC TO DC A generator (alternator) creates an alternating current (AC) because the current changes polarity during the alternator's rotation. However, a battery cannot "store" alternating current; therefore, this alternating current is changed to direct current (DC) by diodes inside the alternator. Diodes are one-way electrical check valves that permit current to flow in only one direction.

GENERATOR CONSTRUCTION

HOUSING A generator is constructed using a two-piece cast aluminum housing. Aluminum is used because of its lightweight, nonmagnetic properties, and heat transfer properties needed to help keep the generator cool. A front ball bearing is pressed into the front housing, called the **drive-end housing**, to provide the support and friction reduction necessary for the belt-driven rotor assembly. The rear housing, or the **slip-ring-end housing**, usually contains either a roller bearing or ball bearing support for the rotor and mounting for the brushes, diodes, and internal voltage regulator (if so equipped). ● **SEE FIGURES 17–1 AND 17–2.**

GENERATOR OVERRUNNING PULLEYS

PURPOSE AND FUNCTION Many generators are equipped with an **overrunning alternator pulley (OAP)**, also called an *overrunning clutch pulley* or an *alternator clutch*

FIGURE 17–3 An OAP on a Chevrolet Corvette generator.

TECH TIP

Generator Horsepower and Engine Operation

Many technicians are asked how much power certain accessories require. A 100-ampere generator requires about 2 horsepower from the engine. One horsepower is equal to 746 watts. Watts are calculated by multiplying amperes times volts.

$$\textbf{Power in watts} = 100\text{ A} \times 14.5\text{ V} = 1{,}450\text{ W}$$
$$1\text{ hp} = 746\text{ W}$$

Therefore, 1,450 watts is about 2 horsepower.

Allowing about 20% for mechanical and electrical losses adds another 0.4 horsepower. Therefore, when someone asks how much power it takes to produce 100 amperes from an generator, the answer is 2.4 horsepower.

Many generators delay the electrical load to prevent the engine from stumbling when a heavy electrical load is applied. The voltage regulator or vehicle computer is capable of gradually increasing the output of the alternator over a period of several minutes. Even though 2 horsepower does not sound like much, a sudden demand for 2 horsepower from an idling engine can cause the engine to run rough or stall. The difference in part numbers of various generators is often an indication of the time interval over which the load is applied. Therefore, using the wrong replacement generator could cause the engine to stall!

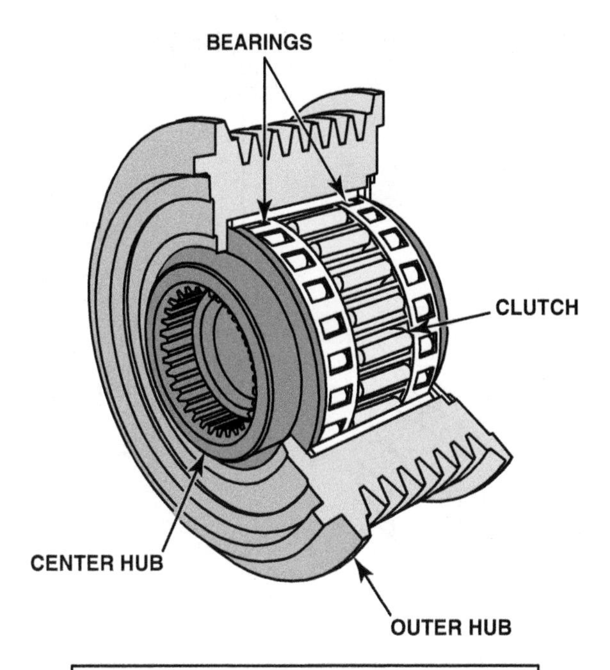

FIGURE 17–4 An exploded view of an overrunning alternator pulley showing all of the internal parts.

pulley. The purpose of this pulley is to help eliminate noise and vibration in the accessory drive belt system, especially when the engine is at idle speed. At idle, engine impulses are transmitted to the alternator through the accessory drive belt. The mass of the rotor of the alternator tends to want to keep spinning, but the engine crankshaft speeds up and slows down slightly due to the power impulses. Using a one-way clutch in the alternator pulley allows the belt to apply power to the alternator in only one direction, thereby reducing fluctuations in the belt. ● **SEE FIGURES 17–3 AND 17–4.**

A conventional drive pulley attaches to the alternator (rotor) shaft with a nut and lock washer. In the overrunning clutch pulley, the inner race of the clutch acts as the nut as it screws on to the shaft. Special tools are required to remove and install this type of pulley.

Another type of alternator pulley uses a dampener spring inside, plus a one-way clutch. These units have the following names:

- **Isolating Decoupler Pulley (IDP)**
- **Active Alternator Pulley (AAP)**
- **Alternator Decoupler Pulley (ADP)**
- **Alternator Overrunning Decoupler Pulley**
- **Overrunning Alternator Dampener (OAD)** (most common term)

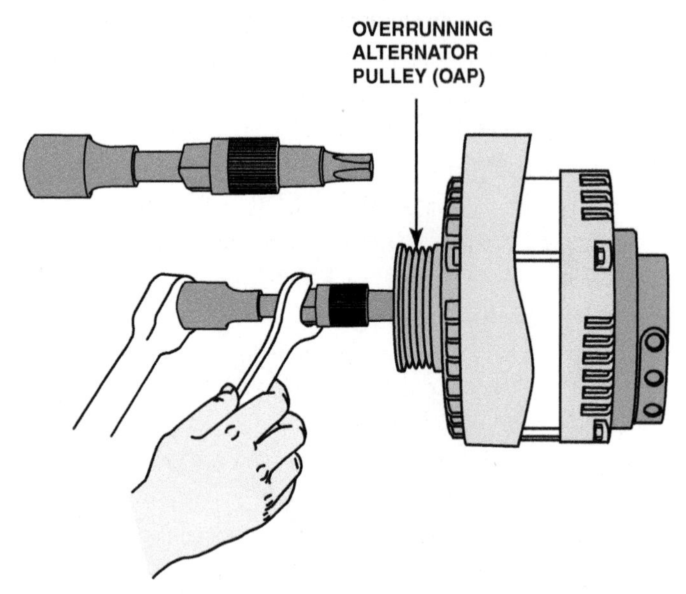

FIGURE 17–5 A special tool is needed to remove and install overrunning alternator pulleys or dampeners.

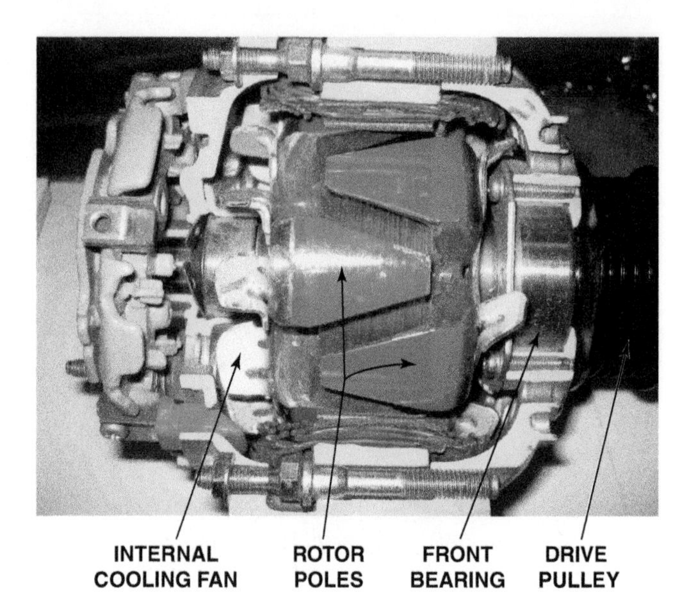

FIGURE 17–6 A cutaway of an alternator, showing the rotor and cooling fan that is used to force air through the unit to remove the heat created when it is charging the battery and supplying electrical power for the vehicle.

FREQUENTLY ASKED QUESTION

Can I Install an OAP or an OAD to My Alternator?

Usually, no. An alternator needs to be equipped with the proper shaft to allow the installation of an OAP or OAD. This also means that a conventional pulley often cannot be used to replace a defective overrunning alternator pulley or dampener with a conventional pulley. Check service information for the exact procedure to follow.

TECH TIP

Always Check the OAP or OAD First

Overrunning alternator pulleys and overrunning alternator dampeners can fail. The most common factor is the one-way clutch. If it fails, it can freewheel and not power the alternator or it can lock up and not provide the dampening as designed. If the charging system is not working, the OAP or OAD could be the cause, rather than a fault in the alternator itself.

In most cases, the entire alternator assembly will be replaced because each OAP or OAD is unique for each application and both require special tools to remove and replace.

● **SEE FIGURE 17–5.**

OAP or OAD pulleys are primarily used on vehicles equipped with diesel engines or on luxury vehicles where noise and vibration need to be kept at a minimum. Both are designed to:

- Reduce accessory drive belt noise
- Improve the life of the accessory drive belt
- Improve fuel economy by allowing the engine to be operated at a low idle speed

GENERATOR COMPONENTS AND OPERATION

ROTOR CONSTRUCTION The **rotor** is the rotating part of the generator and is driven by the accessory drive belt. The rotor creates the magnetic field of the generator and produces a current by electromagnetic induction in the stationary stator windings. The rotor is constructed of many turns of copper wire coated with a varnish insulation wound over an iron core. The iron core is attached to the rotor shaft.

At both ends of the rotor windings are heavy-gauge metal plates bent over the windings with triangular fingers called **claw poles**. These pole fingers do not touch, but alternate or interlace, as shown in ● **FIGURE 17–6**.

HOW ROTORS CREATE MAGNETIC FIELDS The two ends of the rotor winding are connected to the rotor's slip rings. Current for the rotor flows from the battery into one brush that

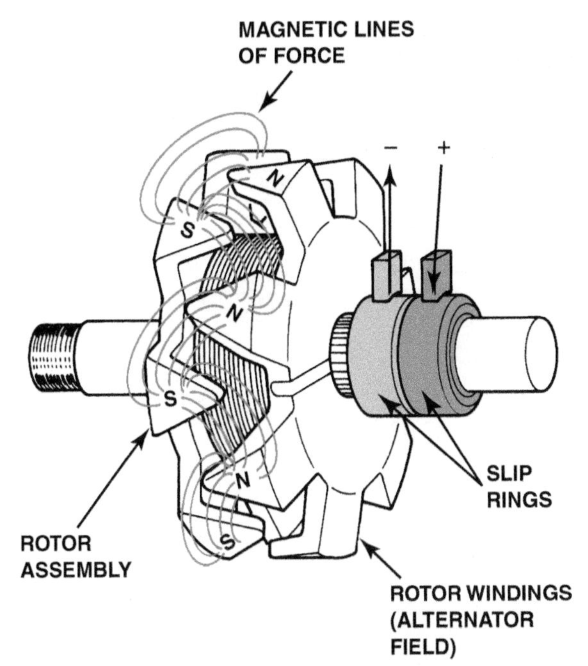

FIGURE 17–7 Rotor assembly of a typical generator. Current through the slip rings causes the "fingers" of the rotor to become alternating north and south magnetic poles. As the rotor revolves, these magnetic lines of force induce a current in the stator windings.

rides on one of the slip rings, then flows through the rotor winding, then exits the rotor through the other slip ring and brush. One brush is considered to be the "positive" brush and one is considered to be the "negative" or "ground" brush. The voltage regulator is connected to either the positive or the negative brush and controls the field current through the rotor that controls the output of the generator.

If current flows through the rotor windings, the metal pole pieces at each end of the rotor become electromagnets. Whether a north or a south pole magnet is created depends on the *direction* in which the wire coil is wound. Because the pole pieces are attached to each end of the rotor, one pole piece will be a north pole magnet. The other pole piece is on the opposite end of the rotor and therefore is viewed as being wound in the opposite direction, creating a south pole. Therefore, the rotor fingers are alternating north and south magnetic poles. The magnetic fields are created between the alternating pole piece fingers. These individual magnetic fields produce a current by electromagnetic induction in the stationary stator windings. ● **SEE FIGURE 17–7.**

ROTOR CURRENT The current necessary for the field (rotor) windings is conducted through slip rings with carbon brushes. The maximum rated alternator output in amperes depends on the number and gauge of the rotor windings. Substituting rotors from one generator to another can greatly affect maximum output. Many commercially rebuilt generators are tested and then display a sticker to indicate their tested output. The original rating stamped on the housing is then ground off.

The current for the field is controlled by the voltage regulator and is conducted to the slip rings through carbon brushes. The brushes conduct only the field current, which is usually between 2 and 5 amperes.

STATOR CONSTRUCTION The **stator** consists of the stationary coil windings inside the generator. The stator is supported between the two halves of the generator housing, with three copper wire windings that are wound on a laminated metal core.

As the rotor revolves, its moving magnetic field induces a current in the stator windings. ● **SEE FIGURE 17–8.**

DIODES **Diodes** are constructed of a semiconductor material (usually silicon) and operate as a one-way electrical check valve that permits the current to flow in only one direction. Generators use a minimum of six diodes (one positive and one negative set for each of the three stator windings) to convert alternating current to direct current.

Diodes used in alternators are included in a single part called a rectifier, or *rectifier bridge*. A rectifier not only includes the diodes but also the cooling fins and connections for the stator windings and the voltage regulator. ● **SEE FIGURE 17–9.**

DIODE TRIO Some generators are equipped with a diode trio that supplies current to the brushes from the stator windings. A diode trio uses three diodes, in one housing, with one diode for each of the three stator windings and then one output terminal.

HOW A GENERATOR WORKS

FIELD CURRENT IS PRODUCED A rotor inside a generator is turned by a belt and drive pulley which are turned by the engine. The magnetic field of the rotor generates a current in the stator windings by electromagnetic induction. ● **SEE FIGURE 17–10.**

Field current flowing through the slip rings to the rotor creates an alternating north and south pole on the rotor, with a magnetic field between each finger of the rotor.

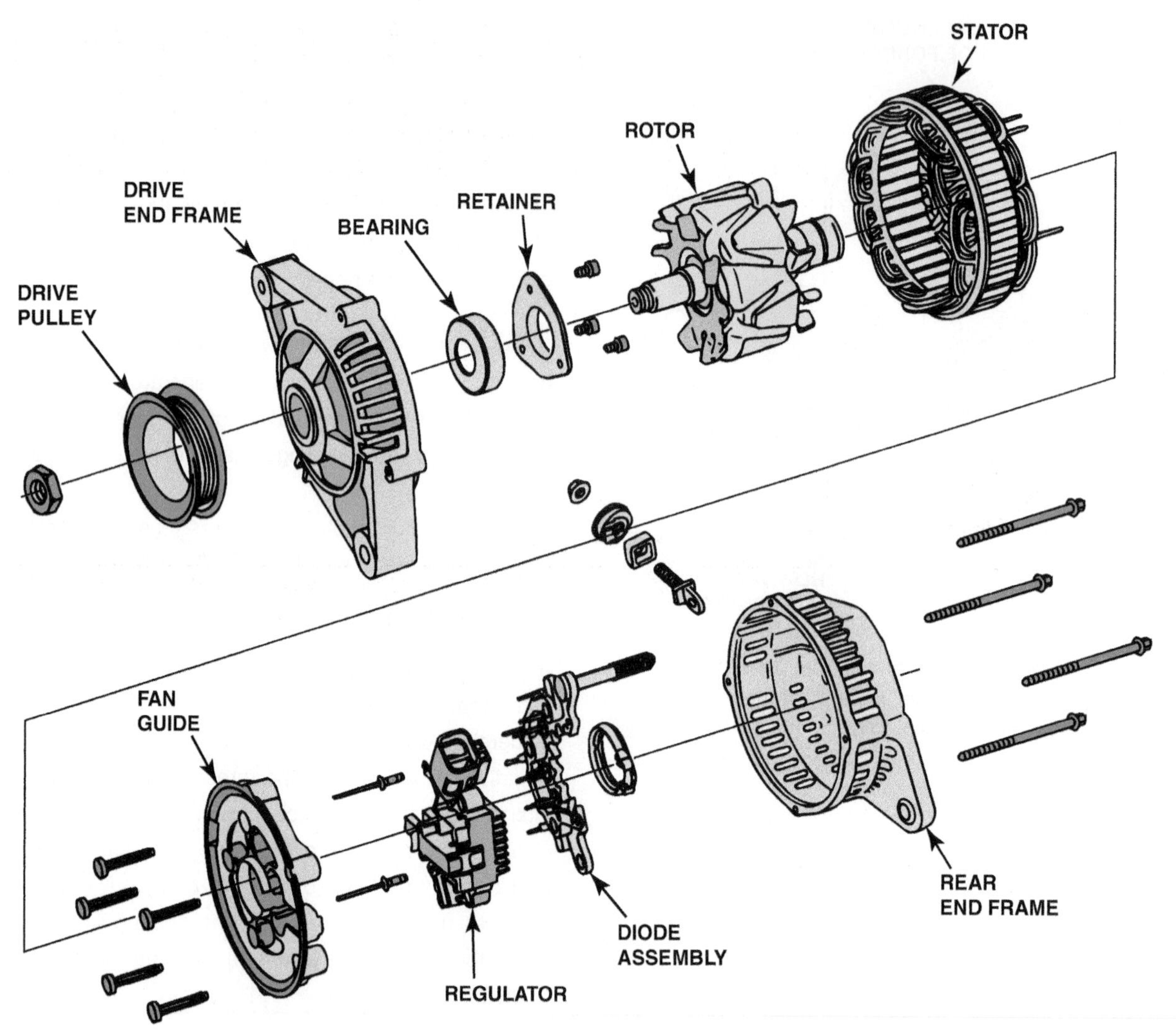

FIGURE 17–8 An exploded view of a typical alternator (generator) showing all of its internal parts including the stator windings.

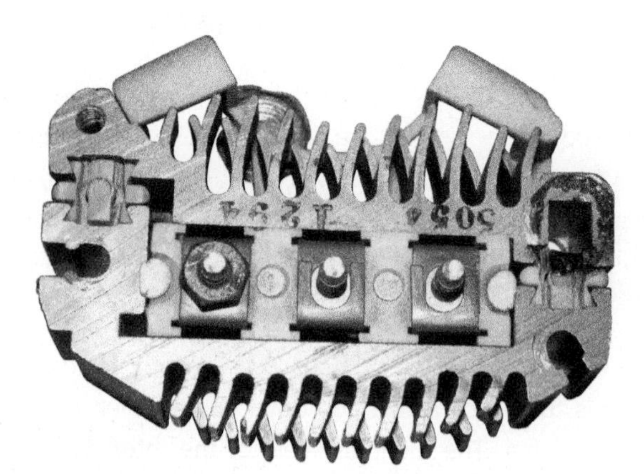

FIGURE 17–9 A rectifier usually includes six diodes in one assembly and is used to rectify AC voltage from the stator windings into DC voltage suitable for use by the battery and electrical devices in the vehicle.

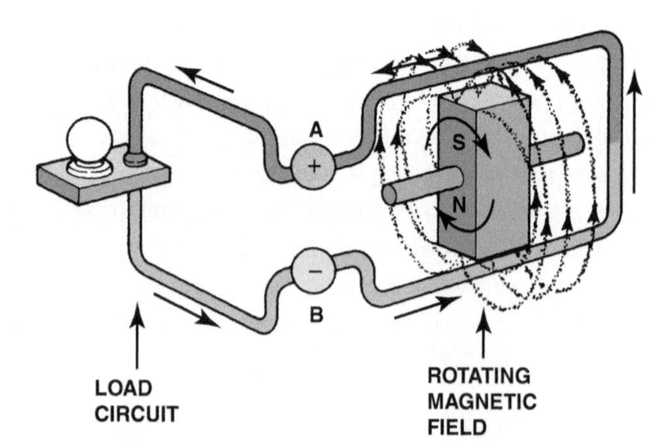

FIGURE 17–10 Magnetic lines of force cutting across a conductor induce a voltage and current in the conductor.

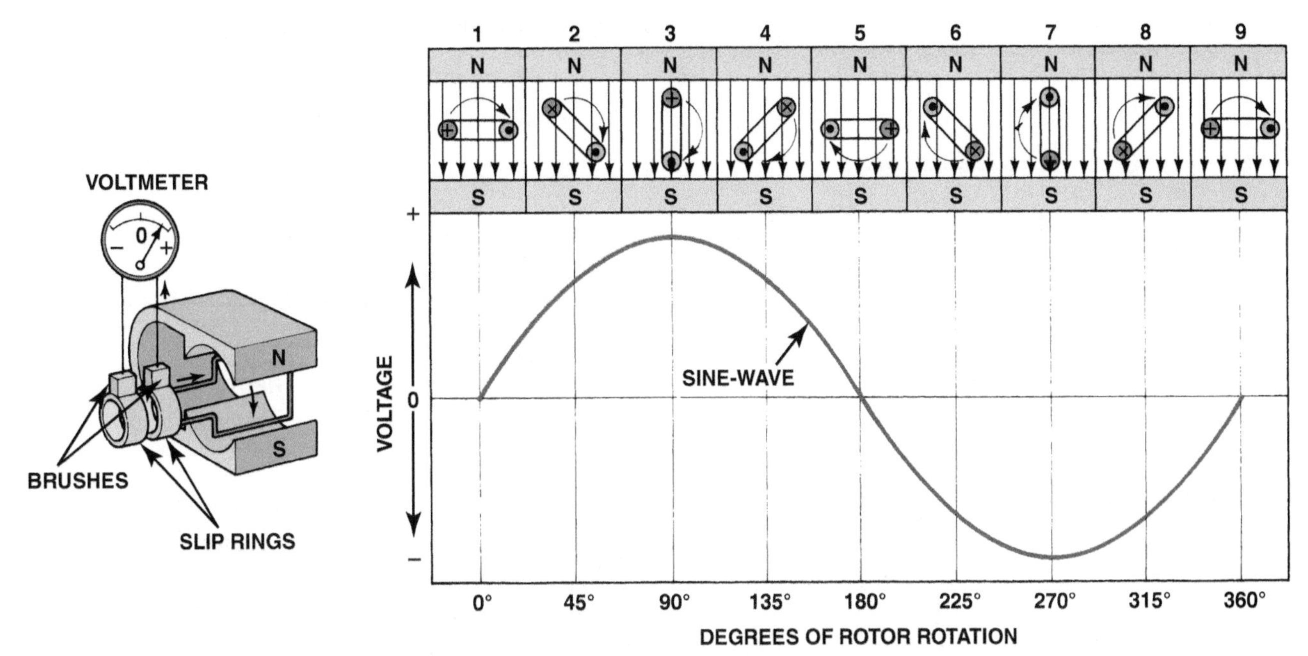

FIGURE 17–11 A sine wave (shaped like the letter *S* on its side) voltage curve is created by one revolution of a winding as it rotates in a magnetic field.

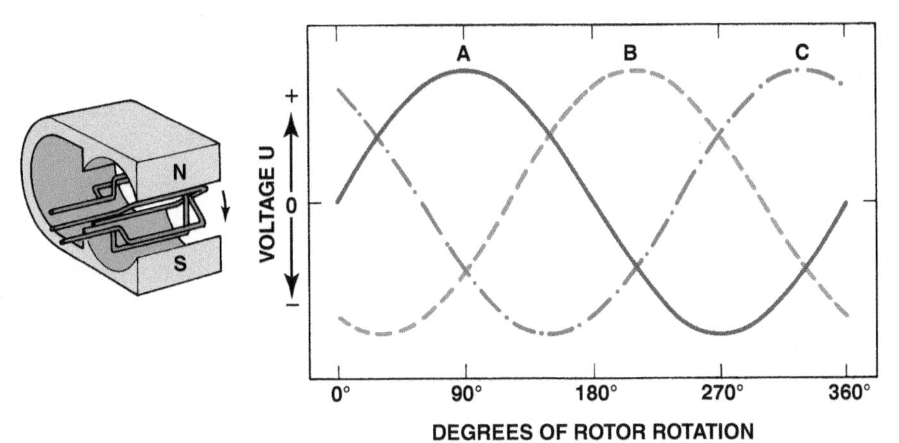

FIGURE 17–12 When three windings (A, B, and C) are present in a stator, the resulting current generation is represented by the three sine waves. The voltages are 120 degrees out of phase. The connection of the individual phases produces a three-phase alternating voltage.

CURRENT IS INDUCED IN THE STATOR The induced current in the stator windings is an alternating current because of the alternating magnetic field of the rotor. The induced current starts to increase as the magnetic field starts to induce current in each winding of the stator. The current then peaks when the magnetic field is the strongest and starts to decrease as the magnetic field moves away from the stator winding. Therefore, the current generated is described as being of a sine wave or alternating current pattern. ● **SEE FIGURE 17–11.**

As the rotor continues to rotate, this sine wave current is induced in each of the three windings of the stator.

Because each of the three windings generates a sine wave current, as shown in ● **SEE FIGURE 17–12**, the resulting currents combine to form a three-phase voltage output.

The current induced in the stator windings connects to diodes (one-way electrical check valves) that permit the alternator output current to flow in only one direction. All alternators contain six diodes, one pair (a positive and a negative diode) for each of the three stator windings. Some alternators contain eight diodes with another pair connected to the center connection of a wye-type stator.

WYE-CONNECTED STATORS The Y (pronounced "wye" and generally so written) type or star pattern is the most commonly used alternator stator winding connection. ● **SEE FIGURE 17–13.**

The output current with a wye-type stator connection is constant over a broad alternator speed range.

Current is induced in each winding by electromagnetic induction from the rotating magnetic fields of the rotor. In a wye-type stator connection, the currents must combine because two windings are always connected in series. ● **SEE FIGURE 17–14.**

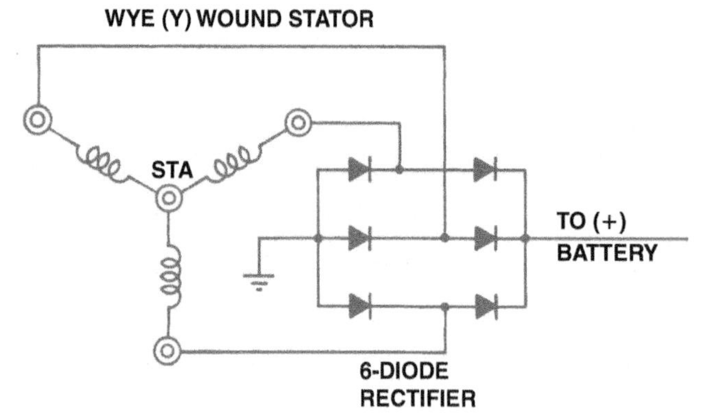

FIGURE 17–13 Wye-connected stator winding.

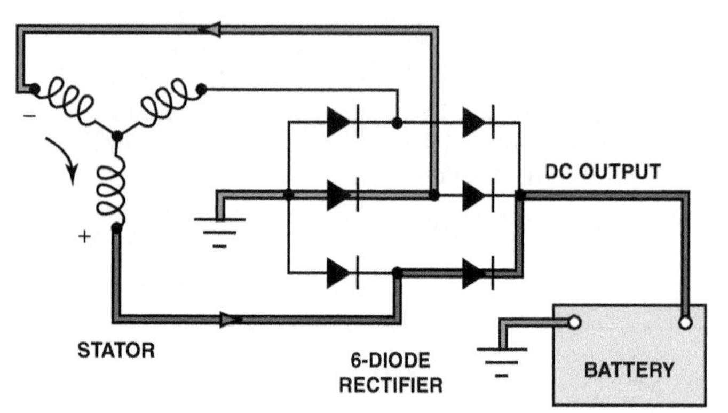

FIGURE 17–14 As the magnetic field, created in the rotor, cuts across the windings of the stator, a current is induced. Notice that the current path includes passing through one positive (+) diode on the way to the battery and one negative (−) diode as a complete circuit is completed through the rectifier and stator.

The current produced in each winding is added to the other windings' current and then flows through the diodes to the generator output terminal. One-half of the current produced is available at the neutral junction (usually labeled "STA" for stator). The voltage at this center point is used by some alternator manufacturers to control the charge indicator light or is used by the voltage regulator to control the rotor field current.

DELTA-CONNECTED STATORS The **delta winding** is connected in a triangular shape. Delta is a Greek letter shaped like a triangle. ● **SEE FIGURE 17–15.**

Current induced in each winding flows to the diodes in a parallel circuit. More current can flow through two parallel circuits than can flow through a series circuit (as in a wye-type stator connection).

Delta-connected stators are used on generators where high output at high-generator revolutions per minute (RPM) is required. The delta-connected generator can produce 73% more current than the same generator with wye-type stator connections. For example, if an generator with a wye-connected

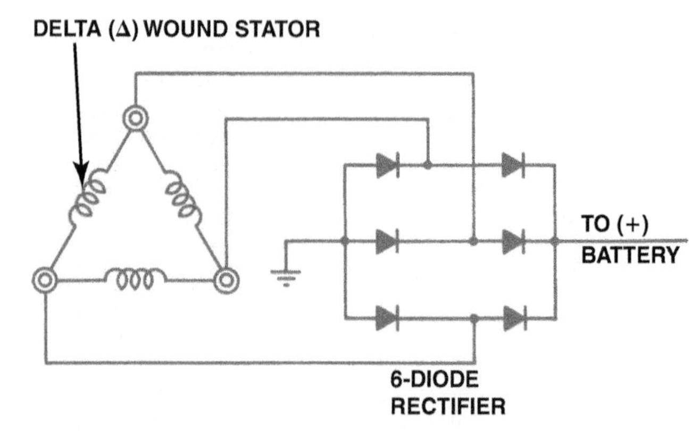

FIGURE 17–15 Delta-connected stator winding.

stator can produce 55 amperes, the *same* generator with delta-connected stator windings can produce 73% more current, or 95 amperes ($55 \times 1.73 = 95$). The delta-connected generator, however, produces lower current at low speed and must be operated at high speed to produce its maximum output.

GENERATOR OUTPUT FACTORS

The output voltage and current of a generator depend on the following factors:

1. **Speed of rotation.** Generator output is increased with rotational speed up to the generator maximum possible ampere output. Generators normally rotate at a speed two to three times faster than engine speed, depending on the relative pulley sizes used for the belt drive. For example, if an engine is operating at 5000 RPM, the generator will be rotating at about 15,000 RPM.
2. **Number of conductors.** A high-output generator contains more turns of wire in the stator windings. Stator winding connections (whether wye or delta) also affect the maximum generator output. ● **SEE FIGURE 17–16** for an example of a stator that has six rather than three windings, which greatly increases the amperage output of the generator.
3. **Strength of the magnetic field.** If the magnetic field is strong, a high output is possible because the current generated by electromagnetic induction is dependent on the number of magnetic lines of force that are cut.
   a. The strength of the magnetic field can be increased by increasing the number of turns of conductor wire wound on the rotor. A higher output generator rotor has more turns of wire than a generator rotor with a low rated output.

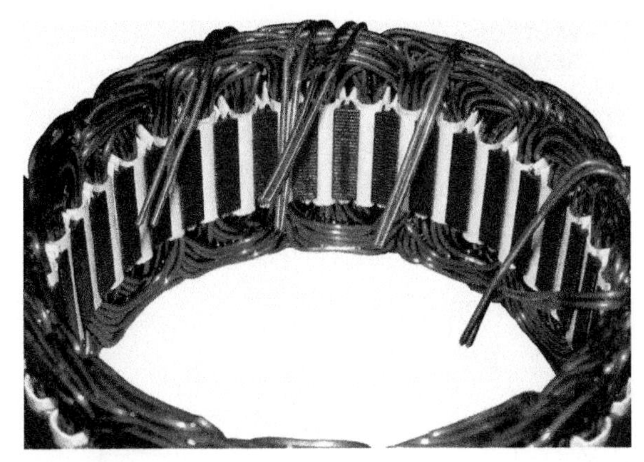

FIGURE 17–16 A stator assembly with six, rather than the normal three, windings.

b. The strength of the magnetic field also depends on the current through the field coil (rotor). Because magnetic field strength is measured in ampere-turns, the greater the amperage or the number of turns, or both, the greater the generator output.

GENERATOR VOLTAGE REGULATION

PRINCIPLES An automotive generator must be able to produce electrical pressure (voltage) higher than battery voltage to charge the battery. Excessively high voltage can damage the battery, electrical components, and the lights of a vehicle. Basic principles include the following:

- If no (zero) amperes of current existed throughout the field coil of the rotor generator output would be zero because without field current a magnetic field does not exist.
- The field current required by most automotive generators is less than 3 amperes. It is the *control* of the *field* current that controls the output of the generator.
- Current for the rotor flows from the battery positive post, through the rotor positive brush, into the rotor field winding, and exits the rotor winding through the rotor ground brush. Most voltage regulators control field current by controlling the amount of field current through the ground brush.
- The voltage regulator simply opens the field circuit if the voltage reaches a predetermined level, then closes the field circuit again as necessary to maintain the correct charging voltage. ● **SEE FIGURE 17–17.**
- The electronic circuit of some voltage regulators cycle between 10 and 7,000 times per second as needed to accurately control the field current through the rotor; others (newer GM models) use pulse width modulation (PWM) at a rate of 400 Hz to control the generator output.

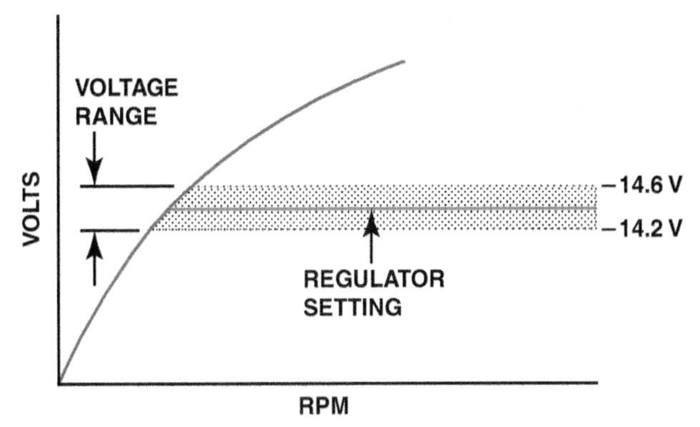

FIGURE 17–17 Typical voltage regulator range.

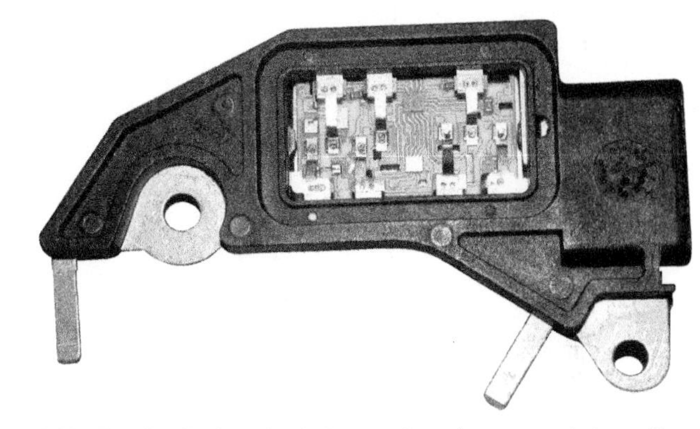

FIGURE 17–18 A typical electronic voltage regulator with the cover removed showing the circuits inside. (A CS series regulator is shown)

REGULATOR OPERATION

- The control of the field current is accomplished by opening and closing the *ground* side or the positive feed side of the field circuit through the rotor.
- The zener diode is a major electronic component that makes voltage regulation possible. A zener diode blocks current flow until a specific voltage is reached, then it permits current to flow. Generator voltage from the stator and diodes is first sent through a thermistor, which changes resistance with temperature, and then to a zener diode. When the upper-limit voltage is reached, the zener diode conducts current to a transistor, which then opens the field (rotor) circuit. The electronics are usually housed in a separate part inside the generator. ● **SEE FIGURES 17–18 AND 17–19.**

BATTERY CONDITION AND CHARGING VOLTAGE

If the automotive battery is discharged, its voltage will be lower than the voltage of a fully charged battery. The generator will supply charging current, but it may not reach the maximum charging voltage. For example, if a vehicle is jump started and run at a fast idle (2000 RPM), the charging voltage may be only 12 volts. In this case, the following may occur.

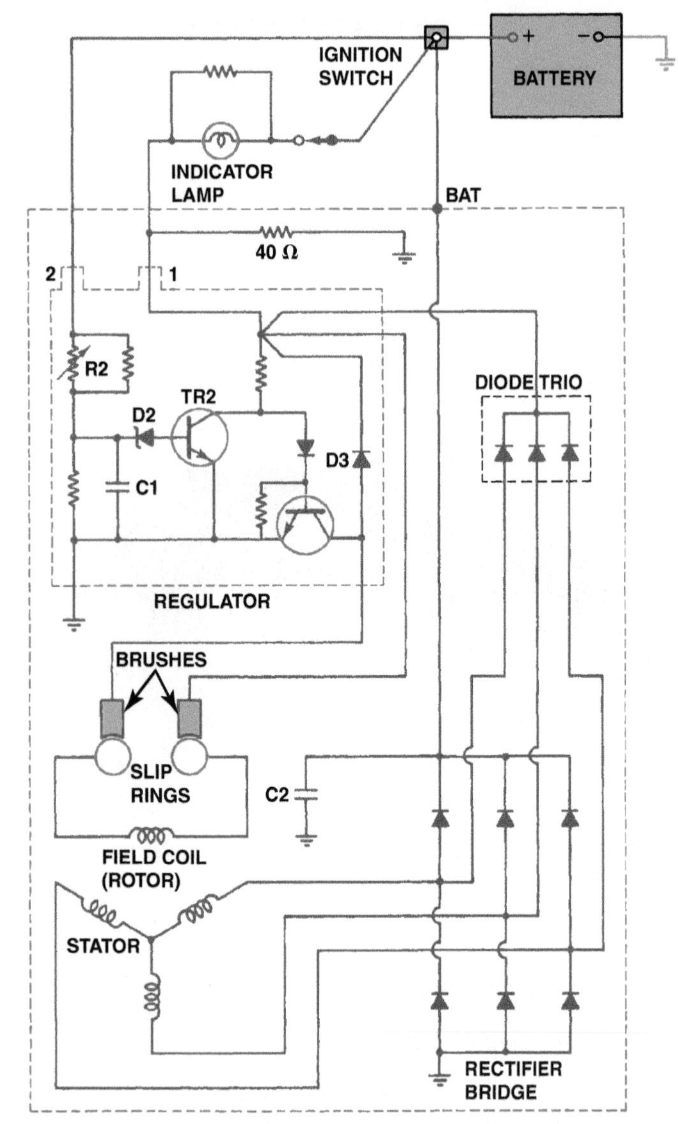

FIGURE 17–19 Typical General Motors SI-style generator with an integral voltage regulator. Voltage present at terminal 2 is used to reverse bias the zener diode (D2) that controls TR2. The positive brush is fed by the ignition current (terminal I) plus current from the diode trio.

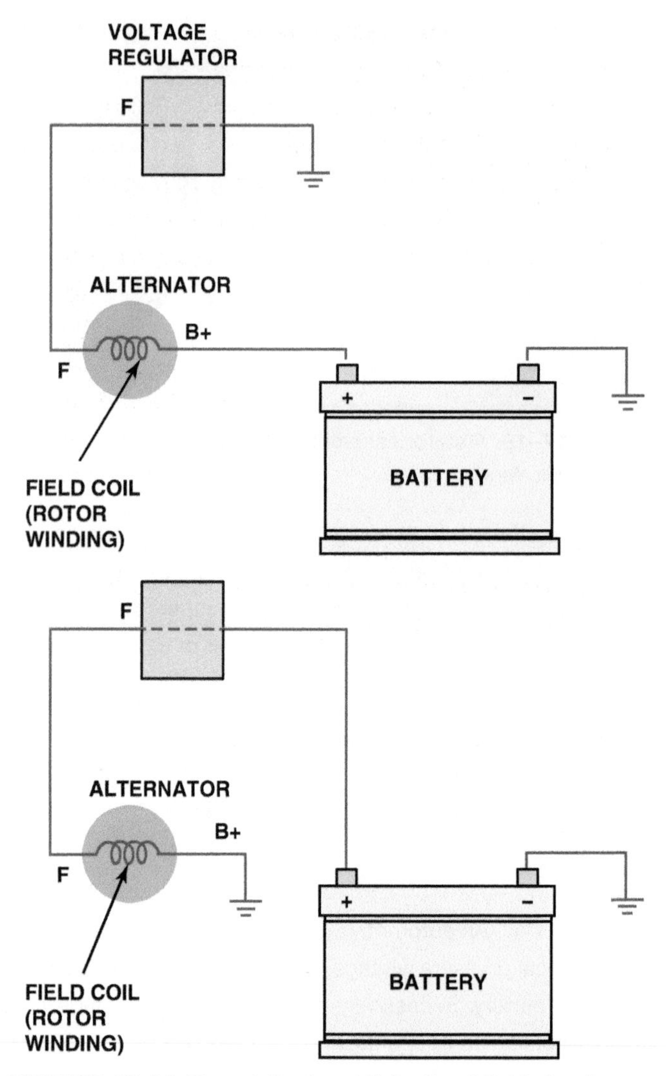

FIGURE 17–20 Type A (top) and B (bottom) field circuits.

- As the battery becomes charged and the battery voltage increases, the charging voltage will also increase, until the voltage regulator limit is reached.
- Then the voltage regulator will start to control the charging voltage. A good, but discharged, battery should be able to convert into chemical energy all the current the generator can produce. As long as generator voltage is higher than battery voltage, current will flow from the alternator (high voltage) to the battery (lower voltage).
- Therefore, if a voltmeter is connected to a discharged battery with the engine running, it may indicate charging voltage that is lower than normally acceptable.

FREQUENTLY ASKED QUESTION

What are A and B field circuits?

Older DC generators and early AC generators (alternators) all used an external regulator to control the current through the generator field coils. These are called Type A or Type B circuits, depending on how the field of the generator is energized.

A type A circuit has voltage applied to the field coils and the regulator toggles the fields to ground for voltage control.

A type B circuit has the field coils grounded internally and the regulator toggles (sends) voltage to the fields for voltage control. **SEE FIGURE 17–20.**

In other words, the condition and voltage of the battery *do* determine the charging rate of the alternator. It is often stated that the battery is the true "voltage regulator" and that the voltage regulator simply acts as the upper-limit voltage control. This is the reason why all charging system testing *must* be performed with a reliable and known to be good battery, at least 75% charged, to be assured of accurate test results. If a discharged battery is used during charging system testing, tests could mistakenly indicate a defective generator and/or voltage regulator.

TEMPERATURE COMPENSATION All voltage regulators (mechanical or electronic) provide a method for increasing the charging voltage slightly at low temperatures and for lowering the charging voltage at high temperatures. A battery requires a higher charging voltage at low temperatures because of the resistance to chemical reaction changes. However, the battery would be overcharged if the charging voltage were not reduced during warm weather. Electronic voltage regulators use a temperature-sensitive resistor in the regulator circuit. This resistor, called a **thermistor**, provides lower resistance as the temperature increases. A thermistor is used in the electronic circuits of the voltage regulator to control charging voltage over a wide range of underhood temperatures.

NOTE: Voltmeter test results may vary according to temperature. Charging voltage tested at 32°F (0°C) will be higher than for the same vehicle tested at 80°F (27°C) because of the temperature-compensation factors built into voltage regulators.

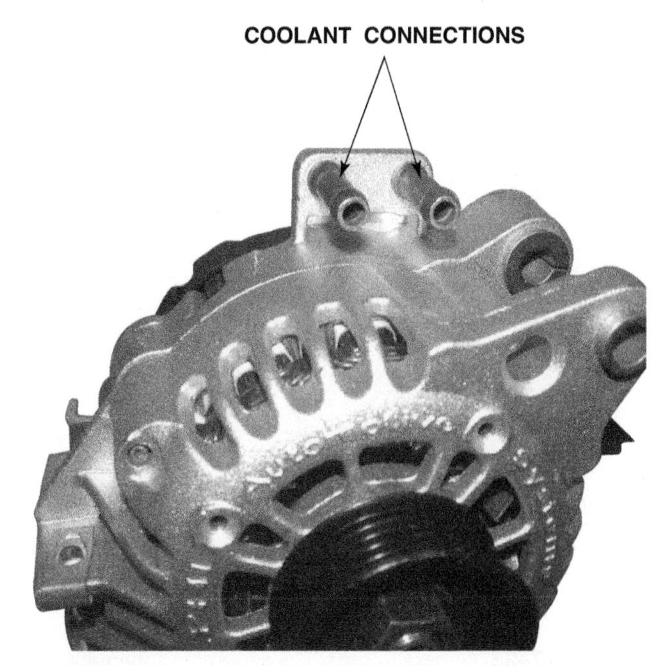

FIGURE 17–21 A liquid-cooled generator showing the hose connections where coolant from the engine flows through the rear frame of the generator.

GENERATOR COOLING

Generators create heat during normal operation and this heat must be removed to protect the components inside, especially the diodes and voltage regulator. The types of cooling include:

- External fan
- Internal fan(s)
- Both an external fan and an internal fan
- Coolant cooled (SEE FIGURE 17–21.)

CS GENERATORS

Beginning in the mid-1980s, General Motors introduced a smaller, yet higher-output, series of generators, the CS (charging system) series. Following the letters CS are numbers indicating the outside diameter of the stator laminations. **SEE FIGURE 17–22.**

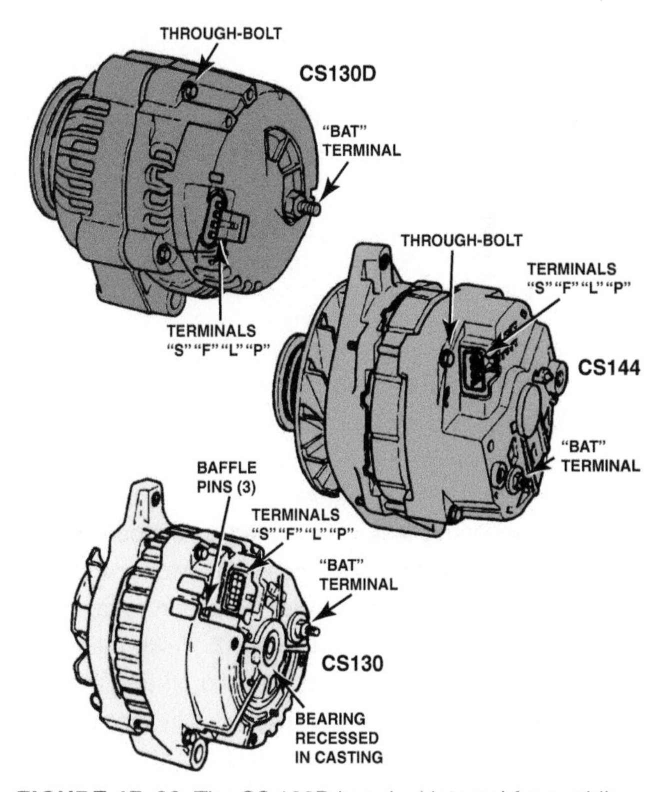

FIGURE 17–22 The CS 130D has dual internal fans, while other generators may have one external fan or both an external and internal fan. (Courtesy of General Motors)

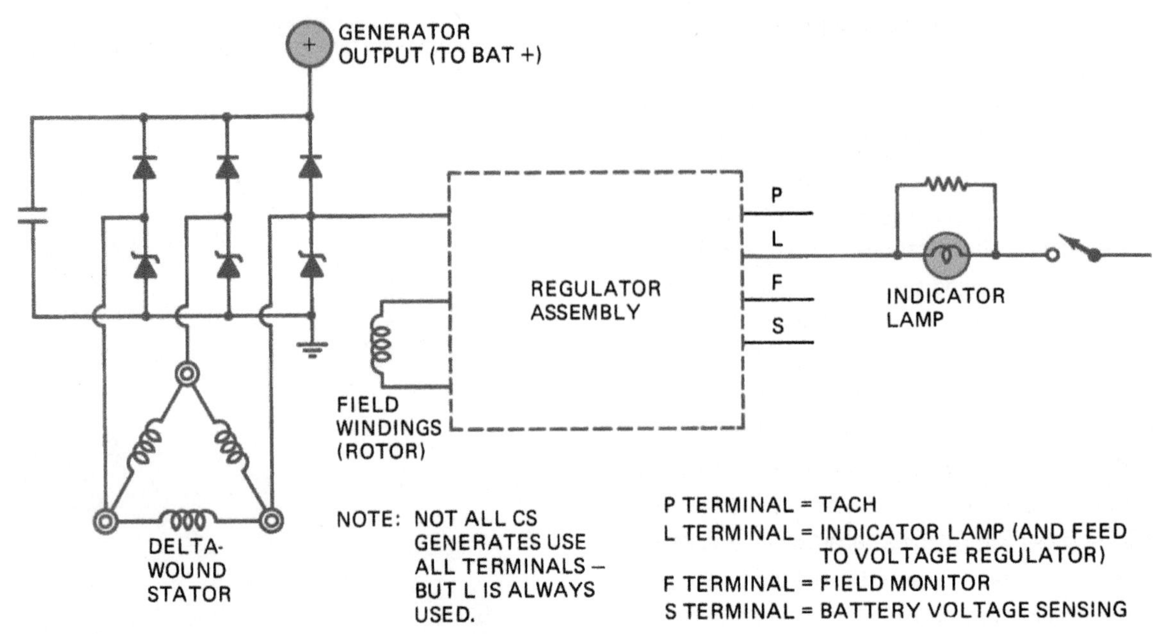

FIGURE 17–23 General Motors CS generator schematic. Notice the use of zener diodes in the rectifier to help control any high-voltage surges that could affect delicate computer circuits. If a high-voltage surge does occur, the zener diode(s) will be reverse biased and the potentially harmful voltage will be safely conducted to ground. Voltage must be present at the L terminal (12 volts in this application) to allow the generator to start charging.

Typical sizes, designations, and output include the following:

CS 121, 5-SI 74 A

CS 130, 9-SI 105 A

CS 144, 17-SI 120 A

These generators feature two cooling fans, one internal or both internal (indicated by the suffix "D") and terminals designed to permit connection to an on-board computer through terminals L and F.

The reduced size generators also feature ball bearings front and rear. The voltage is controlled by the body control module (BCM) or engine control module (ECM), depending on the year and model. On older vehicles the internal voltage regulator alone controls output. ● **SEE FIGURE 17–23**. The voltage regulator switches the field voltage on and off at a fixed frequency of about 400 Hz. Voltage is controlled by varying the on versus off time of the field current.

NOTE: Voltage must be present at the "L" terminal before the generator will begin to charge. This is usually either 12 volts or 5 volts, depending on the vehicle. Refer to service information (SI) for the vehicle being serviced.

COMPUTER-CONTROLLED GENERATORS

TYPES OF SYSTEMS Computers can interface with the charging system in three ways:

1. The computer can *activate* the charging system by turning on and off the field current to the rotor. In other words, the computer, usually the powertrain control module (PCM), controls the field current to the rotor.
2. The computer can *monitor* the operation of the generator and increase engine speed if needed during conditions when a heavy load is demanded by the generator.
3. The computer can *control* the generator by controlling generator output to match the needs of the electrical system. This system detects the electrical needs of the vehicle and commands the generator to charge only when needed to improve fuel economy.

COMPUTER CONTROL Computer control of the charging system has the following advantages:

1. The computer controls the field of the generator, which can pulse it on or off as needed for maximum efficiency, thereby saving fuel.

FIGURE 17–24 A Hall-effect current sensor attached to the positive battery cable is used as part of the EPM system.

COMMAND DUTY CYCLE	ALTERNATOR OUTPUT VOLTAGE
10%	11.0 V
20%	11.6 V
30%	12.1 V
40%	12.7 V
50%	13.3 V
60%	13.8 V
70%	14.4 V
80%	14.9 V
90%	15.5 V

CHART 17–1

The output voltage is controlled by varying the duty cycle as controlled by the PCM.

2. Engine idle can also be improved by turning on the generator slowly, rather than all at once, if an electrical load is switched on, such as the air-conditioning system.
3. Most computers can also reduce the load on the electrical system if the demand exceeds the capacity of the charging system by reducing fan speed, shutting off rear window defoggers, or increasing engine speed to cause the generator to increase the amperage output.

NOTE: A commanded higher-than-normal idle speed may be the result of the computer compensating for an abnormal electrical load. This higher idle speed could indicate a defective battery or other electrical system faults.

4. The computer can monitor the charging system and set diagnostic trouble codes (DTCs) if a fault is detected. Many systems allow the service technician to control the charging of the generator using a scan tool.
5. Because the charging system is computer controlled, it can be checked using a scan tool. Some vehicle systems allow the scan tool to activate the generator field and then monitor the output to help detect fault locations. Always follow the vehicle manufacturer's diagnostic procedure.

GM REGULATED VOLTAGE CONTROL SYSTEMS*

Regulated Voltage Control (RVC) systems, also called **electrical power management (EPM)** systems, regulate the generator's output voltage to support all electrical functions, ensure optimum performance, and improve battery life. RVC systems improve fuel economy, extend battery life, increase switch life, enhance lamp longevity, and communicate the state of the charging system to the driver.

The RVC system controls the charging system to prevent the battery from undercharging at low engine revolutions per minute (RPM), and overcharging at high engine RPM. This is accomplished by monitoring the amount of current flowing out of and into the battery. Except in the Generation I system, a Hall-effect sensor attached to the negative or positive battery cable is used to measure the current. ● **SEE FIGURE 17–24.**

The objective of the RVC system is to maintain the battery at an 80% state of charge (SOC). The engine control module (ECM) controls the generator by changing the on-time of the current through the rotor. ● **SEE FIGURE 17–25**. The on-time, called **duty cycle**, varies from 5% to 95%. ● **SEE CHART 17–1.**

There are two types of RVC systems, stand-alone RVC (SARVC) and up-integrated RVC. RVC systems consist of the following components: battery current sensor, body control module (BCM), engine control module (ECM), powertrain control module (PCM), instrument panel cluster (IPC), and generator battery control module.

NOTE: Each type of RVC system may utilize a different combination of components depending on the system generation.

*General Motors Course # 16041.09W.

FIGURE 17–25 The amount of time current is flowing through the field (rotor) determines the generator output.

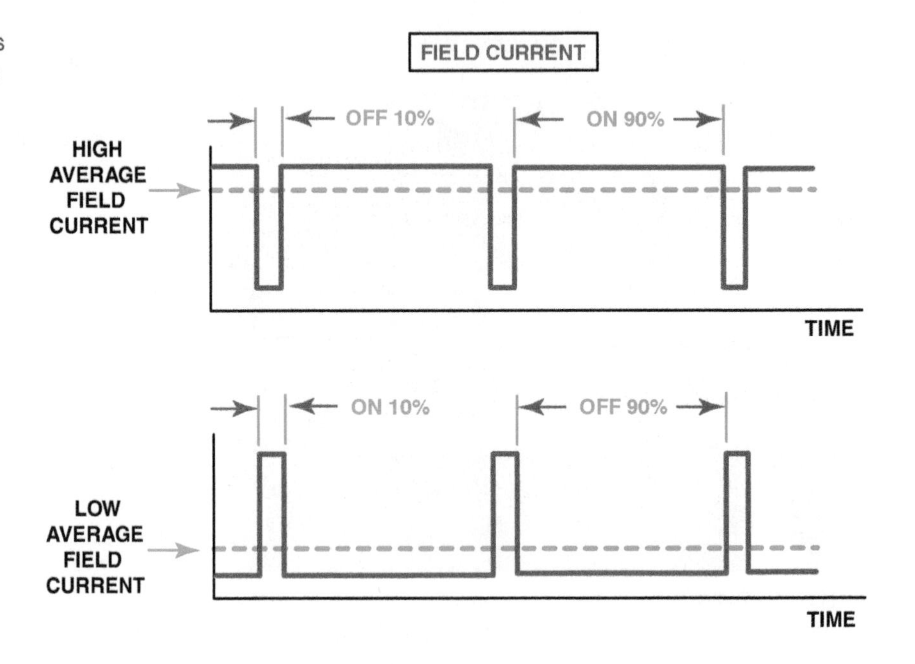

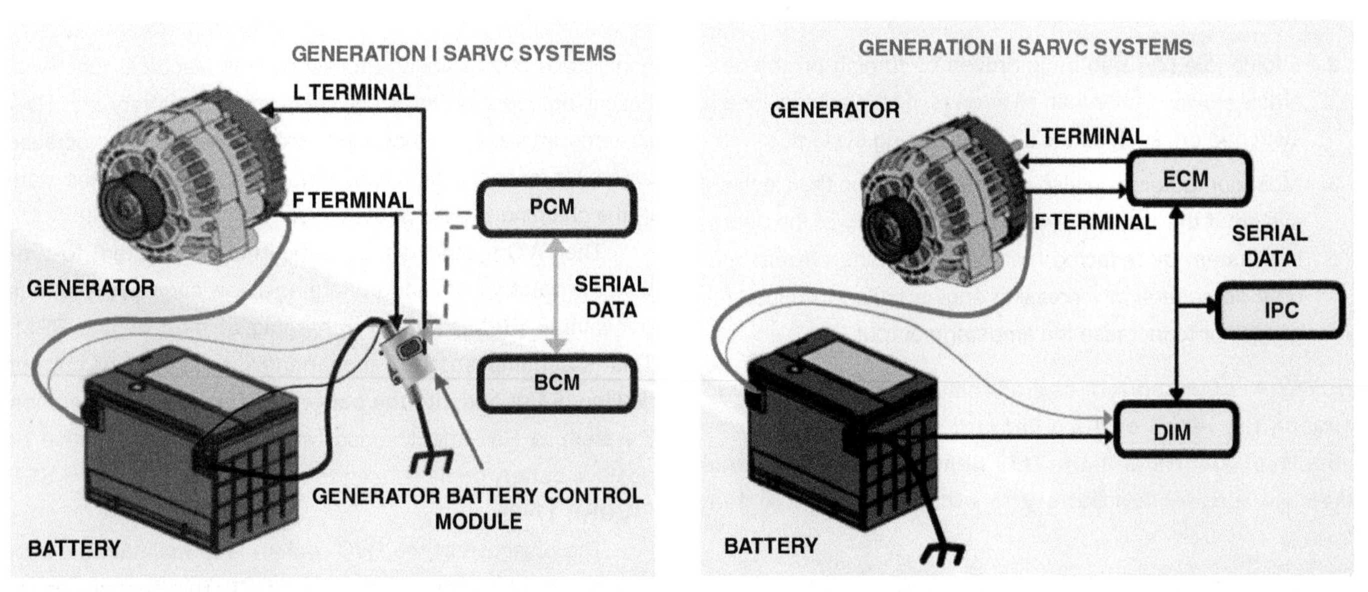

FIGURE 17–26 Generation I and II regulated voltage control (RVC) systems. The Generation II system does not use a current sensor. (Courtesy of General Motors)

There are two generations of SARVC systems, Generations I and II. ● **SEE FIGURE 17–26.**

- Generation I SARVC systems use the generator battery control module and current sensor component to interact with the BCM. In Generation I SARVC systems, the BCM communicates with the system components. The generator battery control module controls the generator voltage output and communicates with the PCM.
- Generation II SARVC systems do not use the generator battery control module or current sensor component. In generation II SARVC systems, the ECM, dash integration module (DIM), and the IPC communicate to control the generator voltage output.

There are two generations of up-integrated regulated voltage control (RVC) systems, Generations III and IV. Up-integrated systems use a battery current sensor to receive accurate voltage measurements from the battery positive terminal. ● **SEE FIGURE 17–27.**

- In Generation III up-integrated systems, the generator, ECM, battery current sensor, IPM, and the IPC communicate to control the generator voltage output.
- In Generation IV up-integrated systems, the generator, PCM, battery current sensor, BCM, and the IPC communicate to control the generator voltage output.

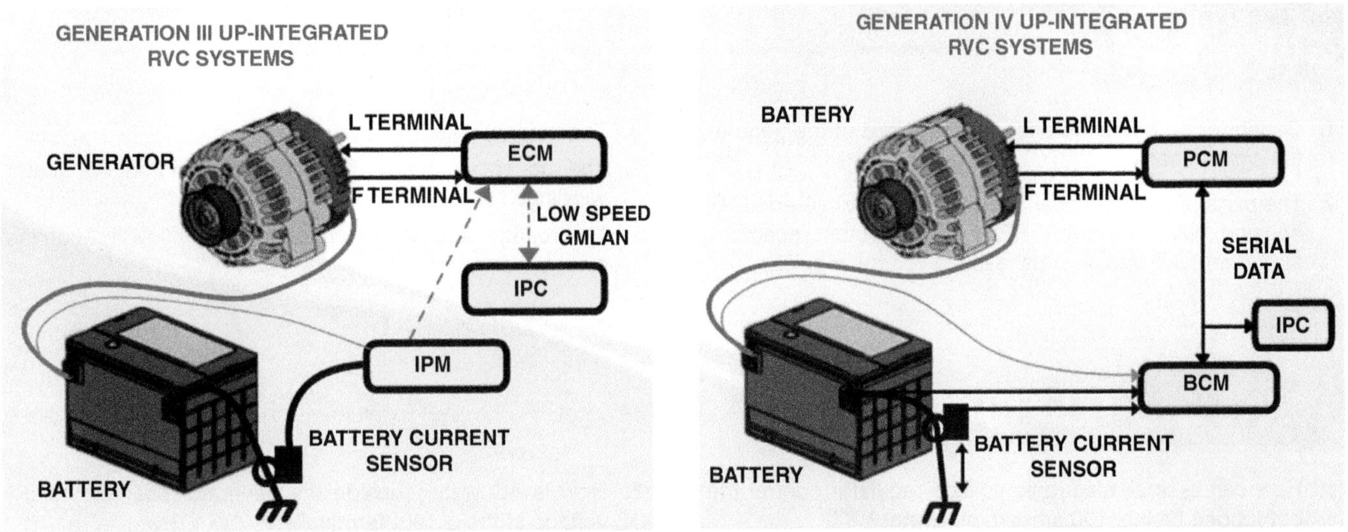

FIGURE 17–27 Generation III and IV up-integrated RVC systems both use a Hall-effect current sensor. (Courtesy of General Motors)

MODES OF OPERATION Most systems have these six modes of operation.

1. **Charge mode.** The charge mode is activated when any of the following occurs:
   - Electric cooling fans are on high speed.
   - Rear window defogger is on.
   - Battery state of charge (SOC) is less than 80%.
   - Outside (ambient) temperature is less than 32°F (0°C).
2. **Fuel economy mode.** This mode reduces the load on the engine from the generator for maximum fuel economy. This mode is activated when the following conditions are met:
   - Ambient temperature is above 32°F (0°C).
   - The SOC of the battery is 80% or higher.
   - The cooling fans and rear defogger are off.

   The target voltage is 13 volts and will return to the charge mode, if needed.
3. **Voltage reduction mode.** This mode is commanded to reduce the stress on the battery during low-load conditions. This mode is activated when the following conditions are met:
   - Ambient temperature is above 32°F (0°C).
   - Battery discharge rate is less than 7 amperes.
   - Rear defogger is off.
   - Cooling fans are on low or off.
   - Target voltage is limited to 12.7 volts.

TECH TIP

The Voltage Display Can Be a Customer Concern

A customer may complain that the voltmeter reading on the dash fluctuates up and down. This may be normal as the computer-controlled charging system commands various modes of operation based on the operating conditions. Follow the vehicle manufacturer's recommended procedures to verify proper operation.

4. **Start-up mode.** This mode is selected after engine start and commands a charging voltage of 14.5 volts for 30 seconds. After 30 seconds, the mode is changed depending on conditions.
5. **Battery sulfation mode.** This mode is commanded if the output voltage is less than 13.2 volts for 45 minutes, which can indicate that sulfated plates could be the cause. The target voltage is 13.9 to 15.5 volts for three minutes. After three minutes, the system returns to another mode based on conditions.
6. **Headlight mode.** This mode is selected when the headlights are on and the target voltage is 14.5 volts.

SUMMARY

1. Generator output is increased if the speed of the generator is increased.
2. The parts of a typical alternator include the drive-end (DE) housing, slip-ring-end (SRE) housing, rotor assembly, stator, rectifier bridge, brushes, and voltage regulator.
3. The magnetic field is created in the rotor.
4. The generator output current is created in the stator windings.
5. The voltage regulator controls the current flow through the rotor winding.

REVIEW QUESTIONS

1. How can a small electronic voltage regulator control the output of a typical 100 ampere generator?
2. What are the component parts of a typical generator?
3. How is the computer used to control an generator?
4. Why do voltage regulators include temperature compensation?
5. How is AC voltage inside the generator changed to DC voltage at the output terminal?
6. What is the purpose of an OAP or OAD?

CHAPTER QUIZ

1. Technician A says that the diodes regulate the generator output voltage. Technician B says that the field current can be computer controlled. Which technician is correct?
   - **a.** Technician A only
   - **b.** Technician B only
   - **c.** Both Technicians A and B
   - **d.** Neither Technician A nor B
2. A magnetic field is created in the ________ in an generator (AC alternator).
   - **a.** Stator
   - **b.** Diodes
   - **c.** Rotor
   - **d.** Drive-end frame
3. The voltage regulator controls current through the ________.
   - **a.** Generator brushes
   - **b.** Rotor
   - **c.** Generator field
   - **d.** All of the above
4. Technician A says that two diodes are required for each stator winding lead. Technician B says that diodes change alternating current into direct current. Which technician is correct?
   - **a.** Technician A only
   - **b.** Technician B only
   - **c.** Both Technicians A and B
   - **d.** Neither Technician A nor B
5. The generator output current is produced in the ________.
   - **a.** Stator
   - **b.** Rotor
   - **c.** Brushes
   - **d.** Diodes (rectifier bridge)
6. Generator brushes are constructed from ________.
   - **a.** Copper
   - **b.** Aluminum
   - **c.** Carbon
   - **d.** Silver-copper alloy
7. How much current flows through the generator brushes?
   - **a.** All of the generator output flows through the brushes
   - **b.** 25 to 35 amperes, depending on the vehicle
   - **c.** 10 to 15 amperes
   - **d.** 2 to 5 amperes
8. Technician A says that an alternator overrunning pulley is used to reduce vibration and noise. Technician B says that an overrunning alternator pulley or dampener uses a one-way clutch. Which technician is correct?
   - **a.** Technician A only
   - **b.** Technician B only
   - **c.** Both Technicians A and B
   - **d.** Neither Technician A nor B
9. Operating an generator in a vehicle with a defective battery can harm the ________.
   - **a.** Diodes (rectifier bridge)
   - **b.** Stator
   - **c.** Voltage regulator
   - **d.** Brushes
10. Technician A says that a wye-wound stator produces more maximum output than the same generator equipped with a delta-wound stator. Technician B says that a generator equipped with a delta-wound stator produces more maximum output than a wye-wound stator. Which technician is correct?
    - **a.** Technician A only
    - **b.** Technician B only
    - **c.** Both Technicians A and B
    - **d.** Neither Technician A nor B

chapter 18

CHARGING SYSTEM DIAGNOSIS AND SERVICE

LEARNING OBJECTIVES

After studying this chapter, the reader will be able to:

1. Discuss the various methods to test the charging system.
2. Discuss the generator output test.
3. Explain how to disassemble a generator and test its component parts.

This chapter will help you prepare for the ASE Electrical/Electronic Systems (A6) certification test content area "D" (Charging System Diagnosis and Repair).

KEY TERMS

GM STC OBJECTIVES

GM Service Technical College topics covered in this chapter are as follows:

1. GM charging systems, overview, operation, and diagnosis.
2. Identify charging system characteristics.
3. Identify methods in using the Tech 2 or other scan tool in performing diagnostic procedures.
4. Identify charging system diagnosis.

FIGURE 18–1 The digital multimeter should be set to read DC volts, with the red lead connected to the positive (+) battery terminal and the black meter lead connected to the negative (−) battery terminal.

? FREQUENTLY ASKED QUESTION

What Is a Full-Fielding Test?

Full fielding is a procedure used on older noncomputerized vehicles for bypassing the voltage regulator that could be used to determine if the generator is capable of producing its designed output. This test is no longer performed for the following reasons:

- The voltage regulator is built into the generator, therefore requiring that the entire assembly be replaced even if just the regulator is defective.
- When the regulator is bypassed, the generator can produce a high voltage (over 100 volts in some cases), which could damage all of the electronic circuits in the vehicle.

Always follow the vehicle manufacturer's recommended testing procedures.

CHARGING SYSTEM TESTING AND SERVICE

BATTERY STATE OF CHARGE The charging system can be tested as part of a routine vehicle inspection or to determine the reason for a no-charge or reduced charging circuit performance. The battery *must* be at least 75% charged before testing the generator and the charging system. A weak or defective battery will cause inaccurate test results. If in doubt, replace the battery with a known good shop battery for testing.

CHARGING VOLTAGE TEST The **charging voltage test** is the easiest way to check the charging system voltage at the battery. Use a digital multimeter to check the voltage, as follows:

STEP 1 Select DC volts.

STEP 2 Connect the red meter lead to the positive (+) terminal of the battery and the black meter lead to the negative (−) terminal of the battery.

NOTE: The polarity of the meter leads is not too important when using a digital multimeter. If the meter leads are connected backward on the battery, the resulting readout will simply have a negative (−) sign in front of the voltage reading.

STEP 3 Start the engine and increase the engine speed to about 2000 RPM (fast idle) and record the charging voltage. ● **SEE FIGURE 18–1.**

Specifications for charging voltage = 13.5 to 15 volts

- If the voltage is too high, check that the generator is properly grounded.
- If the voltage is lower than specifications, then there is a fault with the wiring or the generator.
- If the wiring and the connections are okay, then additional testing is required to help pinpoint the root cause. Replacement of the generator and/or battery is often required if the charging voltage is not within factory specifications.

SCAN TESTING THE CHARGING CIRCUIT Most vehicles that use a computer-controlled charging system can be diagnosed using a scan tool. Not only can the charging voltage be monitored, but also in many vehicles, the field circuit can be controlled and the output voltage monitored to check that the system is operating correctly. ● **SEE FIGURE 18–2.**

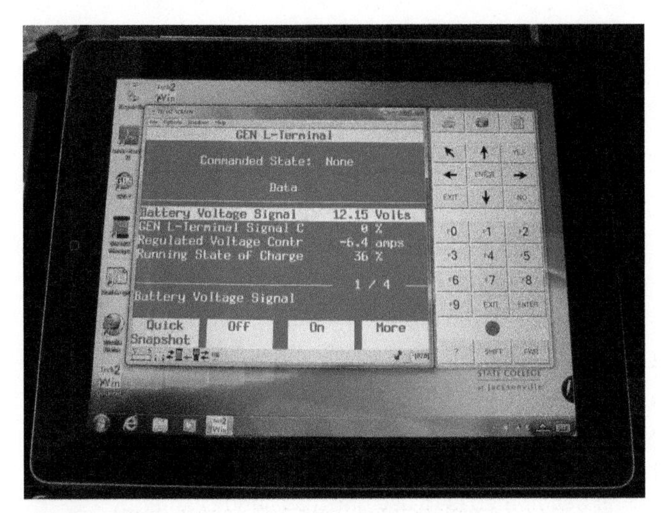

FIGURE 18–2 A scan tool can be used to diagnose charging system problems. (Courtesy of Jeffrey Rehkopf)

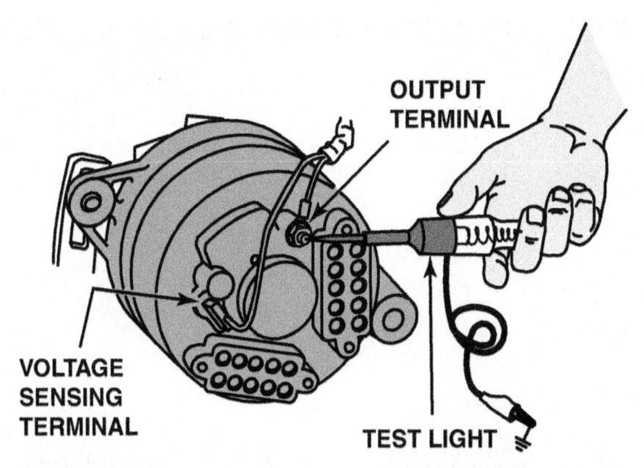

FIGURE 18–3 Before replacing a generator, the wise technician checks that battery voltage is present at the output and battery voltage sense terminals. if no voltage is detected, then there is a fault in the wiring.

TECH TIP

Use a Test Light to Check for a Defective Fusible Link

Most generators use a fusible link or mega fuse between the output terminal and the positive (+) terminal of the battery. If this fusible link or fuse is defective (blown), then the charging system will not operate at all. Many generators have been replaced repeatedly because of a blown fusible link that was not discovered until later. A quick and easy test to check if the fusible link is okay is to touch a test light to the output terminal. With the other end of the test light attached to a good ground, the fusible link or mega fuse is okay if the light comes on. This test confirms that the circuit between the generator and the battery has continuity. ● **SEE FIGURE 18–3.**

FIGURE 18–4 This accessory drive belt is worn and requires replacement. Newer belts are made from ethylene propylene diene monomer (EPDM). This rubber does not crack like older belts and may not show wear even though the ribs do wear and can cause slippage.

DRIVE BELT INSPECTION AND ADJUSTMENT

BELT VISUAL INSPECTION It is generally recommended that all belts be inspected regularly and replaced as needed. Replace any serpentine belt that has more than three cracks in any one rib that appears in a 3 inch span. Check service information for the specified procedure and recommended replacement interval. ● **SEE FIGURE 18–4.**

BELT TENSION MEASUREMENT If the vehicle does not use a belt tensioner, then a belt tension gauge is needed to achieve the specified belt tension. Install the belt and operate the engine with all of the accessories turned on to "run-in" the belt for at least five minutes. Adjust the tension of the accessory drive belt to factory specifications or use the following table for an example of the proper tension based on the size of the belt.

There are four ways that vehicle manufacturers specify that the belt tension is within factory specifications:

1. **Belt tension gauge.** A belt tension gauge is needed to determine if it is at the specified belt tension. Install the belt and operate the engine with all of the accessories

FIGURE 18–5 Check service information for the exact marks where the tensioner should be located for proper belt tension.

SERPENTINE BELTS	
NUMBER OF RIBS USED	**TENSION RANGE (LB)**
3	45–60
4	60–80
5	75–100
6	90–125
7	105–145
V-BELTS	
V-BELT TOP WIDTH (in.)	**TENSION RANGE (LB)**
1/4	45–65
5/16	60–85
25/64	85–115
31/64	105–145

CHART 18–1

Typical belt tension for various widths of belts. Tension is the force needed to depress the belt as displayed on a belt tension gauge.

turned on to "run-in" the belt for at least five minutes. Adjust the tension of the accessory drive belt to factory specifications, or see ● **CHART 18–1** for an example of the proper tension based on the size of the belt.

2. **Marks on a tensioner.** Many tensioners have marks that indicate the normal operating tension range for the accessory drive belt. Check service information for the preferred location of the tensioner mark. ● **SEE FIGURE 18–5.**
3. **Torque wrench reading.** Some vehicle manufacturers specify that a beam-type torque wrench be used to determine the torque needed to rotate the tensioner. If the torque reading is below specifications, the tensioner must be replaced.

TECH TIP

The Hand Cleaner Trick

Lower-than-normal generator output could be the result of a loose or slipping drive belt. All belts (V and serpentine multigroove) use an interference angle between the angle of the Vs of the belt and the angle of the Vs on the pulley. As the belt wears, the interference angles are worn off of both edges of the belt. As a result, the belt may start to slip and make a squealing sound even if tensioned properly.

A common trick used to determine if the noise is belt related is to use grit-type hand cleaner or scouring powder. With the engine off, sprinkle some powder onto the pulley side of the belt. Start the engine. The excess powder will fly into the air, so get away from under the hood when the engine starts. If the belts are now quieter, you know that it was the glazed belt that made the noise.

The noise can sound exactly like a noisy bearing. Therefore, before you start removing and replacing parts, try the hand cleaner trick.

Often, the grit from the hand cleaner will remove the glaze from the belt and the noise will not return. However, if the belt is worn or loose, the noise will return and the belt should be replaced. A fast, alternative method to see if the noise is from the belt is to spray water from a squirt bottle at the belt with the engine running. If the noise stops, the belt is the cause of the noise. The water quickly evaporates and, therefore, unlike the gritty hand cleaner, water simply finds the problem—it does not provide a short-term fix.

4. **Deflection.** Depress the belt between the two pulleys that are the farthest apart; the flex or deflection should be 1/2 inch (13 mm).

AC RIPPLE VOLTAGE CHECK

PRINCIPLES A good generator should produce very little AC voltage or current output. It is the purpose of the diodes in the generator to rectify or convert most AC voltage into DC voltage. While it is normal to measure some AC voltage from

FIGURE 18–6 This overrunning generator dampener (OAD) is longer than an overrunning generator pulley (OGP) because it contains a dampener spring as well as a one-way clutch. Be sure to check that it locks in one direction.

TECH TIP

Check the Overrunning Clutch

If low or no generator output is found, remove the generator drive belt and check the overrunning generator pulley (OGP) or overrunning generator dampener (OAD) for proper operation. Both types of overrunning clutches use a one-way clutch. Therefore, the pulley should freewheel in one direction and rotate the generator rotor when rotated in the opposite direction. ● **SEE FIGURE 18–6.**

a generator, excessive AC voltage, called AC ripple, is undesirable and indicates a fault with the rectifier diodes or stator windings inside the generator.

TESTING AC RIPPLE VOLTAGE The procedure to check for **AC ripple voltage** includes the following steps:

STEP 1 Set the digital meter to read AC volts.

STEP 2 Start the engine and operate it at 2000 RPM (fast idle).

STEP 3 Connect the voltmeter leads to the positive and negative battery terminals.

STEP 4 Turn on the headlights to provide an electrical load on the generator.

NOTE: A more accurate reading can be obtained by touching the meter lead to the output or "battery" terminal of the generator. ● **SEE FIGURE 18–7.**

The results should be interpreted as follows: If the rectifier diodes and stator are good, the voltmeter should read *less* than 400 millivolts (0.4 volt) AC. If the reading is over 500 millivolts (0.5 volt) AC, the rectifier diodes or stator are defective.

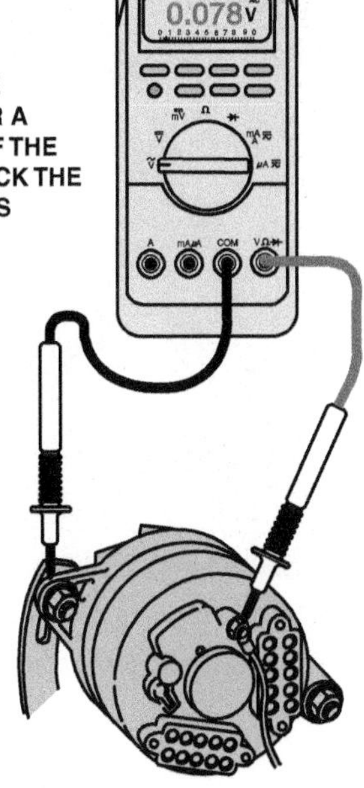

FIGURE 18–7 Testing AC ripple at the output terminal of the generator is more accurate than testing at the battery due to the resistance of the wiring between the generator and the battery. The reading shown on the meter, set to AC volts, is only 78 millivolts (0.078 volt), far below what the reading would be if a diode were defective.

NOTE: Many conductance testers, such as Midtronic and Snap-On, automatically test for AC ripple.

TESTING AC RIPPLE CURRENT

All generators should create direct current (DC) if the diodes and stator windings are functioning correctly. A mini clamp-on meter capable of measuring AC amperes can be used to check the generator. A good generator should produce less than 10% of its rated amperage output in AC ripple amperes. For example, a generator rated at 100 amperes should not produce more than 10 amperes AC ripple ($100 \times 10\% = 10$). It is normal for a good generator to produce 3 or 4 amperes of AC ripple current to the battery. Only if the AC ripple current exceeds 10% of the rating of the generator should the generator be repaired or replaced.

TEST PROCEDURE To measure the AC current to the battery, perform the following steps.

STEP 1 Start the engine and turn on the lights to create an electrical load on the generator.

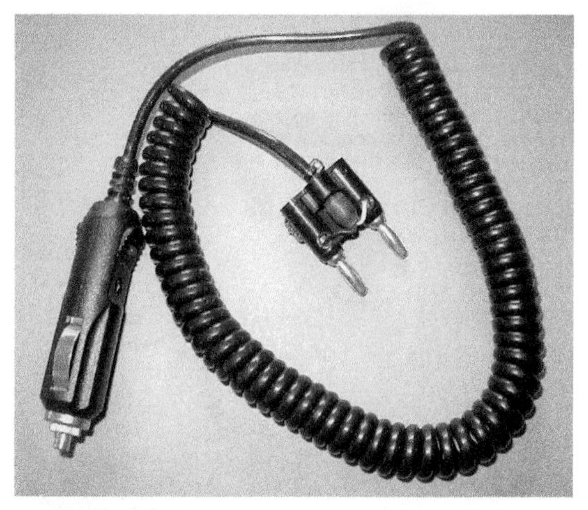

FIGURE 18–8 Charging system voltage can be easily checked at the lighter plug by connecting a lighter plug to the voltmeter through a double banana plug.

FIGURE 18–9 A mini clamp-on meter can be used to measure generator output as shown here (105.2 amperes). Then the meter can be used to check AC current ripple by selecting AC amps on the rotary dial. AC ripple current should be less than 10% of the DC current output.

TECH TIP

The Lighter Plug Trick

Battery voltage measurements can be read through the lighter socket. Simply construct a test tool using a lighter plug at one end of a length of two-conductor wire and the other end connected to a double banana plug. The double banana plug will fit most meters in the common (COM) terminal and the volt terminal of the meter. This is handy to use while road-testing the vehicle under real-life conditions. Both DC voltage and AC ripple voltage can be measured. ● **SEE FIGURE 18–8.**

STEP 2 Using a mini clamp-on digital multimeter, place the clamp around either all of the positive (+) battery cables or all of the negative (−) battery cables.

An AC/DC current clamp adapter can also be used with a conventional digital multimeter set on the DC millivolts scale.

STEP 3 To check for AC current ripple, switch the meter to read AC amperes and record the reading. Read the meter display.

STEP 4 The results should be within 10% of the specified generator rating. A reading of greater than 10 amperes AC indicates defective generator diodes. ● **SEE FIGURE 18–9.**

CHARGING SYSTEM VOLTAGE DROP TESTING

GENERATOR WIRING For the proper operation of any charging system, there must be good electrical connections between the battery positive terminal and the generator output terminal. The generator must also be properly grounded to the engine block.

Many manufacturers of vehicles run the lead from the output terminal of the generator to other connectors or junction blocks that are electrically connected to the positive terminal of the battery. If there is high resistance (a high voltage drop) in these connections or in the wiring itself, the battery will not be properly charged.

VOLTAGE DROP TEST PROCEDURE When there is a suspected charging system problem (with or without a charge indicator light on), simply follow these steps to measure the voltage drop of the insulated (power-side) charging circuit.

STEP 1 Start the engine and run it at a fast idle (about 2000 engine RPM).

STEP 2 Turn on the headlights to ensure an electrical load on the charging system.

STEP 3 Using any voltmeter set to read DC volts, connect the positive test lead (red) to the output terminal of the generator. Attach the negative test lead (black) to the positive post of the battery.

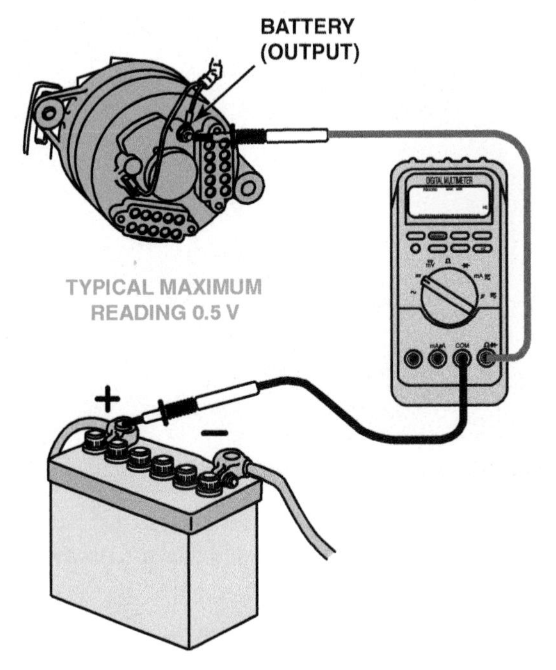

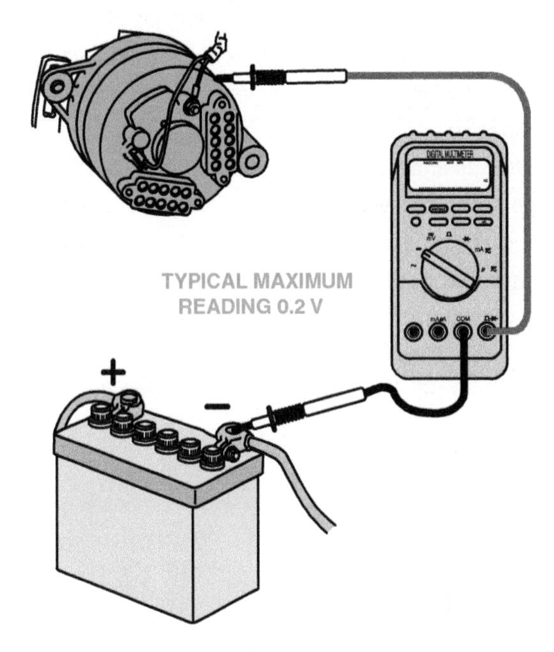

FIGURE 18–10 Voltmeter hookup to test the voltage drop of the charging circuit.

The results should be interpreted as follows:

1. If there is less than a 0.5 volt (500 millivolt) reading, then all wiring and connections are satisfactory.
2. If the voltmeter reads higher than 0.5 volt, there is excessive resistance (voltage drop) between the generator output terminal and the positive terminal of the battery.
3. If the voltmeter reads battery voltage (or close to battery voltage), there is an open circuit between the battery and the generator output terminal.

To determine whether the generator is correctly grounded, maintain the engine speed at 2000 RPM with the headlights on. Connect the positive voltmeter lead to the case of the generator and the negative voltmeter lead to the negative terminal of the battery. The voltmeter should read less than 0.2 volt (200 millivolts) if the generator is properly grounded. If the reading is over 0.2 volt, connect one end of an auxiliary ground wire to the case of the generator and the other end to a good engine ground. ● **SEE FIGURE 18–10.**

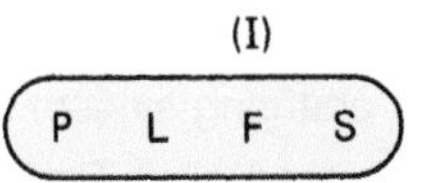

FIGURE 18–11 Typical GM CS generator wiring plug identification. Note that terminal F is sometimes labeled "I" on some generators.

GENERAL MOTORS CS SERIES AC GENERATORS

A General Motors CS series generator requires only two wires to operate–the battery (BAT feed and the wire to the L terminal. The CS series generators are designed to operate as stand-alone generators or be controlled by a vehicle computer system. ● **SEE FIGURE 18–11** for terminal identification.

The P terminal is connected directly to the stator winding that produces about one-half of the system alternating current and is used as a tachometer signal. *P* is used because it is

TECH TIP

Use a Fused Jumper Wire as a Diagnostic Tool

When diagnosing a generator charging problem, try using a fused jumper wire to connect the positive and negative terminals of the generator directly to the positive and negative terminals of the battery. If a definite improvement is noticed, the problem is in the wiring of the vehicle. High resistance, due to corroded connections or loose grounds, can cause low generator output, repeated regulator failures, slow cranking, and discharged batteries. A voltage drop test of the charging system can also be used to locate excessive resistance (high voltage drop) in the charging circuit, but using a fused jumper wire is often faster and easier.

FIGURE 18–12 A typical tester used to test batteries as well as the cranking and charging system. Always follow the operating instructions.

an abbreviation for "phase", an output for the stator phases. Terminal S is the sensing terminal for true battery voltage. Terminal F (for *feedback*) is the computer-sensing terminal. The computer monitors this terminal, sets trouble codes, and alerts the driver if there is a charging system malfunction. The letter *I* is sometimes used instead of *F*. The L terminal (for lamp) is used to turn on or control the voltage regulator, thus initiating charging system operation.

DIAGNOSING CS SERIES GENERATORS If the charge indicator is on in the dash, unplug the connector at the generator (which can have up to four wires). Start the engine and observe the dash charge light. If the light is still on, there is a short to ground in the L wire circuit between the generator and the dash. If the light is out, check for voltage at L terminal of the connector.

If there is voltage at the L terminal (remember, the connector is still unplugged), the problem is in the generator, if charging is not occurring.

If there is no voltage available at the L terminal, apply a voltage through a standard test light to the L terminal of the generator. This supplies the power for the regulator. (This is not full-fielding the generator, only supplying power to the regulator.) If the generator output is now normal, the problem is in the wiring to the L terminal of the generator. Check all fuses, all fusible links, and the charge indicator light.

GENERATOR OUTPUT TEST

PRELIMINARY CHECKS A generator output test measures the current (amperes) of the generator. A charging circuit may be able to produce correct charging circuit voltage, but not be able to produce adequate amperage output. If in doubt about charging system output, first check the condition of the generator drive belt. With the engine off, attempt to rotate the fan of the generator by hand. Replace or tighten the drive belt if the generator fan can be rotated this way.

CARBON PILE TEST PROCEDURE A carbon pile tester uses plates of carbon to create an electrical load. A carbon pile test is used to load test a battery and/or a generator. ● **SEE FIGURE 18–12**.

The testing procedure for generator output is as follows:

STEP 1 Connect the starting and charging test leads according to the manufacturer's instructions, which usually include installing the amp clamp around the output wire near the generator.

STEP 2 Turn off all electrical accessories to be sure that the tester is measuring the true output of the generator.

STEP 3 Start the engine and operate it at 2000 RPM (fast idle). Turn the load increase control slowly to obtain the highest reading on the ammeter scale. Do not allow the voltage to drop below 12.6 volts. Note the ampere reading.

STEP 4 Add 5 to 7 amperes to the reading because this amount of current is used by the ignition system to operate the engine.

STEP 5 Compare the output reading to factory specifications. The rated output may be stamped on the generator or can be found in service information.

CAUTION: *NEVER* disconnect a battery cable with the engine running. All vehicle manufacturers warn not to do this, because this was an old test, before generators, to see if a generator could supply current to operate the ignition system without a battery. When a battery cable is removed, the generator (or PCM) will lose the battery voltage sense signal. Without a battery voltage sense circuit, the generator will do one of two things, depending on the make and model of vehicle.

- **The generator output can exceed 100 volts. This high voltage may not only damage the generator but also electrical components in the vehicle, including the PCM and all electronic devices.**
- **The generator stops charging as a fail safe measure to protect the generator and all of the electronics in the vehicle from being damaged due to excessively high voltage.**

MINIMUM REQUIRED GENERATOR OUTPUT

PURPOSE All charging systems must be able to supply the electrical demands of the electrical system. If lights and accessories are used constantly and the generator cannot supply the necessary ampere output, the battery will be drained. To determine the minimum electrical load requirements, connect an inductive ammeter probe around either battery cable or the generator output cable. ● **SEE FIGURE 18–13.**

NOTE: If using an inductive pickup ammeter, be certain that the pickup is over *all* the wires leaving the battery terminal.

Failure to include the small body ground wire from the negative battery terminal to the body or the small positive wire (if testing from the positive side) will *greatly decrease* the current flow readings.

PROCEDURE After connecting an ammeter correctly in the battery circuit, continue as follows:

1. Start the engine and operate to about 2000 RPM (fast idle).
2. Turn the heat selector to air-conditioning (if the vehicle is so equipped).

FIGURE 18–13 The best place to install a charging system tester amp probe is around the generator output terminal wire, as shown in the figure.

TECH TIP

Bigger Is Not Always Better

Many technicians are asked to install a higher output generator to allow the use of emergency equipment or other high-amperage equipment such as a high-wattage sound system.

Although many higher output units can be physically installed, it is important not to forget to upgrade the wiring and the fusible link(s) in the generator circuit. Failure to upgrade the wiring could lead to overheating. The usual failure locations are at junctions or electrical connectors.

3. Turn the blower motor to high speed.
4. Turn the headlights on bright.
5. Turn on the rear defogger.
6. Turn on the windshield wipers.
7. Turn on any other accessories that may be used continuously (do not operate the horn, power door locks, or other units that are not used for more than a few seconds).
8. Observe the ammeter. The current indicated is the electrical load that the generator is able to exceed to keep the battery fully charged.

TEST RESULTS The minimum acceptable generator output is 5 amperes greater than the accessory load. A negative (discharge) reading indicates that the generator is not capable of supplying the current (amperes) that may be needed.

FIGURE 18–14 Replacing a generator is not always as easy as it is from a Buick with a 3800 V-6, where the generator is easy to access. Many generators are difficult to access, and require the removal of other components.

GENERATOR REMOVAL

After diagnosis of the charging system has determined that there is a fault with the generator, it must be removed safely from the vehicle. Always check service information for the exact procedure to follow on the vehicle being serviced. A typical removal procedure includes the following steps.

STEP 1 Before disconnecting the negative battery cable, use a test light or a voltmeter and check for battery voltage at the output terminal of the generator. A complete circuit must exist between the generator and the battery. If there is no voltage at the generator output terminal, check for a blown fusible link or other electrical circuit fault.

STEP 2 Disconnect the negative (−) terminal from the battery. (Use a memory saver to maintain radio, memory seats, and other functions.)

STEP 3 Remove the accessory drive belt that drives the generator.

STEP 4 Remove electrical wiring, fasteners, spacers, and brackets, as necessary, and remove the generator from the vehicle. ● **SEE FIGURE 18–14.**

GENERATOR DISASSEMBLY

NOTE: The generator is normally replaced with a new or remanufactured unit rather than being repaired by the technician. The process is shown here and at the end of the chapter for information purposes.

FIGURE 18–15 Always mark the case of the generator before disassembly to make it easy for correct reassembly.

TECH TIP

The Sniff Test

When checking for the root cause of a generator failure, one test that a technician could do is to sniff (smell) the generator. If the generator smells like a dead rat (rancid smell), the stator windings have been overheated by trying to charge a discharged or defective battery. If the battery voltage is continuously low, the voltage regulator will continue supplying full-field current to the generator. The voltage regulator is designed to cycle on and off to maintain a narrow charging system voltage range.

If the battery voltage is continually below the cutoff point of the voltage regulator, the generator is continually producing current in the stator windings. This constant charging can often overheat the stator and burn the insulating varnish covering the stator windings. If the generator fails the sniff test, the technician should replace the stator and other generator components that are found to be defective *and* replace or recharge and test the battery.

DISASSEMBLY PROCEDURE

STEP 1 Mark the case with a scratch or with chalk to ensure proper reassembly of the generator case. ● **SEE FIGURE 18–15.**

STEP 2 After the through bolts have been removed, carefully separate the two halves. The stator windings must stay with the rear case. When this happens, the brushes and springs will fall out.

STEP 3 Remove the rectifier assembly and voltage regulator.

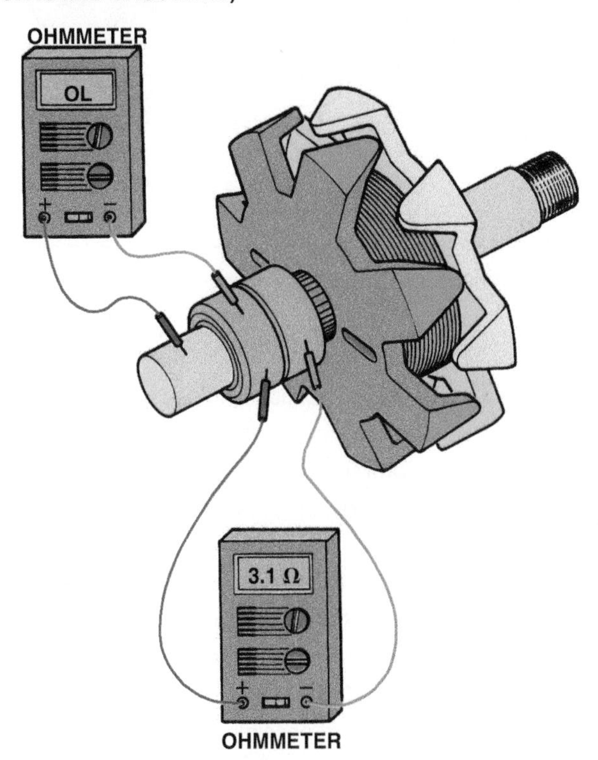

FIGURE 18–16 Testing a generator rotor using an ohmmeter.

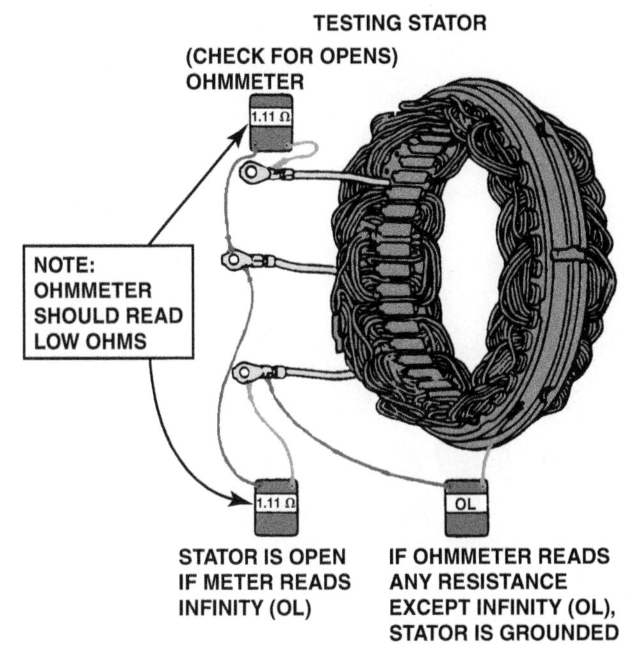

FIGURE 18–17 If the ohmmeter reads infinity between any two of the three stator windings, the stator is open and, therefore, defective. The ohmmeter should read infinity between any stator lead and the steel laminations. If the reading is less than infinity, the stator is grounded. Stator windings cannot be tested if shorted because the normal resistance is very low.

ROTOR TESTING The slip rings on the rotor should be smooth and round (within 0.002 inch of being perfectly round).

- If grooved, the slip rings can be machined to provide a suitable surface for the brushes. Do not machine beyond the minimum slip-ring dimension as specified by the manufacturer.
- If the slip rings are discolored or dirty, they can be cleaned with 400-grit or fine emery (polishing) cloth. The rotor must be turned while being cleaned to prevent flat spots on the slip rings.
- Measure the resistance between the slip rings using an ohmmeter. Typical resistance values and results include the following:

1. The resistance measured between either slip ring and the steel rotor shaft should be infinity (OL). If there is continuity, then the rotor is shorted to ground.
2. Rotor resistance range is normally between 2.4 and 6 ohms.
3. If the resistance is below specification, the rotor is shorted.
4. If the resistance is above specification, the rotor connections are corroded or open.

If the rotor is found to be bad, it must be replaced or repaired at a specialized shop. ● **SEE FIGURE 18–16.**

NOTE: The cost of a replacement rotor may exceed the cost of an entire rebuilt generator. Be certain, however, that the rebuilt generator is rated at the same output as the original or higher.

STATOR TESTING The stator must be disconnected from the diodes (rectifiers) before testing. Because all three windings of the stator are electrically connected (either wye or delta), an ohmmeter can be used to check a stator.

- There should be low resistance at all three stator leads (continuity).
- There should *not* be continuity (in other words, there should be a meter reading of infinity ohms) when the stator is tested between any stator lead and the metal stator core.
- If there is continuity, the stator is shorted-to-ground and must be repaired or replaced. ● **SEE FIGURE 18–17.**

NOTE: Because the resistance is very low for a normal stator, it is generally *not* possible to test for a *shorted* (copper-to-copper) stator. A shorted stator will, however, greatly reduce generator output. An ohmmeter cannot detect an open stator if the stator is delta wound. The ohmmeter will still indicate low resistance because all three windings are electrically connected.

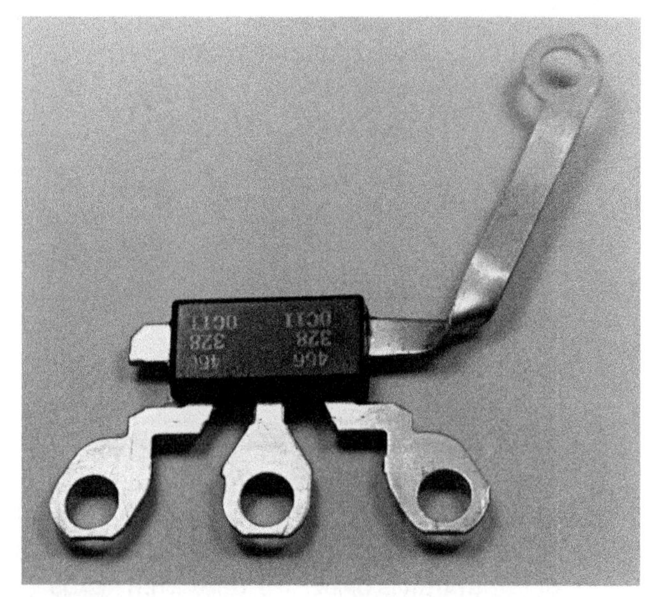

FIGURE 18–18 Typical diode trio. If one leg of a diode trio is open, the generator may produce close to normal output, but the charge indicator light on the dash will be on dimly.

TESTING THE DIODE TRIO Many generators are equipped with a diode trio. A diode is an electrical one-way check valve that permits current to flow in only one direction. Because *trio* means "three," a diode trio is three diodes connected together. ● **SEE FIGURE 18–18.**

The diode trio is connected to all three stator windings. The current generated in the stator flows through the diode trio to the internal voltage regulator. The diode trio is designed to supply current for the field (rotor) and turns off the charge indicator light when the generator voltage equals or exceeds the battery voltage. If one of the three diodes in the diode trio is defective (usually open), the generator may produce close-to-normal output; however, the charge indicator light will be "on" dimly.

A diode trio should be tested with a digital multimeter. The meter should be set to the diode-check position. The multimeter should indicate 0.5 to 0.7 volt (500 to 700 millivolts) one way and OL (overlimit) after reversing the test leads and touching all three connectors of the diode trio.

TESTING THE RECTIFIER

TERMINOLOGY The rectifier assembly usually is equipped with six diodes including three positive diodes and three negative diodes (one positive and one negative for each winding of the stator).

FIGURE 18–19 A typical rectifier bridge that contains all six diodes in one replaceable assembly.

METER SETUP The rectifier(s) (diodes) should be tested using a multimeter that is set to "diode check" position on the digital multimeter (DMM).

Because a diode (rectifier) should allow current to flow in only one direction, each diode should be tested to determine if the diode allows current flow in one direction and blocks current flow in the opposite direction. To test some generator diodes, it may be necessary to unsolder the stator connections. ● **SEE FIGURE 18–19.**

Accurate testing is not possible unless the diodes are separated electrically from other generator components.

TESTING PROCEDURE Connect the leads to the leads of the diode (pigtail and housing of the rectifier bridge). Read the meter. Reverse the test leads. A good diode should have high resistance (OL) one way (reverse bias) and low voltage drop of 0.5 to 0.7 volt (500 to 700 millivolts) the other way (forward bias).

RESULTS Open or shorted diodes must be replaced. Most generators group or combine all positive and all negative diodes in one replaceable rectifier component.

REASSEMBLING THE GENERATOR

BRUSH HOLDER REPLACEMENT Generator carbon brushes often last for many years and require no scheduled maintenance. The life of the generator brushes is extended because they conduct only the field (rotor) current, which is normally only 2 to 5 amperes. The generator brushes should be inspected when the generator is disassembled and should be replaced when worn to less than 1/2 inch long. Brushes are commonly purchased assembled together in a brush holder.

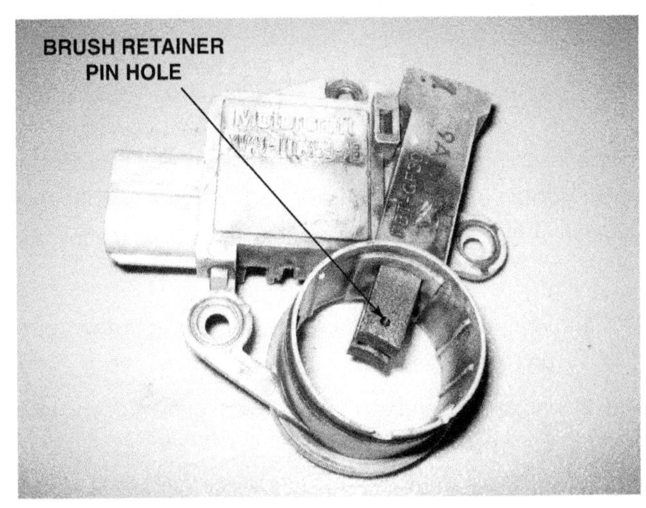

FIGURE 18–20 A brush holder assembly with new brushes installed. The holes in the brushes are used to hold the brushes up in the holder when it is installed in the generator. After the rotor has been installed, the retaining pin is removed which allows the brushes to contact the slip rings of the rotor.

After the brushes are installed (usually retained by two or three screws) and the rotor is installed in the generator housing, a brush retainer pin can be pulled out through an access hole in the rear of the generator, allowing the brushes to be pressed against the slip rings by the brush springs. ● **SEE FIGURE 18–20**.

BEARING SERVICE AND REPLACEMENT The bearings of a generator must be able to support the rotor and reduce friction. A generator must be able to rotate at up to 15,000 RPM and withstand the forces created by the drive belt. The front bearing is usually a ball bearing type and the rear can be either a smaller roller or ball bearing.

The old or defective bearing can sometimes be pushed out of the front housing and the replacement pushed in by applying pressure with a socket or pipe against the outer edge of the bearing (outer race). Replacement bearings are usually prelubricated and seated. Many generator front bearings must be removed from the rotor using a special puller.

GENERATOR ASSEMBLY After testing or servicing, the generator rectifier(s), regulator, stator, and brush holder must be reassembled using the following steps.

STEP 1 If the brushes are internally mounted, insert a wire through the holes in the brush holder to hold the brushes against the springs.

STEP 2 Install the rotor and front-end frame in proper alignment with the mark made on the outside of the generator housing. Install the through bolts. Before removing the wire pin holding the brushes, spin the generator pulley. If the generator is noisy or not rotating freely, the generator can easily be disassembled again to check for the cause. After making certain the generator is free to rotate, remove the brush holder pin and spin the generator again by hand. The noise level may be slightly higher with the brushes released onto the slip rings.

STEP 3 Generators should be tested on a bench tester, if available, before they are reinstalled on a vehicle. When installing the generator on the vehicle, be certain that all mounting bolts and nuts are tight. The battery terminal should be covered with a plastic or rubber protective cap to help prevent accidental shorting to ground, which could seriously damage the generator.

REMANUFACTURED GENERATORS

Remanufactured or rebuilt generators are totally disassembled and rebuilt. Even though there are many smaller rebuilders who may not replace all worn parts, the major national remanufacturers *totally* remanufacture the generator. Old generators (called **cores**) are totally disassembled and cleaned. Both bearings are replaced and all components are tested. Rotors are rewound to original specifications if required. The rotor windings are not counted but are rewound on the rotor "spool," using the correct-gauge copper wire, to the *weight* specified by the original manufacturer. New slip rings are replaced as required, soldered to the rotor spool windings, and machined. The rotors are also balanced and measured to ensure that the outside diameter of the rotor meets specifications. An undersized rotor will produce less generator output because the field must be close to the stator windings for maximum output. Bridge rectifiers are replaced, if required. Every generator is then assembled and tested for proper output, boxed, and shipped to a warehouse. Individual parts stores (called jobbers) purchase parts from various regional or local warehouses.

The Two-Minute Generator Repair

A Chevrolet pickup truck was brought to a shop for routine service. The customer stated that the battery required a jump start after a weekend of sitting. The technician tested the battery and the charging system voltage using a small handheld digital multimeter. The battery voltage was 12.4 volts (about 75% charged), but the charging voltage was also 12.4 volts at 2000 RPM. Because normal charging voltage should be 13.5 to 15 volts, it was obvious that the charging system was not operating correctly.

The technician checked the dash and found that the "charge" light was not on. Before removing the generator for service, the technician checked the wiring connection on the generator. When the connector was removed, it was discovered to be rusty. After the contacts were cleaned, the charging system was restored to normal operation. The technician had learned that the simple things should always be checked first before tearing into a big or expensive repair.

GENERATOR INSTALLATION

Before installing a replacement generator, check service information for the exact procedure to follow for the vehicle being serviced. A typical installation procedure includes the following steps:

STEP 1 Verify that the replacement generator is the correct unit for the vehicle.

STEP 2 Install the generator wiring on the generator and install the generator.

STEP 3 Check the condition of the drive belt and replace, if necessary. Install the drive belt over the drive pulley.

STEP 4 Properly tension the drive belt.

STEP 5 Tighten all fasteners to factory specifications.

STEP 6 Double-check that all fasteners are correctly tightened and remove all tools from the engine compartment area.

STEP 7 Reconnect the negative battery cable.

STEP 8 Start the engine and verify proper charging circuit operation.

1 Before the generator is disassembled, it is spin tested and connected to a scope to check for possible defective components.

2 The scope pattern shows that the voltage output is far from being a normal pattern. This pattern indicates serious faults in the rectifier diodes.

3 The first step is to remove the drive pulley. This rebuilder is using an electric impact wrench to accomplish the task.

4 Carefully inspect the drive galley for damage of embedded rubber from the drive belt. The slightest fault can cause a vibration, noise, or possible damage to the generator.

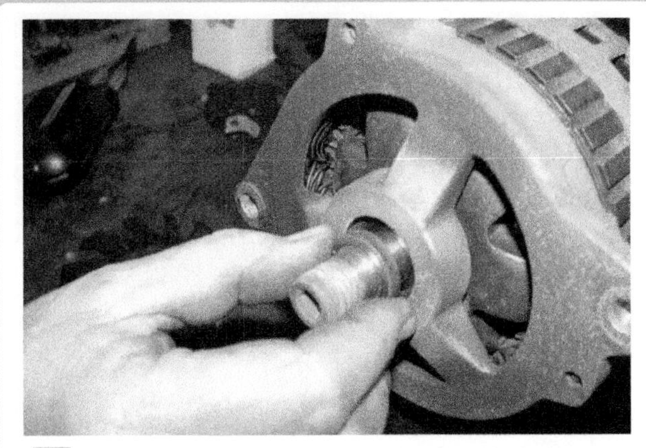

5 Remove the external fan (if equipped) and then the spacers as shown.

6 Next pop off the plastic cover (shield) covering the stator/rectifier connection.

CONTINUED ▶

7 Using a diagonal cutter, cut the weld to separate the stator from the rectifier.

8 After removing the bolts the end housing and stator can be separated from the rear (slip-ring-end) housing.

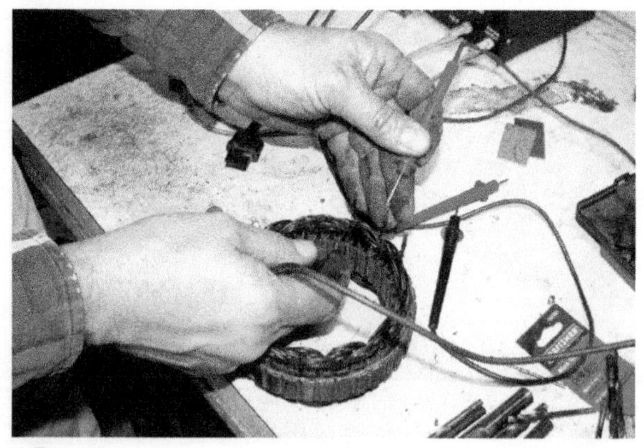

9 The stator is checked by visual inspection for discoloration or other physical damage, and then checked with an ohmmeter to see if the windings are shorted-to-ground.

10 The front bearing is removed from the drive-end housing using a press.

11 A view of the slip-ring-end (SRE) housing showing the black plastic shield, which helps direct air flow across the rectifier.

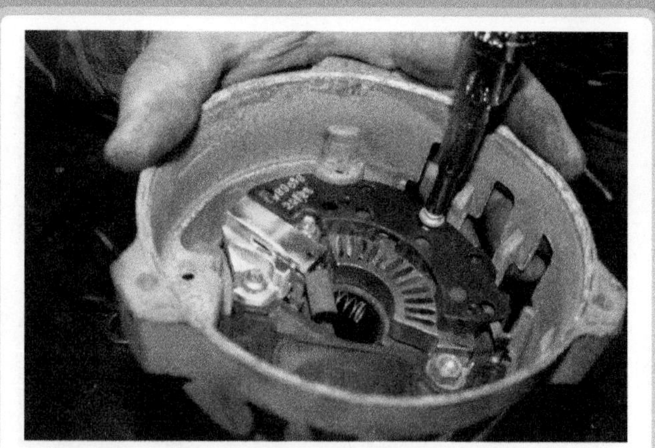

12 After the shield has been removed, the rectifier, regulator, and brush holder assembly can be removed by removing the retaining screws.

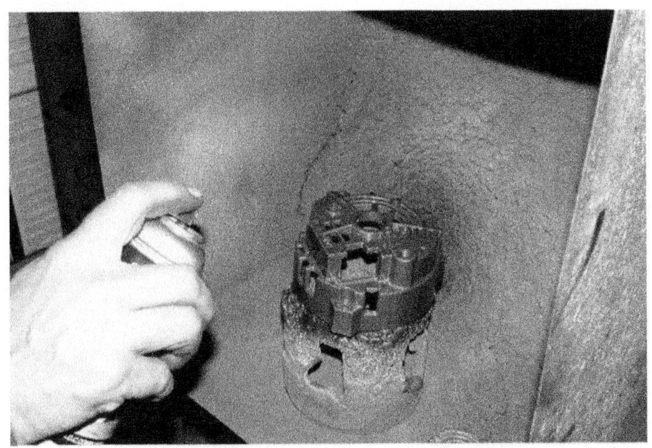

13 This rebuilder is painting the housing using a high-quality industrial grade spray paint to make the rebuilt generator look like new.

14 The slip rings on the rotor are being machined on a lathe.

15 The rotor is being tested using an ohmmeter. The specifications for the resistance between the slip rings on the CS-130 are 2.2 to 3.5 ohms.

16 The rotor is also tested between the slip ring and the rotor shaft. This reading should be infinity.

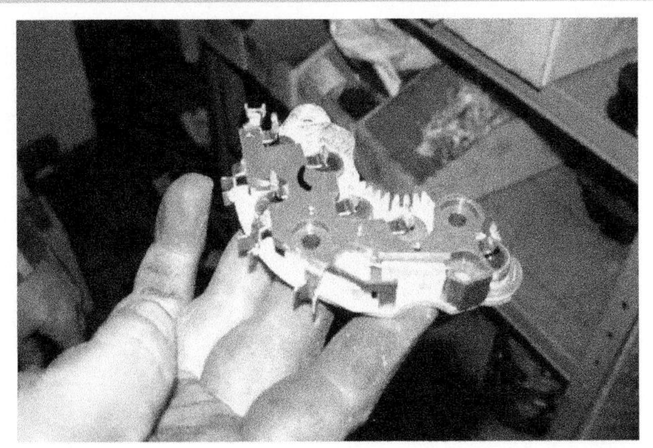

17 A new rectifier. This replacement unit is significantly different than the original but is designed to replace the original unit and meets the original factory specifications.

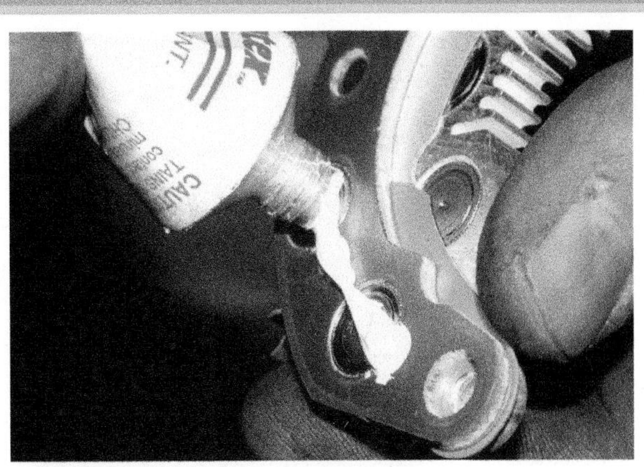

18 Silicone heat transfer compound is applied to the heat sink of the new rectifier.

CONTINUED ▶

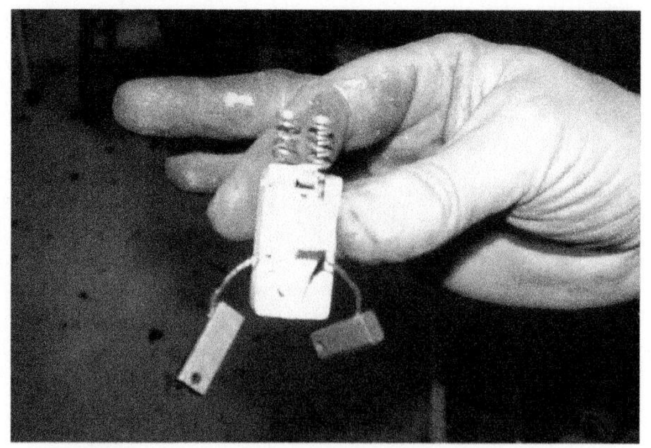

19 Replacement brushes and springs are assembled into the brush holder.

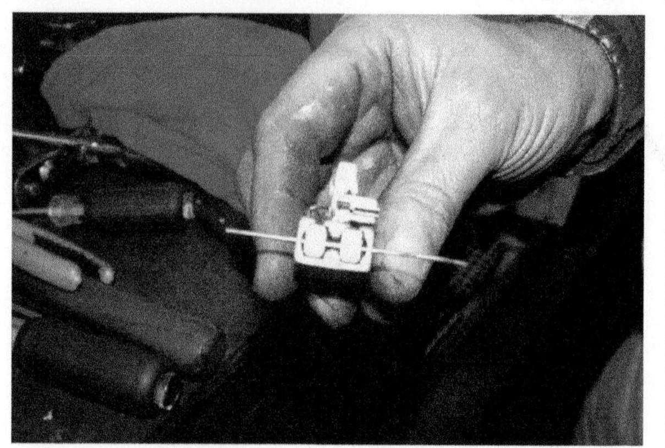

20 The brushes are pushed into the brush holder and retained by a straight wire, which extends through the rear housing of the generator. This wire is then pulled out when the unit is assembled.

21 Here is what the CS generator looks like after installing the new brush holder assembly, rectifier bridge, and voltage regulator.

22 The junction between the rectifier bridge and the voltage regulator is soldered.

23 The plastic deflector shield is snapped back into location using a blunt chisel and a hammer. This shield directs the airflow from the fan over the rectifier bridge and voltage regulator.

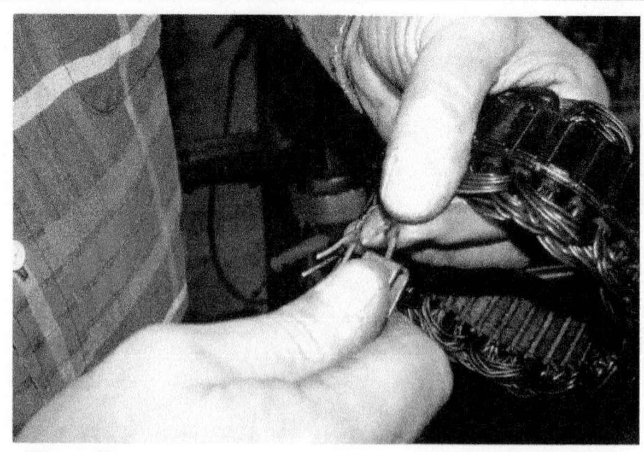

24 Before the stator windings can be soldered to the rectifier bridge, the varnish insulation is removed from the ends of the leads.

25 After the stator has been inserted into the rear housing the stator leads are soldered to the copper lugs of the rectifier bridge.

26 New bearings are installed. A spacer is placed between the bearing and the slip rings to help prevent the possibility that the bearing could move on the shaft and short against the slip ring.

27 The slip-ring-end (SRE) housing is aligned with the marks made during disassembly and is pressed into the drive-end (DE) housing.

28 The retaining bolts, which are threaded into the drive-end housing from the back of the generator are installed.

29 The external fan and drive pulley are installed and the retaining nut is tightened on the rotor shaft.

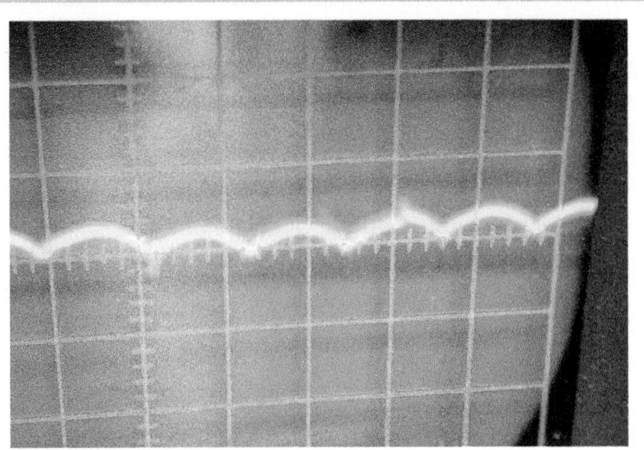

30 The scope pattern shows that the diodes and stator are functioning correctly and voltage check indicates that the voltage regulator is also functioning correctly.

SUMMARY

1. Charging system testing requires that the battery be at least 75% charged to be assured of accurate test results. Normal charging voltage (at 2000 engine RPM) is 13.5 to 15 volts.
2. To check for excessive resistance in the wiring between the generator and the battery, a voltage drop test should be performed.
3. Generators do not produce their maximum rated output unless required by circuit demands. Therefore, to test for maximum generator output, the battery must be loaded to force the generator to produce its maximum output.
4. Each generator should be marked across its case before disassembly to ensure proper clock position during reassembly. After disassembly, all generator internal components should be tested using an ohmmeter. The following components should be tested:
   a. Stator
   b. Rotor
   c. Diodes
   d. Diode trio (if the generator is so equipped)
   e. Bearings
   f. Brushes (should be more than 1/2 inch long)

REVIEW QUESTIONS

1. How does a technician test the voltage drop of the charging circuit?
2. How does a technician measure the amperage output of a generator?
3. What tests can be performed to determine whether a diode or stator is defective before removing the generator from the vehicle?

CHAPTER QUIZ

1. To check the charging voltage, connect a digital multimeter (DMM) to the positive (+) and the negative (−) terminals of the battery and select ________.
   a. DC volts
   b. AC volts
   c. DC amps
   d. AC amps
2. To check for ripple voltage from the generator, connect a digital multimeter (DMM) and select ________.
   a. DC volts
   b. AC volts
   c. DC amps
   d. AC amps
3. The maximum allowable alternating current (AC) in amperes that is being sent to the battery from the generator is ________.
   a. 0.4 ampere
   b. 1 to 3 amperes
   c. 3 to 4 amperes
   d. 10% of the rated output of the generator
4. Why should the lights be turned on when checking for ripple voltage or alternating current from the generator?
   a. To warm the battery
   b. To check that the battery is fully charged
   c. To create an electrical load for the generator
   d. To test the battery before conducting other tests
5. An acceptable charging circuit voltage on a 12 volt system is ________.
   a. 13.5 to 15 volts
   b. 12.6 to 15.6 volts
   c. 12 to 14 volts
   d. 14.9 to 16.1 volts
6. Technician A says that the computer can be used to control the output of the generator by controlling the field current. Technician B says that voltage regulators control the generator output by controlling the field current through the rotor. Which technician is correct?
   a. Technician A only
   b. Technician B only
   c. Both Technicians A and B
   d. Neither Technician A nor B
7. Technician A says that a voltage drop test of the charging circuit should only be performed when current is flowing through the circuit. Technician B says to connect the leads of a voltmeter to the positive and negative terminals of the battery to measure the voltage drop of the charging system. Which technician is correct?
   a. Technician A only
   b. Technician B only
   c. Both Technicians A and B
   d. Neither Technician A nor B

8. When testing a generator rotor, if an ohmmeter shows zero ohms with one meter lead attached to the slip rings and the other meter lead touching the rotor shaft, the rotor is ________.
   a. Okay (normal)
   b. Defective (shorted-to-ground)
   c. Defective (shorted-to-voltage)
   d. Okay (rotor windings are open)

9. A generator diode is being tested using a digital multimeter set to the diode-check position. A good diode will read ________ if the leads are connected one way across the diode and ________ if the leads are reversed.
   a. 300/300
   b. 0.475/0.475
   c. OL/OL
   d. 0.551/OL

10. A generator could test as producing lower-than-normal output, yet be okay, if the ________.
   a. Battery is weak or defective
   b. Engine speed is not high enough during testing
   c. Drive belt is loose or slipping
   d. All of the above

chapter 19

LIGHTING AND SIGNALING CIRCUITS

LEARNING OBJECTIVES

After studying this chapter, the reader will be able to:

1. Describe how an exterior lighting systems works.
2. Read and interpret a bulb chart.
3. Discuss the operation of brake lights and turn signals.
4. Inspect, replace, and aim headlights and bulbs.
5. Discuss troubleshooting procedures for lighting and signaling circuits.

This chapter will help you prepare for the ASE Electrical/Electronic Systems (A6) certification test content area "E" (Lighting System Diagnosis and Repair).

KEY TERMS

AFS 286
Brake lights 275
Candlepower 272
CHMSL 275
Color shift 286
Composite headlight 284
Courtesy lights 291
DRL 289
HID 284
Kelvin (K) 285
LED 276
Rheostat 281
Trade number 271
Troxler effect 295
Xenon headlights 285

 STC OBJECTIVES

GM Service Technical College topics covered in this chapter are as follows:

1. Determine system operation by utilizing a description of operation and wiring schematic.
2. Diagnose a dome light circuit.

INTRODUCTION

The vehicle has many different lighting and signaling systems, each with its own specific components and operating characteristics. The major light-related circuits and systems covered include the following:

- Exterior lighting
- Headlights (halogen, HID, and LED)
- Bulb trade numbers
- Brake lights
- Turn signals and flasher units
- Courtesy lights
- Light-dimming rearview mirrors

FIGURE 19–1 A dual-filament (3157, left) bulb contains both a low-intensity filament for taillights or parking lights and a high-intensity filament for brake lights and turn signals. Bulbs come in a variety of shapes and sizes. The numbers shown are the trade numbers. (Courtesy of Jeffrey Rehkopf)

EXTERIOR LIGHTING

HEADLIGHT SWITCH CONTROL Exterior lighting is controlled by the headlight switch, which is connected directly to the battery on most vehicles. Therefore, if the light switch is left on manually, the lights could drain the battery. Older headlight switches contained a built-in circuit breaker. If excessive current flows through the headlight circuit, the circuit breaker will momentarily open the circuit, then close it again. The result is headlights that flicker on and off rapidly. This feature allows the headlights to function, as a safety measure, in spite of current overload.

The headlight switch controls the following lights on most vehicles, usually through a module:

1. Headlights
2. Taillights
3. Side-marker lights
4. Front parking lights
5. Dash lights
6. Interior (dome) light(s)

COMPUTER-CONTROLLED LIGHTS Because these lights can easily drain the battery if accidentally left on, many newer vehicles control these lights through computer modules. The computer module keeps track of the time the lights are on and can turn them off if the time is excessive. The computer can control either the power side or the ground side of the circuit.

For example, a typical computer-controlled lighting system usually includes the following steps:

STEP 1 The driver depresses or rotates the headlight switch.

STEP 2 The signal from the headlight switch is sent to the nearest control module.

STEP 3 The control module then sends a request to the headlight control module to turn on the headlights as well as the front park and side-marker lights.

Through the data BUS, the rear control module receives the lights on signal and turns on the lights at the rear of the vehicle.

STEP 4 All modules monitor current flow through the circuit and will turn on a bulb failure warning light if it detects an open bulb or a fault in the circuit.

STEP 5 After the ignition has been turned off, the modules will turn off the lights after a time delay to prevent the battery from being drained.

BULB NUMBERS

TRADE NUMBER The number used on automotive bulbs is called the bulb **trade number**, as recorded with the American National Standards Institute (ANSI). The number is the same regardless of the manufacturer. ● **SEE FIGURE 19–1**.

CANDLEPOWER The trade number also identifies the size, shape, number of filaments, and amount of light produced, measured in **candlepower**. For example, the 1156 bulb, commonly used for backup lights, is 32 candlepower. A 194 bulb, commonly used for dash or side-marker lights, is rated at only 2 candlepower. The amount of light produced by a bulb is determined by the resistance of the filament wire, which also affects the amount of current (in amperes) required by the bulb.

It is important that the correct trade number of bulb always be used for replacement to prevent circuit or component damage. The correct replacement bulb for a vehicle is usually listed in the owner or service manual. ● **REFER TO CHART 19–1** for a listing of common bulbs and their specifications used in most vehicles.

BULB NUMBER	FILAMENTS	AMPERAGE LOW/HIGH	WATTAGE LOW/HIGH	CANDLEPOWER LOW/HIGH
Headlights				
1255/H1	1	4.58	55.00	129.00
1255/H3	1	4.58	55.00	121.00
6024	2	2.73/4.69	35.00/60.00	27,000/35,000
6054	2	2.73/5.08	35.00/65.00	35,000/40,000
9003	2	4.58/5.00	55.00/60.00	72.00/120.00
9004	2	3.52/5.08	45.00/65.00	56.00/95.00
9005	1	5.08	65.00	136.00
9006	1	4.30	55.00	80.00
9007	2	4.30/5.08	55.00/65.00	80.00/107.00
9008	2	4.30/5.08	55.00/65.00	80.00/107.00
9011	1	5.08	65.00	163.50
Headlights (HID–Xenon)				
D2R	Air Gap	0.41	35.00	222.75
D2S	Air Gap	0.41	35.00	254.57
Taillights, Stop, and Turn Lamps				
1156	1	2.10	26.88	32.00
1157	2	0.59/2.10	8.26/26.88	3.00/32.00
2057	2	0.49/2.10	6.86/26.88	2.00/32.00
3057	2	6.72/26.88	0.48/2.10	1.50/24.00
3155	1	1.60	20.48	21.00
3157	2	0.59/2.10	8.26/26.88	2.20/24.00
4157	2	0.59/2.10	8.26/26.88	3.00/32.00
7440	1	1.75	21.00	36.60
7443	2	0.42/1.75	5.00/21.00	2.80/36.60
17131	1	0.33	4.00	2.80
17635	1	1.75	21.00	37.00
17916	2	0.42/1.75	5.00/21.00	1.20/35.00
Parking, Daytime Running Lamps				
24	1	0.24	3.36	2.00
67	1	0.59	7.97	4.00
168	1	0.35	4.90	3.00
194	1	0.27	3.78	2.00
889	1	3.90	49.92	43.00
912	1	1.00	12.80	12.00
916	1	0.54	7.29	2.00
1034	2	0.59/1.80	8.26/23.04	3.00/32.00
1156	1	2.10	26.88	32.00
1157	2	0.59/2.10	8.26/26.88	3.00/32.00
2040	1	0.63	8.00	10.50
2057	2	0.49/2.10	6.86/26.88	1.50/24.00
2357	2	0.59/2.23	8.26/28.54	3.00/40.00
3157	2	0.59/2.10	8.26/26.88	3.00/32.00
3357	2	0.59/2.23	8.26/28.54	3.00/40.00
3457	2	0.59/2.23	8.26/28.51	3.00/40.00
3496	2	0.66/2.24	8.00/27.00	3.00/45.00
3652	1	0.42	5.00	6.00
4114	2	0.59/2/23	8.26/31.20	3.00/32.00
4157	2	0.59/2.10	8.26/26/88	3.00/32.00
7443	2	0.42/1.75	5.00/21.00	2.80/36.60
17131	1	0.33	4.00	2.80
17171	1	0.42	5.00	4.00
17177	1	0.42	5.00	4.00
17311	1	0.83	10.00	10.00
17916	2	0.42/1.75	5.00/21.00	1.20/35.00
68161	1	0.50	6.00	10.00
Center High-Mounted Stop Lamp (CHMSL)				
70	1	0.15	2.10	1.50
168	1	0.35	4.90	3.00
175	1	0.58	8.12	5.00
211-2	1	0.97	12.42	12.00
577	1	1.40	17.92	21.00
579	1	0.80	10.20	9.00
889	1	3.90	49.92	43.00
891	1	0.63	8.00	11.00
906	1	0.69	8.97	6.00
912	1	1.00	12.80	12.00
921	1	1.40	17.92	21.00
922	1	0.98	12.54	15.00
1141	1	1.44	18.43	21.00
1156	1	2.10	26.88	32.00

CHART 19–1

Bulb chart sorted by typical applications. Check the owner's manual, service information, or a bulb manufacturer's application chart for the exact bulb to use.

CONTINUED

BULB NUMBER	FILAMENTS	AMPERAGE LOW/HIGH	WATTAGE LOW/HIGH	CANDLEPOWER LOW/HIGH
2723	1	0.20	2.40	1.50
3155	1	1.60	20.48	21.00
3156	1	2.10	26.88	32.00
3497	1	2.24	27.00	45.00
7440	1	1.75	21.00	36.60
17177	1	0.42	5.00	4.00
17635	1	1.75	21.00	37.00
License Plate, Glove Box, Dome, Side Marker, Trunk, Map, Ashtray, Step/Courtesy, and Underhood				
37	1	0.09	1.26	0.50
67	1	0.59	7.97	4.00
74	1	0.10	1.40	.070
98	1	0.62	8.06	6.00
105	1	1.00	12.80	12.00
124	1	0.27	3.78	1.50
161	1	0.19	2.66	1.00
168	1	0.35	4.90	3.00
192	1	0.33	4.29	3.00
194	1	0.27	3.78	2.00
211-1	1	0.968	12.40	12.00
212-2	1	0.74	9.99	6.00
214-2	1	0.52	7.02	4.00
293	1	0.33	4.62	2.00
561	1	0.97	12.42	12.00
562	1	0.74	9.99	6.00
578	1	0.78	9.98	9.00
579	1	0.80	10.20	9.00
PC579	1	0.80	10.20	9.00
906	1	0.69	8.97	6.00
912	1	1.00	12.80	12.00
917	1	1.20	14.40	10.00
921	1	1.40	17.92	21.00
1003	1	0.94	12.03	15.00
1155	1	0.59	7.97	4.00
1210/H2	1	8.33	100.00	239.00
1210/H3	1	8.33	100.00	192.00
1445	1	0.14	2.02	0.70
1891	1	0.24	3.36	2.00
1895	1	0.27	3.78	2.00
3652	1	0.42	5.00	6.00
11005	1	0.39	5.07	4.00
11006	1	0.24	3.36	2.00
12100	1	0.77	10.01	9.55
13050	1	0.38	4.94	3.00
17036	1	0.10	1.20	0.48

CONTINUED

BULB NUMBER	FILAMENTS	AMPERAGE LOW/HIGH	WATTAGE LOW/HIGH	CANDLEPOWER LOW/HIGH
17097	1	0.25	3.00	1.76
17131	1	0.33	4.00	2.80
17177	1	0.42	5.00	4.00
17314	1	0.83	10.00	8.00
17916	2	0.42/1.75	5.00/21.00	1.20/35.00
47830	1	0.39	5.00	6.70
Instrument Panel				
37	1	0.09	1.26	0.50
73	1	0.08	1.12	0.30
74	1	0.10	1.40	0.70
PC74	1	0.10	1.40	0.70
PC118	1	0.12	1.68	0.70
124	1	0.27	3.78	1.50
158	1	0.24	3.36	2.00
161	1	0.19	2.66	1.00
192	1	0.33	4.29	3.00
194	1	0.27	3.78	2.00
PC194	1	0.27	3.78	2.00
PC195	1	0.27	3.78	1.80
1210/H1	1	8.33	100.00	217.00
1210/H3	1	8.33	100.00	192.00
17037	1	0.10	1.20	0.48
17097	1	0.25	3.00	1.76
17314	1	0.83	10.00	8.00
Backup, Cornering, and Fog/Driving Lamps				
67	1	0.59	7.97	4.00
579	1	0.80	10.20	9.00
880	1	2.10	26.88	43.00
881	1	2.10	26.88	43.00
885	1	3.90	49.92	100.00
886	1	3.90	49.92	100.00
893	1	2.93	37.50	75.00
896	1	2.93	37.50	75.00
898	1	2.93	37.50	60.00
899	1	2.93	37.50	60.00
921	1	1.40	17.92	21.00
1073	1	1.80	23.04	32.00
1156	1	2.10	26.88	32.00
1157	2	0.59/2.10	8.26/26.88	3.00/32.00
1210/H1	1	8.33	100.00	217.00
1255/H1	1	4.58	55.00	129.00
1255/H3	1	4.58	55.00	121.00
1255/H11	1	4.17	55.00	107.00
2057	2	0.49/2.10	6.86/26.88	1.50/24.00
3057	2	0.48/2.10	6.72/26.88	2.00/32.00

CONTINUED

BULB NUMBER	FILAMENTS	AMPERAGE LOW/HIGH	WATTAGE LOW/HIGH	CANDLEPOWER LOW/HIGH
3155	1	1.60	20.48	21.00
3156	1	2.10	26.88	32.00
3157	2	0.59/2.10	8.26/26.88	3.00/32.00
4157	2	0.59/2.10	8.26/26/88	3.00/32.00
7440	1	1.75	21.00	36.00
9003	2	4.58/5.00	55.00/60.00	72.00/120.00
9006	1	4.30	55.00	80.00
9145	1	3.52	45.00	65.00
17635	1	1.75	21.00	37.00

FIGURE 19–2 Bulbs that have the same trade number have the same operating voltage and wattage. The NA means that the bulb uses a natural amber glass ampoule with clear turn signal lenses.

BULB NUMBER SUFFIXES Many bulbs have suffixes that indicate some feature of the bulb, while keeping the same size and light output specifications.

Typical bulb suffixes include:

- NA: natural amber (amber glass)
- A: amber (painted glass)
- HD: heavy duty
- LL: long life
- IF: inside frosted
- R: red
- B: blue
- G: green

● **SEE FIGURE 19–2.**

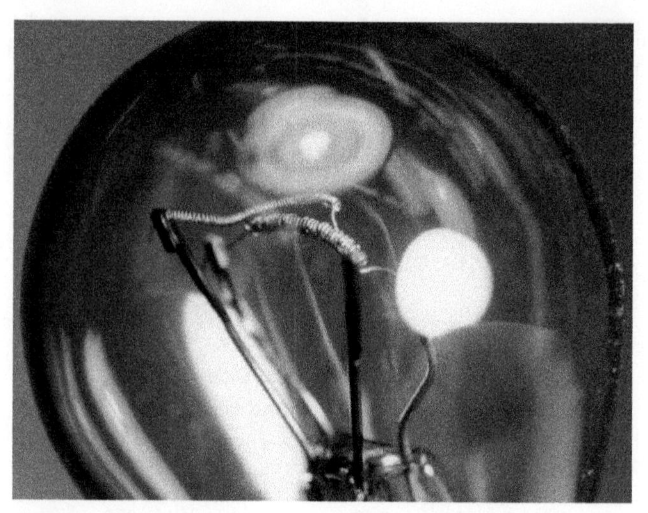

FIGURE 19–3 Close-up a 2057 dual-filament (double-contact) bulb that failed. Notice that the top filament broke from its mounting and melted onto the lower filament. This bulb caused the dash lights to come on whenever the brakes were applied.

REAL WORLD FIX

Weird Problem—Easy Solution

A General Motors minivan had the following electrical problems:

- The turn signals flashed rapidly on the left side.
- With the ignition key off, the lights-on warning chime sounded if the brake pedal was depressed.
- When the brake pedal was depressed, the dome light came on.

All of these problems were caused by one defective 2057 dual-filament bulb, as shown in ● **FIGURE 19–3.**

Apparently, the two filaments were electrically connected when one filament broke and then welded to the other filament. This caused the electrical current to feed back from the brake light filament into the taillight circuit, causing all the problems.

TESTING BULBS Bulbs can be tested using two basic tests.

1. Perform a visual inspection of any bulb. Many faults, such as a shorted filament, corroded connector, or water, can cause weird problems that are often thought to be wiring issues.

● **SEE FIGURES 19–4 AND 19–5.**

FIGURE 19–4 Corrosion caused the two terminals of this dual-filament bulb to be electrically connected.

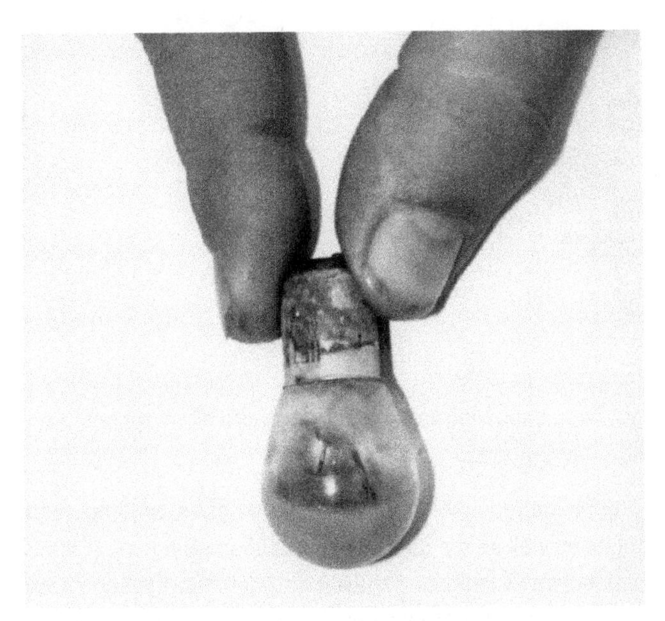

FIGURE 19–5 Often the best diagnosis is a thorough visual inspection. This bulb was found to be filled with water, which caused problems.

2. Bulbs can be tested using an ohmmeter and checking the resistance of the filament(s). Most bulbs will read low resistance, between 0.5 and 20 ohms, at room temperature depending on the bulb. Test results include:
   - **Normal resistance.** The bulb is good. Check both filaments if it is a two-filament bulb. ● **SEE FIGURE 19–6.**
   - **Zero ohms.** It is unlikely but possible for the bulb filament to be shorted.
   - **OL (electrically open).** The reading indicates that the bulb filament is broken.

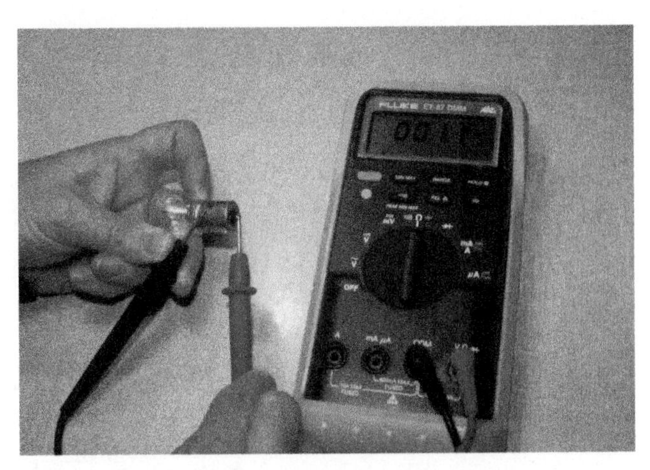

FIGURE 19–6 This single-filament bulb is being tested with a digital multimeter set to read resistance in ohms. The reading of 1.1 ohms is the resistance of the bulb when cold. As soon as current flows through the filament, the resistance increases about 10 times. It is the initial surge of current flowing through the filament when the bulb is cool that causes many bulbs to fail in cold weather as a result of the reduced resistance. As the temperature increases, the resistance increases.

BRAKE LIGHTS

OPERATION, NON-BCM CONTROLLED **Brake lights**, also called stop lights or stop lamps, use the high-intensity filament of a double-filament bulb. (The low-intensity filament is for the taillights.) When the brakes are applied, the brake switch is closed and the brake lamps light. The brake switch receives current from a fuse that is hot all the time. The brake light switch is a normally open (N.O.) switch, but is closed when the driver depresses the brake pedal. Since 1986, all vehicles sold in the United States have a third brake light commonly referred to as the **center high-mounted stop light (CHMSL)**. ● **SEE FIGURE 19–7.**

The brake switch is also used as an input switch (signal) for the following:

1. Cruise control (deactivates when the brake pedal is depressed)
2. Antilock brakes (ABS)
3. Brake shift interlock (prevents shifting from park position unless the brake pedal is depressed)

STOP LAMPS, BCM-CONTROLLED On computer controlled stop lamps the body control module (BCM) receives inputs and then turns the stop lamps on or off. The brake pedal position (BPP) sensor is used to sense the action of the driver application of the brake pedal. The BPP sensor provides an analog voltage

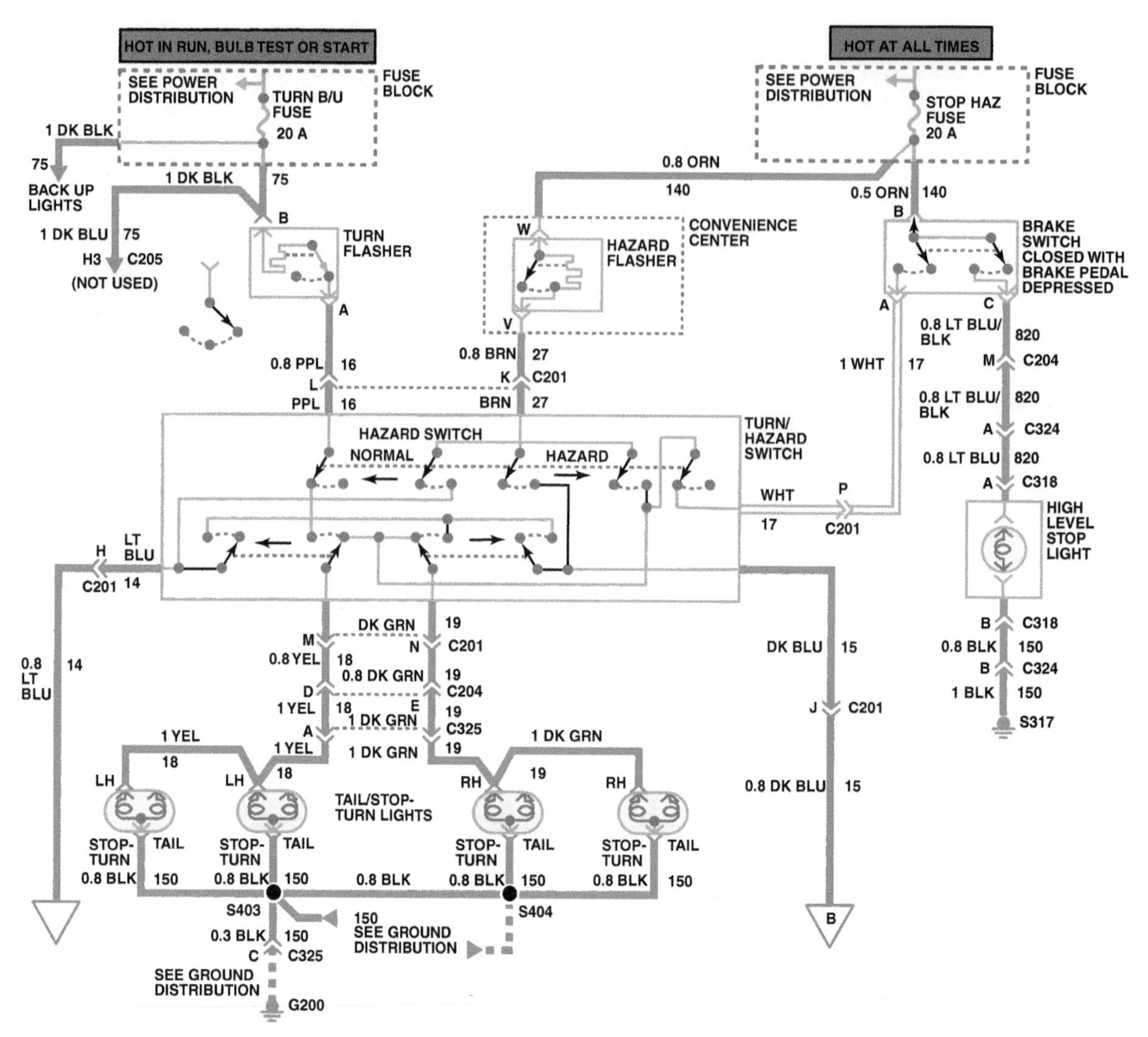

FIGURE 19–7 Typical non-computer controlled brake light and taillight circuit showing the brake switch and all of the related circuit components.

signal that will increase as the brake pedal is applied. The BCM provides a low reference signal and a 5-volt reference voltage to the BPP sensor. When the variable signal reaches a voltage threshold indicating the brakes have been applied, the BCM will apply battery voltage to the left and right stop lamp control circuits as well as the center high mounted stop lamp (CHMSL) control circuit illuminating the left and right stop lamps and the CHMSL. ● **SEE FIGURE 19–8.**

FREQUENTLY ASKED QUESTION

Why Are LEDs Used for Brake Lights?

Light-emitting diode (LED) brake lights are frequently used for high-mounted stop lamps (CHMSLs) for the following reasons.

1. **Faster illumination.** An LED will light up to 200 ms faster than an incandescent bulb, which requires some time to heat the filament before it is hot enough to create light. This faster illumination can mean the difference in stopping distances at 60 mph (100 km/h) by about 18 ft (6 m) due to the reduced reaction time for the driver of the vehicle behind.
2. **Longer service life.** LEDs are solid-state devices that do not use a filament to create light. As a result, they are less susceptible to vibration and will often last the life of the vehicle.

NOTE: Aftermarket replacement LED bulbs that are used to replace conventional bulbs may require the use of a different type of flasher unit due to the reduced current draw of the LED bulbs. ● **SEE FIGURE 19–9.**

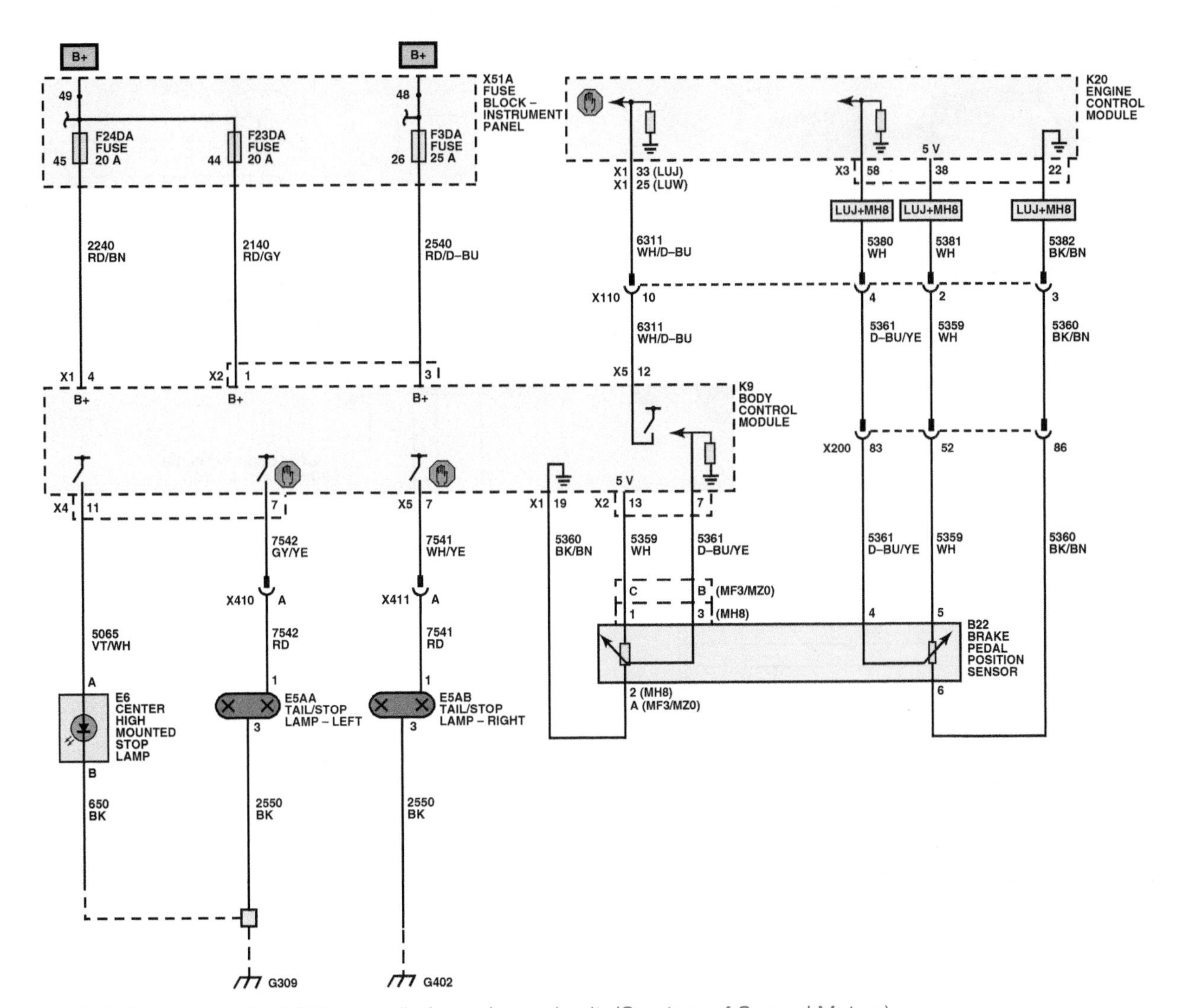

FIGURE 19–8 Schematic of a BCM-controlled stop lamp circuit. (Courtesy of General Motors)

FIGURE 19–9 A replacement LED taillight bulb is constructed of many small, individual light-emitting diodes.

TURN SIGNALS

OPERATION The turn signal circuit is supplied power from the ignition switch and operated by a lever and switch. ● **SEE FIGURE 19–10.**

When the turn signal switch is moved in either direction, the corresponding turn signal lamps receive current through the flasher unit. The flasher unit causes the current to start and stop as the turn signal lamp flashes on and off with the interrupted current.

ONE-FILAMENT STOP/TURN BULBS In many vehicles, the stop and turn signals are both provided by one filament. When the turn signal switch is turned on (closed), the filament receives interrupted current through the flasher unit. When the brakes are applied, the current first flows to the turn signal

FIGURE 19–10 The turn signal switch is located under the air bag clock spring and steering wheel. The cancel pin turns off the signal after a turn. (Courtesy of Jeffrey Rehkopf)

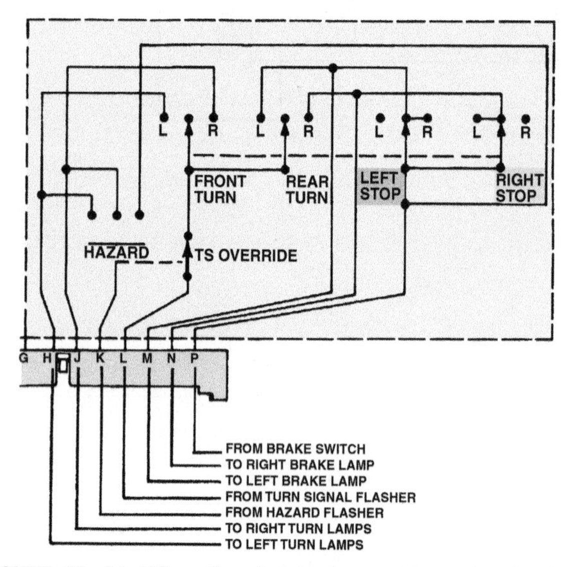

FIGURE 19–11 When the stop lamps and turn signals share a common bulb filament, stop light current flows through the turn signal switch.

switch, except for the high-mounted stop, which is fed directly from the brake switch. If neither turn signal is on, then current through the turn signal switch flows to both rear brake lights. If the turn signal switch is operated (turned to either left or right), current flows through the flasher unit on the side that was selected and directly to the brake lamp on the opposite side. If the brake pedal is not depressed, then current flows through the flasher and only to one side. **SEE FIGURE 19–11.**

Moving the lever up or down completes the circuit through the flasher unit and to the appropriate turn signal lamps. A turn signal switch includes cams and springs that cancel the signal after the turn has been completed. As the steering wheel

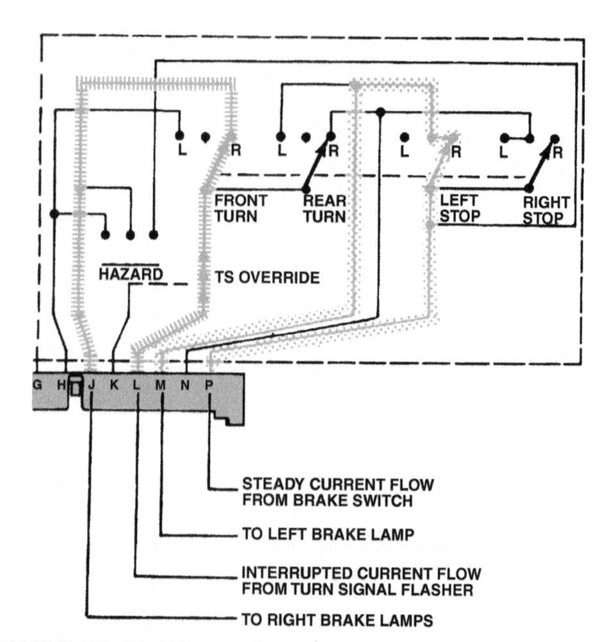

FIGURE 19–12 When a right turn in signaled, the turn signal switch contacts send flasher current to the right-hand filament and brake switch current to the left-hand filament.

is turned in the signaled direction and then returns to its normal position, the cams and springs cause the turn signal switch contacts to open and break the circuit.

TWO-FILAMENT STOP/TURN BULBS In systems using separate filaments for the stop and turn lamps, the brake and turn signal switches are not connected. If the vehicle uses the same filament for both purposes, then brake switch current is routed through contacts within the turn signal switch. By linking certain contacts, the bulbs can receive either brake switch current or flasher current, depending upon which direction is being signaled. For example, **FIGURE 19–12** shows current flow through the switch when the brake switch is closed and a right turn is signaled.

Steady current through the brake switch is sent to the left brake lamp. Interrupted current from the turn signal is sent to the right turn lamps.

FLASHER UNITS A turn signal flasher unit is a metal or plastic can containing a switch that opens and closes the turn signal circuit. Vehicles can be equipped with many different types of flasher units. **SEE FIGURE 19–13.**

- **DOT flashers.** The turn signal flasher is designed to transmit the current to light the front and rear bulbs on only one side at a time. The U.S. Department of Transportation (DOT) regulation requires that the driver be alerted when a turn signal bulb is not working. This is achieved by using a series-type flasher unit. The flasher

FIGURE 19–13 Three styles of flasher units.

unit requires current flow through two bulbs (one in the front and one in the rear) in order to flash. A bimetal flasher is often used in this application. The operation of this flasher is current sensitive, which means that the flasher will stop flashing when one of the light bulbs is out and that it will flash at a faster rate when adding additional load, such as a trailer. The turn signal lamp current is passed through the bimetal element and causes heating. When the element is hot enough, the bimetal distorts, opening the contacts and turning off the lamps. After the bimetal cools, it returns to the original shape, closing the contacts and turning on the lamps again. This sequence is repeated until the load is removed. If one bulb burns out, the turn signal indicator lamp on the dash will remain lit. The flasher will not flash because there is not enough current flow through the one remaining bulb to cause the flasher to become heated enough to open.

- **Hybrid flashers.** The hybrid flashers have an electronic flasher control circuit to operate the internal electromechanical relay and are commonly called a flasher relay. This type of flasher has a stable electronic timing circuitry that enables a wide operating voltage and temperature range. The hybrid flasher has a lamp current-sensing circuit which will cause the flash rate to double when a bulb is burned out.
- **Solid-state flashers.** The solid-state flashers have an internal electronic circuit for timing and solid-state power output devices for load switching. Solid-state units cause the turn indicator to flash rapidly if a bulb is burned out. These electronic flashers are compatible with older systems that can be used as replacement units.
- **Hazard warning flasher.** Some older vehicles also have a hazard warning flasher with the primary function of causing both the left and right turn signal lamps to flash when the hazard warning switch is activated.

FREQUENTLY ASKED QUESTION

Where Is the Flasher Unit?

Many newer vehicles do not use a flasher unit. On many vehicles, such as on many 2006+ General Motors vehicles, the turn signal switch is an input to the body control module (BCM). The BCM sends a signal through the data lines to the lighting module(s) to flash the lights. The BCM also sends a signal to the radio which sends a clicking sound to the driver's side speaker even if the radio is off.

FREQUENTLY ASKED QUESTION

Why Does the Side-Marker Light Flash Alternately?

A question that service technicians are asked frequently is why the side-marker light goes out alternately when the turn signal is on, and is on when the turn signal is off. Some vehicle owners think that there is a fault with the vehicle, but this is normal operation. The side-marker light goes out when the lights are on and the turn signal is flashing because there are 12 volts on both sides of the bulb (see points X and Y in ● **FIGURE 19–14**).

Normally, the side-marker light gets its ground through the turn signal bulb.

- **Combination turn signal and hazard warning flasher.** The combination flasher is a device that combines the functions of a turn signal flasher and a hazard warning flasher into one package, which often uses three electrical terminals.

TURN SIGNAL LAMPS, BCM-CONTROLLED Refer to ● **FIGURE 19–15**. Ground is applied at all times to the turn signal/multifunction switch. The turn signal lamps may only be activated with the ignition switch in the ON or START positions. When the turn signal/multifunction switch is placed in either the TURN RIGHT or TURN LEFT position, ground is applied to the body control module (BCM) through either the right turn or left turn signal switch circuit. The BCM responds to the turn signal switch input by applying a pulsating voltage to the front and rear turn signal lamps through their respective control circuits.

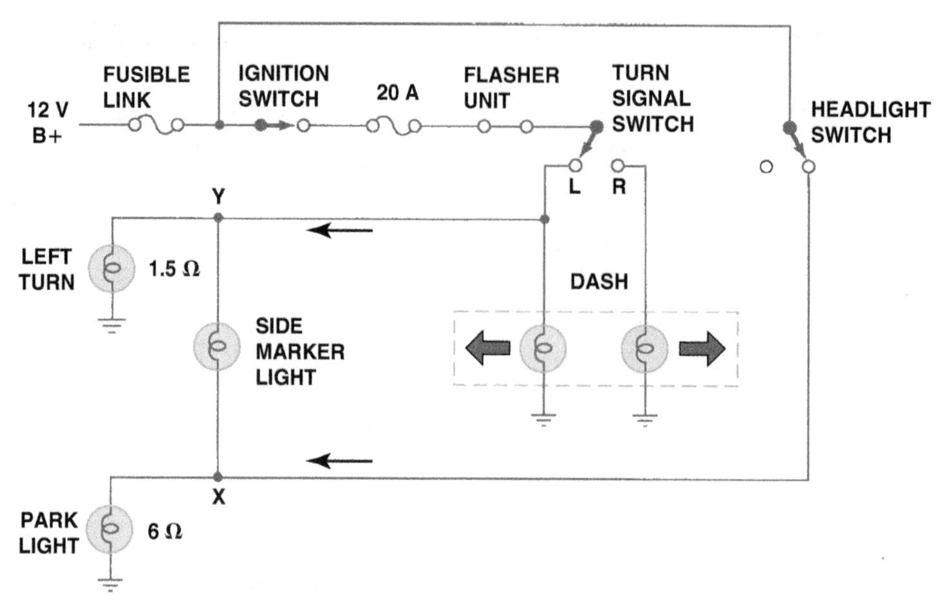

FIGURE 19–14 The side-marker light goes out whenever there is voltage at both points X and Y. These opposing voltages stop current flow through the side-marker light. The left turn light and left park light are actually the same bulb (usually 2057) and are shown separately to help explain how the side-marker light works on many vehicles.

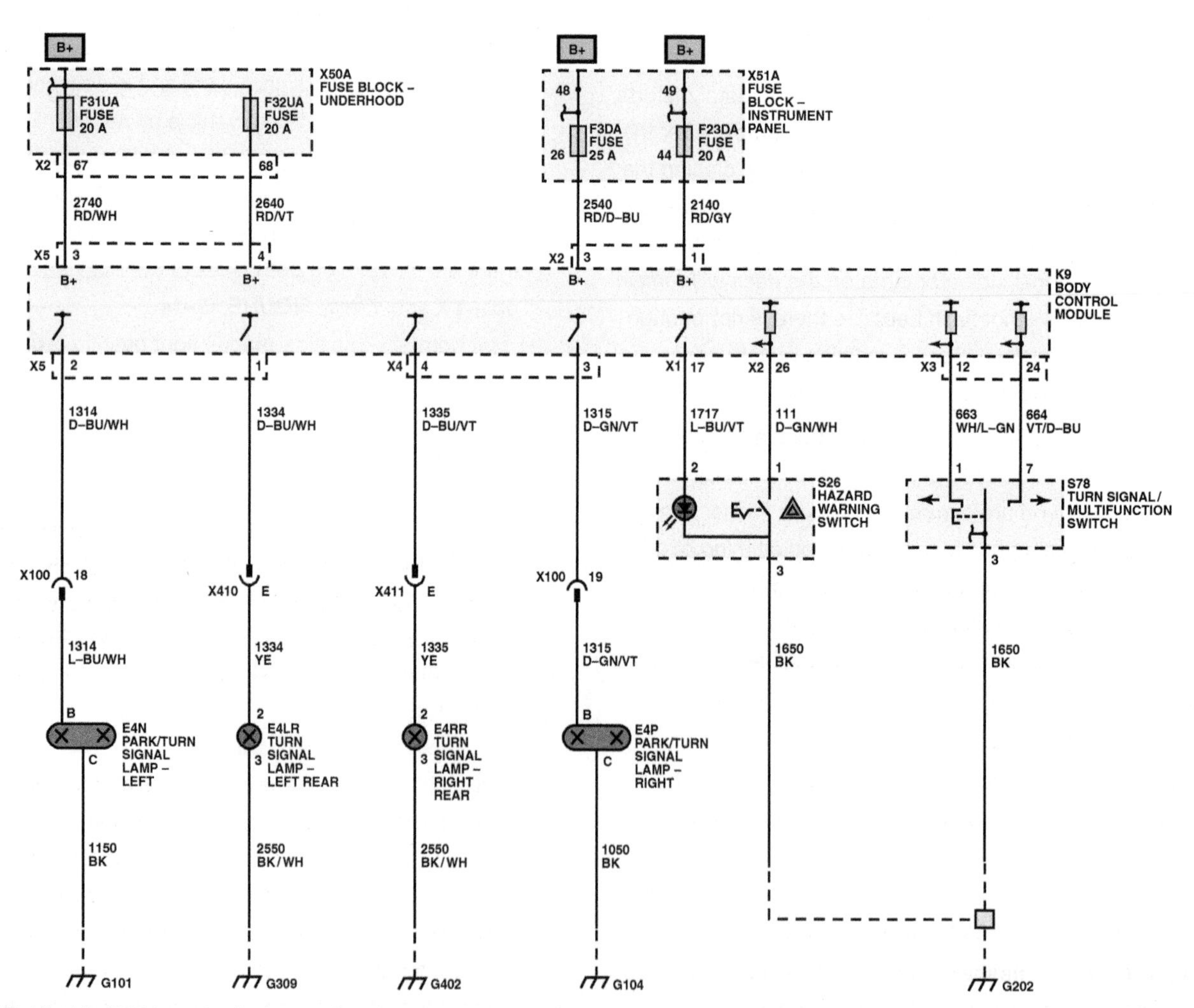

FIGURE 19–15 BCM-controlled turn signals schematic. Notice that there is no flasher unit. (Courtesy of General Motors)

When a turn signal request is received by the BCM, a serial data message is sent to the instrument panel cluster (IPC), requesting the respective turn signal indicator to be pulsed ON and OFF. The turn signal "clicking" sound is produced by either the instrument cluster of the left radio speaker.

HAZARD LAMPS, BCM-CONTROLLED The hazard flashers may be activated in any power mode. The hazard switch signal circuit is momentarily grounded when the hazard switch is pressed. The body control module responds to the hazard switch signal input by supplying battery voltage to all four turn signal lamps in an ON and OFF duty cycle. When the hazard switch is activated, the BCM sends a serial data message to the instrument panel cluster, requesting both turn signal indicators to be cycled ON and OFF.

NOTE: BCM-controlled systems do not have flasher units.

HEADLIGHTS AND PARKING LIGHTS

HEADLIGHT SWITCHES The headlight switch operates the exterior and interior lights of most vehicles. On noncomputer-controlled lighting systems, the headlight switch is connected directly to the battery through a fusible link, and has continuous power or is "hot" all the time. A circuit breaker is built into most older model headlight switches to protect the headlight circuit. ● **SEE FIGURE 19–16**.

The headlight switch may include the following:

- The interior dash lights can often be dimmed manually by rotating the headlight switch knob or by another rotary knob that controls a variable resistor (called a **rheostat**). The rheostat drops the voltage sent to the dash lights. Whenever there is a voltage drop (increased resistance), there is heat. A coiled resistance wire is built into a ceramic housing that is designed to insulate the rest of the switch from the heat and allow heat to escape.
- The headlight switch also contains a built-in circuit breaker that will rapidly turn the headlights on and off in the event of a short circuit. This prevents a total loss of headlights. If the headlights are rapidly flashing on and off, check the entire headlight circuit for possible shorts. The circuit breaker controls only the headlights. The other lights controlled by the headlight switch (taillights, dash lights, and parking lights) are fused separately. Flashing headlights may also be caused by a failure in the built-in circuit breaker, requiring replacement of the switch assembly.

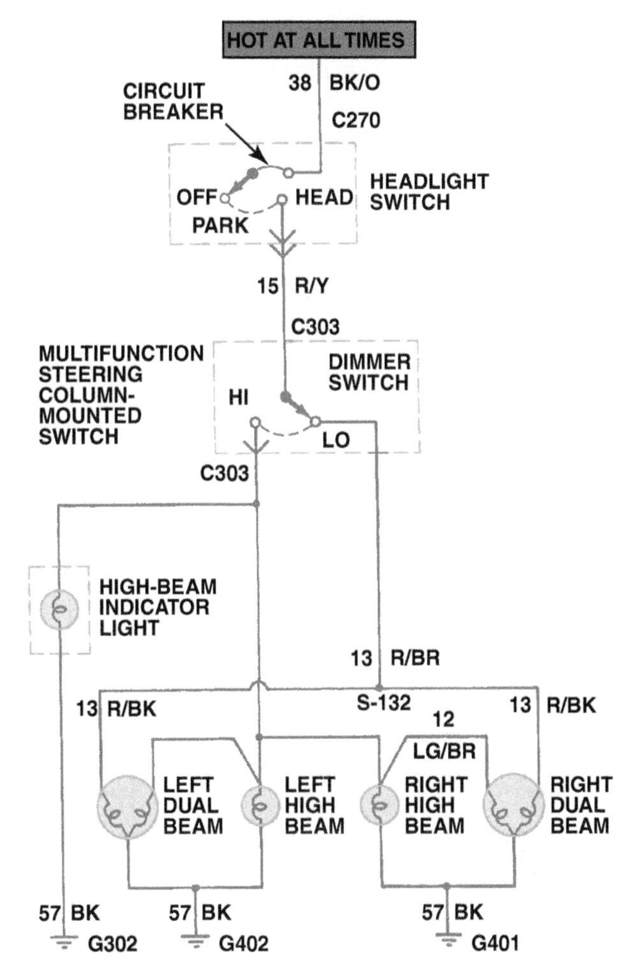

FIGURE 19–16 Typical headlight circuit diagram. Note that the headlight switch is represented by a dotted outline indicating that other circuits (such as dash lights) also operate from the switch.

PARK, TAIL, AND LICENSE LAMPS When the headlamp switch is placed in the HEAD or PARK position, ground is applied to the park lamp switch ON signal circuit to the body control module (BCM). The BCM responds by applying voltage to the park lamps, tail lamps, and license lamps control circuits, illuminating the park, tail, and license lamps. ● **SEE FIGURE 19–17**.

AUTOMATIC HEADLIGHTS Computer-controlled lights use a light sensor that signals the computer when to turn on the headlights. The sensor is mounted on the dashboard or mirror. Often these systems have a driver-adjusted sensitivity control that allows for the lights to be turned on at various levels of light. Most systems also have a computer module control over the time that the lights remain on after the ignition has been turned off and the last door has been closed. A scan tool is often needed to change this time delay.

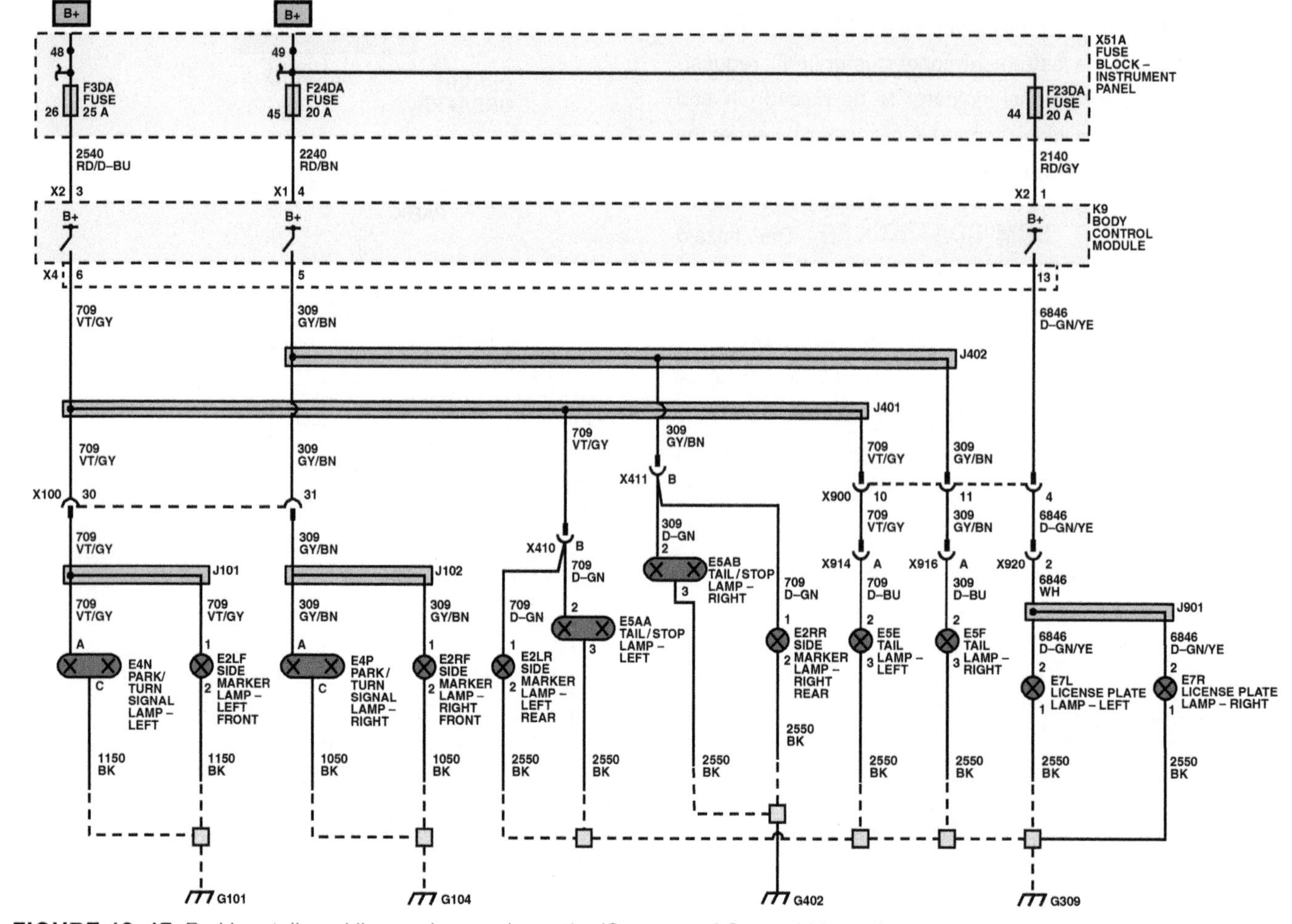

FIGURE 19–17 Parking, tail, and license lamp schematic. (Courtesy of General Motors)

BCM-CONTROLLED LOW BEAM HEADLAMPS The body control module (BCM) monitors three signal circuits from the headlamp switch. When the headlamp switch is in the "AUTO" position, all three signal circuits are open. When placed in the AUTO position, the BCM monitors inputs from the ambient light sensor to determine if headlamps are required or if the daytime running lamps will be activated, based on outside lighting conditions.

When the headlamp switch is placed in the OFF position, the headlamp switch "headlamps OFF" signal circuit is grounded, indicating to the BCM that the exterior lamps should be turned OFF. With the headlamp switch in the "PARK" position, the headlamp switch park lamps ON signal circuit is grounded, indicating that the park lamps have been requested.

LOW BEAM OPERATION When the headlamp switch is placed in the "HEADLAMP" position, both the headlamp switch park lamps ON signal circuit and the headlamp switch headlamps ON signal circuit are grounded. The BCM responds to the inputs by illuminating the park lamps and headlamps. When the low beam headlamps are requested, the BCM applies B+ to both low beam headlamp control circuits illuminating the low beam headlamps. ● **SEE FIGURE 19–18.**

The BCM will also command the low beam headlamps ON during daylight conditions when the following conditions are met:

- Headlamp switch in the AUTO position
- Windshield wipers ON
- Vehicle in any gear but PARK—automatic transmission
- Vehicle in motion—manual transmission

When the BCM commands the low beam headlamps ON, the driver will notice that the interior backlighting for the instrument cluster and the various other switches dim to the level of brightness selected by the instrument panel dimmer switch.

HIGH BEAM OPERATION When the low beam headlamps are ON and the turn signal/multifunction switch is placed in the high beam position, ground is applied to the BCM through the high beam signal circuit. The BCM responds to the high beam request by applying ground to the high beam relay control

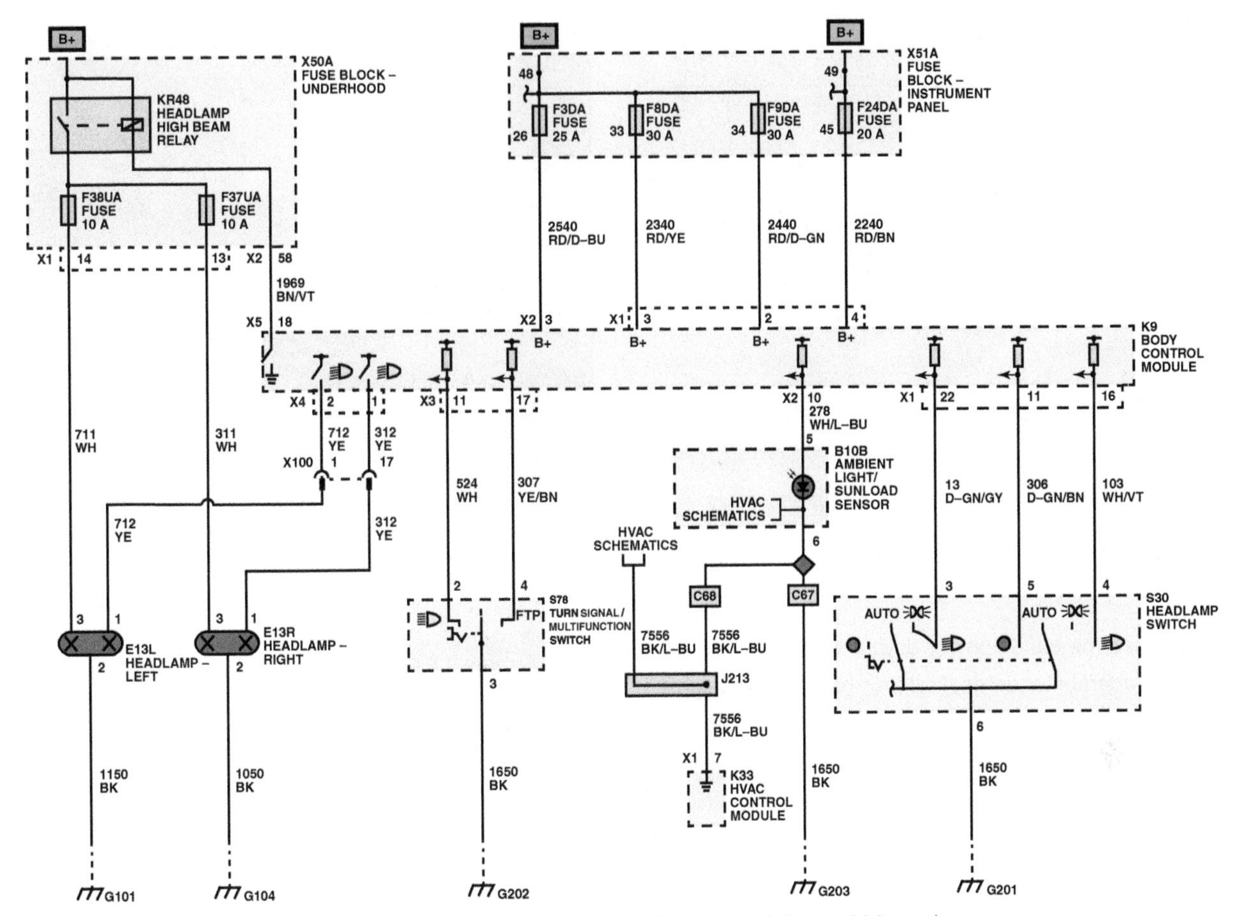

FIGURE 19–18 A typical BCM-controlled headlamps schematic. (Courtesy of General Motors)

circuit which energizes the high beam relay. With the high beam relay energized, the switch contacts close allowing battery voltage to flow through the left and right high beam fuses to the high beam control circuits illuminating the left and right high beam headlamps. Each headlamp is permanently grounded at ground G101 for the left headlamp and ground G104 for the right headlamp.

SEALED BEAM HEADLIGHTS A sealed beam headlight consists of a sealed glass or plastic assembly containing the bulb, a reflective surface, and prism lenses to properly focus the light beam. Low-beam headlights contain two filaments and three electrical terminals.

- One for low beam
- One for high beam
- Common ground

High-beam headlights contain only one filament and two terminals. Because low-beam headlights also contain a high-beam filament, the entire headlight assembly must be replaced if either filament is defective. ● **SEE FIGURE 19–19.**

FIGURE 19–19 A typical four-headlight system using sealed beam headlights.

HALOGEN SEALED BEAM HEADLIGHTS Halogen sealed beam headlights are brighter and more expensive than normal headlights. Because of their extra brightness, it is common practice to have only two headlights on at any

FIGURE 19–20 A typical composite headlamp assembly. The lens, housing, and bulb sockets are usually included as a complete assembly.

FIGURE 19–21 Handle a halogen bulb by the base to prevent the skin's oil from getting on the glass.

one time, because the candlepower output would exceed the maximum U.S. federal standards if all four halogen headlights were on. Therefore, before trying to repair the problem that only two of the four lamps are on, check the owner or shop manual for proper operation.

CAUTION: Do not attempt to wire all headlights together. The extra current flow could overheat the wiring from the headlight switch through the dimmer switch and to the headlights. The overloaded circuit could cause a fire.

COMPOSITE HEADLIGHTS **Composite headlights** are constructed using a replaceable bulb and a fixed lens cover that is part of the vehicle. Composite headlights are the result of changes in the aerodynamic styling of vehicles where sealed beam lamps could no longer be used. ● **SEE FIGURE 19–20.**

The replaceable bulbs are usually bright halogen bulbs. Halogen bulbs get very hot during operation, between 500°F and 1,300°F (260°C and 700°C). It is important never to touch the glass of any halogen bulb with bare fingers because the natural oils of the skin on the glass bulb can cause the bulb to break when it heats during normal operation.

TECH TIP

Diagnose Bulb Failure

Halogen bulbs can fail for various reasons. Some causes for halogen bulb failure and their indications are as follows:

- **Gray color.** Low voltage to bulb (check for corroded socket or connector)
- **White (cloudy) color.** Indication of an air leak
- **Broken filament.** Usually caused by excessive vibration
- **Blistered glass.** Indication that someone has touched the glass

NOTE: ***Never touch the glass (called the ampoule) of any halogen bulb.*** **The oils from your fingers can cause unequal heating of the glass during operation, leading to a shorter-than-normal service life.** ● **SEE FIGURE 19–21.**

HIGH-INTENSITY DISCHARGE HEADLIGHTS

PARTS AND OPERATION **High-intensity discharge (HID)** headlights produce a distinctive blue-white light that is crisper, clearer, and brighter than light produced by a halogen headlight.

High-intensity discharge lamps do not use a filament like conventional electrical bulbs, but contain two electrodes about 0.2 inch (5 mm) apart. A high-voltage pulse is sent to the bulb which arcs across the tips of electrodes producing light.

It creates light from an electrical discharge between two electrodes in a gas-filled arc tube. It produces twice the light with less electrical input than conventional halogen bulbs.

The HID lighting system consists of the discharge arc source, igniter, ballast, and headlight assembly. ● **SEE FIGURE 19–22.**

The two electrodes are contained in a tiny quartz capsule filled with xenon gas, mercury, and metal halide salts. HID

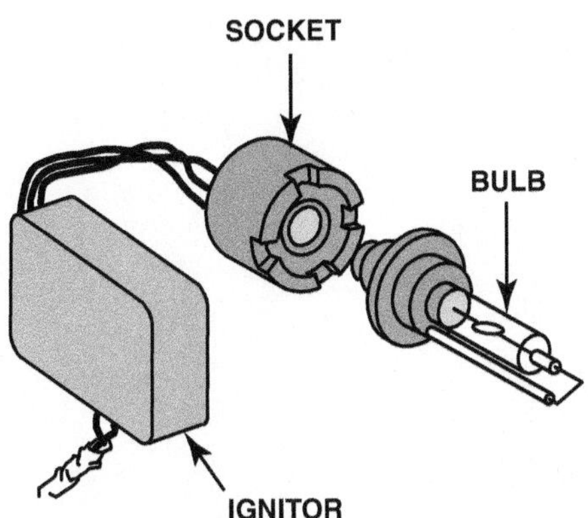

FIGURE 19–22 The igniter contains the ballast and transformer needed to provide high-voltage pulses to the arc tube bulb.

> **FREQUENTLY ASKED QUESTION**
>
> **What Is the Difference between the Temperature of the Light and the Brightness of the Light?**
>
> The temperature of the light indicates the color of the light. The brightness of the light is measured in lumens. A standard 100 watt incandescent light bulb emits about 1,700 lumens. A typical halogen headlight bulb produces about 2,000 lumens, and a typical HID bulb produces about 2,800 lumens.

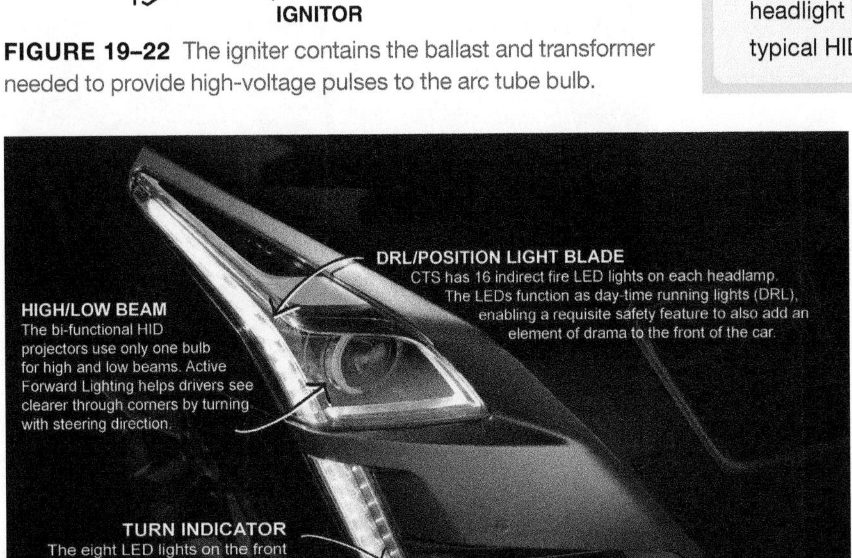

FIGURE 19–23 HID (xenon) headlights are part of the headlamp module on this Cadillac. (Courtesy of General Motors)

headlights are also called **xenon headlights**. The lights and support electronics are expensive, but they should last the life of the vehicle unless physically damaged.

HID headlights produce a white light giving the lamp a blue-white color. The color of light is expressed in temperature using the Kelvin scale. **Kelvin (K)** temperature is the Celsius temperature plus 273 degrees. Typical color temperatures include:

- Daylight: 5,400°K
- HID: 4,100°K
- Halogen: 3,200°K
- Incandescent (tungsten): 2,800°K

● **SEE FIGURE 19–23.**

The HID ballast is powered by 12 volts from the headlight switch on the body control module. The HID headlights operate in three stages or states.

1. Start-up or stroke state
2. Run-up state
3. Steady state

START-UP OR STROKE STATE When the headlight switch is turned to the on position, the ballast may draw up to 20 amperes at 12 volts. The ballast sends multiple high-voltage pulses to the arc tube to start the arc inside the bulb. The voltage provided by the ballast during the start-up state ranges from −600 volts to +600 volts, which is increased by a transformer to about 25,000 volts. The increased voltage is used to create an arc between the electrodes in the bulb.

RUN-UP STATE After the arc is established, the ballast provides a higher than steady state voltage to the arc tube to keep the bulb illuminated. On a cold bulb, this state could last as long as 40 seconds. On a hot bulb, the run-up state may last only 15 seconds. The current requirements during the run-up state are about 360 volts from the ballast and a power level of about 75 watts.

STEADY STATE The steady state phase begins when the power requirement of the bulb drops to 35 watts. The ballast provides a minimum of 55 volts to the bulb during steady state operation.

BI-XENON HEADLIGHTS Some vehicles are equipped with bi-xenon headlights, which use a shutter to block some of the light during low-beam operation and then mechanically move to expose more of the light from the bulb for high-beam operation. Because xenon lights are relatively slow to start working, vehicles equipped with bi-xenon headlights use two halogen lights for the "flash-to-pass" feature.

FAILURE SYMPTOMS The following symptoms indicate bulb failure:

- A light flickers
- Lights go out (caused when the ballast assembly detects repeated bulb restrikes)
- Color changes to a dim pink glow

Bulb failures are often intermittent and difficult to repeat. However, bulb failure is likely if the symptoms get worse over time. Always follow the vehicle manufacturer's recommended testing and service procedures.

DIAGNOSIS AND SERVICE High-intensity discharge headlights will change slightly in color with age. This **color shift** is usually not noticeable unless one headlight arc tube assembly has been replaced due to a collision repair, and then the difference in color may be noticeable. The difference in color will gradually change as the arc tube ages and should not be too noticeable by most customers. If the arc tube assembly is near the end of its life, it may not light immediately if it is turned off and then back on immediately. This test is called a "hot restrike" and if it fails, a replacement arc tube assembly may be needed or there is another fault, such as a poor electrical connection, that should be checked.

WARNING

Always adhere to all warnings because the high-voltage output of the ballast assembly can cause personal injury or death.

LED HEADLIGHTS

Some vehicles, including several Lexus models, use LED headlights either as standard equipment (Lexus LS600h) or optional. ● **SEE FIGURE 19–24.**

Advantages include:

- Long service life
- Reduced electrical power required

FIGURE 19–24 LED headlights usually require multiple units to provide the needed light. (Courtesy of General Motors)

Disadvantages include:

- High cost
- Many small LEDs required to create the necessary light output

HEADLIGHT AIMING

According to U.S. federal law, all headlights, regardless of shape, must be able to be aimed using headlight aiming equipment. Older vehicles equipped with sealed beam headlights used a headlight aiming system that attached to the headlight itself. ● **SEE FIGURES 19–25 AND 19–26.** Also see the photo sequence on headlight aiming at the end of the chapter.

ADAPTIVE FORWARD LIGHTING SYSTEM

PARTS AND OPERATION A system that mechanically moves the headlights to follow the direction of the front wheels is called **adaptive (**or **advanced) forward light system,** or **AFS**. The AFS provides a wider range of visibility during cornering. The headlights are usually capable of rotating from 5 to 15 degrees left or right, depending on the direction of the turn. Vehicles that use AFS include many General Motors models, usually as an extra cost option. ● **SEE FIGURE 19–27.**

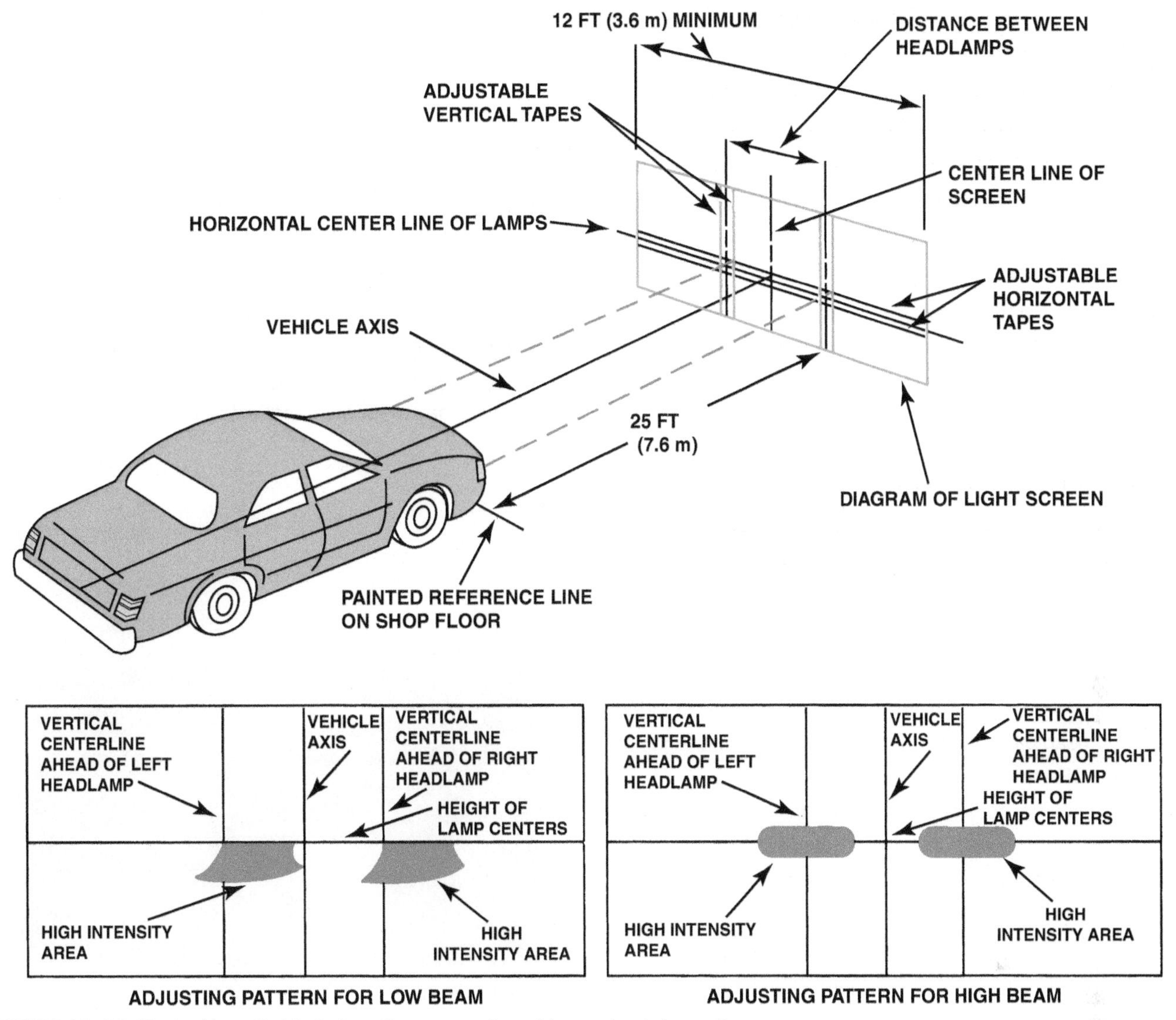

FIGURE 19–25 Typical headlight aiming diagram as found in service information.

FIGURE 19–26 Many composite headlights have a built-in bubble level to make aiming easy and accurate.

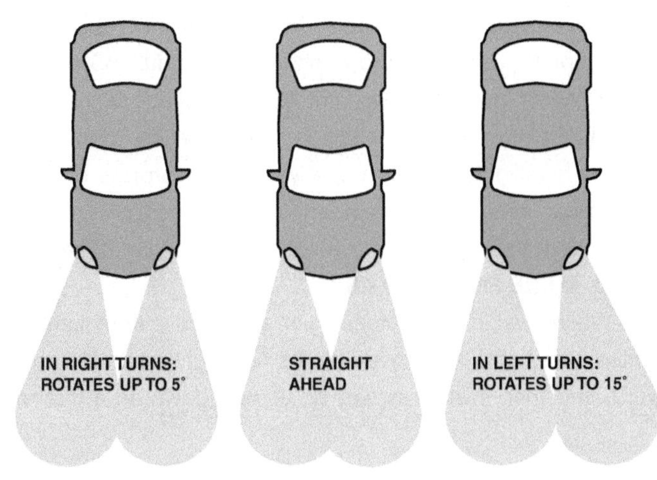

FIGURE 19–27 Adaptive front lighting systems rotate the low-beam headlight in the direction of travel. Operation of the left headlight is indicated above.

NOTE: These angles are reversed on vehicles sold in countries that drive on the left side of the road, such as Great Britain, Japan, Australia, and New Zealand.

The vehicle has to be moving above a predetermined speed, usually above 20 mph (30 km/h) for full movement, and the lights stop moving when the speed drops below about 3 mph (5 km/h).

The specific components of the AFS are:

- **Headlamp Control Module—**Is fully functional when ignition is in RUN position with an operating voltage range of 10.5–16 volt.
- **Right Headlamp Motor—**Moves the headlamp a maximum of 15° right or maximum of 5° left, as commanded by headlamp control module.
- **Left Headlamp Motor—**Moves the headlamp a maximum of 15° left or maximum of 5° right, as commanded by headlamp control module.

Auxiliary components used by AFS are the body control module (BCM), electronic brake control module (EBCM), engine control module (ECM), and transmission control module (TCM). The headlamp control module activates when ignition moves from CRANK to RUN.

The headlamp control module calculates vehicle speed, transmission gear selection, power mode, and headlamp switch position from serial data messages received from BCM, EBCM, ECM, and TCM. Data is sent to the headlamp control module via GMLAN. The direction the headlamps move is determined by steering wheel angle (data sent by EBCM) and is limited by steering angles of + / − 90°. There is a "dead zone" of about + / − 10° near steering center in which AFS will not move headlamps.

Depending on data received, headlamp control module activates motors to move headlamps right or left. Movement of headlamps is restricted at low vehicle speeds; full movement doesn't occur in vehicle speeds of 30 mph and above.

AFS is often used in addition to self-leveling motors so that the headlights remain properly aimed regardless of how the vehicle is loaded. Without self-leveling, headlights would shine higher than normal if the rear of the vehicle is heavily loaded. ● **SEE FIGURE 19–28.**

When a vehicle is equipped with an adaptive front lighting system, the lights are moved by the headlight controller outward, and then inward as well as up and down as a test of the system. This action is quite noticeable to the driver, and is normal operation of the system.

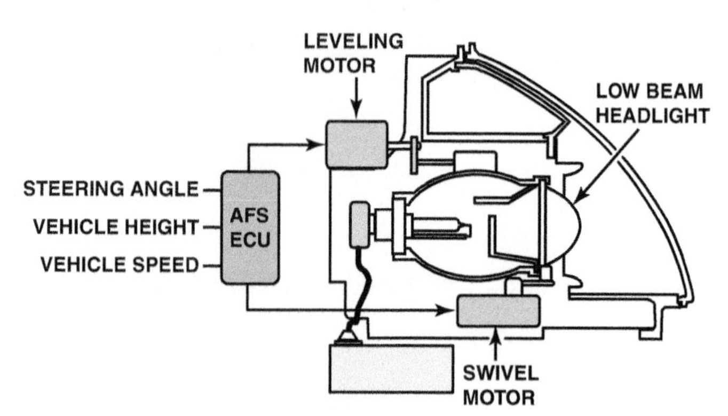

FIGURE 19–28 Some adaptive forward lighting systems use two motors, one for the up and down movement and the other for rotating the low-beam headlight to the left and right.

FIGURE 19–29 Typical dash-mounted switch that allows the driver to disable the front lighting system.

DIAGNOSIS AND SERVICE The first step when diagnosing an AFS fault is to perform the following visual inspection:

- Start by checking that the AFS is switched on. Most AFS headlights are equipped with a switch that allows the driver to turn the system on and off. ● **SEE FIGURE 19–29.**
- Check that the system performs a self-test during start-up.

- Verify that both low-beam and high-beam lights function correctly. The system may be disabled if a fault with one of the headlights is detected.
- Use a scan tool to test for any AFS-related diagnostic trouble codes. Some systems allow the AFS to be checked and operated using a scan tool.

Always follow the recommended testing and service procedures as specified by the vehicle manufacturer in service information.

TECH TIP

Checking a Dome Light Can Be Confusing

If a technician checks a dome light with a test light, both sides of the bulb will "turn on the light" if the bulb is good. This will be true if the system's "ground switched" doors are closed and the bulb is good. This confuses many technicians because they do not realize that the ground will not be sensed unless the door is open.

DAYTIME RUNNING LIGHTS

PURPOSE AND FUNCTION **Daytime running lights (DRLs)** involve operation of the following:

- Front parking lights
- Separate DRL lamps
- Headlights (usually at reduced current and voltage) when the vehicle is running

Canada has required daytime running lights on all new vehicles since 1990. Studies have shown that DRLs have reduced accidents where used.

Daytime running lights primarily use a control module that turns on either the low- or high-beam headlights or separate daytime running lights. The lights on some vehicles come on when the engine starts. Other vehicles will turn on the lamps when the engine is running but delay their operation until a signal from the vehicle speed sensor indicates that the vehicle is moving.

To avoid having the lights on during servicing, some systems will turn off the headlights when the parking brake is applied and the ignition switch is cycled off then back on. Others will only light the headlights when the vehicle is in a drive gear. ● **SEE FIGURE 19–30.**

CAUTION: Most factory daytime running lights operate the headlights at reduced intensity. These are *not* designed to be used at night. Normal intensity of the headlights (and operation of the other external lamps) is actuated by turning on the headlights as usual.

DAYTIME RUNNING LAMPS (DRL), BCM-CONTROLLED

The daytime running lamps (DRL) will illuminate the right and left low beam headlamps continuously. The DRLs will operate when the following conditions are met:

- The ignition is in the RUN or CRANK position
- The shift lever is out of the PARK position for vehicles equipped with automatic transmissions or the parking brake is released for vehicles with manual transmissions
- The low and high beam headlamps are OFF

Refer to FIGURE 19–18 for operation. The ambient light sensor is used to monitor outside lighting conditions. The ambient light sensor provides a voltage signal that will vary between 0.2 and 4.9 volts depending on outside lighting conditions. The body control module (BCM) provides a 5-volt reference signal to the ambient light sensor which is permanently grounded at G203. The BCM monitors the ambient light sensor signal circuit to determine if outside lighting conditions are correct for either daytime running lights or automatic lamp control (ALC) when the headlamp switch is in the AUTO position. In daylight conditions the BCM will command the low beam headlamps ON. Any function or condition that turns on the headlamps will cancel DRL operation.

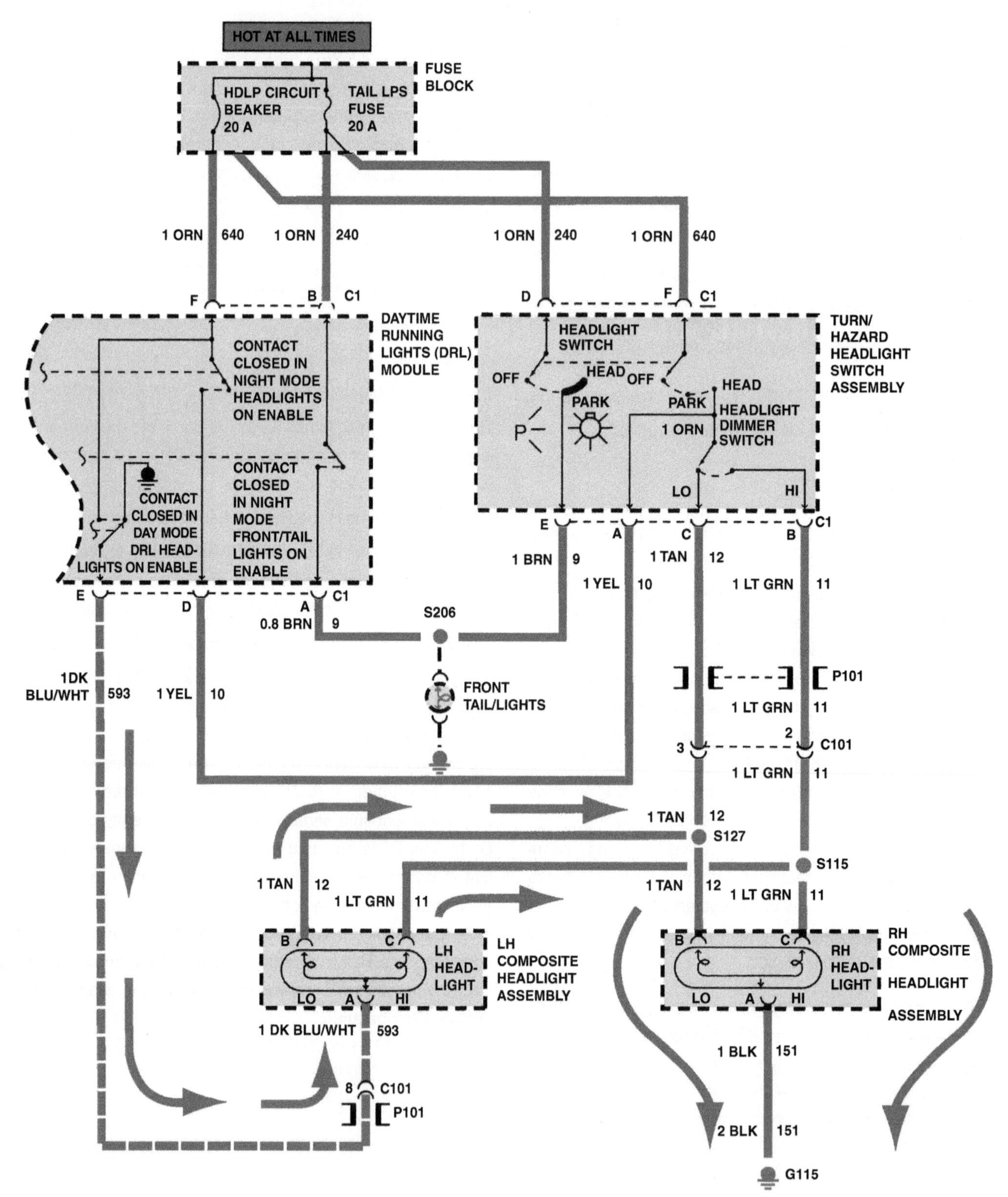

FIGURE 19–30 Typical daytime running light (DRL) circuit. Follow the arrows from the DRL module through both headlights. Notice that the left and right headlights are connected in series, resulting in increased resistance, less current flow, and dimmer than normal lighting. When the normal headlights are turned on, both headlights receive full battery voltage, with the left headlight grounding through the DRL module.

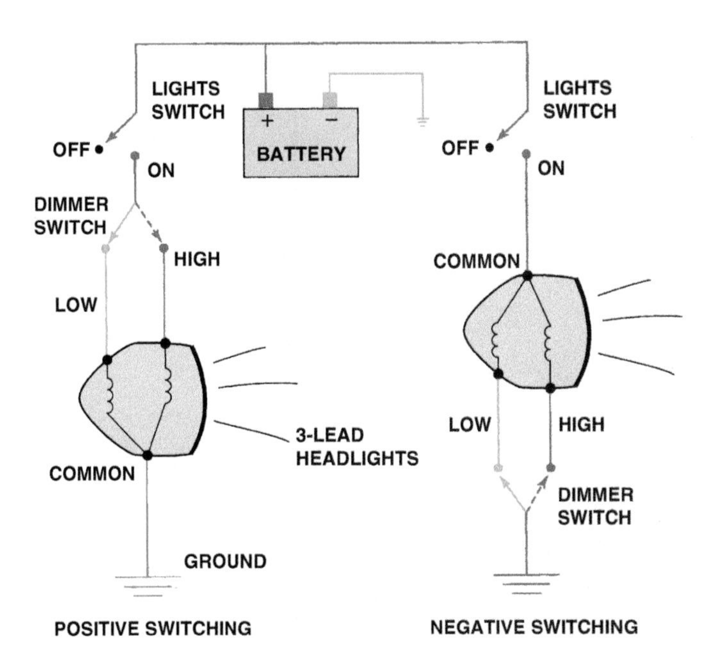

FIGURE 19–31 Most vehicles use positive switching of the high- and low-beam headlights. Notice that both filaments share the same ground connection. Some vehicles use negative switching and place the dimmer switch between the filaments and the ground.

DIMMER SWITCHES

On older vehicles the headlight switch controls the power or hot side of the headlight circuit. The current is then sent to the dimmer switch, which allows current to flow to either the high-beam or the low-beam filament of the headlight bulb, as shown in ● **FIGURE 19–31**.

An indicator light illuminates on the dash when the high beams are selected.

Newer vehicles use the BCM to control high and low beam operation. The turn signal/multifunction switch is an input to the BCM, either a ground or voltage signal, to the high beam signal circuit in the BCM. The BCM then turns on the high beam headlights through the high beam relay. Refer to ● **FIGURE 19–32**.

INSTRUMENT PANEL DIMMING, BCM CONTROLLED

The I/P dimmer switch controls are located on the headlamp switch assembly and are used to increase and decrease the brightness of the interior backlighting components. When the I/P dimmer switch is placed in a desired brightness position, the body control module (BCM) receives a signal from the I/P dimmer switch and responds by applying a pulse width modulated (PWM) voltage to the hazard switch light emitting diode (LED) backlighting control circuit, illuminating the LED to the desired level of brightness. ● **SEE FIGURE 19–33**.

COURTESY LIGHTS

Courtesy light is a generic term primarily used for interior lights, including overhead (dome) and under-the-dash (courtesy) lights. These interior lights are controlled by operating switches located in the doorjambs or door latch assembly of the vehicle doors or by a switch on the dash. ● **SEE FIGURE 19–34**.

Most newer vehicles operate the interior lights through the vehicle computer or through an electronic module. Because the exact wiring and operation of these units differ, consult the service information for the exact model of the vehicle being serviced. ● **SEE FIGURE 19–35**.

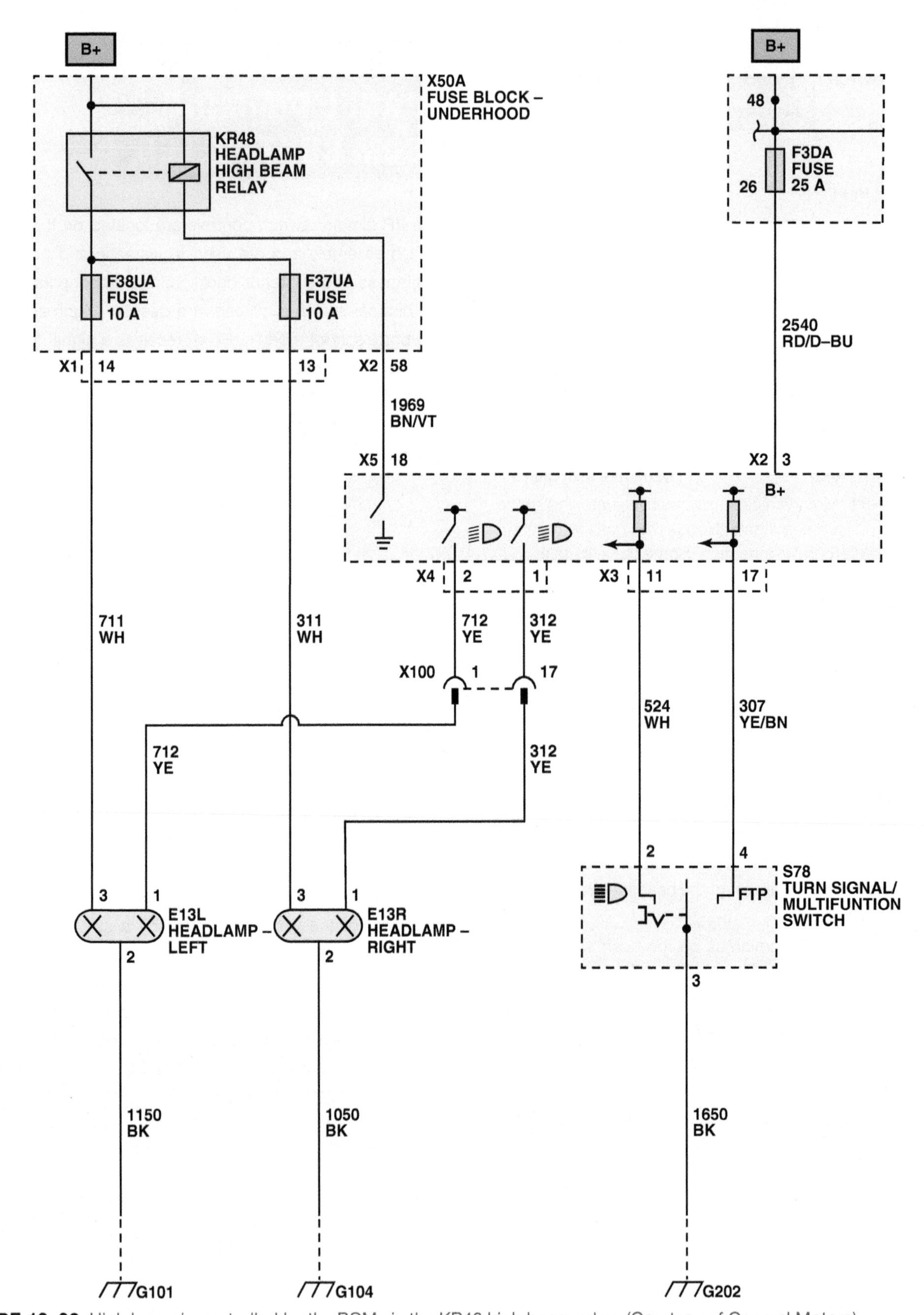

FIGURE 19–32 High beam is controlled by the BCM via the KR48 high beam relay. (Courtesy of General Motors)

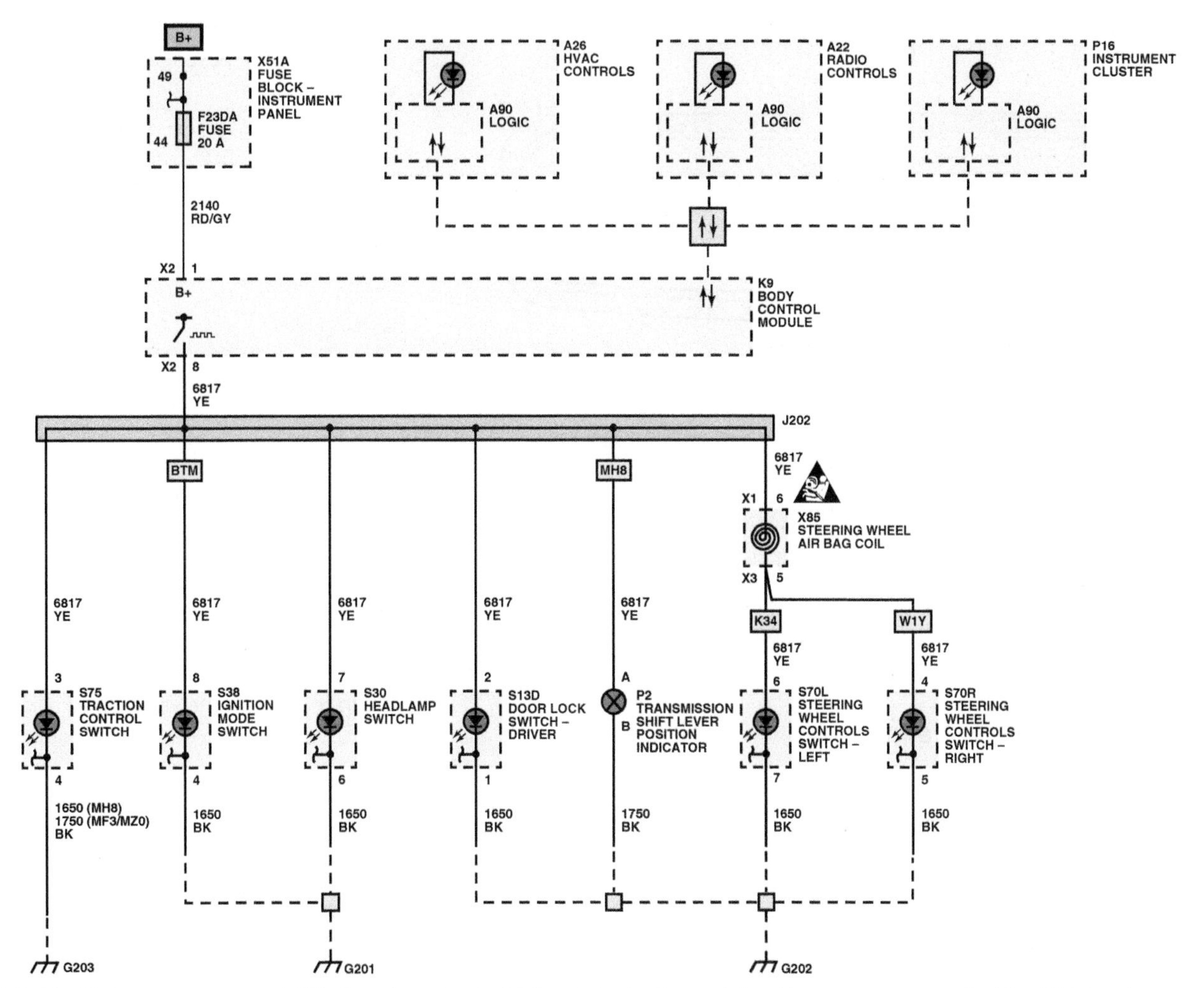

FIGURE 19–33 Instrument panel (IP) dimming circuit. LEDs at the bottom of the diagram are controlled by PWM voltage; brightness at the top is requested by the BCM over GM LAN. (Courtesy of General Motors)

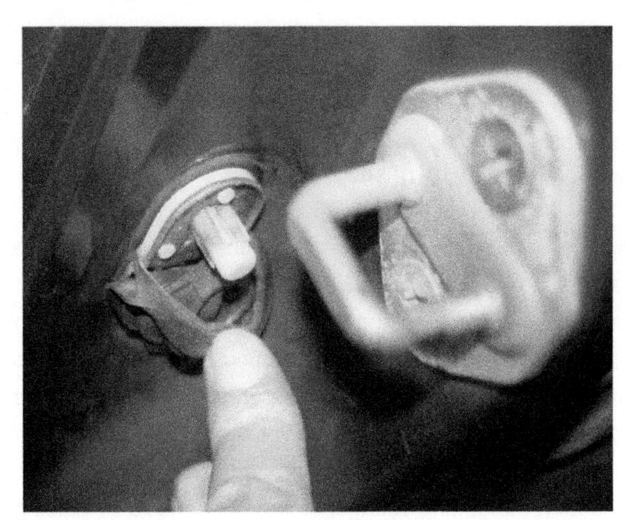

FIGURE 19–34 A typical courtesy light doorjamb switch. Newer vehicles use the door switch as an input to the vehicle computer and the computer turns the interior lights on or off. By placing the lights under the control of the computer, the vehicle engineers have the opportunity to delay the lights after the door is closed and to shut them off after a period of time to avoid draining the battery.

FIGURE 19–35 On many vehicles the door lock motor and door switches are part of the door latch assembly. The switches are inputs to the door module or body control module. In this example D is ground; C is Key Switch signal; B is Door Ajar signal; A is Interior Lamp Switch Signal. (Courtesy of Jeffrey Rehkopf)

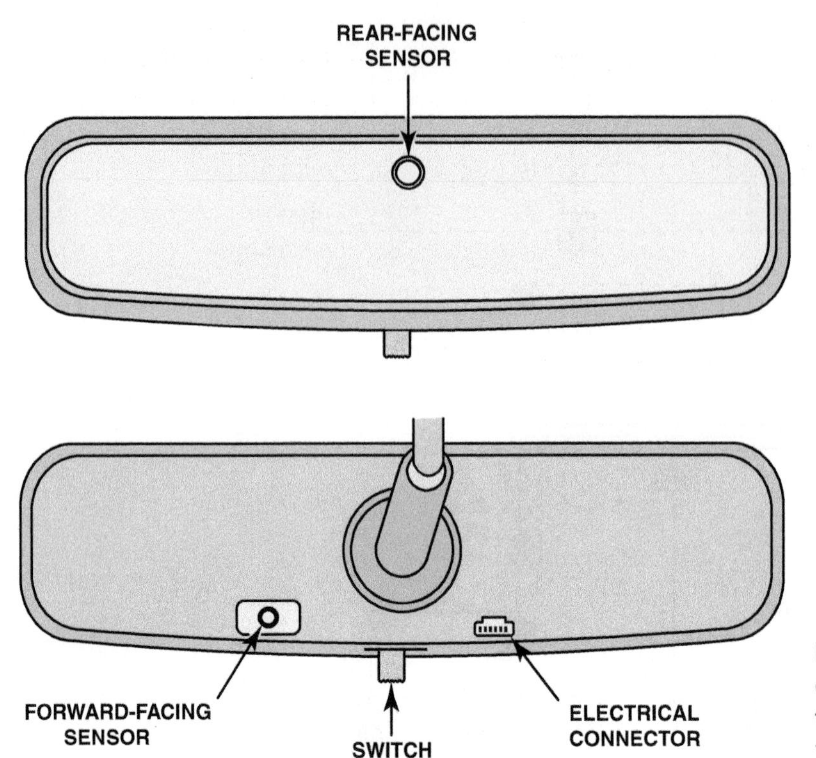

FIGURE 19–36 An automatic dimming mirror compares the amount of light toward the front of the vehicle to the rear of the vehicle and allies a voltage to cause the gel to darken the mirror.

ILLUMINATED ENTRY

Some vehicles are equipped with illuminated entry, meaning the interior lights are turned on for a given amount of time when the outside door handle is operated while the doors are locked. Most vehicles equipped with illuminated entry also light the exterior door keyhole. Vehicles equipped with body computers use the input from the key fob remote to "wake up" the power supply for the body computer.

AUTOMATIC DIMMING MIRRORS

PARTS AND OPERATION Automatic dimming mirrors use electrochromic technology to dim the mirror in proportion to the amount of headlight glare from other vehicles at the rear. The electrochromic technology developed by Gentex Corporation uses a gel that changes with light between two pieces of glass. One piece of glass acts as a reflector and the other has a transparent (clear) electrically conductive coating. The inside rearview mirror also has a forward-facing light sensor that is used to detect darkness and signal the rearward-facing sensor to begin to check for excessive glare from headlights behind the vehicle. The rearward-facing sensor sends a voltage to the electrochromic gel in the mirror that is in proportion to the amount of glare detected. The mirror dims in proportion to the glare and then becomes like a standard rearview mirror when the glare is no longer detected. If automatic dimming mirrors are used on the exterior, the sensors in the interior mirror and electronics are used to control both the interior and exterior mirrors. ● **SEE FIGURE 19–36.**

DIAGNOSIS AND SERVICE If a customer concern states that the mirrors do not dim when exposed to bright headlights from the vehicle behind, the cause could be sensors or the mirror itself. Be sure that the mirror is getting electrical power. Most automotive dimming mirrors have a green light to indicate the presence of electrical power. If no voltage is found at the mirror, follow standard troubleshooting procedures to find the cause. If the mirror is getting voltage, start the diagnosis by placing a strip of tape over the forward-facing light sensor. Turn the ignition key on, engine off (KOEO), and observe the operation of

FREQUENTLY ASKED QUESTION

What Is the Troxler Effect?

The **Troxler effect**, also called *Troxler fading,* is a visual effect where an image remains on the retina of the eye for a short time after the image has been removed. The effect was discovered in 1804 by Igney Paul Vital Troxler (1780–1866), a Swiss physician. Because of the Troxler effect, headlight glare can remain on the retina of the eye and create a blind spot. At night, this fading away of the bright light from the vehicle in the rear reflected by the rearview mirror can cause a hazard.

TECH TIP

The Weirder the Problem, the More Likely It Is a Poor Ground Connection

Bad grounds are often the cause for feedback or lamps operating at full or partial brilliance. At first the problem looks weird because often the switch for the lights that are on dimly is not even turned on. When an electrical device is operating and it lacks a proper ground connection, the current will try to find ground and will often cause other circuits to work. Check all grounds before replacing parts.

the mirror when a flashlight or trouble light is directed onto the mirror. If the mirror reacts and dims, the forward-facing sensor is defective. Most often, the entire mirror assembly has to be replaced if any sensor or mirror faults are found.

One typical fault with automatic dimming mirrors is a crack can occur in the mirror assembly, allowing the gel to escape from between the two layers of glass. This gel can drip onto the dash or center console and harm these surfaces. The mirror should be replaced at the first sign of any gel leakage.

LIGHTING SYSTEM DIAGNOSIS

Diagnosing any faults in the lighting and signaling systems usually includes the following steps:

STEP 1 Verify the customer concern.

STEP 2 Perform a visual inspection, checking for collision damage or other possible causes that would affect the operation of the lighting circuit.

STEP 3 Connect a factory or enhanced scan tool with bidirectional control of the computer modules to check for proper operation of the affected lighting circuit.

STEP 4 Follow the diagnostic procedure as found in service information to determine the root cause of the problem.

LIGHTING SYSTEM SYMPTOM GUIDE

The following list will assist technicians in troubleshooting lighting systems.

Problem	Possible Causes and/or Solutions
One headlight dim	1. Poor ground connection on body 2. Corroded connector
One headlight out (low or high beam)	1. Burned out headlight filament (Check the headlight with an ohmmeter. There should be a low-ohm reading between the power-side connection and the ground terminal of the bulb.) 2. Open circuit (no 12 volts to the bulb)
Both high- and low-beam headlights out	1. Burned out bulbs (Check for voltage at the wiring connector to the headlights for a possible open circuit to the headlights or open [defective] dimmer switch.) 2. Open circuit (no 12 volts to the bulb)
All headlights inoperative	1. Burned out filaments in all headlights (Check for excessive charging system voltage.) 2. Defective dimmer switch 3. Defective headlight switch
Slow turn signal operation	1. Defective flasher unit 2. High resistance in sockets or ground wire connections 3. Incorrect bulb numbers
Turn signals operating on one side only	1. Burned out bulb on affected side 2. Poor ground connection or defective socket on affected side 3. Incorrect bulb number on affected side 4. Defective turn signal switch
Interior light(s) inoperative	1. Burned out bulb(s) 2. Open in the power-side circuit (blown fuse) 3. Open in doorjamb switch(es)
Interior lights on all the time	1. Shorted doorjamb switch 2. Shorted control switch
Brake lights inoperative	1. Defective brake switch 2. Defective turn signal switch 3. Burned out brake light bulbs 4. Open circuit or poor ground connection 5. Blown fuse
Hazard warning lights inoperative	1. Defective hazard flasher unit 2. Open in hazard circuit 3. Blown fuse 4. Defective hazard switch
Hazard warning lights blinking too rapidly	1. Incorrect flasher unit 2. Shorted wiring to front or rear lights 3. Incorrect bulb numbers

1 The driver noticed that the taillight fault indicator (icon) on the dash was on any time the lights were on.

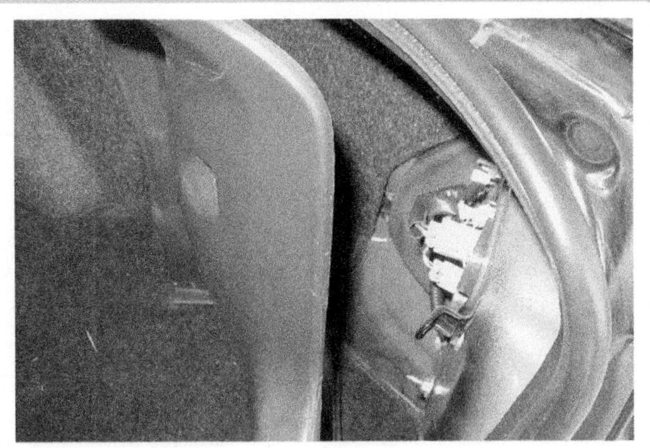

2 A visual inspection at the rear of the vehicle indicated that the right rear taillight bulb did not light. Removing a few screws from the plastic cover revealed the taillight assembly.

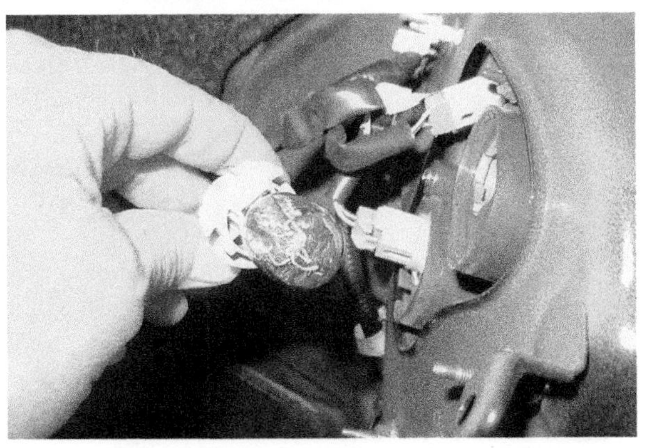

3 The bulb socket is removed from the taillight assembly by gently twisting the base of the bulb counterclockwise.

4 The bulb is removed from the socket by gently grasping the bulb and pulling the bulb straight out of the socket. Many bulbs required that you rotate the bulb 90° (1/4 turn) to release the retaining bulbs.

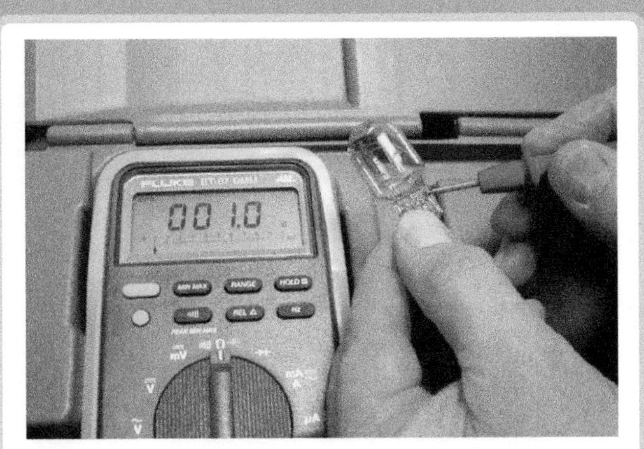

5 The new 7443 replacement bulb is being checked with an ohmmeter to be sure that it is okay before it is installed in the vehicle.

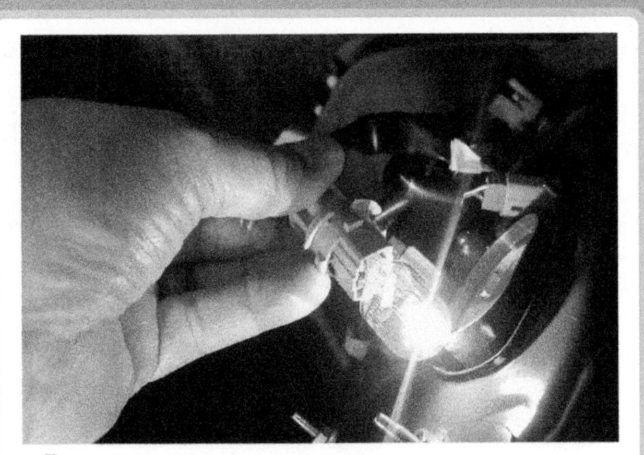

6 The replacement bulb in inserted into the taillight socket and the lights are turned on to verify proper operation before putting the components back together.

OPTICAL HEADLIGHT AIMING

1 Before checking the vehicle for headlight aim, be sure that all the tires are at the correct inflation pressure, and that the suspension is in good working condition.

2 The headlight aim equipment will have to be adjusted for the slope of the floor in the service bay. Start the process by turning on the laser light generator on the side of the aimer body.

3 Place a yardstick or measuring tape vertically in front of the center of the front wheel, noting the height of the laser beam.

4 Move the yardstick to the center of the rear wheel and measure the height of the laser beam at this point. The height at the front and rear wheels should be the same.

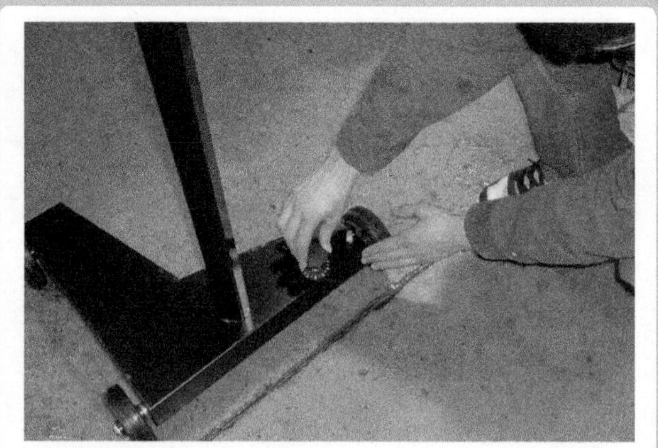

5 If the laser beam height measurements are not the same, the floor slope of the aiming equipment must be adjusted. Turn the floor slope knob until the measurements are equal.

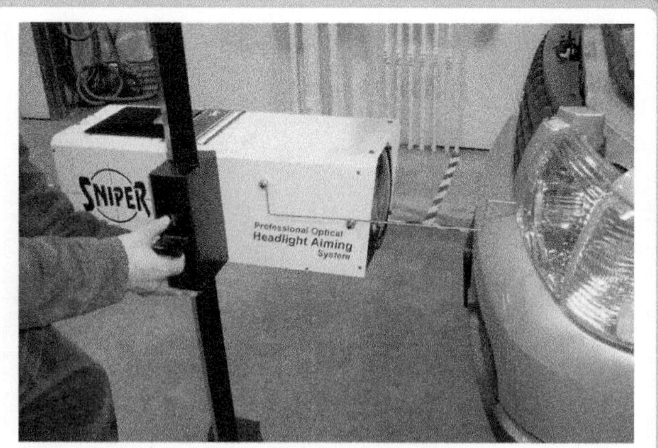

6 Place the aimer in front of the headlight to be checked, at a distance of 10 to 14 inches (25 to 35 cm). Use the aiming pointer to adjust the height of the aimer to the middle of the headlight.

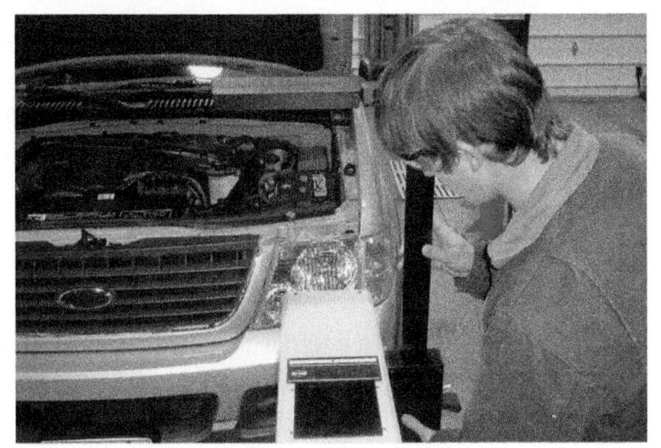

7 Align the aimer horizontally, using the pointer to place the aimer at the center of the headlight.

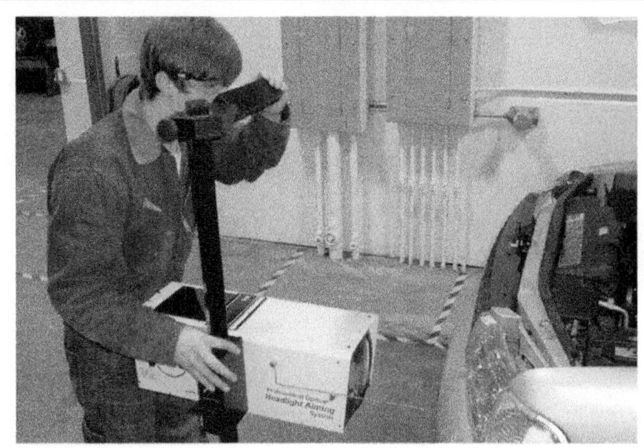

8 Lateral alignment (aligning the body of the aimer with the body of the vehicle) is done by looking through the upper visor. The line in the upper visor is aligned with symmetrical points on the vehicle body.

9 Turn on the vehicle headlights, being sure to select the correct beam position for the headlight to be aimed.

10 View the light beam through the aimer window. The position of the light pattern will be different for high and low beams.

11 If the first headlight is aimed adequately, move the aimer to the headlight on the opposite side of the vehicle. Follow the previous steps to position the aimer accurately.

12 If adjustment is required, move the headlight adjusting screws using a special tool or a 1/4 inch drive ratchet/socket combination. Watch the light beam through the aimer window to verify the adjustment.

SUMMARY

1. Automotive bulbs are identified by trade numbers.
2. The trade number is the same regardless of manufacturer for the exact same bulb specification.
3. Daytime running lights (DRLs) are used on many vehicles.
4. High-intensity discharge (HID) headlights are brighter and have a blue tint.
5. Turn signal flashers come in many different types and construction.

REVIEW QUESTIONS

1. Why should the exact same trade number of bulb be used as a replacement?
2. Why is it important to avoid touching a halogen bulb with your fingers?
3. How do you diagnose a turn signal operating problem?
4. How do you aim headlights on a vehicle equipped with aerodynamic-style headlights?

CHAPTER QUIZ

1. Technician A says that the bulb trade number is the same for all bulbs of the same size. Technician B says that a dual-filament bulb has different candlepower ratings for each filament. Which technician is correct?
   - **a.** Technician A only
   - **b.** Technician B only
   - **c.** Both Technicians A and B
   - **d.** Neither Technician A nor B
2. Two technicians are discussing flasher units. Technician A says that only a DOT-approved flasher unit should be used for turn signals. Technician B says that a parallel (variable-load) flasher will function for turn signal usage, although it will not warn the driver if a bulb burns out. Which technician is correct?
   - **a.** Technician A only
   - **b.** Technician B only
   - **c.** Both Technicians A and B
   - **d.** Neither Technician A nor B
3. Interior overhead lights (dome lights) are operated by doorjamb switches that ________.
   - **a.** Complete the power side of the circuit
   - **b.** Complete the ground side of the circuit
   - **c.** Move the bulb(s) into contact with the power and ground
   - **d.** Either a or b depending on application
4. Electrical feedback is usually a result of ________.
   - **a.** Too high a voltage in a circuit
   - **b.** Too much current (in amperes) in a circuit
   - **c.** Lack of a proper ground
   - **d.** Both a and b
5. According to Chart 19–1, which bulb is brightest?
   - **a.** 194
   - **b.** 168
   - **c.** 194NA
   - **d.** 1157
6. If a 1157 bulb were to be installed in a left front parking light socket instead of a 2057 bulb, what would be the most likely result?
   - **a.** The left turn signal would flash faster.
   - **b.** The left turn signal would flash slower.
   - **c.** The left parking light would be slightly brighter.
   - **d.** The left parking light would be slightly dimmer.
7. A technician replaced a 1157NA with a 1157A bulb. Which is the most likely result?
   - **a.** The bulb is brighter because the 1157A candlepower is higher.
   - **b.** The amber color of the bulb is a different shade.
   - **c.** The bulb is dimmer because the 1157A candlepower is lower.
   - **d.** Both b and c
8. A customer complained that every time he turned on his vehicle's lights, the left-side turn signal indicator light on the dash remained on. The most likely cause is a ________.
   - **a.** Poor ground to the parking light (or taillight) bulb on the *left* side
   - **b.** Poor ground to the parking light (or taillight) bulb on the *right* side, causing current to flow to the left-side lights
   - **c.** Defective (open) parking light (or taillight) bulb on the left side
   - **d.** Both a and c
9. A defective taillight or front park light bulb could cause the ________.
   - **a.** Turn signal indicator on the dash to light when the lights are turned on
   - **b.** Dash lights to come on when the brake lights are on
   - **c.** Lights-on warning chime to sound if the brake pedal is depressed
   - **d.** All of the above
10. A defective brake switch could prevent proper operation of the ________.
    - **a.** Cruise control
    - **b.** ABS brakes
    - **c.** Shift interlock
    - **d.** All of the above

chapter 20

DRIVER INFORMATION AND NAVIGATION SYSTEMS

LEARNING OBJECTIVES

After studying this chapter, the reader will be able to:

1. Identify the meaning of dash warning symbols.
2. Explain the operation of electronic speedometers and electronic odometers.
3. Describe how a navigation system works.
4. Explain the operation and diagnosis of OnStar, backup camera, and backup sensor.
5. Describe how to troubleshoot malfunctioning dash instruments.

This chapter will help you prepare for the ASE Electrical/Electronic Systems (A6) certification test content area "F" (Gauges, Warning Devices, and Driver Information System Diagnosis and Repair).

KEY TERMS

Backup camera 319
brake fluid level sensor 306
EEPROM 313
GPS 315
HUD 313
IP 307
LDWS 320
NVRAM 313
PM generator 311
Pressure differential switch 306
RPA 319
red brake warning lamp (RBWL) 306

GM STC OBJECTIVES

GM Service Technical College topics covered in this chapter are as follows:

1. Describe what OnStar is. (19040.37D1)
2. Discuss the general features of the OnStar system.
3. Identify the components, operation, and diagnostic and service procedures of the Rear Vision Camera (RVC) system.
4. Identify the components, operation, and diagnostic and service procedures of the Lane Departure System.
5. Identify the characteristics, components, operation, and diagnostic and service procedures of the Rear Parking Assist System.

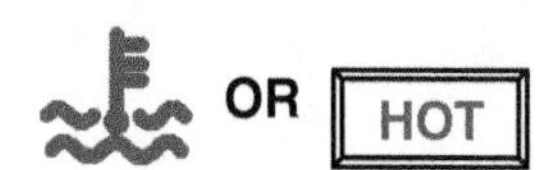

FIGURE 20–1 Engine coolant temperature is too high.

FIGURE 20–2 Engine oil pressure too low.

FIGURE 20–3 Water detected in fuel. Notice to drain the water from the fuel filter assembly on a vehicle equipped with a diesel engine.

FIGURE 20–4 Maintenance required. This usually means that the engine oil is scheduled to be changed or other routine service items need to be replaced or checked.

DASH WARNING SYMBOLS

PURPOSE AND FUNCTION All vehicles are equipped with warning lights that are often confusing to drivers. Because many vehicles are sold throughout the world, symbols instead of words are being used as warning lights. The dash warning lights are often called *telltale* lights as they are used to notify the driver of a situation or fault.

BULB TEST When the ignition is first turned on, all of the warning lights come on as part of a self-test and to help the driver or technician spot any warning light that may be burned out. Technicians or drivers who are familiar with what lights should light may be able to determine if one or more warning lights are not on when the ignition is first turned on. Most factory scan tools can be used to command all of the warning lights on to help determine if one is not working.

ENGINE FAULT WARNING Engine fault warning lights include the following:

- **Engine coolant temperature.** This warning lamp should come on when the ignition is first turned on as a bulb check and if the coolant temperature reaches 248°F to 258°F (120°C to 126°C), depending on the make and model of the vehicle. ● **SEE FIGURE 20–1.**

  If the engine coolant temperature warning lamp comes on while driving, perform the following in an attempt to reduce the temperature.

  1. Turn off the air-conditioning.
  2. Turn on the heater.
  3. If the hot light remains on, drive to a safe location and shut off the engine and allow it to cool to help avoid serious engine damage.

- **Engine oil pressure.** This warning lamp should light when the ignition is first turned on as a bulb check; or if the engine oil pressure light comes on when driving, perform the following:

  1. Pull off the road as soon as possible.
  2. Shut off the engine.
  3. Check the oil level.
  4. Do not drive the vehicle with the engine oil light on or severe engine damage can occur.

  ● **SEE FIGURE 20–2.**

- **Water in diesel fuel warning.** This warning lamp will light when the ignition is first turned on as a bulb check and if water is detected in the diesel fuel. This lamp is only used or operational in vehicles equipped with a diesel engine. If the water in diesel fuel warning lamp comes on, do the following:

  1. Remove the water using the built-in drain, usually part of the fuel filter.
  2. Check service information for the exact procedure to follow.

  ● **SEE FIGURE 20–3.**

- **Maintenance required warning.** The maintenance required lamp comes on when the ignition is first turned on as a bulb check and if the vehicle requires service. The service required could include:

  1. Oil and oil filter change
  2. Tire rotation
  3. Inspection

  Check service information for the exact service required. ● **SEE FIGURE 20–4.**

- **Malfunction indicator lamp (MIL),** also called a **check engine** or **service engine soon (SES) light**. This warning lamp comes on when the ignition is first turned on as a

FIGURE 20–5 Malfunction indicator lamp (MIL), also called a check engine light. The light means the engine control computer has detected a fault.

FIGURE 20–6 Charging system fault detected.

bulb test and then only if a fault in the powertrain control module (PCM) has been detected. If the MIL comes on when driving, it is not necessary to stop the vehicle, but the cause for why the warning lamp came on should be determined as soon as possible to avoid harming the engine or engine control systems. The MIL could come on if any of the following has been detected.

1. A sensor or actuator is electrically open or shorted.
2. A sensor is out of range for expected values.
3. An emission control system failure occurs, such as a loose gas cap.

If the MIL is on, a diagnostic trouble code has been set. Use a scan tool to retrieve the code(s) and follow service information for the exact procedure to follow. ● **SEE FIGURE 20–5.**

ELECTRICAL SYSTEM-RELATED WARNING LIGHTS

- **Charging system fault.** This warning lamp will come on when the ignition is first turned on as a bulb check and if a fault in the charging system has been detected. The lamp could include a fault with any of the following:

  1. Battery state of charge (SOC), electrical connections, or the battery itself
  2. Alternator or related wiring

  ● **SEE FIGURE 20–6.**

  If the charge system warning lamp comes on, continue to drive until it is safe to pull over. The vehicle can usually be driven for several miles using battery power alone.

Check the following by visible inspection.

1. Alternator drive belt
2. Loose or corroded electrical connections at the battery
3. Loose or corroded wiring to the alternator
4. Defective alternator

FIGURE 20–7 Fasten safety belt warning light.

FIGURE 20–8 Fault detected in the supplemental restraint (airbag) system.

FIGURE 20–9 Fault detected in base brake system.

SAFETY-RELATED WARNING LIGHTS Safety-related warning lamps include the following:

- **Safety belt warning lamp.** The safety belt warning lamp will light and sound an alarm to notify the driver if the driver's side or passenger's side safety belt is not fastened. It is also used to indicate a fault in the safety belt circuit. Check service information for the exact procedure to follow if the safety belt warning light remains on even when the belts are fastened. ● **SEE FIGURE 20–7.**
- **Airbag warning lamp.** The airbag warning lamp comes on and flashes when the ignition is first turned on as part of a self-test of the system. If the airbag warning lamp remains on after the self-test, then the airbag controller has detected a fault. Check service information for the exact procedure to follow if the airbag warning lamp is on. ● **SEE FIGURE 20–8.**

  NOTE: The passenger side airbag light may indicate that it is on or off, depending if there is a passenger or an object heavy enough to trigger the seat sensor. Electronic devices placed on the seat can also affect the system.

- **Red brake fault warning light.** All vehicles are equipped with a red brake warning (RBW) lamp that lights if a fault in the base (hydraulic) brake system is detected. Three types of sensors are used to light this warning light.
  1. A brake fluid level sensor located in the master cylinder brake fluid reservoir
  2. A pressure switch located in the pressure differential switch, which detects a difference in pressure between the front and rear or diagonal brake systems
  3. The parking brake could be applied. ● **SEE FIGURE 20–9.**

FIGURE 20–10 Brake light bulb failure detected.

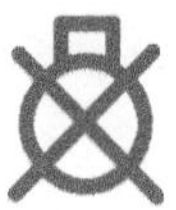

FIGURE 20–11 Exterior light bulb failure detected.

FIGURE 20–12 Worn brake pads or linings detected.

FIGURE 20–13 Fault detected in antilock brake system.

If the red brake warning light comes on, do not drive the vehicle until the cause is determined and corrected.

- **Brake light bulb failure.** Some vehicles are able to detect if a brake light is burned out. The warning lamp will warn the driver when a situation like this occurs. ● **SEE FIGURE 20–10.**
- **Exterior light bulb failure.** Many vehicles use the body control module (BCM) to monitor current flow through all of the exterior lights and therefore can detect if a bulb is not working. ● **SEE FIGURE 20–11.**
- **Worn brake pads.** Some vehicles are equipped with sensors built into the disc brake pads that are used to trigger a dash warning light. The warning light often comes on when the ignition is first turned on as a bulb check and then goes out. If the brake pad warning lamp is on, check service information for the exact service procedure to follow. ● **SEE FIGURE 20–12.**
- **Antilock brake system (ABS) fault.** The amber antilock brake system warning light comes on if the ABS controller detects a fault in the antilock braking system. Examples of what could trigger the warning light include:
  1. Defective wheel speed sensor
  2. Low brake fluid level in the hydraulic control unit assembly
  3. Electrical fault detected anywhere in the system

  ● **SEE FIGURE 20–13.**

  If the amber ABS warning lamp is on, it is safe to drive the vehicle, but the antilock portion may not function.

FIGURE 20–14 Low tire pressure detected.

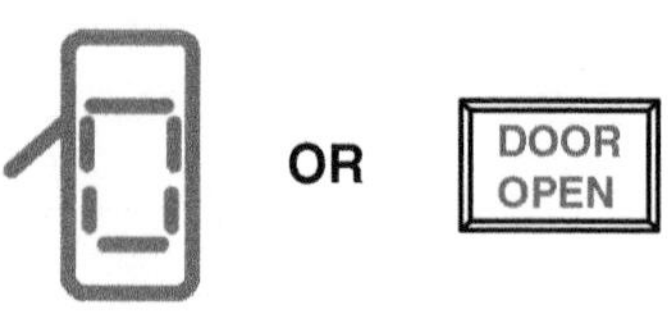

FIGURE 20–15 Door open or ajar.

FIGURE 20–16 Windshield washer fluid low.

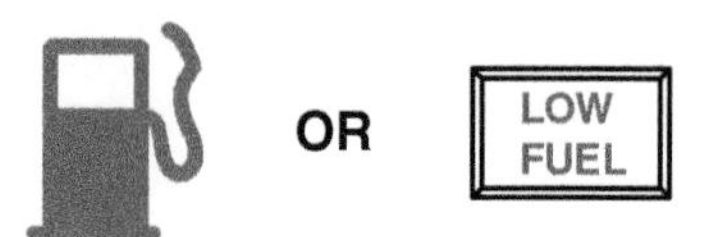

FIGURE 20–17 Low fuel level.

TECH TIP

Check the Spare

Some vehicles that are equipped with a full-size spare tire also have a sensor in the spare. If the warning lamp is on and all four tires are properly inflated, check the spare.

- **Low tire pressure warning.** A tire pressure monitoring system (TPMS) warns if the inflation pressure of a tire has decreased by 25% (about 8 psi). If the warning lamp or message of a low tire is displayed, check the tire pressures before driving. If the inflation pressure is low, repair or replace the tire. ● **SEE FIGURE 20–14.**

DRIVER INFORMATION SYSTEM

- **Door open or ajar warning light.** If a door is open or ajar, a warning light is used to notify the driver. Check and close all doors and tailgates before driving. ● **SEE FIGURE 20–15.**
- **Windshield washer fluid low.** A sensor in the windshield washer fluid reservoir is used to turn on the low washer fluid warning lamp. ● **SEE FIGURE 20–16.**
- **Low fuel warning.** A low fuel indicator light is used to warn the driver that the fuel level is low. In most vehicles, the light comes on when there is between 1 and 3 gallons (3.8 and 11 liters) of fuel remaining. ● **SEE FIGURE 20–17.**

FIGURE 20–18 Headlights on.

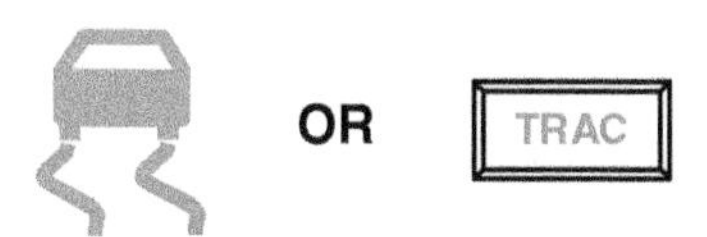

FIGURE 20–19 Low traction detected. Traction control system is functioning to restore traction (usually flashes when actively working to restore traction).

VSC

FIGURE 20–20 Vehicle stability control system either off or working if flashing.

TRAC
OFF

FIGURE 20–21 Traction control system has been turned off.

- **Headlights on light.** This dash indicator lights whenever the headlights are on. ● **SEE FIGURE 20–18.**

  NOTE: This light may or may not indicate that the headlights are on if the headlight switch is set to the automatic position.

- **Low traction detected.** On a vehicle equipped with a traction control system (TCS), a dash indicator light is flashed whenever the system is working to restore traction. If the low traction warning light is flashing, reduce the rate of acceleration to help the system restore traction of the drive wheels with the road surface. ● **SEE FIGURE 20–19.**
- **Electronic stability control.** If a vehicle is equipped with electronic stability control (ESC), also called vehicle stability control (VSC), the dash indicator lamp will flash if the system is trying to restore vehicle stability. ● **SEE FIGURE 20–20.**
- **Traction off.** If the traction control system (TCS) is turned off by the driver, an indicator lamp lights to help remind the driver that this system has been turned off and will not be able to restore traction when lost. The system reverts to on, when the ignition is turned off, and then back on as the traction off button is depressed. ● **SEE FIGURE 20–21.**

CRUISE

FIGURE 20–22 Indicates that the cruise control is on and able to maintain vehicle speed if set.

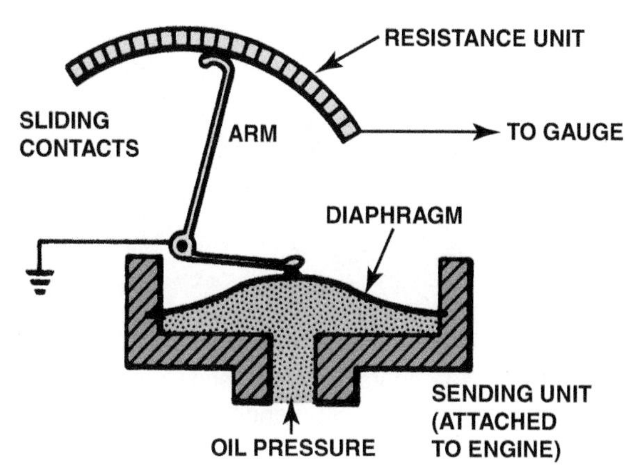

FIGURE 20–23 A typical oil pressure sending unit provides a varying amount of resistance as engine oil pressure changes. The output from the sensor is a variable voltage.

- **Cruise indicator lamp.** Most vehicles are equipped with a switch that turns on the cruise control. The cruise (speed) control system does not work unless it has been turned on to help prevent accidental engagement. When the cruise control has been turned on, the cruise indicator light is on. ● **SEE FIGURE 20–22.**

OIL PRESSURE WARNING DEVICES

OPERATION The oil pressure lamp operates through use of an oil pressure sensor unit, which is screwed into the engine block, and grounds the electrical circuit and lights the dash warning lamp in the event of low oil pressure, that is, 3 to 7 psi (20 to 50 kilopascals [kPa]). Normal oil pressure is generally between 10 and 60 psi (70 and 400 kPa). Some vehicles are equipped with variable voltage oil pressure sensors rather than a simple pressure switch. ● **SEE FIGURE 20–23.**

OIL PRESSURE LAMP DIAGNOSIS To test the operation of the oil pressure warning circuit, unplug the wire from the oil pressure sending unit, usually located near the oil filter, with the ignition switch on. With the wire disconnected from the sending unit, the warning lamp should be off. If the wire is touched to a ground, the warning lamp should be on. If there is *any* doubt of the operation of the oil pressure warning lamp,

FIGURE 20–24 A temperature gauge showing normal operating temperature between 180°F and 215°F, depending on the specific vehicle and engine.

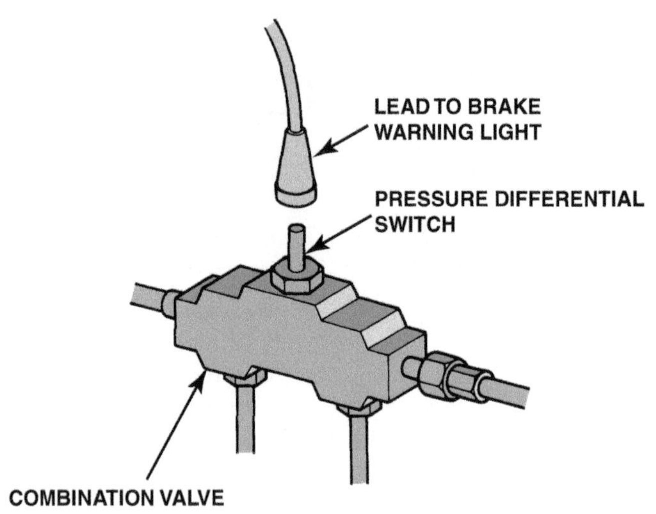

FIGURE 20–25 The pressure differential switch grounds the warning lamp when there is loss of pressure in one part of the brake hydraulic system.

always check the actual engine oil pressure using a gauge that can be screwed into the opening that is left after unscrewing the oil pressure sending unit. For removing the sending unit, special sockets are available at most auto parts stores, or a 1 inch or 1 1/16 inches 6-point socket may be used for most units.

REAL WORLD FIX

The Low Oil Pressure Story

After replacing valve cover gaskets on a Chevrolet V-8, the technician discovered that the oil pressure warning lamp was on. After checking the oil level and finding everything else okay, the technician discovered a wire pinched under the valve cover.

The wire went to the oil pressure sending unit. The edge of the valve cover had cut through the insulation and caused the current from the oil lamp to go to ground through the engine. Normally the oil lamp comes on when the sending unit grounds the wire from the lamp.

The technician freed the pinched wire and covered the cut with silicone sealant to prevent corrosion damage.

TEMPERATURE LAMP/ GAUGE DIAGNOSIS

The "hot" lamp, or engine coolant overheat warning lamp, warns the driver whenever the engine coolant temperature is between 248°F and 258°F (120°C and 126°C). This temperature is slightly below the boiling point of the coolant in a properly operating cooling system. The temperature sensor on older models was separate from the sensor used by the engine computer. However, most vehicles now use the engine coolant temperature (ECT) sensor for engine temperature gauge operation. To test this sensor, use a scan tool to verify proper engine temperature and follow the vehicle manufacturer's recommended testing procedures. ● **SEE FIGURE 20–24**.

BRAKE WARNING LAMP

DIFFERENTIAL SWITCH A **pressure-differential switch** was used on vehicles built after 1967 to warn the driver of a loss of pressure in one of the two separate hydraulic systems by lighting the dashboard **red brake warning lamp (RBWL)**. This system was used until it was replaced with the brake fluid level sensors used on today's vehicles. ● **SEE FIGURE 20–25**.

The dash warning lamp is often the same lamp as that used to warn the driver that the parking brake is on. The warning lamp is usually operated by using the parking brake lever or brake hydraulic pressure switch to complete the ground for the warning lamp circuit. If the warning lamp is on, first check if the parking brake is fully released.

FLUID LEVEL SENSOR Most master cylinders use a **brake fluid level sensor** or switch in the master cylinder reservoir.

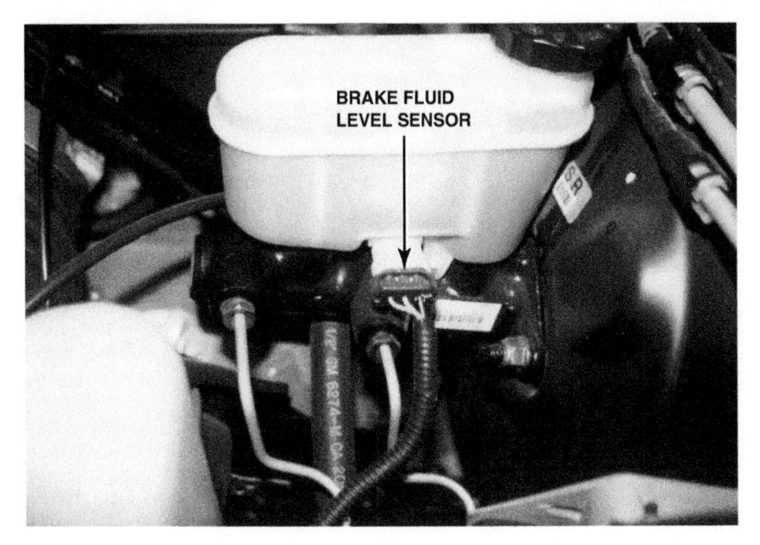

FIGURE 20–26 The red brake warning lamp can be turned on if the brake fluid level is low.

This sensor will light the red brake warning lamp on the dash if low brake fluid level is detected. A float-type sensor or a magnetic reed switch is commonly used and provides a completed electrical circuit when the brake fluid level is low. ● **SEE FIGURE 20–26**.

DIAGNOSING A RED "BRAKE" WARNING LAMP

Activation of the red brake dash warning lamp can be for any one of several reasons:

1. **Parking Brake "On."** The same dash warning lamp is used to warn the driver that the parking brake is on.
2. **Low Brake Fluid.** This lights the red dash warning lamp on vehicles equipped with a master cylinder reservoir brake fluid level switch.
3. **Unequal Brake Pressure.** The pressure-differential switch is used on older vehicles with a front/rear brake split system to warn the driver whenever there is low brake pressure to either the front or rear brakes.
4. **Worn brake pads.** Vehicles equipped with electronic pad wear sensors can turn on the light when the pads wear out.

ANALOG DASH INSTRUMENTS

An analog display uses a needle to show the value, whereas a digital display uses numbers. Analog electromagnetic dash instruments use small electromagnetic coils that are connected to a sending unit for such things as fuel level, water temperature, and oil pressure. The sensors are the same regardless of the type of display used. The resistance of the sensor varies with what is being measured. ● **SEE FIGURE 20–27** for typical electromagnetic fuel gauge operation.

NETWORK COMMUNICATION

DESCRIPTION Many instrument panels are operated by electronic control units that communicate with the engine control computer for engine data such as revolutions per minute (RPM) and engine temperature. These electronic **instrument panels (IPs)** use the voltage changes from variable-resistance sensors, such as that of the fuel gauge, to determine fuel level. Therefore, even though the sensor in the fuel tank is the same, the display itself may be computer controlled. The data is transmitted to the instrument cluster as well as to the powertrain control module through serial data lines. Because all sensor inputs are interconnected, the technician should always follow the factory recommended diagnostic procedures. ● **SEE FIGURE 20–28** on page 309.

ELECTRONIC ANALOG GAUGES

DESCRIPTION Most analog dash displays since the early 1990s are electronically or computer controlled. The sensors may be the same, but the sensor information is sent to the body or vehicle computer through a data BUS, and then

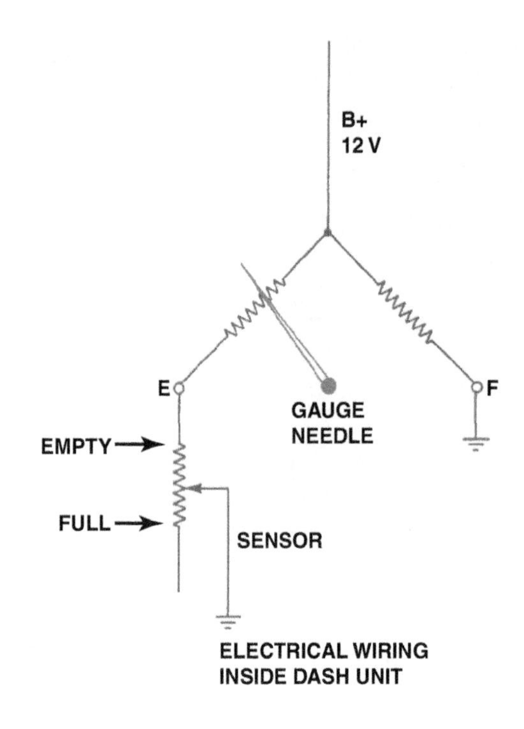

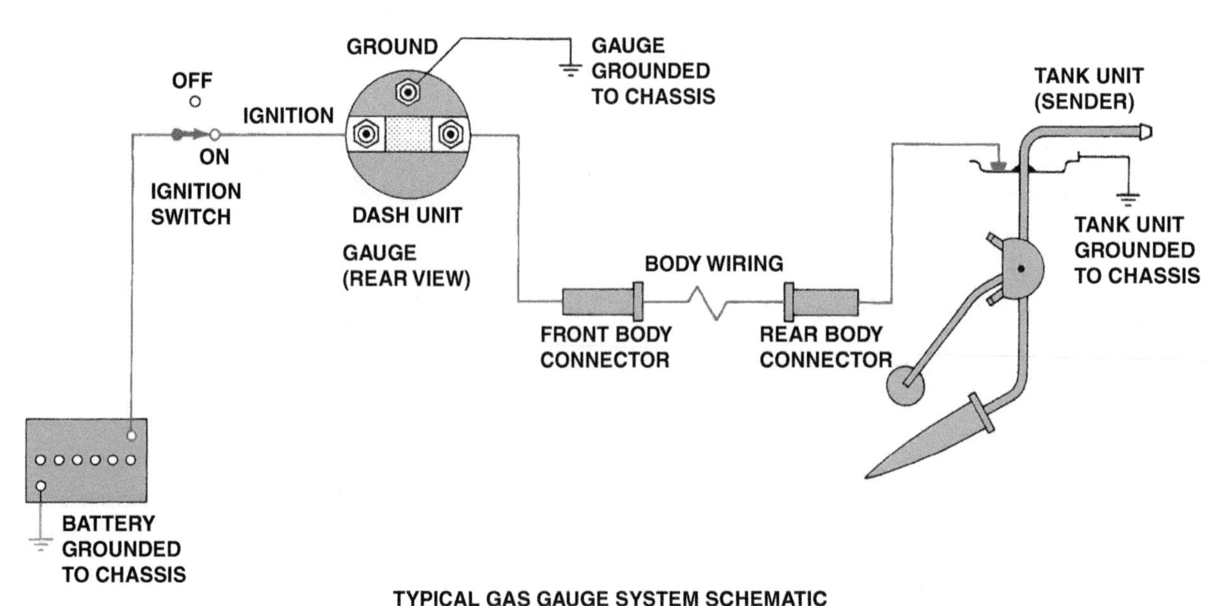

FIGURE 20–27 Electromagnetic fuel gauge wiring. If the sensor wire is unplugged and grounded, the needle should point to "E" (empty). If the sensor wire is unplugged and held away from ground, the needle should point to "F" (full).

the computer controls current through small electromagnets or stepper motors that move the needle of the gauge. ● **SEE FIGURES 20–29 AND 20–30.**

NOTE: Many electronic gauge clusters are checked at key on where the dash display needles will be commanded to 1/4, 1/2, 3/4, and full positions before returning to their normal readings. This self-test allows the service technician to check the operation of each individual gauge, even though replacing the entire instrument panel cluster is usually necessary to repair an inoperative gauge.

DIAGNOSIS The dash electronic circuits are often too complex to show on a wiring diagram. Instead, all related electronic circuits are simply indicated as a solid box with "electronic

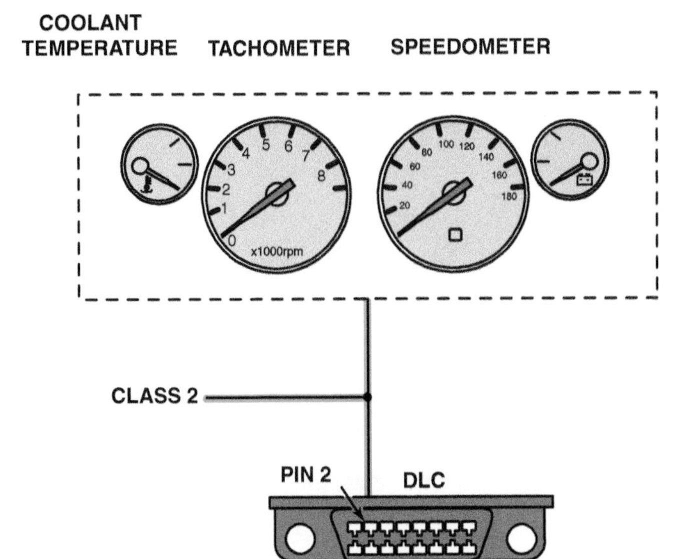

FIGURE 20–28 A typical instrument display uses data from the sensors over serial data lines to the individual gauges.

(a)

(b)

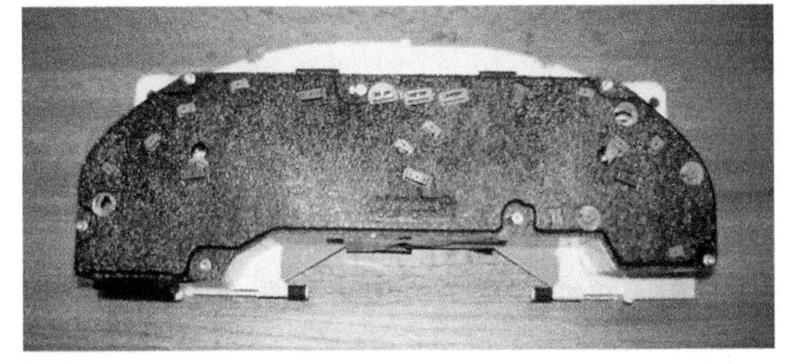

(c)

FIGURE 20–29 (a) View of the vehicle dash with the instrument cluster removed. Sometimes the dash instruments can be serviced by removing the padded dash cover (crash pad) to gain access to the rear of the dash. (b) The front view of the electronic analog dash display. (c) The rear view of the dash display showing that there are a few bulbs that can be serviced, but otherwise the unit is serviced as an assembly.

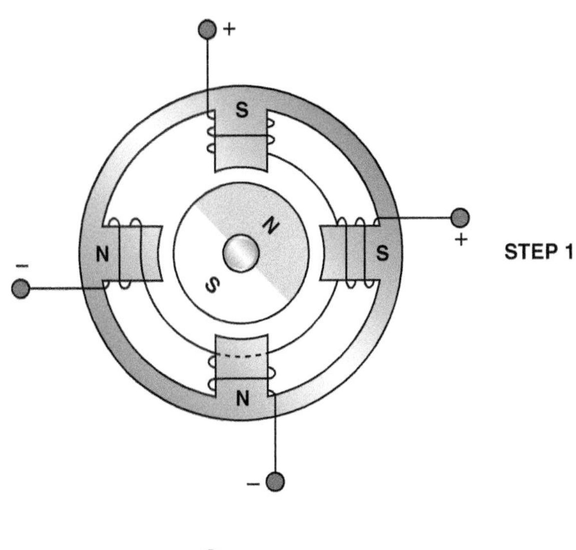

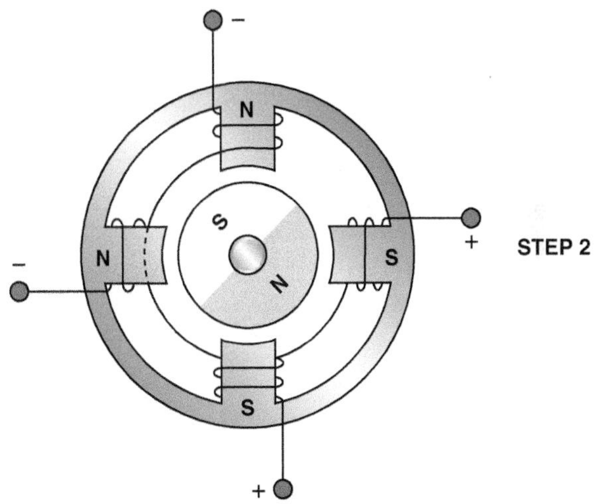

FIGURE 20–30 Most stepper motors use four wires which are pulsed by the computer to rotate the armature in steps.

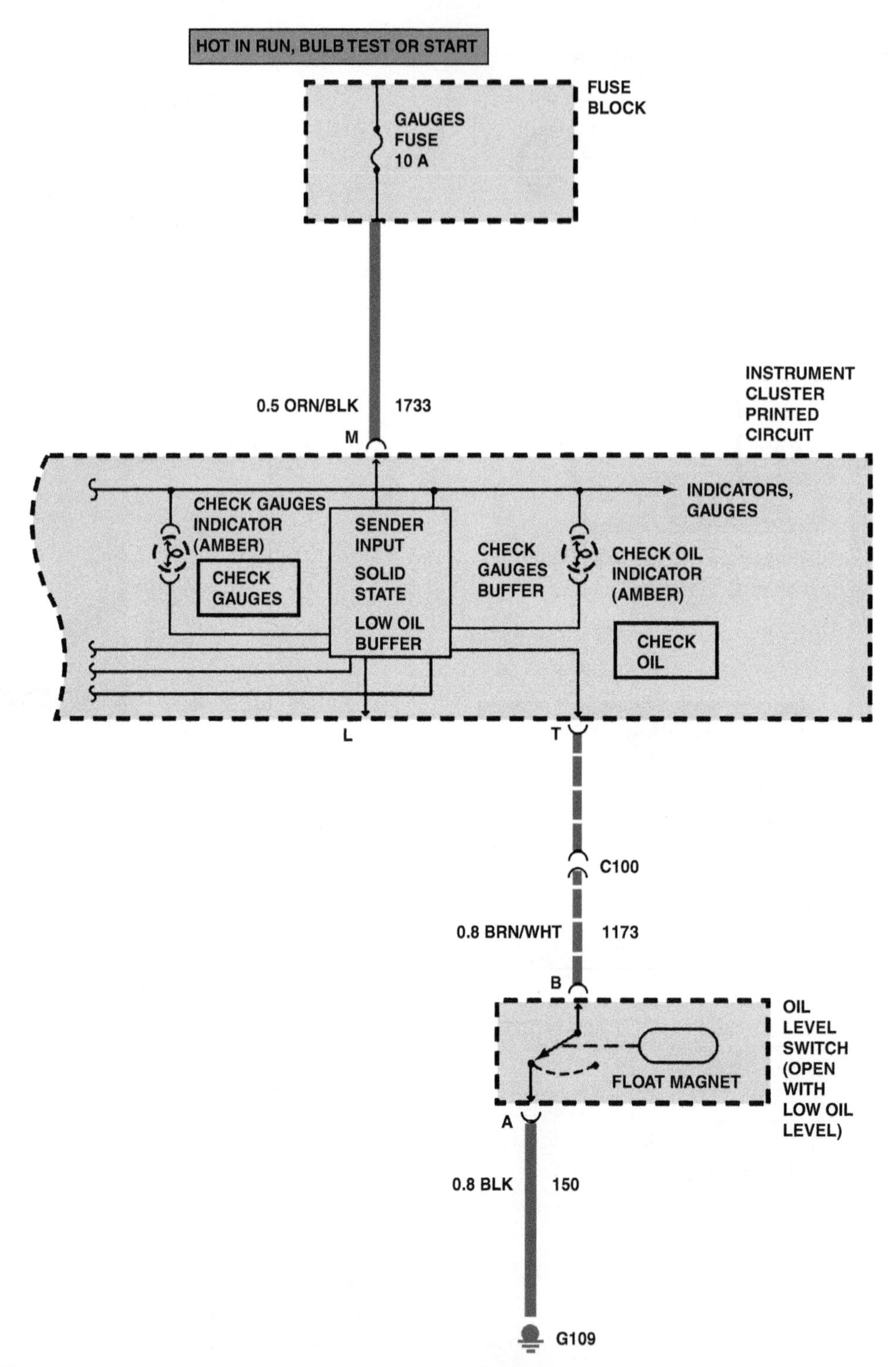

FIGURE 20–31 The ground for the "check oil" indicator lamp is controlled by the electronic low-oil buffer. Even though this buffer is connected to an oil level sensor, the buffer also takes into consideration the amount of time the engine has been stopped and the temperature of the engine. The only way to properly diagnose a problem with this circuit is to use the procedures specified by the vehicle manufacturer.

module" printed on the diagram. A scan tool is needed to diagnosis the operation of a computer-controlled analog dash instrument display.

Note that the grounding for the "check oil" dash indicator lamp is accomplished through an electronic buffer. The exact conditions, such as amount of time since the ignition was shut off, are unknown to the technician. To correctly diagnose problems with this type of circuit, technicians must read, understand, and follow the written diagnostic procedures specified by the vehicle manufacturer. **SEE FIGURE 20–31.**

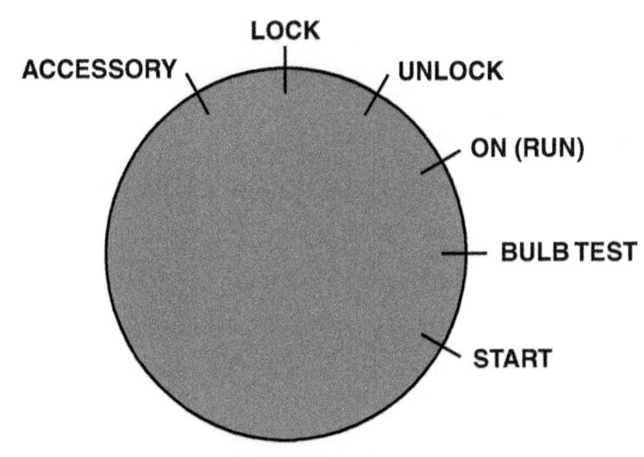

FIGURE 20–32 Typical ignition switch positions. Notice the bulb check position between "on" (run) and "start." These inputs are often just voltage signals to the body control module and can be checked using a scan tool.

FIGURE 20–33 Many newer vehicles place the ignition switch on the dash and incorporate antitheft controls. Note the location of the accessory position.

TECH TIP

The Bulb Test

Many ignition switches have six positions. Notice the bulb test position (between "on" and "start"). When the ignition is turned to "on" (run), some dash warning lamps are illuminated. When the bulb test position is reached, additional dash warning lamps often are lighted. Technicians use this ignition switch position to check the operation of fuses that protect various circuits. Dash warning lamps are not all powered by the same fuses. If an electrical component or circuit does not work, the power side (fuse) can be quickly checked by observing the operation of the dash lamps that share a common fuse with the problem circuit. Consult a wiring diagram for fuse information on the exact circuit being tested. ● **SEE FIGURES 20–32 AND 20–33.**

FIGURE 20–34 A vehicle speed sensor located in the extension housing of the transmission. Some vehicles use the wheel speed sensors for vehicle speed information.

ELECTRONIC SPEEDOMETERS

OPERATION Electronic dash displays ordinarily use an electric vehicle speed sensor and a reluctor on the output shaft of the transmission. These speed sensors contain a permanent magnet and generate a voltage in proportion to the vehicle speed. These speed sensors are commonly called **permanent magnet (PM) generators**. ● **SEE FIGURE 20–34.**

The output of a PM generator speed sensor is an AC voltage that varies in frequency and amplitude with increasing vehicle speed. The PM generator speed signal is sent to the instrument cluster electronic circuits. These specialized electronic circuits include a buffer amplifier circuit that converts the variable sine wave voltage from the speed sensor to an on/off signal that can be used by other electronic circuits to indicate a vehicle's speed. The vehicle speed is then displayed by either an electronic needle-type speedometer or by numbers on a digital display.

ODOMETERS An odometer is a dash display that indicates the total miles traveled by the vehicle. Some dash displays also

(a)

(b)

FIGURE 20–35 (a) Older odometers are mechanical and are operated by a stepper motor. (b) Many vehicles are equipped with an electronic odometer.

include a trip odometer that can be reset and used to record total miles traveled on a trip or the distance traveled between fuel stops. Electronic dash displays can use either an electrically driven mechanical odometer or a digital display odometer to indicate miles traveled. On mechanical type odometers, a small electric motor, called a stepper motor, is used to turn the number wheels of a mechanical-style odometer. A pulsed voltage is fed to this stepper motor, which moves in relation to the miles traveled. ● **SEE FIGURE 20–35.**

Digital odometers use LED, LCD, or VTF displays to indicate miles traveled. Because total miles must be retained when the ignition is turned off or the battery is disconnected, a special electronic chip must be used that will retain the miles traveled.

TECH TIP

The Soldering Gun Trick

Diagnosing problems with digital or electronic dash instruments can be difficult. Replacement parts generally are expensive and usually not returnable if installed in the vehicle. A popular trick that helps isolate the problem is to use a soldering gun near the PM generator.

A PM generator contains a coil of wire wound around a magnet. As the transmission reluctor revolves a voltage is produced. It is the *frequency* of this voltage that the dash (or engine) computer uses to calculate vehicle speed.

A soldering gun plugged into 110 volts AC will provide a strong *varying* magnetic field around the soldering gun. This magnetic field is constantly changing at the rate of 60 cycles per second. This frequency of the magnetic field induces a voltage in the windings of the PM generator. This induced voltage at 60 hertz (Hz) is converted by the computer circuits to a miles per hour (mph) reading on the dash.

To test the electronic speedometer, turn the ignition to "on" (engine off) and hold a soldering gun near the PM generator.

CAUTION: The soldering gun tip can get hot, so hold it away from wiring or other components that may be damaged by the hot tip.

If the PM generator, wiring, computer, and dash are okay, the speedometer should register a speed, usually 54 miles per hour (87 km/h). If the speedometer does not work when the vehicle is driven, the problem is in the PM generator drive.

If the speedometer does not register a speed when the soldering gun is used, the problem could be caused by the following:

1. Defective PM generator (check the windings with an ohmmeter)
2. Defective (open or shorted) wiring from the PM generator to the computer
3. Defective computer or dash circuit

These special chips are called **nonvolatile random-access memory (NVRAM)**. *Nonvolatile* means that the information stored in the electronic chip is not lost when electrical power is removed. Some vehicles use a chip called **electronically erasable programmable read-only memory (EEPROM)**. Most digital odometers can read up to 999,999.9 miles or kilometers (km), and then the display indicates error. If the chip is damaged or exposed to static electricity, it may fail to operate and "error" may appear.

SPEEDOMETER/ODOMETER SERVICE If the speedometer and odometer fail to operate, check the following:

- The speed sensor should be the first item checked. With the vehicle safely raised off the ground and supported, check vehicle speed using a scan tool. If a scan tool is not available, disconnect the wires from the speed sensor near the output shaft of the transmission. Connect a multimeter set on AC volts to the terminals of the speed sensor and rotate the drive wheels with the transmission in neutral. A good speed sensor should indicate approximately 2 volts AC if the drive wheels are rotated by hand.
- If the speed sensor is working, check the wiring from the speed sensor to the dash cluster. If the wiring is good, the instrument panel (IP) should be sent to a specialty repair facility.
- If the speedometer operates correctly but the mechanical odometer does not work, the odometer stepper motor, the number wheel assembly, or the circuit controlling the stepper motor is defective. If the digital odometer does not operate but the speedometer operates correctly, then the dash cluster must be removed and sent to a specialized repair facility. A replacement chip is available only through authorized sources; if the odometer chip is defective, the original number of miles must be programmed into the replacement chip.

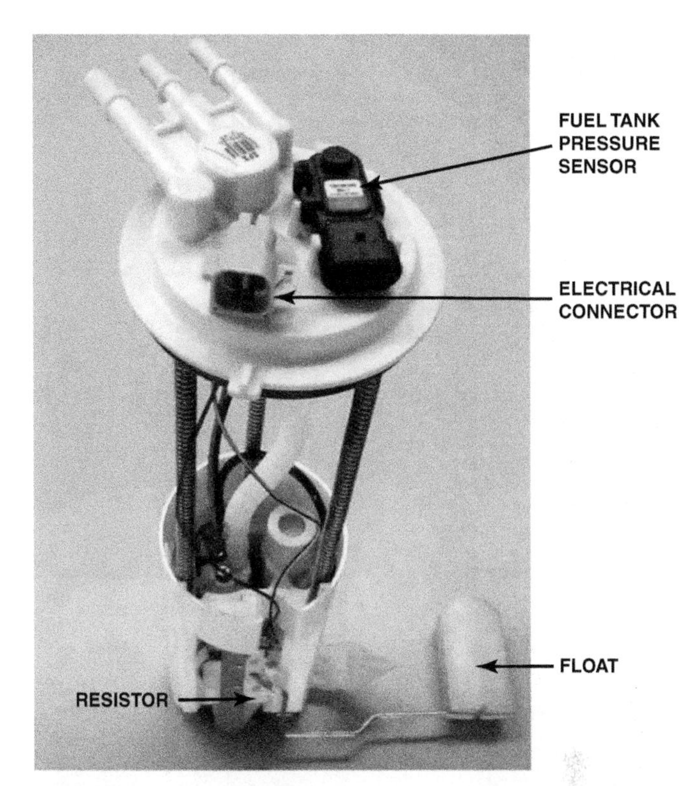

FIGURE 20–36 A fuel tank module assembly that contains the fuel pump and fuel level sensor in one assembly.

ELECTRONIC FUEL LEVEL GAUGES

OPERATION Electronic fuel level gauges ordinarily use the same fuel tank sending unit as that used on conventional fuel gauges. The tank unit consists of a float attached to a variable resistor. As the fuel level changes, the resistance of the sending unit changes. As the resistance of the tank unit changes, the dash-mounted gauge also changes. The only difference between a digital fuel level gauge and a conventional needle type is in the display. Digital fuel level gauges can be either numerical (indicating gallons or liters remaining in the tank) or a bar graph display. **SEE FIGURE 20–36.**

The diagnosis of a problem is the same as that described earlier for conventional fuel gauges. If the tests indicate that the dash unit is defective, usually the *entire* dash gauge assembly must be replaced.

HEAD-UP DISPLAY

The **head-up display (HUD)** is a supplemental display that projects the vehicle speed and sometimes other data, such as turn signal information, onto the windshield. The projected image looks as if it is some distance ahead, making it easy for the driver to see without having to refocus on a closer dash display. **SEE FIGURES 20–37 AND 20–38.**

The head-up display can also have the brightness controlled on most vehicles that use this type of display. The HUD unit is installed in the instrument panel (IP) and uses a mirror to project vehicle information onto the inside surface of the windshield. **SEE FIGURE 20–39.**

FIGURE 20–37 A typical head-up display showing vehicle speed and engine RPM. (Courtesy of General Motors)

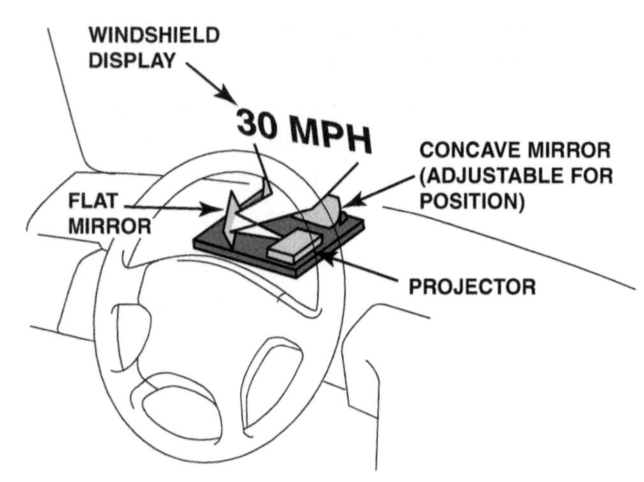

FIGURE 20–39 A typical head-up display (HUD) unit.

FIGURE 20–38 The dash-mounted control for the head-up display on this Cadillac allows the driver to move the image up and down on the windshield for best viewing.

FIGURE 20–40 A night vision camera behind the grille of a Cadillac.

Follow the vehicle manufacturer's recommended diagnostic and testing procedures if any faults are found with the head-up display.

NIGHT VISION

PARTS AND OPERATION Night vision systems use a camera that is capable of observing objects in the dark to assist the driver while driving at night. The primary night viewing illumination devices are the headlights. The night vision option uses a head-up display (HUD) to improve the vision of the driver beyond the scope of the headlights. Using a HUD display allows the driver to keep eyes on the road and hands on the wheel for maximum safety.

Besides the head-up display, the night vision camera uses a special thermal imaging or infrared technology. The camera is mounted behind the grill in the front of the vehicle. ● **SEE FIGURE 20–40.**

The camera creates pictures based on the heat energy emitted by objects rather than from light reflected on an object as in a normal optical camera. The image looks like a black and white photo negative when hot objects (higher thermal energy) appear light or white, and cool objects appear dark or black. Other parts of the night vision system include:

- **On/off and dimming switch.** This allows the driver to adjust the brightness of the display and to turn it on or off as needed.
- **Up/down switch.** The night vision HUD system has an electric tilt adjust motor that allows the driver to adjust the image up or down on the windshield within a certain image.

CAUTION: Becoming accustomed to night vision can be difficult and may take several nights to get used to looking at the head-up display.

DIAGNOSIS AND SERVICE The first step when diagnosing a fault with the night vision system is to verify the concern. Check the owner manual or service information for proper operation. For example, the Cadillac night vision system requires the following actions to function.

1. The ignition has to be in the on (run) position.
2. The Twilight Sentinel photo cell must indicate that it is dark.
3. The headlights must be on.
4. The switch for the night vision system must be on and the brightness adjusted so the image is properly displayed.

The night vision system uses a camera in the front of the vehicle that is protected from road debris by a grille. However, small stones or other debris can get past the grille and damage the lens of the camera. If the camera is damaged, it must be replaced as an assembly because no separate parts are available. Always follow the vehicle manufacturer's recommended testing and servicing procedures.

NAVIGATION AND GPS

PURPOSE AND FUNCTION The **global positioning system (GPS)** uses 24 satellites in orbit around the earth to provide signals for navigation devices. GPS is funded and controlled by the U.S. Department of Defense (DOD). While the system can be used by anyone with a GPS receiver, it was designed for and is operated by the U.S. military. ● **SEE FIGURE 20–41.**

BACKGROUND The current global positioning system was developed after a civilian airplane from Korean Airlines, Flight 007, was shot down as it flew over Soviet territory in 1983. The system became fully operational in 1991. Civilians were granted use of GPS that same year, but with less accuracy than the system used by the military.

Until 2000, the nonmilitary use of GPS was purposely degraded by a computer program called selection availability (S/A) built into the satellite transmission signals. After 2000, the S/A has been officially turned off, allowing nonmilitary users more accurate position information from the GPS receivers.

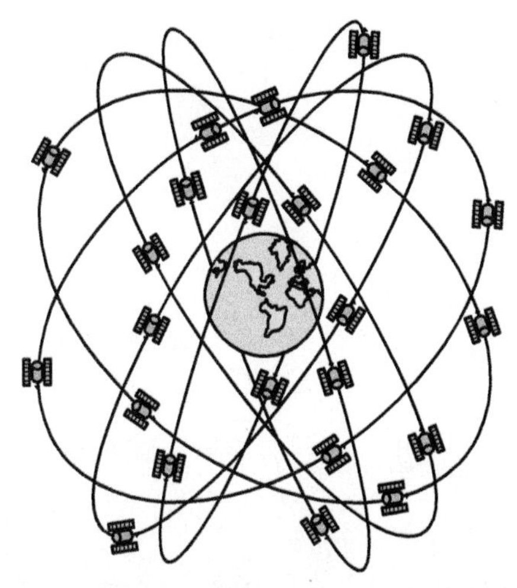

FIGURE 20–41 Global positioning systems use 24 satellites in high earth orbit whose signals are picked up by navigation systems. The navigation system computer then calculates the location based on the position of the satellite overhead.

FREQUENTLY ASKED QUESTION

Does the Government Know Where I Am?

No. The navigation system uses signals from the satellites and uses the signals from three or more to determine position. If the vehicle is equipped with OnStar, then the vehicle position can be monitored by the use of the cellular telephone link to OnStar call centers. Unless the vehicle has a cellular phone connection to the outside world, the only people who will know the location of the vehicle are the persons inside the vehicle viewing the navigation screen.

NAVIGATION SYSTEM PARTS AND OPERATION Navigation systems use the GPS satellites for basic location information. The navigation controller located in the rear of the vehicle uses other sensors, including a digitized map to display the location of the vehicle.

- **GPS satellite signals.** These signals from at least three satellites are needed to locate the vehicle.

FIGURE 20–42 A typical GPS display screen showing the location of the vehicle.

TECH TIP

Window Tinting Can Hurt GPS Reception

Most factory-installed navigation systems use a GPS antenna inside the rear back glass or under the rear package shelf. If a metalized window tint is applied to the rear glass, the signal strength from the GPS satellites can be reduced. If the customer concern includes inaccurate or nonfunctioning navigation, check for window tint.

- **Yaw sensor.** This sensor is often used inside the navigation unit to detect movement of the vehicle during cornering. This sensor is also called a "g" sensor because it measures force; 1 g is the force of gravity.
- **Vehicle speed sensor.** This sensor input is used by the navigation controller to determine the speed and distance the vehicle travels. This information is compiled and compared to the digital map and GPS satellite inputs to locate the vehicle.
- **Audio output/input.** Voice-activated factory units use a built-in microphone at the center top of the windshield and the audio speakers speech output.

Navigation systems include the following components.

1. Screen display ● **SEE FIGURE 20–42.**
2. GPS antenna

TECH TIP

Touch Screen Tip

Most vehicle navigation systems use a touch screen for use by the driver (or passenger) to input information or other on-screen prompts. Some touch screens use infrared beams projected from the top and bottom plus across the screen to form a grid. The system detects where on the screen a finger is located by the location of the beams that are cut. Do not push harder on the display if the unit does not respond, or the display unit may get damaged. If no response is detected when lightly depressing the screen, rotate the finger to cause the infrared beams to be cut.

3. Navigation control unit, usually with map information stored on different media formats, depending on the navigation radio. The possible media formats are:
   - DVD/CD
   - HDD (hard disc drive)
   - SD card
   - Compact flash card

The map data includes street names and the following information:

1. Points of interest (POI), including automated teller machines (ATMs), restaurants, schools, colleges, museums, shopping, and airports, as well as vehicle dealer locations
2. Business addresses and telephone numbers, including hotels and restaurants (If the telephone number is listed in the business telephone book, it can usually be displayed on the navigation screen. If the telephone number of the business is known, the location can be displayed.)

NOTE: Private residences or cellular telephone numbers are not included in the database of telephone numbers stored on the navigation system DVD.

3. Turn-by-turn directions to addresses that are selected by:
   - NPoints of interest (POI)
   - NTyped in using a keyboard shown on the display

The navigation unit then often allows the user to select the fastest way to the destination, as well as the shortest way, or assists in how to avoid toll roads. ● **SEE FIGURE 20–43.**

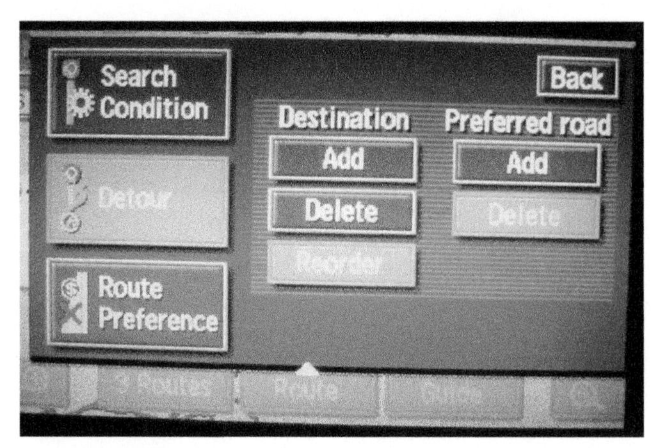

FIGURE 20–43 A typical navigation display showing various options. Some systems do not allow access to these functions if the vehicle is in gear and/or moving.

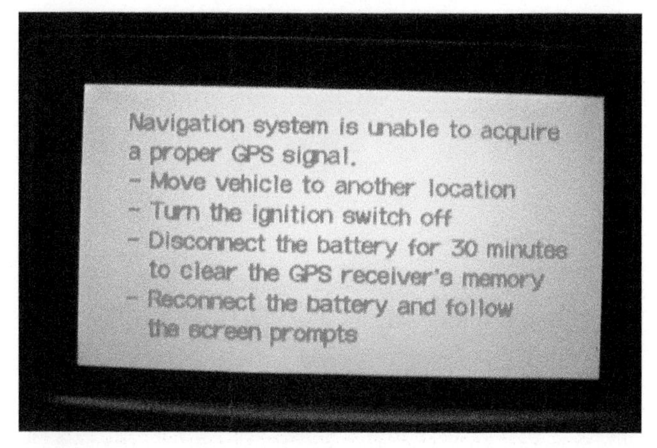

FIGURE 20–44 A screen display of a navigation system that is unable to acquire usable signals from GPS satellites.

DIAGNOSIS AND SERVICE For the correct functioning of the navigation system, three inputs are needed.

- Location
- Direction
- Speed

The navigation system uses the GPS satellite and map data to determine a location. Direction and speed are determined by the navigation computer from inputs from the satellite, plus the yaw sensor and vehicle speed sensor. The following symptoms may occur and be a customer complaint. Knowing how the system malfunctions helps to determine the most likely cause.

- If the vehicle icon jumps down the road, a fault with the vehicle speed (VS) sensor input is usually indicated.
- If the icon rotates on the screen, but the vehicle is not being driven in circles, a fault with the yaw sensor or yaw sensor input to the navigation controller is likely.
- If the icon goes off course and shows the vehicle on a road that it is not on, a fault with the GPS antenna is the most common reason for this situation.

Sometimes the navigation system itself will display a warning that views from the satellite are not being received. Always follow the displayed instructions. SEE FIGURE 20–44.

? FREQUENTLY ASKED QUESTION

What Is Navigation-Enhanced Climate Control?

Some vehicles use data from the navigation system to help control the automatic climate control system. Data about the location of the vehicle includes:

- **Time and date.** This information allows the automatic climate control system to determine where the sun is located.
- **Direction of travel.** The navigation system can also help the climate control system determine the direction of travel.

As a result of the input from the navigation system, along with the various other sensors in the vehicle, the automatic climate control system can more accurately control cabin temperature. For example, if the vehicle was traveling south in the late afternoon in July, the climate control system could assume that the passenger side of the vehicle would be warmed more by the sun than the driver's side and could increase the airflow to the passenger side to help compensate for the additional solar heating.

ONSTAR

PARTS AND OPERATION OnStar is a system that includes the following functions:

1. Cellular telephone
2. Global positioning antenna and computer

OnStar is standard or optional on most General Motors vehicles and selected other brands and models, to help the driver in an emergency or to provide other services. The cellular telephone is used to communicate with the driver from advisors at service centers. The advisor at the service center is able to see the location of the vehicle as transmitted from the GPS antenna and computer system in the vehicle on a display. OnStar does not display the location of the vehicle to the driver unless the vehicle is also equipped with a navigation system.

FIGURE 20–45 The three-button OnStar control is located on the inside rearview mirror. The left button (telephone handset icon) is pushed if a hands-free cellular call is to be made. The center button is depressed to contact an OnStar advisor and the right emergency button is used to request that help be sent to the vehicle's location.

Unlike most navigation systems, the OnStar system requires a monthly fee. OnStar was first introduced in 1996 as an option on some Cadillac models. Early versions used a handheld cellular telephone while later units used a group of three buttons mounted on the inside rearview mirror and a hands-free cellular telephone. ● **SEE FIGURE 20–45.**

The first version used analog cellular service while later versions used a dual mode (analog and digital) service until 2007. Since 2007, all OnStar systems use digital cellular service, which means that older systems that were analog only need to be upgraded.

The OnStar system includes the following features, which can vary depending on the level of service desired and cost per month:

- **Automatic notification of airbag deployment.** If the airbag is deployed, the advisor is notified immediately and attempts to call the vehicle. If there is no reply, or if the occupants report an emergency, the advisor will contact emergency services and give them the location of the vehicle.
- **Emergency services.** If the red button is pushed, OnStar immediately locates the vehicle and contacts the nearest emergency service agency.
- **Stolen vehicle location assistance.** If a vehicle is reported stolen, a call center advisor can track the vehicle.
- **Remote door unlock.** An OnStar advisor can send a cellular telephone message to the vehicle to unlock the vehicle if needed.
- **Roadside assistance.** When called, an OnStar advisor can locate a towing company or locate a provider who can bring gasoline or change a flat tire.
- **Accident assistance.** An OnStar advisor is able to help with the best way to handle an accident. The advisor can supply a step-by-step checklist of the things that should be done plus call the insurance company, if desired.
- **Remote horn and lights.** The OnStar system is tied into the lights and horn circuits so an advisor can activate them if requested to help the owner locate the vehicle in a parking lot or garage.
- **Vehicle diagnosis.** Because the OnStar system is tied to the PCM, an OnStar advisor can help with diagnosis if there is a fault detected. The system works as follows:
  - The malfunction indicator light (MIL) (check engine) comes on to warn the driver that a fault has been detected.
  - The driver can depress the OnStar button to talk to an advisor and ask for a diagnosis.
  - The OnStar advisor will send a signal to the vehicle requesting the status from the powertrain control module (PCM), as well as the controller for the antilock brakes and the airbag module.
  - The vehicle then sends any diagnostic trouble codes to the advisor. The advisor can then inform the driver about the importance of the problem and give advice as to how to resolve the problem.

DIAGNOSIS AND SERVICE The OnStar system can fail to meet the needs of the customer if any of the following conditions occur:

1. Lack of cellular telephone service in the area
2. Poor global positioning system (GPS) signals, which can prevent an OnStar advisor from determining the position of the vehicle
3. Transport of the vehicle by truck or ferry so that it is out of contact with the GPS satellite in order for an advisor to properly track the vehicle

If all of the above are okay and the problem still exists, follow service information diagnostic and repair procedures. If a new vehicle communication interface module (VCIM) is installed in the vehicle, the electronic serial number (ESN) must be tied to the vehicle. Follow service information instructions for the exact procedures to follow.

FIGURE 20–46 A typical view displayed on the navigation screen from the backup camera.

FIGURE 20–47 A typical fisheye-type backup camera usually located near the center on the rear of the vehicle near the license plate.

BACKUP CAMERA

PARTS AND OPERATION A **backup camera** is used to display the area at the rear of the vehicle in a screen display on the dash when the gear selector is placed in reverse. Backup cameras are also called *reversing cameras* or *rearview cameras.*

Backup cameras are different from normal cameras because the image displayed on the dash is flipped so it is a mirror image of the scene at the rear of the vehicle. This reversing of the image is needed because the driver and the camera are facing in opposite directions. Backup cameras were first used in large vehicles with limited rearward visibility, such as motor homes. Many vehicles equipped with navigation systems today include a backup camera for added safety while reversing. ● **SEE FIGURE 20–46.**

The backup camera contains a wide-angle or fisheye lens to give the largest viewing area. Most backup cameras are pointed downward so that objects on the ground, as well as walls, are displayed. ● **SEE FIGURE 20–47.**

DIAGNOSIS AND SERVICE Faults in the backup camera system can be related to the camera itself, the display, or the connecting wiring. The main input to the display unit comes from the transmission range switch which signals the backup camera when the transmission is shifted into reverse.

To check the transmission range switch, perform the following:

1. Check if the backup (reverse) lights function when the gear selector is placed in reverse with the key on, engine off (KOEO).
2. Check that the transmission/transaxle is fully engaged in reverse when the selector is placed in reverse.

Most of the other diagnosis involves visual inspection, including:

1. Checking the backup camera for damage.
2. Checking the screen display for proper operation.
3. Checking that the wiring from the rear camera to the body is not cut or damaged.

Always follow the vehicle manufacturer's recommended diagnosis and repair procedures.

FRONT/REAR PARK ASSIST

COMPONENTS Backup sensors are used to warn the driver if there is an object behind the vehicle while reversing. The system used in General Motors vehicles is called **rear park assist (RPA)**, or Front/Rear Park Assist (if equipped) and includes the following components:

- Ultrasonic object sensors built into the rear bumper assembly
- A display with three lights usually located inside the vehicle above the rear window and visible to the driver in the rearview mirror
- An electronic control module that uses an input from the transmission range switch and lights the warning lamps needed when the vehicle gear selector is in reverse

FIGURE 20–48 A typical backup sensor display located above the rear window inside the vehicle. The warning lights are visible in the inside rearview mirror.

 TECH TIP

Check for Repainted Bumper

The ultrasonic sensors embedded in the bumper are sensitive to paint thickness because the paint covers the sensors. If the system does not seem to be responding to objects, and if the bumper has been repainted, measure the paint thickness using a nonferrous paint thickness gauge. The maximum allowable paint thickness is 6 mils (0.006 inch or 0.15 millimeter).

OPERATION The three-light display includes two amber lights and one red light. The following lights are displayed depending on the distance from the rear bumper.

- One amber lamp will light when the vehicle is in reverse and traveling at less than 3 miles per hour (5 km/h) and the sensors detect an object 40 to 60 inches (102 to 152 cm) from the rear bumper. A chime or audio signal also sounds once when an object is detected, to warn the driver to look at the rear parking assist display. ● **SEE FIGURE 20–48.**
- Two amber lamps light when the distance between the rear bumper and an object is between 20 and 40 inches (50 and 100 cm) and the chime will sound again.
- Two amber lamps and the red lamp light and the chime sounds continuously when the distance between the rear bumper and the object is between 11 and 20 inches (28 and 50 cm).

FIGURE 20–49 The small round buttons in the rear bumper are ultrasonic sensors used to sense distance to an object.

If the distance between the rear bumper and the object is less than 11 inches (28 cm), all indicator lamps flash and the chime will sound continuously.

The ultrasonic sensors embedded in the rear bumper "fire" individually every 150 milliseconds (27 times per second). ● **SEE FIGURE 20–49.**

The sensors fire and then receive a return signal and arm to fire again in sequence from the left sensor to the right sensor.

Each sensor has the following three wires.

1. An 8 volt supply wire from the RPA module, used to power the sensor
2. A reference low or ground wire
3. A signal line, used to send and receive commands to and from the RPA module

DIAGNOSIS The rear parking assist control module is capable of detecting faults and storing diagnostic trouble codes (DTCs). If a fault has been detected by the control module, the red lamp flashes and the system is disabled. Follow service information diagnostic procedures because the rear parking assist module cannot usually be accessed using a scan tool. Most systems use the warning lights to indicate trouble codes.

LANE DEPARTURE WARNING SYSTEM

PARTS AND OPERATION The **lane departure warning system (LDWS)** uses cameras to detect if the vehicle is crossing over lane marking lines on the pavement. Some systems use two cameras, one mounted on each outside rearview

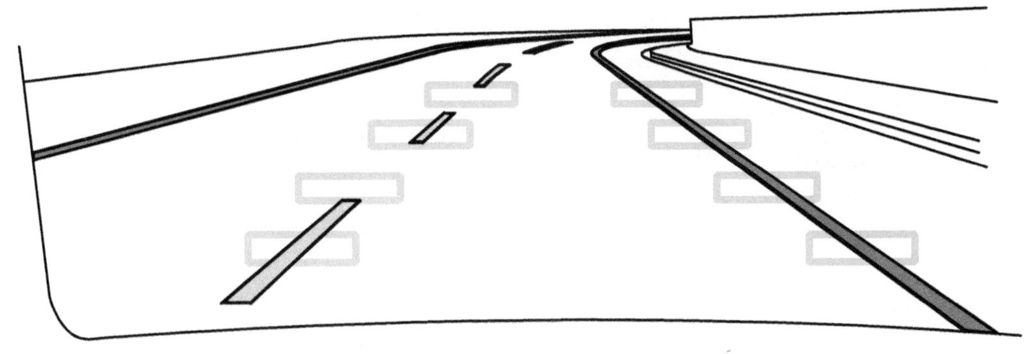

FIGURE 20–50 A lane departure warning system often uses cameras to sense the road lines and warns the driver if the vehicle is not staying within the lane, unless the turn signal is on.

mirror. Some systems use infrared sensors located under the front bumper to monitor the lane markings on the road surface.

The system names also vary according to vehicle manufacturer, including:

Honda/Acura: lane keep assist system (LKAS)

Toyota/Lexus: lane monitoring system (LMS)

General Motors: lane departure warning (LDW)

Ford: lane departure warning (LDW)

Nissan/Infiniti: lane departure prevention (LDP) system

If the cameras detect that the vehicle is starting to cross over a lane dividing line, a warning chime will sound or a vibrating mechanism mounted in the driver's seat cushion is triggered on the side where the departure is being detected. This warning will not occur if the turn signal is on in the same direction as detected. ● **SEE FIGURE 20–50.**

DIAGNOSIS AND SERVICE Before attempting to service or repair a lane departure warning system fault, check service information for an explanation on how the system is supposed to work. If the system is not working as designed, perform a visual inspection of the sensors or cameras, checking for damage from road debris or evidence of body damage, which could affect the sensors. After a visual inspection, follow the vehicle manufacturer's recommended diagnosis procedures to locate and repair the fault in the system.

ELECTRONIC DASH INSTRUMENT DIAGNOSIS AND TROUBLESHOOTING

If one or more electronic dash gauges do not work correctly, first check the WOW display that lights all segments to full brilliance whenever the ignition switch is first switched on. If *all* segments of the display do *not* operate, then the entire electronic cluster must be replaced in most cases. If all segments operate during the WOW display but do not function correctly afterward, the problem is most often a defective sensor or defective wiring to the sensor.

All dash instruments except the voltmeter use a variable-resistance unit as a sensor for the system being monitored. Most new-vehicle dealers are required to purchase essential test equipment, including a test unit that permits the technician to insert various fixed-resistance values in the suspected circuit. For example, if a 45 ohm resistance is put into the fuel gauge circuit that reads from 0 to 90 ohms, a properly operating dash unit should indicate one-half tank. The same tester can produce a fixed signal to test the operation of the speedometer and tachometer. If this type of special test equipment is not available, the electronic dash instruments can be tested using the following procedure.

TECH TIP

Keep Stock Overall Tire Diameter

Whenever larger (or smaller) wheels or tires are installed, the speedometer and odometer calibration are also thrown off. This can be summarized as follows:

- **Larger diameter tires.** The speed showing on the speedometer is slower than the actual speed. The odometer reading will show fewer miles than actual.
- **Smaller diameter tires.** The speed showing on the speedometer is faster than the actual speed. The odometer reading will show more miles than actual.

General Motors trucks can be recalibrated with a recalibration kit (1988-1991) or with a replacement controller assembly called a digital ratio adapter controller (DRAC) located under the dash. It may be possible to recalibrate the speedometer and odometer on earlier models, before 1988, or vehicles that use speedometer cables by replacing the drive gear in the transmission. Check service information for the procedure on the vehicle being serviced.

1. With the ignition switched off, unplug the wire(s) from the sensor for the function being tested. For example, if the oil pressure gauge is not functioning correctly, unplug the wire connector at the oil pressure sending unit.
2. With the sensor wire unplugged, turn the ignition switch on and wait until the WOW display stops. The display for the affected unit should show either fully lighted segments or no lighted segments, depending on the make of the vehicle and the type of sensor.
3. Turn the ignition switch off. Connect the sensor wire lead to ground and turn the ignition switch on. After the WOW display, the display should be the opposite (either fully on or fully off) of the results in step 2.

TESTING RESULTS If the electronic display functions fully on and fully off with the sensor unplugged and then grounded, the problem is a defective sensor. If the electronic display fails to function fully on and fully off when the sensor wire(s) are opened and grounded, the problem is usually in the wiring from the sensor to the electronic dash or it is a defective electronic cluster.

CAUTION: Whenever working on or *near* any type of electronic dash display, always wear a wire attached to your wrist (wrist strap) connected to a good body ground to prevent damaging the electronic dash with static electricity.

MAINTENANCE REMINDER LAMPS

Maintenance reminder lamps indicate that the oil should be changed or that other service is required. There are numerous ways to extinguish a maintenance reminder lamp. Some require the use of a special tool. Always check the owner manual or service information for the exact procedure for the vehicle being serviced. For example, to reset the oil service reminder light on many General Motors vehicles, you have to perform the following:

STEP 1 Turn the ignition key on (engine off).

STEP 2 Depress the accelerator pedal three times and hold it down on the fourth.

STEP 3 When the reminder light flashes, release the accelerator pedal.

STEP 4 Turn the ignition key to the off position.

STEP 5 Start the engine and the light should be off.

SUMMARY

1. Most digital and analog (needle-type) dash gauges use variable-resistance sensors.
2. Dash warning lamps are called telltale lamps.
3. Many electronically operated or computer-operated dash indicators require that a service manual be used to perform accurate diagnosis.
4. Permanent magnet (PM) generators produce an AC signal and are used for vehicle speed and wheel speed sensors.
5. Navigation systems and warning systems are part of the driver information system on many vehicles.

REVIEW QUESTIONS

1. How does a stepper motor analog dash gauge work?
2. What are LED, LCD, VTF, and CRT dash displays? Describe each.
3. How do you diagnose a problem with a red brake warning lamp?
4. How do you test the dash unit of a fuel gauge?
5. How does a navigation system determine the location of the vehicle?

CHAPTER QUIZ

1. Two technicians are discussing a fuel gauge on a General Motors vehicle. Technician A says that if the ground wire connection to the fuel tank sending unit becomes rusty or corroded, the fuel gauge will read lower than normal. Technician B says that if the power lead to the fuel tank sending unit is disconnected from the tank unit and grounded (ignition on), the fuel gauge should go to empty. Which technician is correct?
   - **a.** Technician A only
   - **b.** Technician B only
   - **c.** Both Technicians A and B
   - **d.** Neither Technician A nor B
2. If an oil pressure warning lamp on a General Motors vehicle is on all the time, yet the engine oil pressure is normal, the problem could be a _______.
   - **a.** Defective (shorted) oil pressure sending unit (sensor)
   - **b.** Defective (open) oil pressure sending unit (sensor)
   - **c.** Wire shorted-to-ground between the sending unit (sensor) and the dash warning lamp
   - **d.** Both a and c
3. When the oil pressure drops to between 3 and 7 psi, the oil pressure lamp lights by _______.
   - **a.** Opening the circuit
   - **b.** Shorting the circuit
   - **c.** Grounding the circuit
   - **d.** Conducting current to the dash lamp by oil
4. A brake warning lamp on the dash remains on whenever the ignition is on. If the wire to the pressure differential switch (usually a part of a combination valve or built into the master cylinder) is unplugged, the dash lamp goes out. Technician A says that this is an indication of a fault in the hydraulic brake system. Technician B says that the problem is probably due to a stuck parking brake cable switch. Which technician is correct?
   - **a.** Technician A only
   - **b.** Technician B only
   - **c.** Both Technicians A and B
   - **d.** Neither Technician A nor B
5. A customer complains that every time the lights are turned on in the vehicle, the dash display dims. What is the most probable explanation?
   - **a.** Normal behavior for LED dash displays
   - **b.** Normal behavior for VTF dash displays
   - **c.** Poor ground in lighting circuit causing a voltage drop to the dash lamps
   - **d.** Feedback problem most likely caused by a short-to-voltage between the headlights and dash display
6. Technician A says that LCDs may be slow to work at low temperatures. Technician B says that an LCD dash display can be damaged if pressure is exerted on the front of the display during cleaning. Which technician is correct?
   - **a.** Technician A only
   - **b.** Technician B only
   - **c.** Both Technicians A and B
   - **d.** Neither Technician A nor B
7. Technician A says that backup sensors use LEDs to detect objects. Technician B says that a backup sensor will not work correctly if the paint is thicker than 0.006 inch. Which technician is correct?
   - **a.** Technician A only
   - **b.** Technician B only
   - **c.** Both Technicians A and B
   - **d.** Neither Technician A nor B
8. Technician A says that metal-type tinting can affect the navigation system. Technician B says most navigation systems require a monthly payment for use of the GPS satellite. Which technician is correct?
   - **a.** Technician A only
   - **b.** Technician B only
   - **c.** Both Technicians A and B
   - **d.** Neither Technician A nor B
9. Technician A says that the data displayed on the dash can come from the engine computer. Technician B says that the entire dash assembly may have to be replaced even if just one unit fails. Which technician is correct?
   - **a.** Technician A only
   - **b.** Technician B only
   - **c.** Both Technicians A and B
   - **d.** Neither Technician A nor B
10. How does changing the size of the tires affect the speedometer reading?
    - **a.** A smaller diameter tire causes the speedometer to read faster than actual speed and more than actual mileage on the odometer.
    - **b.** A smaller diameter tire causes the speedometer to read slower than the actual speed and less than the actual mileage on the odometer.
    - **c.** A larger diameter tire causes the speedometer to read faster than the actual speed and more than the actual mileage on the odometer.
    - **d.** A larger diameter tire causes the speedometer to read slower than the actual speed and more than the actual mileage on the odometer.

chapter 21

HORN, WIPER, AND BLOWER MOTOR CIRCUITS

LEARNING OBJECTIVES

After studying this chapter, the reader will be able to:

1. Describe how the horn operates, and diagnose faulty horn operation.
2. Explain the testing and diagnosis of windshield wipers and windshield washers.
3. Explain the operation and diagnosis of a blower motor.

This chapter will help you prepare for the ASE Electrical/Electronic Systems (A6) certification test content area "G" (Horn and Wiper/Washer Diagnosis and Repair) and content area "H" (Accessories Diagnosis and Repair).

KEY TERMS

Horns 325
Pulse wipers 329
Rain sense wipers 334
Series-wound field 327
Shunt field 327
Variable-delay wipers 329
Windshield wipers 327

GM STC OBJECTIVES

GM Service Technical College topics covered in this chapter are as follows:

1. Operation and characteristics of horn, wiper, and blower circuits as used on General Motors vehicles.

HORNS

PURPOSE AND FUNCTION **Horns** are electric devices that emit a loud sound used to alert other drivers or persons in the area. Horns are manufactured in several different tones ranging from 1,800 to 3,550 Hz. Vehicle manufacturers select from various horn tones for a particular vehicle sound. ● **SEE FIGURE 21–1.**

When two horns are used, each has a different tone when operated separately, yet the sound combines when both are operated.

HORN CIRCUITS Automotive horns usually operate on full battery voltage wired from the battery, through a fuse, switch, and then to the horns. Most vehicles use a horn relay. On older vehicles, the horn button on the steering wheel or column completes a circuit to ground that closes a relay, and the heavy current flow required by the horn then travels from the relay to the horn. Without a horn relay, the high current of the horns must flow through the steering wheel horn switch.

On most current models, the horn relay is controlled by the body control module (BCM). The horn switch is an input to the BCM, and the BCM controls the horn relay. In this way, the BCM can also "beep" the horn when the vehicle is locked or unlocked, using the key fob remote. ● **SEE FIGURE 21–2.**

HORN OPERATION A vehicle horn is an actuator that converts an electrical signal to sound. The horn circuit includes an armature (a coil of wire) and contacts that are attached to a diaphragm. When energized, the armature causes the diaphragm to move up, which then opens a set of contact points that deenergize the armature circuit. As the diaphragm moves down, the contact points close, reenergize the armature circuit, and the diaphragm moves up again. This rapid opening and closing of the contact points causes the diaphragm to vibrate at an audible frequency. The sound created by the diaphragm is magnified as it travels through a trumpet attached to the diaphragm chamber. Most horn systems typically use one or two horns, but some have up to four. Those with multiple horns use both high- and low-pitch units to achieve a harmonious tone. Only a high-pitched unit is used in single-horn applications. The horn assembly is marked with an "H" or "L" for pitch identification.

HORN SYSTEM DIAGNOSIS There are four types of horn failure:

- No horn operation
- Intermittent operation
- Constant operation
- Weak or low volume sound

FIGURE 21–1 Two horns are used on this vehicle. Many vehicles use only one horn, often hidden underneath the vehicle.

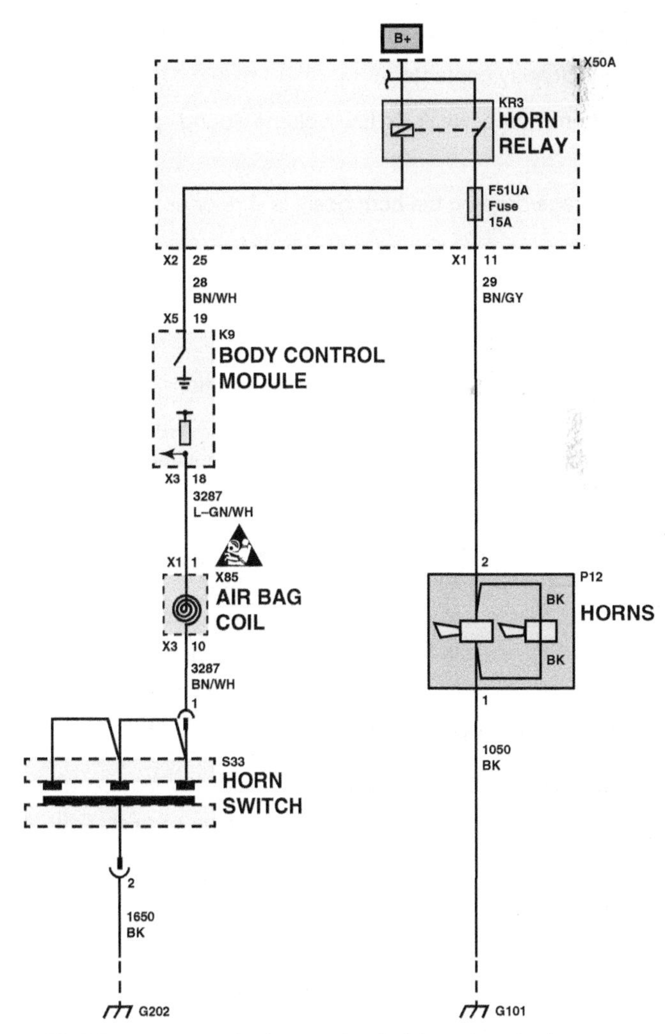

FIGURE 21–2 A typical horn circuit. Note that the horn switch grounds a circuit inside the body control module and then the BCM provides a ground for the relay. (Courtesy of General Motors)

If a horn does not operate at all, check for the following:

- Burned fuse or fusible link
- Open circuit
- Defective horn
- Faulty relay
- Defective horn switch
- Poor ground (horn mounting)
- Corroded or rusted electrical connector

If a horn operates intermittently, check for the following:

- Loose contact at the switch
- Loose, frayed, or broken wires
- Defective relay

If the horn sounds continuously and cannot be shut off, check the following:

- Horn switch contacts stuck closed
- Short-to-ground on the relay control circuit
- Stuck relay contacts

If the horn has a weak or low volume sound, check for the following:

- Voltage drop on the horn positive wire or connection
- A loose or corroded ground connection
- Horn damaged or flooded with water or mud

HORN SERVICE When a horn malfunctions, circuit tests are made to determine if the horn, relay, switch, or wiring is the source of the failure. Typically, a digital multimeter (DMM) is used to perform voltage drop and continuity checks to isolate the failure.

- **Switch and relay.** A momentary contact switch is used to sound the horn. The horn switch is mounted to the steering wheel in the center of the steering column on some models and is part of a multifunction switch installed on the steering column.

  CAUTION: If steering wheel needs to be removed for diagnosis or repair of the horn circuit, follow service information procedures for disarming the airbag circuit prior to removing the steering wheel, and for the specified test equipment to use.

  On most late-model vehicles, the horn relay is located in a centralized power distribution center along with other relays, circuit breakers, and fuses. The horn relay bolts onto an inner fender or the bulkhead in the engine compartment of older vehicles. Check the relay to determine if the coil is being energized and if current passes through the power circuit when the horn switch is depressed.

  Obtain an electrical schematic of the horn circuit and use a voltmeter to test input, output, and control voltage.

- **Circuit testing.** Circuit testing involves the following steps:

STEP 1 Make sure the fuse or fusible link is good before attempting to troubleshoot the circuit.

STEP 2 Check that the ground connections for the horn are clean and tight. Most horns ground to the chassis through the mounting bolts. High ground circuit resistance due to corrosion, road dirt, or loose fasteners may cause no, or intermittent, horn operation.

STEP 3 On a system with a relay, test the power output circuit and the control circuit. Check for voltage available at the horn, voltage available at the relay, and continuity through the switch. When no relay is used, there are two wires leading to the horn switch, and a connection to the steering wheel is made with a double contact slip ring. Test points on this system are similar to those of a system with a relay, but there is no control circuit.

STEP 4 On BCM-controlled horns, use a scan tool special function feature to command the BCM to sound the horn. The scan tool can turn the relay one and off, allowing testing of the circuit.

HORN REPLACEMENT Horns are generally mounted on the radiator core support by bolts and nuts or sheet metal screws. It may be necessary to remove the grille or other parts to access the horn mounting screws. If a replacement horn is required, attempt to use a horn of the same tone as the original. The tone is usually indicated by a number or letter stamped on the body of the horn. To replace a horn, simply remove the fasteners and lift the old horn from its mounting bracket.

Clean the attachment area on the mounting bracket and chassis before installing the new horn. Some models use a corrosion-resistant mounting bolt to ensure a ground connection. ● **SEE FIGURE 21–3.**

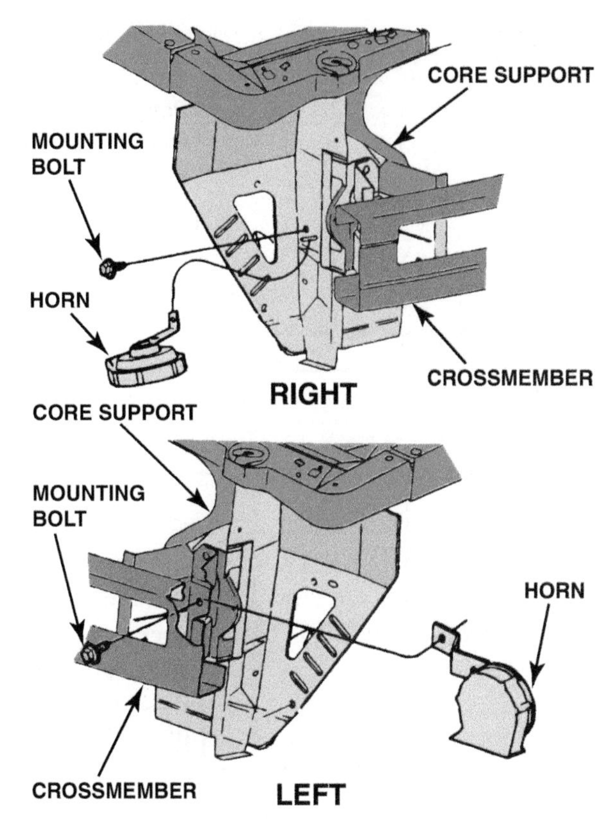

FIGURE 21–3 Horns typically mount to the radiator core support or bracket at the front of the vehicle.

WINDSHIELD WIPER AND WASHER SYSTEM

PURPOSE AND FUNCTION **Windshield wipers** are used to keep the viewing area of the windshield clean of rain. Windshield wiper systems and circuits vary greatly between manufacturers as well as between models. Some vehicles combine the windshield wiper and windshield washer functions into a single system. Many minivans and sport utility vehicles (SUVs) also have a rear window wiper and washer system that works independently of the windshield system. In spite of the design differences, all windshield and rear window wiper and washer systems operate in a similar fashion.

COMPUTER CONTROLLED Most wipers since the 1990s have used the body computer to control the actual operation of the wiper. The wiper controls are simply a command to the computer. The computer may also turn on the headlights whenever the wipers are on, which is the law in some states. ● **SEE FIGURE 21–4**.

WIPER AND WASHER COMPONENTS A typical combination of wiper and washer system consists of the following:

- Wiper motor
- Gearbox
- Wiper arms and linkage
- Washer pump
- Hoses and jets (nozzles)
- Fluid reservoir
- Combination switch
- Wiring and electrical connectors
- Electronic control module

The motor and gearbox assembly is wired to the wiper switch on the instrument panel or steering column or to the wiper control module. ● **SEE FIGURE 21–5**.

Some systems use either a one- or two-speed wiper motor, whereas others have a variable-speed motor.

WINDSHIELD WIPER MOTORS The windshield wipers ordinarily use a special two-speed electric motor. Most are compound-wound motors, a motor type, which provides for two different speeds.

- **Series-wound field**
- **Shunt field**

One speed is achieved in the series wound field and the other speed in the shunt wound field. The wiper switch provides the necessary electrical connections for either motor speed. Switches in the mechanical wiper motor assembly provide the necessary operation for "parking" and "concealing" of the wipers. ● **SEE FIGURE 21–6** for a typical wiper motor assembly.

- **Wiper motor operation.** Most wiper motors use a permanent magnet motor with a low speed + brush and a high speed + brush. The brushes connect the battery to the internal windings of the motor, and the two brushes provide for two different motor speeds.

  The ground brush is directly opposite the low-speed brush. The high-speed brush is off to the side of the low-speed brush. When current flows through the high-speed brush, there are fewer turns on the armature between

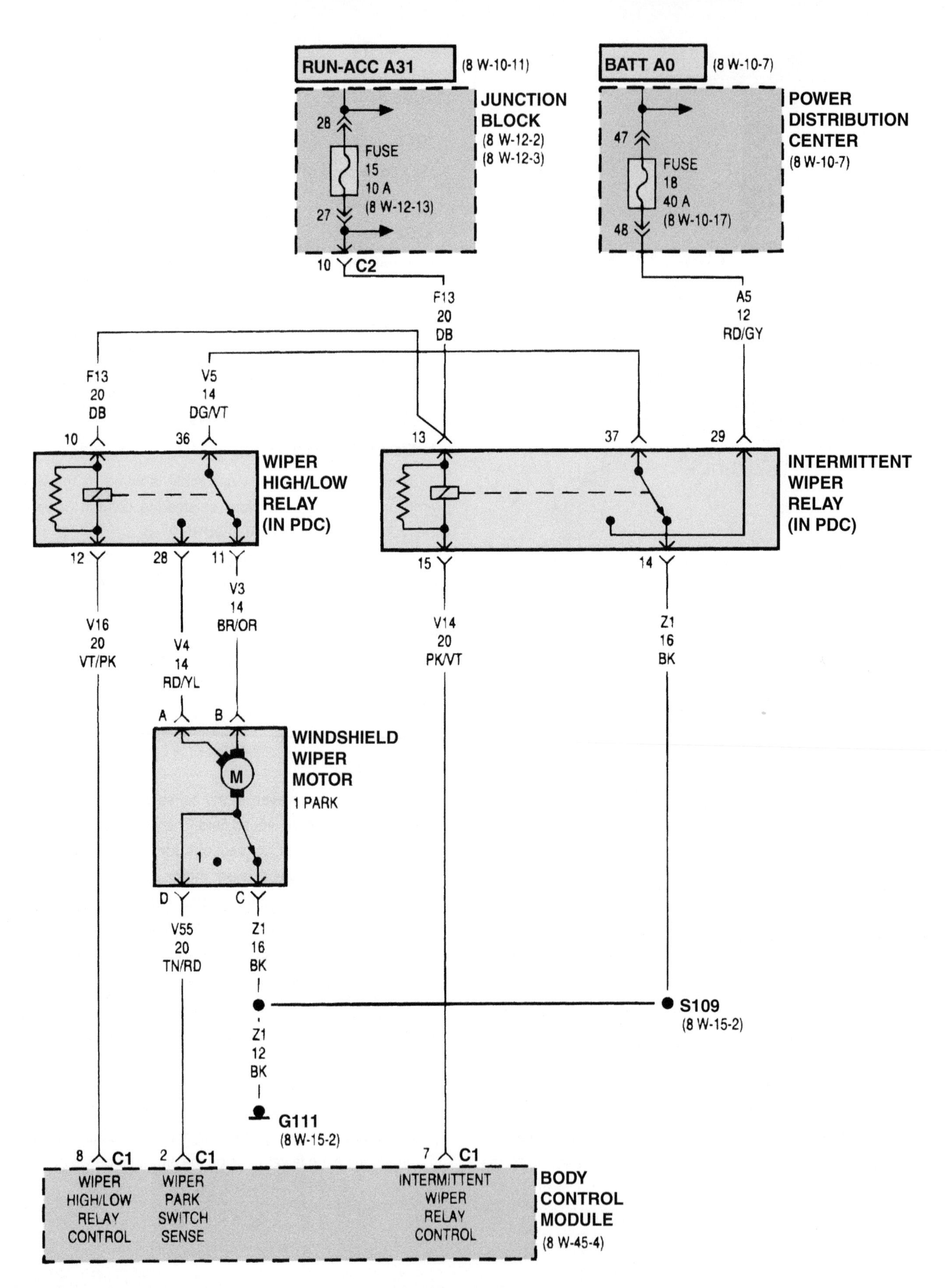

FIGURE 21–4 A circuit diagram is necessary to troubleshoot a windshield wiper problem.

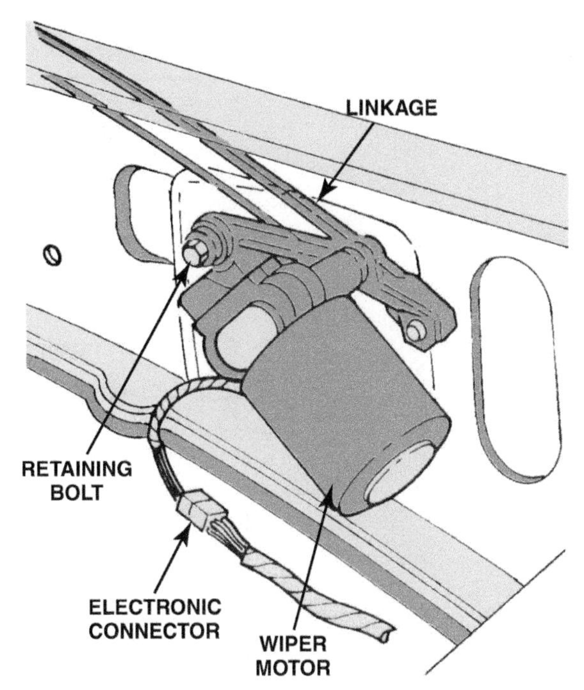

FIGURE 21–5 The motor and linkage bolt to the body and connect to the switch with a wiring harness.

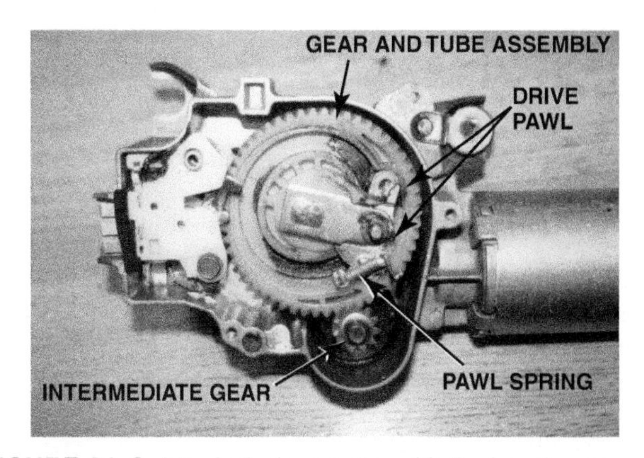

FIGURE 21–6 A typical wiper motor with the housing cover removed. The motor itself has a worm gear on the shaft that turns the small intermediate gear, which then rotates the gear and tube assembly, which rotates the crank arm (not shown) that connects to the wiper linkage.

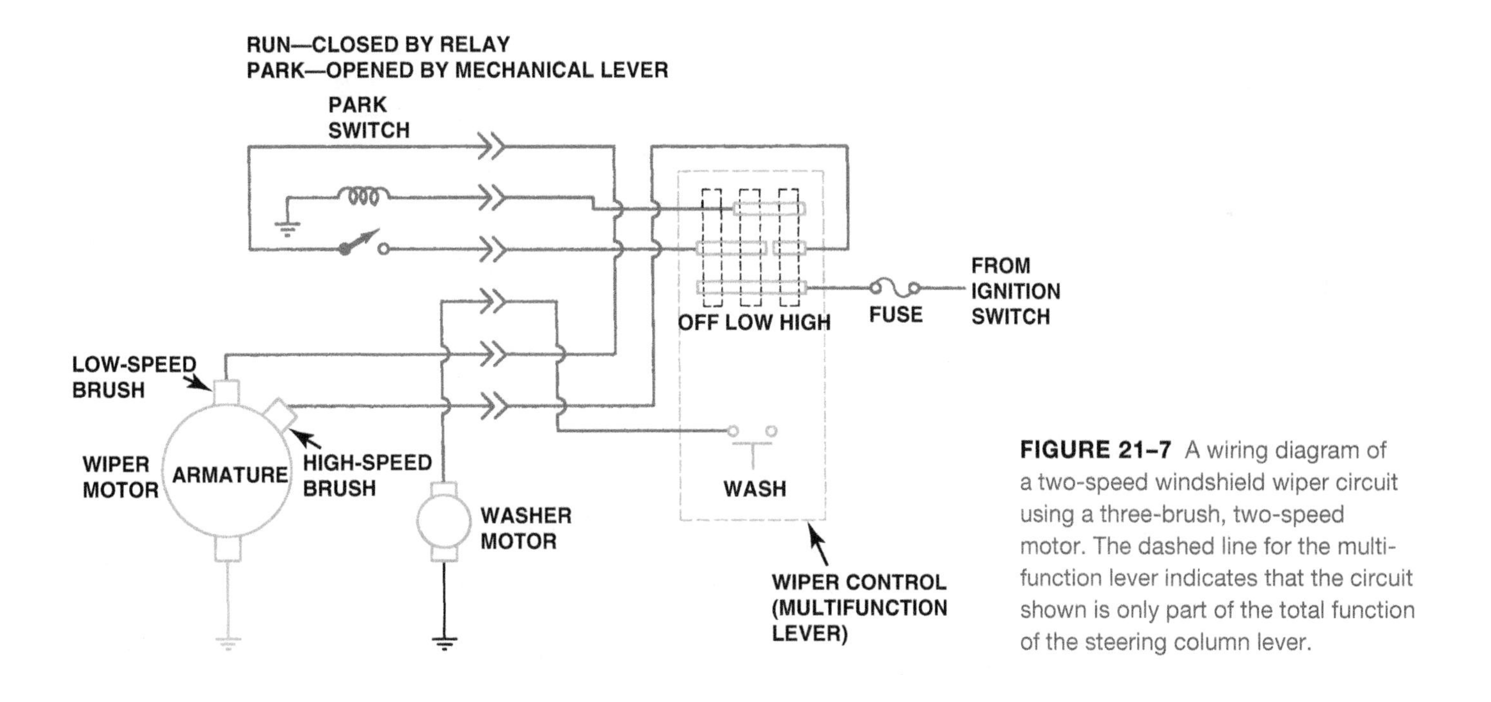

FIGURE 21–7 A wiring diagram of a two-speed windshield wiper circuit using a three-brush, two-speed motor. The dashed line for the multifunction lever indicates that the circuit shown is only part of the total function of the steering column lever.

the hot and ground brushes, and therefore the resistance is less. With less resistance, more current flows and the armature revolves faster. ● **SEE FIGURES 21–7 AND 21–8.**

- **Variable wipers.** The **variable-delay wipers** (also called **pulse wipers**) use an electronic circuit with a variable resistor that controls the time of the charge and discharge of a capacitor. The charging and discharging of the capacitor controls the circuit for the operation of the wiper motor. ● **SEE FIGURE 21–9.**

HIDDEN WIPERS Some vehicles are equipped with wipers that become hidden when turned off. These wipers are also

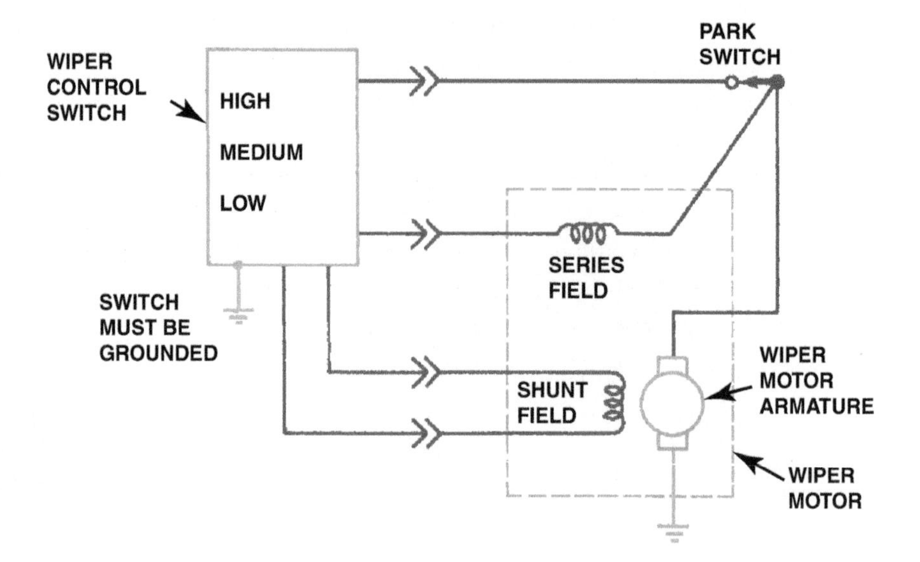

FIGURE 21–8 A wiring diagram of a three-speed windshield wiper circuit using a two-brush motor, but both a series-wound and a shunt field coil.

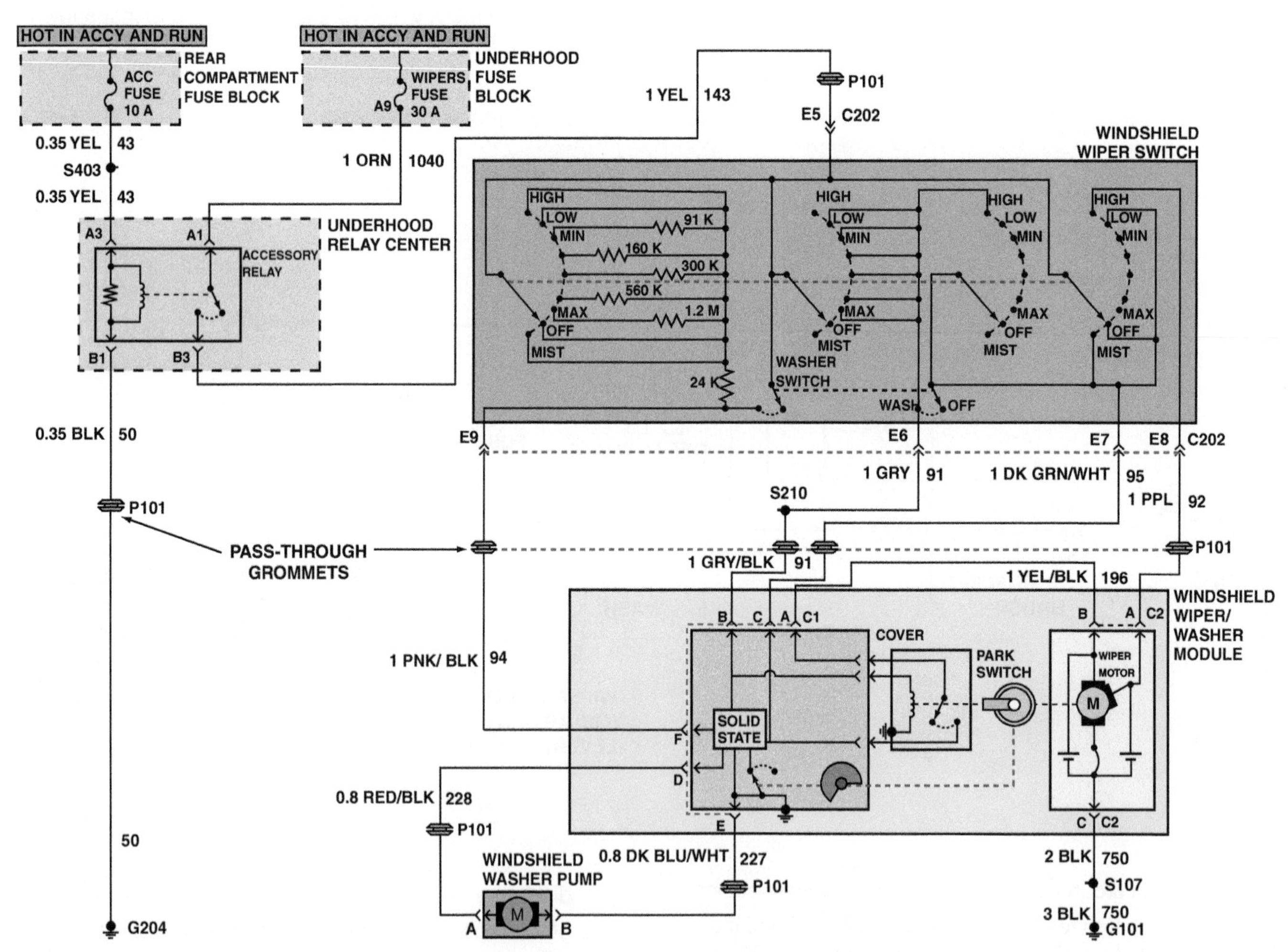

FIGURE 21–9 A variable pulse rate windshield wiper circuit. Notice that the wiring travels from the passenger compartment through pass-through grommets to the underhood area.

called *depressed wipers.* The gearbox has an additional linkage arm to provide depressed parking for hidden wipers. This link extends to move the wipers into the park position when the motor turns in reverse of operating direction. With depressed park, the motor assembly includes an internal park switch. The park switch completes a circuit to reverse armature polarity in the motor when the windshield wiper switch is turned off. The park circuit opens once the wiper arms are in the park position. Instead of a depressed park feature, some systems simply extend the cleaning arc below the level of the hood line.

PULSE WIPE SYSTEMS Windshield wipers may also incorporate a delay, or intermittent operation, a feature commonly called "pulse wipe." The length of the delay, or the frequency of the intermittent operation, is adjustable on some systems. Pulse wipe systems may rely on simple electrical controls, such as a variable-resistance switch, or be controlled electronically through a control module.

With any electronic control system, it is important to follow the diagnosis and test procedures recommended by the manufacturer for that specific vehicle.

A typical pulse, or interval, wiper system uses either a governor or a solid-state module that contains either a variable resistor or rheostat and capacitor. The module connects into the electrical circuitry between the wiper switch and wiper motor. The variable resistor or rheostat controls the length of the interval between wiper pulses. A solid-state pulse wipe timer regulates the control circuit of the pulse relay to direct current to the motor at the prescribed interval. ● **SEE FIGURE 21–10.**

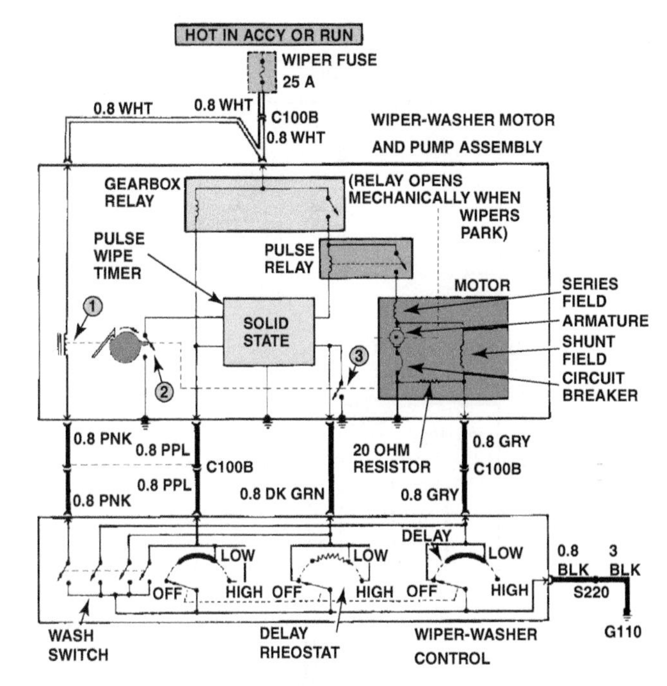

① RATCHET RELEASE SOLENOID
(OPERATED WHEN WASH SWITCH DEPRESSED)
② WASHER OVERRIDE SWITCH
(CLOSED DURING WASH CYCLE)
③ HOLDING SWITCH
(OPEN AT THE END OF EACH SWEEP)

FIGURE 21–10 Circuit diagram of a rheostat-controlled, electronically timed interval wiper.

WINDSHIELD WIPER DIAGNOSIS Windshield wiper failure may be the result of an electrical fault or a mechanical problem, such as binding linkage. Generally, if the wipers operate at one speed setting but not another, the problem is electrical.

To determine if there is an electrical or mechanical problem, access the motor assembly and disconnect the wiper arm linkage from the motor and gearbox. Depending on the type of vehicle, this procedure may involve the following:

- Removing body trim panels from the covered areas at the base of the windshield to gain access to the linkage connectors
- Switching the motor on to each speed (If the motor operates at all speeds, the problem is mechanical. If the motor still does not operate, the problem is electrical.)

If the wiper motor does not run at all, check for the following:

- Grounded or inoperative switch
- Defective motor
- Circuit wiring fault
- Poor electrical ground connection

If the motor operates but the wipers do not, check for the following:

- Stripped gears in the gearbox or stripped linkage connection
- Loose or separated motor-to-gearbox connection
- Loose linkage to the motor connection

If the motor does not shut off, check for the following:

- Defective park switch inside the motor
- Defective wiper switch
- Poor ground connection at the wiper switch

WINDSHIELD WIPER TESTING When the wiper motor does not operate with the linkage disconnected, perform the following steps to determine the fault. ● **SEE FIGURE 21–11.**

To test the wiper system, perform the following steps:

STEP 1 Refer to the circuit diagram or a connector pin chart for the vehicle being serviced to determine the test points for voltage measurements.

STEP 2 Switch the ignition on and set the wiper switch to a speed at which the motor does not operate.

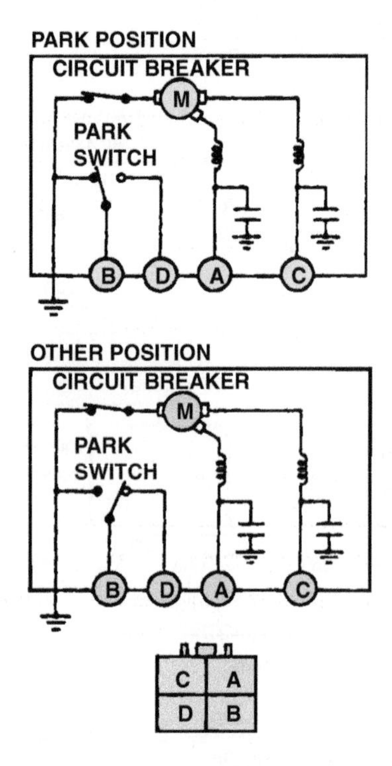

TERMINAL	OPERATION SPEED
C	LOW
A	HIGH

FIGURE 21–11 A wiper motor connector pin chart.

STEP 3 Check for battery voltage available at the appropriate wiper motor terminal for the selected speed. If voltage is available to the motor, an internal motor problem is indicated. No voltage available indicates a switch or circuit failure.

STEP 4 Check for proper ground connections.

STEP 5 Check that battery voltage is available at the motor side of the wiper switch. If battery voltage is available, the circuit is open between the switch and motor. No voltage available indicates either a faulty switch or a power supply problem.

STEP 6. Check for battery voltage available at the power input side of the wiper switch. If voltage is available, the switch is defective. Replace the switch. No voltage available to the switch indicates a circuit problem between the battery and switch.

WINDSHIELD WIPER SERVICE Wiper motors are replaced if defective. The motor usually mounts on the bulkhead (firewall). Bulkhead-mounted units are accessible from under the hood, while the cowl panel needs to be removed to service a motor mounted in the cowl. ● **SEE FIGURE 21–12.**

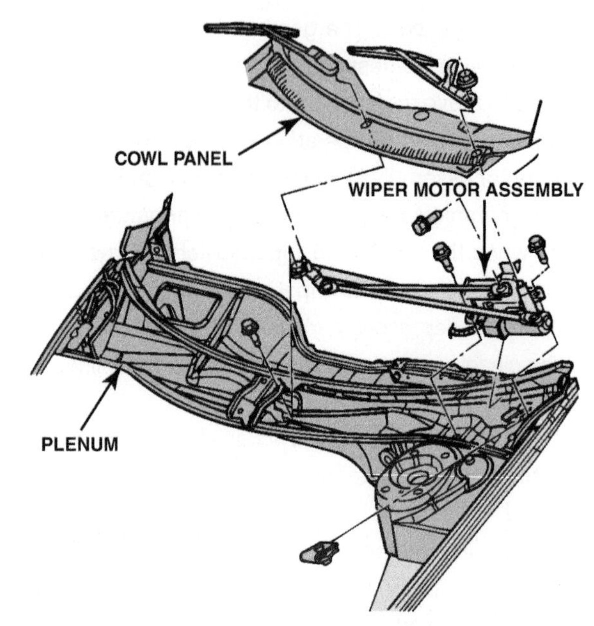

FIGURE 21–12 The wiper motor and linkage mount under the cowl panel on many vehicles.

FREQUENTLY ASKED QUESTION

How Do Wipers Park?

Some vehicles have wiper arms that park lower than the normal operating position so that they are hidden below the hood when not in operation. This is called a *depressed park position.* When the wiper motor is turned off, the park switch allows the motor to continue to turn until the wiper arms reach the bottom edge of the windshield. Then the park switch reverses the current flow through the wiper motor, which makes a partial revolution in the opposite direction. The wiper linkage pulls the wiper arms down below the level of the hood and the park switch is opened, stopping the wiper motor.

After gaining access to the motor, removal is simply a matter of disconnecting the linkage, unplugging the electrical connectors, and unbolting the motor. Move the wiper linkage through its full travel by hand to check for any binding before installing the new motor.

Rear window wiper motors are generally located inside the rear door panel of station wagons, or the rear hatch panel on vehicles with a hatchback or liftgate. ● **SEE FIGURE 21–13.**

After removing the trim panel covering the motor, replacement is essentially the same as replacing the front wiper motor.

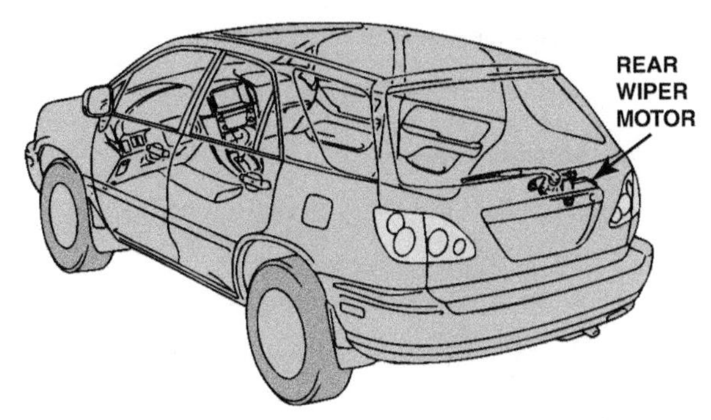

FIGURE 21–13 A single wiper arm mounts directly to the motor on most rear wiper applications.

Wiper control switches are installed either on the steering column or on the instrument panel.

Steering column wiper switches, which are operated by controls on the end of a switch stalk (usually called a *multifunction switch*), require partial disassembly of the steering column for replacement.

WINDSHIELD WASHER OPERATION Most vehicles use a positive-displacement or centrifugal-type washer pump located in the washer reservoir. A momentary contact switch, which is often part of a steering column-mounted combination switch assembly, energizes the washer pump. Washer pump switches are installed either on the steering column or on the instrument panel. The nozzles can be located on the bulkhead or in the hood depending on the vehicle.

WINDSHIELD WASHER DIAGNOSIS Inoperative windshield washers may be caused by the following:

- Blown fuse or open circuit
- Empty reservoir
- Clogged nozzle
- Broken, pinched, or clogged hose
- Loose or broken wire
- Blocked reservoir screen
- Leaking reservoir
- Defective pump

To diagnose the washer system, follow service information procedures that usually include the following steps:

STEP 1 To quick check any washer system, make sure the reservoir has fluid and is not frozen, and then disconnect the pump hose and operate the washer switch.

TECH TIP

Use a Scan Tool to Check Accessories

Most vehicles built since 2000 can have the lighting and accessory circuits checked using a scan tool. A technician can use the following:

- Factory scan tool, such as:
  - Tech 2, Tech2Win and Multiple Diagnostic Interface, or Global Diagnostic System 2 (GDS2)
  - DRB III or Star Scan or Star Mobile or WiTech (Chrysler-Jeep vehicles)
  - New Generation Star or IDS (Ford)
  - TIS Tech Stream (Toyota/Lexus)
- Enhanced aftermarket scan tool that has body bidirectional control capability, including:
  - Snap-on Modis, Solus, or Verus
  - OTC Genisys
  - Autoengenuity

Using a bidirectional scan tool allows the technician to command the operation of electrical accessories such as windows, lights, and wipers. If the circuit operates correctly when commanded by the scan tool and does not function using the switch(es), follow service information instructions to diagnose the switch circuits.

NOTE: Always use good-quality windshield washer fluid from a closed container to prevent contaminated fluid from damaging the washer pump. Radiator antifreeze (ethylene glycol) should never be used in any windshield wiper system.

● **SEE FIGURE 21–14.**

STEP 2 If fluid squirts from the pump, the delivery system is at fault, not the motor, switch, or circuitry.

STEP 3 If no fluid squirts from the pump, the problem is most likely a circuit failure, defective pump, or faulty switch.

STEP 4 A clogged reservoir screen also may be preventing fluid from entering the pump.

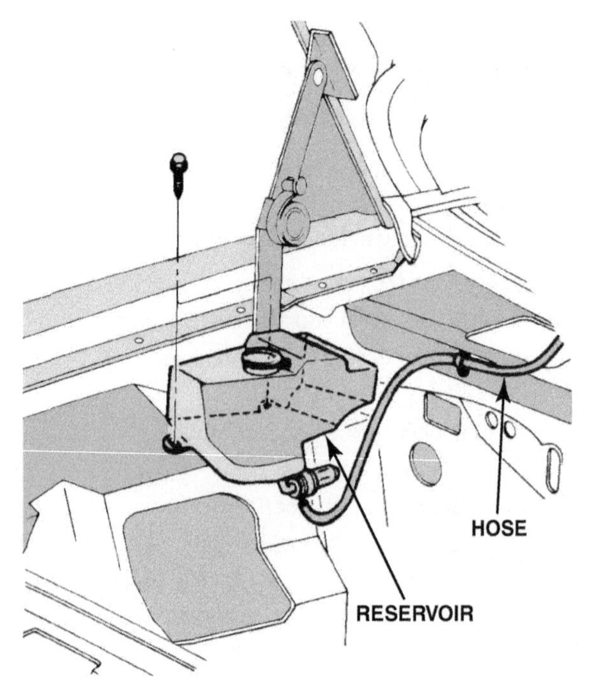

FIGURE 21–14 Disconnect the hose at the pump and operate the switch to check a washer pump.

WINDSHIELD WASHER SERVICE When a fluid delivery problem is indicated, check for the following:

- Blocked, pinched, broken, or disconnected hose
- Clogged nozzles
- Blocked washer pump outlet

If the pump motor does not operate, check for battery voltage available at the pump while operating the washer switch. If voltage is available and the pump does not run, check for continuity on the pump ground circuit. If there is no voltage drop on the ground circuit, replace the pump motor.

If battery voltage is not available at the motor, check for power through the washer switch. If voltage is available at and through the switch, there is a problem in the wiring between the switch and pump. Perform voltage drop tests to locate the fault. Repair the wiring as needed and retest.

Washer motors are not repairable and are simply replaced if defective. Centrifugal or positive-displacement pumps are located on or inside the washer reservoir tank or cover and secured with a retaining ring or nut. ● **SEE FIGURE 21–15.**

RAIN SENSE WIPER SYSTEM

PARTS AND OPERATION **Rain sense wiper** systems use a sensor located at the top of the windshield on the inside to detect rain droplets. The rain sensor itself is about the size of a wristwatch. It is actually a light sensor that uses infrared light beams to detect water through light refraction when droplets contact the windshield.

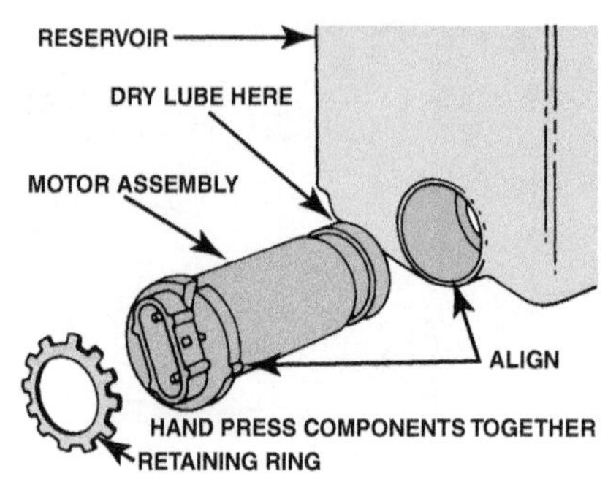

FIGURE 21–15 Washer pumps usually install into the reservoir and are held in place with a retaining ring.

Using advanced software, the sensor takes a reading once every 40 milliseconds. The sensor measures droplet size and frequency and sends a signal to the BCM that determines how slow or fast the wipers move. ● **SEE FIGURES 21–16 AND 21–17.**

The wiper switch can be left on the sense position all of the time, and if no rain is sensed, the wipers will not swipe. Unlike the traditional "intermittent" wiper settings, adjustment of the wipers adjusts the sensitivity of the sensor rather than changing the time between wipes. The speed at which the vehicle is traveling also factors in wipe frequency.

The system also works in the dark. While the human eye perceives total darkness, a raindrop still refracts some amount of light. The programming of the rain sensor changes sensitivity in darkness to trigger more wipes, reducing windshield glare.

DIAGNOSIS AND SERVICE If there is a complaint about the rain sense wipers not functioning correctly, check the owner manual to be sure that they are properly set and adjusted. Also, verify that the windshield wipers are functioning correctly on all speeds before diagnosing the rain sensor circuits. Always follow the vehicle manufacturer's recommended diagnosis and testing procedures.

BLOWER MOTOR

PURPOSE AND FUNCTION The same blower motor moves air inside the vehicle for the following:

1. Air-conditioning
2. Heat
3. Defrosting
4. Defogging
5. Venting of the passenger compartment

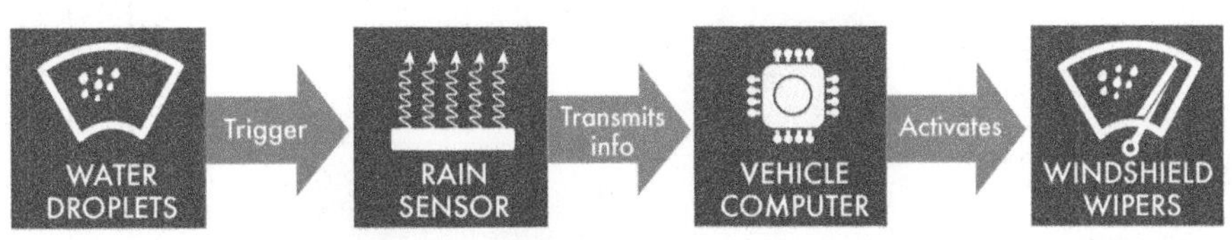

FIGURE 21–16 The rain sensing system adjusts wipers according to the amount of rain on the windshield. (Courtesy of General Motors)

FIGURE 21–17 The rains sensor mounts behind the inside rearview mirror. (Courtesy of General Motors)

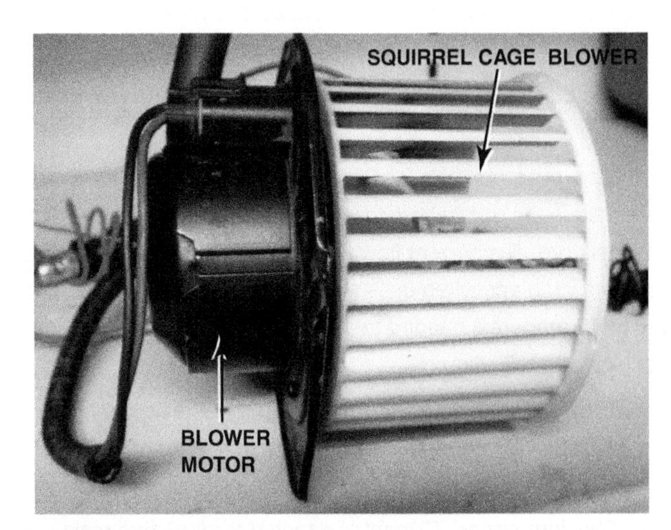

FIGURE 21–18 A squirrel cage blower motor. A replacement blower motor usually does not come equipped with the squirrel cage blower, so it has to be switched from the old motor.

The motor turns a squirrel cage-type fan. A squirrel cage-type fan is able to move air without creating a lot of noise. The fan switch controls the path that the current follows to the blower motor. ● **SEE FIGURE 21–18.**

PARTS AND OPERATION The motor is usually a permanent magnet, one-speed motor that operates at its maximum speed with full battery voltage. The switch gets current from the fuse panel with the ignition switch on, and then directs full

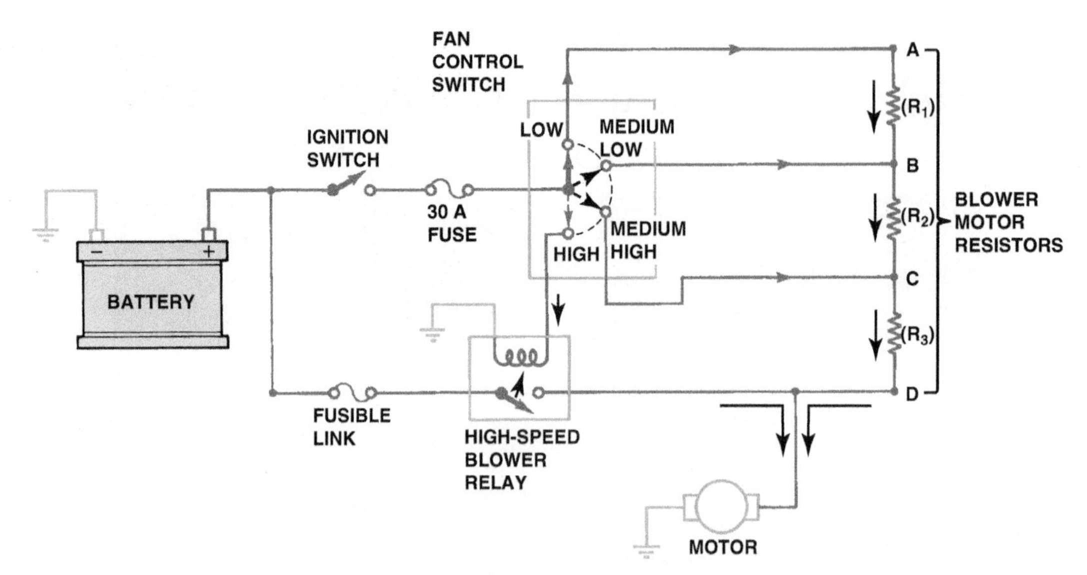

FIGURE 21–19 A typical blower motor circuit with four speeds. The three lowest fan speeds (low, medium low, and medium high) use the blower motor resistors to drop the voltage to the motor and reduce current to the motor. On high, the resistors are bypassed. The "high" position on the fan switch energizes a relay, which supplies the current for the blower on high through a fusible link.

battery voltage to the blower motor for high speed and to the blower motor through resistors for lower speeds.

VARIABLE SPEED CONTROL The fan switch controls the path of current through a resistor pack to obtain different fan speeds of the blower motor. The electrical path can be:

- Full battery voltage for high-speed operation
- Through one or more resistors to reduce the voltage and the current to the blower motor, which then rotates at a slower speed

The resistors are located near the blower motor and mounted in the duct where the airflow from the blower can cool the resistors. The current flow through the resistor is controlled by the switch and often uses a relay to carry the heavy current (10 to 12 amperes) needed to power the fan. Normal operation includes the following:

- **Low speed.** Current flows through three resistors in series to drop the voltage to about 4 volts and 4 amperes.
- **Medium speed.** Current is directed through two resistors in series to lower the voltage to about 6 volts and 6 amperes.
- **Medium-high speed.** Current is directed through one resistor resulting in a voltage of about 9 volts and 9 amperes.

FIGURE 21–20 A typical blower motor resistor pack used to control blower motor speed. Some blower motor resistors are flat and look like a credit card and are called "credit card resistors."

- **High speed.** Full battery voltage, usually through a relay, is applied to the blower motor resulting in a current of about 12 amperes.

● **SEE FIGURES 21–19 AND 21–20.**

NOTE: Some vehicles place the blower motor resistors on the ground side of the motor circuit. The location of the resistors does not affect the operation because they are connected in series.

Some blower motors are electronically controlled by the body control module (BCM) and include electronic circuits to achieve a variable speed. ● **SEE FIGURE 21–21.**

FIGURE 21–21 A brushless DC motor that uses the body computer to control the speed. (Courtesy of Sammy's Auto Service, Inc.)

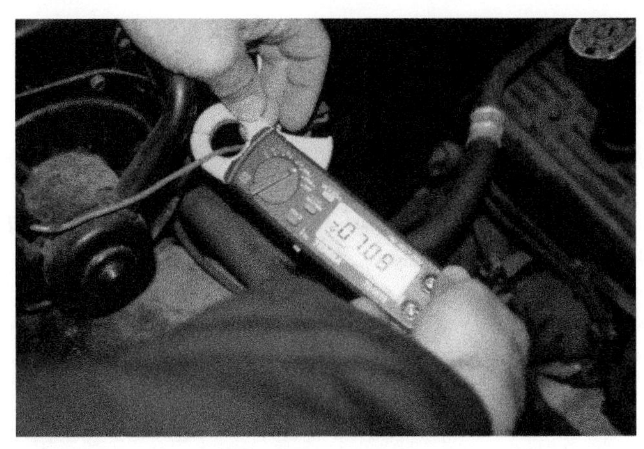

FIGURE 21–22 Using a mini AC/DC clamp-on multimeter to measure the current drawn by a blower motor.

BLOWER MOTOR DIAGNOSIS If the blower motor does not operate at any speed, the problem could be any of the following:

1. Defective ground wire or ground wire connection
2. Defective blower motor (not repairable; must be replaced)
3. Open circuit in the power-side circuit, including fuse, wiring, or fan switch

If the blower works on lower speeds but not on high speed, the problem is usually an inline fuse or high-speed relay that controls the heavy current flow for high-speed operation. The high-speed fuse or relay usually fails as a result of internal blower motor bushing wear, which causes excessive resistance to motor rotation. At slow blower speeds, the resistance is not as noticeable and the blower operates normally. The blower motor is a sealed unit, and if defective, must be replaced as a unit. The squirrel cage fan usually needs to be removed from the old motor and attached to the replacement motor. If the blower motor operates normally at high speed but not at any of the lower speeds, the problem could be melted wire resistors or a defective switch.

The blower motor can be tested using a clamp-on DC ammeter. ● **SEE FIGURE 21–22.**

Most blower motors do not draw more than 15 amperes on high speed. A worn or defective motor usually draws more current than normal and could damage the blower motor resistors or blow a fuse if not replaced.

TECH TIP

The 20 Ampere Fuse Test

Most blower motors operate at about 12 amperes on high speed. If the bushings (bearings) on the armature of the motor become worn or dry, the motor turns more slowly. Because a motor also produces counterelectromotive force (CEMF) as it spins, a slower-turning motor will actually draw more amperes than a fast-spinning motor.

If a blower motor draws too many amperes, the resistors or the electronic circuit controlling the blower motor can fail. Testing the actual current draw of the motor is sometimes difficult because the amperage often exceeds the permissible amount for most digital meters.

One test recommended by General Motors Co. is to unplug the power lead to the motor (retain the ground on the motor) and use a fused jumper lead with one end connected to the battery's positive terminal and the other end to the motor terminal. Use a 20 amperes fuse in the test lead, and operate the motor for several minutes. If the blower motor is drawing more than 20 amperes, the fuse will blow. Some experts recommend using a 15 amperes fuse. If the 15 amperes fuse blows and the 20 amperes fuse does not, then you know the approximate blower motor current draw.

HORN, WIPER, AND BLOWER MOTOR SYMPTOM GUIDE

The following list will assist technicians in troubleshooting electrical accessory systems:

Blower Motor Problem	Possible Causes and/or Solutions
Blower motor does not operate.	1. Blown fuse 2. Poor ground connection on blower motor 3. Defective motor (Use a fused jumper wire connected between the positive terminal of the battery and the blower motor power lead connection [lead disconnected] to check for blower motor operation.) 4. Defective control switch 5. Resistor block open or defective blower motor control module
Blower motor operates only on high speed.	1. Open in the resistors located in the air box near the blower motor 2. Stuck or defective high-speed relay 3. Defective blower motor control switch
Blower motor operates in lower speed(s) only, no high speed.	1. Defective high-speed relay or blower high-speed fuse **NOTE: If the high-speed fuse blows a second time, check the current drawn by the motor and replace the blower motor if the current draw is above specifications. Check for possible normal operation if the rear window defogger is not in operation; some vehicles electrically prevent simultaneous operation of the high-speed blower and rear window defogger to help reduce the electrical loads.**
Windshield Wiper or Washer Problem	**Possible Causes and/or Solutions**
Windshield wipers are inoperative.	1. Blown fuse 2. Poor ground on the wiper motor or the control switch 3. Defective motor or linkage problem
Windshield wipers operate on high speed or low speed only.	1. Defective switch 2. Defective motor assembly 3. Poor ground on the wiper control switch
Windshield washers are inoperative.	1. Defective switch 2. Empty reservoir or clogged lines or discharge nozzles 3. Poor ground on the washer pump motor
Horn Problem	**Possible Causes and/or Solutions**
Horn(s) are inoperative.	1. Poor ground on horn(s) 2. Defective relay (if used); open circuit in the steering column 3. Defective horn (Use a fused jumper wire connected between the positive terminal of the battery and the horn [horn wire disconnected] to check for proper operation of the horn.)
Horn(s) produce low volume or wrong sound.	1. Poor ground at horn 2. Incorrect frequency of horn
Horn blows all the time.	1. Stuck horn relay (if used) 2. Short-to-ground in the wire to the horn button

SUMMARY

1. Horn frequency can range from 1,800 to 3,550 Hz.
2. Most horn circuits use a relay, and the current through the relay coil is controlled by the horn switch.
3. Most windshield wipers use a three-brush, two-speed motor.
4. Windshield washer diagnosis includes checking the pump both electrically and mechanically for proper operation.
5. Many blower motors use resistors wired in series to control blower motor speed.
6. A good blower motor should draw less than 20 amperes.

REVIEW QUESTIONS

1. What are the three types of horn failure?
2. How is the horn switch used to operate the horn?
3. How do you determine if a windshield wiper problem is electrical or mechanical?
4. Why does a defective blower motor draw more current (amperes) than a good motor?

CHAPTER QUIZ

1. Technician A says that a defective high-speed blower motor relay could prevent high-speed blower operation, yet allows normal operation at low speeds. Technician B says that a defective (open) blower motor resistor can prevent low-speed blower operation, yet permit normal high-speed operation. Which technician is correct?
   - **a.** Technician A only
   - **b.** Technician B only
   - **c.** Both Technicians A and B
   - **d.** Neither Technician A nor B
2. To determine if a windshield wiper problem is electrical or mechanical, the service technician should _______.
   - **a.** Disconnect the linkage arm from the windshield wiper motor and operate the windshield wiper
   - **b.** Check to see if the fuse is blown
   - **c.** Check the condition of the wiper blades
   - **d.** Check the washer fluid for contamination
3. A weak-sounding horn is being diagnosed. Technician A says that a poor ground connector at the horn itself can be the cause. Technician B says an open relay can be the cause. Which technician is correct?
   - **a.** Technician A only
   - **b.** Technician B only
   - **c.** Both Technicians A and B
   - **d.** Neither Technician A nor B
4. What controls the operation of a pulse wiper system?
   - **a.** Resistor that controls current flow to the wiper motor
   - **b.** Solid-state (electronic) module
   - **c.** Variable-speed gem set
   - **d.** Transistor
5. Which pitch horn is used for a single horn application?
   - **a.** High pitch
   - **b.** Low pitch
6. The horn switch on the steering wheel on a vehicle that uses a horn relay _______.
   - **a.** Sends electrical power to the horns
   - **b.** Provides the ground circuit for the horn
   - **c.** Grounds the horn relay coil
   - **d.** Provides power (12 volts) to the horn relay
7. A rain sense wiper system uses a rain sensor that is usually mounted _______.
   - **a.** Behind the grille
   - **b.** Outside of the windshield at the top
   - **c.** Behind the inside rearview mirror
   - **d.** On the roof
8. Technician A says a blower motor can be tested using a fused jumper lead. Technician B says a blower motor can be tested using a clamp-on ammeter. Which technician is correct?
   - **a.** Technician A only
   - **b.** Technician B only
   - **c.** Both Technicians A and B
   - **d.** Neither Technician A nor B
9. A defective blower motor draws more current than a good motor because the _______.
   - **a.** Speed of the motor increases
   - **b.** CEMF decreases
   - **c.** Airflow slows down, which decreases the cooling of the motor
   - **d.** Both a and c
10. Windshield washer pumps can be damaged if _______.
    - **a.** Pure water is used in freezing weather
    - **b.** Contaminated windshield washer fluid is used
    - **c.** Ethylene glycol (antifreeze) is used
    - **d.** All of the above

chapter 22

ACCESSORY CIRCUITS

LEARNING OBJECTIVES

After studying this chapter, the reader will be able to:

1. Explain how cruise control operates and how to troubleshoot the circuit.
2. Discuss how to test a heated rear window defogger circuit and rear window heating grids.
3. Describe how power windows and power seats operate.
4. Diagnose incorrect electric lock and keyless entry operation, and determine necessary action.
5. Explain how a antitheft system works, and diagnose faulty operation.

This chapter will help you prepare for the ASE Electrical/Electronic Systems (A6) certification test content area "H" (Accessories Diagnosis and Repair).

KEY TERMS

Adjustable pedals 357
Backlight 346
Center high-mounted stop light (CHMSL) 343
Control wires 352
Cruise control 341
Direction wires 352
Electric adjustable pedals (EAP) 357
HomeLink 348
Independent switches 349
Key fob 358
Lockout switch 349
Lumbar 353
Master control switch 349
Peltier effect 355
Permanent magnet electric motors 349
Rubber coupling 353
Screw jack assembly 353
Throttle actuator control (TAC) 343
Window regulator 350

GM STC OBJECTIVES

GM Service Technical College topics covered in this chapter are as follows:

1. Characteristics, components, and operation of keyless entry and security systems.
2. Characteristics, components, and operation of the content theft deterrent systems.
3. Diagnostic strategies and service considerations for keyless entry and security systems.

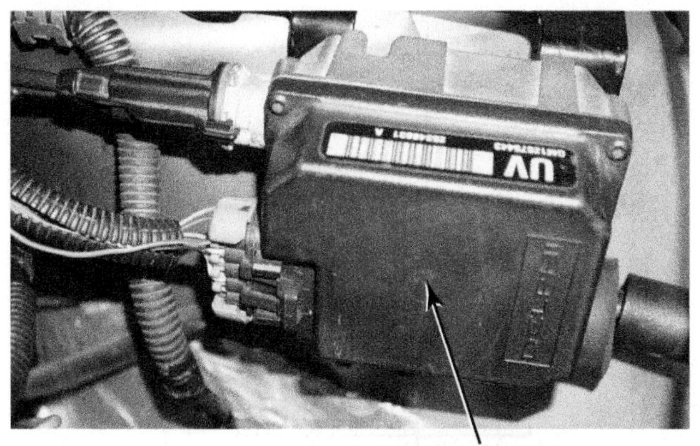

FIGURE 22–1 This cruise control servo unit has an electrical connection with wires that go to the cruise control module or the vehicle computer, depending on the vehicle. An electric servo motor and cable connects to the throttle linkage to maintain the preset speed.

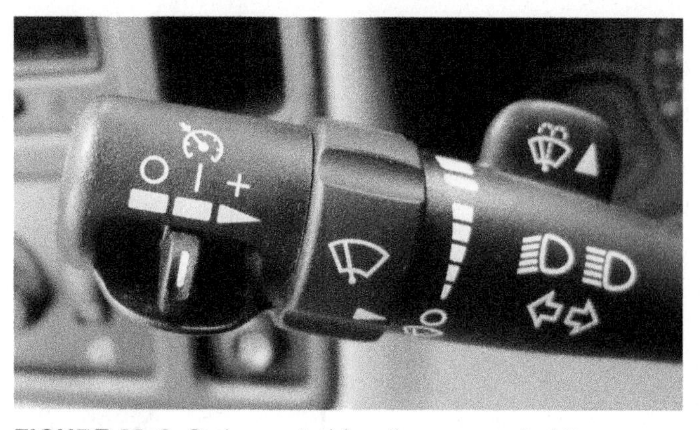

FIGURE 22–2 Cruise control functions are part of the multifunction switch. (Courtesy of Jeffrey Rehkopf)

WARNING

Most vehicle manufacturers warn in the owner manual that cruise control should not be used when it is raining or if the roads are slippery. Cruise control systems operate the throttle, and, if the drive wheels start to hydroplane, the vehicle slows, causing the cruise control unit to accelerate the engine. When the engine is accelerated and the drive wheels are on a slippery road surface, vehicle stability will be lost and might possibly cause a crash.

CRUISE CONTROL

PARTS INVOLVED **Cruise control** (also called *speed control*) is a combination of electrical and mechanical components designed to maintain a constant, set vehicle speed without driver pressure on the accelerator pedal. Major components of a typical cruise control system include the following:

1. **Servo unit.** The servo unit attaches to the throttle linkage through a cable or chain.

    The servo unit controls the movement of the throttle by receiving a controlled amount of vacuum from a control module. ● **SEE FIGURE 22–1.**

    Some systems use a stepper motor and do not use engine vacuum.
2. **Computer or cruise control module.** This unit receives inputs from the brake switch, throttle position (TP) sensor, and vehicle speed sensor. It operates the solenoids or stepper motor to maintain the set speed.
3. **Speed set control.** A speed set control is a switch or control located on the steering column, steering wheel, dash, or console. Many cruise control units feature coast, accelerate, and resume functions. ● **SEE FIGURE 22–2.**
4. **Safety release switches.** When the brake pedal is depressed, the cruise control system is disengaged through use of an electrical or vacuum switch, usually located on the brake pedal bracket. Both electrical and vacuum releases are used to be certain that the cruise control system is released, even in the event of failure of one of the release switches.

CRUISE CONTROL OPERATION A typical cruise control system can be set only if the vehicle speed is 30 mph or more. In a noncomputer-operated system, the transducer contains a low-speed electrical switch that closes when the speed-sensing section of the transducer senses a speed exceeding the minimum engagement speed.

When the set button is depressed on the cruise control, solenoid values on the servo unit allow engine vacuum to be applied to one side of the diaphragm, which is attached to the throttle plate of the engine through a cable or linkage. The servo unit usually contains two solenoids to control the opening and closing of the throttle.

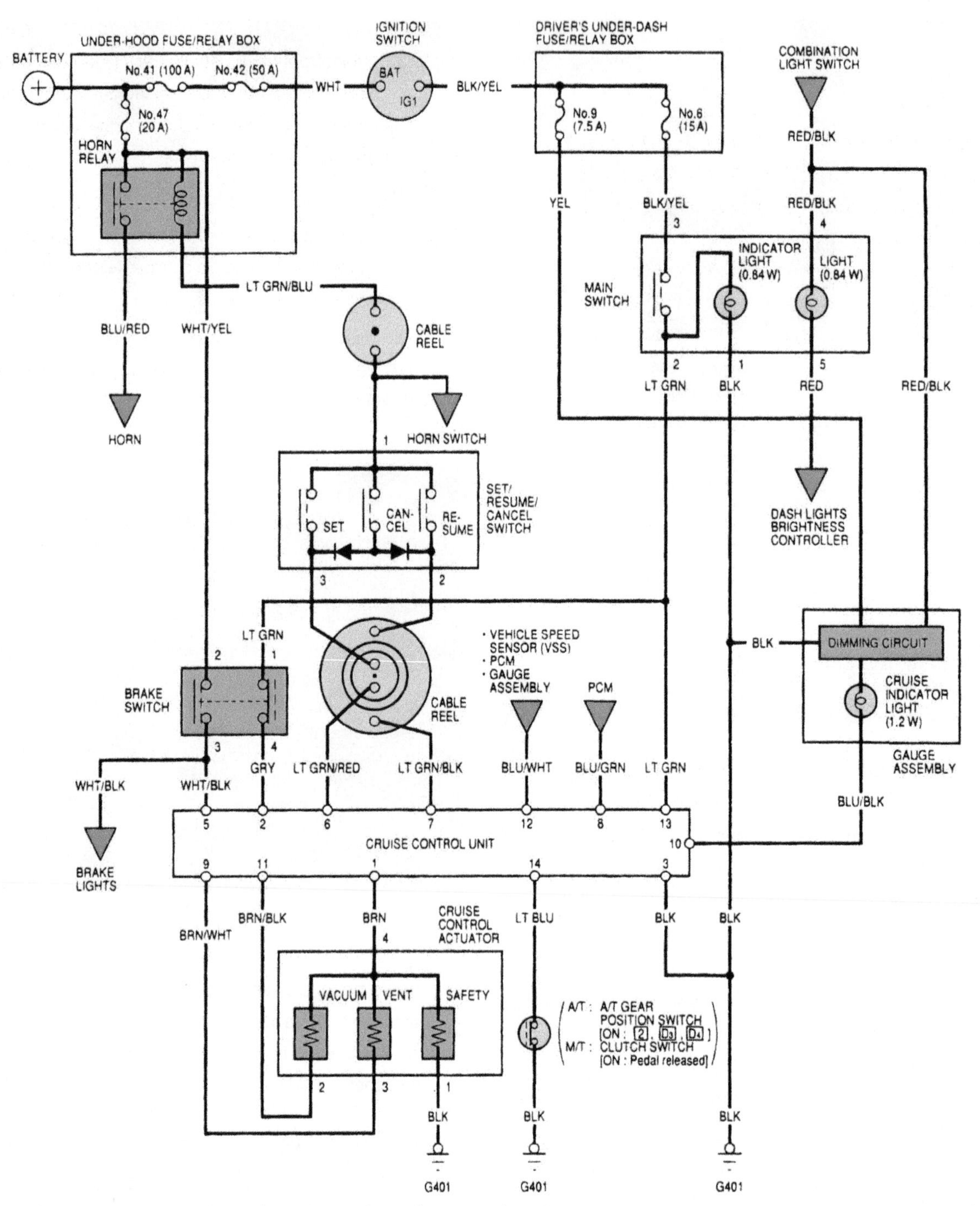

FIGURE 22–3 Circuit diagram of a typical electronic cruise control system.

- One solenoid opens and closes to control the passage, which allows engine vacuum to be applied to the diaphragm of the servo unit, increasing the throttle opening.
- One solenoid bleeds air back into the sensor chamber to reduce the throttle opening.

The throttle position (TP) sensor or a position sensor, inside the servo unit, sends the throttle position information to the cruise control module.

Most computer-controlled cruise control systems use the vehicle's speed sensor input to the engine control computer for speed reference. Computer-controlled cruise control units also use servo units for throttle control, control switches for driver control of cruise control functions, and both electrical and vacuum brake pedal release switches. ● **SEE FIGURE 22–3.**

TECH TIP

Bump Problems

Cruise control problem diagnosis can involve a complex series of checks and tests. The troubleshooting procedures vary among manufacturers (and year), so a technician should always consult a service manual for the exact vehicle being serviced. However, every cruise control system uses a brake safety switch and, if the vehicle has manual transmission, a clutch safety switch. The purpose of these safety switches is to ensure that the cruise control system is disabled if the brakes or the clutch is applied. Some systems use redundant brake pedal safety switches, one electrical to cut off power to the system and the other a vacuum switch used to bleed vacuum from the actuating unit.

If the cruise control "cuts out" or disengages itself while traveling over bumpy roads, the most common cause is a misadjusted brake (and/or clutch) safety switch(es). Often, a simple readjustment of these safety switches will cure the intermittent cruise control disengagement problems.

CAUTION: Always follow the manufacturer's recommended safety switch adjustment procedures. If the brake safety switch(es) is misadjusted, it could keep pressure applied to the master brake cylinder, resulting in severe damage to the braking system.

TECH TIP

Check the Third Brake Light

On some older General Motors vehicles, the cruise control will not work if the third brake light is out. This third brake light is called the **center high-mounted stop light (CHMSL)**. Always check the brake lights first if the cruise control does not work on a General Motors vehicle.

TROUBLESHOOTING CRUISE CONTROL Cruise control system troubleshooting is usually performed using the step-by-step procedure as specified by the vehicle manufacturer.

The usual steps in the diagnosis of an inoperative or incorrectly operating mechanical-type cruise control include the following:

STEP 1 Use a factory or enhanced scan tool to retrieve any cruise control diagnostic trouble codes (DTCs). Perform bidirectional testing if possible using the scan tool.

STEP 2 Check that the cruise control fuse is not blown and that the cruise control dash light is on when the cruise control is turned on.

STEP 3 Check for proper operation of the brake and/or clutch switch.

STEP 4 Inspect the throttle cable and linkage between the sensor unit and the throttle plate for proper operation without binding or sticking.

STEP 5 Check the vacuum hoses for cracks or other faults.

STEP 6 Check that the vacuum servo unit (if equipped), using a hand-operated vacuum pump, can hold vacuum without leaking.

STEP 7 Check the servo solenoids for proper operation, including a resistance measurement check.

ELECTRONIC THROTTLE CRUISE CONTROL

PARTS AND OPERATION Most newer vehicles are equipped with a **throttle actuator control (TAC)** system. Vehicles equipped with such a system do not use throttle actuators for the cruise control. The ETC system operates the throttle under all engine operating conditions. An ETC system uses a DC electric motor to move the throttle plate that is spring loaded to a partially open position. The motor actually closes the throttle at idle against spring pressure. The spring-loaded position is the default position and results in a high idle speed. The powertrain control module (PCM) uses the input signals from the *accelerator pedal position (APP)* sensor to determine the desired throttle position. The PCM then commands the throttle to the necessary position of the throttle plate. ● **SEE FIGURE 22–4.**

FIGURE 22–4 A typical throttle actuator with the protective covers removed.

FIGURE 22–5 A trailer icon lights on the dash of this Cadillac when the transmission trailer towing mode is selected.

TECH TIP

Use Trailer Tow Mode

Some customers complain that when using cruise control while driving in hilly or mountainous areas, the speed of the vehicle will sometimes go 5 to 8 mph below the set speed. The automatic transmission then downshifts, the engine speed increases, and the vehicle returns to the set speed. To help avoid the slowdown and rapid acceleration, ask the customer to select the trailer towing position. When this mode is selected, the automatic transmission downshifts almost as soon as the vehicle speed starts to decrease. This results in a smoother operation and is less noticeable to both the driver and passengers. ● **SEE FIGURE 22–5.**

The cruise control on a vehicle equipped with an electronic throttle control system consists of a switch to set the desired speed. The PCM receives the vehicle speed information from the vehicle speed (VS) sensor and uses the ETC system to maintain the set speed.

DIAGNOSIS AND SERVICE Any fault in the APP sensor or ETC system will disable the cruise control function. Always follow the specified troubleshooting procedures, which will usually include the use of a scan tool to properly diagnose the ETC system.

RADAR CRUISE CONTROL

PURPOSE AND FUNCTION The purpose of a radar cruise control system is to give the driver more control over the vehicle by keeping an assured clear distance behind the vehicle in front. If the vehicle in front slows, the radar cruise control detects the slowing vehicle and automatically reduces the speed of the vehicle to keep a safe distance. Then if the vehicle speeds up, the radar cruise control also allows the vehicle to increase to the preset speed. This makes driving in congested areas easier and less tiring.

TERMINOLOGY Depending on the manufacturer, radar cruise control is also referred to as the following:

- **Adaptive cruise control** (Audi, Ford, General Motors, and Hyundai)
- **Dynamic cruise control** (BMW, Toyota/Lexus)
- **Active cruise control** (Mini Cooper, BMW)
- **Autonomous cruise control** (Mercedes)

It uses forward-looking radar to sense the distance to the vehicle in front and maintains an assured clear distance. This type of cruise control system works within the following conditions.

1. Speeds from 20 to 100 mph (30 to 161 km/h)
2. Designed to detect objects as far away as 500 ft (150 m)

The cruise control system is able to sense both distance and relative speed. ● **SEE FIGURE 22–6.**

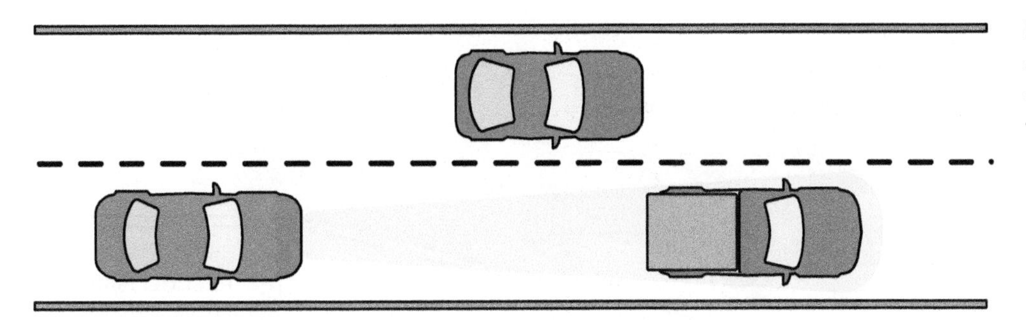

FIGURE 22–6 Radar cruise control uses sensors to keep the distance the same even when traffic slows ahead.

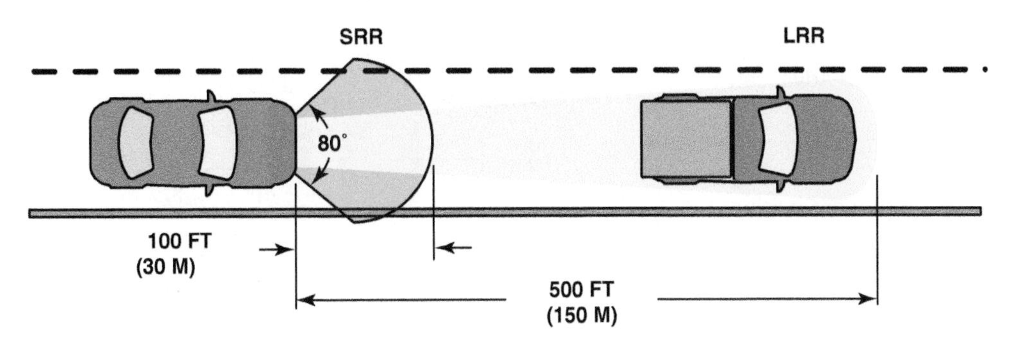

FIGURE 22–7 Most radar cruise control systems use radar, both long and short range. Some systems use optical or infrared cameras to detect objects.

FREQUENTLY ASKED QUESTION

Will Radar Cruise Control Set Off My Radar Detector?

It is doubtful. The radar used for radar cruise control systems operates on frequencies that are not detectable by police radar detector units. Cruise control radar works on the following frequencies:

- 76 to 77 GHz (long range)
- 24 GHz (short range)

The frequencies used for the various types of police radar include the following:

- X-band: 8 to 12 GHz
- K-band: 24 GHz
- Ka-band: 33 to 36 GHz

The only time there may be interference is when the radar cruise control, as part of a precollision system, starts to use short-range radar (SRR) in the 24 GHz frequency. This would trigger the radar detector but would be an unlikely event and just before a possible collision with a vehicle coming toward you.

PARTS AND OPERATION Radar cruise control systems use long-range radar (LRR) to detect faraway objects in front of the moving vehicle. Some systems use a short-range radar (SRR) and/or infrared (IR) or optical cameras to detect distances for when the distance between the moving vehicle and another vehicle in front is reduced. ● SEE FIGURE 22–7.

The radar frequencies include the following:

- 76 to 77 GHz (long-range radar)
- 24 GHz (short-range radar)

PRE-COLLISION SYSTEM

PURPOSE AND FUNCTION The purpose and function of a pre-collision system is to monitor the road ahead and prepare to avoid a collision, and to protect the driver and passengers. A pre-collision makes use of the following vehicle systems:

1. The long-range and short-range radar or detection systems used by a radar cruise control system to detect objects in front of the vehicle
2. Antilock brake system (ABS)
3. Adaptive (radar) cruise control
4. Brake assist system

TERMINOLOGY Pre-collision systems can be called by various names depending on the make of the vehicle. Some commonly used names for a pre-collision or pre-crash system include the following:

- **Ford/Lincoln:** Collision Warning with Brake Support
- **Honda/Acura:** Collision Mitigation Brake System (CMBS)
- **Mercedes-Benz:** Pre-Safe or Attention Assist

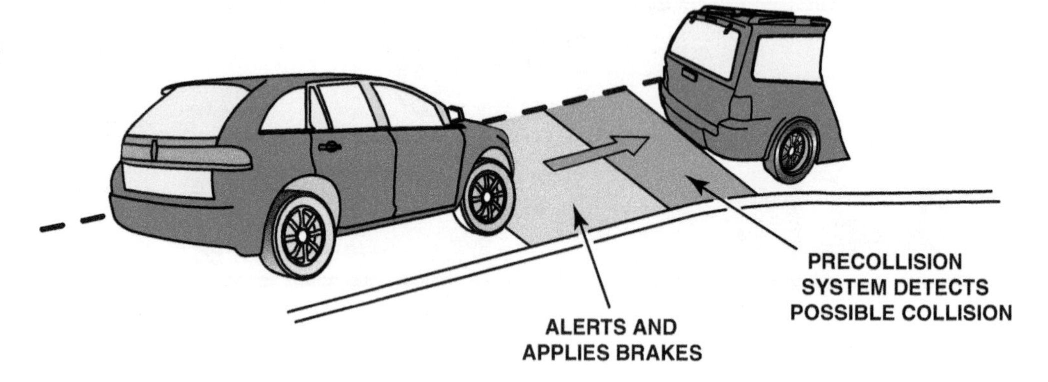

FIGURE 22–8 A precollision system is designed to prevent a collision first and then interacts to prepare for a collision if needed.

- **Toyota/Lexus:** Pre-Collision System (PCS) or Advanced Pre-Collision System (APCS)
- **General Motors:** Pre-Collision System (PCS)
- **Volvo:** Collision Warning with Brake Support or Collision Warning with Brake Assist

OPERATION The system functions by monitoring objects in front of the vehicle and can act to avoid a collision by the following actions:

- Sounds an alarm
- Flashes a warning lamp
- Applies the brakes and brings the vehicle to a full stop (if needed), if the driver does not react

SEE FIGURE 22–8.

If the system is unable to prevent a collision, the system will perform the following actions:

1. Apply the brakes full force to reduce vehicle speed as much as possible.
2. Close all windows and the sunroof to prevent the occupants from being ejected from the vehicle.
3. Move the seats to an upright position.
4. Raise the headrest (if electrically powered).
5. Pretension the seat belts.
6. Airbags and seat belt tensioners function as designed during the collision.

HEATED REAR WINDOW DEFOGGERS

PARTS AND OPERATION An electrically heated rear window defogger system uses an electrical grid baked on the glass that warms the glass to about 85°F (29°C) and clears it of fog or frost. The rear window is also called a **backlight**. The rear window defogger system is controlled by a driver-operated switch and a timer relay. **SEE FIGURE 22–9.**

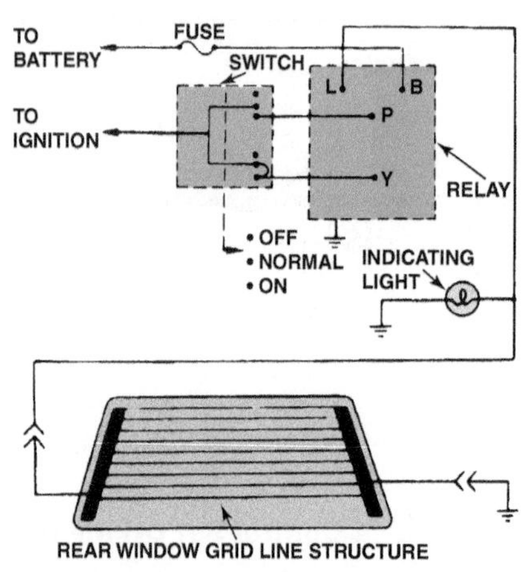

FIGURE 22–9 A switch and relay control current through the heating grid of a rear window defogger.

The timer relay is necessary because the window grid can draw up to 30 amperes, and continued operation would put a strain on the battery and the charging system. Generally, the timer relay permits current to flow through the rear window grid for only 10 minutes. If the window is still not clear of fog after 10 minutes, the driver can turn the defogger on again, but after the first 10 minutes any additional defogger operation is limited to 5 minutes.

PRECAUTIONS Electric grid-type rear window defoggers can be damaged easily by careless cleaning or scraping of the inside of the rear window glass. Short, broken sections of the rear window grid can be repaired using a special epoxy-based electrically conductive material. If more than one section is damaged or if the damaged grid length is greater than approximately 1.5 inches (3.8 cm), a replacement rear window glass may be required to restore proper defogger operation.

The electrical current through the grids depends, in part, on the temperature of the conductor grids. As the temperature decreases, the resistance of the grids decreases and the current flow increases, helping to warm the rear glass. As

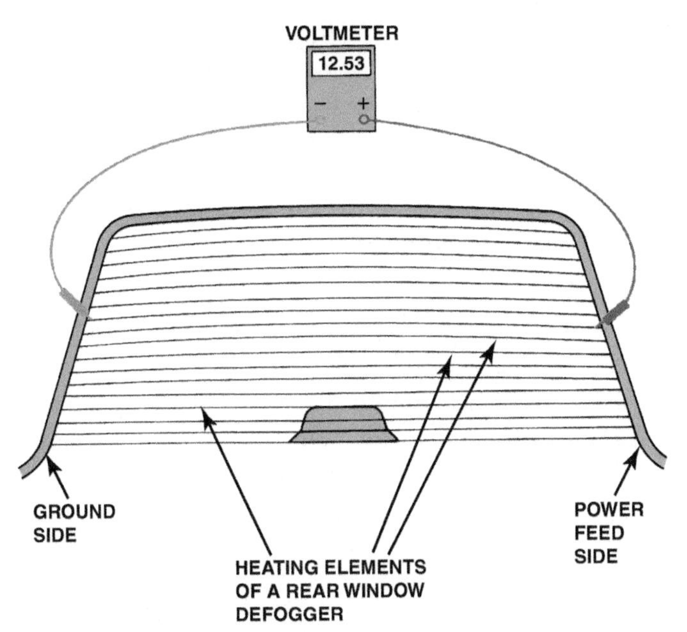

FIGURE 22–10 A rear window defogger electrical grid can be tested using a voltmeter to check for a decreasing voltage as the meter lead is moved from the power side toward the ground side. As the voltmeter positive lead is moved along the grid (on the inside of the vehicle), the voltmeter reading should steadily decrease as the meter approaches the ground side of the grid.

the temperature of the glass increases, the resistance of the conductor grids increases and the current flow decreases. Therefore, the defogger system tends to self-regulate the electrical current requirements to match the need for defogging.

NOTE: Some vehicles use the wire grid of the rear window defogger as the radio antenna. Therefore, if the grid is damaged, radio reception can also be affected.

HEATED REAR WINDOW DEFOGGER DIAGNOSIS

Troubleshooting a nonfunctioning rear window defogger unit involves using a test light or a voltmeter to check for voltage to the grid. If no voltage is present at the rear window, check for voltage at the switch and relay timer assembly. A poor ground connection on the opposite side of the grid from the power side can also cause the rear defogger not to operate. Because most defogger circuits use an indicator light switch and a relay timer, it is possible to have the indicator light on, even if the wires are disconnected at the rear window grid. A voltmeter can be used to test the operation of the rear window defogger grid. ● **SEE FIGURE 22–10.**

With the negative test terminal attached to a good body ground, carefully probe the grid conductors. There should be a decreasing voltage reading as the probe is moved from the power ("hot") side of the grid toward the ground side of the grid.

REPAIR OR REPLACEMENT If there is a broken grid wire, it can be repaired using an electrically conductive substance available in a repair kit.

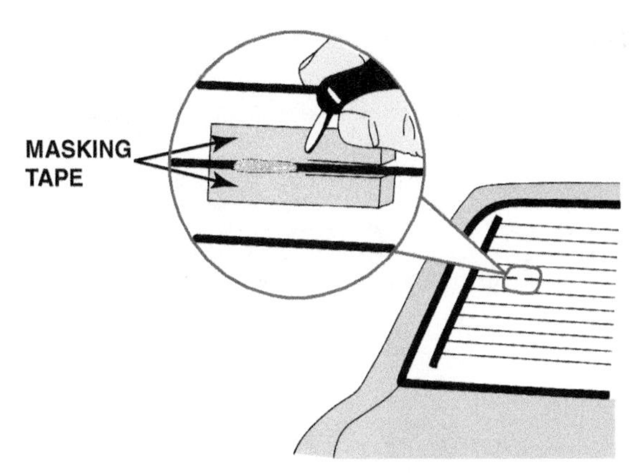

FIGURE 22–11 The typical repair material contains conductive silver-filled polymer, which dries in 10 minutes and is usable in 30 minutes.

TECH TIP

The Breath Test

It is difficult to test for the proper operation of all grids of a rear window defogger unless the rear window happens to be covered with fog. A common trick that works is to turn on the rear defogger and exhale onto the outside of the rear window glass. In a manner similar to that of people cleaning eyeglasses with their breath, this procedure produces a temporary fog on the glass so that all sections of the rear grids can quickly be checked for proper operation.

Most vehicle manufacturers recommend that grid wire less than 2 inches (5 cm) long be repaired. If a bad section is longer than 2 inches, the entire rear window will need to be replaced. ● **SEE FIGURE 22–11.**

HEATED MIRRORS

PURPOSE AND FUNCTION The purpose and function of heated outside mirrors is to heat the surface of the mirror, which evaporates moisture on the surface. The heat helps keep ice and fog off the mirrors, to allow for better driver visibility.

PARTS AND OPERATION Heated outside mirrors are often tied into the same electrical circuit as the rear window defogger. Therefore, when the rear defogger is turned on, the heating grid on the backside of the mirror is also turned on. Some vehicles use a switch for each mirror.

FIGURE 22–12 Typical HomeLink garage door opener buttons. Notice that three different units can be controlled from the vehicle using the HomeLink system.

DIAGNOSIS The first step in any diagnosis procedure is to verify the customer concern. Check the owner's manual or service information for the proper method to use to turn on the heated mirrors.

NOTE: Heated mirrors are not designed to melt snow or a thick layer of ice.

If a fault has been detected, follow service information instructions for the exact procedure to follow. If the mirror itself is found to be defective, it is usually replaced as an assembly instead of being repaired.

HOMELINK GARAGE DOOR OPENER

OPERATION **HomeLink** is a device installed in many vehicles that duplicates the radio-frequency code of the original garage door opener. The frequency range which HomeLink is able to operate is 288 to 418 MHz. The typical vehicle garage door opening system has three buttons that can be used to operate one or more of the following devices.

1. Garage doors equipped with a radio transmitter electric opener
2. Gates
3. Entry door locks
4. Lighting or small appliances

The devices include both fixed-frequency devices, usually older units, and rolling (encrypted) code devices. ● **SEE FIGURE 22–12**.

PROGRAMMING A VEHICLE GARAGE DOOR OPENER

When a vehicle is purchased, it must be programmed using the transmitter for the garage door opener or other device.

NOTE: The HomeLink garage door opening controller can only be programmed by using a transmitter. If an automatic garage door system does not have a remote transmitter, HomeLink cannot be programmed.

Normally, the customer is responsible for programming the HomeLink to the garage door opener. However, some customers may find that help is needed from the service department. The steps that are usually involved in programming HomeLink in the vehicle to the garage door opener are as follows:

STEP 1 Unplug the garage door opener during programming to prevent it from being cycled on and off, which could damage the motor.

STEP 2 Check that the frequency of the handheld transmitter is between 288 and 418 MHz.

STEP 3 Install new batteries in the transmitter to be assured of a strong signal being transmitted to the HomeLink module in the vehicle.

STEP 4 Turn the ignition on, engine off (KOEO).

STEP 5 While holding the transmitter 4 to 6 inches away from the HomeLink button, press and hold the HomeLink button while pressing and releasing the handheld transmitter every two seconds. Continue pressing and releasing the transmitter until the indicator light near the HomeLink button changes from slow blink to a rapid flash.

STEP 6 Verify that the vehicle garage door system (HomeLink) button has been programmed. Press and hold the garage door button. If the indicator light blinks rapidly for two seconds and then comes on steady, the system has been successfully programmed using a rolling code design. If the indicator light is on steady, then it has been successfully programmed to a fixed-frequency device.

DIAGNOSIS AND SERVICE If a fault occurs with the HomeLink system, first verify that the garage door opener is functioning correctly. Also, check if the garage door opener remote control is capable of operating the door. Repair the garage door opener system as needed.

If the problem still exists, attempt reprogramming the HomeLink vehicle system, being sure that the remote has a fresh battery.

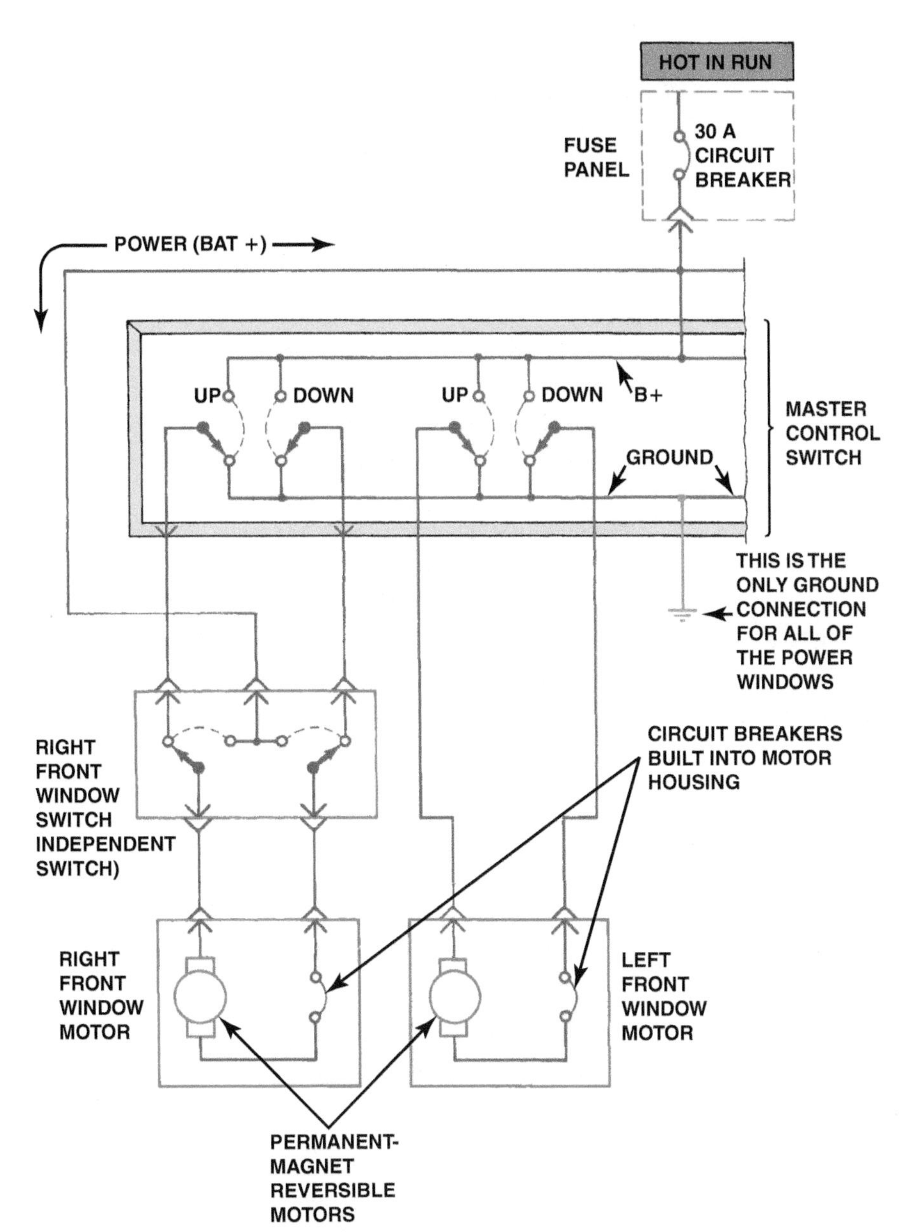

FIGURE 22–13 A typical power window circuit using PM motors. Control of the direction of window operation is achieved by directing the polarity of the current through the nongrounded motors. The only ground for the entire system is located at the master control (driver's side) switch assembly.

POWER WINDOWS

SWITCHES AND CONTROLS Power windows use electric motors to raise and lower door glass. They can be operated by both a **master control switch** located beside the driver and additional **independent switches** for each electric window. Some power window systems use a **lockout switch** located on the driver's controls to prevent operation of the power windows from the independent switches. Power windows are designed to operate only with the ignition switch in the on (run) position, although some manufacturers use a time delay for accessory power after the ignition switch is turned off. This feature permits the driver and passengers an opportunity to close all windows or operate other accessories for about 10 minutes or until a vehicle door is opened after the ignition has been turned off. This feature is often called *retained accessory power*.

POWER WINDOW MOTORS Most power window systems use **permanent magnet (PM) electric motors**. It is possible to run a PM motor in the reverse direction simply by reversing the polarity of the two wires going to the motor. Most power window motors do not require that the motor be grounded to the body (door) of the vehicle. The ground for all the power windows is most often centralized near the driver's master control switch. The up-and-down motion of the individual window motors is controlled by double-pole, double-throw (DPDT) switches. These DPDT switches have five contacts and permit battery voltage to be applied to the power window motor, as well as reverse the polarity and direction of the motor. Each motor is protected by an electronic circuit breaker. These circuit breakers are built into the motor assembly and are not a separate replaceable part.
● **SEE FIGURE 22–13**.

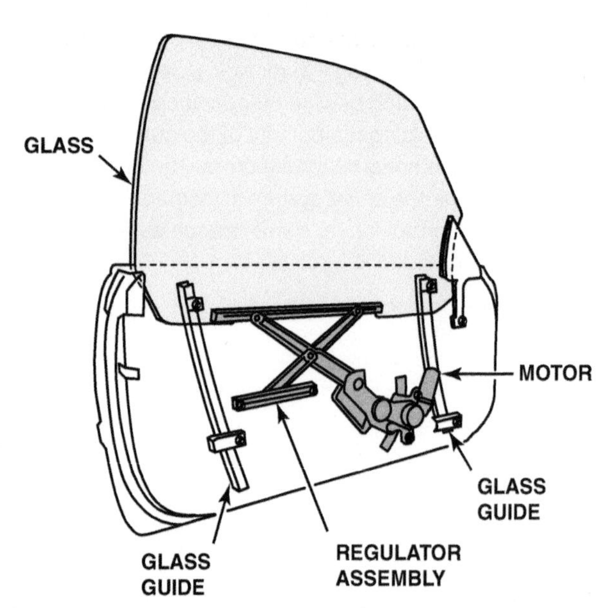

FIGURE 22–14 An electric motor and a regulator assembly raise and lower the glass on a power window.

The power window motors rotate a mechanism called a **window regulator**. The window regulator is attached to the door glass and controls opening and closing of the glass. Door glass adjustments such as glass tilt and upper and lower stops are usually the same for both power and manual windows. ● **SEE FIGURE 22–14.**

AUTO DOWN/UP FEATURES Many power windows are equipped with an auto down feature that allows windows to be lowered all of the way if the control switch is moved to a detent or held down for longer than 0.3 second. The window will then move down all the way to the bottom and then the motor stops.

Many vehicles are equipped with the auto up feature that allows the driver to raise the driver's side or all windows in some cases, with just one push of the button. A sensor in the window motor circuit measures the current through the motor. The circuit is opened if the window touches an object, such as a hand or finger. When the window reaches the top or hits an object, the current through the window motor increases. When the upper limit amperage draw is reached, the motor circuit is opened and the window either stops or reverses.

Most newer power windows use network communications modules to operate the power windows, and the switches are simply voltage signals to the module which supplies current to the individual window motors. ● **SEE FIGURE 22–15.**

The main components of a BCM-controlled window system are:

- Driver door module (DDM) (up to 2006, depending on model)
- Driver door switch (DDS) (2007 and later, depending on model)
- Passenger door module (PDM) (up to 2006, depending on model)
- Passenger door switch (PDS) (2007 and later, depending on model)
- Body control module (BCM)
- Class 2 data (depending on model)
- GM LAN data (depending on model)

TECH TIP

Programming Auto Down Power Windows

Many vehicles are equipped with automatic operation that can cause the window to go all the way down (or up) if the switch is depressed beyond a certain point or held for a fraction of a second. Sometimes this feature is lost if the battery in the vehicle has been disconnected. Although this programming procedure can vary depending on the make and model, many times the window(s) can be reprogrammed without using a scan tool by simply depressing and holding the down button for 10 seconds. If the vehicle is equipped with an auto up feature, repeat the procedure by holding the button up for 10 seconds. Always check exact service information for the vehicle being serviced.

The drivers door module or switch assembly contains power window switches for each door, which are integral components. Closing any of the normally open, rocker type switches provides the DDM (DDS) with a request for power window operation. The switches for the two front windows may have three positions UP, Down and Express Down, while the switches for the rear windows usually have only two positions, UP and Down. Each of these switches and their positions is a direct input into the DDM or DDS.

Upon receiving a request for power window operation, the DDM (DDS), also transmits a Class 2 or GM LAN message to the body control module (BCM) indicating the switch and it's changed position. The BCM can then command window operation of the appropriate window.

The two rear passenger doors have there own switches. Each of the switches contain a power window up and down relay. A switch activation alone can control the up and down functions of the rear windows or the BCM, upon receiving a Class 2 message from the DDM, can control the set of relays which will activate the rear window motors. ● **SEE FIGURE 22–16.**

FIGURE 22–15 A master power window control panel with the buttons and the cover removed.

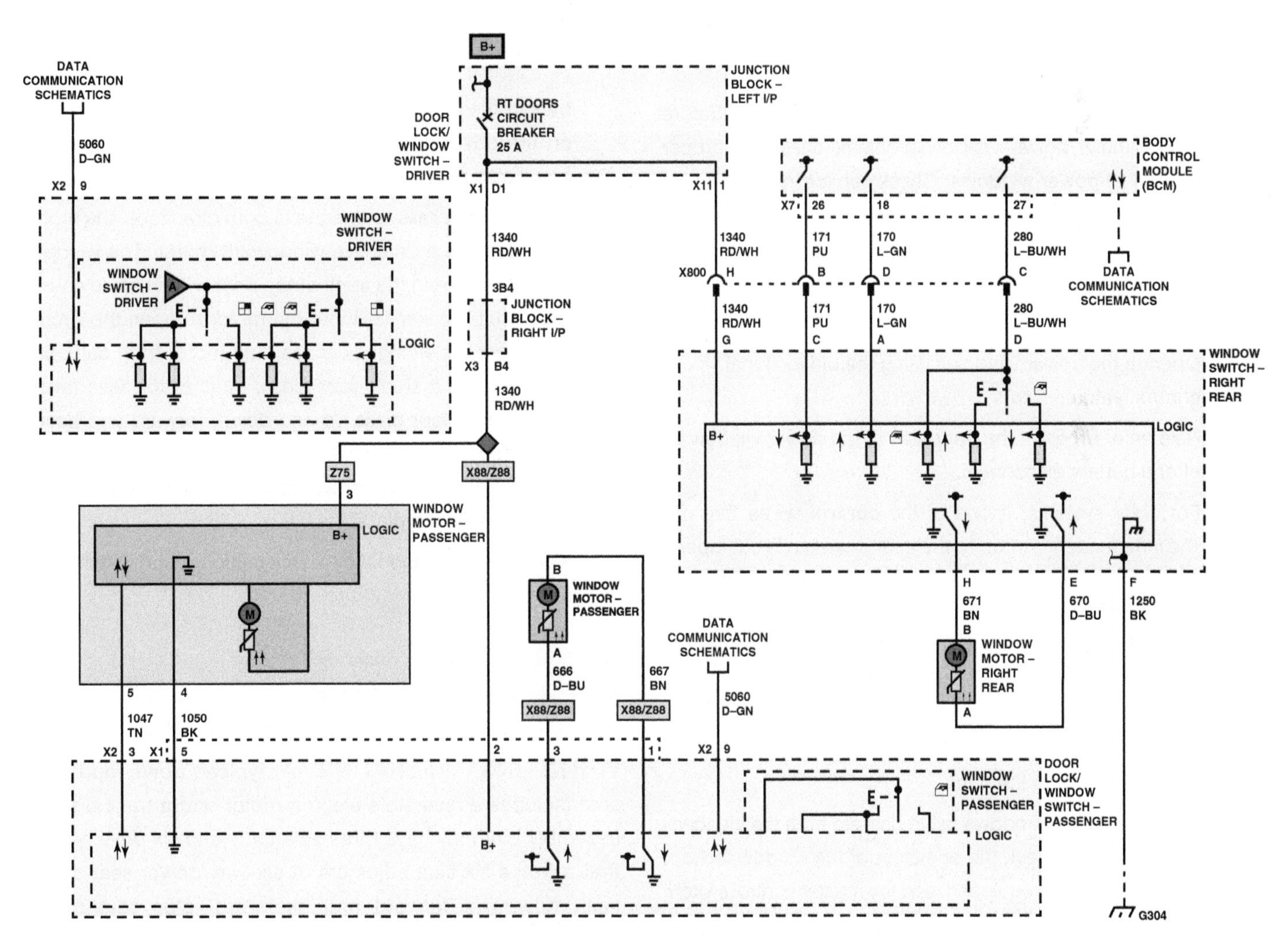

FIGURE 22–16 The power windows can be operated directly by the related switch or by way of a data connection. This diagram shows two vehicle options, one with express-down windows (Z75) or without express-down (X88/Z88). (Courtesy of General Motors)

FIGURE 22–17 A power seat uses electric motors under the seat, which drive cables that extend to operate screw jacks (up and down) or gears to move the seat forward and back.

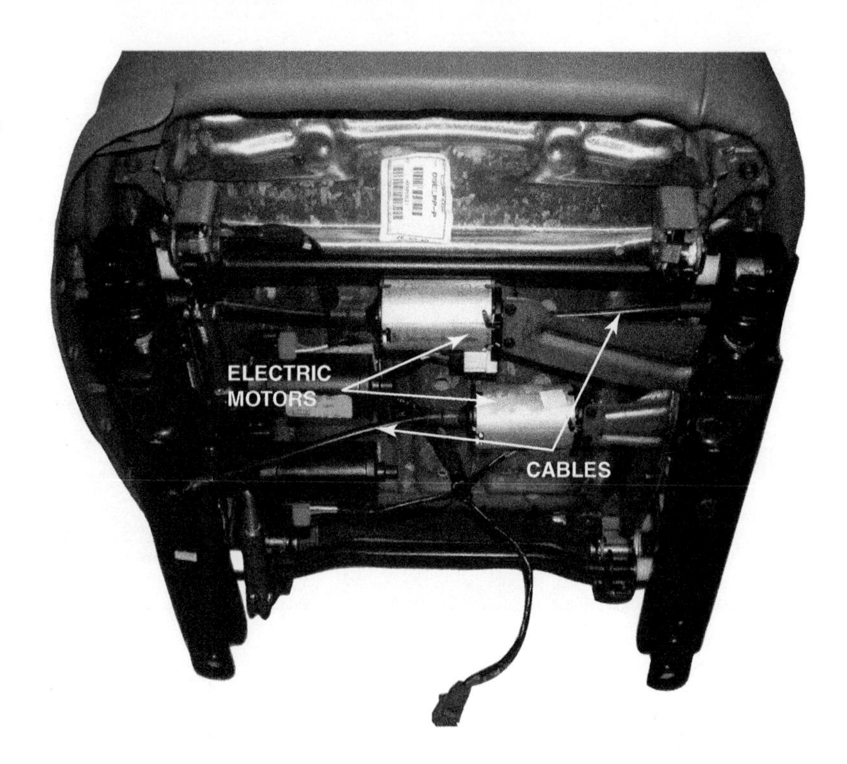

TROUBLESHOOTING POWER WINDOWS Before troubleshooting a power window problem, check for proper operation of all power windows. Check service information for the exact procedure to follow. In a newer system, a scan tool can be used to perform the following:

- Check for B (body) or U (network) diagnostic trouble codes (DTCs).
- Operate the power windows using the bidirectional control feature.
- Relearn or program the operation of the power windows after a battery disconnect.

For older systems, if one of the **control wires** that run from the independent switch to the master switch is cut (open), the power window may operate in only one direction. The window may go down but not up, or vice versa. However, if one of the **direction wires** that run from the independent switch to the motor is cut (open), the window will not operate in either direction. The direction wires and the motor must be electrically connected to permit operation and change of direction of the electric lift motor in the door.

1. If *both* rear door windows fail to operate from the independent switches, check the operation of the window lockout (if the vehicle is so equipped) and the master control switch.
2. If one window can move in one direction only, check for continuity in the control wires (wires between the independent control switch and the master control switch).
3. If *all* windows fail to work or fail to work occasionally, check, clean, and tighten the ground wire(s) located either behind the driver's interior door panel or under the dash on the driver's side. A defective fuse or circuit breaker could also cause all the windows to fail to operate.
4. If one window fails to operate in both directions, the problem could be a defective window lift motor. The window could be stuck in the track of the door, which could cause the circuit breaker built into the motor to open the circuit to protect the wiring, switches, and motor from damage. To check for a stuck door glass, attempt to move (even slightly) the door glass up and down, forward and back, and side to side. If the window glass can move slightly in all directions, the power window motor should be able to at least move the glass.
5. Always refer to and follow service information when diagnosing power window circuits.

POWER SEATS

PARTS AND OPERATION A typical power-operated seat includes a reversible electric motor and a transmission assembly that may have three solenoids and six *drive cables* that turn the six seat adjusters. A six-way power seat offers seat movement forward and backward, plus seat cushion movement up and down at the front and the rear. The drive cables are similar to speedometer cables because they rotate inside a cable housing and connect the power output of the seat transmission to a gear or screw jack assembly that moves the seat. ● **SEE FIGURE 22–17.**

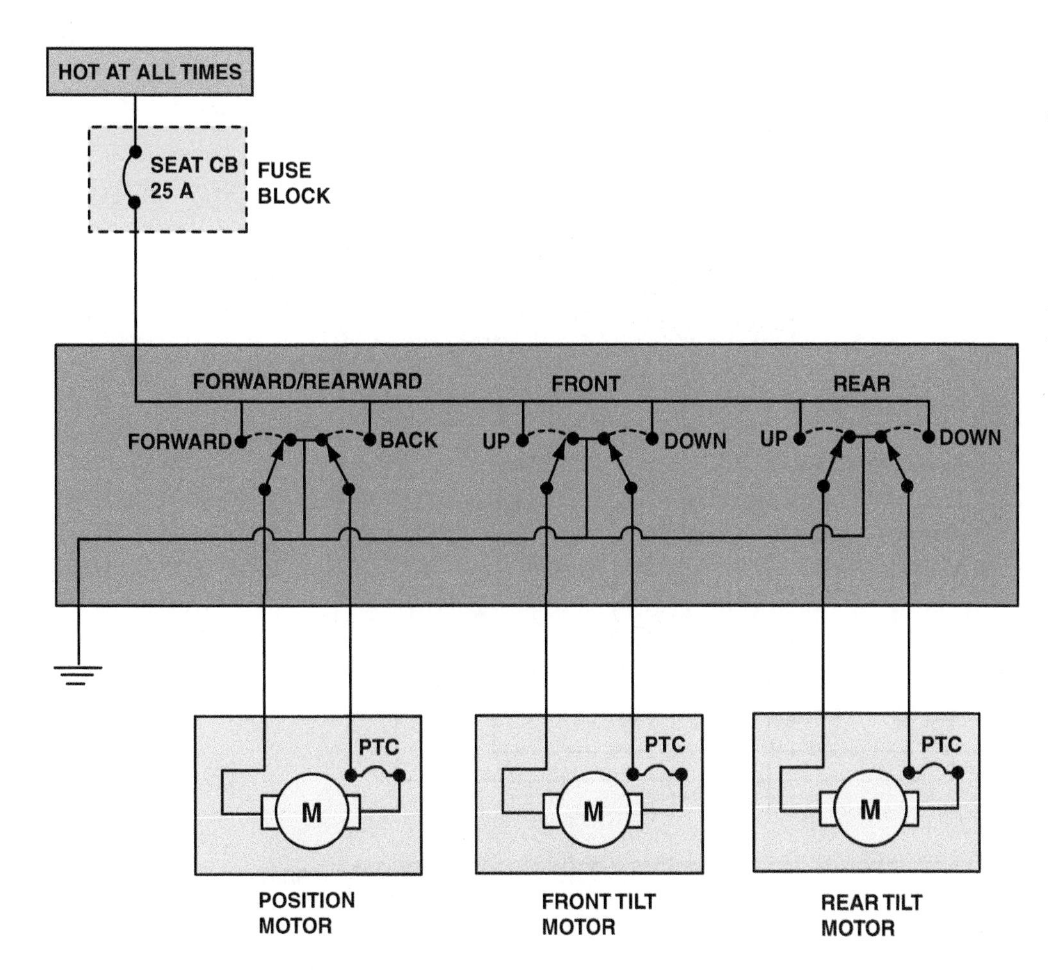

FIGURE 22–18 A typical power seat circuit diagram. Notice that each motor has a built-in electronic (solid-state) PTC circuit protector. The seat control switch can change the direction in which the motor(s) runs by reversing the direction in which the current flows through the motor.

A **screw jack assembly** is often called a *gear nut*. It is used to move the front or back of the seat cushion up and down.

A **rubber coupling**, usually located between the electric motor and the transmission, prevents electric motor damage in the event of a jammed seat. This coupling is designed to prevent motor damage.

Most power seats use a permanent magnet motor that can be reversed by simply reversing the polarity of the current sent to the motor by the seat switch. ● **SEE FIGURE 22–18.**

POWER SEAT MOTOR(S) Most PM motors have a built-in circuit breaker or PTC circuit protector to protect the motor from overheating. Many Ford power seat motors use three separate armatures inside one large permanent magnet field housing. Some power seats use a series-wound electric motor with two separate field coils, one field coil for each direction of rotation. This type of power seat motor typically uses a relay to control the direction of current from the seat switch to the corresponding field coil of the seat motor. This type of power seat can be identified by the "click" heard when the seat switch is changed from up to down or front to back, or vice versa. The click is the sound of the relay switching the field coil current. Some power seats use as many as eight separate PM motors that operate all functions of the seat, including headrest height, seat length, and side bolsters, in addition to the usual six-way power seat functions.

NOTE: Some power seats use a small air pump to inflate a bag (or bags) in the lower part of the back of the seat, called the **lumbar**, because it supports the lumbar section of the spine. The lumbar section of the seat can also be changed, using a lever or knob that the driver can move to change the seat section for the lower back.

MEMORY SEAT Memory seats use a potentiometer to sense the position of the seat. The seat position can be programmed into the body control module (BCM) or memory seat module and stored by position number 1, 2, or 3. The driver pushes the desired button and the seat moves to the stored position. ● **SEE FIGURE 22–19.**

TROUBLESHOOTING POWER SEATS Power seats are usually wired from the fuse panel so they can be operated without having to turn the ignition switch to on (run). If a power seat does not operate or make any noise, the circuit breaker (or fuse,

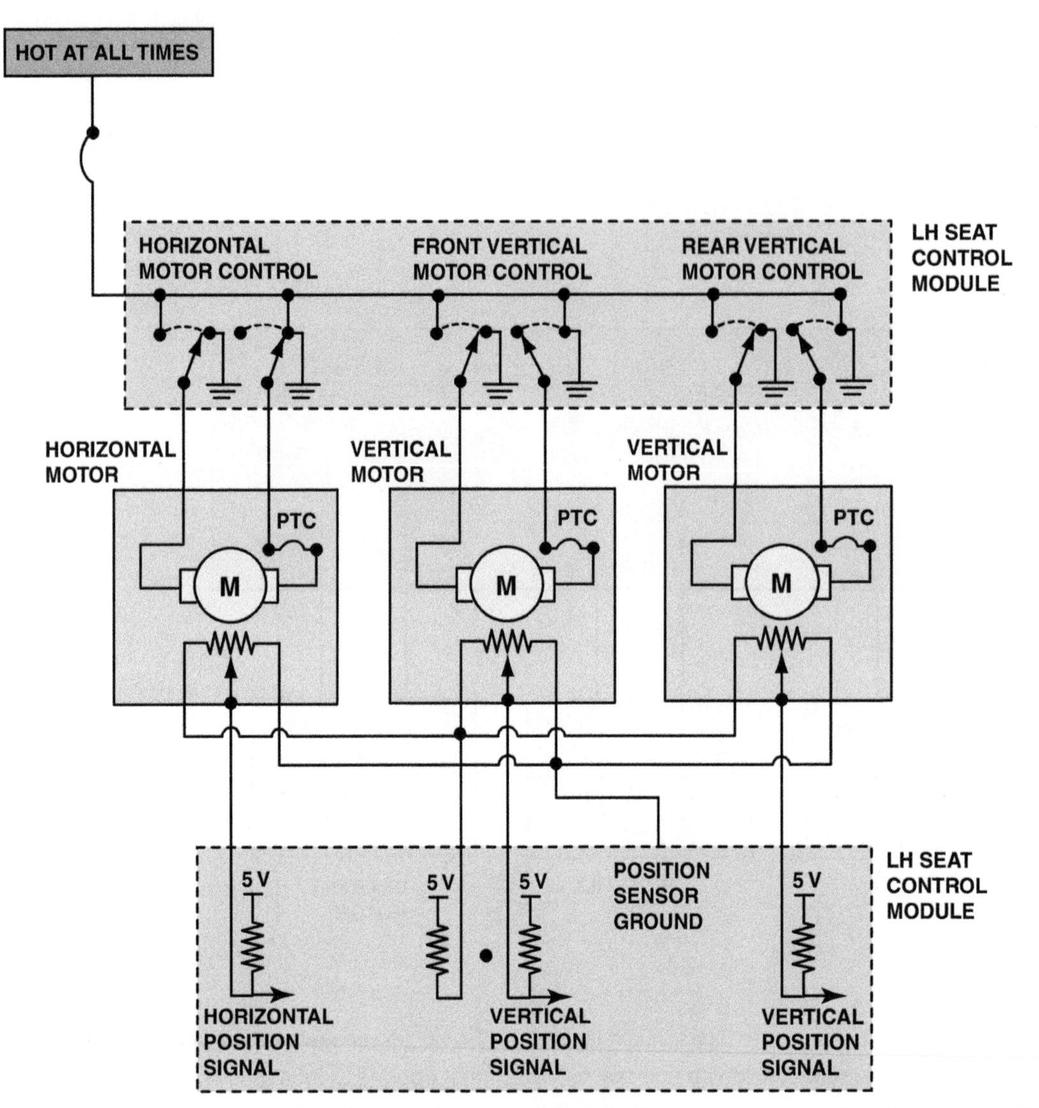

FIGURE 22–19 A typical memory seat module showing the three-wire potentiometer used to determine seat position.

if the vehicle is so equipped) should be checked first. The steps usually include the following:

STEP 1 Check service information for the exact procedure to follow when diagnosing power seats. If the seat relay clicks, the circuit breaker is functioning, but the relay or electric motor may be defective.

STEP 2 Remove the screws or clips that retain the controls to the inner door panel or seat and check for voltage at the seat control.

STEP 3 Check the ground connection(s) at the transmission and clutch control solenoids (if equipped). The solenoids must be properly grounded to the vehicle body for the power seat circuit to operate.

If the power seat motor runs but does not move the seat, the most likely fault is a worn or defective rubber clutch sleeve between the electric seat motor and the transmission.

Easy Exit Seat Programming

Some vehicles are equipped with memory seats that allow the seat to move rearward when the ignition is turned off to allow easy exit from the vehicle. Vehicles equipped with this feature include an *exit/entry* button that is used to program the desired exit/entry position of the seat for each of two drivers.

If the vehicle is not equipped with this feature and only one driver primarily uses the vehicle, the second memory position can be programmed for easy exit and entry. Simply set position 1 to the desired seat position and position 2 to the entry/exit position. Then, when exiting the vehicle, press memory 2 to allow easy exit and easy entry the next time. Press memory 1 when in the vehicle to return the seat memory to the desired driving position.

TECH TIP

What Every Driver Should Know About Power Seats

Power seats use an electric motor or motors to move the position of the seat. These electric motors turn small cables that operate mechanisms that move the seat. *Never* place rags, newspapers, or any other object under a power seat. Even ice scrapers can get caught between moving parts of the seat and can often cause serious damage or jamming of the power seat.

If the seat relay clicks but the seat motor does not operate, the problem is usually a defective seat motor or defective wiring between the motor and the relay. If the power seat uses a motor relay, the motor has a double reverse-wound field for reversing the motor direction. This type of electric motor must be properly grounded. Permanent magnet motors do not require grounding for operation.

NOTE: Power seats are often difficult to service because of restricted working room. If the entire seat cannot be removed from the vehicle because the track bolts are covered, attempt to remove the seat from the top of the power seat assembly. These bolts are almost always accessible regardless of seat position.

ELECTRICALLY HEATED SEATS

PARTS AND OPERATION Heated seats use electric heating elements in the seat bottom, as well as in the seat back in many vehicles. The heating element is designed to warm the seat and/or back of the seat to about 100°F (37°C) or close to normal body temperature (98.6°F). Many heated seats also include a high-position or a variable temperature setting, so the temperature of the seats can therefore be as high as 110°F (44°C).

A temperature sensor in the seat cushion is used to regulate the temperature. The sensor is a variable resistor, which changes with temperature and is used as an input signal to a heated seat control module. The heated seat module uses the seat temperature input, as well as the input from the high–low (or variable) temperature control, to turn the current on or off to the heating element in the seat. Some vehicles are equipped with heated seats in both the rear and the front seats.

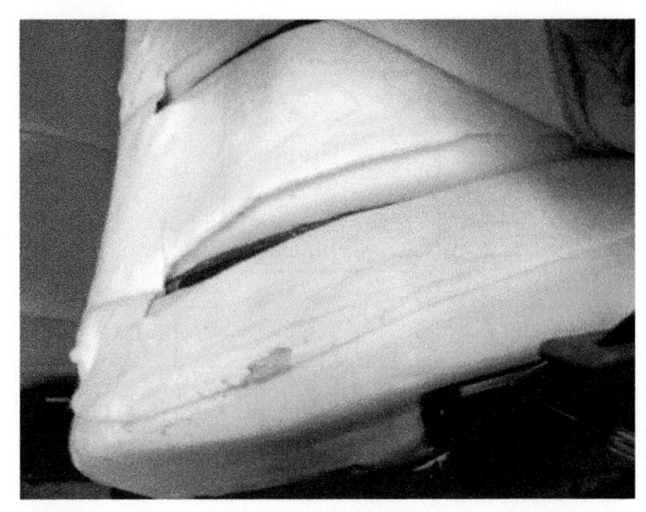

FIGURE 22–20 The heating element of a heated seat is a replaceable part, but service requires that the upholstery be removed. The yellow part is the seat foam material and the entire white cover is the replaceable heating element. This is then covered by the seat material.

DIAGNOSIS AND SERVICE When diagnosing a heated seat concern, start by verifying that the switch is in the on position and that the temperature of the seat is below normal body temperature. Using service information, check for power and ground at the control module and to the heating element in the seat. Most vehicle manufacturers recommend replacing the entire heating element if it is defective. ● **SEE FIGURE 22–20.**

HEATED AND COOLED SEATS

PARTS AND OPERATION Most electrically heated and cooled seats use a thermoelectric device (TED) located under the seat cushion and seat back. The thermoelectric device consists of positive and negative connections between two ceramic plates. Each ceramic plate has copper fins to allow the transfer of heat to air passing over the device and directed into the seat cushion. The thermoelectric device uses the **Peltier effect**, named after the inventor, Jean C. A. Peltier, a French clockmaker. When electrical current flows through the module, one side is heated and the other side is cooled. Reversing the polarity of the current changes the side to be heated. ● **SEE FIGURE 22–21.**

Most vehicles equipped with heated and cooled seats use two modules per seat, one for the seat cushion and one for the

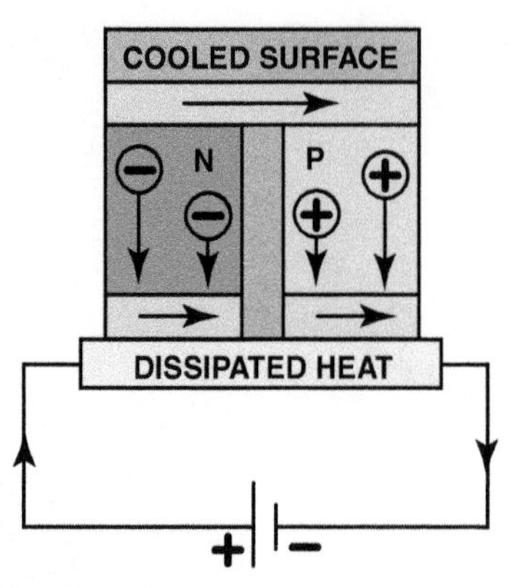

FIGURE 22–21 A Peltier effect device is capable of heating or cooling, depending on the polarity of the applied current.

TECH TIP

Check the Seat Filter

Heated and cooled seats often use a filter to trap dirt and debris to help keep the air passages clean. If a customer complains of a slow heating or cooling of the seat, check the air filter and replace or clean as necessary. Check service information for the exact location of the seat filter and for instructions on how to remove and/or replace it.

seat back. When the heated and cooled seats are turned on, air is forced through a filter and then through the thermoelectric modules. The air is then directed through passages in the foam of the seat cushion and seat back. Each thermoelectric device has a temperature sensor, called a thermistor. The control module uses sensors to determine the temperature of the fins in the thermoelectric device so the controller can maintain the set temperature.

DIAGNOSIS AND SERVICE The first step in any diagnosis is to verify that the heated–cooled seat system is not functioning. Check the owner's manual or service information for the specified procedures. If the system works partially, check the air filter, usually located under the seat for each thermoelectric device. A partially clogged filter can restrict airflow and reduce the heating or cooling effect. If the system control indicator light is not on or the system does not work at all, check for power and ground at the thermoelectric devices. Always follow the vehicle manufacturer's recommended diagnosis and service procedures.

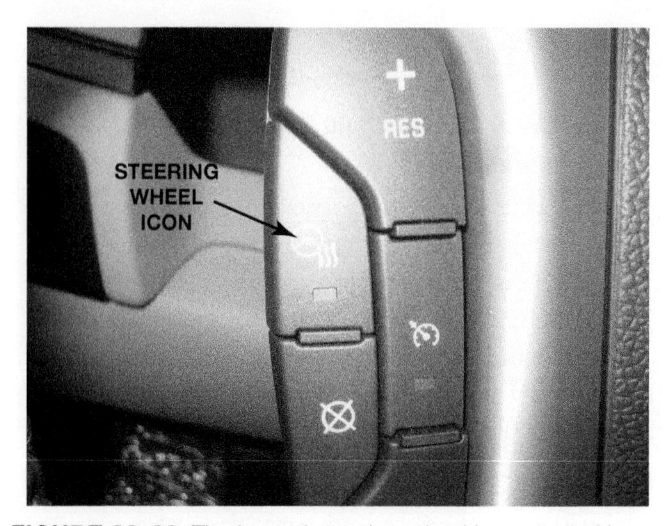

FIGURE 22–22 The heated steering wheel is controlled by a switch on the steering wheel in this vehicle.

HEATED STEERING WHEEL

PARTS INVOLVED A heated steering wheel usually consists of the following components:

- Steering wheel with a built-in heater in the rim
- Heated steering wheel control switch
- Heated steering wheel control module

OPERATION When the steering wheel heater control switch is turned on, a signal is sent to the control module and electrical current flows through the heating element in the rim of the steering wheel. ● **SEE FIGURE 22–22.**

The system remains on until the ignition switch is turned off or the driver turns off the control switch. The temperature of the steering wheel is usually calibrated to stay at about 90°F (32°C), and it requires three to four minutes to reach that temperature depending on the outside temperature.

DIAGNOSIS AND SERVICE Diagnosis of a heated steering wheel starts with verifying that the heated steering wheel is not working as designed.

NOTE: Most heated steering wheels do not work if the temperature inside the vehicle is about 90°F (32°C) or higher.

If the heated steering wheel is not working, follow the service information testing procedures which would include a check of the following:

1. Check the heated steering wheel control switch for proper operation. This is usually done by checking for voltage at

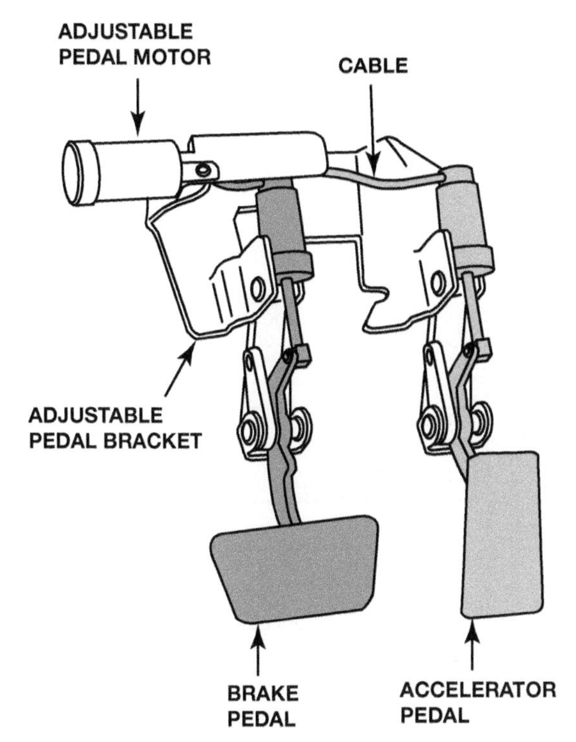

FIGURE 22–23 A typical adjustable pedal assembly. Both the accelerator and the brake pedal can be moved forward and rearward by using the adjustable pedal position switch.

both terminals of the switch. If voltage is available at only one of the two terminals of the switch and the switch has been turned on and off, an open (defective) switch is indicated.

2. Check for voltage and ground at the terminals leading to the heating element. If voltage is available at the heating element and the ground has less than 0.2 volt drop to a good chassis ground, the heating element is defective. The entire steering wheel has to be replaced if the element is defective.

Always follow the vehicle manufacturer's recommended diagnosis and testing procedures.

ADJUSTABLE PEDALS

PURPOSE AND FUNCTION **Adjustable pedals**, also called **electric adjustable pedals (EAP)**, place the brake pedal and the accelerator pedal on movable brackets that are motor operated. A typical adjustable pedal system includes the following components:

- **Adjustable pedal position switch.** Allows the driver to position the pedals
- **Adjustable pedal assembly.** Includes the motor, threaded adjustment rods, and a pedal position sensor

● **SEE FIGURE 22–23.**

FIGURE 22–24 Electrically folded mirror in the folded position.

FIGURE 22–25 The electric mirror control is located on the driver's side door panel on this Cadillac Escalade.

REAL WORLD FIX

The Case of the Haunted Mirrors

The owner complained that while driving, either one or the other outside mirror would fold in without any button being depressed. Unable to verify the customer concern, the service technician looked at the owner's manual to find out exactly how the mirrors were supposed to work. In the manual, a caution statement said that if the mirror is electrically folded inward and then manually pushed out, the mirror will not lock into position. The power folding mirrors must be electrically cycled outward, using the mirror switches to lock them in position. After cycling both mirrors inward and outward electrically, the problem was solved. ● **SEE FIGURES 22–24 AND 22–25.**

TECH TIP

Check the Remote

The memory function may be programmed to a particular key fob remote, which would command the adjustable pedals to move to the position set in memory. Always check both remote settings before attempting to repair a problem that may not be a problem.

The position of the pedals, as well as the position of the seat system, is usually included as part of the memory seat function and can be set for two or more drivers.

DIAGNOSIS AND SERVICE The first step when there is a customer concern about the functioning of the adjustable pedals is to verify that the unit is not working as designed. Check the owner manual or service information for the proper operation. Follow the vehicle manufacturer's recommended troubleshooting procedure. Many diagnostic procedures include the use of a factory scan tool with bidirectional control capabilities to test this system.

OUTSIDE FOLDING MIRRORS

Mirrors that can be electrically folded inward are a popular feature, especially on larger sport utility vehicles. A control inside is used to fold both mirrors inward when needed, such as when entering a garage or close parking spot. For diagnosis and servicing of outside folding mirrors, check service information for details.

ELECTRIC POWER DOOR LOCKS

Electric power door locks use a permanent magnet (PM) reversible motor to lock or unlock all vehicle door locks from a control switch or switches.

The electric motor uses a built-in circuit breaker and operates the lock-activating rod. PM reversible motors do not require grounding because, as with power windows, the motor control is determined by the polarity of the current through the two motor wires. ● **SEE FIGURE 22–26.**

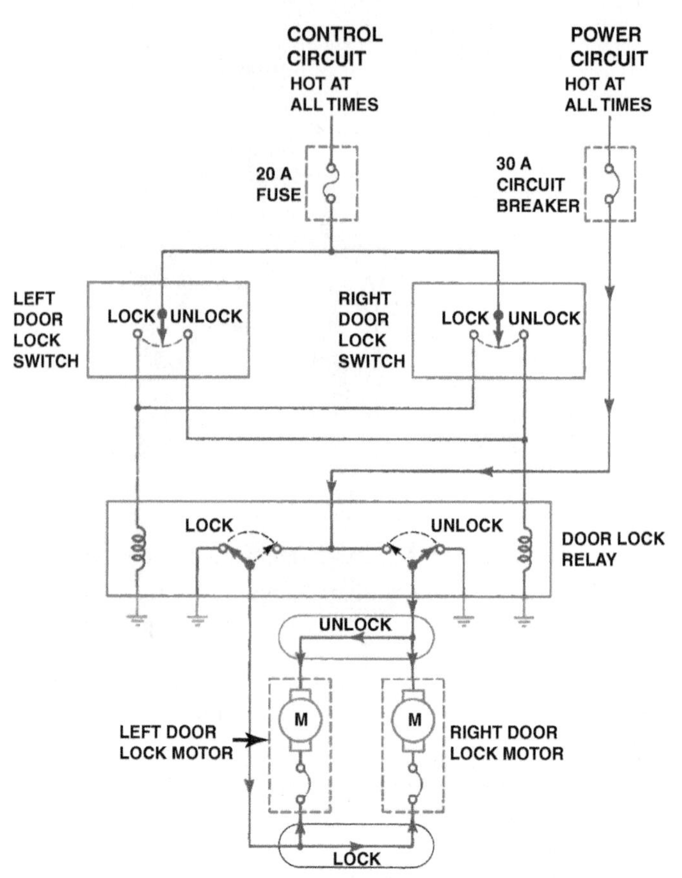

FIGURE 22–26 A typical electric power door lock circuit diagram. Note that the control circuit is protected by a fuse, whereas the power circuit is protected by a circuit breaker. As with the operation of power windows, power door locks typically use reversible permanent magnet (PM) nongrounded electric motors. These motors are geared mechanically to the lock–unlock mechanism.

Some two-door vehicles do *not* use a power door lock relay because the current flow for only two PM motors can be handled through the door lock switches. However, most four-door vehicles and vans with power locks on rear and side doors use a relay to control the current flow necessary to operate four or more power door lock motors. The door lock relay is controlled by the door lock switch and is commonly the location of the one and only *ground* connection for the entire door lock circuit.

KEYLESS ENTRY

Even though some vehicles use a keypad located on the outside of the door, most keyless entry systems use a wireless transmitter built into the key or key fob. A **key fob** is a decorative tab or item on a key chain. ● **SEE FIGURE 22–27.**

FIGURE 22–27 A key fob remote with the cover removed showing the replaceable battery.

The transmitter broadcasts a signal that is received by the electronic control module, which is generally mounted in the trunk or under the instrument panel. ● SEE FIGURE 22–28.

The electronic control unit sends a voltage signal to the door lock actuator(s) located in the doors. Generally, if the transmitter unlock button is depressed once, only the driver's door is unlocked. If the unlock button is depressed twice, then all doors unlock.

ROLLING CODE RESET PROCEDURE Many keyless remote systems use a rolling code type of transmitter and receiver. In a conventional system, the transmitter emits a certain fixed frequency, which is received by the vehicle control module. This single frequency can be intercepted and rebroadcast to open the vehicle.

A rolling code type of transmitter emits a different frequency every time the transmitter button is depressed and then rolls over to another frequency so that it cannot be intercepted. Both the transmitter and the receiver must be kept in synchronized order so that the remote will function correctly.

If the transmitter is depressed when it is out of range from the vehicle, the proper frequency may not be recognized by the receiver, which did not roll over to the new frequency when the transmitter was depressed. If the transmitter does not work, try to resynchronize the transmitter to the receiver by depressing and holding both the lock and the unlock button for 10 seconds when within range of the receiver.

KEYLESS ENTRY DIAGNOSIS A small battery powers the transmitter, and a weak battery is a common cause of remote power locks failing to operate. If the keyless entry

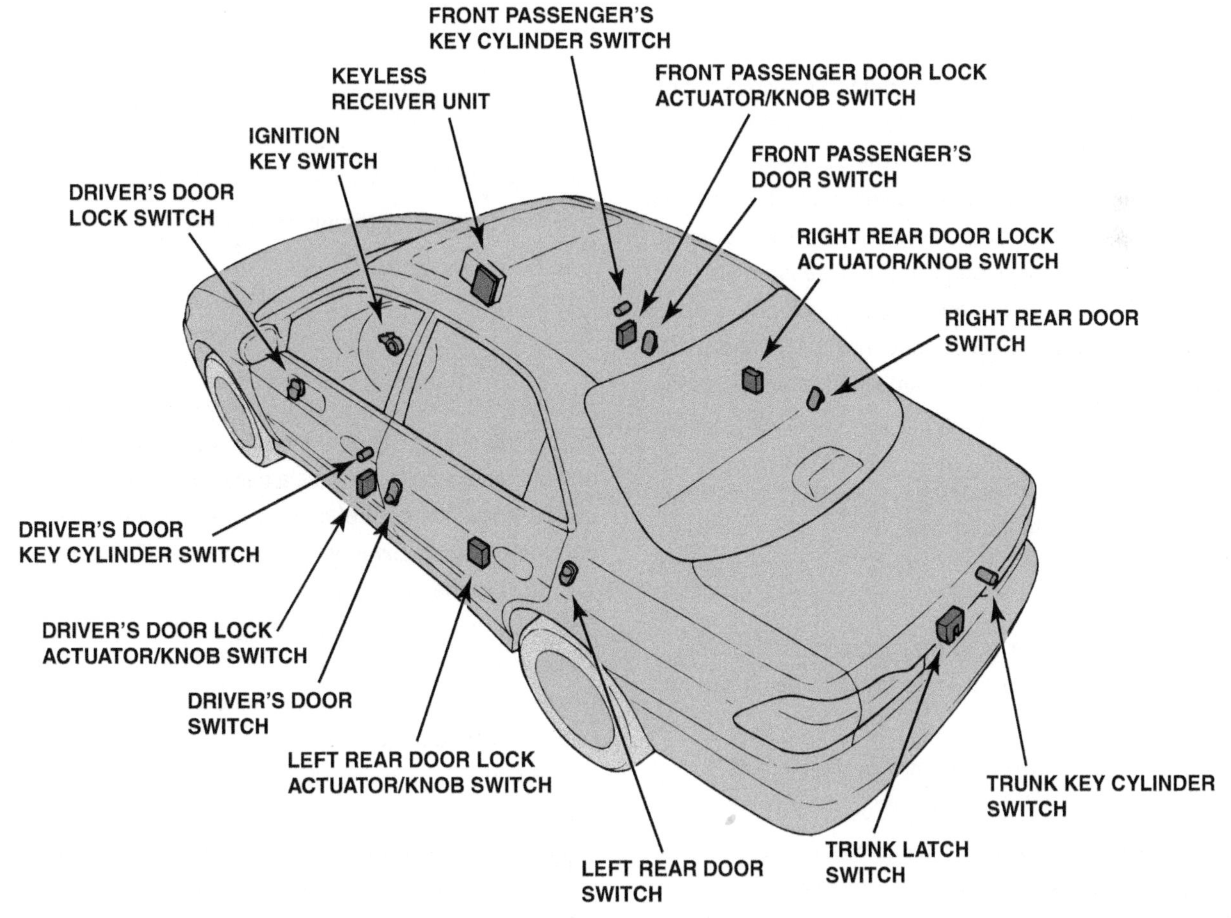

FIGURE 22–28 A typical vehicle showing the location of the various components of the remote keyless entry system.

system fails to operate after the transmitter battery has been replaced, check the following items:

- Mechanical binding in the door lock
- Low vehicle battery voltage
- Blown fuse
- Open circuit to the control module
- Defective control module
- Defective transmitter

PROGRAMMING A NEW REMOTE If a new or additional remote transmitter is to be used, it must be programmed to the vehicle. The programming procedure varies and may require the use of a scan tool. Check service information for the exact procedure to follow. ● **SEE CHART 22–1**.

ANTITHEFT SYSTEMS

PARTS AND OPERATION Antitheft devices flash lights or sound an alarm if the vehicle is broken into or vandalized. In addition to the alarm, some systems prevent the engine from starting by disabling the starter, ignition, or fuel system once the antitheft device is activated. Others permit the engine to start, but then disable it after several seconds. Switches in the doorjambs, trunk, and hood provide an input signal to the control module should an undesirable entry occur on a typical system. Some antitheft systems are more complex and also have electronic sensors that trigger the alarm if there is a change in battery current draw, a violent vehicle motion, or if glass is broken. These sensors also provide an input signal to the control module, which may be a separate antitheft unit or incorporated into the PCM or BCM. ● **SEE FIGURE 22–29** for an example of a shock sensor used in an antitheft alarm system.

ANTITHEFT SYSTEM DIAGNOSIS Most factory-installed antitheft systems are integrated with several other circuits to form a complex, multiple-circuit system. The major steps are as follows:

1. It is essential to have accurate diagrams, specifications, and test procedures for the specific model being serviced.
2. The easiest way to reduce circuit complexity is to use the wiring diagram to break the entire system into its subcircuits, then check only those related to the problem.
3. If any step indicates that a subcircuit is not complete, check the power source, ground, components, and wiring in that subcircuit.

FIGURE 22–29 A shock sensor used in alarm and antitheft systems. If the vehicle is moved, the magnet will move relative to the coil, inducing a small voltage that will trigger the alarm.

Many systems use a computer chip in the plastic part of the key. Most systems are electronically regulated and have a self-diagnostic program. This self-diagnostic program is generally accessed and activated using a scan tool. Diagnostic and test procedures are similar as for any of the other electronic control systems used on the vehicle.

ANTITHEFT SYSTEM TESTING AND SERVICE Before performing any diagnostic checks, make sure that all of the following electrical devices function correctly:

- Parking and low-beam headlights
- Dome and courtesy lights
- Horn
- Electric door locks

Circuit information from these devices often provides basic inputs to the control module. If a problem is detected in any of these circuits, such as a missing signal or a signal that is out of range, the control module disables the antitheft system and may record a diagnostic trouble code (DTC).

If all of the previously mentioned devices are operational, check all the circuits leading to the antitheft control module. Make sure all switches are in their normal or off positions. Doorjamb switches complete the ground circuit when a door is opened. ● **SEE FIGURE 22–30**.

Frequently, corrosion that builds up on the switch contacts prevents the switch from operating properly. Conduct voltage drop tests to isolate faulty components and circuit problems. Repair as needed and retest to confirm that the system is operational. Follow procedures from the manufacturer to clear

MAKE/MODEL	NOTES	PROCEDURE
Buick Rendezvous Lucerne LaCrosse **Chevrolet** Blazer Impala Monte Carlo Uplander **Pontiac** Grand Prix Montana **Saturn** Relay	A scan tool is required. A total of four transmitters can be learned. All transmitters to be programmed must be programmed at the same time. Activating program mode erases previously learned codes.	1. Install a scan tool and access the BCM Special Functions, Lift Gate Module (LGM), or Module Setup; Program Key Fobs menu. 2. Press the start key on the scan tool. 3. Press and hold both the lock and unlock buttons on the first transmitter. Within 5 to 10 seconds the scan tool will report that the transmitter is programmed. 4. Repeat step 3 to program up to four transmitters. 5. Turn off and remove the scan tool to exit programming mode.
Buick Rainier **Cadillac** Escalade **Chevrolet** C/K Trucks Suburban Tahoe Trailblazer **Saab** 9-7 (some)	Fobs can also be programmed with a scan tool. All fobs to be used must be programmed at the same time. The first fob learned will be fob 1 and fob 2 will be the second fob learned.	1. Enter the vehicle and close all the doors. 2. Insert the key into the ignition lock. 3. Press and hold the door unlock switch, turn the ignition on, off, and then release the unlock switch. 4. The door locks will cycle one time to confirm programming mode. 5. Press and hold the lock and unlock buttons on the key fob for about 15 seconds. 6. The locks will cycle once when the fob has been learned. 7. Repeat steps 5 and 6 to program any additional fobs. 8. Turn the ignition key to run, to exit the programming mode.
Cadillac CTS SRX	All programmed key fobs will be erased. All transmitters to be programmed must be relearned during this procedure. Up to four fobs can be programmed. The first to be learned will be fob 1 and the second to be learned will be fob 2.	1. Install the scan tool and turn the ignition on. 2. Navigate to the Body, RFA (or RCDLR), Special Functions; Program Key Fobs menu. 3. Follow the directions on the scan tool to program the transmitters.
Cadillac Deville Seville **Pontiac** Bonneville Grand Am	Up to four transmitters can be programmed. All fobs to be used must be programmed at the same time. The first fob learned will be fob 1 and the second that is learned will be fob 2.	1. Install a scan tool and turn on the ignition. 2. Navigate to the Remote Function Actuator (RFA) module: Special function, Program Key Fobs menu to activate program mode. 3. The doors will lock and unlock to indicate programming mode. 4. Press and hold the lock and unlock buttons on the fob. The door locks will cycle to indicate the fob has been learned. 5. Repeat step 4 for any additional fobs. 6. To exit programming mode, turn off and remove the scan tool.

CHART 22-1

CONTINUED

Remote keyless programming steps for popular vehicles. Procedures may also apply to similar vehicles by the same manufacturer. Always refer to service information for specific vehicles.

MAKE/MODEL	NOTES	PROCEDURE
Cadillac STS XLR **Chevrolet** Corvette	A scan tool can also be used to program key fobs. This procedure will take 30 minutes to complete. All programmed key fobs will be erased. All transmitters to be programmed must be relearned during this procedure. Up to four fobs can be programmed. The first to be learned will be fob 1 and the second to be learned will be fob 2.	1. Start with the vehicle off. 2. Place the fob to be learned in the console pocket with the buttons facing forward. 3. Insert the vehicle key into the driver's door lock cylinder and cycle the key five times within 5 seconds. The DIC will display "OFF/ACC TO LEARN." 4. Press the OFF/ACC part of the ignition button. 5. The DIC will display "WAIT 10 MINUTES," then count down to zero, 1 minute at a time. The display will change to "OFF/ACC TO LEARN." 6. Repeat steps 4 and 5 two more times for a total of 30 minutes. 7. When the DIC displays "OFF/ACC TO LEARN" for the fourth time, press the OFF/ACC button again; the DIC will display "READY FOR FOB 1." 8. When fob 1 has been learned, a beep will be heard and the DIC will display "READY FOR FOB 2." 9. Remove fob 1 from the pocket and insert fob 2. A beep will be heard when that fob has been learned. 10. Repeat steps 8 and 9 for additional fobs. 11. To exit programming, press the OFF/ACC portion of the ignition button.
Chevrolet Cavalier Equinox Malibu SSR S/T Trucks **Saab** 9–7 (some models) **Saturn** Vue	A scan tool is required. Up to four transmitters can be programmed. On vehicles with personalization features, the transmitters are numbered 1 and 2. The first transmitter programmed will become driver 1 and the second will become driver 2.	1. Install the scan tool and navigate to the BCM or RFA menu, Special Functions; select Program Key Fobs. 2. Select Add/Replace Key Fob to program a new or additional fob. 3. Select Clear Memory and Program All Fobs option to replace all fobs or to recode driver 1 and driver 2 fobs. 4. Follow the scan tool instructions to complete the programming.
Chevrolet Venture van GM "U" vans	All fobs to be used must be programmed at the same time. Up to four transmitters can be programmed. If the BCM displays DTCs in step 5, they may have to be resolved before programming can continue.	1. With the ignition key out of the ignition, remove the BCM PRGRM fuse from the passenger side fuse block. 2. Enter the vehicle and close all doors. 3. Insert the key and turn the ignition to ACC. 4. The seat belt indicator and chime will activate two, three, or four times, depending on the type of BCM in the vehicle. 5. Turn the key off and then back to ACC within 1 second. If the BCM has any stored DTCs, they will be displayed by the chime and belt indicator at this time. 6. Open and close any door. The chime will sound to indicate programming mode. 7. Press and hold the fob lock and unlock buttons for about 14 seconds. The BCM will sound the chime when the fob has been learned. 8. Repeat step 7 for up to four total transmitters. 9. After programming, remove the ignition key and replace the BCM PRGRM fuse.

CONTINUED

MAKE/MODEL	NOTES	PROCEDURE
Pontiac G6 **Saturn** Ion L300	A scan tool is used to program key fobs. Up to four transmitters can be programmed. If any key fob is programmed, all fobs must be programmed at the same time. On vehicles with personalization features, the transmitters are numbered 1 and 2. The first transmitter programmed will become driver 1 and the second will become driver 2.	1. Install the scan tool and navigate to the Program Key Fobs menu. 2. Select the number of fobs to be programmed. 3. Press and hold the lock and unlock buttons on the first fob to be programmed. The locks should cycle to indicate okay. NOTE: This fob becomes driver 1 key fob. 4. Repeat step 3 for the second fob. This fob becomes driver 2 key fob. 5. Repeat step 3 for any other key fobs to be programmed. 6. Turn off and remove the scan tool to exit programming.
Saab 9-2	Up to four transmitters can be programmed.	1. Sit in the driver's seat and close all doors. 2. Open and close the driver's door. 3. Turn the ignition switch from "on" to lock, 10 times within 15 seconds. The horn will chirp to indicate programming mode. 4. Open and close the driver's door. 5. Press any button on the fob to be programmed. 6. The horn will chirp two times to indicate that the transmitter has been learned. 7. Repeat steps 4, 5, and 6 for any additional transmitters. 8. To exit from programming mode, remove the key from the ignition. The horn should chirp three times to confirm.

CONTINUED

DTC records and then run the self-diagnostic program to verify repairs. Some system diagnostic procedures specify the use of special testers. ● **SEE FIGURE 22–31**.

● **SEE CHART 22–2** for programming procedures for selected vehicles.

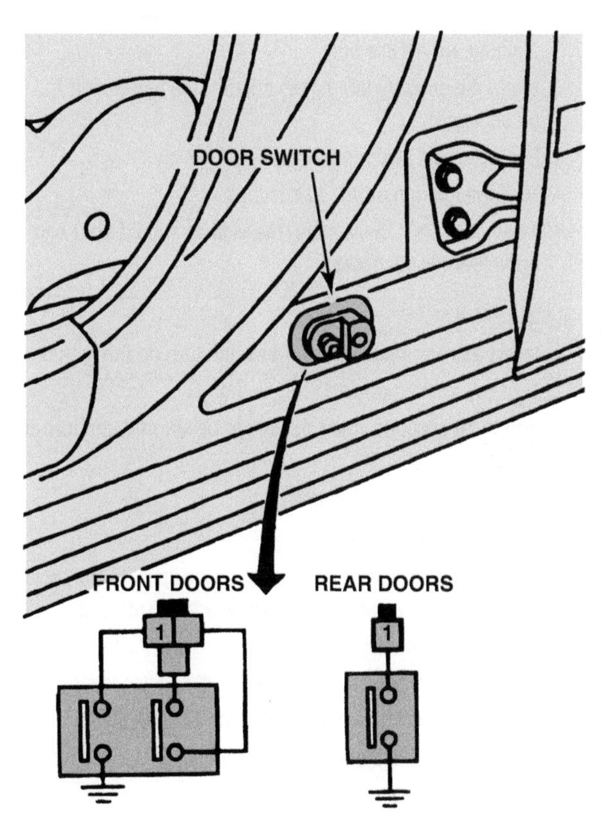

FIGURE 22–30 Door switches, which complete the ground circuit with the door open, are a common source of high resistance.

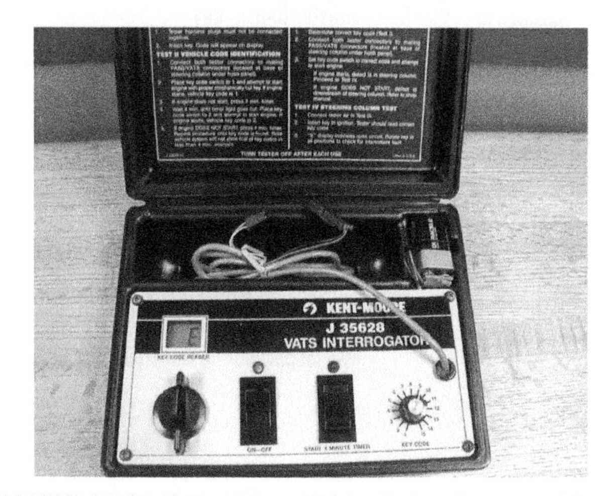

FIGURE 22–31 A special tool is needed to diagnose older General Motors VATS security system and special keys that contain a resistor pellet.

MAKE/MODEL	NOTES	PROCEDURES
General Motors Passkey Passkey II (except vehicles with BCM)	The Passkey decoder will learn the first pellet read when the decoder module is first installed. This learned value cannot be changed. A Passkey Interrogator special tool is needed to read key pellet resistance when replacing keys. The tool will read out a code number related to the pellet resistance. PELLET CODE — RESISTANCE 1 — 402 2 — 523 3 — 681 4 — 887 5 — 1,130 6 — 1,470 7 — 1,870 8 — 2,370 9 — 3,010 10 — 3,740 11 — 4,750 12 — 6,040 13 — 7,500 14 — 9,530 15 — 11,800	NEW DECODER MODULE 1. Install the new decoder module. 2. Insert the key and start the vehicle to program the pellet code into the new module. DUPLICATE KEY 1. Use the Interrogator tool to read the existing key code. 2. Obtain a key with the matching pellet code and cut the key to match the original key. LOST KEY 1. The Interrogator tool must be used to determine the stored code. 2. Cut a blank key so that the ignition can be turned. 3. Access the lock cylinder 2 wire connector and connect it to the Interrogator. 4. Alternately select each of the 15 code positions on the Interrogator until the vehicle starts. This is then the correct pellet code. 5. Obtain the correct coded key and cut it to fit.
General Motors Passkey II (vehicles with BCM)	On vehicles with a body control module (BCM), the Passkey II pellet code is stored in the BCM. The BCM can learn the pellet code of a replacement key using a scan tool or this procedure. Make sure that the battery is fully charged. If the learning procedure is not successful, check the system for codes and repair.	1. Insert the key to be learned and turn the ignition on. Leave the switch on for 11 minutes. The security lamp will be on or flashing during this time. 2. When the security lamp goes off, turn the ignition off for 30 seconds. 3. Repeat step 1 two more times. 4. Turn the ignition off for 30 seconds. 5. Attempt to start the vehicle. The vehicle should start and run if the learn is successful.
General Motors Passkey III Passkey III+	Quick-Learn requires at least one programmed master (black) key. Keys can be learned with a scan tool. If no programmed master key is available, the 30 minute Auto Learn procedure must be used. Auto Learn procedure will erase all learned keys. Make sure that the battery is fully charged. On vehicles with a driver information center (DIC), a "STARTING DISABLED DUE TO THEFT" message will display during the 10 minute timer.	QUICK LEARN 1. Insert a programmed master key and turn on the ignition. 2. Turn the ignition off and remove the key. 3. Within 10 seconds insert the key to be learned and turn the ignition on. 4. The key is now programmed. 30 MINUTE AUTO LEARN 1. Insert the new master key and turn on the ignition. The security lamp should be on and then turn it off after 10 minutes. 2. Turn the ignition off for 5 seconds. 3. Repeat steps 1 and 2 two more times (30 minutes total). 4. From the off position, turn on and start the vehicle. 5. The vehicle should start and run, indicating the key has been learned.

CHART 22–2

CONTINUED

Immobilizer or vehicle theft deterrent key learn procedures for some popular vehicles.

MAKE/MODEL	NOTES	PROCEDURES
General Motors Passlock (early systems)	Passlock systems do not have coded keys. Replacement or new keys do not have to be learned. Early Passlock systems pass an "R" code to the instrument cluster and then the IPC sends a password on to the PCM. Perform this procedure if replacing the instrument cluster, lock cylinder, or PCM.	1. After parts are installed, attempt to start the vehicle. 2. The vehicle should start and stall. 3. Leave the key on and wait until the flashing theft lamp stays on steady. 4. Attempt to start the vehicle again. It should start and continue to run. 5. The theft lamp should flash for 10 seconds and then go out to indicate the password has been learned.
General Motors Passlock (later models)	Replacement or additional keys do not have to be learned. Programming is necessary if the Passlock sensor, BCM, or PCM has been replaced. A scan tool can also be used to program the Passlock system. ● **SEE FIGURE 22–32**.	1. Turn the ignition on and attempt to start the vehicle. 2. The vehicle will not start. Release the key to on. Wait about 10 minutes for the security lamp to go off. 3. Turn off the ignition for 5 seconds. 4. Repeat steps 1 through 3 two more times. 5. For a fourth time turn, the key on and start the vehicle. The vehicle should start and run, indicating that the lock code has been learned.

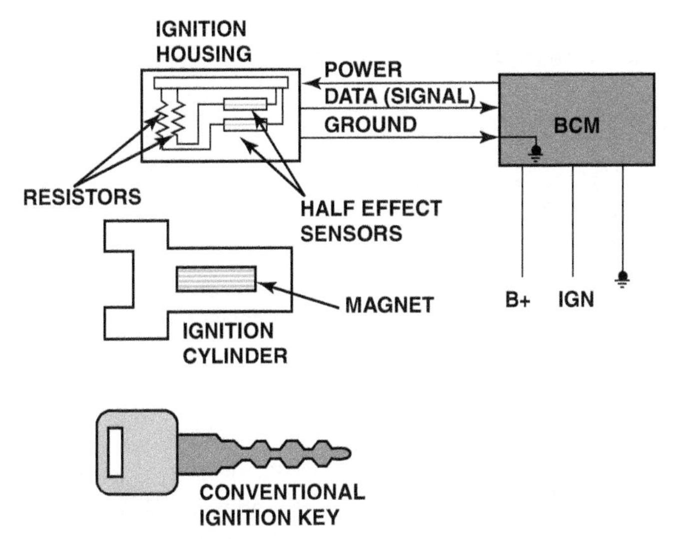

FIGURE 22–32 The Passlock series of General Motors security systems uses a conventional key. The magnet is located in the ignition lock cylinder and triggers the Hall-effect sensors.

ELECTRICAL ACCESSORY SYMPTOM GUIDE

Cruise Control Problem	Possible Causes and/or Solutions
Cruise (speed) control is inoperative.	1. Blown fuse 2. Defective or misadjusted electrical or vacuum safety switch near the brake pedal arm 3. Lack of engine vacuum to servo or transducer 4. Defective transducer; defective speed control switch
Cruise (speed) control speed is incorrect or variable.	1. Misadjusted activation cable 2. Defective or pinched vacuum hose 3. Misadjustment of transducer

Power Windows Problem	Possible Causes and/or Solutions
Power windows are inoperative.	1. Defective (blown) fuse (circuit breaker) 2. Defective relay (if used) 3. Poor ground for master control switch 4. Poor connections at switch(es) or motor(s) 5. Open circuit (usually near the master control switch) 6. Defective lockout switch
One power window is inoperative.	1. Defective motor; defective or open control switch 2. Open or loose wiring to the switch or the motor
Only one power window can be operated from the master switch.	1. Poor connection or open circuit in the control wire(s)

Power Seats Problem	Possible Causes and/or Solutions
Power seats are inoperative, no click or noise.	1. Defective circuit breaker 2. Poor ground at the switch or relay (if used) 3. Open in the wiring between the switch and relay (if used); defective switch 4. Defective solenoid(s) or wiring 5. Defective door switch
Power seats are inoperative, click is heard.	1. "Flex" in the cables from the motor(s) to check for motor operation (If flex is felt, the motor is trying to operate the gear nut or the screw jack.) 2. Binding or obstruction 3. Defective motor (The click is generally the relay sound.) 4. Defective solenoid(s) or wiring to the solenoid(s)
All power seat functions are operative except one.	1. Defective motor 2. Defective solenoid or wiring to the solenoid

Electric Power Door Lock Problem	Possible Causes and/or Solutions
Power door locks are inoperative.	1. Defective circuit breaker, fuse, or wiring to the switch or relay (if used) 2. Defective relay (if used); defective switch 3. Defective door lock solenoid or ground for solenoid (if solenoid operated) 4. Open in the wiring to the door lock solenoid or the motor 5. Mechanical obstruction of the door lock mechanism
Only one door lock is inoperative.	1. Defective switch; poor ground on the solenoid (if solenoid operated) 2. Defective door lock solenoid or motor; poor electrical connection at the motor or solenoid

Rear Window Defogger Problem	Possible Causes and/or Solutions
Rear window defogger is inoperative.	1. Proper operation by performing breath test and/or voltmeter (Check at the power side of the rear window grid.); defective relay or timer assembly 2. Defective switch 3. Open ground connection at the rear window grid **NOTE: If there is an open circuit (power side or ground side), the dash indicator light will still operate in most cases.**
Rear window defogger cleans only a portion of the rear window.	1. Broken grid wire(s) or poor electrical connections at either the power side or the ground side of the wire grid

DOOR PANEL REMOVAL

1 Looking at the door panel, there appears to be no visible fasteners.

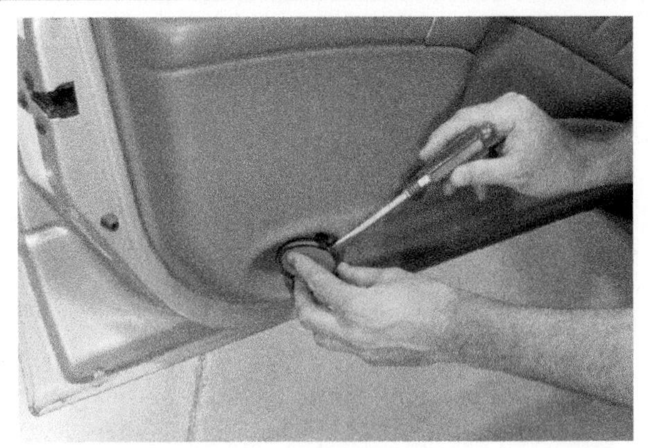

2 Gently prying at the edge of the light shows that it snaps in place and can be easily removed.

3 Under the red “door open” warning light is a fastener.

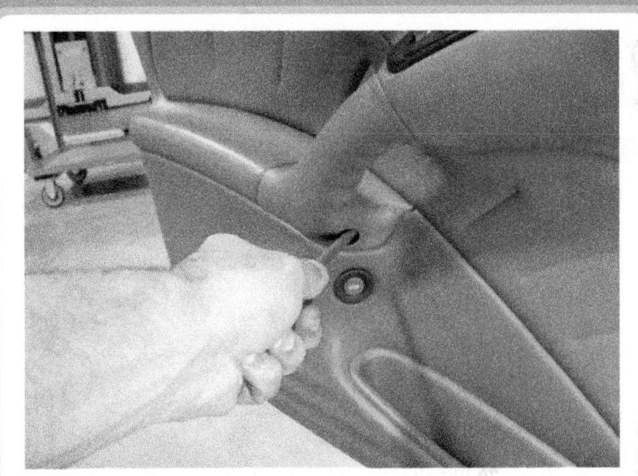

4 Another screw is found under the armrest.

5 A screw is removed from the bezel around the interior door handle.

6 The electric control panel is held in by clips.

CONTINUED ▶

7 Another screw is found after the control panel is removed.

8 The panel beside the outside mirror is removed by gently prying.

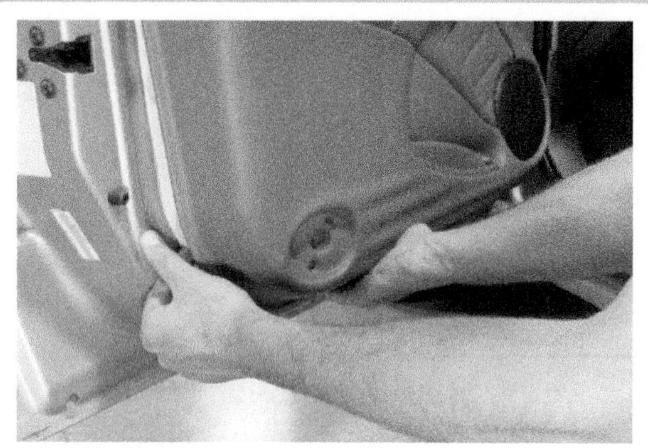

9 A gentle tug and the door panel is removed.

10 The sound-deadening material also acts as a moisture barrier and would need to be removed to gain access to the components inside the door.

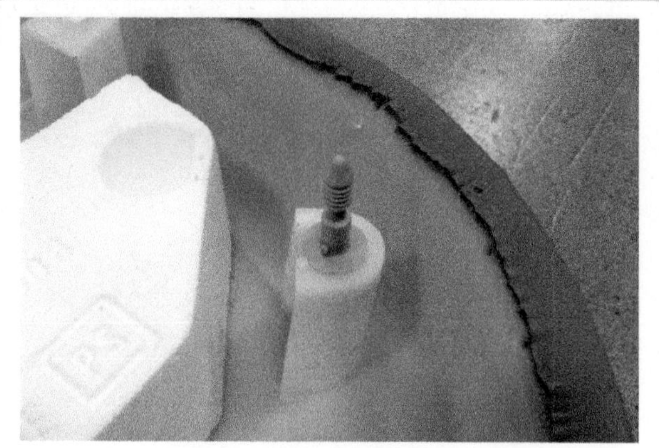

11 Carefully inspect the door panel clips before reinstalling the door panel.

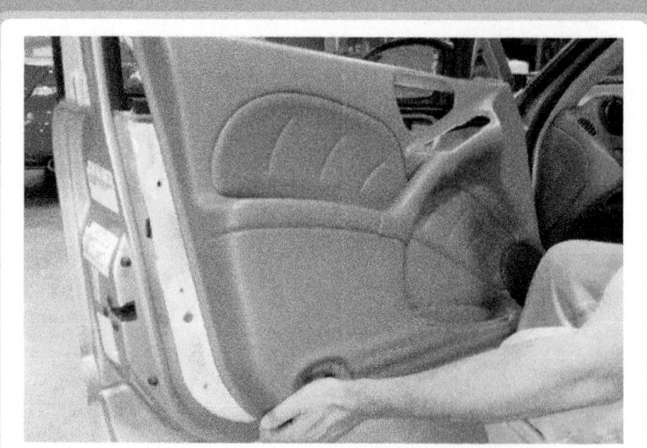

12 Align and press the door panel clips into the openings and reinstall all of the fasteners and components.

SUMMARY

1. Most power windows and power door locks use a permanent magnet motor that has a built-in circuit breaker and is reversible. The control switches and relays direct the current through the motors.
2. The current flow through a rear window defogger is often self-regulating. As the temperature of the grid increases, its resistance increases, reducing current flow. Some rear window defoggers are also used as radio antennas.
3. Radar cruise control systems use many of the same components as the precollision system.
4. Remote keyless entry systems use a wireless transmitter built into the key fob to operate the power door lock.
5. Factory antitheft systems must function properly to allow the engine to crank and/or start.

REVIEW QUESTIONS

1. How do power door locks on a four-door vehicle function with only one ground wire connection?
2. How does a rear window defogger regulate how much current flows through the grids based on temperature?
3. What is the usual procedure to follow to resynchronize a remote keyless entry transmitter?
4. How do heated and cooled seats operate?

CHAPTER QUIZ

1. The owner of a vehicle equipped with cruise control complains that the cruise control often stops working when driving over rough or bumpy pavement. Technician A says the brake switch may be out of adjustment. Technician B says a defective servo unit is the most likely cause. Which technician is correct?
   - **a.** Technician A only
   - **b.** Technician B only
   - **c.** Both Technicians A and B
   - **d.** Neither Technician A nor B
2. Technician A says that the cruise control on a vehicle that uses an electronic throttle control (ETC) system uses a servo to move the throttle. Technician B says that the cruise control on a vehicle with ETC uses the APP sensor to set the speed. Which technician is correct?
   - **a.** Technician A only
   - **b.** Technician B only
   - **c.** Both Technicians A and B
   - **d.** Neither Technician A nor B
3. All power windows fail to operate from the independent switches but all power windows operate from the master switch. Technician A says the window lockout switch may be on. Technician B says the power window relay could be defective. Which technician is correct?
   - **a.** Technician A only
   - **b.** Technician B only
   - **c.** Both Technicians A and B
   - **d.** Neither Technician A nor B
4. Technician A says that a defective ground connection at the master control switch (driver's side) could cause the failure of all power windows. Technician B says that if *one* control wire is disconnected, all windows will fail to operate. Which technician is correct?
   - **a.** Technician A only
   - **b.** Technician B only
   - **c.** Both Technicians A and B
   - **d.** Neither Technician A nor B
5. A typical radar cruise control system uses ________.
   - **a.** Long-range radar (LRR)
   - **b.** Short-range radar (SRR)
   - **c.** Electronic throttle control system to control vehicle speed
   - **d.** All of the above
6. When checking the operation of a rear window defogger with a voltmeter, ________.
   - **a.** The voltmeter should be set to read AC volts
   - **b.** The voltmeter should read close to battery voltage anywhere along the grid
   - **c.** Voltage should be available anytime at the power side of the grid because the control circuit just completes the ground side of the heater grid circuit
   - **d.** The voltmeter should indicate decreasing voltage when the grid is tested across the width of the glass
7. PM motors used in power windows, mirrors, and seats can be reversed by ________.
   - **a.** Sending current to a reversed field coil
   - **b.** Reversing the polarity of the current to the motor
   - **c.** Using a reverse relay circuit
   - **d.** Using a relay and a two-way clutch

8. If only one power door lock is inoperative, a possible cause is a ________.
   a. Poor ground connection at the power door lock relay
   b. Defective door lock motor (or solenoid)
   c. Defective (open) circuit breaker for the power circuit
   d. Defective (open) fuse for the control circuit
9. A keyless remote control stops working. Technician A says the battery in the remote could be dead. Technician B says that the key fob may have to be resynchronized. Which technician is correct?
   a. Technician A only
   b. Technician B only
   c. Both Technicians A and B
   d. Neither Technician A nor B
10. Two technicians are discussing antitheft systems. Technician A says that some systems require a special key. Technician B says that some systems use a computer chip in the key. Which technician is correct?
    a. Technician A only
    b. Technician B only
    c. Both Technicians A and B
    d. Neither Technician A nor B

chapter 23

SAFETY BELTS AND AIRBAGS

LEARNING OBJECTIVES

After studying this chapter, the reader will be able to:

1. Diagnose and repair faulty safety belts and retractors.
2. Explain the operation of front airbags.
3. Describe the procedures to diagnose and repair common faults in airbag systems.
4. Disarm and enable the airbag system for vehicle service.
5. Explain how the passenger presence system works.

This chapter will help you prepare for the ASE Electrical/Electronic Systems (A6) certification test content area "H" (Accessories Diagnosis and Repair).

KEY TERMS

GM STC OBJECTIVES

GM Service Technical College topics covered in this chapter are as follows:

1. Diagnose concerns related to air bag system frontal inflator module operation.
2. Diagnose concerns related to air bag system side impact inflator module operation.
3. Diagnose concerns related to air bag system seat belt pretensioner operation.
4. Diagnose concerns related to air bag system by using the SIR Driver/Passenger Load tool.

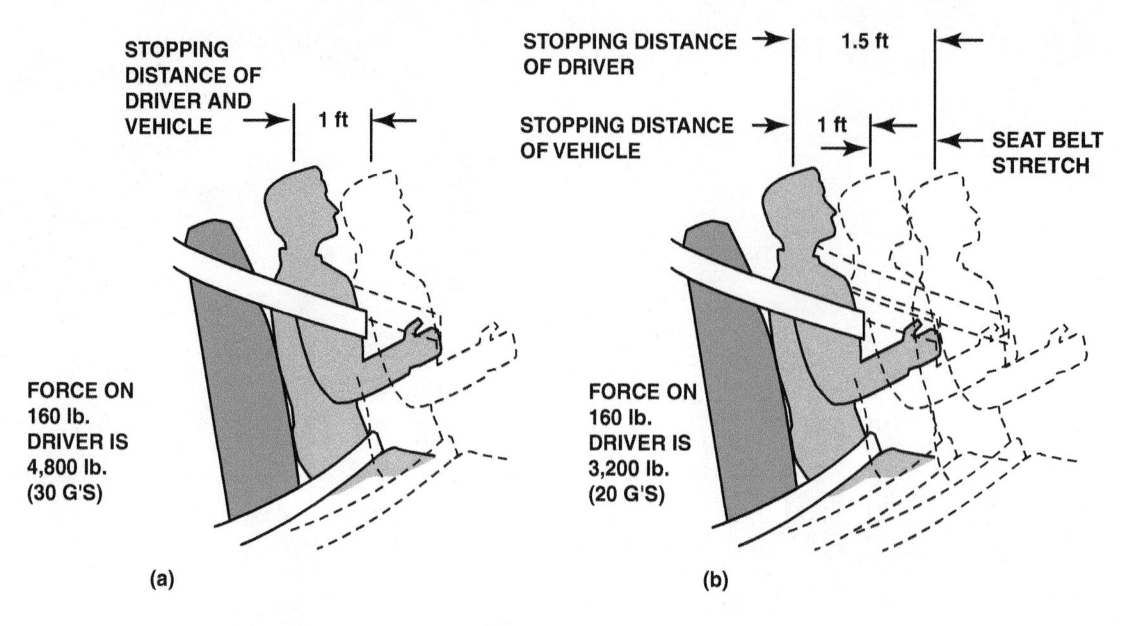

FIGURE 23–1 (a) Safety belts are the primary restraint system. (b) During a collision the stretching of the safety belt slows the impact to help reduce bodily injury.

SAFETY BELTS AND RETRACTORS

SAFETY BELTS Safety belts are used to keep the driver and passengers secured to the vehicle in the event of a collision. Most safety belts include three-point support and are constructed of nylon webbing about 2 inches (5 cm) wide. The three support points include two points on either side of the seat for the belt over the lap and one crossing over the upper torso, which is attached to the "B" pillar or seat back. Every crash consists of three types of collisions.

Collision 1: The vehicle strikes another vehicle or object.

Collision 2: The driver and/or passengers hit objects inside the vehicle if unbelted.

Collision 3: The internal organs of the body hit other organs or bones, which causes internal injuries.

If a safety belt is being worn, the belt stretches, absorbing a lot of the impact, thereby preventing collision with other objects in the vehicle and reducing internal injuries. ● **SEE FIGURE 23–1**.

BELT RETRACTORS Safety belts are also equipped with one of the following types of retractors.

- Nonlocking retractors, which are used primarily on recoiling
- Emergency locking retractors, which lock the position of the safety belt in the event of a collision or rollover
- Emergency and web speed-sensitive retractors, which allow freedom of movement for the driver and passenger but lock if the vehicle is accelerating too fast or if the vehicle is decelerating too fast

● **SEE FIGURE 23–2** for an example of an inertia-type seat belt locking mechanism.

SAFETY BELT LIGHTS AND CHIMES All late-model vehicles are equipped with a safety belt warning light on the dash and a chime that sounds if the belt is not fastened. ● **SEE FIGURE 23–3**.

Some vehicles will intermittently flash the reminder light and sound a chime until the driver and sometimes the front passenger fasten their safety belts.

PRETENSIONERS A **pretensioner** is an explosive (pyrotechnic) device that is part of the seat belt retractor assembly and tightens the seat belt as the airbag is being deployed. The purpose of the pretensioning device is to force the occupant back into position against the seat back and to remove any slack in the seat belt. ● **SEE FIGURE 23–4**.

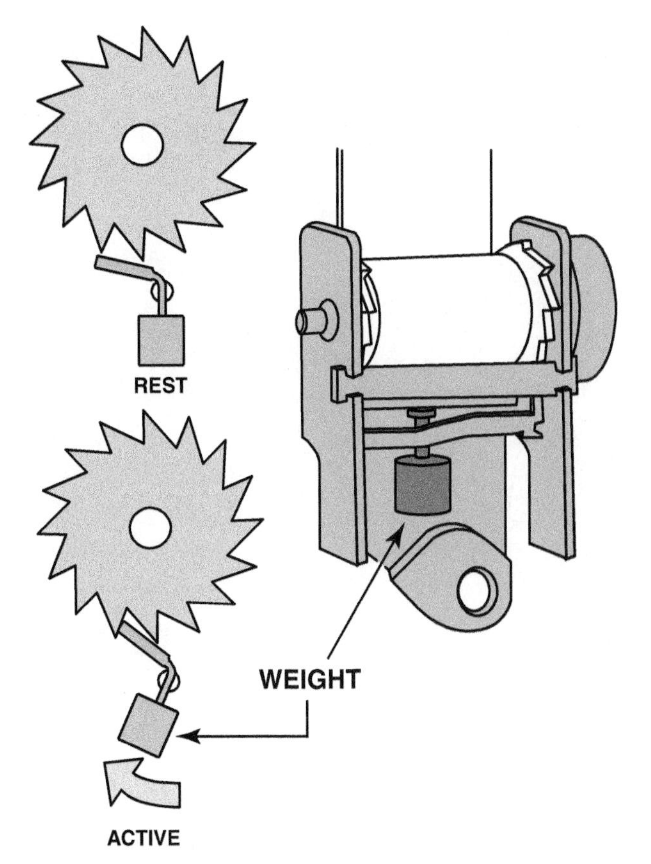

FIGURE 23–2 Most safety belts have an inertia-type mechanism that locks the belt in the event of rapid movement.

FIGURE 23–3 A typical safety belt warning light.

CAUTION: The seat belt pretensioner assemblies must be replaced in the event of an airbag deployment. Always follow the vehicle manufacturer's recommended service procedure. Pretensioners are explosive devices that could be ignited if voltage is applied to the terminals. Do not use a jumper wire or powered test light around the wiring near the seat belt latch wiring. Always follow the vehicle manufacturer's recommended test procedures.

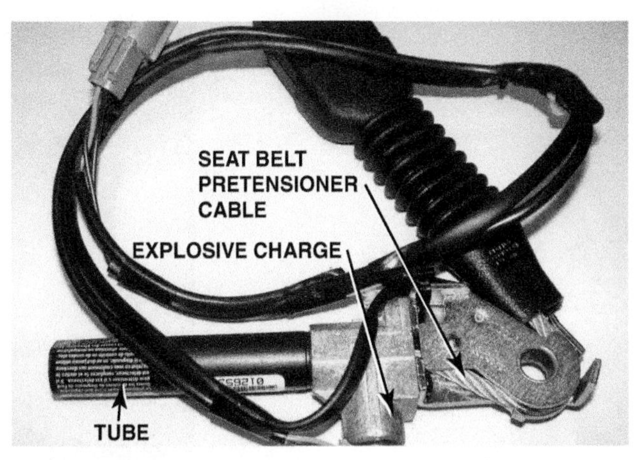

FIGURE 23–4 A small explosive charge in the pretensioner forces the end of the seat belt down the tube, which removes any slack in the seat belt.

FRONT AIRBAGS

PURPOSE AND FUNCTION **Airbag** passive restraints are designed to cushion the driver (or passenger, if the passenger side is so equipped) during a frontal collision. The system consists of one or more nylon bags folded up in compartments located in the steering wheel, dashboard, interior panels, or side pillars of the vehicle. ● **SEE FIGURE 23–5**.

During a crash of sufficient force, pressurized gas instantly fills the airbag and then deploys out of the storage compartment to protect the occupant from serious injury. These airbag systems may be known by many different names, including the following:

1. **Supplemental restraint system (SRS)**
2. **Supplemental inflatable restraints (SIR)**
3. **Supplemental air restraints (SAR)**

Most airbags are designed to supplement the safety belts in the event of a collision, and front airbags are meant to be deployed only in the event of a frontal impact within 30 degrees of center. Front (driver and passenger side) airbag systems are *not* designed to inflate during side or rear impact. The force required to deploy a typical airbag is approximately equal to the force of a vehicle hitting a wall at over 10 mph (16 km/h).

The force required to trigger the sensors within the system prevents accidental deployment if curbs are hit or the brakes are rapidly applied. The system requires a substantial force to deploy the airbag to help prevent accidental inflation.

FIGURE 23–5 Today's vehicles have air bag protection for driver and passengers, protecting from frontal and side impacts. (Courtesy of General Motors)

PARTS INVOLVED ● **SEE FIGURE 23–6** for an overall view of the parts included in a typical airbag system.

The parts include:

1. Sensors
2. Airbag (inflator) module
3. Clockspring wire coil in the steering column
4. Control module
5. Wiring and connectors

OPERATION To cause inflation, the following events must occur.

- To cause a deployment of the airbag, two sensors must be triggered at the same time. The **arming sensor** is used to provide electrical power, and a *forward* or *discriminating sensor* is used to provide the ground connection.
- The arming sensor provides the electrical power to the airbag heating unit, called a **squib**, inside the inflator module.
- The squib uses electrical power and converts it into heat for ignition of the propellant used to inflate the airbag.
- Before the airbag can inflate, however, the squib circuit also must have a ground provided by the forward or the discriminating sensor. In other words, two sensors (arming and forward sensors) *must* be triggered *at the same time* before the airbag will be deployed. ● **SEE FIGURE 23–7**.

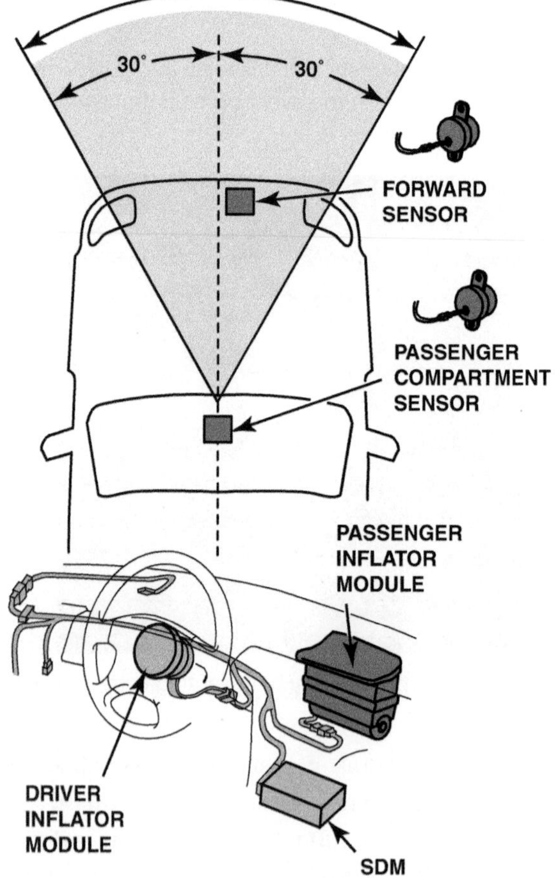

FIGURE 23–6 A typical airbag system showing many of the components. The SDM is the "sensing and diagnostic module" and includes the arming sensor as well as the electronics that keep checking the circuits for continuity and the capacitors that are discharged to deploy the airbags.

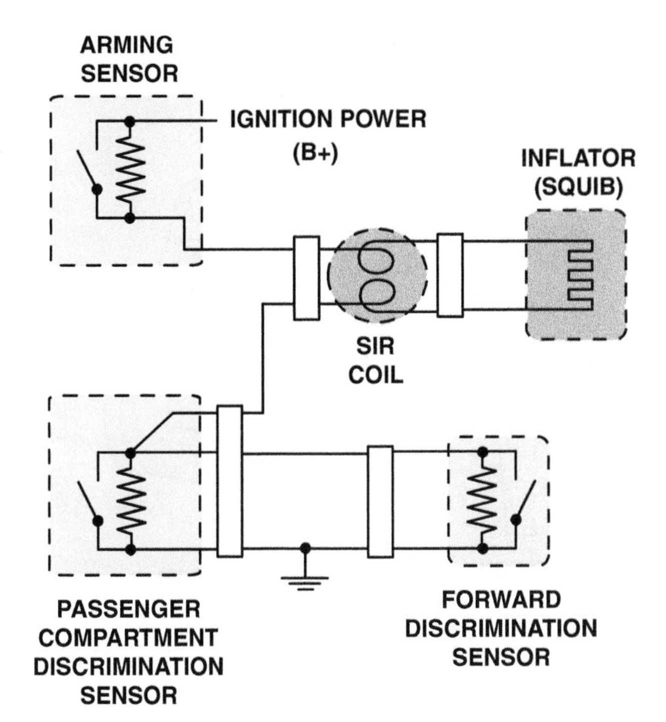

FIGURE 23–7 A simplified airbag deployment circuit. Note that both the arming sensor and at least one of the discriminating sensors must be activated at the same time. The arming sensor provides the power, and either one of the discriminating sensors can provide the ground for the circuit.

TYPES OF AIRBAG INFLATORS There are two different types of inflators used in airbags.

1. **Solid fuel.** This type uses sodium azide pellets and, when ignited, generates a large quantity of nitrogen gas that quickly inflates the airbag. This was the first type used and is still commonly used in driver and passenger side airbag inflator modules. ● **SEE FIGURE 23–8.** The squib is the electrical heating element used to ignite the gas-generating material, usually sodium azide. It requires about 2 amperes of current to heat the heating element and ignite the inflator.
2. **Compressed gas.** Commonly used in passenger side airbags and roof-mounted systems, the compressed gas system uses a canister filled with argon gas, plus a small percentage of helium at 3,000 PSI (435 kPa). A small igniter ruptures a burst disc to release the gas when energized. The compressed gas inflators are long cylinders that can be installed inside the instrument panel, seat back, door panel, or along any side rail or pillar of the vehicle. ● **SEE FIGURE 23–9.**

Once the inflator is ignited, the nylon bag quickly inflates (in about 30 milliseconds or 0.030 seconds) with nitrogen gas generated by the inflator. During an actual frontal collision

FIGURE 23–8 The inflator module is being removed from the airbag housing. The squib, inside the inflator module, is the heating element that ignites the pyrotechnic gas generator that rapidly produces nitrogen gas to fill the airbag.

FIGURE 23–9 This shows a deployed side curtain airbag on a training vehicle.

accident, the driver is being thrown forward by the driver's own momentum toward the steering wheel. The strong nylon bag inflates at the same time. Personal injury is reduced by the spreading of the stopping force over the entire upper-body region. The normal collapsible steering column remains in operation and collapses in a collision when equipped with an airbag system. The bag is equipped with two large side vents that allow the bag to deflate immediately after inflation, once the bag has cushioned the occupant in a collision.

TIMELINE FOR AIRBAG DEPLOYMENT Following are the times necessary for an airbag deployment in milliseconds (each millisecond is equal to 0.001 second or 1/1,000 of a second).

1. Collision occurs: 0.0 millisecond
2. Sensors detect collision: 16 milliseconds (0.016 second)

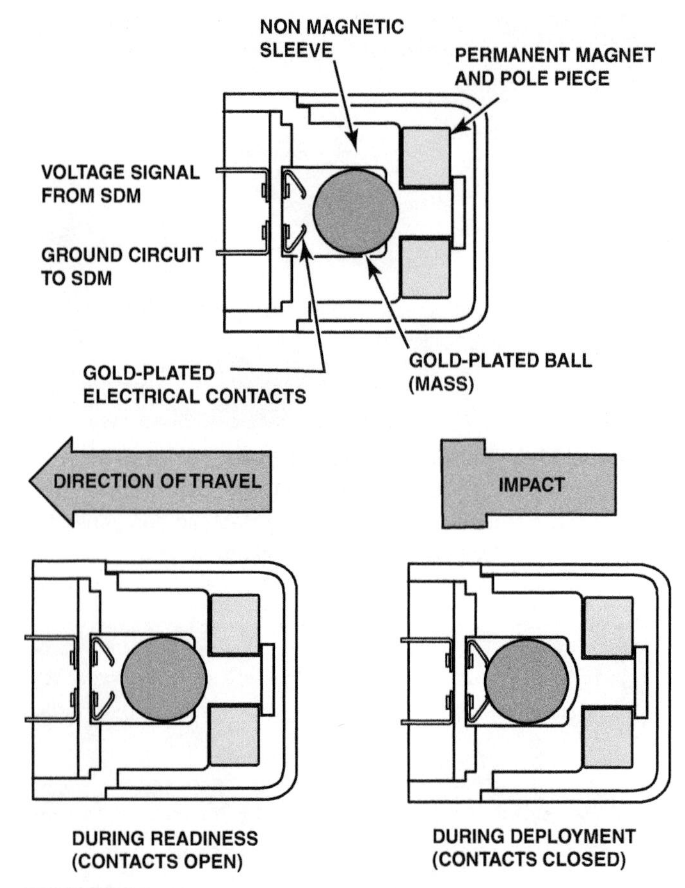

FIGURE 23–10 An airbag magnetic sensor.

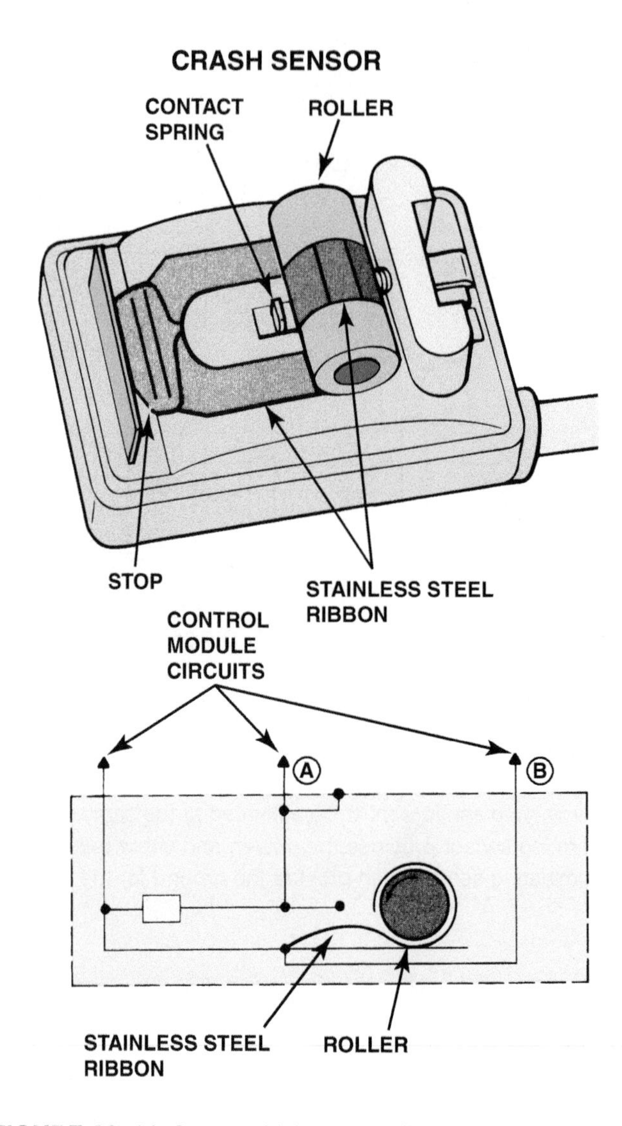

FIGURE 23–11 Some vehicles use a ribbon-type crash sensor.

3. Airbag is deployed and seam cover rips: 40 milliseconds (0.040 second)
4. Airbag is fully inflated: 100 milliseconds (0.100 second)
5. Airbag deflated: 250 milliseconds (0.250 second)

In other words, an airbag deployment occurs and is over in about a quarter of a second.

SENSOR OPERATION All three sensors are basically switches that complete an electrical circuit when activated. The sensors are similar in construction and operation, and the *location* of the sensor determines its name. All airbag sensors are rigidly mounted to the vehicle and *must* be mounted with the arrow pointing toward the front of the vehicle to ensure that the sensor can detect rapid forward deceleration.

There are three basic styles (designs) of airbag sensors.

1. **Magnetically retained gold-plated ball sensor.** This sensor uses a permanent magnet to hold a gold-plated steel ball away from two gold-plated electrical contacts. ● **SEE FIGURE 23–10.**

   If the vehicle (and the sensor) stops rapidly enough, the steel ball is released from the magnet because the inertia force of the crash was sufficient to overcome the magnetic pull on the ball and then makes contact with the two gold-plated electrodes. The steel ball only remains in contact with the electrodes for a relatively short time because the steel ball is drawn back into contact with the magnet.

2. **Rolled up stainless-steel ribbon-type sensor.** This sensor is housed in an airtight package with nitrogen gas inside to prevent harmful corrosion of the sensor parts. If the vehicle (and the sensor) stops rapidly, the stainless-steel roll "unrolls" and contacts the two gold-plated contacts. Once the force is stopped, the stainless-steel roll will roll back into its original shape. ● **SEE FIGURE 23–11.**

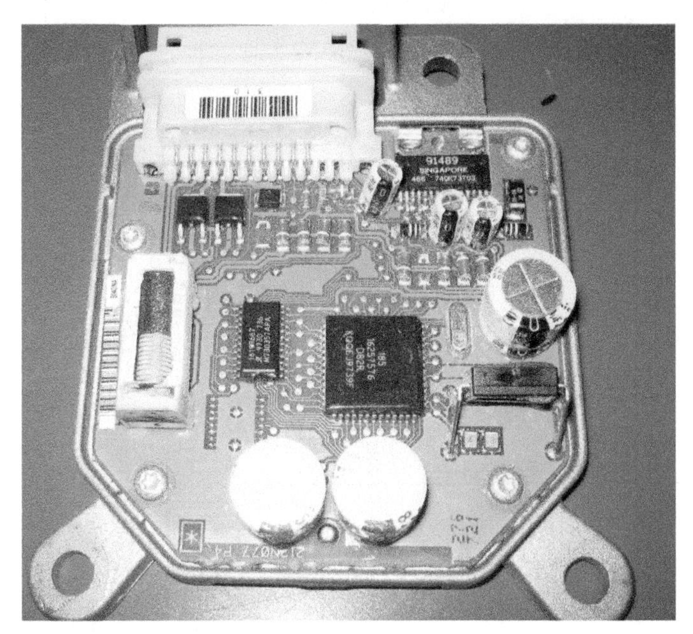

FIGURE 23–12 A sensing and diagnostic module that includes an accelerometer.

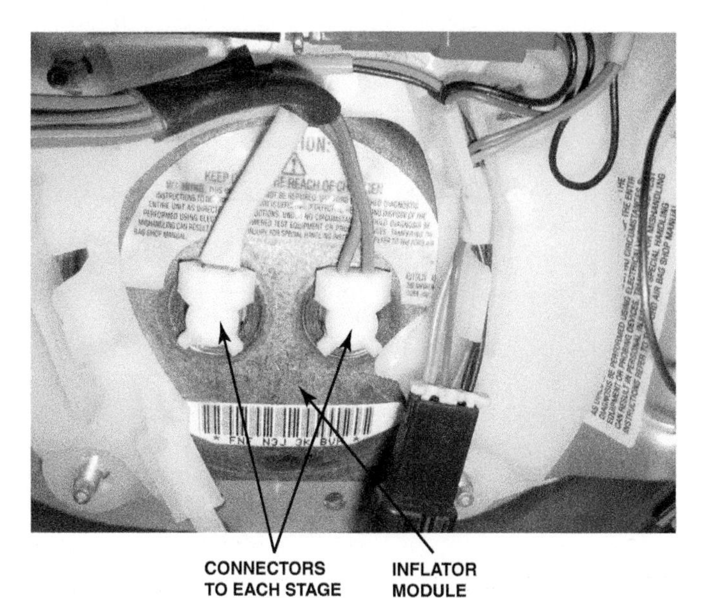

FIGURE 23–13 A driver's side airbag showing two inflator connectors. One is for the lower force inflator and the other is for the higher force inflator. Either can be ignited or both at the same time if the deceleration sensor detects a severe impact.

3. **Integral sensor.** Some vehicles use electronic **deceleration sensors** built into the inflator module, called **integral sensors**. For example, General Motors uses the term *sensing and diagnostic module* (*SDM*) to describe their integrated sensor/module assembly. These units contain an accelerometer-type sensor that measures the rate of deceleration and, through computer logic, determines if the airbags should be deployed. ● **SEE FIGURE 23–12.**

TWO-STAGE AIRBAGS Two-stage airbags, often called advanced airbags or smart airbags, use an accelerometer-type of sensor to detect force of the impact. This type of sensor measures the actual amount of deceleration rate of the vehicle and is used to determine whether one or both elements of a two-stage airbag should be deployed.

- **Low-stage deployment.** This lower force deployment is used if the accelerometer detects a low-speed crash.
- **High-stage deployment.** This stage is used if the accelerometer detects a higher speed crash or a more rapid deceleration rate.
- **Both low- and high-stage deployment.** Under severe high-speed crashes, both stages can be deployed. ● **SEE FIGURE 23–13.**

WIRING Wiring and connectors are very important for proper identification and long life. Airbag-related circuits have the following features.

- All electrical wiring connectors and conduit for airbags are colored yellow.
- To ensure proper electrical connection to the inflator module in the steering wheel, a coil assembly is used in the steering column. This coil is a ribbon of copper wires that operates much like a window shade when the steering wheel is rotated. As the steering wheel is rotated, this coil, usually called a **clockspring**, prevents the lack of continuity between the sensors and the inflator assembly that might result from a horn-ring type of sliding conductor. ● **SEE FIGURE 23–14.**
- Inside the yellow plastic airbag connectors are gold-plated terminals that are used to prevent corrosion.

Most airbag systems also contain a diagnostic unit that often includes an auxiliary power supply, which is used to provide the current to inflate the airbag if the battery is disconnected from the vehicle during a collision. This auxiliary power supply normally uses capacitors that are discharged through the squib of the inflation module. When the ignition is turned off, these capacitors are discharged. Therefore, after a few minutes an airbag system will not deploy if the vehicle is hit while parked.

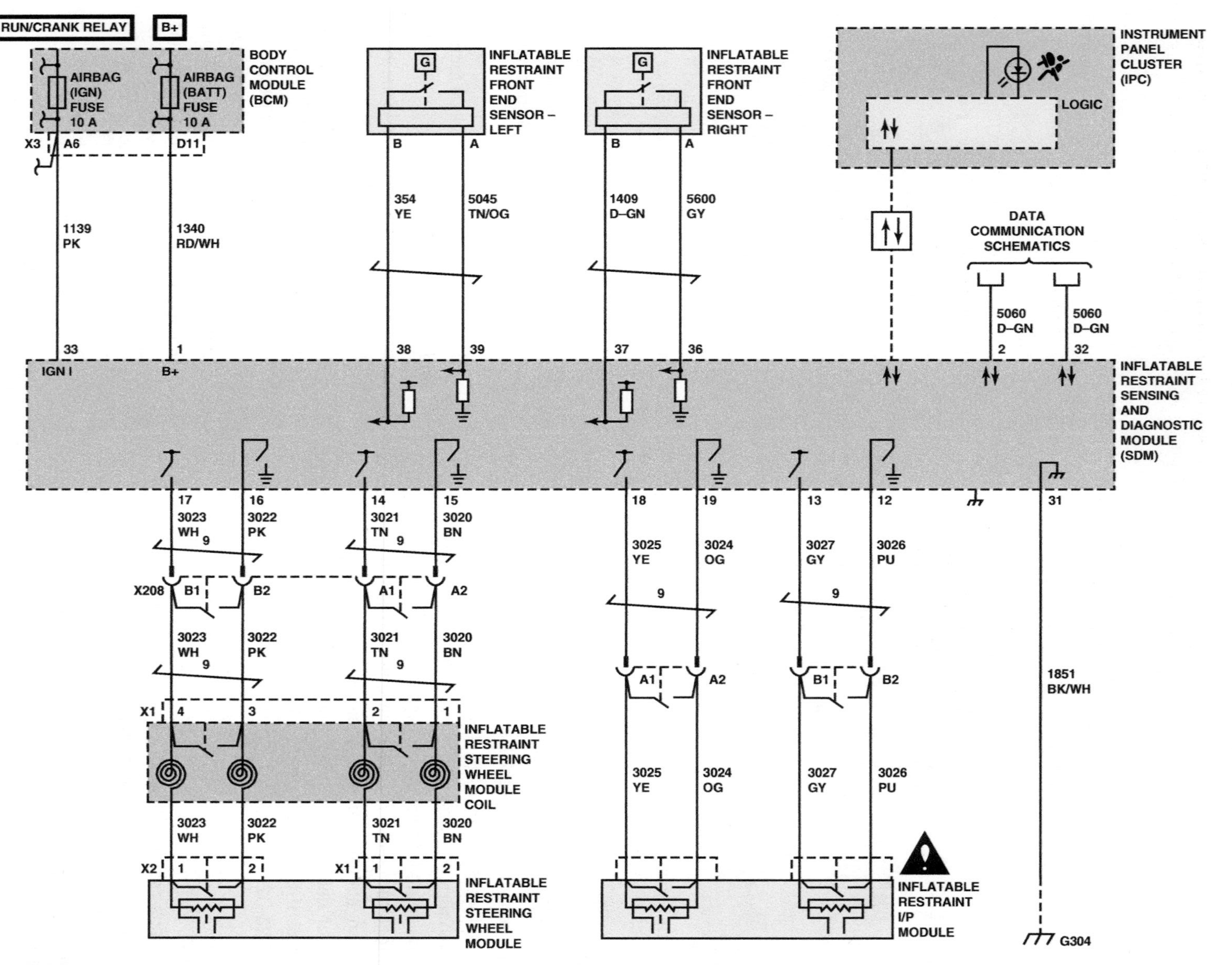

FIGURE 23–14 The airbag sensing and diagnostic module (SDM) is linked to the vehicle data system and the instrument panel. Notice the airbag wire connecting the module to the airbag through the clockspring (coil). (Courtesy of General Motors)

FIGURE 23–15 A typical seat (side) airbag that deploys from the side of the seat.

SAFETY TIP

Dual-Stage Airbag Caution

Many vehicles are equipped with **dual-stage airbags** (two-stage airbags) that actually contain two separate inflators, one for less severe crashes and one for higher speed collisions. These systems are sometimes called smart airbag systems because the accelerometer-type sensor used can detect how severe the impact is and deploy one or both stages. If one stage is deployed, the other stage is still active and could be accidentally deployed. A service technician cannot tell by looking at the airbag whether both stages have deployed. Always handle a deployed airbag as if it has not been deployed and take all precautions necessary to keep any voltage source from getting close to the inflator module terminals.

SEAT AND SIDE CURTAIN AIRBAGS

SEAT AIRBAGS Side and/or *curtain airbags* use a variety of sensors to determine if they need to be deployed. Side airbags are mounted in one of two general locations.

- In the side bolster of the seat (**SEE FIGURE 23–15.**)
- In the door panel

Most side airbag sensors use an electronic accelerometer to detect when to deploy the airbags, which are usually mounted to the bottom of the left and right "B" pillars (where the front doors latch) behind a trim panel on the inside of the vehicle.

SIDE CURTAIN AIRBAGS Side curtain airbags are usually deployed by a module based on input from many different sensors, including a lateral acceleration sensor and wheel speed sensors. For example, in one system used by Ford, the ABS controller commands that the brakes on one side of the vehicle be applied, using down pressure while monitoring the wheel speed sensors. If the wheels slow down with little brake pressure, the controller assumes that the vehicle could roll over, thereby deploying the side curtain airbags.

KNEE AIRBAGS Some vehicles are equipped with knee airbags usually on the driver's side. Use caution if working under the dash and always follow the vehicle manufacturer's specified service procedures.

AIRBAG DIAGNOSIS TOOLS AND EQUIPMENT

SELF-TEST PROCEDURE The electrical portion of airbag systems is constantly checked by the circuits within the airbag-energizing power unit or through the airbag controller. The electrical airbag components are monitored by applying a small-signal voltage from the airbag controller through the various sensors and components. Each component and sensor uses a resistor in parallel with the load or open sensor switch for use by the diagnostic signals. If continuity exists, the testing circuits will measure a small voltage drop. If an open or short circuit occurs, a dash warning light is lighted and a possible diagnostic trouble code (DTC) is stored. Follow exact manufacturer's recommended procedures for accessing and erasing airbag diagnostic trouble codes.

Diagnosis and service of airbag systems usually require some or all of the following items.

- Digital multimeter (DMM)
- Airbag simulator, often called a load tool
- Scan tool
- Shorting bar or shorting connector(s)
- Airbag system tester
- Vehicle-specific test harness
- Special wire repair tools or connectors, such as crimp-and-seal weatherproof connectors
  SEE FIGURE 23–16.

FIGURE 23–16 An airbag diagnostic tester. Included in the plastic box are electrical connectors and a load tool that substitutes for the inflator module during troubleshooting.

TECH TIP

Pocket the Ignition Key to Be Safe

When replacing any steering gear such as a rack-and-pinion steering unit, be sure that no one accidentally turns the steering wheel. If the steering wheel is turned without being connected to the steering gear, the airbag wire coil (clockspring) can become off center. This can cause the wiring to break when the steering wheel is rotated after the steering gear has been replaced. To help prevent this from occurring, simply remove the ignition key from the ignition and keep it in your pocket while servicing the steering gear.

CAUTION: Most vehicle manufacturers specify that the negative battery terminal be removed when testing or working around airbags. Be aware that a memory saver device used to keep the computer and radio memory alive can supply enough electrical power to deploy an airbag.

PRECAUTIONS Take the following precautions when working with or around airbags.

1. Always follow all precautions and warning stickers on vehicles equipped with airbags.
2. Maintain a safe working distance from all airbags to help prevent the possibility of personal injury in the unlikely event of an unintentional airbag deployment.
   - Side impact airbag: 5 inches (13 cm) distance
   - Driver front airbag: 10 inches (25 cm) distance
   - Passenger front airbag: 20 inches (50 cm) distance
3. In the event of a collision in which the bag(s) is deployed, the inflator module *and* all sensors usually must be replaced to ensure proper future operation of the system.
4. Avoid using a self-powered test light around the yellow airbag wiring. Even though it is highly unlikely, a self-powered test light could provide the necessary current to accidentally set off the inflator module and cause an airbag deployment.
5. Use care when handling the inflator module section when it is removed from the steering wheel. Always hold the inflator away from your body.
6. If handling a deployed inflator module, always wear gloves and safety glasses to avoid the possibility of skin irritation from the sodium hydroxide dust, which is used as a lubricant on the bag(s), that remains after deployment.
7. Never jar or strike a sensor. The contacts inside the sensor may be damaged, preventing the proper operation of the airbag system in the event of a collision.
8. When mounting a sensor in a vehicle, make certain that the arrow on the sensor is pointing toward the front of the vehicle. Also be certain that the sensor is securely mounted.

FREQUENTLY ASKED QUESTION

What Are Smart Airbags?

Smart airbags use the information from sensors to determine the level of deployment. Sensors used include:

- **Vehicle speed (VS) sensors.** This type of sensor has a major effect on the intensity of a collision. The higher the speed is, the greater the amount of impact force.
- **Seat belt fastened switch.** If the seat belt is fastened, as determined by the seat belt buckle switch, the airbag system will deploy accordingly. If the driver or passenger is not wearing a seat belt, the airbag system will deploy with greater force compared to when the seat belt is being worn.
- **Passenger seat sensor.** The sensor in the seat on the passenger's side determines the force of deployment. If there is not a passenger detected, the passenger side airbag will not deploy on the vehicle equipped with a passenger seat sensor system.

AIRBAG SYSTEM SERVICE

DIS-ARMING The airbags should be dis-armed (temporarily disconnected), whenever performing service work on any of the following locations.

- Steering wheel
- Dash or instrument panel
- Glove box (instrument panel storage compartment)

Check service information for the exact procedure, which usually includes the following steps.

STEP 1 Disconnect the negative battery cable.

STEP 2 Remove the airbag fuse and wait the specified amount of time.

STEP 3 (Some vehicles) Disconnect the yellow electrical connector located at the base of the steering column to disable the driver's side airbag.

STEP 4 (Some vehicles) Disconnect the yellow electrical connector for the passenger side airbag.

This procedure is called "disabling airbags" in most service information. Always follow the vehicle manufacturer's specified procedures.

DIAGNOSTIC AND SERVICE PROCEDURE Airbag system components and their location in the vehicle vary according to system design, but the basic principles of testing are the same as for other electrical circuits. Use service information to determine how the circuit is designed and the correct sequence of tests to be followed.

- Some airbag systems require the use of special testers. The built-in safety circuits of such testers prevent accidental deployment of the airbag.
- If such a tester is not available, follow the recommended alternative test procedures specified by the manufacturer.
- Access the self-diagnostic system and check for diagnostic trouble code (DTC) records.
- The scan tool is needed to access the data stream on most systems.

SELF-DIAGNOSIS All airbag systems can detect system electrical faults, and if found will disable the system and notify the driver through an airbag warning lamp in the instrument cluster. Depending on circuit design, a system fault may cause the warning lamp to fail to illuminate, remain lit continuously, or flash. Some systems use a tone generator that produces an audible warning when a system fault occurs or if the warning lamp is inoperative.

FREQUENTLY ASKED QUESTION

Why Change Knee Bolsters If Switching to Larger Wheels?

Larger wheels and tires can be installed on vehicles, but the powertrain control module (PCM) needs to be reprogrammed so the speedometer and other systems that are affected by a change in wheel/tire size can work effectively. When 20 inches wheels are installed on General Motors trucks or sport utility vehicles (SUVs), GM specifies that replacement knee bolsters be installed. Knee bolsters are the padded area located on the lower part of the dash where a driver or passenger's knees would hit in the event of a front collision. The reason for the need to replace the knee bolsters is to maintain the crash testing results. The larger 20 inches wheels would tend to be forced further into the passenger compartment in the event of a front-end collision. Therefore to maintain the frontal crash rating standard, the larger knee bolsters are required.

WARNING: Failure to perform the specified changes when changing wheels and tires could result in the vehicle not being able to provide occupant protection as designed by the crash test star rating that the vehicle originally achieved.

The warning lamp should illuminate with the ignition key on and engine off as a bulb check. If not, the diagnostic module is likely disabling the system. If the airbag warning light remains on, the airbags may or may not be disabled, depending on the specific vehicle and the fault detected. Some warning lamp circuits have a timer that extinguishes the lamp after a few seconds. The airbag system generally does not require service unless there is a failed component. However, a steering wheel-mounted airbag module is routinely removed and replaced in order to service switches and other column-mounted devices.

DRIVER SIDE AIRBAG MODULE REPLACEMENT

For the specific model being serviced, carefully follow the procedures provided by the vehicle manufacturer to disable and remove the airbag module. Failure to do so may result in serious injury and extensive damage to the vehicle. Replacing a

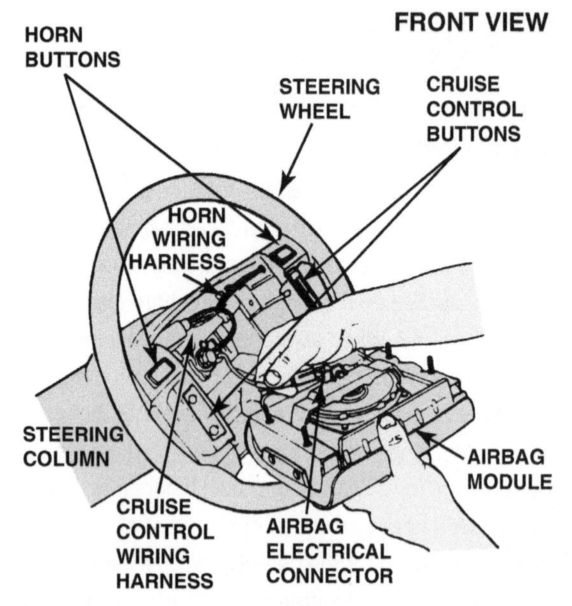

FIGURE 23–17 After disconnecting the battery and the yellow connector at the base of the steering column, the airbag inflator module can be removed from the steering wheel and the yellow airbag electrical connector at the inflator module disconnected.

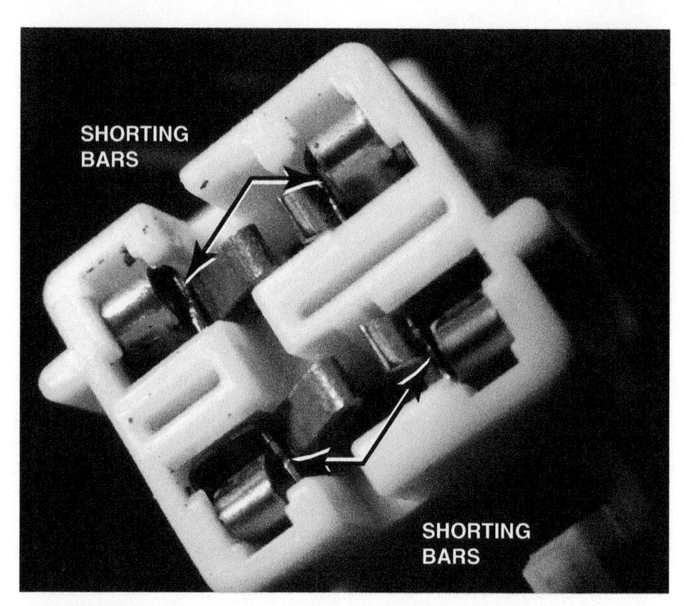

FIGURE 23–18 Shorting bars are used in most airbag connectors. These spring-loaded clips short across both terminals of an airbag connector when it is disconnected to help prevent accidental deployment of the airbag. If electrical power was applied to the terminals, the shorting bars would simply provide a low-resistance path to the other terminal and not allow current to flow past the connector. The mating part of the connector has a tapered piece that spreads apart the shorting bars when the connector is reconnected.

discharged airbag is costly. The following procedure reviews the basic steps for removing an airbag module. Do not substitute these general instructions for the specific procedure recommended by the manufacturer.

1. Turn the steering wheel until the front wheels are positioned straight ahead. Some components on the steering column are removed only when the front wheels are straight.
2. Switch the ignition off and disconnect the negative battery cable, which cuts power to the airbag module.
3. Once the battery is disconnected, wait as long as recommended by the manufacturer before continuing. When in doubt, wait at least 10 minutes to make sure the capacitor is completely discharged.
4. Loosen and remove the nuts or screws that hold the airbag module in place. On some vehicles, these fasteners are located on the back of the steering wheel. On other vehicles, they are located on each side of the steering wheel. The fasteners may be concealed with plastic finishing covers that must be pried off with a small screwdriver to access them.
5. Carefully lift the airbag module from the steering wheel and disconnect the electrical connector. Connector location varies: Some are below the steering wheel behind a plastic trim cover; others are at the top of the column under the module. ● **SEE FIGURES 23–17 AND 23–18.**
6. Store the module pad side up in a safe place where it will not be disturbed or damaged while the vehicle is being serviced. Do not attempt to disassemble the airbag module. If the airbag is defective, replace the entire assembly.

When installing the airbag module, make sure the clockspring is correctly positioned to ensure module-to-steering-column continuity. ● **SEE FIGURE 23–19.**

Always route the wiring exactly as it was before removal. Also, make sure the module seats completely into the steering wheel. Secure the assembly using new fasteners, if specified.

SAFETY WHEN MANUALLY DEPLOYING AIRBAGS

Airbag modules cannot be disposed of unless they are deployed. Do the following to prevent injury when manually deploying an airbag.

- When possible, deploy the airbag outside of the vehicle. Follow the vehicle manufacturer's recommendations.
- Follow the vehicle manufacturer's procedures and equipment recommendations.

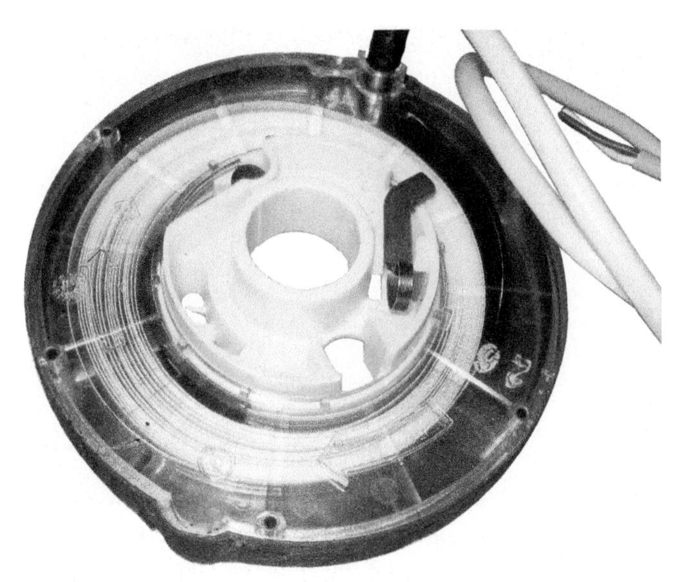

FIGURE 23–19 An airbag clockspring showing the flat conductor wire. It must be properly positioned to ensure proper operation.

- Wear the proper hearing and eye protection.
- Deploy the airbag with the trim cover facing up.
- Stay at least 20 ft (6 m) from the airbag. (Use long jumper wires attached to the wiring and routed outside the vehicle to a battery.)
- Allow the airbag module to cool.

SEE FIGURE 23–20.

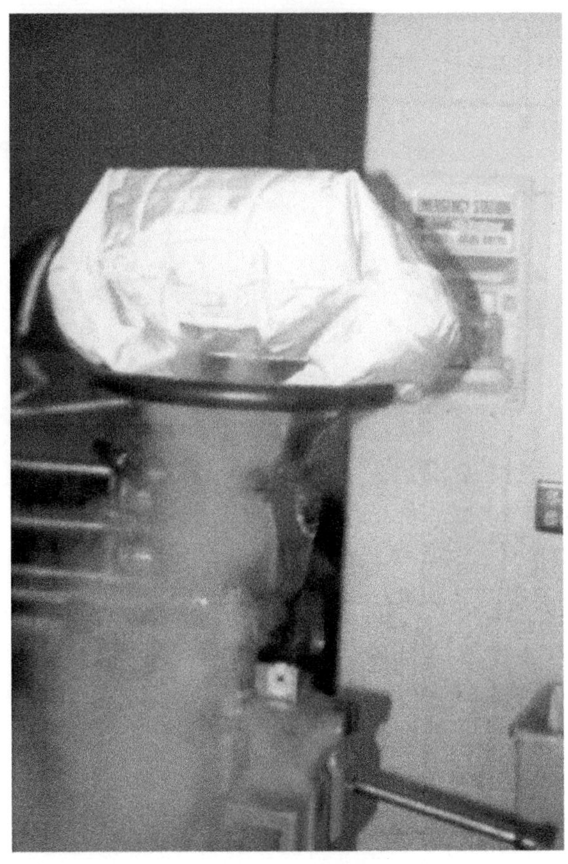

FIGURE 23–20 An airbag being deployed as part of a demonstration in an automotive laboratory.

OCCUPANT DETECTION SYSTEMS

PURPOSE AND FUNCTION The U.S. Federal Motor Vehicle Safety Standard 208 (FMVSS) specifies that the passenger side airbag be disabled or deployed with reduced force under the following conditions. This system is referred to as an **occupant detection system (ODS)** or the **passenger presence system (PPS)**.

- When there is no weight on the seat and no seat belt is fastened, the passenger side airbag will not deploy and the passenger airbag light should be off. SEE FIGURE 23–21.
- The passenger side airbag will be disabled and the disabled airbag light will be "on" only if at least 10 to 37 lb (4.5 to 17 kg) is on the passenger seat, which would generally represent a seated child.

FIGURE 23–21 A dash warning lamp will light if the passenger side airbag is off because no passenger was detected by the seat sensor.

FIGURE 23–22 The passenger side airbag "on" lamp will light if a passenger is detected on the passenger seat.

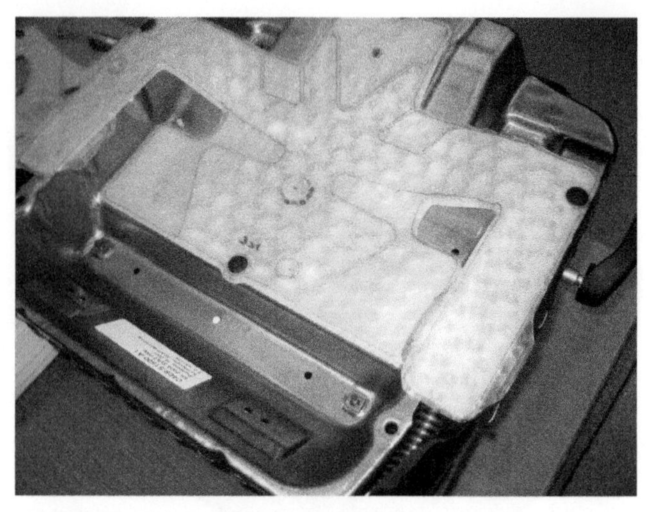

FIGURE 23–23 A gel-filled (bladder-type) occupant detection sensor showing the pressure sensor and wiring.

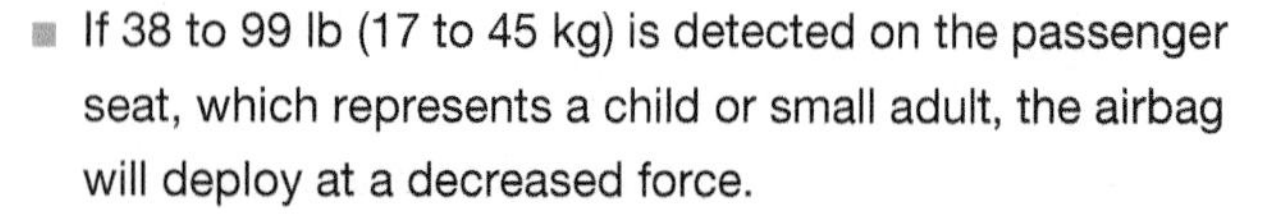

- If 38 to 99 lb (17 to 45 kg) is detected on the passenger seat, which represents a child or small adult, the airbag will deploy at a decreased force.
- If 99 lb (45 kg) or more is detected on the passenger seat, the airbag will deploy at full force, depending on the severity of the crash, speed of the vehicle, and other factors which may result in the airbag deploying at a reduced force.

  ● **SEE FIGURE 23–22.**

FIGURE 23–24 A resistor-type occupant detection sensor. The weight of the passenger strains these resistors, which are attached to the seat, thereby signaling to the module the weight of the occupant.

TYPE OF SEAT SENSOR The passenger presence system (PPS) uses one of three types of sensors.

- **Gel-filled bladder sensor.** This type of occupant sensor uses a silicone-filled bag that has a pressure sensor attached. The weight of the passenger is measured by the pressure sensor, which sends a voltage signal to the module controlling the airbag deployment. A safety belt tension sensor is also used with a gel-filled bladder system to monitor the tension on the belt. The module then uses the information from both the bladder and the seat belt sensor to determine if a tightened belt may be used to restrain a child seat. ● **SEE FIGURE 23–23.**
- **Capacitive strip sensors.** This type of occupant sensor uses several flexible conductive metal strips under the seat cushion. These sensor strips transmit and receive a low-level electric field, which changes due to the weight of the front passenger seat occupant. The module determines the weight of the occupant based on the sensor values.
- **Force-sensing resistor sensors.** This type of occupant sensor uses resistors, which change their resistance based on the stress that is applied. These resistors are part of the seat structure, and the module can determine the weight of the occupant based on the change in the resistance of the sensors. ● **SEE FIGURE 23–24.**

CAUTION: Because the resistors are part of the seat structure, it is very important that all seat fasteners be torqued to factory specifications to ensure proper operation of the occupant detection system. A *seat track position (STP) sensor* is used by the airbag controller to determine the position of the seat. If the seat is too close to the airbag, the controller may disable the airbag.

DIAGNOSING OCCUPANT DETECTION SYSTEMS A fault in the system may cause the passenger side airbag light to turn on when there is no weight on the seat. A scan tool is often used to check or calibrate the seat, which must be empty, by commanding the module to rezero the seat sensor. Some systems use a unit that has various weights along with a scan tool to calibrate and diagnose the occupant detection system. ● **SEE FIGURE 23–25.**

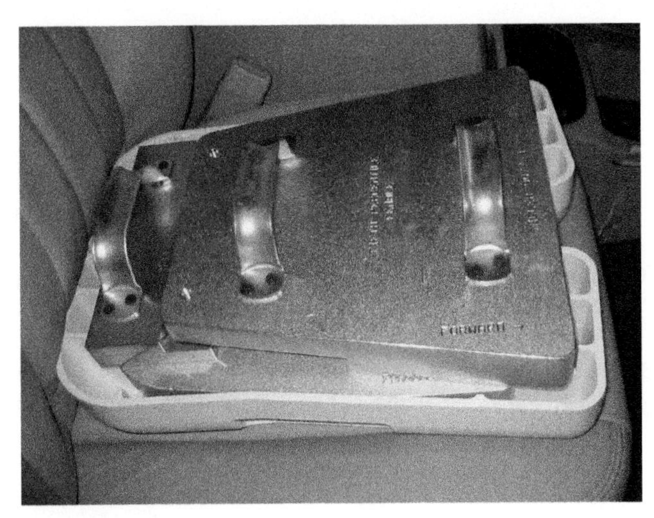

FIGURE 23–25 A test weight is used to calibrate the occupant detection system on this vehicle.

TECH TIP

Aggressive Driving and OnStar

If a vehicle equipped with the OnStar system is being driven aggressively and the electronic stability control system has to intercede to keep the vehicle under control, OnStar may call the vehicle to see if there has been an accident. The need for a call from OnStar usually will be determined if the accelerometer registers slightly over 1 g-force, which could be achieved while driving on a race track.

EVENT DATA RECORDERS

PARTS AND OPERATION As part of the airbag controller on many vehicles, the **event data recorder (EDR)** is used to record parameters just before and slightly after an airbag deployment. The following parameters are recorded.

- Vehicle speed
- Brake on/off
- Seat belt fastened
- G-forces as measured by the accelerometer

Unlike an airplane event data recorder, a vehicle unit is not a separate unit and does not record voice conversations and does not include all crash parameters. This means that additional crash data, such as skid marks and physical evidence at the crash site, will be needed to fully reconstruct the incident.

The EDR is embedded into the airbag controller and receives data from many sources and at varying sample rates. The data is constantly being stored in a memory buffer and not recorded into the EPROM unless an airbag deployment has been commanded. The combined data is known as an *event file.* The airbag is commanded on, based on input mainly from the accelerometer sensor. This sensor, usually built into the airbag controller, is located inside the vehicle. The accelerometer calculates the rate of change of the speed of the vehicle. This determines the acceleration rate and is used to predict if that rate is high enough to deploy the frontal airbags. The airbags will be deployed if the threshold g-value is exceeded. The passenger side airbag will also be deployed unless it is suppressed by either of the following:

- No passenger is detected.
- The passenger side airbag switch is off.

DATA EXTRACTION Data extraction from the event data recorder in the airbag controller can only be achieved using a piece of equipment known as the Crash Data Retrieval System, manufactured by Vetronics Corporation. This is the only authorized method for retrieving event files and only certain organizations are allowed access to the data. These groups or organizations include:

- Original equipment manufacturer's representatives
- National Highway Traffic Safety Administration
- Law enforcement agencies
- Accident reconstruction companies

Crash data retrieval must only be done by a trained crash data retrieval (CDR) technician or analyst. A technician undergoes specialized training and must pass an examination. An analyst must attend additional training beyond that of a technician to achieve CDR analyst certification.

SUMMARY

1. Airbags use a sensor(s) to determine if the rate of deceleration is enough to cause bodily harm.
2. All airbag electrical connectors and conduit are yellow and all electrical terminals are gold plated to protect against corrosion.
3. Always follow the manufacturer's procedure for disabling the airbag system prior to any work performed on the system.
4. Frontal airbags only operate within 30 degrees from center and do not deploy in the event of a rollover, side, or rear collision.
5. Two sensors must be triggered at the same time for an airbag deployment to occur. Many newer systems use an accelerometer-type crash sensor that actually measures the amount of deceleration.
6. Pretensioners are explosive (pyrotechnic) devices that remove the slack from the seat belt and help position the occupant.
7. Occupant detection systems use sensors in the seat to determine whether the airbag will be deployed and with full or reduced force.

REVIEW QUESTIONS

1. What are the safety precautions to follow when working around an airbag?
2. What sensor(s) must be triggered for an airbag deployment?
3. How should deployed inflation modules be handled?
4. What is the purpose of pretensioners?

CHAPTER QUIZ

1. A vehicle is being repaired after an airbag deployment. Technician A says that the inflator module should be handled as if it is still live. Technician B says rubber gloves should be worn to prevent skin irritation. Which technician is correct?
   - **a.** Technician A only
   - **b.** Technician B only
   - **c.** Both Technicians A and B
   - **d.** Neither Technician A nor B
2. A seat belt pretensioner is ________.
   - **a.** A device that contains an explosive charge
   - **b.** Used to remove slack from the seat belt in the event of a collision
   - **c.** Used to force the occupant back into position against the seat back in the event of a collision
   - **d.** All of the above
3. What conducts power and ground to the driver's side airbag?
   - **a.** Twisted-pair wires
   - **b.** Clockspring
   - **c.** Carbon contact and brass surface plate on the steering column
   - **d.** Magnetic reed switch
4. Two technicians are discussing dual-stage airbags. Technician A says that a deployed airbag is safe to handle regardless of which stage caused the deployment of the airbag. Technician B says that both stages ignite, but at different speeds depending on the speed of the vehicle. Which technician is correct?
   - **a.** Technician A only
   - **b.** Technician B only
   - **c.** Both Technicians A and B
   - **d.** Neither Technician A nor B
5. Where are shorting bars used?
   - **a.** In pretensioners
   - **b.** At the connectors for airbags
   - **c.** In the crash sensors
   - **d.** In the airbag controller
6. Technician A says that a deployed airbag can be repacked, reused, and reinstalled in the vehicle. Technician B says that a deployed airbag should be discarded and replaced with an entire new assembly. Which technician is correct?
   - **a.** Technician A only
   - **b.** Technician B only
   - **c.** Both Technicians A and B
   - **d.** Neither Technician A nor B
7. What color are the airbag electrical connectors and conduit?
   - **a.** Blue
   - **b.** Red
   - **c.** Yellow
   - **d.** Orange
8. Driver and/or passenger front airbags will only deploy if a collision occurs how many degrees from straight ahead?
   - **a.** 10 degrees
   - **b.** 30 degrees
   - **c.** 60 degrees
   - **d.** 90 degrees
9. How many sensors must be triggered at the same time to cause an airbag deployment?
   - **a.** One
   - **b.** Two
   - **c.** Three
   - **d.** Four
10. The electrical terminals used for airbag systems are unique because they are ________.
    - **a.** Solid copper
    - **b.** Tin-plated heavy-gauge steel
    - **c.** Silver plated
    - **d.** Gold plated

chapter 24

AUDIO SYSTEM OPERATION AND DIAGNOSIS

LEARNING OBJECTIVES

After studying this chapter, the reader will be able to:

1. Describe how AM, FM, and satellite radio work.
2. Describe antennas and their diagnosis.
3. Discuss the purpose, function, and types of speakers.
4. Discuss crossovers and voice recognition systems.
5. Explain how Bluetooth systems work.
6. List causes and corrections of radio noise and interference.

This chapter will help you prepare for the ASE Electrical/Electronic Systems (A6) certification test content area "H" (Accessories Diagnosis and Repair).

GM STC OBJECTIVES

GM Service Technical College topics covered in this chapter are as follows:

1. Perform diagnosis of audio system concerns and isolate the root cause.
2. Perform diagnosis of XM radio systems concerns and isolate the root cause.
3. Perform satellite antenna diagnosis for GPS and XM reception concerns.
4. Use the DMM for speaker voltage testing.

KEY TERMS

AUDIO FUNDAMENTALS

INTRODUCTION The audio system of today's vehicles is a complex combination of antenna system, receiver, amplifier, and speakers, all designed to provide living room-type music reproduction while the vehicle is traveling in city traffic or at highway speed.

Audio systems produce audible sounds and include the following:

- Radio (AM, FM, and satellite)
- Antenna systems that are used to capture electronic energy broadcast to radios
- Speaker systems
- Aftermarket enhancement devices that increase the sound energy output of an audio system
- Diagnosis of audio-related problems

Many audio-related problems can be addressed and repaired by a service technician.

TYPES OF ENERGY There are two types of energy that affect audio systems.

- **Electromagnetic energy or radio waves.** Antennas capture the radio waves which are then sent to the radio or receiver to be amplified.
- **Acoustical energy, usually called sound.** Radios and receivers amplify the radio wave signals and drive speakers which reproduce the original sound as transmitted by radio waves.

  SEE FIGURE 24–1.

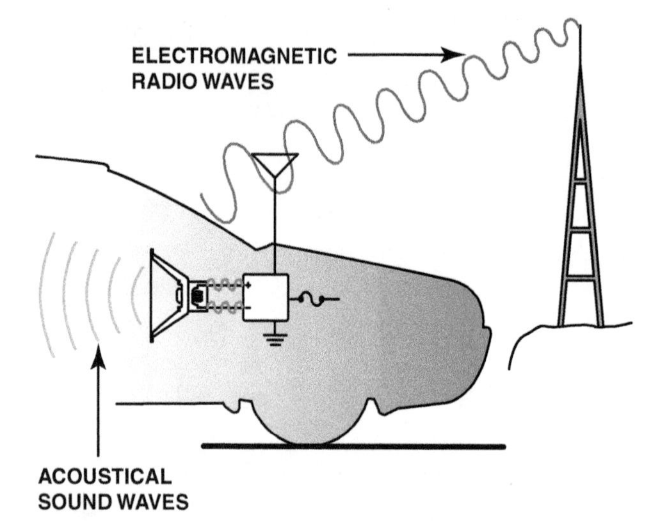

FIGURE 24–1 Audio systems use both electromagnetic radio waves and sound waves to reproduce sound inside the vehicle.

TERMINOLOGY Radio waves travel at approximately the speed of light (186,282,000 miles per second) and are electromagnetic. Radio waves are measured in two ways, wavelength and frequency. A radio wave has a series of high points and low points. A wavelength is the time and distance between two consecutive points, either high or low. A wavelength is measured in meters. **Frequency**, also known as **radio frequency (RF)**, is the number of times a particular waveform repeats itself in a given amount of time, and is measured in **hertz (Hz)**. A signal with a frequency of 1 Hz is one radio wavelength per second. Radio frequencies are measured in kilohertz (kHz), thousands of wavelengths per second, and megahertz (MHz), millions of wavelengths per second. **SEE FIGURE 24–2.**

- The higher the frequency, the shorter the wavelength.
- The lower the frequency, the longer the wavelength.

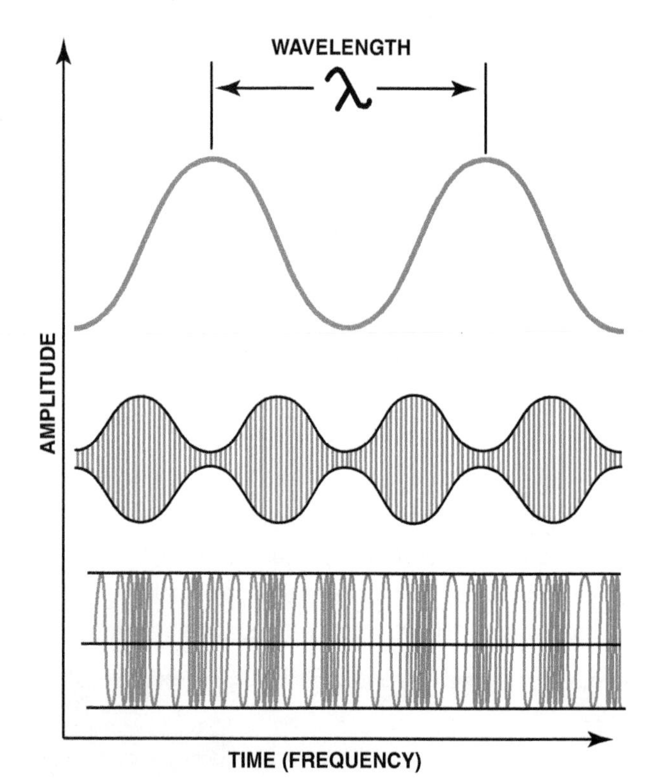

FIGURE 24–2 The relationship among wavelength, frequency, and amplitude.

A longer wavelength can travel a further distance than a shorter wavelength. Therefore, lower frequencies provide better reception at further distances.

- AM radio frequencies range from 530 to 1,710 kHz.
- FM radio frequencies range from 87.9 to 107.9 MHz.

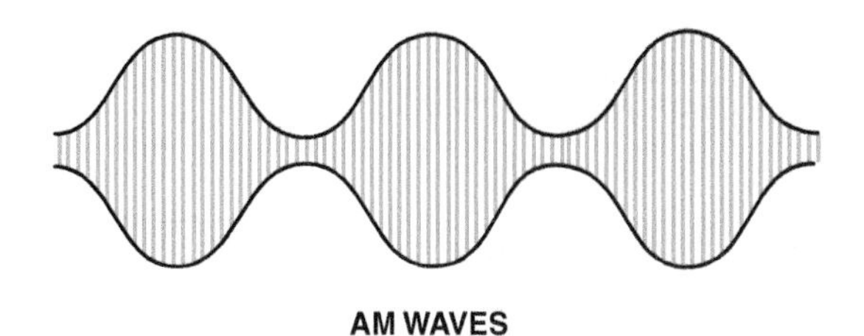

FIGURE 24–3 The amplitude changes in AM broadcasting.

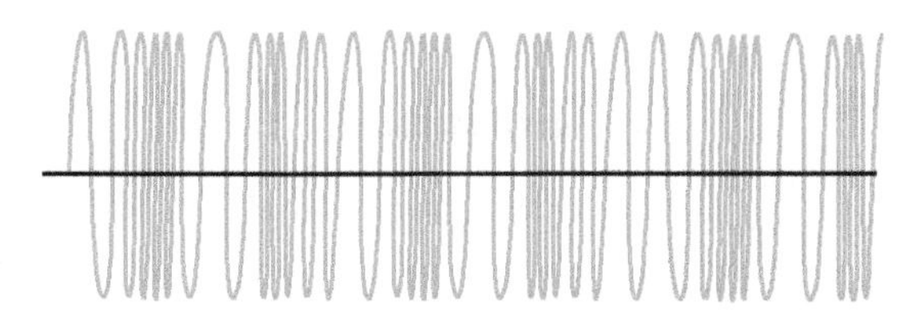

FIGURE 24–4 The frequency changes in FM broadcasting and the amplitude remains constant.

MODULATION **Modulation** is the term used to describe when information is added to a constant frequency. The base radio frequency used for RF is called the *carrier wave.* A carrier is a radio wave that is changed to carry information. The two types of modulation are:

- **Amplitude modulation (AM)**
- **Frequency modulation (FM)**

AM waves are radio waves that have amplitude that can be varied, transmitted, and detected by a receiver. Amplitude is the height of the wave as graphed on an oscilloscope. ● **SEE FIGURE 24–3.**

FM waves are also radio waves that have a frequency that can be varied, transmitted, and detected by a receiver. This type of modulation changes the number of cycles per second, or frequency, to carry the information. ● **SEE FIGURE 24–4.**

RADIO WAVE TRANSMISSION More than one signal can be carried by a radio wave. This process is called *sideband operation.* Sideband frequencies are measured in kilohertz. The amount of the signal above the assigned frequency is referred to as the upper sideband. The amount of the signal below the assigned frequency is called the lower sideband. This capability allows radio signals to carry stereo broadcasts. Stereo broadcasts use the upper sideband to carry one channel of the stereo signal, and the lower sideband to carry the other channel. When the signal is decoded by the radio, these two signals become the right and left channels. ● **SEE FIGURE 24–5.**

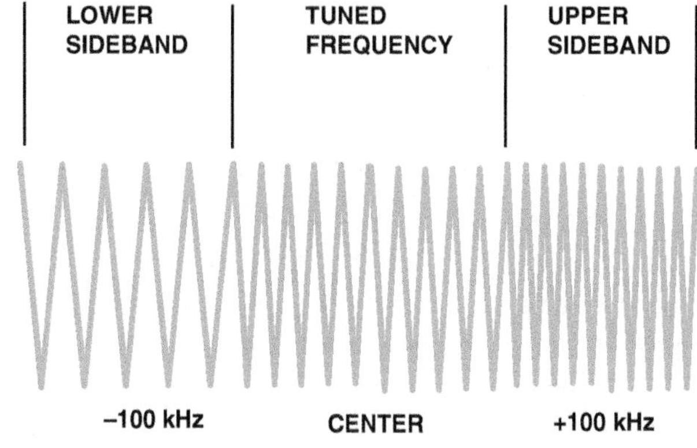

FIGURE 24–5 Using upper and lower sidebands allows stereo to be broadcast. The receiver separates the signals to provide left and right channels.

? FREQUENTLY ASKED QUESTION

What Does a "Capture" Problem Mean?

A capture problem affects only FM reception and means that the receiver is playing more than one station if two stations are broadcasting at the same frequency. Most radios capture the stronger signal and block the weaker signal. However, if the stronger signal is weakened due to being blocked by buildings or mountains, the weaker signal will then be used. When this occurs, it will sound as if the radio is changing stations by itself. This is not a fault with the radio, but simply a rare occurrence with FM radio.

NOISE Because radio waves are a form of electromagnetic energy, other forms of energy can impact them. For example, a bolt of lightning generates broad radio-frequency bandwidths known as radio-frequency interference (RFI). RFI is one type of electromagnetic interference (EMI) and is the frequency that interferes with radio transmission.

AM CHARACTERISTICS AM radio reception can be achieved over long distances from the transmitter because the waves can bounce off the ionosphere, usually at night. Even during the day, the AM signals can be picked up some distance from the transmitter. AM radio reception depends on a good antenna. If there is a fault in the antenna circuit, AM reception is affected the most.

FM CHARACTERISTICS Because FM waves have a high RF and a short wavelength, they travel only a short distance. The waves cannot follow the shape of the earth but instead travel in a straight line from the transmitter to the receiver. FM waves will travel through the ionosphere and into space and do not reflect back to earth like AM waves.

MULTIPATH Multipath is caused by reflected, refracted, or line of sight signals reaching an antenna at different times. Multipath results from the radio receiving two signals to process on the same frequency. This causes an echo effect in the speakers. *Flutter,* or *picket fencing* as it is sometimes called, is caused by the blocking of part of the FM signal. This blocking causes a weakening of the signal resulting in only part of the signal getting to the antenna, causing an on-again off-again radio sound. Flutter also occurs when the transmitter and the receiving antenna are far apart.

RADIOS AND RECEIVERS

The antenna receives the radio wave where it is converted into very weak fluctuating electrical current. This current travels along the antenna lead-in to the radio that amplifies the signal and sends the new signal to the speakers where it is converted into acoustical energy.

Most late-model radios and receivers use five input/output circuits.

1. **Power.** Usually a constant 12-volt feed to keep the internal clock alive.
2. **Ground.** This is the lowest voltage in the circuit and connects indirectly to the negative terminal of the battery.
3. **Serial data.** Used to turn the unit on and off and provide other functions such as steering wheel control operation.
4. **Antenna input.** From one or more antennas.
5. **Speaker outputs.** These wires connect the receiver to the speakers or as an input to an amplifier.

ANTENNAS

TYPES OF ANTENNAS The typical radio electromagnetic energy from the broadcast antenna induces a signal in the antenna that is very small, only about 25 microvolts AC (0.000025 VAC) in strength. The radio contains amplifier circuits that increase the received signal strength into usable information.

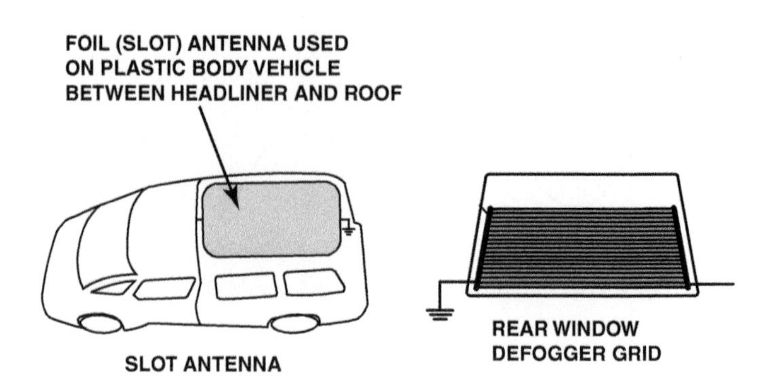

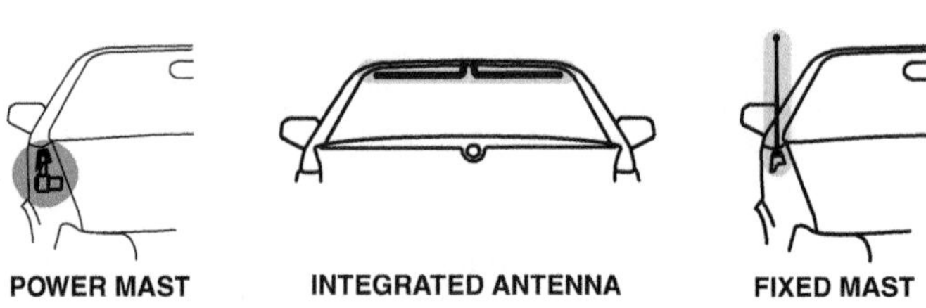

FIGURE 24–6 The five types of antennas used on General Motors vehicles include the slot antenna, fixed mast antenna, rear window defogger grid antenna, a powered mast antenna, and an integrated antenna.

For example, the five types of antennas used on vehicles include the following:

- **Slot antenna.** The slot antenna is concealed in the roof of some plastic body vehicles such as older General Motors plastic body vans. This antenna is surrounded by metal on a Mylar sheet.
- **Rear window defogger grid.** This type of system uses the heating wires to receive the signals and special circuitry to separate the RF from the DC heater circuit.
- **Powered mast.** These antennas are controlled by the radio. When the radio is turned on, the antenna is raised; when the radio is shut off, the antenna is retracted. The antenna system consists of an antenna mast and a drive motor controlled by the radio "on" signal through a relay.
- **Fixed mast antenna.** This antenna offers the best overall performance currently available. The mast is simply a vertical rod. Mast antennas are typically located on the fender or rear quarter panel of the vehicle.
- **Integrated antenna.** This type of antenna is sandwiched in the windshield and an appliqué on the rear window glass. The antenna in the rear window is the primary antenna and receives both AM and FM signals. The secondary antenna is located in the front windshield typically on the passenger side of the vehicle. This antenna receives only FM signals.

SEE FIGURE 24–6.

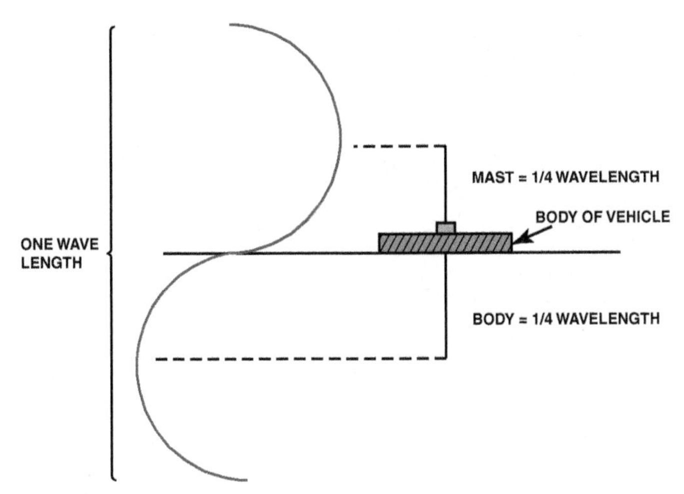

FIGURE 24–7 The ground plane is actually one-half of the antenna.

FREQUENTLY ASKED QUESTION

What Is a Ground Plane?

Antennas designed to pick up the electromagnetic energy that is broadcast through the air to the transmitting antenna are usually one-half wavelength high, and the other half of the wavelength is the **ground plane**. This one-half wavelength in the ground plane is literally underground.

For ideal reception, the receiving antenna should also be the same as the wavelength of the signal. Because this length is not practical, a design compromise uses the length of the antenna as one-fourth of the wavelength; in addition, the body of the vehicle itself is one-fourth of the wavelength. The body of the vehicle, therefore, becomes the ground plane. ● **SEE FIGURE 24–7.**

Any faulty condition in the ground plane circuit will cause the ground plane to lose effectiveness, such as:

- Loose or corroded battery cable terminals
- Acid buildup on battery cables
- Engine grounds with high resistance
- Loss of antenna or audio system grounds
- Defective alternator, causing an AC ripple exceeding 50 millivolts (0.050 volt)

ANTENNA HEIGHT The antenna collects all radio-frequency signals. An AM radio operates best with as long an antenna as possible, but FM reception is best when the antenna height is exactly 31 inches (79 cm). Most fixed-length antennas are, therefore, exactly this height. Even the horizontal section of a windshield antenna is 31 inches (79 cm) long.

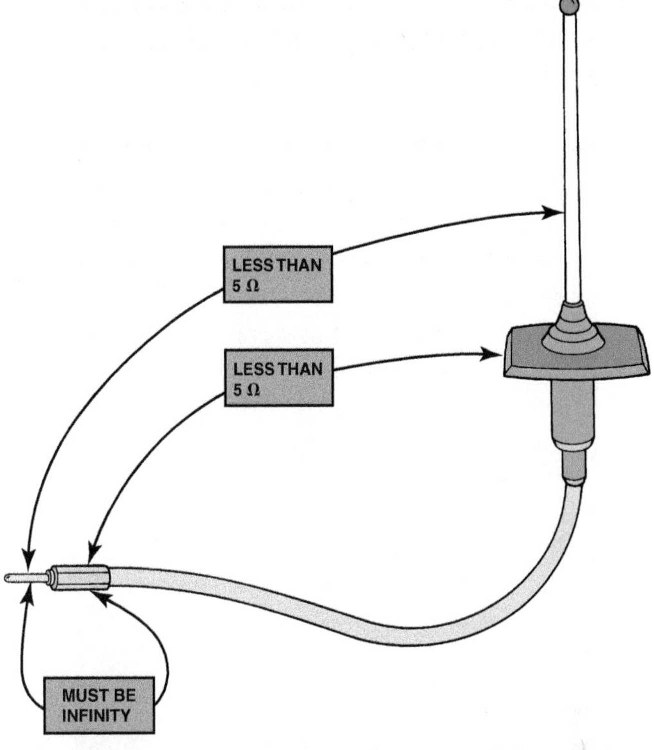

FIGURE 24–8 If all ohmmeter readings are satisfactory, the antenna is good.

A defective antenna will:

- Greatly affect AM radio reception
- May affect FM radio reception

ANTENNA DIAGNOSIS

ANTENNA TESTING If the antenna or lead-in cable is broken (open), FM reception will be heard but may be weak, and there will be *no* AM reception. An ohmmeter should read infinity between the center antenna lead and the antenna case. For proper reception and lack of noise, the case of the antenna must be properly grounded to the vehicle body. ● **SEE FIGURE 24–8.**

POWER ANTENNA TESTING AND SERVICE Most power antennas use a circuit breaker and a relay to power a reversible, permanent magnet (PM) electric motor that moves a nylon cord attached to the antenna mast. Some vehicles have a dash-mounted control that can regulate antenna mast

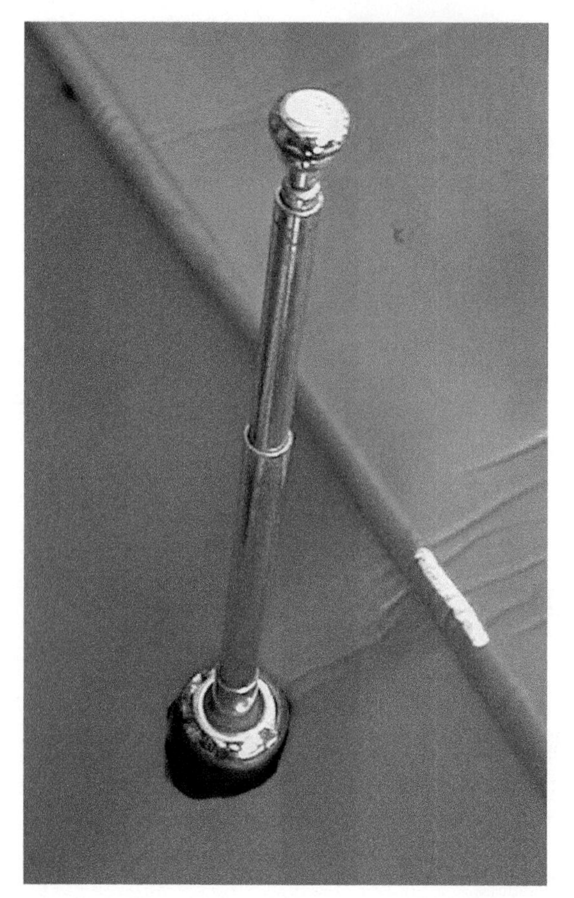

FIGURE 24–9 Cutting a small hole in a fender cover helps to protect the vehicle when replacing or servicing an antenna.

TECH TIP

The Hole in the Fender Cover Trick

A common repair is to replace the mast of a power antenna. To help prevent the possibility of causing damage to the body or paint of the vehicle, cut a hole in a fender cover and place it over the antenna. ● **SEE FIGURE 24–9.**

If a wrench or tool slips during the removal or installation process, the body of the vehicle will be protected.

height and/or operation, whereas many operate automatically when the radio is turned on and off. The power antenna assembly is usually mounted between the outer and inner front fender or in the rear quarter panel. The unit contains the motor, a spool for the cord, and upper- and lower-limit switches. The power antenna mast is tested in the same way

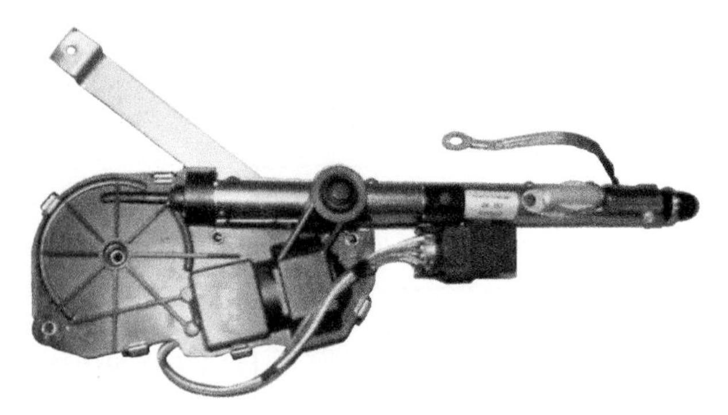

FIGURE 24–10 A typical power antenna assembly. Note the braided ground wire used to ensure that the antenna has a good ground plane.

as a fixed-mast antenna. (An infinity reading should be noted on an ohmmeter when the antenna is tested between the center antenna terminal and the housing or ground.) Except in the case of cleaning or mast replacement, most power antennas are either replaced as a unit or repaired by specialty shops. ● **SEE FIGURE 24–10.**

Making certain that the drain holes in the motor housing are not plugged with undercoating, leaves, or dirt can prevent many power antenna problems. All power antennas should be kept clean by wiping the mast with a soft cloth and lightly oiling with light oil such as WD-40 or a similar grade oil.

SPEAKERS

PURPOSE AND FUNCTION The purpose of any **speaker** is to reproduce the original sound as accurately as possible. Speakers are also called *loudspeakers.* The human ear is capable of hearing sounds from a very low frequency of 20 Hz (cycles per seconds) to as high as 20,000 Hz. No one speaker is capable of reproducing sound over such a wide frequency range. ● **SEE FIGURE 24–11.**

Good-quality speakers are the key to a proper sounding radio or sound system. Replacement speakers should be securely mounted and wired according to the correct *polarity.* ● **SEE FIGURE 24–12.**

IMPEDANCE MATCHING All speakers used on the same radio or amplifier should have the same internal coil resistance, called **impedance**. If unequal-impedance speakers are used, sound quality may be reduced and serious damage to the radio may result. ● **SEE FIGURE 24–13.**

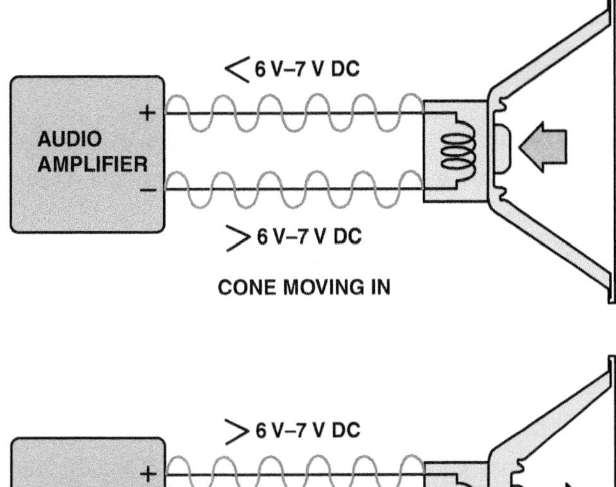

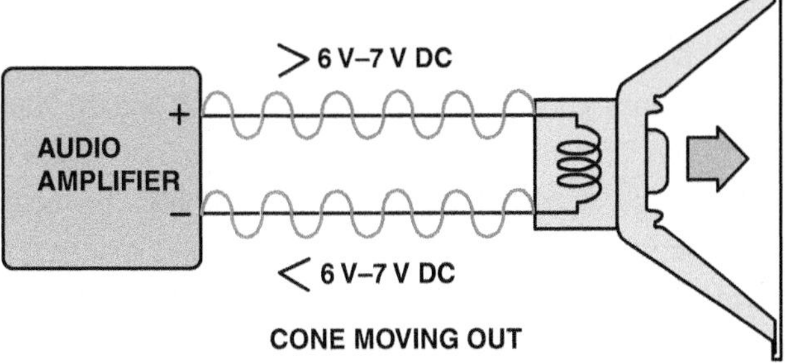

FIGURE 24–11 Between 6 and 7 volts is applied to each speaker terminal, and the audio amplifier then increases the voltage on one terminal and at the same time decreases the voltage on the other terminal causing the speaker cone to move. The moving cone then moves the air, causing sound.

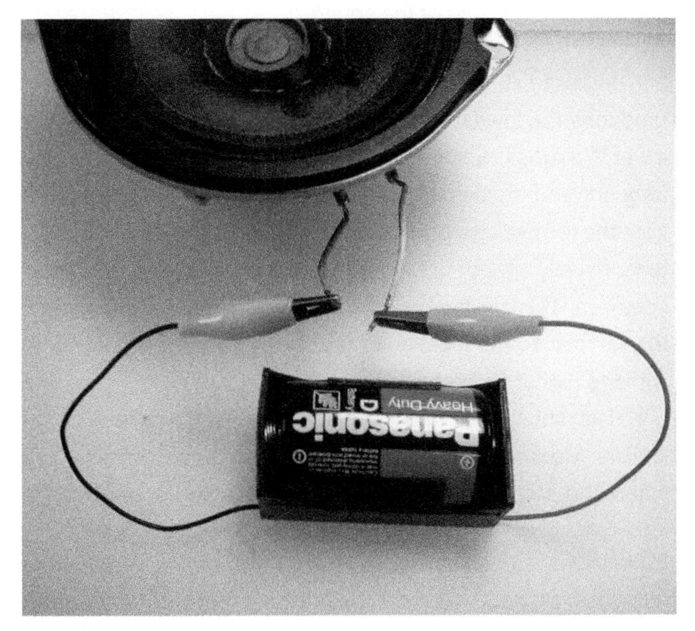

FIGURE 24–12 A typical automotive speaker with two terminals. The polarity of the speakers can be identified by looking at the wiring diagram in the service manual or by using a 1.5-volt battery to check. When the battery positive is applied to the positive terminal of the speaker, the cone will move outward. When the battery leads are reversed, the speaker cone will move inward.

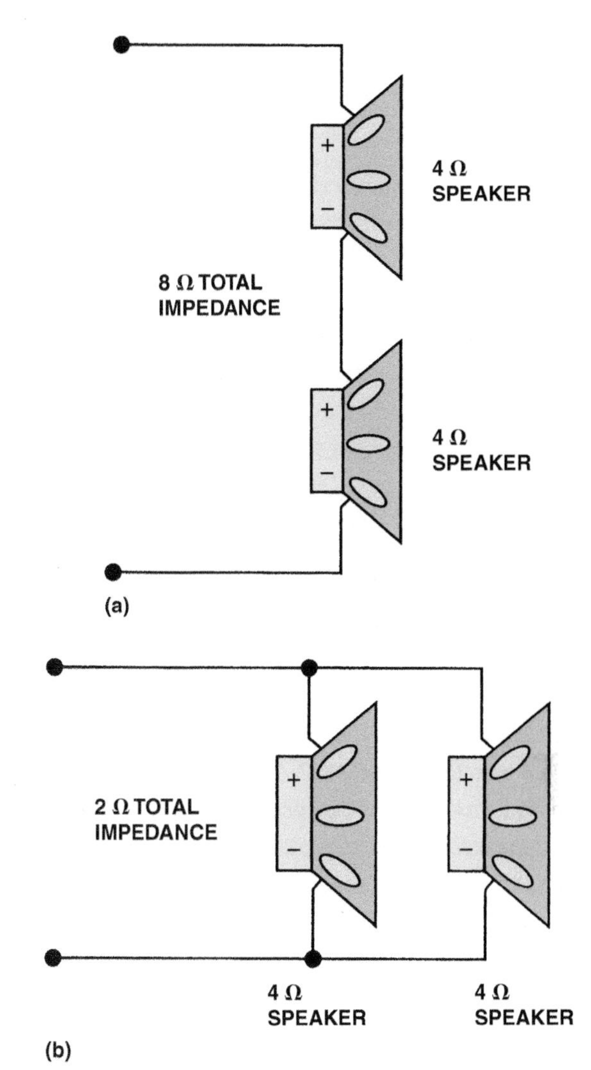

FIGURE 24–13 (a) Two 4 ohm speakers connected in series result in total impedance of 8 ohms. (b) Two 4 ohm speakers connected in parallel result in total impedance of 2 ohms.

All speakers should have the same impedance. For example, if two 4 ohm speakers are being used for the rear and they are connected in parallel, the total impedance is 2 ohms.

$$R_T = \frac{\textbf{4 Ω (impedance of each speaker)}}{\textbf{2 (number of speakers in parallel)}} = \textbf{2 ohms}$$

The front speakers should also represent a 2 ohm load from the radio or amplifier. See the following example.

Two front speakers: each 2 ohms

Two rear speakers: each 8 ohms

Solution: Connect the front speakers in series (connect the positive [+] of one speaker to the negative [−] of the other) for a total impedance of 4 ohms (2 Ω + 2 Ω = 4 Ω). Connect the two rear speakers in parallel (connect the positive [+] of each speaker together and the negative [−] of each speaker together) for a total impedance of 4 ohms (8 Ω ÷ 2 = 4 Ω).

 TECH TIP

Skin Effect

When a high-frequency signal (AC voltage) is transmitted through a wire, the majority of it travels on the outside surface of the wire. This characteristic is called skin effect. The higher the frequency is, the closer to the outer surface the signal moves. To increase audio system output, most experts recommend the use of wire that has many strands of very fine wire to increase the surface area or the skin area of the conductor. Therefore, most aftermarket speaker wires are stranded with many small-diameter copper strands.

SPEAKER WIRING The wire used for speakers should be as large a wire (as low an AWG gauge number) as is practical in order to be assured that full power is reaching the speakers. Typical "speaker wire" is about 22 gauge (0.35 mm^2), yet tests conducted by audio engineers have concluded that increasing the wire gauge—up to 4 gauge (19 mm^2) or larger—greatly increases sound quality. All wiring connections should be soldered after making certain that all speaker connections have the correct polarity.

CAUTION: Be careful when installing additional audio equipment on a General Motors vehicle system that uses a two-wire speaker connection called a floating ground system. Other systems run only one power (hot) lead to each speaker and ground the other speaker lead to the body of the vehicle.

This arrangement helps prevent interference and static that could occur if these components were connected to a chassis (vehicle) ground. If the components are chassis grounded, there may be a difference in the voltage potential (voltage); this condition is called a *ground loop.*

CAUTION: Regardless of radio speaker connections used, *never* operate any radio without the speakers connected, or the speaker driver section of the radio may be damaged as a result of the open speaker circuit.

SPEAKER TYPES

INTRODUCTION No one speaker is capable of reproducing sound over such a wide frequency range. Therefore, speakers are available in three basic types.

1. Tweeters are for high-frequency ranges.
2. Midrange are for mid-frequency ranges.
3. Woofers and subwoofers are for low-frequency ranges.

TWEETER A **tweeter** is a speaker designed to reproduce high-frequency sounds, usually between 4,000 and 20,000 Hz (4 and 20 kHz). Tweeters are very directional. This means that the human ear is most likely to be able to detect the location of the speaker while listening to music. This also means that a tweeter should be mounted in the vehicle where the sound can be directed in line of sight to the listener. Tweeters are usually mounted on the inside door near the top, windshield "A" pillar, or similar locations.

MIDRANGE A midrange speaker is designed and manufactured to be able to best reproduce sounds in the middle of the human hearing range, from 400 to 5,000 Hz. Most people are sensitive to the sound produced by these midrange speakers. These speakers are also directional in that the listener can usually locate the source of the sound.

SUBWOOFER A **subwoofer**, sometimes called a *woofer,* produces the lowest frequency of sounds, usually 125 Hz and lower. A *midbass* speaker may also be used to reproduce those frequencies between 100 and 500 Hz. Low-frequency sounds from these speakers are *not* directional. This means that the listener usually cannot detect the source of the sound from these speakers. The low-frequency sounds seem to be everywhere in the vehicle, so the location of the speakers is not as critical as with the higher frequency speakers.

The subwoofer can be placed almost anywhere in the vehicle. Most subwoofers are mounted in the rear of the vehicle where there is more room for the larger subwoofer speakers.

SPEAKER FREQUENCY RESPONSE Frequency response is how a speaker responds to a range of frequencies. A typical frequency response for a midrange speaker may be 500 to 4,000 Hz.

WARNING

Hearing loss is possible if exposed to loud sounds. According to noise experts (audiologists), hearing protection should be used whenever the following occurs.

1. You must raise your voice to be heard by others next to you.
2. You cannot hear someone else speaking who is less than 3 ft (1 m) away.
3. You are operating power equipment, such as a lawnmower.

FREQUENTLY ASKED QUESTION

What Is a Bass Blocker?

A bass blocker is a capacitor and coil assembly that effectively blocks low frequencies. A bass blocker is normally used to block low frequencies being sent to the smaller front speakers. Using a bass blocker allows the smaller front speakers to more efficiently reproduce the midrange and high-range frequency sounds.

SOUND LEVELS

DECIBEL SCALE A **decibel (dB)** is a measure of sound power, and it is the faintest sound a human can hear in the midband frequencies. The dB scale is not linear (straight line) but logarithmic, meaning that a small change in the dB reading results in a large change in volume of noise. An increase of 10 dB in sound pressure is equal to doubling the perceived volume. Therefore, a small difference in dB rating means a big difference in the sound volume of the speaker.

EXAMPLES Some examples of decibel sound levels include:

- Quiet, faint — 30 dB: whisper, quiet library; 40 dB: quiet room
- Moderate — 50 dB: moderate range sound; 60 dB: normal conversation
- Loud — 70 dB: vacuum cleaner, city traffic; 80 dB: busy noisy traffic, vacuum cleaner
- Extremely loud — 90 dB: lawnmower, shop tools; 100 dB: chain saw, air drill
- Hearing loss possible — 110 dB: loud rock music

CROSSOVERS

DEFINITION A **crossover** is designed to separate the frequency of a sound and send a certain frequency range, such as low-bass sounds, to a woofer designed to reproduce these low-frequency sounds. There are two types of crossovers: passive and active.

PASSIVE CROSSOVER A passive crossover does not use an external power source. Rather it uses a coil and a capacitor to block certain frequencies that a particular type of speaker cannot handle and allow just those frequencies that it can handle to be applied to the speaker. For example, a 6.6 millihenry coil and a 200 microfarad capacitor can effectively pass 100 Hz frequency sound to a large 10 inches subwoofer. This type of passive crossover is called a **low-pass filter**, because it passes (transfers) only the low-frequency sounds to the speaker and blocks all other frequencies. A **high-pass filter** is used to transfer higher frequency (over 100 Hz) to smaller speakers.

ACTIVE CROSSOVER **Active crossovers** use an external power source and produce superior performance. An active crossover is also called an *electronic crossover* or *crossover network*. These units include many powered filters and are considerably more expensive than passive crossovers. Two amplifiers are necessary to fully benefit from an active crossover. One amplifier is for the higher frequencies and midrange and the other amplifier is for the subwoofers. If you are on a budget and plan to use just one amplifier, then use passive crossover. If you can afford to use two or more amplifiers, then consider using the electronic (active) crossover. ● **SEE FIGURE 24–14** for an example of crossovers used in factory-installed systems.

AFTERMARKET SOUND SYSTEMS

POWER AND GROUND UPGRADES If adding an amplifier and additional audio components, be sure to include the needed power and ground connections. These upgrades can include:

- A separate battery for the audio system
- An inline fuse near the battery to protect the wiring and the components

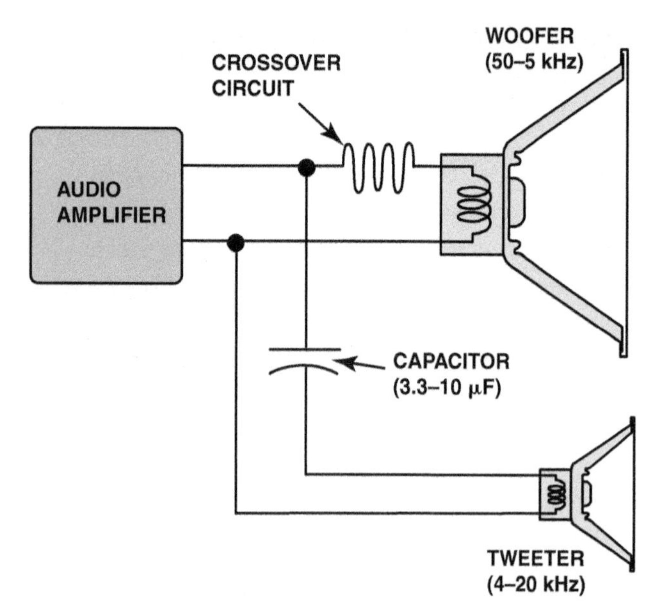

FIGURE 24–14 Crossovers are used in audio systems to send high-frequency sounds to the small (tweeter) speakers and low-frequency sounds to larger (woofer) speakers.

FIGURE 24–15 Two capacitors connected in parallel provide the necessary current flow to power large subwoofer speakers.

- Wiring that is properly sized to the amperage draw of the system and the length of wire (The higher the output wattage, the greater the amperage required and the larger the wire gauge needed. The longer the distance between the battery and the components, the larger the wire gauge needed for best performance.)
- Ground wires at least the same gauge as the power side wiring (Some experts recommend using extra ground wires for best performance.)

Read, understand, and follow all instructions that come with audio system components.

POWERLINE CAPACITOR A **powerline capacitor**, also called a **stiffening capacitor**, refers to a large capacitor (often abbreviated CAP) of 0.25 farad or larger connected to an amplifier power wire. The purpose and function of this capacitor is to provide the electrical reserve energy needed by the amplifier to provide deep bass notes. **SEE FIGURE 24–15.**

Battery power is often slow to respond; and when the amplifier attempts to draw a large amount of current, the capacitor will try to stabilize the voltage level at the amplifier by discharging stored current as needed.

A rule of thumb is to connect a capacitor with a capacity of 1 farad for each 1,000 watts of amplifier power. **SEE CHART 24–1.**

CAPACITOR INSTALLATION A powerline capacitor connects to the power leads between the inline fuse and the amplifier. **SEE FIGURE 24–16.**

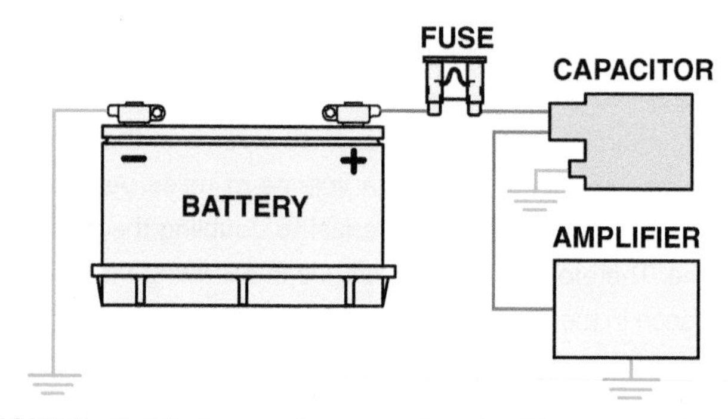

FIGURE 24–16 A powerline capacitor should be connected through the power wire to the amplifier as shown. When the amplifier requires more electrical power (watts) than the battery can supply, the capacitor will discharge into the amplifier and supply the necessary current for the fraction of a second it is needed by the amplifier. At other times when the capacitor is not needed, it draws current from the battery to keep it charged.

POWERLINE CAPACITOR USAGE GUIDE

WATTS (AMPLIFIER)	RECOMMENDED CAPACITOR IN FARADS (MICROFARADS)
100 W	0.10 farad (100,000 μF)
200 W	0.20 farad (200,000 μF)
250 W	0.25 farad (250,000 μF)
500 W	0.50 farad (500,000 μF)
750 W	0.75 farad (750,000 μF)
1,000 W	1.00 farad (1,000,000 μF)

CHART 24–1

The rating of the capacitor needed to upgrade an audio system is directly related to the wattage of the system.

FREQUENTLY ASKED QUESTION

What Do the Amplifier Specifications Mean?

RMS power	**RMS** means root-mean-square and is the rating that indicates how much power the amplifier is capable of producing continuously.
RMS power at 2 ohms	This specification in watts indicates how much power the amplifier delivers into a 2 ohm speaker load. This 2 ohm load is achieved by wiring two 4 ohm speakers in parallel or by using 2 ohm speakers.
Peak power	Peak power is the maximum wattage an amplifier can deliver in a short burst during a musical peak.
THD	**Total harmonic distortion (THD)** represents the amount of change of the signal as it is being amplified. The lower the number, the better the amplifier (e.g., a 0.01% rating is better than a 0.07% rating).
Signal-to-noise ratio	This specification is measured in decibels (dB) and compares the strength of the signal with the level of the background noise (hiss). A higher volume indicates less background noise (e.g., a 105 dB rating is better than a 100 dB rating).

If the capacitor were connected to the circuit as shown without "precharging," the capacitor would draw so much current that it would blow the inline fuse. To safely connect a large capacitor, it must be *precharged.* To precharge the capacitor, follow these steps.

STEP 1 Connect the negative (2) terminal of the capacitor to a good chassis ground.

STEP 2 Insert an automotive 12 volts light bulb, such as a headlight or parking light, between the positive (1) terminal of the capacitor and the positive terminal of the battery. The light will come on as the capacitor is being charged and then go out when the capacitor is fully charged.

STEP 3 Disconnect the light from the capacitor, and then connect the power lead to the capacitor. The capacitor is now fully charged and ready to provide the extra power necessary to supplement battery power to the amplifier.

FIGURE 24–17 Voice commands can be used to control many functions, including navigation systems, climate control, telephone, and radio.

VOICE RECOGNITION

PARTS AND OPERATION **Voice recognition** is an expanding technology. It allows the driver of a vehicle to perform tasks, such as locate an address in a navigation system by using voice commands rather than buttons. In the past, users had to say the exact words to make it work such as the following examples listed from an owner manual for a vehicle equipped with a voice-actuated navigation system.

"Go home"

"Repeat guidance"

"Nearest ATM"

The problem with these simple voice commands was that the exact wording had to be spoken. The voice recognition software would compare the voice command to a specific list of words or phrases stored in the system in order for a match to occur. Newer systems recognize speech patterns and take action based on learned patterns. Voice recognition can be used for the following functions.

1. Navigation system operation (**SEE FIGURE 24–17**.)
2. Sound system operation
3. Climate control system operation
4. Telephone dialing and other related functions (**SEE FIGURE 24–18**.)

A microphone is usually placed in the driver's side sun visor or in the overhead console in the center top portion of the windshield area.

FIGURE 24–18 The voice command icon on the steering wheel of a Cadillac.

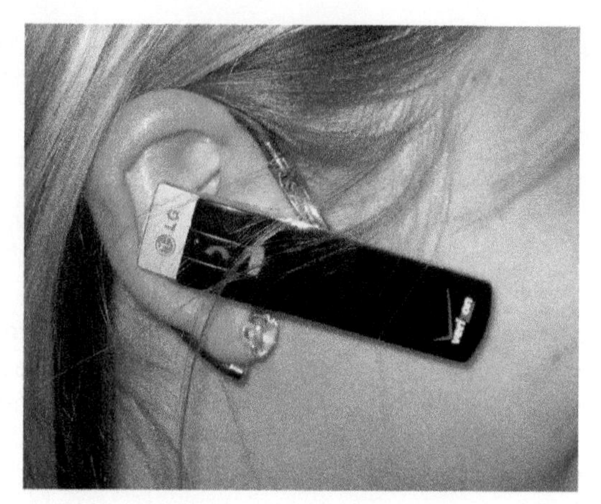

FIGURE 24–19 Bluetooth earpiece that contains a microphone and speaker unit that is paired to a cellular phone. The telephone has to be within 33 ft (10 m) of the earpiece.

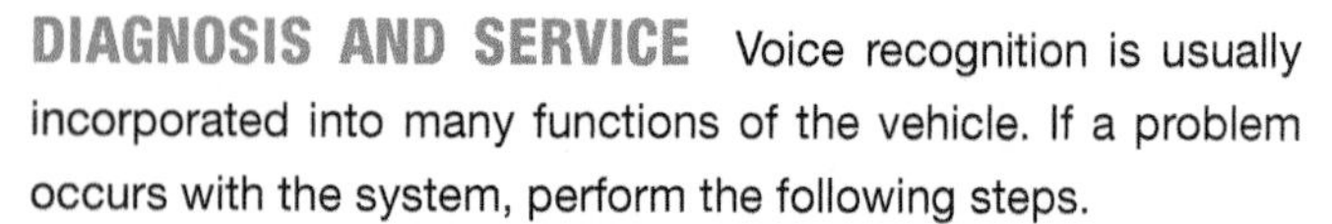

DIAGNOSIS AND SERVICE Voice recognition is usually incorporated into many functions of the vehicle. If a problem occurs with the system, perform the following steps.

1. Verify the customer complaint (concern). Check the owner manual or service information for the proper voice commands and verify that the system is not functioning correctly.
2. Check for any aftermarket accessories that may interfere or were converted to components used by the voice recognition system, such as remote start units, MP3 players, or any other electrical component.
3. Check for stored diagnostic trouble codes (DTCs) using a scan tool.
4. Follow the recommended troubleshooting procedures as stated in service information.

BLUETOOTH

Bluetooth is a (radio frequency) standard for short-range communications. The range of a typical Bluetooth device is 33 ft (10 m) and it operates in the ISM (industrial, scientific, and medical) band between 2.4000 and 2.4835 MHz.

Bluetooth is a wireless standard that works on two levels.

- It provides physical communication using low power, requiring only about 1 milliwatt (1/1,000 of a watt) of electrical power, making it suitable for use with small handheld or portable devices, such as an ear-mounted speaker/microphone.
- It provides a standard protocol for how bits of data are sent and received.

? FREQUENTLY ASKED QUESTION

Where Did Bluetooth Get Its Name?

The early adopters of the standard used the term "Bluetooth," and they named it for Harold Bluetooth, the king of Denmark in the late 900s. The king was able to unite Denmark and part of Norway into a single kingdom.

The Bluetooth standard is an advantage because it is wireless, low cost, and automatic. The automotive use of Bluetooth technology is in the operation of a cellular telephone being tied into the vehicle. The vehicle allows the use of hands-free telephone usage. A vehicle that is Bluetooth telephone equipped has the following components.

- A Bluetooth receiver can be built into the navigation or existing sound system.
- A microphone allows the driver to use voice commands as well as telephone conversations from the vehicle to the cell via Bluetooth wireless connections.

Many cell phones are equipped with Bluetooth which may allow the caller to use an ear-mounted microphone and speaker. ● **SEE FIGURE 24–19.**

If the vehicle and the cell phone are equipped with Bluetooth, the speaker and microphone can be used as a hands-free telephone when the phone is in the vehicle. The cell phone can be activated in the vehicle by using voice commands.

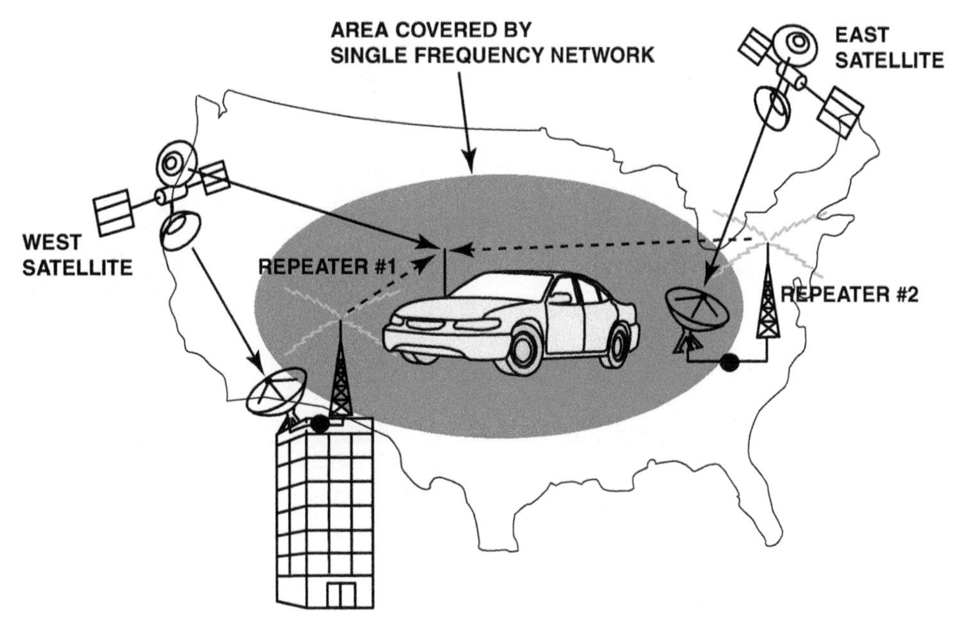

FIGURE 24–20 SDARS uses satellites and repeater stations to broadcast radio.

SATELLITE RADIO

PARTS AND OPERATION Satellite radio, also called **Satellite Digital Audio Radio Services** or **SDARS**, is a fee-based system that uses satellites to broadcast high-quality radio. SDARS broadcasts on the S-band of 2.1320 to 2.345 GHz.

SIRIUS/XM RADIO Sirius/XM radio is standard equipment but is optional in most vehicles. XM radio uses two satellites launched in 2001 called Rock (XM-2) and Roll (XM-1) in a geosynchronous orbit above North America. Two replacement satellites, Rhythm (XM-3) and Blues (XM-4), were launched in 2006. Sirius and XM radio combined in 2008 and now share some programming. The two types of satellite radios use different protocols and, therefore, require separate radios unless a combination unit is purchased.

RECEPTION Reception from satellites can be affected by tall buildings and mountains. To help ensure consistent reception, both SDARS providers do the following:

- Include in the radio itself a buffer circuit that can store several seconds of broadcasts to provide service when traveling out of a service area
- Provide land-based repeater stations in most cities (**SEE FIGURE 24–20.**)

FREQUENTLY ASKED QUESTION

Can Two Bluetooth Telephones Be Used in a Vehicle?

Usually. In order to use two telephones, the second phone needs to be given a name. When both telephones enter the vehicle, check which one is recognized. Say "phone status" and the system will tell you to which telephone the system is responding. If it is not the one you want, simply say, "next phone," and it will move to the other one.

ANTENNA To be able to receive satellite radio, the antenna needs to be able to receive signals from both the satellite and the repeater stations located in many large cities. There are various types and shapes of antennas, including those shown in **FIGURES 24–21 AND 24–22.**

DIAGNOSIS AND SERVICE The first step in any diagnosis is to verify the customer complaint (concern). If no satellite service is being received, first check with the customer to verify that the monthly service fee has been paid and the account is up to date. If poor reception is the cause, carefully check the antenna for damage or faults with the lead-in wire. The antennas must be installed on a metal surface to provide the proper ground plane.

FIGURE 24–21 An aftermarket XM radio antenna mounted on the rear deck lid. The deck lid acts as the ground plane for the antenna.

FIGURE 24–22 A shark-fin-type factory antenna used for both XM and OnStar.

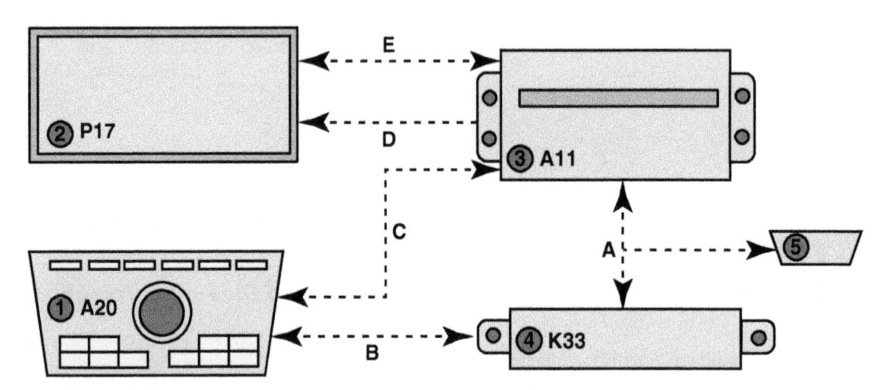

FIGURE 24–23 Diagram of an Infotainment system. (1) Radio/HVAC Controls; (2) Info Display Module; (3) Radio; (4) HVAC Control Module; (5) Data Link Connector; (A) GMLAN; (B) Local Interconnect Network (LIN); (C) CAN Graphical Interface; (D) Digital Video; (E) Touch Screen Serial Data. (Courtesy of General Motors)

For all other satellite radio fault problems, check service information for the exact tests and procedures. Always follow the factory recommended procedures. Check the following websites for additional information.

- www.xmradio.com
- www.sirius.com
- www.siriusxm.com

INFOTAINMENT SYSTEMS

General Motors installs a variety of radio, navigation, and other entertainment and information systems, depending on the vehicle and trim level. It is beyond the scope of this text to detail each system; however, the current trend is to integrate the systems into a packaged system, with operations controlled by a central screen. ● **SEE FIGURE 24–23.**

GM collectively calls these systems **Infotainment** systems, while each division has a system name for the uplevel vehicles.

- For Chevrolet the system is called MyLink.
- For Buick and GMC the system is called IntelliLink.
- Cadillac has the CUE (Cadillac User Experience) system.

The systems use a 7 inch or 8 inch touch screen and an icon-based interface. Each infotainment system can be controlled by using the faceplate knobs and buttons, touch screen, and voice recognition (activated through the audio steering wheel controls).

The standard color radio option offers entry/mid-level radio features. The uplevel connected color radio option is equipped with the MyLink or IntelliLink system. ● **SEE FIGURE 24–24.**

FIGURE 24–24 The MyLink (IntelliLink) screen has many options. (Courtesy of General Motors)

The MyLink system has provisions for the owner to stream music and other programs via a smartphone and the Internet, and includes these features:

- AM/FM/XM/CD/MP3 capability
- USB input
- AUX input
- iPod™/mass storage device support
- Bluetooth hands-free phone and Bluetooth audio streaming
- Voice Recognition control
- Full color touch screen
- Upgradeable software by USB

ADVANCED CONNECTIVITY The Connected Color Radio offers the following advanced connectivity features and benefits.

Bluetooth—Instead of just being able to wirelessly connect a cell phone to the vehicle to make phone calls, Bluetooth acts almost like a USB connection that is wireless. It enables access to features on a smartphone such as contact lists and music. It also allows data (music, voice, information) to stream over the wireless Bluetooth connection. The Bluetooth hardware, antenna, and functionality are part of the radio.

Bluetooth Streaming Audio—Bluetooth Streaming Audio allows pairing a smartphone, feature phone, or device that supports streaming audio and then playing music wirelessly through the Bluetooth connection.

Enhanced Voice Recognition—Short commands can be given through the hands-free push to talk button and the radio system will respond. Voice recognition can be set up to respond to English, Spanish, French, and German. To use the enhanced voice recognition, simply press the push to talk button on the steering wheel. The radio will say, "Please say a command," followed by a beep. After the beep, the user tells the radio what to do.

Smartphone App Integration—Initially, this feature will function only with iPhone, Android, or Blackberry smartphones. It uses the phone's data connection to access Internet radio applications (apps) like Pandora® and Stitcher™. The apps reside on the radio and are accessed from the touch screen or from the voice recognition system. They get the data they need through the Bluetooth or USB (for iPhone) connection accessing the phone's data connection. ● **SEE FIGURE 24–25.**

NOTE Customers should be advised that when connected, they are using their smartphone's data plan.

Gracenote—Gracenote technology embedded into the CCR radio helps manage and navigate the USB device music collection. When a USB device is connected to the radio, Gracenote identifies the music collection and delivers the correct album, artist name, genres, and cover art on the screen. Gracenote will try to fill in any missing information. Gracenote database updates will be available online in these sites: www.chevrolet.com/mylink, www.buick.com/intellilink, and www.gmc.com/intellilink.

USB Connectivity—An iPod™, an MP3 Player, or a USB flash drive can be plugged into the USB port. The USB icon displays when a USB device is connected.

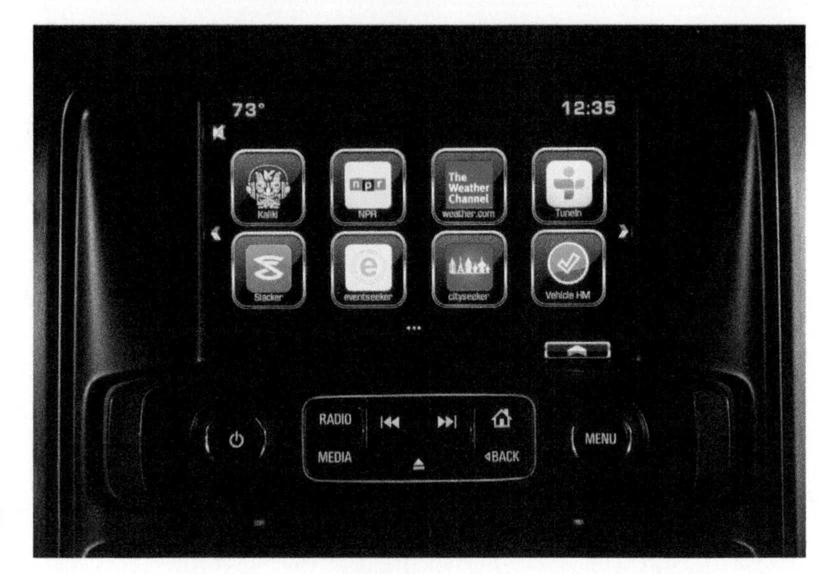

FIGURE 24–25 Various smartphone apps can be installed to the MyLink system. (Courtesy of General Motors)

CUSTOMER UPGRADEABILITY In addition to the standard software update via TIS2Web, the new Connected Color Radio is upgradeable by the owner using the USB connection. Downloads are Web-based using a USB flash drive, and allow existing features to be updated and new features to be added.

Customers can perform some updates themselves; others may require a trip to the dealership. Customers will be informed when they can or cannot perform updates themselves.

The updates must be saved to a USB drive, which is taken to the vehicle and plugged into the USB port. The engine must be running until the update is completed.

INTERNET RADIO SERVICES Currently, two embedded applications reside in the radio—Pandora Internet Radio and Stitcher SmartRadio. The Pandora and Stitcher applications on the radio are used to remotely control each application on the smartphone.

Pandora Internet Radio is a free personalized Internet radio service that streams radio stations based on favorite artists or genres (a slight delay may occur when loading a song or changing a station). Pandora is not available in Canada.*

NOTE: The ability to stream Pandora or Stitcher via the radio is directly dependent on the smartphone data connection to the cellular tower/Wi-Fi hotspot. If the data connection is poor or intermittent, music streaming will be adversely affected.

*General Motors TechLink, April 2012.

INFOTAINMENT SYSTEM COMPONENTS AND OPERATION

MEDIA ORIENTED SYSTEMS TRANSPORT (MOST)**

At the core of the latest Infotainment systems is the **Media Oriented Systems Transport (MOST)** bus, a high-speed multimedia network technology. The serial MOST bus is a ring-type network and uses synchronous data communication to transmit audio, video, data, and control information among any of the devices attached. ● **SEE FIGURE 24–26**.

The MOST bus can use a fiber optic or two-wire structure to communicate among these devices. GM systems are wired. The Remote Radio Receiver, Human Machine Interface Module, Remote Optical Drive, Audio Amplifier, and the Instrument Panel Cluster all communicate on the MOST bus.

The MOST enable circuit is used to wake the network and trigger network diagnostics. Any component on the MOST bus may assert the enable circuit, but communications are initiated by the MOST bus master. Some functions of the MOST bus master are as follows:

- Responsible for normal wake-up and initialization of communication on the network
- Receives vehicle power state information from the vehicle power mode master. The MOST bus master uses this information to control the power state of the Infotainment system

**General Motors Service Information document # 3289492.

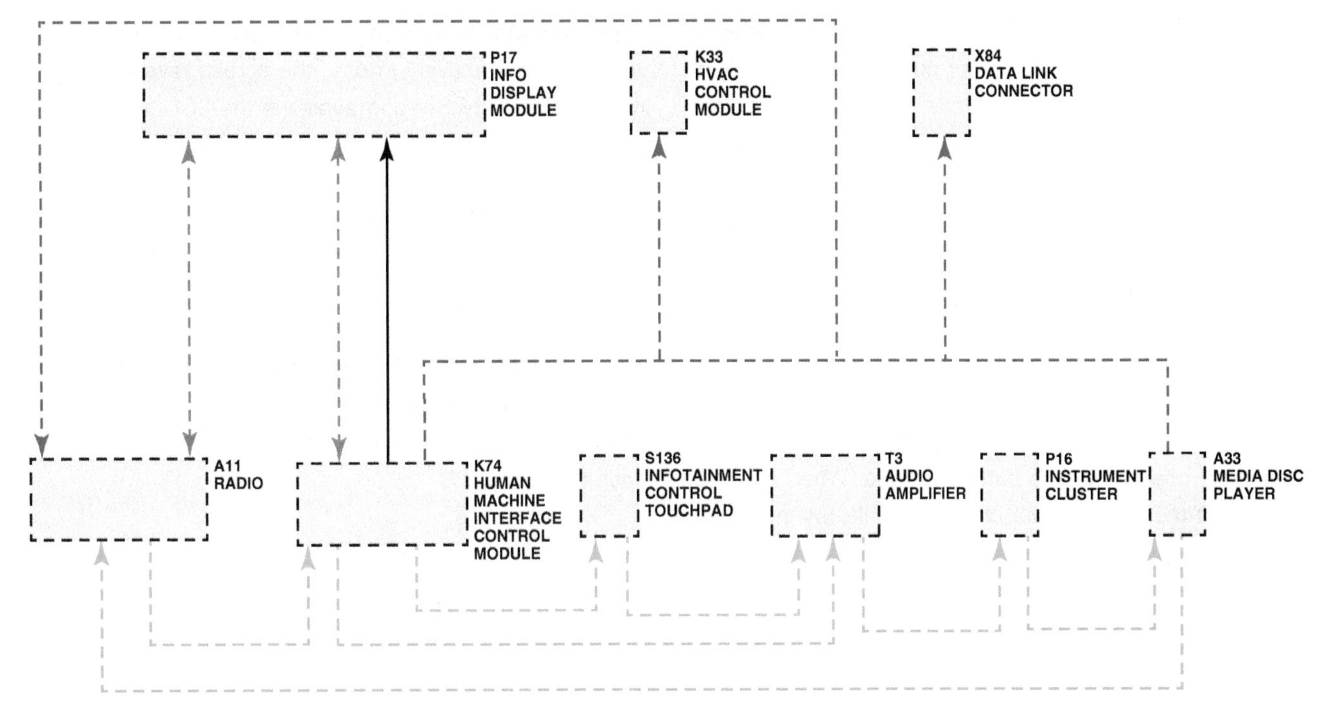

FIGURE 24–26 This block diagram shows the Infotainment system and its related communication networks. The MOST network is shown in green; the LIN is shown in blue; and the serial network (GM LAN) is shown in red. (Courtesy of General Motors)

- Responsible for maintaining known good network configuration
- Reports MOST bus errors/DTCs

LOCAL INTERCONNECT NETWORK (LIN) The **Local Interconnect Network (LIN)** bus is a single-wire communication system. This bus is used to exchange information between a master control module and other smart devices which provide supporting functionality.

The Remote Radio Receiver, Human Machine Interface Module, Information Display, Infotainment Controls, and the Multifunction Controls all communicate on the LIN bus.

GMLAN The Remote Radio Receiver, Audio Amplifier, and the Human Machine Interface Module communicate with other components and systems in the vehicle via GMLAN.

SYSTEM COMPONENTS

REMOTE RADIO RECEIVER The radio is the MOST BUS master. The radio also communicates with other components and systems within the vehicle via GMLAN. The remote radio receiver is responsible for receiving all broadcast audio bands. Broadcast signals from AM, FM, and XM bands are transmitted to the radio via the vehicle antenna systems.

RADIO POWER The radio receives battery power and ground from the vehicle harness. The power mode master of the vehicle provides system power mode to the radio via serial data messages. The power mode master determines the system power mode by processing power mode information from ignition switch inputs. Serial data power modes supported by the radio are OFF, ACCESSORY, RUN, and CRANK REQUEST.

RADIO AUDIO OUTPUTS When equipped with an external amplifier, the radio outputs audio signals digitally over the MOST bus. When not equipped with an external amplifier, the radio outputs audio directly to the speakers.

HUMAN MACHINE INTERFACE CONTROL MODULE The human machine interface module is responsible for the following: Video for the infotainment display, Bluetooth®, USB, memory card reader, and speech recognition functions.

The human machine interface module communicates with the info display module via the LIN bus for control information, touch communications, and dimming level. Digital video data is sent to the display through a dedicated video cable.

MEDIA DISC PLAYER The media disc player receives control information and outputs digital audio over the MOST bus. The media disc player receives battery power and ground from the vehicle harness.

AUDIO AMPLIFIER (IF EQUIPPED) The purpose of the amplifier is to increase the power of a voltage or current signal. A fused battery voltage circuit provides the main amplifier power. The audio amplifier is part of the MOST network and receives audio signals and control information from the MOST bus.

The output signal of an amplifier may consist of the same frequencies as the input signal or it may consist of only a portion of the frequencies as in the case of a subwoofer or mid-range speaker. The audio amplifier amplifies the signal and sends it to the appropriate speakers.

Each of the audio output channel circuits (+) and (−) at the audio amplifier outputs has a DC bias voltage that is approximately one half of the battery voltage. When using a DMM, each of the audio output channel circuits will measure approximately 6.5 V DC. The audio being played on the system is produced by a varying AC voltage that is centered around the DC bias voltage on the same circuit. The AC voltage is what causes the speaker cone to move and produce sound.

The frequency (Hz) of the AC voltage signal is directly related to the frequency of the input (audio source playing) to the audio system. Both the DC bias voltage and the AC voltage signals are needed for the audio system to properly produce sound. Both the DC bias voltage and the AC voltage signals are needed for the audio system to properly produce sound.

SPEAKER OPERATION Speakers turn electrical energy into mechanical energy to move air, using a permanent magnet and an electromagnet. The electromagnet is energized when the radio or amplifier (if equipped) delivers current to the voice coil on the speaker. The voice coil will form a north and south pole that will cause the voice coil and the speaker cone to move in relation to the permanent magnet. The current delivered to the speaker is rapidly changing alternating current (A/C). This causes the speaker cone to move in two directions producing sound.

INFOTAINMENT CONTROLS AND DISPLAY The infotainment display and controls are a separate component from the radio, combined into an assembly. The assembly contains the control knobs and buttons for all audio and HVAC functions and the information display. The assembly is supplied battery voltage and ground from the vehicle harness.

Control information, touch communications, and dimming level for the display are communicated via a LIN serial data circuit to the human machine interface control module.

The human machine interface control module sends the display digital video data for on-screen display through a dedicated video cable.

The information display provides a feedback on the touch screen and certain controls. Buttons pulse (vibrate) when pressed to affirm that the command is being carried out. When not actively in use, the screen reverts to minimal images. Proximity Sensing awakens the LCD screen when a hand approaches it. The controls communicate via a LIN serial data circuit with the remote radio receiver.

HVAC data for controls and status indicators is communicated between the HVAC controls and the HVAC control module with a separate LIN serial data circuit. HVAC status screen information from the HVAC control module is transmitted to the radio on the GMLAN serial data circuit. The radio communicates the desired screen information to the human machine interface module to be sent to the information display using the video data circuits.**

DIAGNOSIS System diagnosis is explained in the appropriate service information section. A scan tool is essential (GDES2 or a Tech 2) when diagnosing these systems. In addition, a service tool is available to help with diagnosis. The Multi-Media Interface Tester (MIT), EL-50334-20, is designed to perform pass or fail testing on the Bluetooth®, USB, and auxiliary jack connections.

The MIT outputs four distinct audio files to test the Bluetooth, AUX/Line-In, and USB functions of the audio system. The operation of each test function is confirmed by a confirmation message played back through the vehicle's audio system. ● **SEE FIGURE 24–27.**

RADIO INTERFERENCE

DEFINITION Radio interference is caused by variations in voltage in the powerline or may be picked up by the antenna. A "whine" that increases in frequency with increasing engine speed is usually referred to as an **alternator whine** and is eliminated by installing a radio choke or a filter capacitor in the power feed wire to the radio. ● **SEE FIGURE 24–28.**

CAPACITOR USAGE Ignition noise is usually a raspy sound that varies with the speed of the engine. This noise is usually eliminated by the installation of a capacitor on the positive side of the ignition coil. The capacitor should be connected to the power feed wire to either the radio or the amplifier, or both. The capacitor *has* to be grounded. A capacitor allows AC interference to pass through to ground while blocking the flow of DC current. Use a 470 μF, 50 volt electrolytic capacitor, which is readily available from most radio supply stores. A special coaxial capacitor can also be used in the powerline. ● **SEE FIGURE 24–29.**

**General Motors Service Information document # 3289492.

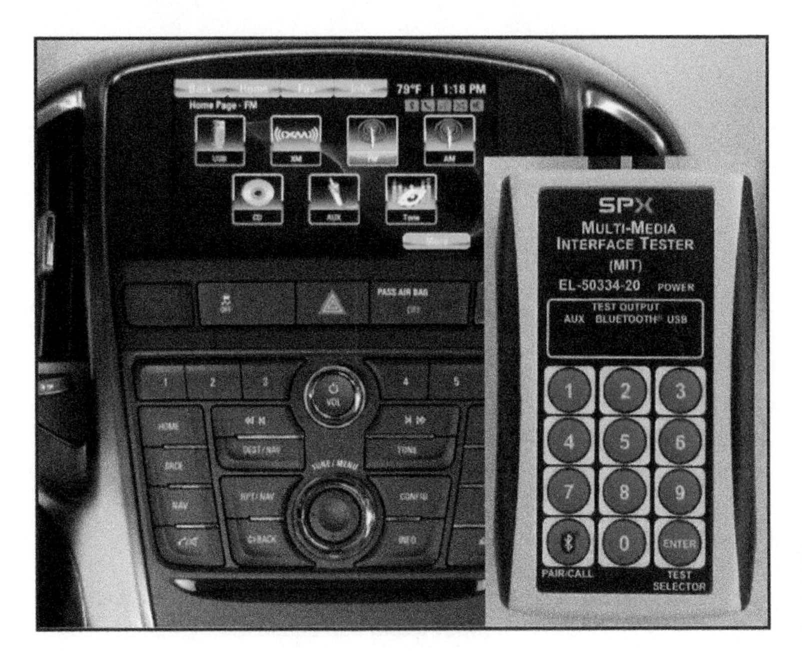

FIGURE 24–27 Special tools are available to help in diagnosis of the Infotainment system. (Courtesy of General Motors)

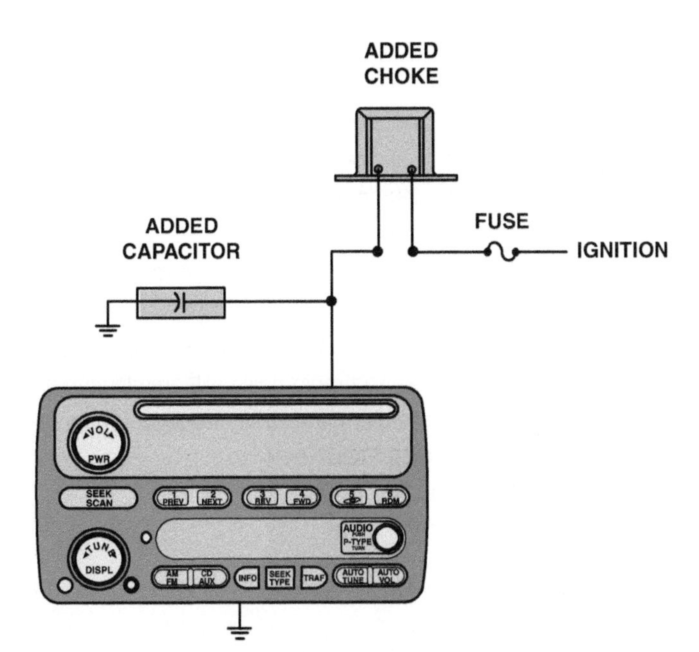

FIGURE 24–28 A radio choke and/or a capacitor can be installed in the power feed lead to any radio, amplifier, or equalizer.

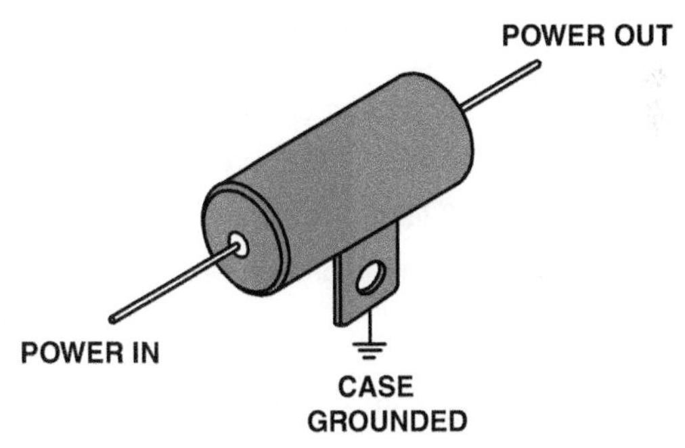

FIGURE 24–29 A coaxial capacitor, like this one, can be installed in the power feed wire to the blower motor to eliminate interference caused by the blower motor.

RADIO CHOKE A **radio choke**, which is a coil of wire, can also be used to reduce or eliminate radio interference. Again, the radio choke is installed in the power feed wire to the radio equipment. Radio interference being picked up by the antenna can best be eliminated by stopping the source of the interference and making certain that all units containing a coil, such as electric motors, have a capacitor or diode attached to the power-side wire.

BRAIDED GROUND WIRE Using a braided ground wire is usually specified when electrical noise is a concern. The radio-frequency signals travel on the surface of a conductor rather than through the core of the wire. A braided ground strap is used because the overlapped wires short out any radio-frequency signals traveling on the surface.

FREQUENTLY ASKED QUESTION

What Does ESN Mean?

ESN means electronic serial number. This is necessary information to know when reviewing satellite radio subscriptions. Each radio has its own unique ESN, often found on a label at the back or bottom of the unit. It is also often shown on scan tools or test equipment designed to help diagnose faults in the units.

TECH TIP

The Separate Battery Trick

Whenever diagnosing sound system interference, try running separate 14 gauge wire(s) from the sound system power lead and ground to a separate battery outside of the vehicle. If the noise is still heard, the interference is *not* due to an alternator diode or other source in the wiring of the vehicle.

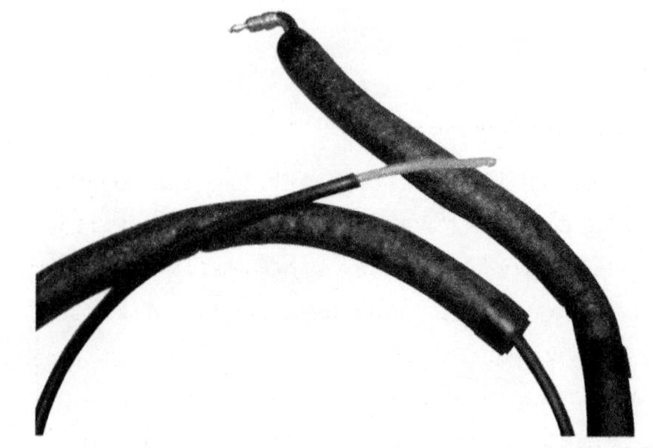

FIGURE 24–30 A "sniffer" can be made from an old antenna lead-in cable by removing about 3 inches of the outer shielding from the end. Plug the lead-in cable into the antenna input of the radio and tune the radio to a weak station. Move the end of the antenna wire around the vehicle dash area. The sniffer is used to locate components that may not be properly shielded or grounded and can cause radio interference through the case (housing) of the radio itself.

AUDIO NOISE SUMMARY In summary:

- Radio noise can be broadcast or caused by noise (voltage variations) in the power circuit to the radio.
- Most radio interference complaints come when someone installs an amplifier, power booster, equalizer, or other radio accessory.
- *A major cause of this interference is the variation in voltage through the ground circuit wires. To prevent or reduce this interference, make sure all ground connections are clean and tight.*
- Placing a capacitor in the ground circuit also may be beneficial.

CAUTION: Amplifiers sold to boost the range or power of an antenna often increase the level of interference and radio noise to a level that disturbs the driver.

Capacitor and/or radio chokes are the most commonly used components. Two or more capacitors can be connected in parallel to increase the capacity of the original capacitor. A "sniffer" can be used to locate the source of the radio noise. A sniffer is a length of antenna wire with a few inches of insulation removed from the antenna end. The sniffer is attached to the antenna input terminal of the radio, and the radio is turned on and set to a weak station. The other end of the sniffer is then moved around areas of the dash to locate where the source of the interference originates. The radio noise will greatly increase if the end of the sniffer comes close to where electromagnetic leakage is occurring. ● **SEE FIGURE 24–30.**

● **SEE CHART 24–2.**

AUDIO NOISE CONTROL SYMPTOM CHART		
NOISE SOURCE	**WHAT IT SOUNDS LIKE**	**WHAT TO TRY**
Alternator	A whine whose pitch changes with engine speed	Install a capacitor to a ground at the alternator output
Ignition	Ticking that changes with engine speed	Use a sniffer to further localize the source of the problem
Turn signals	Popping in time with the turn signals	Install a capacitor across the turn signal flasher
Brake lights	Popping whenever the brake pedal is depressed	Install a capacitor across the brake light switch contacts
Blower motor	Ticking in time with the blower motor	Install a capacitor to ground at the motor hot lead
Dash lamp dimmer	A buzzy whine whose pitch changes with the dimmer setting	Install a capacitor to ground at the dimmer hot lead
Horn switch	Popping when the horn is sounded	Install a capacitor between the hot lead and horn lead at the horn relay
Horn	Buzzing synchronized with the horn	Install a capacitor to ground at each horn hot lead
Amplifier power supply	A buzz, not affected by engine speed	Ground the amplifier chassis using a braided ground strap

CHART 24–2

Radio noise can have various causes, and knowing where or when the noise occurs helps pin down the location.

REAL WORLD FIX

Lightning Damage

A radio failed to work in a vehicle that was outside during a thunderstorm. The technician checked the fuses and verified that power was reaching the radio. Then the technician noticed the antenna. It had been struck by lightning. Obviously, the high voltage from the lightning strike traveled to the radio receiver and damaged the circuits. Both the radio and the antenna were replaced to correct the problem. ● **SEE FIGURE 24–31**.

FIGURE 24–31 The tip of this antenna was struck by lightning.

REAL WORLD FIX

The General Motors Security Radio Problem

A customer replaced the battery in a General Motors vehicle and now the radio display shows "LOC." This means that the radio is locked and there is a customer code stored in the radio.

Other displays and their meaning include:

"InOP"	This display indicates that too many incorrect codes have been entered and the radio must be kept powered for one hour and the ignition turned on before any more attempts can be made.
"SEC"	This display means there is a customer's code stored and the radio is unlocked, secured, and operable.
"---"	This means there is no customer code stored and the radio is unlocked.
"REP"	This means the customer's code has been entered once and the radio now is asking that the code be repeated to verify it was entered correctly the first time.

To unlock the radio, the technician used the following steps (the code number being used is 4321).

STEP 1 Press the "HR" (hour) button: "000" is displayed.

STEP 2 Set the first two digits using the hour button: "4300" is displayed.

STEP 3 Set the last two digits of the code using the "MIN" (minutes) button: "4321" is displayed.

STEP 4 Press the AM-FM button to enter the code. The radio is unlocked and the clock displays "1:00."

Thankfully, the owner had the security code. If the owner had lost the code, the technician would have to secure a scrambled factory backup code from the radio and then call a toll-free number to obtain another code for the customer. The code will only be given to authorized dealers or repair facilities.

SUMMARY

1. Radios receive AM (amplitude modulation) and FM (frequency modulation) signals that are broadcast through the air.
2. The radio antenna is used to induce a very small voltage signal as an input into the radio from the electromagnetic energy via the broadcast station.
3. AM requires an antenna whereas FM may be heard from a radio without an antenna.
4. Speakers reproduce the original sound, and the impedance of all speakers should be equally matched.

5. Crossovers are used to block certain frequencies to allow each type of speaker to perform its job better. A low-pass filter is used to block high-frequency sounds being sent to large woofer speakers, and a high-pass filter blocks low-frequency sounds being sent to tweeters.

6. Radio interference can be caused by many different things, such as a defective alternator, a fault in the ignition system, a fault in a relay or solenoid, or a poor electrical ground connection.

REVIEW QUESTIONS

1. Why do AM signals travel farther than FM signals?
2. What are the purpose and function of the ground plane?
3. How do you match the impedance of speakers?
4. What two items may need to be added to the wiring of a vehicle to control or reduce radio noise?

CHAPTER QUIZ

1. Technician A says that a radio can receive AM signals, but not FM signals, if the antenna is defective. Technician B says that a good antenna should give a reading of about 500 ohms when tested with an ohmmeter between the center antenna wire and ground. Which technician is correct?
   - **a.** Technician A only
   - **b.** Technician B only
   - **c.** Both Technicians A and B
   - **d.** Neither Technician A nor B
2. An antenna lead-in wire should have how many ohms of resistance between the center terminal and the grounded outer covering?
   - **a.** Less than 5 ohms
   - **b.** 5 to 50 ohms
   - **c.** 300 to 500 ohms
   - **d.** Infinity (OL)
3. Technician A says that a braided ground wire is best to use for audio equipment to help reduce interference. Technician B says to use insulated 14 gauge or larger ground wire to reduce interference. Which technician is correct?
   - **a.** Technician A only
   - **b.** Technician B only
   - **c.** Both Technicians A and B
   - **d.** Neither Technician A nor B
4. What maintenance should be performed to a power antenna to help keep it working correctly?
   - **a.** Remove it from the vehicle and lubricate the gears and cable.
   - **b.** Clean the mast with a soft cloth and lubricate with a light oil.
   - **c.** Disassemble the mast and pack the mast with silicone grease (or equal).
   - **d.** Loosen and then retighten the retaining nut.
5. If two 4 ohm speakers are connected in parallel, meaning positive (+) to positive (+) and negative (−) to negative (−), the total impedance will be ________.
   - **a.** 8 ohms
   - **b.** 4 ohms
   - **c.** 2 ohms
   - **d.** 1 ohm
6. If two 4 ohm speakers are connected in series, meaning the positive (+) of one speaker connected to the negative (−) of the other speaker, the total impedance will be ________.
   - **a.** 8 ohms
   - **b.** 5 ohms
   - **c.** 4 ohms
   - **d.** 1 ohm
7. An aftermarket satellite radio has poor reception. Technician A says that a lack of a proper ground plane on the antenna could be the cause. Technician B says that mountains or tall buildings can interfere with reception. Which technician is correct?
   - **a.** Technician A only
   - **b.** Technician B only
   - **c.** Both Technicians A and B
   - **d.** Neither Technician A nor B
8. 100,000 μF means ________.
   - **a.** 0.10 farad
   - **b.** 0.01 farad
   - **c.** 0.001 farad
   - **d.** 0.0001 farad
9. A radio choke is actually a ________.
   - **a.** Resistor
   - **b.** Capacitor
   - **c.** Coil (inductor)
   - **d.** Transistor
10. What device passes AC interference to ground and blocks DC voltage, and is used to control radio interference?
    - **a.** Resistor
    - **b.** Capacitor
    - **c.** Coil (inductor)
    - **d.** Transistor

chapter 25

GENERAL MOTORS HYBRID VEHICLES

LEARNING OBJECTIVES

After studying this chapter, the reader will be able to:

1. Identify General Motors hybrid electric and extended range electric vehicles.
2. Describe how the parallel hybrid truck system works.
3. Describe the features and operating characteristics of the Saturn, Chevrolet, and Buick mild hybrids, and two-mode hybrid vehicles.
4. Describe how the Chevrolet VOLT works.

KEY TERMS

Auxiliary power outlet (APO) 411
Belt alternator starter (BAS) 412
Drive motor/Generator Control Module (DMCM) 419
Electric machine (EM) 410
Electro-hydraulic power steering (EHPS) 411
eAssist 414
Flywheel alternator starter (FAS) 410
Parallel hybrid truck (PHT) 410
Valve-regulated lead-acid (VRLA) 411

GM STC OBJECTIVES

GM Service Technical College topics covered in this chapter are as follows:

1. Characteristics of a hybrid vehicle (08530.10W).
2. Characteristics and modes of operation of a parallel hybrid vehicle.
3. Characteristics and modes of operation of a series/parallel hybrid vehicle.
4. Applications that use hybrid technology.

CHEVROLET/GMC PARALLEL HYBRID TRUCK

INTRODUCTION Chevrolet/GMC parallel hybrid truck is a mild hybrid that was built from 2004 until 2008. It looked like a conventional extended cab pickup and was sold throughout the United States. It used four batteries.

- A conventional 12-volt flooded-type lead-acid battery under the hood, which is used for all of the 12-volt accessories.
- Three 12-volt valve-regulated lead-acid (VRLA), also called absorbed glass mat (AGM), batteries connected in series located under the rear seat, which provides 36 volts for the electric traction motor.

Hybrid components in the truck are warranted for eight years/100,000 miles (160,000 km), in addition to the standard three year/60,000 km warranty. It is a mild hybrid design and is not able to be propelled using battery power alone.

BACKGROUND The overall goal of the **parallel hybrid truck (PHT)** was to maximize fuel economy on vehicles with relatively high fuel consumption and high-potential sales volume.

The Vortec 5300 V-8 engine delivers 295 horsepower (220 kW) and 335 lb-ft (463 N-m) of torque, which is the same as its non-hybrid counterpart.

Instead of a conventional starter motor and alternator, parallel hybrid pickups use a compact 14-kW electric induction motor or starter generator integrated in between the engine and transmission. The starter generator, which operates from the 36-volt battery pack, provides fast, quiet starting power and allows automatic engine stops/starts to conserve fuel. It also performs the following functions:

- Smoothes out torsional vibrations (driveline surges)
- Generates electrical current to charge the batteries and run auxiliary power outlets
- Provides coast-down regenerative braking, as an aid to fuel economy

STARTER-ALTERNATOR The pickup trucks have what General Motors calls a **flywheel alternator starter (FAS)** hybrid system. A three-phase AC induction electric motor, called an **electric machine (EM)** by General Motors, was selected due to low cost, simple controls, and its ability to perform over the entire engine operating range. The electric machine was designed as two assemblies.

- A rotor assembly
- A stator assembly

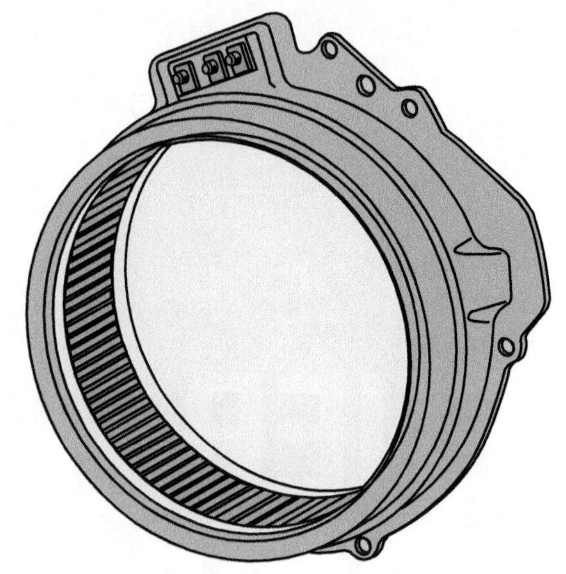

FIGURE 25–1 Stator assembly used on the General Motors parallel hybrid truck.

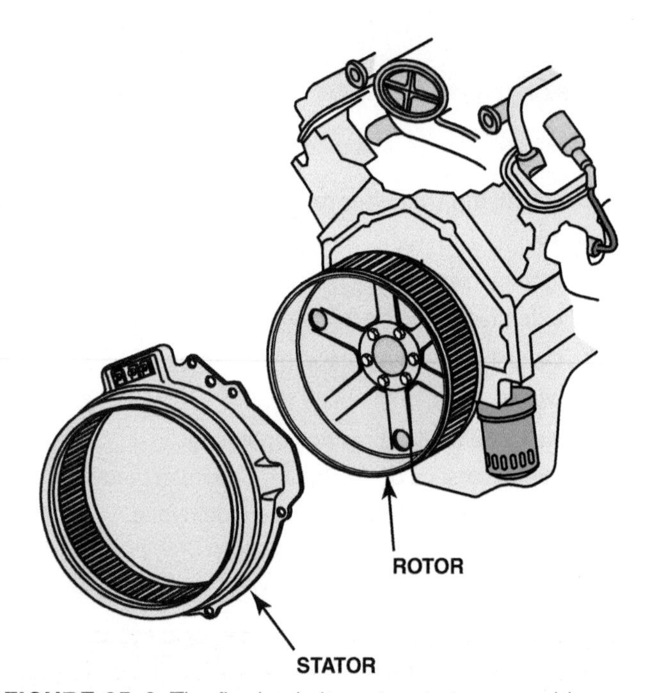

FIGURE 25–2 The flywheel alternator starter assembly.

The stator is located around the rotor and supported by a stamped assembly positioned by existing dowels on the rear face of the engine block and clamped between the engine and transmission. ● **SEE FIGURE 25–1.**

The rotor assembly consists of an electric rotor assembled onto a rotor hub that is bolted to the engine crankshaft and wrapped around the torque converter. ● **SEE FIGURE 25–2.**

When the trucks are slowing down or come to a stop, the fuel is shut off and the engine stops. This saves fuel and when the driver releases the brake pedal, the flywheel alternator starter, taking electrical power from the 36-volt batteries, is

FIGURE 25–3 The three VRLA batteries are located under the rear seat. Note the clear plastic tubes used to the event that pressure in the batteries opens the ventvalve(s).

FIGURE 25–4 The double outlets in the bed of the truck are covered with a spring-loaded rubber-sealed cover to keep out water and dirt.

? FREQUENTLY ASKED QUESTION

Does the Electric Motor Smooth Out the Pulsations of the Engine?

Yes. Most hybrid electric vehicles use the computer to control the electric motor to help reduce variations in crankshaft speeds that normally occur at low speeds. This variation in crankshaft speed is due to the firing impulses and is most noticeable to a driver when the torque converter clutch is engaged at low vehicle speeds, and is often described as "chuggle" or "surge." The PCM commands the electric motor to apply a counter torque to the crankshaft to smooth out these disturbances.

used to restart the engine. The electric motor is not capable of propelling the truck forward at lower speeds using battery and electric motor power alone.

BATTERIES Three **valve-regulated lead-acid (VRLA)** batteries are used in the GM parallel hybrid truck to provide a 36-volt nominal voltage and a charging voltage of 42 volts. ● SEE FIGURE 25–3.

IDLE STOP The engine is turned off as the vehicle slows to a stop and remains stopped until the brake pedal is released. This saves fuel and when the driver releases the brake pedal, the flywheel alternator starter, taking electrical power from the 36-volt batteries, is used to restart the engine. The vehicle may not enter idle stop mode under certain conditions, such as when the charge of the 36-volt battery pack is low.

AUXILIARY POWER OUTLETS Parallel hybrid trucks have power outlets located under the rear seat of the cab and in the pickup bed. These two duplex power outlets are called the **Auxiliary Power Outlet (APO)** system.

The APO system provides up to 2,400 watts of 120-volt AC power. With these power outlets, most auxiliary electrical equipment and devices with a maximum limit of 2,400 watts can be plugged in.

The APO features ground-fault-protected 120-volt, 20-amp (2,400 W) outlets (a duplex outlet under the rear seat in the cab, and a duplex outlet in the truck bed), which can be used to power heavy-duty tools and equipment when used in "normal" mode. ● **SEE FIGURE 25–4.**

ELECTROHYDRAULIC POWER STEERING A 42-volt **electrohydraulic power steering (EHPS)** pump replaces the traditional belt-driven unit. This system allows the operation of the power steering when the engine is off during idle stop conditions. On the Chevrolet Silverado hybrid truck, the EHPS module controls the power steering motor, which has the function of providing hydraulic power to the brake booster and the steering gear.

A secondary function includes the ability to improve fuel economy by operating on a demand basis and the ability to provide speed-dependent variable-effort steering.

FIGURE 25–5 An electrohydraulic power steering assembly on a Chevrolet hybrid pickup truck.

The EHPS module controls the EHPS power pack, which is an integrated assembly consisting of the following components:

- Electric motor
- Hydraulic pump
- Fluid reservoir
- Reservoir cap
- Fluid level sensor
- Electronic controller
- Electrical connectors

● **SEE FIGURE 25–5.**

PARALLEL HYBRID TRUCK SYSTEMS

HVAC SYSTEM Passenger compartment cooling needs during typical engine auto stop events (statistically, 90% of stops are less than 30 seconds) are met by careful management of refrigerant capacity already in the system prior to the stop. Longer stops result in automatic engine restarts. Passenger compartment heating during auto stop periods is accomplished by a simple electric auxiliary coolant pump plumbed in series with the heater core. The auxiliary pump maintains the flow of hot coolant through the heater core during engine idle stop mode.

ELECTRIC MACHINE COOLING Coolant for the electric machine (electric motor) is taken from the engine water jacket, and is routed around the stator through a water channel created by the two stampings used as a carrier for the stator, and returned to the suction side of the engine coolant system. This plumbing blocks the flow of coolant through the electric machine until the engine thermostat opens so as not to delay engine warm up.

GENERAL MOTORS MILD (ASSIST) HYBRIDS

PURPOSE AND FUNCTION The purpose of the mild hybrid design used in mid-size General Motors vehicles is to provide many of the benefits of a hybrid electric vehicle with the least amount of changes to the base vehicle as possible.

The mild (assist) GM hybrid system is found on some Saturn and Chevrolet vehicles.

PARTS AND OPERATION The Saturn VUE and Aura and the Chevrolet Malibu hybrid systems include:

- Idle stop feature (start-stop)
- Regenerative braking

These functions are made possible by using a **Belt Alternator Starter (BAS)** system. ● **SEE FIGURE 25–6.**

The large alternator, which is also used to start the gasoline engine during restarts after idle stop, is driven by a wider-than-normal accessory drive belt. A dual-tensioner assembly for the motor/generator allows for the transfer of a small amount of torque to the drive system for very brief periods of time. The assembly combines a hydraulic strut tensioner and a friction-damped rotary tensioner on a common pivoting arm to control the bi-directional loads (motoring and generating). ● **SEE FIGURE 25–7.**

The gasoline engine is started using the conventional starter motor when first started. After the engine is running, the hybrid assist system can then provide the following:

1. Electrically motored creep at startup
2. Light power assist during acceleration
3. Light regenerative mode during deceleration

The system consists of the following six elements:

1. The electric motor/generator unit that replaces the alternator, and is capable of 115 lb-ft (156 N-m) of auto-start torque
2. Engine coolant-cooled power electronics that control the motor/generator unit and provide 12-volt vehicle accessory power
3. A 36-volt NiMH hybrid battery pack capable of delivering and receiving more than 10 kW of peak power. The battery pack is behind the rear seat
4. An engine control module

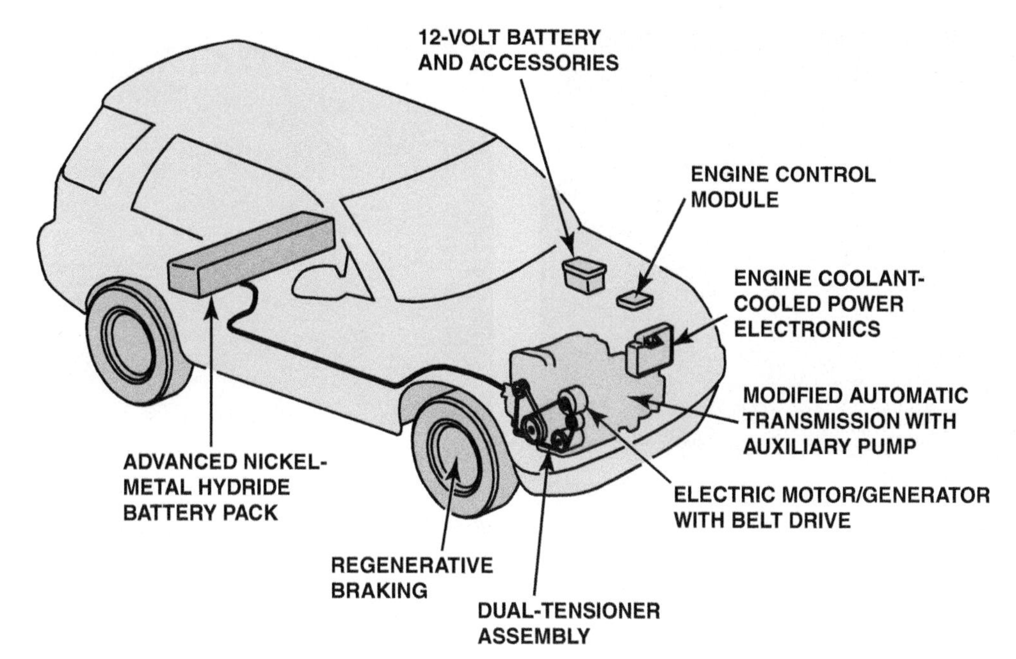

FIGURE 25–6 An overall view of the components and their locations in a Saturn VUE hybrid electric vehicle.

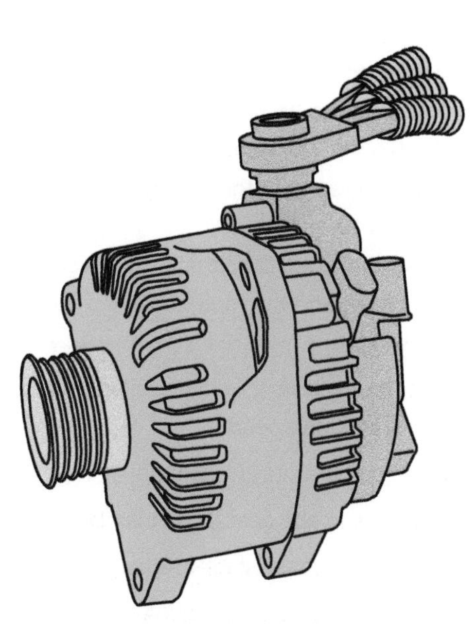

FIGURE 25–7 The motor/generator assembly used in the Saturn VUE and Chevrolet Malibu hybrid vehicle.

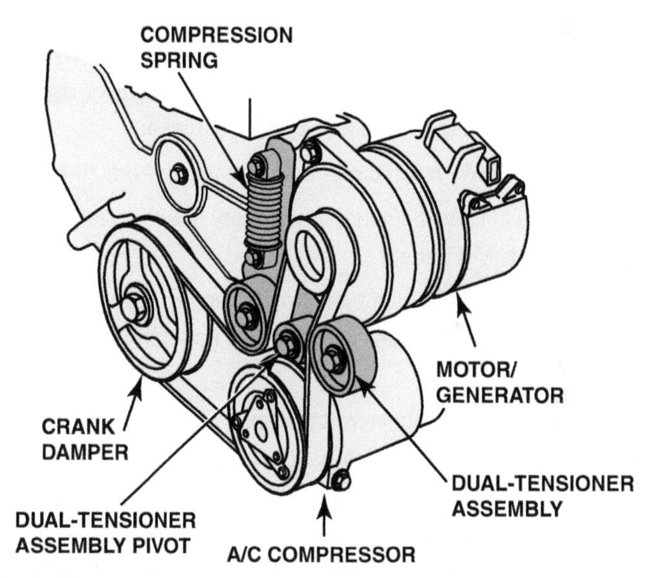

FIGURE 25–8 The Saturn VUE engine showing the location of the accessories and the motor/generator. The dual-tensioner assembly that controls the motoring (help to power the vehicle) and generating loads.

5. An engine accessory drive using a dual-tensioner assembly and aramid cord belt that enables transfer of starting/propelling and generating torque (● **SEE FIGURE 25–8**).
6. Hybrid-enabled Hydra-Matic 4T45-E electronically controlled four-speed automatic transaxle that includes an auxiliary oil pump and unique hybrid controls

The mild hybrid enables early fuel cutoff to the engine during deceleration and shuts off the engine at idle (idle stop mode). Regenerative braking and optimized charging combined with an energy storage system further enhance fuel economy while maintaining all vehicle accessories and air-conditioning systems during the periods when the engine is temporarily shut off.

The BAS systems indicate a range of improvement in fuel economy of around 10% to 12%.

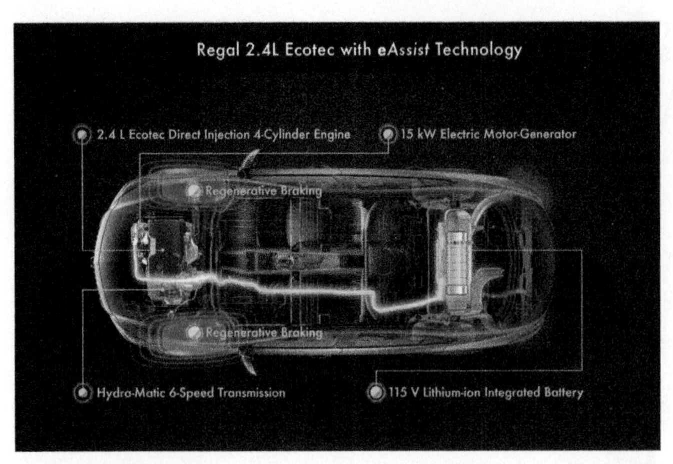

FIGURE 25–9 Buick Regal, showing the eAssist components and locations. (Courtesy of General Motors)

FIGURE 25–10 The driver information center (DIC) show the operating mode of the eAssist system. (Courtesy of General Motors)

BUICK/CHEVROLET eASSIST

The BAS system used in the Chevrolet and Saturn mild hybrids was updated in 2011 to include the following changes used in the 2012+ Chevrolet Malibu and Buick LaCrosse:

1. Uses a 15-kW electric motor-generator up from 10 kW
2. Uses a 115-volt lithium-ion battery pack instead of a 36-volt NiMH battery
3. The motor-generator can provide up to 15 horsepower and 79 lb-ft of torque to the 2.4 liter Ecotech four-cylinder engine

The use of the higher voltage and more powerful motor-generator allows the archive a boost to acceleration and an improvement in fuel economy.

OPERATION Combined with a 2.4 liter Ecotech four-cylinder, the **eAssist** system (RPO HP6) uses a lithium-ion high-voltage battery, electric induction motor-generator, and regenerative braking to provide added vehicle performance when needed while also enhancing fuel economy. ● **SEE FIGURE 25–9.**

The high-voltage battery system is designed to provide power assistance to the internal combustion engine, rather than store energy for all-electric propulsion. While in fuel shut-off mode, the induction motor-generator unit continues spinning along with the engine to provide regenerative braking, torque smoothing, and immediate take-off power when the driver presses on the accelerator.

As the vehicle comes to a stop, the induction motor-generator unit spins the engine, bringing it to a smooth stop, properly positioned for an immediate restart when the driver releases the brake pedal or applies the accelerator pedal.

NOTE: The regenerative braking process requires unique transmission shifting as the vehicle slows down.

AUTO STOP While driving, when the brake is applied and the vehicle comes to a complete stop, the engine may turn off. This is referred to as an auto stop (the tachometer gauge will read AUTO STOP). The engine restarts immediately when the brake pedal is released or the accelerator pedal is applied. ● **SEE FIGURE 25–10.**

There are several conditions that may cause the engine to remain running or restart when the vehicle is stopped. These include the following:

- The engine, transmission, or high-voltage battery has not reached operating temperature
- The outside temperature is less than –4°F (–20°C).
- The shift lever is in any gear other than Drive (D).
- The high-voltage battery state-of-charge is low.
- The climate control system requires the engine to run based on the climate control or defog setting. Using the eco air-conditioning mode will result in more frequent and longer auto stops.
- The auto stop time is greater than two minutes.
- The malfunction indicator lamp (MIL) is illuminated.

When on a moderate or steep grade, the hill-assist system captures brake pressure to help the driver more comfortably accelerate from a stop, which reduces the tendency of the vehicle to roll backward with the engine in shutdown mode.

FIGURE 25–11 This Cadillac Escalade, like many hybrids built by General Motors, have many emblems showing that it is a hybrid electric vehicle.

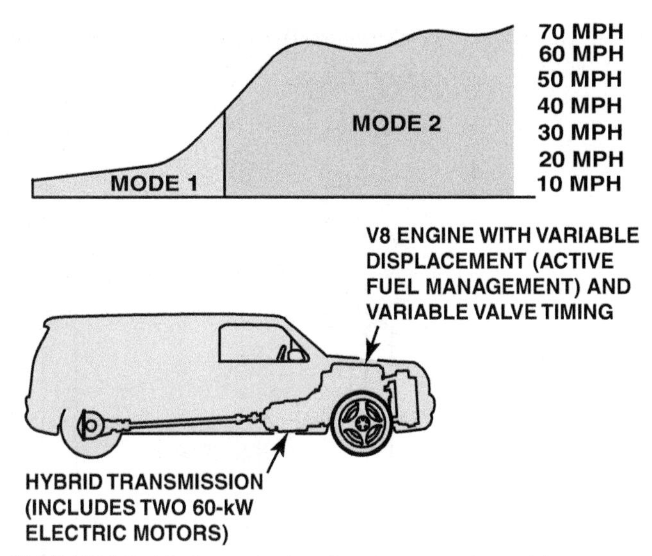

FIGURE 25–12 A graph showing the operation of the two-mode hybrid vehicle. At lower speeds, the vehicle is capable of being propelled using electrical power alone and assist at higher speeds.

eASSIST COMPONENTS

Motor/Generator The liquid-cooled electric induction motor/generator is mounted to the engine in place of the 12-volt generator to provide both motor assist and electric-generating functions through a unique engine belt-drive system. The motor-generator is connected to the crankshaft pulley using a specially designed serpentine belt and drive belt tensioner.

Generator and Battery Controls The eAssist generator control and battery module assembly contains the high-voltage lithium-ion battery pack, the integrated power inverter, and 14 volt power module. It is located in the forward area of the trunk compartment and weighs about 65 pounds (29 kg). An electric fan cools the generator control and battery module, drawing air from a vent located in the package tray, behind the rear seat.

Transmission The Hydra-Matic 6T40 transmission is designed to enhance powertrain efficiency. Significant internal transmission changes to clutch controls and hardware provide reduced spin losses while improving shift response and time.

An auxiliary, electric-driven transmission oil pump has been added to keep the transmission fluid flowing and clutches applied when the engine shuts down during an auto stop.

The added electric power provided by the eAssist system also allows for higher transmission gearing, improving steady-state efficiency without impacting acceleration performance or driveability. The system's ability to provide electric assist at cruising speeds allows the driver to accelerate lightly or ascend mild grades without the transmission downshifting.*

GENERAL MOTORS TWO-MODE HYBRID

BACKGROUND The two-mode system was a joint development by General Motors, Chrysler, and BMW. The two-mode hybrid system is a full (strong) hybrid and is capable of propelling the vehicle on battery power alone. The two-mode system is used in the following vehicles:

- BMW 7 series hybrid (2011+)
- Cadillac Escalade (2008+) ● **SEE FIGURE 25–11.**
- Chevrolet Tahoe (2008+)
- Chrysler Aspen (2009)
- Dodge Durango (2009)
- GMC Yukon (2008+)

● **SEE FIGURE 25–12.**

The first mode provides fuel-saving capability in low-speed, stop-and-go driving with a combination of full electric propulsion and engine power. In the first mode, the vehicle can operate in three ways:

- Electric power only
- Engine power only
- Any combination of engine and electric power

*General Motors Techlink, July 2011.

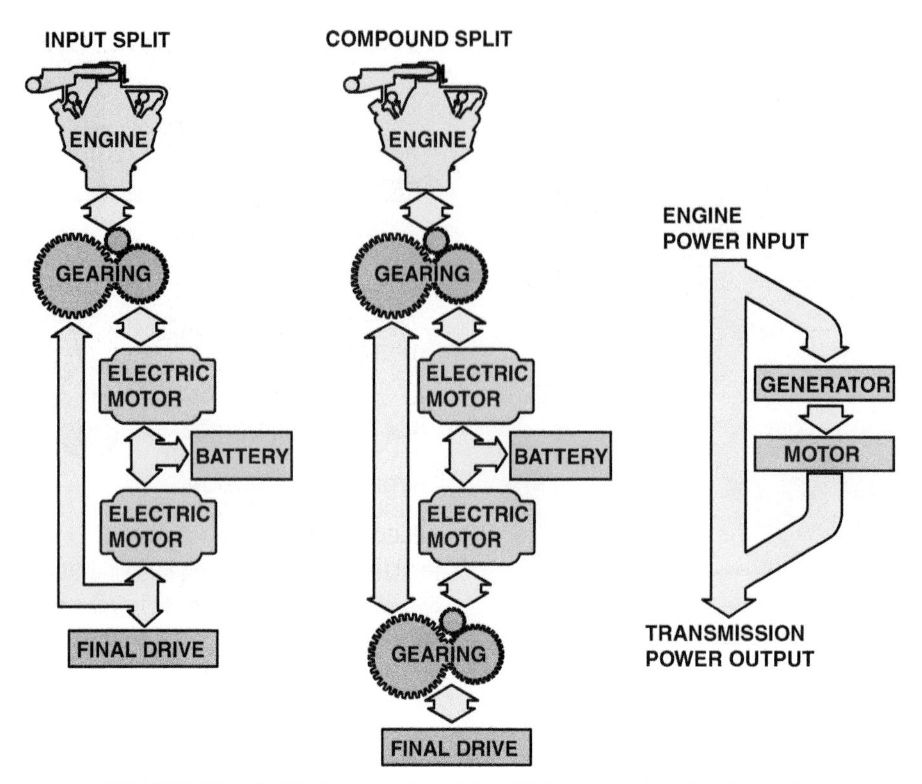

FIGURE 25–13 The two modes of the GM two-mode hybrid vehicle.

The second mode is used primarily at highway speeds to optimize fuel economy, while providing full engine power when conditions demand it, such as trailer towing or climbing steep grades. ● **SEE FIGURE 25–13.**

POWERTRAIN The two-mode hybrid features two 60-kW motors in the hybrid drivetrain, a 300-volt battery pack, and a V-8 engine. The BMW 7 series two-mode hybrid uses a twin-turbocharged 4.4 liter V-8 with an 8-speed automatic transmission. A two-mode hybrid electric vehicle is capable of increasing the fuel economy by about 25%, depending on the type of driving conditions. Like all hybrids, the two-mode combines the power of a gasoline engine with that of electric motors and includes:

- Regenerative braking that captures kinetic energy that would otherwise be lost
- Idle stop

Existing full hybrid systems have a single mode of operation, using a single planetary gearset to split engine power to drive the wheels or charge the battery. ● **SEE FIGURE 25–14.**

These systems are effective at low speeds because they can move the vehicle without running the gasoline engine. ● **SEE FIGURE 25–15.**

FIGURE 25–14 The two-mode General Motors hybrid encloses both electric motor-generators inside the transmission case.

At higher speeds, when the engine is needed, using the electric motors has much less benefit. As a result, sending power through electric motors and a variable transmission is roughly 20% less efficient than driving the vehicle through a purely mechanical power path, using gears.

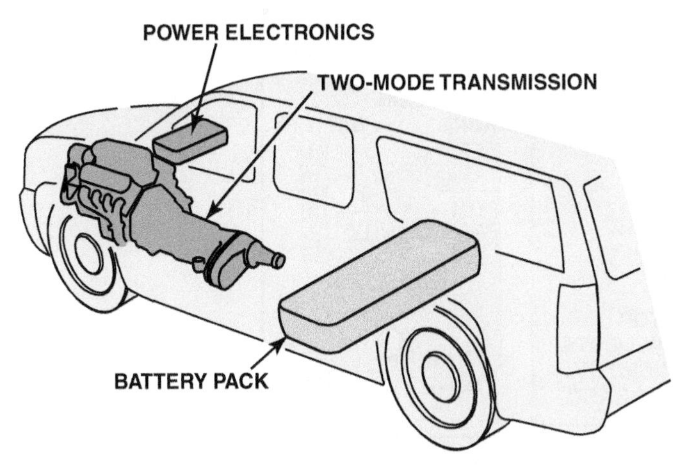

FIGURE 25–15 The two-mode General Motors vehicle is a full (strong) hybrid capable of propelling the vehicle using battery power alone.

FIGURE 25–16 Cutaway view of the two-mode General Motors transmission.

COMPONENTS The components of the two-mode transmission include the following:

- Two 60-kW electric motor/generators assemblies
- Three planetary gearsets (one is located in front of motor/generator A; another is located between the two motor/generators; and the last is located behind motor/generator B)
- Four wet plate clutches (two friction [rotating] and two mechanical [stationary] clutch assemblies)

  ● **SEE FIGURE 25–16**

TWO MODES OF OPERATION

- The *first mode* is for accelerating from standstill to second gear. At low speed and light load, the vehicle can move with
  - Either of the two electric motors alone
  - The internal combustion engine (ICE) alone
  - Or a combination of the two (electric motor and/or ICE)

FIGURE 25–17 The high-voltage NiMH battery pack is located underneath rear seat in the GM two-mode hybrid vehicle. The high-voltage disconnect plug is located on the passenger side of the battery pack.

In this mode, the engine (if running) can be shut off under certain conditions and everything will continue to operate on electric power alone. The hybrid system can restart the ICE at any time as needed. One of the motor/generators operates as a generator to charge the high-voltage battery and the other works as a motor to assist in propelling the vehicle.

The *second mode* takes the vehicle from second gear through to overdrive. At higher loads and speeds, the ICE always runs. In the second mode, the motor/generators and planetary gearsets are used to keep torque and horsepower at a maximum. As the vehicle speed increases, various combinations of the four fixed ratio planetary gears engage and/or disengage to multiply engine torque, allowing one or the other of the motor/generators to perform as a generator to charge the high-voltage battery.

BATTERIES Two-mode hybrid electric vehicles use two batteries.

1. **Auxiliary Battery** A conventional 12-volt flooded-type lead-acid battery located under the hood. The BMW uses an absorbed glass mat (AGM) 12-volt battery located under the hood. This auxiliary battery is used to power the electronics and the accessories in the vehicle. The 12-volt battery is kept charged by the use of a DC-to-DC converter that converts the 300 volts DC from the high-voltage battery pack to 12 volts DC for the auxiliary battery.
2. **High-Voltage Battery** A 300-volt nickel-metal halide (NiMH) battery pack located under the rear seat. ● **SEE FIGURE 25–17.**

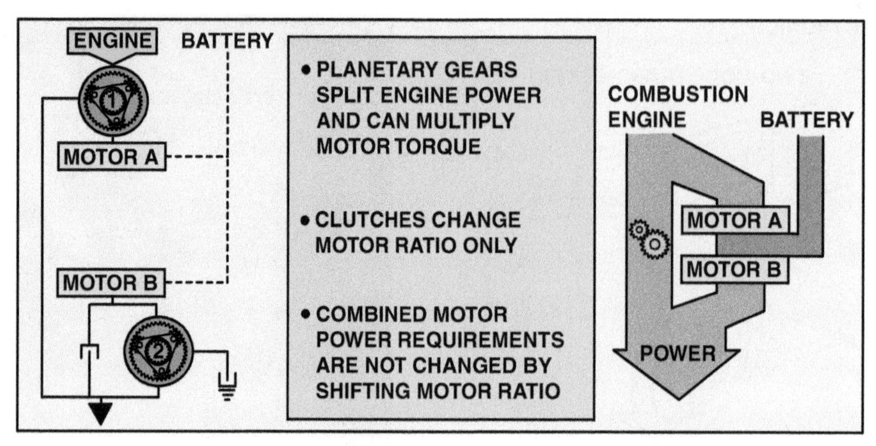

FIGURE 25–18 The gasoline engine can power motor A to charge the batteries and help propel the vehicle.

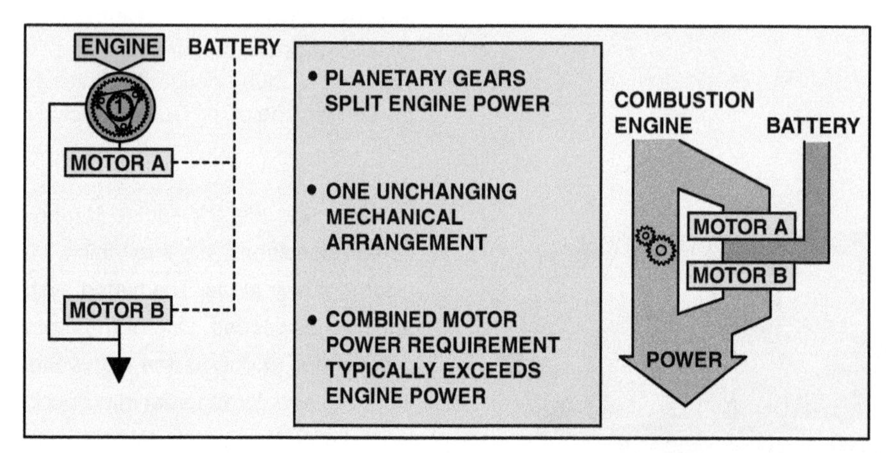

FIGURE 25–19 The high-voltage battery current can be fed to both electric motors to propel the vehicle.

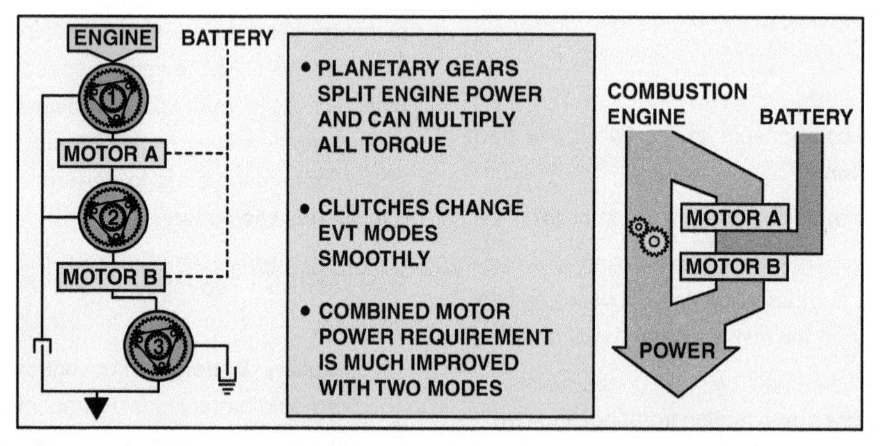

FIGURE 25–20 The electric motors can be used to assist the gasoline engine to provide additional torque for rapid acceleration.

The high-voltage battery is charged by the motor/generators inside the transmission. The motors are used to propel the vehicle and to charge the high-voltage battery during regenerative braking.

TWO-MODE OPERATION The two-mode system also has an electrically variable transmission but adds two planetary gearsets (which multiply the torque from the power source). This arrangement provides two operating modes for the electric motors. ● **SEE FIGURES 25–18 THROUGH 25–21.**

The system adds two clutches, which can engage different gear combinations to provide an overlay of four fixed (not variable) gear ratios. To the driver, the gear ratios feel and function like the stepped gear changes of an automatic

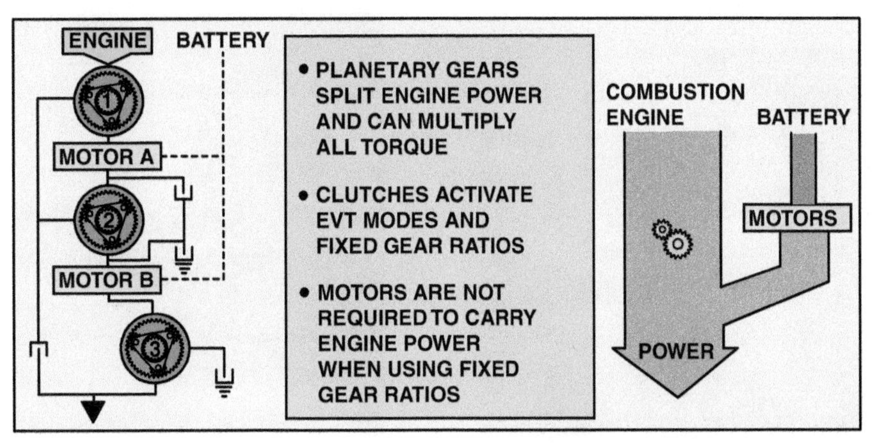

FIGURE 25–21 The gasoline engine alone can be used to power the vehicle.

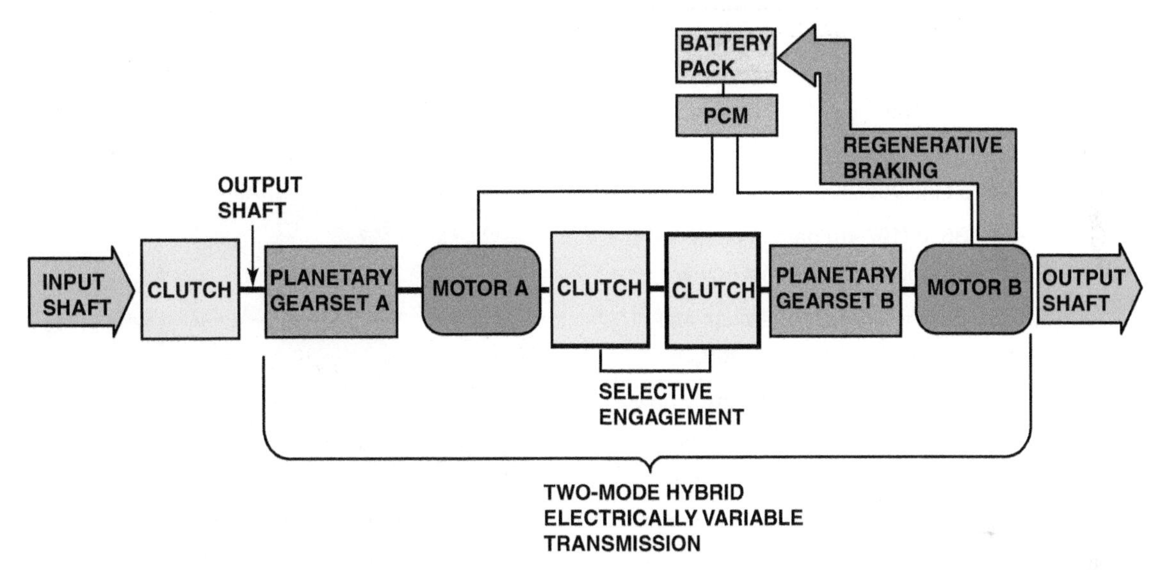

FIGURE 25–22 The two-mode hybrid also operates on electric power alone during deceleration (regenerative braking).

transmission. Because the torque is multiplied by the gears, the electric motors can be much smaller and fit in the same housing as the transmission. ● **SEE FIGURE 25–22.**

SERVICE Besides the normal routine service that any vehicle requires, the two-mode hybrid uses a separate cooling system that may require service. This cooling is for the **Drive Motor/ Generator Control Module (DMCM)** system. The DMCM cooling system includes:

- The DMCM coolant surge tank,
- DMCM surge tank pressure cap,
- DMCM cooling pumps, and
- Hybrid cooling radiator and the Drive Motor/Generator Control Module (DMCM).

The DMCM cooling system uses a 50–50 pre-mixed DEX-COOL™ coolant and de-ionized water.

TECH TIP

Raise the Hood to Keep the Engine Running

When the hood is raised, the engine will remain running and will not try to enter into the idle stop mode. This makes it easy when service work that requires the engine to be kept running is being performed.

CHEVROLET VOLT

GENERAL DESCRIPTION The 2011+ Chevrolet Volt is an extended range electric vehicle based on the Chevrolet Cruze platform. ● **SEE FIGURE 25–23.**

FIGURE 25–23 The Chevrolet Volt is a small four-door vehicle that is based on the Chevrolet Cruze platform.

FIGURE 25–24 The high-voltage battery pack is housed between the seat under the center console and under the rear seat. The orange high-voltage disconnect plug is shown in this cutaway model and is located under the console package tray.

Major features of the Volt include:

- A 346-volt lithium-ion 435 lb (197 kg) battery pack that has 16 kW-h (16,000 watt-hours) of electrical power.
- A 1.4 liter four-cylinder gasoline engine (premium fuel required) that produces 84 HP and 92 lb-ft of torque.
- The Volt is capable of being powered using battery power alone for 25 to 50 miles depending on conditions and driving style.
- The gasoline engine is started only to extend the range once the state-of-charge of the battery pack is reduced to 25% to 35%.

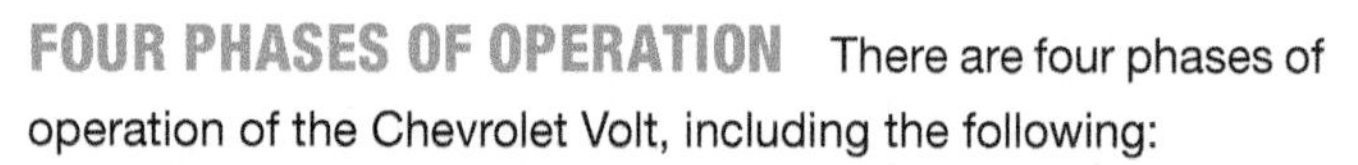

FOUR PHASES OF OPERATION There are four phases of operation of the Chevrolet Volt, including the following:

1. In the low-speed electric vehicle mode the vehicle is powered up to about 40 MPH (65 km/h) using electrical power from the high-voltage battery to the larger of the two electric motors inside the transaxle assembly.
2. Above about 40 MPH, the second smaller electric motor starts to help power the vehicle, which reduces the speed of the larger electric motor. The two motors connect to the planetary gearset inside the transaxle and their combined torque is applied to the drive wheels.
3. The vehicle operates using electric power alone until the state-of-charge (SOC) of the high-voltage battery drops to about 30%. At this time the gasoline engine will start, which drives the smaller electric motor so it becomes a generator to keep the high-voltage battery at about 30% SOC (25% to 35%).

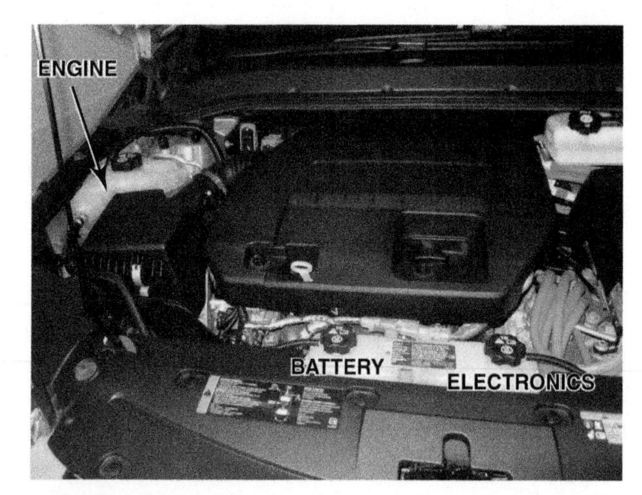

FIGURE 25–25 The engine cooling system uses a 20 PSI pressure cap and the cooling systems for the battery and the electronic use 5 PSI caps.

4. At highway speeds, the gasoline engine helps power the vehicle. Using the gasoline engine to help power the vehicle at higher speeds allows the large electric motor to operate slower, thereby increasing its efficiency.

VOLT HIGH-VOLTAGE BATTERY PACK The high-voltage battery pack in the Chevrolet Volt is located under the center console and under the rear seat. ● **SEE FIGURE 25–24.**

The battery pack is kept cool by using a separate cooling system. The system has a coolant pump and reservoir with a 5 PSI pressure cap. ● **SEE FIGURE 25–25.**

When the high-voltage battery pack drops to about 30% SOC, the gasoline engine is turned on to maintain a certain

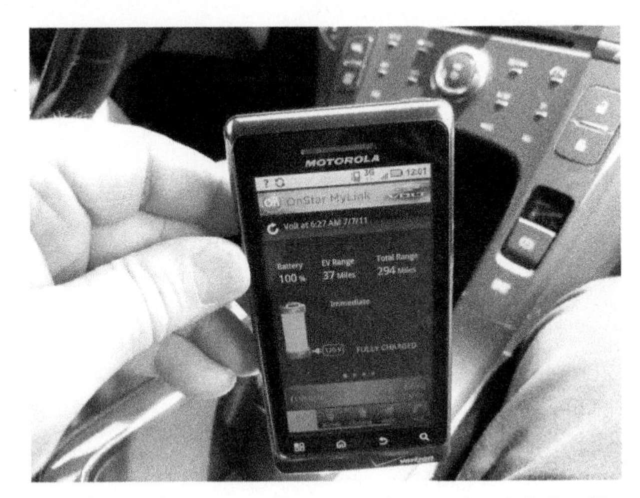

FIGURE 25–26 A smart phone can be used to link On-Star in the Chevrolet Volt, which can send state-of-charge and estimated range information.

level of battery charge. However, the engine is not able to fully charge the high-voltage battery pack. To fully charge the high-voltage battery, the vehicle must be plugged into to an electrical outlet. ● **SEE FIGURE 25–26** (smart phone photo).

Normal battery usage includes two modes.

- **Charge Depleting Mode.** During the charge depleting mode, the vehicle is operating in electric vehicle (EV) mode only. All power needed to propel the vehicle, and for heating and cooling is provided by the high-voltage battery. The range of EV mode can vary from 25 to 50 miles depending on the weather and driving conditions.
- **Charge Sustaining Mode.** In charge sustaining mode, the gasoline engine is cycled on and off as needed to keep the battery state-of-charge at the 25% to 35% level. The driver has no control over when the engine starts or stops. ● **SEE FIGURE 25–27.**

ENGINE MAINTENANCE MODE Engine operation in the Chevrolet Volt is carefully monitored because the engine may not run for many days or even weeks if driven in electric vehicle mode. The engine maintenance mode is needed because

- Fuel can age and old fuel can cause problems with proper engine operation.
- The engine needs to be kept lubricated to prevent rust and corrosion of internal engine parts.

To address these concerns the controller is capable of performing the following functions:

1. **Fuel weathering.** The powertrain control module keeps track of the amount of fuel used during each drive cycle and calculates the percentage of new and old fuel that is in the fuel tank. If the age of the fuel is determined to be a concern, the driver will be notified that the engine will likely be started to use some of the fuel. It may require several drive cycles for the fuel to be used so that even though the vehicle is used only in electric vehicle mode, it will use some gasoline to maintain fresh fuel in the system. This condition will be rarely used and the gasoline life monitor will attempt to keep the fuel in the tank so that it is less than a year old.
2. **Engine Lubrication.** To help keep the mechanical parts of the gasoline engine at peak operating efficiency, the powertrain control module will perform the following actions if the engine has not run for an extended period.
    - Power the actuators such as the electronic throttle control (no engine operation)
    - Engine spin without fuel
    - Engine spin with fuel and spark (too lean to cause the engine to actually start)
    - Engine start to purge contaminates from the oil and condensate from the exhaust

These modes will be very rare and used only if the vehicle is used in electric vehicle mode for many days or weeks.

CHARGING THE VOLT The Chevrolet Volt comes equipped with a charging cable that is stored in the rear compartment which can be used to recharge the vehicle from a 110- to 120-volt outlet (Level 1). ● **SEE FIGURE 25–28.**

The standard connector built into the left front fender of the vehicle can also be used to charge the vehicle from a 220- to 240-volt (Level 2) charging station or outlet.

For technical service information on General Motors hybrid electric vehicles, visit *www.gmtechinfo.com*.

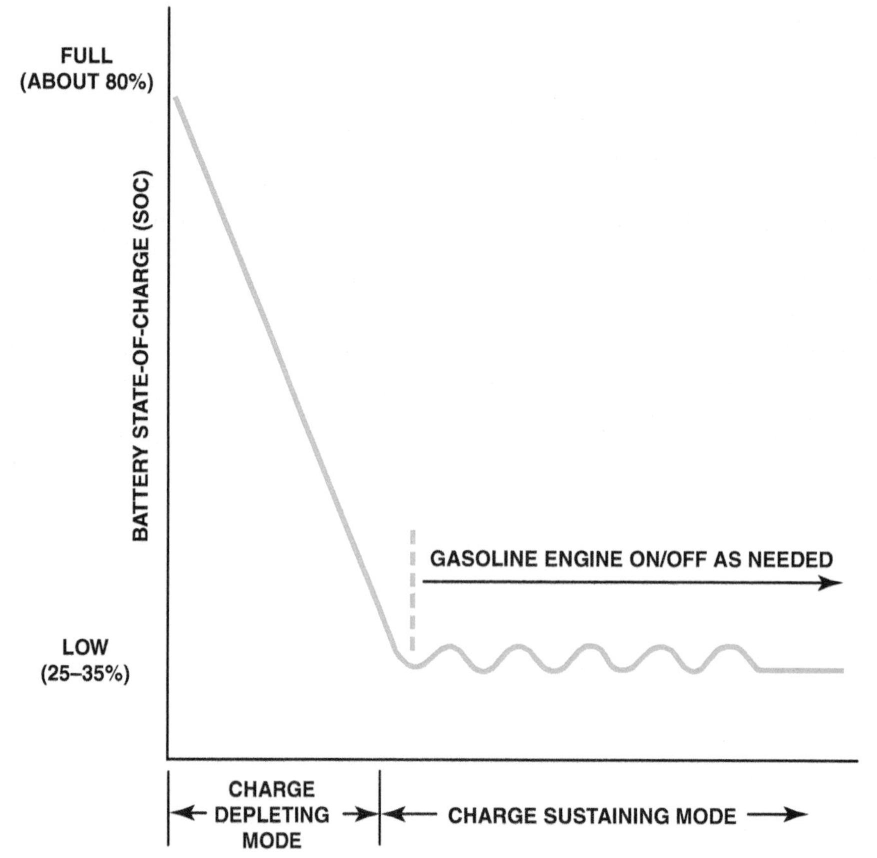

FIGURE 25–27 After the Volt has been charged it uses the electrical power stored in the high-voltage battery to propel the vehicle and provide heating and cooling for 25 to 50 miles (40 to 80 km). Then the gasoline engine starts and maintains the SOC between 25% and 35%. The gasoline engine cannot fully charge the high-voltage batteries but rather the vehicle has to be plugged in to provide a higher SOC level.

(a)

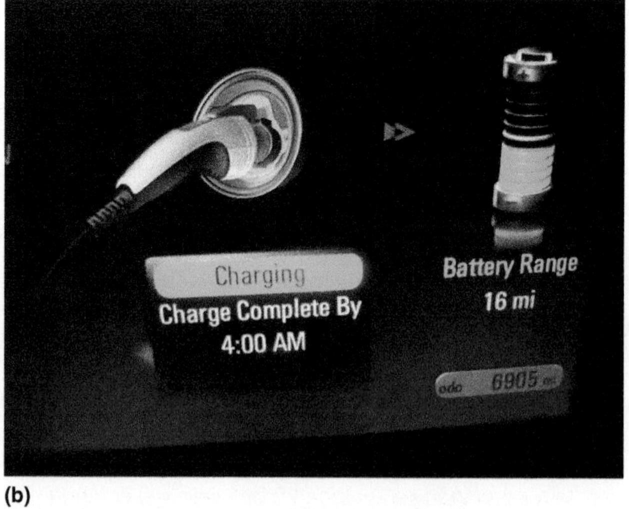

(b)

FIGURE 25–28 (a) The Chevrolet Volt is charged using a standard SAE 1772 connector using either 110 or 220 volts. (b) After connecting the charging plug, a light on the top of the dash turns green and the dash display shows the estimated time when the high-voltage battery will be fully charged and the estimated current range using battery power alone.

SUMMARY

1. The General Motors parallel hybrid truck (PHT) uses three conventional VRLA 12-volt batteries connected in series to provide 36 volts (42 volts charging).
2. The flywheel alternator starter (FAS) is used to start the gasoline engine, provide torque smoothing, and generate electrical energy for the 120-volt outlets.
3. The PHT has four auxiliary power outlets (APOs)—two at the base of the rear seat and two in the bed of the truck.
4. The Saturn, Buick, and Chevrolet Malibu hybrid vehicles use a Belt Alternator Starter (BAS) system and 36-volt NiMH or 115-volt Li-ion battery packs.
5. The General Motors two-mode hybrid electric vehicle uses two 60-kW electric motors inside the transmission and can power the vehicle using electrical energy alone.
6. The two-mode hybrid vehicle is a full (strong) hybrid and uses a 300-volt NiMH battery pack located under the rear seat.
7. The Chevrolet Volt is an extended range electric vehicle that uses a 16-kW lithium-ion battery and a four-cylinder gasoline engine to help propel the vehicle and to keep the high-voltage battery at about 25% to 35% state-of-charge to extend the driving range of the vehicle.

REVIEW QUESTIONS

1. How is the General Motors parallel hybrid truck able to supply 120 volts AC to the auxiliary power outlet?
2. What type of hybrid vehicle is the GM PHT truck, and what are its capabilities?
3. How does the electrohydraulic power steering (EHPS) work?
4. How does the Belt Alternator Starter (BAS) system work?
5. How does the two-mode hybrid system work?

CHAPTER QUIZ

1. The batteries used in the General Motors parallel hybrid truck are ______________.
   a. 300-volt NiMH
   b. Three 12-volt valve-regulated lead-acid
   c. 36-volt NiMH
   d. 144-volt D-cell-size NiMH
2. In a General Motors parallel hybrid truck, where is the electric motor (FAS)?
   a. Attached to the crankshaft at the rear of the engine
   b. Belt driven at the front of the engine
   c. Inside the transmission assembly
   d. At the output of the transmission
3. Which statement(s) is/are true about the auxiliary power outlets (APOs)?
   a. Two outlets are inside the cab of the truck
   b. Two outlets are in the bed of the truck
   c. Each outlet provides 120 volts AC
   d. All of the above
4. The General Motors mild (assist) hybrid system uses what type of motor/generator?
   a. Induction type motor located between the engine and the transmission
   b. A belt operated motor/generator
   c. A motor/generator located inside the transmission/transaxle
   d. A conventional but larger than normal alternator and starter
5. How is the high-voltage battery recharged in a Chevrolet Volt?
   a. By plugging the vehicle into an electrical outlet
   b. By running the engine
   c. Can be charged using the gasoline engine or by plugging it into an outlet
   d. By using a special high-voltage battery charger at a shop or dealer only
6. What type of battery pack is used in the Saturn VUE and Chevrolet Malibu BAS hybrid?
   a. 300-volt NiMH
   b. Three 12-volt valve-regulated lead-acid
   c. 36-volt NiMH
   d. 144-volt D-cell-size NiMH
7. The Saturn VUE and Chevrolet Malibu hybrid electric vehicles (BAS) are capable of which of the following?
   a. Idle stop
   b. Regenerative braking
   c. Powering the vehicle from a stop using electricpower only
   d. Both a and b are correct

8. Where are the electric motors in the two-mode hybrid vehicle?
   a. Belt driven at the front of the engine
   b. Inside the transmission housing
   c. At the rear of the engine
   d. At the rear (output shaft) of the transmission

9. The batteries used in the GM two-mode hybrid vehicles are ______________.
   a. 300-volt NiMH
   b. Three 12-volt valve-regulated lead-acid
   c. 36-volt NiMH
   d. 144-volt D-cell-size NiMH

10. Two technicians are discussing the two-mode hybrid vehicle. Technician A says that the two modes are idle stop and regenerative braking. Technician B says that the trucks are full (strong) hybrids capable of operating on battery power alone. Which technician is correct?
   a. Technician A only
   b. Technician B only
   c. Both technicians A and B
   d. Neither technician A nor B

chapter 26

HEATING AND AIR-CONDITIONING PRINCIPLES

LEARNING OBJECTIVES

After studying this chapter, the reader should be able to:

1. Discuss the changes of states of matter.
2. Discuss the effect of heat and temperature on matter.
3. Discuss the two types of humidity.
4. Explain heating and cooling load.
5. Explain the three ways in which heat flows.
6. Describe the air-conditioning process.
7. Explain the purpose of an HVAC system.

This chapter will help you prepare for the ASE Heating and Air Conditioning (A7) certification test content area "A" (A/C System Service, Diagnosis, and Repair).

KEY TERMS

GM STC OBJECTIVES

GM Service Technical College topics covered in this chapter are as follows:

1. The difference between heat and temperature. (01510.01W)
2. The effects of heat, temperature changes, and latent heat.
3. The definitions of heat flow, radiation, conduction, and convection.
4. The effects of pressure and boiling points.
5. The behaviors of refrigerants and coolants.

FIGURE 26–1 Water is a substance that can be found naturally in solid, liquid, and vapor states.

FREQUENTLY ASKED QUESTION

Why Is Liquid Sprayed from a Can Cold?

If a pressurized can of liquid is sprayed continuously, the can becomes cold, and so does the liquid being sprayed. The can becomes cold because the pressure in the can is reduced while spraying, allowing the liquid propellant inside the can to boil and absorb heat. The propellant vapor is further cooled as it decompresses when it hits the open air. Rapid decompression results in a rapid temperature drop.

INTRODUCTION

PURPOSE AND FUNCTION The **heating, ventilation, and air-conditioning (HVAC)** system of an automobile is designed to provide comfort for the driver and passengers. It is intended to maintain in-vehicle temperature and humidity within a range that is comfortable for the people inside and provide fresh, clean air. The air-conditioning system transfers the heat from inside the vehicle and moves it to the outside of the vehicle. The heater is needed in cold climates to prevent freezing or death.

PRINCIPLES INVOLVED On earth, matter is found in one of three different phases or states:

1. Solid
2. Liquid
3. Vapor (gas)

The state depends upon the nature of the substance, the temperature, and the pressure or force exerted on it. Water occurs naturally in all three states: solid ice, liquid water, and water vapor, depending upon the temperature and pressure. ● **SEE FIGURE 26–1**.

CHANGES OF STATE A solid is a substance that cannot be compressed and has strong resistance to flow. The molecules of a solid attract each other strongly, and resist changes in volume and shape.

- A substance is solid at any temperature below its melting point. Melting point is a characteristic of the substance, and is related to the temperature at which a solid turns to liquid. For water, the melting point is 32°F (0°C), which means that changes can be observed between liquid water and ice under normal weather conditions.
- A liquid is a substance that cannot be compressed. A substance in a liquid state has a fixed volume, but no definite shape.
- The boiling point is the temperature at which a liquid substance turns to vapor. For water at normal sea level conditions, the boiling point is 212°F (100°C). A vapor is a substance that can be easily compressed, has no resistance to flow, and no fixed volume. Since a vapor flows, it is considered a fluid just like liquids are.

A vapor condenses to liquid if the temperature falls below the vaporizing temperature. Again, the difference is simply whether heat is being added or taken away. Boiling point and condensation point temperatures are not fixed because they vary with pressure.

HEAT AND TEMPERATURE Molecules in a substance tend to vibrate rapidly in all directions, and this disorganized energy is called heat. The intensity of vibration depends on how much kinetic energy, or energy of motion, the atom or molecule contains. Heat and temperature are not the same.

- **Temperature** is the measure of the level of energy. Temperature is measured in degrees.
- **Heat** is measured in the metric unit called **calorie** and expresses the amount of heat needed to raise the temperature of one gram of water one degree Celsius. Heat is also measured in **British Thermal Units (BTU)**. One BTU is the heat required to raise the temperature of one pound of water 1°F at sea level. One BTU equals 252 calories.

FIGURE 26–2 The extra heat required to change a standard amount of water at its boiling point to vapor is called latent heat of vaporization.

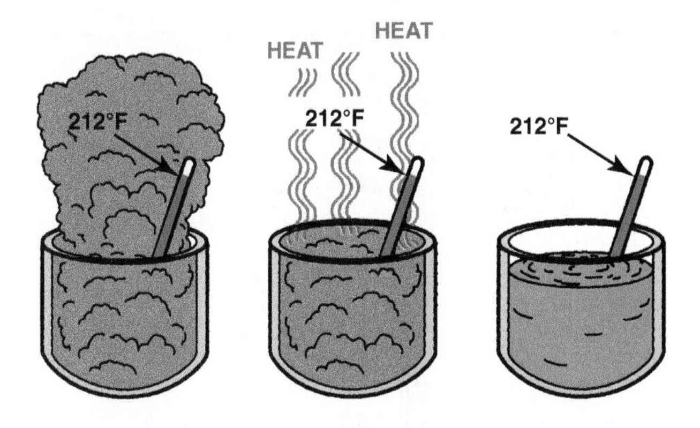

FIGURE 26–3 The latent heat of vaporization that water vapor stores is released when the vapor condenses to a liquid. The temperature stays the same.

SENSIBLE HEAT Sensible heat makes sense because it can be felt and measured on a thermometer. If there is 1 lb. of water at 40°F and 1 BTU of heat is added to it, the temperature will increase to 41°F. Adding another BTU of heat will increase the temperature to 42°F and adding another 170 BTU (212-42) will increase the temperature to 212°F, the boiling point.

LATENT HEAT **Latent heat** is the "extra" heat that is needed to transform a substance from one state to another. Imagine that a solid or a liquid is being heated on a stove. When the solid reaches its melting point, or the liquid reaches its boiling point, their temperatures stop rising. The solid begins to melt, and the liquid begins to boil. This occurs without any sensible change in temperature, even though heat is still being applied from the burner. The water in the container on the stove boils at a temperature of 212°F (100°C) at sea level, for as long as any liquid water remains. As heat is further added to the water, heat will be used in changing the state of the liquid to a vapor. This extra, hidden amount of energy necessary to change the state of a substance is called latent heat. ● **SEE FIGURES 26–2 AND 26–3.**

Latent heat is important in the operation of an air-conditioning system because the cooling effect is derived from changing the state of liquid refrigerant to vapor. The liquid refrigerant absorbs the latent heat of vaporization, making the air cooler. The cooler air is then blown into the passenger compartment.

TEMPERATURE, VOLUME, AND PRESSURE OF A VAPOR Unlike a solid, vapor has no fixed volume. Increasing the temperature of a vapor, while keeping the volume confined in the same space, increases the pressure. This happens as the vibrating vapor molecules collide more and more energetically with the walls of the container. Conversely, decreasing the temperature decreases the pressure. This relationship between temperature and pressure in vapor is why a can of nonflammable refrigerant can explode when heated by a flame—the pressure buildup inside the can will eventually exceed the can's ability to contain the pressure. Increasing the pressure by compressing vapor increases the temperature. Decreasing the pressure by permitting the vapor to expand decreases the temperature.

HEAT INTENSITY Intensity of heat is important to us because if it is too cold, humans feel uncomfortable and is measured in degrees. Extremely cold temperatures can cause frostbite and hypothermia. The other end of the scale can also be uncomfortable and may cause heat stress and dehydration. Humans have a temperature **comfort zone** somewhere between 68°F and 78°F (20°C and 26°C). This comfort zone varies among individuals. ● **SEE FIGURE 26–4.**

RULES OF HEAT TRANSFER Heating and air conditioning must follow the basic rules of heat transfer. An understanding of these rules helps greatly in understanding the systems.

- Heat always flows from hot to cold (from higher level of energy to lower level of energy). ● **SEE FIGURE 26–5.**
- To warm a person or item, heat must be added.
- To cool a person or item, heat must be removed.
- A large amount of heat is absorbed when a liquid changes state to vapor.
- A large amount of heat is released when a vapor changes state to a liquid.
- Compressing a gas concentrates the heat and increases the temperature.

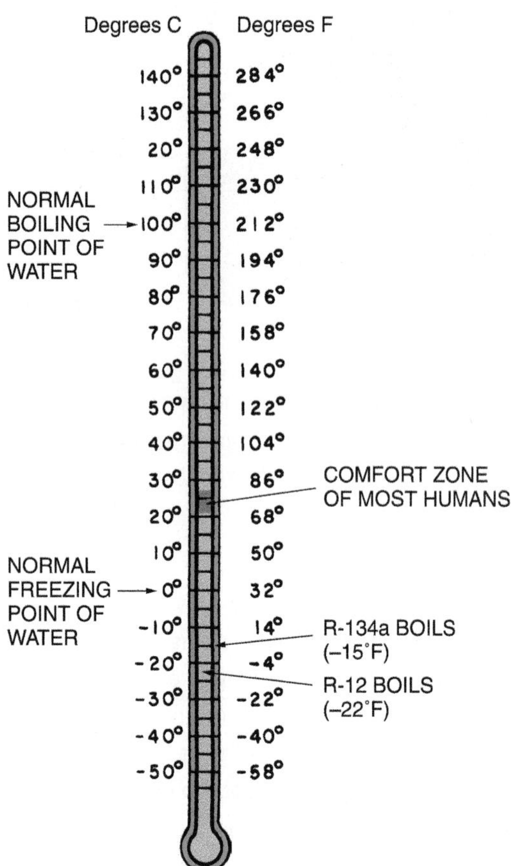

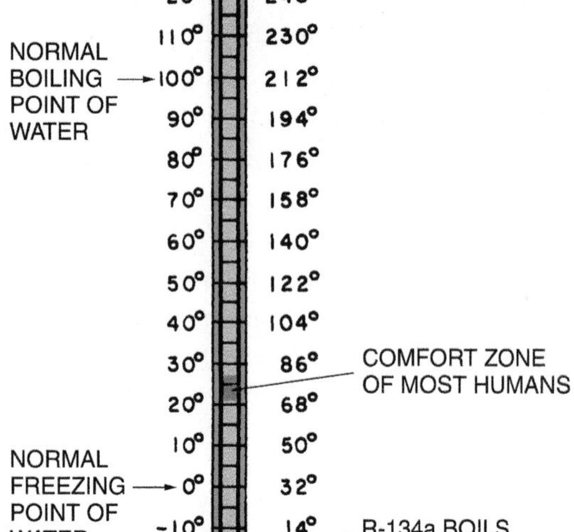

FIGURE 26–4 Heat intensity is measured using a thermometer. The two common measuring scales, Celsius and Fahrenheit, are shown here. This thermometer is also marked with water freezing and boiling and refrigerant boiling temperatures.

TECH TIP

Quick and Easy Temperature Conversion

Temperature in service information and on scan tools are often expressed in degrees Celsius, which is often confusing to those used to temperature expressed in Fahrenheit degrees. A quick and easy way to get an *approximate* conversion is to take the degrees in Celsius, double it, and add 25.

For example,

Celsius × 2 + 25 = approximate Fahrenheit degrees:

0°C × 2 = 0 + 25 = 25°F (actual = 32°F)
10°C × 2 = 20 + 25 = 45°F (actual = 50°F)
15°C × 2 = 30 + 25 = 55°F (actual = 59°F)
20°C × 2 = 40 + 25 = 65°F (actual = 68°F)
25°C × 2 = 50 + 25 = 75°F (actual = 77°F)
30°C × 2 = 60 + 25 = 85°F (actual = 86°F)
35°C × 2 = 70 + 25 = 95°F (actual = 95°F)
40°C × 2 = 80 + 25 = 105°F (actual = 104°F)
45°C × 2 = 90 + 25 = 115°F (actual = 113°F)
50°C × 2 = 100 + 25 = 125°F (actual = 122°F)

The simplest way to convert between the Fahrenheit and Celsius scales accurately is to use a conversion chart or use an app on a smart phone.

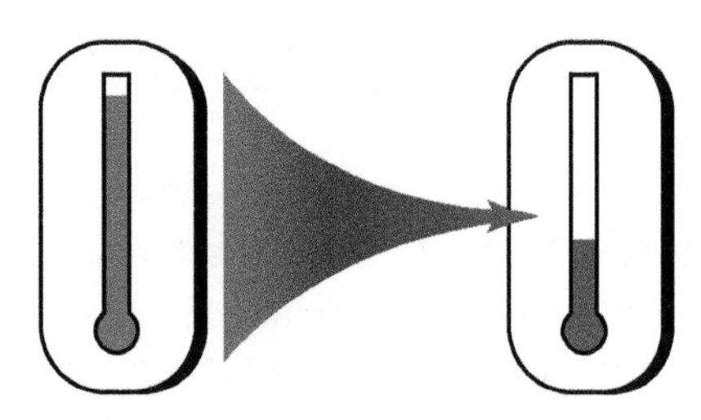

HOT TRAVELS TO COLD UNTIL THE TEMPERATURES EQUAL.

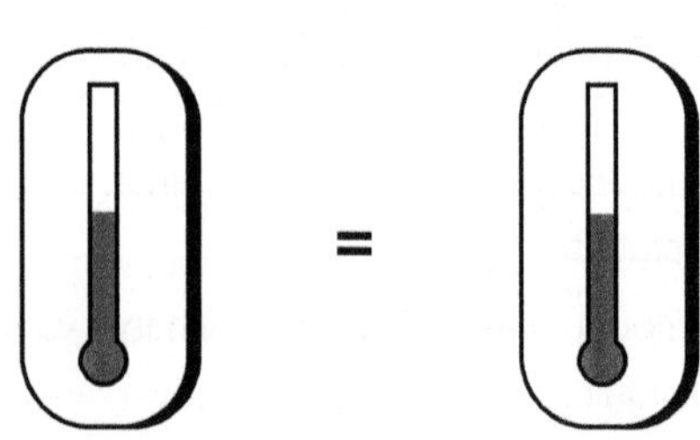

FIGURE 26–5 Heat travels from higher temperature (higher energy level) to lower temperature (lower energy level).

HUMIDITY Humidity refers to water vapor present in the air. The level of humidity depends upon the amount of water vapor present and the temperature of the air. The amount of water vapor in the air tends to be higher near lakes or the ocean, because more water is available to evaporate from their surfaces. In desert areas with little open water, the amount of water vapor in the air tends to be low.

- **Absolute humidity** is the mass of water vapor in a given volume of air.
- **Relative humidity (RH)** is the percentage of how much moisture is present in the air compared to how much moisture the air is capable of holding at that temperature.

Relative humidity is commonly measured with a *hygrometer* or a *psychrometer*. A hygrometer depends on a sensitive element that expands and contracts, based on the humidity. Hygrometers typically resemble a clock, with the scale reading from 0% to 100% relative humidity. ● **SEE FIGURE 26–6.**

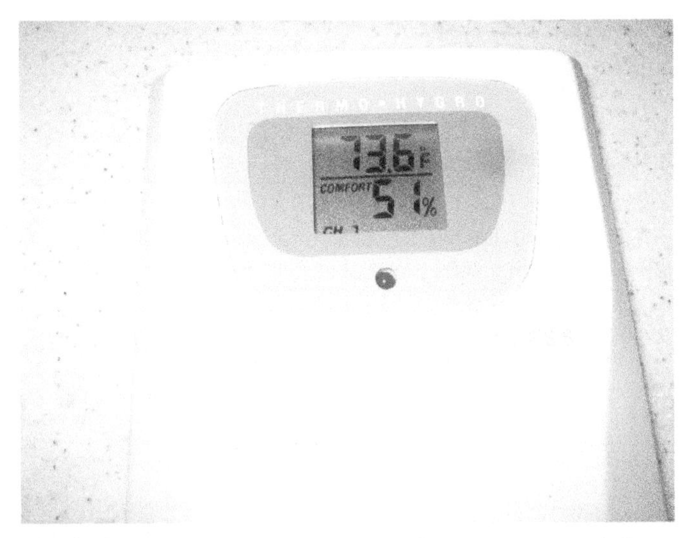

FIGURE 26–6 A combination meter that measures and displays both the temperature and the humidity is useful to use when working on air-conditioning systems.

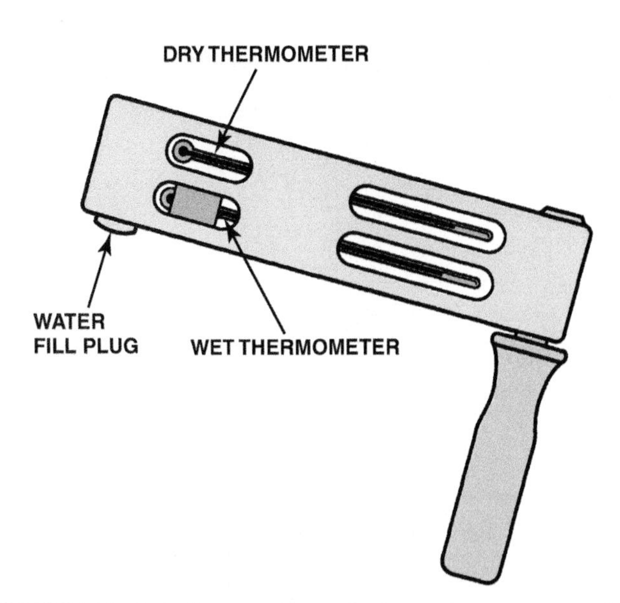

FIGURE 26–7 A sling psychrometer is used to measure relative humidity.

HEATING AND COOLING LOAD

HEATING LOAD **Heating load** is the term used when additional heat is needed. The actual load is the number of BTUs or calories of heat energy that must be added. In a home or office, burning fuel is the usual way to generate heat using coal, gas, or oil as a fuel. In most vehicles, the heat is provided by the heated coolant from the engine cooling system. This coolant is typically at a temperature of 190°F to 205°F (88°C to 98°C) when the engine reaches its normal operating temperature. ● **SEE FIGURE 26–8**.

In most vehicles, heated coolant is circulated through a heat exchanger, called a **heater core**. Air is circulated through the heater core, where it absorbs heat. Then it is blown into the passenger compartment, where the heat travels on to warm the car interior and occupants. The air from the blower motor moves the heat from the heater core to the passengers.

COOLING WITH ICE One way to move heat, called **cooling load**, is with a block of ice. A substantial amount of latent heat is required to change the state of the solid ice into a liquid:

- 144 BTU per lb. (80 calories per gram).
- A 50-lb. block of ice represents 50 × 144, or 7,200 BTU, of cooling power when it changes from 50 lb. of solid at 32°F to 50 lb. of liquid at 32°F.

FREQUENTLY ASKED QUESTION

What Is a Sling Pyschrometer?

A psychrometer is a measuring instrument used to measure relative humidity. It uses two thermometers, one of which has the bulb covered in a cotton wick soaked in distilled water from a built-in reservoir. The wick keeps the bulb of the "wet thermometer" wet so that it can be cooled by evaporation. Sling psychrometers are spun round in the air a certain number of times. Water evaporates from the cotton wick at a rate inversely proportional to the relative humidity of the air.

- Faster if the humidity is low
- Slower if the humidity is high.

The "dry thermometer" measures the air temperature.

- The higher the relative humidity, the closer the readings of the two thermometers.
- The lower the humidity, the greater the difference in temperature of the two thermometers.

The different temperatures indicated by the wet and dry thermometers are compared to values given in a chart, which gives the relative humidity. While a sling psychrometer is still used, most technicians use an electronic instrument to measure relative humidity. ● **SEE FIGURE 26–7**.

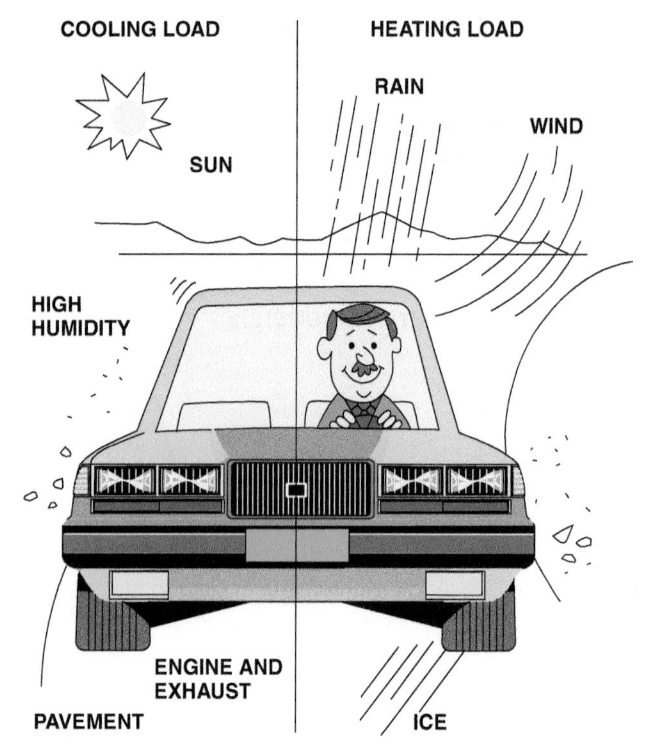

FIGURE 26–8 Winter presents a heat load where heat must be added for comfort (right). Summer presents a cooling load.

In the early days of air conditioning, the term *ton* was commonly used. A ton of air conditioning was the amount of heat it took to melt a ton of ice: 2,000 × 144, or 288,000 BTU. ● **SEE FIGURE 26–9**.

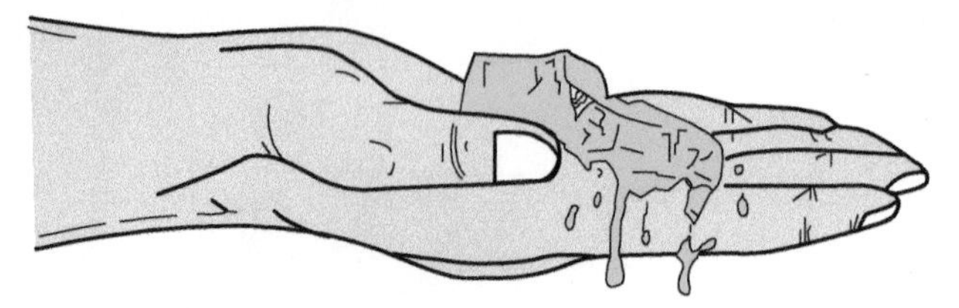

FIGURE 26–9 Ice has a cooling effect because of latent heat of fusion which means that it absorbs heat as it melts.

FIGURE 26–10 At one time, evaporative coolers were used to cool car interiors. Air forced through a water-wetted mesh produces evaporation and a cooling effect.

EVAPORATIVE COOLING A method of cooling that works well in areas of low humidity is evaporation of water, commonly called **evaporative cooling**. If water is spread thinly over the extremely large area of a meshed cooler pad and air is blown across it, the water evaporates. For every pound of water that evaporates, 970 BTU (540 calories per gram) of heat is absorbed. This is the latent heat of evaporation, just as when it is boiled. This is a natural process and uses only the energy required by the blower to circulate the air through the cooler pads and on to the space to be cooled. Disadvantages of evaporative coolers, often called "swamp coolers" includes:

- increases the relative humidly
- not effective in areas of high humidity because the water does not evaporate rapidly enough to be efficient.

At one time, window-mounted evaporative coolers were used in cars. They were not very popular because they were unattractive and worked well only in dry areas. ● **SEE FIGURE 26–10**.

MECHANICAL COOLING A third way to handle a cooling load is by the use of **mechanical refrigeration**, which is called air conditioning. This system also uses evaporation of a liquid and the large amount of heat required for evaporation. The refrigerant boils so that it changes from liquid to gas, but it is condensed back to gas using an engine or electrically powered compressor to move the refrigerant and to increase its pressure in the system. ● **SEE FIGURE 26–11**.

HEAT MOVEMENT

Heat can travel through one or more of three paths as it moves from hot to cold:

1. conduction
2. convection
3. radiation. ● **SEE FIGURE 26–12**.

CONDUCTION The simplest heat movement method is conduction, by which heat travels through a medium such as a solid or liquid, moving from one molecule of the material to the next. For

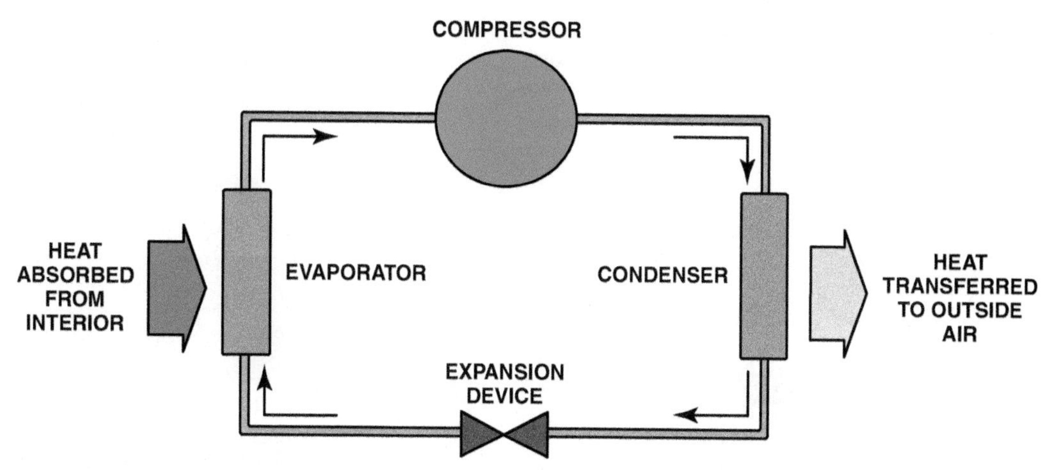

FIGURE 26–11 Heat, from in-vehicle cabin air, causes the refrigerant to boil in the evaporator (left). The compressor increases the pressure and moves refrigerant vapor to the condenser, where the heat is transferred to ambient air. This also causes the vapor to return to liquid form.

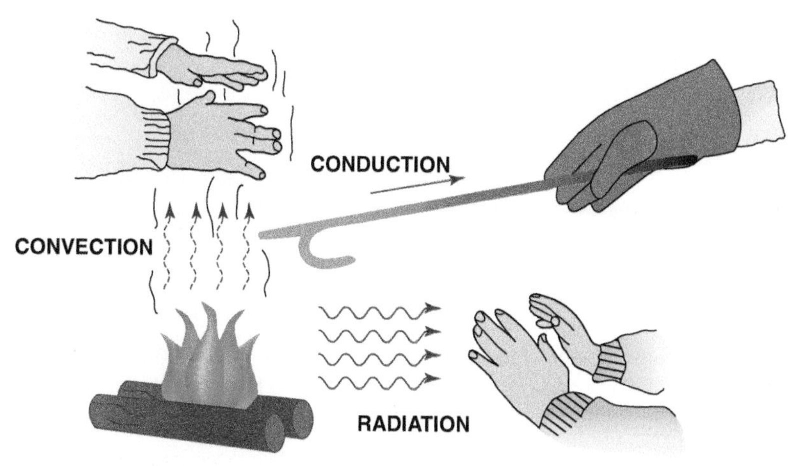

FIGURE 26–12 Heat can be moved from the source by convection, conduction, or radiation.

example, if one end of wire is heated, the heat will travel through the material itself and will be transferred through to the other end of the wire. Some materials (most of the metals) are good heat conductors. Copper and aluminum are among the best of the commonly used metals, so most heat exchangers (radiators, evaporators, and condensers) use copper or aluminum.

- Some materials, such as wood, are poor heat conductors.
- Some materials, such as Styrofoam, conduct heat so poorly that they are called insulators. Most good insulators incorporate a lot of air or gaseous material in their structure because air is a poor conductor of heat.

CONVECTION Convection is a process of transferring heat by moving the heated medium, usually air or a liquid. An example of convection is the engine cooling system. Coolant is heated in the water jackets next to the cylinders and combustion chambers. Then the coolant is pumped to the radiator, where the heat is transferred to the air traveling through the radiator. Convection also occurs in the interior of the vehicle when air is circulated past the driver and passengers to pick up heat and moved to the evaporator, where the heat is transferred to the evaporator fins. The evaporator fins are cooler, so heat is transferred easily.

RADIATION Heat can travel through heat rays and pass from one location to another without warming the air through which it passes. The best example of this is the heat from the sun, which passes through cold space and warms our planet and everything it shines on. Radiant heat can pass from any warmer object through air to any cooler object. It is affected by the color and texture of the heat emitter, where the heat leaves, and the collector, where the heat is absorbed. Dark, rough surfaces make better heat emitters and collectors than light-colored, smooth surfaces.

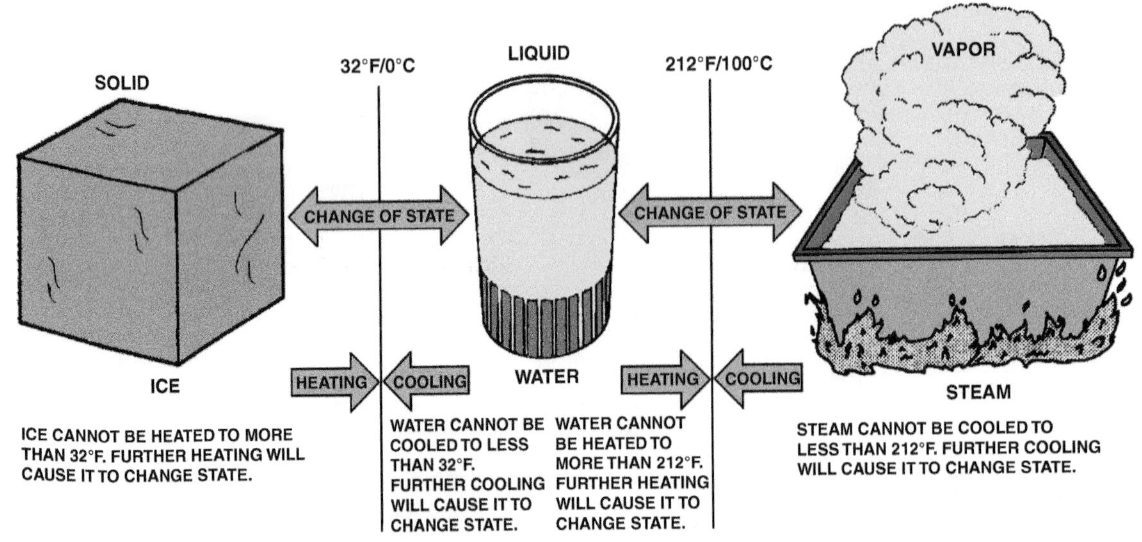

FIGURE 26–13 Matter can change state by adding or removing heat.

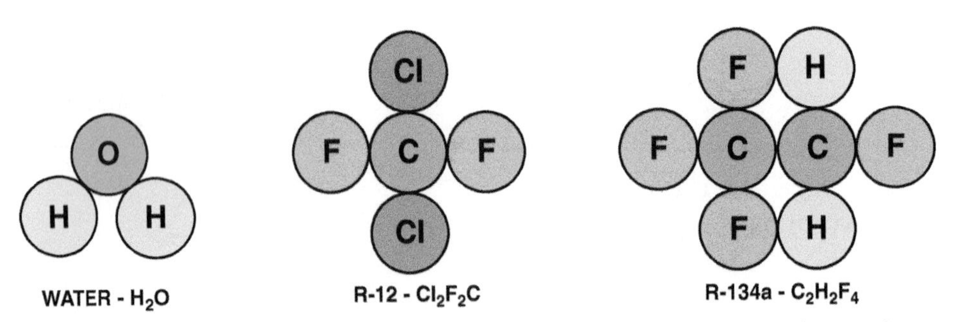

FIGURE 26–14 A water molecule contains two oxygen atoms and one hydrogen atom. R-12 is a combination of one carbon, two chlorine, and two fluorine atoms. R-134a is a combination of two carbon, four fluorine, and two hydrogen atoms.

NOTE: At one time, California Highway Patrol cars were painted all black. Painting the tops white benefited the patrol officers by lowering the in-vehicle temperature significantly.

AIR-CONDITIONING PROCESS

CHANGING THE STATE OF THE REFRIGERANT The air-conditioning process works using a fluid, called *refrigerant,* which continuously changes state from liquid to gas and back to liquid.

Most states of matter can be changed from one state to another by adding or removing heat. ● **SEE FIGURE 26–13.**

Molecules are the building blocks for all things that can be seen or felt. Molecules are combinations of atoms, which are in turn made up of electrons, neutrons, and protons. The protons are in the center, or nucleus, of the atom and the electrons travel in an orbit around them. There are about 100 basic elements or atoms, each having a different atomic number, that combine with other elements to make the many, varied molecules. The atomic number of an element is based on the number of electrons and protons in that element. The periodic table of elements seen in most chemistry laboratories shows the relationship of these elements.

Water molecules, for example, are called H_2O. This is a combination of a single oxygen atom and two hydrogen atoms. Hydrogen has an atomic number of 1 (1 proton and 1 electron), and oxygen has an atomic number of 8 (8 electrons and 8 protons). ● **SEE FIGURE 26–14**. The three states of water are well known and include:

1. Solid ice
2. Liquid water
3. Vapor (gaseous).

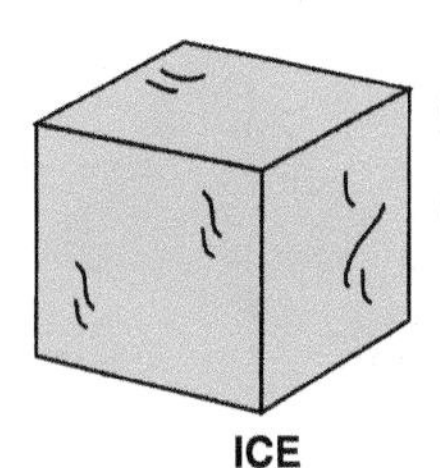

FIGURE 26–15 Ice is a solid form of water with a low temperature and slow molecular action.

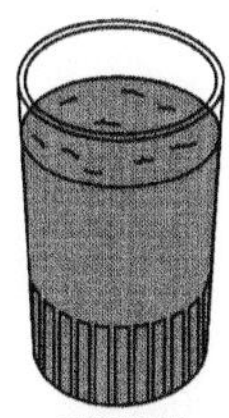

FIGURE 26–16 Water is warmer than ice and can flow to take the shape of any container.

SOLID Solid matter has a definite shape and substance. Solids exert pressure in only one direction, and that is downward because of gravity. For example, ice is the solid form of water, and will hold its shape, and is cold. Water is normally a solid at temperatures below 32°F (0°C), which is the normal freezing point. The electrons in the molecule's atoms are still orbiting around the protons, but the movement has been slowed because much of the heat energy has been removed. ● **SEE FIGURE 26–15.**

LIQUID Adding heat to most solids causes them to reach their melting point. It is the same material, but heat energy has broken the molecular bond and the matter becomes fluid. Fluid has no shape and it takes the shape of its container. Liquids can flow through a pipe or hose and can be pumped such as by the air-conditioning compressor.

Water is normally a liquid between 32°F and 212°F (0°C and 100°C). The molecules are same as ice, but heat energy has increased the movement of the electrons. ● **SEE FIGURE 26–16.**

GAS When heat is added to most liquids, it produces gas as the liquids boil. It is the same material, but the heat energy has broken the molecular bonds still further so that the molecules have no shape at all and have expanded so much that they have very little weight. A gas molecule exerts pressure in every direction. Gases can also be pumped through hoses and pipes, making them easy to move through the air-conditioning (A/C) system.

At temperatures above 212°F (100°C), water normally boils to become a gas, called *steam*. Again, the molecules are the same as water or ice, but heat energy has greatly increased molecular movement. ● **SEE FIGURE 26–17.**

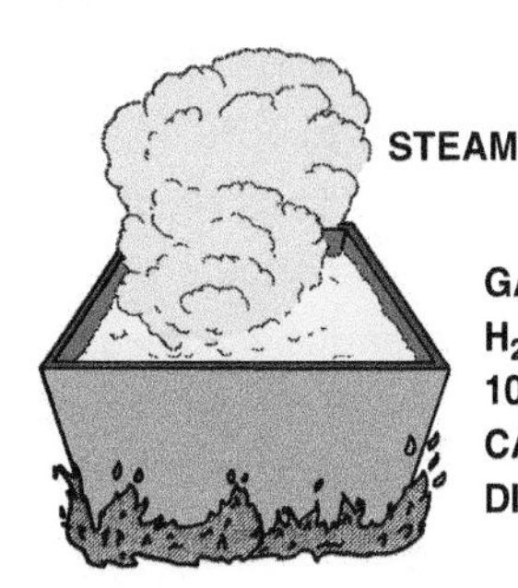

FIGURE 26–17 Adding heat to water produces steam, the gas state, with a much freer molecular action.

PURPOSE OF AN HVAC SYSTEM

CONTROL OF TEMPERATURE The goal in heating and air conditioning is to maintain a comfortable in-vehicle temperature and humidity. This is affected by the size of the vehicle, the number of passengers, and the amount of glass area, to name only a few variables. The internal body temperature of humans is about 98.6°F (37°C), which seems odd when our most comfortable temperature is 68°F to 78°F (20°C to 26°C). This means that in the summer heat must be continually given off to be comfortable, but in the winter suitable clothing is needed to maintain warmth. Body comfort is also affected by radiant heat. In the winter, the sun warms the body. Solar engineers are working on ways to control this heat flow as the amount of glass area of a vehicle increases.

The velocity of air past our bodies is another factor in human comfort. Air movement is an important part of heating and air-conditioning systems.

CONTROL OF HUMIDITY Humid cold air feels much colder than dry air at the same temperature.

- Humid hot air slows down our natural body cooling system (evaporation of perspiration), so it can make a day feel much hotter.
- Air that is too dry also tends to make people feel uncomfortable.
- As with temperature, a range of humidity that most people feel comfortable in a relative humidly of about 45% to 55%.

As the air-conditioning system operates, it dehumidifies (removes moisture) from the air. Water vapor condenses on the cold evaporator fins just as it would on a glass holding a cold drink. This condensed water then drops off the evaporator and runs out the drain at the bottom of the evaporator case. In-vehicle humidity is reduced to about 40% to 45% on even the most humid days if the A/C is operated long enough. A good example of this dehumidification process occurs when a vehicle's A/C is operated on cold days when the windows are fogged up. It usually takes only a short time to dry the air and remove the fog from the windows. ● **SEE FIGURE 26–18.**

CLEANLINESS A side effect of air conditioning is the cleaning of the air coming into the car as it passes through the cooling ductwork. The act of cooling and dehumidifying air at the A/C evaporator causes water droplets to form on the evaporator fins. Dust and other contaminants in the air that come into contact with these droplets become trapped and are flushed out of the system as the water drops drain from the evaporator. Most recent vehicles use a **cabin filter** in the A/C and heating systems to clean the air by trapping dust and pollen particles before they enter the passenger compartment.

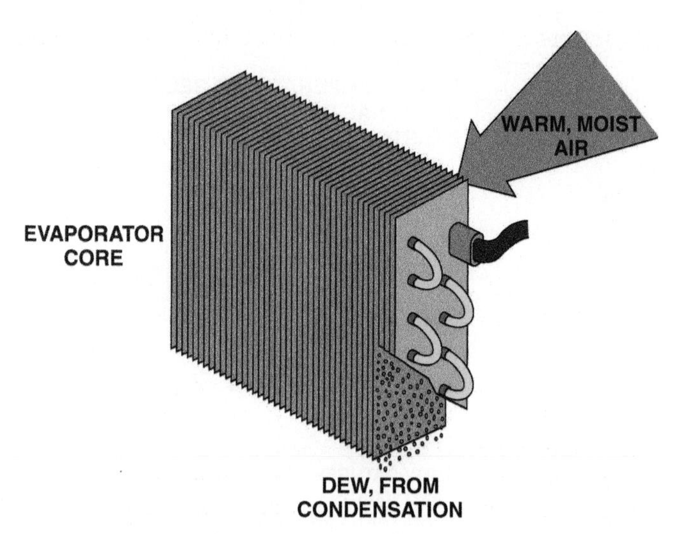

FIGURE 26–18 When air comes into contact with the cold evaporator, excess moisture forms dew. This condensed moisture leaves the car through the evaporator drain.

SUMMARY

1. Heat is moved into or out of the passenger compartment to obtain a good comfort level.
2. Heat intensity is measured using the Fahrenheit or Celsius scales, and heat quantity is measured using calories and BTU.
3. The comfort zone of most humans is between 68°F and 78°F (20°C and 26°C) and 45% to 50% humidity.
4. A/C systems reduce humidity by removing moisture (water) from the air.
5. HVAC systems clean air because particles are caught by moisture on the evaporator and by filters.

REVIEW QUESTIONS

1. What does the abbreviation "HVAC" stand for?
2. What are the three states of matter?
3. What is the difference between heat and temperature?
4. How is relative humidity measured?
5. How does heat move?

CHAPTER QUIZ

1. The three different phases or states of matter include ________.
   a. Solid, water, and steam
   b. Ice, liquid, and gas
   c. Solid, liquid, and gas
   d. Liquid, water, and steam
2. An air-conditioning system cools the interior of the vehicle by ________.
   a. Moving the heat from inside the vehicle to outside the vehicle
   b. Blowing cold air into the interior
   c. Moving heat from inside of the vehicle to the engine cooling system
   d. Using the engine to move air

3. Twenty degrees Celsius is about how many degrees Fahrenheit?
   a. 25
   b. 45
   c. 65
   d. 85
4. Heat intensity is measured in _______.
   a. BTUs
   b. Degrees
   c. RH
   d. Calories
5. A psychrometer measures _______.
   a. Temperature
   b. Relative humidity
   c. Amount of heat
   d. Radiation
6. A BTU is a measure of _______.
   a. Temperature
   b. Relative humidity
   c. Amount of heat
   d. Radiation
7. Heat transferred through the air is called _______.
   a. Radiation
   b. Insulation
   c. Convection
   d. Conduction
8. Air-conditioning process works through a fluid, called a _______ that continuously changes state from liquid to gas and back to liquid.
   a. Element
   b. Conductor
   c. Insulator
   d. Refrigerant
9. Humans prefer temperatures that are between _______ and _______.
   a. 55°F; 65°F
   b. 60°F; 70°F
   c. 68°F; 78°F
   d. 76°F; 86°F
10. Humans prefer relative humidity that is between _______ and _______.
    a. 10%; 20%
    b. 20%; 30%
    c. 30%; 40%
    d. 45%; 55%

chapter 27

THE REFRIGERATION CYCLE

LEARNING OBJECTIVES

After studying this chapter, the reader should be able to:

1. Explain how the A/C system works.
2. Identify the low and high side of an A/C system.
3. Explain the purpose and function of evaporators in an A/C system.
4. Explain the purpose and function of thermal expansion valves and orifice tube systems.
5. Explain the purpose and function of condensers.

This chapter will help you prepare for the ASE Heating and Air Conditioning (A7) certification test content area "A" (A/C System Service, Diagnosis, and Repair).

KEY TERMS

Compressor 437
Condenser 437
Evaporator 437
Flooded 441
High side 438
Liquid line 438
Low side 438
Orifice tube (OT) 439
Overcharge 444
Pressure sensor 446
Pressure switch 445
Receiver–drier 442
Refrigeration cycle 437
Starved 441
Sub-cooling 443
Suction line 445
Superheat 445
Thermal expansion valve (TXV) 439
Thermistor 445
Undercharge 444

GM STC OBJECTIVES

GM Service Technical College topics covered in this chapter are as follows:

1. The high and low pressure sides of an air-conditioning system (01510.08W).
2. The purpose and operation of an evaporator.
3. The purpose of a compressor.
4. The purpose and operation of a condenser.
5. The purpose, types, and operation of refrigerant metering devices.
6. The purpose, types, and operation of storage-dehydrators.

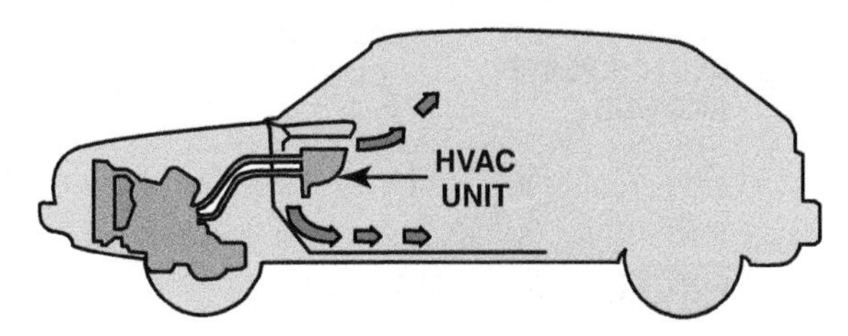

FIGURE 27–1 Air is circulated through the A/C and heating system and the vehicle to either add or remove heat.

BASIC PRINCIPLES

Automotive A/C systems operate on the principle of moving heat from inside to outside of the vehicle. Heat travels from a higher temperature (higher energy level) to a lower temperature (lower energy level).

- The flow of a refrigerant through the system is called the **refrigeration cycle** and is used to cool the interior of the vehicle.
- The heating system transfers heat from the engine's cooling system to the passenger compartment. ● **SEE FIGURE 27–1.**

REFRIGERANT MOVEMENT All automotive air-conditioning systems are closed and sealed. A refrigerant is circulated through the system by a **compressor** that is usually powered by the engine through an accessory drive belt. Older systems used CFC-12 as a refrigerant, commonly referred to as *R-12* or by its DuPont trade name of *Freon*. Starting in the early 1990s, most vehicle manufacturers now use HFC 134a (R-134a), a refrigerant that is less harmful to the atmosphere. The basic principle of the refrigeration cycle is that as a liquid changes into a gas, heat is absorbed. The heat that is absorbed by an automotive air-conditioning system is the heat from inside the vehicle.

HOW THE A/C SYSTEM WORKS The air-conditioning (A/C) system works as follows:

1. High-pressure liquid refrigerant flows through an expansion device or restriction, which controls the amount of refrigerant that is allowed to pass through.
2. When the high-pressure liquid passes through the expansion device, the pressure drops. This causes the liquid refrigerant to evaporate in a small radiator-type unit called the **evaporator**. When the refrigerant evaporates, it absorbs heat (latent heat) when changing from a liquid to a gas. As the heat is absorbed by the refrigerant, the evaporator becomes cold. ● **SEE FIGURE 27–2.**

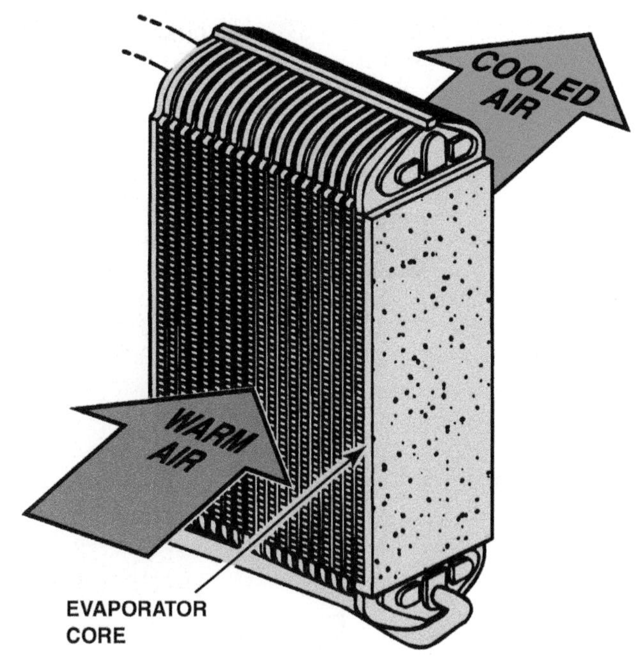

FIGURE 27–2 The evaporator removes heat from the air that enters a vehicle by transferring it to the vaporizing refrigerant.

3. After the refrigerant has evaporated into a *low-pressure gas* in the evaporator, it flows into the engine-driven compressor. The compressor compresses the low-pressure refrigerant gas into a high-temperature, high-pressure gas and forces it through the system. ● **SEE FIGURE 27–3.**
4. This high-pressure gas flows into the condenser located in front of the cooling system radiator. The **condenser** looks like another radiator, and its purpose and function is to remove heat from the high-pressure gas. In the condenser, the high-pressure gas changes (condenses) to form a high-pressure liquid as the latent heat from the refrigerant is released to the air. ● **SEE FIGURE 27–4.**
5. The high-pressure liquid then flows to the expansion device, which controls the amount of refrigerant that is allowed to pass through and meters the flow into the evaporator. When the high pressure of the liquid passes through the expansion device, the pressure drops and causes the refrigerant to vaporize, starting the cycle all over again.
6. Air is blown through the evaporator by the blower motor. The air is cooled as heat is removed from the air and transferred to the refrigerant in the evaporator. This cooled air is then directed inside the passenger compartment through vents.

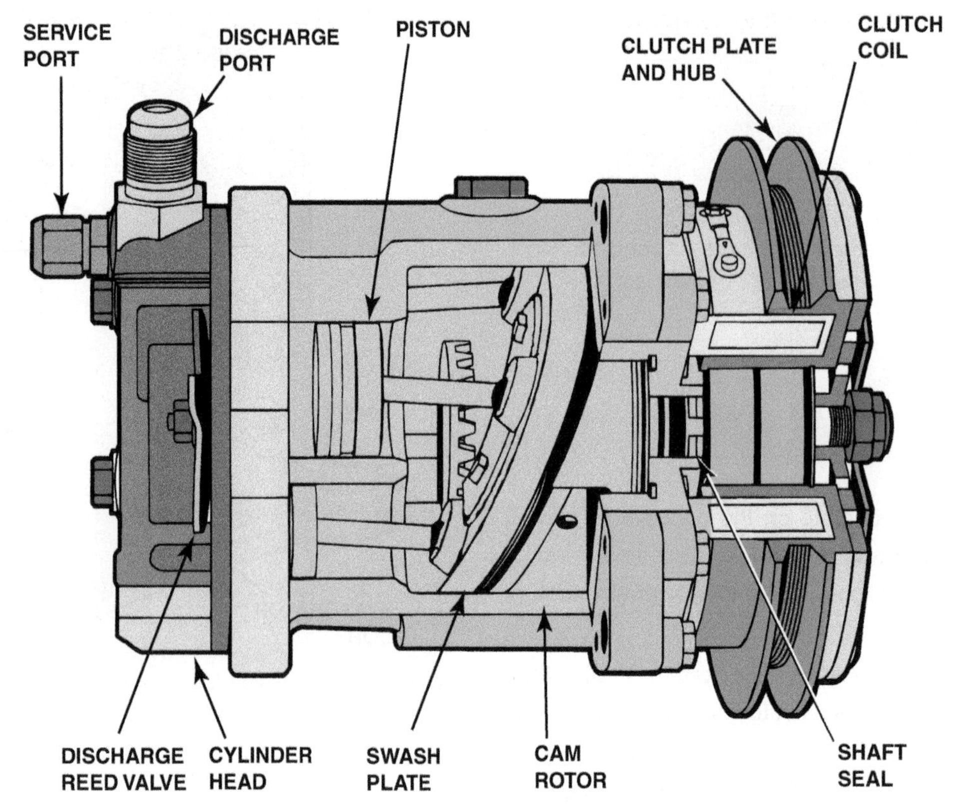

FIGURE 27–3 The compressor provides the mechanical force needed to pressurize the refrigerant.

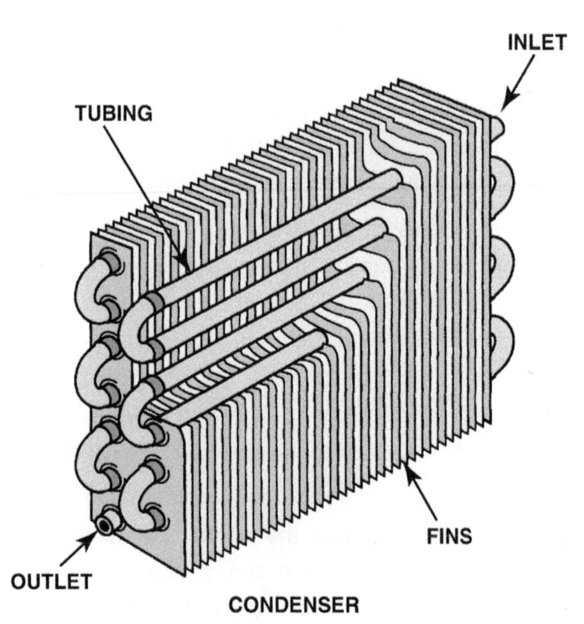

FIGURE 27–4 The condenser changes the refrigerant vapor into a liquid by transferring heat from the refrigerant to the air stream that flows between the condenser fins.

HIGH- AND LOW-SIDE IDENTIFICATION A/C systems can be divided into two parts:

1. **Low side**—it has low pressure and temperature. The low side begins at the expansion device and ends at the compressor. The refrigerant boils or evaporates in the low side.

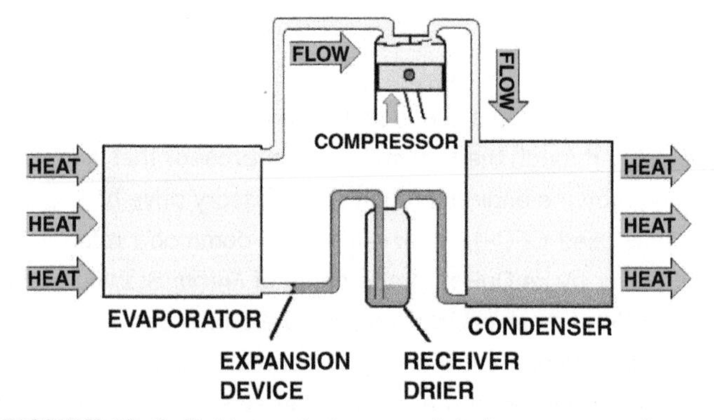

FIGURE 27–5 Refrigerant changes state to a vapor as it absorbs heat in the low side and into a liquid as it loses heat in the high side.

2. **High side**—it has higher pressures and temperatures. The high side begins at the compressor and ends at the expansion device. The refrigerant condenses in the high side. The high side line after the condenser is often called the **liquid line**. ● **SEE FIGURE 27–5.**

In an operating system, the low and high sides can be identified by:

- **Pressure**—A pressure gauge shows low pressure in the low side and high pressure in the high side.
- **Sight**—On high-humidity days, the cold low-side tubing often collects water droplets and may even frost.

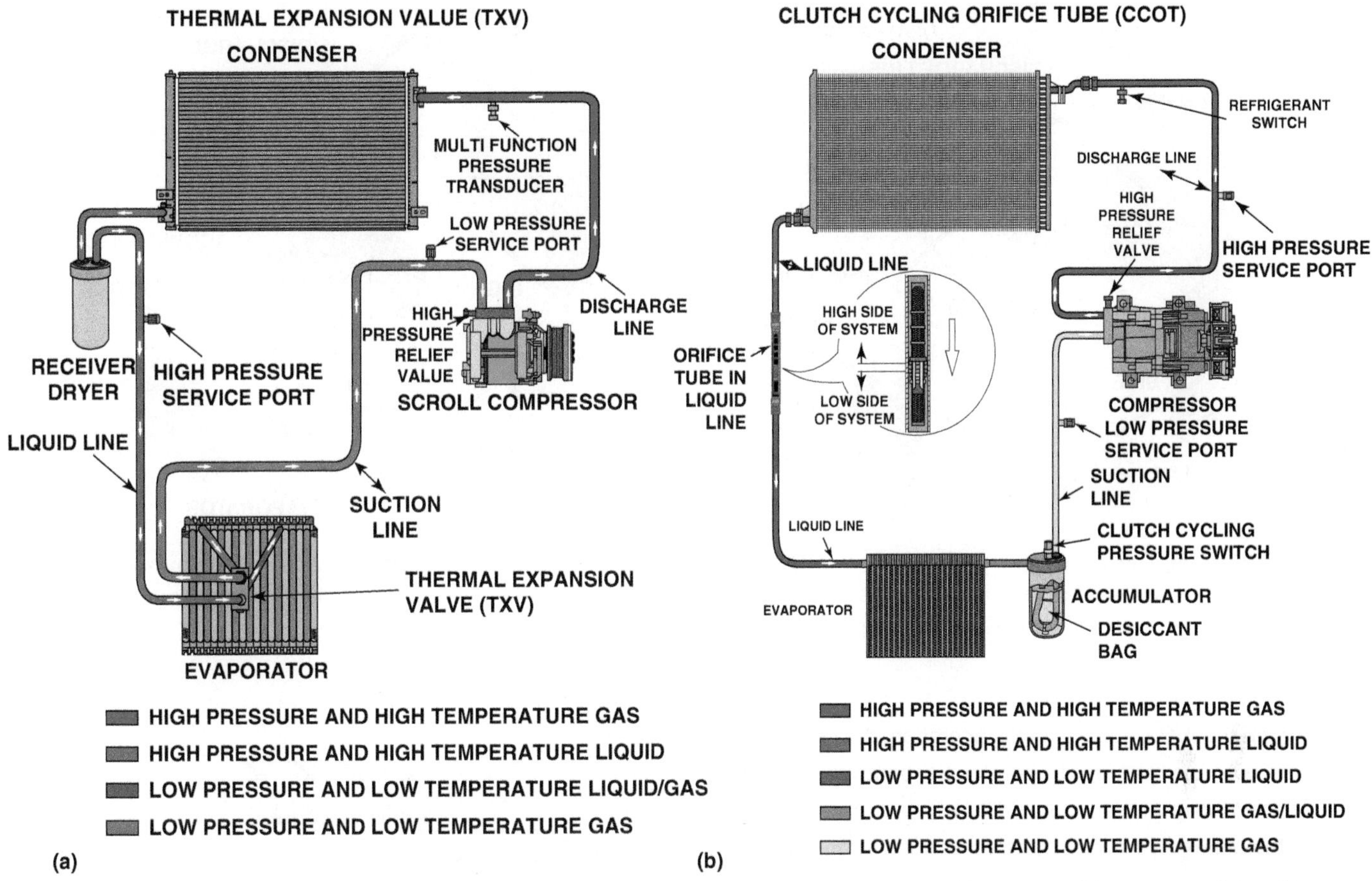

FIGURE 27–6 Automotive A/C systems are either a TXV system with a receiver–drier (a) or an OT system with an accumulator. (b) Various compressors are used with both systems.

FREQUENTLY ASKED QUESTION

How Does the Inside of the Vehicle Get Cooled?

The underlying principle involved in air conditioning or refrigeration is that "cold attracts heat." Therefore, a cool evaporator attracts the hot air inside the vehicle. Heat always travels toward cold and when the hot air passes through the cold evaporator, the heat is absorbed by the cold evaporator, which lowers the temperature of the air. The cooled air is then forced into the passenger compartment by the blower through the air-conditioning vents.

- **Temperature**—The low side is cool to cold, and the high side is hot.
- **Tubing size**—Low-side tubes and hoses are larger (vapor), and high-side tubes and hoses are smaller (liquid).

TYPES OF EXPANSION DEVICES Automotive A/C systems are of two types:

- **Orifice tube (OT)** systems—Orifice tube systems are also called *cycling clutch orifice tube (CCOT)* and *fixed orifice tube (FOT)* systems.
- **Thermal expansion valve (TXV)** systems. TXV systems are now being used by most vehicle manufacturers mainly due to the reduction in the amount of refrigerant required in this type of system. SEE FIGURE 27–6.

LOW-SIDE OPERATION The low side begins at the refrigerant expansion or flow metering device, either a TXV or an OT, which produces a pressure drop. The low side ends at the compressor, which increases the pressure.

These components are part of the low-side:

TXV or OT
Evaporator
Accumulator (OT systems)
Compressor suction port
Larger, low pressure lines and hoses

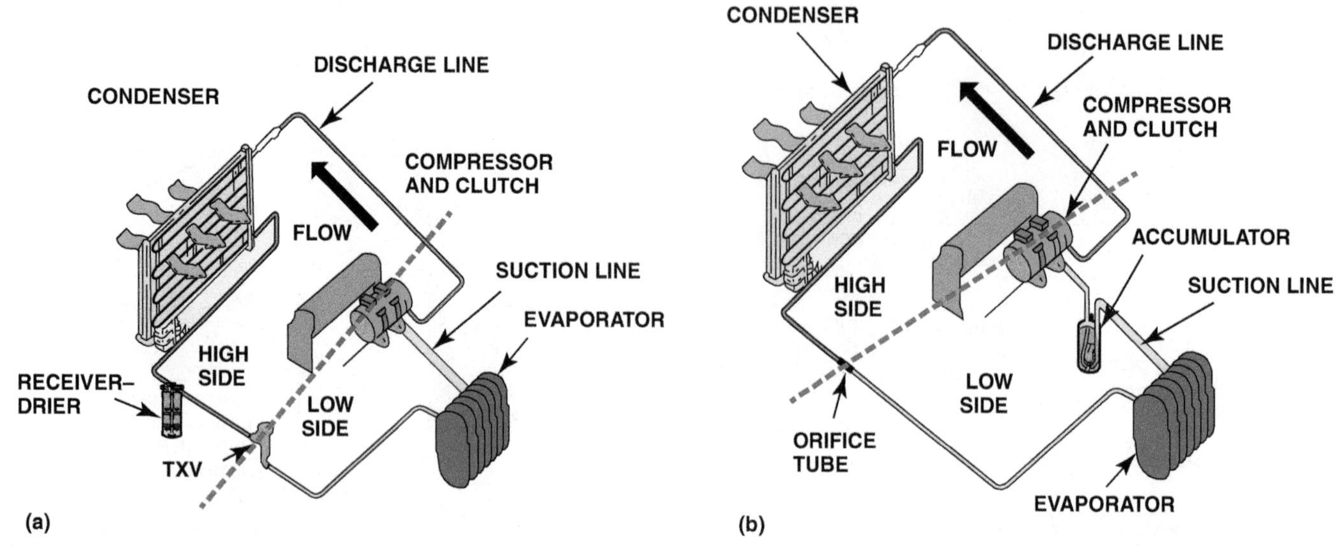

FIGURE 27–7 (a) The low side begins at the TXV or OT and includes the evaporator and suction line to the compressor. (b) The OT system includes an accumulator. The area above the red line represents high pressure and high temperature. The area below the red line represents lower temperature and lower pressures.

When the A/C system is in full operation, the goal of most systems is to maintain an evaporator temperature just above the freezing point of water, 32°F (0°C). This temperature produces the greatest heat exchange without ice formation on the evaporator fins (evaporator icing reduces the heat transfer). ● **SEE FIGURE 27–7**.

HIGH SIDE OPERATION The high side begins at the compressor and the compressor outlet (discharge) port. The high side ends at the metering device (TXV or OT). These components are part of the high side:

Compressor and discharge port

Condenser

Receiver–drier (TXV systems)

Smaller, high pressure lines and hoses

The high pressure vapor is pumped to the condenser where it releases heat, causing the refrigerant to change into a liquid. Liquid refrigerant then passes through the metering device to start the cycle all over again. ● **SEE FIGURE 27–8**.

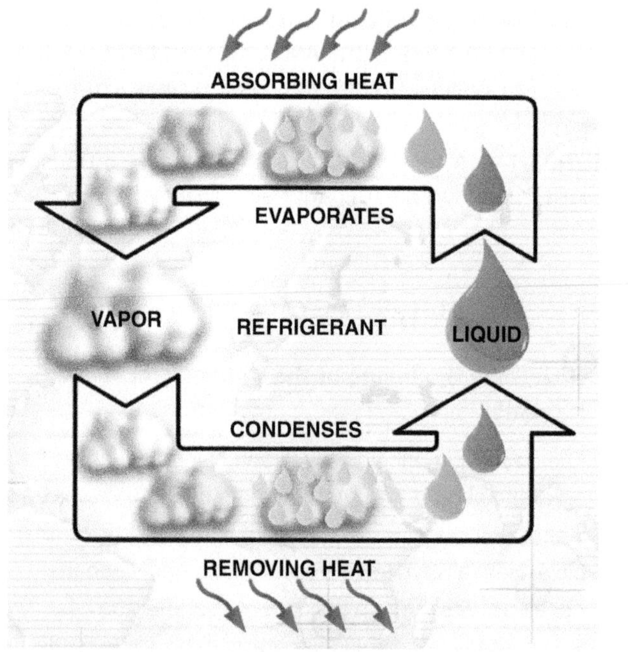

FIGURE 27–8 The heat inside the vehicle boils the refrigerant inside the evaporator. The heat is then contained in the vapor as latent heat. The compressor raises the temperature of the vapor. The refrigerant is then cooled by the outside air, condensing back to a liquid. The liquid enters the evaporator where the cycle repeats. (Courtesy of General Motors)

EVAPORATORS

PURPOSE AND FUNCTION The evaporator is the heat exchanger and absorbs heat from the passenger compartment. The refrigerant enters the evaporator as a liquid spray or mist, leaving an area of a few hundred pounds per square inch (PSI) and passing through a small orifice into an area of about 30 PSI. Like most heat exchangers, a well-designed evaporator has a large amount of surface area in contact with the refrigerant and the air from the passenger compartment. The heat from the air

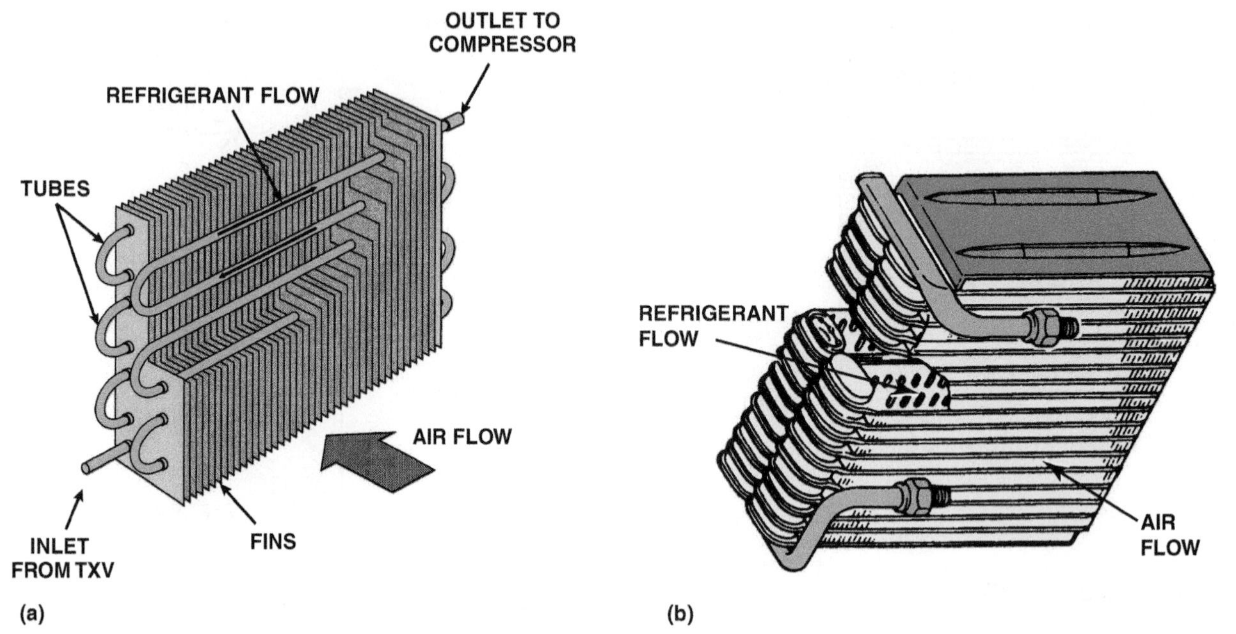

FIGURE 27–9 (a) A tube-and-fin and (b) a plate evaporator. Each type has a large contact area for heat to leave the air and enter the refrigerant.

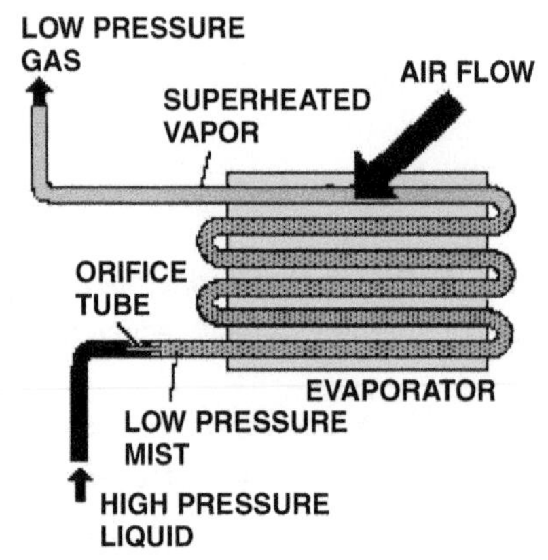

FIGURE 27–10 As liquid refrigerant enters the evaporator, the boiling point will try to drop to as low as 32°F (0°C) because of the drop in pressure. The cold temperature causes the refrigerant to absorb heat from the air circulated through the evaporator.

causes the refrigerant to boil and turn into vapor and the cooler air is returned to the passenger compartment. ● **SEE FIGURE 27–9.**

The cold temperature in the evaporator is produced by boiling the refrigerant. Refrigerants have very low boiling points, well below 0°F (–18°C); when the liquid boils, it absorbs a large amount of heat, called the *latent heat of vaporization*. To produce cooling, liquid refrigerant must enter and boil inside the evaporator. The amount of heat an evaporator absorbs is directly related to the amount of liquid refrigerant that boils inside it. ● **SEE FIGURE 27–10.**

A properly operating evaporator has a temperature just above 32°F (0°C), and refrigerant pressure is directly related to temperature. R-134a has a 27-PSI (186-kPa) pressure at 32°F (0°C). Abnormal temperatures and pressures indicate that something is wrong, such as the evaporator might have too much or too little refrigerant.

- **Starved evaporator—**An evaporator that has a low pressure but a temperature that is too warm is called "starved," which means that not enough refrigerant is entering to produce the desired cooling effect. A starved evaporator is usually the result of a restriction at or before the expansion device or an undercharge of refrigerant.
- **Flooded evaporator—**If more refrigerant enters the evaporator than can boil, the evaporator floods. In this case the pressure is higher than normal. A flooded evaporator is usually the result of having too much refrigerant in the system. ● **SEE FIGURE 27–11.**

THERMAL EXPANSION VALVES

TERMINOLOGY A thermal expansion valve (TXV) is a variable valve that changes the size of the valve opening in response to the cooling load of the evaporator. A TXV is controlled by evaporator temperature and pressure so that it opens to flow as much refrigerant as possible when a lot of cooling is needed and all of the refrigerant must boil in the evaporator. Most TXVs are calibrated so that the outlet temperature is a few

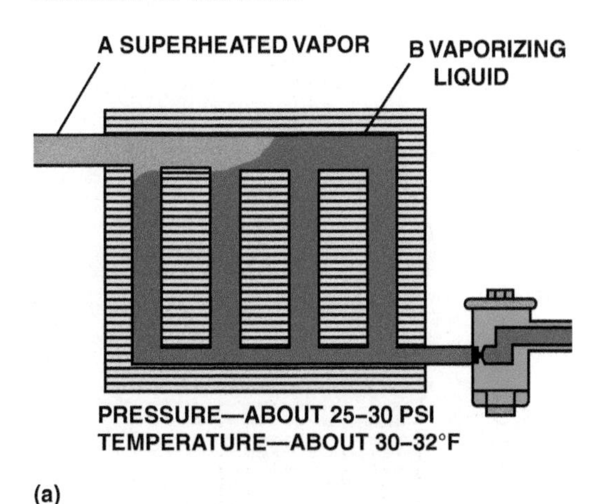

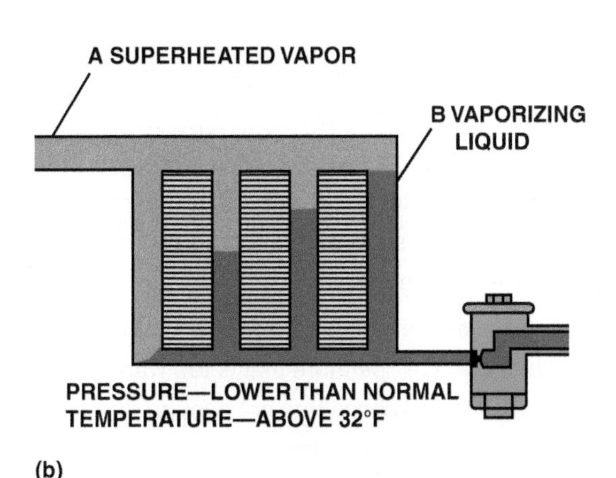

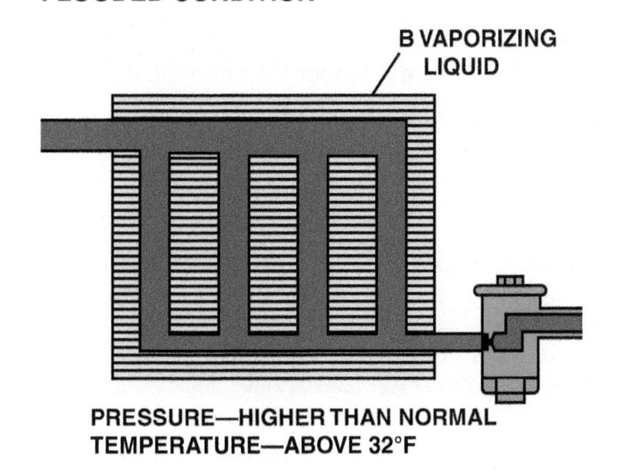

FIGURE 27–11 (a) If proper amount of refrigerant enters the evaporator, it has a slight superheat as it leaves. (b) A starved condition, in which not enough refrigerant enters the evaporator, does not produce as much cooling. (c) If too much refrigerant enters, the evaporator floods because the refrigerant will not at all boil.

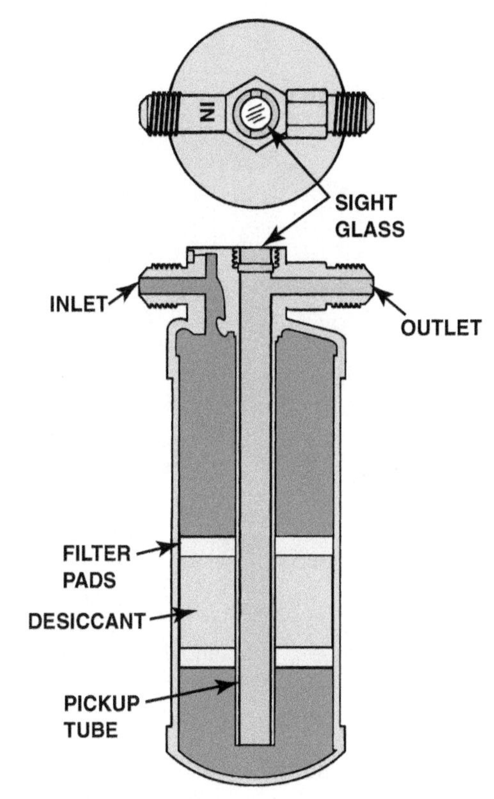

FIGURE 27–12 Expansion-valve systems store excess refrigerant in a receiver–drier, which is located in the high-side liquid section of the system.

degrees above the inlet pressure and temperature. When there is a lower cooling load, the TXV must reduce the flow.

RECEIVER–DRIER A **receiver–drier** is used in the high side of a TXV system. It contains a desiccant to remove moisture and provides a storage chamber for liquid refrigerant. Most receiver–driers also contain a filter to trap debris that might plug the TXV.

NOTE: Older receiver–driers have a sight glass in the outlet line so the refrigerant flow can be checked to see if it is all liquid or contains bubbles. A receiver–drier should be about half full of liquid, so vapor bubbles are an indication of an undercharge. A sight glass is not used in most R-134a systems because the refrigerant has a cloudy appearance in a properly charged system.
● **SEE FIGURE 27–12.**

ORIFICE TUBE SYSTEMS

PURPOSE AND FUNCTION An orifice tube is a restriction in the liquid line that forces the refrigerant to expand as it passes through the small opening (orifice). When the refrigerant expands, the temperature of the refrigerant drops and starts

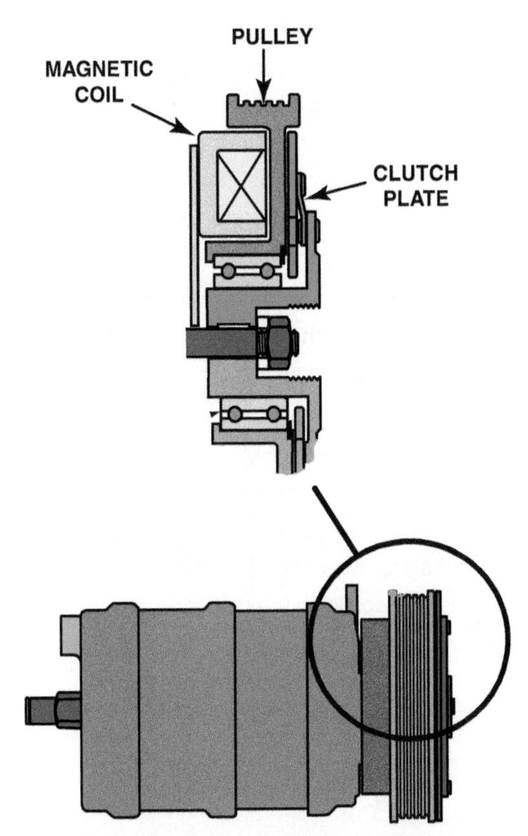

FIGURE 27–13 The compressor clutch allows the compressor to cycle off and on to control evaporator temperature and to shut the system off.

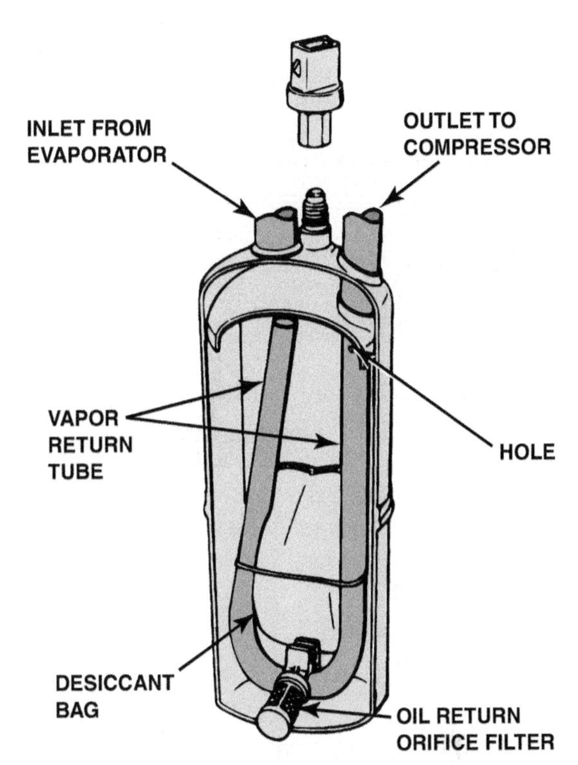

FIGURE 27–14 Expansion-valve systems store excess refrigerant in a receiver–drier, which is located in the high-side liquid section of the system, whereas orifice tube systems store excess refrigerant in an accumulator (shown here) located in the low-side vapor section of the system.

to evaporate in the evaporator. There are two basic designs of orifice tube system, including:

1. A cycling clutch (CC) system that disengages the compressor and shuts the system off when the evaporator temperature or pressure drops below freezing point. Most A/C compressors are driven by a belt from the engine through an electromagnetic clutch. ● **SEE FIGURE 27–13.**
2. A variable displacement compressor that is used to control evaporator pressure and temperature by controlling the amount of refrigerant passing through the orifice tube.

ACCUMULATOR A receiver–drier is used with a thermal expansion valve (TXV) system and is located in the high-pressure (liquid) side of the system between the condenser and the expansion valve. An accumulator is used in the orifice tube system and is located in the low side of the system between the evaporator and the compressor. The accumulator is needed in an orifice tube system because it prevents any liquid refrigerant from entering the compressor, which would destroy the compressor. It traps and holds any liquid refrigerant that leaves the evaporator. Both assemblies contain a desiccant to remove any moisture that may be in the system to help prevent possible acid formation, which can decrease component life. ● **SEE FIGURE 27–14.**

FREQUENTLY ASKED QUESTION

What Is Sub-Cooling?

The term **sub-cooling** refers to a liquid existing at a temperature below its normal condensation temperature. The condenser removes heat and changes a high-pressure vapor into a high-pressure liquid. As the superheated (high-pressure) gas is pushed into the condenser, the temperature is reduced. The refrigerant does not start to change state until the temperature reaches what is called its *saturated pressure-temperature*. At saturation pressure-temperature point, the change of state becomes latent heat (invisible or hidden heat). The temperatures of the liquid and the vapor will stay the same until the temperature of the refrigerant starts to drop. Temperature of the refrigerant will start to drop once 98% to 99% of the refrigerant becomes liquid.

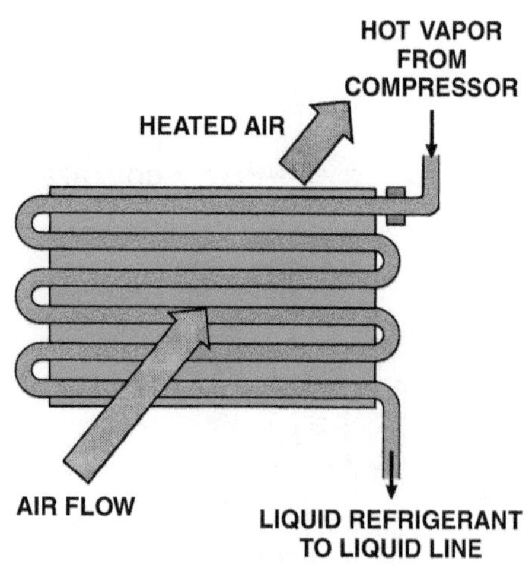

FIGURE 27–15 A condenser is a heat exchanger that transfers heat from the refrigerant to the air flowing through it.

CONDENSERS

PURPOSE AND FUNCTION The condenser, like the evaporator, is a heat exchanger. Low-pressure refrigerant vapor is compressed by the compressor into a high-temperature, high-pressure vapor. This vapor then passes into the condenser where air passing over the condenser cools the refrigerant and causes it to condense into a high-temperature liquid. The refrigerant enters the top of the condenser as a hot vapor and leaves from the bottom as a cooler liquid. ● **SEE FIGURE 27–15.**

REFRIGERANT CHARGE LEVEL

For an A/C system to work properly, there should be a constant flow of liquid refrigerant through the TXV or OT. While operating, the following occurs:

- The evaporator contains a refrigerant mist in the first two-thirds to three-fourths of its volume, with vapor in the remaining portion.
- The condenser contains a condensing vapor in the upper portion, with liquid in the bottom passages.
- The line connecting the condenser to the expansion device is filled with liquid.
- The accumulator is about half full of liquid so that liquid refrigerant does not enter the compressor.

Most recent A/C systems have improved efficiency and reduced the size of some components so they can operate with smaller charge volumes. At one time, the refrigerant capacity of many domestic systems was in the 3-lb. to 4-lb. (1.4 to 1.8 kg) range. However, recently most systems hold 1.0 to 2.5 lb. (0.5 to 1.1 kg) of refrigerant. Some of the new systems have a capacity of less than a pound (0.43 kg). With this reduced volume, an accurate charge amount is more critical.

- **Undercharge—**If the volume of liquid drops so that vapor bubbles pass through the TXV or OT, the system is **undercharged** and its cooling effectiveness is reduced.
- **Overcharge—**If an excessive amount of refrigerant is put into a system then the excess volume partially fills the condenser as a liquid and reduces its effective volume. This is called an **overcharge** and causes abnormally high pressures, especially in the high side, and poor cooling at the evaporator.

EVAPORATOR ICING CONTROLS

PURPOSE AND FUNCTION Most A/C systems operate at maximum capacity when it is necessary to cool the vehicle. Compressor size (displacement) and the sizes of the evaporator and condenser determine cooling power and are designed to cool the vehicle and its passengers on a hot day. Vehicle size and glass area, compressor displacement and operating speed, number of passengers, ambient temperature, and vehicle speed are all design parameters that are considered during the initial design of the A/C and heating systems. Some systems are designed to cool a vehicle with the engine at idle speed and the compressor running at its slowest speed. Newer systems are made as small as practical to reduce HVAC system size and vehicle weight for improved fuel economy.

As the vehicle cools down, the cooling load on the evaporator drops, and its temperature also drops. As mentioned earlier, the minimum temperature for an evaporator is 32°F, the point at which water freezes and ice and frost form. There are several ways of preventing evaporator icing, including the following:

- Cycling the compressor clutch
- Controlling evaporator pressure so it does not drop below 30 PSI
- Reducing the displacement of the compressor by using a variable displacement compressor.

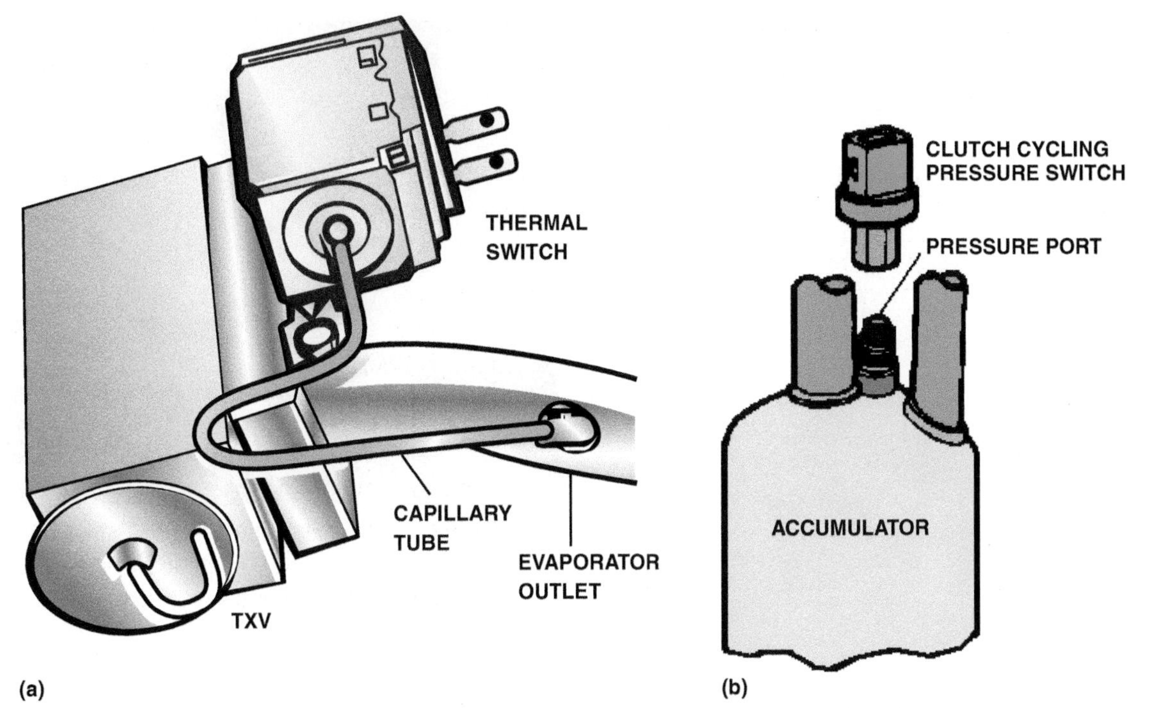

FIGURE 27–16 (a) Most TXV systems use a thermal switch to cycle the compressor out when the evaporator gets too cold. (b) Most orifice tube systems use a pressure switch to cycle the compressor out when the low-side pressure drops too low.

> **? FREQUENTLY ASKED QUESTION**
>
> **What Is "Superheat"?**
>
> **Superheat** is the amount of heat added to the refrigerant after it has changed from liquid to vapor. Superheat is usually measured as the actual temperature difference between the boiling point of the refrigerant at the inlet and at the outlet of the evaporator. Typical values for superheat in an evaporator are between 4°F and 16°F (3°C and 10°C). Superheat is important because it ensures that all (or almost all) of the refrigerant vaporizes before leaving the evaporator.

Early A/C systems used a temperature-controlled switch mounted in the airstream from the evaporator. This thermal switch, also called an *icing switch* or *defrost switch*, was set to open and stop the current flow to the clutch when the temperature drops below 32°F (0°C) and reclose when there is a temperature increase of about 10 Fahrenheit degrees (6 Celsius degrees). This causes a pressure increase of about 10 to 20 PSI that, in turn, produces sufficient temperature rise to melt any frost or ice on the fins.

THERMISTOR Some newer systems use a **thermistor**, which is a solid-state device that changes its electrical resistance in inverse relationship to its temperature. When the temperature increases, the resistance of the thermistor decreases. It is used as an input to an electronic control module (ECM) to provide the actual evaporator temperature control.

PRESSURE SWITCHES Many orifice tube systems use a two-wire **pressure switch** mounted in the accumulator or the **suction line** to the compressor. The evaporator temperature and pressure are closely linked and they drop together. As the *cutout pressure switch* senses the pressure dropping below a certain point (about 30 PSI), it opens to stop the compressor. The pressure switch recloses when the pressure increases, with the *cut-in pressure* being about 42 to 49 PSI, depending on the vehicle. With either of these systems, if ice and frost start to form because the evaporator gets too cold, the ice and frost melt during the off part of the cycle. ● **SEE FIGURE 27–16**.

Most recent vehicles use a **pressure sensor** in place of a pressure switch. The resistance of the sensor changes in direct relation to the pressure. It is an input to a PCM/ECM used for compressor clutch, cooling fan, and idle-speed control as well as low-pressure and high-pressure protection. ● **SEE FIGURE 27–17**.

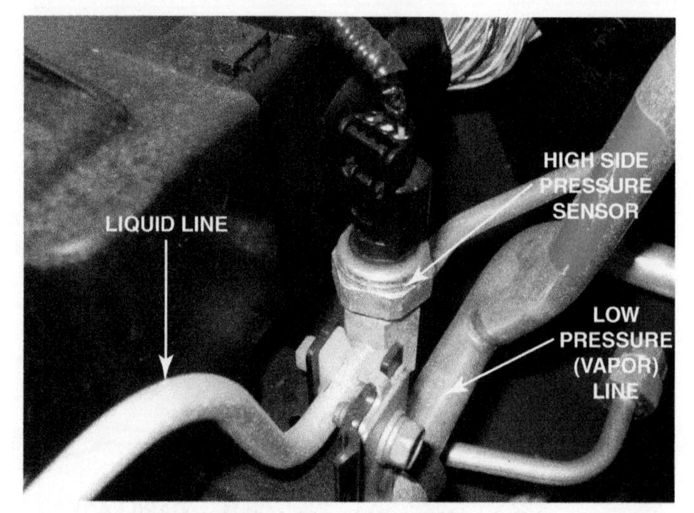

FIGURE 27–17 A typical three-wire pressure sensor used on the high side (vapor) line. The three wires are the voltage supply (usually 5 volts), ground, and signal wire.

SUMMARY

1. Automotive A/C systems operate on the principle of moving heat from inside to outside of the vehicle.
2. All automotive air-conditioning systems are closed and sealed. A refrigerant is circulated through the system by a compressor that is usually powered by the engine through an accessory drive belt.
3. The liquid refrigerant evaporates in a small radiator-type unit called the evaporator.
4. After the refrigerant has evaporated into a low-pressure gas in the evaporator, the refrigerant flows into the engine-driven compressor.
5. From the compressor, high-pressure gas flows into the condenser located in front of the cooling system radiator.
6. Automotive A/C systems are either orifice tube (OT) systems or thermal expansion valve (TXV) systems.
7. A receiver–drier is used with a TXV system and is located in the high pressure (liquid) side of the system.
8. An accumulator is used in orifice tube system and is located in the low side of the system between the evaporator and the compressor.

REVIEW QUESTIONS

1. How does the air conditioning cool the inside of the vehicle?
2. What are the major components of the refrigeration cycle?
3. How does the refrigeration system work?
4. What is the difference between a pressure switch and a pressure sensor?

CHAPTER QUIZ

1. Heat travels from ________.
   a. Hot to cooler
   b. Cold to warmer
   c. Either from hot to cool or the other way depending on the weather
   d. From outside the vehicle to inside the vehicle
2. The refrigerant is circulated through the system by a ________.
   a. Condenser
   b. Evaporator
   c. Compressor
   d. Thermal expansion valve or orifice tube

3. When the refrigerant evaporates, it absorbs heat when it changes from liquid to gas. In which unit does this occur?
   a. Condenser
   b. Evaporator
   c. Compressor
   d. Thermal expansion valve or orifice tube
4. Which unit contains a desiccant to remove moisture from the system?
   a. Receiver–drier
   b. Accumulator
   c. Evaporator
   d. Both a and b
5. The low side of the refrigeration cycle means ________.
   a. It has low pressure and temperature
   b. It begins at the compressor and ends at the expansion device
   c. That the refrigerant condenses
   d. It is called the liquid line
6. The high side of the refrigeration cycle means it ________.
   a. Has low pressure and temperature
   b. Begins at the expansion device and ends at the compressor
   c. Has higher pressures and temperatures
   d. The refrigerant boils or evaporates
7. A ________ is a variable valve that changes the size of the valve opening in response to the cooling load of the evaporator.
   a. Orifice tube
   b. Compressor
   c. Thermal expansion valve (TXV)
   d. Receiver–drier
8. Pressures are controlled in an orifice tube (OT) system by ________.
   a. Using a variable valve
   b. Cycling an electromagnetic compressor clutch on and off as needed.
   c. Using a variable displacement compressor
   d. Either b or c
9. Which condition can cause the air-conditioning system to produce less-than normal cooling?
   a. Overcharged with refrigerant
   b. Undercharged with refrigerant
   c. Either under- or overcharged
   d. Superheat condition
10. The condenser ________.
   a. Is a heat exchanger
   b. Is a device where air passing over it cools the refrigerant and causes it to change into a high-temperature liquid.
   c. Transfers heat from the refrigerant to the air flowing
   d. All of the above

chapter 28

AIR-CONDITIONING COMPRESSORS AND SERVICE

LEARNING OBJECTIVES

After studying this chapter, the reader should be able to:

1. State the different types of A/C compressors.
2. Discuss the parts and operation of compressor clutches.
3. Discuss compressor valves and switches.
4. Explain A/C compressor diagnosis and service.

This chapter will help you prepare for the ASE Heating and Air Conditioning (A7) certification test content area "B" (Refrigeration System Diagnosis, and Repair).

KEY TERMS

GM STC OBJECTIVE

GM Service Technical College topic covered in this chapter is as follows:

1. The purpose, types, controlling components, and operation of a compressor (01510.08W).

FIXED DISPLACEMENT COMPRESSOR

VARIABLE DISPLACEMENT COMPRESSOR

SCROLL COMPRESSOR

FIGURE 28–1 General Motors uses different types of compressors, depending on the make and model of vehicle. (Courtesy of General Motors)

COMPRESSORS

PURPOSE AND FUNCTION The air-conditioning compressor can be thought of as a pump that circulates refrigerant. It has to work against the restriction of the thermal expansion valve (TXV) or orifice tube (OT). The pressure must be increased to the point where refrigerant temperature is above ambient air temperature and there is enough heat transfer at the condenser to get rid of all the heat absorbed in the evaporator. Most A/C compressors are driven by a belt and pulley from the engine. ● **SEE FIGURE 28–1**.

TYPES OF A/C COMPRESSORS There are many types of A/C compressors used on vehicles, including the following:

- **Piston compressors**—Most older automotive compressors used a crankshaft, similar to a small gasoline engine, and a reciprocating-piston type. Newer piston compressors use a swash or wobble plate.
- **Vane compressors**—Vane compressor has vanes that contact the rotor housing at each end, and they slide to make a seal at each end as the rotor turns.
- **Scroll compressors**—Scroll compressors require rather complex machining to achieve constant sealing between the fixed and movable scrolls.

PISTON COMPRESSORS

PISTON COMPRESSOR OPERATION A piston compressor moves the pistons up and down in a cylinder to produce pumping action and controls the refrigerant flow with two sets of **reed valves**.

- The downward or **suction stroke** of the piston causes the refrigerant to flow from the compressor suction cavity to push the suction reed open and fill the cylinder. The suction cavity is connected to the evaporator so it contains refrigerant vapor at evaporator pressure.
- An upward stroke, or **discharge stroke**, of the piston generates pressure to force the refrigerant through the discharge reed into the discharge chamber and on to the condenser. The discharge pressure becomes high-side pressure. ● **SEE FIGURE 28–2**.

Piston compressors have some disadvantages; chief among them is the high inertial loads that result from moving a piston at a rather high speed, bringing it to a stop, moving it at a high speed in the opposite direction, bringing it to a stop, and so on. This movement produces vibrations and severe stress on moving parts.

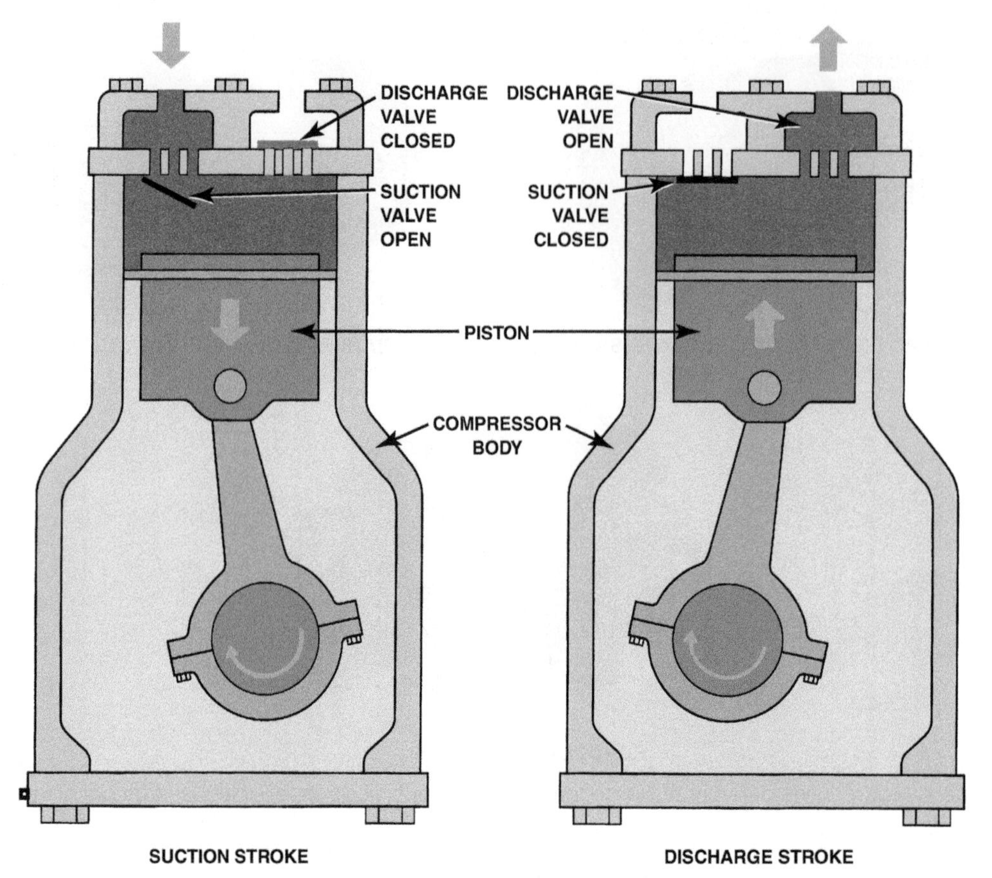

FIGURE 28–2 In a piston compressor, when moving downward, the piston creates a drop in pressure inside the cylinder. The resulting difference in pressure allows the suction valve to open. Refrigerant then flows into the cylinder. When the piston moves upward on discharge stroke, the pressure closes the intake valve and forces the refrigerant out the discharge valve.

COAXIAL SWASH-PLATE COMPRESSORS Coaxial swash-plate compressors drive the pistons through a swash plate, which is attached to the driveshaft. The swash plate is mounted at an angle so it will wobble and cause the reciprocating action of the pistons. ● **SEE FIGURE 28–3.**

- The swash plate revolves with the shaft with each piston having a pair of bearings that can pivot as the swash plate slides through them.
- The pistons are double ended so that each end can pump, and the pistons are arranged parallel to and around the driveshaft. This is called a *coaxial* arrangement.
- One driveshaft revolution causes each piston end to move through a complete pumping cycle. The most common arrangement is three double pistons making a 6-cylinder compressor and a 10-cylinder using five pistons. ● **SEE FIGURE 28–4.**

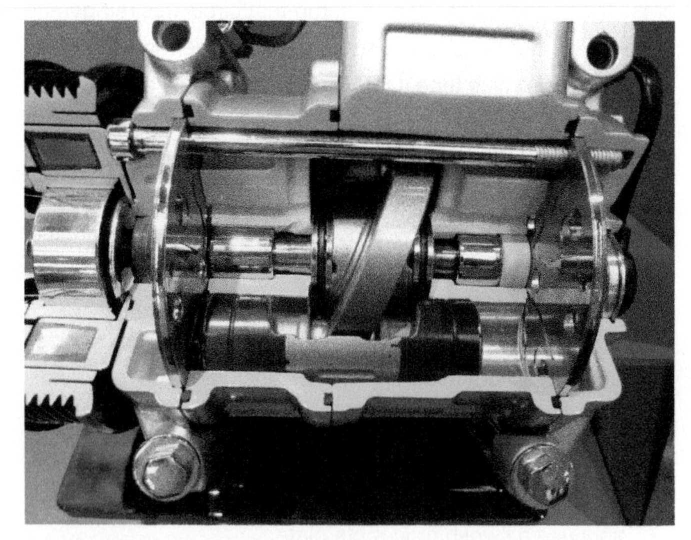

FIGURE 28–3 Cutaway picture of a coaxial swash-plate compressor. (Courtesy of Jeffrey Rehkopf)

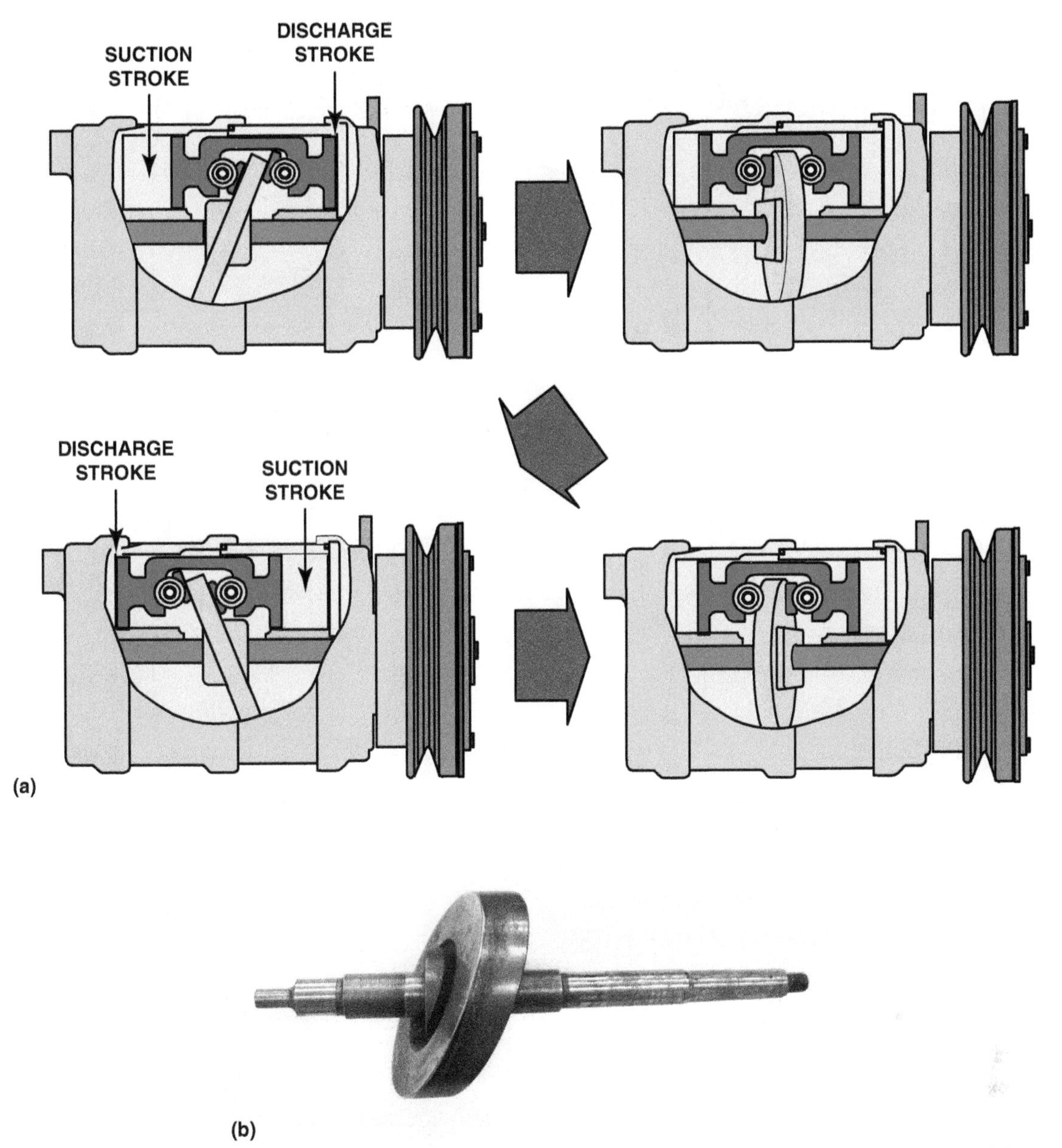

FIGURE 28–4 (a) One of the double pistons of a swash-plate compressor as it move through a pumping cycle. (b) A compressor shaft with the swash plate.

A swash-plate compressor must have passages to transfer refrigerant between the suction and discharge chambers at each end of the compressor. The suction crossover passage is usually designed so that it can provide lubrication to internal moving parts.

General Motors uses the following compressor designations:

- Delco Air: DA
- Harrison: H
- Harrison Radiator: HA
- Harrison Redesigned: HR
- High Efficiency: HE
- Truck: HT
- Upgraded HT: HU

Nippondenso also manufactures 6-cylinder and 10-cylinder coaxial compressors. This design is used as OEM equipment by the Chrysler Corporation, the Ford Motor Company, and other vehicle manufacturers around the world. Denso coaxial compressors, like those of other manufacturers, use a four-part aluminum body with two cylinder assemblies, rear head, and front head sealed by O-rings. Either a single-key drive or a splined drive is used between the clutch drive plate and the compressor shaft.

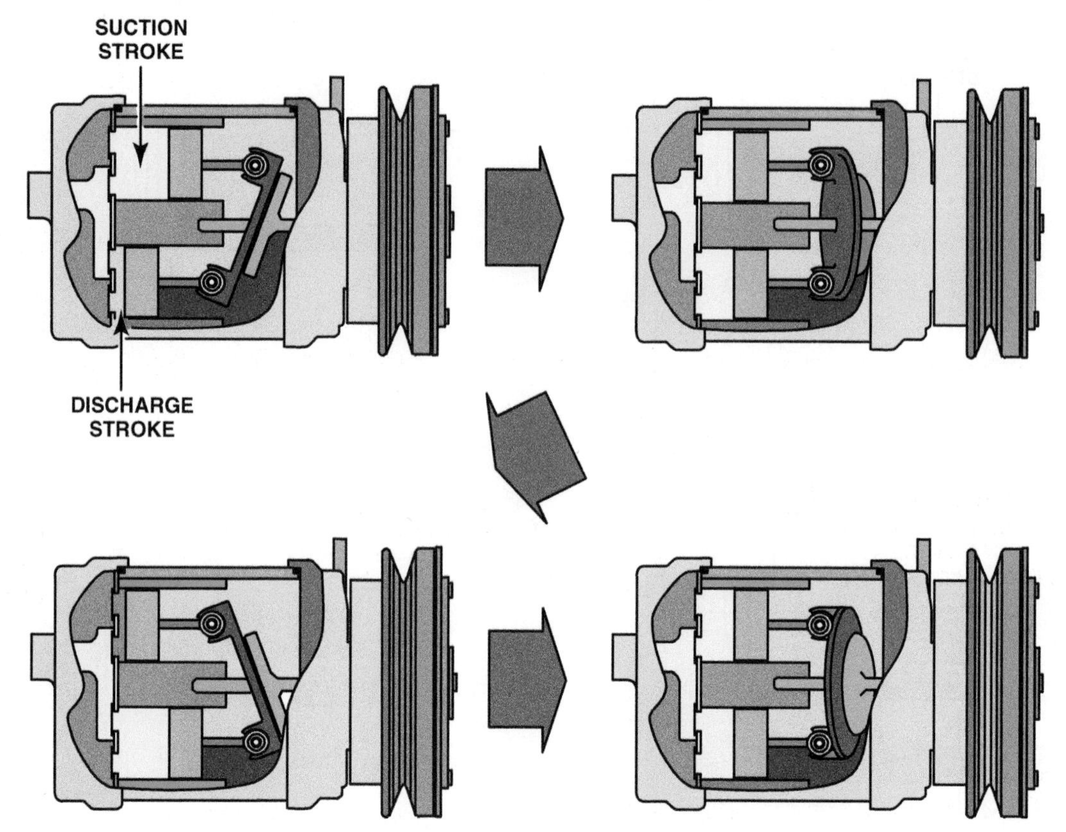

FIGURE 28–5 Two of the pistons of a wobble-plate compressor as they move through a pumping cycle.

Several other Japanese manufacturers have produced 6-cylinder and 10-cylinder swash-plate compressors. These manufacturers include:

- Calsonic
- Hitachi
- Mitsubishi
- Nihon Radiator
- Seltec
- Zexel (formerly Diesel Kiki)

COAXIAL WOBBLE-PLATE COMPRESSORS Wobble-plate compressors drive the pistons through an angle plate that looks somewhat like a swash plate, but the wobble plate does not rotate and drives single pistons through piston rods. Wobble-plate compressors commonly use five or seven cylinders. ● **SEE FIGURE 28–5.**

VARIABLE DISPLACEMENT WOBBLE-PLATE COMPRESSORS A variable displacement compressor provides smooth operation with no clutch cycling, a constant 32°F (0°C) evaporator, and the most efficiency. Changing the compressor displacement is the major control for preventing evaporator icing. ● **SEE FIGURE 28–6.**

FIGURE 28–6 A variable displacement compressor. With the wobble plate almost flat, the pistons are at minimum stroke. (Courtesy of Jeffrey Rehkopf)

- This design includes a large compressor that can pump enough refrigerant to meet high cooling loads, and it reduces the displacement and pumping capacity of the compressor to match the needs of the evaporator as the evaporator cools.

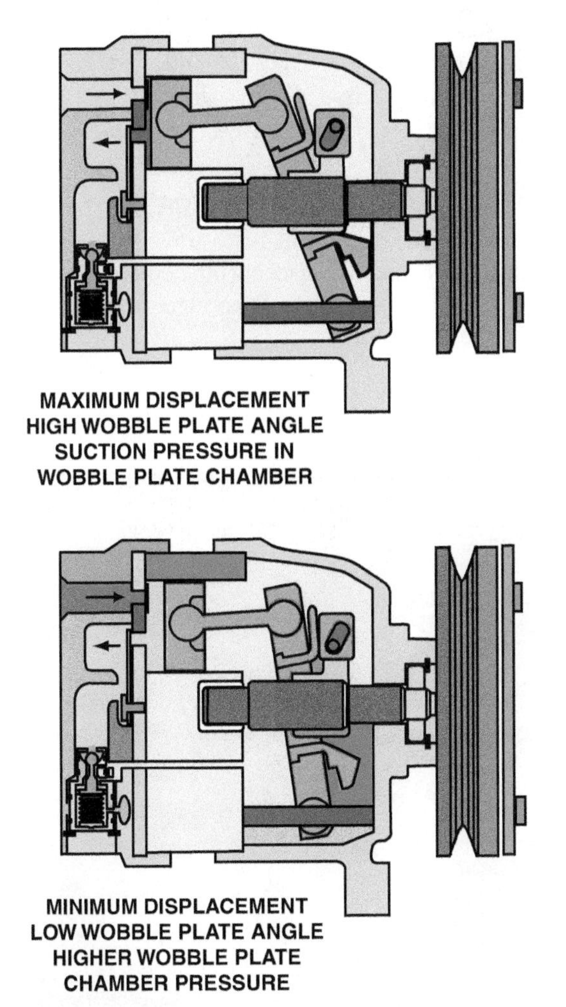

FIGURE 28–7 A variable displacement compressor can change the angle of the wobble plate and piston stroke. This angle is changed by a control valve that senses evaporator pressure, which in turn changes wobble chamber pressure.

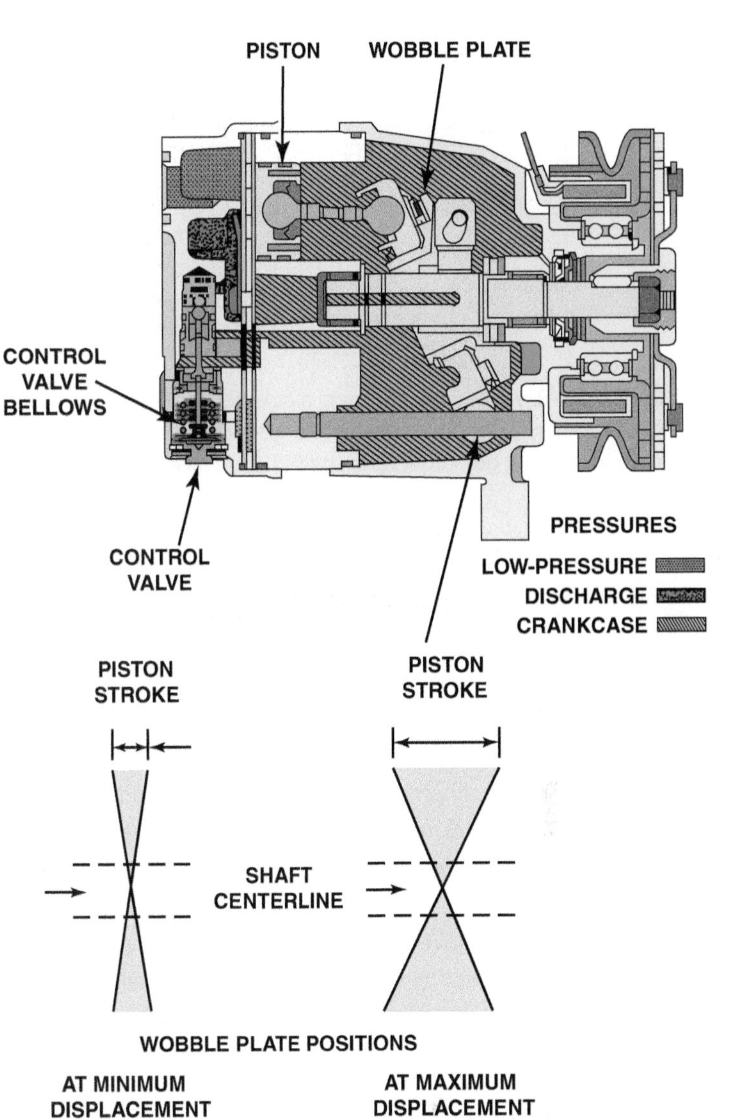

FIGURE 28–8 When the evaporator cools and low-side pressure drops, the piston stroke of a variable displacement compressor is reduced so that compressor output matches the cooling load.

- When there is a low cooling load at the evaporator, the wobble plate is moved to a less angled position. Some designs can reduce wobble-plate angle to 1% or 2% of the maximum stroke angle. This feature makes the compressor more efficient by reducing the drive load when it is not needed and also eliminates the need to cycle the compressor off and on. ● **SEE FIGURE 28–7.**

Wobble-plate angle is determined by the relative pressure at each end of the piston, and the angle is controlled by changing the pressure in the crankcase.

- When cooling load calls for high output and maximum displacement, crankcase pressure is kept low, and the wobble plate is at its maximum angle. The control valve bleeds crankcase pressure into the compressor suction cavity to lower the pressure.
- When cooling demand lessens, the control valve closes the bleed to the suction cavity and opens a passage between the discharge cavity and the crankcase, raising the pressure. Increasing crankcase pressure raises the pressure on the bottom side of the pistons and causes the wobble plate to move to low angle, reducing displacement. A typical variable compressor has a displacement of 0.6 cu. inch to 9.2 cu. inch (10 cc to 151 cc). When maximum cooling is needed, the displacement is 9.2 cu. inch, and this can drop to as low as 0.6 cu. inch as needed to keep the evaporator pressure above the freezing point or at the pressure to produce the desired outlet air temperature. Compressor displacement is adjusted by changing the pressure inside the crank/piston chamber using a pressure or electronically controlled valve. ● **SEE FIGURE 28–8.**

FIGURE 28–9 As the rotor turns in a counterclockwise direction, the vanes move in and out to follow the contour of the housing. This action forms chambers that get larger at the suction ports and smaller at the discharge ports. Evaporator pressure fills the chambers as they get larger, and the reducing size forces the refrigerant into the high side.

VANE COMPRESSORS

CONSTRUCTION The vanes of these compressors are mounted in a rotor that runs inside a round, eccentric, or a somewhat elliptical, chamber. The vanes slide in and out of the rotor as their outer end follows the shape of the chamber. Compressors with a round, eccentric chamber have one pumping action per vane per revolution. Compressors with an elliptical housing have two pumping actions per vane per revolution. This type of compressor is sometimes called *balanced* because there is a pressure chamber on each side of the rotor.

OPERATION As the rotor turns, in one or two areas, the chamber behind the vane increases in size. This area has a port connected to the suction cavity. The following vane traps the refrigerant and forms a chamber as it passes by the suction port. The trapped refrigerant is carried around to the discharge port. In this location, the chamber size gets smaller which increases the gas pressure and forces it into the high side. Vane compressors have the advantage of being very compact and vibration free. ● **SEE FIGURE 28–9.**

FIGURE 28–10 Cutaway of a scroll compressor showing the fixed scroll (green) and the movable scroll (gray). (Courtesy of Jeffrey Rehkopf)

SCROLL COMPRESSORS

CONSTRUCTION Scroll compressors use two major components:

1. **Fixed scroll**—The fixed scroll is attached to the compressor housing.
2. **Movable scroll**—The movable scroll is mounted over an eccentric bushing and counterweight on the crankshaft. It does not rotate, but it moves in an orbit relative to the stationary scroll and as a result is also called an *orbiting piston compressor*.

Both scrolls have a spiral shape that forms one side of the pumping chamber. ● **SEE FIGURE 28–10.**

OPERATION As the scroll orbits, it forms a pumping chamber that is open at the outer end. This chamber is moved to the center by the scroll's action as the pressure is increased. Two or three chambers are present at the same time. The outer ends of the scrolls are open to the suction port, and the inner ends connect to the discharge port. ● **SEE FIGURE 28–11.**

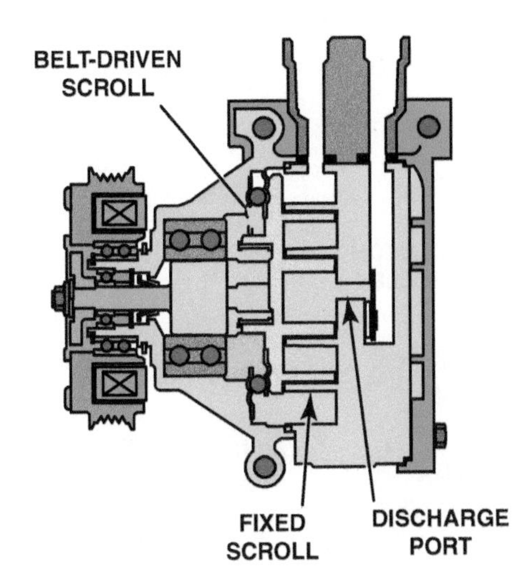

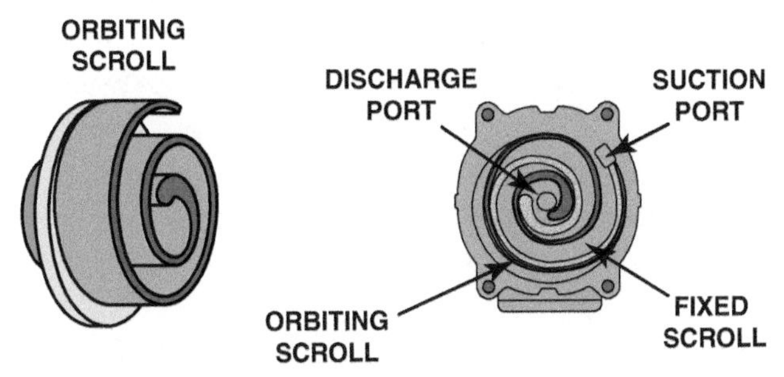

FIGURE 28–11 As the orbital scroll moves, it forms pumping chambers/gas pockets that start at the suction port and forces the refrigerant to the discharge port at the center.

ADVANTAGES A scroll compressor has the advantage of having very smooth operation and low-engagement torque that allows the use of a small clutch. A scroll compressor can also be driven at higher revolutions per minute (RPM) than other designs, so that a smaller drive pulley is used. This compressor design is also much more efficient than the other compressor styles when it is operated at the design speed, which is an advantage for vehicles that tend to run most of the time at cruising speed.

DISADVANTAGES The one disadvantage of a scroll compressor is that it is more expensive to manufacture and therefore costs more than a piston-type compressor.

COMPRESSOR CLUTCHES

PURPOSE AND FUNCTION Electromagnetic clutches allow the compressor to be turned on and off. The clutch uses a coil of wire where a magnetic field is generated when electrical current flows through it. The magnetic field pulls the drive plate against the rotating pulley to drive the compressor. ● **SEE FIGURE 28–12.**

FIGURE 28–12 The clutch assembly mounts on the front of the compressor and drives the compressor input shaft. (Courtesy of Jeffrey Rehkopf)

PARTS AND OPERATION Magnetic clutches include:

1. The clutch coil and pulley are both mounted on an extension from the front of the compressor housing. The pulley and its bearing are mounted on an extension of the front head. This placement allows the side load of the drive belt to be absorbed by the pulley bearing and compressor housing. It also allows easier servicing of individual clutch parts.
2. The **drive plate** is attached to the compressor shaft. The drive plate is also called an *a clutch pulley,* and the pulley is also called a *rotor*. ● **SEE FIGURE 28–13.**

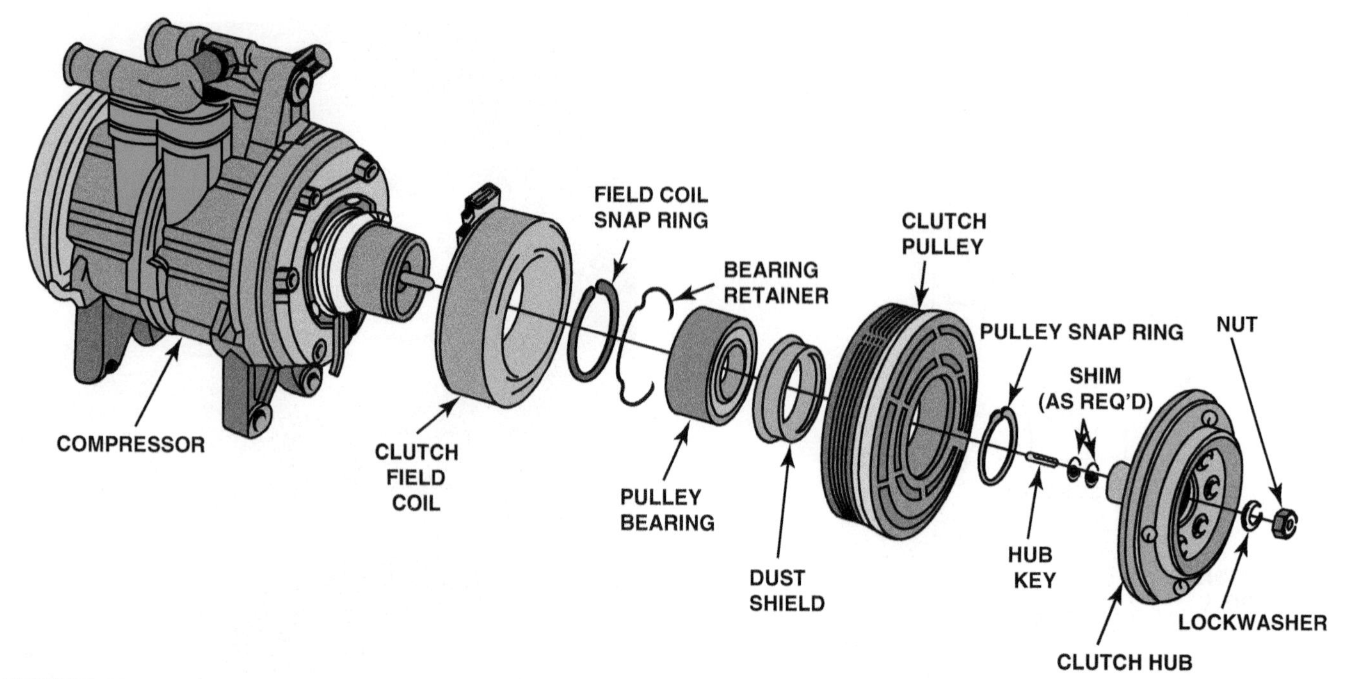

FIGURE 28–13 The electromagnetic clutch assembly includes the clutch field coil, where the magnetic field is created; the clutch pulley, which rides on the pulley bearing; and the clutch hub, which is attached to the input shaft of the compressor. The small shims are added or deleted as needed to adjust the air gap between the clutch hub and the clutch pulley.

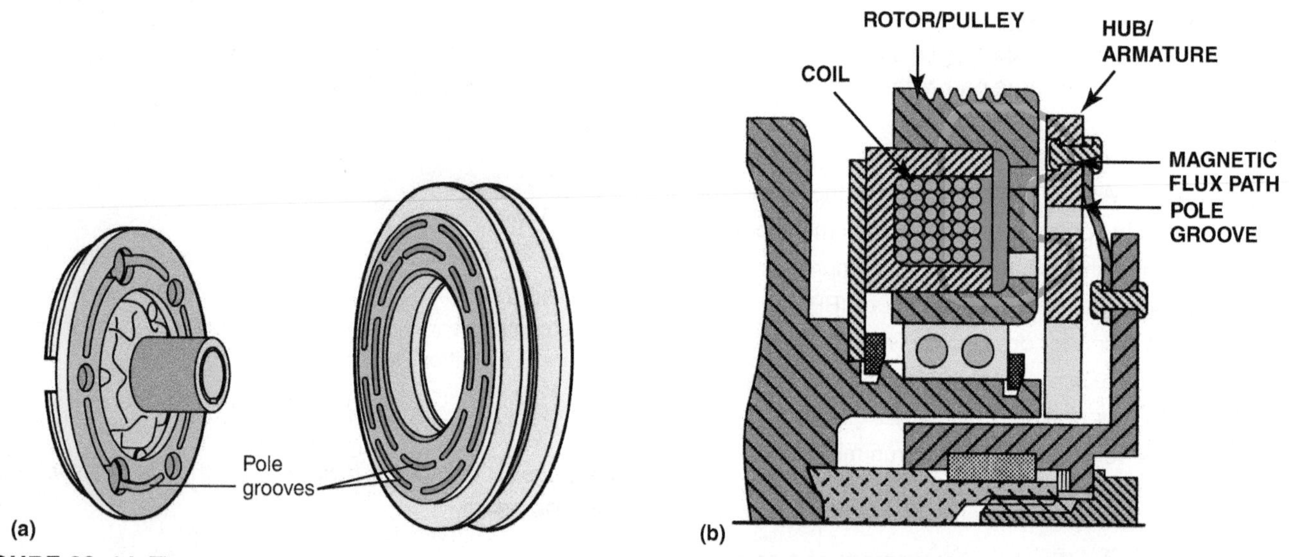

FIGURE 28–14 The magnetic flux path is from the coil and through the metal of the rotor and clutch hub. When it meets a pole groove, it travels from the hub to the rotor or vice versa, which increases the clutch holding power.

Some design factors used to increase holding power include the following:

- The number of *flux poles*, which are the slots in the face of the clutch armature (the greater the number of slots, the stronger the magnetic hold). ● **SEE FIGURE 28–14**.
- The diameter of the rotor and armature (the larger the diameter, the greater the holding power).
- The use of copper or aluminum in the clutch coil winding (copper produces about 20% greater torque capacity).
- The current draw of the coil (the lower the resistance of the coil, the greater the current draw and the stronger the magnetic field that is created).

DAMPER DRIVES Some recent vehicles use a clutchless **damper drive**, electronic-controlled, variable displacement

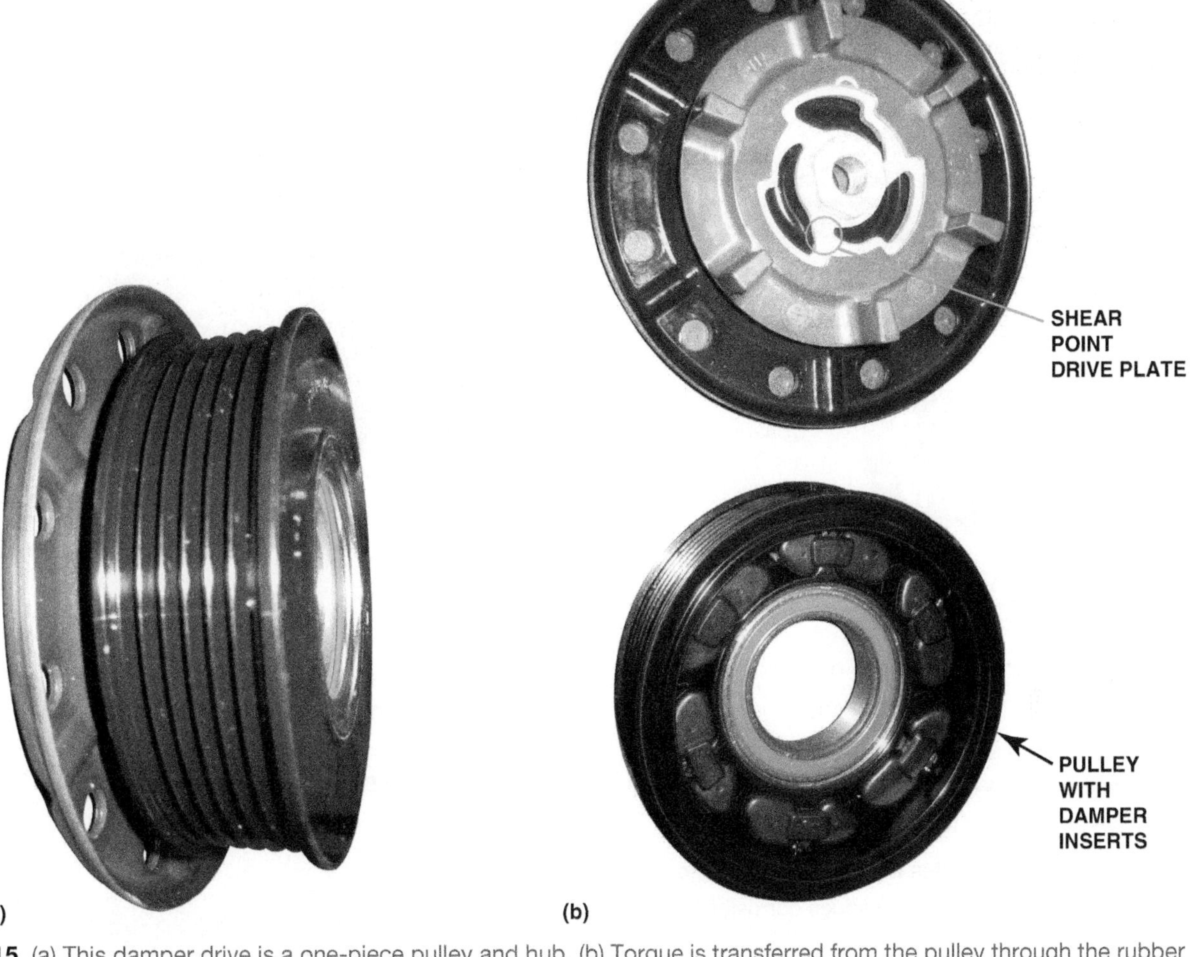

FIGURE 28–15 (a) This damper drive is a one-piece pulley and hub. (b) Torque is transferred from the pulley through the rubber damper inserts (bottom), and the drive plate uses torque-limiting fingers that will shear if the compressor should seize (top).

compressor. The pulley always drives the compressor through a rubber portion that dampens rotating engine pulsations.

One variable displacement compressor that uses a damper drive is electronically controlled to go to minimum displacement of 2% output when A/C is not used. This displacement requires very little power and is enough to circulate oil through the moving parts. The pulley drive plate includes a metal or rubber shear portion that can break to protect the drive belt in case the compressor should fail and lock up. ● **SEE FIGURE 28–15.**

Damper drives cannot be cycled for evaporator temperature control. The compressor displacement control valve responds to electrical signals from the control module, and this controls evaporator temperature to deliver the desired outlet temperature and prevent evaporator freeze-up. ● **SEE FIGURE 28–16.**

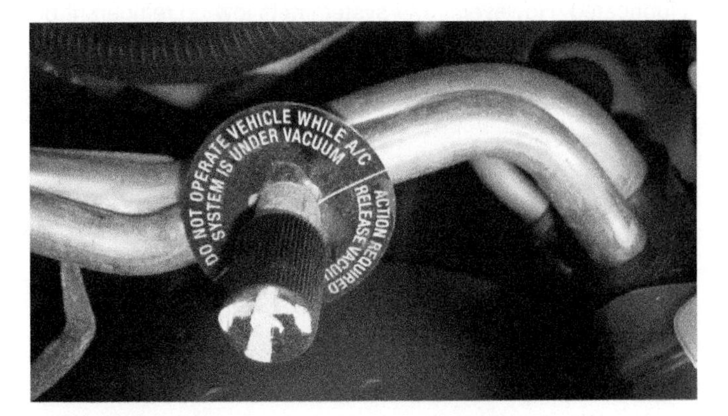

FIGURE 28–16 This tag on the service port indicates a damper-drive compressor that can be damaged if the engine is run without refrigerant in the system.

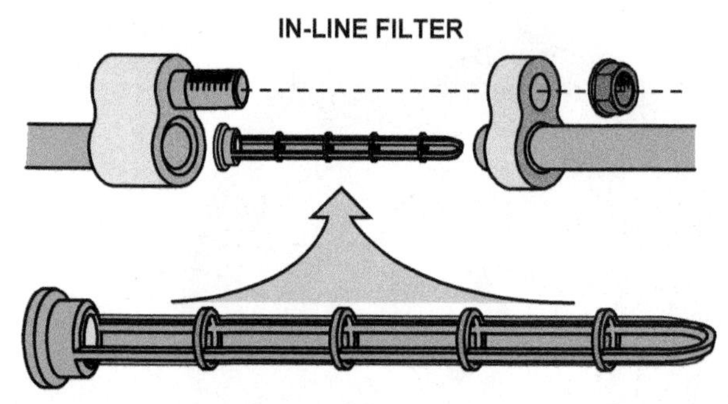

FIGURE 28–17 This filter, about the size of an orifice tube, is installed in the suction line by the vehicle manufacturer.

COMPRESSOR LUBRICATION

NEED FOR LUBRICATION Refrigerant oil serves several purposes, including the following:

- It lubricates the moving parts of the compressor to reduce friction and prevents wear.
- Refrigerant oil also helps *seal* the compressor shaft seal, the insides of the hoses, and various connections between the parts to reduce refrigerant leakage.
- In addition, it lubricates the TXV and coats the metal parts inside the system to reduce corrosion.

HOW OIL IS CIRCULATED The refrigerant oil used in R-134a system does not mix with the refrigerant but instead moves through the system simply with the movement of the refrigerant. If the system has the proper refrigerant charge and the proper amount of oil in the system, then the compressor is lubricated. However, if the system gets low on refrigerant or does not have enough oil in the system, then compressor wear and eventual failure will likely occur.

FILTERS AND MUFFLERS

Inline filters are available from aftermarket sources for installation in the liquid line between the condenser and the OT or TXV and are designed to filter the refrigerant to stop debris from plugging the expansion device.

General Motors is installing inline filters, called a *refrigerant or expansion valve filter*, in the liquid line on some TXV systems. This filter is about the same size and shape as an orifice tube, installed in a connector in the liquid line, and is secured by a joint or reduced/dimpled line section. ● **SEE FIGURE 28–17.**

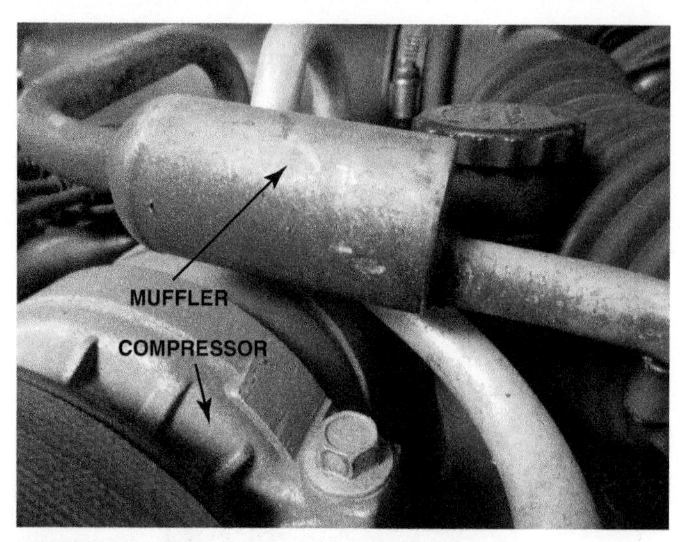

FIGURE 28–18 A muffler is a simple expansion or pulsation chamber with baffles inside the unit to help smooth compressor pressure pulses.

Mufflers are installed in the discharge or suction line of some systems. These mufflers are usually a simple baffled cylinder and are used to dampen the pumping noise of the compressor. ● **SEE FIGURE 28–18.**

COMPRESSOR VALVES AND SWITCHES

PRESSURE RELIEF VALVES Excessive high-side pressure can produce compressor damage and a potential safety hazard if the system should rupture. Compressors are controlled and protected with several different types of pressure relief valves. Many early systems contain a *high-pressure relief* or *release valve,* or release valve which was mounted on the compressor or at some location in the high side.

- R-12 relief valves are set to release pressure at 440 to 550 PSI (3,000 to 3,800 kPa).
- R-134a valves are a little higher at 500 to 600 PSI (3,450 to 4,130 kPa).
- A relief valve is spring-loaded so excessive pressure will open the valve, and as soon as the excess pressure is released, the valve will reclose. ● **SEE FIGURE 28–19.**

NOTE: Newer systems are designed to release the clutch and shut the system off if pressures get too high to avoid venting refrigerant into the atmosphere.

FIGURE 28–19 A high-pressure relief valve contains a strong spring that keeps the valve closed unless high-side pressure (from the left) forces it open. The valve then closes when the pressure drops.

SWITCHES ON THE COMPRESSOR Many compressors contain one or more of the following:

- A low-pressure switch
- A high-pressure switch
- A low- and/or high-pressure sensor

Switches can be connected to ports, leading to either the suction or discharge cavities in the compressor. These switches are usually used in circuits either to protect the compressor or system from damage or as sensors for the engine control module. ● SEE FIGURE 28–20.

COMPRESSOR CONTROL SWITCHES Various electrical switches are used in A/C systems to prevent evaporator icing, protect the compressor, and control fan motors. Control switches can be located anywhere in the system, including at the following positions:

- Compressor discharge
- Compressor suction cavities
- Receiver–drier
- Accumulator. ● SEE FIGURE 28–21.

COMPRESSOR SPEED SENSOR Some vehicles use an A/C compressor speed (RPM) sensor so the ECM will know if the compressor is running, and by comparing the compressor and engine speed signals, the ECM can determine if the compressor clutch is slipping excessively. This system is often called a *belt lock* or *belt protection system.* It prevents a locked-up compressor from destroying the engine drive belt, which, in turn, can cause engine overheating or loss of power steering. If the ECM detects an excessive speed difference for more than a few seconds, it will turn the compressor off. Check service information for the vehicle being serviced to determine if the compressor has a speed (RPM) sensor and if so, where it is located and how to check it for proper operation.

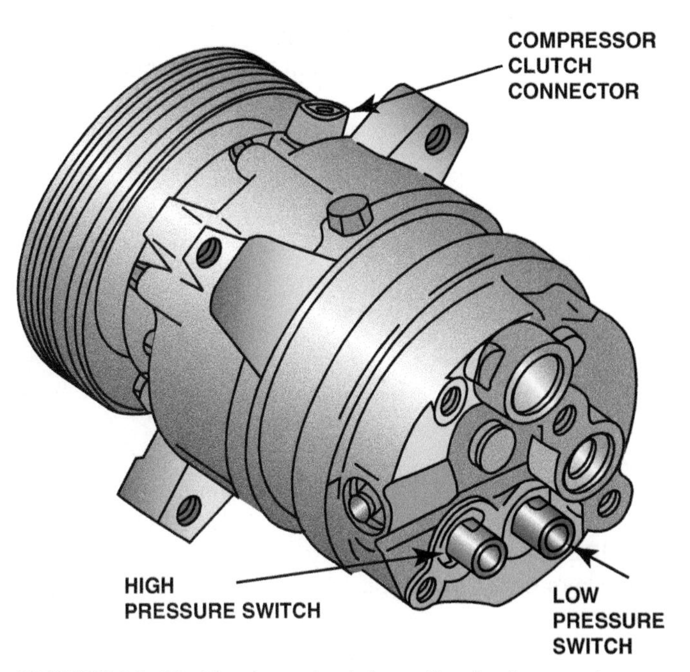

FIGURE 28–20 Check service information for the exact purpose and function of each of the switches located on the compressor because they can vary according to make, model, and year of manufacture of vehicle and can also vary as to what compressor is used.

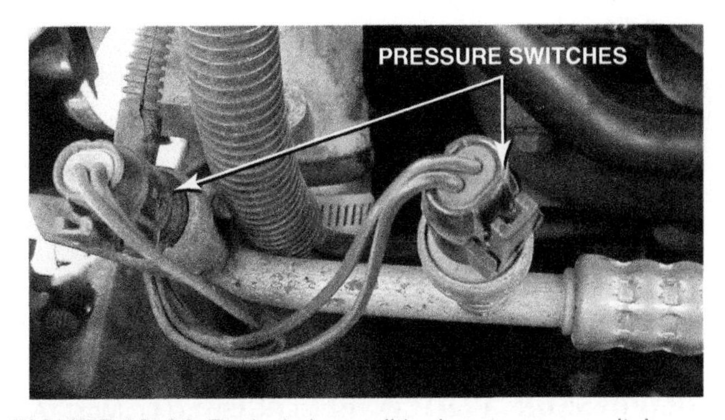

FIGURE 28–21 Typical air-conditioning pressure switches. Check service information to determine the purpose and function of each switch for the vehicle being inspected.

A/C COMPRESSOR DIAGNOSIS AND SERVICE

COMPRESSOR CLUTCH DIAGNOSIS A faulty compressor or compressor clutch is indicated if the following conditions are observed:

- The high- and low-side pressures are too close, within 50 PSI (345 kPa).

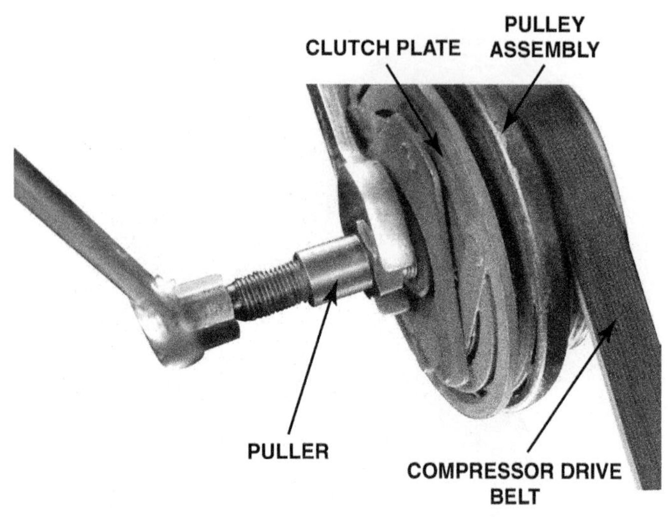

FIGURE 28–22 After removing the retaining nut from the A/C compressor shaft, a special puller is used to remove the compressor clutch plate (hub).

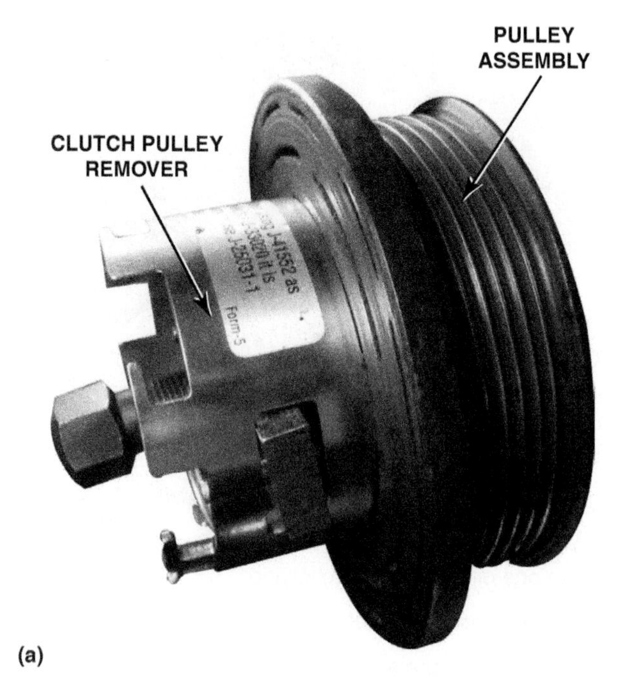

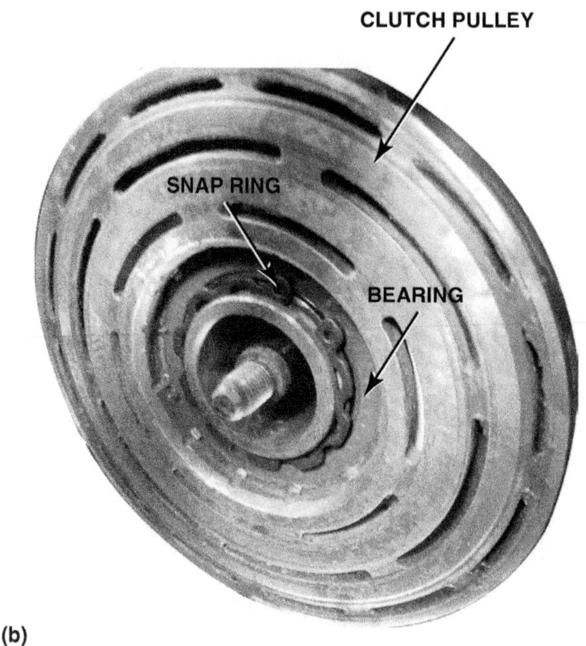

FIGURE 28–23 (a) The pulley assembly is removed using a special puller. (b) The pulley assembly includes the bearing which may or may not be a replaceable part, depending on the compressor.

- It cannot produce a high-side pressure of 350 PSI (2,400 kPa) or greater. This test usually requires disconnecting the fan(s) or blocking condenser airflow.
- There is visible damage to the compressor, clutch, or pulley.
- The compressor shaft rotates freely with no resistance.
- Shaft rotation is rough or harsh.
- There is free play when shaft rotation is reversed.
- The clutch has too much or too little air gap.
- The clutch does not apply or release.
- The pulley rotation is rough or with too much free play.

COMPRESSOR CLUTCH REPLACEMENT Most A/C compressor clutches are three-part assemblies with a separate drive hub (armature), rotor pulley, and coil. To remove a clutch assembly, check service information for the exact procedure to follow. Most specified procedures include the following steps:

STEP 1 Remove the locknut or bolt from the compressor shaft. A clutch hub wrench is often required to keep the hub from turning. Some compressors do not use a locknut.

STEP 2 Use the correct tool to pull the hub from the compressor shaft. ● **SEE FIGURE 28–22.**

STEP 3 A special puller is required on most compressors but the rotor pulley can be slid off some compressors, such as the Nippondenso compressors. ● **SEE FIGURE 28–23.**

To reinstall a clutch assembly, perform the following steps:

STEP 1 Install the coil, making sure that the anti-rotation pins and holes are aligned and the wire connector is in the correct position. The coil must be pressed in place on some compressors.

STEP 2 Install the coil retaining ring.

FIGURE 28–24 Air gap of the clutch is adjusted on some A/C compressors by using thin metal washers called shims. Adding a shim increases the gap and deleting a shim decreases the gap.

FIGURE 28–25 On some compressors, the air gap is adjusted by pressing the plate to the correct position.

STEP 3 Install the rotor pulley and replace the retainer ring. Some retainer rings have a beveled face and this side must face away from the pulley. Test this installation by rotating the rotor pulley and it must rotate freely, with no interference.

STEP 4 On some compressors, install the adjusting shims onto the shaft, install the drive key, and align and install the hub. ● **SEE FIGURE 28–24.**
On many GM compressors, the hub must be pulled onto the shaft. It should be pulled on just far enough to get the correct air gap. ● **SEE FIGURE 28–25.**

STEP 5 If used, install the shaft locknut or bolt and tighten it to the correct torque.

FIGURE 28–26 The shaft seal must keep refrigerant from escaping out the front of the compressor. Most compressors have an oil flow routed to them to reduce wear and improve the sealing action.

STEP 6 Check the clutch air gap at three locations around the clutch. The clearance should be within specifications at all three points. If it is too wide or too narrow, readjust the air gap. If it is too wide at one point and too narrow at another, replace the hub.

COMPRESSOR SHAFT SEAL REPLACEMENT Every compressor has a seal that keeps refrigerant from escaping through the opening where the pulley driveshaft enters. ● **SEE FIGURE 28–26.**

Many compressors use a rotating **seal cartridge** attached to the shaft and a stationary **seal seat** attached to the front of the compressor housing. The compressor has one or two flats on the shaft so that the seal cartridge is positively driven.

A gasket or rubber O-ring is used so that the seal seat makes a gas-tight seal at the housing, and the seat has an extremely smooth sealing face. The carbon-material sealing member is spring loaded so that its smooth face makes tight contact with the seal seat. The cartridge also uses a rubber O-ring or molded rubber unit to seal the carbon to the driveshaft. Another important part of the seal is the compressor oil, which lubricates the surfaces and forms the final seal between the sealing surfaces. ● **SEE FIGURE 28–27.**

A ceramic material has replaced cast iron for the seal seat in some compressors. Ceramic is not affected by water or acids, which can cause rust, corrosion, or etching of the iron seats. Ceramic seats are easy to identify because they are white instead of gray.

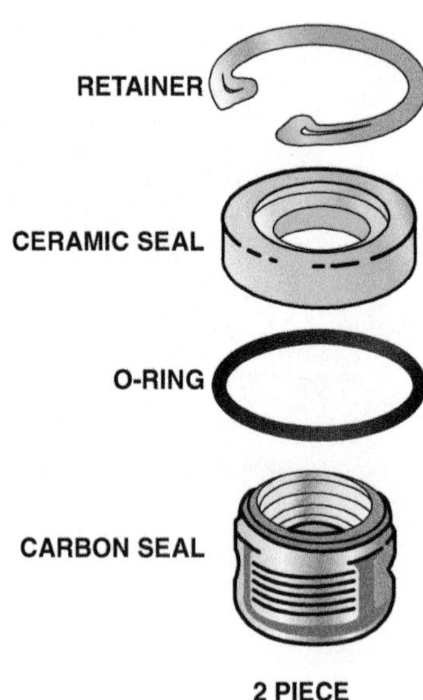

FIGURE 28–27 Many compressors use a two-piece seal with a rotating carbon seal and a stationary seal.

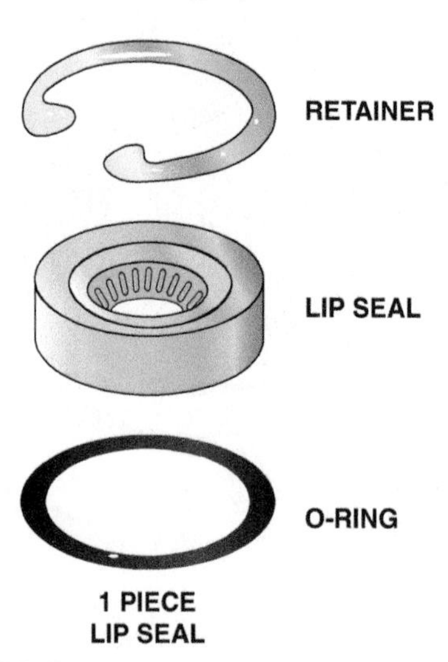

FIGURE 28–28 Some newer compressors use a stationary lip seal that seals against the rotating shaft.

Recent compressor designs use a **lip seal**. The lip of the seal is made from Teflon and rides against a perfectly smooth portion of the driveshaft. Some shaft seals use double seating lips. The outer shell of the seal fits into a recess in the compressor housing and is sealed using a rubber O-ring. Gas pressure in the compressor ensures a tight fit between the seal lip and the shaft. ● **SEE FIGURE 28–28.**

With most compressors, the shaft seal is removed from the front of the compressor after the clutch plate has been removed. Seals should not be reused and new seal parts should always be installed. Special tools are required for most seal replacements. These tools are designed to grip the seal cartridge or seat so they can be quickly pulled out or slid back into the proper position. Always follow the specified procedure found in service information.

COMPRESSOR DRIVE BELT SERVICE Engine drive belts should be checked periodically for damage and proper tension. If a belt shows excessive wear, severe glazing, rubber breakdown, or frayed cords, it should be replaced. The automatic tensioner on some vehicles includes a belt stretch indicator. If belt stretch of more than 1% is indicated, the belt should be replaced.

Belt slippage is caused by either a worn tensioner or worn belt. A belt that is too tight causes an excessively high load on the bearings of the components driven by the belt. Traditionally, belt tension is checked by pushing the center of the belt inward and then pulling it outward while noticing how much the belt is able to be deflected. Most manufacturers use a total movement of 0.5 inch as a maximum distance. Most manufacturers recommend using a belt tension gauge that is hooked onto the belt and uses a scale to show tension.

Quick and Easy Belt Noise Test

With the engine running at idle speed, use a spray bottle and squirt some water on the belt and listen for a noise change. If the noise increases, there is a belt tension problem. If the noise decreases but then returns, there is a belt alignment problem.

To remove and replace a drive belt using an automatic tensioner, perform the following steps:

STEP 1 Note the routing of the belt as per the under hood decal or in-service information.

STEP 2 Relieve the belt tensioner and slip the belt off the pulleys. A wrench can be used for this procedure on most tensioners. ● **SEE FIGURE 28–29.**

STEP 3 Remove the belt, and with the belt off, spin the pulleys to make sure that they are clean, not worn, and rotate freely.

STEP 4 Install the belt on some pulleys, rotate the tensioner, slide the belt into the proper position, and release the tensioner.

STEP 5 Check to ensure proper belt placement on each pulley.

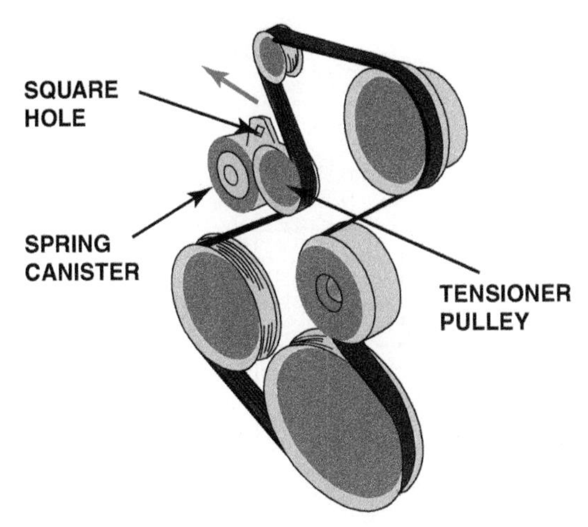

FIGURE 28–29 Moving the automatic tensioner outward allows the serpentine belt to be removed from the pulleys.

FREQUENTLY ASKED QUESTION

What Is a Stretch Belt?

Starting in 2007, some vehicles use a stretch-fit, multi-rib belt without a tensioner. The elastic nature of the belt allows it to be stretched to install it over the pulleys, and the stretch provides the tension to keep it from slipping. A special tool or strap is required to install a stretch belt, and some manufacturers advise to cut the belt to remove it. ● **SEE FIGURE 28–30**.

Cut the old belt to remove it. Place the new belt in position on the pulley that is hardest to get to. If the second pulley has holes through it:

- Install the belt onto the pulley, and secure it to the pulley using a zip tie.
- Rotate the engine by hand far enough so the pulley rotates and pulls the belt into position.
- Cut the zip tie.

If the second pulley does not have holes, a special tool will be required:

- Place the special tool into position on the second pulley.
- Start the belt onto the pulley.
- Rotate the engine by hand far enough so the pulley rotates and pulls the belt into position.
- Remove the special tool.

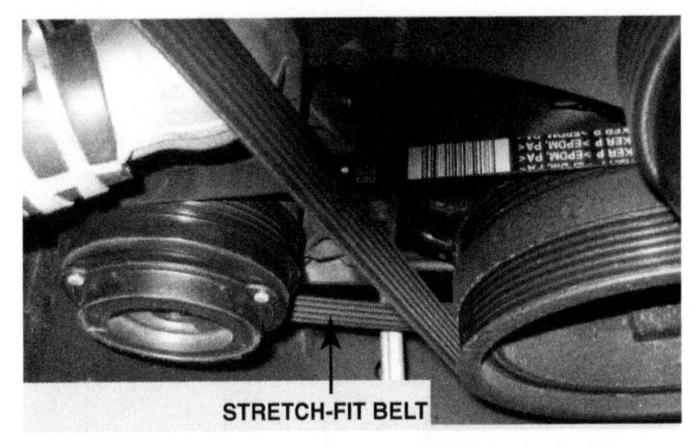

FIGURE 28–30 A stretch-fit belt is identified by the lack of an idler or method of adjusting belt tension.

Start the engine and check for proper belt operation.

BELT TENSIONER SERVICE The tensioner is designed to keep the belt tight enough so it does not slip but not so tight that the belt or bearings in the driven components will fail early. It must also dampen the tensioner arm to stop excess motion/bouncing and align the pulley to the belt. Tensioners can fail, and at least one source recommends installing a new tensioner when the belt is replaced.

To check belt tensioner operation, perform the following steps:

STEP 1 With the engine running at idle speed, observe any tensioner movement, and there should be a rather gentle motion. If it appears to bounce back and forth a large amount, the dampener portion is probably worn out.

STEP 2 Stop the engine and move the tensioner through its travel. It should move smoothly against the spring pressure with no catches or free portions.

STEP 3 With the pulley unloaded, check for free play or rough bearing operation. Next, spin the bearing and it should rotate smoothly for two or three revolutions. Also, check for excess arm motion at the pivot bushing.

CLUTCH ELECTRICAL SYSTEM* The compressor clutch is engaged by an electromagnetic field. One side of the electromagnet is grounded and the other side receives power from the A/C compressor relay. The relay is usually controlled

*Adapted from General Motors Service Information, Document # 1962851.

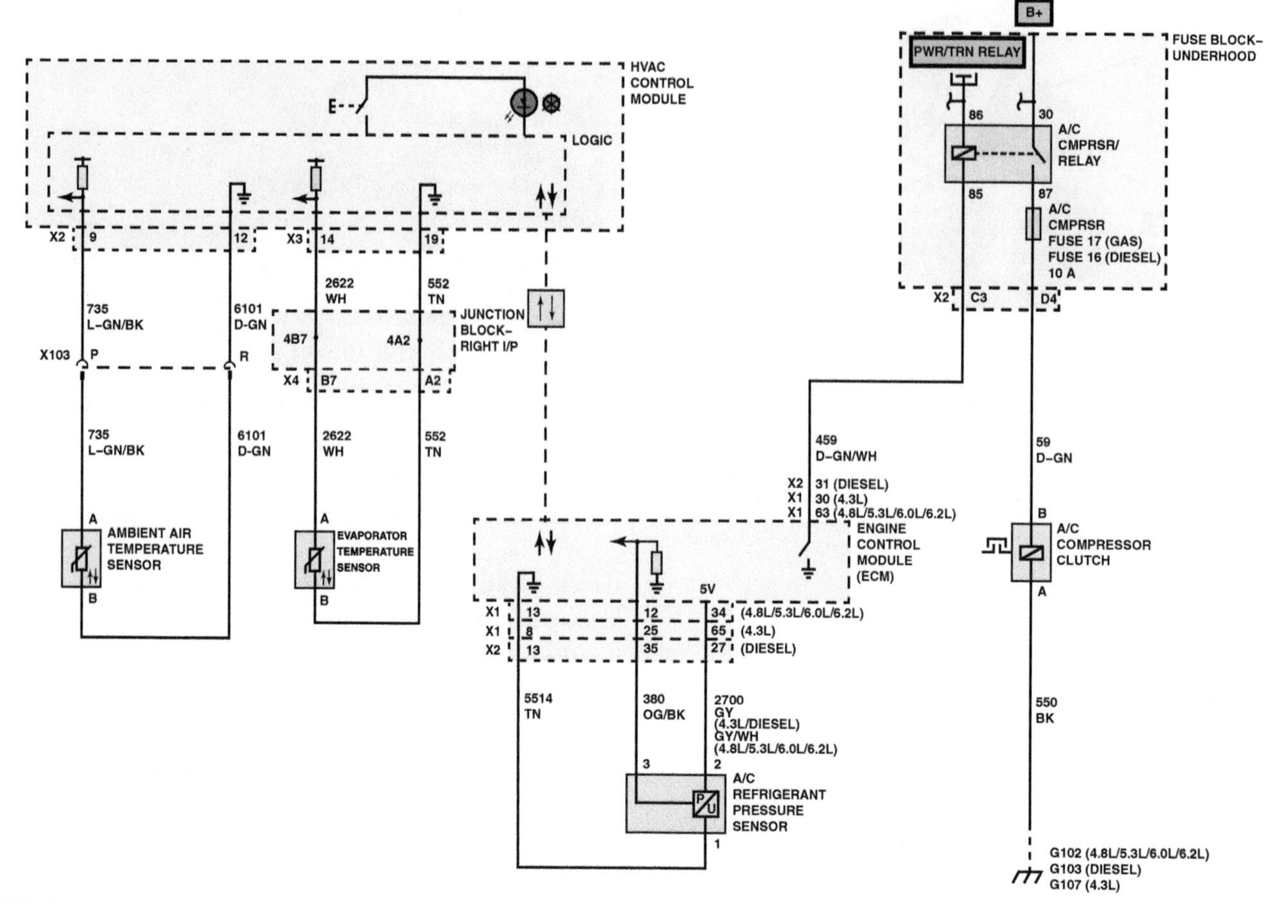

FIGURE 28–31 This diagram shows the components of the compressor clutch electrical system. (Courtesy of General Motors)

by the vehicle engine control module (ECM) but that will depend on the make and model. Always refer to service information for detailed operation. ● **SEE FIGURE 28–31**.

CIRCUIT/SYSTEM VERIFICATION Before the clutch will engage, a number of parameters must be met. To verify correct operation of the compressor clutch control system, follow these steps:

1. Verify the A/C system has the correct refrigerant charge.
2. Verify the scan tool A/C High Side Pressure Sensor parameter is between 2957 and 151 kPa (22–429 psi).
3. Verify the scan tool Evaporator Temperature Sensor parameter is greater than 2°C (34°F).
4. With the engine running, press the A/C request switch. The A/C compressor clutch should engage.
5. Place the mode switch in the defrost position. The A/C compressor clutch should engage.

If the A/C compressor clutch does not engage, refer to Circuit/System Testing.

CIRCUIT/SYSTEM TESTING

1. With the ignition OFF, disconnect the A/C relay.
2. With the ignition ON, verify that a test lamp does not illuminate between the control circuit terminal 85 and ground. If the test lamp illuminates, test the control circuit for a short to voltage.
3. Verify that a test lamp illuminates between the ignition circuit terminal 86 and ground.
4. Verify that a test lamp illuminates between the relay B+ circuit terminal 30 and ground.
5. Disconnect the harness connector at the A/C clutch coil. Test for less than 1 ohm between the ground circuit terminal A and ground.
6. Connect the harness connector at the A/C clutch coil.

FIGURE 28–32 The decal on this compressor identifies the type (SDB709) and the serial number. Note also that it uses a seven-groove, multi-V clutch, four mounting bolts, and vertical-pad service ports at the side.

7. Connect a 10-amp fused jumper wire between the relay B+ circuit terminal 30 and control circuit terminal 87. Verify the A/C clutch is activated. If the A/C clutch does not activate, test the wiring to the clutch for an open/high resistance. If the circuit tests normal, test or replace the A/C clutch.
8. Connect a test lamp between the ignition circuit terminal 86 and the control circuit terminal 85. Command the A/C relay output function ON and OFF with a scan tool. The test lamp should turn ON and OFF when changing between the commanded states.
   - If the test lamp is always ON, test the control circuit for a short to ground. If the circuit tests normal, replace the ECM.
   - If the test lamp is always OFF, test the control circuit for a short to voltage, or an open/high resistance. If the circuit tests normal, replace the ECM.
   - If all circuits test normal, test or replace the A/C clutch relay.

REPLACEMENT COMPRESSORS

IDENTIFY THE UNIT Replacement compressors are available as new or rebuilt units, and proper identification is made from the vehicle make, model, and engine size. Then, if needed, proper identification is made by the old compressor make and model. At times, a failed compressor is replaced with a different compressor make and model if the mounting points, clutch diameter and belt position, and line fittings are the same. ● **SEE FIGURE 28–32.**

OIL CHARGE Having proper amount of oil in the system during compressor replacement is an important factor. Too much oil in a system can reduce system performance, and too little oil can cause early compressor failure. Many compressors are equipped with *shipping oil* that is not intended for long-term lubrication. Be sure to read the information that usually comes with the new compressor and follow the directions. Most manufacturers recommend draining all the oil from the old compressor and measuring the amount drained. ● **SEE FIGURE 28–33.**

If the amount is within certain limits, such as 3 oz. to 5 oz., the same amount of new oil is then added to the new compressor. Always follow the manufacturer's recommendations or processes. A typical example includes:

- If the amount drained was below 3 oz., then 3 oz. of new oil is added.
- If the amount drained is more than 5 oz., then 5 oz. of new oil is added.

Replacing a compressor usually includes the following steps:

STEP 1 Adjust the oil level in the compressor as instructed by the vehicle or compressor manufacturer.

STEP 2 Install the compressor on the engine and replace the mounting bolts.

STEP 3 Install the drive belt, adjust the belt tension, and tighten all mounting bolts to the correct torque.

STEP 4 Using new gaskets or O-rings, connect the discharge and suction lines. Then evacuate, charge, and check for leaks at all fittings.

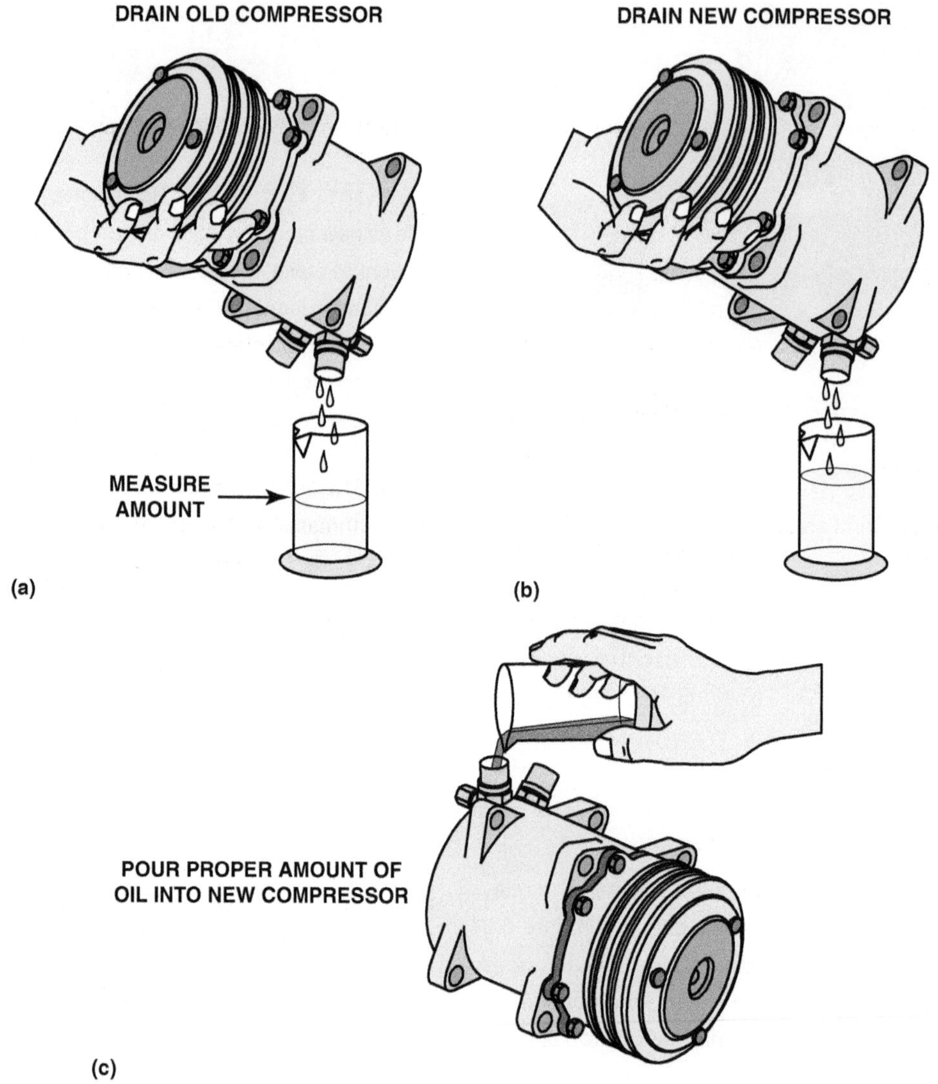

DRAIN OLD COMPRESSOR

DRAIN NEW COMPRESSOR

MEASURE AMOUNT

(a)

(b)

POUR PROPER AMOUNT OF OIL INTO NEW COMPRESSOR

(c)

FIGURE 28–33 (a) The oil should be drained from the old compressor; rotate the compressor shaft and the compressor to help the draining. (b) Drain the oil from the new compressor. (c) Pour the same amount of oil drained from the old compressor or the amount specified by the compressor manufacturer of the proper oil into the new compressor.

SUMMARY

1. Various compressor types and models are used with vehicle A/C systems.
2. Compressor models can use a variety of clutch and pulley designs.
3. Some variable displacement compressors use a damper pulley in place of a clutch.
4. Compressors are lubricated by oil that is circulated by the refrigerant.

REVIEW QUESTIONS

1. What type compressors are used in air-conditioning systems?
2. How does a coaxial swash-plate compressor work?
3. What is the difference between a vane compressor and a radial compressor?
4. How does an electromagnetic compressor clutch work?
5. How is the air gap of a compressor clutch adjusted?

CHAPTER QUIZ

1. In a piston-type compressor, the pistons are driven by a ________.
   a. Crankshaft
   b. Swash plate
   c. Wobble plate
   d. All of the above depending on the compressor
2. Most piston compressors use what type of valves?
   a. Poppet
   b. Reed
   c. Sliding
   d. On-off gate-type valve
3. The basic designs of A/C compressors include ________, ________, and ________?
   a. Piston, vane, and scroll
   b. Radial, piston, and vane
   c. Balanced, piston, and scroll
   d. Fixed scroll, moveable scroll, and piston
4. Two technicians are discussing variable displacement piston compressors. Technician A says that the wobble plate is moved to the high-angle position for maximum output when the cooling load is high. Technician B says that the wobble-plate angle is controlled by the pressure in the crankcase of the compressor. Which technician is correct?
   a. Technician A only
   b. Technician B only
   c. Both technicians A and B
   d. Neither A nor B
5. Wobble-plate compressors commonly use how many cylinders?
   a. Two
   b. Three
   c. Four
   d. Five or seven
6. A clutchless damper drive is used with what type of A/C compressor?
   a. Electronic-controlled, variable displacement compressor
   b. Scroll compressor
   c. Vane-type compressor
   d. Orbiting piston compressor
7. How is the A/C compressor lubricated in an R-134a system?
   a. Uses a sump to hold 3 oz. to 5 oz. of refrigerant oil
   b. The refrigerant oil mixes with the refrigerant
   c. The oil moves through the system simply with the movement of the refrigerant
   d. Most compressors are a sealed unit with their own oil
8. Switches located on the compressor itself may include ________.
   a. A low-pressure switch
   b. High-pressure switch
   c. Low- and/or high-pressure sensor
   d. Any or all of the above
9. A compressor speed (belt lock) sensor is used by the HVAC ECM to ________.
   a. Prevent a locked-up compressor from destroying the engine drive belt
   b. Determine if the compressor is running, and by comparing the compressor and engine speed signals
   c. Vary the displacement of the compressor based on compressor speed
   d. Both a and b
10. A stretch-type A/C compressor belt is removed using ________.
    a. A special tool
    b. A knife (cut it)
    c. By removing the compressor
    d. By removing the tensioner

REFRIGERANTS AND REFRIGERANT OILS

LEARNING OBJECTIVES

After studying this chapter, the reader should be able to:

1. Discuss the depletion of the ozone layer and the resulting issues of global warming.
2. Explain the impact of legislative laws on automotive A/C systems.
3. Discuss identifying refrigerants and proper storage container.
4. State the changes considered for future refrigerants.
5. Explain refrigerant safety precautions.
6. Discuss the different types and viscosities of refrigerant oils.

This chapter will help you prepare for the ASE Heating and Air Conditioning (A7) certification test content area "B" (Refrigeration System Diagnosis, and Repair).

KEY TERMS

Clean Air Act 471
Environmental Protection Agency (EPA) 471
Global warming 471
Global warming potential (GWP) 471
Greenhouse effect 471
Greenhouse gases (GHGs) 471
Hygroscopic 477
Ozone 470
Ozone depletion potential (ODP) 470
Section 609 471
Significant new alternatives policy (SNAP) 472
Stratosphere 469
Total environmental warming impact (TEWI) 472

GM STC OBJECTIVES

GM Service Technical College topics covered in this chapter are as follows:

1. The behaviors of refrigerants and coolants (01510.01W).
2. The effects of refrigerants and coolants.
3. The purpose and process of refrigerant recovery (01510.10W).

FIGURE 29–1 Large 30 pound containers of R-134a are light blue for easy identification.

REFRIGERANTS

Refrigerants are colorless and odorless compounds. Usually, the only way to know that they are present in a container is how the container feels when it is picked up or shaken. On manifold gauge sets equipped with a sight glass, bubbles can be seen in the clear liquid as it passes by the glass.

Refrigerants are commonly available in several sizes of containers including:

- Small cans of 12 to 14 oz. (400 g)—at one time, this was 15 oz. of R-12 and 1 oz. of can for a total of 1 lb and
- Larger drums or canisters of 15 or 30 lb. (6.8 or 13.6 kg). Small containers of R-12 can be purchased only by certified technicians; however, there are no national restrictions on purchase of R-134a.

Refrigerant containers are color coded:

- R-12 containers are white
- R-22 containers are green
- R-134a containers are light blue. ● **SEE FIGURE 29–1**.

Refrigerant containers are usually disposable. These containers should be evacuated into a recovery unit, marked empty, and properly disposed of when they are emptied. The

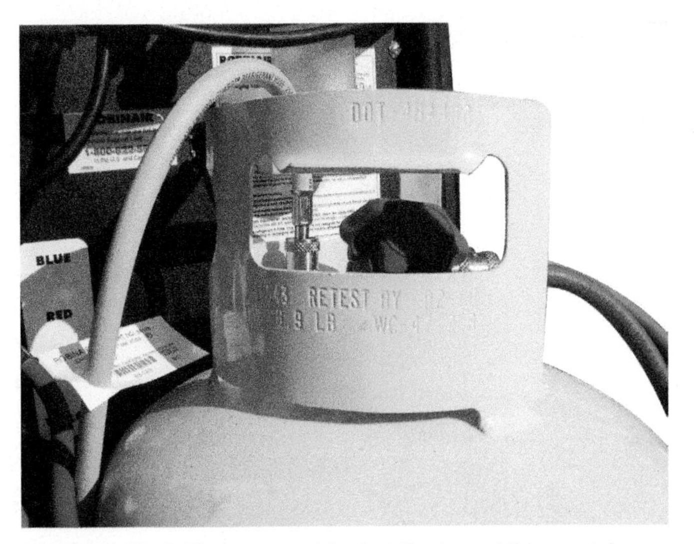

FIGURE 29–2 The stamped text at the top of this container reads "DOT-4BA400."

TECH TIP

Think of a Propane Tank

An effective way of telling how much propane there is in the container used in the barbecue grill is by pouring hot water over the container. The heat will warm the top of the container, cause some of the liquid to boil, and increase container pressure, and if felt along the container, the liquid will cool the lower part. The liquid fills the container to the point of temperature change. This same procedure can be used to estimate the level of refrigerant in the container.

storage containers for recycled refrigerant must be approved by the DOT and carry the proper marking to show this. ● **SEE FIGURE 29–2**.

ENVIRONMENTAL ISSUES

TERMINOLOGY Planet Earth is unique in many ways. It has an atmosphere that contains a percentage of oxygen high enough to allow mammals, including humans, to live in it. This atmosphere extends outward from Earth for about 31 miles (50 km). The upper layer of the atmosphere is called the **stratosphere**, and it begins about 7 to 10 miles (11 to 16 km)

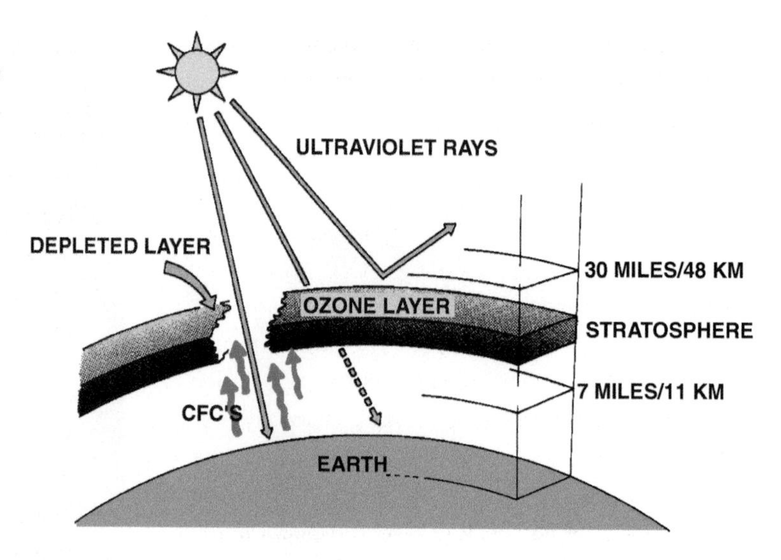

FIGURE 29–3 Depletion of the ozone layer allows more ultraviolet radiation from the sun to reach Earth's surface.

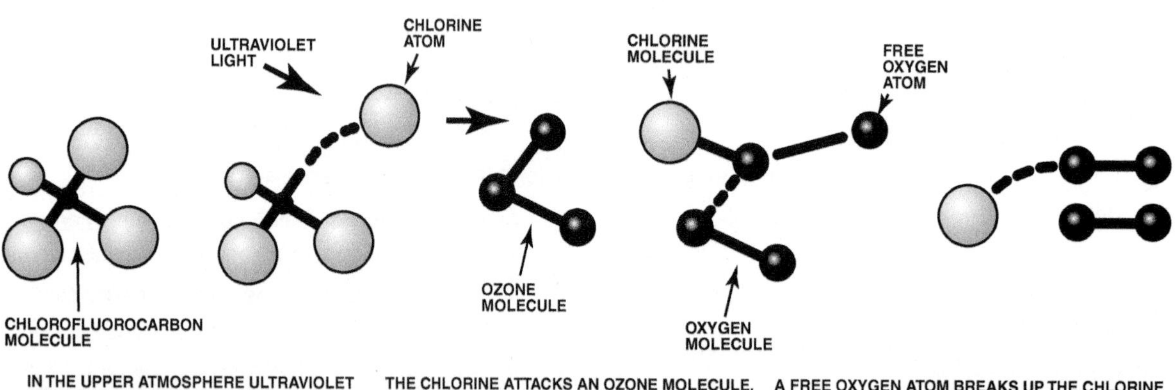

FIGURE 29–4 Chlorofluorocarbon molecules break apart in the atmosphere.

up and extends to the outer limits. A layer of **ozone** (O_3) extends around the Earth in the stratosphere. **SEE FIGURE 29–3.**

The ozone layer is important to us because it blocks the ultraviolet wavelengths of light generated by the sun. Ultraviolet rays can be very harmful to our way of life. In humans, an excess of these rays can cause an increase in skin cancer and cataract of the eyes, as well as damage to our immune system. These same problems can affect many animals. Ultraviolet rays can also damage plants and vegetables. This damage probably also extends to plankton and larvae in the sea, the base of the food chain for sea animals.

In the late 1900s, it was determined that the ozone layer is getting much thinner, and large holes are being created in it (mostly near the South Pole). The ozone layer is not providing the same protection from ultraviolet (UV) rays as it once did. It has been determined that

1. The breakup or depletion of the ozone layer is caused by human-made chemical pollution.
2. One of the major ozone-depleting chemicals is chlorine.
3. One of the major sources of chlorine in the atmosphere is R-12.

OZONE DEPLETION A chlorine atom from a *chlorinated fluorocarbon (CFC)* such as R-12 can travel into the stratosphere if it escapes or is released. There, under the effects of ice clouds and sunlight, it can combine with one of the oxygen atoms of an ozone molecule to form chlorine monoxide and an ordinary oxygen molecule, O_2. This destroys that ozone molecule. The chlorine can then break away and attack other ozone molecules. It is believed that 1 chlorine atom can destroy 10,000 to 100,000 ozone molecules. **SEE FIGURE 29–4.**

- CFCs do the most damage and have an **ozone depletion potential (ODP)** of 1.
- Hydrochlorofluorocarbon (HCFC), such as R-22, have an ODP around 0.01–0.1, so they have a lesser effect on the ozone layer.

- Since HFCs, such as R-134a, contain no chlorine and have an ODP of 0 (zero), they have no detrimental effect on the ozone layer.

GLOBAL WARMING AND GREENHOUSE GASES

Another area of concern is a layer of gases that is causing a **greenhouse effect**. This gas layer traps heat at the Earth's surface and lower atmosphere, and it is increasing the temperature of our living area. This is called **global warming**. CFC and hydrocarbon (HC) gases such as butane and propane are considered **greenhouse gases (GHGs)**. **Global warming potential (GWP)** compares the ability of different gases to trap heat, and this is based on the heat-absorbing ability of each gas. The lower the GWP number, the lower the global warming potential.

FREQUENTLY ASKED QUESTION

How Are Refrigerants Named?

The procedure for determining the refrigerant number is rather tedious and is of no value to the refrigerant technician. It is included in this text to reduce some misconceptions and wild stories that have passed through the industry.

The system is based on the number of carbon, hydrogen, and fluorine atoms in the refrigerant molecule. It is a four-character system with the letter "a" added to indicate a nonsymmetrical or asymmetrical molecule. The most common method of numbering system is:

C (MINUS 1), H (PLUS 1), and F.

R-12 has 1 C (1−1=0), O H (0+1=1), and 2 F (2), to get 012, or 12.

Look at the R-12 and R-134a molecules in ● **FIGURE 29–5**.

- The R-12 molecule is symmetrical, which means that both sides are the same.
- The R-134a molecule is asymmetrical (the reason for the "a" in the name), which means that the left and right sides are different, so an *a* is added to it. There is an HFC-134 molecule that is symmetrical, but it is not used in automotive refrigerant systems. Letters further down the alphabet indicate greater changes in the molecule.

Technically, the name should be CFC-12, HFC-134a, or HCFC-22 to indicate that the refrigerant is a chlorofluorocarbon, hydrofluorocarbon, or hydrochlorofluorocarbon. The chemical name for CFC-12, dichlorodifluoromethane, indicates that the molecule has two (di) chlorine atoms and two fluorine atoms, and the suffix methane indicates there is one carbon atom. The prefixes, di = 2, tri = 3, tetra = 4, penta = 5, and so on indicate the number of atoms. With carbon-based molecules, single-C molecules are methanes, double-C molecules are ethanes, three-C molecules are propanes, and so on. HFC-134a, tetrafluoroethane, has four (tetra) F atoms, two H (not mentioned in the name), and the suffix ethane, meaning there are two carbon atoms.

This explanation is not complete because chemists use other designations to show bonds within the molecule and different number combinations for blend refrigerants. This numbering system is based on the carbon chain that the molecule is built around. In this system, HFC-134a is 1, 1, 1, 2 tetrafluoroethane.

LEGISLATION

CLEAN AIR ACT At a conference in Montreal, Canada, in 1987, the United States, along with 22 other countries, agreed to limit the production of ozone-depleting chemicals. This agreement is referred to as the Montreal Protocol. In 1990, President Bush (senior) signed the **Clean Air Act**, which phased out the production of CFCs in the United States by the year 2000. R-12 production in the United States ceased at the end of 1995.

SECTION 609 **Section 609** is a portion of the Clean Air Act that places certain requirements on the *mobile vehicle air conditioning (MVAC)* service field. Important portions of this section require the following:

Effective January 1, 1992

- Technicians who repair or service automotive A/C systems shall be properly trained and certified and use approved refrigerant recovery and recycling equipment.
- Recovery and recycling equipment must be properly approved.
- Each shop that performs A/C service on motor vehicles shall certify to the **Environmental Protection Agency (EPA)** that it is using approved recycling equipment and that only properly trained and certified technicians are using this equipment.

Effective November 15, 1992

- Sales of small containers of R-12 (less than 20 lb) are restricted to certified technicians.

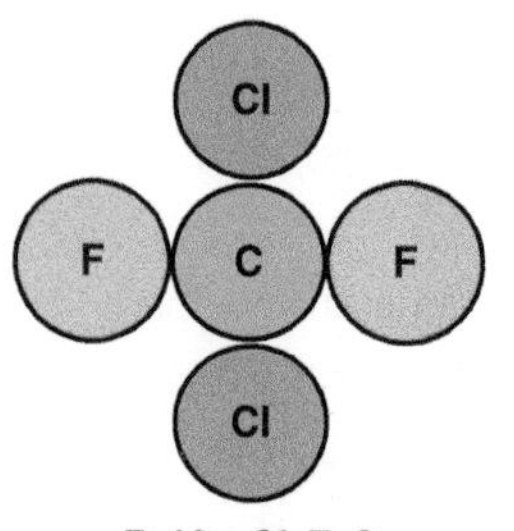

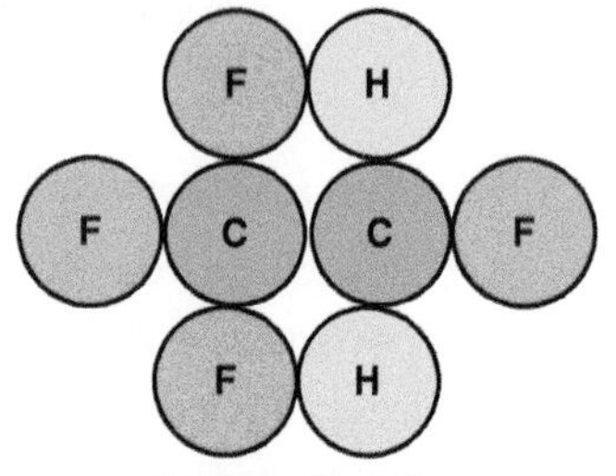

R-12 - Cl_2F_2C R-134a - $C_2H_2F_4$

FIGURE 29–5 R-12 is a combination of one carbon, two chlorine, and two fluorine atoms. R-134a is a combination of two carbon and four fluorine atoms.

Effective November 14, 1994

- MVAC technicians must be certified to purchase EPA-acceptable blend refrigerants.

Effective January 29, 1998

- Equipment used to recover or recycle R-134a must meet SAE standards.
- MVAC service technicians must be certified to handle non-ozone-depleting refrigerants (including R-134a).

Under the Montreal Protocol, phaseout of HCFC began in 2004 (a 35% reduction). This reduction began in 2003 in the United States. Manufacture of HCFC-22 was stopped in 2010 in the United States and other developed countries.

EUROPEAN R-134A PHASE-OUT The concern about climate change has caused the European Community to try to reduce greenhouse gas (GHG) emissions. Part of this legislation is to require that refrigerants must have a GWP of less than 150. Some refrigerants and their GWP include the following:

- R-134a has a GWP of 1,300 (a rather high potential contributor for climate change).
- R-1234yf has a GWP of 4. ● **SEE FIGURE 29–6**.
- R-744—Carbon dioxide (CO_2) has a GWP of 1

This rule was originally to take effect in 2011, with R-134a banned from new vehicle HVAC systems.

KYOTO PROTOCOL There is still a major concern about the global warming effect from escaping refrigerants. Representatives from many nations met in Kyoto, Japan, in 1997 and established the Kyoto Protocol to address these concerns. The **Total Environmental Warming Impact (TEWI)** index was developed, which rates the impact of various refrigerants along with the energy required to perform the cooling operation. In the future, it is hoped that each nation will reduce its negative impact on the environment to zero in order to reduce global change to zero. All factors concerning energy use and emissions will be considered. The United States has not signed the Kyoto Protocol.

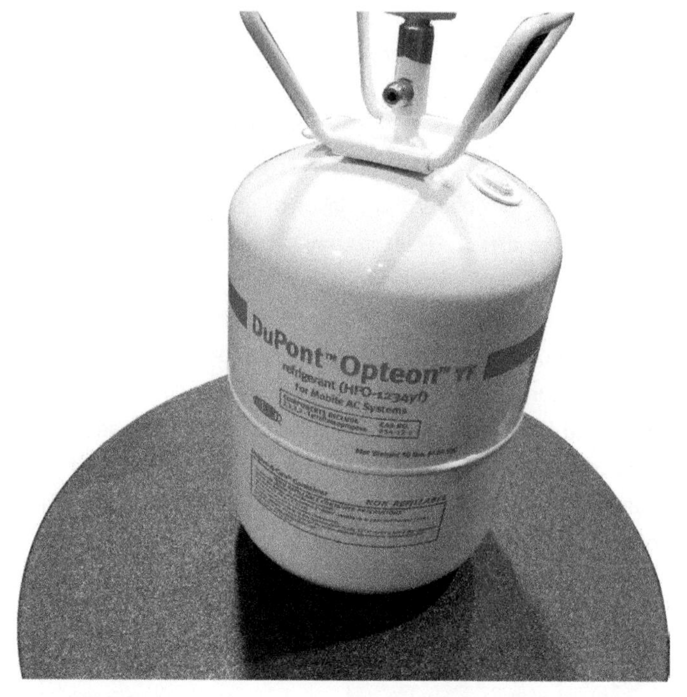

FIGURE 29–6 R-1234yf refrigerant is sold in white containers with a red stripe.

ALTERNATE REFRIGERANTS

SNAP PROGRAM Section 612 of the 1990 Clean Air Act established the **Significant New Alternatives Policy (SNAP)** program to determine acceptable replacements for Class I and Class II chemicals. Class I chemicals include CFCs, and Class II chemicals are HCFCs. SNAP is administered by the EPA and identifies refrigerants that are acceptable from their ozone-depleting potential, global warming potential, flammability, and toxicity characteristics. Alternate refrigerants are not tested on their refrigeration quality, but only on their human health and environmental risks.

R-134a is the only alternate product for an R-12 system that has addressed the SAE retrofit documents (SAE J1657, SAE J1658, SAE J1659, and SAE J1662). These standards are designed to ensure long-term, trouble-free operation of the A/C system. An alternate refrigerant can only be used under the following conditions:

- Each refrigerant must have its own unique set of fittings, and all ports not converted must be permanently disabled.
- Each refrigerant must have a label with a unique color that specifies pertinent information.

R-1234yf REFRIGERANT PROPERTIES		R-134a
TEMPERATURE DEGREES C	PRESSURE KPA	PRESS. KPA
−20	149.9	
−10	220.3	97
0	314.0	188
10	435.6	310
20	590.0	471
30	782.3	655
40	1017.8	917
50	1302.3	1222
60	1641.9	1590
70	2043.4	2048
80	2515.7	2522

R-1234yf PRESSURES ARE HIGHER THAN R-134a REFRIGERANTS WITHIN THIS RANGE (−20 to 40)

R-1234yf PRESSURES ALIGN, THEN TREND LOWER THAN R-134a REFRIGERANTS WITHIN THIS RANGE (50 to 80)

FIGURE 29–7 This chart compares R-1234yf pressures to R-134a pressures. The differing pressures require components that are different, also. (Courtesy of General Motors)

- All original refrigerants must be removed before charging with the new refrigerant.
- With blends that contain HCFC-22, hoses must be replaced with less permeable barrier hoses.
- With systems that include a high-pressure release device, a high-pressure shutoff switch must be installed.
- Blends containing HCFC-22 will be phased out in the near future.

COUNTERFEIT AND BOOTLEG REFRIGERANTS With the high cost and limited availability of R-12, a refrigerant buyer has to be careful about where it is purchased from. Refrigerant is being illegally imported, and some of it is contaminated. There are stories of counterfeit refrigerant being sold in containers marked the same as those of a reputable, domestic manufacturer. Other stories tell of containers that contain nothing but water with air pressure. Refrigerant should be purchased only from a known source that handles reputable products.

FUTURE REFRIGERANTS

TEWI In order to meet the requirements of **Total Environment Warming Impact (TEWI)**, several possible changes are being considered as a new refrigerant:

- R-1234yf
- R-744—Carbon dioxide (CO_2)

R-1234YF R-1234yf has operating characteristics that are very similar to R-134a, with a much lower GWP and an ODP of zero.

While a R-1234yf refrigerant system operates similarly to R-134a systems, the refrigerant is NOT compatible with other refrigeration systems. This new refrigerant utilizes a different lubricant, which is also not compatible with other systems.

R-1234yf REFRIGERANT SERVICE The R-1234yf refrigerant requires the use of specific recharge, recovery, and recycling equipment, with an integrated refrigerant identifier. A dedicated contaminated refrigerant reclaiming machine is also required. Specific leak testing procedures are mandatory. In order to ensure that the correct equipment is used, R-1234yf has unique service fittings.

R-1234yf REFRIGERANT PROPERTIES The vapor pressure of R-1234yf is slightly different than that of R-134a, at both low temperatures and at high temperatures. At lower temperatures (below 40°C/104°F), the vapor pressure of R-1234yf is significantly higher than that of R-134a. The result of this refrigerant property is that the higher "low side pressures" are normal, and a certified evaporator (designed to withstand higher "low side" pressures) must be used on R-1234yf systems.

At vapor pressures above 40°C (104°F) the R-1234yf trends slightly lower than R-134a. This property results in lower high side pressures being normal, but producing a less efficient refrigeration cycle. ● **SEE FIGURE 29–7**.

In order to improve efficiency, an integrated heat exchanger is utilized to raise high side temperatures. The

FIGURE 29–8 The label on this vehicle shows that CO_2 (R744) is being used as the refrigerant.

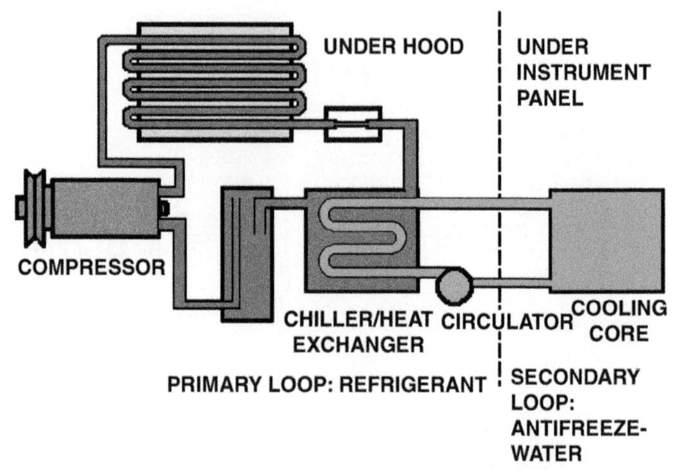

FIGURE 29–9 A secondary loop A/C system keeps the potentially dangerous or flammable refrigerant out of the passenger compartment by using a chiller/heat exchanger to cool an antifreeze and water mixture. This fluid then transfers heat from the cooling core in the air distribution section to the chiller.

integrated heat exchanger is incorporated into the lines and transfers heat into the refrigerant vapor traveling into the suction side of the compressor, as well as lowering the temperature of the liquid refrigerant entering the thermostatic expansion valve.

There is concern about possible cross-contamination with either R-134a or an HC:

- As little as 5% of R-134a can increase the pressure in a recovery cylinder high enough to cause an "air purge" and this can empty the entire cylinder in the attempt to bleed off air.
- As little as 2% HC with 2% air will increase flammability to greater-than-limits established by the refrigerant manufacturers.

R-744 A carbon dioxide (CO_2) system (R-744) is similar to the present-day systems, but it requires extremely high pressure, 7 to 10 times that of an R-134a system. This pressure puts a large added load on the compressor, heat exchanger, and refrigerant lines. The extremely high pressure requires more power from the engine to operate the compressor and this extra energy causes reduced fuel economy. At this time, a CO_2 system is up to 40% less efficient than an R-134a system and can cost about 30% more to produce. ● **SEE FIGURE 29–8.**

SECONDARY LOOP SYSTEMS Systems that use potentially hazardous refrigerants will probably use a secondary loop. The CO_2 or HC portion of the system will be entirely under the hood. A heat exchanger will provide a very cold fluid to connect with a liquid-to-air heat exchanger that will replace the evaporator. These systems might also use refrigerant-leak-sensing devices and a method of venting/dumping the refrigerant into one of the fender wells to prevent a buildup of refrigerant in the passenger area. ● **SEE FIGURE 29–9.**

FREQUENTLY ASKED QUESTION

Will R-134a Systems be Required to be Retrofitted to R-1234yf?

No. Currently there is no mandate to retrofit existing systems with R-1234yf. If a vehicle came from the factory with R-1234yf, then this refrigerant, of course, should be used when servicing the vehicle and will require a special machine. R-134a is and will be the primary refrigerant for most vehicles and there is no need or legal requirement to replace it with any other refrigerant at this time.

REFRIGERANT SAFETY PRECAUTIONS

POTENTIAL HAZARDS Refrigerants should be handled only by trained and certified technicians because of potential safety hazards:

- Physiological reaction
- Asphyxiation (this means that the refrigerant does not contain oxygen, so it cannot sustain life if breathed)
- Frostbite and blindness
- Poisoning
- Combustion
- Explosion of storage containers

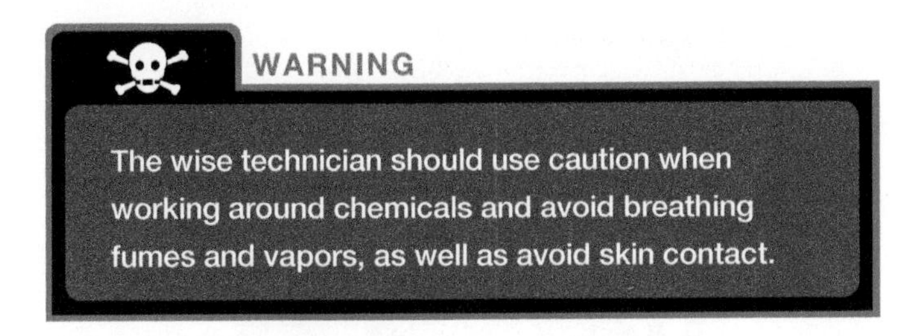

PRECAUTIONS Precautions when working with refrigerants include:

- Wear safety goggles or a clear face shield and protective clothing (gloves) when working with refrigerants. If refrigerant splashes into your eyes, blindness can occur. If refrigerant splashes into your eyes or onto your skin, do not rub that body part but instead flush it with cool, clean water to restore the temperature. Place sterile gauze over the eye to keep it clean and get professional medical attention immediately.
- Always be in a well-ventilated shop area when working with refrigerants and avoid small, enclosed areas. Refrigerants do not contain oxygen and are heavier than air. If they are released into a confined area, they fill the lower space, forcing air and its oxygen upward. Any humans or animals that breathe refrigerants can be asphyxiated, and lack of oxygen can cause loss of consciousness or death. In case of accidental release of refrigerant into the atmosphere move immediately to an area with adequate ventilation.
- If liquid refrigerant is splashed onto the skin or into the eyes of a human or animal, it immediately boils and absorbs heat from the body part it is in direct contact with. The temperature of the area is reduced to the low boiling point of the refrigerant, which is cold enough to freeze that body part.
- If a CFC such as R-12 or R-22 comes into contact with a flame or heated metal, a poisonous gas similar to phosgene is formed. This can occur while using a flame-type leak detector, if refrigerant is drawn into a running engine, or even if it is drawn through burning tobacco. An indication that a poisonous gas is forming is a bitter taste.
- Several flammable refrigerants have been marketed, and even though they have been banned and are illegal, they still show up. A mixture of more than 2% hydrocarbon (butane, isobutane, or propane) is considered flammable and about 4 oz. in a vehicle interior can become an explosive mixture. R-134a can become combustible at higher pressures if mixed with air. Air should not be used to flush an R-134a system because of the remote chance of a fire or explosion.

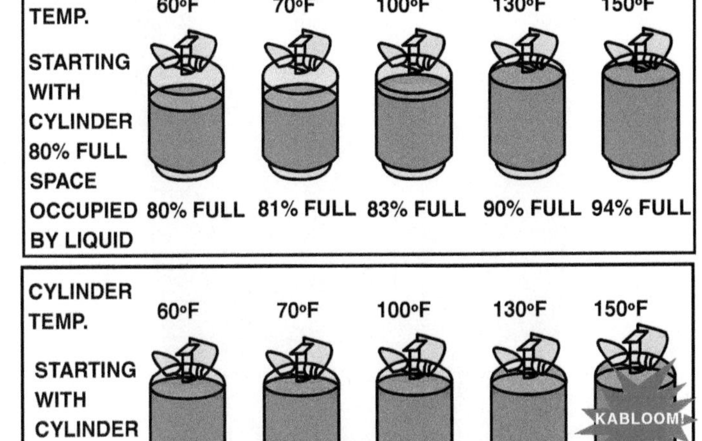

FIGURE 29–10 When recovering refrigerant, the container should be filled to a maximum of about 80%.

- When refrigerant containers are filled, room is left for expansion and the container is marked with its critical temperature, the maximum that it should be subjected to. Refrigerant containers are designed to contain the refrigerant under pressures encountered in normal working and storage conditions. Container pressure is about the same as the vapor pressure for that refrigerant up to a certain temperature point, which is where the liquid has expanded to fill the entire container. Beyond this point, any further expansion of the liquid generates very high hydraulic pressures that will rupture the container. The chance of container rupture is generally low unless the container is overfilled or overheated. ● **SEE FIGURE 29–10.**

REFRIGERANT OILS

PURPOSE AND FUNCTION Refrigerant oil serves several purposes, the most important being to *lubricate* the moving parts of the compressor to reduce friction and prevent wear. Refrigerant oil also helps *seal* the compressor shaft seal, the insides of the hoses, and various connections between the parts to reduce refrigerant leakage. In addition, it lubricates the TXV and coats the metal parts inside the system to reduce corrosion.

FIGURE 29–11 A container of refrigerant, which is sold as a replacement for R-12 or R-134a and contains R-134a but also butane, a flammable gas.

FIGURE 29–12 PAG oil is the type of refrigerant oil specified for use in most R-134a systems and the "150" is the viscosity.

FREQUENTLY ASKED QUESTION

What Is an HC Refrigerant?

A hydrocarbon (HC) refrigerant is butane/propane blend that is very flammable. It can be used and will function in present-day A/C systems, but it is too dangerous to use. At the present time, the EPA and 19 states have banned flammable refrigerants. A leak or rupture in the evaporator could easily result in a vehicle explosion. Some hydrocarbon refrigerants are sold to do-it-yourselfers (DIY) and while it can work in either an R-12 or an R-134a system, it would contaminate the system and make future repairs of the system much more difficult. ● **SEE FIGURE 29–11.**

The oil used in a system must be completely compatible with all the materials in the system.

TYPES OF REFRIGERANT OILS There are three types of refrigerant oils:

1. **Mineral oil** is used in R-12 systems.
2. **PAG (polyalkylene glycol)** oils are used in most R-134a systems. ● **SEE FIGURE 29–12.**
3. **POE (polyol ester)** oils are used in a few R-134a systems, and either ester or PAG oils are used in R-12 systems retrofitted to R-134a.

PAG and POE are synthetic, human-made oils. Neither PAG nor POE is soluble in R-134a, but they are easily moved by the refrigerant flow so they travel through the entire A/C system.

Most compressor remanufacturers consider PAG oil a better lubricant than ester oil. There are over 30 varieties of refrigerant oils. whose use depends on the requirements of a particular compressor or system. Piston, scroll, and vane compressors often require a different oil viscosity because of the different operating characteristics and use internal pressures to circulate the oil.

Compressors normally contain a certain amount of oil (often just 2 to 8 oz.). The oil level can only be checked by removing the compressor and draining all of the oil, then measuring how much oil came out. This is normally done when a compressor is replaced or when major service is performed on the system.

VISCOSITY OF REFRIGERANT OILS Viscosity is a measurement of how thick the oil is and how easily it flows. The oil must be thick enough so that moving parts float on an oil film. Oil must also be fluid enough to flow into the tiny spaces between parts. Compressor manufacturers normally specify the oil type and their viscosity for each type of compressor. The viscosity is determined by the International Standards Organization (ISO) viscosities and is measured at 40°C. Refrigerant oils are commonly available in several viscosities, including:

- **46**—Called low viscosity—typically used in Ford vehicles
- **100**—Called medium to high viscosity—typically used in Chrysler vehicles
- **150**—Called ultra-high viscosity—typically used in General Motors vehicles

NOTE: Always check the under hood decal or service information for the exact oil and viscosity to use. ● SEE FIGURE 29–13.

HYGROSCOPIC It is important to keep oil containers closed so that the oil does not absorb water from the atmosphere. A mineral oil can absorb about 0.005% water by weight. PAG or ester oil is very **hygroscopic**, which means that it can absorb moisture directly from the air up to about 2% to 6%. Some synthetic oils undergo hydrolysis if exposed to too much water and revert back to their original components: acid and alcohol.

OIL FLOW IN THE SYSTEM When a system operates, oil is either absorbed or pushed by the refrigerant and moves through the system. It does not stay in any particular component, but a certain amount of oil can be expected in each component. Oil can also move while a system is shut off because of temperature changes of its various parts.

FIGURE 29–13 This under hood decal gives a part number (A 001 989 08 03) instead of the type and viscosity of refrigerant oil. Service information was needed to determine that it was PAG 46.

SUMMARY

1. Refrigerants escaping into the atmosphere can have detrimental effects on the ozone layer and also increase climate change and global warming.
2. The Clean Air Act places requirements for technicians to follow when servicing mobile A/C systems.
3. The SNAP rule limits which refrigerants can be used in a system.
4. Safety precautions should be followed when handling refrigerants.
5. Refrigerant oils are available in different viscosities and the specified oil must be used when servicing an air-conditioning system.

REVIEW QUESTIONS

1. What do the colors of the containers mean?
2. How can a refrigerant cause the breakup or depletion of the ozone layer?
3. What does the Clean Air Act mandate when it comes to servicing air-conditioning systems?
4. What is a SNAP refrigerant?
5. What precautions are necessary when working with refrigerants?
6. What does hygroscopic mean when referring to refrigerant oils?

CHAPTER QUIZ

1. R-134a is in what color container?
   a. Red
   b. Light blue
   c. White
   d. Green
2. Where is the ozone layer?
   a. Near ground level
   b. In the stratosphere
   c. In the clouds
   d. Between the ground and the clouds
3. The lower the GWP number, the ______________.
   a. Lower the global warming potential
   b. Higher the global warming potential
   c. Closer it is to the ground
   d. The more poisonous
4. What section of the Clean Air Act concerns work on air-conditioning systems in vehicles?
   a. 847
   b. 609
   c. 777
   d. 103a

5. Which refrigerant has the lowest Global warming potential (GWP)?
   a. R-744
   b. R-134a
   c. R-1234yf
   d. R-12
6. What is true about a SNAP refrigerant?
   a. Requires unique set of fittings, and all ports not converted must be permanently disabled.
   b. Each refrigerant must have a label with a unique color that specifies pertinent information.
   c. All original refrigerants must be removed before charging with the new refrigerant.
   d. All of the above
7. The letter "a" in R-134a means ____________?
   a. Automatic
   b. Automotive
   c. Asymmetrical
   d. Atmosphere
8. Precautions when working with refrigerants include ____________.
   a. Wear safety goggles or a clear face shield and protective clothing (gloves) when working with refrigerants
   b. Always be in a well-ventilated shop area when working with refrigerants and avoid small, enclosed areas
   c. When refrigerant containers are filled, reserve room is left for expansion and the container is marked with its critical temperature
   d. All of the above
9. What type of refrigerant oil is used in R-134a systems?
   a. POE
   b. PAG
   c. Mineral
   d. Either a or b depending on the vehicle
10. Why is it important to keep refrigerant oil containers closed?
    a. To keep dirt out
    b. To keep it from evaporating
    c. To keep it from absorbing moisture from the air
    d. It can leak out by traveling upward through an open container

chapter 30

A/C SYSTEM COMPONENTS, OPERATION, AND SERVICE

LEARNING OBJECTIVES

After studying this chapter, the reader will be able to:

1. Discuss the purpose and function of compressors and condensers.
2. Describe the operation of thermal expansion valves.
3. Explain the construction and operation of orifice tubes.
4. Explain the purpose and function of evaporators and accumulators.
5. Discuss the use of lines and hoses in refrigeration.
6. Describe electrical switches and evaporator temperature controls used in A/C systems.

Explain component replacement procedures.

This chapter will help you prepare for the ASE Heating and Air Conditioning (A7) certification test content area "B" (Refrigeration System Diagnosis, and Repair).

KEY TERMS

Barrier hoses 490
Captive O-ring 493
Discharge line 490
H-block 483
Liquid line 490
Sight glass 490
Suction line 490
Thermistor 495
Transducer 495

GM STC OBJECTIVES

GM Service Technical College topics covered in this chapter are as follows:

1. The purpose and operation of an evaporator (01510.08W).
2. The purpose, types, controlling components, and operation of a compressor.
3. The purpose and operation of a condenser.
4. The purpose and types of fans.
5. The purpose, types, and operation of refrigerant metering devices.
6. The purpose, types, and operation of storage-dehydrators.
7. The purpose and types of refrigerant hoses, lines, and oils.
8. Basic air-conditioning service procedures.

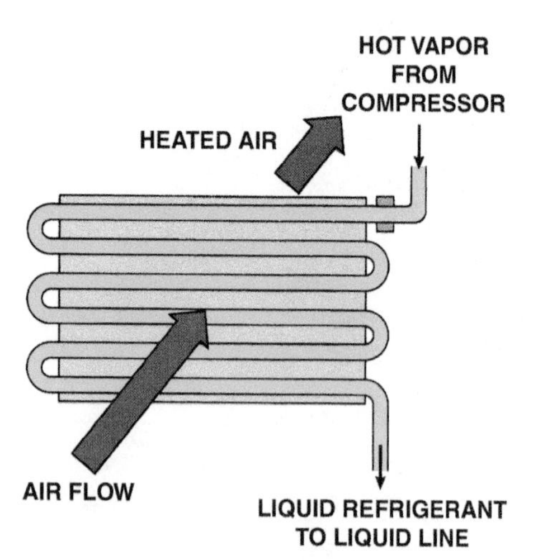

FIGURE 30–1 A condenser is a heat exchanger that transfers heat from the refrigerant to the air flowing through it.

FIGURE 30–2 Cooling fans can either be electric or engine driven. Some vehicles may have more than one fan. (Courtesy of General Motors)

BACKGROUND

Automotive air-conditioning components have been evolving steadily since the introduction of air conditioning (A/C) in vehicles in 1940. Air conditioning in the early days was a very expensive option available only in luxury cars. Today, A/C is standard in many different models and uses many different types of systems.

COMPRESSORS

The compressor is the pump in the system that circulates the refrigerant. The pressure must be increased until the refrigerant temperature is above ambient air temperature so the condenser can get rid of all the heat absorbed in the evaporator. See Chapter 28 for details on compressors.

CONDENSERS

PURPOSE AND FUNCTION The condenser is a heat exchanger that is used to get rid of the heat removed from the passenger compartment. The condenser cools the hot refrigerant vapors, which while passing through the condensing tubes condense into high pressure liquid. The condenser of most vehicles is mounted in front of the radiator into which warm air is forced through when the vehicle is moving forward. ● **SEE FIGURE 30–1**.

On most vehicles, the condenser is part of a *cooling module* that can combine the following:

- Condenser
- Radiator
- Automatic transmission fluid cooler
- Power steering pump oil coolers
- An intercooler for engine intake air on turbocharged engines

Air flow through the condenser is increased by the use of a cooling fan.

- Many rear-wheel-drive (RWD) vehicles use a fan driven by a fan clutch mounted on the water pump. Some RWD vehicles use both an engine-driven fan with fan clutch and an electric fan. ● **SEE FIGURE 30–2**.
- Front-wheel-drive (FWD) vehicles normally use an electric motor to drive the fan(s).

This motor is controlled by relays that are controlled by the powertrain control module (PCM) using temperature information from the engine coolant temperature (ECT) sensor.

CONDENSER CONSTRUCTION Automotive condensers are heat exchangers that are made in several designs, including the following:

- Older *tube-and-fin* condensers are merely a tube bent back and forth into a *serpentine shape* with fins attached. After the tubes are pressed through the fins, return bends or manifolds are used to connect the tubes to give the desired flow pattern.

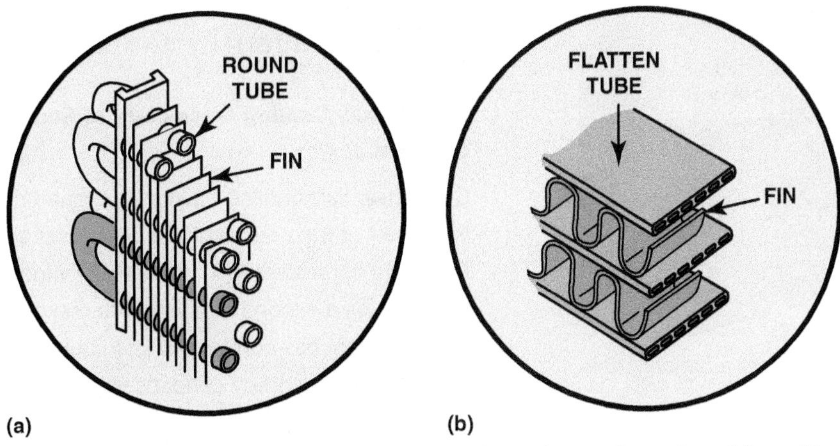

FIGURE 30–3 (a) A tube-and-fin condenser is made up of a series of fins with the round tubes passing through them. (b) An extruded tube condenser uses flat tubes with the fins attached between them. Flat tube condensers can use either parallel or serpentine flow.

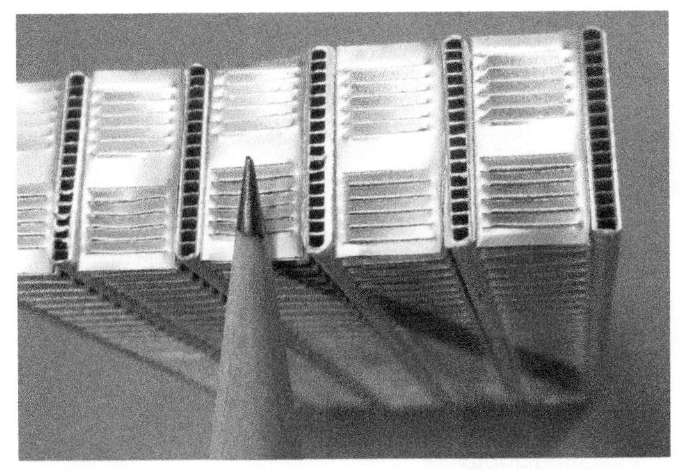

FIGURE 30–4 Notice the size of the passages in a condenser cross section compared to the point of a pencil. The passages are the small holes running vertically and the pencil is pointing to the aluminum fins that are used to transfer the heat from the passages to the air.

- Newer condensers use a *flattened tube* with fins between the tubes. The tubing is formed in either *serpentine* or *parallel-flow* arrangement. **SEE FIGURE 30–3.**
- Many condensers use a flattened, extruded aluminum tube that is divided into small refrigerant passages, 13–16 passages in a tube that is 0.875 inch (24 mm) wide. **SEE FIGURE 30–4.**
- The numerous fins provide the large amount of contact area needed with the airstream. Newer flat-tube, parallel-flow condensers are more efficient and transfer heat much better and therefore allows for a smaller and lighter component.
- Condensers can be constructed using either a parallel flow or serpentine flow design. **SEE FIGURE 30–5.**

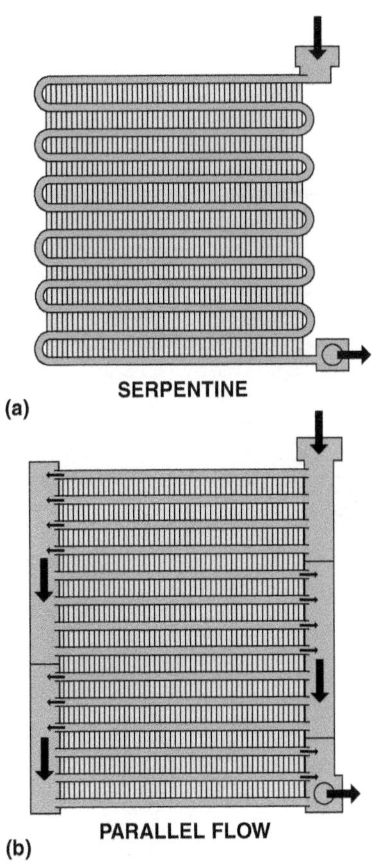

FIGURE 30–5 (a) The refrigerant follows a winding path through a serpentine condenser. (b) The refrigerant follows a back-and-forth path through a parallel-flow condenser.

As the gas condenses to a liquid, the volume is reduced because a gas has about 1,000 times the volume of the same liquid. As the latent heat of condensation is transferred to the airstream, the refrigerant vapor transforms into a liquid state. **SEE FIGURE 30–6.**

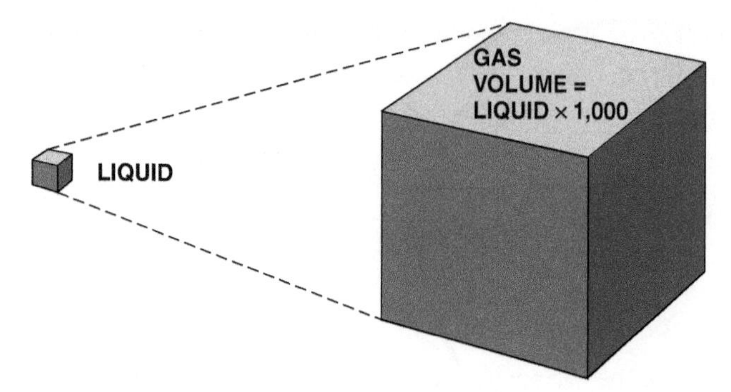

FIGURE 30–6 The volume of gas that enters a condenser is about 1,000 times the volume of liquid leaving it.

FREQUENTLY ASKED QUESTION

Why Is Sub-Cooling Necessary in Some Condensers?

Condenser sub-cooling makes sure that there is a liquid seal at the bottom of the condenser so the liquid line or receiver will not have any vapors. Some vehicles use a second condenser to make sure that the refrigerant has condensed into liquid before it exits. These secondary condensers can be called as:

- *Dual condenser*
- *Secondary condenser*
- *Sub-condenser*
- *Modulator.* ● **SEE FIGURE 30–7.**

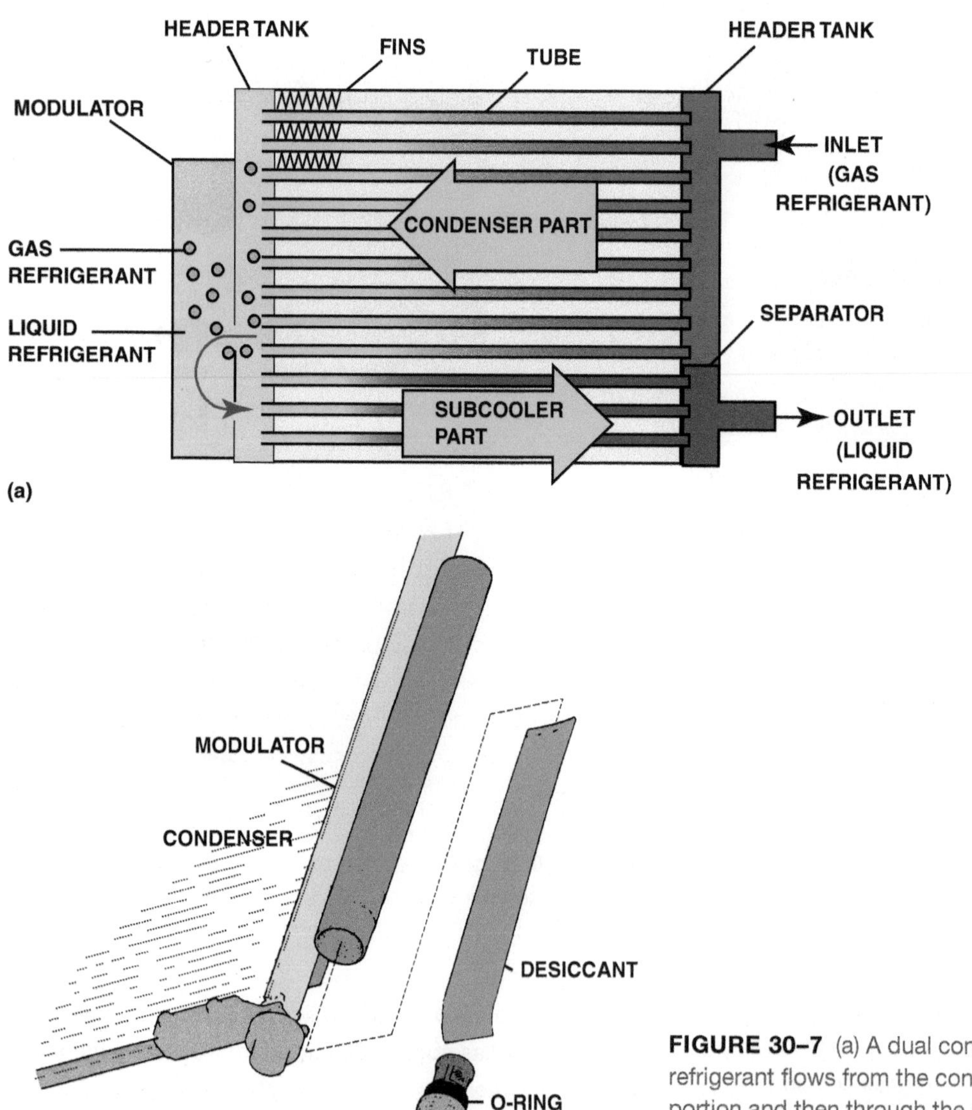

FIGURE 30–7 (a) A dual condenser is a condenser where the refrigerant flows from the condenser portion through the modulator portion and then through the sub-cooling part. (b) The modulator is built as part of the condenser and often includes a removable plug that allows desiccant replacement.

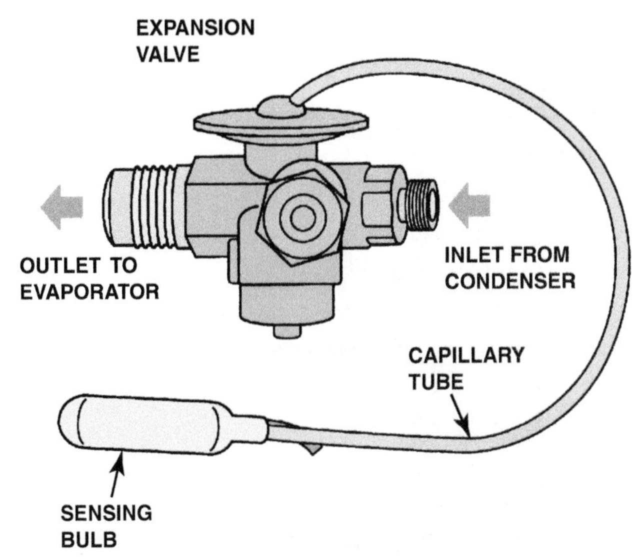

FIGURE 30–8 A typical expansion valve uses an inlet and outlet attachment for the evaporator, and a temperature-sensing bulb that is attached to the evaporator outlet tube.

THERMAL EXPANSION VALVES

PURPOSE AND FUNCTION A thermal expansion valve (TXV) senses both temperature and pressure and controls the flow of refrigerant into the evaporator. ● **SEE FIGURE 30–8.**

ADVANTAGES The advantages of a TXV system over an orifice tube system include:

1. The TXV can maintain a low superheat to ensure that the majority of the evaporator surface is being used, resulting in higher efficiency.
2. Requires a smaller refrigerant charge.

DISADVANTAGES The disadvantages of a TXV system over an orifice tube system include:

1. The system costs more to produce
2. There are more moving parts

CONSTRUCTION AND OPERATION In most TXV systems, the evaporator outlet is connected to the compressor inlet by the suction line with an internal diameter (ID) of about 5/8 inch or 3/4 inch (16 mm or 19 mm). The key to the operation of the expansion valve is the variable orifice. In these systems, the outlet from the high-pressure side to the low-pressure side is a variable-diameter hole. A pintle valve is a ball-and-seat valve used to increase or decrease the size of the opening. The expansion valve uses the pintle valve to control how rapidly refrigerant enters the evaporator. The expansion valve controls the refrigerant flow in response to the temperature of the evaporator outlet, measured by the remotely mounted sensing bulb and capillary tube.

- At room temperature, a TXV is open because the gas pressure in the capillary tube exerts a rather high pressure, greater than the spring pressure.
- As the system cools, the temperature of the sensing bulb and capillary tube drops, and the pressure on the diaphragm drops with it.
- When the pressure at the diaphragm drops below the spring pressure, the valve closes.
- If an excess amount of refrigerant enters the evaporator, high refrigerant pressure can act through the balance passage to close the TXV. ● **SEE FIGURE 30–9.**

INTERNALLY AND EXTERNALLY BALANCED TXVS There is also a passage that allows evaporator pressure to act on the both sides of the diaphragm in opposition to thermal bulb pressure.

- In an *internally balanced* TXV, this passage is open to evaporator inlet pressure.
- In an *externally balanced* TXV, this passage connects to a length of small tubing that connects to the evaporator tailpipe. Externally balanced valves are used on some larger evaporators to give better response. ● **SEE FIGURE 30–10.**

TYPES OF TXVS Thermal expansion valves come in many types. Most of the early valves threaded onto the evaporator inlet, and the liquid line threaded onto the valve. The thermal bulb, or end of the capillary tube, was clamped onto the evaporator tailpipe or inserted into a well in it. The thermal bulb has to be clamped tightly and should be well insulated so that it can transmit an accurate temperature signal. If used, the external equalizer line is threaded onto a fitting on the tailpipe. Most TXVs have a small, very fine screen at their inlet. This screen traps debris that can plug the valve, and it can be removed for cleaning or installing a replacement.

H-BLOCKS The thermal expansion block valve is another type of TXV. It is called an **H-block** or an *H-valve.* The H-block design uses four passages and controls the refrigerant flow using opposing pressures. These valves are connected with both the liquid and suction lines and the evaporator inlet and

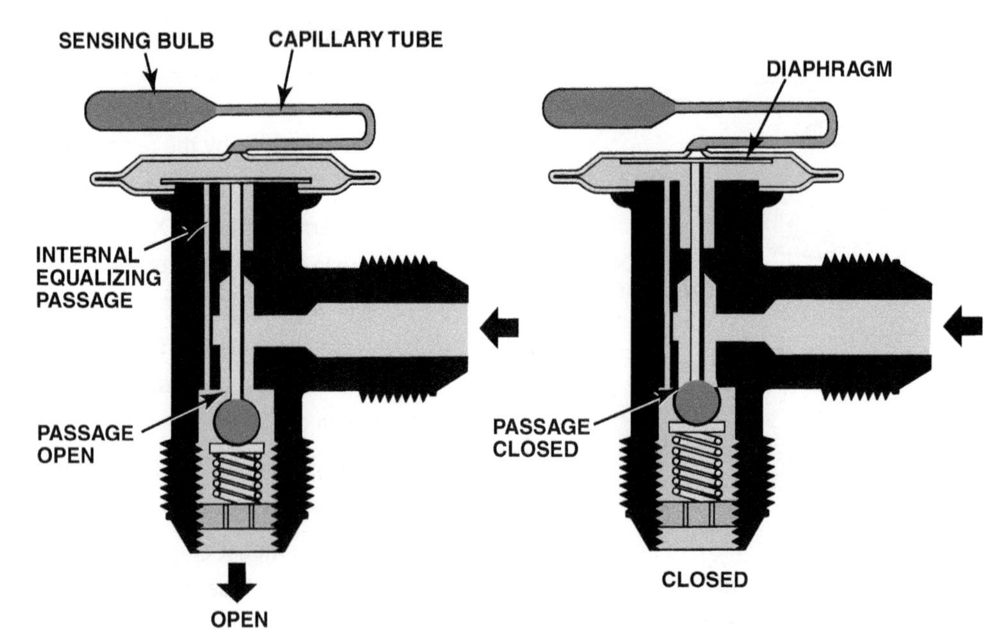

FIGURE 30–9 Pressure from the capillary tube pushes on the spring-loaded diaphragm to open the expansion valve. As the pressure in the capillary tube contracts, the reduced pressure on the diaphragm allows the valve to close.

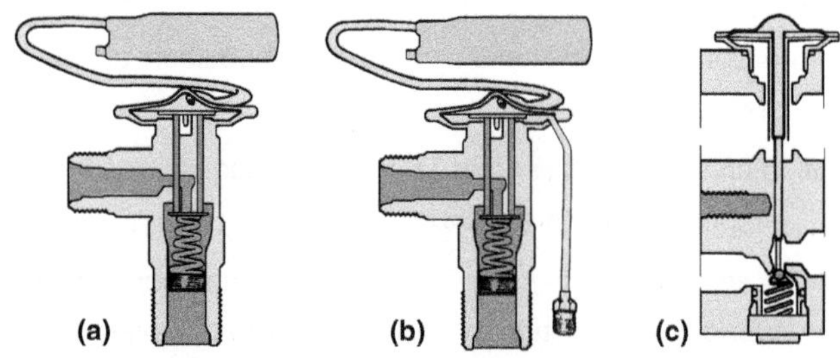

FIGURE 30–10 (a) An internal equalized TXV has two large connectors for the liquid line and evaporator. (b) An external equalized TXV has an additional smaller line to connect to the evaporator outlet. (c) Block-type TXV has four openings that connect to the evaporator, liquid line, and suction line.

FREQUENTLY ASKED QUESTION

How Is the Compressor Kept Lubricated?

The expansion valve opens and closes to control the refrigerant flow in response to the temperature of the evaporator outlet. To make sure that the compressor receives the oil it needs for lubrication when the valve is closed, a slot is cut in the ball seat inside the expansion valve to permit a small amount of refrigerant and oil to pass through to the evaporator and then to the compressor. ● **SEE FIGURE 30–11.**

outlet. Some block valves use threaded fittings, and some are bolted between manifolds and are sealed using O-rings. ● **SEE FIGURE 30–12.**

ORIFICE TUBE SYSTEMS

TERMINOLOGY An orifice tube (OT), also called an *expansion tube,* is a fixed-diameter orifice that the refrigerant must flow through. The diameter varies between systems and is about 1/16 inch (0.065 inch or 1.6 mm). ● **SEE FIGURE 30–13.**

ADVANTAGES The advantages of an orifice tube system over a TXV system include:

1. An orifice tube system is much simpler and cheaper to produce than a TXV system.
2. In an orifice tube system, the only moving part in the system is the A/C compressor.

DISADVANTAGES The disadvantages of an orifice tube system over a TXV system include:

1. An orifice tube system cannot respond to evaporator temperature. At times of low cooling loads, the orifice tube allows too much refrigerant to flow and floods the evaporator with liquid.
2. Another disadvantage is that an orifice tube system requires more refrigerant than a TXV system.

CONSTRUCTION AND OPERATION Most orifice tubes are a thin brass tube that is a couple of inches long and has a plastic filter screen around it. This tube is sized to flow the

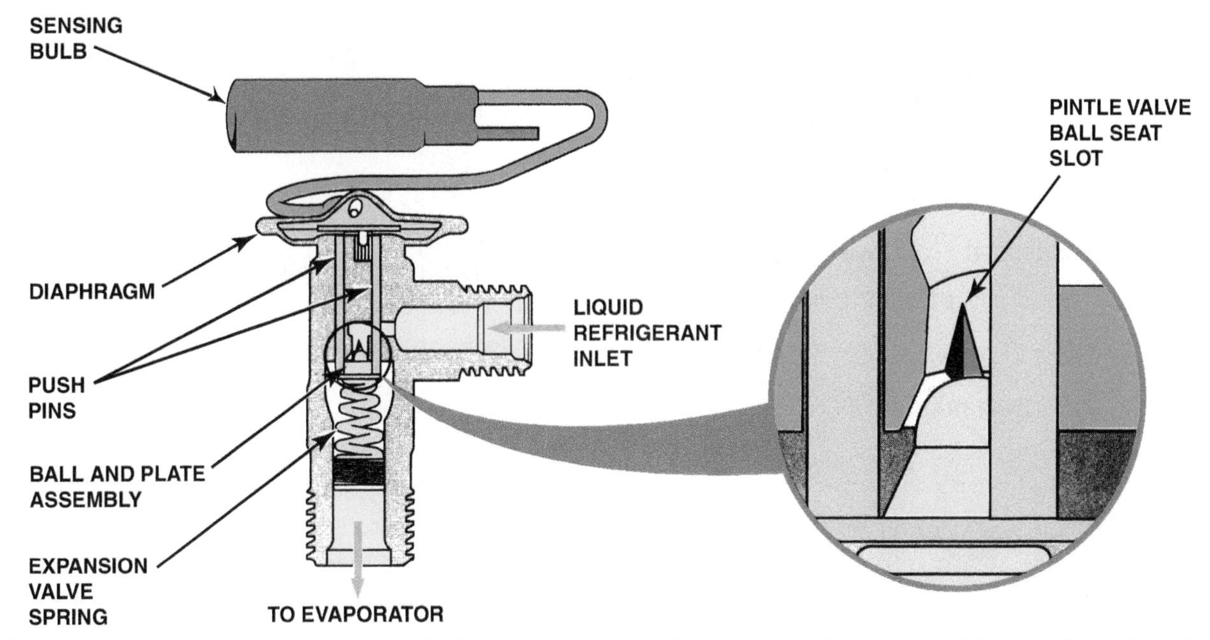

FIGURE 30–11 A slot cut in the ball seat inside the expansion valve permits refrigerant and oil to pass through at all times, even when the valve is closed.

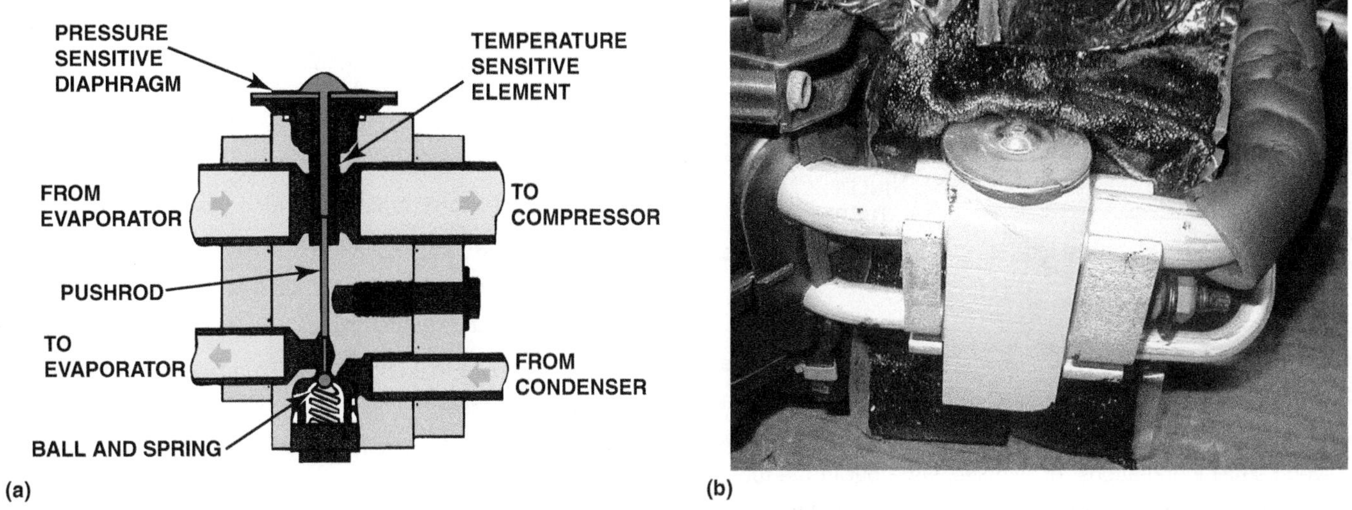

FIGURE 30–12 (a) An H-valve (H-block) combines the temperature sensing and pressure-regulating functions into a single assembly. (b) An H-valve removed from the vehicle for a better view.

proper amount of refrigerant into the evaporator for maximum cooling loads. Some orifice tubes use a filter made up of many small plastic beads in place of the screen. The orifice tube floods the evaporator during light cooling loads, so a low-side accumulator is always used with an orifice. The flow through an orifice tube is also affected by pressure, and excessive high-side pressure can cause evaporator pressure and temperature to become too high.

ORIFICE TUBE SIZES Manufacturers color-code orifice tubes to identify the car make and model for which a tube is made. There are at least eight different sizes of orifice tubes that have similar appearance, but the size ranges from 0.047 inch to 0.072 inch (1.19 mm to 1.83 mm). An R-134a system uses a larger orifice tube than a similar R-12 system.

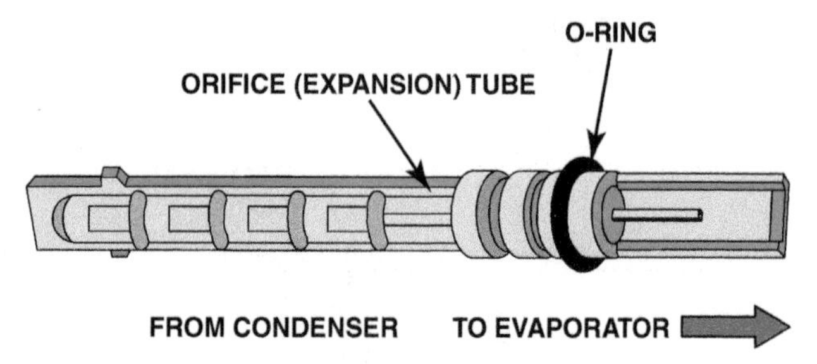

FIGURE 30–13 A typical orifice tube. The refrigerant flow is from the left toward the right.

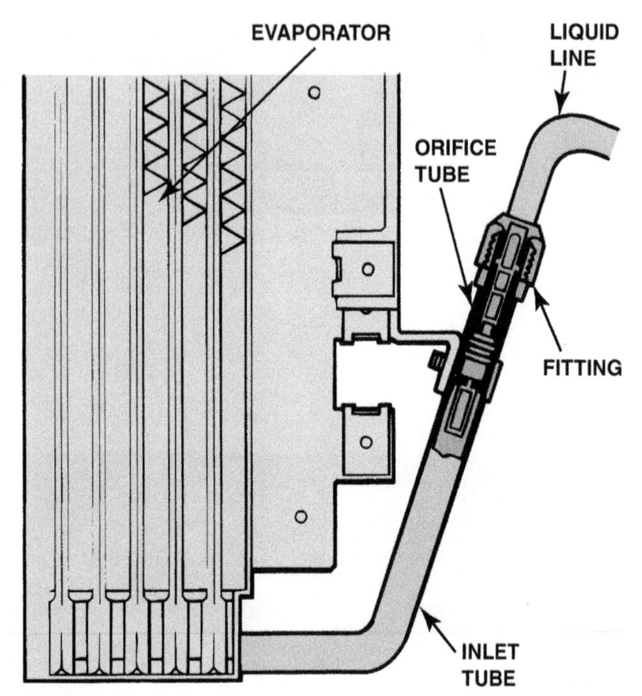

FIGURE 30–14 The orifice tube is usually located at the inlet tube to the evaporator.

ORIFICE TUBE LOCATION Some orifice tubes are placed into the evaporator inlet tube. An O-ring was fitted around it to stop refrigerant from flowing past the outside of the orifice tube. Several small indentations, or dimples, were put in the tubing wall to keep the orifice tube from moving too far into the evaporator. ● **SEE FIGURE 30–14.**

Many newer vehicles place the orifice tube in the liquid line farther away from the evaporator, close to the condenser outlet. This is done because of complaints of hissing noises that occur after the vehicle is shut off and high-side pressure bleeds down through the orifice tube.

VARIABLE ORIFICE VALVES (VOVS) Most orifice tubes have a fixed-size orifice that is sized for the proper refrigerant flow to produce maximum A/C performance during 50-MPH to 60-MPH (80 km/h to 100 km/h) operation. At lower engine speeds, the orifice size is too large. The orifice size should produce the high-side pressure for proper condenser action and the pressure drop into the low side to produce proper evaporator pressure and temperature. Vehicles that spend considerable time idling commonly have poor A/C performance and experience short compressor life.

TECH TIP

How to Locate the Orifice Tube

In an operating system, the orifice tube position can be located by finding the point where the liquid line changes temperature from hot to cold. If the system is not operating, the dimples in the line show the orifice tube location.

One type uses a variable orifice valve (VOV) that senses temperature to change the size of the orifice. A bimetal coil spring senses the temperature of the liquid refrigerant, and when the temperature increases, the spring moves the variable port to a closed position. This increases the restriction at the VOV and reduces the flow to the evaporator. ● **SEE FIGURE 30–15.**

A VOV is similar to a fixed orifice tube but contains an internal valve. Installation of the VOV is simply a matter of removing the orifice tube and installing the VOV. Always follow the instructions for the variable orifice valve if installing it in place of a fixed orifice tube.

ELECTRONIC ORIFICE TUBE A few General Motors vehicles used an electronic orifice tube for a year or two. It uses a solenoid to change orifice diameter.

- It has an orifice diameter of 0.062 inch (1.6 mm) when the solenoid is off (de-energized).
- This increases to 0.080 inch (2.0 mm) when the solenoid is on (energized).

The solenoid is controlled by an ECM, or electronic control module, that uses primarily the following three inputs:

1. Vehicle speed (VS) sensor
2. Engine RPM
3. High-side pressure

Energizing the solenoid increases the orifice size to reduce high-side pressure. ● **SEE FIGURE 30–16.**

EVAPORATORS

PURPOSE AND FUNCTION An evaporator, sometimes called the *evaporator core*, is a heat exchanger. The purpose and function of the evaporator is to remove heat from the air

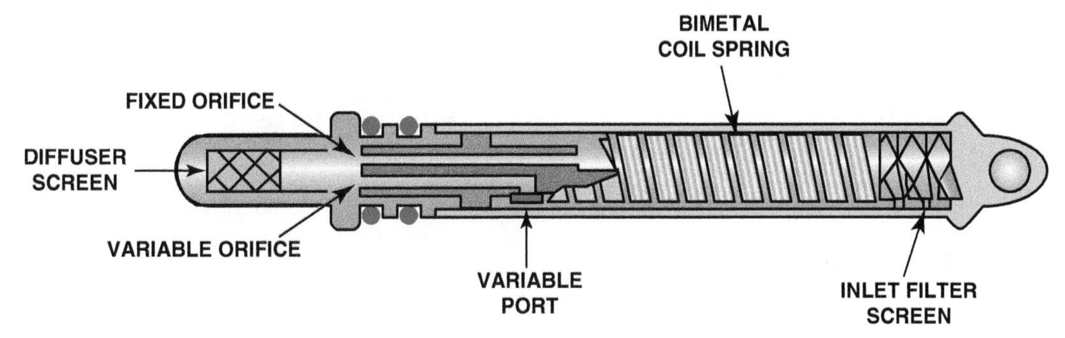

FIGURE 30–15 A VOV uses a bimetal coil spring to sense the temperature of the refrigerant. A higher temperature will cause the coil to expand and partially close the variable port to increase the restriction.

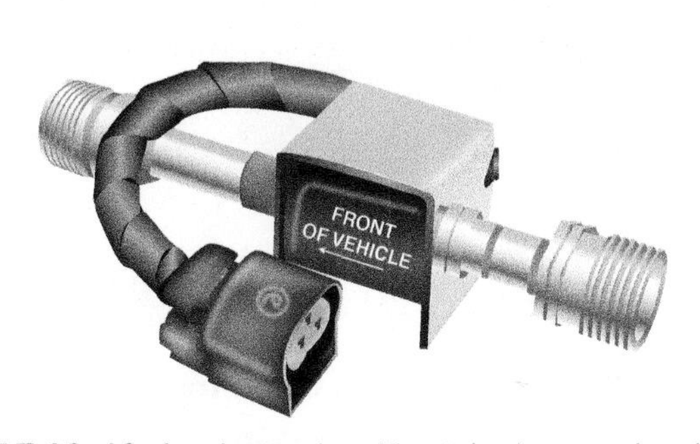

FIGURE 30–16 An electronic orifice tube has a solenoid so it can change orifice size.

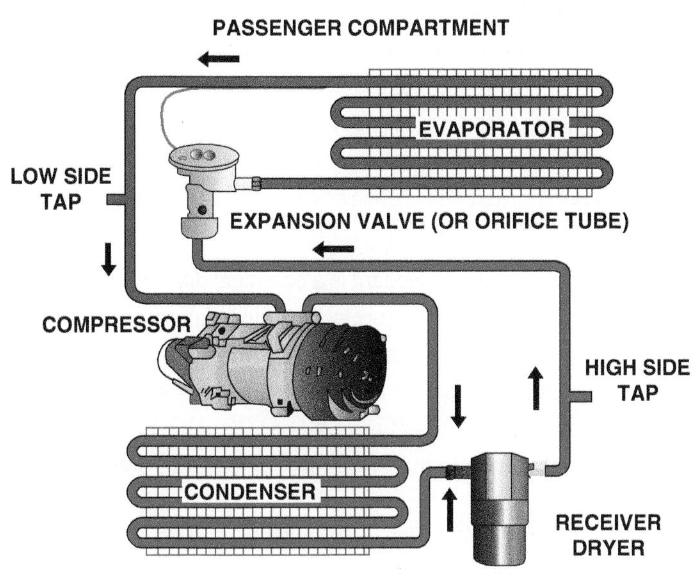

FIGURE 30–17 The evaporator is part of the low pressure side of the refrigeration cycle and is used to transfer the heat from inside the vehicle to the refrigerant flowing through the internal tubes.

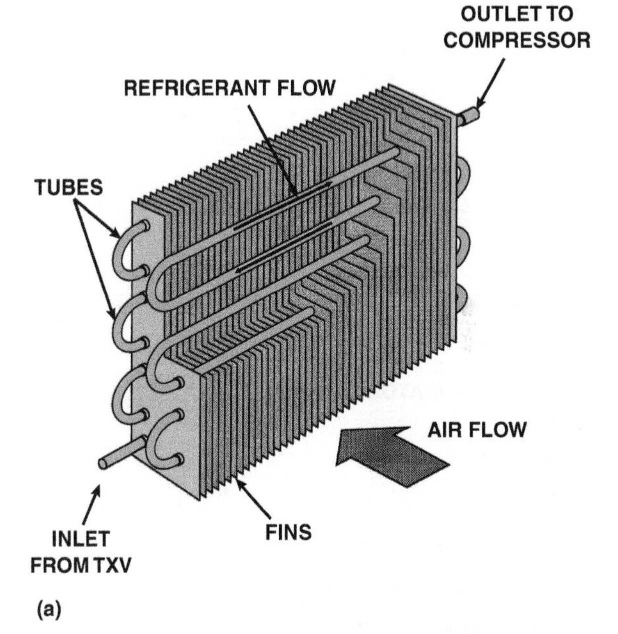

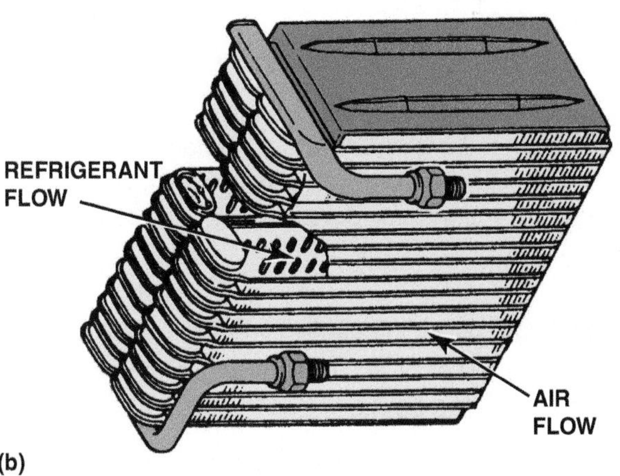

FIGURE 30–18 (a) An older design tube-and-fin evaporator. (b) A plate evaporator. Each type has a large contact area for heat to leave the air and enter the refrigerant.

being forced through it to cool the inside of the vehicle. **SEE FIGURE 30–17.**

CONSTRUCTION Most evaporators are a series of plates sandwiched together to form both the refrigerant and air passages. **SEE FIGURE 30–18.**

Evaporators are normally made from aluminum because it has good thermal properties and is lightweight. Evaporators have at least two line connections:

1. The smaller inlet line (liquid line) connects to the TXV or orifice tube.

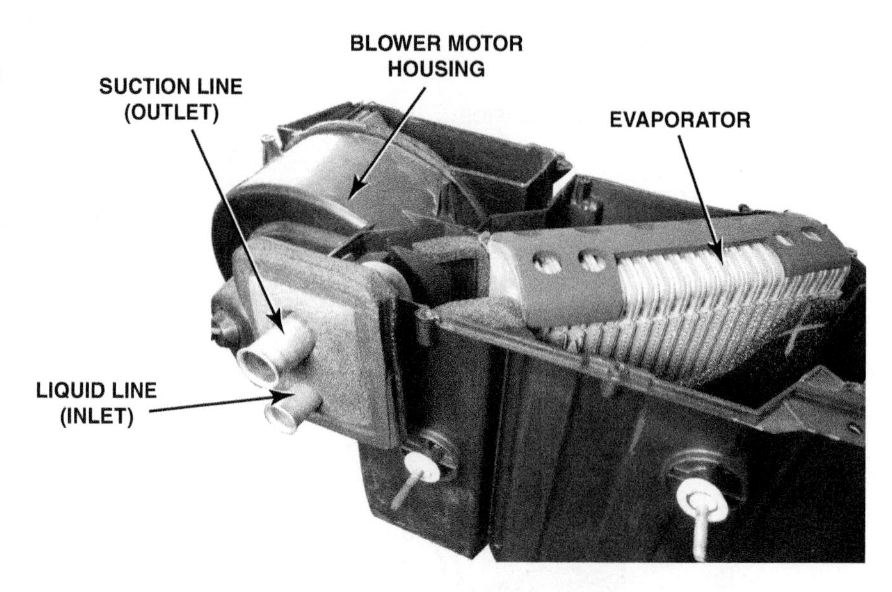

FIGURE 30–19 An evaporator in the HVAC case showing the liquid line inlet tube from the TXV or orifice tube and suction outlet line to the compressor or accumulator.

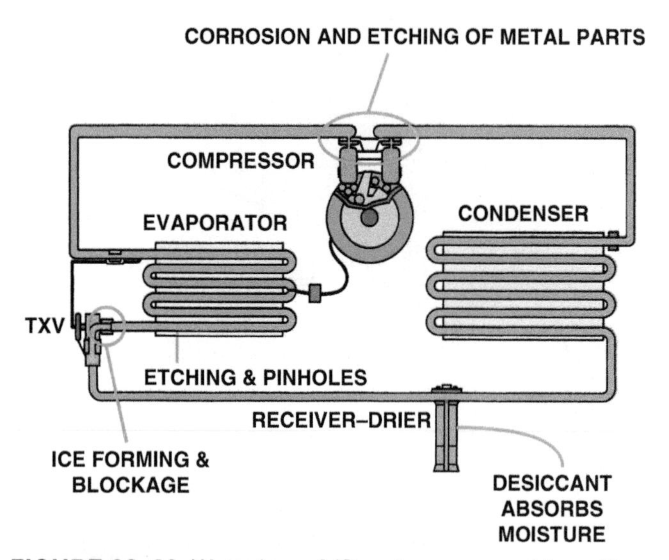

FIGURE 30–20 Water in an A/C system can combine with the refrigerant to form acid. These acids can etch and dissolve components, causing corrosion of metal parts, and ice blockage at the expansion device.

2. The larger outlet line connects to the suction line and to the compressor or accumulator on orifice tube systems. ● **SEE FIGURE 30–19.**

RECEIVER–DRIERS AND ACCUMULATORS

PURPOSE AND FUNCTION The purpose of refrigerant storage is to compensate for volume changes due to temperature change or refrigerant loss. Desiccant is needed to remove moisture or water, which can cause rusting or corrosion.

> **? FREQUENTLY ASKED QUESTION**
>
> **Where Does the Moisture Come from Inside a Sealed Air-Conditioning System?**
>
> Small amounts of moisture may remain after sealing the system and the desiccant is there to absorb this left-over moisture. Also, during operation, the pressures in the system change and moisture and air can be drawn into the system through microscopic openings in the rubber hoses, compressor shaft seal, Schrader valves, and O-rings.

DESICCANT The desiccant is a chemical drying agent called "molecular sieve" and its job is to remove all traces of water vapor from a system. Water can mix with refrigerant to form acids, which cause rust and corrosion of metal parts. Water can also freeze and form ice at the TXV or OT, which can block the flow of refrigerant into the evaporator. ● **SEE FIGURE 30–20.**

ACCUMULATORS The accumulator serves three major functions.

1. Prevents liquid refrigerant from passing to the compressor
2. Holds the desiccant, which helps remove moisture from the system
3. Holds a reserve of refrigerant

An accumulator is a container that holds about 1 quart (1 L) or less in volume. The inlet line from the evaporator enters near the top of the accumulator and then drops downward to the bottom, and then exits the accumulator at the top. This routing of the outlet tube separates the refrigerant vapor at the top from the liquid at the bottom so that only vapor will leave

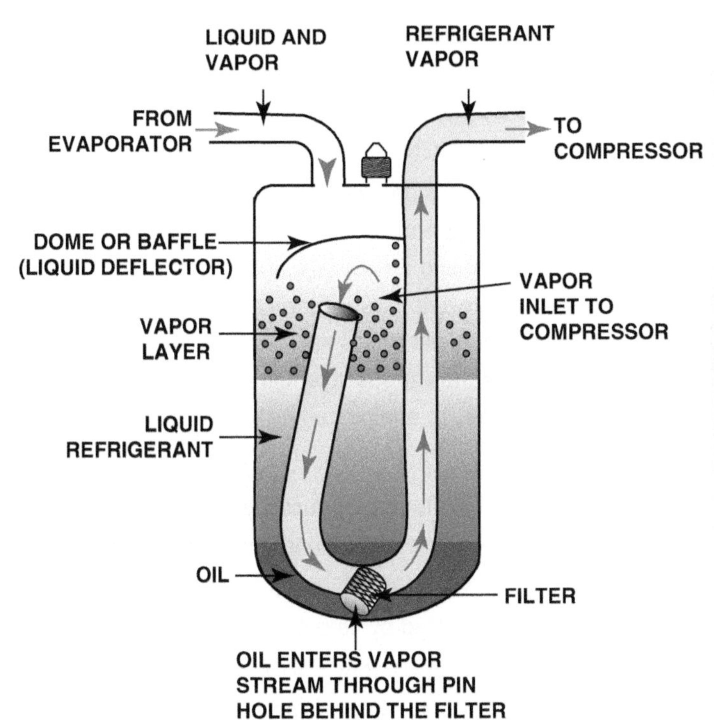

FIGURE 30–21 Accumulators are designed so that vapors from the top move to the compressor. They contain desiccant to absorb water from the refrigerant and many include a fitting for low-side pressure and the clutch cycling switch.

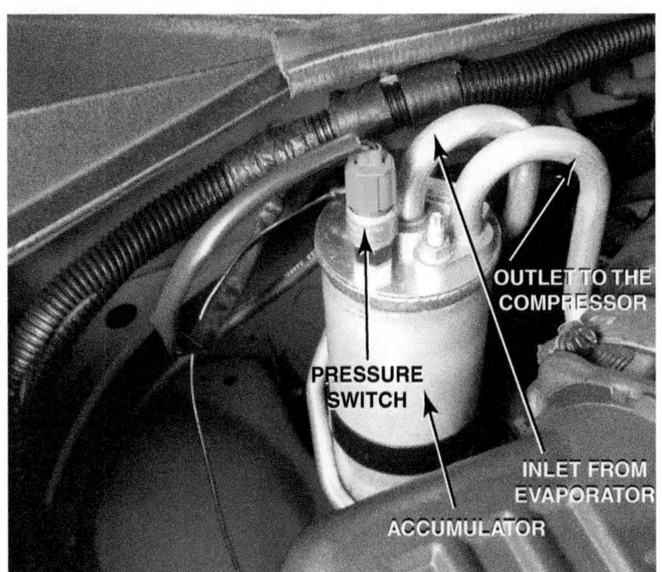

FIGURE 30–22 This accumulator has an accumulator tube that connects the evaporator outlet and a suction line that connects to the compressor inlet.

the accumulator. A small opening for oil bleed is at the lowermost point and this opening usually has a filter so debris will not block it. A small amount of liquid refrigerant and oil also leaves through the bleed hole. The oil ensures that the compressor is lubricated. ● **SEE FIGURE 30–21**.

The accumulator of some R-134a systems has an insulating jacket to help reduce the heat absorption and lower the air temperature at the discharge ducts. The accumulator or receiver–drier is normally replaced if a system has been opened to the atmosphere for a period of time because the desiccant is probably saturated with moisture. It is also standard practice to replace the accumulator or receiver–drier whenever major service work is done on a system, especially if the compressor is replaced.

The accumulator inlet on many vehicles is connected directly to the evaporator, whereas on other vehicles it is mounted separately with a metal tube or rubber hose to connect it to the evaporator. The outlet of the accumulator is connected to the compressor inlet through the flexible suction line. ● **SEE FIGURE 30–22**.

FREQUENTLY ASKED QUESTIONS

Are All Desiccants the Same?

No. Desiccant is usually contained in a cloth bag inside the accumulator or receiver–drier. A desiccant variety referred to as XH-5 has commonly been used with R-12 in automotive systems. XH-5 has the ability to absorb about 1% of its weight in water, but this chemical suffers damage when it absorbs fluorine from R-134a, and it begins to decompose when it comes into contact with R-134a and PAG oil. Other desiccant types (XH-7 and XH-9) are compatible with R-134a and PAG. The receiver–drier or accumulator used in an R-134a system should use XH-7 or XH-9 desiccant. All new accumulators and receiver–driers contain either XH-7 or XH-9 desiccant. An accumulator or receiver–drier should contain the desiccant type and amount to suit the A/C system that it is designed for and is therefore required to be replaced when retrofitting a system from R-12 to R-134a.

RECEIVER–DRIER The receiver–drier is used with TXV systems and is normally found in the high-side liquid line, somewhere between the condenser and the TXV. The receiver–drier inlet directs incoming refrigerant into the container so that the refrigerant passes through a filter pad and the desiccant. The outlet, often called a pickup tube, begins near the bottom and usually exits at the top. A fine mesh filter screen is used at the inner opening to stop any debris from passing out of the receiver–drier and on to the TXV. The receiver–drier is usually about half full of liquid refrigerant during system operation. ● **SEE FIGURE 30–23**.

A receiver–drier can also include a pressure relief plug, and/or switch. The relief plug releases excess pressure and the pressure switch is used to sense high-side pressure, preventing compressor operation if the pressure is either too low or too high.

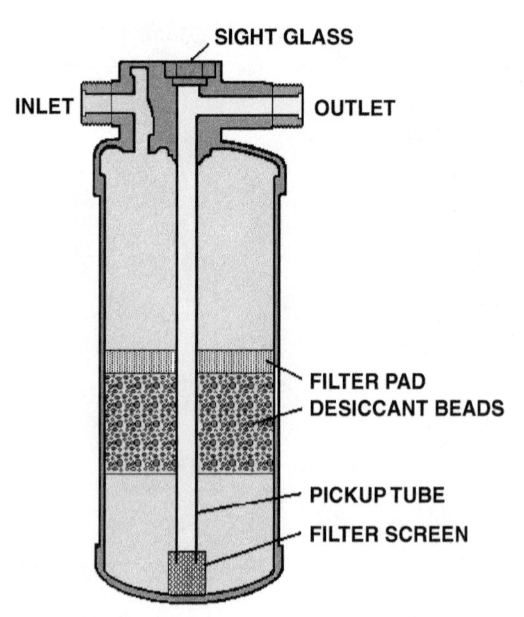

FIGURE 30–23 A cutaway view of a receiver–drier showing the filter pads and desiccant.

? FREQUENTLY ASKED QUESTIONS

What is a sight glass?

On R12 systems the receiver–drier can contain a sight glass which allows observation of the refrigerant flow as it leaves the receiver–drier. In a properly operating system he flow of R-12 Is normally invisible (except on very hot days) because R-12 is clear, and the oil is dissolved in the liquid refrigerant. Bubbles or foam in the sight glass can indicate abnormal operation. R-134a appears cloudy in a sight glass, so sight glasses are not used in R-134a systems.

LINES AND HOSES

PURPOSE AND FUNCTION The various system components must be interconnected so that refrigerant can circulate through the system. The components are connected using hoses and tubing (also called pipes).

HOSE CONSTRUCTION Both flexible rubber and rigid metal tubing are used to link the components. The connections to the compressor must be flexible to allow for engine and compressor movement. ● **SEE FIGURE 30–24.**

- Early R-12 hoses were made from rubber with one or two layers of braided reinforcing material. Over time refrigerants could permeate most of these flexible hose materials and slowly escape from the system. This design hose worked with R-12, but R-134a (smaller molecule) would permeate (leak through) R-12 hoses.
- Refrigerant hoses designed for R-134a are made with one or two nonpermeable inner layers with internal reinforcement and an outer layer for protection. The nonpermeable nylon layer forms a leak-proof barrier which is why these hoses are called **barrier hoses**. The materials for the various layers are developed to keep refrigerant loss to a minimum.

METAL LINES Metal tubing (usually aluminum) is used in many systems to connect stationary components, such as a condenser, to the receiver–drier or OT.

- Metal tubing is sized by its outside diameter (OD).
- Pipe and hose are sized by the inside diameter (ID) and these sizes are often nominal (approximate) sizes.

A number sizing is often used for refrigerant hose and fittings, with the most popular sizes being #6, #8, #10, and #12—

- #6 equals 5/16 inch (7.9 mm),
- #8 equals 13/32 inch (10.3 mm),
- #10 equals 1/2 inch (12.7 mm), and
- #12 equals 5/8 inch (15.9 mm).

Although metal aluminum tubing does not have permeation problems, corrosion caused by battery spillage or water can create holes in the tubing and produce leakage.

The lines in a system are named for their function or what they contain. Starting at the compressor, the **discharge line**, sometimes called the *hot gas line,* connects the compressor to the condenser inlet. The **liquid line** connects the condenser outlet to the receiver–drier and TXV or OT. A TXV system can have two liquid lines, one on each side of the receiver–drier. The **suction line** connects the evaporator outlet to the accumulator or compressor and has the largest diameter because it transfers a low-pressure vapor. ● **SEE FIGURE 30–25.**

HOSE SIZES

- The suction line has an ID of 1/2 inch or 5/8 inch (12.7 mm to 15.9 mm) (a #10 or #12 hose). The liquid line has the smallest diameter, usually an ID of 5/16 inch (7.9 mm) (#6 hose).
- The discharge line has an ID of 13/32 inch or 1/2 inch (10.3 mm or 12.7 mm) (#8 or # 10 hose). Metric sizes are also used. ● **SEE FIGURE 30–26.**

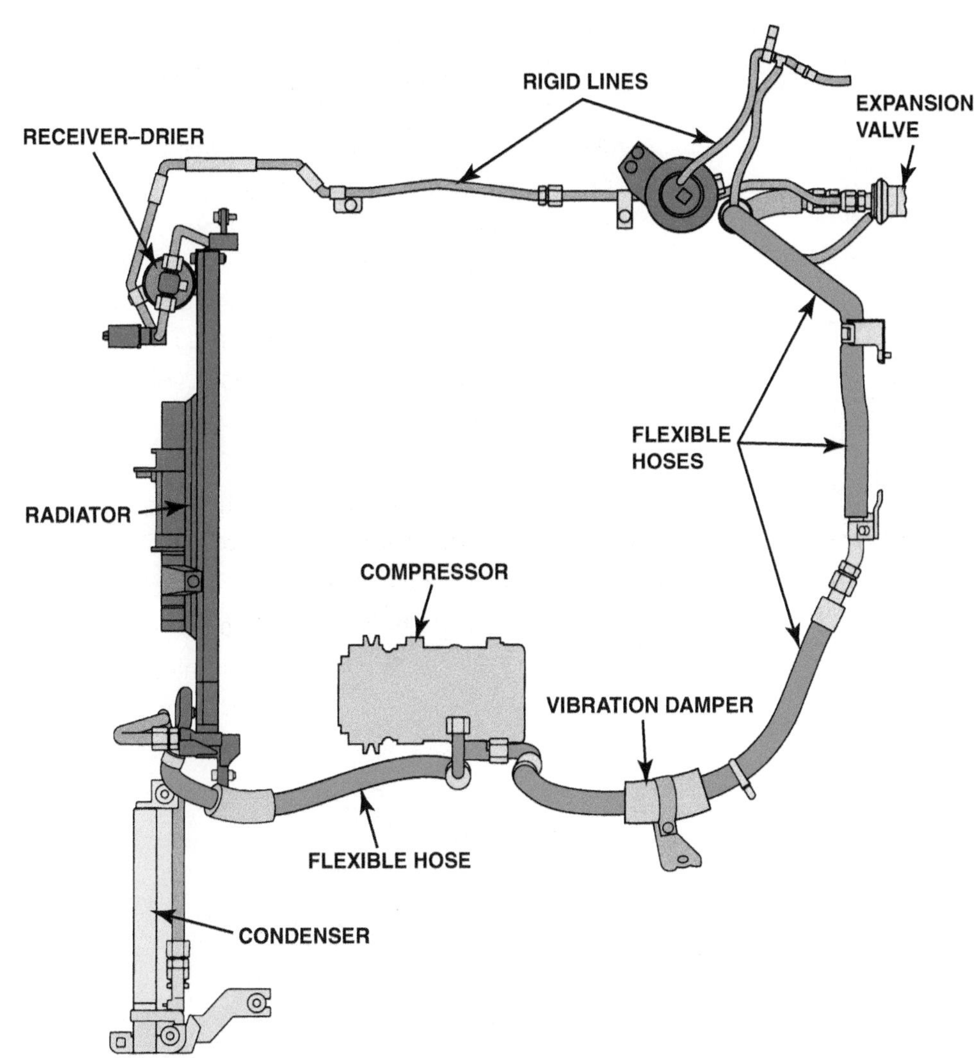

Parts shown are not in their normal position and some parts are missing, but it does show the locations of the rigid and flexible refrigerant lines.

FIGURE 30–24 Rigid lines and flexible hoses are used throughout the air-conditioning system. The line to and from the compressor must be flexible because it is attached to the engine, which moves on its mounts during normal vehicle operation.

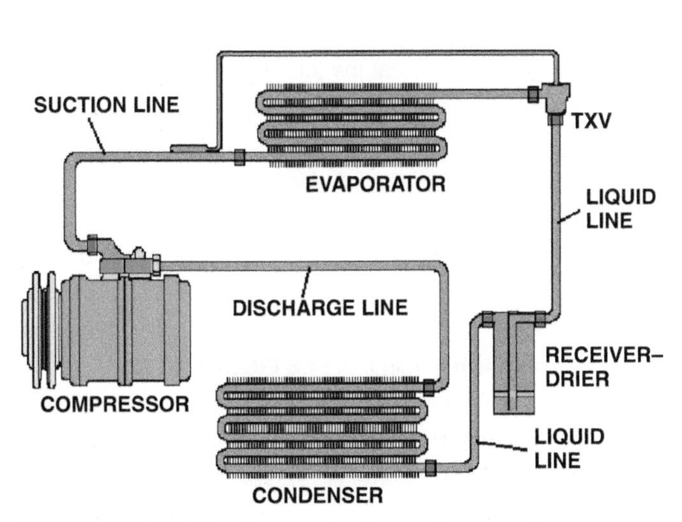

FIGURE 30–25 The three major hoses/lines are the discharge, liquid, and suction lines. A system can have two liquid lines.

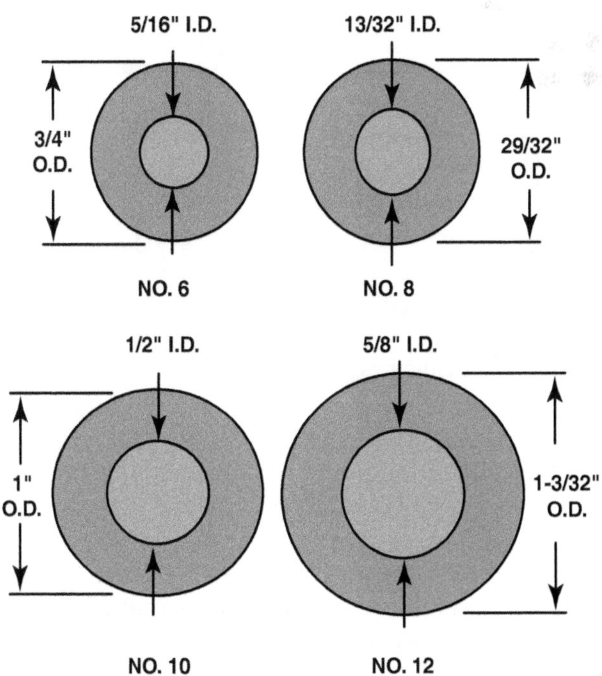

FIGURE 30–26 Most systems use three of these four refrigerant hose sizes. Metal tubing is sized by its outside diameter (OD) and hoses are sized by the inside diameter (ID).

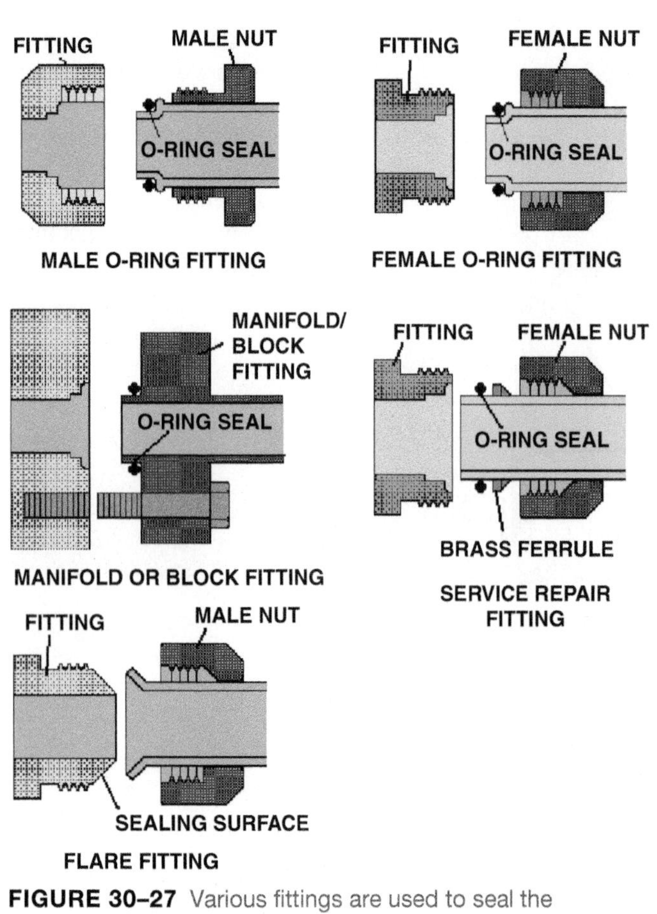

FIGURE 30–27 Various fittings are used to seal the refrigerant line connections. The service fitting is used for metal line repairs or to insert an inline filter.

SPRING LOCK COUPLER

GARTER SPRING

MALE FITTING

FEMALE FITTING

PLASTIC COUPLER

QUICK-CONNECT COUPLER

FIGURE 30–28 Spring-lock and quick-connect couplers are merely pushed together, and a garter spring or plastic cage holds them coupled.

CONNECTIONS The lines and hoses are connected to the major components using fittings of several different styles. ● **SEE FIGURE 30–27.**

These fittings allow the lines to be disconnected and are designed to keep refrigerant leakage to a minimum. Most new fittings use an O-ring seal that can be replaced during service. A variety of quick-connect couplers are used to hold two lines together and seal the joint between them. Some of these are held together by one of at least two styles of plastic couplers. One style was held together using a garter spring and was called a *spring-lock fitting.* ● **SEE FIGURES 30–28 AND 30–29.**

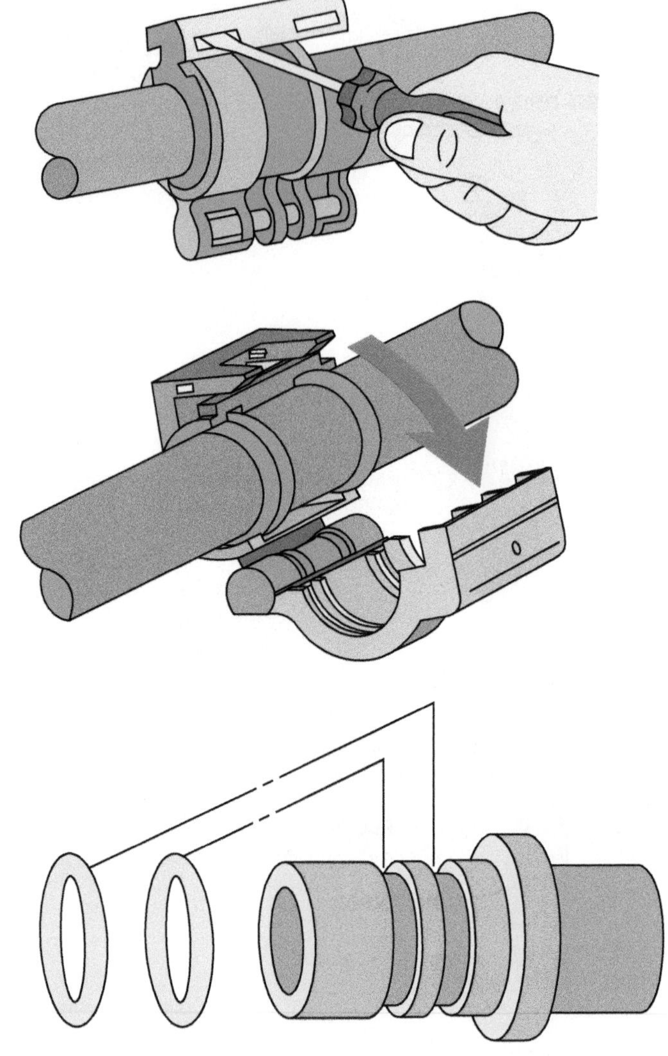

FIGURE 30–29 This plastic retainer snaps over retaining ridges on the metal A/C lines. (Courtesy of General Motors)

Spring-lock fittings are connected by pushing one line over the other until the garter spring moves into position. A special tool is inserted into the fitting to expand the garter spring and release the fitting. ● **SEE FIGURE 30–30.**

A spring-lock fitting is a type of quick-disconnect fitting. The female portion of this fitting has a flare-like ridge at the end and is gripped by a garter spring when connected. One or two O-rings form the seal between the two fitting parts. These O-rings must be resilient enough to compensate for slight movement between the two parts. ● **SEE FIGURE 30–31.**

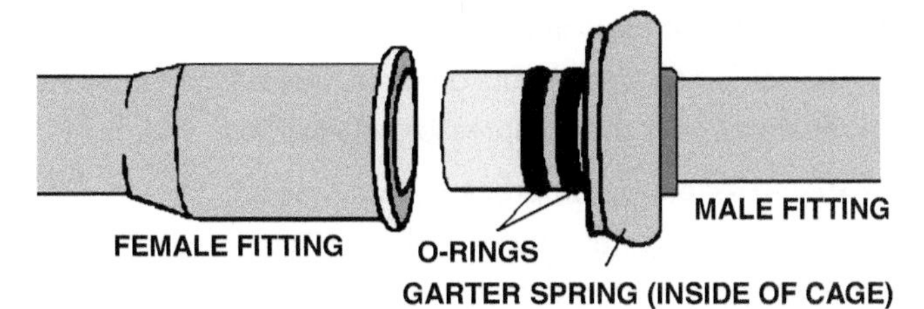

FIGURE 30–30 A spring-lock fitting is a type of quick disconnect fitting that is sealed by two O-rings and held together by a garter spring. A special tool is required to expand the garter spring to release the fitting.

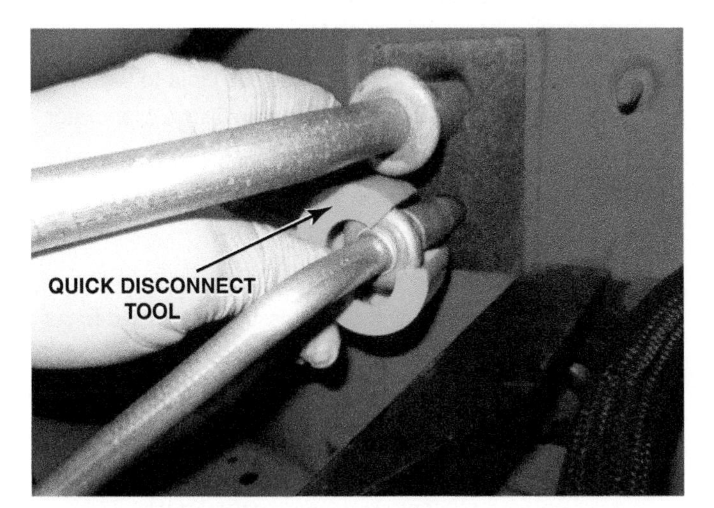

FIGURE 30–31 A quick disconnect tool being used to remove the connections at the evaporator.

O-RINGS O-ring fittings squeeze a rubber O-ring between the two parts being connected to make the seal. In most cases, the O-ring is slid over the end of the metal tube and is located by a raised metal ring or bead. Some manufacturers use a **captive O-ring**, also called *captured O-ring*, which locates the O-ring more positively in a shallow groove. Captive O-rings use a larger-diameter cross section than standard O-rings. Some designs use two O-rings for a better seal.

O-rings made from Chloroprene or Neoprene are is commonly used in R-12 systems, but these materials are not compatible with R-134a. High-grade nitrile O-rings are made from highly saturated nitrile (HSN or HNBR), often tinted blue or green, are used in R-134a systems. There is no color standard for O-rings but most manufacturers use the following color designations:

- Black indicates nitrile or neoprene.
- Blue indicates neoprene or nitrile.
- Green indicates HNBR and are generally used in R-134a systems. ● **SEE FIGURE 30–32.**

FIGURE 30–32 Having an assortment of O-rings makes it easier for the service technician to be able to use the correct ones during a repair.

ELECTRICAL SWITCHES AND EVAPORATOR TEMPERATURE CONTROLS

CONTROL SWITCHES Various electrical switches are used in A/C systems to prevent evaporator icing, protect the compressor, and control fan motor. Control switches can be located anywhere in the system, such as at the

- compressor discharge
- suction cavities
- receiver–drier
- accumulator

Some recent systems use a variable displacement compressor to prevent icing. A valve that senses evaporator pressure is used in the compressor, and compressor displacement is reduced in response to that pressure.

At one time, the A/C electrical circuit was rather simple.

- A typical circuit connects the evaporator blower motor and compressor clutch at one master switch.
- The power to the compressor clutch passes through a temperature switch that opens the circuit to cycle the clutch when the evaporator gets too cold.
- The power to the blower motor passes through a speed control switch so that the blower speed can be changed. ● **FIGURE 30–33.**

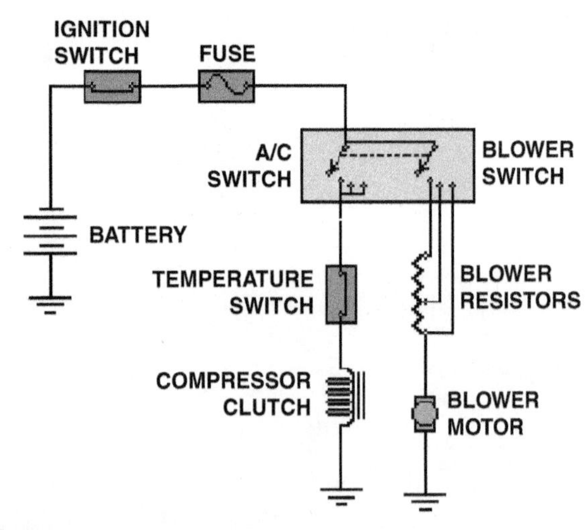

FIGURE 30–33 Many early A/C systems used a simple electrical circuit.

Today various switches, sensors, and relays are used. A sensor is usually an input to an electronic control module (ECM) or body control module (BCM). Many sensors provide a variable signal, so the ECM/BCM will know the actual temperature or pressure at a given point. A relay is essentially a magnetic switch that is controlled by another switch. A relay is used to control a greater amount of current than a switch can handle. Relays also allow computer control modules to control electrical circuits. Any one A/C system will have some of these, but not all of the following, depending on the type of system and manufacturer:

- **A/C clutch relay.** Controls current to compressor clutch and is used to turn the compressor on or off
- **Ambient sensor or switch.** Senses outside temperature and is designed to prevent compressor operation when ambient temperature is below a certain point, about 35°F to 40°F (2°C to 4°C)
- **Compressor high-pressure sensor or switch.** Mounted in compressor discharge cavity and senses high-side pressure and is used to cut out the compressor clutch if pressure is too high or too low, or provides a signal to another device if pressure is too high
- **Compressor low-pressure sensor or switch.** Mounted in compressor suction cavity, this switch or sensor detects low-side pressure and is used to cut out the compressor clutch if pressure is too low
- **Compressor RPM sensor.** Provides input to the ECM/BCM that the compressor is running. The ECM/BCM will cut out compressor if RPM is low indicating belt slippage or impending compressor lockup. (Used by some manufacturers)
- **Compressor superheat sensor or switch.** Function is similar to compressor low-pressure switch
- **Compressor high-temperature switch.** Mounted on the compressor and shuts off the compressor if the temperature of the compressor gets too hot
- **Power steering compressor cutoff switch.** Mounted at power steering gear, this switch senses pressure in that system and is used to stop the compressor when the pressure in the power steering rises to a high level which increases the load on the engine. The A/C compressor clutch is then disengaged to help remove some of the load from the engine so it does not stall
- **Engine coolant temperature (ECT).** Mounted near the engine thermostat and senses engine temperature for ECM/BCM and is also used to turn on cooling fan(s)
- **Evaporator pressure sensor.** Provides input to the ECM/BCM as to the operating pressure in the evaporator
- **Evaporator temperature sensor.** Mounted at the evaporator, it senses temperature and is used to cycle compressor clutch to prevent icing
- **High-pressure cutout switch.** Mounted at receiver–drier or liquid line, it senses high-side pressure and is used to cut out the compressor clutch if high-side pressure is too high
- **High-temperature cutoff sensor or switch.** Mounted at the condenser outlet, this switch or sensor measures condenser temperature and is used to cut out the compressor clutch if the temperature is too high
- **Low-pressure cutout sensor or switch.** Mounted in receiver–drier or liquid line, it senses low pressure and is used to cut out the compressor clutch if the refrigerant charge becomes too low.
- **Master switch.** Mounted at control head, this switch can be operated by the driver to turn the system on or off
- **Pressure cycling switch.** Mounted at accumulator, it senses low pressure and cycles the compressor clutch to prevent evaporator icing
- **Thermostatic cycling switch.** Mounted at evaporator, it senses air temperature and cycles the compressor clutch to prevent evaporator icing
- **Trinary pressure switch.** Mounted at receiver–drier, this switch senses high-side pressure and cuts out compressor clutch if pressure is too high or too low. It can also be used to control radiator shutters or fan motor
- **Blower relay.** Can be used to turn blower on or off and provide high blower speed
- **Clutch cutoff relay.** Can be used to interrupt compressor clutch

- **Condenser fan relay.** Used to turn condenser fan motor on or off
- **Radiator fan relay.** Used to turn radiator fan motor on or off

THERMOSTATIC CYCLING SWITCH Evaporator temperature in a cycling clutch system is sometimes controlled by either a thermostatic (thermal) switch or pressure switch. A bellows switch uses a capillary tube inserted into the evaporator fins. As with a TXV, the gas pressure in the capillary tube is exerted on the bellows. In a thermal switch, the bellows acts on a set of contact points.

- A warm evaporator produces a higher bellows pressure, which keeps the points closed.
- A cold evaporator reduces the pressure, which causes the points to open.

A bimetal switch has a contact arm that is laminated from two metals with very different thermal expansion rates.

- When the switch is warm, the metals expand to close the contact points.
- When the switch cools, the metals in the arm contract to open the contacts.

Both of these, switch styles are calibrated so they are closed at temperatures above 32°F (0°C) and open at temperatures below 32°F (0°C). When the switch opens, the compressor clutch is turned off. This causes the evaporator to begin warming, and after a few degrees of temperature increase, the switch closes to cycle the compressor in again. Recent systems use thermistors to sense temperature instead of thermostatic cycling switches.

PRESSURE CYCLING SWITCHES A pressure cycling switch is mounted to sense low-side pressure, usually on the accumulator. Ice begins to form when evaporator pressure drops below 30 PSI in an R-12 system or slightly less in an R-134a system.

- When evaporator pressure is above 30 PSI, the switch contacts close so the compressor will operate.
- When the pressure drops below 30 PSI, the switch opens and cycles the compressor out.
- After the compressor stops, evaporator pressure will increase and after an increase of about 10 to 20 PSI, the pressure switch contacts close to cycle the compressor in again. This switch also prevents the compressor from operating if the pressure is too low from a refrigerant loss. ● **SEE FIGURE 30–34.**

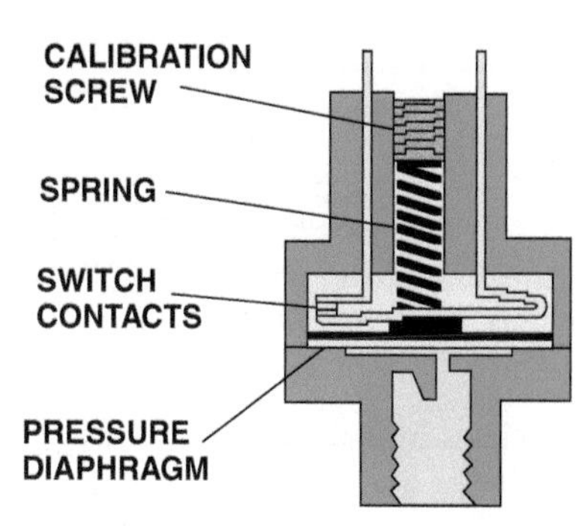

FIGURE 30–34 A pressure switch is either on or off. The contacts are closed by gas pressure on the diaphragm; they are opened by the spring.

LOW-PRESSURE SWITCH Many systems include a low-pressure switch, located in a low-pressure line or the accumulator to prevent compressor operation if there is loss of refrigerant. The compressor is lubricated by the oil that is mixed with and circulates with the refrigerant. A loss of refrigerant will also mean a loss of lubricating oil and subsequent compressor failure.

THERMISTORS AND TRANSDUCERS Recent systems use solid-state sensors that do not use switch contacts. The most common sensors are thermistors and transducers and these provide a variable output instead of just being on or off like a switch.

- A **thermistor** is commonly used to sense temperatures. It is basically an electrical resistor that changes resistance in inverse relationship to its temperature. Most automotive thermistors are of the negative temperature coefficient (NTC) type. The resistance will increase as the temperature drops and vice versa.
- A **transducer** senses pressure and changes a variable pressure signal into a variable electrical signal. A transducer can let the ECM/BCM know the actual pressure in the low- and/or high side of the system. Sensors provide an electrical signal to a control module that, in turn, controls compressor clutch or condenser fan operation.

REAR A/C SYSTEMS

PARTS AND OPERATION Some larger vehicles (vans and small buses) have dual heat and A/C assemblies, with the rear unit mounted in a rear side panel or in the roof. The rear A/C unit

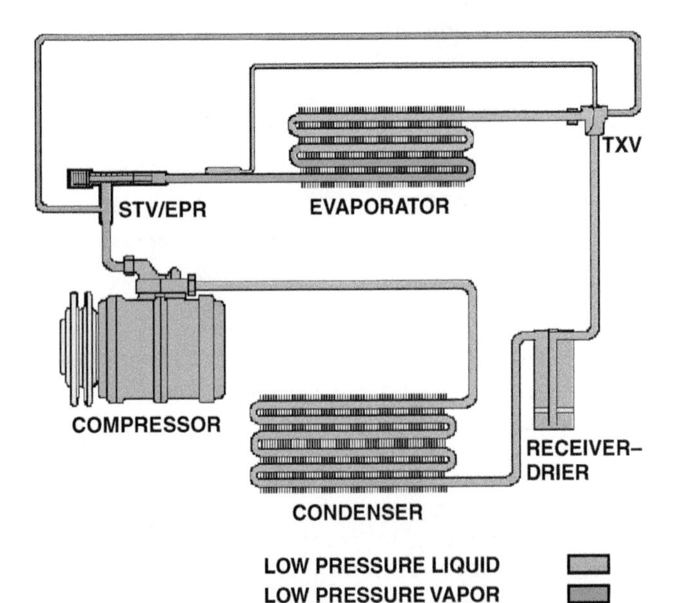

FIGURE 30–35 A suction throttling valve (STV) prevents evaporator pressure from dropping below 30 PSI, and this keeps ice from forming on the evaporator.

> **? FREQUENTLY ASKED QUESTION**
>
> **What Do STV, POA, and EPR Valves Do?**
>
> Some older systems used a valve to control evaporator pressure to prevent icing. These valves include the following:
>
> - Suction throttling valves (STVs)
> - Pilot-operated absolutes (POAs)
> - Evaporator pressure regulators (EPRs)
>
> Some import vehicles used an EPR valve through the late 1990s. These valves were mounted at the evaporator outlet, the compressor inlet, or somewhere between. Most of them sense evaporator pressure and when the pressure starts to drop below a certain point, the valve closes down to restrict refrigerant flow to the compressor. A system that uses a suction throttling valve maintains an almost constant evaporator temperature of 32°F (0°C) without cycling the clutch. ● **SEE FIGURE 30–35.**

consists of an evaporator and a TXV that operates in parallel flow with the front unit. ● **SEE FIGURE 30–36.**

Tee fittings are placed in the liquid and suction lines so that refrigerant can flow through both units, with the flow through the rear unit dependent on the cooling load. Many dual systems use an orifice tube to control the refrigerant flow into the front evaporator and a TXV with the rear evaporator. The rear evaporator is normally mounted in an assembly that includes a blower, heater core, and doors to control the air temperature and where the air returns to the passenger compartment.

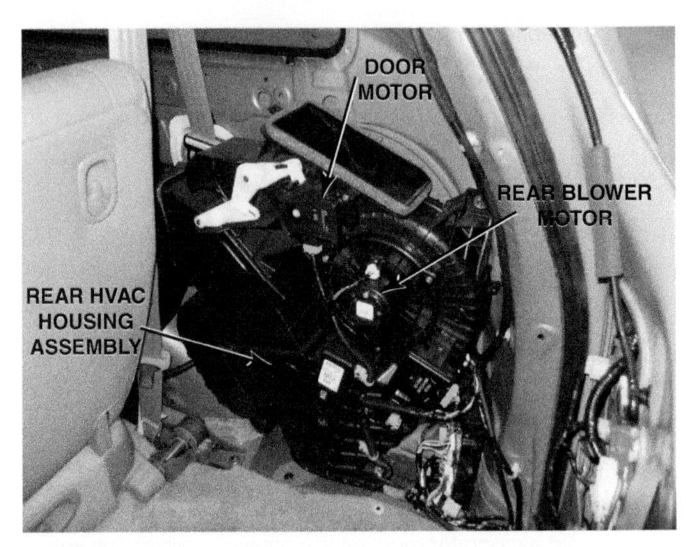

FIGURE 30–36 The rear HVAC module assembly used on a minivan.

CAUTION: The rear TXV shuts off the flow through the rear evaporator when the rear system is shut off, but a potential problem is created. A TXV does not ensure a complete shutoff, and refrigerant can leak through the valve. Because the blower is shut off, the evaporator will chill, and liquid refrigerant can puddle in this portion of the suction line. The puddle can flow down the suction line and slug the compressor, causing a knock and possible damage. To help prevent this from occurring, ask the customer to operate the rear air-conditioning unit regularly. Some systems use a solenoid valve to stop any flow through the rear system.

COMPONENT REPLACEMENT PROCEDURES

FITTING REPAIR If the fitting is tight and still leaks, it must be taken apart and inspected for damage and a new O-ring must be installed. Torque specifications for the various line fittings are provided by the vehicle manufacturer. Always use two wrenches when servicing fittings.

LINE REPLACEMENT A faulty hose or metal line is often repaired by replacing it with a new or repaired line. This normally involves disconnecting each end of the line at a fitting. Some hoses, like the suction and discharge, are combined with

FIGURE 30–37 The line to the condenser from the compressor includes a flange mount with an O-ring.

another line, making a double connection at the compressor, condenser, or evaporator. A manifold-type connector sealed with a pair of O-rings is commonly used at these connections. Some aftermarket suppliers can custom make hoses with fittings. ● **SEE FIGURE 30–37**.

A/C COMPONENT REPLACEMENT Inside-out failure is caused by acids inside of the system, and the accumulator or

FREQUENTLY ASKED QUESTION

Can a Refrigerant Hose Be Repaired?

Sometimes. A hose can sometimes be repaired by replacing one of the ends or by cutting out a damaged section and splicing it back together. Most OEM hoses are connected to the line fitting with a captive, *beadlock ferrule* so the hose is gripped by both the line connector and the ferrule.

- New service fittings with a captive metal ferrule and a new clamping method have been developed to allow field repair and makeup of barrier hoses. These fittings are commonly called *bubble-style crimp*. ● **SEE FIGURE 30–38**.
- The most common fitting styles are female and male O-rings (determined by the nut threads), female and male flares, male insert fittings, and spring-lock styles. These fittings are usually available for the four common hose sizes and in straight, 45° bend, and 90° bend shapes.

When repairing a line or hose, always follow the instructions that came with the repair kit to be sure of a proper and leak-free repair.

(a)

(c)

(b)

FIGURE 30–38 (a) When making a bubble-style crimp on the metal shell, dies of the correct size are installed in the tool. (b) The shell is crimped by tightening the tool drive bolt. (c) A finished crimp.

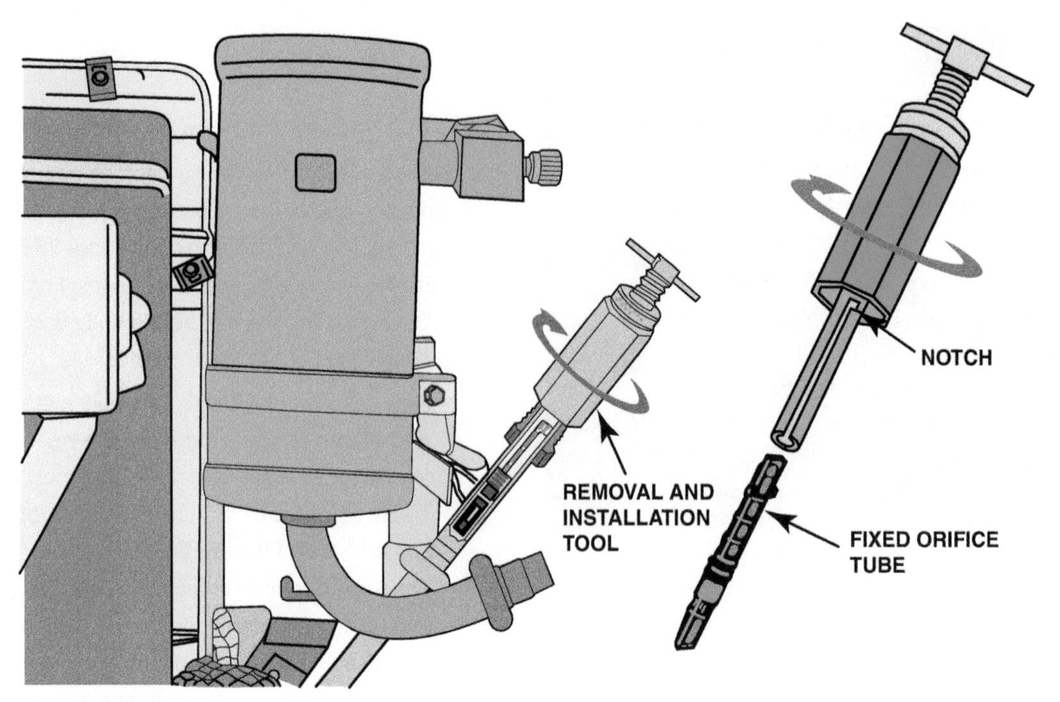

FIGURE 30–39 Needle-nose pliers can sometimes be used to remove an orifice tube, but it requires this special tool in most cases.

COMPONENT	FLUID OZ.	CC
Accumulator	2	60
Condenser	1	30
Evaporator	2	60
Each hose	0.3	10
Receiver–drier	0.5	15

CHART 30–1

When an A/C component is removed, a certain amount of oil is also removed. These are typical amounts. Check service information for the exact amount to add to each component if it is replaced.

receiver–drier must be replaced and the refrigerant recycled to remove these acids.

- When the compressor, evaporator, or TXV is replaced, the receiver–drier or accumulator should also be replaced. These failures were possibly caused by system contamination. This contamination will probably have loaded the desiccant capacity with moisture.
- It is good practice to keep the plastic caps in place on the new components until just before installation to keep as much moisture out of the system as possible.
- Removal of an accumulator, condenser, evaporator, or receiver–drier also removes a certain amount of oil from the system, and new oil should be added to the new part. The actual amount is usually specified in service information. ● **SEE CHART 30–1.**

If faulty, a major A/C component (the accumulator, condenser, evaporator, OT, receiver–drier, or TXV) is repaired by replacing it with a new one. At one time, replacement of these components was relatively easy because there was rather good access, except when working with evaporators. In most cases, getting the evaporator case out of the vehicle is tedious and time consuming, sometimes requiring the vehicle or evaporator case to be cut. Many technicians will not change an evaporator without consulting service information for the exact procedure to follow.

To remove and replace a major A/C component, locate, read, understand, and follow the specified repair procedure found in service information, which usually includes the following steps:

STEP 1 Recover the refrigerant from the system.

STEP 2 Disconnect and cap the refrigerant lines to the component.

STEP 3 Disconnect any mounting brackets and wires connected to the component and remove it.

STEP 4 Install the new part and attach any mounting brackets and wires.

STEP 5 Remove the line caps, pour the proper amount of the correct refrigerant oil into the component or line, and connect the refrigerant lines.

STEP 6 Evacuate and recharge the system and test for leaks.

ORIFICE TUBE REPLACEMENT A special puller that attaches to the orifice tube or needle-nose pliers is normally used to remove it. ● **SEE FIGURE 30–39.**

TECH TIP

Because It Fits, Does Not Mean It Is Correct!

Many air-conditioning systems use orifice tubes that look similar if not identical. They are usually color-coded for identification. Always use the recommended orifice tube for the vehicle you are servicing. Some examples of the various colors and sizes available include:

Make, Color, Orifice Size (Inches)

Chrysler, purple, 0.0605
Ford, red, 0.0605
Ford, orange, 0.0560
Ford, brown, 0.0470
Ford, green, 0.0505
GM, yellow, 0.0605. ● **SEE FIGURE 30–40.**

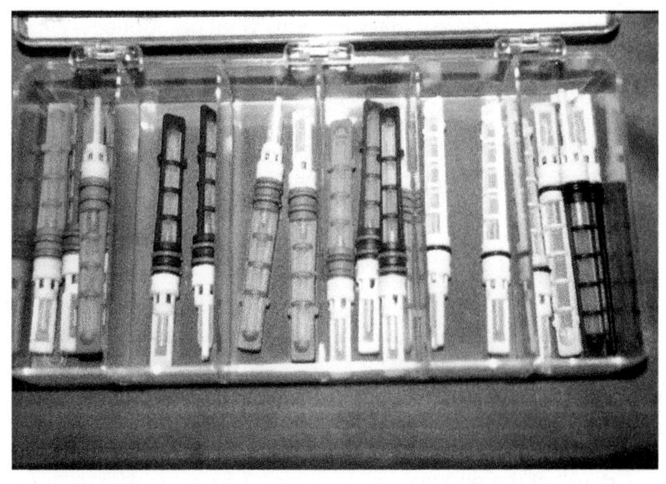

FIGURE 30–40 An assortment of orifice tubes in a plastic case with each size and color listed for easy access.

TXV REPLACEMENT When replacing a TXV (other than a H-block-type TXV), the thermal bulb must be securely attached to the evaporator outlet tube. This area must be clean to ensure good heat transfer. After attaching the thermal bulb, it must be wrapped with insulating refrigerant tape. This thick, pliable tape is used to keep outside heat from reaching the thermal bulb.

SUMMARY

1. The major condenser types are round or flat tube and serpentine or parallel flow.
2. TXVs are used with receiver–driers that are located in the high side.
3. Orifice tubes divide the high side from the low side and are used with accumulators located in the low side.
4. Various line fitting types are used to connect the components.
5. Various switches, sensors, and controls are used to control compressor, blower, and fan operation.

REVIEW QUESTIONS

1. What components are often included as part of a cooling module?
2. Why is sub-cooling necessary in some condensers?
3. What are the advantages and disadvantages of a thermal expansion valve system?
4. What are the advantages and disadvantages of an orifice tube system?
5. What is the difference between an accumulator and a receiver–drier?
6. Why does refrigerant oil need to be added to a new component such as a condenser?
7. What is the difference between a switch and a sensor?

CHAPTER QUIZ

1. A cooling module usually includes what components?
   a. Radiator
   b. Automatic transmission fluid cooler
   c. A/C condenser
   d. All of the above
2. As the latent heat of condensation is transferred to the airstream in the condenser, the refrigerant changes into __________.
   a. Liquid
   b. Vapor
   c. Gas
   d. Solid
3. The advantages of a TXV system over an orifice tube system include __________.
   a. Simpler and cheaper to produce
   b. Requires a smaller refrigerant charge
   c. Fewer moving parts
   d. All of the above
4. The advantages of an orifice tube system over a TXV system include __________.
   a. Simpler and cheaper to produce
   b. Requires a smaller refrigerant charge
   c. More moving parts
   d. All of the above
5. The orifice tube is usually located at the inlet tube to the __________.
   a. Condenser
   b. Compressor
   c. Evaporator
   d. Receiver–drier
6. Two technicians are discussing A/C systems. Technician A says that the TXV system uses a receiver–drier mounted in the suction line. Technician B says that the orifice tube system uses an accumulator in the suction line. Which technician is correct?
   a. Technician A only
   b. Technician B only
   c. Both Technicians A and B
   d. Neither Technician A nor B
7. The liquid line __________.
   a. Connects the condenser to the receiver–drier and TXV or OT
   b. Is smaller than the suction line
   c. Is larger in diameter than the suction line
   d. Both a and b
8. O-rings used in R-134a systems are usually __________.
   a. Green
   b. Black
   c. Red
   d. Orange
9. Various electrical switches used in A/C systems __________.
   a. Prevent evaporator icing
   b. Protect the compressor
   c. Control fan motors
   d. All of the above
10. Rear air-conditioning systems use __________.
    a. An orifice tube to control the refrigerant flow into the front evaporator and a TXV with the rear evaporator
    b. An orifice tube to control the refrigerant flow into the front evaporator and the rear evaporator
    c. A TXV to control the refrigerant flow into the front evaporator and an orifice tube with the rear evaporator
    d. A TXV to control the refrigerant flow into the front evaporator and the rear evaporator

chapter 31

AIR MANAGEMENT SYSTEM

LEARNING OBJECTIVES

After studying this chapter, the reader should be able to:

1. Prepare for the ASE Heating and Air Conditioning (A7) certification test content area "B" (Refrigeration System Component Diagnosis and Repair).
2. Discuss the different components of an air management system.
3. Explain airflow control and air temperature control in an A/C system.
4. Discuss plenum and control doors.
5. Explain nonelectrical and electronic HVAC controls.

KEY TERMS

Air doors 502
Air inlet 502
Air management system 502
Cabin filter 505
Dual-zone 508
Inside air 502
Mode door 502
Outside air 502
Plenum 506
Recirculation 502
Temperature-blend door 502
Vacuum actuator 508

GM STC OBJECTIVES

GM Service Technical College topics covered in this chapter are as follows:

1. Air flow in the heating system. (01510.01W)
2. Air flow in an air-conditioning system.
3. The ventilation system.
4. The purpose and components of air distribution ductwork. (01510.05W)
5. The purpose and types of air doors/valves in a single zone air ventilation system.
6. The purpose and components of a manual HVAC control panel.

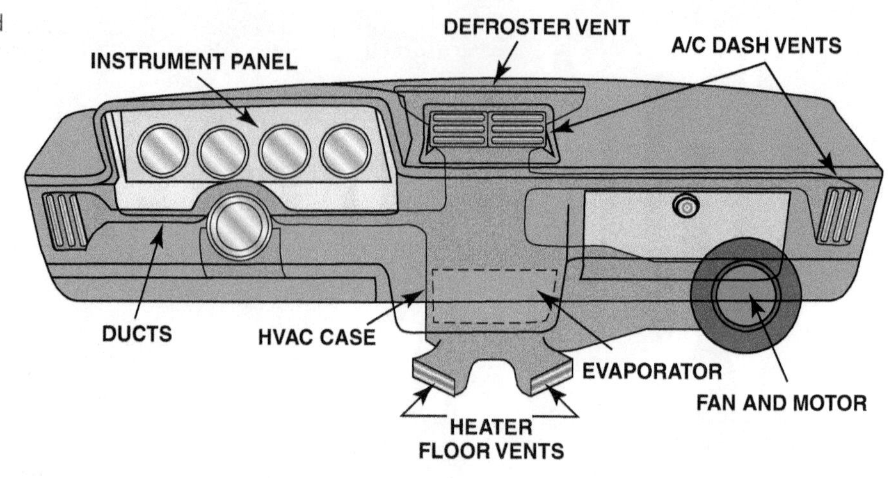

FIGURE 31–1 The HVAC airflow is directed toward the windshield, dash or floor vents, or combinations depending on the system settings.

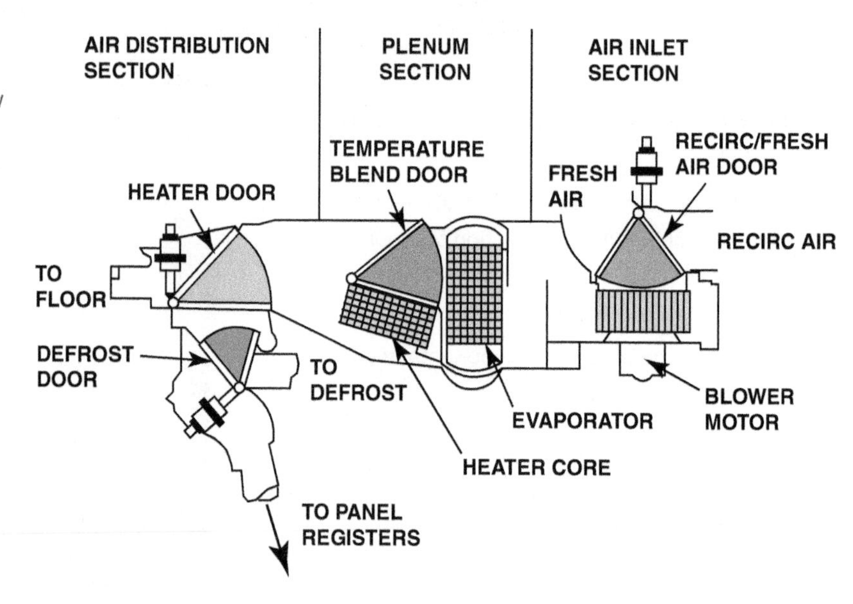

FIGURE 31–2 The three major portions of the A/C and heat system are air inlet, plenum, and air distribution. The shaded portions show the paths of the four control doors.

INTRODUCTION

TERMINOLOGY A system that contains the HVAC plenum, ducts, and **air doors** is called the **air management system**, or *air distribution system,* and controls the airflow to the passenger compartment. ● **SEE FIGURE 31–1**.

Air flows into the case that contains the evaporator and heater core from two possible inlets:

1. **Outside air**, often called *fresh air*
2. **Inside air**, usually called **recirculation**.

Proper temperature control to enhance passenger comfort during heating should maintain an air temperature in the footwell about 7°F to 14°F (4°C to 8°C) above the temperature around the upper body. This is done by directing the heated airflow to the floor. During A/C operation, the upper body should be cooler, so the airflow is directed to the instrument panel registers. Airflow is usually controlled by three or more doors, which are called *flap doors* or valves by some manufacturers. The three doors include:

- **Air inlet** door—used to select outside or inside air inlet
- **Temperature-blend door**—used to adjust air temperature
- **Mode door**—used to select air discharge location ● **SEE FIGURE 31–2**.

The use of these components allows the system to provide airflow under the following conditions:

1. Fresh outside air or recirculated air
2. Air conditioning
3. Defrost
4. Heat.

HEAT For heat position, the following can be done:

- Temperature set to the desired setting
- Air intake—select outside air for faster heating

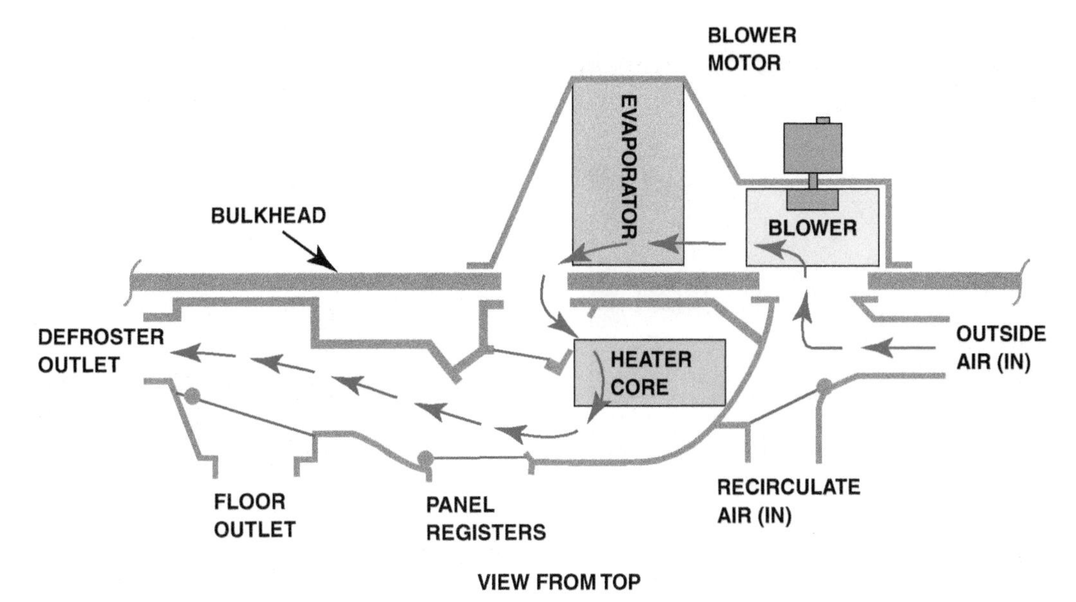

FIGURE 31–3 In the defog or defrost mode position, the air is directed through the evaporator to remove the moisture from the air before being sent through the heater core to warm the air.

- Air conditioning set to off
- Set airflow to floor
- Fan speed to desired speed

AIR CONDITIONING For air-conditioning position, the following can be done:

- Temperature set to the desired setting
- Air intake—set to outside air or recirculation under high humidity conditions.
- Airflow—select dash vents (also called panel vents)
- Air conditioning set to on
- Fan speed set to desired speed

VENTILATION For ventilation position, the following can be done:

- Temperature set to desired temperature
- Air intake—select outside air
- Airflow—set to dash (panel) vents
- Air conditioning set to off
- Fan speed set to desired speed

DEFOGGING OR DEFROSTING THE INSIDE OF THE WINDSHIELD For defogging or defrosting position, the following can be done:

- Temperature set to high temperature
- Air intake set to outside air
- Airflow set to windshield (defrost)
- Fan speed set to desired speed. ● SEE FIGURE 31–3.

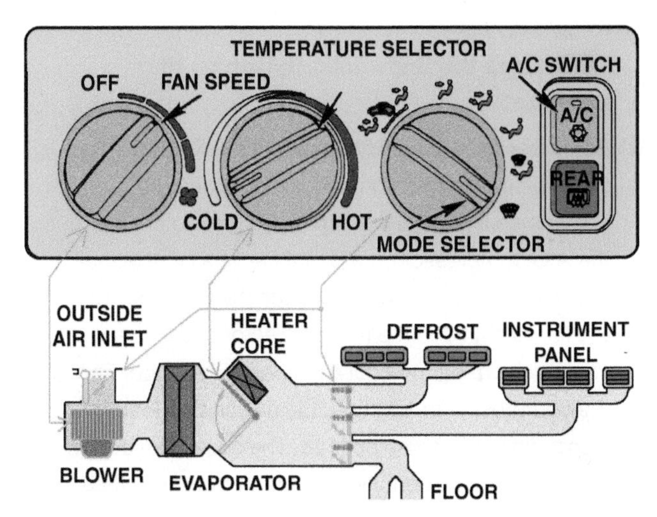

FIGURE 31–4 Most HVAC control heads include a control for turning the system on and setting the mode of operation, a control for adjusting the temperature, and a control for the fan speed.

HVAC CONTROL HEAD The HVAC control head or panel is mounted in the instrument panel cluster or console. The control head includes the following controls:

- System on and off
- Outside or recirculated air
- Mode position (floor, vent, or defrost)
- Temperature desired
- Blower speed

The control head is connected to various parts through electrical connections, vacuum connections, mechanical cables, or a combination of these. ● SEE FIGURE 31–4.

FIGURE 31–5 The A/C compressor is turned on or off by depressing the "snow flake" button on the dash.

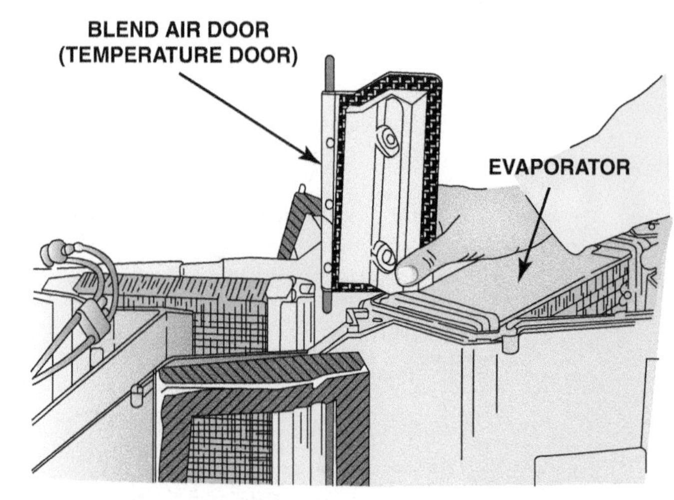

FIGURE 31–6 Many air control doors swing on their upper and lower pivots (in red).

FREQUENTLY ASKED QUESTION

What Does the Snowflake Button on the Dash Do?

Some people, such as those who drive vehicles that are equipped with automatic climate control systems, sometimes find it hard to figure out how to engage the A/C compressor on a rental car or a vehicle that they have not driven before. Often the driver will turn the fan to high and the mode selector to the dash vent position, but no cool air is being delivered. For the compressor to function, the button that looks like a snowflake has to be pushed. The snowflake button is actually the air conditioning on/off button. ● **SEE FIGURE 31–5**.

AIRFLOW CONTROL

OPERATION The amount or volume of HVAC air is controlled by blower speed. A multispeed blower is used to force air through the ductwork when the vehicle is moving at low speeds or to increase the airflow at any speed. At highway speeds, most systems have a natural airflow from ram air pressure created by the forward movement of the vehicle. This is the pressure generated at the base of the windshield by the speed of the vehicle. This airflow is improved in some vehicles by outlet registers placed in low-pressure areas toward the rear of the vehicle. Higher speeds move more air. The inlet and outlet directions of the airflow and the discharge air temperature are controlled by swinging, sliding, or rotating doors.

TYPES OF DOORS Most systems use a flap door that swings about 45° to 90°. Swinging doors are very simple and require little maintenance. ● **SEE FIGURE 31–6**.

Another design uses a *rotary door*. This pan-shaped door has openings at the side and edge, and the door rotates about 100° to one of four different positions. Each position directs airflow to the desired outlet(s).

The space under the instrument panel is very crowded, and some vehicles use a smaller HVAC case using a *sliding mode door* also called a *rolling door*. Some of these door designs roll up, similar to a window shade, and unroll to block a passage. ● **SEE FIGURE 31–7**.

AIR TEMPERATURE CONTROL

TEMPERATURE CONTROL USING AIRFLOW Most HVAC systems are considered *reheat* systems in that the incoming air is chilled as it passes through the evaporator. The air is then heated as part or all of the flow passes through the heater core so it reaches the desired in-vehicle temperature. ● **SEE FIGURE 31–8**.

TEMPERATURE CONTROL USING A VARIABLE COMPRESSOR Some vehicles use an electronically controlled variable displacement compressor with an air temperature sensor after the evaporator. Compressor displacement is adjusted to cool the evaporator just enough to cool the air to

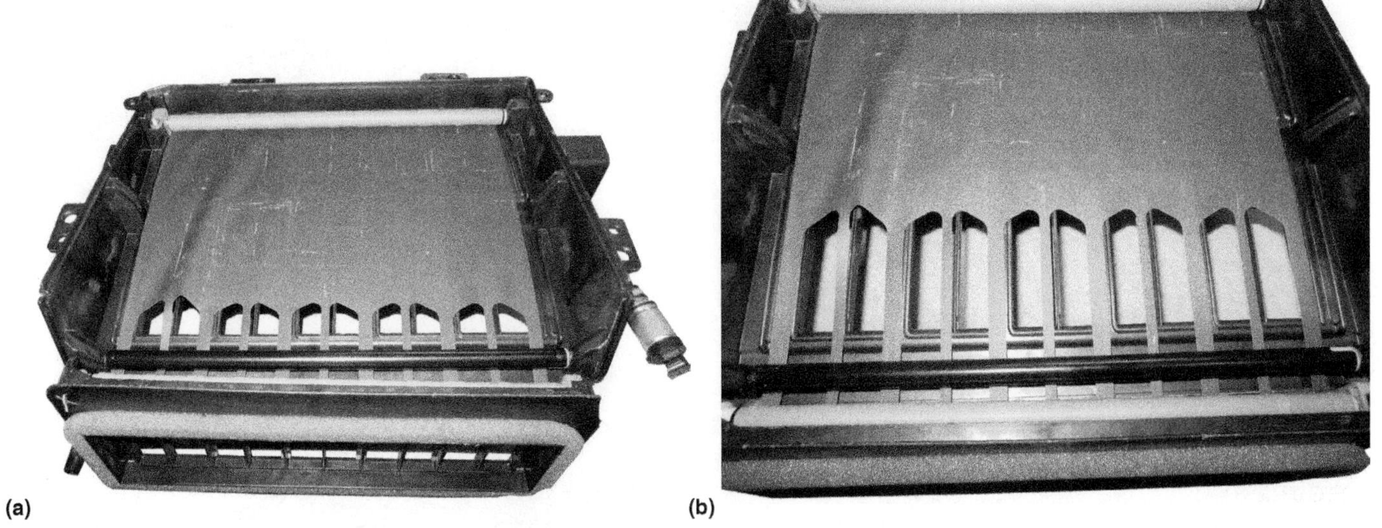

FIGURE 31–7 (a) A typical rolling-door type HVAC door that is shown almost fully closed. (b) The same door shown about half open.

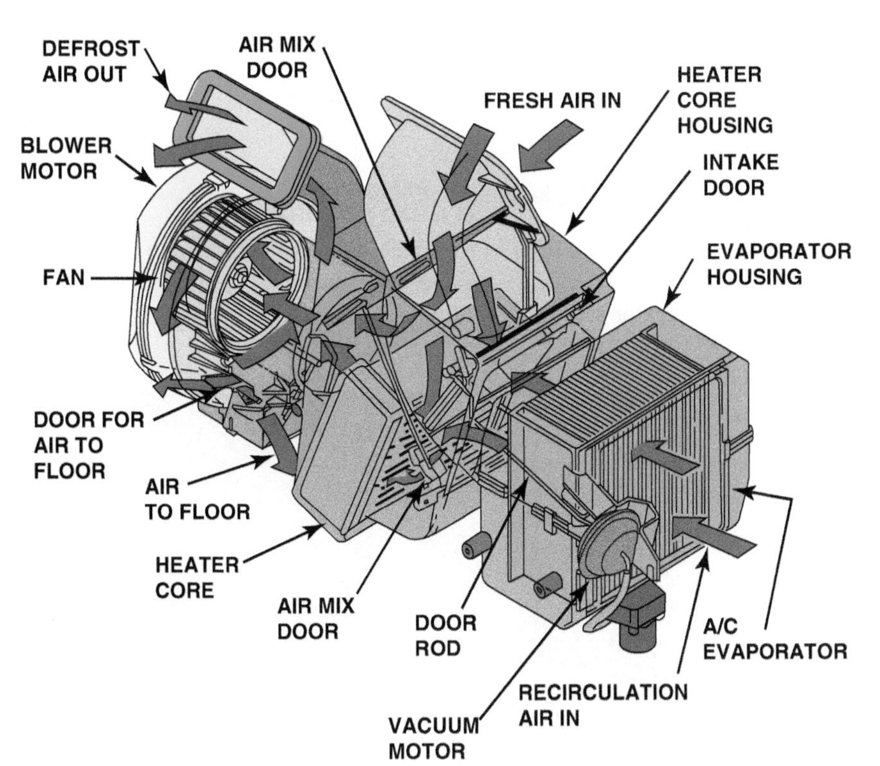

FIGURE 31–8 The blower motor forces air to flow through the A/C evaporator to remove moisture from the air before it is sent through the heater core where the air is heated before being directed to the defrost and floor vents.

the desired temperature. Cooling the incoming air no more than necessary reduces compressor load on the engine and helps to improve fuel economy.

AIR FILTRATION

TERMINOLOGY Most HVAC systems include a **cabin filter** to remove small dust or pollen particles from the incoming airstream. This filter is also called a/an

- *HVAC air filter*
- *Interior ventilation filter*
- *Micron filter*
- *Particulate filter*
- *Pollen filter*

These filters require periodic replacement. If they are not serviced properly, they will cause an airflow reduction when plugged. ● **SEE FIGURE 31–9**.

FIGURE 31–9 A cabin filter being removed from behind the glove compartment. The dark color is part of the filter and is activated charcoal used to help remove odors.

TYPES OF CABIN FILTERS There are two types of filter media:

1. **Particle filters.** Particle filters remove solid particles such as dust, soot, spores, and pollen using a special paper or nonwoven fleece material; they can trap particles that are about 3 microns or larger.
2. **Adsorption filters.** Adsorption filters remove noxious gases and odors using an activated charcoal media with the charcoal layer between layers of filter media. These two filter types can be combined into a two-stage filter. The filter media can have an electrostatic charge to make it more efficient.

CASES AND DUCTS

The evaporator and heater housing is molded from reinforced plastic and contains the following components:

- Evaporator
- Heater core
- Blower motor
- Most of the air control doors

The housing, called a **plenum**, is connected to the air inlets and outlets using formed plastic. These parts are required to contain and direct airflow, reduce noise, keep outside water and debris from entering, and isolate engine fumes and noises. ● **SEE FIGURE 31–10.**

Air can enter the duct system from either the plenum chamber in front of the vehicle's windshield (outside air) or from the *recirc* (short for *recirculation*) or return register (inside air). The return register is often positioned below the right end of the instrument panel. (The right and left sides of the vehicle are always described as seen by the driver.) The outside air plenum often includes a screen to keep leaves and other large debris from entering with the air. ● **SEE FIGURE 31–11.**

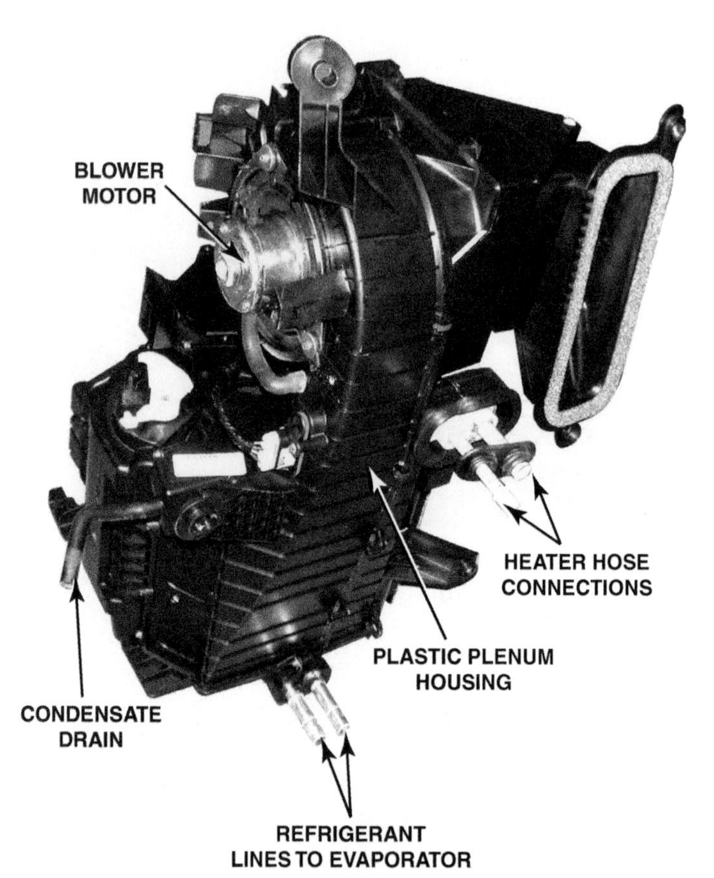

FIGURE 31–10 A typical HVAC housing that often has to be removed from the vehicle as an assembly to get access to the heater core and evaporator.

FIGURE 31–11 The air inlet to the HVAC system is usually at the base of the windshield and covered with a plastic screen (grille) to help keep debris such as leaves from entering the system.

TECH TIP

Keep the Air Screen Clean

The outside air inlet screen must be kept in good condition to prevent debris and small animals from entering the HVAC case. Leaves and pine needles can enter, decay, and mold. Mice have been known to enter and build nests and/or die. Any of these conditions can create a bad smell and are very difficult to clean.

PLENUM AND CONTROL DOORS

AIR INLET CONTROL DOOR The **air inlet control** door is also called

- *Fresh air door*
- *Recirculation door*
- *Outside air door*

This door is normally positioned so it allows airflow from one source while it shuts off the other. It can be positioned to allow

- Fresh air to enter while shutting off the recirculation opening
- Air to return or recirculate from inside the vehicle while shutting off fresh air
- A mix of fresh air and return air.

In many newer vehicles, the door is set to the fresh air position in all function lever positions except off, max heat, and max A/C. Max A/C and max heat settings position the door to recirculate in-vehicle air.

NOTE: In some vehicles, the recirculation door blocks most of the outside air and allows 80% of the air to be recirculated from the passenger compartment. About 20% of the air entering the passenger compartment is outside air to help keep the air in the passenger compartment fresh and keep the CO_2 levels low.

TEMPERATURE-BLEND DOOR Most systems position the evaporator so all air must pass through it. This allows removal of moisture by the evaporator's cold temperature. Many systems operate the A/C when defrost is selected to dry

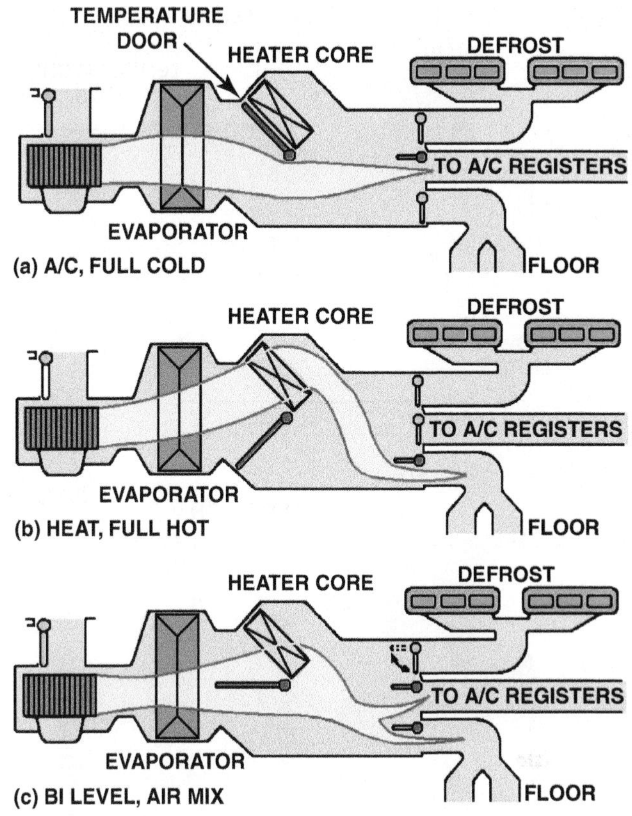

FIGURE 31–12 (a) The temperature and mode doors swing to direct all of the cool air past the heater core, (b) through the core to become hot, (c) or to blend hot and cool air.

the air. The heater core is placed downstream so that air can be routed either through or around it using one or two doors to control this airflow. The door used to control interior temperature is called the temperature-blend door. Some other names for this door include:

- *Air-mix door*
- *Temperature door*
- *Blend door*
- *Diverter door*
- *Bypass door.* ● **SEE FIGURE 31–12.**

The temperature-blend door is connected to the temperature knob or lever at the control head using a mechanical cable or an electric actuator. When the temperature lever is set to the coldest setting, the temperature-blend door routes all air so it bypasses the heater core. This causes the air entering the passenger compartment to be the coldest, coming straight from the evaporator. When the temperature lever is set to the hottest setting, the temperature-blend door routes all air through the heater core, and heated air goes to the passenger compartment. Setting the control lever to somewhere between cold and hot will mix or *temper* cold and hot air, allowing the driver to adjust the temperature to whatever is desired.

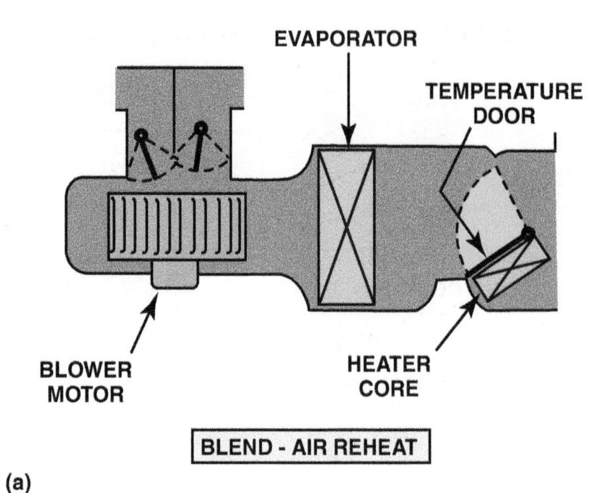

FIGURE 31–13 (a) In a blend-air system, all of the air is cooled. Then some of it is reheated and blended with the cool air to get the right temperature. (b) In a reheat system, all of the air is cooled and then reheated to the correct temperature.

In the past in some vehicles, the evaporator was placed or stacked right next to the heater core. These systems controlled the air temperature by regulating the amount of reheat at the heater core. ● **SEE FIGURE 31–13.**

AIR DISTRIBUTION AND OUTLETS Air from the plenum can flow into one or two of three outlet paths:

1. The A/C registers (vents) in the face of the instrument panel
2. The defroster registers at the base of the windshield
3. The heater outlets at the floor under the instrument panel.

Most vehicles include two ducts under the front seats or in the center console to transfer air to the rear seat area. Most vehicles also direct airflow to the side windows to defog the side windows. ● **SEE FIGURE 31–14.**

Airflow to these ducts is controlled by one or more mode doors controlled by the function lever or buttons. Mode doors are also called *function*, *floor-defrost*, and *panel-defrost* doors. Mode/function control sets the doors as follows:

- **A/C:** in-dash registers with outside air inlet
- **Max A/C:** in-dash registers with recirculation
- **Heat:** floor level with outside air inlet
- **Max Heat:** floor level with recirculation
- **Bi-level:** both in-dash and floor discharge
- **Defrost:** windshield registers

In many systems, a small amount of air is directed to the defroster ducts when in the heat mode, and while in defrost mode, a small amount of air goes to the floor level.

DUAL-ZONE SYSTEMS **Dual-zone** systems allow the driver and passenger to select different temperature settings. The temperature choices can be as much as 30°F (16°C) different. Dual-zone systems split the duct and airflow past the heater core and use two air mix valves or doors with each air mix valve/temperature door controlled by a separate actuator.

NONELECTRICAL HVAC CONTROLS

CABLE-OPERATED SYSTEMS Mechanical systems are the least expensive. Most early control heads used purely mechanical operation for the doors, and one or more cables connected the function lever to the air inlet and mode doors. The temperature lever was also connected to the temperature-blend door by another cable. These mechanical levers were rather simple and usually trouble free, but they had some disadvantages. They tended to bind and could require a good deal of effort to operate.

NOTE: The stiff wire cable may be called a Bowden cable in service information.

VACUUM-OPERATED SYSTEMS Many vehicles use **vacuum actuators**, sometimes called *vacuum motors,* to operate the air inlet and mode doors. ● **SEE FIGURE 31–15.**

The doors are controlled by a vacuum valve that is operated by the control head. Vacuum controls operate more easily than cables, and vacuum hoses are much easier to route through congested areas than cables. ● **SEE FIGURE 31–16.**

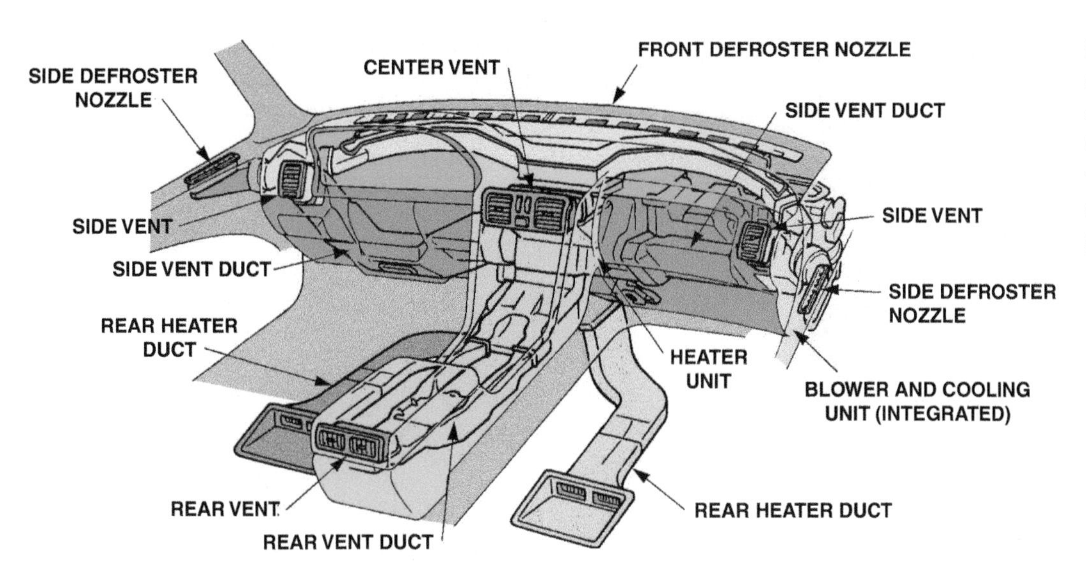

FIGURE 31–14 Ducts are placed in the center console or on the floor under the front seats to provide heated and cooled air to the rear seat passengers.

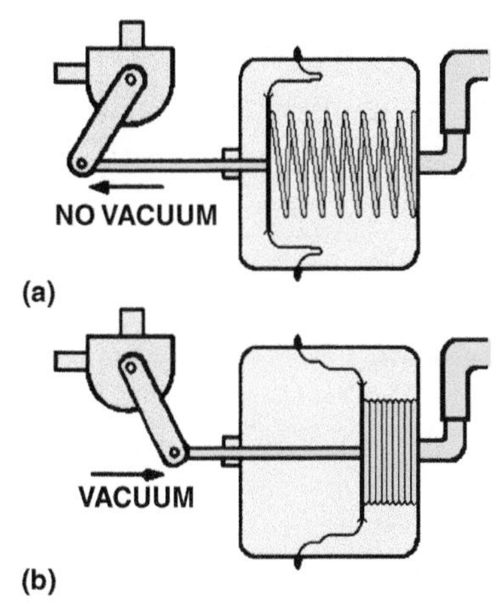

FIGURE 31–15 (a) With no vacuum signal, the spring extends the actuator shaft to place the door in a certain position. (b) A vacuum signal pulls the shaft inward and moves the door to the other position.

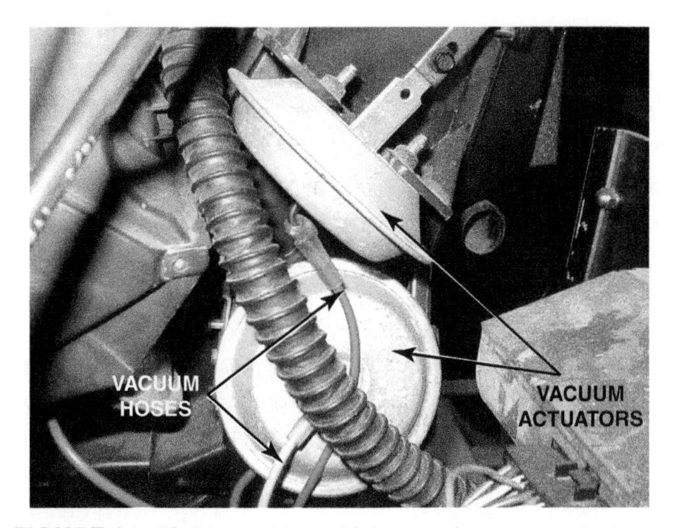

FIGURE 31–16 Many older vehicles used vacuum actuators to move the HVAC doors. When vacuum actuators operate, they alter the air–fuel mixture in the engine. Because vacuum controls affect engine operation and therefore emissions, recent vehicles use electric control systems.

TECH TIP

Defrost All the Time? Check the Vacuum

A common problem with older vehicles that use vacuum actuators involves airflow from the defroster ducts even though the selector lever is in other positions. The defrost setting is the default position in the event of a failure with the vacuum supply. The defrost position is used because it is the *safest* position. For safety, the windshield must remain free from frost. Heat is also supplied to the passenger compartments not only through defrost ducts but also through the heater vents at floor level. If the airflow is mostly directed to the windshield, check under the hood for a broken, disconnected, or missing vacuum hose. Check the vacuum reserve container for cracks or rust (if metal) that could prevent the container from holding vacuum. Check all vacuum hose connections at the intake manifold and trace each carefully, inspecting for cracks, splits, or softened areas that may indicate a problem.

NOTE: This problem of incorrect airflow inside the vehicle often occurs after another service procedure has been performed, such as air filter or cabin filter replacement. The movement of the technician's body and arms can cause a hose to be pulled loose or a vacuum fitting to break without the service technician being aware that anything wrong has occurred.

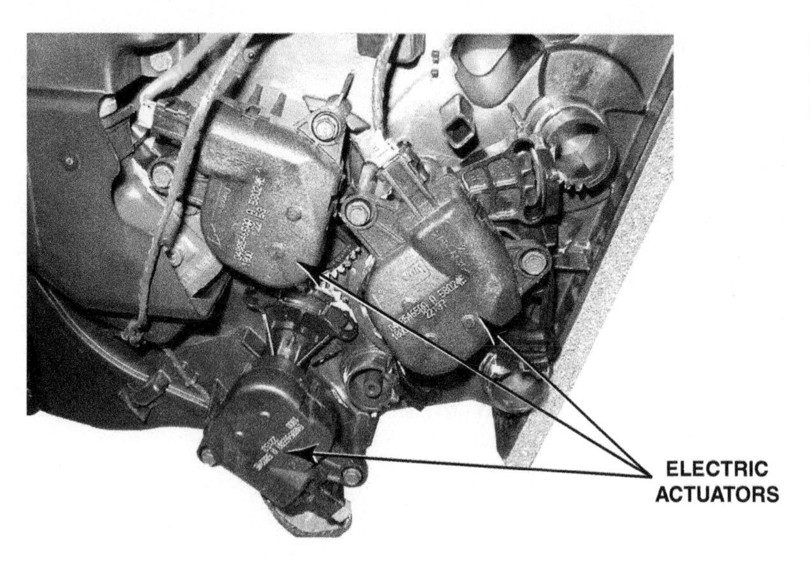

FIGURE 31–17 Three compact, electric actuators/ servomotors operate the doors in this part of the HVAC case.

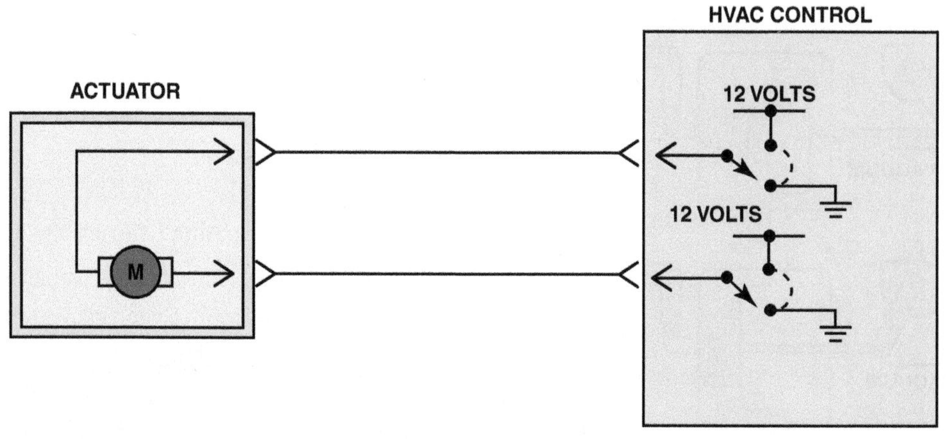

FIGURE 31–18 A two-wire HVAC electronic actuator where the direction of rotation is controlled by the HVAC control head or module, which changes the direction of rotation by changing the polarity of the power and ground connection at the motor.

ELECTRONIC HVAC CONTROLS

OVERVIEW Most recent vehicles use electrical function switches at the HVAC control head. These are often called *electromechanical controls.* These switches operate a group of solenoid valves that control the vacuum flow to the vacuum motors or use an electric actuator (motor) to operate the air distribution and temperature-blend doors. ● **SEE FIGURE 31–17**.

TWO-WIRE ACTUATORS A typical two-wire actuator rotates when electric impulses are sent to the brushes by the HVAC control head. The direction of rotation, and therefore the movement of the HVAC door position, is changed by changing the wire that is pulsed with power and the other brush is then connected to ground. ● **SEE FIGURE 31–18.**

THREE-WIRE ACTUATORS A typical three-wire actuator uses a power, ground, and an input signal wire from the HVAC control module. There is a module (logic chip) inside the motor assembly that receives a 0- to 5-volt signal from the HVAC control module. When the actuator gets a 0-volt signal from the control module, it rotates in one direction and when it receives a 5-volt signal it rotates in the opposite direction. If the motor receives a 2.5-volt signal, the motor stops rotating. ● **SEE FIGURE 31–19.**

FIVE-WIRE ACTUATORS A five-wire actuator uses two wires to power the motor (power and ground) and three wires for a potentiometer that is used to signal the HVAC control module of the motor's location. The potentiometer may be a separate gear-driven part attached to the motor or a part of the printed circuit board where a slider moves across resistive paint to create the potentiometer signal voltage. ● **SEE FIGURE 31–20.**

Most HVAC control modules convert the potentiometer signal voltage to a binary number ranging from 0 to 255, which

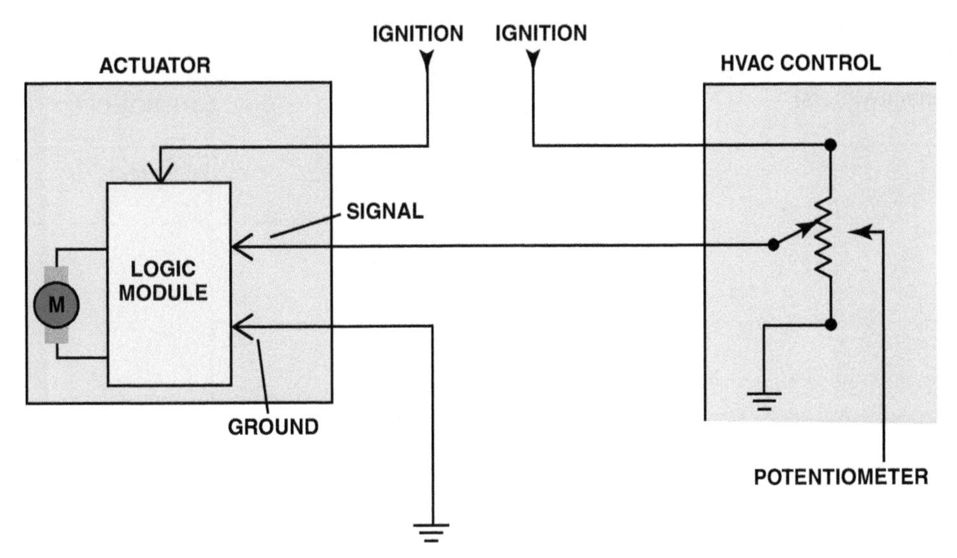

FIGURE 31–19 Three-wire actuators include a logic chip inside the motor assembly. The HVAC control module then sends a 0- to 5-volt signal to the motor assembly to control the direction of rotation.

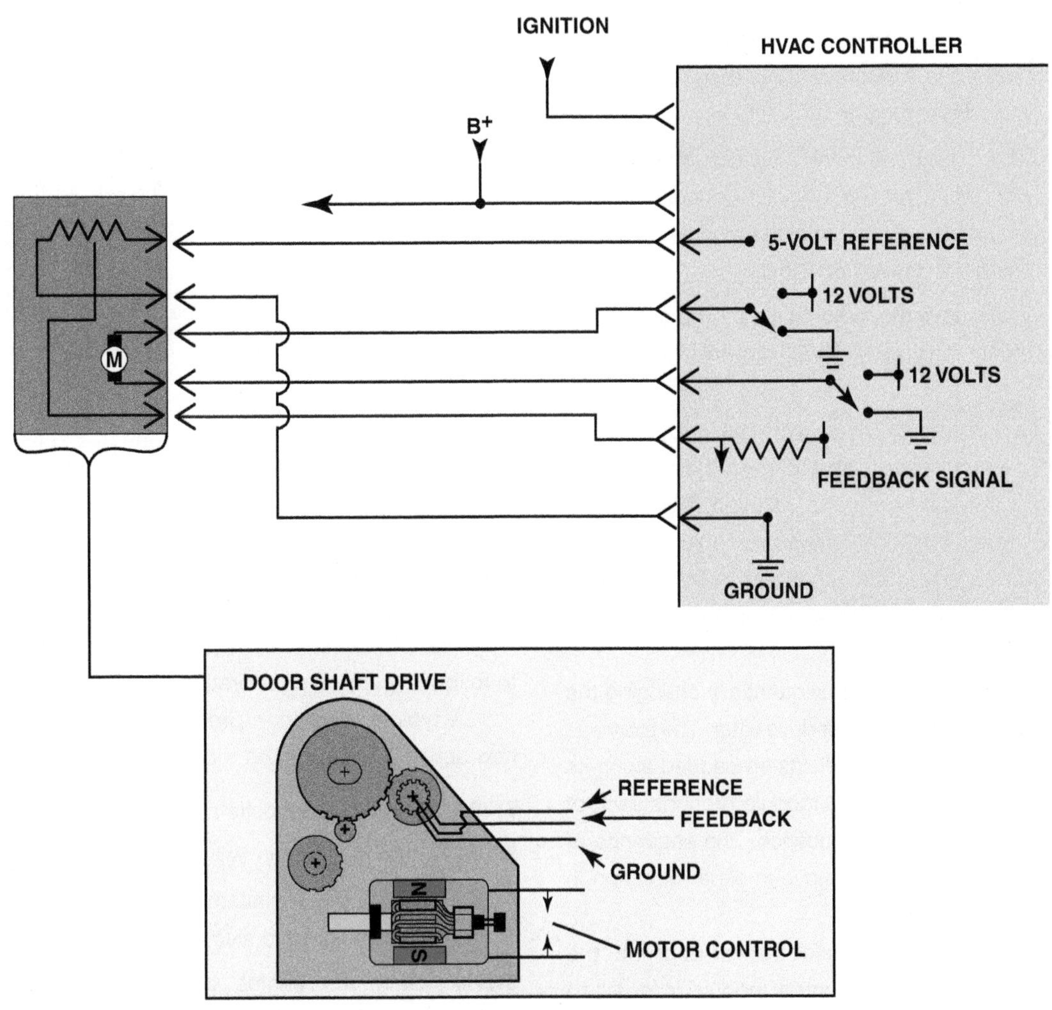

FIGURE 31–20 A typical five-wire HVAC actuator showing the two wires used to power the motor and the three wires used for the motor position potentiometer.

FIGURE 31–21 The stepper motor coils are grounded sequentially to rotate the motor armature one way or the other. (Courtesy of General Motors)

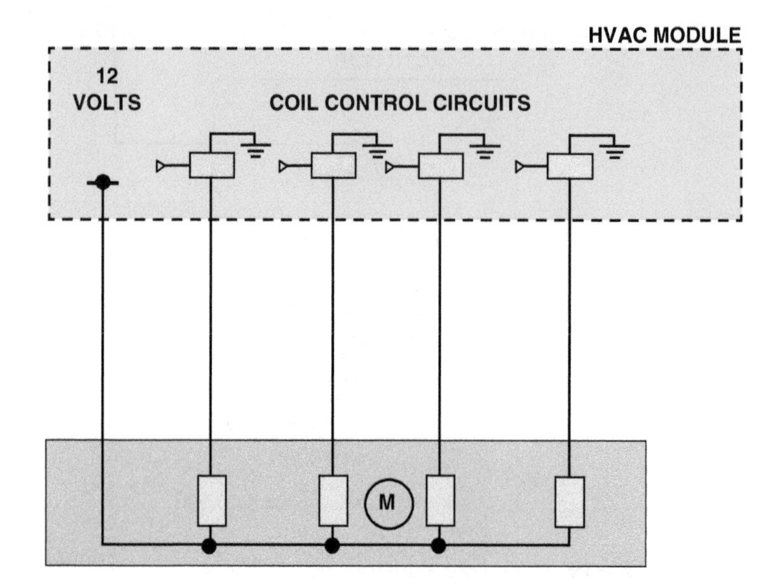

is often seen on a scan tool display. The HVAC control module monitors the feedback signal to determine the actual location of the door and to determine if the door is stuck. If the motor position falls outside of the expected range, there can be two things that can occur, depending on the vehicle.

1. The controller, usually the HVAC control module, will drive the motor until it reaches its desired location. This action can result a ticking sound being heard as the motor attempts to reach the desired position.
2. The controller will move the motor to a default position and then stop working, making the HVAC controls inoperative.

STEPPER MOTOR ACTUATORS With the introduction of the Global A vehicle architecture, an additional type of actuator, referred to as a stepper type motor, is used. The actuator has a 5-wire connector; the HVAC control module supplies a 12-volt reference voltage to the stepper motor and energizes the four stepper motor coils with a pulsed ground signal. ● **SEE FIGURE 31–21.**

The stepper motor operates by sequentially changing the magnetic fields that surround the inner drive rotor. The rotor has permanent magnets with alternating fields embedded along its axis. To create movement, the HVAC module grounds each of the four coils in a predetermined sequence. The sequence of magnetic fields causes the rotor to move in either direction in stepped increments.

The stepper motor puts the related flap or door into the position calculated by the HVAC control module in order to reach the selected position. The null point of the stepper motor is calibrated when the stepper motor is new. When the stepper motor is calibrated, the HVAC control module can drive the applicable coil to reach exactly the desired position of the flap or door. ● **SEE FIGURE 31–22.**

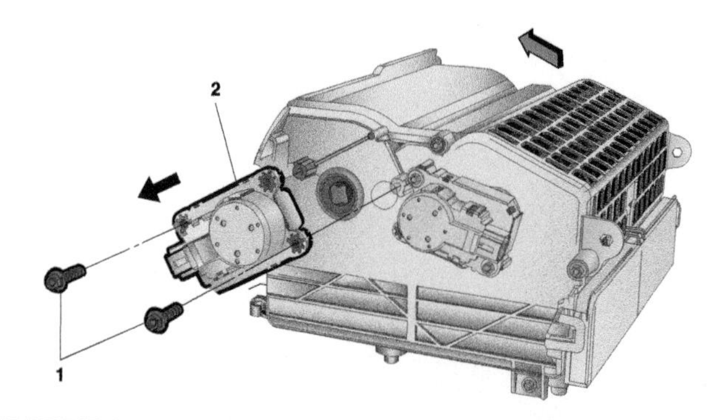

FIGURE 31–22 After replacement the stepper motor actuator must be calibrated by the HVAC control module. (Courtesy of General Motors)

ACTUATOR CALIBRATION PROCEDURES An HVAC actuator may need to be calibrated after the actuator has been replaced. Check service information for the specified procedure to follow to perform a calibration if needed.

A typical calibration procedure to use when installing a new actuator using a scan tool includes the following steps:

STEP 1 Clear all diagnostic trouble codes (DTCs)

STEP 2 Turn the ignition switch to the off position

STEP 3 Install the replacement actuator and reconnect all mechanical and electrical connections

STEP 4 Start the engine and select motor recalibration program on the scan tool under the HVAC Special Functions menu

STEP 5 Verify that no diagnostic trouble codes have been set.

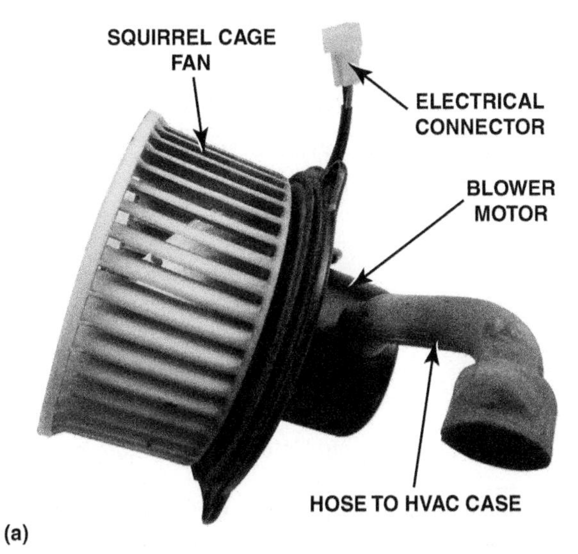

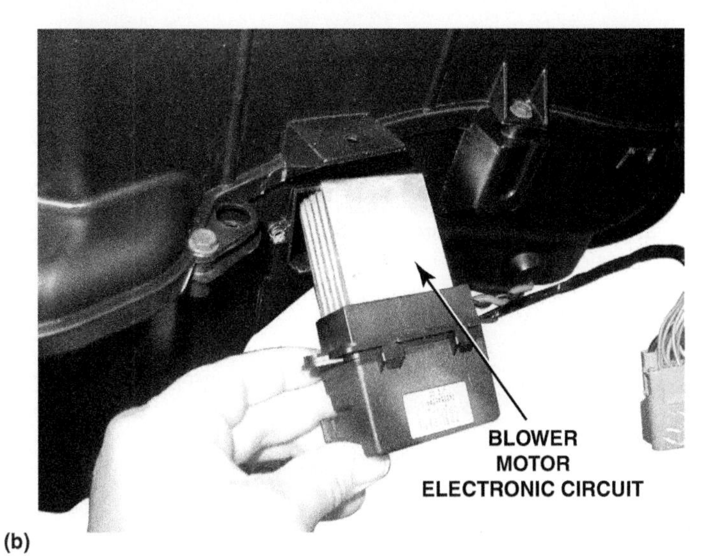

FIGURE 31–23 (a) A typical blower motor assembly with a squirrel-cage fan attached. The hose to the HVAC case is used to bring clean cabin air, instead of dirty outside air, to cool the motor. (b) Blower motor speed is controlled through an electronic circuit (shown) or through a resistor pack.

CAUTION: Do not operate an actuator prior to installation to "test it" because many actuators operate until they are stopped mechanically by the door being fully open or fully closed. If operated using a battery and jumper wires without being installed, the actuator can be moved beyond its normal range of motion and will not operate correctly, if at all, when placed into the HVAC housing.

Alternate Method (without Scan Tool):

NOTE: Do not adjust any controls on the HVAC control module while the module is self-calibrating. If interrupted, improper HVAC performance will result.

STEP 1 Clear all DTCs.

STEP 2 Place the ignition switch to the OFF position.

STEP 3 Install the replacement HVAC actuator.

STEP 4 Connect all previously disconnected components.

STEP 5 Remove the HVAC control module fuse for a minimum of 10 s.

STEP 6 Reinstall the HVAC control module fuse.

STEP 7 Turn the ignition ON and wait 40 s for the HVAC control module to self-calibrate.

BLOWER MOTOR CONTROL

Blower speed control in many of these systems is through a multiposition electrical switch and a group of resistors or electronic controls. The position of the switch determines the amount of resistance in the blower circuit and therefore the speed of the motor. Electronically controlled systems use a pulse-width-modulated (PWM) voltage supply to control the blower motor, which switches the motor off and on, up to 40,000 times per second. Increasing the length of the "on time" produces higher speeds. ● **SEE FIGURE 31–23.**

NOTE: Refer to Chapter 21 for more information on the blower motor.

SUMMARY

1. Air flows into the case that contains the evaporator and heater core from two possible inlets:
   - Outside air, often called fresh air
   - Inside air, usually called recirculation air.
2. The HVAC case contains a blower, A/C evaporator, the heater core, and doors to control the air temperature and flow.
3. The control head allows the driver to change blower speed, adjust the temperature, turn A/C on or off, and direct the airflow.
4. Control heads transfer motion to the HVAC case through cables, vacuum control, or electronics and electric motors.
5. Most HVAC systems include a filter to remove particles and odors from the air.

REVIEW QUESTIONS

1. What do the three main HVAC doors do?
2. In the heating mode, why is heat directed toward the floor?
3. Where is the air directed when the control panel is set to defog/defrost position?
4. What is meant by a "reheat" system?
5. Where does the air enter the vehicle when outside air is selected?

CHAPTER QUIZ

1. "Outside air" is also called ________ air.
   - **a.** Recirculation
   - **b.** Fresh
   - **c.** Plenum
   - **d.** Mode
2. The ________ door is used to select air discharge location such as floor, vent, or defrost.
   - **a.** Mode
   - **b.** Recirculation
   - **c.** Fresh air
   - **d.** Any of the above
3. What does the snowflake button on the dash do?
   - **a.** Selects defrost
   - **b.** Selects defog
   - **c.** Turns on the A/C compressor
   - **d.** Selects fresh outside air
4. The ram air that enters the vehicle from the outside enters the HVAC system from ________.
   - **a.** Beside the headlight
   - **b.** Behind the grille
   - **c.** The side vents
   - **d.** The base of the windshield
5. Most systems use a flap door that swings about ________ degrees.
   - **a.** 10 to 25
   - **b.** 25 to 40
   - **c.** 45 to 90
   - **d.** 90 to 120
6. The air inlet control door is also called ________.
   - **a.** Fresh air door
   - **b.** Recirculation door
   - **c.** Outside air door
   - **d.** Any of the above
7. The door used to control interior temperature is called the ________ door.
   - **a.** Temperature-blend
   - **b.** Recirculation
   - **c.** Outside (fresh) air
   - **d.** Mode
8. A five-wire actuator uses two wires to power the motor (power and ground) and three wires for ________.
   - **a.** Direction control
   - **b.** A feedback potentiometer
   - **c.** Static electricity protection
   - **d.** Redundant control
9. What is a "reheat" system?
   - **a.** A system that uses two heater cores.
   - **b.** A system that cools the air through the evaporator then heats the air though airflow though the heater core.
   - **c.** A system that uses both engine coolant and an electronic heater to heat the air.
   - **d.** Any of the above depending on the make and model of vehicle.
10. Most recent HVAC actuators are recalibrated after replacement by ________.
    - **a.** Using a battery and jumper wires
    - **b.** Using a scan tool or allow it to self-learn after being installed
    - **c.** Turning the ignition switch to the off position
    - **d.** None of the above

chapter 32

COOLING SYSTEM OPERATION AND DIAGNOSIS

LEARNING OBJECTIVES

After studying this chapter, the reader should be able to:

1. Describe how coolant flows through an engine.
2. Discuss the operation of the thermostat.
3. Explain the purpose and function of the radiator pressure cap.
4. Describe the operation and service of water pumps.
5. Discuss how to diagnose cooling system problems.

This chapter will help you prepare for ASE certification test A7 Heating and Air Conditioning content area "C" (Heating and Engine Cooling Systems Diagnosis and Repair).

KEY TERMS

Bar 524
Bleed holes 526
Bypass 518
Centrifugal pump 525
Coolant recovery system 524
Cooling fins 521
Core tubes 521
Impeller 525
Parallel flow system 526
Reverse cooling 525
Scroll 525
Series flow system 526
Series-parallel flow system 526
Silicone coupling 528
Steam slits 526
Surge tank 524
Thermostatic spring 528

GM STC OBJECTIVES

GM Service Technical College topics covered in this chapter are as follows:

1. The components of the engine cooling system. (00510.01W-R2)
2. Normal and abnormal engine coolant.
3. Engine coolant system service inspection procedures.
4. Perform cooling system tests (pressure, combustion leakage, and temperature); determine needed repairs.
5. Test and replace thermostat.

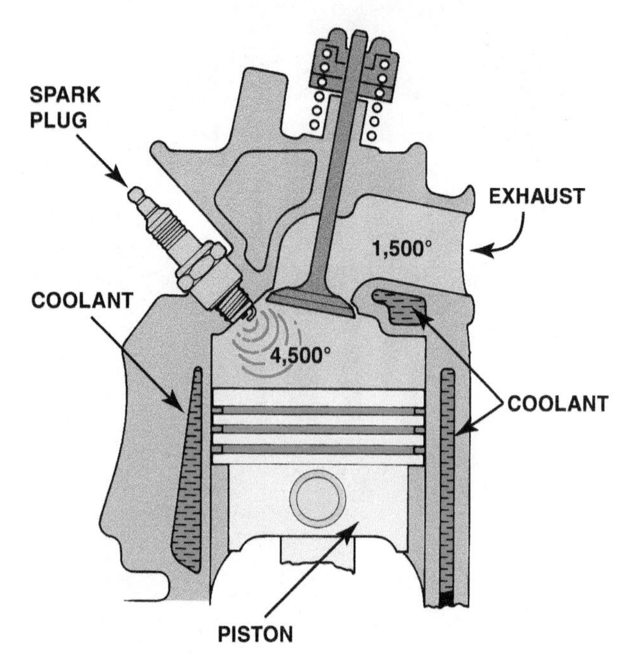

FIGURE 32–1 Typical combustion and exhaust temperatures.

TECH TIP

Overheating Can Be Expensive

A faulty cooling system seems to be a major cause of engine failure. Engine rebuilders often have nightmares about seeing their rebuilt engine placed back in service in a vehicle with a clogged radiator. Most engine technicians routinely replace the water pump and all hoses after an engine overhaul or repair. The radiator should also be checked for leaks and proper flow whenever the engine is repaired or replaced. Overheating is one of the most common causes of engine failure.

COOLING SYSTEM

PURPOSE AND FUNCTION Satisfactory cooling system operation depends on the design and operating conditions of the system. The design is based on heat output of the engine, radiator size, type of coolant, size of water pump (coolant pump), type of fan, thermostat, and system pressure. The cooling system must allow the engine to warm up to the required operating temperature as rapidly as possible and then maintain that temperature.

Peak combustion temperatures in the engine run from 4,000°F to 6,000°F (2,200°C to 3,300°C). The combustion temperatures will *average* between 1,200°F and 1,700°F (650°C and 925°C). Continued temperatures as high as this would weaken engine parts, so heat must be removed from the engine. The cooling system keeps the head and cylinder walls at a temperature that is within the range for maximum efficiency. The cooling system removes about one-third of the heat created in the engine. Another third escapes to the exhaust system. ● **SEE FIGURE 32–1.**

LOW-TEMPERATURE ENGINE PROBLEMS Engine operating temperatures must be above a minimum temperature for proper engine operation. If the coolant temperature does not reach the specified temperature as determined by the thermostat, then the following engine-related faults can occur.

- A P0128 diagnostic trouble code (DTC) can be set. This code indicates "coolant temperature below thermostat regulating temperature," which is usually caused by a defective thermostat staying open or partially open.
- Moisture created during the combustion process can condense and flow into the oil. *For each gallon of fuel used, moisture equal to a gallon of water is produced.* The condensed moisture combines with unburned hydrocarbons and additives to form carbonic acid, sulfuric acid, nitric acid, hydrobromic acid, and hydrochloric acid.

To reduce cold engine problems and to help start engines in cold climates, most manufacturers offer block heaters as an option. These block heaters are plugged into household current (110 volts AC) and the heating element warms the coolant.

HIGH-TEMPERATURE ENGINE PROBLEMS Maximum temperature limits are required to protect the engine. Higher than normal temperatures can cause the following engine-related issues.

- High temperatures will oxidize the engine oil producing hard carbon and varnish. The varnish will cause the hydraulic valve lifter plungers to stick. Higher than normal temperatures will also cause the oil to become thinner (lower viscosity than normal). Thinned oil will also get into the combustion chamber by going past the piston rings and through valve guides to cause excessive oil consumption.
- The combustion process is very sensitive to temperature. High coolant temperatures raise the combustion temperatures to a point that may cause detonation (also called spark knock or ping) to occur.

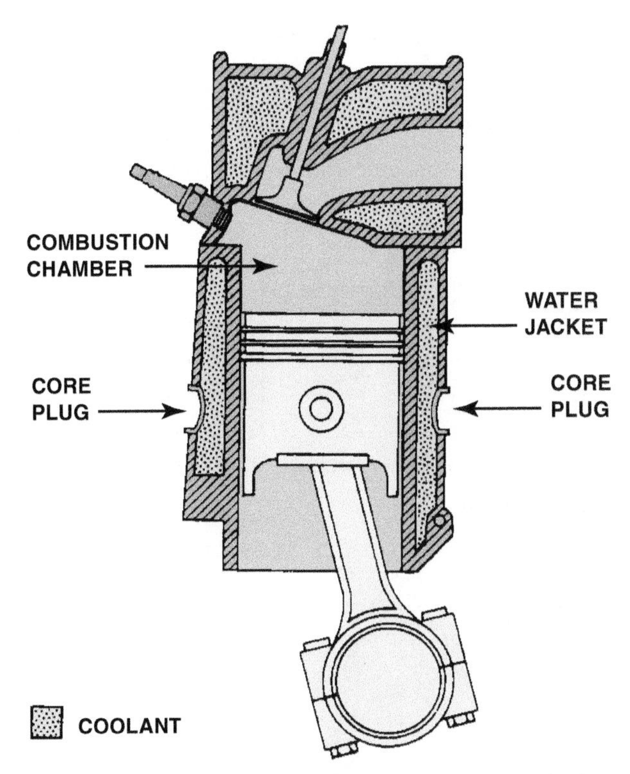

FIGURE 32–2 Coolant circulates through the water jackets in the engine block and cylinder head.

COOLING SYSTEM OPERATION

PURPOSE AND FUNCTION Coolant flows through the engine, where it picks up heat. It then flows to the radiator, where the heat is given up to the outside air. The coolant continually recirculates through the cooling system, as illustrated in ● **FIGURES 32–2 AND 32–3.**

COOLING SYSTEM OPERATION The temperature of the coolant rises as much as 15°F (8°C) as it goes through the engine and cools as it goes through the radiator. *The coolant flow rate may be as high as 1 gallon (4 liters) per minute for each horsepower the engine produces.*

Hot coolant comes out of the thermostat housing on the top of the engine on most engines. The engine coolant outlet is connected to the radiator by the upper radiator hose and clamps. The coolant in the radiator is cooled by air flowing through the radiator. As the coolant moves through the radiator, it cools. The cooler coolant leaves the radiator through an outlet and the lower radiator hose and then flows to the inlet side of the water pump, where it is recirculated through the engine.

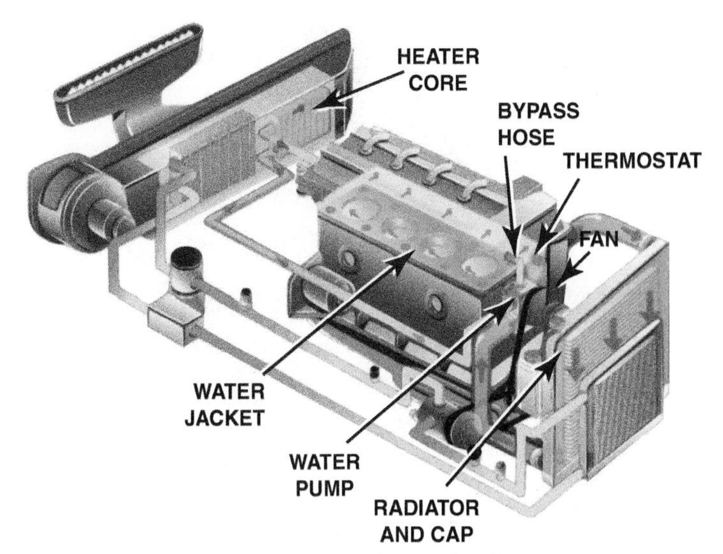

FIGURE 32–3 Coolant flow through a typical engine cooling system. (Courtesy of General Motors)

NOTE: Some engine designs such as General Motor's 4.8, 5.3, 5.7, and 6.0 liter V-8s place the thermostat on the inlet side of the water pump. As the cooled coolant hits the thermostat, the thermostat closes until the coolant temperature again causes it to open. Placing the thermostat in the inlet side of the water pump therefore reduces the rapid temperature changes that could cause stress in the engine, especially if aluminum heads are used with a cast iron block.

Radiators are designed for the maximum rate of heat transfer using minimum space. Cooling airflow through the radiator is aided by a belt- or electric motor–driven cooling fan.

THERMOSTATS

PURPOSE AND FUNCTION There is a normal operating temperature range between low-temperature and high-temperature extremes. The thermostat controls the minimum normal temperature. The thermostat is a temperature-controlled valve placed at the engine coolant outlet on most engines.

THERMOSTAT OPERATION An encapsulated wax-based plastic pellet heat sensor is located on the engine side of the thermostatic valve. As the engine warms, heat causes the wax pellet to expand. ● **SEE FIGURE 32–4.**

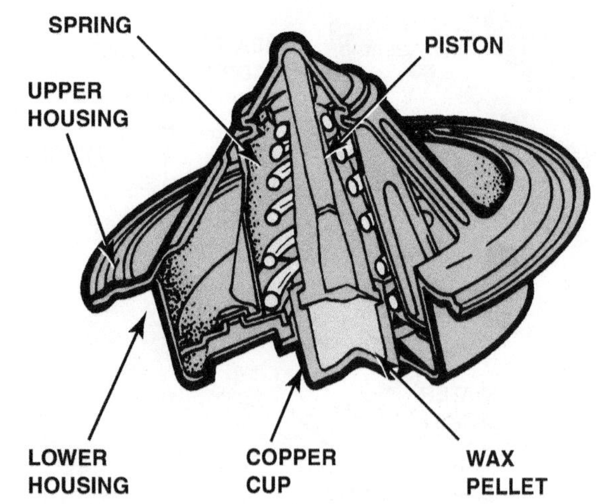

FIGURE 32–4 A cross section of a typical wax-actuated thermostat showing the position of the wax pellet and spring.

A mechanical link, connected to the heat sensor, opens the thermostat valve. As the thermostat begins to open, it allows some coolant to flow to the radiator, where it is cooled. The remaining part of the coolant continues to flow through the bypass, thereby bypassing the thermostat and flowing back through the engine. ● **SEE FIGURE 32–5.**

The rated temperature of the thermostat indicates the temperature at which the thermostat starts to open. The thermostat is fully open at about 20°F higher than its opening temperature. ● **SEE CHART 32–1.**

If the radiator, water pump, and coolant passages are functioning correctly, the engine should always be operating within the opening and fully open temperature range of the thermostat. ● **SEE FIGURE 32–6.**

NOTE: A bypass around the closed thermostat allows a small part of the coolant to circulate within the engine during warm-up. It is a small passage that leads from the engine side of the thermostat to the inlet side of the water pump. It allows some coolant to bypass the thermostat even when the thermostat is open. The bypass may be cast or drilled into the engine and pump parts. ● SEE FIGURES 32–7 AND 32–8.

The bypass aids in uniform engine warm-up. Its operation eliminates hot spots and prevents the building of excessive coolant pressure in the engine when the thermostat is closed.

ELECTRIC THERMOSTAT On some engines the thermostat has an electric heating grid that is controlled by the engine control module (ECM). Both the coolant and the heater grid, depending on the load conditions of the engine, can heat the wax pellet. The heater grid is pulse-width modulated (PWM) to ground by the ECM; this results in more efficient fuel consumption and reduced emissions in city driving and low-speed cruising. ● **SEE FIGURE 32–9.**

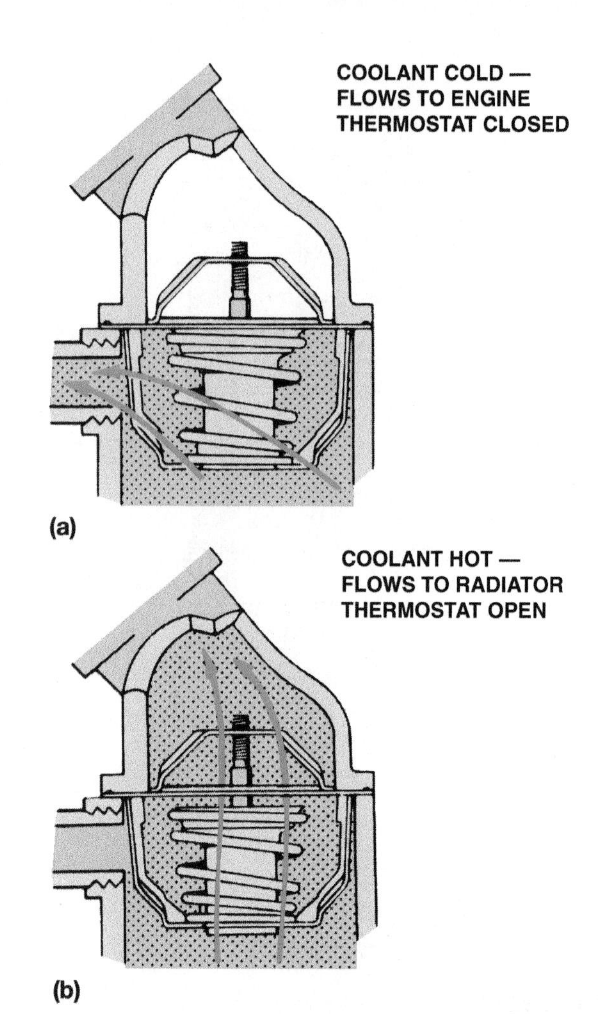

FIGURE 32–5 (a) When the engine is cold, the coolant flows through the bypass. (b) When the thermostat opens, the coolant can flow to the radiator.

THERMOSTAT TEMPERATURE RATING	STARTS TO OPEN	FULLY OPEN
180°F	180°F	200°F
195°F	195°F	215°F

CHART 32–1

The temperature of the coolant depends on the rating of the thermostat.

THERMOSTAT TESTING There are three basic methods used to check the operation of the thermostat.

1. **Hot water method.** If the thermostat is removed from the vehicle and is closed, insert a 0.015 inch (0.4 mm) feeler gauge in the opening so that the thermostat will hang on the feeler gauge. The thermostat should then be suspended by the feeler gauge in a container of water or

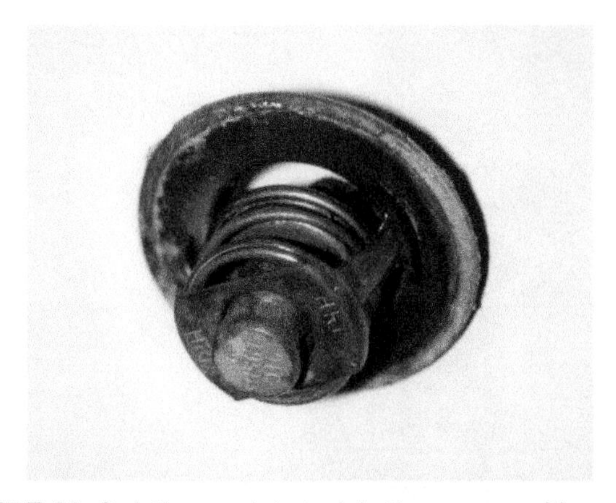

FIGURE 32–6 A thermostat stuck in the open position caused the engine to operate too cold. If a thermostat is stuck closed, this can cause the engine to overheat.

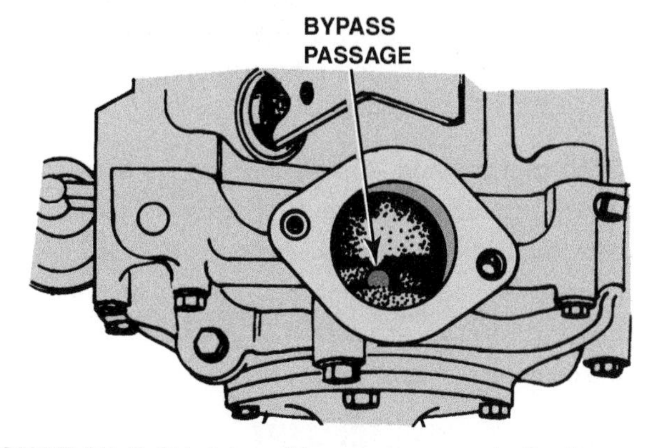

FIGURE 32–7 This internal bypass passage in the thermostat housing directs cold coolant to the water pump.

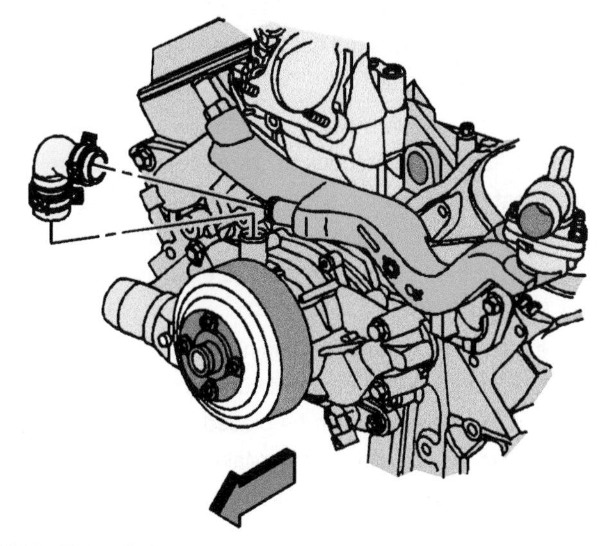

FIGURE 32–8 The water pump may have an external bypass tube and hose. (Courtesy of General Motors)

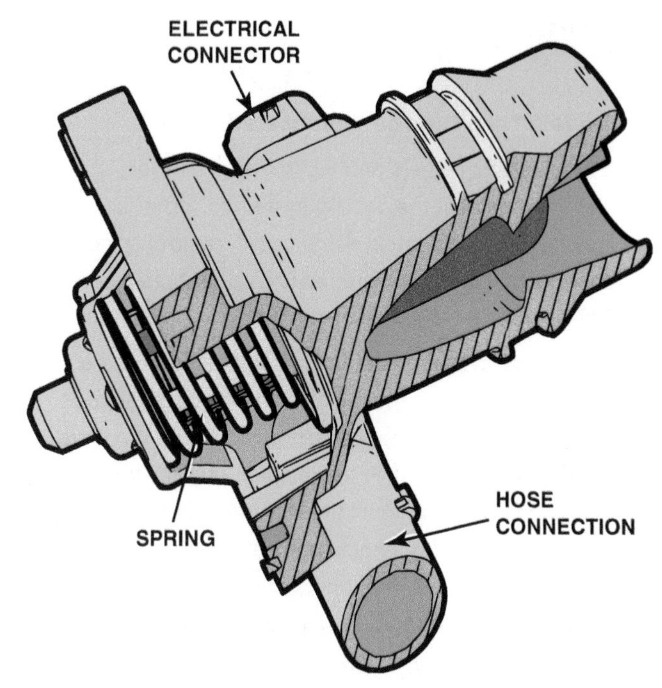

FIGURE 32–9 The electric thermostat is warmed by the coolant or an electric heater. (Courtesy of General Motors)

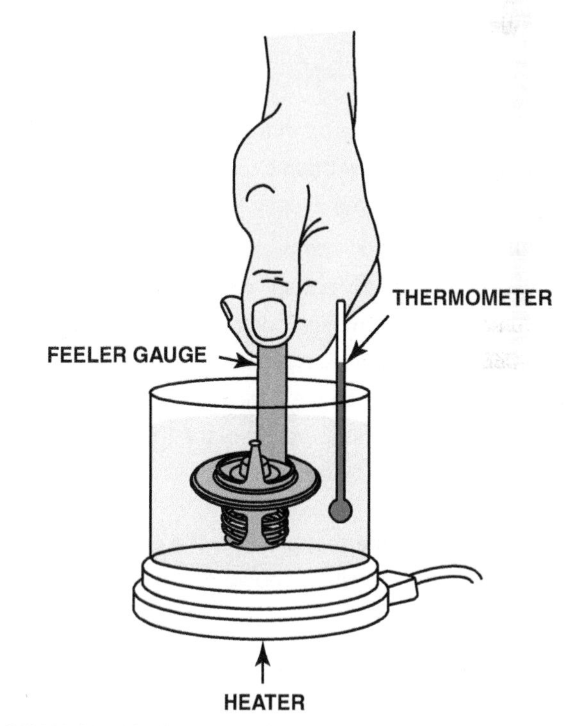

FIGURE 32–10 Checking the opening temperature of a thermostat.

coolant along with a thermometer. The container should be heated until the thermostat opens enough to release and fall from the feeler gauge. The temperature at which the thermostat falls is the opening temperature of the thermostat. If it is within 5°F (4°C) of the temperature stamped on the thermostat, the thermostat is satisfactory for use. If the temperature difference is greater, the thermostat should be replaced. ● **SEE FIGURE 32–10.**

2. **Infrared thermometer method.** An infrared thermometer (also called a pyrometer) can be used to measure the temperature of the coolant near the thermostat. The area on the engine side of the thermostat should be at the

TECH TIP

Do Not Take Out the Thermostat!

Some vehicle owners and technicians remove the thermostat in the cooling system to "cure" an overheating problem. In some cases, removing the thermostat can *cause* overheating rather than stop it. This is true for three reasons.

1. Without a thermostat the coolant can flow more quickly through the radiator. The thermostat adds some restriction to the coolant flow, and therefore keeps the coolant in the radiator longer. This also allows additional time for the heat transfer between the hot engine parts and the coolant. The presence of the thermostat thus ensures a greater reduction in the coolant temperature before it returns to the engine.
2. Heat transfer is greater with a greater difference between the coolant temperature and air temperature. Therefore, when coolant flow rate is increased (no thermostat), the temperature difference is reduced.
3. Without the restriction of the thermostat, much of the coolant flow often bypasses the radiator entirely and returns directly to the engine.

If overheating is a problem, removing the thermostat will usually not solve the problem. Remember, the thermostat controls the temperature of the engine coolant by opening at a certain temperature and closing when the temperature falls below the minimum rated temperature of the thermostat.

highest temperature that exists in the engine. A properly operating cooling system should cause the pyrometer to read as follows:

- As the engine warms, the temperature reaches near thermostat opening temperature.
- As the thermostat opens, the temperature drops just as the thermostat opens, sending coolant to the radiator.
- As the thermostat cycles, the temperature should range between the opening temperature of the thermostat and 20°F (11°C) above the opening temperature.

NOTE: If the temperature rises higher than 20°F (11°C) above the opening temperature of the thermostat, inspect the cooling system for a restriction or low coolant flow. A clogged radiator could also cause the excessive temperature rise.

FIGURE 32–11 Some thermostats are an integral part of the housing. This thermostat and radiator hose housing is serviced as an assembly. Some thermostats snap into the engine radiator fill tube underneath the pressure cap.

3. **Scan tool method.** A scan tool can be used on many vehicles to read the actual temperature of the coolant as detected by the engine coolant temperature (ECT) sensor. Although the sensor or the wiring to and from the sensor may be defective, at least the scan tool can indicate what the computer "thinks" is the engine coolant temperature.

THERMOSTAT REPLACEMENT Two important things about a thermostat include the following:

1. An overheating engine *may* result from a faulty thermostat.
2. An engine that does not get warm enough *always* indicates a faulty thermostat.

To replace the thermostat, coolant will have to be drained from the radiator drain petcock to lower the coolant level below the thermostat. It is not necessary to completely drain the system. The hose should be removed from the thermostat housing neck and then the housing removed to expose the thermostat. ● **SEE FIGURES 32–11 AND 32–12.**

The gasket flanges of the engine and thermostat housing should be cleaned, and the gasket surface of the housing must be flat. The thermostat should be placed in the engine with the sensing pellet *toward* the engine. Make sure that the thermostat position is correct, and install the thermostat housing with a new gasket or O-ring.

CAUTION: Failure to set the thermostat into the recessed groove will cause the housing to become tilted when tightened. If this happens and the housing bolts are tightened, the housing will usually crack, creating a leak.

The upper hose should then be installed and the system refilled. Install the correct size of radiator hose clamp. ● **SEE FIGURES 32–13 AND 32–14.**

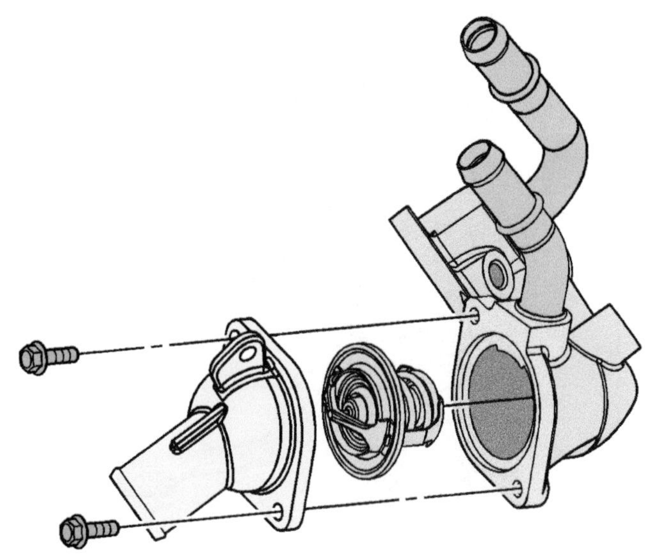

FIGURE 32–12 The thermostat can be removed after removing the bolts and cover. (Courtesy of General Motors)

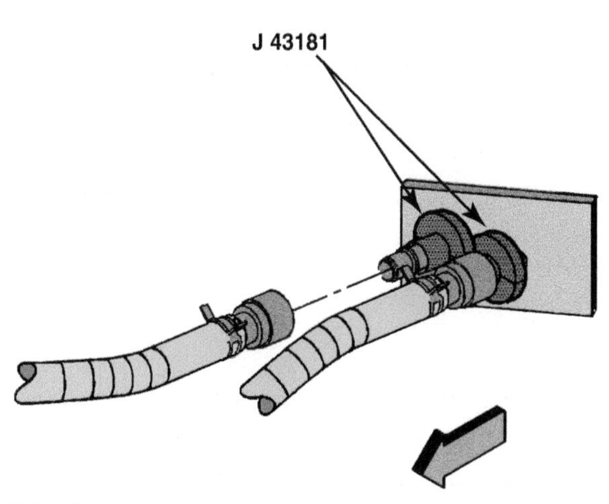

FIGURE 32–14 Some coolant hoses may have a quick-connect fitting that requires a special tool to disconnect. (Courtesy of General Motors)

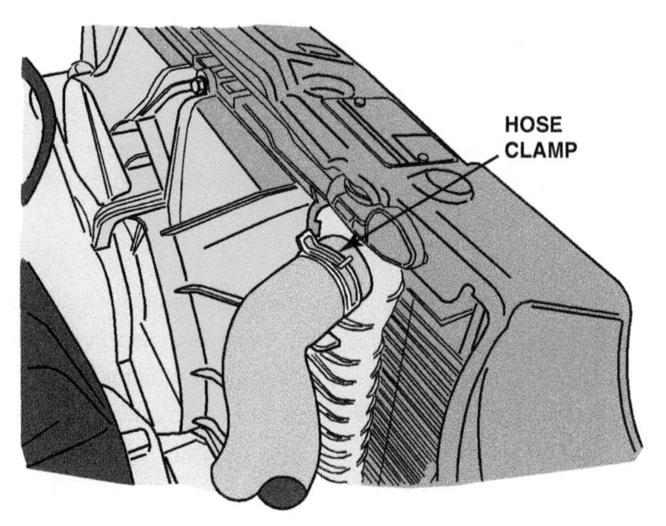

FIGURE 32–13 A spring-type hose clamp, as used on many General Motors vehicles. (Courtesy of General Motors)

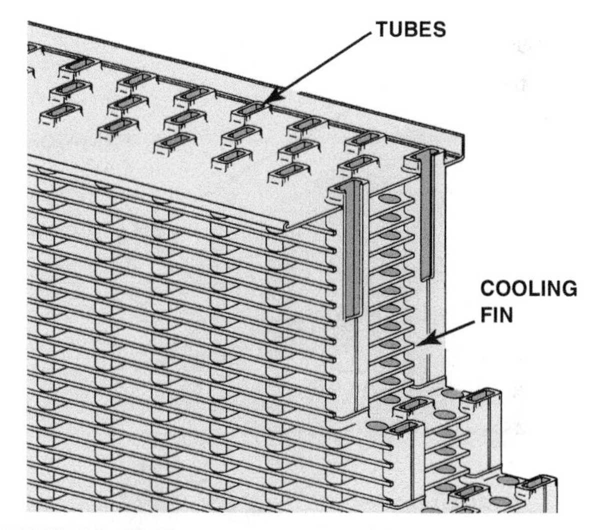

FIGURE 32–15 The tubes and fins of the radiator core.

RADIATORS

TYPES The two types of radiator cores in common use in most vehicles are:

- Serpentine fin core
- Plate fin core

In each of these types, the coolant flows through oval-shaped **core tubes**. Heat is transferred through the tube wall and soldered joint to **cooling fins**. The fins are exposed to the air that flows through the radiator, which removes heat from the radiator and carries it away. **SEE FIGURES 32–15 AND 32–16**.

Older automobile radiators were made from yellow brass. Since the 1980s, most radiators have been made from aluminum with nylon-reinforced plastic side tanks. These materials are corrosion resistant, have good heat transferability, and are easily formed.

Core tubes are made from 0.0045 to 0.012 inch (0.1 to 0.3 mm) thicksheet of brass or aluminum, using the thinnest possible materials for each application. The metal is rolled into round tubes and the joints are sealed with a locking seam.

The two basic designs of radiators include:

1. **Down-flow radiators.** This design was used mostly in older vehicles, where the coolant entered the radiator at the top and flowed downward, exiting the radiator at the bottom.
2. **Cross-flow radiators.** Most radiators use a cross-flow design, where the coolant flows from one side of the radiator to the opposite side.

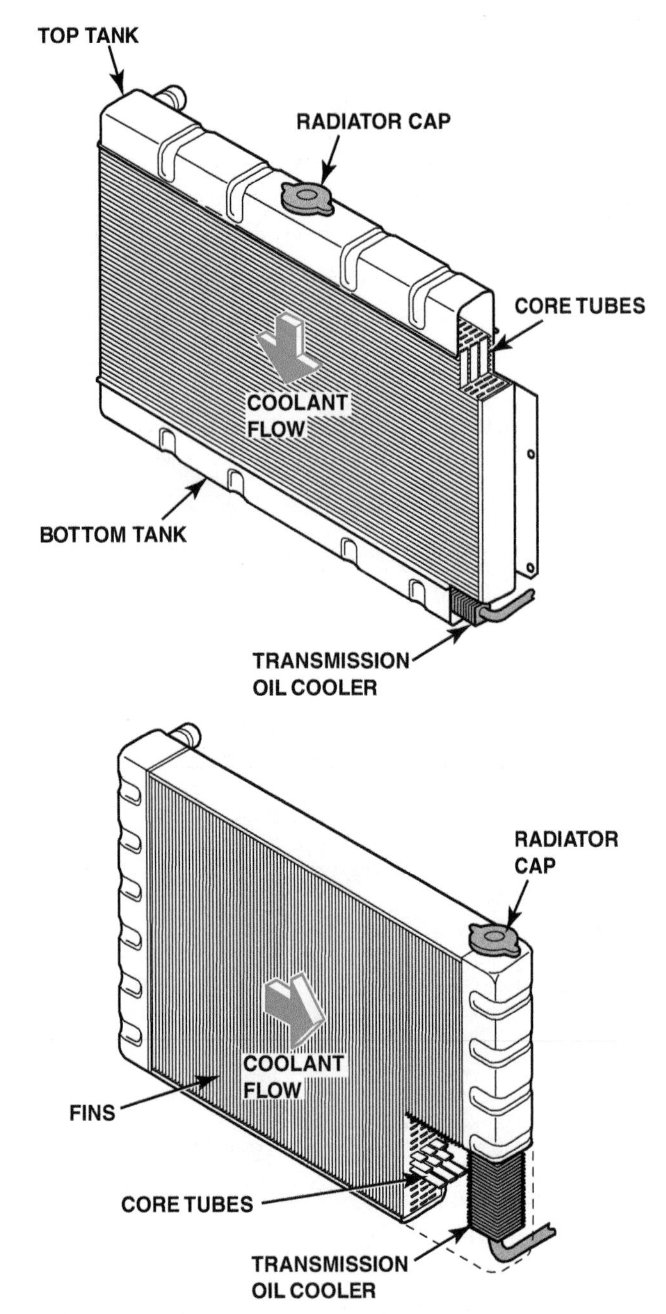

FIGURE 32–16 A radiator may be either a down-flow or a cross-flow type.

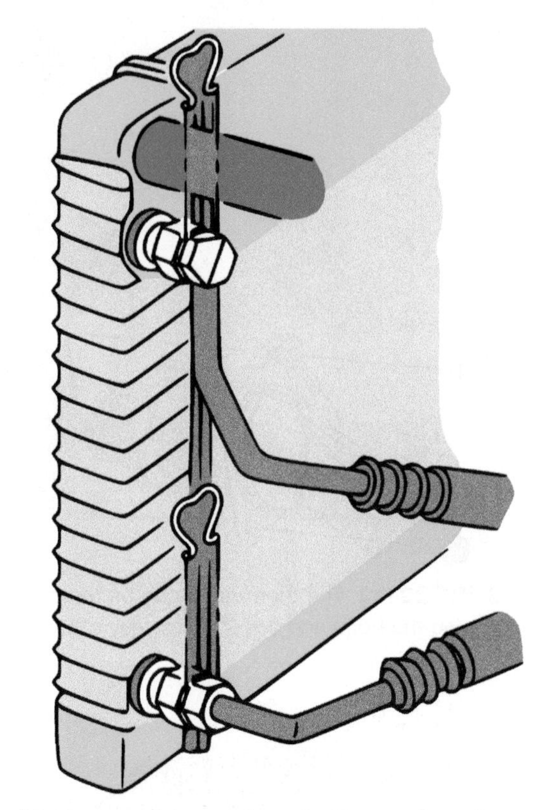

FIGURE 32–17 Many vehicles equipped with an automatic transmission use a transmission fluid cooler installed in one of the radiator tanks.

HOW RADIATORS WORK The main limitation of heat transfer in a cooling system is in the transfer from the radiator to the air. Heat transfers from the water to the fins as much as seven times faster than heat transfers from the fins to the air, assuming equal surface exposure. The radiator must be capable of removing an amount of heat energy approximately equal to the heat energy of the power produced by the engine. *Each horsepower is equivalent to 42 BTUs (10,800 calories) per minute.* As the engine power is increased, the heat-removing requirement of the cooling system is also increased.

With a given frontal area, radiator capacity may be increased by increasing the core thickness, packing more material into the same volume, or both. The radiator capacity may also be increased by placing a shroud around the fan so that more air will be pulled through the radiator.

NOTE: The lower air dam in the front of the vehicle is used to help direct the air through the radiator. If this air dam is broken or missing, the engine may overheat, especially during highway driving due to the reduced airflow through the radiator.

When a transmission oil cooler is used in the radiator, it is placed in the outlet tank, where the coolant has the lowest temperature. ● **SEE FIGURE 32–17.**

PRESSURE CAPS

OPERATION On many radiators the filler neck is fitted with a pressure cap. ● **SEE FIGURE 32–18**. The cap has a spring-loaded valve that closes the cooling system vent. This causes cooling pressure to build up to the pressure setting of the cap. At this point, the valve will release the excess

FIGURE 32–18 Many radiator caps are mounted on the radiator.

TECH TIP

Working Better Under Pressure

A problem that sometimes occurs with a high-pressure cooling system involves the water pump. For the pump to function, the inlet side of the pump must have a lower pressure than its outlet side. If inlet pressure is lowered too much, the coolant at the pump inlet can boil, producing vapor. The pump will then spin the coolant vapors and not pump coolant. This condition is called *pump cavitation.* Therefore, a radiator cap could be the cause of an overheating problem. A pump will not pump enough coolant if not kept under the proper pressure for preventing vaporization of the coolant.

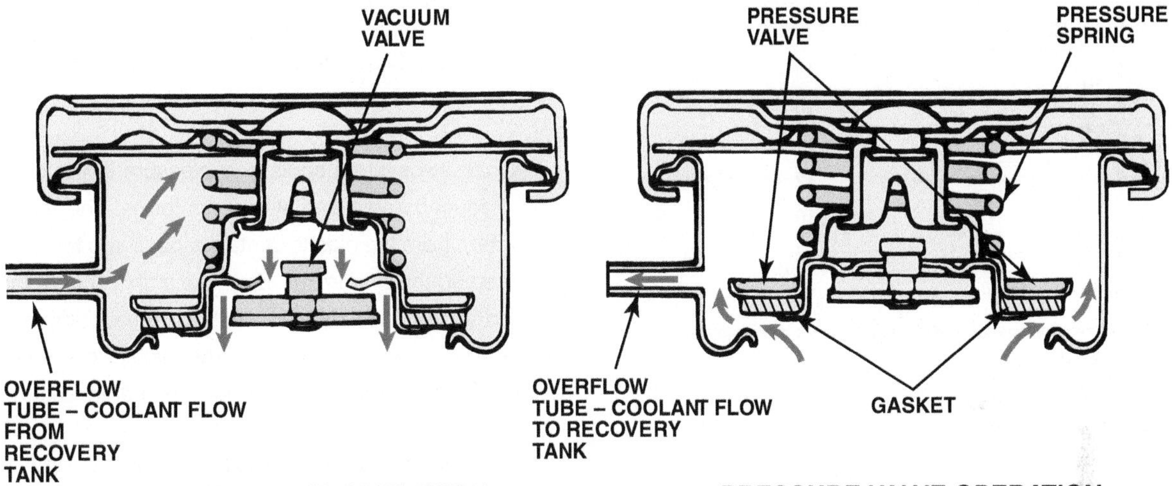

FIGURE 32–19 The pressure valve maintains the system pressure and allows excess pressure to vent. The vacuum valve allows coolant to return to the system from the recovery tank.

pressure to prevent system damage. Engine cooling systems are pressurized to raise the boiling temperature of the coolant.

- *The boiling temperature will increase by approximately 3°F (1.6°C) for each pound of increase in pressure.*
- At sea level, water will boil at 212°F (100°C). With a 15 PSI (100 kPa) pressure cap, water will boil at 257°F (125°C), which is a maximum operating temperature for an engine.

FUNCTIONS The specified coolant system temperature serves two functions:

1. It allows the engine to run at an efficient temperature, close to 200°F (93°C), with no danger of boiling the coolant.
2. The higher the coolant temperature, the more heat the cooling system can transfer. The heat transferred by the cooling system is proportional to the temperature difference between the coolant and the outside air. This characteristic has led to the design of small, high-pressure radiators that are capable of handling large quantities of heat. For proper cooling, the system must have the right pressure cap correctly installed.

A vacuum valve is part of the pressure cap and is used to allow coolant to flow back into the radiator when the coolant cools down and contracts. ● **SEE FIGURE 32–19**.

NOTE: The proper operation of the pressure cap is especially important at high altitudes. The boiling point of water is lowered by about 1°F for every 550 ft increase in altitude. Therefore, in Denver, Colorado (altitude 5,280 ft), the boiling point of water is about 202°F, and at the top of Pike's Peak in Colorado (14,110 ft) water boils at 186°F.

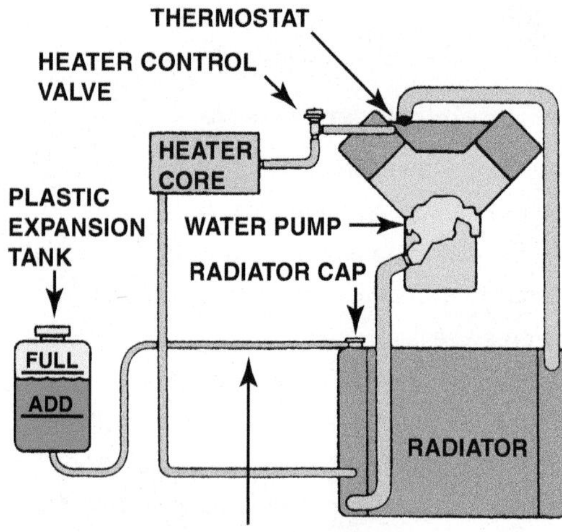

FIGURE 32–20 The level in the coolant recovery system raises and lowers with engine temperature.

BAR OR ATMOSPHERES	POUNDS PER SQUARE INCH (PSI)
1.1	16
1.0	15
0.9	13
0.8	12
0.7	10
0.6	9
0.5	7

CHART 32–2

Comparison showing the metric pressure as shown on the top of the cap to pounds per square inch (PSI).

METRIC RADIATOR CAPS According to the *SAE Handbook,* all radiator caps must indicate their nominal (normal) pressure rating. Most original equipment radiator caps are rated at about 14 to 16 PSI (97 to 110 kPa).

However, some vehicles use radiator pressure indicated in a unit called a **bar**. One bar is the pressure of the atmosphere at sea level, or about 14.7 PSI. The conversions in ● **CHART 32–2** can be used when replacing a radiator cap, to make certain it matches the pressure rating of the original.

COOLANT RECOVERY SYSTEMS

PURPOSE AND FUNCTION Excess pressure usually forces some coolant from the system through an overflow. Most cooling systems connect the overflow to a plastic reservoir to hold excess coolant while the system is hot. ● **SEE FIGURE 32–20.**

FIGURE 32–21 Some vehicles use a surge tank, which is located at the highest level of the cooling system, with a radiator cap. (Courtesy of Jeffrey Rehkopf)

When the system cools, the pressure in the cooling system is reduced and a partial vacuum forms. This vacuum pulls the coolant from the plastic container back into the cooling system, keeping the system full. Because of this action, the system is called a **coolant recovery system**. A vacuum valve allows coolant to reenter the system as the system cools so that the radiator parts will not collapse under the partial vacuum.

SURGE TANK Some vehicles use a **surge tank**, which is located at the highest level of the cooling system and holds about 1 quart (1 liter) of coolant. A hose attaches to the bottom of the surge tank to the inlet side of the water pump. A smaller bleed hose attaches to the side of the surge tank to the highest point of the radiator. The bleed line allows some coolant circulation through the surge tank, and air in the system will rise below the radiator cap and be forced from the system if the pressure in the system exceeds the rating of the radiator cap. ● **SEE FIGURE 32–21.**

WATER PUMPS

OPERATION The water pump (also called a coolant pump) is driven by one of two methods.

- Crankshaft belt
- Camshaft

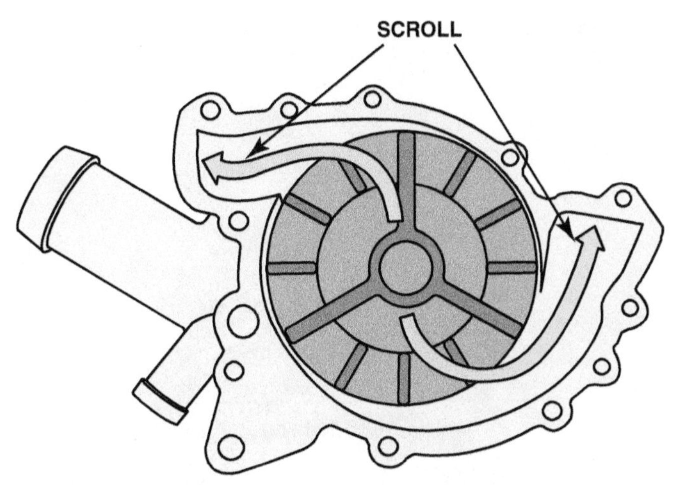

FIGURE 32–22 Coolant flow through the impeller and scroll of a coolant pump for a V-type engine.

FIGURE 32–23 A demonstration engine running on a stand, showing the amount of coolant flow that actually occurs through the cooling system.

Coolant recirculates from the radiator to the engine and back to the radiator. Low-temperature coolant leaves the radiator by the bottom outlet. It is pumped into the warm engine block, where it picks up some heat. From the block, the warm coolant flows to the hot cylinder head, where it picks up more heat.

NOTE: Some engines use **reverse cooling**. This means that the coolant flows from the radiator to the cylinder head(s) before flowing to the engine block.

Water pumps are not positive displacement pumps. The water pump is a **centrifugal pump** that can move a large volume of coolant without increasing the pressure of the coolant. The pump pulls coolant in at the center of the **impeller**. Centrifugal force throws the coolant outward so that it is discharged at the impeller tips. ● **SEE FIGURE 32–22**.

FIGURE 32–24 This severely corroded water pump could not circulate enough coolant to keep the engine cool. As a result, the engine overheated and blew a head gasket.

FREQUENTLY ASKED QUESTION

How Much Coolant Can a Water Pump Move?

A typical water pump can move a maximum of about 7,500 gallons (28,000 liters) of coolant per hour, or recirculate the coolant in the engine over 20 times per minute. This means that a water pump could be used to empty a typical private swimming pool in an hour! The slower the engine speed, the less power is consumed by the water pump. However, even at 35 mph (56 km/h), the typical water pump still moves about 2,000 gallons (7,500 liters) per hour or 0.5 gallon (2 liters) per second! ● **SEE FIGURE 32–23**.

As engine speeds increase, more heat is produced by the engine and more cooling capacity is required. The pump impeller speed increases as the engine speed increases to provide extra coolant flow at the very time it is needed.

The coolant leaving the pump impeller is fed through a **scroll**. The scroll is a smoothly curved passage that changes the fluid flow direction with minimum loss in velocity. The scroll is connected to the front of the engine so as to direct the coolant into the engine block. On V-type engines, two outlets are often used, one for each cylinder bank. Occasionally, diverters are necessary in the water pump scroll to equalize coolant flow between the cylinder banks of a V-type engine in order to equalize the cooling.

WATER PUMP SERVICE A worn impeller on a water pump can reduce the amount of coolant flow through the engine. ● **SEE FIGURE 32–24**.

FIGURE 32–25 The bleed weep hole in the water pump allows coolant to leak out of the pump and not be forced into the bearing. If the bearing failed, more serious damage could result.

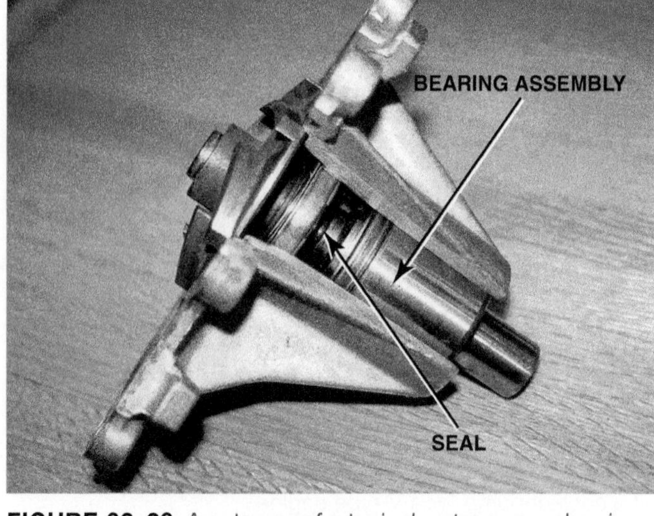

FIGURE 32–26 A cutaway of a typical water pump showing the long bearing assembly and the seal. The weep hole is located between the seal and the bearing. If the seal fails, then coolant flows out of the weep hole to prevent the coolant from damaging the bearing.

TECH TIP

Release the Belt Tension Before Checking a Water Pump

The technician should release water pump belt tension before checking for water pump bearing looseness. To test a water pump bearing, it is normal to check the fan for movement; however, if the drive belt is tight, any looseness in the bearing will not be felt.

If the seal of the water pump fails, then coolant will leak out of the weep hole. The hole allows coolant to escape without getting trapped and forced into the water pump bearing assembly. ● **SEE FIGURE 32–25**.

The hole allows coolant to escape without getting trapped and forced into the water pump bearing assembly.

If the bearing is defective, then the pump will usually be noisy and will have to be replaced. Before replacing a water pump that has failed because of a loose or noisy bearing, check all of the following:

1. Drive belt tension
2. Bent fan
3. Fan for balance

If the water pump drive belt is too tight, then excessive force may be exerted against the pump bearing. If the cooling fan is bent or out of balance, then the resulting vibration can damage the water pump bearing. ● **SEE FIGURE 32–26**.

COOLANT FLOW IN THE ENGINE

TYPES OF SYSTEMS Coolant flows through the engine in one of the following ways.

- **Parallel flow system.** In the **parallel flow system**, coolant flows into the block under pressure and then crosses the head gasket to the head through main coolant passages beside *each* cylinder.
- **Series flow system.** In the **series flow system**, the coolant flows around all the cylinders on each bank. All the coolant flows to the *rear* of the block, where large main coolant passages allow the coolant to flow across the head gasket. The coolant then enters the rear of the heads. In the heads, the coolant flows forward to a crossover passage on the intake manifold outlet at the *highest point* in the engine cooling passage. This is usually located at the front of the engine. The outlet is either on the heads or in the intake manifold.
- **Series-parallel flow system.** Some engines use a combination of these two coolant flow systems and call it a **series-parallel flow system**. Any steam that develops will go directly to the top of the radiator. In series flow systems, **bleed holes** or **steam slits** in the gasket, block, and head perform the function of letting out the steam.

FIGURE 32–27 A Chevrolet V-8 block that shows the large coolant holes and the smaller gas vent or bleed holes that must match the head gasket when the engine is assembled.

COOLANT FLOW AND HEAD GASKET DESIGN Most V-type engines use cylinder heads that are interchangeable side to side, but not all engines. Therefore, based on the design of the cooling system and flow through the engine, it is very important to double check that the cylinder head is matched to the block and that the head gasket is installed correctly (end for end) so that all of the cooling passages are open to allow the proper flow of coolant through the system. ● **SEE FIGURE 32–27.**

ENGINE OIL COOLER The coolant can also be directed through an oil cooler to cool the oil during high-temperature conditions. The oil cooler also helps to warm the engine oil when the engine is first started in cold weather. Because the coolant usually reaches operating temperature before the oil, the cooler can also heat the cold engine oil so it reaches normal operating temperature more quickly, thereby helping to reduce engine wear. ● **SEE FIGURE 32–28.**

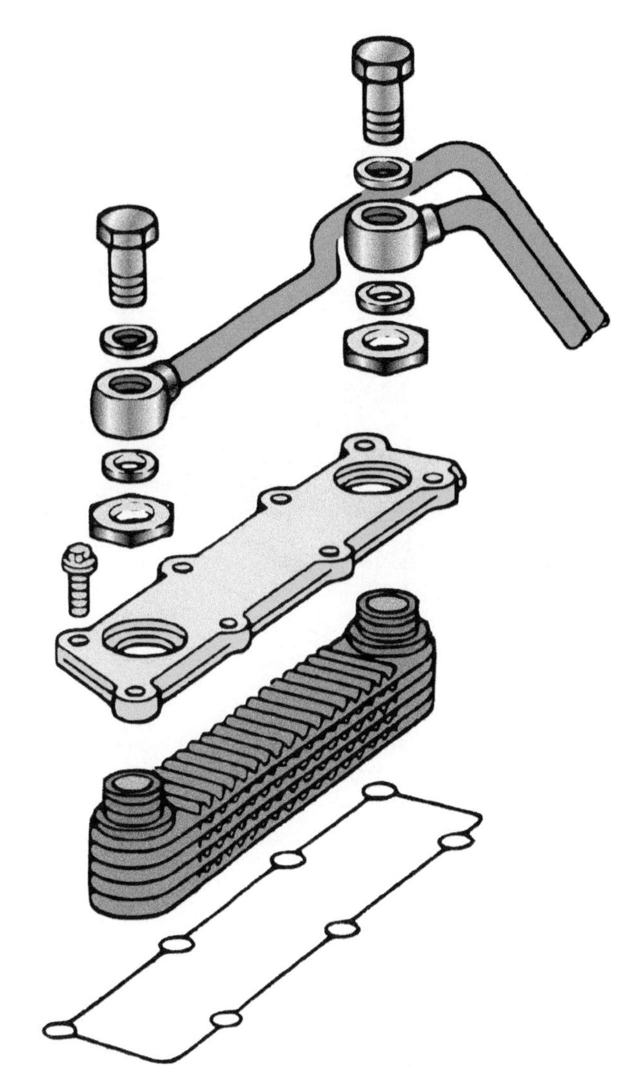

FIGURE 32–28 This engine oil cooler mounts in the water jacket of the engine block. (Courtesy of General Motors)

COOLING FANS

ELECTRONICALLY CONTROLLED COOLING FAN Two types of electric cooling fans used on many engines include:

- One two-speed cooling fan
- Two cooling fans (one for normal cooling and one for high heat conditions)

The PCM commands low-speed fans on under the following conditions:

- ECT exceeds approximately 223°F (106°C).
- A/C refrigerant pressure exceeds 190 PSI (1,310 kPa).
- After the vehicle is shut off, the engine coolant temperature at key-off is greater than 284°F (140°C) and system voltage is more than 12 volts. The fan(s) will stay on for approximately three minutes.

The PCM commands the high-speed fan on under the following conditions:

- ECT reaches 230°F (110°C).
- A/C refrigerant pressure exceeds 240 PSI (1,655 kPa).
- Certain diagnostic trouble codes (DTCs) set.

To prevent a fan from cycling on and off excessively at idle, the fan may not turn off until the ignition switch is moved to the off position or the vehicle speed exceeds approximately 10 mph (16 km/h).

Many rear-wheel-drive vehicles and all transverse engines drive the fan with an electric motor. ● **SEE FIGURE 32–29.**

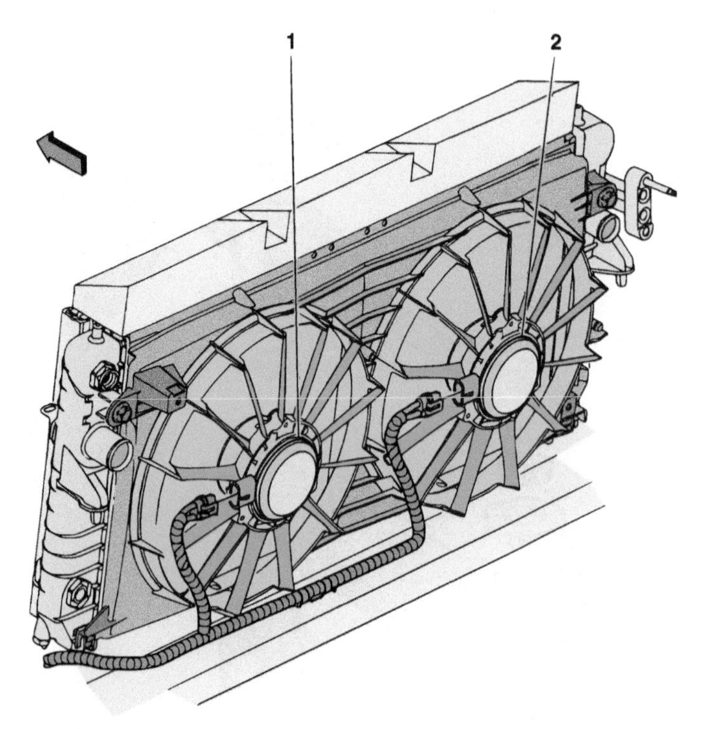

FIGURE 32–29 These cooling fans can be individually controlled by the powertrain control module (PCM). (Courtesy of General Motors)

FIGURE 32–30 A typical engine-driven thermostatic spring cooling fan.

NOTE: Most electric cooling fans are computer controlled. To save energy, the cooling fans are turned off whenever the vehicle is traveling faster than 35 mph (55 km/h). The ram air caused by the vehicle speed is enough to keep the radiator cool. Of course, if the computer senses that the temperature is still too high, the computer will turn on the cooling fan, to "high," if possible, in an attempt to cool the engine to avoid severe engine damage.

THERMOSTATIC FANS On some rear-wheel-drive vehicles, a thermostatic cooling fan is driven by a belt from the crankshaft. It turns faster as the engine turns faster. Generally, the engine is required to produce more power at higher speeds. Therefore, the cooling system will also transfer more heat. Increased fan speed aids in the required cooling. Engine heat also becomes critical at low engine speeds in traffic where the vehicle moves slowly. The thermostatic fan is designed so that it uses little power at high engine speeds and minimizes noise. Three types of thermostatic fans include:

1. **Silicone coupling.** The **silicone coupling** fan drive is mounted between the drive pulley and the fan.

WARNING

Some electric cooling fans can come on after the engine is off without warning. Always keep hands and fingers away from the cooling fan blades unless the electrical connector has been disconnected to prevent the fan from coming on. Always follow all warnings and cautions.

NOTE: When diagnosing an overheating problem, look carefully at the cooling fan. If silicone is leaking, then the fan may not be able to function correctly and should be replaced.

2. **Thermostatic spring.** A second type of thermal fan has a **thermostatic spring** added to the silicone coupling fan drive. The thermostatic spring operates a valve that allows the fan to freewheel when the radiator is cold. As the radiator warms to about 150°F (65°C), the air hitting the thermostatic spring will cause the spring to change its shape. The new shape of the spring opens a valve that allows the drive to operate like the silicone coupling drive. When the engine is very cold, the fan may operate at high speeds for a short time until the drive fluid warms slightly. The silicone fluid will then flow into a reservoir to let the fan speed drop to idle. ● **SEE FIGURE 32–30**.

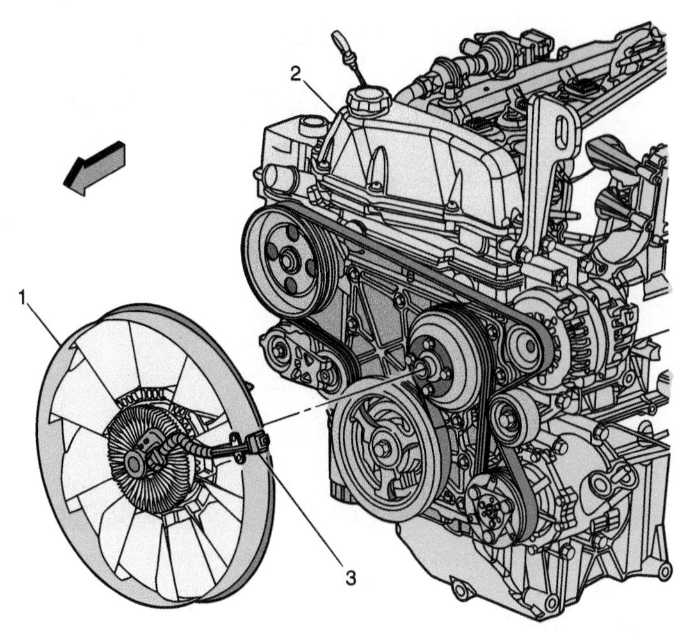

FIGURE 32–31 An electronically controlled fan hub showing the fan (1), the engine (2), and the electrical connector (3). (Courtesy of General Motors)

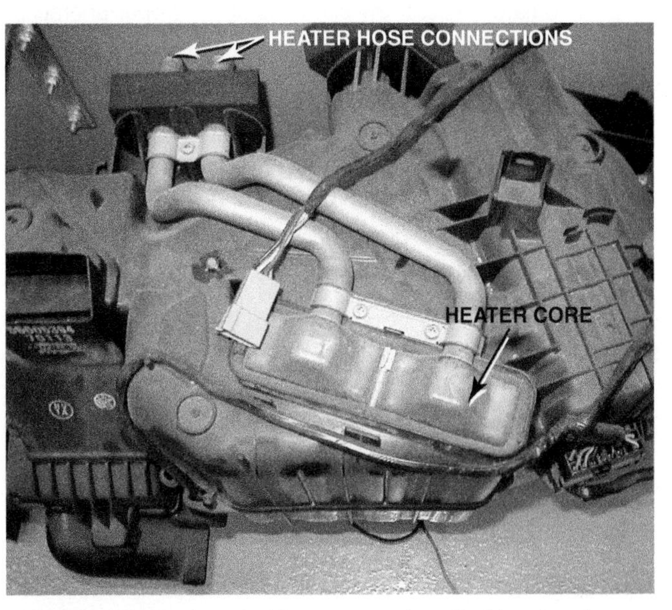

FIGURE 32–32 A typical heater core installed in a heating, ventilation, and air-conditioning (HVAC) housing assembly.

TECH TIP

Be Sure to Always Use a Fan Shroud

A fan shroud forces the fan to draw air through the radiator. If a fan shroud is not used, then air is drawn from around the fan and will reduce the airflow through the radiator. Many overheating problems are a result of not replacing the factory shroud after engine work or body repair work to the front of the vehicle.

3. Another version of the silicone fan coupling uses an internal solenoid to control the flow of the viscous fluid within the fan hub. The solenoid is controlled by a relay, which in turn is controlled by the vehicle's powertrain control module (PCM). ● **SEE FIGURE 32–31**.

The fan is designed to move enough air at the lowest fan speed to cool the engine when it is at its highest coolant temperature. The fan shroud is used to increase the cooling system efficiency. The fan or fan clutch may be bolted or threaded onto the water pump fan hub. ● **SEE FIGURE 32–31**.

HEATER CORES

PURPOSE AND FUNCTION Most of the heat absorbed from the engine by the cooling system is wasted. Some of this heat, however, is recovered by the vehicle heater. Heated coolant is passed through tubes in the small core of the heater. Air is passed across the heater fins and is then sent to the passenger compartment. In some vehicles, the heater and air conditioning work in series to maintain vehicle compartment temperature. ● **SEE FIGURE 32–32**.

HEATER PROBLEM DIAGNOSIS When the heater does not produce the desired amount of heat, many owners and technicians replace the thermostat before doing any other troubleshooting. It is true that a defective thermostat is the reason for the *engine* not to reach normal operating temperature, but there are many other causes besides a defective thermostat that can result in lack of heat from the heater. To determine the exact cause, follow this procedure.

STEP 1 After the engine has been operated, feel the upper radiator hose. If the engine is up to proper operating temperature, the upper radiator hose should be too hot to hold. The hose should also be pressurized.

a. If the hose is not hot enough, replace the thermostat.

b. If the hose is not pressurized, test or replace the radiator pressure cap if it will not hold the specified pressure.

c. If okay, see step 2.

STEP 2 With the engine running, feel both heater hoses. (The heater should be set to the maximum heat position.) Both hoses should be too hot to hold. If both hoses are warm (not hot) or cool, check the heater control valve for proper operation (if equipped). If one hose is hot and the other (return) is just warm or cool, then remove both hoses from the heater core or engine and flush the heater core with water from a garden hose.

STEP 3 If both heater hoses are hot and there is still a lack of heating concern, then the fault is most likely due to an airflow blend door malfunction. Check service information for the exact procedure to follow.

NOTE: Heat from the heater that "comes and goes" is most likely the result of low coolant level. Usually with the engine at idle, there is enough coolant flow through the heater. At higher engine speeds, however, the lack of coolant through the heads and block prevents sufficient flow through the heater.

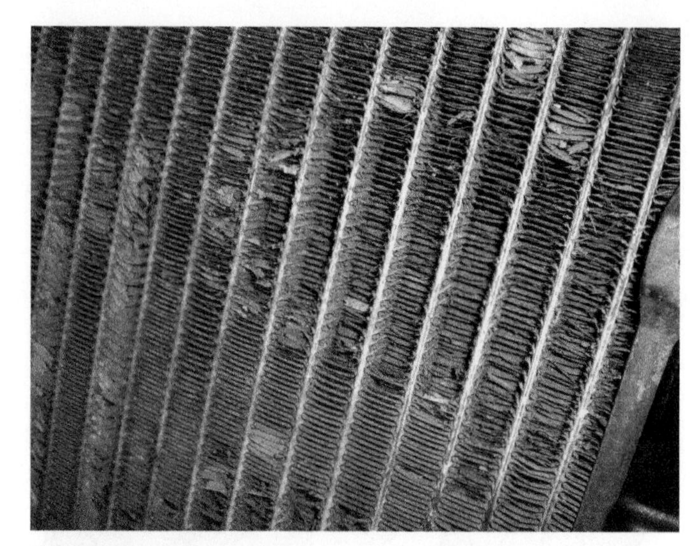

FIGURE 32–33 A heavily corroded radiator from a vehicle that was overheating. A visual inspection discovered that the corrosion had eaten away many of the cooling fins, yet did not leak. This radiator was replaced and it solved the overheating problem.

COOLING SYSTEM TESTING

VISUAL INSPECTION Many cooling system faults can be found by performing a thorough visual inspection. Items that can be inspected visually include:

- Water pump drive belt for tension or faults
- Cooling fan for faults
- Heater and radiator hoses for condition and leaks
- Coolant overflow or surge tank coolant level
- Evidence of coolant loss
- Radiator condition ● **SEE FIGURE 32–33**.

PRESSURE TESTING Pressure testing using a hand-operated pressure tester is a quick and easy cooling system test. The radiator cap is removed (engine cold!) and the tester is attached in the place of the radiator cap. By operating the plunger on the pump, the entire cooling system is pressurized. ● **SEE FIGURE 32–34**.

FIGURE 32–34 Pressure testing the cooling system. A typical hand-operated pressure tester applies pressure equal to the radiator cap pressure. The pressure should hold; if it drops, this indicates a leak somewhere in the cooling system. An adapter is needed for vehicles that use a threaded pressure cap.

CAUTION: Do not pump up the pressure beyond that specified by the vehicle manufacturer. Most systems should not be pressurized beyond 14 PSI (100 kPa). If a greater pressure is used, then it may cause the water pump, radiator, heater core, or hoses to fail.

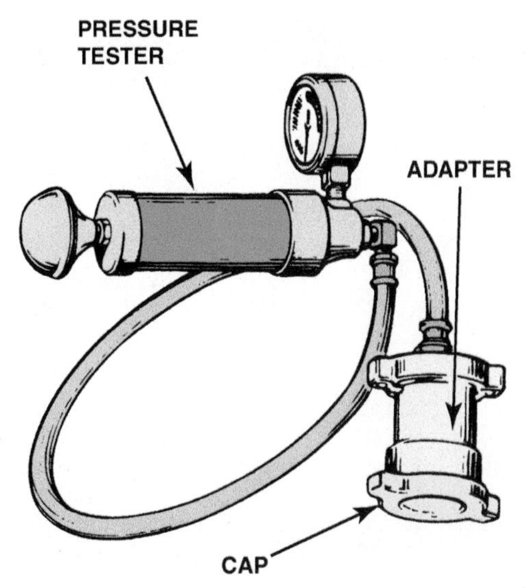

FIGURE 32–35 The pressure cap should be checked for proper operation using a pressure tester as part of the cooling system diagnosis.

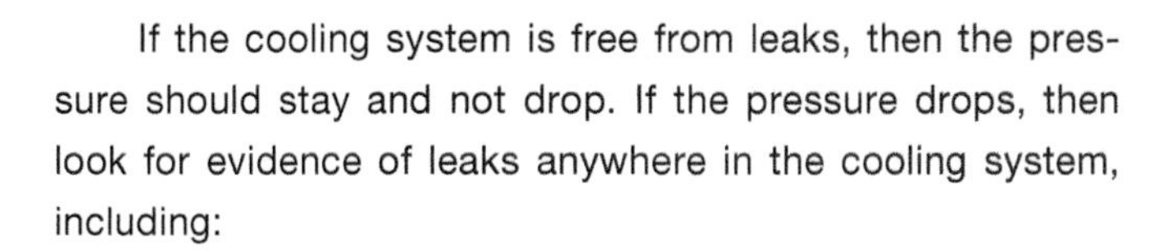

If the cooling system is free from leaks, then the pressure should stay and not drop. If the pressure drops, then look for evidence of leaks anywhere in the cooling system, including:

1. Heater hoses
2. Radiator hoses
3. Radiator
4. Heater core
5. Cylinder head
6. Core plugs in the side of the block or cylinder head

Pressure testing should be performed whenever there is a leak or suspected leak. The pressure tester can also be used to test the radiator cap. An adapter is used to connect the pressure tester to the radiator cap. Replace any cap that will not hold pressure. ● **SEE FIGURE 32–35**.

COOLANT DYE LEAK TESTING One of the best methods to check for a coolant leak is to use a fluorescent dye in the coolant, one that is specifically designed for coolant. Operate the vehicle with the dye in the coolant until the engine reaches normal operating temperature. Use a black light to inspect all areas of the cooling system. When there is a leak, it will be easy to spot because the dye in the coolant will be seen as bright green. ● **SEE FIGURE 32–36**.

FIGURE 32–36 Use dye specifically made for coolant when checking for leaks using a black light.

FIGURE 32–37 When an engine overheats, often the coolant overflow container boils.

COOLANT TEMPERATURE WARNING LIGHT

PURPOSE AND FUNCTION Most vehicles are equipped with a heat sensor for the engine operating temperature indicator light. If the warning light comes on during driving (or the temperature gauge goes into the red danger zone), then the coolant temperature is about 250°F to 258°F (120°C to 126°C), which is still *below* the boiling point of the coolant (assuming a properly operating pressure cap and system). ● **SEE FIGURE 32–37**.

PRECAUTIONS If the coolant temperature warning light comes on, then follow these steps.

STEP 1 Shut off the air conditioning and turn on the heater. The heater will help rid the engine of extra heat. Set the blower speed to high.

STEP 2 If possible, shut the engine off and let it cool. (This may take over an hour.)

STEP 3 Never remove the radiator cap when the engine is hot.

STEP 4 Do *not* continue to drive with the hot light on, or serious damage to your engine could result.

STEP 5 If the engine does not feel or smell hot, it is possible that the problem is a faulty hot light sensor or gauge. Continue to drive, but to be safe, stop occasionally and check for any evidence of overheating or coolant loss.

COMMON CAUSES OF OVERHEATING Overheating can be caused by defects in the cooling system, such as the following:

1. Low coolant level
2. Plugged, dirty, or blocked radiator
3. Defective fan clutch or electric fan
4. Incorrect ignition timing (if adjustable)
5. Low engine oil level
6. Broken fan drive belt
7. Defective radiator cap
8. Dragging brakes
9. Frozen coolant (in freezing weather)
10. Defective thermostat
11. Defective water pump (the impeller slipping on the shaft internally)
12. Blocked cooling passages in the block or cylinder head(s)

COOLING SYSTEM INSPECTION

COOLANT LEVEL The cooling system is one of the most maintenance-free systems in the engine. Normal maintenance involves an occasional check on the coolant level. It should also include a visual inspection for signs of coolant system leaks and for the condition of the coolant hoses and fan drive belts.

CAUTION: The coolant level should only be checked when the engine is cool. Removing the pressure cap from a hot engine will release the cooling system pressure while the coolant temperature is above its atmospheric boiling temperature. When the cap is removed, the pressure will instantly drop to atmospheric pressure level, causing the coolant to boil immediately. Vapors from the boiling liquid will blow coolant from the system. Coolant will be lost, and someone may be injured or burned by the high-temperature coolant that is blown out of the filler opening.

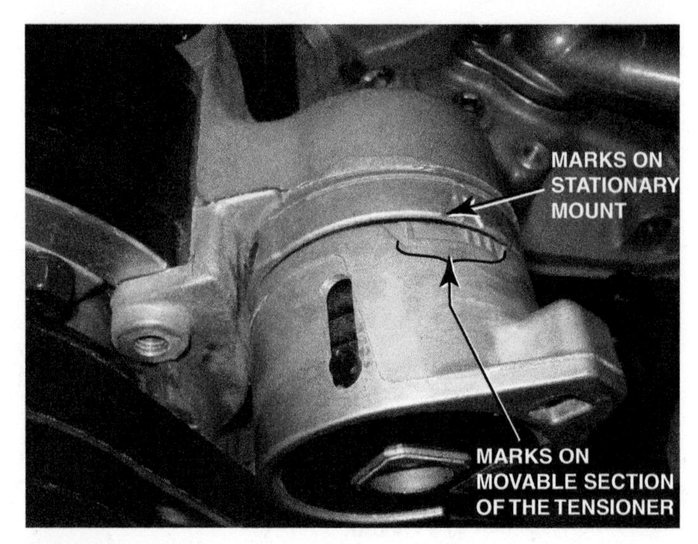

FIGURE 32–38 Typical marks on an accessory drive belt tensioner.

ACCESSORY DRIVE BELT TENSION Drive belt condition and proper installation are important for the proper operation of the cooling system.

There are a number of ways vehicle manufacturers specify that the belt tension is within factory specifications.

1. **Belt tension gauge.** A belt tension gauge is needed to achieve the specified belt tension. Install the belt and operate the engine with all of the accessories turned on, to run in the belt for at least five minutes. Adjust the tension of the accessory drive belt to factory specifications or use ● **CHART 32–3** for an example of the proper tension based on the size of the belt. Replace any serpentine belt that has more than three cracks in any one rib that appears in a 3 inch span.
2. **Marks on the tensioner.** Many tensioners have marks that indicate the normal operating tension range for the accessory drive belt. Check service information for the location of the tensioner mark. ● **SEE FIGURE 32–38.**
3. **Torque wrench reading.** Some vehicle manufacturers specify that a beam-type torque wrench be used to determine the torque needed to rotate the tensioner. If the torque reading is below specifications, the tensioner must be replaced.

NUMBER OF RIBS USED	TENSION RANGE (LB.)
3	45 – 60
4	60 – 80
5	75 – 100
6	90 – 125
7	105 – 145

CHART 32–3

The number of ribs determines the tension range of the belt.

TECH TIP

The Water Spray Trick

Lower-than-normal alternator output could be the result of a loose or slipping drive belt. All belts (V and serpentine multigroove) use an interference angle between the angle of the Vs of the belt and the angle of the Vs on the pulley. A belt wears this interference angle off the edges of the V of the belt. As a result, the belt may start to slip and make a squealing sound even if tensioned properly.

A common trick to determine if the noise is from the belt is to spray water from a squirt bottle at the belt with the engine running. If the noise stops, the belt is the cause of the noise. The water quickly evaporates and therefore, water just finds the problem—it does not provide a short-term fix.

4. **Deflection.** Depress the belt between the two pulleys that are the farthest apart and the flex or deflection should be about 1/2 inch.

COOLING SYSTEM SERVICE

FLUSHING COOLANT Flushing the cooling system includes the following steps:

STEP 1 Drain the system (dispose of the old coolant correctly).

STEP 2 Fill the system with clean water and flushing/cleaning chemical.

STEP 3 Start the engine until it reaches operating temperature with the heater on.

STEP 4 Drain the system and fill with clean water.

FIGURE 32–39 Many vehicle manufacturers recommend that the bleeder valve be opened whenever refilling the cooling system. (Courtesy of Jeffrey Rehkopf)

STEP 5 Repeat until drain water runs clear (any remaining flush agent will upset pH).

STEP 6 Fill the system with 50/50 antifreeze/water mix or premixed coolant.

STEP 7 Start the engine until it reaches operating temperature with the heater on.

STEP 8 Adjust coolant level as needed.

Bleeding the air out of the cooling system is important because air can prevent proper operation of the heater and can cause the engine to overheat. ● **SEE FIGURE 32–39.**

In most systems, small air pockets can occur. The engine must be thoroughly warmed to open the thermostat. This allows full coolant flow to remove the air pockets. The heater must also be turned to full heat. On some vehicles a special tool can be used to remove all of the air from the system. Always refer to service information for the proper procedure. ● **SEE FIGURE 32–40.**

NOTE: The Vac-n-Fill system is a required tool when servicing the Chevrolet Volt and some other GM vehicles. The Volt high voltage battery cooling system MUST be filled using the Vac-n-Fill equipment.

COOLANT EXCHANGE MACHINE Many coolant exchange machines are able to perform one or more of the following operations.

- Exchange old coolant with new coolant
- Flush the cooling system
- Pressure or vacuum check the cooling system for leaks

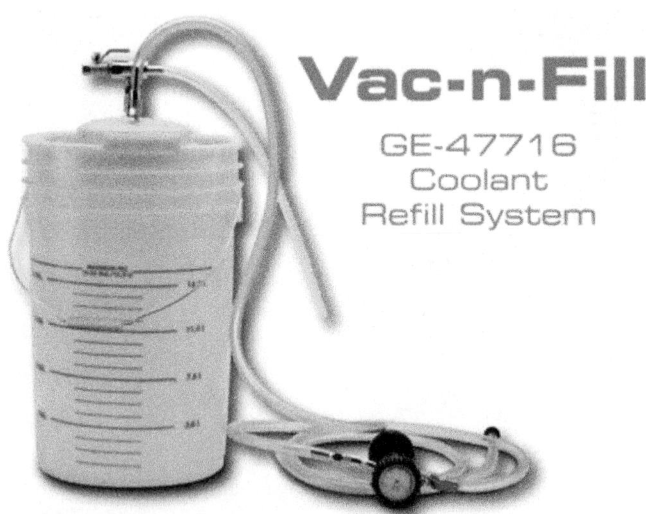

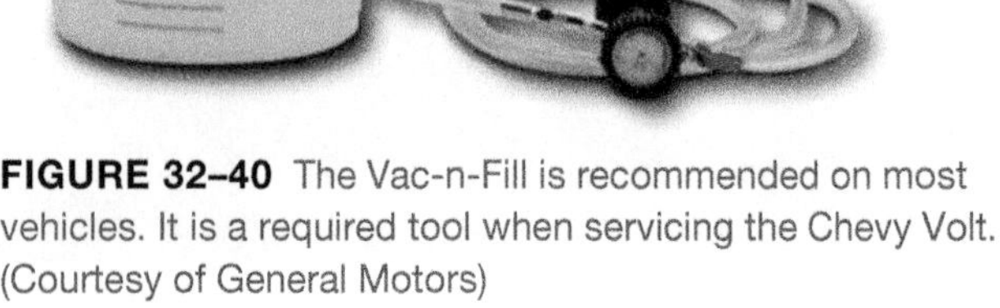

FIGURE 32–40 The Vac-n-Fill is recommended on most vehicles. It is a required tool when servicing the Chevy Volt. (Courtesy of General Motors)

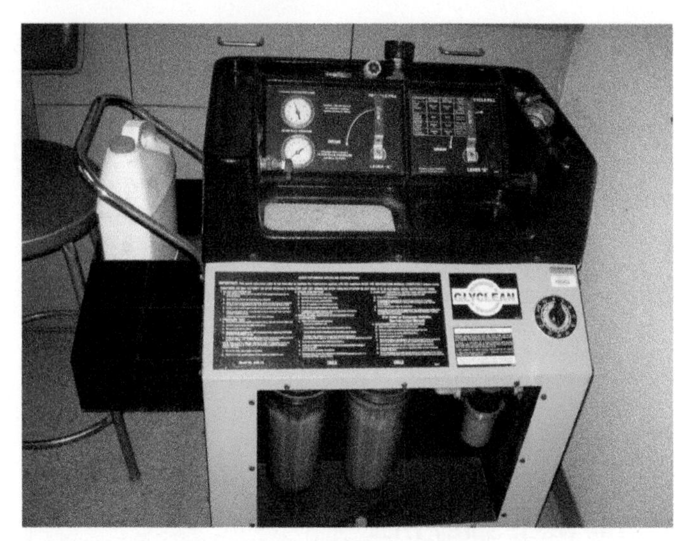

FIGURE 32–41 Using a coolant exchange machine helps eliminate the problem of air getting into the system which can cause overheating or lack of heat due to air pockets getting trapped in the system.

TECH TIP

Always Replace the Pressure Cap

Replace the old radiator cap with a new cap with the same pressure rating. The cap can be located on the following:

1. Radiator
2. Coolant recovery reservoir
3. Upper radiator hose

WARNING

Never remove a pressure cap from a hot engine. When the pressure is removed from the system, the coolant will immediately boil and will expand upward, throwing scalding coolant in all directions. Hot coolant can cause serious burns.

The use of a coolant exchange machine pulls a vacuum on the cooling system which helps illuminate air pockets from forming during coolant replacement. If an air pocket were to occur, then the following symptoms may occur:

1. **Lack of heat from the heater.** Air rises and can form in the heater core, which will prevent coolant from flowing.
2. **Overheating.** The engine can overheat due to the lack of proper coolant flow through the system.

Always follow the operating instructions for the coolant exchange machine being used. ● **SEE FIGURE 32–41.**

HOSE INSPECTION Coolant system hoses are critical to engine cooling. As the hoses get old, they become either soft or brittle and sometimes swell in diameter. Their condition depends on their material and on the engine service conditions. If a hose breaks while the engine is running, all coolant will be lost. A hose should be replaced any time it appears to be abnormal. ● **SEE FIGURE 32–42.**

NOTE: To make hose removal easier and to avoid possible damage to the radiator, use a utility knife and slit the hose lengthwise. Then simply peel the hose off.

The hose and hose clamp should be positioned so that the clamp is close to the bead on the neck. This is especially important on aluminum hose necks to avoid corrosion. When the hoses are in place and the drain petcock is closed, the cooling system can be refilled with the correct coolant mixture.

DISPOSING OF USED COOLANT Used coolant drained from vehicles should be disposed of according to state or local laws. Some communities permit draining into the sewer. Ethylene glycol will easily biodegrade. There could be problems with groundwater contamination, however, if coolant is spilled on open ground. Check with recycling companies authorized by local or state governments for the exact method recommended for disposal in your area.

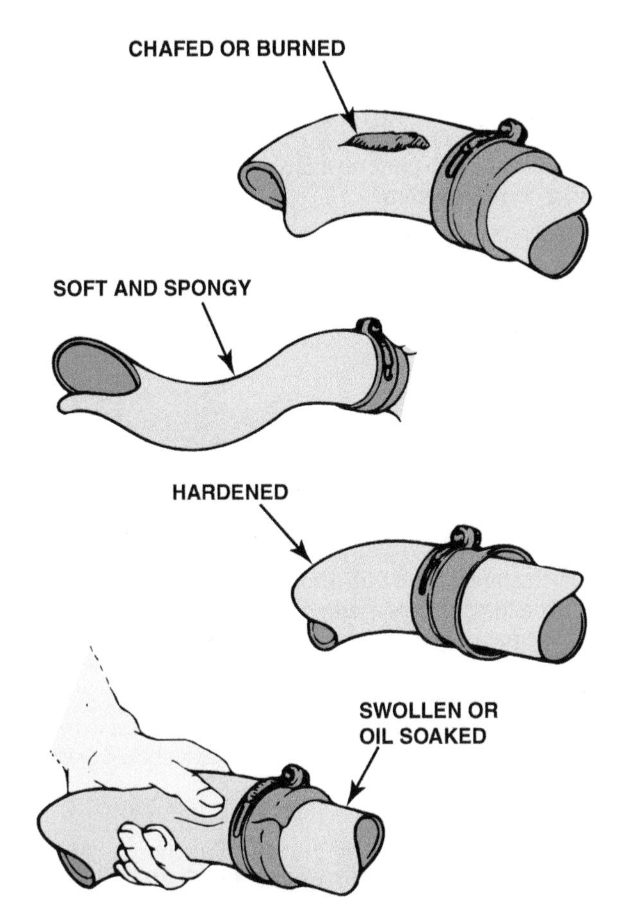

FIGURE 32–42 All cooling system hoses should be checked for wear or damage.

CLEANING THE RADIATOR EXTERIOR Overheating can result from exterior and interior radiator plugging. External plugging is caused by dirt and insects. This type of plugging can be seen if you look straight through the radiator while a light is held behind it. It is most likely to occur on off-road vehicles. The plugged exterior of the radiator core can usually be cleaned with water pressure from a hose. The water is aimed at the *engine side* of the radiator. The water should flow freely through the core at all locations. If this does not clean the core, the radiator should be removed for cleaning at a radiator shop.

FIGURE 32–43 The top 3/8 inch hose is designed for oil and similar liquids, whereas the 3/8 inch hose below is labeled "heater hose" and is designed for coolant.

TECH TIP

Always Use Heater Hoses Designed for Coolant

Many heater hoses are sizes that can also be used for other purposes such as oil lines. Always check and use hose that states it is designed for heater or cooling system use. ● **SEE FIGURE 32–43.**

TECH TIP

Quick and Easy Cooling System Problem Diagnosis

1. If overheating occurs in slow stop-and-go traffic, the usual cause is low airflow through the radiator. Check for airflow blockages or cooling fan malfunction.
2. If overheating occurs at highway speeds, the cause is usually a radiator or coolant circulation problem. Check for a restricted or clogged radiator.

SUMMARY

1. The purpose and function of the cooling system is to maintain proper engine operating temperature.
2. The thermostat controls engine coolant temperature by opening at its rated opening temperature to allow coolant to flow through the radiator.
3. Coolant fans are designed to draw air through the radiator to aid in the heat transfer process, drawing the heat from the coolant and transferring it to the outside air through the radiator.
4. The cooling system should be tested for leaks using a hand-operated pressure pump.
5. Water pumps are usually engine driven and circulate coolant through the engine and the radiator when the thermostat opens.
6. Coolant flows through the radiator hoses to and from the engine and through heater hoses to send heated coolant to the heater core in the passenger compartment.

REVIEW QUESTIONS

1. What is normal operating coolant temperature?
2. Explain the flow of coolant through the engine and radiator.
3. Why is a cooling system pressurized?
4. What is the purpose of the coolant system bypass?
5. Describe how to perform a drain, flush, and refill procedure on a cooling system.
6. Explain the operation of a thermostatic cooling fan.
7. Describe how to diagnose a heater problem.
8. What are 10 common causes of overheating?

CHAPTER QUIZ

1. The upper radiator collapses when the engine cools. What is the most likely cause?
   a. Defective upper radiator hose
   b. Missing spring from the upper radiator hose, which is used to keep it from collapsing
   c. Defective thermostat
   d. Defective pressure cap
2. What can be done to prevent air from getting trapped in the cooling system when the coolant is replaced?
   a. Pour the coolant into the radiator slowly.
   b. Use a coolant exchange machine that draws a vacuum on the system.
   c. Open the air bleeder valves while adding coolant.
   d. Either b or c
3. Heat transfer is improved from the coolant to the air when the ____________.
   a. Temperature difference is great
   b. Temperature difference is small
   c. Coolant is 95% antifreeze
   d. Both a and c
4. A water pump is a positive displacement type of pump.
   a. True
   b. False
5. Water pumps ____________.
   a. Work only at idle and low speeds and are disengaged at higher speeds
   b. Use engine oil as a lubricant and coolant
   c. Are driven by the engine crankshaft or camshaft
   d. Disengage during freezing weather to prevent radiator failure
6. What diagnostic trouble code (DTC) could be set if the thermostat is defective?
   a. P0300
   b. P0171
   c. P0440
   d. P0128
7. Which statement is *true* about thermostats?
   a. The temperature marked on the thermostat is the temperature at which the thermostat should be fully open.
   b. Thermostats often cause overheating.
   c. The temperature marked on the thermostat is the temperature at which the thermostat should start to open.
   d. Both a and b
8. Technician A says that the radiator should always be inspected for leaks and proper flow before installing a rebuilt engine. Technician B says that overheating during slow city driving can only be due to a defective electric cooling fan. Which technician is correct?
   a. Technician A only
   b. Technician B only
   c. Both Technicians A and B
   d. Neither Technician A nor B
9. A customer complains that the heater works sometimes, but sometimes only cold air comes out while driving. Technician A says that the water pump is defective. Technician B says that the cooling system could be low on coolant. Which technician is correct?
   a. Technician A only
   b. Technician B only
   c. Both Technicians A and B
   d. Neither Technician A nor B
10. The normal operating temperature (coolant temperature) of an engine equipped with a 195°F thermostat is ____________.
    a. 175°F to 195°F
    b. 185°F to 205°F
    c. 195°F to 215°F
    d. 175°F to 215°F

chapter 33

HEATING SYSTEM OPERATION AND DIAGNOSIS

LEARNING OBJECTIVES

After studying this chapter, the reader should be able to:

1. Discuss the operation of heating systems.
2. Discuss the diagnosis of heating systems.
3. Explain the operation of electrically heated seats.
4. Explain the operation of heated and cooled seats.
5. Explain the operation of heated steering wheel.

This chapter will help you prepare for the ASE Heating and Air Conditioning (A7) certification test content area "C" (Heating and Engine Cooling Systems Diagnosis, and Repair).

KEY TERMS

Cellular 539
Control valve 539
Heater core 538
Inlet hose 539
Outlet hose 539
Peltier effect 541
Thermoelectric device (TED) 541

GM STC OBJECTIVES

GM Service Technical College topics covered in this chapter are as follows:

1. Air flow in the heating system. (01510.01W)
2. The purpose and components of the heating system. (01510.05W)
3. The procedure for replacing a heater core. (01510.10W)

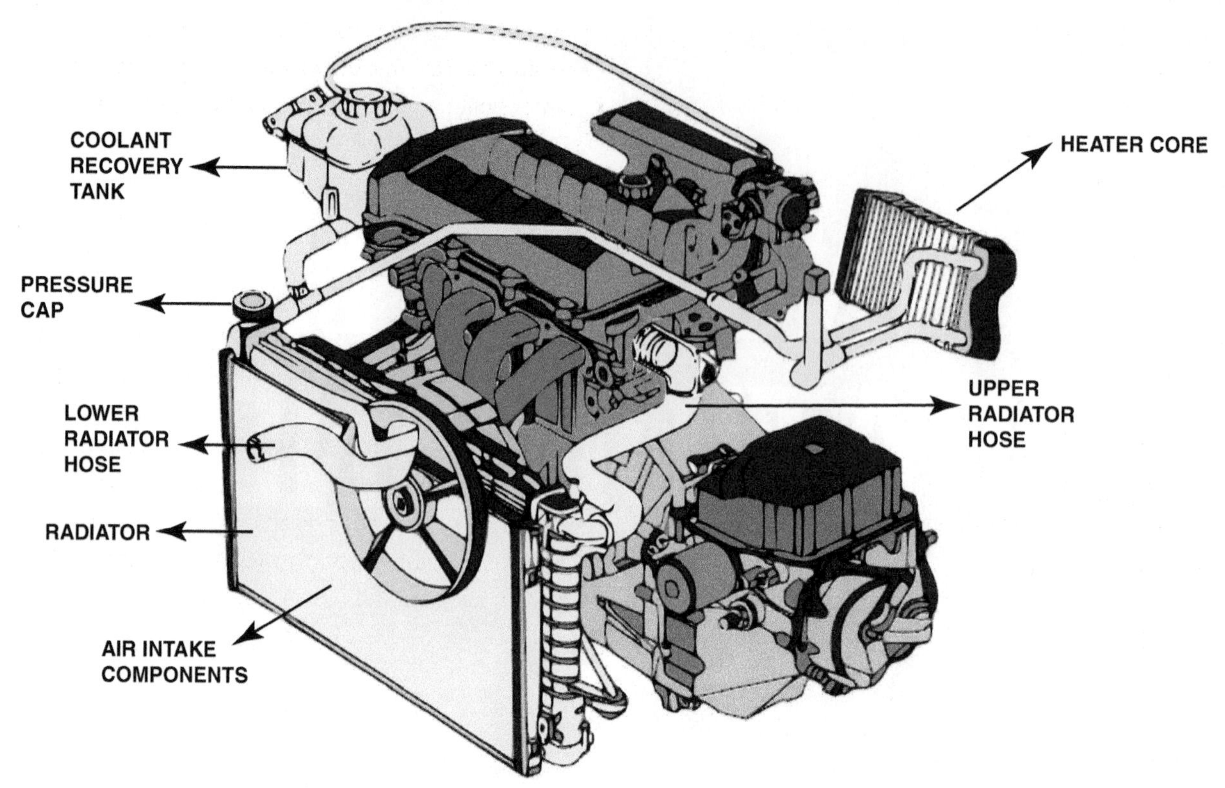

FIGURE 33–1 The main parts of a vehicle's heating system are the heater core, blower, heater hoses, and, in some cases, a heater control valve. (Courtesy of General Motors)

HEATING SYSTEM

PURPOSE AND FUNCTION The heating system uses heat that would normally be wasted (removed by the radiator) to warm the interior of the vehicle. The heating system is made up of the heater core, hoses, engine, and/or electric water pump, and, in some systems, a control valve as well as a blower motor to create the needed airflow into the passenger compartment. ● **SEE FIGURE 33–1**.

The Federal Motor Vehicle Safety Standard (FMVSS) 103 requires that every vehicle sold in the United States must be able to defrost or defog certain areas of the windshield in a specified amount of time. Most HVAC systems will default to defrost if there is a failure in the control part of the system. The defrost/defog system in most vehicles uses the heater and diverts the airflow to the base of the windshield.

HEATER CORE The **heater core** is a heat exchanger much like the condenser, evaporator, and radiator. Heat transfers from the coolant, to the fins, and to the air passing through the core. As with other heat exchangers, there is a large area of

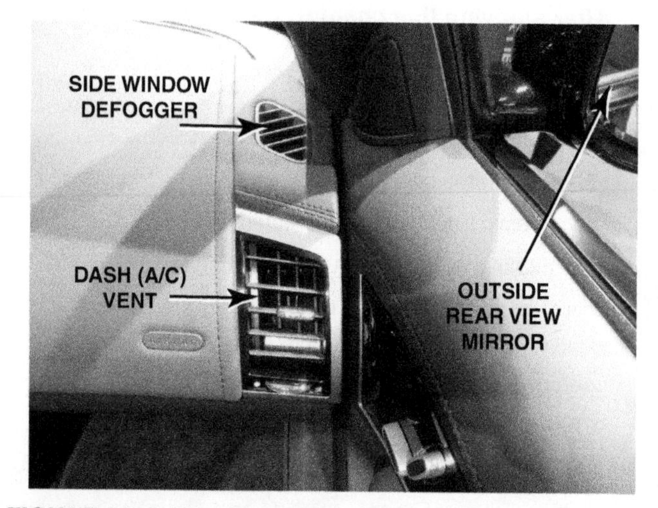

FIGURE 33–2 The side window defogger, also called a side window demister, directs air to the side window to clear the area so the side view mirror can be seen by the driver.

FREQUENTLY ASKED QUESTION

What Are Side Window Defoggers?

Most vehicles direct airflow from the defroster discharge vents to the side windows to defog them. This helps the driver to see through the side window and observe the side mirror to improve safety. ● **SEE FIGURE 33–2**.

FIGURE 33–3 The heater core is like a small radiator that transfers heat from the coolant into the passenger compartment. (Courtesy of Jeffrey Rehkopf)

fin-to-air contact to allow sufficient heat transfer and airflow. ● **SEE FIGURE 33–3.**

Most heater cores use a **cellular** form of construction that is somewhat like a plate-type evaporator. The tubes are made by joining two corrugated brass or aluminum plates, and the corrugated fins are attached between pairs of tubes. In most cores, the tanks at the ends of the core serve as manifolds to direct the flow back and forth through the core. Many systems use a smaller diameter for the heater inlet than for the outlet. This allows an easier exit of the coolant and reduces the pressure inside the core, which in turn reduces the possibility of leaks.

OPERATION The **inlet hose** to the heater core connects to an outlet fitting near the engine thermostat, or an area of the engine with the hottest coolant. The **outlet hose** from the heater core is connected near the inlet of the water pump, which is the area with the lowest coolant pressure. When the engine runs, coolant flows through the engine's water jackets, past the thermostat, and through the heater core. The heated coolant warms the heater core and the air passing through it.

NOTE: Some vehicles have a **control valve** in the heater inlet hose that allows coolant flow to be shut off when MAX cooling is selected to keep hot coolant from flowing through the heater core. Most new vehicles do not use a control valve, and the temperature of the air to the passenger compartment is controlled by an air temperature blend door in the heater plenum.

FLOW RESTRICTORS Some systems include a restrictor to slow the coolant velocity as it passes through the heater core. The restrictor can be part of the manifold fitting or the inlet heater hose assembly. The major purpose of the restrictor is to slow the flow rate in order to reduce noise and to allow for a better heat transfer to the fins and for a better heat transfer from the coolant to the air.

DUAL HEATING SYSTEMS

Larger vehicles with rear A/C systems include a heater in the rear unit. These rear units include a heater core and temperature-blend door. The heater hoses include a tee fitting in each hose, so heated coolant can flow through either or both heater cores. Some manufacturers include a water valve in the rear heater core so that hot coolant can be kept out of the core during A/C operation to improve A/C efficiency. The heater operation on these units is the same as that for a front unit.

HEATER DIAGNOSIS

VISUAL INSPECTION The diagnosis of a heater problem or concern should start with a visual inspection. The following items should be checked or tested:

1. **Coolant level and condition**—Low coolant level can cause a lack of heat from the heater. Low coolant level can also cause occasional loss of heat. The coolant should also be tested to make sure that it provides the needed freezing protection and pH.
2. **Water pump drive belt condition and proper tension** Drive belt condition and proper installation are important for the proper operation of the cooling system. There are four ways vehicle manufacturers specify that the belt tension is within factory specifications.
   - **Belt tension gauge.** A belt tension gauge is needed to achieve the specified belt tension. Install the belt and operate the engine with all of the accessories turned on, to run in the belt for at least five minutes. Adjust the tension of the accessory drive belt to factory specifications. Replace any serpentine belt that has more than three cracks in any one rib that appears in a 3 inch span.

WARNING

Do not remove the radiator cap when the engine is hot. Allow the vehicle to sit several hours before removing the pressure cap to check the radiator coolant level.

FIGURE 33–4 A special wrench being used to remove the tension from the accessory drive belt so it can be removed.

TECH TIP

Always Follow the Vehicle Manufacture's Recommended Procedure

The cooling system will not function correctly if air is not released (burped) from the system after a refill. On some vehicles the radiator cap is on the overflow bottle and is not the highest point. The manufacturer may have a very specific bleed procedure or the air will NEVER come out. If the system has had the coolant replaced and there is a complaint for lack of heat from the heater, then it is possible that air is trapped in the heater core.

- **Marks on the tensioner.** Many tensioners have marks that indicate the normal operating tension range for the accessory drive belt. Check service information for the location of the tensioner mark. ● **SEE FIGURE 33–4.**
- **Torque wrench reading.** Some vehicle manufacturers specify that a beam-type torque wrench be used to determine the torque needed to rotate the tensioner. If the torque reading is below specifications, the tensioner must be replaced.
- **Deflection.** Depress the belt between the two pulleys that are the farthest apart and the flex or deflection should be 1/2 inch.

NOTE: The water pump itself can have a slipping or corroded impeller, which can cause a reduction in the amount of coolant that is circulated.

FIGURE 33–5 When the plastic tool is placed in to the grooves of a good belt (left), the gauge sits above the surface. When the gauge tool is placed in the grooves of a belt that is worn and requires replacement (right), the gauge is below the surface of the ribs of the belt.

TECH TIP

Use a Belt Wear Gauge

Gates Rubber Company offers a free belt wear gauge that can be used to measure the depth of the curves in a serpentine belt. Wear cannot be seen but can result in a drive belt slipping, especially on belts used to drive AC compressors on hot days. ● **SEE FIGURE 33–5.**

3. **To check a radiator or condenser for possible clogged or restricted areas—**Carefully touch the outside of the unit with your hand. Any cool spots indicate that the radiator or condenser is clogged in that cool area.
4. **Check the temperature of the heater hoses—**Both should be hot to the touch. If the supply hose is hotter than the return hose, this indicates that the heater core could be partially clogged. Check that the control valve, if equipped, is open by checking to see that the temperature is the same on both sides of the valve.
5. **Check for proper airflow across the heater core—**If the airflow is blocked by leaves or debris, this can reduce the amount of heat being delivered to the passenger compartment. Check that the cabin filter is clean and not restricted.
6. **Pressure test the cooling system—**Pressure testing the cooling system insures that the system is leak-free and that the pressure cap is working as designed.

SCAN TOOL DIAGNOSIS Using a factory or factory-level aftermarket scan tool, perform the following:

- Use a scan tool and check for any stored or pending diagnostic trouble codes, especially those that pertain to the HVAC controls.

TECH TIP

Hot/Cold/Hot/Cold Heater Diagnosis

A common customer complaint is a lack of heat from the heater but only at idle, even though there seems to be plenty of heat when the engine is operating at highway speeds. This is a classic symptom of low coolant level. The heater core is usually mounted higher than the water jackets/passages in the engine and low coolant levels allow circulation when idling, but do not allow heater operation at higher engine speeds. This is because the velocity of the coolant drops at idle, putting air into the highest point (heater core) resulting in no heat at idle. The lower-than-normal coolant level in the radiator prevents enough flow to supply the heater core.

- Check that the coolant temperature as measured by the engine coolant temperature (ECT) sensor is correct, usually 180°F to 200°F (82°C to 93°C). Check service information for the specified coolant temperature for the vehicle being checked.
- Perform a bidirectional control of the blend door and blower motor speed, if possible, to confirm that these are operating correctly.

ELECTRICALLY HEATED SEATS

PARTS AND OPERATION Heated seats use electric heating elements in the seat bottom, as well as in the seat back in many vehicles. The heating element is designed to warm the seat and/or back of the seat to about 100°F (37°C) or close to normal body temperature (98.6°F). Many heated seats also include a high position or a variable temperature setting, so the temperature of the seats can therefore be as high as 110°F (44°C). A temperature sensor in the seat cushion is used to regulate the temperature. The sensor is a variable resistor that changes with temperature and is used as an input signal to a heated seat control module. The heated seat module uses the seat temperature input, as well as the input from the high–low (or variable) temperature control, to turn the current on or off to the heating element in the seat. Some vehicles are equipped with heated seats in both the rear and the front.

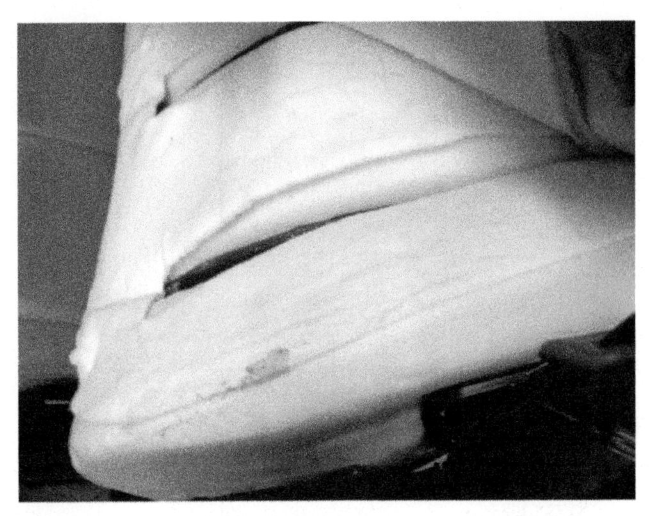

FIGURE 33–6 The heating element of a heated seat is a replaceable part, but service requires that the upholstery be removed. The yellow part is the seat foam material and the entire white cover is the replaceable heating element. This is then covered by the seat material.

DIAGNOSIS AND SERVICE After verifying that the heated seats are not functioning as designed, check that the switch is in the on position and that the temperature of the seat is below normal body temperature. Using service information, check for power and ground at the control module and to the heating element in the seat. Most vehicle manufacturers recommend replacing the entire heating element if it is defective. ● **SEE FIGURE 33–6.**

HEATED AND COOLED SEATS

PARTS AND OPERATION Most electrically heated and cooled seats use a **thermoelectric device (TED)** located under the seat cushion and seat back. The thermoelectric device consists of positive and negative connections between two ceramic plates. Each ceramic plate has copper fins to allow the transfer of heat to air passing over the device and directed into the seat cushion. The thermoelectric device uses the **Peltier effect**, named after the inventor, Jean C. A. Peltier, a French clockmaker. When electrical current flows through the module, one side is heated and the other side is cooled. Reversing the polarity of the current changes the side being heated. ● **SEE FIGURE 33–7.**

Most vehicles equipped with heated and cooled seats use two modules per seat, one for the seat cushion and one for the seat back. When the heated and cooled seats are turned on, air is forced through a filter and then through the thermoelectric

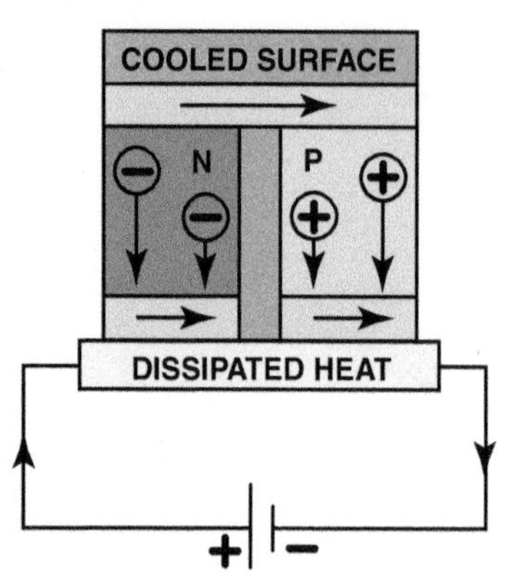

FIGURE 33–7 A Peltier effect device is capable of heating or cooling, depending on the polarity of the applied current.

modules. The air is then directed through passages in the foam of the seat cushion and seat back. Each thermoelectric device has a temperature sensor, called a thermistor. The control module uses sensors to determine the temperature of the fins in the thermoelectric device so the controller can maintain the set temperature.

DIAGNOSIS AND SERVICE The first step in any diagnosis is to verify that the heated–cooled seat system is not functioning. Check the owner's manual or service information for the correct operating instructions and specified diagnostic procedures.

HEATED STEERING WHEEL

PARTS INVOLVED A heated steering wheel usually consists of the following components.

- Steering wheel with a built-in heater in the rim
- Heated steering wheel control switch
- Heated steering wheel control module

OPERATION When the steering wheel heater control switch is turned on, a signal is sent to the control module and electrical current flows through the heating element in the rim of the steering wheel. ● **SEE FIGURE 33–8.**

The system remains on until the ignition switch is turned off or the driver turns off the control switch. The temperature of the steering wheel is usually calibrated to stay at about 90°F (32°C), and it requires three to four minutes to reach that temperature depending on the outside temperature.

DIAGNOSIS AND SERVICE Diagnosis of a heated steering wheel starts with verifying that it is not working as designed.

FIGURE 33–8 The heated steering wheel is controlled by a switch on the steering wheel in this vehicle.

TECH TIP

Check the Seat Filter

Heated and cooled seats often use a filter to trap dirt and debris to help keep the air passages clean. If a customer complains of a slow heating or cooling of the seat, check the air filter and replace or clean as necessary. Check service information for the exact location of the seat filter and for instructions on how to remove and/or replace it.

NOTE: Most heated steering wheels do not work if the temperature inside the vehicle is about 90°F (32°C) or higher.

If the heated steering wheel is not working, follow the service information testing procedures, which includes a check of the following:

1. Check the heated steering wheel control switch for proper operation. This is usually done by checking for voltage at both terminals of the switch. If voltage is available at only one of the two terminals of the switch and the switch has been turned on and off, an open (defective) switch is indicated.
2. Check for voltage and ground at the terminals leading to the heating element. If voltage is available at the heating element and the ground has less than 0.2 volt drop to a good chassis ground, the heating element is defective. The entire steering wheel has to be replaced if the element is defective. Always follow the vehicle manufacturer's recommended diagnosis and testing procedures.

LACK OF HEAT CONCERN DIAGNOSIS

1 The diagnosis of a lack of heat from the heater started by checking the temperature of the heater hoses. The return hose was colder than the inlet hose.

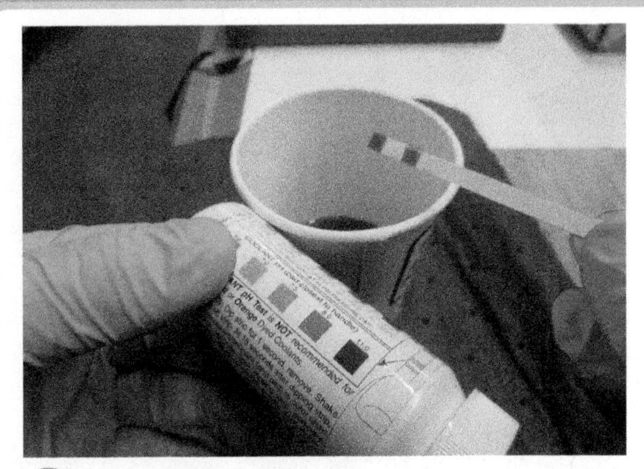

2 A test strip was used to check the coolant pH and freezing point. The coolant did not look fresh, but the test strip results were normal.

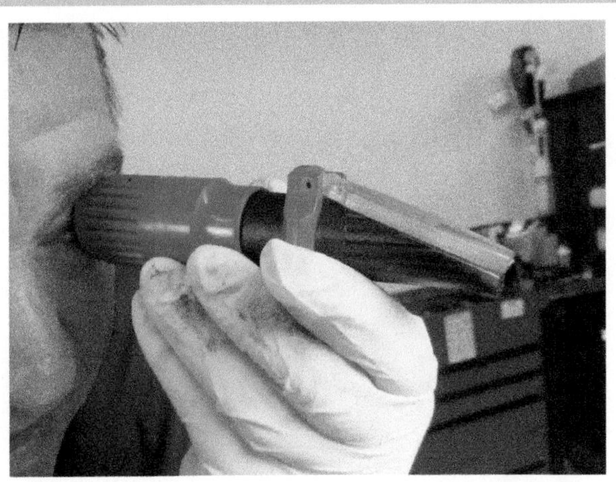

3 The refractometer test confirmed that the freezing protection was –32°F (–36°C).

4 The heater core was flushed and the coolant flow through the hoses seemed to be normal.

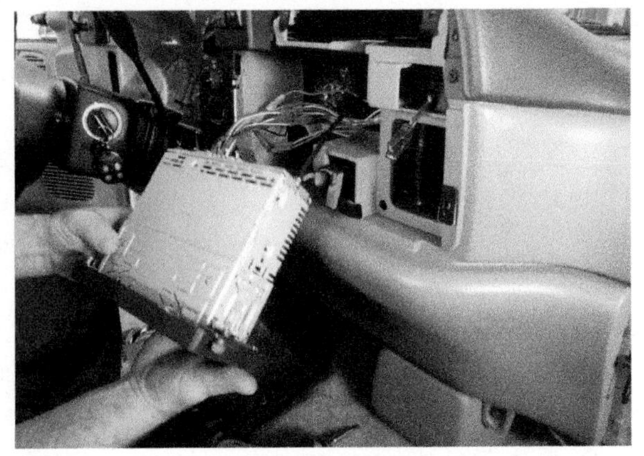

5 The heater core had to be removed and this required that the dash assembly be removed. Starting to remove the dash components to get access to the heater core.

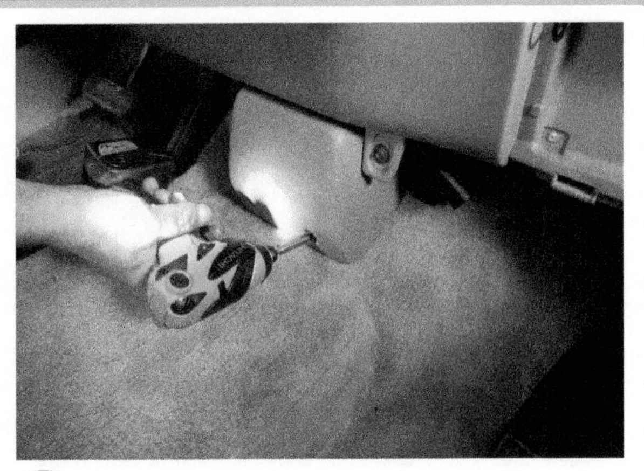

6 The fasteners used to retain the dash to the bulkhead being removed.

CONTINUED ▶

7 The A/C system was evacuated because it was necessary to remove HVAC module which contained the evaporator.

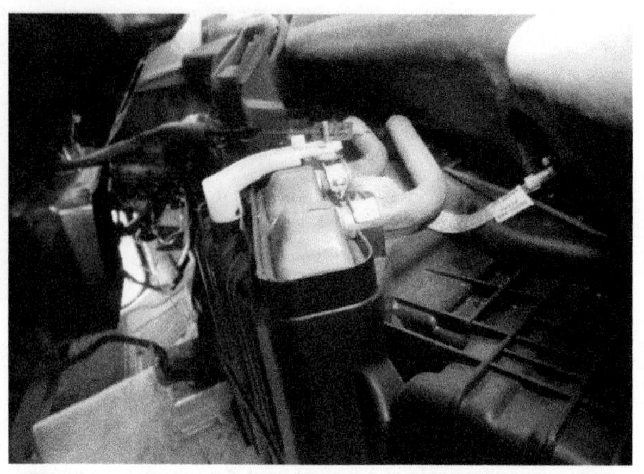

8 The dash was pulled back and the heater core became visible.

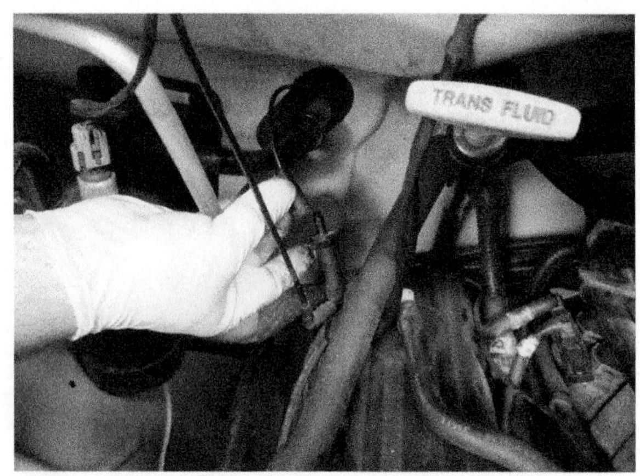

9 The vacuum connection for the HVAC vacuum actuators was disconnected from under the hood.

10 The AC hose connections to the evaporator were removed.

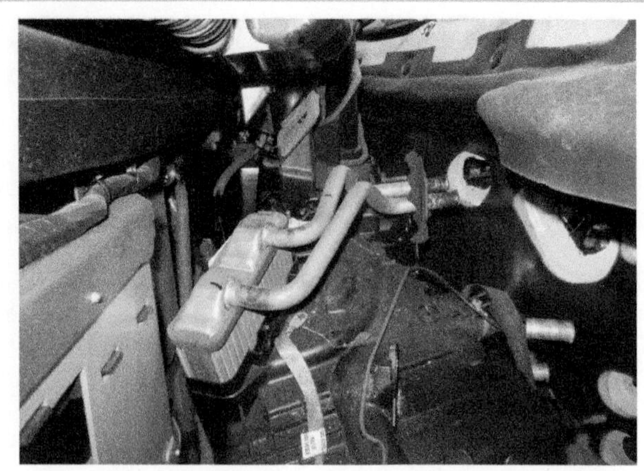

11 The heater core being removed from inside the vehicle.

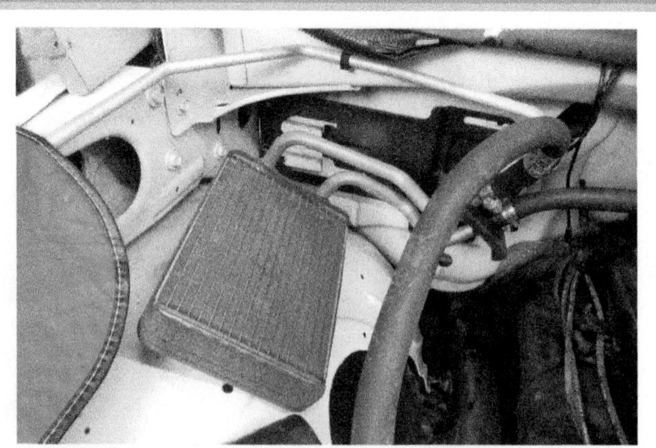

12 The heater core was removed from the inside of the vehicle and connected to the heater hoses and then the engine was started. The heater core felt hot across the entire surface.

13 However, when an infrared temperature gun was used, the temperature measured about 20 degrees different depending on where it was measured.

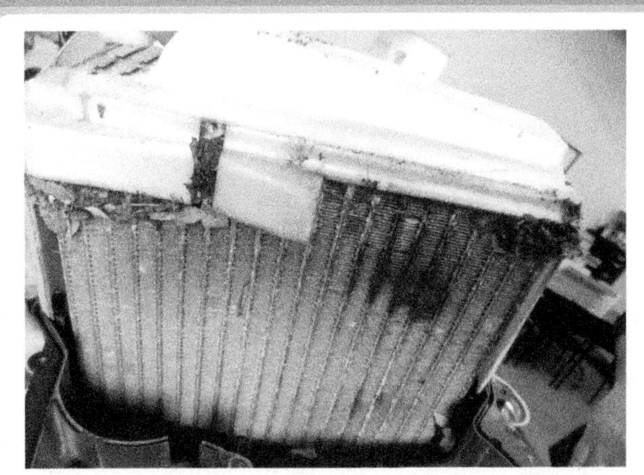

14 The evaporator had signs that it was leaking as indicted by the oil seen on the surface. It was replaced at the same time as the heater core.

15 The replacement heater core was made of aluminum instead of brass like the original but it fit correctly.

16 Fresh coolant was added to a coolant exchange machine. Conventional green (IAT) coolant was used because it had been converted to this coolant during a previous service.

17 The coolant was replaced using the exchange machine, which also purged any air from the system. The heater worked great and was very hot and the vehicle owner was pleased.

18 The end cap was removed from the old heater core and connected to a water hose. Only about half of the tubes were flowing water confirming that this heater core did require replacement.

SUMMARY

1. Hot coolant from the engine cooling system provides the heat for heater output.
2. The heater core resembles a small radiator, and it is connected to the engine by a pair of hoses.
3. Some vehicles use a coolant flow valve to control heater output temperature.
4. Heated seats can use either a resistance type heater located under the seat covering or a thermoelectric (TE) device.
5. Heater diagnosis includes verifying the problem, visual inspections and scan tool usage.

REVIEW QUESTIONS

1. What are the purpose and function of the heating system?
2. How does a thermoelectric (TE) device work?
3. What steps are included in performing a visual inspection of the heating system?
4. How can a scan tool be used to check the operation of the heater?

CHAPTER QUIZ

1. Heat is transferred from the hot coolant flowing through the ____________ to warm the air flowing through the fins of the heater core.
   - **a.** Water tubes
   - **b.** Hoses
   - **c.** Radiator
   - **d.** Blower motor
2. The heater control valve, if used, shuts off the coolant flow to the heater core when which of the following setting is selected?
   - **a.** Defrost
   - **b.** MAX cooling
   - **c.** Recirculation
   - **d.** Bi-level (both dash vents and floor vents)
3. The heating element is designed to warm the seat and/or back of the seat to about ____________.
   - **a.** 50°F (10°C)
   - **b.** 70°F (21°C)
   - **c.** 90°F (32°C)
   - **d.** 100°F (38°C)
4. A typical vehicle heater works by ____________.
   - **a.** Electrically controlling using a thermostat
   - **b.** Making the A/C system operate in reverse using the evaporator as the heater
   - **c.** Transferring heat from the coolant, to the fins, and to the air passing through the core.
   - **d.** Using hot coolant from the engine to heat the evaporator
5. How does a thermoelectric device (TED) work?
   - **a.** It uses an electric current passed through coolant
   - **b.** With a type of electric resistance heating
   - **c.** When electrical current flows through the module, one side is heated and the other side is cooled.
   - **d.** Using a temperature-controlled thermostat in the cooling system to control coolant temperature.
6. What is the purpose of a flow restrictor, if used?
   - **a.** It slows the coolant velocity as it passes through the heater core.
   - **b.** To reduce noise
   - **c.** It allows for a better heat transfer to the fins/air
   - **d.** All of the above
7. Coolant is hottest in which area?
   - **a.** Bottom of the radiator
   - **b.** Near the thermostat
   - **c.** At the overflow container
   - **d.** Near the outlet of the water pump
8. A seat filter is used with what type of system?
   - **a.** Most cabin heating systems
   - **b.** Heated and cooled seats
   - **c.** All heated seats
   - **d.** Heated steering wheels to keep the air around the driver free from dust and pollen
9. If the cooling system is low on coolant, what is a common symptom?
   - **a.** The cooling system will not supply any coolant to the heater core, so there will be complaint of lack of heat
   - **b.** Hot air comes from the heater at all times due to an overheated engine
   - **c.** Lack of heat from the heater, only when the vehicle is at idle, and plenty of heat when the engine is operating at highway speeds.
   - **d.** None of the above
10. The water pump drive belt should be checked for ____________.
    - **a.** Proper tension
    - **b.** Condition
    - **c.** Wear using a gauge
    - **d.** All of the above

chapter 34

AUTOMATIC TEMPERATURE CONTROL SYSTEMS

LEARNING OBJECTIVES

After studying this chapter, the reader should be able to:

1. Discuss the purpose and function of automatic temperature control (ATC) systems.
2. Discuss the sensors used in ATC systems.
3. State the need for airflow control.
4. Discuss the purpose of automatic HVAC controls.
5. Discuss how to diagnose the electrical ATC system faults.
6. Explain the automatic climatic control diagnostic procedure.
7. Explain the types of actuators in ATC systems.

GM STC OBJECTIVES

GM Service Technical College topics covered in this chapter are as follows:

1. Identify the operation of the manual and automatic HVAC control system. (11044.05W1)
2. Identify the operation of the manual and automatic dual zone HVAC control system.

KEY TERMS

FIGURE 34–1 The automatic A/C controls on this vehicle feature dual-zone temperature control (2014 GMC). (Courtesy of General Motors)

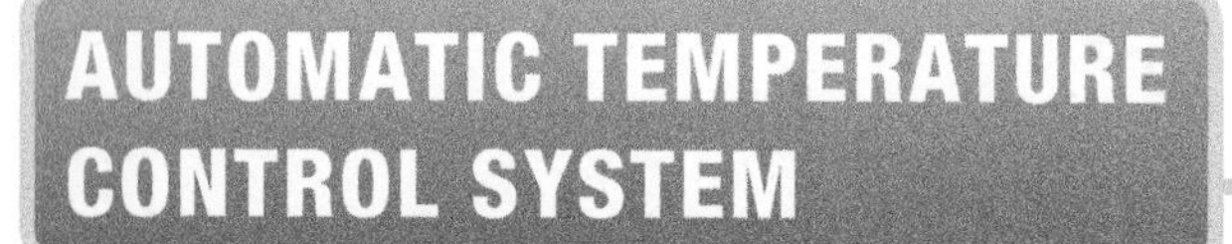

AUTOMATIC TEMPERATURE CONTROL SYSTEM

PURPOSE AND FUNCTION The purpose and function of the **heating, ventilation, and air conditioning (HVAC)** system are to provide comfortable temperature and humidity levels inside the passenger compartment. Proper temperature control to enhance passenger comfort during heating should maintain air temperature at the foot level about 7°F to 14°F (4°C to 8°C) above the temperature around the upper body. This is accomplished by directing the heated airflow to the floor. During A/C operation, the upper body should be cooler, so the airflow is directed to the instrument panel registers. Most new HVAC systems use an electronic control head and some use electronic blower speed control, radiator fan, and control of compressor displacement.

OPERATION **Automatic temperature control (ATC) System** is also called **automatic climatic control system** or **automatic air-conditioning system**. With an automatic temperature control system (ATC), the driver can turn the automatic controls on or off and select the desired temperature. The ATC will adjust the following:

- Blower speed
- Temperature door
- Air inlet door
- Mode door to achieve the proper temperature
- Control the air-conditioning compressor operation. ● **SEE FIGURE 34–1**.

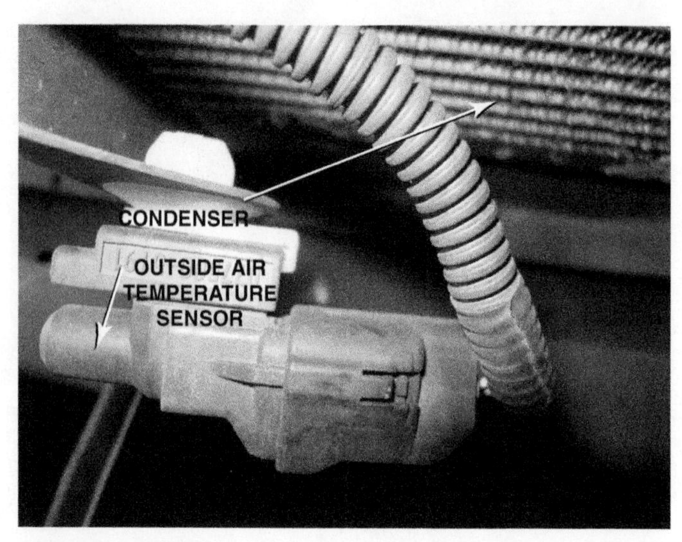

FIGURE 34–2 The outside air temperature sensor is mounted behind the grill, in front of the A/C condenser on this vehicle.

SENSORS

PURPOSE AND FUNCTION The purpose of using sensors is to provide information to the HVAC controller regarding the conditions outside as well as inside the vehicle. Sensors provide data about the following:

- Different conditions that can affect temperature conditions within the vehicle.
- Temperature setting at the control head.
- Operation of the A/C system.
- Operation of the engine.

Pressure and temperature sensors are used to determine the condition in the air-conditioning system so that the controller can

- Provide the most efficient use of energy.
- Provide a comfortable interior environment for the driver and passengers.
- Reduce the load on the engine and electrical system as much as possible to improve fuel economy.

OUTSIDE AIR TEMPERATURE SENSOR The **outside air temperature (OAT) sensor**, also called the **ambient temperature sensor**, measures outside air temperature and is often mounted at the radiator shroud or in the area behind the front grill. ● **SEE FIGURE 34–2**.

An ambient sensor is a thermistor that is mounted in the air stream passing through the front of the vehicle or the airflow entering the HVAC case.

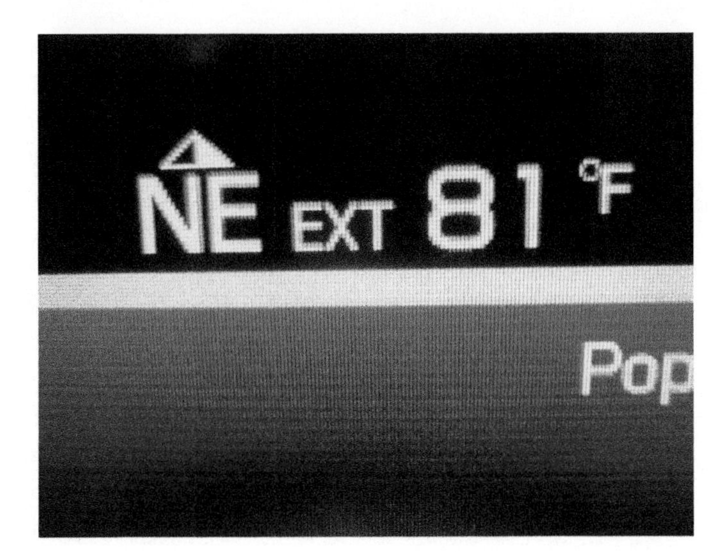

FIGURE 34–3 The outside air temperature in displayed on the navigation screen on this vehicle and uses the information from the outside air temperature sensor.

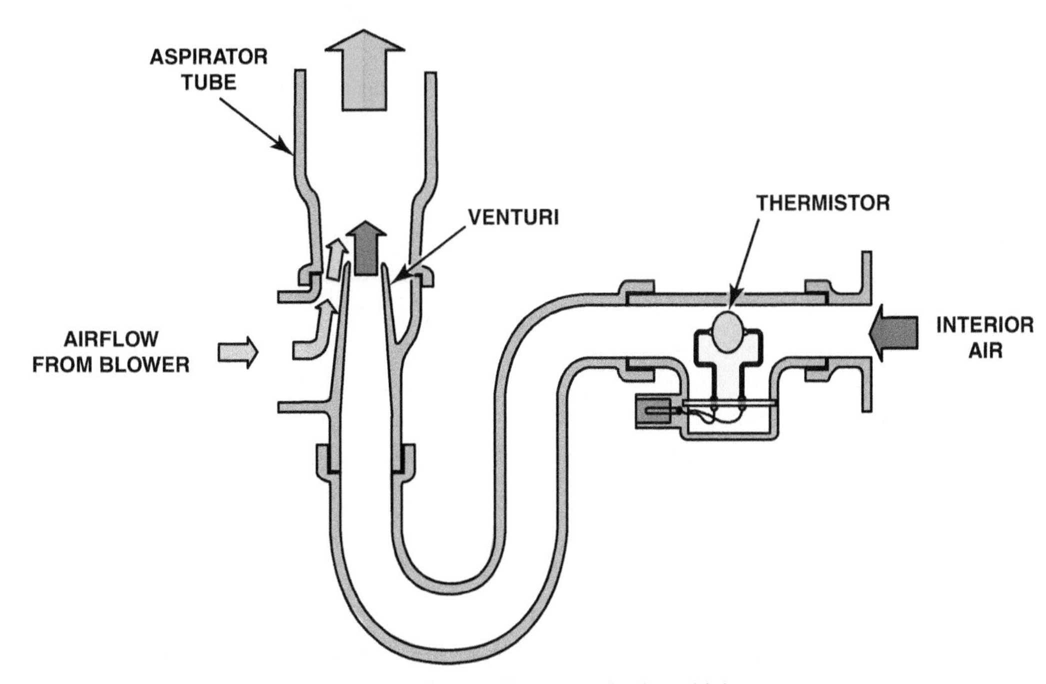

FIGURE 34–4 The airflow from the blower causes airflow to flow past the in-vehicle temperature sensor.

- Most automotive thermistors are of the **negative temperature coefficient (NTC)** type; the resistance changes in an inverse or negative relationship with temperature.
- The resistance is low when the temperature is high.
- This sensor is used to determine the outside air temperature that is displayed on the dash. ● **SEE FIGURE 34–3.**

IN-VEHICLE TEMPERATURE SENSOR The **in-vehicle temperature sensor** is often mounted behind the instrument panel, and a set of holes or a small grill allows air to pass by it. Air from the blower motor passes through the venturi of the aspirator and pulls in-vehicle air past the thermistor. This tube is called an aspirator tube and is connected to the blower housing. Blower operation produces airflow through the aspirator and past the sensor. ● **SEE FIGURE 34–4.**

DISCHARGE AIR TEMPERATURE SENSOR The **discharge air temperature (DAT)** sensor is used to measure the temperature of the air leaving the dash vents.

- This discharge air temperature can be used to determine the proper temperature position or control compressor output.

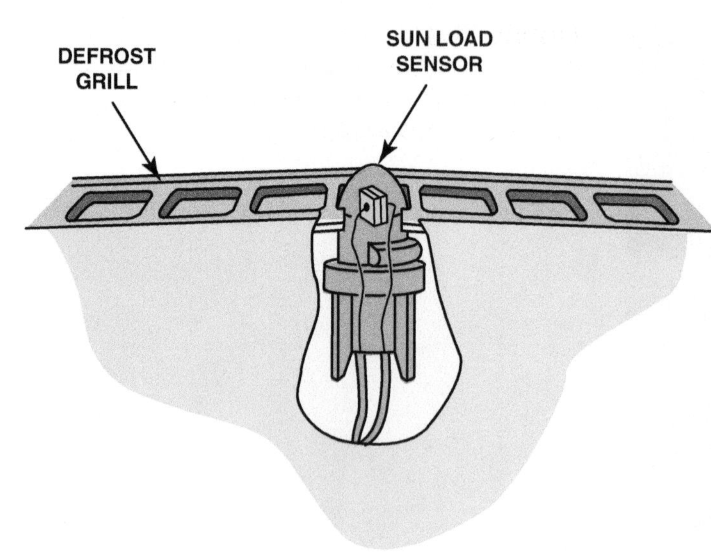

FIGURE 34–5 Sun load sensors are usually located at the top of the instrument panel.

- This sensor reading is used by the controller, which compares the discharge temperature reading with the temperature set by the driver. The control module calculates the temperature difference and automatically adjusts the HVAC system to minimize this temperature difference.

EVAPORATOR TEMPERATURE SENSOR The **evaporator temperature (EVT) sensor** is used to measure the temperature of the air leaving the evaporator. Evaporator air temperature can be used to determine the proper temperature position or control compressor output. The long probe of this sensor passes between the evaporator fins. It provides an evaporator temperature input to the HVAC controller.

SUN LOAD SENSOR The **sun load sensor** (also called a *solar sensor*) is normally mounted on top of the instrument panel and is used to measure radiant heat load that might cause an increase of the in-vehicle temperature. Bright sunshine provides a signal to the HVAC electronic control module (ECM) that things are going to get hotter. Sun load sensors are photodiodes that change their electrical conductivity based on the level of sunlight striking the sensor.

A typical sun load sensor uses a 5-volt operating voltage and produces an output signal voltage based on the intensity of the light.

- Dark = 0.3 volts
- Bright sunshine = 3.0 volts. ● **SEE FIGURE 34–5.**

INFRARED SENSORS **Infrared (IR) sensors** are mounted in the control head or overhead in the headliner. They monitor the surface temperature of the head and chest area of the driver and passenger. The system will adjust the temperature and position of the airflow depending on the temperature they measure and the settings on the control head. The detector does not emit an infrared beam but instead measures the infrared rays that are emitted from the person's body.

COMPRESSOR SPEED SENSOR The air-conditioning (A/C) **compressor speed sensor**, also called a *lock* or *belt lock* sensor, is used so the ECM will know if the compressor is running, and by comparing the compressor and engine speed signals, the ECM can determine if the compressor clutch or drive belt is slipping excessively. It prevents a locked-up compressor from destroying the engine drive belt, which, in turn, can cause engine overheating or loss of power steering. If the ECM detects an excessive speed differential for more than a few seconds, it will turn the compressor off. The sensor can be either a magnetic type or a Hall-effect sensor. Check service information for details about the sensor for the vehicle being serviced.

ENGINE COOLANT TEMPERATURE SENSOR The **engine coolant temperature (ECT) sensor** is a thermistor and measures the temperature of the engine coolant and is usually located near the engine thermostat. The ECT sensor is used to keep the system from turning on the heater before the coolant is warmed up, which is often called *cold engine lockout.* ● **SEE FIGURE 34–6.**

PRESSURE TRANSDUCERS A **pressure transducer** can be used in the low- and/or high-pressure refrigerant line. The transducer converts the system pressure into an electrical signal that allows the ECM to monitor pressure. The pressure signal can be used by the controller to:

- Cycle the compressor to prevent evaporator freeze-up.
- Change orifice tube size.
- Change the compressor displacement.
- Shut off the compressor or speedup cooling fan operation because of high pressures.
- Prevent compressor operation if the refrigerant level is low or empty.

The pressures of the refrigerant systems can be checked by looking at the scan tool scan data of a factory or factory-level scan tool. ● **SEE CHART 34–1** for some sample pressures and what they could indicate.

AIR QUALITY SENSORS Some systems use an *air quality sensor*, which detects hydrocarbons (HC) or ozone (O_3). HC can come from vehicle engine exhaust or decaying animal material and often produces offensive odors. Ozone is an irritant to

FIGURE 34–6 The engine coolant temperature (ECT) sensor is located in the engine water jacket near the thermostat, cylinder head, or engine block (Courtesy of General Motors)

LOW SIDE PRESSURE	HIGH SIDE PRESSURE	CONDITION
25–35 PSI	170–200 PSI	Normal operation
Low	Low	Low refrigerant charge level
Low	High	Restriction in high-side line
High	High	System is overcharged
High	Low	Restriction in the low-side line

CHART 34–1

Sample refrigerant system pressures and possible causes as shown from the pressure sensors and displayed on a scan tool. Check service information for the exact procedures to follow if the pressures are not correct.

the respiratory system. When the system detects an air quality issue, it automatically switches to using mostly inside air, about 80%, and reduces the amount of outside air entering the system to about 20% from the normal 80% level. This reduces the amount of outside air that is being brought into the passenger compartment until the system detects healthy outside air.

RELATIVE HUMIDITY SENSOR A few vehicles use a **relative humidity (RH) sensor** to determine the level of in-vehicle humidity. High RH increases the cooling load. A relative humidity sensor uses the capacitance change of a polymer thin film capacitor to detect the relative amount of moisture in the air. A dielectric polymer layer absorbs water molecules through a thin metal electrode and causes capacitance change proportional to relative humidity. The thin polymer layer reacts very fast, usually in less than 5 seconds to 90% of the final value of relative humidity. The sensor responds to the full range from 0% to 100% relative humidity. The output of the sensor is sent to the HVAC controller, which uses the information to control the air inlet door and the air-conditioning compressor operation to achieve the desired level of humidity (20% to 40%) in the passenger compartment.

FIGURE 34–7 Some automatic HVAC system use the information from the factory navigation system to fine tune the interior temperature and airflow needs based on location and the direction of travel.

GPS SENSOR A few vehicles that are equipped with a global positioning system (GPS) for navigation will have a sun position strategy that tracks the angle of the sunlight entering the vehicle. Cooler in-vehicle temperatures are required if the vehicle is positioned so sunlight enters through the windshield or side windows. ● **SEE FIGURE 34–7**.

FIGURE 34–8 Three electric actuators can be easily seen on this demonstration unit. However, accessing these actuators in a vehicle can be difficult.

ACTUATORS

PURPOSE AND FUNCTION HVAC actuators are electric or mechanical devices that move doors to provide the needed airflow at the correct time and location. Actuators include:

- Electric motors that operate the air/temperature doors, also called *flaps*.
- Feedback circuits that provide position information to the HVAC controller.

TYPES OF ACTUATORS An actuator is a part that moves the vanes or valves. Actuators used in air-conditioning systems are either electric or vacuum operated and includes these different types. ● **SEE FIGURE 34–8.**

1. **Dual-Position Actuator.** A **dual-position actuator** is able to move either open or closed. An example of this type of actuator is the recirculation door, which can be either open or closed.
2. **Three-Position Actuator.** A **three-position actuator** is able to provide three air door positions, such as the bi-level door, which could allow defrost only, floor only, or a mixture of the two.
3. **Variable-Position Actuator.** A **variable-position actuator** is capable of positioning a valve in any position. The variable position actuators may be a stepper-motor type or use a feedback potentiometer to detect the actual position of the door or valve. ● **SEE FIGURE 34–9.**

ELECTRIC ACTUATOR MOTORS Electric actuator motors are used to move air doors. Electric door actuators can be either

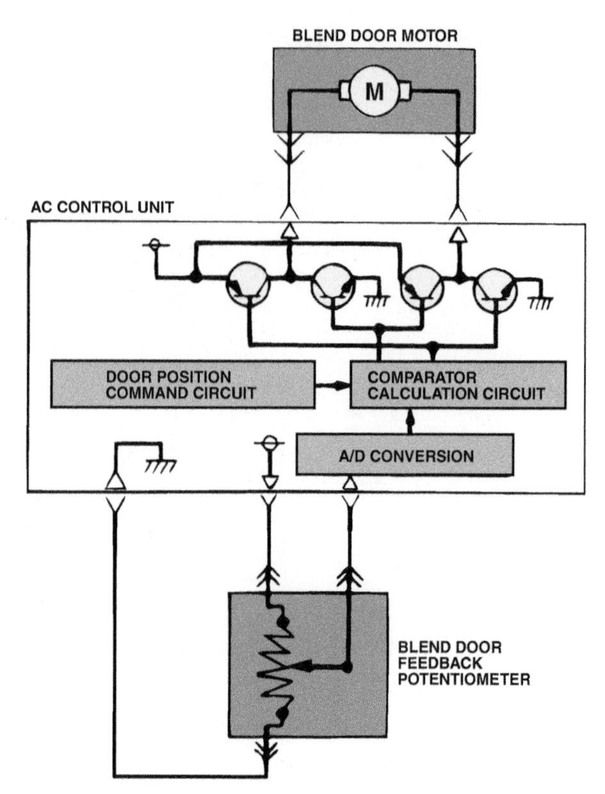

FIGURE 34–9 The feedback circuit signals the AC control unit with the blend door position.

continuous-position or two-position units (open or closed). Variable-position actuators can stop anywhere in their range and need a feedback circuit so the ECM will know their position. The temperature-blend door is operated by an electric servomotor (continuous-position actuator) that can move the door to any position called for to produce an air mix of the desired temperature. Some systems have the ability to count the actuator motor commutator segments so it is able to determine how far the motor revolves. A small current-flow reduction occurs as the space between the commutator bars passes under the brushes of a DC motor. This system needs no feedback circuit, but a calibration procedure must be performed if a motor or HVAC controller is replaced. Newer systems have an output from the ECM to each of the controlled actuators or outputs.

- **Temperature Control Actuators.** The heater core is placed downstream from the evaporator in the airflow so that air can be routed either through or around it and one or two doors are used to control this airflow. This door is usually called the **temperature-blend door**. Some other names that are used depending on the vehicle manufacturer include the following:
  - *air mix door*
  - *temperature door*
  - *blend door*
  - *diverter door*
  - *bypass door*

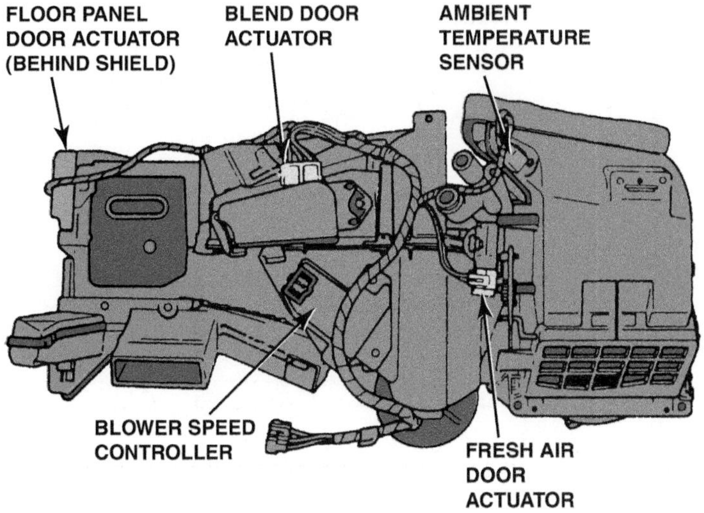

FLOOR PANEL DOOR ACTUATOR (BEHIND SHIELD)

BLEND DOOR ACTUATOR

AMBIENT TEMPERATURE SENSOR

BLOWER SPEED CONTROLLER

FRESH AIR DOOR ACTUATOR

FIGURE 34–10 A typical HVAC system showing some of the airflow door locations.

- **Mode control actuators.** Some vehicles include ducts to transfer air to the rear seat area, and some vehicles include ducts to demist the vehicle's side windows with warm air. Airflow to these ducts is controlled by one or more **mode doors** controlled by the function lever or buttons. Mode doors are also called *function, floor-defrost,* or *panel-defrost* doors. Mode/function control sets the doors as follows:
- **A/C:** in-dash registers with outside air inlet
- **Max A/C:** in-dash registers with recirculation
- **Heat:** floor level with outside air inlet
- **Max Heat:** floor level with recirculation
- **Bi-level:** both in-dash and floor discharge
- **Defrost:** windshield registers

Many control heads also provide for in-between settings, which combine some of these operations. In many systems, a small amount of air is directed to the defroster ducts when in the heat mode, and while in defrost mode a small amount of air goes to the floor level.

- **Inlet air control door actuators.** Air can enter the duct system from either the plenum chamber in front of the windshield (outside air) or return register (inside air). The return register is often positioned below the right end of the instrument panel. (The right and left sides of the vehicle are always described as seen by the driver.) The *inlet air control door* can also be called the
- *fresh air door,*
- *recirculation door, or*
- *outside air door*

This door is normally positioned so it allows airflow from one source while it shuts off the other. It can be positioned to allow outside air to enter

- While shutting off the recirculation opening.
- To allow air to return or recirculate from inside the vehicle while shutting off outside (fresh) air.
- To allow a mix of outside air and return air.

In most vehicles, the door is set to the outside air position in all function lever positions except off, max heat, and max A/C. Max A/C and max heat settings position the door to recirculate in-vehicle air. ● **SEE FIGURE 34–10**.

DUAL-ZONE SYSTEMS In many vehicles, the HVAC system is capable of supplying discharge air of more than one temperature to different areas in the vehicle. This type of system is usually referred to as a **Dual-Zone System** and allows the driver and front seat passenger to set their own desired temperature. Dual-zone systems contain two separately commanded temperature-blend doors. The temperature difference between the driver and front seat passenger can be up to 30°F (16°C), usually with settings between 60°F (16°C) minimum and 90°F (32°C) maximum. ● **SEE FIGURE 34–11**.

Tri-Zone and *Quad-Zone* systems are usually found in passenger vans, sport utility vehicles, and luxury cars, which allow the passengers in the rear of the vehicle to control the temperature at their location.

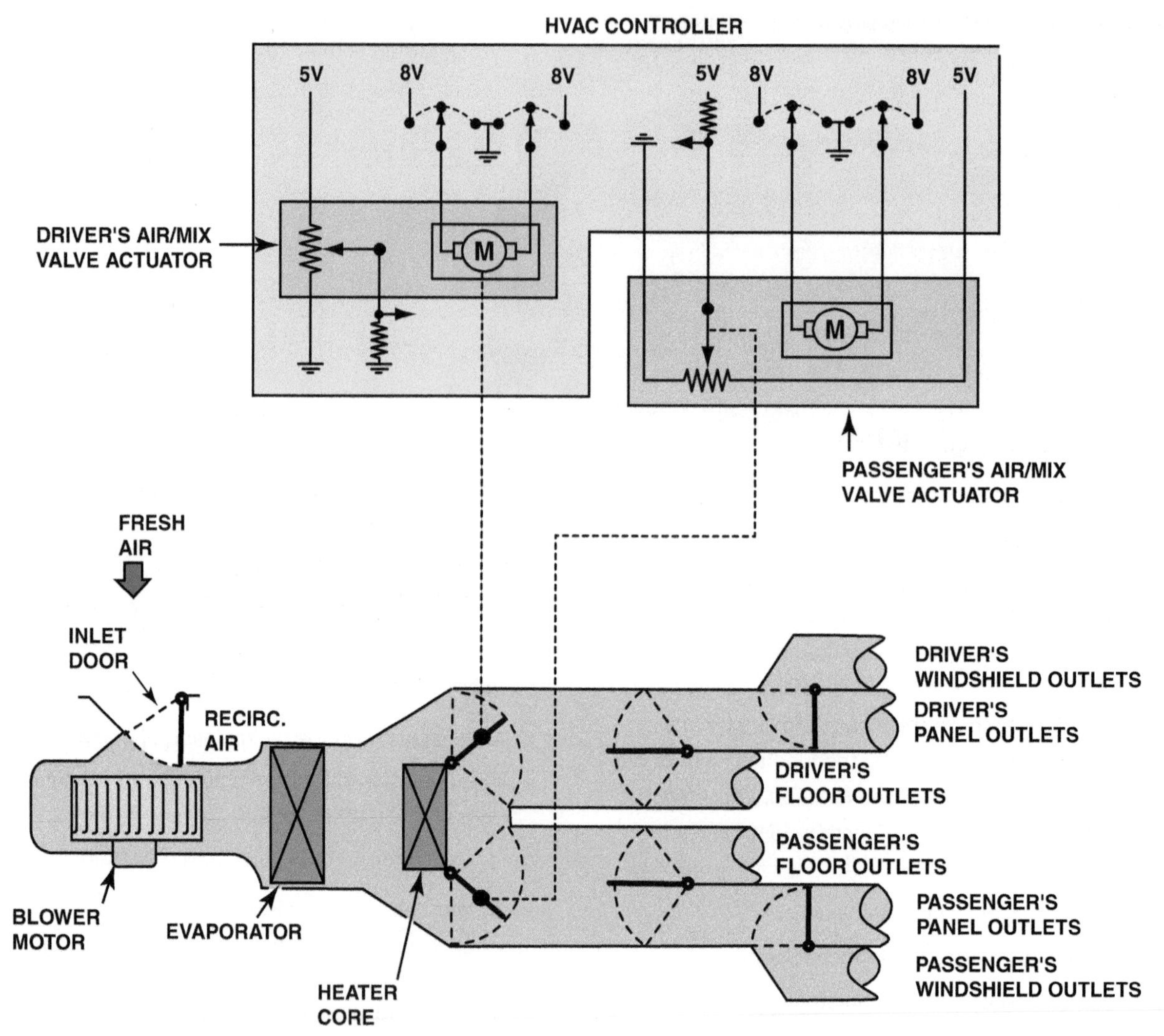

FIGURE 34–11 A dual-climate control system showing the airflow and how it splits.

AUTOMATIC HVAC CONTROLS

OPERATION The HVAC control head or panel is mounted in the instrument cluster. The control head provides the switches and levers needed to control the different aspects of the heating and A/C system, which include:

- HVAC system on and off
- A/C on/off
- Outside or recirculated air
- A/C, defrost, or heating mode
- Temperature desired
- Blower speed

COMPRESSOR CONTROLS Most air-conditioning compressors use an electromagnetic clutch. A coil of wire inside the clutch creates a strong magnetic field that when activated connects the input shaft of the compressor to the drive pulley.

Some systems may connect one or more switches in series with the compressor clutch so that all have to be functioning before the compressor clutch can be engaged. A low- and high-pressure switch or sensor may also be an input to the PCM or HVAC controller for use in controlling the compressor.

- **Low-pressure switch:** This pressure switch is electrically closed only if there is 8 PSI to 24 PSI (55 kPa to 165 kPa) of refrigerant pressure. This amount of pressure means that the system is sufficiently charged to provide lubrication for the compressor.

- **High-pressure switch:** This pressure switch is located in the high-pressure side of the A/C system. If the pressure exceeds a certain level, typically 375 PSI (2,600 kPa), the pressure switch opens, thereby preventing possible damage to the air-conditioning system due to excessively high pressure.
- **A/C relay:** The relay supplies power to the compressor.

CONTROL MODULE The control module used for automatic climatic control systems can be referred by various terms depending on the exact make and model of vehicle. Some commonly used terms include the following:

- ECM (Electronic control module)
- BCM (Body control module)
- HVAC control module (often is built into the smart control head)
- HVAC controller or programmer

The control modules are programmed to open or close circuits to the actuators based on the values of the various sensors. Although it is unable to handle the electric current for devices such as the compressor clutch or blower motor, the control module can operate relays. These relays in turn control the electric devices. Units that use small current flows, such as a light-emitting diode (LED) or digital display, can operate directly from the control module.

An ECM is programmed with various strategies to suit the requirements of the particular vehicle. For example, when A/C is requested by the driver of a vehicle with a relatively small engine, the ECM will probably increase the engine idle RPM. On this same vehicle, the ECM will probably shut off the compressor clutch during wide-open throttle (WOT).

- On other vehicles, the ECM might shut off the compressor clutch at very high speeds to prevent the compressor from spinning too fast.
- In some cases, part of the A/C and heat operating strategy is built into the control head and are called **smart control heads**. In some cases, door operating strategy is built into the door operating motors, and these are called **smart motors**.
- Many control modules are programmed to run a test sequence, called *self-diagnosis*, at start-up (when the ignition key is turned on). If improper electrical values are found, the ECM indicates a failure often by blinking the A/C indicator light at the control head. Some ECMs also monitor the system during operation and indicate a failure or stop the compressor if there is a problem.

FREQUENTLY ASKED QUESTION

Why Is the Blower Speed So High?

This question is often asked by passengers when riding in a vehicle equipped with automatic climatic control. The controller does command a high blower speed if:

- The outside temperature is low and the engine coolant temperature is hot enough to provide heat. The high blower speed is used to warm the passenger compartment as quickly as possible then when the temperature has reached the preset level, then the blower speed is reduced to maintain the preset temperature.
- The outside temperature is hot and the air-conditioning compressor is working to provide cooling. The high speed blower is used to circulate air through the evaporator in an attempt to cool the passenger compartment as quickly as possible. Once the temperature reaches close to the preset temperature, the blower speed is reduced to keep the temperature steady.

One system, for example, notes the frequency of clutch cycling that indicates a low refrigerant charge level. If there are too many clutch cycles during a certain time period, the control module will shut off the compressor and set a diagnostic trouble code (DTC). ● **SEE FIGURE 34–12**.

AUTOMATIC CLIMATIC CONTROL DIAGNOSIS

DIAGNOSTIC PROCEDURE If a fault occurs in the automatic climatic system, check service information for the specified procedure to follow for the vehicle being checked. Most vehicle service information includes the following steps:

STEP 1 Verify the customer concern. Check that the customer is operating the system correctly. For example, most automatic systems cannot detect when the defrosters are needed so most systems have a control that turns on the defroster(s).

STEP 2 Perform a thorough visual inspection of the heating and cooling system for any obvious faults.

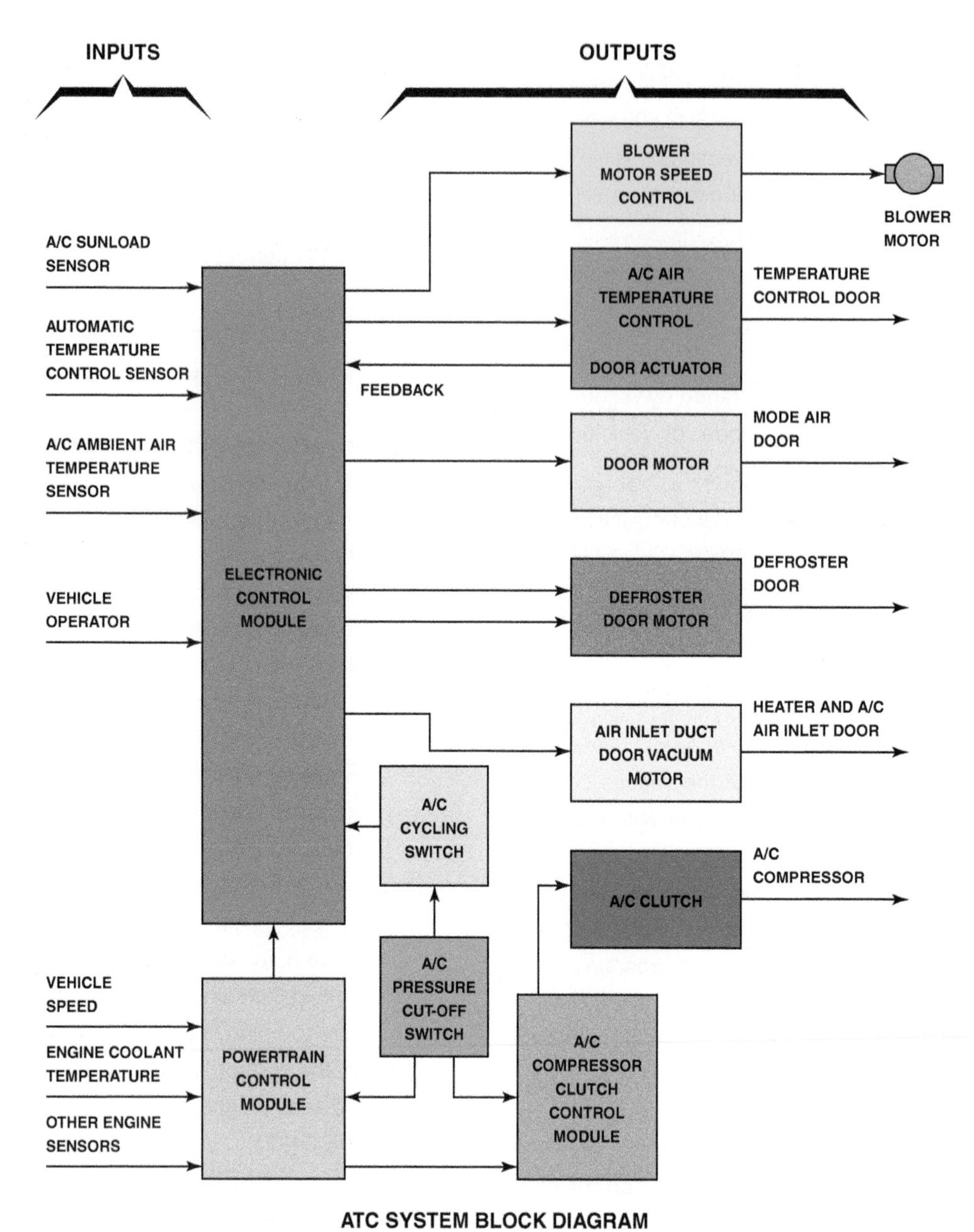

FIGURE 34–12 A block diagram showing the inputs to the electronic control assembly and the outputs; note that some of the outputs have feedback to the ECM.

STEP 3 Use a factory scan tool or a factory level aftermarket scan tool and check for diagnostic trouble codes (DTCs).

NOTE: Older systems can be accessed using the buttons on the control panel and then the diagnostic trouble codes and system data are displayed. Check service information for the exact procedures to follow for the vehicle being serviced.

STEP 4 If there are stored diagnostic trouble codes, follow service information instructions for diagnosing the system. ● **SEE CHART 34–2** for sample diagnostic trouble codes (DTCs).

STEP 5 If there are no stored diagnostic trouble codes, check scan tool data for possible fault areas in the system.

SCAN TOOL BIDIRECTIONAL CONTROL Scan tools are the most important tool for any diagnostic work on all vehicles.

The scan tool (Tech 2 or GDS 2) is designed to provide bidirectional capability, which allows the service technician the opportunity to operate components using the scan tool, thereby confirming that the component is able to work when commanded.

ATC-RELATED DIAGNOSTIC TROUBLE CODES	
BODY DIAGNOSTIC TROUBLE CODE (DTC)	DESCRIPTION
B0126	Right Panel Discharge Temperature Fault
B0130	Air Temperature/Mode Door Actuator Malfunction
B0131	Right Heater Discharge Temperature Fault
B0145	Auxiliary HAVC Actuator Circuit
B0159	Outside Air Temperature Sensor Circuit Range/Performance
B0160/B0162	Ambient Air Temperature Sensor Circuit
B0164	Passenger Compartment Temperature Sensor #1 (Single Sensor or LH) Circuit Range/Performance
B0169	In-Vehicle Temp Sensor Failure (passenger—not used)
B0174	Output Air Temperature Sensor #1 (Upper Single or LH) Circuit Range/ Performance
B0179	Output Air Temperature Sensor #2 (Lower Single or LH) Circuit Range/ Performance
B0183	Sun Load Sensor Circuit
B0184	Solar Load Sensor #1 Circuit Range (sun load)
B0188	Sun Load Sensor Circuit
B0189	Solar Load Sensor #2 Circuit Range (sun load)
B0229	HVAC Actuator Circuit
B0248	Mode Door Inoperative Error
B0249	Heater/Defrost/AC Door Range Error
B0263	HVAC Actuator Circuit
B0268	Air/Inlet Door Inoperative Error
B0269	Air Inlet Door Range Error
B0408	Temperature Control #1 (Main/Front) Circuit Malfunction
B0409	Air Mix Door #1 Range Error
B0414	Air Temperature/Mode Door Actuator Malfunction
B0418	HVAC Actuator Circuit
B0419	Air Mix Door #2 Range Error
B0423	Air Mix Door #2 Inoperative Error
B0424	Air Temperature/Mode Door Actuator Malfunction
B0428	Air Mix Door #3 Inoperative Error

CHART 34–2

Sample automatic climatic control diagnostic trouble codes.

SENSOR	TYPICAL VALUE
Inside air temperature sensor	−40°F–120°F (−40°C–49°C)
Ambient air temperature sensor	−40°F–120°F (−40°C–49°C)
Engine coolant temperature (ECT) sensor	40°F–250°F (−40°C–121°C)
Sun load Sensor	0.3 volts (dark) 3.0 volts (bright)
Evaporator temperature sensor	Usually 34°F–44°F (1°C–7°C)
Relative Humidity sensor	0%–100%

CHART 34–3

Typical sensors and values that may be displayed on a scan tool. Check service information for the exact specifications for the vehicle being serviced.

FREQUENTLY ASKED QUESTION

What Are the Symptoms of a Broken Blend Door?

Blend doors can fail and cause the following symptoms:

- Clicking noise from the actuator motor assembly as it tries to move a broken door.
- Outlet temperature can change from hot to cold or from cold to hot at any time especially when cornering because the broken door is being forced one way or the other due to the movement of the vehicle.
- A change in the temperature when the fan speed is changed. The air movement can move the broken blend door into another position which can change the vent temperature.

If any of these symptoms are occurring, then a replacement blend door is required.

ATC components that may be able to be controlled or checked using a scan tool include:

- Blower speed control (faster and slower to check operation)
- Command the position of airflow doors to check for proper operation and to check for proper airflow from the vents and ducts
- Values of all sensors
- Pressures of the refrigerant in the high and low sides of the systems

Typical scan tool data and their meaning are shown in **CHART 34–3**.

SUMMARY

1. Automatic temperature control (ATC) systems use sensors to detect the conditions both inside and outside the vehicle.
2. The sensors used include the following:
   - Sun load sensor
   - Evaporator temperature sensor
   - Ambient air temperature (outside air temperature) sensor
   - In-vehicle temperature sensor
   - Infrared Sensors
   - Engine coolant temperature (ECT) sensor
3. Pressure transducers detect the pressures in the refrigerant system.
4. Some systems use a compressor speed sensor to detect if the drive belt or compressor clutch is slipping.
5. The heater core is placed downstream from the evaporator in the airflow so that air can be routed either through or around it and one or two doors are used to control this airflow. This door is called the *temperature-blend door.*
6. In many vehicles, the HVAC system is capable of supplying discharge air of more than one temperature to different areas in the vehicle. This type of system is usually referred to as a *Dual-Zone Climate Control System.*
7. The diagnostic steps include:
   - Verifying the customer concern
   - Performing a thorough vial inspection
   - Retrieveing diagnostic trouble codes (DTCs)
   - Checking the data as displayed on a scan tool to determine what sensors or actuators are at fault

REVIEW QUESTIONS

1. What sensors are used in a typical automatic temperature control (ATC) system?
2. Why is a feedback potentiometer used on an electric actuator?
3. What is the purpose of the aspirator tube in the in-vehicle temperature sensor section?

CHAPTER QUIZ

1. What is the purpose of a compressor speed sensor?
   a. To allow the ECM to control compressor speed
   b. So that the ECM can control high-side pressure
   c. So that the ECM can control low-side pressure
   d. To let the ECM know if the compressor is turning
2. Which sensor is also called the ambient air temperature sensor?
   a. Outside air temperature (OAT)
   b. In-vehicle temperature
   c. Discharge air temperature
   d. Evaporator outlet temperature
3. What is the most common type of sun load sensor?
   a. Potentiometer
   b. Negative temperature coefficient (NTC) thermistor
   c. Photodiode
   d. Positive temperature coefficient (PTC) thermistor
4. An actuator can be capable of how many position(s)?
   a. Two
   b. Three
   c. Variable
   d. All of the above
5. A dual-zone A/C system allows different ____________ settings for the driver and passenger.
   a. Fan speed
   b. Temperature
   c. Compressor
   d. Vent
6. Which sensor might use an aspirator tube?
   a. In-vehicle temperature
   b. Outside air temperature (OAT)
   c. Discharge air temperature
   d. Evaporator outlet temperature
7. Technician A says that the sun load sensor controls the headlights. Technician B says that the sun load sensor is located behind the sun visor. Which technician is correct?
   a. Technician A only
   b. Technician B only
   c. Both technicians A and B
   d. Neither technician A nor B
8. Automatic temperature control (ATC) diagnosis includes ____________.
   a. Verifying customer concern
   b. Performing a visual inspection
   c. Checking for stored diagnostic trouble codes
   d. All of the above
9. The control module used for automatic climatic control systems is called ____________.
   a. ECM
   b. BCM
   c. HVAC control module
   d. Any of the above depending on the make and model of vehicle
10. A feedback potentiometer is used to ____________.
    a. Provide feedback to the driver as to where the controls are set
    b. Provide feedback to the controller as to the location of a door or valve
    c. Give temperature information about the outside air temperature to the dash display
    d. Any of the above depending on the exact make and model of vehicle

chapter 35

HYBRID AND ELECTRIC VEHICLE HVAC SYSTEMS

LEARNING OBJECTIVES

After studying this chapter, the reader will be able to:

1. Prepare for the ASE Heating and Air Conditioning (A7) certification test content area "A" (A/C System Service, Diagnosis, and Repair).
2. Explain the basic operation of the air-conditioning system used in hybrid electric vehicles.
3. Discuss the types of compressors used in a hybrid electric vehicle.
4. Explain the components and modes of operation of a coolant heat storage system.
5. Describe the parts and operation of cabin heating systems.
6. Explain the effect of heat on the electrical/electronic systems of a hybrid electric vehicle.

KEY TERMS

Battery chiller 568
Extended-range electric vehicles (EREV) 567
Internal combustion engine (ICE) 560
POE (polyol ester) 561
PTC heate 563

GM STC OBJECTIVES

GM Service Technical College topic covered in this chapter are as follows:

1. Identify the characteristics and operation of HVAC systems as applied to General Motors hybrid and extended-range electric vehicles.

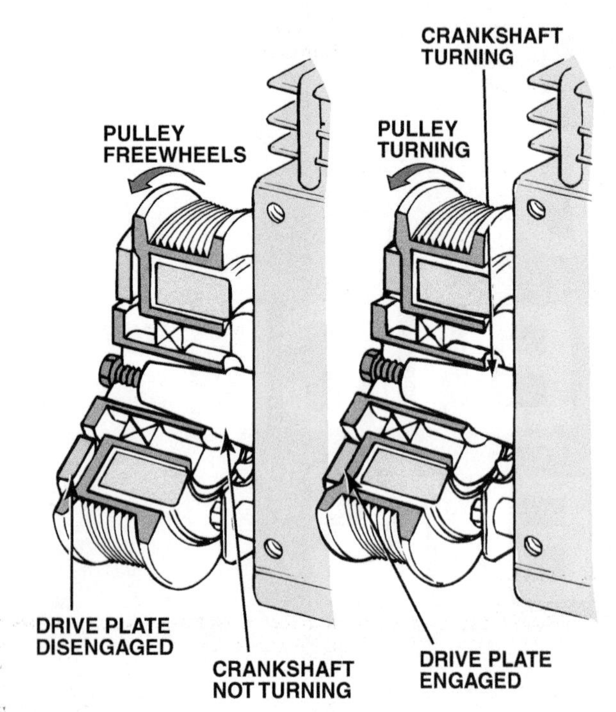

FIGURE 35–1 The A/C compressor clutch allows the compressor to engage and disengage as necessary while the ICE continues to run.

HYBRID VEHICLE AIR-CONDITIONING SYSTEMS

BASIC OPERATION The fundamental purpose of any air-conditioning system is to absorb heat in one location and then dissipate (move) that heat to another location. Control of the air-conditioning system is the operation of the compressor, as well as the airflow across the evaporator and condenser. The types of compressors used in hybrid electric vehicles include:

- Belt-driven by the **internal combustion engine (ICE)** accessory drive. The belt drive often uses an electrically operated clutch, which allows the compressor pulley to disconnect from the compressor and stop refrigerant flow in the system while the ICE continues to run. ● **SEE FIGURE 35–1**.
- Dual compressor where a large section of it is driven by the internal combustion engine and another smaller section driven by a high-voltage electric motor to provide cooling during idle stop (start/stop) operation.
- Electric-motor powered compressor used in all electric vehicles and many hybrid electric vehicles.

In most situations, fresh air is brought in from outside the vehicle and then is heated or cooled before being sent to the appropriate vents. It is also possible to draw air from the passenger compartment itself, when the system is placed in the recirculation mode.

The fresh air coming into the vehicle is sometimes sent through a cabin filter first, in order to remove particulate matter and prevent clogging of the evaporator. All the incoming air must pass through the A/C evaporator core after leaving the blower motor. ● **SEE FIGURE 35–2**.

In defrost mode, the A/C compressor is activated and the evaporator core is cooled to the point where any humidity in the air will condense on the evaporator and then be drained outside the vehicle. When in defrost mode, the air leaving the A/C evaporator core can be sent through the heater core to raise its temperature. This warm air is sent to the defrost outlets on the driver's and the passenger side of the windshield.

The temperature of the air is controlled by the position of the air blend door, as it either directs varying amounts of air over the heater core or bypasses it completely. On most systems, when the controls are in the heat position (mode), the A/C compressor is turned off and the evaporator operates at ambient temperature. This means that any temperature change of the incoming air is now controlled only by the blend door as it directs the air across the heater core. The heater core is part of the ICE cooling system, and hot coolant is circulated through it by the water pump.

When the system is placed in the *A/C mode,* the A/C compressor is engaged and the blower motor circulates air over the evaporator.

- The cool, dehumidified air is then sent to the blend door, where the air can bypass the heater core completely when maximum cooling effect is required.
- If warmer air is desirable, its temperature can be increased by changing the position of the blend door so that some of the air passes through the heater core on its way to the distribution ducts.
- The final air temperature is achieved by blending the heated air from the heater core with the unheated air.

HYBRID ELECTRIC VEHICLE COMPRESSORS

CONVENTIONAL COMPRESSORS Many hybrid vehicles use an electrically driven compressor, which allows compressor and A/C operation with the engine at idle stop. Hybrids using belt-driven compressors use a 12-volt electromagnetic

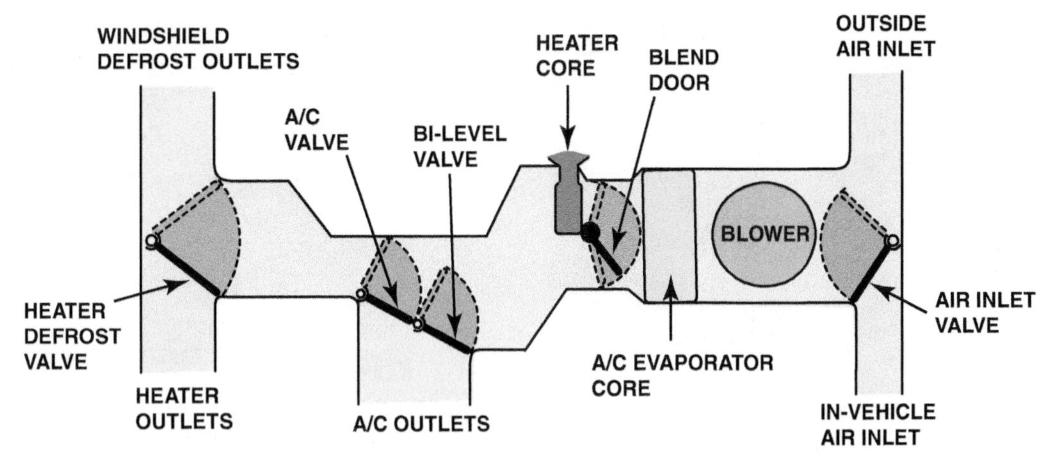

FIGURE 35–2 A basic air distribution system. Air can enter the system from outside the vehicle, or from the passenger compartment while in the recirculation mode.

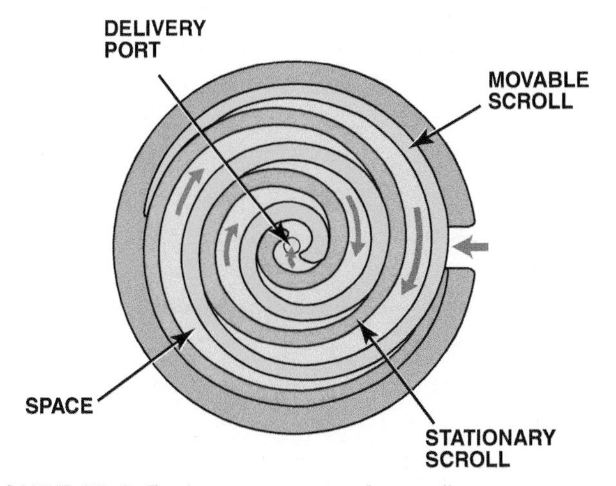

FIGURE 35–3 Basic components of a scroll compressor. Note the "pockets" of refrigerant that occupy the spaces labeled with red arrows.

clutch, similar to non hybrid vehicles. When this type is used on a hybrid electric vehicle, the idle stop (start/stop) function is disabled so that cooling can continue when the vehicle is stopped. The most commonly used compressor design in hybrid electric vehicles is the scroll compressor. ● **SEE FIGURE 35–3**.

DUAL COMPRESSORS A few hybrid vehicles use a dual compressor that is belt- (85%) and electrically (15%) driven. The electrically driven portion of a Honda hybrid electric vehicle compressor operates at more than 150 volts (depending on the vehicle), 3-phase AC and provides cooling during idle stop.

- One part is a 4.6 cu. inch (75 cc) belt-driven compressor that will stop when the engine stops
- The other part is a smaller 0.9 cu. inch (15 cc) electric compressor that runs during idle-auto-stop to keep the A/C operating.

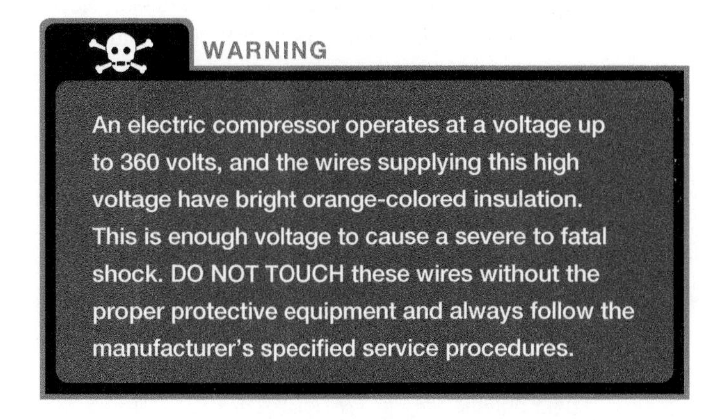

Some hybrid vehicles use an A/C system much like other vehicles, but run the engine constantly when the A/C is turned on. Many hybrid vehicles use an electrically driven compressor that allows compressor and A/C operation with the engine at idle stop. ● **SEE FIGURES 35–4 AND 35–5**.

ELECTRIC COMPRESSORS Electric vehicles do not have an internal combustion engine, and hybrid vehicles do not always run the engine continuously during vehicle operation. An electric compressor operates from the high-voltage (HV) battery pack and can therefore provide cooling under all conditions if the vehicle is moving or has stopped. An electric compressor combines a scroll compressor with a DC or AC electric motor. ● **SEE FIGURE 35–6**.

The electric motor is cycled on and off to produce the desired cooling. This unit can be mounted at any convenient place on the vehicle since it does not need to be connected to an engine.

REFRIGERANT OIL Hybrid vehicle A/C systems that use electric-driven compressors must use **POE (polyol ester)** oil, unlike all other compressors. PAG (polyalkylene glycol) oil,

FIGURE 35–4 Hybrid electric vehicle A/C compressor. Note that this unit is primarily belt driven but has a high-voltage electric motor built in to allow A/C system operation during idle stop. The orange cable indicates that the wires carry high voltage. Always follow the safety warning precautions as specified by the vehicle manufacturer.

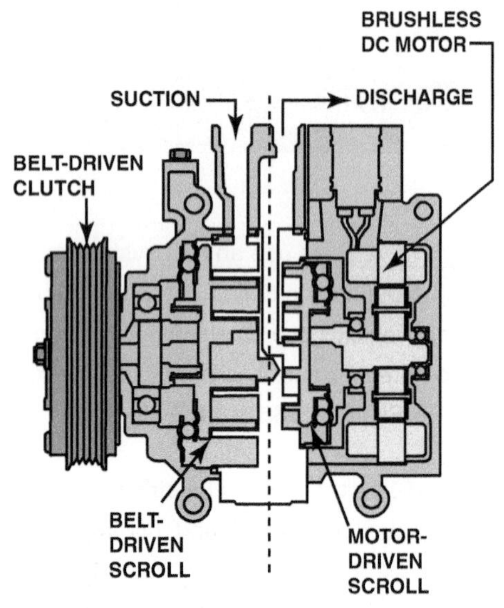

FIGURE 35–5 This hybrid vehicle A/C compressor uses two scrolls with one being belt driven (left) and the other driven by a brushless DC motor (right).

which is used in non hybrid vehicles, is slightly conductive and can cause deterioration of the insulation on the windings of the compressor motor. This can cause the compressor to become electrically conductive, which can result in electrical leakage. This leakage can potentially be hazardous during future service. ● **SEE FIGURE 35–7**.

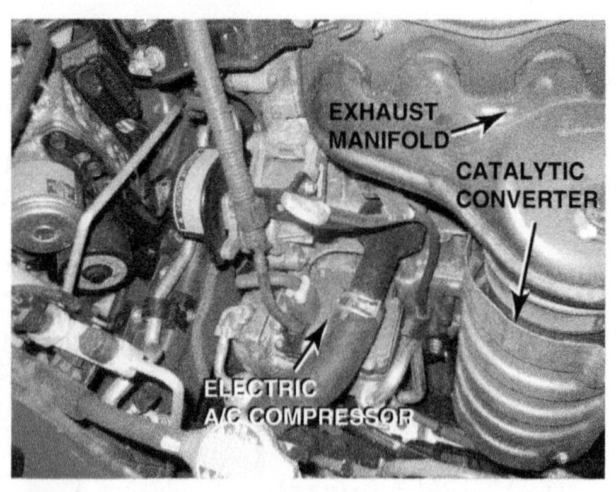

FIGURE 35–6 The compressor is mounted to the engine block in the same location as a conventional engine-driven compressor. This mounting helps reduce noise and vibration.

FIGURE 35–7 Specific A/C compressor oil designed for use in Honda hybrid vehicle air-conditioning systems.

CABIN HEATING SYSTEMS

PARTS AND OPERATION Heater cores are built similar to radiators, but are much smaller and often have the inlet and outlet pipes located at the same end of the assembly. The heater core is located in the passenger compartment, inside the plenum chamber (air distribution box) for the vehicle's heater and air-conditioning components. ● **SEE FIGURE 35–8**.

Air entering the plenum chamber must first pass through the A/C evaporator core, and then is directed either through or around the heater core by the blend door (*air mix valve*). The temperature of the air can be adjusted by changing the position of

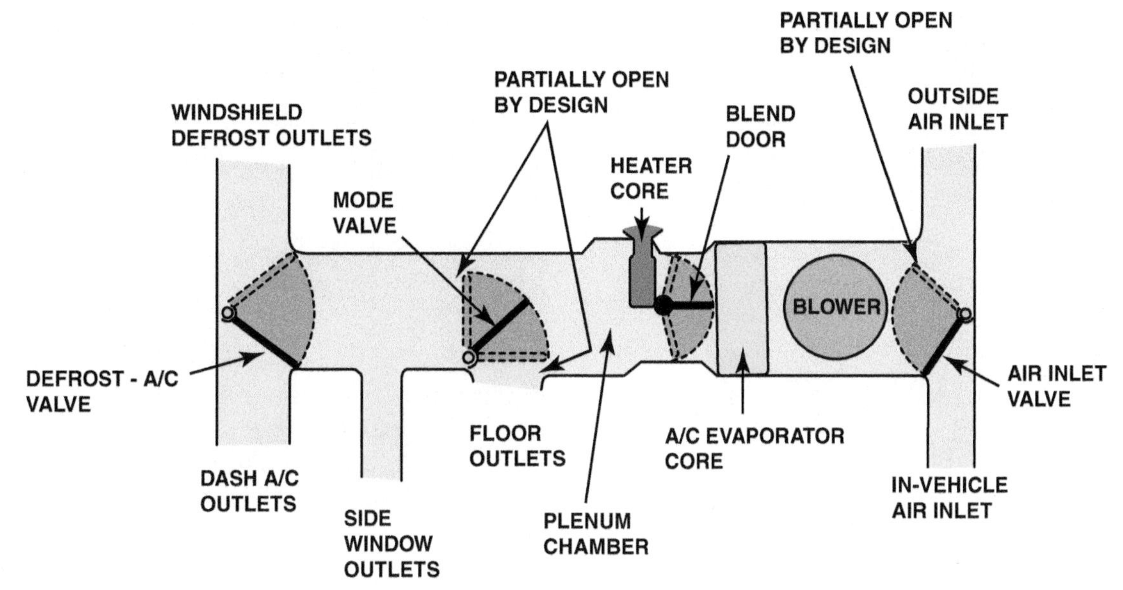

FIGURE 35–8 The heater core is located in the plenum chamber. Air temperature in this system is controlled by the position of the air mix valve (blend door).

the blend door and the percentage of air that is sent through the heater core. Once the air leaves the heater core, it is blended with any air that has bypassed it and then is sent to its destination through a series of mode doors. These direct the air to specific areas of the vehicle, depending on the mode that is selected at the time. For instance, the defrost mode will require that the air be directed toward the windshield outlets. In the heat mode, however, the air can be sent to the instrument panel outlets and/or the floor vents depending on make and model.

ENGINE-OFF HEATER OPERATION Some vehicles, including some hybrid and non hybrids, have a feature that allows the heater to keep operating when the engine is turned off. In cold climates, this keeps the vehicle warm during start/stop operation. The main component is an electric pump, called an *auxiliary water pump*, along with a bypass valve. ● **SEE FIGURE 35–9**.

When the heater is on and the engine is off, the bypass valve closes, and the pump operates to maintain a hot coolant flow through the heater core. When the heater is on with the engine running, the pump is shut off, and the bypass valve is open to allow normal heater operation.

PTC HEATERS Another approach used to heat the interior on a hybrid electric vehicle is to use **PTC heaters** built into the heater core itself. Positive temperature coefficient (PTC) refers to the tendency of a conductor to increase its electrical resistance as its temperature increases. PTC heaters convert electrical energy into heat, and this is used to boost heat to the passenger compartment. ● **SEE FIGURE 35–10**.

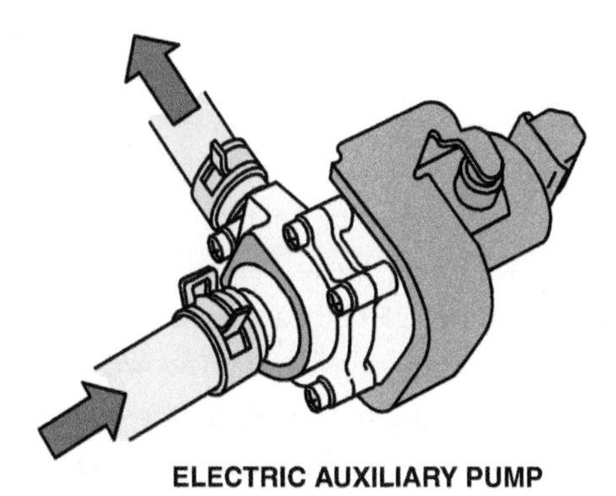

FIGURE 35–9 The electric auxiliary pump is turned on whenever cabin heating is requested and the ICE is in idle stop mode. The pump is plumbed in series with the hoses running to the vehicle heater core.

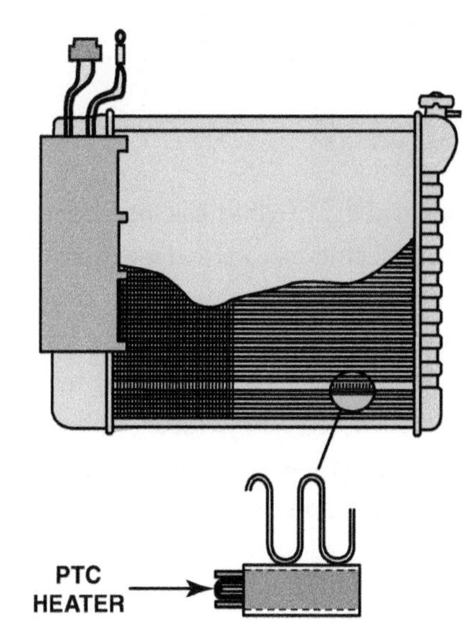

FIGURE 35–10 PTC heaters can be located on the heater core itself to help boost heat to the passenger compartment when coolant temperature is low.

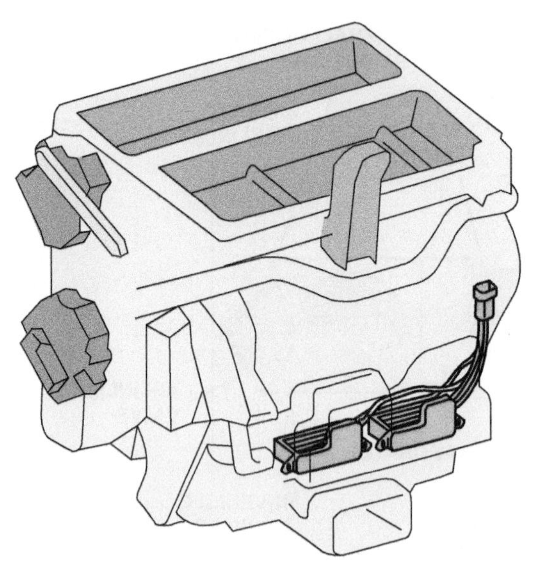

FIGURE 35–11 Two PTC heaters are located in the footwell air ducts in this example. These are energized when the coolant temperature is low and MAX HOT is requested in the FOOT or FOOT/DEF modes.

PTC heaters can also be located in the air ducts in the form of a honeycomb-shaped grid. Air that is leaving the plenum chamber passes through these heaters before it enters the passenger compartment. PTC heaters are located in the heater core, as well as the footwell air ducts. The A/C electronic control unit turns on the PTC heaters when the coolant temperature is low and MAX HOT is requested. ● **SEE FIGURE 35–11**.

HEV ELECTRIC MOTOR AND CONTROL SYSTEM COOLING

NEED FOR COOLING Hybrid electric vehicles are unique in that they have electric motors and electronic controls that are not found in vehicles with conventional drivetrains. These components are designed to operate under heavy load with high current and voltage demands, so they tend to generate excessive heat during vehicle operation. Special auxiliary are incorporated into hybrid electric vehicles to prevent overheating of these critical components.

EFFECTS OF HEAT ON THE ELECTRICAL/ ELECTRONIC SYSTEM Electronic components operate more efficiently as their temperature decreases and can suffer permanent damage if overheated. All hybrid electric vehicles have cooling systems for their motors and motor controls, and some utilize air cooling to remove excess heat from these components. ● **SEE FIGURE 35–12**.

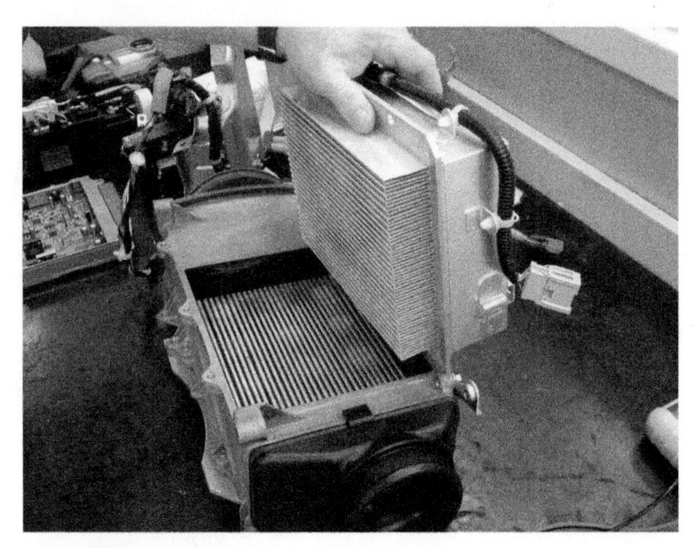

FIGURE 35–12 The motor and HV battery electronics on some hybrid vehicles are air cooled. Note the cooling fins for the modules.

Many hybrid electric vehicles use a liquid cooling system for their motors and motor controls.

SYSTEM CONSTRUCTION The liquid cooling systems used for the motors and motor controls on hybrid electric vehicles have much in common with conventional ICE cooling systems. There is a separate expansion tank that acts as a coolant reservoir for the system, and the coolant is often the same type that is used for the ICE cooling system. A radiator is used to dissipate excess heat and is located at the front of the vehicle. Some designs may have the radiator incorporated into the ICE radiator, or it may be separate. ● **SEE FIGURE 35–13**.

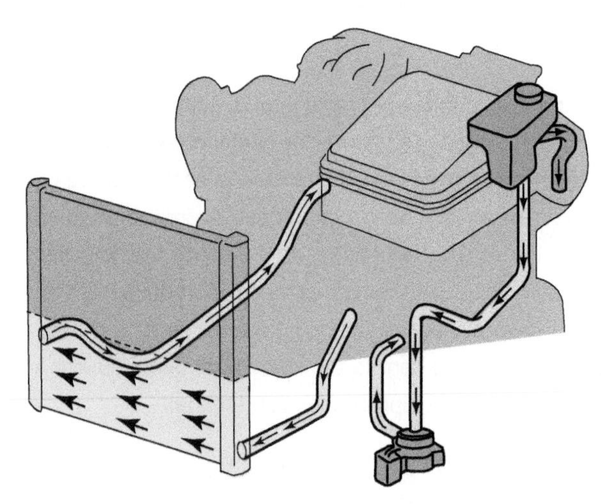

FIGURE 35–13 The electric motors and the motor controls are cooled using a separate cooling system. This system uses a radiator that is integral with the ICE cooling system radiator.

FIGURE 35–14 This transaxle has cooling passages for both of the motor-generators. Note the coolant pipes (two on each assembly) on the lower left of this photograph.

FREQUENTLY ASKED QUESTION

Why Isn't the ICE Cooling System Used to Cool HEV Motors and Motor Controls?

Most internal combustion engine (ICE) cooling systems operate at over 200°F (93°C) and use a thermostat to control the cooling system temperature. For maximum efficiency, it is important that the engine operate as close to this temperature at all times. Electric motors and the motor controls, however, tend to operate more efficiently at lower temperatures. The engine cooling system runs too hot to allow these components to operate at peak efficiency, so a separate low-temperature system is often used.

A low-voltage electric water pump is used to circulate the coolant, and it is often configured to run whenever the vehicle is in operation. The coolant is circulated through the various components in the system, which could include the following:

- Electric motor-generator(s)
- DC–DC converter
- Inverter
- Transmission oil cooler
- Other high-load control modules and the high-voltage batteries in some cases, although these often have their own separate system to warm and cool the battery pack

FREQUENTLY ASKED QUESTION

What Coolant Is Used in a Motor/Electronics Cooling System?

Most hybrid electric vehicle manufacturers specify that the same coolant used in the vehicle's ICE (engine) be used for cooling the motors and electronics. For instance, Toyota specifies that its Super Long Life Coolant be used in the second-generation Prius ICE cooling system as well as the inverter cooling system.

A thermostat is not used in these systems. In some HEV systems, the electric pump will circulate the coolant whenever the key is on. In other systems, peak temperature is controlled by turning on the pump, and then the fan, at progressively higher speeds as the coolant temperature rises. Some hybrid transaxles have cooling passages located near both of the electric motor-generators and these are part of the inverter cooling system. Coolant hoses connect to fittings on the transmission case and direct coolant in and out of each of the motor-generator assemblies. ● **SEE FIGURE 35–14.**

TWO-MODE HVAC SYSTEM OVERVIEW

CAUTION: The HVAC system contains a high-voltage compressor. Service performed on or around the compressor and the high-voltage components requires disabling the high-voltage system. Therefore, proper safety precautions are REQUIRED before servicing these systems. These precautions apply to all high-voltage HVAC systems.

- Required PPE—When disabling the high-voltage system, ALWAYS use the required personal protective equipment (PPE). Required PPE includes class 0 insulation gloves and safety glasses with side shields. Recommended PPE includes rubber-soled shoes and nonsynthetic clothing.
- Class 0 Insulation Gloves—Always perform a visual inspection of the class 0 insulation gloves before each use. It is important to look at the date stamp on each glove and make sure the gloves are certified up to date. Each glove must be recertified every six months to ensure proper glove protection. Refer to service information for proper glove inspection procedures.

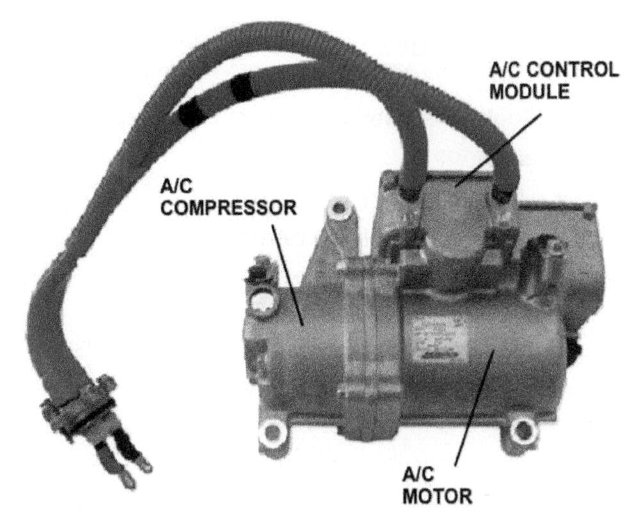

FIGURE 35–15 The two-mode compressor receives DC voltage from the HV battery and the ACCM converts this to AC voltage. (Courtesy of General Motors)

- Shop Hazards—Make sure your work environment is safe by checking for potential shop hazards, as well as removing any conductive materials you might be wearing or may encounter.

AIR-CONDITIONING COMPRESSOR The two-mode hybrid vehicle uses an all-electric-driven air-conditioning compressor that is bussed DC power directly from the 300-volt power inverter module (PIM). The compressor contains high-voltage energy whenever the hybrid battery high-voltage contactor relays are closed. The compressor is directly coupled to a three-phase electric motor which receives power from the air-conditioning control module (ACCM). ● **SEE FIGURE 35–15.**

The two-mode compressor uses a smaller capacity (30 cc) scroll-type pump than a conventional compressor because the ICE does not drive the compressor. A conventional vehicle system cycles the compressor to control system pressure. However, on a two-mode system, compressor speed can be controlled electrically to achieve proper system pressure.

AIR-CONDITIONING CONTROL MODULE The ACCM inverts 300 volts direct current (DC) from the hybrid battery to 300 volts alternating current three-phase energy to power the compressor motor. The 12-volt side of the ACCM controls air-conditioning compressor speed using information from the engine control module (ECM), hybrid powertrain control module (HPCM), and electronic climate control (ECC) module over low-speed communication lines. Note that the ACCM is part of the HV compressor assembly and is not serviced separately.

COMPRESSOR OPERATION While the compressor is operating, it creates a unique sound which may be heard during auto stop and engine ON events. When ECC blower speed increases, compressor speed also increases to accommodate system pressure and needs. In other words, compressor speeds may follow ECC blower speeds to match system pressure and temperature needs, masking compressor noise.

COMPRESSOR REPLACEMENT AND SERVICE The air-conditioning compressor is self contained and replaced as a unit. However, the pressure relief valve, low voltage connector, PIM connector O-ring, and PIM connector captured bolts are air-conditioning compressor components that can all be replaced separately.

POE OIL AND O-RINGS The two-mode hybrid air-conditioning compressor uses standard R134A refrigerant, but a different type of oil than conventional compressors. Because the oil and refrigerant are accessible to the electric motor, any moisture within the system can cause high-voltage electrical isolation concerns.

The two-mode hybrid compressor uses polyol ester (POE) oil and is required in order to keep the high-voltage system isolated from the chassis. If the wrong oil is used, or moisture enters the system, the hybrid system may detect a loss of isolation. Only POE oil should be used in this system and the recharging equipment must be flushed properly before any service is performed on the system.

NOTE: When moisture is absorbed in the POE oil, the oil may become acidic and possibly cause compressor failure. Refer to service information for proper servicing procedures.

Different types of O-rings and hoses are utilized in two-mode vehicles because of the POE oil used in the system. If standard O-rings or hoses are used, swelling may occur, creating a leak in the system. Be sure to use the correct O-ring or hose when repairing the system. Refer to service information for correct O-ring part numbers and procedures. When assembling threaded connections, use mineral oil just as you would on a conventional system.

ACCUMULATOR The two-mode air-conditioning system needs to remain as moisture free as possible to maintain system isolation. Therefore, the air-conditioning system utilizes an accumulator, which contains approximately three times more desiccant than accumulators used in conventional systems.

LOW-SIDE PRESSURE TRANSDUCER Another component which makes the two-mode air-conditioning system different from a conventional air-conditioning system is the use of

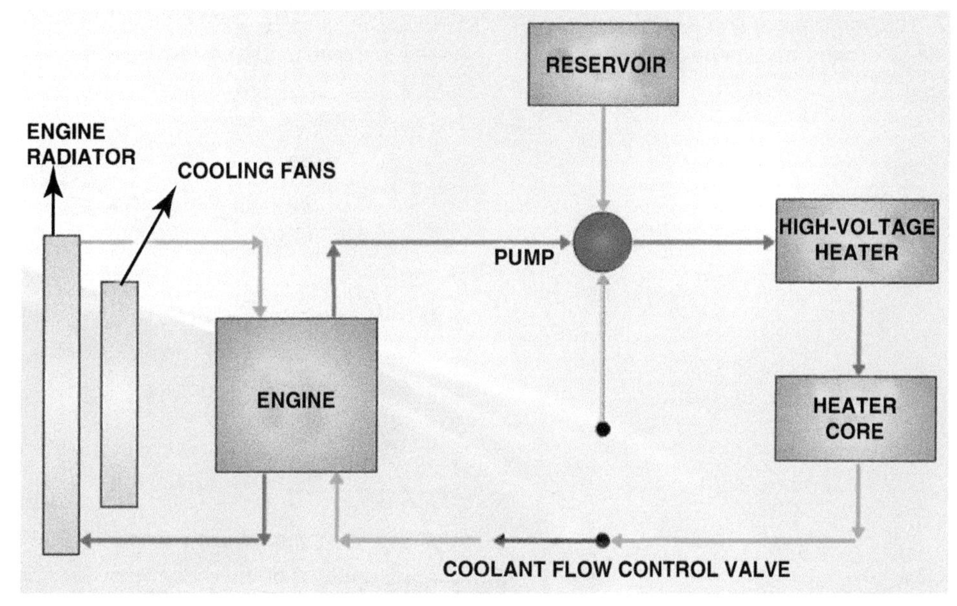

FIGURE 35–16 A block diagram of the EREV heating system. (Courtesy of General Motors)

a low-side pressure transducer instead of a pressure switch. A scan tool can read both high- and low-side pressures without raising the hood of the two-mode vehicle. Always verify readings by connecting gauges to the system.

TWO-MODE HEATING SYSTEM

The two-mode hybrid is equipped with an electric circulation pump. The electric circulation pump supplies the cabin heater core with engine coolant to maintain heater system demand while the vehicle is in auto stop or electric drive mode. If the engine coolant temperature drops too low due to extended auto stop, or electric drive mode, the engine will restart to maintain its specified coolant temperature level.

The electric circulation pump is located in the passenger side of the engine compartment, near the front of the engine. This 12-volt electric circulation pump is controlled by the ECC module, through a relay which is located in the underhood hybrid bussed electrical center (HBEC).*

*Adapted from General Motors Centre of Learning Course # 1844.01W-R2.

EXTENDED-RANGE ELECTRIC VEHICLE HVAC SYSTEM

Chevrolet Volt/Cadillac ELR **extended-range electric vehicles (EREV)** include a single or dual-zone, automatic climate control heating, ventilation, and air conditioning (HVAC) system that has been modified from conventional vehicle HVAC systems. The system includes components that provide heating and air-conditioning functions whether the internal combustion engine is operating or not. The unique components include:

- High-voltage heater
- High-voltage A/C compressor
- Drive motor battery coolant cooler (chiller)

HEATING SYSTEM The heating system on the extended-range electric vehicle uses the engine or the high-voltage heater to warm the coolant that supplies heat for the cabin. When the internal combustion engine is running, the engine provides the required heat. When the ICE is not running, the high-voltage heater warms the coolant. The high-voltage heater provides different levels of heat depending on the requested set temperature. ● **SEE FIGURE 35–16.**

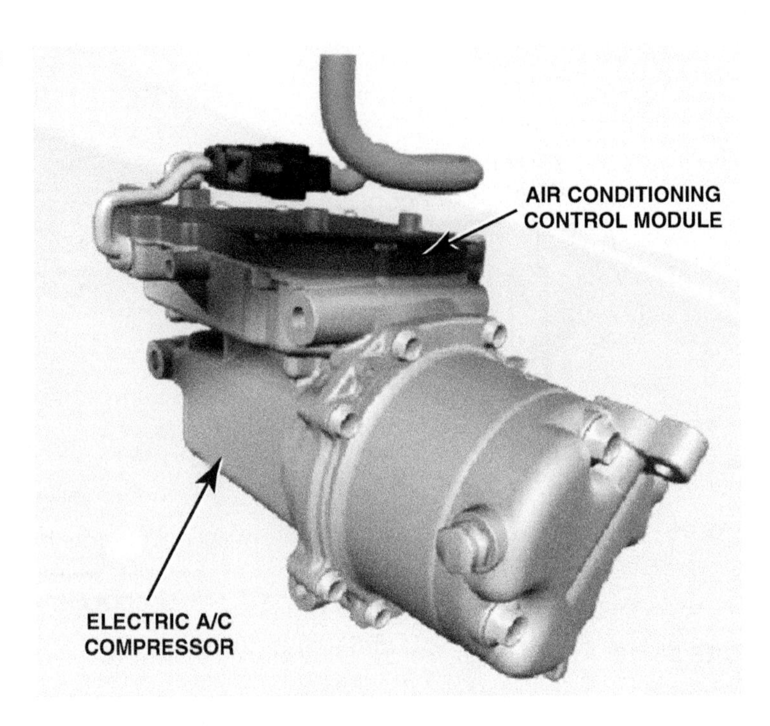

FIGURE 35–17 The compressor and ACCM are replaced as a unit. (Courtesy of General Motors)

AIR-CONDITIONING SYSTEM The R-134A air-conditioning system on the extended-range electric vehicle is similar to conventional vehicles with the exception of an electronic variable compressor and a drive motor battery coolant cooler, also called a **battery chiller**. The compressor is driven by a high-voltage electric motor and provides the necessary refrigerant pressures to meet cooling demand. The drive motor battery chiller works in parallel with the HVAC refrigerant system and provides the same function as an additional evaporator core by extracting heat from the high-voltage battery cooling system.

EREV AIR-CONDITIONING COMPRESSOR The electric A/C compressor is motor-driven rather than belt-driven, enabling the compressor to operate independently from the engine. The A/C compressor receives 360-volts DC from the high-voltage battery. An internal inverter changes the DC voltage to three-phase alternating current (AC). The heating, ventilation, and air conditioning or HVAC control module receives sensor inputs and calculates the actions required to accomplish an A/C request. The module varies the compressor motor speed up to 10,000 revolutions per minute (RPM) to maintain the requested cooling level. The compressor is not cycled on and off but is electronically variable. ● **SEE FIGURE 35–17**.

EREV RECEIVER-DEHYDRATOR The receiver-dehydrator is a storage unit for refrigerant that is attached to the condenser and ensures the thermal expansion valve (TXV) receives only liquid refrigerant. The receiver-dehydrator contains desiccant that absorbs moisture from the refrigerant. The extended-range electric vehicle's receiver-dehydrator contains two to three times more desiccant than a conventional A/C system. The system requires an increased amount of desiccant because the unique oil in the system absorbs a greater amount of moisture. The desiccant cartridge is replaceable. Refer to service information for the procedure.

EVAPORATOR CORE AND CONDENSER The evaporator core is similar to those found on conventional vehicles. Refrigerant entering the evaporator is regulated by a TXV. The condenser, located behind the high-voltage battery radiator, enables the transfer of heat from the refrigerant to the passing air. A refrigerant filter screen is located at the inlet of the condenser to trap debris in the event of a compressor failure.

DRIVE MOTOR BATTERY COOLANT COOLER The drive motor battery coolant cooler (battery chiller) is a fluid to fluid heat exchanger that cools the high-voltage battery. The cooler receives liquid refrigerant through a TXV and absorbs heat from the high-voltage battery coolant. ● **SEE FIGURE 35–18.**

LUBRICANT The electric A/C compressor requires POE oil. POE oil does not conduct electricity, which ensures isolation of the high-voltage compressor from the vehicle chassis. However, POE oil is more hygroscopic than polyalkylene glycol (PAG) oil, requiring 2–3 times more desiccant in the system.

SYSTEM OPERATION The operation of the A/C system is similar to a conventional A/C system. However, the A/C compressor contains an internal module that is commanded by the

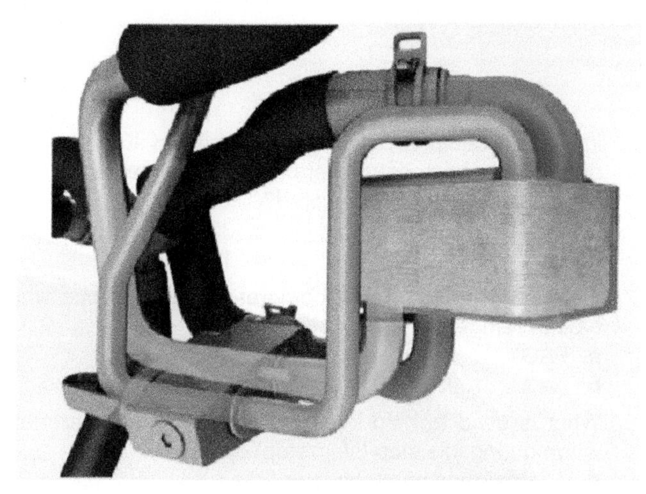

FIGURE 35–18 The battery chiller uses refrigerant to help cool the HV battery. (Courtesy of General Motors)

hybrid powertrain control module 2 (HPCM 2) through serial data communication. The engine control module and HPCM 2 monitor sensors to determine the optimal compressor motor speed. Similar to a conventional evaporator core, the battery chiller uses a TXV and refrigerant to cool the high-voltage battery coolant. Several plates inside the battery chiller separate the battery coolant and refrigerant. The refrigerant also passes through a second TXV into the cabin evaporator core for cabin cooling.*

NOTE: The battery cooling system circulates a premixed Dexcool®, which is a 50/50 mixture of Dexcool and deionized water. Deionized water is required for high-voltage isolation and to prevent corrosion from effecting heat sink performance. Always use premixed coolant and never use tap water in the battery coolant system.

SERVICE PROCEDURES

Most hybrid electric vehicle HVAC-related service procedures are the same as a conventional vehicle except for the following precautions:

- If recovering the refrigerant, use a hybrid-specific refrigerant recovery, recycling, and recharge (RRR) machine.
- If removing a compressor that is powered by high voltage and has orange cables running to it, disable the HV system by removing the high-voltage disconnect (service) plug before work is begun.
- Always follow the vehicle-specific service procedures.
- Read, understand, and follow all safety notices and precautions.

SUMMARY

1. The purpose of the ICE cooling system is to bring the ICE up to optimum temperature as quickly as possible and to maintain that temperature under all operating conditions.
2. A thermostat is used to maintain the optimum coolant temperature in the ICE cooling system, however, thermostats are not normally used in EV- and hybrid-dedicated electronics cooling systems.
3. Electric auxiliary pumps are used to circulate coolant in the heating system when an HEV enters idle stop mode.
4. The coolant heat storage system is used to limit vehicle emissions during cold starts.
5. Most HEVs use scroll compressors in their air-conditioning systems.
6. Some HEVs use A/C compressors with electric drive or a combination belt-electric drive mechanism.
7. PTC heaters are used to provide supplemental heat in HEV heating systems.
8. Most hybrid electric vehicle manufacturers specify that the same coolant used in the vehicle's ICE (engine) be used for cooling the motors and electronics.

REVIEW QUESTIONS

1. How does a coolant heat storage system work?
2. Why is an HEV motor/electronics cooling system separate from that of the ICE?
3. What is the function of a PTC heater, and why is it used in an HEV heating system?

*Adapted from General Motors Centre of Learning Course # 18420.05W-R2.

CHAPTER QUIZ

1. A coolant heat storage tank can keep coolant warm for a maximum of ____________ day(s).
   a. One
   b. Two
   c. Three
   d. Four
2. The coolant heat storage system is being discussed. Technician A says that the water valve is driven by an electric motor. Technician B says that the storage tank has its own electric water pump. Which technician is correct?
   a. Technician A only
   b. Technician B only
   c. Both Technicians A and B
   d. Neither Technician A nor B
3. Technician A says that PTC heaters can be built into a conventional heater core assembly. Technician B says that a PTC heater's electrical resistance will decrease as its temperature increases. Which technician is correct?
   a. Technician A only
   b. Technician B only
   c. Both Technicians A and B
   d. Neither Technician A nor B
4. All of the following statements about hybrid electric vehicle A/C compressors are true, *except* ____________.
   a. Most are reciprocating piston designs
   b. Some use a belt drive along with an electric motor
   c. Some use only an electric motor without a belt drive
   d. Nonconductive refrigeration oil must be used with A/C compressors utilizing an electric drive motor
5. What is the color of the plastic conduit used over high-voltage wiring to an electric A/C compressor?
   a. Blue
   b. Orange
   c. Yellow
   d. Red
6. In a dual compressor that uses both an ICE-powered part and an electrical-power part, what percentage of the capacity is electrically powered?
   a. 85%
   b. 55%
   c. 35%
   d. 15%
7. Electrically powered A/C compressors should use what type of refrigerant oil?
   a. PAO
   b. PAG
   c. POE
   d. Polyalkylene glycol
8. What is used to help keep the passenger compartment warm during idle stop (start/stop) operation?
   a. A larger than normal heater core
   b. An auxiliary electric water pump
   c. Operating the A/C in reverse to provide heating instead of cooling
   d. A larger than usual water jackets in the ICE
9. In most hybrid electric vehicles, the air entering the plenum chamber passes through the ____________.
   a. Heater core
   b. Cabin filter
   c. Evaporator
   d. Recirculation door
10. What coolant is used in a motor/electronics cooling system?
    a. Special electrically insulted coolant
    b. DEX-COOL
    c. The same coolant that is used in the ICE cooling system
    d. Either a or c depending on the vehicle

chapter 36

REFRIGERANT RECOVERY, RECYCLING, AND RECHARGING

LEARNING OBJECTIVES

After studying this chapter, the reader should be able to:

1. Explain the steps involved in the service and repair of A/C systems.
2. Discuss the procedure for identifying refrigerants in an A/C system.
3. Explain the procedure for refrigerant recovery in A/C systems.
4. Explain the procedure for recycling refrigerant in A/C systems.
5. Discuss the purpose of flushing an A/C system.
6. Explain the procedure for evacuating an A/C system.
7. Discuss the procedure for recharging an A/C system.
8. Explain how to retrofit a R-12 system to a R-134a system.
9. Explain the purpose of sealants and stop leaks.

This chapter will help you prepare for the ASE Heating and Air Conditioning (A7) certification test content area "A" (A/C System Service, Diagnosis, and Repair).

KEY TERMS

GM STC OBJECTIVES

GM Service Technical College topics covered in this chapter are as follows:

1. Recover refrigerant from a vehicle A/C system.
2. Evacuate a vehicle A/C system.
3. Recharge a vehicle A/C system.

CLEAN AIR ACT

SECTION 609 The U.S. Clean Air Act has placed a group of requirements on A/C service. These requirements can be viewed at www.epa.gov/ozone/title6/609

Section 609 of the Act gives EPA the power to enforce the following requirements:

- Preventing the release or venting of CFC-12 or HFC-134a.
- Technicians are required to use approved equipment to recover and recycle CFC-12 and HFC 134a.
- Technicians who repair or service CFC-12 and HFC-134a systems must be trained and certified by an EPA-approved organization. Visit www.ase.com or www.macsw.org for details on how to become a certified air-conditioning technician.
- Service shops must maintain records of refrigerant transfer and technician certification.
- Service shops must certify to EPA that they have and are properly using approved refrigerant recovery equipment.
- The sales of small cans of CFC-12 (R-12) can be made only to certified technicians.
- CFC-12 equipment can be permanently converted to HFC-134a, but must meet SAE standard J2210.
- CFC-12 systems can be retrofitted to use a SNAP refrigerant.
- EPA can assess civil penalties and fines for violations of these requirements.

REQUIREMENTS A major part of the Clean Air Act requires recycling of R-12 and R-134a instead of releasing/venting the refrigerant to the atmosphere, which is banned by Section 609. This makes economic sense. At one time, new (also called virgin) R-12 was inexpensive, well under $1.00 per pound. When R-12 was inexpensive, it was standard practice to simply vent the contents of a system to the atmosphere when service was needed. Also, many people kept adding R-12 to a system rather than going through the trouble and expense of repairing a leak. This was called *topping off a system.* It is socially irresponsible not to repair any fixable leaks. Small containers of R-12 are no longer available to do-it-yourselfers.

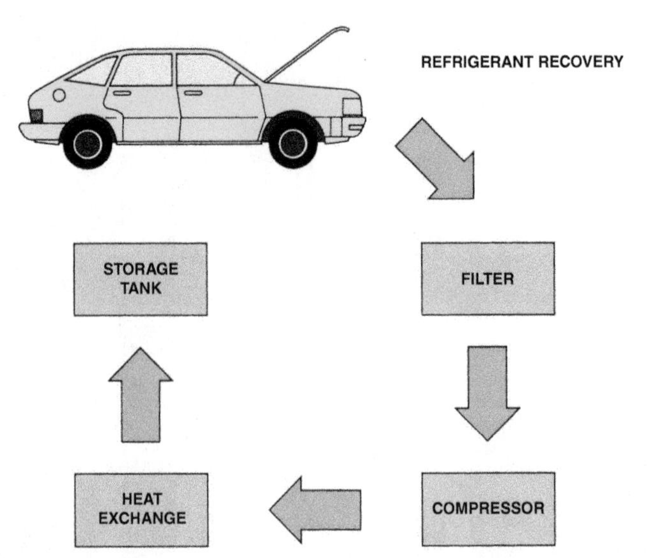

FIGURE 36–1 A recovery unit removes refrigerant vapor from the vehicle. Then it filters the refrigerant before compressing it so it condenses and can be stored as a liquid in the storage tank.

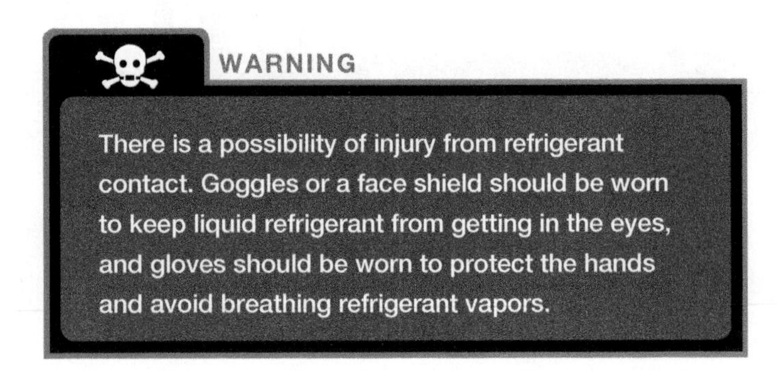

A/C SERVICE OPERATIONS

Air conditioning service usually begins by identifying what refrigerant or other chemicals may be in the system and the recovery of whatever refrigerant is left in the system. ● **SEE FIGURE 36–1.**

Recovery means that all refrigerant is removed from a system so it can be stored in a container in liquid form. This refrigerant will be recycled into the same or another A/C system.

Recycling is the process of removing moisture (water), oil, and noncondensable gases (air) from the recovered R-12 or R-134a so it meets the standards of new refrigerant. Recycled refrigerant should be at least 98% pure. R-134a can also be recovered and recycled, but separate equipment, dedicated to R-134a, is required. Recycled R-12 or R-134a can be used in the same way as new refrigerant.

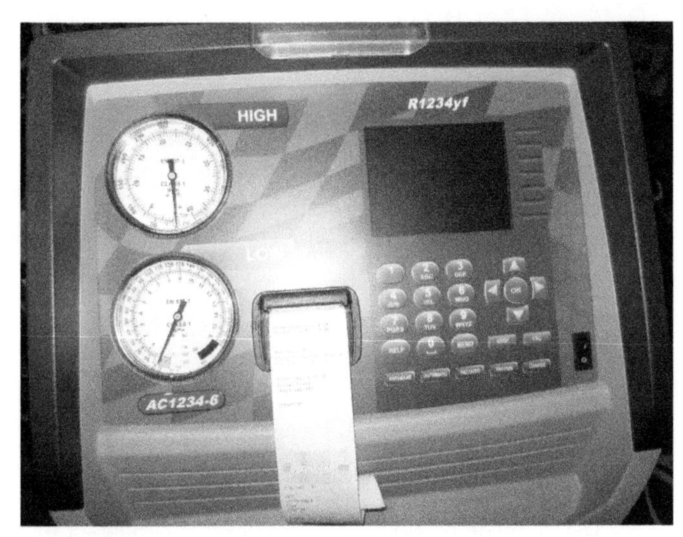

FIGURE 36–2 An RRR machine designed for use with systems that use R-1234yf. The machine used must match the refrigerant used in the vehicle.

FREQUENTLY ASKED QUESTION

What Is an AC Virus?

Recovering an A/C system which is contaminated by a substance such as R-22 or a hydrocarbon into the recovery container that is half full of R-134a will ruin the R-134a in the container. If the container is then used to service another vehicle, the contaminated refrigerant will contaminate the other system(s) as well. Many compare a system with contaminated refrigerant to a virus that can sometimes spread from person to person at a very rapid rate.

SAE RRR MACHINE STANDARDS **Refrigerant recovery, recycling, and recharging (RRR)** equipment must be certified to meet SAE J2788, which has more stringent requirements starting in 2006 over the previous SAE J2210 standard (1991–2005). Under the old SAE J2210 standard, recovery efficiency and charge accuracy were never requirements. Some features of the SAE 2788 units include:

- The new SAE J-2788 standard requires that all R-134a service equipment manufactured after January 1, 2008, must recover 95% of the refrigerant and recharge to within an accuracy of 1/2 ounce.
- The old J2210 standard claimed to be accurate to only within 0.88 oz (which is a 3% error on a system that holds 2 pounds of refrigerant or a 7% error on a system that holds 14 oz).

FIGURE 36–3 A typical refrigerant identification machine. The readout indicates what kind of refrigerant is in the system. If a blend or some other contaminated refrigerant is discovered, it should be recovered and stored in a separate container to keep it from contaminating fresh refrigerant.

- Machines compliant with standard SAE 2788 will remove more of the old refrigerant during the recovery process and provide more accurate refrigerant measurement during charging.
- Recovery equipment is available as a single unit or in combination with a recycling unit. Recovery-only units are simpler and less expensive than equipment that handle both recovery and recycling. ● **SEE FIGURE 36–2.**

NOTE: These are U.S. standards. Refer to Canadian or other applicable refrigerant handling standards for the country in which the vehicle is being serviced.

SERVICE AND REPAIR OF A/C SYSTEMS

STEPS INVOLVED The service and repair of heating and A/C systems consists of the following steps:

- Identifying the refrigerant in a system using an **identifier**. ● **SEE FIGURE 36–3.**
- Recovery and disposal of contaminated refrigerant.
- Evacuation of the system.
- Repair of the system as needed, including the replacement of system components.
- Maintaining proper oil level when components are serviced.
- Checking the oil level in a compressor or system.

Several important facts when servicing an A/C system include the following:

- A/C systems are designed to operate using a specific amount of a particular refrigerant.
- A/C systems are designed to operate using a specific amount of particular refrigerant oil.
- Adding any other chemical into a system can create a chemical problem that can cause system damage or failure. Most original equipment manufacturer (OEM) and aftermarket suppliers consider any chemical other than the refrigerant and oil to be a foreign material.

REFRIGERANT CONTAMINATION At one time, contaminated refrigerant was rare, and the majority of the problems were caused by air from improper service procedures. Today, there are a variety of sealants, blends, hydrocarbons, and other refrigerants, along with recycled refrigerants available to both repair facilities and do-it-your (DIY) consumers that make the potential for contamination a much greater problem. The A/C industry defines refrigerant as:

- 98% and better is considered pure
- Less than 97% pure is contaminated
- Between 97% and 98% is questionable

Purity standards (SAE J1991 and J2099 and ARI 700-88) permit a small amount of contamination with new (virgin) and recycled refrigerant.

NONCONDENSABLE GASES (NCG) Any refrigerant that contains **noncondensable gases (NCG)**, usually air, or with 2% or greater of another material (foreign refrigerant), is considered contaminated. If the contamination is greater than 5%, problems such as excessive high-side pressures and incorrect clutch cycling rate result. Other unseen problems of oil breakdown, seal deterioration, or compressor wear can also occur. NCG contamination can be reduced by recycling the refrigerant, purging the air out, or diluting the contaminated refrigerant with pure refrigerant. Contamination from a foreign refrigerant requires that the refrigerant be sent off for reclaiming or disposal.

SEALANT CONTAMINATION HVAC technicians believe that contamination is caused by chemicals that are intended to seal refrigerant leaks. There are two general types of sealants:

- A chemical that promotes swelling of rubber O-rings and seals
- An epoxy/epoxy-like, one- or two-part, moisture-cure polymer

TECH TIP

Always Check for Sealant

It is possible for an A/C system to contain a sealant, and this sealant can damage a refrigerant identifier and/or a service unit when that refrigerant in the system is recovered. It is highly recommended to test for a sealant before starting any other refrigerant service. The only sure way to completely remove a sealant from a system is to replace every component that contains refrigerant.

A simple and relatively inexpensive sealant identifier is Quick Detect made by Neutronics. Recovering refrigerant that contains a sealant might cause damage to the recovery unit. There have been cases where sealant has cured inside the machine, causing loss of the machine or expensive repairs.

REFRIGERANT IDENTIFICATION

TYPES OF IDENTIFIERS To determine what the correct refrigerant is for the system being serviced, inspect the service fittings and the refrigerant label. There are two types of refrigerant identifiers.

1. A *go-no-go identifier* (accuracy to about 90% or better), indicates the purity of the refrigerant with lights that indicate pass or fail.
2. A *diagnostic identifier* (accuracy to 98% or better) displays the percentage by weight of R-12, R-134a, R-22, hydrocarbons, and air (displayed as NCG). If the contaminant is air, the refrigerant can be safely recovered and recycled for reuse. If the contaminant is another refrigerant or a hydrocarbon, then a special recovery procedure is required and the recovered mixture needs to be sent off for disposal or recycling. Recycling machines cannot remove a foreign refrigerant, only air, water, oil, or particulates. Newer identifiers include a printer port, so hard-copy readout can be printed for the customer's use. ● **SEE FIGURE 36–4**.

STEPS TO IDENTIFY THE REFRIGERANT To identify the refrigerant in a system, perform the following steps:

STEP 1 The first step on all identifiers is CALIBRATION, because most units do not allow it to be connected to the low side line until the identifier shows what to do on the display.

FIGURE 36–4 A typical printout showing that the system has 100% R-1234yf refrigerant in the system.

STEP 2 With the engine and the system shut off, connect the identifier to the low-side service port using the correct hose assembly for the system's refrigerant type.

STEP 3 Check the filter on the identifier for the incoming gas because it will show a color change when it needs to be changed.

STEP 4 Allow a gas sample to enter the unit. Some units include a warning device to make sure liquid refrigerant does not enter it. Many units include a warning device to indicate a flammable refrigerant.

STEP 5 Read the display to determine the nature of the refrigerant. Some units allow printing of the results at that time. If the refrigerant is good or is contaminated with air, it can be safely recovered and recycled.

STEP 6 When the analysis is complete, some units display instructions to disconnect the sampling hose and then bleed out the gas that was sampled. If the hose is not disconnected when prompted, the identifier can bleed all of the refrigerant from the system.

TECH TIP

Check New Refrigerant to Be Sure

Some shops are using the diagnostic-style identifier as a quality-control check of new refrigerant they purchase. Badly contaminated new (virgin) refrigerant is also being found, some of which is manufactured and being imported illegally. Contaminated refrigerant should not be charged into a system. An identifier cannot identify a blend directly, but can identify the refrigerants contained in the blend.

REFRIGERANT RECOVERY

EQUIPMENT NEEDED Most recovery units are made as part of the service equipment and contain two service hoses (a low and high side) that are connected to the system service ports and low- and high-side pressure gauges. For R134a systems, confirm that the service connector knurled knob is turned fully counterclockwise. Then slide the quick connect collar up the hose and press the fitting onto the system service fitting and release the collar. It should "snap into place" and not be able to be removed. Slowly tighten the knurled knob until fully opened.

- Remove the caps (blue for the low side and red for the high side) from the service fittings and connect the A/C manifold gauge.
- The connector has a sliding collar and a knurled knob. Slide the collar up the hose, press the fitting onto the orifice and release the collar, and then hand-tighten the knurled knob.

NOTE: Some recovery units are connected to the center hose (yellow hose) of a manifold gauge set and are used for adding refrigerant or evacuating the system.

Blends or contaminated mixtures should be recovered using a different, separate machine. The hoses should be equipped with shutoff valves within 12 inches of the ends and have end fittings to match the refrigerant used.

A wise technician is careful to check the service history of the vehicle. Many refuse to service a vehicle if there is a chance that it contains contaminated refrigerant. Recovery units contain the following:

- An oil separator
- A compressor-like pump
- A condenser-like heat exchanger

RECOVERY OPERATION Recovery units draw refrigerant vapor out of a system and convert it into liquid for storage. Recovery units weigh the amount of refrigerant that is recovered, which tells the technician if all refrigerant from a fully charged system has been recovered or whether the system was fully charged. Oil removed and separated during the recovery

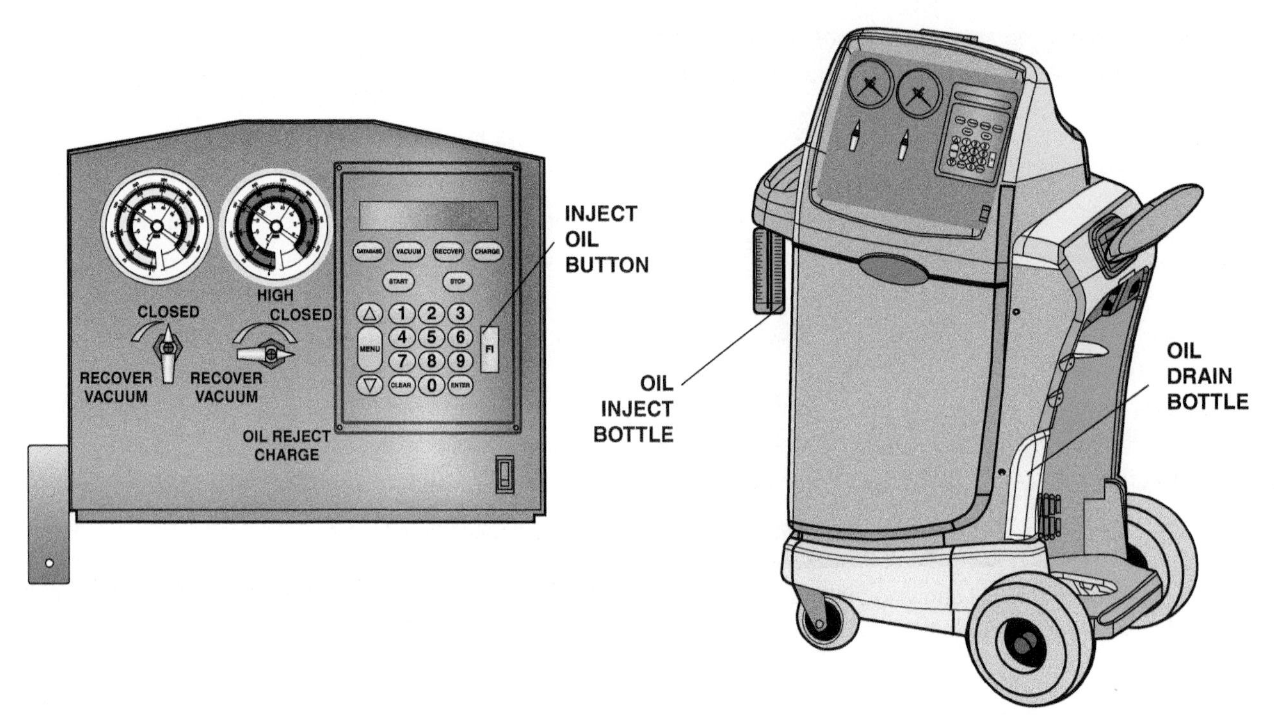

FIGURE 36–5 During the recovery process, oil from the system is separated into a container so the technician will know how much oil was removed.

process is usually drained into a measuring cup and noted so that this amount of new oil can be replaced in the system as it is recharged. Oil recovery also gives an indication of the oil volume in the system. If no oil is recovered, then the system was likely low on oil. If an excessive amount of oil is recovered, the system probably had too much oil.

Some refrigerant in a system is absorbed in the oil and does not leave the oil immediately when a system is emptied. This trapped refrigerant *out-gases* or boils out of the oil later, after the pressure has been removed. Recovery units shut off automatically after the main refrigerant charge has been removed and the system drops into a slight vacuum. To completely remove the refrigerant, run the normal recovery procedure and then recheck the pressure after a 5-minute wait. If the pressure has increased, restart the recovery process.

STORAGE CONTAINER CERTIFICATION The storage container for recovered refrigerant must be approved by the Department of Transportation (DOT) and carry the letters *DOT* and the certification numbers. The tank must also include the date for tank retesting. Recovery tanks must be inspected and certified to be in good condition every 5 years. Some rules to follow for refrigerant recovery storage tanks include:

- Do not reuse disposable cylinders.
- Make sure that the recovery cylinder is labeled for the refrigerant being recovered.
- Make sure the *tank certification* is in order and the cylinder retest date has not expired.
- Inspect the cylinder for damage and do not fill if damaged.

STEPS TO RECOVER REFRIGERANT To recover the refrigerant from a system, include the following steps:

STEP 1 Identify the refrigerant in the system.

STEP 2 Make sure the hoses have the proper shutoff valves and are compatible with the refrigerant in the system.

STEP 3 Connect the recovery unit to the system or to the center hose of the manifold gauge set, following the directions of the manufacturer.

STEP 4 Open the required valves and turn the machine on to start the recovery process, following the directions of the machine's manufacturer.

STEP 5 Continue the recovery until the machine shuts off or the pressure reading has dropped into a vacuum.

STEP 6 Verify completion of recovery by shutting off all valves and watching the system pressure. If pressure rises above 0 PSI within 5 minutes, repeat steps 4 and 5 to recover the remaining refrigerant.

STEP 7 Drain, measure, and record the amount of oil removed from the system with the refrigerant and dispose of properly. This amount of new oil should be added during the recharging process. ● **SEE FIGURE 36–5.**

The system can now be repaired. Present standards and good work habits require that recovered refrigerant be recycled before reuse, even if it is to be returned to the system from which it was recovered.

SMALL CANS At one time, refrigerants were commonly sold in small cans that contained 12 oz or 15 oz of refrigerant. Refrigerant is available in larger containers, usually 30 pounds, and larger, and these are commonly used by service shops. Small cans are commonly used by untrained do-it-yourself (DIY) vehicle owners.

There are concerns that DIY service often results in the wasting or release of a substantial amount of unused refrigerant for several reasons:

- A tendency to not repair refrigerant leaks and instead just add additional refrigerant to the leaking system.
- Partially charged systems are topped off, so they probably end up over- or undercharged.
- A small can may not have the correct amount of refrigerant for the system, so charging is often stopped before the can is empty.
- The remaining refrigerant in the can will probably leak out over time.

RECOVERING CONTAMINATED REFRIGERANT Special procedures should be followed to remove a contaminated or a blend refrigerant from a system. Do not use an R-12 or R-134a recovery machine to remove a contaminated or a blend refrigerant. One alternative is to convert an older R12 machine and gauge set to recover refrigerant blends, but this equipment should be used only for blends from that point on unless it is decontaminated after each use. If the mixture contains more than 4% hydrocarbon, it should be considered explosive, and this mixture should not be recovered using an electric-powered machine. Air-powered recovery machines are available and can be safely used for explosive mixtures.

Hydrocarbon refrigerants, blends, and unknown mixtures should be recovered into containers clearly labeled "CONTAMINATED REFRIGERANT." The proper color for these containers is gray with a yellow top. Recovery of contaminated refrigerant using a recovery machine is essentially the same as recovering uncontaminated refrigerant except that the contaminated material must be sent off for disposal or off-site recycling.

RECYCLING REFRIGERANT

QUALITY STANDARDS Most recycling units pump the recovered refrigerant through a very fine filter to remove foreign particles, past a desiccant to remove water, and through an oil separator to remove any excess oil. Air is removed by venting it, using the noncondensable purge, from the top of the liquid refrigerant. Recycled refrigerant must meet the same purity standards as new (virgin) refrigerant including:

- Less than 15 ppm moisture
- Less than 4000 ppm oil
- Less than 330 ppm air

Some machines have a sight glass equipped with a moisture indicator so that the operator can tell when the moisture has been removed.

SAE 2788 recovery units are designed to stop operation if a filter or desiccant change is needed. Some units can perform the recycling process in a single pass from the storage container through the cleaning process and back to the storage container. These machines often complete the recycling process while the system is being evacuated. Others require several passes, and the recovery process continues to circulate the refrigerant as long as necessary. ● **SEE FIGURE 36–6.**

The recycling machine is dedicated to a particular type of refrigerant, and a recycled blend can only be recharged back into the vehicle it came from or another vehicle from the same fleet.

PROCEDURE FOR RECYCLING To recycle a refrigerant, perform the following steps:

STEP 1 Open the valves or perform the programming steps required by the machine manufacturer and turn on the machine.

STEP 2 The machine operates until excess foreign particles and water have been removed or for a programmed length of time and then shuts off. Check the moisture indicator to ensure that the refrigerant is dry. If the machine does not shut off in the proper amount of time, its internal filters or desiccant probably require service. ● **SEE FIGURE 36–7.**

NOTE: Sometimes all of the air will not be removed in one recycling pass. Repeat the recycling as needed to remove all of the excess air.

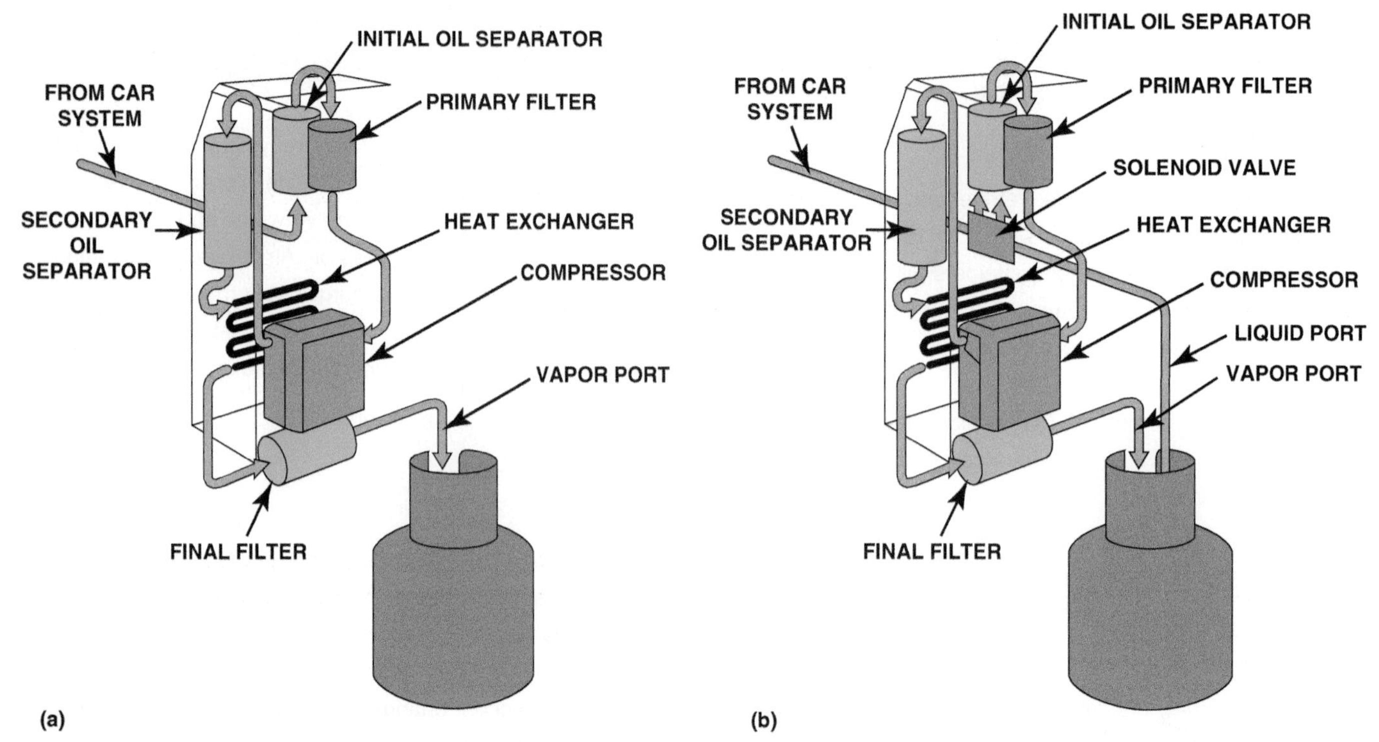

FIGURE 36–6 A single-pass recycling machine (a) cleans and filters the refrigerant as it is being recovered. A multipass machine (b) recovers the refrigerant in one operation and then cycles the refrigerant through filters and separators in another operation.

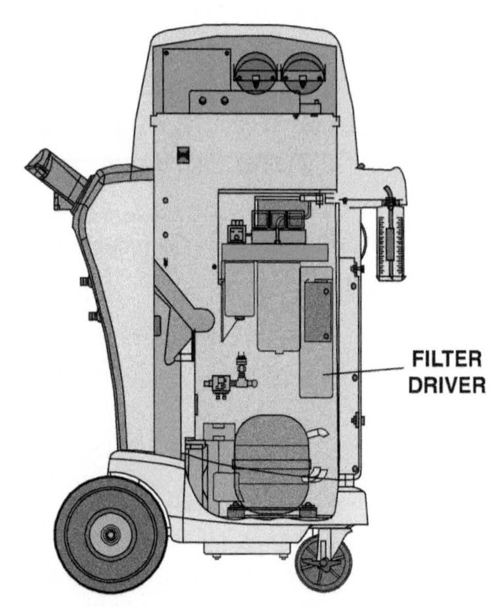

FIGURE 36–7 Recycling machines have a filter and desiccant that must be replaced after a certain amount of use.

AIR CONTAMINATION CHECKS After recovery is completed, check for excess air in the refrigerant can be done by looking at the *pressure–temperature (PT)* relationship. This is best done after the temperature of the refrigerant has stabilized to the level it was at the start of the workday.

To check the PT relationship, perform the following steps:

STEP 1 Keep the storage container at a temperature above 65°F (21°C) and away from direct sunlight for 12 hours.

STEP 2 Read the pressure in the container using a calibrated pressure gauge with 1 PSI increments.

STEP 3 Read the temperature of the air next to the container.

STEP 4 Compare the pressure and temperature readings. ● **SEE CHART 36–1**.

TEST RESULTS

- If the pressure for a particular temperature is less than that given in the table, the refrigerant does not contain an excess amount of air and is considered uncontaminated.
- If the pressure is greater than that given in the table, slowly vent or purge gas from the top of the container (red valve) until the pressure drops below that given in the table.
- If the pressure does not drop, the refrigerant must be recycled or sent off for disposal or recovery.
- Some recycling machines can purge air from the system during recycling using either a manual or an automatic process. If recycling is attempted with contaminated refrigerants or some blends, the entire mix can be

TEMPERATURE °F	PRESSURE (PSIG) R-12	PRESSURE (PSIG) R-134a
65	74	69
66	75	70
67	76	71
68	78	73
69	79	74
70	80	76
71	82	77
72	83	79
73	84	80
74	86	82
75	87	83
76	88	84
77	90	85
78	92	88
79	94	90
80	96	91
81	98	93
82	99	95
83	100	96
84	101	98
85	102	100
86	103	102
87	105	103
88	107	105
89	108	107
90	110	109
91	111	111
92	113	113
93	115	115
94	116	117
95	118	118
96	120	120
97	122	122
98	124	125
99	125	127
100	127	129
101	129	131
102	130	133
103	132	135
104	134	137
105	136	139
106	138	142
107	140	144
108	142	146
109	144	149
110	146	151

CHART 36–1

Pressure/temperature chart used to determine if there is air trapped in the refrigerant container.

TECH TIP

Watch Out for "Black Death"

Debris from a failed compressor is often a fine, black material called *goo* or **black death** by many technicians. This mixture of small particles of aluminum and Teflon from the piston rings of the compressor, mixed with overheated refrigerant oil, will be trapped between the compressor discharge port and the orifice tube. It is very difficult to flush this out, especially from the condenser.

discharged as the machine attempts to purge air. With refrigerant that is severely contaminated with air, it is thought that some of the excess air is contained in the liquid, not just sitting on top of it, making it much more difficult to purge. If this is the case, make a partial air purge and then let the PT relationship stabilize over a period of time and recheck.

FLUSHING AN A/C SYSTEM

PURPOSE **Flushing** is using a liquid or a gas to clean the inside passages of an air-conditioning system. When a compressor fails, it usually sends solid compressor particles into the high and possibly the low sides, which can plug the condenser passages and orifice tube. It is always a good practice to check the orifice tube or TXV filter screen for debris when there is a compressor failure and to flush the system, install a filter, or both before completing the job.

Flushing is done by pumping a liquid material through the passages in a reverse or normal flow direction. A gas, such as shop air or nitrogen, is not really effective in removing solid material. Flushing agents can be a commercial flushing solution, liquid refrigerant, or very lightweight ester oil. However, ester oil has very little or no solvent action, and remaining ester oil might be difficult to remove from the system.

NOTE: Manufacturers may not warranty a replacement compressor if the accumulator and orifice tube or receiver–drier is not also replaced.

TECH TIP

The Coffee Filter Trick

Some technicians place a paper coffee filter to catch the debris in the flush material going to the catch container. This allows them to inspect the material being flushed from the system and to determine the effectiveness of the flushing process.

CAUTIONS AND PRECAUTIONS No flushing operation will remove 100% of metal debris from a failed compressor. The very small metal particles can be stopped only by a filter.

- A compressor cannot be flushed.
- An accumulator or receiver–drier is not flushed and is simply replaced.
- If the evaporator is flushed, it is simply done to remove excessive refrigerant oil that may be trapped in the system during previous service.
- Most experts believe that the OT or TXV screen will filter out larger debris.
- Most flushing is done on the high side, from the compressor connection of the discharge line to the OT or receiver–drier fitting of the liquid line, or vice versa.

Approved flush solvents for A/C systems are based on HFC refrigerants that have a fairly high boiling point such as Genesolv SF, which has a boiling point of 59°F (15°C). Technicians using flushing chemicals have a professional responsibility to:

- Obtain, read, and understand the MSDS (SDS) for that chemical.
- Learn how that chemical works and how it should be used.
- Determine if that chemical will do no harm to the system and the materials like gaskets and seals.
- Determine how to remove that chemical from the system and how to properly recycle or dispose of it.

Flushing is intended to remove the following:

- Contaminated polyalkylene glycol (PAG) oil
- Desiccant, following a desiccant bag failure
- Overcharge of PAG oil
- Refrigerant contamination

General Motors recommends flushing with liquid R-134a, using the ACR 2000 or GE-48800 Cool Tech service equipment. Check the appropriate service information for specific recommendations.

FIGURE 36–8 The Robinair Cool Tech # GE-48800 is the recommended RRR equipment for General Motors dealerships. (Courtesy of General Motors)

Be careful when choosing a flushing solvent, and always follow the recommendations of the manufacturer. The solvent cleanses the system of impurities while removing some debris. It may also absorb moisture that may be in the system. Each solvent should be supplied with a safety data sheet. Be sure to read this information, and then use the product as directed.

Select the proper flushing agent for the A/C system being serviced. Consult the equipment instructions and add the recommended quantity of the flushing agent to fill the tank of the machine. The type of equipment being used will determine which components may be flushed. ● **SEE FIGURES 36–8 AND 36–9.**

Exact procedures vary by equipment and system, so it is important to follow both the machine operating instructions and the vehicle service manual procedures. With any type of flushing equipment, the refrigerant must be recovered from the system before connecting the machine.

FIGURE 36–9 This adapter kit is used with the GE-48800 for system flushing. (Courtesy of General Motors)

SETUP

1. Recover the refrigerant from the system.
2. Close service coupler valves and disconnect hoses from vehicle access ports.
3. Close the valve on the external source tank.

NOTE: During this procedure, up to 12 lbs. of refrigerant is charged into the vehicle A/C system. If the flushing cycle is stopped before it is complete and the external source valve is open, the unit automatically adds refrigerant to the internal storage vessel, and there will be no room to recover the refrigerant used for flushing.

4. Remove the A/C system orifice tube, and reconnect the fittings to create a bypass. ● **SEE FIGURE 36–10**. If the system has a TXV, remove it and install the H-block pass-through adapter into the system.
5. Disconnect the compressor hose block at the rear of the compressor.
6. Attach the compressor block adapter (from the flushing kit) to the system side of the compressor block. ● **SEE FIGURE 36–11**. Also, remove and bypass the accumulator or receiver–drier, if specified by the manufacturer.
7. Referring to service information, configure the block connectors to provide forward- or back-flushing of the refrigerant, which flows from the unit through the red high-side connection hose.

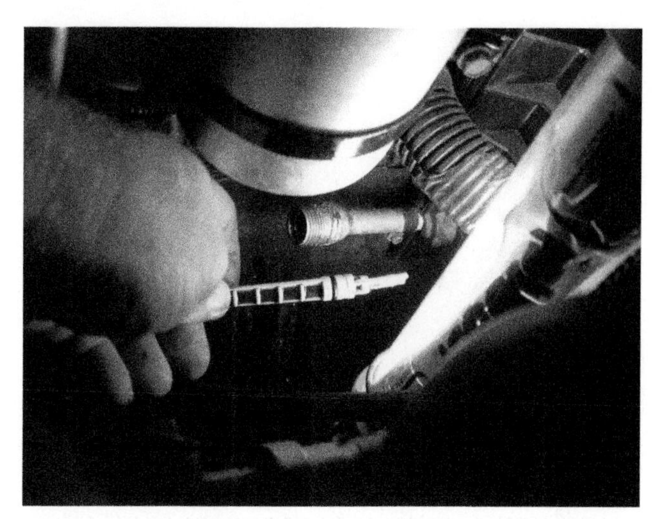

FIGURE 36–10 Remove the orifice tube and reconnect the lines without the tube. (Courtesy of Jeffrey Rehkopf)

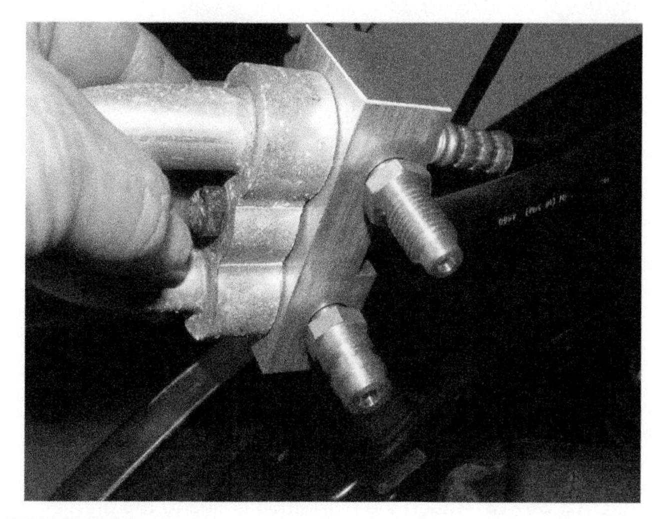

FIGURE 36–11 Remove the lines from the compressor and install the bridge block. (Courtesy of Jeffrey Rehkopf)

TECH TIP

General Motors recommends reverse flushing only in the case of a desiccant bag failure or compressor failure. Forward flushing is recommended for excessive or contaminated oil, or for any other reason.

8. Open the red service coupler.
9. Verify that a flushing filter is correctly installed in the flushing filter housing. If the filter has a check valve, remove it first. ● **SEE FIGURE 36–12**.

FIGURE 36–12 Put a new, clean filter in the filter housing. (Courtesy of Jeffrey Rehkopf)

FIGURE 36–13 Attach the ACR 2000 hoses to the bridge block and the filter housing. (Courtesy of Jeffrey Rehkopf)

10. Connect the filter housing to the desired return side of the adapter block and to the blue low-side hose. Open the blue service coupler and then open the isolation valve on the filter housing hose. ● **SEE FIGURE 36–13.**

FLUSHING PROCEDURE

1. Following the prompts on the screen, perform the flushing procedure. Refer to service information for details on forward and reverse flushing. The ACR 2000 refers to the flushing procedure as an "Oil Flush" procedure on the menu. The preset flushing time is 10 minutes. ● **SEE FIGURE 36–14.**
2. The unit will automatically run the vacuum pump for five minutes to remove air from the A/C system.
3. The unit flushes the system for the designated length of time and then enters a recovery mode.

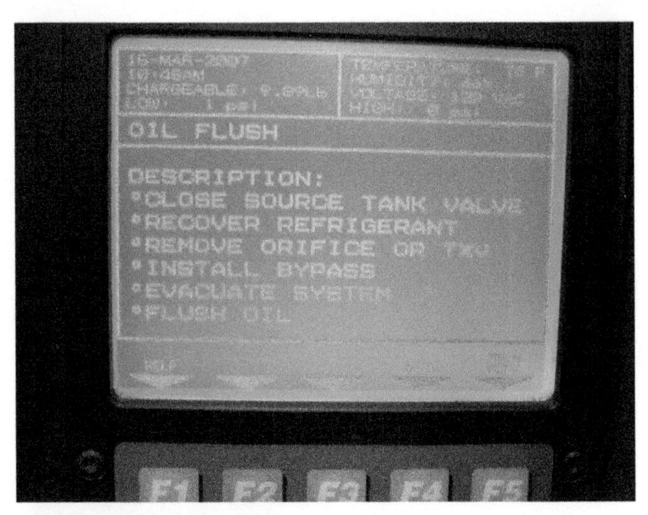

FIGURE 36–14 Follow the prompts to flush the system. (Courtesy of Jeffrey Rehkopf)

TECH TIP

Although the flushing time is 10 minutes, it takes considerably longer to clear the vehicle of liquid R-134a. Be patient and do not interrupt the process. It may take up to a half hour to finish.

NOTE: If the external flushing filter gets plugged, the unit displays FLUSH FILTER BLOCKED RETRY? Press START / YES to retry. Press STOP / NO to end the flush process and recover the refrigerant. With the refrigerant recovered, the filter can be serviced and the process repeated.

4. Oil that has been collected drains into the graduated oil drain bottle. Remove the bottle and measure the oil.
5. When the unit displays FLUSHING COMPLETE, close service couplers, remove hoses, and reassemble the vehicle's A/C system to its original state. This includes a new accumulator or receiver–drier, a new compressor, new O-rings, and new expansion device (TXV or orifice tube).
6. Open the valve on the source tank.
7. Evacuate and recharge the vehicle according to the service information instructions.

NOTE: It is important that the correct amount and type of oil be added to the system. After flushing, the only oil in the system will be the small amount in the compressor. This process removes all of the oil from the other components.

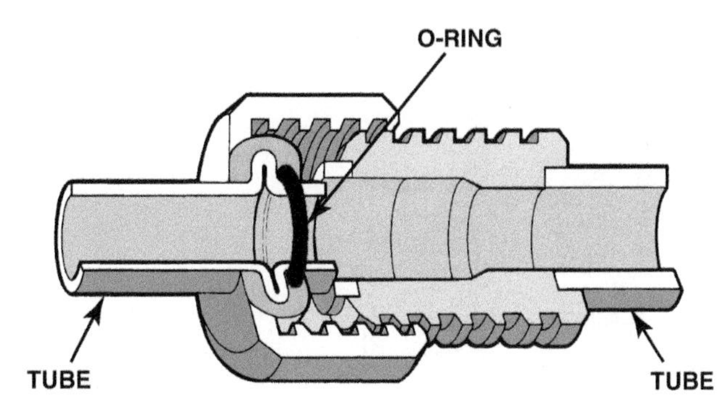

FIGURE 36–15 O-rings are usually made of neoprene rubber or highly saturated nitriles (HSN) to withstand high temperatures and flexing. O-rings should be changed during a retrofit procedure.

REPLACING COMPONENTS

GENERAL GUIDELINES After all refrigerant has been removed from the system, repairs can be accomplished. For example, the evaporator can be removed from the vehicle and replaced. Seal all openings of the system during service to prevent moisture and contaminants from entering. If the system has been opened to the atmosphere for a length of time (over 24 hours), most experts recommend replacing the drier to help prevent the possibility of damaging moisture being trapped in the system. After all repairs are completed, the system should then be evacuated.

NOTE: Be sure to follow all instructions regarding the amount of oil that needs to be added to the system if components have been replaced.

O-RINGS The O-ring seal is part of a fitting that holds the ends of two refrigerant lines or hoses together inside a connector. ● **SEE FIGURE 36–15.**

The O-ring forms the seal between the lines or hoses and the connector. The O-rings usually are made of highly saturated nitriles (HSN) or neoprene rubber and remain flexible over a wide range of temperatures. The O-ring must be lubricated with clean refrigerant oil before assembly to ensure a good seal. ● **SEE FIGURES 36–16 AND 36–17.**

TIGHTENING CONNECTIONS All threaded connections should be started by hand to avoid cross-threading. Two wrenches, one on each threaded fitting, should be used to loosen or tighten O-ring fittings. Using two wrenches is often called using "double-wrenches" or a "backup wrench" and is specified to be used by most vehicle manufacturers.

SERVICE FITTINGS AND CAPS Each type of refrigerant has its own unique service fitting. This prevents accidental use of the wrong service equipment and/or the introduction of the wrong refrigerant. Service valves are found almost anywhere on the system. They may be located on the receiver–drier, accumulator, compressor, muffler, or in the lines themselves. All service valves should have plastic coverings called *service caps*. ● **SEE FIGURE 36–18.**

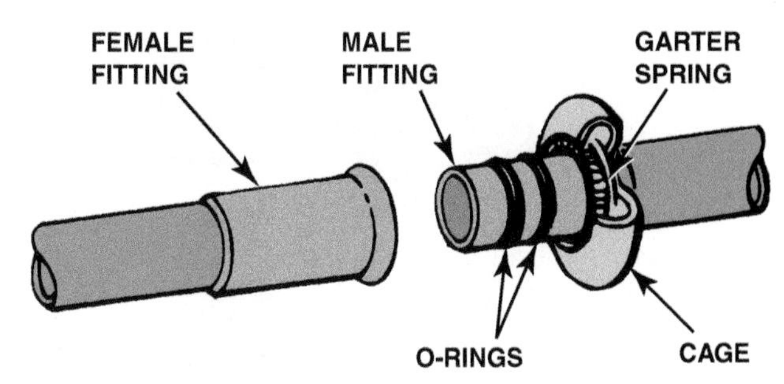

FIGURE 36–16 A typical spring-lock coupling.

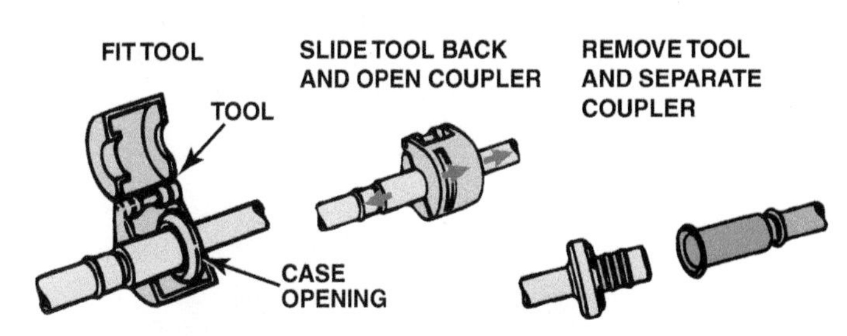

FIGURE 36–17 A special tool is needed to remove and install this spring-lock coupling.

Along with preventing dirt from entering the system, service valve caps have O-rings which become the primary seal if a valve leaks.

Always reattach the caps after any service has been performed, and replace them if missing. Another built-in precaution is the refrigerant cut-off valve, which keeps the refrigerant in the service hose instead of allowing it to vent to the atmosphere.

SCHRADER VALVES A Schrader valve is similar to a tire valve. Internal pressure holds Schrader valves closed. There is also a small spring to keep the valve seated if the internal pressure becomes insufficient. When the service connection is made, the depressor in the end of the service hose or service coupling pushes on a small pin inside the valve which opens the valve. The valve opens only when the service line connection is nearly complete, preventing contamination of the system or the unnecessary release of refrigerant. Many experts recommend that the Schrader valves be replaced whenever the system is open after recovery of refrigerant. ● **SEE FIGURE 36–19.**

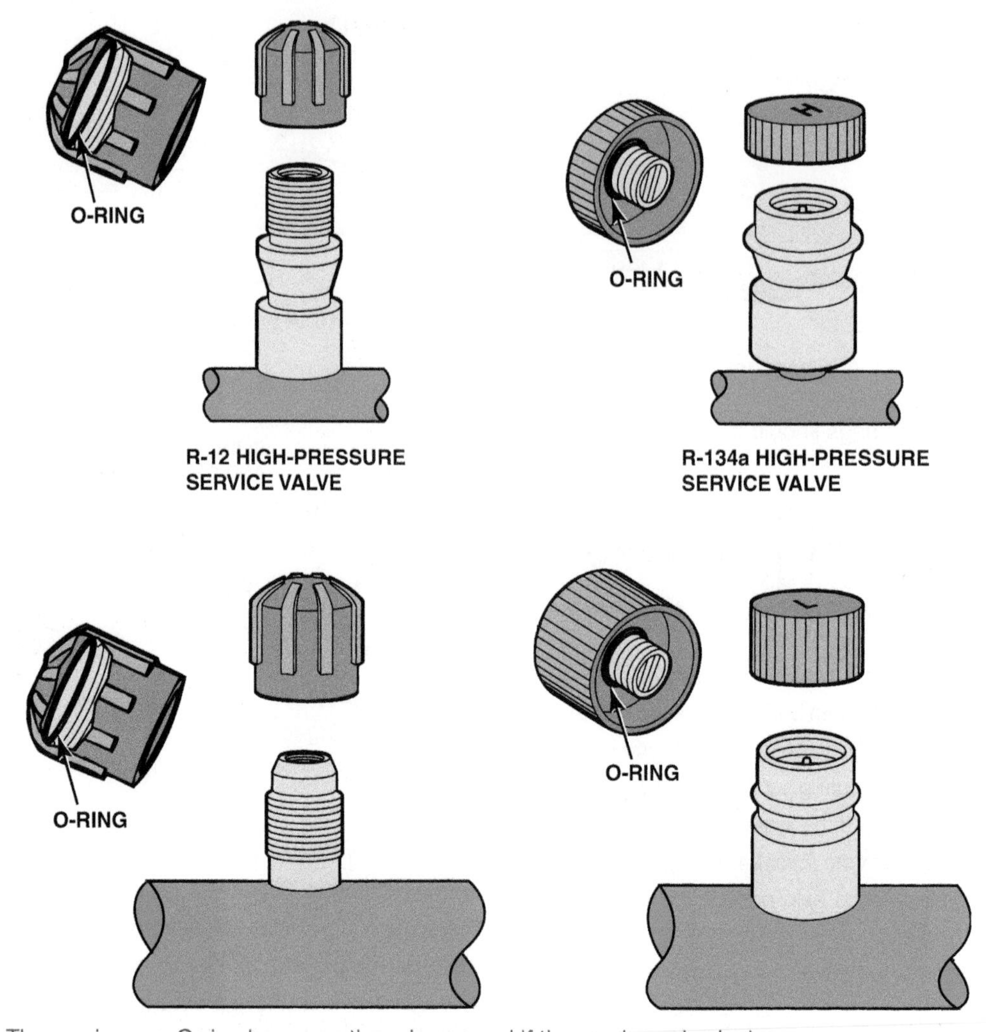

FIGURE 36–18 The service cap O-ring becomes the primary seal if the service valve leaks.

IN-LINE FILTER

PURPOSE AND FUNCTION If a compressor fails, debris in the form of metal particles often travels downstream through the condenser and is trapped in the receiver–drier or orifice tube, where it can cause plugging. Some compressor failures send debris into the suction line. The technician can install an in-line filter in either the high side or low side to trap any debris still in the system to protect the new compressor. They are available with connections to fit different line sizes. Some filters include a new orifice tube. ● **SEE FIGURE 36–20**.

INSTALLING AN IN-LINE FILTER To install an in-line filter, perform the following steps:

STEP 1 Recover the refrigerant from the system.

STEP 2 Select the location for the filter. The flexibility of hose allows a great deal of freedom. With metal tubing, locate a straight section of tubing slightly longer than the filter's connections.

STEP 3 Evacuate and recharge the system after installation of the filter.

NOTE: Most aftermarket air-conditioning filters have a directional arrow that indicates refrigerant flow direction.

REFRIGERANT OIL

ADDING OIL When it is determined that additional oil must be added, several types of low-cost injectors can be used to push oil into a charged or empty system. The proper amount of

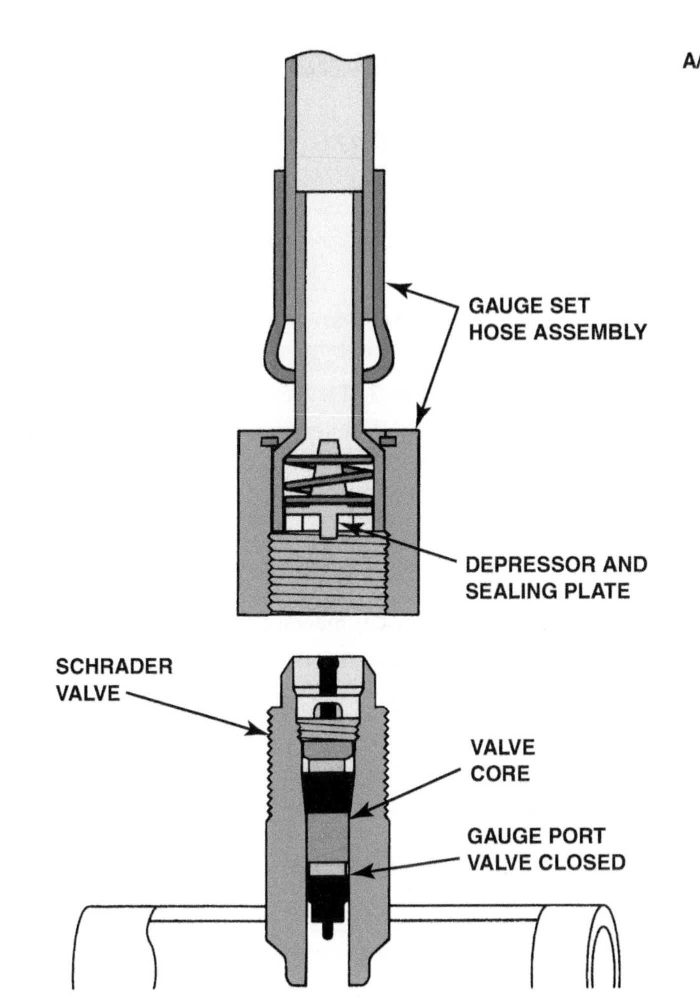

FIGURE 36–19 A depressor pin on the gauge set opens the Schrader valve when the connection is almost completely tightened. This prevents accidental refrigerant discharge.

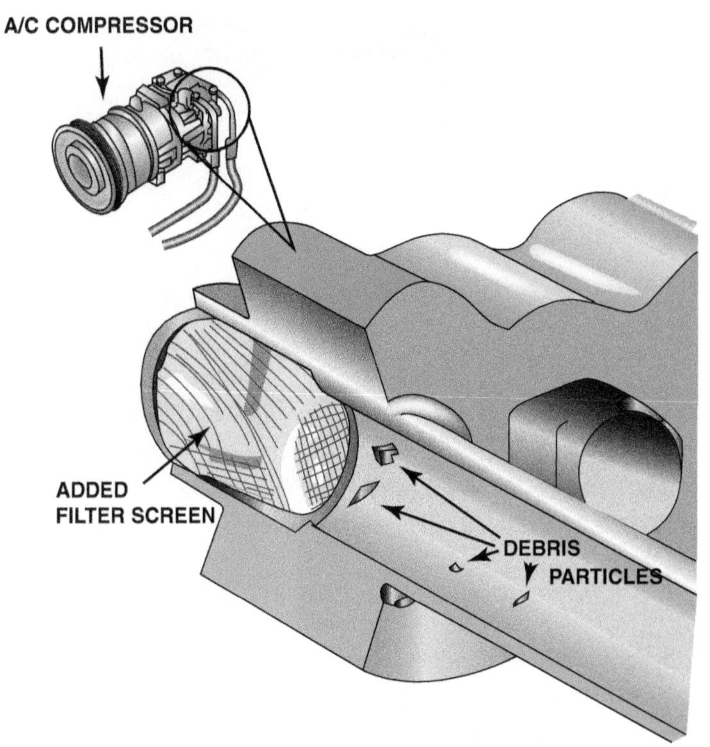

FIGURE 36–20 A typical aftermarket filter is installed in the suction line at the entrance to the compressor that is designed to catch any debris that could harm the compressor.

oil can be kept in a system by adding oil to replace the amount of oil removed when a component is replaced. This is usually only a small amount. Compressor replacement requires a slightly larger amount. ● **SEE FIGURE 36–21**.

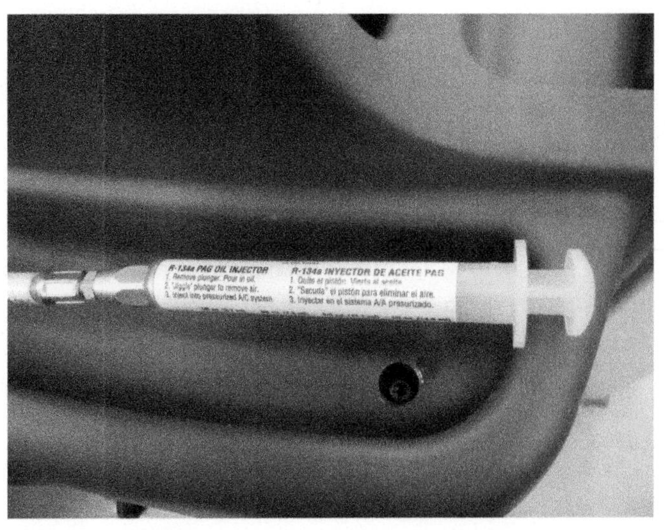

FIGURE 36–21 A variety of oil injectors are available for purchase. Some can be used while the system is under a vacuum, and some can force oil into a charged system.

EVACUATING THE SYSTEM

PURPOSE OF EVACUATING THE SYSTEM After a system has been repaired, all of the air and moisture that might have entered must be removed. **Evacuation** means that a vacuum will be applied to the system to remove the air and to vaporize any moisture that may be in the system. Although water boils at 212°F (100°C) at sea level, it can boil at much lower temperatures when the pressure is reduced. In other words, if a vacuum is applied to the air-conditioning system, the low pressure will cause any trapped moisture in the system to vaporize (boil). This water vapor is then removed from the system through the vacuum pump and released into the

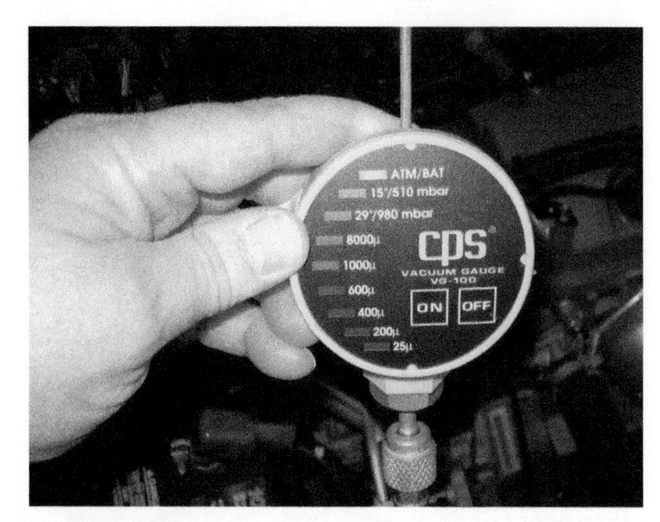

FIGURE 36–22 An air-conditioning vacuum gauge that reads in microns.

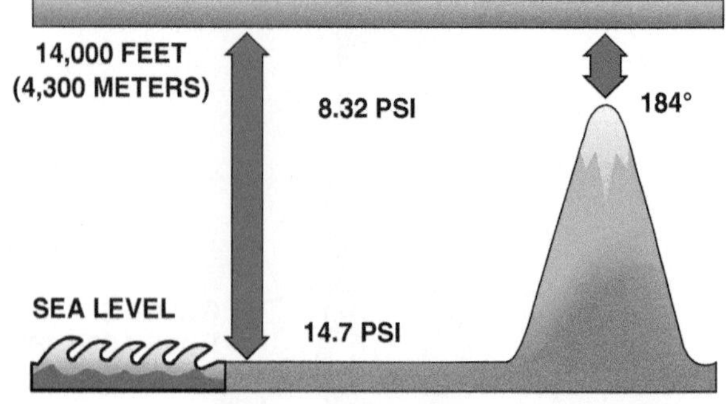

FIGURE 36–23 At sea level, water boils at 212°F (100°C), whereas at the top of Pike's Peak, at 14,000 feet, water boils at 184°F (84°C) because the atmospheric pressure is reduced from 14.7 PSI to just 8.32 PSI.

atmosphere. It is important to evacuate the system to at least 29 inch Hg of vacuum for at least 45 minutes to be assured that all of the moisture has been removed. For best results, the vacuum should be higher than 29.2 inch Hg. The higher the vacuum and the longer that it is allowed to evacuate, the better.

MICRON RATING Vacuum pumps are rated by both **cubic feet per minute (cfm)** and **micron** ratings.

- The cfm rating is the volume it can pump.
- The micron rating is how deep a vacuum the pump can pull.

An automotive A/C system requires about a 1.2-cfm to 1.5-cfm rating, whereas a larger system used in a bus or truck needs about a 5-cfm to 6-cfm vacuum pump. Using a vacuum pump that is too small requires a much longer evacuation time, which causes excess wear on the vacuum pump.

- A perfect vacuum is 29.92 inch Hg, or 0 microns (a micron is equal to one-millionth of a meter).
- A good vacuum pump will pull a system down to under 500 microns (29.90 inch Hg). This high vacuum drops the boiling point of water to around 0°F (–18°C). ● **SEE FIGURE 36–22**.

VACUUM PUMP SERVICE A vacuum pump must be maintained with the proper oil level and periodic oil changes to ensure proper operation. If service is neglected, it will not pump to its design capabilities. Some modern electronic units flash a warning if the oil has not been changed at the proper interval (about 10 hours).

New refrigerant service machines are equipped with oil-less vacuum pumps. These units eliminate the need to service the vacuum pump. To run without lubricating oil, the pump uses two high-temperature plastic pistons operating in anodized aluminum cylinders with a Teflon-type cup seal. The piston and connecting rod are one piece. The crankshaft and connecting rod bearings use a lubricated and sealed ball bearing.

EVACUATION TIME It is accepted practice to evacuate an automotive system for at least 45 minutes. Vacuum readings are affected by altitude and the drop in atmospheric pressure that occurs.

- At 1,000 feet (305 m) above sea level, a complete vacuum is 28.92 inch Hg, about 1 inch Hg less than at sea level.
- This pressure will decrease at a rate of about 1 inch Hg per 1,000-foot elevation increase. ● **SEE FIGURE 36–23**.

EVACUATION PROCEDURE To evacuate a system using a manifold gauge set and portable vacuum pump, perform the following steps:

STEP 1 Connect the manifold service hoses to the system service ports if necessary but these are normally still connected from the recovery process. There should be no or very little pressure in the system.

STEP 2 Open both manifold valves completely (and the vacuum pump valve if there is one) and start the vacuum pump. An air discharge should be noticed from the vacuum pump as well and a drop in gauge pressures. ● **SEE FIGURE 36–24**.

FIGURE 36–24 The timer was set on this RRR machine to 60 minutes so the technician can be doing other service work while the A/C system was being evacuated.

FREQUENTLY ASKED QUESTION

What Is a Micron?

A typical vacuum gauge reads in inches of Mercury (inch Hg) and the recommended vacuum level needed to remove moisture from the system is a vacuum of 29.2 inch Hg. or lower. However, many experts recommend using a vacuum gauge that measures the amount of air remaining in the system rather than just the vacuum.

- A micron is one millionth of a meter and there are about 760,000 microns of air at atmospheric pressure.
- The lower the pressure, the lower the number of microns of air.
- A vacuum reading of 29.72 inch Hg. is about 5,000 microns.
- Many experts recommend that the micron level be 500 or less for best results.

This is particularly important when evacuating a dual-climate control system where two evaporators are used and there are long lengths of refrigerant lines. **SEE CHART 36–2**.

TEMPERATURE F°/C°	INCH HG	MICRONS*	PRESSURE (PSI)
212/100	0.00	759,068	14.7
205/96	5.00	536,000	12.3
194/90	9.81	525,526	10.2
176/80	16.0	355,092	6.9
158/70	20.8	233,680	4.5
140/60	24.1	149,352	2.9
122/50	26.4	92,456	1.8
104/40	27.9	55,118	1.1
86/30	28.8	31,750	0.6
80/27	29.0	25,400	0.5
76/24	29.1	22,860	0.4
72/22	29.2	20,320	0.4
69/21	29.3	17,780	0.3
64/18	29.4	15,240	0.3
59/15	29.5	12,700	0.2
53/12	29.6	10,160	0.2
45/7	29.7	7,620	0.1
32/0	29.8	4,572	0.1
21/–6	29.9	2,540	0.1
6/–14	29.9	1,270	0.1
–24/–31	30.0	254	0.0
–35/–37	30.0	127	0.0

**The remaining pressure in the system in microns.*

CHART 36–2

Boiling temperature of water at converted pressures.

STEP 3 Check the gauge pressures periodically. After about 5 minutes, the pressure should be lower than 20 inch Hg. A leak is usually the cause if the pressure has not dropped this low. Confirm a leak by closing all valves, shutting off the vacuum pump, and watching the pressure. If it steadily increases, there is a leak that must be located and repaired before continuing.

NOTE: Some vehicles with rear A/C use an electric solenoid valve to block flow through the rear unit when it is turned off. This solenoid should be activated so it will open and allow a more complete evacuation of the system.

STEP 4 Continue evacuating until 500 microns is reached or for the desired length of time, close all valves, shut off the vacuum pump, and note the low-side pressure.

STEP 5 After 5 minutes, recheck the low-side pressure. If the vacuum is held steady, the system is good and ready to be recharged. If the low-side pressure increases, a possible leak is indicated.

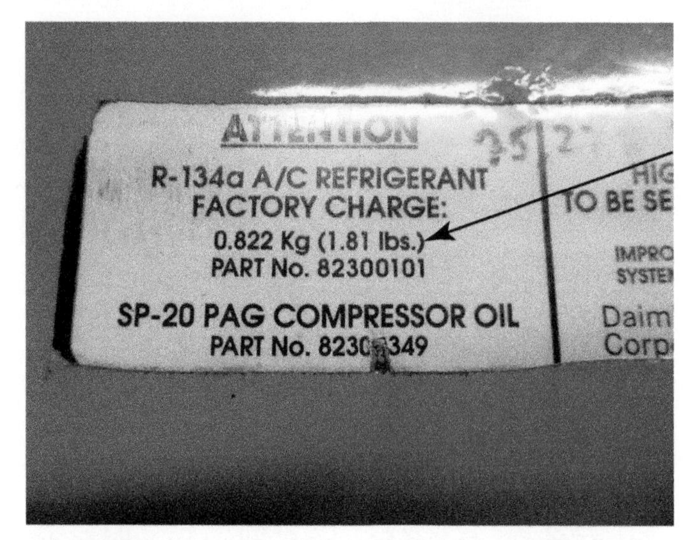

FIGURE 36–25 The underhood decal states that this vehicle requires 1.81 pounds (0.822 Kg) of R-134a refrigerant.

FIGURE 36–26 The calibrated scale on the RRR machine displayed the exact amount of specified refrigerant that was added to the system (1.81 lb.).

When using an A/C service unit, evacuation is often simply a matter of flipping a switch, assuming the service hoses are still connected from the recovery operation. Most recent machines use electric solenoids to control the flow inside the machine, and starting the evacuation process opens the solenoids needed for this process. Some also include a microprocessor that can be programmed to run the vacuum pump for the desired length of time. Some older charging stations are purely mechanical and on these units, the proper valves must be opened as the vacuum pump is started.

TECH TIP

Check the Scale for Accuracy

Most charging scales can be calibrated to ensure accuracy. If a known amount of weight is placed on the scale, the weight indicator should show the correct amount of charge. If a test weight is not available use:

- 9 pennies weigh 1 ounce
- 27 pennies weigh 3 ounces

RECHARGING THE SYSTEM

PROCEDURES After the system has been evacuated, it can be recharged with the correct amount of new or recycled refrigerant. Charge level is system specific, and as the volume of HVAC systems has been downsized, the amount of refrigerant has become critical. The charge level is normally found on a specification decal fastened to some location under the hood and includes both the type and volume of refrigerant used. The decal is attached to the compressor on some older systems. If the decal is missing or illegible, charge specifications are also given in service information from the vehicle manufacturer or aftermarket A/C component suppliers. ● **SEE FIGURE 36–25.**

RECOMMENDED CHARGING PROCEDURE The best method to charge a system is to start with an empty system and charge the specific amount of refrigerant as measured by accurate scales. Most shops use larger refrigerant containers, in the 30-lb size, so the amount to be charged into the system must be measured. The most commonly used A/C service machines use charging cylinders and electronic scales. These can be either individual portable units or parts of a charging station.

Current service units use electronic scales that can be programmed for the desired charge level. The refrigerant container is placed on the scale, and a hose is used to connect its valve to the scale. The operator then programs in the charge volume desired and starts the charge process. When the proper amount of refrigerant has left the container, an electric solenoid in the unit shuts off the refrigerant flow. These units can also be operated manually, with the operator holding down a button or switch until the desired amount of refrigerant has left the container. ● **SEE FIGURE 36–26.**

Moving the refrigerant into the system requires that the charging container pressure be greater than the system pressure. Because the process begins with the system in a vacuum, the first portion goes rather quickly, but the first 1/2 lb or so fills the internal volume and starts generating pressure. As refrigerant boils and leaves the container, it cools the remaining refrigerant and causes a pressure drop. Many charging stations include heaters, which are similar to an electric blanket, to raise the internal pressure of the refrigerant container to help force refrigerant into the system. When heaters are used, the system does not need to be operated. The pressure difference can also be increased by starting the A/C system so the low-side pressure drops and then charging only into the low side.

With the system running, caution should be exercised if charging liquid into it to avoid **slugging** the compressor, which is allowing liquid refrigerant to enter a running compressor which can cause severe damage.

- If the container is upright, gas will exit.
- If the container is upside down, liquid will exit.

To charge a system using a charging station, perform the following steps:

STEP 1 Enter the specified amount of refrigerant into the charging scale or into the charging cylinder unit. This charge process begins with the system still under a vacuum, with the manifold valves closed.

STEP 2 Follow the machine instructions for the unit being used.

STEP 3 When the charge volume has entered the system, close the necessary valves.

STEP 4 Start the A/C system, let it run until the pressures stabilize, and note system pressures.

RETROFITTING R-134a INTO AN R-12 SYSTEM

PURPOSE **Retrofitting** is the changing of a system from one refrigerant to another and is normally a repair-driven operation and not done until absolutely necessary. All experts agree that if a system was designed for R-12, then R-12 should be used in it when it requires service, even though some systems cool better after changing to R-134a.

SECTION 609, EPA RETROFIT REQUIREMENTS For a proper retrofit, this section of the Clean Air Act requires:

- The technician must be Section 609 certified.
- Recover the original refrigerant using EPA-approved equipment.
- Replace service fittings to match new refrigerant.
- Install new label to cover the old one.
- Ensure that the system has barrier hoses.
- Ensure that the system has a high-pressure shutoff switch.

REFRIGERANT CHOICE FOR RETROFIT Any EPA SNAP-approved refrigerant can be used to replace R-12. All of them require a similar retrofit procedure. Some of the points to consider when choosing the refrigerant include the following:

- It must be EPA approved and have unique service fittings and label.
- R-134a is the refrigerant used in every new vehicle and is EPA SNAP-approved refrigerant.
- Unique service, recovery, and charging equipment are required for each refrigerant. Most shops have R-12 and R-134a equipment.
- Most major compressor manufacturers and builders will not warranty a compressor that failed if a blend refrigerant was used in it.
- Blends are outlawed in some areas, including at least one Canadian province.

All vehicle manufacturers, the Mobile Air Conditioning Society (MACS), and most service shops prefer using R-134a for retrofits. It is relatively inexpensive, used in all new vehicles (so the service equipment is fairly common), readily available, and, because it is a single-compound refrigerant, can be recycled in service shops.

RECOMMENDED GUIDELINES When retrofitting, at least 99% of the R-12 must be removed from the system and as much of the mineral oil as practical. A recovery unit removes the R-12 vapor and some of the oil and measures the amount of refrigerant and oil removed. Complete R-12 removal can be difficult because of the amount absorbed by any remaining oil.

FIGURE 36–27 When a system is retrofitted from R-12 (CFC-12) to R-134a (HFC-134a), the proper service fittings have to be used to help assure that cross-contamination does not occur.

- Oil removal also helps remove R-12 from the system. Oil removal is more difficult because it must be drained out, flushed out, or pulled out with the R-12. Complete draining of some components requires removal of the component, which greatly increases the labor costs. Any remaining mineral oil will probably gel and settle at the bottom of the accumulator or evaporator.
- R-134a is lighter than R-12, so a system should be charged to about 80% to 90% the amount of the R-12 capacity. Some aftermarket sources have made R-134a retrofit charge capacities available. To figure the new charge level, simply multiply the R-12 charge level by 0.8. For example, if the R-12 charge is 2 pounds (32 oz), $32 \times 0.8 = 27$ oz. General Motors guidelines for retrofitting its vehicles recommend multiplying the R-12 charge level by 0.9 and subtracting 0.25 pounds (4 oz). Using this procedure, $32 \times 0.9 = 29 - 4 = 25$, slightly less than the 27 oz (80% level).

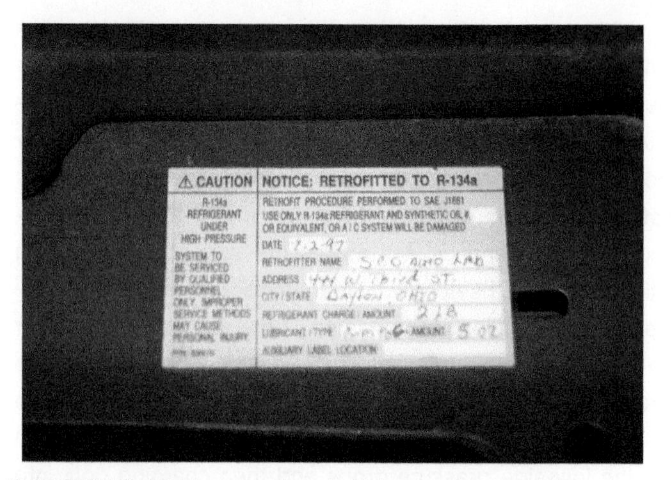

FIGURE 36–28 An underhood sticker is also installed indicating that the system was retrofitted to HFC-134a and when it was done and by whom.

RETROFIT KITS Both vehicle and aftermarket manufacturers produce kits that include the parts needed to convert particular systems. Vehicle manufacturer kits are for Type I retrofit, and most aftermarket sources provide Type II kits. Depending on the particular make and model, a kit can include the following:

- A sticker to identify that it is an R-134a system, along with the charge level. ● **SEE FIGURE 36–27**
- R-134a-type service fittings to be permanently installed over the existing service fittings
- Replacement hoses with the R-134a service fittings
- Replacement O-rings
- Ester or PAG oil
- Replacement receiver–drier or accumulator with XH-7 or XH-9 desiccant
- Replacement system switches calibrated for R-134a pressures
- Replacement TXV or OT calibrated for R-134a pressures
- High-pressure cutoff switch

The R-12 service fittings must either be converted to R-134a service fittings or permanently capped to make them unusable. Normally, they are converted using a **conversion fitting**. A conversion fitting is threaded over the R-12 fitting, using the same Schrader valve, and a thread-lock adhesive locks it in place. It is a good practice to replace the old Schrader valves with ones that are compatible with R-134a. ● **SEE FIGURE 36–28.**

TECH TIP

Orifice Tube Size

When retrofitting an OT system, it is a good practice to replace the orifice tube with one that is about 0.010 inch (0.026 mm) smaller than the one used in the R-12 system. R-134a is lighter than R-12, and the R-12 orifice tube will allow too much R-134a to flow. This can drop high-side pressure and increase low-side pressure.

RETROFIT PROCEDURE To retrofit a system, perform the following steps:

STEP 1 Visually inspect the system to ensure good condition, install a gauge set, and operate the system to bring it up to operating temperatures. Check for proper operation and note any needed repairs. Record the high-side pressure for later comparison.

STEP 2 Recover the R-12 from the system and remove as much oil-dissolved R-12 as possible.

STEP 3 Make any repairs to the system to cure problems that were found in step 1.

STEP 4 If the compressor failed, remove the failed compressor, flush the system and/or install a high-side filter, and install the replacement compressor along with a new accumulator or receiver–drier.

STEP 5 Check the system to determine whether a high-pressure relief valve is used. If it has one, a high-pressure cutoff switch must be installed to stop the compressor before pressure relief valve release pressures occur. The switch is installed so it senses high-side pressure and is wired into the clutch circuit or relay so it can interrupt clutch operation.

STEP 6 If directed, replace the receiver–drier or accumulator.

STEP 7 Replace any line-fitting O-rings on connections that were disturbed, or as directed.

STEP 8 Replace any switches and valves as directed.

STEP 9 Add the proper type and amount of oil—ester or PAG—into the compressor oil fill port or suction port. If the accumulator or receiver–driver is replaced, pour part of the oil into the inlet port.

STEP 10 Install the R-134a service fittings. Any old Schrader valves that remain in service should be replaced with new R-134a valves.

STEP 11 Fill out and install the identifying decal to properly identify the system. The old label must be rendered unreadable.

STEP 12 Connect a vacuum pump to the system and pull a minimum vacuum of 29 inch Hg (500 microns) for at least 30 minutes to evacuate the system.

STEP 13 Recharge the system using R-134a. Charge the system with 80% to 90% of the specified amount of R-12.

STEP 14 Operate the system and check for proper operation, paying careful attention to the high-side pressure.

STEP 15 Test for leaks.

SEALANTS AND STOP LEAKS

PURPOSE Sealants and stop leaks are included in many do-it-yourself automotive air-conditioning charging products and are designed to plug or stop small refrigerant leaks. Most HVAC technicians dislike stop leaks as they are considered an inadequate or temporary repair method, and the best repair is to actually fix the leak.

SEALANT CONCERNS Sealants and stop leaks will not always be successful in stopping leaks and may damage the A/C system. Stop leaks also have a reputation of plugging up small orifices that might reduce the proper flow of refrigerant or oil. There is also a question of whether the stop-leak material will contaminate the rather delicate chemical balance within an A/C system. In spite of these misgivings, sealants have become popular with do-it-yourselfers.

TYPES There are two types of sealants:

1. *Type I* is a rubber "conditioner," often called a "seal swell." It contains a hydrocarbon or alcohol that softens rubber that has hardened and causes it to swell. Type I sealants are generally low cost and will cause little harm and are often available combined in a small can of refrigerant. Other sealants are single- or two-part chemicals that are moisture-activated and remain fluid until in contact with moisture.

2. *Type II* is a stop-leak chemical that hardens when it contacts moisture as it escapes, which results in sealing the leak. Type II sealants have hardened and caused plugging inside a system and service machines. The sealant is purchased as a kit that includes the sealant with a valve and short adapter hose. The absence of moisture inside the A/C system along with the heat and pressure prevent the sealant from hardening inside the system. ● **SEE FIGURE 36–29**.

CAUTION: Refrigerant service machines have experienced problems of lines plugging up and solenoids sticking after recovering refrigerant from systems containing sealants. One compressor manufacturer will not warranty compressors if sealant residue is found in them.

FIGURE 36–29 Use A/C stop leak with caution.

SUMMARY

1. The first refrigerant service operation is to identify the refrigerant in the system and check to make sure that it does not contain a sealant.
2. Evacuation is the process of removing the refrigerant from the system and storing it for reuse.
3. Recharging is the process of charging the system with the specified amount of refrigerant.
4. Sealants should be detected before a system is evacuated to protect the recovery machine from damage and to prevent the sealant from being added to the refrigerant in the RRR machine.
5. Retrofitting is the process of changing the refrigerant from one refrigerant to another, usually from R-12 to R-134a.
6. The refrigerant is recovered from a system so that service operations can be performed.
7. Recycling removes foreign particles, water, and air from refrigerant.

REVIEW QUESTIONS

1. What does the EPA Section 609 act affect?
2. What is considered to be a noncondensable gas (NCG)?
3. What are the steps to recover refrigerant?
4. What are the criteria for recycled refrigerant?
5. Why do A/C experts recommend using a micron meter to measure the vacuum in the system?
6. What are the steps required to be performed when retrofitting an R-12 system to R-134a?

CHAPTER QUIZ

1. Section 609 of the Act gives EPA the power to enforce which of the following requirements?
   a. Preventing the release or venting of CFC-12 or HFC-134a
   b. Technicians are required to use approved equipment to recover and recycle CFC-12 and HFC 134a
   c. Technicians who repair or service CFC-12 and HFC-134a systems must be trained and certified by an EPA-approved organization
   d. All of the above
2. What is the SAE specification number of an RRR machine that meets the 2006 and newer standards?
   a. J-2788
   b. J-2210
   c. J-1930
   d. J-0560
3. The storage container for recovered refrigerant must be approved by ____________.
   a. EPA
   b. OSHA
   c. DOT
   d. MSDS
4. Why should service technicians check a system for the presence of sealant before recovering the refrigerant?
   a. It can harm the recovery machine.
   b. It can contaminate the refrigerant that is stored in the recovery container.
   c. It can damage a refrigerant identifier.
   d. All of the above.
5. Present standards and good work habits require that recovered refrigerant be ____________.
   a. Sent out to be discarded at an EPA-certified facility
   b. Recycled
   c. Sent back to the manufacturer
   d. Kept for 90 days
6. A container of R-134a is tested at a pressure of 115 PSI at an ambient temperature of 75°F. This means ____________.
   a. This is normal pressure for R-134a at that temperature
   b. That there is R-12 in the container
   c. Air is in the refrigerant
   d. That there is sealant in the container
7. What component cannot be flushed?
   a. The compressor
   b. The condenser
   c. The receiver–drier
   d. None of the above
8. What is called "black death"?
   a. Dried refrigerant
   b. Teflon from the compressor piston rings
   c. Sealant
   d. Desiccant
9. When retrofitting a system from R-12 to R-134a, what is required to be done?
   a. Install a label
   b. Convert the fittings
   c. Replace the receiver–drier
   d. All of the above
10. Most experts recommend that the system be drawn down to how many microns?
    a. 500 or lower
    b. 10,000 or higher
    c. 243
    d. 127

chapter 37

A/C SYSTEM DIAGNOSIS AND REPAIR

LEARNING OBJECTIVES

After studying this chapter, the reader should be able to:

1. Describe the eight-step diagnostic procedure for an A/C system.
2. Explain how to perform a visual inspection of an A/C system.
3. Discuss how to perform an A/C performance test.
4. Describe how to determine the root cause of the problem in an A/C system.

This chapter will help you prepare for the ASE Heating and Air Conditioning (A7) certification test content area "A" (A/C System Service, Diagnosis, and Repair).

KEY TERMS

After-blow module 603
Body control module (BCM) 599
Diagnostic trouble codes (DTCs) 598
Electronic evaporator dryer (EED) 603
Groundout 604
Technical service bulletin (TSB) 599
Visual inspection 596

GM STC OBJECTIVES

GM Service Technical College topics covered in this chapter are as follows:

1. Perform a system performance/function test on the vehicle's A/C system.
2. Leak-check an A/C system (electronic detector).
3. Leak-check an A/C system (trace dye/UV light).

THE DIAGNOSTIC PROCEDURE

STEPS INVOLVED When diagnosing a heating and air-conditioning system problem, most vehicle manufacturers recommend that the following steps be performed.

STEP 1 **Verify the Customer Complaint (concern).** Some times the customer does not understand how the system is supposed to work or does not explain the fault clearly. Verifying the fault also means that the technician can verify that the problem has been corrected after the service procedure has been performed.

STEP 2 **Perform a Thorough Visual Inspection.** Heating and air-conditioning problems are often found by looking carefully at all of the components, checking for obvious faults or damage due to an accident or road debris.

STEP 3 **Check for Diagnostic Trouble Codes (DTCs).** Many heating and air-conditioning systems use sensors and actuators, which are computer-controlled and will set diagnostic trouble codes in the event of component failure.

STEP 4 **Check for Related Technical Service Bulletins (TSBs).** If there has been a technical service bulletin released to solve a known problem, it saves a lot of time to know what to do rather than spend a lot of time trying to find and correct a customer concern.

STEP 5 **Perform an A/C Performance Test.** An A/C performance test is used to determine how well the system is able to remove heat from inside the vehicle and move it to the outside of the vehicle.

STEP 6 **Determine the Root Cause.** Perform pressure and temperature measurements to help determine the root cause of the customer concern.

STEP 7 **Repair the System.** Replace or repair the components that are defective or are no longer working as designed and recharge the system according to factory specifications.

STEP 8 **Verify the Repair.** Drive the vehicle under similar conditions that caused the customer to complain and verify that the concern has been corrected.

TECH TIP

"Evaporator Dandruff"

Many evaporators have a chemical coating to prevent bacterial growth that can cause bad odors. The coating can flake off and can sometimes be seen as dust specks on the top of the dash where it flows out of the defroster ducts and lands on the dash. This is commonly called "evaporator dandruff." The only fix for this condition is to replace the evaporator but because this usually involves the removal of the entire dash assembly, most vehicle owners simply ignore the dust on the dash.

STEP 1 VERIFY THE CUSTOMER CONCERN

UNDERSTAND THE EXACT FAULT The customer concern needs to be addressed and if possible, when the problem occurs if the problem is intermittent, identify when it occurs. Sometimes the A/C system is functioning normally for the conditions which could include any of the following:

- Higher than normal outside air (ambient) temperature
- High humidity level
- A new vehicle that is larger or has more glass area than the customer's previous vehicle
- A new vehicle that is black or dark in color compared their previous vehicle that was a lighter color such as white or silver. A light-colored vehicle reflects light and the heat from the sun instead of being absorbed as with a dark-colored vehicle.

Other needed information include the following:

1. What recent service work was performed on the vehicle? (Previous repairs could have somehow affected the HVAC system.)
2. Has the vehicle been in storage? (Animals like to build nests in HVAC ducts and this would affect A/C operation and airflow.)

FREQUENTLY ASKED QUESTION

What Does "Short Cycling" Mean?

If the system is low on refrigerant, it will short cycle, or rapidly cycle on and off. This is the result of the compressor pulling the refrigerant out of the low side quickly to open either the cycling switch or the low-pressure switch. With the compressor off, the flow into the evaporator raises the pressure enough to reclose the switch and restart the compressor. With a normal charge, the low-side pressure should be 15 PSI to 35 PSI and the clutch should be on for 45 seconds to 90 seconds and be off for only about 15 seconds to 30 seconds.

3. Has the vehicle been involved in a collision? (Hidden damage could cause faults in the HVAC system yet may not be visible from the outside.)
4. Has the A/C system been serviced before? (This may indicate that the system has sealant or other impurities in the system.)
5. The service adviser or shop owner should ask the customer to state their concern in their own words such as:
   - "Does not seem to cool as fast as last year"
   - " Airflow through the driver's side vent appears to be blocked"
   - "The fan noise is high but I can't feel much cool air coming from the vents"

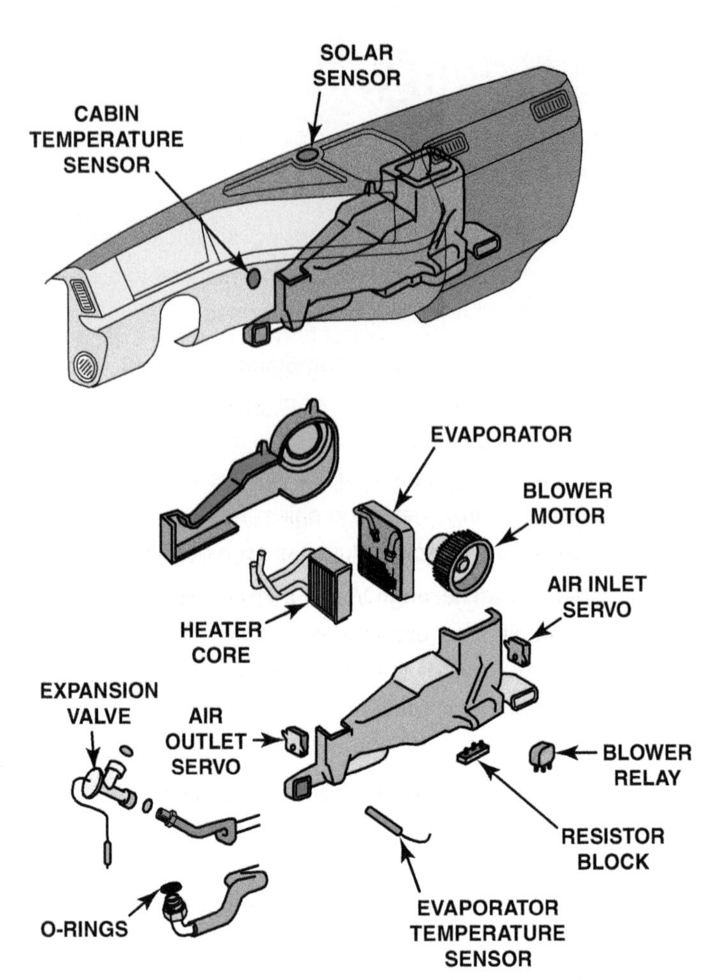

FIGURE 37–1 A visual inspection checks all of the visible, underhood components for possible wear or damage. The underdash components are checked for noise and proper airflow.

STEP 2 VISUAL INSPECTIONS

UNDERHOOD CHECKS Most vehicle manufacturers specify that the first step after the customer concern has been verified is that a **visual inspection** be performed. A visual inspection of the underhood items includes the following:

1. Check the condition of the A/C compressor drive belt.
2. Check the tension of the A/C compressor drive belt and the automatic tensioner.
3. Inspect the refrigerant hoses and lines for signs of oily residue and damage. Oil residue with caked-on dirt indicates a probable leak. Check that each of the A/C service ports is capped.
4. While checking the hoses and lines, determine if the system uses a thermal expansion valve system (TXV) or an orifice tube (OT) system and if a variable displacement compressor is used. The compressor shape and model number are used for identification.
5. Check that the compressor mounting bolts are tight.
6. Check to make sure that the air gap of the A/C compressor clutch is correct.
7. Check the electrical wires to the clutch, blower motor, and any A/C switches for good, tight connections, and possible damage. ● **SEE FIGURE 37–1**.
8. Check the condition of any vacuum hoses between the intake manifold and bulkhead, if equipped.
9. Check the faces of the condenser and radiator core for restriction to airflow caused by debris and clean as needed.

TECH TIP

Broken Condenser Line?—Check the Engine Mounts!

Most air-conditioning systems use aluminum and flexible rubber lines between the compressor and the condenser. Because the compressor is mounted on and driven by the engine and the condenser is mounted to the body, these lines can break if the engine mounts are defective. The rubber hoses attached between the aluminum fittings of the compressor and condenser are designed to absorb normal engine movement. Worn engine mounts would allow the engine to move too much. Aluminum lines cannot be flexed without crushing and breaking. Therefore, the wise technician will carefully inspect and replace any and all worn engine mounts if a broken aluminum condenser line is discovered to prevent a premature failure of a replacement condenser.

IN-VEHICLE CHECKS With the engine off, the in-vehicle checks include:

1. Operate the blower switch through its various speeds while listening to the fan and motor for unusual noises. Note that some systems do not have blower operation unless the heat or A/C controls are on or unless the engine is running.
2. Move the temperature lever of mechanically operated doors to both ends of its travel. It should move smoothly and stop before making contact at the ends. An early or late stop indicates that adjustment is needed.

ENGINE RUNNING CHECKS With the engine running, the underhood checks include the following:

1. Make sure the compressor clutch is engaged and the compressor is running. Listen for any signs of improper compressor operation.
2. Turn off the A/C clutch to make sure it releases smoothly. With the clutch released, listen for proper clutch bearing operation.
3. Feel the temperature of the A/C lines and hoses. Be cautious on the high side because the lines should be warm to hot, with the temperature increasing. All the lines on the low side should be cool to cold, with the temperature getting colder. ● **SEE FIGURE 37–2**.

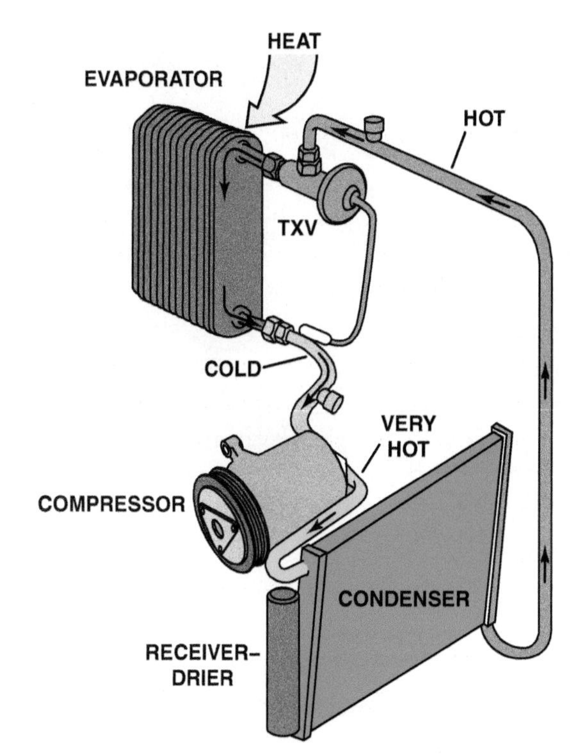

FIGURE 37–2 When a system is operating properly, the suction line to the compressor should be cool, and the discharge line should be hot to very hot. The liquid lines should also be hot.

TECH TIP

Listen for the Thunk

On some vehicles, a "thunk" noise occurs as the temperature door contacts the stop. In an area with cold winters, it may "thunk" at the full heat stop. If in an area with hot summers, it may "thunk" at the full cool stop.

4. Feel the temperature of the heater hoses. If the engine is at operating temperature, both hoses should feel hot. Expect a temperature difference on heater systems with a coolant flow control valve.
5. Check the engine cooling fan operation (if running). The fan should be turning smoothly, with good airflow.
6. Check the evaporator drain. After a few minutes of operation on a humid day there should be a small puddle of water under the evaporator area and drops of water coming from the drain. With some vehicles, the drain is routed through a frame member. Check the service information for the location of the drain.

NOTE: In areas with very low humidity, water may not drain out.

FIGURE 37–3 Check the condensate drain to be sure it is open if there is water on the carpet on the passenger side.

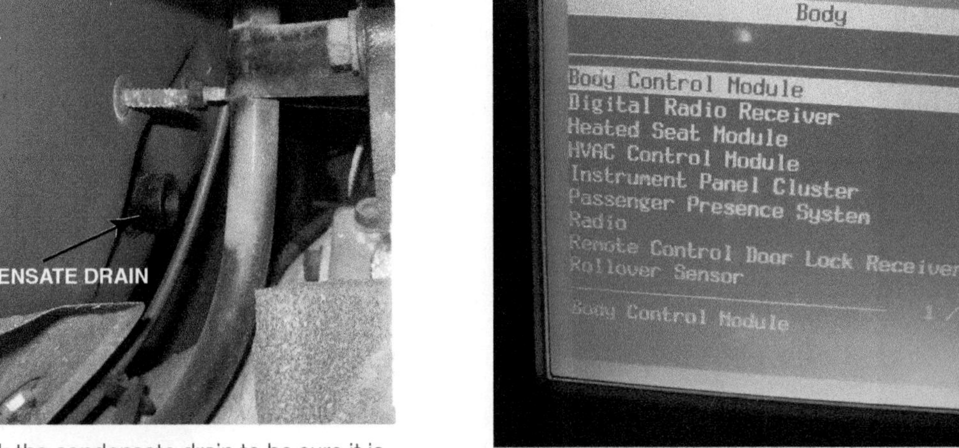

FIGURE 37–4 Under "Body" on most scan tools, select "HVAC Control Module" for access to the HVAC-related data and diagnostic trouble codes.

TECH TIP

Water on the Carpet? Check the Evaporator Water Drain

If the evaporator water drip tube becomes clogged with mud, leaves, or debris, water will build up inside the evaporator housing and spill out onto the carpet on the passenger side. Customers often think that the windshield or door seals are leaking. Most evaporator water drains are not visible unless the vehicle is hoisted. ● **SEE FIGURE 37–3**.

7. With the engine still running, check inside the vehicle for overall operation. By now the A/C system should be delivering cool to cold air from the instrument panel registers. Refer to performance check for actual expected temperatures. Moving the temperature control should cause a temperature change. Changing the function control to heat should move the air discharge to floor level.

TECH TIP

Look for DTCs in "Body" and "Chassis"

Whenever diagnosing a customer concern with the HVAC system, check for diagnostic trouble codes (DTCs) under chassis and body systems and do not just look under engines. Therefore, a global or generic scan tool that can read only "P" codes is not suitable for diagnosing an HVAC system. Engine or emission control-type codes are "P" codes, whereas module communications are "U" codes. These are most often found when looking for DTCs under chassis or body systems. Chassis-related codes are labeled "C" and body system-related codes are labeled "B" codes and these can cause an HVAC issue if they affect a sensor that is used by the system. ● **SEE FIGURE 37–4**.

STEP 3 CHECK DIAGNOSTIC TROUBLE CODES

Check for any stored **diagnostic trouble codes (DTCs)**. While most HVAC-related trouble codes will be "B," "C," or "U" codes, there are some that are Powertrain-related P-codes including:

- P0645 A/C Clutch Relay Control Circuit
- P0646 A/C Clutch Relay Control Circuit Low
- P0647 A/C Clutch Relay Control Circuit High
- P0691 Fan 1 Control Circuit Low
- P0692 Fan 1 Control Circuit High
- P0693 Fan 2 Control Circuit Low
- P0694 Fan 2 Control Circuit High

These P-codes can be read using global (generic) scan tools that are available at low cost. However, to get access to the codes set for most of the HVAC-related faults requires the use of a factory scan tool or an enhanced factory-level aftermarket scan tool.

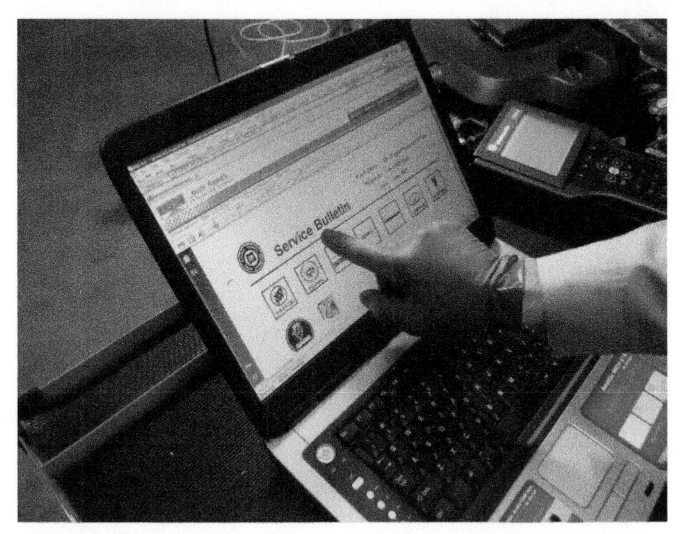

FIGURE 37–5 After checking for stored diagnostic trouble codes (DTCs), the wise technician checks service information for any technical service bulletins that may relate to the vehicle being serviced.

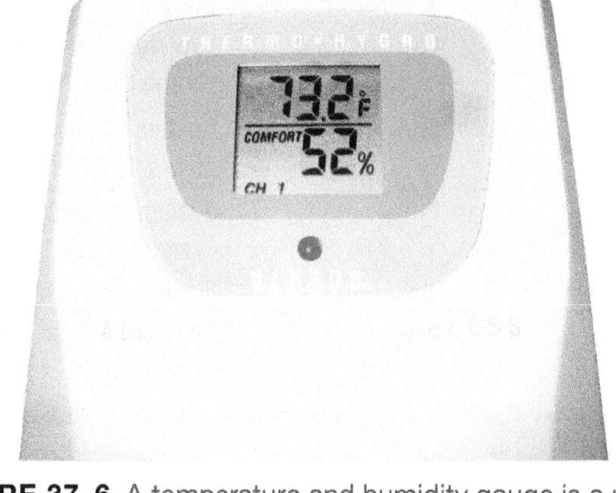

FIGURE 37–6 A temperature and humidity gauge is a useful tool for air-conditioning work. The higher the relative humidity, the more difficult it is for the air-conditioning system to lower the temperature inside the vehicle.

Many HVAC systems are controlled by the **body control module (BCM)**. The air conditioning-related diagnostic trouble codes are usually found under "Body" or "Chassis" and if set, will NOT light the "check engine" warning light. The check engine light is called the *malfunction indicator light (MIL)* and is commanded on if there is a fault that can affect emissions. Because HVAC-related faults often do not affect emissions, a DTC will be stored and may light another warning lamp or a message on the dash, such as the "A/C on" lamp will flash to indicate that a fault has been detected with the HVAC system.

STEP 4 CHECK FOR TECHNICAL SERVICE BULLETINS

Technical service bulletins (TSBs) are issued by vehicle and aftermarket manufacturers to inform technicians of a situation or technical problem and give the corrective steps and a list of parts needed to solve the problem. Any diagnostic trouble codes should be retrieved before looking at the technical service bulletins because many bulletins include what DTCs may or may not be present. TSBs are often released by new vehicle manufacturers to dealership service departments. ● **SEE FIGURE 37–5.**

While some of these TSBs concern minor problems covering few vehicles, many contain very helpful solutions to hard-to-find problems that cover many vehicles. TSBs can also be purchased through aftermarket companies that are licensed and available on a Web site. Visit the National Automotive Service Task Force (NASTF) Web site (www.NASTF.org) for a list of the Web addresses for all vehicle manufacturers' sites where the full text of TSBs can be purchased directly. Factory TSBs can often save the technician many hours of troubleshooting.

STEP 5 PERFORM A/C PERFORMANCE TEST

PROCEDURE An A/C performance test is used to determine if the system is capable of performing as designed. Most A/C performance tests include the following steps:

STEP 1 Install pressure gauges to the service ports.

STEP 2 Start the engine, set the parking brake, and raise the idle to 2000 RPM.

STEP 3 Measure ambient temperature 3 inches (80 mm) in front of the condenser and measure the humidity. ● **SEE FIGURE 37–6.**

STEP 4 Place a thermometer in the air conditioner center vent.

STEP 5 Set the air conditioner for maximum cooling.

STEP 6 Open the doors and set the blower speed to high, which applies the maximum load on the system.

LOW SIDE	HIGH SIDE	CONDITION (POSSIBLE CAUSE)
25 PSI–35 PSI	170 PSI–200 PSI	Normal
Low	Low	Low refrigerant charge level
Low	High	Restriction in high-side line
High	High	System is overcharged. Expansion valve stuck open
High	Low	Restriction in the low-side line

CHART 37–1

Pressure gauge readings and possible causes.

AMBIENT AIR TEMPERATURE	HUMIDITY	LOW-SIDE PRESSURE (PSI)	HIGH-SIDE PRESSURE (PSI)
70°F/21°C	Low	25–30	140–190
	High	28–35	165–220
80°F/27°C	Low	26–33	150–200
	High	30–36	190–260
90°F/32°C	Low	31–37	170–220
	High	37–45	210–290
100°F/38°C	Low	35–44	195–245
	High	38–48	230–320
110°F/43°C	Low	40–50	235–285
	High	42-52	260–350

CHART 37–2

The average R-134a pressure–temperature readings during a performance test. The high-side pressure of R-12 systems will be lower at higher temperatures.

STEP 7 Allow the system to operate for another five minutes before recording the gauge readings.

- Check the sight glass (if equipped). Clear? Bubbles? Foam?
- Check the A/C lines for frosting:
  - Low-Side Lines: Frosted (indicates low refrigerant level and should be corrected before continuing the test).
  - High-Side Lines: Frosted (indicates a restriction where the frost begins and should be corrected before continuing the test).

SYSTEM TEST RESULTS Wait several minutes to allow the system to reach maximum output and observe the thermometer.

- If 35°F to 45°F (2°C to 7°C) or 30°F (15°C) cooler than the outside air temperature, the system is functioning normally.
- If over 45°F (7°C) or over 30°F (15°C) warmer than the outside air temperature, continue with pressure gauge testing.

If pressures are okay and the sight glass is clear, but vent temperature is high, check for a blend door or heater control valve problem, or look for a possible system oil overcharge. If pressures vary from specifications, perform further tests to locate the problem. ● **SEE CHART 37–1**.

STEP 6 DETERMINE THE ROOT CAUSE

TEMPERATURE AND PRESSURE MEASUREMENTS

Temperature and pressure are directly related in A/C systems. As the ambient temperature increases, the high-side pressure must also increase to have a heat transfer at the condenser. The temperature of the vapor must be higher than the ambient temperature to allow enough heat to be removed for condensation. Also, higher ambient temperatures, and high humidity, usually mean a higher heat load on the evaporator. This means a larger quantity of heat has to be removed at the condenser.

- The **high-side pressure** is directly related to the amount of heat that needs to be removed and the heat transfer at the condenser.
- **Low-side pressure** indicates the boiling point of the temperature of the evaporator. If the pressure is too high, the boiling point of the refrigerant and temperature of the evaporator are too high. Low-side pressure that is too low indicates the evaporator is too cold and may ice over (freeze), or that there is not enough boiling refrigerant in the evaporator to remove an adequate amount of heat. ● **SEE CHART 37–2**.

Insufficient heat transfer at the condenser is usually the cause of high-side pressure that is too high. The number one cause of poor heat transfer is lack of airflow across the

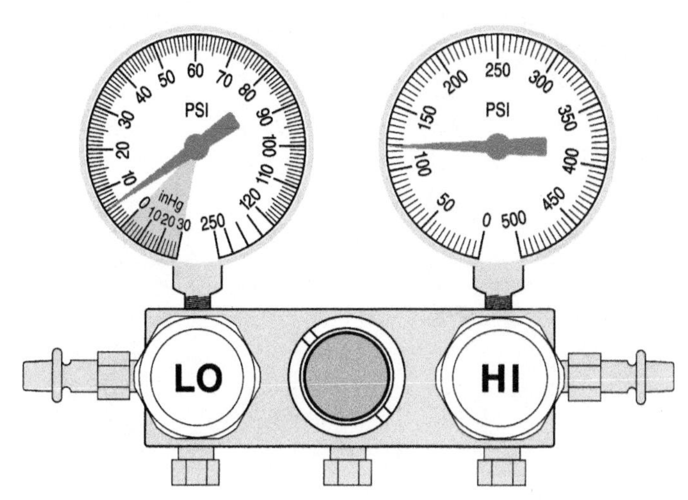

FIGURE 37–7 When both low- and high-side pressures are low, the system is undercharged with refrigerant.

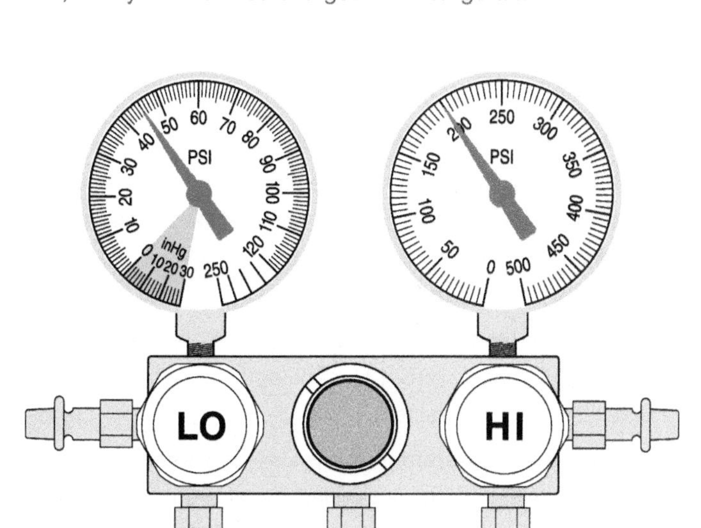

FIGURE 37–8 Both low- and high-side pressures higher than normal indicate that the system is overcharged with refrigerant.

condenser. The vehicle is dependent upon portable fans to move enough air when testing in the shop. It may be necessary to drive the vehicle at 30 MPH (48 km/h) to get the ram air necessary to determine if lack of airflow is the reason for the poor heat transfer.

Another cause of excessive high-side pressure is contamination with a different refrigerant. Mixing R-12 and R-134a raises the condensing pressure of the mixture. If the system is contaminated with R22, the pressures can become extremely high. At 150°F, the pressure of R-12 is 235 PSI (1,620 kPa), R-134a is 263 PSI (1,813 kPa), and R-22 is 381 PSI (2,627 kPa). This is an important reason to use a refrigerant identifier. ● **SEE FIGURES 37–7** through **37–9**.

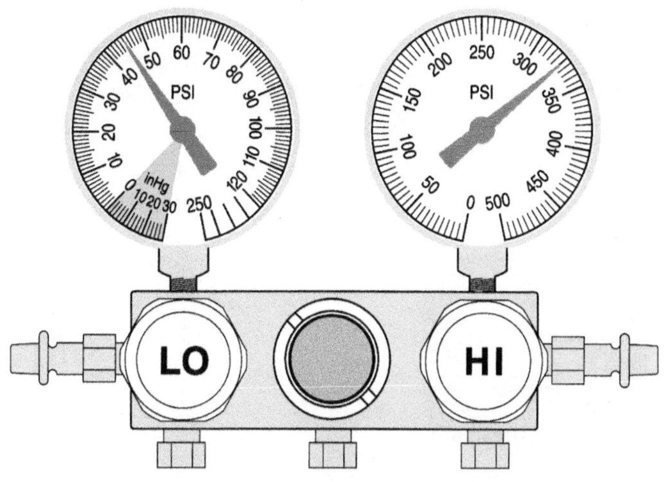

FIGURE 37–9 Lack of proper airflow across the condenser is usually the cause of high low-side and higher high-side pressures.

TECH TIP

The Paper Test

To determine if there is adequate airflow through a condenser, many technicians place a sheet of paper or a dollar bill in front of the condenser when the cooling fans are operating. With the engine running at idle speed, the bill should stick to the condenser.

TEMPERATURE DIFFERENCES ACROSS COMPONENTS

Test for heat transfer at the evaporator and the condenser.

- **Temperature change across the evaporator**—The temperature difference between the inlet and outlet of the evaporator should be same or within 5° (−5° to +5°). If the temperature at the evaporator outlet is more than 6° different than the inlet temperature, then the system is overcharged.
- **Temperature change across the condenser**—The temperature difference between the inlet and outlet of the condenser should be 20°F to 50°F (10°C to 30°C).
  - If the temperature difference is higher than 50°, then the system is undercharged or has a restriction.
  - If the temperature difference is less than 19°, then the system is overcharged.

FIGURE 37–10 A clogged orifice tube.

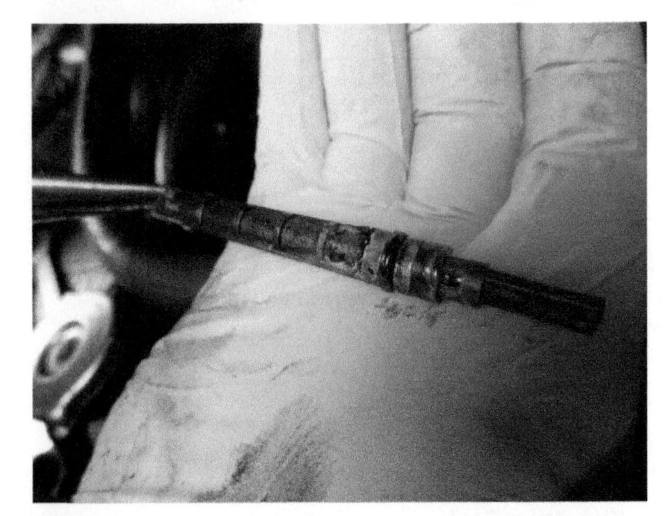

FIGURE 37–11 A partially restricted orifice tube should be replaced if discovered during service.

TECH TIP

The Clogged Orifice Tube Test

A clogged orifice tube is a common air-conditioning system failure. When the orifice tube becomes clogged, it blocks the flow of refrigerant through the evaporator, which causes a reduced cooling of the passenger compartment. To check for a possible restriction in the system, follow these easy steps:

STEP 1 Connect the A/C pressure gauge to both low and high-side pressure fittings.

STEP 2 Operate the A/C system for 5 minutes to 10 minutes.

STEP 3 Shut off the A/C system and watch the pressure gauges. If the pressures do not equalize quickly, then there is a restriction in the system. Normal time needed to equalize is often 15 seconds to 30 seconds, depending on the amount of refrigerant. ● **SEE FIGURES 37–10 AND 37–11.**

NOTE: To locate a restriction anywhere in the system, feel along the system lines. The restriction exists at the point of greatest temperature difference. "Frosting" is a good indication of a restriction.

FIGURE 37–12 A black light being used to look for refrigerant leaks after a fluorescent dye was injected into or added to the system. The dye will glow a bright yellow-green color at the point of the leak.

LEAK DETECTION If the A/C system is low on a charge of refrigerant, the sources of the leak should be found and corrected. Look for oily areas that are formed when refrigerant leaks and some refrigerant oil is lost. It is this oil that indicates a refrigerant leak. The two preferred methods of leak detection include:

- **Electronic leak detector.** Many of these units can detect both CFC-12 and HFC-134a. The detector will sound a tone if a leak is detected.
- **Dye in the refrigerant.** A dye is added to some refrigerant to help the technician visually spot a leak in the refrigerant system. This method works well except for leaks in the evaporator, which are usually not visible. ● **SEE FIGURE 37–12.**

ADDRESSING SYSTEM ODORS Some systems develop a musty, moldy smell, which is not really a fault of the system. Some sources classify these odors into two types:

- "Dirty socks/gym locker" odor, which has an organic cause
- "Refrigerator, cement, or dusty room" odor, which is caused by chemicals

The organic odor problem is most common in areas with high relative humidity, and it is caused by mildew-type fungus growth on the evaporator and in the evaporator plenum. Modern evaporators have more fins that are closer together, and they tend to trap more moisture and bacterial growth. The cool surface of the evaporator collects moisture as it dehumidifies the air, and most of this moisture runs out the bottom of the case. After a vehicle is shut off, the moist surface of the evaporator warms up, and this warm, wet area becomes an ideal environment for fungus and bacteria growth. A coating is

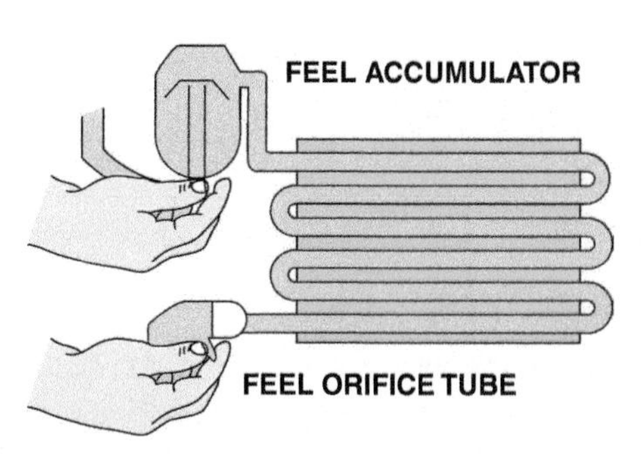

FIGURE 37–13 If the system is fully charged, the outlet temperature of the line leaving the evaporator should be about the same as the temperature of the line entering the evaporator after the expansion valve or orifice tube. In a fully to slightly undercharged orifice tube system, the bottom of the accumulator is as cold as the line just downstream from the orifice tube. A warmer accumulator indicates an undercharge.

TECH TIP

The Touch, Feel Test

A quick-and-easy test to check the state of charge of an orifice tube system is to use one hand and touch the evaporator side of the orifice tube. Touch your other hand to the inlet to the accumulator. The following conditions can be determined by noticing the temperature of these two locations. ● **SEE FIGURE 37–13.**

- **Normal operation**—both temperatures about the same
- **Undercharged condition**—accumulator temperature higher (warmer) than the orifice tube temperature

Just remember: High pressure means that the temperature of the component or line will also be high (hot). Low pressure means that the temperature of the component or line will also be low (cold). For example, the inlet to the compressor (low pressure) should always be cool, whereas the outlet of the compressor (high pressure) should always be hot.

applied to many evaporators to speed up water runoff, which helps dry the evaporator and reduce bacterial growth. Airborne bacteria also collect on this surface, and if the surface stays moist, these bacteria will live and grow, creating the unpleasant smell. Then when the air conditioning or blower motor is turned on, that smell blows through the vehicle.

TECH TIP

Leak-Testing the Evaporator

A quick-and-easy test to check whether the evaporator is leaking refrigerant is to remove the blower motor resistor pack. The blower motor resistor pack is almost always located directly "downstream" and near the blower motor. Removing the blower motor resistor pack gives access to the area near the evaporator. Inserting the probe of a leak detector into this open area allows the detector to test the air close to the evaporator and bend the detector hose to check the LOWER part of the evaporator box, because refrigerant will fall to the lower point. If the vehicle does not use a blower motor resistor or if it is difficult to access, hoist the vehicle and insert the sniffer probe in the condensate tube.

CAUTION: This is not recommended if the system has been run recently. If water is drawn into the electronic leak detector, it will be destroyed. If checking at the condensate line, make sure that the tube is dry and that there is no condensation ready to drip out of the tube.

Several companies market chemicals, essentially fungicides, to kill the bacterial growth, or detergents to clean the evaporator core. Some of these chemicals and a procedure to use them have been approved by vehicle manufacturers. These chemicals are sprayed into the ductwork or onto the evaporator fins. ● **SEE FIGURE 37–14.**

Some manufacturers install an **electronic evaporator dryer (EED)**, also called an **after-blow module**. In some vehicles, the A/C control module is programmed to operate the blower for a drying cycle after the car is shut off. This device turns on the blower (with the ignition switched off for 30 minutes to 50 minutes) and lets it run long enough to dry off the evaporator after the vehicle has been shut off. One system waits 10 minutes, runs the blower on high for 10 seconds, shuts down for another 10 minutes, and then repeats the 10-second operation and 10-minute pause for 10 cycles. The object is to blow the moist air out of the evaporator without discharging the battery. EED modules are available for vehicles with either ground or B+ side blower switch systems.

(a)

(b)

FIGURE 37–14 (a) A spray can of mold and mildew eliminator available at local parts stores and used on the evaporator before installing it into the vehicle is good insurance against HVAC-related odors. (b) Spraying the clear liquid mold and mildew eliminator onto the surface of the replacement evaporator prior to installation.

TECH TIP

Hot/Cold Sides

A complaint of uneven air discharge temperature from the instrument panel registers (cold on one side and warm on the other) can be caused by a low charge level. This will cause some parts of the evaporator to be cold while others are warm. It is possible for the air from the cold side to flow to a single register.

TECH TIP

The Radio "POP" Trick

Many air-conditioning compressor clutch circuits contain a diode that is used to suppress the high-voltage spike that is generated whenever the compressor clutch coil is disengaged (turned off). If this diode were to fail, a high voltage (up to 400 volts) could damage sensitive electronic components in the vehicle including the electronic air-conditioning compressor clutch control unit (if so equipped). Another thing that can occur is that the radio, other modules, or vehicle networks will intermittently turn off and then back on again because of the high-voltage spike.

To check this diode, tune the radio to a weak AM station near 1400 Hz and cycle the air-conditioning compressor on and off. If a "pop" noise is heard from the radio speaker(s), then the diode is defective and must be replaced.

NOTE: While some A/C compressor diodes can be replaced separately, some of these air-conditioning compressor clutch diodes are part of the wiring harness or connector assembly. ● **SEE FIGURE 37–15.**

ADDRESSING NOISE ISSUES The A/C system is the potential source for several noise problems, and the compressor and clutch are the main culprits.

- A moaning or growling noise is a relatively low-frequency noise caused by something moving slowly.
- A whine or squeal is a high-pitch and high-frequency noise produced by something moving rapidly.

Another problem can be **groundout**, in which a vibrating metal A/C line contacts another surface (flat surfaces produce the most noise). Groundout problems also occur when the exhaust pipes make metal-to-metal contact with the vehicle body.

(a)

(b)

FIGURE 37–15 The A/C compressor diode (if used) may be located in the vehicle fuse box or as part of the clutch electrical connector. (Courtesy of Jeffrey Rehkopf)

STEP 7 REPAIR THE SYSTEM

PURPOSE OF ANY REPAIR The purpose of any repair is to restore the system to like-new condition. Typical repairs include the replacement of components such as the accumulator, compressor, evaporator, condenser, and refrigerant lines/hoses.

TYPICAL REPAIRS Many repairs require the proper identification, recovery, evacuation, and charging of the refrigerant. After the repairs, the system should function and operate as designed and the repair should be performed according to established industry and vehicle manufacturer's recommendations. All safety precautions should be adhered to and the refrigerant handled according to federal and local laws.

STEP 8 VERIFY THE REPAIR

OPERATE THE SYSTEM After the repairs or service procedures have been performed, verify that the system is working as designed. If needed, drive the vehicle under the same conditions that it was when the customer concern was corrected to verify the repair. Document the work order and return the vehicle to the customer in clean condition.

LACK OF COOLING DIAGNOSIS AND REPAIR

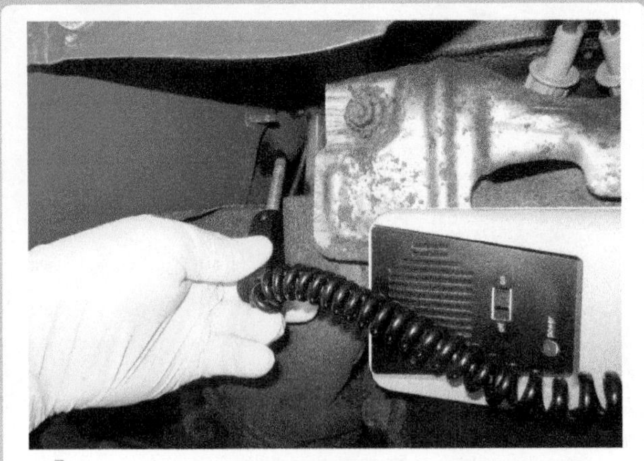

1 The lack of cooling was diagnosed using a leak detector which showed the evaporator was leaking when tested at the condensate drain.

2 The recovery process was started by connecting an RRR machine to the high-side fitting.

3 The low-side fitting was connected and the recovery was begun.

4 Very little refrigerant was left in the system (0.1 lb.) but it was recovered.

5 The lines to the accumulator were removed.

6 The lines to the evaporator were disconnected using a quick connect tool to release the fitting.

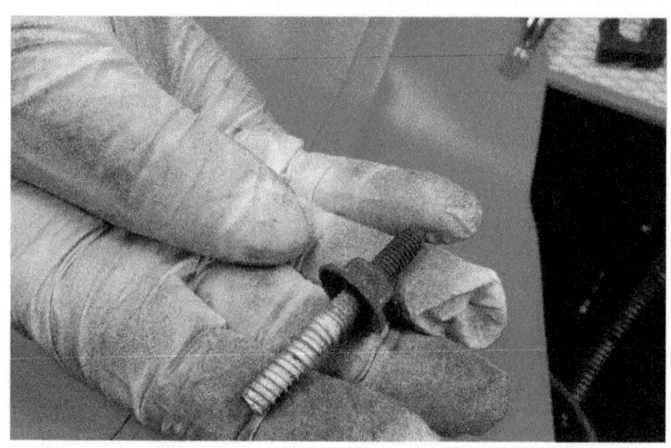

7 The retaining nuts were removed that held the HVAC case to the bulkhead from under the hood. In one case, the entire stud came out instead of just the nut.

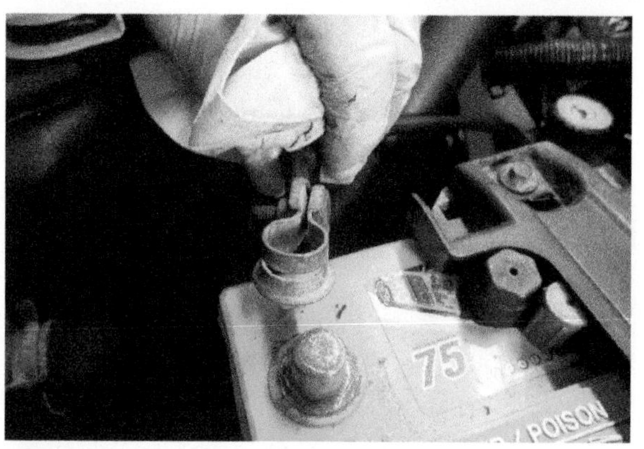

8 Before going into the passenger compartment, the negative battery cable was disconnected from the battery.

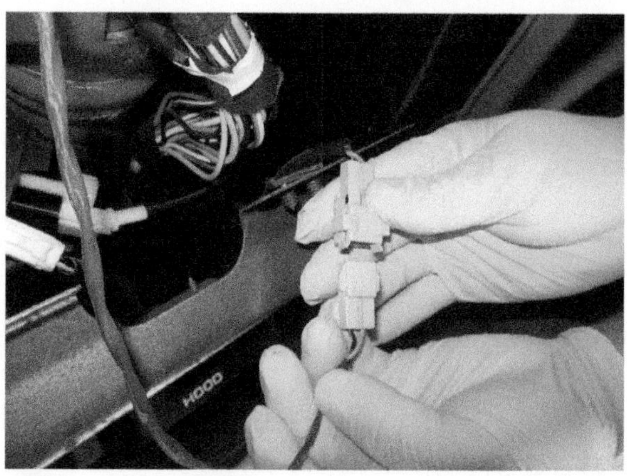

9 The dash assembly is being removed and this process includes disconnecting the airbag wiring connectors.

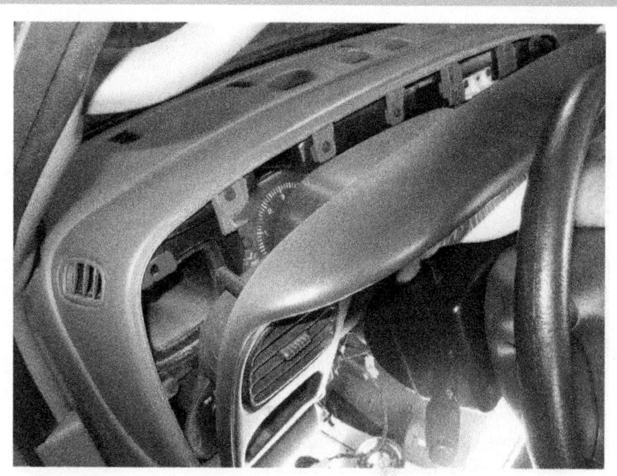

10 The driver/passenger side of the dash is removed to gain access to the additional fasteners.

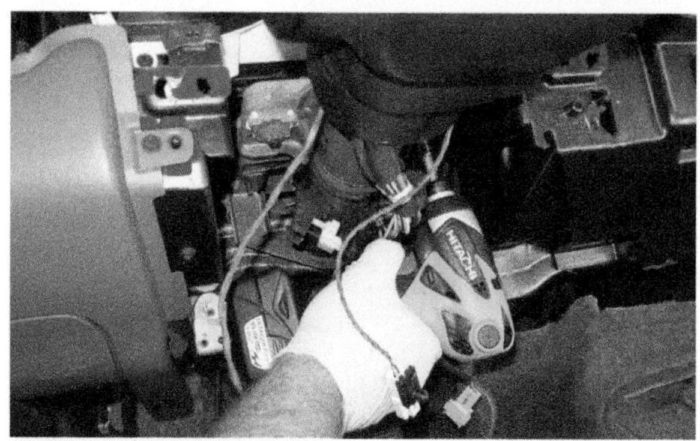

11 The two retaining fasteners that hold the steering column to the dash are removed and the steering column lowered into the seat, which is protected using a fender cover.

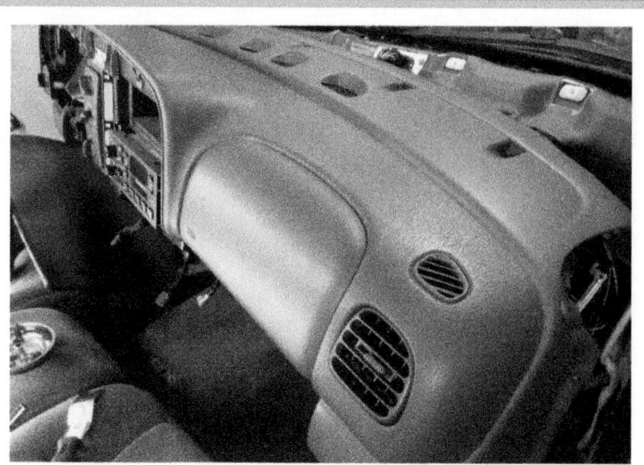

12 The entire dash assembly is being gently removed.

CONTINUED ▶

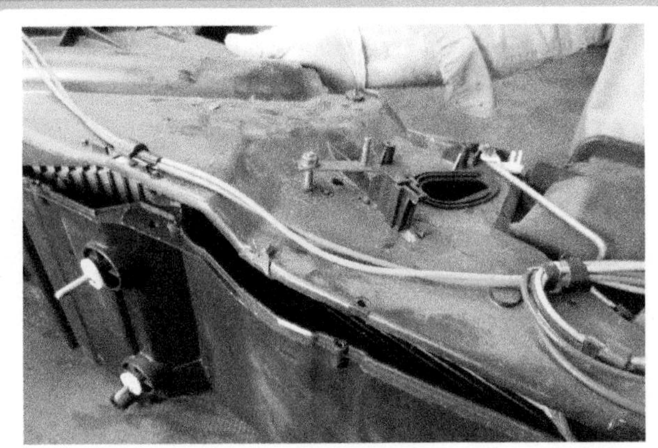

13 The HVAC case is being disassembled after being removed from the vehicle.

14 The evaporator is removed and it shows signs of leaking. The root cause of the lack of cooling has been found.

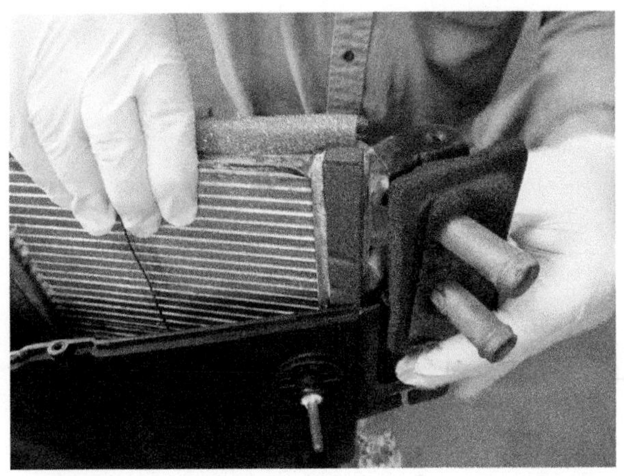

15 The heater core was also removed and replaced at the same time as insurance against possible future leaks from this unit.

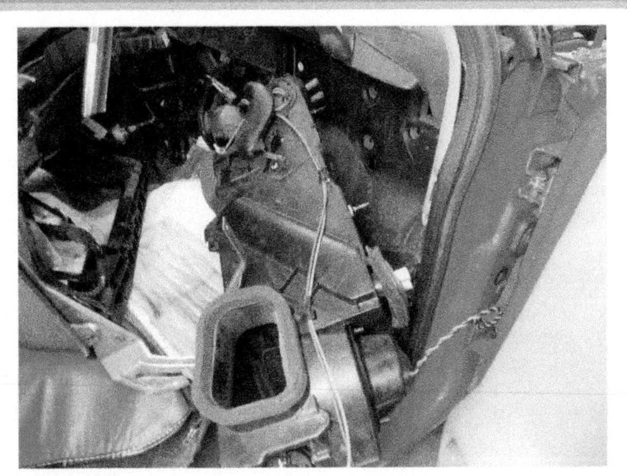

16 The reassembled HVAC case is now being installed back into the vehicle.

17 A new accumulator is installed.

18 The system was evacuated and charged to the specified amount and the pressure gauge and temperature reading indicated that the system was restored to like-new performance.

SUMMARY

1. When diagnosing a heating and air-conditioning system problem, most vehicle manufacturers recommend that the following steps be performed.

   STEP 1 Verify the Customer Complaint (concern).

   STEP 2 Perform a Thorough Visual Inspection.

   STEP 3 Check for Diagnostic Trouble Codes (DTCs).

   STEP 4 Check for Related Technical Service Bulletins (TSBs).

   STEP 5 Perform an A/C Performance Test.

   STEP 6 Determine the Root Cause.

   STEP 7 Repair the System.

   STEP 8 Verify the Repair.

2. Sometimes the A/C system is functioning normally for the conditions which could include any of the following:
   - Higher than normal outside air (ambient) temperature
   - High humidity level

REVIEW QUESTIONS

1. What are the steps that a technician should follow when diagnosing and repairing an air-conditioning-related problem?
2. What should be checked as part of a visual inspection?
3. Why should the diagnostic trouble codes be checked before checking for technical service bulletins?
4. What does the temperature difference across the condenser and evaporator tell the technician?
5. How can leaks in the refrigeration system be detected?

CHAPTER QUIZ

1. When should the service technician check for technical service bulletins?
   a. As soon as the work order is received with the customer complaint shown.
   b. Before performing a visual inspection.
   c. After checking for any stored diagnostic trouble codes.
   d. After making the repair.
2. Clear water is seen dripping from beneath the vehicle on the passenger side. This means _____________.
   a. Normal operation
   b. Possible clogged evaporator
   c. Possible restricted condenser
   d. Receiver–drier is at the end of its normal life
3. What should the technician use when checking for diagnostic trouble codes?
   a. Any global (generic) scan tool that can read body and/or chassis codes.
   b. Any enhanced scan tool that can read body and/or chassis codes.
   c. Only a factory scan tool can be used.
   d. A paper clip because it is needed to read codes.
4. What could be wrong if the A/C compressor clutch cycles on and off rapidly?
   a. Defective A/C compressor
   b. Defective A/C compressor clutch
   c. Low of refrigerant charge
   d. Shorted compressor clutch diode
5. The low-side pressure is low and the high-side pressure is higher than normal. What is the most likely cause?
   a. Low refrigerant charge level.
   b. System is overcharged. Expansion valve stuck open.
   c. Restriction in the low-side line.
   d. Restriction in high-side line.
6. The temperature difference between the inlet and outlet of the condenser should be _____________.
   a. Equal
   b. 10°F to 30°F
   c. 20°F to 50°F
   d. Greater than 60°
7. Most A/C performance tests include _____________.
   a. Start the engine, set the parking brake, and raise the idle to 2000 RPM
   b. Measure ambient temperature 3 inches (80 mm) in front of the condenser and measure the humidity
   c. Open the doors and set the blower speed to high, which applies the maximum load on the system
   d. All of the above
8. Several different methods of leak detection are available including _____________.
   a. Visual inspection looking for refrigerant oil stains
   b. Dye
   c. Using a leak detector
   d. Any of the above
9. A "pop" is heard from the radio speaker(s) when the system is operating with the air conditioning on. What is the most likely cause?
   a. Normal operation
   b. A defective compressor clutch diode
   c. A restricted evaporator
   d. A worn A/C compressor
10. A musty, moldy smell in the air-conditioning system is usually due to _____________.
    a. Mildew-type fungus growth
    b. A restricted evaporator
    c. A clogged condenser
    d. A weak or defective AC compressor

INDEX

Note: The letter 'f' following locators refers to figures respectively

B

C

D

F

G

H

P

R

S

T

U

V